~~216~~ $12

FUNDAMENTAL AND PHYSICAL CONSTANTS

quantity	symbol	U.S.	SI
Charge			
electron	e		-1.6022×10^{-19} C
proton	p		$+1.6022 \times 10^{-19}$ C
Density			
air [STP] [32°F (0°C)]		0.0805 lbm/ft^3	1.29 kg/m^3
air [70°F (20°C), 1 atm]		0.0749 lbm/ft^3	1.20 kg/m^3
earth [mean]		345 lbm/ft^3	5520 kg/m^3
mercury		849 lbm/ft^3	1.360×10^4 kg/m^3
seawater		64.0 lbm/ft^3	1025 kg/m^3
water [mean]		62.4 lbm/ft^3	1000 kg/m^3
Distance [mean]			
earth radius		2.09×10^7 ft	6.370×10^6 m
earth-moon separation		1.26×10^9 ft	3.84×10^8 m
earth-sun separation		4.89×10^{11} ft	1.49×10^{11} m
moon radius		5.71×10^6 ft	1.74×10^6 m
sun radius		2.28×10^9 ft	6.96×10^8 m
first Bohr radius	a_0	1.736×10^{-10} ft	5.292×10^{-11} m
Gravitational Acceleration			
earth [mean]	g	32.174 (32.2) ft/sec^2	9.8067 (9.81) m/s^2
moon [mean]		5.47 ft/sec^2	1.67 m/s^2
Mass			
atomic mass unit	u	3.66×10^{-27} lbm	1.6606×10^{-27} kg
earth		1.32×10^{25} lbm	6.00×10^{24} kg
electron [rest]	m_e	2.008×10^{-30} lbm	9.109×10^{-31} kg
moon		1.623×10^{23} lbm	7.36×10^{22} kg
neutron [rest]	m_n	3.693×10^{-27} lbm	1.675×10^{-27} kg
proton [rest]	m_p	3.688×10^{-27} lbm	1.673×10^{-27} kg
sun		4.387×10^{30} lbm	1.99×10^{30} kg
Pressure, atmospheric		14.696 (14.7) lbf/in^2	1.0133×10^5 Pa
Temperature, standard		32°F (492°R)	0°C (273K)
absolute zero		−459.67°F (0°R)	−273.16°C (0K)
triple point, water		32.02°F, 0.0888 psia	0.01109°C, 0.6123 kPa
Velocity			
earth escape		3.67×10^4 ft/sec	1.12×10^4 m/s
light [vacuum]	c	9.84×10^8 ft/sec	2.9979 (3.00) $\times 10^8$ m/s
sound [air, STP]	a	1090 ft/sec	331 m/s
[air, 70°F (20°C)]		1130 ft/sec	344 m/s
Volume, molal ideal gas [STP]		359 ft^3/lbmol	22.41 m^3/kmol
Fundamental Constants			
Avogadro's number	N_A		6.022×10^{23} mol^{-1}
Bohr magneton	μ_B		9.2732×10^{-24} J/T
Boltzmann constant	k	5.65×10^{-24} ft-lbf/°R	1.38065×10^{-23} J/K
Faraday constant	F		96 487 C/mol
gravitational constant	g_c	32.174 lbm-ft/lbf-sec^2	n.a.
gravitational constant	G	3.440×10^{-8} ft^4/lbf-sec^4	6.674×10^{-11} $N{\cdot}m^2/kg^2$
nuclear magneton	μ_N		5.050×10^{-27} J/T
permeability of a vacuum	μ_0		1.2566×10^{-6} N/A^2 (H/m)
permittivity of a vacuum	ϵ_0		8.854×10^{-12} $C^2/N{\cdot}m^2$ (F/m)
Planck's constant	h		6.6256×10^{-34} J·s
Rydberg constant	R_∞		1.097×10^7 m^{-1}
specific gas constant, air	R	53.35 ft-lbf/lbm-°R	287.03 J/kg·K
Stefan-Boltzmann constant	σ	1.713×10^{-9} Btu/ft^2-hr-$°R^4$	5.670×10^{-8} $W/m^2{\cdot}K^4$
universal gas constant	R^*	1545.35 ft-lbf/lbmol-°R	8314.47 J/kmol·K
	R^*	1.986 Btu/lbmol-°R	0.08206 atm·L/mol·K

quantity	symbol	value
Mathematical Constants		
Archimedes' number (pi)	π	3.14159 26536
base of natural logs	e	2.71828 18285
Euler constant	γ	0.57721 56649

Power Reference Manual
for the Electrical and Computer PE Exam

Second Edition

John A. Camara, PE

Professional Publications, Inc. • Belmont, California

Benefit by Registering This Book with PPI

- Get book updates and corrections.
- Hear the latest exam news.
- Obtain exclusive exam tips and strategies.
- Receive special discounts.

Register your book at **ppi2pass.com/register**.

Report Errors and View Corrections for This Book

PPI is grateful to every reader who notifies us of a possible error. Your feedback allows us to improve the quality and accuracy of our products. You can report errata and view corrections at **ppi2pass.com/errata**.

POWER REFERENCE MANUAL FOR THE ELECTRICAL AND COMPUTER PE EXAM

Second Edition

Current printing of this edition: 2

Printing History

date	edition number	printing number	update
Mar 2014	1	3	Minor corrections.
Mar 2016	2	1	New edition. Code updates. Additional content. Copyright update.
Aug 2017	2	2	Minor corrections. Minor cover updates.

Printed in the United States of America.

PPI
1250 Fifth Avenue
Belmont, CA 94002
(650) 593-9119
ppi2pass.com

ISBN: 978-1-59126-502-3

Library of Congress Control Number: 2015954486

F E D C B A

Topics

Mathematics
Basic Theory
Field Theory
Circuit Theory
Generation
Distribution
System Analysis
Protection and Safety
Machinery and Devices
Electronics
Special Applications
Measurement/ Instrumentation
Electrical Materials
Codes and Standards
Professional Practice
Support Material

Where do I find practice problems to test what I've learned in this Reference Manual?

The *Power Reference Manual* provides a knowledge base that will prepare you for the Electrical and Computer PE Power exam. But there's no better way to exercise your skills than to practice solving problems. To simplify your preparation, please consider *Power Practice Problems for the Electrical and Computer PE Exam.* This publication provides you with more than 600 practice problems, each with a complete, step-by-step solution.

Table of Contents

Topic XIV: Codes and Standards

Topic XV: Professional Practice

Topic XVI: Support Material

Appendices Table of Contents

Preface

To pass the Electrical and Computer Power PE exam administered by the National Council of Examiners for Engineering and Surveying (NCEES), you need to understand both the concepts and the practical applications of electrical power engineering. Therefore, my goal for *Power Reference Manual for the Electrical and Computer PE Exam* is to clearly explain the concepts outlined by the NCEES exam specifications, as well as present examples that show how these concepts are applied.

For this second edition, I updated all material related to and dependent on the 2014 edition of the *National Electrical Code* (NEC) and the 2012 edition of the *National Electrical and Safety Code* (NESC). I also wrote significant new content on the *IEEE 3000 Standards Collection™ for Industrial and Commercial Power Systems*, formerly known as the *IEEE Color Books*, as well as on workplace electrical safety, personal protective equipment (PPE), and transformer types. The existing chapter on the NEC was expanded to include new content on special occupancies, special equipment, special conditions, and communication systems. I've also included four new appendices: one on PPE, two on ambient temperature correction factors, and one on plug and receptacle configurations.

FOR WHOM THIS BOOK IS WRITTEN

The *Power Reference Manual* is written for you. It is written to be a reference at all times in your career, whether you are an undergraduate student or an experienced engineer. If you are an exam candidate, the *Power Reference Manual* is an efficient resource for exploring exam topics systematically and exhaustively. If you are a practicing power engineer, it functions as a comprehensive reference, discussing all aspects of this field in a realistic manner, and incorporating the most common formulas and data. Finally, for the engineering student, it presents a thorough review of the fundamentals of electrical engineering.

The scope of this book is beyond that of the Power PE exam. When the exam is complete and you are a licensed electrical engineer, you will need a resource for electrical engineering information, either in your work or simply to satisfy the curiosity that makes us thinking humans. In both the exam and your career, I hope this book serves you well.

LOOKING FORWARD TO THE FUTURE

Future editions of this book will be very much shaped by what you and others want to see in it. I am braced for the influx of comments and suggestions from those readers who (1) want more topics, (2) want more detail in existing topics, and (3) think that continuing to include some topics is just plain lame. Computer hardware and programming change quickly, as does the grand unified (atomic) theory of everything. I expect several chapters to generate some kind of debate. Should you find an error in this book, know that it is mine and that I regret it. Beyond that, I hope two things happen. First, please let me know about the error by using the error reporting form on the PPI website (**ppi2pass.com/errata**). Second, I hope you learn something from the error—I know I will! I appreciate suggestions for improvement, additional questions, and recommendations for expansion so that new editions or similar texts will better meet the needs of future examinees. Read carefully, prepare well, and you will triumph over the exam!

John A. Camara, PE

Acknowledgments

It is with enduring gratitude that I thank Michael R. Lindeburg, PE, who, with the sixth edition of the *Electrical Engineering Reference Manual*, allowed me to realize my dream of authoring a significant engineering text. In taking the reins from the title's previous author, the late Raymond Yarbrough, I was entrusted with the responsibility of shaping that text as the NCEES Electrical and Computer PE exam and the engineering profession evolved. It is my hope that the *Power Reference Manual for the Electrical and Computer PE Exam*—which is based on the *Electrical Engineering Reference Manual*—will match the comprehensive and cohesive quality of the best of PPI's family of engineering books, which I have admired for three decades.

The second edition of the *Power Reference Manual* would not have been possible without the professionalism and meticulous attention to detail shown by Ellen Nordman, lead editor. In addition, my thanks go out to the wonderfully responsive team at PPI, including Serena Cavanaugh, associate acquisitions editor; Sam Webster, editorial manager; Nicole Evans, associate project manager; Thomas Bliss, David Chu, Sierra Cirimelli-Low, Tyler Hayes, Tracy Katz, Julia Lopez, Scott Marley, Heather Turbeville, and Ian A. Walker, copy editors; Tom Bergstrom, production associate and technical illustrator; Kate Hayes, production associate; Cathy Schrott, production services manager, who keeps it all moving forward; and Sarah Hubbard, director of Product Development and Implementation, whose guidance from the beginning has been most appreciated.

Gregg Wagener, PE, reviewed each of the chapters of the first edition of this book, providing valuable insight, correcting numerous errors, and in general, teaching me a great deal along the way. Thank you to James Mirabile, PE, and Nanzhu Zhang, who technically reviewed the new content for the second edition.

Michael R. Lindeburg, PE, provided Chapters 1–14 on mathematics, as well as the chapters on economic analysis, law, ethics, and engineering licensure. He also provided sections of the illumination chapter. These contributions saved me considerable time, and I am grateful to use material of this quality. John Goularte, Jr., taught me firsthand the applications of biomedical engineering, and the real-world practicality of this topic is attributable to him.

Additionally, the following individuals contributed in a variety of ways, from thoughtful reviews, criticisms, and corrections, to suggestions for additional material. I am grateful for their efforts, which resulted in enhanced coverage of the material and enriched my knowledge of each topic. Thank you to Chris Iannello, PhD; Thomas P. Barrera, PhD; Robert G. Henriksen, PhD; Glenn G. Butcher, DCS; Chen J. Chang, MSE; and Aubrey Clements, BSEE.

The knowledge expressed in this book represents years of training, instruction, and self-study in electrical, electronic, mechanical, nuclear, marine, space, and a variety of other types of engineering—all of which I find fascinating. I would not understand any of it were it not for the teachers, instructors, mentors, family, and friends who have spent countless hours with me and from whom I have learned so much. A few of them include Mary Avila, Capt. C. E. Ellis, Claude Estes, Capt. Karl Hasslinger, Jerome Herbeck, Jack Hunnicutt, Ralph Loya, Harry Lynch, Harold Mackey, Michael O'Neal, Mark Richwine, Charles Taylor, Abigail Thyarks, and Jim Triguerio. There are others who have touched my life in special ways: Jim, Marla, Lora, Todd, Tom, and Rick. There are many others, though here they will remain nameless, to whom I am indebted. Thanks.

A very special thanks to my son, Jac Camara, whose annotated periodic table became the basis for the Electrical Materials chapter. Additionally, our discussions about quantum mechanics and a variety of other topics helped stir the intellectual curiosity I thought was fading.

Thanks also to my daughter, Cassiopeia, who has been a constant source of inspiration, challenge, and pride.

Thanks to my lady, Shelly, for providing insight into the intricacies of the NEC and patiently missing long hours together during the update of this edition. Also, thanks to Wyatt Wade, who is teaching me the "new math."

Finally, thanks to my mother, Arlene, who made me a better person.

John A. Camara, PE

Acknowledgments

Codes and References

The information that was used to write and update this book was based on the exam specifications at the time of publication. However, as with engineering practice itself, the PE exam is not always based on the most current codes or cutting-edge technology. Similarly, codes, standards, and regulations adopted by state and local agencies often lag issuance by several years. It is likely that the codes that are most current, the codes that you use in practice, and the codes that are the basis of your exam will all be different. However, differences between code editions typically minimally affect the technical accuracy of this book, and the methodology presented remains valid. For more information about the variety of codes related to electrical engineering, refer to the following organizations and their websites.

American National Standards Institute (ansi.org)
Electronic Components Industry Association (ecianow.org)
Federal Communications Commission (fcc.gov)
Institute of Electrical and Electronics Engineers (ieee.org)
International Organization for Standardization (iso.org)
International Society of Automation (isa.org)
National Electrical Manufacturers Association (nema.org)
National Fire Protection Association (nfpa.org)

The PPI website (**ppi2pass.com/eefaq**) provides the dates and editions of the codes, standards, and regulations on which NCEES has announced the PE exams are based. It is your responsibility to find out which codes are relevant to your exam.

The minimum recommended library of materials to bring to the Electrical and Computer PE Power exam consists of this book, any applicable code books, a standard handbook of electrical engineering, and one or two textbooks that cover fundamental circuit theory (both electrical and electronic).

CODES AND STANDARDS

47 CFR 73: *Code of Federal Regulations*, "Title 47—Telecommunication, Part 73—Radio Broadcast Rules," 2014. Office of the Federal Register National Archives and Records Administration, Washington, DC. (Communications.)

IEEE/ASTM SI 10: *American National Standard for Metric Practice*, 2010. ASTM International, West Conshohocken, PA. (Metric.)

IEEE Std 141 (IEEE Red Book): *IEEE Recommended Practice for Electric Power Distribution for Industrial Plants*, 1993. The Institute of Electrical and Electronics Engineers, Inc., New York, NY.

IEEE Std 142 (IEEE Green Book): *IEEE Recommended Practice for Grounding of Industrial and Commercial Power Systems*, 2007. The Institute of Electrical and Electronics Engineers, Inc., New York, NY.

IEEE Std 241 (IEEE Gray Book): *IEEE Recommended Practice for Electrical Power Systems in Commercial Buildings*, 1990. The Institute of Electrical and Electronics Engineers, Inc., New York, NY.

IEEE Std 242 (IEEE Buff Book): *IEEE Recommended Practice for Protection and Coordination of Industrial and Commercial Power Systems*, 2001. The Institute of Electrical and Electronics Engineers, Inc., New York, NY.

IEEE Std 399 (IEEE Brown Book): *IEEE Recommended Practice for Industrial and Commercial Power Systems Analysis*, 1997. The Institute of Electrical and Electronics Engineers, Inc., New York, NY.

IEEE Std 446 (IEEE Orange Book): *IEEE Recommended Practice for Emergency and Standby Power Systems for Industrial and Commercial Applications*, 1995. The Institute of Electrical and Electronics Engineers, Inc., New York, NY.

IEEE Std 493 (IEEE Gold Book): *IEEE Recommended Practice for the Design of Reliable Industrial and Commercial Power Systems*, 2007. The Institute of Electrical and Electronics Engineers, Inc., New York, NY.

IEEE Std 551 (IEEE Violet Book): *IEEE Recommended Practice for Calculating Short-Circuit Currents in Industrial and Commercial Power Systems*, 2006. The Institute of Electrical and Electronics Engineers, Inc., New York, NY.

IEEE Std 602 (IEEE White Book): *IEEE Recommended Practice for Electric Systems in Health Care Facilities*, 2007. The Institute of Electrical and Electronics Engineers, Inc., New York, NY.

IEEE Std 739 (IEEE Bronze Book): *IEEE Recommended Practice for Energy Management in Industrial and Commercial Facilities*, 1995. The Institute of Electrical and Electronics Engineers, Inc., New York, NY.

IEEE Std 902 (IEEE Yellow Book): *IEEE Guide for Maintenance, Operation, and Safety of Industrial and Commercial Power Systems*, 1998. The Institute of Electrical and Electronics Engineers, Inc., New York, NY.

IEEE Std 1015 (IEEE Blue Book): *IEEE Recommended Practice for Applying Low-Voltage Circuit Breakers Used in Industrial and Commercial Power Systems*, 2006. The Institute of Electrical and Electronics Engineers, Inc., New York, NY.

IEEE Std 1100 (IEEE Emerald Book): *IEEE Recommended Practice for Powering and Grounding Electronic Equipment*, 2005. The Institute of Electrical and Electronics Engineers, Inc., New York, NY.

NEC (NFPA 70): *National Electrical Code*, 2014. National Fire Protection Association, Quincy, MA. (Power.)

NESC (IEEE C2): *2012 National Electrical Safety Code*, 2012. The Institute of Electrical and Electronics Engineers, Inc., New York, NY. (Power.)

REFERENCES

Anthony, Michael A. *NEC Answers*. New York, NY: McGraw-Hill. (*National Electrical Code* example applications textbook.)

Bronzino, Joseph D. *The Biomedical Engineering Handbook*. Boca Raton, FL: CRC Press. (Electrical and electronics handbook.)

Chemical Rubber Company. *CRC Standard Mathematical Tables and Formulae*. Boca Raton, FL: CRC Press. (General engineering reference.)

Croft, Terrell and Wilford I. Summers. *American Electricians' Handbook*. New York, NY: McGraw-Hill. (Power handbook.)

Earley, Mark W. et al. *National Electrical Code Handbook*, 2014 ed. Quincy, MA: National Fire Protection Association. (Power handbook.)

Fink, Donald G. and H. Wayne Beaty. *Standard Handbook for Electrical Engineers*. New York, NY: McGraw-Hill. (Power and electrical and electronics handbook.)

Grainger, John J. and William D. Stevenson, Jr. *Power System Analysis*. New York, NY: McGraw-Hill. (Power textbook.)

Horowitz, Stanley H. and Arun G. Phadke. *Power System Relaying*. Chichester, West Sussex: John Wiley & Sons, Ltd. (Power protection textbook.)

Huray, Paul G. *Maxwell's Equations*. Hoboken, NJ: John Wiley & Sons, Inc. (Power and electrical and electronics textbook.)

Jaeger, Richard C. and Travis Blalock. *Microelectronic Circuit Design*. New York, NY: McGraw-Hill Education. (Electronic fundamentals textbook.)

Lee, William C.Y. *Wireless and Cellular Telecommunications*. New York, NY: McGraw-Hill. (Electrical and electronics handbook.)

Marne, David J. *National Electrical Safety Code (NESC) 2012 Handbook*. New York, NY: McGraw-Hill Professional. (Power handbook.)

McMillan, Gregory K. and Douglas Considine. *Process/Industrial Instruments and Controls Handbook*. New York, NY: McGraw-Hill Professional. (Power and electrical and electronics handbook.)

Millman, Jacob and Arvin Grabel. *Microelectronics*. New York, NY: McGraw-Hill. (Electronic fundamentals textbook.)

Mitra, Sanjit K. *An Introduction to Digital and Analog Integrated Circuits and Applications*. New York, NY: Harper & Row. (Digital circuit fundamentals textbook.)

Parker, Sybil P., ed. *McGraw-Hill Dictionary of Scientific and Technical Terms*. New York, NY: McGraw-Hill. (General engineering reference.)

Plonus, Martin A. *Applied Electromagnetics*. New York, NY: McGraw-Hill. (Electromagnetic theory textbook.)

Rea, Mark S., ed. *The IESNA Lighting Handbook: Reference & Applications*. New York, NY: Illuminating Engineering Society of North America. (Power handbook.)

Shackelford, James F. and William Alexander, eds. *CRC Materials Science and Engineering Handbook*. Boca Raton, FL: CRC Press, Inc. (General engineering handbook.)

Van Valkenburg, M.E. and B.K. Kinariwala. *Linear Circuits*. Englewood Cliffs, NJ: Prentice-Hall. (AC/DC fundamentals textbook.)

Wildi, Theodore and Perry R. McNeill. *Electrical Power Technology*. New York, NY: John Wiley & Sons (Power theory and application textbook.)

Introduction

PART 1: HOW YOU CAN USE THIS BOOK

QUICKSTART

If you are the type of reader who desires information distilled into concise bits and you are ready to begin your study, then read this section and get started!

- This book takes complex topics and parses them into condensed, readable, understandable chunks.
- The National Council of Examiners for Engineering and Surveying (NCEES) Electrical and Computer—Power PE exam consists of problems that require (on average) six minutes to solve. The exam problems focus on small portions of complex electrical engineering topics. Your study should reflect this.
- Once the exam is over, this book will remain a useful reference you can use to review basic and complex power engineering concepts that you may encounter in your professional practice.

Because the chapters in this book are independent, you can focus your study on your weakest areas first. Review the table of contents to familiarize yourself with the book, then use Table 1 in this Introduction as a guide for focusing your study. This book's index is a great starting place for searching for specific information and for planning your review of the topics covered by the NCEES specifications. The mathematics, theory, and specialty topics, while not directly listed in the NCEES requirements, supply fundamental background on topics that may be necessary for problem solving on the exam. The appendices are succinct references for common electrical information. I have included the Professional chapters to refine your professional engineering knowledge and round out your knowledge base.

Commence study immediately. Six months is not an unreasonable time frame to prepare for the exam, though your learning curve will vary depending upon your background. Start with your weakest areas and work toward your strongest. Review every week.

Work problems three to five times a week. The key to success on the exam is to work as many problems as possible. Browse PPI's website for additional resources, such as this book's companion, *Power Practice Problems for the Electrical and Computer PE Exam*, as well as *Power Practice Exams for the Electrical and Computer PE Exam*, and PPI's Exam Cafe.

To minimize the time spent searching for often-used formulas and data, prepare a one-page summary of all the important formulas and information in each subject area. You can then use these summaries during the exam instead of searching for them in this book. You may want to consider *Power Quick Reference for the Electrical and Computer PE Exam*, which PPI has published for this type of use.

This book focuses on the basics that are the building blocks of engineering knowledge. This fundamental knowledge is useful regardless of the periodic changes to the NCEES exam requirements. The topic names might change, but the knowledge necessary to solve the types of problems necessary to pass the exam remains the same. Understanding the basics allows you to solve complex problems, which can always be divided into simpler, individual concepts. A thorough review of the basics, terminology, and problems in this text and the *Power Practice Problems* companion book will enable you to solve the variety of problems that might appear on the PE exam or in your professional practice. Continue to study until the exam date. Enjoy the learning process and revel in the knowledge gained.

If you are the type of reader who desires a more thorough review and wants a deeper understanding of the PE exam before commencing your studies, read the rest of the Introduction.

ABOUT THIS BOOK

While selecting and preparing the content in this book, I followed the NCEES Electrical and Computer—Power PE exam specifications. I have endeavored throughout the text to include the exact terminology used by these specifications.

While I wrote the book with the NCEES specifications in mind, the content of the *Power Reference Manual* is not limited to the specifications. This book also contains the fundamental building blocks you need to build a complete understanding of power engineering. In each chapter, I explore the mathematical, theoretical, and practical applications of the chapter's topic. You can focus your study on any or all of these types of applications.

The content of the *Power Reference Manual* and its companion book, *Power Practice Problems for the*

Electrical and Computer PE Exam, are guided by these concepts. As a result, you will find relatively easy, moderately difficult, and difficult examples in this book. The relatively easy examples and concepts are meant to remind you of the fundamentals. Studying these examples will help bring you closer to the level of knowledge you had readily available when college afforded you the opportunity to concentrate on these subjects daily. The moderately difficult examples can be solved in approximately six minutes and are meant to simulate the examination. The difficult examples and topics are meant to explore more complex areas of study and demonstrate how difficult problems might be broken into simpler pieces.

Each chapter builds on the knowledge gained from previous chapters and will refer to them, or even repeat some concepts in a slightly different manner. Finally, references to standards or to specific idiosyncrasies described in the footnotes are meant to provide jumping-off points for further study that will round off your knowledge. If you encounter these subjects on the exam, they will likely quote from the reference or standard and ask for an interpretation or calculation based on the quoted material.

FOR THE PE CANDIDATE

While you are preparing for the Power exam, the following suggestions may help.

- Concentrate primarily on the fundamentals of electrical engineering. Take time to review the principles of any problem you find difficult.
- Use the index extensively. Every significant term, law, theorem, and concept has been indexed. If you do not recognize a term, look for it in the index and read the related text for a more comprehensive overview. The index is the quickest way to find material in an exam situation, so you should familiarize yourself with it before exam day. General terms or concepts, which are covered in a variety of subject areas, may not be indexed, though the specifics of the concepts will be listed.
- Some subjects appear in more than one chapter. Use the index liberally and learn all there is to know about a particular subject.
- Know which subjects in this book are covered on the PE exam, and which are provided as support material. Chapters 1–25 and 58–62 are supportive and do not cover exam subjects. They do, however, provide background for the other chapters, enhance understanding of fundamentals, and are useful after the exam for quick reference. (The exam subjects known at the time of this printing are provided in Table 1. These are subject to change; the most up-to-date information can be found on PPI's website at **ppi2pass.com/eefaq**.)
- Become intimately familiar with this book. This means knowing the order of the chapters, the approximate location of important figures and tables, what appendices are available, and so on.
- After reviewing a chapter of this book, solve each of the related problems in the companion book, *Power Practice Problems for the Electrical and Computer PE Exam*. Practice in problem solving is the key to success on the exam. It is a good idea to begin with problems that fall in your weakest areas and work toward your stronger areas.
- Use the answers to the practice problems to check your work. If your answer does not match the given solution, study the solution to determine why. The majority of this book uses SI units, except where customary U.S. (English) units are the standard. Some equations in the U.S. system use the term g/g_c. For calculations at standard gravity, 1.00 is the numerical value of this fraction. Therefore, it is necessary to incorporate this quantity only in problems with a nonstandard gravity, or when being meticulous with units.
- Use Table 3 as your guide to focusing your exam preparation. Table 3 is found at the end of this Introduction.
- Practice by completing as many problems as possible. Remember that you will have an average of only six minutes per problem on the exam, so detailed design work is not possible. Rather, it is more likely that you will need to solve a portion of a design problem, complete a piece of an analysis puzzle, or demonstrate understanding of general requirements or standard practice.

IF YOU ARE A PRACTICING ENGINEER OR AN ENGINEERING STUDENT

If you are a practicing engineer or an engineering student and you have obtained this book as a general-reference handbook, the following suggestions and information may be of assistance.

- Both the index and table of contents are extensive and are meant as the starting point for any information desired. Every significant term, law, theorem, and concept has been indexed in every conceivable way—forwards and backwards.
- Chapters are grouped into interrelated topics, each offering a broad overview of a given area of electrical engineering. If you want a quick introduction to a given topic, refer to that portion of the book.
- The mathematics chapters can be used as quick reviews when working any engineering problems.
- The Advanced Engineering Mathematics chapter provides, in simple but accurate terms, the

Table 1 Detailed Analysis of Tested Subjects

ELECTRICAL AND COMPUTER—POWER EXAMINATION

General Power Engineering (24 questions)

Measurement and instrumentation (6 questions): instrument transformers; wattmeters; VOM metering; insulation testing; ground resistance testing

Special applications (8 questions): lightning and surge protection; reliability; illumination engineering; demand and energy management/calculations; engineering economics

Codes and standards (10 questions): *National Electrical Code*; *National Electrical Safety Code*; electric shock and burns

Circuit Analysis (16 questions)

Analysis (9 questions): three-phase circuit analysis; symmetrical components; per unit analysis; phasor diagrams

Devices and power electronic circuits (7 questions): battery characteristics and ratings; power supplies; relays, switches, and PLCs; variable-speed drives

Rotating Machines and Electromagnetic Devices (16 questions)

Rotating machines (10 questions): synchronous machines; induction machines; generator/motor applications; equivalent circuits; speed-torque characteristics; motor starting

Electromagnetic devices (6 questions): transformers; reactors; testing

Transmission and Distribution (24 questions)

System analysis (10 questions): voltage drop; voltage regulation; power factor correction and voltage support; power quality; fault current analysis; grounding; transformer connections; transmission line models

Power system performance (6 questions): power flow; load sharing: parallel generators or transformers; power system stability

Protection (8 questions): overcurrent protection; protective relaying; protective devices; coordination

information necessary to understand the importance and use of various advanced mathematical concepts. This should allow you to more fully comprehend and appreciate professional journals or engineering-focused presentations that use such concepts without resorting to a complete review of undergraduate math texts.

- The Advanced Engineering Mathematics chapter is also meant as the starting point for delving more deeply into advanced mathematical concepts by providing the names and defining the variables of frequently used mathematical formulas. This is done to help you through those situations where you know you need more mathematical information, but you do not know where to begin, what problems to ask, or what mathematical concept is involved.
- The appendices are meant as quick references for common electrical information.

IF YOU ARE AN INSTRUCTOR

If you are teaching a review course for the PE exam without the benefit of recent, firsthand exam experience, you can use the material in this book as a guide to prepare your lectures. Your students should use the book as a textbook. You should emphasize the subjects in each chapter and avoid subjects that were omitted. Subjects that do not appear in this book have rarely, if ever, appeared on the PE exam.

In an attempt to over-prepare students, I suggest assigning homework with practice problems that are more difficult and more varied than actual exam problems. It is often more efficient to cover several procedural steps in one practice problem than to ask simple "one liners" or definition problems.

"Capacity assignment" should be the goal in a successful review course. To do all the homework for some chapters requires approximately 15 to 20 hours of preparation per week. One student may be able to put in 20 hours, while another may only be able to put in 10 hours; assigning the full 20 hours of homework ensures that each student is working to his or her capacity. After the PE exam, all students will honestly say that they could not have prepared any more than they did in your course.

Instead of individually grading students' homework, permit students to make use of existing solutions to check their own work and learn approaches to the problems in their homework set. This will allow you to address special needs or questions written on the assignments, rather than simply marking the work as right or wrong. Solved practice problems are available in the companion *Power Practice Problems*.

Lecture coverage of some exam subjects is necessarily brief; other subjects are not covered at all. These omissions are intentional; they are not the result of scheduling. Why? First, time is not on your side in a review course. Second, some subjects rarely appear on the exam. Third, there are some subjects that are not well-received by the students, because they are not subjects that are pertinent to all students' objectives. Unless you have six months in which to teach your PE review course, your students' time can be better spent covering topics that apply to the majority of examinees. Table 1 lists the number of problems on the exam that will be devoted to a given topic. Using this and the depth of coverage in this book, you can decide which topics to focus on in your review course.

In organizing the review course format outlined in Table 2, I have tried to order the subjects in a logical,

Table 2 *Typical PE Review Course Format*

session	subject covered	chapter(s)
1	Introduction: PE Exam; Mathematics; Professional Practice	Introduction, 1–15, 58–62
2	Basic Theory and DC Circuits	16–19, 26
3	Field Theory and AC Circuits	20, 21, 27
4	Linear Circuits	29–31
5	Field Theory and Transformers	22, 23, 28
6	Three-Phase Electricity	34
7	Power Generation and Distribution	33, 35, 36
8	Power Transformers and Transmission Lines	37, 38, 39
9	Power System Analysis	40, 41
10	Protection and Safety	42
11	Rotating DC Machinery	43
12	Rotating AC Machinery	44
13	Electronics	45
14	Codes and Standards	56, 57

progressive manner, keeping my eye on "playing the high-probability subjects." Early in the course, I cover those subjects that everyone needs to learn. I leave the more specific subjects to the end.

Use the formulas, illustrations, and tables of data in the text to minimize chalkboard time. Avoid proofs and emphasize problem solving. The more problems solved, the higher the chances of exam success.

The practice problems in the companion book, *Power Practice Problems for the Electrical and Computer PE Exam*, range from quick-answer to in-depth problems. The more in-depth problems can be used as homework or timed problems.

One possible review course format is outlined in Table 2. The sessions are weekly, with the course lasting three and a half months. Most sessions could be covered in three hours of lecture. All the skipped chapters and any related practice problems are presented as floating assignments to be made up in the students' "free time."

I strongly believe in exposing students to a realistic sample exam, but since the review course usually ends only a few days before the real PE exam, I hesitate to make students sit for several hours in the late evening to take an in-class exam. Instead, I recommend distributing and assigning a take-home sample exam at the first meeting of the review course.

Caution your students not to peek at the sample exam prior to taking it. Looking at the sample exam or otherwise using it to direct their review will produce unwarranted specialization in subjects contained in the sample exam. For the sample exam to be realistic practice, it needs to mimic as much as possible the challenges of the actual exam.

There are many ways to organize a PE review course, depending on your available time, budget, intended audience, facilities, and enthusiasm. However, all good course formats have the same result: students struggle with the workload during the course, and then they breeze through the exam afterward.

PART 2: EVERYTHING YOU EVER WANTED TO KNOW ABOUT THE PE EXAM

WHAT IS THE FORMAT OF THE PE EXAM?

The NCEES PE exam in electrical and computer engineering consists of two four-hour sessions separated by a one-hour lunch period. The exam is an open-book exam. All problems are multiple-choice. Upon registering for the exam, examinees choose to work one of three exam modules: Computer Exam, Electrical and Electronics Exam, or Power Exam. Each exam consists of 80 problems (40 in the morning and 40 in the afternoon) that test knowledge in the areas specified. Examinees must work all 80 problems in the exam of their choice.

For more information on NCEES and the exams necessary to become a licensed electrical engineer, refer to Chap. 62, Engineering Licensing in the United States.

WHAT SUBJECTS ARE ON THE PE EXAM?

NCEES has published a description of subjects on the exam. Irrespective of the published exam structure, the exact number of problems that will appear in each subject area cannot be predicted reliably.

There is no guarantee that any single subject will occur in any quantity. One of the reasons for this is that some of the problems span several disciplines. You might consider a pump selection problem to come from the subject of fluids, while NCEES might categorize it as engineering economics.

WHAT IS THE TYPICAL PROBLEM FORMAT?

Almost all of the problems are stand-alone—that is, they are completely independent. However, NCEES allows that some sets of problems may start with a statement of a "situation" that will apply to (typically) two to five following problems. Such grouped problems are increasingly rare, however.

Each of the problems will have four answer options labeled "A," "B," "C," and "D." If the answer options are numerical, they will be displayed in increasing value. One of the answer options is correct (or will be "most nearly correct," as described in the following section). The remaining answer options are incorrect and may consist of one or more "logical distractors," the term used by NCEES to designate incorrect options that look correct.

NCEES intends the problems to be unrelated. Problems are independent or start with new given data. A mistake on one of the problems should not cause you to get a subsequent problem wrong. However, considerable time may be required to repeat previous calculations with a new set of given data.

HOW MUCH "LOOK UP" IS REQUIRED ON THE EXAM?

Since the problems are multiple-choice in design, all required data will appear in the situation statement. Since the exam would be unfair if it was possible to arrive at an incorrect answer after making valid assumptions or using plausible data, you will not generally be required to come up with numerical data that might affect your success on the problem. There will also be superfluous information in the majority of problems.

WHAT DOES "MOST NEARLY" REALLY MEAN?

One of the more disquieting aspects of these problems is that the available answer options are seldom exact. Answer options generally have only two or three significant digits. Exam problems ask, "Which answer option is most nearly the correct value?" or they instruct you to complete the sentence, "The value is approximately . . ." A lot of self-confidence is required to move on to the next problem when you do not find an exact match for the answer you calculated, or if you have had to split the difference because no available answer option is close.

NCEES has described it like this:

> Many of the problems on NCEES exams require calculations to arrive at a numerical answer. Depending on the method of calculation used, it is very possible that examinees working correctly will arrive at a range of answers. The phrase "most nearly" is used to accommodate answers that have been derived correctly but that may be slightly different from the correct answer option given on the exam. You should use good engineering judgment when selecting your choice of answer. For example, if the problem asks you to calculate an electrical current or determine the load on a beam, you should literally select the answer option that is most nearly what you calculated, regardless of whether it is more or less than your calculated value. However, if the problem asks you to select a fuse or circuit breaker to protect against a calculated current or to size a beam to carry a load, you should select an answer option that will safely carry the current or load. Typically, this requires selecting a value that is closest to but larger than the current or load.

The difference is significant. Suppose you are asked to calculate "most nearly" the current drawn by a given electrical load. Suppose, also, that the result of the calculation is 11 A. If the provided answer options were (A) 5 A, (B) 10 A, (C) 15 A, and (D) 20 A, you would go with answer option (B), because this is most nearly what you calculated. If, however, you were asked to select the breaker that provides the required overcurrent protection for this load, you would select answer option (C).

HOW MUCH MATHEMATICS IS NEEDED FOR THE EXAM?

Generally, only simple algebra, trigonometry, and geometry are needed on the PE exam. You will need to use the trigonometric, logarithm, square root, and similar buttons on your calculator. There is no need to use any other method for these functions. For functions that I recommend you program into your calculator, see the section "What about Calculators?"

There are no pure mathematics problems (algebra, geometry, trigonometry, etc.) on the exam. However, you will need to apply your knowledge of these subjects to the exam problems. Except for simple quadratic equations, you will probably not need to find the roots of higher-order equations. Occasionally, it will be convenient to use the equation-solving capability of an advanced calculator. However, other solution methods will always exist.

There is little or no use of calculus on the exam. Rarely, you may need to take a simple derivative to find a maximum or minimum of some function. Even rarer is the need to integrate to find an average or moment of inertia.

There is essentially no need to solve differential equations. On the exam, there are simple linear circuits and components that can be solved or explained with differential equations. Often these applications, and others that use such mathematics, can be solved by other means.

Basic statistical analysis of observed data may be necessary. Statistical calculations are generally limited to finding means, medians, standard deviations, variances, percentiles, and confidence limits. The only population distribution you need to be familiar with is the normal curve. Probability, reliability, hypothesis testing, and statistical quality control are not explicit exam subjects.

The Power exam is concerned with numerical answers, not with proofs or derivations. You will not be asked to prove or derive formulas.

Occasionally, a calculation may require an iterative solution method. Generally, there is no need to complete more than two iterations. You will not need to program your calculator to obtain an "exact" answer, nor will you generally need to use complex numerical methods.

HOW ABOUT ENGINEERING ECONOMICS?

For most of the early years of engineering licensing, problems on engineering economics appeared frequently on the exams. This is no longer the case. However, in its outline of exam subjects, NCEES notes: "Some problems may require knowledge of engineering economics." What this means is that engineering economics might appear in several problems on the exam, or the subject might be totally absent. While the degree of engineering economics knowledge may have decreased somewhat, the basic economic concepts (e.g., time value of money, present worth, nonannual compounding, comparison of alternatives, etc.) are still valid test subjects.

WHAT ABOUT PROFESSIONALISM AND ETHICS?

For many decades, NCEES has considered adding professionalism and ethics problems to the PE exam. However, these subjects are not part of the test outline, and there has yet to be an ethics problem in the exam.

IS THE EXAM TRICKY?

Other than providing superfluous data, the PE exam is not a "tricky exam." The exam does not overtly try to get you to fail. Examinees manage to fail on a regular basis with perfectly straightforward problems. The exam problems are difficult in their own right. NCEES does not need to provide misleading or conflicting statements. However, you will find that commonly made mistakes are represented in the available answer options. Thus, the alternative answers (known as distractors) will be logical.

Problems are generally practical, dealing with common and plausible situations that you might experience in your job. Circuit design is restricted to connecting electrical components to create a desired outcome. You will not be asked to design a new IC chip or build an equivalent vacuum tube circuit.

DOES NCEES WRITE EXAM PROBLEMS AROUND THIS BOOK?

Only NCEES knows what NCEES uses to write its exam problems. However, it is irrelevant, because this book is not intended to (1) be everything you need to pass the exam, (2) expose exam secrets or exam problems, or (3) help you pass when you do not deserve to pass. NCEES knows about this book, but worrying about NCEES writing exam problems based on information that is or is not in this book means you are placing too much dependency on this book. This book, for example, will teach you how to use aspects of many standards and codes. Expecting that this book will replace those standards and codes is unrealistic. This book will provide instruction in certain principles. Expecting that you will not need to learn anything else is unrealistic. This book presents many facts, definitions, and numerical values. Expecting that you will not need to know other facts, definitions, and numerical values is unrealistic. What NCEES uses to write exam problems will not have any effect on what you need to do to prepare for the exam.

WHAT MAKES THE PROBLEMS DIFFICULT?

Some problems are difficult because the pertinent theory is not obvious. There may be only one acceptable procedure, and it may be heuristic (or defined by a code) such that nothing else will be acceptable.

Some problems are difficult because the data needed is hard to find. Some data just is not available unless you happen to have brought the right reference book. For example, building requirements concerning electricity are in the *National Electrical Code*; without the code and without knowledge of where given information is located within the code, certain problems will be nearly impossible to solve.

Some problems are difficult because they defy the imagination. Some problems are difficult because the computational time is lengthy. Some problems are difficult because the terminology is obscure, and you just do not know what the terms mean. This can happen in almost any subject.

DOES THE PE EXAM USE SI UNITS?

The PE exam primarily uses the SI system. A few topics utilize the customary U.S. system where they correspond to commonly accepted industry standards. For example, problems involving the *National Electrical Code* were, until recently, given in customary U.S. units with SI provided in parenthesis. Even in this topic, SI is now foremost and customary U.S units are now in parentheses.

WHY DOES NCEES REUSE SOME PROBLEMS?

NCEES reuses some of the more reliable problems from each exam. The percentage of repeat problems is not high—no more than 25% of the exam. NCEES repeats problems in order to equate the performance of one group of examinees with the performance of an earlier group. The repeated problems are known as *equaters*, and together, they are known as the *equating subtest.*

Occasionally, a new problem appears on the exam that very few of the examinees do well on. Usually, the reason for this is that the subject is too obscure or the problem is too difficult. Also, there have been cases where a low percentage of the examinees get the answer correct because the problem was inadvertently stated in a poor or confusing manner. Problems that everyone gets correct are also considered defective.

NCEES tracks the usage and "success" of each of the exam problems. "Rogue" problems are not repeated without modification. This is one of the reasons historical analysis of problem types should not be used as the basis of your review.

DOES NCEES USE THE EXAM TO PRE-TEST FUTURE PROBLEMS?

NCEES does not use the PE exam to "pre-test" or qualify future problems. (It does use this procedure on the FE exam, however.) All of the problems you work will contribute toward your final score.

ARE THE EXAMPLE PROBLEMS IN THIS BOOK REPRESENTATIVE OF THE EXAM?

The example problems in this book are intended to be instructional and informative. They were written to illustrate how their respective concepts can be implemented. Example problems are not intended to represent exam problems or provide guidance on what you should study.

ARE THE PRACTICE PROBLEMS REPRESENTATIVE OF THE EXAM?

The practice problems in the companion book, *Power Practice Problems for the Computer and Electrical PE Exam*, cover the most likely exam subjects. Some of the practice problems are multiple-choice, and some require free-format solutions. However, they are generally more comprehensive and complex than actual exam problems, regardless of their formats. Some of the practice problems are marked "*Time limit: one hour.*" Compared to the six-minute problems on the PE exam, such one-hour problems are considerably more time-consuming.

All of the practice problems are original. Since NCEES does not release old exams, none of the practice problems are actual exam problems.

WHAT REFERENCE MATERIAL IS PERMITTED IN THE EXAM?

The PE exam is an open-book exam. Most states do not have any limits on the numbers and types of books you can use. Personal notes in a three-ring binder and other semipermanent covers can usually be used.

Some states use a "shake test" to eliminate loose papers from binders. Make sure that nothing escapes from your binders when they are inverted and shaken.

The references you bring into the exam room in the morning do not have to be the same as the references you use in the afternoon. However, you cannot share books with other examinees during the exam.

A few states do not permit collections of solved problems such as *Schaum's Outlines* series, sample exams, and solutions manuals. A few states maintain a formal list of banned books.

Strictly speaking, loose paper and scratch pads are not permitted in the examination room. Certain types of preprinted graphs and logarithmically scaled graph papers (which are almost never needed) should be three-hole punched and brought in a three-ring binder. An exception to this restriction may be made for laminated and oversize charts, graphs, and tables that are commonly needed for particular types of problems.

It is a good idea to check with your state board to learn what is and is not permitted in the examination room. PPI provides information about each state board on the PPI website (**ppi2pass.com/faqs/state-boards**).

HOW MANY BOOKS SHOULD YOU BRING?

Except for codes and standards, you should not need many books in the examination room. The trouble is, you cannot know in advance which ones you will need. That is the reason why many examinees show up with boxes and boxes of books. Since this book is not a substitute for your own experience and knowledge, without a doubt, there are things that you will need that are not in this book. But there are not so many that you need to bring your company's entire library. The exam is very fast-paced. You will not have time to use books with which you are not thoroughly familiar. The exam does not require you to know obscure solution methods or to use difficult-to-find data. You will not need articles printed in an industry magazine; you will not need doctoral theses or industry proceedings; and you will not need to know about recent industry events.

So, it really is unnecessary to bring a large quantity of books with you. Essential books are identified in the Codes and References section of this book, and you should be able to decide which support you need for the areas in which you intend to work. This book and five to 10 other references of your choice should be sufficient for most of the problems you answer.

MAY TABS BE PLACED ON PAGES?

It is common to tab pages in your books in an effort to reduce the time required to locate useful sections. Inasmuch as some states consider Post-it® notes to be "loose paper," your tabs should be of the more permanent variety. Although you can purchase tabs with gummed attachment points, it is also possible simply to use transparent tape to secure the Post-its you have already placed in your books.

CAN YOU WRITE AND MARK IN YOUR BOOKS?

During your preparation, you may write anything you want, anywhere in your books, including this one. You can use pencil, pen, or highlighter in order to further your understanding of the content. However, during the exam, you must avoid the appearance of taking notes about the exam. This means that you should write only on the scratch paper that is provided. During the exam, other than drawing a line across a wide table of numbers, or using your pencil to follow a line on a graph, you should not write in your books.

WHAT ABOUT CALCULATORS?

The exam requires the use of a scientific calculator. However, it may not be obvious that you should bring a spare calculator with you to the exam. It is always unfortunate when an examinee is not able to finish because his or her calculator was dropped or stolen or stopped working for some unknown reason.

To protect the integrity of its exams, NCEES has banned communicating and text-editing calculators from the exam site. NCEES provides a list of calculator models acceptable for use during the exam. Calculators not included in the list are not permitted. Check the current list of permissible devices at the PPI website (**ppi2pass.com/calculators**). Contact your state board to determine if nomographs and specialty slide rules are permitted.

The exam has not been optimized for any particular brand or type of calculator. In fact, for most calculations, a $15 scientific calculator will produce results as satisfactory as those from a $200 calculator. There are definite benefits to having built-in statistical functions, graphing, unit-conversion, and equation-solving capabilities. However, these benefits are not so great as to give anyone an unfair advantage.

It is essential that a calculator used for the PE exam have the following functions.

- trigonometric and inverse trigonometric functions
- hyperbolic and inverse hyperbolic functions
- π
- $\sqrt{x}$ and x^2
- both common and natural logarithms
- y^x and e^x

For maximum speed and utility, your calculator should also have or be programmed for the following functions.

- converting between polar (phasor) and rectangular vectors
- performing simultaneous linear equations
- performing rms calculations
- simplifying Smith chart calculations
- calculating factors for economic analysis problems
- calculating determinants of matrices
- linear regression
- calculating factors for economic analysis problems

You may not share calculators with other examinees. Be sure to take your calculator with you whenever you leave the exam room for any length of time.

Laptop and tablet computers, and electronic readers are not permitted in the examination room. Their use has been considered, but no states actually permit them. However, considering the nature of the exam problems, it is very unlikely that these devices would provide any advantage.

ARE MOBILE DEVICES PERMITTED?

You may not possess or use a cell phone, laptop or tablet computer, electronic reader, or other communications or text-messaging device during the exam, regardless of whether it is on. You will not be frisked upon entrance to the exam, but should a proctor discover that you are in possession of a communications device, you should expect to be politely excluded from the remainder of the exam.

HOW YOU SHOULD GUESS

There is no deduction for incorrect answers, so guessing is encouraged. However, since NCEES produces defensible licensing exams, there is no pattern to the placement of correct responses. Since the quantitative responses are sequenced according to increasing values, the placement of a correct answer among other numerical distractors is a function of the distractors, not of some statistical normalizing routine. Therefore, it is irrelevant whether you choose all "A," "B," "C," or "D" when you get into guessing mode during the last minute or two of the exam period.

The proper way to guess is as an engineer. You should use your knowledge of the subject to eliminate illogical answer options. Illogical answer options are those that violate good engineering principles, that are outside normal operating ranges, or that require extraordinary assumptions. Of course, this requires you to have some basic understanding of the subject in the first place. Otherwise, you go back to random guessing. That is the reason that the minimum passing score is higher than 25%.

You will not get any points using the "test-taking skills" that helped you in college. You will not be able to eliminate any [verb] answer options from "Which [noun] ..." problems. You will not find problems with options of the "more than 50" and "less than 50" variety. You will not find one answer option among the four that has a different number of significant digits, or has a verb in a different tense, or has some singular/plural discrepancy with the stem. The distractors will always match the stem, and they will be logical.

HOW IS THE EXAM GRADED AND SCORED?

The maximum number of points you can earn on the Power PE exam is 80. The minimum number of points for passing (referred to by NCEES as the *cut score*) varies from exam to exam. The cut score is determined through a rational procedure, without the benefit of knowing examinees' performance on the exam. That is, the exam is not graded on a curve. The cut score is selected based on what you are expected to know, not based on passing a certain percentage of engineers.

Each of the problems is worth one point. Grading is straightforward, since a computer grades your score sheet. Either you get the problem right or you do not. If you mark two or more answers for the same problem, no credit is given for the problem.

You will receive the results of your exam from your state board (not NCEES) online through your MyNCEES account or by mail, depending on your state. Eight to ten weeks will pass before NCEES releases the results to the state boards. However, the state boards take varying amounts of additional time before notifying examinees. You should allow three to four months for notification.

Your score may or may not be revealed to you, depending on your state's procedure. Even if the score is reported to you, it may have been scaled or normalized to 100%. It may be difficult to determine whether the reported score is out of 80 points or is out of 100%.

HOW IS THE CUT SCORE ESTABLISHED?

The raw cut score may be established by NCEES before or after the exam is administered. Final adjustments may be made following the exam date.

NCEES uses a process known as the *Modified Angoff* procedure to establish the cut score. This procedure starts with a small group (the cut score panel) of professional engineers and educators selected by NCEES. Each individual in the group reviews each problem and makes an estimate of its difficulty. Specifically, each individual estimates the number of minimally qualified engineers out of a hundred examinees who should know the correct answer to the problem. (This is equivalent to predicting the percentage of minimally qualified engineers who will answer correctly.)

Next, the panel assembles, and the estimates for each problem are openly compared and discussed. Eventually, a consensus value is obtained for each. When the panel has established a consensus value for every problem, the values are summed and divided by 100 to establish the cut score.

Various minor adjustments can be made to account for examinee population (as characterized by the average performance on any equater problems) and any flawed problems. Rarely, security breaches result in compromised problems or exams. How equater problems, exam flaws, and security issues affect examinee performance is not released by NCEES to the public.

CHEATING AND EXAM SUBVERSION

There are not very many ways to cheat on an open-book test. The proctors are well trained in spotting the few ways that do exist. It goes without saying that you should not talk to other examinees in the room, nor should you pass notes back and forth. You should not write anything into your books or take notes on the content of the exam. You should not use your cell phone. The number of people who are released to use the restroom may be limited to prevent discussions.

NCEES regularly reuses problems that have appeared on previous exams. Therefore, exam integrity is a serious issue with NCEES, which goes to great lengths to make sure nobody copies the problems. You may not keep your exam booklet or scratch paper, enter the text of problems into your calculator, or copy problems into your own material.

NCEES has become increasingly unforgiving about loss of its intellectual property. NCEES routinely prosecutes violators and seeks financial redress for loss of its exam problems, as well as invalidating any engineering license you may have earned by taking one of its exams while engaging in prohibited activities. Your state board may impose additional restrictions on your right to retake any exam if you are convicted of such activities. In addition to tracking down the sources of any exam problem compilations that it becomes aware of, NCEES is also aggressive in pursuing and prosecuting examinees who disclose the contents of the exam in online forum and "chat" environments. Your constitutional rights to free speech and expression will not protect you from civil prosecution for violating the nondisclosure agreement that NCEES requires you to sign before taking the exam. If you wish to participate in a dialogue about a particular exam subject, you must do so in such a manner that does not violate the essence of your nondisclosure agreement. This requires decoupling your discussion from the exam and reframing the problem to avoid any exam particulars.

The proctors are concerned about exam subversion, which generally means activity that might invalidate the exam or the examination process. The most common form of exam subversion involves trying to copy exam problems for future use. However, in their zeal to enforce and protect, proctors have shown unforgiving intolerance of otherwise minor infractions such as using your own pencil, using a calculator not on the approved list, possessing a cell phone, or continuing to write for even an instant after "pencils down" is called. For such infractions, you should expect to have the results of your exam invalidated, and all of your pleas and arguments in favor of forgiveness to be ignored. Even worse, since you will summarily be considered to have cheated, your state board will most likely prohibit you from retaking the exam for a number of exam cycles. There is no mercy built into the NCEES and state board procedures.

PART 3: HOW TO PREPARE FOR AND PASS THE POWER PE EXAM

WHAT SHOULD YOU STUDY?

The exam covers many diverse subjects. Strictly speaking, you do not have to study every subject on the exam in order to pass. However, the more subjects you study, the more you will improve your chances of passing. You should use Table 3 to help you plan which subjects you are going to study, and when. You should decide early in the preparation process which subjects you are going to study. The strategy you select will depend on your background. The following are the four most common strategies.

A broad approach is the key to success for examinees who have recently completed their academic studies. This strategy is to review the fundamentals in a broad range of undergraduate subjects (which means studying all or most of the chapters in this book). The exam includes enough fundamentals problems to make this strategy worthwhile. Overall, it is the best approach.

Engineers who have little time for preparation tend to concentrate on the subject areas in which they hope to find the most problems. By studying the list of exam subjects, some have been able to focus on those subjects that will give them the highest probability of finding enough problems that they can answer. This strategy works as long as the exam cooperates and has enough of the types of problems they need. Too often, though, examinees who pick and choose subjects to review cannot find enough problems to complete the exam.

Engineers who have been away from classroom work for a long time tend to concentrate on the subjects in which they have had extensive experience, in the hope that the exam will feature lots of problems in those subjects. This method is seldom successful.

Some engineers plan on modeling their solutions from similar problems they have found in textbooks, collections of solutions, and old exams. These engineers often spend a lot of time compiling and indexing the example and sample problem types in all of their books. This is not a legitimate preparation method, and it is almost never successful.

DO YOU NEED A CLASSROOM REVIEW COURSE?

Approximately 60% of first-time PE examinees take an instructor-led review course of some form. Live classroom and internet courses of various types, as well as previously recorded lessons, are available for some or all of the exam topics. Live courses and instructor-moderated internet courses provide several significant advantages over self-directed study, some of which may apply to you.

- A course structures and paces your review. It ensures that you keep going forward without getting bogged down in one subject.
- A course focuses you on a limited amount of material. Without a course, you might not know which subjects to study.
- A course provides you with the problems you need to solve. You will not have to spend time looking for them.
- A course spoon-feeds you the material. You may not need to read the book!
- The course instructor can answer your questions when you are stuck.

You probably already know if any of these advantages apply to you. A review course will be less valuable if you are thorough, self-motivated, and highly disciplined.

HOW LONG SHOULD YOU STUDY?

We have all heard stories of the person who did not crack a book until the week before the exam and still passed it with flying colors. Yes, these people really exist. However, I am not one of them, and you probably are not either. In fact, I am convinced that these people are as rare as the ones who have taken the exam five times and still have not passed it.

A thorough review takes approximately 300 hours. Most of this time is spent solving problems. Some of it may be spent in class; some is spent at home. Some examinees spread this time over a year. Others try to cram it all into two months. Most classroom review courses last for three or four months. The best time to start studying will depend on how much time you can spend per week.

WHAT THE WELL-HEELED ELECTRICAL ENGINEER SHOULD BEGIN ACCUMULATING

There are many references and resources that you should begin to assemble for review and for use in the exam.

It is unlikely that you could pass the PE exam without accumulating other books and resources. There certainly is not much margin for error if you show up with only one book. There are many problems that require knowledge, data, and experience that are presented and described only in codes, standards, and references dedicated to a single subject. You would have to be truly lucky to go in "bare," find the right mix of problems, and pass.

Few examinees are able to accumulate all of the references needed to support the exam's entire body of knowledge. The accumulation process is too expensive and time-consuming, and the sources are too diverse. Like purchasing an insurance policy, what you end up with will be more a function of your budget than of your needs. In some cases, one book will satisfy several needs.

The books and other items listed in Codes and References are regularly cited by examinees as being particularly useful to them. This listing only includes the major "named" books that have become standard references in the industry. These books are in addition to any textbooks or resources that you might choose to bring.

ADDITIONAL REVIEW MATERIAL

In addition to this book and its companion, *Power Practice Problems for the Electrical and Computer Engineering PE Exam*, PPI can provide you with many targeted references and study aids, some of which are listed here. All of these resources have stood the test of time, which means that examinees continually report their usefulness and that PPI keeps them up to date.

- *Power Practice Exams for the Electrical and Computer PE Exam.*
- *Engineering Unit Conversions*
- Electrical PE Exam Cafe

For a complete listing of available sample problems, other books, calculator software, videos, and other exam-related products, please go to PPI's website at **ppi2pass.com/electrical**.

WHAT YOU WILL NOT NEED

Generally, people bring too many things to the exam. One general rule is that you should not bring books that you have not looked at during your review. If you did not need a book while doing the example problems in this book, you will not need it during the exam.

There are some other things that you will not need.

- Books on basic and introductory subjects: You will not need books that cover trigonometry, geometry, or calculus.
- Books on non-exam subjects: Such subjects as materials science, statics, dynamics, mechanics of materials, drafting, history, the English language, geography, and philosophy are not part of the exam.
- Books on mathematical analysis, numerical analysis, or extensive mathematics tabulations.
- Obscure books and materials: Books that are in foreign languages, doctoral theses, and papers presented at technical societies will not be needed during the exam.
- Old textbooks or obsolete, rare, and ancient books: NCEES exam committees are aware of which textbooks are in use. Material that is available only in out-of-print publications and old editions will not be used.
- Handbooks in other disciplines: You probably will not need a civil, mechanical, or industrial engineering handbook.
- Crafts- and trades-oriented books: The exam does not expect you to have detailed knowledge of trades or manufacturing operations.
- Manufacturer's literature and catalogs: No part of the exam requires you to be familiar with products that are proprietary to any manufacturer.
- U.S. government publications: With the exceptions of the publications mentioned and referenced in this book, no government publications are required for the PE exam.
- Your state's laws: The PE exam is a national exam. Nothing unique to your state will appear on it. (However, federal legislation affecting engineers, particularly in environmental areas, is fair game.)

SHOULD YOU LOOK FOR OLD EXAMS?

The traditional approach to preparing for standardized tests includes working sample tests. However, NCEES does not release old tests or problems after they are used. Therefore, there are no official problems or tests available from legitimate sources. NCEES publishes booklets of sample problems and solutions to illustrate the format of the exam. However, these problems have been compiled from various previous exams, and the resulting publication is not a true "old exam." Furthermore, NCEES sometimes constructs its sample problems books from problems that have been pulled from active use for various reasons, including poor performance. Such marginal problems, while accurately reflecting the format of the exam, are not always representative of actual exam subjects.

WHAT SHOULD YOU MEMORIZE?

You get lucky here, because it is not necessary to memorize anything. The exam is open-book, so you can look up any procedure, formula, or piece of information that you need. You can speed up your problem-solving response time significantly if you do not have to look up the conversion from W to ft-lbf/sec, or the definition of the sine of an angle, but you do not even have to memorize these kinds of things. As you work practice problems in the companion book, *Power Practice Problems for the Electrical and Computer PE Exam*, you will automatically memorize the things that you come across more than a few times.

DO YOU NEED A REVIEW SCHEDULE?

It is important that you develop and adhere to a review outline and schedule. Once you have decided which subjects you are going to study, you can allocate the available time to those subjects in a manner that makes sense to you. If you are not taking a classroom review course (where the order of preparation is determined by the lectures), you should make an outline of subjects for self-study to use for scheduling your preparation. A fill-in-the-dates schedule is provided in Table 3 at the end of this introduction.

A SIMPLE PLANNING SUGGESTION

Designate some location (a drawer, a corner, a cardboard box, or even a paper shopping bag left on the floor) as your "exam catchall." Use your catchall during the months before the exam when you have revelations about things you should bring with you. For example, you might realize that the plastic ruler marked off in tenths of an inch that is normally kept in the kitchen junk drawer can help you. Or, you might decide that a certain book is particularly valuable, that it would be nice to have dental floss after lunch, or that large rubber bands and clips are useful for holding books open.

It is not actually necessary to put these treasured items in the catchall during your preparation. You can, of course, if it is convenient. But if these items will have other functions during the time before the exam, at least write yourself a note and put the note into the catchall. When you go to pack your exam kit a few days before the exam, you can transfer some items immediately, and the notes will be your reminders for the other items that are back in the kitchen drawer.

HOW YOU CAN MAKE YOUR REVIEW REALISTIC

In the exam, you must be able to recall solution procedures, formulas, and important data quickly. You must remain sharp for eight hours or more. When you played a sport back in school, your coach tried to put you in game-related situations. Preparing for the PE exam is not much different from preparing for a big game. Some part of your preparation should be realistic and representative of the exam environment.

There are several things you can do to make your review more representative. For example, if you gather most of your review resources (i.e., books) in advance and try to use them exclusively during your review, you will become more familiar with them. (Of course, you can also add to or change your references if you find inadequacies.)

Learning to use your time wisely is one of the most important lessons you can learn during your review. You will undoubtedly encounter problems that end up taking much longer than you expected. In some instances, you will cause your own delays by spending too much time looking through books for things you need (or just by looking for the books themselves!). Other times, the problems will entail too much work. Learn to recognize these situations so that you can make an intelligent decision about skipping such problems in the exam.

WHAT TO DO A FEW DAYS BEFORE THE EXAM

There are a few things you should do a week or so before the exam. You should arrange for childcare and transportation. Since the exam does not always start or end at the designated time, make sure that your childcare and transportation arrangements are flexible.

Check PPI's website for last-minute updates and errata to any PPI books you might have and are bringing to the exam.

If you have not already done so, read the "Advice from Examinees" section of PPI's website.

If it is convenient, visit the exam location in order to find the building, parking areas, exam room, and restrooms. If it is not convenient, you may find driving directions and/or site maps online.

Take the battery cover off your calculator and check to make sure you are bringing the correct size replacement batteries. Some calculators require a different kind of battery for their "permanent" memories. Put the cover back on and secure it with a piece of masking tape. Write your name on the tape to identify your calculator.

If your spare calculator is not the same as your primary calculator, spend a few minutes familiarizing yourself with how it works. In particular, you should verify that your spare calculator is functional.

PREPARE YOUR CAR

[ ] Gather snow chains, a shovel, and tarp to kneel on while installing chains.
[ ] Check tire pressure.
[ ] Check your spare tire.
[ ] Check for tire installation tools.
[ ] Verify that you have the vehicle manual.
[ ] Check fluid levels (oil, gas, water, brake fluid, transmission fluid, window-washing solution).
[ ] Fill up with gas.

[ ] Check battery and charge if necessary.
[ ] Know something about your fuse system (where fuses are, how to replace them, etc.).
[ ] Assemble all required maps.
[ ] Fix anything that might slow you down (missing wiper blades, etc.).
[ ] Check your taillights.
[ ] Affix the current DMV vehicle registration tag.
[ ] Fix anything that might get you pulled over on the way to the exam (burned-out taillight or headlight, broken lenses, bald tires, missing license plate, noisy muffler).
[ ] Treat the inside windows with anti-fog solution.
[ ] Put a roll of paper towels in the back seat.
[ ] Gather exact change for any bridge tolls or toll roads.
[ ] Find and install your electronic toll tag, if applicable.
[ ] Put $20 in your glove box.
[ ] Check for current vehicle registration and proof of insurance.
[ ] Locate a spare door and ignition key.
[ ] Find your AAA or other roadside-assistance cards and phone numbers.
[ ] Plan alternate routes.

PREPARE YOUR EXAM KITS

Second in importance to your scholastic preparation is the preparation of your two exam kits. The first kit consists of a bag, box (plastic milk crates hold up better than cardboard in the rain), or wheeled travel suitcase containing items to be brought with you into the exam room.

[ ] your exam authorization notice
[ ] current, signed government-issued photographic identification (e.g., driver's license, not a student ID card)
[ ] this book
[ ] other textbooks and reference books
[ ] regular dictionary
[ ] review course notes in a three-ring binder
[ ] cardboard boxes or plastic milk crates to use as bookcases
[ ] primary calculator
[ ] spare calculator
[ ] instruction booklets for your calculators
[ ] extra calculator batteries
[ ] two straightedges (e.g., ruler, scale, triangle, protractor)
[ ] protractor
[ ] scissors
[ ] stapler
[ ] transparent tape
[ ] magnifying glass
[ ] small (jeweler's) screwdriver for fixing your glasses or for removing batteries from your calculator
[ ] unobtrusive (quiet) snacks or candies, already unwrapped
[ ] two small plastic bottles of water
[ ] travel pack of tissue (keep in your pocket)
[ ] handkerchief
[ ] headache remedy
[ ] personal medication
[ ] $5.00 in assorted coinage
[ ] spare contact lenses and multipurpose contact lens cleaning solution
[ ] backup reading glasses (no case)
[ ] eye drops
[ ] light, comfortable sweater
[ ] loose shoes or slippers
[ ] cushion for your chair
[ ] earplugs
[ ] wristwatch with alarm
[ ] several large trash bags ("raincoats" for your boxes of books)
[ ] roll of paper towels
[ ] wire coat hanger (to hang up your jacket or to get back into your car in an emergency)
[ ] extra set of car keys on a string around your neck

The second kit consists of the following items and should be left in a separate bag or box in your car in case it is needed.

[ ] copy of your exam authorization notice
[ ] light lunch
[ ] beverage in thermos or can
[ ] sunglasses
[ ] extra pair of prescription glasses
[ ] raincoat, boots, gloves, hat, and umbrella
[ ] street map of the exam area
[ ] parking permit
[ ] battery-powered desk lamp
[ ] your cell phone
[ ] length of rope

The following items cannot be used during the exam and should be left at home.

[ ] personal pencils and erasers (NCEES distributes mechanical pencils at the exam.)
[ ] fountain pens
[ ] radio, CD player, MP3 player, or other media player
[ ] battery charger
[ ] extension cords
[ ] scratch paper
[ ] notepads
[ ] drafting compass
[ ] circular ("wheel") slide rules

PREPARE FOR THE WORST

All of the occurrences listed in this section have happened to examinees. Granted, you cannot prepare for every eventuality. But, even though each of these occurrences taken individually is a low-probability event, taken together, they are worth considering in advance.

- Imagine getting a flat tire, getting stuck in traffic, or running out of gas on the way to the exam.
- Imagine rain and snow as you are carrying your cardboard boxes of books into the exam room.
- Imagine arriving late. Can you get into the exam without having to make two trips from your car?
- Imagine having to park two blocks from the exam site. How are you going to get everything to the exam room? Can you actually carry everything that far? Could you use a furniture dolly, a supermarket basket, or perhaps a helpmate?
- Imagine a Star Trek convention, square-dancing contest, construction, or an auction in the next room.
- Imagine a site without any heat, with poor lighting, or with sunlight streaming directly into your eyes.
- Imagine a hard folding chair and a table with one short leg.
- Imagine a site next to an airport with frequent takeoffs, or next to a construction site with a pile driver, or next to the NHRA State Championship.
- Imagine a seat where someone nearby chews gum with an open mouth; taps his pencil or drums her fingers; or wheezes, coughs, and sneezes for eight hours.
- Imagine the distraction of someone crying or of proctors evicting yelling and screaming examinees who have been found cheating.
- Imagine the tragedy of another examinee's serious medical emergency.
- Imagine a delay of an hour while you wait for someone to unlock the building, turn on the heat, or wait for the head proctor to bring instructions.
- Imagine a power outage occurring sometime during the exam.
- Imagine a proctor who (a) tells you that one of your favorite books cannot be used in the exam, (b) accuses you of cheating, or (c) calls "time up" without giving you any warning.
- Imagine not being able to get your lunch out of your car or find a restaurant.
- Imagine getting sick or nervous in the exam.
- Imagine someone stealing your calculator during lunch.

WHAT TO DO THE DAY BEFORE THE EXAM

Take the day before the exam off from work to relax. Do not cram. A good night's sleep is the best way to start the exam. If you live a considerable distance from the exam site, consider getting a hotel room in which to spend the night.

Practice setting up your exam work environment. Carry your boxes to the kitchen table. Arrange your "bookcases" and supplies. Decide what stays on the floor in boxes and what gets an "honored position" on the tabletop.

Use your checklist to make sure you have everything. Make sure your exam kits are packed and ready to go. Wrap your boxes in plastic bags in case it is raining when you carry them from the car to the examination room.

Calculate your wake-up time and set the alarms on two bedroom clocks. Select and lay out your clothing items. (Dress in layers.) Select and lay out your breakfast items.

If it is going to be hot on exam day, put your (plastic) bottles of water in the freezer.

Make sure you have gas in your car and money in your wallet.

WHAT TO DO THE DAY OF THE EXAM

Turn off the quarterly and hourly alerts on your wristwatch. Leave your cell phone in the car. If you must bring it, set it on silent mode. Bring a morning newspaper.

You should arrive at least 30 minutes before the exam starts. This will allow time for finding a convenient parking place, bringing your materials to the exam room, making room and seating changes, and calming down. Be prepared, though, to find that the exam room is not open or ready at the designated time.

Once you have arranged the materials around you on your table, take out your morning newspaper and look cool. (Only nervous people work crossword puzzles.)

WHAT TO DO DURING THE EXAM

All of the procedures typically associated with timed, proctored, machine-graded assessment tests will be in effect when you take the PE exam.

The proctors will distribute the exam booklets and answer sheets if they are not already on your tables. However, you should not open the booklets until instructed to do so. You may read the information on the front and back covers, and you should write your name in any appropriate blank spaces.

Listen carefully to everything the proctors say. Do not ask your proctors any engineering problems. Even if they are knowledgeable in engineering, they will not be permitted to answer your problems.

Answers to problems are recorded on an answer sheet contained in the test booklet. The proctors will guide you through the process of putting your name and other biographical information on this sheet when the time comes, which will take approximately 15 minutes. You will be given the full four hours to answer problems. Time to initialize the answer sheet is not part of your four hours.

The common suggestions to "completely fill the bubbles and erase completely" apply here. NCEES provides each examinee with a mechanical pencil with HB lead. The use of ballpoint pens and felt-tip markers is prohibited.

If you finish the exam early and there are still more than 30 minutes remaining, you will be permitted to leave the room. If you finish less than 30 minutes before the end of the exam, you may be required to remain until the end. This is done to be considerate of the people who are still working.

Be prepared to stop working immediately when the proctors call "pencils down" or "time is up." Continuing to work for even a few seconds will completely invalidate your exam.

When you leave, you must return your examination booklet. You may not keep the examination booklet for later review.

If there are any problems that you think were flawed, in error, or unsolvable, ask a proctor for a "reporting form" on which you can submit your comments. Follow your proctor's advice in preparing this document.

WHAT ABOUT EATING AND DRINKING IN THE EXAMINATION ROOM?

The official rule is probably the same in every state: no eating or drinking during the exam. That makes sense, for a number of reasons. Some examination sites do not want (or do not permit) stains and messes. Others do not want crumbs to attract ants and rodents. Your table partners do not want spills or smells. Nobody wants the distractions. Your proctors cannot give you a new exam booklet when the first one is ruined with coffee.

How this rule is administered varies from site to site and from proctor to proctor. Some proctors enforce the letter of law, threatening to evict you from the examination room when they see you chewing gum. Others may permit you to have bottled water, as long as you store the bottles on the floor where any spills will not harm what is on the table. No one is going to let you crack peanuts while you work on the exam, but I cannot see anyone complaining about a hard candy melting away in your mouth. You will just have to find out when you get there.

HOW TO SOLVE MULTIPLE-CHOICE PROBLEMS

When you begin each session of the exam, observe the following suggestions.

- Use only the pencil provided.
- Do not spend an inordinate amount of time on any single problem. If you have not answered a problem in a reasonable amount of time, make a note of it and move on.
- Set your wristwatch alarm for five minutes before the end of each four-hour session, and use that remaining time to guess at all of the remaining problems. Odds are that you will be successful with about 25% of your guesses, and these points will more than make up for the few points that you might earn by working during the last five minutes.
- Make mental notes about any problem for which you cannot find a correct response, that appears to have two correct responses, or that you believe has some technical flaw. Errors in the exam are rare, but they do occur. Such errors are usually discovered during the scoring process and discounted from the exam, so it is not necessary to tell your proctor, but be sure to mark the one best answer before moving on.
- Make sure all of your responses on the answer sheet are dark, and completely fill the bubbles.

SOLVE PROBLEMS CAREFULLY

Many points are lost to carelessness. Keep the following items in mind when you are solving practice problems. Hopefully, these suggestions will be automatic in the exam.

[ ] Did you recheck your mathematical equations?
[ ] Do the units cancel out in your calculations?
[ ] Did you convert between radius and diameter?
[ ] Did you recheck all data obtained from other sources, tables, and figures?
[ ] Did you convert to the requested units?

SHOULD YOU TALK TO OTHER EXAMINEES AFTER THE EXAM?

The jury is out on this problem. People react quite differently to the examination experience. Some people are energized. Most are exhausted. Some people need to unwind by talking with other examinees, describing every detail of their experience, and dissecting every exam problem. Other people need lots of quiet space, and prefer just to get into a hot tub to soak and sulk. Most engineers, apparently, are in this latter category.

Since everyone who took the exam has seen it, you will not be violating your "oath of silence" if you talk about the details with other examinees immediately after the exam. It is difficult not to ask how someone else approached a problem that had you completely stumped. However, keep in mind that it is very disquieting to think you answered a problem correctly, only to have someone tell you where you went wrong.

To ensure you do not violate the nondisclosure agreement you signed before taking the exam, make sure you do not discuss any exam particulars with people who have not also taken the exam.

AFTER THE EXAM

Yes, there is something to do after the exam. Most people return home, throw their exam "kits" into the corner, and collapse. A week later, when they can bear to think about the experience again, they start integrating their exam kits back into their normal lives. The calculators go back into the desk, the books go back on the shelves, the $5.00 in change goes back into the piggy bank, and all of the miscellaneous stuff you brought with you to the exam is put back wherever it came.

Here is what I suggest you do as soon as you get home.

- [ ] Thank your spouse and children for helping you during your preparation.
- [ ] Take any paperwork you received on exam day out of your pocket, purse, or wallet. Put this inside your *Power Reference Manual.*
- [ ] Reflect on any statements regarding exam secrecy to which you signed your agreement in the exam.
- [ ] Call your employer and tell him/her that you need to take a mental health day off on Monday.

A few days later, when you can face the world again, do the following.

- [ ] Make notes about anything you would do differently if you had to take the exam over again.
- [ ] Consolidate all of your application paperwork, correspondence to/from your state, and any paperwork that you received on exam day.
- [ ] If you took a live review course, call or email the instructor (or write a note) to say, "Thanks."
- [ ] Return any books you borrowed.
- [ ] Write thank-you notes to all of the people who wrote letters of recommendation or reference for you.
- [ ] Find and read the chapter in this book that covers ethics. There were no ethics problems on your PE exam, but it does not make any difference. Ethical behavior is expected of a PE in any case. Spend a few minutes reflecting on how your performance (obligations, attitude, presentation, behavior, appearance, etc.) might be about to change once you are licensed. Consider how you are going to be a role model for others around you.
- [ ] Put all of your review books, binders, and notes someplace where they will be out of sight.

FINALLY

By the time you have "undone" all of your preparations, you might have thought of a few things that could help future examinees. If you have any sage comments about how to prepare, any suggestions about what to do in or bring to the exam, any comments on how to improve this book, or any funny anecdotes about your experience, I hope you will share these with me. By this time, you will be the "expert," and I will be your biggest fan.

AND THEN, THERE'S THE WAIT . . .

Waiting for the exam results is its own form of mental torture.

Yes, I know the exam is 100% multiple-choice, and grading should be almost instantaneous. But, you are going to wait, nevertheless. There are many reasons for the delay.

Although the actual machine grading "only takes seconds," consider the following facts: (a) NCEES prepares multiple exams for each administration, in case one becomes unusable (i.e., is inappropriately released) before the exam date. (b) Since the actual version of the exam used is not known until after it is finally given, the cut score determination occurs after the exam date.

I would not be surprised to hear that NCEES receives dozens, if not hundreds, of claims from well-meaning examinees who were 100% certain that the exams they took were seriously flawed to some degree—that there was not a correct answer for such-and-such problem, that there were two answers for such-and-such problem, or even, perhaps, that such-and-such problem was missing from their examination booklet altogether. Each of these claims must be considered as a potential adjustment to the cut score.

Then, the exams must actually be graded. Since grading nearly 50,000 exams (counting all the FE and PE exams) requires specialized equipment, software, and training not normally possessed by the average employee, as well as time to do the work (also not normally possessed by the average employee), grading is invariably outsourced.

Outsourced grading cannot begin until all of the states have returned their score sheets to NCEES and NCEES has sorted, separated, organized, and consolidated the score sheets into whatever "secret sauce sequence" is best. During grading, some of the score sheets "pop out" with any number of abnormalities that demand manual scoring.

After the individual exams are scored, the results are analyzed in a variety of ways. Some of the analysis looks at passing rates by such delineators as degree, major, university, site, and state. Part of the analysis looks for similarities between physically adjacent examinees (to look for cheating). Part of the analysis looks for examination sites that have statistically abnormal group performance. And, some of the analysis looks for exam problems that have a disproportionate fraction of successful or unsuccessful examinees. Anyway, you get the idea: Grading is not merely a matter of putting your examination booklet and answer sheet in an electronic reader. All of these steps have to be completed for 100% of the examinees before any results can be delivered.

Once NCEES has graded your test and notified your state, when you hear about it depends on when the work is done by your state. Some states have to approve the results at a board meeting; others prepare the certificates before sending out notifications. Some states are more computerized than others. Some states have 50 examinees, while others have 10,000. Some states are shut down by blizzards and hurricanes; others are administratively challenged—understaffed, inadequately trained, or over budget.

There is no pattern to the public release of results. None. The exam results are not released to all states simultaneously. (The states with the fewest examinees often receive their results soonest.) They are not released by discipline. They are not released alphabetically by state or examinee name. The examinees who did not pass are not notified first (or last). Your coworker might receive his or her notification today, and you might be waiting another three weeks for yours.

Some states post the names of the successful examinees, or unsuccessful examinees, or both on their official state websites before the results go out. Others update their websites after the results go out. Some states do not list much of anything on their websites.

Remember, too, that the size or thickness of the envelope you receive from your state does not mean anything. Some states send a big congratulations package and certificate. Others send a big package with a new application to repeat the exam. Some states send a postcard. Some send a one-page letter. Some states send you an invoice for your license fees. (Ahh, what a welcome bill!) You just have to open it to find out.

AND WHEN YOU PASS ...

[ ] Celebrate.
[ ] Notify the people who wrote letters of recommendation or reference for you.
[ ] Read "FAQs about What Happens After You Pass the PE Exam" on PPI's website.
[ ] Ask your employer for a raise.
[ ] Tell the folks at PPI (who have been rootin' for you all along) the good news.

Table 3 *Schedule for Self-Study*

chapter number	subject	date to start	date to finish
1	Systems of Units		
2	Energy, Work, and Power		
3	Engineering Drawing Practice		
4	Algebra		
5	Linear Algebra		
6	Vectors		
7	Trigonometry		
8	Analytic Geometry		
9	Differential Calculus		
10	Integral Calculus		
11	Differential Equations		
12	Probability and Statistical Analysis of Data		
13	Computer Mathematics		
14	Numerical Analysis		
15	Advanced Engineering Mathematics		
16	Electromagnetic Theory		
17	Electronic Theory		
18	Communication Theory		
19	Acoustic and Transducer Theory		
20	Electrostatics		
21	Electrostatic Fields		
22	Magnetostatics		
23	Magnetostatic Fields		
24	Electrodynamics		
25	Maxwell's Equations		
26	DC Circuit Fundamentals		
27	AC Circuit Fundamentals		
28	Transformers		
29	Linear Circuit Analysis		
30	Transient Analysis		
31	Time Response		
32	Frequency Response		
33	Generation Systems		
34	Three-Phase Electricity and Power		
35	Batteries, Fuel Cells, and Power Supplies		
36	Power Distribution		
37	Power Transformers		
38	Transmission Lines		
39	The Smith Chart		
40	Power System Analysis		
41	Analysis of Control Systems		

chapter number	subject	date to start	date to finish
42	Protection and Safety		
43	Rotating DC Machinery		
44	Rotating AC Machinery		
45	Electronics Fundamentals		
46	Junction Transistors		
47	Field Effect Transistors		
48	Electrical and Electronic Devices		
49	Digital Logic		
50	Lightning Protection and Grounding		
51	Illumination		
52	Power System Management		
53	Measurement and Instrumentation		
54	Electrical Materials		
55	Biomedical Electrical Engineering		
56	National Electrical Code		
57	National Electrical Safety Code		
58	Engineering Economic Analysis		
59	Engineering Law		
60	Engineering Ethics		
61	Electrical Engineering Frontiers		
62	Engineering Licensing in the United States		

Topic I: Mathematics

Chapter

1. Systems of Units
2. Energy, Work, and Power
3. Engineering Drawing Practice
4. Algebra
5. Linear Algebra
6. Vectors
7. Trigonometry
8. Analytic Geometry
9. Differential Calculus
10. Integral Calculus
11. Differential Equations
12. Probability and Statistical Analysis of Data
13. Computer Mathematics
14. Numerical Analysis
15. Advanced Engineering Mathematics

1 Systems of Units

1. INTRODUCTION

The purpose of this chapter is to eliminate some of the confusion regarding the many units available for each engineering variable. In particular, an effort has been made to clarify the use of the so-called English systems, which for years have used the *pound* unit both for force and mass—a practice that has resulted in confusion even for those familiar with it.

2. COMMON UNITS OF MASS

The choice of a mass unit is the major factor in determining which system of units will be used in solving a problem. It is obvious that one will not easily end up with a force in pounds if the rest of the problem is stated in meters and kilograms. Actually, the choice of a mass unit determines more than whether a conversion factor will be necessary to convert from one system to another (e.g., between SI and English units). An inappropriate choice of a mass unit may actually require a conversion factor *within* the system of units.

The common units of mass are the gram, pound, kilogram, and slug.[1] There is nothing mysterious about these units. All represent different quantities of matter, as Fig. 1.1 illustrates. In particular, note that the pound and slug do not represent the same quantity of matter.[2]

Figure 1.1 Common Units of Mass

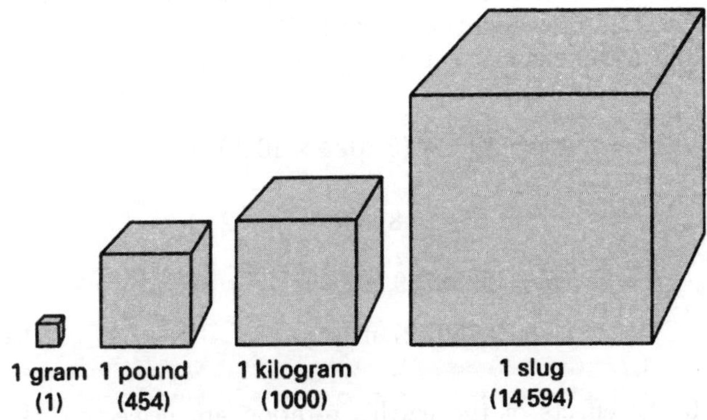

3. MASS AND WEIGHT

In SI, *kilograms* are used for mass and *newtons* for weight (force). The units are different, and there is no confusion between the variables. However, for years the term *pound* has been used for both mass and weight. This usage has obscured the distinction between the two: mass is a constant property of an object; weight varies with the gravitational field. Even the conventional use of the abbreviations *lbm* and *lbf* (to distinguish between pounds-mass and pounds-force) has not helped eliminate the confusion.

It is true that an object with a mass of one pound will have an earthly weight of one pound, but this is true only on the earth. The weight of the same object will be much less on the moon. Therefore, care must be taken when working with mass and force in the same problem.

The relationship that converts mass to weight is familiar to every engineering student.

$$W = mg \qquad 1.1$$

Equation 1.1 illustrates that an object's weight will depend on the local acceleration of gravity as well as the object's mass. The mass will be constant, but gravity will depend on location. Mass and weight are not the same.

[1]Normally, one does not distinguish between a unit and a multiple of that unit, as is done here with the gram and the kilogram. However, these two units actually are bases for different consistent systems.

[2]A slug is approximately equal to 32.1740 pounds-mass.

4. ACCELERATION OF GRAVITY

Gravitational acceleration on the earth's surface is usually taken as 32.2 ft/sec^2 or 9.81 m/s^2. These values are rounded from the more precise values of 32.1740 ft/sec^2 and 9.806 65 m/s^2. However, the need for greater accuracy must be evaluated on a problem-by-problem basis. Usually, three significant digits are adequate, since gravitational acceleration is not constant anyway but is affected by location (primarily latitude and altitude) and major geographical features.

The term *standard gravity*, g_0, is derived from the acceleration at essentially any point at sea level and approximately 45° N latitude. If additional accuracy is needed, the gravitational acceleration can be calculated from Eq. 1.2. This equation neglects the effects of large land and water masses. ϕ is the latitude in degrees.

$$g_{\text{surface}} = g'\Big(1 + (5.3024 \times 10^{-3})\sin^2\phi - (5.8 \times 10^{-6})\sin^2 2\phi\Big)$$

$$g' = 32.08769\ \text{ft/sec}^2 = 9.780\,327\ \text{m/s}^2 \qquad 1.2$$

If the effects of the earth's rotation are neglected, the gravitational acceleration at an altitude h above the earth's surface is given by Eq. 1.3. r_e is the earth's radius.

$$g_h = g_{\text{surface}}\left(\frac{r_e}{r_e + h}\right)^2$$

$$r_e = 3959\ \text{mi} = 6.378\,1 \times 10^6\ \text{m} \qquad 1.3$$

5. CONSISTENT SYSTEMS OF UNITS

A set of units used in a calculation is said to be *consistent* if no conversion factors are needed.[3] For example, a moment is calculated as the product of a force and a lever arm length.

$$M = Fr \qquad 1.4$$

A calculation using Eq. 1.4 would be consistent if M was in newton-meters, F was in newtons, and r was in meters. The calculation would be inconsistent if M was in ft-kips, F was in kips, and r was in inches (because a conversion factor of 1/12 would be required).

The concept of a consistent calculation can be extended to a system of units. A *consistent system of units* is one in which no conversion factors are needed for any calculation. For example, Newton's second law of motion can be written without conversion factors. Newton's second law simply states that the force required to accelerate an object is proportional to the acceleration of the object. The constant of proportionality is the object's mass.

$$F = ma \qquad 1.5$$

Equation 1.5 is $F = ma$, not $F = Wa/g$ or $F = ma/g_c$. Equation 1.5 is consistent: it requires no conversion factors. This means that in a consistent system where conversion factors are not used, once the units of m and a have been selected, the units of F are fixed. This has the effect of establishing units of work and energy, power, fluid properties, and so on.

The decision to work with a consistent set of units is desirable but not necessary. Problems in fluid flow and thermodynamics are routinely solved in the United States with inconsistent units. This causes no more of a problem than working with inches and feet when calculating a moment. It is necessary only to use the proper conversion factors.

6. THE ENGLISH ENGINEERING SYSTEM[4]

Through common and widespread use, pounds-mass (lbm) and pounds-force (lbf) have become the standard units for mass and force in the *English Engineering System*. The English Engineering System is used in subjects in this book where the units of that system remain in widespread use. Otherwise, SI (metric) units are used. Where instructive, both systems are used.

There are subjects in the United States in which the use of pounds for mass is firmly entrenched. For example, most thermodynamics, fluid flow, and heat transfer problems have traditionally been solved using the units of lbm/ft^3 for density, Btu/lbm for enthalpy, and Btu/lbm-°F for specific heat. Unfortunately, some equations contain both lbm-related and lbf-related variables, as does the steady flow conservation of energy equation, which combines enthalpy in Btu/lbm with pressure in lbf/ft^2.

The units of pounds-mass and pounds-force are as different as the units of gallons and feet, and they cannot be canceled. A mass conversion factor, g_c, is needed to make the equations containing lbf and lbm dimensionally consistent. This factor is known as the *gravitational constant* and has a value of 32.1740 lbm-ft/lbf-sec^2. The numerical value is the same as the standard acceleration

[3]The terms *homogeneous* and *coherent* are also used to describe a consistent set of units.

[4]The name "English Engineering System" is often used to refer to the system of units used in the United States, as well as the system of units formerly used in the United Kingdom. The system is properly known as the United States Customary System but is sometimes referred to as the Imperial System or British Imperial System. It differs from the English Engineering System in that its use of units preceded the standardization of the English System in 1824. The United Kingdom completed its shift to SI units, commonly known as the metric system, around 2009. The United States remains the only industrialized nation not to have officially shifted to the metric system, though the Omnibus Trade and Competitiveness Act of 1988 designates "the metric system of measurement as the preferred system of weights and measures for United States trade and commerce."

of gravity, but g_c is not the local gravitational acceleration, g.[5] g_c is a conversion constant, just as 12.0 is the conversion factor between feet and inches.

The English Engineering System is an inconsistent system as defined according to Newton's second law. $F = ma$ cannot be written if lbf, lbm, and ft/sec^2 are the units used. The g_c term must be included.

$$F \text{ in lbf} = \frac{(m \text{ in lbm})\left(a \text{ in } \frac{\text{ft}}{\text{sec}^2}\right)}{g_c \text{ in } \frac{\text{lbm-ft}}{\text{lbf-sec}^2}} \qquad 1.6$$

In Eq. 1.6, g_c does more than "fix the units." Since g_c has a numerical value of 32.1740, it actually changes the calculation numerically. A force of 1.0 pound will not accelerate a 1.0-pound mass at the rate of 1.0 ft/sec^2.

In the English Engineering System, work and energy are typically measured in ft-lbf (mechanical systems) or in British thermal units, Btu (thermal and fluid systems). One Btu is equal to 778.17 ft-lbf.

Example 1.1

Calculate the weight in lbf of a 1.00 lbm object in a gravitational field of 27.5 ft/sec^2.

Solution

From Eq. 1.6,

$$F = \frac{ma}{g_c} = \frac{(1.00 \text{ lbm})\left(27.5 \ \frac{\text{ft}}{\text{sec}^2}\right)}{32.2 \ \frac{\text{ft-lbm}}{\text{lbf-sec}^2}}$$

$$= 0.854 \text{ lbf}$$

7. OTHER FORMULAS AFFECTED BY INCONSISTENCY

It is not a significant burden to include g_c in a calculation, but it may be difficult to remember when g_c should be used. Knowing when to include the gravitational constant can be learned through repeated exposure to the formulas in which it is needed, but it is safer to carry the units along in every calculation.

The following is a representative (but not exhaustive) listing of formulas that require the g_c term. In all cases, it is assumed that the standard English Engineering System units will be used.

- kinetic energy

$$E = \frac{mv^2}{2g_c} \quad \text{(in ft-lbf)} \qquad 1.7$$

- potential energy

$$= \frac{mgz}{g_c} \quad \text{(in ft-lbf)} \qquad 1.8$$

- pressure at a depth

$$p = \frac{\rho g h}{g_c} \quad \text{(in lbf/ft}^2\text{)} \qquad 1.9$$

Example 1.2

A rocket that has a mass of 4000 lbm travels at 27,000 ft/sec. What is its kinetic energy in ft-lbf?

Solution

From Eq. 1.7,

$$E_k = \frac{mv^2}{2g_c} = \frac{(4000 \text{ lbm})\left(27{,}000 \ \frac{\text{ft}}{\text{sec}}\right)^2}{(2)\left(32.2 \ \frac{\text{lbm-ft}}{\text{lbf-sec}^2}\right)}$$

$$= 4.53 \times 10^{10} \text{ ft-lbf}$$

8. WEIGHT AND WEIGHT DENSITY

Weight, W, is a force exerted on an object due to its placement in a gravitational field. If a consistent set of units is used, Eq. 1.1 can be used to calculate the weight of a mass. In the English Engineering System, however, Eq. 1.10 must be used.

$$W = \frac{mg}{g_c} \qquad 1.10$$

Both sides of Eq. 1.10 can be divided by the volume of an object to derive the *weight density*, γ, of the object. Equation 1.11 illustrates that the weight density (in units of lbf/ft^3) can also be calculated by multiplying the mass density (in units of lbm/ft^3) by g/g_c. Since g and g_c usually have the same numerical values, the only effect of Eq. 1.12 is to change the units of density.

$$\frac{W}{V} = \left(\frac{m}{V}\right)\left(\frac{g}{g_c}\right) \qquad 1.11$$

$$\gamma = \frac{W}{V} = \left(\frac{m}{V}\right)\left(\frac{g}{g_c}\right) = \frac{\rho g}{g_c} \qquad 1.12$$

Weight does not occupy volume. Only mass has volume. The concept of weight density has evolved to simplify certain calculations, particularly fluid calculations. For example, pressure at a depth is calculated from Eq. 1.13. (Compare this with Eq. 1.9.)

$$p = \gamma h \qquad 1.13$$

[5]It is acceptable (and recommended) that g_c be rounded to the same number of significant digits as g. Therefore, a value of 32.2 for g_c would typically be used.

9. THE ENGLISH GRAVITATIONAL SYSTEM

Not all English systems are inconsistent. Pounds can still be used as the unit of force as long as pounds are not used as the unit of mass. Such is the case with the consistent *English Gravitational System.*

If acceleration is given in ft/sec^2, the units of mass for a consistent system of units can be determined from Newton's second law. The combination of units in Eq. 1.14 is known as a *slug.* g_c is not needed at all since this system is consistent. It would be needed only to convert slugs to another mass unit.

$$\text{units of } m = \frac{\text{units of } F}{\text{units of } a} = \frac{\text{lbf}}{\frac{\text{ft}}{\text{sec}^2}} = \frac{\text{lbf-sec}^2}{\text{ft}} \quad 1.14$$

Slugs and pounds-mass are not the same, as Fig. 1.1 illustrates. However, both are units for the same quantity: mass. Equation 1.15 will convert between slugs and pounds-mass.

$$\text{no. of slugs} = \frac{\text{no. of lbm}}{g_c} \quad 1.15$$

The number of slugs is not derived by dividing the number of pounds-mass by the local gravity. g_c is used regardless of the local gravity. The conversion between feet and inches is not dependent on local gravity; neither is the conversion between slugs and pounds-mass.

Since the English Gravitational System is consistent, Eq. 1.16 can be used to calculate weight. The local gravitational acceleration is used.

$$W \text{ in lbf} = (m \text{ in slugs})\left(g \text{ in } \frac{\text{ft}}{\text{sec}^2}\right) \quad 1.16$$

10. THE ABSOLUTE ENGLISH SYSTEM

The obscure *Absolute English System* takes the approach that mass must have units of pounds-mass (lbm) and the units of force can be derived from Newton's second law. The units for F cannot be simplified any more than they are in Eq. 1.17. This particular combination of units is known as a *poundal.*[6] A poundal is not the same as a pound.

$$\begin{aligned}\text{units of } F &= (\text{units of } m)(\text{units of } a) \\ &= (\text{lbm})\left(\frac{\text{ft}}{\text{sec}^2}\right) \\ &= \frac{\text{lbm-ft}}{\text{sec}^2}\end{aligned} \quad 1.17$$

Poundals have not seen widespread use in the United States. The English Gravitational System (using slugs for mass) has greatly eclipsed the Absolute English System in popularity. Both are consistent systems, but there seems to be little need for poundals in modern engineering. Figure 1.2 shows the poundal in comparison to other common units of force.

[6]A poundal is equal to 0.03108 pounds-force.

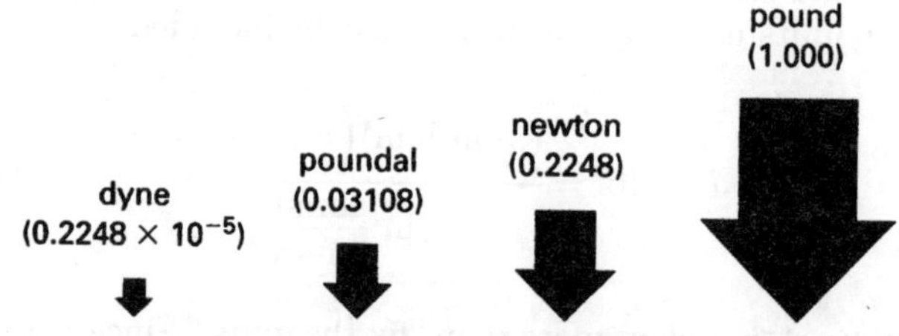

***Figure 1.2** Common Force Units*

11. METRIC SYSTEMS OF UNITS

Strictly speaking, a *metric system* is any system of units that is based on meters or parts of meters. This broad definition includes *mks systems* (based on meters, kilograms, and seconds) as well as *cgs systems* (based on centimeters, grams, and seconds).

Metric systems avoid the pounds-mass versus pounds-force ambiguity in two ways. First, a unit of weight is not established at all. All quantities of matter are specified as mass. Second, force and mass units do not share a common name.

The term *metric system* is not explicit enough to define which units are to be used for any given variable. For example, within the cgs system there is variation in how certain electrical and magnetic quantities are represented (resulting in the ESU and EMU systems). Also, within the mks system, it is common practice in some industries to use kilocalories as the unit of thermal energy, while the SI unit for thermal energy is the joule. This shows a lack of uniformity even within the metricated engineering community.[7]

The "metric" parts of this book use SI, which is the most developed and codified of the so-called metric systems.[8] There will be occasional variances with local engineering custom, but it is difficult to anticipate such variances within a book that must itself be consistent.

12. THE cgs SYSTEM

The cgs system is used widely by chemists and physicists. It is named for the three primary units used to construct its derived variables: the centimeter, the gram, and the second.

[7]In the "field test" of the metric system conducted over the past 200 years, other conventions are to use kilograms-force (kgf) instead of newtons and kgf/cm^2 for pressure (instead of pascals).

[8]SI units are an outgrowth of the *General Conference of Weights and Measures*, an international treaty organization that established the *Système International d'Unités* (*International System of Units*) in 1960. The United States subscribed to this treaty in 1975.

When Newton's second law is written in the cgs system, the following combination of units results.

$$\begin{aligned}\text{units of force} &= (m \text{ in g})\left(a \text{ in } \frac{\text{cm}}{\text{s}^2}\right)\\ &= \text{g}\cdot\text{cm/s}^2\end{aligned} \qquad 1.18$$

This combination of units for force is known as a *dyne.* Energy variables in the cgs system have units of dyne·cm or, equivalently, g·cm^2/s^2. This combination is known as an *erg.* There is no uniformly accepted unit of power in the cgs system, although calories per second is frequently used.

The fundamental volume unit in the cgs system is the cubic centimeter (cc). Since this is the same volume as one thousandth of a liter, units of milliliters (mL) are also used.

13. SI UNITS (THE mks SYSTEM)

SI units comprise an *mks system* (so named because it uses the meter, kilogram, and second as base units). All other units are derived from the base units, which are completely listed in Table 1.1. This system is fully consistent, and there is only one recognized unit for each physical quantity (variable).

Table 1.1 *SI Base Units*

quantity	name	symbol
length	meter	m
mass	kilogram	kg
time	second	s
electric current	ampere	A
temperature	kelvin	K
amount of substance	mole	mol
luminous intensity	candela	cd

Two types of units are used: base units and derived units. The *base units* are dependent only on accepted standards or reproducible phenomena. The *derived units* (Table 1.2 and Table 1.3) are made up of combinations of base units. Prior to 1995, radians and steradians were classified as *supplementary units.*

In addition, there is a set of non-SI units that may be used. This concession is primarily due to the significance and widespread acceptance of these units. Use of the non-SI units listed in Table 1.4 will usually create an inconsistent expression requiring conversion factors.

The SI unit of force can be derived from Newton's second law. This combination of units for force is known as a *newton.*

$$\begin{aligned}\text{units of force} &= (m \text{ in kg})\left(a \text{ in } \frac{\text{m}}{\text{s}^2}\right)\\ &= \text{kg}\cdot\text{m/s}^2\end{aligned} \qquad 1.19$$

Table 1.2 *Some SI Derived Units with Special Names*

quantity	name	symbol	expressed in terms of other units
frequency	hertz	Hz	1/s
force	newton	N	kg·m/s^2
pressure, stress	pascal	Pa	N/m^2
energy, work, quantity of heat	joule	J	N·m
power, radiant flux	watt	W	J/s
quantity of electricity, electric charge	coulomb	C	
electric potential, potential difference, electromotive force	volt	V	W/A
electric capacitance	farad	F	C/V
electric resistance	ohm	Ω	V/A
electric conductance	siemens	S	A/V
magnetic flux	weber	Wb	V·s
magnetic flux density	tesla	T	Wb/m^2
inductance	henry	H	Wb/A
luminous flux	lumen	lm	
illuminance	lux	lx	lm/m^2
plane angle	radian	rad	–
solid angle	steradian	sr	–

Energy variables in SI units have units of N·m or, equivalently, kg·m^2/s^2. Both of these combinations are known as a *joule.* The units of power are joules per second, equivalent to a *watt.*

Example 1.3

A 10 kg block hangs from a cable. What is the tension in the cable? (Standard gravity equals 9.81 m/s^2.)

Solution

$$\begin{aligned}F &= mg\\ &= (10\ \text{kg})\left(9.81\ \frac{\text{m}}{\text{s}^2}\right)\\ &= 98.1\ \text{kg}\cdot\text{m/s}^2 \quad (98.1\ \text{N})\end{aligned}$$

Example 1.4

A 10 kg block is raised vertically 3 m. What is the change in potential energy?

Solution

$$\begin{aligned}\Delta E_p &= mg\Delta h\\ &= (10\ \text{kg})\left(9.81\ \frac{\text{m}}{\text{s}^2}\right)(3\ \text{m})\\ &= 294\ \text{kg}\cdot\text{m}^2/\text{s}^2 \quad (294\ \text{J})\end{aligned}$$

Mathematics

Table 1.3 Some SI Derived Units

quantity	description	symbol
area	square meter	m^2
volume	cubic meter	m^3
speed—linear	meter per second	m/s
speed—angular	radian per second	rad/s
acceleration—linear	meter per second squared	m/s^2
acceleration—angular	radian per second squared	rad/s^2
density, mass density	kilogram per cubic meter	kg/m^3
concentration (of amount of substance)	mole per cubic meter	mol/m^3
specific volume	cubic meter per kilogram	m^3/kg
luminance	candela per square meter	cd/m^2
absolute viscosity	pascal second	Pa·s
kinematic viscosity	square meters per second	m^2/s
moment of force	newton meter	N·m
surface tension	newton per meter	N/m
heat flux density, irradiance	watt per square meter	W/m^2
heat capacity, entropy	joule per kelvin	J/K
specific heat capacity, specific entropy	joule per kilogram kelvin	J/kg·K
specific energy	joule per kilogram	J/kg
thermal conductivity	watt per meter kelvin	W/m·K
energy density	joule per cubic meter	J/m^3
electric field strength	volt per meter	V/m
electric charge density	coulomb per cubic meter	C/m^3
surface density of charge, flux density	coulomb per square meter	C/m^2
permittivity	farad per meter	F/m
current density	ampere per square meter	A/m^2
magnetic field strength	ampere per meter	A/m
permeability	henry per meter	H/m
molar energy	joule per mole	J/mol
molar entropy, molar heat capacity	joule per mole kelvin	J/mol·K
radiant intensity	watt per steradian	W/sr

14. RULES FOR USING SI UNITS

In addition to having standardized units, the set of SI units also has rigid syntax rules for writing the units and combinations of units. Each unit is abbreviated with a specific symbol. The following rules for writing and combining these symbols should be adhered to.

Table 1.4 Acceptable Non-SI Units

quantity	unit name	symbol or abbreviation	relationship to SI unit
area	hectare	ha	1 ha = 10 000 m^2
energy	kilowatt-hour	kW·h	1 kW·h = 3.6 MJ
mass	metric ton[a]	t	1 t = 1000 kg
plane angle	degree (of arc)	°	1° = 0.017 453 rad
speed of rotation	revolution per minute	r/min	1 r/min = $2\pi/60$ rad/s
temperature interval	degree Celsius	°C	1°C = 1K
time	minute	min	1 min = 60 s
	hour	h	1 h = 3600 s
	day (mean solar)	d	1 d = 86 400 s
	year (calendar)	a	1 a = 31 536 000 s
velocity	kilometer per hour	km/h	1 km/h = 0.278 m/s
volume	liter[b]	L	1 L = 0.001 m^3

[a]The international name for metric ton is *tonne*. The metric ton is equal to the *megagram* (Mg).

[b]The international symbol for liter is the lowercase l, which can be easily confused with the numeral 1. Several English-speaking countries have adopted the script ℓ or uppercase L (as does this book) as a symbol for liter in order to avoid any misinterpretation.

- The expressions for derived units in symbolic form are obtained by using the mathematical signs of multiplication and division. For example, units of velocity are m/s. Units of torque are N·m (not N-m or Nm).
- Scaling of most units is done in multiples of 1000.
- The symbols are always printed in roman type, regardless of the type used in the rest of the text. The only exception to this is in the use of the symbol for liter, where the use of the lower case "el" (l) may be confused with the numeral one (1). In this case, "liter" should be written out in full, or the script ℓ or L used. (L is used in this book.)
- Symbols are not pluralized: 45 kg (not 45 kgs).
- A period after a symbol is not used, except when the symbol occurs at the end of a sentence.
- When symbols consist of letters, there must always be a full space between the quantity and the symbols: 45 kg (not 45kg). However, for planar angle designations, no space is left: 42°12′45″ (not 42° 12′ 45″).
- All symbols are written in lowercase, except when the unit is derived from a proper name: m for meter, s for second, A for ampere, Wb for weber, N for newton, W for watt.
- Prefixes are printed without spacing between the prefix and the unit symbol (e.g., km is the symbol for kilometer). Table 1.5 lists common prefixes, their symbols, and their values.
- In text, when no number is involved, the unit should be spelled out. Example: Carpet is sold by the square meter, not by the m^2.

*Table 1.5 SI Prefixes**

prefix	symbol	value
exa	E	10^{18}
peta	P	10^{15}
tera	T	10^{12}
giga	G	10^{9}
mega	M	10^{6}
kilo	k	10^{3}
hecto	h	10^{2}
deka (or deca)	da	10^{1}
deci	d	10^{-1}
centi	c	10^{-2}
milli	m	10^{-3}
micro	μ	10^{-6}
nano	n	10^{-9}
pico	p	10^{-12}
femto	f	10^{-15}
atto	a	10^{-18}

*There is no "B" (billion) prefix. In fact, the word "billion" means 10^9 in the United States but 10^{12} in most other countries. This unfortunate ambiguity is handled by avoiding the use of the term "billion."

- Where a decimal fraction of a unit is used, a zero should always be placed before the decimal marker: 0.45 kg (not .45 kg). This practice draws attention to the decimal marker and helps avoid errors of scale.

- A practice in some countries is to use a comma as a decimal marker, while the practice in North America, the United Kingdom, and some other countries is to use a period as the decimal marker. Furthermore, in some countries that use the decimal comma, a period is frequently used to divide long numbers into groups of three. Because of these differing practices, spaces must be used instead of commas to separate long lines of digits into easily readable blocks of three digits with respect to the decimal marker: 32 453.246 072 5. A space (half-space preferred) is optional with a four-digit number: 1 234 or 1234.

- The word *ton* has multiple meanings. In the United States and Canada, the *short ton* of 2000 lbm (907.18 kg; 8896.44 N) is used. Previously, for commerce within the United Kingdom, a *long ton* of 2240 lbm (1016.05 kg) was used. A *metric ton* (or, *tonne*) is 1000 kg (10^6 Mg; 2205 lbm). In air conditioning industries, a *ton of refrigeration* is equivalent to a cooling rate of 200 Btu/min (12,000 Btu/hr). For explosives, a *ton of explosive power* is approximately the energy given off by 2000 lbm of TNT, standardized by international convention as 10^9 cal (1 Gcal; 4.184 GJ) for both "ton" and "tonne" designations. Various definitions are used in shipping, where *freight ton* (*measurement ton* or MTON) refers to volume, usually 40 ft^3. Many other specialty definitions are also in use.

15. PRIMARY DIMENSIONS

Regardless of the system of units chosen, each variable representing a physical quantity will have the same *primary dimensions.* For example, velocity may be expressed in miles per hour (mph) or meters per second (m/s), but both units have dimensions of length per unit time. Length and time are two of the primary dimensions, as neither can be broken down into more basic dimensions. The concept of primary dimensions is useful when converting little-used variables between different systems of units, as well as in correlating experimental results (i.e., dimensional analysis).

There are three different sets of primary dimensions in use.[9] In the *$ML\theta T$ system*, the primary dimensions are mass (M), length (L), time (θ), and temperature (T). All symbols are uppercase. In order to avoid confusion between time and temperature, the Greek letter theta is used for time.[10]

All other physical quantities can be derived from these primary dimensions.[11] For example, work in SI units has units of N·m. Since a newton is a kg·m/s^2, the primary dimensions of work are ML^2/θ^2. The primary dimensions for many important engineering variables are shown in Table 1.6. If it is more convenient to stay with traditional English units, it may be more desirable to work in the *$FML\theta TQ$* system (sometimes called the *engineering dimensional system*). This system adds the primary dimensions of force (F) and heat (Q). Work (ft-lbf in the English system) has the primary dimensions of FL. (Compare this with the primary dimensions for work in the *$ML\theta T$* system.) Thermodynamic variables are similarly simplified.

Dimensional analysis will be more conveniently carried out when one of the four-dimension systems ($ML\theta T$ or $FL\theta T$) is used. Whether the $ML\theta T$, $FL\theta T$, or $FML\theta TQ$ system is used depends on what is being derived and who will be using it, and whether or not a consistent set of variables is desired. Conversion constants such as g_c and J will almost certainly be required if the $ML\theta T$ system is used to generate variables for use in the English systems. It is also much more convenient to use the $FML\theta TQ$ system when working in the fields of thermodynamics, fluid flow, heat transfer, and so on.

16. DIMENSIONLESS GROUPS

A *dimensionless group* is derived as a ratio of two forces or other quantities. Considerable use of dimensionless groups is made in certain subjects, notably fluid

[9]One of these, the $FL\theta T$ system, is not discussed here.

[10]This is the most common usage. There is a lack of consistency in the engineering world about the symbols for the primary dimensions in dimensional analysis. Some writers use t for time instead of θ. Some use H for heat instead of Q. And, in the worst mix-up of all, some have reversed the use of T and θ.

[11]A *primary dimension* is the same as a *base unit* in the SI set of units. The SI units add several other base units, as shown in Table 1.1, to deal with variables that are difficult to derive in terms of the four primary base units.

Mathematics

Table 1.6 Dimensions of Common Variables

variable (common symbol)	dimensional system $ML\theta T$	$FL\theta T$	$FMLT\theta Q$
mass (m)	M	$F\theta^2/L$	M
force (F)	ML/θ^2	F	F
length (L)	L	L	L
time (θ or t)	θ	θ	θ
temperature (T)	T	T	T
work (W)	ML^2/θ^2	FL	FL
heat (Q)	ML^2/θ^2	FL	Q
acceleration (a)	L/θ^2	L/θ^2	L/θ^2
frequency (n or f)	$1/\theta$	$1/\theta$	$1/\theta$
area (A)	L^2	L^2	L^2
coefficient of thermal expansion (β)	$1/T$	$1/T$	$1/T$
density (ρ)	M/L^3	$F\theta^2/L^4$	M/L^3
dimensional constant (g_c)	1.0	1.0	$ML/\theta^2 F$
specific heat at constant pressure (c_p); at constant volume (c_v)	$L^2/\theta^2 T$	$L^2/\theta^2 T$	Q/MT
heat transfer coefficient (h); overall (U)	$M/\theta^3 T$	$F/\theta LT$	$Q/\theta L^2 T$
power (P)	ML^2/θ^3	FL/θ	FL/θ
heat flow rate ($\dot{Q}$)	ML^2/θ^3	FL/θ	Q/θ
kinematic viscosity (ν)	L^2/θ	L^2/θ	L^2/θ
mass flow rate ($\dot{m}$)	M/θ	$F\theta/L$	M/θ
mechanical equivalent of heat (J)	–	–	FL/Q
pressure (p)	$M/L\theta^2$	F/L^2	F/L^2
surface tension (σ)	M/θ^2	F/L	F/L
angular velocity (ω)	$1/\theta$	$1/\theta$	$1/\theta$
volumetric flow rate ($\dot{m}/\rho = \dot{V}$)	L^3/θ	L^3/θ	L^3/θ
conductivity (k)	$ML/\theta^3 T$	$F/\theta T$	$Q/L\theta T$
thermal diffusivity (α)	L^2/θ	L^2/θ	L^2/θ
velocity (v)	L/θ	L/θ	L/θ
viscosity, absolute (μ)	$M/L\theta$	$F\theta/L^2$	$F\theta/L^2$
volume (V)	L^3	L^3	L^3

mechanics and heat transfer. For example, the Reynolds number, Mach number, and Froude number are used to distinguish between distinctly different flow regimes in pipe flow, compressible flow, and open channel flow, respectively.

Table 1.7 contains information about the most common dimensionless groups used in fluid mechanics and heat transfer.

17. LINEAL AND BOARD FOOT MEASUREMENTS

The term *lineal* is often mistaken as a typographical error for *linear*. Although "lineal" has its own specific meaning very different from "linear," the two are often used interchangeably by engineers.[12] The adjective *lineal* is often encountered in the building trade (e.g., 12 lineal feet of lumber), where the term is used to distinguish it from board feet measurement.

A *board foot* (abbreviated bd-ft) is not a measure of length. Rather, it is a measure of volume used with lumber. Specifically, a board foot is equal to 144 in^3 (2.36×10^{-3} m^3). The name is derived from the volume of a board 1 foot square and 1 inch thick. In that sense, it is parallel in concept to the acre-foot. Since lumber cost is directly related to lumber weight and volume, the board foot unit is used in determining the overall lumber cost.

18. DIMENSIONAL ANALYSIS

Dimensional analysis is a means of obtaining an equation that describes some phenomenon without understanding the mechanism of the phenomenon. The most serious limitation is the need to know beforehand which variables influence the phenomenon. Once these are known or assumed, dimensional analysis can be applied by a routine procedure.

The first step is to select a system of primary dimensions. (See Sec. 1.15.) Usually the $ML\theta T$ system is used, although this choice may require the use of g_c and J in the final results.

The second step is to write a functional relationship between the dependent variable and the independent variable, x_i.

$$y = f(x_1, x_2, \ldots, x_m) \qquad 1.20$$

This function can be expressed as an exponentiated series. The C_1, a_i, b_i, ..., z_i in Eq. 1.21 are unknown constants.

$$y = C_1 x_1^{a_1} x_2^{b_1} x_3^{c_1} \cdots x_m^{z_1} + C_2 x_1^{a_2} x_2^{b_2} x_3^{c_2} \cdots x_m^{z_2} + \cdots \qquad 1.21$$

The key to solving Eq. 1.21 is that each term on the right-hand side must have the same dimensions as y. Simultaneous equations are used to determine some of the a_i, b_i, c_i, and z_i. Experimental data are required to determine the C_i and remaining exponents. In most analyses, it is assumed that the $C_i = 0$ for $i \geq 2$.

Since this method requires working with m different variables and n different independent dimensional quantities (such as M, L, θ, and T), an easier method is desirable. One simplification is to combine the m variables into dimensionless groups called *pi-groups*. (See Table 1.6.)

[12] *Lineal* is best used when discussing a line of succession (e.g., a lineal descendant of a particular person). *Linear* is best used when discussing length (e.g., a linear dimension of a room).

Table 1.7 *Common Dimensionless Groups*

name	symbol	formula	interpretation
Biot number	Bi	$\frac{hL}{k_s}$	$\frac{\text{surface conductance}}{\text{internal conduction of solid}}$
Cauchy number	Ca	$\frac{\mathrm{v}^2}{\frac{B_s}{\rho}} = \frac{\mathrm{v}^2}{a^2}$	$\frac{\text{inertia force}}{\text{compressive force}} = \text{Mach number}^2$
Eckert number	Ec	$\frac{\mathrm{v}^2}{2c_p\Delta T}$	$\frac{\text{temperature rise due to energy conversion}}{\text{temperature difference}}$
Eötvös number	Eo	$\frac{\rho g L^2}{\sigma}$	$\frac{\text{buoyancy}}{\text{surface tension}}$
Euler number	Eu	$\frac{\Delta p}{\rho \mathrm{v}^2}$	$\frac{\text{pressure force}}{\text{inertia force}}$
Fourier number	Fo	$\frac{kt}{\rho c_p L^2} = \frac{\alpha t}{L^2}$	$\frac{\text{rate of conduction of heat}}{\text{rate of storage of energy}}$
Froude number*	Fr	$\frac{\mathrm{v}^2}{gL}$	$\frac{\text{inertia force}}{\text{gravity force}}$
Graetz number*	Gz	$\left(\frac{D}{L}\right)\left(\frac{\mathrm{v}\rho c_p D}{k}\right)$	$\frac{(\text{Re})(\text{Pr})}{L/D}$ $\frac{\text{heat transfer by convection in entrance region}}{\text{heat transfer by conduction}}$
Grashof number*	Gr	$\frac{g\beta\Delta T L^3}{\nu^2}$	$\frac{\text{buoyancy force}}{\text{viscous force}}$
Knudsen number	Kn	$\frac{\lambda}{L}$	$\frac{\text{mean free path of molecules}}{\text{characteristic length of object}}$
Lewis number*	Le	$\frac{\alpha}{D_c}$	$\frac{\text{thermal diffusivity}}{\text{molecular diffusivity}}$
Mach number	M	$\frac{\mathrm{v}}{a}$	$\frac{\text{macroscopic velocity}}{\text{speed of sound}}$
Nusselt number	Nu	$\frac{hL}{k}$	$\frac{\text{temperature gradient at wall}}{\text{overall temperature difference}}$
Péclet number	Pé	$\frac{\mathrm{v}\rho c_p D}{k}$	$(\text{Re})(\text{Pr})$ $\frac{\text{heat transfer by convection}}{\text{heat transfer by conduction}}$
Prandtl number	Pr	$\frac{\mu c_p}{k} = \frac{\nu}{\alpha}$	$\frac{\text{diffusion of momentum}}{\text{diffusion of heat}}$
Reynolds number	Re	$\frac{\rho \mathrm{v} L}{\mu} = \frac{\mathrm{v}L}{\nu}$	$\frac{\text{inertia force}}{\text{viscous force}}$
Schmidt number	Sc	$\frac{\mu}{\rho D_c} = \frac{\nu}{D_c}$	$\frac{\text{diffusion of momentum}}{\text{diffusion of mass}}$
Sherwood number*	Sh	$\frac{k_D L}{D_c}$	$\frac{\text{mass diffusivity}}{\text{molecular diffusivity}}$
Stanton number	St	$\frac{h}{\mathrm{v}\rho c_p} = \frac{h}{c_p G}$	$\frac{\text{heat transfer at wall}}{\text{energy transported by stream}}$
Stokes number	Sk	$\frac{\Delta p L}{\mu \mathrm{v}}$	$\frac{\text{pressure force}}{\text{viscous force}}$
Strouhal number*	Sl	$\frac{L}{t\mathrm{v}} = \frac{L\omega}{\mathrm{v}}$	$\frac{\text{frequency of vibration}}{\text{characteristic frequency}}$
Weber number	We	$\frac{\rho \mathrm{v}^2 L}{\sigma}$	$\frac{\text{inertia force}}{\text{surface tension force}}$

*Multiple definitions exist, most often the square or square root of the formula shown.

If these dimensionless groups are represented by π_1, $\pi_2, \pi_3, \ldots, \pi_k$, the equation expressing the relationship between the variables is given by the *Buckingham π-theorem.*

$$f(\pi_1, \pi_2, \pi_3, \ldots, \pi_k) = 0 \qquad 1.22$$

$$k = m - n \qquad 1.23$$

The dimensionless pi-groups are usually found from the m variables according to an intuitive process.

Example 1.5

A solid sphere rolls down a submerged incline. Find an equation for the velocity, v.

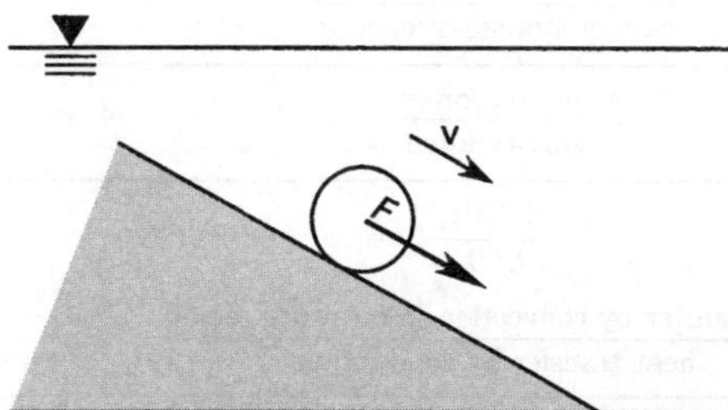

Solution

Assume that the velocity depends on the force, F, due to gravity, the diameter of the sphere, D, the density of the fluid, ρ, and the viscosity of the fluid, μ.

$$\mathrm{v} = f(F, D, \rho, \mu) = CF^aD^b\rho^c\mu^d$$

This equation can be written in terms of the primary dimensions of the variables.

$$\frac{L}{\theta} = C\left(\frac{ML}{\theta^2}\right)^a L^b\left(\frac{M}{L^3}\right)^c\left(\frac{M}{L\theta}\right)^d$$

Since L on the left-hand side has an implied exponent of one, a necessary condition is

$$1 = a + b - 3c - d \quad (L)$$

Similarly, the other necessary conditions are

$$-1 = -2a - d \quad (\theta)$$
$$0 = a + c + d \quad (M)$$

Solving simultaneously yields

$$b = -1$$
$$c = a - 1$$
$$d = 1 - 2a$$
$$\mathrm{v} = CF^aD^{-1}\rho^{a-1}\mu^{1-2a}$$
$$= C\left(\frac{\mu}{D\rho}\right)\left(\frac{F\rho}{\mu^2}\right)^a$$

C and a must be determined experimentally.

Energy, Work, and Power

Nomenclature

c	specific heat	Btu/lbm-°F	J/kg·°C
C	molar specific heat	Btu/lbmol-°F	J/kmol·°C
E	energy	ft-lbf	J
F	force	lbf	N
g	gravitational acceleration, 32.2 (9.81)	ft/sec^2	m/s^2
g_c	gravitational constant, 32.2	lbm-ft/lbf-sec^2	n.a.
h	height	ft	m
hp	horsepower	ft-lbf/sec	kW
I	mass moment of inertia	lbm-ft^2	kg·m^2
J	Joule's constant, 778.2	ft-lbf/Btu	–
k	spring constant	lbf/ft	N/m
m	mass	lbm	kg
$\dot{m}$	mass flow rate	lbm/sec	kg/s
MW	molecular weight	lbm/lbmol	kg/kmol
p	pressure	lbf/ft^2	Pa
P	power	ft-lbf/sec	kW
q	heat	Btu/lbm	J/kg
Q	heat	Btu	J
r	radius	ft	m
R	resistance	Ω	Ω
s	distance	ft	m
t	time	sec	s
T	temperature	°F	°C
T	torque	ft-lbf	N·m
u	specific energy	ft-lbf/lbm	J/kg
U	internal energy	Btu	J
v	velocity	ft/sec	m/s
W	work	ft-lbf	J
x	displacement	ft	m

Symbols

δ	displacement	ft	m
η	efficiency	–	–
θ	angular position	rad	rad
ρ	mass density	lbm/ft^3	kg/m^3
v	specific volume	ft^3/lbm	m^3/kg
ϕ	angle	deg	deg
ω	angular velocity	rad/sec	rad/s

Subscripts

f	frictional
p	constant pressure
v	constant volume

1. ENERGY OF A MASS

The *energy* of a mass represents the capacity of the mass to do work. Such energy can be stored and released. There are many forms that it can take, including mechanical, thermal, electrical, and magnetic energies. Energy is a positive, scalar quantity (although the change in energy can be either positive or negative).

The total energy of a body can be calculated from its mass, m, and its *specific energy*, u (i.e., the energy per unit mass).[1]

$$E = mu \qquad 2.1$$

Typical units of mechanical energy are foot-pounds and joules. (A joule is equivalent to the units of N·m and kg·m^2/s^2.) In traditional English-unit countries, the *British thermal unit* (Btu) is used for thermal energy, whereas the kilocalorie (kcal) is still used in some applications in SI countries. *Joule's constant*, or the *Joule equivalent* (778.17 ft-lbf/Btu, usually shortened to 778, three significant digits), is used to convert between English mechanical and thermal energy units.

$$\text{energy in Btu} = \frac{\text{energy in ft-lbf}}{J} \qquad 2.2$$

Two other units of large amounts of energy are the therm and the quad. A *therm* is 10^5 Btu (1.055×10^8 J). A *quad* is equal to a quadrillion (10^{15}) Btu. This is 1.055×10^{18} J, or roughly the energy contained in 200 million barrels of oil.

[1]The use of the symbols E and u for energy is not consistent in the engineering field.

2. LAW OF CONSERVATION OF ENERGY

The *law of conservation of energy* says that energy cannot be created or destroyed. However, energy can be converted into different forms. Therefore, the sum of all energy forms is constant.

$$\sum E = \text{constant} \qquad 2.3$$

3. WORK

Work, *W*, is the act of changing the energy of a particle, body, or system. For a mechanical system, *external work* is work done by an external force, whereas *internal work* is done by an internal force. Work is a signed, scalar quantity. Typical units are inch-pounds, foot-pounds, and joules. Mechanical work is seldom expressed in British thermal units or kilocalories. (See Chap. 52 for more information on the units of work.)

For a mechanical system, work is positive when a force acts in the direction of motion and helps a body move from one location to another. Work is negative when a force acts to oppose motion. (Friction, for example, always opposes the direction of motion and can do only negative work.) The work done on a body by more than one force can be found by superposition.

From a thermodynamic standpoint, work is positive if a particle or body does work on its surroundings. Work is negative if the surroundings do work on the object (e.g., blowing up a balloon represents negative work to the balloon). Although this may be a difficult concept to grasp, it is consistent with the conservation of energy, because the sum of negative work and the positive energy increase is zero (i.e., no net energy change in the system).[2]

The work performed by a variable force or torque is calculated from the dot products of Eq. 2.4 and Eq. 2.5.

$$W_{\text{variable force}} = \int \mathbf{F} \cdot d\mathbf{s} \quad \text{[linear systems]} \qquad 2.4$$

$$W_{\text{variable torque}} = \int \mathbf{T} \cdot d\boldsymbol{\theta} \quad \text{[rotational systems]} \qquad 2.5$$

The work done by a force or torque of constant magnitude is

$$W_{\text{constant force}} = \mathbf{F} \cdot \mathbf{s} = Fs\cos\phi \quad \text{[linear systems]} \qquad 2.6$$

$$W_{\text{constant torque}} = \mathbf{T} \cdot \boldsymbol{\theta} = Fr\theta\cos\phi \quad \text{[rotational systems]} \qquad 2.7$$

The nonvector forms illustrate that only the component of force or torque in the direction of motion contributes to work. (See Eq. 2.6 and Eq. 2.7.)

[2]This is just a partial statement of the *first law of thermodynamics*.

Common applications of the work done by a constant force are frictional work and gravitational work. (See Fig. 2.1.) The work to move an object a distance s against a frictional force of F_f is

$$W_{\text{friction}} = F_f s \qquad 2.8$$

Figure 2.1 *Work of a Constant Force*

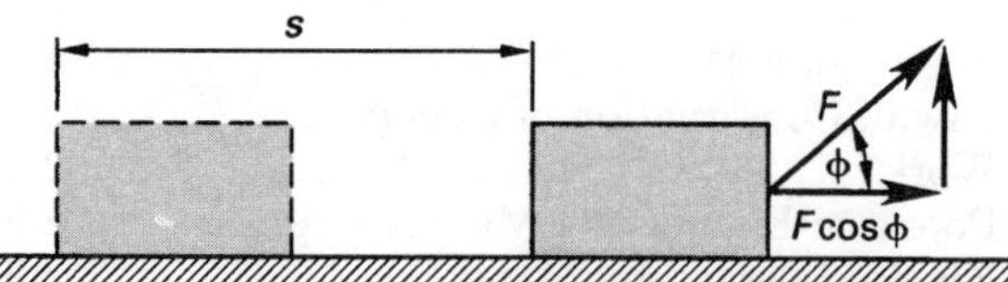

The sign of the work depends on the direction of the motion. The work done *by* gravity when a mass m changes in elevation from h_1 to h_2 is

$$W_{\text{gravity}} = mg(h_2 - h_1) \quad \text{[SI]} \qquad 2.9(a)$$

$$W_{\text{gravity}} = \left(\frac{mg}{g_c}\right)(h_2 - h_1) \quad \text{[U.S.]} \qquad 2.9(b)$$

The sign of the work depends on the direction of the motion. The work done by or on a *linear spring* whose length or deflection changes from δ_1 to δ_2 is given by Eq. 2.10.[3] It does not make any difference whether the spring is a compression spring or an extension spring.

$$W_{\text{spring}} = \tfrac{1}{2}k(\delta_2^2 - \delta_1^2) \qquad 2.10$$

Example 2.1

A lawn mower engine is started by pulling a cord wrapped around a sheave. The sheave radius is 8.0 cm. The cord is wrapped around the sheave two times. If a constant tension of 90 N is maintained in the cord during starting, what work is done on the engine?

Solution

The starting torque on the engine is

$$T = Fr = (90\ \text{N})\left(\frac{8.0\ \text{cm}}{100\ \frac{\text{cm}}{\text{m}}}\right) = 7.2\ \text{N·m}$$

The cord wraps around the sheave two times; therefore, $(2)(2\pi) = 12.6$ rad. From Eq. 2.7, the work done by a constant torque is

$$W = T\theta = (7.2\ \text{N·m})(12.6\ \text{rad}) = 90.7\ \text{J}$$

[3]A *linear spring* is one for which the linear relationship $F = kx$ is valid.

Example 2.2

A 200 lbm crate is pushed 25 ft at constant velocity across a warehouse floor. There is a frictional force of 60 lbf between the crate and floor. What work is done by the frictional force on the crate?

Solution

From Eq. 2.8,

$$\begin{aligned} W_{\text{friction}} &= -F_f s = (60 \text{ lbf})(25 \text{ ft}) \\ &= -1500 \text{ ft-lbf} \end{aligned}$$

The work is opposite to the direction of motion.

4. POTENTIAL ENERGY OF A MASS

Potential energy (*gravitational energy*) is a form of mechanical energy possessed by a body due to its relative position in a gravitational field. Potential energy is lost when the elevation of a body decreases. The lost potential energy usually is converted to kinetic energy or heat.

$$E_{\text{potential}} = mgh \quad \text{[SI]} \quad 2.11(a)$$

$$E_{\text{potential}} = \frac{mgh}{g_c} \quad \text{[U.S.]} \quad 2.11(b)$$

In the absence of friction and other nonconservative forces, the change in potential energy of a body is equal to the work required to change the elevation of the body.

$$W = \Delta E_{\text{potential}} \quad 2.12$$

5. KINETIC ENERGY OF A MASS

Kinetic energy is a form of mechanical energy associated with a moving or rotating body. The kinetic energy of a body moving with instantaneous linear velocity, v, is

$$E_{\text{kinetic}} = \tfrac{1}{2}m\text{v}^2 \quad \text{[SI]} \quad 2.13(a)$$

$$E_{\text{kinetic}} = \frac{m\text{v}^2}{2g_c} \quad \text{[U.S.]} \quad 2.13(b)$$

A body can also have rotational kinetic energy.

$$E_{\text{rotational}} = \tfrac{1}{2}I\omega^2 \quad \text{[SI]} \quad 2.14(a)$$

$$E_{\text{rotational}} = \frac{I\omega^2}{2g_c} \quad \text{[U.S.]} \quad 2.14(b)$$

According to the *work-energy principle* (see Sec. 2.9), the kinetic energy is equal to the work necessary to initially accelerate a stationary body or to bring a moving body to rest.

$$W = \Delta E_{\text{kinetic}} \quad 2.15$$

Example 2.3

A solid disk flywheel ($I = 200$ kg·m^2) is rotating with a speed of 900 rpm. What is its rotational kinetic energy?

Solution

The angular rotational velocity is

$$\omega = \frac{\left(900 \ \frac{\text{rev}}{\text{min}}\right)\left(2\pi \ \frac{\text{rad}}{\text{rev}}\right)}{60 \ \frac{\text{s}}{\text{min}}} = 94.25 \text{ rad/s}$$

From Eq. 2.14, the rotational kinetic energy is

$$\begin{aligned} E &= \tfrac{1}{2}I\omega^2 = \left(\tfrac{1}{2}\right)(200 \text{ kg·m}^2)\left(94.25 \ \frac{\text{rad}}{\text{s}}\right)^2 \\ &= 888 \times 10^3 \text{ J} \quad (888 \text{ kJ}) \end{aligned}$$

6. SPRING ENERGY

A spring is an energy storage device because the spring has the ability to perform work. In a perfect spring, the amount of energy stored is equal to the work required to compress the spring initially. The stored spring energy does not depend on the mass of the spring. Given a spring with spring constant (stiffness) k, the *spring energy* is

$$E_{\text{spring}} = \tfrac{1}{2}k\delta^2 \quad 2.16$$

Example 2.4

A body of mass m falls from height h onto a massless, simply supported beam. The mass adheres to the beam. If the beam has a lateral stiffness k, what will be the deflection, δ, of the beam?

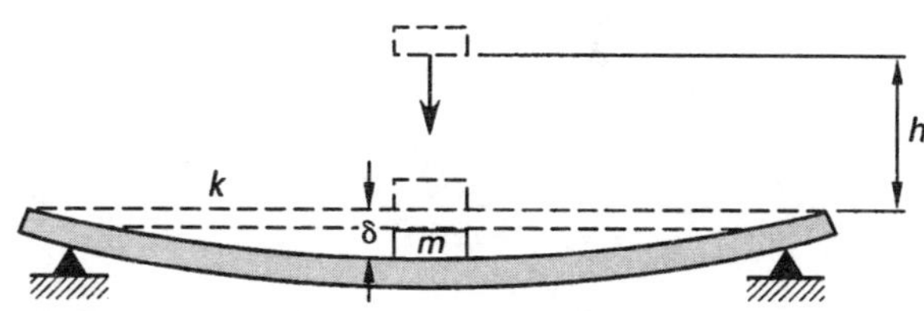

Solution

The initial energy of the system consists of only the potential energy of the body. Using consistent units, the change in potential energy is

$$E = mg(h + \delta) \quad \text{[consistent units]}$$

All of this energy is stored in the spring. Therefore, from Eq. 2.16,

$$\tfrac{1}{2}k\delta^2 = mg(h+\delta)$$

Solving for the deflection,

$$\delta = \frac{mg \pm \sqrt{mg(2hk+mg)}}{k}$$

7. PRESSURE ENERGY OF A MASS

Since work is done in increasing the pressure of a system (e.g., work is done in blowing up a balloon), mechanical energy can be stored in pressure form. This is known as *pressure energy, static energy, flow energy, flow work,* and *p-V work (energy)*. For a system of pressurized mass m, the pressure energy is

$$E_{\text{flow}} = \frac{mp}{\rho} = mpv \quad [v = \text{specific volume}] \qquad 2.17$$

8. INTERNAL ENERGY OF A MASS

The property of internal energy is encountered primarily in thermodynamics problems. Typical units are British thermal units, joules, and kilocalories. The total internal energy, usually given the symbol U, of a body increases when the body's temperature increases.[4] In the absence of any work done on or by the body, the change in internal energy is equal to the heat flow, Q, into the body. Q is positive if the heat flow is into the body and negative otherwise.

$$U_2 - U_1 = Q \qquad 2.18$$

An increase in internal energy is needed to cause a rise in temperature. Different substances differ in the quantity of heat needed to produce a given temperature increase. The heat, Q, required to change the temperature of a mass, m, by an amount, ΔT, is called the *specific heat* (*heat capacity*) of the substance, c.

$$Q = mc\Delta T \qquad 2.19$$

$$c = \frac{Q}{m\Delta T} \qquad 2.20$$

Because specific heats of solids and liquids are slightly temperature dependent, the mean specific heats are used when evaluating processes covering a large temperature range.

The lowercase c implies that the units are Btu/lbm-°F or J/kg·°C. Typical values of specific heat are given in Table 2.1. The *molar specific heat*, designated by the symbol C, has units of Btu/lbmol-°F or J/kmol·°C.

$$C = (\text{MW}) \times c \qquad 2.21$$

For gases, the specific heat depends on the type of process during which the heat exchange occurs. Specific heats for constant-volume and constant-pressure processes are designated by c_v and c_p, respectively.

$$Q = mc_v\Delta T \quad [\text{constant-volume process}] \qquad 2.22$$

$$Q = mc_p\Delta T \quad [\text{constant-pressure process}] \qquad 2.23$$

Values of c_p and c_v for solids and liquids are essentially the same. However, the designation c_p is often encountered for solids and liquids. Table 2.1 gives values of c_p for selected liquids and solids.

Table 2.1 *Approximate Specific Heats of Selected Liquids and Solids**

	c_p	
substance	Btu/lbm-°F	kJ/kg·°C
aluminum, pure	0.23	0.96
aluminum, 2024-T4	0.2	0.84
ammonia	1.16	4.86
asbestos	0.20	0.84
benzene	0.41	1.72
brass, red	0.093	0.39
bronze	0.082	0.34
concrete	0.21	0.88
copper, pure	0.094	0.39
Freon-12	0.24	1.00
gasoline	0.53	2.20
glass	0.18	0.75
gold, pure	0.031	0.13
ice	0.49	2.05
iron, pure	0.11	0.46
iron, cast (4% C)	0.10	0.42
lead, pure	0.031	0.13
magnesium, pure	0.24	1.00
mercury	0.033	0.14
oil, light hydrocarbon	0.5	2.09
silver, pure	0.06	0.25
steel, 1010	0.10	0.42
steel, stainless 301	0.11	0.46
tin, pure	0.055	0.23
titanium, pure	0.13	0.54
tungsten, pure	0.032	0.13
water	1.0	4.19
wood (typical)	0.6	2.50
zinc, pure	0.088	0.37

(Multiply Btu/lbm-°F by 4.1868 to obtain kJ/kg·°C.)
*Values in Btu/lbm-°F are the same as cal/g·°C. Values in kJ/kg·°C are the same as kJ/kg·K.

9. WORK-ENERGY PRINCIPLE

Since energy can be neither created nor destroyed, external work performed on a conservative system goes into

[4]The *thermal energy*, represented by the body's enthalpy, is the sum of internal and pressure energies.

changing the system's total energy. This is known as the *work-energy principle* (or *principle of work and energy*).

$$W = \Delta E = E_2 - E_1 \qquad 2.24$$

Generally, the term *work-energy principle* is limited to use with mechanical energy problems (i.e., conversion of work into kinetic or potential energies). When energy is limited to kinetic energy, the work-energy principle is a direct consequence of Newton's second law but is valid only for inertial reference systems.

By directly relating forces, displacements, and velocities, the work-energy principle introduces some simplifications into many mechanical problems.

- It is not necessary to calculate or know the acceleration of a body to calculate the work performed on it.
- Forces that do not contribute to work (e.g., are normal to the direction of motion) are irrelevant.
- Only scalar quantities are involved.
- It is not necessary to individually analyze the particles or component parts in a complex system.

Example 2.5

A 4000 kg elevator starts from rest, accelerates uniformly to a constant speed of 2.0 m/s, and then decelerates uniformly to a stop 20 m above its initial position. Neglecting friction and other losses, what work was done on the elevator?

Solution

By the work-energy principle, the work done on the elevator is equal to the change in the elevator's energy. Since the initial and final kinetic energies are zero, the only mechanical energy change is the potential energy change.

Taking the initial elevation of the elevator as the reference (i.e., $h_1 = 0$), and combining Eq. 2.9(a) and Eq. 2.24,

$$\begin{aligned} W &= E_{2,\text{potential}} - E_{1,\text{potential}} = mg(h_2 - h_1) \\ &= (4000 \text{ kg})\left(9.81 \ \frac{\text{m}}{\text{s}^2}\right)(20 \text{ m}) \\ &= 785 \times 10^3 \text{ J} \quad (785 \text{ kJ}) \end{aligned}$$

10. CONVERSION BETWEEN ENERGY FORMS

Conversion of one form of energy into another does not violate the conservation of energy law. However, most problems involving conversion of energy are really just special cases of the work-energy principle. For example, consider a falling body that is acted upon by a gravitational force. The conversion of potential energy into kinetic energy can be interpreted as equating the work done by the constant gravitational force to the change in kinetic energy.

In general terms, *Joule's law* states that one energy form can be converted without loss into another. There are two specific formulations of Joule's law. As related to electricity, $P = I^2R = V^2/R$ is the common formulation of Joule's law. As related to thermodynamics and ideal gases, Joule's law states that "the change in internal energy of an ideal gas is a function of the temperature change, not of the volume." This latter form can also be stated more formally as "at constant temperature, the internal energy of a gas approaches a finite value that is independent of the volume as the pressure goes to zero."

Example 2.6

A 2 lbm projectile is launched straight up with an initial velocity of 700 ft/sec. Neglecting air friction, calculate the (a) kinetic energy immediately after launch, (b) kinetic energy at maximum height, (c) potential energy at maximum height, (d) total energy at an elevation where the velocity has dropped to 300 ft/sec, and (e) maximum height attained.

Solution

(a) From Eq. 2.13, the kinetic energy is

$$\begin{aligned} E_{\text{kinetic}} &= \frac{mv^2}{2g_c} = \frac{(2 \text{ lbm})\left(700 \ \frac{\text{ft}}{\text{sec}}\right)^2}{(2)\left(32.2 \ \frac{\text{lbm-ft}}{\text{lbf-sec}^2}\right)} \\ &= 15{,}217 \text{ ft-lbf} \end{aligned}$$

(b) The velocity is zero at the maximum height. Therefore, the kinetic energy is zero.

(c) At the maximum height, all of the kinetic energy has been converted into potential energy. Therefore, the potential energy is 15,217 ft-lbf.

(d) Although some of the kinetic energy has been transformed into potential energy, the total energy is still 15,217 ft-lbf.

(e) Since all of the kinetic energy has been converted into potential energy, the maximum height can be found from Eq. 2.11.

$$E_{\text{potential}} = \frac{mgh}{g_c}$$

$$15{,}217 \text{ ft-lbf} = \frac{(2 \text{ lbm})\left(32.2 \ \frac{\text{ft}}{\text{sec}^2}\right)h}{32.2 \ \frac{\text{lbm-ft}}{\text{lbf-sec}^2}}$$

$$h = 7609 \text{ ft}$$

Example 2.7

A 4500 kg ore car rolls down an incline and passes point A traveling at 1.2 m/s. The ore car is stopped by a spring bumper that compresses 0.6 m. A constant friction force of 220 N acts on the ore car at all times. What is the spring constant?

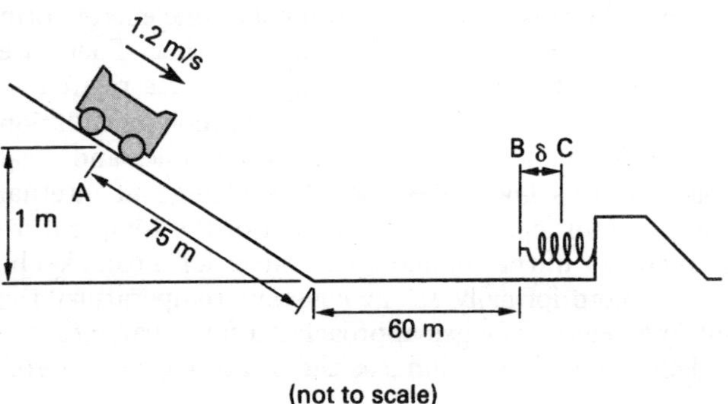

Solution

The car's total energy at point A is the sum of the kinetic and potential energies.

$$\begin{aligned} E_{\text{total,A}} &= E_{\text{kinetic}} + E_{\text{potential}} = \tfrac{1}{2}mv^2 + mgh \\ &= \left(\tfrac{1}{2}\right)(4500\ \text{kg})\left(1.2\ \frac{\text{m}}{\text{s}}\right)^2 \\ &\quad + (4500\ \text{kg})\left(9.81\ \frac{\text{m}}{\text{s}^2}\right)(1\ \text{m}) \\ &= 47\,385\ \text{J} \end{aligned}$$

At point B, the potential energy has been converted into additional kinetic energy. However, except for friction, the total energy is the same as at point A. Since the frictional force does negative work, the total energy remaining at point B is

$$\begin{aligned} E_{\text{total,B}} &= E_{\text{total,A}} - W_{\text{friction}} \\ &= 47\,385\ \text{J} - (220\ \text{N})(75\ \text{m} + 60\ \text{m}) \\ &= 17\,685\ \text{J} \end{aligned}$$

At point C, the maximum compression point, the remaining energy has gone into compressing the spring a distance $\delta = 0.6$ m and performing a small amount of frictional work.

$$\begin{aligned} E_{\text{total,B}} = E_{\text{total,C}} &= W_{\text{spring}} + W_{\text{friction}} \\ &= \tfrac{1}{2}k\delta^2 + F_f\delta \\ 17\,685\ \text{J} &= \tfrac{1}{2}k(0.6\ \text{m})^2 + (220\ \text{N})(0.6\ \text{m}) \end{aligned}$$

The spring constant can be determined directly.

$$k = 97\,520\ \text{N/m} \quad (97.5\ \text{kN/m})$$

11. POWER

Power is the amount of work done per unit time. It is a scalar quantity. (Although power is calculated from two vectors, the vector dot-product operation is seldom needed.) Typical basic units of power are ft-lbf/sec and watts (J/s), although *horsepower* is widely used. Table 2.2 can be used to convert units of power.

$$P = \frac{W}{\Delta t} \qquad \text{2.25}$$

For a body acted upon by a force or torque, the instantaneous power can be calculated from the velocity.

$$P = F\text{v} \quad \text{[linear systems]} \qquad \text{2.26}$$

$$P = T\omega \quad \text{[rotational systems]} \qquad \text{2.27}$$

For a fluid flowing at a rate of $\dot{m}$, the unit of time is already incorporated into the flow rate (e.g., lbm/sec). If the fluid experiences a specific energy change of Δu, the power generated or dissipated will be

$$P = \dot{m}\Delta u \qquad \text{2.28}$$

Table 2.2 *Useful Power Conversion Formulas*

1 hp	= 550 ft-lbf/sec
	= 33,000 ft-lbf/min
	= 0.7457 kW
	= 0.7068 Btu/sec
1 kW	= 737.6 ft-lbf/sec
	= 44,250 ft-lbf/min
	= 1.341 hp
	= 0.9483 Btu/sec
1 Btu/sec	= 778.17 ft-lbf/sec
	= 46,680 ft-lbf/min
	= 1.415 hp

Example 2.8

When traveling at 100 km/h, a car supplies a constant horizontal force of 50 N to the hitch of a trailer. What tractive power (in horsepower) is required for the trailer alone?

Solution

From Eq. 2.26, the power being generated is

$$\begin{aligned} P = F\text{v} &= \frac{(50\ \text{N})\left(100\ \frac{\text{km}}{\text{h}}\right)\left(1000\ \frac{\text{m}}{\text{km}}\right)}{\left(60\ \frac{\text{s}}{\text{min}}\right)\left(60\ \frac{\text{min}}{\text{h}}\right)\left(1000\ \frac{\text{W}}{\text{kW}}\right)} \\ &= 1.389\ \text{kW} \end{aligned}$$

Using a conversion from Table 2.2, the horsepower is

$$P = \left(1.341\ \frac{\text{hp}}{\text{kW}}\right)(1.389\ \text{kW}) = 1.86\ \text{hp}$$

12. EFFICIENCY

For energy-using systems (such as cars, electrical motors, elevators, etc.), the *energy-use efficiency*, η, of a system is the ratio of an ideal property to an actual property. The property used is commonly work, power, or, for thermodynamics problems, heat. When the rate of work is constant, either work or power can be used to calculate the efficiency. Otherwise, power should be used. Except in rare instances, the numerator and denominator of the ratio must have the same units.[5]

$$\eta = \frac{P_{\text{ideal}}}{P_{\text{actual}}} \quad [P_{\text{actual}} \geq P_{\text{ideal}}] \qquad 2.29$$

For energy-producing systems (such as electrical generators, prime movers, and hydroelectric plants), the *energy-production efficiency* is

$$\eta = \frac{P_{\text{actual}}}{P_{\text{ideal}}} \quad [P_{\text{ideal}} \geq P_{\text{actual}}] \qquad 2.30$$

The efficiency of an *ideal machine* is 1.0 (100%). However, all *real machines* have efficiencies of less than 1.0.

[5]The *energy-efficiency ratio* used to evaluate refrigerators, air conditioners, and heat pumps, for example, has units of Btu per watt-hour (Btu/W-hr).

3 Engineering Drawing Practice

1. INTRODUCTION[1]

This chapter is designed as a brief introduction to the reading of drawings and schematics to which the electrical engineer may be exposed. As many mechanical-electrical interfaces exist, Sec. 3.2 through Sec. 3.11 present the basic conventions used in such drawings. Section 3.13 is an overview of the electrical symbols used in schematics.[2] The various symbols are discussed throughout this book.

2. NORMAL VIEWS OF LINES AND PLANES

A *normal view* of a line is a perpendicular projection of the line onto a viewing plane parallel to the line. In the normal view, all points of the line are equidistant from the observer. Therefore, the true length of a line is viewed and can be measured.

Generally, however, a line will be viewed from an oblique position and will appear shorter than it actually is. The normal view can be constructed by drawing an auxiliary view from the orthographic view.[3] (See Sec. 3.6.)

Similarly, a normal view of a plane figure is a perpendicular projection of the figure onto a viewing plane parallel to the plane of the figure. All points of the plane are equidistant from the observer. Therefore, the true size and shape of any figure in the plane can be determined.

3. INTERSECTING AND PERPENDICULAR LINES

A single orthographic view is not sufficient to determine whether two lines intersect. However, if two or more views show the lines as having the same common point (i.e., crossing at the same position in space), then the lines intersect. In Fig. 3.1, the subscripts F and T refer to front and top views, respectively.

Figure 3.1 Intersecting and Nonintersecting Lines

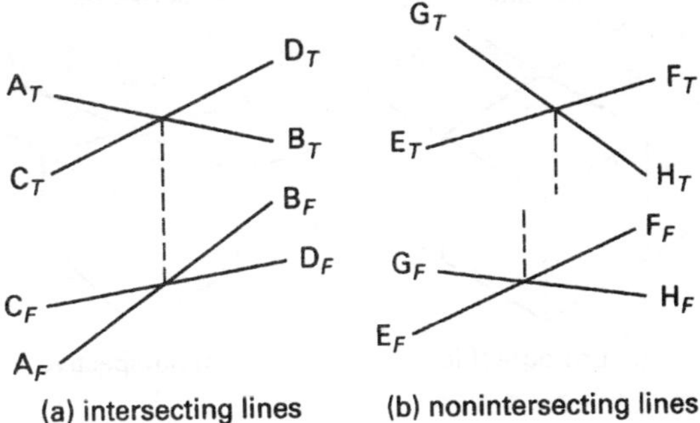

According to the *perpendicular line principle*, two perpendicular lines appear perpendicular only in a normal view of either one or both of the lines. Conversely, if two lines appear perpendicular in any view, the lines are perpendicular only if the view is a normal view of one or both of the lines.

4. TYPES OF VIEWS

Objects can be illustrated in several different ways depending on the number of views, the angle of observation, and the degree of artistic latitude taken for the purpose of simplifying the drawing process.[4] Table 3.1 categorizes the types of views.

The different types of views are easily distinguished by their *projectors* (i.e., projections of parallel lines on the object). For a cube, there are three sets of projectors corresponding to the three perpendicular axes. In an *orthographic* (*orthogonal*) *view*, the projectors are

[1]This chapter is not meant to show "how to do it" as much as it is to present the conventions and symbols of engineering drawing.

[2]A drawing is the surface portrayal of a form or figure in line. A schematic is a presentation of the element-by-element relationship of all parts of a system.

[3]The technique for constructing a normal view is covered in engineering drafting texts.

[4]The omission of perspective from a drawing is an example of a step taken to simplify the drawing process.

parallel. In a *perspective* (*central*) *view*, some or all of the projectors converge to a point. Figure 3.2 illustrates the orthographic and perspective views of a block.

Table 3.1 *Types of Views of Objects*

- orthographic views
  - principal views
  - auxiliary views
  - oblique views
    - cavalier projection
    - cabinet projection
    - clinographic projection
    - axonometric views
      - isometric
      - dimetric
      - trimetric
- perspective views
  - parallel perspective
  - angular perspective

Figure 3.2 *Orthographic and Perspective Views of a Block*

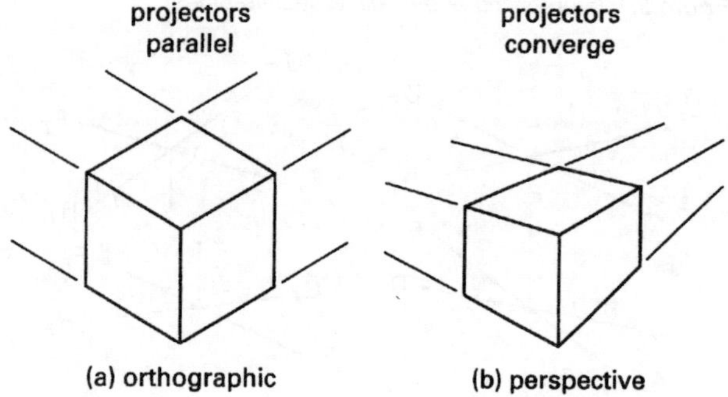

5. PRINCIPAL (ORTHOGRAPHIC) VIEWS

In a *principal view* (also known as a *planar view*), one of the sets of projectors is normal to the view. That is, one of the planes of the object is seen in a normal view. The other two sets of projectors are orthogonal and are usually oriented horizontally and vertically on the paper. Because background details of an object may not be visible in a principal view, it is necessary to have at least three principal views to completely illustrate a symmetrical object. At most, six principal views will be needed to illustrate complex objects.

The relative positions of the six views have been standardized and are shown in Fig. 3.3, which also defines the *width* (also known as *depth*), *height*, and *length* of the object. The views that are not needed to illustrate features or provide dimensions (i.e., *redundant views*) can be omitted. The usual combination selected consists of the top, front, and right side views.

It is common to refer to the front, side, and back views as *elevations* and to the top and bottom views as *plan views*. These terms are not absolute since any plane can be selected as the front.

Figure 3.3 *Positions of Standard Orthographic Views*

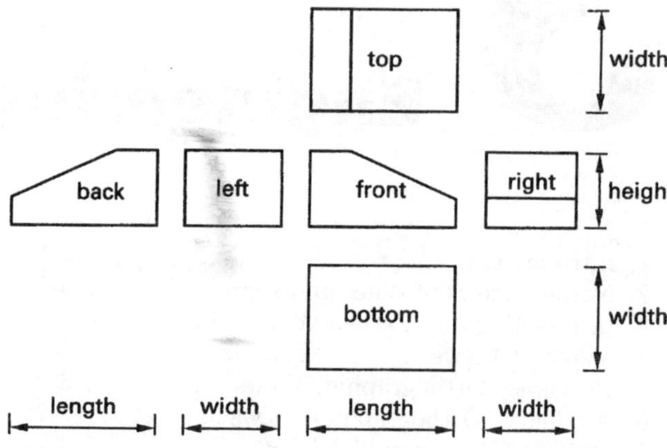

6. AUXILIARY (ORTHOGRAPHIC) VIEWS

An *auxiliary view* is needed when an object has an inclined plane or curved feature or when there are more details than can be shown in the six principal views. As with the other orthographic views, the auxiliary view is a normal (face-on) view of the inclined plane. Figure 3.4 illustrates an auxiliary view.

Figure 3.4 *Auxiliary View*

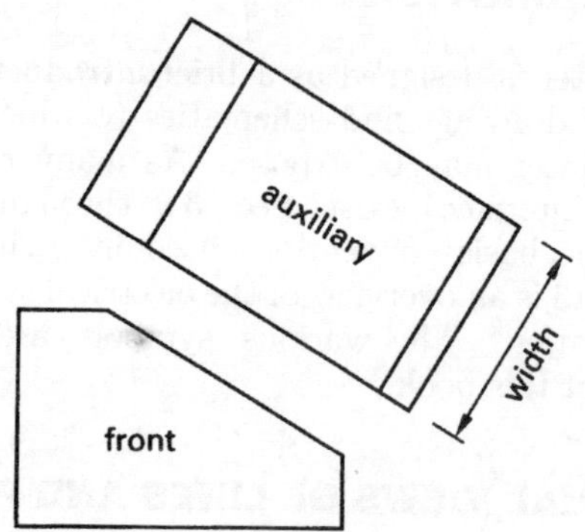

The projectors in an auxiliary view are perpendicular to only one of the directions in which a principal view is observed. Accordingly, only one of the three dimensions of width, height, and depth can be measured (scaled). In a *profile auxiliary view*, the object's width can be measured. In a *horizontal auxiliary view* (*auxiliary elevation*), the object's height can be measured. In a *frontal auxiliary view*, the depth of the object can be measured.

7. OBLIQUE (ORTHOGRAPHIC) VIEWS

If the object is turned so that three principal planes are visible, it can be completely illustrated by a single *oblique view*.[5] In an oblique view, the direction from

[5]Oblique views are not unique in this capability—perspective drawings share it. Oblique and perspective drawings are known as *pictorial drawings* because they give depth to the object by illustrating it in three dimensions.

which the object is observed is not (necessarily) parallel to any of the directions from which principal and auxiliary views are observed.

In two common methods of oblique illustration, one of the view planes coincides with an orthographic view plane. Two of the drawing axes are at right angles to each other; one of these is vertical, and the other (the *oblique axis*) is oriented at 30° or 45° (originally chosen to coincide with standard drawing triangles). The ratio of scales used for the horizontal, vertical, and oblique axes can be 1:1:1 or 1:1:$^1/_2$. The latter ratio helps to overcome the visual distortion due to the absence of perspective in the oblique direction.

Cavalier (45° oblique axis and 1:1:1 scale ratio) and *cabinet* (45° oblique axis and 1:1:$^1/_2$ scale ratio) *projections* are the two common types of oblique views that incorporate one of the orthographic views. If an angle of 9.5° is used (as in illustrating crystalline lattice structures), the technique is known as *clinographic projection.* Figure 3.5 illustrates cavalier and cabinet oblique drawings.

Figure 3.5 *Cavalier and Cabinet Oblique Drawings*

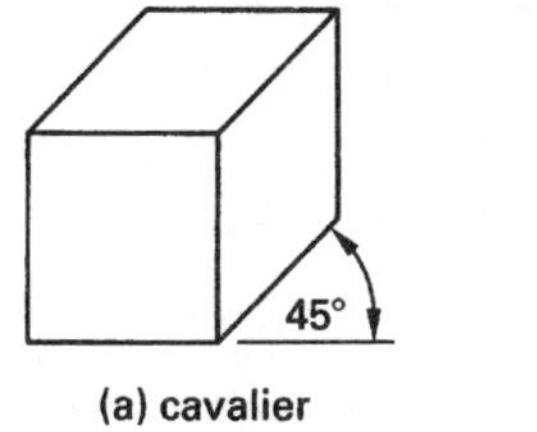

(a) cavalier

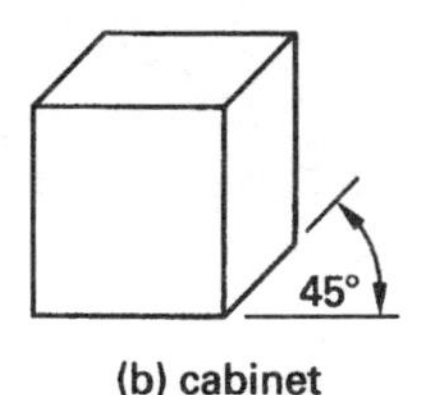

(b) cabinet

8. AXONOMETRIC (ORTHOGRAPHIC OBLIQUE) VIEWS

In axonometric views, the view plane is not parallel to any of the principal orthographic planes. Figure 3.6 illustrates types of axonometric views. Axonometric views and axonometric drawings are not the same. In a *view (projection)*, one or more of the face lengths is foreshortened. In a *drawing*, the lengths are drawn full length, resulting in a distorted illustration. Table 3.2 lists the proper ratios that should be observed.

In an *isometric view*, the three projectors intersect at equal angles (120°) with the plane. This simplifies construction with standard 30° drawing triangles. All of the faces are foreshortened an equal amount, to $\sqrt{2/3}$, or approximately 81.6% of the true length. In a *dimetric view*, two of the projectors intersect at equal angles, and only two of the faces are equally reduced in length. In a *trimetric view*, all three intersection angles are different, and all three faces are reduced different amounts.

Table 3.2 *Axonometric Foreshortening*

view	projector intersection angles	proper ratio of sides
isometric	120°, 120°, 120°	0.82:0.82:0.82
dimetric	131°25′, 131°25′, 97°10′	1:1:$^1/_2$
	103°38′, 103°38′, 152°44′	$^3/_4$:$^3/_4$:1
trimetric	102°28′, 144°16′, 113°16′	1:$^2/_3$:$^7/_8$
	138°14′, 114°46′, 107°	1:$^3/_4$:$^7/_8$

Figure 3.6 *Types of Axonometric Views*

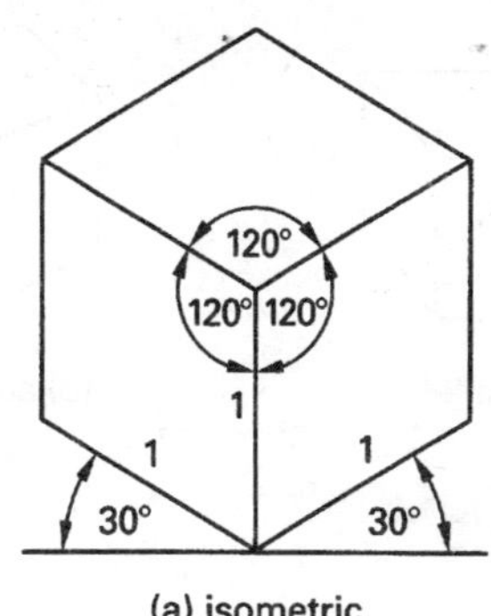

(a) isometric

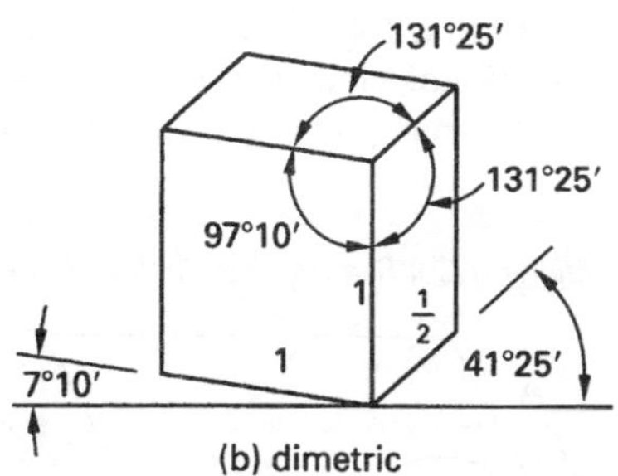

(b) dimetric

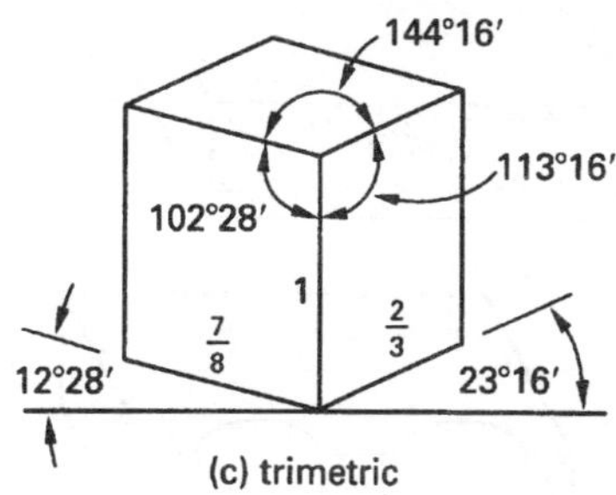

(c) trimetric

9. PERSPECTIVE VIEWS

In a *perspective view*, one or more sets of projectors converge to a fixed point known as the *center of vision.* In the *parallel perspective*, all vertical lines remain vertical in the picture; all horizontal frontal lines remain horizontal. Therefore, one face is parallel to the observer and only one set of projectors converges. In the *angular perspective*, two sets of projectors converge. In the little-used *oblique perspective*, all three sets of projectors converge. Figure 3.7 illustrates types of perspective views.

Mathematics

Figure 3.7 Types of Perspective Views

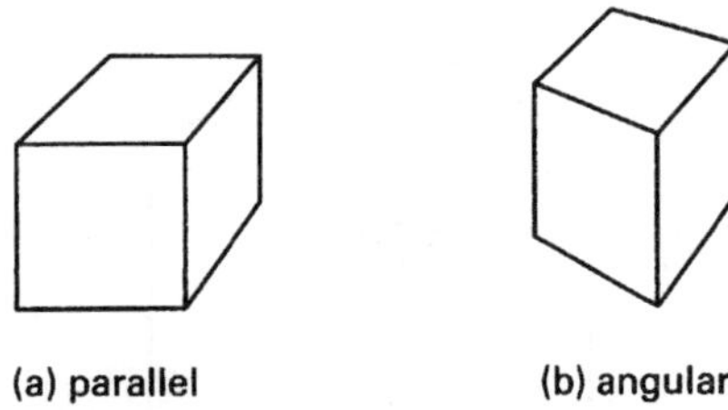

10. SECTIONS

A *section* is an imaginary cut taken through an object to reveal the shape or interior construction.[6] Figure 3.8 illustrates the standard symbol for a *sectioning cut* and the resulting sectional view. Section arrows are perpendicular to the cutting plane and indicate the viewing direction.

Figure 3.8 Sectioning Cut Symbol and Sectional View

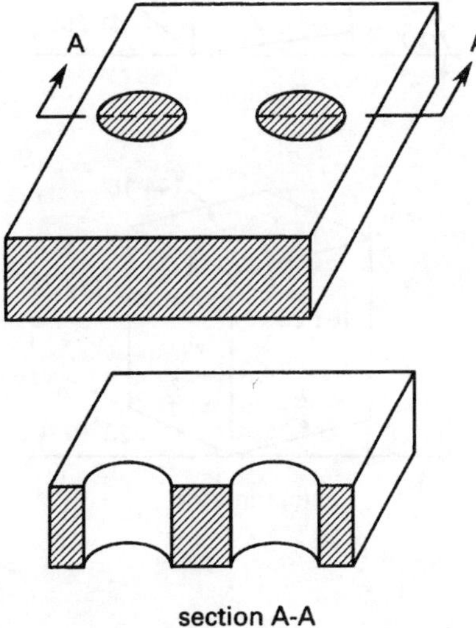

11. TOLERANCES

The *tolerance* for a dimension is the total permissible variation or difference between the acceptable limits. The tolerance for a dimension can be specified in two ways: either as a general rule in the title block (e.g., ±0.001 in unless otherwise specified) or as specific limits that are given with each dimension (e.g., 2.575 in ±0.005 in).

12. SURFACE FINISH

ANSI B46.1 specifies surface finish by a combination of parameters.[7] The basic symbol for designating these factors is shown in Fig. 3.9. In the symbol, A is the maximum *roughness height index*, B is the optional minimum *roughness height*, C is the peak-to-valley *waviness height*, D is the optional peak-to-valley *waviness spacing* (*width*) rating, E is the optional *roughness width cutoff* (*roughness sampling length*), F is the *lay*, and G is the *roughness width*. Unless minimums are specified, all parameters are maximum allowable values, and all lesser values are permitted.

Since the roughness varies, the waviness height is an arithmetic average within a sampled square, and the designation A is known as the *roughness weight*, R_a.[8] Values are normally given in microns (μm) or microinches (μin) in SI or customary U.S. units, respectively. A value for the roughness width cutoff of 0.80 mm (0.03 in) is assumed when E is not specified. Other standard values in common use are 0.25 mm (0.010 in) and 0.08 mm (0.003 in). The lay symbol, F, can be = (parallel to indicated surface), ⊥ (perpendicular), C (circular), M (multidirectional), P (pitted), R (radial), or X (crosshatch).

If a small circle is placed at the A position, no machining is allowed and only cast, forged, die-cast, injection-molded, and other unfinished surfaces are acceptable.

Figure 3.9 Surface Finish Designations

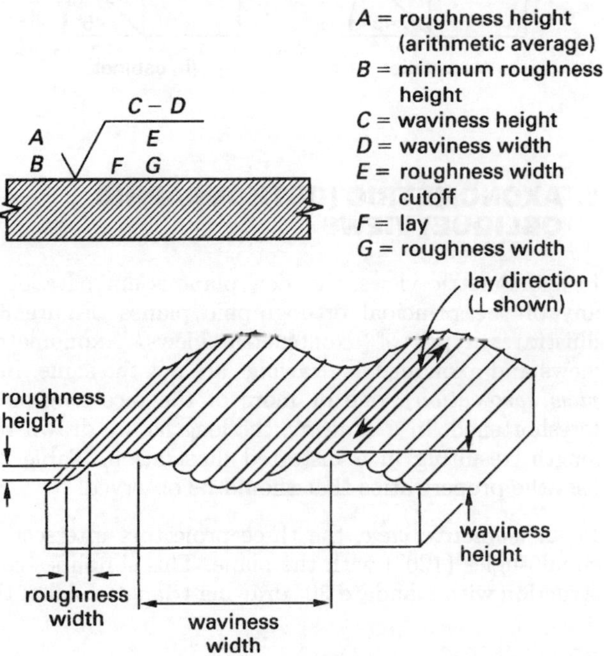

[6]The term *section* is also used to mean a *cross section*—a slice of finite but negligible thickness that is taken from an object to show the cross section or interior construction at the plane of the slice.

[7]Specification does not indicate appearance (i.e., color, luster) or performance (i.e., hardness, corrosion resistance, microstructure).

[8]The symbol R_a is the same as the AA (arithmetic average) and CLA (centerline average) terms used in other (and earlier) standards.

13. ELECTRICAL SYMBOLS

Basic vacuum tube symbols are shown in Table 3.3. Symbols for common types of basic circuit elements (e.g., resistors, inductors, and capacitors) are shown in Table 3.4.

Table 3.3 *Vacuum Tube Symbols*

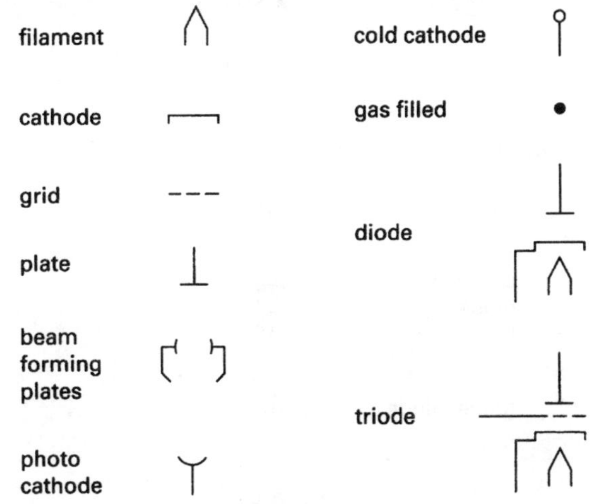

Table 3.4 *Basic Circuit Element Symbols*

resistors:
- fixed
- variable
- fusible
- light dependent
- temperature dependent ($t°$)

inductors*:
- fixed air core
- fixed iron core (magnetic core)
- variable

*Both the symbols and are used, with the former more commonly used to represent transformers.

capacitors:
- unpolarized
- polarized
- variable or tuning
- ganged

One-line electrical diagrams, often associated with power equipment or transmission line representations, use a set of standard symbols. These symbols are present on electrical and electronic schematics, especially those for ground connections, connections in general, and electrical rotating machinery. Table 3.5 shows some of these symbols.

The basic symbols for semiconductor devices are displayed in Table 3.6.

Always refer to the legend of a particular drawing or schematic to ensure understanding of the symbols utilized.

Table 3.5 *One-Line Electrical Diagram Symbols*

protective devices:
- protective gap
- fuse
- thermal overload device
- magnetic overload

circuit breakers:
- air
- drawout type

winding symbols:
- single phase
- two phase
- three-phase wye (or star)
- three-phase wye (grounded)
- three-phase delta

connections:
- common tie
- ground
- chassis ground

Table 3.6 *Semiconductor Symbols*

p-region on *n*-region

[*p*] [*n*]

or

[*p*] [*n*]

n-region on *p*-region

[*n*] [*p*]

or

[*n*] [*p*]

transistors:

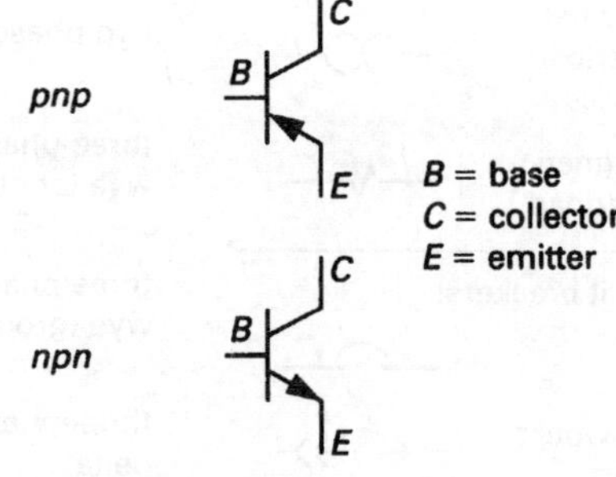

field effect transistors (FETs)*:

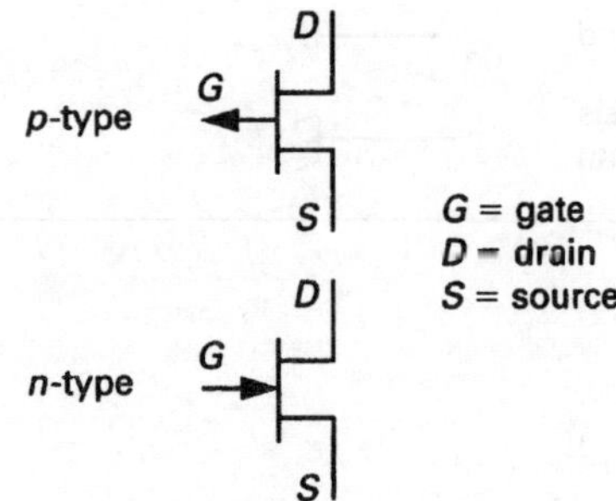

*The gate and source can be shown closer together to aid in differentiating the source and the drain.

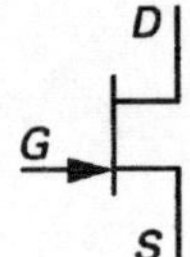

unijunction transistors (UJTs):

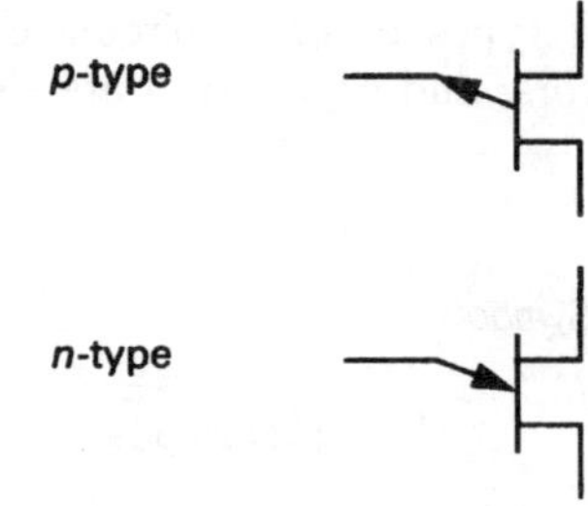

n-type

two-terminal devices:

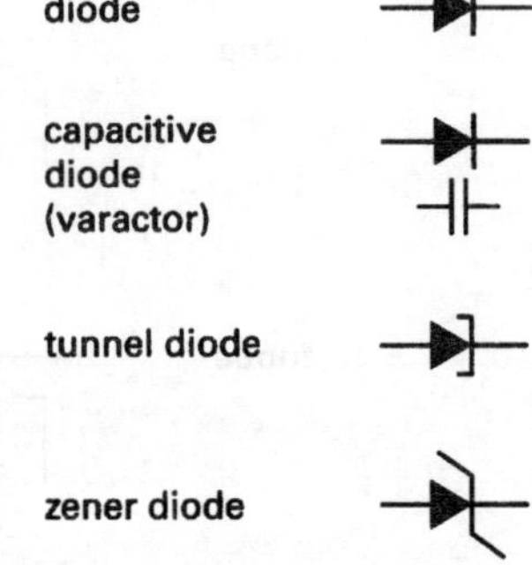

three-terminal device:

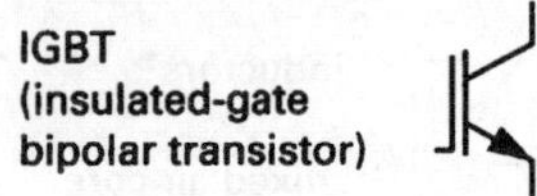

photodiode:

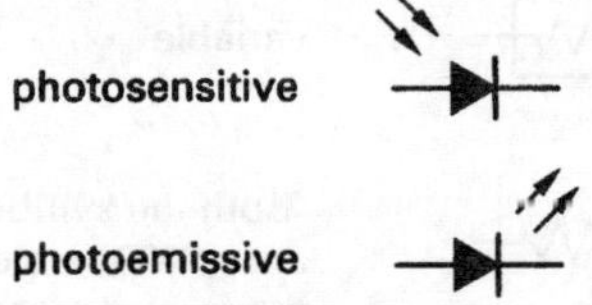

miscellaneous devices:

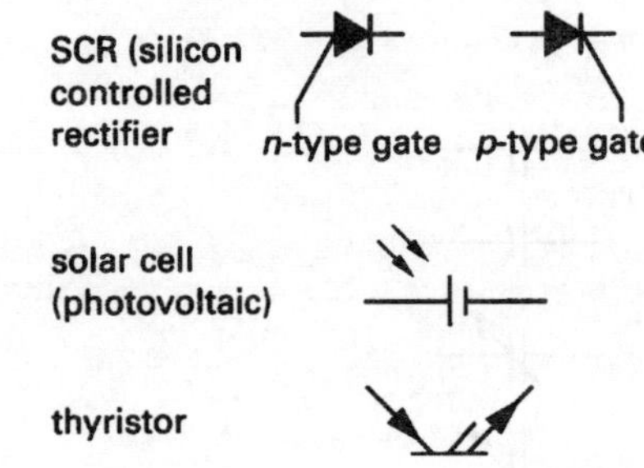

4 Algebra

1. INTRODUCTION

Engineers working in design and analysis encounter mathematical problems on a daily basis. Although algebra and simple trigonometry are often sufficient for routine calculations, there are many instances when certain advanced subjects are needed. This chapter and the following, in addition to supporting the calculations used in other chapters, consolidate the mathematical concepts most often needed by engineers.

2. SYMBOLS USED IN THIS BOOK

Many symbols, letters, and Greek characters are used to represent variables in the formulas used throughout this book. These symbols and characters are defined in the nomenclature section of each chapter. However, some of the other symbols in this book are listed in Table 4.2.

3. GREEK ALPHABET

Table 4.1 lists the Greek alphabet.

Table 4.1 The Greek Alphabet

A	α	alpha	N	ν	nu
B	β	beta	Ξ	ξ	xi
Γ	γ	gamma	O	o	omicron
Δ	δ	delta	Π	π	pi
E	ϵ	epsilon	P	ρ	rho
Z	ζ	zeta	Σ	σ	sigma
H	η	eta	T	τ	tau
Θ	θ	theta	Υ	υ	upsilon
I	ι	iota	Φ	ϕ	phi
K	κ	kappa	X	χ	chi
Λ	λ	lambda	Ψ	ψ	psi
M	μ	mu	Ω	ω	omega

4. TYPES OF NUMBERS

The *numbering system* consists of three types of numbers: real, imaginary, and complex. *Real numbers*, in turn, consist of rational numbers and irrational numbers. *Rational real numbers* are numbers that can be written as the ratio of two integers (e.g., 4, $^2/_5$, and $^1/_3$).[1] *Irrational real numbers* are nonterminating, nonrepeating numbers that cannot be expressed as the ratio of two integers (e.g., π and $\sqrt{2}$). Real numbers can be positive or negative.

Imaginary numbers are square roots of negative numbers. The symbols i and j are both used to represent the square root of -1.[2] For example, $\sqrt{-5} = \sqrt{5}\sqrt{-1} = \sqrt{5}i$. *Complex numbers* consist of combinations of real and imaginary numbers (e.g., $3 - 7i$).

5. SIGNIFICANT DIGITS

The significant digits in a number include the leftmost, nonzero digits to the rightmost digit written. Final answers from computations should be rounded off to the number of decimal places justified by the data.

[1]Notice that 0.3333333 is a nonterminating number, but as it can be expressed as a ratio of two integers (i.e., 1/3), it is a rational number.

[2]The symbol j is used to represent the square root of -1 in electrical calculations to avoid confusion with the current variable, i.

Mathematics

Table 4.2 Symbols Used in This Book

symbol	name	use	example
Σ	sigma	series summation	$\sum_{i=1}^{3} x_i = x_1 + x_2 + x_3$
π	pi	3.1415926...	$P = \pi D$
e	base of natural logs	2.71828...	
Π	pi	series multiplication	$\prod_{i=1}^{3} x_i = x_1 x_2 x_3$
Δ	delta	change in quantity	$\Delta h = h_2 - h_1$
$\overline{}$	over bar	average value	$\overline{x}$
$\dot{}$	over dot	per unit time	$\dot{m}$ = mass flowing per second
!	factorial[a]		$x! = x(x-1)(x-2)\cdots(2)(1)$
$\vert\;\vert$	absolute value[b]		$\vert -3 \vert = +3$
$\sim$	similarity		$\Delta ABC \sim \Delta DEF$
$\approx$	approximately equal to		$x \approx 1.5$
$\cong$	congruency		$ST \cong UV$
$\propto$	proportional to		$x \propto y$
$\equiv$	equivalent to		$a + bi \equiv re^{i\theta}$
∞	infinity		$x \to \infty$
log	base-10 logarithm		$\log 5.74$
ln	natural logarithm		$\ln 5.74$
exp	exponential power		$\exp(x) = e^x$
rms	root-mean-square		$V_{rms} = \sqrt{\frac{1}{n}\sum_{i=1}^{n} V_i^2}$
$\angle$	phasor or angle		$\angle 53°$

[a]*Zero factorial* (0!) is frequently encountered in the form of $(n-n)!$ when calculating permutations and combinations. Zero factorial is defined as 1.
[b]The notation abs(x) is also used to indicate the absolute value.

The answer can be no more accurate than the least accurate number in the data. Of course, rounding should be done on final calculation results only. It should not be done on interim results.

There are two ways that significant digits can affect calculations. For the operations of multiplication and division, the final answer is rounded to the number of significant digits in the least significant multiplicand, divisor, or dividend. So, $2.0 \times 13.2 = 26$ since the first multiplicand (2.0) has two significant digits only.

For the operations of addition and subtraction, the final answer is rounded to the position of the least significant digit in the addenda, minuend, or subtrahend. So, $2.0 + 13.2 = 15.2$ because both addenda are significant to the tenth's position; but $2 + 13.4 = 15$ since the 2 is significant only in the ones' position.

The multiplication rule should not be used for addition or subtraction, as this can result in strange answers. For example, it would be incorrect to round 1700 + 0.1 to 2000 simply because 0.1 has only one significant digit. Table 4.3 gives examples of significant digits.

When rounding to fewer digits than available, the *American National Standard for Metric Practice* (IEEE/ASTM SI 10) dictates that the last digit retained should be rounded up if the number is odd and rounded down if the number is even. For example, 2.55 and 2.65 rounded to two significant digits would both be 2.6. This allows for a balance in the rounding of numbers, whereas always rounding five up or down would not.

Table 4.3 Examples of Significant Digits

number as written	number of significant digits	implied range
341	3	340.5–341.5
34.1	3	34.05–34.15
0.00341	3	0.003405–0.003415
341×10^7	3	340.5×10^7–341.5×10^7
3.41×10^{-2}	3	3.405×10^{-2}–3.415×10^{-2}
3410	3	3405–3415
3410*	4	3409.5–3410.5
341.0	4	340.95–341.05

*It is permitted to write "3410." to distinguish the number from its three-significant-digit form, although this is rarely done.

6. EQUATIONS

An *equation* is a mathematical statement of equality, such as $5 = 3 + 2$. *Algebraic equations* are written in terms of *variables*. In the equation $y = x^2 + 3$, the value of variable y depends on the value of variable x. Therefore, y is the *dependent variable* and x is the *independent variable*. The dependency of y on x is clearer when the equation is written in *functional form*: $y = f(x)$.

A *parametric equation* uses one or more independent variables (*parameters*) to describe a function.[3] For example, the parameter θ can be used to write the parametric equations of a unit circle.

$$x = \cos\theta \tag{4.1}$$

$$y = \sin\theta \tag{4.2}$$

A unit circle can also be described by a *nonparametric equation*.[4]

$$x^2 + y^2 = 1 \tag{4.3}$$

7. FUNDAMENTAL ALGEBRAIC LAWS

Algebra provides the rules that allow complex mathematical relationships to be expanded or condensed. Algebraic laws may be applied to complex numbers, variables, and real numbers. The general rules for changing the form of a mathematical relationship are given as follows.

- commutative law for addition:

$$A + B = B + A \tag{4.4}$$

- commutative law for multiplication:

$$AB = BA \tag{4.5}$$

- associative law for addition:

$$A + (B + C) = (A + B) + C \tag{4.6}$$

- associative law for multiplication:

$$A(BC) = (AB)C \tag{4.7}$$

- distributive law:

$$A(B + C) = AB + AC \tag{4.8}$$

8. POLYNOMIALS

A *polynomial* is a rational expression—usually the sum of several variable terms known as *monomials*—that does not involve division. The *degree of the polynomial* is the highest power to which a variable in the expression is raised. The following *standard polynomial forms* are useful when trying to find the roots of an equation.

$$(a + b)(a - b) = a^2 - b^2 \tag{4.9}$$

$$(a \pm b)^2 = a^2 \pm 2ab + b^2 \tag{4.10}$$

$$(a \pm b)^3 = a^3 \pm 3a^2b + 3ab^2 \pm b^3 \tag{4.11}$$

$$(a^3 \pm b^3) = (a \pm b)(a^2 \mp ab + b^2) \tag{4.12}$$

$$(a^n - b^n) = (a - b)\begin{pmatrix} a^{n-1} + a^{n-2}b + a^{n-3}b^2 \\ + \cdots + b^{n-1} \end{pmatrix}$$

$$[n \text{ is any positive integer}] \tag{4.13}$$

$$(a^n + b^n) = (a + b)\begin{pmatrix} a^{n-1} - a^{n-2}b + a^{n-3}b^2 \\ - \cdots + b^{n-1} \end{pmatrix}$$

$$[n \text{ is any positive odd integer}] \tag{4.14}$$

The *binomial theorem* defines a polynomial of the form $(a + b)^n$.

$$(a+b)^n = \underset{[i=0]}{a^n} + \underset{[i=1]}{na^{n-1}b} + \underset{[i=2]}{C_2a^{n-2}b^2} + \cdots$$

$$+ C_ia^{n-i}b^i + \cdots + nab^{n-1} + b^n \tag{4.15}$$

$$C_i = \frac{n!}{i!(n-i)!} \quad [i = 0, 1, 2, \ldots, n] \tag{4.16}$$

The coefficients of the expansion can be determined quickly from *Pascal's triangle*—each entry is the sum of the two entries directly above it. (See Fig. 4.1.)

Figure 4.1 *Pascal's Triangle*

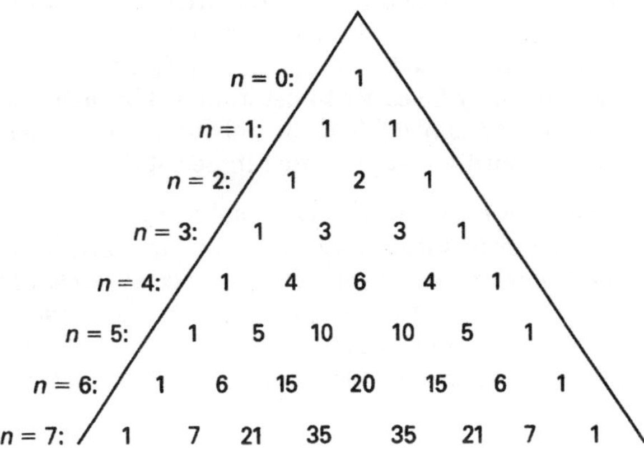

9. ROOTS OF QUADRATIC EQUATIONS

A *quadratic equation* is an equation of the general form $ax^2 + bx + c = 0$ $[a \neq 0]$. The *roots*, x_1 and x_2, of the equation are the two values of x that satisfy it.

$$x_1, x_2 = \frac{-b \pm \sqrt{b^2 - 4ac}}{2a} \tag{4.17}$$

$$x_1 + x_2 = -\frac{b}{a} \tag{4.18}$$

$$x_1x_2 = \frac{c}{a} \tag{4.19}$$

[3]As used in this section, there is no difference between a parameter and an independent variable. However, the term *parameter* is also used as a descriptive measurement that determines or characterizes the form, size, or content of a function. For example, the radius is a parameter of a circle, and mean and variance are parameters of a probability distribution. Once these parameters are specified, the function is completely defined.

[4]Since only the coordinate variables are used, this equation is also said to be in *Cartesian equation form*.

The types of roots of the equation can be determined from the *discriminant* (i.e., the quantity under the radical in Eq. 4.17).

- If $b^2 - 4ac > 0$, the roots are real and unequal.
- If $b^2 - 4ac = 0$, the roots are real and equal. This is known as a *double root.*
- If $b^2 - 4ac < 0$, the roots are complex and unequal.

10. ROOTS OF GENERAL POLYNOMIALS

It is difficult to find roots of cubic and higher-degree polynomials because few general techniques exist. *Cardano's formula* (*method*) uses closed-form equations to laboriously calculate roots for general cubic (3rd degree) polynomials. Compromise methods are used when solutions are needed on the spot.

- *inspection and trial and error:* Finding roots by inspection is equivalent to making reasonable guesses about the roots and substituting into the polynomial.
- *graphing:* If the value of a polynomial $f(x)$ is calculated and plotted for different values of x, an approximate value of a root can be determined as the value of x at which the plot crosses the x-axis.
- *factoring:* If at least one root (say, $x = r$) of a polynomial $f(x)$ is known, the quantity $x - r$ can be factored out of $f(x)$ by long division. The resulting quotient will be lower by one degree, and the remaining roots may be easier to determine. This method is particularly applicable if the polynomial is in one of the standard forms presented in Sec. 4.8.
- *special cases:* Certain polynomial forms can be simplified by substitution or solved by standard formulas if they are recognized as being special cases. (The standard solution to the quadratic equation is such a special case.) For example, $ax^4 + bx^2 + c = 0$ can be reduced to a polynomial of degree 2 if the substitution $u = x^2$ is made.
- *numerical methods:* If an approximate value of a root is known, numerical methods (bisection method, Newton's method, etc.) can be used to refine the value. The more efficient techniques are too complex to be performed by hand.

11. EXTRANEOUS ROOTS

With simple equalities, it may appear possible to derive roots by basic algebraic manipulations.[5] However, multiplying each side of an equality by a power of a variable may introduce *extraneous roots.* Such roots do not satisfy the original equation even though they are derived according to the rules of algebra. Checking a calculated root is always a good idea, but is particularly necessary if the equation has been multiplied by one of its own variables.

Example 4.1

Use algebraic operations to determine a value that satisfies the following equation. Determine if the value is a valid or extraneous root.

$$\sqrt{x-2} = \sqrt{x} + 2$$

Solution

Square both sides.

$$x - 2 = x + 4\sqrt{x} + 4$$

Subtract x from each side, and combine the constants.

$$4\sqrt{x} = -6$$

Solve for x.

$$x = \left(\frac{-6}{4}\right)^2 = \frac{9}{4}$$

Substitute $x = 9/4$ into the original equation.

$$\sqrt{\frac{9}{4} - 2} = \sqrt{\frac{9}{4}} + 2$$

$$\frac{1}{2} = \frac{7}{2}$$

Since the equality is not established, $x = 9/4$ is an extraneous root.

12. DESCARTES' RULE OF SIGNS

Descartes' rule of signs determines the maximum number of positive (and negative) real roots that a polynomial will have by counting the number of sign reversals (i.e., changes in sign from one term to the next) in the polynomial. The polynomial $f(x) = 0$ must have real coefficients and must be arranged in terms of descending powers of x.

- The number of positive roots of the polynomial equation $f(x) = 0$ will not exceed the number of sign reversals.
- The difference between the number of sign reversals and the number of positive roots is an even number.
- The number of negative roots of the polynomial equation $f(x) = 0$ will not exceed the number of sign reversals in the polynomial $f(-x)$.
- The difference between the number of sign reversals in $f(-x)$ and the number of negative roots is an even number.

[5]In this sentence, *equality* means a combination of two expressions containing an equal sign. Any two expressions can be linked in this manner, even those that are not actually equal. For example, the expressions for two nonintersecting ellipses can be equated even though there is no intersection point. Finding extraneous roots is more likely when the underlying equality is false to begin with.

Example 4.2

Determine the possible numbers of positive and negative roots that satisfy the following polynomial equation.

$$4x^5 - 5x^4 + 3x^3 - 8x^2 - 2x + 3 = 0$$

Solution

There are four sign reversals, so up to four positive roots exist. To keep the difference between the number of positive roots and the number of sign reversals an even number, the number of positive real roots is limited to zero, two, and four.

Substituting $-x$ for x in the polynomial results in

$$-4x^5 - 5x^4 - 3x^3 - 8x^2 + 2x + 3 = 0$$

There is only one sign reversal, so the number of negative roots cannot exceed one. There must be exactly one negative real root in order to keep the difference to an even number (zero in this case).

13. RULES FOR EXPONENTS AND RADICALS

In the expression $b^n = a$, b is known as the *base* and n is the *exponent* or *power*. In Eq. 4.20 through Eq. 4.33, a, b, m, and n are any real numbers with limitations listed.

$$b^0 = 1 \quad [b \neq 0] \qquad 4.20$$

$$b^1 = b \qquad 4.21$$

$$b^{-n} = \frac{1}{b^n} = \left(\frac{1}{b}\right)^n \quad [b \neq 0] \qquad 4.22$$

$$\left(\frac{a}{b}\right)^n = \frac{a^n}{b^n} \quad [b \neq 0] \qquad 4.23$$

$$(ab)^n = a^n b^n \qquad 4.24$$

$$b^{m/n} = \sqrt[n]{b^m} = (\sqrt[n]{b})^m \qquad 4.25$$

$$(b^n)^m = b^{nm} \qquad 4.26$$

$$b^m b^n = b^{m+n} \qquad 4.27$$

$$\frac{b^m}{b^n} = b^{m-n} \quad [b \neq 0] \qquad 4.28$$

$$\sqrt[n]{b} = b^{1/n} \qquad 4.29$$

$$(\sqrt[n]{b})^n = (b^{1/n})^n = b \qquad 4.30$$

$$\sqrt[n]{ab} = \sqrt[n]{a}\sqrt[n]{b} = a^{1/n}b^{1/n} = (ab)^{1/n} \qquad 4.31$$

$$\sqrt[n]{\frac{a}{b}} = \frac{\sqrt[n]{a}}{\sqrt[n]{b}} = \left(\frac{a}{b}\right)^{1/n} \quad [b \neq 0] \qquad 4.32$$

$$\sqrt[m]{\sqrt[n]{b}} = \sqrt[mn]{b} = b^{1/mn} \qquad 4.33$$

14. LOGARITHMS

Logarithms can be considered to be exponents. For example, the exponent n in the expression $b^n = a$ is the logarithm of a to the base b. Therefore, the two expressions $\log_b a = n$ and $b^n = a$ are equivalent.

The base for *common logs* is 10. Usually, "log" will be written when common logs are desired, although "$\log_{10}$" appears occasionally. The base for *natural* (*Napierian*) *logs* is 2.71828..., a number which is given the symbol e. When natural logs are desired, usually "ln" will be written, although "$\log_e$" is also used.

Most logarithms will contain an integer part (the *characteristic*) and a decimal part (the *mantissa*). The common and natural logarithms of any number less than one are negative. If the number is greater than one, its common and natural logarithms are positive. Although the logarithm may be negative, the mantissa is always positive. For negative logarithms, the characteristic is found by expressing the logarithm as the sum of a negative characteristic and a positive mantissa.

For common logarithms of numbers greater than one, the characteristics will be positive and equal to one less than the number of digits in front of the decimal. If the number is less than one, the characteristic will be negative and equal to one more than the number of zeros immediately following the decimal point.

If a negative logarithm is to be used in a calculation, it must first be converted to *operational form* by adding the characteristic and mantissa. The operational form should be used in all calculations and is the form displayed by scientific calculators.

The logarithm of a negative number is a complex number.

Example 4.3

Use logarithm tables to determine the operational form of $\log_{10} 0.05$.

Solution

Since the number is less than one and there is one leading zero, the characteristic is found by observation to be −2. From a book of logarithm tables, the mantissa of 5.0 is 0.699. Two ways of expressing the combination of mantissa and characteristic are used.

method 1: $\bar{2}.699$

method 2: $8.699 - 10$

The operational form of this logarithm is $-2 + 0.699 = -1.301$.

15. LOGARITHM IDENTITIES

Prior to the widespread availability of calculating devices, logarithm identities were used to solve complex calculations by reducing the solution method to table look-up, addition, and subtraction. Logarithm identities are still useful in simplifying expressions containing exponentials and other logarithms. In Eq. 4.34 through Eq. 4.45, $a \neq 1$, $b \neq 1$, $x > 0$, and $y > 0$.

$$\log_b b = 1 \quad 4.34$$

$$\log_b 1 = 0 \quad 4.35$$

$$\log_b b^n = n \quad 4.36$$

$$\log x^a = a \log x \quad 4.37$$

$$\log \sqrt[n]{x} = \log x^{1/n} = \frac{\log x}{n} \quad 4.38$$

$$b^{n \log_b x} = x^n = \text{antilog}(n \log_b x) \quad 4.39$$

$$b^{\log_b x/n} = x^{1/n} \quad 4.40$$

$$\log xy = \log x + \log y \quad 4.41$$

$$\log \frac{x}{y} = \log x - \log y \quad 4.42$$

$$\log_a x = \log_b x \log_a b \quad 4.43$$

$$\ln x = \ln 10 \log_{10} x \approx 2.3026 \log_{10} x \quad 4.44$$

$$\log_{10} x = \log_{10} \ln x \, e \approx 0.4343 \ln x \quad 4.45$$

Example 4.4

The surviving fraction, f, of a radioactive isotope is given by $f = e^{-0.005t}$. For what value of t will the surviving percentage be 7%?

Solution

$$f = 0.07 = e^{-0.005t}$$

Take the natural log of both sides.

$$\ln 0.07 = \ln e^{-0.005t}$$

From Eq. 4.36, $\ln e^x = x$. Therefore,

$$-2.66 = -0.005t$$

$$t = 532$$

16. PARTIAL FRACTIONS

The method of *partial fractions* is used to transform a proper polynomial fraction of two polynomials into a sum of simpler expressions, a procedure known as *resolution*.[6,7] The technique can be considered to be the act of "unadding" a sum to obtain all of the addends.

Suppose $H(x)$ is a proper polynomial fraction of the form $P(x)/Q(x)$. The object of the resolution is to determine the partial fractions u_1/v_1, u_2/v_2, and so on, such that

$$H(x) = \frac{P(x)}{Q(x)} = \frac{u_1}{v_1} + \frac{u_2}{v_2} + \frac{u_3}{v_3} + \cdots \quad 4.46$$

The form of the denominator polynomial $Q(x)$ will be the main factor in determining the form of the partial fractions. The task of finding the u_i and v_i is simplified by categorizing the possible forms of $Q(x)$.

case 1: $Q(x)$ factors into n different linear terms.

$$Q(x) = (x - a_1)(x - a_2)\cdots(x - a_n) \quad 4.47$$

Then,

$$H(x) = \sum_{i=1}^{n} \frac{A_i}{x - a_i} \quad 4.48$$

case 2: $Q(x)$ factors into n identical linear terms.

$$Q(x) = (x - a)(x - a)\cdots(x - a) \quad 4.49$$

Then,

$$H(x) = \sum_{i=1}^{n} \frac{A_i}{(x - a)^i} \quad 4.50$$

case 3: $Q(x)$ factors into n different quadratic terms, $x^2 + p_i x + q_i$.

Then,

$$H(x) = \sum_{i=1}^{n} \frac{A_i x + B_i}{x^2 + p_i x + q_i} \quad 4.51$$

case 4: $Q(x)$ factors into n identical quadratic terms, $x^2 + px + q$.

Then,

$$H(x) = \sum_{i=1}^{n} \frac{A_i x + B_i}{(x^2 + px + q)^i} \quad 4.52$$

Once the general forms of the partial fractions have been determined from inspection, the *method of undetermined coefficients* is used. The partial fractions are all cross-multiplied to obtain $Q(x)$ as the denominator, and the coefficients are found by equating $P(x)$ and the cross-multiplied numerator.

Example 4.5

Resolve $H(x)$ into partial fractions.

$$H(x) = \frac{x^2 + 2x + 3}{x^4 + x^3 + 2x^2}$$

[6]To be a *proper polynomial fraction*, the degree of the numerator must be less than the degree of the denominator. If the polynomial fraction is improper, the denominator can be divided into the numerator to obtain whole and fractional polynomials. The method of partial fractions can then be used to reduce the fractional polynomial.

[7]This technique is particularly useful for calculating integrals and inverse Laplace transforms in subsequent chapters.

Solution

Here, $Q(x) = x^4 + x^3 + 2x^2$ factors into $x^2(x^2 + x + 2)$. This is a combination of cases 2 and 3.

$$H(x) = \frac{A_1}{x} + \frac{A_2}{x^2} + \frac{A_3 + A_4 x}{x^2 + x + 2}$$

Cross multiplying to obtain a common denominator yields

$$\frac{(A_1 + A_4)x^3 + (A_1 + A_2 + A_3)x^2 + (2A_1 + A_2)x + 2A_2}{x^4 + x^3 + 2x^2}$$

Since the original numerator is known, the following simultaneous equations result.

$$\begin{aligned} A_1 + A_4 &= 0 \\ A_1 + A_2 + A_3 &= 1 \\ 2A_1 + A_2 &= 2 \\ 2A_2 &= 3 \end{aligned}$$

The solutions are $A_1 = 0.25$; $A_2 = 1.50$; $A_3 = -0.75$; and $A_4 = -0.25$.

$$H(x) = \frac{1}{4x} + \frac{3}{2x^2} - \frac{x+3}{4(x^2 + x + 2)}$$

17. SIMULTANEOUS LINEAR EQUATIONS

A *linear equation* with n variables is a polynomial of degree 1 describing a geometric shape in n-space. A *homogeneous linear equation* is one that has no constant term, and a *nonhomogeneous linear equation* has a constant term.

A solution to a set of simultaneous linear equations represents the intersection point of the geometric shapes in n-space. For example, if the equations are limited to two variables (e.g., $y = 4x - 5$), they describe straight lines. The solution to two simultaneous linear equations in 2-space is the point where the two lines intersect. The set of the two equations is said to be a *consistent system* when there is such an intersection.[8]

Simultaneous equations do not always have unique solutions, and some have none at all. In addition to crossing in 2-space, lines can be parallel or they can be the same line expressed in a different equation format (i.e., dependent equations). In some cases, parallelism and dependency can be determined by inspection. In most cases, however, matrix and other advanced methods must be used to determine whether a solution exists. A set of linear equations with no simultaneous solution is known as an *inconsistent system*.

[8]A homogeneous system always has at least one solution: the *trivial solution*, in which all variables have a value of zero.

Several methods exist for solving linear equations simultaneously by hand.[9]

- *graphing:* The equations are plotted and the intersection point is read from the graph. This method is possible only with two-dimensional problems.
- *substitution:* An equation is rearranged so that one variable is expressed as a combination of the other variables. The expression is then substituted into the remaining equations wherever the selected variable appears.
- *reduction:* All terms in the equations are multiplied by constants chosen to eliminate one or more variables when the equations are added or subtracted. The remaining sum can then be solved for the other variables. This method is also known as *eliminating the unknowns*.
- *Cramer's rule:* This is a procedure in linear algebra that calculates determinants of the original coefficient matrix **A** and of the n matrices resulting from the systematic replacement of column **A** by the constant matrix **B**.

Example 4.6

Solve the following set of linear equations by (a) substitution and (b) reduction.

$$2x + 3y = 12 \quad \text{[Eq. I]}$$
$$3x + 4y = 8 \quad \text{[Eq. II]}$$

Solution

(a) From Eq. I, solve for variable x.

$$x = 6 - 1.5y \quad \text{[Eq. III]}$$

Substitute $6 - 1.5y$ into Eq. II wherever x appears.

$$\begin{aligned} (3)(6 - 1.5y) + 4y &= 8 \\ 18 - 4.5y + 4y &= 8 \\ y &= 20 \end{aligned}$$

Substitute 20 for y in Eq. III.

$$x = 6 - (1.5)(20) = -24$$

The solution $(-24, 20)$ should be checked to verify that it satisfies both original equations.

(b) Eliminate variable x by multiplying Eq. I by 3 and Eq. II by 2.

$$3 \times \text{Eq. I: } 6x + 9y = 36 \quad \text{[Eq. I']}$$
$$2 \times \text{Eq. II: } 6x + 8y = 16 \quad \text{[Eq. II']}$$

Subtract Eq. II′ from Eq. I′.

$$y = 20 \quad \text{[Eq. I'} - \text{Eq. II']}$$

[9]Other methods exist, but they require a computer.

Substitute $y = 20$ into Eq. I′.

$$6x + (9)(20) = 36$$

$$x = -24$$

The solution (−24, 20) should be checked to verify that it satisfies both original equations.

18. COMPLEX NUMBERS

A *complex number*, **Z**, is a combination of real and imaginary numbers. When expressed as a sum (e.g., $a + bi$), the complex number is said to be in *rectangular* or *trigonometric form*. The complex number can be plotted on the real-imaginary coordinate system known as the *complex plane*, as illustrated in Fig. 4.2.

Figure 4.2 *A Complex Number in the Complex Plane*

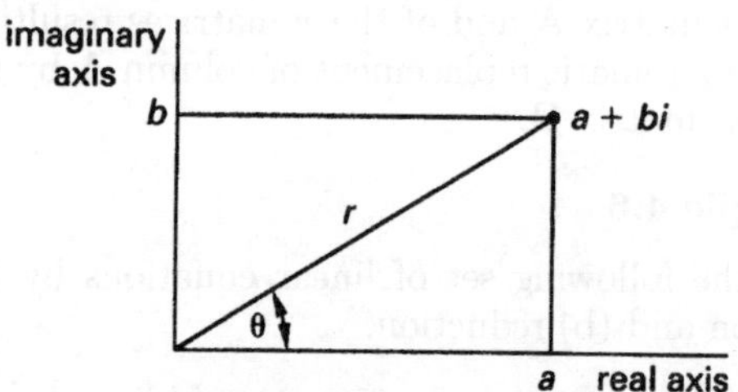

The complex number $\mathbf{Z} = a + bi$ can also be expressed in *exponential form*.[10] The quantity r is known as the *modulus* of **Z**; θ is the *argument*.

$$a + bi \equiv re^{i\theta} \qquad 4.53$$

$$r = \text{mod}\,\mathbf{Z} = \sqrt{a^2 + b^2} \qquad 4.54$$

$$\theta = \arg \mathbf{Z} = \arctan\frac{b}{a} \qquad 4.55$$

Similarly, the *phasor form* (also known as the *polar form*) is

$$\mathbf{Z} = r\angle\theta \qquad 4.56$$

The *rectangular form* can be determined from r and θ.

$$a = r\cos\theta \qquad 4.57$$

$$b = r\sin\theta \qquad 4.58$$

$$\mathbf{Z} = a + bi = r\cos\theta + ir\sin\theta$$

$$= r(\cos\theta + i\sin\theta) \qquad 4.59$$

The *cis form* is a shorthand method of writing a complex number in rectangular (trigonometric) form.

$$a + bi = r(\cos\theta + i\sin\theta) = r\,\text{cis}\,\theta \qquad 4.60$$

[10]The terms *polar form*, *phasor form*, and *exponential form* are all used somewhat interchangeably.

Euler's equation, as shown in Eq. 4.61, expresses the equality of complex numbers in exponential and trigonometric form.

$$e^{i\theta} = \cos\theta + i\sin\theta \qquad 4.61$$

Related expressions are

$$e^{-i\theta} = \cos\theta - i\sin\theta \qquad 4.62$$

$$\cos\theta = \frac{e^{i\theta} + e^{-i\theta}}{2} \qquad 4.63$$

$$\sin\theta = \frac{e^{i\theta} - e^{-i\theta}}{2i} \qquad 4.64$$

Example 4.7

What is the exponential form of the complex number $\mathbf{Z} = 3 + 4i$?

Solution

$$r = \sqrt{a^2 + b^2} = \sqrt{3^2 + 4^2} = \sqrt{25}$$
$$= 5$$

$$\theta = \arctan\frac{b}{a} = \arctan\frac{4}{3} = 0.927 \text{ rad}$$

$$\mathbf{Z} = re^{i\theta} = 5e^{i(0.927)}$$

19. OPERATIONS ON COMPLEX NUMBERS

Most algebraic operations (addition, multiplication, exponentiation, etc.) work with complex numbers, but notable exceptions are the inequality operators. The concept of one complex number being less than or greater than another complex number is meaningless.

When adding two complex numbers, real parts are added to real parts, and imaginary parts are added to imaginary parts.

$$(a_1 + ib_1) + (a_2 + ib_2) = (a_1 + a_2) + i(b_1 + b_2) \qquad 4.65$$

$$(a_1 + ib_1) - (a_2 + ib_2) = (a_1 - a_2) + i(b_1 - b_2) \qquad 4.66$$

Multiplication of two complex numbers in rectangular form is accomplished by the use of the algebraic distributive law and the definition $i^2 = -1$.

Division of complex numbers in rectangular form requires use of the *complex conjugate*. The complex conjugate of the complex number $(a + bi)$ is $(a - bi)$. By multiplying the numerator and the denominator by the complex conjugate, the denominator will be converted to the real number $a^2 + b^2$. This technique is known as *rationalizing* the denominator and is illustrated in Ex. 4.7(c).

Multiplication and division are often more convenient when the complex numbers are in exponential or phasor forms, as Eq. 4.67 and Eq. 4.68 show.

$$(r_1 e^{i\theta_1})(r_2 e^{i\theta_2}) = r_1 r_2 e^{i(\theta_1+\theta_2)} \qquad 4.67$$

$$\frac{r_1 e^{i\theta_1}}{r_2 e^{i\theta_2}} = \left(\frac{r_1}{r_2}\right) e^{i(\theta_1-\theta_2)} \qquad 4.68$$

Taking powers and roots of complex numbers requires *de Moivre's theorem*, Eq. 4.69 and Eq. 4.70.

$$\mathbf{Z}^n = (re^{i\theta})^n = r^n e^{in\theta} \qquad 4.69$$

$$\sqrt[n]{\mathbf{Z}} = (re^{i\theta})^{1/n} = \sqrt[n]{r} e^{i(\theta+k360^\circ/n)} \quad [k = 0, 1, 2, \ldots, n-1] \qquad 4.70$$

Example 4.8

Perform the following complex arithmetic.

(a) $(3+4i)+(2+i)$

(b) $(7+2i)(5-3i)$

(c) $\dfrac{2+3i}{4-5i}$

Solution

(a)
$$(3+4i)+(2+i) = (3+2)+(4+1)i = 5+5i$$

(b)
$$\begin{aligned}(7+2i)(5-3i) &= (7)(5)-(7)(3i)+(2i)(5) - (2i)(3i)\\ &= 35-21i+10i-6i^2\\ &= 35-21i+10i-(6)(-1)\\ &= 41-11i\end{aligned}$$

(c) Multiply the numerator and denominator by the complex conjugate of the denominator.

$$\frac{2+3i}{4-5i} = \frac{(2+3i)(4+5i)}{(4-5i)(4+5i)} = \frac{-7+22i}{(4)^2+(5)^2} = \frac{-7}{41} + i\frac{22}{41}$$

20. LIMITS

A *limit* (*limiting value*) is the value a function approaches when an independent variable approaches a target value. (See Fig. 4.3.) For example, suppose the value of $y = x^2$ is desired as x approaches 5. This could be written as

$$\lim_{x\to 5} x^2 \qquad 4.71$$

The power of limit theory is wasted on simple calculations such as this but is appreciated when the function is undefined at the target value. The object of limit theory

Figure 4.3 *Existence of Limits*

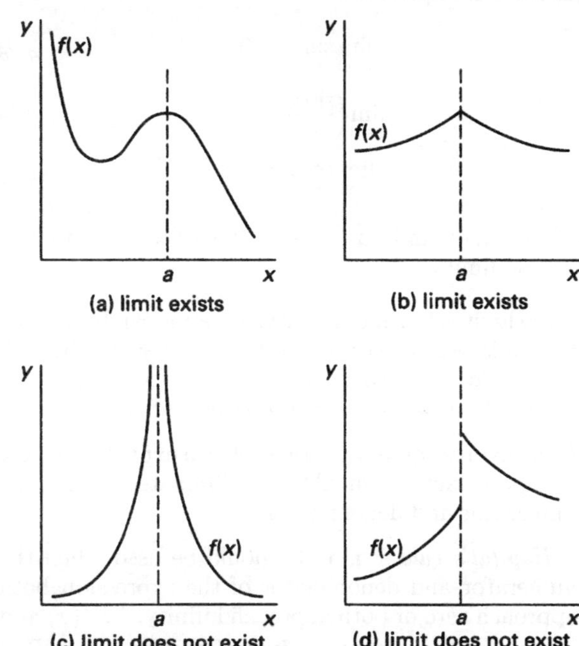

is to determine the limit without having to evaluate the function at the target. The general case of a limit evaluated as x approaches the target value a is written as

$$\lim_{x\to a} f(x) \qquad 4.72$$

It is not necessary for the actual value $f(a)$ to exist for the limit to be calculated. The function $f(x)$ may be undefined at point a. However, it is necessary that $f(x)$ be defined on both sides of point a for the limit to exist. If $f(x)$ is undefined on one side, or if $f(x)$ is discontinuous at $x = a$ (as in Fig. 4.3(c) and Fig. 4.3(d)), the limit does not exist.

The following theorems can be used to simplify expressions when calculating limits.

$$\lim_{x\to a} x = a \qquad 4.73$$

$$\lim_{x\to a} (mx+b) = ma+b \qquad 4.74$$

$$\lim_{x\to a} b = b \qquad 4.75$$

$$\lim_{x\to a} (kF(x)) = k \lim_{x\to a} F(x) \qquad 4.76$$

$$\lim_{x\to a}\left(F_1(x)\begin{Bmatrix}+\\-\\\times\\\div\end{Bmatrix}F_2(x)\right) = \lim_{x\to a}(F_1(x))\begin{Bmatrix}+\\-\\\times\\\div\end{Bmatrix}\lim_{x\to a}(F_2(x)) \qquad 4.77$$

The following identities can be used to simplify limits of trigonometric expressions.

$$\lim_{x\to 0} \sin x = 0 \quad 4.78$$

$$\lim_{x\to 0}\left(\frac{\sin x}{x}\right) = 1 \quad 4.79$$

$$\lim_{x\to 0} \cos x = 1 \quad 4.80$$

The following standard methods (tricks) can be used to determine limits.

- If the limit is taken to infinity, all terms can be divided by the largest power of x in the expression. This will leave at least one constant. Any quantity divided by a power of x vanishes as x approaches infinity.
- If the expression is a quotient of two expressions, any common factors should be eliminated from the numerator and denominator.
- *L'Hôpital's rule*, Eq. 4.81, should be used when the numerator and denominator of the expression both approach zero or both approach infinity.[11] $P^k(x)$ and $Q^k(x)$ are the kth derivatives of the functions $P(x)$ and $Q(x)$, respectively. (L'Hôpital's rule can be applied repeatedly as required.)

$$\lim_{x\to a}\left(\frac{P(x)}{Q(x)}\right) = \lim_{x\to a}\left(\frac{P^k(x)}{Q^k(x)}\right) \quad 4.81$$

Example 4.9

Evaluate the following limits.

(a) $\lim_{x\to 3}\left(\frac{x^3-27}{x^2-9}\right)$

(b) $\lim_{x\to \infty}\left(\frac{3x-2}{4x+3}\right)$

(c) $\lim_{x\to 2}\left(\frac{x^2+x-6}{x^2-3x+2}\right)$

Solution

(a) Factor the numerator and denominator. (L'Hôpital's rule can also be used.)

$$\begin{aligned}\lim_{x\to 3}\left(\frac{x^3-27}{x^2-9}\right) &= \lim_{x\to 3}\left(\frac{(x-3)(x^2+3x+9)}{(x-3)(x+3)}\right)\\ &= \lim_{x\to 3}\left(\frac{x^2+3x+9}{x+3}\right)\\ &= \frac{(3)^2+(3)(3)+9}{3+3}\\ &= 9/2\end{aligned}$$

[11] L'Hôpital's rule should not be used when only the denominator approaches zero. In that case, the limit approaches infinity regardless of the numerator.

(b) Divide through by the largest power of x. (L'Hôpital's rule can also be used.)

$$\begin{aligned}\lim_{x\to \infty}\left(\frac{3x-2}{4x+3}\right) &= \lim_{x\to \infty}\left(\frac{3-\frac{2}{x}}{4+\frac{3}{x}}\right)\\ &= \frac{3-\frac{2}{\infty}}{4+\frac{3}{\infty}} = \frac{3-0}{4+0}\\ &= 3/4\end{aligned}$$

(c) Use L'Hôpital's rule. (Factoring can also be used.) Take the first derivative of the numerator and denominator.

$$\begin{aligned}\lim_{x\to 2}\left(\frac{x^2+x-6}{x^2-3x+2}\right) &= \lim_{x\to 2}\left(\frac{2x+1}{2x-3}\right) = \frac{(2)(2)+1}{(2)(2)-3}\\ &= \frac{5}{1}\\ &= 5\end{aligned}$$

21. SEQUENCES AND PROGRESSIONS

A *sequence*, $\{A\}$, is an ordered *progression* of numbers, a_i, such as 1, 4, 9, 16, 25, ... The *terms* in a sequence can be all positive, all negative, or of alternating signs. a_n is known as the *general term* of the sequence.

$$\{A\} = \{a_1, a_2, a_3, \ldots, a_n\} \quad 4.82$$

A sequence is said to *diverge* (i.e., be *divergent*) if the terms approach infinity or if the terms fail to approach any finite value, and it is said to *converge* (i.e., be *convergent*) if the terms approach any finite value (including zero). That is, the sequence converges if the limit defined by Eq. 4.83 exists.

$$\lim_{n\to\infty} a_n \begin{cases}\text{converges if the limit is finite}\\ \text{diverges if the limit is infinite}\\ \quad\text{or does not exist}\end{cases} \quad 4.83$$

The main task associated with a sequence is determining the next (or the general) term. If several terms of a sequence are known, the next (unknown) term must usually be found by intuitively determining the pattern of the sequence. In some cases, though, the method of *Rth-order differences* can be used to determine the next term. This method consists of subtracting each term from the following term to obtain a set of differences. If the differences are not all equal, the next order of differences can be calculated.

Example 4.10

What is the general term of the sequence $\{A\}$?

$$\{A\} = \left\{3,\ \frac{9}{2},\ \frac{27}{6},\ \frac{81}{24},\ldots\right\}$$

Solution

The solution is purely intuitive. The numerator is recognized as a power series based on the number 3. The denominator is recognized as the factorial sequence. The general term is

$$a_n = \frac{3^n}{n!}$$

Example 4.11

Find the sixth term in the sequence $\{A\} = \{7,\ 16,\ 29,\ 46,\ 67,\ a_6\}$.

Solution

The sixth term is not intuitively obvious, so the method of *R*th-order differences is tried. The pattern is not obvious from the first order differences, but the second order differences are all 4.

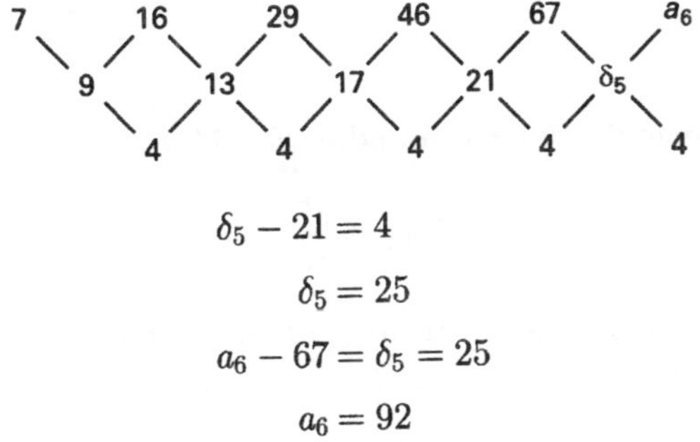

$$\begin{aligned} \delta_5 - 21 &= 4 \\ \delta_5 &= 25 \\ a_6 - 67 = \delta_5 &= 25 \\ a_6 &= 92 \end{aligned}$$

Example 4.12

Does the sequence with general term e^n/n converge or diverge?

Solution

See if the limit exists.

$$\lim_{n\to\infty}\left(\frac{e^n}{n}\right) = \frac{\infty}{\infty}$$

Since ∞/∞ is inconclusive, apply L'Hôpital's rule. Take the derivative of both the numerator and the denominator with respect to n.

$$\lim_{n\to\infty}\left(\frac{e^n}{1}\right) = \frac{\infty}{1} = \infty$$

The sequence diverges.

22. STANDARD SEQUENCES

There are four standard sequences: the geometric, arithmetic, harmonic, and *p*-sequence.

- *geometric sequence:* The geometric sequence converges for $-1 < r \leq 1$ and diverges otherwise. a is known as the *first term*; r is known as the *common ratio.*

$$a_n = ar^{n-1} \quad \left[\begin{array}{c} a \text{ is a constant} \\ n = 1,\ 2,\ 3,\ldots,\ \infty \end{array}\right] \qquad 4.84$$

example: $\{1, 2, 4, 8, 16, 32\}\quad (a = 1,\ \ r = 2)$

- *arithmetic sequence:* The arithmetic sequence always diverges.

$$a_n = a + (n-1)d \quad \left[\begin{array}{c} a \text{ and } d \text{ are constants} \\ n = 1,\ 2,\ 3,\ldots,\ \infty \end{array}\right] \qquad 4.85$$

example: $\{2, 7, 12, 17, 22, 27\}\quad (a = 2,\ \ d = 5)$

- *harmonic sequence:* The harmonic sequence always converges.

$$a_n = \frac{1}{n} \quad [n = 1,\ 2,\ 3,\ldots,\ \infty] \qquad 4.86$$

example: $\{1,\ 1/2,\ 1/3,\ 1/4,\ 1/5,\ 1/6\}$

- *p-sequence:* The *p*-sequence converges if $p \geq 0$ and diverges if $p < 0$. (This is different from the *p*-series whose convergence depends on the sum of its terms.)

$$a_n = \frac{1}{n^p} \quad [n = 1,\ 2,\ 3,\ldots,\ \infty] \qquad 4.87$$

example: $\{1,\ 1/4,\ 1/9,\ 1/16,\ 1/25,\ 1/36\}\quad (p = 2)$

23. SERIES

A *series* is the sum of terms in a sequence. There are two types of series. A *finite series* has a finite number of terms, and an *infinite series* has an infinite number of terms.[12] The main tasks associated with series are determining the sum of the terms and whether the series converges. A series is said to *converge* (be *convergent*) if the sum, S_n, of its term exists.[13] A finite series is always convergent.

[12]The term *infinite series* does not imply the sum is infinite.

[13]This is different from the definition of convergence for a sequence where only the last term was evaluated.

Mathematics

The performance of a series based on standard sequences (defined in Sec. 4.22) is well known.

- *geometric series:*

$$S_n = \sum_{i=1}^{n} ar^{i-1} = \frac{a(1-r^n)}{1-r} \quad \text{[finite series]} \tag{4.88}$$

$$S_n = \sum_{i=1}^{\infty} ar^{i-1} = \frac{a}{1-r} \quad \left[\begin{array}{c}\text{infinite series} \\ -1 < r < 1\end{array}\right] \tag{4.89}$$

- *arithmetic series:* The infinite series diverges unless $a = d = 0$.

$$\begin{aligned} S_n &= \sum_{i=1}^{n} \left(a + (i-1)d\right) \\ &= \frac{n\left(2a + (n-1)d\right)}{2} \quad \text{[finite series]} \end{aligned} \tag{4.90}$$

- *harmonic series:* The infinite series diverges.
- *p-series:* The infinite series diverges if $p \le 1$. The infinite series converges if $p > 1$. (This is different from the *p*-sequence whose convergence depends only on the last term.)

24. TESTS FOR SERIES CONVERGENCE

It is obvious that all finite series (i.e., series having a finite number of terms) converge. That is, the sum, S_n, defined by Eq. 4.91 exists.

$$S_n = \sum_{i=1}^{n} a_i \tag{4.91}$$

Convergence of an infinite series can be determined by taking the limit of the sum. If the limit exists, the series converges; otherwise, it diverges.

$$\lim_{n\to\infty} S_n = \lim_{n\to\infty} \sum_{i=1}^{n} a_i \tag{4.92}$$

In most cases, the expression for the general term a_n will be known, but there will be no simple expression for the sum S_n. Therefore, Eq. 4.92 cannot be used to determine convergence. It is helpful, but not conclusive, to look at the limit of the general term. If the limit, as defined in Eq. 4.93, is nonzero, the series diverges. If the limit equals zero, the series may either converge or diverge. Additional testing is needed in that case.

$$\lim_{n\to\infty} a_n \begin{cases} = 0 & \text{inconclusive} \\ \neq 0 & \text{diverges} \end{cases} \tag{4.93}$$

Two tests can be used independently or after Eq. 4.93 has proven inconclusive: the ratio and comparison tests.

The *ratio test* calculates the limit of the ratio of two consecutive terms.

$$\lim_{n\to\infty} \frac{a_{n+1}}{a_n} \begin{cases} < 1 & \text{converges} \\ = 1 & \text{inconclusive} \\ > 1 & \text{diverges} \end{cases} \tag{4.94}$$

The *comparison test* is an indirect method of determining convergence of an unknown series. It compares a standard series (geometric and *p*-series are commonly used) against the unknown series. If all terms in a positive standard series are smaller than the terms in the unknown series and the standard series diverges, the unknown series must also diverge. Similarly, if all terms in the standard series are larger than the terms in the unknown series and the standard series converges, then the unknown series also converges.

In mathematical terms, if A and B are both series of positive terms such that $a_n < b_n$ for all values of n, then (a) B diverges if A diverges, and (b) A converges if B converges.

Example 4.13

Does the infinite series A converge or diverge?

$$A = 3 + \frac{9}{2} + \frac{27}{6} + \frac{81}{24} + \cdots$$

Solution

The general term was found in Ex. 4.9 to be

$$a_n = \frac{3^n}{n!}$$

Since limits of factorials are not easily determined, use the ratio test.

$$\begin{aligned} \lim_{n\to\infty} \left(\frac{a_{n+1}}{a_n}\right) &= \lim_{n\to\infty} \left(\frac{\frac{3^{n+1}}{(n+1)!}}{\frac{3^n}{n!}}\right) = \lim_{n\to\infty} \left(\frac{3}{n+1}\right) = \frac{3}{\infty} \\ &= 0 \end{aligned}$$

Since the limit is less than 1, the infinite series converges.

Example 4.14

Does the infinite series A converge or diverge?

$$A = 2 + \frac{3}{4} + \frac{4}{9} + \frac{5}{16} + \cdots$$

Solution

By observation, the general term is

$$a_n = \frac{1+n}{n^2}$$

The general term can be expanded by partial fractions to

$$a_n = \frac{1}{n} + \frac{1}{n^2}$$

However, $1/n$ is the harmonic series. Since the harmonic series is divergent and this series is larger than the harmonic series (by the term $1/n^2$), this series also diverges.

25. SERIES OF ALTERNATING SIGNS[14]

Some series contain both positive and negative terms. The ratio and comparison tests can both be used to determine if a series with alternating signs converges. If a series containing all positive terms converges, then the same series with some negative terms also converges. Therefore, the all-positive series should be tested for convergence. If the all-positive series converges, the original series is said to be *absolutely convergent.* (If the all-positive series diverges and the original series converges, the original series is said to be *conditionally convergent.*)

[14]This terminology is commonly used even though it is not necessary that the signs strictly alternate.

Alternatively, the ratio test can be used with the absolute value of the ratio. The same criteria apply.

$$\lim_{n\to\infty}\left|\frac{a_{n+1}}{a_n}\right| \begin{cases} <1 & \text{converges} \\ =1 & \text{inconclusive} \\ >1 & \text{diverges} \end{cases} \qquad 4.95$$

5 Linear Algebra

1. MATRICES

A *matrix* is an ordered set of *entries* (*elements*) arranged rectangularly and set off by brackets.[1] The entries can be variables or numbers. A matrix by itself has no particular value—it is merely a convenient method of representing a set of numbers.

The size of a matrix is given by the number of rows and columns, and the nomenclature $m \times n$ is used for a matrix with m rows and n columns. For a *square matrix*, the number of rows and columns will be the same, a quantity known as the *order* of the matrix.

Bold uppercase letters are used to represent matrices, while lowercase letters represent the entries. For example, a_{23} would be the entry in the second row and third column of matrix **A**.

$$\mathbf{A} = \begin{bmatrix} a_{11} & a_{12} & a_{13} \\ a_{21} & a_{22} & a_{23} \\ a_{31} & a_{32} & a_{33} \end{bmatrix}$$

A *submatrix* is the matrix that remains when selected rows or columns are removed from the original matrix.[2] For example, for matrix **A**, the submatrix remaining after the second row and second column have been removed is

$$\begin{bmatrix} a_{11} & a_{13} \\ a_{31} & a_{33} \end{bmatrix}$$

An *augmented matrix* results when the original matrix is extended by repeating one or more of its rows or columns or by adding rows and columns from another matrix. For example, for the matrix **A**, the augmented matrix created by repeating the first and second columns is

$$\left[\begin{array}{ccc|cc} a_{11} & a_{12} & a_{13} & a_{11} & a_{12} \\ a_{21} & a_{22} & a_{23} & a_{21} & a_{22} \\ a_{31} & a_{32} & a_{33} & a_{31} & a_{32} \end{array}\right]$$

2. SPECIAL TYPES OF MATRICES

Certain types of matrices are given special designations.

- *cofactor matrix:* the matrix formed when every entry is replaced by the cofactor (see Sec. 5.4) of that entry
- *column matrix:* a matrix with only one column
- *complex matrix:* a matrix with complex number entries
- *diagonal matrix:* a square matrix with all zero entries except for the a_{ij} for which $i = j$
- *echelon matrix:* a matrix in which the number of zeros preceding the first nonzero entry of a row increases row by row until only zero rows remain. A *row-reduced echelon matrix* is an echelon matrix in which the first nonzero entry in each row is a 1 and all other entries in the columns are zero.
- *identity matrix:* a diagonal (square) matrix with all nonzero entries equal to 1, usually designated as **I**, having the property that **AI** = **IA** = **A**
- *null matrix:* the same as a zero matrix
- *row matrix:* a matrix with only one row
- *scalar matrix:*[3] a diagonal (square) matrix with all diagonal entries equal to some scalar k

[1]The term *array* is synonymous with *matrix*, although the former is more likely to be used in computer applications.

[2]By definition, a matrix is a submatrix of itself.

[3]Although the term *complex matrix* means a matrix with complex entries, the term *scalar matrix* means more than a matrix with scalar entries.

- *singular matrix:* a matrix whose determinant is zero (see Sec. 5.10)
- *skew symmetric matrix:* a square matrix whose transpose (see Sec. 5.9) is equal to the negative of itself (i.e., $\mathbf{A} = -\mathbf{A}^t$)
- *square matrix:* a matrix with the same number of rows and columns (i.e., $m = n$)
- *symmetric(al) matrix:* a square matrix whose transpose is equal to itself (i.e., $\mathbf{A}^t = \mathbf{A}$), which occurs only when $a_{ij} = a_{ji}$
- *triangular matrix:* a square matrix with zeros in all positions above or below the diagonal
- *unit matrix:* the same as the identity matrix
- *zero matrix:* a matrix with all zero entries

Figure 5.1 shows examples of special matrices.

Figure 5.1 *Examples of Special Matrices*

$$\begin{bmatrix} 9 & 0 & 0 & 0 \\ 0 & -6 & 0 & 0 \\ 0 & 0 & 1 & 0 \\ 0 & 0 & 0 & 5 \end{bmatrix} \quad \begin{bmatrix} 2 & 18 & 2 & 18 \\ 0 & 0 & 1 & 9 \\ 0 & 0 & 0 & 9 \\ 0 & 0 & 0 & 0 \end{bmatrix} \quad \begin{bmatrix} 1 & 9 & 0 & 0 \\ 0 & 0 & 1 & 0 \\ 0 & 0 & 0 & 1 \\ 0 & 0 & 0 & 0 \end{bmatrix}$$

(a) diagonal (b) echelon (c) row-reduced echelon

$$\begin{bmatrix} 1 & 0 & 0 & 0 \\ 0 & 1 & 0 & 0 \\ 0 & 0 & 1 & 0 \\ 0 & 0 & 0 & 1 \end{bmatrix} \quad \begin{bmatrix} 3 & 0 & 0 & 0 \\ 0 & 3 & 0 & 0 \\ 0 & 0 & 3 & 0 \\ 0 & 0 & 0 & 3 \end{bmatrix} \quad \begin{bmatrix} 2 & 0 & 0 & 0 \\ 7 & 6 & 0 & 0 \\ 9 & 1 & 1 & 0 \\ 8 & 0 & 4 & 5 \end{bmatrix}$$

(d) identity (e) scalar (f) triangular

3. ROW EQUIVALENT MATRICES

A matrix **B** is said to be *row equivalent* to a matrix **A** if it is obtained by a finite sequence of *elementary row operations* on **A**:

- interchanging the ith and jth rows
- multiplying the ith row by a nonzero scalar
- replacing the ith row by the sum of the original ith row and k times the jth row

However, two matrices that are row equivalent as defined do not necessarily have the same determinants. (See Sec. 5.5.)

Gauss-Jordan elimination is the process of using these elementary row operations to row-reduce a matrix to echelon or row-reduced echelon forms, as illustrated in Ex. 5.8. When a matrix has been converted to a row-reduced echelon matrix, it is said to be in *row canonical form.* The phrases *row-reduced echelon form* and *row canonical form* are synonymous.

4. MINORS AND COFACTORS

Minors and cofactors are determinants of submatrices associated with particular entries in the original square matrix. The *minor* of entry a_{ij} is the determinant of a submatrix resulting from the elimination of the single row i and the single column j. For example, the minor corresponding to entry a_{12} in a 3 × 3 matrix **A** is the determinant of the matrix created by eliminating row 1 and column 2.

$$\text{minor of } a_{12} = \begin{vmatrix} a_{21} & a_{23} \\ a_{31} & a_{33} \end{vmatrix} \qquad 5.1$$

The *cofactor* of entry a_{ij} is the minor of a_{ij} multiplied by either +1 or −1, depending on the position of the entry. (That is, the cofactor either exactly equals the minor or it differs only in sign.) The sign is determined according to the following positional matrix.[4]

$$\begin{bmatrix} +1 & -1 & +1 & \cdots \\ -1 & +1 & -1 & \cdots \\ +1 & -1 & +1 & \cdots \\ \vdots & \vdots & \vdots & \end{bmatrix}$$

For example, the cofactor of entry a_{12} in matrix **A** (described in Sec. 5.4) is

$$\text{cofactor of } a_{12} = -\begin{vmatrix} a_{21} & a_{23} \\ a_{31} & a_{33} \end{vmatrix} \qquad 5.2$$

Example 5.1

What is the cofactor corresponding to the −3 entry in the following matrix?

$$\mathbf{A} = \begin{bmatrix} 2 & 9 & 1 \\ -3 & 4 & 0 \\ 7 & 5 & 9 \end{bmatrix}$$

Solution

The minor's submatrix is created by eliminating the row and column of the −3 entry.

$$\mathbf{M} = \begin{bmatrix} 9 & 1 \\ 5 & 9 \end{bmatrix}$$

The minor is the determinant of **M**.

$$|\mathbf{M}| = (9)(9) - (5)(1) = 76$$

The sign corresponding to the −3 position is negative. Therefore, the cofactor is −76.

[4]The sign of the cofactor a_{ij} is positive if $(i + j)$ is even and is negative if $(i + j)$ is odd.

5. DETERMINANTS

A *determinant* is a scalar calculated from a square matrix. The determinant of matrix **A** can be represented as $D\{\mathbf{A}\}$, $\text{Det}(\mathbf{A})$, $\Delta\mathbf{A}$, or $|\mathbf{A}|$.[5] The following rules can be used to simplify the calculation of determinants.

- If **A** has a row or column of zeros, the determinant is zero.
- If **A** has two identical rows or columns, the determinant is zero.
- If **B** is obtained from **A** by adding a multiple of a row (column) to another row (column) in **A**, then $|\mathbf{B}| = |\mathbf{A}|$.
- If **A** is triangular, the determinant is equal to the product of the diagonal entries.
- If **B** is obtained from **A** by multiplying one row or column in **A** by a scalar k, then $|\mathbf{B}| = k|\mathbf{A}|$.
- If **B** is obtained from the $n \times n$ matrix **A** by multiplying by the scalar matrix k, then $|k\mathbf{A}| = k^n|\mathbf{A}|$.
- If **B** is obtained from **A** by switching two rows or columns in **A**, then $|\mathbf{B}| = -|\mathbf{A}|$.

Calculation of determinants is laborious for all but the smallest or simplest of matrices. For a 2 × 2 matrix, the formula used to calculate the determinant is easy to remember.

$$\mathbf{A} = \begin{bmatrix} a & b \\ c & d \end{bmatrix}$$

$$|\mathbf{A}| = \begin{vmatrix} a & b \\ c & d \end{vmatrix} = ad - bc \qquad 5.3$$

Two methods are commonly used for calculating the determinant of 3 × 3 matrices by hand. The first uses an augmented matrix constructed from the original matrix and the first two columns (as shown in Sec. 5.1).[6] The determinant is calculated as the sum of the products in the left-to-right downward diagonals less the sum of the products in the left-to-right upward diagonals.

$$\mathbf{A} = \begin{bmatrix} a & b & c \\ d & e & f \\ g & h & i \end{bmatrix}$$

augmented $\mathbf{A}$ =

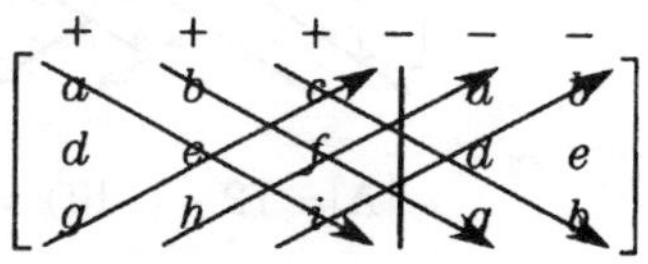
+ + + − − −
a b c a b
d e f d e
g h i g h

5.4

$$|\mathbf{A}| = aei + bfg + cdh - gec - hfa - idb \qquad 5.5$$

The second method of calculating the determinant is somewhat slower than the first for a 3 × 3 matrix but illustrates the method that must be used to calculate determinants of 4 × 4 and larger matrices. This method is known as *expansion by cofactors.* One row (column) is selected as the base row (column). The selection is arbitrary, but the number of calculations required to obtain the determinant can be minimized by choosing the row (column) with the most zeros. The determinant is equal to the sum of the products of the entries in the base row (column) and their corresponding cofactors.

$$\mathbf{A} = \begin{bmatrix} a & b & c \\ d & e & f \\ g & h & i \end{bmatrix}$$

$$|\mathbf{A}| = a\begin{vmatrix} e & f \\ h & i \end{vmatrix} - d\begin{vmatrix} b & c \\ h & i \end{vmatrix} + g\begin{vmatrix} b & c \\ e & f \end{vmatrix} \qquad 5.6$$

Example 5.2

Calculate the determinant of matrix **A** (a) by cofactor expansion, and (b) by the augmented matrix method.

$$\mathbf{A} = \begin{bmatrix} 2 & 3 & -4 \\ 3 & -1 & -2 \\ 4 & -7 & -6 \end{bmatrix}$$

Solution

(a) Since there are no zero entries, it does not matter which row or column is chosen as the base. Choose the first column as the base.

$$\begin{aligned} |\mathbf{A}| &= 2\begin{vmatrix} -1 & -2 \\ -7 & -6 \end{vmatrix} - 3\begin{vmatrix} 3 & -4 \\ -7 & -6 \end{vmatrix} + 4\begin{vmatrix} 3 & -4 \\ -1 & -2 \end{vmatrix} \\ &= (2)(6 - 14) - (3)(-18 - 28) + (4)(-6 - 4) \\ &= 82 \end{aligned}$$

[5]The vertical bars should not be confused with the square brackets used to set off a matrix, nor with absolute value.

[6]It is not actually necessary to construct the augmented matrix, but doing so helps avoid errors.

(b)

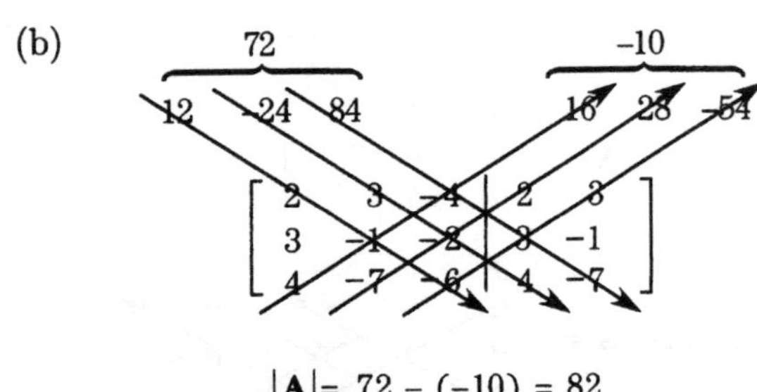

$|\mathbf{A}| = 72 - (-10) = 82$

6. MATRIX ALGEBRA[7]

Matrix algebra differs somewhat from standard algebra.

- *equality:* Two matrices, **A** and **B**, are equal only if they have the same numbers of rows and columns *and* if all corresponding entries are equal.
- *inequality:* The > and < operators are not used in matrix algebra.
- *commutative law of addition:*

$$\mathbf{A} + \mathbf{B} = \mathbf{B} + \mathbf{A} \qquad 5.7$$

- *associative law of addition:*

$$\mathbf{A} + (\mathbf{B} + \mathbf{C}) = (\mathbf{A} + \mathbf{B}) + \mathbf{C} \qquad 5.8$$

- *associative law of multiplication:*

$$(\mathbf{AB})\mathbf{C} = \mathbf{A}(\mathbf{BC}) \qquad 5.9$$

- *left distributive law:*

$$\mathbf{A}(\mathbf{B} + \mathbf{C}) = \mathbf{AB} + \mathbf{AC} \qquad 5.10$$

- *right distributive law:*

$$(\mathbf{B} + \mathbf{C})\mathbf{A} = \mathbf{BA} + \mathbf{CA} \qquad 5.11$$

- *scalar multiplication:*

$$k(\mathbf{AB}) = (k\mathbf{A})\mathbf{B} = \mathbf{A}(k\mathbf{B}) \qquad 5.12$$

It is important to recognize that, except for trivial and special cases, matrix multiplication is not commutative. That is,

$$\mathbf{AB} \neq \mathbf{BA}$$

7. MATRIX ADDITION AND SUBTRACTION

Addition and subtraction of two matrices is possible only if both matrices have the same numbers of rows and columns (i.e., order). They are accomplished by adding or subtracting the corresponding entries of the two matrices.

[7]Since matrices are used to simplify the presentation and solution of sets of linear equations, matrix algebra is also known as *linear algebra*.

8. MATRIX MULTIPLICATION

A matrix can be multiplied by a scalar, an operation known as *scalar multiplication*, in which case all entries of the matrix are multiplied by that scalar. For example, for the 2 × 2 matrix **A**,

$$k\mathbf{A} = \begin{bmatrix} ka_{11} & ka_{12} \\ ka_{21} & ka_{22} \end{bmatrix}$$

A matrix can be multiplied by another matrix, but only if the left-hand matrix has the same number of columns as the right-hand matrix has rows. *Matrix multiplication* occurs by multiplying the elements in each left-hand matrix row by the entries in each right-hand matrix column, adding the products, and placing the sum at the intersection point of the participating row and column.

Matrix division can only be accomplished by multiplying by the inverse of the denominator matrix. There is no specific division operation in matrix algebra.

Example 5.3

Determine the product matrix **C**.

$$\mathbf{C} = \begin{bmatrix} 1 & 4 & 3 \\ 5 & 2 & 6 \end{bmatrix} \begin{bmatrix} 7 & 12 \\ 11 & 8 \\ 9 & 10 \end{bmatrix}$$

Solution

The left-hand matrix has three columns, and the right-hand matrix has three rows. Therefore, the two matrices can be multiplied.

The first row of the left-hand matrix and the first column of the right-hand matrix are worked with first. The corresponding entries are multiplied, and the products are summed.

$$c_{11} = (1)(7) + (4)(11) + (3)(9) = 78$$

The intersection of the top row and left column is the entry in the upper left-hand corner of the matrix **C**.

The remaining entries are calculated similarly.

$$c_{12} = (1)(12) + (4)(8) + (3)(10) = 74$$
$$c_{21} = (5)(7) + (2)(11) + (6)(9) = 111$$
$$c_{22} = (5)(12) + (2)(8) + (6)(10) = 136$$

The product matrix is

$$\mathbf{C} = \begin{bmatrix} 78 & 74 \\ 111 & 136 \end{bmatrix}$$

9. TRANSPOSE

The *transpose*, $\mathbf{A}^t$, of an $m \times n$ matrix $\mathbf{A}$ is an $n \times m$ matrix constructed by taking the ith row and making it the ith column. The diagonal is unchanged. For example,

$$\mathbf{A} = \begin{bmatrix} 1 & 6 & 9 \\ 2 & 3 & 4 \\ 7 & 1 & 5 \end{bmatrix}$$

$$\mathbf{A}^t = \begin{bmatrix} 1 & 2 & 7 \\ 6 & 3 & 1 \\ 9 & 4 & 5 \end{bmatrix}$$

Transpose operations have the following characteristics.

$$(\mathbf{A}^t)^t = \mathbf{A} \quad \text{5.13}$$

$$(k\mathbf{A})^t = k(\mathbf{A}^t) \quad \text{5.14}$$

$$\mathbf{I}^t = \mathbf{I} \quad \text{5.15}$$

$$(\mathbf{AB})^t = \mathbf{B}^t\mathbf{A}^t \quad \text{5.16}$$

$$(\mathbf{A}+\mathbf{B})^t = \mathbf{A}^t + \mathbf{B}^t \quad \text{5.17}$$

$$|\mathbf{A}^t| = |\mathbf{A}| \quad \text{5.18}$$

10. SINGULARITY AND RANK

A *singular matrix* is one whose determinant is zero. Similarly, a *nonsingular* matrix is one whose determinant is nonzero.

The *rank* of a matrix is the maximum number of linearly independent row or column vectors.[8] A matrix has rank r if it has at least one nonsingular square submatrix of order r but has no nonsingular square submatrix of order more than r. While the submatrix must be square (in order to calculate the determinant), the original matrix need not be.

The rank of an $m \times n$ matrix will be, at most, the smaller of m and n. The rank of a null matrix is zero. The ranks of a matrix and its transpose are the same. If a matrix is in echelon form, the rank will be equal to the number of rows containing at least one nonzero entry. For a 3×3 matrix, the rank can either be 3 (if it is nonsingular), 2 (if any one of its 2×2 submatrices is nonsingular), 1 (if it and all 2×2 submatrices are singular), or 0 (if it is null).

The determination of rank is laborious if done by hand. Either the matrix is reduced to echelon form by using elementary row operations, or exhaustive enumeration is used to create the submatrices and many determinants are calculated. If a matrix has more rows than columns and row-reduction is used, the work required to put the matrix in echelon form can be reduced by working with the transpose of the original matrix.

[8]The *row rank* and *column rank* are the same.

Example 5.4

What is the rank of matrix $\mathbf{A}$?

$$\mathbf{A} = \begin{bmatrix} 1 & -2 & -1 \\ -3 & 3 & 0 \\ 2 & 2 & 4 \end{bmatrix}$$

Solution

Matrix $\mathbf{A}$ is singular because $|\mathbf{A}| = 0$. However, there is at least one 2×2 nonsingular submatrix:

$$\begin{vmatrix} 1 & -2 \\ -3 & 3 \end{vmatrix} = (1)(3) - (-3)(-2) = -3$$

Therefore, the rank is 2.

Example 5.5

Determine the rank of matrix $\mathbf{A}$ by reducing it to echelon form.

$$\mathbf{A} = \begin{bmatrix} 7 & 4 & 9 & 1 \\ 0 & 2 & -5 & 3 \\ 0 & 4 & -10 & 6 \end{bmatrix}$$

Solution

By inspection, the matrix can be row-reduced by subtracting two times the second row from the third row. The matrix cannot be further reduced. Since there are two nonzero rows, the rank is 2.

$$\begin{bmatrix} 7 & 4 & 9 & 1 \\ 0 & 2 & -5 & 3 \\ 0 & 0 & 0 & 0 \end{bmatrix}$$

11. CLASSICAL ADJOINT

The *classical adjoint* is the transpose of the cofactor matrix. The resulting matrix can be designated as $\mathbf{A}_{\text{adj}}$, adj$\{\mathbf{A}\}$, or $\mathbf{A}^{\text{adj}}$.

Example 5.6

What is the classical adjoint of matrix $\mathbf{A}$?

$$\mathbf{A} = \begin{bmatrix} 2 & 3 & -4 \\ 0 & -4 & 2 \\ 1 & -1 & 5 \end{bmatrix}$$

Solution

The matrix of cofactors is

$$\begin{bmatrix} -18 & 2 & 4 \\ -11 & 14 & 5 \\ -10 & -4 & -8 \end{bmatrix}$$

The transpose of the matrix of cofactors is

$$\mathbf{A}_{\text{adj}} = \begin{bmatrix} -18 & -11 & -10 \\ 2 & 14 & -4 \\ 4 & 5 & -8 \end{bmatrix}$$

12. INVERSE

The product of a matrix $\mathbf{A}$ and its inverse, $\mathbf{A}^{-1}$, is the identity matrix, $\mathbf{I}$. Only square matrices have inverses, but not all square matrices are invertible. A matrix has an inverse if and only if it is nonsingular (i.e., its determinant is nonzero).

$$\mathbf{AA}^{-1} = \mathbf{A}^{-1}\mathbf{A} = \mathbf{I} \qquad 5.19$$

$$(\mathbf{AB})^{-1} = \mathbf{B}^{-1}\mathbf{A}^{-1} \qquad 5.20$$

The inverse of a 2×2 matrix is easily determined by formula.

$$\mathbf{A} = \begin{bmatrix} a & b \\ c & d \end{bmatrix}$$

$$\mathbf{A}^{-1} = \frac{\begin{bmatrix} d & -b \\ -c & a \end{bmatrix}}{|\mathbf{A}|} \qquad 5.21$$

For a 3×3 or larger matrix, the inverse is determined by dividing every entry in the classical adjoint by the determinant of the original matrix.

$$\mathbf{A}^{-1} = \frac{\mathbf{A}_{\text{adj}}}{|\mathbf{A}|} \qquad 5.22$$

Example 5.7

What is the inverse of matrix $\mathbf{A}$?

$$\mathbf{A} = \begin{bmatrix} 4 & 5 \\ 2 & 3 \end{bmatrix}$$

Solution

The determinant is calculated as

$$|\mathbf{A}| = (4)(3) - (2)(5) = 2$$

Using Eq. 5.22, the inverse is

$$\mathbf{A}^{-1} = \frac{\begin{bmatrix} 3 & -5 \\ -2 & 4 \end{bmatrix}}{2} = \begin{bmatrix} \frac{3}{2} & -\frac{5}{2} \\ -1 & 2 \end{bmatrix}$$

Check.

$$\mathbf{AA}^{-1} = \begin{bmatrix} 4 & 5 \\ 2 & 3 \end{bmatrix}\begin{bmatrix} \frac{3}{2} & -\frac{5}{2} \\ -1 & 2 \end{bmatrix} = \begin{bmatrix} 6-5 & -10+10 \\ 3-3 & -5+6 \end{bmatrix}$$

$$= \begin{bmatrix} 1 & 0 \\ 0 & 1 \end{bmatrix} = \mathbf{I} \quad \text{[OK]}$$

13. WRITING SIMULTANEOUS LINEAR EQUATIONS IN MATRIX FORM

Matrices are used to simplify the presentation and solution of sets of simultaneous linear equations. For example, the following three methods of presenting simultaneous linear equations are equivalent:

$$a_{11}x_1 + a_{12}x_2 = b_1$$
$$a_{21}x_1 + a_{22}x_2 = b_2$$

$$\begin{bmatrix} a_{11} & a_{12} \\ a_{21} & a_{22} \end{bmatrix}\begin{bmatrix} x_1 \\ x_2 \end{bmatrix} = \begin{bmatrix} b_1 \\ b_2 \end{bmatrix}$$

$$\mathbf{AX} = \mathbf{B}$$

In the second and third representations, $\mathbf{A}$ is known as the *coefficient matrix*, $\mathbf{X}$ as the *variable matrix*, and $\mathbf{B}$ as the *constant matrix*.

Not all systems of simultaneous equations have solutions, and those that do may not have unique solutions. The existence of a solution can be determined by calculating the determinant of the coefficient matrix. These rules are summarized in Table 5.1.

- If the system of linear equations is homogeneous (i.e., $\mathbf{B}$ is a zero matrix) and $|\mathbf{A}|$ is zero, there are an infinite number of solutions.
- If the system is homogeneous and $|\mathbf{A}|$ is nonzero, only the trivial solution exists.
- If the system of linear equations is nonhomogeneous (i.e., $\mathbf{B}$ is not a zero matrix) and $|\mathbf{A}|$ is nonzero, there is a unique solution to the set of simultaneous equations.
- If $|\mathbf{A}|$ is zero, a nonhomogeneous system of simultaneous equations may still have a solution. The requirement is that the determinants of all substitutional matrices (see Sec. 5.14) are zero, in which case there will be an infinite number of solutions. Otherwise, no solution exists.

Table 5.1 Solution Existence Rules for Simultaneous Equations

	$\mathbf{B}=0$	$\mathbf{B}\neq 0$
$\lvert\mathbf{A}\rvert=0$	infinite number of solutions (linearly dependent equations)	either an infinite number of solutions or no solution at all
$\lvert\mathbf{A}\rvert\neq 0$	trivial solution only ($x_i=0$)	unique nonzero solution

14. SOLVING SIMULTANEOUS LINEAR EQUATIONS

Gauss-Jordan elimination can be used to obtain the solution to a set of simultaneous linear equations. The coefficient matrix is augmented by the constant matrix. Then, elementary row operations are used to reduce the coefficient matrix to canonical form. All of the operations performed on the coefficient matrix are performed on the constant matrix. The variable values that satisfy the simultaneous equations will be the entries in the constant matrix when the coefficient matrix is in canonical form.

Determinants are used to calculate the solution to linear simultaneous equations through a procedure known as *Cramer's rule.*

The procedure is to calculate determinants of the original coefficient matrix **A** and of the n matrices resulting from the systematic replacement of a column in **A** by the constant matrix **B**. For a system of three equations in three unknowns, there are three substitutional matrices, $\mathbf{A}_1$, $\mathbf{A}_2$, and $\mathbf{A}_3$, as well as the original coefficient matrix, for a total of four matrices whose determinants must be calculated.

The values of the unknowns that simultaneously satisfy all of the linear equations are

$$x_1 = \frac{|\mathbf{A}_1|}{|\mathbf{A}|} \qquad 5.23$$

$$x_2 = \frac{|\mathbf{A}_2|}{|\mathbf{A}|} \qquad 5.24$$

$$x_3 = \frac{|\mathbf{A}_3|}{|\mathbf{A}|} \qquad 5.25$$

Example 5.8

Use Gauss-Jordan elimination to solve the following system of simultaneous equations.

$$2x+3y-4z=1$$
$$3x-y-2z=4$$
$$4x-7y-6z=-7$$

Solution

The augmented matrix is created by appending the constant matrix to the coefficient matrix.

$$\left[\begin{array}{ccc|c} 2 & 3 & -4 & 1 \\ 3 & -1 & -2 & 4 \\ 4 & -7 & -6 & -7 \end{array}\right]$$

Elementary row operations are used to reduce the coefficient matrix to canonical form. For example, two times the first row is subtracted from the third row. This step obtains the 0 needed in the a_{31} position.

$$\left[\begin{array}{ccc|c} 2 & 3 & -4 & 1 \\ 3 & -1 & -2 & 4 \\ 0 & -13 & 2 & -9 \end{array}\right]$$

This process continues until the following form is obtained.

$$\left[\begin{array}{ccc|c} 1 & 0 & 0 & 3 \\ 0 & 1 & 0 & 1 \\ 0 & 0 & 1 & 2 \end{array}\right]$$

$x=3$, $y=1$, and $z=2$ satisfy this system of equations.

Example 5.9

Use Cramer's rule to solve the following system of simultaneous equations.

$$2x+3y-4z=1$$
$$3x-y-2z=4$$
$$4x-7y-6z=-7$$

Solution

The determinant of the coefficient matrix is

$$|\mathbf{A}| = \begin{vmatrix} 2 & 3 & -4 \\ 3 & -1 & -2 \\ 4 & -7 & -6 \end{vmatrix} = 82$$

The determinants of the substitutional matrices are

$$|\mathbf{A}_1| = \begin{vmatrix} 1 & 3 & -4 \\ 4 & -1 & -2 \\ -7 & -7 & -6 \end{vmatrix} = 246$$

$$|\mathbf{A}_2| = \begin{vmatrix} 2 & 1 & -4 \\ 3 & 4 & -2 \\ 4 & -7 & -6 \end{vmatrix} = 82$$

$$|\mathbf{A}_3| = \begin{vmatrix} 2 & 3 & 1 \\ 3 & -1 & 4 \\ 4 & -7 & -7 \end{vmatrix} = 164$$

The values of x, y, and z that will satisfy the linear equations are

$$x = \frac{246}{82} = 3$$

$$y = \frac{82}{82} = 1$$

$$z = \frac{164}{82} = 2$$

15. EIGENVALUES AND EIGENVECTORS

Eigenvalues and eigenvectors (also known as *characteristic values* and *characteristic vectors*) of a square matrix **A** are the scalars k and matrices **X** such that

$$\mathbf{AX} = k\mathbf{X} \qquad 5.26$$

The scalar k is an eigenvalue of **A** if and only if the matrix $(k\mathbf{I} - \mathbf{A})$ is singular; that is, if $|k\mathbf{I} - \mathbf{A}| = 0$. This equation is called the *characteristic equation* of the matrix **A**. When expanded, the determinant is called the *characteristic polynomial.* The method of using the characteristic polynomial to find eigenvalues and eigenvectors is illustrated in Ex. 5.10.

If all of the eigenvalues are unique (i.e., nonrepeating), then Eq. 5.27 is valid.

$$[k\mathbf{I} - \mathbf{A}]\mathbf{X} = 0 \qquad 5.27$$

Example 5.10

Find the eigenvalues and nonzero eigenvectors of the matrix **A**.

$$\mathbf{A} = \begin{bmatrix} 2 & 4 \\ 6 & 4 \end{bmatrix}$$

Solution

$$k\mathbf{I} - \mathbf{A} = \begin{bmatrix} k & 0 \\ 0 & k \end{bmatrix} - \begin{bmatrix} 2 & 4 \\ 6 & 4 \end{bmatrix} = \begin{bmatrix} k-2 & -4 \\ -6 & k-4 \end{bmatrix}$$

The characteristic polynomial is found by setting the determinant $|k\mathbf{I} - \mathbf{A}|$ equal to zero.

$$(k-2)(k-4) - (-6)(-4) = 0$$
$$k^2 - 6k - 16 = (k-8)(k+2) = 0$$

The roots of the characteristic polynomial are $k = +8$ and $k = -2$. These are the eigenvalues of **A**.

Substituting $k = 8$,

$$k\mathbf{I} - \mathbf{A} = \begin{bmatrix} 8-2 & -4 \\ -6 & 8-4 \end{bmatrix} = \begin{bmatrix} 6 & -4 \\ -6 & 4 \end{bmatrix}$$

The resulting system can be interpreted as the linear equation $6x_1 - 4x_2 = 0$. The values of x that satisfy this equation define the eigenvector. An eigenvector **X** associated with the eigenvalue +8 is

$$\mathbf{X} = \begin{bmatrix} x_1 \\ x_2 \end{bmatrix} = \begin{bmatrix} 4 \\ 6 \end{bmatrix}$$

All other eigenvectors for this eigenvalue are multiples of **X**. Normally **X** is reduced to smallest integers.

$$\mathbf{X} = \begin{bmatrix} 2 \\ 3 \end{bmatrix}$$

Similarly, the eigenvector associated with the eigenvalue −2 is

$$\mathbf{X} = \begin{bmatrix} x_1 \\ x_2 \end{bmatrix} = \begin{bmatrix} +4 \\ -4 \end{bmatrix}$$

Reducing this to smallest integers gives

$$\mathbf{X} = \begin{bmatrix} +1 \\ -1 \end{bmatrix}$$

6 Vectors

1. INTRODUCTION

A physical property or quantity can be described by a scalar, vector, or tensor. A *scalar* has only magnitude. Knowing its value is sufficient to define a scalar. Mass, enthalpy, density, and speed are examples of scalars.

Force, momentum, displacement, and velocity are examples of vectors. A *vector* is a directed straight line with a specific magnitude and is specified completely by its direction (consisting of the vector's *angular orientation* and its *sense*) and magnitude. A vector's *point of application* (*terminal point*) is not needed to define the vector.[1] Two vectors with the same direction and magnitude are said to be *equal vectors* even though their *lines of action* may be different.[2]

A vector can be designated by a boldface variable (as in this book) or as a combination of the variable and some other symbol. For example, the notations $\mathbf{V}$, $\overline{V}$, $\hat{V}$, $\vec{V}$, and $\underline{V}$ are used by different authorities to represent vectors. In this book, the magnitude of a vector can be designated by either $|\mathbf{V}|$ or V (italic but not bold).

Stress, dielectric constant, and magnetic susceptibility are examples of tensors. A *tensor* has magnitude in a specific direction but the direction is not unique. Tensors are frequently associated with *anisotropic materials* that have different properties in different directions. A tensor in three-dimensional space is defined by nine components, compared with the three that are required to define vectors. These components are written in matrix form. Stress, σ, at a point, for example, would be defined by the following tensor matrix.

$$\sigma \equiv \begin{bmatrix} \sigma_{xx} & \sigma_{xy} & \sigma_{xz} \\ \sigma_{yx} & \sigma_{yy} & \sigma_{yz} \\ \sigma_{zx} & \sigma_{zy} & \sigma_{zz} \end{bmatrix}$$

2. VECTORS IN n-SPACE

In some cases, a vector, $\mathbf{V}$, will be designated by its two endpoints in n-dimensional vector space. The usual vector space is three-dimensional force-space. Usually, one of the points will be the origin, in which case the vector is said to be "based at the origin," "origin-based," or "zero-based."[3] If one of the endpoints is the origin, specifying a terminal point P would represent a force directed from the origin to point P.

If a coordinate system is superimposed on the vector space, a vector can be specified in terms of the n coordinates of its two endpoints. The magnitude of the vector $\mathbf{V}$ is the distance in vector space between the two points, as given by Eq. 6.1. Similarly, the direction is defined by the angle the vector makes with one of the axes. Figure 6.1 illustrates a vector in two dimensions.

$$|\mathbf{V}| = \sqrt{(x_2 - x_1)^2 + (y_2 - y_1)^2} \qquad 6.1$$

$$\phi = \arctan \frac{y_2 - y_1}{x_2 - x_1} \qquad 6.2$$

Figure 6.1 Vector in Two-Dimensional Space

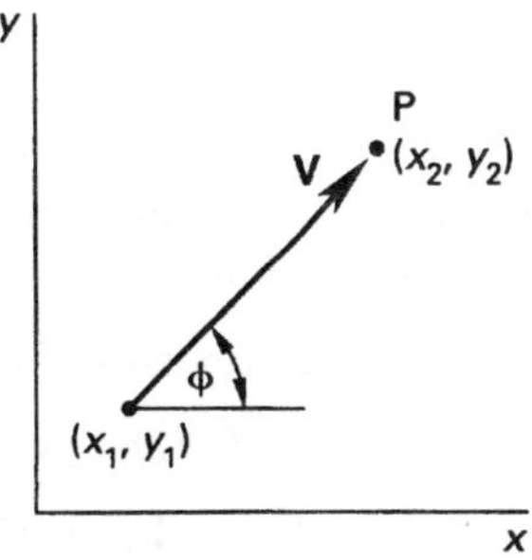

[1]A vector that is constrained to act at or through a certain point is a *bound vector (fixed vector)*. A *sliding vector (transmissible vector)* can be applied anywhere along its line of action. A *free vector* is not constrained and can be applied at any point in space.

[2]A distinction is sometimes made between equal vectors and equivalent vectors. *Equivalent vectors* produce the same effect but are not necessarily equal.

[3]Any vector directed from P_1 to P_2 can be transformed into a zero-based vector by subtracting the coordinates of point P_1 from the coordinates of terminal point P_2. The transformed vector will be equivalent to the original vector.

The *components* of a vector are the projections of the vector on the coordinate axes. (For a zero-based vector, the components and the coordinates of the endpoint are the same.) Simple trigonometric principles are used to resolve a vector into its components. A vector reconstructed from its components is known as a *resultant vector*. Figure 6.2 shows the location of direction angles.

$$V_x = |\mathbf{V}|\cos\phi_x \qquad 6.3$$

$$V_y = |\mathbf{V}|\cos\phi_y \qquad 6.4$$

$$V_z = |\mathbf{V}|\cos\phi_z \qquad 6.5$$

$$|\mathbf{V}| = \sqrt{V_x^2 + V_y^2 + V_z^2} \qquad 6.6$$

In Eq. 6.3 through Eq. 6.5, ϕ_x, ϕ_y, and ϕ_z are the *direction angles*—the angles between the vector and the x-, y-, and z-axes, respectively. The cosines of these angles are known as *direction cosines*. The sum of the squares of the direction cosines is equal to 1.

$$\cos^2\phi_x + \cos^2\phi_y + \cos^2\phi_z = 1 \qquad 6.7$$

Figure 6.2 *Direction Angles of a Vector*

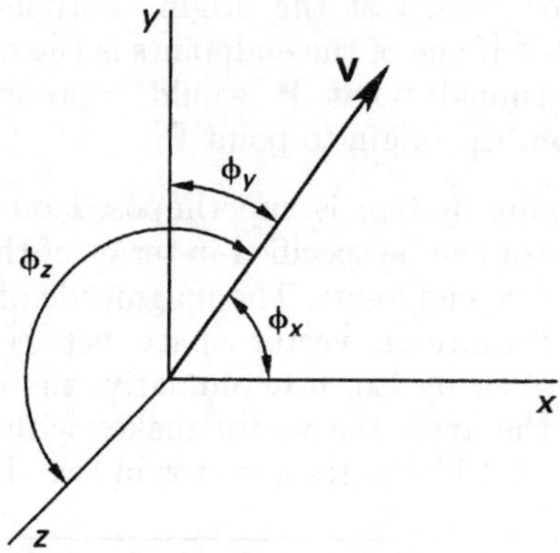

3. UNIT VECTORS

Unit vectors are vectors with unit magnitudes (i.e., magnitudes of 1). They are represented in the same notation as other vectors. (Unit vectors in this book are written in boldface type.) Although they can have any direction, the standard unit vectors (the *Cartesian unit vectors* **i**, **j**, and **k**) have the directions of the x-, y-, and z-coordinate axes and constitute the *Cartesian triad*, as illustrated in Fig. 6.3.

A vector **V** can be written in terms of unit vectors and its components.

$$\mathbf{V} = |\mathbf{V}|\mathbf{a} = V_x\mathbf{i} + V_y\mathbf{j} + V_z\mathbf{k} \qquad 6.8$$

Figure 6.3 *Cartesian Unit Vectors*

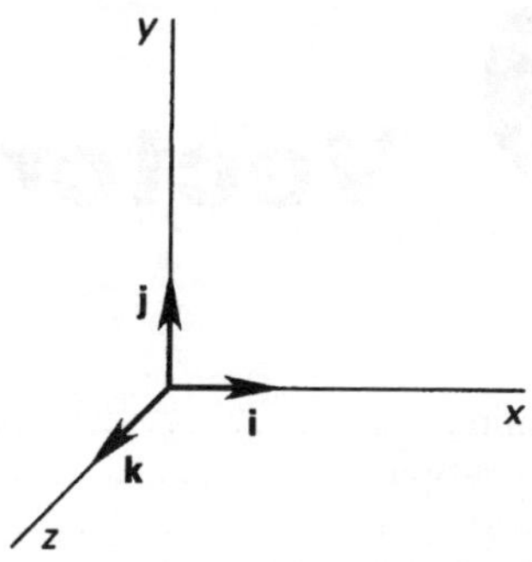

The unit vector, **a**, has the same direction as the vector **V** but has a length of 1. This unit vector is calculated by dividing the original vector, **V**, by its magnitude, |**V**|.

$$\mathbf{a} = \frac{\mathbf{V}}{|\mathbf{V}|} = \frac{V_x\mathbf{i} + V_y\mathbf{j} + V_z\mathbf{k}}{\sqrt{V_x^2 + V_y^2 + V_z^2}} \qquad 6.9$$

4. VECTOR REPRESENTATION

The most common method of representing a vector is by writing it in *rectangular form*—a vector sum of its orthogonal components. In rectangular form, each of the orthogonal components has the same units as the resultant vector.

$$\mathbf{A} \equiv A_x\mathbf{i} + A_y\mathbf{j} + A_z\mathbf{k} \quad \text{[three dimensions]}$$

However, the vector is also completely defined by its magnitude and associated angle. These two quantities can be written together in *phasor form*, sometimes referred to as *polar form*.

$$\mathbf{A} \equiv |\mathbf{A}|\angle\phi = A\angle\phi$$

5. CONVERSION BETWEEN SYSTEMS

The choice of the **ijk** triad may be convenient but is arbitrary. A vector can be expressed in terms of any other set of unit vectors, **uvw**.

$$\mathbf{V} = V_x\mathbf{i} + V_y\mathbf{j} + V_z\mathbf{k} = V'_x\mathbf{u} + V'_y\mathbf{v} + V'_z\mathbf{w} \qquad 6.10$$

The two representations are related.

$$V'_x = \mathbf{V}\cdot\mathbf{u} = (\mathbf{i}\cdot\mathbf{u})V_x + (\mathbf{j}\cdot\mathbf{u})V_y + (\mathbf{k}\cdot\mathbf{u})V_z \qquad 6.11$$

$$V'_y = \mathbf{V}\cdot\mathbf{v} = (\mathbf{i}\cdot\mathbf{v})V_x + (\mathbf{j}\cdot\mathbf{v})V_y + (\mathbf{k}\cdot\mathbf{v})V_z \qquad 6.12$$

$$V'_z = \mathbf{V}\cdot\mathbf{w} = (\mathbf{i}\cdot\mathbf{w})V_x + (\mathbf{j}\cdot\mathbf{w})V_y + (\mathbf{k}\cdot\mathbf{w})V_z \qquad 6.13$$

Equation 6.11 through Eq. 6.13 can be expressed in matrix form. The dot products are known as the *coefficients of transformation*, and the matrix containing them is the *transformation matrix*.

$$\begin{pmatrix} V'_x \\ V'_y \\ V'_z \end{pmatrix} = \begin{pmatrix} \mathbf{i}\cdot\mathbf{u} & \mathbf{j}\cdot\mathbf{u} & \mathbf{k}\cdot\mathbf{u} \\ \mathbf{i}\cdot\mathbf{v} & \mathbf{j}\cdot\mathbf{v} & \mathbf{k}\cdot\mathbf{v} \\ \mathbf{i}\cdot\mathbf{w} & \mathbf{j}\cdot\mathbf{w} & \mathbf{k}\cdot\mathbf{w} \end{pmatrix} \begin{pmatrix} V_x \\ V_y \\ V_z \end{pmatrix} \qquad 6.14$$

6. VECTOR ADDITION

Addition of two vectors by the *polygon method* is accomplished by placing the tail of the second vector at the head (tip) of the first. The sum (i.e., the *resultant vector*) is a vector extending from the tail of the first vector to the head of the second, as shown in Fig. 6.4. Alternatively, the two vectors can be considered as the two sides of a parallelogram, while the sum represents the diagonal. This is known as addition by the *parallelogram method.*

Figure 6.4 *Addition of Two Vectors*

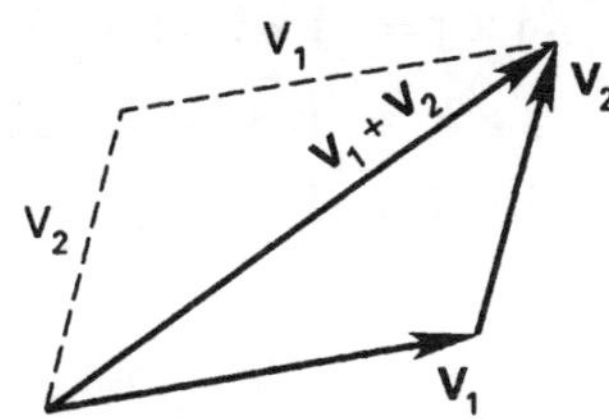

The components of the resultant vector are the sums of the components of the added vectors (that is, $V_{1x}+V_{2x}$, $V_{1y}+V_{2y}$, $V_{1z}+V_{2z}$).

Vector addition is both commutative and associative.

$$\mathbf{V}_1+\mathbf{V}_2=\mathbf{V}_2+\mathbf{V}_1 \qquad 6.15$$

$$\mathbf{V}_1+(\mathbf{V}_2+\mathbf{V}_3)=(\mathbf{V}_1+\mathbf{V}_2)+\mathbf{V}_3 \qquad 6.16$$

7. MULTIPLICATION BY A SCALAR

A vector, **V**, can be multiplied by a scalar, *c*. If the original vector is represented by its components, each of the components is multiplied by *c*.

$$c\mathbf{V}=c|\mathbf{V}|\mathbf{a}=cV_x\mathbf{i}+cV_y\mathbf{j}+cV_z\mathbf{k} \qquad 6.17$$

Scalar multiplication is distributive.

$$c(\mathbf{V}_1+\mathbf{V}_2)=c\mathbf{V}_1+c\mathbf{V}_2 \qquad 6.18$$

8. VECTOR DOT PRODUCT

The *dot product* (*scalar product*), $\mathbf{V}_1\cdot\mathbf{V}_2$, of two vectors is a scalar that is proportional to the length of the projection of the first vector onto the second vector, as illustrated in Fig. 6.5.[4]

Figure 6.5 *Vector Dot Product*

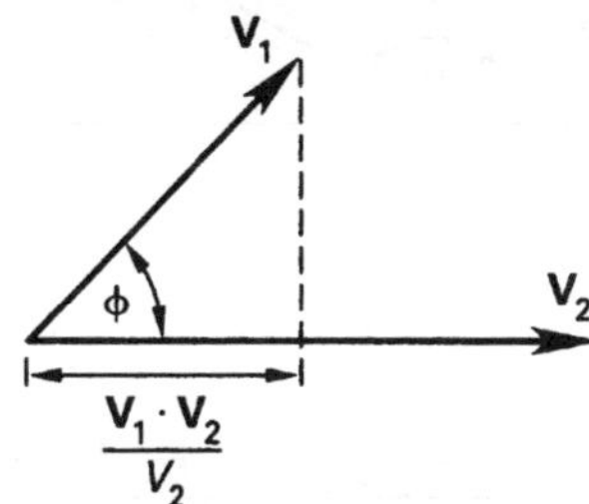

The dot product is commutative and distributive.

$$\mathbf{V}_1\cdot\mathbf{V}_2=\mathbf{V}_2\cdot\mathbf{V}_1 \qquad 6.19$$

$$\mathbf{V}_1\cdot(\mathbf{V}_2+\mathbf{V}_3)=\mathbf{V}_1\cdot\mathbf{V}_2+\mathbf{V}_1\cdot\mathbf{V}_3 \qquad 6.20$$

The dot product can be calculated in two ways, as Eq. 6.21 indicates. ϕ is limited to 180° and is the angle between the two vectors.

$$\begin{aligned}\mathbf{V}_1\cdot\mathbf{V}_2&=|\mathbf{V}_1|\,|\mathbf{V}_2|\cos\phi \\ &=V_{1x}V_{2x}+V_{1y}V_{2y}+V_{1z}V_{2z}\end{aligned} \qquad 6.21$$

When Eq. 6.21 is solved for the angle between the two vectors, ϕ, it is known as the *Cauchy-Schwartz theorem.*

$$\cos\phi=\frac{V_{1x}V_{2x}+V_{1y}V_{2y}+V_{1z}V_{2z}}{|\mathbf{V}_1|\,|\mathbf{V}_2|} \qquad 6.22$$

The dot product can be used to determine whether a vector is a unit vector and to show that two vectors are orthogonal (perpendicular). For any unit vector, **u**,

$$\mathbf{u}\cdot\mathbf{u}=1 \qquad 6.23$$

For two non-null orthogonal vectors,

$$\mathbf{V}_1\cdot\mathbf{V}_2=0 \qquad 6.24$$

Equation 6.23 and Eq. 6.24 can be extended to the Cartesian unit vectors.

$$\mathbf{i}\cdot\mathbf{i}=1 \qquad 6.25$$

$$\mathbf{j}\cdot\mathbf{j}=1 \qquad 6.26$$

$$\mathbf{k}\cdot\mathbf{k}=1 \qquad 6.27$$

$$\mathbf{i}\cdot\mathbf{j}=0 \qquad 6.28$$

$$\mathbf{i}\cdot\mathbf{k}=0 \qquad 6.29$$

$$\mathbf{j}\cdot\mathbf{k}=0 \qquad 6.30$$

[4]The dot product is also written in parentheses without a dot, that is, $(\mathbf{V}_1\mathbf{V}_2)$.

Mathematics

Example 6.1

What is the angle between the zero-based vectors $\mathbf{V}_1 = (-\sqrt{3}, 1)$ and $\mathbf{V}_2 = (2\sqrt{3}, 2)$?

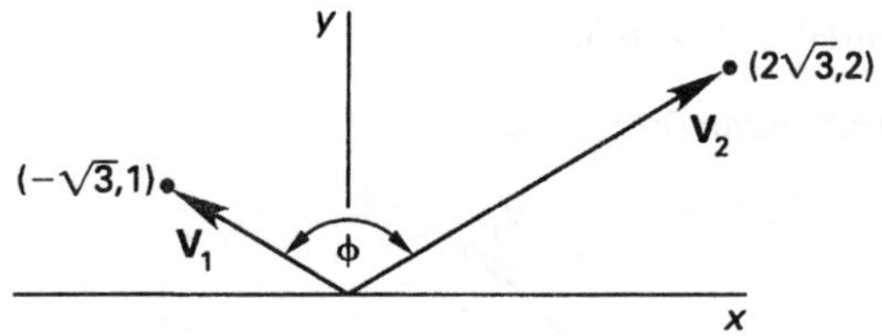

Solution

From Eq. 6.22,

$$\begin{aligned}\cos\phi &= \frac{V_{1x}V_{2x} + V_{1y}V_{2y}}{|\mathbf{V}_1|\,|\mathbf{V}_2|} = \frac{V_{1x}V_{2x} + V_{1y}V_{2y}}{\sqrt{V_{1x}^2 + V_{1y}^2}\sqrt{V_{2x}^2 + V_{2y}^2}} \\ &= \frac{(-\sqrt{3})(2\sqrt{3}) + (1)(2)}{\sqrt{(-\sqrt{3})^2 + (1)^2}\sqrt{(2\sqrt{3})^2 + (2)^2}} \\ &= -\frac{1}{2} \\ \phi &= \arccos\left(-\frac{1}{2}\right) \\ &= 120°\end{aligned}$$

9. VECTOR CROSS PRODUCT

The *cross product* (*vector product*), $\mathbf{V}_1 \times \mathbf{V}_2$, of two vectors is a vector that is orthogonal (perpendicular) to the plane of the two vectors.[5] The unit vector representation of the cross product can be calculated as a third-order determinant. Figure 6.6 illustrates the vector cross product.

$$\mathbf{V}_1 \times \mathbf{V}_2 = \begin{vmatrix} \mathbf{i} & V_{1x} & V_{2x} \\ \mathbf{j} & V_{1y} & V_{2y} \\ \mathbf{k} & V_{1z} & V_{2z} \end{vmatrix} \qquad 6.31$$

The direction of the cross-product vector corresponds to the direction a right-hand screw would progress if vectors $\mathbf{V}_1$ and $\mathbf{V}_2$ are placed tail-to-tail in the plane they define and $\mathbf{V}_1$ is rotated into $\mathbf{V}_2$. The direction can also be found from the *right-hand rule.*

Figure 6.6 *Vector Cross Product*

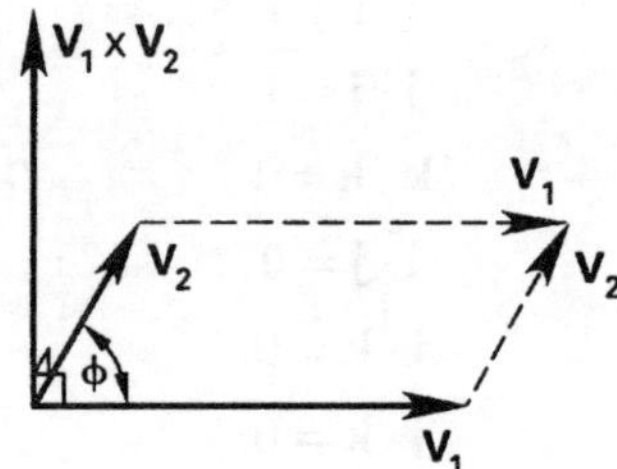

[5]The cross product is also written in square brackets without a cross, that is, $[\mathbf{V}_1\mathbf{V}_2]$.

The magnitude of the cross product can be determined from Eq. 6.32, in which ϕ is the angle between the two vectors and is limited to 180°. The magnitude corresponds to the area of a parallelogram that has $\mathbf{V}_1$ and $\mathbf{V}_2$ as two of its sides.

$$|\mathbf{V}_1 \times \mathbf{V}_2| = |\mathbf{V}_1||\mathbf{V}_2|\sin\phi \qquad 6.32$$

Vector cross multiplication is distributive but not commutative.

$$\mathbf{V}_1 \times \mathbf{V}_2 = -(\mathbf{V}_2 \times \mathbf{V}_1) \qquad 6.33$$

$$c(\mathbf{V}_1 \times \mathbf{V}_2) = (c\mathbf{V}_1) \times \mathbf{V}_2 = \mathbf{V}_1 \times (c\mathbf{V}_2) \qquad 6.34$$

$$\mathbf{V}_1 \times (\mathbf{V}_2 + \mathbf{V}_3) = \mathbf{V}_1 \times \mathbf{V}_2 + \mathbf{V}_1 \times \mathbf{V}_3 \qquad 6.35$$

If the two vectors are parallel, their cross product will be zero.

$$\mathbf{i}\times\mathbf{i} = \mathbf{j}\times\mathbf{j} = \mathbf{k}\times\mathbf{k} = 0 \qquad 6.36$$

Equation 6.31 and Eq. 6.33 can be extended to the unit vectors.

$$\mathbf{i}\times\mathbf{j} = -\mathbf{j}\times\mathbf{i} = \mathbf{k} \qquad 6.37$$

$$\mathbf{j}\times\mathbf{k} = -\mathbf{k}\times\mathbf{j} = \mathbf{i} \qquad 6.38$$

$$\mathbf{k}\times\mathbf{i} = -\mathbf{i}\times\mathbf{k} = \mathbf{j} \qquad 6.39$$

Example 6.2

Find a unit vector orthogonal to $\mathbf{V}_1 = \mathbf{i} - \mathbf{j} + 2\mathbf{k}$ and $\mathbf{V}_2 = 3\mathbf{j} - \mathbf{k}$.

Solution

The cross product is a vector orthogonal to $\mathbf{V}_1$ and $\mathbf{V}_2$.

$$\begin{aligned}\mathbf{V}_1 \times \mathbf{V}_2 &= \begin{vmatrix} \mathbf{i} & 1 & 0 \\ \mathbf{j} & 1 & 3 \\ \mathbf{k} & 2 & -1 \end{vmatrix} \\ &= -5\mathbf{i} + \mathbf{j} + 3\mathbf{k}\end{aligned}$$

Check to see whether this is a unit vector.

$$|\mathbf{V}_1 \times \mathbf{V}_2| = \sqrt{(-5)^2 + (1)^2 + (3)^2} = \sqrt{35}$$

Since its length is $\sqrt{35}$, the vector must be divided by $\sqrt{35}$ to obtain a unit vector.

$$\mathbf{a} = \frac{-5\mathbf{i} + \mathbf{j} + 3\mathbf{k}}{\sqrt{35}}$$

10. MIXED TRIPLE PRODUCT

The *mixed triple product* (*triple scalar product* or just *triple product*) of three vectors is a scalar quantity representing the volume of a parallelepiped with the three vectors making up the sides. It is calculated as a determinant. Since Eq. 6.40 can be negative, the absolute value must be used to obtain the volume in that case. Figure 6.7 shows a mixed triple product.

$$\mathbf{V}_1 \cdot (\mathbf{V}_2 \times \mathbf{V}_3) = \begin{vmatrix} V_{1x} & V_{1y} & V_{1z} \\ V_{2x} & V_{2y} & V_{2z} \\ V_{3x} & V_{3y} & V_{3z} \end{vmatrix} \qquad 6.40$$

Figure 6.7 *Vector Mixed Triple Product*

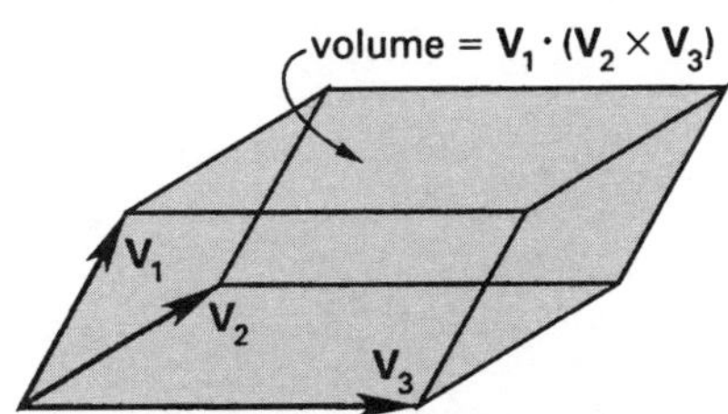

The mixed triple product has the property of *circular permutation*, as defined by Eq. 6.41.

$$\mathbf{V}_1 \cdot (\mathbf{V}_2 \times \mathbf{V}_3) = (\mathbf{V}_1 \times \mathbf{V}_2) \cdot \mathbf{V}_3 \qquad 6.41$$

11. VECTOR TRIPLE PRODUCT

The *vector triple product* is a vector defined by Eq. 6.42. The quantities in parentheses on the right-hand side are scalars.

$$\mathbf{V}_1 \times (\mathbf{V}_2 \times \mathbf{V}_3) = (\mathbf{V}_1 \cdot \mathbf{V}_3)\mathbf{V}_2 - (\mathbf{V}_1 \cdot \mathbf{V}_2)\mathbf{V}_3 \qquad 6.42$$

12. VECTOR FUNCTIONS

A vector can be a function of another parameter. For example, a vector **V** is a function of variable t when its V_x, V_y, and V_z are functions of t.

$$\mathbf{V}(t) = (2t - 3)\mathbf{i} + (t^2 + 1)\mathbf{j} + (-7t + 5)\mathbf{k} \qquad 6.43$$

When the functions of t are differentiated (or integrated) with respect to t, the vector itself is differentiated (integrated).[6]

$$\frac{d\mathbf{V}(t)}{dt} = \frac{dV_x}{dt}\mathbf{i} + \frac{dV_y}{dt}\mathbf{j} + \frac{dV_z}{dt}\mathbf{k} \qquad 6.44$$

Similarly, the integral of the vector is

$$\int \mathbf{V}(t)\,dt = \mathbf{i}\int V_x\,dt + \mathbf{j}\int V_y\,dt + \mathbf{k}\int V_z\,dt \qquad 6.45$$

[6]This is particularly valuable when converting among position, velocity, and acceleration vectors.

7 Trigonometry

1. DEGREES AND RADIANS

Degrees and *radians* are two units for measuring angles. One complete circle is divided into 360 degrees (written 360°) or 2π radians (abbreviated *rad*).[1] The conversions between degrees and radians are

multiply	by	to obtain
radians	$\frac{180}{\pi}$	degrees
degrees	$\frac{\pi}{180}$	radians

The number of radians in an angle, θ, corresponds to two times the area within a circular sector with arc length θ and a radius of one, as shown in Fig. 7.1. Alternatively, the area of a sector with central angle θ radians is $\theta/2$ for a *unit circle* (i.e., a circle with a radius of one unit).

2. PLANE ANGLES

A *plane angle* (usually referred to as just an *angle*) consists of two intersecting lines and an intersection point known as the *vertex*. The angle can be referred to by a capital letter representing the vertex (e.g., B in Fig. 7.2), a letter representing the angular measure (e.g., B or β), or by three capital letters, where the middle letter is the vertex and the other two letters are two points on different lines, and either the symbol ∠ or ∢ (e.g., ∢ ABC).

[1]The abbreviation *rad* is also used to represent *radiation absorbed dose*, a measure of radiation exposure.

Figure 7.1 Radians and Area of Unit Circle

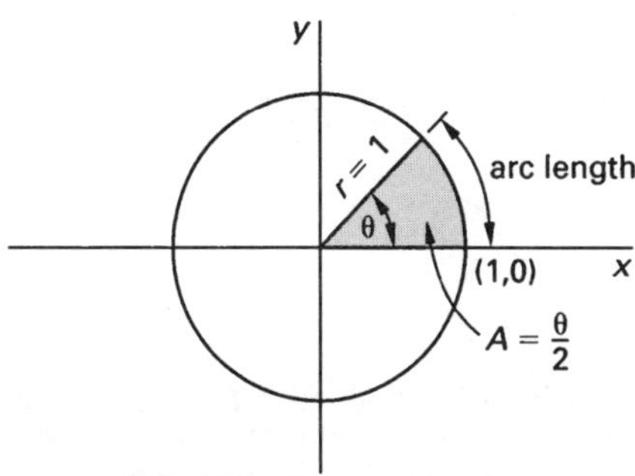

Figure 7.2 Angle

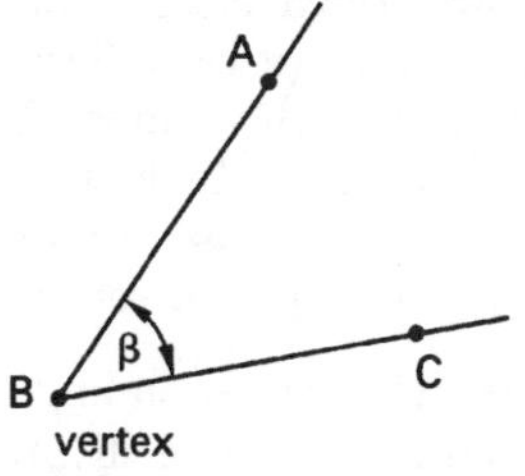

The angle between two intersecting lines generally is understood to be the smaller angle created.[2] Angles have been classified as follows.

- *acute angle:* an angle less than 90° ($\pi/2$ rad)
- *obtuse angle:* an angle more than 90° ($\pi/2$ rad) but less than 180° (π rad)
- *reflex angle:* an angle more than 180° (π rad) but less than 360° (2π rad)
- *related angle:* an angle that differs from another by some multiple of 90° ($\pi/2$ rad)
- *right angle:* an angle equal to 90° ($\pi/2$ rad)
- *straight angle:* an angle equal to 180° (π rad), that is, a straight line

Complementary angles are two angles whose sum is 90° ($\pi/2$ rad). *Supplementary angles* are two angles whose

[2]In books on geometry, the term *ray* is used instead of *line*.

sum is 180° (π rad). *Adjacent angles* share a common vertex and one (the interior) side. Adjacent angles are supplementary only if their exterior sides form a straight line.

Vertical angles are the two angles with a common vertex and with sides made up by two intersecting straight lines, as shown in Fig. 7.3. Vertical angles are equal.

Figure 7.3 Vertical Angles

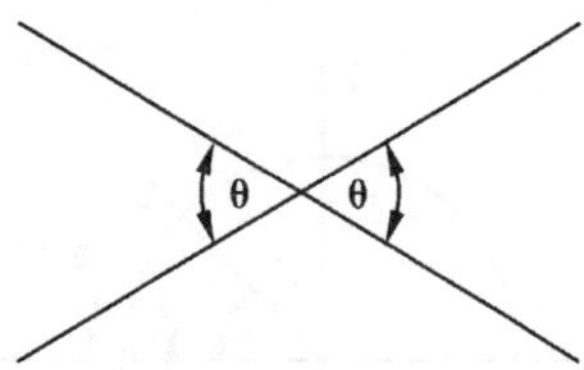

Angle of elevation and *angle of depression* are surveying terms referring to the angle above and below the horizontal plane of the observer, respectively.

3. TRIANGLES

A *triangle* is a three-sided closed polygon with three angles whose sum is 180° (π rad). Triangles are identified by their vertices and the symbol Δ (e.g., ΔABC in Fig. 7.4). A side is designated by its two endpoints (e.g., AB in Fig. 7.4) or by a lowercase letter corresponding to the capital letter of the opposite vertex (e.g., c).

In *similar triangles*, the corresponding angles are equal and the corresponding sides are in proportion. (Since there are only two independent angles in a triangle, showing that two angles of one triangle are equal to two angles of the other triangle is sufficient to show similarity.) The symbol for similarity is ~. In Fig. 7.4, ΔABC ~ ΔDEF (i.e., ΔABC is similar to ΔDEF).

Figure 7.4 Similar Triangles

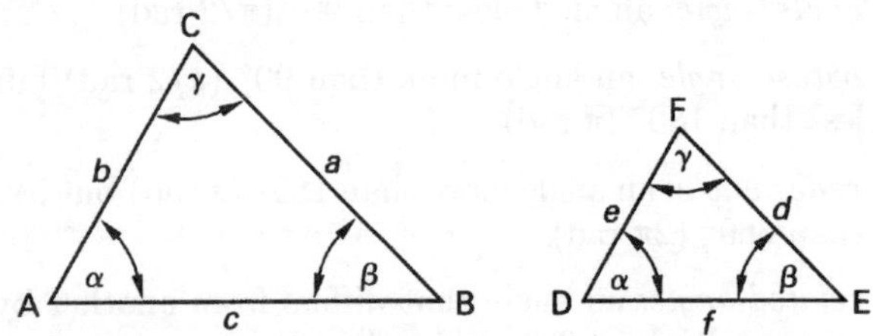

4. RIGHT TRIANGLES

A *right triangle* is a triangle in which one of the angles is 90° (π/2 rad). The remaining two angles are complementary. If one of the acute angles is chosen as the reference, the sides forming the right angle are known as the *adjacent side*, x, and the *opposite side*, y. The longest side is known as the *hypotenuse*, r. The *Pythagorean theorem* relates the lengths of these sides.

$$x^2 + y^2 = r^2 \qquad 7.1$$

In certain cases, the lengths of unknown sides of right triangles can be determined by inspection.[3] This occurs when the lengths of the sides are in the ratios of 3:4:5, 1:1:$\sqrt{2}$, 1:$\sqrt{3}$:2, and 5:12:13. Figure 7.5 illustrates a 3:4:5 triangle.

Figure 7.5 3:4:5 Right Triangle

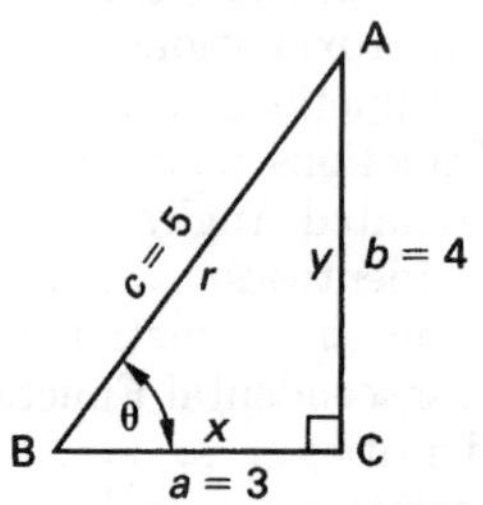

5. CIRCULAR TRANSCENDENTAL FUNCTIONS

The *circular transcendental functions* (usually referred to as the *transcendental functions, trigonometric functions*, or *functions of an angle*) are calculated from the sides of a right triangle. Equation 7.2 through Eq. 7.7 refer to Fig. 7.5.

$$\text{sine: } \sin\theta = \frac{y}{r} = \frac{\text{opposite}}{\text{hypotenuse}} \qquad 7.2$$

$$\text{cosine: } \cos\theta = \frac{x}{r} = \frac{\text{adjacent}}{\text{hypotenuse}} \qquad 7.3$$

$$\text{tangent: } \tan\theta = \frac{y}{x} = \frac{\text{opposite}}{\text{adjacent}} \qquad 7.4$$

$$\text{cotangent: } \cot\theta = \frac{x}{y} = \frac{\text{adjacent}}{\text{opposite}} \qquad 7.5$$

$$\text{secant: } \sec\theta = \frac{r}{x} = \frac{\text{hypotenuse}}{\text{adjacent}} \qquad 7.6$$

$$\text{cosecant: } \csc\theta = \frac{r}{y} = \frac{\text{hypotenuse}}{\text{opposite}} \qquad 7.7$$

Three of the transcendental functions are reciprocals of the others. However, while the tangent and cotangent

[3]These cases are almost always contrived examples. There is nothing intrinsic in nature to cause the formation of triangles with these proportions.

functions are reciprocals of each other, the sine and cosine functions are not.

$$\cot\theta = \frac{1}{\tan\theta} \qquad 7.8$$

$$\sec\theta = \frac{1}{\cos\theta} \qquad 7.9$$

$$\csc\theta = \frac{1}{\sin\theta} \qquad 7.10$$

The trigonometric functions correspond to the lengths of various line segments in a right triangle with a unit hypotenuse. Figure 7.6 shows such a triangle inscribed in a unit circle.

Figure 7.6 *Trigonometric Functions in a Unit Circle*

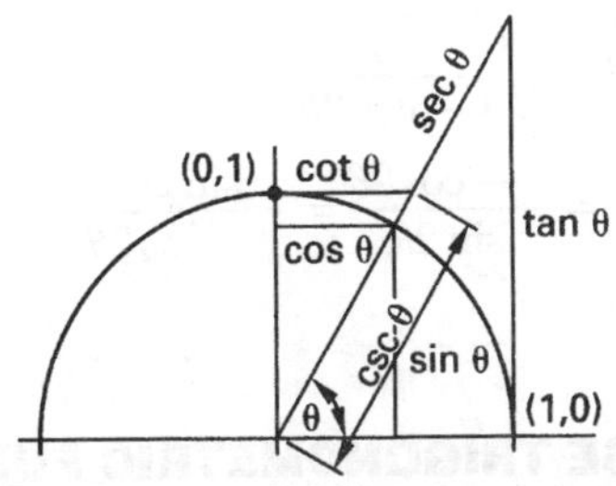

6. SMALL ANGLE APPROXIMATIONS

When an angle is very small, the hypotenuse and adjacent sides are essentially equal in length, and certain approximations can be made. (The angle θ must be expressed in radians in Eq. 7.11 and Eq. 7.12.)

$$\sin\theta \approx \tan\theta \approx \theta\big|_{\theta < 10^\circ\ (0.175\ \text{rad})} \qquad 7.11$$

$$\cos\theta \approx 1\big|_{\theta < 5^\circ\ (0.0873\ \text{rad})} \qquad 7.12$$

7. GRAPHS OF THE FUNCTIONS

Figure 7.7 illustrates the periodicity of the sine, cosine, and tangent functions.[4]

Figure 7.7 *Graphs of Sine, Cosine, and Tangent Functions*

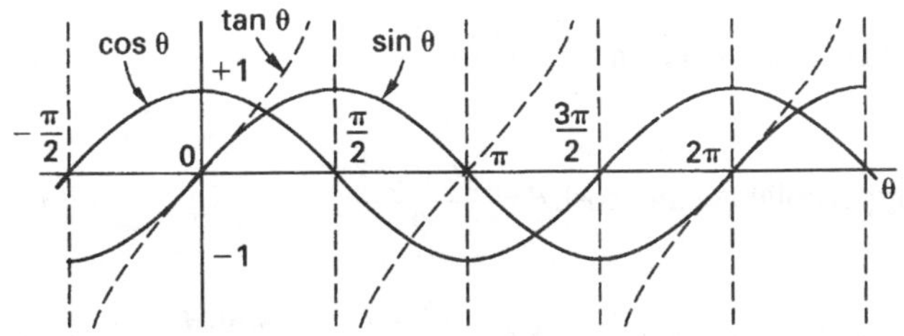

[4]The remaining functions, being reciprocals of these three functions, are also periodic.

8. SIGNS OF THE FUNCTIONS

Table 7.1 shows how the sine, cosine, and tangent functions vary in sign with different values of θ. All three functions are positive for angles $0° \le \theta \le 90°$ ($0 \le \theta \le \pi/2$ rad), but only the sine is positive for angles $90° < \theta \le 180°$ ($\pi/2$ rad $< \theta \le \pi$ rad). The concept of quadrants is used to summarize the signs of the functions: angles up to 90° ($\pi/2$ rad) are in quadrant I, between 90° and 180° ($\pi/2$ and π rad) are in quadrant II, and so on.

Table 7.1 *Signs of the Functions by Quadrant*

	quadrant			
function	I	II	III	IV
sine	+	+	−	−
cosine	+	−	−	+
tangent	+	−	+	−

9. FUNCTIONS OF RELATED ANGLES

Figure 7.7 shows that the sine, cosine, and tangent curves are symmetrical with respect to the horizontal axis. Furthermore, portions of the curves are symmetrical with respect to a vertical axis. The values of the sine and cosine functions repeat every 360° (2π rad), and the absolute values repeat every 180° (π rad). This can be written as

$$\sin(\theta + 180°) = -\sin\theta \qquad 7.13$$

Similarly, the tangent function repeats every 180° (π rad), and its absolute value repeats every 90° ($\pi/2$ rad).

Table 7.2 summarizes the functions of the related angles.

Table 7.2 *Functions of Related Angles*

$f(\theta)$	$-\theta$	$90°-\theta$	$90°+\theta$	$180°-\theta$	$180°+\theta$
sin	$-\sin\theta$	$\cos\theta$	$\cos\theta$	$\sin\theta$	$-\sin\theta$
cos	$\cos\theta$	$\sin\theta$	$-\sin\theta$	$-\cos\theta$	$-\cos\theta$
tan	$-\tan\theta$	$\cot\theta$	$-\cot\theta$	$-\tan\theta$	$\tan\theta$

10. TRIGONOMETRIC IDENTITIES

There are many relationships between trigonometric functions. For example, Eq. 7.14 through Eq. 7.16 are well known.

$$\sin^2\theta + \cos^2\theta = 1 \qquad 7.14$$

$$1 + \tan^2\theta = \sec^2\theta \qquad 7.15$$

$$1 + \cot^2\theta = \csc^2\theta \qquad 7.16$$

Mathematics

Other relatively common identities are listed as follows.[5]

- *double-angle formulas*

$$\sin 2\theta = 2\sin\theta\cos\theta = \frac{2\tan\theta}{1+\tan^2\theta} \quad 7.17$$

$$\cos 2\theta = \cos^2\theta - \sin^2\theta = 1 - 2\sin^2\theta$$

$$= 2\cos^2\theta - 1 = \frac{1-\tan^2\theta}{1+\tan^2\theta} \quad 7.18$$

$$\tan 2\theta = \frac{2\tan\theta}{1-\tan^2\theta} \quad 7.19$$

$$\cot 2\theta = \frac{\cot^2\theta - 1}{2\cot\theta} \quad 7.20$$

- *two-angle formulas*

$$\sin(\theta \pm \phi) = \sin\theta\cos\phi \pm \cos\theta\sin\phi \quad 7.21$$

$$\cos(\theta \pm \phi) = \cos\theta\cos\phi \mp \sin\theta\sin\phi \quad 7.22$$

$$\tan(\theta \pm \phi) = \frac{\tan\theta \pm \tan\phi}{1 \mp \tan\theta\tan\phi} \quad 7.23$$

$$\cot(\theta \pm \phi) = \frac{\cot\phi\cot\theta \mp 1}{\cot\phi \pm \cot\theta} \quad 7.24$$

- *half-angle formulas* ($\theta < 180°$)

$$\sin\frac{\theta}{2} = \sqrt{\frac{1-\cos\theta}{2}} \quad 7.25$$

$$\cos\frac{\theta}{2} = \sqrt{\frac{1+\cos\theta}{2}} \quad 7.26$$

$$\tan\frac{\theta}{2} = \sqrt{\frac{1-\cos\theta}{1+\cos\theta}} = \frac{\sin\theta}{1+\cos\theta} = \frac{1-\cos\theta}{\sin\theta} \quad 7.27$$

- *miscellaneous formulas* ($\theta < 90°$)

$$\sin\theta = 2\sin\frac{\theta}{2}\cos\frac{\theta}{2} \quad 7.28$$

$$\sin\theta = \sqrt{\frac{1-\cos 2\theta}{2}} \quad 7.29$$

$$\cos\theta = \cos^2\frac{\theta}{2} - \sin^2\frac{\theta}{2} \quad 7.30$$

$$\cos\theta = \sqrt{\frac{1+\cos 2\theta}{2}} \quad 7.31$$

$$\tan\theta = \frac{2\tan\frac{\theta}{2}}{1-\tan^2\frac{\theta}{2}}$$

$$= \frac{2\sin\frac{\theta}{2}\cos\frac{\theta}{2}}{\cos^2\frac{\theta}{2} - \sin^2\frac{\theta}{2}} \quad 7.32$$

$$\tan\theta = \sqrt{\frac{1-\cos 2\theta}{1+\cos 2\theta}}$$

$$= \frac{\sin 2\theta}{1+\cos 2\theta} = \frac{1-\cos 2\theta}{\sin 2\theta} \quad 7.33$$

$$\cot\theta = \frac{\cot^2\frac{\theta}{2} - 1}{2\cot\frac{\theta}{2}}$$

$$= \frac{\cos^2\frac{\theta}{2} - \sin^2\frac{\theta}{2}}{2\sin\frac{\theta}{2}\cos\frac{\theta}{2}} \quad 7.34$$

$$\cot\theta = \sqrt{\frac{1+\cos 2\theta}{1-\cos 2\theta}}$$

$$= \frac{1+\cos 2\theta}{\sin 2\theta} = \frac{\sin 2\theta}{1-\cos 2\theta} \quad 7.35$$

11. INVERSE TRIGONOMETRIC FUNCTIONS

Finding an angle from a known trigonometric function is a common operation known as an *inverse trigonometric operation.* The inverse function can be designated by adding "inverse," "arc-," or the superscript −1 to the name of the function. For example,

$$\text{inverse}\sin 0.5 = \arcsin 0.5 = \sin^{-1} 0.5 = 30°$$

12. HYPERBOLIC TRANSCENDENTAL FUNCTIONS

Hyperbolic transcendental functions (normally referred to as *hyperbolic functions*) are specific equations containing combinations of the terms e^θ and $e^{-\theta}$. These combinations appear regularly in certain types of problems (e.g., analysis of cables and heat transfer from fins) and are given specific names and symbols to simplify presentation.[6]

$$\text{hyperbolic sine: } \sinh\theta = \frac{e^\theta - e^{-\theta}}{2} \quad 7.36$$

$$\text{hyperbolic cosine: } \cosh\theta = \frac{e^\theta + e^{-\theta}}{2} \quad 7.37$$

$$\text{hyperbolic tangent: } \tanh\theta = \frac{e^\theta - e^{-\theta}}{e^\theta + e^{-\theta}} = \frac{\sinh\theta}{\cosh\theta} \quad 7.38$$

[5]It is an idiosyncrasy of the trade that these formulas are conventionally referred to as *formulas*, not *identities*.

[6]The hyperbolic sine and cosine functions are pronounced (by some) as "sinch" and "cosh," respectively.

hyperbolic cotangent: $\coth\theta = \frac{e^\theta + e^{-\theta}}{e^\theta - e^{-\theta}} = \frac{\cosh\theta}{\sinh\theta}$ 7.39

hyperbolic secant: $\text{sech}\,\theta = \frac{2}{e^\theta + e^{-\theta}} = \frac{1}{\cosh\theta}$ 7.40

hyperbolic cosecant: $\text{csch}\,\theta = \frac{2}{e^\theta - e^{-\theta}} = \frac{1}{\sinh\theta}$ 7.41

Hyperbolic functions cannot be related to a right triangle, but they are related to a rectangular (equilateral) hyperbola, as shown in Fig. 7.8. The shaded area has a value of $\theta/2$ and is sometimes given the units of *hyperbolic radians.*

$$\sinh\theta = \frac{y}{a} \quad 7.42$$

$$\cosh\theta = \frac{x}{a} \quad 7.43$$

$$\tanh\theta = \frac{y}{x} \quad 7.44$$

$$\coth\theta = \frac{x}{y} \quad 7.45$$

$$\text{sech}\,\theta = \frac{a}{x} \quad 7.46$$

$$\text{csch}\,\theta = \frac{a}{y} \quad 7.47$$

Figure 7.8 *Equilateral Hyperbola and Hyperbolic Functions*

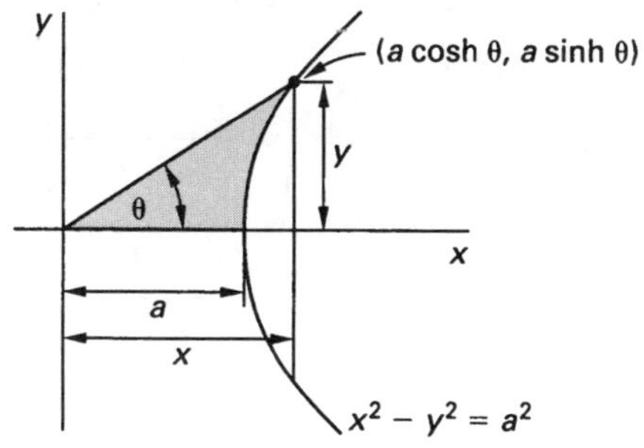

13. HYPERBOLIC IDENTITIES

The hyperbolic identities are different from the standard trigonometric identities. Some of the most important identities are presented as follows.

$$\cosh^2\theta - \sinh^2\theta = 1 \quad 7.48$$

$$1 - \tanh^2\theta = \text{sech}^2\,\theta \quad 7.49$$

$$1 - \coth^2\theta = -\text{csch}^2\,\theta \quad 7.50$$

$$\cosh\theta + \sinh\theta = e^\theta \quad 7.51$$

$$\cosh\theta - \sinh\theta = e^{-\theta} \quad 7.52$$

$$\sinh(\theta \pm \phi) = \sinh\theta\cosh\phi \pm \cosh\theta\sinh\phi \quad 7.53$$

$$\cosh(\theta \pm \phi) = \cosh\theta\cosh\phi \pm \sinh\theta\sinh\phi \quad 7.54$$

$$\tanh(\theta \pm \phi) = \frac{\tanh\theta \pm \tanh\phi}{1 \pm \tanh\theta\tanh\phi} \quad 7.55$$

14. GENERAL TRIANGLES

A *general triangle* (also known as an *oblique triangle*) is one that is not specifically a right triangle, as shown in Fig. 7.9. Equation 7.56 calculates the area of a general triangle.

$$\text{area} = \tfrac{1}{2}ab\sin C = \tfrac{1}{2}bc\sin A = \tfrac{1}{2}ca\sin B \quad 7.56$$

The *law of sines*, as shown in Eq. 7.57, relates the sides and the sines of the angles.

$$\frac{\sin A}{a} = \frac{\sin B}{b} = \frac{\sin C}{c} \quad 7.57$$

The *law of cosines* relates the cosine of an angle to an opposite side. (Equation 7.58 can be extended to the two remaining sides.)

$$a^2 = b^2 + c^2 - 2bc\cos A \quad 7.58$$

The *law of tangents* relates the sum and difference of two sides. (Equation 7.59 can be extended to the two remaining sides.)

$$\frac{a-b}{a+b} = \frac{\tan\left(\frac{A-B}{2}\right)}{\tan\left(\frac{A+B}{2}\right)} \quad 7.59$$

Figure 7.9 *General Triangle*

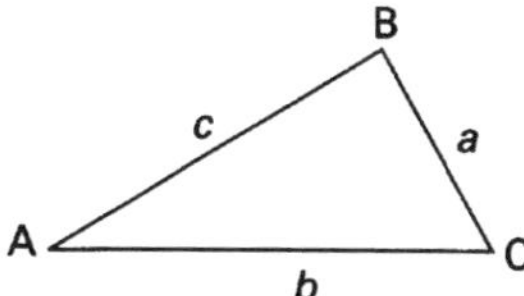

15. SPHERICAL TRIGONOMETRY

A *spherical triangle* is a triangle that has been drawn on the surface of a sphere, as shown in Fig. 7.10. The *trihedral angle* O–ABC is formed when the vertices A, B, and C are joined to the center of the sphere. The *face angles* (BOC, COA, and AOB in Fig. 7.10) are used to measure the sides (a, b, and c in Fig. 7.10). The *vertex angles* are A, B, and C. Angles are used to measure both vertex angles and sides.

Figure 7.10 *Spherical Triangle ABC*

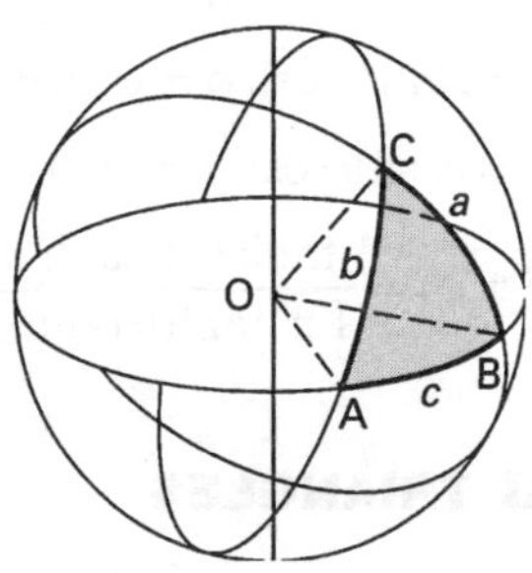

The following rules are valid for spherical triangles for which each side and angle is less than 180°.

- The sum of the three vertex angles is greater than 180° and less than 540°.

$$180° < \mathrm{A} + \mathrm{B} + \mathrm{C} < 540° \qquad 7.60$$

- The sum of any two sides is greater than the third side.
- The sum of the three sides is less than 360°.

$$0° < a + b + c < 360° \qquad 7.61$$

- If the two sides are equal, the corresponding angles opposite are equal, and the converse is also true.
- If two sides are unequal, the corresponding angles opposite are unequal. The greater angle is opposite the greater side.

The *spherical excess*, ϵ, is the amount by which the sum of the vertex angles exceeds 180°. The *spherical defect*, *d*, is the amount by which the sum of the sides differs from 360°.

$$\epsilon = \mathrm{A} + \mathrm{B} + \mathrm{C} - 180° \qquad 7.62$$

$$d = 360° - (a + b + c) \qquad 7.63$$

There are many trigonometric identities that define the relationships between angles in a spherical triangle. Some of the more common identities are presented as follows.

- *law of sines*

$$\frac{\sin \mathrm{A}}{\sin a} = \frac{\sin \mathrm{B}}{\sin b} = \frac{\sin \mathrm{C}}{\sin c} \qquad 7.64$$

- *first law of cosines*

$$\cos a = \cos b \cos c + \sin b \sin c \cos \mathrm{A} \qquad 7.65$$

- *second law of cosines*

$$\cos \mathrm{A} = -\cos \mathrm{B} \cos \mathrm{C} + \sin \mathrm{B} \sin \mathrm{C} \cos a \qquad 7.66$$

16. SOLID ANGLES

A *solid angle*, ω, is a measure of the angle subtended at the vertex of a cone, as shown in Fig. 7.11. The solid angle has units of *steradians* (abbreviated *sr*). A steradian is the solid angle subtended at the center of a unit sphere (i.e., a sphere with a radius of one) by a unit area on its surface. Since the surface area of a sphere of radius r is r^2 times the surface area of a unit sphere, the solid angle is equal to the area cut out by the cone divided by r^2.

$$\omega = \frac{\text{surface area}}{r^2} \qquad 7.67$$

Figure 7.11 *Solid Angle*

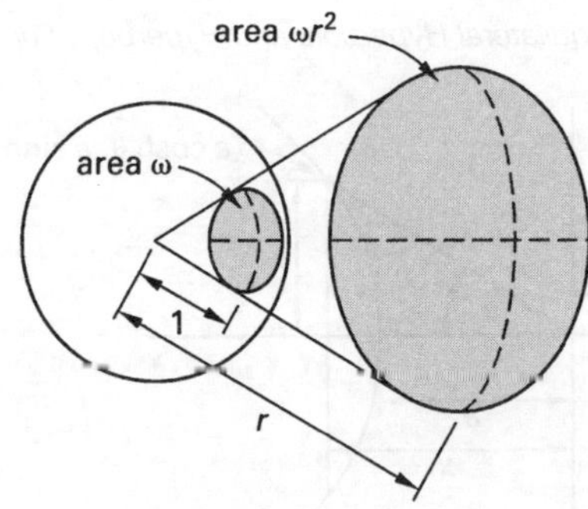

Analytic Geometry

1. MENSURATION OF REGULAR SHAPES

The dimensions, perimeter, area, and other geometric properties constitute the *mensuration* (i.e., the measurements) of a geometric shape. Appendix 8.A and App. 8.B contain formulas and tables used to calculate these properties.

Example 8.1

In the study of open channel fluid flow, the hydraulic radius is defined as the ratio of flow area to wetted perimeter. What is the hydraulic radius of a 6 in inside diameter pipe filled to a depth of 2 in?

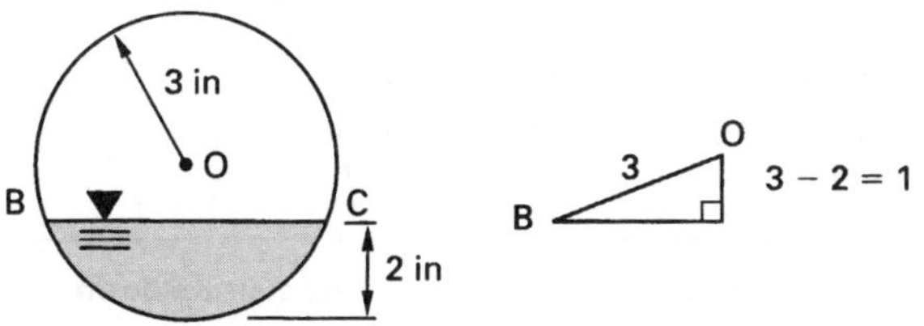

Solution

Points O, B, and C constitute a circular segment and are used to find the central angle of the circular segment.

$$\tfrac{1}{2}\angle\text{BOC} = \arccos\tfrac{1}{3} = 70.53^\circ$$

$$\phi = \angle\text{BOC} = \frac{(2)(70.53^\circ)(2\pi)}{360^\circ} = 2.462 \text{ rad}$$

From App. 8.A, the area in flow and arc length are

$$\begin{aligned} A &= \tfrac{1}{2}r^2(\phi - \sin\phi) \\ &= \left(\tfrac{1}{2}\right)(3 \text{ in})^2\big(2.462 \text{ rad} - \sin(2.462 \text{ rad})\big) \\ &= 8.251 \text{ in}^2 \end{aligned}$$

$$\begin{aligned} s &= r\phi \\ &= (3 \text{ in})(2.462 \text{ rad}) \\ &= 7.386 \text{ in} \end{aligned}$$

The hydraulic radius is

$$\begin{aligned} r_h &= \frac{A}{s} = \frac{8.251 \text{ in}^2}{7.386 \text{ in}} \\ &= 1.12 \text{ in} \end{aligned}$$

2. AREAS WITH IRREGULAR BOUNDARIES

Areas of sections with irregular boundaries (such as creek banks) cannot be determined precisely, and approximation methods must be used. If the irregular side can be divided into a series of cells of equal width, either the trapezoidal rule or Simpson's rule can be used. Figure 8.1 shows an example of an irregular area.

Figure 8.1 *Irregular Areas*

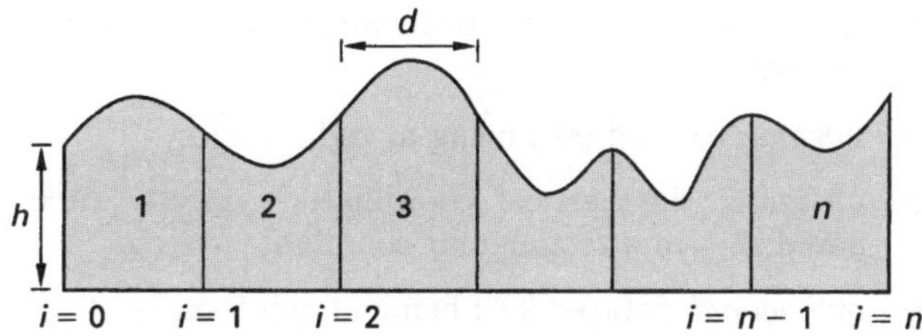

If the irregular side of each cell is fairly straight, the *trapezoidal rule* is appropriate.

$$A = \frac{d}{2}\left(h_0 + h_n + 2\sum_{i=1}^{n-1} h_i\right) \qquad 8.1$$

If the irregular side of each cell is curved (parabolic), *Simpson's rule* should be used. (n must be even to use Simpson's rule.)

$$A = \frac{d}{3}\left(h_0 + h_n + 4\sum_{\substack{i \text{ odd} \\ i=1}}^{n-1} h_i + 2\sum_{\substack{i \text{ even} \\ i=2}}^{n-2} h_i\right) \qquad 8.2$$

3. GEOMETRIC DEFINITIONS

The following terms are used in this book to describe the relationship or orientation of one geometric figure to another. Figure 8.2 illustrates some of the following geometric definitions.

- *abscissa:* the horizontal coordinate, typically designated as x in a rectangular coordinate system
- *asymptote:* a straight line that is approached but not intersected by a curved line
- *asymptotic:* approaching the slope of another line; attaining the slope of another line in the limit
- *center:* a point equidistant from all other points
- *collinear:* falling on the same line
- *concave:* curved inward (in the direction indicated)[1]
- *convex:* curved outward (in the direction indicated)
- *convex hull:* a closed figure whose surface is convex everywhere
- *coplanar:* falling on the same plane
- *inflection point:* a point where the second derivative changes sign or the curve changes from concave to convex; also known as a *point of contraflexure*
- *locus of points:* a set or collection of points having some common property and being so infinitely close together as to be indistinguishable from a line
- *node:* a point on a line from which other lines enter or leave
- *normal:* rotated 90°; being at right angles
- *ordinate:* the vertical coordinate, typically designated as y in a rectangular coordinate system
- *orthogonal:* rotated 90°; being at right angles
- *saddle point:* a point in three-dimensional space where all adjacent points are higher in one direction (the direction of the saddle) and lower in an orthogonal direction (the direction of the sides)
- *tangent point:* having equal slopes at a common point

[1]This is easily remembered since one must go inside to explore a cave.

Figure 8.2 Geometric Definitions

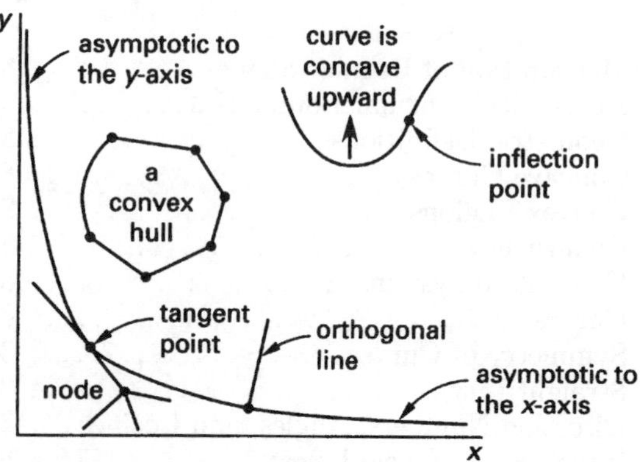

4. CONCAVE CURVES

Concavity is a term that is applied to curved lines. A *concave up curve* is one whose function's first derivative increases continuously from negative to positive values. Straight lines drawn tangent to concave up curves are all below the curve. The graph of such a function may be thought of as being able to "hold water."

The first derivative of a *concave down curve* decreases continuously from positive to negative. A graph of a concave down function may be thought of as "spilling water." (See Fig. 8.3.)

Figure 8.3 Concave Curves

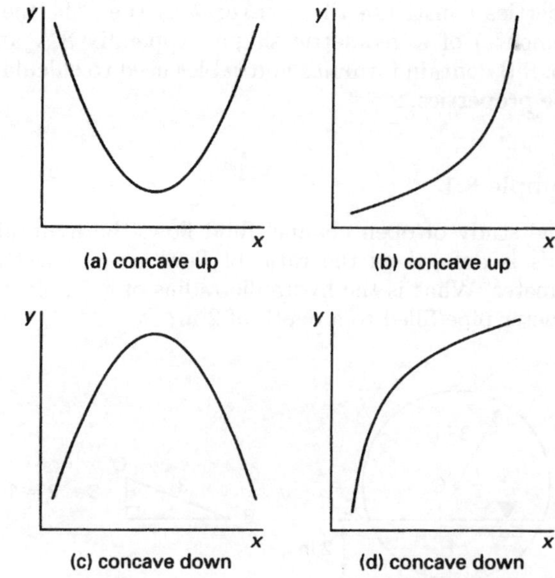

5. CONVEX REGIONS

Convexity is a term that is applied to sets and regions.[2] It plays an important role in many mathematics subjects. A set or multidimensional region is *convex* if it contains the line segment joining any two of its points; that is, if a straight line is drawn connecting any two points in a convex region, that line will lie entirely within the region. For example, the interior of a parabola is a convex region, as is a solid sphere. The *void* or *null region* (i.e., an empty set of points), single points, and straight lines are convex sets. A convex region bounded by separate, connected line segments is known as a *convex hull*. (See Fig. 8.4.)

Figure 8.4 Convexity

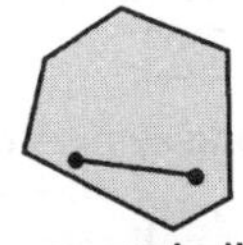

Within a convex region, a local maximum is also the global maximum. Similarly, a local minimum is also the global minimum. The intersection of two convex regions is also convex.

6. CONGRUENCY

Congruence in geometric figures is analogous to *equality* in algebraic expressions. Congruent line segments are segments that have the same length. Congruent angles have the same angular measure. Congruent triangles have the same vertex angles and side lengths.

In general, *congruency*, indicated by the symbol $\cong$, means that there is one-to-one correspondence between all points on two objects. This correspondence is defined by the *mapping function* or *isometry*, which can be a translation, rotation, or reflection. Since the identity function is a valid mapping function, every geometric shape is congruent to itself.

Two congruent objects can be in different spaces. For example, a triangular area in three-dimensional space can be mapped into a triangle in two-dimensional space.

7. COORDINATE SYSTEMS

The manner in which a geometric figure is described depends on the coordinate system that is used. The three-dimensional system (also known as the *rectangular coordinate system* and *Cartesian coordinate system*) with its x-, y-, and z-coordinates is the most commonly used in engineering. Table 8.1 summarizes the components needed to specify a point in the various coordinate systems. Figure 8.5 illustrates the use of and conversion between the coordinate systems.

Table 8.1 Components of Coordinate Systems

name	dimensions	components
rectangular	2	x, y
rectangular	3	x, y, z
polar	2	r, θ
cylindrical	3	r, θ, z
spherical	3	r, θ, ϕ

Figure 8.5 Different Coordinate Systems

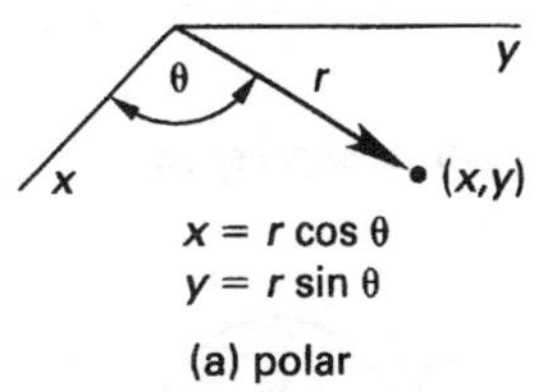

(a) polar

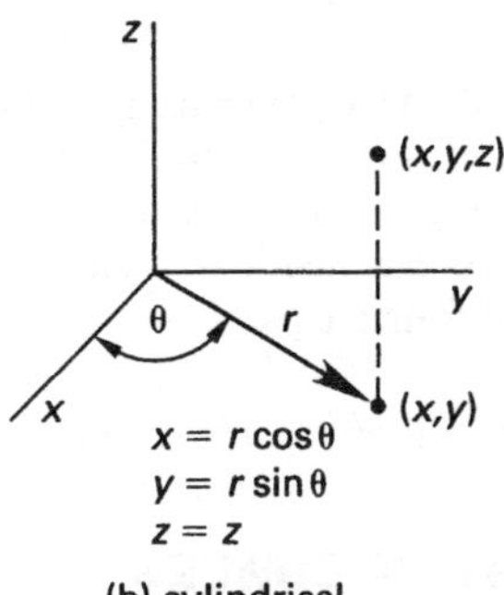

(b) cylindrical

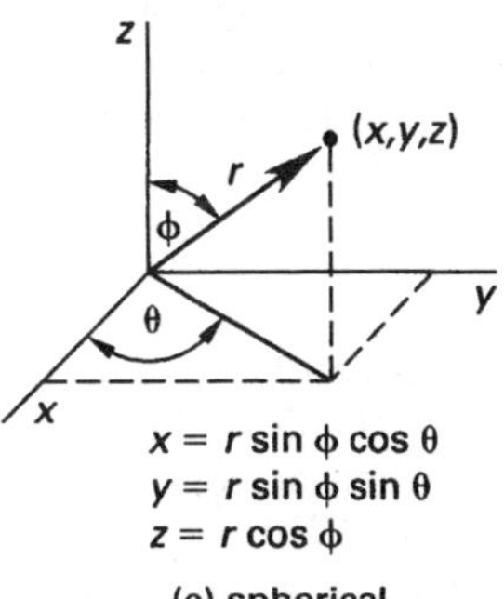

(c) spherical

8. CURVES

A *curve* (commonly called a *line*) is a function over a finite or infinite range of the independent variable. When a curve is drawn in two- or three-dimensional space, it is known as a *graph of the curve*. It may or may not be possible to describe the curve mathematically. The *degree of a curve* is the highest exponent in

[2]It is tempting to define regions that fail the convexity test as being "concave." However, it is more proper to define such regions as "nonconvex." In any case, it is important to recognize that convexity depends on the reference point: an observer within a sphere will see the spherical boundary as convex; an observer outside the sphere may see the boundary as nonconvex.

the function. For example, Eq. 8.3 is a fourth-degree curve.

$$f(x) = 2x^4 + 7x^3 + 6x^2 + 3x + 9 = 0 \quad 8.3$$

An *ordinary cycloid* ("wheel line") is a curve traced out by a point on the rim of a wheel that rolls without slipping. (See Fig. 8.6.) Cycloids that start with the tracing point down (i.e., on the x-axis) are described in parametric form by Eq. 8.4 and Eq. 8.5 and in rectangular form by Eq. 8.6. In Eq. 8.4 through Eq. 8.6, using the minus sign results in a bottom (downward) cusp at the origin; using the plus sign results in a vertex (trough) at the origin and a top (upward) cusp.

$$x = r(\theta \pm \sin\theta) \quad 8.4$$

$$y = r(1 \pm \cos\theta) \quad 8.5$$

$$x = r \arccos\frac{r - y}{r} \pm \sqrt{2ry - y^2} \quad 8.6$$

Figure 8.6 Cycloid (cusp at origin shown)

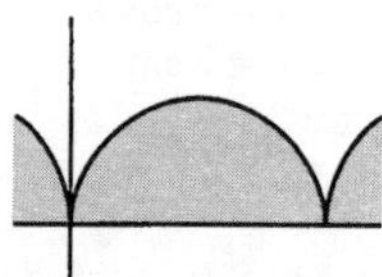

An *epicycloid* is a curve generated by a point on the rim of a wheel that rolls on the outside of a circle. A *hypocycloid* is a curve generated by a point on the rim of a wheel that rolls on the inside of a circle. The equation of a hypocycloid of four cusps is

$$x^{2/3} + y^{2/3} = r^{2/3} \quad [\text{4 cusps}] \quad 8.7$$

9. SYMMETRY OF CURVES

Two points, P and Q, are symmetrical with respect to a line if the line is a perpendicular bisector of the line segment PQ. If the graph of a curve is unchanged when y is replaced with $-y$, the curve is symmetrical with respect to the x-axis. If the curve is unchanged when x is replaced with $-x$, the curve is symmetrical with respect to the y-axis.

Repeating waveforms can be symmetrical with respect to the y-axis. A curve $f(x)$ is said to have *even symmetry* if $f(x) = f(-x)$. (Alternatively, $f(x)$ is said to be a *symmetrical function.*) With even symmetry, the function to the left of $x = 0$ is a reflection of the function to the right of $x = 0$. (In effect, the y-axis is a mirror.) The cosine curve is an example of a curve with even symmetry.

A curve is said to have *odd symmetry* if $f(x) = -f(-x)$. (Alternatively, $f(x)$ is said to be an *asymmetrical function.*[3]) The sine curve is an example of a curve with odd symmetry.

A curve is said to have *rotational symmetry* (*half-wave symmetry*) if $f(x) = -f(x + \pi)$.[4] Curves of this type are identical except for a sign reversal on alternate half-cycles. (See Fig. 8.7.)

Table 8.2 describes the type of function resulting from the combination of two functions.

Figure 8.7 Waveform Symmetry

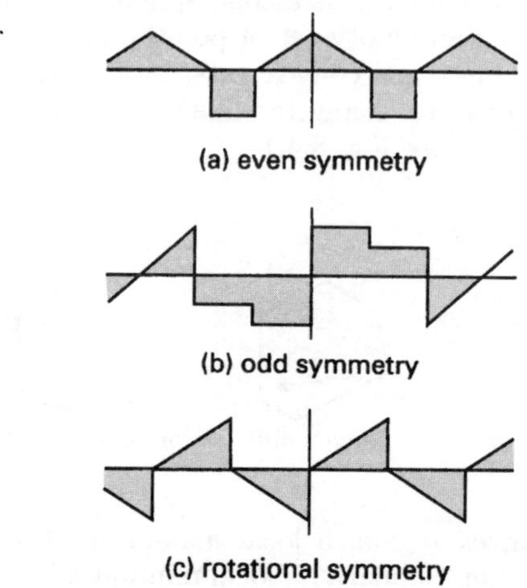

Table 8.2 Combinations of Functions

	operation			
	+	−	×	÷
$f_1(x)$ even, $f_2(x)$ even	even	even	even	even
$f_1(x)$ odd, $f_2(x)$ odd	odd	odd	even	even
$f_1(x)$ even, $f_2(x)$ odd	neither	neither	odd	odd

10. STRAIGHT LINES

Figure 8.8 illustrates a straight line in two-dimensional space. The *slope* of the line is m, the *y-intercept* is b, and the *x-intercept* is a. The equation of the line can be represented in several forms. The procedure for finding the equation depends on the form chosen to represent the line. In general, the procedure involves substituting one or more known points on the line into the equation in order to determine the coefficients.

Figure 8.8 Straight Line

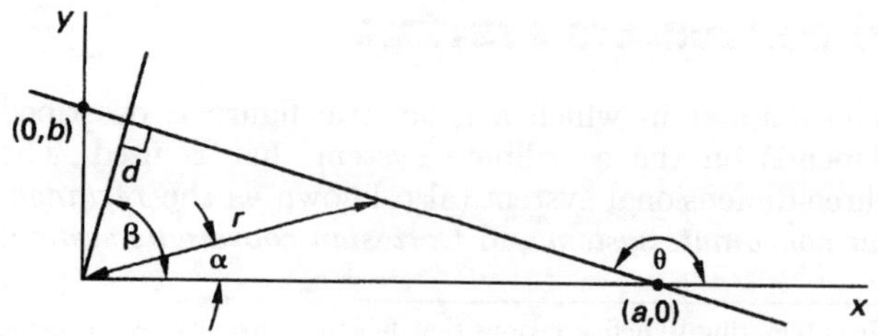

[3]Although they have the same meaning, the semantics of "even symmetry" and "asymmetrical function" are contradictory.

[4]The symbol π represents half of a full cycle of the waveform, not the value 3.141...

- *general form*

$$Ax+By+C=0 \quad 8.8$$

$$A=-mB \quad 8.9$$

$$B=\frac{-C}{b} \quad 8.10$$

$$C=-aA=-bB \quad 8.11$$

- *slope-intercept form*

$$y=mx+b \quad 8.12$$

$$m=\frac{-A}{B}=\tan\theta=\frac{y_2-y_1}{x_2-x_1} \quad 8.13$$

$$b=\frac{-C}{B} \quad 8.14$$

$$a=\frac{-C}{A} \quad 8.15$$

- *point-slope form*

$$y-y_1=m(x-x_1) \quad 8.16$$

- *intercept form*

$$\frac{x}{a}+\frac{y}{b}=1 \quad 8.17$$

- *two-point form*

$$\frac{y-y_1}{x-x_1}=\frac{y_2-y_1}{x_2-x_1} \quad 8.18$$

- *normal form*

$$x\cos\beta+y\sin\beta-d=0 \quad 8.19$$

(d and β are constants; x and y are variables.)

- *polar form*

$$r=\frac{d}{\cos(\beta-\alpha)} \quad 8.20$$

(d and β are constants; r and α are variables.)

11. DIRECTION NUMBERS, ANGLES, AND COSINES

Given a directed line from (x_1, y_1, z_1) to (x_2, y_2, z_2), the *direction numbers* are

$$L=x_2-x_1 \quad 8.21$$

$$M=y_2-y_1 \quad 8.22$$

$$N=z_2-z_1 \quad 8.23$$

The distance between two points is

$$d=\sqrt{L^2+M^2+N^2} \quad 8.24$$

The *direction cosines* are

$$\cos\alpha=\frac{L}{d} \quad 8.25$$

$$\cos\beta=\frac{M}{d} \quad 8.26$$

$$\cos\gamma=\frac{N}{d} \quad 8.27$$

Note that

$$\cos^2\alpha+\cos^2\beta+\cos^2\gamma=1 \quad 8.28$$

The *direction angles* are the angles between the axes and the lines. They are found from the inverse functions of the direction cosines.

$$\alpha=\arccos\frac{L}{d} \quad 8.29$$

$$\beta=\arccos\frac{M}{d} \quad 8.30$$

$$\gamma=\arccos\frac{N}{d} \quad 8.31$$

The direction cosines can be used to write the equation of the straight line in terms of the unit vectors. The line **R** would be defined as

$$\mathbf{R}=d(\mathbf{i}\cos\alpha+\mathbf{j}\cos\beta+\mathbf{k}\cos\gamma) \quad 8.32$$

Similarly, the line may be written in terms of its direction numbers.

$$\mathbf{R}=L\mathbf{i}+M\mathbf{j}+N\mathbf{k} \quad 8.33$$

Example 8.2

A line passes through the points (4, 7, 9) and (0, 1, 6). Write the equation of the line in terms of its (a) direction numbers and (b) direction cosines.

Solution

(a) The direction numbers are

$$L=4-0=4$$

$$M=7-1=6$$

$$N=9-6=3$$

Using Eq. 8.33,

$$\mathbf{R}=4\mathbf{i}+6\mathbf{j}+3\mathbf{k}$$

(b) The distance between the two points is

$$d=\sqrt{(4)^2+(6)^2+(3)^2}=7.81$$

The line in terms of its direction cosines is

$$\mathbf{R}=\frac{4\mathbf{i}+6\mathbf{j}+3\mathbf{k}}{7.81}$$

$$=0.512\mathbf{i}+0.768\mathbf{j}+0.384\mathbf{k}$$

12. INTERSECTION OF TWO LINES

The intersection of two lines is a point. The location of the intersection point can be determined by setting the two equations equal and solving them in terms of a common variable. Alternatively, Eq. 8.34 and Eq. 8.35 can be used to calculate the coordinates of the intersection point.

$$x = \frac{B_2C_1 - B_1C_2}{A_2B_1 - A_1B_2} \quad \text{8.34}$$

$$y = \frac{A_1C_2 - A_2C_1}{A_2B_1 - A_1B_2} \quad \text{8.35}$$

13. PLANES

A *plane* in three-dimensional space (see Fig. 8.9) is completely determined by one of the following:

- three noncollinear points
- two nonparallel vectors $\mathbf{V}_1$ and $\mathbf{V}_2$ and their intersection point P_0
- a point P_0 and a vector, $\mathbf{N}$, normal to the plane (i.e., the *normal vector*)

Figure 8.9 *Plane in Three-Dimensional Space*

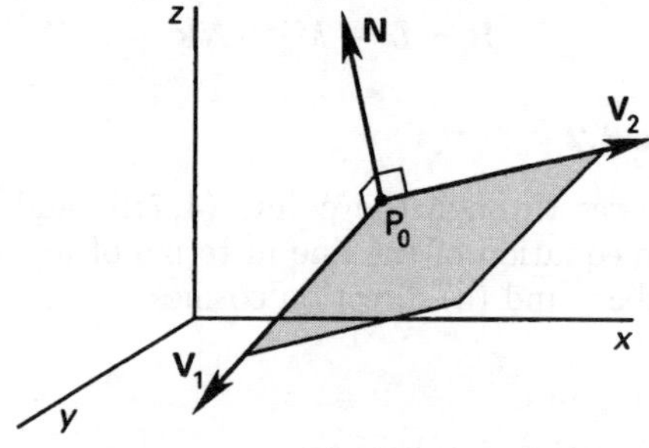

The plane can be specified mathematically in one of two ways: in rectangular form or as a parametric equation. The general form is

$$A(x - x_0) + B(y - y_0) + C(z - z_0) = 0 \quad \text{8.36}$$

x_0, y_0, and z_0 are the coordinates of the intersection point of any two vectors in the plane. The coefficients A, B, and C are the same as the coefficients of the normal vector, $\mathbf{N}$.

$$\mathbf{N} = \mathbf{V}_1 \times \mathbf{V}_2 = A\mathbf{i} + B\mathbf{j} + C\mathbf{k} \quad \text{8.37}$$

Equation 8.36 can be simplified as follows.

$$Ax + By + Cz + D = 0 \quad \text{8.38}$$

$$D = -(Ax_0 + By_0 + Cz_0) \quad \text{8.39}$$

The following procedure can be used to determine the equation of a plane from three noncollinear points, P_1, P_2, and P_3, or from a normal vector and a single point.

step 1: (If the normal vector is known, go to step 3.) Determine the equations of the vectors $\mathbf{V}_1$ and $\mathbf{V}_2$ from two pairs of the points. For example, determine $\mathbf{V}_1$ from points P_1 and P_2, and determine $\mathbf{V}_2$ from P_1 and P_3. Express the vectors in the form $A\mathbf{i} + B\mathbf{j} + C\mathbf{k}$.

$$\mathbf{V}_1 = (x_2 - x_1)\mathbf{i} + (y_2 - y_1)\mathbf{j} + (z_2 - z_1)\mathbf{k} \quad \text{8.40}$$

$$\mathbf{V}_2 = (x_3 - x_1)\mathbf{i} + (y_3 - y_1)\mathbf{j} + (z_3 - z_1)\mathbf{k} \quad \text{8.41}$$

step 2: Find the normal vector, N, as the cross product of the two vectors.

$$\mathbf{N} = \mathbf{V}_1 \times \mathbf{V}_2 = \begin{vmatrix} \mathbf{i} & (x_2 - x_1) & (x_3 - x_1) \\ \mathbf{j} & (y_2 - y_1) & (y_3 - y_1) \\ \mathbf{k} & (z_2 - z_1) & (z_3 - z_1) \end{vmatrix} \quad \text{8.42}$$

step 3: Write the general equation of the plane in rectangular form (see Eq. 8.36) using the coefficients A, B, and C from the normal vector and any one of the three points as P_0.

The parametric equations of a plane also can be written as a linear combination of the components of two vectors in the plane. Referring to Fig. 8.9, the two known vectors are

$$\mathbf{V}_1 = V_{1x}\mathbf{i} + V_{1y}\mathbf{j} + V_{1z}\mathbf{k} \quad \text{8.43}$$

$$\mathbf{V}_2 = V_{2x}\mathbf{i} + V_{2y}\mathbf{j} + V_{2z}\mathbf{k} \quad \text{8.44}$$

If s and t are scalars, the coordinates of each point in the plane can be written as Eq. 8.45 through Eq. 8.47. These are the parametric equations of the plane.

$$x = x_0 + sV_{1x} + tV_{2x} \quad \text{8.45}$$

$$y = y_0 + sV_{1y} + tV_{2y} \quad \text{8.46}$$

$$z = z_0 + sV_{1z} + tV_{2z} \quad \text{8.47}$$

Example 8.3

The following points are coplanar.

$$P_1 = (2, 1, -4)$$
$$P_2 = (4, -2, -3)$$
$$P_3 = (2, 3, -8)$$

Determine the equation of the plane in (a) general form and (b) parametric form.

Solution

(a) Use the first two points to find a vector, $\mathbf{V}_1$.

$$\begin{aligned}\mathbf{V}_1 &= (x_2 - x_1)\mathbf{i} + (y_2 - y_1)\mathbf{j} + (z_2 - z_1)\mathbf{k}\\ &= (4-2)\mathbf{i} + (-2-1)\mathbf{j} + \big(-3-(-4)\big)\mathbf{k}\\ &= 2\mathbf{i} - 3\mathbf{j} + 1\mathbf{k}\end{aligned}$$

Similarly, use the first and third points to find $\mathbf{V}_2$.

$$\begin{aligned}\mathbf{V}_2 &= (x_3 - x_1)\mathbf{i} + (y_3 - y_1)\mathbf{j} + (z_3 - z_1)\mathbf{k}\\ &= (2-2)\mathbf{i} + (3-1)\mathbf{j} + \big(-8-(-4)\big)\mathbf{k}\\ &= 0\mathbf{i} + 2\mathbf{j} - 4\mathbf{k}\end{aligned}$$

From Eq. 8.42, determine the normal vector as a determinant.

$$\mathbf{N} = \begin{vmatrix} \mathbf{i} & 2 & 0\\ \mathbf{j} & -3 & 2\\ \mathbf{k} & 1 & -4 \end{vmatrix}$$

Expand the determinant across the top row.

$$\begin{aligned}\mathbf{N} &= \mathbf{i}(12-2) - 2(-4\mathbf{j} - 2\mathbf{k})\\ &= 10\mathbf{i} + 8\mathbf{j} + 4\mathbf{k}\end{aligned}$$

The rectangular form of the equation of the plane uses the same constants as in the normal vector. Use the first point and write the equation of the plane in the form of Eq. 8.36.

$$(10)(x-2) + (8)(y-1) + (4)(z+4) = 0$$

The three constant terms can be combined by using Eq. 8.39.

$$D = -\big((10)(2) + (8)(1) + (4)(-4)\big) = -12$$

The equation of the plane is

$$10x + 8y + 4z - 12 = 0$$

(b) The parametric equations based on the first point and for any values of s and t are

$$\begin{aligned}x &= 2 + 2s + 0t\\ y &= 1 - 3s + 2t\\ z &= -4 + 1s - 4t\end{aligned}$$

The scalars s and t are not unique. Two of the three coordinates can also be chosen as the parameters. Dividing the rectangular form of the plane's equation by 4 to isolate z results in an alternate set of parametric equations.

$$\begin{aligned}x &= x\\ y &= y\\ z &= 3 - 2.5x - 2y\end{aligned}$$

14. DISTANCES BETWEEN GEOMETRIC FIGURES

The smallest distance, d, between various geometric figures is given by the following equations.

- between two points in (x, y, z) format:

$$d = \sqrt{(x_2 - x_1)^2 + (y_2 - y_1)^2 + (z_2 - z_1)^2} \qquad 8.48$$

- between a point (x_0, y_0) and a line $Ax + By + C = 0$:

$$d = \frac{|Ax_0 + By_0 + C|}{\sqrt{A^2 + B^2}} \qquad 8.49$$

- between a point (x_0, y_0, z_0) and a plane $Ax + By + Cz + D = 0$:

$$d = \frac{|Ax_0 + By_0 + Cz_0 + D|}{\sqrt{A^2 + B^2 + C^2}} \qquad 8.50$$

- between two parallel lines $Ax + By + C = 0$:

$$d = \left|\frac{|C_2|}{\sqrt{A_2^2 + B_2^2}} - \frac{|C_1|}{\sqrt{A_1^2 + B_1^2}}\right| \qquad 8.51$$

Example 8.4

What is the minimum distance between the line $y = 2x + 3$ and the origin $(0,0)$?

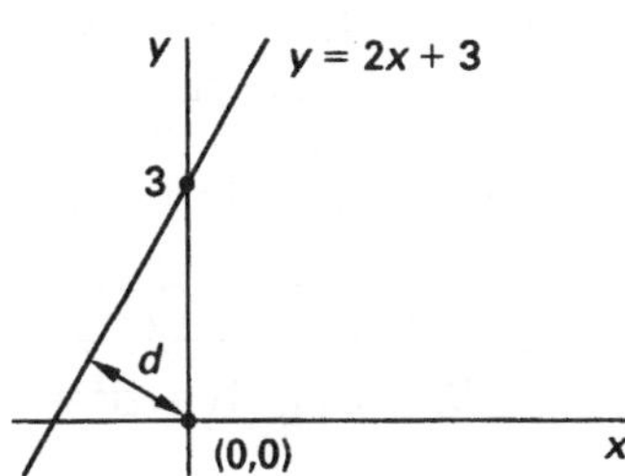

Solution

Put the equation in general form.

$$Ax + By + C = 2x - y + 3 = 0$$

Use Eq. 8.49 with $(x, y) = (0,0)$.

$$d = \frac{|Ax + By + C|}{\sqrt{A^2 + B^2}} = \frac{(2)(0) + (-1)(0) + 3}{\sqrt{(2)^2 + (-1)^2}}$$
$$= \frac{3}{\sqrt{5}}$$

15. ANGLES BETWEEN GEOMETRIC FIGURES

The angle, ϕ, between various geometric figures is given by the following equations.

- between two lines in $Ax + By + C = 0$, $y = mx + b$, or direction angle formats:

$$\phi = \arctan\left(\frac{A_1 B_2 - A_2 B_1}{A_1 A_2 + B_1 B_2}\right) \tag{8.52}$$

$$\phi = \arctan\left(\frac{m_2 - m_1}{1 + m_1 m_2}\right) \tag{8.53}$$

$$\phi = |\arctan(m_1) - \arctan(m_2)| \tag{8.54}$$

$$\phi = \arccos\left(\frac{L_1 L_2 + M_1 M_2 + N_1 N_2}{d_1 d_2}\right) \tag{8.55}$$

$$\phi = \arccos\left(\begin{array}{c}\cos\alpha_1 \cos\alpha_2 + \cos\beta_1 \cos\beta_2 \\ + \cos\gamma_1 \cos\gamma_2\end{array}\right) \tag{8.56}$$

If the lines are parallel, then $\phi = 0$.

$$\frac{A_1}{A_2} = \frac{B_1}{B_2} \tag{8.57}$$

$$m_1 = m_2 \tag{8.58}$$

$$\alpha_1 = \alpha_2;\ \beta_1 = \beta_2;\ \gamma_1 = \gamma_2 \tag{8.59}$$

If the lines are perpendicular, then $\phi = 90°$.

$$A_1 A_2 = -B_1 B_2 \tag{8.60}$$

$$m_1 = \frac{-1}{m_2} \tag{8.61}$$

$$\alpha_1 + \alpha_2 = \beta_1 + \beta_2 = \gamma_1 + \gamma_2 = 90° \tag{8.62}$$

- between two planes in $A\mathbf{i} + B\mathbf{j} + C\mathbf{k} = 0$ format, the coefficients A, B, and C are the same as the coefficients for the normal vector. (See Eq. 8.37.) ϕ is equal to the angle between the two normal vectors.

$$\cos\phi = \frac{|A_1 A_2 + B_1 B_2 + C_1 C_2|}{\sqrt{A_1^2 + B_1^2 + C_1^2}\sqrt{A_2^2 + B_2^2 + C_2^2}} \tag{8.63}$$

Example 8.5

Use Eq. 8.52, Eq. 8.53, and Eq. 8.54 to find the angle between the lines.

$$y = -0.577x + 2$$
$$y = +0.577x - 5$$

Solution

Write both equations in general form.

$$-0.577x - y + 2 = 0$$
$$0.577x - y - 5 = 0$$

(a) From Eq. 8.52,

$$\phi = \arctan\left(\frac{A_1 B_2 - A_2 B_1}{A_1 A_2 + B_1 B_2}\right)$$
$$= \arctan\left(\frac{(-0.577)(-1) - (0.577)(-1)}{(-0.577)(0.577) + (-1)(-1)}\right)$$
$$= 60°$$

(b) Use Eq. 8.53.

$$\phi = \arctan\left(\frac{m_2 - m_1}{1 + m_1 m_2}\right)$$
$$= \arctan\left(\frac{0.577 - (-0.577)}{1 + (0.577)(-0.577)}\right)$$
$$= 60°$$

(c) Use Eq. 8.54.

$$\phi = |\arctan(m_1) - \arctan(m_2)|$$
$$= |\arctan(-0.577) - \arctan(0.577)|$$
$$= |-30° - 30°|$$
$$= 60°$$

16. CONIC SECTIONS

A *conic section* is any one of several curves produced by passing a plane through a cone as shown in Fig. 8.10. If α is the angle between the vertical axis and the cutting plane and β is the cone generating angle, Eq. 8.64 gives the *eccentricity*, ϵ, of the conic section. Values of the eccentricity are given in Fig. 8.10.

$$\epsilon = \frac{\cos\alpha}{\cos\beta} \tag{8.64}$$

All conic sections are described by second-degree polynomials (i.e., *quadratic equations*) of the following form.[5]

$$Ax^2 + Bxy + Cy^2 + Dx + Ey + F = 0 \tag{8.65}$$

This is the *general form*, which allows the figure axes to be at any angle relative to the coordinate axes. The *standard forms* presented in the following sections pertain to figures whose axes coincide with the coordinate axes, thereby eliminating certain terms of the general equation.

Figure 8.11 can be used to determine which conic section is described by the quadratic function. The quantity $B^2 - 4AC$ is known as the *discriminant.* Figure 8.11

[5]One or more straight lines are produced when the cutting plane passes through the cone's vertex. Straight lines can be considered to be quadratic functions without second-degree terms.

determines only the type of conic section; it does not determine whether the conic section is degenerate (e.g., a circle with a negative radius).

Figure 8.10 Conic Sections Produced by Cutting Planes

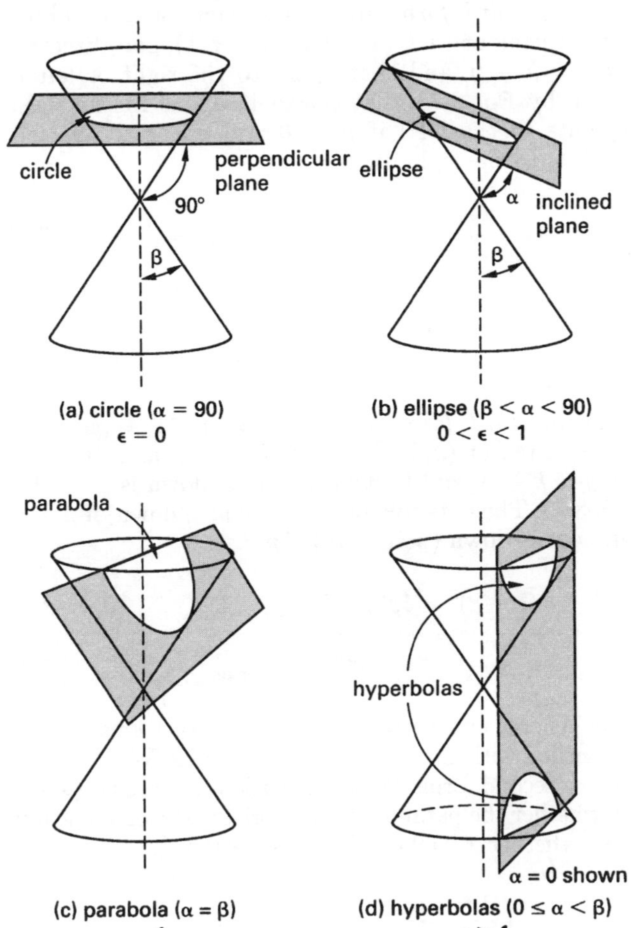

Figure 8.11 Determining Conic Sections from Quadratic Equations

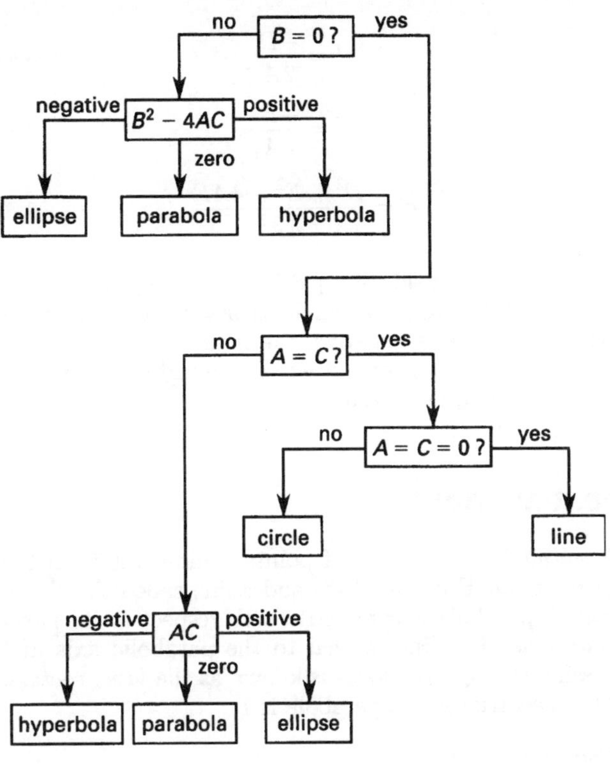

Example 8.6

What geometric figures are described by the following equations?

(a) $4y^2 - 12y + 16x + 41 = 0$

(b) $x^2 - 10xy + y^2 + x + y + 1 = 0$

(c) $x^2 + 4y^2 + 2x - 8y + 1 = 0$

(d) $x^2 + y^2 - 6x + 8y + 20 = 0$

Solution

(a) Referring to Fig. 8.11, $B = 0$ since there is no xy term, $A = 0$ since there is no x^2 term, and $AC = (0)(4) = 0$. This is a parabola.

(b) $B \neq 0$; $B^2 - 4AC = (-10)^2 - (4)(1)(1) = +96$. This is a hyperbola.

(c) $B = 0$; $A \neq C$; $AC = (1)(4) = +4$. This is an ellipse.

(d) $B = 0$; $A = C$; $A = C = 1$ ($\neq 0$). This is a circle.

17. CIRCLE

The general form of the equation of a circle, as illustrated in Fig. 8.12, is

$$Ax^2 + Ay^2 + Dx + Ey + F = 0 \quad 8.66$$

Figure 8.12 Circle

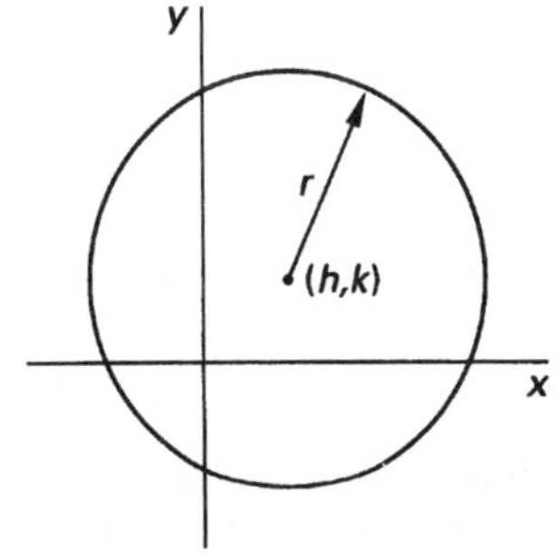

The *center-radius form* of the equation of a circle with radius r and center at (h, k) is

$$(x - h)^2 + (y - k)^2 = r^2 \quad 8.67$$

The two forms can be converted by use of Eq. 8.68 through Eq. 8.70.

$$h = \frac{-D}{2A} \quad 8.68$$

$$k = \frac{-E}{2A} \quad 8.69$$

$$r^2 = \frac{D^2 + E^2 - 4AF}{4A^2} \quad 8.70$$

If the right-hand side of Eq. 8.70 is positive, the figure is a circle. If it is zero, the circle shrinks to a point. If the right-hand side is negative, the figure is imaginary. A *degenerate circle* is one in which the right-hand side is less than or equal to zero.

18. PARABOLA

A *parabola* is the locus of points equidistant from the *focus* (point F in Fig. 8.13) and a line called the *directrix*. A parabola is symmetric with respect to its *parabolic axis*. The line normal to the parabolic axis and passing through the focus is known as the *latus rectum*. The eccentricity of a parabola is 1.

Figure 8.13 Parabola

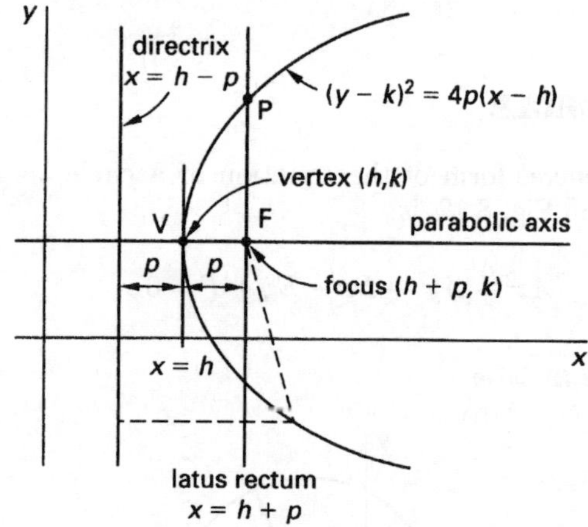

There are two common types of parabolas in the Cartesian plane—those that open right and left, and those that open up and down. Equation 8.65 is the general form of the equation of a parabola. With Eq. 8.71, the parabola points horizontally to the right if $CD > 0$ and to the left if $CD < 0$. With Eq. 8.72, the parabola points vertically up if $AE > 0$ and down if $AE < 0$.

$$Cy^2 + Dx + Ey + F = 0 \Big|_{\substack{C, D \neq 0 \\ \text{opens horizontally}}} \quad 8.71$$

$$Ax^2 + Dx + Ey + F = 0 \Big|_{\substack{A, E \neq 0 \\ \text{opens vertically}}} \quad 8.72$$

The *standard form* of the equation of a parabola with vertex at (h, k), focus at $(h + p, k)$, and directrix at $x = h - p$, and that opens to the right or left is given by Eq. 8.73. The parabola opens to the right (points to the left) if $p > 0$ and opens to the left (points to the right) if $p < 0$.

$$(y - k)^2 = 4p(x - h)\Big|_{\text{opens horizontally}} \quad 8.73$$

$$y^2 = 4px \Big|_{\substack{\text{vertex at origin} \\ h = k = 0}} \quad 8.74$$

The *standard form* of the equation of a parabola with vertex at (h, k), focus at $(h, k + p)$, and directrix at $y = k - p$, and that opens up or down is given by Eq. 8.75. The parabola opens up (points down) if $p > 0$ and opens down (points up) if $p < 0$.

$$(x - h)^2 = 4p(y - k)\Big|_{\text{opens vertically}} \quad 8.75$$

$$x^2 = 4py\Big|_{\text{vertex at origin}} \quad 8.76$$

The general and vertex forms of the equations can be reconciled with Eq. 8.77 through Eq. 8.79. Whether the first or second forms of these equations are used depends on whether the parabola opens horizontally or vertically (i.e., whether $A = 0$ or $C = 0$), respectively.

$$h = \begin{cases} \dfrac{E^2 - 4CF}{4CD} & \text{[opens horizontally]} \\[2ex] \dfrac{-D}{2A} & \text{[opens vertically]} \end{cases} \quad 8.77$$

$$k = \begin{cases} \dfrac{-E}{2C} & \text{[opens horizontally]} \\[2ex] \dfrac{D^2 - 4AF}{4AE} & \text{[opens vertically]} \end{cases} \quad 8.78$$

$$p = \begin{cases} \dfrac{-D}{4C} & \text{[opens horizontally]} \\[2ex] \dfrac{-E}{4A} & \text{[opens vertically]} \end{cases} \quad 8.79$$

19. ELLIPSE

An *ellipse* has two foci separated along the *major axis* by a distance $2c$. The line perpendicular to the major axis passing through the center of the ellipse is the *minor axis*. The lines perpendicular to the major axis passing through the foci are the *latus recta*. The distance between the two vertices is $2a$. The ellipse is the locus of points such that the sum of the distances from the two foci is $2a$. Referring to Fig. 8.14,

$$F_1P + PF_2 = 2a \qquad 8.80$$

Figure 8.14 *Ellipse*

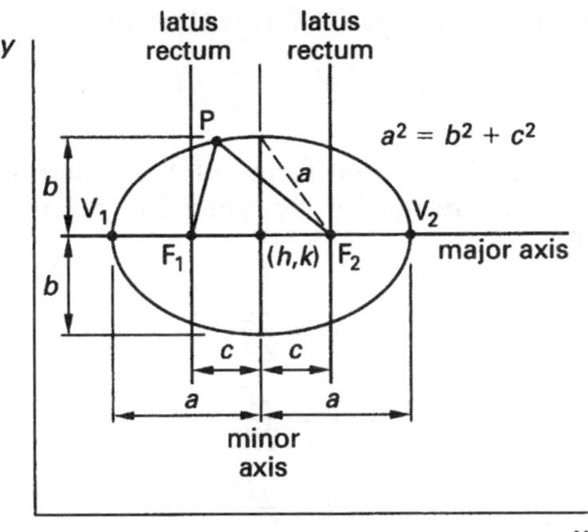

Equation 8.81 is the standard equation used for an ellipse with axes parallel to the coordinate axes, while Eq. 8.65 is the general form. F is not independent of A, C, D, and E for the ellipse.

$$Ax^2 + Cy^2 + Dx + Ey + F = 0\Big|_{\substack{AC>0\\ A\neq C}} \qquad 8.81$$

Equation 8.82 gives the standard form of the equation of an ellipse centered at (h, k). Distances a and b are known as the *semimajor distance* and *semiminor distance*, respectively.

$$\frac{(x-h)^2}{a^2} + \frac{(y-k)^2}{b^2} = 1 \qquad 8.82$$

The distance between the two foci is $2c$.

$$2c = 2\sqrt{a^2 - b^2} \qquad 8.83$$

The *aspect ratio* of the ellipse is

$$\text{aspect ratio} = \frac{a}{b} \qquad 8.84$$

The *eccentricity*, ϵ, of the ellipse is always less than 1. If the eccentricity is zero, the figure is a circle (another form of a *degenerative ellipse*).

$$\epsilon = \frac{\sqrt{a^2 - b^2}}{a} < 1 \qquad 8.85$$

The standard and center forms of the equations of an ellipse can be reconciled by using Eq. 8.86 through Eq. 8.89.

$$h = \frac{-D}{2A} \qquad 8.86$$

$$k = \frac{-E}{2C} \qquad 8.87$$

$$a = \sqrt{C} \qquad 8.88$$

$$b = \sqrt{A} \qquad 8.89$$

20. HYPERBOLA

A *hyperbola* has two foci separated along the *transverse axis* by a distance $2c$. Lines perpendicular to the transverse axis passing through the foci are the *conjugate axes*. The distance between the two vertices is $2a$, and the distance along a conjugate axis passing through each vertex between two points on the asymptotes is $2b$. The hyperbola is the locus of points such that the difference in distances from the two foci is $2a$. Referring to Fig. 8.15,

$$F_2P - PF_1 = 2a \qquad 8.90$$

Figure 8.15 *Hyperbola*

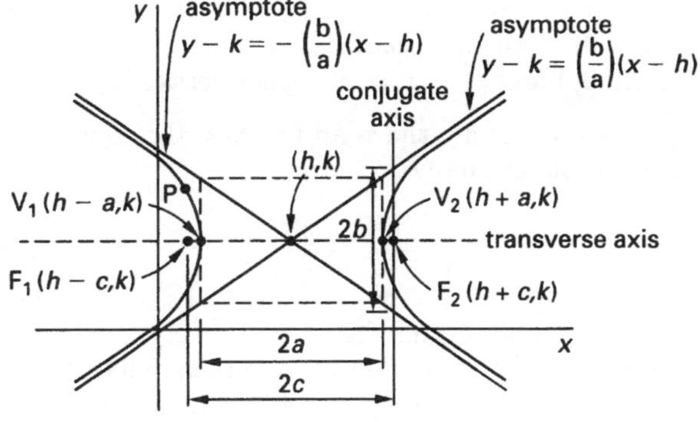

Equation 8.91 is the standard equation of a hyperbola. Coefficients A and C have opposite signs.

$$Ax^2 + Cy^2 + Dx + Ey + F = 0\big|_{AC<0} \qquad 8.91$$

Equation 8.92 gives the standard form of the equation of a hyperbola centered at (h, k) and opening to the left and right.

$$\frac{(x-h)^2}{a^2} - \frac{(y-k)^2}{b^2} = 1\Big|_{\text{opens horizontally}} \qquad 8.92$$

Equation 8.93 gives the standard form of the equation of a hyperbola that is centered at (h, k) and is opening up and down.

$$\frac{(y-k)^2}{a^2} - \frac{(y-k)^2}{b^2} = 1 \Big|_{\text{opens vertically}} \quad 8.93$$

The distance between the two foci is $2c$.

$$2c = 2\sqrt{a^2 + b^2} \quad 8.94$$

The *eccentricity*, ϵ, of the hyperbola is calculated from Eq. 8.95 and is always greater than 1.

$$\epsilon = \frac{c}{a} = \frac{\sqrt{a^2 + b^2}}{a} > 1 \quad 8.95$$

The hyperbola is asymptotic to the lines given by Eq. 8.96 and Eq. 8.97.

$$y = \pm\frac{b}{a}(x-h) + k \Big|_{\text{opens horizontally}} \quad 8.96$$

$$y = \pm\frac{a}{b}(x-h) + k \Big|_{\text{opens vertically}} \quad 8.97$$

For a *rectangular (equilateral) hyperbola*, the asymptotes are perpendicular, $a = b$, $c = \sqrt{2}a$, and the eccentricity is $\epsilon = \sqrt{2}$. If the hyperbola is centered at the origin (i.e., $h = k = 0$), then the equations are $x^2 - y^2 = a^2$ (opens horizontally) and $y^2 - x^2 = a^2$ (opens vertically).

If the asymptotes are the x- and y-axes, the equation of the hyperbola is simply

$$xy = \pm\frac{a^2}{2} \quad 8.98$$

The general and center forms of the equations of a hyperbola can be reconciled by using Eq. 8.99 through Eq. 8.103. Whether the hyperbola opens left and right or up and down depends on whether M/A or M/C is positive, respectively, where M is defined by Eq. 8.99.

$$M = \frac{D^2}{4A} + \frac{E^2}{4C} - F \quad 8.99$$

$$h = \frac{-D}{2A} \quad 8.100$$

$$k = \frac{-E}{2C} \quad 8.101$$

$$a = \begin{cases} \sqrt{-C} & \text{[opens horizontally]} \\ \sqrt{-A} & \text{[opens vertically]} \end{cases} \quad 8.102$$

$$b = \begin{cases} \sqrt{A} & \text{[opens horizontally]} \\ \sqrt{C} & \text{[opens vertically]} \end{cases} \quad 8.103$$

21. SPHERE

Equation 8.104 is the general equation of a sphere. The coefficient A cannot be zero.

$$Ax^2 + Ay^2 + Az^2 + Bx + Cy + Dz + E = 0 \quad 8.104$$

Equation 8.105 gives the standard form of the equation of a sphere centered at (h, k, l) with radius r.

$$(x-h)^2 + (y-k)^2 + (z-l)^2 = r^2 \quad 8.105$$

The general and center forms of the equations of a sphere can be reconciled by using Eq. 8.106 through Eq. 8.109.

$$h = \frac{-B}{2A} \quad 8.106$$

$$k = \frac{-C}{2A} \quad 8.107$$

$$l = \frac{-D}{2A} \quad 8.108$$

$$r = \sqrt{\frac{B^2 + C^2 + D^2}{4A^2} - \frac{E}{A}} \quad 8.109$$

22. HELIX

A *helix* is a curve generated by a point moving on, around, and along a cylinder such that the distance the point moves parallel to the cylindrical axis is proportional to the angle of rotation about that axis. (See Fig. 8.16.) For a cylinder of radius r, Eq. 8.110 through Eq. 8.112 define the three-dimensional positions of points along the helix. The quantity $2\pi k$ is the *pitch* of the helix.

$$x = r\cos\theta \quad 8.110$$

$$y = r\sin\theta \quad 8.111$$

$$z = k\theta \quad 8.112$$

Figure 8.16 Helix

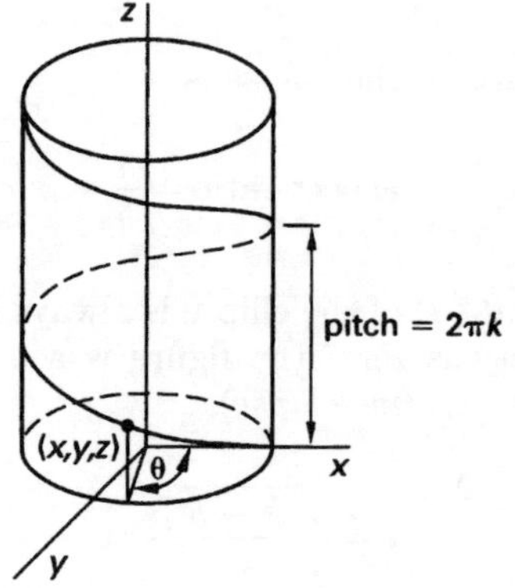

Differential Calculus

1. DERIVATIVE OF A FUNCTION

In most cases, it is possible to transform a continuous function, $f(x_1, x_2, x_3, \ldots)$, of one or more independent variables into a derivative function.[1] In simple cases, the *derivative* can be interpreted as the slope (tangent or rate of change) of the curve described by the original function. Since the slope of the curve depends on x, the derivative function will also depend on x. The derivative, $f'(x)$, of a function $f(x)$ is defined mathematically by Eq. 9.1. However, limit theory is seldom needed to actually calculate derivatives.

$$f'(x) = \lim_{\Delta x \to 0} \frac{\Delta f(x)}{\Delta x} \qquad 9.1$$

The derivative of a function $f(x)$, also known as the *first derivative*, is written in various ways, including

$$f'(x), \frac{df(x)}{dx}, \frac{df}{dx}, \mathbf{D}f(x), \mathbf{D}_x f(x), \dot{f}(x), sf(s)$$

A *second derivative* may exist if the derivative operation is performed on the first derivative—that is, a derivative is taken of a derivative function. This is written as

$$f''(x), \frac{d^2 f(x)}{dx^2}, \frac{d^2 f}{dx^2}, \mathbf{D}^2 f(x), \mathbf{D}_x^2 f(x), \ddot{f}(x), s^2 f(s)$$

[1]A function, $f(x)$, of one independent variable, x, is used in this section to simplify the discussion. Although the derivative is taken with respect to x, the independent variable can be anything.

Newton's notation (e.g., $\dot{m}$ and $\ddot{x}$) is generally only used with functions of time.

A *regular* (*analytic* or *holomorphic*) *function* possesses a derivative. A point at which a function's derivative is undefined is called a *singular point*, as Fig. 9.1 illustrates.

Figure 9.1 *Derivatives and Singular Points*

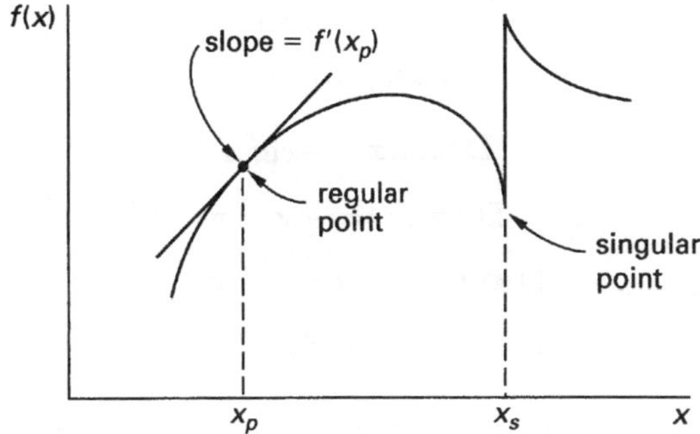

2. ELEMENTARY DERIVATIVE OPERATIONS

Equation 9.2 through Eq. 9.5 summarize the elementary derivative operations on polynomials and exponentials. Equation 9.2 and Eq. 9.3 are particularly useful. (a, n, and k represent constants. $f(x)$ and $g(x)$ are functions of x.)

$$\mathbf{D}k = 0 \qquad 9.2$$

$$\mathbf{D}x^n = nx^{n-1} \qquad 9.3$$

$$\mathbf{D}\ln x = \frac{1}{x} \qquad 9.4$$

$$\mathbf{D}e^{ax} = ae^{ax} \qquad 9.5$$

Equation 9.6 through Eq. 9.17 summarize the elementary derivative operations on transcendental (trigonometric) functions.

$$\mathbf{D}\sin x = \cos x \qquad 9.6$$

$$\mathbf{D}\cos x = -\sin x \qquad 9.7$$

$$\mathbf{D}\tan x = \sec^2 x \qquad 9.8$$

$$\mathbf{D}\cot x = -\csc^2 x \qquad 9.9$$

$$\mathbf{D}\sec x = \sec x \tan x \qquad 9.10$$

$$\mathbf{D}\csc x = -\csc x \cot x \qquad 9.11$$

$$\mathbf{D}\arcsin x = \frac{1}{\sqrt{1-x^2}} \qquad 9.12$$

$$\mathbf{D}\arccos x = -\mathbf{D}\arcsin x \qquad 9.13$$

$$\mathbf{D}\arctan x = \frac{1}{1+x^2} \qquad 9.14$$

$$\mathbf{D}\operatorname{arccot} x = -\mathbf{D}\arctan x \qquad 9.15$$

$$\mathbf{D}\operatorname{arcsec} x = \frac{1}{x\sqrt{x^2-1}} \qquad 9.16$$

$$\mathbf{D}\operatorname{arccsc} x = -\mathbf{D}\operatorname{arcsec} x \qquad 9.17$$

Equation 9.18 through Eq. 9.23 summarize the elementary derivative operations on hyperbolic transcendental functions. Derivatives of hyperbolic functions are not completely analogous to those of the regular transcendental functions.

$$\mathbf{D}\sinh x = \cosh x \qquad 9.18$$

$$\mathbf{D}\cosh x = \sinh x \qquad 9.19$$

$$\mathbf{D}\tanh x = \operatorname{sech}^2 x \qquad 9.20$$

$$\mathbf{D}\coth x = -\operatorname{csch}^2 x \qquad 9.21$$

$$\mathbf{D}\operatorname{sech} x = -\operatorname{sech} x \tanh x \qquad 9.22$$

$$\mathbf{D}\operatorname{csch} x = -\operatorname{csch} x \coth x \qquad 9.23$$

Equation 9.24 through Eq. 9.29 summarize the elementary derivative operations on functions and combinations of functions.

$$\mathbf{D}kf(x) = k\mathbf{D}f(x) \qquad 9.24$$

$$\mathbf{D}\big(f(x) \pm g(x)\big) = \mathbf{D}f(x) \pm \mathbf{D}g(x) \qquad 9.25$$

$$\mathbf{D}\big(f(x)\cdot g(x)\big) = f(x)\mathbf{D}g(x) + g(x)\mathbf{D}f(x) \qquad 9.26$$

$$\mathbf{D}\left(\frac{f(x)}{g(x)}\right) = \frac{g(x)\mathbf{D}f(x) - f(x)\mathbf{D}g(x)}{\big(g(x)\big)^2} \qquad 9.27$$

$$\mathbf{D}\big(f(x)\big)^n = n\big(f(x)\big)^{n-1}\mathbf{D}f(x) \qquad 9.28$$

$$\mathbf{D}f\big(g(x)\big) = \mathbf{D}_g f(g)\mathbf{D}_x g(x) \qquad 9.29$$

Example 9.1

What is the slope at $x = 3$ of the curve $f(x) = x^3 - 2x$?

Solution

The derivative function found from Eq. 9.3 determines the slope.

$$f'(x) = 3x^2 - 2$$

The slope at $x = 3$ is

$$f'(3) = (3)(3)^2 - 2 = 25$$

Example 9.2

What are the derivatives of the following functions?

(a) $f(x) = 5\sqrt[3]{x^5}$

(b) $f(x) = \sin x \cos^2 x$

(c) $f(x) = \ln(\cos e^x)$

Solution

(a) Using Eq. 9.3 and Eq. 9.24,

$$\begin{aligned} f'(x) &= 5\mathbf{D}\sqrt[3]{x^5} = 5\mathbf{D}(x^5)^{1/3} \\ &= (5)\left(\tfrac{1}{3}\right)(x^5)^{-2/3}\mathbf{D}x^5 \\ &= (5)\left(\tfrac{1}{3}\right)(x^5)^{-2/3}(5)(x^4) \\ &= 25x^{2/3}/3 \end{aligned}$$

(b) Using Eq. 9.26,

$$\begin{aligned} f'(x) &= \sin x\mathbf{D}\cos^2 x + \cos^2 x\mathbf{D}\sin x \\ &= (\sin x)(2\cos x)(\mathbf{D}\cos x) + \cos^2 x\cos x \\ &= (\sin x)(2\cos x)(-\sin x) + \cos^2 x\cos x \\ &= -2\sin^2 x\cos x + \cos^3 x \end{aligned}$$

(c) Using Eq. 9.29,

$$\begin{aligned} f'(x) &= \left(\frac{1}{\cos e^x}\right)\mathbf{D}\cos e^x \\ &= \left(\frac{1}{\cos e^x}\right)(-\sin e^x)\mathbf{D}e^x \\ &= \left(\frac{-\sin e^x}{\cos e^x}\right)e^x \\ &= -e^x\tan e^x \end{aligned}$$

3. CRITICAL POINTS

Derivatives are used to locate the local *critical points* of functions of one variable—that is, *extreme points* (also known as *maximum* and *minimum* points) as well as the *inflection points* (*points of contraflexure*). The plurals *extrema*, *maxima*, and *minima* are used without the word "points." These points are illustrated in Fig. 9.2. There is usually an inflection point between two adjacent local extrema.

The first derivative is calculated to determine the locations of possible critical points. The second derivative is calculated to determine whether a particular point is a local maximum, minimum, or inflection point, according to the following conditions. With this method, no distinction is made between local and global extrema. Therefore, the extrema should be compared with the function values at the endpoints of the interval, as

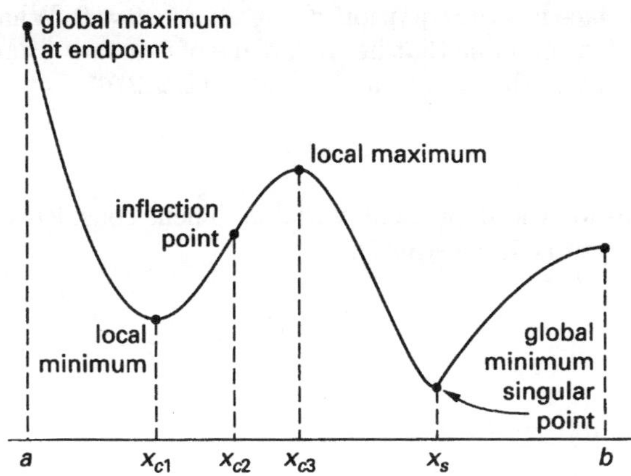

Figure 9.2 Extreme and Inflection Points

illustrated in Ex. 9.3.[2] Generally, $f'(x) \neq 0$ at an inflection point.

$$f'(x_c) = 0 \text{ at any extreme point, } x_c \quad 9.30$$

$$f''(x_c) < 0 \text{ at a maximum point} \quad 9.31$$

$$f''(x_c) > 0 \text{ at a minimum point} \quad 9.32$$

$$f''(x_c) = 0 \text{ at an inflection point} \quad 9.33$$

Example 9.3

Find the global extrema of the function $f(x)$ on the interval $[-2, +2]$.

$$f(x) = x^3 + x^2 - x + 1$$

Solution

The first derivative is

$$f'(x) = 3x^2 + 2x - 1$$

Since the first derivative is zero at extreme points, set $f'(x)$ equal to zero and solve for the roots of the quadratic equation.

$$3x^2 + 2x - 1 = (3x - 1)(x + 1) = 0$$

The roots are $x_1 = {}^1\!/_3$, $x_2 = -1$. These are the locations of the two local extrema.

The second derivative is

$$f''(x) = 6x + 2$$

[2]It is also necessary to check the values of the function at singular points (i.e., points where the derivative does not exist).

Substituting x_1 and x_2 into $f''(x)$,

$$f''(x_1) = (6)\left(\frac{1}{3}\right) + 2 = 4$$

$$f''(x_2) = (6)(-1) + 2 = -4$$

Therefore, x_1 is a local minimum point (because $f''(x_1)$ is positive), and x_2 is a local maximum point (because $f''(x_2)$ is negative). The inflection point between these two extrema is found by setting $f''(x)$ equal to zero.

$$f''(x) = 6x + 2 = 0 \text{ or } x = -\tfrac{1}{3}$$

Since the question asked for the global extreme points, it is necessary to compare the values of $f(x)$ at the local extrema with the values at the endpoints.

$$f(-2) = -1$$

$$f(-1) = 2$$

$$f(\tfrac{1}{3}) = 22/27$$

$$f(2) = 11$$

Therefore, the actual global extrema are the endpoints.

4. DERIVATIVES OF PARAMETRIC EQUATIONS

The derivative of a function $f(x_1, x_2, \ldots, x_n)$ can be calculated from the derivatives of the parametric equations $f_1(s), f_2(s), \ldots, f_n(s)$. The derivative will be expressed in terms of the parameter, s, unless the derivatives of the parametric equations can be expressed explicitly in terms of the independent variables.

Example 9.4

A circle is expressed parametrically by the equations

$$x = 5\cos\theta$$

$$y = 5\sin\theta$$

Express the derivative dy/dx (a) as a function of the parameter θ and (b) as a function of x and y.

Solution

(a) Taking the derivative of each parametric equation with respect to θ,

$$\frac{dx}{d\theta} = -5\sin\theta$$

$$\frac{dy}{d\theta} = 5\cos\theta$$

Then,

$$\frac{dy}{dx} = \frac{\frac{dy}{d\theta}}{\frac{dx}{d\theta}} = \frac{5\cos\theta}{-5\sin\theta} = -\cot\theta$$

(b) The derivatives of the parametric equations are closely related to the original parametric equations.

$$\frac{dx}{d\theta} = -5\sin\theta = -y$$

$$\frac{dy}{d\theta} = 5\cos\theta = x$$

$$\frac{dy}{dx} = \frac{\frac{dy}{d\theta}}{\frac{dx}{d\theta}} = \frac{-x}{y}$$

5. PARTIAL DIFFERENTIATION

Derivatives can be taken with respect to only one independent variable at a time. For example, $f'(x)$ is the derivative of $f(x)$ and is taken with respect to the independent variable x. If a function, $f(x_1, x_2, x_3, \ldots)$, has more than one independent variable, a *partial derivative* can be found, but only with respect to one of the independent variables. All other variables are treated as constants. Symbols for a partial derivative of f taken with respect to variable x are $\partial f/\partial x$ and $f_x(x, y)$.

The geometric interpretation of a partial derivative $\partial f/\partial x$ is the slope of a line tangent to the surface (a sphere, ellipsoid, etc.) described by the function when all variables except x are held constant. In three-dimensional space with a function described by $z = f(x, y)$, the partial derivative $\partial f/\partial x$ (equivalent to $\partial z/\partial x$) is the slope of the line tangent to the surface in a plane of constant y. Similarly, the partial derivative $\partial f/\partial y$ (equivalent to $\partial z/\partial y$) is the slope of the line tangent to the surface in a plane of constant x.

Example 9.5

What is the partial derivative $\partial z/\partial x$ of the following function?

$$z = 3x^2 - 6y^2 + xy + 5y - 9$$

Solution

The partial derivative with respect to x is found by considering all variables other than x to be constants.

$$\frac{\partial z}{\partial x} = 6x - 0 + y + 0 - 0 = 6x + y$$

Example 9.6

A surface has the equation $x^2 + y^2 + z^2 - 9 = 0$. What is the slope of a line that lies in a plane of constant y and is tangent to the surface at $(x, y, z) = (1, 2, 2)$?[3]

Solution

Solve for the dependent variable. Then, consider variable y to be a constant.

$$z = \sqrt{9 - x^2 - y^2}$$

$$\begin{aligned}\frac{\partial z}{\partial x} &= \frac{\partial(9 - x^2 - y^2)^{1/2}}{\partial x} \\ &= \left(\frac{1}{2}\right)(9 - x^2 - y^2)^{-1/2}\left(\frac{\partial(9 - x^2 - y^2)}{\partial x}\right) \\ &= \left(\frac{1}{2}\right)(9 - x^2 - y^2)^{-1/2}(-2x) \\ &= \frac{-x}{\sqrt{9 - x^2 - y^2}}\end{aligned}$$

At the point $(1, 2, 2)$, $x = 1$ and $y = 2$.

$$\left.\frac{\partial z}{\partial x}\right|_{(1,2,2)} = \frac{-1}{\sqrt{9 - (1)^2 - (2)^2}} = -\frac{1}{2}$$

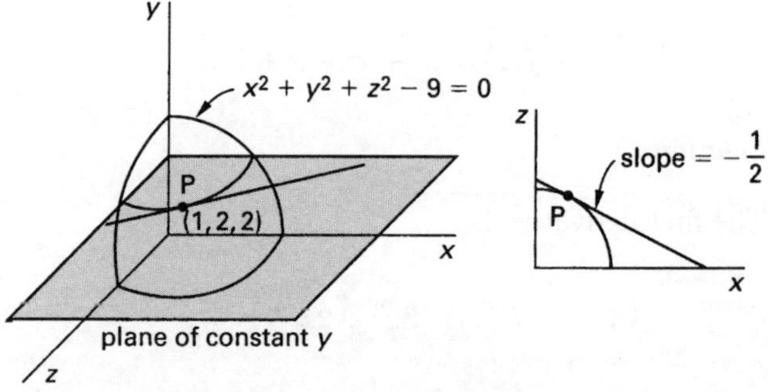

6. IMPLICIT DIFFERENTIATION

When a relationship between n variables cannot be manipulated to yield an explicit function of $n - 1$ independent variables, that relationship implicitly defines the nth variable. Finding the derivative of the implicit variable with respect to any other independent variable is known as *implicit differentiation*.

An implicit derivative is the quotient of two partial derivatives. The two partial derivatives are chosen so that dividing one by the other eliminates a common differential. For example, if z cannot be explicitly

[3]Although only implied, it is required that the point actually be on the surface (i.e., it must satisfy the equation $f(x, y, z) = 0$).

extracted from $f(x,y,z)=0$, the partial derivatives $\partial z/\partial x$ and $\partial z/\partial y$ can still be found as follows.

$$\frac{dz}{dx}=\frac{-\frac{\partial f}{\partial x}}{\frac{\partial f}{\partial z}} \quad 9.34$$

$$\frac{dz}{dy}=\frac{-\frac{\partial f}{\partial y}}{\frac{\partial f}{\partial z}} \quad 9.35$$

Example 9.7

Find the derivative dy/dx of

$$f(x,y)=x^2+xy+y^3$$

Solution

Implicit differentiation is required because x cannot be extracted from $f(x,y)$.

$$\frac{\partial f}{\partial x}=2x+y$$

$$\frac{\partial f}{\partial y}=x+3y^2$$

$$\frac{dy}{dx}=\frac{-\frac{\partial f}{\partial x}}{\frac{\partial f}{\partial y}}=\frac{-(2x+y)}{x+3y^2}$$

Example 9.8

Solve Ex. 9.6 using implicit differentiation.

Solution

$$f(x,y,z)=x^2+y^2+z^2-9=0$$

$$\frac{\partial f}{\partial x}=2x$$

$$\frac{\partial f}{\partial z}=2z$$

$$\frac{\partial z}{\partial x}=\frac{-\frac{\partial f}{\partial x}}{\frac{\partial f}{\partial z}}=\frac{-2x}{2z}=-\frac{x}{z}$$

At the point $(1,2,2)$,

$$\frac{\partial z}{\partial x}=-\frac{1}{2}$$

7. TANGENT PLANE FUNCTION

Partial derivatives can be used to find the equation of a plane tangent to a three-dimensional surface defined by $f(x,y,z)=0$ at some point, P_0.

$$\begin{aligned}T(x_0,y_0,z_0) &= (x-x_0)\left.\frac{\partial f(x,y,z)}{\partial x}\right|_{P_0} \\ &\quad +(y-y_0)\left.\frac{\partial f(x,y,z)}{\partial y}\right|_{P_0} \\ &\quad +(z-z_0)\left.\frac{\partial f(x,y,z)}{\partial z}\right|_{P_0} \\ &= 0\end{aligned} \quad 9.36$$

The coefficients of x, y, and z are the same as the coefficients of **i**, **j**, and **k** of the normal vector at point P_0. (See Sec. 9.10.)

Example 9.9

What is the equation of the plane that is tangent to the surface defined by $f(x,y,z)=4x^2+y^2-16z=0$ at the point $(2,4,2)$?

Solution

Calculate the partial derivatives and substitute the coordinates of the point.

$$\left.\frac{\partial f(x,y,z)}{\partial x}\right|_{P_0}=8x|_{(2,4,2)}=(8)(2)=16$$

$$\left.\frac{\partial f(x,y,z)}{\partial y}\right|_{P_0}=2y|_{(2,4,2)}=(2)(4)=8$$

$$\left.\frac{\partial f(x,y,z)}{\partial z}\right|_{P_0}=-16|_{(2,4,2)}=-16$$

$$\begin{aligned}T(2,4,2) &= (16)(x-2)+(8)(y-4)-(16)(z-2) \\ &= 16x+8y-16z-32\end{aligned}$$

Substitute into Eq. 9.36, and divide both sides by 8.

$$2x+y-2z-4=0$$

8. GRADIENT VECTOR

The slope of a function is the change in one variable with respect to a distance in a chosen direction. Usually, the direction is parallel to a coordinate axis. However, the maximum slope at a point on a surface may not be in a direction parallel to one of the coordinate axes.

The *gradient vector function* $\nabla f(x, y, z)$ (pronounced "del f") gives the maximum rate of change of the function $f(x, y, z)$.

$$\nabla f(x,y,z) = \frac{\partial f(x,y,z)}{\partial x}\mathbf{i} + \frac{\partial f(x,y,z)}{\partial y}\mathbf{j} + \frac{\partial f(x,y,z)}{\partial z}\mathbf{k} \quad \text{9.37}$$

Example 9.10

A two-dimensional function is defined by

$$f(x,y) = 2x^2 - y^2 + 3x - y$$

(a) What is the gradient vector for this function? (b) What is the direction of the line passing through the point (1, −2) that has a maximum slope? (c) What is the maximum slope at the point (1, −2)?

Solution

(a) It is necessary to calculate two partial derivatives in order to use Eq. 9.37.

$$\frac{\partial f(x,y)}{\partial x} = 4x + 3$$

$$\frac{\partial f(x,y)}{\partial y} = -2y - 1$$

$$\nabla f(x,y) = (4x+3)\mathbf{i} + (-2y-1)\mathbf{j}$$

(b) Find the direction of the line passing through (1, −2) with maximum slope by inserting $x = 1$ and $y = -2$ into the gradient vector function.

$$\begin{aligned}\mathbf{V} &= \Big((4)(1)+3\Big)\mathbf{i} + \Big((-2)(-2)-1\Big)\mathbf{j} \\ &= 7\mathbf{i} + 3\mathbf{j}\end{aligned}$$

(c) The magnitude of the slope is

$$|\mathbf{V}| = \sqrt{(7)^2 + (3)^2} = 7.62$$

9. DIRECTIONAL DERIVATIVE

Unlike the gradient vector (covered in Sec. 9.8), which calculates the maximum rate of change of a function, the *directional derivative*, indicated by $\nabla_u f(x,y,z)$, $D_u f(x,y,z)$, or $f'_u(x,y,z)$, gives the rate of change in the direction of a given vector, $\mathbf{u}$ or $\mathbf{U}$. The subscript u implies that the direction vector is a unit vector, but it does not need to be, as only the direction cosines are calculated from it.

$$\begin{aligned}\nabla_u f(x,y,z) = &\left(\frac{\partial f(x,y,z)}{\partial x}\right)\cos\alpha \\ &+ \left(\frac{\partial f(x,y,z)}{\partial y}\right)\cos\beta \\ &+ \left(\frac{\partial f(x,y,z)}{\partial z}\right)\cos\gamma\end{aligned} \quad \text{9.38}$$

$$\mathbf{U} = U_x\mathbf{i} + U_y\mathbf{j} + U_z\mathbf{k} \quad \text{9.39}$$

$$\cos\alpha = \frac{U_x}{|\mathbf{U}|} = \frac{U_x}{\sqrt{U_x^2 + U_y^2 + U_z^2}} \quad \text{9.40}$$

$$\cos\beta = \frac{U_y}{|\mathbf{U}|} \quad \text{9.41}$$

$$\cos\gamma = \frac{U_z}{|\mathbf{U}|} \quad \text{9.42}$$

Example 9.11

What is the rate of change of $f(x,y) = 3x^2 + xy - 2y^2$ at the point (1, −2) in the direction $4\mathbf{i} + 3\mathbf{j}$?

Solution

The direction cosines are given by Eq. 9.40 and Eq. 9.41.

$$\cos\alpha = \frac{U_x}{|\mathbf{U}|} = \frac{4}{\sqrt{(4)^2 + (3)^2}} = \frac{4}{5}$$

$$\cos\beta = \frac{U_y}{|\mathbf{U}|} = \frac{3}{5}$$

The partial derivatives are

$$\frac{\partial f(x,y)}{\partial x} = 6x + y$$

$$\frac{\partial f(x,y)}{\partial y} = x - 4y$$

The directional derivative is given by Eq. 9.38.

$$\nabla_u f(x,y) = \left(\tfrac{4}{5}\right)(6x+y) + \left(\tfrac{3}{5}\right)(x-4y)$$

Substituting the given values of $x = 1$ and $y = -2$,

$$\begin{aligned}\nabla_u f(1,-2) &= \left(\tfrac{4}{5}\right)\Big((6)(1)-2\Big) + \left(\tfrac{3}{5}\right)\Big(1-(4)(-2)\Big) \\ &= \frac{43}{5} \quad (8.6)\end{aligned}$$

10. NORMAL LINE VECTOR

Partial derivatives can be used to find the vector normal to a three-dimensional surface defined by $f(x,y,z) = 0$

at some point P_0. The coefficients of **i**, **j**, and **k** are the same as the coefficients of x, y, and z calculated for the equation of the tangent plane at point P_0. (See Sec. 9.7.)

$$\mathbf{N} = \left.\frac{\partial f(x,y,z)}{\partial x}\right|_{P_0}\mathbf{i} + \left.\frac{\partial f(x,y,z)}{\partial y}\right|_{P_0}\mathbf{j} + \left.\frac{\partial f(x,y,z)}{\partial z}\right|_{P_0}\mathbf{k} \qquad 9.43$$

Example 9.12

What is the vector normal to the surface of $f(x,y,z) = 4x^2 + y^2 - 16z = 0$ at the point $(2,4,2)$?

Solution

The equation of the tangent plane at this point was calculated in Ex. 9.9 to be

$$T(2,4,2) = 2x + y - 2z - 4 = 0$$

A vector that is normal to the tangent plane through this point is

$$\mathbf{N} = 2\mathbf{i} + \mathbf{j} - 2\mathbf{k}$$

11. DIVERGENCE OF A VECTOR FIELD

The *divergence*, div **F**, of a vector field $\mathbf{F}(x,y,z)$ is a scalar function defined by Eq. 9.45 and Eq. 9.46.[4] The divergence of **F** can be interpreted as the *accumulation* of flux (i.e., a flowing substance) in a small region (i.e., at a point). One of the uses of the divergence is to determine whether flow (represented in direction and magnitude by **F**) is compressible. Flow is incompressible if div **F** = 0, since the substance is not accumulating.

$$\mathbf{F} = P(x,y,z)\mathbf{i} + Q(x,y,z)\mathbf{j} + R(x,y,z)\mathbf{k} \qquad 9.44$$

$$\operatorname{div}\mathbf{F} = \frac{\partial P}{\partial x} + \frac{\partial Q}{\partial y} + \frac{\partial R}{\partial z} \qquad 9.45$$

It may be easier to calculate divergence from Eq. 9.46.

$$\operatorname{div}\mathbf{F} = \nabla\cdot\mathbf{F} \qquad 9.46$$

The vector del operator, ∇, is defined as

$$\nabla = \frac{\partial}{\partial x}\mathbf{i} + \frac{\partial}{\partial y}\mathbf{j} + \frac{\partial}{\partial z}\mathbf{k} \qquad 9.47$$

If there is no divergence, then the dot product calculated in Eq. 9.46 is zero.

Example 9.13

Calculate the divergence of the following vector function.

$$\mathbf{F}(x,y,z) = xz\mathbf{i} + e^x y\mathbf{j} + 7x^3 y\mathbf{k}$$

Solution

From Eq. 9.45,

$$\operatorname{div}\mathbf{F} = \frac{\partial}{\partial x}xz + \frac{\partial}{\partial y}e^x y + \frac{\partial}{\partial z}7x^3 y = z + e^x + 0$$
$$= z + e^x$$

12. CURL OF A VECTOR FIELD

The *curl*, curl **F**, of a vector field $\mathbf{F}(x,y,z)$ is a vector field defined by Eq. 9.49 and Eq. 9.50. The curl **F** can be interpreted as the *vorticity* per unit area of flux (i.e., a flowing substance) in a small region (i.e., at a point). One of the uses of the curl is to determine whether flow (represented in direction and magnitude by **F**) is rotational. Flow is irrotational if curl **F** = 0.

$$\mathbf{F} = P(x,y,z)\mathbf{i} + Q(x,y,z)\mathbf{j} + R(x,y,z)\mathbf{k} \qquad 9.48$$

$$\operatorname{curl}\mathbf{F} = \left(\frac{\partial R}{\partial y} - \frac{\partial Q}{\partial z}\right)\mathbf{i} + \left(\frac{\partial P}{\partial z} - \frac{\partial R}{\partial x}\right)\mathbf{j} + \left(\frac{\partial Q}{\partial x} - \frac{\partial P}{\partial y}\right)\mathbf{k} \qquad 9.49$$

It may be easier to calculate the curl from Eq. 9.50. (The vector del operator, ∇, was defined in Eq. 9.47.)

$$\operatorname{curl}\mathbf{F} = \nabla\times\mathbf{F} = \begin{vmatrix} \mathbf{i} & \mathbf{j} & \mathbf{k} \\ \frac{\partial}{\partial x} & \frac{\partial}{\partial y} & \frac{\partial}{\partial z} \\ P(x,y,z) & Q(x,y,z) & R(x,y,z) \end{vmatrix} \qquad 9.50$$

If the velocity vector is **V**, then the vorticity is

$$\omega = \nabla\times\mathbf{V} = \omega_x\mathbf{i} + \omega_y\mathbf{j} + \omega_z\mathbf{k} \qquad 9.51$$

The *circulation* is the line integral of the velocity, **V**, along a closed curve.

$$\Gamma = \oint V\cdot ds = \oint \omega\cdot dA \qquad 9.52$$

Example 9.14

Calculate the curl of the following vector function.

$$\mathbf{F}(x,y,z) = 3x^2\mathbf{i} + 7e^x y\mathbf{j}$$

Solution

Using Eq. 9.50,

$$\operatorname{curl}\mathbf{F} = \begin{vmatrix} \mathbf{i} & \mathbf{j} & \mathbf{k} \\ \frac{\partial}{\partial x} & \frac{\partial}{\partial y} & \frac{\partial}{\partial z} \\ 3x^2 & 7e^x y & 0 \end{vmatrix}$$

[4]A bold letter, **F**, is used to indicate that the vector is a function of x, y, and z.

Expand the determinant across the top row.

$$\mathbf{i}\left(\frac{\partial}{\partial y}(0)-\frac{\partial}{\partial z}(7e^xy)\right)-\mathbf{j}\left(\frac{\partial}{\partial x}(0)-\frac{\partial}{\partial z}(3x^2)\right)$$
$$+\mathbf{k}\left(\frac{\partial}{\partial x}(7e^xy)-\frac{\partial}{\partial y}(3x^2)\right)$$
$$=\mathbf{i}(0-0)-\mathbf{j}(0-0)+\mathbf{k}(7e^xy-0)$$
$$=7e^xy\mathbf{k}$$

13. TAYLOR'S FORMULA

Taylor's formula (series) can be used to expand a function around a point (i.e., approximate the function at one point based on the function's value at another point). The approximation consists of a series, each term composed of a derivative of the original function and a polynomial. Using Taylor's formula requires that the original function be continuous in the interval $[a, b]$ and have the required number of derivatives. To expand a function, $f(x)$, around a point, a, in order to obtain $f(b)$, Taylor's formula is

$$f(b)=f(a)+\frac{f'(a)}{1!}(b-a)+\frac{f''(a)}{2!}(b-a)^2+\cdots+\frac{f^n(a)}{n!}(b-a)^n+R_n(b) \qquad 9.53$$

In Eq. 9.53, the expression f^n designates the nth derivative of the function $f(x)$. To be a useful approximation, two requirements must be met: (1) point a must be relatively close to point b, and (2) the function and its derivatives must be known or easy to calculate. The last term, $R_n(b)$, is the uncalculated remainder after n derivatives. It is the difference between the exact and approximate values. By using enough terms, the remainder can be made arbitrarily small. That is, $R_n(b)$ approaches zero as n approaches infinity.

It can be shown that the remainder term can be calculated from Eq. 9.54, where c is some number in the interval $[a, b]$. With certain functions, the constant c can be completely determined. In most cases, however, it is possible only to calculate an upper bound on the remainder from Eq. 9.55. M_n is the maximum (positive) value of $f^{n+1}(x)$ on the interval $[a, b]$.

$$R_n(b)=\frac{f^{n+1}(c)}{(n+1)!}(b-a)^{n+1} \qquad 9.54$$

$$|R_n(b)|\le M_n\frac{|(b-a)^{n+1}|}{(n+1)!} \qquad 9.55$$

14. MACLAURIN POWER APPROXIMATIONS

If $a=0$ in the Taylor series, Eq. 9.53 is known as the *Maclaurin series*. The Maclaurin series can be used to approximate functions at some value of x between 0 and 1. The following common approximations may be referred to as Maclaurin series, Taylor series, or power series approximations.

$$\sin x\approx x-\frac{x^3}{3!}+\frac{x^5}{5!}-\frac{x^7}{7!}+\cdots+(-1)^n\frac{x^{2n+1}}{(2n+1)!} \qquad 9.56$$

$$\cos x\approx 1-\frac{x^2}{2!}+\frac{x^4}{4!}-\frac{x^6}{6!}+\cdots+(-1)^n\frac{x^{2n}}{(2n)!} \qquad 9.57$$

$$\sinh x\approx x+\frac{x^3}{3!}+\frac{x^5}{5!}+\frac{x^7}{7!}+\cdots+\frac{x^{2n+1}}{(2n+1)!} \qquad 9.58$$

$$\cosh x\approx 1+\frac{x^2}{2!}+\frac{x^4}{4!}+\frac{x^6}{6!}+\cdots+\frac{x^{2n}}{(2n)!} \qquad 9.59$$

$$e^x\approx 1+x+\frac{x^2}{2!}+\frac{x^3}{3!}+\cdots+\frac{x^n}{n!} \qquad 9.60$$

$$\ln(1+x)\approx x-\frac{x^2}{2}+\frac{x^3}{3}-\frac{x^4}{4}+\cdots+(-1)^{n+1}\frac{x^n}{n} \qquad 9.61$$

$$\frac{1}{1-x}\approx 1+x+x^2+x^3+\cdots+x^n \qquad 9.62$$

10 Integral Calculus

1. INTEGRATION

Integration is the inverse operation of differentiation. For that reason, *indefinite integrals* are sometimes referred to as *antiderivatives*.[1] Although expressions can be functions of several variables, integrals can only be taken with respect to one variable at a time. The *differential term* (dx in Eq. 10.1) indicates that variable. In Eq. 10.1, the function $f'(x)$ is the *integrand*, and x is the variable of integration.

$$\int f'(x)dx = f(x) + C \qquad 10.1$$

While most of a function, $f(x)$, can be "recovered" through integration of its derivative, $f'(x)$, a constant term will be lost. This is because the derivative of a constant term vanishes (i.e., is zero), leaving nothing to recover from. A *constant of integration*, C, is added to the integral to recognize the possibility of such a constant term.

2. ELEMENTARY OPERATIONS

Equation 10.2 through Eq. 10.8 summarize the elementary integration operations on polynomials and exponentials.[2]

Equation 10.2 and Eq. 10.3 are particularly useful. (C and k represent constants. $f(x)$ and $g(x)$ are functions of x.)

$$\int k\,dx = kx + C \qquad 10.2$$

$$\int x^m\,dx = \frac{x^{m+1}}{m+1} + C \quad [m \neq -1] \qquad 10.3$$

$$\int \frac{1}{x}\,dx = \ln|x| + C \qquad 10.4$$

$$\int e^{kx}\,dx = \frac{e^{kx}}{k} + C \qquad 10.5$$

$$\int xe^{kx}\,dx = \frac{e^{kx}(kx-1)}{k^2} + C \qquad 10.6$$

$$\int k^{ax}\,dx = \frac{k^{ax}}{a\ln k} + C \qquad 10.7$$

$$\int \ln x\,dx = x\ln x - x + C \qquad 10.8$$

Equation 10.9 through Eq. 10.20 summarize the elementary integration operations on transcendental functions

$$\int \sin x\,dx = -\cos x + C \qquad 10.9$$

$$\int \cos x\,dx = \sin x + C \qquad 10.10$$

$$\int \tan x\,dx = \ln|\sec x| + C = -\ln|\cos x| + C \qquad 10.11$$

$$\int \cot x\,dx = \ln|\sin x| + C \qquad 10.12$$

$$\int \sec x\,dx = \ln|\sec x + \tan x| + C = \ln\left|\tan\left(\frac{x}{2} + \frac{\pi}{4}\right)\right| + C \qquad 10.13$$

$$\int \csc x\,dx = \ln|\csc x - \cot x| + C = \ln\left|\tan\frac{x}{2}\right| + C \qquad 10.14$$

$$\int \frac{dx}{k^2 + x^2} = \frac{1}{k}\arctan\frac{x}{k} + C \qquad 10.15$$

[1]The difference between an indefinite and definite integral (covered in Sec. 10.7) is simple: An *indefinite integral* is a function, while a *definite integral* is a number.

[2]More extensive listings, known as *tables of integrals*, are widely available. (See App. 10.A.)

$$\int \frac{dx}{\sqrt{k^2 - x^2}} = \arcsin \frac{x}{k} + C \quad [k^2 > x^2] \qquad 10.16$$

$$\int \frac{dx}{x\sqrt{x^2 - k^2}} = \frac{1}{k} \operatorname{arcsec} \frac{x}{k} + C \quad [x^2 > k^2] \qquad 10.17$$

$$\int \sin^2 x\, dx = \tfrac{1}{2}x - \tfrac{1}{4}\sin 2x + C \qquad 10.18$$

$$\int \cos^2 x\, dx = \tfrac{1}{2}x - \tfrac{1}{4}\sin 2x + C \qquad 10.19$$

$$\int \tan^2 x\, dx = \tan x - x + C \qquad 10.20$$

Equation 10.21 through Eq. 10.26 summarize the elementary integration operations on hyperbolic transcendental functions. Integrals of hyperbolic functions are not completely analogous to those of the regular transcendental functions.

$$\int \sinh x\, dx = \cosh x + C \qquad 10.21$$

$$\int \cosh x\, dx = \sinh x + C \qquad 10.22$$

$$\int \tanh x\, dx = \ln|\cosh x| + C \qquad 10.23$$

$$\int \coth x\, dx = \ln|\sinh x| + C \qquad 10.24$$

$$\int \operatorname{sech} x\, dx = \arctan(\sinh x) + C \qquad 10.25$$

$$\int \operatorname{csch} x\, dx = \ln\left|\tanh \frac{x}{2}\right| + C \qquad 10.26$$

Equation 10.27 through Eq. 10.30 summarize the elementary integration operations on functions and combinations of functions

$$\int kf(x)dx = k\int f(x)dx \qquad 10.27$$

$$\int (f(x) + g(x))\, dx = \int f(x)dx + \int g(x)dx \qquad 10.28$$

$$\int \frac{f'(x)}{f(x)} dx = \ln|f(x)| + C \qquad 10.29$$

$$\begin{aligned} \int f(x)dg(x) &= f(x)\int dg(x) - \int g(x)df(x) + C \\ &= f(x)g(x) - \int g(x)df(x) + C \end{aligned} \qquad 10.30$$

Example 10.1

Find the integral with respect to x of

$$3x^2 + \tfrac{1}{3}x - 7$$

Solution

This is a polynomial function, and Eq. 10.3 can be applied to each of the three terms.

$$\int (3x^2 + \tfrac{1}{3}x - 7)\, dx = x^3 + \tfrac{1}{6}x^2 - 7x + C$$

3. INTEGRATION BY PARTS

Equation 10.30, repeated here, is known as *integration by parts*. $f(x)$ and $g(x)$ are functions. The use of this method is illustrated by Ex. 10.2.

$$\int f(x)dg(x) = f(x)g(x) - \int g(x)df(x) + C \qquad 10.31$$

Example 10.2

Find the following integral.

$$\int x^2 e^x\, dx$$

Solution

x^2e^x is factored into two parts so that integration by parts can be used.

$$\begin{aligned} f(x) &= x^2 \\ dg(x) &= e^x\, dx \\ df(x) &= 2x\, dx \end{aligned}$$

$$g(x) = \int dg(x) = \int e^x\, dx = e^x$$

From Eq. 10.31, disregarding the constant of integration (which cannot be evaluated),

$$\int f(x)dg(x) = f(x)g(x) - \int g(x)df(x)$$

$$\int x^2 e^x\, dx = x^2 e^x - \int e^x(2x)\, dx$$

The second term is also factored into two parts, and integration by parts is used again. This time,

$$\begin{aligned} f(x) &= x \\ dg(x) &= e^x\, dx \\ df(x) &= dx \end{aligned}$$

$$g(x) = \int dg(x) = \int e^x\, dx = e^x$$

From Eq. 10.31,

$$\begin{aligned}\int 2xe^x\,dx &= 2\int xe^x\,dx = 2\left(xe^x - \int e^x\,dx\right)\\ &= 2(xe^x - e^x)\end{aligned}$$

Then, the complete integral is

$$\begin{aligned}\int x^2e^x\,dx &= x^2e^x - 2(xe^x - e^x) + C\\ &= e^x(x^2 - 2x + 2) + C\end{aligned}$$

4. SEPARATION OF TERMS

Equation 10.28 shows that the integral of a sum of terms is equal to a sum of integrals. This technique is known as *separation of terms.* In many cases, terms are easily separated. In other cases, the technique of *partial fractions* can be used to obtain individual terms. These techniques are illustrated by Ex. 10.3 and Ex. 10.4.

Example 10.3

Find the following integral.

$$\int \frac{(2x^2+3)^2}{x}\,dx$$

Solution

$$\begin{aligned}\int \frac{(2x^2+3)^2}{x}\,dx &= \int \frac{4x^4+12x^2+9}{x}\,dx\\ &= \int \left(4x^3 + 12x + \frac{9}{x}\right)dx\\ &= x^4 + 6x^2 + 9\ln|x| + C\end{aligned}$$

Example 10.4

Find the following integral.

$$\int \frac{3x+2}{3x-2}\,dx$$

Solution

The integrand is larger than 1, so use long division to simplify it.

$$\begin{array}{r|l} & \text{1 rem 4} \quad \left(1 + \dfrac{4}{3x-2}\right)\\ 3x-2 & 3x+2\\ & 3x-2\\ \hline & \text{4 remainder}\end{array}$$

$$\begin{aligned}\int \frac{3x+2}{3x-2}\,dx &= \int \left(1 + \frac{4}{3x-2}\right)dx\\ &= \int dx + \int \frac{4}{3x-2}\,dx\\ &= x + \tfrac{4}{3}\text{in}\,|(3x-2)| + C\end{aligned}$$

5. DOUBLE AND HIGHER-ORDER INTEGRALS

A function can be successively integrated. (This is analogous to successive differentiation.) A function that is integrated twice is known as a *double integral*; if integrated three times, it is a *triple integral*, and so on. Double and triple integrals are used to calculate areas and volumes, respectively.

The successive integrations do not need to be with respect to the same variable. Variables not included in the integration are treated as constants.

There are several notations used for a multiple integral, particularly when the product of length differentials represents a differential area or volume. A double integral (i.e., two successive integrations) can be represented by one of the following notations.

$$\iint f(x,y)\,dx\,dy, \quad \int_{R^2} f(x,y)\,dx\,dy,$$

$$\text{or} \iint_{R^2} f(x,y)\,dA$$

A triple integral can be represented by one of the following notations.

$$\iiint f(x,y,z)\,dx\,dy\,dz, \quad \int_{R^3} f(x,y,z)\,dx\,dy\,dz,$$

$$\text{or} \iiint_{R^3} f(x,y,z)\,dV$$

Example 10.5

Find the following double integral.

$$\iint (x^2 + y^3x)\,dx\,dy$$

Solution

$$\int (x^2 + y^3x)\,dx = \tfrac{1}{3}x^3 + \tfrac{1}{2}y^3x^2 + C_1$$

$$\int \left(\tfrac{1}{3}x^3 + \tfrac{1}{2}y^3x^2 + C_1\right)dy = \tfrac{1}{3}yx^3 + \tfrac{1}{8}y^4x^2 + C_1y + C_2$$

So,

$$\iint (x^2 + y^3x)\,dx\,dy = \tfrac{1}{3}yx^3 + \tfrac{1}{8}y^4x^2 + C_1y + C_2$$

6. INITIAL VALUES

The constant of integration, C, can be found only if the value of the function $f(x)$ is known for some value of x_0. The value $f(x_0)$ is known as an *initial value* or *initial condition*. To completely define a function, as many initial values, $f(x_0)$, $f'(x_1)$, $f''(x_2)$, and so on, as there are integrations are needed. x_0, x_1, x_2, and so on, can be, but do not have to be, the same.

Example 10.6

It is known that $f(x) = 4$ when $x = 2$ (i.e., the initial value is $f(2) = 4$). Find the original function.

$$\int (3x^3 - 7x)\,dx$$

Solution

The function is

$$\begin{aligned} f(x) &= \int (3x^3 - 7x)\,dx \\ &= \tfrac{3}{4}x^4 - \tfrac{7}{2}x^2 + C \end{aligned}$$

Substituting the initial value determines C.

$$\begin{aligned} 4 &= \left(\tfrac{3}{4}\right)(2)^4 - \left(\tfrac{7}{2}\right)(2)^2 + C \\ 4 &= 12 - 14 + C \\ C &= 6 \end{aligned}$$

The function is

$$f(x) = \tfrac{3}{4}x^4 - \tfrac{7}{2}x^2 + 6$$

7. DEFINITE INTEGRALS

A *definite integral* is restricted to a specific range of the independent variable. (Unrestricted integrals of the types shown in all preceding examples are known as *indefinite integrals*.) A definite integral restricted to the region bounded by *lower* and *upper limits* (also known as *bounds*), x_1 and x_2, is written as

$$\int_{x_1}^{x_2} f(x)\,dx$$

Equation 10.32 indicates how definite integrals are evaluated. It is known as the *fundamental theorem of calculus*.

$$\int_{x_1}^{x_2} f'(x)\,dx = f(x)\Big|_{x_1}^{x_2} = f(x_2) - f(x_1) \qquad 10.32$$

A common use of a definite integral is the calculation of work performed by a force, F, that moves an object from position x_1 to x_2. The force can be constant or a function of x.

$$W = \int_{x_1}^{x_2} F\,dx \qquad 10.33$$

Example 10.7

Evaluate the following definite integral.

$$\int_{\pi/4}^{\pi/3} \sin x\,dx$$

Solution

From Eq. 10.32,

$$\begin{aligned} \int_{\pi/4}^{\pi/3} \sin x\,dx &= -\cos x\Big|_{\pi/4}^{\pi/3} = -\cos\frac{\pi}{3} - \left(-\cos\frac{\pi}{4}\right) \\ &= -0.5 - (-0.707) \\ &= 0.207 \end{aligned}$$

8. AVERAGE VALUE

The average value of a function $f(x)$ that is integrable over the interval $[a, b]$ is

$$\text{average value} = \frac{1}{b-a}\int_a^b f(x)\,dx \qquad 10.34$$

9. AREA

Equation 10.35 calculates the area, A, bounded by $x = a$, $x = b$, $f_1(x)$ above, and $f_2(x)$ below. ($f_2(x) = 0$ if the area is bounded by the x-axis.) This is illustrated in Fig. 10.1.

$$A = \int_a^b \big(f_1(x) - f_2(x)\big)\,dx \qquad 10.35$$

Figure 10.1 *Area Between Two Curves*

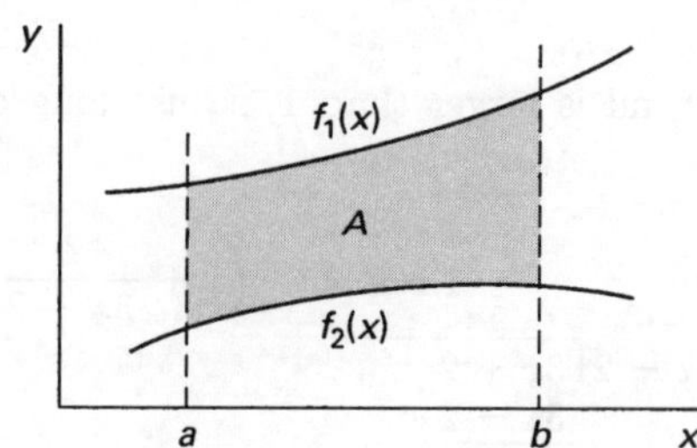

Example 10.8

Find the area between the x-axis and the parabola $y = x^2$ in the interval $[0, 4]$.

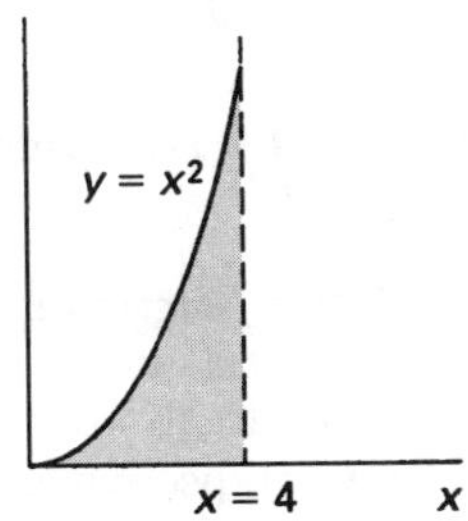

Solution

Referring to Eq. 10.35,

$$f_1(x) = x^2$$
$$f_2(x) = 0$$

$$A = \int_a^b \big(f_1(x) - f_2(x)\big)\,dx = \int_0^4 x^2\,dx$$
$$= \frac{x^3}{3}\Big|_0^4$$
$$= 64/3$$

10. ARC LENGTH

Equation 10.36 gives the length of a curve defined by $f(x)$ whose derivative exists in the interval $[a, b]$.

$$\text{length} = \int_a^b \sqrt{1 + \big(f'(x)\big)^2}\,dx \qquad 10.36$$

11. PAPPUS' THEOREMS[3]

The first and second theorems of Pappus are:[4]

- *first theorem:* Given a curve, C, that does not intersect the y-axis, the area of the *surface of revolution* generated by revolving C around the y-axis is equal to the product of the length of the curve and the circumference of the circle traced by the centroid of curve C.

$$A = \text{length} \times \text{circumference}$$
$$= \text{length} \times 2\pi \times \text{radius} \qquad 10.37$$

- *second theorem:* Given a plane region, R, that does not intersect the y-axis, the *volume of revolution* generated by revolving R around the y-axis is equal to the product of the area and the circumference of the circle traced by the centroid of area R.

$$V = \text{area} \times \text{circumference}$$
$$= \text{area} \times 2\pi \times \text{radius} \qquad 10.38$$

12. SURFACE OF REVOLUTION

The surface area obtained by rotating $f(x)$ about the x-axis is

$$A = 2\pi \int_{x=a}^{x=b} f(x)\sqrt{1 + \big(f'(x)\big)^2}\,dx \qquad 10.39$$

The surface area obtained by rotating $f(y)$ about the y-axis is

$$A = 2\pi \int_{y=c}^{y=d} f(y)\sqrt{1 + \big(f'(y)\big)^2}\,dy \qquad 10.40$$

Example 10.9

The curve $f(x) = \frac{1}{2}x$ over the region $x = [0, 4]$ is rotated about the x-axis. What is the surface of revolution?

Solution

The surface of revolution is

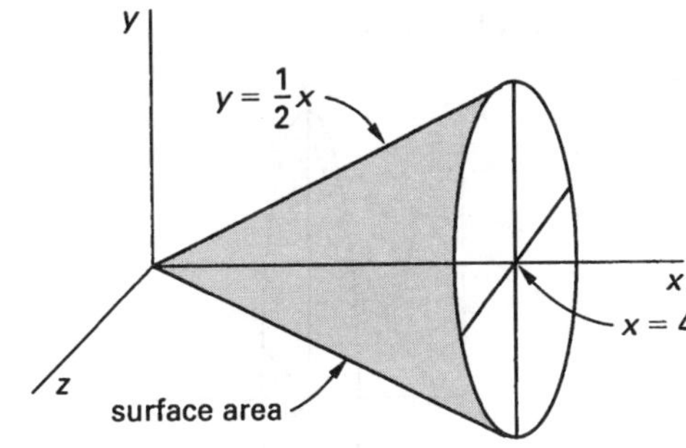

Since $f(x) = \frac{1}{2}x$, $f'(x) = 1/2$. From Eq. 10.39, the area is

$$A = 2\pi \int_{x=a}^{x=b} f(x)\sqrt{1 + \big(f'(x)\big)^2}\,dx$$
$$= 2\pi \int_0^4 \tfrac{1}{2}x\sqrt{1 + \left(\tfrac{1}{2}\right)^2}\,dx$$
$$= \frac{\sqrt{5}}{2}\pi \int_0^4 x\,dx$$
$$= \frac{\sqrt{5}}{2}\pi \frac{x^2}{2}\Big|_0^4$$
$$= \frac{\sqrt{5}}{2}\pi\left(\frac{(4)^2 - (0)^2}{2}\right)$$
$$= 4\sqrt{5}\pi$$

[3]This section is an introduction to surfaces and volumes of revolution but does not involve integration.
[4]Some authorities call the first theorem the second, and vice versa.

13. VOLUME OF REVOLUTION

The volume obtained by rotating $f(x)$ about the x-axis is given by Eq. 10.41. $f^2(x)$ is the square of the function, not the second derivative. Equation 10.41 is known as the *method of discs.*

$$V = \pi \int_{x=a}^{x=b} f^2(x)dx \qquad 10.41$$

The volume obtained by rotating $f(x)$ about the y-axis can be found from Eq. 10.41 (i.e., using the method of discs) by rewriting the limits and equation in terms of y, or alternatively, the *method of shells* can be used, resulting in the second form of Eq. 10.42.

$$\begin{aligned} V &= \pi \int_{y=c}^{y=d} f^2(y)dy \\ &= 2\pi \int_{x=a}^{x=b} xf(x)dx \end{aligned} \qquad 10.42$$

Example 10.10

The curve $f(x) = x^2$ over the region $x = [0, 4]$ is rotated about the x-axis. What is the volume of revolution?

Solution

The volume of revolution is

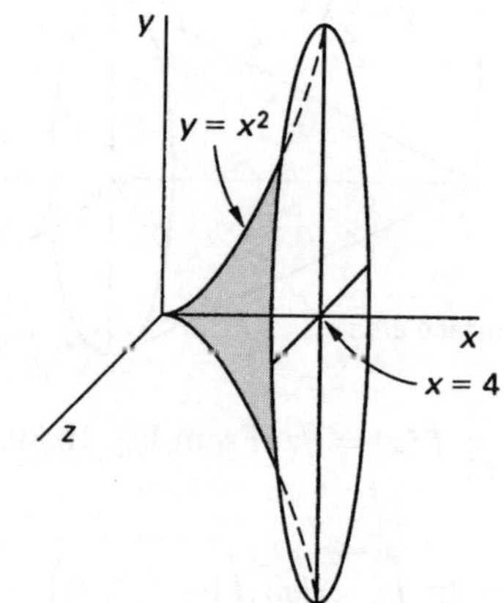

$$\begin{aligned} V &= \pi \int_a^b f^2(x)dx = \pi \int_0^4 (x^2)^2 dx = \pi \frac{x^5}{5}\Big|_0^4 \\ &= \pi\left(\frac{1024}{5} - 0\right) \\ &= 204.8\pi \end{aligned}$$

14. MOMENTS OF A FUNCTION

The *first moment of a function* is a concept used in finding centroids and centers of gravity. Equation 10.43 and Eq. 10.44 are for one- and two-dimensional problems, respectively. It is the exponent of x (1 in this case) that gives the moment its name.

$$\text{first moment} = \int xf(x)dx \qquad 10.43$$

$$\text{first moment} = \iint xf(x, y)dx\,dy \qquad 10.44$$

The *second moment of a function* is a concept used in finding moments of inertia with respect to an axis. Equation 10.45 and Eq. 10.46 are for two- and three-dimensional problems, respectively. Second moments with respect to other axes are analogous.

$$(\text{second moment})_x = \iint y^2 f(x, y)dy\,dx \qquad 10.45$$

$$(\text{second moment})_x = \iiint (y^2 + z^2)f(x, y, z)dy\,dz\,dx \qquad 10.46$$

15. FOURIER SERIES

Any periodic waveform can be written as the sum of an infinite number of sinusoidal terms, known as *harmonic terms* (i.e., an infinite series). Such a sum of terms is known as a *Fourier series*, and the process of finding the terms is *Fourier analysis.* (Extracting the original waveform from the series is known as *Fourier inversion.*) Since most series converge rapidly, it is possible to obtain a good approximation to the original waveform with a limited number of sinusoidal terms.

Fourier's theorem is Eq. 10.47.[5] The object of a Fourier analysis is to determine the coefficients a_n and b_n. The constant a_0 can often be determined by inspection since it is the average value of the waveform.

$$\begin{aligned} f(t) &= a_0 + a_1 \cos \omega t + a_2 \cos 2\omega t + \cdots \\ &\quad + b_1 \sin \omega t + b_2 \sin 2\omega t + \cdots \end{aligned} \qquad 10.47$$

ω is the *natural (fundamental) frequency* of the waveform. It depends on the actual waveform period, T.

$$\omega = \frac{2\pi}{T} \qquad 10.48$$

To simplify the analysis, the time domain can be normalized to the radian scale. The normalized scale is obtained by dividing all frequencies by ω. Then, the Fourier series becomes

$$\begin{aligned} f(t) &= a_0 + a_1 \cos t + a_2 \cos 2t + \cdots \\ &\quad + b_1 \sin t + b_2 \sin 2t + \cdots \end{aligned} \qquad 10.49$$

[5]The independent variable used in this section is t, since Fourier analysis is most frequently used in the time domain.

The coefficients a_n and b_n are found from the following relationships.[6]

$$a_0 = \frac{1}{2\pi}\int_0^{2\pi} f(t)dt$$
$$= \frac{1}{T}\int_0^{T} f(t)dt \qquad 10.50$$

$$a_n = \frac{1}{\pi}\int_0^{2\pi} f(t)\cos nt\, dt$$
$$= \frac{2}{T}\int_0^{T} f(t)\cos nt\, dt \quad [n \geq 1] \qquad 10.51$$

$$b_n = \frac{1}{\pi}\int_0^{2\pi} f(t)\sin nt\, dt$$
$$= \frac{2}{T}\int_0^{T} f(t)\sin nt\, dt \quad [n \geq 1] \qquad 10.52$$

While Eq. 10.51 and Eq. 10.52 are always valid, the work of integrating and finding a_n and b_n can be greatly simplified if the waveform is recognized as being symmetrical. Table 10.1 summarizes the simplifications.[7]

Example 10.11

Find the first four terms of a Fourier series that approximates the repetitive step function illustrated.

$$f(t) = \begin{Bmatrix} 1 & 0 < t < \pi \\ 0 & \pi < t < 2\pi \end{Bmatrix}$$

f(t)
1
π
2π
t

Solution

From Eq. 10.50,

$$a_0 = \frac{1}{2\pi}\int_0^{\pi} (1)dt + \frac{1}{2\pi}\int_{\pi}^{2\pi} (0)dt = \frac{1}{2}$$

This value of 1/2 corresponds to the average value of $f(t)$. It could have been found by observation.

$$a_1 = \frac{1}{\pi}\int_0^{\pi} (1)\cos t\, dt + \frac{1}{\pi}\int_{\pi}^{2\pi} (0)\cos t\, dt = \frac{1}{\pi}\sin t\Big|_0^{\pi} + 0$$
$$= 0$$

In general,

$$a_n = \frac{1}{\pi}\frac{\sin nt}{n}\Big|_0^{\pi} = 0$$

$$b_1 = \frac{1}{\pi}\int_0^{\pi} (1)\sin t\, dt + \frac{1}{\pi}\int_{\pi}^{2\pi} (0)\sin t\, dt$$
$$= \left(\frac{1}{\pi}\right)\left(-\cos t\Big|_0^{\pi}\right) = 2/\pi$$

Table 10.1 *Fourier Analysis Simplifications for Symmetrical Waveforms*

	even symmetry $f(-t) = f(t)$	odd symmetry $f(-t) = -f(t)$
full-wave symmetry* $f(t + 2\pi) = f(t)$	$b_n = 0$ [all n]	$a_0 = 0$ $a_n = 0$ [all n]
$\lvert A_2\rvert = \lvert A_1\rvert$ $\lvert A_{total}\rvert = \lvert A_1\rvert$ *any repeating waveform	$a_n = \frac{1}{\pi}\int_0^{2\pi} f(t)\cos nt\, dt$ [all n]	$b_n = \frac{1}{\pi}\int_0^{2\pi} f(t)\sin nt\, dt$ [all n]
half-wave symmetry* $f(t + \pi) = -f(t)$	$a_n = 0$ [even n] $b_n = 0$ [all n]	$a_0 = 0$ $a_n = 0$ [all n] $b_n = 0$ [even n]
$\lvert A_2\rvert = \lvert A_1\rvert$ $\lvert A_{total}\rvert = 2\lvert A_1\rvert$ *same as rotational symmetry	$a_n = \frac{2}{\pi}\int_0^{\pi} f(t)\cos nt\, dt$ [odd n]	$b_n = \frac{2}{\pi}\int_0^{\pi} f(t)\sin nt\, dt$ [odd n]
quarter-wave symmetry $f(t + \pi) = -f(t)$	$a_0 = 0$ $a_n = 0$ [even n] $b_n = 0$ [all n]	$a_0 = 0$ $a_n = 0$ [all n] $b_n = 0$ [even n]
$\lvert A_2\rvert = \lvert A_1\rvert$ $\lvert A_{total}\rvert = 4\lvert A_1\rvert$	$a_n = \frac{4}{\pi}\int_0^{\frac{\pi}{2}} f(t)\cos nt\, dt$ [odd n]	$b_n = \frac{4}{\pi}\int_0^{\frac{\pi}{2}} f(t)\sin nt\, dt$ [odd n]

[6]Several valid forms of the Fourier series are used. The one shown, Eq. 10.50, is consistent with that used by NCEES. Other forms remove the fraction 1/2 embedded in the a_0 term in Eq. 10.50 and instead show it specifically as $\frac{1}{2}a_0$ in an equation such as Eq. 10.47. Additionally, other forms show the interval of integration as $-\pi$ to $+\pi$. Because of the periodicity of the integrands, the interval of integration may be replaced by any other interval of length 2π, or, if the period is other than 2π, by the interval for the period T.

[7]The Fourier series of an even function is called a *Fourier cosine series.* The Fourier series of an odd function is called a *Fourier sine series.* If both the cosine and sine terms are present in a given function, the Fourier series may be written with coefficients labeled as c_n that are combinations of the a_n and b_n coefficients.

In general,

$$b_n = \left(\frac{1}{\pi}\right)\left(\frac{-\cos nt}{n}\bigg|_0^\pi\right) = \begin{cases} 0 & \text{for } n \text{ even} \\ \dfrac{2}{\pi n} & \text{for } n \text{ odd} \end{cases}$$

The series is

$$f(t) = \frac{1}{2} + \frac{2}{\pi}(\sin t + \tfrac{1}{3}\sin 3t + \tfrac{1}{5}\sin 5t + \cdots)$$

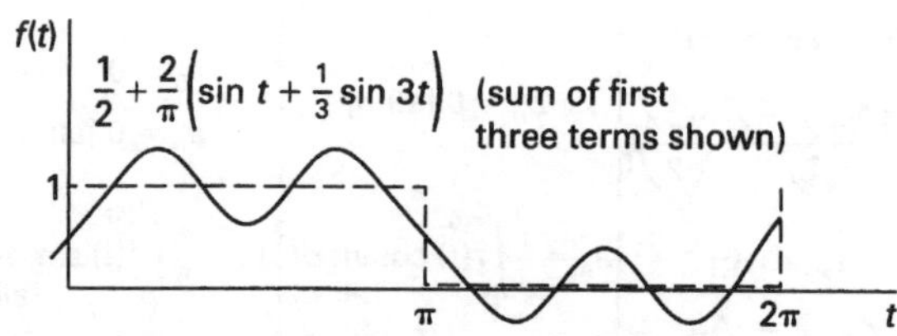

16. FAST FOURIER TRANSFORMS

Many mathematical operations are needed to implement a true Fourier transform. While the terms of a Fourier series might be slowly derived by integration, a faster method is needed to analyze real-time data. The *fast Fourier transform* (FFT) is a computer algorithm implemented in *spectrum analyzers* (*signal analyzers* or *FFT analyzers*) and replaces integration and multiplication operations with table look-ups and additions.[8]

Since the complexity of the transform is reduced, the transformation occurs more quickly, enabling efficient analysis of waveforms with little or no periodicity.[9]

Using a spectrum analyzer requires choosing the frequency band (e.g., 0–20 kHz) to be monitored. (This step automatically selects the sampling period. The lower the frequencies sampled, the longer the sampling period.) If they are not fixed by the analyzer, the numbers of time-dependent input variable samples (e.g., 1024) and frequency-dependent output variable values (e.g., 400) are chosen.[10] There are half as many frequency lines as data points because each line contains two pieces of information—real (amplitude) and imaginary (phase). The *resolution* of the resulting frequency analysis is

$$\text{resolution} = \frac{\text{frequency bandwidth}}{\text{no. of output variable values}} \qquad 10.53$$

[8]*Spectrum analysis*, also known as *frequency analysis*, *signature analysis*, and *time-series analysis*, develops a relationship (usually graphical) between some property (e.g., amplitude or phase shift) and frequency.

[9]Hours and days of manual computations are compressed into milliseconds.

[10]Two samples per time-dependent cycle (at the maximum frequency) is the lower theoretical limit for sampling, but the practical minimum rate is approximately 2.5 samples per cycle. This will ensure that *alias components* (i.e., low-level frequency signals) do not show up in the frequency band of interest.

17. INTEGRAL FUNCTIONS

Integrals that cannot be evaluated as finite combinations of elementary functions are called *integral functions*. These functions are evaluated by series expansion. Some of the more common functions are listed as follows.[11,12]

- *integral sine function:*

$$\mathrm{Si}(x) = \int_0^x \frac{\sin x}{x}\,dx = x - \frac{x^3}{3\cdot 3!} + \frac{x^5}{5\cdot 5!} - \frac{x^7}{7\cdot 7!} + \cdots \qquad 10.54$$

- *integral cosine function:*

$$\mathrm{Ci}(x) = \int_{-\infty}^x \frac{\cos x}{x}\,dx = -\int_x^\infty \frac{\cos x}{x}\,dx = C_E + \ln x - \frac{x^2}{2\cdot 2!} + \frac{x^4}{4\cdot 4!} - \cdots \qquad 10.55$$

- *integral exponential function:*

$$\mathrm{Ei}(x) = \int_{-\infty}^x \frac{e^x}{x}\,dx = -\int_{-x}^\infty \frac{e^{-x}}{x}\,dx = C_E + \ln x + x + \frac{x^2}{2\cdot 2!} + \frac{x^3}{3\cdot 3!} + \cdots \qquad 10.56$$

- *error function:*

$$\mathrm{erf}(x) = \frac{2}{\sqrt{\pi}}\int_0^x e^{-x^2}\,dx = \left(\frac{2}{\sqrt{\pi}}\right)\left(\frac{x}{1\cdot 0!} - \frac{x^3}{3\cdot 1!} + \frac{x^5}{5\cdot 2!} - \frac{x^7}{7\cdot 3!} + \cdots\right) \qquad 10.57$$

- *Fresnel integrals:*

$$C(x) = \int_0^x \cos(t^2)\,dt \qquad 10.58$$

$$S(x) = \int_0^x \sin(t^2)\,dt \qquad 10.59$$

[11]Other integral functions include the gamma function and elliptic integral.

[12]C_E in Eq. 10.55 and Eq. 10.56 is *Euler's constant*.

$$C_E = \int_{+\infty}^0 e^{-x}\ln x\,dx = \lim_{m\to\infty}\left(1 + \frac{1}{2} + \frac{1}{3} + \cdots + \frac{1}{m} - \ln m\right) = 0.577215665$$

11 Differential Equations

1. TYPES OF DIFFERENTIAL EQUATIONS

A *differential equation* is a mathematical expression combining a function (e.g., $y = f(x)$) and one or more of its derivatives. The *order* of a differential equation is the highest derivative in it. *First-order differential equations* contain only first derivatives of the function, *second-order differential equations* contain second derivatives (and may contain first derivatives as well), and so on.

A *linear differential equation* can be written as a sum of products of multipliers of the function and its derivatives. If the multipliers are scalars, the differential equation is said to have *constant coefficients.* If the function or one of its derivatives is raised to some power (other than one) or is embedded in another function (e.g., y embedded in $\sin y$ or e^y), the equation is said to be *nonlinear.*

Each term of a *homogeneous differential equation* contains either the function (y) or one of its derivatives; that is, the sum of derivative terms is equal to zero. In a *nonhomogeneous differential equation,* the sum of derivative terms is equal to a nonzero *forcing function* of the independent variable (e.g., $g(x)$). In order to solve a nonhomogeneous equation, it is often necessary to solve the homogeneous equation first. The homogeneous equation corresponding to a nonhomogeneous equation is known as a *reduced equation* or *complementary equation.*

The following examples illustrate the types of differential equations.

$y' - 7y = 0$	homogeneous, first-order linear, with constant coefficients
$y'' - 2y' + 8y = \sin 2x$	nonhomogeneous, second-order linear, with constant coefficients
$y'' - (x^2 - 1)y^2 = \sin 4x$	nonhomogeneous, second-order, nonlinear

An *auxiliary equation* (also called the *characteristic equation*) can be written for a homogeneous linear differential equation with constant coefficients, regardless of order. This auxiliary equation is simply the polynomial formed by replacing all derivatives with variables raised to the power of their respective derivatives.

The purpose of solving a differential equation is to derive an expression for the function in terms of the independent variable. The expression does not need to be explicit in the function, but there can be no derivatives in the expression. Since, in the simplest cases, solving a differential equation is equivalent to finding an indefinite integral, it is not surprising that *constants of integration* must be evaluated from knowledge of how the system behaves. Additional data are known as *initial values*, and any problem that includes them is known as an *initial value problem.*[1]

Most differential equations require lengthy solutions and are not efficiently solved by hand. However, several types are fairly simple and are presented in this chapter.

[1]The term *initial* implies that time is the independent variable. While this may explain the origin of the term, initial value problems are not limited to the time domain. A *boundary value problem* is similar, except that the data come from different points. For example, additional data in the form $y(x_0)$ and $y'(x_0)$ or $y(x_0)$ and $y'(x_1)$ that need to be simultaneously satisfied constitute an initial value problem. Data of the form $y(x_0)$ and $y(x_1)$ constitute a boundary value problem. Until solved, it is difficult to know whether a boundary value problem has zero, one, or more than one solution.

Example 11.1

Write the complementary differential equation for the following nonhomogeneous differential equation.

$$y'' + 6y' + 9y = e^{-14x} \sin 5x$$

Solution

The complementary equation is found by eliminating the forcing function, $e^{-14x} \sin 5x$.

$$y'' + 6y' + 9y = 0$$

Example 11.2

Write the auxiliary equation to the following differential equation.

$$y'' + 4y' + y = 0$$

Solution

Replacing each derivative with a polynomial term whose degree equals the original order, the auxiliary equation is

$$r^2 + 4r + 1 = 0$$

2. HOMOGENEOUS, FIRST-ORDER LINEAR DIFFERENTIAL EQUATIONS WITH CONSTANT COEFFICIENTS

A homogeneous, first-order linear differential equation with constant coefficients will have the general form of Eq. 11.1.

$$y' + ky = 0 \tag{11.1}$$

The auxiliary equation is $r + k = 0$ and it has a root of $r = -k$. Eq. 11.2 is the solution.

$$y = Ae^{rx} = Ae^{-kx} \tag{11.2}$$

If the initial condition is known to be $y(0) = y_0$, the solution is

$$y = y_0 e^{-kx} \tag{11.3}$$

3. FIRST-ORDER LINEAR DIFFERENTIAL EQUATIONS

A first-order linear differential equation has the general form of Eq. 11.4. $p(x)$ and $g(x)$ can be constants or any function of x (but not of y). However, if $p(x)$ is a constant and $g(x)$ is zero, it is easier to solve the equation as shown in Sec. 11.2.

$$y' + p(x)y = g(x) \tag{11.4}$$

The *integrating factor* (which is usually a function) to this differential equation is

$$u(x) = \exp\left(\int p(x)\,dx\right) \tag{11.5}$$

The closed-form solution to Eq. 11.4 is

$$y = \frac{1}{u(x)}\left(\int u(x)g(x)\,dx + C\right) \tag{11.6}$$

For the special case where $p(x)$ and $g(x)$ are both constants, Eq. 11.4 becomes

$$y' + ay = b \tag{11.7}$$

If the initial condition is $y(0) = y_0$, then the solution to Eq. 11.7 is

$$y = \frac{b}{a}(1 - e^{-ax}) + y_0 e^{-ax} \tag{11.8}$$

Example 11.3

Find a solution to the following differential equation.

$$y' - y = 2xe^{2x} \quad y(0) = 1$$

Solution

This is a first-order linear equation with $p(x) = -1$ and $g(x) = 2xe^{2x}$. The integrating factor is

$$u(x) = \exp\left(\int p(x)\,dx\right) = \exp\left(\int -1\,dx\right) = e^{-x}$$

The solution is given by Eq. 11.6.

$$\begin{aligned} y &= \frac{1}{u(x)}\left(\int u(x)g(x)\,dx + C\right) \\ &= \frac{1}{e^{-x}}\left(\int e^{-x}2xe^{2x}\,dx + C\right) \\ &= e^{x}\left(2\int xe^{x}\,dx + C\right) \\ &= e^{x}(2xe^{x} - 2e^{x} + C) \\ &= e^{x}(2e^{x}(x-1) + C) \end{aligned}$$

From the initial condition,

$$\begin{aligned} y(0) &= 1 \\ e^0((2)(e^0)(0-1) + C) &= 1 \\ 1((2)(1)(-1) + C) &= 1 \end{aligned}$$

Therefore, $C = 3$. The complete solution is

$$y = e^{x}(2e^{x}(x-1) + 3)$$

4. FIRST-ORDER SEPARABLE DIFFERENTIAL EQUATIONS

First-order separable differential equations can be placed in the form of Eq. 11.9. For clarity and convenience, y' is written as dy/dx.

$$m(x) + n(y)\frac{dy}{dx} = 0 \qquad 11.9$$

Equation 11.9 can be placed in the form of Eq. 11.10, both sides of which are easily integrated. An initial value will establish the constant of integration.

$$m(x)dx = -n(y)dy \qquad 11.10$$

5. FIRST-ORDER EXACT DIFFERENTIAL EQUATIONS

A *first-order exact differential equation* has the form

$$f_x(x, y) + f_y(x, y)y' = 0 \qquad 11.11$$

Notice that $f_x(x, y)$ is the exact derivative of $f(x, y)$ with respect to x, and $f_y(x, y)$ is the exact derivative of $f(x, y)$ with respect to y. The solution is

$$f(x, y) - C = 0 \qquad 11.12$$

6. HOMOGENEOUS, SECOND-ORDER LINEAR DIFFERENTIAL EQUATIONS WITH CONSTANT COEFFICIENTS

Homogeneous second-order linear differential equations with constant coefficients have the form of Eq. 11.13. They are most easily solved by finding the two roots of the auxiliary equation. (See Eq. 11.14.)

$$y'' + k_1y' + k_2y = 0 \qquad 11.13$$

$$r^2 + k_1r + k_2 = 0 \qquad 11.14$$

There are three cases. If the two roots of Eq. 11.14 are real and different, the solution is

$$y = A_1e^{r_1x} + A_2e^{r_2x} \qquad 11.15$$

If the two roots are real and the same, the solution is

$$y = A_1e^{rx} + A_2xe^{rx} \qquad 11.16$$

$$r = \frac{-k_1}{2} \qquad 11.17$$

If the two roots are imaginary, they will be of the form $(\alpha + i\omega)$ and $(\alpha - i\omega)$, and the solution is

$$y = A_1e^{\alpha x}\cos\omega x + A_2e^{\alpha x}\sin\omega x \qquad 11.18$$

In all three cases, A_1 and A_2 must be found from the two initial conditions.

Example 11.4

Solve the following differential equation.

$$y'' + 6y' + 9y = 0$$

$$y(0) = 0 \qquad y'(0) = 1$$

Solution

The auxiliary equation is

$$r^2 + 6r + 9 = 0$$

$$(r + 3)(r + 3) = 0$$

The roots to the auxiliary equation are $r_1 = r_2 = -3$. Therefore, the solution has the form of Eq. 11.16.

$$y = A_1e^{-3x} + A_2xe^{-3x}$$

The first initial condition is

$$y(0) = 0$$

$$A_1e^0 + A_2(0)e^0 = 0$$

$$A_1 + 0 = 0$$

$$A_1 = 0$$

To use the second initial condition, the derivative of the equation is needed. Making use of the known fact that $A_1 = 0$,

$$y' = \frac{d}{dx}A_2xe^{-3x} = -3A_2xe^{-3x} + A_2e^{-3x}$$

Using the second initial condition,

$$y'(0) = 1$$

$$-3A_2(0)e^0 + A_2e^0 = 1$$

$$0 + A_2 = 1$$

$$A_2 = 1$$

The solution is

$$y = xe^{-3x}$$

7. NONHOMOGENEOUS DIFFERENTIAL EQUATIONS

A nonhomogeneous equation has the form of Eq. 11.19. $f(x)$ is known as the *forcing function.*

$$y'' + p(x)y' + q(x)y = f(x) \qquad 11.19$$

The solution to Eq. 11.19 is the sum of two equations. The *complementary solution*, y_c, solves the complementary (i.e., homogeneous) problem. The *particular solution*, y_p, is any specific solution to the nonhomogeneous Eq. 11.19 that is known or can be found. Initial values are used to evaluate any unknown coefficients in the complementary solution *after* y_c and y_p have been combined. (The particular solution will not have any unknown coefficients.)

$$y = y_c + y_p \tag{11.20}$$

Two methods are available for finding a particular solution. The *method of undetermined coefficients*, as presented here, can be used only when $p(x)$ and $q(x)$ are constant coefficients and $f(x)$ takes on one of the forms in Table 11.1.

The particular solution can be read from Table 11.1 if the forcing function is of one of the forms given. Of course, the coefficients A_i and B_i are not known—these are the *undetermined coefficients.* The exponent s is the smallest nonnegative number (and will be 0, 1, or 2), which ensures that no term in the particular solution, y_p, is also a solution to the complementary equation, y_c. s must be determined prior to proceeding with the solution procedure.

Once y_p (including s) is known, it is differentiated to obtain y'_p and y''_p, and all three functions are substituted into the original nonhomogeneous equation. The resulting equation is rearranged to match the forcing function, $f(x)$, and the unknown coefficients are determined, usually by solving simultaneous equations.

If the forcing function, $f(x)$, is more complex than the forms shown in Table 11.1, or if either $p(x)$ or $q(x)$ is a function of x, the method of *variation of parameters* should be used. This complex and time-consuming method is not covered in this book.

Table 11.1 *Particular Solutions**

form of $f(x)$	form of y_p
$P_n(x) = a_0x^n + a_1x^{n-1} + \cdots + a_n$	$x^s\begin{pmatrix} A_0x^n + A_1x^{n-1} \\ + \cdots + A_n \end{pmatrix}$
$P_n(x)e^{\alpha x}$	$x^s\begin{pmatrix} A_0x^n + A_1x^{n-1} \\ + \cdots + A_n \end{pmatrix}e^{\alpha x}$
$P_n(x)e^{\alpha x}\begin{Bmatrix} \sin \omega x \\ \cos \omega x \end{Bmatrix}$	$x^s\begin{pmatrix} \begin{pmatrix} A_0x^n + A_1x^{n-1} \\ + \cdots + A_n \end{pmatrix} \\ \times e^{\alpha x}\cos \omega x \\ + \begin{pmatrix} B_0x^n + B_1x^{n-1} \\ + \cdots + B_n \end{pmatrix} \\ \times e^{\alpha x}\sin \omega x \end{pmatrix}$

*$P_n(x)$ is a polynomial of degree n.

Example 11.5

Solve the following nonhomogeneous differential equation.

$$y'' + 2y' + y = e^x \cos x$$

Solution

step 1: Find the solution to the complementary (homogeneous) differential equation.

$$y'' + 2y' + y = 0$$

Since this is a differential equation with constant coefficients, write the auxiliary equation.

$$r^2 + 2r + 1 = 0$$

The auxiliary equation factors in $(r + 1)^2 = 0$ with two identical roots at $r = -1$. Therefore, the solution to the homogeneous differential equation is

$$y_c(x) = C_1e^{-x} + C_2xe^{-x}$$

step 2: Use Table 11.1 to determine the form of a particular solution. Since the forcing function has the form $P_n(x)e^{\alpha x}\cos \omega x$ with $P_n(x) = 1$ (equivalent to $n = 0$), $\alpha = 1$, and $\omega = 1$, the particular solution has the form

$$y_p(x) = x^s(Ae^x \cos x + Be^x \sin x)$$

step 3: Determine the value of s. Check to see if any of the terms in $y_p(x)$ will themselves solve the homogeneous equation. Try $Ae^x \cos x$ first.

$$\frac{d}{dx}(Ae^x \cos x) = Ae^x \cos x - Ae^x \sin x$$

$$\frac{d^2}{dx^2}(Ae^x \cos x) = -2Ae^x \sin x$$

Substitute these quantities into the homogeneous equation.

$$y'' + 2y' + y = 0$$

$$-2Ae^x \sin x + 2Ae^x \cos x$$

$$-2Ae^x \sin x + Ae^x \cos x = 0$$

$$3Ae^x \cos x - 4Ae^x \sin x = 0$$

Disregarding the trivial (i.e., $A = 0$) solution, $Ae^x \cos x$ does not solve the homogeneous equation.

Next, try $Be^x \sin x$.

$$\frac{d}{dx}(Be^x \sin x) = Be^x \cos x + Be^x \sin x$$

$$\frac{d^2}{dx^2}(Be^x \sin x) = 2Be^x \cos x$$

Substitute these quantities into the homogeneous equation.

$$y'' + 2y' + y = 0$$

$$\begin{aligned} 2Be^x \cos x + 2Be^x \cos x & \\ + 2Be^x \sin x + Be^x \sin x &= 0 \\ 3Be^x \sin x + 4Be^x \cos x &= 0 \end{aligned}$$

Disregarding the trivial ($B = 0$) case, $Be^x \sin x$ does not solve the homogeneous equation.

Since none of the terms in $y_p(x)$ solve the homogeneous equation, $s = 0$, and a particular solution has the form

$$y_p(x) = Ae^x \cos x + Be^x \sin x$$

step 4: Use the method of unknown coefficients to determine A and B in the particular solution. Drawing on the previous steps, substitute the quantities derived from the particular solution into the nonhomogeneous equation.

$$y'' + 2y' + y = e^x \cos x$$

$$\begin{aligned} -2Ae^x \sin x + 2Be^x \cos x & \\ + 2Ae^x \cos x - 2Ae^x \sin x & \\ + 2Be^x \cos x + 2Be^x \sin x & \\ + Ae^x \cos x + Be^x \sin x &= e^x \cos x \end{aligned}$$

Combining terms,

$$\begin{aligned} (-4A + 3B)e^x \sin x + (3A + 4B)e^x \cos x & \\ = e^x \cos x & \end{aligned}$$

Equating the coefficients of like terms on either side of the equal sign results in the following simultaneous equations.

$$-4A + 3B = 0$$

$$3A + 4B = 1$$

The solution to these equations is

$$A = \frac{3}{25}$$

$$B = \frac{4}{25}$$

A particular solution is

$$y_p(x) = \left(\tfrac{3}{25}\right)(e^x \cos x) + \left(\tfrac{4}{25}\right)(e^x \sin x)$$

step 5: Write the general solution.

$$\begin{aligned} y(x) &= y_c(x) + y_p(x) \\ &= C_1 e^{-x} + C_2 x e^{-x} + \left(\tfrac{3}{25}\right)(e^x \cos x) \\ &\quad + \left(\tfrac{4}{25}\right)(e^x \sin x) \end{aligned}$$

The values of C_1 and C_2 would be determined at this time if initial conditions were known.

8. NAMED DIFFERENTIAL EQUATIONS

Some differential equations with specific forms are named after the individuals who developed solution techniques for them.

- *Bessel equation of order ν*

$$x^2 y'' + xy' + (x^2 - \nu^2)y = 0 \qquad 11.21$$

- *Cauchy equation*

$$\begin{aligned} a_0 x^n \frac{d^n y}{dx^n} + a_1 x^{n-1} \frac{d^{n-1} y}{dx^{n-1}} + \cdots & \\ + a_{n-1} x \frac{dy}{dx} + a_n y = f(x) & \qquad 11.22 \end{aligned}$$

- *Euler equation*

$$x^2 y'' + \alpha x y' + \beta y = 0 \qquad 11.23$$

- *Gauss' hypergeometric equation*

$$x(1 - x)y'' + (c - (a + b + 1)x)y' - aby = 0 \qquad 11.24$$

- *Legendre equation of order λ*

$$(1 - x^2)y'' - 2xy' + \lambda(\lambda + 1)y = 0 \quad [-1 < x < 1] \qquad 11.25$$

9. LAPLACE TRANSFORMS

Traditional methods of solving nonhomogeneous differential equations by hand are usually difficult and/or time consuming. *Laplace transforms* can be used to reduce many solution procedures to simple algebra.

Every mathematical function, $f(t)$, for which Eq. 11.26 exists has a Laplace transform, written as $\mathcal{L}(f)$ or $F(s)$. The transform is written in the s-domain, regardless of the independent variable in the original function.[2] (The variable s is equivalent to a derivative operator, although it may be handled in the equations as a simple

[2]It is traditional to write the original function as a function of the independent variable t rather than x. However, Laplace transforms are not limited to functions of time.

variable.) Equation 11.26 converts a function into a Laplace transform.

$$\mathcal{L}(f(t)) = F(s) = \int_0^{\infty} e^{-st} f(t)\,dt \qquad 11.26$$

Equation 11.26 is not often needed because tables of transforms are readily available. (Appendix 11.A contains some of the most common transforms.)

Extracting a function from its transform is the *inverse Laplace transform* operation. Although other methods exist, this operation is almost always done by finding the transform in a set of tables.[3]

$$f(t) = \mathcal{L}^{-1}(F(s)) \qquad 11.27$$

Example 11.6

Find the Laplace transform of the following function.

$$f(t) = e^{at} \qquad [s > a]$$

Solution

Applying Eq. 11.26,

$$\mathcal{L}(e^{at}) = \int_0^{\infty} e^{-st} e^{at}\,dt = \int_0^{\infty} e^{-(s-a)t}\,dt = -\frac{e^{-(s-a)t}}{s-a}\bigg|_0^{\infty}$$

$$= \frac{1}{s-a} \qquad [s > a]$$

10. STEP AND IMPULSE FUNCTIONS

Many forcing functions are sinusoidal or exponential in nature; others, however, can only be represented by a step or impulse function. A *unit step function*, u_t, is a function describing the disturbance of magnitude 1 that is not present before time t but is suddenly there after time t. A step of magnitude 5 at time $t = 3$ would be represented as $5u_3$. (The notation $5u(t-3)$ is used in some books.)

The *unit impulse function*, δ_t, is a function describing a disturbance of magnitude 1 that is applied and removed so quickly as to be instantaneous. An impulse of magnitude 5 at time 3 would be represented by $5\delta_3$. (The notation $5\delta(t-3)$ is used in some books.)

Example 11.7

What is the notation for a forcing function of magnitude 6 that is applied at $t = 2$ and that is completely removed at $t = 7$?

Solution

The notation is $f(t) = 6(u_2 - u_7)$.

[3]Other methods include integration in the complex plane, convolution, and simplification by partial fractions.

Example 11.8

Find the Laplace transform of u_0, a unit step at $t = 0$.

$$f(t) = 0 \text{ for } t < 0$$

$$f(t) = 1 \text{ for } t \geq 0$$

Solution

Since the Laplace transform is an integral that starts at $t = 0$, the value of $f(t)$ prior to $t = 0$ is irrelevant.

$$\mathcal{L}(u_0) = \int_0^{\infty} e^{-st}(1)\,dt = -\frac{e^{-st}}{s}\bigg|_0^{\infty} = 0 - \frac{-1}{s}$$

$$= \frac{1}{s}$$

11. ALGEBRA OF LAPLACE TRANSFORMS

Equations containing Laplace transforms can be simplified by applying the following principles.

- *linearity theorem* (c is a constant.)

$$\mathcal{L}(cf(t)) = c\mathcal{L}(f(t)) = cF(s) \qquad 11.28$$

- *superposition theorem* ($f(t)$ and $g(t)$ are different functions.)

$$\mathcal{L}(f(t) \pm g(t)) = \mathcal{L}(f(t)) \pm \mathcal{L}(g(t))$$
$$= F(s) \pm G(s) \qquad 11.29$$

- *time-shifting theorem* (*delay theorem*)

$$\mathcal{L}(f(t-b)u_b) = e^{-bs}F(s) \qquad 11.30$$

- *Laplace transform of a derivative*

$$\mathcal{L}(f^n(t)) = -f^{n-1}(0) - sf^{n-2}(0) - \cdots$$
$$- s^{n-1}f(0) + s^n F(s) \qquad 11.31$$

- *other properties*

$$\mathcal{L}\left(\int_0^t f(u)\,du\right) = \frac{1}{s}F(s) \qquad 11.32$$

$$\mathcal{L}(tf(t)) = -\frac{dF}{ds} \qquad 11.33$$

$$\mathcal{L}\left(\frac{1}{t}f(t)\right) = \int_0^{\infty} F(u)\,du \qquad 11.34$$

12. CONVOLUTION INTEGRAL

A complex Laplace transform, $F(s)$, will often be recognized as the product of two other transforms, $F_1(s)$ and $F_2(s)$, whose corresponding functions $f_1(t)$ and $f_2(t)$ are known. Unfortunately, Laplace transforms cannot be computed with ordinary multiplication. That is, $f(t) \neq f_1(t)f_2(t)$ even though $F(s) = F_1(s)F_2(s)$.

However, it is possible to extract $f(t)$ from its *convolution*, $h(t)$, as calculated from either of the *convolution integrals* in Eq. 11.35. This process is demonstrated in Ex. 11.9. χ is a dummy variable.

$$\begin{aligned} f(t) &= \mathcal{L}^{-1}(F_1(s)F_2(s)) \\ &= \int_0^t f_1(t-\chi)f_2(\chi)d\chi \\ &= \int_0^t f_1(\chi)f_2(t-\chi)d\chi \end{aligned} \qquad 11.35$$

Example 11.9

Use the convolution integral to find the inverse transform of

$$F(s) = \frac{3}{s^2(s^2+9)}$$

Solution

$F(s)$ can be factored as

$$F_1(s)F_2(s) = \left(\frac{1}{s^2}\right)\left(\frac{3}{s^2+9}\right)$$

As the inverse transforms of $F_1(s)$ and $F_2(s)$ are $f_1(t) = t$ and $f_2(t) = \sin 3t$, respectively, the convolution integral from Eq. 11.35 is

$$\begin{aligned} f(t) &= \int_0^t (t-\chi)\sin 3\chi d\chi \\ &= \int_0^t (t\sin 3\chi - \chi \sin 3\chi)d\chi \\ &= t\int_0^t \sin 3\chi d\chi - \int_0^t \chi \sin 3\chi d\chi \end{aligned}$$

Expand using integration by parts.

$$\begin{aligned} f(t) &= -\tfrac{1}{3}t\cos 3\chi + \tfrac{1}{3}\chi\cos 3\chi - \tfrac{1}{9}\sin 3\chi\Big|_0^t \\ &= \frac{3t - \sin 3t}{9} \end{aligned}$$

13. USING LAPLACE TRANSFORMS

Any nonhomogeneous linear differential equation with constant coefficients can be solved with the following procedure, which reduces the solution to simple algebra. A complete table of transforms simplifies or eliminates step 5.

step 1: Put the differential equation in standard form (i.e., isolate the y'' term).

$$y'' + k_1y' + k_2y = f(t) \qquad 11.36$$

step 2: Take the Laplace transform of both sides. Use the linearity and superposition theorems. (See Eq. 11.28 and Eq. 11.29.)

$$\mathcal{L}(y'') + k_1\mathcal{L}(y') + k_2\mathcal{L}(y) = \mathcal{L}(f(t)) \qquad 11.37$$

step 3: Use Eq. 11.38 and Eq. 11.39 to expand the equation. (These are specific forms of Eq. 11.31.) Use a table to evaluate the transform of the forcing function.

$$\mathcal{L}(y'') = s^2\mathcal{L}(y) - sy(0) - y'(0) \qquad 11.38$$

$$\mathcal{L}(y') = s\mathcal{L}(y) - y(0) \qquad 11.39$$

step 4: Use algebra to solve for $\mathcal{L}(y)$.

step 5: If needed, use partial fractions to simplify the expression for $\mathcal{L}(y)$.

step 6: Take the inverse transform to find $y(t)$.

$$y(t) = \mathcal{L}^{-1}(\mathcal{L}(y)) \qquad 11.40$$

Example 11.10

Find $y(t)$ for the following differential equation.

$$y'' + 2y' + 2y = \cos t$$

$$y(0) = 1 \qquad y'(0) = 0$$

Solution

step 1: The equation is already in standard form.

step 2: $\mathcal{L}(y'') + 2\mathcal{L}(y') + 2\mathcal{L}(y) = \mathcal{L}(\cos t)$

step 3: Use Eq. 11.38 and Eq. 11.39. Use App. 11.A to find the transform of $\cos t$.

$$s^2\mathcal{L}(y) - sy(0) - y'(0) + 2s\mathcal{L}(y) - 2y(0) + 2\mathcal{L}(y) = \frac{s}{s^2+1}$$

But, $y(0) = 1$ and $y'(0) = 0$.

$$s^2\mathcal{L}(y) - s + 2s\mathcal{L}(y) - 2 + 2\mathcal{L}(y) = \frac{s}{s^2+1}$$

step 4: Combine terms and solve for $\mathcal{L}(y)$.

$$\mathcal{L}(y)(s^2 + 2s + 2) - s - 2 = \frac{s}{s^2+1}$$

$$\begin{aligned} \mathcal{L}(y) &= \frac{\frac{s}{s^2+1} + s + 2}{s^2 + 2s + 2} \\ &= \frac{s^3 + 2s^2 + 2s + 2}{(s^2+1)(s^2+2s+2)} \end{aligned}$$

step 5: Expand the expression for $\mathcal{L}(y)$ by partial fractions.

$$\begin{aligned}\mathcal{L}(y) &= \frac{s^3+2s^2+2s+2}{(s^2+1)(s^2+2s+2)} \\ &= \frac{A_1 s+B_1}{s^2+1}+\frac{A_2 s+B_2}{s^2+2s+2} \\ &= \frac{\begin{array}{c}s^3(A_1+A_2)+s^2(2A_1+B_1+B_2)\\ +\,s(2A_1+2B_1+A_2)+(2B_1+B_2)\end{array}}{(s^2+1)(s^2+2s+2)}\end{aligned}$$

The following simultaneous equations result.

$$\begin{aligned} A_1 + A_2 &= 1 \\ 2A_1 + B_1 + B_2 &= 2 \\ 2A_1 + A_2 + 2B_1 &= 2 \\ 2B_1 + B_2 &= 2 \end{aligned}$$

These equations have the solutions $A_1 = 1/5$, $A_2 = 4/5$, $B_1 = 2/5$, and $B_2 = 6/5$.

step 6: Refer to App. 11.A and take the inverse transforms. (The numerator of the second term is rewritten from $(4s + 6)$ to $((4s + 4) + 2)$.)

$$\begin{aligned} y &= \mathcal{L}^{-1}\Big(\mathcal{L}(y)\Big) \\ &= \mathcal{L}^{-1}\left(\frac{\left(\frac{1}{5}\right)(s+2)}{s^2+1}+\frac{\left(\frac{1}{5}\right)(4s+6)}{s^2+2s+2}\right) \\ &= \left(\tfrac{1}{5}\right)\left(\begin{array}{l}\mathcal{L}^{-1}\left(\dfrac{s}{s^2+1}\right)+2\mathcal{L}^{-1}\left(\dfrac{1}{s^2+1}\right) \\ \quad +4\mathcal{L}^{-1}\left(\dfrac{s-(-1)}{\left(s-(-1)\right)^2+1}\right) \\ \quad +2\mathcal{L}^{-1}\left(\dfrac{1}{\left(s-(-1)\right)^2+1}\right)\end{array}\right) \\ &= \left(\tfrac{1}{5}\right)\left(\cos t + 2\sin t + 4e^{-t}\cos t + 2e^{-t}\sin t\right) \end{aligned}$$

14. THIRD- AND HIGHER-ORDER LINEAR DIFFERENTIAL EQUATIONS WITH CONSTANT COEFFICIENTS

The solutions of third- and higher-order linear differential equations with constant coefficients are extensions of the solutions for second-order equations of this type. Specifically, if an equation is homogeneous, the auxiliary equation is written and its roots are found. If the equation is nonhomogeneous, Laplace transforms can be used to simplify the solution.

Consider the following homogeneous differential equation with constant coefficients.

$$y^n + k_1 y^{n-1} + \cdots + k_{n-1}y' + k_n y = 0 \qquad 11.41$$

The auxiliary equation to Eq. 11.41 is

$$r^n + k_1 r^{n-1} + \cdots + k_{n-1}r + k_n = 0 \qquad 11.42$$

For each real and distinct root r, the solution contains the term

$$y = Ae^{rx} \qquad 11.43$$

For each real root r that repeats m times, the solution contains the term

$$y = (A_1 + A_2 x + A_3 x^2 + \cdots + A_m x^{m-1})e^{rx} \qquad 11.44$$

For each pair of complex roots of the form $r = \alpha \pm i\omega$ the solution contains the terms

$$y = e^{\alpha x}(A_1 \sin \omega x + A_2 \cos \omega x) \qquad 11.45$$

15. APPLICATION: ENGINEERING SYSTEMS

There is a wide variety of engineering systems (mechanical, electrical, fluid flow, heat transfer, and so on) whose behavior is described by linear differential equations with constant coefficients.

16. APPLICATION: MIXING

A typical mixing problem involves a liquid-filled tank. The liquid may initially be pure or contain some solute. Liquid (either pure or as a solution) enters the tank at a known rate. A drain may be present to remove thoroughly mixed liquid. The concentration of the solution (or, equivalently, the amount of solute in the tank) at some given time is generally unknown. (See Fig. 11.1.)

Figure 11.1 *Fluid Mixture Problem*

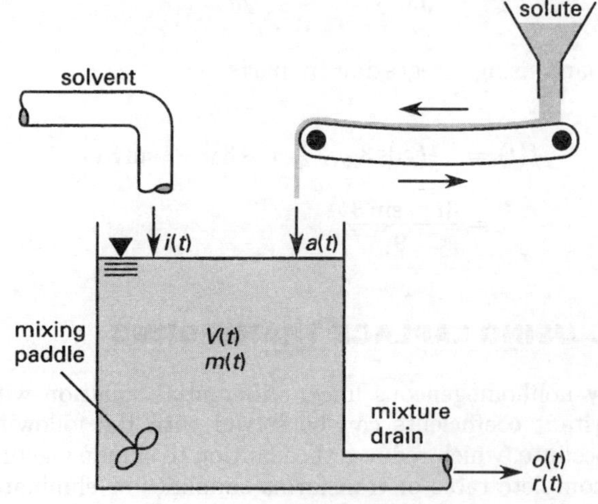

If $m(t)$ is the mass of solute in the tank at time t, the rate of solute change will be $m'(t)$. If the solute is being added at the rate of $a(t)$ and being removed at the rate of $r(t)$, the rate of change is

$$\begin{aligned} m'(t) &= \text{rate of addition} - \text{rate of removal} \\ &= a(t) - r(t) \end{aligned} \qquad 11.46$$

The rate of solute addition $a(t)$ must be known and, in fact, may be constant or zero. However, $r(t)$ depends on the concentration, $c(t)$, of the mixture and volumetric flow rates at time t. If $o(t)$ is the volumetric flow rate out of the tank, then

$$r(t) = c(t)o(t) \qquad 11.47$$

However, the concentration depends on the mass of solute in the tank at time t. Recognizing that the volume, $V(t)$, of the liquid in the tank may be changing with time,

$$c(t) = \frac{m(t)}{V(t)} \qquad 11.48$$

The differential equation describing this problem is

$$m'(t) = a(t) - \frac{m(t)o(t)}{V(t)} \qquad 11.49$$

Example 11.11

A tank contains 100 gal of pure water at the beginning of an experiment. Pure water flows into the tank at a rate of 1 gal/min. Brine containing 1/4 lbm of salt per gallon enters the tank from a second source at a rate of 1 gal/min. A perfectly mixed solution drains from the tank at a rate of 2 gal/min. How much salt is in the tank 8 min after the experiment begins?

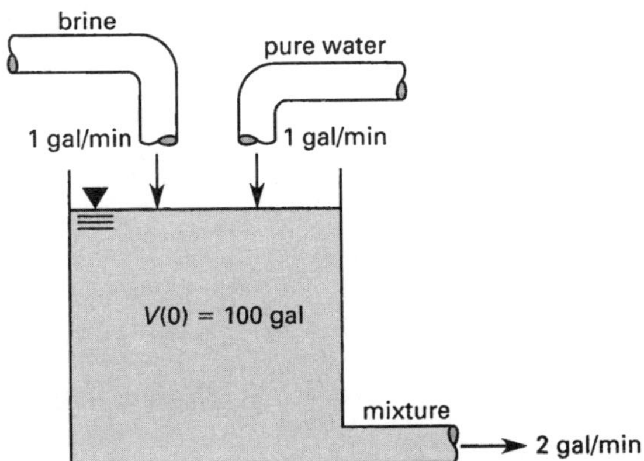

Solution

Let $m(t)$ represent the mass of salt in the tank at time t. 0.25 lbm of salt enters the tank per minute (that is, $a(t)$ = 0.25 lbm/min). The salt removal rate depends on the concentration in the tank. That is,

$$\begin{aligned} r(t) = o(t)c(t) &= \left(2\ \frac{\text{gal}}{\text{min}}\right)\left(\frac{m(t)}{100\ \text{gal}}\right) \\ &= \left(0.02\ \frac{1}{\text{min}}\right)m(t) \end{aligned}$$

From Eq. 11.46, the rate of change of salt in the tank is

$$\begin{aligned} m'(t) &= a(t) - r(t) \\ &= 0.25\ \frac{\text{lbm}}{\text{min}} - \left(0.02\ \frac{1}{\text{min}}\right)m(t) \end{aligned}$$

$$m'(t) + \left(0.02\ \frac{1}{\text{min}}\right)m(t) = 0.25\ \text{lbm/min}$$

This is a first-order linear differential equation of the form of Eq. 11.7. Since the initial condition is $m(0) = 0$, the solution is

$$\begin{aligned} m(t) &= \left(\frac{0.25\ \frac{\text{lbm}}{\text{min}}}{0.02\ \frac{1}{\text{min}}}\right)\left(1 - e^{-\left(0.02\ \frac{1}{\text{min}}\right)t}\right) \\ &= (12.5\ \text{lbm})\left(1 - e^{-\left(0.01\ \frac{1}{\text{min}}\right)t}\right) \end{aligned}$$

At $t = 8$,

$$\begin{aligned} m(t) &= (12.5\ \text{lbm})\left(1 - e^{-\left(0.02\ \frac{1}{\text{min}}\right)(8\ \text{min})}\right) \\ &= (12.5\ \text{lbm})(1 - 0.852) \\ &= 1.85\ \text{lbm} \end{aligned}$$

17. APPLICATION: EXPONENTIAL GROWTH AND DECAY

Equation 11.50 describes the behavior of a substance whose quantity, $m(t)$, changes at a rate proportional to the quantity present. The constant of proportionality, k, will be negative for decay (e.g., radioactive decay) and positive for growth (e.g., compound interest).

$$m'(t) = km(t) \qquad 11.50$$

$$m'(t) - km(t) = 0 \qquad 11.51$$

If the initial quantity of substance is $m(0) = m_0$, then Eq. 11.51 has the solution

$$m(t) = m_0 e^{kt} \qquad 11.52$$

If $m(t)$ is known for some time t, the constant of proportionality is

$$k = \frac{1}{t}\ln\left(\frac{m(t)}{m_0}\right) \qquad 11.53$$

For the case of a decay, the *half-life*, $t_{1/2}$, is the time at which only half of the substance remains. The relationship between k and $t_{1/2}$ is

$$kt_{1/2} = \ln\left(\tfrac{1}{2}\right) = -0.693 \qquad 11.54$$

18. APPLICATION: EPIDEMICS

During an epidemic in a population of n people, the density of sick (contaminated, contagious, affected, etc.) individuals is $\rho_s(t) = s(t)/n$, where $s(t)$ is the number of sick individuals at a given time, t. Similarly, the density of well (uncontaminated, unaffected, susceptible, etc.) individuals is $\rho_w(t) = w(t)/n$, where $w(t)$ is the number of well individuals. Assuming there is no quarantine, the population size is constant, individuals move about freely, and sickness does not limit the activities of individuals, the rate of contagion, $\rho_s'(t)$, will be $k\rho_s(t)\rho_w(t)$, where k is a proportionality constant.

$$\rho_s'(t) = k\rho_s(t)\rho_w(t) = k\rho_s(t)(1-\rho_s(t)) \qquad 11.55$$

This is a separable differential equation that has the solution

$$\rho_s(t) = \frac{\rho_s(0)}{\rho_s(0) + (1-\rho_s(0))e^{-kt}} \qquad 11.56$$

19. APPLICATION: SURFACE TEMPERATURE

Newton's law of cooling states that the surface temperature, T, of a cooling object changes at a rate proportional to the difference between the surface and ambient temperatures. The constant k is a positive number.

$$T'(t) = -k(T(t) - T_{\text{ambient}}) \quad [k>0] \qquad 11.57$$

$$T'(t) + kT(t) - kT_{\text{ambient}} = 0 \quad [k>0] \qquad 11.58$$

This first-order linear differential equation with constant coefficients has the following solution (from Eq. 11.8).

$$T(t) = T_{\text{ambient}} + (T(0) - T_{\text{ambient}})e^{-kt} \qquad 11.59$$

If the temperature is known at some time t, the constant k can be found from Eq. 11.60.

$$k = \frac{-1}{t}\ln\left(\frac{T(t) - T_{\text{ambient}}}{T(0) - T_{\text{ambient}}}\right) \qquad 11.60$$

20. APPLICATION: EVAPORATION

The mass of liquid evaporated from a liquid surface is proportional to the exposed surface area. Since quantity, mass, and remaining volume are all proportional, the differential equation is

$$\frac{dV}{dt} = -kA \qquad 11.61$$

For a spherical drop of radius r, Eq. 11.61 reduces to

$$\frac{dr}{dt} = -k \qquad 11.62$$

$$r(t) = r(0) - kt \qquad 11.63$$

For a cube with sides of length s, Eq. 11.61 reduces to

$$\frac{ds}{dt} = -2k \qquad 11.64$$

$$s(t) = s(0) - 2kt \qquad 11.65$$

12 Probability and Statistical Analysis of Data

1. SET THEORY

A *set* (usually designated by a capital letter) is a population or collection of individual items known as *elements* or *members*. The *null set*, $\emptyset$, is empty (i.e., contains no members). If A and B are two sets, A is a *subset* of B if every member in A is also in B. A is a *proper subset* of B if B consists of more than the elements in A. These relationships are denoted as follows.

$$A \subseteq B \quad \text{[subset]}$$

$$A \subset B \quad \text{[proper subset]}$$

The *universal set*, U, is one from which other sets draw their members. If A is a subset of U, then A' (also designated as A^{-1}, $\tilde{A}$, $-A$, and $\overline{A}$) is the *complement* of A and consists of all elements in U that are not in A. This is illustrated in a *Venn diagram* in Fig. 12.1(a).

The *union of two sets*, denoted by $A \cup B$ and shown in Fig. 12.1(b), is the set of all elements that are either in A or B or both. The *intersection of two sets*, denoted by $A \cap B$ and shown in Fig. 12.1(c), is the set of all elements that belong to both A and B. If $A \cap B = \emptyset$, A and B are said to be *disjoint sets.*

Figure 12.1 *Venn Diagrams*

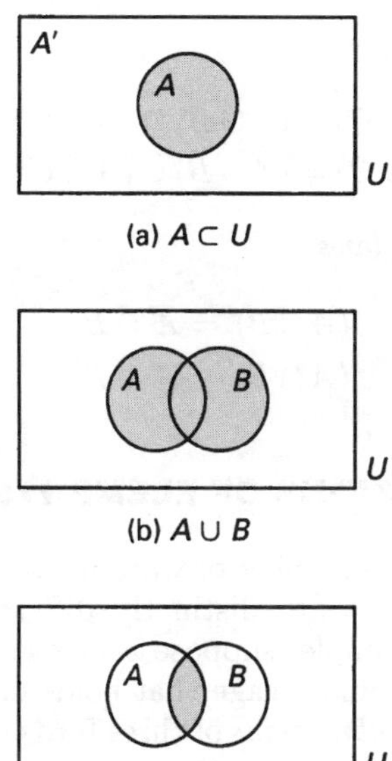

If A, B, and C are subsets of the universal set, the following laws apply.

- *identity laws*

$$A \cup \emptyset = A \qquad 12.1$$

$$A \cup U = U \qquad 12.2$$

$$A \cap \emptyset = \emptyset \qquad 12.3$$

$$A \cap U = A \qquad 12.4$$

- *idempotent laws*

$$A \cup A = A \qquad 12.5$$

$$A \cap A = A \qquad 12.6$$

- *complement laws*

$$A \cup A' = U \qquad 12.7$$
$$(A')' = A \qquad 12.8$$
$$A \cap A' = \emptyset \qquad 12.9$$
$$U' = \emptyset \qquad 12.10$$

- *commutative laws*

$$A \cup B = B \cup A \qquad 12.11$$
$$A \cap B = B \cap A \qquad 12.12$$

- *associative laws*

$$(A \cup B) \cup C = A \cup (B \cup C) \qquad 12.13$$
$$(A \cap B) \cap C = A \cap (B \cap C) \qquad 12.14$$

- *distributive laws*

$$A \cup (B \cap C) = (A \cup B) \cap (A \cup C) \qquad 12.15$$
$$A \cap (B \cup C) = (A \cap B) \cup (A \cap C) \qquad 12.16$$

- *de Morgan's laws*

$$(A \cup B)' = A' \cap B' \qquad 12.17$$
$$(A \cap B)' = A' \cup B' \qquad 12.18$$

2. COMBINATIONS OF ELEMENTS

There are a finite number of ways in which n elements can be combined into distinctly different groups of r items. For example, suppose a farmer has a hen, a rooster, a duck, and a cage that holds only two birds. The possible *combinations* of three birds taken two at a time are (hen, rooster), (hen, duck), and (rooster, duck). The birds in the cage will not remain stationary, and the combination (rooster, hen) is not distinctly different from (hen, rooster). That is, the groups are not *order conscious.*

The number of combinations of n items taken r at a time is written $C(n,r)$, C_r^n, ${}_nC_r$, or $\binom{n}{r}$ (pronounced "n choose r") and given by Eq. 12.19. It is sometimes referred to as the *binomial coefficient.*

$$\binom{n}{r} = C(n,r) = \frac{n!}{(n-r)!r!} \quad [\text{for } r \le n] \qquad 12.19$$

Example 12.1

Six people are on a sinking yacht. There are four life jackets. How many combinations of survivors are there?

Solution

The groups are not order-conscious. From Eq. 12.19,

$$C(6,4) = \frac{n!}{(n-r)!r!} = \frac{6!}{(6-4)!4!} = \frac{6\cdot5\cdot4\cdot3\cdot2\cdot1}{(2\cdot1)(4\cdot3\cdot2\cdot1)}$$
$$= 15$$

3. PERMUTATIONS

An order-conscious subset of r items taken from a set of n items is the *permutation* $P(n,r)$, also written P_r^n and ${}_nP_r$. The permutation is order conscious because the arrangement of two items (say a_i and b_i) as a_ib_i is different from the arrangement b_ia_i. The number of permutations is

$$P(n,r) = \frac{n!}{(n-r)!} \quad [\text{for } r \le n] \qquad 12.20$$

If groups of the entire set of n items are being enumerated, the number of permutations of n items taken n at a time is

$$P(n,n) = \frac{n!}{(n-n)!} = \frac{n!}{0!} = n! \qquad 12.21$$

A *ring permutation* is a special case of n items taken n at a time. There is no identifiable beginning or end, and the number of permutations is divided by n.

$$P_{\text{ring}}(n,n) = \frac{P(n,n)}{n} = (n-1)! \qquad 12.22$$

Example 12.2

A pianist knows four pieces but will have enough stage time to play only three of them. Pieces played in a different order constitute a different program. How many different programs can be arranged?

Solution

The groups are order conscious. From Eq. 12.20,

$$P(4,3) = \frac{n!}{(n-r)!} = \frac{4!}{(4-3)!} = \frac{4\cdot3\cdot2\cdot1}{1} = 24$$

Example 12.3

Seven diplomats from different countries enter a circular room. The only furnishings are seven chairs arranged around a circular table. How many ways are there of arranging the diplomats?

Solution

All seven diplomats must be seated, so the groups are permutations of seven objects taken seven at a time. Since there is no head chair, the groups are ring permutations. From Eq. 12.22,

$$P_{\text{ring}}(7,7) = (7-1)! = 6\cdot5\cdot4\cdot3\cdot2\cdot1$$
$$= 720$$

4. PROBABILITY THEORY

The act of conducting an experiment (trial) or taking a measurement is known as *sampling. Probability theory* determines the relative likelihood that a particular event will occur. An *event*, e, is one of the possible outcomes of the *trial*. Taken together, all of the possible events constitute a finite *sample space*, $E = [e_1, e_2, \ldots, e_n]$. The trial is drawn from the *population* or *universe*. Populations can be finite or infinite in size.

Events can be numerical or nonnumerical, discrete or continuous, and dependent or independent. An example of a nonnumerical event is getting tails on a coin toss. The number from a roll of a die is a discrete numerical event. The measured diameter of a bolt produced from an automatic screw machine is a numerical event. Since the diameter can (within reasonable limits) take on any value, its measured value is a continuous numerical event.

An event is *independent* if its outcome is unaffected by previous outcomes (i.e., previous runs of the experiment) and *dependent* otherwise. Whether or not an event is independent depends on the population size and how the sampling is conducted. Sampling (a trial) from an infinite population is implicitly independent. When the population is finite, *sampling with replacement* produces independent events, while *sampling without replacement* changes the population and produces dependent events.

The terms *success* and *failure* are loosely used in probability theory to designate obtaining and not obtaining, respectively, the tested-for condition. "Failure" is not the same as a *null event* (i.e., one that has a zero probability of occurrence).

The *probability* of event e_1 occurring is designated as $p\{e_1\}$ and is calculated as the ratio of the total number of ways the event can occur to the total number of outcomes in the sample space.

Example 12.4

There are 380 students in a rural school—200 girls and 180 boys. One student is chosen at random and is checked for gender and height. (a) Define and categorize the population. (b) Define and categorize the sample space. (c) Define the trials. (d) Define and categorize the events. (e) In determining the probability that the student chosen is a boy, define success and failure. (f) What is the probability that the student is a boy?

Solution

(a) The population consists of 380 students and is finite.

(b) In determining the gender of the student, the sample space consists of the two outcomes $E = [\text{girl, boy}]$. This sample space is nonnumerical and discrete. In determining the height, the sample space consists of a range of values and is numerical and continuous.

(c) The trial is the actual sampling (i.e., the determination of gender and height).

(d) The events are the outcomes of the trials (i.e., the gender and height of the student). These events are independent if each student returns to the population prior to the random selection of the next student; otherwise, the events are dependent.

(e) The event is a success if the student is a boy and is a failure otherwise.

(f) From the definition of probability,

$$p\{\text{boy}\} = \frac{\text{no. of boys}}{\text{no. of students}} = \frac{180}{380} = 0.47$$

5. JOINT PROBABILITY

Joint probability rules specify the probability of a combination of events. If n mutually exclusive events from the set E have probabilities $p\{e_i\}$, the probability of any one of these events occurring in a given trial is the sum of the individual probabilities. The events in Eq. 12.23 come from a single sample space and are linked by the word *or*.

$$p\{e_1 \text{ or } e_2 \text{ or } \cdots \text{ or } e_k\} = p\{e_1\} + p\{e_2\} + \cdots + p\{e_k\} \qquad 12.23$$

When given two independent sets of events, E and G, Eq. 12.24 will give the probability that either event e_i or g_i, or both, will occur. The events in Eq. 12.24 come from two different sample spaces and are linked by the word *or*.

$$p\{e_i \text{ or } g_i\} = p\{e_i\} + p\{g_i\} - p\{e_i\}p\{g_i\} \qquad 12.24$$

When given two independent sets of events, E and G, Eq. 12.25 will give the probability that events e_i and g_i will both occur. The events in Eq. 12.25 come from two different sample spaces and are linked by the word *and*.

$$p\{e_i \text{ and } g_i\} = p\{e_i\}p\{g_i\} \qquad 12.25$$

Example 12.5

A bowl contains five white balls, two red balls, and three green balls. What is the probability of getting either a white ball or a red ball in one draw from the bowl?

Solution

Since the two possible events are mutually exclusive and come from the same sample space, Eq. 12.23 can be used.

$$p\{\text{white or red}\} = p\{\text{white}\} + p\{\text{red}\} = \frac{5}{10} + \frac{2}{10} = \frac{7}{10}$$

Example 12.6

One bowl contains five white balls, two red balls, and three green balls. Another bowl contains three yellow balls and seven black balls. What is the probability of getting a red ball from the first bowl and a yellow ball from the second bowl in one draw from each bowl?

Solution

Eq. 12.25 can be used because the two trials are independent.

$$\begin{aligned} p\{\text{red and yellow}\} &= p\{\text{red}\}p\{\text{yellow}\} \\ &= \left(\frac{2}{10}\right)\left(\frac{3}{10}\right) \\ &= \frac{6}{100} \end{aligned}$$

6. COMPLEMENTARY PROBABILITIES

The probability of an event occurring is equal to one minus the probability of the event not occurring. This is known as *complementary probability*.

$$p\{e_i\} = 1 - p\{\text{not } e_i\} \qquad 12.26$$

Equation 12.26 can be used to simplify some probability calculations. Specifically, calculation of the probability of numerical events being "greater than" or "less than" or quantities being "at least" a certain number can often be simplified by calculating the probability of complementary event.

Example 12.7

A fair coin is tossed five times.[1] What is the probability of getting at least one tail?

Solution

The probability of getting at least one tail in five tosses could be calculated as

$$\begin{aligned} p\{\text{at least 1 tail}\} &= p\{\text{1 tail}\} + p\{\text{2 tails}\} \\ &\quad + p\{\text{3 tails}\} + p\{\text{4 tails}\} \\ &\quad + p\{\text{5 tails}\} \end{aligned}$$

However, it is easier to calculate the complementary probability of getting no tails (i.e., getting all heads).

$$\begin{aligned} p\{\text{at least 1 tail}\} &= 1 - p\{\text{0 tails}\} \\ &= 1 - (0.5)^5 \\ &= 0.96875 \end{aligned}$$

7. CONDITIONAL PROBABILITY

Given two dependent sets of events, E and G, the probability that event e_k will occur given the fact that the dependent event g has already occurred is written as $p\{e_k|g\}$ and given by *Bayes' theorem*, Eq. 12.27.

$$p\{e_k|g\} = \frac{p\{e_k \text{ and } g\}}{p\{g\}} = \frac{p\{g|e_k\}p\{e_k\}}{\sum_{i=1}^{n} p\{g|e_i\}p\{e_i\}} \qquad 12.27$$

[1]It makes no difference whether one coin is tossed five times or five coins are each tossed once.

8. PROBABILITY DENSITY FUNCTIONS

A *density function* is a nonnegative function whose integral taken over the entire range of the independent variable is unity. A *probability density function* (PDF) is a mathematical formula that gives the probability of a discrete numerical event. A *numerical event* is an occurrence that can be described (usually) by an integer. For example, 27 cars passing through a bridge toll booth in an hour is a discrete numerical event. Figure 12.2 shows a graph of a typical probability density function.

A probability density function, $f(x)$, gives the probability that discrete event x will occur. That is, $p\{x\} = f(x)$. Important discrete probability density functions are the binomial, hypergeometric, and Poisson distributions.

Figure 12.2 Probability Density Function

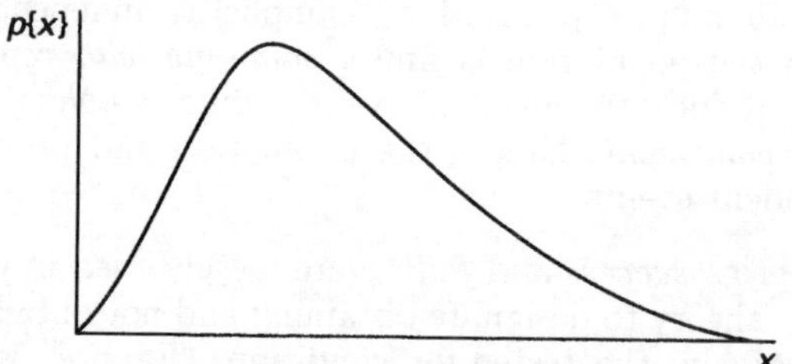

9. BINOMIAL DISTRIBUTION

The *binomial probability density function* is used when all outcomes can be categorized as either successes or failures. The probability of success in a single trial is $\hat{p}$, and the probability of failure is the complement, $\hat{q} = 1 - \hat{p}$. The population is assumed to be infinite in size so that sampling does not change the values of $\hat{p}$ and $\hat{q}$. (The binomial distribution can also be used with finite populations when sampling with replacement.)

Equation 12.28 gives the probability of x successes in n independent *successive trials*. The quantity $\binom{n}{x}$ is the binomial coefficient, identical to the number of combinations of n items taken x at a time.

$$p\{x\} = f(x) = \binom{n}{x}\hat{p}^x\hat{q}^{n-x} \qquad 12.28$$

$$\binom{n}{x} = \frac{n!}{(n-x)!x!} \qquad 12.29$$

Equation 12.28 is a true (discrete) distribution, taking on values for each integer value up to n. The mean, μ, and variance, σ^2 (see Sec. 12.22 and Sec. 12.23), of this distribution are

$$\mu = n\hat{p} \qquad 12.30$$

$$\sigma^2 = n\hat{p}\hat{q} \qquad 12.31$$

Example 12.8

Five percent of a large batch of high-strength steel bolts purchased for bridge construction are defective. (a) If seven bolts are randomly sampled, what is the probability that exactly three will be defective? (b) What is the probability that two or more bolts will be defective?

Solution

(a) The bolts are either defective or not, so the binomial distribution can be applied.

$$\hat{p} = 0.05 \quad [\text{success} = \text{defective}]$$

$$\hat{q} = 1 - 0.05 = 0.95 \quad [\text{failure} = \text{not defective}]$$

From Eq. 12.28,

$$\begin{aligned} p\{3\} = f(3) &= \binom{n}{x}\hat{p}^x\hat{q}^{n-x} = \binom{7}{3}\hat{p}^3\hat{q}^{7-3} \\ &= \left(\frac{7\cdot6\cdot5\cdot4\cdot3\cdot2\cdot1}{4\cdot3\cdot2\cdot1\cdot3\cdot2\cdot1}\right)(0.05)^3(0.95)^4 \\ &= 0.00356 \end{aligned}$$

(b) The probability that two or more bolts will be defective could be calculated as

$$p\{x \geq 2\} = p\{2\} + p\{3\} + p\{4\} + p\{5\} + p\{6\} + p\{7\}$$

This method would require six probability calculations. It is easier to use the complement of the desired probability.

$$p\{x \geq 2\} = 1 - p\{x \leq 1\} = 1 - (p\{0\} + p\{1\})$$

$$p\{0\} = \binom{n}{x}\hat{p}^x\hat{q}^{n-x} = \binom{7}{0}(0.05)^0(0.95)^7 = (0.95)^7$$

$$\begin{aligned} p\{1\} &= \binom{n}{x}\hat{p}^x\hat{q}^{n-x} = \binom{7}{1}(0.05)^1(0.95)^6 \\ &= (7)(0.05)(0.95)^6 \end{aligned}$$

$$\begin{aligned} p\{x \geq 2\} &= 1 - \left((0.95)^7 + (7)(0.05)(0.95)^6\right) \\ &= 1 - (0.6983 + 0.2573) \\ &= 0.0444 \end{aligned}$$

10. HYPERGEOMETRIC DISTRIBUTION

Probabilities associated with sampling from a finite population without replacement are calculated from the *hypergeometric distribution.* If a population of finite size M contains K items with a given characteristic (e.g., red color, defective construction), then the probability of finding x items with that characteristic in a sample of n items is

$$p\{x\} = f(x) = \frac{\binom{K}{x}\binom{M-K}{n-x}}{\binom{M}{n}} \quad [\text{for } x \leq n] \qquad 12.32$$

11. MULTIPLE HYPERGEOMETRIC DISTRIBUTION

Sampling without replacement from finite populations containing several different types of items is handled by the *multiple hypergeometric distribution.* If a population of finite size M contains K_i items of type i (such that $\Sigma K_i = M$), the probability of finding x_1 items of type 1, x_2 items of type 2, and so on, in a sample size n (such that $\Sigma x_i = n$) is

$$p\{x_1, x_2, x_3, \ldots\} = \frac{\binom{K_1}{x_1}\binom{K_2}{x_2}\binom{K_3}{x_3}\cdots}{\binom{M}{n}} \qquad 12.33$$

12. POISSON DISTRIBUTION

Certain events occur relatively infrequently but at a relatively regular rate. The probability of such an event occurring is given by the *Poisson distribution.* Suppose an event occurs, on the average, λ times per period. The probability that the event will occur x times per period is

$$p\{x\} = f(x) = \frac{e^{-\lambda}\lambda^x}{x!} \quad [\lambda > 0] \qquad 12.34$$

λ is both the mean and the variance of the Poisson distribution.

Example 12.9

The number of customers arriving at a hamburger stand in the next period is a Poisson distribution having a mean of eight. What is the probability that exactly six customers will arrive in the next period?

Solution

$\lambda = 8$ and $x = 6$. From Eq. 12.34,

$$p\{6\} = \frac{e^{-\lambda}\lambda^x}{x!} = \frac{e^{-8}(8)^6}{6!} = 0.122$$

13. CONTINUOUS DISTRIBUTION FUNCTIONS

Most numerical events are *continuously distributed* and are not constrained to discrete or integer values. For example, the resistance of a 10% 1 Ω resistor may be any value between 0.9 Ω and 1.1 Ω. The probability of an exact numerical event is zero for continuously distributed variables. That is, there is no chance that a numerical event will be *exactly* x.[2] It is possible to determine only the probability that a numerical event will be less than x, greater than x, or between the values of x_1 and x_2, but not exactly equal to x.

While an expression, $f(x)$, for a probability density function can be written, it is used to derive the *continuous distribution function* (CDF), $F(x_0)$, which gives the probability of numerical event x_0 or less occurring, as illustrated in Fig. 12.3.

$$p\{X < x_0\} = F(x_0) = \int_0^{x_0} f(x)dx \qquad 12.35$$

$$f(x) = \frac{dF(x)}{dx} \qquad 12.36$$

Figure 12.3 *Continuous Distribution Function*

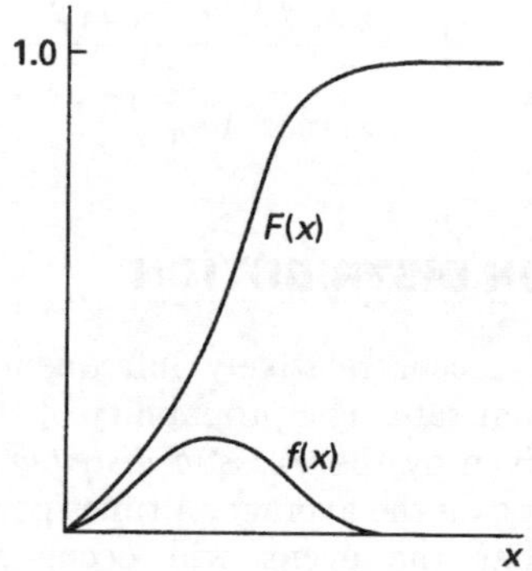

14. EXPONENTIAL DISTRIBUTION

The continuous *exponential distribution* is given by its probability density and continuous distribution functions.

$$f(x) = \lambda e^{-\lambda x} \qquad 12.37$$

$$p\{X < x\} = F(x) = 1 - e^{-\lambda x} \qquad 12.38$$

The mean and variance of the exponential distribution are

$$\mu = \frac{1}{\lambda} \qquad 12.39$$

$$\sigma^2 = \frac{1}{\lambda^2} \qquad 12.40$$

15. NORMAL DISTRIBUTION

The *normal distribution* (*Gaussian distribution*) is a symmetrical distribution commonly referred to as the *bell-shaped curve*, which represents the distribution of outcomes of many experiments, processes, and phenomena. (See Fig. 12.4.) The probability density and continuous distribution functions for the normal distribution with mean μ and variance σ^2 are

$$f(x) = \frac{e^{-\frac{1}{2}\left(\frac{x-\mu}{\sigma}\right)^2}}{\sigma\sqrt{2\pi}} \quad [-\infty < x < +\infty] \qquad 12.41$$

$$p\{\mu < X < x_0\} = F(x_0) = \frac{1}{\sigma\sqrt{2\pi}}\int_0^{x_0} e^{-\frac{1}{2}\left(\frac{x-\mu}{\sigma}\right)^2} dx \qquad 12.42$$

Figure 12.4 *Normal Distribution*

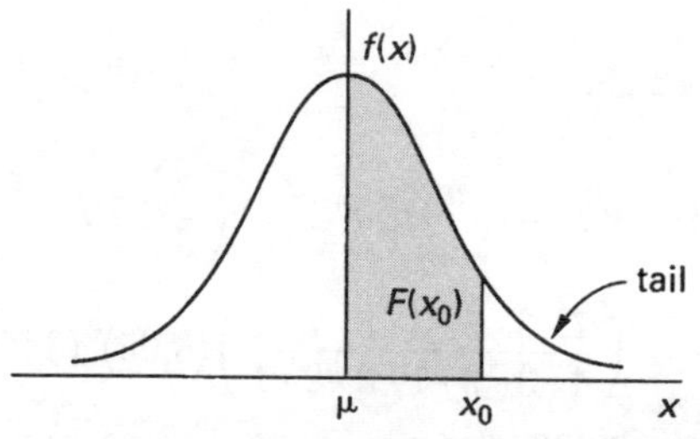

Since $f(x)$ is difficult to integrate, Eq. 12.41 is seldom used directly, and a *standard normal table* is used instead. (See App. 12.A.) The standard normal table is based on a normal distribution with a mean of zero and a standard deviation of 1. Since the range of values from an experiment or phenomenon will not generally correspond to the standard normal table, a value, x_0, must be converted to a *standard normal value*, z. In Eq. 12.43, μ and σ are the mean and standard deviation, respectively, of the distribution from which x_0 comes. For all practical purposes, all normal distributions are completely bounded by $\mu \pm 3\sigma$.

$$z = \frac{x_0 - \mu}{\sigma} \qquad 12.43$$

Numbers in the standard normal table, as given by App. 12.A, are the probabilities of the normalized x being between zero and z and represent the areas under the curve up to point z. When x is less than μ, z will be negative. However, the curve is symmetrical, so the table value corresponding to positive z can be used. The probability of x being greater than z is the complement of the table value. The curve area past point z is known as the *tail of the curve.*

The *error function*, erf(x), and its complement, erfc(x), are defined by Eq. 12.44 and Eq. 12.45. The error function is used to determine the probable error of a measurement.

$$\text{erf}(x_0) = \frac{2}{\sqrt{\pi}}\int_0^{x_0} e^{-x^2} dx \qquad 12.44$$

$$\text{erfc}(x_0) = 1 - \text{erf}(x_0) \qquad 12.45$$

[2]It is important to understand the rationale behind this statement. Since the variable can take on any value and has an infinite number of significant digits, we can infinitely continue to increase the precision of the value. For example, the probability is zero that a resistance will be exactly 1 Ω because the resistance is really 1.03 or 1.0260008 or 1.02600080005, and so on.

Mathematics

Example 12.10

The mass, m, of a particular hand-laid fiberglass (Fibreglas™) part is normally distributed with a mean of 66 kg and a standard deviation of 5 kg. (a) What percent of the parts will have a mass less than 72 kg? (b) What percent of the parts will have a mass in excess of 72 kg? (c) What percent of the parts will have a mass between 61 kg and 72 kg?

Solution

(a) The 72 kg value must be normalized, so use Eq. 12.43. The standard normal variable is

$$z = \frac{x-\mu}{\sigma} = \frac{72\,\text{kg} - 66\,\text{kg}}{5\,\text{kg}} = 1.2$$

Reading from App. 12.A, the area under the normal curve is 0.3849. This represents the probability of the mass, m, being between 66 kg and 72 kg (i.e., z being between 0 and 1.2). However, the probability of the mass being less than 66 kg is also needed. Since the curve is symmetrical, this probability is 0.5. Therefore,

$$p\{m < 72\,\text{kg}\} = p\{z < 1.2\} = 0.5 + 0.3849 = 0.8849$$

(b) The probability of the mass exceeding 72 kg is the area under the tail past point z.

$$p\{m > 72\,\text{kg}\} = p\{z > 1.2\} = 0.5 - 0.3849 = 0.1151$$

(c) The standard normal variable corresponding to $m = 61$ kg is

$$z = \frac{x-\mu}{\sigma} = \frac{61\,\text{kg} - 66\,\text{kg}}{5\,\text{kg}} = -1$$

Since the two masses are on opposite sides of the mean, the probability will have to be determined in two parts.

$$\begin{aligned} p\{61 < m < 72\} &= p\{61 < m < 66\} + p\{66 < m < 72\} \\ &= p\{-1 < z < 0\} + p\{0 < z < 1.2\} \\ &= 0.3413 + 0.3849 \\ &= 0.7262 \end{aligned}$$

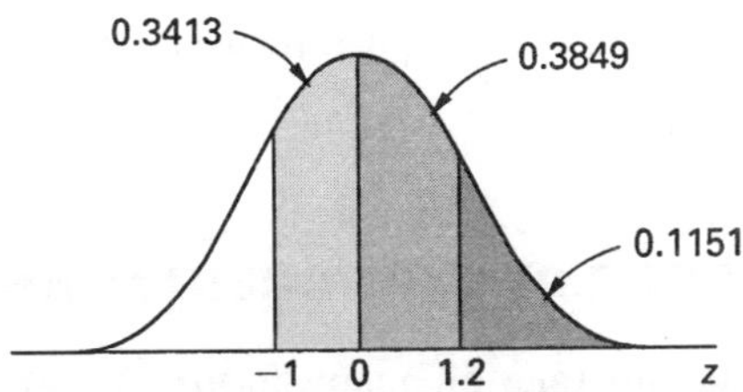

16. APPLICATION: RELIABILITY

Introduction

Reliability, $R\{t\}$, is the probability that an item will continue to operate satisfactorily up to time t. The *bathtub distribution*, Fig. 12.5, is often used to model the probability of failure of an item (or, the number of failures from a large population of items) as a function of time. Items initially fail at a high rate, a phenomenon known as *infant mortality*. For the majority of the operating time, known as the *steady-state operation*, the failure rate is constant (i.e., is due to random causes). After a long period of time, the items begin to deteriorate and the failure rate increases. (No mathematical distribution describes all three of these phases simultaneously.)

Figure 12.5 *Bathtub Reliability Curve*

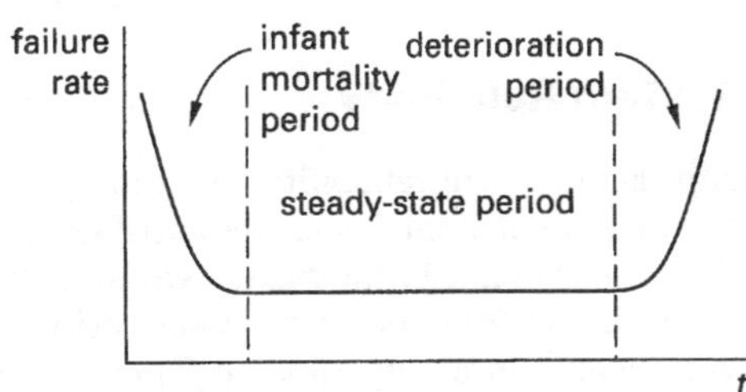

The *hazard function*, $z\{t\}$, represents the *conditional probability of failure*—the probability of failure in the next time interval, given that no failure has occurred thus far.[3]

$$z\{t\} = \frac{f(t)}{R(t)} = \frac{\frac{dF(t)}{dt}}{1 - F(t)} \qquad 12.46$$

Exponential Reliability

Steady-state reliability is often described by the *negative exponential distribution*. This assumption is appropriate whenever an item fails only by random causes and does not experience deterioration during its life. The parameter λ is related to the *mean time to failure* (MTTF) of the item.[4]

$$R\{t\} = e^{-\lambda t} = e^{-t/\text{MTTF}} \qquad 12.47$$

$$\lambda = \frac{1}{\text{MTTF}} \qquad 12.48$$

Equation 12.47 and the exponential continuous distribution function, Eq. 12.38, are complementary.

$$R\{t\} = 1 - F(t) = 1 - \left(1 - e^{-\lambda t}\right) = e^{-\lambda t} \qquad 12.49$$

The hazard function for the negative exponential distribution is

$$z\{t\} = \lambda \qquad 12.50$$

Therefore, the hazard function for exponential reliability is constant and does not depend on t (i.e., on the age of the item). In other words, the expected future life of an item is independent of the previous history (length of operation). This lack of memory is consistent with the

[3]The symbol $z\{t\}$ is traditionally used for the hazard function and is not related to the standard normal variable.

[4]The term "mean time *between* failures" is improper. However, the term *mean time before failure* (MTBF) is acceptable.

assumption that only random causes contribute to failure during steady-state operations. And since random causes are unlikely discrete events, their probability of occurrence can be represented by a Poisson distribution with mean λ. That is, the probability of having x failures in any given period is

$$p\{x\} = \frac{e^{-\lambda}\lambda^x}{x!} \qquad 12.51$$

Serial System Reliability

In the analysis of system reliability, the binary variable X_i is defined as one if item i operates satisfactorily and zero if otherwise. Similarly, the binary variable Φ is one only if the entire system operates satisfactorily. Therefore, Φ will depend on a *performance function* containing the X_i.

A *serial system* is one for which all items must operate correctly for the system to operate. Each item has its own reliability, R_i. For a serial system of n items, the performance function is

$$\Phi = X_1 X_2 X_3 \cdots X_n = \min(X_i) \qquad 12.52$$

The probability of a serial system operating correctly is

$$p\{\Phi = 1\} = R_{\text{serial system}} = R_1 R_2 R_3 \cdots R_n \qquad 12.53$$

Parallel System Reliability

A *parallel system* with n items will fail only if all n items fail. Such a system is said to be *redundant* to the nth degree. Using redundancy, a highly reliable system can be produced from components with relatively low individual reliabilities.

The performance function of a redundant system is

$$\begin{aligned}\Phi &= 1 - (1 - X_1)(1 - X_2)(1 - X_3) \cdots (1 - X_n) \\ &= \max(X_i)\end{aligned} \qquad 12.54$$

The reliability of the parallel system is

$$\begin{aligned}R &= p\{\Phi = 1\} \\ &= 1 - (1 - R_1)(1 - R_2)(1 - R_3) \cdots (1 - R_n)\end{aligned} \qquad 12.55$$

Example 12.11

The reliability of an item is exponentially distributed with mean time to failure (MTTF) of 1000 hr. What is the probability that the item will not have failed after 1200 hr of operation?

Solution

The probability of not having failed before time t is the reliability. From Eq. 12.48 and Eq. 12.49,

$$\lambda = \frac{1}{\text{MTTF}} = \frac{1}{1000 \text{ hr}} = 0.001 \text{ hr}^{-1}$$

$$R\{1200\} = e^{-\lambda t} = e^{(-0.001 \text{ hr}^{-1})(1200 \text{ hr})} = 0.3$$

Example 12.12

What are the reliabilities of the following systems?

(a)

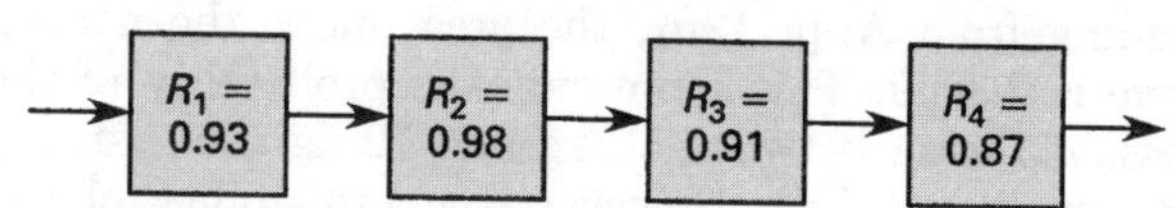

(b)

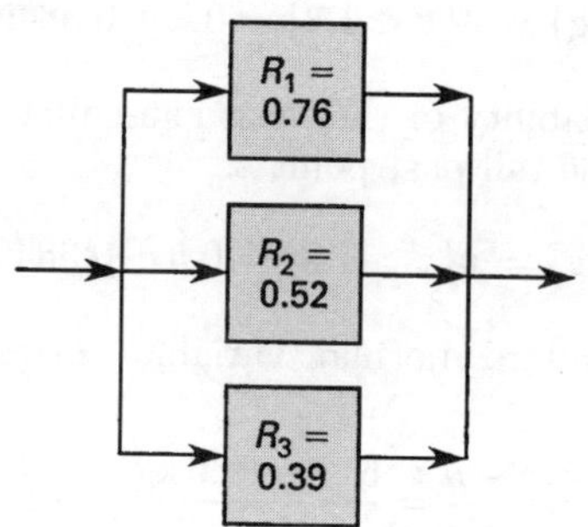

Solution

(a) This is a serial system. From Eq. 12.53,

$$\begin{aligned}R &= R_1 R_2 R_3 R_4 = (0.93)(0.98)(0.91)(0.87) \\ &= 0.72\end{aligned}$$

(b) This is a parallel system. From Eq. 12.55,

$$\begin{aligned}R &= 1 - (1 - R_1)(1 - R_2)(1 - R_3) \\ &= 1 - (1 - 0.76)(1 - 0.52)(1 - 0.39) \\ &= 0.93\end{aligned}$$

17. ANALYSIS OF EXPERIMENTAL DATA

Experiments can take on many forms. An experiment might consist of measuring the mass of one cubic foot of concrete or measuring the speed of a car on a roadway. Generally, such experiments are performed more than once to increase the precision and accuracy of the results.

Both systematic and random variations in the process being measured will cause the observations to vary, and the experiment would not be expected to yield the same result each time it was performed. Eventually, a collection of experimental outcomes (observations) will be available for analysis.

The *frequency distribution* is a systematic method for ordering the observations from small to large, according to some convenient numerical characteristic. The *step interval* should be chosen so that the data are presented in a meaningful manner. If there are too many intervals, many of them will have zero frequencies; if there are too few intervals, the frequency distribution will have little value. Generally, 10 to 15 intervals are used.

Once the frequency distribution is complete, it can be represented graphically as a *histogram*. The procedure in drawing a histogram is to mark off the interval limits (also known as *class limits*) on a number line and then draw contiguous bars with lengths that are proportional to the frequencies in the intervals and that are centered on the midpoints of their respective intervals. The continuous nature of the data can be depicted by a *frequency polygon*. The number or percentage of observations that occur up to and including some value can be shown in a *cumulative frequency table*.

Example 12.13

The number of cars that travel through an intersection between 12 noon and 1 p.m. is measured for 30 consecutive working days. The results of the 30 observations are

79, 66, 72, 70, 68, 66, 68, 76, 73, 71, 74, 70, 71, 69, 67, 74, 70, 68, 69, 64, 75, 70, 68, 69, 64, 69, 62, 63, 63, 61

(a) What are the frequency and cumulative distributions? (Use a distribution interval of two cars per hour.) (b) Draw the histogram. (Use a cell size of two cars per hour.) (c) Draw the frequency polygon. (d) Graph the cumulative frequency distribution.

Solution

(a)

cars per hour	frequency	cumulative frequency	cumulative percent
60–61	1	1	3
62–63	3	4	13
64–65	2	6	20
66–67	3	9	30
68–69	8	17	57
70–71	6	23	77
72–73	2	25	83
74–75	3	28	93
76–77	1	29	97
78–79	1	30	100

(b)

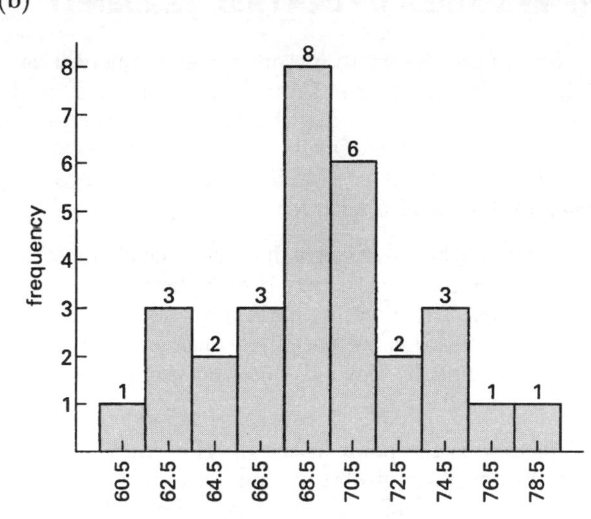

(c)

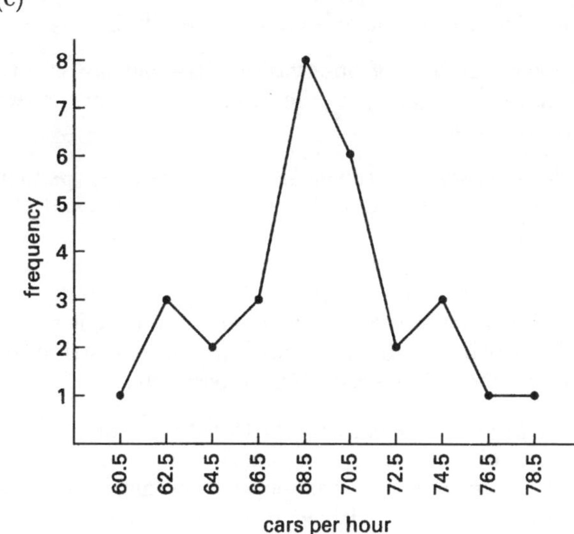

(d)

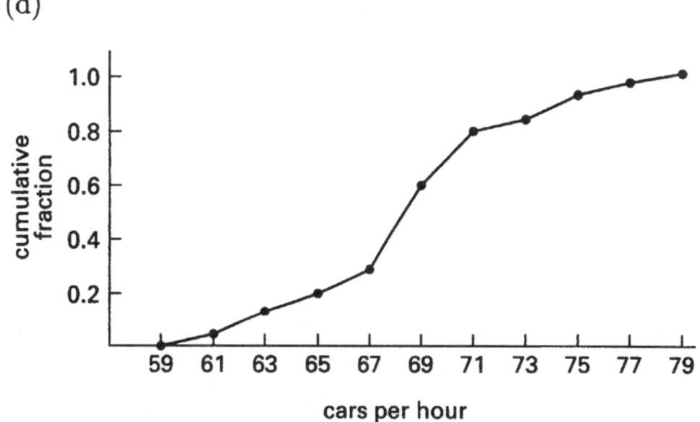

18. MEASURES OF CENTRAL TENDENCY

It is often unnecessary to present the experimental data in their entirety, either in tabular or graphical form. In such cases, the data and distribution can be represented by various parameters. One type of parameter is a measure of *central tendency.* Mode, median, and mean are measures of central tendency.

The *mode* is the observed value that occurs most frequently. The mode may vary greatly between series of observations. Therefore, its main use is as a quick measure of the central value since little or no computation is required to find it. Beyond this, the usefulness of the mode is limited.

The *median* is the point in the distribution that partitions the total set of observations into two parts containing equal numbers of observations. It is not influenced by the extremity of scores on either side of the distribution. The median is found by counting up (from either end of the frequency distribution) until half of the observations have been accounted for.

For even numbers of observations, the median is estimated as some value (i.e., the average) between the two center observations.

Similar in concept to the median are *percentiles* (*percentile ranks*), *quartiles*, and *deciles.* The median could also have been called the *50th percentile* observation. Similarly, the 80th percentile would be the observed value (e.g., the number of cars per hour) for which the cumulative frequency was 80%. The quartile and decile points on the distribution divide the observations or distribution into segments of 25% and 10%, respectively.

The *arithmetic mean* is the arithmetic average of the observations. The sample mean, $\overline{x}$, can be used as an unbiased estimator of the population mean, μ. The *mean* may be found without ordering the data (as was necessary to find the mode and median). The mean can be found from the following formula.

$$\overline{x} = \left(\frac{1}{n}\right)(x_1 + x_2 + \cdots + x_n) = \frac{\sum x_i}{n} \qquad 12.56$$

The *geometric mean* is used occasionally when it is necessary to average ratios. The geometric mean is calculated as

$$\text{geometric mean} = \sqrt[n]{x_1 x_2 x_3 \cdots x_n} \quad [x_i > 0] \qquad 12.57$$

The *harmonic mean* is defined as

$$\text{harmonic mean} = \frac{n}{\frac{1}{x_1} + \frac{1}{x_2} + \cdots + \frac{1}{x_n}} \qquad 12.58$$

The *root-mean-squared* (rms) *value* of a series of observations is defined as

$$x_{\text{rms}} = \sqrt{\frac{\sum x_i^2}{n}} \qquad 12.59$$

Example 12.14

Find the mode, median, and arithmetic mean of the distribution represented by the data given in Ex. 12.13.

Solution

First, resequence the observations in increasing order.

61, 62, 63, 63, 64, 64, 66, 66, 67, 68, 68, 68, 68, 69, 69, 69, 69, 70, 70, 70, 70, 71, 71, 72, 73, 74, 74, 75, 76, 79

The mode is the interval 68–69, since this interval has the highest frequency. If 68.5 is taken as the interval center, then 68.5 would be the mode.

The 15th and 16th observations are both 69, so the median is

$$\frac{69 + 69}{2} = 69$$

The mean can be found from the raw data or from the grouped data using the interval center as the assumed observation value. Using the raw data,

$$\overline{x} = \frac{\sum x}{n} = \frac{2069}{30} = 68.97$$

19. MEASURES OF DISPERSION

The simplest statistical parameter that describes the variability in observed data is the *range.* The range is found by subtracting the smallest value from the largest. Since the range is influenced by extreme (low probability) observations, its use as a measure of variability is limited.

The *standard deviation* is a better estimate of variability because it considers every observation. That is, in Eq. 12.60, N is the total population size, not the sample size, n.

$$\sigma = \sqrt{\frac{\sum (x_i - \mu)^2}{N}} = \sqrt{\frac{\sum x_i^2}{N} - \mu^2} \qquad 12.60$$

The standard deviation of a sample (particularly a small sample) is a biased (i.e., is not a good) estimator of the population standard deviation. An *unbiased estimator* of the population standard deviation is the *sample standard deviation, s*.[5]

$$s = \sqrt{\frac{\sum (x_i - \overline{x})^2}{n-1}} = \sqrt{\frac{\sum x_i^2 - \frac{\left(\sum x_i\right)^2}{n}}{n-1}} \qquad 12.61$$

If the sample standard deviation, s, is known, the standard deviation of the sample, σ_{sample}, can be calculated.

$$\sigma_{\text{sample}} = s\sqrt{\frac{n-1}{n}} \qquad 12.62$$

The *variance* is the square of the standard deviation. Since there are two standard deviations, there are two variances. The *variance of the sample* is σ^2, and the *sample variance* is s^2.

The *relative dispersion* is defined as a measure of dispersion divided by a measure of central tendency. The *coefficient of variation* is a relative dispersion calculated from the sample standard deviation and the mean.

$$\text{coefficient of variation} = \frac{s}{\overline{x}} \qquad 12.63$$

Skewness is a measure of a frequency distribution's lack of symmetry.

$$\text{skewness} = \frac{\overline{x} - \text{mode}}{s} \approx \frac{3(\overline{x} - \text{median})}{s} \qquad 12.64$$

Example 12.15

For the data given in Ex. 12.13, calculate (a) the sample range, (b) the standard deviation of the sample, (c) an unbiased estimator of the population standard deviation, (d) the variance of the sample, and (e) the sample variance.

Solution

$$\sum x_i = 2069$$

$$\left(\sum x_i\right)^2 = (2069)^2 = 4{,}280{,}761$$

$$\sum x_i^2 = 143{,}225$$

$$n = 30$$

$$\overline{x} = \frac{2069}{30} = 68.967$$

(a)

$$R = x_{\text{max}} - x_{\text{min}} = 79 - 61 = 18$$

(b)

$$\sigma = \sqrt{\frac{\sum x_i^2}{n} - (\overline{x})^2} = \sqrt{\frac{143{,}225}{30} - \left(\frac{2069}{30}\right)^2} = 4.215$$

(c)

$$s = \sqrt{\frac{\sum x_i^2 - \frac{\left(\sum x_i\right)^2}{n}}{n-1}} = \sqrt{\frac{143{,}225 - \frac{4{,}280{,}761}{30}}{29}} = 4.287$$

(d)

$$\sigma^2 = 17.77$$

(e)

$$s^2 = 18.38$$

20. CENTRAL LIMIT THEOREM

Measuring a sample of n items from a population with mean μ and standard deviation σ is the general concept of an experiment. The sample mean, $\overline{x}$, is one of the parameters that can be derived from the experiment. This experiment can be repeated k times, yielding a set of averages $(\overline{x}_1, \overline{x}_2, \ldots, \overline{x}_k)$. The k numbers in the set themselves represent samples from distributions of averages. The average of averages, $\overline{\overline{x}}$, and sample standard deviation of averages, $s_{\overline{x}}$ (known as the *standard error of the mean*), can be calculated.

The *central limit theorem* characterizes the distribution of the sample averages. The theorem can be stated in several ways, but the essential elements are the following points.

1. The averages, $\overline{x}_i$, are normally distributed variables, even if the original data from which they are calculated are not normally distributed.
2. The grand average, $\overline{\overline{x}}$ (i.e., the average of the averages), approaches and is an unbiased estimator of μ.

$$\mu \approx \overline{\overline{x}} \qquad 12.65$$

[5]There is a subtle yet significant difference between *standard deviation of the sample*, σ (obtained from Eq. 12.60 for a finite sample drawn from a larger population) and the *sample standard deviation*, s (obtained from Eq. 12.61). While σ can be calculated, it has no significance or use as an estimator. It is true that the difference between σ and s approaches zero when the sample size, n, is large, but this convergence does nothing to legitimize the use of σ as an estimator of the true standard deviation. (Some people say "large" is 30, others say 50 or 100.)

The standard deviation of the original distribution, σ, is much larger than the standard error of the mean.

$$\sigma \approx \sqrt{n} s_{\overline{x}} \quad 12.66$$

21. CONFIDENCE LEVEL

The results of experiments are seldom correct 100% of the time. Recognizing this, researchers accept a certain probability of being wrong. In order to minimize this probability, the experiment is repeated several times. The number of repetitions depends on the desired level of confidence in the results.

If the results have a 5% probability of being wrong, the *confidence level*, C, is 95% that the results are correct, in which case the results are said to be *significant*. If the results have only a 1% probability of being wrong, the confidence level is 99%, and the results are said to be *highly significant*. Other confidence levels (90%, 99.5%, etc.) are used as appropriate.

22. APPLICATION: CONFIDENCE LIMITS

As a consequence of the central limit theorem, sample means of n items taken from a normal distribution with mean μ and standard deviation σ will be normally distributed with mean μ and variance σ^2/n. The probability that any given average, $\overline{x}$, exceeds some value, L, is

$$p\{\overline{x} > L\} = p\left\{ z > \left| \frac{L-\mu}{\frac{\sigma}{\sqrt{n}}} \right| \right\} \quad 12.67$$

L is the *confidence limit* for the confidence level $1 - p\{\overline{x} > L\}$ (expressed as a percent). Values of z are read directly from the standard normal table. As an example, $z = 1.645$ for a 95% confidence level since only 5% of the curve is above that z in the upper tail. Similar values are given in Table 12.1. This is known as a *one-tail confidence limit* because all of the probability is given to one side of the variation.

Table 12.1 *Values of z for Various Confidence Levels*

confidence level, C	one-tail limit, z	two-tail limit, z
90%	1.28	1.645
95%	1.645	1.96
97.5%	1.96	2.17
99%	2.33	2.575
99.5%	2.575	2.81
99.75%	2.81	3.00

With *two-tail confidence limits*, the probability is split between the two sides of variation. There will be upper and lower confidence limits, UCL and LCL, respectively.

$$p\{\text{LCL} < \overline{x} < \text{UCL}\} = p\left\{ \frac{\text{LCL}-\mu}{\frac{\sigma}{\sqrt{n}}} < z < \frac{\text{UCL}-\mu}{\frac{\sigma}{\sqrt{n}}} \right\} \quad 12.68$$

23. APPLICATION: BASIC HYPOTHESIS TESTING

A *hypothesis test* is a procedure that answers the question, "Did these data come from [a particular type of] distribution?" There are many types of tests, depending on the distribution and parameter being evaluated. The simplest hypothesis test determines whether an average value obtained from n repetitions of an experiment could have come from a population with known mean, μ, and standard deviation, σ. A practical application of this question is whether a manufacturing process has changed from what it used to be or should be. Of course, the answer (i.e., "yes" or "no") cannot be given with absolute certainty—there will be a confidence level associated with the answer.

The following procedure is used to determine whether the average of n measurements can be assumed (with a given confidence level) to have come from a known population.

step 1: Assume random sampling from a normal population.

step 2: Choose the desired confidence level, C.

step 3: Decide on a one-tail or two-tail test. If the hypothesis being tested is that the average has or has not *increased* or *decreased*, choose a one-tail test. If the hypothesis being tested is that the average has or has not *changed*, choose a two-tail test.

step 4: Use Table 12.1 or the standard normal table to determine the z-value corresponding to the confidence level and number of tails.

step 5: Calculate the actual standard normal variable, z'.

$$z' = \frac{\overline{x}-\mu}{\frac{\sigma}{\sqrt{n}}} \quad 12.69$$

step 6: If $z' \geq z$, the average can be assumed (with confidence level C) to have come from a different distribution.

Example 12.16

When it is operating properly, a cement plant has a daily production rate that is normally distributed with a mean of 880 tons/day and a standard deviation of 21 tons/day. During an analysis period, the output is measured on 50 consecutive days, and the mean output is found to be 871 tons/day. With a 95% confidence level, determine whether the plant is operating properly.

Solution

step 1: Given.

step 2: $C = 0.95$ is given.

step 3: Since a specific direction in the variation is not given (i.e., the example does not ask whether the average has decreased), use a two-tail hypothesis test.

step 4: From Table 12.1, $z = 1.96$.

step 5: From Eq. 12.69,

$$z' = \left|\frac{\bar{x} - \mu}{\frac{\sigma}{\sqrt{n}}}\right| = \left|\frac{871 - 880}{\frac{21}{\sqrt{50}}}\right| = 3.03$$

Since $3.03 > 1.96$, the distributions are not the same. There is at least a 95% probability that the plant is not operating correctly.

24. APPLICATION: STATISTICAL PROCESS CONTROL

All manufacturing processes contain variation due to random and nonrandom causes. Random variation cannot be eliminated. *Statistical process control* (SPC) is the act of monitoring and adjusting the performance of a process to detect and eliminate nonrandom variation.

Statistical process control is based on taking regular (hourly, daily, etc.) samples of n items and calculating the mean, $\bar{x}$, and range, R, of the sample. To simplify the calculations, the range is used as a measure of the dispersion. These two parameters are graphed on their respective x-bar and *R-control charts*, as shown in Fig. 12.6.[6] Confidence limits are drawn at $\pm 3\sigma/\sqrt{n}$. From a statistical standpoint, the control chart tests a hypothesis each time a point is plotted. When a point falls outside these limits, there is a 99.75% probability that the process is out of control. Until a point exceeds the control limits, no action is taken.[7]

25. MEASURES OF EXPERIMENTAL ADEQUACY

An experiment is said to be *accurate* if it is unaffected by experimental error. In this case, *error* is not synonymous with *mistake*, but rather includes all variations not within the experimenter's control.

[6]Other charts (e.g., the *sigma chart*, *p-chart*, and *c-chart*) are less common but are used as required.

[7]Other indications that a correction may be required are seven measurements on one side of the average and seven consecutively increasing measurements. Rules such as these detect shifts and trends.

Figure 12.6 Typical Statistical Process Control Charts

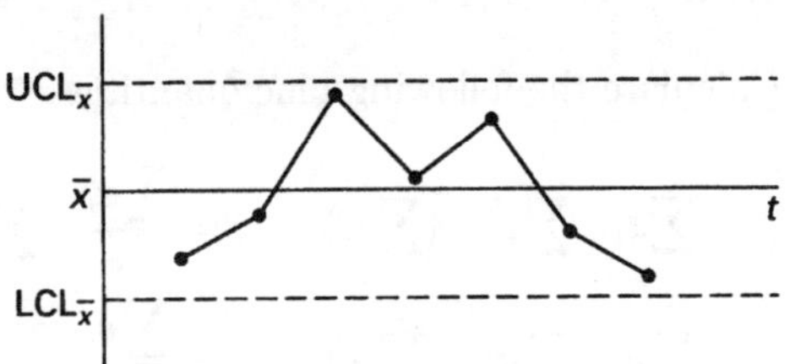

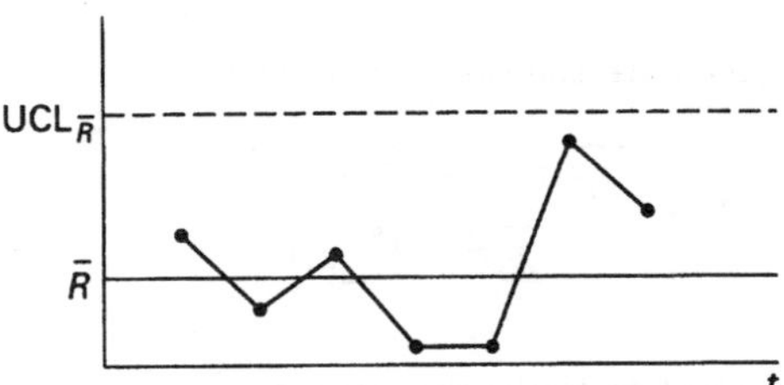

For example, suppose a gun is aimed at a point on a target and five shots are fired. The mean distance from the point of impact to the sight in point is a measure of the alignment accuracy between the barrel and sights. The difference between the actual value and the experimental value is known as ***bias***.

Precision is not synonymous with accuracy. Precision is concerned with the repeatability of the experimental results. If an experiment is repeated with identical results, the experiment is said to be precise.

The average distance of each impact from the centroid of the impact group is a measure of the precision of the experiment. It is possible to have a highly precise experiment with a large bias.

Most of the techniques applied to experiments in order to improve the accuracy (i.e., reduce bias) of the experimental results (e.g., repeating the experiment, refining the experimental methods, or reducing variability) actually increase the precision.

Sometimes the word *reliability* is used with regard to the precision of an experiment. In this case, a "reliable estimate" is used in the same sense as a "precise estimate."

Stability and *insensitivity* are synonymous terms. A stable experiment will be insensitive to minor changes in the experimental parameters. For example, suppose the centroid of a bullet group is 2.1 in from the target point at 65°F and 2.3 in away at 80°F. The sensitivity of the experiment to temperature change would be

$$\text{sensitivity} = \frac{\Delta x}{\Delta T} = \frac{2.3\text{ in} - 2.1\text{ in}}{80°\text{F} - 65°\text{F}} = 0.0133\text{ in/°F}$$

26. LINEAR REGRESSION

If it is necessary to draw a straight line ($y = mx + b$) through n data points $(x_1, y_1), (x_2, y_2), \ldots, (x_n, y_n)$, the following method based on the *method of least squares* can be used.

step 1: Calculate the following nine quantities.

$$\sum x_i \quad \sum x_i^2 \quad \left(\sum x_i\right)^2 \quad \overline{x} = \frac{\sum x_i}{n} \quad \sum x_i y_i$$

$$\sum y_i \quad \sum y_i^2 \quad \left(\sum y_i\right)^2 \quad \overline{y} = \frac{\sum y_i}{n}$$

step 2: Calculate the slope, m, of the line.

$$m = \frac{n\sum(x_i y_i) - \left(\sum x_i\right)\left(\sum y_i\right)}{n\sum x_i^2 - \left(\sum x_i\right)^2} \qquad 12.70$$

step 3: Calculate the y-intercept, b.

$$b = \overline{y} - m\overline{x} \qquad 12.71$$

step 4: To determine the goodness of fit, calculate the correlation coefficient, r.

$$r = \frac{n\sum(x_i y_i) - \left(\sum x_i\right)\left(\sum y_i\right)}{\sqrt{\left(n\sum x_i^2 - \left(\sum x_i\right)^2\right)\left(n\sum y_i^2 - \left(\sum y_i\right)^2\right)}} \qquad 12.72$$

If m is positive, r will be positive; if m is negative, r will be negative. As a general rule, if the absolute value of r exceeds 0.85, the fit is good; otherwise, the fit is poor. r equals 1.0 if the fit is a perfect straight line.

A low value of r does not eliminate the possibility of a nonlinear relationship existing between x and y. It is possible that the data describe a parabolic, logarithmic, or other nonlinear relationship. (Usually this will be apparent if the data are graphed.) It may be necessary to convert one or both variables to new variables by taking squares, square roots, cubes, or logarithms, to name a few of the possibilities, in order to obtain a linear relationship. The apparent shape of the line through the data will give a clue to the type of variable transformation that is required. The curves in Fig. 12.7 may be used as guides to some of the simpler variable transformations.

Figure 12.8 illustrates several common problems encountered in trying to fit and evaluate curves from experimental data. Figure 12.8(a) shows a graph of clustered data with several extreme points. There will be moderate correlation due to the weighting of the extreme points, although there is little actual correlation at low values of the variables. The extreme data should be excluded, or the range should be extended by obtaining more data.

Figure 12.8(b) shows that good correlation exists in general, but extreme points are missed and the overall correlation is moderate. If the results within the small

Figure 12.7 *Nonlinear Data Curves*

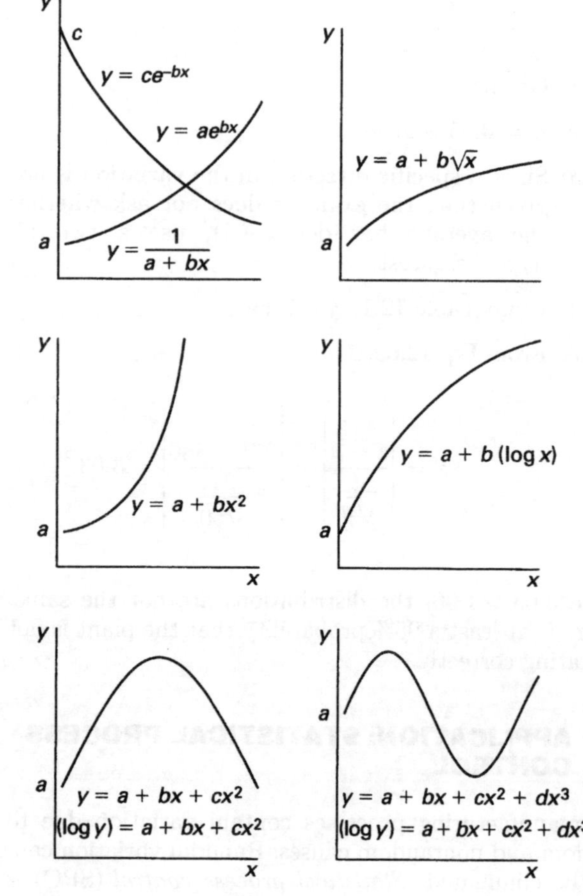

Figure 12.8 *Common Regression Difficulties*

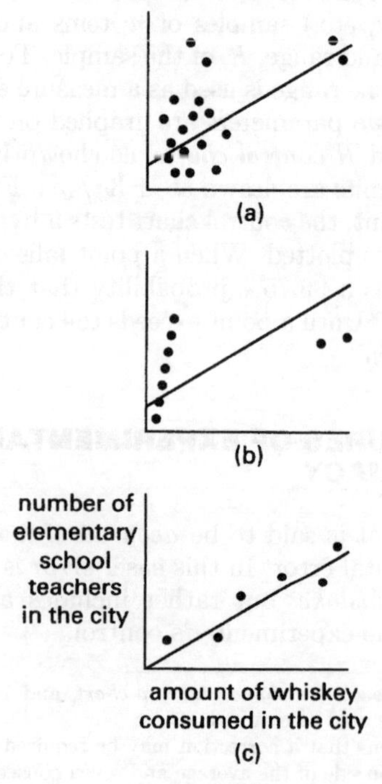

linear range can be used, the extreme points should be excluded. Otherwise, additional data points are needed, and curvilinear relationships should be investigated.

Figure 12.8(c) illustrates the problem of drawing conclusions of cause and effect. There may be a predictable relationship between variables, but that does not imply a cause and effect relationship. In the case shown, both variables are functions of a third variable, the city population. But there is no direct relationship between the plotted variables.

Example 12.17

An experiment is performed in which the dependent variable y is measured against the independent variable x. The results are as follows.

x	y
1.2	0.602
4.7	5.107
8.3	6.984
20.9	10.031

(a) What is the least squares straight line equation that represents this data? (b) What is the correlation coefficient?

Solution

(a)

$$\sum x_i = 35.1$$

$$\sum y_i = 22.72$$

$$\sum x_i^2 = 529.23$$

$$\sum y_i^2 = 175.84$$

$$\left(\sum x_i\right)^2 = 1232.01$$

$$\left(\sum y_i\right)^2 = 516.38$$

$$\overline{x} = 8.775$$

$$\overline{y} = 5.681$$

$$\sum x_i y_i = 292.34$$

$$n = 4$$

From Eq. 12.70, the slope is

$$m = \frac{(4)(292.34) - (35.1)(22.72)}{(4)(529.23) - (35.1)^2} = 0.42$$

From Eq. 12.71, the y-intercept is

$$b = 5.681 - (0.42)(8.775) = 2.0$$

The equation of the line is

$$y = 0.42x + 2.0$$

(b) From Eq. 12.72, the correlation coefficient is

$$r = \frac{(4)(292.34) - (35.1)(22.72)}{\sqrt{\begin{array}{l}\big((4)(529.23) - 1232.01\big) \\ \quad \times \big((4)(175.84) - 516.38\big)\end{array}}} = 0.914$$

Example 12.18

Repeat Ex. 12.17 assuming the relationship between the variables is nonlinear.

Solution

The first step is to graph the data. Because the graph has the appearance of the fourth case, it can be assumed that the relationship between the variables has the form of $y = a + b \log x$. Therefore, the variable change $z = \log x$ is made, resulting in the following set of data.

z	y
0.0792	0.602
0.672	5.107
0.919	6.984
1.32	10.031

If the regression analysis is performed on this set of data, the resulting equation and correlation coefficient are

$$y = -0.036 + 7.65z$$

$$r = 0.999$$

This is a very good fit. The relationship between the variable x and y is approximately

$$y = -0.036 + 7.65 \log x$$

13 Computer Mathematics

1. POSITIONAL NUMBERING SYSTEMS

A *base-b number*, N_b, is made up of individual *digits*. In a *positional numbering system*, the position of a digit in the number determines that digit's contribution to the total value of the number. Specifically, the position of the digit determines the power to which the *base* (also known as the *radix*), b, is raised. For *decimal numbers*, the radix is 10, hence the description *base-10 numbers*.

$$(a_n a_{n-1} \cdots a_2 a_1 a_0)_b = a_n b^n + a_{n-1} b^{n-1} + \cdots + a_2 b^2 + a_1 b + a_0 \qquad 13.1$$

The leftmost digit, a_n, contributes the greatest to the number's magnitude and is known as the *most significant digit* (MSD). The rightmost digit, a_0, contributes the least and is known as the *least significant digit* (LSD).

2. CONVERTING BASE-*b* NUMBERS TO BASE-10

Equation 13.1 converts base-b numbers to base-10 numbers. The calculation of the right-hand side of Eq. 13.1 is performed in the base-10 arithmetic and is known as the *expansion method.*[1]

[1]Equation 13.1 works with any base number. The *double-dabble (double and add) method* is a specialized method of converting from base-2 to base-10 numbers.

Converting base-b numbers (i.e., decimals) to base-10 is similar to converting whole numbers and is accomplished by Eq. 13.2.

$$(0.a_1 a_2 \cdots a_m)_b = a_1 b^{-1} + a_2 b^{-2} + \cdots + a_m b^{-m} \qquad 13.2$$

3. CONVERTING BASE-10 NUMBERS TO BASE-*b*

The *remainder method* is used to convert base-10 numbers to base-b numbers. This method consists of successive divisions by the base, b, until the quotient is zero. The base-b number is found by taking the remainders in the reverse order from which they were found. This method is illustrated in Ex. 13.1 and Ex. 13.3.

Converting a base-10 fraction to base-b requires multiplication of the base-10 fraction and subsequent fractional parts by the base. The base-b fraction is formed from the integer parts of the products taken in the same order in which they were determined. This is illustrated in Ex. 13.2(d).

4. BINARY NUMBER SYSTEM

There are only two *binary digits (bits)* in the *binary number system:* zero and one.[2] Thus, all binary numbers consist of strings of bits (i.e., zeros and ones). The leftmost bit is known as the *most significant bit* (MSB), and the rightmost bit is the *least significant bit* (LSB).

As with digits from other numbering systems, bits can be added, subtracted, multiplied, and divided, although only digits 0 and 1 are allowed in the results. The rules of bit addition are

$$0+0=0$$

$$0+1=1$$

$$1+0=1$$

$$1+1=0 \text{ carry } 1$$

Example 13.1

(a) Convert $(1011)_2$ to base-10. (b) Convert $(75)_{10}$ to base-2.

[2]Alternatively, the binary states may be called *true* and *false, on* and *off, high* and *low*, or *positive* and *negative*.

Solution

(a) Using Eq. 13.1 with $b = 2$,

$$(1)(2)^3 + (0)(2)^2 + (1)(2)^1 + 1 = 11$$

(b) Use the remainder method. (See Sec. 13.3.)

$$75 \div 2 = 37 \text{ remainder } 1$$
$$37 \div 2 = 18 \text{ remainder } 1$$
$$18 \div 2 = 9 \text{ remainder } 0$$
$$9 \div 2 = 4 \text{ remainder } 1$$
$$4 \div 2 = 2 \text{ remainder } 0$$
$$2 \div 2 = 1 \text{ remainder } 0$$
$$1 \div 2 = 0 \text{ remainder } 1$$

The binary representation of $(75)_{10}$ is $(1001011)_2$.

5. OCTAL NUMBER SYSTEM

The *octal (base-8) system* is one of the alternatives to working with long binary numbers. Only the digits 0 through 7 are used. The rules for addition in the octal system are the same as for the decimal system except that the digits 8 and 9 do not exist. For example,

$$7 + 1 = 6 + 2 = 5 + 3 = (10)_8$$
$$7 + 2 = 6 + 3 = 5 + 4 = (11)_8$$
$$7 + 3 = 6 + 4 = 5 + 5 = (12)_8$$

Example 13.2

Perform the following operations.

(a) $(2)_8 + (5)_8$

(b) $(7)_8 + (6)_8$

(c) Convert $(75)_{10}$ to base-8.

(d) Convert $(0.14)_{10}$ to base-8.

(e) Convert $(13)_8$ to base-10.

(f) Convert $(27.52)_8$ to base-10.

Solution

(a) The sum of 2 and 5 in base-10 is 7, which is less than 8 and, therefore, is a valid number in the octal system. The answer is $(7)_8$.

(b) The sum of 7 and 6 in base-10 is 13, which is greater than 8 (and, therefore, needs to be converted). Using the remainder method (see Sec. 13.3),

$$13 \div 8 = 1 \text{ remainder } 5$$
$$1 \div 8 = 0 \text{ remainder } 1$$

The answer is $(15)_8$.

(c) Use the remainder method (see Sec. 13.3).

$$75 \div 8 = 9 \text{ remainder } 3$$
$$9 \div 8 = 1 \text{ remainder } 1$$
$$1 \div 8 = 0 \text{ remainder } 1$$

The answer is $(113)_8$.

(d) Refer to Sec. 13.3.

$$0.14 \times 8 = 1.12$$
$$0.12 \times 8 = 0.96$$
$$0.96 \times 8 = 7.68$$
$$0.68 \times 8 = 5.44$$
$$0.44 \times 8 = \text{etc.}$$

The answer, $(0.1075\ldots)_8$, is constructed from the integer parts of the products.

(e) Use Eq. 13.1.

$$(1)(8) + 3 = (11)_{10}$$

(f) Use Eq. 13.1 and Eq. 13.2.

$$\begin{aligned}(2)(8)^1 &+ (7)(8)^0 \\ &+ (5)(8)^{-1} + (2)(8)^{-2} = 16 + 7 + \frac{5}{8} + \frac{2}{64} \\ &= (23.656)_{10}\end{aligned}$$

6. HEXADECIMAL NUMBER SYSTEM

The *hexadecimal (base-16) system* is a shorthand method of representing the value of four binary digits at a time.[3] Since 16 distinctly different characters are needed, the capital letters A through F are used to represent the decimal numbers 10 through 15. The progression of hexadecimal numbers is illustrated in Table 13.1.

Example 13.3

(a) Convert $(4D3)_{16}$ to base-10. (b) Convert $(1475)_{10}$ to base-16. (c) Convert $(0.8)_{10}$ to base-16.

Solution

(a) The hexadecimal number D is 13 in base-10. Using Eq. 13.1,

$$(4)(16)^2 + (13)(16)^1 + 3 = (1235)_{10}$$

[3]The term *hex number* is often heard.

Table 13.1 Binary, Octal, Decimal, and Hexadecimal Equivalents

binary	octal	decimal	hexadecimal
0	0	0	0
1	1	1	1
10	2	2	2
11	3	3	3
100	4	4	4
101	5	5	5
110	6	6	6
111	7	7	7
1000	10	8	8
1001	11	9	9
1010	12	10	A
1011	13	11	B
1100	14	12	C
1101	15	13	D
1110	16	14	E
1111	17	15	F
10000	20	16	10

(b) Use the remainder method. (See Sec. 13.3.)

$$1475 \div 16 = 92 \text{ remainder } 3$$
$$92 \div 16 = 5 \text{ remainder } 12$$
$$5 \div 16 = 0 \text{ remainder } 5$$

Since $(12)_{10}$ is $(C)_{16}$, (or hex C), the answer is $(5C3)_{16}$.

(c) Refer to Sec. 13.3.

$$0.8 \times 16 = 12.8$$
$$0.8 \times 16 = 12.8$$
$$0.8 \times 16 = \text{etc.}$$

Since $(12)_{10} = (C)_{16}$, the answer is $(0.CCCCC\ldots)_{16}$.

7. CONVERSIONS AMONG BINARY, OCTAL, AND HEXADECIMAL NUMBERS

The octal system is closely related to the binary system since $(2)^3 = 8$. Conversion from a binary to an octal number is accomplished directly by starting at the LSB (right-hand bit) and grouping the bits in threes. Each group of three bits corresponds to an octal digit. Similarly, each digit in an octal number generates three bits in the equivalent binary number.

Conversion from a binary to a hexadecimal number starts by grouping the bits (starting at the LSB) into fours. Each group of four bits corresponds to a hexadecimal digit. Similarly, each digit in a hexadecimal number generates four bits in the equivalent binary number.

Conversion between octal and hexadecimal numbers is easiest when the number is first converted to a binary number.

Example 13.4

(a) Convert $(5431)_8$ to base-2. (b) Convert $(1001011)_2$ to base-8. (c) Convert $(1011111101111001)_2$ to base-16.

Solution

(a) Convert each octal digit to binary digits.

$$(5)_8 = (101)_2$$
$$(4)_8 = (100)_2$$
$$(3)_8 = (011)_2$$
$$(1)_8 = (001)_2$$

The answer is $(101100011001)_2$.

(b) Group the bits into threes starting at the LSB.

1 001 011

Convert these groups into their octal equivalents.

$$(1)_2 = (1)_8$$
$$(001)_2 = (1)_8$$
$$(011)_2 = (3)_8$$

The answer is $(113)_8$.

(c) Group the bits into fours starting at the LSB.

1011 1111 0111 1001

Convert these groups into their hexadecimal equivalents.

$$(1011)_2 = (B)_{16}$$
$$(1111)_2 = (F)_{16}$$
$$(0111)_2 = (7)_{16}$$
$$(1001)_2 = (9)_{16}$$

The answer is $(BF79)_{16}$.

8. COMPLEMENT OF A NUMBER

The *complement*, N^*, of a number, N, depends on the machine (computer, calculator, etc.) being used. Assuming that the machine has a maximum number, n, of digits per integer number stored, the b's and $(b-1)$'s complements are

$$N_b^* = b^n - N \quad 13.3$$
$$N_{b-1}^* = N_b^* - 1 \quad 13.4$$

For a machine that works in base-10 arithmetic, the *ten's* and *nine's complements* are

$$N_{10}^* = 10^n - N \quad 13.5$$
$$N_9^* = N_{10}^* - 1 \quad 13.6$$

For a machine that works in base-2 arithmetic, the *two's* and *one's complements* are

$$N_2^* = 2^n - N \qquad 13.7$$

$$N_1^* = N_2^* - 1 \qquad 13.8$$

9. APPLICATION OF COMPLEMENTS TO COMPUTER ARITHMETIC

Equation 13.9 and Eq. 13.10 are the practical applications of complements to computer arithmetic.

$$(N^*)^* = N \qquad 13.9$$

$$M - N = M + N^* \qquad 13.10$$

The binary one's complement is easily found by switching all of the ones and zeros to zeros and ones, respectively. It can be combined with a technique known as *end-around carry* to perform subtraction. End-around carry is the addition of the *overflow bit* to the sum of N and its ones complement.

Example 13.5

(a) Simulate the operation of a base-10 machine with a capacity of four digits per number and calculate the difference $(18)_{10} - (6)_{10}$ with ten's complements.

(b) Simulate the operation of a base-2 machine with a capacity of five digits per number and calculate the difference $(01101)_2 - (01010)_2$ with two's complements.

(c) Solve part (b) with a one's complement and end-around carry.

Solution

(a) The ten's complement of 6 is

$$(6)_{10}^* = (10)^4 - 6 = 10{,}000 - 6 = 9994$$

Using Eq. 13.10,

$$18 - 6 = 18 + (6)_{10}^* = 18 + 9994 = 10{,}012$$

However, the machine has a maximum capacity of four digits. Therefore, the leading 1 is dropped, leaving 0012 as the answer.

(b) The two's complement of $(01010)_2$ is

$$\begin{aligned} N_2^* &= (2)^5 - N = (32)_{10} - N \\ &= (100000)_2 - (01010)_2 \\ &= (10110)_2 \end{aligned}$$

From Eq. 13.10,

$$\begin{aligned} (01101)_2 - (01010)_2 &= (01101)_2 + (10110)_2 \\ &= (100011)_2 \end{aligned}$$

Since the machine has a capacity of only five bits, the leftmost bit is dropped, leaving $(00011)_2$ as the difference.

(c) The one's complement is found by reversing all the digits.

$$(01010)_1^* = (10101)_2$$

Adding the one's complement,

$$(01101)_2 + (10101)_2 = (100010)_2$$

The leading bit is the overflow bit, which is removed and added to give the difference.

$$(00010)_2 + (1)_2 = (00011)_2$$

10. COMPUTER REPRESENTATION OF NEGATIVE NUMBERS

On paper, a minus sign indicates a negative number. This representation is not possible in a machine. Hence, one of the n digits, usually the MSB, is reserved for sign representation. (This reduces the machine's capacity to represent numbers to $n - 1$ bits per number.) It is arbitrary whether the sign bit is 1 or 0 for negative numbers as long as the MSB is different for positive and negative numbers.

The one's complement is ideal for forming a negative number since it automatically reverses the MSB. For example, $(00011)_2$ is a five-bit representation of decimal 3. The ones complement is $(11100)_2$, which is recognized as a negative number because the MSB is 1.

Example 13.6

Simulate the operation of a six-digit binary machine that uses one's complements for negative numbers.

(a) What is the machine representation of $(-27)_{10}$? (b) What is the decimal equivalent of the two's complement of $(-27)_{10}$? (c) What is the decimal equivalent of the one's complement of $(-27)_{10}$?

Solution

(a) $(27)_{10} = (011011)_2$. The negative of this number is the same as the one's complement: $(100100)_2$.

(b) The two's complement is one more than the one's complement. (See Eq. 13.7 and Eq. 13.8.) Therefore, the two's complement is

$$(100100)_2 + 1 = (100101)_2$$

This represents $(-26)_{10}$.

(c) From Eq. 13.9, the complement of a complement of a number is the original number. Therefore, the decimal equivalent is -27.

11. BOOLEAN ALGEBRA FUNDAMENTAL POSTULATES

Boolean algebra is a system of mathematics that deals with the algebraic treatment of logic. When the variables of the system are limited to only one of two values, the system is called two-valued. *Switching algebra* is the term often used to describe two-valued Boolean algebra. The basic postulates of switching algebra follow.

- Postulate 1: A Boolean variable x has two possible exclusive values, 0 and 1. That is,

$$\text{if } x = 0 \text{ then } x \neq 1, \text{ and}$$
$$\text{if } x = 1 \text{ then } x \neq 0$$

- Postulate 2: The NOT operation, "$\overline{x}$" or x', is defined as

$$\overline{0} = 1 \quad \overline{1} = 0$$

- Postulate 3: The logical operations AND ($\wedge$ or $\cdot$), and OR ($\vee$ or $+$) are defined as

$$0 \cdot 0 = 0 \quad 0 + 0 = 0$$
$$0 \cdot 1 = 0 \quad 0 + 1 = 1$$
$$1 \cdot 0 = 0 \quad 1 + 0 = 1$$
$$1 \cdot 1 = 1 \quad 1 + 1 = 1$$

Sometimes the symbol for the AND operation is dropped completely and $x_1 \cdot x_2$ is represented as x_1x_2.

12. BOOLEAN ALGEBRA LAWS

The process of reducing logical expressions to simpler forms is aided by the use of the following relationships, or laws, of switching algebra.[4] In the following, x can represent a single variable or a general logical function.

- Special properties of 0 and 1

$$0 + x = x \quad 0 \cdot x = 0$$
$$1 + x = 1 \quad 1 \cdot x = x$$

- Idempotence laws

$$x + x = x \quad x \cdot x = x$$

- Complementation laws

$$x + \overline{x} = 1 \quad x \cdot \overline{x} = 0$$

- Involution

$$\overline{(\overline{x})} = x$$

- Commutative laws

$$x + y = y + x \quad x \cdot y = y \cdot x$$

- Associative laws

$$x + (y + z) = (x + y) + z$$
$$x \cdot (y \cdot z) = (x \cdot y) \cdot z$$

- Distributive laws

$$x \cdot (y + z) = (x \cdot y) + (x \cdot z)$$
$$x + (y \cdot z) = (x + y) \cdot (x + z)$$

- Absorption laws

$$x + (x \cdot y) = x \qquad x \cdot (x + y) = x$$
$$x + (\overline{x} \cdot y) = x + y \quad x \cdot (\overline{x} + y) = x \cdot y$$

13. BOOLEAN ALGEBRA THEOREMS

The simplification of logical expressions is further enhanced by a set of relationships that exist as theorems, although one takes the name of a law. These theorems follow.

- Simplification theorems

$$xy + x\overline{y} = x \quad (x + y)(x + \overline{y}) = x$$

The AND symbol is often assumed between parentheses.

$$x + xy = x \qquad x(x + y) = x$$
$$(x + \overline{y})y = xy \quad x\overline{y} + y = x + y$$

- de Morgan's laws

$$\overline{(x + y + z + \cdots)} = \overline{x}\,\overline{y}\,\overline{z}\cdots$$
$$\overline{(xyz\cdots)} = \overline{x} + \overline{y} + \overline{z}\cdots$$
$$\overline{\big(f(x_1, x_2, x_3, \cdots, x_n, 0, 1, +, \cdot)\big)}$$
$$= f(\overline{x_1}, \overline{x_2}, \overline{x_3}, \cdots, \overline{x_n}, 1, 0, \cdot, +)$$

- Duality

$$(x + y + z + \cdots)^{\mathrm{D}} = xyz\cdots$$
$$(xyz\cdots)^{\mathrm{D}} = x + y + z + \cdots$$
$$\big(f(x_1, x_2, x_3, \cdots, x_n, 0, 1, +, \cdot)\big)^{\mathrm{D}}$$
$$= f(x_1, x_2, x_3, \cdots, x_n, 1, 0, \cdot, +)$$

- Consensus theorem

$$xy + yz + \overline{x}z = xy + \overline{x}z$$
$$(x + y)(y + z)(\overline{x} + z) = (x + y)(\overline{x} + z)$$
$$(x + y)(\overline{x} + z) = xz + \overline{x}y$$

[4]The reduction of such expressions is desired to minimize circuit complexity and costs when realizing a given expression with electronic logic circuits.

14 Numerical Analysis

1. NUMERICAL METHODS

Although the roots of second-degree polynomials are easily found by a variety of methods (by factoring, completing the square, or using the quadratic equation), easy methods of solving cubic and higher-order equations exist only for specialized cases. However, cubic and higher-order equations occur frequently in engineering, and they are difficult to factor. Trial and error solutions, including graphing, are usually satisfactory for finding only the general region in which the root occurs.

Numerical analysis is a general subject that covers, among other things, iterative methods for evaluating roots to equations. The most efficient numerical methods are too complex to present and, in any case, work by hand. However, some of the simpler methods are presented here. Except in critical problems that must be solved in real time, a few extra calculator or computer iterations will make no difference.[1]

2. FINDING ROOTS: BISECTION METHOD

The *bisection method* is an iterative method that "brackets" (also known as "straddles") an interval containing the *root* or *zero* of a particular equation.[2] The size of the interval is halved after each iteration. As the method's name suggests, the best estimate of the root after any iteration is the midpoint of the interval. The maximum error is half the interval length. The procedure continues until the size of the maximum error is "acceptable."[3]

The disadvantages of the bisection method are (a) the slowness in converging to the root, (b) the need to know the interval containing the root before starting, and (c) the inability to determine the existence of or find other real roots in the starting interval.

The bisection method starts with two values of the independent variable, $x = L_0$ and $x = R_0$, which straddle a root. Since the function passes through zero at a root, $f(L_0)$ and $f(R_0)$ will have opposite signs. The following algorithm describes the remainder of the bisection method.

Let n be the iteration number. Then, for $n = 0, 1, 2, \ldots$, perform the following steps until sufficient accuracy is attained.

step 1: Set $m = \frac{1}{2}(L_n + R_n)$.

step 2: Calculate $f(m)$.

step 3: If $f(L_n)f(m) \leq 0$, set $L_{n+1} = L_n$ and $R_{n+1} = m$. Otherwise, set $L_{n+1} = m$ and $R_{n+1} = R_n$.

step 4: $f(x)$ has at least one root in the interval $[L_{n+1}, R_{n+1}]$. The estimated value of that root, x^*, is

$$x^* \approx \tfrac{1}{2}(L_{n+1} + R_{n+1})$$

The maximum error is $\frac{1}{2}(R_{n+1} - L_{n+1})$.

Example 14.1

Use two iterations of the bisection method to find a root of

$$f(x) = x^3 - 2x - 7$$

Solution

The first step is to find L_0 and R_0, which are the values of x that straddle a root and have opposite signs. A table can be made and values of $f(x)$ calculated for random values of x.

x	−2	−1	0	+1	+2	+3
$f(x)$	−11	−6	−7	−8	−3	+14

Since $f(x)$ changes sign between $x = 2$ and $x = 3$, $L_0 = 2$ and $R_0 = 3$.

First iteration, $n = 0$:

$$m = \tfrac{1}{2}(L_n + R_n) = \left(\tfrac{1}{2}\right)(2 + 3) = 2.5$$

$$f(2.5) = x^3 - 2x - 7 = (2.5)^3 - (2)(2.5) - 7 = 3.625$$

[1]Most advanced hand-held calculators have "root finder" functions that use numerical methods to iteratively solve equations.
[2]The equation does not have to be a pure polynomial. The bisection method requires only that the equation be defined and determinable at all points in the interval.
[3]The bisection method is not a closed method. Unless the root actually falls on the midpoint of one iteration's interval, the method continues indefinitely. Eventually, the magnitude of the maximum error is small enough not to matter.

Since $f(2.5)$ is positive, a root must exist in the interval [2, 2.5]. Therefore, $L_1 = 2$ and $R_1 = 2.5$. At this point, the best estimate of the root is

$$x^* \approx \left(\tfrac{1}{2}\right)(2 + 2.5) = 2.25$$

The maximum error is $\left(\tfrac{1}{2}\right)(2.5 - 2) = 0.25$.

Second iteration, $n = 1$:

$$m = \left(\tfrac{1}{2}\right)(2 + 2.5) = 2.25$$

$$f(2.25) = (2.25)^3 - (2)(2.25) - 7 = -0.1094$$

Since $f(2.25)$ is negative, a root must exist in the interval [2.25, 2.5]. Therefore, $L_2 = 2.25$ and $R_2 = 2.5$. The best estimate of the root is

$$x^* \approx \tfrac{1}{2}(L_{n+1} + R_{n+1}) \approx \left(\tfrac{1}{2}\right)(2.25 + 2.5) = 2.375$$

The maximum error is $\left(\tfrac{1}{2}\right)(2.5 - 2.25) = 0.125$.

3. FINDING ROOTS: NEWTON'S METHOD

Many other methods have been developed to overcome one or more of the disadvantages of the bisection method. These methods have their own disadvantages.[4]

Newton's method is a particular form of *fixed-point iteration*. In this sense, "fixed point" is often used as a synonym for "root" or "zero." However, fixed-point iterations get their name from functions with the characteristic property $x = g(x)$ such that the limit of $g(x)$ is the fixed point (i.e., is the root).

All fixed-point techniques require a starting point. Preferably, the starting point will be close to the actual root.[5] And, while Newton's method converges quickly, it requires the function to be continuously differentiable.

Newton's method algorithm is simple. At each iteration (n = 0, 1, 2, etc.), Eq. 14.1 estimates the root. The maximum error is determined by looking at how much the estimate changes after each iteration. If the change between the previous and current estimates (representing the magnitude of error in the estimate) is too large, the current estimate is used as the independent variable for the subsequent iteration.[6]

$$x_{n+1} = g(x_n) = x_n - \frac{f(x_n)}{f'(x_n)} \qquad 14.1$$

[4]The *regula falsi (false position) method* converges faster than the bisection method but is unable to specify a small interval containing the root. The *secant method* is prone to round-off errors and gives no indication of the remaining distance to the root.

[5]Theoretically, the only penalty for choosing a starting point too far away from the root will be a slower convergence to the root.

[6]Actually, the theory defining the maximum error is more definite than this. For example, for a large enough value of n, the error decreases approximately linearly. Therefore, the consecutive values of x_n converge linearly to the root as well.

Example 14.2

Solve Ex. 14.1 using two iterations of Newton's method. Use $x_0 = 2$.

Solution

The function and its first derivative are

$$f(x) = x^3 - 2x - 7$$

$$f'(x) = 3x^2 - 2$$

First iteration, $n = 0$:

$$x_0 = 2$$

$$f(x_0) = f(2) = (2)^3 - (2)(2) - 7 = -3$$

$$f'(x_0) = f'(2) = (3)(2)^2 - 2 = 10$$

$$x_1 = x_0 - \frac{f(x_0)}{f'(x_0)} = 2 - \frac{-3}{10} = 2.3$$

Second iteration, $n = 1$:

$$x_1 = 2.3$$

$$f(x_1) = (2.3)^3 - (2)(2.3) - 7 = 0.567$$

$$f'(x_1) = (3)(2.3)^2 - 2 = 13.87$$

$$x_2 = x_1 - \frac{f(x_1)}{f'(x_1)} = 2.3 - \frac{0.567}{13.87} = 2.259$$

4. NONLINEAR INTERPOLATION: LAGRANGIAN INTERPOLATING POLYNOMIAL

Interpolating between two points of known data is common in engineering. Primarily due to its simplicity and speed, straight-line interpolation is used most often. Even if more than two points on the curve are explicitly known, they are not used. Since straight-line interpolation ignores all but two of the points on the curve, it ignores the effects of curvature.

A more powerful technique that accounts for the curvature is the *Lagrangian interpolating polynomial*. This method uses an nth degree parabola (polynomial) as the interpolating curve.[7] This method requires that $f(x)$ be continuous and real-valued on the interval

[7]The Lagrangian interpolating polynomial reduces to straight-line interpolation if only two points are used.

$[x_0, x_n]$ and that $n + 1$ values of $f(x)$ are known corresponding to $x_0, x_1, x_2, \ldots, x_n$.

The procedure for calculating $f(x)$ at some intermediate point x^* starts by calculating the Lagrangian interpolating polynomial for each known point.

$$L_k(x^*) = \prod_{\substack{i=0\\ i\neq k}}^{n} \frac{x^* - x_i}{x_k - x_i} \qquad 14.2$$

The value of $f(x)$ at x^* is calculated from Eq. 14.3.

$$f(x^*) = \sum_{k=0}^{n} f(x_k) L_k(x^*) \qquad 14.3$$

The Lagrangian interpolating polynomial has two primary disadvantages. The first is that a large number of additions and multiplications are needed.[8] The second is that the method does not indicate how many interpolating points should be (or should have been) used. Other interpolating methods have been developed that overcome these disadvantages.[9]

Example 14.3

A real-valued function has the following values.

$$f(1) = 3.5709$$

$$f(4) = 3.5727$$

$$f(6) = 3.5751$$

Use the Lagrangian interpolating polynomial to determine the value of the function at 3.5.

Solution

The procedure for applying Eq. 14.2, the Lagrangian interpolating polynomial, is illustrated in tabular form. Notice that the term corresponding to $i = k$ is omitted from the product.

$k = 0$: $\quad i = 0 \quad i = 1 \quad i = 2$

$$L_0(3.5) = \left(\frac{3.5-1}{1-1}\right)\left(\frac{3.5-4}{1-4}\right)\left(\frac{3.5-6}{1-6}\right) = 0.08333$$

(the first factor is crossed out)

$k = 1$:

$$L_1(3.5) = \left(\frac{3.5-1}{4-1}\right)\left(\frac{3.5-4}{4-4}\right)\left(\frac{3.5-6}{4-6}\right) = 1.04167$$

(the second factor is crossed out)

$k = 2$:

$$L_2(3.5) = \left(\frac{3.5-1}{6-1}\right)\left(\frac{3.5-4}{6-4}\right)\left(\frac{3.5-6}{6-6}\right) = -0.12500$$

(the third factor is crossed out)

Equation 14.3 is used to calculate the estimate.

$$\begin{aligned} f(3.5) &= (3.5709)(0.08333) + (3.5727)(1.04167) \\ &\quad + (3.5751)(-0.12500) \\ &= 3.57225 \end{aligned}$$

5. NONLINEAR INTERPOLATION: NEWTON'S INTERPOLATING POLYNOMIAL

Newton's form of the interpolating polynomial is more efficient than the Lagrangian method of interpolating between known points.[10] Given $n + 1$ known points for $f(x)$, the *Newton form of the interpolating polynomial* is

$$f(x^*) = \sum_{i=0}^{n} \left(f[x_0, x_1, \ldots, x_i] \prod_{j=0}^{i-1} (x^* - x_j) \right) \qquad 14.4$$

$f[x_0, x_1, \ldots, x_i]$ is known as the ith *divided difference.*

$$f[x_0, x_1, \ldots, x_i] = \sum_{k=0}^{i} \left(\frac{f(x_k)}{\begin{array}{c}(x_k - x_0)\cdots(x_k - x_{k-1}) \\ \times (x_k - x_{k+1}) \cdots (x_k - x_i)\end{array}} \right) \qquad 14.5$$

It is necessary to define the following two terms.

$$f[x_0] = f(x_0) \qquad 14.6$$

$$\prod (x^* - x_j) = 1 \quad [i=0] \qquad 14.7$$

Example 14.4

Repeat Ex. 14.3 using Newton's form of the interpolating polynomial.

[8]As with the numerical methods for finding roots previously discussed, the number of calculations probably will not be an issue if the work is performed by a calculator or computer.

[9]Other common methods for performing interpolation include the *Newton form* and *divided difference table.*

[10]In this case, "efficiency" relates to the ease in adding new known points without having to repeat all previous calculations.

Solution

Since there are $n + 1 = 3$ data points, $n = 2$. Evaluate the terms for $i = 0$–2.

$i = 0$:

$$f[x_0]\prod_{j=0}^{-1}(x^* - x_j) = f(x_0)(1) = f(x_0)$$

$i = 1$:

$$f[x_0, x_1]\prod_{j=0}^{0}(x^* - x_j) = f[x_0, x_1](x^* - x_0)$$

$$f[x_0, x_1] = \frac{f(x_0)}{x_0 - x_1} + \frac{f(x_1)}{x_1 - x_0}$$

$i = 2$:

$$f[x_0, x_1, x_2]\prod_{j=0}^{1}(x^* - x_j) = f[x_0, x_1, x_2](x^* - x_0) \times (x^* - x_1)$$

$$f[x_0, x_1, x_2] = \frac{f(x_0)}{(x_0 - x_1)(x_0 - x_2)} + \frac{f(x_1)}{(x_1 - x_0)(x_1 - x_2)} + \frac{f(x_2)}{(x_2 - x_0)(x_2 - x_1)}$$

Substitute known values.

$$\begin{aligned} f(3.5) &= 3.5709 + \left(\frac{3.5709}{1-4} + \frac{3.5727}{4-1}\right)(3.5-1) \\ &\quad + \left(\frac{3.5709}{(1-4)(1-6)} + \frac{3.5727}{(4-1)(4-6)} + \frac{3.5751}{(6-1)(6-4)}\right) \\ &\quad \times (3.5-1)(3.5-4) \\ &= 3.57225 \end{aligned}$$

This answer is the same as that determined in Ex. 14.3.

15 Advanced Engineering Mathematics

1. OVERVIEW

This chapter emphasizes advanced engineering mathematics not covered in earlier chapters. A grasp of such material is necessary in order to understand various aspects of electrical engineering, especially at the design and theoretical levels. The province of the engineer in mathematics, however, resides in the recognition of the form of, and restrictions on, the solution.

2. POWER SERIES METHOD: HOMOGENEOUS LINEAR DIFFERENTIAL EQUATIONS

When the coefficients of a homogeneous linear differential equation are a function of x, the methods explained in Chap. 11 are no longer applicable.[1] Further, the solutions may be *nonelementary functions*.[2] Equations of this type include Bessel's equation and Legendre's equation.[3]

- Bessel Equation of Order ν

$$x^2y'' + xy' + (x^2 - \nu^2)y = 0 \qquad 15.1$$

- Legendre Equation of Order λ

$$(1 - x^2)y'' - 2xy' + \lambda(\lambda + 1)y = 0 \quad [-1 < x < 1] \qquad 15.2$$

The solutions of such equations are power series; the method of obtaining them is the *power series method*. A power series is of the following form.[4]

$$\sum_{m=0}^{\infty} c_m(x-a)^m = c_0 + c_1(x-a) + c_2(x-a)^2 + \cdots \qquad 15.3$$

The constants c_0, c_1, ... are the *coefficients* of the series. The constant a is called the *center*. The variable is x.

A common power series is the Maclaurin series.

$$\frac{1}{1-x} = \sum_{m=0}^{\infty} x^m = 1 + x + x^2 + \cdots \quad \left[\begin{array}{l}\text{If } |x| < 1, \text{ then this power series}\\ \text{is known as a geometric series.}\end{array}\right] \qquad 15.4$$

[1]Equations discussed in this chapter are *ordinary differential equations* (ODEs) as are most equations in Chap. 11. The term "ordinary" distinguishes these equations from *partial differential equations* (PDEs).
[2]*Elementary functions* are those that can be formed by a finite number of mathematical operations (addition, subtraction, multiplication, and division) from algebraic functions and exponential, logarithmic, and trigonometric functions. *Nonelementary functions* are those that cannot be so formed. Do not confuse nonelementary functions with transcendental functions. *Transcendental functions* are nonalgebraic functions, for example, exponential, logarithmic, and trigonometric.
[3]Bessel and Legendre equations and their associated polynomials occur in many engineering applications. As a single example, they are used in approximating the response of filter circuits.
[4]A power series does not include negative or fractional powers of x.

Mathematics

$$e^x = \sum_{m=0}^{\infty} \frac{x^m}{m!}$$

$$= 1 + x + \frac{x^2}{2!} + \frac{x^3}{3!} + \cdots \qquad 15.5$$

$$\cos x = \sum_{m=0}^{\infty} \frac{(-1)^m x^{2m}}{2m!}$$

$$= 1 - \frac{x^2}{2!} + \frac{x^4}{4!} - \cdots \qquad 15.6$$

$$\sin x = \sum_{m=0}^{\infty} \frac{(-1)^m x^{2m+1}}{(2m+1)!}$$

$$= x - \frac{x^3}{3!} + \frac{x^5}{5!} - \cdots \qquad 15.7$$

The steps to the power series method are as follows.

step 1: Given a differential equation, represent the given functions in the equation by a power series in powers of x (or $x-a$, if solutions of this type are desired). If the functions are polynomials, as is often the case in practical applications, nothing need be done in this first step.

step 2: Assume a solution in the form of a power series.

step 3: Substitute this series and the series obtained by termwise differentiation into the original equation.

step 4: Collect like powers of x and equate the sum of their coefficients to zero.

step 5: Determine the unknown coefficients in the assumed power series solution successively.

step 6: Substitute the values of the unknown coefficients into the assumed power series solution.

Example 15.1

Solve the following homogeneous, first-order linear ordinary differential equation with constant coefficients.[5]

$$y' + ky = 0$$

Solution

step 1: Represent the given functions in the equation by a power series in powers of x. Because the coefficients are constant (i.e., no functions are used in the equation), this step is not required.

step 2: Assume a solution in the form of a power series.

$$y(x) = c_0 + c_1 x + c_2 x^2 + \cdots = \sum_{m=0}^{\infty} c_m x^m$$

[5]The solution to this differential equation was found using other methods in Sec. 11.2. The power series method is used on this simple differential equation to clarify the method.

step 3: Substitute this series and the series obtained by termwise differentiation into the original equation. Differentiation of $y(x)$ termwise results in the following equation.

$$y'(x) = 0 + c_1 + 2c_2 x + 3c_3 x^2 + \cdots$$

Substituting into the original equation gives

$$y' + ky = 0$$

$$(0 + c_1 + 2c_2 x + 3c_3 x^2 + \cdots) + k(c_0 + c_1 x + c_2 x^2 + \cdots) = 0$$

step 4: Collect like powers of x and equate the sum of their coefficients to zero.

$$(c_1 + kc_0) + (2c_2 + kc_1)x + (3c_3 + kc_2)x^2 + \cdots = 0$$

For this equation to be true, the sum of the coefficients of each power of x must be zero. That is,

$$c_1 + kc_0 = 0$$
$$2c_2 + kc_1 = 0$$
$$3c_3 + kc_2 = 0 \cdots$$

step 5: Determine the unknown coefficients in the assumed power series successively. That is, using the results of step 4, determine the coefficients in terms of c_0, which remains arbitrary.

$$c_1 = -kc_0$$

$$c_2 = \frac{-k}{2} c_1 = \frac{k^2}{2!} c_0$$

$$c_3 = \frac{-k}{3} c_2 = \frac{-k^3}{3!} c_0 \ \ldots$$

step 6: Substitute these values into the assumed power series solution from step 2.

$$y(x) = c_0 + c_1 x + c_2 x^2 + \cdots = \sum_{m=0}^{\infty} c_m x^m$$

$$= c_0 + (-kc_0)x + \left(\frac{k^2}{2!} c_0\right) x^2 + \left(\frac{-k^3}{3!} c_0\right) x^3 + \cdots$$

$$= c_0 \left(1 - kx + \frac{k^2}{2!} x^2 - \frac{k^3}{3!} x^3 + \cdots\right)$$

For clarification, let $-kx = z$.

$$y(z) = c_0 \left(1 + z + \frac{z^2}{2!} + \frac{z^3}{3!} + \cdots\right)$$

The terms in parentheses represent the Maclaurin series for e^z. The solution is

$$y(z) = c_0 e^z$$

or

$$y(x) = c_0 e^{-kx}$$

The initial conditions can be used to determine the value of the constant c_0.

3. EXTENDED POWER SERIES METHOD: HOMOGENEOUS LINEAR DIFFERENTIAL EQUATIONS

If a negative power of x exists in an equation—in other words, if the equation is not analytic at $x = 0$—the *extended power series method* is used. (In simpler terms, with a negative power of x, x is in the denominator; the equation cannot be defined as $x = 0$ because this involves division by zero. This is also referenced as a *singularity.*) The extended power series method can be used on any differential equation of the following form.

$$y'' + \left(\frac{a(x)}{x}\right)y' + \left(\frac{b(x)}{x^2}\right)y = 0 \qquad 15.8$$

The Cauchy or Euler equation, Eq. 15.9, is of this form.[6]

$$x^2y'' + axy' + by = 0 \qquad 15.9$$

Rearranging to show that the equation is not analytic at $x=0$ (i.e., dividing by x^2 and allowing a and b to be functions of x) changes Eq. 15.9 into the form of Eq. 15.8.

Gauss' hypergeometric equation, Eq. 15.10, can also be solved using this method.[7]

$$x(1-x)y'' + \Big(c - (a+b+1)x\Big)y' - aby = 0 \qquad 15.10$$

In Eq. 15.8, it is assumed that $a(x)$ and $b(x)$ are analytic at $x=0$; therefore, at least one solution can be represented in the form[8]

$$\begin{aligned} y(x) &= x^r \sum_{m=0}^{\infty} c_m x^m \\ &= x^r(c_0 + c_1x + c_2x^2 + \cdots) \end{aligned} \qquad 15.11$$

[6]This form is equivalent to that given in Chap. 11.
[7]This equation applies to probability and statistics.
[8]Using the form of the solution shown, the method is called the *Frobenius method.*

The exponent r may be any real or complex number and is chosen so that $c_0 \neq 0$.[9] Expanding $a(x)$ and $b(x)$ into a power series, differentiating Eq. 15.11, substituting into Eq. 15.8, and equating the sum of the coefficients of each power of x to zero yields the *indicial equation*, from the coefficients associated with x^r.

$$r^2 + (a_0 - 1)r + b_0 = 0 \qquad 15.12$$

Equation 15.12 will yield a basis of solutions. One of the solutions will be of the form of Eq. 15.11; the other form depends on the roots of the indicial equation. There are three possible cases depending on these roots.

case 1: distinct roots r_1 and r_2, which do not differ by an integer[10]

$$y_1(x) = x^{r_1}(c_0 + c_1x + c_2x^2 + \cdots)$$
$$y_2(x) = x^{r_2}(d_0 + d_1x + d_2x^2 + \cdots)$$

case 2: double root, r

$$y_1(x) = x^r(c_0 + c_1x + c_2x^2 + \cdots)$$
$$r = \frac{1-a_0}{2}$$
$$y_2(x) = y_1(x)\ln x + x^r \sum_{m=1}^{\infty} A_m x^m$$

The a_0 in the equation for $y_1(x)$ is from the indicial equation, Eq. 15.12.

case 3: distinct roots, r_1 and r_2, that differ by an integer, for example, 1, 2, 3, ...

$$y_1(x) = x^{r_1}(c_0 + c_1x + c_2x^2 + \cdots)$$
$$y_2(x) = k_p y_1(x)\ln x + x^{r_2} \sum_{m=0}^{\infty} c_m x^m \quad [x > 0]$$

4. GENERAL METHOD FOR SOLVING NONHOMOGENEOUS DIFFERENTIAL EQUATIONS

In Chap. 11, the method of undetermined coefficients was used to find a particular solution to nonhomogeneous equations. It was limited to simpler differential equations in which the coefficients $p(x)$ and $q(x)$ are constants and the forcing function $f(x)$ was of a particular form. The method of *variation of parameters* is completely general but more complicated, and it removes all such restrictions.

[9]Restraining r to be a nonnegative integer would make the entire expression a power series, as covered in Sec. 15.2 and Sec. 15.3.
[10]This includes complex conjugate roots.

Consider the nonhomogeneous second-order linear differential equation, Eq. 15.13.[11]

$$y'' + p(x)y' + q(x)y = f(x) \qquad 15.13$$

The homogeneous portion of Eq. 15.13 has a solution of the following form.

$$y_h(x) = c_1y_1(x) + c_2y_2(x) \qquad 15.14$$

The coefficients c_1 and c_2 are selected to be functions $u(x)$ and $v(x)$ that make the following a particular solution to Eq. 15.13.[12]

$$y_p(x) = u(x)y_1(x) + v(x)y_2(x) \qquad 15.15$$

The parameters $u(x)$ and $v(x)$ are then determined in terms of $y_1(x)$ and $y_2(x)$ by rearranging the particular solution, Eq. 15.15. The result is substituted into the homogeneous solution, Eq. 15.14. The associated first and second derivatives are determined. The results are substituted into Eq. 15.13. The result is a particular solution of the following completely general form.

$$\begin{aligned} y_p(x) &= -y_1(x)\int \frac{y_2(x)f(x)}{W}dx \\ &\quad + y_2(x)\int \frac{y_1(x)f(x)}{W}dx \end{aligned} \qquad 15.16$$

The term W is called the Wronskian of y_1 and y_2 and is given by

$$W = y_1(x)y_2'(x) - y_1'(x)y_2(x) \qquad 15.17$$

The parameters $u(x)$ and $v(x)$ are

$$u(x) = -\int \frac{y_2(x)f(x)}{W}dx \qquad 15.18$$

$$v(x) = \int \frac{y_1(x)f(x)}{W}dx \qquad 15.19$$

The overall result is that the combination of Eq. 15.14 and Eq. 15.16 constitute a general solution to Eq. 15.13, that is, to any nonhomogeneous differential equation.

Example 15.2

Solve the following differential equation.

$$y'' + y = \sec x$$

Solution

The homogeneous portion of the equation is

$$y'' + y = 0$$

The characteristic equation is[13]

$$\begin{aligned} r^2 + 0r + 1 &= 0 \\ r^2 &= -1 \\ r &= \pm\sqrt{-1} = \pm j \end{aligned}$$

The homogeneous solution is

$$y_h(x) = c_1 \cos x + c_2 \sin x \qquad \text{[I]}$$

From this,

$$y_1 = \cos x \text{ and } y_2 = \sin x$$

From Eq. 15.17, the Wronskian is

$$\begin{aligned} W\big(y_1(x), y_2(x)\big) &= y_1(x)y_2'(x) - y_1'(x)y_2(x) \\ &= \cos x \cos x - (-\sin x)\sin x \\ &= \cos^2 x + \sin^2 x = 1 \end{aligned}$$

Substitute into Eq. 15.16 to determine the particular solution.

$$\begin{aligned} y_p(x) &= -y_1(x)\int \frac{y_2(x)f(x)}{W}dx \\ &\quad + y_2(x)\int \frac{y_1(x)f(x)}{W}dx \\ &= -\cos x\int \frac{\sin x \sec x}{1}dx \\ &\quad + \sin x\int \frac{\cos x \sec x}{1}dx \end{aligned}$$

Substitute the definition of the secant and then integrate.

$$\begin{aligned} y_p(x) &= -\cos x\int \sin x\left(\frac{1}{\cos x}\right)dx \\ &\quad + \sin x\int \cos x\left(\frac{1}{\cos x}\right)dx \\ &= -\cos x\int \tan x dx + \sin x\int dx \\ &= -\cos x(-\ln|\cos x|) + (\sin x)(x) \\ &= \cos x \ln|\cos x| + x\sin x \end{aligned} \qquad \text{[II]}$$

[11]The functions p, q, and f are continuous on an open interval I.
[12]This is defined on the same open interval I mentioned in the previous footnote. Also, recall that since $j = 1\angle 90°$, $j^2 = 1\angle 180° = -1$, making $j = \sqrt{-1}$.

[13]See Chap. 11.

Combine the homogeneous (or complementary) solution I with the particular solution II to obtain the general solution.

$$\begin{aligned} y(x) &= y_h(x) + y_p(x) \\ &= c_1\cos x + c_2\sin x + \cos x\ \ln|\cos x| + x\sin x \\ &= (c_1 + \ \ln|\cos x|)\cos x + (c_2 + x)\sin x \end{aligned}$$

5. ORTHOGONALITY OF FUNCTIONS

Let $g_m(x)$ and $g_n(x)$ be two real-valued functions defined on an interval $a \leq x \leq b$. Further, define the integral of the two as

$$(g_m, g_n) = \int_a^b g_m(x)g_n(x)dx \qquad 15.20$$

The functions g_m and g_n are *orthogonal* on the interval $a \leq x \leq b$ if the integral in Eq. 15.20 equals zero.[14] That is,

$$\begin{aligned} (g_m, g_n) &= \int_a^b g_m(x)g_n(x)dx \\ &= 0 \quad [m \neq n] \end{aligned} \qquad 15.21$$

The nonnegative square root of (g_m, g_m) is called the *norm* of g_m.

$$\|g_m\| = \sqrt{(g_m, g_m)} = \sqrt{\int_a^b g_m^2(x)dx} \qquad 15.22$$

6. LAPLACE TRANSFORM: INITIAL AND FINAL VALUE THEOREMS

Laplace transforms simplify the solution of nonhomogeneous differential equations. Additionally, the initial conditions are automatically included in the analysis.[15]

$$\begin{aligned} f(t) &= L^{-1}\{F(s)\} \\ &= \frac{1}{2\pi j}\int_{c-j\infty}^{c+j\infty} F(s)e^{st}ds \end{aligned} \qquad 15.23$$

Equation 15.23 can be used to obtain the inverse transform (i.e., moves from the s domain back into the time domain), although the use of transform tables is more common. Further, the exact inverse transform may not be necessary. To understand how the equation, that is, the circuit, reacts at $t=0$, use the *initial value theorem*, Eq. 15.24.[16]

$$f(0) = \lim_{s\to\infty} sF(s) \qquad 15.24$$

The *final value theorem*, Eq. 15.25, can be used to determine the value of the function in the time domain as time approaches infinity.[17]

$$f(\infty) = \lim_{s\to 0} sF(s) \qquad 15.25$$

Example 15.3

The Laplace transform of a given voltage is shown. Evaluate the voltage at $t=0$.

$$V(s) = \frac{4s}{(s^2+9)(s^2+1)}$$

Solution

Expand the denominator and divide the numerator and denominator by s^2.

$$V(s) = \frac{4s}{s^4 + 10s^2 + 9} = \frac{\frac{4}{s}}{s^2 + 10 + \frac{9}{s^2}}$$

Apply the initial value theorem, Eq. 15.24.

$$\begin{aligned} v(0) &= \lim_{s\to\infty} sV(s) \\ &= \lim_{s\to\infty} s\left(\frac{\frac{4}{s}}{s^2 + 10 + \frac{9}{s^2}}\right) \\ &= \lim_{s\to\infty}\left(\frac{4}{s^2 + 10 + \frac{9}{s^2}}\right) \\ &= 0 \end{aligned}$$

The voltage function associated with the given voltage is

$$v(t) = \sin t \sin 2t$$

The value of $v(t)$ is indeed zero at $t=0$ as $\sin(0)=0$, confirming the answer.

[14] Both Legendre polynomials and Bessel functions are orthogonal. Important orthogonal sets of functions arise as solutions to linear second-order differential equations.

[15] The terms $f(0)$ and $f'(0)$ are required when transforming a second-order differential, thereby immediately placing the initial conditions into use.

[16] $f(0)$ is more technically $f(0^+)$, that is, the value at zero when approached from the positive side. This is based on the normally used Laplace transform being defined from zero to infinity.

[17] Technically, Eq. 15.25 should be $\lim_{t\to\infty} f(t) = \lim_{s\to 0} sF(s)$.

Example 15.4

Given the voltage in the frequency domain (or s domain), as shown, what is the value of the voltage at $t=\infty$?[18]

$$V(s) = \frac{1}{s(s+1)^2}$$

Solution

Expand the denominator.

$$\begin{aligned} V(s) &= \frac{1}{s(s+1)^2} = \frac{1}{s(s+1)(s+1)} = \frac{1}{s(s^2+2s+1)} \\ &= \frac{1}{s^3+2s^2+s} \end{aligned}$$

Apply the final value theorem, Eq. 15.25.

$$\begin{aligned} v(\infty) &= \lim_{s\to 0} sV(s) \\ &= \lim_{s\to 0} s\left(\frac{1}{s^3+2s^2+s}\right) \\ &= \lim_{s\to 0} \left(\frac{s}{s^3+2s^2+s}\right) \\ &= \lim_{s\to 0} \left(\frac{1}{s^2+2s+1}\right) = 1 \end{aligned}$$

The voltage function associated with the given voltage is

$$v(t) = 1 - e^{-t} - te^{-t}$$

The value of $v(t)$ as $t \to \infty$ can be seen to be equal to 1 from this equation, confirming the solution.[19]

7. HALF-RANGE EXPANSIONS

In some engineering problems, a function $f(t)$ defined over a finite interval may be more readily manipulated and a solution more easily obtained if the function is represented as a Fourier series. (See Sec. 15.8.)

Given a function $f(t)$ defined on some interval $0 \le t \le l$ or $0 \le t \le T/2$ where T is the period,[20] the *even periodic extension* of $f(t)$ of period $T=2l$ is given by the Fourier cosine series.

$$f(t) = a_0 + \sum a_n \cos \frac{n\pi t}{l} \qquad \text{15.26}$$

[18]Asking for the value of the voltage at $t=\infty$ is the equivalent of asking for the steady-state value.
[19]As t approaches ∞, the value of s approaches 0. In other words, as the function, or circuit, approaches a steady-state condition, the number of oscillations (frequency) approaches zero.
[20]The definition of the interval means the period is $T/2=l$ or $T=2l$.

The coefficients of the series are given by the following.

$$a_0 = \frac{1}{l}\int_0^l f(t)\,dt \qquad \text{15.27}$$

$$a_n = \frac{2}{l}\int_0^l f(t)\cos\frac{n\pi t}{l}\,dt \quad [n=1,2,3,\ldots] \qquad \text{15.28}$$

The *odd periodic extension* of $f(t)$ on the same interval is given by the Fourier sine series.

$$f(t) = \sum_{n=1}^{\infty} b_n \sin\frac{n\pi t}{l} \qquad \text{15.29}$$

The coefficients are

$$b_n = \frac{2}{l}\int_0^l f(t)\sin\frac{n\pi t}{l}\,dt \quad [n=1,2,3,\ldots] \qquad \text{15.30}$$

The series represented by Eq. 15.26 and Eq. 15.29 along with their associated coefficients are called the *half-range expansions* of the given function $f(t)$. Figure 15.1 shows a given function $f(t)$ and its half-range expansions. In the half-range expansion in Fig. 15.1, $f_1(t) = f(t)$ on the interval $0 \le t \le l$. Also, $f_2(t) = f(t)$ on this same interval.

Figure 15.1 Periodic Extensions

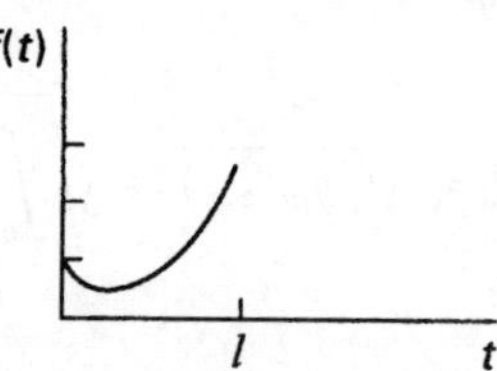

(a) given function $f(t)$

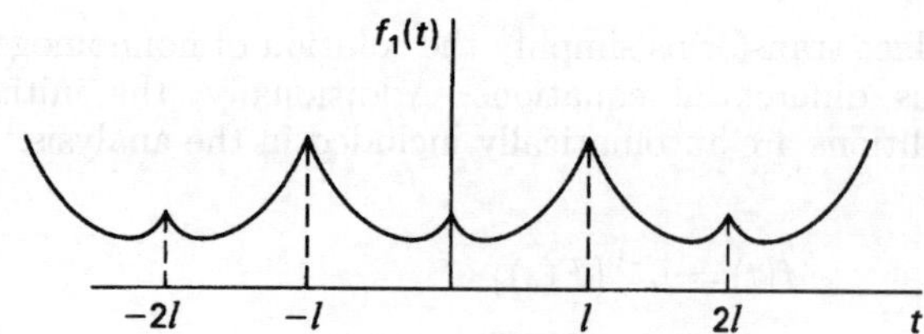

(b) even periodic extension of $f(t)$

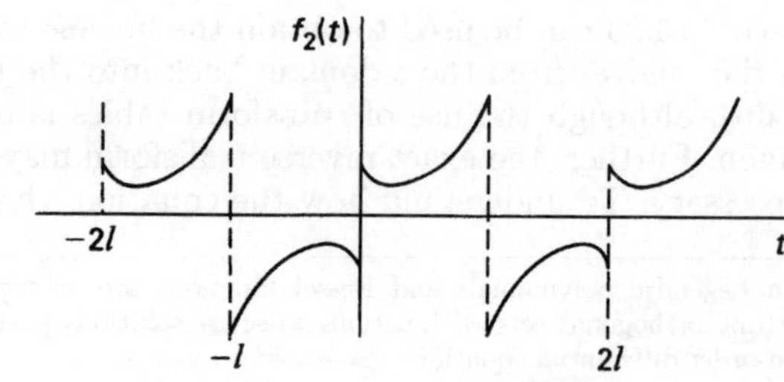

(c) odd periodic extension of $f(t)$

Summing the infinite series from the graphs in Fig. 15.1 leaves the original function $f(t)$ in place, as was given in Eq. 15.29. This is especially apparent in Fig. 15.1(c).

8. PARTIAL DIFFERENTIAL EQUATIONS: SEPARATION OF VARIABLES

A *partial differential equation* involves one or more partial derivatives of an unknown function of two or more variables. As before, the order of the equation is the order of the highest derivative. The equation is *linear* if the dependent variable and its partial derivatives are of the first degree, that is, not raised to a power greater than one. The equation is defined as *homogeneous* if each term consists of a dependent variable or one of its derivatives.

Some important second-order partial differential equations follow. (In the following equations, c is a constant, t is time, and u represents the unknown function. All are shown in Cartesian coordinates. Additionally, the constant c is shown as c^2 to indicate a positive quantity in the physical world.)

One-dimensional wave equation:[21]

$$\frac{\partial^2 u}{\partial t^2} = c^2 \frac{\partial^2 u}{\partial x^2} \qquad 15.31$$

One-dimensional heat equation:

$$\frac{\partial u}{\partial t} = c^2 \frac{\partial^2 u}{\partial x^2} \qquad 15.32$$

Two-dimensional Laplace equation:

$$\frac{\partial^2 u}{\partial x^2} + \frac{\partial^2 u}{\partial y^2} = 0 \qquad 15.33$$

Two-dimensional Poisson equation:

$$\frac{\partial^2 u}{\partial x^2} + \frac{\partial^2 u}{\partial y^2} = f(x, y) \qquad 15.34$$

Three-dimensional Laplace equation:[22]

$$\frac{\partial^2 u}{\partial x^2} + \frac{\partial^2 u}{\partial y^2} + \frac{\partial^2 u}{\partial z^2} = 0 \qquad 15.35$$

The equations are solved based on the conditions at a boundary, that is, the *boundary conditions*. Those equations having time as one of the variables may already have the conditions at $t=0$ (the *initial conditions*) given or known, and they are solved as constrained by those conditions.

An important aspect of the solutions to linear partial differential equations is the following fundamental theorem. If u_1 and u_2 are any of the solutions of a linear homogeneous partial differential equation in some region, then another solution is

$$u = c_1 u_1 + c_2 u_2 \qquad 15.36$$

The terms c_1 and c_2 represent constants.

The following method of solving such problems is termed the *separation of variables.*[23]

step 1: Apply the separation of variables to obtain ordinary differential equations.[24]

step 2: Determine solutions that satisfy the boundary conditions.

step 3: Compose the solutions such that the initial conditions are satisfied.

Example 15.5

Solve the one-dimensional wave equation with the given boundary and initial conditions.

$$\frac{\partial^2 u}{\partial t^2} = c^2 \left(\frac{\partial^2 u}{\partial x^2}\right)$$

The boundary conditions are that the function $u(x, t)$ equals zero at $x=0$ and $x=l$.[25] This can be expressed in mathematical terms as

$$u(0, t) = 0 \qquad [\text{I}]$$

$$u(l, t) = 0 \qquad [\text{II}]$$

Further, this function must take on the values in conditions I and II for all t. Let the initial deflection be $f(x)$ and the initial velocity be $g(x)$.

$$u(x, 0) = f(x) \qquad [\text{III}]$$

$$\left.\frac{\partial u}{\partial t}\right|_{t=0} = g(x) \qquad [\text{IV}]$$

Solution

step 1: Apply the separation of variables to obtain ordinary differential equations. Separating variables yields an unknown function of the following form.

$$u(x, t) = F(x)G(t)$$

[21]A solution to this one-dimensional case is given in Ex. 15.5. Solutions to the two-dimensional case in Cartesian coordinates involve double Fourier series. Solutions to the two-dimensional case in polar coordinates involve Bessel functions.

[22]The theory of solutions to this equation is called *potential theory.*

[23]The separation of variables method is also called the *product method.*

[24]Recall that ordinary differential equations are those that do not involve partial differentiation.

[25]A vibrating string secured at each end would provide such conditions, as would a confined electromagnetic wave.

From this function,

$$\frac{\partial^2 u(x,t)}{\partial t^2} = F(x)\ddot{G}(t)$$

$$\frac{\partial^2 u(x,t)}{\partial x^2} = F''(x)G(t)$$

The dots represent differentiation with respect to t. The primes represent differentiation with respect to x. Insert these results into the original wave equation.

$$\frac{\partial^2 u}{\partial t^2} = c^2\left(\frac{\partial^2 u}{\partial x^2}\right)$$

$$F(x)\ddot{G}(t) = c^2 F''(x)G(t)$$

$$\frac{F''(x)}{F(x)} = \frac{\ddot{G}(t)}{c^2 G(t)}$$

The variables t and x are now separated. Further, the result must be equal to a constant, say k. This can be deduced from the fact that changing the variable t changes only one side of the equation, and vice versa for the variable x.

$$\frac{F''(x)}{F(x)} = \frac{\ddot{G}(t)}{c^2 G(t)} = k$$

This leads to two linear ordinary differential equations (ODEs).

$$F''(x) - kF(x) = 0$$

$$\ddot{G}(t) - kc^2 G(t) = 0$$

step 2: Determine solutions based on the boundary conditions. Starting with the function $F(x)$, the boundary conditions in I and II were

$$u(0,t) = 0 = F(0)G(t)$$

$$u(l,t) = 0 = F(l)G(t)$$

Letting $G(t) \equiv 0$ gives $u(x,t) \equiv 0$, which is trivial. Therefore, $G(t) \neq 0$. This means that

$$F(0) = 0 \text{ and } F(l) = 0$$

If $k = 0$, the general solution of the ODE for the function $F(x)$ is of the form $F(x) = ax + b$.[26] Because $F(0) = 0$, b must be zero. Because $F(l) = 0$, a must be zero. Hence, $F(x) \equiv 0$ giving $u(x,t) \equiv 0$, all of which results in a trivial solution.

Let k be some positive value, $k = \mu^2$. The solution is now of the following form.

$$F(x) = Ae^{\mu x} + Be^{-\mu x}$$

But again, applying the boundary conditions makes $A = 0$ and $B = 0$. Therefore, $F(x) \equiv 0$ and $u(x,t) \equiv 0$.

To have a useful solution, k must be negative, say $k = -p^2$. This provides a solution of the following form.

$$F(x) = A\cos px + B\sin px$$

$F(0) = 0$ makes $A = 0$. The second boundary condition yields

$$F(l) = 0 = B\sin pl$$

Because $B = 0$ would give $F(x) \equiv 0$ and $u(x,t) \equiv 0$, $\sin pl$ must be equal to zero. That is,

$$pl = n\pi \text{ or } p = \frac{n\pi}{l} \quad [n = 0, 1, 2, \ldots]$$

Letting $B = 1$, an infinite number of solutions are of the following form.

$$F_n(x) = \sin\frac{n\pi}{l}x \quad [n = 1, 2, 3, \ldots]$$

Because k is restricted to negative values, that is, $k = -p^2$, the ODE for the function containing the t variable is

$$\ddot{G}(t) - kc^2 G(t) = 0$$

$$\ddot{G}(t) + p^2c^2 G(t) = 0$$

$$\ddot{G}(t) + \left(\frac{n\pi}{l}\right)^2 c^2 G(t) = 0$$

$$\ddot{G}(t) + \lambda_n^2 G(t) = 0 \quad \left[\lambda_n = \left(\frac{n\pi}{l}\right)c\right]$$

The solution is therefore[27]

$$G_n(t) = B_n\cos\lambda_n t + B_n^*\sin\lambda_n t$$

Writing the function $u_n(x,t) = F_n(t)G_n(t)$ gives

$$u(x,t) = \sin\frac{n\pi}{l}x \times (B_n\cos\lambda_n t + B_n^*\sin\lambda_n t) \quad [n = 1, 2, 3, \ldots] \qquad \text{[V]}$$

[26]The characteristic equation results in a double root of zero. See Chap. 11 for the method used on homogeneous, second-order linear differential equations with constant coefficients.

[27]The symbols B_n and B_n^* are commonly used as b_n in representations of Fourier sine series. The symbol B_n^* is not to be confused with that of a complex number.

This function satisfies the boundary conditions given in Eq. I and Eq. II. The functions are *eigenfunctions*, or characteristic functions. The values for λ_n are *eigenvalues*, or characteristic values. The function u_n represents harmonic motion of a frequency $\lambda_n/2\pi = cn/2l$ cycles per unit time. The *fundamental mode* occurs when $n=1$. Each mode n has $n-1$ nodes, that is, points that do not move.[28] Some fundamental nodes are shown.

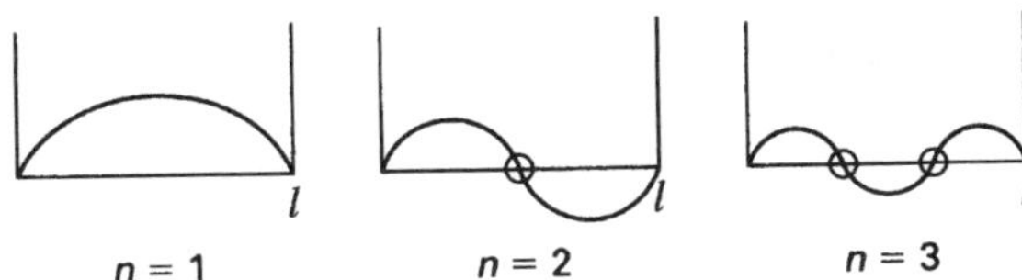

step 3: Compose the solution such that the initial conditions are satisfied.

A single solution of the form of Eq. V will not, in general, satisfy both initial conditions III and IV. Based on the fundamental theorem given by Eq. 15.36, the sum of the solutions is a solution. Therefore, the following combination of solutions to Eq. V provides a solution to the one-dimensional wave equation.

$$u(x,t) = \sum_{n=1}^{\infty} u_n(x,t)$$

$$= \sum \sin \frac{n\pi}{l} x$$

$$(B_n \cos \lambda_n t + B_n^* \sin \lambda_n t) \quad \text{[VI]}$$

Applying the initial condition III to Eq. VI gives

$$u(x,0) = \sum_{n=1}^{\infty} B_n \sin \frac{n\pi}{l} x = f(x) \quad \text{[VII]}$$

To make Eq. VI satisfy initial condition III, B_n must be chosen as the half-range expansion of $f(x)$, that is, the Fourier sine series of $f(x)$.[29]

$$B_n = \frac{2}{l} \int_0^l f(x) \sin \frac{n\pi}{l} x \, dx$$

$$[n = 1, 2, 3, \ldots] \quad \text{[VIII]}$$

By making this choice for B_n, Eq. VII becomes valid and Eq. VI satisfies initial condition III.

One item now remains: make Eq. VI satisfy initial condition IV. Start by taking the derivative of Eq. VI with respect to time and applying condition IV. This gives the following equation.

$$\left.\frac{\partial u}{\partial t}\right|_{t=0}$$

$$= \sum_{n=1}^{\infty} \sin \frac{n\pi}{l} x \left.\begin{pmatrix} -B_n \lambda_n \sin \lambda_n t \\ + B_n^* \lambda_n \cos \lambda_n t \end{pmatrix}\right|_{t=0}$$

$$= \sum_{n=1}^{\infty} B_n^* \lambda_n \sin \frac{n\pi}{l} x = g(x) \quad \text{[IX]}$$

Using the same logic, and recalling that $\lambda_n = cn\pi/l$ gives

$$B_n^* \lambda_n = \frac{2}{l} \int_0^l g(x) \sin \frac{n\pi}{l} x \, dx$$

$$B_n^* = \frac{2}{cn\pi} \int_0^l g(x) \sin \frac{n\pi}{l} x \, dx \quad \text{[X]}$$

With B_n^* equal to the condition given in Eq. X, Eq. VI satisfies initial condition IV. All the conditions are now met.

Eq. VI is a solution to the one-dimensional wave equation with the coefficients B_n and B_n^* defined by Eq. VIII and Eq. IX.[30]

9. SPECIAL FUNCTIONS

Several different special functions of use occur as *named equations* with different forms and symbols.

Unit Step Function

Symbols: u_t; $S_k(t)$; $H(t-k)$

This function may be used to represent what occurs when a switch is turned on.

$$u_t = \begin{cases} 0 & 0 < t < k \\ 1 & t > k \end{cases} \quad \text{15.37}$$

[28]This is not counting the constrained endpoints.

[29]The Fourier sine series results in an odd extension of $f(x)$. In other words, the original function $f(x)$ is extended in the coefficient B_n which, when summed, leaves only $f(x)$. See the graphs in Sec. 15.7.

[30]Also, the series represented in Eq. VI must converge, and the series obtained by differentiating Eq. VI twice (termwise) with respect to x and t must converge and sum to $\partial^2 u/\partial x^2$ and $\partial^2 u/\partial t^2$. Other solution methods are possible. One is called the *D'Alembert* solution. Another is using a variation of parameters and Laplace transforms. Also, Laplace transforms can change the partial differential equation into an ordinary differential equation that can be solved using standard methods.

Equation 15.37 mathematically represents what is shown in Fig. 15.2. The symbol $H(t-k)$ is called the *Heaviside step function.*[31]

Figure 15.2 *Unit Step Function*

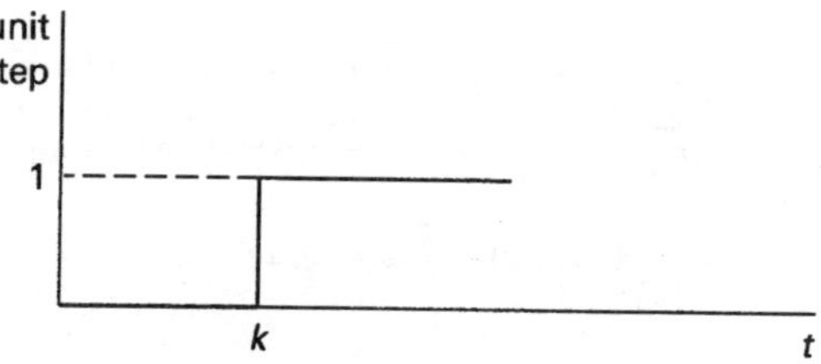

Unit Finite Impulse Function

$$I(h, t-t_0) = H(t-t_0) - H\big(t-(t_0+h)\big) \quad 15.38$$

$$I(h, t-t_0) = \begin{cases} 0 & t < t_0 \text{ and } t > (t_0+h) \\ 1 & t_0 < t < (t_0+h) \end{cases} \quad 15.39$$

The *unit finite impulse function* is shown graphically in Fig. 15.3.

Of special importance in Fig. 15.3 part (c) is that the unit finite impulse function, that is, the shaded region, represents an area equal to 1 for any $h > 0$. This area represents a rectangle with the base equal to $t_0 - (t_0+h) = h$ and the height equal to $1/h$; therefore, the area is $h(1/h) = 1$. The impulse is the integral of force $(1/h)$ over the interval of time (h).

Delta Function; Dirac Function; Impulse Function; Unit Impulse Function

Symbols: δt; $\delta(t-t_0)$

If the limit as $h \rightarrow 0$ is taken on the unit finite impulse function, the result is the *delta*, or *Dirac function.*[32] This function is sometimes called the *impulse function* or *unit impulse function* and not always differentiated from the unit finite impulse function.

$$\lim_{h\rightarrow 0} I(h, t-t_0) \equiv \delta(t-t_0) = \begin{cases} 0 & t \neq t_0 \\ \infty & t = t_0 \end{cases} \quad 15.40$$

This is shown graphically in Fig. 15.4.[33]

Figure 15.3 *Unit Finite Impulse Function*

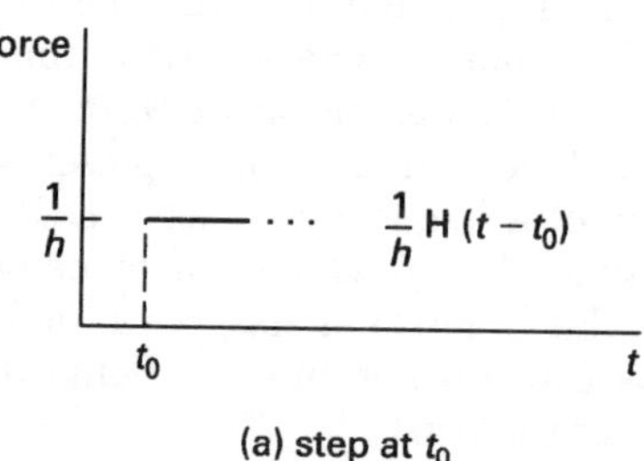

(a) step at t_0

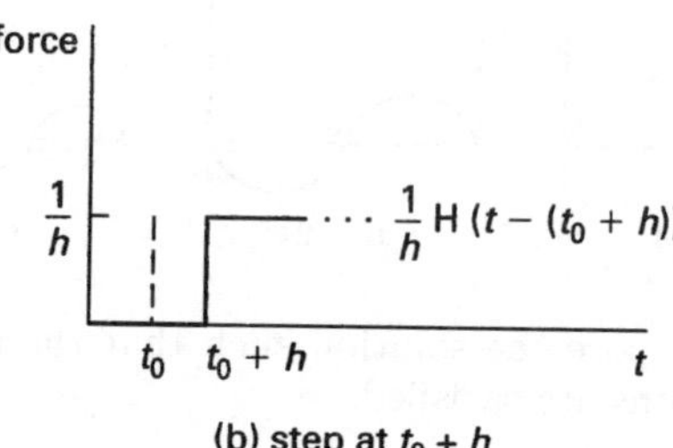

(b) step at $t_0 + h$

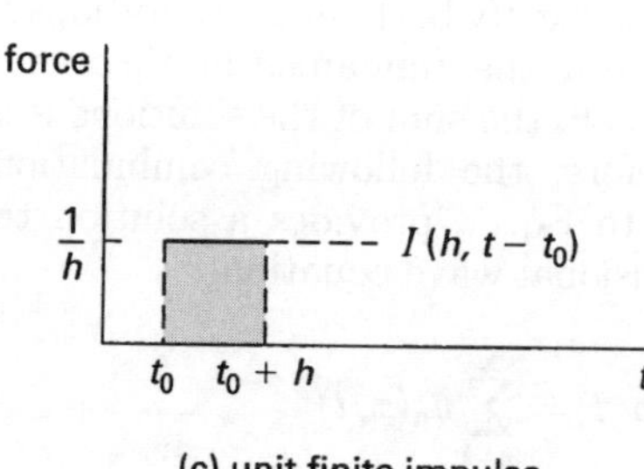

(c) unit finite impulse

Figure 15.4 *Delta Function*

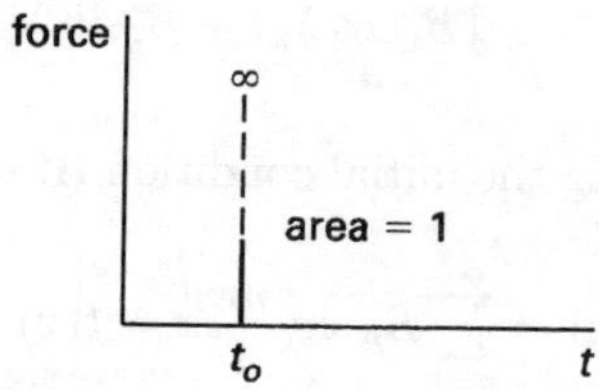

The delta function is strictly defined only in terms of integration. The delta function displays a property called *sifting* that is given by Eq. 15.41.[34]

$$\int_{-\infty}^{+\infty} \delta(t-t_0) f(t)\, dt = f(t_0) \quad 15.41$$

[31]Oliver Heaviside was an English electrical engineer whose treatment of the derivative process (represented by D) in algebraic terms led to the development of the Laplace transform.

[32]Paul Dirac was an English physicist whose theory of negative energy holes predicted the existence of positrons.

[33]There is much more to the delta function. The mathematics shown is not strictly true. The delta function is not a proper function and is therefore shown as equivalent to (≡) the limit.

[34]The impulse function is a powerful analytical tool. For example, electrical signals can be represented as weighted sums of shifted impulses—essentially this can be the interpretation of Eq. 15.41. Using this interpretation, the response of a *linear time invariant* (LTI) system can be analyzed for any arbitrary input, including *discrete* inputs, that is, those inputs that are not continuous. (Recall that the term "linear" indicates that superposition can be used; "time invariant" means a time shift in the input signal causes an identical time shift in the output signal.) Using the principle of superposition, one can then obtain the output of an electrical circuit to any input. Using the unit finite impulse function, one can do the same for continuous time systems.

Gamma Function

Symbols: $\Gamma(\alpha)$; $(\alpha-1)!$

Also known as the *factorial function*, the *gamma function* is defined by the following integral.[35]

$$\Gamma(\alpha)=\int_0^{\infty} e^{-t}t^{(\alpha-1)}\,dt \quad [\alpha>0] \qquad 15.42$$

The gamma function is shown graphically in Fig. 15.5.

Figure 15.5 *Gamma Function*

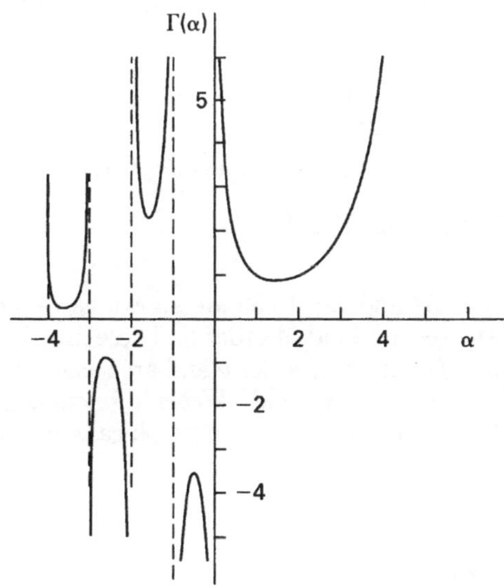

The important functional relation of the gamma function is

$$\Gamma(\alpha+1)=\alpha\Gamma(\alpha) \qquad 15.43$$

Example 15.6

Find $\Gamma(2)$.

Solution

Apply the definition given by Eq. 15.42.

$$\Gamma(\alpha)=\int_0^{\infty} e^{-t}t^{(\alpha-1)}\,dt \quad [\alpha>1]$$

$$\Gamma(2)=\int_0^{\infty} e^{-t}t^{(2-1)}\,dt$$

$$=\int_0^{\infty} e^{-t}t\,dt$$

From a mathematical handbook of integrals, the result of the previous equation is

$$\Gamma(2)=\frac{e^{-t}}{(-1)^2}(-t-1)\Big|_0^{\infty}$$

$$=e^{-t}(-t-1)\Big|_0^{\infty}$$

$$=\frac{-(t+1)}{e^t}\Big|_0^{\infty}$$

Use L'Hôpital's rule to determine the value at infinity.

$$\frac{-(t+1)}{e^t}\Big|_0^{\infty}=0-\left(\frac{-(0+1)}{e^0}\right)=1$$

This result could also have been accomplished using the functional relation of Eq. 15.43 as shown.

$$\Gamma(\alpha+1)=\alpha\Gamma(\alpha)$$

$$\Gamma(2)=\Gamma(1+1)=1\Gamma(1)=1$$

Finally, the factorial symbol could have been used.[36]

$$\Gamma(\alpha)=(\alpha-1)!$$

$$\Gamma(2)=(2-1)!=1!=1$$

A table of gamma function values is given in App. 15.A.

Error Function and Complementary Error Function

The error function, also called the *probability integral*, was defined in Chap. 10 as

$$\operatorname{erf}(x)=\frac{2}{\sqrt{\pi}}\int_0^{x} e^{-t^2}\,dt \qquad 15.44$$

$2/\sqrt{\pi}$ normalizes the function so that $\operatorname{erf}(\infty)=1$. By contrast, the *complementary error function* is defined as

$$\operatorname{erfc}(x)=1-\operatorname{erf}(x)=\frac{2}{\sqrt{\pi}}\int_x^{\infty} e^{-t^2}\,dt \qquad 15.45$$

The error function is shown graphically in Fig. 15.6.

Bessel Functions of the First Kind

A Bessel differential equation of order ν is given by

$$x^2y''+xy'+(x^2-\nu^2)y=0 \qquad 15.46$$

A particular solution is given by the Bessel function of the first kind of order ν. (This result is obtained using the power series method.)

$$J_\nu(x)=x^\nu\sum_{m=0}^{\infty}\frac{(-1)^m x^{2m}}{2^{2m+\nu}m!\Gamma(\nu+m+1)} \qquad 15.47$$

[35]Also defined in terms of the gamma function are the *beta function* and *incomplete gamma functions*.

[36]An important value of gamma is $\Gamma(1/2)=\sqrt{\pi}$.

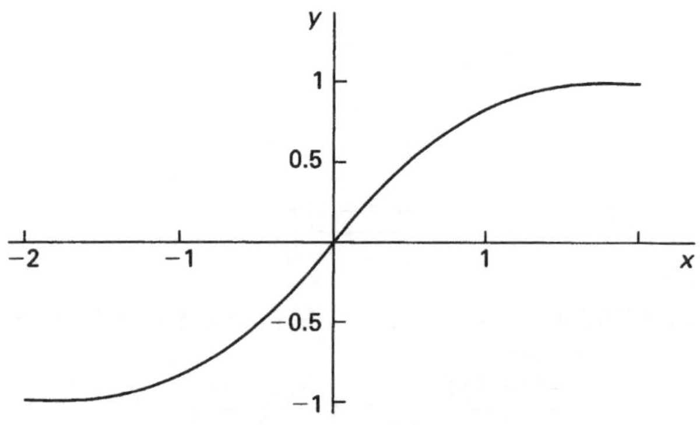

Figure 15.6 Error Function

Integer values of ν are often denoted by n. The functions J_0 and J_1 are important in many engineering problems. These functions are shown graphically in Fig. 15.7.

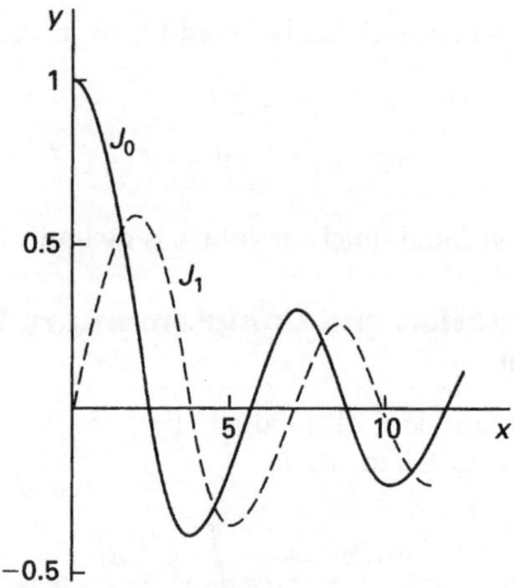

Figure 15.7 Bessel Functions of the First Kind

A table of values for $J_0(x)$ and $J_1(x)$ is given in App. 15.B.

Bessel Functions of the Second Kind

If the value of ν or n is zero in a Bessel differential equation, and dividing by the variable x, the result is

$$xy'' + y' + xy = 0 \qquad 15.48$$

The standard particular solution is known as the *Bessel function* of the second kind of order zero and is given by the following equation. This is also called *Neumann's function* of order zero.

$$Y_0(x) = \frac{2}{\pi}\left(\begin{array}{c} J_0(x)\left(\ln\frac{x}{2}+\gamma\right) \\ +\sum\left(\frac{(-1)^{m-1}h_m}{2^{2m}(m!)^2}\right)x^{2m}\end{array}\right) \qquad 15.49$$

The constant γ is Euler's constant (also given as C_E, C, and $\ln\gamma$) which is defined as

$$\begin{aligned}\gamma &= \lim_{s\to\infty}\left(1+\frac{1}{2}+\cdots+\frac{1}{s}-\ln s\right) \\ &\approx 0.577\end{aligned} \qquad 15.50$$

The variable h_m is defined as

$$h_m = 1+\frac{1}{2}+\cdots+\frac{1}{m} \quad [m=1,2,3\ldots] \qquad 15.51$$

The second standard solution to Eq. 15.48 for all ν is given by

$$\begin{aligned}Y_\nu(x) &= \left(\frac{1}{\sin\nu\pi}\right) \\ &\quad\times\left(J_\nu(x)\cos\nu\pi - J_{-\nu}(x)\right)\end{aligned} \qquad 15.52$$

$$Y_n(x) = \lim_{\nu\to n} Y_\nu(x) \qquad 15.53$$

Equation 15.52 and Eq. 15.53 are known as *Bessel functions* of the second kind of order ν. These are also called *Neumann's functions of order* ν and are sometimes given as $N_\nu(x)$. They are also called *Weber's functions* in some texts. These functions are shown graphically in Fig. 15.8.

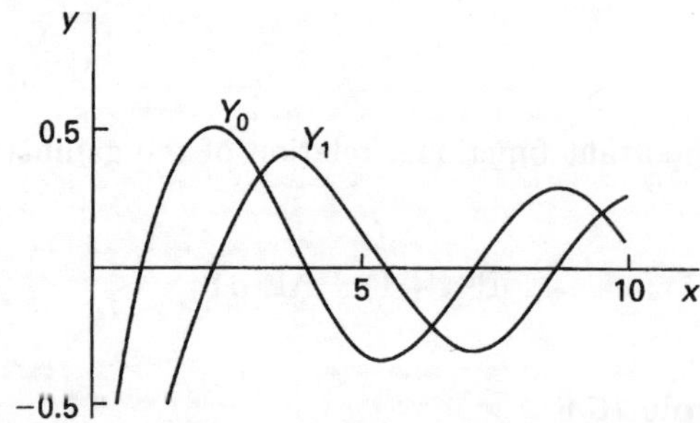

Figure 15.8 Bessel Functions of the Second Kind

Bessel Functions of the Third Kind

A general solution to Bessel's equation for all values of ν is given by Eq. 15.54.

$$y(x) = c_1 J_\nu(x) + c_2 Y_\nu(x) \qquad 15.54$$

Sometimes a solution needs to be complex for real values of x. In such cases, solutions take on the following forms.

$$H_\nu^{(1)}(x) = J_\nu(x) + jY_\nu(x) \qquad 15.55$$

$$H_\nu^{(2)}(x) = J_\nu(x) - jY_\nu(x) \qquad 15.56$$

These linearly independent functions are called *Bessel functions* of the third kind of order ν. They are also called the *first and second Hankel functions* of order ν.

Hyperbolic Bessel Functions

Hyperbolic Bessel functions are given by Eq. 15.57.

$$I_m(x) = j^{-m} J_m(jx) \qquad 15.57$$

10. CONTINUOUS-TIME SYSTEMS: FOURIER SERIES AND TRANSFORMS

Fourier series are used to analyze periodic signals. Fourier integrals, also called Fourier transforms, are used to analyze aperiodic signals. The power of Fourier analysis is that if the input signal can be expressed in terms of periodic complex exponentials or sinusoids, then the output can be expressed in the same form, simplifying analysis. Further, representing otherwise intractable equations as a periodic Fourier series allows for analysis in ordinary and partial differential equations.

Example 15.7

Show that the response of a linear time-invariant system (LTI) to a complex exponential input is the same complex exponential with a change in amplitude.[37] In other words, show[38]

$$e^{st} \rightarrow H(s)e^{st}$$

The term $H(s)$ is the complex amplitude factor and in general will be a function of the complex variable s.

Solution

Assume an input signal of the form $x(t) = e^{st}$ provided to an LTI circuit with impulse response $h(t)$.[39]

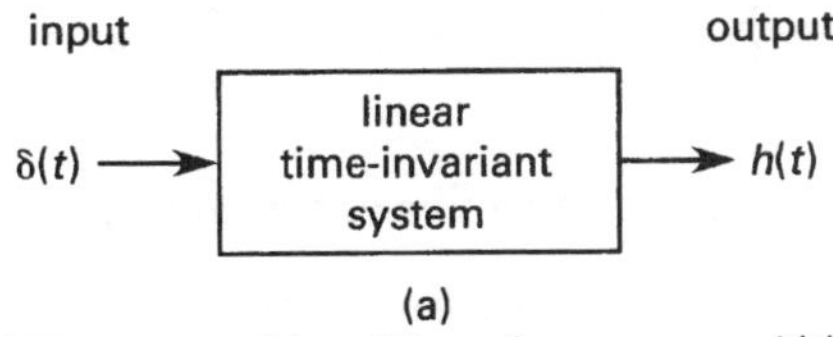

(a)
LTI system with unit impulse response $h(t)$

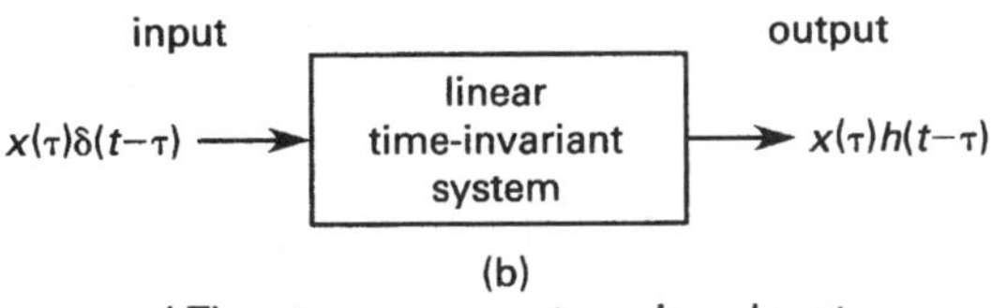

(b)
LTI system response to an impulse at time τ with amplitude $x(\tau)$

Writing the output $y(t)$ in terms of the convolution integral gives the following equation. (Recall that the convolution integral takes two functions defined in the s domain, in this case the impulse and the input signal $x(t)$, and gives the output in the time domain t directly without determining the inverse Laplace transform.)

$$y(t) = \int_{-\infty}^{+\infty} h(t-\tau)x(\tau)\,d\tau = \int_{-\infty}^{+\infty} h(\tau)e^{-s\tau}e^{st}\,d\tau$$

Because the term e^{st} does not affect the integral, it can be moved outside the integral.

$$y(t) = e^{st}\int_{-\infty}^{+\infty} h(\tau)e^{-s\tau}\,d\tau$$

This can be rewritten as

$$y(t) = e^{st}H(s) = H(s)e^{st}$$

$$H(s) = \int_{-\infty}^{+\infty} h(\tau)e^{-s\tau}\,d\tau$$

Any complex exponential is an eigenfunction of an LTI system. The constant $H(s)$ for a specified value s is the eigenvalue of the eigenfunction e^{st}.[40]

Consider an input signal of the form

$$x(t) = a_1e^{s_1t} + a_2e^{s_2t} + a_3e^{s_3t}$$

The response to each term separately is

$$a_1e^{s_1t} \rightarrow a_1H(s_1)e^{s_1t}$$
$$a_2e^{s_2t} \rightarrow a_2H(s_2)e^{s_2t}$$
$$a_3e^{s_3t} \rightarrow a_3H(s_3)e^{s_3t}$$

Because, by the definition of "linear," superposition applies, the output $y(t)$ is

$$y(t) = a_1H(s_1)e^{s_1t} + a_2H(s_2)e^{s_2t} + a_3H(s_3)e^{s_3t}$$

In general terms,

$$\text{input} \rightarrow \text{output}$$
$$\sum_k a_ke^{s_kt} \rightarrow \sum_k a_kH(s_k)e^{s_kt}$$

In general, s is complex and of the form $s = a + j\omega$. If s is restricted to be equal to $j\omega$, that is, $s = j\omega$, then the exponentials are of the form $e^{j\omega t}$ and Fourier transforms are used. If s is allowed to be of the more general form $s = a + j\omega$, then Laplace transforms are used.

[37]"Linear" indicates that superposition applies. "Time invariant" indicates that a shift in the input signal $x(t)$ to $x(t-\tau)$ causes a shift in the output signal $y(t)$ to $y(t-\tau)$. When such a shift occurs in an equation, the equation is referred to as a *difference equation* and, as will be shown, represents a discrete time system.

[38]This is equivalent to showing the same thing for sinusoids, because from Euler's relation $e^{j\omega t} = \cos\omega t + j\sin\omega t$.

[39]The impulse response $h(t)$ occurs due to an input $\delta(t)$. This is simply a notation, which can vary considerably.

[40]The value $H(s)$ is determined by the electrical circuit and represents the circuit's response to an impulse $\delta(t-\tau)$. $H(s)$, with $s = j\omega$, is the Fourier transform of the response of the circuit to an impulse signal. Generally written as $H(\omega)$, it is termed the *frequency response* of the system.

11. CONTINUOUS-TIME SYSTEMS: FOURIER SERIES REPRESENTATION OF A PERIODIC SIGNAL

A *continuous-time signal* is one in which the independent variable is continuous, that is, the signal is defined for a continuum of values. Figure 15.9 shows an example of such a signal.

Figure 15.9 *Continuous-Time Periodic Signal*

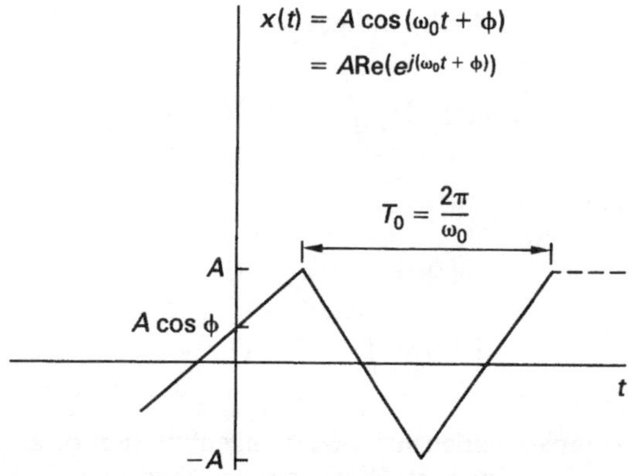

In Fig. 15.9, the values T_0 and ω_0 represent the *fundamental period* and *fundamental frequency*, respectively.

That is, they represent the minimum positive, nonzero value for which $x(t) = x(t+T)$ is true for a periodic function.

The Fourier series representation for any continuous time periodic signal is

$$x(t) = \sum_{k=-\infty}^{+\infty} a_k e^{jk\omega_0 t} \qquad 15.58$$

$$a_k = \frac{1}{T_0}\int_{T_0} x(t)e^{-jk\omega_0 t}dt \qquad 15.59$$

There are other representations of a Fourier series. (See Sec. 10.15 for an example.) This complex exponential form is very useful in electrical analysis. It is called both the *complex Fourier series* and the *exponential Fourier series*.

In Fourier series representation, the input term $x(t)$ is real and has the form of a cosine function.[41] When $k=0$, the value of $x(t)$ represents the DC or constant value.

Fourier analysis is the determination of coefficients using the *analysis equation*, Eq. 15.59, to obtain the result. The output $y(t)$ for a circuit is the input $x(t)$ represented as a Fourier series (see Eq. 15.58 and Eq. 15.59) multiplied by an amplitude factor $H(s)$. The amplitude factor $H(s)$ is determined by the circuit through which the given input (i.e., the signal) is processed.[42] The amplitude factor $H(s)$ is the circuit's response, $h(t-\tau)$, to an impulse, $\delta(t-\delta)$.[43] Some properties of the Fourier series for periodic signals are given in App. 15.C.

Example 15.8

Represent $x(t)$ from Fig. 15.9 as a Fourier series.

Solution

One could apply the analysis equation given in Eq. 15.59. However, in this simple case, it is easier to expand the equation in terms of complex exponentials and identify the Fourier series coefficients by inspection. Specifically, the $\cos(\omega_0 t+\phi)$ function can be expressed as

$$\cos(\omega_0 t+\phi) = \tfrac{1}{2}e^{j(\omega_0 t+\phi)} + \tfrac{1}{2}e^{-j(\omega_0 t+\phi)}$$

In the first term $k=1$. Therefore,

$$a_k = a_1 = \tfrac{1}{2}$$

In the second term $k=-1$. Therefore,

$$a_k = a_{-1} = \tfrac{1}{2}$$

For all other values of k, $a_k = 0$. The function in terms of a Fourier series can be represented as

$$x(t) = \sum_{-\infty}^{+\infty} a_k e^{jk(\omega_0 t+\phi)} = \tfrac{1}{2}e^{j(\omega_0 t+\phi)} + \tfrac{1}{2}e^{-j(\omega_0 t+\phi)}$$

The shift of $+\phi$ is the equivalent of multiplying by $e^{j\phi}$.

To confirm the result, use Euler's relation to expand the Fourier series and obtain the original equation. That is,

$$\begin{aligned} x(t) &= \tfrac{1}{2}e^{j(\omega_0 t+\phi)} + \tfrac{1}{2}e^{-j(\omega_0 t+\phi)} \\ &= \tfrac{1}{2}\Big(\cos(\omega_0 t+\phi) + j\sin(\omega_0 t+\phi)\Big) \\ &\quad + \tfrac{1}{2}\begin{pmatrix} \cos\Big(-(\omega_0 t+\phi)\Big) \\ +j\sin\Big(-(\omega_0 t+\phi)\Big) \end{pmatrix} \end{aligned}$$

[41]This is because the summation is from $-\infty$ to $+\infty$ and $a_k = a_{-k}$ (sometimes written $a_k = a^*_{-k}$), so the imaginary terms cancel. Hint: Use Euler's relation and trigonometric identities on the first few terms as a proof.

[42]This may seem rather circular. In different terms, if one represents the input as a Fourier series, that is, as Eq. 15.58, and then replaces the a_k with the Fourier coefficients (also called the *spectral coefficients*) as given by Eq. 15.59, the output is obtained by multiplying by the Fourier transform of the impulse input signal (given by $H(s)$ earlier in this section). The s is restricted to $s = j\omega$ when using the Fourier transform.

[43]When using the Fourier transform, the s is limited to $j\omega$ and $H(s)$ is the Fourier transform of $h(t-\tau)$. If s has both real and imaginary parts, that is, $s = a + j\omega$, $H(s)$ is the Laplace transform of $h(t-\tau)$.

Applying the trigonometric identities $\sin(-\alpha) = -\sin\alpha$ and $\cos(-\alpha) = \cos\alpha$ gives

$$\begin{aligned} x(t) &= \tfrac{1}{2}\Big(\cos(\omega_0 t + \phi) + j\sin(\omega_0 t + \phi)\Big) \\ &\quad + \tfrac{1}{2}\Big(\cos(\omega_0 t + \phi) - j\sin(\omega_0 t + \phi)\Big) \\ &= \cos(\omega_0 t + \phi) \end{aligned}$$

12. CONTINUOUS-TIME SYSTEMS: FOURIER TRANSFORM OF AN APERIODIC SIGNAL

Many signals of interest to an electrical engineer are *aperiodic*. The Fourier series representation can still be utilized if one considers the aperiodic signal as the limit of a periodic signal as the period becomes arbitrarily large.[44] The result is[45,46]

$$x(t) = \frac{1}{2\pi}\int_{-\infty}^{+\infty} X(\omega)e^{j\omega t}\,d\omega \qquad 15.60(a)$$

$$x(t) = \int_{-\infty}^{+\infty} X(f)e^{j2\pi ft}\,df \qquad 15.60(b)$$

$$X(\omega) = \int_{-\infty}^{+\infty} x(t)e^{-j\omega t}\,dt \qquad 15.61(a)$$

$$X(f) = \int_{-\infty}^{+\infty} x(t)e^{-j2\pi ft}\,dt \qquad 15.61(b)$$

Equation 15.60 and Eq. 15.61 are referred to as a *Fourier transform pair.* Equation 15.61 is the *Fourier transform* or *Fourier integral.* Equation 15.60 is the *inverse Fourier transform* that is the *synthesis equation* of the pair.[47]

Some properties of the Fourier transform for aperiodic signals are given in App. 15.D. Some Fourier transform pairs are given in App. 15.E.

The arrangement of aperiodic signals is analogous to that of periodic functions. Equation 15.60 correlates with Eq. 15.58. The complex exponentials formed by the synthesis equations have amplitudes represented by the coefficients a_k given by Eq. 15.59 in the periodic case, representing a discrete set of harmonically related frequencies $k\omega_0$ with $k = 0, \pm1, \pm2, \ldots$ For aperiodic signals, Eq. 15.60 represents the linear combination of complex exponentials with amplitudes $(X(\omega)/2\pi)\,d\omega$, with $X(\omega)$ given by Eq. 15.61. Examples of Fourier transformations are shown in Fig. 15.10. (Obviously, the sine and cosine functions shown in the figure are not aperiodic, but rather, periodic. The Fourier transform of a periodic signal with Fourier coefficients can be thought of as a train of impulses occurring at harmonically related frequencies for which the area of the impulse at the kth harmonic frequency ($k\omega_0$) is 2π times the kth Fourier series coefficient a_k. All the calculations in the aperiodic section apply.)

Figure 15.10 *Fourier Transformations*

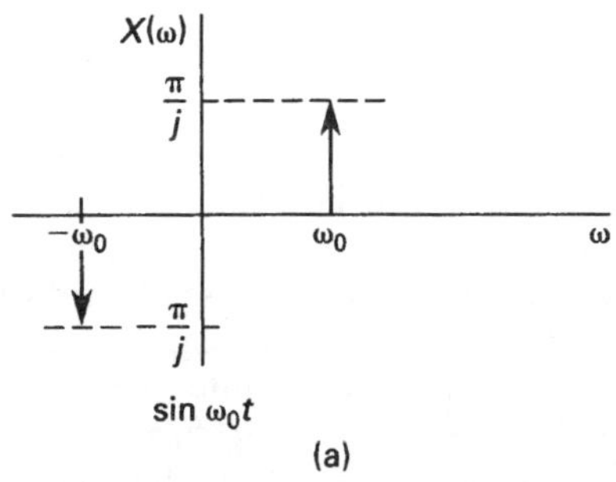

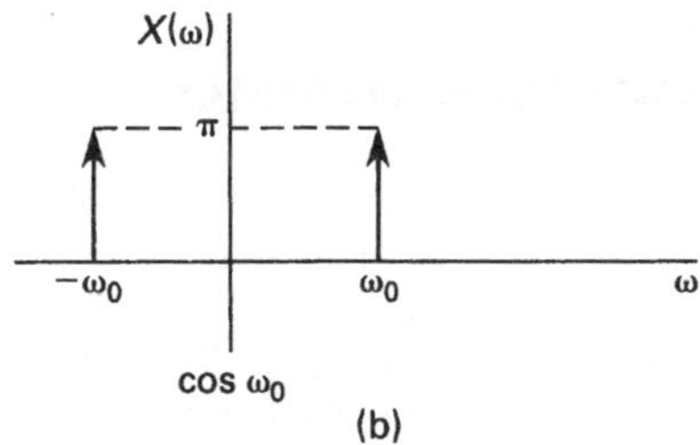

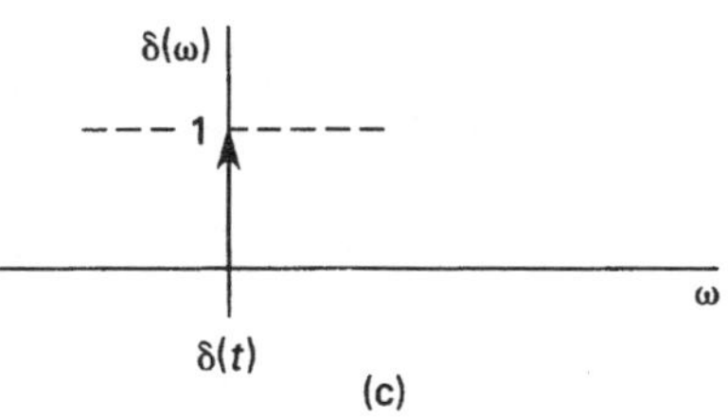

Fourier transforms are useful for the analysis of continuous-time systems that are linear time-invariant. (The broad range of circuitry using resistors, capacitors, and inductors are linear time-invariant.) The output signals of such circuits are related proportionally to the input signals by an amplitude factor.

In mathematical terms, an input signal $x(t)$ can be represented by an infinite number of pulses, summed as shown in Eq. 15.62.

$$x(t) = \int_{-\infty}^{+\infty} x(\tau)\delta(t-\tau)\,d\tau \qquad 15.62$$

[44]This is equivalent to saying $\omega_0 \to 0$ because $\omega_0 = 2\pi/T_0$. In other words, no periodicity exists.

[45]The derivation of most of the formulas used in this book will not be given unless they enhance the understanding of a fundamental principle.

[46]Equation 15.60(a) and Eq. 15.61(a) are not symmetrical, but they are used in electrical engineering and in physics. Mathematicians prefer the symmetrical versions of Eq. 15.60(b) and Eq. 15.61(b).

[47]This is because the inverse Fourier transform decomposes a signal into a linear combination of complex exponentials similar to the function of Eq. 15.58.

The output signal is then given by the following equation.[48]

$$y(t) = \int_{-\infty}^{+\infty} x(\tau)h(t-\tau)d\tau \qquad 15.63$$

Equation 15.63 is of the form of a convolution integral (see Sec. 15.13). Using data from App. 15.D, Eq. 15.63 can be written as[49]

$$\begin{aligned} y(t) &= \int_{-\infty}^{+\infty} x(\tau)h(t-\tau)d\tau \\ &= \mathcal{F}^{-1}\big(X(\omega)H(\omega)\big) \end{aligned} \qquad 15.64$$

Herein lies the power of the Fourier transform: the often difficult convolution can be completely avoided. Convert the input signal $x(t)$ into the frequency domain as $X(\omega)$. Convert the impulse response $h(t)$ into $H(\omega)$. The convolution of $x(t)$ and $h(t)$ is simply the product of $X(\omega)$ and $H(\omega)$, and the output $y(t)$ is realized by applying the inverse Fourier transform to the result. In other words, a convolution in the time domain is multiplication in the frequency domain.

13. CONVOLUTION INTEGRAL

The convolution property of a given transform T is one that gives the inverse transform $T^{-1}\{X(\tau)H(\tau)\}$ directly in terms of the original functions $x(t)$ and $h(t)$.[50] This was evident in Eq. 15.64 where the inverse transform $T^{-1}\{X(\tau)H(\tau)\}$, that is, $y(t)$, was given in terms of the original functions $x(t)$ and $h(t)$.[51] A generalized form of the *convolution integral* is[52]

$$\begin{aligned} f(t) * g(t) &= \int_a^b f(\tau)g(t-\tau)d\tau \\ &= T^{-1}\{F(s)G(s)\} \end{aligned} \qquad 15.65$$

When the convolution property is invoked for a Fourier transform, the limits of integration a and b become $-\infty$ and $+\infty$, respectively. Also, s becomes ω (or $j\omega$ in some texts) and $G(\omega)$ represents $H(\omega)$. ($H(\omega)$ is the Fourier transform of the system impulse response, or the frequency response, for LTI circuits as discussed in previous sections of this chapter.) When invoked for a Laplace transformation, a and b become 0 and $+\infty$, respectively, and s is generalized to real and imaginary components, $s = \sigma + j\omega$. (This essentially gives the properties of magnitude and phase.)

Convolution properties follow. The variable t has been dropped in Eq. 15.66 through Eq. 15.69 as convolution can be used with more than one variable.

$$f * g = g * f \quad \text{[commutative law]} \qquad 15.66$$

$$f * (g_1 + g_2) = f * g_1 + f * g_2 \quad \text{[distributive law]} \qquad 15.67$$

$$(f * g) * v = f * (g * v) \quad \text{[associative law]} \qquad 15.68$$

$$f * 0 = 0 * f = 0 \qquad 15.69$$

Convolution is used to filter a signal linearly. It is used in electrical engineering (e.g., signal processing) because of its relationship to Fourier transforms. Specifically, the convolution of two functions in the time domain equals the multiplication of the transforms in the frequency domain, as shown in Eq. 15.65. In many problems, this property simplifies calculations. *Correlation* is similar to convolution in mathematical form, and is used to display the interdependence or association of two variables. It also is used in signal processing (e.g., image processing), probability and statistics, and areas of engineering requiring such applications, for example, communications. The autocorrelation function of a signal $f(t)$ is

$$\phi_{fg}(t) = \int_{-\infty}^{+\infty} f(t+\tau)g(\tau)d\tau \qquad 15.70$$

14. LAPLACE TRANSFORMS

The Laplace transform, whose properties and uses are given in Chap. 11, is repeated here for convenience.

$$\mathcal{L}(f(t)) = F(s) = \int_0^{\infty} e^{-st}f(t)dt \qquad 15.71$$

Equation 15.71 is more specifically termed the *unilateral Laplace transformation.* The primary use of this transform is in the solution of linear constant-coefficient differential equations with initial conditions. (For an example, see Sec. 11.9.)

[48]Here no change in the input signal representation occurs. When using the Fourier series of $x(t)$ in the process of obtaining $y(t)$, such a change does occur. Though the transform method is more straightforward, the Fourier series representation has numerous applications and was therefore presented in depth in Sec. 15.11.

[49]Some texts write $X(\omega)$ and $H(\omega)$ as $X(j\omega)$ and $H(j\omega)$. The j is normally understood to be present.

[50]The dummy variable τ is used to represent the transform and can be any transform: Fourier, Fourier sine, Fourier cosine, Laplace, and others.

[51]It is often easier to find the result using convolution rather than applying the inverse transform to $X(\omega)H(\omega)$.

[52]A convolution integral is a function that expresses the amount of overlap that occurs when one function overlaps with (shifts over) another. It represents a blending of the functions.

By contrast, the *bilateral Laplace transform* of a general signal $f(t)$ is given by Eq. 15.72.

$$\mathcal{L}(f(t)) = F(s) = \int_{-\infty}^{+\infty} e^{-st} f(t)\,dt \qquad 15.72$$

The unilateral transform is more commonly encountered. The term unilateral is not often used. The differentiation between the two types is simply shown in the limits of integration. From the bilateral Laplace transform, the relationship between it and the Fourier transform is easily seen.

$$F(s)\big|_{s=j\omega} = \mathcal{F}\big(f(t)\big) \qquad 15.73$$

If s is not purely imaginary, the relationship is given mathematically as

$$F(\sigma + j\omega) = \int_{-\infty}^{+\infty} \left(f(t)e^{-\sigma t}\right) e^{-j\omega t}\,dt \qquad 15.74$$

In other words, the Laplace transform of a function with $s = \sigma + j\omega$ is the Fourier transform of $f(t)$ multiplied by the real exponential $e^{-\sigma t}$. Properties of the Laplace transform as well as some common transforms themselves are given in App. 15.D and App. 15.G. These appendices include the *region of convergence* (ROC or R in some texts), that is, the restrictions necessary to make the transform valid, which will become important when the frequency response of electrical circuits is considered.

15. DISCRETE-TIME SYSTEMS: FOURIER SERIES AND TRANSFORMS[53]

A *discrete-time signal* is one in which the independent variable is defined only at discrete times. Figure 15.11 shows an example of such a signal.

Figure 15.11 *Discrete-Time Periodic Signal*

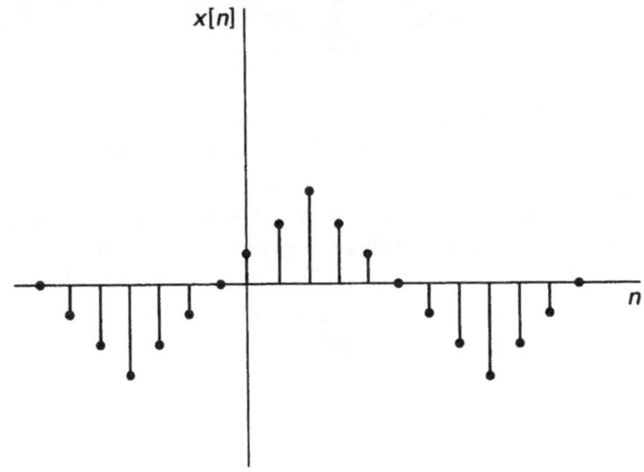

n replaces t as the variable in order to distinguish between the continuous-time and discrete-time cases. Additionally, brackets are used, $[n]$, to further distinguish discrete-time systems.

The response of an LTI system to inputs that can be represented as complex exponential sequences is analogous to continuous-time systems.[54] That is, in general terms,

$$x[n] = \sum_k a_k z_k^n \rightarrow y[n] = \sum_k a_k H[z_k] z_k^n \qquad 15.75$$
$$\text{input} \rightarrow \text{output}$$

Here $H[z_k]$, instead of being the Fourier or Laplace transform of the system impulse response $h(t)$, is the Fourier or the z-transform of the discrete time system impulse response $h[n]$. When $z = e^{j\Omega}$ with Ω real, $H(z_k)$ is the Fourier transform. When $z = e^{\sigma + j\Omega}$, $H(z_k)$ is the z-transform.[55]

16. DISCRETE-TIME SYSTEMS: FOURIER SERIES REPRESENTATION OF A PERIODIC SIGNAL

Similar to the continuous-time case, the Fourier series representation of a discrete-time periodic signal, of period N, is given by the following synthesis and analysis equations. (The symbol $\langle N\rangle$ indicates an interval of length N.)

$$X[n] = \sum_{k=\langle N\rangle} a_k e^{jk(2\pi/N)n} \qquad 15.76$$

$$a_k = \frac{1}{N} \sum_{n=\langle N\rangle} x[n] e^{-jk(2\pi/N)n} \qquad 15.77$$

The important difference in the Fourier series representation in this discrete-time case as compared to the continuous-time case is that it is finite (over the interval N).[56]

17. DISCRETE-TIME SYSTEMS: FOURIER TRANSFORM OF AN APERIODIC SIGNAL

Applying an analogous procedure to that used for continuous-time systems, that is, constructing a periodic signal from the aperiodic signal and then taking the limit as the period approaches infinity, gives the following.

$$x[n] = \frac{1}{2\pi} \int_{2\pi} x[\Omega] e^{j\Omega n}\,d\Omega \qquad 15.78$$

[53]These systems, and their associated Fourier series and transforms, are sometimes called *finite* rather than discrete.

[54]See the example in Sec. 15.10 for the form of the proof, as only the results will be presented here.

[55]This discrete transform, the z-transform, is discussed in Sec. 15.19. It is the discrete-time counterpart to the Laplace transform for continuous-time cases.

[56]This finiteness is important because it is applied in the FFT or Fast Fourier Transform. The FFT algorithm is well-suited to computers allowing efficient analysis of signals using discrete methods, which prior to computers had been too time consuming.

$$X(\Omega) = \sum_{n=-\infty}^{+\infty} x[n]e^{-j\Omega n} \qquad 15.79$$

The function given by Eq. 15.79 is called the *discrete-time Fourier transform* and the two equations together, Eq. 15.78 and Eq. 15.79, are termed the *Fourier transform pair.* As before, Eq. 15.78 is the synthesis equation and Eq. 15.79 is the analysis equation. The properties of the discrete-time Fourier series (used with periodic signals) are given in App. 15.H. Discrete-time Fourier transforms (used with aperiodic signals) are given in App. 15.I.

18. CONVOLUTION SUM

The *convolution sum,* also known as the *superposition sum,* is the discrete-time counterpart of the continuous-time convolution integral. The convolution sum is given by Eq. 15.80.

$$\begin{aligned} y[n] &= \sum_{k=-\infty}^{+\infty} x[k]h[n-k] \\ &= \sum_{k=-\infty}^{+\infty} h[k]x[n-k] \\ &= x[n] * h[n] \end{aligned} \qquad 15.80$$

The summation is the convolution of the sequences $x[n]$ and $h[n]$, which is represented by $x[n] * h[n]$. The equation represents the response of a discrete, *linear, time-invariant* (LTI) *system* to any arbitrary input based on the system's response to a unit impulse. This result for discrete systems is similar to that for continuous systems. That is, an LTI system is characterized completely by its impulse responses.

19. z-TRANSFORMS

The z-transform is the discrete-time counterpart to the Laplace transform. Also, it represents the generalization of the Fourier transform for discrete-time systems; that is, it is the transform to use when the value of s in a discrete-time system is $s = \sigma + j\Omega$.[57] The z-transform of a sequence is defined as[58]

$$X(z) = \sum_{n=-\infty}^{+\infty} x[n]z^{-n} \qquad 15.81$$

[57]This means that s has both real and imaginary parts, that is, a magnitude and a phase.

[58]This is the *bilateral z-transform.* The *unilateral z-transform* is summed from $n = 0$ to $n = \infty$ much as the unilateral Laplace transform commonly used. Also, again similar to the unilateral Laplace transform, the unilateral z-transform is useful in solving for the output of circuits with a response given by constant coefficient difference equations with nonzero initial conditions.

The variable z is complex. The relationship between the z-transform and the discrete-time Fourier transform is given by Eq. 15.82.

$$X(z)\big|_{z=e^{j\Omega}} = F(x[n]) \qquad 15.82$$

Some properties of the z-transform are given in App. 15.J. Some common z-transforms are given in App. 15.K.

20. TRANSFORMATION OF INTEGRALS

A *transform* takes on the following form.[59]

$$T\{f\} = \int_a^b f(k)k(x,\tau)dx = f(\tau) \qquad 15.83$$

Essentially one integrates out a variable, x in this case, and places the function in another *space* or *domain,* that of τ in this case. The symbol $T\{f\}$ indicates the transform of the function and $k(x,\tau)$ is referred to as the *kernel.* For the Fourier transform the kernel is $e^{j\omega_0}\tau$. For the Laplace transform, the kernel is $e^{-s\tau}$. Many others exist. A transform places the function in another *function space,* for example, k, ω, s, and so on.

By contrast, the *transformation of integrals* changes the way a physical system is viewed and has important consequences in electrical engineering. Such transformations follow.

Gauss' Divergence Theorem

$$\iint_{A(V)} \mathbf{F}\cdot d\mathbf{A} = \iiint_{V(A)} \nabla\cdot\mathbf{F}dV \qquad 15.84$$

An arbitrary vector field is represented by $\mathbf{F}$. The symbol $A(V)$ indicates the surface area over the volume of concern. The $d\mathbf{A}$ indicates the normal unit differential surface area, and $V(A)$ is the volume bounded by the surface A. Equation 15.84 says the net flux through a closed surface equals the summation of the scalar sources inside; that is, it equals the divergence of the field integrated over the volume. The symbols

$$\iint_{A(V)} \text{ and } \iiint_{V(A)}$$

are sometimes represented as

$$\int_s \text{ or } \oint_S \text{ or } \iint_S \text{ or } \oint\!\!\oint_S$$

and

$$\int_V \text{ or } \iiint_V$$

[59]This is a general linear integral transformation; that is, superposition is applicable.

Stokes' Theorem

Stokes' theorem is analogous to Gauss' theorem. It relates the curl of a vector field **F** inside a closed line, or contour, to the circulation—that is, the curl of **F** along that contour. See Fig. 15.12 for an illustration of the theorem.[60]

Figure 15.12 Curl and Circulation of Vector Field

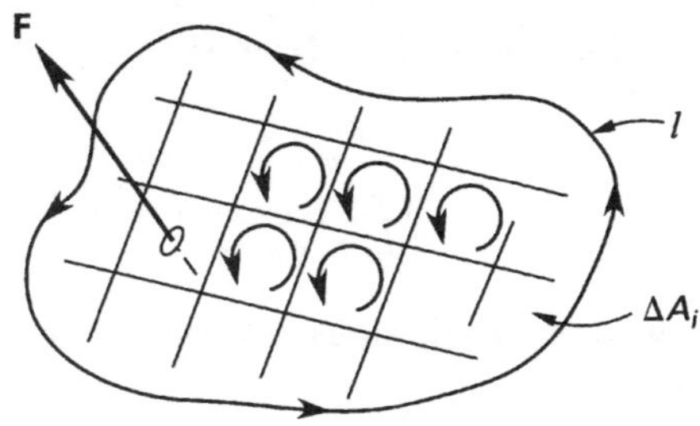

In mathematical terms, Stokes' theorem is

$$\oint_{l(A)} \mathbf{F} \cdot d\mathbf{l} = \iint_{A(l)} \nabla \times \mathbf{F} \cdot d\mathbf{A} \qquad 15.85$$

The symbols

$$\oint_{l(A)} \text{ and } \iint_{A(l)}$$

are sometimes represented as

$$\int_C \text{ and } \int_S$$

C indicates a *contour* or *line integral,* and S indicates a *surface integral.*

Green's Theorem

Green's theorem is actually a special case of Stokes' theorem with the vector **F** in the same plane as the surface A. If $g(x, y)$ and $f(x, y)$ are functions defined in this plane, say x and y, then Green's theorem in mathematical terms is

$$\oint_{l(A)} (f(x,y)\,dx + g(x,y)\,dy) = \iint_{A(l)} \left(\frac{\partial g}{\partial x} - \frac{\partial f}{\partial y}\right) dx\,dy \qquad 15.86$$

This theorem transforms double integrals, that is, surface integrals, into line integrals. This is similar to Gauss' theorem that transformed volume integrals (triple integrals) into area integrals (double integrals). Green's theorem is sometimes useful in simplifying the calculation of the integral.

21. LINE INTEGRALS

A line integral is a generalization of the concept of a definite integral. For example, a definite integral is given by the following.

$$\int_a^b f(x)\,dx \qquad 15.87$$

Let this integral apply to the curve in Fig. 15.13.

Figure 15.13 Simple Curve

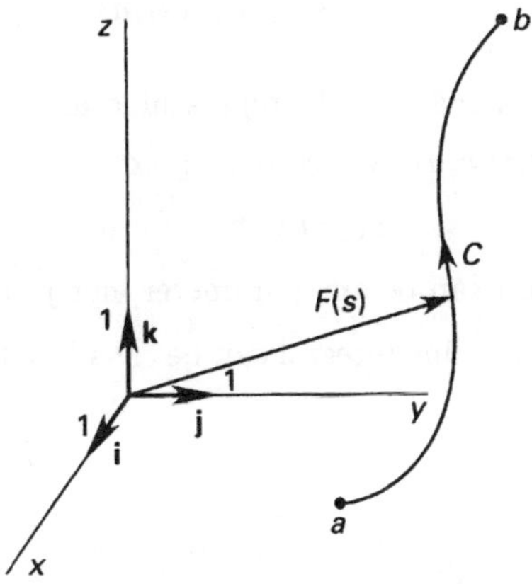

Because curve C is *simple*, that is, it has no double points, it may be represented in Cartesian coordinates as[61]

$$a \le s \le b$$

$$\mathbf{r}(s) = x(s)\mathbf{i} + y(s)\mathbf{j} + z(s)\mathbf{k} \qquad 15.88$$

The *line integral*[62] of f along C from a to b is

$$\int_C f(x,y,z)\,ds \qquad 15.89$$

The curve C is called the *path of integration.*

If the line integral is over a closed path as illustrated in Fig. 15.14, the integral would be written as

$$\oint_C f(x,y,z)\,ds \qquad 15.90$$

[60]A paddle wheel's rate of rotation can be thought of as measuring the curl of a turbulent fluid of velocity **F** with the curl in the direction of the paddle wheel axis, $\mathbf{n}dA = d\mathbf{A}$.

[61]Another way of defining a smooth curve is to say it has unique tangents at all points within the curve of the line.

[62]Probably a better name would be *curve integral* rather than line integral, but this term is not commonly used.

Figure 15.14 Closed Path

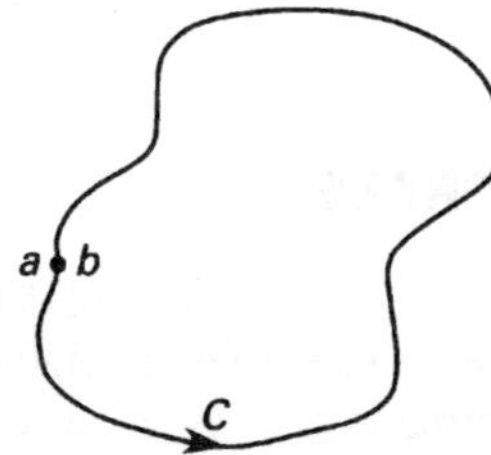

Line integrals are evaluated by changing them into definite integrals. The key is to represent the path of integration C in the desired form. One possible form is

$$\int_C f(x, y, z)\,ds = \int_a^b f\big(x(s), y(s), z(s)\big)\,ds \qquad 15.91$$

The value of $\mathbf{r}$ may also be represented as

$$\mathbf{r}(t) = x(t)\mathbf{i} + y(t)\mathbf{j} + z(t)\mathbf{k}$$
$$t_0 \le t \le t_1 \qquad 15.92$$

The variable t can be any parameter, not just time.

In this case, the line integral can be calculated using the following.

$$\int_C f(x, y, z)\,ds = \int_{t_0}^{t_1} f\big(x(t), y(t), z(t)\big)\frac{ds}{dt}\,dt \qquad 15.93$$

In this equation,

$$\frac{ds}{dt} = \sqrt{\dot{\mathbf{r}} \cdot \dot{\mathbf{r}}} = \sqrt{\dot{x}^2 + \dot{y}^2 + \dot{z}^2} \qquad 15.94$$

Example 15.9

Integrate the function $f(x, y) = 3xy^2$ over the portion of the circle from a to b as shown. The direction of C determines the positive direction and the integration path.

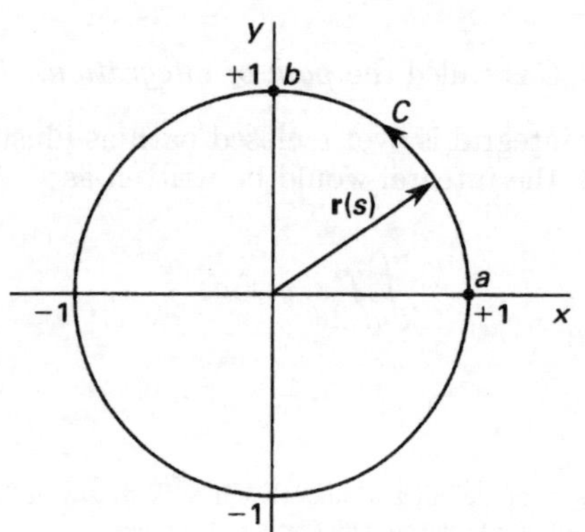

Solution

The value of $\mathbf{r}(s)$ can be seen as

$$\mathbf{r}(s) = \cos s\mathbf{i} + \sin s\mathbf{j}$$

Using Eq. 15.88 gives

$$\int_C f(x, y)\,ds = \int_a^b f\big(x(s), y(s)\big)\,ds$$

$$\int_C 3xy^2\,ds = \int_0^{\pi/2} 3\cos s\sin^2 s\,ds = 3\int_0^{\pi/2} \cos s\sin^2 s\,ds$$

Using a change of variables, let $u = \sin s$. The integration becomes[63]

$$3\int_0^{\pi/2} \cos s\,\sin^2 s\,ds = 3\int_0^1 u^2\,du = 3\frac{u^3}{3}\Big|_0^1 = 3\left(\tfrac{1}{3} - \tfrac{0}{3}\right) = 1$$

Example 15.10

Consider a particle upon which a variable force $\mathbf{p}$ acts along the curve C. The work done by the force $\mathbf{p}$ is given by the line integral

$$w = \int_C \mathbf{p}\cdot d\mathbf{r}$$

Show that this equals the gain in kinetic energy.

Solution

First,

$$d\mathbf{r} = \frac{d\mathbf{r}}{dt}\,dt = \mathbf{v}\,dt$$

Substituting gives

$$w = \int_{t_0}^{t_1} \mathbf{p}\cdot\mathbf{v}\,dt$$

The values t_0 and t_1 are the initial and final values of dt. Newton's second law states that

$$\mathbf{p} = m\frac{d^2\mathbf{r}}{dt^2} = m\frac{d\mathbf{v}}{dt}$$

[63]At this point, one could choose to determine the integral from standard math tables.

The variable m represents the mass. Substituting a second time gives

$$\begin{aligned} w &= \int_{t_0}^{t_1} \mathbf{p}\cdot\mathbf{v}dt = \int_{t_0}^{t_1}\left(m\frac{d\mathbf{v}}{dt}\right)\cdot\mathbf{v}dt \\ &= \int_{t_0}^{t_1}\frac{d}{dt}(m\mathbf{v}\cdot\mathbf{v})dt \\ &= \int_{t_0}^{t_1} dm|\mathbf{v}|^2 \\ &= \frac{m}{2}|\mathbf{v}|^2\Big|_{t_0}^{t_1} \end{aligned}$$

This is the desired result.

The value of a line integral of a given function depends not only on the endpoints but also on the shape of the integration path.

22. COMPLEX NUMBERS

Two additional important items not covered in Chap. 4 regarding complex numbers are now considered.

Let any complex number z be represented by

$$z = x + jy \qquad 15.95$$

The z is not bolded as in Chap. 4 to avoid confusion with vector variables. Although it is somewhat similar, z is a point and not a direction. Further, j is used instead of i to represent $\sqrt{-1}$ to avoid confusion with the variable for current.

The complex conjugate is

$$\overline{z} = z^* = x - jy \qquad 15.96$$

The real portion (Re or $\mathcal{R}$) of a complex number z is

$$\text{Re}(z) = x = \frac{1}{2}(z + z^*) \qquad 15.97$$

The imaginary part (Im or $\mathcal{I}$) of a complex numberz is

$$\text{Im}(z) = y = \frac{1}{2j}(z - z^*) \qquad 15.98$$

Also, the familiar formulas for real trigonometric functions hold for complex values.

Example 15.11

Express the Euler formula for complex values in terms of real functions. The Euler formula is

$$e^{j\theta} = \cos\theta + j\sin\theta$$

Solution

If θ is complex, and representing θ as z (which is standard in electrical engineering symbology), gives

$$e^{jz} = \cos z + j\sin z$$

Using a trigonometric identity, the individual terms are

$$\begin{aligned} \cos z &= \cos(x + jy) \\ &= \cos x\cos jy - \sin x\sin jy \\ \sin z &= \sin(x + jy) \\ &= \sin x\cos jy + \cos x\sin jy \end{aligned}$$

Because the familiar formulas apply, the hyperbolic cosine and sine for a complex number are

$$\begin{aligned} \cos z &= \frac{1}{2}(e^{jz} + e^{-jz}) \\ \sin z &= \frac{1}{2j}(e^{jz} - e^{-jz}) \end{aligned}$$

This allows the following representation.

$$\begin{aligned} \cos jy &= \tfrac{1}{2}(e^{j(jy)} + e^{-j(jy)}) \\ &= \tfrac{1}{2}(e^{-y} + e^{+y}) \\ &= \cosh y \end{aligned}$$

Also,

$$\begin{aligned} \sin jy &= \frac{1}{2j}\left(e^{j(jy)} - e^{-j(jy)}\right) \\ &= \frac{1}{2j}(e^{-y} - e^{+y}) \\ &= j\sinh y \end{aligned}$$

Substituting gives

$$\begin{aligned} \cos z &= \cos(x + jy) \\ &= \cos x\cosh y - j\sin x\sinh y \\ \sin z &= \sin(x + jy) \\ &= \sin x\cosh y + j\cos x\sinh y \end{aligned}$$

It is this real representation of the terms that is useful in the actual calculation of $\cos z$ and $\sin z$.

The theory of *complex analytic functions* can be useful in the study of potential theory in electrostatics. The development of such functions will not be presented but is mentioned to place them in the context of electrical engineering. The results of such calculations are shown as force field lines, or potential lines, in field theory.

Topic II: Basic Theory

Chapter

16. Electromagnetic Theory
17. Electronic Theory
18. Communication Theory
19. Acoustic and Transducer Theory

16 Electromagnetic Theory

Nomenclature

$\mathbf{a}$	acceleration	m/s^2
$\mathbf{a}$	unit vector	–
A, $\mathbf{A}$	area	m^2
B, $\mathbf{B}$	magnetic flux density	T (Wb/m^2)
c	speed of light, 2.9979×10^8	m/s
C	capacitance	F
d	distance	m
D, $\mathbf{D}$	electric flux density	C/m^2
e^-	charge on an electron, -1.6022×10^{-19}	C
E, $\mathbf{E}$	electric field strength	V/m
F, $\mathbf{F}$	force	N
G	conductance	S (A/V)
G	universal gravitational constant, 6.672×10^{-11}	$N{\cdot}m^2/kg^2$
h	Planck's constant, 6.6256×10^{-34}	J·s
H, $\mathbf{H}$	magnetic field strength	A/m
i	variable current	A
I, $\mathbf{I}$	constant or rms current	A
J, $\mathbf{J}$	current density	A/m^2
k	electrostatic proportionality constant, 8.987×10^9	$N{\cdot}m^2/C^2$
l, $\mathbf{l}$	length	m
ℓ	azimuthal quantum number	–
L	inductance	H
m	mass	kg
m	spatial quantum number	–
m_s, s	spin quantum number	–
M, $\mathbf{M}$	magnetization	A/m
M	mutual induction	H
n	principal quantum number	–
N	number of turns	–
p	pole strength	Wb (N/T)
p^+	charge on a proton, 1.6022×10^{-19}	C
$\mathbf{P}$	polarization	C/m^2
q, Q	charge	C
r, $\mathbf{r}$	radius	m
R	ratio	–
R	resistance	Ω
$\mathcal{R}$	reluctance	A/Wb
s, $\mathbf{s}$	surface area	m^2
S	elastance	1/F
t	thickness	m
t	time	s
u	energy density	J/m^3
U	energy	J
v	variable voltage	V
v, $\mathbf{v}$	velocity[1]	m/s
V	constant or rms voltage	V
V_{volume}	volume	m^3

[1]The term "velocity" is often used in text where it would be more technically correct to use the term "speed." *Velocity* is a vector quantity with speed as its magnitude.

Symbols

Γ	reciprocal inductance	1/H
ϵ	permittivity	$C^2/N{\cdot}m^2$ (F/m)
ϵ_0	free-space permittivity, 8.854×10^{-12}	$C^2/N{\cdot}m^2$ (F/m)
θ	angle	degrees
Λ	flux linkage	Wb
μ	mobility	$m^2/V{\cdot}s$
μ	permeability	N/A^2 (H/m)
μ_0	free-space permeability, 1.2566×10^{-6}	N/A^2 (H/m)
ρ	charge density	C/m^3
σ	conductivity	S·m
ϕ, Φ	magnetic flux	Wb
χ	susceptibility	–
ψ, Ψ	electric flux	C
ω	angular velocity	rad/s

Subscripts

1–2	from 1 to 2; or effect on 2 caused by the field of 1
ave	average
c	conduction
ci	cast iron
d	displacement drift
e	electric
l	line
m	magnetic
m	mean
r	relative

PART 1: FUNDAMENTALS

As a general rule, electrical effects emanate from static charges, while magnetic effects emanate from charges in motion. To be more precise, *electricity* is any manifestation of energy conversion of charge carriers that results in forces in the direction of motion of those charge carriers.[2] *Magnetism* is any manifestation of the kinetic energy of charge carriers that arises from forces or produces forces in the direction perpendicular to the motion of those charge carriers. The two were long thought to be separate phenomena, but were united by Maxwell's equations.[3] *Electromagnetism* deals with control of the average movement of charges, in generally linear elements, and usually at considerable power levels. For comparison, see the description of electronics in Sec. 17.1.

The charges of interest are the electron and the proton, designated as e^- and p^+. The mass, charge, and charge-to-mass ratios for the proton and electron are given in Table 16.1. The SI unit of charge is the coulomb (C). The charge of one electron is sometimes referred to as one *electrostatic unit* (esu). One C is approximately equal to 6.24×10^{18} esu. One mole of electrons is referred to as a *faraday*. One faraday is approximately equal to 96 500 C.

Table 16.1 *Properties of the Electron and Proton*

	electron	proton
mass at rest, kg	9.1096×10^{-31}	1.6726×10^{-27}
charge, C	-1.6022×10^{-19}	$+1.6022 \times 10^{-19}$
charge-to-mass ratio, C/kg	1.7588×10^{11}	9.5791×10^{7}

The *extranuclear structure* of an atom, that is, its electrical or electronic structure, determines the defining characteristics of that atom. Orbiting electrons are arranged in shells designated *K, L, M, N, O, P,* and *Q*. The inner shell electrons are tightly bound and interact only with high-energy particles, such as gamma rays. The outer shells in complex atoms—those with numerous neutrons and protons, and thus numerous electrons—tend to be loosely bound. It is these outer shells that determine the electrical and chemical properties of the elements. The energy of the orbiting electron is determined by four quantum numbers: the *principle quantum number* (n), the *azimuthal quantum number* (ℓ), the *spatial quantum number* (m), and the *spin quantum number* (m_s or s).[4,5]

When electrons are in close proximity, such as in a crystalline solid, nearby atoms affect their behavior and the electron's energy is no longer uniquely determined. Indeed, the single energy level of an electron in a free atom is spread into a band, or range, of energy levels.[6] The *conduction band* is the range of energy states in a solid in which electrons can move freely (that is, they can effect transitions between energy levels). In other words, the band is not full of *conduction electrons.* When every energy level is full, the material is known as an *insulator* or a *dielectric.*

[2]The forces produce displacement, velocity, or acceleration.

[3]Special relativity also confirms the link between electricity and magnetism in that it shows that electric and magnetic fields appear or vanish depending upon the motion of the observer. Further, it was the Lorentz transformation that required invariance in Maxwell's equations; that is, electric and magnetic fields must remain in the same form for stationary and moving systems. This requirement is necessary because, without it, electrical and optical phenomena would differ depending on one's motion. It is important to understand that, while the experimental results do differ depending on the motion of the observer, the fundamental phenomena remain unchanged. For example, Maxwell's equations describing electromagnetism remain in the same form. The Lorentz transformation was the genesis of Einstein's special theory of relativity.

[4]The azimuthal quantum number, ℓ, specifies the angular orbital momentum. Electrons whose value of ℓ is 0, 1, 2, and 3 are referred to as the *s*, *p*, *d*, and *f* electrons (for historical reasons related to the "picture" these electrons produced during experiments).

[5]The spin has two values, $\pm 1/2$, and is equal to $h/2\pi$ where h is Planck's constant.

[6]This band of energy levels allows what is referred to in some texts as an *electron cloud* to exist. The theory is by no means universal. Consult electrical conduction theory texts for more information.

Table 16.2 Common Conductors[a]

material	gage[b]	diameter (in)	area[c] (circular mils)	nominal DC resistance (ohms/1000 ft at 68°F (20°C))
aluminum[d]	AWG 8	0.1285	16,510	1.030
copper[e]	AWG 8	0.1285	16,510	0.6533
steel[f]	BWG 8	0.165	27,239[g]	3.28[h]

[a]The English Engineering System is used in this table, as it represents the system most commonly used in this area of electrical engineering.
[b]AWG stands for American Wire Gage, the usual standard for nonferrous wires, rods, and plates. BWG stands for Birmingham Wire Gage, the usual standard for galvanized iron and steel wire.
[c]One circular mil is a unit having the area corresponding to the area of a circle with a diameter of 0.001 in.
[d]This information is for bare, solid, all-aluminum hard-drawn wire and is taken from ASTM International's Standard, ASTM B230/B230M.
[e]This value is considered a trade maximum. It depends on the specific process used when manufacturing the wire. ASTM requirements do exist for the resistance of copper for various processes.
[f]The type of steel referenced here is used for telephone and telegraph wire.
[g]The area in circular mils varies slightly depending upon the construction, that is, the number of strands and arrangement.
[h]This number is referenced to 1000 ft here for convenience in comparison with the other conductors. Transmission cables and telephone or telegraph wires are often referenced to longer lengths, as in ohms/mile.

PART 2: ELECTRIC AND MAGNETIC PHENOMENA

1. ELECTROMAGNETIC EFFECTS

Electromagnetic effects are caused by the dynamic behavior of elementary particles, that is, the electron and the proton, due to their mass and charge. These effects can differ greatly from free-space effects, depending upon the material in which the particles find themselves. There are three major classes of electromagnetic materials.

- *Conductors or semiconductors* are materials through which charges flow more or less easily. Some common conductors are given in Table 16.2. The properties of widely used semiconductors are given in Table 16.3. Good conductors have a conductivity range of 1×10^3 S/m to 6×10^3 S/m.
- *Dielectrics or insulators* are materials that inhibit the passages of charges. Some common insulators are given in Table 16.4. Insulators have a conductivity range of 10^{-18} S/m up to approximately 10^{-4} S/m. The determining factors for the conductivity of insulators are the application and the voltage stress (that is, the magnitude of the voltage being insulated).
- *Magnetic materials* are materials in which the motion of the charges produces perpendicular effects, that is, enhanced transverse effects. A very small portion of the elements in the periodic table exhibit such effects. These materials are given in Table 16.5.

Chapter 54 further discusses the properties of electrical materials. This information is necessary to understanding how electromagnetic and related theories can be applied to practical ends.

Table 16.3 Common Semiconductors

semiconductor	symbol	periodic table group
silicon	Si	IV
germanium	Ge	IV
gallium arsenide	GaAs	III and V
diamond and graphite	–	–
selenium	Se	VI
silicon carbide	SiC	IV
indium antimonide	InSb	III and V

Table 16.4 Common Insulators

insulating material
polyvinyl chloride (PVC)
butyl rubber
neoprene
plastics
ceramics
insulating gasses and liquids

Table 16.5 Magnetic Materials

magnetic material	symbol	atomic number
iron	Fe	26
cobalt	Co	27
nickel	Ni	28

2. CONDUCTION EFFECTS

Conduction effects occur in systems having mobile charges when an electric force is applied. The charges move in the direction of the applied force. As mentioned in Part 1, conduction effects result from the loosely bound outer shell electrons of atoms. In very strong conductors, electrons can be released and move under the application of small chemical, thermal, or other forces.

3. DIELECTRIC PHENOMENA

When a system contains bound charges, the electrons still move under the application of an electric force, but only to a limited extent. This process is called the *dielectric phenomenon*. It results in a current termed the *displacement current*, which is related to the *electric flux density*, **D**, in the dielectric material. In turn, the electric flux density, **D**, is proportional to the *electric field strength* or *intensity*, **E**.[7] Heat losses occur in dielectrics due to hysteresis effects at very high frequencies. Hysteresis effects occur in magnetic materials at much lower frequencies.

Dielectrics can sustain only a certain field strength before they break down and conduction occurs. This maximum field strength is called the *dielectric strength*, and its unit of measure is volts per meter of thickness of the material tested. The results vary, depending upon thickness and the environmental conditions of the test. For the same material, thin dielectrics break down at lower field strengths than thick dielectrics.

4. MAGNETIC PHENOMENA

Magnetic phenomena are caused by the directed motion of charges. The effects occur perpendicular to this motion. In magnetic materials, the phenomena are associated with the orientation of the orbiting electrons and, to some extent, the spin of the electrons. Magnetic phenomena occur in a relatively small number of materials whose outer shell orbits fill prior to the inner shells. This occurs because the outer shell orbit configuration is actually a lower energy configuration than the inner shell.

Several types of magnetism exist.

- *Ferromagnetism* is produced by the exchange of forces between atomic moments. This type of magnetism produces strongly magnetic materials with high permeability. Large clusters of atoms group together to form *magnetic domains*, each of which has the same atomic moment alignment within the domain. Ferromagnetism, named for iron, is the common magnetism one associates with horseshoes or toy magnets. See Fig. 16.1 for an example of a crystalline structure with domains and Fig. 16.2 for an example of magnetization curves. Figure 16.3 shows the spin arrangements of several types of magnetism.

- *Paramagnetism* is produced by the orbital or spin moments of the electrons, or both. The atomic alignment is minimal, that is, there is little or no domain formation, and as a result, paramagnetic materials are not strongly magnetic.

- *Antiferromagnetism* is produced by the exchange of forces between atomic moments. Antiferromagnetic materials have an antiparallel arrangement of equal spins. Their magnetism is of a similar strength to that of paramagnetic materials.

- *Diamagnetism* is produced by electron spins in antiparallel pairs in closed electron shells. Diamagnetic materials are weakly repulsive to external magnetic fields. These materials possess an internal magnetic field that opposes any externally applied magnetic field.

- *Ferrimagnetism* is produced by the moment that results from the combination of two antiferromagnetic lattices. Ferrimagnetic materials contain two kinds of magnetic ions, arranged antiparallel but with unequal spins.

Figure 16.1 *Magnetic Domains*

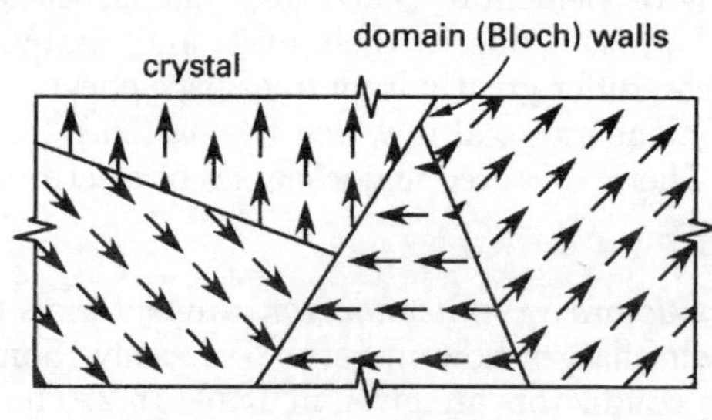

Figure 16.2 *Magnetization Curves*

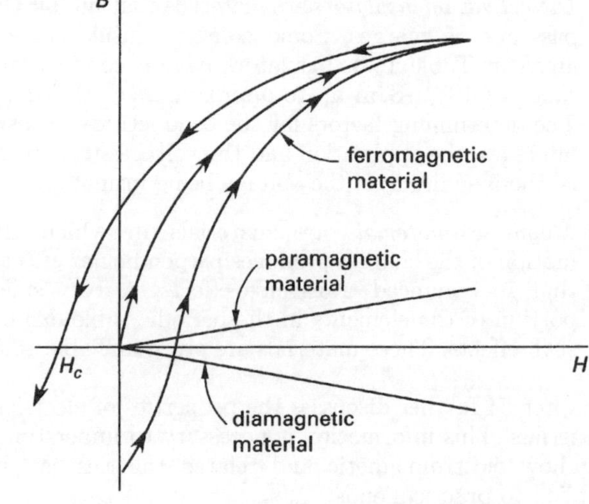

[7]Both of these are discussed more fully in Sec. 16.11.

Figure 16.3 Magnetic Spin Arrangements

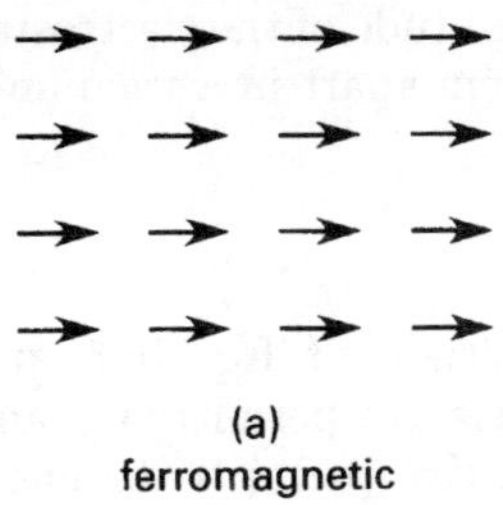

(a)
ferromagnetic

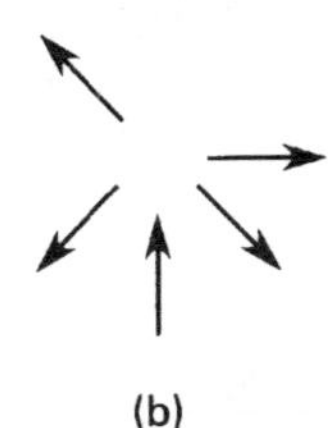

(b)
paramagnetic

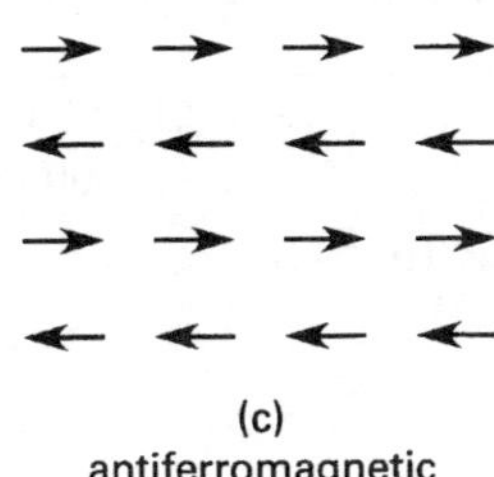

(c)
antiferromagnetic

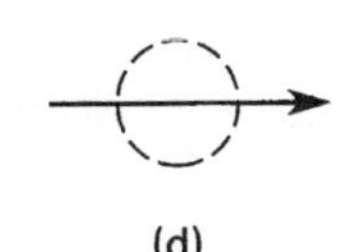

(d)
diamagnetic*

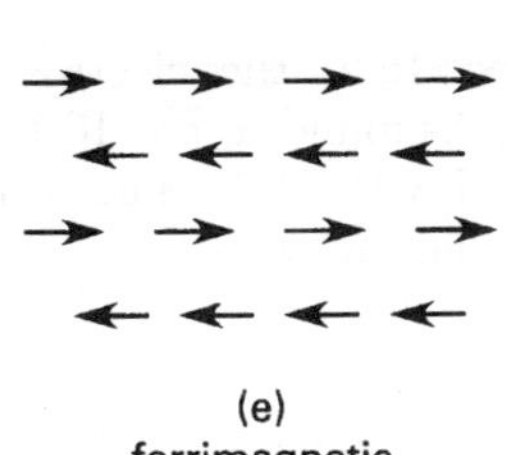

(e)
ferrimagnetic

*The circle represents the internal magnetic field.

5. THERMOELECTRIC PHENOMENA

There are three significant relations between electricity and heat of importance to the electrical engineer. These are the Seebeck effect, the Thomson effect, and the Peltier effect. Each will be explained in terms of its applications. For now, the important point is that heat is capable of providing the energy to release and transport electrons.

PART 3: ELECTROSTATICS

As the name implies, electrostatics is the field of electrical engineering that deals with the electrical effects of stationary charges. An important concept in this field is that all electrical phenomena are expressible in terms of the masses and charges of the elementary particles involved.

6. ELECTRIC CHARGES

An electrically neutral material has equal numbers of positive and negative charge carriers. Equation 16.1 describes bodies in this unelectrified state.

$$Q^+ = Q^- = 0 \quad \textit{16.1}$$

Only integer numbers of charge carriers exist. Therefore, the transport of an integer number n charge carriers to a body from another body means that

$$Q^+ = np^+ \text{ and } Q^- = ne^- \quad \textit{16.2}$$

Like charges exhibit an electric repulsion. Unlike charges exhibit an electric attraction. An attraction also exists between the masses of any two charged particles. The force associated with the gravitational attraction is very small in comparison to the force of the electrical attraction. Therefore, the gravitational force can be ignored except when the charges are accelerated.

Static electricity formed by positive charges, such as that produced by rubbing silk on a glass rod, is called *vitreous static electricity.* When the static electricity is formed by negative charges, such as that produced by rubbing fur on a rubber, amber, or plastic rod, it is termed *resinous static electricity.*

Example 16.1

What is the gravitational force of attraction between two electrons located 1 cm apart?

Solution

Use Newton's law of gravitation.

$$F = \frac{Gm_1m_2}{r^2}$$

Newton's universal constant or *gravitational constant,* G, is equal to 6.672×10^{-11} N·m^2/kg^2. The mass of the electrons (taken from Table 16.1) and the 1 cm distance between the electrons are substituted into the equation, giving the following.

$$F = \frac{Gm_1m_2}{r^2} = \frac{Gm_e^2}{r^2}$$

$$= \frac{\left(6.672 \times 10^{-11} \frac{\text{N·m}^2}{\text{kg}^2}\right)(9.1096 \times 10^{-31}\ \text{kg})^2}{(1\ \text{cm})^2\left(\frac{1\ \text{m}}{100\ \text{cm}}\right)^2}$$

$$= 5.537 \times 10^{-67}\ \text{N}$$

7. COULOMB'S LAW

Coulomb's law states that the force between charges at rest in an isotropic medium is proportional to the product of the charges and inversely proportional to both the square of the distance between them and the dielectric coefficient of the medium. The force acts in a straight line between the centers of the charges.[8] Equation 16.3 describes Coulomb's law in mathematical terms.

$$\mathbf{F}_{1\text{-}2} = \frac{Q_1Q_2}{4\pi\epsilon r_{1\text{-}2}^2}\mathbf{a}_{r_{1\text{-}2}} \qquad 16.3$$

The vector $\mathbf{a}_{r_{1\text{-}2}}$ is a unit vector having a direction along a straight line connecting Q_1 and Q_2. The term ϵ represents the *dielectric coefficient*, more commonly called the *permittivity*. The 4π term occurs in many electric field calculations and, for this reason, is included in this *rationalized* form of Coulomb's law in order to cancel terms in integrals over spheres and make calculations easier. This is perhaps more readily seen when Coulomb's law is written in terms of the electric field, **E**.

$$\mathbf{F}_{1\text{-}2} = \mathbf{F}_2 = Q_2\mathbf{E}_1 \qquad 16.4$$

Coulomb's law is also stated in its Gaussian SI form as follows.

$$\mathbf{F}_{1\text{-}2} = \frac{kQ_1Q_2}{\epsilon_r r^2}\mathbf{a}_{1\text{-}2} \qquad 16.5$$

The term ϵ_r represents the *relative permittivity* of the dielectric. ($\epsilon_r = 1.0$ for a vacuum.) The constant k has an approximate value of 8.987×10^9 N·m²/C². In Eq. 16.3 through Eq. 16.5 the force is measured in newtons, the distance in meters, and the charge in coulombs.

[8]In Eq. 16.3, the unit vector, and thus the force, is *from 1 to 2*. The force direction is the direction charge 2 will move due to the electric field of charge 1. See Eq. 16.4. The notation often varies. The unit vector $\mathbf{a}_{r_{1\text{-}2}}$ (see Eq. 16.3) or $a_{1\text{-}2}$ (see Eq. 16.5) is sometimes written $\mathbf{a}_r$, $\mathbf{a}_{12}$, or even $\mathbf{a}_{21}$. The $\mathbf{a}_{21}$ notation is used as a reminder that, although the vector is *from 1 to 2*, it is determined mathematically by subtracting the position elements of 2 from 1; for example, $x_2 - x_1$, $y_2 - y_1$, and $z_2 - z_1$. (The unit vector is then determined by dividing the distance between the two points.) At times, the notation 1-2 merely indicates the connection between the points and one must determine the correct direction from other information.

Example 16.2

What is the magnitude of the electrostatic force between two electrons 1 cm apart in a vacuum?

Solution

Take the magnitude of Eq. 16.5 and substitute the values for the relative permittivity and k given in this section. The relative permittivity has no units. As the term "relative" suggests, it is a comparison. This is discussed more fully in Sec. 16.9.

$$\mathbf{F}_{1\text{-}2} = k\left(\frac{Q_1Q_2}{\epsilon_r r^2}\right)\mathbf{a}_{1\text{-}2}$$

$$|\mathbf{F}_{1\text{-}2}| = F = k\left(\frac{Q_1Q_2}{\epsilon_r r^2}\right)$$

$$= \frac{\left(8.987 \times 10^9 \frac{\text{N·m}^2}{\text{C}^2}\right)(-1.6022 \times 10^{-19}\ \text{C})^2}{(1)(1\ \text{cm})^2\left(\frac{1\ \text{m}}{100\ \text{cm}}\right)^2}$$

$$= 2.307 \times 10^{-24}\ \text{N}$$

Example 16.3

What is the ratio of the electrostatic force to the gravitational force between two electrons 1 cm apart in a vacuum?

Solution

As found in Ex. 16.2, the electrostatic repulsion is 2.307×10^{-24} N. As found in Ex. 16.1, the gravitational attraction is 5.537×10^{-67} N. The ratio of electrostatic force to gravitational force is

$$R = \frac{2.307 \times 10^{-24}\ \text{N}}{5.537 \times 10^{-67}\ \text{N}}$$

$$= 4.166 \times 10^{42}$$

The electrostatic force that these elementary particles experience is 42 powers of ten greater than the gravitational force of attraction between them. Clearly, the gravitational attraction between such particles can be ignored in all but the most exacting calculations.

Coulomb's law implies that the principle of superposition applies for electric charges. This means that a system of charges is a linear system. The total resultant effect at any point within the system can be determined from the vector sum of the individual components.

Example 16.4

Three point charges in a vacuum are arranged in a straight line. Find the force on point charge A.

A 3 m B 4 m C

400 μC −200 μC 800 μC

Solution

Because all three charges are in line, vector analysis is not needed. The force on point charge A is the sum of the individual forces from the other two point charges. Use the Gaussian form of Coulomb's law.

From Eq. 16.5,

$$\begin{aligned} F_{\text{A-B\&C}} &= F_{\text{B\&C-A}} \\ &= F_{\text{A-B}} + F_{\text{A-C}} = \frac{kQ_\text{A}Q_\text{B}}{r_{\text{A-B}}^2} + \frac{kQ_\text{A}Q_\text{C}}{r_{\text{A-C}}^2} \\ &= 8.987 \times 10^9 \ \frac{\text{N·m}^2}{\text{C}^2} \\ &\quad \times \left(\frac{(400 \times 10^{-6}\ \text{C})(-200 \times 10^{-6}\ \text{C})}{(3\ \text{m})^2} + \frac{(400 \times 10^{-6}\ \text{C})(800 \times 10^{-6}\ \text{C})}{(7\ \text{m})^2} \right) \\ &= -79.9\ \text{N} + 58.7\ \text{N} \\ &= -21.2\ \text{N} \end{aligned}$$

Because the sign is negative, the force on A is toward B and C.

Example 16.5

Determine the magnitude and direction of the force on point charge A caused by point charges B and C.

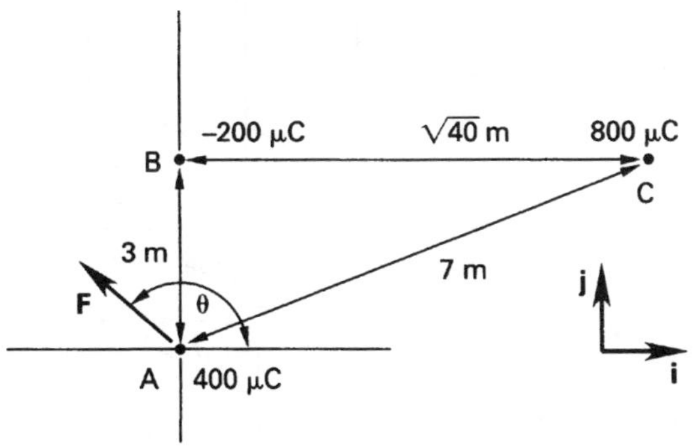

Solution

The charges and distances are the same as in Ex. 16.4.

$$F_{\text{B-A}} = F_{\text{A-B}} = -79.9\ \text{N} \quad \text{[attractive]}$$

$$F_{\text{C-A}} = F_{\text{A-C}} = -58.7\ \text{N} \quad \text{[repulsive]}$$

Determine the unit vectors for these forces. Let point A correspond to the origin. The unit vectors ($\mathbf{a}_{\text{B-A}}$ and $\mathbf{a}_{\text{C-A}}$) from B and C to A are determined as

$$\begin{aligned} \mathbf{a}_{\text{B-A}} &= \frac{(x_\text{A} - x_\text{B})\mathbf{i} + (y_\text{A} - y_\text{B})\mathbf{j}}{\sqrt{(x_\text{A} - x_\text{B})^2 + (y_\text{A} - y_\text{B})^2}} \\ &= \frac{(0-0)\mathbf{i} + (0-3)\mathbf{j}}{\sqrt{(0-0)^2 + (0-3)^2}} \\ &= -\mathbf{j} \end{aligned}$$

$$\begin{aligned} \mathbf{a}_{\text{C-A}} &= \frac{(0 - \sqrt{40})\mathbf{i} + (0-3)\mathbf{j}}{\sqrt{(0-\sqrt{40})^2 + (0-3)^2}} \\ &= -0.904\mathbf{i} - 0.429\mathbf{j} \end{aligned}$$

The total force is

$$\begin{aligned} &F_{\text{B-A}}\mathbf{a}_{\text{B-A}} + F_{\text{C-A}}\mathbf{a}_{\text{C-A}} \\ &\quad = (-79.9\ \text{N})(-\mathbf{j}) + (58.7\ \text{N})(-0.904\mathbf{i} - 0.429\mathbf{j}) \\ &\quad = -53.1\mathbf{i}\ \text{N} + 54.7\mathbf{j}\ \text{N} \end{aligned}$$

The magnitude of the force is

$$|\text{F}_{\text{A-B\&C}}| = \sqrt{(-53.1\ \text{N})^2 + (54.7\ \text{N})^2} = 76.2\ \text{N}$$

The direction (counterclockwise angle with respect to the horizontal) is

$$\theta = 180° - \arctan \frac{54.7\ \text{N}}{53.1\ \text{N}} = 180° - 45.9° = 134.1°$$

8. ELECTRIC FIELDS

The *electric field*, **E**, with units of V/m, is the region in which an electric charge exerts a measurable force on another charge, above and beyond the gravitational force that the two charges exert on each other. It is a *vector force field* because a test charge placed in the field will experience a force in a given direction and with a specific magnitude. The direction of the field is along flux lines. These lines, ideally, represent the direction of the force on an infinitely small, isolated positive test charge.[9] The actual direction is represented by the unit vector **a**. Equation 16.6 represents the *electric field strength*, or *electric field intensity*, in a substance with permittivity ϵ at a distance r from a point charge Q.

$$\mathbf{E} = \frac{Q}{4\pi\epsilon r^2}\mathbf{a} \qquad 16.6$$

Coulomb forces obey the principle of linear superposition in free space and in many dielectrics. Therefore, the electric field strength or intensity acting on a charge Q

[9]The positive test charge is arbitrary, but universal. The charge would ideally have to be infinitely small in order to prevent its own electric field from distorting the original field.

Basic Theory

located at point 1 as a result of charges $Q_1, Q_2, \ldots, Q_i$ at distances r_{i1} from point 1 can be represented as the following sum. The vector a_{ri1} is the unit vector with direction from Q_i to point 1.

$$\mathbf{E}_1 = \sum_{i=1}^{i} \frac{Q_i}{4\pi\epsilon r_{i1}^2}\mathbf{a}_{r_{i1}} \tag{16.7}$$

Additionally, the electric field intensity at any point p produced by charges $Q_1, Q_2, \ldots, Q_n$ at distances $r_1, r_2, \ldots, r_n$ from the point is the vector sum of the field strengths produced by the charges individually.

If one is dealing with charges distributed uniformly throughout space, the electric field strength at a distance r that is large when compared to dV_{volume} is given by Eq. 16.8.

$$\mathbf{E}_1 = \int_{V_{\text{volume}}} \frac{\rho\, dV_{\text{volume}}}{4\pi\epsilon r_{i1}^2}\mathbf{a}_{r_{i1}} \tag{16.8}$$

The unit vector $\mathbf{a}_{ri1}$ is directed from dv to point 1, at which $\mathbf{E}$ is evaluated. The unit vector, the distance, r, and the charge density, ρ, are usually functions of the variable of integration. The total charge, Q, in the volume V_{volume} that has the charge density ρ is

$$Q = \int_{V_{\text{volume}}} \rho\, dV_{\text{volume}} \tag{16.9}$$

The electric fields discussed to this point have been radial in nature. Not all fields are radial. As seen in Eq. 16.8, the electric field depends on the orientation of the surface or volume that contains the charge. A *uniform electric field*, commonly used in sensing instruments and other important applications, can be created between two flat plates, as shown in Fig. 16.4.

Figure 16.4 *Uniform Electric Field*

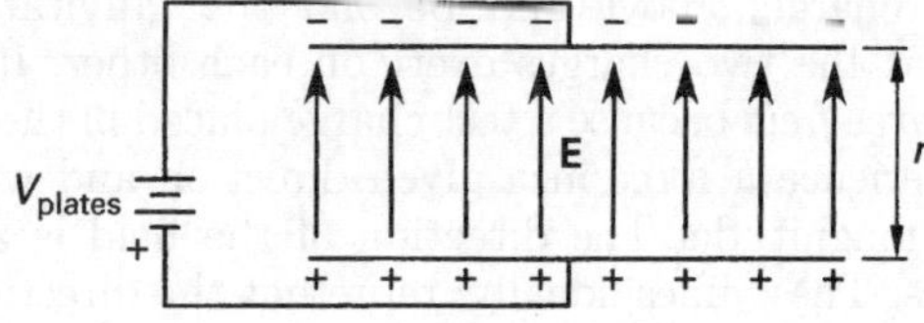

With a potential difference of V volts between the flat plates separated by distance r, the electric field strength is

$$E_{\text{uniform}} = \frac{V_{\text{plates}}}{r} \tag{16.10}$$

The field strengths, or intensities, for several configurations are given in Table 16.6.

An electric field is an abstraction that allows one to visualize how charges in different locations within the field will experience a force. A *line of electric force* is a curve drawn so that it has the direction of the force acting on the test charge, (i.e., it is in the direction of the electric field). A *tube of force* is created by drawing lines of force through the boundary of any closed curve. The lines form a tubular surface not cut by any other lines of force.

Table 16.6 *Electric Fields and Capacitance for Various Configurations*

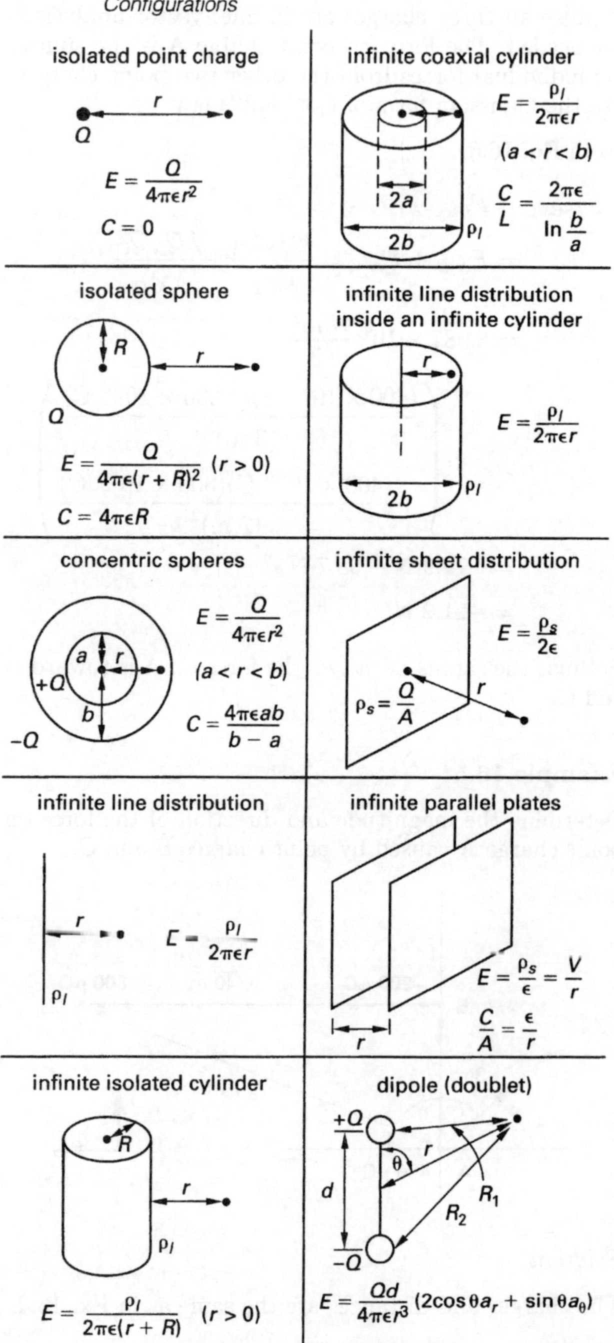

9. PERMITTIVITY AND SUSCEPTIBILITY

The forces between electric charges depend upon the environment in which they are located. The maximum coulombic force occurs in free space. In all other materials, the coulombic force is reduced. Another way of viewing this phenomenon is to say that the electric flux does not pass equally through all materials.[10] In fact, it cannot pass through conductive material at all and is diminished to varying degrees by various dielectrics.

The *permittivity of free space*, that is, a vacuum, ϵ_0, may be considered to be representative of the dielectric properties of free space. In actuality, it is a factor that provides consistent units for Coulomb's law when units of force, charge, and distance are arbitrary. The product of the permittivity of free space and the *relative permittivity* of a medium gives the *permittivity* of that medium.

$$\epsilon = \epsilon_0 \epsilon_r \qquad 16.11$$

In general, the relative permittivity is a complex number. Except for very high frequencies, the imaginary component, or quadrature component, can be ignored. The real component is often termed the *dielectric constant.* Because the value of ϵ_r depends on frequency as well as other factors, the term *dielectric coefficient* is more appropriate. The relative permittivity can also be viewed in terms of capacitance, as shown in Eq. 16.12.

$$\epsilon_r = \frac{C_{\text{with dielectric}}}{C_{\text{vacuum}}} \qquad 16.12$$

Typical relative permittivities are given in Table 16.7.

The permittivity can also be expressed as follows.

$$\epsilon = \epsilon_0 (1 + \chi_e) \qquad 16.13$$

The term χ_e represents the electric susceptibility of a given dielectric. It is a numeric measure of the polarization or displacement of electrons in the atoms or molecules of the dielectric.

The units for permittivity are $C^2/N{\cdot}m^2$ or F/m. The permittivity of free space, or vacuum, ϵ_0, in SI units is $8.854 \times 10^{-12}\ C^2/N{\cdot}m^2$ (F/m).

Example 16.6

A certain capacitor is constructed of two square parallel plates with air as the dielectric. If a teflon insert is used in place of the air, by what factor does the capacitance increase?

Solution

The capacitance is directly proportional to the permittivity and is given by the following formula.

$$C = \frac{\epsilon A}{r} = \frac{\epsilon_0 \epsilon_r A}{r}$$

Because the example asks for the factor by which the capacitance increases, not the actual value of the capacitance, one need merely compare the permittivity of teflon to that of air. Using data from Table 16.7,

$$\frac{\epsilon_{r,\text{teflon}}}{\epsilon_{r,\text{air}}} = \frac{2.0}{1.00059} = 2.0$$

One could use the upper bound of the permittivity for the teflon, in which case the factor by which the capacitance increases would be approximately 2.2.

Table 16.7 Typical Relative Permittivities (20°C and 1 atm)

material	ϵ_r	material	ϵ_r
acetone	21.3	mineral oil	2.24
air	1.00059	mylar	2.8–3.5
alcohol	16–31	olive oil	3.11
amber	2.9	paper	2.0–2.6
asbestos paper	2.7	paper (kraft)	3.5
asphalt	2.7	paraffin	1.9–2.5
bakelite	3.5–10	polyethylene	2.25
benzene	2.284	polystyrene	2.6
carbon dioxide	1.001	porcelain	5.7–6.8
carbon tetrachloride	2.238	quartz	5
		rock	≈ 5
castor oil	4.7	rubber	2.3–5.0
diamond	16.5	shellac	2.7–3.7
glass	5–10	silicon oil	2.2–2.7
glycerine	56.2	slate	6.6–7.4
hydrogen	1.003	sulfur	3.6–4.2
lucite	3.4	teflon	2.0–2.2
marble	8.3	vacuum	1.000
methanol	22	water	80.37
mica	2.5–8	wood	2.5–7.7

10. ELECTRIC FLUX

Flux is defined as the electric or magnetic lines of force in a region. The *electric flux*, Ψ, is a scalar field representing these imaginary lines of force. Flux lines leave or enter any surface at right angles to that surface, and the orientations of the electric field lines and the electric flux lines always coincide. By convention, the flux lines are directed outward from the positive charge and inward toward the negative charge, or to infinity if no negative charge exists in the area, as shown in Fig. 16.5. In Fig. 16.5, only a few representative lines of flux are shown. Also, if the charge on a hollow sphere is all the same sign, the flux outside this sphere is the same as the flux that would exist from a point charge at the center. Therefore, for the purposes of drawing flux lines and calculating forces, a hollow charged sphere can be replaced by a point charge at the center.

[10]The concept of flux is another abstraction used to help one visualize electric interactions. It is discussed more fully in Sec 16.10.

Figure 16.5 Flux Lines

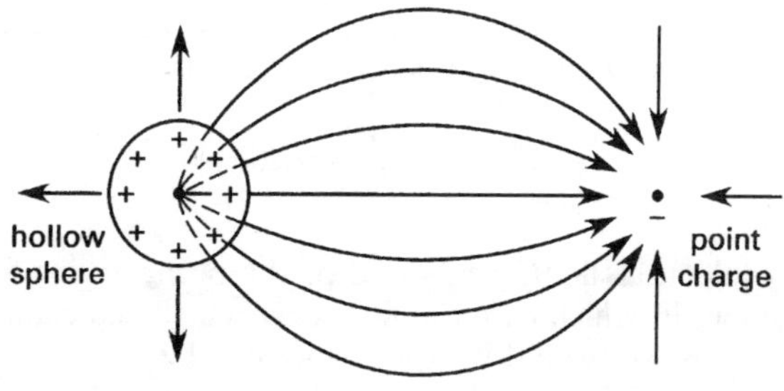

The flux, Ψ, is numerically equal to the charge. That is, by definition, one coulomb of electric charge gives rise to one coulomb of electric flux.

$$\Psi = Q \qquad 16.14$$

The electric flux is also called the *displacement flux*. This is because the flux is a quantity associated with the amount of bound charge, Q, displaced in a dielectric material that is subject to an electric field. This terminology can be understood by referring to Fig. 16.6.

Figure 16.6 Displacement Flux Experiment

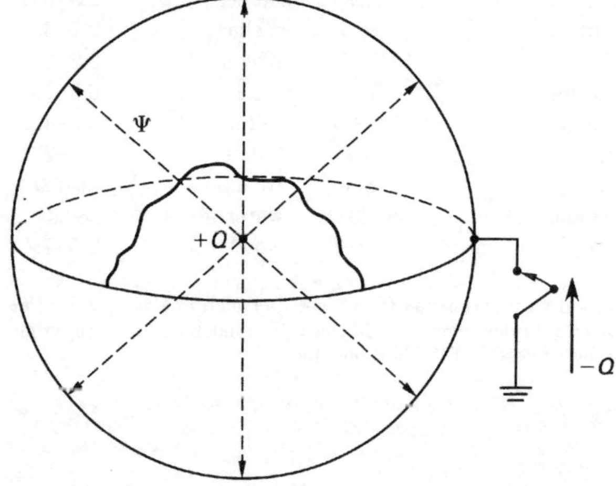

Early experimenters enclosed a fixed positive charge inside a spherical conducting shell as shown in Fig. 16.6. When the switch was closed, a charge of $-Q$, equal in magnitude to the enclosed positive charge, was found on the shell and was believed to be caused by the transient flow of negative charge induced by the flux from the $+Q$. In other words, the positive charge *displaced* the $-Q$ charge and forced it onto the surface. The scalar electric flux field, Ψ, is not directly measurable as is the electric field, **E**. Instead it is inferred from such experiments.

11. ELECTRIC FLUX DENSITY

The *electric flux density*, **D**, is also referred to as the *displacement density* or simply the *displacement*. It is also loosely referred to as the *vector flux density*. It is a vector quantity that, like the electric flux, Ψ, cannot be directly measured. It is given in Eq. 16.15. Unlike the electric field strength, **E**, the electric flux density or displacement, **D**, is independent of the medium.[11]

$$\mathbf{D} = \epsilon\mathbf{E} = \epsilon_0\epsilon_r\mathbf{E} \qquad 16.15$$

The magnitude of **D**, or D, is also referred to as the *flux density*. It is the number of flux lines per unit area perpendicular to the flux. It is a scalar quantity with units of C/m^2.

$$D = |\mathbf{D}| = \frac{\psi}{A} = \frac{Q}{A} = \sigma \qquad 16.16$$

When the medium is nonisotropic, the permittivity, ϵ, is a tensor and the displacement, **D**, and electric field strength, **E**, are not necessarily in the same direction.

Many terms with differing units are used to refer to "flux density," such as ρ_l in units of C/m and ρ_V in units of C/m^3. This confusion is avoided in texts by referring to such terms as ρ_l and ρ_V as charge densities.[12] Care must be exercised when interpreting the phrase "flux density."

12. GAUSS' LAW FOR ELECTROSTATICS

Gauss' law for electrostatics is a necessary consequence of Coulomb's law. Gauss' law states that the amount of electric flux, Ψ, passing through any closed surface is proportional to (and in our rationalized system of SI units, equal to) the total charge, Q, contained within the surface. The surface is called a *Gaussian surface*. Gauss' law in a variety of forms is given in Eq. 16.17, Eq. 16.18, and Eq. 16.19. **ds** is an element of the surface area with magnitude ds and direction normal to **ds**, and θ is the angle between the normal to the surface area and **D**.

$$\begin{aligned}\Psi &= \iint D\,dA = \iint \sigma\,dA \\ &= Q \quad [d\mathbf{A} \text{ parallel to } \mathbf{D}]\end{aligned} \qquad 16.17$$

$$\begin{aligned}\Psi &= \iint \mathbf{D}\cdot d\mathbf{s} = \iint D\cos\theta\,ds \\ &= Q \quad [\text{arbitrary surface}]\end{aligned} \qquad 16.18$$

$$\Psi = \iiint \rho\,dv = Q \quad [\text{arbitrary volume}] \qquad 16.19$$

[11]This is because **E** is multiplied by the permittivity. See Eq. 16.6 by way of explanation, noting the permittivity in the denominator.

[12]This is consistent with the electric flux, Ψ, being numerically equal to the charge. Another way of considering this is to realize that a rationalized set of units (the SI system) that makes the flux equal to the charge is being used. If the system were not rationalized, the flux would merely be proportional to the charge.

Because the electric field strength or intensity, **E**, is related to the displacement or electric flux density, **D**, by the permittivity, Gauss' law as given in Eq. 16.17 can also be stated mathematically as

$$\Psi = \iint \epsilon E dA = Q \quad [d\mathbf{A} \text{ parallel to } \mathbf{E}] \qquad 16.20$$

Special Gaussian surfaces are used in highly symmetrical charge configurations to simplify calculations with Gauss' law. Such surfaces meet the following criteria.

1. The surface is closed.
2. At each point on the chosen surface, **D**, is either normal or tangential to the surface.
3. The flux density, or displacement, magnitude, D, is sectionally constant over the portion of the surface where **D** is normal.

   Requirement (1) is a general requirement of Gauss' law. Requirement (2) completely eliminates integrals and makes the calculation simple. Requirement (3) allows D to be removed from the integral.

Example 16.7

Given a point charge of constant value Q, what is the flux density, D, at a distance r from the charge?

Solution

First, arbitrarily construct a sphere around the point charge as shown.

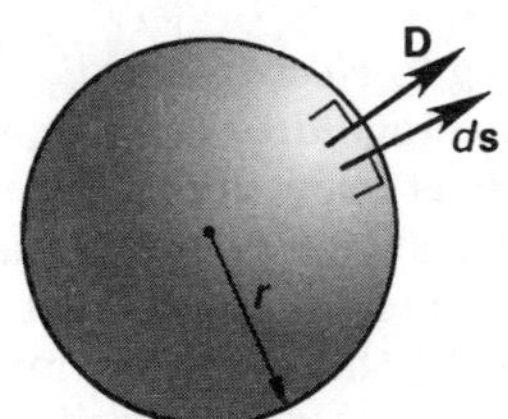

A sphere surrounding a point charge meets all the requirements of a special Gaussian surface. Using Eq. 16.18 gives the following.

$$\Psi = \iint \mathbf{D} \cdot d\mathbf{s} = \iint D \cos\theta ds = Q$$

Because the lines of flux emanate radially outward from the point charge, **D** and ds are parallel and $\cos\theta = 1$. Additionally, the magnitude of the flux density, D, is constant at a given r. This results in the following manipulations.

$$Q = \oint D \cos\theta ds = \oint D(1) ds$$
$$= D \oint ds$$

The surface area of a sphere is given by $4\pi r^2$.

$$Q = D \oint ds = D4\pi r^2$$

$$D = \frac{Q}{4\pi r^2}$$

Using the same logic as in Ex. 16.7 on a uniform line charge of density ρ_l of length l at a distance of r gives the following equation for displacement magnitude.

$$D = \frac{\rho_l}{2\pi r} \qquad 16.21$$

13. CAPACITANCE AND ELASTANCE

If charges Q^+ and Q^- exist on conductors separated by free space or by a dielectric, the mutual attraction "stores" the charges, holding them in place. The charges produce a stress in the dielectric measured by the electric field strength, **E**, and a potential difference exists between the plates. *Capacitance* is the property of a system of conductors and dielectrics that permits this storage of charges.

If a potential difference is applied across two conductors, charges will build up, creating an electric field between the conductors. The amount of the charge, Q, is proportional to the applied voltage. The proportionality constant is called the *capacitance* and has units of *farads*, F. This is shown in Eq. 16.22. A farad is a large capacitance and very few capacitors are constructed with a value of even 1.0 F. Most circuit elements have capacitances in the range of μF (10^{-6}) or pF (10^{-12}).

$$Q = CV \qquad 16.22(a)$$

$$C = \frac{Q}{V} \qquad 16.22(b)$$

The *elastance*, S, is the reciprocal of the capacitance.

$$S = \frac{1}{C} = \frac{V}{Q} = \frac{\frac{U}{Q}}{Q} \qquad 16.23$$

As shown in Eq. 16.23, the elastance can also be defined as the amount of energy, U, required per charge, Q, to transfer a unit charge between two conductors separated by a dielectric.

14. CAPACITORS

A *capacitor* (once called a *condenser*) is a device that stores electric charge. It consists of conductors separated by a dielectric. A common type of capacitor consists of two parallel plates of equal area A separated by a distance r. This type of capacitor is called a *parallel plate capacitor* and its capacitance is determined by Eq. 16.24.

$$C = \frac{\epsilon A}{r} \qquad 16.24$$

Equation 16.24 ignores the effects of *fringing*, as do many calculations involving capacitors. An example of fringing is shown in Fig. 16.7. The formula for the capacitance of other configurations is given in Table 16.6.

Assuming the plates in a capacitor begin with the same potential ($V=0$), the total energy, U (in J), in the electric field of a capacitor with capacitance C charged to a potential V is given by Eq. 16.25.

$$U = \tfrac{1}{2}CV^2 = \tfrac{1}{2}VQ = \tfrac{1}{2}\frac{Q^2}{C} \qquad 16.25$$

For n capacitors in parallel, the total capacitance is

$$C_{\text{total}} = C_1 + C_2 + C_3 + \cdots + C_n \qquad 16.26$$

For n capacitors in series, the total capacitance is

$$\frac{1}{C_{\text{total}}} = \frac{1}{C_1} + \frac{1}{C_2} + \frac{1}{C_3} + \cdots + \frac{1}{C_n} \qquad 16.27$$

Figure 16.7 Fringing

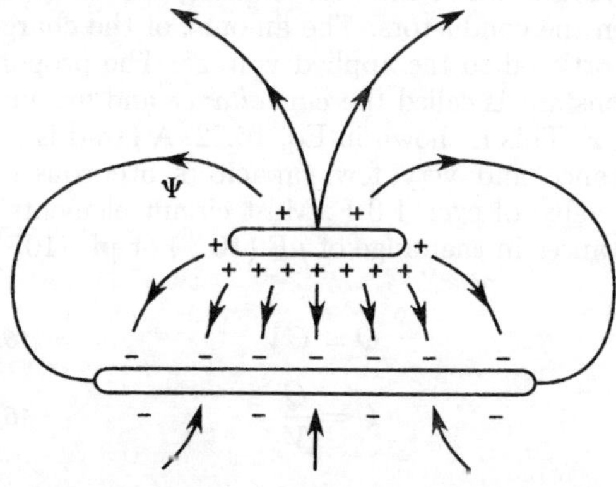

15. ENERGY DENSITY IN AN ELECTRIC FIELD

In a rectangular dielectric solid of surface area A and thickness t with a potential difference of V applied to either surface, the average energy (in J) is given by

$$\begin{aligned} U_{\text{ave}} &= \tfrac{1}{2}V_{\text{voltage}}Q = \tfrac{1}{2}(Et)(DA) = \tfrac{1}{2}EDV_{\text{volume}} \\ &= \tfrac{1}{2}\epsilon E^2 V_{\text{volume}} \\ &= \tfrac{1}{2}\frac{D^2}{\epsilon} V_{\text{volume}} \end{aligned} \qquad 16.28$$

Equation 16.28 in all its forms assumes that **D** and **E** are in the same direction. If not, the dot product must be used to determine the energy.

The average energy density at any point in this electric field is given in Eq. 16.29.

$$u_{\text{ave}} = \frac{U_{\text{ave}}}{V_{\text{volume}}} = \tfrac{1}{2}\epsilon E^2 = \tfrac{1}{2}\frac{D^2}{\epsilon} \qquad 16.29$$

PART 4: ELECTROKINETICS

16. SPEED AND MOBILITY OF CHARGE CARRIERS

A positive free charge moving in an electric field through a vacuum is shown in Fig. 16.8. Such a particle would experience a force given by the following.[13]

$$\mathbf{F} = Q\mathbf{E} = m\mathbf{a} \qquad 16.30$$

Figure 16.8 Free Charge in the Electric Field of a Vacuum

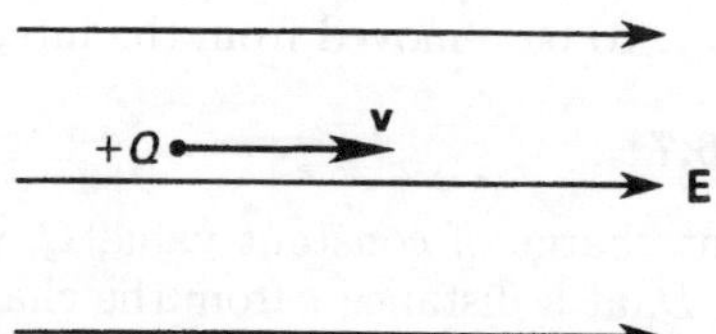

The force the particle experiences is unopposed and results in the constant acceleration of the particle as long as it remains in the electric field.

In metals, according to *electron-gas theory*, the electrons instead reach an average *drift velocity*, $\mathbf{v}_d$, because they undergo collisions as they move through the electric field. These collisions are between the electrons and the atoms within the crystalline structure of the metal and are caused by the thermal vibrations of the individual atoms.[14] The average drift velocity is

$$\begin{aligned} \mathbf{v}_d &= \int \mathbf{a}dt = \int \frac{\mathbf{F}}{m}dt = \frac{Q}{m}t_m\mathbf{E} \\ &= \mu\mathbf{E} \end{aligned} \qquad 16.31$$

The term t_m is the mean free time between collisions and m is the mass of the particle.[15] The mobility of the moving charge is given by μ. Because the particles in the metal undergo an increase in vibratory motion as the temperature increases, the mobility varies with the temperature—and with the crystalline structure of the

[13]This is identical to Eq. 16.4 without the subscripts. Nominally, subscripts need only be used to avoid confusion when multiple fields or charges are involved.

[14]The thermal vibrations can be treated in a quantum mechanical fashion as discrete particles called *phonons*.

[15]The variation in the overall field within the metal is caused by the periodic variations established by the lattice atoms. The field variations can result in collisions, accounted for in some formulas by an *effective electron mass*, m_e^*. Only the thermal vibrations must be determined when such a mass is used.

material. As the temperature increases, the mobility decreases, resulting in a lower drift velocity and a lower current. In circuit analysis this phenomenon is called *resistivity*. The resistivity is the reciprocal of the conductivity.

17. CURRENT[16]

An *electric current* is the movement of charges past a particular reference point or through a specified surface. The symbol I is generally used for constant currents while i is used for time-varying currents. Current is measured in *ampères* (A), commonly called amps. 1 A = 1 C/s. By convention, the current moves in the direction a positive charge would move (from the positive terminal to the negative terminal), that is, opposite to the flow of electrons. *Ohm's law* relates current to voltage and resistance as shown in Eq. 16.32.

$$V = IR \qquad 16.32$$

For simple DC circuits, this relationship is adequate and the current is defined by Eq. 16.33.[17]

$$I = \frac{dQ}{dt} \qquad 16.33$$

However, when the charges exist in a liquid or a gas, as in the field of electronic engineering, or when the charges are both positive and negative and have different characteristics, as in semiconductors, Eq. 16.33 is inadequate and one must be more specific in defining current. As a result, the *current density vector*, **J**, with units of A/m^2 is more frequently utilized in electromagnetic theory.

The *total current* is given by the sum of the convection current, the displacement current, and the conduction current. The convection and conduction currents are considered true currents and the displacement current a virtual current. Unless otherwise stated, the "current" will normally be the conduction current, but care must be used when determining what "current" is being referenced.

18. CONVECTION CURRENT[18]

In a material containing a volume V of mobile charges with a charge density of ρ moving across a surface area A with an average speed of $\mathrm{v}_{\mathrm{ave}}$, the *convection current* will be given by Eq. 16.34.

$$I = \rho A \mathrm{v}_{\mathrm{ave}} = \left(\frac{Q}{V}\right) A \mathrm{v}_{\mathrm{ave}} = \left(\frac{Q}{Al}\right) A\left(\frac{l}{t}\right) = \frac{Q}{t} \qquad 16.34$$

In terms of the current density vector, the convection current is given by Eq. 16.35 and can be visualized as shown in Fig. 16.9.

$$\mathbf{J} = \rho \mathbf{v}_d \qquad 16.35$$

Figure 16.9 *Convection Current*

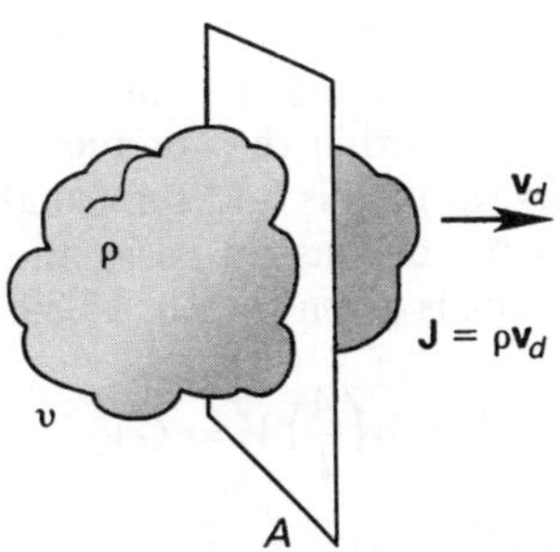

Convection, as the name implies, involves the diffusion of charges in a medium, as opposed to the movement of charges under the influence of an electric field. Although not of general use in electrical engineering, the concept of a convection current is sometimes useful in electromagnetic field theory.

19. DISPLACEMENT CURRENT[19]

When the electric field varies with time, a current can be made to flow in a dielectric material. That is, charges can be displaced within a dielectric by the electric fields of the charges moving outside the borders of the dielectric. This is because the electric fields of the charged particles extend well beyond the dimensions of the charges themselves. The current so created is called the *displacement current*, i_d, and is given by Eq. 16.36.

$$i_d = \frac{d\psi}{dt} \qquad 16.36$$

The displacement current in terms of the current density vector, **J**, is given by Eq. 16.37.

$$\mathbf{J}_d = \frac{\delta \mathbf{D}}{\delta t} \qquad 16.37$$

The displacement current, i_d, through any given surface can be obtained by integration of the normal component

[16]Compare an electric field's induced current to a magnetic field's induced voltage in Sec. 16.33.

[17]The use of the capital I indicates that the speed of the charges referred to is constant. If the speed is not constant then the current varies with time and i would be used as the appropriate symbol. In practical terms, this means that currents in DC circuits (unidirectional current flow) in which the current magnitude varies could be labeled i.

[18]Compare convection current with the movement of plasma (see Sec. 17.3).

[19]Compare displacement current to the displacement magnetic effect in Sec. 16.35.

of the displacement current density vector, $\mathbf{J}_d$, over the surface as in Eq. 16.38.

$$i_d = \frac{d}{d_t}\int_A \mathbf{D}\cdot d\mathbf{A} \qquad 16.38$$

The displacement current is equal to zero if the electric flux is constant with respect to time. That is, to cause a current to "flow" through a dielectric without the benefit of a direct connection to a conductor (as in the case of a capacitor), the electric flux must be changing with respect to time.[20]

20. CONDUCTION CURRENT[21]

A *conduction current* occurs in the presence of an electric field within a conductor. The magnitude of this current is given by the number of mobile charges transferred per unit time within the conductor. The charges may be electrons (in metals and semiconductors), holes (in semiconductors), or ions (in gases and electrolytes). Conduction current is given by Eq. 16.39.

$$I_c = \sigma\left(\frac{A}{l}\right)V = GV \qquad 16.39$$

The symbol σ indicates the conductivity of the material and is a property of a unit volume of the material. The area of the conductor is given by A and the potential difference by V. The length represented by l is the length over which the potential is applied. The quantity G is the conductance of the specified conductor and depends on the dimensions of the material used.

The conduction current in terms of the conduction current density vector, $\mathbf{J}_c$, is given by Eq. 16.40.

$$\mathbf{J}_c = \rho\mathbf{v}_d \qquad 16.40$$

This is identical to Eq. 16.35. Applying Eq. 16.31, specifically $\mathbf{v}_d = \mu\mathbf{E}$, puts Eq. 16.40 in the more widely recognized form shown in Eq. 16.41.

$$\mathbf{J} = \sigma\mathbf{E} \qquad 16.41$$

Equation 16.41 is called the *point form of Ohm's law.*

PART 5: MAGNETOSTATICS

21. MAGNETIC POLES

Mobile charges set in motion by an electric field carry their own electric fields with them. The subsequent three-dimensional field can be resolved into longitudinal and transverse components, referenced to the direction of motion. The transverse fields produce magnetic forces between moving charges. This force also acts between the conductors in which the charges move. It is this force that constitutes the basis for motor and generator theory in electric power engineering. Moving charges represented by the current, i, and the magnetic field established by their motion are shown in Fig. 16.10(a). Another completely valid view of the same situation is shown in Fig. 16.10(b). If the charges move with a uniform motion, the field is referred to as *magnetostatic.* If the charges move nonuniformly, the field is referred to as *magnetokinetic.*

Figure 16.10 *Magnetic Fields of Moving Charges*

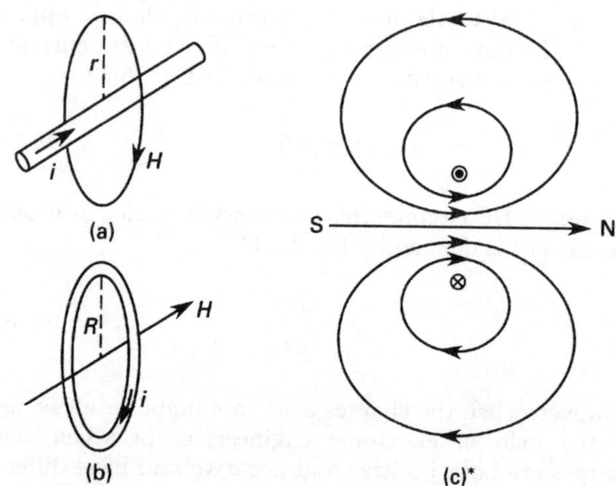

*The symbol "⊗" indicates the current flows into the page. The symbol "⊙" indicates flow out of the page. That is, the dot indicates the head of an arrow and the "⊗" represents the tail. Assume positive current flow. This correlates with the *right-hand rule* that states that placing the thumb in the direction of the current flow and curling the fingers inward gives the direction of the magenetic field.

A magnetic field can exist only with two equal and opposite poles, referred to as the *north pole* and the *south pole.*[22] The combination of the north and south pole linked together is referred to as a *dipole.* The magnetic *dipole moment, m,* depends on the pole strength, p, and the distance between the poles.

$$m = pd \qquad 16.42$$

The magnetic moment is a mathematical construct. Because the poles cannot be separated for measurement, the distance, d, is not measurable. Nevertheless, the magnetic moment is an important and fundamental quantity, and it can be measured. Magnetic moments are measured in terms of *Bohr magnetons.* In ferromagnetic material, a Bohr magneton is the moment

[20]The charges do not flow through the dielectric but are instead displaced within the dielectric. Also, this displacement ceases if the electric flux within the dielectric no longer changes.

[21]Compare conduction current with the principles of the magnetic circuit in Sec. 16.32.

[22]An electric charge can exist as a single charged object. Magnetic poles cannot exist as single north or south poles. (Mathematically there is no reason for the nonexistence of such a *magnetic monopole,* but none has yet been found.)

produced by one unpaired electron, which is equivalent to 9.24×10^{-27} A·m^2.

Figure 16.10(c) represents a loop of current creating a magnetic field. As the diameter of the current loop becomes infinitely small, the field approaches that of a magnetic dipole. A magnetic dipole field is shown in Fig. 16.11.

Figure 16.11 *Magnetic Dipole Field*

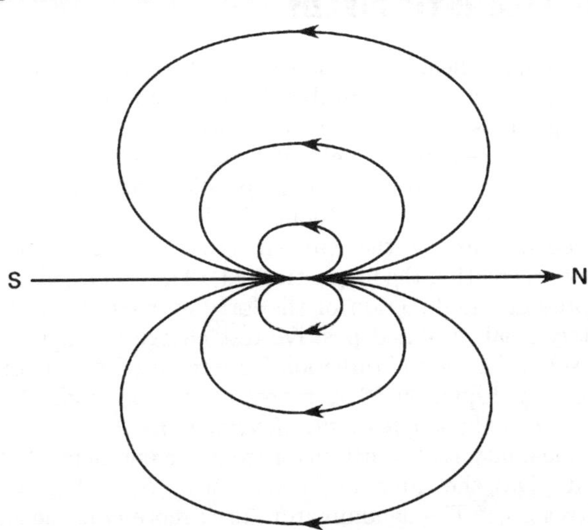

The infinitely small loop of current can be viewed as a single electron moving in an orbit. Hence, the magnetic moments of individual atoms and the usefulness of the Bohr magneton.[23]

For ease of presentation, and because in magnetic materials many groups of atoms tend to align themselves into domains, individual electrons or loops of current are not usually shown. Instead they are more often shown as in Fig. 16.12, which illustrates the standard convention that lines of *magnetic flux* are directed outward from the north pole (that is, the *magnetic source*), and inward at the south pole (that is, the *magnetic sink*).

Figure 16.12 *Typical Magnetic Field*

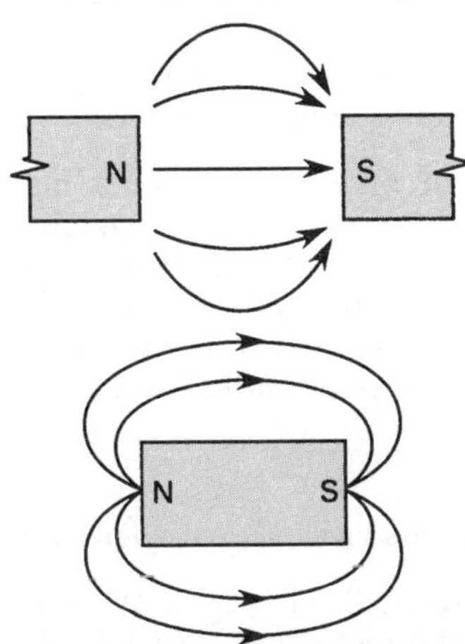

[23]It was Ampère who suggested over a hundred years ago that the magnetism observed in materials was caused by "molecular currents."

22. BIOT-SAVART LAW

The Biot-Savart law is basic to magnetostatics and magnetokinetics just as Coulomb's law is basic to electrostatics and electrokinetics. The law, in conjunction with Ampère's law, relates the force $\mathbf{F}_{1\text{-}2}$ between two current elements as in Eq. 16.43.

$$\mathbf{F}_{1\text{-}2} = \frac{\mu}{4\pi}\left(\frac{(I_2 d\mathbf{l}_2) \times (I_1 d\mathbf{l}_1) \times (\mathbf{r}_{1\text{-}2})}{r_{1\text{-}2}^2}\right) \qquad 16.43$$

The term μ is the magnetic permeability and $\mathbf{r}_{1\text{-}2}$ is a unit vector from the point represented by differential current element $d\mathbf{l}_1$ to $d\mathbf{l}_2$.[24] Both current elements are in the magnetic field of the other. For parallel currents Eq. 16.43 becomes

$$F_{1\text{-}2} = \frac{\mu}{4\pi r^2} I_1 dl_1 I_2 dl_2 \qquad 16.44$$

Because differential elements of current cannot be isolated, Eq. 16.43 and Eq. 16.44 are of use only when integrated over the entire path of the currents I_1 and I_2. The situation described by Eq. 16.43 and Eq. 16.44 is shown in Fig. 16.13.

Figure 16.13 *Magnetic Force Between Current Elements*

The magnetic force between two moving charges is given by Eq. 16.45. A specific case is shown in Fig. 16.13.

$$\mathbf{F}_{1\text{-}2} = \frac{\mu}{4\pi}\left(\frac{Q_1 Q_2}{r_{1\text{-}2}^2}\right)\mathbf{v}_2 \times (\mathbf{v}_1 \times \mathbf{r}_{1\text{-}2}) \qquad 16.45$$

If the charges are moving parallel to one another, the maximum magnetic force is

$$F = \frac{\mu}{4\pi r^2} Q_1 Q_2 \mathrm{v}^2 \qquad 16.46$$

[24]The magnetic permeability, μ, can only be used where one is operating over a linear region of the *BH* curve. If not, the magnetic permeability of free space, μ_0, should be used in its place. For hard ferromagnetic materials, **B** and **M** are directly related and μ need not be used.

Basic Theory

Comparison of Magnetic and Electric Force

The maximum coulombic force between moving charges in free space can be deduced from Eq. 16.3 to be

$$F_e = \frac{Q_1 Q_2}{4\pi\epsilon_0 r^2} \qquad 16.47$$

The maximum magnetic force between moving charges in free spaces can be deduced from Eq. 16.45 to be

$$F_m = \frac{\mu_0 Q_1 Q_2 \text{v}^2}{4\pi r^2} \qquad 16.48$$

Comparing the maximum magnetic force F_m to the maximum electric force F_e gives

$$\frac{F_m}{F_e} = \epsilon_0 \mu_0 \text{v}^2 \qquad 16.49$$

Consider Eq. 16.49. In electrical engineering practice the velocities of electrons are much less than the velocity of light, therefore the magnetic force is much weaker than the electric force. Additionally, the speed of light squared is fundamentally related to magnetic permeability and the electric permittivity as follows.

$$c^2 = \frac{1}{\mu_0 \epsilon_0} \qquad 16.50$$

Finally, the magnetic force can be viewed as the electric force multiplied by the factor $(\text{v}/c)^2$. That is, the magnetic force is a result of charges in relative motion.

$$F_m = F_e \left(\frac{\text{v}}{c}\right)^2 \qquad 16.51$$

The magnetic and electric forces on two reference charges moving in parallel are shown in Fig. 16.14.

Figure 16.14 Magnetic and Electric Forces on Moving Charges

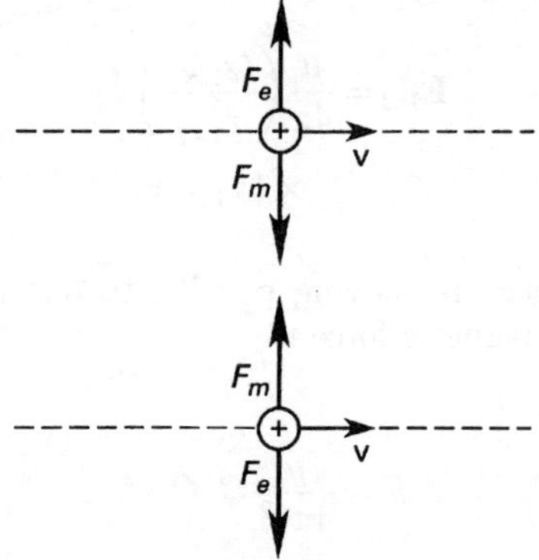

Magnetic Superposition

Magnetic forces occur as a result of the interaction of a pair of current-carrying conductors or the equivalent. One sets up the magnetic field and the other exhibits a force reaction to this same magnetic field. In free space and most nonmagnetic materials, the magnitude of the magnetic field is proportional to the total current that sets up the field as long as the permeability of the medium is constant. However, this is not true for iron and other ferromagnetic materials. Therefore, in general, the principle of linear superposition does not apply to current-carrying conductors in the vicinity of ferromagnetic materials.

23. MAGNETIC FIELDS

A magnetic field is a region in which a moving electric charge exerts a measurable force on another moving charge. It is a vector force field because a moving test charge placed in the field will experience a force in a given direction and with a specific magnitude. The direction of the field is along the flux lines, which are perpendicular to the current (that is, the moving charges creating the magnetic field). These lines, ideally, represent the direction of the force exerted on an infinitely small, isolated positive test charge moving with speed v. The actual direction is represented by the unit vector **a**. Equation 16.52 represents the *strength of the B-field*, or the *magnetic flux density*, in a substance with permeability μ at a distance r from current element $d\mathbf{l}_1$ that gives the direction of one turn, N, of the total current, I.[25] The denominator, $4\pi r^2$, represents the surface area of a sphere centered on the current element $d\mathbf{l}$ whose surface just touches the point in space where the **B**-field is being evaluated. Specific configurations may be evaluated with Eq. 16.52 by changing **B** to $d\mathbf{B}$ and l to dl and integrating.

$$\mathbf{B} = \frac{\mu I l}{4\pi r^2}\mathbf{a} \qquad 16.52$$

B should not be called the magnetic field strength. Magnetic field strength is designated by **H**.[26] Although **H**, which will be defined in Sec. 16.26, is directly related to the current I (as **E** is directly related to the charge), the fundamental field is **B**, as **B** enters into the force relations (as **E** enters into the force relations in electrokinetics). **B** is analogous with **E**, and **H** is analogous with **D**. This is true though a review of the units in each case would seem to indicate otherwise.

The amount of *magnetic flux* in a magnetic field is symbolized by Φ or by the commonly used ϕ, both measured in webers (Wb). The magnetic flux density, **B**, is measured in teslas (T). One T is equivalent to 1 Wb/m^2. **B** is also known as the *magnetic induction*, the *intensity of magnetization*, and the *dipole moment*

[25] The **B** field has two sources. One is the current, and the other is the magnetization, **M**.

[26] For historical reasons, the concept of the **B** field in terms of magnetic flux density developed first, prior to the concept of a field or field strength **H**. The names are opposites of those used when discussing electric quantities. Nevertheless, **B** is analogous to **E**, and **H** is analogous to **D**.

per unit volume.[27] **B** is given in terms of the flux in Eq. 16.53 and Eq. 16.54.

$$B = |\mathbf{B}| = \frac{\phi}{A} \quad \text{16.53}$$

$$\mathbf{B} = \frac{\phi}{A}\mathbf{a} \quad \text{16.54}$$

24. PERMEABILITY AND SUSCEPTIBILITY

The behavior and magnitude of magnetic forces between moving charges depend on the environment in which the moving charges are located. The largest magnetic forces occur in ferromagnetic materials. Another way of viewing this phenomenon is to say that magnetic flux does not pass equally through all materials. In fact, diamagnetic material repels magnets.

The *permeability of free space* (free space is a vacuum), μ_0, represents the magnetic properties of free space. However, it is better regarded as a constant relating the units in mechanical and electromagnetic systems. 1.2566×10^{-6} N/A^2 = $4\pi \times 10^{-7}$ N/A^2 (H/m) is the value for μ_0 in the SI system. The product of the permeability of free space and the *relative permeability* of a medium gives the *permeability* of the medium.

$$\mu = \mu_0 \mu_r \quad \text{16.55}$$

Permeability is often defined in terms of the magnetic flux density and the magnetic field strength necessary to create it, as in Eq. 16.56.

$$\mu = \frac{B}{H} \quad \text{16.56}$$

When **B** and **H** are not parallel, the permeability, μ, is a tensor. In general, when losses occur within a material, the permeability becomes a complex number and curves that show the variations in the real and imaginary terms are called the *magnetic spectrum* or *permeability spectrum* of the material.

The permeability can also be expressed as follows.

$$\mu = \mu_0(1 + \chi_m) \quad \text{16.57}$$

The term χ_m is a dimensionless quantity called the *magnetic susceptibility*. Magnetic susceptibility is a measure of the alignment in a magnetic material. It is also the ratio of the magnetization, M, to the external applied magnetic field, H.

$$\chi = \frac{M}{H} \quad \text{16.58}$$

For diamagnetic and paramagnetic materials, the susceptibility is essentially constant. For ferromagnetic materials, the susceptibility is nonlinear with respect to the applied field, as shown in Fig. 16.2.

[27]The word "induction" stems from an earlier era. When an unmagnetized piece of iron was brought near a magnet, magnetic poles were said to be "induced" in the iron.

25. MAGNETIC FLUX

The magnetic flux, Ψ_m, or more correctly, Φ, is defined as the surface integral of the normal component of the magnetic flux density, **B**, over an area, A, and is given by Eq. 16.59.

$$\Phi = \Psi_m = \iint \mathbf{B} \cdot d\mathbf{A} \quad \text{16.59}$$

It is also useful to define the flux, ϕ, as the magnetic lines of force in a region.[28] In this case, the flux is given by Eq. 16.60, with N representing the number of complete turns of closely spaced conductors of length l carrying current I. The term NI in Eq. 16.60 is sometimes referred to as the *magnetomotive force* (mmf or preferably F_m).

$$\phi = \mu N I l = BA \quad \text{16.60}$$

The orientation of the magnetic field lines and magnetic flux lines depends on the properties of the material, specifically the permeability. By convention, the flux lines exit the north pole as given by the right-hand rule.[29] The magnetic field lines always form closed loops. They do not terminate on anything but themselves (see Fig. 16.10 and Fig. 16.11). Since no magnetic charges exist, the magnetic flux on a closed sphere (or any closed surface) is equal to zero.

26. MAGNETIC FIELD STRENGTH

The *magnetic field strength*, **H**, in units of A/m, is derived from the magnetic flux density.

$$\mathbf{H} = \frac{1}{\mu}\mathbf{B} \quad \text{16.61}$$

The magnetic field strength is a vector point function whose curl is the current density. Because it is proportional to the magnetic flux density in regions where no magnetized substances exist, and because the *cgs* system units define **H** in terms of lines of flux per unit area, **H** is referred to as the magnetic flux density in texts on magnetic theory. This explains its correlation with the electric flux density, **D**.[30] When using the SI system of

[28]The symbol for the flux, Ψ_m, does not necessarily have to change. It is shown here as commonly used, ϕ. The standard symbol for magnetic flux is Φ. The additional symbols are used to alert the reader to the range of symbols that may be encountered.

[29]In the case of straight wire, the thumb indicates the current direction and the fingers curl in the direction of the magnetic field. For a coil of wire, the fingers curl in the direction of the positive current flow and the thumb points in the direction of the magnetic field direction. The thumb points toward the north pole.

[30]There are other reasons that **H** is analogous to **D**. For instance, **D** is used to avoid direct reference to polarization charges and the polarization, **P**. It is then easier to work with **E** and **D**. **H** is used to avoid direct reference to the magnetization, **M**. It is then easier to work with **B** and **H**.

units, **H** should be referred to only as the magnetic field strength.

The magnetic field and inductance for various configurations are given in Table 16.8.

Example 16.8

A long wire with a radius of 0.001 m carries 30 A of current. What are the magnetic field strength and the magnetic flux density 0.01 m from the surface of the wire? (This is roughly equivalent to asking what are the magnetic field strength and flux density at the inside surface of a conduit that contains a 10 AWG copper wire which is carrying the maximum current allowed by the *National Electrical Code.*)

Solution

From Table 16.8, the magnetic field strength is

$$H = \frac{I}{2\pi r} = \frac{30 \text{ A}}{2\pi(0.001 \text{ m} + 0.01 \text{ m})} = 434.06 \text{ A/m}$$

From Eq. 16.56, using the permeability of free space, the flux density is

$$B = \mu H = \left(4\pi \times 10^{-7} \ \frac{\text{Wb}}{\text{A·m}}\right)\left(434.06 \ \frac{\text{A}}{\text{m}}\right) = 5.45 \times 10^{-4} \text{ T}$$

27. GAUSS' LAW FOR MAGNETIC FLUX

A law similar to Gauss' law for electric flux can be written for magnetic flux. (See Sec. 16.12.) Such a law states that the amount of flux passing through any closed surface is equal to zero.

$$\Phi = \psi_m = \iint \mathbf{B} \cdot d\mathbf{A} = 0 \qquad 16.62$$

The physical significance of Eq. 16.62 is that magnetic flux lines form continuous closed loops. Additionally, this indicates that the divergence of any magnetic field is zero. In other words, no magnetic charge exists. The law refers to magnetic flux rather than magnetostatics as it is also applicable to time-varying magnetic fields (that is, magnetokinetics).

28. INDUCTANCE AND RECIPROCAL INDUCTANCE

When charges are in motion, a current is said to be flowing and a magnetic field is established. When the current varies with time, the generated magnetic field opposes the current change and in doing so "stores" energy in the magnetic field. *Inductance, L*, is the property of a system of conductors and circuits that permits this storage.

***Table 16.8** Magnetic Field and Inductance for Various Configurations*

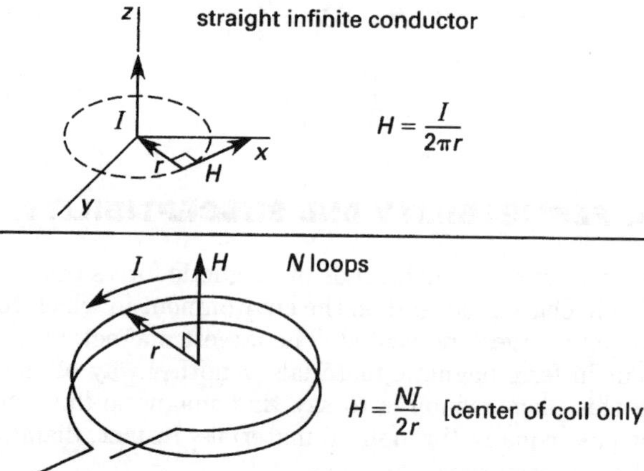

infinite cylindrical coil helix (solenoid)

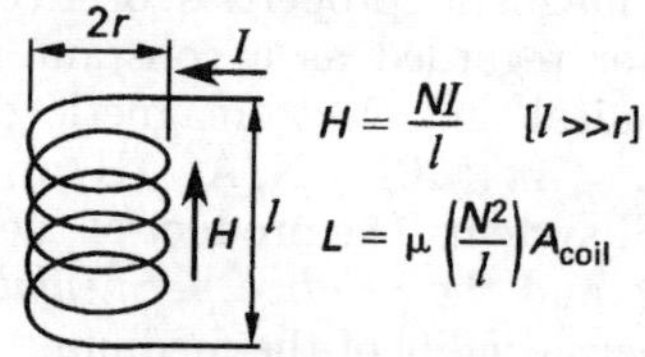

$H = \frac{NI}{l}$ [$l >> r$]

$L = \mu\left(\frac{N^2}{l}\right)A_{\text{coil}}$

torus (toroidal coil)

l, mean, R, r, μ, N_{turns}, I

$H = \frac{NI}{2\pi R}$ [$r << R$]

$L = \frac{\mu N^2 A_{\text{core}}}{l_{\text{mean}}} = \frac{\mu N^2 r^2}{2R}$

coaxial cable (high frequencies)

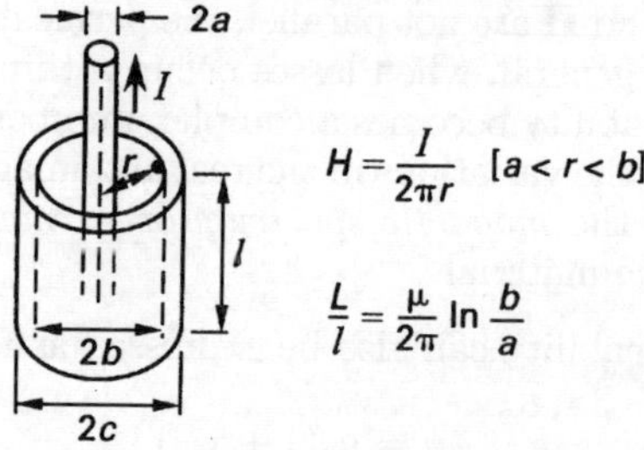

parallel transmission lines (high frequencies)

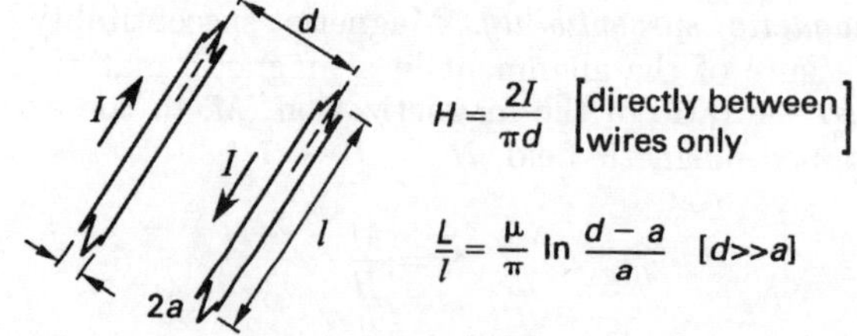

Reprinted with permission from *Core Engineering Concepts for Students and Professionals*, by Michael R. Lindeburg, PE, copyright 2010, by Professional Publications, Inc.

If a time-varying current is applied to a conductor, usually in the form of a coil of wire, the magnetic field builds up, producing a potential difference across the ends of the conductor. The amount of potential, V, is proportional to the magnitude of the rate of change of current. The proportionality constant is called the *inductance*, L, and is measured in *henries*, H. This is shown in Eq. 16.63.[31]

$$\text{v} = L\frac{di}{dt} \tag{16.63}$$

The *reciprocal inductance*, Γ, a seldom used term, is the reciprocal of the inductance.

$$\Gamma = \frac{1}{L} \tag{16.64}$$

The type of inductance just described is also called *self-inductance*. It can also be described in terms of *flux linkage*, Λ (measured in Wb). Flux linkage is defined as the lines of flux that link the entire magnetic circuit. For example, in Fig. 16.15 the "linked flux" is the flux through the center of the solenoid that links the current flows of all the wire turns (the small flux loops around each turn of wire are ignored). Using this definition, the inductance would be given by Eq. 16.65.

$$L = \frac{\Lambda}{I} = \frac{N\phi}{I} \tag{16.65}$$

Mutual inductance, M, vital to transformer theory, is the term used to describe the fact that a changing current in one circuit induces an electromotive force in another. It is given by Eq. 16.66 where Λ represents the flux links in one winding and I is the current in the other winding.

$$M = \frac{\delta\Lambda}{\delta I} \tag{16.66}$$

Figure 16.15 *Magnetic Flux Links in a Solenoid*

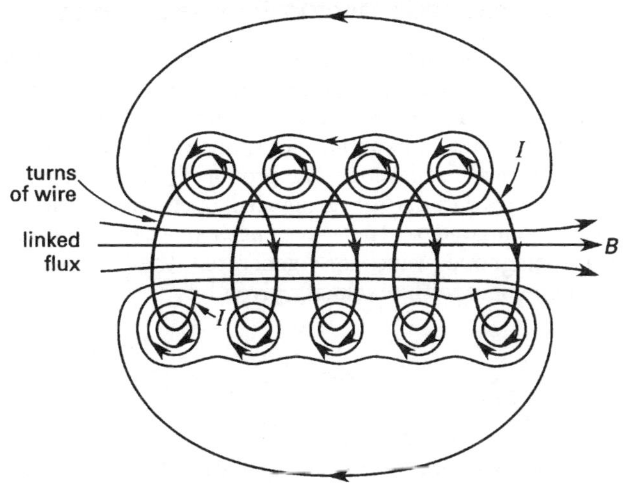

[31]This equation technically belongs in Part 6 of this chapter. It is often stated as the definition of inductance and is used here for that reason.

29. INDUCTORS

An *inductor* is a device that stores magnetic energy. It typically consists of coils of wire with a core material in the center, which can be air. A common type of inductor consists of N turns of wire with radius r and an iron core center with permeability μ. This type of inductor is called a *solenoid*, and its inductance is determined by Eq. 16.67.

$$L = \mu\frac{N^2 A_{\text{coil}}}{l} \tag{16.67}$$

The formula for the inductance of other configurations is given in Table 16.8.

The total energy, U (in joules), in the magnetic field of an inductor with inductance L and current I is

$$U = \tfrac{1}{2}LI^2 \tag{16.68}$$

For inductors in series, the total inductance is

$$L_{\text{total}} = L_1 + L_2 + L_3 + \cdots + L_n \tag{16.69}$$

The reciprocal of the total inductance for inductors in parallel is

$$\frac{1}{L_{\text{total}}} = \frac{1}{L_1} + \frac{1}{L_2} + \frac{1}{L_3} + \cdots + \frac{1}{L_n} \tag{16.70}$$

30. ENERGY DENSITY IN A MAGNETIC FIELD

The average energy (in joules) in a magnetic field is

$$U_{\text{ave}} = \tfrac{1}{2}\phi NI \tag{16.71}$$

The average energy density at any point in this magnetic field is

$$u_{\text{ave}} = \frac{U_{\text{ave}}}{V_{\text{volume}}} = \tfrac{1}{2}BH = \tfrac{1}{2}\mu H^2 = \tfrac{1}{2}\frac{B^2}{\mu} \tag{16.72}$$

The average energy stored in inductors that are mutually coupled is given by Eq. 16.73. Here M is the mutual inductance (with units of henries) and I_1 and I_2 represent the currents in the two inductors.

$$U_{\text{ave}} = MI_1I_2 \tag{16.73}$$

PART 6: MAGNETOKINETICS

31. SPEED AND DIRECTION OF CHARGE CARRIERS

A positive free charge moving with velocity, **v**, in a magnetic field experiences a force given by Eq. 16.74.

$$\mathbf{F} = Q\mathbf{v} \times \mathbf{B} \qquad 16.74$$

The magnetic force can change a charge's direction but not its kinetic energy. This is in contrast to the force the charge experiences in an electric field, which does work on the charge and therefore changes its kinetic energy.[32]

If there are many charges flowing through a conductor in a magnetic field, that is, if a current-carrying conductor exists in an external magnetic field, the differential force equation is

$$d\mathbf{F} = (dq)(\mathbf{v} \times \mathbf{B}) = (I\,dt)(\mathbf{v} \times \mathbf{B}) = I(d\mathbf{l} \times \mathbf{B}) \qquad 16.75$$

In Eq. 16.75, dl is a unit length in the direction of conventional (i.e., positive) current flow I. If the conductor is straight and the magnetic field is constant along the length of the conductor, a version of *Ampère's law* results.

$$F = NIBl\sin\theta \qquad 16.76$$

N represents the number of turns of current-carrying conductor in the magnetic field and θ is the angle between the conductor (i.e., the current element $d\mathbf{l}$) and the magnetic field, **B**. If these are at right angles, which is often the case, the force is represented by

$$F = NIBl \qquad 16.77$$

The situations in Eq. 16.75 through Eq. 16.77 are illustrated in Fig. 16.16.

Care must be taken to properly interpret directions in the figure. If an external magnetic force is applied in the direction shown, the conductor has a velocity in the same direction. Applying Eq. 16.74 then gives the direction of force on the charge carriers within the conductor —in this case, in the direction of the current shown. This is the basis of generator theory. If instead a current is applied to a loop of wire, of which the conductor represents one section, the velocity in Eq. 16.74 is in the direction of the charge carriers, that is, the direction of the current flow shown. In this case, the force given by Eq. 16.74 is in a direction opposite to the force arrow shown. This is the basis of motor theory.

[32]Consider that if a charge is not in motion with respect to a magnetic field, it cannot "see" the magnetic field and therefore experiences no change in kinetic energy. A charge stationary with respect to an electric field can "see" the field and therefore experiences a force that changes its kinetic energy.

Figure 16.16 *Conductor in a Magnetic Field*

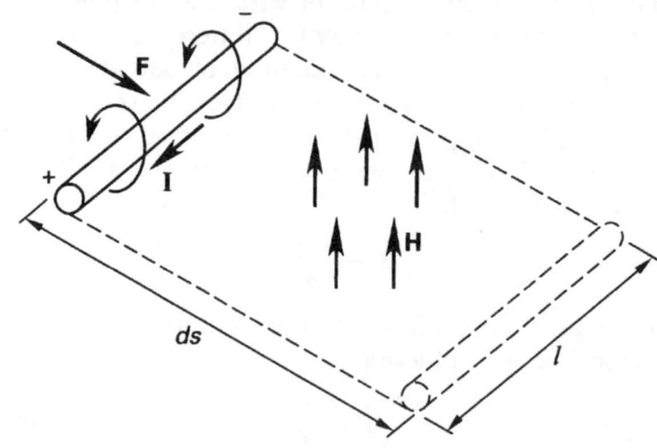

If a magnetic field, **B**, is uniform within a region and a charge has an initial velocity perpendicular to this field, the resulting path of the particle is a circle of radius r. The magnitude of the force is determined by Eq. 16.74. Such a force would be directed toward the center of the circle. The centripetal acceleration is of a magnitude $\omega^2 r = \mathrm{v}^2/r$. Applying Newton's second law gives the following for the radius.

$$|Q|\,\mathrm{v}B = m\frac{\mathrm{v}^2}{r} \qquad 16.78$$

Rearranging Eq. 16.78 gives

$$r = \frac{m\mathrm{v}}{|Q|\,B} \qquad 16.79$$

When both electric and magnetic fields are in a given region at the same time, the force is given by the *Lorentz force equation*, Eq. 16.80.

$$\mathbf{F} = Q(\mathbf{E} + \mathbf{v} \times \mathbf{B}) \qquad 16.80$$

Electromagnetic Oscillations and Waves

A time-varying electric field produces time-varying currents that generate magnetic fields. If the electromagnetic variations are restricted to a small region, such as within a circuit, no spatial variations of the field occur and electrical oscillations are produced. If the electromagnetic variations are generated over large areas of space, compared to the wavelength of the oscillation, electromagnetic waves are generated that propagate through space with a velocity given by

$$\mathrm{v} = \frac{1}{\sqrt{\epsilon\mu}} \qquad 16.81$$

When a conduction current generates the electromagnetic field, the magnetic field strength, **H**, is proportional to the electric field strength, **E**, and an induction field is set up, keeping the energy within the circuit. If the electromagnetic field is generated by a displacement current, the magnetic field strength, **H**, is directly proportional to the rate of change of the electric field strength, $d\mathbf{E}/dt$. A radiation field is thereby created, and energy can leave the system (i.e., be radiated into space).

Electric and Magnetic Energy Conversion

Applying the law of conservation of energy to a lossless, energy-storing system in which charges move in directed nonuniform motion means the sum of the instantaneous values of electric and magnetic energy must remain constant. Changes in the electric energy necessarily mean changes in the magnetic energy as given in Eq. 16.82.

$$du_e = -du_m \qquad 16.82$$

32. VOLTAGE AND THE MAGNETIC CIRCUIT

An *electric potential* or *voltage* is the work done on a unit charge to bring it from some specified reference point to another point. The symbol V is generally used for constant voltages and v is used for time-varying systems. The unit is the *volt* with 1 V = 1 J/C. By convention, current flows from the positive terminal on a voltage source to the negative outside the source and from negative to positive inside the source. Also by convention, the current flows from the positive terminal to the negative terminal through a resistor.[33] This situation is illustrated in Fig. 16.17 along with an analogous *magnetic circuit.*

Figure 16.17 *Magnetic-Electric Circuit Analogy*

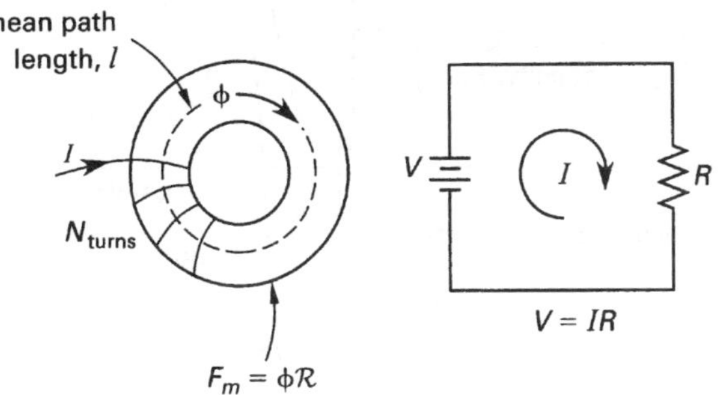

In a magnetic circuit, the flux density is proportional to the *magnetomotive force* (mmf), F_m, given in Eq. 16.83.

$$F_m = IN \qquad 16.83$$

The magnetomotive force is in units of amp-turns. Turns, like revolutions or radians, are for clarification only and can be disregarded in equation manipulations.

[33]These directions are based on conventional current flow and are arbitrary. As long as one remains consistent when choosing reference directions for both voltage and current, the correct results will be obtained.

The *reluctance*, $\mathcal{R}$, measured in A/Wb, is analogous to electrical resistance and is given by

$$\mathcal{R} = \frac{l}{\mu A} \qquad 16.84$$

The permeability depends on the flux density. Therefore, iterations between Eq. 16.84 and Eq. 16.83 may be needed if either the reluctance or flux density is an unknown.

The magnetic equation that correlates with Ohm's law in electric circuits is

$$F_m = Hl = \phi\mathcal{R} \qquad 16.85$$

Example 16.9

A 0.001 m slice is taken from a cast-iron toroidal coil ($\mu_r = 2000$) with a 0.35 m mean toroidal diameter and a 0.07 m core diameter. A steady unknown current flows through 300 turns of wire wrapped around the coil, producing a constant flux density in the core and air gap. What current is required to establish a flux of 10 mWb across the air gap?

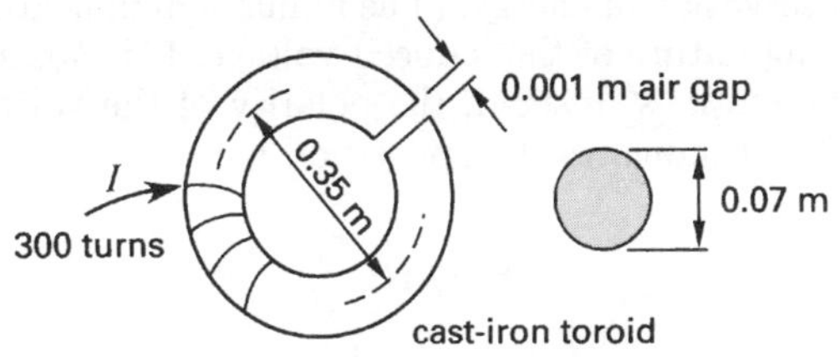

Solution

The mean path length through the cast iron is

$$l_{ci} = \pi(0.35 \text{ m}) - 0.001 \text{ m} = 1.0986 \text{ m}$$

The cross-sectional area of the flux path is

$$A = \frac{\pi}{4}d^2 = \frac{\pi}{4}(0.07 \text{ m})^2 = 0.003848 \text{ m}^2$$

The total reluctance is the sum of the reluctances of the cast iron and air paths. From Eq. 16.84,

$$\begin{aligned}\mathcal{R} &= \sum\frac{l}{\mu A} = \frac{1}{\mu_0 A}\sum\frac{l}{\mu_r}\\ &= \frac{1}{\left(4\pi\times 10^{-7}\,\frac{\text{Wb}}{\text{A·m}}\right)(0.003848 \text{ m}^2)}\\ &\quad\times\left(\frac{1.0986 \text{ m}}{2000} + \frac{0.001 \text{ m}}{1}\right)\\ &= 3.204\times 10^5 \text{ A/Wb}\end{aligned}$$

The flux, ϕ, is given as 10^{-2} Wb. From Eq. 16.83, the magnetomotive force is

$$F_m = IN = I\,300$$

Basic Theory

From Eq. 16.83 and Eq. 16.85, the current is

$$I = \frac{\phi \mathcal{R}}{N} = \frac{(10^{-2}\ \text{Wb})\left(3.203 \times 10^5\ \frac{\text{A}}{\text{Wb}}\right)}{300\ \text{turns}}$$
$$= 10.68\ \text{A}$$

33. MAGNETIC FIELD-INDUCED VOLTAGE

When a conductor "cuts" lines of magnetic flux at a given rate, a voltage called the *induced electromotance*, or *electromotive force* (emf) is produced in the conductor, as illustrated in Fig. 16.16. The magnitude of such an *electromagnetic induction* is given in Eq. 16.86, which is the mathematical statement of *Faraday's law*. Faraday's law states that the induced voltage is proportional to the time rate of change of the magnetic flux linked with the circuit. The minus sign is mandated by *Lenz's law*, which states that the current flow in a conductor caused by an induced voltage moves in a direction that creates a magnetic flux that opposes the magnetic flux inducing the voltage. More succinctly, the direction of the induced emf is such as to oppose any change in current. This is a consequence of the law of conservation of energy. (The minus sign indicates this opposing nature of the induced voltage. It is not shown in all equations. Instead, the polarity of the voltage in the circuit indicates the opposition.)

$$V = -N\frac{d\phi}{dt} \qquad 16.86$$

Because the magnetic flux linkage is given by $\Lambda = N\phi$, and the induced voltage is the time rate of change of these linkages, the induced voltage can be written in terms of the velocity of the conductors.

$$V = N\frac{d\phi}{dt} = NBl\frac{ds}{dt} = NBlv \qquad 16.87$$

In both Eq. 16.86 and Eq. 16.87, N is the number of conductors, l is the length of the conductor in the magnetic field, and ds is the differential distance traveled by the conductor through the field (see Fig. 16.16). Equation 16.86 is the *flux-changing method* of generating a voltage. Equation 16.87 is the *flux-cutting method* of generating a voltage.

34. EDDY CURRENTS

Eddy currents, also known as *Foucault currents*, occur as a result of induced voltages in a conducting medium by a varying magnetic field. Because the currents so induced cannot leave the conducting medium, they "deflect" off the edges to form loops, or circulating currents. Such currents result in Joule heating of the conducting medium and can result in considerable losses if not adequately addressed in the design of the system.

35. DISPLACEMENT CURRENT MAGNETIC EFFECT

The displacement current explained in Sec. 16.19 occurs only in dielectrics and insulators. The conduction electrons in these materials are tightly bound and cannot move freely through the material. Nevertheless, the displacement current creates a magnetic field intensity, **H**, proportional to the time rate of change of the electric field intensity, **E**, producing the displacement current. (By contrast, in a conduction current, **H** and **E** are in time phase.) Because **H** is proportional to $d\mathbf{E}/dt$, and **E** is proportional to $d\mathbf{H}/dt$, it is possible to generate electromagnetic waves in nonconducting material.

36. MAGNETIC HYSTERESIS

Plotting the values of the magnetic flux density, **B**, for various values of the magnetic field strength, or magnetizing force, **H**, for a ferromagnetic material reveals a nonlinear relationship. This is illustrated in Fig. 16.18. The phenomenon causing values of **B** to lag behind **H** so that the fields differ in magnitude is called *hysteresis*. The loop traces a complete cycle of **BH** values and is termed the *hysteresis loop*.

Figure 16.18 *Magnetic Hysteresis*

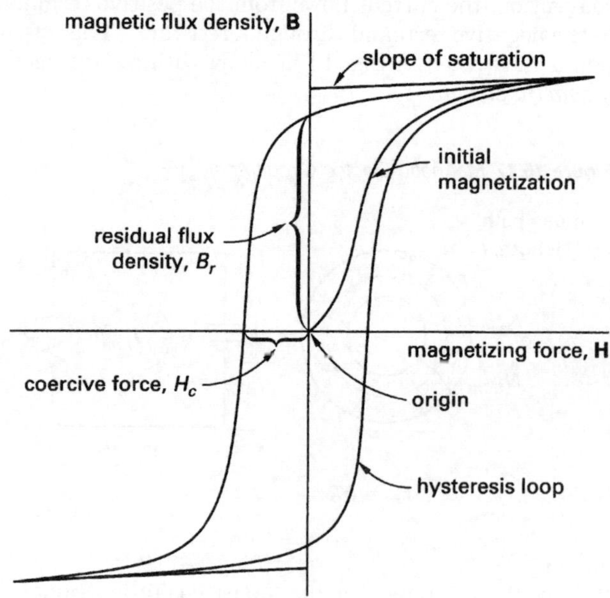

17 Electronic Theory

Nomenclature

A_0	constant	–
B, **B**	magnetic flux density	T (Wb/m^2)
C	constant	cm^{-6}
D	diffusion constant	m^2/s
e^-	electron charge	C
E, **E**	electric field strength	V/m (N/C)
E	energy	J or eV
E_c	conduction-band energy edge	J or eV
E_v	valence band energy edge	J or eV
f	frequency	Hz
F, **F**	force	N
h	Planck's constant, 6.6256×10^{-34}	J·s
J, **J**	current density	A/m^2
k	Boltzmann's constant, 1.3807×10^{-23}	J/K
m	mass	kg
m_e	electron mass	kg
n	electron carrier density	m^{-3}
N_c	effective conduction-band density state	cm^{-3}
N_v	effective valence-band density state	cm^{-3}
p	hole carrier density	m^{-3}
p^+	proton charge	C
q	electric unit charge, 1.6022×10^{-19}	C
Q	charge	C
r	radius	m
T	temperature	K
U	energy	J
v, **v**	velocity[1]	m/s
V	constant or rms voltage	V

Symbols

ϵ_0	free-space permittivity, 8.854×10^{-12}	C^2/N·m^2
λ	wavelength	m
μ	mobility	m^2/V·s
ν	frequency	rad/s
ρ	charge density	C/m^3
ρ_e	electron charge density	electrons/m^3
σ	conductivity	S/m
τ	mean lifetime	s
φ	work function[2]	J
ω	angular velocity	rad/s

Subscripts

0	initial or rest
d	drift
e	electron
g	gap
h	hole
i	intrinsic
n	electrons
p	holes
T	temperature
$\parallel$	parallel
$\perp$	perpendicular

[1]The term "velocity" is often used in text where it would be more technically correct to use the term "speed." Velocity is a vector quantity with speed as its magnitude.

1. FUNDAMENTALS

Electronic theory is the study of subatomic particles, such as electrons, in materials other than metals. It is concerned with the release, transport, control, and collection of charge carriers in a vacuum, in gases, and in semiconductors. While the charge carriers are usually electrons, they can be holes (a concept used to describe the absence of an electron), positive ions, or negative ions.

The fundamentals of electronic engineering are identical to those for electricity and magnetism. Because of this, many of the formulas used in this chapter will be repetitious. Their form and symbology may differ slightly, as often occurs in electrical engineering texts; this is purposefully done to expose readers to variations they may encounter. The principles and results are consistent. All the electromagnetic theory presented in Chap. 16 is applicable to electronic theory as well.

The primary difference between electronics and electromagnetism is in their applications. Electronic engineering allows for much greater control over the instantaneous, rather than the average, movement of charge carriers. Additionally, the control of these carriers can be extremely rapid. Electronic devices are generally nonlinear. That is, the output terminals can provide an amplified version of the current, voltage, or power supplied to the input terminals. The devices tend to be

[2]The units for the work function are often electron volts, eV. One eV is equivalent to 1.6022×10^{-19} J.

smaller than their electromagnetic counterparts and use considerably less energy. The field of electronic engineering is focused on *active electron devices*, those devices requiring an external power source in order to maintain their terminals at suitable operating voltages and currents. Electronics is generally more dependent on the material properties of a medium, such as composition and electronic structure, as well as the mode of electric conduction.

2. CHARGES IN A VACUUM

A charge Q in a vacuum experiencing an electric field intensity $\mathbf{E}$ (with units of V/m) will undergo an acceleration in the direction of the field. The force is

$$\mathbf{F}_{\|} = Q\mathbf{E} \tag{17.1}$$

If the charge is in a magnetic field of flux density $\mathbf{B}$ (with units of T) and velocity $\mathbf{v}$ in m/s, it will experience a transverse force of

$$\mathbf{F}_{\perp} = Q\mathbf{v} \times \mathbf{B} \tag{17.2}$$

The direction of motion of the particle will be a circle if the magnetic field is uniform and the particle enters it at right angles. The radius of the circle is

$$r = \frac{m\mathrm{v}}{QB} \tag{17.3}$$

If the charge is in an electromagnetic field, the total force is

$$\mathbf{F} = \mathbf{F}_{\|} + \mathbf{F}_{\perp} = Q(\mathbf{E} + (\mathbf{v} \times \mathbf{B})) \tag{17.4}$$

The energy of a charge Q in a vacuum, moving in an electric field $\mathbf{E}$ from point 1 to point 2, will experience a change in energy U (in units of J)[3] of

$$U = Q\int_{x_1}^{x_2} \mathbf{E}\cdot d\mathbf{x} = Q(V_1 - V_2) \tag{17.5}$$

Vacuum Tubes

A *vacuum tube* is a device in which electrons move through a vacuum or gaseous medium in a gas-tight envelope. In a basic vacuum tube, called a *triode*, the electrons are emitted from a heated *cathode* and move within the tube to the *plate* by an intervening *grid* that is used for amplification.

Tubes dominated electronic technology in the first half of the twentieth century as semiconductors do today. The release of electrons from the surface of a metal, which is the basis of the vacuum tube, remains an important area of study in electronic applications in space and in the construction of metal-semiconductor junctions.

[3]The magnetic field acts transverse to the direction of motion and can therefore only change the direction, not the energy.

Work Function

The *work function*, φ, is a measure of the energy required to remove an electron from the surface of a metal. It is usually given in terms of electron volts, eV. It is specifically the minimum energy necessary to move an electron from the *Fermi level* of a metal (an energy level below which all available energy levels are filled, and above which all are empty, at 0K) to infinity, that is, the vacuum level. This energy level must be reached in order to move electrons from semiconductor devices into the metal conductors that constitute the remainder of an electrical circuit. It is also the energy level of importance in the design of optical electronic devices.

In photosensitive electronic devices an incoming photon provides the energy to release an electron and make it available to the circuit, that is, free it so that it may move under the influence of an electric field. This phenomenon whereby a short wavelength photon interacts with an atom and releases an electron is called the *photoelectric effect*. The photon energy required to release an electron is

$$h\nu = \varphi + \tfrac{1}{2}m\mathrm{v}^2 \tag{17.6}$$

The term h is *Planck's constant* and ν is the *threshold frequency*. If the photon does not have the minimum energy required, no electron is released.

3. CHARGES IN LIQUIDS AND GASES

In any substance, such as a liquid or a gas, where the motion of a charged particle is hindered by surrounding particles, an average velocity called the *drift velocity*, $\mathbf{v}_d$, is established. The drift velocity is dependent on the electric field intensity, $\mathbf{E}$, and the *mobility*, μ, of the charge carrier as shown in Eq. 17.7.

$$\mathbf{v}_d = \mu\mathbf{E} \tag{17.7}$$

Mobility is measured in $\mathrm{m^2/V{\cdot}s}$. The concept of drift velocity in liquids and gases is illustrated in Fig. 17.1.

Conductivity

The mobility of the charge carriers is of special importance in electronic engineering, more so than in electrical engineering, although the principles are identical. The mobility is proportional to the conductivity.

$$\sigma = \rho\mu \tag{17.8}$$

The conductivity, σ, is measured in siemens per meter, S/m. The charge density is in units of $\mathrm{C/m^3}$. Metals have approximately 10^{28} atoms/m^3 with one to two electrons free to move, while silicon semiconductors have 10^{16} intrinsic charge carriers per cubic meter. Concentrations in gases vary widely with the type of gas and the conditions under which the gas is held.

Figure 17.1 *Drift Velocity in Liquids and Gases*

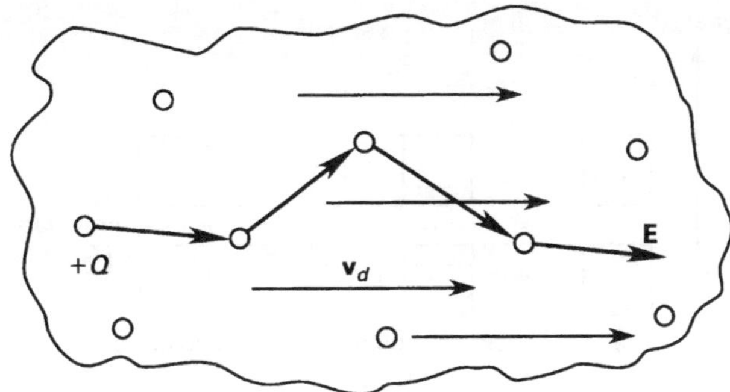

Figure 17.2 *Conductivity Formulas*

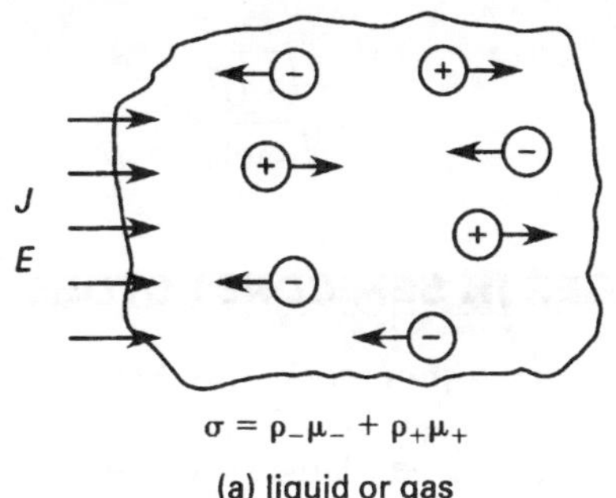

$\sigma = \rho_-\mu_- + \rho_+\mu_+$

(a) liquid or gas

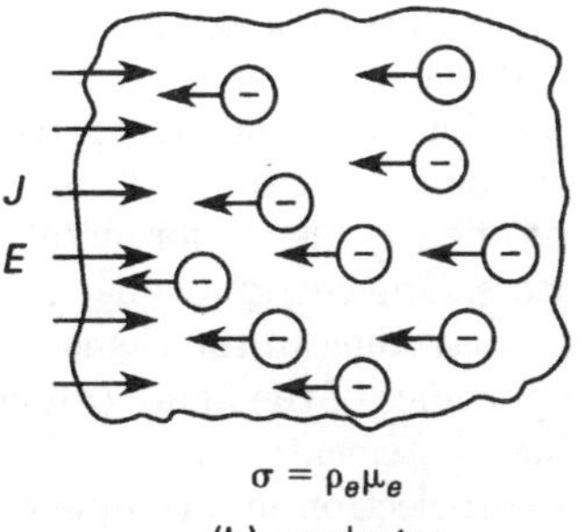

$\sigma = \rho_e\mu_e$

(b) conductor

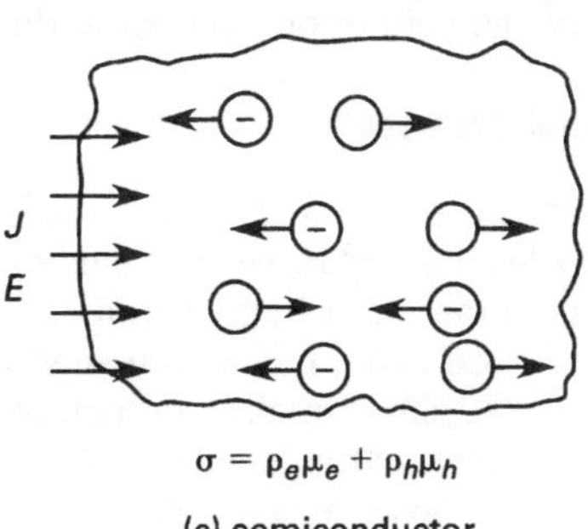

$\sigma = \rho_e\mu_e + \rho_h\mu_h$

(c) semiconductor

In metallic conductors, the conductivity is often defined by the point form of Ohm's law and given by

$$\sigma = \frac{|\mathbf{J}|}{|\mathbf{E}|} = \frac{J}{E} \tag{17.9}$$

Conductivity is more complex for liquids and gases. In general, both positive and negative ions may be present, some singly charged and some doubly charged, possibly with different masses. The conductivity must account for all these factors. If one assumes that the negative and positive ions are alike, the formula for conductivity in liquids and gases has only two terms. See Fig. 17.2 for the formula and a comparison to the formulas for conductors and semiconductors.

Example 17.1

A particular 10 AWG copper wire has a conductivity of 58×10^6 S/m. Assuming 10^{28} conduction electrons per cubic meter, what is the mobility of the electrons in the copper?

Solution

From Eq. 17.8, the conductivity is

$$\sigma = \rho\mu$$

Rearranging gives

$$\mu = \frac{\sigma}{\rho}$$

The conductivity is given. The charge density is

$$\begin{aligned}\rho &= \left(10^{28}\ \frac{\text{electron}}{\text{m}^3}\right)\left(-1.6022 \times 10^{-19}\ \frac{\text{C}}{\text{electron}}\right)\\ &= -1.6022 \times 10^{9}\ \text{C/m}^3\end{aligned}$$

Substituting gives

$$\begin{aligned}\mu &= \frac{\sigma}{\rho} = \frac{58 \times 10^6\ \frac{\text{S}}{\text{m}}}{-1.6022 \times 10^9\ \frac{\text{C}}{\text{m}^3}}\\ &= -3.6 \times 10^{-2}\ \text{S}\cdot\text{m}^2/\text{C} \quad (-3.6 \times 10^{-2}\ \text{m}^2/\text{V}\cdot\text{s})\end{aligned}$$

In metallic conductors, the charge carriers are electrons that move in a direction opposite that of the electric field. The charge density, ρ, and the mobility, μ, are both negative, leaving the conductivity positive. By convention, the electrons moving in one direction are treated as positive charges moving in the opposite direction, and ρ and μ are reported as positive.

Plasmas

Plasma, often viewed as a fourth state of matter, can be considered a highly ionized gas. Plasmas are highly conductive, with high concentrations of roughly equal numbers of positive and negative ions and a low voltage gradient. The negative charge carriers tend to be electrons that vibrate longitudinally about their mean position with a frequency of

$$f = \frac{1}{2\pi}\sqrt{\frac{\rho_e Q^2}{\epsilon_0 m_e}} \tag{17.10}$$

Let λ be the wavelength of oscillation of the frequency given by Eq. 17.10. Then the velocity of the electrons is

$$\mathrm{v} = \lambda\sqrt{\frac{\rho_e Q^2}{\epsilon_0 m_e}} \qquad 17.11$$

4. CHARGES IN SEMICONDUCTORS

The conductivity is proportional to the concentration of free charge carriers (see Eq. 17.8). In a good conductor, this concentration is very large ($\approx 10^{28}$ electrons/m^3). In an insulator, it is very small ($\approx 10^7$ electrons/m^3). Semiconductor charge carrier concentrations vary between these approximate values. The semiconducting nature of these types of materials originates not so much from the number of carriers but from the arrangement of allowed energy states into bands. Figure 17.3 illustrates the general arrangement of energy levels for metals, semiconductors, and insulators. The energy bands and gaps are not to scale and vary according to the element or compound utilized. The importance of the figure can be summed up by the following statement: an electric field cannot impart energy to an electron in a completely filled band. The electron must be able to absorb energy and move into a new allowable energy level in order to move under the influence of the field or of a potential difference.

Energy-Band Theory[4]

Electrons exist in *core shells* (i.e., completely filled allowed energy levels) and in *valence shells* (i.e., allowed energy levels, some of which remain vacant). For example, the carbon atom has a core shell that is completely filled, the K shell (in electron configuration parlance, $1s^2$). The carbon atom's valence shell, the L shell, has only four of the possible eight electrons it could hold (in electron configuration parlance, $2s^2 2p^2$). When individual carbon atoms are brought together in a crystalline structure, the valence electrons are exposed to the potential energy changes of nearby atoms as illustrated in Fig. 17.4.

Because of this periodic potential, each energy level in the outermost shell splits into a number of discrete levels.[5] This is illustrated in Fig. 17.5. The periodic potential is usually accounted for in electronic equations by the use of an *effective mass* for electrons and holes, m_e^* and m_h^*.

At the atomic spacing *d*, atoms behave as individual atoms with discrete energy levels centered around the K and L shells. At atomic spacing *c*, the valence band splits, with two states in a lower band and six in an upper. At atomic spacing *b*, the bands overlap. Discrete energy levels still exist, but now an electron can move from low in the band to high in the band in small discrete steps of energy called *quanta*.[6] At atomic spacing *a*, the original L valence shell has been split into two distinct bands of

Figure 17.3 *Energy Band Diagrams*

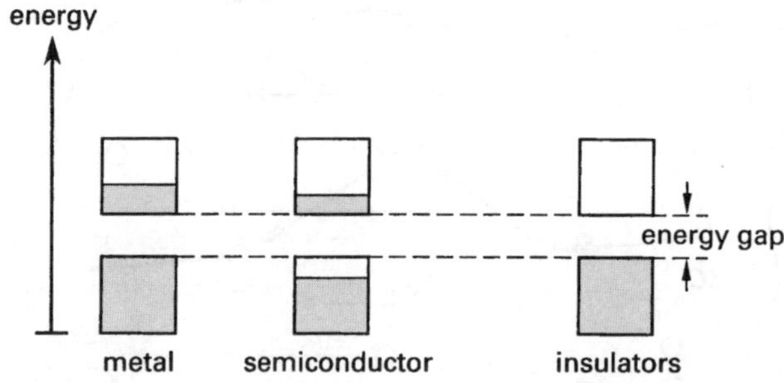

Figure 17.4 *Electron Potential Energy Variation in a One-Dimensional Crystal*

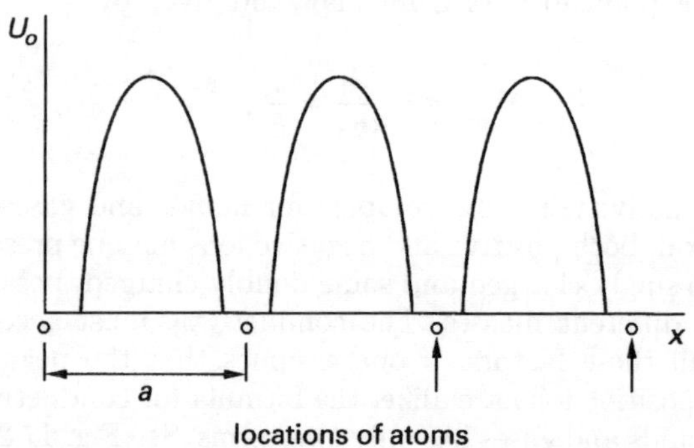

Figure 17.5 *Energy Bands of Carbon as a Function of Atomic Spacing*

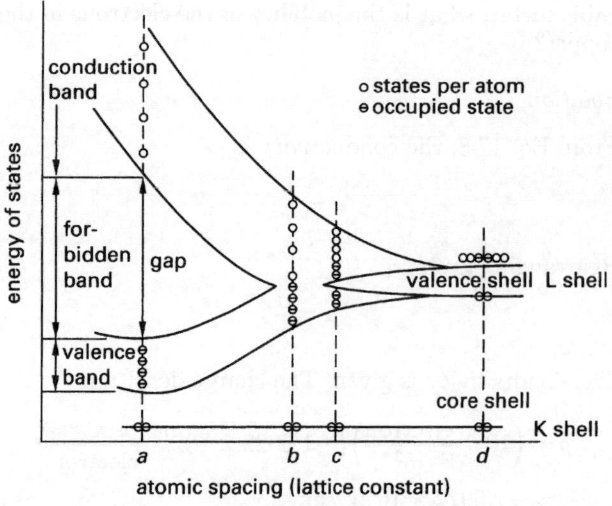

[4]Quantum theory plays the dominant role in explaining energy-band theory. Electrical engineering is primarily concerned with the results.

[5]In quantum mechanical terms, this occurs because the eigenfunctions of each atom obtained from a solution of Schrödinger's equation overlap those of neighboring atoms. This phenomenon is similar to the resonant double peak that occurs when two LC tuning circuits with identical resonant frequencies are brought close together.

[6]At atomic spacing *c* and *d*, the electrons can also move in discrete energy levels, but of several quanta.

energy. The lower retains the name *valence band* and the upper is called the *conduction band.* These bands are normally shown as in Fig. 17.6.

Figure 17.6 *Multiple Energy Bands*

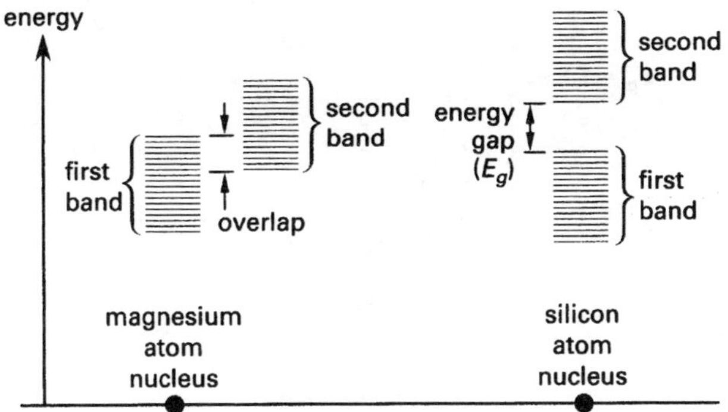

Intrinsic and Extrinsic Semiconductors

Silicon and germanium are the bases of most semiconductor devices. They each possess four valence electrons, leaving four unfilled electron positions. Their crystalline lattice structure uses covalent bonding, that is, a sharing of the electrons, to "complete" the outer shell of eight electrons. (The sharing of electrons does not actually fill the remaining positions but does bond the individual atoms into the crystalline structure.) This situation is illustrated symbolically in Fig. 17.7. In the figure, only silicon atoms are present. When a semiconductor is pure, it is called an *intrinsic semiconductor.*

The energy the structure possesses simply due to its temperature is sometimes adequate to break a covalent bond and free an electron. The amount of energy needed to free an electron is the necessary energy to overcome the energy gap, E_g, and raise the electron to the conduction band. In this manner, a *hole* is created in the valence band and a free electron in the conduction band, as illustrated in Fig. 17.8. The hole and the free electron are both *carriers.* That is, they both contribute significantly to the current upon the application of a voltage. This type of carrier generation is termed *thermal carrier generation.* The current so generated is the *intrinsic conduction.* The intrinsic carrier density, n_i, is given by Eq. 17.12.

$$n_i^2 = Ce^{-E_g/kT} \qquad 17.12$$

As Eq. 17.12 illustrates, the intrinsic concentration is a function of temperature. The constant, C, takes into account the effective masses of electrons and holes and is actually constant only at thermal equilibrium.[7] If temperature varies, the constant C can be replaced with A_0T^3, where A_0 is constant and T is the absolute temperature. The Boltzmann constant, k, is given in units consistent with energy gap units, 8.62×10^{-5} eV/K. The

[7]The constant C is the product of the *effective density states* N_c and N_v (see App. 17.A).

Figure 17.7 *Crystal Structure of Silicon*

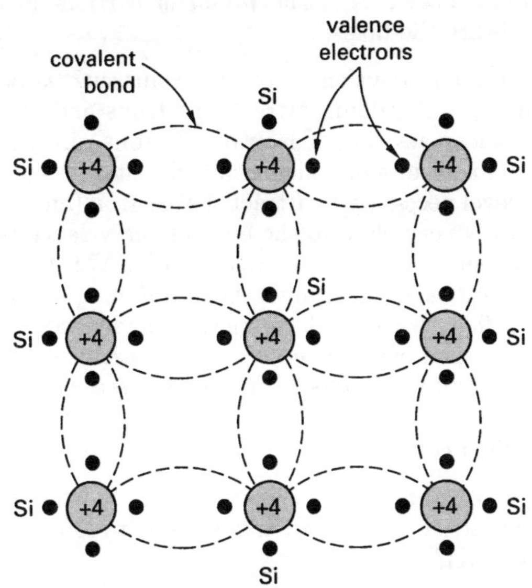

Figure 17.8 *Silicon Crystal with Broken Covalent Bond*

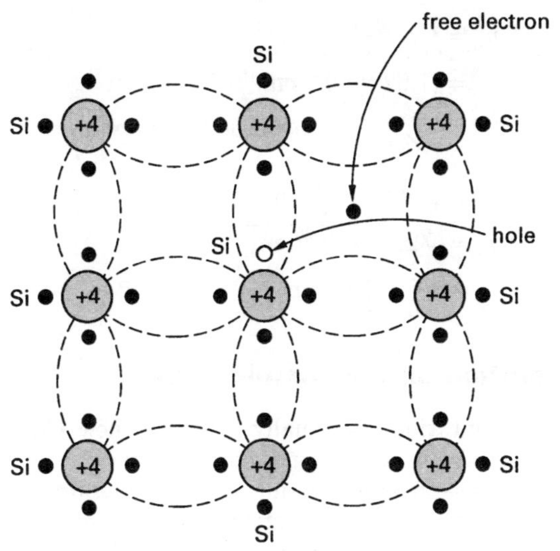

energy gap in silicon is approximately 1.1 eV, and in germanium is approximately 0.8 eV.

Silicon and germanium atoms are tetravalent (i.e., they have four electrons in the valence band). The conductivity of semiconductors composed of these elements can be increased by *doping,* that is, adding impurities to the mix. When a semiconductor is doped, it is called an *extrinsic semiconductor.*

The doping of silicon with pentavalent substitutional impurities such as phosphorus, arsenic, or antimony results in a higher electron density. This is because the fifth electron on the impurity atom is easily ionized, because its energy level is very close to the edge of the conduction-band energy, E_c. This situation is illustrated in Fig. 17.9(a). Such *dopants* are called *donors.* Any such donor mix is called *n-type,* because the *majority carriers* are electrons. Regions doped with donors are sometimes labeled n^+. The *minority carriers* in such materials are the holes.

The addition of trivalent impurity atoms such as boron, aluminum, and gallium attracts electrons and thereby creates vacancies (i.e., a lack of electrons—in essence, holes). This results in a higher hole density, because the energy level necessary to attract the electron (and create the hole) is very close to the edge of the valence band, E_v. This situation is illustrated in Fig. 17.9(b). Such dopants are called *acceptors*. Any such acceptor mix is called *p-type*, because the majority carriers are holes. Regions doped with acceptors are sometimes labeled p^+. The minority carriers in such materials are electrons.

Example 17.2

Given a constant, C, of 7.62×10^{38} cm^{-6}, what is the intrinsic carrier concentration in silicon at room temperature (300K)?

Solution

The intrinsic carrier concentration is given by Eq. 17.12.

$$\begin{aligned} n_i^2 &= Ce^{-E_g/kT} \\ &= (7.62 \times 10^{38}\ \text{cm}^{-6}) \\ &\quad \times \left(e^{-1.1\ \text{eV}/(8.62\times10^{-5}\ \text{eV/K})(300\text{K})}\right) \\ &= 2.56 \times 10^{20}\ \text{cm}^{-6} \\ n_i &= \sqrt{2.56 \times 10^{20}\ \text{cm}^{-6}} \\ &= 1.6 \times 10^{10}\ \text{cm}^{-3} \end{aligned}$$

Generation and Recombination

Thermal agitation continually generates new electron-hole pairs within a semiconductor.[8] However, electron-hole pairs disappear as a result of *recombination*. This is the process whereby free electrons fall into empty covalent bonds, filling the holes and reducing the number of charge carriers. The average electron or hole lives a certain mean time, τ_n for electrons and τ_p for holes.[9] These are important parameters in the design of semiconductor devices.

Law of Mass Action

Adding n-type impurities decreases the number of holes below the number existing in an intrinsic semiconductor; conversely, adding p-type impurities decreases the number of free electrons. Theoretical analysis indicates that, under thermal equilibrium, the product of free negative and positive (hole) concentrations remains constant regardless of the amount of doping. This is a form of the *law of mass action* and is given by

$$np = n_i^2 \qquad 17.13$$

The result of this law is that the doping of an intrinsic semiconductor increases the conductivity and results in a conductor that utilizes either predominately electrons (n-type) or predominately holes (p-type) as carriers. The intrinsic concentration is a function of temperature (see Eq. 17.12).

Figure 17.9 *Energy Band Diagrams of n-type and p-type Semiconductors*

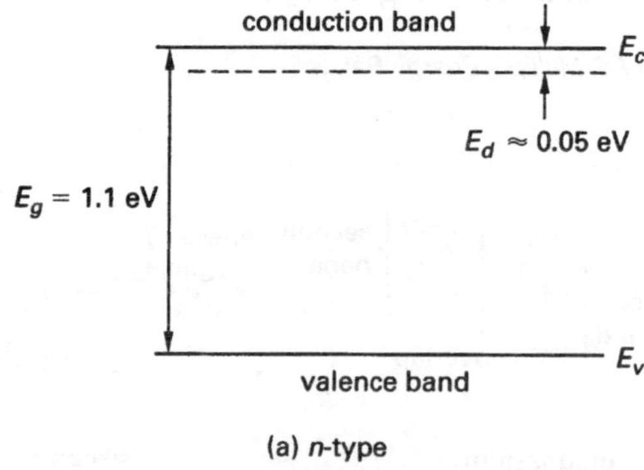

(a) n-type

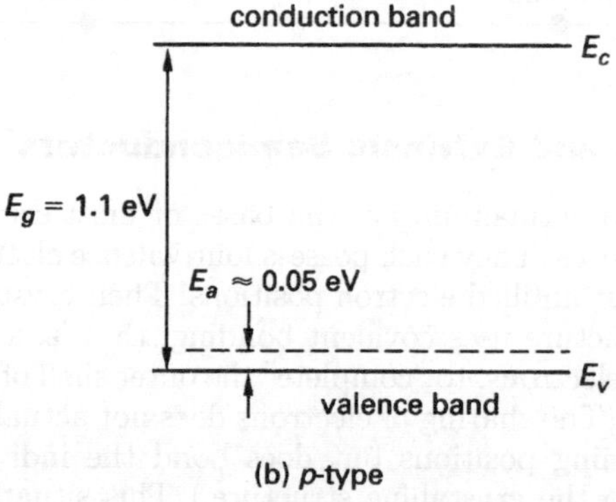

(b) p-type

Diffusion

Whenever a concentration gradient of any kind exists, a flow will occur in an attempt to equalize the concentration. This phenomenon is called *diffusion* and is governed by a *diffusion coefficient*, D, measured in units of m^2/s. In a semiconductor, when an n-type region is placed in contact with a p-type region, a *diffusion current* flows to equalize the concentration of electrons and holes. This diffusion current has no analog in metals. The magnitude of the diffusion current density is

$$J = -qD\frac{d\rho}{dx} \qquad 17.14$$

The minus sign indicates that the current flow is in a direction opposite the concentration gradient. This is shown in Fig. 17.10.

[8]Any form of energy that can be absorbed by the semiconductor, such as illumination, radiation, and acoustic energy, can create electron-hole pairs.

[9]Do not confuse these terms with relaxation time that occurs for conduction and uses identical symbols.

Figure 17.10 Diffusion Current Flow

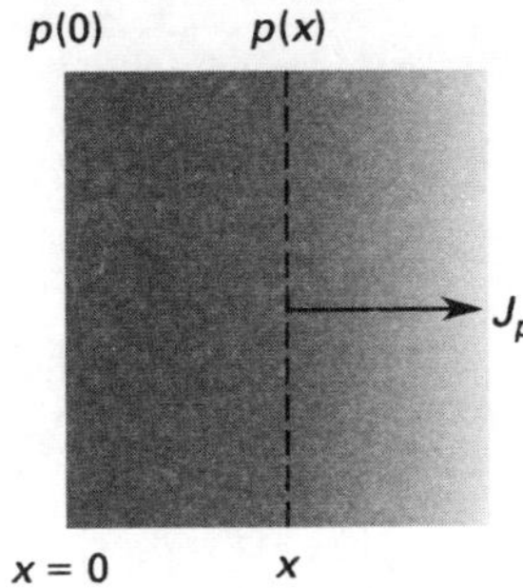

The diffusion and mobility of the electron and hole carriers are governed by the *Einstein equation*, Eq. 17.15.

$$V_T = \frac{D_p}{\mu_p} = \frac{D_n}{\mu_n} \qquad 17.15$$

The term V_T is the *volt-equivalent of temperature* and is defined by Eq. 17.16.

$$V_T = \frac{kT}{q} \qquad 17.16$$

Boltzmann's constant in Eq. 17.16 must be in units of J/K and the temperature, T, is in units of K. The unit of charge, q, is 1.6022×10^{-19} C.

The diffusion current will continue to flow until an electric field is established that opposes the flow. When this occurs, a region is created between the two semiconductor types that is devoid of free charge carriers. This region is called the *space-charge region*, because the region is space containing a charge but no free carriers. This region is also called the *depletion region* or *transition region*. The space-charge region is illustrated in Fig. 17.11 and forms the basis for the operation and control of all semiconductor devices.

A partial listing of the properties of silicon, germanium, and gallium arsenide is given in App. 17.A.

Example 17.3

What is the volt-equivalent of temperature at 300K?

Solution

From Eq. 17.16, the volt-equivalent of temperature is

$$\begin{aligned} V_T &= \frac{kT}{q} \\ &= \frac{\left(1.3807 \times 10^{-23}\ \frac{\text{J}}{\text{K}}\right)(300\text{K})}{1.6022 \times 10^{-19}\ \text{C}} \\ &= 2.58 \times 10^{-2}\ \text{V} \end{aligned}$$

Figure 17.11 pn Junction Properties

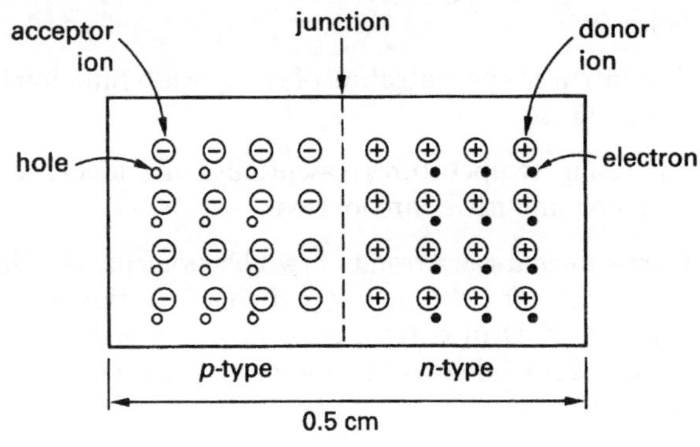

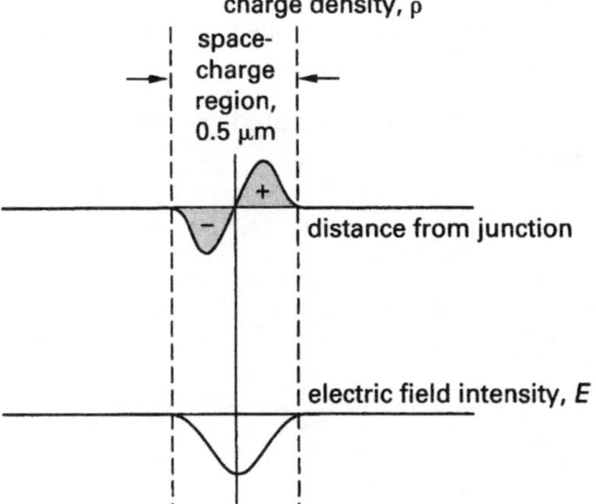

Example 17.4

If the mobility of electrons in silicon is 1.5 cm²/V·s at room temperature (300K), what is the value of the diffusion coefficient, D?

Solution

From Eq. 17.15, the Einstein relationship gives

$$V_T = \frac{D_n}{\mu_n}$$

Rearranging gives

$$\begin{aligned} D_n &= V_T \mu_n \\ &= (0.0258\ \text{V})\left(1.5\ \frac{\text{cm}^2}{\text{V·s}}\right)\left(\frac{1\ \text{m}}{100\ \text{cm}}\right)^2 \\ &= 3.87 \times 10^{-6}\ \text{m}^2/\text{s} \end{aligned}$$

Summary of Fundamental Semiconductor Principles

The principles of semiconductor behavior are summarized as follows.

- Two types of charge carriers exist: electrons and holes.
- A semiconductor may have donor impurities and, as a result, mobile charge carriers that are primarily electrons.

- A semiconductor may have acceptor impurities and, as a result, mobile charge carriers that are primarily holes.
- The intrinsic concentration of carriers is a function of temperature.
- At room temperature, essentially all donor and acceptor impurities are ionized.
- Current occurs as a result of two phenomena: (1) Diffusion current is by means of a concentration gradient and (2) Drift current is by means of an external electric field (identical to the conduction current in metals).
- Charge carriers are continuously generated thermally and simultaneously disappear, by means of recombination.
- Across an open circuited *pn* junction, a potential exists.

18 Communication Theory

Nomenclature

A	alphabet or message size	–
BW	bandwidth	Hz
$c(t)$	carrier signal	–
f	frequency	Hz
$f(t)$	time domain function	–
$F(\omega)$	frequency domain function	–
$H(X)$	information entropy	–
i	variable current	A
I	constant or rms current	A
$I(x_i)$	information content of x_i	–
k	Boltzmann's constant, 1.3807×10^{-23}	J/K
n	number of carriers	–
$p(x_i)$	probability of element x_i	–
P	power	W
q	charge per carrier	C
r	ratio	dB
R	resistance	Ω
t	time	s
T	temperature	K
T_S	sampling period	s
v	variable voltage	V
V	constant or rms voltage	V

Symbols

Δt	rise time or time duration	s
$\Delta\Omega$	equivalent rectangular bandwidth	Hz
τ_d	delay time	s
ω	angular frequency	rad/s

Subscripts

c	center or carrier
I	information (signal)
m	message (signal)
n	noise
R	received or receiver
s	signal
S	sampling

1. FUNDAMENTALS

The purpose of any communication system is to deliver *information* represented as a set of data, $\{I\}$, over a given *channel*, C, to the desired *destination* or *receiver*, R. The information comes from a *source* and is usually transmitted as electric signals controlled by the sender. This is illustrated in Fig. 18.1.

Figure 18.1 Communication System

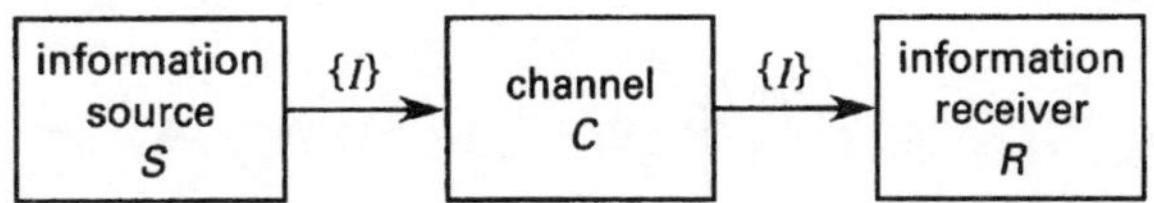

Ideally, the information received would match the information sent. However, because of channel limitations and noise, this may not be the case. Instead, a corrupted version of the information, $\{I^*\}$, may be received. The design objectives in any communication system are to minimize channel distortion and noise and to maximize the amount and rate of the information sent. As in most engineering problems, a trade-off exists. Sending more information at a greater rate increases distortion and error.

Some forms of information are better suited for transmission over given channels than others. Therefore, encoders are used at the source to change the information set, $\{I\}$, into a more suitable form of information, $\{I_E\}$.[1] This is illustrated by Fig. 18.2.

Analog and Digital Signals

An *analog signal* operates with variables that are continuously measurable. An analog signal has an infinite number of values, although a given instrument may not have the resolution necessary to determine them. An example of such a signal is the human voice. The receiver, the human ear, does not always hear or resolve the entire signal, but the continuously varying values are in the voice's waveform.

[1]The information is encoded in the sense that its format is changed to suit the medium of the channel, not for security purposes. Encoding for security purposes is termed *encrypting*.

Figure 18.2 Communication System with Encoding and Decoding

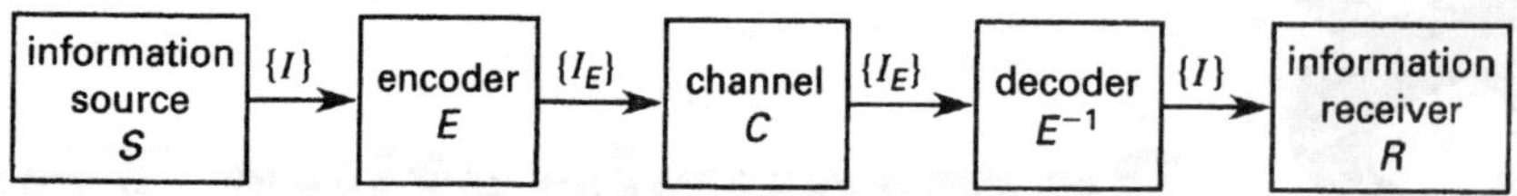

A *digital signal* operates with nominally discontinuous, or discrete, variables that can vary in frequency, amplitude, or polarity. Computer inputs and computer information processing utilize digital signals.

It is often necessary in electrical engineering to convert an analog signal into a digital signal. To do so, the analog signal must be sampled. Figure 18.3 illustrates the sampling of a portion of an analog signal, part (a), in the time domain. The sampling of impulse pulses, part (b), results in a digital signal whose individual impulse pulses vary in amplitude, part (c).

Figure 18.3 Sampling

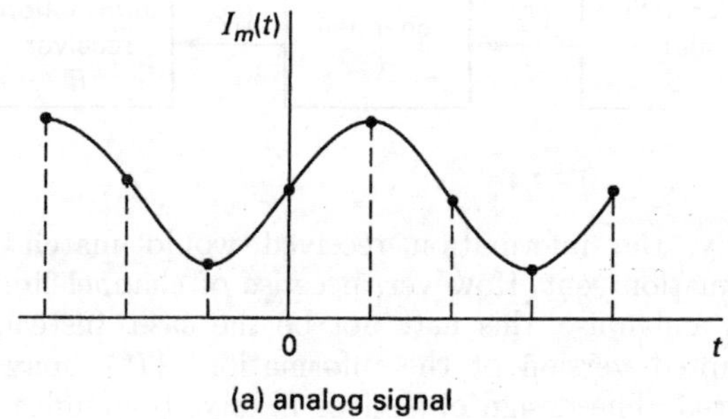

(a) analog signal

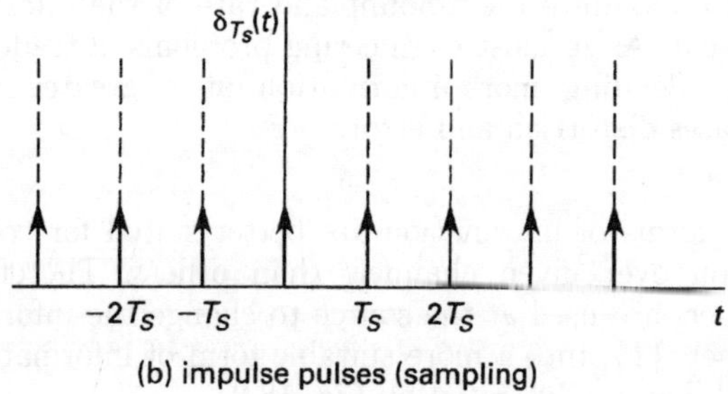

(b) impulse pulses (sampling)

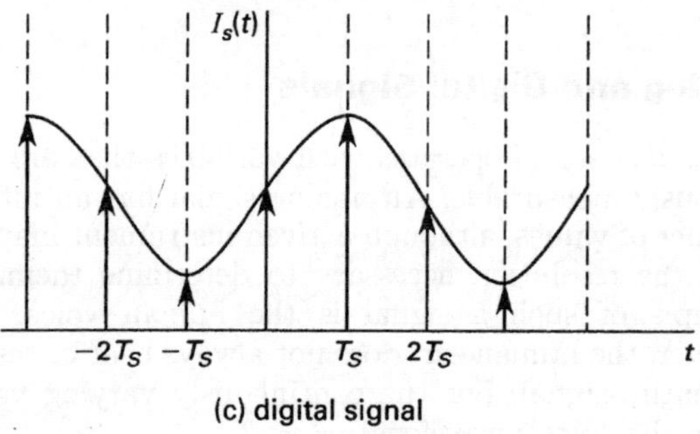

(c) digital signal

The *sampling theorem* is the basis for conversion of an analog signal to a digital signal. A signal that is sampled the minimum number of times while retaining all the properties of the original in the sampled signal is called an *ideal sampled signal.* The rate that achieves this ideal sampling is called the *Nyquist rate*, f_S, and is given by Eq. 18.1.[2]

$$f_S = \frac{1}{T_S} = 2f_I \tag{18.1}$$

The term f_I is the frequency of the original information signal (sometimes called the *message signal*). The *Nyquist interval*, T_S, is the sampling period corresponding to the Nyquist rate and is given by

$$T_S = \frac{1}{2f_I} \tag{18.2}$$

Example 18.1

In general, telephone line voice signals are limited to about 3300 Hz. What is the minimum sampling rate required in order to retain the original signal?

Solution

The minimum sampling rate is given by the Nyquist rate, Eq. 18.1. Applying this equation,

$$\begin{aligned} f_S &= 2f_I \\ &= (2)(3300\ \text{Hz}) \\ &= 6600\ \text{Hz} \end{aligned}$$

Example 18.2

What is the Nyquist interval for a wideband channel with a bandwidth of 4000 Hz?

Solution

The Nyquist interval is the period of sampling corresponding to the Nyquist rate. That is, it is the period corresponding to twice the information frequency. Using Eq. 18.2 and substituting gives

$$\begin{aligned} T_S &= \frac{1}{2f_I} = \frac{1}{(2)(4000\ \text{Hz})} \\ &= 1.250 \times 10^{-4}\ \text{s} \end{aligned}$$

[2]In actual practice the sampling rate is slightly higher than the Nyquist criteria to avoid *aliasing*, or *foldover*, which distorts the original signal. Also, sampling rates higher than the Nyquist rate are required to properly sample a sinusoidal signal because if one happened to sample when the sinusoid was equal to zero, all subsequent samples would be zero.

Basic Signal Theory

Consider the arbitrary nonnegative function, $f(t)$, shown in Fig. 18.4(a), which is the impulse response of a particular linear circuit. The time that $f(t)$ is appreciably different from zero is the *duration* of $f(t)$. When used in terms of the system, the duration is called the *rise time* or *response time* and is given the symbol Δt. The term τ_d is called the *delay time* of the system. The integral of the pulse represents the step-function response to the impulse $f(t)$ (see Fig. 18.4(b)).

Figure 18.4 *Signal Duration and Delay*

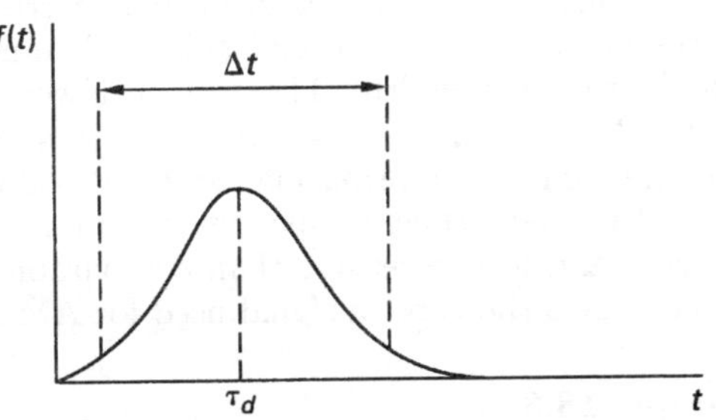

(a) impulse response of a linear system

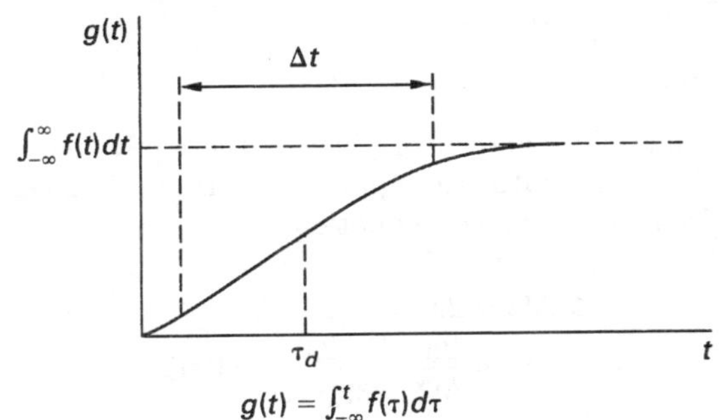

(b) integral of the impulse response

Because the function $f(t)$ is nonnegative, a new nonnegative function with unit area can be defined. This function is

$$\frac{f(t)}{\int_{-\infty}^{+\infty} f(t)\,dt} \qquad 18.3$$

Equation 18.3 and its frequency domain counterpart, that is, its Fourier transform counterpart, are used to define an equivalent rectangular signal of bandwidth, $\Delta\Omega$, and height, $F(\omega)$. This is illustrated by the dashed lines in Fig. 18.5. The same process is followed in the time domain and yields an equivalent rectangular duration, Δt.

The *bandwidth* is defined by the frequency range over which $F(\omega)$ is appreciably greater than zero. The duration in the time domain is directly related to the bandwidth in the frequency domain. Furthermore, the two

Figure 18.5 *Equivalent Rectangular Bandwidth*

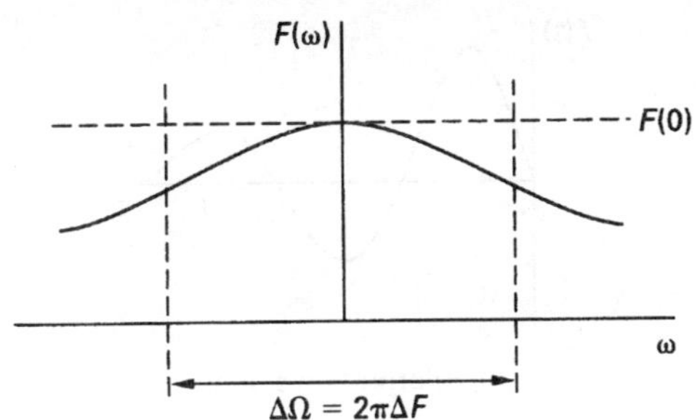

are related in a fundamental way by various *uncertainty relationships*. The primary uncertainty relationship is

$$\Delta t \Delta\Omega = 2\pi \qquad 18.4$$

The second moment uncertainty relationship, which relates the energy, is

$$\Delta t_2 \Delta\Omega_2 \geq \frac{1}{2} \qquad 18.5$$

Many other uncertainty relationships exist.

Signals, such as the function $f(t)$, are modified prior to being placed in a communication channel for transmission. Signals containing the desired information are placed on, or mixed with, a carrier signal and then transmitted to the receiver. The situation is illustrated in mathematical terms in Fig. 18.6.

Figure 18.6 *Basic Signal Processing: Functional Representation*

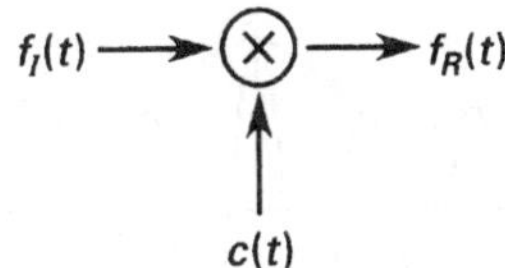

The processing indicated in the figure takes place in the information source, *S*, of Fig. 18.1 or in the encoder, *E*, of Fig. 18.2. Consider a specific signal. Let the information signal, $f_I(t)$, be given by Fig. 18.7(a). When mixed with a carrier signal, $c(t)$, of Fig. 18.7(b), the resultant signal, $f_R(t)$, is given as Fig. 18.7(c).[3]

If the amplitude of the carrier signal is sufficient, the resultant signal, $f_R(t)$, does not cross the x-axis and is represented as shown in Fig. 18.8.

The carrier is centered at zero on the frequency axis. One *sideband* is centered at ω_c and the other at $-\omega_c$. In each sideband, the frequencies above the center frequency $|\omega_c|$ are termed the *upper sideband* and those below constitute the *lower sideband*. If the information

[3]The specific signal shown in Fig. 18.7(c) represents $f_R(t) \approx f(t)\cos \omega_c t$ and is called a *suppressed-carrier amplitude modulation system*.

Figure 18.7 Basis Signal Processing: Graphical Representation

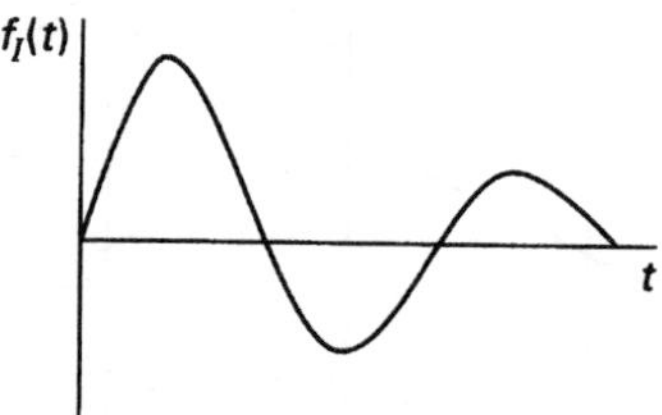

(a) information signal

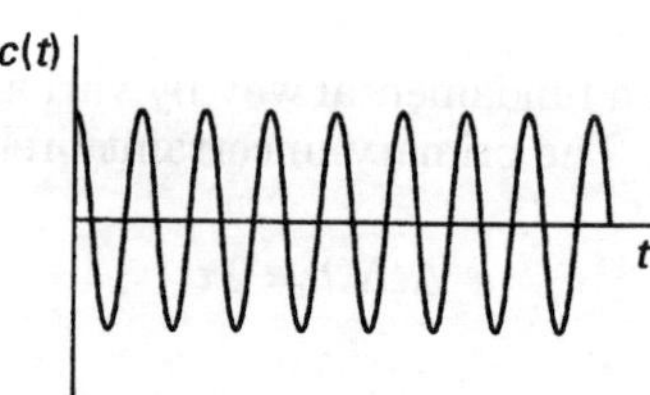

(b) carrier signal

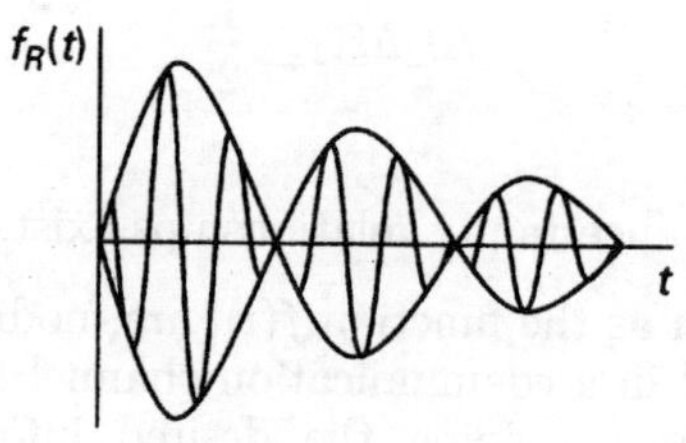

(c) resultant signal

Figure 18.8 Double Sideband AM Signal and Frequency Spectrum

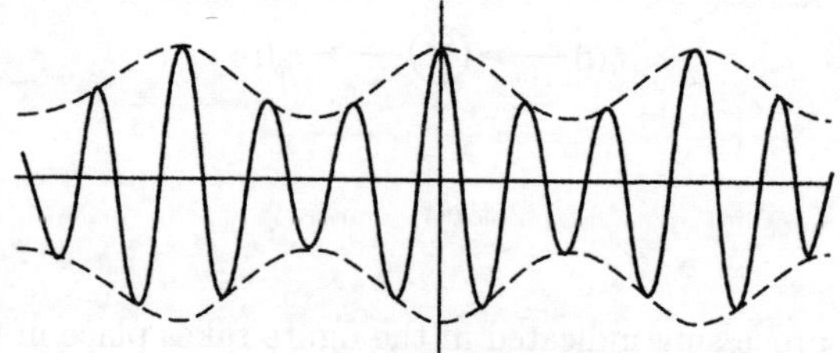

(a) double sideband AM signal

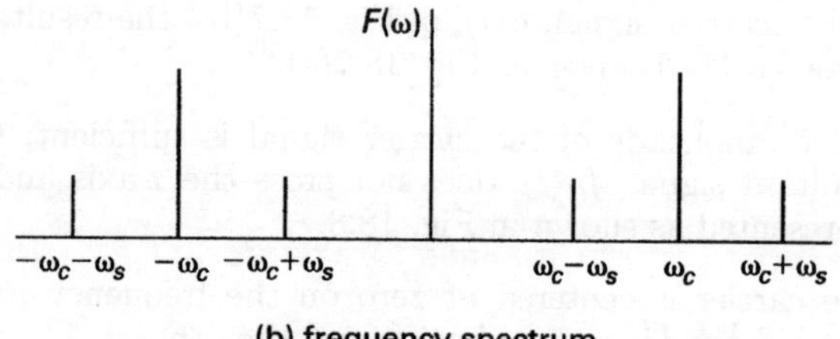

(b) frequency spectrum

signal is not sinusoidal, the frequency spectrum spreads as shown in Fig. 18.9.

Information signals, or message sets, are modified in order to transmit information over a given channel with

Figure 18.9 Spreading of Frequency Spectrum

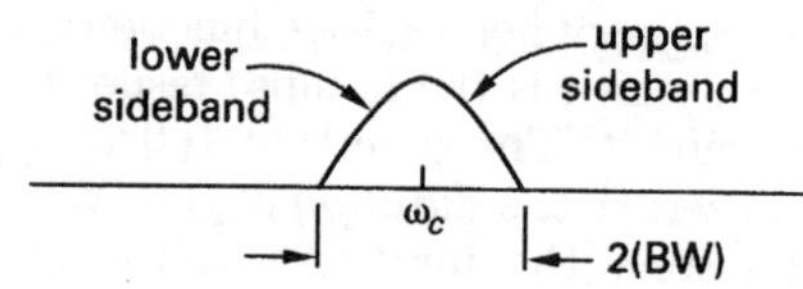

less error. Additionally, a carrier makes it easier to demodulate the signal at the receiver. However, the information desired, that is, the original signal, is completely contained within the upper or lower sideband of a single sideband. The transmitted carrier simply absorbs power and the second sideband unnecessarily expands the bandwidth. The former lowers the efficiency. The latter makes poor use of the frequency spectrum, meaning less information can be passed in a given bandwidth. Nevertheless, such a system is simple to construct and less expensive than comparable systems. In fact, this is the type of signal used for AM radio.

Example 18.3

If the bandwidth of a narrowband, or telegraph-grade, communication system is 300 Hz, what is the response time?

Solution

Assuming the bandwidth given is the equivalent rectangular bandwidth, the applicable formula is given by Eq. 18.4. Applying this equation,

$$\Delta t \Delta \Omega = 2\pi$$

$$\Delta t = \frac{2\pi}{\Delta \Omega} = \frac{2\pi}{300\ \text{Hz}} = 0.02\ \text{s}$$

Information Entropy

Information theory quantifies the communication process. To model a communication system, the information must be predictable. It must be possible for each piece of information to be represented by a *probability of occurrence*, $p(x_i)$. The more probable a particular piece of information, the less *self-information*, $I(x_i)$, a message contains. The information content is defined by Eq. 18.6.

$$I(x_i) = \log \frac{1}{p(x_i)} = -\log p(x_i) \qquad 18.6$$

The choice of the base for the logarithm is arbitrary, but is normally chosen as 2. In this case the unit of self-information, $I(x_i)$, is the common *binary digit* or *bit.* If base-e is used the unit is called the *nat.* (Base-e is convenient for its analytical properties when integration and differentiation are realized with electric circuitry.) Base-10 (rarely used) would result in the unit called a *hartley.*

Example 18.4

What is the information content of a message with only two possible values, zero and one, of equal probability?

Solution

Assume base 2. Using Eq. 18.6, and noting that equal probability means a 0.5 chance of a zero and a 0.5 chance of a one, gives

$$I(x_i) = \log_2 \frac{1}{p(x_i)} = \log_2 \frac{1}{0.5} = 1 \text{ bit}$$

(Unless otherwise told, assume base 2. Also, the bit is not a unit in the true sense.)

Example 18.5

The probability of occurrence of a given message in a computer system is 1.5259×10^{-5}. What is the information content of a message?

Solution

Use Eq. 18.6 and substitute the given information.

$$\begin{aligned} I(x_i) &= \log_2 \frac{1}{p(x_i)} = \log_2 \frac{1}{1.529 \times 10^{-5}} \\ &= 16 \text{ bits} \end{aligned}$$

One of the more important statistical properties of the information variable $I(x_i)$ is its mean or average value. In information theory, the average value of a message is called its *entropy* and is given the symbol $H(X)$. The entropy is defined by Eq. 18.7.

$$\begin{aligned} H(X) &= \sum_{i=1}^{A} p(x_i) I(x_i) \\ &= -\sum_{i=1}^{A} p(x_i) \log_2 p(x_i) \end{aligned} \qquad 18.7$$

The term A is the alphabet size, that is, the total number of elements in the message set. Properly interpreted, Eq. 18.7 is identical to entropy in statistical mechanics or thermodynamics. In statistical mechanics, entropy is a measure of the disorder of the system.

Similarly, in information theory, entropy is a measure of the uncertainty associated with a message. The units of entropy are the *bit*, the *nat*, and the *hartley* and are sometimes given as bits/message, nats/message, or hartleys/message.

The following properties emerge for a binary system.

- $H(X) \geq 0$.
- $H(X) = 0$ only for $p = 0$ or 1.
- $H(X)$ is a maximum at $p = 1/2$.

In the general case, for an alphabet size of A, the entropy has the following properties.

- $H(X) \geq 0$.
- $H(X) = 0$ if and only if all the probabilities are zero except for one, which is unity.
- $H(X) \leq \log_b A$.
- $H(X) = \log_b A$ if and only if all the probabilities are equal so that $p(x_i) = 1/A$ for all i.

Example 18.6

What is the entropy of a binary system with probabilities p and $1 - p$?

Solution

The entropy is given by Eq. 18.7. The alphabet size is two. That is, the value of the elements can be either a zero or a one in a binary system.

$$\begin{aligned} H(X) &= -\sum_{i=1}^{A} p(x_i) \log_2 p(x_i) \\ &= -\sum_{i=1}^{2} p(x_i) \log_2 p(x_i) \\ &= -\Big(p \log_2 p + (1-p) \log_2 (1-p)\Big) \\ &= -\Big((0.5) \log_2 0.5 + (1-0.5) \log_2 (1-0.5)\Big) \\ &= 1 \text{ bit/message} \end{aligned}$$

More enlightening than the mathematical answer in Ex. 18.6 is the plot of $H(X)$ versus the probability.

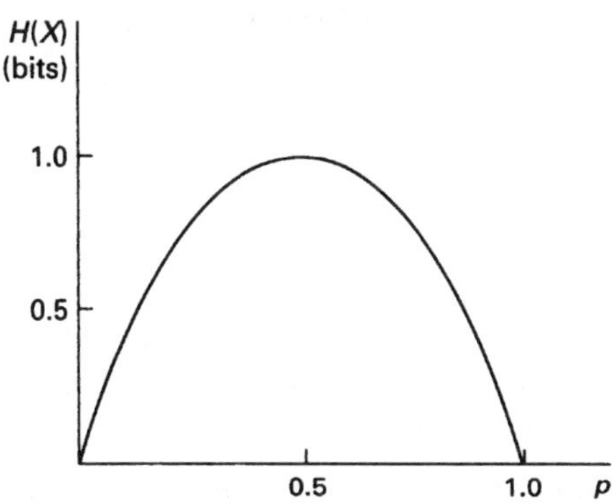

Example 18.7

A certain communication system consists of three possible messages with probabilities of 1/2, 1/4, and 1/4. What is the entropy of the message?

Solution

The information content of the probability 1/2 message is as follows.

From Eq. 18.6,

$$I(x_i) = \log_2 \frac{1}{p(x_i)} = \log_2 \frac{1}{0.25} = 1 \text{ bit}$$

The information content of the probability 1/4 message is as follows.

From Eq. 18.6,

$$I(x_i) = \log_2 \frac{1}{p(x_i)} = \log_2 \frac{1}{0.25} = 2 \text{ bits}$$

From Eq. 18.7, the entropy is

$$H(X) = -\sum_{i=1}^{A} p(x_i) \log_2 p(x_i)$$

$$= \sum_{i=1}^{3} p(x_i) I(x_i)$$

$$= \left(\frac{1}{2}\right)(1 \text{ bit}) + \left(\frac{1}{4}\right)(2 \text{ bits}) + \left(\frac{1}{4}\right)(2 \text{ bits})$$

$$= 1.5 \text{ bits/message}$$

Decibels

The use of a unit known as the *decibel*, dB, is common in communications theory and electrical engineering in general. The decibel is a unit that describes the ratio of two powers or intensities. In acoustics, it also describes the ratio of a power to a reference power. In mathematical terms, the decibel is

$$r = 10 \log_{10} \frac{P_2}{P_1} \qquad 18.8$$

When the ratio is negative, a power loss is indicated. Because the decibel is a log function, decibel gain or loss in the successive stages of a communication system can be added algebraically.

When the ratio is taken with respect to 1 watt, that is, when $P_1 = 1$ W, the ratio is commonly called the *decibel watt* and abbreviated dBW. When the ratio is taken with respect to 1 milliwatt, that is, when $P_1 = 1 \times 10^{-3}$ W, the ratio is termed the *decibels above 1 milliwatt* and abbreviated dBm. Several other decibel ratios are used in electrical engineering and care must be taken to ensure that the correct reference power level is used.

The decibel is also used as a ratio of currents and voltages, though technically this should only be done if the impedances of the measured quantities are equal. This is often ignored, and the decibel for currents or voltages is given by Eq. 18.9

$$r = 10 \log \left(\frac{I_2}{I_1}\right)^2 = 10 \log \left(\frac{V_2}{V_1}\right)^2 = 20 \log \frac{I_2}{I_1}$$

$$= 20 \log \frac{V_2}{V_1} \qquad 18.9$$

2. CHANNELS AND BANDS

As shown in Fig. 18.1, all communication systems use a channel. The *channel* is the medium that connects the information source to the destination, or receiver. The media may be metallic conductors such as copper, specialized cables such as coaxial cable or fiber optic cable, waveguides (devices that guide the propagation of electromagnetic waves using the physical construction of the guide), the atmosphere or, in the case of satellites, space itself.

Channels are classified according to bandwidth. Narrowband channels use bandwidths of 300 Hz or less. Voice-grade channels use 300 Hz to 4 kHz (the audible frequency range of the human voice is approximately 300–3400 Hz). Wideband channels use 4 kHz and greater bandwidths.[4]

Bands represent a somewhat arbitrary range of electromagnetic-wave frequencies between definite limits. Though somewhat arbitrary, the properties of the bands differ, creating different uses for each of the bands. For example, VLF (very low frequency) transmissions form surface waves that travel great distances and are able to penetrate water to a certain degree, making them useful for submarine communications. Standard AM radio uses the MF band while FM radio uses the VHF band. Television utilizes the VHF and UHF bands. Satellites, with their need for high data rates and minimal environmental interference, as well as their ability to utilize line-of-sight communication from Earth, function in the UHF and SHF bands. The bands are shown in Table 18.1. The electromagnetic spectrum is shown in App. 18.A.

3. CODING

In designing a communication system, one attempts to minimize the probability of information error, or bit error, and at the same time minimize the length of the message. Coding theorems guarantee the existence of codes that allow transmission within a channel's capacity at an arbitrarily small error rate. However, to do so, especially in the presence of noise, may require excessively long messages—that is, a low data rate—or raising the signal power above acceptable rates. To

[4]The channel capacity is determined by the type of channel and is given by the *Shannon coding theorem*. The results of this theorem are among the most important in information theory and communications engineering.

Table 18.1 Frequency Bands

symbol	frequency range	frequency band
VLF	3–30 kHz	very low
LF	30–300 kHz	low
MF	300–3000 kHz	medium
HF	3–30 MHz	high
VHF	30–300 MHz	very high
UHF	300–3000 MHz	ultra high
SHF	3–30 GHz	super high
EHF	30–300 GHz	extremely high

overcome this difficulty, *error-correcting codes* have been developed to correct some of the errors that occur during the transmission of an information signal through a noisy channel. A simple example is a parity check code. An example of an *even parity* check code for detecting single errors is shown in Table 18.2. This code determines the total number of ones or zeros in the message and assigns a one if the total is odd and a zero if the total is even.

Table 18.2 Even Parity Check Code

information	check digit	transmitted signal
000	0	0000
001	1	1001
010	1	1010
011	0	0011
100	1	1100
101	0	0101
110	0	0110
111	1	1111

4. NOISE

Noise or *interference* is any unwanted input to a communication system. Such input causes meaningless or erroneous information that must be ignored or removed. Noise results from random processes and can be classified in mathematical terms, but it is more useful to classify random processes as either *natural noise* or *artificial noise*. However, one mathematical classification is worthy of mention. *White noise* or *additive white Gaussian noise* is defined as a waveform in which all observations are jointly Gaussian and mutually independent. The power spectral density is constant with respect to frequency variation. That is, the noise is random and unaffected by changes in frequency. (Such noise cannot technically exist, but the concept is usefully applied to a system with a constant power spectral density over a frequency range wider than the bandwidth.)

Natural Sources of Noise

Thermal noise, or *Johnson noise*, is generated by the random motion of electrons in conductors and semiconductors. The mean squared value of the thermal noise voltage is given by Eq. 18.10.

$$\overline{v_n^2} = 4kTR(\text{BW}) \qquad 18.10$$

The bandwidth, BW, is the frequency bandwidth of the monitoring equipment. The resistance of the circuit or element is given by R in ohms. The temperature, T, is in kelvins. Because the voltage is squared, Eq. 18.10 is proportional to the average power generated by the noise. Thermal noise is constant from DC values to very high frequencies and is considered white noise.

Shot noise, sometimes called *Schottky noise*, is caused by electrons moving across a potential barrier in a random way. It was originally defined for vacuum tubes as the voltage developed because of random variations in the number and velocity of electrons emitted by a heated cathode. Current can be defined as

$$I = q\frac{dn}{dt} \qquad 18.11$$

q represents the charge per carrier and dn/dt is the number of carriers passing through the potential per second. Because of the random thermal velocities of the electrons, the value of dn/dt fluctuates. If the average current is given by I and the total current by $i(t)$, then the noise current is

$$i_n(t) = i(t) - I \qquad 18.12$$

Assuming the process is completely random, all frequencies are equally likely. The mean squared value of the *shot noise current* is

$$\overline{i_n^2} = 2qI(\text{BW}) \qquad 18.13$$

This noise causes sputtering or popping sounds in radio receivers and snow effects in televisions.

Defect noise describes a number of related phenomena exhibited by noise voltages across terminals of devices when DC currents pass through them. The magnitude depends on the surface imperfections of the terminals. Such noise is also called *flicker noise*, *contact noise*, *excess noise*, *current noise*, and *1/f noise*.

Example 18.8

What is the noise voltage of a 10,000 Ω resistor operating in a 4000 Hz frequency bandwidth at room temperature?

Solution

Use Eq. 18.10. Room temperature is 300K.

$$\begin{aligned}\overline{v_n^2} &= 4kTR(\text{BW}) \\ &= (4)\left(1.38\times10^{-23}\ \frac{\text{J}}{\text{K}}\right) \\ &\quad\times(300\text{K})(10\times10^{3}\ \Omega)(4000\ \text{Hz}) \\ &= 6.62\times10^{-13}\ \text{V}^2\end{aligned}$$

Solve for the voltage directly.

$$\begin{aligned}\overline{v_n^2} &= 6.62\times10^{-13}\ \text{V}^2 \\ v_n &= \sqrt{6.62\times10^{-13}\ \text{V}^2} \\ &= 8.14\times10^{-7}\ \text{V}\end{aligned}$$

Artificial Noise

The natural noise explained in Sec. 18.4 is fundamental in nature in that the proximate cause of the noise is the random, noncontinuous nature of electric charge under the influence of a potential. When a noise is generated by electrical or electronic equipment, which is in principle under human control, the noise is classified as *artificial* or *environmental noise*. Artificial noise may be grouped into three main categories.

Interference noise results when a radio station or television channel interferes with another. This may be caused by poor antenna design or unexpected atmospheric effects. This also includes *crosstalk* between communication links caused by multipath propagation or reflection.

Hum is a periodic signal arising from power lines. It occurs at the power supply frequency or its harmonics (60 Hz or 120 Hz in the United States). In transformer iron cores with loose laminations, the frequency of the hum is twice the supply frequency and is related to the alignment of magnetic domains in the material. Hum can be eliminated by proper construction, filtering, and shielding.

Impulse noise describes numerous phenomena. Impulse noises are often modeled as low-frequency shot processes, that is, as the superposition of a small number of large impulses. Sources may include the ignition noise from automobiles, corona noise from high-voltage transmission lines, switching noise, and lightning. Impulse noises are usually best handled with limiters or blankers that protect the system during the occurrence of the noise and return it to normal when the noisy period has passed.

5. MODULATION

Modulation is the process whereby some parameter of one wave is altered by some parameter of another wave. A particular case of modulation is illustrated in Fig. 18.7 and Fig. 18.8. Modulation is necessary because in many instances it may be inefficient or impossible to transmit a given spectrum through the required channel. In addition, modulation is used for frequency translation of a signal in order to maximize the efficiency of an antenna or move the signal to an assigned frequency band. Modulation makes signal processing, particularly the recovery of the signal information, easier. It can also be used to enhance *multiplexing*, a process whereby multiple signals are transmitted within the same channel. Figure 18.2 shows the block diagram for a communication system with the modulator replaced by an encoder. In fact, modulation is a form of encoding. In addition, the encoding covered in Sec. 18.3 would still occur alongside any modulation.

Signals can be modulated in amplitude or frequency. (AM stands for amplitude modulation and FM for frequency modulation.) Both sidebands or a single sideband can be transmitted. The original carrier wave may or may not be present.

19 Acoustic and Transducer Theory

Nomenclature

B	bulk modulus	Pa
c	specific heat	J/kg·K
C	capacitance	F
E	modulus of elasticity	Pa
f	frequency	Hz
I	intensity	W/m^2
L	inductance	H
L_p	sound pressure level	dB
m	mass per area	kg/m^2
M	total mass	kg
MW	molecular weight (mass)[1]	kg/kmol
p	pressure	Pa
R	resistance	Ω
R	specific gas constant for air, 287	J/kg·K
R^*	universal gas constant, 8314.3	J/kmol·K
T	period	s
T	temperature	K
v	velocity	m/s
V	voltage	V

Symbols

β	sound intensity level	dB
γ	isentropic exponent	–
γ	ratio of specific heats	–
η	energy density	J/m^3
λ	wavelength	m
ρ	density	kg/m^3

Subscripts

0	reference
p	pressure
s	sound
v	volume

[1]This is also called the *kilogram molecular weight*. It is truly a mass, not a weight. The *gram molecular weight*, measured in g/mol·K, is also common. Care should be exercised when applying the units.

1. INTRODUCTION

Sound is an alteration of properties, such as pressure, particle displacement or velocity, or density, in an elastic medium. Sound is a wave phenomenon, like electromagnetic radiation, but with two important fundamental differences. First, sound is transmitted as a *longitudinal wave*, also known as a *compression wave*, that causes alternating compression and expansion (*rarefaction*), while electromagnetic waves are transverse waves. Second, sound waves require a medium through which to travel. Electromagnetic waves do not; they can travel through a vacuum.

For visualization purposes, sound waves are often represented as transverse waves as shown in Fig. 19.1. The *wavelength* is the distance between successive compressions or expansions.

Figure 19.1 *Longitudinal Waves*

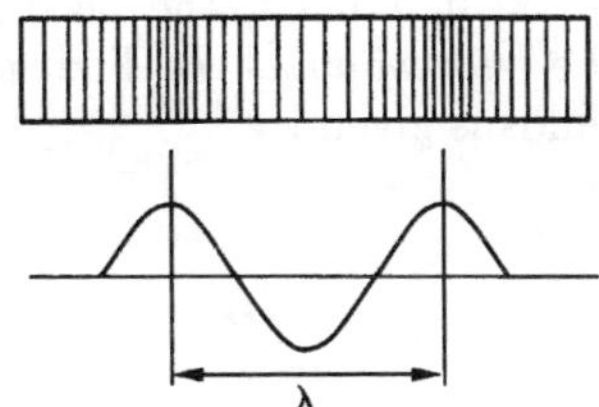

The usual relationships among wavelength, propagation velocity, frequency, and period remain valid.

$$\mathrm{v}_s = f\lambda \qquad 19.1$$

$$T = \frac{1}{f} \qquad 19.2$$

Sound waves traveling through a medium change the energy level of the medium. This energy is called *acoustic energy* and represents the difference between the total energy of the system and the system energy without the sound waves present. Electrical engineering is concerned with the manipulation and use of this energy. The basic interface between acoustic energy and electrical engineering is illustrated in Fig. 19.2.

A *transducer* provides the interface and is the device that provides a usable output in response to a specified *measurand.* The measurand is a physical quantity, property, or condition to be measured.[2] The transducer can be *self-generating,* in which case no external excitation power is required. Such transduction elements are termed *passive elements,* whereas those that require excitation power are *active elements.*

Figure 19.2 *Acoustic/Electrical Interface*

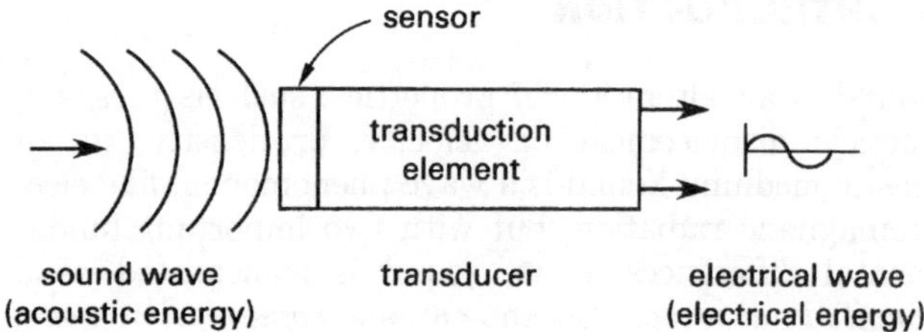

2. SOUND WAVE PROPAGATION VELOCITY

The propagation velocity of sound, v_s, is commonly called the *speed of sound* and *sonic velocity.* The sound velocity depends on the compressibility of the medium through which the sound wave is traveling. This occurs because the sound wave physically moves atoms in the medium, that is, places them closer together (compression) or farther apart (rarefaction), as it propagates. In solids, the compressibility is accounted for by the *elastic modulus, E.* In liquids, the elastic modulus is referred to as the *bulk modulus, B.* The speed of sound in solids and liquids is given by

$$v_s = \sqrt{\frac{E}{\rho}} \qquad 19.3$$

For ideal gases, the propagation velocity is

$$v_s = \sqrt{\frac{\gamma R^* T}{\text{MW}}} = \sqrt{\gamma R T} \qquad 19.4$$

In thermodynamics, the term γ is the ratio of the specific heat at a constant pressure to the specific heat at a constant volume, c_p/c_v, and is called the *isentropic exponent.*

The speed of sound in a few common materials is given in Table 19.1. Various values of the universal gas constant are given in Table 19.2. When the universal gas constant is divided by the molecular weight, that is, R^*/MW, it becomes the *specific gas constant, R.*

[2]These definitions are paraphrased from ANSI MC6.1.

Table 19.1 *Approximate Speeds of Sound (at one atmospheric pressure ≈ 10^5 Pa)*

material	speed of sound (m/s)
air	330 @ 0°C
aluminum	4990
carbon dioxide	260 @ 0°C
hydrogen	970 @ 0°C
steel	5150
water	1490 @ 20°C

Table 19.2 *Values of the Universal Gas Constant, R**

units in SI and metric systems
8.3143 kJ/kmol·K
8314.3 J/kmol·K
0.08206 atm·L/mol·K
1.986 cal/mol·K
8.3143 J/mol·K
82.06 atm·cm^3/mol·K
0.08206 atm·m^3/kmol·K
8314.3 kg·m^2/s^2·kmol·K
8314.3 m^3·Pa/kmol·K
8.314 × 10^7 erg/mol·K

Example 19.1

What is the speed of sound in air at 293K, given an isentropic exponent of 1.4?

Solution

Use Eq. 19.4.

$$v_s = \sqrt{\frac{\gamma R^* T}{\text{MW}}}$$

The universal gas constant is 8.3143 J/mol·K. The molecular weight of air is approximately 29 g/mol. (This assumes 78% N_2, 21% O_2, and 1% Ar.) Substituting gives

$$\begin{aligned} v_s &= \sqrt{\frac{\gamma R^* T}{\text{MW}}} \\ &= \sqrt{\frac{(1.4)\left(8.3143\ \frac{\text{J}}{\text{mol·K}}\right)(293\text{K})\left(10^3\ \frac{\text{g}}{\text{kg}}\right)}{\left(29\ \frac{\text{g}}{\text{mol}}\right)}} \\ &= 343\ \text{m/s} \end{aligned}$$

3. ENERGY AND INTENSITY OF SOUND WAVES

A sound wave, or acoustic energy, manifests itself in the displacement and motion of the medium through which it travels. One could measure the energy of the medium in terms of this displacement or motion. However, the most widely used and measurable property of sound waves is *sound pressure*, the fluctuation above and below the equilibrium pressure (or reference pressure), p_0, of the medium in which the sound wave travels. The energy density, η, is

$$\eta = \tfrac{1}{2}\frac{p_0^2}{\rho \text{v}^2} \qquad 19.5$$

Like many other items in electrical engineering, the *intensity* of the sound wave, which is the power delivered per unit of wave front, is the value of concern. The intensity is

$$I = \eta \text{v} = \tfrac{1}{2}\frac{p_0^2}{\rho \text{v}} \qquad 19.6$$

The human ear can sense sound wave intensities from approximately 10^{-12} W/m^2 up to the pain threshold of 1 W/m^2. Because this encompasses a 12 decade range, the intensities are normally measured in decibels, dB, compared to a reference intensity, as shown in Eq. 19.7.

$$\beta = 10 \log \frac{I}{I_0} \qquad 19.7$$

Because the pressures that sound waves create are the normally measured quantities, the *sound-pressure level*, L_p, as defined in Eq. 19.8 is normally used in lieu of Eq. 19.7.

$$L_p = 20 \log \frac{p}{p_0} \qquad 19.8$$

The defined reference pressure for sounds in air is 20 μPa. This correlates with 0 dB, which is the threshold for human hearing.

Example 19.2

The density of air is approximately 1.3 kg/m^3. What pressure is expected at a transducer's sensor for a sound intensity level of 1 W/m^2 in air at 273K?

Solution

Use Eq. 19.6.

$$I = \tfrac{1}{2}\frac{p_0^2}{\rho \text{v}}$$

From Table 19.1, the velocity (speed) of sound in air at 273K (0°C) is 330 m/s. The density of air is given. Rearranging to solve for the equilibrium pressure and substituting gives

$$I = \tfrac{1}{2}\frac{p_0^2}{\rho \text{v}}$$

$$\begin{aligned} p_0 = \sqrt{2I\rho \text{v}} &= \sqrt{(2)\left(1\ \frac{\text{W}}{\text{m}^2}\right)\left(1.3\ \frac{\text{kg}}{\text{m}^3}\right)\left(330\ \frac{\text{m}}{\text{s}}\right)} \\ &= 29.3 \text{ Pa} \end{aligned}$$

The small change in pressure caused by a sound wave is superimposed on normal atmospheric pressure of approximately 10^5 Pa. Though small, this pressure results in a sensation of pain for most people. Any audio system design should use this pressure as an absolute maximum.

4. TRANSDUCTION PRINCIPLES

The transduction principle of a given transducer determines nearly all its other characteristics. There are three *self-generating* transduction types: photovoltaic, piezoelectric, and electromagnetic. All other types require the use of an external excitation power source.

In *photovoltaic transduction*, light is directed onto the junction of two dissimilar metals, generating a voltage. This type of transduction is used primarily in optical sensors. It can also be used with the measurand controlling a mechanical-displacement shutter that varies the intensity of the built-in light source. *Piezoelectric transduction* occurs because certain crystals generate an electrostatic charge or potential when placed in compression or tension or by bending forces. In *electromagnetic transduction*, the measurand is converted into a voltage by a change in magnetic flux that occurs when magnetic material moves relative to a coil with a ferrous core. These self-generating types of transduction are illustrated in Fig. 19.3.

In *capacitive transduction*, the measurand is converted into a change in capacitance. This can be realized by moving the electrodes of a given capacitor or changing the position of the dielectric. *Inductive transduction* is the change in self-inductance of a single coil. *Photoconductive transduction* converts the measurand into a conductance change within a semiconductor material caused by a change in the incident illumination. This is similar to photovoltaic transduction (see Fig. 19.4 to compare).

In *reluctive transduction*, the measurand is converted to an AC voltage change caused by a change in the reluctance path between two or more coils. Reluctance in a magnetic circuit is analogous to resistance in an electrical circuit. This principle also applies to differential transformers and inductance bridges. *Potentiometric transduction* converts the measurand into a position change of a moveable contact on a resistance element. This changes the end-to-end resistance relative to the wiper-arm position controlled by the measurand. *Resistive transduction* is similar to potentiometric transduction except that the change in resistance is brought about by heating or cooling, mechanical stresses, or other external forces. *Strain-gauge transduction* is a special case of resistive transduction generated by a resistance change, caused by a strain, in two of the four arms of a Wheatstone bridge. Here, however, the output is always given by the bridge output voltage.

Figure 19.3 *Self-Generating Transducers*

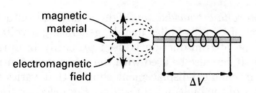

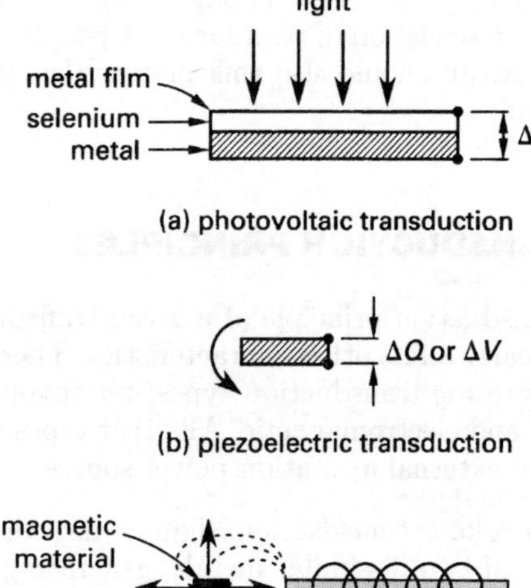

Figure 19.4 *Photoconductive Transduction*

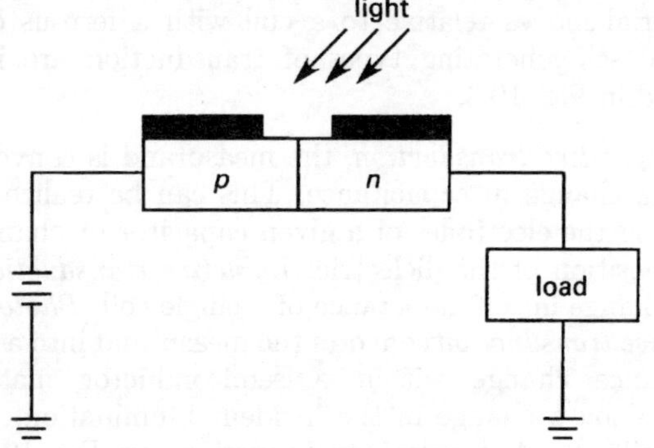

5. EQUIVALENT CIRCUIT ELEMENTS

In electrical engineering, analogs have been developed for the basic elements of acoustic and mechanical systems. These analogs are illustrated in Fig. 19.5.

Figure 19.5 *Basic Elements of Electrical, Acoustic, and Mechanical Systems*

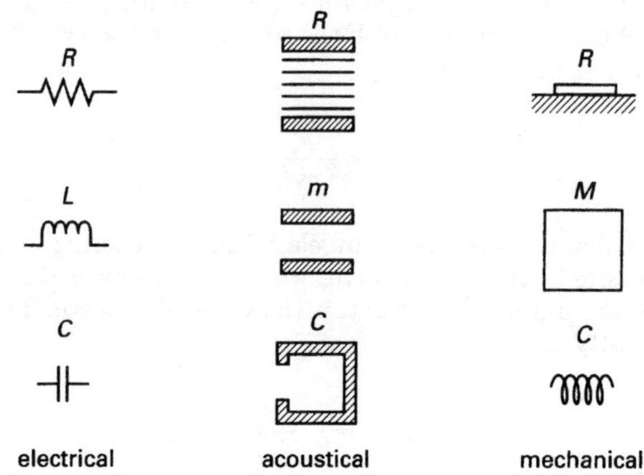

Resistance in acoustic systems can be defined as air friction or viscosity effects in narrow slots. In mechanical systems, resistance is analogous to friction. Inductance in acoustic systems is mass per unit area in a constriction or narrow opening, while in a mechanical system it is the mass itself. Capacitance in an acoustic system is the inverse stiffness of an enclosed volume of air when subjected to piston-like action. It is proportional to the volume enclosed. In mechanical systems, a spring is analogous to capacitance.

Topic III: Field Theory

Chapter

20 Electrostatics

Nomenclature

a	unit vector	–
d, **d**	distance	m
D, **D**	electric flux density	C/m^2
E, **E**	electric field strength	N/C (V/m)
F, **F**	force	N
l, **l**, L	length	m
p, **p**	dipole moment	C·m
q	electric unit charge, 1.6022×10^{-19}	C
Q	charge	C
ΔQ	infinitesimal test charge	C
r, **r**, R	radius or distance	m
s, S	surface area	m^2
T, **T**	torque	N·m
U	energy	J
v	variable voltage	V
v	volume	m^3
V	constant or rms voltage	V
ΔV	potential difference	V
W	work	J

Symbols

ϵ	permittivity	C^2/N·m^2 (F/m)
ϵ_0	free-space permittivity, 8.854×10^{-12}	C^2/N·m^2 (F/m)
ρ	charge density	C/m^3
ρ_l	line charge density	C/m
ρ_s	surface charge density	C/m^2
ρ_v	volume charge density	C/m^3
Ψ	electric flux	C

Subscripts

1–2	from 1 to 2; or effect on 2 caused by the field of 1
e	electric
l	line
r	radial
s	surface

PART 1: ELECTROSTATIC FIELDS

A *field* is an entity that acts as an intermediary in interactions between particles. It is distributed over part or all of space and is a function of space coordinates and time (if the field is not static). An electric field is fundamental in nature, causing a charged body to be attracted or repelled by another charged body. The electric field, **E**, represents the force and direction a positive test charge would experience if placed in the field. It is measured in N/C, which is equivalent to V/m. The electric field strength is given by[1]

$$\mathbf{E}_1 = \frac{\mathbf{F}_{1-2}}{\Delta Q_2} = \left(\frac{Q_1}{4\pi\epsilon r^2}\right)\mathbf{a} \qquad 20.1$$

The unit vector, **a**, points in the direction the positive test charge will move at the point of concern within the field, that is, along the lines of flux. The electric field is a vector force field in that a charge placed within the field experiences a force in a specific direction. The force field for a single charged particle is illustrated in Fig. 20.1.

Figure 20.1 *Electric Field of a Positive Charge*

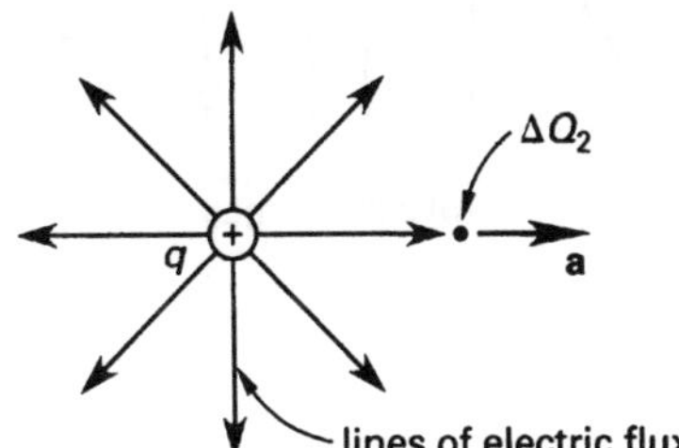

By convention, lines of flux are said to originate at positive charges and terminate at negative charges. Within an electric field, like charges repel and unlike charges attract. Therefore, a positive sign on vector **E** indicates repulsion and a negative sign indicates attraction.

[1]The electric field is defined in terms of the force per unit positive test charge. The rightmost equation is actually the realization of this definition for a point charge.

Field Theory

The force fields of multiple charges interact and the principle of superposition is applicable. The interaction of two unlike charges and of two like charges is illustrated in Fig. 20.2.

Figure 20.2 *Electric Field Interaction*

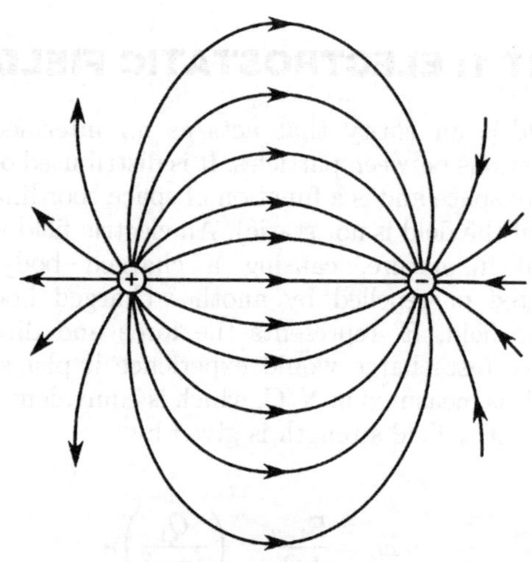

(a) unlike charges

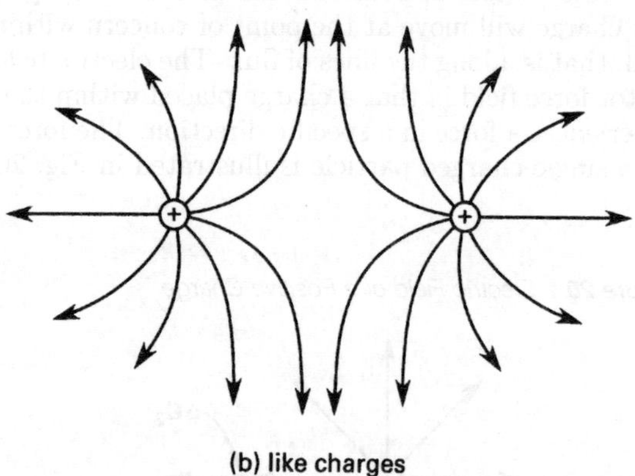

(b) like charges

Equation 20.1 was written in terms of the test charge. Normally an electric field is understood to be the field as it would exist without the test charge present. Further, subscripts are unnecessary in simple configurations. Therefore, in general, the force on a charged object (or distribution of charged objects) is

$$\mathbf{F} = Q\mathbf{E} \qquad 20.2$$

The lines of electric flux comprise a scalar field normally given the symbol Ψ. The flux is equal in magnitude to the charge enclosed. By definition, one coulomb of electric charge gives rise to one coulomb of electric flux. The electric flux density, **D**, is a vector field with units of flux per unit area or C/m². This is not a measurable field. It is used in electrical engineering to remove the effect of the medium, as can be seen by its defining equation.

$$\mathbf{D} = \epsilon\mathbf{E} \qquad 20.3$$

1. POINT CHARGE

The electric field of a point charge was given in Eq. 20.1. It is repeated here with the unit vector specifically showing the radial spreading of the electric field. The electric field of a point charge follows the ***inverse-square law*** and is normally written in spherical coordinates.

$$\mathbf{E} = \left(\frac{Q}{4\pi\epsilon r^2}\right)\mathbf{a}_r \qquad 20.4$$

The potential, V, associated with a point charge is given by Eq. 20.5. The potential is a ***work field*** and is always a scalar. It represents the energy required to transport a test charge from one point to another in an electric field. The unit of potential (v or V) is the volt, V, which is equivalent to J/C.

$$V = \frac{Q}{4\pi\epsilon r} \qquad 20.5$$

Example 20.1

A positive charge is located at the origin of a Cartesian coordinate system. What is the electric field at point (2, 2, 3) (distance in units of meters) with no dielectric present?

Solution

The situation is as shown.

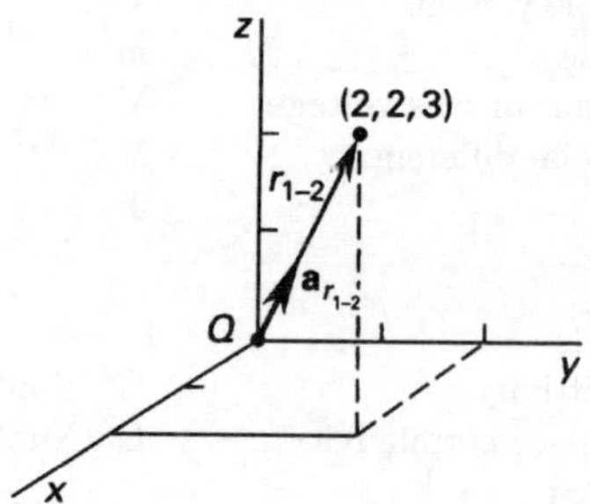

Using Eq. 20.1 and noting that point 1 of the equation is the origin and point 2 is the point (2, 2, 3) gives

$$\mathbf{E}_1 = \left(\frac{Q_1}{4\pi\epsilon r_{1-2}^2}\right)\mathbf{a}_{r_{1-2}}$$

Because no dielectric is present, $\epsilon = \epsilon_0 = 8.854 \times 10^{-12}$ C²/N·m². The distance r_{1-2} is

$$r_{1-2} = \sqrt{(2\text{ m} - 0)^2 + (2\text{ m} - 0)^2 + (3\text{ m} - 0)^2} = 4.1\text{ m}$$

The vector $\mathbf{r}_{1-2} = 2\mathbf{a}_x + 2\mathbf{a}_y + 3\mathbf{a}_z$. Therefore the unit vector in the direction of $\mathbf{r}_{1-2}$ is given by

$$\begin{aligned}\mathbf{a}_{r_{1-2}} &= \frac{\mathbf{r}_{1-2}}{r_{1-2}} = \frac{2\mathbf{a}_x + 2\mathbf{a}_y + 3\mathbf{a}_z}{4.1 \text{ m}} \\ &= 0.49\mathbf{a}_x + 0.49\mathbf{a}_y + 0.73\mathbf{a}_z\end{aligned}$$

The magnitude of the electric field is

$$\begin{aligned}E_1 &= \frac{Q_1}{4\pi\epsilon_0 r_{1-2}^2} \\ &= \frac{1.6022 \times 10^{-19} \text{ C}}{4\pi\left(8.854 \times 10^{-12} \frac{\text{C}^2}{\text{N}\cdot\text{m}^2}\right)(4.1 \text{ m})^2} \\ &= 8.6 \times 10^{-11} \text{ N/C}\end{aligned}$$

Therefore, the electric field is

$$\begin{aligned}\mathbf{E}_1 &= E_1 \mathbf{a}_{r_{1-2}} \\ &= \left(8.6 \times 10^{-11} \frac{\text{N}}{\text{C}}\right)(0.49\mathbf{a}_x + 0.49\mathbf{a}_y + 0.73\mathbf{a}_z)\end{aligned}$$

2. DIPOLE CHARGE

A *dipole* consists of equal and opposite point charges separated by a small distance *d*. Substances whose molecules are dipoles are referred to as *polar substances.* Water is the most significant of these substances. The two hydrogens are separated by 105° with the oxygen atom at the vertex. A dipole is illustrated in Fig. 20.3.

Figure 20.3 Dipole

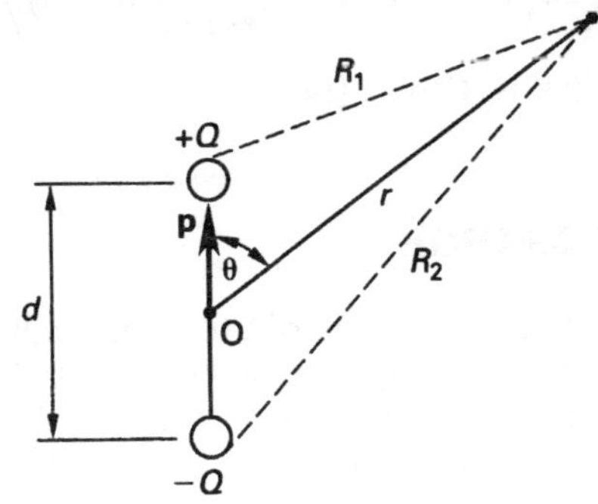

The *dipole moment* is defined by Eq. 20.6.[2]

$$\mathbf{p} = Q\mathbf{d} \qquad 20.6$$

The electric field of a dipole, given in spherical coordinates, is

$$\mathbf{E} = \left(\frac{Qd}{4\pi\epsilon r^3}\right)(2\cos\theta \mathbf{a}_r + \sin\theta \mathbf{a}_\theta) \qquad 20.7$$

[2]The charge, *q*, used in the dipole moment equation is equal to the *unit positive charge*, often called simply the positive charge. This is done because a unit positive charge is the arbitrary reference. In any work (energy) equation involving a dipole, a 2 multiplies the result to account for both charges.

The potential arising from two charges at a distance *r* from the dipole's center of symmetry is[3]

$$V = \frac{Q(R_2 - R_1)}{4\pi\epsilon R_1 R_2} \approx \frac{Qd\cos\theta}{4\pi\epsilon r^2} \qquad 20.8$$

A dipole in an electric field will experience a torque that tends to align the dipole moment, **p**, with the field, **E**. This is shown in Fig. 20.4. The torque is given by Eq. 20.9 and Eq. 20.10.

$$\mathbf{T} = \mathbf{p} \times \mathbf{E} \qquad 20.9$$

$$\text{T} = dF\sin\theta = dQE\sin\theta \qquad 20.10$$

Figure 20.4 Dipole in an Electric Field

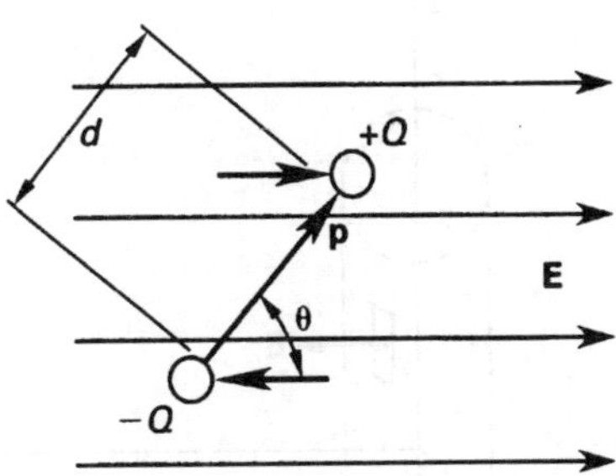

The minimum and maximum energy configurations for a dipole in the presence of an electric field are shown in Fig. 20.5.

Figure 20.5 Dipole Energy Configurations

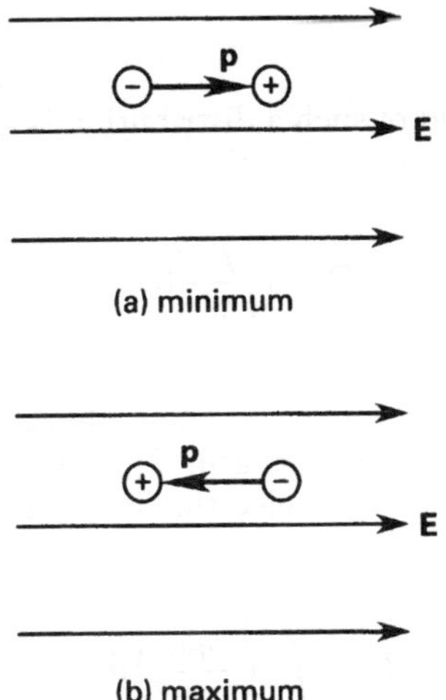

3. COULOMB'S LAW

So far, the equations in this chapter have described the effect of the two charges in the dipole on a third test charge within their electric field. The force between the

[3]When $r \gg d$, the distances R_1 and R_2 from the two dipole charges to the point (r, θ) are essentially the same.

Field Theory

two charges in the dipole, or between any two charges, is described by Coulomb's law and is defined by

$$\mathbf{F} = \left(\frac{Q_1 Q_2}{4\pi\epsilon r^2}\right)\mathbf{a} \qquad 20.11$$

4. LINE CHARGE

Another useful standard charge configuration is that of an infinite, straight-line distribution of charges. This distribution is illustrated in Fig. 20.6.

Figure 20.6 *Infinite Line Charge*

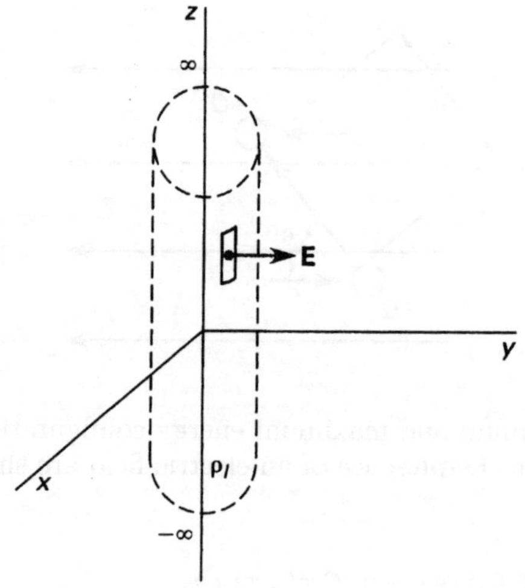

The electric field of such a distribution is given in cylindrical coordinates as

$$\mathbf{E} = \left(\frac{\rho_l}{2\pi\epsilon r}\right)\mathbf{a}_r \qquad 20.12$$

If the line is not an infinite line charge but instead exists over some finite length, L, then a line integral must be utilized. The electric field for a finite line charge density is

$$\mathbf{E} = \int_L \left(\frac{\rho_l \mathbf{a}_r}{4\pi\epsilon r^2}\right) dl \qquad 20.13$$

Example 20.2

An infinite uniform line charge in free space has a charge density, ρ_l, of 10 nC/m and lies along the x-axis of an arbitrary Cartesian coordinate system using meters as the unit of distance. What is the electric field strength, **E**, at point (2, 2, 3)?

Solution

The situation is as shown in the figure.

The field is constant with respect to x. Charges in the y-z plane must be determined in cylindrical coordinates.

$$r = \sqrt{(2\text{ m})^2 + (3\text{ m})^2} = 3.6\text{ m}$$

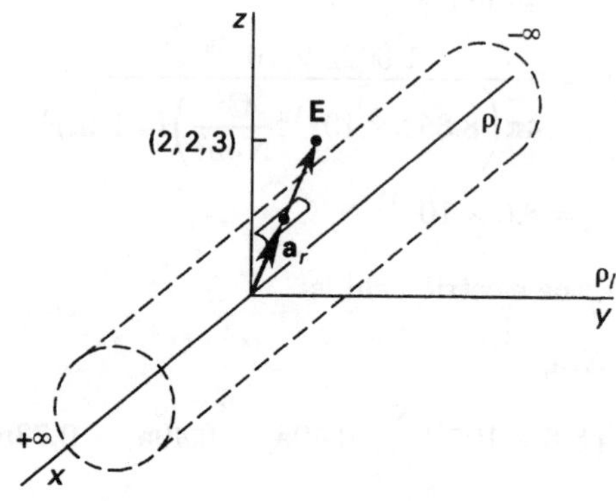

The electric field is

$$\mathbf{E} = \left(\frac{\rho_l}{2\pi\epsilon_0 r}\right)\mathbf{a}_r$$

$$= \left(\frac{10\times 10^{-9}\ \frac{\text{C}}{\text{m}}}{2\pi\left(8.854\times 10^{-12}\ \frac{\text{C}^2}{\text{N}\cdot\text{m}^2}\right)(3.6\text{ m})}\right)\mathbf{a}_r$$

$$= (49.9\text{ N/C})\mathbf{a}_r \quad ((49.9\text{ V/m})\mathbf{a}_r)$$

5. SHEET CHARGE

Consider electric charges distributed uniformly over an infinite plane with density ρ_s as shown in Fig. 20.7. Such a charge density produces an electric field given by Eq. 20.14. The term $\mathbf{a}_n$ represents the unit normal vector to the surface, S.

$$\mathbf{E} = \left(\frac{\rho_s}{2\epsilon}\right)\mathbf{a}_n \qquad 20.14$$

This type of charge configuration is also known as a *surface charge* or *sheet charge*. For a finite surface charge that exists in more than one plane as shown in Fig. 20.8, the electric field is given by the surface integral shown in Eq. 20.15.

$$\mathbf{E} = \int_S \left(\frac{\rho_s \mathbf{a}_r}{4\pi\epsilon r^2}\right) ds \qquad 20.15$$

Figure 20.7 Infinite Surface Charge (viewed edge-on)

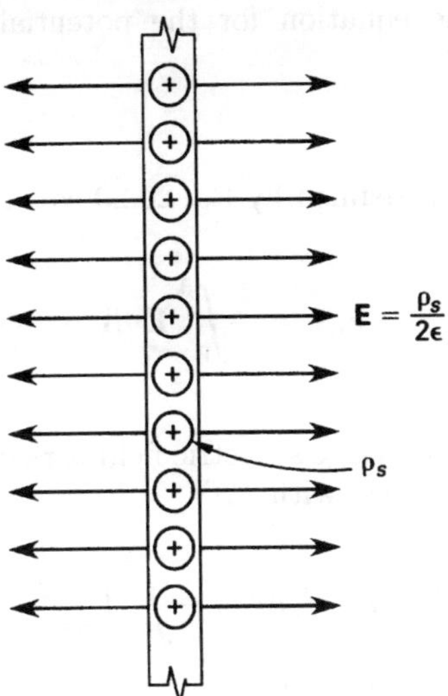

Figure 20.8 Sheet Charge

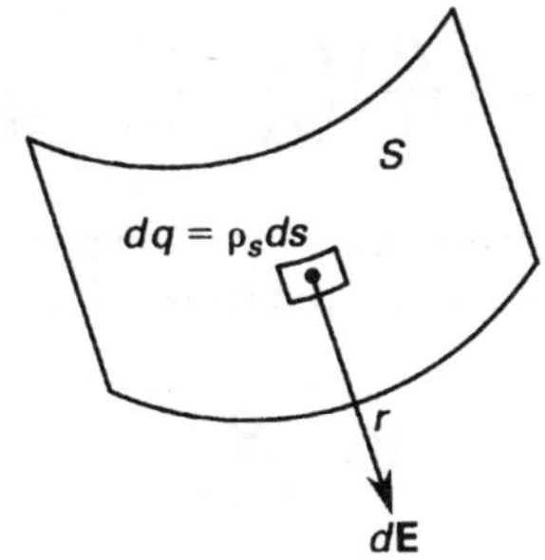

PART 2: DIVERGENCE

The main indicators of a changing vector field are the *divergence* and the *curl*. The curl is best studied in terms of magnetic fields. The divergence is a scalar that bears a similarity to the derivative of a function. In terms of any vector field, **A**, the *divergence theorem*, or *Gauss' divergence theorem*, is

$$\oint_S \mathbf{A}\cdot d\mathbf{s} = \int_V (\nabla\cdot\mathbf{A})\,dv \qquad 20.16$$

If the vector field is the electric flux density, **D**, then the divergence theorem relates the flux emitted over a closed surface to the divergence of the flux density vector from the same volume, that is, the outward flow of the flux per unit volume. The term $\nabla\cdot\mathbf{D}$ is defined as the divergence of **D**. The term ∇ is called the *del operator*.[4]

$$\begin{aligned}\text{div}\,\mathbf{D} &= \nabla\cdot\mathbf{D}\\ &= \left(\frac{\partial}{\partial x}\right)\mathbf{x} + \left(\frac{\partial}{\partial y}\right)\mathbf{y} + \left(\frac{\partial}{\partial z}\right)\mathbf{z}\cdot(D_x + D_y + D_z)\\ &= \frac{\partial D_x}{\partial x} + \frac{\partial D_y}{\partial y} + \frac{\partial D_z}{\partial z}\end{aligned} \qquad 20.17$$

From Gauss' law for electrostatics, the divergence of the electric flux density is

$$\text{div}\,\mathbf{D} = \nabla\cdot\mathbf{D} = \rho \qquad 20.18$$

Equation 20.18 is the differential form of Gauss' law given in integral form in Sec. 16.12.

Example 20.3

In the space charge region on the *n*-side of a *pn* junction, the electric flux density is

$$\mathbf{D} = 8x^2\mathbf{x} + 2\mathbf{y} + 2\mathbf{z}$$

What is the charge density in this region?

Solution

The charge density (in units of C/m^3) can be determined from Eq. 20.18.

$$\begin{aligned}\rho = \nabla\cdot\mathbf{D} &= \left(\frac{\partial D_x}{\partial x}\right)\mathbf{x} + \left(\frac{\partial D_y}{\partial y}\right)\mathbf{y} + \left(\frac{\partial D_z}{\partial z}\right)\mathbf{z}\\ &= \left(\frac{\partial(8x^2)}{\partial x}\right)\mathbf{x} + \left(\frac{\partial(2)}{\partial y}\right)\mathbf{y} + \left(\frac{\partial(2)}{\partial z}\right)\mathbf{z}\\ &= 16x + 0 + 0\\ &= 16x\ \text{C/m}^3\end{aligned}$$

PART 3: POTENTIAL

The term "potential" was defined in Sec. 20.1 as the work that must be done against an electric field to bring a unit charge from an arbitrary reference point to the point in question. The reference point is normally infinity. For practical purposes, the surface of the earth or another large conductor is the reference point. The potential at a given point does not depend on the charge placed at the point, but rather on the charge or charge distribution that creates the electric field. The units of potential are volts (V). One volt represents one joule (J) of work expended moving one coulomb (C) of charge from one point to another (J/C). The potential is given in terms of the electric field as

$$\mathbf{E} = -\nabla V \qquad 20.19$$

[4]The del operator is defined only in the Cartesian coordinate system. It is used in the cylindrical and spherical coordinate systems, but care must be taken to ensure correct results.

Field Theory

The negative gradient of the potential is equal to the electric field. The minus sign is required, since positive charges moving in the direction of the electric field move from higher to lower potentials. That is, the potential decreases as the distance from the source of the electric field increases.

Because the potential is the work done per unit positive test charge in moving the charge from, say, point B to point A, the potential can be represented in integral form as follows.

$$V_{\mathrm{BA}} = \frac{W}{\Delta Q} = -\int_{\mathrm{B}}^{\mathrm{A}} \mathbf{E}\cdot d\mathbf{l} \qquad \text{20.20}$$

The potential is at point A with respect to point B and is the work done in moving the positive test charge from B to A. The potential difference between two points is often represented as ΔV (not to be confused with the gradient of the potential, ∇V).

It is often useful to represent the potential in yet another way. Gauss' law was stated in Eq. 20.18 in terms of the electric flux density, **D**. Restating this equation in terms of the electric field, **E**, and assuming the permittivity is a constant, Gauss' law becomes[5]

$$\operatorname{div} \mathbf{E} = \nabla\cdot\mathbf{E} = \frac{\rho}{\epsilon} \qquad \text{20.21}$$

Substituting Eq. 20.19 into Eq. 20.21 gives the following useful result and introduces the Laplacian operator, ∇^2.

$$\nabla^2 v = -\frac{\rho}{\epsilon} \qquad \text{20.22}$$

Equation 20.22 is called *Poisson's equation*. When there are no charges present, the right-hand side of Eq. 20.22 is zero and the equation is called *Laplace's equation*. The Laplacian operator for Cartesian coordinates is

$$\nabla^2 = \frac{\partial^2}{\partial x^2} + \frac{\partial^2}{\partial y^2} + \frac{\partial^2}{\partial z^2} \qquad \text{20.23}$$

Table 20.1 shows the relationships (proportionalities) between the electric field, the potential, and the distance, r, from the charge or charge distribution.

***Table 20.1** Electric Field and Potential versus Distance*

charge distribution		electric field		potential
point charge	$\propto$	$\frac{1}{r^2}$	$\propto$	$\frac{1}{r}$
line charge	$\propto$	$\frac{1}{r}$	$\propto$	$\ln r$
surface charge	$\propto$	constant	$\propto$	r

[5]The permittivity assumption is valid for linear, homogeneous, and isotropic mediums—a wide class of useful materials.

Example 20.4

Determine the equation for the potential of a point charge.

Solution

The potential is defined by Eq. 20.20 as

$$V_{\mathrm{BA}} = -\int_{\mathrm{B}}^{\mathrm{A}} \mathbf{E}\cdot d\mathbf{l}$$

Because a point charge's electric field is radial, the term $\mathbf{E}\cdot d\mathbf{l}$ can be replaced with $E_r dr$.

$$\begin{aligned} V_{\mathrm{BA}} &= -\int_{\mathrm{B}}^{\mathrm{A}} E_r\,dr = -\int_{\mathrm{B}}^{\mathrm{A}} \left(\frac{Q}{4\pi\epsilon r^2}\right) dr \\ &= \frac{-Q}{4\pi\epsilon}\int_{\mathrm{B}}^{\mathrm{A}} \frac{1}{r^2}\,dr \\ &= \left(\frac{Q}{4\pi\epsilon}\right)\left(\frac{1}{r_{\mathrm{A}}} - \frac{1}{r_{\mathrm{B}}}\right) \end{aligned}$$

The normal reference point is infinity.

$$V_{\mathrm{A}-\infty} = \left(\frac{Q}{4\pi\epsilon}\right)\left(\frac{1}{r_{\mathrm{A}}} - \frac{1}{\infty}\right) = \frac{Q}{4\pi\epsilon r_{\mathrm{A}}}$$

The subscripts are normally not shown, and free space is assumed unless otherwise indicated. Therefore,

$$V = \frac{Q}{4\pi\epsilon_0 r}$$

(Equation 20.19 could have been used as the starting point as well.)

PART 4: WORK AND ENERGY

Work is force moving an object over a distance. In an electric field, a positive charge will experience a force tending to move it in the direction of the electric field. Work, dW, is accomplished only by the force in the direction of motion. When a force is applied in a direction other than the direction of motion, the charge will change direction but no energy will be transferred to the charge. This is why magnetic fields change the direction of moving charges but transfer no energy.

Let the direction of motion be given by $d\mathbf{l}$. Let θ be the angle between the direction of motion, $d\mathbf{l}$ (the actual path of the charge), and the direction of the electric field, **E**. The situation for an arbitrary path is shown in Fig. 20.9.[6]

[6]External forces not shown would be required to make the positive charge follow the indicated path.

Figure 20.9 *Positive Charge in an Electric Field (arbitrary path)*

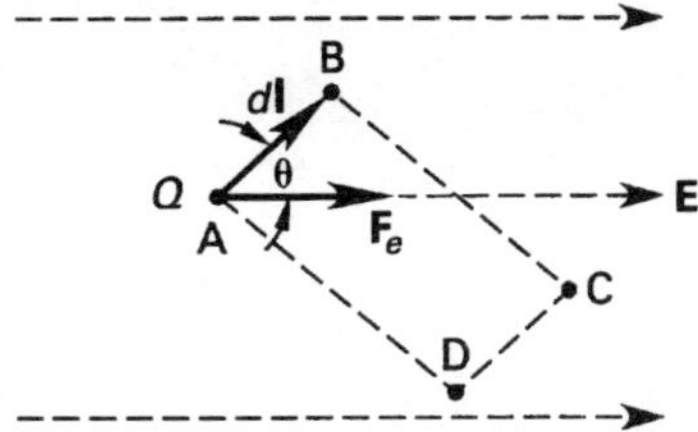

The differential work done by the electric field is

$$dW = -Q\mathbf{E}\cdot d\mathbf{l} = -QE\cos\theta dl \qquad 20.24$$

The total work done in moving a charge from point A to point B in an electric field is

$$W_{AB} = -\int_A^B Q\mathbf{E}\cdot d\mathbf{l}$$

$$= -\int_A^B QE\cos\theta dl \qquad 20.25$$

The work is positive if an external force is required to move the charge (as in bringing two repulsive charges closer together or moving a charge against the force of the electric field). Work is negative if the electric field does the work; for example, allowing attracting charges to come together or allowing a positive charge to move in the direction of the electric field. Positive work indicates that the system requires an energy input and negative work indicates work output. No work is done in moving a charge perpendicular to a field, that is, across the lines of electric flux or anywhere along the lines of an equipotential surface.

When Eq. 20.25 is calculated for all the charges in a given charge configuration creating an electrostatic field, the result is the total energy, U, stored in the field. The energy is "stored" in the volume of the electrostatic field and is given by the following expressions.

$$U = \tfrac{1}{2}\int \rho V dv = \tfrac{1}{2}\int \mathbf{D}\cdot\mathbf{E} dv = \tfrac{1}{2}\int \epsilon E^2 dv$$

$$= \tfrac{1}{2}\int \frac{D^2}{\epsilon} dv \qquad 20.26$$

An electrostatic field is a *conservative field.* No work is done in moving a charge around a closed path within an electrostatic field, such as path ABCDA in Fig. 20.9. Further, the work done in moving a charge between any two points within the field is independent of the path.

This means the integral of the electric field around a closed path equals zero as shown in Eq. 20.26.

$$W = -\oint Q\mathbf{E}\cdot d\mathbf{l} = 0 \qquad 20.27$$

The result given by Eq. 20.26 is true for electrostatic fields and must be modified for time varying fields.

Example 20.5

Neglecting fringing, what is the total energy stored in a parallel plate capacitor with a capacitance of 50 μF and a plate voltage difference of 120 V?

Solution

Neglecting fringing, the capacitor's electric field is identical to the field of infinite parallel plates and is uniform. That is, from Table 16.6,

$$E = \frac{V}{r}$$

Substituting and noting that the field is constant and can be removed from inside the integral gives

$$U = \tfrac{1}{2}\int \epsilon E^2 dv$$

$$U = \tfrac{1}{2}\epsilon E^2 \int dv = \tfrac{1}{2}\epsilon\left(\frac{V}{r}\right)^2 \int dv$$

$$= \tfrac{1}{2}\frac{\epsilon V^2}{r^2}\int dv$$

The capacitance is given by

$$C = \frac{\epsilon A}{r}$$

Substituting,

$$U = \tfrac{1}{2}\frac{\epsilon V^2}{r^2}\int dv = \tfrac{1}{2}\frac{CV^2}{rA}\int dv$$

The volume equals the area of the plates, A, multiplied by the distance between them, r.

$$U = \tfrac{1}{2}\left(\frac{CV^2}{rA}\right)(Ar) = \tfrac{1}{2}CV^2$$

Substituting the given information gives

$$U = \tfrac{1}{2}(50\times 10^{-6}\ \text{F})(120\ \text{V})^2$$

$$= 0.36\ \text{J}$$

Field Theory

21 Electrostatic Fields

Nomenclature

A	area	m^2
d, $\mathbf{d}$	distance	m
D, $\mathbf{D}$	electric flux density	C/m^2
E, $\mathbf{E}$	electric field strength	N/C (V/m)
f	frequency	Hz
J, $\mathbf{J}$	current density	A/m^2
p, $\mathbf{p}$	dipole moment	C·m
P, $\mathbf{P}$	polarization	C/m^2
q	electric unit charge, 1.6022×10^{-19}	C
Q	charge	C
t	time	s
v	volume	m^3
W	work	J

Symbols

ϵ	permittivity	$C^2/N{\cdot}m^2$ or F/m
ϵ_0	free-space permittivity, 8.854×10^{-12}	$C^2/N{\cdot}m^2$
θ	angle	degrees
ρ_{sp}	polarization surface charge density	C/m^2
σ	conductivity	S/m
χ	susceptibility	–
ω	angular frequency	rad/s

Subscripts

0	free space
c	conduction
d	displacement
D	density
e	electric
i	induced or internal
r	relative
sp	surface polarization

1. POLARIZATION

When an electric field is present in a dielectric, the bound electric charges tend to align in the lowest energy configuration possible. In doing so, they increase the flux density, $\mathbf{D}$. The alignment is referred to as the *polarization* of the dielectric and is given the symbol $\mathbf{P}$. The polarization is the total dipole moment, $\mathbf{p}$, per unit volume, v.

$$\mathbf{P} = \frac{\mathbf{p}}{v} \qquad 21.1$$

The flux density is

$$\mathbf{D} = \epsilon_0 \mathbf{E} + \mathbf{P} \qquad 21.2$$

Polarization increases the flux density above the level that would exist under free-space conditions with the same electric field intensity (the polarization of free space equals zero). This occurs even when permanent dipole molecules are not initially present in the dielectric. This can be understood by considering that when an atom is placed within the influence of an electric field,the positive protons tend to move toward the direction of the electric field and the electrons tend to move toward the direction opposite the electric field. In this manner, a dipole is set up within the atom. See Fig. 21.1 for an illustration of this concept. The charge in such induced dipoles is usually not a complete unit charge (1.6022×10^{-19} C) but is instead some fraction thereof, because the charges are not completely separate. The dipole is analogous to spring energy in mechanical systems. The work done in the distortion of the atom is recoverable upon removal of the electric field.

Figure 21.1 *Induced Dipole Moment*

(a) neutral atom or molecule

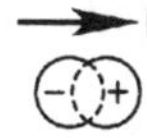

(b) application of an external electric field

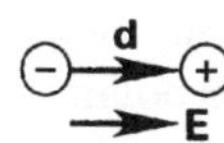

(c) equivalent dipole representation

The electric field and the polarization can have different directions, though this is often not the case. In isotropic, linear material, $\mathbf{E}$ and $\mathbf{P}$ are parallel at each point.

$$\mathbf{P} = \chi_e \epsilon_0 \mathbf{E} \qquad 21.3$$

The term *isotropic* indicates that the medium's properties are independent of direction. The term *linear* indicates that the permittivity, ϵ, is constant regardless of the magnitude of the charge present. The term χ_e is the electric susceptibility. This dimensionless term measures the ease of polarization in a material. For an isotropic linear material, Eq. 21.2 can be rewritten as[1]

$$\mathbf{D} = \epsilon_0(1 + \chi_e)\mathbf{E} \qquad 21.4$$

The relative permittivity can be represented by Eq. 21.5.

$$\epsilon_r = 1 + \chi_e \qquad 21.5$$

In free space, then, because no dielectrics are present, no polarization occurs. The electric flux density, **D**, is

$$\mathbf{D} = \epsilon_0\mathbf{E} \qquad 21.6$$

When a dielectric is present, the flux density is increased and is given by Eq. 21.2. Importantly, Eq. 21.2 is often written without direct reference to polarization, but rather in terms of the relative permittivity. The relationship is shown explicitly in Eq. 21.7.

$$\mathbf{D} = \epsilon\mathbf{E} = \epsilon_0\epsilon_r\mathbf{E} = \epsilon_0\mathbf{E} + \mathbf{P} \qquad 21.7$$

Example 21.1

Water in vapor form has a fractional charge of approximately 5×10^{-20} C and a dipole distance of approximately 10^{-10} m. What is the maximum possible polarization for one mole of water molecules at standard temperature and pressure (STP)?

Solution

The maximum polarization is

$$P = \frac{p}{v}$$

Determine the dipole moment.

$$\begin{aligned} p &= qd = (5 \times 10^{-20}\ \text{C})(10^{-10}\ \text{m}) \\ &= 5 \times 10^{-30}\ \text{C·m} \end{aligned}$$

At STP, the volume of 1 mol of an ideal gas is 22.4 L or 0.0224 m^3. The maximum polarization is

$$\begin{aligned} P &= \frac{p}{v} \\ &= \frac{5 \times 10^{-30}\ \text{C·m}}{0.0224\ \text{m}^3} \\ &= 2.2 \times 10^{-28}\ \text{C/m}^2 \end{aligned}$$

[1]If the material is anisotropic or nonlinear, or both, χ_e is a tensor.

Example 21.2

What are the magnitudes of **P** and **D** for a dielectric in an electric field with a strength of $E = 38.8 \times 10^3$ V/m with a relative permittivity of 4.1 C^2/N·m^2?

Solution

First,

$$P = \chi_e\epsilon_0 E$$

The susceptibility is found from

$$\begin{aligned} \epsilon_r &= 1 + \chi_e \\ \chi_e &= \epsilon_r - 1 = 4.1 - 1 = 3.1 \end{aligned}$$

So,

$$\begin{aligned} P &= \chi_e\epsilon_0 E \\ &= (3.1)\left(8.854 \times 10^{-12}\ \frac{\text{C}^2}{\text{N·m}^2}\right)\left(38.8 \times 10^3\ \frac{\text{V}}{\text{m}}\right) \\ &= 1.06 \times 10^{-6}\ \text{C/m}^2 \end{aligned}$$

Second,

$$\begin{aligned} D &= \epsilon_0\epsilon_r E \\ &= \left(8.854 \times 10^{-12}\ \frac{\text{C}^2}{\text{N·m}^2}\right)\left(4.1\ \frac{\text{C}^2}{\text{N·m}^2}\right) \\ &\quad \times \left(38.8 \times 10^3\ \frac{\text{V}}{\text{m}}\right) \\ &= 1.41 \times 10^{-6}\ \text{C/m}^2 \end{aligned}$$

Example 21.3

An electric field strength is measured as 0.20×10^6 V/m. If the electric flux density is required to be 8×10^{-6} C/m^2 by the design, what is the minimum susceptibility required of the dielectric?

Solution

The flux density is given. Using this value,

$$\begin{aligned} D &= \epsilon_0(1 + \chi_e)E \\ \chi_e &= \frac{D}{\epsilon_0 E} - 1 \\ &= \frac{8 \times 10^{-6}\ \frac{\text{C}}{\text{m}^2}}{\left(8.854 \times 10^{-12}\ \frac{\text{C}^2}{\text{N·m}^2}\right)\left(0.20 \times 10^6\ \frac{\text{V}}{\text{m}}\right)} - 1 \\ &= 3.5 \end{aligned}$$

2. POLARIZED FIELDS

If a dipole is present, it can be aligned either partially or fully with the electric field. An example of such an alignment is illustrated in Fig. 21.2.

Figure 21.2 *Dipole Alignment About a Free Positive Charge*

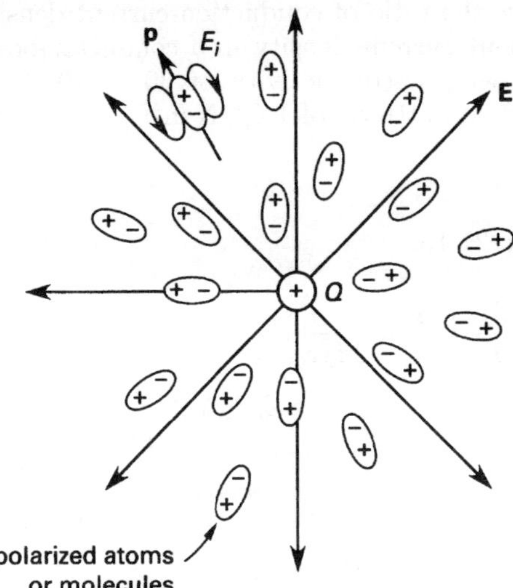

If a molecule contains a permanent dipole moment, it is called a *polar molecule.* Two common examples are water and salt (NaCl). If the molecule has no dipole moment in the absence of an electric field, it is called a *nonpolar molecule.* In this case, the dipole moment can be induced as explained in Sec. 21.1 and shown in Fig. 21.1. Yet a third type of dielectric exists that exhibits permanent polarization in the absence of an electric field but does so only after an induced polarization has taken place. Some waxes, for instance, are polarized when molten and exposed to an electric field. After cooling, some of the induced polarization remains. This special case of aligned polar molecules occurs below a temperature called the *Curie point.* Such materials are analogous to permanent magnets and are called ferroelectrics. Ferroelectrics exhibit the properties of electrical hysteresis.

A dielectric with its dipole moments aligned produces an internal electric field, E_i, that opposes the original electric field, E_0, that created it. The energy expended in the alignment results in a dielectric, or dipole, electric field, E_d. This situation is illustrated in Fig. 21.3(a). The net result is that a surface charge appears at either end of the dielectric as shown in Fig. 21.3(b). The electric flux density is increased above the level that would exist in free space by an amount equal to the polarization as indicated by Eq. 21.2. This can be inferred from the units used to measure electric flux density, C/m^2. The charges per unit area have been increased above the level that exists in free space. An important point to remember is that the net charge in the dielectric remains zero. Also, the polarization surface charge is a real accumulation of charge. It does contribute to the internal and external electric fields. It does not, however, enter into boundary conditions when solving electrical engineering problems. Furthermore, the divergence of **D** depends on only the free charge density.

Figure 21.3 *Dielectric Electric Fields and Surface Charge*

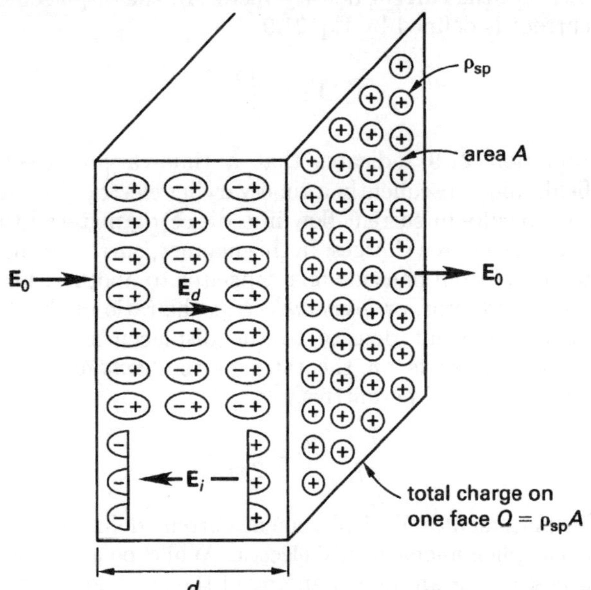

(a) dielectric electric field, $\mathbf{E}_d$

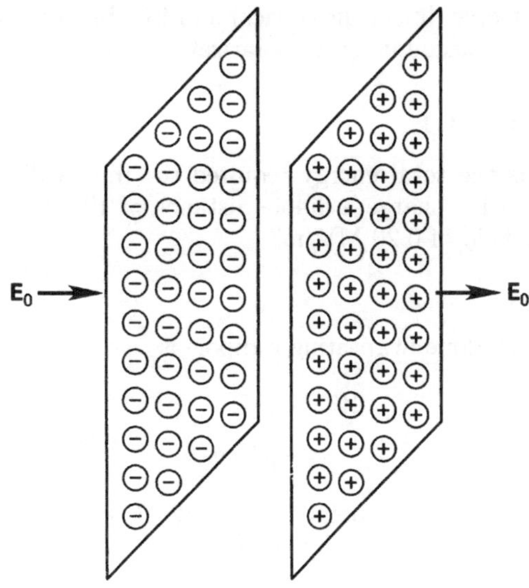

(b) dielectric surface charge

3. ELECTRIC DISPLACEMENT

The electric flux density, **D**, is also called the displacement density. It represents the displacement of charges within a dielectric. The motion of the charges within the dielectric solid is not translational, as it is for conduction current. Instead, the dipole moments rotate under the influence of the electric field (see Fig. 20.4). If the angle $\pi/2$ is chosen as the arbitrary zero energy reference, the work of rotating a dipole is

$$W = -\mathbf{p}\cdot\mathbf{E} \qquad 21.8$$

This motion of charges is the *displacement current*. In terms of the current density vector, **J**, the displacement current is defined by Eq. 21.9.

$$\mathbf{J}_D = \frac{\delta \mathbf{D}}{\delta t} \qquad 21.9$$

Equation 21.9 indicates that a time-varying electric field, and subsequently a time-varying electric flux density, results in current flow in a dielectric material that contains no free charges. In this manner, current is made to "flow" from one side of a capacitor to another. For a poor conductor, or *lossy dielectric*, the ratio of the conduction current density to the displacement current density depends on the ratio of the *conductivity*, σ, and the overall permittivity, ϵ.

$$\frac{J_c}{J_d} = \frac{\sigma}{\omega\epsilon} = \frac{\sigma}{2\pi f \epsilon_0 \epsilon_r} \qquad 21.10$$

Displacement and displacement current define the electrical phenomena in a dielectric. While polarization is necessary for an understanding of the situation, it complicates calculations. For this reason, in practice with simple dielectrics, work is done directly with **D** and not **P** by using the relative permittivity as shown in Eq. 21.7. The relative permittivity is easily determined (for isotropic, linear dielectrics) and for this reason, can be used to account for all polarization effects.

Example 21.4

What is the total energy required to "flip" a dipole of unit positive charge a dipole distance of 10^{-10} m in an electric field of 0.20 MV/m?

Solution

The work done in rotating a dipole is

$$\begin{aligned} W &= -\mathbf{p}\cdot\mathbf{E} \\ &= -pE\cos\theta \end{aligned}$$

The zero reference point is $\theta = \pi/2$. The total energy required to rotate the dipole 90° is

$$|W_{90°}| = pE$$

To flip the dipole (i.e., to rotate it 180°) requires double that amount of energy.

$$|W_{180°}| = 2pE$$

Substituting,

$$\begin{aligned} W_{\text{total}} &= 2pE = 2qdE \\ &= (2)(1.6022 \times 10^{-19}\ \text{C})(10^{-10}\ \text{m}) \\ &\quad \times \left(0.20 \times 10^6\ \frac{\text{V}}{\text{m}}\right) \\ &= 6.4 \times 10^{-24}\ \text{J} \end{aligned}$$

Example 21.5

What is the ratio of conduction current density to displacement current density in a conductor operating at 60 Hz with a conductivity of 30 × 10^6 S/m and a relative permittivity of 1 $\text{C}^2/\text{N}\cdot\text{m}^2$?

Solution

Use Eq. 21.10.

$$\begin{aligned} \frac{J_c}{J_d} &= \frac{\sigma}{\omega\epsilon} = \frac{\sigma}{2\pi f \epsilon_0 \epsilon_r} \\ &= \frac{30 \times 10^6\ \frac{\text{S}}{\text{m}}}{2\pi(60\ \text{Hz})\left(8.854 \times 10^{-12}\ \frac{\text{C}^2}{\text{N}\cdot\text{m}^2}\right)(1)} \\ &= 9.0 \times 10^{15} \end{aligned}$$

The displacement current is a minor effect in conductors, and only needs to be considered at high frequencies.

4. POLARIZATION IN DIELECTRICS VERSUS CONDUCTORS

In a metal conductor, free conduction charges are mobile and move to the surface of the metal under the influence of an electrostatic field. Because the charges are plentiful and mobile, the surface charge builds up until the induced electric field cancels the applied electric field. Consequently, the electric field in the interior of a metal is zero. That is, the polarization of a conductor by the external electric field is complete.

For dielectrics, on the other hand, the charges are bound and can only move a small fraction of an atomic diameter. The induced electric field, E_i, partially cancels the applied electric field. The result is an internal electric field in the dielectric that points in the same direction as the applied field (for isotropic, linear dielectrics). Furthermore, the induced charges on the surface of the dielectric are not equal in magnitude to the induced charges on metallic surfaces. (This could occur only if ϵ_r were infinite.) In short, a dielectric in an external electric field becomes polarized, but not to such an extent that the interior electric field goes to zero.

22 Magnetostatics

Nomenclature

a	unit vector	–
A	area	m^2
A	vector magnetic potential	Wb/m (T·m)
B, **B**	magnetic flux density	T (Wb/m^2)
c	speed of light (vacuum), 3.00×10^8	m/s
d, **d**	distance	m
E, **E**	electric field strength	V/m
F, **F**	force	N
h	Planck's constant (6.6256×10^{-34})	J·s
H, **H**	magnetic field strength	A/m
I, **I**	constant or rms current	A
J, **J**	current density	A/m^2
K, **K**	sheet current	A
l, **l**	length	m
m, **m**	magnetic moment	A·m^2 (J/T)
m	mass	kg
M	magnetization	A/m
n	unit vector	–
N	number of turns	–
p	pole strength	N/(A/m) (J/A or Wb)
p_m	magnetic dipole moment	A·m^2 (J/T)
p	electric dipole moment	C·m
q	electric unit charge, 1.6022×10^{-19}	C
q_m	magnetic charge	A·m (N/T)
Q	charge	C
r, **r**	radius or distance	m
s, **s**	surface area	m^2
S	total surface area	m^2
T	torque	N·m
U, **U**	magnetic potential	A
v, **v**	velocity[1]	m/s
V	constant or rms voltage	V
V_m	magnetomotance	A
V	volume	m^3
W	work	J

Symbols

ϵ	permittivity	C^2/N·m^2 (F/m)
θ	angle	degrees
μ	permeability	N/A^2 (H/m)
μ_0	free-space permeability, 1.2566×10^{-6}	N/A^2 (H/m)
ρ	charge density	C/m^3
Φ, ϕ	magnetic flux	Wb (N/(A/m) or J/A)
ω	angular frequency	rad/s

Subscripts

0	initial
1–2	from 1 to 2; or effect on 2 caused by the field of 1
a	applied
e	electrical or electron
f	final
m	magnetic
p	pole
r	relative
V	per unit volume

PART 1: THE MAGNETIC FIELD

A *field* is an entity that acts as an intermediary in interactions between moving charges. It is distributed over part or all of space and is a function of space coordinates and time (if the field is not static). A *magnetic field* is elementary in nature. Charges moving through the field experience a force. The strength of the **B**-field is the electromotive force (V or J/C) produced when one turn of a loop of area A is linked in 1 sec.[2] It is measured in Wb/m^2, equivalent to V·s/m^2. The strength of the **B**-field is

$$\mathbf{B} = \left(\frac{\mu I l}{4\pi r^2}\right)\mathbf{a} \qquad 22.1$$

[1]The term "velocity" is often used in text where it would be more technically correct to use the term "speed." Velocity is a vector quantity with speed as its magnitude.

[2]The term *linked* indicates that the magnetic flux is passed through the area of concern—in this case, the area of the loop.

The unit vector, **a**, points in the direction of the **B**-field, along the lines of flux. It is perpendicular to the direction of the force experienced by the positive moving charges (that is, the current). This occurs because a magnetic force always acts in a direction transverse to the movement of charges. The denominator in Eq. 22.1, $4\pi r^2$, represents the surface area of a sphere centered on the current element dl, whose surface just touches the point in space where the **B**-field is being evaluated. Specific configurations may be evaluated with Eq. 22.1 by changing **B** to $d\mathbf{B}$ and l to dl and integrating. The magnetic field is a vector force field, in that a moving charge within the field experiences a force in a specific direction. The force field for a single line of moving charges is illustrated in Fig. 22.1.

Figure 22.1 *Magnetic Field of a Positive Current*

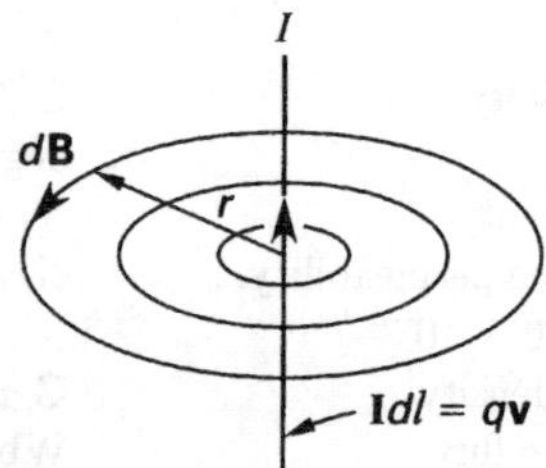

Lines of magnetic flux form continuous loops. By convention, these lines are said to emanate from the north pole of a magnetic dipole and enter the south pole. Like poles repel, and unlike poles attract.

The magnetic force fields of multiple currents interact, and the principle of superposition applies as long as the permeability of the medium is constant. Unfortunately, constant permeability is not exhibited by iron and other ferromagnetic materials, which are widely used around magnetic fields. Therefore, in general, the principle of linear superposition does not apply to current-carrying conductors in the vicinity of ferromagnetic materials.

Equation 22.1 is normally written in terms of the magnetic flux density as shown in Eq. 22.2. Both equations use identical units, Wb/m^2, which in the SI system is a tesla, T. The term **B** is most often called the magnetic flux density.[3]

$$B = \frac{\Phi}{A} \qquad \text{22.2}$$

The force on a current element in a magnetic field is given by the Biot-Savart law in Eq. 16.43. This law can also be viewed in terms of the magnetic field produced by the current elements as explained in Sec. 22.2.

[3]The magnetic flux density, **B**, is the fundamental field in magnetics, and it appears in the force equations. Though the name is reminiscent of the electric flux density, **D**, this is a result of the historical development in the study of magnetics. **B** and **E** are the fundamental fields. **H** and **D** are constructs to aid in the solution of electrical engineering problems.

Magnetic flux, Φ, is defined as the surface integral of the normal component of the flux density, B, over the area.

$$\Phi = \int_A \mathbf{B}\cdot d\mathbf{A} \qquad \text{22.3}$$

It is often more helpful to visualize the magnetic flux as the total quantity of lines of magnetic force set up around a current-carrying conductor. If there are N conductors of length l, the flux is

$$\Phi = \mu N I l = BA \qquad \text{22.4}$$

Example 22.1

A wire carrying a current of 6 A is shown. The current is in the plane of the paper. Determine the magnetic flux density 0.5 m away from a 1 cm segment of the wire at the point indicated.

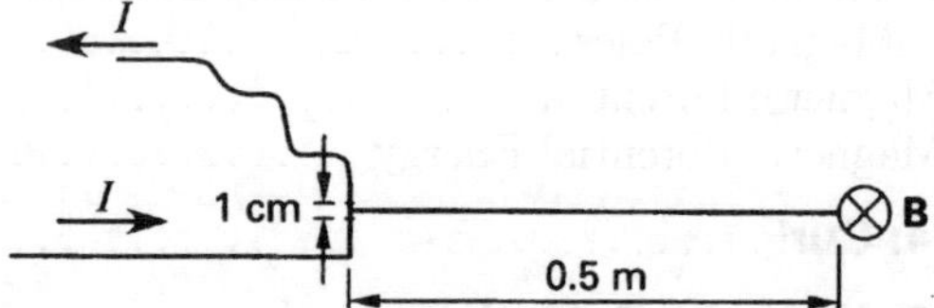

Solution

The magnetic flux density caused by the small segment, dl, 1 cm long is

$$\mathbf{B} = \left(\frac{\mu I l}{4\pi r^2}\right)\mathbf{a}$$

Because the segment is small compared to the distance, r, it is not necessary to integrate. (When to integrate, that is, when to use $d\mathbf{B}$ and dl, depends on the amount of error tolerable in the solution.) Substituting and assuming free-space conditions gives

$$\begin{aligned}\mathbf{B} &= \left(\frac{\mu I l}{4\pi r^2}\right)\mathbf{a} \\ &= \frac{\left(1.2566\times 10^{-6}\ \frac{\text{N}}{\text{A}^2}\right)(6\ \text{A})(1\ \text{cm})\mathbf{a}}{4\pi(0.5\ \text{m})^2\left(100\ \frac{\text{cm}}{\text{m}}\right)} \\ &= (2.4\times 10^{-8}\ \text{T})\ \mathbf{a}\end{aligned}$$

Using the right-hand rule, the direction of the **B**-field is into the paper, as shown.

Example 22.2

A loop of wire in an electric motor approximates a square with four conductors, each of length 0.15 m. Because no other condition is specified, free space is assumed. What is the total flux of the loop if the motor is rated to carry 15 A?

Solution

The applicable equation is

$$\Phi = \mu NIl = BA$$

Substituting known and given values results in

$$\Phi = \mu NIl = \left(1.2566 \times 10^{-6}\ \frac{\text{H}}{\text{m}}\right)(4)(15\text{ A})(0.15\text{ m})$$
$$= 1.1 \times 10^{-5}\text{ Wb}$$

1. BIOT-SAVART LAW

The Biot-Savart law is normally given as relating the force between two current elements as in Sec. 16.22. This is done to emphasize the similarity between electrostatics and magnetostatics and to exhibit the relationship between the Biot-Savart law and Coulomb's law in electromagnetics.

When considering **B** as the flux density, the Biot-Savart law is

$$d\mathbf{B} = \frac{\mu_0}{4\pi}\left(\frac{\mathbf{I} \times \mathbf{r}}{r^2}\right) dl \qquad 22.5$$

The term **r** is the unit vector pointing from the current element $\mathbf{I}dl$ to $d\mathbf{B}$.[4] The magnitude of Eq. 22.5 is given in Eq. 22.6.

$$dB = \frac{\mu_0}{4\pi}\left(\frac{Idl}{r^2}\right)\sin\theta \qquad 22.6$$

Comparing Eq. 22.6 with the equation for an electric field (see Eq. 16.6), the magnetic charge (if one existed) would have the form $q_m = Idl$.

Because the current element $\mathbf{I}dl$ is equivalent to the moving point charge $q\mathbf{v}$, the Biot-Savart law can be written as

$$\mathbf{B} = \frac{\mu_0}{4\pi}\left(\frac{q\mathbf{v} \times \mathbf{r}}{r^2}\right) \qquad 22.7$$

The cross product in Eq. 22.5 and Eq. 22.7 is accounted for when determining the direction of the magnetic field by using the *right-hand rule.* The right-hand rule states that when one's right thumb is pointed in the direction of the positive current flow, the fingers of the right hand curled inward toward the palm give the direction of the magnetic field. The rule can be verified by using the directions shown in Fig. 22.1.

[4]Importantly, the direction of $\mathbf{I}dl$ can be ascribed to the current and written as shown, or it can be ascribed to the differential length and be written as $Id\mathbf{l}$ or $\mathbf{I}d\mathbf{l}$.

2. FORCE ON A MOVING CHARGED PARTICLE

A magnetic field has no effect on stationary charged particles. If, however, the charge Q is moving with velocity **v** in an external magnetic field with flux density **B**, the force on the charge is

$$\mathbf{F} = Q\mathbf{v} \times \mathbf{B} \qquad 22.8$$

The magnetic field referred to in Eq. 22.8 is an external field, not the field created by the moving charged particle. The acceleration of the particle is given by Newton's second law.

The cross product gives the direction of the acceleration at an angle perpendicular to both the velocity **v** and the external magnetic field **B**. A charged particle in a uniform magnetic field will travel in a circular path with a radius and an angular velocity of

$$r = \frac{m\text{v}}{QB} \qquad 22.9$$

$$\omega = \frac{QB}{m} \qquad 22.10$$

Equation 22.8 is sometimes used to define the magnetic field. In doing so, the magnitude of Eq. 22.8 is written as

$$B = \frac{F}{q\text{v}} \qquad 22.11$$

Comparing Eq. 22.11 to Eq. 16.4 shows that just as the electric field can be defined in terms of the force per unit charge that is exerted at a point, the magnetic field can be defined as the force per unit moving charge that is exerted at a point.

Example 22.3

What is the maximum magnitude of the force on an electron moving at 0.1c (i.e., one-tenth the speed of light) in a uniform magnetic field of 5.0×10^{-5} T?

Solution

One-tenth the speed of light is the approximate maximum speed at which relativistic effects can be ignored. The electron is moving at $0.1c = 0.3 \times 10^8$ m/s. (The magnetic field strength is that of the earth's.) The applicable equation is

$$\mathbf{F} = Q\mathbf{v} \times \mathbf{B}$$

Because the maximum magnitude is to be determined, the following adjustments can be made.

$$F = |Q|\text{v}B\sin\theta = |Q|\text{v}B(1) = |Q|\text{v}B$$

Field Theory

Substitute the given values.

$$\begin{aligned} F &= |Q|vB \\ &= (1.6022 \times 10^{-19}\ \text{C})\left(0.3 \times 10^{8}\ \frac{\text{m}}{\text{s}}\right)(5.0 \times 10^{-5}\ \text{T}) \\ &= 2.4 \times 10^{-16}\ \text{N} \end{aligned}$$

3. FORCE ON CURRENT ELEMENTS

A current element $\mathbf{I}dl$ within a magnetic field $\mathbf{B}$ experiences a force of

$$d\mathbf{F} = \mathbf{I}dl \times \mathbf{B} \qquad 22.12$$

Because the term $\mathbf{I}dl$ is a mathematical construct and cannot exist alone but must instead be part of a complete loop or circuit, the force on the entire loop can be obtained from integrating around the loop as shown in Eq. 22.13.

$$\mathbf{F} = \oint \mathbf{I}dl \times \mathbf{B} \qquad 22.13$$

For a straight conductor of length l, the total force on a current-carrying conductor is

$$\mathbf{F} = \mathbf{I}l \times \mathbf{B} \qquad 22.14$$

In terms of magnitude, assuming a uniform magnetic field, Eq. 22.14 can be written as

$$F = lIB\sin\theta \qquad 22.15$$

The direction of the magnetic force can be ascertained by the left-hand *FBI rule* explained in Fig. 22.2.

For a straight, infinitely long conductor, the magnitude of the magnetic field is

$$B = \frac{\mu I}{2\pi r} \qquad 22.16$$

Two straight, infinitely long conductors situated within the magnetic fields of one another experience a force of

$$\frac{\mathbf{F}}{l} = \mathbf{I} \times \mathbf{B} = \left(\frac{\mu I_1 I_2}{2\pi r}\right)\mathbf{a} \qquad 22.17$$

The relationship given by Eq. 22.17 is useful in defining current in terms of force and distance, enabling the measurement of electrical current in terms of familiar mechanical variables. The distance, r, is perpendicular to the wires.

Example 22.4

The force on two parallel wires, approximated as infinite, is measured to be 30×10^{-3} N/m. The wires carry equal currents and repel one another. What are the magnitude and direction of the current in the two wires if the wires are 0.01 m apart?

Solution

The force per unit length is given. Assuming free space, the applicable equation is

$$\frac{\mathbf{F}}{l} = \mathbf{I} \times \mathbf{B} = \left(\frac{\mu I_1 I_2}{2\pi r}\right)\mathbf{a}$$

Determine the magnitude of the current by rearranging the equation to

$$\begin{aligned} I^2 &= \frac{2\pi\left(\frac{F}{l}\right)r}{\mu_0} = \frac{2\pi\left(30 \times 10^{-3}\ \frac{\text{N}}{\text{m}}\right)(0.01\ \text{m})}{1.2566 \times 10^{-6}\ \frac{\text{N}}{\text{A}^2}} \\ &= 1.5 \times 10^3\ \text{A}^2 \\ I &= \sqrt{1.5 \times 10^3\ \text{A}^2} = 38.7\ \text{A} \end{aligned}$$

The direction can be determined in a number of ways, such as by using the right-hand rule or by determining if the magnetic fields aid or oppose each other (that is, if they "link"). Using the equation, and knowing that the force must be in a direction that causes repulsion, the currents must be as shown.

$\mathbf{F}_1 = \mathbf{I}l_1 \times \mathbf{B}_2$ $\quad \mathbf{B}_1 \quad \mathbf{B}_2 \quad$ $\mathbf{F}_2 = \mathbf{I}l_2 \times \mathbf{B}_1$

The magnitude of the current is 38.7 A flowing in opposite directions in the two wires.

4. FORCE ON DISTRIBUTED CURRENT ELEMENTS

Equation 22.8 and Eq. 22.12 can be generalized to the case of force per unit volume.[5] In doing so, the force can be written as

$$\mathbf{F}_V = \mathbf{J} \times \mathbf{B} \qquad 22.18$$

PART 2: LORENTZ FORCE LAW

When an electric field and a magnetic field are present at the same time, the force on a charge is given by the general force law, commonly called the *Lorentz force law.* The Lorentz force combines the electrical and magnetic force and is given in Eq. 22.19.

$$\mathbf{F} = \mathbf{F}_e + \mathbf{F}_m = Q(\mathbf{E} + \mathbf{v} \times \mathbf{B}) \qquad 22.19$$

[5]This is useful for understanding, among other things, plasmas and convection or diffusion currents in semiconductors.

Figure 22.2 Magnetic Force Direction

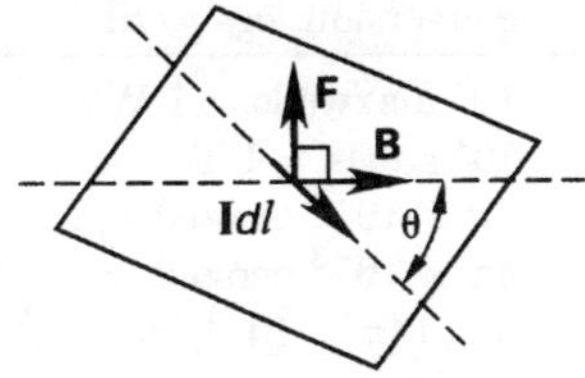

(a) magnetic force direction in terms of Idl

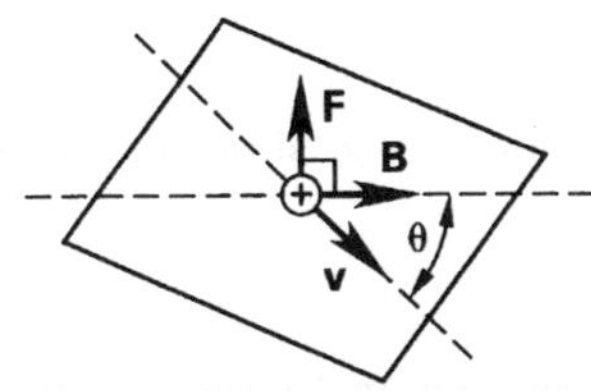

(b) magnetic force direction in terms of individual charge velocities

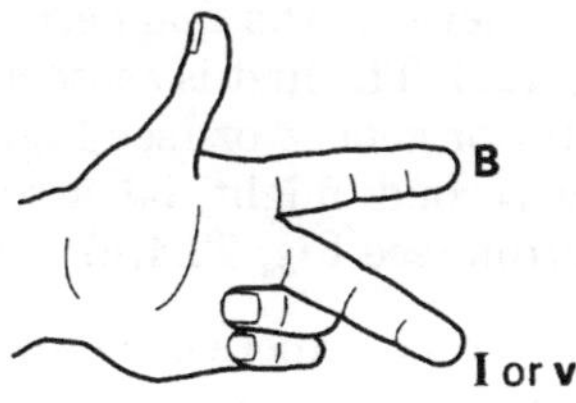

(c) FBI rule

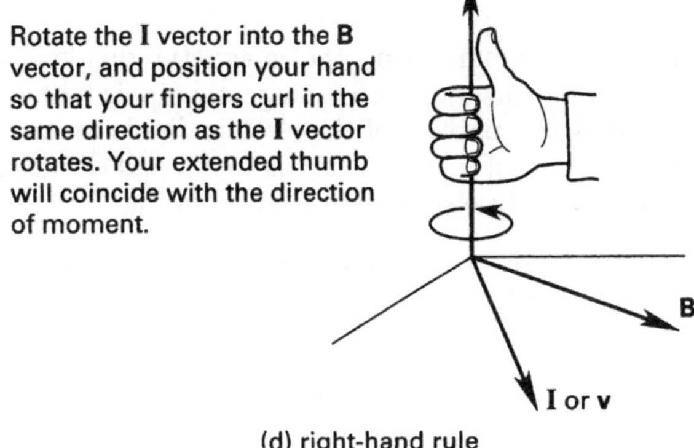

Rotate the I vector into the B vector, and position your hand so that your fingers curl in the same direction as the I vector rotates. Your extended thumb will coincide with the direction of moment.
F
B
I or v

(d) right-hand rule

Similarly, if a charge cloud of density ρ at each point exists in a volume, the Lorentz force is

$$d\mathbf{F} = \rho(\mathbf{E} + \mathbf{v} \times \mathbf{B})dV \qquad 22.20$$

PART 3: TRADITIONAL MAGNETISM

Traditional magnetism was based on concepts similar to those in electrostatics. As a result, the concept of a magnetic charge, q_m, was used, and magnets were described in terms of pole strength, p. The isolated magnetic charge, which does not exist in nature, is called a *monopole*. Nevertheless, the concept of an isolated magnetic charge at the end of a bar magnet is useful in understanding the effects of magnetism. Figure 22.3 illustrates this concept for a bar magnet (magnetic dipole) in an external magnetic field. The force on a magnetic charge is defined as the charge multiplied by the strength of the **B**-field as shown in Eq. 22.21. A magnetic field may, therefore, be defined in the same manner as an electric field; that is, the field represents the force exerted on an infinitesimal magnetic test charge at a point in space $(B = F/q_m)$.

$$\mathbf{F} = q_m\mathbf{B} \qquad 22.21$$

The torque exerted on the magnetic charge in an external magnetic field is

$$\mathbf{T} = q_m\mathbf{d} \times \mathbf{B} \qquad 22.22$$

The magnetic flux density, **B**, is also called the *intensity of magnetization* and the *dipole moment per unit volume*. For a bar magnet of pole strength p, length l, and cross-sectional area A, the magnetic flux density is

$$B = \frac{pl}{lA} = \frac{p}{A} = \frac{\mu m}{V} \qquad 22.23$$

The cgs system of units was the first to be used in the field of magnetism. This resulted in equations whose forms differ significantly from those in the SI system. To ensure consistency, exercise care when using equations in systems other than the SI system. A comparison of the cgs and SI system units for magnetism is given in Table 22.1.

Figure 22.3 Magnetic Charge in a Magnetic Field (Dipole in a Magnetic Field)

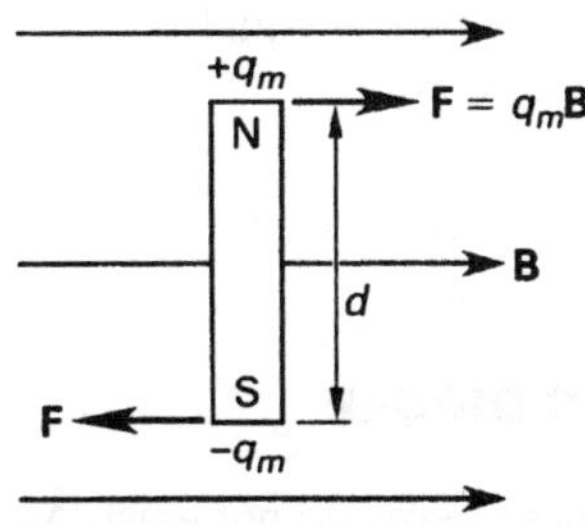

5. MAGNETIC POLE

Defining *pole strength* (not to be confused with the magnetic field strength) provides a method of calculating the force between two permanent magnetic poles analogous to Coulomb's law. The number of *unit-poles* or just *poles*, p, is numerically equal to the flux, ϕ,

Field Theory

Table 22.1 Magnetic Units

quantity	symbol	cgs units	SI units	conversion: cgs to SI
pole strength or flux	p or Φ or ϕ[a]	maxwells[b]	Wb	10^8 maxwells = 1 Wb
flux density	B	gauss[c]	T[d]	10^4 gauss = 1 T
field strength (intensity)	H	oersted	A/m[e]	$4\pi \times 10^{-3}$ oersted = 1 A/m
magnetization[f]	M	oersted	A/m[e]	$4\pi \times 10^{-3}$ oersted = 1 A/m
permeability	μ	gauss/oersted	H/m[g]	$\mu_{cgs}(4\pi \times 10^{-7}) = \mu_{mks}$[h]

[a]The SI symbol for magnetic flux is Φ. The symbol p is used when referring to magnetic poles. The symbol ϕ is a generic flux symbol used in many texts.
[b]A *maxwell* is a line of force.
[c]A gauss is equivalent to lines/cm^2 or maxwells/cm^2.
[d]A tesla is equivalent to Wb/m^2.
[e]A/m is equivalent to N/Wb.
[f]The units used in this row assume the following equation for flux density: $B = \mu_0 H + \mu_0 M$. Other forms of this equation exist. Use care in determining the correct units or in comparing values from different references.
[g]H/m is equivalent to Wb/A·m and Ω·s/m.
[h]The value of μ_{cgs} in the cgs system is 1.

emanating from the end of the magnet. (In the cgs system, the flux is numerically equal to $4\pi p$.) The area of the pole, A_p, and the flux density from the pole, B_p, are related to the pole strength by Eq. 22.24.

$$p = \phi = B_p A_p \qquad 22.24$$

The pole strength, as used here, and the magnetic charge used in Sec. 22.3 are related by Eq. 22.25.

$$q_m = \frac{p}{\mu} \qquad 22.25$$

In the cgs system, the value of the permeability of free space is 1. In free space, the magnetic charge and the pole strength can be used interchangeably.[6] In the SI system, this is not the case, and care must be exercised to ensure consistent units and terminology. In some SI texts, the pole strength is referred to as q_m.

The moving charge of modern magnetism in Eq. 22.11 is the magnetic charge of traditional magnetism in Eq. 22.21. Modern electrical engineering teaches that magnetism is caused by circulating currents (called *amperian currents*) and electron spins. The magnetic charge in such a context is equal to Idl. Equation 22.26 relates these concepts.

$$q_m = qv = Idl \qquad 22.26$$

6. MAGNETIC DIPOLE

Individual magnetic poles do not exist. Magnetic dipoles do exist and are the result of circulating currents. A small loop of circulating current has an associated *magnetic moment*, **m**, whose magnitude depends on the current flow and the area encompassed by that flow as given in Eq. 22.27. The direction of the magnetic moment is perpendicular to the plane of the current loop and is shown as the unit vector **n**.

$$\mathbf{m} = IA\mathbf{n} \qquad 22.27$$

Two views of the magnetic moment are illustrated in Fig. 22.4. The first is associated with a loop of wire in a motor or generator (see Fig. 22.4(a) and Fig. 22.4(b)). The second might be associated with a circulating electron (see Fig. 22.4(c)).

The torque experienced by the magnetic moment is given by Eq. 22.28.

$$\mathbf{T} = \mathbf{m} \times \mathbf{B} \qquad 22.28$$

In traditional magnetism, the magnetic dipole was viewed as a bar magnet with a magnetic dipole moment $p_m = q_m d$. A comparison of the electric dipole moment, the modern view of the magnetic moment, and the traditional magnetic dipole moment is given in Fig. 22.5.

The magnetic moment in terms of current loops, magnetic charge, and pole strength is

$$m = IA = q_m d = \left(\frac{p}{\mu}\right) l \qquad 22.29$$

Example 22.5

Determine the magnitude of the magnetic moment of a single electron at the first *Bohr radius* of 5.292×10^{-11} m if the current of the electron moving in the orbit is equivalent to approximately 1.05×10^{-3} A.

Solution

The magnitude of the magnetic moment, m, is

$$m = IA$$

[6]The permeability, μ, in the cgs system is equivalent to the relative permeability, μ_r, in the mks system.

Figure 22.4 *Magnetic Moment*

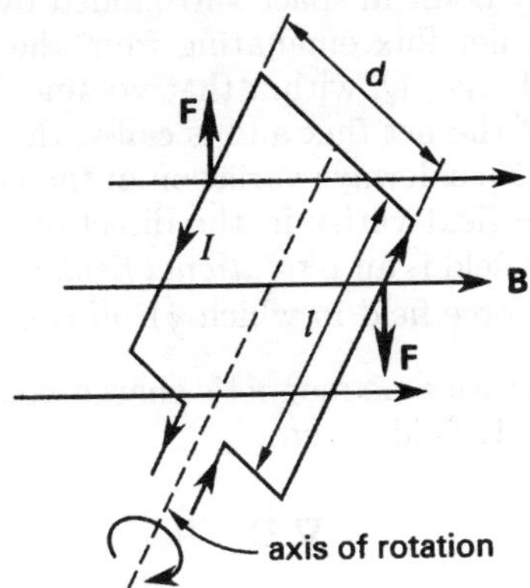

(a) loop of wire carrying current I and encompassing area $A = ld$

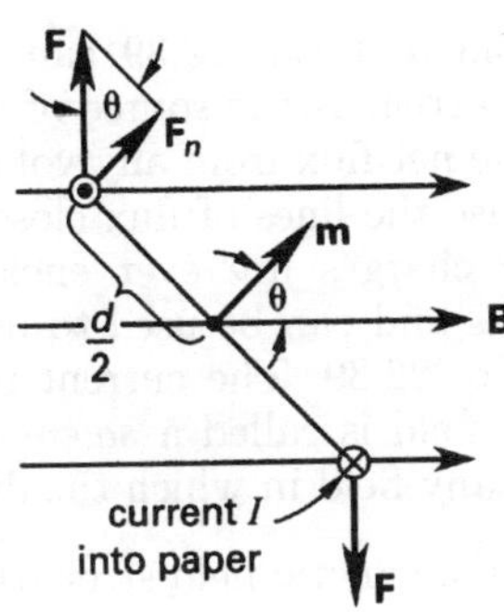

(b) cross section of loop showing the magnetic moment, **m**

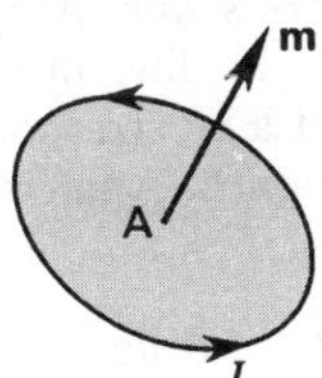

(c) generic view of a magnetic moment

The area correlating to the Bohr radius is

$$A = \pi r^2 = \pi(5.292 \times 10^{-11}\ \text{m})^2 = 8.798 \times 10^{-21}\ \text{m}^2$$

Substituting gives

$$m = IA = (1.05 \times 10^{-3}\ \text{A})(8.798 \times 10^{-21}\ \text{m}^2) = 9.24 \times 10^{-24}\ \text{A·m}^2$$

This is the approximate equivalent of one Bohr magneton, which equals $q_e h/4\pi m_e$.

Figure 22.5 *Dipole Moment Comparison*

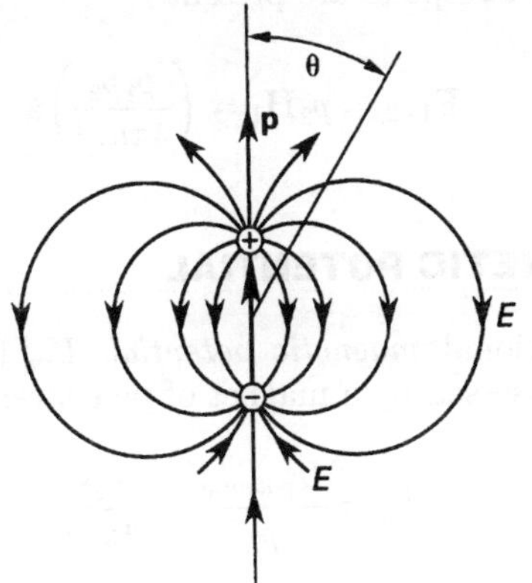

(a) electric dipole moment

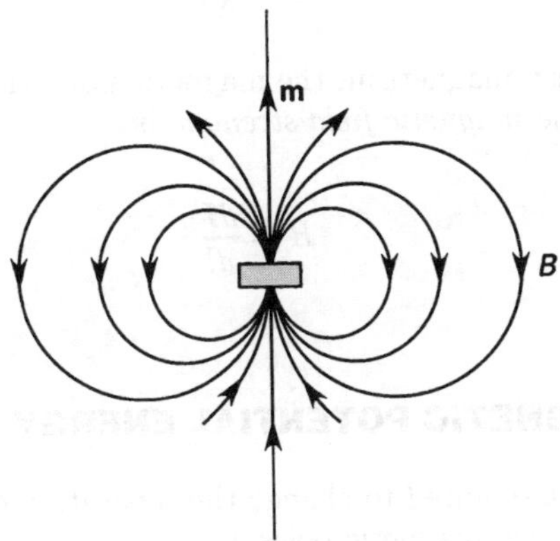

(b) magnetic dipole moment

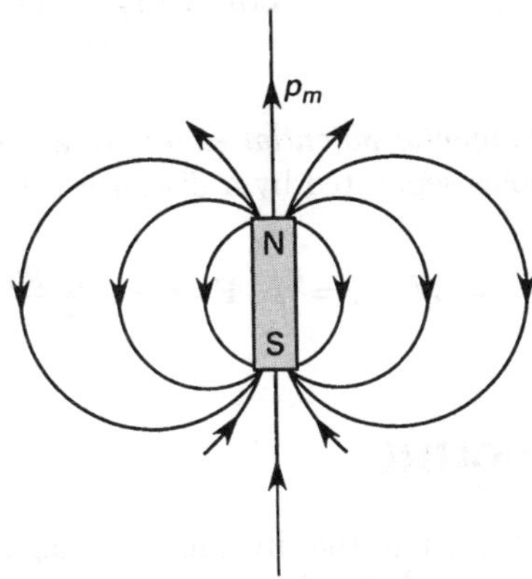

(c) traditional magnetic dipole moment

7. COULOMB'S LAW EQUIVALENT: FORCE BETWEEN MAGNETIC POLES

The force between the pole ends of two magnets separated by a distance, r, written similarly to Coulomb's law, is given by Eq. 22.30. The orientation of the magnets is not important. Vector addition is required if more than two poles are present.

$$\mathbf{F}_{1-2} = p_2\mathbf{H}_1 = \left(\frac{p_1 p_2}{4\pi\mu r^2}\right)\mathbf{a} \qquad 22.30$$

8. MAGNETIC POTENTIAL

The traditional *magnetic potential*, U_m (measured in amps), possessed by a magnet of pole strength, p_A, is

$$U_m = \frac{U_{\text{potential}}}{p_A} = \frac{-p_B}{4\pi\mu r} \qquad 22.31$$

The *magnetic potential gradient* is

$$H = \frac{-\Delta U_m}{\Delta r} \qquad 22.32$$

In modern magnetism, the magnetic potential gradient, called the *magnetic field strength*, is

$$H = \frac{dI}{dl} \qquad 22.33$$

9. MAGNETIC POTENTIAL ENERGY

The *work* required to change the separation distance, r, between two magnetic poles is

$$W = \int_{r_0}^{r_f} \mathbf{F}\cdot d\mathbf{r} = \left(\frac{-p_A p_B}{4\pi\mu}\right)\left(\frac{1}{r_f} - \frac{1}{r_0}\right) \qquad 22.34$$

The total *magnetic potential energy* in a system of two magnetic poles separated by a distance r is

$$U_{\text{potential}} = W_{\infty-r} = \int_{\infty}^{r} \mathbf{F}\cdot d\mathbf{r} = \frac{-p_A p_B}{4\pi\mu r} \qquad 22.35$$

PART 4: CURL

In Chap. 20, Part 3, the divergence of an electric field was found to equal the charge enclosed within the surface bounding the divergence. The applicable equation is repeated here for convenience. Equation 22.36 assumes the permittivity is constant (i.e., the medium is isotropic and linear).

$$\nabla\cdot\mathbf{E} = \frac{\rho}{\epsilon} \qquad 22.36$$

The curl of an electrostatic field is equal to zero.

$$\nabla\times\mathbf{E} = 0 \qquad 22.37$$

Equation 22.36 and Eq. 22.37 can be understood as follows: Any point in space surrounded by a closed surface with a net flux emanating from the volume must contain a charge, Q, within that volume. The charge is the cause of the net flux and is called the *scalar source* of the flux. No sideways variation in the field is present. That is, the field varies in the direction of the source. The electric field is an *irrotational field.* (An irrotational field is any force field in which $\oint\mathbf{F}\cdot dl = 0$.)

In magnetostatics, no magnetic charge exists. The divergence of the **B**-field is zero.

$$\nabla\cdot\mathbf{B} = 0 \qquad 22.38$$

The curl of the **B**-field in free space is

$$\nabla\times\mathbf{B} = \mu_0\mathbf{J} \qquad 22.39$$

Equation 22.38 and Eq. 22.39 can be understood as follows: The current is the source of the flux in a magnetic field. The net flux from any volume, however, will be zero, because the lines of flux close upon themselves (no magnetic charges are ever enclosed). A sidewise variation exists and can be used to measure the source as given by Eq. 22.39. The current is a vector source. The magnetic field is called a *solenoidal field.* (A solenoidal field is any field in which the divergence is zero.)

In terms of the magnetic field strength, Eq. 22.39 is

$$\nabla\times\mathbf{H} = \mathbf{J} \qquad 22.40$$

In integral form, the results of Eq. 22.39 and Eq. 22.40 are known as *Ampère's law.* Ampère's law for steady currents states that the line integral of the magnetic field strength around a closed loop is equal to the current enclosed by the path. Ampère's law is

$$\oint\mathbf{H}\cdot dl = \int_{\text{surface}} \mathbf{J}\cdot ds = I_{\text{enclosed}} \qquad 22.41$$

PART 5: MAGNETIC POTENTIAL

The difference in the magnetic potential between any two points in a magnetic field is given by the product of the magnetic field strength and the length of the path.

$$\int_0^f \mathbf{H}\cdot dl = \mathbf{U}_f - \mathbf{U}_0 \qquad 22.42$$

If the path is closed and no current is contained within it, no potential difference exists, as shown by Eq. 22.43.

$$\oint\mathbf{H}\cdot dl = 0 \qquad 22.43$$

In a closed path that encloses N turns carrying current I, the potential is

$$\oint \mathbf{H}\cdot d\mathbf{l} = NI = V_m \qquad 22.44$$

The term NI, given the symbol V_m, is called the *magnetomotance*. The magnetic potential and the magnetomotance are specified in terms of amperes or ampere-turns.

PART 6: MAGNETIC VECTOR POTENTIAL

The electric field intensity, **E**, was first obtained from known charge configurations. The concept of a potential was then developed, and **E** was determined from the negative gradient of the potential. As a starting point in electrical engineering problems, the potentials on the boundary conductors are known. Then Laplace's equations are used to determine the potential V. Once V is known, **E** and **D** can be determined. Similarly, a *vector magnetic potential*, **A**, can be defined to function in the same manner as electric potential in electric circuits.

To ensure the vector magnetic potential satisfies known conditions in magnetic fields, the following conditions must be met.

$$\nabla \times \mathbf{A} = \mathbf{B} \qquad 22.45$$

$$\nabla\cdot\mathbf{A} = 0 \qquad 22.46$$

The condition required by Eq. 22.46 is called the *gauge condition*, known in magnetostatics as the *Coulomb gauge*. Magnetic vector potential is measured in Wb/m or T·m.

The vector magnetic potential for three standard current configurations—a current filament (see Eq. 22.47), a sheet current (see Eq. 22.48), and a volume current (see Eq. 22.49)—follow.

$$\mathbf{A} = \oint \frac{\mu I d\mathbf{l}}{4\pi r} \qquad 22.47$$

$$\mathbf{A} = \int_S \frac{\mu \mathbf{K} ds}{4\pi r} \qquad 22.48$$

$$\mathbf{A} = \int_V \frac{\mu \mathbf{J} dV}{4\pi r} \qquad 22.49$$

At infinity, the magnetic vector potential is assumed to be zero. Equation 22.47 through Eq. 22.49 cannot be applied if the current distributions are infinite.

PART 7: WORK AND ENERGY

Magnetic forces on moving charge particles and current-carrying conductors result from the magnetic field. Any forces ($\mathbf{F}_a$) used to counter these must be applied from a source external to the magnetic field and be equal and opposite to the force of the field. If motion of the charged particle or current-carrying conductor occurs, the work is given by

$$W = \int_0^f \mathbf{F}_a\cdot d\mathbf{l} \qquad 22.50$$

A positive result from the initial position on the path (l_0) to the final position (l_f) indicates that work was done by an external force on the system; that is, energy was expended against the field. Because the magnetic force is generally not a conservative force field, the entire path of integration must be specified.

23 Magnetostatic Fields

Nomenclature

A	area	m^2
B, **B**	magnetic flux density	T (Wb/m^2)
H, **H**	magnetic field strength	A/m
I	constant or rms current	A
l	length	m
m, **m**	magnetic moment	$A{\cdot}m^2$ (J/T)
M, **M**	magnetization	A/m
N	number of turns	–
p_m	magnetic dipole moment	$A{\cdot}m^2$ (J/T)
P	polarization	C/m^2
Q	charge	C
q_m	magnetic charge	A·m (N/T)
r	radius	m
T	temperature	K
v	volume	m^3

Symbols

μ	permeability	N/A^2 (H/m)
μ_0	free-space permeability 1.2566×10^{-6}	N/A^2 (H/m)
χ	susceptibility	–

Subscripts

0	initial or free space
C	Curie
m	magnetic
r	relative

1. MAGNETIZATION

When a magnetic field is present in a magnetizable material, the magnetic moments tend to align themselves into the lowest energy configuration possible. In doing so, they increase the flux density, **B**. The alignment is referred to as *magnetization* of the material and is given the symbol **M**. The magnetization is the total magnetic moment, **m**, per unit volume, v.

$$\mathbf{M} = \frac{\mathbf{m}}{v} \qquad 23.1$$

The flux density is then given by the following general formula.[1]

$$\mathbf{B} = \mu_0\mathbf{H} + \mathbf{M} \qquad 23.2$$

Magnetization is technically defined as the magnetic polarization divided by the magnetic constant of the system used. In the SI system, the magnetic constant is the permeability of free space. Therefore, the equation becomes

$$\mathbf{B} = \mu_0(\mathbf{H} + \mathbf{M}) = \mu_0\mathbf{H} + \mu_0\mathbf{M} \qquad 23.3$$

The magnetic **B**-field has two sources. The first is the current, I, and the other is the magnetization, **M**. The current is the source of the magnetic field strength, **H**. The magnetic dipoles of individual atoms or molecules (which make up the larger domains) are the source of the magnetization, **M**.

Magnetization increases the flux density in magnetizable materials above the level that would exist under free-space conditions with the same magnetic field strength.[2] This occurs even when the magnetic moments of a material are not initially aligned. When an atom is placed within the influence of a magnetic field, its magnetic moment tends to align with the external magnetic field. The extent to which this occurs depends on the material. This is analogous to spring energy in mechanical systems. The work done in rotating the magnetic moment is recoverable upon removal of the magnetic field. The extent of the recovery again depends on the material.

The magnetic field and the magnetization can have different directions, though this is often not the case on a macroscopic level. In isotropic, linear material, **H** and **M** are parallel and related by

$$\mathbf{M} = \chi_m\mathbf{H} \qquad 23.4$$

The term *isotropic* indicates that the medium's properties are independent of direction. The term *linear* indicates that the permeability, μ, is constant regardless of

[1]Equation 23.2 is of a form similar to that used in electrostatic fields, but it cannot be used in the SI system as written. The study of magnetics developed mainly with the cgs system. In the cgs system $\mu_0 = 1$, but cgs was not a rationalized system, so $\mathbf{B} = \mathbf{H} + 4\pi\mathbf{M}$. (The terms **B**, **H**, and **M** are often parallel, and the equation is, therefore, written as $B = H + 4\pi M$.) The SI system is rationalized, but $\mu_0 = 4\pi \times 10^{-7}$ H/m.

[2]The magnetization, **M**, of free space is zero. A diamagnetic material's magnetization is negative and opposes the applied field. Diamagnetic materials are an exception to the statement in the text.

the magnitude of the current present. The term χ_m is the magnetic susceptibility. This dimensionless term measures the ease of magnetization in a material. For an isotropic linear material, Eq. 23.3 can be rewritten as

$$\mathbf{B} = \mu_0 (1 + \chi_m)\mathbf{H} \quad 23.5$$

Importantly, for ferromagnetic materials (those materials with large permeabilities), Eq. 23.4 and Eq. 23.5 do not hold true. (By contrast, Eq. 23.3 is always valid.) Ferromagnetic materials are nonlinear and exhibit the property of hysteresis. This nonlinearity is easily seen on the *BH* curve in Fig. 23.1, which displays typical hysteresis loops.[3] Because ferromagnetic materials are so important in engineering applications, many permeabilities are defined in the various linear portions of the *BH* curve—and utilized only in those linear regions.

Figure 23.1 *Hysteresis Loops*

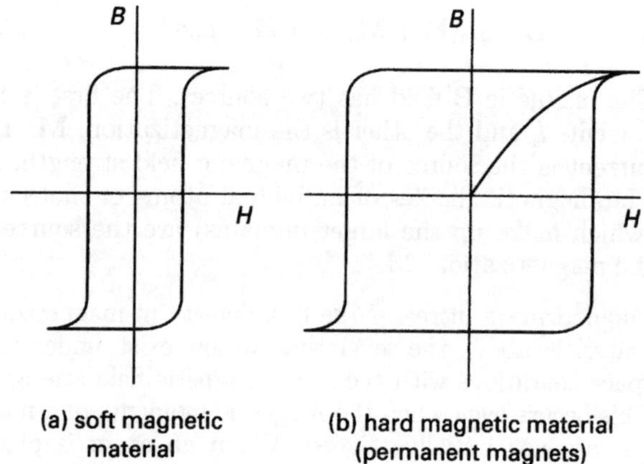

For those magnetic materials that are linear, and for ferromagnetic materials on linear portions of the *BH* curve, the relative permeability is

$$\mu_r = 1 + \chi_m \quad 23.6$$

In free space, no magnetization occurs because no magnetic materials are present, and the magnetic flux density, **B**, is given as

$$\mathbf{B} = \mu_0 \mathbf{H} \quad 23.7$$

If a magnetizable material is present, the flux density is increased and is given by Eq. 23.3. Importantly, Eq. 23.3 is often written without direct reference to the magnetization but rather in terms of the relative permeability. The relationship is shown explicitly in Eq. 23.8.

$$\mathbf{B} = \mu\mathbf{H} = \mu_0\mu_r\mathbf{H} = \mu_0\mathbf{H} + \mu_0\mathbf{M} \quad 23.8$$

[3] *MH* curves are not commonly plotted, though they are identical in shape to *BH* curves.

Example 23.1

The dimensions of a small magnet used in instrumentation are 1 cm × 1 cm × 0.3 cm. The volume is uniformly magnetized and possesses a measured flux density of 3×10^{-2} T and a magnetic moment of 0.01 A·m². What is the magnetization, **M**, of the magnet?

Solution

The magnetization is determined from Eq. 23.1.

$$\mathbf{M} = \frac{\mathbf{m}}{v}$$

The volume is determined from the dimensions.

$$v = (1 \text{ cm})(1 \text{ cm})(0.3 \text{ cm})\left(\frac{1}{100}\ \frac{\text{m}}{\text{cm}}\right)^3 = 0.3 \times 10^{-6} \text{ m}^3$$

Substituting gives

$$\mathbf{M} = \frac{\mathbf{m}}{v} = \frac{0.01 \text{ A·m}^2}{0.3 \times 10^{-6} \text{ m}^3} = 3.3 \times 10^4 \text{ A/m}$$

Example 23.2

What is the magnetic field strength in the magnet described in Ex. 23.1?

Solution

The magnetic field strength, **H**, is related to the flux density and magnetization in all cases.

$$\mathbf{B} = \mu_0(\mathbf{H} + \mathbf{M}) = \mu_0\mathbf{H} + \mu_0\mathbf{M}$$

Rearranging to solve for the desired quantity and substituting the given and calculated information gives

$$\mathbf{B} = \mu_0\mathbf{H} + \mu_0\mathbf{M}$$

Inside a magnet, the **B**-field and magnetization **M** are parallel, while the **H**-field is in opposition. Knowing this, one can change from vector quantities to magnitudes.

$$\begin{aligned} B &= \mu_0 H + \mu_0 M \\ H &= \frac{B - \mu_0 M}{\mu_0} \\ &= \frac{3 \times 10^{-2} \text{ T} - \left(1.2566 \times 10^{-6}\ \frac{\text{H}}{\text{m}}\right)\left(3.3 \times 10^4\ \frac{\text{A}}{\text{m}}\right)}{1.2566 \times 10^{-6}\ \frac{\text{H}}{\text{m}}} \\ &= -9.1 \times 10^3 \text{ A/m} \end{aligned}$$

Consider the following.[4]

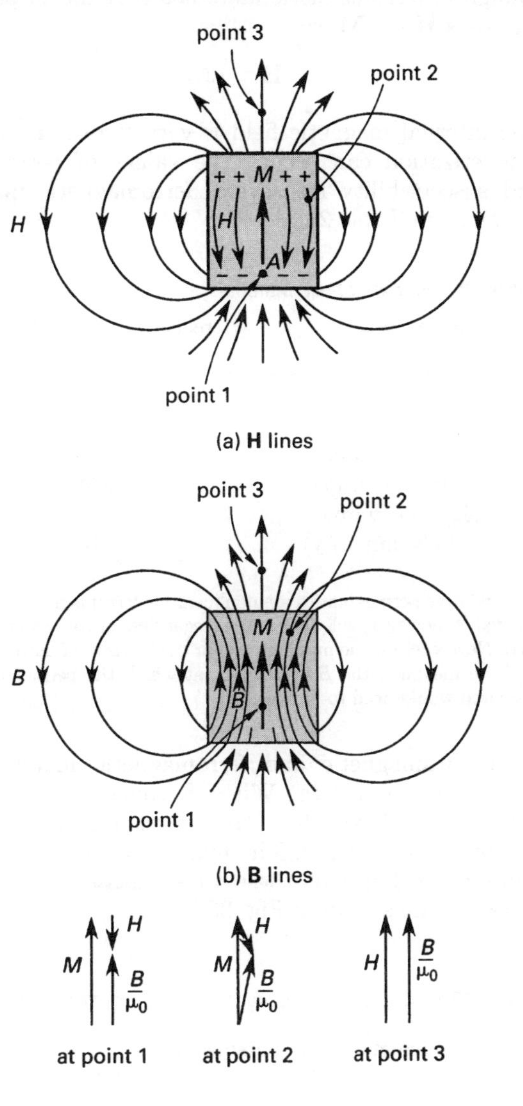

Outside the magnet, $\mathbf{B} = \mu_0\mathbf{H}$ and $\mathbf{M}$ equals zero. $\mathbf{B}$ and $\mathbf{H}$ are therefore parallel outside the magnet.

2. MAGNETIC FIELDS

All materials of concern to electronic engineers contain electrons in motion. In specific orbits, these electrons can be viewed as circulating currents. All materials then contain magnetic dipoles. Such dipoles can fully or partially align with a magnetic field. Parts (a) and (c) of Fig. 23.2 illustrate such an alignment.

[4]More explanation is possible but unnecessary for the majority of magnetic applications in electrical engineering.

***Figure 23.2** Permanent Magnet: Two Theoretical Views*

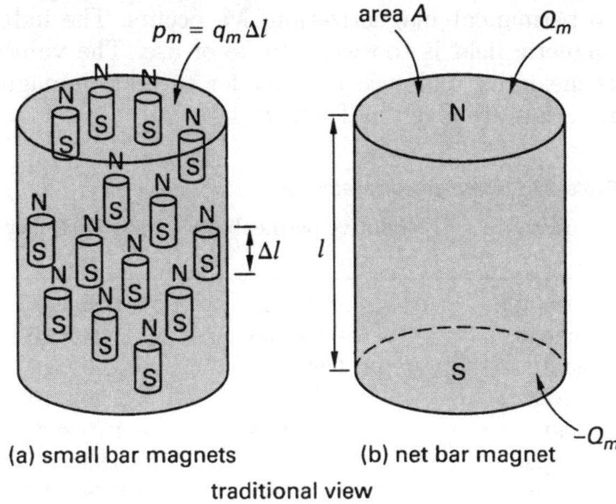

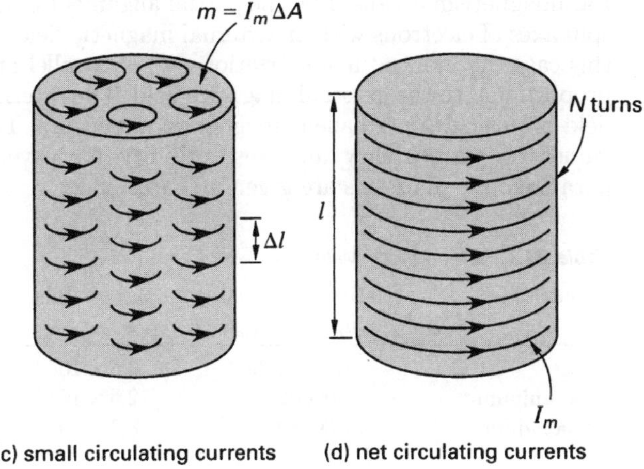

In addition to the movement of the electron in its orbit, the electron exhibits a property known as *spin*. The spin is the equivalent of the rotation of the electron about its axis. This rotation can be viewed as the equivalent of a circulating current and the source of an additional magnetic moment.

The magnetic moment of an atom may take two forms. First, the magnetic moments of all the electrons may be oriented such that they cancel, so that the magnet as a whole has no net magnetic moment. This condition leads to diamagnetism. Second, the cancellation of the electron moments, orbital and spin, may be partial, so that the atom is left with a net magnetic moment. Such an atom is referred to as a *magnetic atom*. This condition leads to several types of magnetism, including para-, ferro-, antiferro-, and ferrimagnetism.

Diamagnetism is exhibited by all materials and is caused by the orbital motion of the electrons. When an external magnetic field is applied, the magnetic moments of orbital pairs of electrons are unbalanced. Currents are induced in the circulating electrons in such a way as to create an internal magnetic field that opposes the external magnetic fields.[5] This is known as *antimagnetism.* No permanent magnetization, **M**, occurs. The induced magnetic field is too weak to be of use. The values of permeability and susceptibility for several diamagnetic materials are given in Table 23.1.

Table 23.1 *Diamagnetic Materials*

substance	relative permeability $\mu_r = 1+\chi_m$	susceptibility χ_m
bismuth	0.99983	-1.66×10^{-4}
mercury	0.999968	-3.20×10^{-5}
gold	0.999964	-3.60×10^{-5}
silver	0.99998	-2.60×10^{-5}
lead	0.999983	-1.70×10^{-5}
copper	0.999991	-0.98×10^{-5}
water	0.999991	-0.88×10^{-5}

Paramagnetism is caused by the partial alignment of the spin axes of electrons with an external magnetic field. In this case the induced magnetization, **M**, is parallel and proportional to the external magnetic field. The internal field is weak. No permanent magnetization occurs. The values of permeability and susceptibility for several paramagnetic materials are given in Table 23.2.

Table 23.2 *Paramagnetic Materials*

substance	relative permeability $\mu_r = 1+\chi_m$	susceptibility χ_m
air	1.00000036	3.6×10^{-7}
aluminum	1.000021	2.5×10^{-5}
palladium	1.00082	8.2×10^{-4}

The most important type of magnetism is ferromagnetism. Ferromagnetism is primarily caused by unpaired electron magnetic moments. The magnetic moments of individual atoms or molecules align themselves within small volumes called *domains.*[6] This effect occurs in iron, nickel, cobalt, and their alloys; in some manganese compounds; and in some rare earth elements. The domains can be completely aligned with weak external magnetic fields, resulting in very large magnetization, **M**. This effect is shown in Fig. 23.2. In part (a) of this figure, the individual magnetic moments (or groups of atoms in domains) are viewed as small bar magnets that have aligned with the external magnetic field (not shown). The net result is shown in part (b). Parts (a) and (b) illustrate the traditional view of magnetism. In the modern view, ferromagnetism is caused by circulating currents (both orbital and spin) as shown in part (c). Because of the cancellation of interior currents, the net result is a sheet of current on the surface of the cylinder as shown in part (d). The magnetization is significant enough in ferromagnetic materials that for all practical purposes $\mathbf{H} \ll \mathbf{M}$, so

$$\mathbf{B} \approx \mu\mathbf{M} \qquad 23.9$$

The internal magnetic field is very strong. Permanent magnetization can occur. The values of permeability and susceptibility for several ferromagnetic materials are given in Table 23.3.

Table 23.3 *Ferromagnetic Materials*

substance	relative permeability* $\mu_r = 1+\chi_m$
cobalt	250
nickel	600
commercial iron (0.2 impurity)	6000
high-purity iron (0.05 impurity)	2×10^5

*The relative permeabilities are nonlinear for ferromagnetic materials. The value, however, will not change, regardless of the system of units used. Nevertheless, the numbers are for comparison of materials only. To have meaning, the *B* or *H* value at which the permeability was measured would need to be specified.

Both paramagnetic and ferromagnetic materials are temperature sensitive. When ferromagnetic materials are heated above the *Curie temperature*, T_C, they become paramagnetic. The magnetization and temperature dependence of these three classes of magnetic materials are shown in Fig. 23.3.

Figure 23.3 *Magnetization Curves and Temperature Dependence*

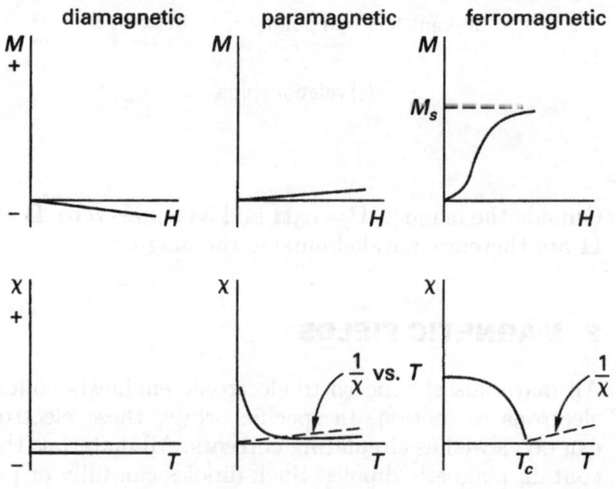

[5]This reaction, best known as *Lenz's law*, is the *counter-electromotive force* (emf) in motors and generators.

[6]Both paramagnetism and ferromagnetism occur as a result of unpaired electron moments. The difference is that ferromagnets exhibit domain formation caused by the quantum mechanical effect of *exchange forces.*

3. AUXILIARY FIELD H

The vector **H**, known as the magnetic field strength, is used to separate the sources of the **B**-field into those caused by external currents and those caused by the magnetization of the material itself. The magnetic field strength, **H**, represents the portion of the flux caused by external currents. The magnetization, **M**, represents the portion of the flux caused by the movement of magnetic dipoles within the material itself.

The magnetic field strength specifies the amount of force per unit flux produced by a conductor of N turns and length l carrying current I. After some manipulation, the magnitude of the magnetic field strength is given by

$$H = \frac{NI}{l} \qquad 23.10$$

Example 23.3

What is the magnetic field strength in a circular electrical device of radius 2.5 cm with 10 turns of wire carrying 30 A?

Solution

The magnetic field strength, H, which does not depend on the material present, is

$$H = \frac{NI}{l}$$

The path length is the circumference of a circle with a radius of 2.5 cm. That is,

$$l = 2\pi r = 2\pi(0.025 \text{ m}) = 0.16 \text{ m}$$

Substituting the given and the calculated gives

$$H = \frac{NI}{l} = \frac{(10)(30 \text{ A})}{0.16 \text{ m}} = 1.9 \times 10^3 \text{ A/m}$$

4. MAGNETIZATION IN NONMAGNETIC AND MAGNETIC MATERIALS

The use of the magnetic field strength, **H**, allows material media to be categorized into three groups. The magnetization vector, **M**, is avoided if possible (as is the polarization vector, **P**) to simplify calculations.

The first group consists of free space and nonmagnetic materials. In this group, the current is the only source of the magnetic field. As the magnetization is zero,

$$\mathbf{B} = \mu_0 \mathbf{H} \qquad 23.11$$

The second group consists of *soft ferromagnetic materials*. Such materials are suitable for linearization. That is, the current can be restricted to operate over a limited portion of the BH curve so that μ is constant. In this case, the governing equation is

$$\mathbf{B} = \mu \mathbf{H} = \mu_0 \mu_r \mathbf{H} \qquad 23.12$$

The magnetization is embedded in the relative permeability term.

The third group consists of the *hard ferromagnetic materials*. These materials are very nonlinear and are suitable for use as permanent magnets. Because of the nonlinearity, the relative permeability term cannot be used. Instead, B and H are given by graphs that are experimentally determined and are typically as shown in Fig. 23.1. Because M is significantly larger than H in such materials, B is related directly to M as indicated in the magnetization curves of Fig. 23.3.

Example 23.4

What magnitude of the magnetic flux density in air is associated with a magnetic field strength of 50 A/m?

Solution

The relative permeability, μ_r, of air (1.00000036) very closely approximates that of free space (1). When air is specified, μ_r can be assumed to be equal to one. Use Eq. 23.11.

$$B = \mu_0 H = \left(1.2566 \times 10^{-6} \ \frac{\text{H}}{\text{m}}\right)\left(50 \ \frac{\text{A}}{\text{m}}\right)$$
$$= 6.3 \times 10^{-5} \text{ T}$$

Example 23.5

What is the flux density if commercial iron is substituted for air in Ex. 23.4?

Solution

Commercial iron has a relative permeability of 6000. From Ex. 23.4, $\mu_0 H = 6.3 \times 10^{-5}$ T. Using these values and Eq. 23.12 gives

$$B_{\text{iron}} = \mu_r B_{\text{air}} = (6000)(6.3 \times 10^{-5} \text{ T}) = 0.38 \text{ T}$$

Field Theory

24 Electrodynamics

Nomenclature

a	unit vector	–
A	area	m^2
AW	atomic weight	g/mol
B, **B**	magnetic flux density	T (Wb/m^2)
c	speed of light, 2.9979×10^8	m/s
E, **E**	electric field strength	V/m
$\mathcal{E}$	electromotive force	V
F	Faraday constant, 96 485	C/mol
F, **F**	force	N
H, **H**	magnetic field strength	A/m
i	variable current	A
I, **I**	constant or rms current	A
J, **J**	current density	A/m^2
l, **l**	length	m
L	self-inductance	H
m	mass	g
N	number of turns	–
P_a, $\mathbf{P}_a$	apparent power	VA
$\mathcal{P}$	average Poynting vector	W/m^2
R	resistance	Ω
s	surface area	m^2
S	Poynting vector	W/m^2
t	time	s
v	velocity	m/s
v, **v**	variable voltage	V
V, **V**	constant or rms voltage	V
V_m	magnetic potential	A or A-turns
W_e	equivalent weight	–
z	valence change	–

Symbols

ϵ	permittivity	$C^2/N{\cdot}m$ (F/m)
μ	permeability	N/A^2 (H/m)
μ_0	free-space permeability, 1.2566×10^{-6}	N/A^2 (H/m)
σ	conductivity	S/m
ϕ	magnetic flux	Wb

Subscripts

ave	average
e	equivalent
g	grams
m	magnetic

1. ELECTROMOTIVE FORCE

It takes energy to drive a charge through a conducting medium. The source of the energy can be electromagnetic, chemical, or mechanical. Whatever the source, the net effect of the energy in any closed circuit is called the *electromotive force*, emf, or *electromotance*, and is given the symbol $\mathcal{E}$. It can be found from the line integral of the force over the path of the circuit.

$$\begin{aligned}\mathcal{E} &= \oint \mathbf{F}\cdot d\mathbf{l} = \oint \frac{\mathbf{J}}{\sigma}\cdot d\mathbf{l} \\ &= \oint \frac{I}{A\sigma}\,dl = I\oint \frac{1}{A\sigma}\,dl = IR\end{aligned} \qquad 24.1$$

The term R represents the total resistance in the external circuit. The force that moves the charges is the electric field, **E**. The electromotive force is measured in volts. In actuality, it is not a force, but the integral of the force per unit charge. The emf can be viewed as energy per unit charge, especially when considering that the unit for emf is the volt. The term electromotive force and the symbol $\mathcal{E}$ are reserved for the sources within an electric circuit, while v and V are used for the remaining components.[1]

Electromotive force can be generated by chemical means as in a battery, mechanically as in a piezoelectric crystal, by heat as in a thermocouple, or by many other methods. The two methods of considerable concern in electrical engineering are dynamic in nature.

The first is the *time-varying magnetic flux method*. The time-varying flux induces a voltage in any conductor present that depends on the number of turns, N, of the conductor and the *flux linkage*, $N\phi$ (the total number of flux lines cut). This is given by Eq. 24.2.

$$v = -N\frac{d\phi}{dt} \qquad 24.2$$

[1]Though this is generally true, in many instances the term *voltage* and its associated symbols (v, V) are used to indicate sources. It is understood that the voltage in those cases represents an electromotive force. Except where confusion may arise, this text will use v and V.

Field Theory

The minus sign is used to indicate that any current caused to flow by an induced electromotive force does so in a direction that generates a magnetic field opposing the flux that induced the emf. This is known as *Lenz's law*. The minus sign is not always shown. In circuit analysis, it is considered understood, though it is still used in equations written to obtain a numeric result. Equation 24.2 is often called *Faraday's law*, though in actuality it is the result of the law for time-varying fields.

Equation 24.2 holds true even when the time-varying magnetic flux in a given circuit element is changing because of the current in the element itself. Therefore, Eq. 24.2 can be written as follows.

$$v = -N\frac{d\phi}{di}\frac{di}{dt} = -L\frac{di}{dt} \qquad 24.3$$

The term L is known as *self-inductance*.

The second dynamic method of inducing an electromotive force is to maintain a constant flux and move the conductors. This is called the *flux-cutting method*. It results in an induced voltage, or electromotive force, of

$$V = -NBlv \qquad 24.4$$

2. FARADAY'S LAW OF ELECTROMAGNETIC INDUCTION

Faraday's law of electromagnetic induction states that the induced voltage is proportional to the rate of change of magnetic flux linked with the circuit. If the rate of change of magnetic flux linked is due to a time rate of change of the magnetic flux, that is, if the circuit (conductor) is stationary, then Faraday's law is

$$\begin{aligned} v &= \oint \mathbf{E}\cdot d\mathbf{l} = -\frac{d}{dt}\int_s \mathbf{B}\cdot ds \\ &= -\int_s \frac{\partial \mathbf{B}}{\partial t}\cdot ds = -\frac{d\phi}{dt} \end{aligned} \qquad 24.5$$

If the rate of change of the magnetic flux linked is due to the motion of the circuit (conductor) and the magnetic field is constant with respect to time, Faraday's law is

$$v = \oint \mathbf{E}\cdot d\mathbf{l} = \oint (\mathbf{v}\times\mathbf{B})\cdot d\mathbf{l} \qquad 24.6$$

The general case of Faraday's law when the magnetic flux is changing and the circuit (conductor) is in motion is

$$\begin{aligned} v &= \oint \mathbf{E}\cdot d\mathbf{l} = \oint (\mathbf{v}\times\mathbf{B})\cdot d\mathbf{l} \\ &= -\int_s \frac{\partial \mathbf{B}}{\partial t}\cdot ds \end{aligned} \qquad 24.7$$

Faraday's law applies to changing magnetic fields. In electrical engineering, the formulas of magnetostatics (Ampère's law, the Biot-Savart law, and others) are used to calculate the fields. While they give results that are only approximately correct, the errors are negligible. Even in the case where the current through a wire is interrupted by a fast-acting switch, or is simply cut, Ampère's law still holds true. Magnetostatic formulas can be used in conjunction with Faraday's law in *quasi-static conditions*. The magnetostatic equations must be abandoned only when the frequencies are extremely high, as in electromagnetic waves and radiation.

3. FARADAY'S LAW OF ELECTROLYSIS

Faraday's law of electrolysis states that the mass of substance deposited at one electrode or liberated at the other varies as in Eq. 24.8. F is the Faraday constant, equal to 96 485 C/mol.[2]

$$m_g = \frac{ItW_e}{F} \qquad 24.8$$

The term W_e is the equivalent weight of an element, the atomic weight in grams divided by the valence change that occurs during electrolysis, z.

$$W_e = \frac{\text{AW}}{z} \qquad 24.9$$

The atomic weight in the case of Faraday's law is given in grams. This is because one *faraday* of electricity will produce one gram of equivalent weight.

Example 24.1

The following electrolysis reaction can occur to copper.

$$\text{Cu}^{+2} + 2\text{e}^- \rightarrow \text{Cu}$$

What current is required to produce five grams of metallic copper (AW = 63.5) from a copper sulfate solution in 1 hr?

Solution

Because the change in the valence of copper is 2, the equivalent weight is

$$W_e = \frac{\text{AW}}{z} = \frac{63.5}{2} = 31.8$$

Rearranging Eq. 24.8 to determine the current required gives

$$m_g = \frac{ItW_e}{F}$$

$$\begin{aligned} I &= \frac{m_gF}{tW_e} = \frac{(5\text{ g})\left(96\,485\ \frac{\text{C}}{\text{mol}}\right)}{(1\text{ hr})\left(3600\ \frac{\text{sec}}{\text{hr}}\right)(31.8)} \\ &= 4.2\text{ A} \end{aligned}$$

[2]An electron has a charge of 1.6022×10^{-19} C. A *faraday* is the charge associated with one mole of electrons. Therefore, a faraday is

$$\begin{aligned} F &= \left(1.6022\times10^{-19}\ \frac{\text{C}}{\text{electron}}\right)(6.022\times10^{23}\text{ electrons}) \\ &= 96\,485\text{ C/mol} \end{aligned}$$

4. POTENTIAL

The electrostatic potential, or voltage, V, is a steady-state phenomenon that is well defined in space and is associated with a conservative electric field. The induced voltage, or potential, of Faraday's law, v, is time dependent and a multivalued function of position. It is associated with a nonconservative field, that of electromotive force.[3] When the potential is associated with a source of energy, the potential is called the electromotive force, or emf.

In magnetic circuits, the potential is called the *magnetomotive force*, mmf, or *magnetomotance*. Like its electric counterpart, mmf is actually not a force, but the line integral of the force around a path. In the magnetic case, it is the energy that moves the flux around the path. Hence, the mmf, or magnetic potential, is

$$\oint \mathbf{H}\cdot d\mathbf{l} = NI = V_m \qquad 24.10$$

5. ENERGY AND MOMENTUM: THE POYNTING VECTOR

Momentum conservation in electrodynamics occurs because the electric and magnetic fields themselves carry momentum. The conservation of energy in electrodynamics is given by *Poynting's theorem* as

$$\frac{dW}{dt} = -\frac{d}{dt}\int_V \tfrac{1}{2}(\epsilon E^2 + \mu H^2)dv - \oint_A (\mathbf{E}\times\mathbf{H})\cdot d\mathbf{A} \qquad 24.11$$

The first integral represents the stored energy in the electric and magnetic fields. The second integral represents the rate at which energy is transferred out of volume V across the surface boundary A. Poynting's theorem says that the work done by the electromotive force on the charges is equal to the decrease in the energy stored in the field, minus the energy that flows out of the surface.

The cross product of **E** and **H** is the Poynting vector, **S** (see Fig. 24.1). It represents the instantaneous power density (in W/m^2) carried by an electromagnetic wave. The direction of **S** is perpendicular to both **E** and **H**, which is normal to the wave front as shown in Fig. 24.1. (In Fig. 24.1, **E** and **B** are shown as the primary fields, consistent with these two vectors representing the fundamental fields.) The Poynting vector is given in Eq. 24.12. In Eq. 24.12, **a** is a unit vector in the direction of propagation of the wave, that is, normal to the wave front.

$$\mathbf{S} = c\epsilon E^2\mathbf{a} = c\mu H^2\mathbf{a} = \mathbf{E}\times\mathbf{H} \qquad 24.12$$

[3]A conservative field is one in which the work done in moving a charge depends only on its initial and final positions and not its path.

Figure 24.1 Poynting Vector

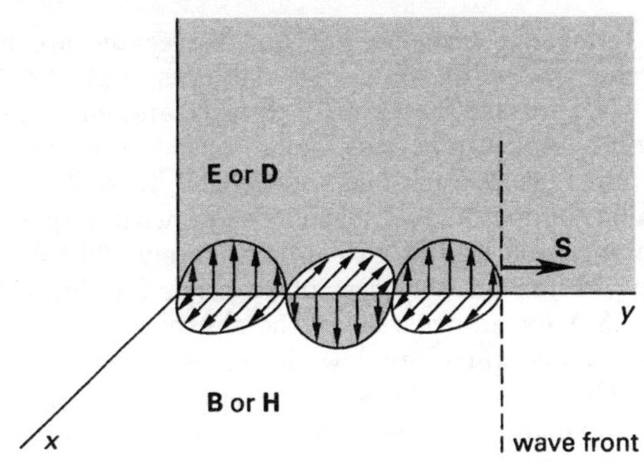

The notation Re(x) stands for the real part of the complex number x. The Poynting vector can be used to solve any problem involving the flow of electric or magnetic energy. The only restriction is that the electric and magnetic fields be related—from the same source. Though Eq. 24.12 represents the instantaneous value of the power density, allowing **E** and **H** to vary dynamically, with a time dependence of $e^{j\omega t}$, leads to a time-average of the Poynting vector, $\mathcal{P}_{\text{ave}}$, given by

$$\mathcal{P}_{\text{ave}} = \tfrac{1}{2}\text{Re}(\mathbf{E}\times\mathbf{H}^*) \qquad 24.13$$

The complex conjugate of **H** is represented by **H***. This follows the results of complex power analysis in which the apparent power is $\mathbf{P}_a = {}^1\!/_2\mathbf{VI}^*$, where the active power is the real part of the apparent power, that is, $\mathcal{P} = {}^1\!/_2\text{Re}\mathbf{VI}^*$.

The Poynting vector, **S**, describes the flow of energy, just as the current density, **J**, describes the flow of current.

6. ELECTROMAGNETIC WAVES

Stationary charges produce an electric field. If the charges are moving with a uniform velocity, a second effect occurs called *magnetism*. If the charges are accelerating, they produce a radiating electromagnetic field capable of transporting energy. The radiating electromagnetic field can be treated as a wave.[4]

The treatment of electrical phenomena as waves dominates electrical engineering fields where the transmission of electromagnetic energy takes place over long distances (often without metallic conductors), such as communications, power transmission, and related areas. In other electrical engineering fields, it is usually more convenient to work with the results, that is, potentials, currents, and resistances.

[4]Electromagnetic waves undergo all of the processes commonly affecting waves, that is, reflection, refraction, diffraction, wave interference, dispersion, and scattering.

7. WAVEGUIDES

The terms *transmission line* and *waveguide* are used interchangeably in some circumstances, but the two types of lines are functionally distinct and suitable for different applications (see Table 24.1).[5] Transmission lines are high power, long wavelength lines. They use metallic conductors to transmit energy with frequencies ranging from 0 Hz (DC) to approximately 500 Hz. (See Chap. 38 for an overview of transmission lines. See App. 18.A for an overview of the electromagnetic spectrum.) *Waveguides* are low power, short wavelength lines. They are used in power generation and distribution supporting subsystems. Waveguides use a variety of conductor types to transmit energy with frequencies ranging from 10^4 Hz (radio) to approximately 10^{15} Hz (optical). See Fig. 24.2 for an overview of waveguide types.[6]

Table 24.1 Transmission Line and Waveguide Comparison

	transmission line	waveguide
conductor number and type	two or more conductors separated by an insulating medium	one enclosed conductor filled with an insulating medium
wave types supported	transverse electromagnetic (TEM) waves	transverse electric (TE) and transverse magnetic (TM) waves
frequency parameters	no cutoff frequency	must operate within the frequency band that allows propagation
losses	significant losses at higher frequencies due to the skin effect	lower losses at high frequencies
power capability	high power capability	low power capability
analysis methods	standard electrical theory and methodology	Maxwell's equations

[5]Since waveguides are designed to carry low power, high frequency waves, power engineers deal with transmission lines more frequently than waveguides.

[6]Waveguide standards include those of the Electronic Industries Alliance (EIA), the International Electrotechnical Commission (IEC), Military Specifications, and the Radio Components Standardization Committee (RCSC), which is one of the original publishers of waveguide standards. EIA waveguides have designations beginning with WR, IEC waveguide designations begin with R, and RCSC waveguide designations begin with WG (waveguide).

Figure 24.2 Waveguide Types

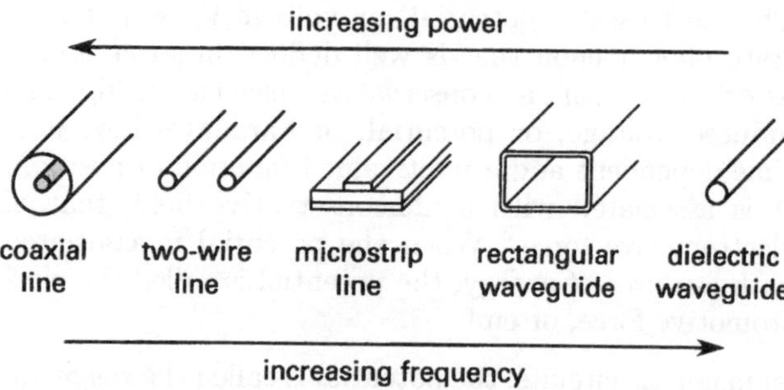

Generally, the opening or width of a waveguide is approximately the same as the distance of the wavelength to be transmitted. However, this is not always the case. For example, the wavelength of a 60 Hz wave is approximately 5×10^6 m, which is outside the transmission range of a waveguide.[7]

[7]At extremely low frequencies (ELFs), the earth-ionosphere boundary is one example of a waveguide for long range communications.

25 Maxwell's Equations

Nomenclature

A	area	m^2
$\mathbf{B}$	magnetic flux density	T or Wb/m^2
c	speed of light, 2.9979×10^8	m/s
$\mathbf{D}$	electric flux density	C/m^2
E, $\mathbf{E}$	electric field strength	V/m
$\mathcal{E}$	electromotive force	V
F_m	magnetomotive force	A or amp-turns
$\mathbf{F}$	force	N
G	conductance	S
H, $\mathbf{H}$	magnetic field strength	A/m
I	constant or rms current	A
$\mathbf{J}$	current density	A/m^2
l, $\mathbf{l}$	length	m
L	inductance	H
$\mathbf{M}$	magnetization	A/m
P_m	permeance	H
$\mathbf{P}$	polarization	C/m^2
Q	charge	C
R	resistance	Ω
$\mathcal{R}$	reluctance	1/H
$\mathbf{s}$	surface area	m^2
t	time	s
T	torque	N·m
U	energy density	J/m^3
v	variable voltage	V
v, $\mathbf{v}$	velocity	m/s
V	constant or rms voltage	V
V	volume	m^3
V_m	magnetic potential	A or amp-turns
W	energy	J
Z	intrinsic impedance	Ω
Z_0	characteristic impedance	Ω

Symbols

ϵ	permittivity	$C^2/N{\cdot}m^2$ or F/m
ϵ_0	free-space permittivity, 8.854×10^{-12}	$C^2/N{\cdot}m^2$ or F/m
μ	permeability	N/A^2 or H/m
μ_0	free-space permeability, 1.2566×10^{-6}	N/A^2 or H/m
ρ	charge density	C/m^3
ρ	resistivity	Ω·m
σ	conductivity	S/m
ϕ	magnetic flux	Wb
χ	susceptibility	–

Subscripts

c	conduction
e	electrical
m	magnetic
r	relative

1. MAXWELL'S EQUATIONS

Maxwell's equations are Gauss', Coulomb's, Faraday's, and Ampère's laws extended by the concept of a displacement current and the notion of a field. Though seldom used in the exact form presented in Table 25.1, they govern all of the electromagnetic phenomena in a medium that is stationary with respect to the coordinate system used.[1] They are valid for linear and nonlinear, isotropic and nonisotropic, homogeneous and nonhomogeneous media for frequencies ranging from zero to microwave frequencies, including many phenomena of light. The general set of Maxwell's equations is given in Table 25.1.

In free space, no charges exist ($\rho = 0$), and no conduction currents exist ($\mathbf{J}_c = 0$). Maxwell's equations then take on the forms given in Table 25.2.

The integral forms of Maxwell's equations are more fundamental in that they display the underlying physical laws. Maxwell's equations describe electromagnetic phenomena on a macroscopic (that is, nonquantum) level. Regions used in the integrals should be larger than atomic dimensions, and time intervals should be long enough to average atomic fluctuations.

2. ELECTROMAGNETIC FIELD VECTORS

The following vectors are electromagnetic field vectors: **E**, **D**, **B**, **H**, and **J**. These vectors define the properties of any medium in terms of the auxiliary equations summarized in Table 25.3.[2]

[1]If the medium is moving, the concepts of relativity must be invoked.
[2]These are auxiliary equations in the sense that they refine terms given in Maxwell's equations.

Table 25.1 Maxwell's Equations

integral form	point form	remarks
$\oint_s \mathbf{D}\cdot d\mathbf{s} = \int_V \rho\, dv$	$\nabla\cdot\mathbf{D} = \rho$	Gauss' law
$\oint_s \mathbf{B}\cdot d\mathbf{s} = 0$	$\nabla\cdot\mathbf{B} = 0$	nonexistence of magnetic monopoles
$\oint \mathbf{E}\cdot d\mathbf{l} = \int_s \left(\frac{-\partial\mathbf{B}}{\partial t}\right)\cdot d\mathbf{s}$	$\nabla\times\mathbf{E} = -\frac{\partial\mathbf{B}}{\partial t}$	Faraday's law
$\oint \mathbf{H}\cdot d\mathbf{l} = \int_s \left(\mathbf{J}_c + \frac{\partial\mathbf{D}}{\partial t}\right)\cdot d\mathbf{s}$	$\nabla\times\mathbf{H} = \mathbf{J}_c + \frac{\partial\mathbf{D}}{dt}$	Ampère's law

Table 25.2 Maxwell's Equations: Free-Space Form

integral form	point form
$\oint_s \mathbf{D}\cdot d\mathbf{s} = 0$	$\nabla\cdot\mathbf{D} = 0$
$\oint_s \mathbf{B}\cdot d\mathbf{s} = 0$	$\nabla\cdot\mathbf{B} = 0$
$\oint \mathbf{E}\cdot d\mathbf{l} = \int_s \left(\frac{-\partial\mathbf{B}}{\partial t}\right)\cdot d\mathbf{s}$	$\nabla\times\mathbf{E} = -\frac{\partial\mathbf{B}}{\partial t}$
$\oint \mathbf{H}\cdot d\mathbf{l} = \int_s \left(\frac{\partial\mathbf{D}}{\partial t}\right)\cdot d\mathbf{s}$	$\nabla\times\mathbf{H} = \frac{\partial\mathbf{D}}{dt}$

Table 25.3 Electromagnetic Field Vector Equations

$$\mathbf{D} = \epsilon\mathbf{E} = \epsilon_0\mathbf{E} + \mathbf{P} = \epsilon_0(1+\chi_e)\mathbf{E}$$
$$\mathbf{B} = \mu\mathbf{H} = \mu_0\mathbf{H} + \mu_0\mathbf{M} = \mu_0(1+\chi_m)\mathbf{H}$$
$$\mathbf{J} = \sigma\mathbf{E} = \rho\mathbf{v}$$

Even though the continuity equation is embedded in Maxwell's equation (as Gauss' and Ampère's laws), for completeness, one should list it separately along with the Lorentz force equation. The *continuity equation* is

$$\nabla\cdot\mathbf{J} = -\frac{\partial\rho}{\partial t} \qquad 25.1$$

The Lorentz force equation is repeated here for convenience.

$$\mathbf{F} = Q(\mathbf{E} + \mathbf{v}\times\mathbf{B}) \qquad 25.2$$

The electromagnetic field vector equations define the electromagnetic properties of the medium. The field vector equations, along with the continuity equation and the Lorentz force equation, are used as the practical applications of Maxwell's equations to the relationships between **E**, **D**, **B**, **H**, and **J**.

3. COMPARISON OF ELECTRIC AND MAGNETIC EQUATIONS

A comparison of electric and magnetic phenomena in terms of circuits is given in Table 25.4.

Table 25.4 Electric and Magnetic Circuit Analogies

electric	magnetic
emf $= V = IR$	mmf $= V_m = \phi\mathcal{R}$
current I	flux ϕ
emf $\mathcal{E}$ or V	mmf V_m
resistance $R = \rho l/A = l/\sigma A$	reluctance $\mathcal{R} = l/\mu A$
resistivity ρ	reluctivity $1/\mu$
conductance $G = 1/R$	permeance $P_m = \mu A/l$
conductivity $\sigma = 1/\rho$	permeability μ

A more extensive comparison of electric and magnetic equations and concepts is given in App. 25.A.

The similarities between electric and magnetic phenomena can also be examined by comparing convection current (see Sec. 16.18) to plasma flow (see Sec. 17.3), conduction current (see Sec. 16.20) to the magnetic circuit (see Sec. 16.32), displacement current (see Sec. 16.19) to the displacement magnetic effect (see Sec. 16.35), and electric field-induced current (see Sec. 16.17) to magnetic field-induced voltage (see Sec. 16.33).

4. CHARACTERISTIC IMPEDANCE

In power and communication engineering, the *characteristic impedance* is the impedance, when connected to the output terminals of a transmission line of any length, that makes the line appear to be infinitely long. Appearing *infinitely long* means that no standing waves exist on the line. Furthermore, the ratio of the voltage and the current is the same for each point along the line.

Characteristic impedance can be measured in a medium or in free space. *Free space*, sometimes referred to as a *classical vacuum*, is an ideal region of space where no particles of matter, no electromagnetic fields, and no gravitational fields exist (i.e., a perfect vacuum).[3] Free space is used as a standard reference medium when it is desirable to consider the properties and behavior of a wave without any boundary effects (e.g., reflection, refraction, or absorption). In free space, waves travel at the speed of light, the superposition principle is

[3]In communications engineering, free space is used to describe a region in the air that is high enough that an electromagnetic wave can travel unaffected by boundary effects, such as those caused by buildings, trees, hills, and the earth.

always exactly true, the permittivity is exactly the electric constant, the permeability is exactly the magnetic constant, and the characteristic impedance is equal to the impedance of free space.

In a medium with known relative permeability and relative permittivity, the *characteristic impedance*, also known as the *intrinsic impedance*, or, in transmission line terminology, the *surge impedance*, is

$$Z = \sqrt{\frac{\mu}{\epsilon}} = \sqrt{\frac{\mu_r \mu_0}{\epsilon_r \epsilon_0}} = Z_0 \sqrt{\frac{\mu_r}{\epsilon_r}} \qquad 25.3$$

The *characteristic impedance of free space*, Z_0, more correctly termed the *characteristic impedance of a vacuum*, is derived from Maxwell's equations and is given by

$$Z_0 = \frac{|\mathbf{E}|}{|\mathbf{H}|} = \frac{E}{H} = \sqrt{\frac{\mu_0}{\epsilon_0}} = \mu_0 c \qquad 25.4$$

The fact that the vacuum of outer space has an impedance does not imply that an electromagnetic wave traveling through it is somehow dissipated. The permeability of a vacuum, ϵ_0, is representative of the inductance, while the permittivity of a vacuum, μ_0, is representative of the capacitance. Therefore, space itself can absorb magnetic and electrical energies, but as in the case of an inductor and a capacitor, the energy is returned to the oscillating electromagnetic wave as it travels.

Example 25.1

What is the characteristic impedance of a vacuum?

Solution

Using Eq. 25.3, the characteristic impedance of a vacuum is

$$Z_0 = \sqrt{\frac{\mu_0}{\epsilon_0}} = \sqrt{\frac{1.2566 \times 10^{-6}\ \frac{\text{H}}{\text{m}}}{8.854 \times 10^{-12}\ \frac{\text{F}}{\text{m}}}}$$

$$= 376.7\ \Omega$$

Field Theory

Topic IV: Circuit Theory

Chapter

26 DC Circuit Fundamentals

Nomenclature

A	area	m^2
d	diameter	m
$\mathcal{E}$	electromotive force	V
G	conductance	S
i	variable current	A
I	constant or rms current	A
l	length	m
n	number of items	–
P	power	W
R	resistance	Ω
S	surface area	m^2
t	time	s
T	temperature	°C
v	variable voltage	V
V	constant or rms voltage	V
VR	voltage regulation	–
W	work	J

Symbols

α	thermal coefficient of resistivity	1/°C
ρ	resistivity	Ω·m
σ	conductivity	S/m

Subscripts

0	initial
cmil	circular mils
Cu	copper
e	equivalent
fl	full load
nl	no load
N	Norton
oc	open circuit
s	source
sc	short circuit
Th	Thevenin

1. VOLTAGE

Voltage is the energy, or work, per unit charge exerted in moving a charge from one position to another. One of the positions is a reference position and is given an arbitrary value of zero. Voltage is also referred to as the *potential difference* to distinguish it from the unit of potential difference, the volt. One volt, V, is the potential difference if one joule of energy moves one coulomb of charge from the reference position to another position. Therefore, 1 V = 1 J/C. Additional equivalent units are W/A, C/F, A/S, and Wb/s.

In practical terms, a potential of one volt is defined as the potential existing between two points of a conducting wire carrying a constant current, I, of one ampere when the power dissipated between these two points is one watt. When the voltage refers to a source of electrical energy, it is termed the *electromotive force* (emf) and sometimes given the symbol $\mathcal{E}$. In a *direct-current* (DC) *circuit*, the voltage, v, may vary in amplitude but not polarity. In many applications, the voltage magnitude, V, is constant as well. That is, it does not vary with time.

2. CURRENT

Current is the amount of charge transported past a given point per unit time. When current is constant or time-invariant, it is given the symbol I.[1] The unit of

[1]The symbol for current is derived from the French word *intensité*.

Circuit Theory

measure for current is the ampere (A), which is equivalent to coulombs per second (C/s). That is, one ampere is equal to the flow of one coulomb of charge past a plane surface, S, in one second. Though current is now known to be electron movement, it was originally viewed as flowing positive charges. This frame-of-reference current is called *conventional current* and is used in this and most other texts. When current refers to the actual flow of electrons, it is termed *electron current.* These concepts are illustrated in Fig. 26.1.

Figure 26.1 *Current Flow*

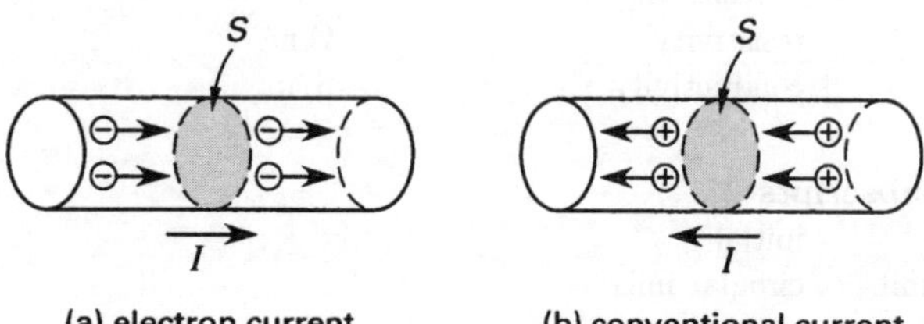

Current does not "flow," though this term is often used in conjunction with current. Additionally, the conservation of charge causes all circuit elements to be electrically neutral, meaning that no net positive or negative charge accumulates on any circuit element.

Example 26.1

A conductor has a constant current of 10 A. How many electrons pass a fixed plane in the conductor in 1 min?

Solution

First, using amperes, determine the charge flowing per minute.

$$10\ \text{A} = \left(10\ \frac{\text{C}}{\text{s}}\right)\left(60\ \frac{\text{s}}{\text{min}}\right) = 600\ \text{C/min}$$

Using the charge of an electron, compute the number of electrons flowing.

$$\frac{600\ \dfrac{\text{C}}{\text{min}}}{1.6022 \times 10^{-19}\ \dfrac{\text{C}}{\text{electron}}} = 3.75 \times 10^{21}\ \text{electrons/min}$$

3. RESISTANCE

Resistance, R, is the property of a circuit element that impedes current flow.[2] It is measured in ohms, Ω. A circuit with zero resistance is a *short circuit,* while a circuit with infinite resistance is an *open circuit.*

Resistors are often constructed from carbon compounds, ceramics, oxides, or coiled wire. Adjustable resistors are known as *potentiometers* and *rheostats.*

[2]Resistance is the real part of *impedance,* which opposes changes in current flow.

The resistance of any substance depends on its *resistivity* (ρ), its length, and its cross-sectional area. Assuming resistivity and cross-sectional area are constant, the resistance of a substance is

$$R = \frac{\rho l}{A} \qquad 26.1$$

Though SI system units are the standard, the units of area, A, are often put in the English Engineering System, especially in many tabulated values in the *National Electrical Code.* The area is given and tabulated for various conductors in *circular mils,* abbreviated cmil. One cmil is the area of a 0.001 in diameter circle. The concept of area in circular mils is represented by Eq. 26.2, Eq. 26.3, and Eq. 26.4.

$$A_{\text{cmil}} = \left(\frac{d_{\text{inches}}}{0.001}\right)^2 \qquad 26.2$$

$$A_{\text{in}^2} = 7.854 \times 10^{-7} \times A_{\text{cmil}} \qquad 26.3$$

$$A_{\text{cm}^2} = 5.067 \times 10^{-6} \times A_{\text{cmil}} \qquad 26.4$$

Resistivity is dependent on temperature. In most conductors, it increases with temperature, because electron movement through the lattice structure becomes increasingly difficult at higher temperatures. The variation of resistivity with temperature is specified by a *thermal coefficient of resistivity,* α, with typical units of 1/°C. The actual resistance or resistivity for a given temperature is calculated with Eq. 26.5 and Eq. 26.6.

$$R = R_0(1 + \alpha\Delta T) \qquad 26.5$$

$$\rho = \rho_0(1 + \alpha\Delta T) \qquad 26.6$$

The thermal coefficients for several common conducting materials are given in Table 26.1.

Example 26.2

What is the resistance of 305 m of 10 AWG (with an area of approximately 10,000 cmil) copper conductor with a resistivity of 2.82×10^{-6} Ω·cm?

Solution

From Eq. 26.1, and using the conversion of Eq. 26.4, the resistance is

$$\begin{aligned} R &= \frac{\rho l}{A} \\ &= \frac{(2.82 \times 10^{-6}\ \Omega\cdot\text{cm})(305\ \text{m})\left(100\ \dfrac{\text{cm}}{\text{m}}\right)}{(10{,}000\ \text{cmil})\left(5.067 \times 10^{-6}\ \dfrac{\text{cm}^2}{\text{cmil}}\right)} \\ &= 1.70\ \Omega \end{aligned}$$

Circuit Theory

Table 26.1 *Approximate Temperature Coefficients of Resistance*[a] *and Percent Conductivities*

material	α (1/°C)	conductivity (%)
aluminum, 99.5% pure	0.00423	63.0
aluminum, 97.5% pure	0.00435	59.8
constantan[b]	0.00001	3.1
copper, IACS (annealed)	0.00402	100.0
copper, pure annealed	0.00428	102.1
copper, hard drawn	0.00402	97.8
gold, 99.9% pure	0.00377	72.6
iron, pure	0.00625	17.5
iron wire, EBB[c]	0.00463	16.2
iron wire, BB[d]	0.00463	13.5
manganin[e]	0.00000	3.41
nickel	0.00622	12.9
platinum, pure	0.00367	14.6
silver, pure annealed	0.00400	108.8
steel wire	0.00463	11.6
tin, pure	0.00440	12.2
zinc, very pure	0.00406	27.7

(Multiply 1/°C by 0.5556 to obtain 1/°F.)

[a]between 0°C and 100°C
[b]58% Cu, 41% Ni, 1% Mn
[c]Extra Best Best grade
[d]Best Best grade
[e]84% Cu, 4% Ni, 12% Mn

4. CONDUCTANCE

The reciprocal of resistivity is *conductivity*, σ, measured in siemens per meter (S/m), often given in siemens per centimeter. The reciprocal of resistance is *conductance*, G, measured in siemens (S). Siemens is the metric unit for the older unit known as the *mho* (1/Ω).

$$\sigma = \frac{1}{\rho} \qquad 26.7$$

$$G = \frac{1}{R} \qquad 26.8$$

The *percent conductivity* is the ratio of a given substance's conductivity to the conductivity of standard IACS (International Annealed Copper Standard) copper, simply called *standard copper*. Alternatively, the percent conductivity is the ratio of standard copper's resistivity to the substance's resistivity.

$$\begin{aligned} \%\ \text{conductivity} &= \frac{\sigma}{\sigma_{\text{Cu}}} \times 100\% \\ &= \frac{\rho_{\text{Cu}}}{\rho} \times 100\% \end{aligned} \qquad 26.9$$

The standard resistivity of copper at 20°C is approximately

$$\begin{aligned} \rho_{\text{Cu,20°C}} &= 1.7241 \times 10^{-6}\ \Omega\text{·cm} \\ &= 0.3403\ \Omega\text{·cmil/cm} \end{aligned} \qquad 26.10$$

The percent conductivities for various substances are given in Table 26.1.

5. OHM'S LAW

Voltage, current, and resistance are related by Ohm's law.[3] The numerical result of Ohm's law is called the *voltage drop* or the *IR drop*.[4]

$$V = IR \qquad 26.11$$

Ohm's law presupposes a *linear circuit*, that is, a circuit consisting of linear elements and linear sources. In mathematical terms, this means that voltage plotted against current will result in a straight line with the slope represented by resistance. A *linear element* is a passive element, such as a resistor, whose performance can be portrayed by a linear voltage-current relationship. A *linear source* is one whose output is proportional to the first power of the voltage or current in the circuit. Many elements are linear or can be represented by equivalent linear circuits over some portion of their operation. Most sources, though not linear, can be represented as ideal sources with resistors in series or parallel to account for the nonlinearity.

6. POWER

If a steady current and voltage produce work, W, in time interval t, the electric power (energy conversion rate) is

$$\begin{aligned} P &= \frac{W}{t} = \frac{VIt}{t} \\ &= VI = \frac{V^2}{R} \end{aligned} \qquad 26.12$$

If the voltage and current vary with time, Eq. 26.12 still applies with these terms expressed as v and i. Equation 26.12 then represents the instantaneous power. The same is true for Eq. 26.13.

Equation 26.13 represents a form of power sometimes referred to as *I squared R* (I^2R) *losses*. Equation 26.13 is also the mathematical statement of *Joule's law of heating effect*.

$$P = I^2R \qquad 26.13$$

[3]This book uses the convention that uppercase letters represent fixed, maximum, effective values or direct current (DC) values; and that lowercase letters represent values that change with time, such as alternating current (AC) values. Direct current (DC) values can change in amplitude over time (but not polarity or direction), and for this reason are sometimes shown as lowercase letters.

[4]It is sometimes helpful to consider electrical problems and theories in terms of their mechanical analogs. The mechanical analogy to Ohm's law in terms of fluid flow is as follows: Voltage is the pressure, current is the flow, and resistance is the head loss caused by friction and restrictions within the system.

7. DECIBELS

Power changes in many circuits range over decades, that is, over several orders of magnitude. As a result, decibels are adopted to express such changes. The use of decibels also has an advantage in that the power gain (or loss) of cascaded stages in series is the sum of the individual stage decibel gains (or losses), making the mathematics easier. Strictly speaking, decibels refer to a power ratio with the denominator arbitrarily chosen as a particular value, P_0. The reference value changes, depending upon the specific area of electrical engineering usage. For example, communications and acoustics use different references.[5] Decibels have come into common usage for referring to voltage and current ratios as well, though an accurate comparison between circuits requires the two terminals being compared to have equivalent resistances.

$$\begin{aligned}\text{ratio (in dB)} &= 10\log_{10}\frac{P}{P_0} \\ &\approx 20\log_{10}\frac{V_2}{V_1} \\ &= 20\log_{10}\frac{I_2}{I_1}\end{aligned} \qquad 26.14$$

Time-varying quantities can also be in the ratio. Care must be taken to compare only instantaneous values, or effective (rms) or time-averaged values to values of the same type.

8. ENERGY SOURCES

Sources of electrical energy include friction between dissimilar substances, contact between dissimilar substances, thermoelectric action (for example, the Thomson, Peltier, and Seebeck effects), the Hall effect, electromagnetic induction, the photoelectric effect, and chemical action. These sources of energy manifest them selves by the potential (V) they generate. When sources of energy are processed through certain electronic circuits, a current source can be created.

An *ideal voltage source* supplies power at a constant voltage, regardless of the current drawn. An *ideal current source* supplies power in terms of a constant current, regardless of the voltage between its terminals. However, real sources have internal resistances that, at higher currents, reduce the available voltage. Consequently, a *real voltage source* cannot maintain a constant voltage when the currents are large. A *real current source* cannot maintain a constant current completely independent of the voltage at its terminals. Real and ideal voltage and current sources are shown in electrical schematic form in Fig. 26.2.

[5]Communications uses many such reference values. One references all power changes to 1 kW. In audio acoustics, the reference is the power necessary to cause the minimum sound pressure level audible to the human ear at 2000 Hz, that is, a pressure of 20 μPa.

Figure 26.2 *Ideal and Real Energy Sources*

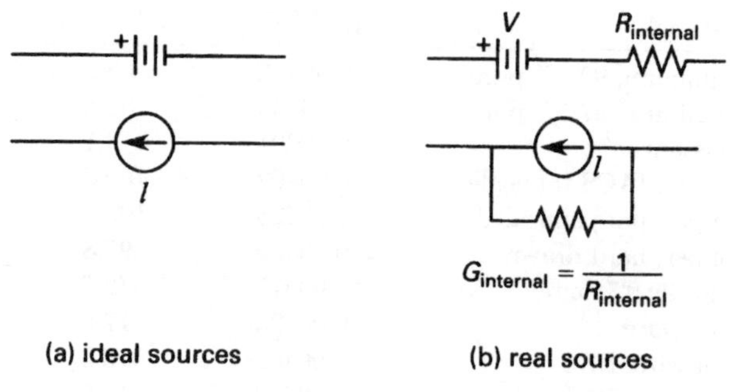

The change in the output of a voltage source is measured as its *voltage regulation*, VR, given by

$$\text{VR} = \frac{V_{\text{nl}} - V_{\text{fl}}}{V_{\text{fl}}} \times 100\% \qquad 26.15$$

Independent sources deliver voltage and current at their rated values regardless of circuit parameters. *Dependent sources* deliver voltage and current at levels determined by a voltage or current somewhere else in the circuit.

9. VOLTAGE SOURCES IN SERIES AND PARALLEL

Voltage sources connected in series, as in Fig. 26.3(a), can be reduced to an equivalent circuit with a single voltage source and a single resistance, as in Fig. 26.3(b). The equivalent voltage and resistance are calculated per Eq. 26.16 and Eq. 26.17.

$$V_e = \sum V_i \qquad 26.16$$

$$R_e = \sum R_i \qquad 26.17$$

Millman's theorem states that n identical sources of voltage V and resistance R connected in parallel, as in Fig. 26.3(c), can be reduced to an equivalent circuit with a single voltage source and a single resistance, as in Fig. 26.3(b). The equivalent voltage and resistance are calculated per Eq. 26.18 and Eq. 26.19.

$$V_e = V \qquad 26.18$$

$$R_e = \frac{R}{n} \qquad 26.19$$

Nonidentical sources can be connected in parallel, but the lower-voltage sources may be "charged" by the higher-voltage sources. That is, the current may enter the positive terminal of the source. A loop current analysis is needed to determine the current direction and magnitude through the voltage sources.

Figure 26.3 *Voltage Sources in Series and Parallel*

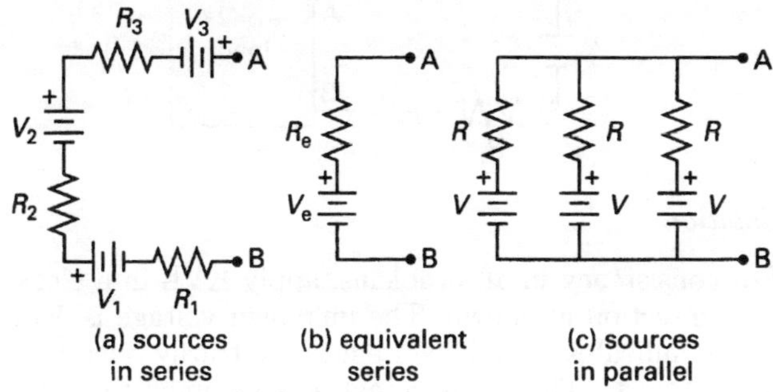

10. CURRENT SOURCES IN SERIES AND PARALLEL

Current sources may be placed in parallel. The circuit is then analyzed using any applicable method to determine the power delivered to the various elements. Current sources of differing magnitude cannot be placed in series without damaging one of them.

11. SOURCE TRANSFORMATIONS

A voltage source of V_s with an internal series resistance of R_s can be represented by an equivalent circuit with a current source of I_s with an internal parallel resistance of the same R_s, and vice versa. Consequently, if Eq. 26.20 is valid, the circuits in Fig. 26.4 are equivalent.

$$V_s = I_s R_s \qquad 26.20$$

Figure 26.4 *Equivalent Sources*

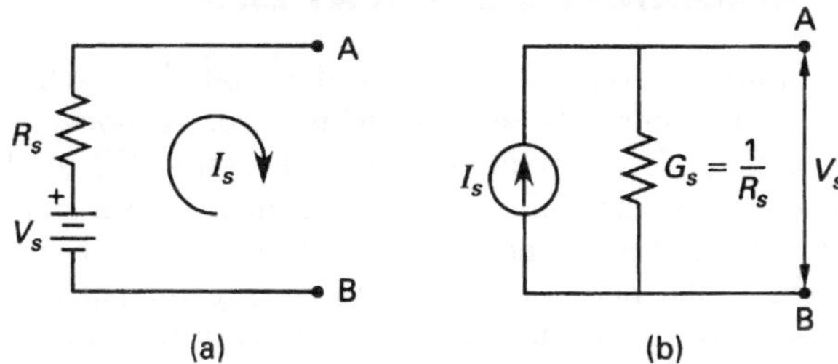

12. MAXIMUM ENERGY TRANSFER

The maximum energy transfer, that is, the maximum power transfer, from a voltage source is attained when the series source resistance (R_s) of Fig. 26.4(a) is reduced to the minimum possible value, with zero being the ideal case. This assumes the load resistance is fixed and the source resistance can be changed. Though this is the ideal situation, it is not often the case. Where the load resistance varies and the source resistance is fixed, maximum power transfer occurs when the load and source resistances are equal.

Similarly, the maximum power transfer from a current source occurs when the parallel source resistance (R_s) of Fig. 26.4(b) is increased to the maximum possible value. This assumes the load resistance is fixed and the source resistance can be changed. In the more common case, where the load resistance varies and the source resistance is fixed, the maximum power transfer occurs when the load and source resistances are equal.

The maximum power transfer occurs in any circuit when the load resistance equals the Norton or Thevenin equivalent resistance.

13. VOLTAGE AND CURRENT DIVIDERS

At times, a source voltage will not be at the required value for the operation of a given circuit. For example, the standard automobile battery voltage is 12 V, while some electronic circuitry uses 8 V for DC biasing. One method of obtaining the required voltage is by use of a circuit referred to as a *voltage divider*. A voltage divider is illustrated in Fig. 26.5(a). The voltage across resistor 2 is given by

$$V_2 = V_s\left(\frac{R_2}{R_1 + R_2}\right) \qquad 26.21$$

If the fraction of the source voltage, V_s, is found as a function of V_2, the result is known as the *gain* or the *voltage-ratio transfer function*. Rearranging Eq. 26.21 to show the specific relationship between R_1 and R_2 gives

$$\frac{V_2}{V_s} = \frac{1}{1 + \frac{R_1}{R_2}} \qquad 26.22$$

An analogous circuit called a *current divider* can be used to produce a specific current. A current divider circuit is shown in Fig. 26.5(b). The current through resistor 2 is

$$I_2 = I_s\left(\frac{R_1}{R_1 + R_2}\right) = I_s\left(\frac{G_2}{G_1 + G_2}\right) \qquad 26.23$$

Figure 26.5 *Divider Circuits*

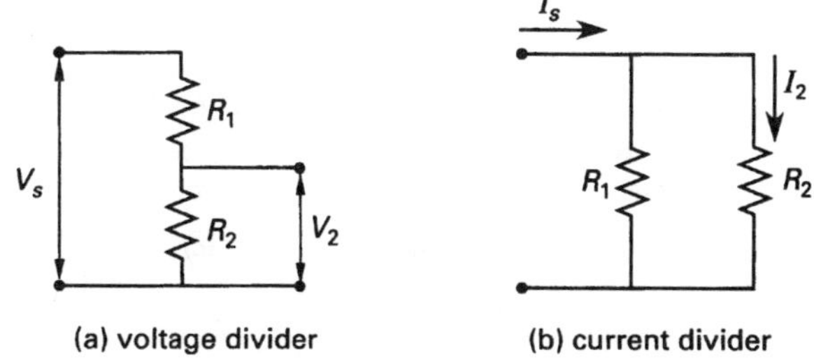

14. KIRCHHOFF'S LAWS

A fundamental principle in the design and analysis of electrical circuits is that the dimensions of the circuit are small. That is, the circuit's dimensions are small when compared to the wavelengths of the electromagnetic quantities that pass within them. Only time variations need be considered, not spatial variations. Consequently, Maxwell's equations, which are partial integro-differential equations, become ordinary integro-differential equations that vary only with time. An *integro-differential equation* contains both integrals and differentials.

The net result is that Gauss' law, in integral form, reduces to Kirchhoff's current law. This occurs because the net charge enclosed in any particular volume of the circuit is zero and, therefore, the sum of the currents at any point must be zero.

Another result is that Faraday's law of electromagnetic induction reduces to Kirchhoff's voltage law. This is because, with a "small" circuit, the surface integral of magnetic flux density is zero; the sum of the voltages around any closed loop must be zero.

15. KIRCHHOFF'S VOLTAGE LAW

Kirchhoff's voltage law (KVL) states that the algebraic sum of the voltages around any closed path within a circuit or network is zero. Some of the voltages may be sources, while others will be voltages caused by current in passive elements. Voltages through the passive elements are often called *voltage drops*. A *voltage rise* may also appear to occur across passive elements in the initial stages of the application of Kirchhoff's voltage law. For resistors, this is only because the direction of the current is arbitrarily chosen. For inductors and capacitors, a voltage rise indicates the release of magnetic or electrical energy. Stated in terms of the voltage rises and drops, KVL is

$$\sum_{\text{loop}} \text{voltage rises} = \sum_{\text{loop}} \text{voltage drops} \qquad 26.24$$

Kirchhoff's voltage law can also be stated as follows: The sum of the voltages around a closed loop must equal zero. The reference directions for voltage rises and drops used in this book are shown in Fig. 26.6.[6]

Figure 26.6 Voltage Reference Directions

rise (a) rise (b) drop (c)

Example 26.3

Consider the following circuit. Determine the expression for V_{AB} across the load in the circuit in the figure.

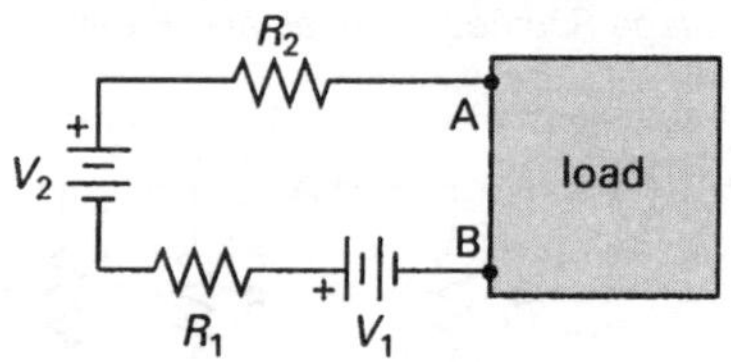

Solution

For consistency in all problems, apply KVL in a clockwise direction as shown. The unknown voltage is V_{AB}, which indicates that the voltage is arbitrarily referenced from A to B. That is, terminal A is assumed to be the high potential (+). Assign polarities to the resistors based on the assumed direction.

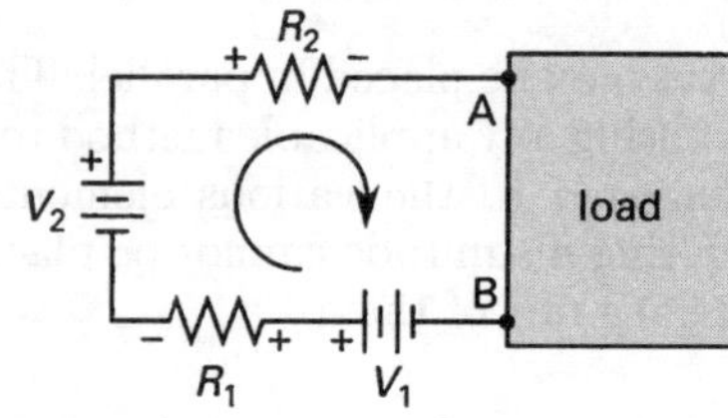

The KVL expression for the loop, clockwise starting at terminal B, is

$$V_1 - V_{R_1} + V_2 - V_{R_2} - V_{AB} = 0$$

Rearrange and solve for V_{AB}.

$$V_{AB} = V_1 - V_{R_1} + V_2 - V_{R_2}$$

16. KIRCHHOFF'S CURRENT LAW

Kirchhoff's current law (KCL) states that the algebraic sum of the currents at a node is zero. A *node* is a connection of two or more circuit elements. When the connection is between two elements, the node is a *simple node*. When the connection is between three or more elements, the node is referred to as a *principal node*. Stated in terms of the currents directed into and out of the node, KCL is

$$\sum_{\text{node}} \text{currents in} = \sum_{\text{node}} \text{currents out} \qquad 26.25$$

Kirchhoff's current law can also be stated as follows: The sum of the currents flowing out of a node must equal zero.[7] The reference directions for positive currents (currents "out") and negative currents (currents "in") used in this book are illustrated in Fig. 26.7.

[6]This alternative statement of Kirchhoff's voltage law and the arbitrary directions chosen are sometimes useful in simplifying the writing of the mathematical equations used in applying KVL.

[7]This alternative statement of Kirchhoff's current law and the arbitrary directions chosen are sometimes useful in simplifying the writing of the mathematical equations used in applying KCL.

Figure 26.7 *Current Reference Directions*

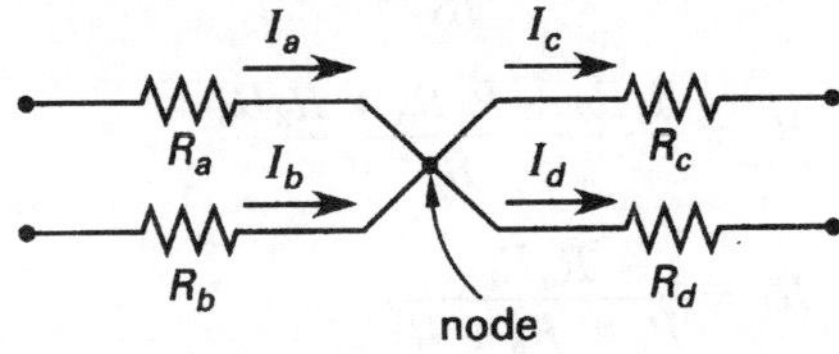

(a) I_a, I_b: negative currents or currents in
(b) I_c, I_d: positive currents or currents out

17. SERIES CIRCUITS

A simple series circuit, as shown in Fig. 26.8, has the following properties (generalized to any number of voltage sources and series resistances).

Figure 26.8 *Series Circuit*

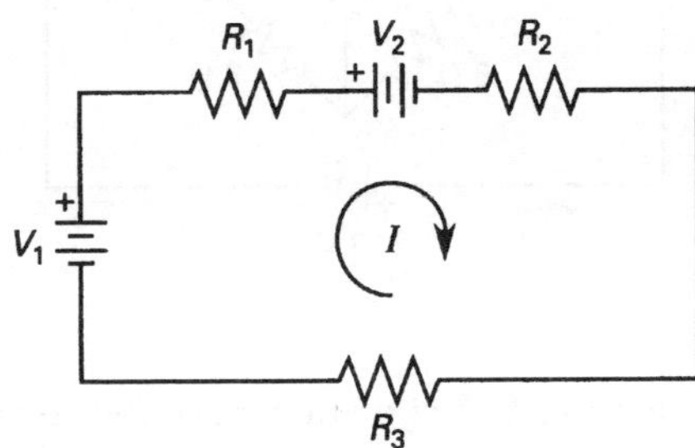

- Current is the same through all the circuit elements.

$$I = I_{R_1} = I_{R_2} = I_{R_3} \cdots = I_{R_n} \qquad 26.26$$

- The equivalent resistance is the sum of the individual resistances.

$$R_e = R_1 + R_2 + R_3 + \cdots + R_n \qquad 26.27$$

- The equivalent applied voltage is the sum of all the voltage sources, with the polarity considered.

$$V_e = \pm V_1 \pm V_2 \pm \cdots \pm V_n \qquad 26.28$$

- The sum of the voltage drops across all circuit elements is equal to the equivalent applied voltage (KVL).

$$V_e = IR_e \qquad 26.29$$

18. PARALLEL CIRCUITS

A simple parallel circuit with only one active source, as shown in Fig. 26.9, has the following properties (generalized to any number of resistors).

Figure 26.9 *Parallel Circuit*

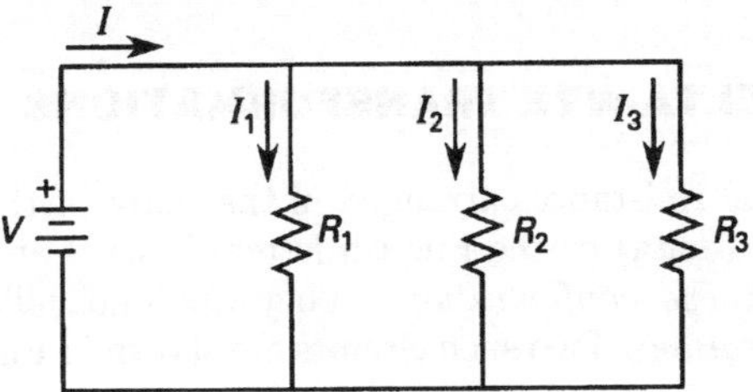

- The voltage across all legs of the circuit is the same.

$$\begin{aligned} V &= V_1 = V_2 = V_3 \cdots = V_n \\ &= I_1R_1 = I_2R_2 = I_3R_3 \cdots = V_n \end{aligned} \qquad 26.30$$

- The reciprocal equivalent resistance is the sum of the reciprocals of the individual resistances. The equivalent conductance is the sum of the individual conductances.

$$\frac{1}{R_e} = \frac{1}{R_1} + \frac{1}{R_2} + \frac{1}{R_3} + \cdots + \frac{1}{R_n} \qquad 26.31(a)$$

$$G_e = G_1 + G_2 + G_3 + \cdots + G_n \qquad 26.31(b)$$

- The total current is the sum of the currents in the individual legs of the circuit (KCL).

$$\begin{aligned} I &= I_1 + I_2 + I_3 + \cdots + I_n \\ &= \frac{V}{R_1} + \frac{V}{R_2} + \frac{V}{R_3} + \cdots + \frac{V}{R_n} \\ &= V(G_1 + G_2 + G_3 + \cdots + G_n) \end{aligned} \qquad 26.32$$

19. ANALYSIS OF COMPLICATED RESISTIVE NETWORKS

The following is a general method that can be used to determine the current flow and voltage drops within a complicated resistive circuit.

step 1: If the circuit is three-dimensional, draw a two-dimensional representation.

step 2: Combine series voltage sources.

step 3: Combine parallel current sources.

step 4: Combine series resistances.

step 5: Combine parallel resistances.

step 6: Repeat steps 2–5 as needed to obtain the current and/or voltage at the desired point, junction, or branch of the circuit. (Do not simplify the circuit beyond what is required to determine the desired quantities.)

Circuit Theory

step 7: If applicable, utilize the delta-wye transformation of Sec. 26.20.

20. DELTA-WYE TRANSFORMATIONS

Electrical resistances arranged in the shape of the Greek letter Δ (delta) or the English letter Y (wye) are known as *delta-wye configurations.* They are also called *pi-T configurations.* Two such circuits are shown in Fig. 26.10.

Figure 26.10 *Delta (Pi)-Wye (T) Configurations*

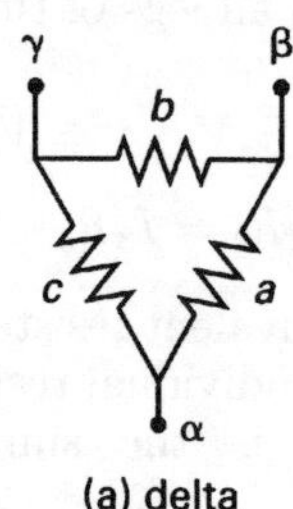

(a) delta

(b) pi

(c) wye

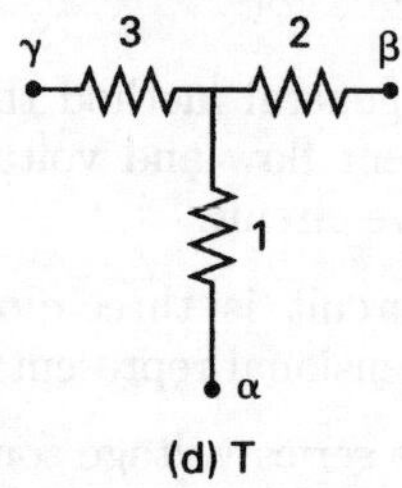

(d) T

The equivalent resistances, which allow transformation between configurations, are

$$R_a = \frac{R_1R_2 + R_1R_3 + R_2R_3}{R_3} \qquad 26.33$$

$$R_b = \frac{R_1R_2 + R_1R_3 + R_2R_3}{R_1} \qquad 26.34$$

$$R_c = \frac{R_1R_2 + R_1R_3 + R_2R_3}{R_2} \qquad 26.35$$

$$R_1 = \frac{R_aR_c}{R_a + R_b + R_c} \qquad 26.36$$

$$R_2 = \frac{R_aR_b}{R_a + R_b + R_c} \qquad 26.37$$

$$R_3 = \frac{R_bR_c}{R_a + R_b + R_c} \qquad 26.38$$

Example 26.4

Simplify the circuit and determine the total current.

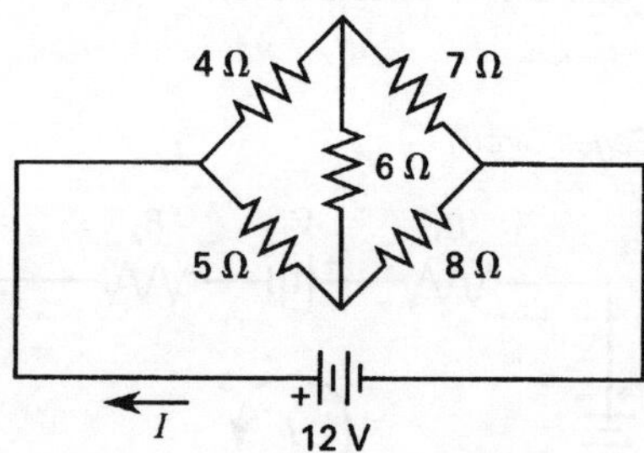

Solution

Convert the 4 Ω-5 Ω-6 Ω delta connection to wye form.

$$\begin{aligned} R_1 &= \frac{R_aR_c}{R_a + R_b + R_c} = \frac{(5\ \Omega)(4\ \Omega)}{5\ \Omega + 6\ \Omega + 4\ \Omega} \\ &= 1.3\ \Omega \end{aligned}$$

$$\begin{aligned} R_2 &= \frac{R_aR_b}{R_a + R_b + R_c} = \frac{(5\ \Omega)(6\ \Omega)}{15\ \Omega} \\ &= 2.0\ \Omega \end{aligned}$$

$$\begin{aligned} R_3 &= \frac{R_bR_c}{R_a + R_b + R_c} = \frac{(6\ \Omega)(4\ \Omega)}{15\ \Omega} \\ &= 1.6\ \Omega \end{aligned}$$

The transformed circuit is

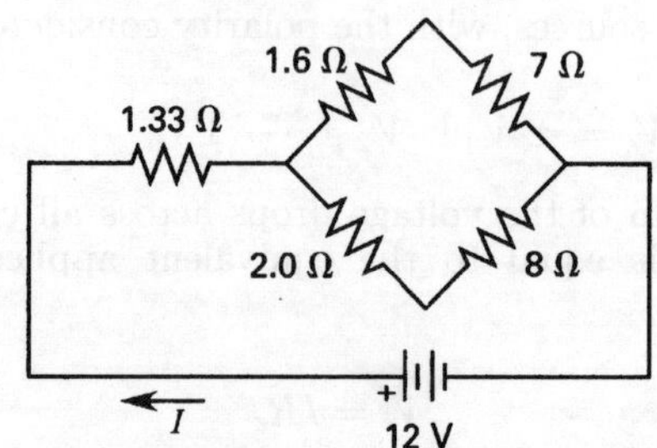

The total equivalent resistance is

$$R_e = 1.33\ \Omega + \frac{1}{\dfrac{1}{1.6\ \Omega + 7\ \Omega} + \dfrac{1}{2\ \Omega + 8\ \Omega}} = 5.95\ \Omega$$

The current is

$$I = \frac{V}{R_e} = \frac{12\ \text{V}}{5.95\ \Omega} = 2.02\ \text{A}$$

21. SUBSTITUTION THEOREM

The *substitution theorem*, also known as the *compensation theorem*, states that any branch in a circuit can be replaced by a substitute branch, as long as the branch voltage and current remain the same, without affecting voltages and currents in any other portion of the circuit.

22. RECIPROCITY THEOREM

In any linear, time-independent circuit with independent current and voltage sources, the ratio of the current in a short circuit in one part of the network to the output of a voltage source in another part of the network is constant—even when the voltage source and the short-circuit positions are interchanged. This principle of *reciprocity* is also applicable to the ratio of the current from a current source and the voltage across an open circuit. Reciprocity between a voltage source and short circuit is illustrated in Fig. 26.11.

Figure 26.11 *Reciprocal Measurements*

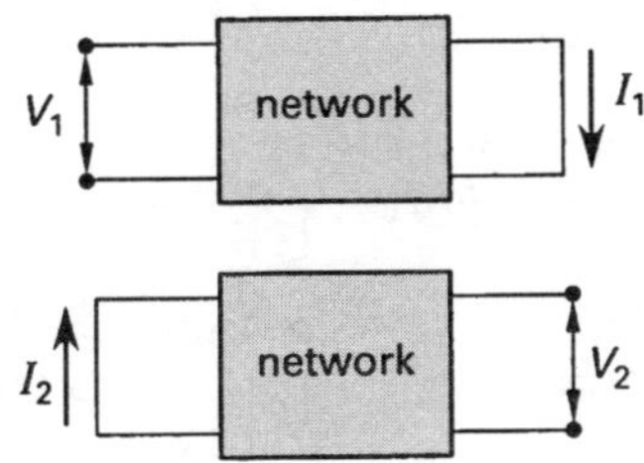

Alternatively, if the network is comprised of linear resistors, the ratio of the applied voltage to the current measured at any point is constant—even when the positions of the source and meter are interchanged. The ratio in both cases is

$$R_{\text{transfer}} = \frac{V_1}{I_1} = \frac{V_2}{I_2} \qquad 26.39$$

The ratio is called the *transfer resistance*. The term *transfer* is used when a given response is determined at a point in a network other than where the driving force is applied. In this case, the voltage is the driving force on one side of the circuit and the current is measured on the other side.

23. SUPERPOSITION THEOREM

The principle of *superposition* is that the response (that is, the voltage across or current through) of a linear circuit element in a network with multiple independent sources is equal to the response obtained if each source is considered individually and the results are summed. The steps involved in determining the desired quantity are as follows.

step 1: Replace all sources except one with their internal resistances.[8] Replace ideal current sources with open circuits. Replace ideal voltage sources with short circuits.

step 2: Compute the desired quantity (either voltage or current) for the element in question attributable to the single source.

step 3: Repeat steps 1 and 2 for each of the sources in turn.

step 4: Sum the calculated values obtained for the current or voltage obtained in step 2. The result is the actual value of the current or voltage in the element for the complete circuit.

Example 26.5

Determine the current through the center leg.

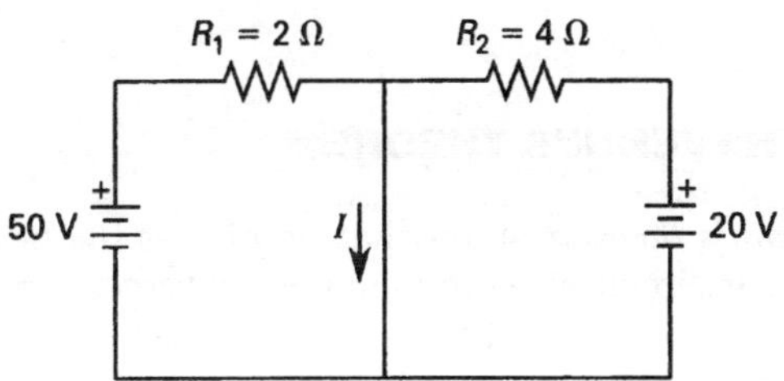

Solution

First, work with the left (50 V) battery. Short out the right (20 V) battery.

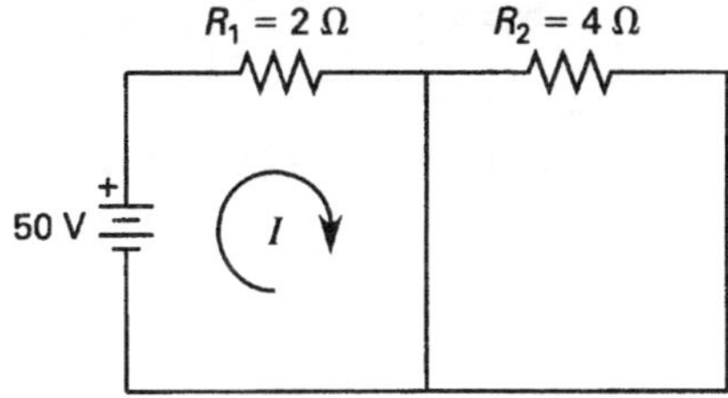

No current will flow through R_2. The equivalent resistance and current are

$$R_e = 2\ \Omega$$

$$I = \frac{V}{R_e} = \frac{50\ \text{V}}{2\ \Omega} = 25\ \text{A}$$

[8]As in all the DC theorems presented in this chapter, this principle also applies to AC circuits or networks. In the case of AC circuits or networks, the term *impedance* would be applicable instead of *resistance*.

Next, work with the right (20 V) battery. Short out the left (50 V) battery.

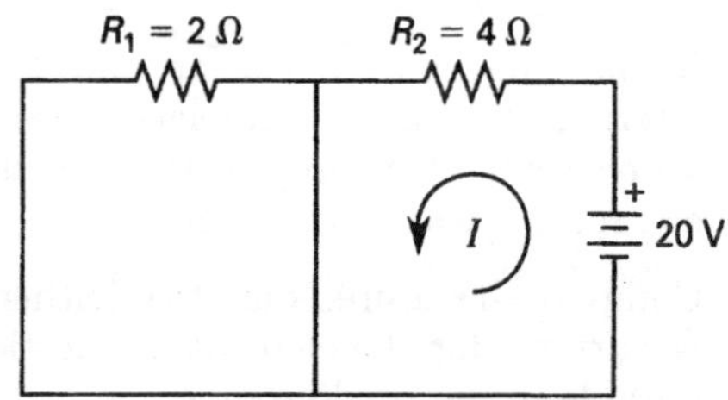

No current will flow through R_1. The equivalent resistance and current are

$$R_e = 4\ \Omega$$

$$I = \frac{V}{R_e} = \frac{20\ \text{V}}{4\ \Omega} = 5\ \text{A}$$

Considering both batteries, the total current flowing is 25 A + 5 A = 30 A.

24. THEVENIN'S THEOREM

Thevenin's theorem states that, insofar as the behavior of a linear circuit at its terminals is concerned, any such circuit can be replaced by a single voltage source, V_{Th}, in series with a single resistance, R_{Th}. The method for determining and utilizing the *Thevenin equivalent circuit*, with designations referring to Fig. 26.12, follows.

Figure 26.12 *Thevenin Equivalent Circuit*

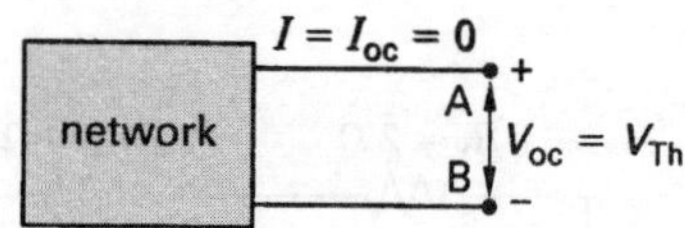

(a) open-circuit measurements

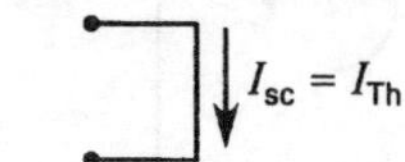

(b) short-circuit measurements

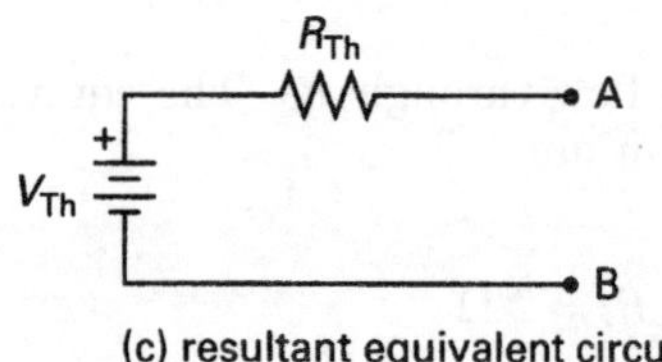

(c) resultant equivalent circuit

step 1: Separate the network that is to be changed into a Thevenin equivalent circuit from its load at two terminals, say A and B.

step 2: Determine the open-circuit voltage, V_{oc}, at terminals A and B.

step 3: Short circuit terminals A and B and determine the current, I_{sc}.

step 4: Calculate the Thevenin equivalent voltage and resistance from the following equations.

$$V_{\text{Th}} = V_{\text{oc}} \qquad \textit{26.40}$$

$$R_{\text{Th}} = \frac{V_{\text{oc}}}{I_{\text{sc}}} \qquad \textit{26.41}$$

step 5: Using the values calculated in Eq. 26.40 and Eq. 26.41, replace the network with the Thevenin equivalent. Reconnect the load at terminals A and B. Determine the desired electrical parameters in the load.

Steps 3 and 4 can be altered by using the following shortcut. Determine the Thevenin equivalent resistance by looking into terminals A and B toward the network with all the power sources altered. Specifically, change independent voltage sources into short circuits and independent current sources into open circuits, then calculate the resistance of the altered network. The resulting resistance is the Thevenin equivalent resistance.

Example 26.6

Determine the Thevenin equivalent circuit that the load resistor, R_{load}, sees.

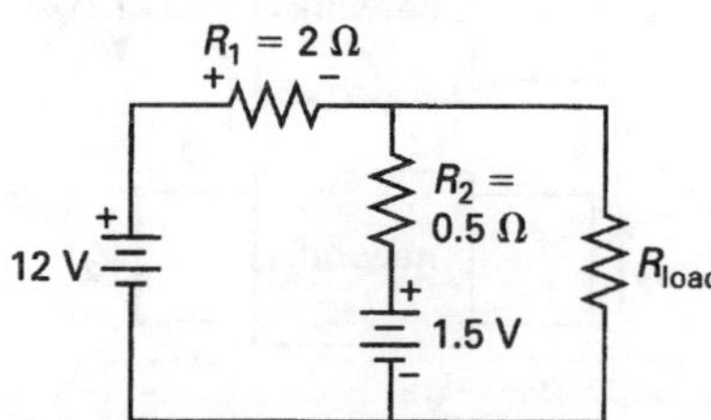

Solution

step 1: Separate the load resistor from the portion of the network to be changed.

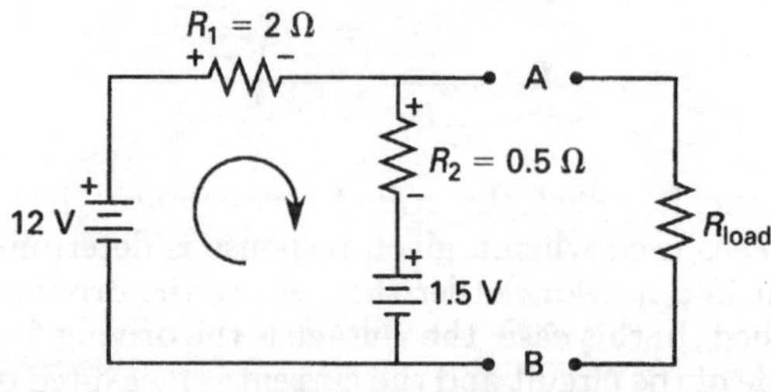

step 2: Apply Kirchhoff's voltage law to find the open-circuit voltage. Starting at terminal B, the equation is

$$12\ \text{V} - V_{R_1} - V_{R_2} - 1.5\ \text{V} = 0$$

$$V_{R_1} + V_{R_2} = 10.5\ \text{V}$$

For clarification, the circuit is redrawn using the result calculated.

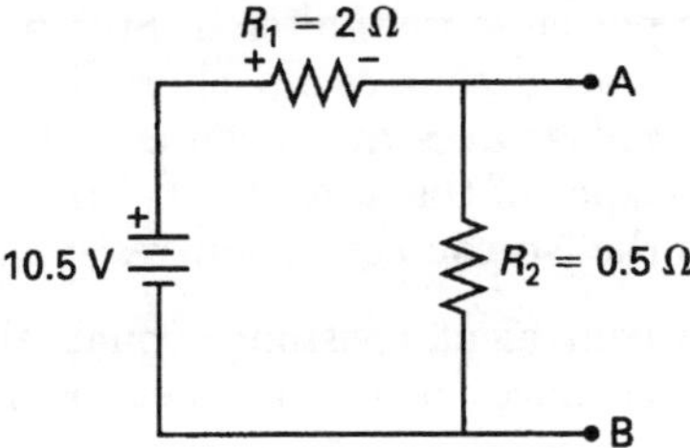

The voltage across R_2 is found using the voltage-divider concept.

$$\begin{aligned} V_{R_2} &= (10.5\ \text{V})\left(\frac{R_2}{R_1+R_2}\right) \\ &= (10.5\ \text{V})\left(\frac{0.5\ \Omega}{2\ \Omega+0.5\ \Omega}\right) \\ &= 2.1\ \text{V} \end{aligned}$$

The circuit branch from A to B can now be illustrated as follows.

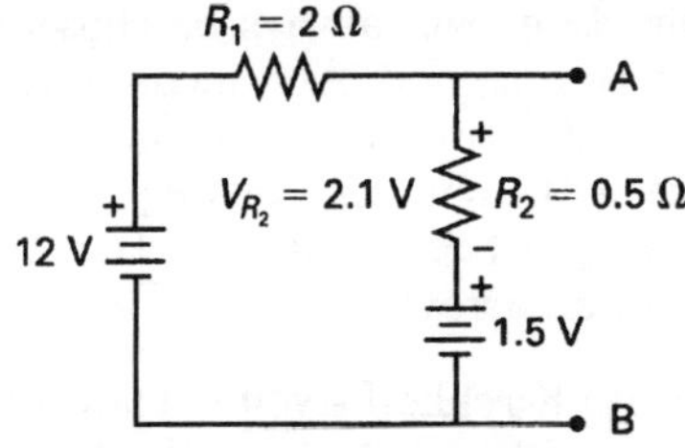

The open-circuit voltage is the voltage across R_2 and the 1.5 V source, so $V_{oc} = 3.6$ V. (This can be found by using KVL around the loop containing the terminals A and B.)

step 3: Short terminals A and B to determine the short-circuit current.

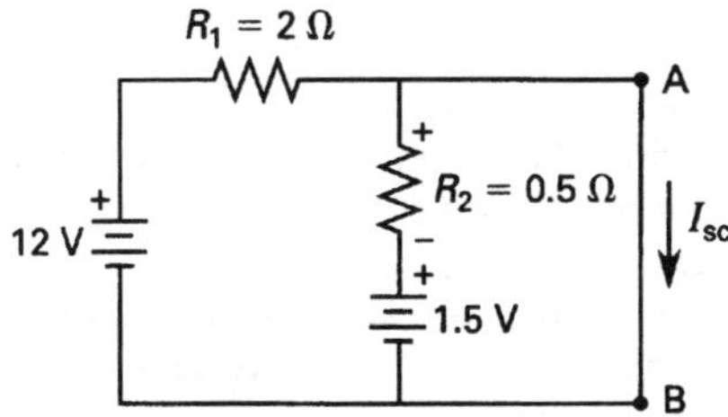

Using either Ohm's law or KVL around the outer loop, determine the short-circuit current contribution from the 12 V source.

$$I_{sc} = \frac{12\ \text{V}}{2\ \Omega} = 6\ \text{A}$$

Using either Ohm's law or KVL around the loop containing the 1.5 V source, R_2, and terminals A and B, determine the short-circuitcurrent contribution from the 1.5 V source.

$$I_{sc} = \frac{1.5\ \text{V}}{0.5\ \Omega} = 3\ \text{A}$$

The total short-circuit current, using the principle of superposition, is $6\ \text{A} + 3\ \text{A} = 9\ \text{A}$.

step 4: Using Eq. 26.40 and Eq. 26.41, calculate the Thevenin equivalent voltage and resistance.

$$V_{\text{Th}} = V_{oc} = 3.6\ \text{V}$$

$$R_{\text{Th}} = \frac{V_{oc}}{I_{sc}} = \frac{3.6\ \text{V}}{9\ \text{A}} = 0.4\ \Omega$$

step 5: Replace the network with the Thevenin equivalent.

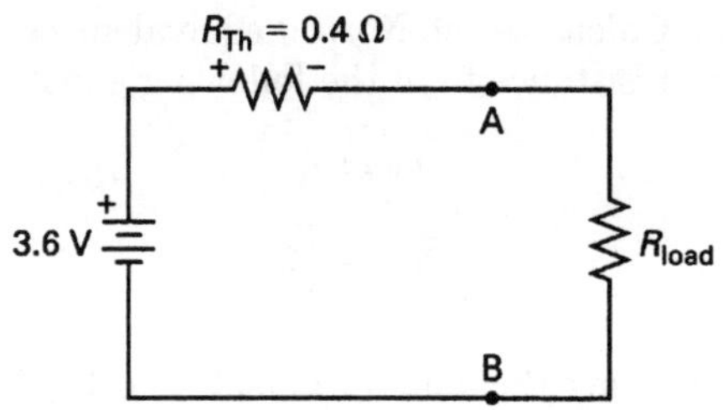

The method outlined is systematic. The shortcut for finding the Thevenin equivalent circuit mentioned earlier in this section would have simplified the solution. For example, short the voltage sources and determine the Thevenin equivalent resistance.

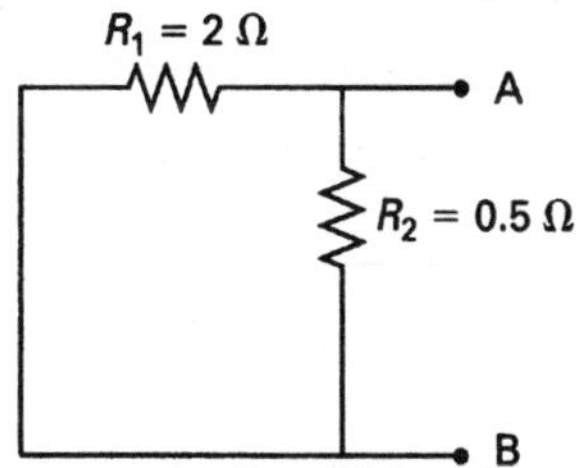

$$R_{\text{Th}} = \frac{R_1 R_2}{R_1 + R_2} = \frac{(2\ \Omega)(0.5\ \Omega)}{2\ \Omega + 0.5\ \Omega} = 0.4\ \Omega$$

The short-circuit current, found using superposition, is 9 A. The Thevenin equivalent voltage can then be found by combining Eq. 26.29 and Eq. 26.41.

$$V_{\text{Th}} = I_{\text{Th}} R_{\text{Th}}$$

Circuit Theory

Substituting the short-circuit current for I_{sc} and the calculated value for R_{Th} results in an identical answer, as required by Thevenin's theorem.

25. NORTON'S THEOREM

Norton's theorem states that, insofar as the behavior of a linear circuit at its terminals is concerned, any such circuit can be replaced by a single current source, I_N, in parallel with a single resistance, R_N. The method for determining and utilizing the *Norton equivalent circuit*, with designations referring to Fig. 26.13, follows.

step 1: Separate the network that is to be changed into a Norton equivalent circuit from its load at two terminals, say A and B.

step 2: Determine the open-circuit voltage, V_{oc}, at terminals A and B.

step 3: Short circuit terminals A and B and determine the current, I_{sc}.

step 4: Calculate the Norton equivalent current and resistance from the following equations.

$$I_N = I_{sc} \quad \text{26.42}$$

$$R_N = \frac{V_{oc}}{I_{sc}} \quad \text{26.43}$$

step 5: Using the values calculated in Eq. 26.42 and Eq. 26.43, replace the network with the Norton equivalent. Reconnect the load at terminals A and B. Determine the desired electrical parameters in the load.

Figure 26.13 Norton Equivalent Circuit

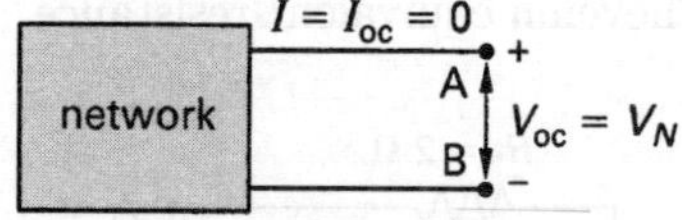

(a) open-circuit measurements

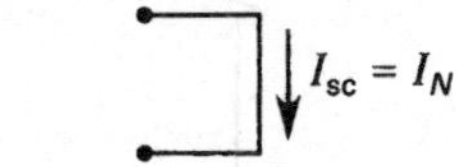

(b) short-circuit measurements

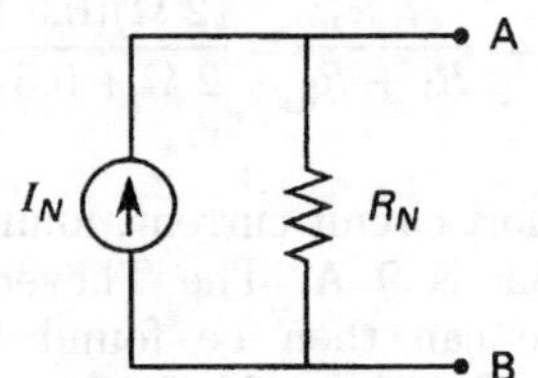

(c) resultant equivalent circuit

Steps 3 and 4 can be altered by using the following shortcut. Determine the Norton equivalent resistance by looking into terminals A and B toward the network with all the power sources altered. Specifically, change independent voltage sources into short circuits and independent current sources into open circuits, then calculate the resistance of the altered network. The resulting resistance is the Norton equivalent resistance.

The Norton equivalent resistance equals the Thevenin equivalent resistance for identical networks.

26. LOOP-CURRENT METHOD

The *loop-current method* is a systematic network-analysis procedure that uses currents as the unknowns. It is also called *mesh analysis* or the *Maxwell loop-current method*. The method uses Kirchhoff's voltage law and is performed on planar networks. It requires $n-1$ simultaneous equations for an n-loop system. The method's steps are as follows.

step 1: Select $n-1$ loops (one loop less than the total number of possible loops).

step 2: Assume current directions for the selected loops. (While this is arbitrary, clockwise directions will always be chosen in this text for consistency. Any incorrectly chosen current direction will result in a negative result when the simultaneous equations are solved in step 4.) Show the direction of the current with an arrow.

step 3: Write Kirchhoff's voltage law for each of the selected loops. Assign polarities based on the direction of the loop current. The voltage of a source is positive when the current flows out of the positive terminal, that is, from the negative terminal to the positive terminal inside the source. The selected direction of the loop current always results in a voltage drop in the resistors of the loop (see Fig. 26.6(c)). Where two loop currents flow through an element, they are summed to determine the voltage drop in that element, using the direction of the current in the loop for which the equation is being written as the positive (i.e., correct) direction. (Any incorrect direction for the loop current will be indicated by a negative sign in the solution in step 4.)

step 4: Solve the $n-1$ equations (from step 3) for the unknown currents.

step 5: If required, determine the actual current in an element by summing the loop currents flowing through the element. (Sum the absolute values to obtain the correct magnitude. The correct direction is given by the loop current with the positive value.)

27. NODE-VOLTAGE METHOD

The *node-voltage method* is a systematic network-analysis procedure that uses voltages as the unknowns. The method uses Kirchhoff's current law. It requires $n-1$ equations for an n-principal node system (equations are not necessary at simple nodes, that is, nodes connecting only two circuit elements). The method's steps are as follows.

step 1: Simplify the circuit, if possible, by combining resistors in series or parallel or by combining current sources in parallel. Identify all nodes. (The minimum total number of equations required will be $n-1$, where n represents the number of principal nodes.)

step 2: Choose one node as the reference node, that is, the node that will be assumed to have ground potential (0 V). (To minimize the number of terms in the equations, pick the node with the largest number of circuit elements as the reference node.)

step 3: Write Kirchhoff's current law for each principal node except the reference node, which is assumed to have a zero potential.

step 4: Solve the $n-1$ equations (from step 3) to determine the unknown voltages.

step 5: If required, use the calculated node voltages to determine any branch current desired.

28. DETERMINATION OF METHOD

When analyzing electrical networks, the method used depends on the circuit elements and their configurations. The loop-current method using Kirchhoff's voltage law is used in circuits without current sources. The node-voltage method using Kirchhoff's current law is used in circuits without voltage sources. When both types of sources are present, one of the following two methods may be used.

method 1: Use each of Kirchhoff's laws, assigning voltages and currents as needed, and substitute any known quantity into the equations as written.

method 2: Use source transformation or source shifting to change the appearance of the circuit so that it contains only the desired sources—voltage sources when using KVL and current sources when using KCL. (Source shifting is manipulating the circuit so that each voltage source has a resistor in series and each current source has a resistor in parallel.)

Use the method that results in the least number of equations. The loop-current method produces $n-1$ equations, where n is the total number of loops. The node-voltage method produces $n-1$ equations, where n is the number of principal nodes. Count the number of loops and the number of principal nodes prior to writing the equations; whichever is least determines the method used.

Additional methods exist, some of which require less work. The advantage of the loop-current and node-voltage methods is that they are systematic and guarantee a solution.

29. PRACTICAL APPLICATION: BATTERIES

In general, a *battery* is defined as a direct-current voltage source made up of one or more units that convert chemical, thermal, nuclear, or solar energy into electrical energy. The most widely used battery type is one that converts the chemical energy contained in its active materials directly into electrical energy by means of an oxidation-reduction reaction.[9] A battery consists of two dissimilar metals, an anode and a cathode, immersed in an electrolyte. The cathode is reduced during the reaction. The transfer of charge is completed within the electrolyte by the flow of ions. This is shown for a single cell in Fig. 26.14. Regardless of the cell or device type, the flow of electrons is always from anode-to-cathode outside and cathode-to-anode inside the cell or device.

Figure 26.14 *Electrochemical Battery (Discharge)*

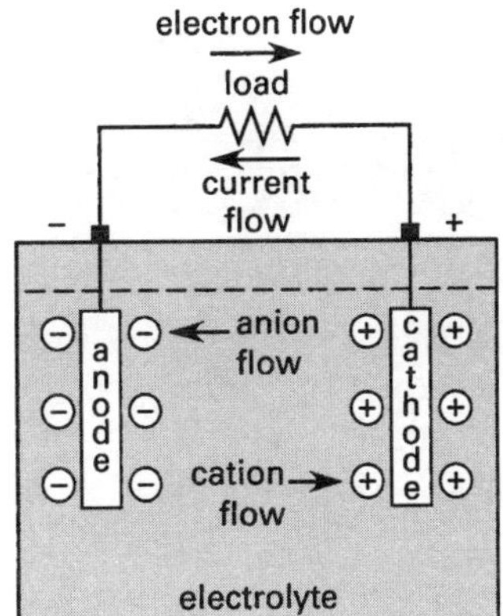

The battery terminals have no absolute voltage value. They have a value relative only to each other, and the terminals and the battery are said to *float.* Usually, one of the terminals is assigned as the reference and given the value of 0 V. This terminal is then the *reference* or *datum* and is said to be *grounded.* If the battery is allowed to float and the circuit is grounded elsewhere, all voltages are assumed to be measured with respect to this ground. The earth is generally regarded as being at 0 V and any circuit tied to earth by an electrical wire is grounded.

[9]In *nonelectrochemical reactions*, the transfer of electrons takes place directly and only heat is involved.

Circuit Theory

The voltage of a battery is determined by the number of cells used. The voltage of the cell is determined by the materials used, as this determines the half-cell oxidation potentials. The capacity of the battery is determined by the amount of materials used and is measured in ampere-hours (A·h). One gram-equivalent weight of material supplies 96,485 C or 26.805 A·h of electric charge.

A *primary battery* is one that uses an electrochemical reaction that is not effciently reversible. An example is the common flashlight battery. This type of battery is also called a *dry cell*, as the electrolyte is a moist paste instead of a liquid solution. A *secondary battery* is rechargeable and has a much higher energy density. An example is the lead-acid storage battery used in automobiles. Such batteries are treated as ideal voltage sources with a series resistance representing internal resistance. Using the specified voltage and resistance, any network containing a battery can be analyzed using the techniques described in this chapter.

27 AC Circuit Fundamentals

Nomenclature

a_n, b_n	Fourier coefficients	–
B	susceptance	S
C	capacitance	F
f	frequency	Hz
G	conductance	S
i	variable current	A
I, $\mathbf{I}$	constant or rms current	A
I^*, $\mathbf{I}^*$	complex conjugate of current	A
L	inductance	H
p	instantaneous power	W
pf	power factor	–
P, $\mathbf{P}$	power	W
Q, $\mathbf{Q}$	reactive power	VAR
R	resistance	Ω
S, $\mathbf{S}$	apparent or complex power	VA
t	time	s
T	period	s
v	variable voltage	V
V, $\mathbf{V}$	constant or rms voltage	V
X	reactance	Ω
Y, $\mathbf{Y}$	admittance	S
Z, $\mathbf{Z}$	impedance	Ω
$\mathbf{Z}$	complex number	–

Symbols

θ	phase angle	rad
ϕ	impedance angle or power angle	rad
ϕ	phase angle difference	rad
ϕ_{pf}	power factor angle	rad
φ	current angle	rad
ω	angular frequency	rad/s

Subscripts

ave	average
C	capacitative or capacitor
e	equivalent
eff	effective
i	imaginary
I	current
L	inductor
m	maximum
p	peak
pf	power factor
R	resistive or resistor
rms	root-mean-square
s	source
thr	threshold
Z	impedance

1. FUNDAMENTALS

Alternating waveforms have currents and voltages that vary with time in a regular and symmetrical manner. Waveform shapes include square, sawtooth, and triangular, along with many variations on these themes. However, for most applications in electrical engineering the variations are sinusoidal in time. In this book, unless otherwise specified, currents and voltages are sinusoidal.[1]

When sinusoidal, the waveform is nearly always referred to as AC, that is, *alternating current*, indicating that the current is produced by the application of a sinusoidal voltage. This means that the flow of electrons changes directions, unlike DC circuits, where the flow of electrons is unidirectional (though the magnitude can change in time).

A circuit is said to be in a *steady-state* condition if the current and voltage time variation is purely constant

[1]Nearly all periodic functions can be represented as the sum of sinusoidal functions, which simplifies the mathematics needed.

(DC) or purely sinusoidal (AC).[2] In this book, unless otherwise specified, all circuits are in a steady-state condition.

All the methods, basic definitions, and equations presented in Chap. 26 involving DC circuits are applicable to AC circuits as well. AC electrical parameters have both magnitudes and angles. Nevertheless, following common practice, phasor notation will be used only if needed to avoid confusion.

2. VOLTAGE

Sinusoidal variables can be expressed in terms of sines or cosines without any loss of generality.[3] A sine waveform is often the standard. If this is the case, Eq. 27.1 gives the value of the instantaneous voltage as a function of time.

$$v(t) = V_m \sin(\omega t + \theta) \qquad 27.1$$

The *maximum value* of the sinusoid is given the symbol V_m and is also known as the *amplitude*. If $v(t)$ is not zero at $t = 0$, the sinusoid is *shifted* and a *phase angle*, θ, must be used, as shown in Fig. 27.1.[4] Also shown in Fig. 27.1 is the *period*, T, which is the time that elapses in one cycle of the sinusoid. The *cycle* is the smallest portion of the sinusoid that repeats.

Figure 27.1 *Sinusoidal Waveform with Phase Angle*

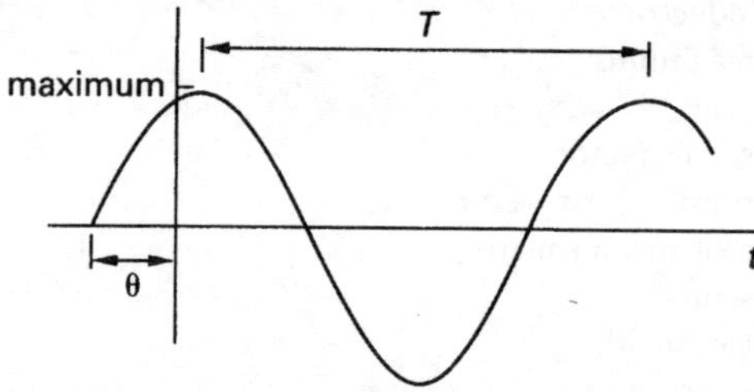

Because the horizontal axis of the voltage in Fig. 27.1 corresponds to time, not distance, the waveform does not have a wavelength. The frequency, f, of the sinusoid is the reciprocal of the period in hertz (Hz). The angular frequency, ω, in rad/s can also be used.

$$f = \frac{1}{T} = \frac{\omega}{2\pi} \qquad 27.2$$

$$\omega = 2\pi f = \frac{2\pi}{T} \qquad 27.3$$

An AC voltage waveform without a phase angle is plotted as a function of differing variables in Fig. 27.2.

[2]Steady-state AC may have a *DC offset*.

[3]The point at which time begins, that is, where $t = 0$, is of no consequence in steady-state AC circuit problems, as the signal repeats itself every cycle. Therefore, it makes no difference whether a sine or cosine waveform is used (though, if one exists, care must be taken to keep the phase angle correct).

[4]The term *phase* is not the same as *phase difference*, which is the difference between corresponding points on two sinusoids of the same frequency.

Figure 27.2 *Sine Wave Plots*

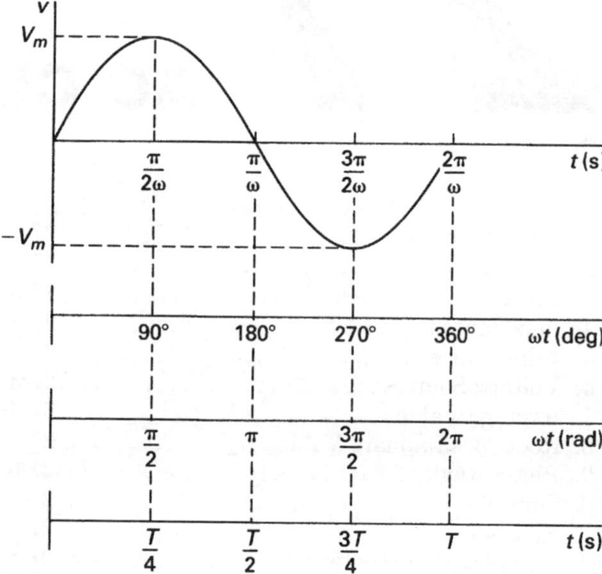

Exponentials, $e^{j\theta}$ and $e^{-j\theta}$, can be combined to produce $\sin\theta$ and $\cos\theta$ terms. As a result, sinusoids can be represented in the following equivalent forms.

- trigonometric: $V_m \sin(\omega t + \theta)$
- exponential: $V_m e^{j\theta}$
- polar or phasor: $V_m \angle\theta$
- rectangular: $V_r + jV_i$

The use of complex exponentials and phasor analysis allows sinusoidal functions to be more easily manipulated mathematically, especially when dealing with derivatives. Exponents are manipulated algebraically, and the resulting sinusoid is then recovered using Euler's relation. This avoids complicated trigonometric mathematics. The angles used in exponentials and with angular frequency, ω, must be in radians. From Euler's relation,

$$e^{j\theta} = \cos\theta + j\sin\theta \qquad 27.4$$

$$j = \sqrt{-1} = 1\angle\frac{\pi}{2} \qquad 27.5$$

Therefore, to regain the sinusoid from the exponential, use Euler's equation, keeping in mind two factors. First, either the real or the imaginary part contains the sinusoid desired, not both.[5] Second, the exponential should not be factored out of any equation until all the derivatives, or integrals, have been taken.

[5]The cosine function could be used as the reference sinusoid, and is in many texts. Over time, many standards have come to use the cosine as the preferred reference. In this text, the sine reference is used when graphing, since this allows the value at $t = 0$ to be zero. Otherwise, the cosine reference will be used. The desired form can be determined from the representation of the voltage (or current) source waveform as either a cosine or a sine function (see Eq. 27.1). Either way, only the phase angle changes.

Circuit Theory

3. CURRENT

Current is the net transfer of electric charge per unit time. When a circuit's driving force, the voltage, is sinusoidal, the resulting current is also (though it may differ by an amount called the *phase angle difference*). The equations and representations in Sec. 27.2, as well as any other representations of sinusoidal waveforms provided in this chapter, apply to current as well as to voltage.

4. IMPEDANCE

Electrical impedance, also known as *complex impedance*, is the total opposition a circuit presents to alternating current. It is equal to the ratio of the complex voltage to the complex current. Impedance, then, is a ratio of phasor quantities and is not itself a function of time. Relating voltage and current in this manner is analogous to Ohm's law, which for AC analysis is referred to as *extended Ohm's law*. Impedance is given the symbol **Z** and is measured in ohms. The three passive circuit elements (resistors, capacitors, and inductors), when used in an AC circuit, are assigned an angle, ϕ, known as the *impedance angle*. This angle corresponds to the phase difference angle produced when a sinusoidal voltage is applied across that element alone.

Impedance is a complex quantity with a magnitude and associated angle. It can be written in *phasor form*—also known as *polar form*—for example, $\mathbf{Z}\angle\phi$, or in *rectangular form* as the complex sum of its resistive (R) and reactive (X) components.

$$\mathbf{Z} \equiv R + jX \qquad 27.6$$

$$R = Z\cos\phi \quad \left[\begin{array}{c}\text{resistive or}\\ \text{real part}\end{array}\right] \qquad 27.7$$

$$X = Z\sin\phi \quad \left[\begin{array}{c}\text{reactive or}\\ \text{imaginary part}\end{array}\right] \qquad 27.8$$

The resistive and reactive components can be combined to form an *impedance triangle*. Such a triangle is shown in Fig. 27.3.

Figure 27.3 *Impedance Triangle*

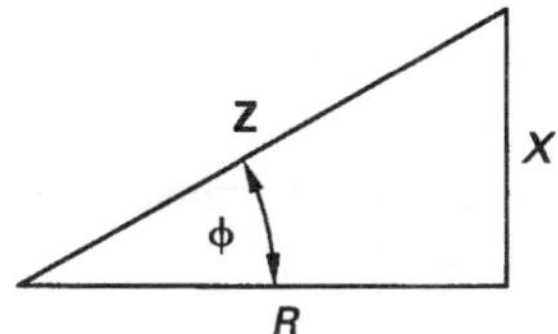

The characteristics of the passive elements in AC circuits, including impedance, are given in Table 27.1.

Table 27.1 *Characteristics of Resistors, Capacitors, and Inductors*

	resistor	capacitor	inductor
value	R (Ω)	C (F)	L (H)
reactance, X	0	$\frac{-1}{\omega C}$	ωL
rectangular impedance, **Z**	$R + j0$	$0 - \frac{j}{\omega C}$	$0 + j\omega L$
phasor impedance, **Z**	$R\angle 0°$	$\frac{1}{\omega C}\angle -90°$	$\omega L\angle 90°$
phase	in-phase	leading	lagging
rectangular admittance, **Y**	$\frac{1}{R} + j0$	$0 + j\omega C$	$0 - \frac{j}{\omega L}$
phasor admittance, **Y**	$\frac{1}{R}\angle 0°$	$\omega C\angle 90°$	$\frac{1}{\omega L}\angle -90°$

5. ADMITTANCE

The reciprocal of impedance is called *admittance*, **Y**. Admittance is useful in analyzing parallel circuits, as admittances can be added directly. The defining equation is

$$\mathbf{Y} = \frac{1}{\mathbf{Z}} = \frac{1}{Z}\angle -\phi \qquad 27.9$$

The reciprocal of the resistive portion of an impedance is known as *conductance*, *G*. The reciprocal of the reactive part is termed the *susceptance*, *B*.

$$G = \frac{1}{R} \qquad 27.10$$

$$B = \frac{1}{X} \qquad 27.11$$

Using these definitions and multiplying by a complex conjugate, admittance can be written in terms of resistance and reactance. Using the same method, impedance can be written in terms of conductance and susceptance. Equation 27.12 and Eq. 27.13 show this and can be used for conversion between admittance and impedance and vice versa.

$$\mathbf{Y} = G + jB = \frac{R}{R^2 + X^2} - j\left(\frac{X}{R^2 + X^2}\right) \qquad 27.12$$

$$\mathbf{Z} = R + jX = \frac{G}{G^2 + B^2} - j\left(\frac{B}{G^2 + B^2}\right) \qquad 27.13$$

6. VOLTAGE SOURCES

The energy for AC voltage sources comes primarily from electromagnetic induction. The concepts of ideal and real sources, as well as regulation, apply to AC sources (see Sec. 26.8). Independent sources deliver voltage and current at their rated values regardless of circuit parameters. Dependent sources, often termed *controlled sources*, deliver voltage and current at levels determined by a voltage or current somewhere else in the circuit. These types of sources occur in electronic circuitry and are also used to model electronic elements, such as transistors. The symbols used for AC voltage sources are shown in Fig. 27.4.

Figure 27.4 AC Voltage Source Symbols

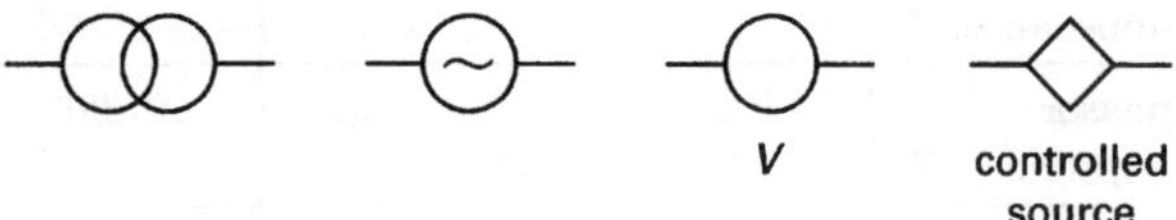

7. AVERAGE VALUE

In purely mathematical terms, the average value of a periodic waveform is the first term of a Fourier series representing the function; that is, it is the zero frequency or DC value. If a function, $f(t)$, repeats itself in a time period T, then the average value of the function is given by Eq. 27.14. In Eq. 27.14, t_1 is any convenient time for evaluating the integral, that is, the time that simplifies the integration. The integral itself is computed over the period.

$$f_{\text{ave}} = \frac{1}{T}\int_{t_1}^{t_1+T} f(t)\,dt \qquad 27.14$$

Integration can be interpreted as the area under the curve of the function. Equation 27.14 divides the net area of the waveform by the period T. This concept is illustrated in Fig. 27.5 and stated in mathematical terms by Eq. 27.15.

$$f_{\text{ave}} = \frac{\text{positive area} - \text{negative area}}{T} \qquad 27.15$$

For the function shown in Fig. 27.5, the area above the axis is called the positive area and the area below is called the negative area. The average value is the net area remaining after the negative area is subtracted from the positive area and the result is divided by the period.

Figure 27.5 Average Value Areas Defined

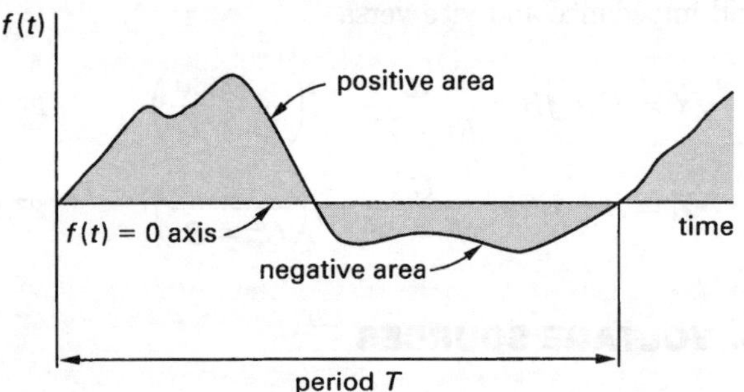

For any periodic voltage, Eq. 27.16 calculates the average value.

$$V_{\text{ave}} = \frac{1}{2\pi}\int_0^{2\pi} v(\theta)\,d\theta = \frac{1}{T}\int_0^T v(t)\,dt \qquad 27.16$$

Any waveform that is symmetrical with respect to the horizontal axis will result in a value of zero for Eq. 27.16. While mathematically correct, the electrical effects of such a voltage occur on both the positive and negative half-cycles. Therefore, the average is instead taken over only half a cycle. This is equivalent to determining the average of a *rectified waveform*, that is, the absolute value of the waveform. The average voltage for a rectified sinusoid is

$$V_{\text{ave}} = \frac{1}{\pi}\int_0^{\pi} v(\theta)\,d\theta = \frac{2V_m}{\pi} \quad \text{[rectified sinusoid]} \qquad 27.17$$

A DC current equal in value to the average value of a rectified AC current produces the same electrolytic effects, such as capacitor charging, plating operations, and ion formation. Nevertheless, not all AC effects can be accounted for using the average value. For instance, in a typical DC meter, the average DC current determines the response of the needle. An AC current sinusoid of equal average magnitude will not result in the same effect since the torque on each half-cycle is in opposite directions, resulting in a net zero effect. Unless the AC signal is rectified, the reading will be zero.

Example 27.1

What is the average value of the pure sinusoid shown?

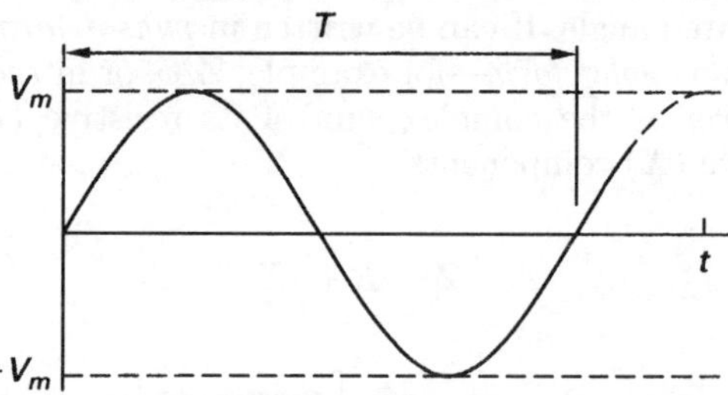

Solution

The sinusoid shown is a sine wave. Substituting into Eq. 27.14 with $t_1 = 0$ gives

$$\begin{aligned} f_{\text{ave}} &= \frac{1}{T}\int_{t_1}^{t_1+T} f(t)\,dt = \frac{1}{T}\int_0^T V_m \sin t\,dt \\ &= \left(\frac{V_m}{T}\right)(-\cos t)\Big|_0^T \\ &= \left(\frac{V_m}{T}\right)(-\cos T + \cos 0) \\ &= \left(\frac{V_m}{T}\right)(-1+1) = 0 \end{aligned}$$

This result is as expected, because a pure sinusoid has equal positive and negative areas.

Example 27.2

Diodes are used to rectify AC waveforms. Real diodes will not pass current until a threshold voltage is reached. As a result, the output is the difference between the sinusoid and the threshold value. This is illustrated in the following figures.

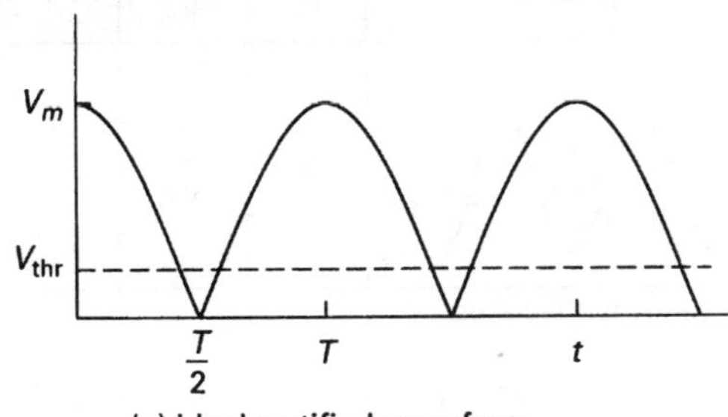

(a) ideal rectified waveform

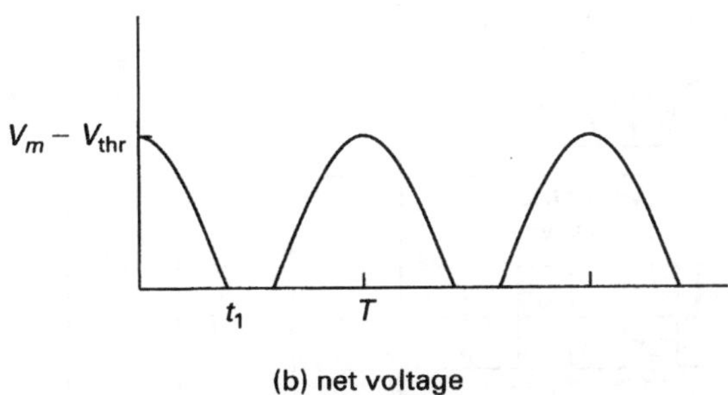

(b) net voltage

What is the expression for the average voltage on the output of a real rectifier diode with a sinusoidal voltage input?

Solution

Because of the symmetry, the average value can be found using half the average determined from Eq. 27.16 by integrating from zero to $T/2$, simplifying the calculation. The waveform is represented by

$$v(t) = \begin{cases} V_m \cos \frac{\pi}{T} t - V_{\text{thr}} & [\text{for } 0 < t < t_1] \\ 0 & [\text{for } t_1 < t < T/2] \end{cases}$$

At $t = t_1$, the following condition exists.

$$V_m \cos \frac{\pi}{T} t_1 = V_{\text{thr}}$$

The equation for one-half the average is

$$\begin{aligned} \tfrac{1}{2} v_{\text{ave}} &= \frac{1}{T} \int_0^{t_1} \left(V_m \cos \frac{\pi}{T} t - V_{\text{thr}} \right) dt \\ &= \left(\frac{1}{T}\right)\left(\frac{T}{\pi}\right) V_m \sin \frac{\pi}{T} t_1 - V_{\text{thr}} \left(\frac{t_1}{T}\right) \end{aligned}$$

Using the equation for the condition at $t = t_1$ and the trigonometric identity $\sin^2\theta + \cos^2\theta = 1$ results in the following equations.

$$\sin \frac{\pi}{T} t_1 = \sqrt{1 - \left(\frac{V_{\text{thr}}}{V_m}\right)^2}$$

$$\frac{t_1}{T} = \frac{1}{\pi} \arccos \frac{V_{\text{thr}}}{V_m}$$

Substituting into the equation for one-half the average and rearranging gives the following final result.

$$\begin{aligned} v_{\text{ave}} &= \frac{2}{\pi} V_m \sqrt{1 - \left(\frac{V_{\text{thr}}}{V_m}\right)^2} \\ &\quad - \frac{2}{\pi} V_{\text{thr}} \arccos \frac{V_{\text{thr}}}{V_m} \end{aligned}$$

The arccos must be expressed in radians. The conclusion drawn is that real diode average values are more complex than values for ideal diode full-wave rectification. However, if ideal diodes are assumed, that is, if the threshold voltage is considered negligible, the following errors are generated.

- 1% error for $V_{\text{thr}}/V_m = 0.0064$
- 2% error for $V_{\text{thr}}/V_m = 0.0128$
- 5% error for $V_{\text{thr}}/V_m = 0.0321$
- 10% error for $V_{\text{thr}}/V_m = 0.0650$

For most practical applications, the ideal diode assumption results in an error of less than 10%.

8. ROOT-MEAN-SQUARE VALUE

In purely mathematical terms, the *effective*, or *root-mean-square* (rms) *value*, of a periodic waveform represented by a function, $f(t)$, which repeats itself in a time period, is given by Eq. 27.18. In Eq. 27.18, t_1 is any convenient time for evaluating the integral, that is, the time that simplifies the integration. The integral itself is computed over the period.

$$f_{\text{rms}}^2 = \frac{1}{T} \int_{t_1}^{t_1+T} f^2(t)\,dt \qquad 27.18$$

For any periodic voltage, Eq. 27.19 calculates the effective or rms value.

$$\begin{aligned} V = V_{\text{eff}} = V_{\text{rms}} &= \sqrt{\frac{1}{2\pi} \int_0^{2\pi} v^2(\theta)\,d\theta} \\ &= \sqrt{\frac{1}{T} \int_0^{T} v^2(t)\,dt} \qquad 27.19 \end{aligned}$$

Normally when a voltage or current variable, V or I, is left unsubscripted, it represents the effective (rms) value.[6] Rectification of the waveform is not necessary to calculate the effective value. (The squaring of the waveform ensures the net result of integration is something other than zero for a sinusoid.) For a sinusoidal waveform, $V = V_m/\sqrt{2} \approx 0.707 V_m$. A DC current of magnitude I produces the same heating effect as an AC current of magnitude I_{eff}.

[6]The value of the standard voltage in the United States, reported as 115–120 V, is an effective value.

Circuit Theory

Table 27.2 gives the characteristics of various commonly encountered alternating waveforms. In this table, two additional terms are introduced. The *form factor* is given by

$$\mathrm{FF} = \frac{V_{\mathrm{eff}}}{V_{\mathrm{ave}}} \qquad 27.20$$

The *crest factor*, CF, also known as the *peak factor* or *amplitude factor*, is

$$\mathrm{CF} = \frac{V_m}{V_{\mathrm{eff}}} \qquad 27.21$$

Nearly all periodic functions can be represented by a Fourier series. A Fourier series can be written as

$$f(t) = \tfrac{1}{2}a_0 + \sum_{n=1}^{\infty} a_n \cos\frac{2\pi n}{T}t + \sum_{n=1}^{\infty} b_n \sin\frac{2\pi n}{T}t \qquad 27.22$$

The average value is the first term, as mentioned in Sec. 27.7. The rms value is

$$f_{\mathrm{rms}} = \sqrt{\left(\tfrac{1}{2}a_0\right)^2 + \tfrac{1}{2}\sum_{n=1}^{\infty}(a_n^2 + b_n^2)} \qquad 27.23$$

Example 27.3

A peak sinusoidal voltage, V_p, of 170 V is connected across a 240 Ω resistor in a lightbulb. What is the power dissipated by the bulb?

Solution

From Table 27.2, the effective voltage is

$$V = \frac{V_p}{\sqrt{2}} = \frac{170\ \mathrm{V}}{\sqrt{2}} = 120.21\ \mathrm{V}$$

The power dissipated is

$$P = \frac{V^2}{R} = \frac{(120.21\ \mathrm{V})^2}{240\ \Omega} = 60.21\ \mathrm{W}$$

Example 27.4

What is the rms value of a constant 3 V signal?

Solution

Using Eq. 27.18, the rms value is

$$v_{\mathrm{rms}}^2 = \frac{1}{T}\int_0^T (3\ \mathrm{V})^2 dt = \left(\frac{1}{T}\right)(9\ \mathrm{V})(T-0)$$

$$v_{\mathrm{rms}} = \sqrt{9\ \mathrm{V}} = 3\ \mathrm{V}$$

Table 27.2 *Characteristics of Alternating Waveforms*

waveform	$\frac{V_{\mathrm{ave}}}{V_m}$	$\frac{V_{\mathrm{rms}}}{V_m}$	FF	CF
V_m	0	$\frac{1}{\sqrt{2}}$	–	$\sqrt{2}$
full-wave rectified sinusoid V_m	$\frac{2}{\pi}$	$\frac{1}{\sqrt{2}}$	$\frac{\pi}{2\sqrt{2}}$	$\sqrt{2}$
half-wave rectified sinusoid V_m	$\frac{1}{\pi}$	$\frac{1}{2}$	$\frac{\pi}{2}$	2
symmetrical square wave V_m, $\frac{T}{2}$	0	1	–	1
unsymmetrical square wave t, $T-t$, V_m, T	$\frac{t}{T}$	$\sqrt{\frac{t}{T}}$	$\sqrt{\frac{T}{t}}$	$\sqrt{\frac{T}{t}}$
sawtooth and symmetrical triangular V_m, $\frac{T}{2}$	0	$\frac{1}{\sqrt{3}}$	–	$\sqrt{3}$
sawtooth and unsymmetrical triangular V_m	$\frac{1}{2}$	$\frac{1}{\sqrt{3}}$	$\frac{2}{\sqrt{3}}$	$\sqrt{3}$

9. PHASE ANGLES

AC circuit inductors and capacitors have the ability to store energy in magnetic and electric fields, respectively. Consequently, while the same shape, the voltage and current waveforms differ by an amount called the *phase angle difference*, ϕ. Ordinarily, the voltage and current sinusoids do not peak at the same time. In a *leading circuit*, the phase angle difference is positive and the current peaks before the voltage (see Fig. 27.6). A leading circuit is termed a *capacitive circuit.* In a *lagging circuit*, the phase angle difference is negative and the current peaks after the voltage. A lagging circuit is termed an *inductive circuit.* These cases are represented mathematically as

$$v(t) = V_m \sin(\omega t + \theta) \quad \text{[reference]} \tag{27.24}$$

$$i(t) = I_m \sin(\omega t + \theta + \phi) \quad \text{[leading]} \tag{27.25}$$

$$i(t) = I_m \sin(\omega t + \theta - \phi) \quad \text{[lagging]} \tag{27.26}$$

Figure 27.6 *Leading Phase Angle Difference*

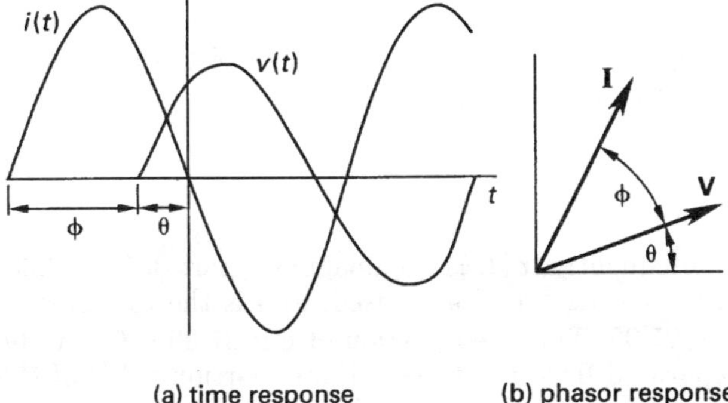

10. SINUSOID

The most important waveform in electrical engineering is the sine function, or a waveform that is sinusoidal in value with respect to time. A sinusoidal waveform has an amplitude (also known as a *magnitude* or *peak value*) that remains constant, such as V_m. The waveform, however, repeats or goes through cycles. From Eq. 27.1, the sinusoid can be characterized by three quantities: magnitude, frequency, and phase angle.[7] The major properties of the sinusoid are covered in Sec. 27.2.

A circuit processes or changes waveforms. This is called *signal processing* or *waveform processing. Signal analysis* is the determination of these waveforms. The sinusoid can be represented as a phasor and used to analyze electrical circuits only if the circuit is in a steady-state condition. The terms AC and DC imply steady-state conditions. For AC circuits, *sinusoidal steady-state* indicates that all voltages and currents within the circuit are sinusoids of the same frequency as the excitation, that is, the driving voltage.

Example 27.5

A reference voltage of $v(t) = 170 \sin \omega t$ is used during an experiment. A measurement of a second voltage, $v_2(t)$ occurs. The second voltage reaches its peak 2.5 ms before the reference voltage and has the same peak value 20 ms later. The peak value is 1.8 times the reference peak. What is the expression for $v_2(t)$?

Solution

Any sinusoidal voltage can be represented as in Eq. 27.1.

$$v(t) = V_m \sin(\omega t + \theta)$$

The peak value (equivalent to V_m) is 1.8 times the reference, or $1.8 \times 170 \text{ V} = 306 \text{ V}$. Calculate the angular frequency using Eq. 27.3 and the given period of 20 ms.

$$\omega = \frac{2\pi}{T} = \frac{2\pi}{20 \text{ ms}} = 314 \text{ rad/s}$$

The phase angle is determined by changing the time, 2.5 ms, into its corresponding angle, θ, as follows.

$$\theta = \omega t = \left(314 \ \frac{\text{rad}}{\text{s}}\right)(2.5 \times 10^{-3} \text{ s}) = 0.79 \text{ rad}$$

Converting 0.79 rad gives a result of 45°. Substituting the calculated values gives the result.

$$v_2(t) = V_m \sin(\omega t + \theta) = 306 \sin(314t + 45°)$$

The result mixes the use of radians (314) and degrees (45). This is often done for clarity. Ensure one or the other is converted prior to calculations. It is often best to deal with all angles in radians.

11. PHASORS

The addition of voltages and currents of the same frequency is simplified mathematically by treating them as phasors.[8] Using a method called the *Steinmetz algorithm,* the following two quantities, introduced in Sec. 27.2, are considered analogs.

$$v(t) = V_m \sin(\omega t + \theta) \text{ and } V = V_m \angle\theta \tag{27.27}$$

Both forms show the magnitude and the phase, but the *phasor,* $V_m\angle\theta$, does not show frequency. In the phasor form, the frequency is implied. The phasor form of Eq. 27.27 is shown in Fig. 27.7.

The phasor is actually a point, but is represented by an arrow of magnitude V_m at an angle θ with respect to a reference, normally taken to be $\theta = 0$. A *reference phasor* is one of known value, usually the driving voltage of a circuit (i.e., the voltage phasor). The reference phasor would be shown in the position of the real axis with θ equal to zero. Phasors are summed using phasor addition, commonly referred to as *vector addition.*[9] The Steinmetz algorithm is illustrated in Fig. 27.8.

In electric circuits with sinusoidal waveforms, an alternate phasor representation is often used. The alternate representation is called the *effective value phasor notation.* In this notation, the rms values of the voltage and

[7] A phase angle of ±90° changes the sine function into a cosine function. Also, frequency refers in this case to the angular frequency, ω, in radians per second. The term frequency also refers to the term *f*, measured in hertz (Hz).

[8] For all circuit elements at the same frequency, the circuit must be in a steady-state condition.

[9] The term "vector addition" is not technically correct because phasors, with the exception of impedance and admittance phasors, rotate with time. The methods, however, are identical.

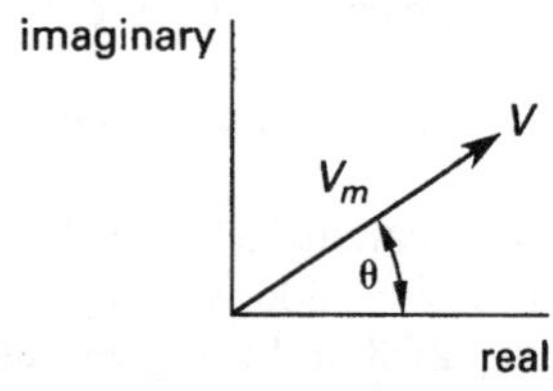

Figure 27.7 Phasor of Magnitude V_m and Angle θ

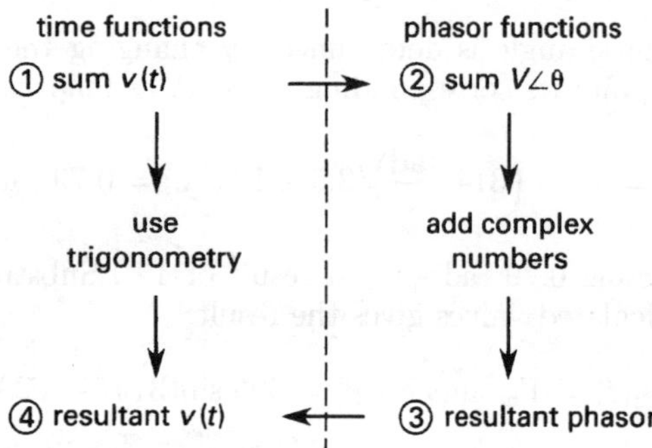

Figure 27.8 Steinmetz Algorithm Steps

current are used instead of the peak values. The angles are also given in degrees. The phasor notation of Eq. 27.27 is then modified to

$$\begin{aligned} V &= V_{\text{rms}}\angle\theta \\ &= \frac{V_m}{\sqrt{2}}\angle\theta \\ &= V\angle\theta \end{aligned} \qquad 27.28$$

The effective value phasor notation represented by the first and second parts of Eq. 27.28 is used by NCEES. When the notation is not specified (for example, when a subscript is not used as in the third part of Eq. 27.28), assume the effective notation. Additionally, many texts do not show phasors in bold font, but instead let the angle symbol indicate that a phasor is being used. Care should be taken to determine the type of notation in use.

Ensure that angles substituted into exponential forms are in radians or errors in mathematical calculations will result.

12. COMPLEX REPRESENTATION

Phasors are plotted in the complex plane. Figure 27.9 shows the relationships among the complex quantities introduced.

In Fig. 27.9, **Z** represents a complex number, not the impedance. Equation 27.1 is repeated here for convenience.

$$v(t) = V_m \sin(\omega t + \theta) \qquad 27.29$$

Figure 27.9 Complex Quantities

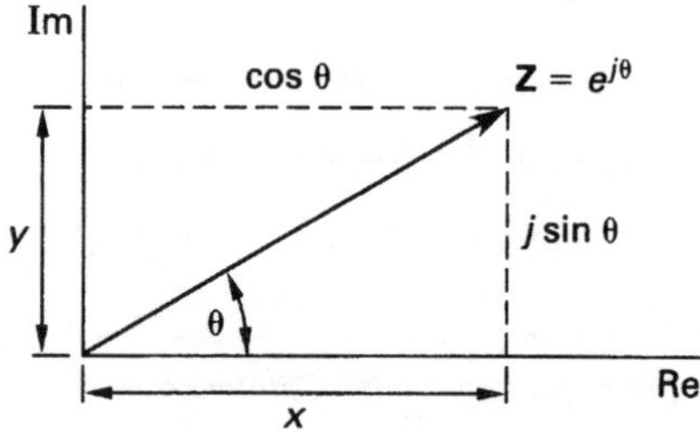

Using Euler's relation, Eq. 27.4, the voltage $v(t)$ can be represented as

$$v(t) = V_m e^{j(\omega t+\theta)} \qquad 27.30$$

Consequently, $v(t)$ is the imaginary part of Eq. 27.30. If the cosine function is used, $v(t)$ is the real part of Eq. 27.30. The fixed portion of Eq. 27.30 ($e^{j\theta}$) can be separated from the time-variable portion ($e^{j\omega t}$) of the function. This is shown in Fig. 27.10(a) and (b). The magnitude of $e^{j\omega t}$ remains equal to one, but the angle increases (rotates counterclockwise) linearly with time. All functions with an $e^{j\omega t}$ term are assumed to rotate counterclockwise with an angular velocity of ω in the complex plane. The voltage can be represented as in Fig. 27.10(c) and

$$v(t) = V_m e^{j\omega t} e^{j\theta} \qquad 27.31$$

If the rotating portion of Eq. 27.31 is assumed to exist, the voltage can be written

$$v(\theta) = V_m e^{j\theta} \qquad 27.32$$

Changing Eq. 27.32 to the phasor form yields

$$\mathbf{V} = V_m\angle\theta \quad \text{or}$$
$$\mathbf{V} = V_{\text{rms}}\angle\theta = \frac{V_m}{\sqrt{2}}\angle\theta = V\angle\theta \qquad 27.33$$

The properties of complex numbers, which can represent voltage, current, or impedance, are summarized in Table 27.3. The designations used in the table are illustrated in Fig. 27.9.

13. RESISTORS

Resistors oppose the movement of electrons. In an *ideal* or *pure resistor*, no inductance or capacitance exists. The magnitude of the impedance is the resistance, R, with units of ohms and an impedance angle of zero. Therefore, voltage and current are in phase in a purely resistive circuit.

$$\mathbf{Z}_R = R\angle 0 = R + j0 \qquad 27.34$$

Figure 27.10 Phasor Rotation in the Complex Plane

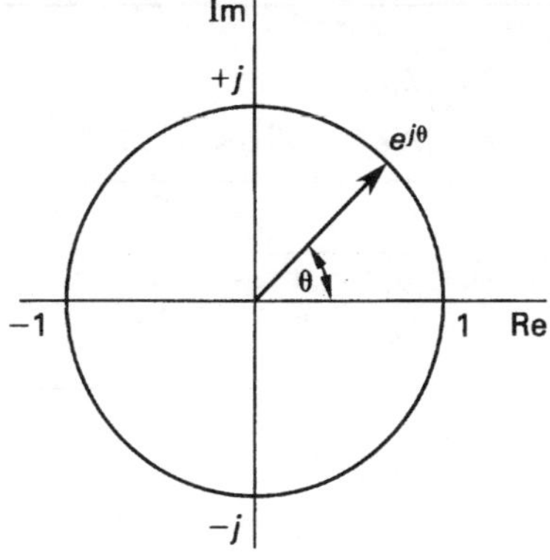

(a) $e^{j\theta}$ for all time

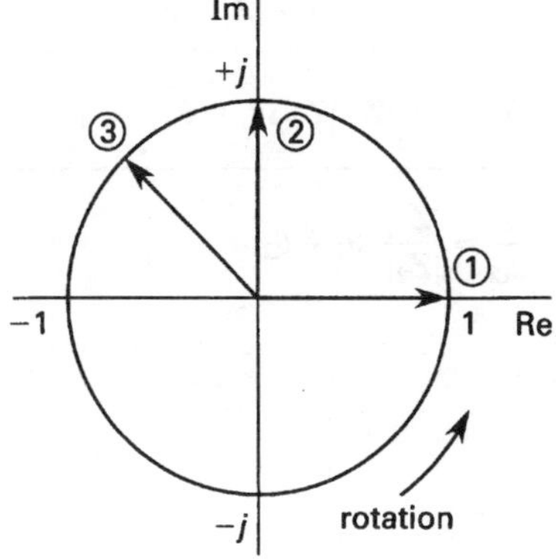

(b) $e^{j(\omega t)}$ for

① $\omega t = 0$

② $\omega t = \frac{\pi}{2}$

③ $\omega t = \frac{3\pi}{4}$

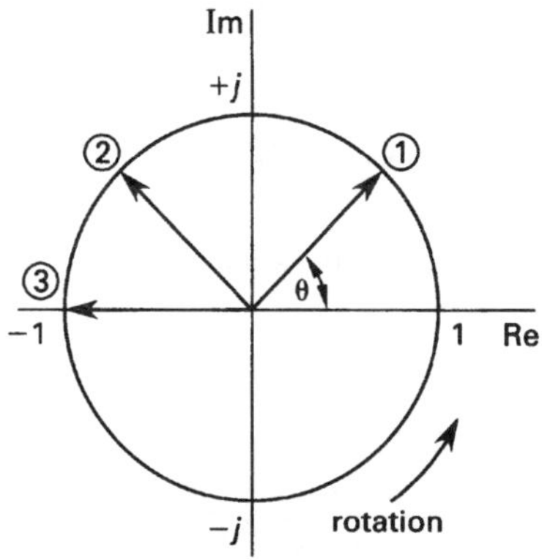

(c) $e^{j(\omega t + \theta)}$ for

① $\omega t = 0$

② $\omega t = \frac{\pi}{2}$

③ $\omega t = \frac{3\pi}{4}$

14. CAPACITORS

Capacitors oppose the movement of electrons by storing energy in an electric field and using this energy to resist changes in voltage over time. Unlike a resistor, an *ideal* or *perfect capacitor* consumes no energy. Equation 27.35 gives the impedance of an ideal capacitor with capacitance C. The magnitude of the impedance is termed the *capacitive reactance*, X_C, with units of ohms and an impedance angle of $-\pi/2(-90°)$. Consequently, the current leads the voltage by 90° in a purely capacitive circuit.[10]

$$\mathbf{Z}_C = X_C \angle -90° = 0 + jX_C \qquad 27.35$$

$$X_C = \frac{-1}{\omega C} = \frac{-1}{2\pi f C} \qquad 27.36$$

15. INDUCTORS

Inductors oppose the movement of electrons by storing energy in a magnetic field and using this energy to resist changes in current over time. Unlike a resistor, an *ideal* or *perfect inductor* consumes no energy. Equation 27.37 gives the impedance of an ideal inductor with inductance L. The magnitude of the impedance is termed the *inductive reactance*, X_L, with units of ohms and an impedance angle of $\pi/2$ (90°). Consequently, the current lags the voltage by 90° in a purely inductive circuit.[11]

$$\mathbf{Z}_L = X_L \angle 90° = 0 + jX_L \qquad 27.37$$

$$X_L = \omega L = 2\pi f L \qquad 27.38$$

16. COMBINING IMPEDANCES

Impedances in combination are like resistors: Impedances in series are added, while the reciprocals of impedances in parallel are added. For series circuits, the resistive and reactive parts of each impedance element are calculated separately and summed. For parallel circuits, the conductance and susceptance of each element are summed. The total impedance is found by a complex addition of the resistive (conductive) and reactive (susceptive) parts. It is convenient to perform the addition in rectangular form.

$$\begin{aligned} Z_e &= \sum Z \\ &= \sqrt{\left(\sum R\right)^2 + \left(\sum X_L - \sum X_C\right)^2} \quad \text{[series]} \end{aligned} \qquad 27.39$$

[10]The impedance angle for a capacitor is negative, but the current phase angle difference is positive—hence the term "leading." This occurs mathematically because the current is obtained by dividing the voltage by the impedance: $\mathbf{I} = \mathbf{V}/\mathbf{Z}$.

[11]The impedance angle for an inductor is positive, but the current phase angle difference is negative—hence the term "lagging." This occurs mathematically because the current is obtained by dividing the voltage by the impedance: $\mathbf{I} = \mathbf{V}/\mathbf{Z}$.

Table 27.3 Properties of Complex Numbers

	rectangular form	polar/exponential form
	$\mathbf{Z} = x + jy$	$\mathbf{Z} = \lvert\mathbf{Z}\rvert\angle\theta$ $\mathbf{Z} = \lvert\mathbf{Z}\rvert e^{j\theta} = \lvert\mathbf{Z}\rvert\cos\theta + j\lvert\mathbf{Z}\rvert\sin\theta$
relationship between forms	$x = \lvert\mathbf{Z}\rvert\cos\theta$ $y = \lvert\mathbf{Z}\rvert\sin\theta$	$\lvert\mathbf{Z}\rvert = \sqrt{x^2 + y^2}$ $\theta = \arctan\frac{y}{x}$
complex conjugate	$\mathbf{Z}^* = x - jy$ $\mathbf{ZZ}^* = (x^2 + y^2) = \lvert z\rvert^2$	$\mathbf{Z}^* = \lvert\mathbf{Z}\rvert e^{-j\theta} = \lvert\mathbf{Z}\rvert\angle -\theta$ $\mathbf{ZZ}^* = (\lvert\mathbf{Z}\rvert e^{j\theta})(\lvert\mathbf{Z}\rvert e^{-j\theta}) = \lvert\mathbf{Z}\rvert^2$
addition	$\mathbf{Z}_1 + \mathbf{Z}_2 = (x_1 + x_2) + j(y_1 + y_2)$	$\mathbf{Z}_1 + \mathbf{Z}_2 = (\lvert\mathbf{Z}_1\rvert\cos\theta_1 + \lvert\mathbf{Z}_2\rvert\cos\theta_2)$ $+ j(\lvert\mathbf{Z}_1\rvert\sin\theta_1 + \lvert\mathbf{Z}_2\rvert\sin\theta_2)$
multiplication	$\mathbf{Z}_1\mathbf{Z}_2 = (x_1x_2 - y_1y_2) + j(x_1y_2 + x_2y_1)$	$\mathbf{Z}_1\mathbf{Z}_2 = \lvert\mathbf{Z}_1\rvert\lvert\mathbf{Z}_2\rvert\angle\theta_1 + \theta_2$
division	$\frac{\mathbf{Z}_1}{\mathbf{Z}_2} = \frac{(x_1x_2 + y_1y_2) + j(x_2y_1 - x_1y_2)}{\lvert\mathbf{Z}_2\rvert^2}$	$\frac{z_1}{z_2} = \frac{\lvert\mathbf{Z}_1\rvert}{\lvert\mathbf{Z}_2\rvert}\angle\theta_1 - \theta_2$

$$\frac{1}{Z_e} = \sum\frac{1}{Z} = Y_e$$

$$= \sqrt{\left(\sum G\right)^2 + \left(\sum B_L - \sum B_C\right)^2} \quad \text{[parallel]} \qquad 27.40$$

The impedance of various series-connected circuit elements is given in App. 27.A. The impedance of various parallel-connected circuit elements is given in App. 27.B.

Example 27.6

Determine the impedance and admittance of the following circuits.

(a)

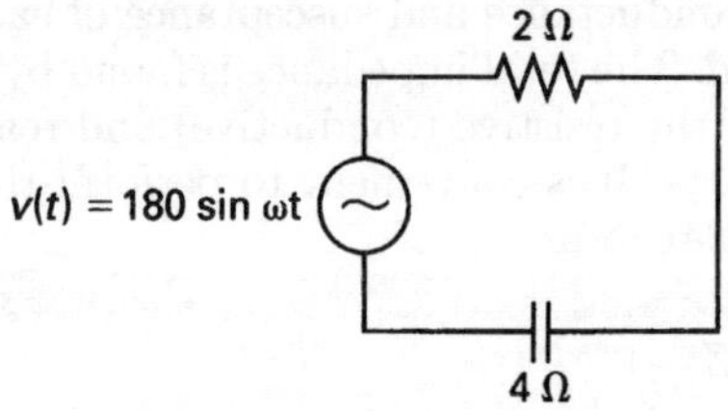

2 Ω

v(t) = 180 sin ωt

4 Ω

(b)

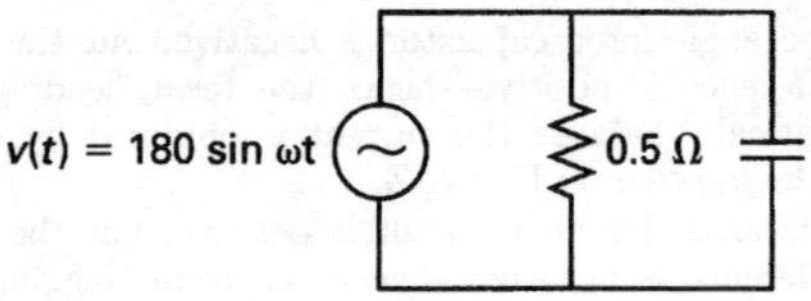

v(t) = 180 sin ωt 0.5 Ω 0.25 Ω

Solution

(a) From Eq. 27.39,

$$Z = \sqrt{R^2 + Z_C^2} = \sqrt{(2\ \Omega)^2 + (4\ \Omega)^2} = 4.47\ \Omega$$

$$\phi = \arctan\frac{X_C}{R} = \arctan\frac{-4\ \Omega}{2\ \Omega} = -63.4°$$

$$\mathbf{Z} = 4.47\ \Omega\angle -63.4°$$

From Eq. 27.9,

$$\mathbf{Y} = \frac{1}{\mathbf{Z}} = \frac{1}{4.47\ \Omega\angle -63.4°} = 0.224\ \text{S}\angle 63.4°$$

(b) Because this is a parallel circuit, work with the admittances.

$$G = \frac{1}{R} = \frac{1}{0.5\ \Omega} = 2\ \text{S}$$

$$B_C = \frac{1}{X_C} = \frac{1}{0.25\ \Omega} = 4\ \text{S}$$

$$Y = \sqrt{G^2 + B_C^2} = \sqrt{(2\ \text{S})^2 + (4\ \text{S})^2} = 4.47\ \text{S}$$

$$\phi = \arctan\frac{B_C}{G} = \arctan\frac{4\ \text{S}}{2\ \text{S}} = 63.4°$$

$$\mathbf{Y} = 4.47\ \text{S}\angle 63.4°$$

$$\mathbf{Z} = \frac{1}{\mathbf{Y}} = \frac{1}{4.47\ \text{S}\angle 63.4°} = 0.224\ \Omega\angle -63.4°$$

17. OHM'S LAW

Ohm's law for AC circuits with linear circuit elements is similar to Ohm's law for DC circuits.[12] The difference is that all the terms are represented as phasors.[13]

$$\mathbf{V} = \mathbf{IZ} \tag{27.41}$$

$$V\angle\theta v = (I\angle\varphi_I)(Z\angle\phi_Z) \tag{27.42}$$

18. POWER

The instantaneous power, p, in a purely resistive circuit is

$$\begin{aligned} p_R(t) &= i(t)v(t) = (I_m \sin \omega t)(V_m \sin \omega t) \\ &= I_m V_m \sin^2 \omega t \\ &= \tfrac{1}{2} I_m V_m - \tfrac{1}{2} I_m V_m \cos 2\omega t \end{aligned} \tag{27.43}$$

The second term of Eq. 27.43 integrates to zero over the period. Therefore, the *average power* dissipated is

$$\begin{aligned} P_R &= \tfrac{1}{2} I_m V_m = \frac{V_m^2}{2R} \\ &= I_{\text{rms}} V_{\text{rms}} = IV \end{aligned} \tag{27.44}$$

Equation 27.28 was used to change the maximum, or peak, values into the more usable rms, or effective, values.

In a purely capacitive circuit, the current leads the voltage by 90°. This allows the current term of Eq. 27.43 to be replaced by a cosine (because the sine and cosine differ by a 90° phase). The instantaneous power in a purely capacitive circuit is

$$\begin{aligned} p_C(t) &= i(t)v(t) \\ &= (I_m \cos \omega t)(V_m \sin \omega t) \\ &= I_m V_m \sin \omega t \cos \omega t \\ &= \tfrac{1}{2} I_m V_m \sin 2\omega t \end{aligned} \tag{27.45}$$

Because the $\sin 2\omega t$ term integrates to zero over the period, the average power is zero. Nevertheless, power is stored during the charging process and returned to the circuit during the discharging process.

$$P_C = 0 \tag{27.46}$$

Similarly, the instantaneous power in an inductor is

$$p_L(t) = -\tfrac{1}{2} I_m V_m \sin 2\omega t \tag{27.47}$$

[12]Ohm's law can be used on *nonlinear devices* (NLD) if the region analyzed is restricted to be approximately linear. When this condition is applied, the analysis is termed *small signal analysis*.

[13]In general, phasors are considered to rotate with time. A secondary definition of a phasor is any quantity that is a complex number. As a result, the impedance is also considered a phasor even though it does not change with time.

Again, the $\sin 2\omega t$ integrates to zero over the period, and the average power is zero. Nevertheless, power is stored during the expansion of the magnetic field and returned to the circuit during the contraction of the magnetic field.

$$P_L = 0 \tag{27.48}$$

19. REAL POWER AND THE POWER FACTOR

In a circuit that contains all three circuit elements (resistors, capacitors, and inductors) or the effects of all three, the average power is calculated from Eq. 27.14 as

$$P_{\text{ave}} = \frac{1}{T}\int_0^T i(t)v(t)\,dt \tag{27.49}$$

Let the generic voltage and current waveforms be represented by Eq. 27.50 and Eq. 27.51. The *current angle* φ is equal to $\theta \pm \phi$.[14]

$$i(t) = I_m \sin(\omega t + \varphi) \tag{27.50}$$

$$v(t) = V_m \sin(\omega t + \theta) \tag{27.51}$$

Substituting Eq. 27.50 and Eq. 27.51 into Eq. 27.49 gives

$$P_{\text{ave}} = \left(\frac{I_m V_m}{2}\right)\cos\phi_{\text{pf}} = I_{\text{rms}} V_{\text{rms}} \cos\phi_{\text{pf}} \tag{27.52}$$

The angle $\phi_{\text{pf}} = |\theta - \varphi|$ is the *power factor angle.* It represents the difference between the voltage and current angles. This difference is the impedance angle, $\pm\phi$. Because $\cos(-\phi) = \cos(+\phi)$, only the absolute value of the impedance angle is used. Because the absolute value is used in the equation, the terms *leading* (for a capacitive circuit) and *lagging* (for an inductive circuit) must be used when describing the power factor. The power factor of a purely resistive circuit equals one; the power factor of a purely reactive circuit equals zero.

Equation 27.52 determines the magnitude of the product of the current and voltage in phase with one another, that is, of a resistive nature. Consequently, Eq. 27.52 determines the power consumed by the resistive elements of a circuit. This average power is called the *real power* or *true power*, and sometimes the *active power.* The quantity $\cos\phi_{\text{pf}}$, or $\cos|\phi|$, is called the *power factor* or *phase factor* and given the symbol pf. The power factor is also equal to the ratio of the real power to the apparent power (see Sec. 27.22). Often, no subscript is used on P when it represents the real power, nor are subscripts used on rms or effective values. The real power is then given by

$$P = IV\cos\phi_{\text{pf}} = IV\text{pf} \tag{27.53}$$

[14]Using a current angle simplifies the derivational mathematics (not shown) and clarifies the definition of the power factor angle given in this section.

It is important to realize that the real power cannot be obtained by multiplying phasors. Doing so results in the addition of the voltage and current angles when the difference is required.

$$P \neq I\angle\varphi V\angle\theta = VI\angle\theta + \varphi \quad 27.54$$

If phasor power is used, the difficulty can be alleviated by using Eq. 27.55.

$$P = \mathrm{Re}(\mathbf{VI}^*) \quad 27.55$$

Equation 27.55 is equivalent to Eq. 27.53.

20. REACTIVE POWER

The *reactive power*, Q, measured in units of VAR, is given by

$$Q = IV\sin\phi_{\text{pf}} \quad 27.56$$

The reactive power, sometimes called the *wattless power*, is the product of the rms values of the current and voltage multiplied by the *quadrature* of the current. The quantity $\sin\phi_{\text{pf}}$ is called the *reactive factor.*[15] The reactive power represents the energy stored in the inductive and capacitive elements of a circuit.

21. APPARENT POWER

The *apparent power*, S, measured in voltamperes, is given by

$$S = IV \quad 27.57$$

The apparent power is the product of the rms values of the voltage and current without regard to the angular relationship between them. The apparent power is representative of the combination of real and reactive power (see Sec. 27.22). As such, not all of the apparent power is dissipated or consumed. Nevertheless, electrical engineers must design systems to adequately handle this power as it exists in the system.

22. COMPLEX POWER AND THE POWER TRIANGLE

The real, reactive, and apparent powers can be related to one another as vectors. The *complex power vector*, **S**, is the vector sum of the *reactive power vector*, **Q**, and the *real power vector*, **P**. The magnitude of **S** is given by Eq. 27.57. The magnitude of **Q** is given by Eq. 27.56. The magnitude of **P** is given by Eq. 27.53. In general, $\mathbf{S} = \mathbf{I}^*\mathbf{V}$ where $\mathbf{I}^*$ is the complex conjugate of the current, that is, the current with the phase difference angle reversed. This convention is arbitrary, but emphasizes that the *power angle*, ϕ, associated with **S** is the same as the overall impedance angle, ϕ, whose magnitude equals the power factor angle, ϕ_{pf}.

The relationship between the magnitudes of these powers is

$$S^2 = P^2 + Q^2 \quad 27.58$$

A drawing of the power vectors in a complex plane is shown in Fig. 27.11 for leading and lagging conditions, and the following relationships are determined from it.

$$P = S\cos\phi \quad 27.59$$

$$Q = S\sin\phi \quad 27.60$$

Figure 27.11 *Power Triangle*

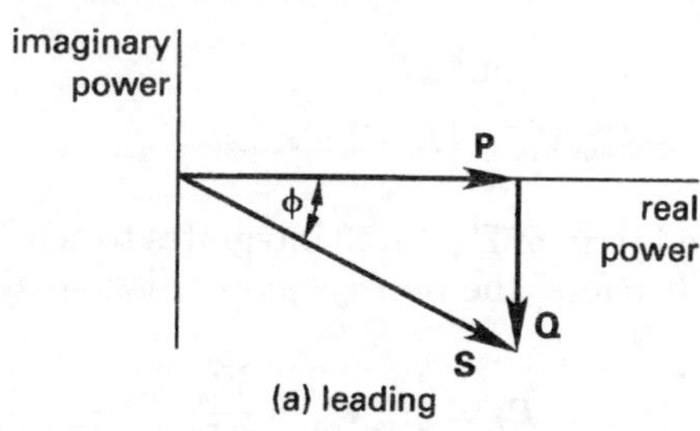

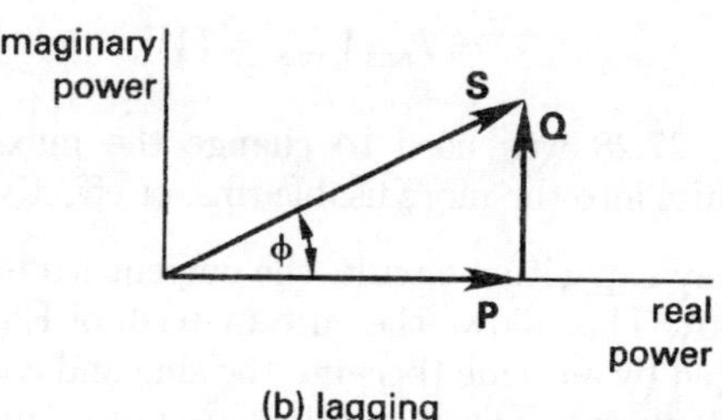

The ratio of reactance, X, to resistance, R, given in Eq. 27.62 and often used in fault analysis and transformer evaluation, can be determined from the power factor, pf, and vice versa. The power factor is the relationship between the current and the voltage. Since this relationship shifts significantly during a short circuit, in fault analysis the short-circuit power factor must be used.[16]

$$\text{pf} = \frac{P}{S} \quad 27.61$$

$$\frac{X}{R} = \tan(\arccos \text{pf}) \quad 27.62$$

Example 27.7

For the circuit shown, find the (a) apparent power, (b) real power, and (c) reactive power; and (d) draw the power triangle.

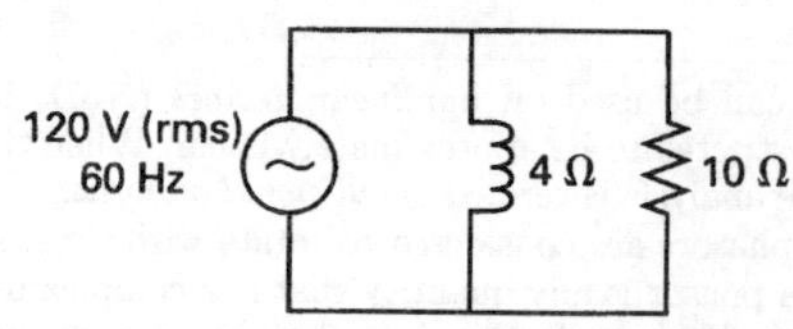

[15]The power factor angle, ϕ_{pf}, is used instead of the absolute value of the impedance angle (as in the power factor) because the sine function results in positive and negative values depending upon the difference between the voltage and current angles. If ϕ_{pf} were not used, $\pm\phi$ would have to be used.

[16]See Chap. 39 for more on the short circuit phenomenon.

Solution

The equivalent impedance is

$$\frac{1}{Z} = \frac{1}{j4\ \Omega} + \frac{1}{10\ \Omega} = -j0.25 + 0.10\ \text{S}$$

$$Z = 3.714\ \Omega$$

$$\phi_Z = \arctan\frac{0.25\ \text{S}}{0.10\ \text{S}} = 68.2^\circ$$

The total current is

$$I = \frac{V}{Z} = \frac{120\ \text{V}\angle 0^\circ}{3.714\ \Omega\angle 68.2^\circ} = 32.31\ \text{A}\angle{-68.2^\circ}$$

(a) The apparent power is

$$S = I^*V = (32.31\ \text{A}\angle 68.2^\circ)(120\ \text{V}\angle 0^\circ) = 3877\ \text{VA}\angle 68.2^\circ$$

(The angle of apparent power is usually not reported.)

(b) The real power is

$$P = \frac{V_R^2}{R} = \frac{(120\ \text{V})^2}{10\ \Omega} = 1440\ \text{W}$$

Alternatively, the real power can be calculated from Eq. 27.59.

$$P = S\cos\phi = (3877\ \text{VA})\cos 68.2^\circ = 1440\ \text{W}$$

(c) The reactive power is

$$Q = \frac{V_L^2}{X_L} = \frac{(120\ \text{V})^2}{4\ \Omega} = 3600\ \text{VAR}$$

Alternatively, the reactive power can be calculated from Eq. 27.60.

$$Q = S\sin\phi = (3877\ \text{VA})\sin 68.2^\circ = 3600\ \text{VAR}$$

(d) The real power is represented by the vector (in rectangular form) 1440 + $j0$. The reactive power is represented by the vector 0 + $j3600$. The apparent power is represented (in phasor form) by the vector $3877\angle 68.2^\circ$. The power triangle is

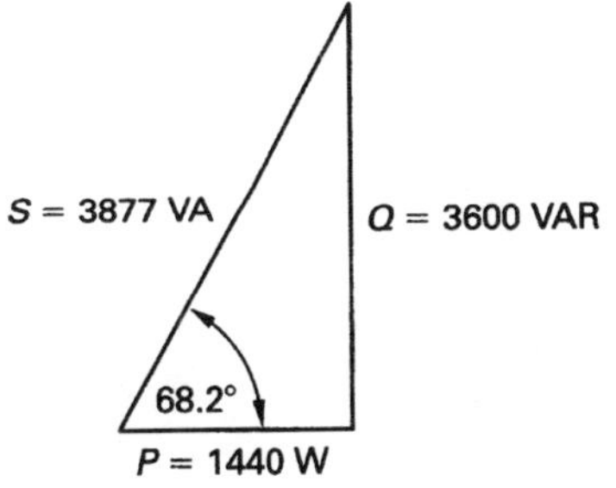

23. MAXIMUM POWER TRANSFER

Assuming a fixed primary impedance, the maximum power condition in an AC circuit is similar to that in a DC circuit. That is, the maximum power is transferred from the source to the load when the impedances match. This occurs when the circuit is in resonance. Resonance is discussed more fully in Chap. 30. The conditions for maximum power transfer, and therefore resonance, are given by

$$R_{\text{load}} = R_s \qquad 27.63$$

$$X_{\text{load}} = -X_s \qquad 27.64$$

24. AC CIRCUIT ANALYSIS

All of the methods and equations presented in Chap. 26, such as Ohm's and Kirchhoff's laws and loop-current and node-voltage methods, can be used to analyze AC circuits as long as complex arithmetic is utilized.

25. COMPARISON OF ANALOG, DISCRETE, AND DIGITAL SIGNALS

A number of different signal types are encountered in AC analysis. Each has distinctive uses and is best suited to distinctive analysis techniques. *Discrete signals* are used in power systems to sample analog values and convert these values into a form that can be used by control circuitry. *Digital signals* are on/off or one/zero signals often used in computers and digital systems that monitor and manipulate control circuitry. Each signal type is shown in Fig. 27.12, and each part of the figure shows a different way of representing the same signal.

An analog signal represents a variable as a continuous measured parameter, as shown in Fig. 27.12(a).[17] A discrete signal is composed of separate and distinct parts. It represents a continuous time-varying signal at discrete time intervals known as the *sampling rate*, as shown in Fig. 27.12(b). The amplitude of the discrete signal is determined by the continuous signal that is sampled. A digital signal makes both the sampling rate (time) and the amplitude discrete, as shown in Fig. 27.12(c).

[17]An analog signal can also be defined as a signal that represents a time-varying feature (variable) of *another* time-varying quantity. A typical example is an AC sinusoidal voltage, which can represent a transducer output, an audio signal, or any of a multitude of sensor signals. In the case of an AC machine, the voltage does not represent another variable per se, though it may be viewed as representing the time-varying magnetic field impressed on the armature or stator. The voltage represents not a "signal" but a continuous voltage variable output directly. The distinction is not usually made, but the term "signal" is used more often when referring to sensor outputs.

Figure 27.12 Representations of a Single AC Signal

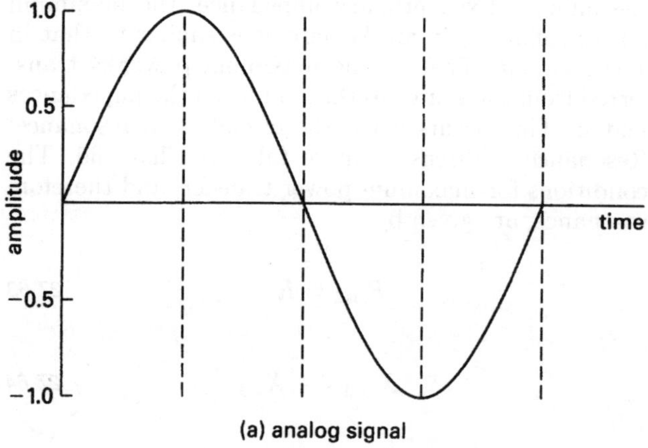

(a) analog signal

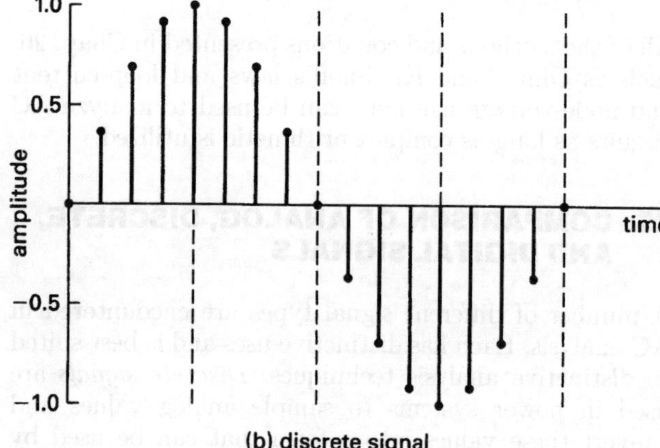

(b) discrete signal

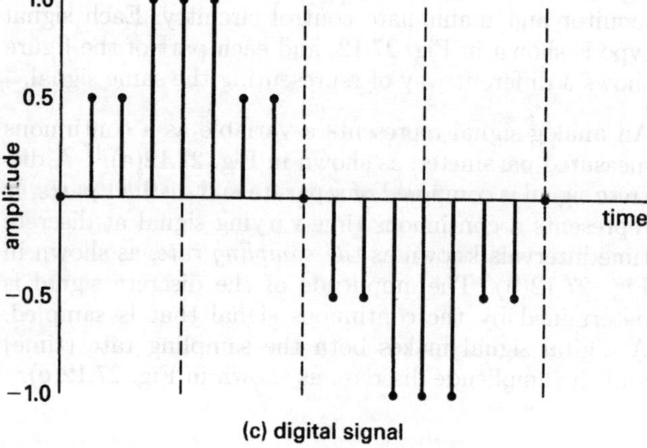

(c) digital signal

28 Transformers

Nomenclature

a	turns ratio	–
B	magnetic flux density	T
C	capacitance	F
f	frequency	Hz
H	magnetic field strength	A/m
i	variable current	A
I	constant or rms current	A
k	coupling coefficient	–
L	inductance or self-inductance	H
M	mutual inductance	H
N	number of items	–
O	origin	–
R	resistance	Ω
t	time	s
v	variable voltage	V
V	constant or rms voltage	V
V_s	effective voltage	V
X	reactance	Ω
Z	impedance	Ω

Symbols

μ	permeability	H/m
Φ	magnetic flux	Wb
ω	angular frequency	rad/s

Subscripts

0	initial
c	coercive or core
C	capacitative or capacitor
ep	effective primary
E	exciting
L	inductor
m	magnetizing, maximum, or mutual
p	primary
r	residual
ref	reflected
s	secondary

1. FUNDAMENTALS

A *transformer* is an electrical device consisting of two or more multiturn coils of wire in close proximity so that their magnetic fields are linked. This linkage allows the transfer of electrical energy from one or more alternating circuits to one or more other alternating circuits by the process of magnetic induction.

Transformers can be used to raise or lower the values of capacitors, inductors, and resistors. They enable the efficient transmission of electrical energy at high voltages over great distances, and then can be used to lower the voltages to safe values for industrial, commercial, and household use.

The principle of the transformer is also the basic principle of induction motors, alternators, and synchronous motors. All these devices are based on the laws of magnetic induction.

2. MAGNETIC COUPLING

Any coil of wire, when energized by a sinusoidal waveform, produces a magnetic flux, Φ, that is also sinusoidal.[1]

$$\Phi(t) = \Phi_m \sin \omega t \qquad 28.1$$

Faraday's law gives the voltage that is induced when a portion of the magnetic flux produced in one coil of wire with N_p turns passes through a second coil of wire with N_s turns. The minus sign is a result of Lenz's law, which states that the induced voltage, and therefore the induced flux, always acts to oppose the voltage (and flux) that created it.

$$v_s(t) = -N_s\left(\frac{d\Phi(t)}{dt}\right) = -N_s\omega\Phi_m \cos \omega t \qquad 28.2$$

The effective voltage is obtained by dividing Eq. 28.2 by the square root of two. The net result is[2]

$$\begin{aligned} V_s(t) &= \frac{-N_s(2\pi f)\Phi_m \cos \omega t}{\sqrt{2}} \\ &= -4.44 N_s f \Phi_m \cos \omega t \end{aligned} \qquad 28.3$$

[1]This ignores harmonic effects in the core of the transformer stemming from variable permeability. These harmonics must be accounted for if inductive interference with communications circuits is a concern.

[2]The capital V_s is used in Eq. 28.3 to indicate the effective voltage value, even though the voltage remains a function of time.

The magnetic flux that passes through both coils is called the *mutual flux*. The magnetic flux that couples only one coil is called the *leakage flux*. The term "leakage" is used because any flux that does not link the coils will not contribute to inducing a voltage in the secondary of the transformer.

The fluxes for a *core transformer* are illustrated in Fig. 28.1. The magnetic flux Φ_{12} is the flux generated by coil 1 that links coil 2. The magnetic flux Φ_{11} is the flux generated by coil 1 that links only coil 1. Similarly, the magnetic flux Φ_{21} is the flux generated by coil 2 that links coil 1, and the magnetic flux Φ_{22} is the flux generated by coil 2 that links only coil 2. Coil 1 and coil 2 are also called the *primary winding* and *secondary winding*, respectively. Magnetic fluxes Φ_{11} and Φ_{22} result in the *self-inductances* of coil 1 and coil 2 (L_p and L_s, respectively).

Figure 28.1 *Fluxes in a Magnetically Coupled Circuit*

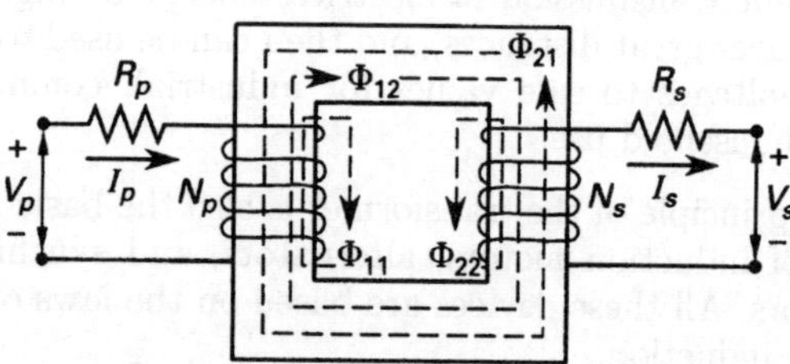

The *mutual inductance*, M, that results from magnetic fluxes Φ_{12} and Φ_{21}, is defined as the ratio of the electromotive force induced in one coil (circuit) to the rate of change in the current in the other coil (circuit). Mutual inductance is the proportionality constant between the induced voltage and the rate of current change.

$$v_p(t) = M\frac{di_s(t)}{dt} \tag{28.4}$$

$$v_s(t) = M\frac{di_p(t)}{dt} \tag{28.5}$$

Just as leakage flux is associated with self-inductance, mutual flux is associated with a *mutual reactance*, which is given by

$$X_m = \omega M \tag{28.6}$$

The induced voltage and generated flux associated with mutual reactance always oppose the flux and current that created them. In Fig. 28.1, for example, Φ_{12} is generated by the current in the primary winding, and results in a voltage in the secondary winding. This secondary voltage causes a current to flow that generates Φ_{21}, which opposes the original flux and generates a voltage drop of $-(I_s)j\omega M$ in the *primary circuit*. By similar reasoning, a voltage drop of $-I_pj\omega M$ is generated in the *secondary circuit* of the transformer as well. This is illustrated in Fig. 28.2(a), which is equivalent to Fig. 28.1, and in Eq. 28.7 and Eq. 28.8, which are obtained by applying Kirchhoff's voltage law (KVL) around the indicated loops.

$$\begin{aligned} V_p &= I_p(R_p + j\omega L_p) - I_sj\omega M \\ &= I_p(R_p + jX_p) - I_sjX_m \\ &= I_pZ_p - I_sjX_m \end{aligned} \tag{28.7}$$

$$\begin{aligned} V_s &= I_s(R_s + j\omega L_s) - I_pj\omega M \\ &= I_s(R_s + jX_s) - I_pjX_m \\ &= I_sZ_s - I_pjX_m \end{aligned} \tag{28.8}$$

Figure 28.2 *Transformer Models*

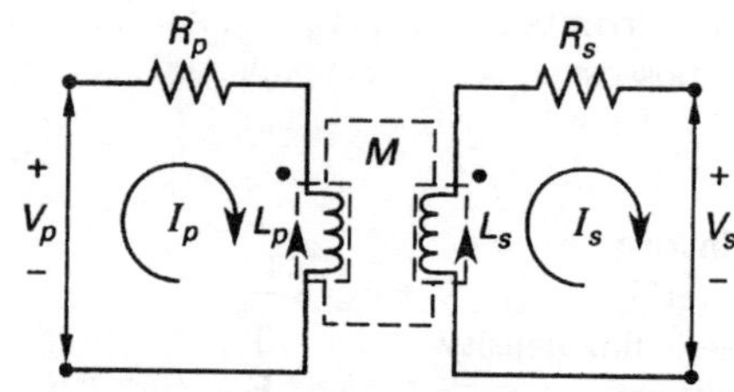

(a) flux linked circuit with mutual inductance

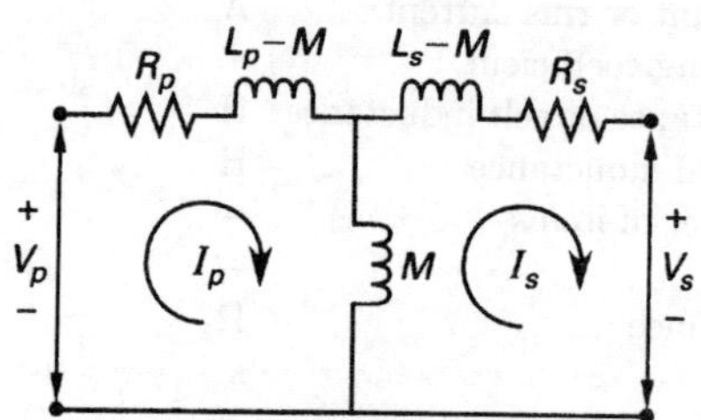

(b) equivalent conductively coupled circuit

The dots shown in Fig. 28.2(a) are used to determine the sign of the mutual inductance when assuming directions of current flow in analysis. When analyzing magnetically coupled circuits, the winding sense, or direction, can be determined using the right-hand rule (see Fig. 28.3). First, select a current direction on the primary side and place a dot where the current enters the winding (see Fig. 28.3(a)). Then use the right-hand rule to determine the current direction necessary for an opposing flux to be generated (see Fig. 28.3(b)). Place a dot where the current leaves this winding.[3] The two dotted terminals are positive simultaneously. With the instantaneous polarity determined, the need for the pictorial representation of the windings shown in Fig. 28.1 is eliminated. The coupled coils can be shown as in Fig. 28.3(c).

[3]This is equivalent to considering the primary current of the transformer to be a load on the source and the secondary current of the transformer to be the source for the secondary current portion of the circuit.

Figure 28.3 *Dot Rule Determination*

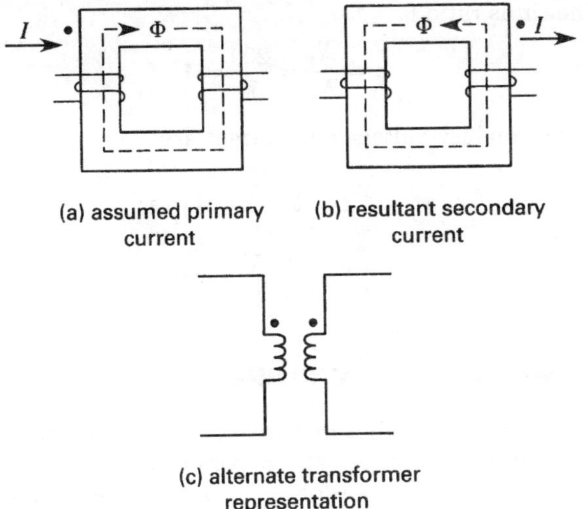

(a) assumed primary current

(b) resultant secondary current

(c) alternate transformer representation

The *dot rule* determines the sign of the mutual reactances, or M-terms.

- When the assumed currents both enter or both leave a pair of coupled coils by the dotted terminals, the sign of the M-terms is the same as that of the inductance terms, or L-terms, in Eq. 28.7 and Eq. 28.8.
- When one current enters by a dotted terminal and the other current leaves by a dotted terminal, the sign of the M-terms is the opposite of that of the L-terms in Eq. 28.7 and Eq. 28.8.

This is shown in Fig. 28.4. The rule is used so that in the application of analysis laws, such as KVL, the resulting equations (see Eq. 28.7 and Eq. 28.8) will be consistent and the mathematical result will be valid for the circuit.

Figure 28.4 *Positive and Negative Mutual Inductances*

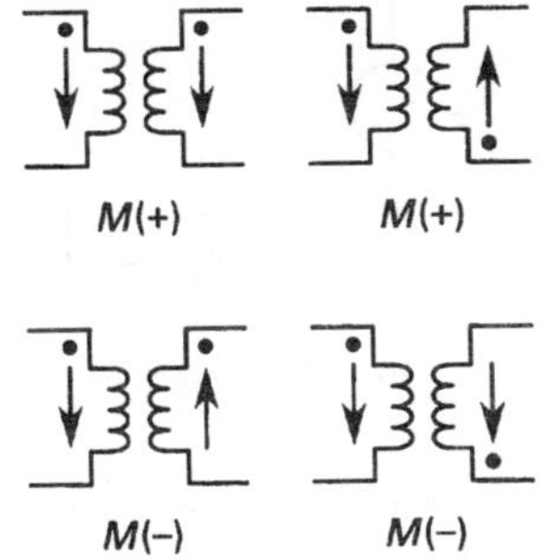

The amount of magnetic coupling is measured by the *coupling coefficient* (*coefficient of coupling*), k, which is the fraction of the total flux that links both coils. The value of k varies from near zero for radio coils to near 1.0 for iron-core transformers.[4]

$$k = \frac{M}{\sqrt{L_pL_s}} = \frac{X_m}{\sqrt{X_pX_s}} \qquad 28.9$$

[4]In terms of the fluxes in Fig. 28.1, the coupling coefficient $k = 1 - \Phi_{11}/(\Phi_{11} + \Phi_{12})$. The high value of the coupling coefficient in iron cores is the main reason for their use in transformers.

3. IDEAL TRANSFORMERS

Transformers, which are magnetically coupled circuits, are used to change voltages, match impedances, and isolate circuits. A primary winding current produces a magnetic flux, which induces a current in the secondary coil. To minimize flux leakage and improve efficiency, transformer coils are wound on magnetic permeable cores that improve the mutual inductance, M (see Sec. 28.2).[5] Two common designs, *core transformers* and *shell transformers*, are shown in Fig. 28.5.

Figure 28.5 *Core and Shell Transformers*

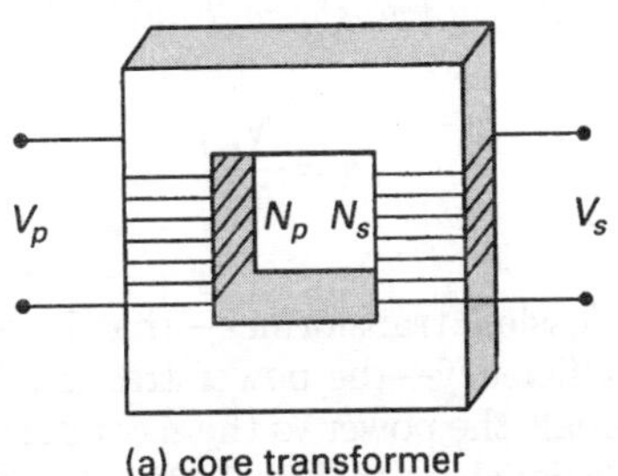

(a) core transformer

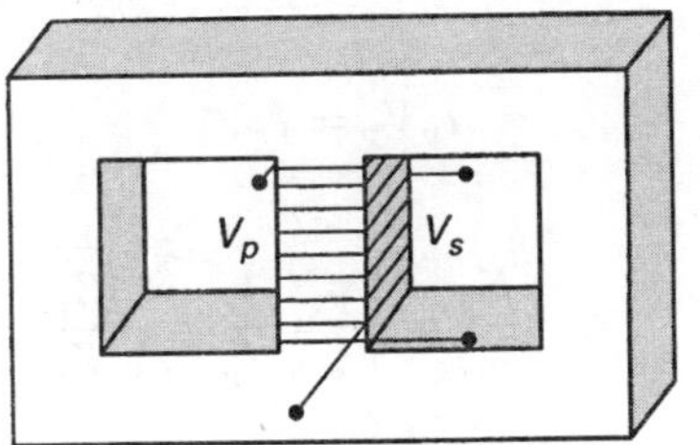

(b) shell transformer

An *ideal transformer* has the following properties.

- The coupling coefficient is unity ($k = 1$); that is, all the flux passes through each coil. There is no leakage flux. So, $|M| = |L_p| = |L_s|$, and $|X_m| = |X_p| = |X_s|$.
- The coils have no resistance; that is, $R_p = R_s = 0\ \Omega$. Combined with the first condition, this means that $Z_p = Z_s$ (see Fig. 28.2(b)).
- All the flux is contained within a path of constant area. This allows the dimensions of the core to be used.
- Core *iron losses* are zero. This indicates that no heating losses exist due to eddy currents or hysteresis loop effects; that is $R_c = 0\,\Omega$ (see Sec. 28.5, Sec. 28.6, and Sec. 28.7).
- The core is infinitely permeable, so no *magnetizing current* (also called the *exciting current*) is needed to create the flux in the core (see Sec. 28.5).

The last two conditions are technically the conditions for an *ideal core*. The transformer and the core can be treated and modeled separately if necessary. The model for an ideal transformer with an ideal core is shown in Fig. 28.6.

[5]An iron core transformer is indicated by one or more lines drawn between the coil symbols on a circuit diagram.

Figure 28.6 Ideal Transformer and Core

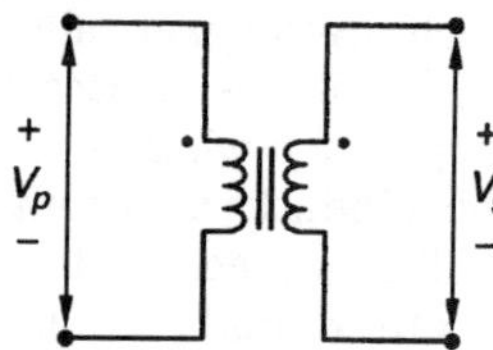

The ratio of the numbers of turns in the primary and secondary windings (N_p and N_s, respectively) is the *turns ratio* (*ratio of transformation*), a.[6]

$$a = \frac{N_p}{N_s} \quad \text{28.10}$$

In an ideal, lossless transformer—that is, a transformer with 100% efficiency—the power transfer from the primary side equals the power to the secondary side. Therefore, the relationships in Eq. 28.11 and Eq. 28.12 are valid.[7]

$$I_p V_p = I_s V_s \quad \text{28.11}$$

$$a = \frac{N_p}{N_s} = \frac{V_p}{V_s} = \frac{I_s}{I_p} = \sqrt{\frac{Z_p}{Z_s}} \quad \text{28.12}$$

If the turns ratio is greater than unity, the transformer decreases the voltage and is called a *step-down transformer*. If the turns ratio is less than unity, the transformer increases the voltage and is called a *step-up transformer*. The assumptions made for ideal transformers are very good approximations for real ones as well, especially for large iron-core transformers.

Example 28.1

An ideal step-down transformer has 200 primary turns and 50 secondary turns in its coils. 440 V are applied across the primary circuit. The secondary load resistance is 5 Ω. What are the secondary voltage and current?

Solution

The turns ratio is

$$a = \frac{N_p}{N_s} = \frac{200}{50} = 4$$

The secondary voltage and current are

$$V_s = \frac{V_p}{a} = \frac{440 \text{ V}}{4} = 110 \text{ V}$$

$$I_s = \frac{V_s}{R_s} = \frac{110 \text{ V}}{5 \ \Omega} = 22 \text{ A}$$

4. IMPEDANCE MATCHING

Matching of impedances between source and load is accomplished to effect the maximum power transfer to the load. The transformer can impedance match. In Fig. 28.7, the impedance in the secondary is

$$Z_s = \frac{V_s}{I_s} \quad \text{28.13}$$

Figure 28.7 Impedance Matching

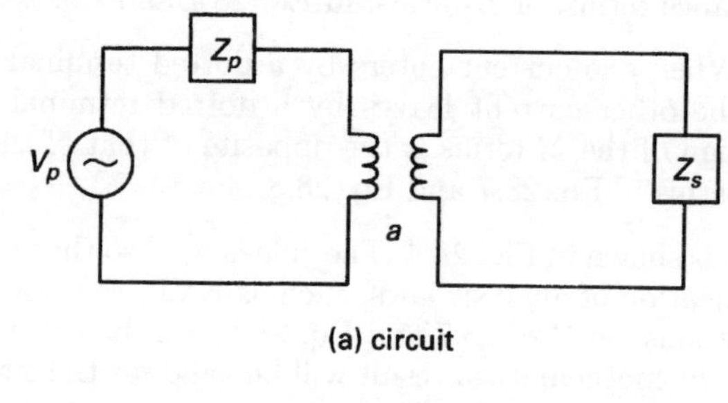

(a) circuit

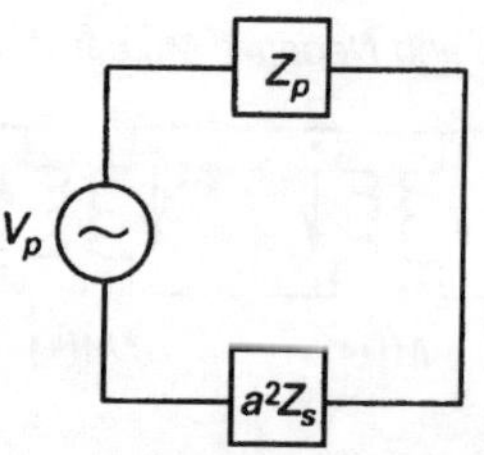

(b) equivalent circuit

The impedance "seen" by the primary side, Z_{ep}, is termed the *effective primary impedance* or the *reflected impedance*, Z_{ref}. The reflected impedance is

$$Z_{ep} = Z_{ref} = \frac{V_p}{I_p} = Z_p + a^2 Z_s \quad \text{28.14}$$

The turns ratio can be found from any of the relationships in Eq. 28.12, or specifically from

$$a = \sqrt{\frac{Z_p}{Z_s}} \quad \text{28.15}$$

[6]Other definitions of the turns ratio are also in use, including the ratio of N_s to N_p (the reciprocal of the value given by Eq. 28.10), the ratio of the two circuits' phase-to-phase values, and the ratio of their line-to-line values. To avoid confusion, some manufacturers give line and phase values instead of a turns ratio. For example, a transformer may be listed as 480/208/120 V wye. This indicates that the primary winding is delta connected and the secondary winding is wye connected (see Chap. 29), the primary line and phase voltage is 480 V, the secondary line voltage is 208 V, and the secondary phase voltage (i.e., phase to neutral) is 120 V.

[7]In an ideal transformer, phasor notation is not used, as all the elements are in phase. This is not the formula for power. The ½ has been dropped as has the complex conjugate of the current (see Chap. 27).

For a fixed primary impedance, the maximum power consumption in the load, that is, the maximum power transfer from the source, takes place when the secondary impedance is the conjugate of the primary impedance. Consequently, the conditions necessary to impedance match are

$$R_p = a^2 R_s \qquad 28.16$$

$$X_p = -a^2 X_s \qquad 28.17$$

Example 28.2

A 15 Ω speaker must be impedance matched to a 10 kΩ amplifier in order to produce the highest quality sound. What is the required turns ratio?

Solution

From Eq. 28.15,

$$a = \sqrt{\frac{Z_p}{Z_s}} = \sqrt{\frac{10 \times 10^3\ \Omega}{15\ \Omega}} = 25.8$$

Example 28.3

The primary of an ideal transformer with a turns ratio of 5 contains a 10 Ω resistor and an 80 mH inductor in series. What is the secondary reactance required to maximize the power transfer in this 60 Hz circuit?

Solution

Equation 28.17 gives the relationship between the reactances.

$$X_p = -a^2 X_s$$

$$X_s = \frac{X_p}{-a^2}$$

The primary reactance is found from

$$X_p = X_L = \omega L = 2\pi f L$$

$$= \frac{2\pi(60\ \text{Hz})(80\ \text{mH})}{1000\ \frac{\text{mH}}{\text{H}}}$$

$$= 30.16\ \Omega$$

Substituting the known and calculated values gives

$$X_s = \frac{X_p}{-a^2} = \frac{30.16\ \Omega}{-(5)^2} = -1.21\ \Omega$$

Example 28.4

For the reactance in Ex. 28.3, what is the value of the capacitance required?

Solution

From Ex. 28.3 and the definition of capacitive reactance,

$$X_s = X_C = -\frac{1}{\omega C}$$

$$C = -\frac{1}{2\pi f X_C} = -\frac{1}{(2\pi)(60\ \text{Hz})(-1.21\ \Omega)}$$

$$= 2.19 \times 10^{-3}\ \text{F}$$

5. REAL TRANSFORMERS

Real transformers have flux leakage, resistance in the coils, and resistance and reactance associated with moving the domains within the core. A model of a real transformer is shown in Fig. 28.8.

Figure 28.8 *Real Transformer Equivalent Circuit*

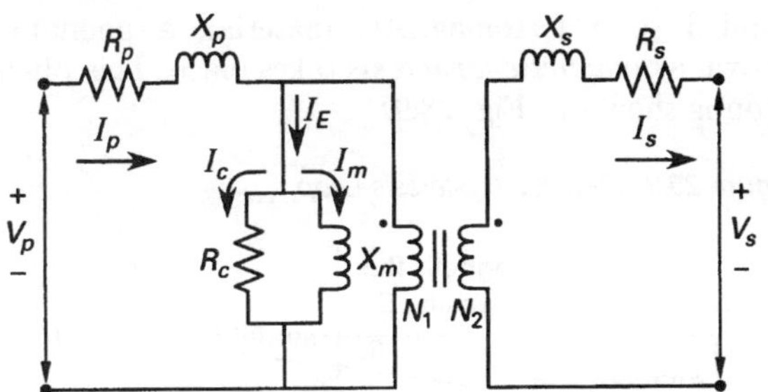

In the model of Fig. 28.8, all the core terms are shown on the primary side. This is arbitrary. Many other equivalent models are possible, though they will all contain these components. Components not yet defined follow.

- *Exciting current*, I_E: The exciting current is the total current needed to produce the magnetic flux in the core.[8]
- *Mutual reactance*, X_m, and *magnetizing current*, I_m: The magnetizing current is the current necessary to generate the flux in the core. It is 90° out of phase with the core current.
- *Core resistance*, R_c, and *core current*, I_c: The core resistance represents iron losses (see Sec. 28.6 and Sec. 28.7). The core current is the current necessary to overcome those losses. This current flows in the primary windings but it is caused by resistive losses physically occurring in the core.

Each leakage flux corresponds to a "loss" of induced voltage in its coil. The loss is considered to be caused

[8]The exciting current flows whether or not the transformer has an attached load. Although not entirely correct, the exciting current is often called the magnetizing current.

Circuit Theory

by an equivalent reactance, X_p or X_s. In terms of the flux, these reactances are

$$X_p = \omega L_p = \frac{V_{X_p}}{I_p} = \frac{4.44 f \Phi_p N_p}{I_p} \qquad 28.18$$

$$X_s = \omega L_s = \frac{V_{X_s}}{I_s} = \frac{4.44 f \Phi_s N_s}{I_s} \qquad 28.19$$

The turns ratio is related to the leakage inductances by

$$a = \sqrt{\frac{L_p}{L_s}} \qquad 28.20$$

6. MAGNETIC HYSTERESIS: BH CURVES

The core of a transformer, especially when specifically designed to increase the coupling, is made from iron. Iron is a ferromagnetic material. When a magnetic field, in this case produced by the coil of the transformer, is applied to a ferromagnetic material, a phenomenon known as *magnetic hysteresis* takes place. This phenomenon is shown in Fig. 28.9.

Figure 28.9 *Magnetic Hysteresis Loop*

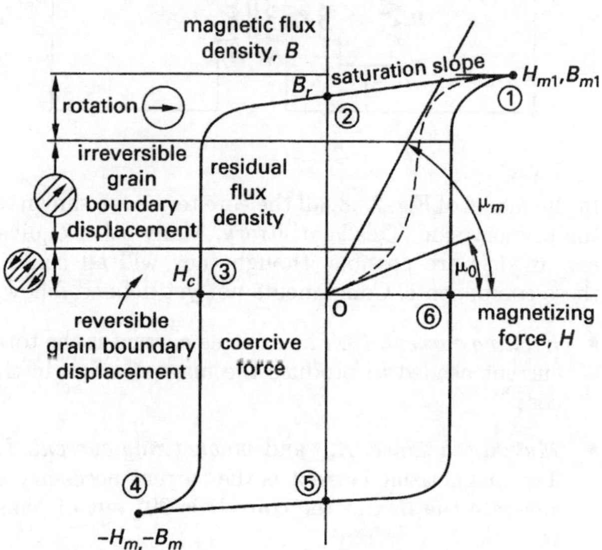

Figure 28.9 plots the magnetic flux density, B, produced for various values of the magnetizing force, H. The relationship is decidedly nonlinear.[9] When a material is first magnetized, the initial BH values occur along the curve from the origin, O, to point 1. This part of the curve, O1, is known as the *initial* or *normal magnetization curve.* The initial slope of O1 is called the *initial permeability*, μ_0 (not to be confused with the permeability of free space, which has the same symbol). The slope of the line from the origin to a point tangent to the normal magnetization curve is called the *maximum permeability*, μ_m. The maximum value of B reached at point 1 is called the *saturation flux density.* If the magnetic field is then decreased, the flux density decreases as well, but not in proportion to the decrease in magnetizing field strength. When H equals zero, at point 2, the amount of magnetic flux density remaining in the core is called the *residual flux density*, B_r. In order to reduce the magnetic field in the core to zero, the magnetizing force must increase in reverse direction of that initially applied. The magnitude of the magnetizing force where B again becomes zero, at point 3, is called the *coercive force*, H_c.

As the magnetizing force continues to increase in the negative direction, the flux density again increases, but in opposite polarity, eventually reaching saturation in this direction at point 4. As the magnetic field is again removed, the magnitude of the magnetic flux density decreases to point 5. A positive value of H (equal to H_c) is required to drive the flux density to zero, at point 6.

As the magnetizing force is increased further, the curve from point 6 to point 1 is followed, completing the cycle but not returning to the origin. The domain, or grain boundary movement, that is the ultimate cause of this curve is shown for the positive portion of the cycle. The phenomenon whereby the values of B lag behind the values of H is called *hysteresis.* The entire loop is called the *hysteresis loop.* The area within the loop represents the energy loss that is due to heating from the movement of the magnetic domains.[10]

7. EDDY CURRENTS

Any conducting mass exposed to variations in magnetic flux contains *eddy currents* (or *Foucault currents*). These currents are a result of magnetic induction caused by the relative motion caused by either variations in the magnetic field or the movement of the mass itself. Such currents result in Joule heating in the mass—in this case, in the core of the transformer. Such an energy loss, in a transformer or in the iron portions of electric machinery, is minimized by dividing the magnetically conducting material into small segments called *laminations*, coated with insulating material. This reduces the length of the path in which current exists.

8. CORE LOSSES

Core losses are defined as the rate of energy conversion into heat in a magnetic material caused by an alternating or pulsing magnetic field. This type of loss is also called the *excitation loss* or *iron loss* (see Sec. 28.5). Core losses are the combination of magnetic hysteresis and eddy current losses.

[9]To this point, all the relationships and equations have assumed linearity in the transformers.

[10]The portion of the energy used to create the flux comes from I_m (see Sec. 28.5). The portion of the energy that is lost as heat comes from I_c (see Sec. 28.5). The current I_c also supplies any eddy current losses (see Sec. 28.7).

29 Linear Circuit Analysis

Nomenclature

A	amplification factor	–
A	area	m^2
b	y-intercept	–
C	capacitance	F
f	frequency	Hz
i	variable current	A
I, $\mathbf{I}$	constant or rms current	A
l	length	m
L	inductance or self-inductance	H
m	slope	–
M	mutual inductance	H
n, N	number of items	–
p	instantaneous power	W
P	power	W
Q	charge	C
r	distance	m
R	resistance	Ω
t	time	s
T	temperature	°C
U	energy	J
v	instantaneous voltage	V
V, $\mathbf{V}$	constant or rms voltage	V
Y	admittance	S
Z, $\mathbf{Z}$	impedance	Ω

Symbols

α	thermal coefficient of resistance	1/°C
ϵ	permittivity	F/m
ϵ_0	free-space permittivity, 8.854×10^{-12}	F/m
ϵ_r	relative permittivity	–
ζ	constant	–
κ	arbitrary constant	–
μ	permeability	H/m
μ_0	free-space permeability, 1.2566×10^{-6}	H/m
ξ	constant	–
ρ	resistivity	Ω·cm
Ψ	magnetic flux	Wb
ω	angular frequency	rad/s

Subscripts

0	initial or free space (vacuum)
20	referenced to 20°C
C	capacitor
e	equivalent
fl	full load
int	internal or source
l	load
L	inductor
m	magnetizing
N	Norton
nl	no load
oc	open circuit
r	ratio or relative
R	resistive or resistor
s	source
sc	short circuit
Th	Thevenin

1. FUNDAMENTALS

Circuit analysis is fundamental to the practice of electrical engineering. Such analysis is based on Kirchhoff's two circuit laws and the values, as well as the fluctuations, of the *circuit variables*, that is, the voltages between terminals and the currents within the network. Circuit theory is nominally divided into determination of *equivalent circuits* and *mathematical analysis* of those equivalent circuits.[1]

[1]In digital systems, the final equivalent circuit is a logic diagram.

Circuit Theory

Equivalent circuits are composed of *lumped elements.* A lumped element is a single element representing a particular electrical property in the entire circuit, or in some portion of the circuit.[2] Equivalent circuit theory is sometimes called *network theory.* The term "network," while often used synonymously with "circuit," is usually reserved for more complicated arrangements of elements. In fact, "circuit" often designates a single loop of a network. Networks are often modeled using ideal elements.

2. IDEAL INDEPENDENT VOLTAGE SOURCES

An *ideal independent voltage source* maintains the voltage at its terminals regardless of the current flowing through the terminals. The symbols for an ideal independent voltage source and its characteristics are shown in Fig. 29.1. The subscript s generally indicates an independent source. The source value can vary with time, but its effective value does not change. That is, the magnitude of its variations in time is without regard to the current. The polarity assigned is arbitrary, but it is often assigned in the direction of positive power flow out of the source. This type of notation is called *source notation.*

Figure 29.1 Ideal Voltage Source

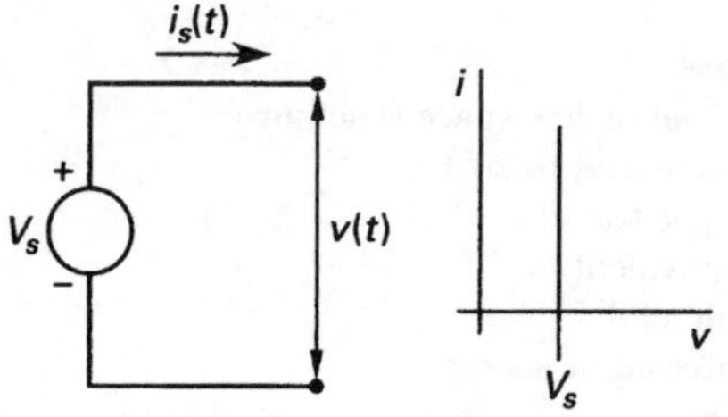

A *real voltage source* cannot maintain the voltage at its terminals when the current becomes too large (in either the positive or the negative direction). The equivalent circuit of a real voltage source is shown in Fig. 29.2. The decrease in voltage as the current increases is measured by the *voltage regulation* given in Eq. 29.1.

$$\text{regulation} = \frac{V_{\text{nl}} - V_{\text{fl}}}{V_{\text{fl}}} \times 100\% \qquad 29.1$$

3. IDEAL INDEPENDENT CURRENT SOURCES

An *ideal independent current source* maintains the current regardless of the voltage at its terminals. The symbols for an ideal independent current source and its characteristics are shown in Fig. 29.3, the subscript s indicating the independent source. The source value can vary with time, but its effective value does not change. That is, the magnitude of its variations in time is without regard to the voltage. The polarity is again arbitrary but is assigned using the source notation.

[2]Electrical properties are distributed throughout elements, though they are concentrated in certain areas. Analyzing such *distributed elements* is difficult. Because of this difficulty, they are represented by lumped parameters. This method is important in the analysis of electrical energy transmission.

Figure 29.2 Real Voltage Source

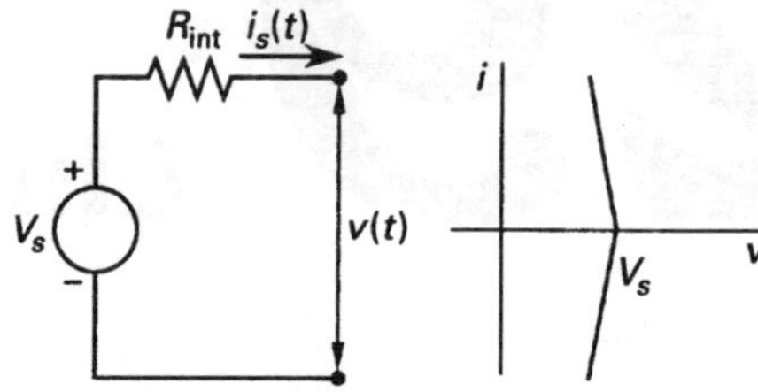

Figure 29.3 Ideal Current Source

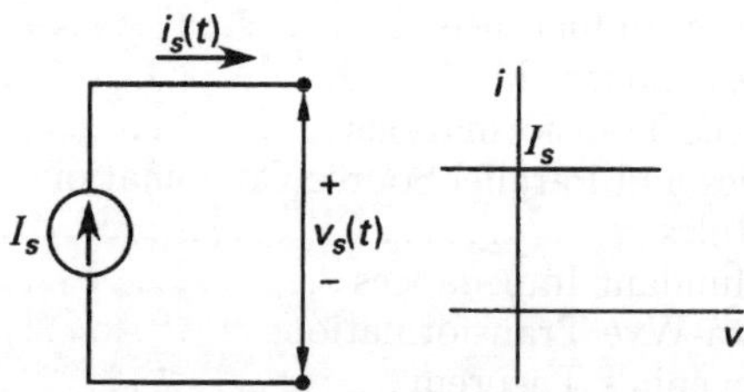

A *real current source* cannot maintain the current at its terminals when the voltage becomes too large (in either the positive or the negative direction). The equivalent circuit of a real current source is shown in Fig. 29.4.

Figure 29.4 Real Current Source

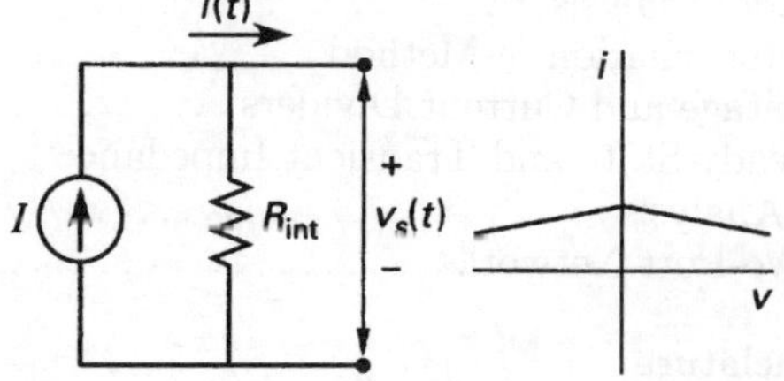

4. IDEAL RESISTORS

An *ideal resistor* maintains the same resistance at its terminals regardless of the power consumed. That is, it is a linear element whose voltage/current relationship is given by Ohm's law.

$$v = iR \qquad 29.2$$

The symbols for an ideal resistor are identical to those for a real resistor and are shown in Fig. 29.5, along with the characteristics of the ideal resistor. The polarity is assigned to make Eq. 29.2, Ohm's law, valid as written. This is sometimes called the *sink notation,* as it corresponds to positive power flow into the resistor.

Figure 29.5 Resistor

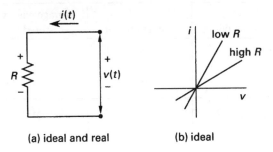

(a) ideal and real (b) ideal

A real resistor undergoes changes in resistance as the temperature changes. The resistance of metals and most alloys increases with temperature while that of carbon and electrolytes decreases. The resistance change is accounted for in circuit design by ensuring the design allows for proper circuit functioning over the expected temperature range (see Sec. 26.3).

5. DEPENDENT SOURCES

Any circuit that is designed to deliver more power to the load than is available at the signal frequency must use an *active device.* An active device is defined as a component capable of amplifying the current or voltage, such as a transistor. An active device uses a source of power other than the input signal in order to accomplish the modification. In order to analyze circuits containing active devices, the devices are replaced by an equivalent circuit of *passive linear elements*, that is, elements that are not sources of energy (such as resistors, inductors, and capacitors), and one or more dependent sources. Such sources may be treated as standard sources using all the linear circuit analysis techniques, with the exception of superposition. Superposition can only be used if the device is limited to operation over a linear region.

6. DEPENDENT VOLTAGE SOURCES

A *dependent voltage source* is one in which the output is controlled by a variable elsewhere in the circuit. For this reason, dependent voltage sources are sometimes called *controlled voltage sources.* The controlling variable can be another voltage, a current, or any other physical quantity, such as light intensity or temperature. The symbols for dependent voltage sources are shown in Fig. 29.6. A *voltage-controlled voltage source* is controlled by a voltage, v, present somewhere else in the circuit. A *current-controlled voltage source* is controlled by a current, i, present somewhere else in the circuit. The terms ζ and ξ are constants.

7. DEPENDENT CURRENT SOURCES

A *dependent current source* is one in which the output is controlled by a variable elsewhere in the circuit. For this reason, dependent current sources are sometimes called

Figure 29.6 Dependent Voltage Sources

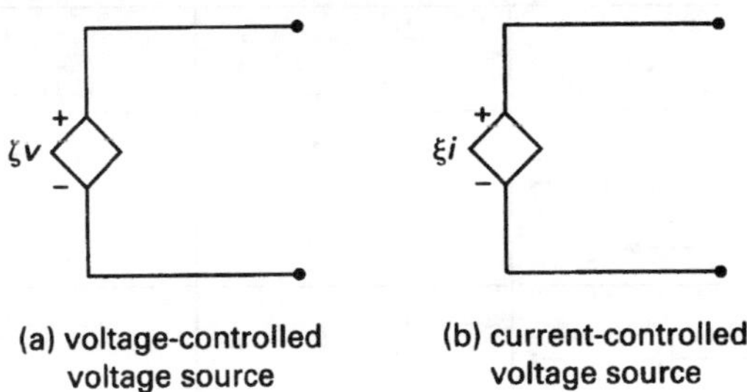

(a) voltage-controlled voltage source (b) current-controlled voltage source

controlled current sources. The controlling variable can be another current, voltage, or any other physical quantity, such as light intensity or temperature. The symbols for dependent current sources are shown in Fig. 29.7. A *voltage-controlled current source* is controlled by a voltage, v, present somewhere else in the circuit. A *current-controlled current source* is controlled by a current, i, present somewhere else in the circuit. The terms ζ and ξ are constants.

Figure 29.7 Dependent Current Sources

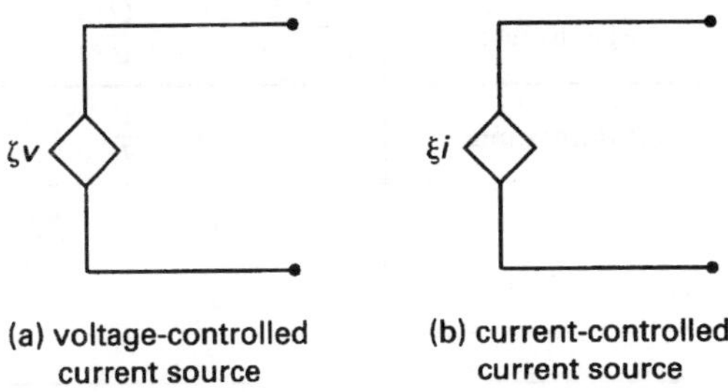

(a) voltage-controlled current source (b) current-controlled current source

8. LINEAR CIRCUIT ELEMENTS

The basic circuit elements are resistance, capacitance, and inductance. Related to the basic element of inductance are mutual inductance and ideal (linear) transformers. A *linear network* is one in which the resistance, capacitance, and inductance are constant with respect to current and voltage, and in which the current or voltage of sources is independent or directly proportional to the other currents and voltages, or their derivatives, in the network. That is, a linear network is one that consists of linear elements and linear sources.[3] Circuit elements or sources are linear if the superposition principle is valid for them, and vice versa. A summary of linear circuit element parameters is given in Table 29.1. The time domain and frequency domain behavior of linear circuit elements, along with their defining equations, are shown in Table 29.2.

9. RESISTANCE

Resistance is the opposition to current flow in a conductor. Technically, it is the opposition that a device or

[3]A linear source is one in which the output is proportional to the first power of a voltage or current in the circuit.

Table 29.1 Linear Circuit Element Parameters

circuit element	voltage	current	instantaneous power	average power	average energy stored
resistor (+ v −, i →)	$v=iR$	$i=\dfrac{v}{R}$	$p=iv=i^2R=\dfrac{v^2}{R}$	$P=IV=I^2R=\dfrac{V^2}{R}$	–
capacitor (+ v −, i →)	$v=\dfrac{1}{C}\int i\,dt+\kappa$	$i=C\dfrac{dv}{dt}$	$p=iv=Cv\dfrac{dv}{dt}$	0	$U=\frac{1}{2}CV^2$
inductor (+ v −, i →)	$v=L\dfrac{di}{dt}$	$i=\dfrac{1}{L}\int v\,dt+\kappa$	$p=iv=Li\dfrac{di}{dt}$	0	$U=\frac{1}{2}LI^2$

Table 29.2 Linear Circuit Parameters, Time and Frequency Domain Representation

parameter	defining equation	time domain	frequency domain[a]
resistance	$R=\dfrac{v}{i}$	$v=iR$	$V=IR$
capacitance	$C=\dfrac{Q}{V}$	$i=C\dfrac{dv}{dt}$	$I=j\omega CV$
self-inductance	$L=\dfrac{\Psi}{I}$	$v=L\dfrac{di}{dt}$	$V=j\omega LI$
mutual inductance[b]	$M_{12}=M_{21}=M=\dfrac{\Psi_{12}}{I_2}=\dfrac{\Psi_{21}}{I_1}$	$v_1=L_1\dfrac{di_1}{dt}+M\dfrac{di_2}{dt}$ $v_2=L_2\dfrac{di_2}{dt}+M\dfrac{di_1}{dt}$	$V_1=j\omega(L_1I_1+MI_2)$ $V_2=j\omega(L_2I_2+MI_1)$

[a]The voltages and currents in the frequency domain column are not shown as vectors; for example, **I** or **V**. They are, however, shown in phasor form. Across any single circuit element (or parameter), the phase angle is determined by the impedance and embodied by the j. If the $j\omega$ (C or L or M) were not shown and instead given as **Z**, the current and voltage would be shown as **I** and **V**, respectively.

[b]Currents I_1 and I_2 (in the mutual inductance row) are in phase. The angle between either I_1 or I_2 and V_1 or V_2 is determined by the impedance angle embodied by the j. Even in a real transformer, this relationship holds, since the equivalent circuit accounts for any phase difference with a magnetizing current, I_m.

material offers to the flow of direct current. In AC circuits, it is the real part of the complex impedance. The resistance of a conductor depends on the resistivity, ρ, of the conducting material. The resistivity is a function of temperature that varies approximately linearly in accordance with Eq. 29.3. A slightly different version of this equation was given in Sec. 26.3. Here, the resistivity is referenced to 20°C and is written ρ_{20}. For copper, ρ_{20} is approximately 1.8×10^{-8} Ω·m. The thermal coefficient of resistivity, also referred to as the *temperature coefficient of resistivity*, is referenced to 20°C and written α_{20}. For copper, α_{20} is approximately 3.9×10^{-3}°C^{-1}. The equation for resistivity at any temperature, T (in °C), then becomes[4]

$$\rho=\rho_{20}\big(1+\alpha_{20}(T-20)\big) \qquad 29.3$$

[4]Copper with a conductivity of 100% will experience a resistance change of approximately 10% for a 25°C (77°F) temperature change.

The resistance of a conductor of length l and cross-sectional area A is

$$R=\frac{\rho l}{A} \qquad 29.4$$

The defining equation for resistance is Ohm's law, or the extended Ohm's law for AC circuits, which relates the linear relationship between the voltage and the current.

$$V=IR \quad \text{or} \quad \mathbf{V}=\mathbf{I}\mathbf{Z}_R \qquad 29.5$$

The power dissipated by the resistor is[5]

$$P=IV=I^2R=\frac{V^2}{R} \qquad 29.6$$

[5]When the term "power" is used, the average power is the referenced quantity. The instantaneous power depends on the time within the given cycle and is written p.

The equivalent resistance of resistors in series is

$$R_e = R_1 + R_2 + \ldots + R_n \quad 29.7$$

The equivalent resistance of resistors in parallel is

$$\frac{1}{R_e} = \frac{1}{R_1} + \frac{1}{R_2} + \ldots + \frac{1}{R_n} \quad 29.8$$

10. CAPACITANCE

Capacitance is a measure of the ability to store charge. Technically, it is the opposition a device or material offers to changing voltage. In a DC circuit, that is, a circuit in which the voltage is constant, the capacitor appears as an open circuit (after any transient condition has subsided). In a DC circuit, if the voltage is zero, the capacitor appears as a short circuit. In an AC circuit, it is the constant of proportionality between the changing voltage and current.

$$i = C\frac{dv}{dt} \quad 29.9$$

Capacitance is a function of the insulating material's *permittivity*, ϵ, between the conducting portions of the capacitor. The permittivity is the product of the *free-space permittivity*, ϵ_0, and the *relative permittivity* or *dielectric constant*, ϵ_r. For free space—that is, a vacuum—the permittivity is 8.854×10^{-12} F/m. The dielectric constant varies, and for many linear materials it has values between 1 and 10.

$$\epsilon = \epsilon_r \epsilon_0 \quad 29.10$$

The capacitance of a simple parallel plate capacitor is given by Eq. 29.11. The term r is the distance between the two parallel plates of equal area, A, and ϵ is the permittivity of the material between the plates.

$$C = \frac{\epsilon A}{r} \quad 29.11$$

The defining equation for capacitance, which relates the amount of charge stored to the voltage across the capacitor terminals, is

$$C = \frac{Q}{V} \quad 29.12$$

The (average) power dissipated by a capacitor is zero. The (average) energy stored in the electric field of a capacitor is

$$U = \tfrac{1}{2}CV^2 = \tfrac{1}{2}QV = \tfrac{1}{2}\left(\frac{Q^2}{C}\right) \quad 29.13$$

The equivalent capacitance of capacitors in series is

$$\frac{1}{C_e} = \frac{1}{C_1} + \frac{1}{C_2} + \ldots + \frac{1}{C_n} \quad 29.14$$

The equivalent capacitance of capacitors in parallel is

$$C_e = C_1 + C_2 + \ldots + C_n \quad 29.15$$

From Eq. 29.9, the voltage cannot change instantaneously, as this would make the current and the power infinite, which clearly cannot be the case. Using this fact, the transient behavior of capacitors can be analyzed. Using this principle is the equivalent of treating a capacitor's arbitrary constant κ in Table 29.1 as the initial voltage, V_0.

Example 29.1

In the circuit shown, the switch has been open for an "extended time," which in such circuits is considered to be a minimum of five time constants. The capacitor has no charge on it at $t=0$ s when the switch is closed. What is the current through the capacitor at the instant the switch is closed?

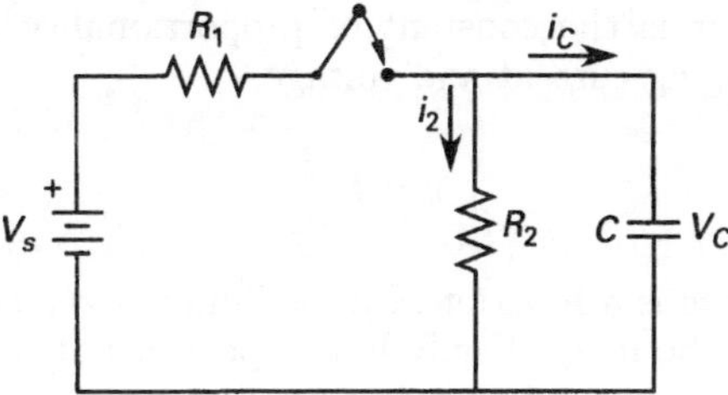

Solution

Because the voltage across the capacitor cannot change instantly, $v(0^-) = v(0^+)$, and the capacitor remains uncharged. The capacitor thus initially acts as a short circuit. All the current flows through the capacitor, bypassing R_2. The current is

$$i_C(0^+) = \frac{V_s}{R_1}$$

The initial voltage across the capacitor, $V_C(0^+)$, is zero.

Example 29.2

Using the circuit given in Ex. 29.1, what is the steady-state voltage across the capacitor?

Solution

The term "steady-state" is another way of saying "an extended period of time," which in such circuits is considered to be a minimum of five time constants. The capacitor will be fully charged and will act as an open circuit. Consequently, $dv/dt=0$ and $i_C=0$. All the current flows through R_2 and is given by

$$I_{R_2} = \frac{V_s}{R_1 + R_2}$$

Because they are in parallel, the voltage across the capacitor is the same as the voltage across R_2. Thus,

$$V_C(\infty) = V_{R_2} = I_2R_2 = \left(\frac{V_s}{R_1+R_2}\right)R_2$$
$$= V_s\left(\frac{R_2}{R_1+R_2}\right)$$

The steady-state current through the capacitor, $I_C(\infty)$, is zero.

11. INDUCTANCE

Inductance is a measure of the ability to store magnetic energy. Technically, it is the opposition a device or material offers to changing current. In a DC circuit, that is, a circuit in which the current is constant, the inductor appears as a short circuit (after any transient condition has subsided). In a DC circuit, if the current is zero, the inductor appears as an open circuit. In AC circuits, it is the constant of proportionality between the changing current and voltage.

$$v = L\frac{di}{dt} \qquad 29.16$$

Inductance is a function of a medium's *permeability*, μ, between the magnetically linked portions of the inductor. The permeability is the product of the *free-space permeability*, μ_0, and the *relative permeability* or *permeability ratio*, μ_r. For free space—that is, a vacuum—the permeability is 1.2566×10^{-6} H/m. The permeability ratio varies. Depending on the material and the flux density, it can have values in the hundreds and sometimes thousands.

$$\mu = \mu_r\mu_0 \qquad 29.17$$

The inductance of a simple toroid shown in Fig. 29.8 is given by Eq. 29.18. The term l is the average flux path distance around the toroid of equal area, A, and μ is the permeability of the medium between the coil turns, N.

$$L = \frac{\mu N^2 A}{l} \qquad 29.18$$

Figure 29.8 *Toroid Inductance Terms*

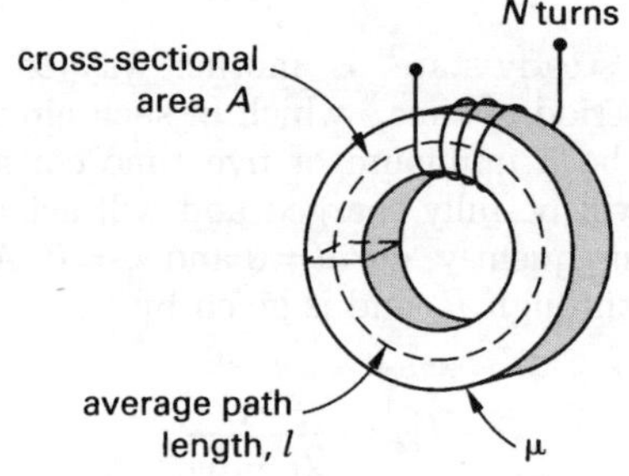

The defining equation for inductance, which relates the amount of flux present to the current flowing through the inductor terminals, is

$$L = \frac{\Psi}{I} \qquad 29.19$$

The (average) power dissipated by an inductor is zero. The (average) energy stored in the magnetic field of an inductor is given by Eq. 29.20. Positive energy indicates energy stored in the inductor. Negative energy indicates energy supplied to the circuit by the inductor.

$$U = \tfrac{1}{2}LI^2 = \tfrac{1}{2}\Psi I = \tfrac{1}{2}\left(\frac{\Psi^2}{L}\right) \qquad 29.20$$

The equivalent inductance of inductors in series is

$$L_e = L_1 + L_2 + \ldots + L_n \qquad 29.21$$

The equivalent inductance of inductors in parallel is

$$\frac{1}{L_e} = \frac{1}{L_1} + \frac{1}{L_2} + \ldots + \frac{1}{L_n} \qquad 29.22$$

From Eq. 29.16, the current cannot change instantaneously, as this would make the voltage and the power infinite, which clearly cannot be the case. Using this fact, the transient behavior of inductors can be analyzed. Using this principle is the equivalent of treating an inductor's arbitrary constant κ in Table 29.1 as the initial current, I_0.

Example 29.3

In the circuit shown, the switch has been open for an extended time, that is, greater than five time constants. The inductor has no associated magnetic field on it at $t = 0$ s when the switch is closed. What is the current through the inductor at the instant the switch is closed?

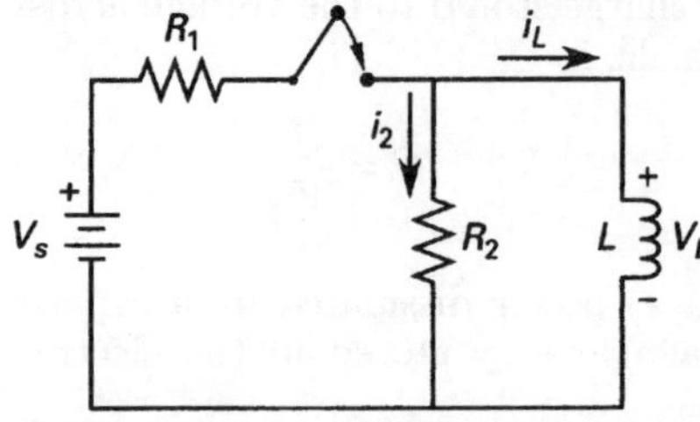

Solution

Because the current across the inductor cannot change instantaneously, $i(0^-) = i(0^+)$ and the inductor remains without a magnetic field. The inductor initially acts as an open circuit. All the current flows through R_2.

$$i_L(0^+) = 0$$

Example 29.4

Using the circuit given in Ex. 29.3, what is the initial voltage across the inductor?

Solution

All the current flows through R_2 and is given by

$$I_{R_2} = \frac{V_s}{R_1 + R_2}$$

Because they are in parallel, the voltage across the inductor is the same as the voltage across R_2. So,

$$V_L(0^+) = V_{R_2} = I_2 R_2 = \left(\frac{V_s}{R_1 + R_2}\right) R_2 = V_s\left(\frac{R_2}{R_1 + R_2}\right)$$

Example 29.5

Using the circuit given in Ex. 29.3, what is the steady-state value of the current through the inductor?

Solution

The term "steady-state" is another way of saying "an extended period of time," which in such circuits is considered to be a minimum of five time constants. The inductor's magnetic field will be at a maximum and the inductor acts as a short circuit. Consequently, $di/dt = 0$, and $v_L = 0$. All the current flows through the inductor, bypassing R_2, and, using Ohm's law, is

$$i_L(\infty) = \frac{V_s}{R_1}$$

12. MUTUAL INDUCTANCE

Mutual inductance is the ratio of the electromotive force induced in one circuit to the rate of change of current in the other circuit. Applicable equations are given in Table 29.2. The concept is explained in Chap. 28. The total energy stored in a system involving mutual inductance is

$$U = \tfrac{1}{2}L_1 I_1^2 + \tfrac{1}{2}L_2 I_2^2 + M I_1 I_2 \qquad \textit{29.23}$$

13. LINEAR SOURCE MODELS

Electric power sources tend to be nonlinear. In order to analyze such sources using circuit analysis, they are modeled as linear sources in combination with linear impedance. Such a source is classified as an *independent source.*

Dependent sources, whose output depends on a parameter elsewhere in the circuit, are associated with active devices, such as transistors. These are not actually sources of power but can be modeled as linear sources in combination with an impedance. The range of operation over which the model is valid must be selected to be approximately linear. Circuit analysis of this type is called *small-signal analysis.*

A third source type is the *transducer* that produces a current or voltage output from a nonelectric input. *Thermocouples* are transducers that produce voltages as functions of temperature. *Photodiodes* are transducers that produce currents as functions of incident illumination level. *Microphones* are transducers that produce voltages as functions of incident sound intensity.

All these sources can be modeled by linear approximation to measured values. Specific models for the electronic components are given in Chap. 53.

Example 29.6

The accompanying illustration shows a representative plot of battery voltage versus load current. (a) What is the linear expression of the battery for the range of operation from 8 A to 12 A? (b) Show the associated model.

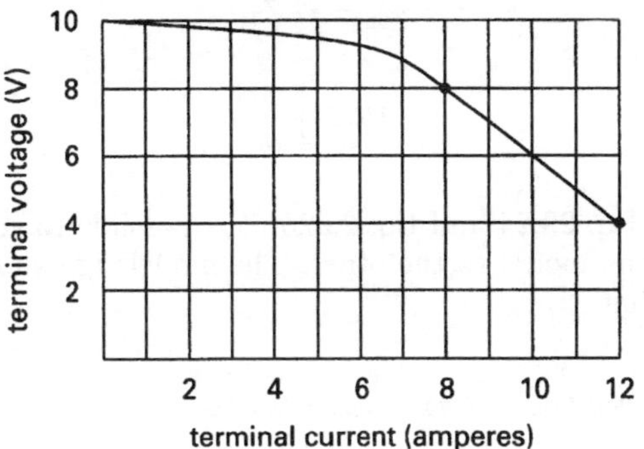

Solution

(a) Using the *two-point method,* equivalent to determining the parameters of a straight line in the formula $y = mx + b$, for the points (8, 8) and (4, 12) gives

$$\frac{I - 8\text{ A}}{A - 8\text{ V}} = \frac{12\text{ A} - 8\text{ A}}{4\text{ V} - 8\text{ V}} = -1\text{ S}$$

(b) Either the voltage or the current may be selected as the dependent variable.

$$V = -I + 16\text{ V} \qquad \text{[I]}$$

$$I = -V + 16\text{ A} \qquad \text{[II]}$$

Equation I is a loop voltage equation with the terminal voltage determined by two sources, one independent (the 16 V) and the other dependent on the current (the −1 S). Equation II is a node current equation with

Circuit Theory

the terminal current determined by two sources, one independent (the 16 A) and the other dependent on the voltage (the −1 S). The resulting models are

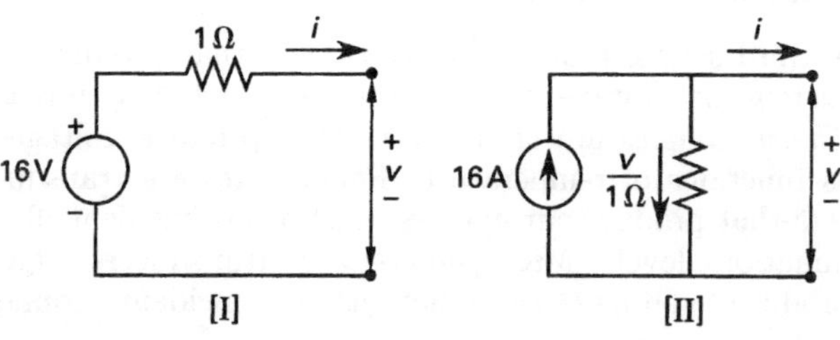

14. SOURCE TRANSFORMATIONS

An electrical source can be modeled as either a voltage source in series with an impedance or a current source in parallel with an impedance. The two models are equivalent in that they have the same terminal voltage/current relationships. Changing from one model to the other is called *source transformation* and is a *circuit reduction* technique. The models' parameters are determined by

$$v_{\text{Th}} = i_N Z_{\text{Th}} \qquad 29.24$$

$$i_N = \frac{v_{\text{Th}}}{Z_{\text{Th}}} \qquad 29.25$$

Using Eq. 29.24 and Eq. 29.25, it is possible to convert from one model to the other. The models are shown in Fig. 29.9.

Figure 29.9 *Source Transformation Models*

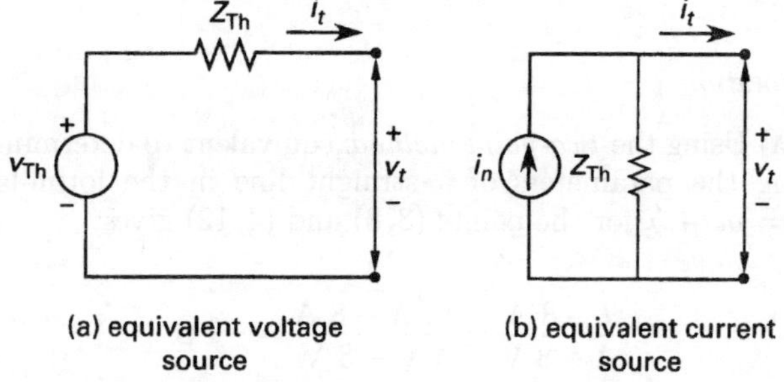

The source transformation is not defined when Z equals zero or when Z approaches ∞. Also, both the open-circuit voltage and the short-circuit current direction must be the same in each model (see Sec. 26.11).

15. SERIES AND PARALLEL SOURCE COMBINATION RULES

One aspect of circuit reduction is the removal of sources that are not required. Section 29.14 referred to ideal current and voltage sources. These can be manipulated, combined, or eliminated as indicated by the following rules.

- An ideal voltage source cannot be placed in parallel with a second ideal voltage source, because the resulting voltage is indeterminate.
- An ideal current source cannot be placed in series with a second ideal current source, because the resulting current is indeterminate.
- An ideal voltage source in parallel with an ideal current source makes the current source redundant. The voltage is set by the voltage source, which absorbs all the current from the current source.
- An ideal current source in series with an ideal voltage source makes the voltage source redundant. The current is set by the current source, which is able to overcome any voltage of the voltage source.
- Ideal voltage sources in series can be combined algebraically using Kirchhoff's voltage law.
- Ideal current sources in parallel can be combined algebraically using Kirchhoff's current law.

The rules above can be used in an iterative manner to reduce a network to a single equivalent source and impedance. This technique is useful during phasor analysis or where a capacitance or inductance is present in a complicated network and the transient current or voltage is of interest. The iterative technique is also useful in the analysis of networks containing nonlinear or active devices.

16. REDUNDANT IMPEDANCES

Circuit reduction occurs when unnecessary impedances are removed according to the following rules.

- Impedance in parallel with an ideal voltage source may be removed. Theoretically, the ideal voltage supplies whatever energy is required to maintain the voltage, so the current flowing through the parallel impedance has no impact on the remainder of the circuit. Care must be exercised when calculating the source current to take into account the current through the parallel impedance.
- Impedance in series with an ideal current source may be removed. Theoretically, the ideal current source supplies whatever energy is required to maintain the current, so the energy loss in the series impedance has no impact on the remainder of the circuit. The voltage across the current source or current source/series impedance combination is determined by the remainder of the circuit.

17. DELTA-WYE TRANSFORMATIONS

Electrical impedances arranged in the shape of the Greek letter delta (Δ) or the English letter Y (wye) are known as *delta-wye configurations*, or *pi-T configurations*. A delta-wye configuration (see Fig. 29.10) can be useful in reducing circuits in power networks.

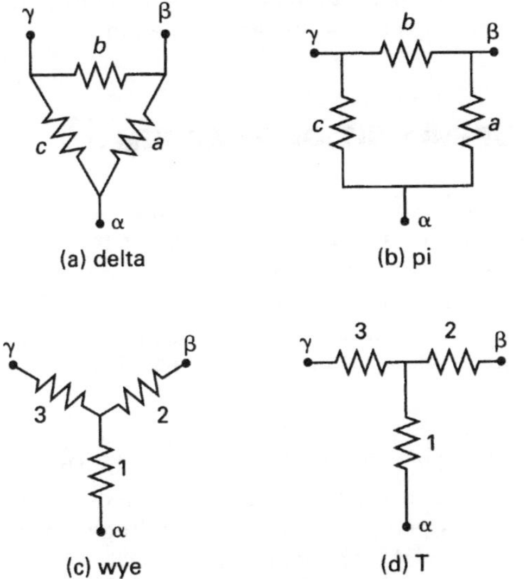

Figure 29.10 *Delta (Pi)-Wye (T) Configurations*

The equivalent impedances, which allow transformation between configurations, are

$$Z_a = \frac{Z_1 Z_2 + Z_1 Z_3 + Z_2 Z_3}{Z_3} \qquad 29.26$$

$$Z_b = \frac{Z_1 Z_2 + Z_1 Z_3 + Z_2 Z_3}{Z_1} \qquad 29.27$$

$$Z_c = \frac{Z_1 Z_2 + Z_1 Z_3 + Z_2 Z_3}{Z_2} \qquad 29.28$$

$$Z_1 = \frac{Z_a Z_c}{Z_a + Z_b + Z_c} \qquad 29.29$$

$$Z_2 = \frac{Z_a Z_b}{Z_a + Z_b + Z_c} \qquad 29.30$$

$$Z_3 = \frac{Z_b Z_c}{Z_a + Z_b + Z_c} \qquad 29.31$$

18. THEVENIN'S THEOREM

Thevenin's theorem states that, insofar as the behavior of a linear circuit at its terminals is concerned, any such circuit can be replaced by a single voltage source, $\mathbf{V}_{\text{Th}}$, in series with a single impedance, $\mathbf{Z}_{\text{Th}}$. The method for determining and utilizing the *Thevenin equivalent circuit*, with designations referring to Fig. 29.11, follows.

step 1: Separate the network that is to be changed into a Thevenin equivalent circuit from its load at two terminals, say A and B.

step 2: Determine the open-circuit voltage, $\mathbf{V}_{\text{oc}}$, at terminals A and B.

step 3: Short-circuit terminals A and B and determine the current, $\mathbf{I}_{\text{sc}}$.

step 4: Calculate the Thevenin equivalent voltage and resistance from the following equations.

$$\mathbf{V}_{\text{Th}} = \mathbf{V}_{\text{oc}} \qquad 29.32$$

$$\mathbf{Z}_{\text{Th}} = \frac{\mathbf{V}_{\text{oc}}}{\mathbf{I}_{\text{sc}}} \qquad 29.33$$

step 5: Using the values calculated in Eq. 29.32 and Eq. 29.33, replace the network with the Thevenin equivalent. Reconnect the load at terminals A and B. Determine the desired electrical parameters in the load.

Steps 3 and 4 can be altered by using the following shortcut. Determine the Thevenin equivalent impedance by looking into terminals A and B toward the network with all the power sources altered. Specifically, change independent voltage sources into short circuits and independent current sources into open circuits. Then calculate the resistance of the altered network. The resulting resistance is the Thevenin equivalent impedance.

Figure 29.11 *Thevenin Equivalent Circuit*

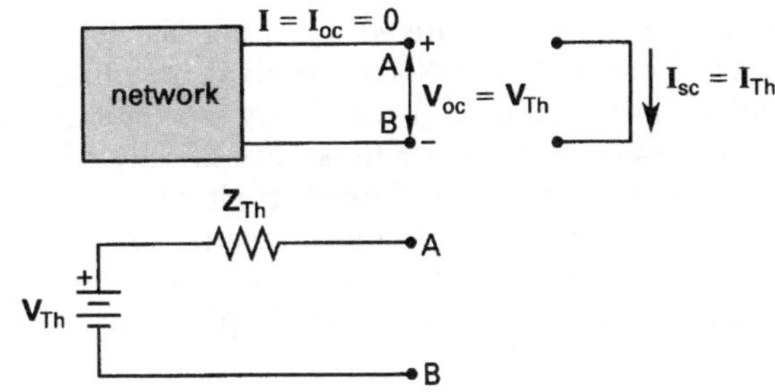

19. NORTON'S THEOREM

Norton's theorem states that, insofar as the behavior of a linear circuit at its terminals is concerned, any such circuit can be replaced by a single current source, $\mathbf{I}_N$, in parallel with a single impedance, $\mathbf{Z}_N$. The method for determining and utilizing the *Norton equivalent circuit*, with designations referring to Fig. 29.12, follows.

step 1: Separate the network that is to be changed into a Norton equivalent circuit from its load at two terminals, say A and B.

step 2: Determine the open-circuit voltage, $\mathbf{V}_{\text{oc}}$, at terminals A and B.

step 3: Short-circuit terminals A and B and determine the current, $\mathbf{I}_{\text{sc}}$.

Figure 29.12 Norton Equivalent Circuit

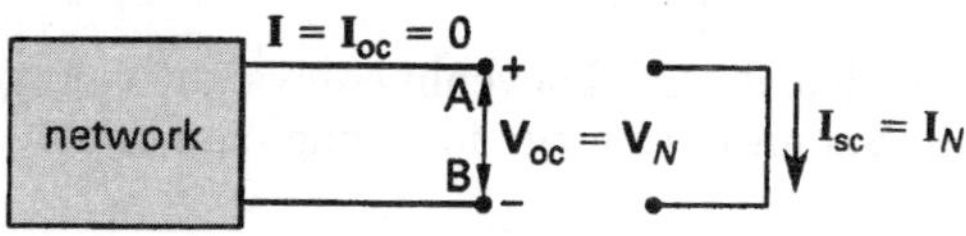

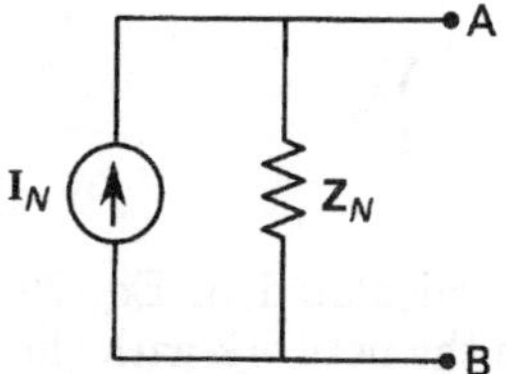

step 4: Calculate the Norton equivalent current and resistance from the following equations.

$$\mathbf{I}_N = \mathbf{I}_{sc} \qquad 29.34$$

$$\mathbf{Z}_N = \frac{\mathbf{V}_{oc}}{\mathbf{I}_{sc}} \qquad 29.35$$

step 5: Using the values calculated in Eq. 29.34 and Eq. 29.35, replace the network with the Norton equivalent. Reconnect the load at terminals A and B. Determine the desired electrical parameters in the load.

Steps 2 and 4 can be altered by using the following shortcut. Determine the Norton equivalent impedance by looking into terminals A and B toward the network with all the power sources altered. Specifically, change independent voltage sources into short circuits and independent current sources into open circuits. Then calculate the resistance of the altered network. The resulting impedance is the Norton equivalent impedance.

The Norton equivalent impedance equals the Thevenin equivalent impedance for identical networks.

20. MAXIMUM POWER TRANSFER THEOREM

The maximum energy, that is, maximum power transfer from a voltage source, occurs when the series source impedance, $\mathbf{Z}_s$, is reduced to the minimum possible value, with zero being the ideal case. This assumes the load impedance is fixed and the source impedance can be changed. Though this is the ideal situation, it is not often the case. Where the load impedance varies and the source impedance is fixed, maximum power transfer occurs when the load and source impedances are complex conjugates. That is, $\mathbf{Z}_l = \mathbf{Z}_s^*$ or $R_{\text{load}} + jX_{\text{load}} = R_s - jX_s$.

Similarly, the maximum power transfer from a current source occurs when the parallel source impedance, $\mathbf{Z}_s$, is increased to the maximum possible value. This assumes the load impedance is fixed and the source impedance can be changed. In the more common case where the load impedance varies and the source impedance is fixed, the maximum power transfer occurs when the load and source impedances are complex conjugates.

The maximum power transfer in any circuit occurs when the load impedance equals the complex conjugate of the Norton or Thevenin equivalent impedance.

21. SUPERPOSITION THEOREM

The principle of *superposition* is that the response of (that is, the voltage across or current through) a linear circuit element in a network with multiple independent sources is equal to the response obtained if each source is considered individually and the results summed. The steps involved in determining the desired quantity follow.

step 1: Replace all sources except one by their internal resistances. Ideal current sources are replaced by open circuits. Ideal voltage sources are replaced by short circuits.

step 2: Compute the desired quantity, either voltage or current caused by the single source, for the element in question.

step 3: Repeat steps 1 and 2 for each of the sources in turn.

step 4: Sum the calculated values obtained for the current or voltage obtained in step 2. The result is the actual value of the current or voltage in the element for the complete circuit.

Superposition is not valid for circuits in which the following conditions exist.

- The capacitors have an initial charge (i.e., an initial voltage) not equal to zero. This principle can be used if the charge is treated as a separate voltage source and the equivalent circuit analyzed.
- The inductors have an initial magnetic field (i.e., an initial current) not equal to zero. This principle can be used if the energy in the magnetic field is treated as a separate current source and this equivalent circuit analyzed.
- Dependent sources are used.

22. MILLER'S THEOREM

A standard two-port network is shown in Fig. 29.13(a). *Miller's theorem* states that if an admittance, Y, is connected between the input and output terminals of a two-port network, as shown in Fig. 29.13(b), the output voltage is linearly dependent on the input voltage. Further, the circuit can be transformed as shown in

Fig. 29.13(c). The transformation equations, with A as a constant that must be determined by independent means, are

$$Y_1 = Y(1 - A) \quad 29.36$$

$$Y_2 = Y\left(1 - \frac{1}{A}\right) \quad 29.37$$

Figure 29.13 *Miller's Theorem*

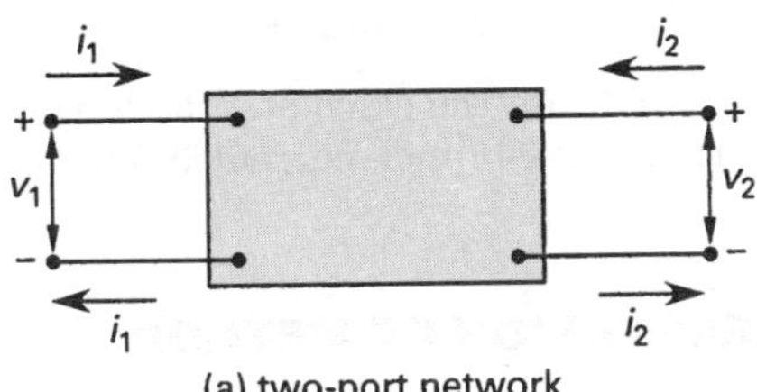

(a) two-port network

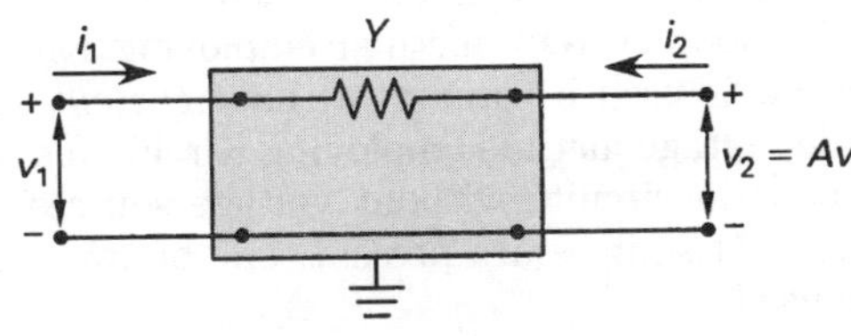

(b) connecting admittance

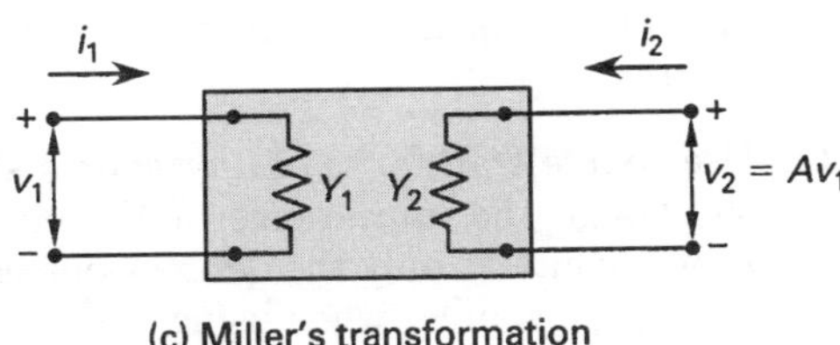

(c) Miller's transformation

Similar formulas can be developed from Eq. 29.36 and Eq. 29.37 for impedances Z_1 and Z_2, though the theorem is not often used in such a manner. Miller's theorem is useful in transistor high-frequency amplifiers, among other applications.

23. KIRCHHOFF'S LAWS

Once an equivalent circuit is determined and reduced to its simplest form, Kirchhoff's voltage and current laws are used to determine the circuit behavior. If the KVL and KCL equations are written in terms of instantaneous quantities, $v(t)$ and $i(t)$, a differential equation results with an order equal to the number of independent energy-storing devices in the circuit. The solution of this equation results in a steady-state component and a transient component, which decays with time. If the equation is written in terms of phasor quantities, **V** and **I**, an algebraic equation results. The solution of this equation alone results in a steady-state component. For an equation with an order higher than two, the most efficient solution method is computer-aided design software.

24. KIRCHHOFF'S VOLTAGE LAW

Kirchhoff's voltage law (KVL) states that the algebraic sum of the phasor voltages around any closed path within a circuit or network is zero. Stated in terms of the voltage rises and drops, the sum of the phasor voltage rises equals the sum of the phasor voltage drops around any closed path within a circuit or network. The method for applying KVL follows.

step 1: Identify the loop.

step 2: Pick a loop direction.[6]

step 3: Assign the loop current in the direction picked in step 2.

step 4: Assign voltage polarities consistent with the loop current direction in step 3.

step 5: Apply KVL to the loop using Ohm's law to express the voltages across each circuit element.

step 6: Solve the equation for the desired quantity.

25. KIRCHHOFF'S CURRENT LAW

Kirchhoff's current law (KCL) states that the algebraic sum of the phasor currents at any node is zero. Stated in terms of the currents out of and into a node, the sum of the currents directed out of a node must equal the sum of the currents directed into the same node. The method for applying KCL follows.

step 1: Identify the nodes and pick a reference or datum node.

step 2: Label the node-to-datum voltage for each unknown node.

step 3: Pick a current direction for each path at every node.[7]

step 4: Apply KCL to the nodes using Ohm's law to express the currents through each circuit branch.

step 5: Solve the equations for the desired quantity.

26. LOOP ANALYSIS

The *loop-current* method is a systematic network-analysis procedure that uses currents as the unknowns. It is

[6]Clockwise is the direction chosen as the standard in this text. Any direction is allowed. The resulting mathematics determines the correct direction of current flow and voltage polarities. Consistency is the key to preventing calculation errors.

[7]Current flow out of the node is the direction chosen as the standard for positive current in this text. Any direction is allowed. The resulting mathematics determines the correct direction of current flow and voltage polarities. Consistency is the key to preventing calculation errors.

also called *mesh analysis* or the *Maxwell loop-current* method. The method uses Kirchhoff's voltage law and is performed on planar networks. It requires $n-1$ simultaneous equations for an n-loop system. The method's steps are as follows.

step 1: Select $n-1$ loops, that is, one loop less than the total number of possible loops.

step 2: Assign current directions for the selected loops. (While this is arbitrary, clockwise directions will always be chosen in this text for consistency.) Any incorrectly chosen current direction will cause a negative result when the simultaneous equations are solved in step 4. Show the direction of the current with an arrow.

step 3: Write Kirchhoff's voltage law for each of the selected loops. Assign polarities based on the direction of the loop current. The voltage of a source is positive when the current flows out of the positive terminal, that is, from the negative terminal to the positive terminal inside the source. The selected direction of the loop current always results in a voltage drop in the resistors of the loop (see Fig. 26.6). Where two loop currents flow through an element, they are summed to determine the voltage drop in that element, using the direction of the current in the loop for which the equation is being written as the positive (i.e., correct) direction. Any incorrect direction for the loop current will be indicated by a negative sign in the solution in step 4.

step 4: Solve the $n-1$ equations from step 3 for the unknown currents.

step 5: If required, determine the actual current in an element by summing the loop currents flowing through the element. Sum the absolute values to obtain the correct magnitude. The correct direction is given by the loop current with the positive value.

27. NODE ANALYSIS

The *node-voltage* method is a systematic network-analysis procedure that uses voltages as the unknowns. The method uses Kirchhoff's current law. It requires $n-1$ equations for an n-principal node system (equations are not necessary at simple nodes, that is, nodes connecting only two circuit elements). The method's steps are as follows.

step 1: Simplify the circuit, if possible, by combining resistors in series or parallel or by combining current sources in parallel. Identify all nodes. The minimum number of equations required will be $n-1$ where n represents the number of principal nodes.

step 2: Choose one node as the reference node, that is, the node that will be assumed to have ground potential (0 V). To minimize the number of terms in the equations, select the node with the largest number of circuit elements to serve as the reference node.

step 3: Write Kirchhoff's current law for each principal node except the reference node, which is assumed to have a zero potential.

step 4: Solve the $n-1$ equations from step 3 to determine the unknown voltages.

step 5: If required, use the calculated node voltages to determine any branch current desired.

28. DETERMINATION OF METHOD

The method used to analyze an electrical network depends on the circuit elements and their configurations. The loop-current method employing Kirchhoff's voltage law is used in circuits without current sources. The node-voltage method employing Kirchhoff's current law is used in circuits without voltage sources. When both types of sources are present, one of two methods may be used.

method 1: Use each of Kirchhoff's laws, assigning voltages and currents as needed, and substitute any known quantity into the equations as written.

method 2: Use source transformation or source shifting to change the appearance of the circuit so that it contains only the desired sources, that is, voltage sources when using KVL and current sources when using KCL. *Source shifting* is manipulating the circuit so that each voltage source has a resistor in series and each current source has a resistor in parallel.

Use the method that results in the least number of equations. The loop-current method produces $n-1$ equations where n is the total number of loops. The node-voltage method produces $n-1$ equations where n is the number of principal nodes. Count the number of loops and the number of principal nodes prior to writing the equations. Whichever is least determines the method used.

Additional methods exist, some of which require less work. The advantage of the loop-current and node-voltage methods is that they are systematic and guarantee a solution.

29. VOLTAGE AND CURRENT DIVIDERS

At times, a source voltage will not be at the required value for the operation of a given circuit. For example, the household voltage of 120 V is too high to properly bias electronic circuitry. One method of obtaining the required voltage without using a transformer is to use a *voltage divider*. Figure 29.14(a) illustrates a voltage divider. The voltage across impedance 2 is

$$\mathbf{V}_2 = \mathbf{V}_s \left(\frac{\mathbf{Z}_2}{\mathbf{Z}_1 + \mathbf{Z}_2} \right) \tag{29.38}$$

Figure 29.14 *Divider Circuits*

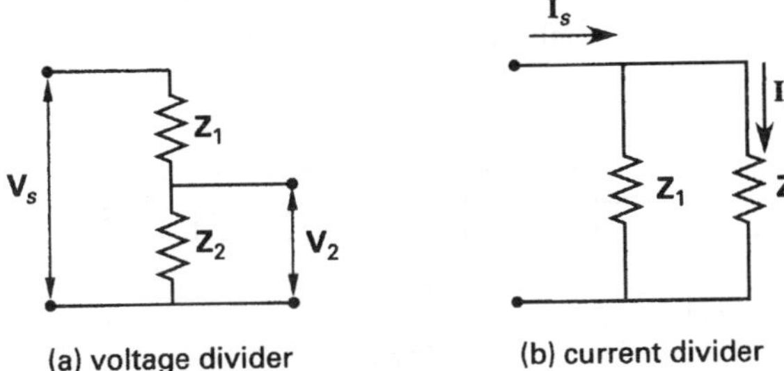

If the fraction of the source voltage, $\mathbf{V}_s$, is found as a function of $\mathbf{V}_2$, the result is known as the *gain* or the *voltage-ratio transfer function*. Rearranging Eq. 29.38 to show the specific relationship between $\mathbf{Z}_1$ and $\mathbf{Z}_2$ gives

$$\frac{\mathbf{V}_2}{\mathbf{V}_s} = \frac{1}{1 + \frac{\mathbf{Z}_1}{\mathbf{Z}_2}} \tag{29.39}$$

A specific current can be obtained through an analogous circuit called a *current divider*, shown in Fig. 29.14(b). The current through impedance 2 is

$$\mathbf{I}_2 = \mathbf{I}_s \left(\frac{\mathbf{Z}_1}{\mathbf{Z}_1 + \mathbf{Z}_2} \right) = \mathbf{I}_s \left(\frac{\mathbf{G}_2}{\mathbf{G}_1 + \mathbf{G}_2} \right) \tag{29.40}$$

30. STEADY-STATE AND TRANSIENT IMPEDANCE ANALYSIS

When the electrical parameters of a circuit do not change with time, the circuit is said to be in a *steady-state* condition. In DC circuits, this condition exists when the magnitude of the parameter is constant. In AC circuits, this condition exists when the frequency is constant. A *transient* is a temporary phenomenon occurring prior to a network reaching steady state.

DC steady-state impedance analysis is based on constant voltages and currents and, therefore, time derivatives of zero. The DC impedances are

- Resistance:

$$Z_R|_{\text{DC}} = R \tag{29.41}$$

- Inductance:

$$v = L\frac{di}{dt} = L(0) = 0 \tag{29.42}$$

$$Z = \frac{v}{i} = \frac{0}{i} = 0 \tag{29.43}$$

$$Z_L|_{\text{DC}} = 0 \quad \text{[short circuit]} \tag{29.44}$$

- Capacitance:

$$i = C\frac{dv}{dt} = C(0) = 0 \tag{29.45}$$

$$Z = \frac{v}{i} = \frac{v}{0} \rightarrow \infty \tag{29.46}$$

$$Z_C|_{\text{DC}} = \infty \quad \text{[open circuit]} \tag{29.47}$$

AC steady-state impedance analysis is based on the phasor form where $df(t)/dt = j\omega f(t)$. The AC steady-state impedances are

- Resistance:

$$Z_R|_{\text{AC}} = R \tag{29.48}$$

- Inductance:

$$v = L\frac{di}{dt} = Lj\omega i \tag{29.49}$$

$$Z = \frac{v}{i} = \frac{Lj\omega i}{i} = j\omega L \tag{29.50}$$

$$Z_L|_{\text{AC}} = j\omega L \tag{29.51}$$

- Capacitance:

$$i = C\frac{dv}{dt} = Cj\omega v \tag{29.52}$$

$$Z = \frac{v}{i} = \frac{v}{Cj\omega v} = \frac{1}{j\omega C} \tag{29.53}$$

$$Z_C|_{\text{AC}} = \frac{1}{j\omega C} \tag{29.54}$$

Transient impedance analysis is based on the phasor form with the complex variable s substituted for $j\omega$. The variable $s = \sigma + j\omega$ and is the same as the Laplace transform variable. The derivative is $df(t)/dt = sf(t)$. The transient impedances are

- Resistance:

$$Z_R = R \tag{29.55}$$

- Inductance:

$$Z_L = sL \tag{29.56}$$

- Capacitance:

$$Z_C = \frac{1}{sC} \tag{29.57}$$

Transient impedances are useful in the stability analysis as well as during transients.

31. TWO-PORT NETWORKS

An electric circuit or network is often used to connect a source to a load, modifying the source energy or information in a given manner as required or desired by the load. If the circuit is such that the current flow into one terminal is equal to the current flow out of a second terminal, the terminal pair is called a *port*. A given network may have any number of ports, the most common being the *two-port network* shown in Fig. 29.15. Four variables exist in this representation: v_1, v_2, i_1, and i_2. The subscript 1 indicates the *input port*, and the subscript 2 indicates the *output port*.

Figure 29.15 *Two-Port Network*

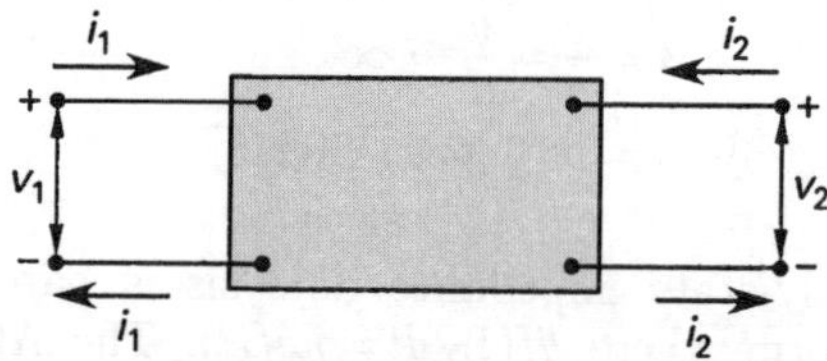

Two-port networks can represent either three- or four-terminal devices, such as transistors or transformers (see Fig. 29.16). The electrical properties of the device within the network are described by a set of parameters identified by double subscripts, the first representing the row and the second the column of a matrix. The *parameter type* is determined by the selection of independent variables for which to solve.

Figure 29.16 *Transformer and Transistor Two-Port Networks*

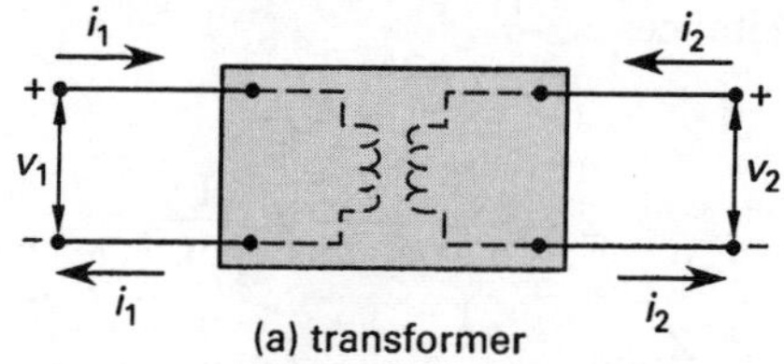

(a) transformer

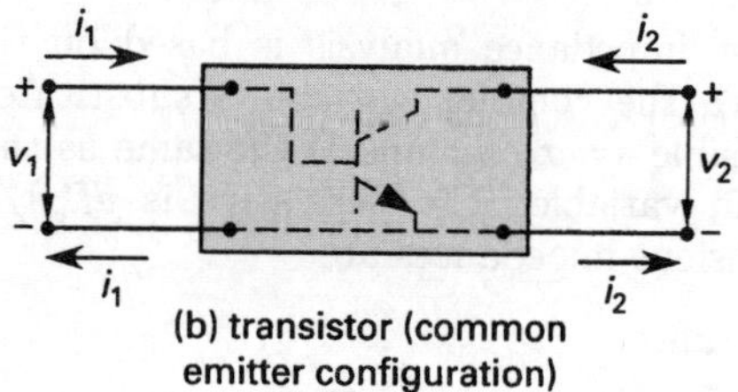

(b) transistor (common emitter configuration)

Open-circuit impedance parameters, or *z*-parameters, occur when the two currents are selected as the independent variables. The *z*-parameter model and *deriving equations* for the individual parameters are shown in Fig. 29.17. The applicable matrix and resulting equations follow.

Figure 29.17 *Impedance Model Parameters*

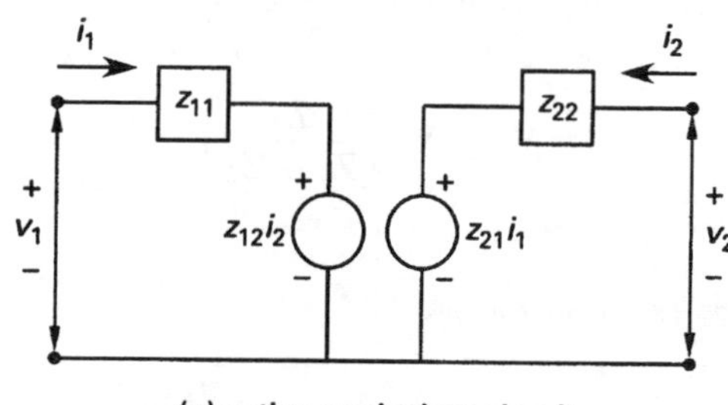

(a) active equivalent circuit

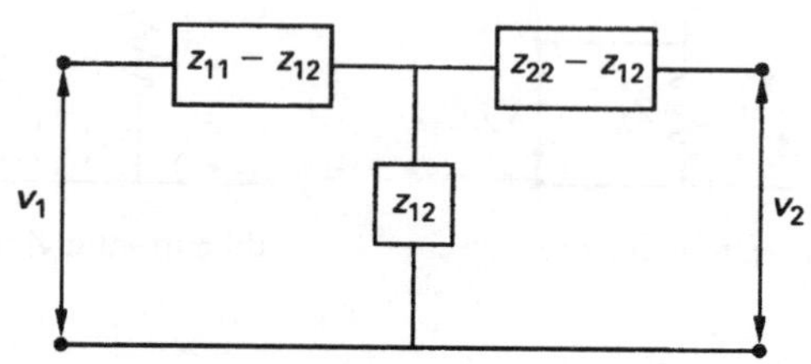

(b) passive T-model equivalent circuit

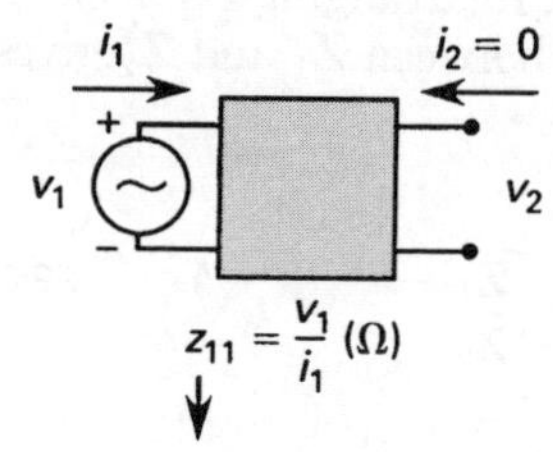

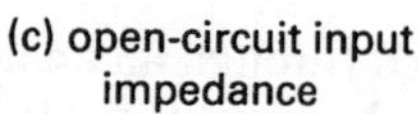

(c) open-circuit input impedance

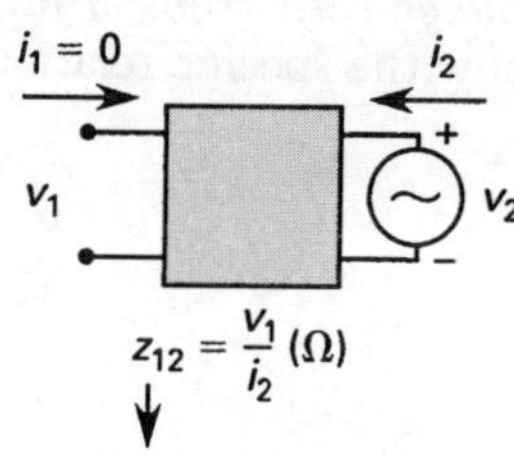

(d) open-circuit reverse transfer impedance

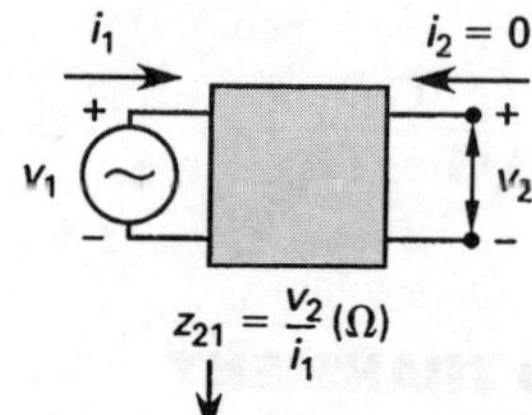

(e) open-circuit forward transfer impedance

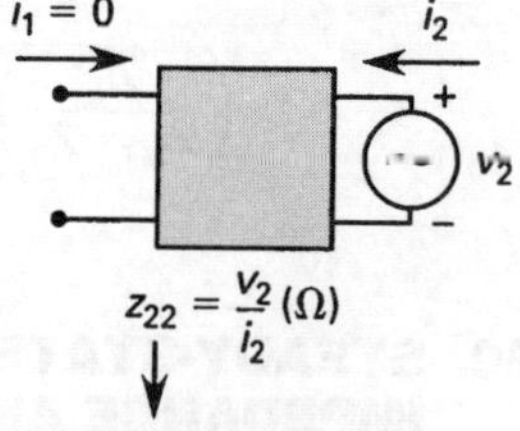

(f) open-circuit output impedance

$$\begin{bmatrix} v_1 \\ v_2 \end{bmatrix} = \begin{bmatrix} z_{11} & z_{12} \\ z_{21} & z_{22} \end{bmatrix} \begin{bmatrix} i_1 \\ i_2 \end{bmatrix} \qquad 29.58$$

$$v_1 = z_{11}i_1 + z_{12}i_2 \qquad 29.59$$

$$v_2 = z_{21}i_1 + z_{22}i_2 \qquad 29.60$$

Short-circuit admittance parameters, or *y*-parameters, occur when the two voltages are selected as the independent variables. The *y*-parameter model and *deriving equations* for the individual parameters are shown in Fig. 29.18. The applicable matrix and resulting equations follow.

$$\begin{bmatrix} i_1 \\ i_2 \end{bmatrix} = \begin{bmatrix} y_{11} & y_{12} \\ y_{21} & y_{22} \end{bmatrix} \begin{bmatrix} v_1 \\ v_2 \end{bmatrix} \qquad 29.61$$

$$i_1 = y_{11}v_1 + y_{12}v_2 \qquad 29.62$$

$$i_2 = y_{21}v_1 + y_{22}v_2 \qquad 29.63$$

Figure 29.18 *Admittance Model Parameters*

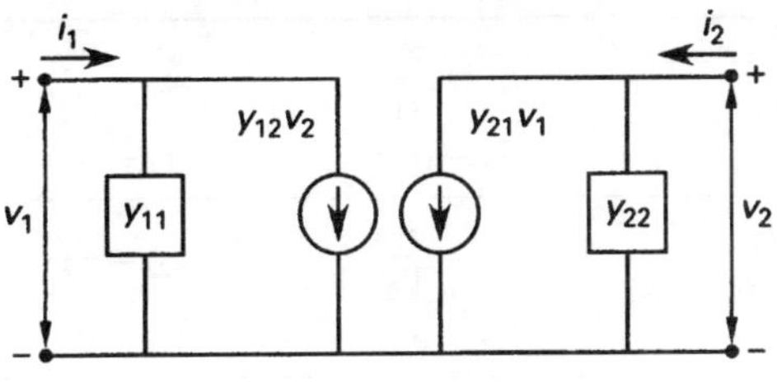

(a) active equivalent circuit

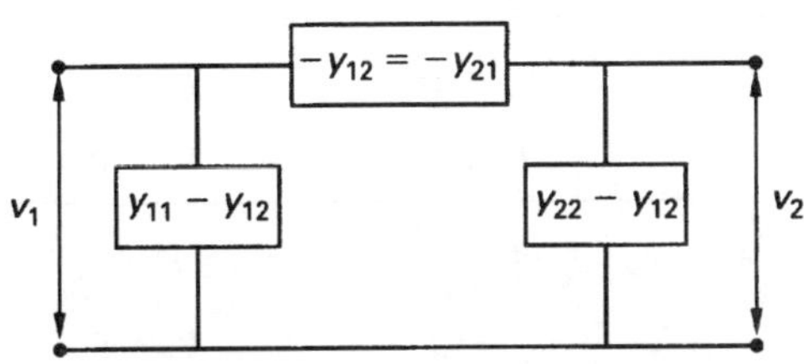

(b) passive π-model equivalent circuit

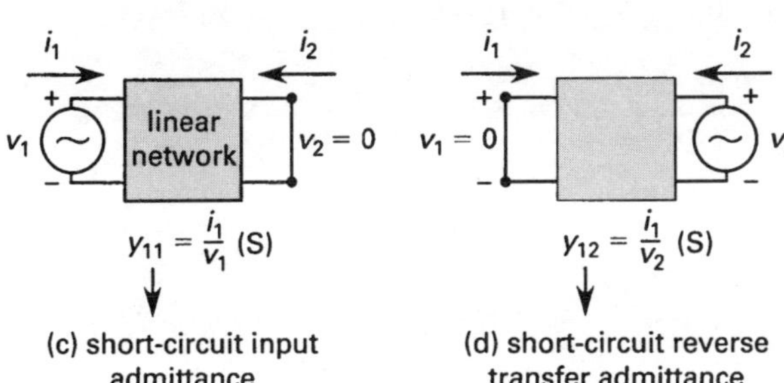

(c) short-circuit input admittance

(d) short-circuit reverse transfer admittance

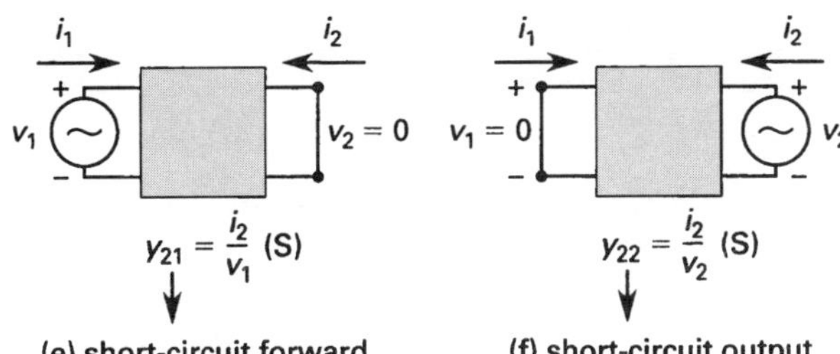

(e) short-circuit forward transfer admittance

(f) short-circuit output admittance

Hybrid parameters, or *h*-parameters, occur when one voltage and one current are selected as the independent variables. Figure 29.19 shows the *h*-parameter model and *deriving equations* for the individual parameters. The applicable matrix and resulting equations follow.

$$\begin{bmatrix} v_1 \\ i_2 \end{bmatrix} = \begin{bmatrix} h_{11} & h_{12} \\ h_{21} & h_{22} \end{bmatrix} \begin{bmatrix} i_1 \\ v_2 \end{bmatrix} \qquad 29.64$$

$$v_1 = h_{11}i_1 + h_{12}v_2 \qquad 29.65$$

$$i_2 = h_{21}i_1 + h_{22}v_2 \qquad 29.66$$

Figure 29.19 *Hybrid Model Parameters*

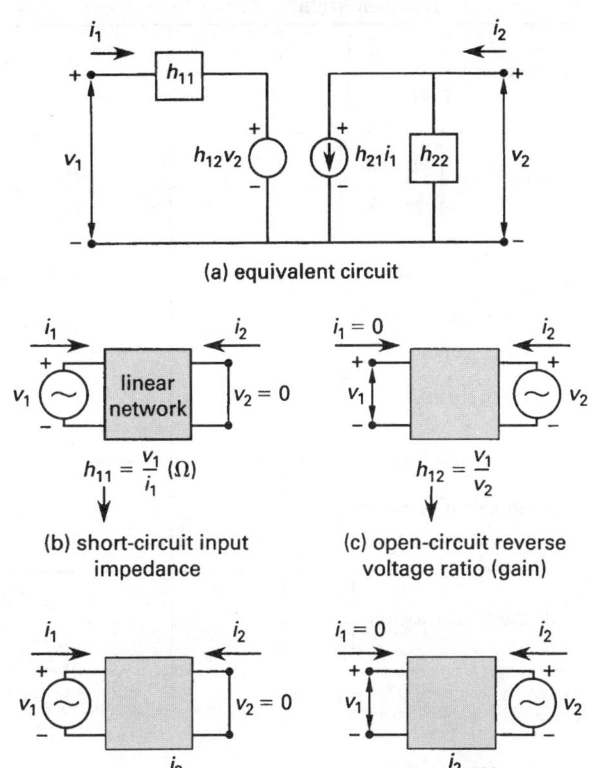

(a) equivalent circuit

(b) short-circuit input impedance

(c) open-circuit reverse voltage ratio (gain)

(d) short-circuit forward current ratio (gain)

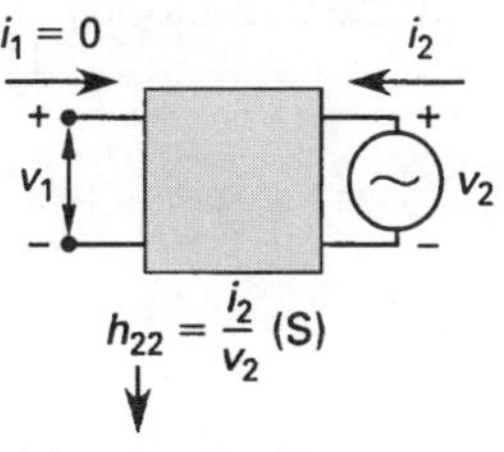

(e) open-circuit output admittance

Inverse hybrid parameters, or *g*-parameters, also occur when one voltage and one current are picked as the independent variables. The applicable matrix and resulting equations follow.

$$\begin{bmatrix} i_1 \\ v_2 \end{bmatrix} = \begin{bmatrix} g_{11} & g_{12} \\ g_{21} & g_{22} \end{bmatrix} \begin{bmatrix} v_1 \\ i_2 \end{bmatrix} \qquad 29.67$$

$$i_1 = g_{11}v_1 + g_{12}i_2 \qquad 29.68$$

$$v_2 = g_{21}v_1 + g_{22}i_2 \qquad 29.69$$

Other parameter representations are possible. Some examples include the *transmission line* or *chain parameters*, and the *inverse transmission line parameters*. The choice of parameter type depends on numerous factors, including whether or not all the parameters exist or are defined for a given network, the mathematical convenience of using a certain set of parameters, and the accuracy of sensitivity of the parameters when considered as part of the overall circuit to which it is connected. The parameters and their deriving equations are summarized in Table 29.3. The formulas for conversion between the parameter types are given in App. 29.A.

Table 29.3 *Two-Port Network Parameters*

representation	deriving equations			
impedance $\begin{bmatrix} z_{11} & z_{12} \\ z_{21} & z_{22} \end{bmatrix}$	$z_{11} = \frac{V_1}{I_1}$ $I_2 = 0$	$z_{12} = \frac{V_1}{I_2}$ $I_1 = 0$	$z_{21} = \frac{V_2}{I_1}$ $I_2 = 0$	$z_{22} = \frac{V_2}{I_2}$ $I_1 = 0$
admittance $\begin{bmatrix} y_{11} & y_{12} \\ y_{21} & y_{22} \end{bmatrix}$	$y_{11} = \frac{I_1}{V_1}$ $V_2 = 0$	$y_{12} = \frac{I_1}{V_2}$ $V_1 = 0$	$y_{21} = \frac{I_2}{V_1}$ $V_2 = 0$	$y_{22} = \frac{I_2}{V_2}$ $V_1 = 0$
hybrid $\begin{bmatrix} h_{11} & h_{12} \\ h_{21} & h_{22} \end{bmatrix}$	$h_{11} = \frac{V_1}{I_1}$ $V_2 = 0$	$h_{12} = \frac{V_1}{V_2}$ $I_1 = 0$	$h_{21} = \frac{I_2}{I_1}$ $V_2 = 0$	$h_{22} = \frac{I_2}{V_2}$ $I_1 = 0$
inverse hybrid $\begin{bmatrix} g_{11} & g_{12} \\ g_{21} & g_{22} \end{bmatrix}$	$g_{11} = \frac{I_1}{V_1}$ $I_2 = 0$	$g_{12} = \frac{I_1}{I_2}$ $V_1 = 0$	$g_{21} = \frac{V_2}{V_1}$ $I_2 = 0$	$g_{22} = \frac{V_2}{I_2}$ $V_1 = 0$
transmission or chain $\begin{bmatrix} A & B \\ C & D \end{bmatrix}$	$A = \frac{V_1}{V_2}$ $I_2 = 0$	$B = \frac{-V_1}{I_2}$ $V_2 = 0$	$C = \frac{I_1}{V_2}$ $I_2 = 0$	$D = \frac{-I_1}{I_2}$ $V_2 = 0$
inverse transmission $\begin{bmatrix} \alpha & \beta \\ \gamma & \delta \end{bmatrix}$	$\alpha = \frac{V_2}{V_1}$ $I_1 = 0$	$\beta = \frac{-V_2}{I_1}$ $V_1 = 0$	$\gamma = \frac{I_2}{V_1}$ $I_1 = 0$	$\delta = \frac{-I_2}{I_1}$ $V_1 = 0$

30 Transient Analysis

Nomenclature

a	constant	–
A	natural response coefficient	–
b	constant	–
BW	bandwidth	Hz
c	constant	–
C	capacitance	F
f	function	–
G	conductance	S
i	variable current	A
I	constant or rms current	A
L	self-inductance	H
N	number	–
P	power	W
q	instantaneous charge	C
Q	quality factor	–
R	resistance	Ω
s	Laplace transform variable	–
t	time	s
v	variable voltage	V
V	constant or rms voltage	V
X	reactance	Ω
Y	admittance	S
Z	impedance	Ω

Symbols

κ	integration constant	–
σ	damping, real part of s	–
τ	time constant	s
ϕ	angle	rad
ϕ	phase difference angle	rad
ω	frequency	rad/s
ω_0	zero resistance frequency	rad/s

Subscripts

0	initial
bat	battery
C	capacitor
d	delay
L	inductor
m	maximum
r	rise
R	resistor
s	source
ss	steady-state
tr	transient

1. FUNDAMENTALS

Whenever a network or circuit undergoes a change, the currents and voltages experience a transitional period during which their properties shift from their former values to their new steady-state values. This time period is called a *transient*. The determination of a circuit's behavior during this period is *transient analysis*. There are three classes of transient problems, all based on the nature of the energy source. These are *DC switching transients*, *AC switching transients*, and *pulse transients*. *Switching transients* occur as a result of a change in topology of the circuit, that is, the physical elements within the circuit change (become connected or disconnected) as a result of the operation of a switch. The switch can be a physical switch or an electrical one, such as a transistor operating in the switching mode. A circuit's response to a switching transient is known as a *step response*.

Pulse transients involve a change in the current or voltage waveform, not in the topology of a circuit. A circuit's response to a pulse transient is known as an *impulse response*.

The energy storage elements in electric circuits are the capacitor and the inductor. When one of these is present, the mathematical representation of the circuit is a first-order differential equation of the form

$$f(t) = b\left(\frac{dx}{dt}\right) + cx \qquad 30.1$$

Either the current or the voltage can be represented as the dependent variable, that is, $f(t)$, in Eq. 30.1. For capacitors, the voltage is used because the energy storage is a function of voltage and a true differential equation results (the capacitor current is an integral equation). For inductors, similar reasoning holds and the current is considered the dependent variable, $f(t)$. The solution to Eq. 30.1 is of the general form

$$x(t) = \kappa + Ae^{-t/\tau} \qquad 30.2$$

See Table 30.1 for the applicable equations for both the inductor and capacitor.

The term A in Eq. 30.2 is the natural response coefficient and is set to match the conditions at some known time in the transient, typically at the onset. The term κ is value-dependent on the forcing function, $f(t)$. The term τ is called the *time constant* and is equal to

$$\tau = \frac{b}{c} \qquad 30.3$$

The time constant for an RC circuit is

$$\tau = RC \qquad 30.4$$

The time constant for an RL circuit is

$$\tau = \frac{L}{R} \qquad 30.5$$

There are two parts to Eq. 30.2, or any solution to a first-order differential equation: the homogeneous (or complementary) solution and the particular solution. The homogeneous solution, that is, the solution where $f(t) = 0$, is called the *natural response* of the circuit and is represented by the exponential term in Eq. 30.2. The decay behavior of a general exponential is shown in Fig. 30.1. The particular solution is called the *forced response*, since it depends on the forcing function, $f(t)$, and is represented by κ in Eq. 30.2.

Figure 30.1 *Exponential Behavior*

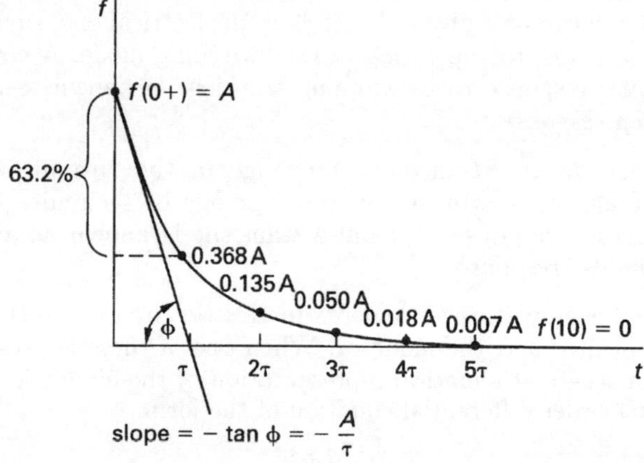

When two independent energy storage elements are present, the mathematical representation of the circuit is a second-order differential equation of the form

$$f(t) = a\left(\frac{d^2x}{dt^2}\right) + b\left(\frac{dx}{dt}\right) + cx \qquad 30.6$$

There are three forms of the solution, depending on the magnitudes of the constants a, b, and c (see Chap. 31).

2. RESISTOR-CAPACITOR CIRCUITS: NATURAL RESPONSE

Consider the generic resistor-capacitor *source-free circuit* shown in Fig. 30.2. When the charged capacitor, C, is connected to the resistor, R, via a complete electrical path, the capacitor will discharge in an attempt to resist the change in voltage. In the process, the capacitor's stored electrical energy is dissipated in the resistor until no further energy remains and no current flows. This gradual decrease is the transient. The voltage follows the form of Fig. 30.1.

Figure 30.2 *Resistor-Capacitor Circuit: Natural Response*

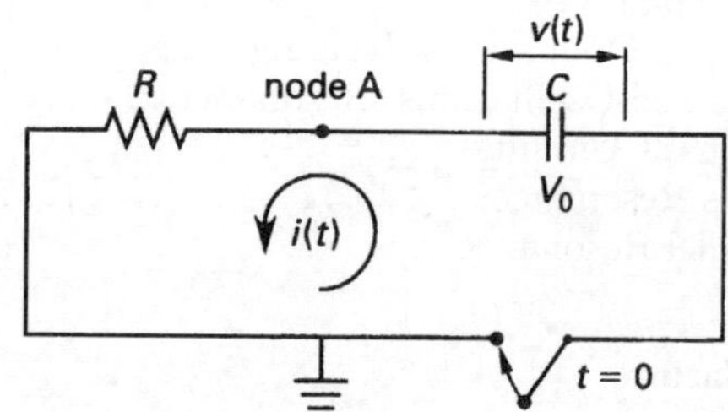

Example 30.1

Determine the formula for the voltage in the circuit of Fig. 30.2 from $t=0$ onward.

Solution

Write Kirchhoff's current law (KCL) for the simple node A between the resistor and the capacitor. The reference node is located at the switch, opposite node A. In keeping with the convention of this book, both currents are assumed to flow out of the node, giving

$$i_C + i_R = 0$$

Substitute the expression for current flow through a capacitor and the Ohm's law expression for current flow in a resistor.

$$C\left(\frac{dv}{dt}\right) + \frac{v}{R} = 0$$

$$\frac{dv}{dt} + \left(\frac{1}{RC}\right)v = 0$$

This is a homogeneous first-order linear differential equation. The solution could be written directly from the information in the mathematics chapters or by knowing that only an exponential form of v can be linearly combined with its derivative. In order to show the process, separate the variables and integrate both sides.

$$\frac{dv}{v} = -\left(\frac{1}{RC}\right)dt$$
$$\int \frac{1}{v}\,dv = \int -\left(\frac{1}{RC}\right)dt$$
$$\ln v = -\frac{1}{RC}t + \kappa$$
$$v(t) = e^{-(1/RC)t+\kappa} = Ae^{-t/RC} = Ae^{-t/\tau}$$

This is the natural response of the system. To determine the constant A, the value of $v(t)$ at some time must be known. Because the initial voltage is V_0, the value of the constant A is

$$v(0) = V_0 = Ae^{-0/\tau} = A$$

The final solution is

$$v(t) = V_0 e^{-t/RC}$$

3. RESISTOR-CAPACITOR CIRCUITS: FORCED RESPONSE

Consider the generic resistor-capacitor circuit shown in Fig. 30.3. When the charged capacitor, C, is connected to the resistor, R, via a complete electrical path, the capacitor will discharge or charge, depending on the magnitude of V_{bat}, in an attempt to resist the change in voltage. In the process, the capacitor's stored electrical energy is either dissipated or enhanced until no further energy change occurs and no current flows. This gradual decrease or increase is the transient. After the passage of time equal to five time constants, 5τ, the final value is within less than 1% of its steady-state value and the transient is considered complete. The voltage follows the form of Fig. 30.1, or its inverse, with the exception that the final steady-state voltage is the voltage of the driving force.

Figure 30.3 *Resistor-Capacitor Circuit: Forced Response*

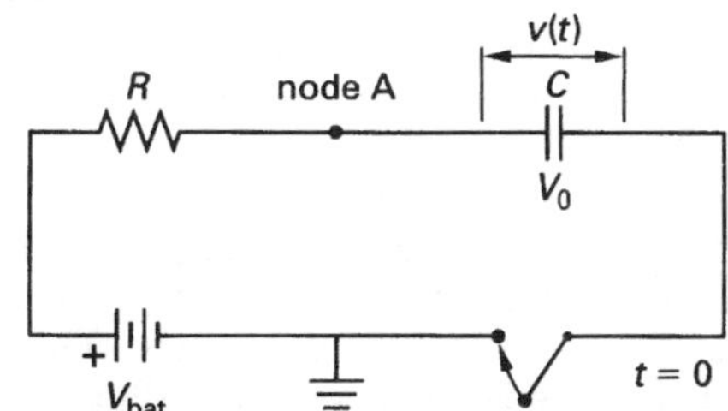

Example 30.2

Determine the formula for the voltage in the circuit of Fig. 30.3 from $t = 0$ onward.

Solution

Write KCL for the simple node A between the resistor and the capacitor. The node between the negative terminal of the battery and the capacitor, opposite node A, is the reference node. Both currents are assumed to flow out of the node in keeping with the convention of this text, giving

$$i_C + i_R = 0$$

Substitute the expression for current flow through a capacitor and the Ohm's law expression for current flow in a resistor.

$$C\left(\frac{dv}{dt}\right) + \frac{v - V_{\text{bat}}}{R} = 0$$
$$\frac{dv}{dt} + \left(\frac{1}{RC}\right)(v - V_{\text{bat}}) = 0$$
$$\frac{dv}{dt} + \left(\frac{1}{RC}\right)v = \left(\frac{1}{RC}\right)V_{\text{bat}}$$

This is a nonhomogeneous first-order linear differential equation. The known constant to the right of the equal sign is called the *forcing* or *driving function*. Such an equation is solved in two steps. The first step is to determine the homogeneous, or complementary, solution, that is, the solution with the forcing function equal to zero. In Ex. 30.1 this was determined to be

$$v_{\text{tr}}(t) = Ae^{-t/RC}$$

This is the natural response of the system and represents the transient portion of the solution. The next step is to determine the particular solution, or forced response, of the system. Because the particular solution is the steady-state solution, the expectation is that the solution would be of the same form as the forcing function. Assuming a constant, κ, as the particular solution and substituting this into the KCL expression for v gives

$$v_{\text{ss}} = \kappa$$
$$\frac{dv}{dt} + \left(\frac{1}{RC}\right)v = \left(\frac{1}{RC}\right)V_{\text{bat}}$$
$$\frac{dv_{\text{ss}}}{dt} + \left(\frac{1}{RC}\right)v_{\text{ss}} = \left(\frac{1}{RC}\right)V_{\text{bat}}$$
$$\frac{d\kappa}{dt} + \left(\frac{1}{RC}\right)\kappa = \left(\frac{1}{RC}\right)V_{\text{bat}}$$
$$0 + \left(\frac{1}{RC}\right)\kappa = \left(\frac{1}{RC}\right)V_{\text{bat}}$$
$$\kappa = V_{\text{bat}}$$

The particular solution is

$$v(\infty) = v_{\text{ss}} = V_{\text{bat}}$$

The total solution is the combination of the complementary (homogeneous) and particular (nonhomogeneous) solutions, that is, the combination of the transient and steady-state solutions. The total solution is

$$v(t) = V_{\text{bat}} + Ae^{-t/RC}$$

The order of the transient and steady-state solutions is unimportant. The exponential term (the transient term) is often listed to the right in an equation.

The natural response coefficient A is again determined by the initial condition $v(0) = V_0$.

$$v(0) = V_0 = V_{\text{bat}} + Ae^{-0/RC} = V_{\text{bat}} + A$$
$$A = V_0 - V_{\text{bat}}$$

The final solution is

$$v(t) = V_{\text{bat}} + (V_0 - V_{\text{bat}})e^{-t/RC}$$

The voltage $v(t)$ is the voltage at node A, that is, the voltage across the capacitor.

4. RESISTOR-INDUCTOR CIRCUITS: NATURAL RESPONSE

Consider the generic resistor-inductor source-free circuit shown in Fig. 30.4. When the inductor, L, whose magnetic field is at a maximum due to initial current, I_0, is connected to the resistor, R, via a complete electrical path, the inductor will discharge in an attempt to resist the change in current. In the process, the inductor's stored magnetic energy is dissipated in the resistor until no further energy remains and no current flows. This gradual decrease is the transient. After the passage of time equal to five time constants, 5τ, the final value is within less than 1% of its steady-state value and the transient is considered complete. The current follows the form of Fig. 30.1.

Figure 30.4 *Resistor-Inductor Circuit: Natural Response*

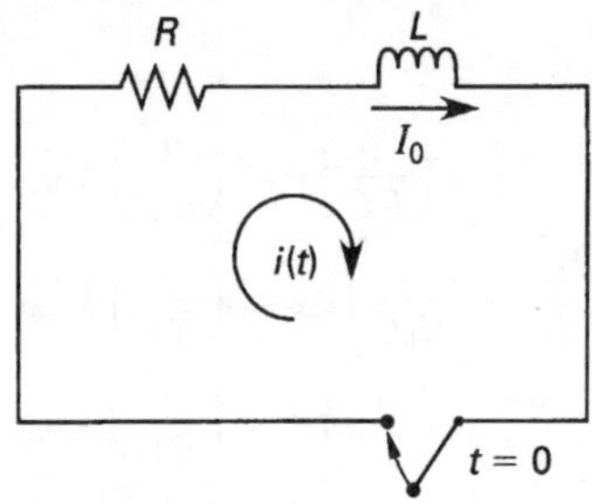

Example 30.3

Determine the formula for the current in the circuit of Fig. 30.4 from $t = 0$ onward.

Solution

Write Kirchhoff's voltage law (KVL) for the loop containing the resistor and the inductor. In keeping with the convention of this book, the reference direction is clockwise, giving

$$v_L + v_R = 0$$

Substitute the expression for voltage across an inductor and the Ohm's law expression for voltage drop in a resistor.

$$L\left(\frac{di}{dt}\right) + iR = 0$$
$$\frac{di}{dt} + \left(\frac{R}{L}\right)i = 0$$

This is a homogeneous first-order linear differential equation. The solution could be written directly from the information in the mathematics chapters or by knowing that only an exponential form of i can be linearly combined with its derivative. The process is similar to that in Ex. 30.1.

$$\frac{di}{i} = -\left(\frac{R}{L}\right)dt$$
$$\int\left(\frac{1}{i}\right)di = \int -\left(\frac{R}{L}\right)dt$$
$$\ln i = -\left(\frac{R}{L}\right)t + \kappa$$
$$i(t) = e^{-(R/L)t+\kappa} = Ae^{-t/(L/R)} = Ae^{-t/\tau}$$

This is the natural response of the system. To determine the constant A, the value of $i(t)$ at some time must be known. Because the initial current is I_0, the value of the constant A is

$$i(0) = I_0 = Ae^{-0/\tau} = A$$

The final solution is

$$i(t) = I_0e^{-t/(L/R)}$$

5. RESISTOR-INDUCTOR CIRCUITS: FORCED RESPONSE

Consider the generic resistor-inductor circuit shown in Fig. 30.5. When the inductor, L, whose magnetic field is at a maximum due to initial current, I_0, is connected to the resistor, R, via a complete electrical path, the inductor's magnetic field will collapse or expand in an attempt to resist the change in current. In the process, the inductor's stored magnetic energy is either dissipated or enhanced until no further energy change occurs and no further current change occurs. This gradual decrease or increase is the transient. After the passage of time equal to five time constants, 5τ, the final value is within less than 1% of its steady-state value and the transient is considered complete. The current follows the form of Fig. 30.1 or its inverse, with the exception that the final steady-state current is the current resulting from the driving force and is determined by the remainder of the circuit.

Figure 30.5 Resistor-Inductor Circuit: Forced Response

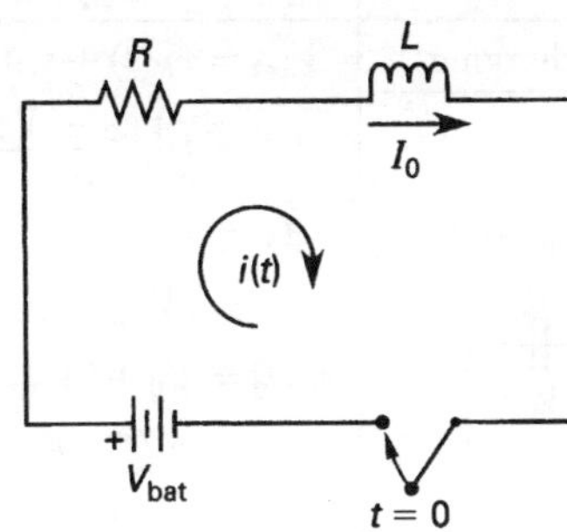

Example 30.4

Determine the formula for the current in the circuit of Fig. 30.5 from $t = 0$ onward.

Solution

Write KVL for the loop containing the resistor and the inductor. In keeping with the convention of this book, the reference direction is clockwise, giving

$$v_L + v_R - V_{\text{bat}} = 0$$

Substitute the expression for voltage across an inductor and the Ohm's law expression for voltage drop in a resistor.

$$L\left(\frac{di}{dt}\right) + iR - V_{\text{bat}} = 0$$

$$\frac{di}{dt} + \left(\frac{R}{L}\right)i = \left(\frac{1}{L}\right)V_{\text{bat}}$$

$$\frac{di}{dt} + \left(\frac{1}{\frac{L}{R}}\right)i = \left(\frac{1}{L}\right)V_{\text{bat}}$$

This is a nonhomogeneous first-order linear differential equation. The known constant to the right of the equal sign is called the *forcing* or *driving function*. Such an equation is solved in two steps. The first step is to determine the particular solution, or forced response, of the system. Because the particular solution is the steady-state solution, the expectation is that the solution will take the same form as the forcing function. Assuming a constant, κ, as the particular solution and substituting this into the KVL expression for i gives

$$i_{\text{ss}} = \kappa$$

$$\frac{di}{dt} + \left(\frac{1}{\frac{L}{R}}\right)i = \left(\frac{1}{L}\right)V_{\text{bat}}$$

$$\frac{di_{\text{ss}}}{dt} + \left(\frac{1}{\frac{L}{R}}\right)i_{\text{ss}} = \left(\frac{1}{L}\right)V_{\text{bat}}$$

$$\frac{d\kappa}{dt} + \left(\frac{1}{\frac{L}{R}}\right)\kappa = \left(\frac{1}{L}\right)V_{\text{bat}}$$

$$0 + \left(\frac{1}{\frac{L}{R}}\right)\kappa = \left(\frac{1}{L}\right)V_{\text{bat}}$$

$$\kappa = \left(\frac{1}{R}\right)V_{\text{bat}}$$

The particular solution is

$$i(\infty) = i_{\text{ss}} = \left(\frac{1}{R}\right)V_{\text{bat}}$$

The next step is to determine the homogeneous, or complementary, solution, that is, the solution with the forcing function equal to zero. In Ex. 30.3 this was determined to be

$$i_{\text{tr}}(t) = Ae^{-t/(L/R)}$$

This is the natural response of the system and represents the transient portion of the solution. The total solution is the combination of the particular (nonhomogeneous) and the complementary (homogeneous) solution, that is, the combination of the steady-state and transient solutions. The total solution is

$$i(t) = \left(\frac{1}{R}\right)V_{\text{bat}} + Ae^{-t/(L/R)}$$

The natural response coefficient A is again determined by the initial condition $i(0) = I_0$. So,

$$i(0) = I_0 = \left(\frac{1}{R}\right)V_{\text{bat}} + Ae^{-0/(L/R)} = \left(\frac{1}{R}\right)V_{\text{bat}} + A$$

$$A = I_0 - \left(\frac{1}{R}\right)V_{\text{bat}}$$

The final solution is

$$i(t) = \left(\frac{1}{R}\right)V_{\text{bat}} + \left(I_0 - \left(\frac{1}{R}\right)V_{\text{bat}}\right)e^{-t/(L/R)}$$

The current $i(t)$ is the current through the inductor.

6. RC AND RL CIRCUITS: SOLUTION METHOD

The method presented in Sec. 30.2 through Sec. 30.5 is valid for any circuit that can be represented by a single equivalent capacitor or inductor and a single equivalent resistor. The solutions were obtained for DC transients.

Table 30.1 is a summary of DC transient responses.[1] AC transients are handled in the same manner except that the forcing function, that is, the source, is represented as a sinusoid. Phasor methods, or sinusoidal methods, are then used to solve the resulting differential equation. If the switch closure were instead treated as a pulse from the source and represented by the unit step function, $u(t)$, the voltage applied at $t=0^+$ would have been $V_{\text{bat}}u(t)$ and the equations would have been handled in the same manner. The steps for solving single resistor and single capacitor/inductor circuits follow.

step 1: Find the steady-state response from the particular solution of the nonhomogeneous equation. The solution will be of the same form as the forcing function.[2] Regardless of transient type, the result will be a steady-state value, κ.

step 2: Determine the transient response from the homogeneous equation or from the known parameters of the time constant. The transient response will take the form $Ae^{-t/\tau}$.

step 3: Sum the steady-state and transient solutions, that is, the particular and complementary solutions, to obtain the total solution. This will take the form $\kappa + Ae^{-t/\tau}$.

step 4: Determine the initial value from the circuit initial conditions. Use this information to calculate the constant A.

step 5: Write the final solution.

This method is used for resistor-capacitor circuits, which are primarily used in electronics. Inductors tend to be large and to change value with both temperature and time. They do not lend themselves well to miniaturization. Capacitors are more easily manufactured on integrated circuits. Inductors are important in power circuits that handle large amounts of current or high voltages and in transformers.

Table 30.1 *Transient Response*

type of circuit	response
series RC, charging $\tau = RC$ $e^{-N} = e^{-t/\tau} = e^{-t/RC}$ R, C, i(t), V_0, V_{bat}, t = 0	$V_{\text{bat}} = v_R(t) + v_C(t)$ $i(t) = \left(\frac{V_{\text{bat}} - V_0}{R}\right)e^{-N}$ $v_R(t) = i(t)R$ $= (V_{\text{bat}} - V_0)e^{-N}$ $v_C(t) = V_0 + (V_{\text{bat}} - V_0) \times (1 - e^{-N})$ $Q_C(t) = C\left(\begin{matrix} V_0 + (V_{\text{bat}} - V_0) \\ \times(1 - e^{-N}) \end{matrix}\right)$
series RC, discharging $\tau = RC$ $e^{-N} = e^{-t/\tau} = e^{-t/RC}$ 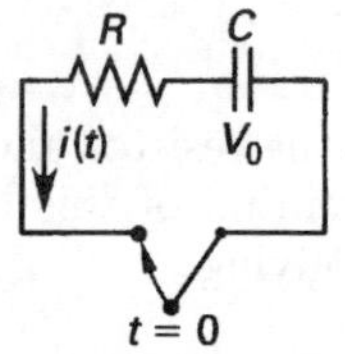 R, C, i(t), V_0, t = 0	$0 = v_R(t) + v_C(t)$ $i(t) = \left(\frac{V_0}{R}\right)e^{-N}$ $v_R(t) = -V_0e^{-N}$ $v_C(t) = V_0e^{-N}$ $Q_C(t) = CV_0e^{-N}$
series RL, charging $\tau = \frac{L}{R}$ $e^{-N} = e^{-t/\tau} = e^{-t/(L/R)}$ 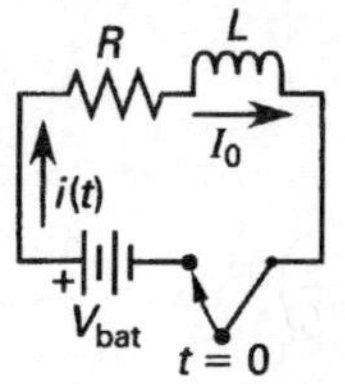 R, L, I_0, i(t), V_{bat}, t = 0	$V_{\text{bat}} = v_R(t) + v_L(t)$ $i(t) = I_0e^{-N} + \left(\frac{V_{\text{bat}}}{R}\right)(1 - e^{-N})$ $v_R(t) = i(t)R$ $= I_0Re^{-N} + V_{\text{bat}}(1 - e^{-N})$ $v_L(t) = (V_{\text{bat}} - I_0R)e^{-N}$
series RL, discharging $\tau = \frac{L}{R}$ $e^{-N} = e^{-t/\tau} = e^{-t/(L/R)}$ 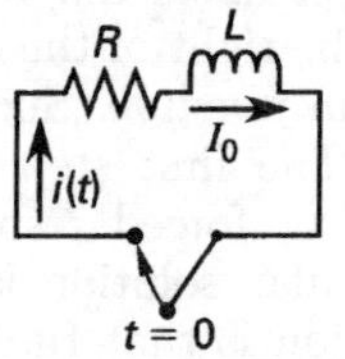 R, L, I_0, i(t), t = 0	$0 = v_R(t) + v_L(t)$ $i(t) = I_0e^{-N}$ $v_R(t) = I_0Re^{-N}$ $v_L(t) = -I_0Re^{-N}$

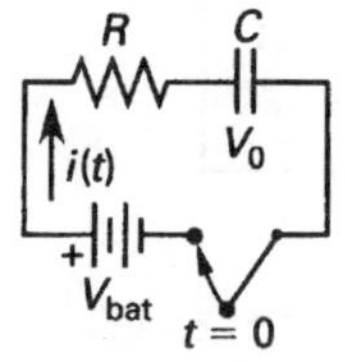

7. RISE TIME

When the input of a capacitive or inductive circuit like the one in Fig. 30.6(a) undergoes a step change, the circuit response is called a *step response.* The step response is illustrated in Fig. 30.6(b). The speed of the response may be quantified in several ways. The time constant, τ, is one possibility. The shorter the time constant, the less time it takes for the circuit to reach a specified value. The value of the time constant is dependent on the circuit. A second possibility is the *rise time*, which is a *defined quantity* applicable to responses in general regardless of the circuit. Let t_{10} be the time a response has reached 10% of its final steady-state value. Let t_{90} be the time a response has reached 90% of its final steady-state value. The rise time, t_r, is then defined as

$$t_r = t_{90} - t_{10} \qquad 30.7$$

[1]The final expression of the response given in the table appears to differ from that given in step 3 of the solution method. The table uses the common form $1 - e^{-t/\tau}$. The forms are equivalent and can be interchanged. Both contain a steady-state and transient component. The steady-state term κ can be found in the $1 - e^{-t/\tau}$ form by letting $t \to \infty$.

[2]If the excitation is constant, the form of the particular solution is a constant and can be determined by simply open-circuiting the terminals of the capacitor and determining the voltage present. This voltage will be the steady-state voltage to which the capacitor will be driven.

Figure 30.6 Rise Time

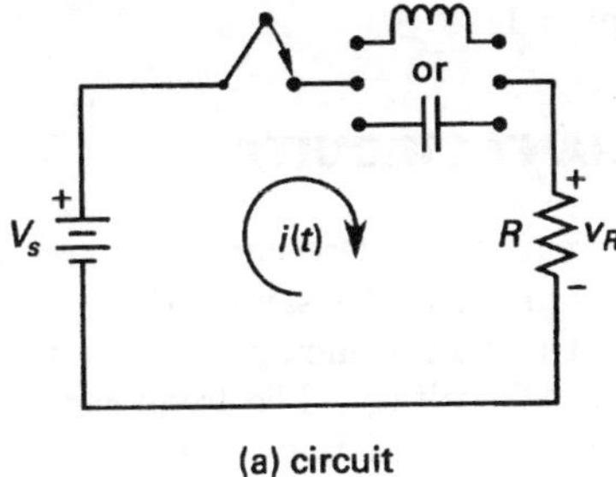

(a) circuit

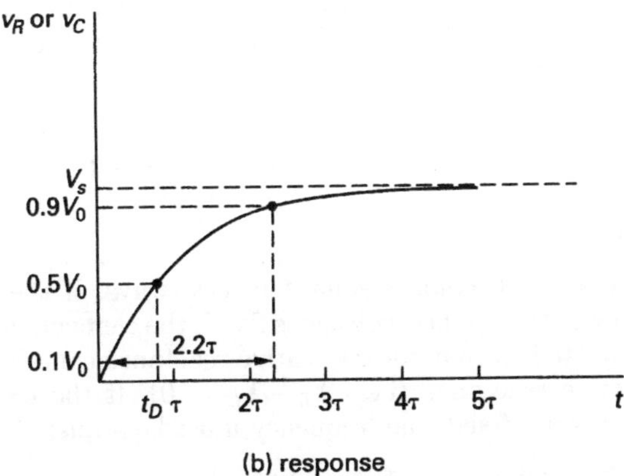

(b) response

The rise time is related to the time constant by Eq. 30.8.

$$t_r = 2.2\tau \qquad 30.8$$

Another defined quantity is the *time delay*, t_d, which is the time for a response to reach 50% of its final steady-state value. The rise and delay times and their relationship to the time constant are shown graphically in Fig. 30.6(b).

8. DAMPED OSCILLATIONS: RINGING

When energized, some circuits undergo voltage or current oscillations that decrease in magnitude over time. Such oscillation is called *ringing*. An example of such a circuit and its response to a switch closure at $t = 0$ is shown in Fig. 30.7. The capacitor is initially charged to V_0 and the inductor has no initial magnetic field. Writing KVL for the circuit yields

$$\frac{1}{C}\int_0^t i\,dt + iR + L\left(\frac{di}{dt}\right) = V_0 \qquad 30.9$$

The current is the rate of change of the charge on the capacitor, that is, $i = dq/dt$. Rearranging mathematically gives

$$\frac{d^2q}{dt^2} + \left(\frac{R}{L}\right)\left(\frac{dq}{dt}\right) + \left(\frac{1}{LC}\right)q = \frac{q_0}{LC} \qquad 30.10$$

Figure 30.7 Ringing Circuit

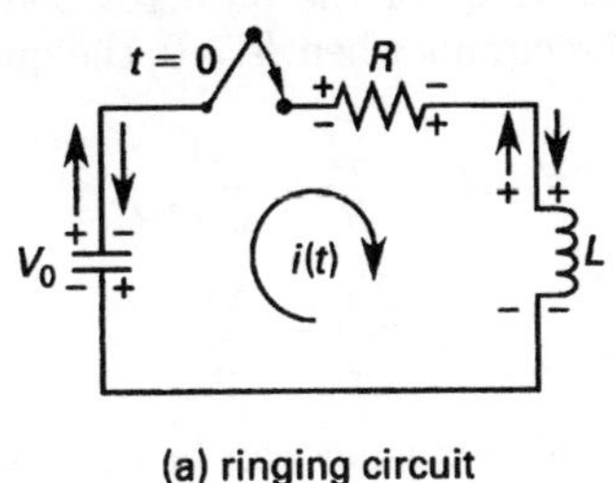

(a) ringing circuit

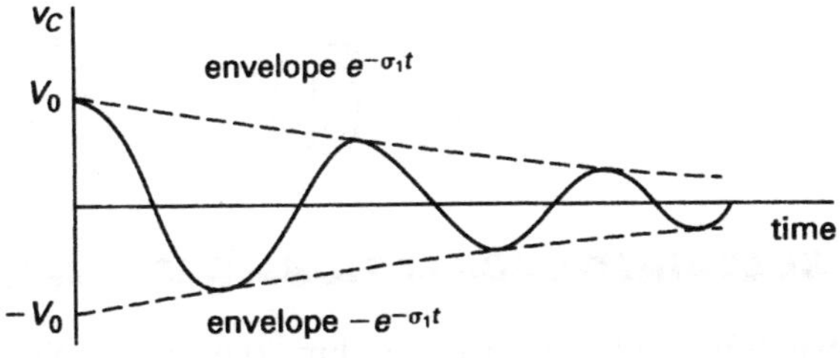

(b) ringing response

Representing this nonhomogeneous second-order differential equation in the *s* domain, appropriate for Laplace transform analysis, results in the following characteristic equation.[3]

$$s^2 + \left(\frac{R}{L}\right)s + \frac{1}{LC} = 0 \qquad 30.11$$

The roots of Eq. 30.11, found from the quadratic equation, take one of three possible forms: (1) real and unequal, (2) real and equal, or (3) complex conjugates.

Ringing occurs for case 3 with roots given by

$$s_1, s_2 = -\sigma_1 \pm j\omega_1 \qquad 30.12$$

The solution to Eq. 30.10 for the ringing circuit is

$$v_C(t) = V_0 e^{-\sigma_1 t}\cos\omega_1 t \qquad 30.13$$

The term σ_1 is called the *damping*. The exponential, $e^{-\sigma_1 t}$, is termed the *envelope* of the waveform. The frequency of the oscillation is given by ω_1. These quantities are illustrated in Fig. 30.7(b). The damping and frequency are found in terms of the circuit parameters from Eq. 30.14 and Eq. 30.15.

$$\sigma_1 = \frac{R}{2L} \qquad 30.14$$

$$\omega_1 = \sqrt{\frac{1}{LC} - \left(\frac{R}{2L}\right)^2} \qquad 30.15$$

[3]A second-order equation should be expected because the circuit contains two independent energy storage devices.

The damping and frequency are also defined in terms of the *quality factor*, Q, of the inductor coil. If ω_0 is the frequency of the circuit when $R = 0$, the quality factor is

$$Q = \frac{\omega_0 L}{R} = \frac{1}{\omega_0 CR} = \frac{X_L}{R} = \frac{X_C}{R} \qquad 30.16$$

The damping and frequency can be defined as

$$\sigma_1 = \frac{\omega_0}{2Q} \qquad 30.17$$

$$\omega_1 = \omega_0 \sqrt{1 - \left(\frac{1}{2Q}\right)^2} \qquad 30.18$$

9. SUSTAINED OSCILLATIONS: RESONANCE

Capacitors and inductors within the same circuit exchange energy. In the ideal, that is, lossless, circuit shown in Fig. 30.8, this exchange can continue indefinitely and is known as a *sustained oscillation.*[4]

Figure 30.8 *Resonant Circuit*

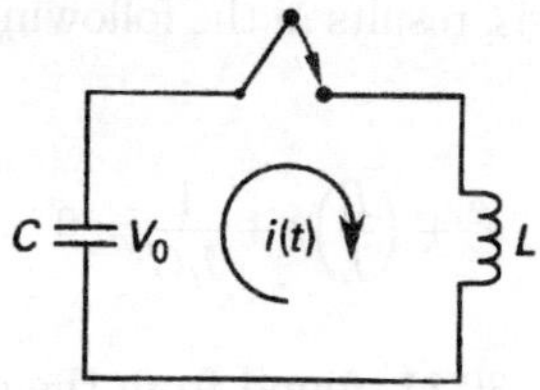

The characteristic equation for the circuit in Fig. 30.8 is

$$s^2 + \frac{1}{LC} = 0 \qquad 30.19$$

The solution is

$$s_1, s_2 = \pm j\sqrt{\frac{1}{LC}} = \pm j\omega_0 \qquad 30.20$$

No damping term, σ_1, exists and the oscillations can continue unfettered. This unimpeded flow of energy naturally occurs at the frequency ω_0 and is called *resonance.*

In a real circuit, resistive components are present and will yield a damping term, σ_1, in the solution to the circuit equation. The result is that an external source of energy is required to make up the loss. Resonance is then defined as a phenomenon of an AC circuit whereby relatively large currents occur near certain frequencies, and a relatively unimpeded oscillation of energy from potential to kinetic occurs. The frequency at which this occurs is called the *resonant frequency.* At the resonant frequency the capacitive reactance equals the inductive reactance, that is, $X_C = X_L$, the current phase angle difference, ϕ, is zero ($\phi = 0$), and the power factor, pf, equals one (pf = 1).

[4]This type of circuit is more than academic and can be realized, or nearly so, with superconducting inductors.

10. RESONANT CIRCUITS

A *resonant circuit* has a zero current phase angle difference. This is equivalent to saying the circuit is purely resistive (i.e., the power factor is equal to one) in its response to an AC voltage. The frequency at which the circuit becomes purely resistive is the *resonant frequency.*

For frequencies below the resonant frequency, a series RLC circuit will be capacitive (leading) in nature; above the resonant frequency, the circuit will be inductive (lagging) in nature.

For frequencies below the resonant frequency, a parallel GLC circuit will be inductive (lagging) in nature; above the resonant frequency, the circuit will be capacitive (leading) in nature.

Circuits can become resonant in two ways. If the frequency of the applied voltage is fixed, the elements must be adjusted so that the capacitive reactance cancels the inductive reactance (i.e., $X_L - X_C = 0$). If the circuit elements are fixed, the frequency must be adjusted.

As Fig. 30.9 and Fig. 30.10 illustrate, a circuit approaches resonant behavior gradually. ω_1 and ω_2 are the *half-power points* (*70% points*, or *3 dB points*) because at those frequencies, the power dissipated in the resistor is half of the power dissipated at the resonant frequency.

$$Z_{\omega 1} = \sqrt{2}R \qquad 30.21$$

$$I_{\omega 1} = \frac{V}{Z_{\omega 1}} = \frac{V}{\sqrt{2}R} = \frac{I_0}{\sqrt{2}} \qquad 30.22$$

$$P_{\omega 1} = I^2 R = \left(\frac{I_0}{\sqrt{2}}\right)^2 R = \tfrac{1}{2}P_0 \qquad 30.23$$

Figure 30.9 *Series Resonance (Band-Pass Filter)*

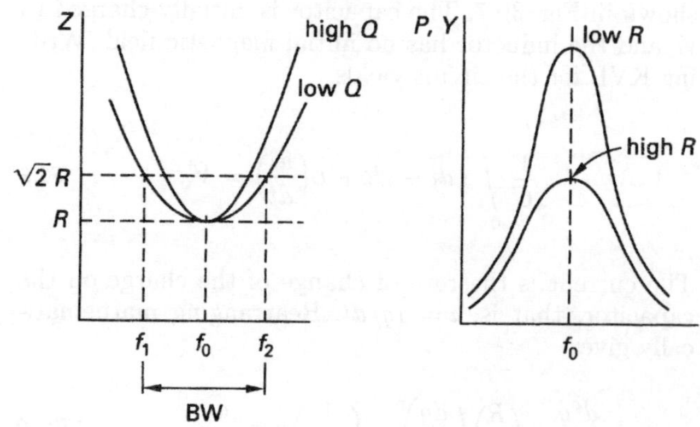

Figure 30.10 *Parallel Resonance (Band-Reject Filter)*

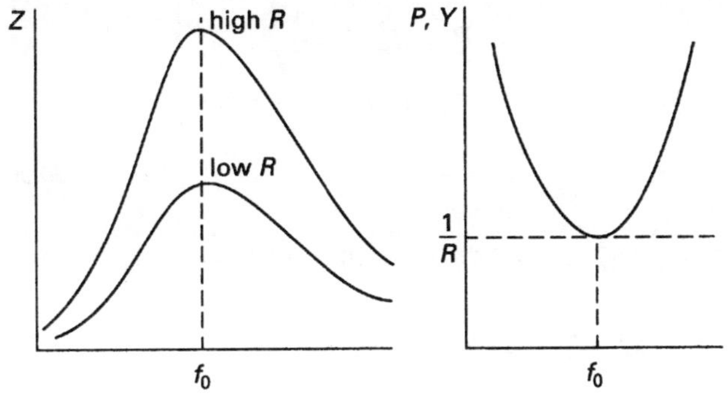

The frequency difference between the half-power points is the *bandwidth*, BW, a measure of circuit selectivity. The smaller the bandwidth, the more selective the circuit.

$$\text{BW} = f_2 - f_1 \quad 30.24$$

The *quality factor*, Q, for a circuit is a dimensionless ratio that compares the reactive energy stored in an inductor each cycle to the resistive energy dissipated.[5] The effect the quality factor has on the frequency characteristic is illustrated by Fig. 30.9.

$$Q = 2\pi\left(\frac{\text{maximum energy stored per cycle}}{\text{energy dissipated per cycle}}\right)$$

$$= \frac{f_0}{(\text{BW})_{\text{Hz}}} = \frac{\omega_0}{(\text{BW})_{\text{rad/s}}}$$

$$= \frac{f_0}{f_2 - f_1} = \frac{\omega_0}{\omega_2 - \omega_1} \quad \left[\begin{array}{l}\text{parallel} \\ \text{or series}\end{array}\right] \quad 30.25$$

Then, the energy stored in the inductor of a series RLC circuit each cycle is

$$U = \frac{I_m^2 L}{2} = I^2 L = Q\left(\frac{I^2 R}{2\pi f_0}\right) \quad 30.26$$

The relationships between the half-power points and quality factor are

$$f_1, f_2 = f_0\left(\sqrt{1 + \frac{1}{4Q^2}} \mp \frac{1}{2Q}\right)$$

$$\approx f_0 \mp \frac{f_0}{2Q} = f_0 \mp \frac{\text{BW}}{2} \quad 30.27$$

Various resonant circuit formulas are tabulated in App. 30.A.

[5]The name *figure of merit* refers to the quality factor calculated from the inductance and internal resistance of a coil.

11. SERIES RESONANCE

In a resonant series RLC circuit,

- impedance is minimum
- impedance equals resistance
- current and voltage are in phase
- current is maximum
- power dissipation is maximum

The total impedance (in rectangular form) of a series RLC circuit is $R + j(X_L - X_C)$. At the resonant frequency, $\omega_0 = 2\pi f_0$,

$$X_L = X_C \quad \text{[at resonance]} \quad 30.28$$

$$\omega_0 L = \frac{1}{\omega_0 C} \quad 30.29$$

$$\omega_0 = 2\pi f_0 = \frac{1}{\sqrt{LC}} \quad 30.30$$

The power dissipation in the resistor is

$$P = \tfrac{1}{2} I_m^2 R = \frac{V_m^2}{2R}$$

$$= I^2 R$$

$$= \frac{V^2}{R} \quad 30.31$$

The quality factor for a series RLC circuit is

$$Q = \frac{X}{R} = \frac{\omega_0 L}{R}$$

$$= \frac{1}{\omega_0 RC} = \frac{1}{R}\sqrt{\frac{L}{C}}$$

$$= \frac{\omega_0}{(\text{BW})_{\text{rad/s}}} = \frac{f_0}{(\text{BW})_{\text{Hz}}}$$

$$= G\omega_0 L = \frac{G}{\omega_0 C} \quad 30.32$$

Example 30.5

A series RLC circuit is connected across a sinusoidal voltage with peak of 20 V. (a) What is the resonant frequency in rad/s? (b) What are the half-power points in rad/s? (c) What is the peak current at resonance? (d) What is the peak voltage across each component at resonance?

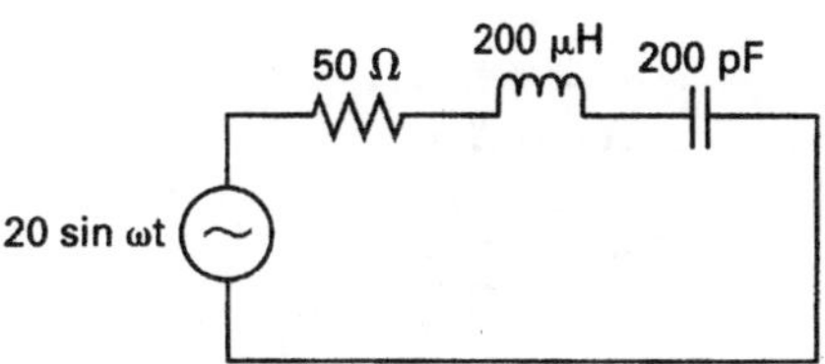

Circuit Theory

Solution

(a) Equation 30.30 gives the resonant frequency.

$$\omega_0 = \frac{1}{\sqrt{LC}} = \frac{1}{\sqrt{(200 \times 10^{-6}\ \text{H})(200 \times 10^{-12}\ \text{F})}}$$
$$= 5 \times 10^6\ \text{rad/s}$$

(b) The half-power points are

$$\omega_1, \omega_2 = \omega_0 \mp \frac{\text{BW}}{2} = \omega_0 \mp \frac{\omega_0}{2Q} = \omega_0 \mp \frac{R}{2L}$$
$$= 5 \times 10^6\ \frac{\text{rad}}{\text{s}} \mp \frac{50\ \Omega}{(2)(200 \times 10^{-6}\ \text{H})}$$
$$= 5.125 \times 10^6\ \text{rad/s},\ 4.875 \times 10^6\ \text{rad/s}$$

(c) The total impedance is the resistance at resonance. The peak resonant current is

$$I_0 = \frac{V_m}{Z_0} = \frac{V_m}{R} = \frac{20\ \text{V}\angle 0°}{50\ \Omega}$$
$$= 0.4\ \text{A}\angle 0°$$

(d) The peak voltages across the components are

$$V_R = I_0 R = (0.4\ \text{A}\angle 0°)(50\ \Omega) = 20\ \text{V}\angle 0°$$
$$V_L = I_0 X_L = I_0 j\omega_0 L$$
$$= j(0.4\ \text{A}\angle 0°)\left(5 \times 10^6\ \frac{\text{rad}}{\text{s}}\right)(200 \times 10^{-6}\ \text{H})$$
$$= 400\ \text{V}\angle 90°$$
$$V_C = I_0 X_C = \frac{I_0}{j\omega_0 C} = \frac{0.4\ \text{A}\angle 0°}{j\left(5 \times 10^6\ \frac{\text{rad}}{\text{s}}\right)(200 \times 10^{-12}\ \text{F})}$$
$$= 400\ \text{V}\angle -90°$$

12. PARALLEL RESONANCE

In a resonant parallel GLC circuit,

- impedance is maximum
- impedance equals resistance
- current and voltage are in phase
- current is minimum
- power dissipation is minimum

The total admittance (in rectangular form) of a parallel GLC circuit is $G + j(B_C - B_L)$. At resonance,

$$X_L = X_C \qquad 30.33$$

$$\omega_0 L = \frac{1}{\omega_0 C} \qquad 30.34$$

$$\omega_0 = 2\pi f_0 = \frac{1}{\sqrt{LC}} \qquad 30.35$$

The power dissipation in the resistor is

$$P = \tfrac{1}{2} I_m^2 R = \frac{V_m^2}{2R}$$
$$= I^2 R$$
$$= \frac{V^2}{R} \qquad 30.36$$

The quality factor for a parallel RLC circuit is

$$Q = \frac{R}{X} = \omega_0 RC = \frac{R}{\omega_0 L}$$
$$= R\sqrt{\frac{C}{L}} = \frac{\omega_0}{(\text{BW})_{\text{rad/s}}} = \frac{f_0}{(\text{BW})_{\text{Hz}}}$$
$$= \frac{\omega_0 C}{G} = \frac{1}{G\omega_0 L} \qquad 30.37$$

Example 30.6

A parallel RLC circuit containing a 10 Ω resistor has a resonant frequency of 1 MHz and a bandwidth of 10 kHz. To what should the resistor be changed in order to increase the bandwidth to 20 kHz without changing the resonant frequency?

Solution

From Eq. 30.25 and Eq. 30.37,

$$Q_{\text{old}} = \frac{f_0}{\text{BW}} = 2\pi f_0 RC$$
$$C = \frac{1}{2\pi R(\text{BW})} = \frac{1}{(2\pi)(10\ \Omega)(10 \times 10^3\ \text{Hz})}$$
$$= 1 \times 10^{-5}/2\pi\ \text{F}$$

The new quality factor is

$$Q_{\text{new}} = \frac{f_0}{\text{BW}}$$
$$= \frac{10^6\ \text{Hz}}{20 \times 10^3\ \text{Hz}}$$
$$= 50$$

From Eq. 30.37, the required resistance is

$$R = \frac{Q}{2\pi f_0 C}$$
$$= \frac{50}{(2\pi)(10^6\ \text{Hz})\left(\frac{1 \times 10^{-5}}{2\pi}\ \text{F}\right)}$$
$$= 5\ \Omega$$

31 Time Response

Nomenclature

a, A	constant	–
A	natural response coefficient	–
b, B	constant	–
c	constant	–
C	capacitance	F
$G(s)$	forward transfer function	–
i	variable current	A
K	error constant, gain constant, or scale factor	–
I, $\mathbf{I}$	constant or rms current	A
L	self-inductance	H
Q	quality factor	–
R	resistance	Ω
s, $\mathbf{s}$	complex frequency	Hz
s	Laplace transform variable	–
s	s-domain variable	–
t	time	s
v	variable voltage	V
V, $\mathbf{V}$	constant or rms voltage	V
Z, $\mathbf{Z}$	impedance	Ω

Symbols

α	damping	Hz
β	oscillation frequency	rad/s
ζ	damping ratio	–
θ	phase angle	rad
κ	steady-state constant	–
σ	neper frequency or napier frequency	Np/s or Hz
τ	time constant	s
ω	angular frequency	rad/s or Hz
ω_0	resonance frequency	rad/s

Subscripts

0	initial or resonant
C	capacitor, compensator, or controller
d	derivative
DC	direct current
i	integral
L	inductor
L/C	inductor or capacitor
n	natural
p	peak
p	proportional
s	source
so	switch open
ss	steady state
tr	transient
Th	Thevenin

1. FUNDAMENTALS

Time domain analysis is analysis that takes place with time as the governing variable. *Sinusoidal analysis* is analysis that occurs when the current and voltage functions are manipulated in their trigonometric form. *Frequency domain analysis* deals with Fourier or Laplace transforms of time functions, which are then functions of frequency. The Fourier transform occurs in k-space which, when k is replaced by ω, is called the frequency domain. The Laplace transform occurs in s-space, also considered the frequency domain. The variable $\mathbf{s}$ is called the *complex frequency* and is given by Eq. 31.1. (In Eq. 31.1, the variable $\mathbf{s}$ is shown in bold to emphasize its phasor properties, which are important in pole-zero analysis.)

$$\mathbf{s} \equiv \sigma + \mathbf{j}\omega \qquad 31.1$$

The variable σ is termed the *neper frequency* and is assigned units of Np/s or Hz. The variable σ represents the ratio of two currents, voltages, or other analogous quantities. The number of nepers is the natural log of this ratio. The term "napier" is sometimes used in place of neper.[1] The Laplace transform and the complex frequency s are used to represent exponential and damped sinusoidal motion in the same manner that ordinary frequency expresses simple harmonic motion. That is, using the variable s allows constants, exponential decay, sinusoids, and damped sinusoids (sinusoids with an

[1]This occurs since the natural logarithm is alternatively called the *Napierian* or *Naperian logarithm.*

Circuit Theory

exponential envelope) to be represented in the same form, Ae^{st} (see Sec. 31.11). Sinusoidal analysis is also considered any analysis using the assumption that the forcing function, the source, is sinusoidal in nature. Sinusoidal analysis in Chap. 27 through Chap. 30 assumed a single frequency. When the frequencies are allowed to vary, the circuit output is called the *frequency analysis* or *frequency response.*

2. FIRST-ORDER ANALYSIS

First-order analysis is performed upon circuits with one energy storage element—that is, one equivalent capacitor or inductor. In circuits containing numerous sources and resistors, the Thevenin or Norton equivalent is taken as seen from the terminals of the single capacitor or inductor. This allows the methods of Chap. 30 to be applied, modified so that the Thevenin voltage, v_{Th}, is used in place of the source voltage and the Thevenin current, i_{Th}, is used as the current through the capacitor and inductor. Figure 31.1(a) represents a generic circuit with a single capacitor or inductor. The source can be either AC or DC. Figure 31.1(b) shows the Thevenin equivalent of the circuit with the switch closed. Figure 31.1(c) shows the new Thevenin equivalent with the switch open. Part (c) of this figure is a convenient way of showing that if the source is eliminated (or changed to a DC source), the resulting currents and voltages will be exponentials (the natural response) and will share a new time constant and possibly new initial currents and voltages.

The differential equation for first-order circuits has the general form

$$f(t) = b\frac{dx}{dt} + cx \quad 31.2$$

The solutions are of the form

$$x(t) = \kappa + Ae^{-t/\tau} \quad 31.3$$

The time constant $\tau = b/c$. κ is a constant, the steady-state component of the solution. The method for determining the complete solution is as follows.

step 1: Determine the Thevenin or Norton equivalent of the circuit as seen from the terminals of the capacitor or inductor.

step 2: Find the steady-state response from the particular solution of the nonhomogeneous equation. The solution will be of the same form as the forcing function.[2] Regardless of transient type, the result will be a steady-state value, κ.

step 3: Determine the transient response from the homogeneous equation or from the known parameters of the time constant. The transient response will be of the form $Ae^{-t/\tau}$.

step 4: Sum the steady-state and transient solutions, that is, the complementary and particular solutions, to obtain the total solution.[3] This will be of the form $\kappa + Ae^{-t/\tau}$.

step 5: Determine the initial value from the circuit initial conditions. Use this information to calculate the constant A.

step 6: Write the final solution.

A capacitive first-order circuit generally uses voltage as the dependent variable. Applying KVL to Fig. 31.1(b) gives

$$v_{\text{Th}} = i_{\text{Th}}R_{\text{Th}} + v_C \quad 31.4$$

Figure 31.1 *First-Order Circuit*

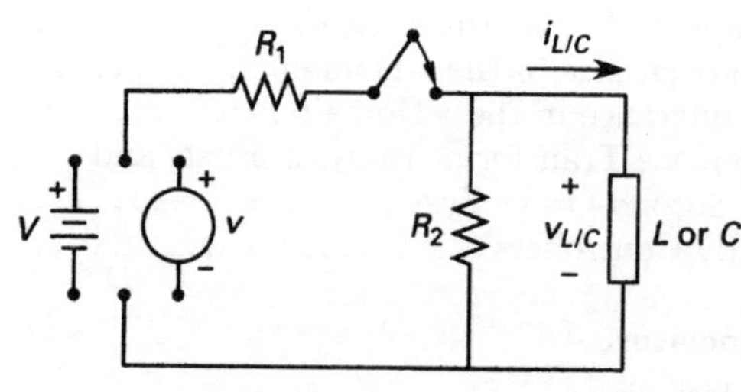

(a) first-order circuit: DC or AC

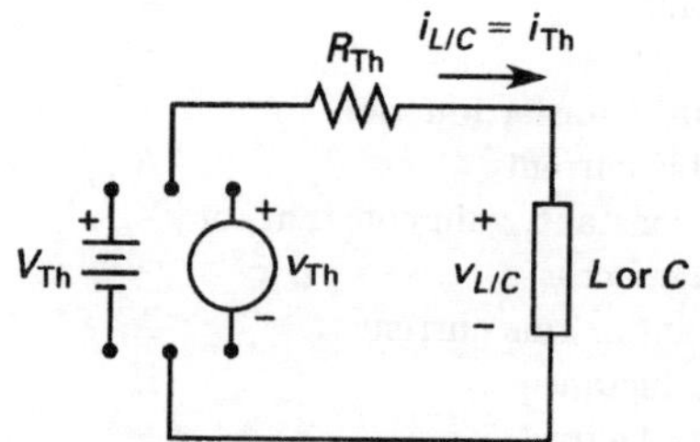

(b) Thevenin equivalent: DC or AC, switch closed

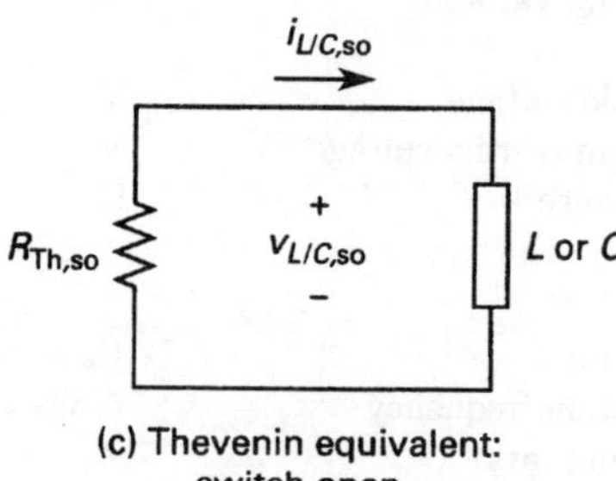

(c) Thevenin equivalent: switch open

[2]If the excitation is constant, the form of the particular solution is a constant and can be determined by simply open-circuiting the terminals of the capacitor and determining the voltage present. This voltage will be the steady-state voltage to which the capacitor will be driven.

[3]The final expression often varies between texts, with some using $1 - e^{-t/\tau}$. The forms are equivalent and can be interchanged. Both contain a steady-state and transient component. The steady-state term κ can be found in the $1 - e^{-t/\tau}$ form by letting $t \rightarrow \infty$.

The Thevenin current is the current flowing through the capacitor, $i_{\text{Th}} = i_C = C(dv_C/dt)$. Substituting gives

$$v_{\text{Th}} = R_{\text{Th}}C\left(\frac{dv_C}{dt}\right) + v_C \qquad 31.5$$

Equation 31.5 is in the general form of a first-order differential equation shown in Eq. 31.2. The solution is

$$v_C(t) = V_{C,\text{ss}} + Ae^{-t/\tau} \qquad 31.6$$

An inductive first-order circuit generally uses current as the dependent variable. Applying KVL to Fig. 31.1(b) gives

$$v_{\text{Th}} = i_{\text{Th}}R_{\text{Th}} + v_L \qquad 31.7$$

The Thevenin current is flowing through the inductor, $i_{\text{Th}} = i_L$. Substituting for the inductor voltage and rearranging gives

$$v_{\text{Th}} = L\left(\frac{di_L}{dt}\right) + R_{\text{Th}}i_L \qquad 31.8$$

Equation 31.8 is in the general form of a first-order differential equation shown in Eq. 31.2. The solution is

$$i_L(t) = I_{L,\text{ss}} + Ae^{-t/\tau} \qquad 31.9$$

3. FIRST-ORDER ANALYSIS: SWITCHING TRANSIENTS

First-order switching transients result from the operation of a switch. The switching mechanism can be an actual switch, relay contacts, or a nonlinear device such as a transistor operating in the saturation or cutoff region. A sudden change in the source results in sudden changes in the voltages and currents within a circuit. A jump in the capacitor voltage requires an impulse current. A jump in the inductor current requires an impulse voltage. Because the conditions in a capacitor or inductor cannot change instantaneously, the post-switched conditions of capacitance, C, or self-inductance, L, are derived from the pre-switched conditions.

Example 31.1[4]

For the network shown, $R_1 = 20\ \Omega$, $R_2 = 10\ \Omega$, $V_{\text{DC}} = 12$ V, and $L = 0.20$ H. The switch has been open for a long time and is closed at $t = 0$. Determine the steady-state response of the network.

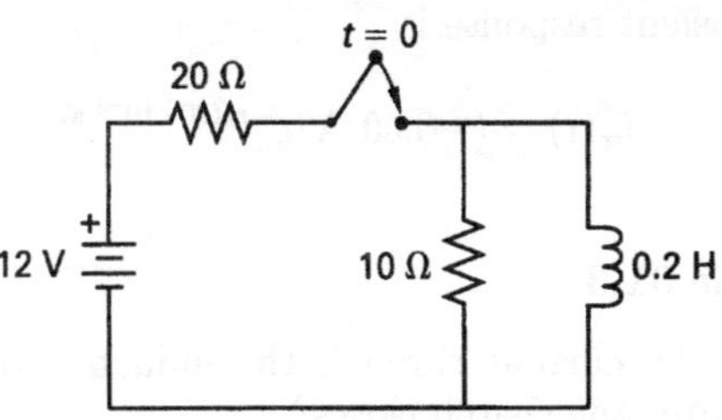

Solution

Using the voltage divider concept, the Thevenin voltage is 4 V. The Thevenin resistance is calculated as seen from the terminals of the inductor, as shown in Fig. 31.1(b). Therefore,

$$\begin{aligned} R_{\text{Th}} &= R_1 \| R_2 = \frac{R_1R_2}{R_1 + R_2} = \frac{(20\ \Omega)(10\ \Omega)}{20\ \Omega + 10\ \Omega} \\ &= 6.67\ \Omega \end{aligned}$$

The steady-state condition is determined from Eq. 31.8, noting that steady-state current is constant $(di/dt = 0)$.

$$\begin{aligned} v_{\text{Th}} &= L\left(\frac{di_L}{dt}\right) + R_{\text{Th}}i_L = (0.2\ \text{H})(0) + (6.67\ \Omega)i_L \\ &= 4\ \text{V} \\ i_{L,\text{ss}} &= \frac{4\ \text{V}}{6.67\ \Omega} = 0.60\ \text{A} \end{aligned}$$

The inductor is a short circuit for steady-state DC conditions, as expected.

Example 31.2

Determine the transient response of the circuit in Ex. 31.1.

Solution

The transient response is of the form $Ae^{-t/\tau}$. The time constant is

$$\tau = \frac{L}{R_{\text{Th}}} = \frac{(0.20\ \text{H})\left(1000\ \frac{\text{s}}{\text{ms}}\right)}{6.67\ \Omega} = 30\ \text{ms}$$

The steady-state current was found to be 0.60 A in Ex. 31.1. Using Eq. 31.9, the total current is

$$i_L(t) = I_{L,\text{ss}} + Ae^{-t/\tau} = 0.60\ \text{A} + Ae^{-t/(30\times10^{-3}\ \text{s})}$$

Because the switch was open for a long time, steady-state conditions are assumed to exist and the initial current in the inductor is zero. That is, $i(0^-) = i(0^+) = 0$. Substituting this information to find the constant A gives

$$\begin{aligned} i_L(0) &= 0.60\ \text{A} + Ae^{-0/(30\times10^{-3}\ \text{s})} = 0.60\ \text{A} + A = 0 \\ A &= -0.60\ \text{A} \end{aligned}$$

[4]Since transient conditions, and a possible AC source, are used in some examples, lowercase letters will be used to designate the voltage and current.

The transient response is

$$i_{tr}(t) = (-0.60\ \text{A})e^{-t/(30\times10^{-3}\ \text{s})}$$

Example 31.3

What is the current through the inductor in Ex. 31.1 120 ms after the switch closes?

Solution

The total solution is the combination of the steady-state and transient responses. From Ex. 31.1 and Ex. 31.2,

$$i_L(t) = 0.60\ \text{A} - (0.60\ \text{A})e^{-t/(30\times10^{-3}\ \text{s})}$$

At 120 ms, or 4τ, after the switch closes, the current is

$$\begin{aligned} i_L(120\ \text{ms}) &= 0.60\ \text{A} - (0.60\ \text{A})e^{(-120\times10^{-3}\ \text{s})/(30\times10^{-3}\ \text{s})} \\ &= 0.60\ \text{A} - (0.60\ \text{A})e^{-4} \\ &= 0.59\ \text{A} \end{aligned}$$

Example 31.4

For the network shown, let $R_1 = 10\ \Omega$, $R_2 = 10\ \Omega$, $V_{DC} = 10$ V, and $C = 50\ \mu$F. The switch has been open for an extended time and is closed at $t = 0$. Determine the steady-state capacitance response.

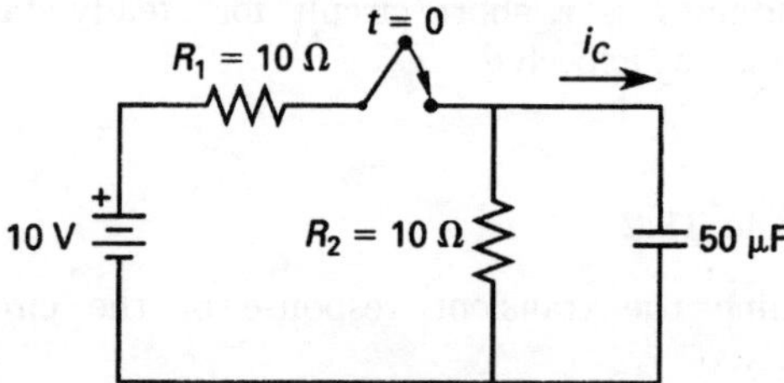

Solution

Using the voltage divider concept, the Thevenin voltage is 5 V. The Thevenin resistance is calculated as seen from the terminals of the capacitor.

$$\begin{aligned} R_{Th} &= R_1 \| R_2 = \frac{R_1 R_2}{R_1 + R_2} = \frac{(10\ \Omega)(10\ \Omega)}{10\ \Omega + 10\ \Omega} \\ &= 5\ \Omega \end{aligned}$$

The steady-state condition is determined from Eq. 31.5, noting that the voltage is constant ($dv/dt = 0$).

$$\begin{aligned} v_{Th} &= R_{Th} C\left(\frac{dv_C}{dt}\right) + v_C \\ &= (5\ \Omega)(50\times10^{-6}\ \text{F})(0) + v_C \\ &= 5\ \text{V} \end{aligned}$$

$$v_{C,ss} = 5\ \text{V}$$

The capacitor is an open circuit for steady-state DC conditions, as expected.

Example 31.5

For the network in Ex. 31.4, the switch opens three time constants into the transient. Determine the response of the network from this time onward.

Solution

After the switch opens, $v_{Th} = 0$ (see Fig. 31.1(c)). Therefore, the steady-state voltage will be zero. The transient response must be of the form $Ae^{-t/\tau}$.

$$v_C(t) = Ae^{-t/\tau} = Ae^{-t/R_{Th}C}$$

Once the switch is open, the Thevenin resistance is

$$R_{Th} = R_2 = 10\ \Omega$$

The time constant is

$$\tau = R_{Th}C = (10\ \Omega)(50\times10^{-6}\ \text{F})\left(10^6\ \frac{\mu\text{s}}{\text{s}}\right) = 500\ \mu\text{s}$$

The total solution is the combination of the steady-state and transient responses, as given in Eq. 31.6.

$$v_C(t) = V_{C,ss} + Ae^{-t/\tau} = 0\ \text{V} + Ae^{-t/(500\times10^{-6}\ \text{s})}$$

In order to determine the constant A, the voltage condition must be known at a certain time. Because the capacitor voltage cannot change instantaneously, $v_C(0^-) = v_C(0^+)$. The switch was opened three time constants into the transient determined by the conditions in Ex. 31.4. The transient response of the circuit in Ex. 31.4 is

$$\begin{aligned} v_C(t) &= Ae^{-t/R_{Th}C} = Ae^{-t/((5\ \Omega)(50\times10^{-6}\ \text{F}))} \\ &= Ae^{-t/250\times10^{-6}\ \text{s}} \end{aligned}$$

The total solution for the circuit in Ex. 31.4, as given by Eq. 31.6, and noting that the initial voltage is zero, is

$$\begin{aligned} v_C(t) &= V_{C,ss} + Ae^{-t/\tau} = 5\ \text{V} + Ae^{-t/(250\times10^{-6}\ \text{s})} \\ v_C(0) &= 5\ \text{V} + Ae^{-0/(250\times10^{-6}\ \text{s})} = 5\ \text{V} + A = 0\ \text{V} \\ A &= -5\ \text{V} \\ v_C(t) &= 5\ \text{V} - 5e^{-t/(250\times10^{-6}\ \text{s})}\ \text{V} \end{aligned}$$

At three time constants into the initial transient,

$$v_C(3\tau) = 5\ \text{V} - 5e^{-((3)(250\times10^{-6}\ \text{s}))/(250\times10^{-6}\ \text{s})}\ \text{V} = 4.75\ \text{V}$$

Use 4.75 V as the initial voltage for the network in Ex. 31.5 to determine the value of A.

$$\begin{aligned} v_C(t) &= V_{C,ss} + Ae^{-t/\tau} = 0\ \text{V} + Ae^{-t/(500\times10^{-6}\ \text{s})} \\ v_C(0^-) &= v_C(0^+) = 4.75\ \text{V} \\ 4.75\ \text{V} &= 0\ \text{V} + Ae^{-0/(500\times10^{-6}\ \text{s})} = A \end{aligned}$$

Circuit Theory

Therefore, the response of the network from the time the switch opens is

$$v_C(t) = V_{C,\text{ss}} + Ae^{-t/\tau} = 0\ \text{V} + 4.75e^{-t/(500\times10^{-6}\ \text{s})}\ \text{V}$$
$$= 4.75e^{-t/(500\times10^{-6}\ \text{s})}\ \text{V}$$

Example 31.6

For the network shown, $R_1 = 10\ \Omega$, $R_2 = 10\ \Omega$, $v =$ 170 V $\sin(377t + 30°)$, and $L = 0.15$ H. Determine the steady-state current with the switch closed.

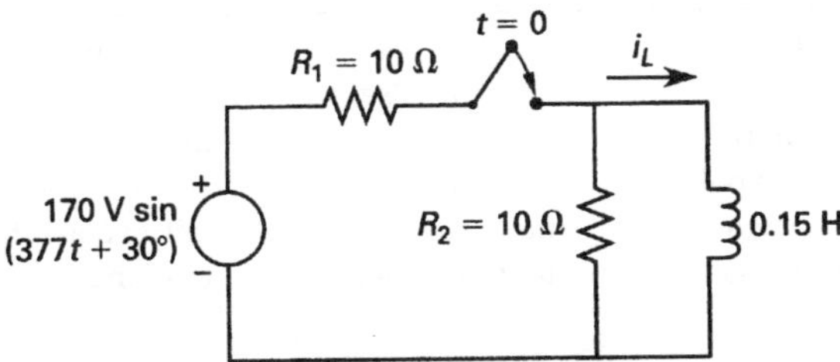

Solution

Using Eq. 31.8 and noting that multiplying by $j\omega$ is equivalent to taking the derivative, gives

$$v_{\text{Th}} = L\left(\frac{di_L}{dt}\right) + R_{\text{Th}}i_L = Lj\omega i_L + R_{\text{Th}}i_L$$
$$i_L = \frac{v_{\text{Th}}}{j\omega L + R_{\text{Th}}}$$

The source is given as sinusoidal. Analysis is the same for sinusoidal sources, but the mathematics is more cumbersome. Instead, use the phasor representation of the sinusoid source and Eq. 31.8.

$$i_L = \frac{v_{\text{Th}}}{R_{\text{Th}} + j\omega L}$$
$$\mathbf{I}_L = \frac{\mathbf{V}_{\text{Th}}}{\mathbf{Z}} = \frac{\mathbf{V}_{\text{Th}}}{\mathbf{Z}_{R_{\text{Th}}} + \mathbf{Z}_L}$$

Write the source voltage in phasor form as 120 V $\angle 30°$. Using the voltage-divider concept, the Thevenin voltage is one-half this value. The impedances of the resistor and inductor are

$$\mathbf{Z}_{R_{\text{Th}}} = 5\ \Omega\angle 0°$$
$$\mathbf{Z}_L = j\omega L = j\left(377\ \frac{\text{rad}}{\text{s}}\right)(0.15\ \text{H}) = 56.55\ \Omega\angle 90°$$
$$\mathbf{Z}_{\text{Th}} = \mathbf{Z}_{R_{\text{Th}}} + \mathbf{Z}_L = (5 + j0) + (0 + j56.55)$$
$$= 5 + j56.55$$
$$= 56.77\ \Omega\angle 84.9°$$

Substitute to determine the steady-state current.

$$\mathbf{I}_L = \frac{\mathbf{V}_{\text{Th}}}{\mathbf{Z}} = \frac{\mathbf{V}_{\text{Th}}}{\mathbf{Z}_{R_{\text{Th}}} + \mathbf{Z}_L} = \frac{60\ \text{V}\angle 30°}{56.77\ \Omega\angle 84.9°}$$
$$= 1.06\ \text{A}\angle{-55°}$$

Convert to the time (sinusoidal) form.

$$\mathbf{I}_L = 1.06\ \text{A}\angle{-55°} = 1.5\ \text{A}\sin(377t - 55°)$$

Effective or rms values are used with phasor quantities. Peak values are used with sinusoidal quantities.[5] Radians and degrees are mixed in the equations strictly for clarity. Radians are necessary for calculations. Transient calculation using sinusoids follows a similar pattern, with phasor analysis being an efficient calculational method. During transients using sinusoids, the exact time of the transient must be specified.

Example 31.7

For the network shown, $R_1 = 10\ \Omega$, $R_2 = 10\ \Omega$, $v =$ 636 V $\cos(2513t + 60°)$, and $C = 1\ \mu$F. Determine the steady-state capacitance voltage with the switch closed.

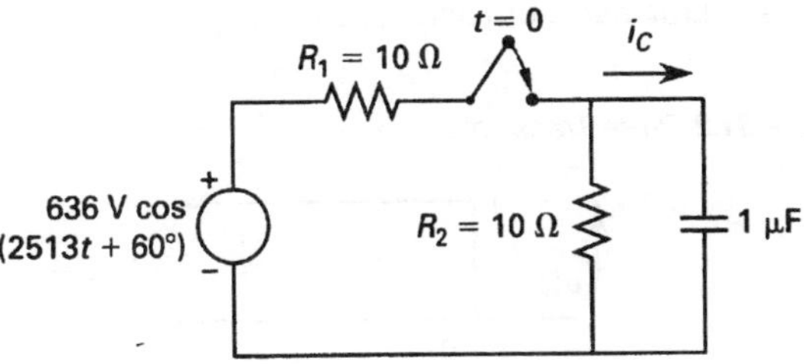

Solution

Using Eq. 31.5 gives

$$v_{\text{Th}} = R_{\text{Th}}C\left(\frac{dv_C}{dt}\right) + v_C$$

Represent the source voltage in phasor form as 450 V$\angle 60°$. Using the voltage divider concept, the Thevenin voltage is one-half this value. Recall that, in the frequency domain, multiplying by $j\omega$ is equivalent to taking the derivative. Combining the two gives

$$v_{\text{Th}} = R_{\text{Th}}C\left(\frac{dv_C}{dt}\right) + v_C$$
$$\mathbf{V}_{\text{Th}} = \mathbf{Z}_{R_{\text{Th}}}Cj\omega\mathbf{V}_C + \mathbf{V}_C = (j\omega\mathbf{Z}_{R_{\text{Th}}}C + 1)\mathbf{V}_C$$
$$\mathbf{V}_C = \frac{\mathbf{V}_{\text{Th}}}{j\omega\mathbf{Z}_{R_{\text{Th}}}C + 1}$$
$$= \frac{225\ \text{V}\angle 60°}{j\left(2513\ \frac{\text{rad}}{\text{s}}\right)(5\ \Omega)(1\times10^{-6}\ \text{F}) + 1}$$
$$= \frac{225\ \text{V}\angle 60°}{1.01\ \Omega\angle 0.72°}$$
$$= 225\ \text{V}\angle 59.28°$$

Transforming into the time domain gives

$$v_{C,\text{ss}}(t) = 318\ \text{V}\cos(2513t + 59°)$$

[5]The representation of phasors as rms quantities and sinusoidal variables as peak quantities is standard in this text. Care should be taken to determine how phasors are being represented in a given document.

Circuit Theory

At the angular frequency given, equivalent to 400 Hz, the capacitance is essentially an open circuit and has little impact on the network. The voltage source in this example was given as the cosine function to emphasize that the choice is arbitrary and does not affect the analysis.

4. FIRST-ORDER ANALYSIS: PULSE TRANSIENTS

A *pulse* is generally defined as a variation in a quantity that is normally constant. When the variation occurs at a single point in time, the pulse is called a *step pulse*. If the magnitude of the step pulse equals one, it is termed a *unit step*. A pulse has a finite duration that is normally brief compared to the time scale of interest. When the duration of the pulse is so short that it can be thought of as infinitesimal, it is called an *impulse*. Figure 31.2 shows a *unit pulse* of each type.

Figure 31.2 *Pulse Transients*

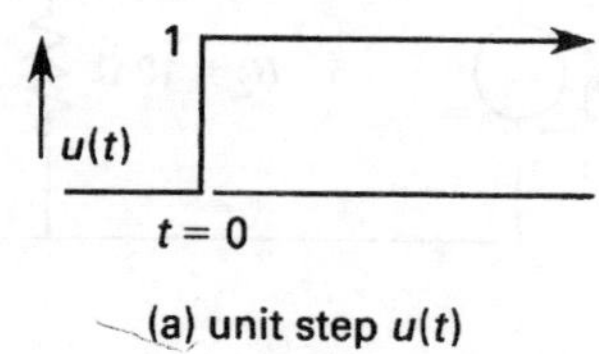

(a) unit step $u(t)$

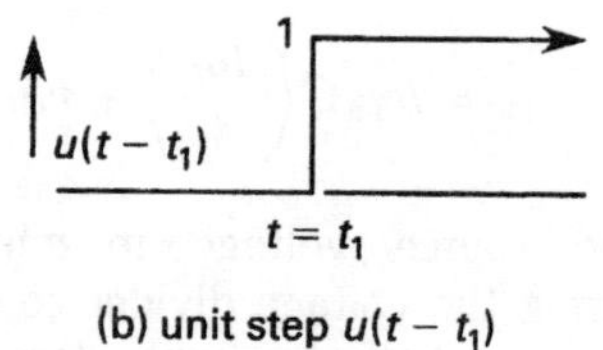

(b) unit step $u(t-t_1)$

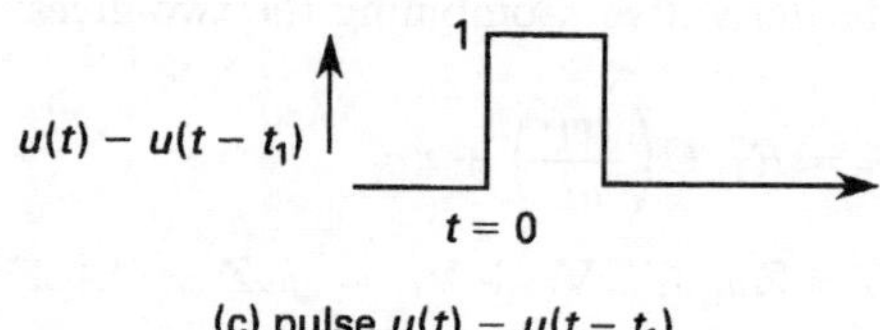

(c) pulse $u(t)-u(t-t_1)$

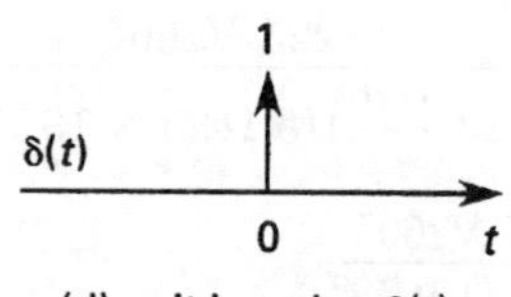

(d) unit impulse $\delta(t)$

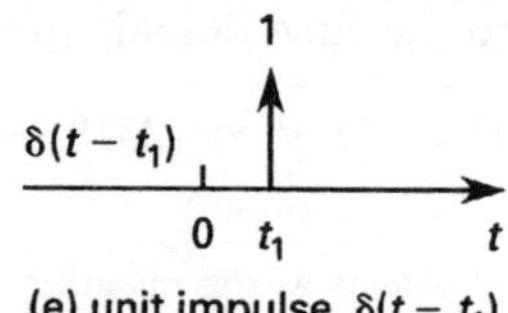

(e) unit impulse $\delta(t-t_1)$

The unit step is defined by Eq. 31.10. The *unit impulse* is the derivative of the unit step function. The relationships between various forcing functions are shown in Table 31.1.

$$u(t-t_1)=0 \quad [t<t_1]$$
$$u(t-t_1)=1 \quad [t>t_1] \qquad \text{31.10}$$

Table 31.1 *Operations on Forcing Functions*

	function when	
function	differentiated	integrated
unit impulse	–	unit step
unit step	unit impulse	unit ramp
unit ramp	unit step	unit parabola
unit parabola	unit ramp	(third degree)
unit exponential	unit exponential	unit exponential
unit sinusoid	unit sinusoid	unit sinusoid

The response of a circuit to a unit step is called a *step response*. The source voltage magnitude, in this case the Thevenin voltage magnitude, is one for a unit step. For a first-order linear circuit like the one shown in Fig. 31.3, the response is determined in terms of the *state variables* of the energy storage device, that is, inductive current or capacitive voltage.[6] If other electrical quantities are required, they can be found from any of the circuit relations. The response equations are

$$v_C(t)=u(t)(1-e^{-t/\tau}) \qquad \text{31.11}$$

$$i_L(t)=u(t)\left(\frac{1}{R_{\text{Th}}}\right)(1-e^{-t/\tau}) \qquad \text{31.12}$$

Figure 31.3 *First-Order Circuit*

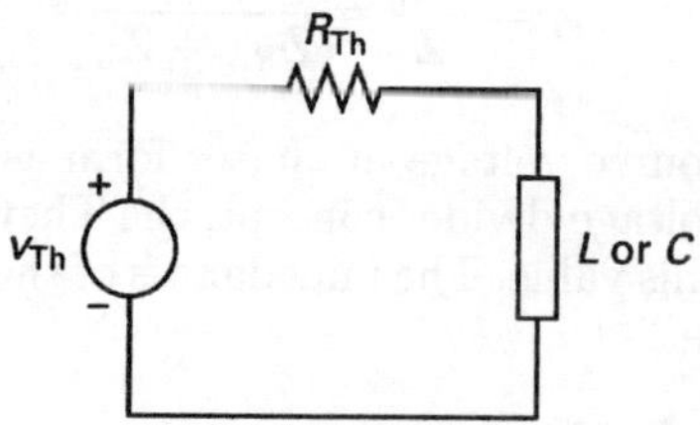

The time constant, τ, for the capacitive circuit is $R_{\text{Th}}C$ and for the inductive circuit is L/R_{Th}.

The step and impulse responses for *RC* and *RL* circuits are shown in Table 31.2. Of importance, the magnitude of the voltage and current sources for a unit step response is one.[7]

[6]A state variable is a variable sufficient to specify the condition of the network and define its future behavior.

[7]The magnitude of one for a unit step is the reason for the apparent nonappearance of the current and voltage in the equations.

Table 31.2 Step and Impulse Responses for First-Order Circuits

circuit	unit step response	unit impulse response
R_{Th}, i_C, v_{Th}, C, v_C	$v_{Th}(t) = u(t)$ $v_C(t) = u(t)\left(1 - e^{-t/(R_{Th}C)}\right)$ $i_C(t) = u(t)\left(\frac{1}{R_{Th}}\right)\left(e^{-t/(R_{Th}C)}\right)$	$v_{Th}(t) = \delta(t)$ $h_v(t) = u(t)\left(\frac{1}{R_{Th}C}\right)\left(e^{-t/(R_{Th}C)}\right)$ $h_i(t) = u(t)\left(-\frac{1}{R_{Th}^2C}\right)\left(e^{-t/(R_{Th}C)}\right)$ $+\,\delta(t)\left(\frac{1}{R_{Th}}\right)$
i_C, i_{Th}, R_{Th}, C, v_C	$i_{Th}(t) = u(t)$ $v_C(t) = u(t)R_{Th}\left(1 - e^{-t/(R_{Th}C)}\right)$ $i_C(t) = u(t) = u(t)\left(e^{-t/(R_{Th}C)}\right)$	$i_{Th}(t) = \delta(t)$ $h_v(t) = u(t)\left(\frac{1}{C}\right)\left(e^{-t/(R_{Th}C)}\right)$ $h_i(t) = u(t)\left(-\frac{1}{R_{Th}C}\right)\left(e^{-t/(R_{Th}C)}\right) + \delta(t)$
R_{Th}, i_C, v_{Th}, L, v_L	$v_{Th}(t) = u(t)$ $v_L(t) = u(t)\left(e^{-t/(L/R_{Th})}\right)$ $i_L(t) = u(t)\left(\frac{1}{R_{Th}}\right)\left(1 - e^{-t/(L/R_{Th})}\right)$	$v_{Th}(t) = \delta(t)$ $h_v(t) = u(t)\left(\frac{R_{Th}}{L}\right)\left(e^{-t/(L/R_{Th})}\right) + \delta(t)$ $h_i(t) = u(t)\left(\frac{1}{L}\right)\left(e^{-t/(L/R_{Th})}\right)$
i_C, i_{Th}, R_{Th}, L, v_L	$i_{Th}(t) = u(t)$ $v_L(t) = u(t)R\left(e^{-t/(L/R_{Th})}\right)$ $i_L(t) = u(t)\left(1 - e^{-t/(L/R_{Th})}\right)$	$i_{Th}(t) = \delta(t)$ $h_v(t) = u(t)\left(-\frac{R_{Th}^2}{L}\right)\left(e^{-t/(L/R_{Th})}\right) + R\delta(t)$ $h_i(t) = u(t)\left(\frac{R_{Th}}{L}\right)\left(e^{-t/(L/R_{Th})}\right)$

Example 31.8

Determine the capacitance voltage in the circuit shown.

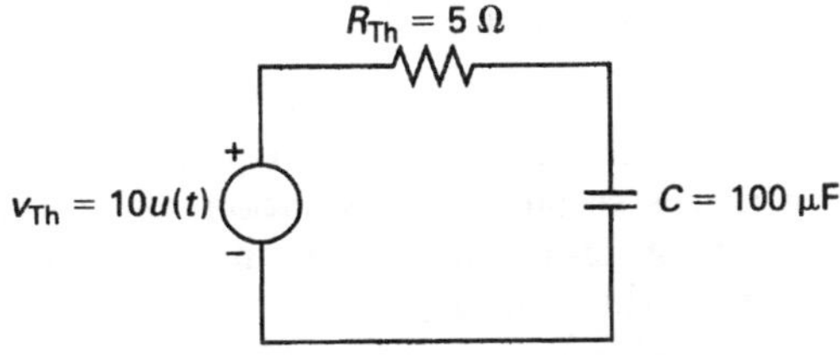

Solution

From Eq. 31.11 or Table 31.2, the form of the capacitance voltage is

$$v_C(t) = u(t)(1 - e^{-t/\tau})$$

The time constant is

$$\tau = R_{Th}C = (5\ \Omega)(100 \times 10^{-6}\ \text{F}) = 500 \times 10^{-6}\ \text{s}$$

The unit step has a magnitude of one. A factor of 10 is required to scale the unit step to the actual value of the Thevenin voltage. The final solution is

$$v_C(t) = 10u(t)\left(1 - e^{-t/(500\times 10^{-6}\ \text{s})}\right)$$

Example 31.9

If the source in Ex. 31.8 outputs two 10 V pulses of 0.010 s duration 0.020 s apart, what is the capacitance voltage?

Solution

From Eq. 31.10 and Fig. 31.2, the voltage source is represented as follows. The brackets in Eq. 31.10 enclose each pulse for clarification.

$$\begin{aligned} v_{\text{Th}}(t) &= (10u(t) - 10u(t - 0.010 \text{ s})) \\ &\quad + (10u(t - 0.030 \text{ s}) - 10u(t - 0.040 \text{ s})) \end{aligned}$$

Using Eq. 31.11 or Table 31.2, scaling by a factor of 10, and applying the principle of superposition gives the total response.

$$\begin{aligned} v_C(t) &= 10u(t)(1 - e^{-t/(500\times10^{-6}\text{ s})}) \\ &\quad - 10u(t - 0.010 \text{ s})(1 - e^{(-t-0.010 \text{ s})/(500\times10^{-6}\text{ s})}) \\ &\quad + 10u(t - 0.030 \text{ s})(1 - e^{(-t-0.030 \text{ s})/(500\times10^{-6}\text{ s})}) \\ &\quad - 10u(t - 0.040 \text{ s})(1 - e^{(-t-0.040 \text{ s})/(500\times10^{-6}\text{ s})}) \end{aligned}$$

5. SECOND-ORDER ANALYSIS

Second-order analysis involves circuits with one equivalent capacitor and one equivalent inductor, that is, two energy storage elements. As in first-order analysis, the method for determining the desired quantity is mathematical analysis in the time domain. Frequency domain analysis, which is also performed in first-order networks, significantly simplifies second-order analysis and is accomplished by transforming either the applicable differential equation or the circuit itself into the *s*-domain.

Laplace transform analysis is then used to determine the desired quantity. (Fourier transform analysis could also be used, but the Laplace transform has the advantage of representing constants, sinusoids, and exponential functions in a single variable, *s*.) Figure 31.4 represents a generic circuit with a single capacitor and inductor.

Figure 31.4 *Second-Order Circuits*

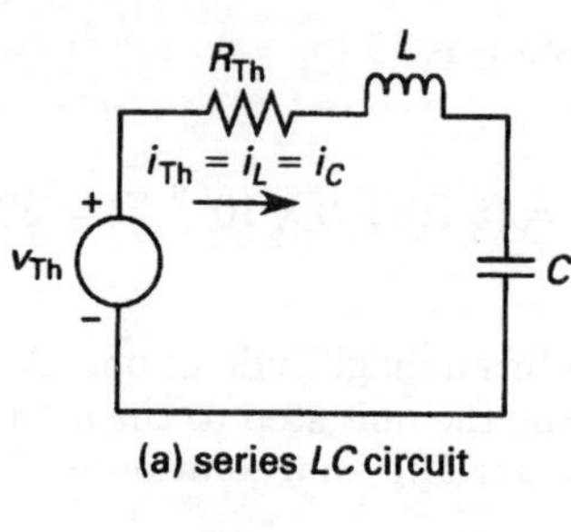

(a) series *LC* circuit

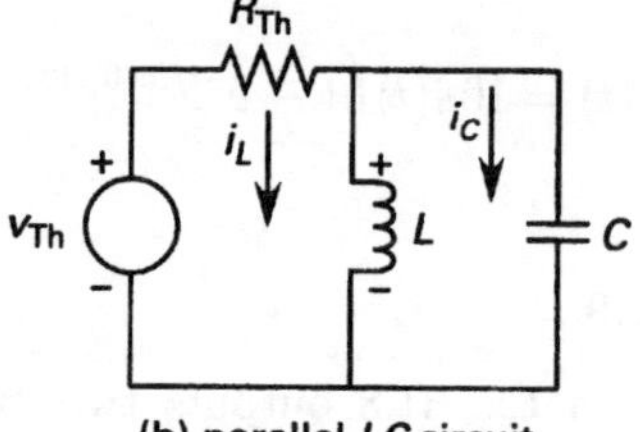

(b) parallel *LC* circuit

The differential equation for second-order circuits takes on the general form

$$f(t) = a\frac{d^2x}{dt^2} + b\frac{dx}{dt} + cx \quad \text{31.13}$$

The associated characteristic equation, written with the roots shown in the *s*-domain, is

$$as^2 + bs + c = 0 \quad \text{31.14}$$

Equation 31.13 can be written in terms of any circuit variable. It is normally written in terms of capacitance voltage or inductance current for reasons identical to those in first-order circuits. The solutions take on one of three forms, depending upon the magnitude of the constants (see Sec. 31.6 through Sec. 31.8). The general response of the three forms is illustrated in Fig. 31.5. *Overdamping* in a linear system is defined as damping over what is required for critical damping. *Critical damping* in a linear system is damping on the threshold between oscillatory and exponential behavior. *Underdamping* is damping that results in oscillatory behavior.

Figure 31.5 *Second-Order Circuit Response to a Step Change*

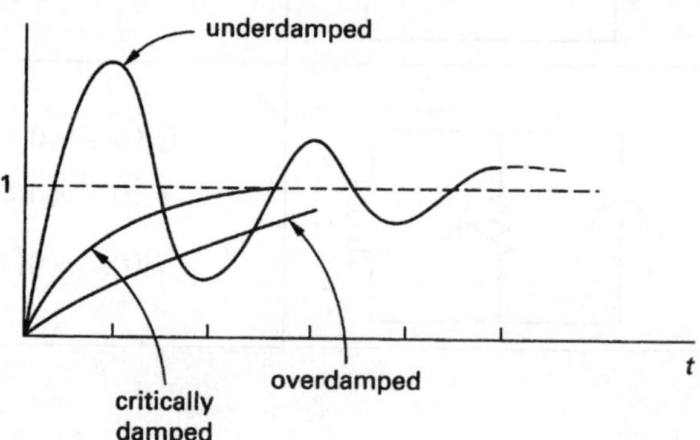

The method for determining the complete solution to second-order networks follows.

step 1: Determine the second-order equation for the circuit to be analyzed by using electrical engineering relationships and laws such as Ohm's law, KVL, KCL, and others.

step 2: Write the characteristic equation associated with the second-order equation. Using the constants of this equation, *a*, *b*, and *c*, determine the form of the solution.

step 3: Write the solution form.

step 4: If required, determine the initial values from the circuit's initial conditions.[8] Determine the steady-state values, x_{ss} and x'_{ss}.

[8]Initial conditions in a circuit are the initial values of the state variable function and its derivative. Two known conditions are required for the solution of a second-order equation.

step 5: Calculate the roots of the characteristic equation.

step 6: Use the information in steps 4 and 5 to calculate the constants in the final solution, A and B. (This involves the solution form in step 3 and its derivative or the direct use of the generic equations for the constants.)

step 7: Write the final solution from the information found in steps 4, 5, and 6.

Steps 4, 5, and 6 are equivalent to performing the following.

step 4′: Find the steady-state response from the particular solution of the nonhomogeneous equation. The solution will be of the same form as the forcing function. Regardless of transient type, the result will be a steady-state value, κ.

step 5′: Determine the transient response from the homogeneous equation. Regardless of the transient type, the transient response will be of the form $Ae^{s_1t} + Be^{s_2t}$. For the critical damped case $s_1 = s_2$, and for the underdamped or oscillatory case the variable s will take on the values $\alpha \pm j\beta$.

step 6′: Sum the steady-state and transient solutions, that is, the complementary and particular solutions, to obtain the total solution. This will be of the form $\kappa + Ae^{s_1t} + Be^{s_2t}$. Calculate the constants in the final solution, A and B.

6. SECOND-ORDER ANALYSIS: OVERDAMPED

If the two roots of Eq. 31.14 are real and different from one another, equivalent to the condition $b^2 > 4ac$, the solution is of the form

$$x(t) = \kappa + Ae^{s_1t} + Be^{s_2t} \qquad 31.15$$

The first term, κ, is a constant, that is, the steady-state component of the solution. This is the particular solution and results from the effect of an energy source or forcing function. The exponential terms represent the natural response of the circuit.[9] The roots of the characteristic equation are

$$s_1 = \frac{-b+\sqrt{b^2-4ac}}{2a} \qquad 31.16$$

$$s_2 = \frac{-b-\sqrt{b^2-4ac}}{2a} \qquad 31.17$$

The coefficients A and B in Eq. 31.15 are found from the initial and steady-state values of x and dx/dt. Because capacitor voltage and inductor current cannot change instantaneously, knowing the conditions immediately prior to any switching transient means knowing the conditions immediately after. By isolating the two energy storage elements and determining the Thevenin equivalent, any second-order circuit can be represented as in Fig. 31.4. In the series RLC circuits of Fig. 31.4(a), the results of applying the initial conditions are

$$v_C(0^+) = V_0 \qquad 31.18$$

$$\frac{dv_C(0^+)}{dt} = \frac{I_0}{C} \qquad 31.19$$

These conditions are a result of the common current path, and $i_C = C(dv_C/dt)$.

For the parallel RLC circuits in Fig. 31.4(b), the results of applying the initial conditions are

$$i_L(0^+) = I_0 \qquad 31.20$$

$$\frac{di_L(0^+)}{dt} = \frac{v_C(0^+)}{L} \qquad 31.21$$

These conditions are a result of the common voltage, and $v_L = L(di_L/dt)$.

The constants A and B are determined by substituting Eq. 31.18 through Eq. 31.21 into Eq. 31.15 and its derivative (multiplying by s is the equivalent of taking the derivative) and then solving for the specific circuit. Equation 31.18 through Eq. 31.21 can be used if network initial conditions are given and circuit element parameters, for example, L and C, are known. The results of such a substitution, in generic terms, are

$$x(0^+) = A + B + x_{ss} \qquad 31.22$$

$$x'(0^+) = s_1A + s_2B + x'_{ss} \qquad 31.23$$

Solving Eq. 31.22 and Eq. 31.23 simultaneously gives the values of the constants.

$$A = \tfrac{1}{2}\left(1+\frac{b}{\sqrt{b^2-4ac}}\right)(x(0^+)-x_{ss}) + \left(\frac{a}{\sqrt{b^2-4ac}}\right)(x'(0^+)-x'_{ss}) \qquad 31.24$$

$$B = \tfrac{1}{2}\left(1-\frac{b}{\sqrt{b^2-4ac}}\right)(x(0^+)-x_{ss}) - \left(\frac{a}{\sqrt{b^2-4ac}}\right)(x'(0^+)-x'_{ss}) \qquad 31.25$$

Example 31.10

Determine the second-order equation for the circuit in Fig. 31.4(a).

Solution

Using KVL and rearranging the result gives

$$v_{Th} = iR_{Th} + L\left(\frac{di}{dt}\right) + \frac{1}{C}\int i\,dt$$

[9]The natural response of an electric circuit is the response that would occur without a forcing function, that is, without an energy source. This portion of the solution goes by numerous names including natural, zero energy, homogeneous, and complementary.

Differentiating gives

$$\frac{dv_{\text{Th}}}{dt} = R_{\text{Th}}\left(\frac{di}{dt}\right) + L\left(\frac{d^2i}{dt^2}\right) + \left(\frac{1}{C}\right)i$$

Rearrange to put into the form of Eq. 31.13.

$$\frac{dv_{\text{Th}}}{dt} = L\left(\frac{d^2i}{dt^2}\right) + R_{\text{Th}}\left(\frac{di}{dt}\right) + \left(\frac{1}{C}\right)i$$

Example 31.11

A certain network is described by the following differential equation.[10]

$$24\text{ V} = 2\left(\frac{d^2v}{dt^2}\right) + 5\left(\frac{dv}{dt}\right) + 2v$$

The initial conditions are $v = 0$ and $dv/dt = 0$. Determine the expression for the voltage for $t \geq 0$.

Solution

Use the method for solving second-order networks outlined in Sec. 31.5.

step 1: Because the second-order equation is given, step 1 is complete.

step 2: The characteristic equation using constants a, b, and c from the second-order equation is

$$2s^2 + 5s + 2 = 0$$

From this, $a = 2$, $b = 5$, $c = 2$, and $b^2 > 4ac$. The network is overdamped.

step 3: The solution form is

$$x(t) = \kappa + Ae^{s_1t} + Be^{s_2t}$$

step 4: The initial conditions are given. The forcing function is a constant 24 V. At steady-state, dv/dt and d^2v/dt^2 equal zero. The steady-state voltages are[11]

$$\begin{aligned}24\text{ V} &= 2\left(\frac{d^2v_{\text{ss}}}{dt^2}\right) + 5\left(\frac{dv_{\text{ss}}}{dt}\right) + 2v_{\text{ss}} \\ &= 0 + 0 + 2v_{\text{ss}} \\ v_{\text{ss}} &= \frac{24}{2} = 12\text{ V}\end{aligned}$$

[10]Dividing the equation by two shows that 12 is the value of the forcing function. Further, the 12 actually represents a step change of the source (Thevenin) voltage at $t = 0$. After the transient ends, the final value of the voltage across the capacitor will be 12 V.

[11]Units are not shown where they might cause confusion with the variables used, such as the voltage v and volts V. Units should be shown in all final solutions.

The steady-state voltage in this case is not a function of time. Therefore,

$$v'_{\text{ss}} = 0\text{ V}$$

step 5: Using Eq. 31.16 and Eq. 31.17, the roots of the characteristic equation are

$$\begin{aligned}s_1 &= \frac{-b + \sqrt{b^2 - 4ac}}{2a} = \frac{-5 + \sqrt{(5)^2 - (4)(2)(2)}}{(2)(2)} \\ &= -1/2\end{aligned}$$

$$\begin{aligned}s_2 &= \frac{-b - \sqrt{b^2 - 4ac}}{2a} = \frac{-(5) - \sqrt{5^2 - (4)(2)(2)}}{(2)(2)} \\ &= -2\end{aligned}$$

step 6: Use Eq. 31.24 and Eq. 31.25 to determine the constants A and B.

$$\begin{aligned}A &= \tfrac{1}{2}\left(1 + \frac{b}{\sqrt{b^2 - 4ac}}\right)(v(0^+) - v_{\text{ss}}) \\ &\quad + \left(\frac{a}{\sqrt{b^2 - 4ac}}\right)(v'(0^+) - v'_{\text{ss}}) \\ &= \left(\frac{1}{2}\right)\left(1 + \frac{5}{\sqrt{(5)^2 - (4)(2)(2)}}\right)(0 - 12) \\ &\quad + \left(\frac{2}{\sqrt{(5)^2 - (4)(2)(2)}}\right)(0 - 0) \\ &= -16\end{aligned}$$

$$\begin{aligned}B &= \tfrac{1}{2}\left(1 - \frac{b}{\sqrt{b^2 - 4ac}}\right)(v(0^+) - v_{\text{ss}}) \\ &\quad - \left(\frac{a}{\sqrt{b^2 - 4ac}}\right)(v'(0^+) - v'_{\text{ss}}) \\ &= \left(\frac{1}{2}\right)\left(1 - \frac{5}{\sqrt{(5)^2 - (4)(2)(2)}}\right)(0 - 12) \\ &\quad - \left(\frac{2}{\sqrt{(5)^2 - (4)(2)(2)}}\right)(0 - 0) \\ &= 4\end{aligned}$$

step 7: Write the final solution.

$$\begin{aligned}v(t) &= \kappa + Ae^{s_1t} + Be^{s_2t} \\ &= 12 - 16e^{(-1/2)t} + 4e^{-2t}\end{aligned}$$

7. SECOND-ORDER ANALYSIS: CRITICALLY DAMPED

If the two roots of Eq. 31.14 are real and the same, equivalent to the condition $b^2 = 4ac$, the solution is of the form

$$x(t) = \kappa + Ae^{st} + Bte^{st} \quad 31.26$$

The first term, κ, is a constant, that is, the steady-state component of the solution. The exponential terms represent the natural response of the circuit. The root of the characteristic equation is

$$s = \frac{-b}{2a} \quad 31.27$$

The coefficients A and B in Eq. 31.26 are found from the initial and steady-state values of x and dx/dt. By the same reasoning used in Sec. 31.6, the initial conditions for a series RLC circuit remain unchanged and are repeated here for convenience.

$$v_C(0^+) = V_0 \quad 31.28$$

$$\frac{dv_C(0^+)}{dt} = \frac{I_0}{C} \quad 31.29$$

Similarly, the initial conditions for parallel RLC circuits are

$$i_L(0^+) = I_0 \quad 31.30$$

$$\frac{di_L(0^+)}{dt} = \frac{v_C(0^+)}{L} \quad 31.31$$

The constants A and B are determined by substituting Eq. 31.28 through Eq. 31.31 into Eq. 31.26 and its derivative (multiplying by s is the equivalent of taking the derivative) and then solving for the specific circuit. Equation 31.28 through Eq. 31.31 can be used if network initial conditions are given and circuit element parameters such as L and C are known. The results of such a substitution, in generic terms, are

$$x(0^+) = A + x_{ss} \quad 31.32$$

$$x'(0^+) = sA + B + x'_{ss} \quad 31.33$$

Solving Eq. 31.32 and Eq. 31.33 simultaneously gives the values of the constants.

$$A = x(0^+) - x_{ss} \quad 31.34$$

$$B = x'(0^+) - x'_{ss} + \left(\frac{b}{2a}\right)A \quad 31.35$$

Example 31.12

Consider the series RLC circuit shown, with $v_{Th} = 10$ V, $R = 80\ \Omega$, $L = 80 \times 10^{-3}$ H, and $C = 50 \times 10^{-6}$ F. The capacitor is initially charged to 5 V and the initial current in the circuit is 187.5×10^{-3} A. Determine the behavior of the circuit, in terms of the capacitance voltage, for $t \geq 0$.

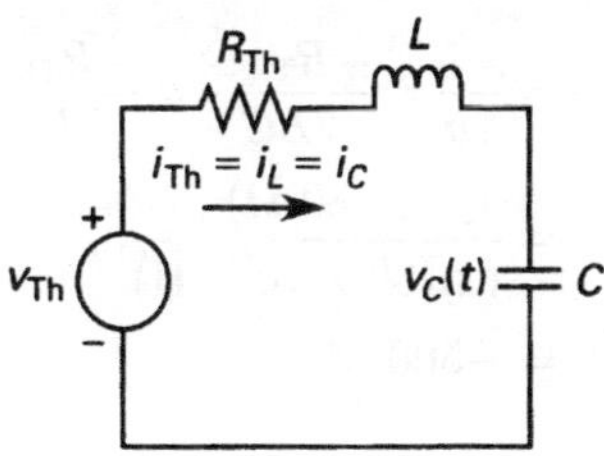

Solution

Using KVL and writing the resulting equation in terms of the capacitance voltage,

$$10\text{ V} = iR_{Th} + L\left(\frac{di}{dt}\right) + v_C$$

Substituting $i = C(dv/dt)$,

$$10\text{ V} = C\left(\frac{dv_C}{dt}\right)R_{Th} + LC\left(\frac{d^2v_C}{dt^2}\right) + v_C$$

$$10\text{ V} = LC\left(\frac{d^2v_C}{dt^2}\right) + R_{Th}C\left(\frac{dv_C}{dt}\right) + v_C$$

The characteristic equation follows.

$$LCs^2 + RCs + 1 = 0$$

Determine the solution form from the relationship between the coefficients of the characteristic equation.

$$\begin{aligned} b^2 &= (RC)^2 = ((80\ \Omega)(50 \times 10^{-6}\text{ F}))^2 = 16 \times 10^{-6} \\ 4ac &= (4)(LC)(1) \\ &= (4)(80 \times 10^{-3}\text{ H})(50 \times 10^{-6}\text{ F})(1) \\ &= 16 \times 10^{-6} \\ b^2 &= 4ac \end{aligned}$$

The circuit is critically damped. The solution is of the form given by Eq. 31.26.

$$v_C(t) = \kappa + Ae^{st} + Bte^{st}$$

The initial conditions are given. The steady-state conditions are found from the results of the KVL analysis.

$$\begin{aligned} 10 &= LC\left(\frac{d^2v_C}{dt^2}\right) + R_{Th}C\left(\frac{dv_C}{dt}\right) + v_C \\ &= LC(0) + R_{Th}C(0) + v_C \end{aligned}$$

$$v_{C,ss} = 10$$

$$v'_{C,ss} = 0$$

Circuit Theory

The roots of the characteristic equation are found from Eq. 31.27 or actual calculation to be[12]

$$s = \frac{-b}{2a} = \frac{-R_{Th}C}{2LC} = \frac{-R_{Th}}{2L}$$
$$= \frac{-(80\ \Omega)}{(2)(80 \times 10^{-3}\ \text{H})}$$
$$= -500$$

Because the initial conditions and circuit parameters are known, instead of using the generic equations (see Eq. 31.32 through Eq. 31.35), use the solution form directly with the initial conditions from Eq. 31.28 and Eq. 31.29.

$$v_C(t) = \kappa + Ae^{st} + Bte^{st}$$
$$v_C(0^+) = 10 + Ae^{s(0)} + B(0)\,e^{s(0)}$$
$$5 = 10 + A$$
$$A = -5$$

The second initial condition results in the following value for B.

$$v_C(t) = k + Ae^{st} + Bte^{st}$$
$$\frac{dv_C(0^+)}{dt} = \frac{I_0}{C} = 0 + Ase^{st} + B(e^{st} + tse^{st})$$
$$\frac{187.5 \times 10^{-3}\ \text{A}}{50 \times 10^{-6}\ \text{F}} = 0 + (-5)(-500)\left(e^{(-500)(0)}\right) + B\left(\begin{array}{c} e^{(-500)(0)} + (0)(-500) \\ \times\, e^{(-500)(0)} \end{array}\right)$$
$$3750 = 2500 + B$$
$$B = 1250$$

The final solution is

$$v_C(t) = \kappa + Ae^{st} + Bte^{st}$$
$$= 10 - 5e^{-500t} + 1250te^{-500t}$$

8. SECOND-ORDER ANALYSIS: UNDERDAMPED

If the two roots of Eq. 31.14 are complex conjugates, equivalent to the condition $b^2 < 4ac$, the solution is of the form[13]

$$x(t) = \kappa + e^{\alpha t}\left(A_e e^{j\beta t} + B_e e^{-j\beta t}\right) \quad \text{31.36}$$

[12]Units are not commonly shown for s. They are inverse seconds.

[13]The subscript e on the constant is a reminder that the constants in Eq. 31.36 and Eq. 31.37 differ.

This can also be represented in the sinusoidal form as

$$x(t) = \kappa + e^{\alpha t}(A\cos\beta t + B\sin\beta t) \quad \text{31.37}$$

The first term, κ, is a constant, that is, the steady-state component of the solution. The exponential terms represent the natural response of the circuit. Equation 31.36 is the ringing response, or damped oscillation, shown in Fig. 30.7(b). The sinusoidal response is enveloped by the exponential, since α is negative.

The variable s used in the overdamped and critically damped cases has been expanded to clarify this behavior and is represented by Eq. 31.38 and Eq. 31.39.[14]

$$s_1 = \alpha + j\beta \quad \text{31.38}$$
$$s_2 = \alpha - j\beta \quad \text{31.39}$$

The roots of the characteristic equation are

$$\alpha = \frac{-b}{2a} \quad \text{31.40}$$
$$\beta = \frac{\sqrt{4ac - b^2}}{2a} \quad \text{31.41}$$

The coefficients A_e and B_e in Eq. 31.36, or A and B in Eq. 31.37, are found from the initial and steady-state values of x and dx/dt. By the same reasoning used in Sec. 31.6, the initial conditions for a series *RLC* circuit remain unchanged and are repeated here for convenience.

$$v_C(0^+) = V_0 \quad \text{31.42}$$
$$\frac{dv_C(0^+)}{dt} = \frac{I_0}{C} \quad \text{31.43}$$

Similarly, the initial conditions for parallel *RLC* circuits are

$$i_L(0^+) = I_0 \quad \text{31.44}$$
$$\frac{di_L(0^+)}{dt} = \frac{v_C(0^+)}{L} \quad \text{31.45}$$

The constants are determined by substituting Eq. 31.42 through Eq. 31.45 into Eq. 31.36 or Eq. 31.37 and their derivatives (multiplying by $j\beta$ is the equivalent of taking the derivative) and then solving for the specific circuit. Equation 31.42 through Eq. 31.45 can be used if network initial conditions are given and circuit element parameters such as L and C are known. The results of such a substitution, in generic terms, are

$$x(0^+) = A + x_{ss} \quad \text{31.46}$$
$$x'(0^+) = \alpha A + \beta B + x'_{ss} \quad \text{31.47}$$

[14]The terminology of α and β is often used in transient analysis. The variable s was split for the underdamped case to clearly show the sinusoids enveloped by the exponentials. The term α is shown in some texts as $-\alpha$. But embedding the negative sign, as in Eq. 31.40, is consistent with the quadratic equation results for the characteristic equation solutions.

Solving Eq. 31.46 and Eq. 31.47 simultaneously gives the values of the constants.

$$A = x(0^+) - x_{ss} \tag{31.48}$$

$$B = \left(\frac{-\alpha}{\beta}\right)A + \left(\frac{1}{\beta}\right)(x'(0^+) - x'_{ss}) \tag{31.49}$$

9. SECOND-ORDER ANALYSIS: PULSE TRANSIENTS

The output of a second-order circuit responding to a step change is illustrated in Fig. 31.5. The electrical parameters defining the response vary depending upon the type of analysis required. (See Sec. 31.10.) The response of a second-order network with zero initial conditions, that is, $x(0) = 0$ and $x'(0)$, to a step change has a wide range of applications. Because a unit step change can be scaled to any specific pulse, it forms the baseline input. Consider Eq. 31.13 with zero initial conditions and the unit step as the input, that is, $f(t) = 1$. The steady-state value of the state variable is

$$x_{ss} = \frac{f(t)}{c} = \frac{1}{c} \tag{31.50}$$

The equations in Sec. 31.6 through Sec. 31.8 are then modified using Eq. 31.50 and the zero initial conditions requirement as follows.[15]

For an overdamped circuit ($b^2 > 4ac$),

$$x(t) = \frac{1}{c} + Ae^{s_1 t} + Be^{s_2 t} \tag{31.51}$$

$$s_1 = \frac{-b + \sqrt{b^2 - 4ac}}{2a} \tag{31.52}$$

$$s_2 = \frac{-b - \sqrt{b^2 - 4ac}}{2a} \tag{31.53}$$

$$A = -\left(\frac{1}{2c}\right)\left(1 + \frac{b}{\sqrt{b^2 - 4ac}}\right) \tag{31.54}$$

$$B = -\left(\frac{1}{2c}\right)\left(1 - \frac{b}{\sqrt{b^2 - 4ac}}\right) \tag{31.55}$$

For a critically damped circuit ($b^2 = 4ac$),

$$x(t) = \frac{1}{c} + Ae^{st} + Bte^{st} \tag{31.56}$$

$$s = \frac{-b}{2a} \tag{31.57}$$

$$A = -\frac{1}{c} \tag{31.58}$$

$$B = -\frac{b}{2ac} \tag{31.59}$$

[15]Because x_{ss} is a constant, x'_{ss}. This condition is also incorporated.

For an underdamped circuit ($b^2 < 4ac$),

$$x(t) = \frac{1}{c} + e^{\alpha t}(A\cos\beta t + B\sin\beta t) \tag{31.60}$$

$$s_1 = \alpha + j\beta \tag{31.61}$$

$$s_2 = \alpha - j\beta \tag{31.62}$$

$$\alpha = \frac{-b}{2a} \tag{31.63}$$

$$\beta = \frac{\sqrt{4ac - b^2}}{2a} \tag{31.64}$$

$$A = -\frac{1}{c} \tag{31.65}$$

$$B = -\frac{\alpha}{\beta c} \tag{31.66}$$

10. HIGHER-ORDER CIRCUITS

A circuit with two or more independent energy storage elements is considered a *higher-order circuit.* Second-order systems dominate the behavior of third- and higher-order systems and can be used to approximate the behavior of such networks. Higher-order analysis takes place primarily in the *s*-domain, using the concept of a complex frequency, *s*. The characteristic equation derived from the network differential equation is described with different terminology depending upon the type of electrical analysis performed. In purely mathematical terms, the characteristic equation is

$$as^2 + bs + c = 0 \tag{31.67}$$

Control systems analysis uses the form given by Eq. 31.68. The *natural frequency*, ω_n, is sometimes written simply as ω. The term ζ is called the *damping ratio.*[16]

$$s^2 + 2\zeta\omega_n s + \omega_n^2 = 0 \tag{31.68}$$

Resonance analysis uses the form given by Eq. 31.69. The resonant frequency is symbolized by ω_0 and the quality factor by Q.

$$s^2 + \left(\frac{\omega_0}{Q}\right)s + \omega_0^2 = 0 \tag{31.69}$$

Transient analysis uses the form given by Eq. 31.70. The *damping* is given the symbol α while the *frequency of oscillation* is β.

$$(s + \alpha)^2 + \beta^2 = 0 \tag{31.70}$$

[16]When $\zeta > 1$, the system is overdamped and will return to equilibrium over an extended period of time but without oscillation. When $\zeta = 1$, the system is critically damped and will return to equilibrium quickly but without oscillation. When $0 < \zeta < 1$, the system is underdamped and will oscillate while returning to equilibrium; as this happens, the amplitude of the oscillation gradually decreases. When $\zeta = 0$, the system is undamped and will oscillate forever at its natural frequency, ω_0. The first three cases are shown in Fig. 31.5.

Circuit Theory

In the transient analysis of Sec. 31.6 through Sec. 31.8, the three cases could have been represented as $\alpha > \omega_0$ (overdamped), $\alpha = \omega_0$ (critically damped), and $\alpha < \omega_0$ (underdamped). The parameters are related as follows.

$$\omega_0^2 = \omega_n^2 = \alpha^2 + \beta^2 \qquad 31.71$$

$$\zeta = \frac{1}{2Q} \qquad 31.72$$

$$\alpha = \frac{\omega_0}{2Q} \qquad 31.73$$

These characteristic equations are the denominators of transfer functions, which are ratios from a network's output point to a network's input point. Their roots, called *poles*, will aid in determining the frequency response of a network without solving the original differential equation. The roots of the characteristic equation determine the natural response of the network, that is, the response without a forcing function (without input). The transfer function and the characteristic equation describe the response of the network independent of input. Any input can then be attached and the output derived by knowing the circuit's transfer function and natural response.[17]

11. COMPLEX FREQUENCY

Electrical circuits can respond to input with a resulting constant output, A, and exponential output, $e^{\alpha t}$, or a sinusoid, $\sin \omega t$.[18] The complex frequency, s, is used to combine the three possibilities and represent each in terms of a single exponential, e^{st}. (See Eq. 31.1 for the definition of complex frequency.)

Consider a signal containing all three possibilities represented by Eq. 31.74, including a phase shift. In such a case, the peak value can be represented by the constant A. (The peak value is used to allow the constant to remain unchanged when in the exponential or sinusoidal form. In the sinusoidal form the peak value must be used to properly represent the function. The form $A\angle\theta$ normally uses an rms value for A, while the exponential form uses a peak value. Care must be taken to ensure the desired form is used.) Such a signal is given by

$$A_p e^{\sigma t} \sin(\omega t + \theta) \qquad 31.74$$

Representing the sinusoid in exponential terms, using the sine term, and dropping the Im prefix as is common practice, allows Eq. 31.74 to be represented as

$$A_p e^{j\theta} e^{(\sigma + j\omega)t} = A_p e^{j\theta} e^{st} \qquad 31.75$$

The e^{st} term will be common to all elements in a network, just as $e^{j\omega t}$ was common to all elements. Only the magnitude A and the angle θ of a given signal need to be carried in calculations. If desired, the e^{st} term is recovered upon completion of analysis, giving the exact form of the response. The variable σ is termed the *neper frequency* and assigned units of Np/s or Hz. The neper frequency σ is always zero or negative in an actual circuit. In Eq. 31.74, for example, if σ were positive, the exponential term would increase without bound, resulting in an infinite response. Clearly, this is not possible. Nevertheless, the σ is considered to carry its sign (in other words, the negative is internal), though the sign is sometimes shown explicitly to clarify the decaying exponential aspect. The variable ω is the angular frequency and is assigned units of rad/s or Hz. A negative value of ω merely implies that the sinusoid carries a negative sign, because $\sin(-\omega t) = -\sin \omega t$.

If $\sigma = 0$ and $\omega = 0$, the signal or input is a constant. If $\sigma < 0$ and $\omega = 0$, the signal represented is an exponential decay. If $\sigma < 0$ and $\omega \neq 0$, the signal is a damped sinusoid.[19] If $\sigma = 0$ and $\omega \neq 0$, the signal is a sinusoid. Table 31.3 shows examples of each of the signal response types based on Eq. 31.75 along with their values in the s-domain. Table 31.4 gives the generalized impedances in the s-domain. Table 31.4 also exhibits the similarity of form between the frequency domain and the s-domain. Working in the s-domain is always possible and simplifies the mathematics. If only a sinusoidal response is desired, upon completion of circuit manipulation, s can be replaced with $j\omega$ and the result will be in the correct form for any computations. Figure 31.6 illustrates how signal response terms plot.

***Table 31.3** Example Signal Representation in the s-Domain*

time function	$A_p e^{j\theta}$ or $A_p\angle\theta$ [a]	s [b]
50	$50\angle 0°$	$0 + j0$
$50e^{-100t}$	$50\angle 0°$	$-100 + j0$ Np/s
$50 \sin(200t - 60°)$	$50\angle -60°$	$0 + j200$ rad/s
$50e^{-100t} \sin(-200t + 60°)$	$50\angle 60°$	$-100 - j200\ s^{-1}$

[a]The format $A\angle\theta$ used in this text normally indicates rms values. Here the peak value is used. If the rms format is used, the values in this column would be divided by $\sqrt{2}$.

[b]A negative frequency $(-\omega)$ value merely implies that the sinusoid carries a negative sign, because $\sin(-\omega t) = -\sin \omega t$ and s would then have a negative j term.

***Table 31.4** Generalized Impedance in the s-Domain*

impedance type	frequency domain value	s-domain value
Z_R	R	R
Z_L	$j\omega L$	sL
Z_C	$\frac{1}{j\omega C}$	$\frac{1}{sC}$

[17]The transfer function gives the steady-state response. The poles provide the transient response.

[18]Even aperiodic signals can be represented in terms of sinusoids. Consequently, these three possibilities represent the full range.

[19]The condition for a damped sinusoid is sometimes given as $\sigma \neq 0$ and $\omega \neq 0$. This is true as long as the value of σ is restricted to negative values. (In real circuits, σ is either zero or negative.) If the negative sign is not carried with the σ but is instead shown in the exponential as $e^{-(\sigma + j\omega)t}$, the condition $\sigma > 0$ and $\omega \neq 0$ represents the damped sinusoid.

Figure 31.6 Signal Response in the s-Domain

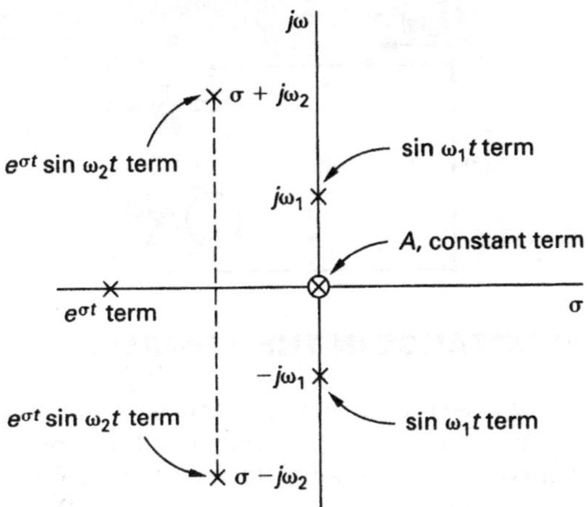

12. LAPLACE TRANSFORM ANALYSIS

The Laplace transform method of analysis moves the state variables, that is, the capacitive voltage and the inductive current, from the time domain into the s-domain. In the s-domain, many forms of excitation, such as the constant, the exponential, and the sinusoid, can be united and easily handled mathematically. The Laplace transform is given by[20]

$$\mathcal{L}\{f(t)\} = \int_{0+}^{\infty} f(t)e^{-st}dt \qquad 31.76$$

The calculation of Eq. 31.76 seldom needs to be accomplished. Numerous tables of Laplace transforms exist (see App. 11.A). Using the data in a Laplace transform table, the properties given by Eq. 31.77 and Eq. 31.78 are used to transform a differential equation representing the behavior of an electrical network.

$$\mathcal{L}\{f(t)\} = F(s) \qquad 31.77$$

$$\mathcal{L}\left\{\frac{d^2f(t)}{dt^2}\right\} = s^2F(s) - sf(0^+) - f'(0^+) \qquad 31.78$$

$$\mathcal{L}\left\{\frac{df(t)}{dt}\right\} = sF(s) - f(0^+) \qquad 31.79$$

Because the energy in a capacitor or inductor cannot change instantaneously, the conditions at $t = 0$ are those at $t = 0^+$. Once the transformation is complete, the inverse Laplace transform can be taken using Eq. 31.80.

$$\mathcal{L}^{-1}\{F(s)\} = f(t) = \frac{1}{2\pi j}\int_{\sigma-j\infty}^{\sigma+j\infty} F(s)e^{st}dt \qquad 31.80$$

[20]Also, taking the Laplace transform of a physical quantity introduces an additional unit of time. For example, $I(s)$ has units of A·s or C. Because the inverse transform removes the extra time unit, such units will be omitted in the s-domain and $I(s)$ is still referred to as the current. The same is done for the voltage, $V(s)$.

The calculation of Eq. 31.80 seldom needs to be accomplished. The same tables used for Laplace transforms contain the inverses. The path of integration of Eq. 31.80 is a straight line parallel to the $j\omega$ axis. All the poles of $F(s)$ lie to the left of this path.

The *classical method* for analyzing electrical networks using Laplace transform analysis follows.

step 1: Write the differential equation describing the circuit behavior from known circuit relationships and laws.

step 2: Transform the differential equation, term by term, into the s-domain using Laplace transform tables (e.g., App. 11.A).

step 3: Solve for the transformed variable, $V(s)$ or $I(s)$, in terms of s.

step 4: Use partial fraction expansion, or any other desired method, to rearrange the terms into those recognizable in Laplace transform tables.

step 5: Equate the coefficients of the powers of s. Solve for the constants.

step 6: Obtain the inverse Laplace transform.

The classical method is useful when the desired quantity is one of the state variables, capacitor voltage or inductor current. If the desired variables are other than the state variables, the *circuit transformation method* is the most direct.

The circuit transformation method follows.

step 1: Transform each individual circuit element into the s-domain. Initial condition sources will exist for energy storage elements. For the capacitor, an initial condition voltage source will be added in series. For the inductor, an initial condition current source will be added in parallel.

step 2: If the desired quantity is one of the variables of an energy storage device (i.e., a state variable), obtain the equivalent Thevenin or Norton form of the circuit as seen from the terminals of the capacitor or inductor.

step 3: Write the differential equation, describing the circuit behavior from known circuit relationships and laws.

step 4: Solve for the desired variable in terms of s.

step 5: Use partial fraction expansion, or any other desired method, to rearrange the terms into those recognizable in Laplace transform tables.

step 6: Equate the coefficients of the powers of s. Solve for the constants.

step 7: Obtain the inverse Laplace transform.

Regardless of the method used, the transformed equations obtained in step 3 are arranged using partial fraction expansion in order to be inversely transformed. This is necessary for transient analysis of first- and second-order systems, because the initial conditions impact the transient. Because the initial conditions do not impact the steady-state value, in sinusoidal analysis, usually called frequency analysis or frequency response, the equations of step 3 reduce to phasor form, so steps 4 through 7 can be skipped and the response analyzed using graphical means. (See Chap. 32.)

13. CAPACITANCE IN THE s-DOMAIN

Capacitors oppose rates of change in voltage. In capacitive circuits, voltage is the state variable and is preferred in system equations. Because the capacitor voltage cannot change instantaneously, $v_C(0+) = v_C(0)$. With this information, the capacitive voltage, $v_C(t)$, and the capacitive current, $C(dv_C(t)/dt)$, are transformed by Eq. 31.81 and Eq. 31.82.

$$\mathcal{L}\{v_C(t)\} = V_C(s) \qquad 31.81$$

$$\begin{aligned}\mathcal{L}\{i_C(t)\} &= \mathcal{L}\left\{C\frac{dv_C(t)}{dt}\right\} \\ &= sCV_C(s) - Cv_C(0)\end{aligned} \qquad 31.82$$

The transformed capacitance must exhibit the properties at its terminals given by Eq. 31.82. Such a model is shown in Fig. 31.7. Using the extended Ohm's law on the current source $sCV_C(s)$, that is, taking the voltage across the source's terminals, $V_C(s)$, and dividing by the current, $sCV_C(s)$, gives the model of Fig. 31.8. Taking the Thevenin equivalent of the model shown in Fig. 31.8 results in a third model given in Fig. 31.9 that shows the initial voltage explicitly. Any of the models is valid; Fig. 31.9 is perhaps the most intuitive.

***Figure 31.7** Capacitor Two-Source Model*

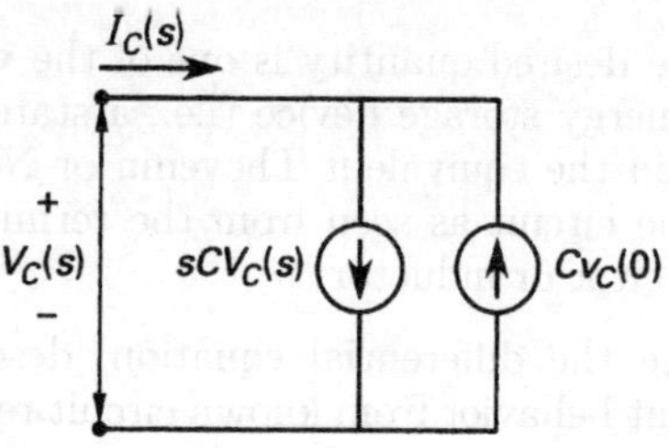

***Figure 31.8** Capacitor Impedance and Initial Current Source Model*

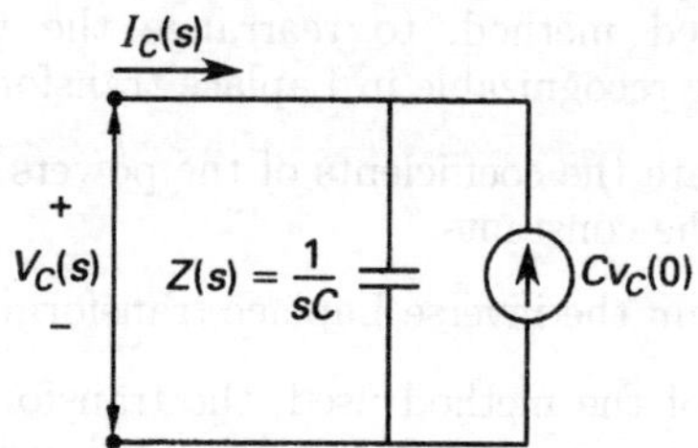

***Figure 31.9** Capacitor Impedance and Initial Voltage Source Model*

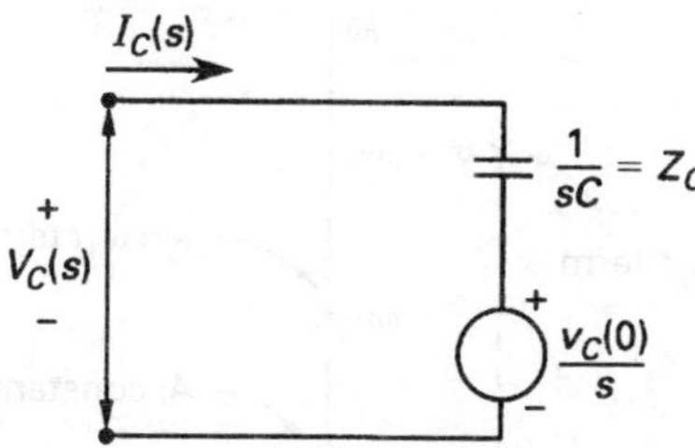

14. INDUCTANCE IN THE s-DOMAIN

Inductors oppose rates of change in current. In inductive circuits, current is the state variable and it is preferred in system equations. Because the inductor current cannot change instantaneously, $i_L(0+) = i_L(0)$. With this information, the inductive current, $i_L(t)$, and the inductive voltage, $L(di_L(t)/dt)$, are transformed by Eq. 31.83 and Eq. 31.84.

$$\mathcal{L}\{i_L(t)\} = I_L(s) \qquad 31.83$$

$$\begin{aligned}\mathcal{L}\{v_L(t)\} &= \mathcal{L}\left\{L\frac{di_L(t)}{dt}\right\} \\ &= sLI_L(s) \quad Li_L(0)\end{aligned} \qquad 31.84$$

The transformed inductance must exhibit the properties at its terminals given by Eq. 31.84. Such a model is shown in Fig. 31.10. Using the extended Ohm's law on the voltage source $sLI_L(s)$, that is, dividing by the current through the source's terminals, $I_L(s)$, gives the model shown in Fig. 31.11. Taking the Norton equivalent of the model in Fig. 31.11 results in a third model given in Fig. 31.12, which shows the initial current explicitly. Any of the models is valid; Fig. 31.12 is perhaps the most intuitive.

***Figure 31.10** Inductor Two-Source Model*

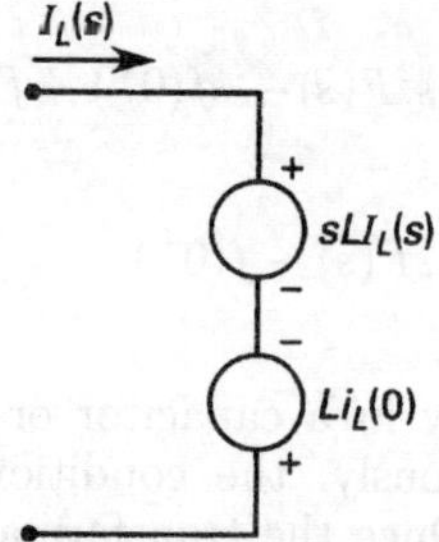

15. LAPLACE TRANSFORM ANALYSIS: FIRST- AND SECOND-ORDER SYSTEMS

Either of the methods in Sec. 31.12 may be applied. If variables other than the capacitor voltage or inductor current are required, the circuit transformation method is the most direct method. The Thevenin form is preferred for capacitive circuits because the final capacitive voltage must equal the Thevenin voltage, thereby acting as a check on the solution. The Norton equivalent is preferred

Figure 31.11 *Inductor Impedance and Initial Voltage Source Model*

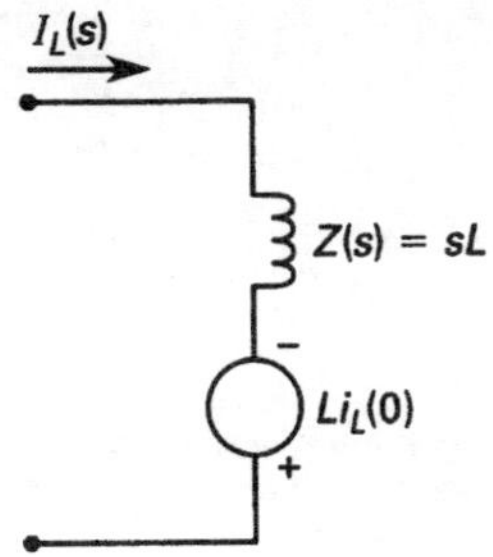

Figure 31.12 *Inductor Impedance and Initial Current Source Model*

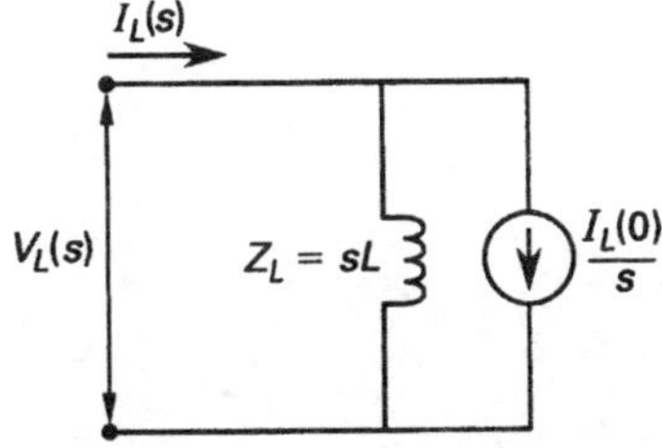

for inductive circuits, because the final inductive current must equal the Norton equivalent current, thereby acting as a check on the solution. Any of the network analysis techniques is applicable when working in the s-domain.

Example 31.13

In the circuit shown, the capacitor has an initial charge of 4.5×10^{-3} C. The switch closes at $t = 0$, applying a steady 120 V source. Determine the current using the Laplace transform method.

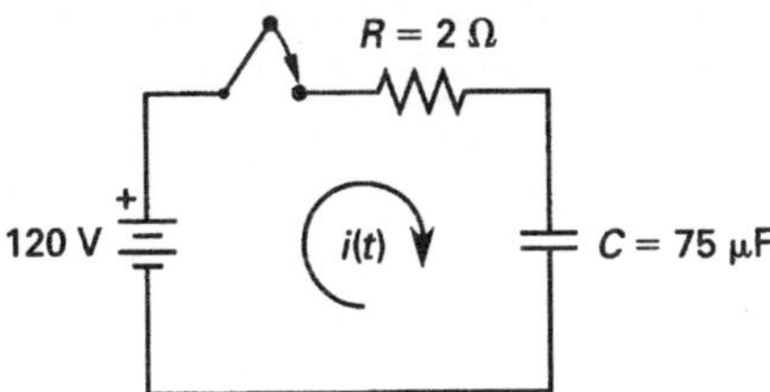

Solution

The classical approach will be used. Using KVL in the direction shown gives

$$V_s - i(t)R - \frac{1}{C}\int i(t)dt = 0$$

Though not stated explicitly, the behavior of the circuit from $t = 0$ onward is the desired result. So,

$$V_s = i(t)R + v_C(0^+) + \frac{1}{C}\int_0^\infty i(t)dt$$

The initial capacitor voltage can be determined from the initial charge, using $C = Q/V$. The constant-source voltage can be considered a unit step at $t = 0$. The initial capacitor voltage is a constant as well. Both transform to the s-domain using $1/s$. (See App. 11.A.) Integration equates to multiplication by $1/s$. Substituting and transforming gives

$$V_s = i(t)R + \frac{Q_0}{C} + \frac{1}{C}\int_0^\infty i(t)dt$$

$$\frac{120\text{ V}}{s} = I(s)(2\ \Omega) + \frac{4.5 \times 10^{-3}\text{ C}}{(75 \times 10^{-6}\text{ F})(s)} + \frac{I(s)}{(75 \times 10^{-6}\text{ F})(s)}$$

$$I(s) = \frac{30\text{ A}}{s + 6.65 \times 10^3}$$

Performing the inverse Laplace transform on $I(s)$ gives

$$i(t) = 30\text{ A}e^{-6.65\times 10^3 t}$$

16. PID CONTROLLERS

A *proportional-integral-derivative controller*, or *PID controller*, can compensate for a wide variety of electrical system responses. A PID controller combines the functions of a proportional controller, an integral controller, and a derivative controller. The proportional controller regulates the magnitude of the difference between the desired and actual levels of a monitored variable, the integral controller regulates the total change in the desired quantity over time, and the derivative controller regulates the rate of change in the monitored variable.

$$G_C(s) = K_p + \frac{K_i}{s} + K_d s \quad \text{[PID controller]} \qquad 31.85$$

In general terms, the proportional gain controls the response time. The integral gain determines the number of oscillations over time (*ringing control*). The derivative gain controls the magnitude of the overshoot. The combination of all three will determine the overall response to any given transient.

Circuit Theory

32 Frequency Response

Nomenclature

A	factor or constant	–
B	constant	–
C	capacitance	F
D	denominator magnitude	–
G	gain	dB
i	instantaneous current	A
I, **I**	effective or DC current	A
K	constant	–
L	self-inductance	H
N	numerator magnitude	–
p	pole	Hz
$r(t)$	response function, time domain	–
$R(s)$	response function, s-domain	–
s	complex frequency	Hz
$T(s)$	transfer or network function	–
v	instantaneous voltage	V
V, **V**	effective or DC voltage	V
X	constant	–
Y	constant	–
z	zero	Hz
Z	impedance	Ω

Symbols

α	angle	rad
β	angle	rad
θ	angle	rad
σ	neper frequency or napier frequency	Np/s or Hz
σ_0	break or corner frequency	Np/s or Hz
ω	angular frequency	rad/s

Subscripts

d	number or denominator
G	gain
n	number or numerator
net	network
p	pole
s	source
z	zero

1. FUNDAMENTALS

In any network with capacitors or inductors or both, transients will occur.[1] Transients can be represented by mathematical circuit models consisting of differential equations. The solutions to these equations depend on a circuit's initial conditions, and the methods outlined in Chap. 31 can be used to obtain these solutions.[2] The Laplace transform method can be used to change differential equation manipulation to algebraic manipulation, and the phasor method can be used to simplify sinusoidal mathematics.

Transient analysis requires the complete analysis of a circuit, including initial conditions, because both the transient and steady-state portions are relevant. The transient portion (natural response) of any output eventually settles on the steady-state (forced response) value determined by the forcing function, that is, the *signal*.[3] *Steady-state analysis* of a circuit, on the other hand, is independent of the circuit's initial condition. Solutions to problems in steady-state analysis can be determined from the circuit characteristics alone. These characteristics are contained within a *transfer function*, $T(s)$, as shown by Eq. 32.1.

$$T(s) = \mathcal{L}\left\{\frac{r(t)}{f(t)}\right\} \quad 32.1$$

The circuit's input signal or forcing function is $f(t)$, and the response to this function is $r(t)$. The transfer function, $T(s)$, is analyzed in the frequency (or s) domain rather than in the time (or t) domain, and this type of analysis is called *frequency response analysis*. Unlike phasor analysis, in which a fixed frequency is assumed, frequency response analysis involves the examination of a range of frequency values. Moreover, frequency response analysis can be used to analyze the effects of *variable frequency inputs* and *spectral inputs*, which have multiple frequencies.

When the signal is sinusoidal (that is, $s = j\omega$), *sinusoidal steady-state analysis* can be used. If the time response to a sinusoidal signal is needed, the signal magnitude, $|F(s)|$, is multiplied by the transfer function

[1]In purely resistive circuits, the response is considered instantaneous.
[2]The arbitrary constant occurring in integration of electrical parameters represents the initial condition.
[3]The natural response is determined from the homogenous solution (i.e., the forcing function is equal to zero) of the differential equation. The forced response is from the particular solution whose form must be that of the forcing function.

magnitude, $|T(s)|$, and their product is the magnitude of the output or response, $|R(s)|$. The angle is found by adding the signal angle, $\angle F(s)$, to the transfer function angle, $\angle T(s)$, which gives the angle of the response, $\angle R(s)$. The result is a phasor that can be changed into sinusoidal form using the known value of $j\omega$. The transfer function contains the circuit's characteristics, so it can be used to determine the output for a variety of inputs.

2. TRANSFER FUNCTION[4]

The term *transfer function* refers to the relationship of one electrical parameter in a network to a second electrical parameter elsewhere in the network.[5] The *network function*, $T_{\text{net}}(s)$, refines this definition to be the ratio of the complex amplitude of an exponential output, $Y(s)$, to the complex amplitude of an exponential input, $X(s)$. If $x(t)$ is Xe^{st} and $y(t)$ is Ye^{st}, and if the linear circuit is made up of lumped elements, the network function can be derived from an input-output differential equation as a rational function of s, and can be written in the following general form.[6]

$$\begin{aligned} T_{\text{net}}(s) &= \frac{Y(s)}{X(s)} \\ &= A\left(\frac{(s-z_1)(s-z_2)\cdots(s-z_n)}{(s-p_1)(s-p_2)\cdots(s-p_d)}\right) \end{aligned} \quad \text{32.2}$$

The complex constants represented as z_1, z_2, and so on are the *zeros* of $T_{\text{net}}(s)$, and are plotted in the s-domain as circles. Zeros are values within a transfer function that result in a *zero response*. A zero response is the minimum a circuit can provide, which is no output. The complex constants represented as p_1, p_2, and so on are the *poles* of $T_{\text{net}}(s)$ and are plotted in the s-domain as X's. Poles are the values within a transfer function that result in an *infinite response*, which is the maximum response a circuit can provide.

Equation 32.2 can be interpreted as the ratio of the response in one part of the s-domain network to the excitation in another part of the network. When s equals any zero, the response of the network will be zero regardless of the excitation. When s equals any pole, the response will be infinite regardless of the excitation. Equation 32.2 is independent of the forcing function. That is, the equation contains the characteristics of the network itself. The response to some source (input) is found by substituting the value of the complex frequency $s=\sigma_s+j\omega_s$ for the source into Eq. 32.2. The variable σ is the *neper frequency* and is assigned units of Np/s or Hz. (See Sec. 32.4.) The variable σ represents the ratio of two currents, voltages, or other analogous quantities. In this case, σ_s represents the ratio of two source frequencies. The properties of the circuit elements themselves are held in the complex zeros and poles, z or p.

[4]In many texts, functions of s are shown with an italic s to avoid confusion with phasors, even though s is a complex number and most complex numbers are shown in bold. When s appears in equations with other complex, bolded variables, s should be treated as a complex number and *not* as a scalar. The zero and pole variables, which are also complex numbers, are shown as italic for consistency.

[5]When the function is unitless—that is, when it is the ratio of voltages, the ratio of currents, and so on—it is called the *transfer ratio*.

[6]This equation is often shown with positive signs. In this case, the zeros are the $-z$ values. The minus signs are shown here to clarify how the s and z values would be combined in the s-domain (see Sec. 32.3).

Example 32.1

Determine the network function of the circuit shown in terms of a current output, $I(s)$, for a given voltage input, $V(s)$.

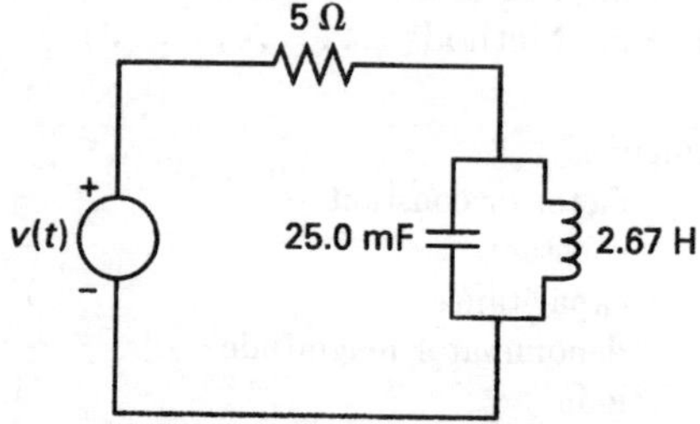

Solution

The network function is

$$T_{\text{net}}(s) = \frac{I(s)}{V(s)} = \frac{1}{Z(s)}$$

Because the calculation will be accomplished in the s-domain, transform the circuit using the properties given in Table. 31.4. The result is

$$Z_C = \frac{1}{sC} = \left(\frac{1}{s(25.0 \text{ mF})}\right)\left(1000 \ \frac{\text{mF}}{\text{F}}\right) = \frac{40 \text{ F}}{s}$$

$$Z_L = sL = s(2.67 \text{ H}) = 2.67s$$

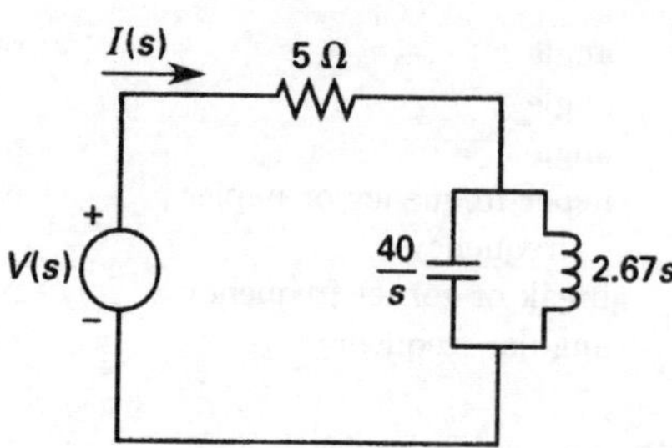

The impedance consists of a resistor in series with the parallel combination of a capacitor and an inductor.

$$Z(s) = \left(5 + \frac{\left(\frac{40}{s}\right)(2.67s)}{\frac{40}{s} + 2.67s}\right) \Omega$$

After manipulating mathematically, the result is

$$Z(s) = \left(\frac{(5)(s^2+8s+15)}{s^2+15}\right)\Omega = \left(\frac{(5)(s+3)(s+5)}{s^2+15}\right)\Omega$$

The network function is

$$T_{\text{net}}(s) = \frac{1}{Z(s)} = 0.2\left(\frac{s^2+15}{(s+3)(s+5)}\right)\Omega$$

Example 32.2

Plot the zeros and poles of the network function in Ex. 32.1.

Solution

The zeros are determined by the roots in the numerator as $s = \pm j\sqrt{15}$, complex conjugate pairs. The poles are determined by the roots in the denominator as $s = -3$ and $s = -5$. Plotting gives the following result.

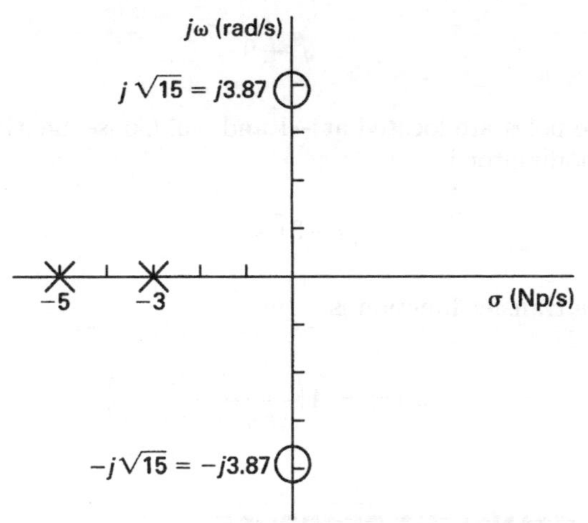

The plot is a complete representation of the network function.

3. STEADY-STATE RESPONSE[7]

The network function is a ratio of phasor quantities, such as $\mathbf{V}_2/\mathbf{V}_1$, $\mathbf{I}_2/\mathbf{I}_1$, $\mathbf{V}_2/\mathbf{I}_1$, and so on. The variables can also be impedances or admittances. As for any phasor quantity, the network function can be expressed as a magnitude at an angle, N or $D\angle\theta$. Each of the zero terms in the numerator and pole terms in the denominator in Eq. 32.2 can be represented as in Eq. 32.3 and Eq. 32.4, respectively.

$$(s - z_n) = N_n\angle\alpha_n \quad 32.3$$

$$(s - p_d) = D_d\angle\beta_d \quad 32.4$$

[7]Functions of s are treated as complex in this section since s has phasor properties.

Substituting Eq. 32.3 and Eq. 32.4 into the generic network equation, Eq. 32.2, provides a mathematical picture of the steady-state response phasors.

$$\begin{aligned} T_{\text{net}}(s) &= A\left(\frac{(N_1\angle\alpha_1)(N_2\angle\alpha_2)\cdots(N_n\angle\alpha_n)}{(D_1\angle\beta_1)(D_2\angle\beta_2)\cdots(D_d\angle\beta_d)}\right) \\ &= A\left(\frac{N_1N_2\cdots N_n}{D_1D_2\cdots D_d}\right)\angle\left(\begin{array}{c}\alpha_1+\alpha_2+\cdots \\ +\alpha_n\end{array}\right. \\ &\quad \left. -(\beta_1+\beta_2+\cdots+\beta_d)\right) \end{aligned} \quad 32.5$$

Equation 32.3 and Eq. 32.4 considered together indicate that the steady-state response of a network to a forcing function input of $s = \sigma + j\omega$ is determined by the lengths of the vectors from the zeros to the s and the vectors from the poles to the s. In addition, the angle is determined by the angles these vectors make with the σ axis.[8] The angles α and β are always between 0° and 90°. Equation 32.5 is a geometric means of determining the network function from the locations of the zeros and poles, without knowing the analytical expression. If the poles and zeros are known, the network function can be written to within a factor of A.

Example 32.3

For the circuit in Ex. 32.1 and the response given in Ex. 32.2, determine the current output for a voltage excitation input of $v = e^{st}$ V where $s = \sqrt{15}$ Np/s.

Solution

Place the input on the pole-zero plot. Draw the vectors from the zeros to the input s. Draw the vectors from the poles to the input s.

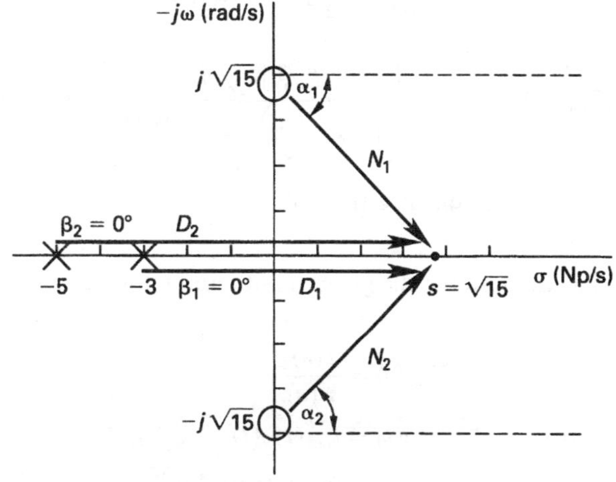

[8]The vectors represent what the circuit must do to respond to the forcing function. The steady-state output of a circuit always takes the form of the forcing or input function.

Circuit Theory

The magnitudes and angles are then calculated. For the zeros,

$$N_1 = N_2 = \sqrt{\left(\sqrt{15}\ \frac{\text{Np}}{\text{s}}\right)^2 + \left(\sqrt{15}\ \frac{\text{Np}}{\text{s}}\right)^2}$$
$$= \sqrt{30}\ \frac{\text{Np}}{\text{s}}$$

$$\alpha_1 = \arctan\left(\frac{-\sqrt{15}\ \frac{\text{Np}}{\text{s}}}{\sqrt{15}\ \frac{\text{Np}}{\text{s}}}\right) = -45°$$

$$\alpha_2 = \arctan\frac{\sqrt{15}\ \frac{\text{Np}}{\text{s}}}{\sqrt{15}\ \frac{\text{Np}}{\text{s}}} = 45°$$

$$\alpha_1 + \alpha_2 = 0°$$

For the poles,

$$D_1 = 3\ \frac{\text{Np}}{\text{s}} + \sqrt{15}\ \frac{\text{Np}}{\text{s}} = 6.87\ \frac{\text{Np}}{\text{s}}$$
$$D_2 = 5\ \frac{\text{Np}}{\text{s}} + \sqrt{15}\ \frac{\text{Np}}{\text{s}} = 8.87\ \frac{\text{Np}}{\text{s}}$$
$$\beta_1 = 0°$$
$$\beta_2 = 0°$$
$$\beta_1 + \beta_2 = 0°$$

The factor A was found in Ex. 32.1. The network function is calculated as

$$T_{\text{net}}(s) = A\left(\frac{N_1 N_2 \cdots N_n}{D_1 D_2 \cdots D_d}\right)\angle(\alpha_1 + \alpha_2 + \cdots + \alpha_n) - (\beta_1 + \beta_2 + \cdots + \beta_d)$$

$$T_{\text{net}}(\sqrt{15}) = (0.2)\left(\frac{(\sqrt{30})(\sqrt{30})}{\left(6.87\ \frac{\text{Np}}{\text{s}}\right)\left(8.87\ \frac{\text{Np}}{\text{s}}\right)}\right)\angle(0° + 0°)$$
$$= 98.46 \times 10^{-3}\angle 0°$$

The response for the given input can be found as

$$T_{\text{net}}(s) = \frac{I(s)}{V(s)}$$
$$i(t) = t_{\text{net}}(t)v(t)$$
$$= 98.46 \times 10^{-3} e^{\sqrt{15}t}$$

The result indicates that as time increases the current becomes infinite, clearly an impossibility. This is consistent with the condition in Sec. 31.10 that σ takes on zero or negative values only.

Example 32.4

Write the expression for the transfer function in the pole-zero plot shown.

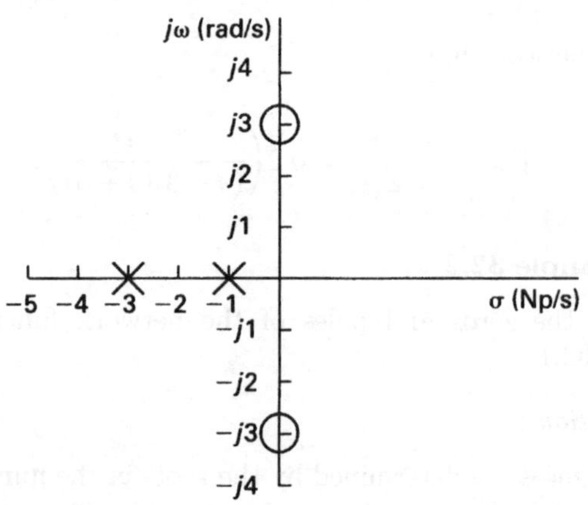

Solution

The zeros are located at $\pm j3$. Consequently, the numerator is

$$s^2 + 9$$

The poles are located at -1 and -3. Consequently, the denominator is

$$(s+3)(s+1)$$

The transfer function is

$$T_{\text{net}}(s) = A\left(\frac{s^2 + 9}{(s+3)(s+1)}\right)$$

4. TRANSIENT RESPONSE

The network function uses the complex frequency to determine the steady-state response. Nevertheless, the circuit characteristics, including the natural response, are embedded in the input equation in the denominator of the network function. Specifically, the transient response is characterized by the poles of the network function.

The real portion of the complex number representing a pole contains the neper frequency, σ, of the circuit for that pole. The *neper frequency*, also called the *napier frequency*, determines the exponential terms in a solution.[9] The exponential terms represent the transient response of a circuit (see Chap. 31). Near a pole, then, the natural response of the circuit elements impacts the overall output of a network, while far away from a pole little effect occurs.

[9]The term "napier" is most commonly known for its association with natural logarithms. It is used here associated with the natural response.

Circuit Theory

Example 32.5

What is the mathematical form of the transient response for the network function in Ex. 32.4?

Solution

The network function in Ex. 32.4 was

$$T_{\text{net}}(s) = A\left(\frac{s^2+9}{(s+3)(s+1)}\right)$$

Because each term is $(s-p)$, the poles occur at $(s-(-p))$. That is, the poles are at -3 and -1. The expanded denominator is

$$s^2+4s+3$$

The coeffcients $a=1$, $b=4$, and $c=3$ indicate that $b^2>4ac$. The transient response is of the form

$$\text{transient response} = Ae^{s_1 t}+Be^{s_2 t}$$

The transient response could be either the current or voltage response, depending upon the ratio represented by the network function. The initial conditions, not included in the network function, would be required to determine the constants A and B.

5. MAGNITUDE AND PHASE PLOTS

A plot of the magnitude and phase of a transfer function in the s-domain provides information regarding the response of that function as frequency varies.[10] Magnitude and phase plots are fundamental to constructing Bode plots.

Consider a transfer function with a single zero at the origin. As only the steady-state response to a sinusoid is of concern, let $s=j\omega$. The transfer function is

$$T(s) = s = s+0 \qquad 32.6$$

$$T(j\omega) = j\omega \qquad 32.7$$

The frequency is allowed to vary as illustrated in Fig. 32.1(a). The magnitude is plotted in Fig. 32.1(b) from $|T(j\omega)|=\omega$. The phase angle variation with frequency is shown in Fig. 32.1(c) from $\alpha=\arctan\omega/0$ or directly from j, which is $1\angle 90°$.

[10]Because the methods apply to all functions in the s-domain, not merely network functions, the term "transfer function" and the symbol $T(s)$ will be used instead of $T_{\text{net}}(s)$.

Figure 32.1 *Single Zero at the Origin*

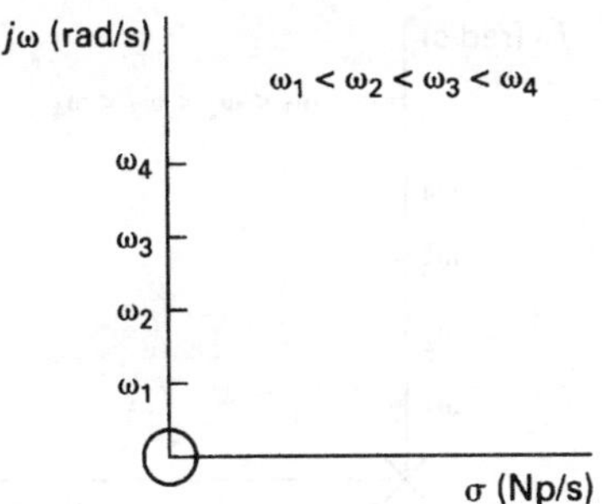

(a) zero at origin, frequency increasing

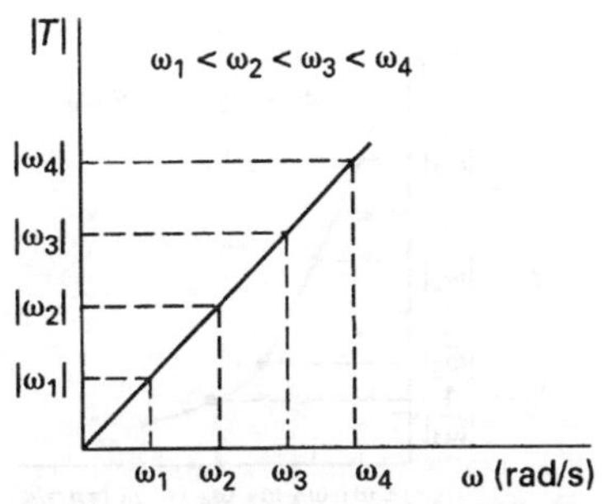

(b) transfer function magnitude, frequency increasing

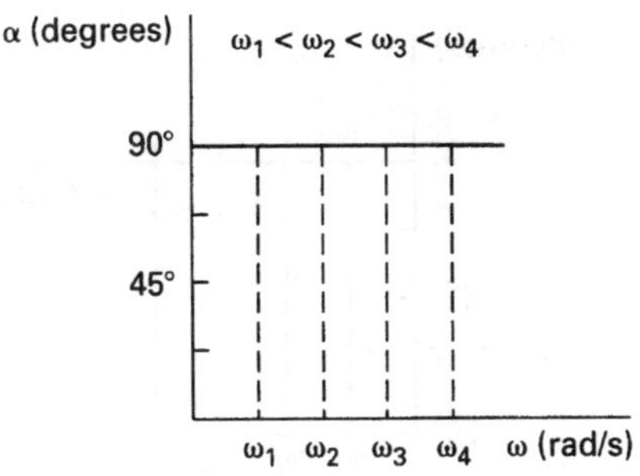

(c) transfer function angle, frequency increasing

Consider a transfer function with a single pole at the origin. As only the steady-state response to a sinusoid is of concern, let $s=j\omega$. The transfer function is

$$T(s) = \frac{1}{s} = \frac{1}{s+0} \qquad 32.8$$

$$T(j\omega) = \frac{1}{j\omega} \qquad 32.9$$

The frequency is allowed to vary as illustrated in Fig. 32.2(a). The magnitude is plotted in Fig. 32.2(b) from $|T(j\omega)|=1/|\omega|$. The phase angle variation with frequency is shown in Fig. 32.2(c) from $\beta=-\arctan\omega/0$ or directly from $1/j$, which is $-j$ or $1\angle{-90°}$.

Figure 32.2 Single Pole at Origin

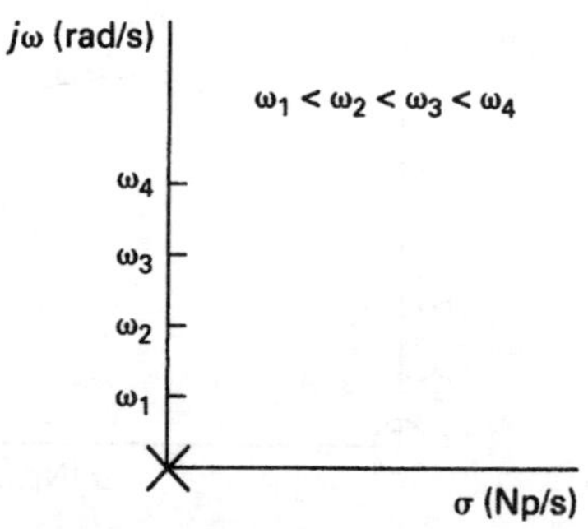

(a) pole at origin, frequency increasing

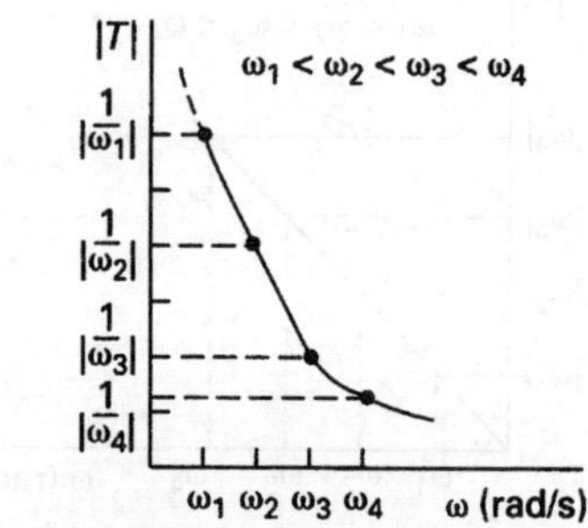

(b) transfer function magnitude, frequency increasing

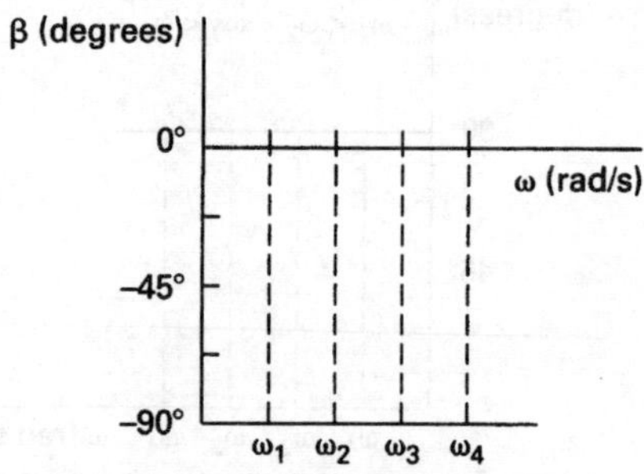

(c) transfer function angle, frequency increasing

Consider a transfer function with a single zero but located on the negative real axis at $-\sigma_n$.[11] Again, only the steady-state response to a sinusoid is of concern, so $s = j\omega$. The transfer function is

$$T(s) = s + \sigma_n \qquad 32.10$$

$$T(j\omega) = j\omega + \sigma_n \qquad 32.11$$

[11]All practical devices will have zeros to the left of the imaginary axis. The restriction of being on the negative real axis will be removed for Bode plots but is used here to clarify the concept of a *break* or *corner frequency* that occurs at zeros and poles, that is, where the real portion of the zero or pole is σ_n or σ_p. The neper frequency, σ, represents the exponential solution or the natural response of the circuit elements that impacts the steady-state response in the area of a zero or a pole.

The frequency is allowed to vary, as illustrated in Fig. 32.3(a).[12] The magnitude is plotted in Fig. 32.3(b) from $|T(j\omega)| = \sqrt{\omega^2 + \sigma_n^2}$. The phase angle variation with frequency is shown in Fig. 32.3(c) from $\alpha = \arctan \omega/\sigma_n$.

Figure 32.3 Single Zero on Negative Real Axis ($-\sigma_n$)

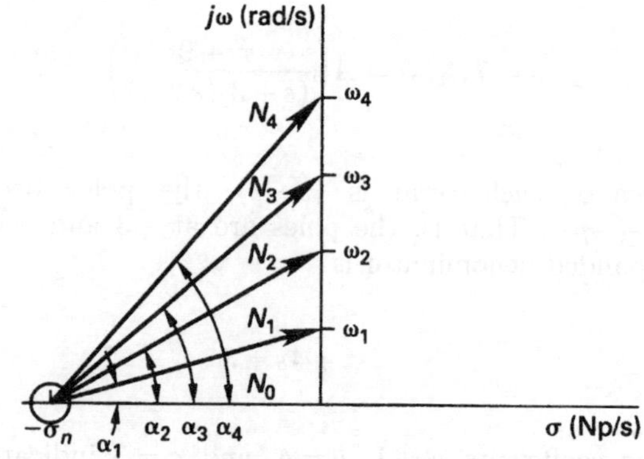

(a) zero on negative real axis, frequency increasing

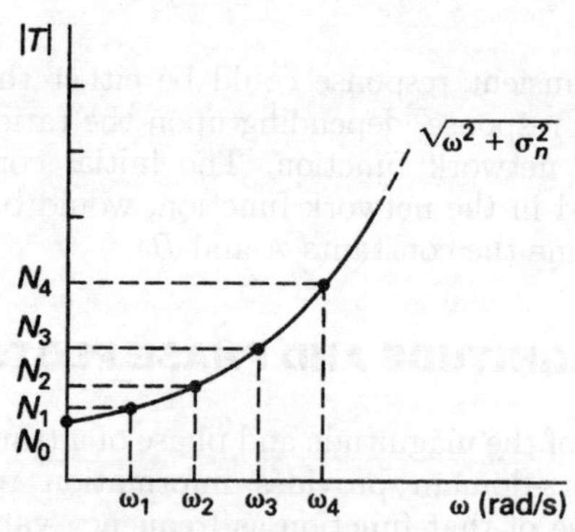

(b) transfer function magnitude, frequency increasing

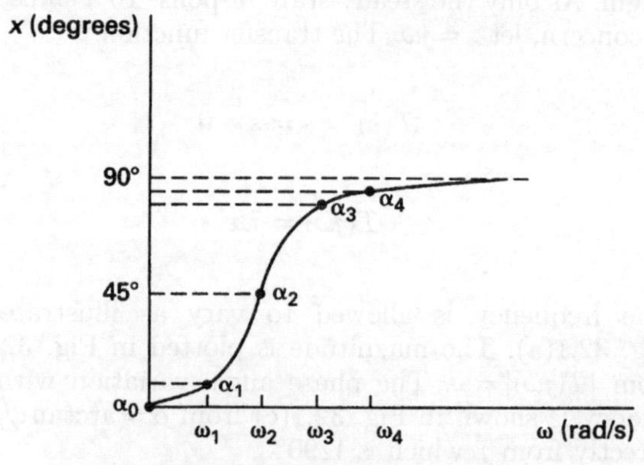

(c) transfer function angle, frequency increasing

[12]The figure is not to scale. The relationships between drawings are approximate, but individual drawings show their true shapes.

Consider a transfer function with a single pole but located on the negative real axis at $-\sigma_n$. Again, only the steady-state response to a sinusoid is of concern, so $s = j\omega$. The transfer function is

$$T(s) = \frac{1}{s + \sigma_d} \qquad 32.12$$

$$T(j\omega) = \frac{1}{j\omega + \sigma_d} \qquad 32.13$$

The frequency is allowed to vary, as illustrated in Fig. 32.4(a).[13] The magnitude is plotted in Fig. 32.4(b) from $|T(jw)| = 1/\sqrt{\omega^2 + \sigma_d^2}$. The phase angle variation with frequency is shown in Fig. 32.4(c) from $\beta = -\arctan \omega/\sigma_d$.

When $\sigma = \omega$, the magnitude of the transfer function for a zero is $\sqrt{\omega^2 + \sigma_n^2} = \sqrt{2}\sigma_n$ and the phase angle is $\alpha = +45°$. When $\sigma = \omega$, the magnitude of the transfer function for a pole is $1/(\sqrt{2}\sigma_d)$ and the angle is $\beta = -45°$. When the transfer function contains numerous zeros and poles, the magnitude is obtained by multiplying the magnitudes of the individual transfer functions, and the phase is obtained by summing the zero angles and subtracting the sum of the pole angles. Equation 32.5 is such a combination, with each $N\angle\alpha$ and $D\angle\beta$ representing the magnitude of one transfer function. The multiplication and division can be replaced by addition and subtraction using logarithmic quantities, guided by Eq. 32.14 and Eq. 32.15.

$$\log xy = \log x + \log y \qquad 32.14$$

$$\log \frac{x}{y} = \log x - \log y \qquad 32.15$$

The magnitude of the transfer function, Eq. 32.5, can be written as

$$\begin{aligned} \log|T(s)| = {} & \log A + \log N_1 + \log N_2 \\ & + \cdots + \log N_n - \log D_1 \\ & - \log D_2 - \cdots - \log D_d \end{aligned} \qquad 32.16$$

6. BODE PLOT PRINCIPLES: MAGNITUDE PLOT

A *Bode plot* or *Bode diagram* is a plot of the gain or phase of an electrical device or network against the frequency. The units of the plot are normally magnitude in decibels versus logarithm of the frequency. The logarithm is useful for algebraic manipulation (see Eq. 32.14 and Eq. 32.15) and the wide frequency range of response. The *decibel*, in units dB, is defined as 10 times the logarithmic ratio of two powers. Nevertheless, it is used for transfer functions in general. Because the I^2 and V^2 are proportional to power, the magnitudes of transfer functions containing the voltage and current are[14]

$$G = 10 \log \left|\frac{I_2}{I_1}\right|^2 = 20 \log \left|\frac{I_2}{I_1}\right| \qquad 32.17$$

$$G = 10 \log \left|\frac{V_2}{V_1}\right|^2 = 20 \log \left|\frac{V_2}{V_1}\right| \qquad 32.18$$

Figure 32.4 *Single Pole on Negative Real Axis ($-\sigma_p$) Plot*

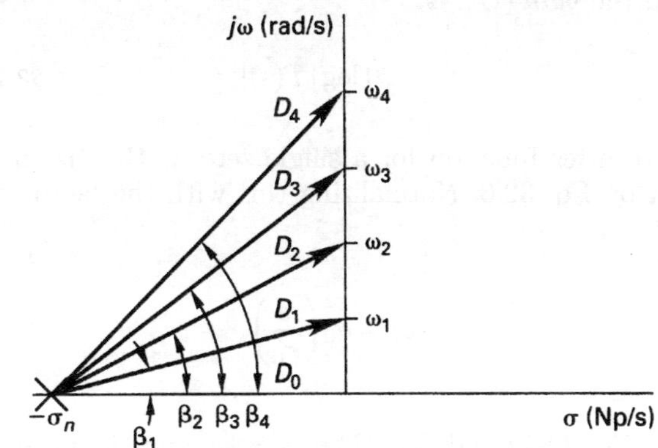

(a) pole on negative real axis, frequency increasing

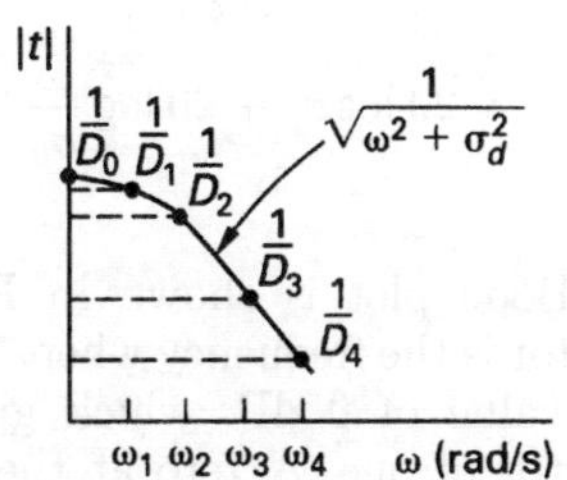

(b) transfer function magnitude, frequency increasing

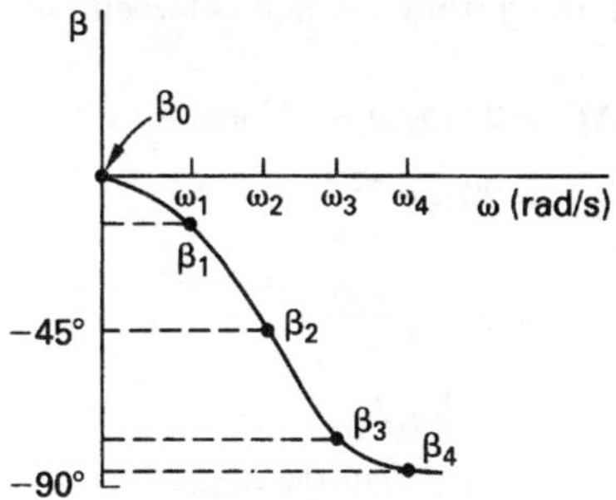

(c) transfer function angle, frequency increasing

[13] The figure is not to scale. The relationships between drawings are approximate, but individual drawings show their true shapes.

[14] To be used in the ratio, the currents or voltages should act on the same impedance. Standard practice ignores this technicality.

Circuit Theory

In general terms, any transfer function's magnitude, called the *gain* (G), is

$$G = 20 \log |T(s)| \qquad 32.19$$

The transfer function for a single zero at the origin is given by Eq. 32.6. Normalizing this with the factor σ_0 gives

$$T(s) = \sigma_0 \left(\frac{s}{\sigma_0} \right) \qquad 32.20$$

Using Eq. 32.19 and Eq. 32.20, $s = j\omega$, and the properties of logarithms from Eq. 32.14 and Eq. 32.15 gives

$$\begin{aligned} G &= 20 \log \left| \sigma_0 \left(\frac{j\omega}{\sigma_0} \right) \right| \\ &= 20 \log \sigma_0 + 20 \log \frac{\omega}{\sigma_0} \end{aligned} \qquad 32.21$$

The resulting Bode plot is shown in Fig. 32.5. The normalizing factor is the frequency where the gain (magnitude) has a value of 0 dB, which gives an actual normalizing factor of one. A zero at the origin results in a line of positive slope passing through $\omega = 1$ or $\log \omega = 0$.[15] The slope is 6 dB/octave or 20 dB/decade. An *octave* is an interval between any two frequencies with a ratio of two to one; for example, $\omega_2 = 2\omega_1$. On the log scale the magnitude change between octaves is

$$\begin{aligned} \Delta G &= 20 \log \omega_2 - 20 \log \omega_1 \\ &= 20 \log \frac{\omega_2}{\omega_1} \\ &= 20 \log \frac{2\omega_1}{\omega_1} \\ &= 20 \log 2 \\ &= 6.02 \approx 6 \text{ dB/octave} \end{aligned} \qquad 32.22$$

Figure 32.5 Bode Plot: Single Zero at Origin

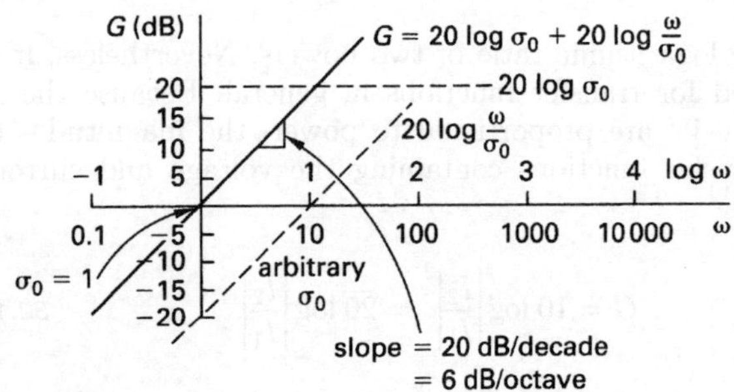

A decade of frequency change alters the magnitude by

$$\begin{aligned} \Delta G &= 20 \log \omega_2 - 20 \log \omega_1 \\ &= 20 \log \frac{\omega_2}{\omega_1} \\ &= 20 \log \frac{10\omega_1}{\omega_1} \\ &= 20 \log 10 \\ &= 20 \text{ dB/decade} \end{aligned} \qquad 32.23$$

The two slopes are equivalent. The constant $20 \log \sigma_0$ does not change the shape of the plot. The constant can be accounted for in one of two ways. First, it can shift the plot upward, as was done in this case. Second, the value of 0 dB could be changed to a new value that is equal to the constant, after plotting the other factors of the gain. The angle for a zero at the origin remains a constant 90°.

The transfer function for a single pole at the origin is given by Eq. 32.8. Normalizing this with the factor σ_0 gives

$$T(s) = \frac{1}{\sigma_0 \left(\frac{s}{\sigma_0} \right)} \qquad 32.24$$

With $s = j\omega$, the gain is

$$\begin{aligned} G &= 20 \log \left| \frac{1}{\sigma_0 \left(\frac{j\omega}{\sigma_0} \right)} \right| \\ &= -20 \log \sigma_0 - 20 \log \frac{\omega}{\sigma_0} \end{aligned} \qquad 32.25$$

The resulting Bode plot is shown in Fig. 32.6. The normalizing factor is the frequency where the gain (magnitude) has a value of 0 dB, which gives an actual normalizing factor of one. A pole at the origin results in a line of negative slope passing through $\omega = 1$ or $\log \omega = 0$.[16] The slope is −6 dB per octave or −20 dB per decade. The constant $-20 \log \sigma_0$ does not change theshape of the plot. In this case, the constant is used to shift the plot downward. Another option is to change the gain coordinates so that the zero value correlates with the value of the constant, after plotting the other terms of the gain. The angle for a pole at the origin remains a constant −90°.

[15]Considering Eq. 32.21 and Fig. 32.5, regardless of where the example normalizing factor, σ_0, is located (in this case, it was at $\sigma_0 = 10$), the final result remains unchanged: a +20 dB/decade line through $\omega = 1$ or $\log \omega = 0$ with an actual σ_0 of 1.

[16]Considering Eq. 32.25 and Fig. 32.6, regardless of where the example normalizing factor, σ_0, is located (in this case, it was at $\sigma_0 = 10$), the final result remains unchanged: a −20 dB per decade line through $\omega = 1$ or $\log \omega = 0$ with an actual σ_0 of 1.

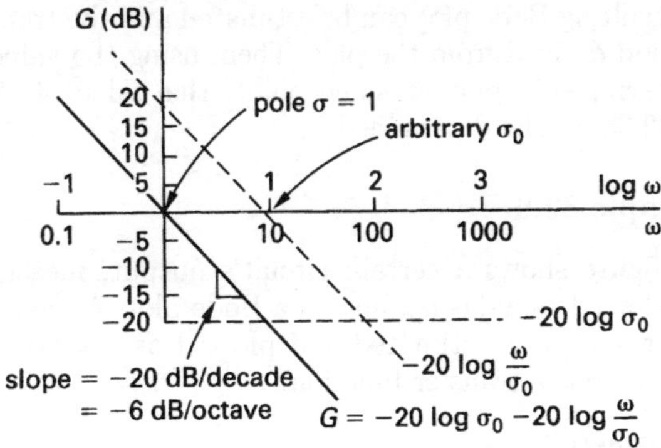

Figure 32.6 *Bode Plot: Single Pole at Origin*

The transfer function for a single zero on the negative real axis at $-\sigma_0$ in the s plane is given by Eq. 32.10. Normalizing this equation gives

$$T(s) = \sigma_0\left(1+\frac{s}{\sigma_0}\right) \qquad \textit{32.26}$$

With $s = j\omega$, the gain is

$$\begin{aligned} G &= 20\log\left|\sigma_0\left(1+\frac{j\omega}{\sigma_0}\right)\right| \\ &= 20\log\sigma_0 + 20\log\left|\left(1+\frac{j\omega}{\sigma_0}\right)\right| \end{aligned} \qquad \textit{32.27}$$

The resulting Bode plot is shown in Fig. 32.7. The constant $20\log\sigma_0$ does not change the shape of the plot but rather shifts it upward or changes the value of the 0 dB to that of the constant (not shown). The normalizing factor is the frequency where the gain (magnitude) has a value of 0 dB. To determine the gain on either side of the normalizing factor, which will now be called the *break frequency* or *corner frequency*, consider the behavior of the asymptotes.[17] At low frequencies, ignoring the constant term of Eq. 32.27 because it only changes the plot's position, the gain is

$$G \approx 20\log\left|\left(1+\frac{0}{\sigma_0}\right)\right| = 0\ \text{dB} \quad [\omega \ll \sigma_0] \qquad \textit{32.28}$$

The low-frequency asymptote is approximated by drawing a line from the break frequency horizontally to the left. At high frequencies, ignoring the constant term of Eq. 32.27, the gain is

$$\begin{aligned} G &\approx 20\log\left|\left(\frac{j\omega}{\sigma_0}\right)\right| \\ &= 20\log\frac{\omega}{\sigma_0} \quad [\omega \gg \sigma_0] \end{aligned} \qquad \textit{32.29}$$

The high-frequency asymptote is approximated by a straight line with a slope of +20 dB/decade. That is,

[17]The break frequency or corner frequency corresponds to the zero or pole location.

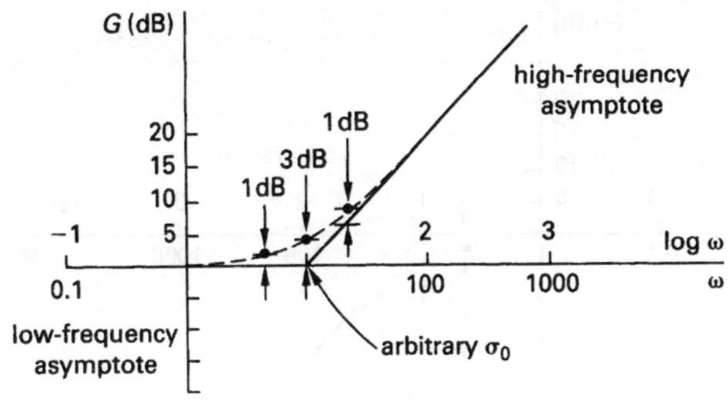

Figure 32.7 *Bode Plot: Single Zero on Negative Real Axis*

the curve "breaks upward" at the break frequency. At the break frequency $\omega = \sigma_0$, the magnitude of the gain is

$$\begin{aligned} G_0 &= 20\log\left|\left(1+\frac{j\omega}{\sigma_0}\right)\right| \\ &= 20\log\left(\sqrt{(1)^2+(1)^2}\right) \\ &= 20\log\sqrt{2} = 3\ \text{dB} \end{aligned} \qquad \textit{32.30}$$

At $\frac{1}{2}\sigma_0$ and $2\sigma_0$, the gain differs by 1 dB from the approximating asymptotes. The exact value of the frequency for a given gain, G_ω, can be determined by rearranging Eq. 32.27 (ignoring the constant factor). The result is

$$\omega = \sigma_0\sqrt{(10)^{G_\omega/10} - 1} \qquad \textit{32.31}$$

While Eq. 32.31 could be used to determine exact values, the asymptotes provide reasonable accuracy and adequate information for most electrical engineering applications. Use of the asymptotic approximations results in an *idealized Bode plot*. Computer programs are available to accurately plot the response of a given circuit should such accuracy become necessary. The angle for a zero on the negative real axis of the s-domain is determined from Eq. 32.26 to be $\arctan\omega/\sigma_0$. At the break frequency, this angle is 45°.

The transfer function for a single pole on the negative real axis at $-\sigma_0$ in the s plane is given by Eq. 32.12. Normalizing the equation and letting $s = j\omega$ results in a gain of

$$\begin{aligned} G &= 20\log\left|\left(\frac{1}{\sigma_0\left(1+\frac{j\omega}{\sigma_0}\right)}\right)\right| \\ &= -20\log\sigma_0 - 20\log\left|\left(1+\frac{j\omega}{\sigma_0}\right)\right| \end{aligned} \qquad \textit{32.32}$$

The plot for the pole is identical to that for a zero, with two exceptions: it has a negative slope and the plot breaks downward at the break frequency as shown in Fig. 32.8.

Figure 32.8 Bode Plot: Single Pole on Negative Real Axis

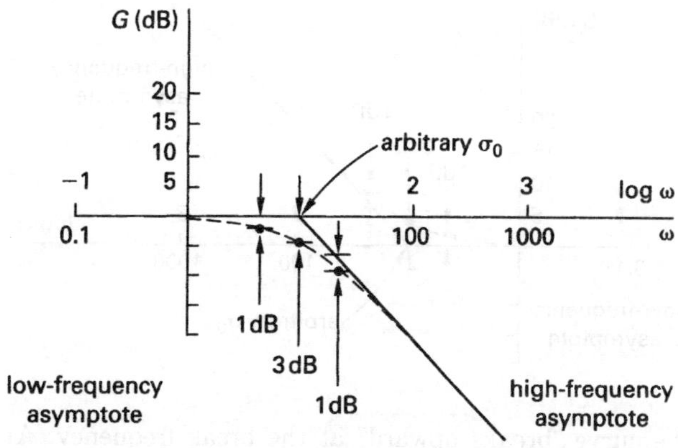

The transfer function for a circuit with numerous zeros and poles can be generalized as

$$T(s) = A\left(\frac{(s+\sigma_{z1})(s+\sigma_{z2}) \cdots \times (s+\sigma_{zn})s^y}{(s+\sigma_{p1})(s+\sigma_{p2}) \cdots \times (s+\sigma_{pd})s^x}\right) \qquad 32.33$$

The term s^y represents zeros at the origin, with y equal to the number of zeros. Zeros at the origin of the s-domain have a normalizing factor of one. The term s^x represents poles at the origin, with x equal to the number of poles. Poles at the origin of the s-domain also have a normalizing factor of one. Changing the form to that useful for Bode plotting,

$$\begin{aligned} G &= 20\log|T(s)| \\ &= 20\log|T(j\omega)| \\ &= 20\log\left(A\left(\frac{\sigma_{z1}\sigma_{z2}\cdots\sigma_{zn}}{\sigma_{p1}\sigma_{p2}\cdots\sigma_{pd}}\right)\right) \\ &\quad + 20\log\left|\left(1+\frac{j\omega}{\sigma_{z1}}\right)\right| + 20\log\left|\left(1+\frac{j\omega}{\sigma_{z2}}\right)\right| + \cdots \\ &\quad + 20\log\left|\left(1+\frac{j\omega}{\sigma_{zn}}\right)\right| + y20\log|j\omega| \\ &\quad - 20\log\left|\left(1+\frac{j\omega}{\sigma_{p1}}\right)\right| - 20\log\left|\left(1+\frac{j\omega}{\sigma_{p2}}\right)\right| - \cdots \\ &\quad - 20\log\left|\left(1+\frac{j\omega}{\sigma_{pd}}\right)\right| - x20\log|j\omega| \end{aligned} \qquad 32.34$$

The normalization used in Eq. 32.34 results in each of the terms being plotted from the break frequency on the $G=0$ dB line. The break frequency (normalizing factor) for the single zeros and poles at the origin is $\omega=1$. (The zeros and poles plot as straight lines and do not actually "break" up or down at the origin. The zero or pole at zero frequency does not get plotted on a Bode plot, because the value is $-\infty$ and $+\infty$.)

The entire plotting process can be reversed. For example, if the output of a circuit is determined experimentally, the resulting Bode plot can be estimated and the transfer function derived from the plot. Then, using the value of the asymptotic plot at some point, the value of A in Eq. 32.33 can be determined.

Example 32.6

The figure shows a certain circuit's output, measured and plotted as a dashed line on a Bode plot. An asymptotic response is estimated and plotted as a solid line. Determine the transfer function.

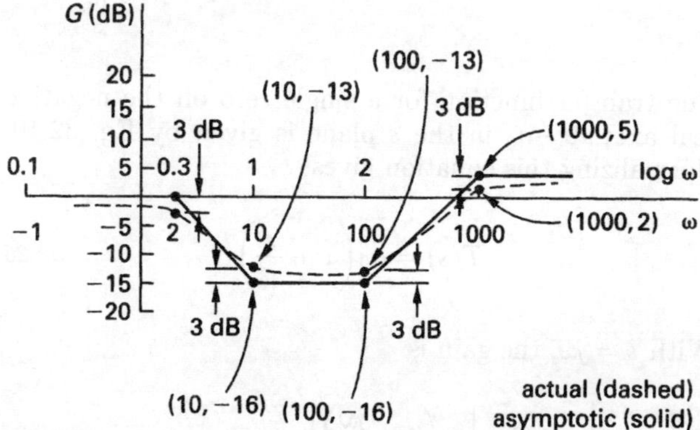

Solution

The transfer function must be in the form of Eq. 32.33. From the plot, the zeros and poles are the points where $G=0$ dB on the frequency axis where the breaks occur. Consequently, the breaks upward at $\omega=10$ and $\omega=100$ indicate zeros. The breaks downward at $\omega=2$ and $\omega=1000$ indicate poles. The transfer function is

$$T(s) = A\left(\frac{(s+10)(s+100)}{(s+2)(s+1000)}\right)$$

Determine A from the value of the asymptotic plot when $s=0$. At $s=0$, the plot indicates a magnitude of 0 dB.

$$G = 20\log|T(0)| = 0 \text{ dB}$$
$$|T(0)| = \text{antilog } 0 = 1$$

Substituting $T(0)=1$ and $s=0$ gives

$$T(s) = A\left(\frac{(s+10)(s+100)}{(s+2)(s+1000)}\right)$$
$$1 = A\left(\frac{(10)(100)}{(2)(1000)}\right)$$
$$A = 2$$

The final solution is

$$T(s) = (2)\left(\frac{(s+10)(s+100)}{(s+2)(s+1000)}\right)$$

7. BODE PLOT PRINCIPLES: PHASE PLOT

The phase of a zero at the origin, such as $s+0$ in the numerator of Eq. 32.33, provides an angle of $+90°$. The phase of a pole at the origin, such as $s+0$ in the denominator of Eq. 32.33, provides an angle of $-90°$. For zeros or poles on the negative real axis of the s-domain, with $s=j\omega$, the angles for each term in Eq. 32.33 are determined from Eq. 32.35 and Eq. 32.36, respectively.

$$\theta_z = \arctan \frac{\omega}{\sigma_{zn}} \quad \text{[radians]} \tag{32.35}$$

$$\theta_p = -\arctan \frac{\omega}{\sigma_{pd}} \quad \text{[radians]} \tag{32.36}$$

At the break frequency, $\omega=\sigma$ for either zeros or poles, and the phase angle is $+45°$ for zeros and $-45°$ for poles. Equation 32.35 and Eq. 32.36 could be used directly to plot the phase angle as a function of frequency. An *idealized Bode phase plot* that is similar to that used for magnitude can be plotted with approximations. Consider Eq. 32.35 and Eq. 32.36 at a frequency $\omega=0.1\sigma$. The absolute value of the zero or pole angle at this frequency is

$$\begin{aligned} |\theta| &= \left| \pm\arctan \frac{0.1\sigma}{\sigma} \right| \\ &= 0.09 \text{ rad} \\ &= 5 \approx 0 \end{aligned} \tag{32.37}$$

This is analogous to the low-frequency asymptote for the magnitude of Eq. 32.28. Consider a frequency one decade higher than the break frequency, $\omega=10\sigma$, for either a zero or a pole. The absolute value of the zero or pole angle at this frequency is

$$\begin{aligned} |\theta| &= \left| \pm\arctan \frac{10\sigma}{\sigma} \right| \\ &= 1.47 \text{ rad} \\ &= 84.3 \approx 90 \end{aligned} \tag{32.38}$$

This is analogous to the high-frequency asymptote for the magnitude of Eq. 32.29. At the break frequency, $\omega=\sigma$ and the angle is 45° for either a zero or a pole. The result is that, from one decade below the break frequency to one decade above, the phase angle for a zero or pole changes from an absolute value of 0° to 90° at a rate of 45° per decade. This approximation is illustrated in Fig. 32.9 for an arbitrary break frequency, that is, arbitrary zero or pole. The approximations used in the idealized phase plot are accurate to within 6°. The phase angles for the entire transfer function must be added as in Eq. 32.5 to obtain the resultant angle. This is done on the Bode plot in a manner analogous to adding the magnitudes.

Figure 32.9 *Bode Phase Plot for a Zero or Pole*

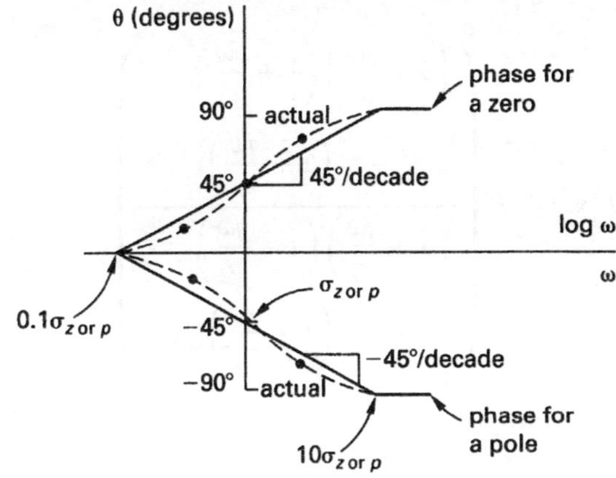

8. BODE PLOT METHODS

The general form of the transfer function of a circuit is[18]

$$T(s) = A\left(\frac{(s+z_1)(s+z_2) \cdots \times (s+z_n)(s)^y}{(s+p_1)(s+p_2) \cdots \times (s+p_d)(s)^x}\right) \tag{32.39}$$

The break frequencies in Eq. 32.39 are the zeros and the poles.[19] The break frequency, better termed the *normalizing factor*, for the zeros at the origin (s^y) and poles at the origin (s^x) is $\omega=1$. As the sinusoidal response is of interest, $s=j\omega$. Using the normalized notation changes Eq. 32.39 to Eq. 32.40.

$$T(j\omega) = A\left(\frac{z_1 z_2 \cdots z_n}{p_1 p_2 \cdots p_d}\right) \times \left(\frac{\left(1+\frac{j\omega}{z_1}\right)\left(1+\frac{j\omega}{z_2}\right) \cdots \times \left(1+\frac{j\omega}{z_n}\right)(j\omega)^y}{\left(1+\frac{j\omega}{p_1}\right)\left(1+\frac{j\omega}{p_2}\right) \cdots \times \left(1+\frac{j\omega}{p_d}\right)(j\omega)^x}\right) \tag{32.40}$$

[18]Transfer functions are often called the *gain* of a circuit. This chapter uses gain to indicate $20\log|T(s)|$.

[19]Using the zero and pole notation removes the restriction of the break frequency occurring only on the negative real axis of the s-domain. The zeros and poles are located at $-z$ and $-p$, respectively. This is the reason the positive sign is used.

Combining the constant term A with the values of the zeros and poles gives a new constant, K.[20]

$$T(j\omega) = K\left(\frac{\left(1+\frac{j\omega}{z_1}\right)\left(1+\frac{j\omega}{z_2}\right)\cdots\times\left(1+\frac{j\omega}{z_n}\right)(j\omega)^y}{\left(1+\frac{j\omega}{p_1}\right)\left(1+\frac{j\omega}{p_2}\right)\cdots\times\left(1+\frac{j\omega}{p_d}\right)(j\omega)^x}\right) \quad 32.41$$

The constant K is

$$K = A\frac{\prod_1^n z_n}{\prod_1^d p_d} \quad 32.42$$

The gain as used in Bode plots is derived from Eq. 32.41 to be

$$G = 20\log K + \sum_{n=1}^{n} 20\log\left|\left(1+\frac{j\omega}{z_n}\right)\right| + 20y\log\omega - \sum_{d=1}^{d} 20\log\left|1+\frac{j\omega}{p_d}\right| - 20x\log\omega \quad 32.43$$

The angle of the gain, θ_G, in degrees is

$$\theta_G = \sum_{n=1}^{n}\arctan\frac{\omega}{z_n} + 90y - \sum_{d=1}^{d}\arctan\frac{\omega}{p_d} - 90x \quad 32.44$$

The overall method for determining the frequency response of a circuit utilizing Bode plots follows.

step 1: Determine the transfer function for the circuit of interest and arrange it in the form of Eq. 32.39 or Eq. 32.41. The break frequencies for each term are the zeros and the poles. Recall that the break frequency for a zero or pole at the origin of the s-domain is $\omega = 1$.

step 2: Plot the break frequency points on the axis of a log frequency scale. That is, plot the points where the magnitude, in decibels, of the transfer function is zero.

step 3: For each zero at the s-domain origin (s^y terms), plot a straight line with a slope of +20 dB per decade through the origin of the Bode plot axis, that is, through $\omega = 1$ ($\log\omega = 0$) and $20\log|T(j\omega)| = 0$.

step 4: For each pole at the s-domain origin (s^x terms), plot a straight line with a slope of −20 dB per decade through the origin of the Bode plot axis, that is, through $\omega = 1$ ($\log\omega = 0$) and $20\log|T(j\omega)| = 0$.

step 5: For all other zeros, draw two asymptotic straight lines from the associated break frequency. The first is of zero slope and extends to lower frequencies. The other has a slope of +20 dB per decade (or 6 dB per octave) and extends to higher frequencies.

step 6: For all other poles, draw two asymptotic straight lines from the associated break frequency. The first is of zero slope and extends to lower frequencies. The other has a slope of −20 dB per decade (or −6 dB per octave) and extends to higher frequencies.

step 7: Sum the asymptotic lines drawn. The result is the idealized Bode plot for the transfer function.

step 8: If greater accuracy is desired, at each break frequency plot a point displaced 3 dB from the intersection of the asymptotic lines. The point is +3 dB from lines that break upward and −3 dB from lines that break downward. At differences of plus or minus one octave from the break frequency, plot a point +1 dB from lines that break upward and −1 dB from lines that break downward. Additional points may be plotted using Eq. 32.31. Join the points in a smooth curve with the ends asymptotic to the idealized Bode plot lines. This smooth curve is the actual plot.

step 9: Determine the value of K from the value of the asymptotic or actual curve at some frequency or by using Eq. 32.42.[21] Using this value of $20\log K$, either shift the plot upward or downward by the amount calculated, or change the 0 dB point on the axis to the value of $20\log K$. In either case, the shape of the Bode plot remains unchanged.

The overall method for determining the phase response of a circuit utilizing Bode plots follows.

step 1: Determine the transfer function for the circuit of interest and arrange it in the form of Eq. 32.39 or Eq. 32.41. The break frequencies for each term are the zeros and the poles. Recall that the break frequency for a zero or pole at the origin of the s-domain is $\omega = 1$.

step 2: Plot the break frequency points on the axis of a log frequency scale. That is, plot the points where the angle $\theta = 0°$.

step 3: For each zero at the s-domain origin (s^y terms), plot a constant angle of +90°.

step 4: For each pole at the s-domain origin (s^x terms), plot a constant angle of −90°.

[20]The constant K is a function of $j\omega$ because of the zeros and poles included.

[21]A pole or zero at the origin in the s-domain has a pole or zero value in Eq. 32.42 of one.

step 5: For all other zeros, plot a point at the break frequency corresponding to +45°. At one decade below the break frequency, plot a point corresponding to 0°. At one decade above the break frequency, plot a point corresponding to +90°. Connect the points. The slope will be +45° per decade.

step 6: For all other poles, plot a point at the break frequency corresponding to −45°. At one decade below the break frequency, plot a point corresponding to 0°. At one decade above the break frequency, plot a point corresponding to −90°. Connect the points. The slope will be −45° per decade.

step 7: Sum the asymptotic lines drawn. The result is an idealized Bode phase plot.

step 8: If greater accuracy is required, additional points may be drawn using Eq. 32.35 for zeros and Eq. 32.36 for poles. Connecting the points with a smooth curve results in the actual plot.

Example 32.7

A certain circuit has the following transfer function.

$$T(s) = \frac{10}{s(s+10)}$$

Draw the associated Bode plot.

Solution

Unless otherwise specified, an idealized Bode plot is assumed. Following the steps in Sec. 32.8 results in the plot shown.

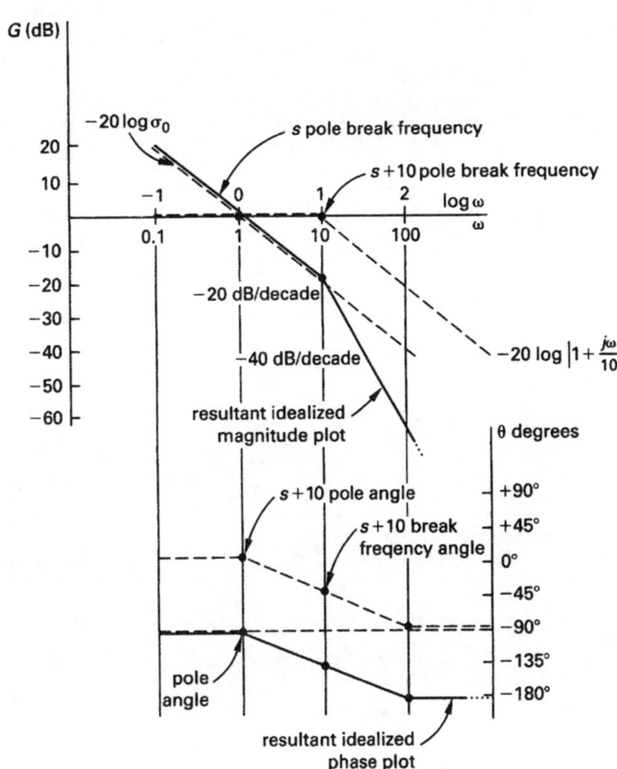

Topic V: Generation

Chapter

33 Generation Systems

Nomenclature

B	magnetic flux density	T
c	speed of light, 2.9979×10^8	m/s
E	energy	J
f	frequency	Hz
f_{droop}	frequency droop	Hz/kW
h	specific enthalpy	kJ/kg
I	effective or DC current	A
m	mass	kg
n_s	synchronous speed	rpm
p	number of poles	–
P	power	kW
Q	heat	J
R	resistance	Ω
s	specific entropy	kJ/kg·K
T	temperature	°C or K
V_{droop}	voltage droop	V/kVAR
W	work	kJ

Symbols

η	efficiency	–
ω	armature angular speed	rad/s

Subscripts

nl	no load
s	synchronous
sys	system

1. FOSSIL FUEL PLANTS

The electric utility industry is one of the largest users of fossil fuels. *Fossil fuels* are those composed of hydrocarbons, such as petroleum, coal, and natural gas. Typical combustion of the fuel produces high-pressure, high-temperature steam.[1] Pressures range from 16.5–24 MPa (2400–3500 psig). A common steam temperature is 540°C (1000°F). The basic thermodynamic cycle is the Rankine cycle shown in Fig. 33.1, along with typical plant components. Turbines used to drive the electrical generators spin at 3600 rpm. The efficiency of the cycle is proportional to the temperature difference between the temperature at the *source*, that is, the steam generator, and the *sink*, that is, the cooling medium of the condenser.

Figure 33.1 *Basic Rankine Cycle*

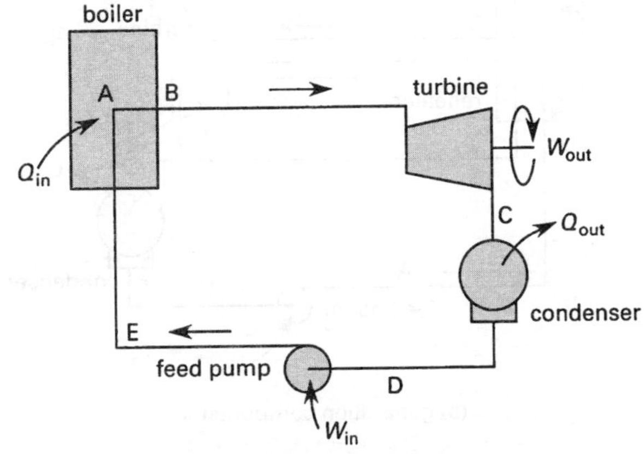

(a) generation components

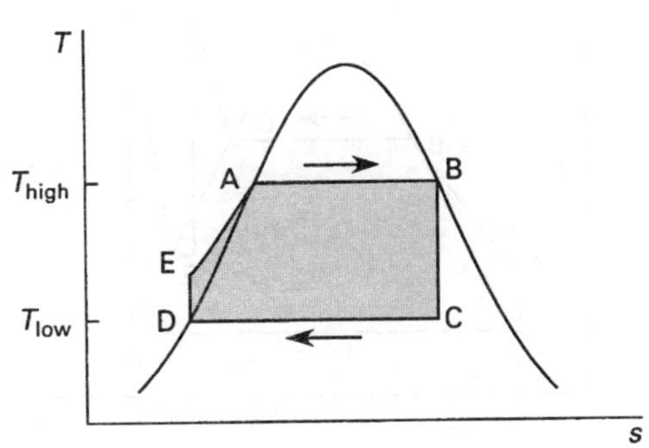

(b) thermodynamic properties

Efficiency improves when the source temperature is raised. Additionally, turbine blade wear is reduced as the moisture content in the expanding steam is reduced. Both are accomplished by adding heat to the steam beyond that required to maintain the steam in the vapor phase, that is, raising the steam temperature above the saturation temperature for the associated pressure, a process known as *superheating*. The maximum practical metallurgical limit on superheating is approximately 625°C (1150°F). In some situations, moisture still forms in the low-pressure turbine stages. This problem is overcome by reheating the steam after partial isentropic

[1]Steam is the most common substance used, but the principles discussed are applicable to vapor cycles in general.

Generation

expansion in the turbine, then allowing the remaining expansion to take place, a process known as *reheating.* A generation plant operating with these modifications is said to use the *reheat cycle.* The steam flow through the plant and the resulting thermodynamic cycle are shown in Fig. 33.2.

Figure 33.2 *Basic Reheat Cycle*

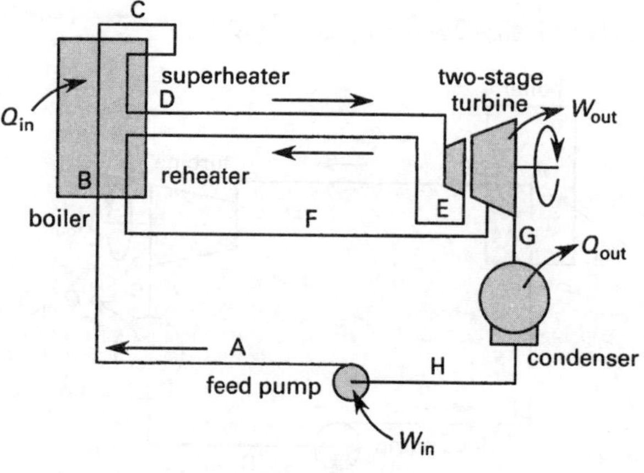

(a) generation components

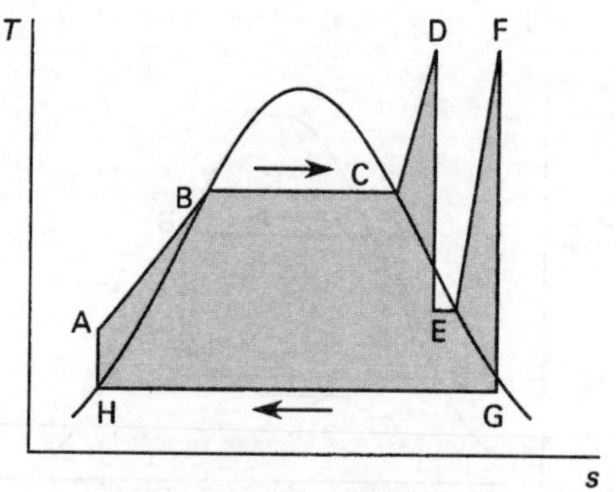

(b) thermodynamic properties

Efficiency can be further increased by minimizing the irreversibilities in the Rankine cycle associated with heating the compressed water to saturation by a finite temperature difference occurring between points E and A in Fig. 33.1 and points A and B in Fig. 33.2. This is accomplished by using heat sources elsewhere in the cycle that have temperatures slightly above that of the compressed liquid. The process is known as *regeneration* and is illustrated in Fig. 33.3. Superheating, reheating, and regeneration are all used to raise the mean effective temperature at which heat is added. Superheating raises the temperature prior to turbine input, reheating raises the temperature prior to low-pressure turbine input, and regeneration raises the temperature of the compressed liquid returning to the boiler, minimizing the heat addition required in the boiler.

Fossil fuels are broadly categorized into solid, liquid, and gaseous types. They may be further classified as natural, manufactured, or by-product. Coal is arguably the most important fossil fuel, because it is widely used and has well-known reserves. The most commonly used types of coal are anthracite, bituminous, subbituminous, and lignite. Environmental concerns involving the burning of coal include the emission of nitrogen oxides, sulfur oxides, particulate matter, and ash. The nitrogen oxides are minimized by control of the fuel-air mixture during combustion and postcombustion reduction using chemical reagents. Sulfur oxides are minimized using a variety of flue-gas desulfurization (FGD) systems. Particulate matter is minimized using electro-static precipitators. Ash is controlled by a variety of means including coal selection, combustion techniques, and various filtration means.

Residual fuel oil is used as the source fuel for combustion in electric generating plants. Residual oil is what remains after lighter hydrocarbons, such as gasoline, have been removed from crude oil. The plant operating principles are similar to those for coal, with the significant difference being the design of the boiler. Environmental concerns include gaseous emissions, removal of oil from environmentally sensitive areas, and oil transportation safety.

Figure 33.3 *Basic Regenerative Cycle*

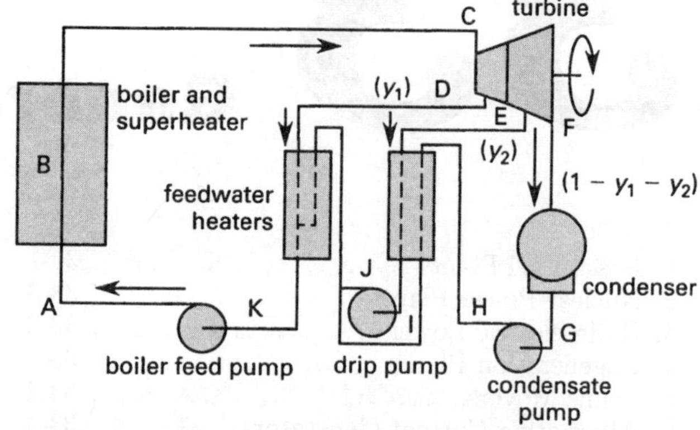

(a) generation components

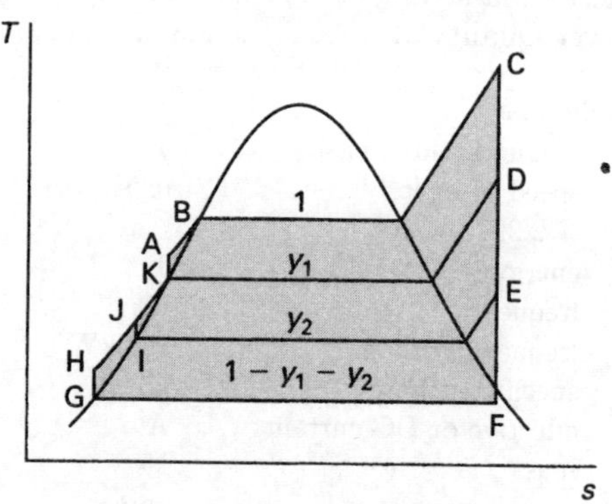

(b) thermodynamic properties

Natural gas is highly valuable for various chemical and space-heating uses. Consequently, its use in large-scale electric power plants is minimal.

2. NUCLEAR POWER PLANTS

The energy process in nuclear power plants is fundamentally different from that of fossil fuel plants. While fossil fuels are burned or undergo combustion, nuclear fuels, normally uranium, change form and in the process release energy. The amount of energy released is given by Einstein's equation.

$$E = mc^2 \qquad 33.1$$

During *fission* a uranium atom is split, by the absorption of a neutron, into two or more elements of lower atomic mass, plus additional neutrons. When the total combined mass of the products is compared to the mass of the reactants, a *mass defect* of approximately 0.215 amu is noted. Substituting this mass defect into Eq. 33.1 results in an energy of approximately 200 MeV/fission (3.2×10^{-11} W·s/fission). *Fusion* combines lighter elements to create an element of higher mass, but it also results in a mass defect and subsequent release of energy. All current commercial electric generation plants are based on the fission process.

Nuclear power plants operate on the same thermodynamic cycles as fossil fuel plants. *Pressurized water reactors* (PWR) are constructed of two loops, a primary and a secondary loop, as shown in Fig. 33.4. The primary system operates in the range of 13.8–17.3 MPa (2000–2500 psia). The temperature of the primary, at approximately 300°C (572°F), is well below the saturation temperature for this pressure range, to prevent boiling and other adverse effects. As a result, the steam side operates at lower temperatures and pressures than in a fossil fuel plant—approximately 8.4 MPa (1230 psia), correlating with the 300°C (572°F) operating temperature on the primary side.

Figure 33.4 *Pressurized Water Reactor*

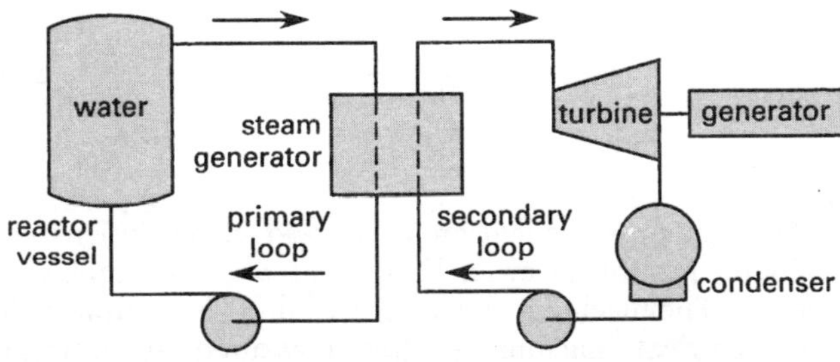

A *boiling water reactor* (BWR) generates steam directly in the core. (See Fig. 33.5.) Such a reactor operates at lower pressures than the typical PWR, usually 6.9 MPa (1000 psia). The steam is passed through various separators and dryers to minimize the moisture content prior to entering the turbine.

Figure 33.5 *Boiling Water Reactor*

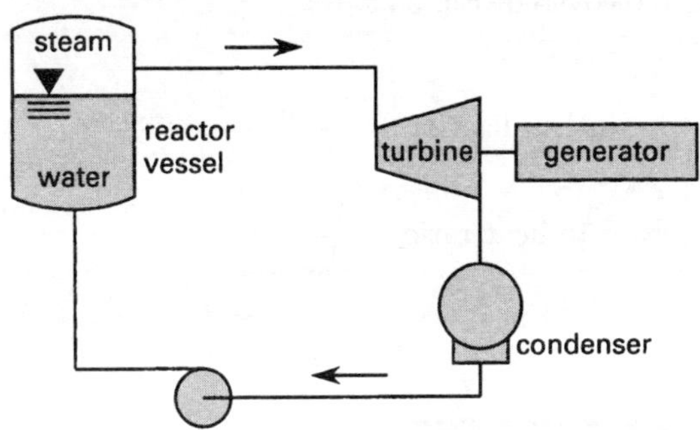

Additional fundamental differences in nuclear power plants compared to fossil fuel plants include a significantly higher thermal energy density, heat generation after shutdown caused by radioactive decay, and minimal environmental emissions. Fossil fuel plants emit carbon dioxide (CO_2), nitrogen dioxide (NO_2), sulfur trioxide (SO_3), particulate matter and, in the case of coal, ash. Nuclear power plants emit small quantities of gaseous products and, when no longer capable of generating power, solid radioactive waste products from the fuel.

3. HYDROELECTRIC POWER

The flow of water from higher elevations to lower elevations changes potential energy to kinetic energy that can be used to drive electric generators directly without the depletion of any fuel. While this eliminates the plant components that deliver heated steam to the turbine and condensed water back to the generator, it does require a source of water large enough to generate a steady output and the ability to control that source. Environmental considerations include the multiple effects of creating large reservoirs of water. Additionally, the multiple uses of the water—for navigation, irrigation, recreation, and flood control—compete with the production of power and significantly impact costs.

4. COGENERATION PLANTS

Cogeneration is the simultaneous on-site generation of electric energy and process steam or heat from the same plant. All power cycles discard a large portion of incoming energy as heat energy. Typical plant efficiencies are on the order of 30%. The other 70% of the energy is rejected to the environment as heat. If this heat is recovered and used for *space heating*, also called *district heating*, or cooling (using an *absorption system*), the process is called a *cogeneration cycle*. If the recovered heat instead vaporizes water in a steam power cycle, the process is called a *combined cycle*. In cogeneration, the recovered heat is used as heat and not converted into mechanical or electrical energy. Two terms are used to measure this recovered energy, neither of which is the same as thermal efficiency. The *fuel utilization* given in Eq. 33.2 is the ratio of useful energy to energy input.

Generation

The *power-to-heat ratio* relates turbine work to the energy recovered, Eq. 33.3.

$$\text{fuel utilization} = \frac{W_{\text{turbine}} + Q_{\text{recovered}}}{Q_{\text{in}}} \qquad 33.2$$

$$\text{power-to-heat ratio} = \frac{W_{\text{turbine}}}{Q_{\text{recovered}}} \qquad 33.3$$

5. PRIME MOVERS

Steam prime movers are of two types: reciprocating and turbine. *Reciprocating prime movers* are used primarily in low-speed (100–400 rpm), high-efficiency applications requiring high starting torque. As such, they are used for auxiliary applications throughout a power plant. *Steam turbines* are the mover of choice for electric power generation. They are variable-speed (1800–25,000 rpm), relatively efficient (85% in large machines), require no internal lubrication, and operate at steam pressures ranging from 34.5 MPa (5000 psig) and 565°C (1050°F) to 1.7 MPa (0.5 in Hg absolute). Most important for electric generation, they can be built in capacities over 1,000,000 kW—more than any other prime mover.

Steam turbines are a series of nozzles in which heat energy is changed to kinetic energy that is then transferred to a rotating wheel or drum and ultimately to an output shaft. *Impulse turbines* use stationary nozzles that drop the pressure with the kinetic energy absorbed in rotating blades operating at approximately constant pressure. *Reaction turbines* drop pressure in both the stationary and rotating portions. The steam turbine types and their associated pressure-velocity diagrams are shown in Fig. 33.6.

All large turbines use multiple stages to improve efficiency. Reaction turbines generally have more stages than impulse designs. A large turbine's first stage is an impulse stage because no pressure drop occurs across the moving blades. This allows for partial-arc admission of the steam. To maximize efficiency, the simple impulse turbine (single row) is the first stage, also called the *control stage*, on heavily loaded reheat turbines. The two-row Rateau impulse turbine is the control stage on small and medium-sized nonreheat turbines, in order to maintain efficiency over a wider range of operating loads. Marine turbines and mechanical drives operating at high numbers of revolutions per minute use single-row impulse first stages in the forward directions but use a single-stage Curtis in the reverse direction to provide the required torque while minimizing windage loss in the forward direction. A Curtis stage absorbs approximately four times the energy of an impulse stage and eight times the energy of a reaction stage.

Losses in a steam turbine include *clearance leakage*, *nozzle leakage*, *rotation* or *windage*, *carryover*, *leaving*, *partial arc*, *supersaturation*, and *moisture loss*. The overall efficiency of a turbine is[2]

$$\eta_{\text{turbine}} = \frac{W_{\text{real}}}{W_{\text{ideal}}} = \frac{h_{\text{in}} - h_{\text{condenser}}}{h_{\text{in}} - h_{\text{out,ideal}}} \qquad 33.4$$

6. ALTERNATING CURRENT GENERATORS

A conductor moving relative to a magnetic field experiences an induced electromotive force or voltage in accordance with Faraday's law. This is called *generator action*. The conductor in which the electromotive force is induced is called the *armature*. For alternating current generators, the armature is physically located on the *stator*, that is, the stationary portion of the generator. The *field* produces the magnetic flux that reacts with the armature. For alternating current generators, the field is located on the *rotor*, that is, the rotating portion of the generator, and is supplied by a direct current that maintains the electromagnetic pole strength, and the output voltage, at the desired value. This arrangement—field on the rotor, armature on the stator—is used primarily because the field current is smaller, easing design requirements for the electric connection, which is commonly made through slip rings and brushes or via brushless exciters. This allows the high-current armature output connections to be made on the stationary portion of the generator. The rotors are either *salient pole* or *cylindrical* as shown in Fig. 33.7. The use of salient pole rotors is mechanically restricted to low rpm applications (approximately 300 rpm). The generation of an AC output voltage using a permanent magnet field is shown in Fig. 33.8.

The speed of rotation of the magnetic field in a synchronous machine is called the *synchronous speed*. Because two poles must pass a given point on the armature in order to complete one cycle (360 electrical degrees), the synchronous speed is

$$n_s = \frac{120f}{p} \qquad 33.5$$

The term n_s is the synchronous speed in revolutions per minute, f is frequency in Hertz, and p is the number of poles.[3] The mechanical distance around the periphery of an electrical machine is often measured in electrical degrees, with 360° the distance between a pole pair.

[2]The ideal outlet enthalpy is determined assuming a constant entropy from the inlet pressure of the turbine to the condenser pressure, that is, a straight line on the Mollier diagram.

[3]The magnetic field is moving on the armature but the armature itself is stationary. Also, a pole is either a single north or a single south. This formula is sometimes written for pole pairs, in which case the factor of 120 becomes 60.

Generation

Figure 33.6 Steam Turbine Types

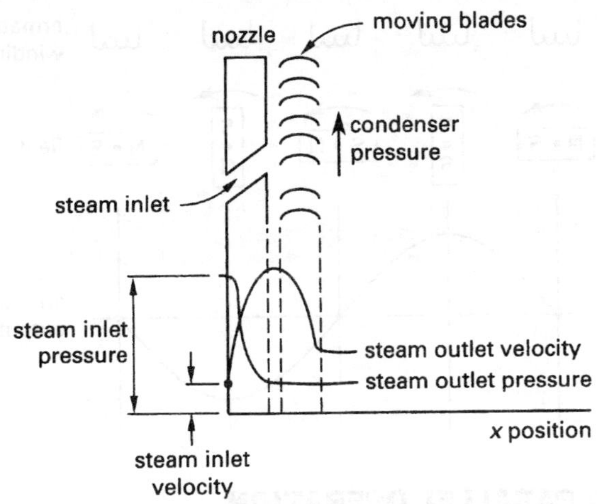

(a) simple impulse turbine

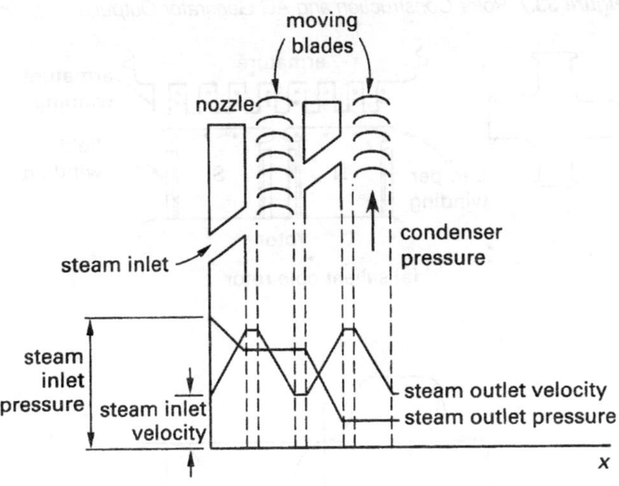

(b) Rateau impulse (pressure staged impulse)

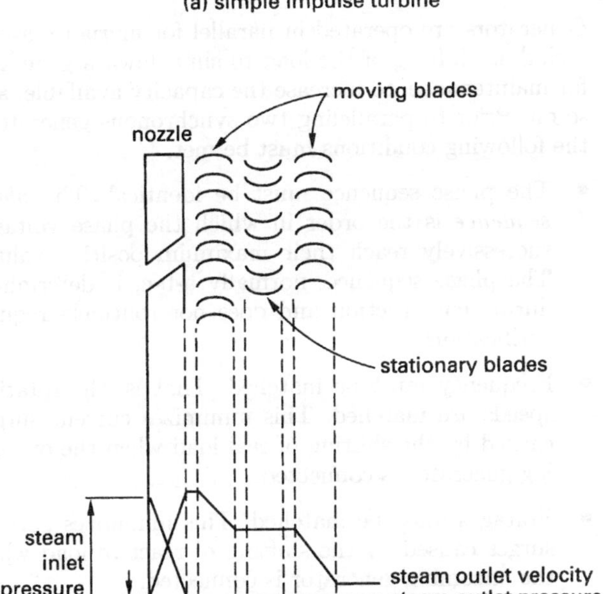

(c) Curtis impulse turbine (velocity staged impulse turbine)

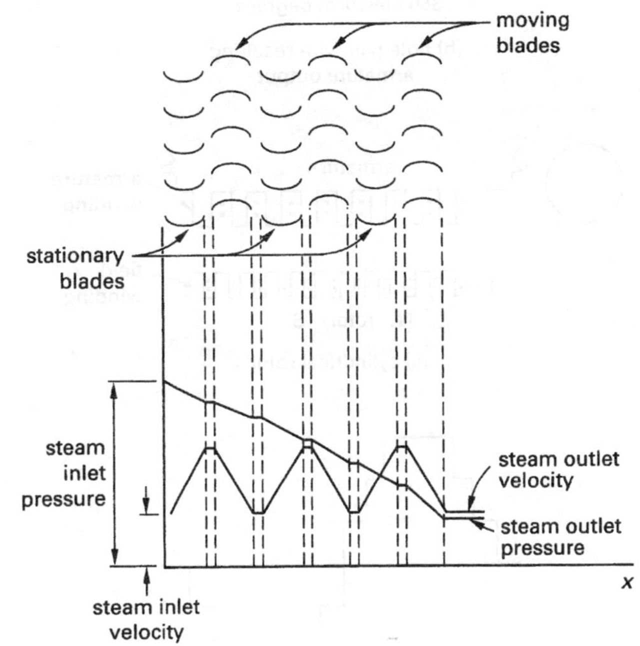

(d) Parsons turbine (reaction turbine)

Because one cycle of a sine wave is 360°, the mechanical and electrical degrees are related by Eq. 33.6.

$$\text{mechanical degrees} = \frac{\text{electrical degrees}}{\text{number of pole pairs}} \qquad 33.6$$

Losses in an AC electrical generator include *windage* and *friction, core, armature copper, field copper,* and *stray* or *load loss.*

Example 33.1

At what speed does the rotor of a four-pole AC generator turn?

Solution

The rotor in an AC generator contains the field, which rotates at synchronous speed. Because the frequency was not stated, assume 60 Hz. Substituting into Eq. 33.5 gives

$$n_s = \frac{120f}{p} = \frac{(120)(60\ \text{Hz})}{4} = 1800\ \text{rpm}$$

Figure 33.7 Rotor Construction and AC Generator Output

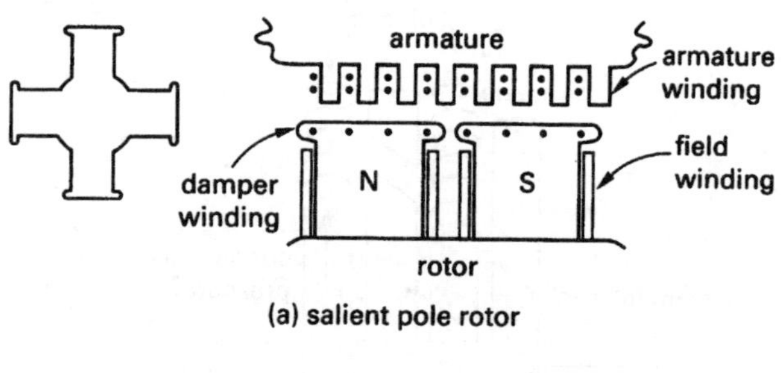

(a) salient pole rotor

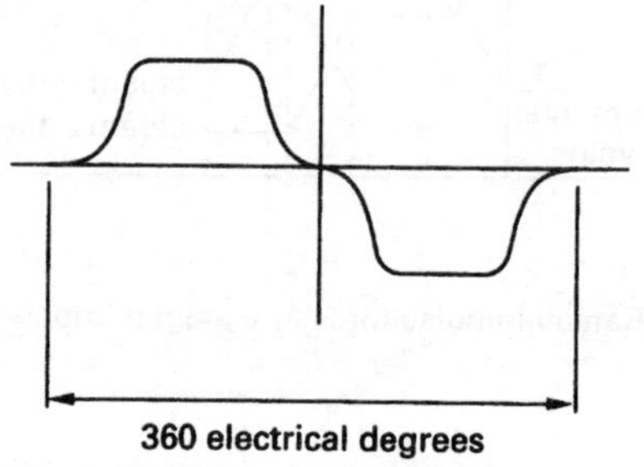

(b) pole pair with resulting armature output

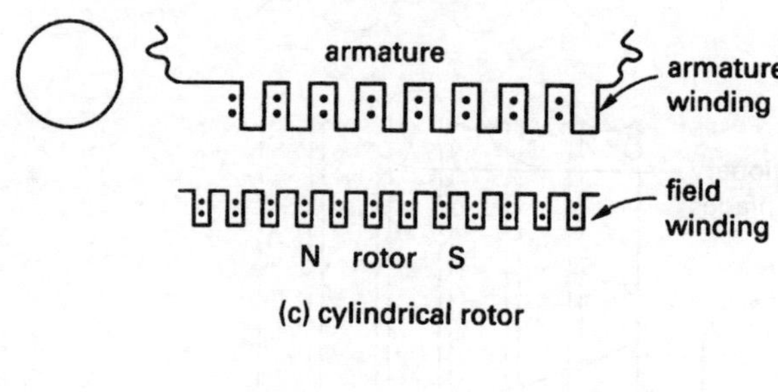

(c) cylindrical rotor

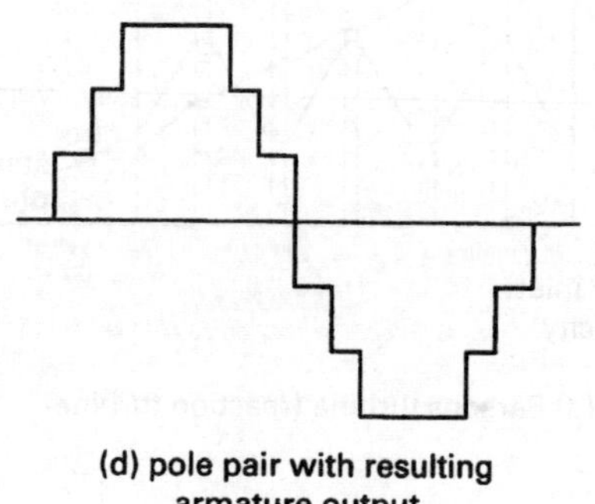

(d) pole pair with resulting armature output

Example 33.2

What is the mechanical spacing, in degrees, between the poles of a 12-pole AC generator?

Solution

The mechanical spacing is the electrical degrees divided by the number of pole pairs. Using Eq. 33.6,

$$\frac{360^\circ}{6} = 60^\circ$$

Figure 33.8 AC Output Generation

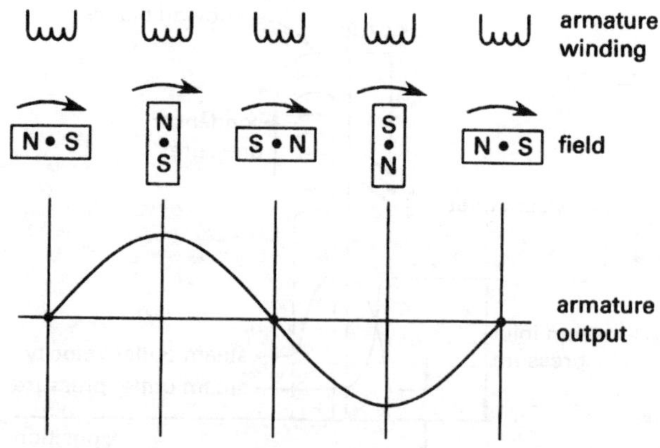

7. PARALLEL OPERATION

Generators are operated in parallel for myriad reasons, including shifting of the load to shut down a generator for maintenance, to increase the capacity available, and so on. Prior to paralleling two synchronous generators, the following conditions must be met.

- The phase sequence must be identical. The *phase sequence* is the order in which the phase voltages successively reach their maximum positive values. The phase sequence, normally a-b-c, is determined during construction and does not routinely require verification.
- Frequency must be matched. That is, the rotation speeds are matched. This minimizes current surges caused by the shifting of real load when the oncoming generator is connected.
- Voltages must be matched. This minimizes current surges caused by the shifting of reactive load when the oncoming generator is connected.

The division of real load among generators operating in parallel is determined by the speed of the generators and the characteristics of the prime mover speed-governing system as shown in Fig. 33.9. The slope of the *speed characteristic line*, also called the *generator* or *frequency droop* (f_{droop}), is determined by the speed-governing system and is constant. The total load is also constant, assuming no loads are started or secured. If the generator's speed is manually changed, that is, if the no-load frequency is adjusted, the associated characteristic line moves up or down and the load is shifted to or from the generator. The resulting frequency of the system (f_{sys}) rises or falls as well. If a load is added to or removed from the system without a change in the generator's no-load frequency, the additional load results in a drop in the system frequency determined by Eq. 33.7.

$$P = \frac{f_{sys} - f_{nl}}{f_{droop}} \qquad 33.7$$

Figure 33.9 Real Load Sharing

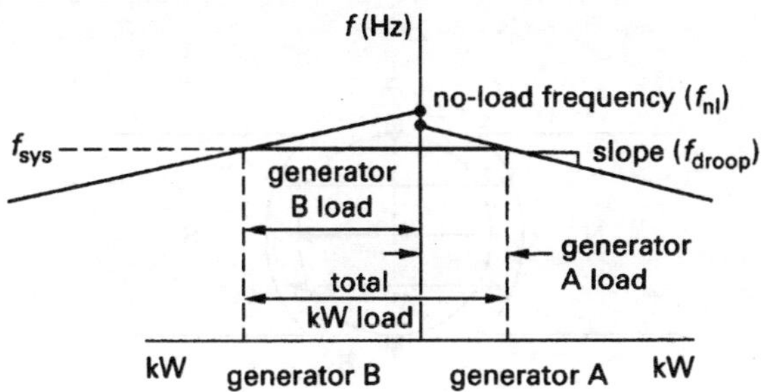

P is the total real power carried by the generator, f_{sys} is the operating frequency of the system, and f_{nl} is the no-load frequency. The frequency droop, f_{droop}, is the slope of the speed characteristic line in units of Hz/kW. Although negative, the frequency droop is often stated as a positive value. If the droop is given or used as a positive value, Eq. 33.7 becomes

$$P = \frac{f_{nl} - f_{sys}}{f_{droop}} \quad 33.8$$

The division of reactive load takes place in the same manner, but is controlled by the voltage regulators. Reactive load sharing is illustrated in Fig. 33.10. The reactive power carried by a generator is determined by Eq. 33.9.

Figure 33.10 Reactive Load Sharing

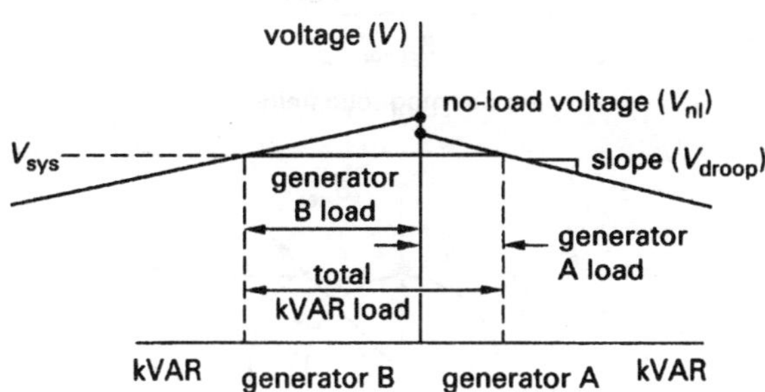

$$Q = \frac{V_{sys} - V_{nl}}{V_{droop}} \quad 33.9$$

If the droop is given or used as a positive value, Eq. 33.9 becomes

$$Q = \frac{V_{nl} - V_{sys}}{V_{droop}} \quad 33.10$$

The slopes of the speed and voltage characteristic lines in Fig. 33.9 and Fig. 33.10 are normally given in percent change of rated frequency or voltage from a no-load to a full-load condition. These are called the *speed regulation* and *voltage regulation* of a generator, respectively.

Example 33.3

A 1000 MW generator has a speed droop of 1%. Rated frequency is 60 Hz. If the generator carries 500 MW at 60 Hz, what is the no-load frequency setpoint of the speed-governing system?

Solution

A speed droop of 1% indicates a change in frequency of 0.01 × 60 Hz from 0 kW to 1000 MW, that is, from no-load to full load. The slope of the speed characteristic line, or curve, is 0.6 Hz/1000 MW. Because the droop is stated as a positive value, Eq. 33.8 is used.

$$P = \frac{f_{nl} - f_{sys}}{f_{droop}}$$

$$\begin{aligned} f_{nl} &= P f_{droop} + f_{sys} \\ &= (500\ \text{MW})\left(\frac{0.6\ \text{Hz}}{1000\ \text{MW}}\right) + 60\ \text{Hz} \\ &= 60.3\ \text{Hz} \end{aligned}$$

8. DIRECT CURRENT GENERATORS

Direct current generators operate under the same principles as AC current generators, except the output is manipulated to provide the desired DC. The definitions of armature and stator remain the same as for AC generators, but their locations are reversed. The field of a DC generator is located on the stator, or stationary portion of the generator. The armature is located on the rotor, or rotating portion of the generator. This arrangement is used primarily because of the need to change the armature output to DC, which is accomplished by the *commutator*. *Commutation* is the process of current reversal in the armature windings that provides direct current to the brushes. Simplified commutation is shown in Fig. 33.11. The physical location of the windings and the electrical connections for a four-coil two-pole generator is shown in Fig. 33.12. The rotors are normally cylindrical. The DC output voltage is a rectified version of that shown in Fig. 33.8.

Figure 33.11 Commutator Action

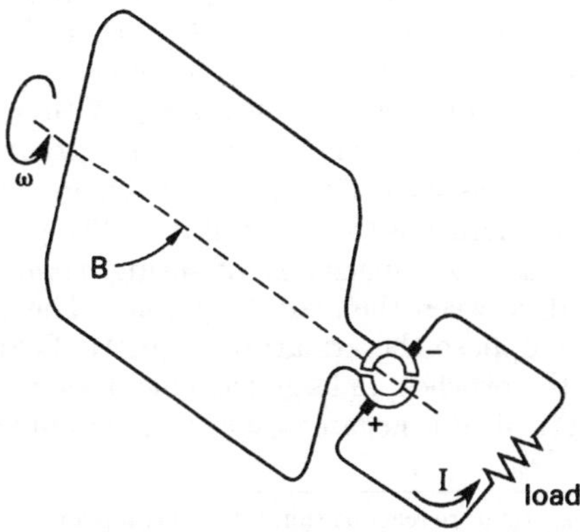

Figure 33.12 *DC Machine Windings*

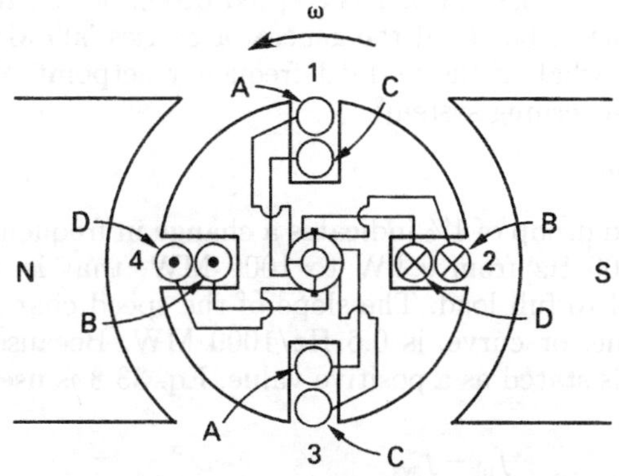

(a) four-coil winding: physical diagram

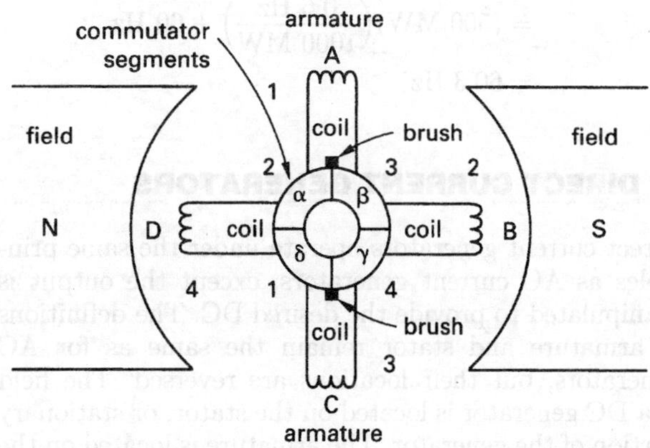

(b) four-coil winding: schematic diagram

Figure 33.13 *DC Generator Armature Flux*

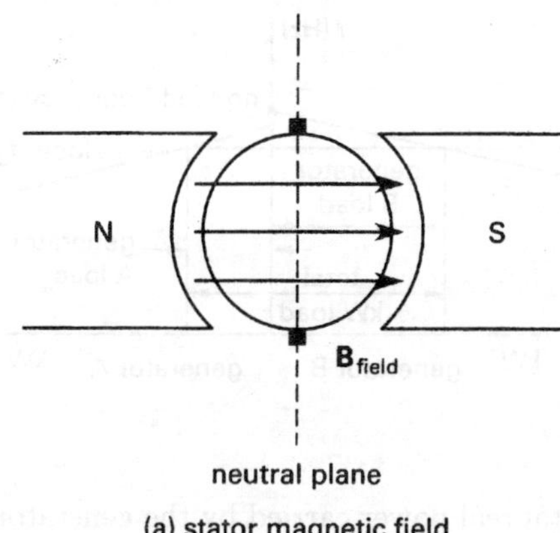

(a) stator magnetic field

(b) armature magnetic field

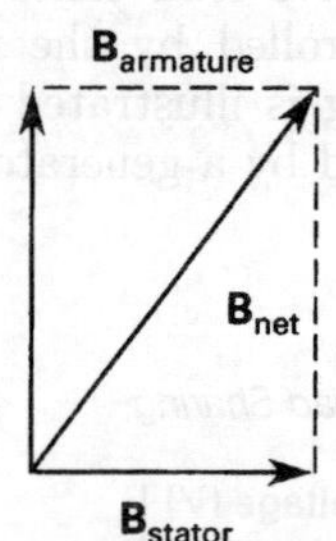

(c) net magnetic field

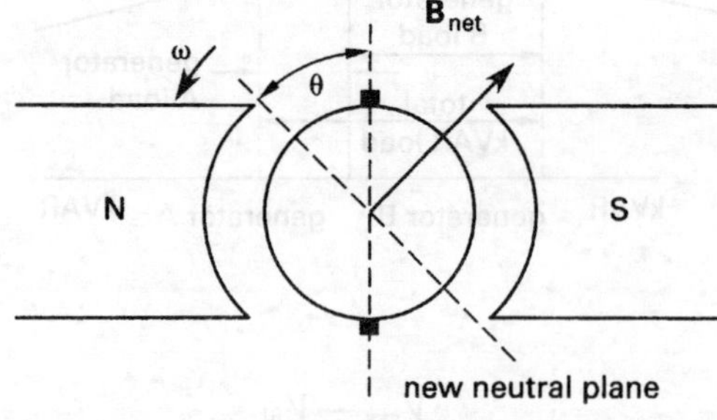

(d) neutral plane shift

Armature reaction is the interaction between the magnetic flux produced by the armature current and the magnetic flux produced by the field current. Armature reaction occurs in both AC and DC generators, lowering the output voltage. The reaction manifests itself by shifting the *neutral plane* as shown in Fig. 33.13.[4]

The neutral plane, also called the *neutral zone*, is the plane where the surface of the armature experiences a magnetic flux density of zero. In DC generators, the shifting of the *neutral plane* results in commutation occurring on coils that have a net voltage. Normally commutation, which shorts the coil connection, occurs when the net voltage on the coil is zero. With a voltage present in the coil, arcing at the brush commutator interface occurs as the coil is shorted. Such arcing lowers brush life and damages the commutator. The problem is solved in most DC machines by using *commutating poles* placed between the main field poles. The commutating poles oppose the armature magnetic field in the vicinity of the brushes. In large DC machines, the armature magnetic field is negated by *compensating windings* electrically in series with the armature windings but physically located in the field pole pieces. These solutions are illustrated in Fig. 33.14.

Losses in a DC electrical generator include windage and friction; core; I^2R or copper losses in the armature; field, compensating, and commutating windings; and *shunt field*, load, and *brush* I^2R and *friction losses.*

[4]In generators, armature reaction shifts the neutral plane in the direction of angular motion. The effect occurs in DC motors as well, but the neutral plane shifts opposite the direction of angular motion.

Figure 33.14 DC Machine Commutating Poles and Compensating Windings

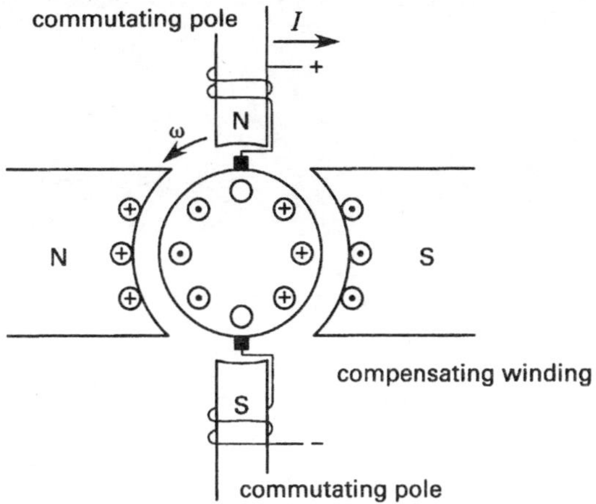

Figure 33.15 Power Disturbances

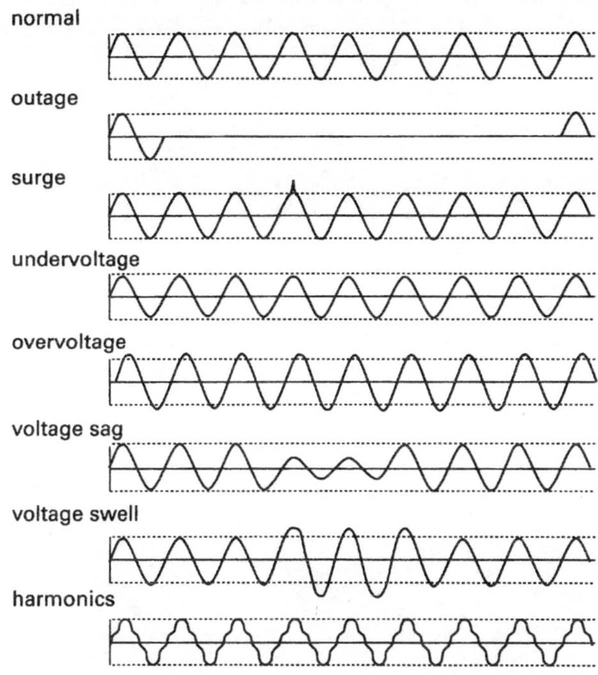

9. ENERGY MANAGEMENT

To ensure safety, security, and reliability, an electric power network must be managed properly. The system designed to accomplish these goals is called the *energy management system* (EMS). The EMS accomplishes four main tasks: generation control, generation scheduling, network analysis, and operator training. The subsystems within the EMS are supervisory control and data acquisition (SCADA), information management, applications, and communications.

The SCADA subsystem gathers data from throughout the network, allows authorized supervisors to control the network, and displays alarms and controls. The information management subsystem stores and controls access to the data required by any portion of the system. The applications subsystem contains the software programs and packages used to accomplish the tasks of the EMS. The communications subsystem generates the connectivity between all the subsystems and the necessary interfaces with operators and outside concerns.

10. POWER QUALITY

Power quality is defined by the equipment supplied. Each piece of electrical equipment has certain power supply requirements, which may vary considerably depending upon purpose and usage. The major areas of power quality follow and are illustrated in Fig. 33.15.

- *Outages*: a complete loss of electrical power. A *blackout* is a complete loss of voltage lasting from one cycle to several days. An outage condition results when faults cause protective devices to function. For example, circuit breakers may be caused to open or fuses to blow.
- *Surge*: a transient with a short duration and high magnitude. The term *spike* is used for peak voltages of 6000 V or more lasting approximately 100 μs to one-half of a cycle. The term *transient* describes a peak voltage of 20,000 V or more, lasting 10 μs to 100 μs. Surges are caused by switching operations or lightning strikes.
- *Undervoltage*: voltage below the rated voltage for a long duration, that is, for several cycles.[5] If the low-voltage condition lasts for an extended time, it is referred to as a *brownout*. Undervoltage conditions result from any number of factors, including loading beyond system capacity, failure of some sources on the grid, and ground faults in the system.
- *Overvoltage*: an increase in the steady-state voltage that lasts for an extended time. It is also called *chronic overvoltage*.[6] Overvoltages are usually caused by improper regulation or voltage regulator failure.
- *Voltage sag*: a drop in voltage. If a low-voltage condition is 80–85% of its rated value for several cycles, the term *sag* or *dip* is used. Sags can result from faults or large starting currents.
- *Voltage swell*: a condition whereby a steady-state rise in voltage occurs for several seconds to approximately one minute.
- *Harmonics*: nonfundamental frequency components of the standard 60 Hz waveform, also called *line noise*. This can occur due to feedback from equipment connected to the line, radio frequency interference, and other electromagnetic sources.

[5]The ANSI standard for service rms voltage low limits is 110–114 V.
[6]The ANSI standard for service rms overvoltage is 126 V.

34 Three-Phase Electricity and Power

Nomenclature

a	ratio of transformation	–
E	electromotive force	V
E	generated voltage	V
I, $\mathbf{I}$	current	A
N	number of turns	–
P	power	W
pf	power factor	–
Q	reactive power	VAR
R	resistance	Ω
S	apparent power	VA
t	time	s
v	variable voltage	V
V	constant or rms voltage	V
V, $\mathbf{V}$	voltage	V
X	reactance	Ω
Z, $\mathbf{Z}$	impedance	Ω

Symbols

η	efficiency	–
ϕ	phase angle difference	rad
ϕ	power factor angle	rad
ω	armature angular speed	rad/s

Subscripts

E	generated velocity
g or gen	generator
l	line
l	line
ll	line-to-line
ln	line-to-neutral
max	maximum
p	phase
pri	primary
pu	per unit
rms	root mean square
sec	secondary
t	total
trans	transmission
V	potential difference between 2 points

1. BENEFITS OF THREE-PHASE POWER

Three-phase energy distribution systems use fewer and smaller conductors and, therefore, are more efficient than multiple single-phase systems providing the same power. Three-phase motors provide a uniform torque, not a pulsating torque as do single-phase motors. Three-phase induction motors do not require additional starting windings or associated switches. When rectified, three-phase voltage has a smoother waveform and less ripple to be filtered.

Using three-phase power results in three currents that tend to cancel each other. Indeed three phases sum to zero in a linear balanced load. This allows the size of the neutral conductor to be reduced or eliminated. Since power transfer for three phases is constant and not pulsating as for a single- or two-phase machine, three-phase machine vibrations are reduced and bearing life is extended.[1] Three-phase systems can produce a magnetic field that rotates in a specified direction, simplifying the design of electric motors.

2. STANDARD NOTATION CONVENTIONS

Waveforms representative of the voltage in a power system are sinusoidal and have constant frequency, unless otherwise stated. Nonsinusoidal waveforms are dealt with by using transforms and transient analysis with varying frequencies. The most common, and mathematically efficient way, to represent such waveforms is with phasors.

Phasor representation is the primary tool for mathematical analysis of electrical systems with the voltage phasor as $\mathbf{V}$ and the current phasor as $\mathbf{I}$, respectively. The magnitude, or the root-mean-square (rms) value, of a

[1]Using two or fewer, or more than three, will result in a situation where the total currents do not cancel one another in a balanced load. Further, there is additional expense involved.

Generation

voltage or current phasor is given by V or I.[2] Maximum values carry a subscript, V_{max}. Impedance, **Z**, is technically not a phasor since it is steady with time. It is, however, a complex number and is treated identically to phasors with regard to notation. Subscripts are added to clarify where necessary.

A generated voltage is given denoted with E and a potential difference between two points with V. The primary (and preferred) method used to indicate potential difference is to use polarity signs, with the symbol "+" used to indicate the higher potential.

A secondary method is to use an arrow between the points, with the tail of the arrow indicating the lower potential.[3] A tertiary method is to use double-subscript notation, with the first subscript indicating the positive polarity.

The symbology for electromotive force, E, is similar to that used in the primary and secondary methods. However, if using the tertiary method, the first subscript will indicate negative polarity. Because of this difference, and to avoid confusion, this chapter uses a double-headed arrow for the potential difference and polarity markings.

Current-flow arrows always indicate the direction of positive current, and current flow is drawn in the same sense as the phase sequence. For example, in illustrations, current is shown as flowing into nodes A, B, and C of either wye or delta connections, and positive-sequence phase currents are shown as flowing from a to b for phase A; b to c for phase B; and c to a for phase C.

Conversion between various representative formats is often required to obtain a desired quantity. For example, voltage expressed as a function of time is given directly by Eq. 34.1. The maximum voltage is given as 169.7 V.[4]

$$v(t) = (169.7\ \text{V})\sin(\omega t + 30°) \qquad 34.1$$

The magnitude (rms) value of the voltage is given by Eq. 34.2. Substituting 169.7 V for V_p, V_{rms} is equal to 120 V.

$$V_{rms} = \frac{V_p}{\sqrt{2}} \qquad 34.2$$

An example using Euler's identity is to change the time representation of the voltage shown in Eq. 34.3 into an effective (rms) phasor representation.

$$v(t) = (169.7\ \text{V})\cos(\omega t + 30°) \qquad 34.3$$

Euler's identity is given by Eq. 34.4.

$$e^{j\theta} = \cos\theta + j\sin\theta \qquad 34.4$$

Either the cosine or the sine can be used as the reference.[5] In this case, the voltage is given in Eq. 34.3. Writing the voltage in terms of Euler's identity, and using the effective phasor notation gives Eq. 34.5.

$$v(t) = \text{Re}\Big((169.7\ \text{V})e^{j(\omega t + 30°)}\Big) \qquad 34.5$$

Separate the voltage into nonrotating and rotating portions. Additionally, separate the maximum value into its rms value and the square root of two as shown in Eq. 34.6.

$$v(t) = \text{Re}\Big((120\ \text{V})e^{j30°}\sqrt{2}e^{j\omega t}\Big) \qquad 34.6$$

Since the rotating portion and the square root of two are common to all the electrical quantities associated with this voltage (consistent with the assumption that only one frequency is present during steady-state conditions, and that effective values are being used), the phasor notation can be written directly from Eq. 34.6, shown in Eq. 34.7.

$$\mathbf{V} = (120\ \text{V})e^{j30°} \qquad 34.7$$

The voltage is given as $\mathbf{V} = 120\ \text{V}\angle 30°$. The angle was carried as 30° throughout, but any calculation using the exponential would need to be accomplished using the radian value $\pi/6$.

3. SINGLE-SUBSCRIPT NOTATION

Single-subscript notation may be used for DC circuits, AC single-phase circuits, and three-phase circuits where confusion cannot occur (for example, where only power calculations occur and directions of currents or polarity of voltages are not required), and for a single phase of a three-phase balanced circuit.

In Fig. 34.1, a single phase of a three-phase circuit is shown. As long as the circuit is balanced, calculations

[2]This text uses bold nomenclature to indicate magnitude. Some texts use $|V|$ or $|I|$, or no notation at all.

[3]This is based on the IEC standard for *Conventions Concerning Electric and Magnetic Circuit*, which is not always followed. Some texts have the head of the arrow at the high or positive potential. Caution should be exercised when reviewing drawings.

[4]The forms of Eq. 34.1 and Eq. 34.3 are not strictly correct mathematically, but this form is commonly seen in electrical engineering practice for convenience. The time-dependent part, ωt, has the units of radians, and the phase angle, 30°, has units of degrees.

[5]The sine reference is often used since the value of the function at $t = 0$ is zero, which correlates with standard graphing techniques. The cosine reference is also often used because it represents the real part of Euler's identity and avoids the j term. Many national and international electrical standards also use cosine as the reference. This text will emphasize the cosine reference unless graphing is involved. (However, both the sine and cosine reference are still in use; only the phase angle changes.) Care should be taken to ensure the desired reference is used, which can be determined from the equation provided. For example, the voltage source is given as a cosine function; therefore, that is the desired reference.

Figure 34.1 Single Phase of a Three-Phase Circuit

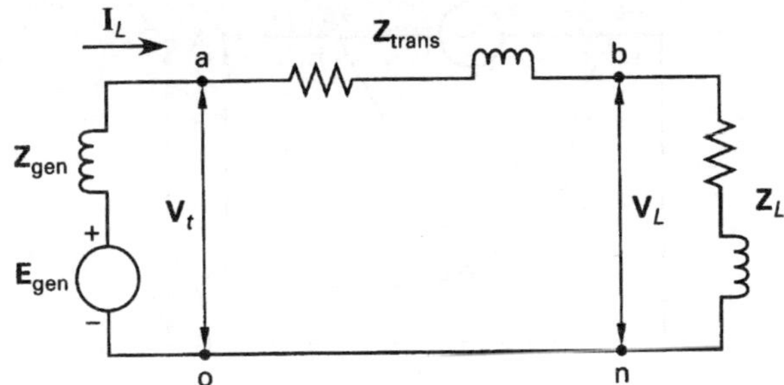

are valid for all three phases. Furthermore, the power calculation on this circuit when multiplied by three results in the total power for the three-phase circuit.

The polarity marks and the direction of the current are required in order to indicate these voltages are phasors. The polarity marks indicate the potential and current flow during the positive half-cycle of the source, in this case, E_g. The potential and current flow change direction in the negative half-cycle, but the power flow direction does not.

The load current, I_L, flows through both the load and the wires, also known as *transmission/distribution lines.* The voltage drop across the transmission line is given by Eq. 34.8.

$$\begin{aligned} \mathbf{V}_{\text{trans}} &= I_{\text{trans}} Z_{\text{trans}} = I_L Z_{\text{trans}} \\ &= \left(\frac{\mathbf{V}_t - \mathbf{V}_L}{\mathbf{Z}_{\text{trans}}}\right)\mathbf{Z}_{\text{trans}} \\ &= \mathbf{V}_t - \mathbf{V}_L \end{aligned} \quad 34.8$$

Using the nodes shown in Fig. 34.1 instead of polarity marks and arrows, and noting that the voltages at nodes a and b can be assumed to be compared to a *reference node* at o (for origin), which could also be called the *neutral node*, n, results in the following instantaneous and phasor single-subscript voltages.

$$v_{\text{a}} = v_t \quad 34.9$$

$$v_{\text{b}} = v_L \quad 34.10$$

$$\mathbf{V}_{\text{a}} = \mathbf{V}_t \quad 34.11$$

$$\mathbf{V}_{\text{b}} = \mathbf{V}_L \quad 34.12$$

$$\mathbf{V}_{\text{trans}} = \mathbf{V}_t - \mathbf{V}_L \quad 34.13$$

4. DOUBLE-SUBSCRIPT NOTATION

Polarity marks and current arrows are not required if double-script notation is used. Three-phase circuits are represented using subscripts to clarify relationships. For example, Eq. 34.13 can be rewritten using the double-subscript notation as Eq. 34.14.

$$\mathbf{V}_{\text{trans}} = \mathbf{V}_{\text{ab}} \quad 34.14$$

In single-subscript notation, when defining a current, the arrow represents the direction of positive current flow. When defining a voltage, a reference node is used, and the voltage at one node is compared to the voltage at another node, though that node was not acknowledged in the representation itself. In double-subscript notation, the voltage of the first subscript is given with respect to the second subscript (i.e., the reference node).

Therefore, the single-subscript load current, $\mathbf{I}_L$, in Fig. 34.1 can be represented as $\mathbf{I}_{\text{ab}}$. The single-subscript voltage $\mathbf{V}_{\text{trans}}$ can be represented as $\mathbf{V}_{\text{ab}}$. The impedance between nodes a and b can be represented as

$$\mathbf{Z}_{\text{trans}} = \frac{\mathbf{V}_{\text{ab}}}{I_{\text{ab}}} = \mathbf{Z}_{\text{ab}} \quad 34.15$$

5. GENERATION OF THREE-PHASE POTENTIAL

The symbolic representation of an AC generator that produces three equal sinusoidal voltages is shown in Fig. 34.2(a). The generated voltage is known as the *phase voltage, V_p,* or *coil voltage.* (Three-phase voltages are almost always stated as effective values.) Primarily because of the location of the windings, the three sinusoids are 120° apart in phase as shown in Fig. 34.2(b). If $\mathbf{V}_{\text{a}}$ is chosen as the reference voltage, then Eq. 34.16 through Eq. 34.18 represent the phasor forms of the three sinusoids. At any moment, the vector sum of these three voltages is zero.

$$\mathbf{V}_{\text{a}} = V_p\angle 0^\circ \quad 34.16$$

$$\mathbf{V}_{\text{b}} = V_p\angle{-120^\circ} \quad 34.17$$

$$\mathbf{V}_{\text{c}} = V_p\angle{-240^\circ} \quad 34.18$$

Equation 34.16 through Eq. 34.18 define an ABC or *positive sequence.* That is, $\mathbf{V}_{\text{a}}$ reaches its peak before $\mathbf{V}_{\text{b}}$, and $\mathbf{V}_{\text{b}}$ peaks before $\mathbf{V}_{\text{c}}$. With a CBA (also written as ACB) or *negative sequence*, obtained by rotating the *field magnet* in the opposite direction, the order of the delivered sinusoids is reversed (i.e., $\mathbf{V}_{\text{c}}$, $\mathbf{V}_{\text{b}}$, $\mathbf{V}_{\text{a}}$).

Although a six-conductor transmission line could be used to transmit the three voltages, it is more efficient to interconnect the windings. The two methods are commonly referred to as *delta* (*mesh*) and *wye* (*star*) connections.

Figure 34.3(a) illustrates delta source connections. The voltage across any two of the lines is known as the *line voltage* (*system voltage*) and is equal to the phase voltage. Any of the coils can be selected as the reference as long as the sequence is maintained. For a positive (ABC) sequence,

$$\mathbf{V}_{\text{CA}} = V_p\angle 0^\circ \quad 34.19$$

$$\mathbf{V}_{\text{AB}} = V_p\angle{-120^\circ} \quad 34.20$$

$$\mathbf{V}_{\text{BC}} = V_p\angle{-240^\circ} \quad 34.21$$

Generation

Figure 34.2 Three-Phase Voltage

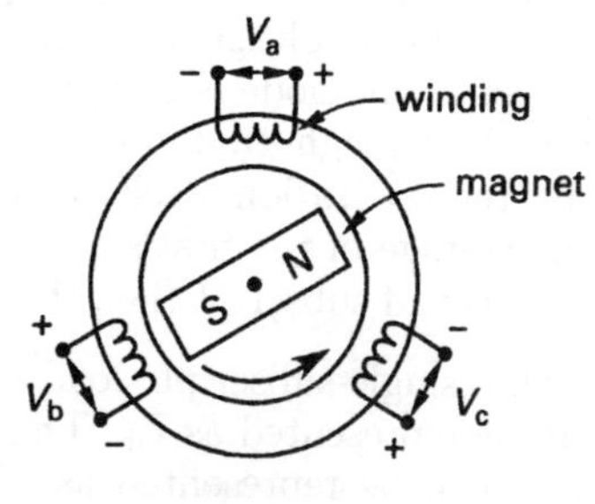

(a) alternator

(b) ABC (positive) sequence

Figure 34.3 Delta and Wye Source Connections

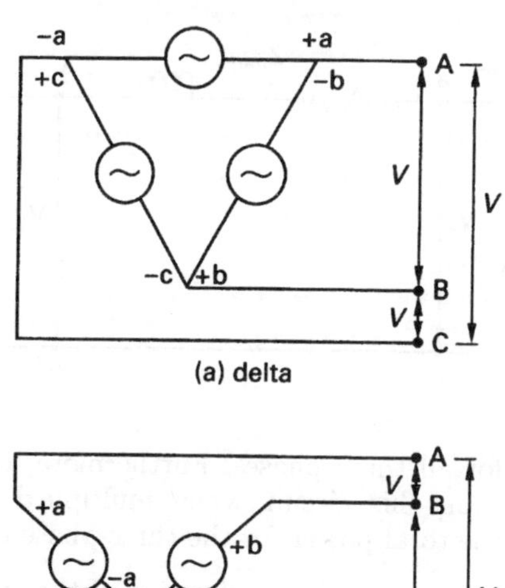

(a) delta

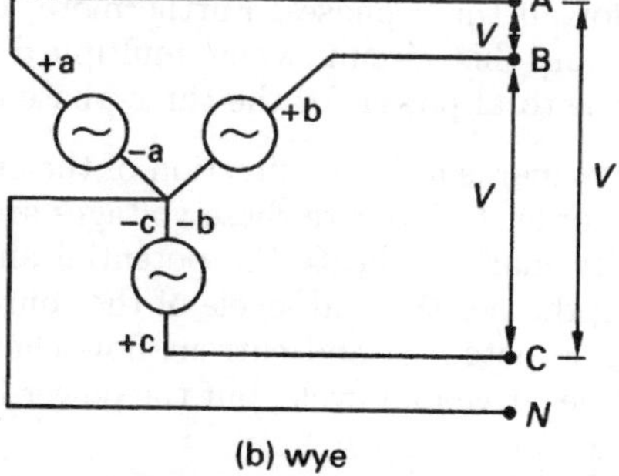

(b) wye

Wye-connected sources are illustrated in Fig. 34.3(b). While the *line-to-neutral voltages* are equal to the phase voltage, the line voltages are greater—$\sqrt{3}$ times the phase voltage. The *grounded wire* (*neutral*) is needed to carry current only if the system is unbalanced.[6] (See Sec. 34.6.) For an ABC sequence, the line voltages are

$$\mathbf{V}_{AB} = \sqrt{3}V_p\angle 30° \quad 34.22$$

$$\mathbf{V}_{BC} = \sqrt{3}V_p\angle -90° \quad 34.23$$

$$\mathbf{V}_{CA} = \sqrt{3}V_p\angle -210° \quad 34.24$$

Although the magnitude of the line voltage depends on whether the generator coils are delta- or wye-connected, each connection results in three equal sinusoidal voltages, each 120° out of phase with the others.

6. DISTRIBUTION SYSTEMS

Three-phase power is delivered by three-wire and four-wire systems. A *four-wire system* consists of three power conductors and a neutral conductor. A *three-wire system* contains only the three power conductors.

Utility power distribution starts with generation. The generator is connected through step-up *subtransmission transformers* that supply *transmission lines*. The actual transmission line voltage depends on the distance between the subtransmission transformers and the user. Distribution *substation transformers* reduce the voltage from the transmission line level to approximately 35 kV. The *primary distribution system* delivers power to *distribution transformers* that reduce voltage still further, to between 120 V and 600 V.

[6]The neutral wire is usually kept to provide for a minor imbalance.

7. BALANCED LOADS

Three impedances are required to fully load a three-phase voltage source. The impedances in a three-phase system are *balanced* when they are identical in magnitude and angle. The voltages and line currents, as well as the real, apparent, and reactive powers, are all identical in a balanced system. Also, the power factor is the same for each phase. Therefore, balanced systems can be analyzed on a per-phase basis. Such calculations are known as *one-line analyses*.

Figure 34.4 illustrates the vector diagram for a balanced delta three-phase system. The phase voltages, **V**, are separated by 120° phase angles, as are the phase currents, **I**. The phase difference angle, ϕ, between a phase voltage and its respective phase current depends on the phase impedance. With delta-connected balanced loads, the phase and line currents differ in phase by 30°.

8. DELTA-CONNECTED LOADS

Figure 34.5 illustrates delta-connected loads.

Figure 34.4 *Positive Sequence (ABC) Balanced Delta Load Vector Diagram*

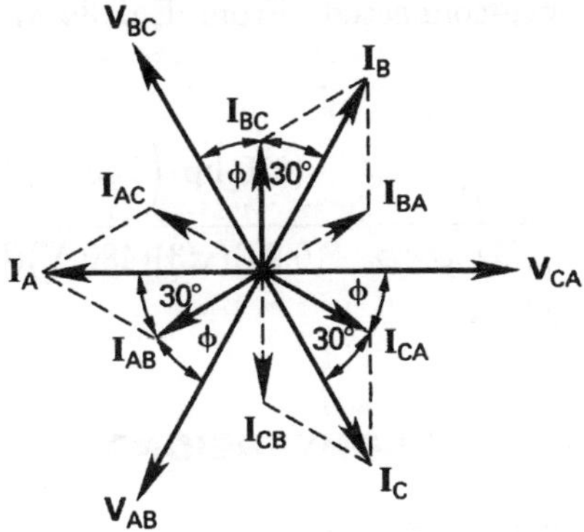

Figure 34.5 *Delta-Connected Loads*

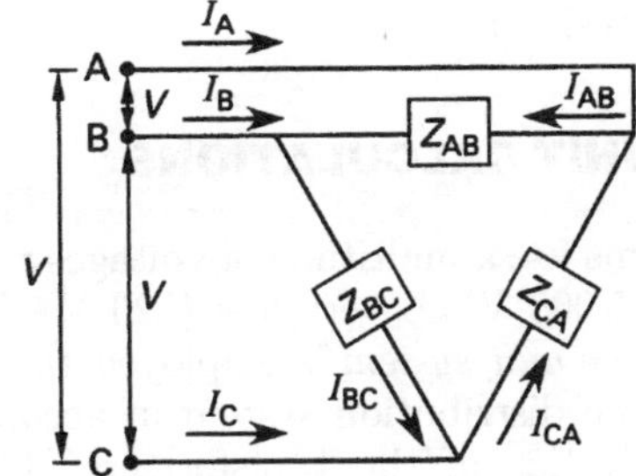

The *phase currents* for a balanced system are calculated from the line voltage (same as the phase voltage). For a positive (ABC) sequence,

$$\mathbf{I}_{AB} = \frac{\mathbf{V}_{AB}}{\mathbf{Z}_{AB}} = \frac{V\angle{-120°}}{Z\angle\phi} = \frac{V}{Z}\angle{-120°} - \phi \qquad 34.25$$

$$\mathbf{I}_{BC} = \frac{\mathbf{V}_{BC}}{\mathbf{Z}_{BC}} = \frac{V}{Z}\angle{-240°} - \phi \qquad 34.26$$

$$\mathbf{I}_{CA} = \frac{\mathbf{V}_{CA}}{\mathbf{Z}_{CA}} = \frac{V}{Z}\angle{-\phi} \qquad 34.27$$

The *line currents* are not the same as the phase currents but are $\sqrt{3}$ times the phase current and displaced $-30°$ in phase from the phase current.

$$|\mathbf{I}_A| = |\mathbf{I}_{AB} - \mathbf{I}_{CA}| = \sqrt{3}I_{AB} \qquad 34.28$$

$$|\mathbf{I}_B| = |\mathbf{I}_{BC} - \mathbf{I}_{AB}| = \sqrt{3}I_{BC} \qquad 34.29$$

$$|\mathbf{I}_C| = |\mathbf{I}_{CA} - \mathbf{I}_{BC}| = \sqrt{3}I_{CA} \qquad 34.30$$

$$I_A = I_B = I_C \quad \text{[balanced]} \qquad 34.31$$

Each impedance in a balanced system dissipates the same real *phase power*, P_p. The total power dissipated is three times the phase power. (This is the same as for wye-connected loads.)

$$\begin{aligned} P_t = 3P_p &= 3V_pI_p\cos|\phi| \\ &= 3V_pI_p\text{pf} \\ &= \sqrt{3}VI\cos\phi \\ &= \sqrt{3}VI\text{pf} \end{aligned} \qquad 34.32$$

Example 34.1

Three identical impedances are connected in delta across a three-phase system with 240 V (rms) line voltages in an ABC sequence. Find the (a) phase current $\mathbf{I}_{AB}$, (b) phase real power, P_p, (c) line current $\mathbf{I}_B$, and (d) total real power.

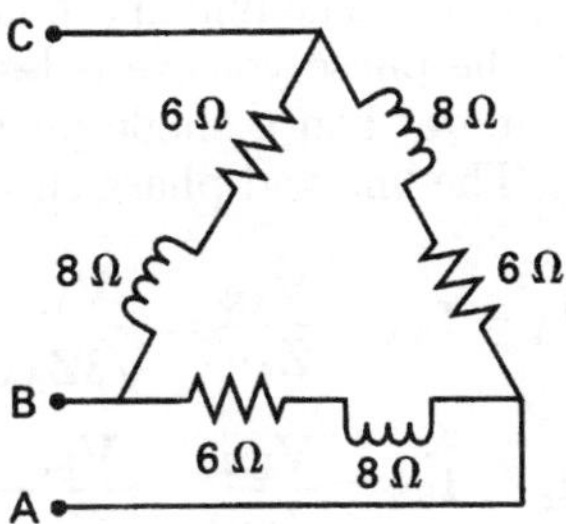

Solution

(a) The phase impedance is

$$\begin{aligned} \mathbf{Z}_p &= \sqrt{R^2 + X_L^2}\,\angle\arctan\frac{X}{R} \\ &= \sqrt{(6\ \Omega)^2 + (8\ \Omega)^2}\,\angle\arctan\frac{8\ \Omega}{6\ \Omega} \\ &= 10\ \Omega\angle 53.13° \end{aligned}$$

The phase current is

$$\mathbf{I}_{AB} = \frac{\mathbf{V}_{AB}}{\mathbf{Z}_{AB}} = \frac{240\ \text{V}\angle{-120°}}{10\ \Omega\angle 53.13°} = 24\ \text{A}\angle{-173.13°}$$

(b) The average phase power is

$$P_p = I_p^2R = (24\ \text{A})^2(6\ \Omega) = 3456\ \text{W}$$

(c) The phase current I_{BC} contributes to the line current.

$$\mathbf{I}_{BC} = \frac{\mathbf{V}_{BC}}{\mathbf{Z}_{BC}} = \frac{240\ \text{V}\angle{-240°}}{10\ \Omega\angle 53.13°} = 24\ \text{A}\angle{-293.13°}$$

$$\begin{aligned} \mathbf{I}_B = \mathbf{I}_{BC} - \mathbf{I}_{AB} &= 24\ \text{A}\angle{-293.13°} \\ &\quad - 24\ \text{A}\angle{-173.13°} \\ &= 41.57\ \text{A}\angle 36.87° \end{aligned}$$

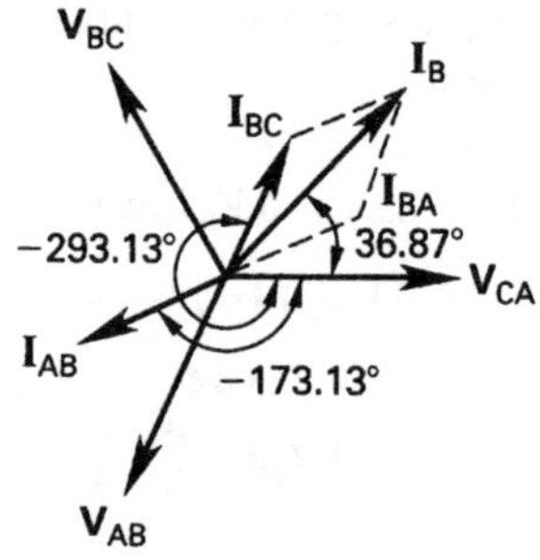

(d) The total power is three times the phase power.

$$P_t = 3P_p = (3)(3456 \text{ W}) = 10{,}368 \text{ W}$$

9. WYE-CONNECTED LOADS

Figure 34.6 illustrates three equal impedances connected in a wye configuration. The line and phase currents are equal. However, the phase voltage is less than the line voltage because more than a single phase is connected across two lines. The line and phase currents are

$$\mathbf{I}_A = \mathbf{I}_{AN} = \frac{\mathbf{V}_{AN}}{\mathbf{Z}_{AN}} = \frac{\mathbf{V}_{AB}}{\sqrt{3}\mathbf{Z}_{AN}} \quad \text{34.33}$$

$$\mathbf{I}_B = \mathbf{I}_{BN} = \frac{\mathbf{V}_{BN}}{\mathbf{Z}_{BN}} = \frac{\mathbf{V}_{BC}}{\sqrt{3}\mathbf{Z}_{BN}} \quad \text{34.34}$$

$$\mathbf{I}_C = \mathbf{I}_{CN} = \frac{\mathbf{V}_{CN}}{\mathbf{Z}_{CN}} = \frac{\mathbf{V}_{CA}}{\sqrt{3}\mathbf{Z}_{CN}} \quad \text{34.35}$$

$$\mathbf{I}_N = 0 \quad \text{[balanced]} \quad \text{34.36}$$

Figure 34.6 *Wye-Connected Loads*

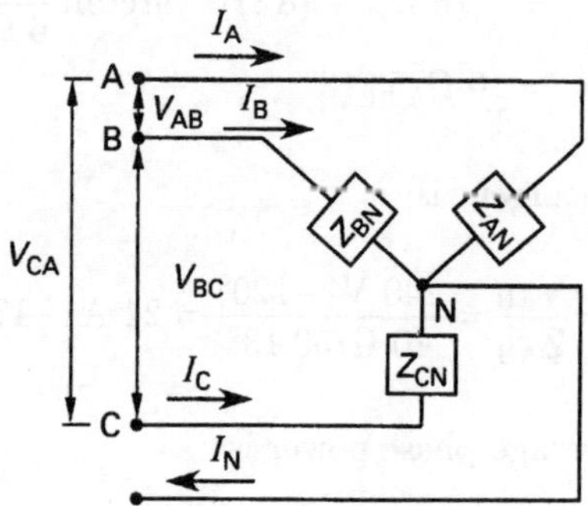

The total power dissipated in a balanced wye-connected system is three times the phase power. (This is the same as for the delta connection.)

$$P_t = 3P_p = 3V_pI_p\cos\phi = \sqrt{3}VI\cos\phi \quad \text{34.37}$$

Example 34.2

A three-phase, 480 V, 250 hp motor has an efficiency of 94% and a power factor of 90%. What is the line current?

Solution

It does not matter whether the motor's windings are delta- or wye-connected. From Eq. 34.37, the line current is

$$\begin{aligned} I &= \frac{P_t}{\eta\sqrt{3}V\cos\phi} = \frac{(250 \text{ hp})\left(745.7 \frac{\text{W}}{\text{hp}}\right)}{(0.94)(\sqrt{3})(480 \text{ V})(0.9)} \\ &= 265 \text{ A} \end{aligned}$$

10. DELTA-WYE CONVERSIONS

It is occasionally convenient to convert a delta system to a wye system and vice versa. The equations in Sec. 26.20, used to convert between delta and wye resistor networks, are applicable if impedances are substituted for resistances.

11. PER-UNIT CALCULATIONS

Power systems use a multitude of voltages ranging from very high (35 000 V) to very low (120 V). To simplify analysis, a *per-unit system* is employed that scales the sections of the distribution system in accordance with the voltage ratios of the transformers. The per-unit system is given mathematically by

$$\text{per unit} = \frac{\text{actual}}{\text{base}} = \frac{\text{percent}}{100} \quad \text{34.38}$$

For a three-phase system, the usual bases are the line voltage (in V) and the total (three-phase) apparent power (in VA) ratings. From these two bases, the base current and impedance can be calculated. Line and phase values may differ in three-phase systems. The relationship between the phase and line quantities for the bases is

$$V_p = \frac{V_l}{\sqrt{3}} \quad \text{34.39}$$

$$S_p = \frac{S_t}{3} \quad \text{34.40}$$

The line-to-phase voltage conversion is necessary in a wye connection only. In a delta connection, the line and phase voltages are equal. Care should be taken when using phase and line quantities because the symbols vary. Phase values are represented here with the subscript p. However, phase values are also given with subscripts ϕ and ln (line-to-neutral). Line values are given with the subscript t (total) here. However, line values are also given with subscripts ll (line-to-line), l (line), 3ϕ, or with no subscript. The per-unit system is represented by Eq. 34.41 through Eq. 34.46. It is presented elsewhere as phase bases for use with single-phase systems. Conversion to a three-phase base, which uses line and total quantities, is accomplished

Generation

with Eq. 34.39 and Eq. 34.40 and is indicated by the subscript 3ϕ in Eq. 34.41 through Eq. 34.46.

$$S_{\text{base}} = S_p = \left(\frac{S}{3}\right)_{3\phi} \quad 34.41$$

$$V_{\text{base}} = V_p = \left(\frac{V}{\sqrt{3}}\right)_{3\phi} \quad 34.42$$

$$I_{\text{base}} = \frac{S_{\text{base}}}{V_{\text{base}}} = \frac{S_p}{V_p} = \left(\frac{S}{\sqrt{3}V}\right)_{3\phi} \quad 34.43$$

$$Z_{\text{base}} = \frac{V_{\text{base}}}{I_{\text{base}}} = \frac{V_p^2}{S_p} = \left(\frac{V^2}{S}\right)_{3\phi} = \left(\frac{V}{\sqrt{3}I}\right)_{3\phi} \quad 34.44$$

$$P_{\text{base}} = P_p = \left(\frac{P}{3}\right)_{3\phi} \quad 34.45$$

$$Q_{\text{base}} = Q_p = \left(\frac{Q}{3}\right)_{3\phi} \quad 34.46$$

The per-unit values are

$$I_{\text{pu}} = \frac{I_{\text{actual}}}{I_{\text{base}}} \quad 34.47$$

$$V_{\text{pu}} = \frac{V_{\text{actual}}}{V_{\text{base}}} \quad 34.48$$

$$Z_{\text{pu}} = \frac{Z_{\text{actual}}}{Z_{\text{base}}} \quad 34.49$$

$$P_{\text{pu}} = \frac{P_{\text{actual}}}{P_{\text{base}}} \quad 34.50$$

$$Q_{\text{pu}} = \frac{Q_{\text{actual}}}{Q_{\text{base}}} \quad 34.51$$

Ohm's law and other circuit analysis methods can be used with the per-unit quantities.

$$V_{\text{pu}} = I_{\text{pu}} Z_{\text{pu}} \quad 34.52$$

All values in a given portion of a power system must be expressed using the same base. The general method for converting from one generic per-unit value, call it χ, to another in a different base is

$$\chi_{\text{pu,new}} = \chi_{\text{pu,old}} \left(\frac{\chi_{\text{base,old}}}{\chi_{\text{base,new}}}\right) \quad 34.53$$

The impedance per-unit value conversion is

$$Z_{\text{pu,new}} = Z_{\text{pu,old}} \left(\frac{V_{\text{base,old}}}{V_{\text{base,new}}}\right)^2 \left(\frac{S_{\text{base,new}}}{S_{\text{base,old}}}\right) \quad 34.54$$

Values of per-unit quantities are initially given or calculated using the rated values of the electrical components (that is, the rating of the generator, transformer, or motor). The per-unit value for a transformer is the same whether referenced to either the low or high side. (Example 34.4 provides evidence for this assertion.) The purpose of Eq. 34.53 and Eq. 34.54 is to adjust to the base value used by the entire system.

When changing from per-unit values to actual values or vice versa (for example, transmission line impedance conversion), the base voltage used must be the voltage value present in the given section of the electrical distribution system for which the results are desired. (Chapter 37 provides details and examples.)

Example 34.3

A wye-connected three-phase device is rated to draw a total of 300 kVA when connected to a line-to-line voltage of 15 kV. The device's per-unit impedance is 0.1414 $+j0.9900$. What is the actual impedance?

Solution

The bases are

$$\begin{aligned} V_{\text{base}} &= V_p = \frac{V}{\sqrt{3}} = \frac{15 \text{ kV}}{\sqrt{3}} \\ &= 8.660 \text{ kV} \\ S_{\text{base}} &= S_{\text{phase}} = \frac{S}{3} = \frac{300 \text{ kVA}}{3} \\ &= 100 \text{ kVA} \end{aligned}$$

The base current is

$$I_{\text{base}} = \frac{S_{\text{base}}}{V_{\text{base}}} = \frac{100 \text{ kVA}}{8.660 \text{ kV}} = 11.55 \text{ A}$$

The base impedance is

$$Z_{\text{base}} = \frac{V_{\text{base}}}{I_{\text{base}}} = \frac{8660 \text{ V}}{11.55 \text{ A}} = 750 \ \Omega$$

From Eq. 34.49, the actual impedance is

$$\begin{aligned} Z_{\text{actual}} &= Z_{\text{pu}} Z_{\text{base}} = (0.1414 + j0.9900)(750 \ \Omega) \\ &= 106.1 + j742.5 \ \Omega \end{aligned}$$

Example 34.4

Review the one-line diagram for a three-phase generator-reactor system in the figure shown. The actual impedance on the low voltage side is $j4.0 \ \Omega$. What is the ratio of the low-side to the high-side per-unit impedance?

Solution

Given the values in the figure, and noting that the impedance on the high voltage side of the transformer must be 16.0 Ω, the quantities in the following table can be calculated. The per-unit values are in the final row.

Table for Solution 34.4

quantity	low voltage side	high voltage side
S_{base}	1200 VA	1200 VA
V_{base}	120 V	240 V
I_{base}	$\frac{S_{\text{base}}}{V_{\text{base}}} = \frac{1200\ \text{VA}}{120\ \text{V}} = 10\ \text{A}$	$\frac{S_{\text{base}}}{V_{\text{base}}} = \frac{1200\ \text{VA}}{240\ \text{V}} = 5\ \text{A}$
Z_{base}	$\frac{V_{\text{base}}^2}{S_{\text{base}}} = \frac{(120\ \text{V})^2}{1200\ \text{VA}} = 12\ \Omega$	$\frac{\mathbf{V}_{\text{base}}^2}{S_{\text{base}}} = \frac{(240\ \text{V})^2}{1200\ \text{VA}} = 48\ \Omega$
$\mathbf{Z}_{\text{pu}}$	$\frac{\mathbf{Z}_{\text{low side}}}{Z_{\text{base}}} = \frac{j4.0\ \Omega}{12\ \Omega} = j0.33\ \text{pu}$	$\frac{\mathbf{Z}_{\text{high side}}}{Z_{\text{base}}} = \frac{j16.0\ \Omega}{48\ \Omega} = j0.33\ \text{pu}$

The ratio of the low-side per-unit value to the high-side per-unit value is 1:1. This indicates that a model of the per-unit impedance for the transformer (single phase) is

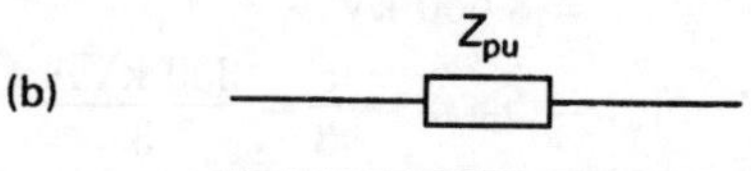

This can be represented in a one-line diagram as

(c) Z_{pu}

Because line quantities are used, this model also works for three-phase systems.

12. UNBALANCED LOADS

The three-phase loads are unequal in an unbalanced system. A fourth conductor, the *neutral conductor*, is required for the line voltages to be constant. Such a system is known as a *four-wire system*. Without the neutral conductor (i.e., in a *three-wire system*), the common point of the load connections is not at zero potential and the voltages across the three impedances vary from the line-to-neutral voltage. The voltage at the common point is known as the *displacement neutral voltage*.

Regardless of whether the line voltages are equal, the line currents are not the same nor do they have a 120° phase difference. Unbalanced systems can be evaluated by computing the phase currents and then applying Kirchhoff's current law (in vector form) to obtain the line currents. The *neutral current* is

$$\mathbf{I}_N = -(\mathbf{I}_\text{A} + \mathbf{I}_\text{B} + \mathbf{I}_\text{C}) \qquad 34.55$$

Example 34.5

Unequal impedances are connected to a 208 V (rms) three-phase system as shown. The sequence is ABC with $\angle V_{\text{CA}} = 0°$. (a) What is the line current $\mathbf{I}_\text{A}$? (b) What is the total real power dissipated?

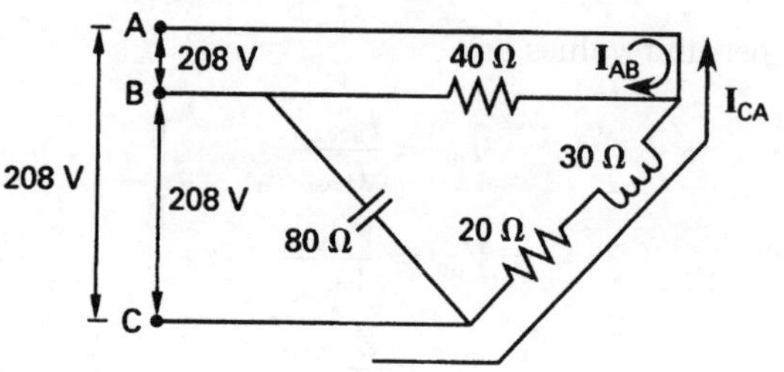

Solution

(a) First, calculate the phase current $\mathbf{I}_{\text{AB}}$.

$$\mathbf{I}_{\text{AB}} = \frac{\mathbf{V}_{\text{AB}}}{\mathbf{Z}_{\text{AB}}} = \frac{208\ \text{V}\angle{-120°}}{40\ \Omega\angle 0°} = 5.20\ \text{A}\angle{-120°}$$

The phase impedance $\mathbf{Z}_{\text{CA}}$ is

$$\begin{aligned}\mathbf{Z}_{\text{CA}} &= \sqrt{R^2 + X_L^2}\ \angle \arctan\frac{X_L}{R} \\ &= \sqrt{(20\ \Omega)^2 + (30\ \Omega)^2}\ \angle \arctan\frac{30\ \Omega}{20\ \Omega} \\ &= 36.06\ \Omega\angle 56.31°\end{aligned}$$

Next, calculate the phase current $\mathbf{I}_{\text{CA}}$.

$$\mathbf{I}_{\text{CA}} = \frac{\mathbf{V}_{\text{CA}}}{\mathbf{Z}_{\text{CA}}} = \frac{208\ \text{V}\angle 0°}{36.06\ \Omega\angle 56.31°} = 5.77\ \text{A}\angle{-56.31}$$

The line current is

$$\begin{aligned}\mathbf{I}_\text{A} &= \mathbf{I}_{\text{AB}} - \mathbf{I}_{\text{CA}} = \mathbf{I}_{\text{AB}} + \mathbf{I}_{\text{AC}} \\ &= 5.20\ \text{A}\angle{-120°} + 5.77\ \text{A}\angle 123.69° \\ &= 5.81\ \text{A}\angle 177°\end{aligned}$$

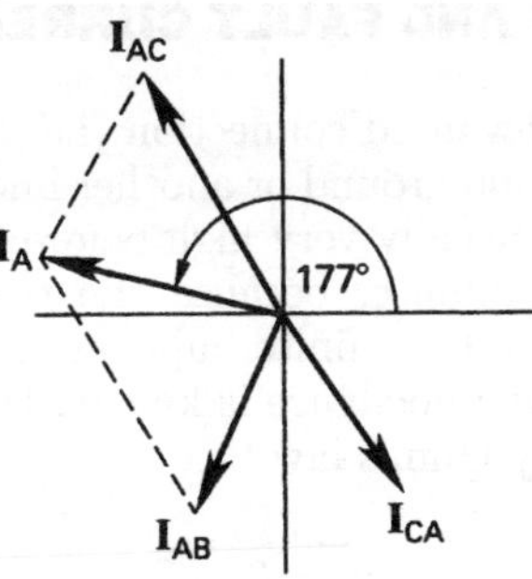

(b) The total power dissipated is

$$P = \sum I^2R = (5.20\ \text{A})^2(40\ \Omega) + (5.77\ \text{A})^2(20\ \Omega) = 1747\ \text{W}$$

13. THREE-PHASE TRANSFORMERS

Transformer banks for three-phase systems have three primary and three secondary windings. Each primary-secondary set is referred to as a *transformer*, as distinguished from the *bank*. Each side can be connected in a delta or wye configuration, making a total of four possible transformer configurations (delta-delta, delta-wye, wye-delta, and wye-wye) as shown in Fig. 34.7. The turns ratio, a, is the same for each winding. Each transformer provides one-third of the total kVA rating, regardless of connection configuration. However, as the figure shows, the secondary voltages and currents depend on the configuration.

$$a = \frac{N_{\text{pri}}}{N_{\text{sec}}} \qquad 34.56$$

Example 34.6

A 240 V (rms) three-phase system drawing 1200 kVA is supplied by a 2400 V (primary-side) transformer bank. Each transformer is connected in a wye-delta (primary-secondary) configuration. What are the (a) ratio of transformation, (b) high-side and low-side winding voltages, (c) high-side and low-side winding currents, and (d) kVA rating?

Solution

This is case (c) in Fig. 34.7.

(a) The primary line voltage is 2400 V.

$$240\ \text{V} = \frac{V}{\sqrt{3}a} = \frac{2400\ \text{V}}{\sqrt{3}a}$$

$$a = \frac{2400\ \text{V}}{(240\ \text{V})(\sqrt{3})} = 5.774$$

Figure 34.7 *Three-Phase Transformer Configurations*

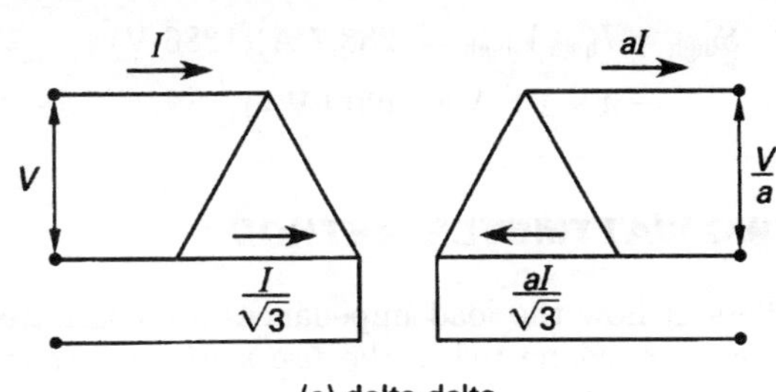

(a) delta-delta

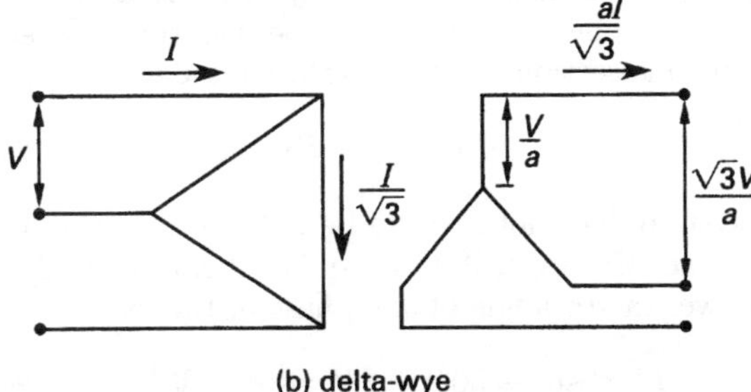

(b) delta-wye

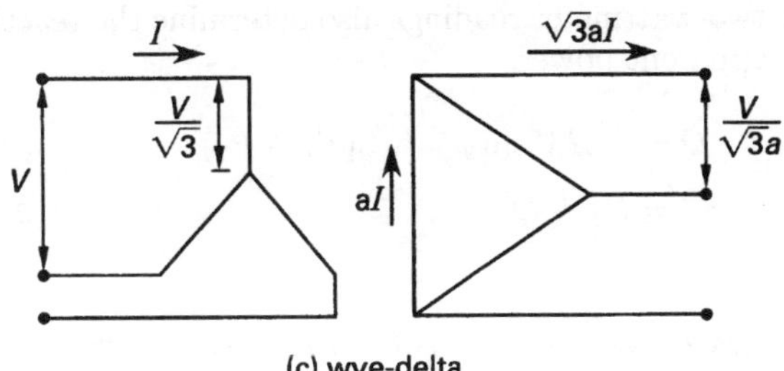

(c) wye-delta

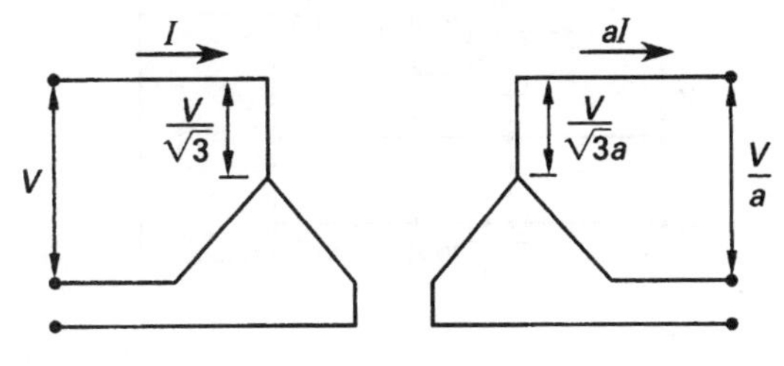

(d) wye-wye

(b) The winding (phase) voltages are

$$V_{\text{high}} = \frac{V}{\sqrt{3}} = \frac{2400\ \text{V}}{\sqrt{3}} = 1386\ \text{V}$$

$$V_{\text{low}} = 240\ \text{V} \quad \text{[given]}$$

(c) The winding (phase) currents are

$$I_{\text{high}} = \frac{S_p}{V_p} = \frac{\sqrt{3}S_t}{3V} = \frac{\sqrt{3}(1200 \times 10^3\ \text{VA})}{(3)(2400\ \text{V})} = 288.7\ \text{A}$$

$$I_{\text{low}} = aI_{\text{high}} = (5.774)(288.7\ \text{A}) = 1667\ \text{A}$$

(d) The transformer winding kVA rating (per phase) is

$$S_{\text{high}} = I_{\text{high}} V_{\text{high}} = (288.7 \text{ A})(1386 \text{ V})$$
$$= 4 \times 10^5 \text{ VA (400 kVA)}$$

14. TWO-WATTMETER METHOD

Regardless of how the load impedances are connected, two *wattmeters* connected in the *two-wattmeter configuration* shown in Fig. 34.8 can be used to determine the total real power in a three-wire system. Because one or both of the power readings can be negative, the additions in this section must be performed algebraically.

$$P_t = P_1 + P_2 \qquad 34.57$$

The system does not need to be balanced for Eq. 34.57 to be used. However, if it is, Eq. 34.58 can be used. ϕ is the power factor angle of one phase of the load.

$$P_t = IV\cos(\phi - 30°) + IV\cos(\phi + 30°) \qquad 34.58$$

The two-wattmeter readings also determine the reactive and apparent power.

$$Q = \sqrt{3}IV\sin\phi = \sqrt{3}(P_1 - P_2) \qquad 34.59$$

$$S^2 = P^2 + Q^2 \qquad 34.60$$

Figure 34.8 Connections for the Two-Wattmeter Method

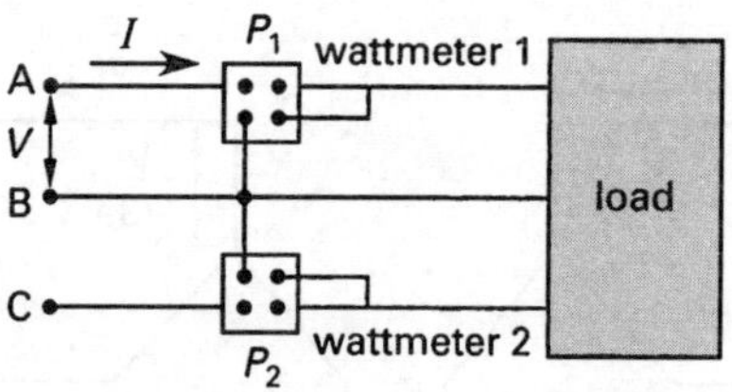

Generation

15. FAULTS AND FAULT CURRENT

A *fault* is an unwanted connection (i.e., a short circuit) between a line and ground or another line. Although the *fault current* is usually very high before circuit breakers trip, it is not infinite, because the transformers and transmission line have finite impedance up to the fault point. If the line impedance is known, the fault current can be found by Ohm's law.[7]

$$V = I_{\text{fault}} Z \qquad 34.61$$

The voltage level during a transient will change, though using the rated value can provide a worst-case condition. Impedance levels change during the three stages of any fault condition. So Eq. 34.61 can, and should, be modified to provide more detailed analysis. Details are given in Chap. 36 and Chap. 40. Additionally, the IEEE Standard 141, titled *IEEE Recommended Practice for Electric Power Distribution for Industrial Plants*, provides detailed information on short-circuit analysis techniques.

[7]Equation 34.61 ignores the transient (DC) current component and the relatively small current flowing in the wire before the fault occurs. Although the fault current has a transient component, it dies out so quickly that it is insignificant.

35 Batteries, Fuel Cells, and Power Supplies

Nomenclature

AW	atomic weight	g/mol
C	capacity	A·h
e^0	standard electrode potential	V
E^0	standard cell potential	V
$\mathcal{E}$	electromotive force	V
I	effective or DC current	A
P	power	W
Q	overcharge	A·h
R	resistance	Ω
V	effective or DC voltage	V
V	volume	L
z	valence or valence change	–

Subscripts

bat	battery
cell	per cell
H_2	hydrogen
int	internal
s	source
t	terminal

1. BATTERY FUNDAMENTALS

A *battery* is a direct-current voltage source that converts chemical, thermal, nuclear, or solar energy into electrical energy. Chemical conversion is used in the majority of batteries and will be the focus of this section. The advantages of a battery as a source of electrical energy are its portability, storage capability, and nonpolluting operation. The disadvantages are its general energy inefficiency and low power capability, though both these disadvantages are lessening as battery technology advances.[1]

A battery may be made up of one or more units called *cells*.[2] A *primary cell* undergoes chemical change in such a manner that one or both electrodes are unusable following discharge. A *secondary cell* undergoes chemical change in a reversible manner and can be recharged following operation. *Shelf life* is the time a battery can be stored without use before its capacity drops to a value that makes it uneconomical to store any further. The *cycle life* is the number of times a battery can be discharged or charged before the output voltage drops to a prescribed value. A *cycle* is a predetermined number of amp-hours (A·h) of discharge or charge.

[1]*Energy efficiency* is the ratio of the discharging energy (in J) to the charging energy (in J). Values between 50% and 70% are common.
[2]The first storage cell was invented by Italian professor Alessandro Volta in 1800.

2. BATTERY THEORY

An *electrolyte* is a compound that, when dissolved in certain solvents (usually water), will conduct electricity. When placed in a solution, an electrolyte dissociates into ions as shown in Fig. 35.1(a). Placing two dissimilar electrodes in the solution results in a migration of the ions according to each ion's affinity. The result is a potential difference between the two electrodes, as shown in Fig. 35.1(b). The magnitude of the potential difference depends on the materials used for the electrodes and is determined by the specific electronic structures, of those materils. Some typical combinations are shown in Table 35.1.

Table 35.1 Cell Voltages

battery type/name	anode (negative electrode)[a]	cathode (positive electrode)[b]	nominal cell voltage (V)
primary/Leclanche	Zn	MnO_2	1.2
primary/ mercury-zinc	Zn	HgO	1.2
primary/ silver-zinc	Zn	AgO	1.6
secondary/ lead-acid	Pb	PbO_2	1.8
secondary/ nickel-cadmium	Cd	NiOOH	1.2
secondary/ silver-zinc	Zn	AgO	1.5
secondary/ nickel-metal hydride	H_2·(metals)	NiOOH	1.2

[a]The anode is positive with respect to the battery, which results in a negatively charged electrode.
[b]The cathode is negative with respect to the battery, which results in a positively charged electrode.

Generation

Figure 35.1 Battery Theory

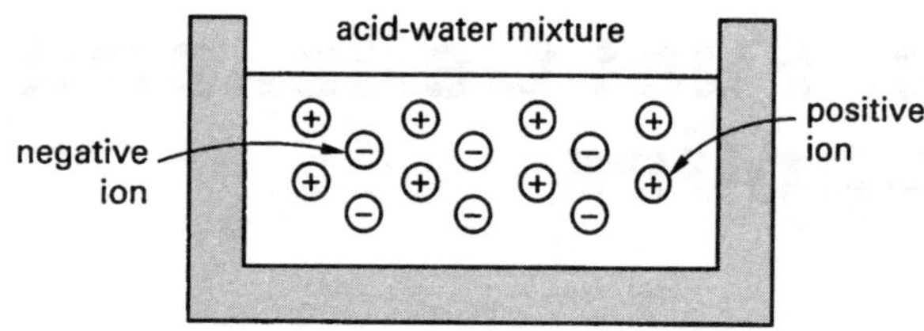

(a) electrolyte dissociation

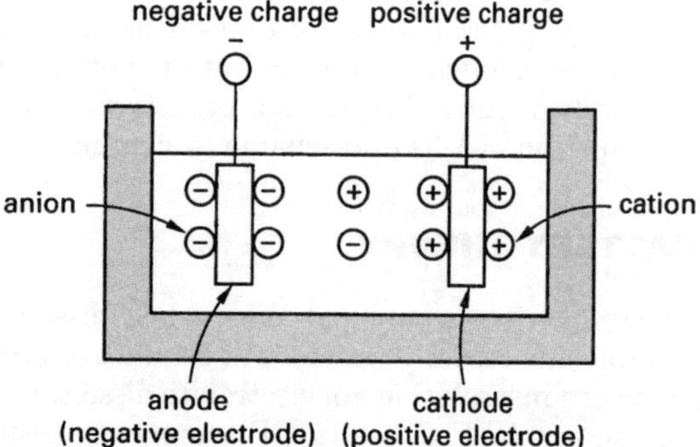

(b) potential due to electrode affinity

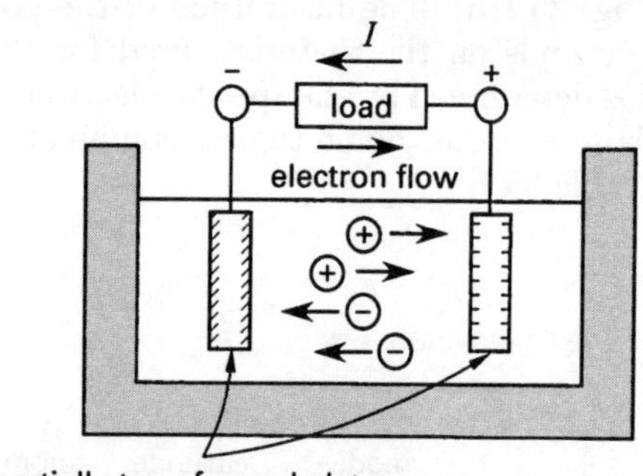

(c) current flow

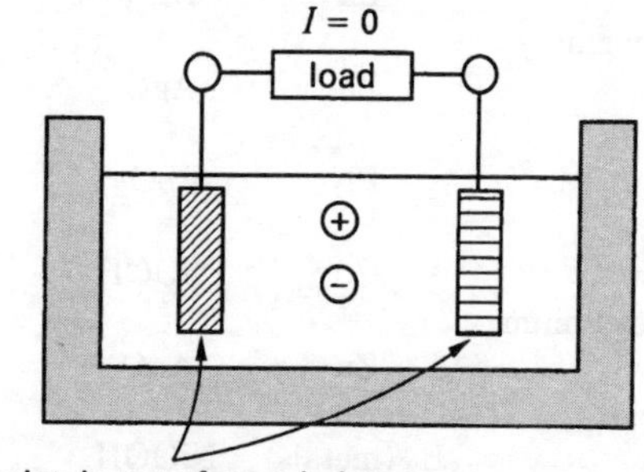

(d) cell discharged (electrode chemical change complete)

The potential difference between two electrodes, or the *standard cell electromotive force (emf)*, E^0, can be calculated from the standard electrode potentials, e^0, using Eq. 35.1.

$$E^0 = e_1^0 - e_2^0 \qquad 35.1$$

Table 35.2 lists standard electrode potentials for common electrode reactions. By convention, this table gives the potentials for reactions based on setting the hydrogen reaction at zero volts as an arbitrary reference. However, the reactions are reversible, and multiplying the tabulated electrode potential by -1 gives the potential for the opposite reaction in which one or more electrons are lost. The larger the potential, the more easily the reaction can occur; a reversible reaction will tend to occur spontaneously in the direction with a positive potential, unless external voltages force the reaction to occur in the opposite direction.

From Table 35.2, then, the available potential from a pair of electrode materials can be found, as well as the direction of the associated reaction. For example, if the electrodes are of lithium and copper, the two electrode reactions would be

$$Li^+ + e^- \rightarrow Li$$

$$Cu^+ + e^- \rightarrow Cu$$

From Table 35.2, when a lithium ion (Li^+) gains an electron to become lithium (Li), the reaction has a potential of –3.05 V. Because this potential is negative, however, this reaction tends to occur spontaneously in the opposite direction, with lithium losing an electron to become a lithium ion. When a copper ion (Cu^+) gains an electron to become copper (Cu), the reaction has a potential of +0.52 V; as this potential is positive, this reaction tends to occur spontaneously in this direction rather than the reverse.

If a copper electrode and a lithium electrode are placed together in solution, with no outside forces acting, the net reaction will be

$$Cu^+ + Li \rightarrow Cu + Li^+$$

Electrons will flow from the lithium to the copper, and the net reaction will have a valence change of one electron and a potential of 3.05 V + 0.52 V = 3.57 V.

The potential is determined entirely by the chemical reactions that result from the materials used. The potential cannot be increased or decreased by changing the amount of material in the electrodes; it can only be changed by using a different combination of materials.

The capacity of the battery is determined by the mass of the materials used. The total *coulombic capacity* (mg/C) or *ampere-hour capacity* (g/A·h) can be estimated from the atomic weight of the element (AW) and the valence change (z), using Eq. 35.2 or Eq. 35.3. However, these equations solve for a theoretical capacity, which a real battery would not achieve; how close a battery's capacity comes to this theoretical value depends on its shape and internal design.

Table 35.2 Standard Reversible Potentials at 25°C

species in equilibrium	e^0 (V)
$Li^+ \rightarrow Li$	−3.05
$K^+ \rightarrow K$	−2.92
$Ba^{++} \rightarrow Ba$	−2.90
$Ca^{++} \rightarrow Ca$	−2.87
$Na^+ \rightarrow Na$	−2.71
$Mg^{++} \rightarrow Mg$	−2.37
$Be^{++} \rightarrow Be$	−1.85
$Al^{+++} \rightarrow Al$	−1.66
$Ti^{++} \rightarrow Ti$	−1.63
$Mn^{++} \rightarrow Mn$	−1.18
$Zn^{++} \rightarrow Zn$	−0.76
$Cr^{+++} \rightarrow Cr$	−0.74
$Ga^{+++} \rightarrow Ga$	−0.52
$Fe^{++} \rightarrow Fe$	−0.44
$Cr^{+++} \rightarrow Cr^{++}$	−0.41
$Cd^{++} \rightarrow Cd$	−0.40
$Ti^{+++} \rightarrow Ti^{++}$	−0.37
$Co^{++} \rightarrow Co$	−0.28
$Ni^{++} \rightarrow Ni$	−0.25
$Cu^{++} \rightarrow Cu^+$	−0.15
$Sn^{++} \rightarrow Sn$	−0.14
$Pb^{++} \rightarrow Pb$	−0.13
$H^+ \rightarrow H$	0
$Cu^{++} \rightarrow Cu$	+0.34
$O_2 \rightarrow OH^-$	+0.40
$Cu^+ \rightarrow Cu$	+0.52
$I_2 \rightarrow 2I^-$	+0.54
$Fe^{+++} \rightarrow Fe^{++}$	+0.77
$Hg_2^{++} \rightarrow 2Hg$	+0.79
$Pd^{++} \rightarrow Pd$	+0.98
$Br_2(l) \rightarrow 2Br^-$	+1.07
$Cr_2O_7^- \rightarrow Cr^{+++}$	+1.33
$Cl_2(g) \rightarrow 2Cl^-$	+1.36
$Ce^{++++} \rightarrow Ce^{+++}$	+1.61
$Au^+ \rightarrow Au$	+1.68
$F_2 \rightarrow 2F^-$	+2.90

$$C_{\text{mg/C}} = \frac{\left(0.010363\ \frac{\text{mol·mg}}{\text{C·g}}\right)(\text{AW})}{z} \quad \text{[coulombic]} \qquad 35.2$$

$$C_{\text{g/A·h}} = \frac{\left(0.037307\ \frac{\text{mol}}{\text{A·h}}\right)(\text{AW})}{z} \quad \text{[ampere-hour]} \qquad 35.3$$

Example 35.1

Copper has an atomic weight of 63.54 g/mol and a valence change of 1. Find the coulombic and ampere-hour capacities of copper.

Solution

Use Eq. 35.2. The coulombic capacity is

$$\begin{aligned} C_{\text{mg/C}} &= \frac{\left(0.010363\ \frac{\text{mol·mg}}{\text{C·g}}\right)(\text{AW})}{z} \\ &= \frac{\left(0.010363\ \frac{\text{mol·mg}}{\text{C·g}}\right)\left(63.54\ \frac{\text{g}}{\text{mol}}\right)}{1} \\ &= 0.66\ \text{mg/C} \end{aligned}$$

Use Eq. 35.3. The ampere-hour capacity is

$$\begin{aligned} C_{\text{g/A·h}} &= \frac{\left(0.037307\ \frac{\text{mol}}{\text{A·h}}\right)(\text{AW})}{z} \\ &= \frac{\left(0.037307\ \frac{\text{mol}}{\text{A·h}}\right)\left(63.54\ \frac{\text{g}}{\text{mol}}\right)}{1} \\ &= 2.4\ \text{g/A·h} \end{aligned}$$

Regardless of the material used and the potential developed in a battery, the chemical reaction that generates current starts when an external load is attached to the terminals. When an external load is attached to the terminals, current flows from the positive electrode (cathode) to the negative electrode (anode) in the external circuit as shown in Fig. 35.1(c). In the internal circuit, current flows from the anode to the cathode. Positive ions (cations) migrate toward the cathode (positive electrode), where they capture incoming electrons and become neutral atoms, transforming the material on the cathode. Negative ions (anions) migrate toward the anode where they lose their electrons by combining with the active material of the anode. When the reactive material on either plate is consumed, the chemical reaction stops and the current ceases as shown in Fig. 35.1(d).

Positive and negative potential on a battery can be referred to in different ways. The anode is positive with respect to the internal circuit (that is, the battery itself). The cathode is negative with respect to the internal circuit (that is, the battery itself). In electrical circuit schematics showing a battery, the power is shown flowing out of the positively charged electrode (the cathode) during discharge, because it is positive with respect to the external circuit. The positive, or conventional, current flows from the anode to the cathode within the battery and from the cathode to the anode in the external circuit (that is, from the negative electrode to the positive electrode within the battery and from the positive to the negative in the external circuit). On an actual battery terminal, however, the anode (negative electrode) is labeled as positive even though it has a negative charge. Alternatively, the anode (negative electrode) can be defined as the terminal at which conventional current enters a battery. This plus and minus convention applies to electrochemical cells.[3]

All batteries possess an internal resistance to current flow that results in a lower output voltage as the current increases. The equivalent circuit for a battery is shown in Fig. 35.2.

Figure 35.2 *Battery Equivalent Circuit*

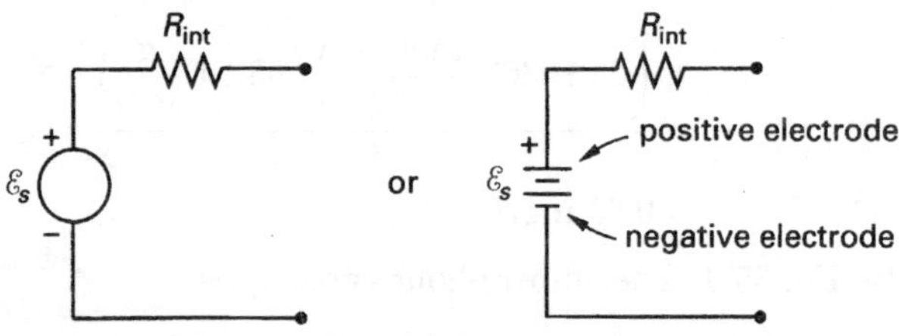

Example 35.2

A 1.5 V battery with an internal resistance of 0.3 Ω is connected to a 1 Ω load. What is the load current?

Solution

The circuit is that of Fig. 35.2 with a 1 Ω resistor attached to the terminals. Using Ohm's law, the current is

$$V = IR$$
$$I = \frac{V}{R} = \frac{V}{R_{\text{int}} + R_{\text{load}}} = \frac{1.5 \text{ V}}{0.3 \ \Omega + 1 \ \Omega}$$
$$= 1.15 \text{ A}$$

Example 35.3

For the battery and load in Ex. 35.2, what is the power consumed by the internal resistance?

Solution

The power is determined by

$$P = I^2R = (1.15 \text{ A})^2(0.3 \ \Omega)$$
$$= 0.397 \text{ W}$$

Example 35.4

For the battery and load in Ex. 35.2, what is the terminal voltage?

Solution

Using KVL around the loop of Fig. 35.2 gives

$$\mathcal{E}_s - IR_{\text{int}} - V_t = 0$$

Using the current from Ex. 35.1 and solving for the terminal voltage gives

$$V_t = \mathcal{E}_s - IR_{\text{int}} = 1.5 \text{ V} - (1.15 \text{ A})(0.3 \ \Omega)$$
$$= 1.15 \text{ V}$$

3. BATTERY TYPES AND CAPACITY

Primary batteries are those that cannot be recharged. Primary batteries that contain no free or liquid electrolyte are called *dry cell batteries*. A *reserve battery* is one in which an essential component is withheld, making the battery inert and permitting long-term storage. A *secondary battery* is one designed so that it can be recharged electrically following discharge. Charging is accomplished by passing a current through the battery in the direction opposite that of the discharge current.

Batteries are also typed according to cell sizes established by the American National Standards Institute (ANSI). Some examples of typical sizes are given in Table 35.3.

As seen in Table 35.3, service capacity of a battery varies. As the current withdrawn is increased, the capacity drops. This occurs primarily because of the battery's internal resistance and the effect of *polarization*. Polarization is the phenomenon whereby, when a cell discharges current, hydrogen is given off at one of the terminals. The hydrogen bubbles act to insulate the electrode, lowering the chemical reaction rate and the current.

Table 35.3 *Cell Size*

cell size	starting drain current (mA)	service capacity (h)
AAA	2	290
	10	45
	20	17
AA	3	350
	15	40
	30	15
B	5	420
	25	65
	50	25
C	5	430
	25	100
	50	40
D	10	500
	50	105
	100	45

[3]The sign convention for electrochemical cells is the opposite of the sign convention for galvanic cells.

Hydrogen production also occurs during battery charging. For a battery that is fully charged, the approximate hydrogen production is given by the empirical formula

$$V_{H_2} \approx 0.42\left(\frac{V_{bat}Q}{V_{cell}}\right) \quad 35.4$$

The term V_{H_2} is the volume of hydrogen produced in liters. Each liter of hydrogen produced correlates to the electrolysis of 0.8 mL of water. The term V_{bat} is the total battery voltage while V_{cell} is the individual cell voltage. The term Q is the amount of overcharge in amp-hours (A·h).

The service capacity combined with the current withdrawn determines the *battery capacity.* Because the capacity varies with current, it is normally measured in amp-hours (A·h) over a standard period. Standard time periods include one, five, eight, and ten hours.

Various battery technologies are shown in Table 35.4.

Example 35.5

A 12 V lead-acid battery with six 2 V cells is left on a charger for 24 h following completion of the charge. Assuming a charging current of 3 A, how much hydrogen is produced?

Solution

From Eq. 35.2, the volume of hydrogen produced is approximately

$$V_{H_2} \approx 0.42\left(\frac{V_{bat}Q}{V_{cell}}\right)$$

The battery voltage is given. The cell voltage is 2 V. The amount of overcharge, Q, is

$$Q = (3\text{ A})(24\text{ h}) = 72\text{ A·h}$$

Substituting gives

$$V_{H_2} \approx 0.42\left(\frac{V_{bat}Q}{V_{cell}}\right) = \left(0.42\ \frac{\text{L}}{\text{A}\cdot\text{h}}\right)\left(\frac{(12\text{ V})(72\text{ A·h})}{2\text{ V}}\right)$$
$$= 181\text{ L}$$

Example 35.6

How much does the capacity of a D-cell battery drop when the drain current is increased from 10 mA to 100 mA?

Solution

From Table 35.3, the capacity in A·h at 10 mA is

$$C = (10 \times 10^{-3}\text{ A})(500\text{ h}) = 5000 \times 10^{-3}\text{ A·h}$$

From Table 35.3, the capacity in A·h at 100 mA is

$$C = (100 \times 10^{-3}\text{ A})(45\text{ h}) = 4500 \times 10^{-3}\text{ A·h}$$

The decrease in capacity is

$$1 - \frac{4500 \times 10^{-3}\text{ A·h}}{5000 \times 10^{-3}\text{ A·h}} = 0.1 \quad (10\%)$$

4. FUEL CELLS

A *fuel cell* is a device that converts chemical energy directly into electrical energy. A fuel cell differs from a battery in that the fuel and oxidant must be continuously supplied and, at least in theory, no heat is produced as a result of the reaction. A fuel cell uses both liquid and gaseous fuels, such as hydrogen, hydrazine, and coal gas. The oxidant is oxygen or air. A hydrogen-oxygen fuel cell is shown in Fig. 35.3.

Figure 35.3 *Hydrogen-Oxygen Fuel Cell*

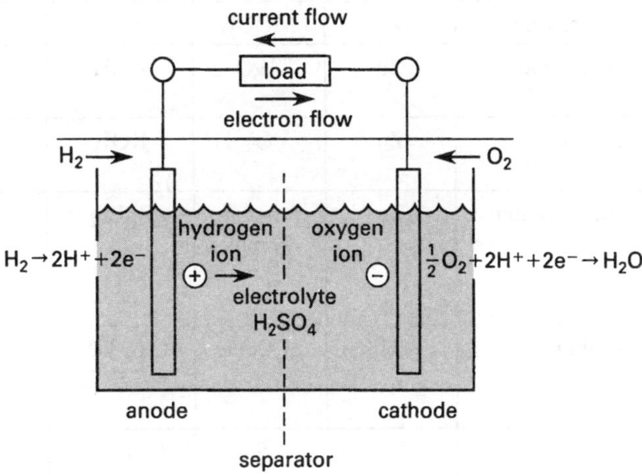

A fuel—in this case, hydrogen—is supplied continuously to an electrode, where it undergoes a catalytic reaction producing electrons and a fuel ion ($2H^+$). The electrons travel through the external circuit, supplying power to the load. The fuel ion migrates through the electrolyte and gas separation barrier to the oxygen electrode. The incoming oxygen reacts with the cathode to produce negative oxygen ions that diffuse in the electrolyte. At the oxidizing electrode's surface, the oxygen, hydrogen ions, and electrons combine through a catalytic reaction. The net result is hydrogen and oxygen combining to form water and produce electrical energy.

Fuel cells are of interest because of their high efficiency, minimal environmental impact, and ability to be constructed in a modular fashion.

5. POWER SUPPLIES

A *power supply* is any source of electrical energy, although the term generally refers to a battery or power line used to supply an electronic circuit with the proper

Table 35.4 Battery Technologies

battery system	negative electrode	positive electrode	electrolyte	nominal voltage (V)	theoretical specific energy (W·h/kg)	practical specific energy (W·h/kg)	practical energy density (W·h/L)	major issues
lead-acid	Pb	PbO_2	H_2SO_4	2.0	252	35	70	heavy, low cycle life, toxic materials
nickel iron	Fe	NiOOH	KOH	1.2	313	45	60	heavy, high maintenance
nickel cadmium	Cd	NiOOH	KOH	1.2	244	50	75	toxic materials, maintenance, cost
nickel hydrogen	H_2	NiOOH	KOH	1.2	434	55	60	cost, high pressure hydrogen, bulky
nickel metal hydride	H (as MH)	NiOOH	KOH	1.2	278–800 (depends on MH)	70	170	cost
nickel zinc	Zn	NiOOH	KOH	1.6	372	60	120	low cycle life
silver zinc	Zn	AgO	KOH	1.9	524	100	180	very expensive, limited life
zinc air	Zn	O_2	KOH	1.1	1320	110	80	low power, limited cycle life, bulky
zinc bromine	Zn	bromine complex	$ZnBr_2$	1.6	450	70	60	low power, hazardous components, bulky
lithium ion	Li	Li_xCoO_2	PC or DMC w/$LiPF_6$	4.0	766	120	200	safety issues, calendar life, cost
sodium sulfur	Na	S	beta alumina	2.0	792	100	>150	high temperature battery, safety, low power electrolyte
sodium nickel chloride	Na	$NiCl_2$	beta alumina	2.5	787	90	>150	high temperature operation, low power

electric voltages and currents for operation.[4] An example is shown in Fig. 35.4.

The purpose of the power supply is to take a given input and shape the output such that the electronic circuitry can operate accurately and without short- or long-term damage. The principles of the component parts are covered in Chap. 28, Chap. 46, and Chap. 51.

6. UNINTERRUPTIBLE POWER SUPPLIES (UPS)

Due to the sensitivity or the importance of a given load, the power to the load must be maintained during interruptions of the normal power source and may need to be *conditioned* to ensure the best possible signal is supplied. That is, the signal should ideally be without distortion. An example of a device that can be used to

[4]Power supplies, along with batteries, relays, switches, programmable logic controllers (PLCs), and variable speed drives, are listed by the NCEES in the exam specifications as a part of the devices and power electronic circuits subtopic under the circuit analysis topic. The principles of these items and others are covered in Chap. 29, Chap. 35, Chap. 44, and Chap. 48.

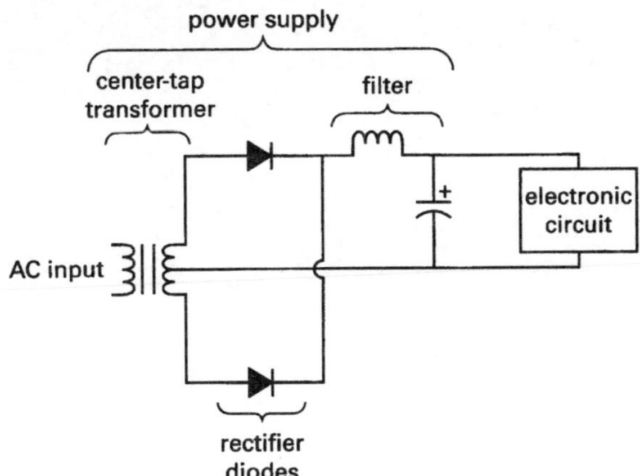

Figure 35.4 Power Supply

provide conditioned power and act as a backup power supply during interruptions in the normal power source is shown in Fig. 35.5.

In Fig. 35.5, the *uninterruptible power supply* (UPS) receives AC input power and rectifies this to DC power. The DC power is used to maintain the charge of a set of batteries whose capacity sets the length of time power can be maintained without the normal power source. Filter circuitry may also be present to smooth distortions of the incoming AC power and provide *conditioned power* to the load. The DC power is then inverted to AC power for use by the load. During normal power failures or excessive distortion (which is sensed by circuitry not shown), the battery immediately supplies the inverter, thereby replacing the incoming source. No interruption occurs to the load, because the battery is on-line at all times. Should a UPS failure itself occur, units are equipped with a bypass used to provide *unconditioned power* directly to the load.

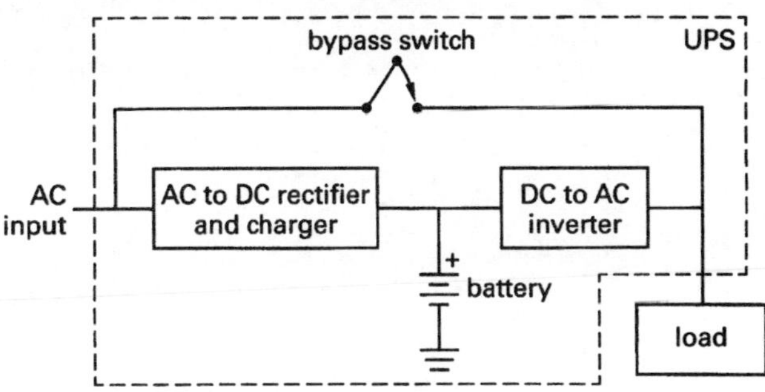

Figure 35.5 Uninterruptible Power Supply Block Diagram

Some units advertised as UPS units are actually *standby power sources*. The main difference can be visualized from Fig. 35.5. If the bypass switch is normally open and the unit is designed to pass power from the rectifier through a conditioning section to an inverter, the unit is a UPS, sometimes called a *conditioned power source* or *conditioned UPS*. If the bypass switch is normally closed and the circuitry is only a battery backup, then the unit is more correctly called a *standby power source*. The associated circuitry does not generally include conditioning filters and the like, and the unit is used as a *battery backup*.

Generation

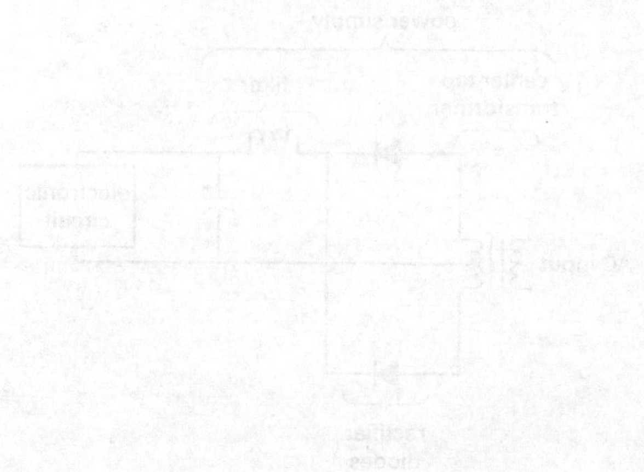

Topic VI: Distribution

Chapter

36. Power Distribution
37. Power Transformers
38. Transmission Lines
39. The Smith Chart

36 Power Distribution

Nomenclature

C	capacitance	F
E	electric field strength	V/m
E'	transient electric field strength	V/m
E''	subtransient electric field strength	V/m
i	variable current	A
I	effective or DC current	A
L	inductance	H
p	pressure	Pa
pf	power factor	–
Q	charge	C
r	radius	m
S	apparent power	VA
v	wind velocity	km/hr
V	effective or DC voltage	V
X	reactance	Ω
X'	transient reactance	V/m
X''	subtransient reactance	V/m
Z, $\mathbf{Z}$	impedance	Ω

Symbols

ϵ	permittivity	F/m
ϵ_0	free-space permittivity, 8.854×10^{-12}	F/m
μ_0	free-space permeability, 1.2566×10^{-6}	H/m
ξ	ratio of radii	–

Subscripts

0	zero sequence
1	positive sequence
2	negative sequence
d	direct or direct axis
dc	DC component
ext	external
f	fault
g	generated or generator
gen1, gen2	generator 1, generator 2
k	short-circuit
l	line or per unit length
L	load
max	maximum
m	motor
n	neutral
p	peak or phase
pu	per unit
q	quadrature
s	synchronous
sc	short circuit
t	terminal

1. FUNDAMENTALS

The *electric distribution system* is the collection of circuitry, high-voltage switchgear, transformers, and related equipment that receives high voltage from the source and delivers it at lower voltages. Its function is to receive power from large bulk sources, that is, generation sources, and distribute it to users at the voltage levels and degrees of reliability required. A hypothetical distribution system one-line diagram is shown in Fig. 36.1. The electric utility may consider the distribution system as that portion of Fig. 36.1 from the distribution substation to the consumer. The symbols used are explained in Fig. 36.2.

One-line diagrams can be used for balanced three-phase systems because the per-phase values are equal. If a neutral line exists, it is not shown. If the diagram is used for *load studies*, the circuit breaker, fuse, and other switching device locations are not shown because they are of no concern. If the diagram is used for *stability studies* or *fault analysis*, the locations and characteristics of circuit breakers, fuses, relays, and other protective devices are shown.

At the *generation level* in Fig. 36.1, the voltage is approximately 6.9 kV. The synchronous generators are shown with their neutrals connected through impedances designed to limit surges in case of a fault in the generator circuit. One or more generators may be attached to a given power bus.

At the *transmission level*, the transmission lines connect the various generators to one another and to the subtransmission lines.

At the *subtransmission level*, the voltage range is 12.5–245 kV. The most commonly used voltages in order of usage are 115 kV, 69 kV, 138 kV, and 230 kV. The subtransmission lines are usually in *grid form*. The grid form allows connections between input buses and various paths to each distribution substation, increasing

Figure 36.1 *Distribution System*

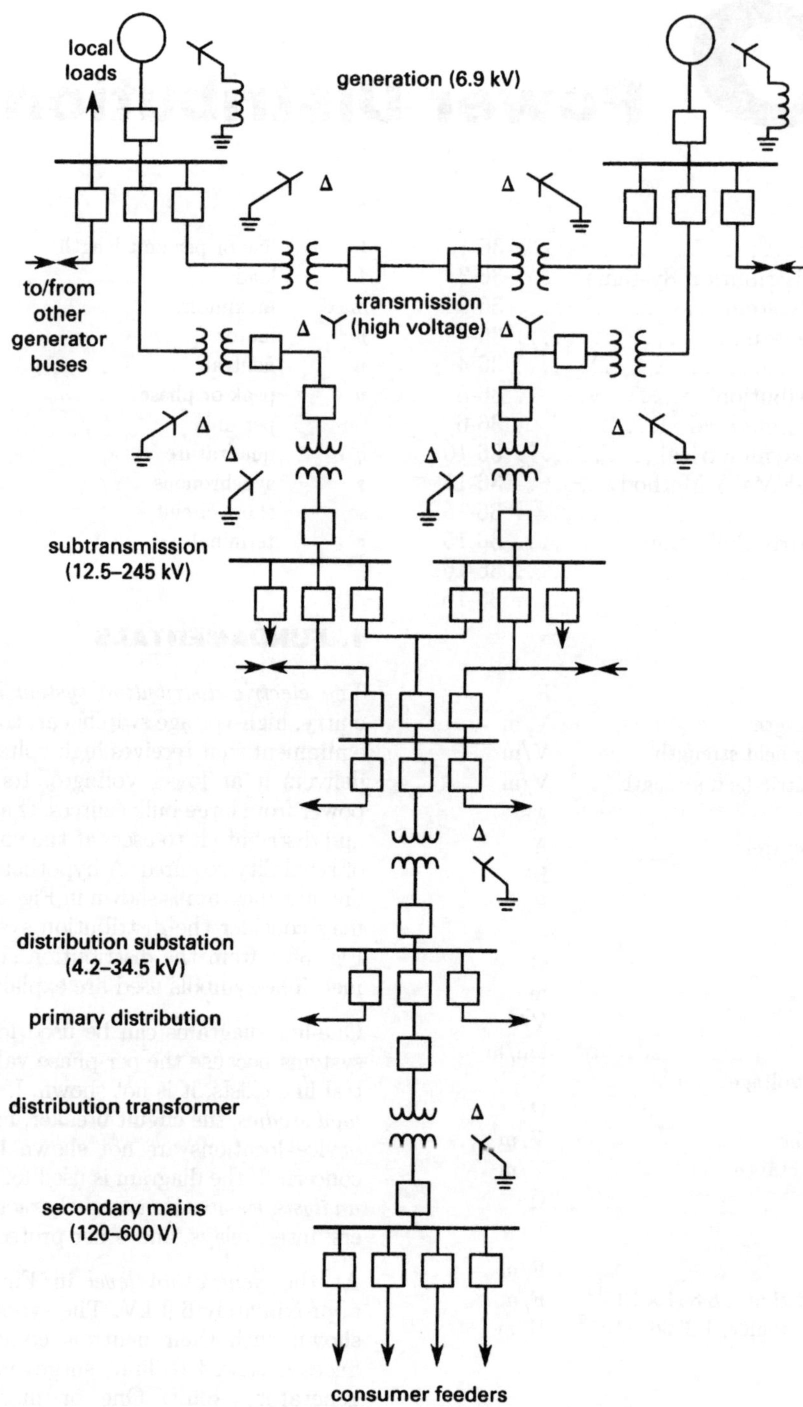

reliability. A minimum of two switchable inputs to each substation input bus is normally used.

At the *primary distribution level*, the voltage range is 4.2–34.5 kV. The most commonly used voltages are 12.5 kV, 25 kV, and 34.5 kV. The actual voltage level is controlled with taps on the substation transformers. The distribution substation supplies several distribution transformers connected in a radial, tree, loop, or grid system. The distribution transformers are mounted on poles, grade-level pads, or underground near the user (a substation may be dedicated to a single large user). Secondary mains operate at a voltage range of 120–600 V.[1]

[1]The ANSI standard for Voltage Range A is 114/228 V to 126/252 V at the service entrance and 108/220 V to 126/252 V at the utilization point.

Distribution

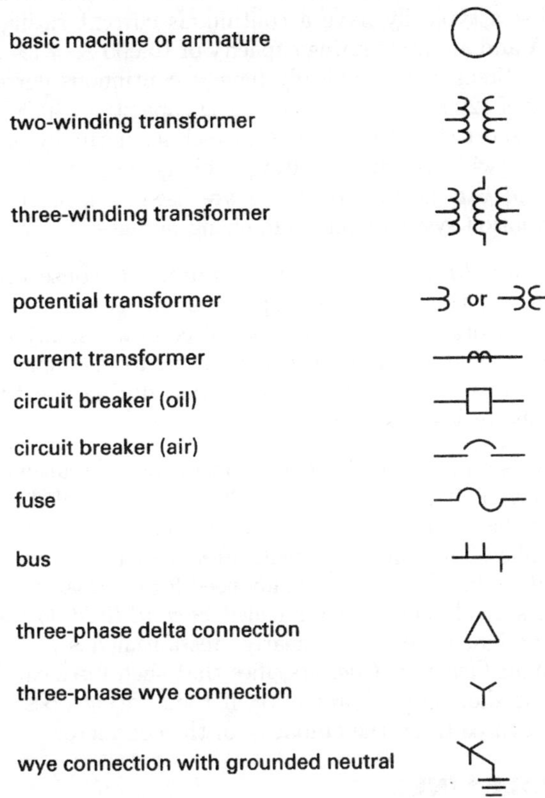

Figure 36.2 *Common Distribution Symbols*

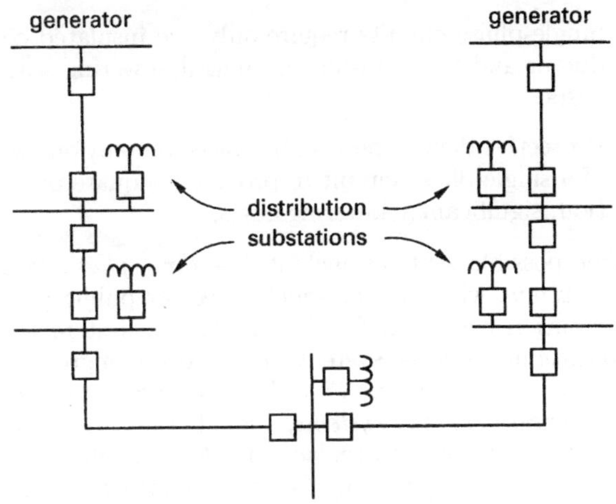

Figure 36.3 *Loop Distribution Pattern*

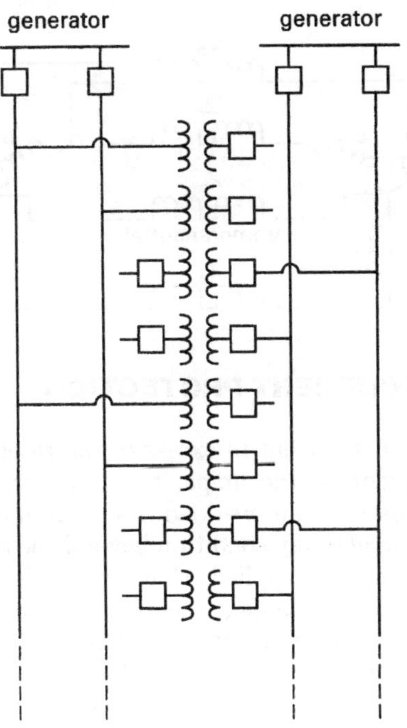

Figure 36.4 *Multiple Distribution Pattern*

2. CLASSIFICATION OF DISTRIBUTION SYSTEMS

Distribution systems are classified in a number of ways, including

- current (AC or DC)
- voltage (120 V, 12.5 kV, 15 kV, 34.5 kV, etc.)
- type of load (residential, commercial, lighting, power, etc.)
- number of conductors (two-wire, three-wire, etc.)
- construction type (overhead or underground)
- connection type

Connection types include radial (seldom used), loop (see Fig. 36.3), network, multiple (see Fig. 36.4), and series (used primarily for street lighting).

3. COMMON-NEUTRAL SYSTEM

In practice, because of economic and operating advantages, almost all distribution systems are three-phase four-wire, common-neutral primary systems like the one shown in Fig. 36.5. The fourth wire acts as a neutral for both the primary and secondary systems (see Fig. 36.1) although one side of the transformer is delta. (The delta connection is used because the third-harmonic voltages generated by the nonlinear nature of the transformer circulate within the delta and do not affect the line currents or voltages.) The neutral is grounded in numerous locations, including at each distribution transformer and to metallic water piping or grounding devices at each service entrance.[2] The fourth wire carries a portion of the unbalanced currents that may exist, while the remainder flows through the earth. Some of the advantages of such a system follow.

- Unbalanced currents flow through the neutral or the earth, ensuring approximately balanced three-phase current flow through the generator. This evens the

[2]The use of nonmetallic water piping requires the use of other grounding techniques described in the *National Electrical Code*.

Distribution

countertorque, minimizes vibration, and extends bearing life.

- Single-phase circuits require only one insulated conductor and the uninsulated neutral, lowering wiring costs.
- Protective devices need to be placed on only one wire of a single-phase circuit to provide adequate protection, significantly lowering costs.

Other possible systems include the *three-phase three-wire system*, which is not widely used for public power distribution except in California. Such a system finds application in marine systems that operate ungrounded to improve reliability. *Two-phase systems* are rarely used. *Direct-current systems*, though they have some advantages, have been replaced by AC systems. *Series systems* are used primarily for street lighting but have been largely replaced by multiple systems.

Figure 36.5 Common-Neutral (Three-Phase Four-Wire) System

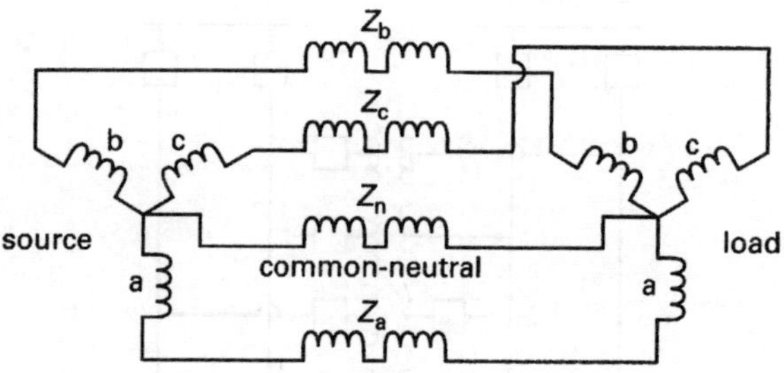

4. OVERCURRENT PROTECTION

Overcurrent is current in excess of the rated current of the equipment or the ampacity of the conductors. It may be caused by an overload, short circuit, or ground fault. *Overload* is any condition beyond the rating of the electrical device supplied that if continued for an extended time will damage the equipment or cause overheating. A short circuit or ground fault is not an overload. The short circuit or ground fault conditions can result in extremely high currents for very short time periods. Overload conditions are longer term. Protective devices include relayed circuit breakers, reclosers, sectionalizers, and fuses.

A *circuit breaker* is an electromagnetic device used to open and close a circuit. The operation of a circuit breaker can be automatic or manual. For overcurrent protection, relays are installed that sense the current and create a corresponding magnetic field whose force is used to trip the breaker if necessary. The trip characteristics of some circuit breakers are adjustable. An *instantaneous trip* is one that occurs without delay. An *inverse time trip* is one in which a delay is deliberately instituted. As the current magnitude increases, the time to trip decreases. The *National Electrical Code* specifies that a circuit breaker is considered to provide adequate short-circuit protection if its rating is no more than six times the ampacity of the conductors.

A *recloser* is a device that opens a circuit instantaneously when a fault occurs but recloses after a short period. Reclosers typically have a continuous current rating of 560 A and an interrupting capacity of 16,000 A or less. A circuit breaker will typically have a continuous current rating of 1200 A and an interrupting capacity of 40,000 A under short-circuit conditions. A recloser normally operates at twice its current rating. This is called the *minimum pickup current.* Reclosers are used to protect the distribution system while minimizing outages.

A *sectionalizer* is a device that counts the consecutive number of times a recloser operates and opens the circuit at a predetermined number of counts, usually two or three. The sectionalizer has no interrupting capacity. It operates to keep the recloser open during the time that the recloser is open.

A *fuse* is an expendable device that opens a circuit when the current becomes excessive. The fuse is a sealed container that contains a conductor surrounded by quartz-sand filler. The conductor melts after the current exceeds a given value. Though normally used for overload conditions, special types of fuses called *current-limiting fuses* are designed to melt nearly instantaneously. The *National Electrical Code* specifies that such fuses can be used as short-circuit protection if their rating does not exceed three times the ampacity of the conductors.

5. POLE LINES

The poles that support overhead electric utility lines are constructed of wood, concrete, steel, and aluminum. Poles are subject to *vertical loading* from the weight of the conducting cable. Normal *transverse loading* occurs when a pole is located at a corner, that is, a point where the geographic direction of the conducting cable changes. Additional vertical and transverse forces account for *wind loading* and *ice loading.* This is accomplished by first calculating the forces using standard mechanical engineering techniques. Then the expected pressures on flat surfaces caused by the wind is determined from

$$p = 71.43\mathrm{v}^2 \qquad 36.1$$

The variable v represents the wind velocity normal to the flat surface in km/h. The pressure is in units of Pa. For curved surfaces, such as a standard pole, the pressure is determined from

$$p = 44.64\mathrm{v}^2 \qquad 36.2$$

Finally, a percentage of this force is added to the result of the mechanical and wind calculations to increase the amount of force that the pole must be designed to withstand. (The force calculations become moment calculations when the pole height is taken into consideration.) The percentage to be added is determined by the extreme wind and icing conditions expected in the area where the pole is to be located. The percentage may vary from 0% to 30%.

Copper conductor used for overhead lines where spans are approximately 61 m (200 ft) or more is the hard-drawn type due to its greater strength. For shorter spans, medium hard-drawn or annealed copper is utilized. For very long spans, aluminum stranded around a steel core is used. This type of cable is called *aluminum conductor steel-reinforced* (ACSR). High-strength aluminum alloys are also used. One type is called *aluminum conductor alloy-reinforced* (ACAR). Another is the *all-aluminum-alloy conductor* (AAAC).

6. UNDERGROUND DISTRIBUTION

Underground installations have increased in popularity, especially in residential districts, primarily for aesthetic reasons, though there are indications that the frequency of faults is lower compared to overhead systems. When faults do occur, however, they are more difficult to access and repair. In a residential district, such a system is called an *underground residential distribution* (URD). Since the development of synthetic insulation such as polyethylene, aluminum conductor is used almost exclusively for underground installations.[3] In copper conductors, standard soft copper is used to improve flexibility. Unlike overhead lines, underground cables must be insulated to ensure the conductor does not contact ground or the cable's external sheath. The insulation also protects against mechanical, chemical, and other effects peculiar to an underground environment.

The insulating material surrounding a conductor is a dielectric. Figure 36.6(a) shows such an arrangement. The insulation thickness must be such that the electric field strength, E, at the surface of the conductor does not break down the insulation. If the cable is too large, it becomes difficult to handle and expensive. If the cable is too small, the dielectric loss becomes large, resulting in overheating of the cable. The optimal thickness is determined by the ratio in Eq. 36.3.

$$\frac{r_2}{r_1} = E \approx 2.718 \qquad 36.3$$

Figure 36.6(a) illustrates a cable with a single dielectric. When more than one dielectric is present, they are arranged to minimize the difference between the maximum and minimum electric field strength across the cable. This arrangement is known as *grading*. *Capacitance grading* is illustrated in Fig. 36.6(b). Given the restriction of a maximum electric field, E_{max}, the operating voltage for a capacitance-graded cable is given by

$$V = E_{\text{max}}\left(r_1 \ln \frac{r_2}{r_1} + r_2 \ln \frac{r_3}{r_2}\right) \qquad 36.4$$

When the dielectric material is separated by coaxial metallic sheaths maintained at a constant voltage level, the grading is called *innersheath grading* (see

[3]Aluminum conductor connections require special attention to ensure adequate contact and avoid corrosion.

Figure 36.6 *Insulated Conductors*

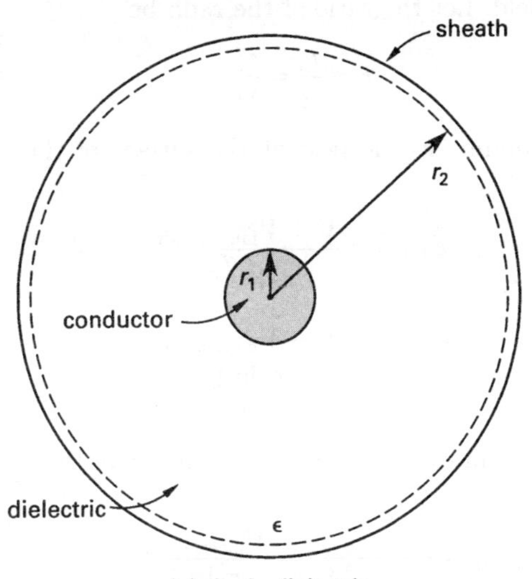

(a) single dielectric

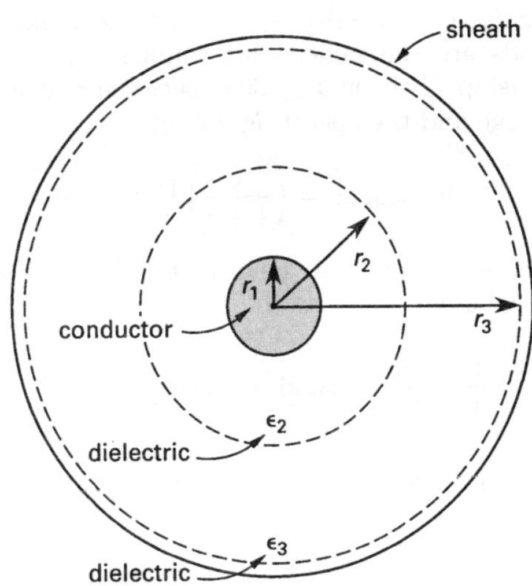

(b) multiple dielectric: capacitance grading

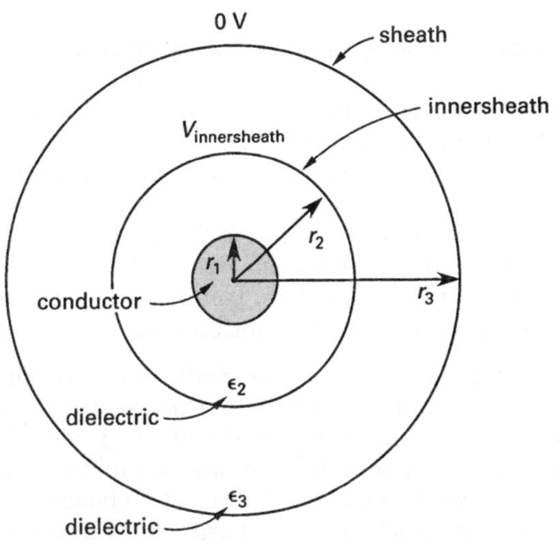

(c) multiple dielectric: innersheath grading

Distribution

Fig. 36.6(c)). The sheathing is used to minimize the electric field. Let the ratio of the radii be

$$\xi = \frac{r_3}{r_2} = \frac{r_2}{r_1} \qquad 36.5$$

The maximum electric field at the surface of the conductor is

$$E_{1,\max} = \frac{V - V_{\text{innersheath}}}{r_1 \ln \frac{r_2}{r_1}} = \frac{V - V_{\text{innersheath}}}{r_1 \ln \xi} \qquad 36.6$$

The maximum electric field at the surface of the innersheath is

$$E_{2,\max} = \frac{V_{\text{innersheath}}}{r_2 \ln \frac{r_3}{r_2}} = \frac{V_{\text{innersheath}}}{r_2 \ln \xi} \qquad 36.7$$

The permittivities are selected so that the maximum electric fields are the same. This condition results in the relationship given in Eq. 36.8 between the innersheath voltage and the operating voltage.

$$V_{\text{innersheath}} = \left(\frac{\xi}{1+\xi}\right) V \qquad 36.8$$

The capacitance per unit length, C_l, for a single-conductor cable is

$$C_l = \frac{Q}{V} = \frac{2\pi\epsilon}{\ln \xi} \quad \text{[farads per meter]} \qquad 36.9$$

The inductance per unit length, L_l, for a single-conductor cable is

$$L_l = \frac{\mu_0}{2\pi} \ln \xi \quad \text{[henries per meter]} \qquad 36.10$$

7. FAULT ANALYSIS: SYMMETRICAL

A *fault* is any defect in a circuit, such as an open circuit, short circuit, or ground. Short-circuit faults, called *shunt faults*, are shown in Fig. 36.7. Open-circuit faults are called *series faults*. Any fault that connects a circuit to ground is termed a *ground fault*. The balanced three-phase short circuit shown in Fig. 36.7(d) is one of the least likely, yet most severe, faults and determines the ratings of the supplying circuit breaker.

A three-phase *symmetrical fault*, such as that in Fig. 36.7(d), can be divided into three stages, as shown in Fig. 36.8. During the *subtransient period*, which lasts only for a few cycles, the current rapidly decreases and the synchronous reactance, X_s, changes to the subtransient reactance, X''_d. There are two reactances: the *direct axis reactance*, X_d, that lags the generator voltage by 90°, and a *quadrature reactance*, X_q, that is

Figure 36.7 *Fault Types, from Most Likely to Least Likely*

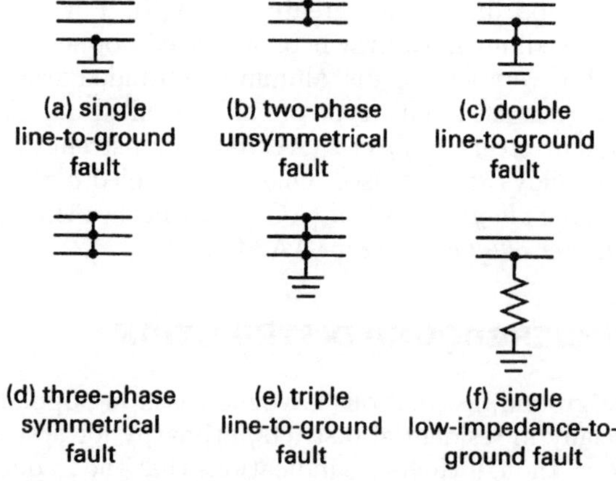

Figure 36.8 *Symmetrical Fault Terminology*

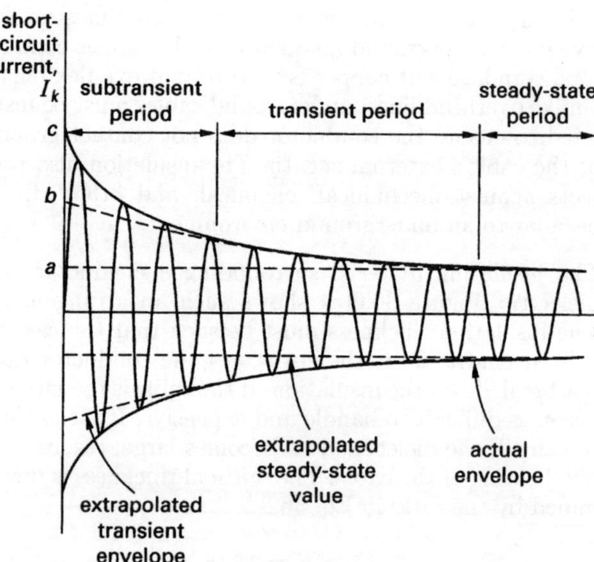

(a) synchronous generator three-phase fault response

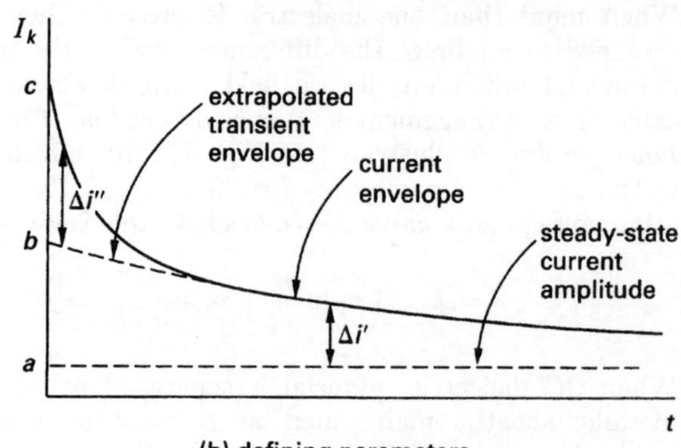

(b) defining parameters

in phase with the generator voltage.[4] The model for a synchronous generator changes from that in Fig. 36.9(a) to that in Fig. 36.9(b). The sudden change in armature current results in a lower armature reactance, but the current through the leakage reactance must remain the same for continuity of energy. For proper modeling, it is necessary to change the voltage, E_g, to E''_g. The voltage E''_g is calculated for the subtransient interval *just prior to the initiation of the fault* using Eq. 36.11.[5]

$$E''_g = V_t + jI_L X''_d \qquad 36.11$$

During the *transient period*, a similar situation exists and the model for the synchronous generator changes from that in Fig. 36.9(b) to that in Fig. 36.9(c). The correct generator voltage for this period is calculated *just prior to the initiation of the fault* using

$$E'_g = V_t + jI_L X'_d \qquad 36.12$$

Both E''_g and E'_g depend on the impedance of the load and the resulting current prior to fault initiation. The reactances associated with the subtransient, transient, and steady-state periods have increasing values; that is, $X''_d < X'_d < X_d$. Typical reactances of three-phase synchronous machines are given in Table 36.1. The corresponding short-circuit currents have decreasing magnitudes; that is, $|I''| > |I'| > |I|$. These currents are the symmetrical rms currents associated with a fault.

In Fig. 36.8, the current at point *c*, I_c, is the *subtransient current*, I'', and is often called the *initial symmetrical rms current*. The *DC component of the fault current* is the exponentially decaying portion—that is, the exponential portion—of the short-circuit current. This DC component and the prospective short-circuit current are shown in Fig. 36.10. The subscript *k*, indicates short-circuit parameters in many standards.[6]

The initial symmetrical rms current (the subtransient current), I'', should not be confused with the initial *symmetrical short-circuit current*, I''_k, shown in Fig. 36.10 and used in IEEE standards and in various short-circuit analysis software programs. I''_k is the maximum rms value of the prospective (available) short-circuit current with the impedance at the zero-time

Figure 36.9 Synchronous Generator Fault Models

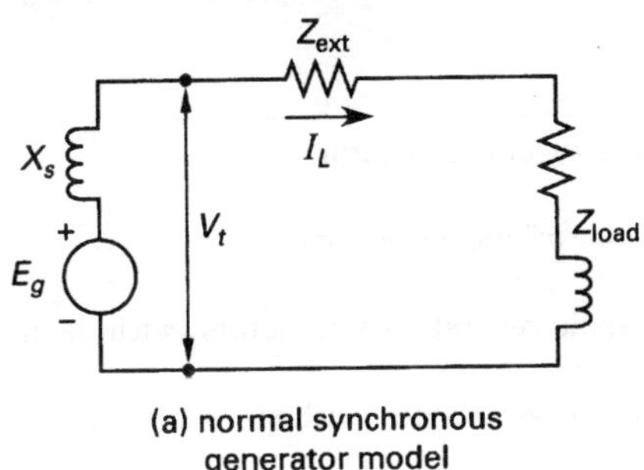

(a) normal synchronous generator model

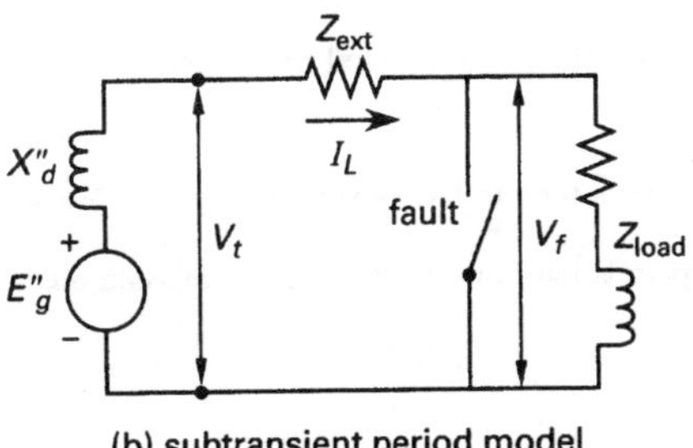

(b) subtransient period model

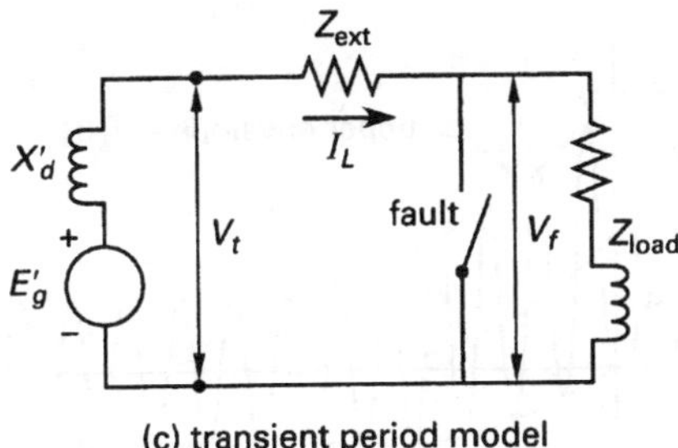

(c) transient period model

value (prior to the fault). The subtransient short-circuit current, I'', is the fault current flowing with the reactance equal to the subtransient reactance (after fault initiation), X''_d.[7] The subtransient reactance is used in determining the *peak short-circuit current*, i_p, shown in Fig. 36.10. This peak current is the maximum possible value of the short-circuit current and is equivalent to the subtransient current, I''.

In a synchronous generator, the subtransient time and reactance depend on the damping in the rotor circuit, the transient time and reactance depend on the damping in the excitation circuits, and the steady-state time and reactance depend on the stator circuit.[8] Most calculations use only the direct axes reactances. Ignoring the quadrature reactances results in an error of approximately 10%, and the values are higher when quadrature reactance is taken into account. Using the quadrature values makes the calculations complex and gains little in accuracy.

[4]During faults, the power factor is low and the quadrature reactance can be ignored.

[5]The general form for the synchronous generator voltage is $E_g = V_t + jIX$.

[6]These standards include IEEE 551, *Recommended Practice for Short-Circuit Calculations in Industrial and Commercial Power Systems* (*IEEE Violet Book*), and IEEE 399, *Recommended Practice for Industrial and Commercial Power Systems Analysis* (*IEEE Brown Book*). These two standards reference international standard IEC 60909-0, *Short-Circuit Currents in Three-Phase A.C. Systems—Part 0: Calculation of Currents*, which uses the *k* subscript used in Fig. 36.10. Slight differences exist between the U.S. standards and the international standard. Both sets of standards, however, are used in short-circuit analysis software, and the terminology used in both may be encountered by a professional engineer.

[7]This is consistent with *Power System Analysis* by Grainger and Stevenson. (See Codes and References.)

[8]The steady-state reactance is the normal generator-synchronous reactance, X_d.

Table 36.1 *Typical Reactances of Three-Phase Synchronous Machines (typical values given above bars, ranges given below bars)*

	X_d (unsaturated)	X_q (rated current)	X'_d (rated voltage)	X''_d (rated voltage)
two-pole turbine generators	1.20	1.16	0.15	0.09
	0.95–1.45	0.92–1.42	0.12–0.21	0.07–0.14
four-pole turbine generators	1.20	1.16	0.23	0.14
	1.00–1.45	0.92–1.42	0.20–0.28	0.12–0.17
salient-pole generators and motors (with dampers)	1.25	0.70	0.30 *	0.20 *
	0.60–1.50	0.40–0.80	0.20–0.50	0.13–0.32
salient-pole generators (without dampers)	1.25	0.70	0.30 *	0.30 *
	0.60–1.50	0.40–0.80	0.20–0.50	0.20–0.50
capacitors (air-coded)	1.85	1.15	0.40	0.27
	1.25–2.20	0.95–1.30	0.36–0.50	0.19–0.30
capacitors (hydrogen-coded at 1/2 psi)	2.20	1.35	0.48	0.32
	1.50–2.65	1.10–1.55	0.36–0.60	0.23–0.36

*High-speed units tend to have low reactance, and low-speed units tend to have high reactance.

Figure 36.10 *IEEE Nomenclature for Three-Phase, Short-Circuit Current*

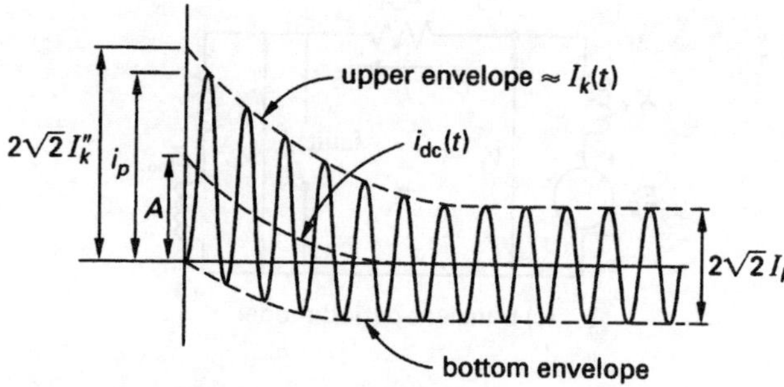

A initial value of aperiodic component
i_{dc} DC (aperiodic) component of short-circuit current
i_p peak short-circuit current
I''_k initial symmetrical short-circuit current
I_k steady-state short-circuit current

The subtransient time period lasts a few cycles and is defined in the IEEE standards as the time it takes for the rapidly changing component of the direct-axis, short-circuit current to decrease to $1/e$, or 0.368, of its initial value. The transient time period is defined in the standards as the time it takes for the slowly changing component of the direct-axis, short-circuit current to decrease to $1/e$, or 0.368, of its initial value.[9] The steady-state period lasts from the end of the transient period until the circuit breaker or other interrupting device acts to remove the fault.

When a synchronous motor is part of a system, it is considered a synchronous generator for fault analysis.[10] Circuit breakers must be able to interrupt fault currents at the rated voltage. The usual fault ratings are in kVA or MVA. The fault rating is the product of the interrupt current capacity and the line-to-line kV rating.

Example 36.1

A synchronous generator and motor are rated for 15 MVA, 13.9 kV, and 25% subtransient reactance. The line impedance connecting the generator and motor is $0.3 + j0.4$ pu, given with the ratings as the base. The motor is operating at 12 MVA, 13 kV, with a 0.8 pf lagging. Determine the per-unit equivalent generated voltage for the motor, $E''_{g,m}$, with the expectation of a fault at the motor terminals.

Solution

The one-line diagram illustrating the situation is

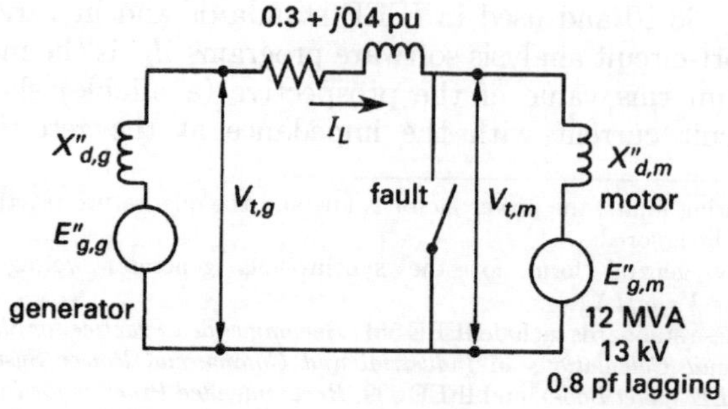

[9]Subtransient times vary with system electrical values, but "a few" cycles is often the range: approximately 0 cycles to 4 cycles, or 0 ms to 66 ms. Transient times last approximately 6 cycles, or from 100 ms to as long as 5 s.

[10]The motor actually becomes a generator when a fault occurs. The spinning motor changes the mechanical energy it possesses into electrical energy, supplying the fault and slowing in the process.

Determine the base quantities for this three-phase system.

$$S_{\text{base}} = S_t = 15 \text{ MVA} \quad (15 \times 10^6 \text{ VA})$$

$$V_{\text{base}} = V_l = 13.9 \text{ kV} \quad (13.9 \times 10^3 \text{ V})$$

$$I_{\text{base}} = \frac{S_{\text{base}}}{\sqrt{3}V_{\text{base}}} = \frac{15 \times 10^6 \text{ VA}}{\sqrt{3}(13.9 \times 10^3 \text{ V})} = 623 \text{ A}$$

$$Z_{\text{base}} = \frac{V_{\text{base}}}{\sqrt{3}I_{\text{base}}} = \frac{13.9 \times 10^3 \text{ V}}{\sqrt{3}(623 \text{ A})} = 12.9 \ \Omega$$

Determine the per-unit quantities of interest, that is, the voltage at the fault site (the terminals of the motor), and the current flowing.

$$V_{\text{pu}} = \frac{V_{\text{actual}}}{V_{\text{base}}} = \frac{13 \times 10^3 \text{ V}}{13.9 \times 10^3 \text{ V}} = 0.935 \text{ pu}$$

$$I_{\text{pu}} = \frac{I_{\text{actual}}}{I_{\text{base}}} = \frac{\frac{12 \times 10^6 \text{ VA}}{\sqrt{3}(13 \times 10^3 \text{ V})}}{0.62 \times 10^3 \text{ A}} = 0.860 \text{ pu}$$

The power factor is 0.8 lagging, so

$$\text{lagging pf} = \cos\theta = 0.8$$
$$\theta = \cos 0.8 = -36.9^\circ$$

The line current, I_l, is

$$I_l = I_{\text{pu}}\angle\theta = 0.860 \text{ pu}\angle{-36.9^\circ} = 0.688 - j0.516 \text{ pu}$$

The equivalent generated voltage for the motor can now be found. Writing the equation for the voltage from the motor terminals to the motor armature gives

$$E''_{g,m} = V_{t,m} - jX''_d I_l = 0.935 \text{ pu} - j(0.25)(0.688 - j0.516 \text{ pu}) = 0.806 - j0.172 \text{ pu} = 0.824 \text{ pu}\angle{-12^\circ}$$

Example 36.2

Determine the per-unit equivalent generated voltage for the generator, $E''_{g,g}$, for the system in Ex. 36.1.

Solution

The motor terminal voltage is known and the line current is known. An equation for the equivalent generated voltage, written from the known quantities at the motor terminals, is

$$E''_{g,g} = V_{t,m} + I_l Z$$

The impedance is from the motor terminals back to the generator, which is the only unknown. From Ex. 36.1, since there is a 25% subtransient reactance, $Z_g = j0.25$ pu.

$$Z = Z_l + Z_g = (0.3 + j0.4 \text{ pu}) + j0.25 \text{ pu} = 0.3 + j0.65 \text{ pu} = 0.715 \text{ pu}\angle 65^\circ$$

Substituting gives

$$E''_{g,g} = V_{t,m} + I_l Z = 0.935 \text{ pu} + (0.860 \text{ pu}\angle{-36.9^\circ})(0.715 \text{ pu}\angle 65^\circ) = 1.5 \text{ pu}\angle 11.1^\circ$$

Example 36.3

Determine the per-unit subtransient period current for the motor in Ex. 36.1.

Solution

A fault at the motor terminals results in a current given by

$$I_m = \frac{E''_{g,m}}{jX''_d} = \frac{0.824 \text{ pu}\angle{-12^\circ}}{0.25 \text{ pu}\angle 90^\circ} = 3.3 \text{ pu}\angle{-102^\circ}$$

Example 36.4

Determine the per-unit subtransient period current for the generator in Ex. 36.1.

Solution

The generator current that flows to the fault is given by

$$I_g = \frac{E''_{g,g}}{Z} = \frac{1.5 \text{ pu}\angle 11.1^\circ}{0.715 \text{ pu}\angle 65^\circ} = 2.09 \text{ pu}\angle{-53.9^\circ}$$

Example 36.5

What is the actual fault current for the circuit in Ex. 36.1?

Solution

The actual fault current is the total current through the fault from the generator and the motor. The synchronous motor acts as a generator to the fault as it gradually loses speed. The subtransient period is too short for the motor to lose a significant amount of speed, so it can be treated as a generator. The total per-unit current is

$$I_{\text{pu}} = I_m + I_g = 3.3 \text{ pu}\angle{-102^\circ} + 2.09 \text{ pu}\angle{-53.9^\circ} = 4.95 \text{ pu}\angle{-83.7^\circ}$$

The actual current is

$$I_{\text{actual}} = I_{\text{pu}} I_{\text{base}} = (4.95 \text{ pu}\angle{-83.7^\circ})(0.62 \text{ kA}) = 3.07 \text{ kA}\angle{-83.7^\circ}$$

Example 36.6

Determine the minimum rating necessary for the motor and generator circuit breakers to protect against the fault described in Ex. 36.1.

Solution

The minimum rating is the base apparent power, S_{base}, multiplied by a factor representing the per-unit current flowing above the base current, $I_{pu,fault}$. For the motor circuit breaker,

$$\begin{aligned} S_{rating} &= I_{pu,fault} S_{base} = (3.3)(15\ \text{MVA}) \\ &= 49.5\ \text{MVA} \end{aligned}$$

For the generator circuit breaker,

$$\begin{aligned} S_{rating} &= I_{pu,fault} S_{base} \\ &= (2.09)(15\ \text{MVA}) \\ &= 31.4\ \text{MVA} \end{aligned}$$

8. FAULT ANALYSIS: UNSYMMETRICAL

Any fault that is not a three-phase short is an *unsymmetrical fault*, also called an *asymmetrical fault*. Examples include the line-to-line and line-to-ground faults shown in Fig. 36.7(b) through Fig. 36.7(f). Such faults are more common than three-phase shorts, but they are more difficult to analyze because they result in uneven phase voltages and currents. Nevertheless, unsymmetrical faults can be analyzed by separating unbalanced phasor components into three sets of symmetrical components as shown in Fig. 36.11.

The three sets of symmetrical components shown in Fig. 36.11 are referred to as the *positive-sequence*, *negative-sequence*, and *zero-sequence* components of the unsymmetrical phasors. The positive-sequence phasors, shown in Fig. 36.12(a), are three equal-magnitude phasors rotating counterclockwise in the sequence a, b, c. Positive-sequence phasors are 120 electrical degrees apart and sum to zero. Positive-sequence components are the ones normally used in electrical engineering. Such components represent balanced three-phase generators, motors, and transformers. The subscript 1 is normally used to indicate the positive sequence.

The negative-sequence phasors, shown in Fig. 36.12(b), are three equal-magnitude phasors rotating counterclockwise in the sequence a, c, b. They can also be represented as a mirror image of the positive-sequence phasors rotating in the clockwise direction. Negative-sequence phasors are 120 electrical degrees apart and sum to zero. The negative-sequence reactance is the average of the subtransient direct and quadrature reactances, as given in Eq. 36.13. The subscript 2 is normally used to indicate the negative sequence.

$$X_2 = \tfrac{1}{2}(X''_d + X''_q) \qquad \textit{36.13}$$

Figure 36.11 *Phasor Diagram: Symmetrical Components of Unbalanced Phasors*

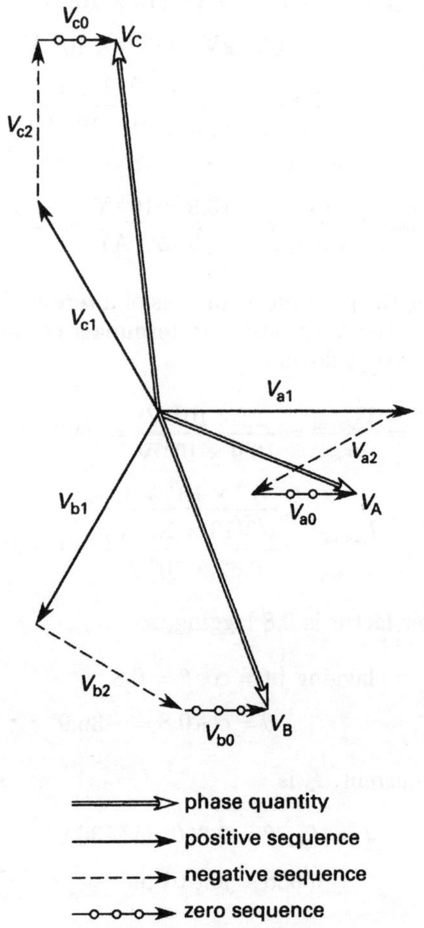

The zero-sequence phasors, shown again in Fig. 36.12(c), are three equal-magnitude phasors coincident in phase sequence and rotating counterclockwise. The subscript 0 is normally used to indicate the zero sequence.

The unsymmetrical phasors of Fig. 36.11 are represented in terms of their symmetrical components by Eq. 36.14 through Eq. 36.16.

$$V_A = V_{a0} + V_{a1} + V_{a2} \qquad \textit{36.14}$$

$$V_B = V_{b0} + V_{b1} + V_{b2} \qquad \textit{36.15}$$

$$V_C = V_{c0} + V_{c1} + V_{c2} \qquad \textit{36.16}$$

Consider the operator a, defined as $1\angle 120°$, a unit vector with an angle of 120°.[11] The properties of a are

$$\begin{aligned} a &= 1\angle 120° = 1 \times e^{j120°} \\ &= -0.5 + j0.866 \end{aligned} \qquad \textit{36.17}$$

$$a^2 = 1\angle 240° = -0.5 - j0.866 = a^* \qquad \textit{36.18}$$

$$a^3 = 1\angle 360° = 1\angle 0° \qquad \textit{36.19}$$

[11]Although a is a vector, it is not generally written as **a**. In this it resembles the operator j, which is equal to $1\angle 90°$ but is rarely written as **j**.

$$a^4 = a \qquad 36.20$$

$$a^5 = a^2 \qquad 36.21$$

$$a^6 = a^3 \qquad 36.22$$

$$1 + a + a^2 = 0 \qquad 36.23$$

Using Eq. 36.14 through Eq. 36.16 and the concept of the phasor **a**, the unsymmetrical components can be represented in terms of a single phase. For example, using phase A gives

$$V_A = V_{a0} + V_{a1} + V_{a2} \qquad 36.24$$

$$V_B = V_{a0} + a^2 V_{a1} + a V_{a2} \qquad 36.25$$

$$V_C = V_{a0} + a V_{a1} + a^2 V_{a2} \qquad 36.26$$

Solving for the sequence components from Eq. 36.24 through Eq. 36.26 gives

$$V_{a0} = \tfrac{1}{3}(V_A + V_B + V_C) \qquad 36.27$$

$$V_{a1} = \tfrac{1}{3}(V_A + a V_B + a^2 V_C) \qquad 36.28$$

$$V_{a2} = \tfrac{1}{3}(V_A + a^2 V_B + a V_C) \qquad 36.29$$

Equations similar to Eq. 36.24 through Eq. 36.29 exist for currents as well. Sequence impedances may also be defined. Impedance through which only positive-sequence currents flow is called a *positive-sequence impedance.* Positive-sequence networks are normally used in electrical engineering, and an example is given in Fig. 36.13(b). *Negative-sequence impedance* is similar to positive-sequence impedance with the reactance given by Eq. 36.13. In addition, a negative-sequence network omits all positive-sequence generators. A sample negative-sequence network is shown in Fig. 36.13(c). *Zero-sequence impedance* is significantly different from positive-sequence impedance. The only machine impedance seen by the zero-sequence impedance is the leakage reactance, X_0. Series reactance is greater than positive-sequence reactance by a factor of 2 to 3.5. A sample zero-sequence network is shown in Fig. 36.13(d). Only a wye-connected load with a grounded neutral permits zero-sequence currents. Only a delta-connected transformer secondary permits zero-sequence currents. Figure 36.14 shows zero-sequence impedances for various configurations. The use of sequence networks simplifies fault calculations.

Figure 36.12 *Components of Unsymmetrical Phasors*

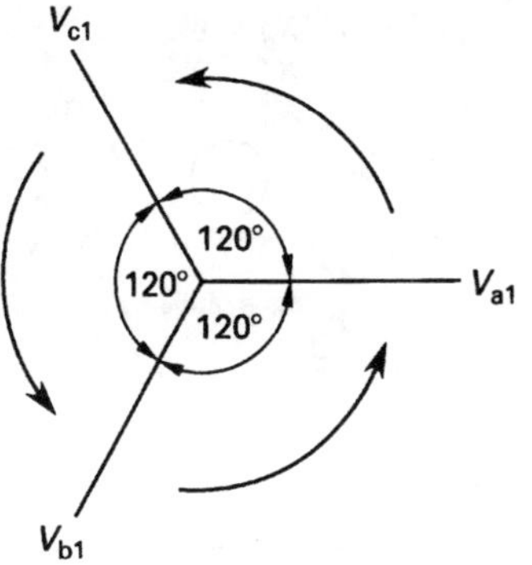

(a) positive sequence

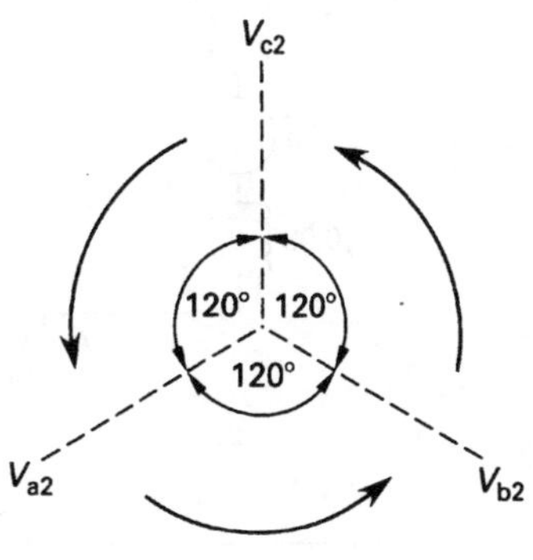

(b) negative sequence

$V_{a0} = V_{b0} = V_{c0}$

(c) zero sequence

9. FAULT ANALYSIS: THE MVA METHOD

The fault analysis techniques in Sec. 36.7 and Sec. 36.8 provide detailed information on the fault, although they also require significant calculation time. Information on those techniques may be found in IEEE Standard 141 titled *IEEE Recommended Practice for Electric Power Distribution for Industrial Plants.* International Standard IEC 60909 also provides guidance for short-circuit analysis that differs from the IEEE guidance and, in some ways, is more detailed. Numerous methods exist, including *ohmic*, *Thevenin*, *E/Z*, and *E/X* methods. Many of these short-circuit analysis methods require the use of computers.

A simplified method, which nevertheless captures the worst-case fault current, is called the *MVA method.* The MVA method doesn't need significant calculation time or software support. This method calculates the fault current that will flow in a system or component for unit voltage supplied from an infinite source. (The infinite source assumption means that the power sources are treated as *reactors*, that is, inductive components with no resistance, and the voltage is steady at the base value throughout.) The MVA method uses the per-unit system, maintains the voltage at a 1 pu value, and calculates the fault (or short-circuit) apparent power, S_{fault} or S_{sc}, and the fault (or short-circuit) current, I_{fault} or I_{sc}. Equation 36.30 and Eq. 36.31 are used to calculate the maximum fault power and fault current.

$$S_{\text{sc}} = \frac{S_{\text{base}}}{Z_{\text{pu}}} \qquad 36.30$$

Figure 36.13 Sample Sequence Networks

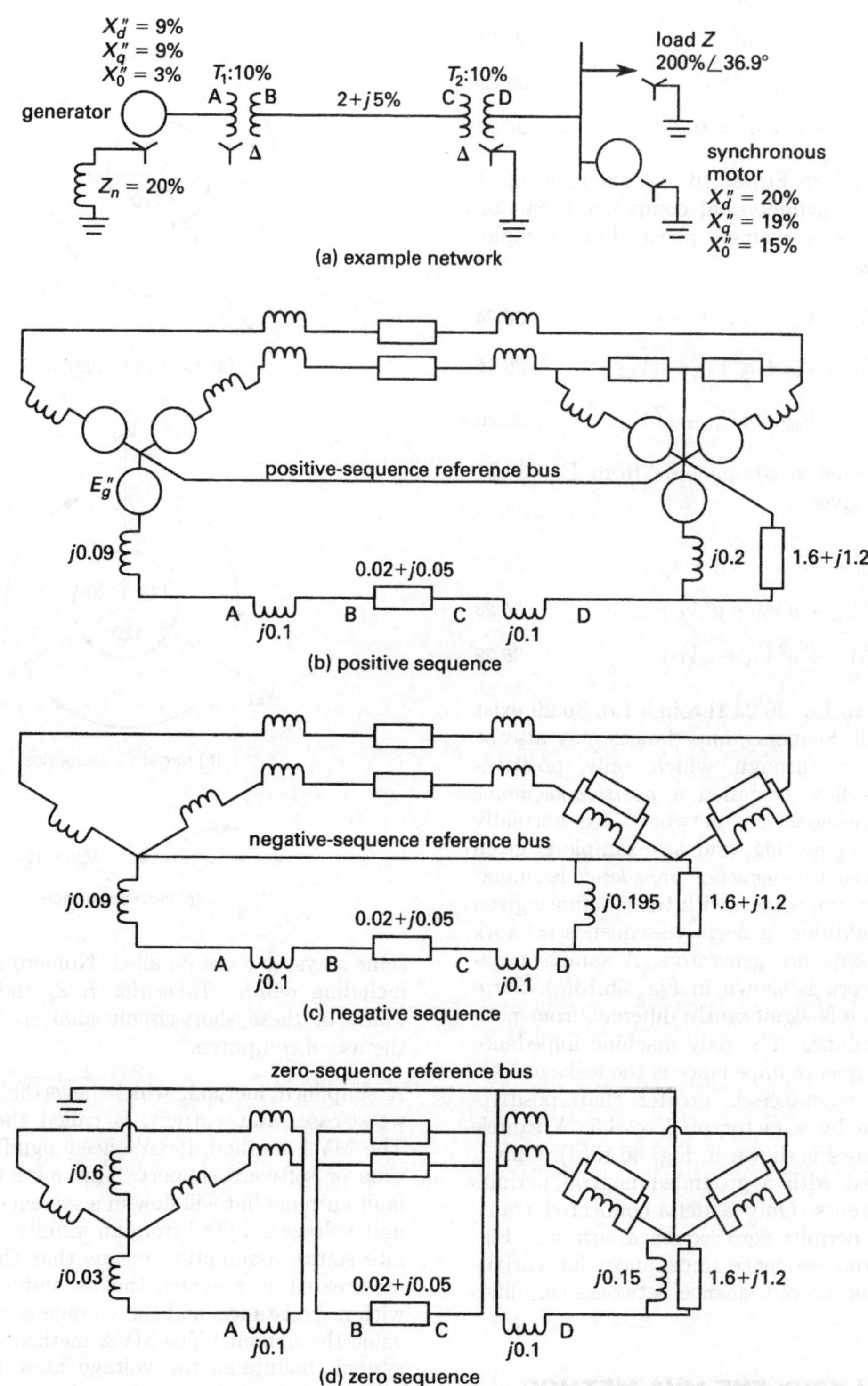

$$I_{sc} = \frac{S_{sc}}{\sqrt{3}V_{base}} \qquad 36.31$$

Variations of this method are used, including one that adds the power sources (i.e., the MVA sources) as vectors. Series power sources are treated as capacitors in series, given that the power transferred to the fault must go through both sources. For example, a generator has to pass its power through an intervening transformer. Parallel power sources are treated as capacitors in parallel, given that power from both is available to the fault.

Motors are treated as generators during short circuit fault analysis because rotational energy is converted to electrical energy and transferred to the fault during such events. Standard practices vary as to the contributions from various size motors.

Figure 36.14 Zero-Sequence Impedances

(a) ungrounded positive-sequence generator

(b) grounded positive-sequence generator

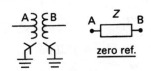

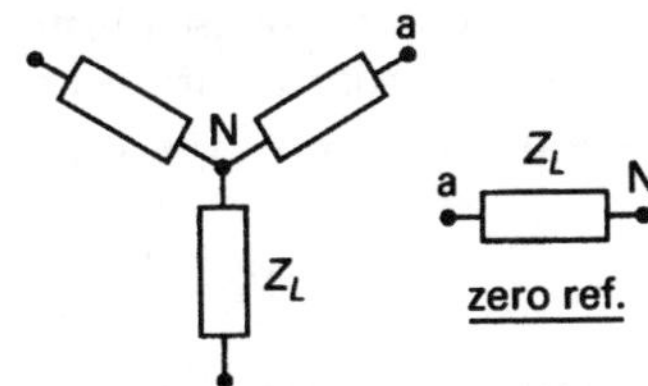

(c) ungrounded wye or delta load

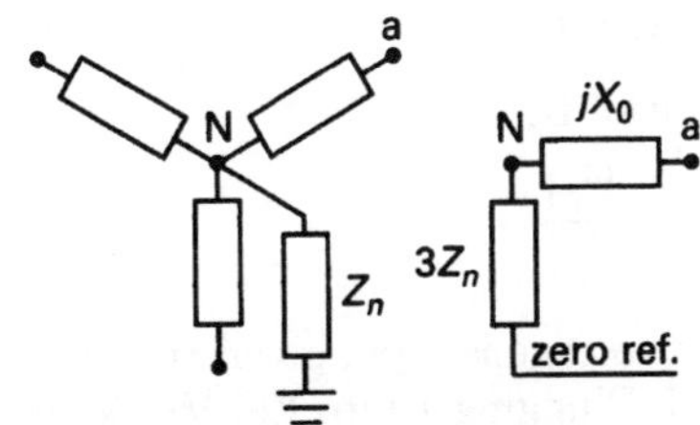

(d) grounded wye load

(e) wye–wye, both neutrals grounded

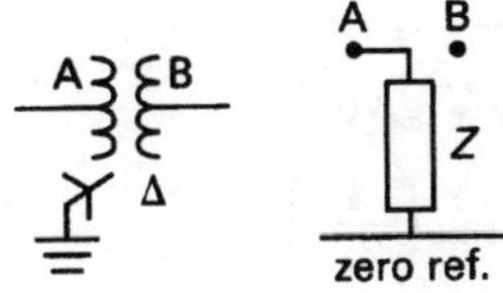

(f) wye–delta, grounded wye neutral

(g) wye–wye, one neutral grounded

(h) wye–delta, ungrounded

(i) wye–wye, ungrounded

(j) delta–delta

Finally, the equations used in this chapter assume symmetrical faults. The power may be calculated based on the positive-, negative-, and zero-sequence impedances, and power may then be combined to provide results for unsymmetrical faults.

Example 36.7

Derive the equation for fault apparent power, S_{sc}, Eq. 36.30.

Solution

The base apparent power is

$$S_{base} = \sqrt{3} I_{base} V_{base} \qquad \text{[I]}$$

Rearranging for the base current gives

$$I_{base} = \frac{S_{base}}{\sqrt{3} V_{base}} \qquad \text{[II]}$$

The fault apparent power and current will use the base voltage, as this provides the worst-case condition and allows for per-unit calculations across transformers. Changing Eq. I and Eq. II to their fault versions gives

$$S_{sc} = \sqrt{3} I_{sc} V_{base} \qquad \text{[III]}$$

$$I_{sc} = \frac{S_{sc}}{\sqrt{3} V_{base}} \qquad \text{[IV]}$$

Equation IV is in the form of Eq. 36.31. The short-circuit (fault) current, I_{sc}, is the actual current that flows in the fault. Using the per-unit system,

$$I_{sc} = I_{actual} = I_{base} I_{pu} \qquad \text{[V]}$$

The per-unit current is given by

$$I_{pu} = \frac{V_{pu}}{Z_{pu}} \qquad \text{[VI]}$$

The per-unit value of the voltage is 1 pu, given that the base voltage is being used. Plugging 1 into Eq. VI yields Eq. VII.

$$I_{pu} = \frac{1}{Z_{pu}} \qquad \text{[VII]}$$

Substituting Eq. VII into Eq. V gives

$$I_{sc} = I_{actual} = I_{base}\left(\frac{1}{Z_{pu}}\right) \qquad \text{[VIII]}$$

Substituting Eq. VIII into Eq. III gives

$$S_{sc} = \sqrt{3}\left(I_{base}\left(\frac{1}{Z_{pu}}\right)\right) V_{base} = \frac{\sqrt{3} I_{base} V_{base}}{Z_{pu}} \qquad \text{[IX]}$$

Eq. IX is used to find the base apparent power in Eq. X.

$$S_{sc} = \frac{\sqrt{3} I_{base} V_{base}}{Z_{pu}} = \frac{S_{base}}{Z_{pu}} \qquad \text{[X]}$$

Equation X is Eq. 36.30, and it completes the derivation.

Example 36.8

Consider the one-line diagram for a three-phase generator-reactor system in the illustration. The power ratings and per-unit impedances based on those ratings are given. Assume a worse-case symmetrical fault. Using the MVA method, what is the maximum short-circuit current at the fault location shown?

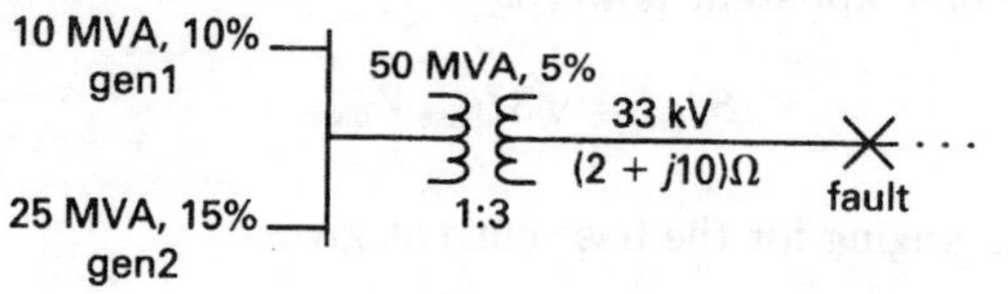

Solution

Using an arbitrary base of 100 MVA for S_{base} and 33 kV for V_{base}, Eq. 34.53 converts to the following reactance values from the given rating base values to the new base. (The reactances, or reactors, are the source of energy to the fault.) A base voltage of 33 kV is used to ensure the proper per-unit value was calculated for the line impedance because the conversion of actual to per-unit values must use the voltage present in the given section of the distribution system. Additionally, the fault is in the 33 kV section of the distribution system, so using 33 kV ensures the correct current calculation for the fault current. Reactance values, X, are used instead of impedance values, Z, because the sources are considered infinite, that is, no resistance is present. Use Eq. 34.53.

$$X_{pu,new} = X_{pu,old}\left(\frac{X_{base,old}}{X_{base,new}}\right)$$

Because no resistance is present, Eq. 34.54 converts to Eq. I.

$$Z_{pu,new} = Z_{pu,old}\left(\frac{S_{base,new}}{S_{base,old}}\right) \qquad \text{[I]}$$

$$\mathbf{Z}_{pu,gen1} = jX_{gen1} = j0.10\left(\frac{100 \text{ MVA}}{10 \text{ MVA}}\right) = j1.0 \text{ pu} \qquad \text{[II]}$$

$$\mathbf{Z}_{pu,gen2} = jX_{gen2} = j0.15\left(\frac{100 \text{ MVA}}{25 \text{ MVA}}\right) = j0.6 \text{ pu} \qquad \text{[III]}$$

$$\mathbf{Z}_{pu,transformer} = jX_{transformer} = j0.05\left(\frac{100 \text{ MVA}}{50 \text{ MVA}}\right) = j0.1 \text{ pu} \qquad \text{[IV]}$$

In some texts, only the reactance, X, is shown. Even so, when used in per-unit calculations, the j is added as a reminder that the value is for a reactive component, and must be added as a vector to any resistive components. Additionally, the impedance, **Z**, may not be shown in the bold format, even though it carries angular information. The form used in Eq. I through Eq. IV is in keeping with IEEE standards.

The impedance of the line is calculated using Eq. 34.49 and a rearranged three-phase version of Eq. 34.44.

$$\mathbf{Z}_{pu} = \frac{\mathbf{Z}_{actual}}{Z_{base}} = \mathbf{Z}_{actual}\left(\frac{S_{base}}{V_{base}^2}\right) = (2\ \Omega + j10\ \Omega)\left(\frac{100 \times 10^6 \text{ VA}}{(33 \times 10^3 \text{ V})^2}\right) = 0.18 \text{ pu} + j0.92 \text{ pu} \qquad \text{[V]}$$

The per-unit values of reactance and impedance do not change as one considers different portions of the distribution system (i.e., each side of the transformer). However, the base voltages do change. The correct base voltage must be used in calculating the per-unit values.

Once complete, the per-unit values may be combined in any way desired, as shown in the next several steps. The results of Eq. I through Eq. V are as shown in the following illustration.

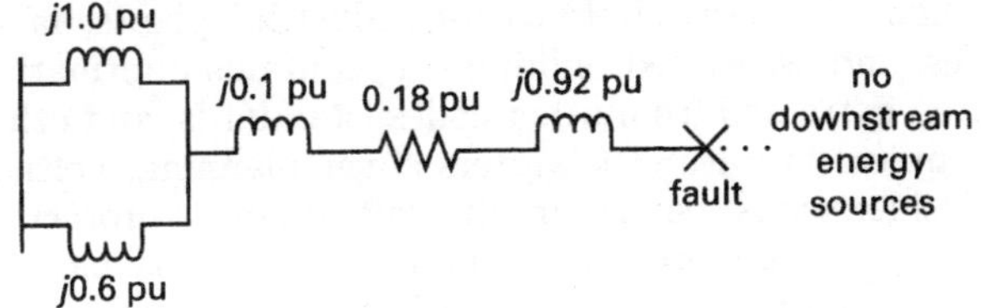

Combining the generator reactances gives

$$\begin{aligned} Z_{\text{total,gen}} &= jX_{\text{total,gen}} \\ &= \frac{(jX_{\text{gen1}})(jX_{\text{gen2}})}{jX_{\text{gen1}} + jX_{\text{gen2}}} \\ &= \frac{(j1.0\text{ pu})(j0.6\text{ pu})}{j1.0\text{ pu} + j0.6\text{ pu}} \\ &= j0.375\text{ pu} \end{aligned} \qquad \text{[VI]}$$

Showing the results of Eq. VI and combining the transformer and line reactances gives

j0.375 0.18 pu j1.02 pu fault

Combining the remaining reactances gives

0.18 pu j1.395 pu fault

The total impedance is

$$\mathbf{Z}_{\text{total}} = (0.18 + j1.395)\text{pu} = 1.41\text{ pu}\angle 83^\circ \qquad \text{[VII]}$$

The angle does not determine the maximum short-circuit current, but it is required for phase calculations and is shown here as clarification only. All the per-unit quantities carry an angle that may be determined if necessary.

Use Eq. 36.30 and the result of Eq. VII to calculate the short-circuit power.

$$\begin{aligned} S_{\text{sc}} &= \frac{S_{\text{base}}}{Z_{\text{pu}}} = \frac{100\text{ MVA}}{1.41\text{ pu}} \\ &= 70.9\text{ MVA} \quad (70.9 \times 10^6\text{ VA}) \end{aligned} \qquad \text{[VIII]}$$

Substitute the result of Eq. VIII into Eq. 36.31 to determine the fault current at the indicated location.

$$\begin{aligned} I_{\text{sc}} &= \frac{S_{\text{sc}}}{\sqrt{3}V_{\text{base}}} = \frac{70.9 \times 10^6\text{ VA}}{\sqrt{3}(33 \times 10^3\text{ V})} \\ &= 1240.4\text{ A} \quad (1240\text{ A}) \end{aligned} \qquad \text{[IX]}$$

10. SMART GRID

Though there is no single definition of a *smart grid*, the term generally applies to broad-based electrical distribution systems using some form of computer-based remote control and automation. The "grid" connects power generation to consumers using substations, transformers, switches, wires, communication infrastructure, and other elements. The "smart" components monitor elements of the system, predict electrical demand, and respond to load deviations and transients to provide continuous power.

The properties of smart grids are guided by Title XIII, "Smart Grid," of the Energy Independence and Security Act of 2007 (EISA), which classifies the following types of projects as smart grid projects.

- projects that optimize the monitoring and control of transmission and distribution, including the use of sensors, communication links, and computer software and systems
- communication infrastructure projects to support the grid
- projects incorporating renewables
- microgrid projects supporting highly reliable and resilient *islanded operation* (i.e., smaller grids operating independently of the main grid or through a single interface with that grid)
- automation projects designed to increase information technology, communications, and overall cyber security

11. IEEE 3000 STANDARDS COLLECTION

The IEEE has established standards to guide the production, distribution, and utilization of electric energy. These IEEE standards are collectively known as the *IEEE 3000 Standards Collection™ for Industrial and Commercial Power Systems*, formerly as the *IEEE Color Books*. Each book focuses on one aspect of electric power and provides a basis for assessing the requirements for a given electrical project. Chapters in this book that cover content related to the individual standards provide a brief overview of each standard, and also are noted as follows.

- IEEE Standard 141, *IEEE Recommended Practice for Electric Power Distribution for Industrial Plants* (*IEEE Red Book*). (See Sec. 36.12.)
- IEEE Standard 142, *IEEE Recommended Practice for Grounding of Industrial and Commercial Power Systems* (*IEEE Green Book*). (See Sec. 50.6.)
- IEEE Standard 241, *IEEE Recommended Practice for Electric Power Systems in Commercial Buildings* (*IEEE Gray Book*). (See Sec. 36.13.)
- IEEE Standard 242, *IEEE Recommended Practice for Protection and Coordination of Industrial and Commercial Power Systems* (*IEEE Buff Book*). (See Sec. 42.18.)

- IEEE Standard 399, *IEEE Recommended Practice for Industrial and Commercial Power Systems Analysis* (*IEEE Brown Book*). (See Sec. 40.11.)
- IEEE Standard 446, *IEEE Recommended Practice for Emergency and Standby Power Systems for Industrial and Commercial Applications* (*IEEE Orange Book*). (See Sec. 52.6.)
- IEEE Standard 493, *IEEE Recommended Practice for the Design of Reliable Industrial and Commercial Power Systems* (*IEEE Gold Book*). (See Sec. 42.19.)
- IEEE Standard 551, *Recommended Practice for Calculating AC Short-Circuit Currents in Industrial and Commercial Power Systems* (*IEEE Violet Book*). (See Sec. 40.12.)
- IEEE Standard 602, *IEEE Recommended Practice for Electric Systems in Health Care Facilities* (*IEEE White Book*). (See Sec. 52.8.)
- IEEE Standard 739, *IEEE Recommended Practice for Energy Management in Industrial and Commercial Facilities* (*IEEE Bronze Book*). (See Sec. 52.7.)
- IEEE Standard 902, *IEEE Guide for Maintenance, Operation, and Safety of Industrial and Commercial Power Systems* (*IEEE Yellow Book*). (See Sec. 42.20.)
- IEEE Standard 1015, *IEEE Recommended Practice for Applying Low-Voltage Circuit Breakers Used in Industrial and Commercial Power Systems* (*IEEE Blue Book*). (See Sec. 42.21.)
- IEEE Standard 1100, *IEEE Recommended Practice for Powering and Grounding Electronic Equipment* (*IEEE Emerald Book*). (See Sec. 50.7.)

12. IEEE RED BOOK

IEEE Standard 141, *IEEE Recommended Practice for Electric Power Distribution for Industrial Plants* (*IEEE Red Book*) compiles the best practices for the design of electric systems for industrial plants, facilities, and associated buildings. It contains information often extracted from other codes, standards, and other technical literature. Chapters cover planning, voltage considerations, short-circuit calculations, protective devices, surge protection, power factors, harmonics, power switching, instrumentation, cable systems, busways, energy management, interface considerations, cost, and power system device numbering.

13. IEEE GRAY BOOK

IEEE Standard 241, *IEEE Recommended Practice for Electric Power Systems in Commercial Buildings* (*IEEE Gray Book*) describes the best practices for the electrical design of commercial buildings. It contains information often extracted from other codes, standards, and other technical literature on the requirements of such structures. Chapters cover load characteristics, voltage considerations, power sources, distribution apparatuses, controllers, services, and equipment vaults. The *IEEE Gray Book* also contains information on electrical equipment rooms, wiring, protection, lighting, space conditioning, automation, expansion, modernization, special occupancy requirements, and energy management.

37 Power Transformers

Nomenclature

a	turns ratio	–
A	ABCD parameter	–
B	magnetic flux density	T
B	susceptance	S
B	ABCD parameter	–
C	ABCD parameter	–
D	ABCD parameter	–
f	frequency	Hz
G	conductance	S
I	effective or DC current	A
k	coupling coefficient	–
L	inductance	H
m	mass	kg
n	Steinmetz exponent	–
N	number of turns	–
P	power	W
pf	power factor	–
R	resistance	Ω
S	apparent power	VA
V	effective or DC voltage	V
VR	voltage regulation	–
X	reactance	Ω
Y	admittance	S
Z	impedance	Ω

Symbols

ω	angular frequency	rad/s

Subscripts

1	primary
2	secondary
at	autotransformer
c	core
Cu	copper
e	eddy current
fl	full load
h	hysteresis
l	line
m	magnetizing
max	maximum
nl	no load
oc	open circuit
O	origin
p	primary
ps	primary to secondary
pu	per unit
s	secondary
sc	short circuit
tw	two-winding transformer
ϕ	phase

1. THEORY

A *transformer* is an electrical device consisting of two or more multiturn coils of wire placed in close proximity in order to link the magnetic field of one to the other. This linkage allows the transfer of electrical energy from one or more alternating circuits to one or more other alternating circuits through the process of magnetic induction. The transformer can be used to raise or lower the value of a capacitor, inductor, or resistor. It enables the efficient transmission of electrical energy at high voltage over great distances, and then is used to lower the voltage to safe values for industrial, commercial, and household use. An exact transformer model with the core parameters referred to the primary side is illustrated in Fig. 37.1.

The parameters are defined as follows.

- R_p is the *primary winding resistance.*
- X_p is the *primary winding reactance.* This is also called the *primary winding leakage reactance.* The *primary winding leakage inductance*, L_p, is sometimes used in the model as well. All the terms represent the inductance not mutually coupled with the secondary winding.
- Z_p is the *primary impedance*, that is, $R_p + jX_p$.

Figure 37.1 Transformer Model

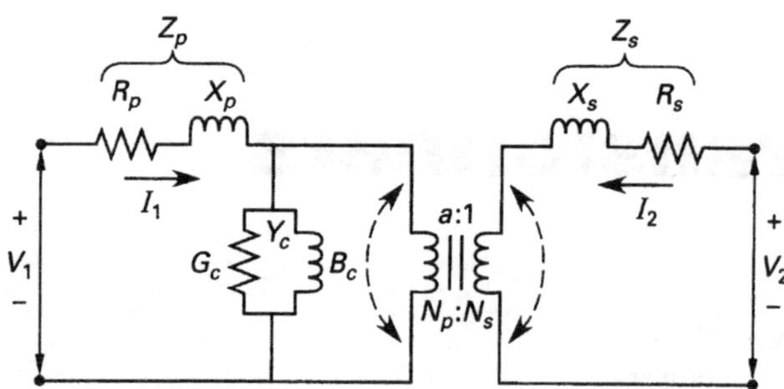

- G_c is the *core conductance*. The reciprocal is sometimes used here and is referred to as the *core resistance*, R_c.
- B_c is the *core susceptance*. The reciprocal is sometimes used here and is referred to as the *core reactance*, X_c. The reciprocal is also called the *core magnetizing reactance*, X_m. The *core inductance*, L_c, is sometimes used in the model as well.
- Y_c is the *core admittance*, that is, $G_c + jB_c$. The reciprocal is the core impedance, Z_c, that is, $R_c + jX_c$.
- The *turns ratio*, N_p/N_s, is given the symbol a, primarily for ease of use in equations. The relationships based on the turns ratio given in Chap. 28, $a = N_p/N_s = V_p/V_s = I_s/I_p = \sqrt{Z_p/Z_s}$, are valid only for ideal transformers, though they are an excellent approximation for large transformers. The relationships are valid for the real transformer in Fig. 37.1, but only if the electrical parameters are referenced to the ideal portion of the real model, shown by the dashed lines.[1]
- X_s is the *secondary winding reactance*. This is also called the *secondary winding leakage reactance*. The *secondary winding leakage inductance*, L_s, is sometimes used in the model as well. All the terms represent the inductance not mutually coupled with the primary winding.
- R_s is the *secondary winding resistance*.
- Z_s is the *secondary impedance*, that is, $R_s + jX_s$.

2. TRANSFORMER RATING

Transformers are rated according to winding voltages, frequency, and kVA rating. Continuous operation at rated values is possible without excessive heat build-up or malfunction of the transformer. The apparent power (measured in kVA) rather than the real power (measured in kW) is used to rate the transformer because the total heating effect depends on the square of the actual current flowing, including that moving in the system between reactive components.

The two main power losses in transformers are *core losses* (*iron losses*) and *copper losses* (*winding losses*). Core losses, represented in Fig. 37.1 by G_c, consist of *eddy current losses* and *hysteresis losses*. Eddy current losses, P_e, are the result of microscopic circulating currents in the iron caused by the magnetic flux passing through. Hysteresis losses, P_h, are caused by the cyclic changes in the magnetic state of the iron. Both these losses are constant and independent of transformer load. That is, they do not vary from no-load to full-load conditions, but they do depend on the mass of iron, m, of the transformer core. Equation 37.1 and Eq. 37.2 can be used to calculate the losses. The maximum flux density, B_{max}, is calculated from Faraday's law. The exponent n is called the *Steinmetz exponent* and varies from 1.5 to 2.5 with a common value of 1.6. The coupling coefficient, k, depends on the transformer design.

$$P_e = k_e B_{\text{max}}^2 f^2 m \qquad 37.1$$

$$P_h = k_h B_{\text{max}}^n f m \qquad 37.2$$

The core losses, ignoring the small line loss caused by R_p, may be approximated from the transformer model in Fig. 37.1 using the following formula.

$$P_c = \frac{V_1^2}{R_c} \qquad 37.3$$

The term R_c is the reciprocal of the core conductance, G_c. Copper losses, P_{Cu}, are caused by the total wire resistance and can be calculated from

$$P_{\text{Cu}} = I^2 R = I_1^2 R_p + I_2^2 R_s \qquad 37.4$$

The *transformer efficiency* is the ratio of the output power to the input power and is at a maximum when the copper losses equal the core losses; that is, when the variable losses equal the constant losses. The *all-day efficiency* is the ratio of energy delivered by a transformer in a 24-hour period to the energy input during the same period.

$$\eta = \frac{P_{\text{out}}}{P_{\text{in}}} = \frac{P_{\text{in}} - \sum P_{\text{losses}}}{P_{\text{in}}} = \frac{P_{\text{out}}}{P_{\text{out}} + P_c + P_{\text{Cu}}} \qquad 37.5$$

Example 37.1

Determine the hysteresis power loss at 60 Hz for a 500 kVA-rated 200 kg iron core transformer with a coupling coefficient of 4×10^{-4} and a Steinmetz exponent of 1.6 in a 1.4 T peak magnetic field.

[1]The ideal transformer model is not used for power transformers because the efficiency, regulation, and power factor must be determined. In this book, the subscripts p and s used for ideal transformers are not used on the current and voltage in the real transformer. They are reserved for electrical parameters of the ideal transformer, that is, the parameters referring to the area of the dashed lines in Fig. 37.1.

Solution

The hysteresis loss is determined from Eq. 37.2.

$$\begin{aligned}P_h &= k_h B_{\text{max}}^n f m = (4 \times 10^{-4})(1.4\ \text{T})^{1.6}(60\ \text{Hz})(200\ \text{kg}) \\ &= 8.22\ \text{W}\end{aligned}$$

Example 37.2

The transformer in Ex. 37.1 uses a primary voltage of 345 kV. If the core resistance is 200 MΩ, estimate the total core power loss.

Solution

The total core losses can be estimated using Eq. 37.3.

$$P_c = \frac{V_1^2}{R_c} = \frac{(345 \times 10^3\ \text{V})^2}{200 \times 10^6\ \Omega} = 595\ \text{W}$$

3. TRANSFORMER CAPACITY

The capacity of a given transformer is determined by its ability to handle the maximum current without changing the properties of the conducting material, which is normally copper, or overheating and breaking down the insulation. In addition to these concerns, each transformer has losses that reduce the available power (see Sec. 28.5).

The capacity is given in terms of the apparent power, S, in kVA. The total capacity of a three-phase transformer is $\sqrt{3}$ times larger than the capacity of a single-phase transformer at the same voltage and current (see Chap. 34).[2]

4. VOLTAGE REGULATION

Voltage regulation is the change in the output voltage from the no-load to the full-load condition.

$$\text{VR} = \frac{V_{\text{nl}} - V_{\text{fl}}}{V_{\text{fl}}} \qquad 37.6$$

The voltage regulation can be expressed in terms of rated quantities. At no load, reflecting the primary voltage to the secondary side gives V_p/a. The rated conditions of the secondary represent full load. The voltage regulation is then

$$\text{VR} = \frac{\frac{V_p}{a} - V_{s,\text{rated}}}{V_{s,\text{rated}}} \qquad 37.7$$

The primary voltage, coming from the large generation source, can be considered constant. That is, the primary side is considered an infinite bus. This assumption is embedded in any regulation equation based on rated values and also in per-unit calculations of the regulation. With output voltage considered to have a 1.0 pu value, the difference between the primary voltage and the output voltage of the transformer secondary in per-unit values is the regulation. That is, to supply the load current and secondary voltage at rated values, the primary voltage must be greater than the secondary. The excess is termed the regulation and is found by using Eq. 37.7 with per-unit values instead of rated values.

Example 37.3

A single-phase transformer rated for 50 kVA operates at a rated secondary voltage of 440 V and 0.8 pf lagging. The primary winding impedance is

$$Z_p = R_p + jX_p = 0.014 + j0.026\ \text{pu}$$

What is the voltage regulation in percent?

Solution

To determine the regulation, the primary voltage is required. Because per-unit impedance is given, the per-unit method will be used. The output voltage, (i.e., the secondary voltage) is 1.0 pu at rated conditions. The output current is 1.0 pu as well. Because the power factor is 0.8 lagging, the current is 1.0 pu$\angle{-36.9°}$ per unit. The impedance in polar form is

$$Z_p = 0.014 + j0.026\ \text{pu} = 0.030\ \text{pu}\angle 61.7°$$

Using the model given in Fig. 37.1 and noting that the per-unit current in the secondary must also flow in the primary, the phase voltage at the primary of the transformer (the voltage at the dashed lines) is

$$\begin{aligned}V_{\text{primary phase}} &= I_{\text{pu}} Z_p \\ &= (1.0\ \text{pu}\angle{-36.9°})(0.030\ \text{pu}\angle 61.7°) \\ &= 0.030\ \text{pu}\angle 24.8°\end{aligned}$$

Reflecting this voltage to the secondary side does not require the turns ratio since per-unit values are in use.[3] Given that the secondary voltage is 1.0 pu, at an angle of 0° since it is the reference, the primary voltage is

$$\begin{aligned}V_{\text{primary reflected}} &= V_{\text{primary phase,pu}} + V_{s,\text{pu}} \\ &= 0.030\ \text{pu}\angle 24.8° + 1.0\ \text{pu}\angle 0° \\ &= 1.027\ \text{pu}\angle 0.7°\end{aligned}$$

[2]Any three-phase circuit power equation includes the factor $\sqrt{3}$, which is approximately 1.73.

[3]The advantage of the per-unit system is that by choosing base primary and secondary voltages related by the turns ratio, which occurs in this example by letting the secondary voltage be the base and relating the primary voltage to it, the transformer is no longer required in the electrical model.

Distribution

The voltage regulation is not concerned with the resulting angle. The regulation is given by

$$\begin{aligned} \text{VR} &= V_{\text{primary reflected,pu}} - V_{\text{rated,pu}} \\ &= 1.027 \text{ pu} - 1.0 \text{ pu} \\ &= 0.027 \text{ pu or } 2.7\% \end{aligned}$$

5. THREE-PHASE TRANSFORMER CONFIGURATIONS

Three-phase transformers can be connected as either delta or wye. When voltage or current ratios are given for three-phase transformers, they are assumed to specify the line conditions, regardless of the connection. If only the turns ratio, a, is given, the line quantities must be calculated. For delta connections, the line and phase voltage and current are given by Eq. 37.8 and Eq. 37.9, respectively.

$$V_l = V_\phi \quad \text{[delta]} \qquad 37.8$$

$$I_l = \sqrt{3} I_\phi \quad \text{[delta]} \qquad 37.9$$

For wye connections, the line and phase voltage and current are given by Eq. 37.10 and Eq. 37.11, respectively.

$$V_l = \sqrt{3} V_\phi \quad \text{[wye]} \qquad 37.10$$

$$I_l = I_\phi \quad \text{[wye]} \qquad 37.11$$

Because either the voltage or the current contains a factor of $\sqrt{3}$, the power for either connection type is given by

$$P = \sqrt{3} I_l V_l (\text{pf}) = 3 I_\phi V_\phi (\text{pf}) \qquad 37.12$$

The line-phase relationships are illustrated in Fig. 37.2 for various connection types.[4] Figure 37.3 shows how the connections would be displayed in an electrical schematic.

Wye connections, with a common neutral, offer economic and operating advantages (see Sec. 36.3). Delta connections allow third-harmonic voltages common to all transformers to circulate within the delta so that line electrical parameters are unaffected. For this reason, transformers normally have at least one delta-connected winding. Additionally, delta connections can suffer a failure of a single phase and still provide 57.7% of their rated load.[5]

[4]The turns ratio, a, is sometimes given as the ratio of the secondary to the primary turns. If this is the case, switch the a from the numerator to the denominator (or from the denominator to the numerator, depending on where a originally appears) to obtain the correct relationship.

[5]When only two phases of a delta connection are purposefully used, it is called an *open delta* or *vee transformer*.

Figure 37.2 Transformer Connections

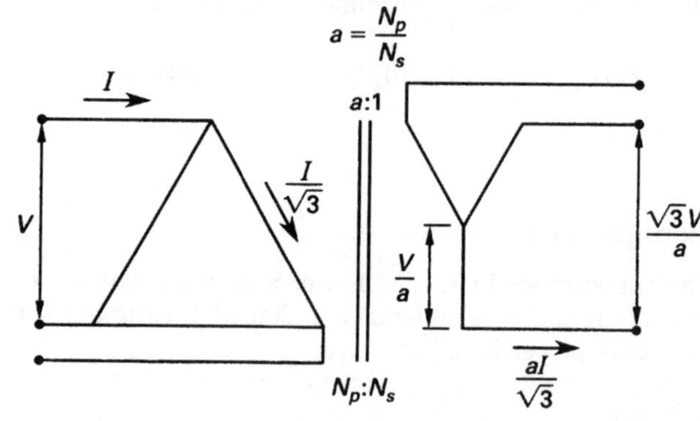

(a) delta-wye connection

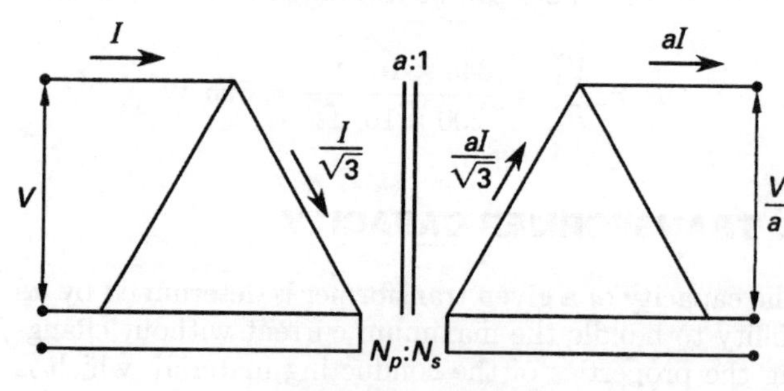

(b) delta-delta connection

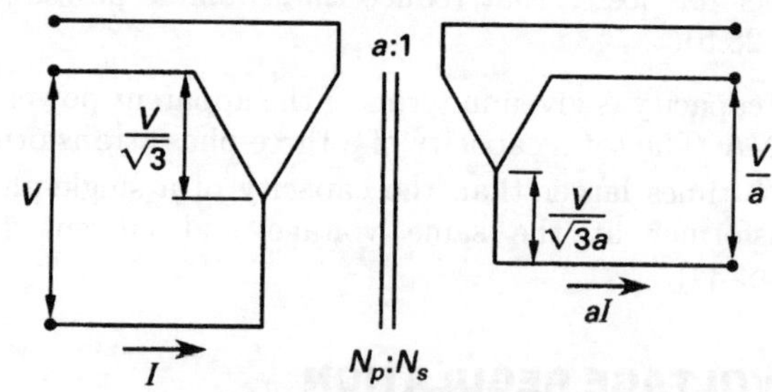

(c) wye-wye connection

Figure 37.3 Standard Transformer Schematic

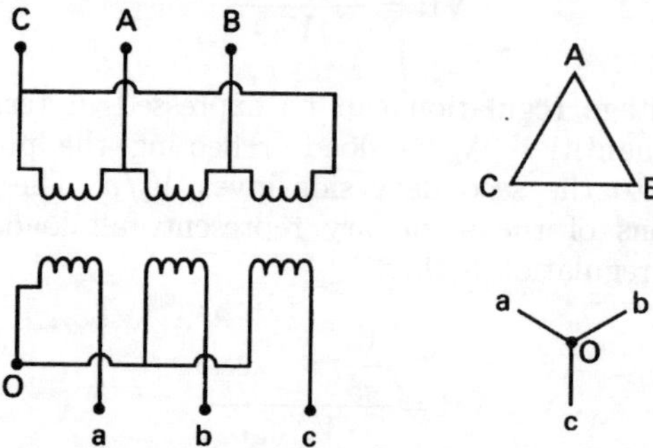

6. TRANSFORMER TESTING

Transformer testing is done to determine the parameters for a real transformer, a model of which is shown in Fig. 37.1. Based on this model, testing determines the rating, efficiency, and equivalent circuit values. Two standardized tests are performed: open-circuit and short-circuit. Open-circuit and short-circuit conditions are established on the low-voltage side of the transformer, which can be either the primary or the secondary side.[6] The primary voltage and current, the secondary voltage and current, and the power are measured in each test. From these values, the transformer's properties are derived. The open-circuit test determines the core parameters and the turns ratio. The short-circuit test determines the winding impedances and verifies the turns ratio. The admittance parameter, Y, accounts for the power loss in the core. The susceptance parameter, B, accounts for energy storage in the core. The resistance parameter, R, accounts for power loss in the windings. The reactance parameter, X, accounts for the leakage (self-inductance) of the primary and secondary windings.

7. OPEN-CIRCUIT TEST

The open-circuit test determines the core parameters and the turns ratio. An open-circuit test is performed by opening the secondary terminals of the transformer. Actually, a voltmeter measures the secondary voltage, V_{2oc}, but the meter has such high resistance it can be considered an open circuit. The rated voltage, V_{1oc}, is applied to the primary terminals. (The rated voltage is used because the core losses are dependent on the magnetic flux density, B_{max}.) The input current, I_{1oc}, is measured as well as the input power, P_{oc}. Because the current flow I_{1oc} is small, the voltage drop across Z_p is negligible. The open-circuit power, P_{oc}, represents the core power loss. The input open-circuit current, I_{1oc}, is the exciting current that maintains the flux in the core. The open-circuit model is shown in Fig. 37.4.

The admittance is

$$Y_c = G_c + jB_c = \frac{I_{1oc}}{V_{1oc}} \qquad 37.13$$

The conductance is

$$G_c = \frac{P_{oc}}{V_{1oc}^2} \qquad 37.14$$

The susceptance is

$$B_c = \frac{1}{X_c} = \frac{-1}{\omega L_c} = -\sqrt{Y_c^2 - G_c^2}$$
$$= \frac{-\sqrt{I_{1oc}^2 V_{1oc}^2 - P_{oc}^2}}{V_{1oc}^2} \qquad 37.15$$

The susceptance is negative $(-1/\omega L)$ for a lagging condition.

[6]Because of this, the terms "high-voltage" (HV) and "low-voltage" (LV) are sometimes used instead of the terms "primary" and "secondary." The terms "high-tension" and "low-tension," synonymous with high- and low-voltage, respectively, are also sometimes used.

Figure 37.4 *Transformer Open-Circuit Test Model*

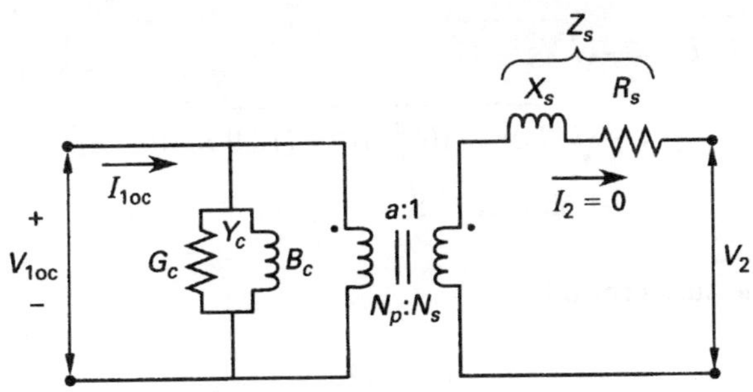

The turns ratio is[7]

$$a_{ps} = \frac{V_{1oc}}{V_{2oc}} \qquad 37.16$$

The power is

$$P_{oc} = V_{1oc}^2 G_c \qquad 37.17$$

The reactive power is

$$Q_{oc} = V_{1oc}^2 B_c \qquad 37.18$$

The apparent power is

$$S_{oc} = V_{1oc}^2 Y_c = V_{1oc} I_{1oc} = \sqrt{P_{oc}^2 + Q_{oc}^2} \qquad 37.19$$

Example 37.4

An open-circuit test is conducted on a 120 V transformer winding. The results of this test are $V_{1oc} = 120$ V, $V_{2oc} = 240$ V, $I_{1oc} = 0.25$ A, and $P_{oc} = 20$ W. Determine the transformer equivalent circuit element parameters given by this data.

Solution

The admittance is

$$Y_c = G_c + jB_c = \frac{I_{1oc}}{V_{1oc}} = \frac{0.25 \text{ A}}{120 \text{ V}}$$
$$= 2.08 \times 10^{-3} \text{ S}$$

The core conductance is

$$G_c = \frac{P_{oc}}{V_{1oc}^2} = \frac{20 \text{ W}}{(120 \text{ V})^2}$$
$$= 1.39 \times 10^{-3} \text{ S}$$

[7]The subscript "ps" is added to clarify the turns ratio as being from the primary to the secondary, as is standard for this text. The turns ratio is sometimes used to indicate the secondary to primary ratio. Care should be taken to determine the definition used in a particular situation.

Distribution

The core susceptance is

$$\begin{aligned} B_c &= -\sqrt{Y_c^2 - G_c^2} \\ &= -\sqrt{(2.08 \times 10^{-3}\ \text{S})^2 - (1.39 \times 10^{-3}\ \text{S})^2} \\ &= -1.55 \times 10^{-3}\ \text{S} \end{aligned}$$

The turns ratio is

$$a_{\text{ps}} = \frac{V_{1\text{oc}}}{V_{2\text{oc}}} = \frac{120\ \text{V}}{240\ \text{V}} = 0.5$$

8. SHORT-CIRCUIT TEST

The short-circuit test determines the winding impedances and verifies the turns ratio. A short-circuit test is performed by shorting the secondary terminals of the transformer. Actually, an ammeter measures the secondary current, $I_{2\text{sc}}$, but the meter has such low resistance that it can be ignored in the analysis of the circuit. A voltage, V_{sc}, is applied to the primary such that the rated current, that is, the volt-amp rating divided by the voltage rating, flows. The voltage will be low because the secondary winding has minimal impedance. The voltage is normally less than 5% of the rated value. Because of this minimal secondary impedance, the core admittance, Y_c, is considered shorted. The effective circuit then consists of the primary impedance in series with the reflected secondary impedance. The input current, $I_{1\text{sc}}$, and output current, $I_{2\text{sc}}$, are measured as well as the input power, P_{sc}. The short-circuit power, P_{sc}, represents the copper loss or I^2R losses in the windings. The short-circuit test model is shown in Fig. 37.5.

Figure 37.5 Transformer Short-Circuit Test Model

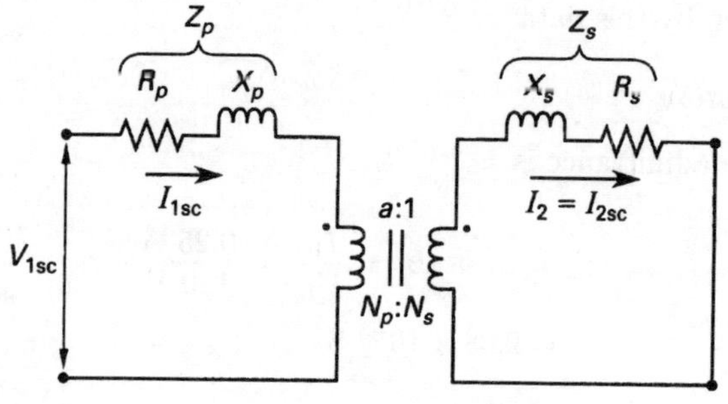

The total impedance, Z, is

$$Z = R_p + jX_p + a_{\text{ps}}^2(R_s + jX_s) = \frac{V_{1\text{sc}}}{I_{1\text{sc}}} \qquad 37.20$$

The total resistance, R, is

$$R = R_p + a_{\text{ps}}^2 R_s = \frac{P_{\text{sc}}}{I_{1\text{sc}}^2} \qquad 37.21$$

To maximize efficiency, transformers are normally designed with R_p equal to $a_{\text{ps}}^2 R_s$. Therefore,

$$R_p = a_{\text{ps}}^2 R_s = \frac{P_{\text{sc}}}{2I_{1\text{sc}}^2} \qquad 37.22$$

The total reactance, X, is

$$X = X_p + a_{\text{ps}}^2 X_s = \frac{\sqrt{I_{\text{sc}}^2 V_{\text{sc}}^2 - P_{\text{sc}}^2}}{I_{1\text{sc}}^2} \qquad 37.23$$

To maximize efficiency, transformers are normally designed with X_p equal to $a_{\text{ps}}^2 X_s$. Therefore,

$$X_p = a_{\text{ps}}^2 X_s = \frac{Q_{\text{sc}}}{2I_{1\text{sc}}^2} \qquad 37.24$$

The turns ratio is

$$a_{\text{ps}} = \frac{I_{2\text{sc}}}{I_{1\text{sc}}} \qquad 37.25$$

The power is

$$P_{\text{sc}} = I_{1\text{sc}}^2 R = I_{1\text{sc}}^2 (R_p + a_{\text{ps}}^2 R_s) \qquad 37.26$$

The reactive power is

$$Q_{\text{sc}} = I_{1\text{sc}}^2 X = I_{1\text{sc}}^2 (X_p + a_{\text{ps}}^2 X_s) \qquad 37.27$$

The apparent power is

$$S_{\text{sc}} = I_{1\text{sc}}^2 Z_{\text{sc}} = V_{1\text{sc}} I_{1\text{sc}} = \sqrt{P_{\text{sc}}^2 + Q_{\text{sc}}^2} \qquad 37.28$$

Example 37.5

A short-circuit test is conducted on a transformer rated at 15 kVA and 1300 primary volts. What is the value of the input current, $I_{1\text{sc}}$?

Solution

Short-circuit tests are conducted at the rated current, that is, $I_{1\text{sc}} = I_{\text{rated}}$. The rated current is found from

$$I_{\text{rated}} = \frac{S_{\text{rated}}}{V_{\text{rated}}} = \frac{15 \times 10^3\ \text{VA}}{1300\ \text{V}} = 11.5\ \text{A}$$

9. ABCD PARAMETERS

ABCD parameters, also known as *transfer* or *chain parameters*, are analytic tools for transmission and distribution problem solving. To use them on transformers, the secondary impedance is reflected to the primary as shown in Fig. 37.6. The result is a two-port network (see Sec. 29.31). The ABCD parameters for any two-port network are

$$V_{\text{in}} = \text{A}V_{\text{out}} - \text{B}I_{\text{out}} \qquad 37.29$$

$$I_{\text{in}} = \text{C}V_{\text{out}} - \text{D}I_{\text{out}} \qquad 37.30$$

Distribution

Figure 37.6 Transformer Two-Port Network

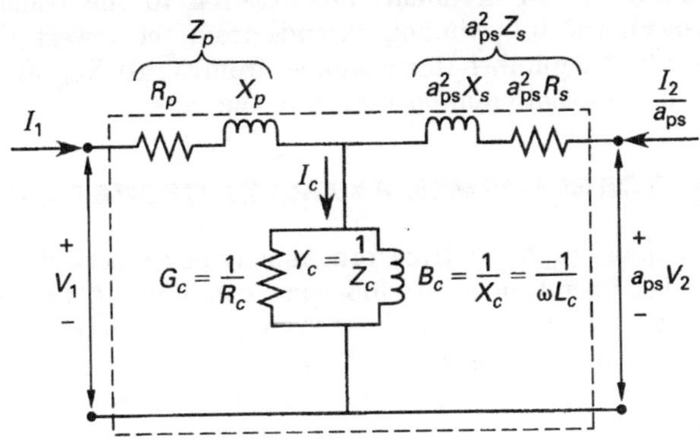

Circuit analysis of the transformer yields Eq. 37.31 through Eq. 37.35, which are used to determine the ABCD parameters.[8]

$$V_1 = a_{ps} V_2 (1 + Z_p Y_c) - \left(\frac{I_2}{a_{ps}}\right)(Z_p + a_{ps}^2 Z_s (1 + Z_p Y_c)) \quad 37.31$$

$$I_1 = a_{ps} V_2 Y_c - \left(\frac{I_2}{a_{ps}}\right)(1 + a_{ps}^2 Z_s Y_c) \quad 37.32$$

$$Z_p = R_p + jX_p \quad 37.33$$

$$Z_s = R_s + jX_s \quad 37.34$$

$$Y_c = G_c + jB_c \quad 37.35$$

To maximize efficiency, transformers are normally designed with Z_p equal to $a_{ps}^2 Z_s$. In this case, the ABCD parameters are given by Eq. 37.36 through Eq. 37.39.

$$\mathrm{A} = 1 + \mathrm{Z}_p Y_c \quad 37.36$$

$$\mathrm{B} = \mathrm{Z}_p(2 + Z_p Y_c) \quad 37.37$$

$$\mathrm{C} = Y_c \quad 37.38$$

$$\mathrm{D} = A \quad 37.39$$

ABCD parameters can be used for two-port networks that are chained together, such as power distribution systems and cascaded amplifiers. For two such cascaded networks, the ABCD parameters are given by Eq. 37.40 through Eq. 37.43.

$$\mathrm{A} = \mathrm{A_1 A_2} + \mathrm{B_1 C_2} \quad 37.40$$

$$\mathrm{B} = \mathrm{A_1 B_2} + \mathrm{B_1 D_2} \quad 37.41$$

$$\mathrm{C} = \mathrm{A_2 C_1} + \mathrm{C_2 D_1} \quad 37.42$$

$$\mathrm{D} = \mathrm{B_2 C_1} + \mathrm{D_1 D_2} \quad 37.43$$

The ABCD parameters for common two-port network types are given in Fig. 37.7.

[8]Care should be used when dealing with the susceptance in Eq. 37.35. Susceptance is negative for an inductive element. Both the susceptance, B_c, and the associated reactance, X_c, carry any negative sign internally.

Figure 37.7 Two-Port Network ABCD Parameters

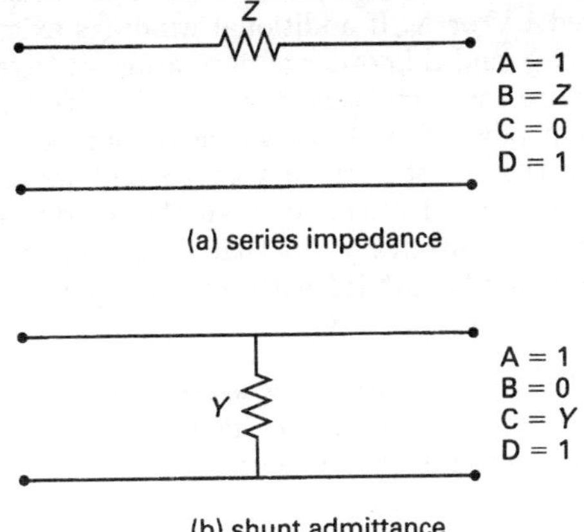

(b) shunt admittance

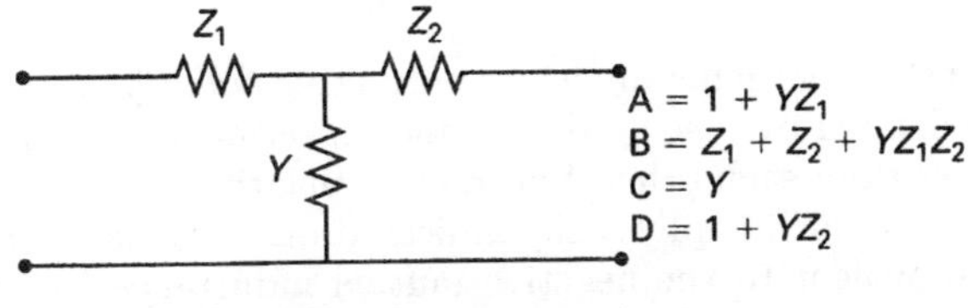

(c) unbalanced tee network

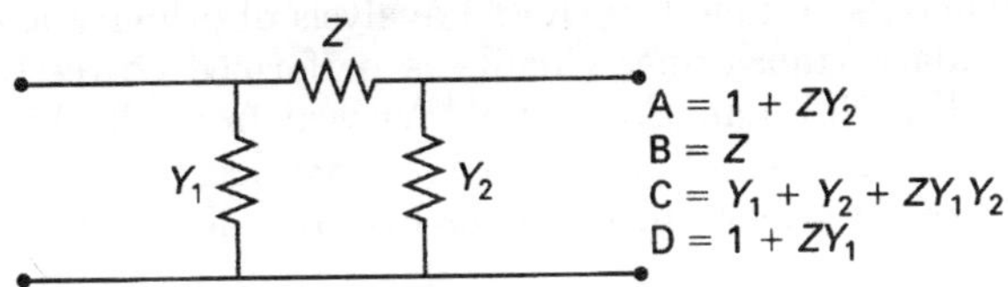

(d) unbalanced pi network

10. TERMINAL MARKING AND POLARITY

The term *connections* as used in regard to transformers refers to the terminal markings and connections of power transformers (as opposed to instrument transformers), distribution transformers, and regulating transformers.[9] A *power transformer* is any transformer that transforms energy between a generator and the distribution primary circuits. *Distribution transformers* transform energy from the primary distribution circuit during transmission to the secondary distribution circuit or to a consumer's service circuit.[10] *Regulating transformers* control the voltage, phase angle, or both to an output circuit and compensate for fluctuations of input voltage or load. *Instrument transformers* isolate and transform voltage or current levels, and are designed to reproduce the voltage or current of the primary circuit in the secondary circuit, with the phase relationship and waveform effectively unchanged.

[9]For more information on markings and connections for transformers, see IEEE Standard C57.12.70, *IEEE Standard for Standard Terminal Markings and Connections for Distribution and Power Transformers*. A related international standard is *Power Transformers—Part 1: General* (IEC 60076-1).

[10]Distribution transformers are generally rated for power in the range of 5 kVA to 500 kVA.

The terminals on the high-voltage side of a transformer are marked HV or H. The terminals on the low-voltage side are marked LV or X. If additional windings exist, they are labeled Y and Z in order of decreasing voltage. If the voltages are the same, then the winding with the highest apparent power (i.e., kVA) rating is marked X, the next Y, and the lowest Z. If the voltages and apparent power ratings are identical between the windings, the designations are arbitrary. In that case, multiple terminals are marked H_1 and H_2, with the corresponding opposite sides marked X_1 and X_2.

The neutral of a wye connection is labeled H_0 or X_0, depending on the location of the neutral (that is, depending on whether the neutral is on the high-voltage side, H, or the low-voltage side, X).

If a two-terminal transformer winding is grounded, the side ground always has the subscript 2, as in H_2 or X_2.

Transformers generally have subtractive polarity, which is indicated by placing the H_1 and X_1 terminals directly across from each other. Subtractive polarity of a transformer (which is physically indicated on a transformer) is equivalent to the negative mutual inductance[11] of a transformer (which is based on theory). Knowing the polarity and the mutual inductance sign ensures that Kirchhoff's voltage law (KVL) analysis of primary and secondary transformer circuits is performed correctly. (See Eq. 28.7 and Eq. 28.8.)[12] Viewed from the high voltage side, the H_1 terminal is always located to the right of the transformer case, as Fig. 37.8 illustrates.[13]

Figure 37.8 Transformer Polarity Types

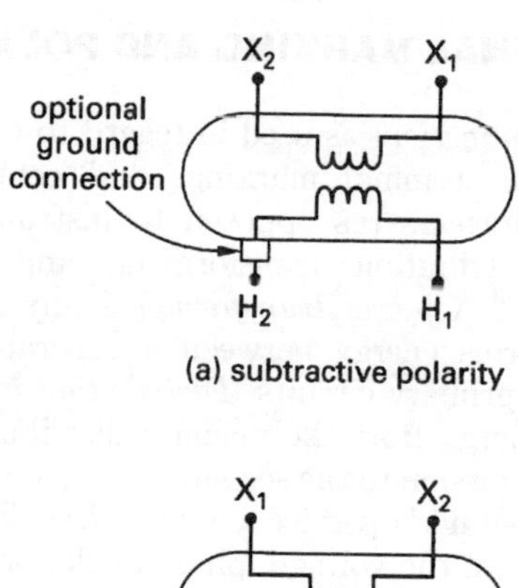

[11]See Sec. 28.2.

[12]The subtractive polarity is the equivalent of a negative value for mutual inductance, M. Regardless of these values, the primary winding is a load to the source and the secondary winding is a source to the load.

[13]The ground is optional and is used as a reference for the voltage.

When a transformer has multiple taps (that is, multiple windings whose terminals are external to the transformer), the full winding extends from the lowest to the highest number (for example, from X_1 to X_5), and the intermediate numbers indicate the taps.

11. TRANSFORMER PARALLEL OPERATION

To expand capacity, transformers may be connected for parallel operation if the following conditions between the two transformers are met: (1) similarly marked terminals must be connected; (2) ground connections must be compatible; and (3) the following values for the transformers must match.

- turns ratios
- primary voltages
- secondary voltages
- resistance and reactance (that is, impedance values)

Properly connected single-phase transformers that are operating in parallel are shown in Fig. 37.9. Three single-phase transformers connected as a three-phase transformer are shown in Fig. 37.10, with the primary side shown at the top of the figure and the secondary at the bottom, as is common practice. If combinations of additive and subtractive polarity are used, the delta and wye connections shown remain the same. That is, the same H and X terminals are connected to one another; only the physical locations of the X_1 and X_2 terminals change.

12. PHASE MARKINGS

Markings on transformers indicate the phase sequence and angular displacement of the transformer. For example, if the high voltage markings are H_1, H_2, and H_3, this indicates that the phase sequence is from winding 1, to winding 2, to winding 3. Such a sequence is known as an *a b c rotation*. The secondary windings will be marked X_1, X_2, and X_3, indicating the corresponding phases for an a-b-c rotation.

The angular displacement is the angle between the high-voltage sides and their corresponding low-voltage sides. It is measured clockwise starting from the line connecting H_1 to the neutral, to the line connecting X_1 to the neutral. The neutral of a delta connection is in the center of the triangle formed by the three windings. The neutral of a wye connection is at the center of the wye, where all connections meet. In a delta-delta or wye-wye connection, the angular displacement is 0°, as shown in Fig. 37.11. In Fig. 37.11(a), the dashed lines indicate the line connecting H_1 to the neutral on the high-voltage side, and connecting X_1 to the neutral on the low-voltage side. Since the dashed lines are parallel to each other and in the same direction, the angular displacement between them is 0°. In Fig. 37.11(b), no dashed lines are used because one winding connection is from H_1 to the neutral, and the other is from X_1 to the

Figure 37.9 Parallel Transformer Connections

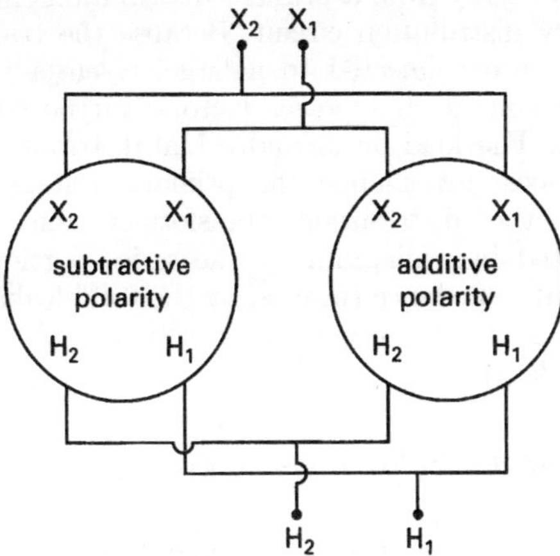

(a) subtractive-additive polarity

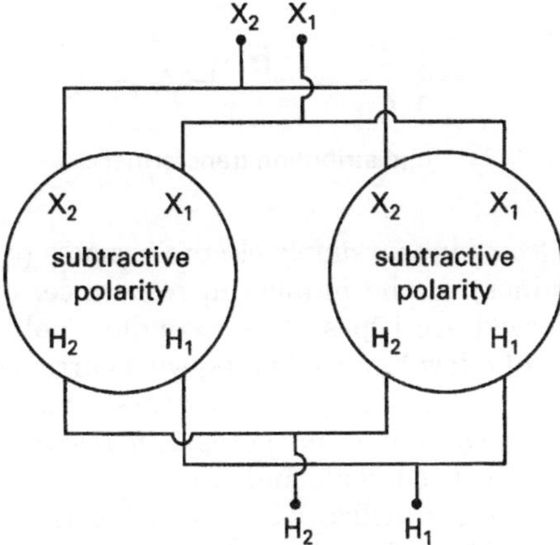

(b) subtractive-subtractive polarity

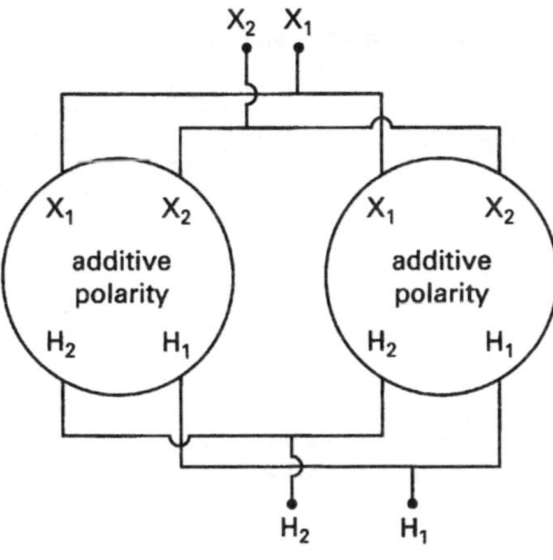

(c) additive-additive polarity

Figure 37.10 Three-Phase Transformer Connections

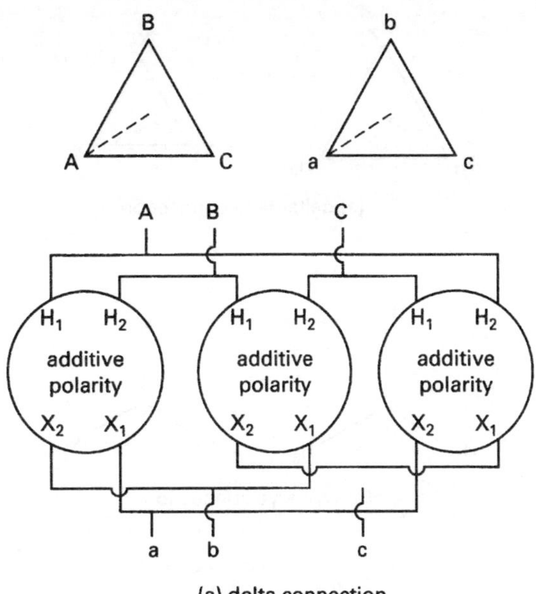

(a) delta connection

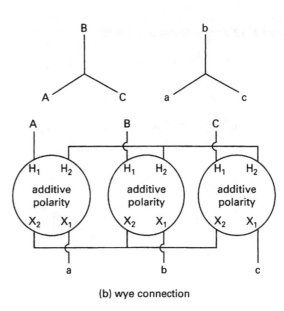

(b) wye connection

neutral. The displacement between these two lines is also 0°. When a transformer is changed from delta to wye or from wye to delta, a 30° displacement is expected, as shown in Fig. 37.12. For delta-wye and wye-delta connections, the displacement between the high-voltage side and the low-voltage side is 30°.[14]

13. GENERAL TRANSFORMER CLASSIFICATIONS

Transformer principles and the analytic methods in this chapter and Chap. 28 are applicable to all types of transformers. The transformer model, which is based

[14]See Chap. 40 for more on wye and delta connections and their associated angle relationships.

Figure 37.11 Angular Displacement of 0°

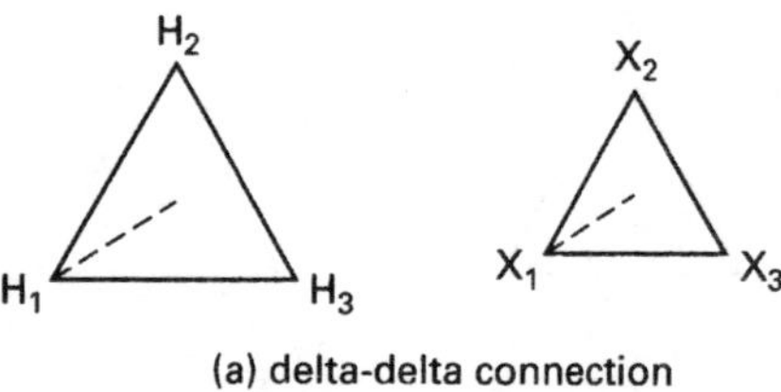

(a) delta-delta connection

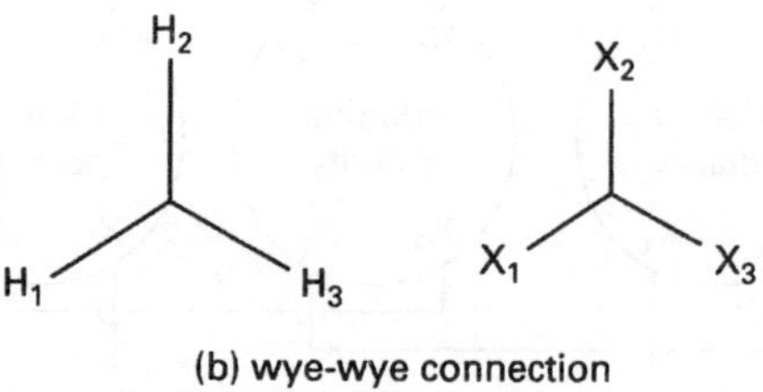

(b) wye-wye connection

Figure 37.12 Angular Displacement of 30°

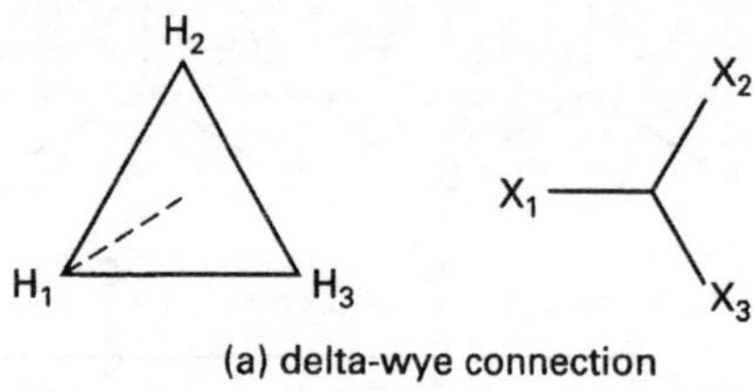

(a) delta-wye connection

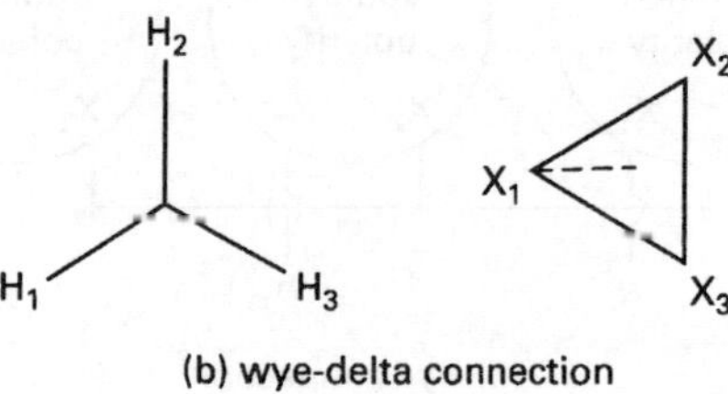

(b) wye-delta connection

on the transformer type and the analysis method used, changes based on the transformer rating and the relative magnitudes of the electrical parameters.

Power transformers are large transformers that transfer energy between a generator and primary distribution circuits. In such large transformers, the total leakage reactance, X, is five times the value of the total resistance, R. As a result, a power transformer can only be accurately represented in a diagram by a reactance as shown in Fig. 37.13(a).

Distribution transformers are transformers that transfer electrical energy from a primary distribution circuit to a secondary distribution circuit. Because the transformer primaries are connected to a large, essentially infinite bus, the voltage and power factors can be considered constant. The load on an individual distribution transformer does not change the primary voltage. Consequently, the distribution transformer can only be represented in a diagram by the series portion of the equivalent impedance (that is, by the total leakage reactance and the total winding resistance, as shown in Fig. 37.13(b)).

Figure 37.13 Special Transformer Models

X

(a) power transformer

Z

(b) distribution transformer

Furnace transformers supply electric furnaces (e.g., welding machines) of the induction, resistance, open-arc, or submerged-arc types. The secondary voltage is in the range of a few hundred volts, with currents as high as 100,000 A. *Grounding transformers* act as a neutral point for grounding purposes. *Instrument transformers* adjust current and voltage to levels that metering devices utilize while isolating the metering circuit from the system being measured.

Typical per-unit values for transformer parameters are given in Table 37.1.

Table 37.1 Typical Transformer Values*

parameter (see Fig. 37.6)	typical per-unit values 3–250 kVA	1–100 MVA
R_p or R_s	0.009–0.005	0.005–0.002
X_p or X_s	0.008–0.025	0.030–0.060
R_c	20–50	100–500
X_c	20–30	30–50
I_c	0.05–0.03	0.03–0.02

*The values for transformers rated between 250 kVA and 1 MVA vary, and their value ranges overlap with the values for transformers rated between 3-250 kVA and 1-100 MVA.

14. BUCK TRANSFORMERS/BOOST TRANSFORMERS/AUTOTRANSFORMERS

An *autotransformer* has only one winding that is common to both the primary and secondary circuits associated with that winding. Current flowing in the output comes directly from the primary winding. (The primary voltage generates a current through the output load.) A portion comes from the inductive process, generating additional voltage and current through the output load. (Primary windings generate a voltage in the secondary windings.)

An autotransformer is also called a *buck* or *boost transformer*, depending on the connection.[15] Connections to the primary and secondary windings are made on one winding either to *boost* the input, which raises the output, or to *buck* the input, which lowers the output. Figure 37.14(a) shows an example of a three-phase autotransformer configured to boost the input, and Fig. 37.14(b) shows one configured to buck the input. In the autotransformer in Fig. 37.14, the single winding has two endpoints and one or more terminals at intermediate points to provide the desired output. Such a transformer can also be automated to move from terminal to terminal in a given time frame in order to start large loads with a minimum starting current.

Figure 37.14 *Autotransformers*

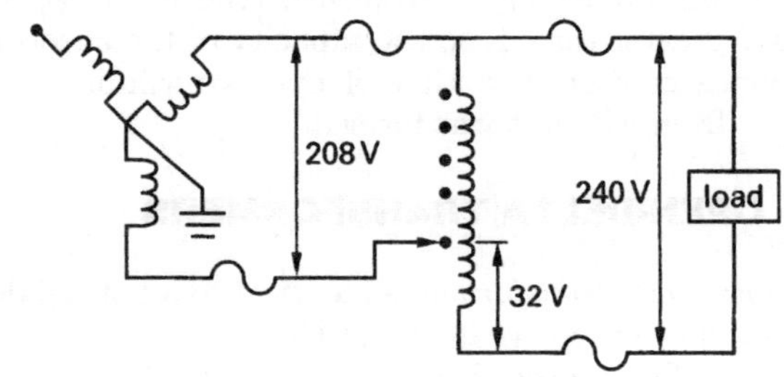

(a) autotransformer for motors, heaters, and air conditioners

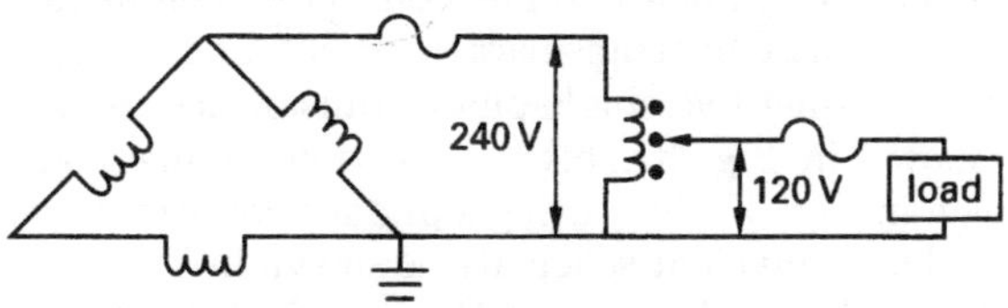

(b) autotransformer for lighting and convenience receptacles

The actual wiring configuration of boost and buck transformers is shown in Fig. 37.15. In terms of apparent power, the capacity of an autotransformer compared to a two-winding transformer is

$$S_{\text{at}} = S_{\text{tw}}\left(1+\frac{N_s}{N_p}\right)$$

$$= S_{\text{tw}}\left(1+\frac{N_2}{N_1}\right) \quad 37.44$$

N_2/N_1 is always greater than one, so the capacity of an autotransformer is always greater than that of a similar two-winding transformer. Generally, the boost (the increase in output voltage over the input voltage) and the buck (the decrease in output voltage under the input voltage) are limited to about 20% of the input.

Figure 37.15 *Autotransformer Connections*

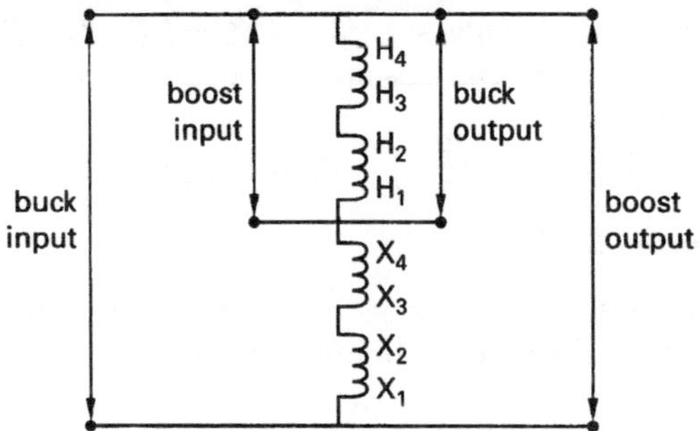

An autotransformer is generally smaller, lighter, and less expensive than an equivalent traditional transformer with two separate windings, because less material is required for its fabrication.[16] One disadvantage of the autotransformer is the lack of electrical isolation between the primary and secondary windings.[17] In addition, if the insulation fails (i.e., if a short occurs), the full output voltage may be applied to the load. Finally, should there be a break in the winding (i.e., an opening), the transformer is then an inductor in series with the load, which under a light load may also result in the full output voltage being applied to the load.

As shown in Fig. 37.14, autotransformers can be used to regulate distribution system voltages to account for losses, or to change voltage from a standard distribution value (208 V or 240 V) to one required by an end user (240 V or 120 V).

Example 37.6

An autotransformer with five evenly spaced taps is used to limit the surge of current to a load, as shown.

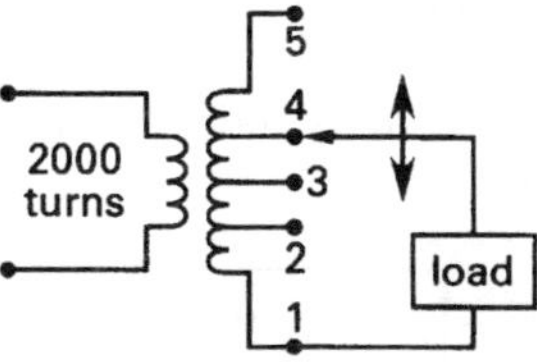

The primary voltage is 220 V. Find the voltage at tap 2 during the startup process.

[15]The term *autotransformer* describes a unit, such as a large motor starter, that is designed to minimize starting transients by using a lower voltage at start-up and then automatically raising the voltage to the rated value. Buck or boost transformers often have taps that must be manually moved and are set once at the desired value for a given installation.

[16]This is true for turns ratios up to about 3:1, after which a standard two-winding transformer is just as economical.

[17]Grounds on the secondary will be seen as a fault by the primary rather than being isolated as in a two-winding transformer. Also, any disturbances on the primary will be felt directly on the secondary loads.

Distribution

Solution

Z_1 is the impedance from the reference point (the ground point) to tap 1. Z_2 is the impedance from tap 1 to tap 2, and so on. Using the concept of a voltage divider from Eq. 29.38, the voltage relationship at tap 2 is

$$V_{\text{load}} = V_1\left(\frac{Z_1+Z_2}{Z_1+Z_2+Z_3+Z_4+Z_5}\right)$$

The taps are evenly spaced, so $Z_1=Z_2=Z_3=Z_4=Z_5$, and the voltage relationship can be simplified.

$$V_{\text{load}} = V_1\left(\frac{2Z}{5Z}\right) = (220\ \text{V})\left(\frac{2}{5}\right) = 88\ \text{V}$$

Example 37.7

A two-winding transformer is shown. The terminals will be connected to create an autotransformer.

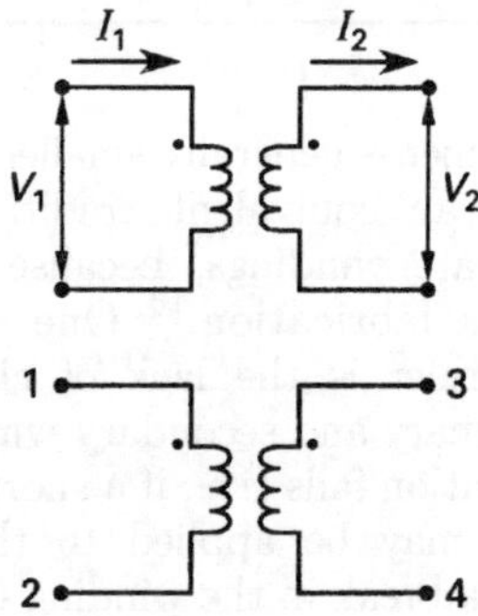

In terms of apparent power, what is the capacity of the autotransformer compared to the original two-winding transformer?

Solution

Connect the two-winding transformer to create an autotransformer.

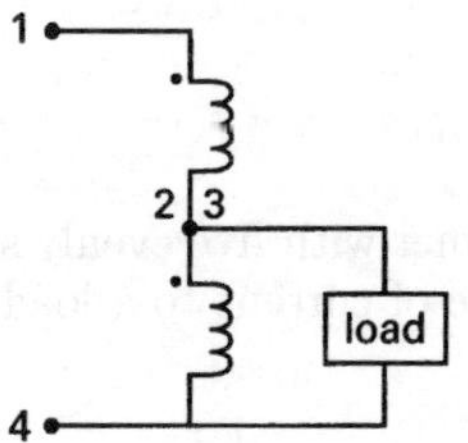

To determine the capacity, label the voltage and current values as shown.

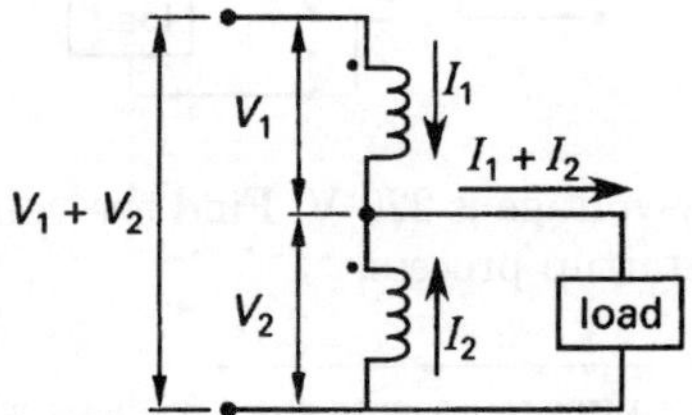

The capacity of the autotransformer is

$$S_{\text{at}} = (V_1+V_2)I_1 = (I_1+I_2)V_2$$

Substitute the turns ratio of the voltage for V_2.

$$\begin{aligned} S_{\text{at}} &= (V_1+V_2)I_1 = I_1V_1 + I_1V_2 \\ &= I_1V_1 + I_1\left(V_1\frac{N_2}{N_1}\right) \\ &= I_1V_1\left(1+\frac{N_2}{N_1}\right) \end{aligned}$$

The capacity of the original two-winding transformer is

$$S_{\text{tw}} = I_1V_1 = I_2V_2$$

Substitute the two-winding capacity into the autotransformer capacity.

$$\begin{aligned} S_{\text{at}} &= I_1V_1\left(1+\frac{N_2}{N_1}\right) \\ &= S_{\text{tw}}\left(1+\frac{N_2}{N_1}\right) \end{aligned}$$

For a step-up voltage transformer, the term N_2/N_1 is always greater than 1, so the capacity of the autotransformer is greater than that of the two-winding transformer from which it was formed.

15. OPEN-DELTA TRANSFORMERS

An *open-delta transformer* is identical to a normal delta transformer except that one of the transformers is not installed. The concept is shown in Fig. 37.16(a) for both the primary and the secondary windings. There is 0° displacement between the primary and secondary voltages for a delta-delta connection. The open-delta connection is used in transmission lines[18] and is typically achieved using two single-phase transformers connected as shown in Fig. 37.16(b). The same connections are shown in Fig. 37.16(c) using standard connection markings. The equivalent schematic is shown in Fig. 37.16(d) and the phasor diagram in Fig. 37.16(e). Even though the connection is open-delta, the primary side is sourced from all three phases. On the secondary side, the voltages are 120° apart, but the full line currents flow through the phases (which does not occur in a normal three-phase connection) and are out of phase with the voltages (by how much depends on the power factor of the load). This results in a reduced capacity of 57.7% for the open-delta transformer, compared to the 66.7% (two-thirds of the original value) that would be expected if one bank of a three-phase transformer bank were removed.[19] The two remaining phases must then carry the current of the missing phase.

[18]Transmission line transformers are known generically as *pots*, *kettles*, or *cans*.

[19]This is also referred to as the 86.6% limit for the capacity of the open-delta transformer. The value of 86.6% is referenced to the full three-phase transformer capacity. The 86.6% limit is determined by the maximum current through a single phase and is independent of the power factor of the load. 86% (referenced to three transformers) of 66.7% (referenced to two transformers) is equal to the stated 57.7% open open-delta capacity. This limit is sometimes called the *utility factor*.

Figure 37.16 Open-Delta Transformer

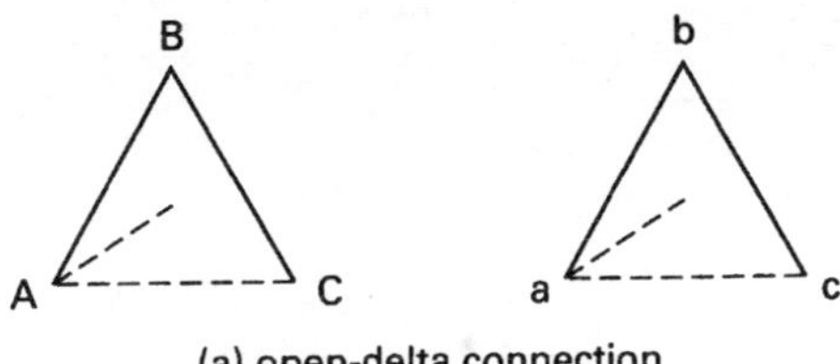

(a) open-delta connection

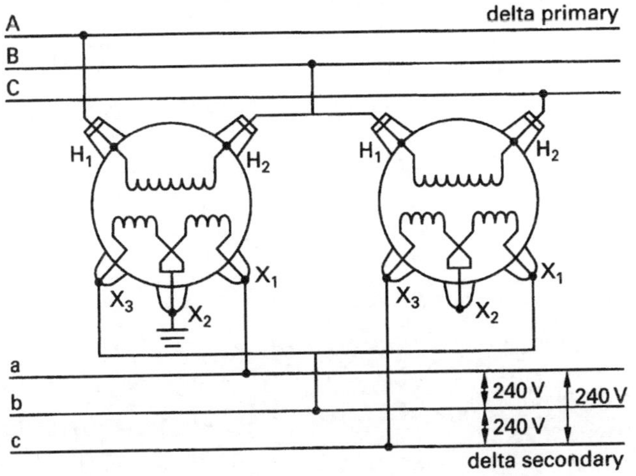

(b) open-delta connection for a transmission line

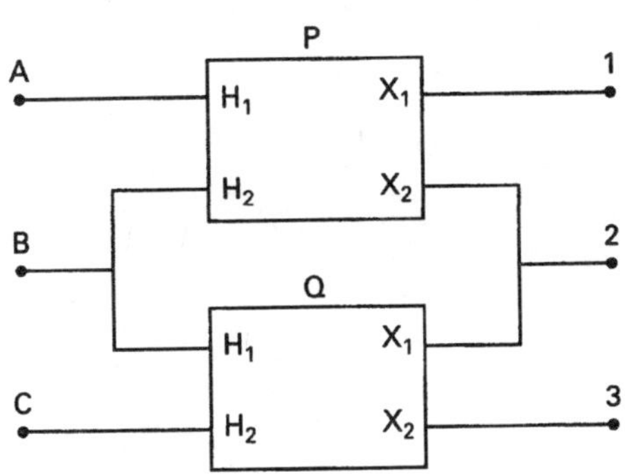

(c) standard open-delta connection

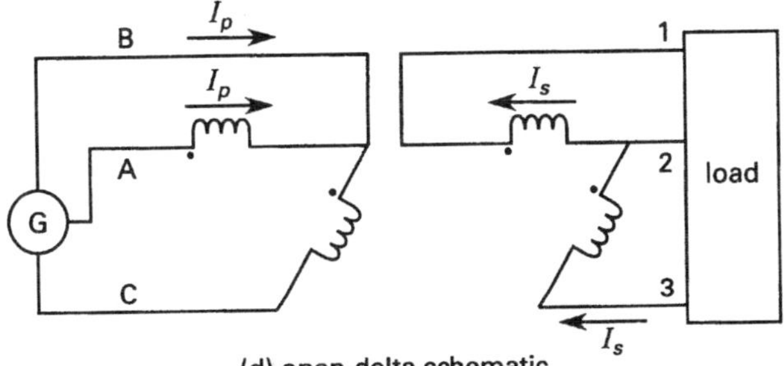

(d) open-delta schematic

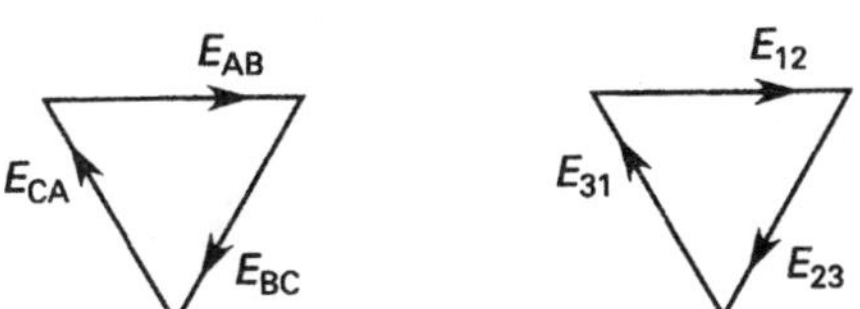

(e) open-delta phasor diagram

Example 37.8

Two single-phase transformers, each rated for 200 kVA and 7200 V/600 V, are connected in an open-delta configuration. What is the maximum capacity of the resulting installed transformer, expressed as a percentage of the installed transformer rating?

Solution

Each transformer is rated for 200 kVA, so the installed transformer rating is

$$(2)(200 \text{ kVA}) = 400 \text{ kVA}$$

However, this is not the maximum capacity. The maximum capacity is limited by the maximum current through a given transformer. The transformer carries the current of the missing phase, so using Eq. 27.57, the current capacity of each secondary transformer is

$$\begin{aligned} S &= IV \\ I &= \frac{S}{V} \\ &= \left(\frac{200 \text{ kVA}}{600 \text{ V}}\right)\left(1000 \frac{\text{VA}}{\text{kVA}}\right) \\ &= 333.33 \text{ A} \end{aligned}$$

The current is limited in any one phase at 333.33 A. The secondary current through lines 1, 2, and 3 is restricted to this level. (See Fig. 37.16(d).) Given this restriction, use Eq. 34.32 to find the maximum capacity.

$$\begin{aligned} S &= \sqrt{3}IV \\ &= \frac{\sqrt{3}(333.33 \text{ A})(600 \text{ V})}{1000 \frac{\text{V}}{\text{kVA}}} \\ &= 346.4 \text{ kVA} \end{aligned}$$

The maximum capacity of the transformer, expressed as a percentage of the installed transformer rating is

$$\begin{aligned} C &= \frac{\text{maximum load}}{\text{installed transformer rating}} \times 100\% \\ &= \left(\frac{346.4 \text{ kVA}}{400 \text{ kVA}}\right) \times 100\% \\ &= 86.6\% \end{aligned}$$

16. ZIGZAG TRANSFORMERS

A *zigzag transformer* is a type of *grounding transformer* that provides a neutral point in an otherwise ungrounded system. The neutral point is then used for the return path for fault currents, for an earth reference point, and for harmonics mitigation (canceling triplet currents). The entire transformer can be used as an autotransformer. A zigzag connection is shown in Fig. 37.17(a), and the zigzag phasor diagram is shown in Fig. 37.17(b).

Distribution

Figure 37.17 *Zigzag Transformer*

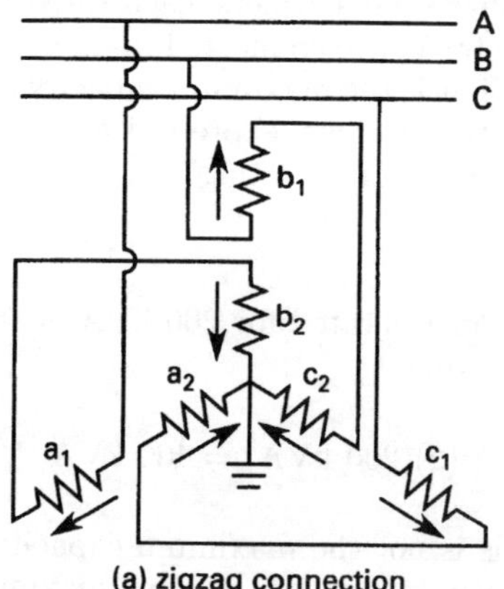

(a) zigzag connection

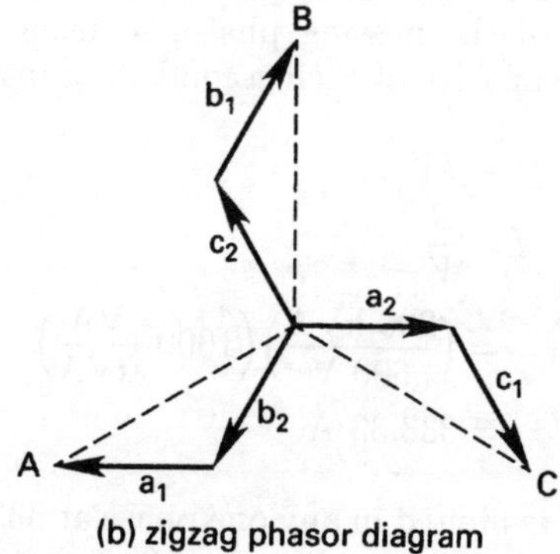

(b) zigzag phasor diagram

38 Transmission Lines

Nomenclature

A	area	m^2
A	ABCD parameter	–
B	susceptance	S
B	ABCD parameter	–
C	capacitance	F
C	ABCD parameter	–
d	distance	m
D	distance	m
D	ABCD parameter	–
E	electric field strength	V/m
f	frequency	Hz
G	conductance	S
GMD	geometric mean distance	m
GMR	geometric mean radius	m
H	magnetic field strength	A/m
I, $\mathbf{I}$	current	A
$\mathbf{J}$	current density	A/m^2
K	correction factor	–
K	skin effect ratio	–
l	length	m
L	inductance	H
M	mutual inductance	H
P	power	W
r	radius	m
R	resistance	Ω
SWR	standing wave ratio	–
t	time	s
T	temperature	°C or K
v	velocity	m/s
V, $\mathbf{V}$	voltage	V
VR	voltage regulation	V
w	width	m
x	distance	m
X	reactance	Ω
Y	admittance	S
Z, $\mathbf{Z}$	impedance	Ω

Symbols

α	attenuation constant	Np/m
α	thermal coefficient of resistance	1/°C
β	phase constant	rad/m
γ	propagation constant	rad/m
Γ	reflection coefficient	–
δ	skin depth	m
ϵ_0	free-space permittivity, 8.854×10^{-12}	F/m
η	efficiency	–
θ	angle	rad
λ	wavelength	m
μ	permeability	H/m
μ_0	free-space permeability, 1.2566×10^{-6}	H/m
μ_r	relative permeability	–
ρ	resistivity	Ω·m
σ	conductivity	S/m
ω	angular velocity	rad/s

Subscripts

0	characteristic
ab	a to b
AC	alternating current
bc	b to c
c	conductor
ca	c to a
C	capacitative
Cu	copper
DC	direct current
e	equivalent
eff	effective
ext	external
fl	full load
int	internal
l	per unit length
L	inductance or inductor
m	mutual
max	maximum
min	minimum
nl	no load
oc	open circuit
P	power transmission

Distribution

R	receiving end
R	resistance
sc	short circuit
S	sending end
w	wave

1. FUNDAMENTALS

A *power transmission line* is a system of conductors that is designed to transfer power from a source to a load. Conductors used in power transmission lines include wires, coaxial cables and lines, and waveguides. A *coaxial cable* or *coaxial line* is one in which the conductor is centered inside, and insulated from, an outer metal tube that acts as a second conductor. Coaxial cables and lines are used at frequencies up to approximately 1 GHz; above 1 GHz, losses become prohibitive and waveguides are used instead. *Waveguides* are devices, such as hollow metal structures of various shapes, that confine electromagnetic waves to particular paths as the waves travel. Waveguides are used at frequencies up to approximately 100 GHz; above 100 GHz, losses become prohibitive and optical guides are used that can operate at frequencies of approximately 2×10^{14} Hz with minimal losses.

The wavelength that is transmitted by a power transmission line must be many times larger than the length of the line itself, or else the power will be radiated into space. For example, a power transmission line operating at 60 Hz will have a wavelength of 5000 km but will rarely have a length greater than 500 km.

Transmission lines that are designed to carry wavelengths that are shorter than or comparable to the length of the line are called *high-frequency transmission lines*. These lines have different properties from power transmission lines and are often used in communication systems for the transmission of signals. Waveguides are used in high-frequency transmission for wavelengths less than approximately 10 cm. A conductor specifically designed with a length comparable to the wavelength is called an *antenna*.

A transmission line guides *transverse electromagnetic* (TEM) *waves*. In TEM waves, the electric and magnetic fields are perpendicular to the direction of propagation. At low frequencies, where the dimensions of the line are small compared to the wavelength, lumped parameters can provide an adequate model of the circuit.[1] This is because the displacement current is small and can be ignored. That is, the storage of electromagnetic energy in the space surrounding the line can be ignored. At high frequencies, the dimensions of the line are large compared to the wavelengths propagated. The displacement current becomes significant. That is, electric and magnetic energy is stored in the space surrounding the line (see Chap. 24 and Chap. 25).

[1]Lumped parameters are single-valued circuit elements that are equivalent to the value in the circuit as a whole, such as a resistor, inductor, or capacitor in a specific position in the circuit that accounts for the line electrical properties as well.

To avoid using field theory to analyze transmission lines, distributed parameters are used along with currents and voltages associated with the electric and magnetic fields. *Distributed parameters* are circuit elements that exist along the entire line but are represented per unit length, which allows the use of common electrical analytic techniques. A transmission line represented by distributed parameters is illustrated in Fig. 38.1.

Figure 38.1 Distributed Parameters

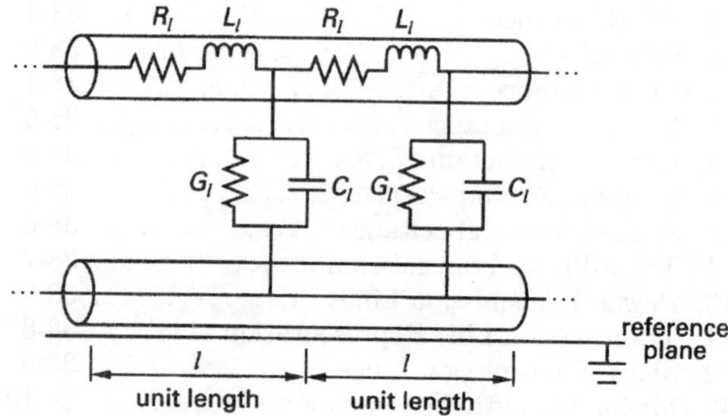

The distributed parameters in Fig. 38.1 are the resistance per unit length, R_l, inductance per unit length, L_l, capacitance per unit length, C_l, and *shunt conductance* per unit length, G_l. The resistance depends on the frequency because of a phenomenon known as the *skin effect*: as frequency rises, current flows closer to the surface of the conductor, which raises the resistance. (The AC resistance at 60 Hz is 5% to 10% higher than the DC resistance.) The *temperature effect* on resistance must also be accounted for in transmission line design. The inductance consists of two terms, the internal inductance, L_{int}, and the external inductance, L_{ext}. The internal inductance depends on the skin effect, decreasing as the frequency increases. The external inductance depends on the geometric arrangement of the conductors. The capacitance also depends on the geometric arrangement of the conductors. The shunt conductance, also called the *line-to-line conductance*, is usually insignificant.

Additional parameters of concern include the characteristic impedance, the standing wave ratio, and the reflection coefficient.[2] The *characteristic impedance* is determined by

$$Z_0 = \sqrt{\frac{Z_l}{Y_l}} \quad [\text{in } \Omega] \qquad 38.1$$

If the transmission line is terminated with the characteristic impedance, no power is reflected back to the source. If the termination impedance is other than Z_0, then the power (signal) from the generator (source) will be partially reflected back to the generator. In wave terminology, the generator's waves will then

[2]The Smith chart (see Chap. 39) is a convenient way to analyze these three parameters for low-loss lines and high-frequency lines.

combine in the transmission line with the reflected waves to form *standing waves*. The *standing wave ratio*, SWR, is the ratio of the maximum to the minimum voltages (currents) encountered along the transmission line. Typically, the SWR is greater than unity. If the terminating impedance matches the characteristic impedance, all the input power provided by the generator will be absorbed by the load and the SWR will equal one. The standing wave ratios are

$$\begin{aligned} \text{SWR} &= \text{VSWR} = \text{CSWR} \\ &= \max \begin{cases} Z_{\text{load}}/Z_0 \\ Z_0/Z_{\text{load}} \end{cases} \end{aligned} \quad 38.2$$

The *reflection coefficient*, Γ, is the ratio of the reflected voltage (current) to the incident voltage (current). The reflection coefficient is

$$\begin{aligned} \Gamma &= \frac{V_{\text{reflected}}}{V_{\text{incident}}} = \frac{I_{\text{reflected}}}{I_{\text{incident}}} \\ &= \left| \frac{Z_{\text{load}} - Z_0}{Z_{\text{load}} + Z_0} \right| \end{aligned} \quad 38.3$$

The fraction of incident power that is reflected back to the source from the load is Γ^2. The relationship between the reflection coefficient and the standing wave ratio is given by Eq. 38.4.

$$\Gamma = \frac{\text{SWR} - 1}{\text{SWR} + 1} \quad 38.4$$

The *velocity of propagation*, v_w, is the velocity at which a wave propagates along a transmission line, given by

$$v_w = \frac{1}{\sqrt{L_l C_l}} \quad 38.5$$

Transmission lines are considered in three lengths.[3] The *short transmission line* is from 0 km to 80 km (0 mi to 50 mi). In a short line, the shunt parameters, G_l and C_l, are insignificant and ignored in the analysis. The *medium transmission line* is from 80 km to 240 km (50 mi to 150 mi). In a medium line, the shunt capacitances are generally lumped in predetermined locations along the line. The *long transmission line* is greater than 240 km (150 mi). A long transmission line is also called a *uniform transmission line* because it has constant distributed parameters along the line from the sending end to the receiving end.

Many transmission line characteristics and formulas are tabulated in English Engineering System units. For this reason, a mixture of metric and English Engineering System units is utilized in this chapter. Additionally, all AC quantities have a magnitude and an angle. However, as is common practice, these items will be shown in phasor notation, or complex notation, only when possible confusion could occur.[4]

2. DC RESISTANCE

Resistance generates the I^2R power loss in transmission lines. Additionally, the IR drop, or voltage drop, affects the voltage regulation. The DC resistance, often called the *characteristic resistance*, R_0, of a conductor of length l, cross-sectional area A, and resistivity ρ is normally given as $\rho l/A$. Because the distributed parameters are given as per-unit-length quantities, the DC resistance per unit length is

$$R_{l,\text{DC}} = R_0 = \frac{\rho}{A} \quad [\text{in } \Omega/\text{m}] \quad 38.6$$

Resistance sensitivity to temperature is calculated from

$$R_{\text{final}} = R_{\text{initial}}(1 + \alpha\Delta T) \quad 38.7$$

Voltage drops are often tabulated for various conductor sizes as shown in Table 38.1.

Table 38.1 *Voltage Drop in Two-Wire DC Circuits (Loop)**

conductor size (AWG)		
copper	approximately equivalent aluminum	voltage drop per 100,000 A·m
6	4	31.3
4	2	19.7
2	1/0	12.4
1/0	3/0	7.80
2/0	4/0	6.19

*Values are calculated for 32°C (90°F) conductors.

Example 38.1

A 50 × 10³ W DC device is located 1.6 km from a motor-generator with a maximum output voltage of 600 V_{DC}. If the lowest voltage the DC device can properly utilize is 575 V, what is the minimum conductor size required?

Solution

The maximum allowable voltage drop is 25 V. To use Table 38.1, the total ampere-meters for this system is required. The maximum current is given by

$$\begin{aligned} P &= IV \\ I &= \frac{P}{V} = \frac{50 \times 10^3 \text{ W}}{575 \text{ V}} = 86.96 \text{ A} \end{aligned}$$

[3]The distances given assume a 60 Hz signal. The engineering methods in each type of transmission line differ. As the frequency increases, the distances of each definition decrease, with the concurrent adjustment in the engineering method used.

[4]Phasor or complex notation usually presents the parameter in bold or with an arrow over the top (**V** or $\vec{V}$).

The total cable distance to reach the site and provide the return path to the generator is

$$D = (2)(1.6 \times 10^3 \text{ m}) = 3.2 \times 10^3 \text{ m}$$

The total ampere-meters is

$$(86.96 \text{ A})(3.2 \times 10^3 \text{ m}) = 278.27 \times 10^3 \text{ A·m}$$

Find the voltage drop to use in the table.

$$\text{allowable voltage drop} = (\text{A·m}_{\text{total}}) \times \left(\frac{\text{voltage drop}}{100{,}000 \text{ A·m}}\right)$$

$$25 \text{ V} = (278.27 \times 10^3 \text{ A·m}) \times \left(\frac{V_{\text{drop}}}{100{,}000 \text{ A·m}}\right)$$

$$V_{\text{drop}} = 8.98 \text{ V}$$

From Table 38.1, the conductor size with an 8.98 voltage drop, or less, is 1/0 for copper or 3/0 for aluminum.

3. SKIN EFFECT

The *skin effect* is the tendency of AC currents to flow near the surface of a conductor. This results in the AC current being restricted to less than the total cross-sectional area of the conductor, resulting in an increased resistance and a decreased internal inductance. The skin effect can be thought of as an electromagnetic phenomenon related to the time required for an external field to penetrate a conductor.[5] It occurs because the surface reactance is smaller than the reactance of other possible current paths within the conductor. The *skin depth*, δ, is the depth beneath the surface of the conductor that is carrying the current at a given frequency caused by electromagnetic waves incident to the surface.[6] Specifically, it is the depth at which the current density is $1/e$, the value of the surface current density. The skin depth for a flat conducting plate is

$$\delta = \frac{1}{\sqrt{\frac{\pi f \mu}{\rho}}} = \frac{1}{\sqrt{\pi f \mu \sigma}} \qquad 38.8$$

The AC resistance per unit length for a flat conducting plate of unit width w is

$$R_{l,\text{AC}} = \frac{\rho}{\delta w} \quad [\text{in } \Omega/\text{m}] \qquad 38.9$$

Conductors are normally round. Equation 38.9 would be applicable only if the skin depth were much less than the radius, r, so that the wire approximated a flat surface. For example, a ratio of $r/\delta > 5.5$ results in less than a 10% error. The actual skin depth for copper conductors is

$$\delta_{\text{Cu}} = \frac{0.066}{\sqrt{f}} \quad [\text{in m}] \qquad 38.10$$

Example 38.2

Determine the copper wire radius in mils where the skin depth equals the radius for the following frequencies: 60 Hz, 10 kHz, 1 MHz.

Solution

First, convert Eq. 38.10 into an equivalent equation based on mils rather than meters.

$$\delta_{\text{Cu}} = \left(\frac{0.066}{\sqrt{f}}\right)\left(\frac{1 \text{ mil}}{0.001 \text{ in}}\right)\left(\frac{1 \text{ in}}{2.54 \text{ cm}}\right)\left(100 \frac{\text{cm}}{\text{m}}\right) = \frac{2.60 \times 10^3}{\sqrt{f}} \text{ mil}$$

Substitute the given frequencies to determine the skin depth.

$$\delta_{\text{Cu}} = \frac{2.60 \times 10^3}{\sqrt{f}} \text{ mil} = \frac{2.60 \times 10^3}{\sqrt{60 \text{ Hz}}} \text{ mil} = 336 \text{ mil}$$

$$\delta_{\text{Cu}} = \frac{2.60 \times 10^3}{\sqrt{f}} \text{ mil} = \frac{2.60 \times 10^3}{\sqrt{10 \times 10^3 \text{ Hz}}} \text{ mil} = 26 \text{ mil}$$

$$\delta_{\text{Cu}} = \frac{2.60 \times 10^3}{\sqrt{f}} \text{ mil} = \frac{2.60 \times 10^3}{\sqrt{1 \times 10^6 \text{ Hz}}} \text{ mil} = 2.6 \text{ mil}$$

The largest solid copper wire is AWG 4/0, that is, AWG 0000.[7] The radius is 230 mil. This indicates that at 60 Hz the electromagnetic wave completely penetrates the conductor. Skin effect occurs at 60 Hz but only for the largest conductors, such as those used for commercial power applications (on the order of 500 kcmil). For most other applications, the 10 kHz frequency mark is often used as the point where skin effect must be accounted for and AC resistance calculated.

4. AC RESISTANCE

Skin effect and the resulting increase in resistance and decrease in internal inductance are often neglected in initial calculations or approximations. In the case of large copper wires at commercial power frequencies, exact calculations must consider the effect.[8] The effect on aluminum conductors is less because aluminum has a higher resistivity. Tables of skin effect ratios are available when the exact value of the AC resistance is required.[9] Table 38.2 gives sample ratios. The skin effect

[5]It is this principle that allows metal shielding to be used to prevent electromagnetic wave penetration of equipment. Conversely, such shielding also prevents the escape of radiated electromagnetic waves.

[6]In generators, the armature current is induced from an electromagnetic wave incident to the surface. This principle applies for generators and any conductor cabling attached to the generator.

[7]Any larger size consists of stranded conductors.

[8]The effect should be accounted for at communications system frequencies as well.

[9]Skin effect ratios are also called *AC/DC resistance ratios*.

Table 38.2 *Skin Effect Ratios*

x	K_R[a]	$K_{L,\text{int}}$[b]
0.0	1.00000	1.00000
0.5	1.00032	0.99984
1.0	1.00519	0.99741
2.0	1.07816	0.96113
4.0	1.67787	0.68632
8.0	3.09445	0.35107
16.0	5.91509	0.17649
32.0	11.56785	0.08835
50.0	17.93032	0.05656
100.0	35.60666	0.02828
∞	∞	0

[a]K_R is the skin effect ratio for AC resistance.
[b]$K_{L,\text{int}}$ is the skin effect ratio for internal inductance.

ratio for the AC resistance, K_R, is used to determine the effective resistance from

$$R_{\text{AC}} = K_R R_{\text{DC}} \tag{38.11}$$

The table is entered using a value of x determined from

$$x = 2\pi r\sqrt{\frac{2f\mu}{\rho}} = \frac{2r\sqrt{2\pi}}{\delta} \tag{38.12}$$

The use of tables is the most accurate practical method of determining the AC resistance. Curves normalized to the DC resistance are sometimes used for intermediate frequencies. The AC resistance can also be found by calculating the effective area of the conductor, that is, the area actually conducting current, and comparing this area to that used by the DC resistance. The DC resistance is given by Eq. 38.6 as ρ/A. Because the resistivity doesn't change, comparing an AC resistance to a DC resistance on the same wire gives

$$\frac{R_{l,\text{AC}}}{R_{l,\text{DC}}} = \frac{A}{A_{\text{eff}}} \tag{38.13}$$

The effective area is

$$A_{\text{eff}} = \pi r^2 - \pi(r-\delta)^2 \tag{38.14}$$

Example 38.3

An AWG 12 copper conductor has a diameter of 81 mil with a DC resistance of approximately 5.4 Ω/km at 25°C. Calculate the resistance per kilometer at 10 kHz with no temperature change.

Solution

Equation 38.13 contains the relationship desired. The skin depth, δ, from Ex. 38.2 is 26 mil. The radius is 40.5 mil. Rearranging Eq. 38.13 and substituting Eq. 38.14 gives

$$\frac{R_{l,\text{AC}}}{R_{l,\text{DC}}} = \frac{A}{A_{\text{eff}}}$$

$$\begin{aligned} R_{l,\text{AC}} &= R_{l,\text{DC}}\left(\frac{A}{A_{\text{eff}}}\right) \\ &= R_{l,\text{DC}}\left(\frac{\pi r^2}{\pi r^2 - \pi(r-\delta)^2}\right) \\ &= R_{l,\text{DC}}\left(\frac{\pi r^2}{\pi r^2\left(1-\left(1-\frac{\delta}{r}\right)^2\right)}\right) \\ &= \left(5.4\ \frac{\Omega}{\text{km}}\right)\left(\frac{1}{1-\left(1-\frac{26\ \text{mil}}{40.5\ \text{mil}}\right)^2}\right) \\ &= 6.2\ \Omega/\text{km} \end{aligned}$$

5. INTERNAL INDUCTANCE

The *internal inductance*, also called the *characteristic inductance*, is that inductance associated with the interior of a solid conductor. That is, the fluxes in the space surrounding the conductor are disregarded. For nonmagnetic materials, the relative permeability, μ_r, is approximately one. The magnetic permeability of a conductor is approximately that of free space; that is, $\mu \approx \mu_0$. The internal inductance per unit length for a conductor is

$$L_{l,\text{int}} = \frac{\mu_0}{8\pi} = 0.5 \times 10^{-7}\ \text{H/m} \tag{38.15}$$

For a single-phase system, two conductors are present, with current flow in opposite directions, and the total internal inductance is twice the value given in Eq. 38.15. The *single-phase internal inductance* is

$$L_{l,\text{int}} = \frac{\mu_0}{4\pi} = 1.0 \times 10^{-7}\ \text{H/m} \tag{38.16}$$

6. EXTERNAL INDUCTANCE

The *external inductance* is that inductance associated with the exterior of a solid conductor. That is, the fluxes in the space surrounding the conductor store the inductive energy. For air, the relative permeability, μ_r, is approximately one. The magnetic permeability of air is approximately that of free space, that is, $\mu \approx \mu_0$. The external inductance per unit length for a single conductor is given by

$$\begin{aligned} L_{l,\text{ext}} &= \frac{\mu_0}{2\pi}\ln\frac{D}{r} \\ &= (2\times 10^{-7})\ln\frac{D}{r} \quad \text{[in H/m]} \end{aligned} \tag{38.17}$$

Distribution

The term D is the distance between the centers of the conductors. The term r is the radius of the conductor. For a single-phase system, two conductors are present, with current flow in opposite directions. The total external inductance is twice the value given in Eq. 38.17. The *single-phase external inductance* is

$$L_{l,\text{ext}} = \frac{\mu_0}{\pi} \ln \frac{D}{r} = (4 \times 10^{-7}) \ln \frac{D}{r} \quad \text{[in H/m]} \qquad 38.18$$

7. SINGLE-PHASE INDUCTANCE

The total inductance of a single-phase system, which consists of two conductors, is the sum of the internal and external inductances.[10]

$$L_l = L_{l,\text{int}} + L_{l,\text{ext}} = \left(\frac{\mu_0}{4\pi}\right)\left(1 + 4 \ln \frac{D}{r}\right) \quad \text{[in H/m]} \qquad 38.19$$

The first term of Eq. 38.19 represents the internal inductance of a solid conductor. The second term represents the inductance caused by external fluxes. Equation 38.19 is simplified using the concept of the *geometric mean radius* (GMR), defined as[11]

$$\text{GMR} = re^{-1/4} \qquad 38.20$$

Substituting the geometric mean radius for the radius in Eq. 38.19 gives the following simplified form for the total single-phase inductance per unit length.

$$L_l = \frac{\mu_0}{\pi} \ln \frac{D}{\text{GMR}} = (4 \times 10^{-7}) \ln \frac{D}{\text{GMR}} \quad \text{[in H/m]} \qquad 38.21$$

Equation 38.21 is equivalent to Eq. 38.19 but represents a different model of the conductor. Specifically, Eq. 38.21 represents an equivalent thin-walled, hollow conductor of radius GMR with no internal flux linkage and, therefore, no internal inductance.

Example 38.4

The wire in Ex. 38.3 is used in a single-phase system operating at 10×10^3 Hz with the lines spaced 0.5 m apart. Ignoring shunt effects, what is the impedance per unit length of line?

Solution

Ignoring the shunt conductance and capacitance gives impedance per unit length as

$$Z_l = R_l + jX_{L,l}$$

The resistance per unit length was determined in Ex. 38.3 to be 6.2×10^{-3} Ω/m. The inductive reactance per unit length is

$$X_{L,l} = 2\pi f L_l$$

The only unknown is the inductance. Using Eq. 38.19 and Eq. 38.20, and noting that a radius of 40.5 mil is 1.029×10^{-3} m, gives

$$\begin{aligned} L_l &= \left(4 \times 10^{-7} \ \frac{\text{H}}{\text{m}}\right) \ln \frac{D}{\text{GMR}} \\ &= \left(4 \times 10^{-7} \ \frac{\text{H}}{\text{m}}\right) \ln \frac{D}{re^{-1/4}} \\ &= \left(4 \times 10^{-7} \ \frac{\text{H}}{\text{m}}\right) \ln \frac{0.5 \text{ m}}{(1.029 \times 10^{-3} \text{ m})e^{-1/4}} \\ &= 2.57 \times 10^{-6} \text{ H/m} \end{aligned}$$

The reactance per unit length is

$$\begin{aligned} X_{L,l} &= 2\pi f L_l \\ &= (2\pi)(10 \times 10^3 \text{ Hz})(2.57 \times 10^{-6} \text{ H/m}) \\ &= 0.161 \ \Omega/\text{m} \end{aligned}$$

The impedance per unit length is

$$Z_l = R_l + jX_{L,l} = 6.2 \times 10^{-3} \ \Omega/\text{m} + j0.161 \ \Omega/\text{m}$$

The inductive reactance has an effect approximately four times greater than the resistance.

8. SINGLE-PHASE CAPACITANCE

The shunt capacitance is the capacitance between the solid conductors. Power transmission line conductors are normally separated by air. For air, the relative permittivity, ϵ_r, is approximately one. Therefore, the permittivity of air is approximately that of free space; that is, $\epsilon \approx \epsilon_0$. The capacitance of a single-phase system, which consists of two conductors, is

$$C_l = \frac{\pi\epsilon_0}{\ln \frac{D}{r}} = \frac{2.78 \times 10^{-11}}{\ln \frac{D}{r}} \quad \text{[in F/m]} \qquad 38.22$$

The term D represents the distance between the centers of the conductors. The term r is the radius of the conductors. For a single-phase system, two conductors are present, with current flow in opposite directions.

[10]The equations used in this text assume that $D \gg r$ and that r is identical in the two conductors, which is generally the case in power transmission lines.

[11]The geometric mean radius is sometimes called the *self geometric mean distance* (GMD).

Example 38.5

A copper AWG 20 wire with a radius of 16 mil is used in a single-phase system at 10×10^3 Hz, with the conductors 0.5 m apart. What is the capacitive reactance per unit length?

Solution

The capacitive reactance per unit length is

$$X_{C,l} = \frac{1}{2\pi f C_l}$$

AWG 20 has a radius of 16 mil or 4.064×10^{-4} m. The capacitance per unit length is

$$\begin{aligned} C_l &= \frac{\pi \epsilon_0}{\ln \frac{D}{r}} \\ &= \frac{\left(\pi 8.854 \times 10^{-12}\ \frac{\text{F}}{\text{m}}\right)}{\ln \frac{0.5\ \text{m}}{4.064 \times 10^{-4}\ \text{m}}} \\ &= 3.91 \times 10^{-12}\ \text{F/m} \end{aligned}$$

The capacitive reactance is

$$\begin{aligned} X_C &= \frac{1}{2\pi f C} \\ &= \frac{1}{(2\pi)(10 \times 10^3\ \text{Hz})(3.91 \times 10^{-12}\ \text{F})} \\ &= 4.07 \times 10^6\ \Omega \\ X_{C,l} &= 4.0 \times 10^6\ \Omega/\text{m} \end{aligned}$$

The capacitive reactance has a much larger effect on the per-unit reactance of the line than does the inductive reactance.

9. THREE-PHASE TRANSMISSION

In three-phase, three-wire, or four-wire balanced transmission, the sum of the instantaneous currents is zero, and the sum of the instantaneous voltages is zero.[12] The conductors are assumed to be equilaterally spaced at distance D. When the conductors are not symmetrically arranged, as is often the case, formulas used for the inductance and capacitance are still valid if the equivalent distance, D_e, is substituted for the distance (see Fig. 38.2(a) and Fig. 38.2(b)).[13]

$$D_e = \sqrt[3]{D_{ab} D_{bc} D_{ca}} \qquad 38.23$$

[12]The three-wire or four-wire balanced assumption allows simplifications in the derivations of the equations in this section.

[13]The equivalent spacing is sometimes called the *mutual geometric mean distance* (D_m or GMD).

Nonsymmetrical spacing causes electrostatic and electromagnetic imbalance among the lines, resulting in unequal phase voltages and currents. In large commercial power transmission systems, the effect is minimal because of the balancing effect of the rotating loads. The effect can be minimized by *transposition* of the lines as illustrated in Fig. 38.2(c). Transposing the lines improves reliability by minimizing the current induced in the event of a fault, reduces power losses, and reduces interference with nearby telecommunications lines.

Figure 38.2 *Transmission Line Spacing*

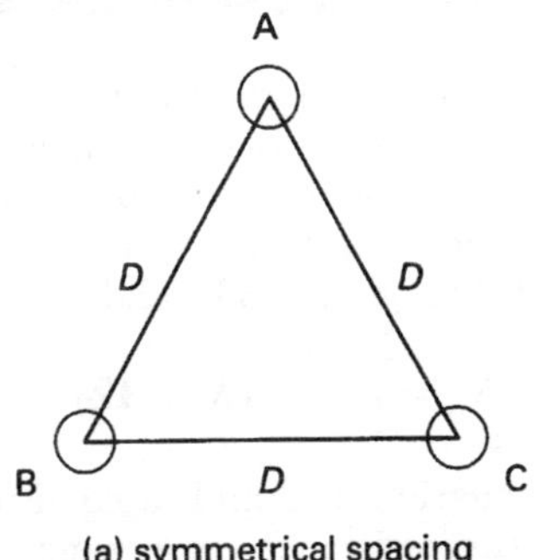

(a) symmetrical spacing

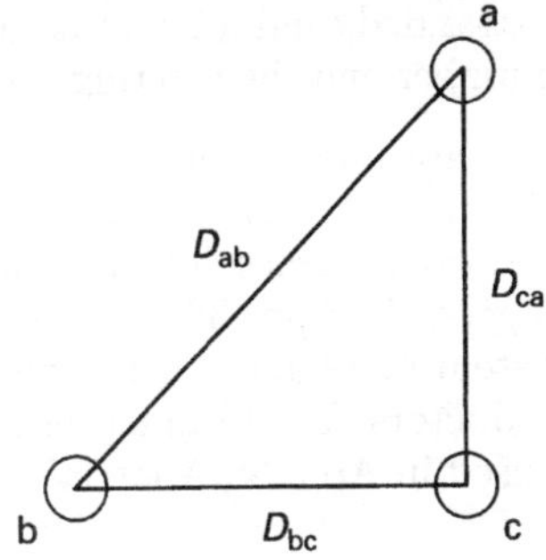

(b) unsymmetrical spacing

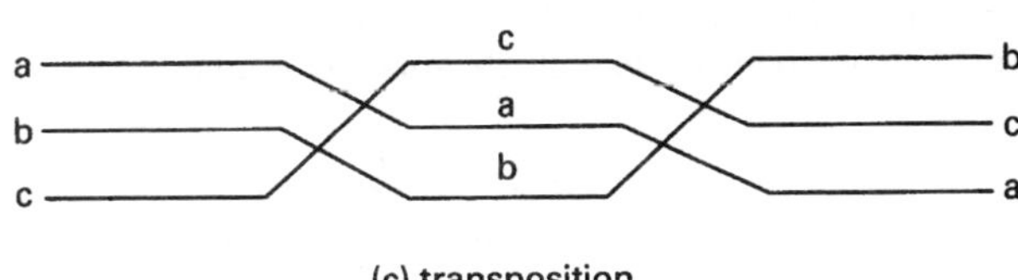

(c) transposition

The per-phase inductance per unit length of a three-phase transmission line is

$$\begin{aligned} L_l &= \frac{\mu_0}{2\pi} \ln \frac{D_e}{\text{GMR}} \\ &= (2 \times 10^{-7}) \ln \frac{D_e}{\text{GMR}} \quad \text{[in H/m]} \end{aligned} \qquad 38.24$$

Mutual inductance, M, also exists, but for symmetrically spaced conductors the total is zero. This equation is similar to Eq. 38.21 for the total inductance of a single-phase system. Specifically, it is one-half the single-phase value. Per-phase values are equivalent to phase-to-neutral values.

Distribution

The per-phase capacitance per unit length of a three-phase transmission line is

$$C_l = \frac{2\pi\epsilon_0}{\ln \frac{D_e}{r}} = \frac{5.56 \times 10^{-11}}{\ln \frac{D_e}{r}} \quad \text{[in F/m]} \qquad 38.25$$

This equation is similar to Eq. 38.22 for the total capacitance of a single-phase system. Specifically, it is twice the single-phase value. Per-phase values are equivalent to phase-to-neutral values.

10. POWER TRANSMISSION LINES

Power transmission lines normally consist of aluminum strands with a steel core or all-aluminum alloy wires. The lines consist of either three or four conductors. The voltage drop in such lines can be calculated using the principles in Sec. 38.3 through Sec. 38.9 and the basic formula.

$$\mathbf{V}_l = \mathbf{I}R_l + j\mathbf{I}X_l = \mathbf{I}\mathbf{Z}_l \qquad 38.26$$

The actual calculation is complicated. Frequency and spacing affect the resistance, inductance, and capacitance. The geometric mean radius given in Eq. 38.20 must be adjusted for stranded conductors, taking into account both the total number and the layering geometry.

Such calculations are unnecessary, because the properties of standard conductors have been tabulated and the conductors given code names for ease of reference. A sample table is given in App. 38.A, which uses English Engineering System units and a 1 ft symmetrical spacing between conductors. The formula for inductive reactance per mile used in App. 38.A is

$$X_L = (2.022 \times 10^{-3}) f \ln \frac{D}{\text{GMR}} \quad [\text{in } \Omega/\text{mi}] \qquad 38.27$$

For line-to-line spacings other than one foot, the correction factor given by Eq. 38.28 must be applied.[14]

$$K_L = 1 + \frac{\ln D}{\ln \frac{1}{\text{GMR}}} \qquad 38.28$$

The formula for capacitive reactance used in App. 38.A is approximately

$$X_C = \frac{1.781}{f} \ln \frac{D}{r} \quad [\text{in M}\Omega\text{-mi}] \qquad 38.29$$

For line-to-line spacings other than one foot, the correction factor given by Eq. 38.30 must be applied.

$$K_C = 1 + \frac{\ln D}{\ln \frac{1}{r}} \qquad 38.30$$

[14]The GMR used in Eq. 38.27, not Eq. 38.20, is the one tabulated. Equation 38.20 is a good approximation of the tabulated value.

The values of D, GMR, and r are in feet when used in Eq. 38.27 through Eq. 38.30. Appendix 38.A can be used for single-phase systems as well. The values for inductive and capacitive reactance, Eq. 38.27 and Eq. 38.29, double for single-phase systems.

Example 38.6

A single-phase system uses merlin conductors spaced 10 ft apart from center to center. The diameter for the merlin conductors is 0.057 ft. What is the total shunt reactance for 10 mi of the system?

Solution

From App. 38.A for merlin conductors, the capacitive reactance for a one-foot spacing is 0.106 MΩ-mi. Since the diameter for the merlin conductors is 0.057 ft, the radius is 0.0285 ft. The correction factor for the 10 ft spacing is

$$K_C = 1 + \frac{\ln D}{\ln \frac{1}{r}} = 1 + \frac{\ln 10}{\ln \frac{1}{0.0285 \text{ ft}}} = 1.647$$

The capacitive reactance for a single conductor is

$$\begin{aligned} X_{C,\text{actual}} &= X_{C,\text{table}} K_C \\ &= (0.106 \text{ M}\Omega\text{-mi})(1.647) \\ &= 0.1746 \text{ M}\Omega\text{-mi} \end{aligned}$$

This is a single-phase system. That is, two conductors are present. The total capacitive reactance is

$$X_{C,\text{total}} = (2)(0.1746 \text{ M}\Omega\text{-mi}) = 0.3492 \text{ M}\Omega\text{-mi}$$

For a 10 mi portion of the system, the capacitive reactance is

$$\begin{aligned} X_C &= \frac{0.3492 \text{ M}\Omega \text{ mi}}{10 \text{ mi}} \\ &= 0.03492 \text{ M}\Omega \end{aligned}$$

11. TRANSMISSION LINE REPRESENTATION

Analysis of transmission lines is accomplished by approximating the line with distributed parameters. The line is represented as a series-parallel combination of electrical components applicable to the length of the transmission system.[15] Three-phase transmission systems are operated in as balanced a configuration as possible, but are seldom equilaterally spaced. Transposition may alleviate this imbalance, but it is seldom used.

[15]The shunt conductance is normally insignificant and is excluded from the models used.

Nevertheless, calculations assume equilateral spacing and transposition. Calculated values closely approximate actual values if the equivalent spacing, D_e, of Eq. 38.23 is used in place of the actual spacing. Given these assumptions, balanced three-phase systems can be analyzed on a per-phase basis using a hypothetical neutral, which contributes no resistance, inductance, or capacitance. Lines with significant imbalances must be analyzed using symmetrical components.

Design of all transmission lines, regardless of length, must account for the voltage regulation and efficiency of power transmission, η_P. In terms of a *sending end* (subscript S), or source, and a *receiving end* (subscript R), or load, these quantities are defined by Eq. 38.31 and Eq. 38.32.

$$\mathrm{VR} = \frac{|V_{R,\mathrm{nl}}| - |V_{R,\mathrm{fl}}|}{|V_{R,\mathrm{fl}}|} \qquad 38.31$$

$$\eta_P = \frac{P_R}{P_S} \qquad 38.32$$

Any transmission line may be represented as a two-port network as shown in Fig. 38.3. The ABCD parameters are sometimes called *generalized circuit constants* and, in general, are complex. (Though complex, common practice is to not show them in complex notation. That is, they are not shown in bold.) The ABCD parameters for the various transmission lines are given in Table 38.3.

Figure 38.3 *Transmission Line Two-Port Network*

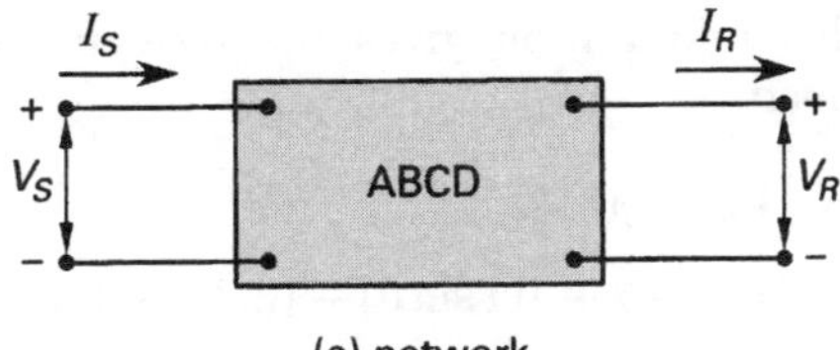

(a) network

$$V_S = AV_R + BI_R$$
$$I_S = CV_R + DI_R$$

(b) equations

$$\begin{bmatrix} V_S \\ I_S \end{bmatrix} = \begin{bmatrix} A & B \\ C & D \end{bmatrix} \begin{bmatrix} V_R \\ I_R \end{bmatrix}$$

(c) matrix form of equations

12. SHORT TRANSMISSION LINES

Short transmission lines are 60 Hz lines that are less than 80 km (50 mi) long. The shunt reactance is excluded from the model because it is much greater than most load impedances and significantly greater than the line impedance. Only the series resistance and inductance are significant as shown in Fig. 38.4.

Figure 38.4 *Short Transmission Line Model*

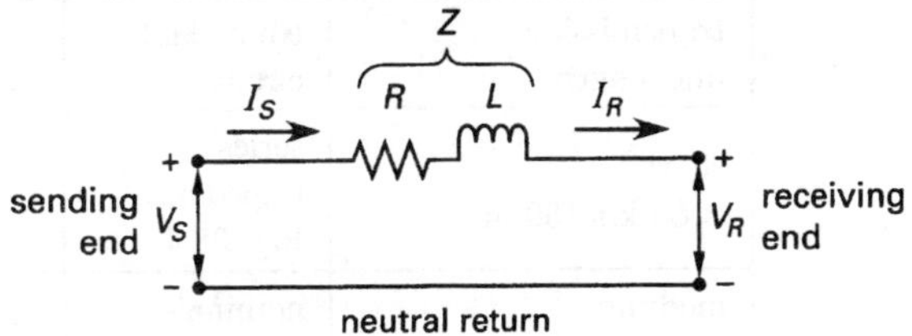

The impedance is given by[16]

$$Z = R + jX_L \qquad 38.33$$

Using the ABCD parameters in Table 38.3 results in Eq. 38.34 and Eq. 38.35.

$$V_S = V_R + I_R Z \qquad 38.34$$

$$I_S = I_R \qquad 38.35$$

Example 38.7

A 30 mi hawk three-phase 60 Hz transmission line has a 10 ft spacing between conductors. The sending-end voltage is 11 kV per phase. The load draws 200 A per phase at 0.8 pf lagging. The entire system is assumed to operate at 25°C. What is the voltage regulation in percent?

Solution

Since it is less than 50 mi in length, this is a short transmission line. The voltage regulation is given by Eq. 38.31. The voltage on the receiving end needs to be determined. The current on the receiving end and its angle (via the power factor) are given. To determine the voltage on the receiving end, the impedance must be calculated. From App. 38.A for a hawk conductor, the uncorrected resistance and reactance are

$$R_{l,\mathrm{AC}} = 0.193\ \Omega/\mathrm{mi}$$
$$X_{L,l} = 0.430\ \Omega/\mathrm{mi}$$

The correction factor for a 10 ft spacing, using the GMR value in App. 38.A, is

$$K_L = 1 + \frac{\ln D}{\ln \frac{1}{\mathrm{GMR}}} = 1 + \frac{\ln 10}{\ln \frac{1}{0.0289}} = 1.6497$$

[16]The quantities represent total amounts but are calculated per unit length, corrected for conductor spacing, and multiplied by the length of the transmission line. The impedance is a per-phase quantity and needs to be multiplied by three to get the line-to-line impedance in a delta-connected load.

Table 38.3 *Per-Phase ABCD Constants for Transmission Lines*

transmission line length	equivalent circuit	A	B	C	D
short <80 km (50 mi)	series impedance Fig. 38.4	1	Z	0	1
medium 80–240 km (50–150 mi)	nominal-T Fig. 38.5(a)	$1+\frac{1}{2}YZ$	$Z(1+\frac{1}{4}YZ)$	Y	$1+\frac{1}{2}YZ$
medium 80–240 km (50–150 mi)	nominal-π Fig. 38.5(b)	$1+\frac{1}{2}YZ$	Z	$Y(1+\frac{1}{4}YZ)$	$1+\frac{1}{2}YZ$
long >240 km (150 mi)	distributed parameters Fig. 38.6	$\cosh \gamma l$	$Z_0 \sinh \gamma l$	$\frac{\sinh \gamma l}{Z_0}$	$\cosh \gamma l$

The corrected reactance is

$$\begin{aligned} X_{L,l(\text{corrected})} &= K_L X_{L,l} \\ &= (1.6497)\left(0.430\ \frac{\Omega}{\text{mi}}\right) \\ &= 0.71\ \Omega/\text{mi} \end{aligned}$$

The impedance for 30 mi of line is

$$\begin{aligned} Z &= (R_{l,\text{AC}} + jX_{L,l})(30\ \text{mi}) \\ &= \left(0.193\ \frac{\Omega}{\text{mi}} + j0.71\ \frac{\Omega}{\text{mi}}\right)(30\ \text{mi}) \\ &= 5.79 + j21.3\ \Omega \\ &= 22.1\ \Omega\angle 74.8^\circ \end{aligned}$$

The power factor is given as 0.8 lagging, therefore the current lags the receiving-end voltage by 36.8°. Because this is referenced to the receiving-end voltage, take V_R to be the reference at 0°. The current is

$$I_R = 200\ \text{A}\angle{-36.8^\circ}$$

The voltage drop across the line at full load is

$$\begin{aligned} V_{\text{drop}} &= I_R Z = (200\ \text{A}\angle{-36.8^\circ})(22.1\ \Omega\angle 74.8^\circ) \\ &= 4.42\ \text{kV}\angle 38.0^\circ \end{aligned}$$

Using the relationship for the sending and receiving ends given by Eq. 38.34 gives

$$\begin{aligned} V_S &= V_R + I_R Z \\ V_R &= V_S - I_R Z = 11\ \text{kV}\angle\theta - 4.42\ \text{kV}\angle 38^\circ \\ |V_R|\ \text{kV}\angle 0^\circ &= |V_R| + j0 \\ &= (11\cos\theta + j11\sin\theta) - (3.48 + j2.72) \end{aligned}$$

The angle θ is the angle of the sending-end voltage with respect to the receiving-end voltage, which was selected as the reference. Because no imaginary portion exists on the left-hand side of the equation, the following must be true.

$$\begin{aligned} 11\sin\theta - 2.72 &= 0 \\ \sin\theta &= \frac{2.72}{11} = 0.247 \end{aligned}$$

Using this information to determine the angle and the term $\cos\theta$ gives

$$\begin{aligned} \theta &= \arcsin 0.247 = 14.3^\circ \\ \cos\theta &= \cos 14.3^\circ = 0.969 \end{aligned}$$

Substituting this information gives the receiving-end voltage at full load.

$$\begin{aligned} |V_R|\ \text{kV}\angle 0^\circ &= |V_R| + j0 \\ &= (11\cos\theta + j11\sin\theta) - (3.48 + j2.72) \\ &= (11)(0.969) + j(11)(0.247) \\ &\quad - 3.48 - j2.72 \\ &= 7.18\ \text{kV}\angle 0^\circ \end{aligned}$$

At a no-load condition, $I_R = 0$ in Eq. 38.34, and the receiving-end voltage equals the sending-end voltage. That is, $V_{R,\text{nl}} = 11$ kV. The regulation, from Eq. 38.31, is

$$\begin{aligned} \text{VR} &= \frac{|V_{R,\text{nl}}| - |V_{R,\text{fl}}|}{|V_{R,\text{fl}}|} = \frac{11\ \text{kV} - 7.18\ \text{kV}}{7.18\ \text{kV}} \\ &= 0.532 \quad (53.2\%) \end{aligned}$$

13. MEDIUM-LENGTH TRANSMISSION LINES

Medium-length transmission lines are 60 Hz lines between 80 km and 240 km (50 mi and 150 mi) long. The shunt reactance is significant enough to be included. The medium-length line is modeled in one of two ways as shown in Fig. 38.5.

Figure 38.5 Medium-Length Transmission Line Models

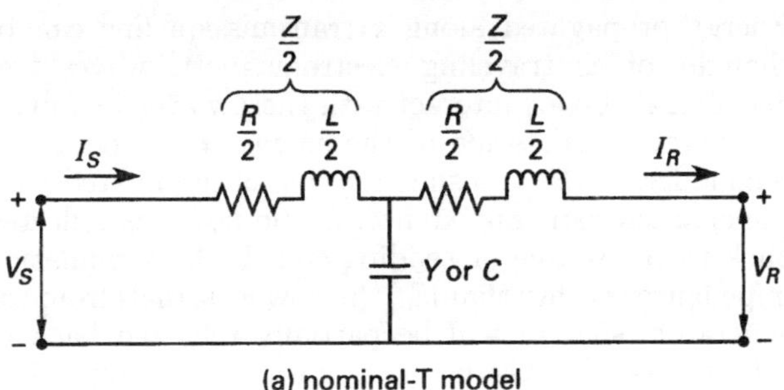

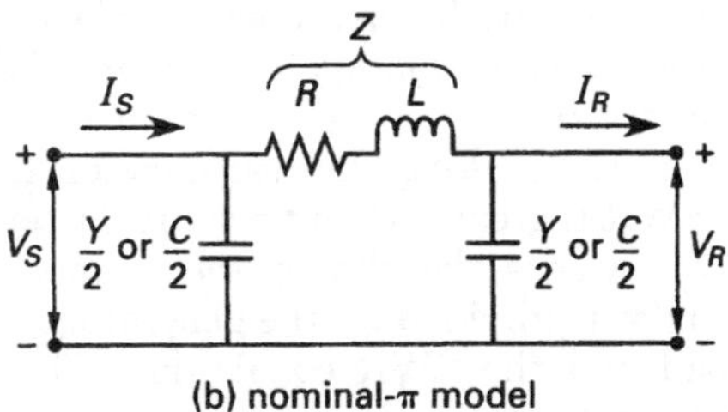

The impedance is[17]

$$Z = R + jX_L \tag{38.36}$$

The admittance is

$$Y = jB_C = \frac{j}{X_C} = -\frac{1}{jX_C} \tag{38.37}$$

Using the ABCD parameters in Table 38.3 gives Eq. 38.38 and Eq. 38.39 for the T approximation.

$$V_S = \left(1 + \tfrac{1}{2}YZ\right)V_R + \left(Z\left(1 + \tfrac{1}{4}YZ\right)\right)I_R \tag{38.38}$$

$$I_S = YV_R + \left(1 + \tfrac{1}{2}YZ\right)I_R \tag{38.39}$$

Using the ABCD parameters in Table 38.3 gives Eq. 38.40 and Eq. 38.41 for the π approximation.[18]

$$V_S = \left(1 + \tfrac{1}{2}YZ\right)V_R + ZI_R \tag{38.40}$$

$$I_S = \left(Y\left(1 + \tfrac{1}{4}YZ\right)\right)V_R + \left(1 + \tfrac{1}{2}YZ\right)I_R \tag{38.41}$$

14. LONG TRANSMISSION LINES

Long transmission lines are 60 Hz lines greater than 240 km (150 mi) long. All the parameters must be represented as distributed parameters, that is, per unit length, as shown in Fig. 38.6. The shunt conductance is

Figure 38.6 Long Transmission Line Model

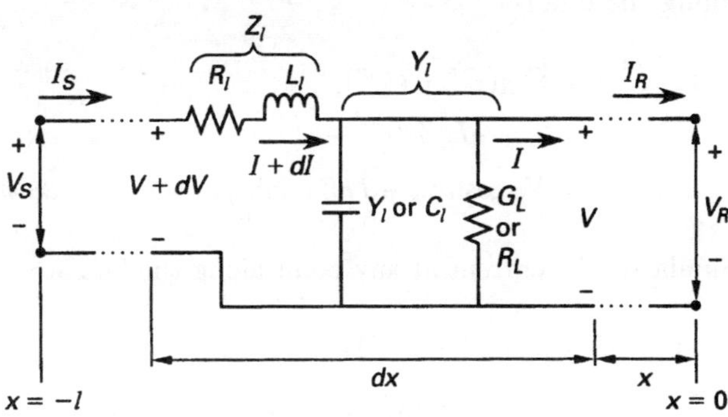

shown for completeness only and is excluded from most calculations, as will be done here.

The impedance per unit length is[19]

$$Z_l = R_l + jX_{L,l} \tag{38.42}$$

The admittance per unit length is

$$\begin{aligned} Y_l &= jB_{C,l} = \frac{j}{X_{C,l}} = -\frac{1}{jX_{C,l}} \\ &= j\omega C_l \end{aligned} \tag{38.43}$$

The voltage at any point along the line of length l in Fig. 38.6 is

$$\frac{\partial^2 V}{\partial x^2} = \gamma^2 V \tag{38.44}$$

The term γ is called the *propagation constant*, measured in radians per meter (mile), and is

$$\begin{aligned} \gamma &= \sqrt{Y_l Z_l} = \sqrt{-B_{C,l}X_{L,l} + jR_l B_{C,l}} \\ &= \alpha + j\beta \end{aligned} \tag{38.45}$$

The magnitude of the propagation constant is

$$|\gamma| = \sqrt{B_{C,l}}\sqrt[4]{R_l^2 + X_{L,l}^2} \tag{38.46}$$

The term α is called the *attenuation constant*, measured in nepers per meter (mile), and is

$$\alpha = |\gamma| \cos\left(\tfrac{1}{2}\arctan\frac{-R_l}{X_{L,l}}\right) \tag{38.47}$$

The term β is called the *phase constant*, measured in radians per meter (mile), and is

$$\beta = |\gamma| \sin\left(\tfrac{1}{2}\arctan\frac{-R_l}{X_{L,l}}\right) \tag{38.48}$$

[17]The quantities represent total amounts but are calculated per unit length, corrected for conductor spacing, and multiplied by the length of the transmission line. Also, the impedance is a per-phase quantity and needs to be multiplied by three to get the line-to-line impedance in a delta-connected load.

[18]The π approximation results in simpler regulation calculations.

[19]The per unit length impedance and admittance are sometimes written with small letters, z and y, to emphasize their per-unit qualities.

Distribution

A solution to Eq. 38.44 giving the voltage at any point along the line is[20]

$$\begin{aligned} V &= \tfrac{1}{2}V_R(e^{\gamma x}+e^{-\gamma x}) \\ &\quad +\tfrac{1}{2}I_RZ_0(e^{\gamma x}-e^{-\gamma x}) \\ &= V_R\cosh\gamma x + I_RZ_0\sinh\gamma x \end{aligned} \qquad 38.49$$

Similarly, the current at any point along the line is

$$\begin{aligned} I &= \tfrac{1}{2}I_R(e^{\gamma x}+e^{-\gamma x}) \\ &\quad +\tfrac{1}{2}\left(\frac{V_R}{Z_0}\right)(e^{\gamma x}-e^{-\gamma x}) \\ &= I_R\cosh\gamma x + \left(\frac{V_R}{Z_0}\right)\sinh\gamma x \end{aligned} \qquad 38.50$$

The term Z_0, called the *characteristic impedance*, is

$$\begin{aligned} Z_0 &= \sqrt{\frac{Z_l}{Y_l}} = \sqrt{\frac{R_l+jX_{L,l}}{jB_{C,l}}} \\ &= \sqrt{X_{L,l}X_{C,l}-jX_{C,l}R_l} \quad \text{[in } \Omega] \end{aligned} \qquad 38.51$$

The magnitude of the characteristic impedance is

$$|Z_0| = \sqrt{X_{C,l}}\sqrt[4]{R_l^2+X_l^2} \qquad 38.52$$

The angle of the characteristic impedance is

$$\angle Z_0 = \tfrac{1}{2}\arctan\frac{-R_l}{X_{L,l}} \qquad 38.53$$

Equation 38.49 and Eq. 38.50 represent *traveling waves.* Specifically, the terms with $e^{-\gamma x}$ represent waves traveling in the positive x-direction. (The distance x is defined as positive to the right, the direction toward the receiving end, that is, toward the direction power flows.) Terms with $e^{\gamma x}$ represent waves traveling in the negative x-direction. The two equations are the value of the two waves superimposed at that point in the line. Because $V = V_S$ (see Eq. 38.49) at $x = -l$ (see Fig. 38.6), and consistent with the ABCD parameters in Table 38.3, the voltage at the sending end is

$$V_S = V_R\cosh\gamma l + I_RZ_0\sinh\gamma l \qquad 38.54$$

Because $I = I_S$ (see Eq. 38.50) at $x = -l$ (see Fig. 38.6), and consistent with the ABCD parameters in Table 38.3, the current at the sending end is

$$I_S = I_R\cosh\gamma l + \frac{V_R}{Z_0}\sinh\gamma l \qquad 38.55$$

The term l in Eq. 38.51 and Eq. 38.52 is the transmission line length.

[20]The solution assumes a sinusoidal steady-state condition. That is, transmission lines are generally AC steady-state networks.

15. REFLECTION COEFFICIENT

Energy propagated along a transmission line can be thought of as traveling electromagnetic waves (see Sec. 38.14). Loads interact with these waves in a manner that absorbs some of the energy and reflects the remainder. If the transmission line is terminated with the characteristic impedance, Z_0, no power is reflected back to the source or sending end. If the termination impedance is other than Z_0, the power (signal) from the generator (source) will be partially reflected back to the generator. Assuming a steady-state sinusoidal source, the generator's waves will then combine in the transmission line with the reflected waves to form *standing waves.* The *standing wave ratio*, SWR, is the ratio of the maximum to the minimum voltages (currents) encountered along the transmission line. Typically, the SWR is greater than unity. If the terminating impedance matches the characteristic impedance, all the input power provided by the generator is absorbed by the load and the SWR equals one. The standing wave ratios are

$$\text{VSWR} = \frac{V_{\text{max}}}{V_{\text{min}}} \qquad 38.56$$

$$\text{ISWR} = \frac{I_{\text{max}}}{I_{\text{min}}} \qquad 38.57$$

The *reflection coefficient*, Γ, is the ratio of the reflected to the incident electric parameter. The voltage reflection coefficient for the load is

$$\Gamma_L = \frac{V_{\text{reflected}}}{V_{\text{incident}}} = \frac{Z_{\text{load}}-Z_0}{Z_{\text{load}}+Z_0} \qquad 38.58$$

The current reflection coefficient is the negative of the reflection coefficient for the voltage. That is,

$$\Gamma_L = \frac{I_{\text{reflected}}}{I_{\text{incident}}} = \frac{Z_0-Z_{\text{load}}}{Z_0+Z_{\text{load}}} \qquad 38.59$$

The fraction of incident power that is reflected back to the source from the load is Γ^2. The relationship between the reflection coefficient and the standing wave ratio is

$$\Gamma = \frac{\text{SWR}-1}{\text{SWR}+1} \qquad 38.60$$

16. TRANSMISSION LINE IMPEDANCE

The equations presented in Sec. 38.1 through Sec. 38.15 assume lossless transmission lines, so that I^2R losses and insulated leakage losses given by G^2V are ignored. In most practical instances, transmission line losses are negligible, so they are commonly ignored. Transmission lines are large in one dimension (the distance or length from source to load) and small in the other two dimensions (the width and height of the conductors and cable configuration). For this reason, the voltage and current waves are directly related to the **E** and **H** fields in the spaces between the conductors. Maxwell's equations

Distribution

can be used to derive the equations for a plane traveling wave.

$$\nabla \times \mathbf{E} = -\frac{\partial \mathbf{B}}{\partial t} \quad 38.61$$

$$\nabla \times \mathbf{H} = \mathbf{J}_c + \frac{\partial \mathbf{D}}{\partial t} \quad 38.62$$

From Eq. 38.61 and Eq. 38.62, as well as Eq. 38.51 through Eq. 38.53 for characteristic impedance, the impedance can be found for any position along a long, lossless transmission line with a sinusoidal source. The impedance along the line at position $x=-d$, where d is a positive distance measured from the receiving end, is given by[21]

$$\begin{aligned} Z(d) &= \frac{V(d)}{I(d)} = Z_0\left(\frac{1+\Gamma e^{-j2\beta d}}{1-\Gamma e^{-j2\beta d}}\right) \\ &= Z_0\left(\frac{Z_R\cos\beta d + jZ_0\sin\beta d}{Z_0\cos\beta d + jZ_R\sin\beta d}\right) \end{aligned} \quad 38.63$$

The term β is the phase constant defined earlier in Sec. 38.14 in terms of the attenuation constant (see Eq. 38.48). β can also be expressed in terms of the velocity of propagation and wavelength.

$$\beta = \frac{\omega}{v_{\text{wave}}} = \frac{2\pi}{\lambda} \quad 38.64$$

Equation 38.63 can be used in place of the Smith chart (see Chap. 39) for finding the impedance at a given point along a transmission line.

A transmission line that is short-circuited at the receiving end has an impedance of $Z_R = 0\ \Omega$ and a voltage of $V_R = 0$ V. Substituting the receiving-end impedance of $0\ \Omega$ into Eq. 38.63 gives

$$Z_{sc} = jZ_0 \tan\beta d \quad 38.65$$

Because no power can be dissipated in a short circuit, the short-circuit impedance is purely reactive.

A transmission line that is open-circuited at the receiving end has an impedance of $Z_R = \infty\ \Omega$, and $I_R = 0$ A. Substituting the receiving-end impedance of $\infty\ \Omega$ into Eq. 38.63 gives

$$Z_{oc} = -jZ_0 \cot\beta d \quad 38.66$$

An open-circuited line is capacitive for $d < \lambda/4$ and inductive for $\lambda/4 < d < \lambda/2$, after which the pattern repeats. So a short-circuited line, which is inductive when $d < \lambda/4$, is equivalent to an open-circuited line that is longer than the short-circuited line by $\lambda/4$ in the range $\lambda/4 < d < \lambda/2$. Figure 38.7 illustrates this concept.

[21]Euler's relation is used to change the exponential terms to sinusoidal terms. The $x=0$ point is at the receiving end of the transmission line and is negative when moving to the left toward a source. Since $x=-d$, Eq. 38.63 is as shown, but d can be considered a positive distance from the receiving end.

Figure 38.7 Input Impedance of Short-Circuited and Open-Circuited Transmission Lines

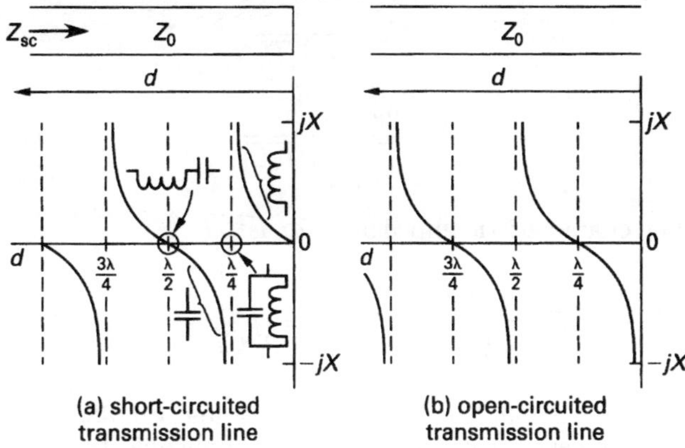

The value of d can be varied to achieve any desired results. Open-circuited and short-circuited stubs can be used on transmission lines to ensure that the impedance of the source-line combination matches the impedance of the receiving-end load. A *stub* is a length of transmission wire that is connected only at one end, with the free end left either open-circuited or short-circuited. The stubs reduce or eliminate reflections and ensure maximum power transfer. The characteristic impedance of a finite transmission line with open-circuit and short-circuit terminations is given by

$$Z_0 = \sqrt{Z_{sc}Z_{oc}} \quad 38.67$$

17. HIGH-FREQUENCY TRANSMISSION LINES

At high frequencies, approximately 1 MHz and higher, wavelengths are shorter and even a few feet of line are treated as a long transmission line. The resistance is normally negligible, specifically, $R \ll \omega L$ and $G \ll \omega C$. Setting the resistance, R, equal to zero simplifies the governing equations. The impedance per unit length is[22]

$$Z_l = jX_{L,l} = j\omega L_l \quad 38.68$$

High-frequency transmission lines use the same admittance per unit length equation as long transmission lines (Eq. 38.43).

The propagation constant, γ, in units of radians per unit length is

$$\begin{aligned} \gamma &= \sqrt{Y_l Z_l} = \sqrt{(j\omega C_l)(j\omega L_l)} = j\omega\sqrt{L_l C_l} \\ &= j\beta \end{aligned} \quad 38.69$$

[22]The per-unit-length impedance and admittance are sometimes written with lowercase letters, z and y, to emphasize their per-unit qualities. In this chapter, the lowercase letters are used for normalized quantities.

Distribution

The term β in Eq. 38.69 is the phase constant. The phase velocity, v_{phase}, and the wavelength are

$$v_{phase} = \frac{1}{\sqrt{L_l C_l}} \quad 38.70$$

$$\lambda = \frac{2\pi}{\beta} = \frac{1}{f\sqrt{L_l C_l}} \quad 38.71$$

The characteristic impedance, Z_0, is[23]

$$Z_0 = \sqrt{\frac{Z_l}{Y_l}} = \sqrt{\frac{j\omega L_l}{j\omega C_l}} = \sqrt{\frac{L_l}{C_l}} \quad 38.72$$

Equation 38.49, Eq. 38.50, Eq. 38.54, and Eq. 38.55 for long power transmission lines are applicable to high-frequency lines when the values from Eq. 38.43, Eq. 38.68, Eq. 38.69, and Eq. 38.72 are substituted.

At high frequencies, however, circuits are designed to minimize reflected waves. This is accomplished by using capacitive compensation to make the load impedance appear to be the characteristic impedance. The expressions for the sending voltage and current are specified in terms of the reflection coefficient, Γ. The voltage traveling in the positive x direction toward the receiving end is associated with

$$V^+ = \tfrac{1}{2}V_R + \tfrac{1}{2}Z_0 I_R \quad 38.73$$

Equation 38.73 is one of the constants from the solution of the wave equation, Eq. 38.74. The other constant is V^-. Both can be determined from the solution to the wave equation given by Eq. 38.49. The positive x direction is defined as the direction of power flow, that is, from the sending end to the receiving end.

$$\frac{\partial^2 V}{\partial x^2} = \gamma^2 V \quad 38.74$$

The terms $\cosh j\beta x = \cos \beta x$ and $\sinh j\beta x = j\sin \beta x$, when combined with Eq. 38.73, allow the sending-end voltage and current to be written in terms of the reflection coefficient.

$$\begin{aligned} V_S &= V^+(e^{j\beta l} + \Gamma e^{-j\beta l}) \\ &= V^+\Big((1+\Gamma)\cos \beta l + j(1-\Gamma)\sin \beta l\Big) \end{aligned} \quad 38.75$$

$$\begin{aligned} I_S &= \left(\frac{V^+}{Z_0}\right)(e^{j\beta l} - \Gamma e^{-j\beta l}) \\ &= \left(\frac{V^+}{Z_0}\right)\Big((1-\Gamma)\cos \beta l + j(1+\Gamma)\sin \beta l\Big) \end{aligned} \quad 38.76$$

[23]The characteristic impedance is the ratio of the voltage to the current on an infinite transmission line. It can be called the *characteristic resistance*, R_0, because the line is lossless. That is, Z_l and Y_l are purely reactive, making Z_0 purely resistive.

The input impedance is V_S/I_S and, in terms of the reflection coefficient, is given by

$$Z_{in} = Z_0\left(\frac{(1+\Gamma)\cos \beta l + j(1-\Gamma)\sin \beta l}{(1-\Gamma)\cos \beta l + j(1+\Gamma)\sin \beta l}\right) \quad 38.77$$

The reflection coefficient is defined in Eq. 38.58. Using this definition, the input impedance can be written in terms of the load impedance, as in Eq. 38.78.

$$Z_{in} = Z_0\left(\frac{Z_{load}\cos \beta l + jZ_0 \sin \beta l}{Z_0 \cos \beta l + jZ_{load}\sin \beta l}\right) \quad 38.78$$

The voltage at any point on a transmission line is a function of the distance from the load because there are two waves traveling on any line, one incident to the load and one reflected from the load. The standing wave ratios defined by Eq. 38.56 and Eq. 38.57 measure the difference between the maximum rms voltage (current) and the minimum rms voltage (current) of the two waves. (The standing wave ratio is a physical variable that can be directly measured.) Where the magnitudes of the waves add, the voltage (current) is at a maximum. Where the magnitudes of the waves subtract, the voltage (current) is at a minimum. The reflection coefficient is defined in terms of the standing wave ratios by Eq. 38.60.

Rearranging Eq. 38.60 to define the voltage standing wave ratio, VSWR, in terms of the reflection coefficient gives

$$\text{VSWR} = \frac{|V^+| + |V^-|}{|V^+| - |V^-|} = \frac{1+|\Gamma|}{1-|\Gamma|} \quad 38.79$$

The VSWR is related to the minimum and maximum impedance by Eq. 38.80 and Eq. 38.81.

$$Z_{max} = Z_0(\text{VSWR}) \quad 38.80$$

$$Z_{min} = \frac{Z_0}{\text{VSWR}} \quad 38.81$$

Example 38.8

A certain transmission line has a characteristic impedance of 50 Ω and a terminating resistance of 100 Ω. What is the reflection coefficient?

Solution

The reflection coefficient is defined by Eq. 38.58 as

$$\Gamma_l = \frac{V_{reflected}}{V_{incident}} = \frac{V_{reflected}}{V_{load}} = \frac{Z_{load} - Z_0}{Z_{load} + Z_0}$$

Substitute the given values.

$$\Gamma_l = \frac{Z_{load} - Z_0}{Z_{load} + Z_0} = \frac{100\ \Omega - 50\ \Omega}{100\ \Omega + 50\ \Omega} = 0.333$$

Example 38.9

For the transmission line described in Ex. 38.8, what is the input impedance if the line is an even number of wavelengths long?

Solution

The transmission line is restricted by the problem statement to be an even number of wavelengths for the frequency carried. In mathematical terms, with n as an integer, this is stated as

$$l = n\lambda = \frac{2\pi n}{\beta}$$

Equation 38.71 is used to change the wavelength, λ, into a function of the phase constant, β. Rearranging gives

$$\beta l = 2\pi n$$

This wavelength restriction causes the sinusoidal terms of the input impedance equation, Eq. 38.78, to be

$$\cos \beta l = 1$$

$$\sin \beta l = 0$$

Substituting these values into Eq. 38.78 gives

$$Z_{\text{in}} = Z_0\left(\frac{Z_{\text{load}}}{Z_0}\right) = Z_{\text{load}} = 100\ \Omega$$

Example 38.10

For the transmission line in Ex. 38.8, what is the voltage standing wave ratio?

Solution

The VSWR can be found by substituting the calculated value of the reflection coefficient (determined in Ex. 38.8) into Eq. 38.79.

$$\text{VSWR} = \frac{1+|\Gamma|}{1-|\Gamma|} = \frac{1+0.333}{1-0.333} = 2$$

Distribution

39 The Smith Chart

Nomenclature

l	length	m
r	normalized resistance	–
R	resistance	Ω
x	distance variable	m
x	normalized reactance	–
X	reactance	Ω
y	normalized admittance	–
z	normalized impedance	–
Z	impedance	Ω

Symbols

β	phase constant	rad/m
Γ	reflection coefficient	–
λ	wavelength	m
ϕ	angle	rad

Subscripts

0	characteristic
c	compensating
cp	compensation point
g	generator
i	imaginary
l	per unit length
L	inductor
min	minimum
r	real
R	receiving end
unk	unknown

1. FUNDAMENTALS

The *Smith chart* is a polar diagram that shows the reflection coefficient (see Eq. 38.3) in terms of normalized resistance and reactance. It was developed around 1940 as a graphical aid in solving transmission line equations. Since then, the solving of such equations has for the most part been taken over by electronic calculators and computer-aided design (CAD) programs,[1] but the Smith chart is still useful for visualizing steady-state conditions on transmission lines. A complete Smith chart is given in App. 39.A.

[1]The Smith chart is based on Eq. 38.58 and Eq. 38.78; these equations are commonly used in place of the Smith chart.

In the Smith chart, curves of constant resistance appear as circles, and curves of constant reactance appear as arcs. Resistance and reactance are normalized with respect to the characteristic impedance, Z_0.

$$z = \frac{Z}{Z_0} = \frac{R+jX}{Z_0} = \frac{1+\Gamma}{1-\Gamma} \qquad 39.1$$

The normalized impedance, z, impedance, Z, resistance, R, reactance, X, and reflection coefficient, Γ, are all functions of distance, and are normally measured with the load at the origin.[2]

In the complex Γ plane, the curves of constant reflection coefficient magnitude, $|\Gamma|$, and of constant voltage standing wave ratio, VSWR (see Eq. 38.2), are circles as shown in Fig. 39.1(a). As these circles are not printed on the Smith chart, their values are obtained from the radial scales shown below the chart in App. 39.A.

Curves of constant normalized resistance, r, are circles as shown in Fig. 39.1(b). The origin of the chart is at the intersection of the $r=1$ circle and the line representing $x=0$. The outermost circle is $r=0$. Arcs of constant normalized reactance, x, are shown as in Fig. 39.1(c). The constant resistance circles of Fig. 39.1(b) and the constant reactance arcs of Fig. 39.1(c) are combined to create the final Smith chart in App. 39.A. Appendix 39.A also includes three circumferential scales. The outer two measure fractions of a wavelength toward the generator (clockwise) or toward the load (counterclockwise). The full circumference represents one-half the wavelength. The inner circumferential scale measures the reflection coefficient angle, $\phi_L = \phi_R - \beta x$.[3]

Any circle drawn around the origin of the Smith chart is called an *impedance locus circle*. At the origin, the impedance locus circle has a radius of zero, which is called the *matched condition*. At this point the total reactance is zero, the load impedance matches the characteristic impedance, making the normalized impedance equal to one, and the reflection coefficient equals zero. Typical values for other electrical conditions are shown in Table 39.1.

[2]Do not confuse the distance x with the normalized reactance x. In some books, the distance is given the symbol d or the normalized reactance is given the symbol χ to avoid confusion. The individual terms in Eq. 39.1 are not listed as functions of distance to avoid the problem in this specific instance.

[3]The x used is distance. When the total distance is from the load (receiving end) to the generator (sending end), the symbol l is used instead.

Figure 39.1 Smith Chart Components

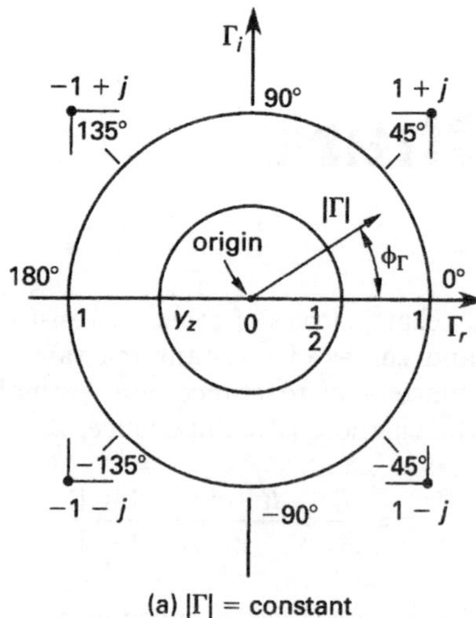

(a) |Γ| = constant

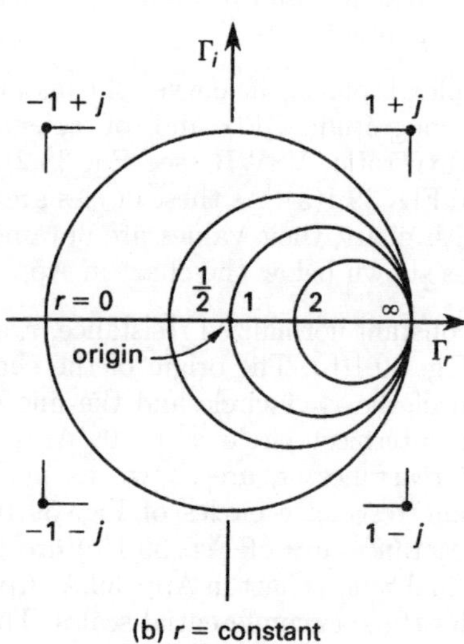

(b) r = constant

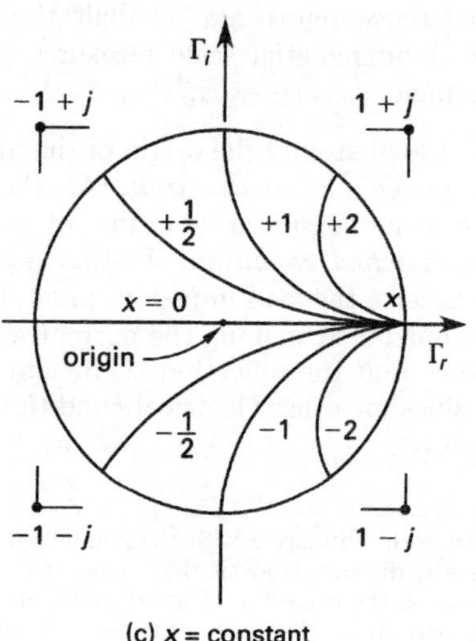

(c) x = constant

Table 39.1 Smith Chart Electrical Conditions

electrical condition	reflection coefficient (Γ)	normalized resistance (r)	normalized reactance (x)
open circuit	$1\angle 0°$	∞ (arbitrary)	arbitrary (∞)
short circuit	$1\angle 180°$	0	0
pure reactance	$1\angle \pm 90°$	0	±1
matched line (pure resistance)	0	1	0

The Smith chart in App. 39.A and the components shown in Fig. 39.1 are arranged to show the real portion of the reflection coefficient along the standard x-axis and the imaginary component along the standard y-axis, subscripted as r and i, respectively. It is also common to rotate the chart 90° clockwise to place the $r=0$ and $x=0$ position—that is, the intersection of the zero resistance circle and the zero reactance line—at the top of the chart as shown in Fig. 39.2.

Figure 39.2 Alternate Smith Chart Positioning

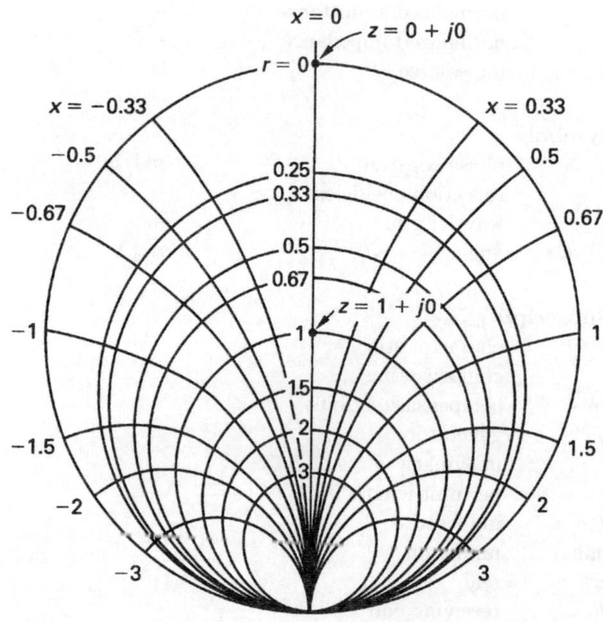

The Smith chart can be used for normalized admittance, y, by considering the r-circles as conductance circles (g-circles) and the x-arcs as susceptance arcs (b-arcs). The angle for the reflection coefficient, Γ, for a given admittance, y, is given by $180° + \phi_\Gamma$, and the point $y = 0 + j0$ represents an open-circuit condition.

The input impedance, compensating reactance, compensating position, reflection coefficient, and standing wave ratio can be graphically determined using the Smith chart.[4] The steps vary and are presented in examples in Sec. 39.3 and Sec. 39.4.

[4]Though the reflection coefficient and standing wave ratio can be determined graphically, App. 39.A substitutes radial scales for the graphical procedure.

Distribution

2. INPUT IMPEDANCE

The general procedure in finding the input impedance is to first determine the normalized load impedance, z_L, using Eq. 39.1. Plot the point z_L, representing the normalized resistance and reactance, at the intersection of the r circle and the x arc. Draw a circle centered at $1 + j0$ (the origin) through the value $r + jx$ corresponding to z_L. This is the impedance locus circle. Draw a radial line from the origin through z_L. Determine the angle βl and the normalized impedance a distance l from the load by following the impedance locus circle clockwise (toward the generator or sending end) through an angle of $2\beta l$.

Example 39.1

A transmission line with an electrical angle, βl, of 6 radians (344°) and a characteristic impedance, z_0, of 50 Ω is terminated with a load of $75 + j25$ Ω. What is the input impedance (i.e., the impedance seen by the generator)?

Solution

step 1: Calculate the normalized load impedance.

$$\begin{aligned} z_L &= \frac{Z_L}{Z_0} = \frac{R_L + jX_L}{Z_0} \\ &= \frac{75 + j25\ \Omega}{50\ \Omega} \\ &= 1.5 + j0.5 \end{aligned}$$

step 2: Plot point $z_L = 1.5 + j0.5$ on the Smith chart.

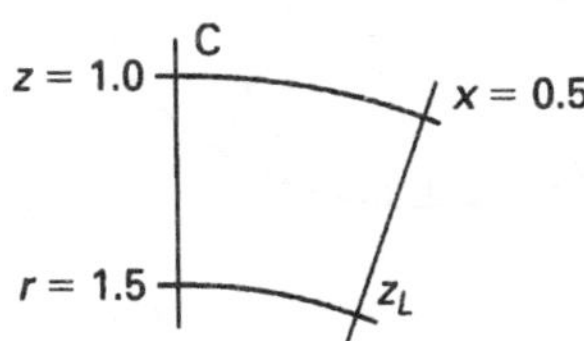

step 3: Draw a circle centered at $1.0 + j0$ (point C), that is, the origin, which passes through point z_L. This is the impedance locus circle.

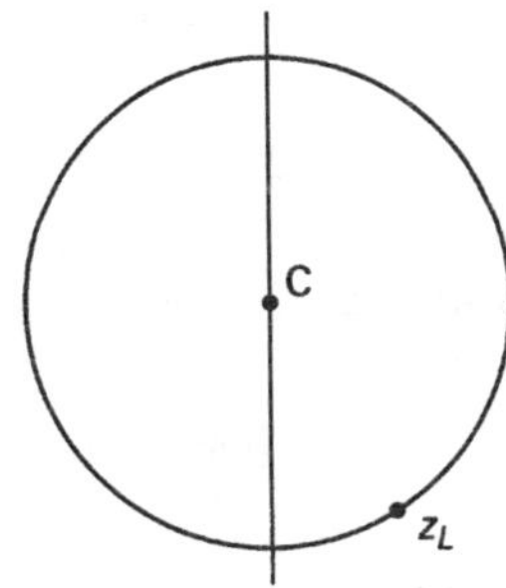

step 4: Extend a line from point C (the origin) through point z_L to the edge of the Smith chart, point L (load point).

step 5: Calculate the angular separation between the source and load points.

$$2\beta l = (2)(344°) = 688°$$

Because this is more than one revolution, subtract the full revolution.

$$688° - 360° = 328°$$

Move clockwise 328° from the load to the generator point.

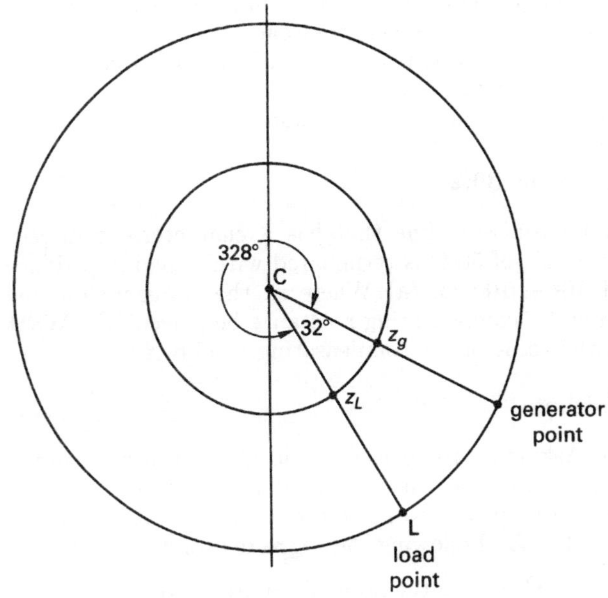

step 6: Read the impedance seen from the generator; that is, the input impedance at the intersection of the generator line and the impedance circle, point z_g.

$$z_g = 1.1 + j0.6$$

step 7: Calculate the input load impedance.

$$\begin{aligned} Z_g &= Z_0 z_g = (50\ \Omega)(1.1 + j0.6) \\ &= 55 + j30\ \Omega \end{aligned}$$

3. COMPENSATION REACTANCE

Compensation is used to make the total reactance of a given transmission line zero. Compensation is also called *impedance matching.* At low frequencies, the matching is accomplished with lumped reactive components (inductors and capacitors). At high frequencies, in order to minimize dissipation losses, a length of open- or short-circuited line is used. Compensation with a single line is called *single-stub matching.* Compensation with two lines is called *double-stub matching.*

The reactance required for compensation can be determined graphically on the Smith chart. The impedance locus circle is set up as before (see Ex. 39.1, steps 1 through 3). The line is traversed toward the source (clockwise) to the intersection of the impedance locus circle and the $r=1$ circle. The intersection is called the *compensation point.* At the compensation point, the impedance is

$$z_{\text{cp}} = 1 + jx \qquad 39.2$$

The compensation reactance is $-x$, which results in a total normalized impedance of $z=1$. Because capacitance is the most desirable compensating reactance for a variety of reasons, the compensation point is always chosen on the positive reactance side of the Smith chart. The positive reactance side of the chart is where $x > 0$, which makes the compensating reactance negative $(x < 0)$.[5]

Example 39.2

A transmission line that has a characteristic impedance, Z_0, of 50 Ω is terminated with a load impedance of $100 + j100$ Ω. (a) Where in the transmission line should a compensating reactance be placed? (b) What is the value of the compensating reactance?

Solution

(a) Asking for the placement of the reactance is equivalent to asking at what electrical angle (βl) the capacitor should be placed to make the impedance at that point equal to Z_0. Following the suggested solution outline,

step 1: Calculate the normalized load impedance.

$$\begin{aligned} z_L &= \frac{Z_L}{Z_0} \\ &= \frac{100 + j100\ \Omega}{50\ \Omega} \\ &= 2 + j2 \end{aligned}$$

step 2: Plot the point z_L on the Smith chart.

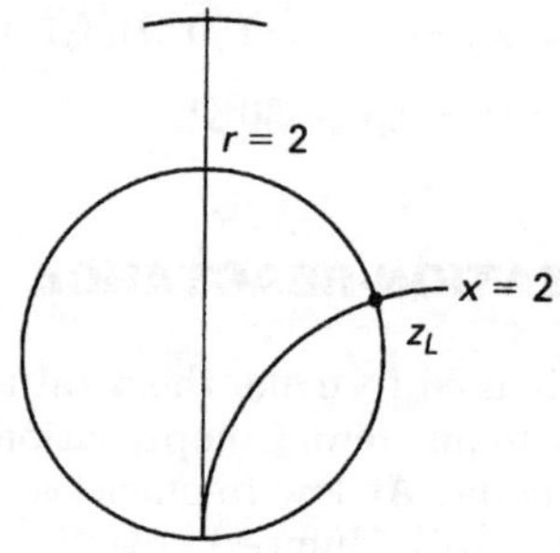

step 3: Draw a circle centered at $1 + j0$, that is, the origin, which passes through point z_L. This is the impedance locus circle.

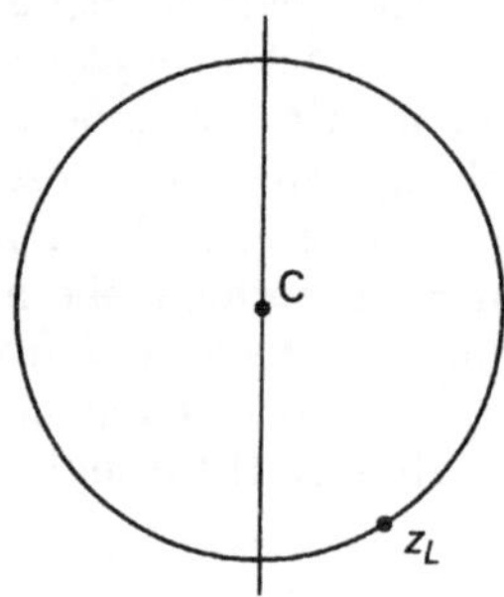

step 4: Extend a line from point C (the origin) through the load point z_L. From the load point, follow the impedance locus circle clockwise, toward the source to point CP (the compensation point), which is the intersection of the impedance locus circle and the $r=1$ circle on the positive reactance side of the Smith chart. (The angle traversed is $2\beta l$.)

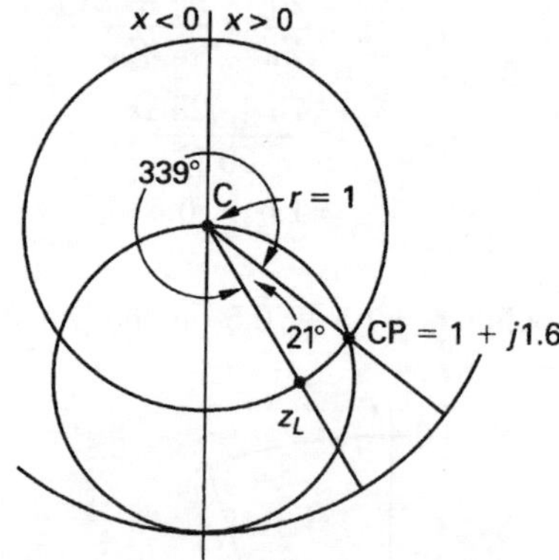

step 5: Measure the angular separation between the compensation and load points. In this case, the value is 21°. Because the electrical angle is always measured from the load toward the source, that is, clockwise, the angular separation is actually 360° − 21° = 339°.

$$2\beta l = 339°$$

The electrical angle (βl) is

$$\beta l = \frac{339°}{2} = 169.5°$$

The compensation point is βl electrical degrees or 169.5°/360° wavelengths (0.471λ) from the load toward the source. The reactance should be placed a distance $l = 0.471\lambda$ from the load

[5]The positive reactance side of the Smith chart is the top if positioned as in App. 39.A, and to the right if positioned as in the examples of this chapter.

toward the source. The distance, l, can also be found using Eq. 38.71.

$$\lambda = \frac{2\pi}{\beta}$$

$$\beta\lambda = 2\pi$$

$$\beta l\lambda = 2\pi l$$

$$l = \left(\frac{\beta l}{2\pi}\right)\lambda = \left(\frac{169.5^\circ}{360^\circ}\right)\lambda = 0.471\lambda$$

step 6: Read the normalized impedance at the compensation point, z_{cp}.

$$z_{cp} = 1 + j1.6$$

(b) The normalized compensating reactance, x_c, must therefore be -1.6. The actual reactance is

$$X_c = Z_0 x_c = (50\ \Omega)(-1.6) = -80\ \Omega$$

4. IMPEDANCE DETERMINATION USING VSWR MEASUREMENTS

The voltage standing wave ratio (VSWR) can be measured using a *slotted line* in which a probe can be placed to determine the maximum and minimum voltages. The distance between a minimum and a maximum is $\lambda/4$ (one-quarter wavelength). The distance between two minimums or two maximums is $\lambda/2$ (one-half wavelength). The first minimum of the VSWR from the load is given the symbol l_{min}. With the load shorted, the distance from the short toward the generator of the first VSWR minimum determines the half-wavelength, which is the position where $\beta l = \pi$.[6] Using the measured VSWR, l_{min}, and the half-wavelength, an unknown impedance on a transmission line can be determined.

The voltage standing wave minimum occurs at the point of minimum impedance. Eq. 38.78 can be written for an unknown impedance, Z_{unk}, at a minimum input impedance as

$$\frac{Z_{in,min}}{Z_0} = \frac{\frac{Z_{unk}}{Z_0} + j\tan\beta l}{1 + j\left(\frac{Z_{unk}}{Z_0}\right)\tan\beta l} \qquad 39.3$$

Eq. 38.81 is rearranged to explicitly show the relationship between the minimum impedance and the VSWR as

$$\frac{Z_{min}}{Z_0} = \frac{1}{\text{VSWR}} = \frac{Z_{in,min}}{Z_0} \qquad 39.4$$

Equation 39.3 and Eq. 39.4 are combined to determine an expression for the unknown impedance in terms of the VSWR, the minimum input impedance, and the minimum distance from the load at which the first minimum of the VSWR occurs.

$$Z_{unk} = Z_0\left(\frac{1 - j(\text{VSWR})\tan\beta l}{\text{VSWR} - j\tan\beta l}\right) \qquad 39.5$$

[6]A short circuit is a minimum and is located at $0 + j0$ (see Table 39.1). A distance of one-half wavelength on the Smith chart is one circumference (see App. 39.A), which places one at $0 + j0$ again—that is, at a minimum. The term βl is the argument of the sinusoidal functions that describe the voltage and will be at a minimum every π radians, or one-half wavelength.

Distribution

Topic VII: System Analysis

Chapter

40 Power System Analysis

Nomenclature

I	constant or rms current	A
$\mathbf{I}^*$	complex conjugate of the current	A
pf	power factor	–
P	power	W
Q	reactive power	VAR
S, $\mathbf{S}$	apparent power	VA
V, $\mathbf{V}$	constant or rms voltage	V
Y	admittance	S
Z	impedance	Ω

Symbols

θ	phase angle	rad
ϕ	power factor angle	rad

Subscripts

a	phase A
b	phase B
c	phase C
l	line
n	neutral
p	phase
pu	per unit
Y	wye
Δ	delta
ϕ	impedance angle

1. INTRODUCTION

Knowledge of the normal operation of power systems and transmission lines in a steady state requires an understanding of the fundamentals of AC circuits, in particular, familiarity with three-phase circuits. Using the concept of sequence currents, the same principles can be applied to analyze abnormal operation. Such abnormal analysis is called *fault analysis*. Both normal and fault analysis techniques use phasor representations of the voltages and currents. The inclusion of these parameters into nodal networks allows them to be solved using matrix algebra and computers.[1]

2. POWER FLOW

The concepts of phasors, complex power, and the power triangle can be found in Chap. 27. The combination of topics is summarized by Eq. 40.1, which is given for a single phase inductive load.

$$\mathbf{S} = \mathbf{VI}^* = P + jQ = VI\cos\theta + jVI\sin\theta \qquad 40.1$$

The direction of power flow can be determined from the sign of the real and reactive power, as indicated in Fig. 40.1. This direction of power flow can also be determined from the angles between the current and voltage as presented in Chap. 42.

Regarding Fig. 40.1,

- $S = P + jQ$ is the input power to the circuit.
- If $P > 0$, the circuit absorbs real power.
- If $P < 0$, the circuit supplies real power.
- If $Q > 0$, the circuit absorbs reactive power (I lags V).
- If $Q < 0$, the circuit supplies reactive power (I leads V).

Figure 40.1 *Power Flow Reference Circuit*

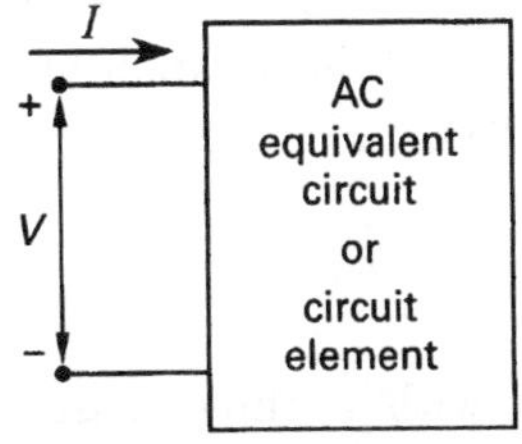

[1]The principles in this chapter can be used to check computer results. Programs such as Mathcad, Simulated Program with Integrated Circuit Emphasis (SPICE), and PSpice can easily perform mathematical analysis of a given design and circuit analysis. Mathcad is a tool for solving, analyzing, and sharing mathematical calculations. SPICE is an industry-standard program used for circuit simulation that can perform DC and AC steady-state and transient analysis, as well as Fourier and Monte Carlo analysis. PSpice is a more limited version of SPICE that can be run on a personal computer with less computational capacity; however, less computational capacity means the results are less accurate.

3. THREE-PHASE CONNECTIONS

Three-phase circuits, both generation and load, are connected with either a wye or delta connection. These connections can be analyzed from either perspective as long as the relationships are maintained. The wye connection is usually shown as Y, and the delta connection is usually shown as Δ. However, the A-to-B-to-C relationship will be represented as shown in Fig. 40.2. The rotation of any given three-phase machine is not shown on these connection diagrams, but rather it depends on the placement of the windings on the machine itself. A phasor diagram shows the rotation, usually with a reference at 0° on what would be the x-axis and assuming counterclockwise rotation.

Figure 40.2 *Wye and Delta Connections*

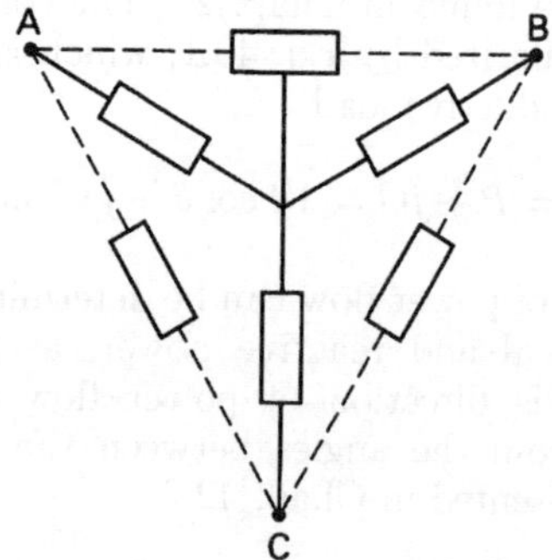

The voltages and currents in balanced three-phase systems during normal operation are always 120° apart. This occurs whether one is referencing phase voltages or currents, or line voltages or currents. As a result, a factor of $\sqrt{3}$ applied to the magnitude and an angle of 30° are commonplace. The result of adding two unit length phasors 120° apart is shown in Fig. 40.3.

Depending upon the type of analysis, the impedance or admittance of a given three-phase system may be required.[2] Conversions are shown in Fig. 40.4. Equation 40.2 summarizes the relationship between delta and wye impedances.

$$Z_Y = \tfrac{1}{3}Z_\Delta \qquad 40.2$$

When presented with a three-phase circuit, assume balanced conditions unless otherwise specified. When presented with voltage, current, and power, assume line-to-line voltage, line current, and total three-phase power unless otherwise specified. The three-phase power is

Figure 40.3 *120° Unity Phasor Addition*

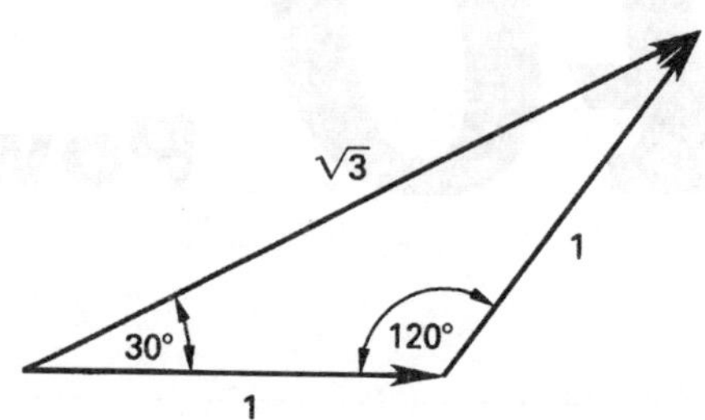

given by Eq. 40.3 and Eq. 40.4, which apply to both delta and wye connections.[3]

$$P = 3V_pI_p\cos\theta = 3V_pI_p\text{pf} \qquad 40.3$$

$$P = \sqrt{3}V_lI_l\cos\theta = \sqrt{3}VI\text{pf} \qquad 40.4$$

The power factor, pf, is the cosine of the angle between the voltage and the current. It is sometimes useful to think of it in terms of Eq. 40.5.

$$\text{pf} = \frac{P}{S} \qquad 40.5$$

The angle between the voltage and current can be calculated from Eq. 40.6.

$$\tan\phi = \frac{Q}{P} \qquad 40.6$$

4. OPERATOR: 120°

The phase displacement of voltages and currents in a balanced three-phase system are always 120° apart; it is useful to have an operator that can perform such counterclockwise rotation.[4] (Recall the operator j that rotates a phasor counterclockwise through 90° and j^2 or -1 results in a rotation of 180°.) To represent the rotation of 120° counterclockwise, the operator a is used. Operator a has the values indicated in Eq. 40.7 through Eq. 40.9.

$$a = 1\angle 120° = 1e^{j2\pi/3} = -0.5 + j0.866 \qquad 40.7$$

$$a^2 = 1\angle 240° = 1e^{j4\pi/3} = -0.5 - j0.866 \qquad 40.8$$

$$a^3 = 1\angle 360° = 1e^{j2\pi} = 1\angle 0° = 1 \qquad 40.9$$

This indicates that $1 + a + a^2 = 0$. Figure 40.5 shows the relationships for the operator a.

[2]Analysis of delta-connected machines can occur as if they were wye-connected, which sometimes simplifies analysis, with a conversion back to delta quantities for the final solution. That is, the neutral does not have to physically exist for circuit analysis relative to the neutral to be valid.

[3]The definitions of the phase voltage, V_p, and the phase current, I_p, in Eq. 40.3 are as follows. If the system is wired wye, V_p and I_p are line-to-neutral values. If the system is wire delta, V_p and I_p are line-to-line values.

[4]This operator greatly simplifies calculations with unsymmetrical components.

System Analysis

Figure 40.4 *Delta-Wye Transformations**

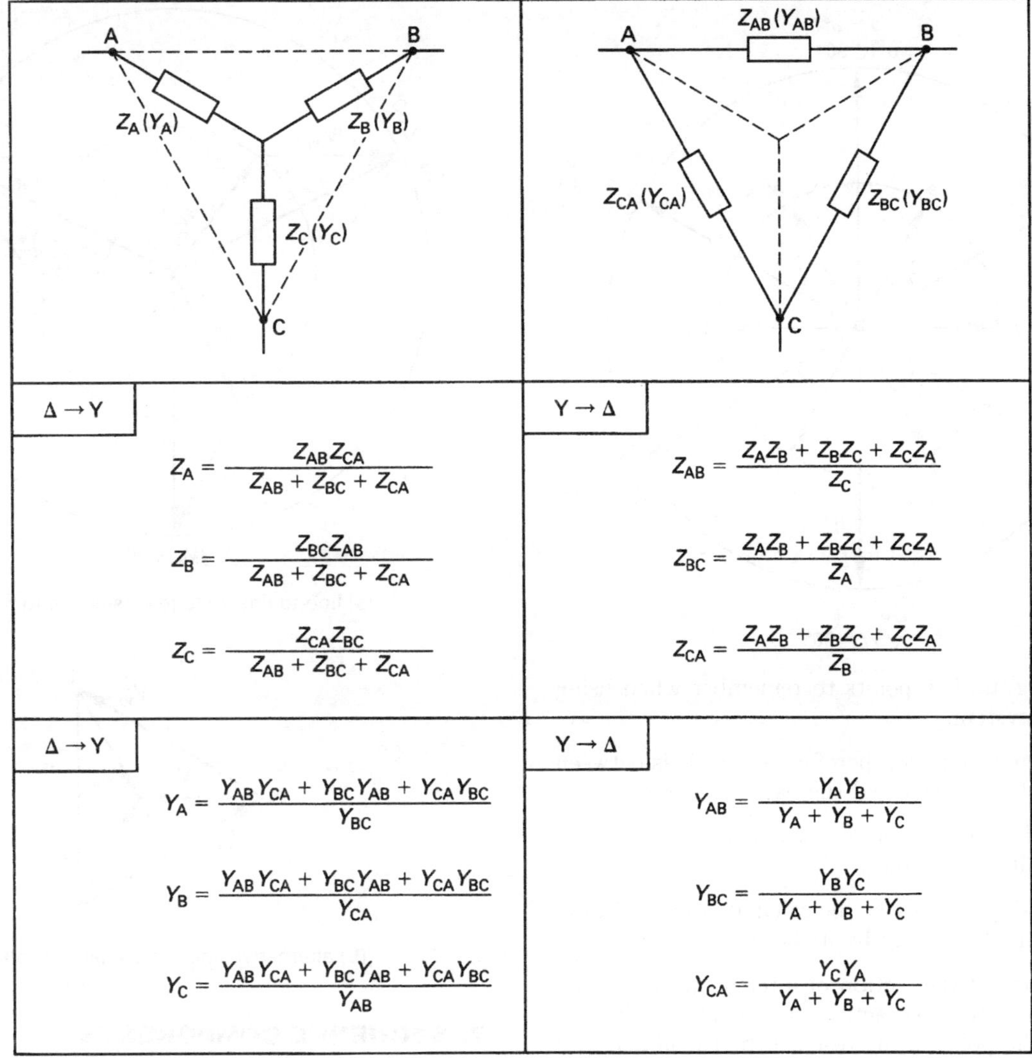

*Admittances and impedances with the same subscripts are reciprocals of one another.

5. BALANCED THREE-PHASE CIRCUIT

In a balanced three-phase circuit, the phase relationships are constant, and therefore, do not depend on the magnitude of the voltage and current values. The relationship for voltages in a balanced wye is shown in Fig. 40.6 (a). An alternate way of showing the voltage phasors is shown in Fig. 40.6 (b).

The relationship for currents in a balanced delta connection is shown in Fig. 40.7 (a). An alternate way of showing the current phasors is shown in Fig. 40.7 (b).

6. PER-UNIT SYSTEM

In a power system with a multitude of voltage levels, the voltage and current changes in the system make calculations more difficult. The per-unit (pu) system simplifies these calculations by establishing four base quantities: power, voltage, current, and impedance. In most cases, the power (in W) and the voltage (in V) are selected as the base. When two of the four base parameters are selected, the other two base quantities can be calculated. The base can be that of the generating system, a major transformer, or completely arbitrary. However, one should use the voltage base in the section of the transmission system that is of interest.

Using all the applicable electrical laws and equations, such as Ohm's law and Kirchhoff's laws, the results are calculated and expressed as a decimal or as a percent value. The product of two per-unit quantities will equal another per-unit quantity. When multiplying two percent values, the result must be divided by 100 to obtain a percent value. The method is described in Sec. 34.8.

Figure 40.5 The 120° Operator

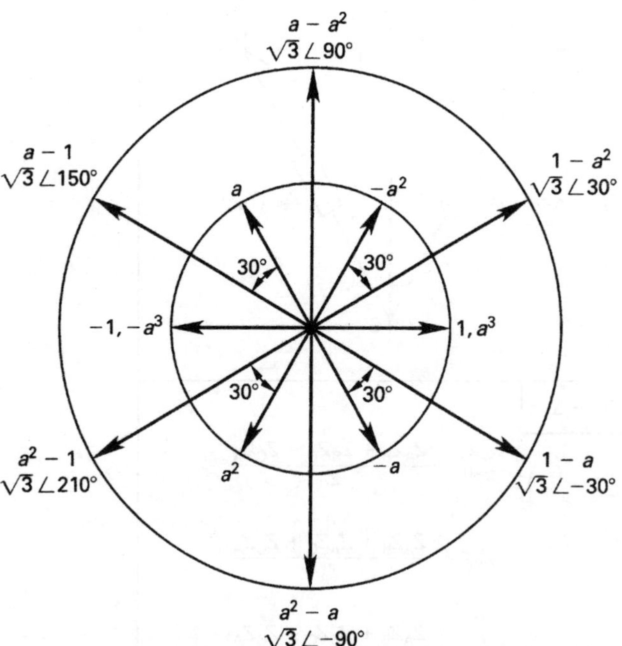

Figure 40.6 Balanced Wye Voltage Relationships

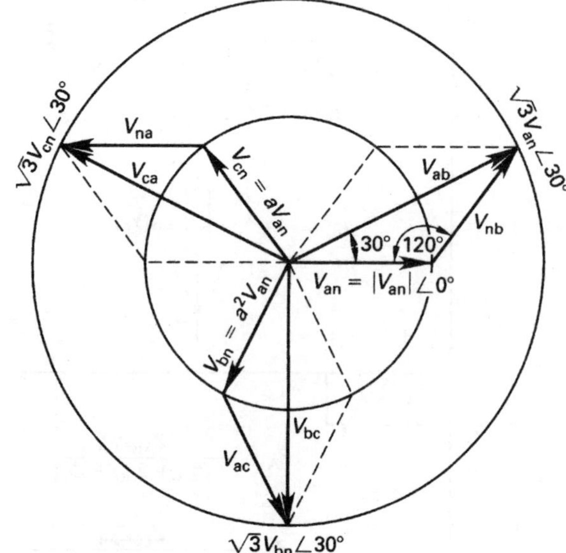

(a) line-to-line voltage versus line-to-neutral

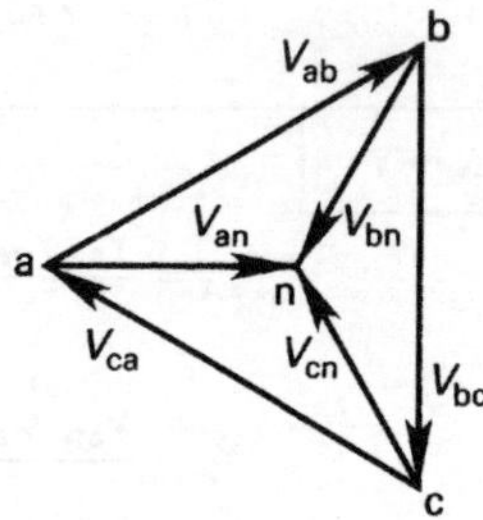

(b) alternative line versus neutral voltages

The following are key points to remember when using the per-unit system.

- The system is unitless; therefore, the ratio is between the actual parameter and the base parameter.
- The symbol "pu" is often used next to the value; however, it is not a unit.
- The actual quantities can be calculated by multiplying a per-unit quantity by its base.
- The voltage of the section of interest must be used when calculating the actual values. Or the impedance to the point of interest can be calculated and then converted to the new base.[5] All subsequent calculations converting to actual values will provide the correct quantities.
- The per-unit system allows one-line drawings of a distribution system to be used which facilitates short-circuit analysis.
- The per-unit impedance of a transformer is identical on either the primary or secondary direction.
- The base can be the total apparent power and line-to-line voltage, or the phase apparent power and phase voltage. Either way, as long as the system is balanced, the per-unit power and per-unit voltage remain equal regardless of the base.

The per-unit equations are summarized in Table 40.1.

[5]If the base impedance in a given section of transmission line is the voltage base in another section, the actual values of current and impedance calculated will be referenced to the incorrect section of the transmission line. For example, if the generators function at 6.9 kV, but the fault is on the 69 kV portion of the line, 69 kV should be used as the base. Or, if this is not done, the impedance can be calculated using the 6.9 kV base and then transformed to the new base. All actual values calculated after the transformation will be correct for that portion of the line.

7. SEQUENCE COMPONENTS

A balanced three-phase system is analyzed using phasors that are equal in magnitude, 120° apart, and in the sequence A-B-C called a *positive sequence.* When a fault occurs, both the magnitude relationship and the phase relationship can change resulting in unbalanced, or unsymmetrical, phasors. Fortescue's theorem states that three unbalanced phasors of a three-phase system can be resolved into three balanced systems of phasors. This allows the analysis of fault conditions with principles and methods used for balanced phasors.[6]

The set of symmetrical components that can be combined to represent unbalanced phasors is shown in Fig. 40.8. The *positive-sequence components* consist of three phasors, 120° apart in phase, and in the same phase sequence as the original phasors. The *negative-sequence phasors* consist of three phasors, 120° apart in phase, and in the opposite phase sequence to that of the original phasors. The *zero-sequence components* consist of three phasors of equal magnitude and zero

[6]Technically, the use of symmetrical components also depends on the voltages and currents being related by constant impedances, making the circuit linear. Therefore, superposition applies.

Figure 40.7 *Balanced Delta Current Relationships*

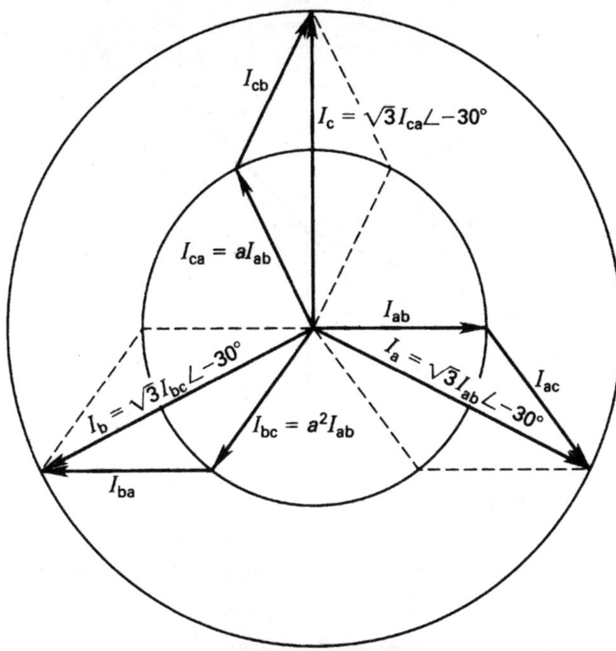

(a) line-to-line currents versus phase currents

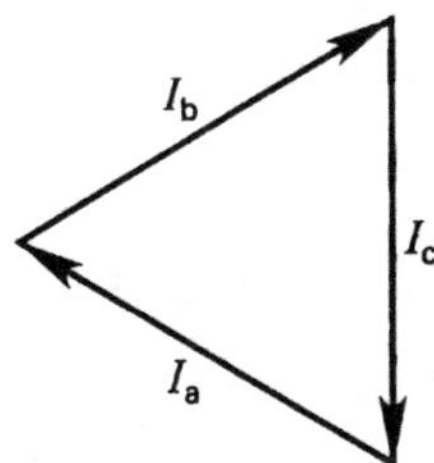

(b) alternative line versus neutral currents

Table 40.1 *Per-Unit Equations*[a,b]

$\text{per unit} = \dfrac{\text{actual}}{\text{base}} = \dfrac{\text{percent}}{100}$	$Z_{\text{base}} = \dfrac{V_{\text{base}}}{I_{\text{base}}} = \dfrac{V_p^2}{S_p} = \left(\dfrac{V^2}{S}\right)_{3\phi} = \left(\dfrac{V}{\sqrt{3}I}\right)_{3\phi}$
$S_{\text{base}} = S_p = \left(\dfrac{S}{3}\right)_{3\phi}$	$P_{\text{base}} = P_p = \left(\dfrac{P}{3}\right)_{3\phi}$
$V_{\text{base}} = V_p = \left(\dfrac{V}{\sqrt{3}}\right)_{3\phi}$	$Q_{\text{base}} = Q_p = \left(\dfrac{Q}{3}\right)_{3\phi}$
$I_{\text{base}} = \dfrac{S_{\text{base}}}{V_{\text{base}}} = \dfrac{S_p}{V_p} = \left(\dfrac{S}{\sqrt{3}V}\right)_{3\phi}$	$\chi_{\text{pu,new}} = \chi_{\text{pu,old}}\left(\dfrac{\chi_{\text{base,old}}}{\chi_{\text{base,new}}}\right)$
$Z_{\text{pu,new}} = Z_{\text{pu,old}}\left(\dfrac{V_{\text{base,old}}}{V_{\text{base,new}}}\right)^2\left(\dfrac{S_{\text{base,new}}}{S_{\text{base,old}}}\right)$	

[a]The equation for $\chi_{\text{pu,new}}$ can only be used to find current, power, or voltage.
[b]The equation for $Z_{\text{pu,new}}$ can only be used to find impedance.

phase displacement from each other. Phasors can be voltage or current phasors. The sequence is indicated by a superscript, zero for zero-sequence, one for positive-sequence, and two for negative-sequence. An example of adding symmetrical components to represent an unbalanced phasor is shown in Fig. 40.8 and Fig. 40.9.

Since each unbalanced phasor is a sum of the symmetrical components, the original phasors are as indicated in Eq. 40.10, Eq. 40.11, and Eq. 40.12.

$$V_a = V_a^{(0)} + V_a^{(1)} + V_a^{(2)} \quad \textit{40.10}$$

$$V_b = V_b^{(0)} + V_b^{(1)} + V_b^{(2)} \quad \textit{40.11}$$

$$V_c = V_c^{(0)} + V_c^{(1)} + V_c^{(2)} \quad \textit{40.12}$$

Using the operator $a = 120°$, each of the sequence components can be represented in terms of the phase A components as indicated for phase B components in Eq. 40.13, Eq. 40.14, and Eq. 40.15, and phase C components in Eq. 40.16, Eq. 40.17, and Eq. 40.18. These equations assume the normal sequence for the system is A-B-C.

$$V_b^{(0)} = V_a^{(0)} \quad \textit{40.13}$$

$$V_b^{(1)} = a^2V_a^{(1)} \quad \textit{40.14}$$

$$V_b^{(2)} = aV_a^{(2)} \quad \textit{40.15}$$

$$V_c^{(0)} = V_a^{(0)} \quad \textit{40.16}$$

$$V_c^{(1)} = aV_a^{(1)} \quad \textit{40.17}$$

$$V_c^{(2)} = a^2V_a^{(2)} \quad \textit{40.18}$$

8. SYMMETRICAL COMPONENTS

Figure 40.9 displays the method for using three sets of symmetrical components to create three unbalanced/unsymmetrical phasors. Equation 40.10, Eq. 40.11, and Eq. 40.12 can be transformed using Eq. 40.13

System Analysis

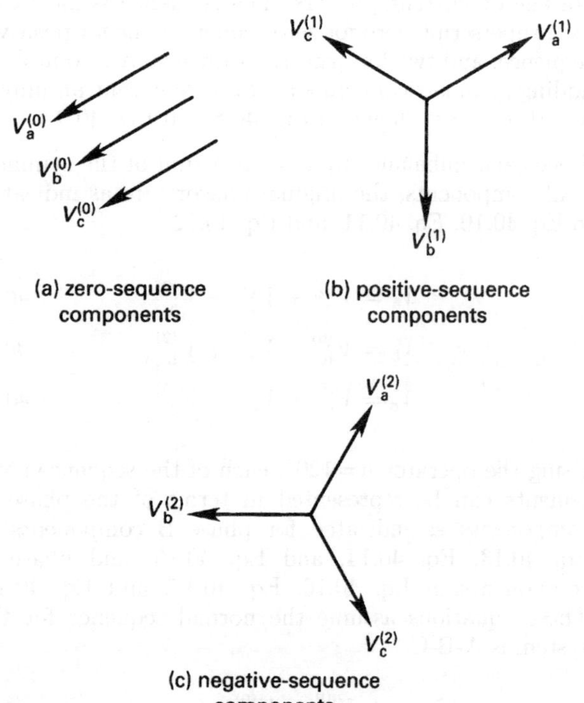

Figure 40.8 Symmetrical Components of Unbalanced Phasors

through Eq. 40.15. The results are shown in Eq. 40.19 through Eq. 40.21.

$$V_a = V_a^{(0)} + V_a^{(1)} + V_a^{(2)} \quad 40.19$$

$$V_b = V_a^{(0)} + a^2 V_a^{(1)} + a V_a^{(2)} \quad 40.20$$

$$V_c = V_a^{(0)} + a V_a^{(1)} + a^2 V_a^{(2)} \quad 40.21$$

The unsymmetrical components can be used to find the symmetrical components using Eq. 40.22, Eq. 40.23, and Eq. 40.24.

$$V_a^{(0)} = \tfrac{1}{3}(V_a + V_b + V_c) \quad 40.22$$

$$V_a^{(1)} = \tfrac{1}{3}(V_a + a V_b + a^2 V_c) \quad 40.23$$

$$V_a^{(2)} = \tfrac{1}{3}(V_a + a^2 V_b + a V_c) \quad 40.24$$

The voltages can be considered the delta connection or phase voltages. If line-to-line voltages (wye connection) are desired, replace V_a, V_b, and V_c with V_{ab}, V_{bc}, and V_{ca}, respectively. If the symmetrical components for phase B are desired, use Eq. 40.13 through Eq. 40.15 to transform Eq. 40.22 through Eq. 40.24. If the symmetrical components for phase C are desired, use Eq. 40.16 through Eq. 40.18 to transform Eq. 40.22 through Eq. 40.24.

In a balanced three-phase system, the sum of the line-to-line voltage phasors will always be zero. Given that, Eq. 40.22 indicates that no zero sequence voltages exist in the line-to-line voltages regardless of the degree of imbalance. The line-to-neutral voltage phasors do not necessarily sum to zero and voltages to neutral may contain zero-sequence components.

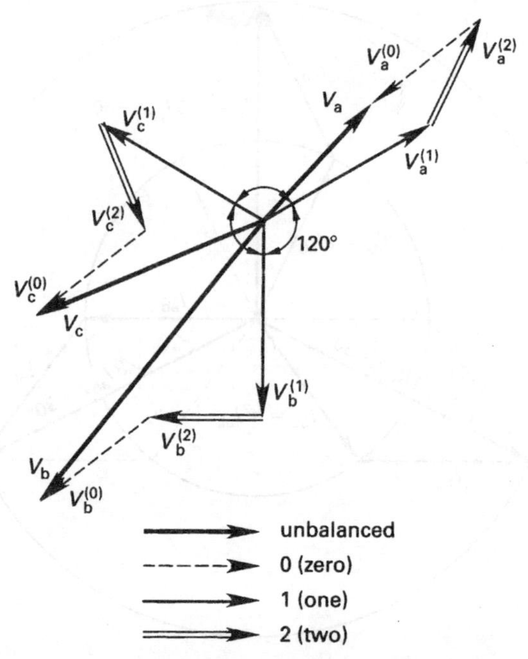

Figure 40.9 Example of Unbalanced Phasor from Symmetrical Phasors

A similar set of equations apply to the current as indicated in Eq. 40.25 through Eq. 40.30.

$$I_a = I_a^{(0)} + I_a^{(1)} + I_a^{(2)} \quad 40.25$$

$$I_b = I_a^{(0)} + a^2 I_a^{(1)} + a I_a^{(2)} \quad 40.26$$

$$I_c = I_a^{(0)} + a I_a^{(1)} + a^2 I_a^{(2)} \quad 40.27$$

$$I_a^{(0)} = \tfrac{1}{3}(I_a + I_b + I_c) \quad 40.28$$

$$I_a^{(1)} = \tfrac{1}{3}(I_a + a I_b + a^2 I_c) \quad 40.29$$

$$I_a^{(2)} = \tfrac{1}{3}(I_a + a^2 I_b + a I_c) \quad 40.30$$

The previous currents can be considered the wye connection or line currents. If the phase currents (delta connection) are desired, replace I_a, I_b, and I_c with I_{ab}, I_{bc}, and I_{ca}, respectively. If the symmetrical components for phase B are desired, use Eq. 40.13 through Eq. 40.15 with currents substituted for the voltages to transform Eq. 40.22 through Eq. 40.24. If the symmetrical components for phase C are desired, then use Eq. 40.16 through Eq. 40.18 with currents substituted for the voltages to transform Eq. 40.22 through Eq. 40.24.

9. SYMMETRICAL WYE AND DELTA CIRCUITS

Three-phase circuits are connected between terminals, usually labeled A, B, and C, and arranged in a delta or wye configuration. The reference is nominally the a-b connection, though it need not be. The equations that follow are applicable regardless of the reference. Symmetrical impedances and the relationships between voltages and currents are shown in Fig. 40.10.

Figure 40.10 *Symmetrical Impedances*

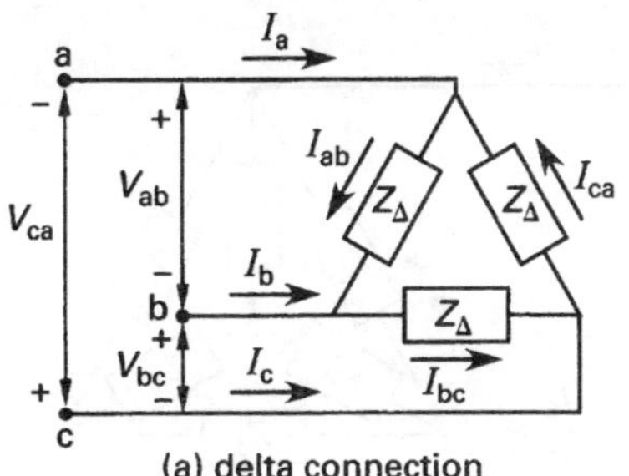

(a) delta connection

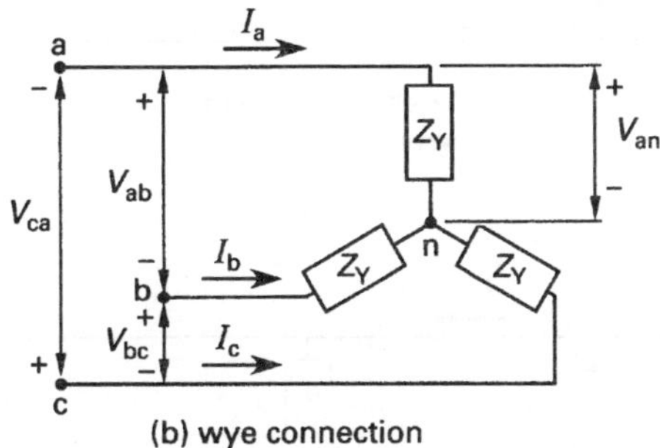

(b) wye connection

In the delta-connected circuit, the currents are given in Eq. 40.31 through Eq. 40.33. The line currents in a delta-connected circuit contain no zero-sequence components.[7]

$$I_a = I_{ab} - I_{ca} \qquad 40.31$$

$$I_b = I_{bc} - I_{ab} \qquad 40.32$$

$$I_c = I_{ca} - I_{bc} \qquad 40.33$$

In a wye-connected circuit, the voltages are calculated using Eq. 40.34 through Eq. 40.36. The line voltages in a wye-connected circuit contain no zero-sequence components.[8]

$$V_{ab} = V_{an} - V_{bn} \qquad 40.34$$

$$V_{bc} = V_{bn} - V_{cn} \qquad 40.35$$

$$V_{ca} = V_{cn} - V_{an} \qquad 40.36$$

The positive- and negative-sequence line (wye) and delta three-phase circuits are shown in Fig. 40.11. The positive- and negative-sequence line-to-neutral (wye) and line-to-line (delta) voltages are shown in Fig. 40.12.

A three-phase circuit can be represented in delta or wye configuration. A delta circuit can be analyzed as a wye-circuit and then converted to delta for the desired values. The conversion, voltage and current relationships, in terms of symmetrical impedances where the $Z_Y = Z_\Delta/3$ is shown in Fig. 40.13.

[7]In delta-connected circuits, zero-sequence currents can flow in the phase currents; however, the currents will flow around the delta, not affecting the line currents.

[8]Recall that the sum of voltages in a wye- or delta-connected three-phase circuit is always zero, and no zero-sequence voltages exist. In wye circuits, there are line-to-neutral zero-sequence voltages, and therefore, there are line-to-neutral zero-sequence currents.

Figure 40.11 *Line and Delta Currents*

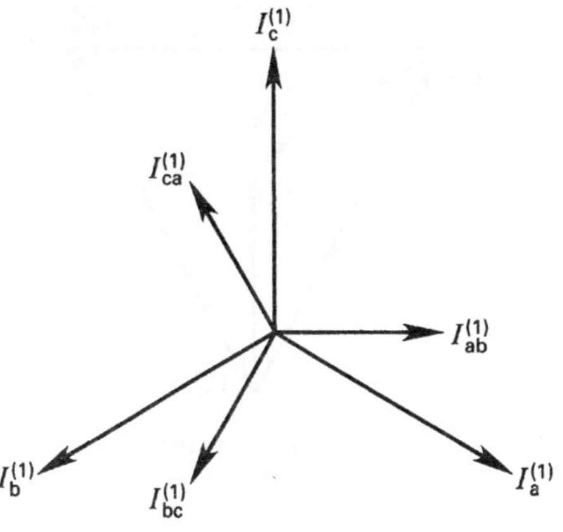

(a) positive-sequence components

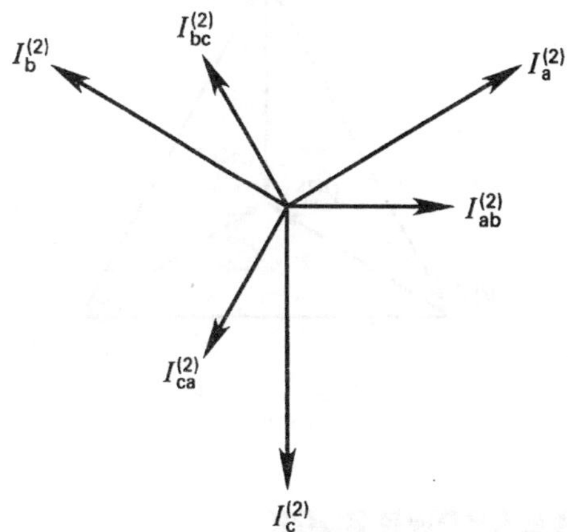

(b) negative-sequence components

10. ELECTROMAGNETIC INTERFERENCE AND COMPATIBILITY

Electromagnetic interference (EMI) is an electromagnetic disturbance that causes the performance of a device, piece of equipment, or system to degrade. If the electromagnetic disturbance comes from an external source, it is called *radiated EMI.* When it comes from an internal source, it is called *conducted EMI.*

Conducted EMI and radiated EMI can each be divided into two types: *continuous wave* (CW) and *transient.* A continuous wave is generally sinusoidal, and its oscillations repeat under steady-state conditions. A transient is a pulse, a damped oscillation, or any other temporary phenomenon resulting in a deviation from steady-state conditions.

A circuit, device, or system is *electromagnetically compatible* if it complies with the EMI standards for a given device that it will operate in or near (i.e., it is not expected to disturb the operation of the device).

Figure 40.12 Line-to-Line and Line-to-Delta Voltages

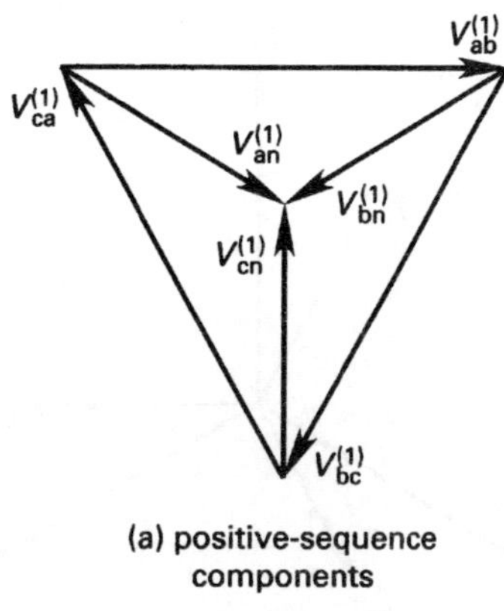

(a) positive-sequence components

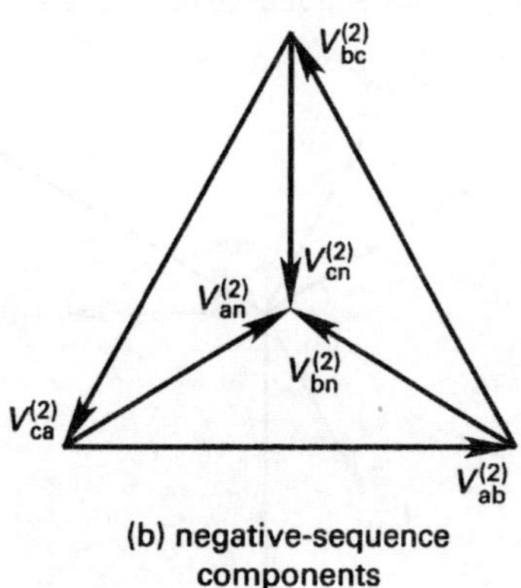

(b) negative-sequence components

Figure 40.13 Symmetrical Delta and Wye Relationships

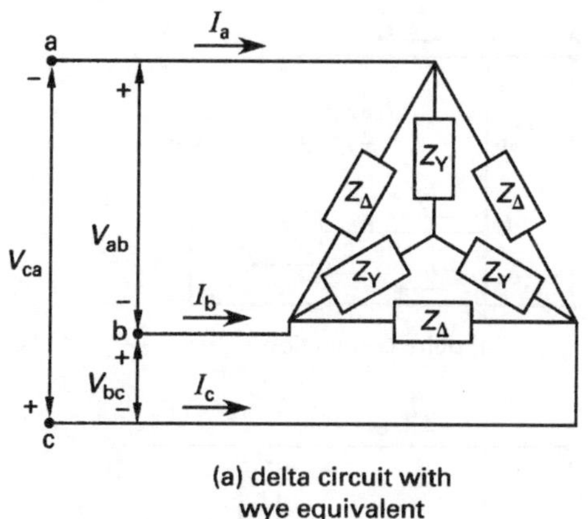

(a) delta circuit with wye equivalent

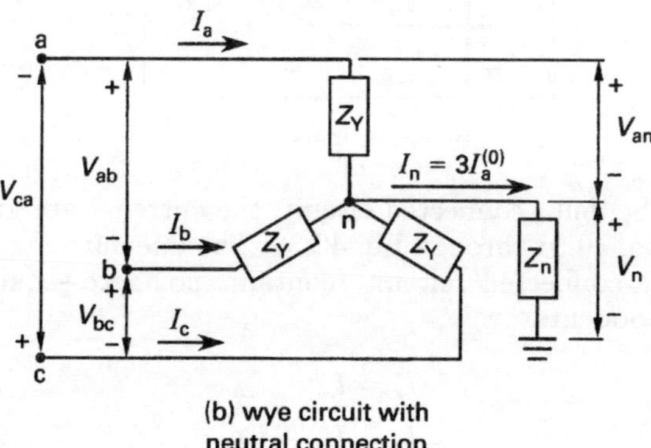

(b) wye circuit with neutral connection

11. IEEE BROWN BOOK

IEEE Standard 399, *IEEE Recommended Practice for Industrial and Commercial Power Systems Analysis* (*IEEE Brown Book*) describes the industry best practices for power system analysis. It contains information, often extracted from other codes, standards, and industry texts, on the use of computer studies for initial and future expansion designs to assist in performance improvement, reliability, and safety. The *IEEE Brown Book* contains chapters on power system analytical procedures; power system modeling; computer solutions to power system problems; load flow; short-circuit, stability, and motor-starting studies; harmonics; switching; reliability; cable ampacity; ground mat; protection coordination; and auxiliary power.

12. IEEE VIOLET BOOK

IEEE Standard 551, *IEEE Recommended Practice for Calculating Short-Circuit Currents in Industrial and Commercial Power Systems* (*IEEE Violet Book*) describes the industry best practices for short-circuit analysis. Some of the information is extracted from other codes, standards, and industry texts. The *IEEE Violet Book* contains chapters on short-circuit currents, calculation techniques, synchronous machines, induction motors, capacitor currents, static converters, ANSI-approved methods, interrupting equipment, unbalanced short-circuit currents, and international standards.

41 Analysis of Control Systems

Nomenclature

A	system matrix	–
$B(s)$	feedback signal	–
B	control matrix	–
C	capacitance	F
C_1, C_2	constant	–
$C(s)$	control output	–
C	ouput matrix	–
$e(t)$	error	–
$E(s)$	error, $\mathcal{L}(e(t))$	–
$f(t)$	forcing function	–
$F(s)$	forcing function, $\mathcal{L}(f(t))$	–
$G(s)$	forward transfer function	–
h	step height	–
$H(s)$	reverse transfer function	–
$i(t)$	current	A
$I(s)$	current, $\mathcal{L}(i(t))$	A
k, K	constant	–
K	gain	–
K	scale factor	–
L	inductance	H
L	length	–
M	fraction overshoot	–
M	gain	–
$M(s)$	manipulated signal	–
n	degrees of freedom, order of the system, or system type	–
N	Nyquist's number	–
$p(t)$	arbitrary function	–
P	number of poles	–
$P(s)$	arbitrary function, $\mathcal{L}(p(t))$	–
Q	quality factor	–
r	real value (root)	–
$r(t)$	time response	–
R	resistance	Ω
$R(s)$	response function, $\mathcal{L}(r(t))$, or reference input	1/s
s	Laplace transform variable or complex frequency variable	–
S	sensitivity	–
t	time	s
$T(s)$	transfer function, $\mathcal{L}(T(t))$	–
$T(t)$	transfer function	–
u	input variable	–
$U(s)$	disturbance	–
U	control vector	–
v	variable voltage	V
V	constant or rms voltage	V
x	state variable	–
X	state vector	–
y	output variable	–
$Y(s)$	output	–
Y	output vector	–
Z	number of zeros	–

Symbols

α	pole-angle	rad
β	zero-angle	rad
ζ	damping ratio	–
τ	time constant	s
ω	natural frequency	rad/s

Subscripts

d	damped or delay
f	forced or feedback
i	input
n	natural
o	output
p	peak or pole-line
r	rise
s	settling
z	zero-line

1. TYPES OF RESPONSE

Natural response (also referred to as *initial condition response, homogeneous response,* and *unforced response*) is the manner in which a system behaves when energy is applied and then subsequently removed. The system is left alone and allowed to do what it would do naturally, without the application of further disturbing forces. In the absence of friction, natural response is characterized by a sinusoidal response function. The response function will contain exponentially decaying sinusoids otherwise.

Forced response is the behavior of a system that is acted upon by a force that is applied periodically. Forced response in the absence of friction is characterized by sinusoidal terms having the same frequency as the forcing function.

Natural and forced responses are present simultaneously in forced systems. The sum of the two responses is the *total response.*[1] This is the reason that differential equations are solved by adding a particular solution to the homogeneous solution. The homogeneous solution corresponds to the natural response; the particular solution corresponds to the forced response.

$$\text{total response} = \text{natural response} + \text{forced response} \qquad 41.1$$

Because the influence of decaying functions disappears after a few cycles, natural response is sometimes referred to as *transient response.* Once the transient response effects have died out, the total response consists entirely of forced terms. This is the *steady-state response.*

2. GRAPHICAL SOLUTION

Graphical solutions are available in some cases, particularly those with homogeneous, step, and sinusoidal inputs. When the system equation is a homogeneous second-order linear differential equation with constant coefficients (see Sec. 11.6) in the form of Eq. 41.2, the natural time response can be determined from Fig. 41.1. The term ω is the natural frequency, ω_n, defined in Sec. 31.10, and ζ is the damping ratio.

$$x'' + 2\zeta\omega x' + \omega^2 x = 0 \qquad 41.2$$

When the system equation is a second-order linear differential equation with constant coefficients and the forcing function is a step of height h (as in Eq. 41.3), the time response can be determined from Fig. 41.2.

$$x'' + 2\zeta\omega x' + \omega^2 x = \omega^2 h \qquad 41.3$$

Figure 41.2 illustrates that a system responding to a step will eventually settle to the steady-state position of the step but that when damping is low ($\zeta < 1$), there will be *overshoot.* Figure 41.3 illustrates this and other parameters of second-order response to a step input. The settling time depends on the tolerance (i.e., the separation of the actual and steady-state responses). The *time delay,* t_d in Fig. 41.3, is the time needed to reach 50% of the steady-state value. The *time constant,* τ, is the time needed to reach approximately 63% of the steady-state value.

Figure 41.1 Natural Response

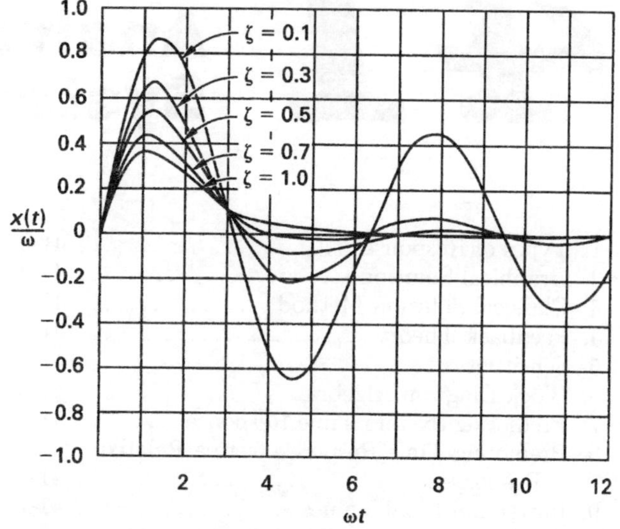

Figure 41.2 Response to a Unit Step

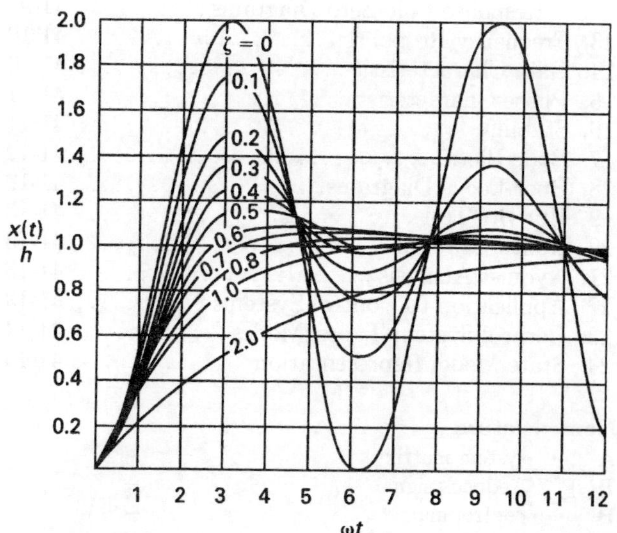

The damped frequency is

$$\omega_d = \omega\sqrt{1-\zeta^2} \qquad 41.4$$

The rise time is

$$t_r = \frac{\pi - \arccos\zeta}{\omega_d} \qquad 41.5$$

[1]This response is a function of time, and, therefore, is referred to as *time response* to distinguish it from *frequency response.* (See Sec. 41.13.)

System Analysis

Figure 41.3 *Second-Order Step Time Response Parameters*

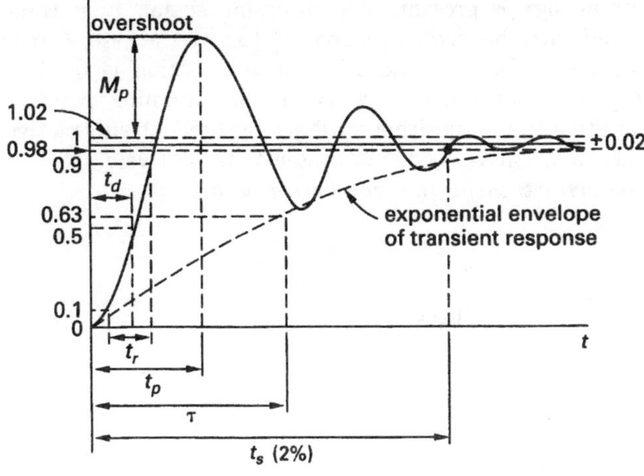

The peak time is

$$t_p = \frac{\pi}{\omega_d} \tag{41.6}$$

The peak gain (fraction overshoot) is

$$M_p = e^{(-\pi\zeta)/\sqrt{1-\zeta^2}} \tag{41.7}$$

The settling time is

$$t_s = \frac{3.91}{\zeta\omega_n} \quad \text{[2\% criterion]} \tag{41.8}$$

$$= \frac{3.00}{\zeta\omega_n} \quad \text{[5\% criterion]} \tag{41.9}$$

The time constant is

$$\tau = \frac{1}{\zeta\omega_n} \tag{41.10}$$

3. CLASSICAL SOLUTION METHOD

A system model can be thought of as a block where the *input signal* is the forcing function, $f(t)$, and the *output signal* is the response function, $r(t)$. The standard analytical method of determining the response of a system from its system equation uses Laplace transforms. Accordingly, the classical solution approach does not determine the output signal or response function directly but derives the *transfer function*, $T(s)$, instead.[2] The transfer function is also known as the *rational function.*

$$T(s) = \frac{\mathcal{L}r(t)}{\mathcal{L}f(t)} \tag{41.11}$$

[2]Strictly speaking, $r(t)/f(t)$ is the *transfer function* and $\mathcal{L}r(t)/\mathcal{L}f(t)$ is the *transform of the transfer function.* However, the distinction is seldom made.

Transfer functions in the s-domain are Laplace transformations of the corresponding time-domain functions. Transforming $r(t)/f(t)$ into $T(s)$ involves more than a simple change of variables. The s symbol can be thought of as a derivative operator; similarly, the integration operator is represented as $1/s$. By convention, Laplace transforms are represented by uppercase letters while operand functions are represented by lowercase letters.

Example 41.1

The following system equation in integrodifferential equation form has been derived for the output current from an electrical network. Convert the system equation to the s-domain.

$$i(t) = \frac{1}{L}\int (v_1 - v_2)\,dt + C\left(\frac{d(v_2 - v_1)}{dt}\right)$$

Solution

Replace all derivative operators by s; replace all integration operators by $1/s$. Replace time domain voltage, $v(t)$, with s-domain voltage, $V(s)$.

$$I(s) = \frac{V_1}{sL} - \frac{V_2}{sL} + sCV_2 - sCV_1$$

Example 41.2

Determine the voltage transfer function for the system shown and draw its black box representation.

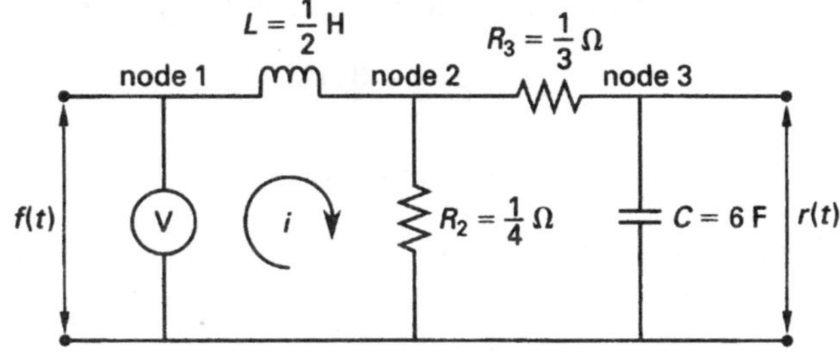

Solution

First, write the system equations. There are two loop currents, so two simultaneous equations are needed. One system equation can be written for each of the three nodes, so there will be one redundant system equation. (At this point, it is not obvious which of the two system equations can be used to derive the transfer function.)

At node 1,

$$i(t) = \frac{1}{L}\int (v_1 - v_2)\,dt = 2\int (v_1 - v_2)\,dt$$

System Analysis

At node 2,

$$0 = \frac{1}{L}\int (v_2 - v_1)dt + \frac{v_2}{R_2} + \frac{v_2 - v_3}{R_3}$$
$$= 2\int (v_2 - v_1)dt + 4v_2 + 3(v_2 - v_3)$$

At node 3,

$$0 = C\left(\frac{dv_3}{dt}\right) + \frac{v_3 - v_2}{R_3}$$
$$= 6\left(\frac{dv_3}{dt}\right) + 3(v_3 - v_2)$$

Next, convert the system equations to the s-domain by substituting s for derivative operations and $1/s$ for integration operations.

For node 1,

$$I(s) = \frac{2(V_1 - V_2)}{s}$$

For node 2,

$$0 = \frac{2(V_2 - V_1)}{s} + 4V_2 + 3(V_2 - V_3)$$

For node 3,

$$0 = 6sV_3 + 3(V_3 - V_2)$$

The transfer function is the ratio of the output to input voltages, V_3/V_1. It does not depend on V_2. The equation for node 1 cannot be used unless $i(t)$ is known, which it (generally) is not. V_2 is eliminated from the equations for nodes 2 and 3. From the second and third nodes,

$$V_2 = \frac{2V_1 + 3sV_3}{2 + 7s}$$
$$= V_3(1 + 2s)$$

The transfer function, $T(s)$, is found by equating these two expressions and solving for V_3/V_1.

$$T(s) = \frac{V_3}{V_1} = \frac{1}{7s^2 + 4s + 1}$$

The black box representation of this system is

$$\longrightarrow \boxed{\frac{1}{7s^2 + 4s + 1}} \longrightarrow$$

4. FEEDBACK THEORY

The output signal is returned as input in a feedback loop (feedback system). A basic feedback system (see Fig. 41.4) consists of two black box units (a *dynamic unit* and a *feedback unit*), a pick-off point (take-off point), and a summing point (*comparator* or *summer*). The summing point is assumed to perform positive addition unless a minus sign is present. The incoming signal, v_i, is combined with the feedback signal, v_f, to give the *error* (*error signal*), e. Whether addition or subtraction is used in Eq. 41.12 depends on whether the summing point is additive (i.e., a positive feedback system) or subtractive (i.e., a negative feedback system), respectively. $E(s)$ is the *error transfer function* (*error gain*).

$$E(s) = \mathcal{L}\big(e(t)\big) = V_i(s) \pm V_f(s)$$
$$= V_i(s) \pm H(s)V_o(s) \qquad 41.12$$

Figure 41.4 Feedback System

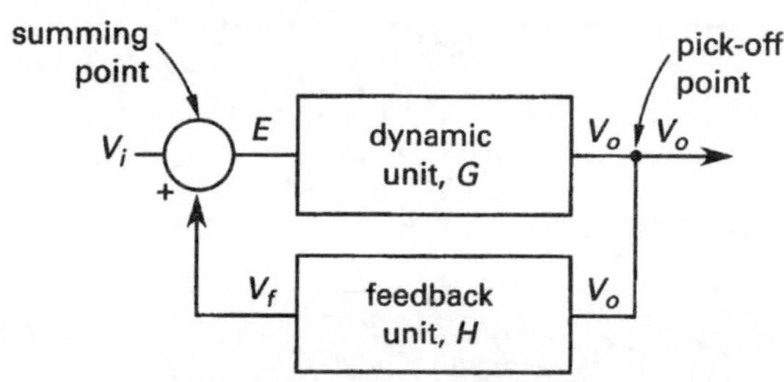

The ratio $E(s)/V_i(s)$ is the *error ratio* (*actuating signal ratio*).

$$\frac{E(s)}{V_i(s)} = \frac{1}{1 + G(s)H(s)} \quad \text{[negative feedback]} \qquad 41.13$$

$$= \frac{1}{1 - G(s)H(s)} \quad \text{[positive feedback]} \qquad 41.14$$

Because the dynamic and feedback units are black boxes, each has an associated transfer function. The transfer function of the dynamic unit is known as the *forward transfer function* (*direct transfer function*), $G(s)$. In most feedback systems—amplifier circuits in particular—the magnitude of the forward transfer function is known as the *forward gain* or *direct gain*. $G(s)$ can be a scalar if the dynamic unit merely scales the error. However, $G(s)$ is normally a complex operator that changes both the magnitude and the phase of the error.

$$V_o(s) = G(s)E(s) \qquad 41.15$$

The pick-off point transmits the output signal, V_o, from the dynamic unit back to the feedback element. The output of the dynamic unit is not reduced by the pickoff point. The transfer function of the feedback unit is the *reverse transfer function* (*feedback transfer function*, *feedback gain*), $H(s)$, which can be a simple magnitude-changing scalar or a phase-shifting function.

$$V_f(s) = H(s)V_o(s) \qquad 41.16$$

System Analysis

The ratio $V_f(s)/V_i(s)$ is the *feedback ratio (primary feedback ratio)*.

$$\frac{V_f(s)}{V_i(s)} = \frac{G(s)H(s)}{1+G(s)H(s)} \quad \text{[negative feedback]} \qquad 41.17$$

$$= \frac{G(s)H(s)}{1-G(s)H(s)} \quad \text{[positive feedback]} \qquad 41.18$$

The *loop transfer function* (*loop gain* or *open-loop transfer function*) is the gain after going around the loop one time, $\pm G(s)H(s)$.

The *overall transfer function* (*closed-loop transfer function, control ratio, system function, closed-loop gain*), $G_{\text{loop}}(s)$, is the overall transfer function of the feedback system. The quantity $1+G(s)H(s)=0$ is the *characteristic equation*. The *order of the system* is the largest exponent of s in the characteristic equation. (This corresponds to the highest-order derivative in the system equation.)

$$G_{\text{loop}}(s) = \frac{V_o(s)}{V_i(s)}$$

$$= \frac{G(s)}{1+G(s)H(s)} \quad \text{[negative feedback]} \qquad 41.19$$

$$= \frac{G(s)}{1-G(s)H(s)} \quad \text{[positive feedback]} \qquad 41.20$$

With positive feedback and $G(s)H(s)$ less than 1.0, G_{loop} will be larger than $G(s)$. This increase in gain is a characteristic of positive feedback systems. As $G(s)H(s)$ approaches 1.0, the closed-loop transfer function increases without bound, which is usually an undesirable effect.

In a negative feedback system, the denominator of Eq. 41.19 is greater than 1.0. Although the closed-loop transfer function will be less than $G(s)$, there may be other desirable effects. Generally, a system with negative feedback is less sensitive to variations in temperature, circuit components values, input signal frequency, and signal noise. Other benefits include distortion reduction, increased stability, and impedance matching.[3]

Example 41.3

A high-gain noninverting amplifier has a gain of 10^6. Feedback is provided by a resistor voltage divider. What is the closed-loop gain?

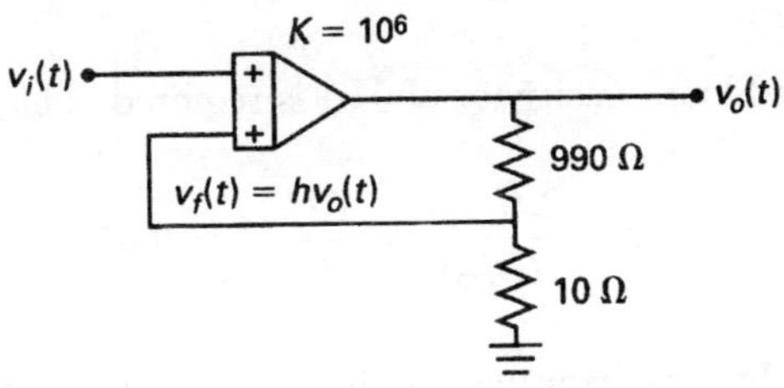

Solution

The fraction of the output signal appearing at the summing point depends on the resistances in the divider circuit.

$$h = \frac{10\ \Omega}{10\ \Omega + 990\ \Omega} = 0.01$$

Because the feedback path only scales the feedback signal, the feedback will be positive. From Eq. 41.20, the closed-loop gain is

$$K_{\text{loop}} = \frac{K}{1-Kh} = \frac{10^6}{1-(10^6)(0.01)} \approx -100$$

5. SENSITIVITY

For large loop gains ($G(s)H(s) \gg 1$) in negative feedback systems, the overall gain is approximately $1/H(s)$. The forward gain is not a factor, and by choosing the correct value of $H(s)$, the output can be made insensitive to variations in $G(s)$.

In general, the *sensitivity*, S, of any variable, A, with respect to changes in another parameter, B, is

$$S_B^A = \frac{d\ln A}{d\ln B} = \frac{\frac{dA}{A}}{\frac{dB}{B}} \approx \left(\frac{\Delta A}{\Delta B}\right)\left(\frac{B}{A}\right) \qquad 41.21$$

The sensitivity of the loop transfer function with respect to the forward transfer function is

$$S_{G(s)}^{G_{\text{loop}}(s)} = \left(\frac{\Delta G_{\text{loop}}(s)}{\Delta G(s)}\right)\left(\frac{G(s)}{G_{\text{loop}}(s)}\right)$$

$$= \frac{1}{1+G(s)H(s)} \quad \left[\begin{array}{c}\text{negative}\\ \text{feedback}\end{array}\right] \qquad 41.22$$

$$= \frac{1}{1-G(s)H(s)} \quad \left[\begin{array}{c}\text{positive}\\ \text{feedback}\end{array}\right] \qquad 41.23$$

Example 41.4

A closed-loop gain of −100 is required from a circuit, and the output signal must not vary by more than ±1%. An amplifier is available, but its output varies by ±20%. How can this amplifier be used?

[3]For circuits to be directly connected in series without affecting their performance, all input impedances must be infinite and all output impedances must be zero.

Solution

A closed-loop sensitivity of 0.01 is required. This means

$$\frac{\Delta v_o}{v_o} = \frac{\Delta G_{\text{loop}} v_i}{G_{\text{loop}} v_i} = \frac{\Delta G_{\text{loop}}}{G_{\text{loop}}} = 0.01$$

Similarly, the existing amplifier has a sensitivity of 20%. This means

$$\frac{\Delta G}{G} = 0.20$$

The ratio of these variations corresponds to the definition of sensitivity as in Eq. 41.21. For positive feedback,

$$\frac{\Delta G_{\text{loop}} G}{\Delta G G_{\text{loop}}} = \frac{0.01}{0.20} = \frac{1}{1 - GH}$$

Solving, $GH = -19$.

Solving for G from Eq. 41.20,

$$G_{\text{loop}} = -100 = \frac{G}{1 - GH} = \frac{G}{1 - (-19)}$$

Solving, $G = -2000$. Finally, solve for H.

$$H = \frac{GH}{G} = \frac{-19}{-2000} = 0.0095$$

6. BLOCK DIAGRAM ALGEBRA

The functions represented by several interconnected black boxes (*cascaded blocks*) can be simplified into a single block operation. Some of the most important simplification rules of block diagram algebra are shown in Fig. 41.5. Case 3 represents the standard feedback model.

Example 41.5

A complex block system is constructed from five blocks and two summing points. What is the overall transfer function?

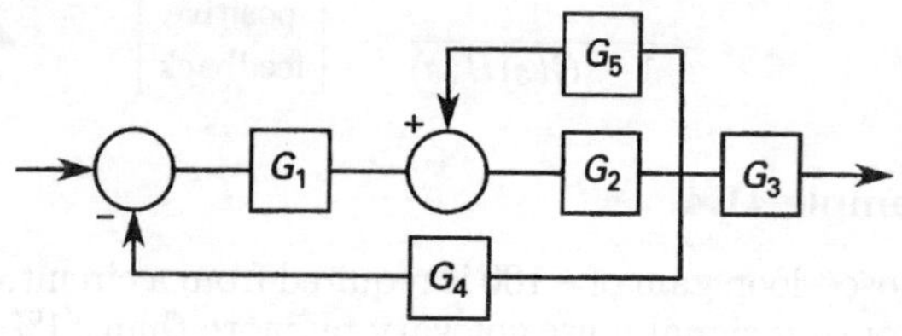

Figure 41.5 *Rules of Simplifying Block Diagrams*

Solution

Use case 5 to move the second summing point back to the first summing point.

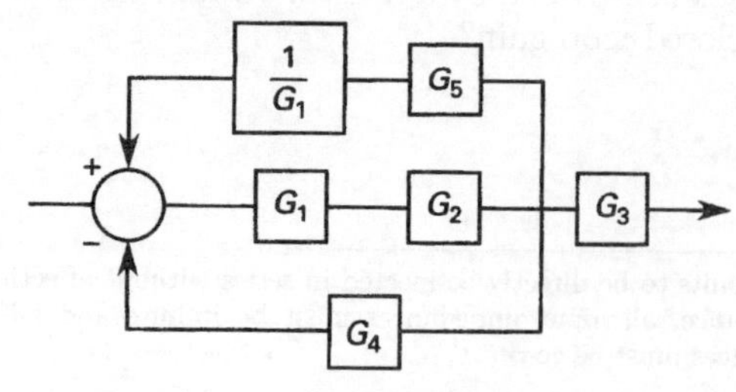

Use case 1 to combine boxes in series.

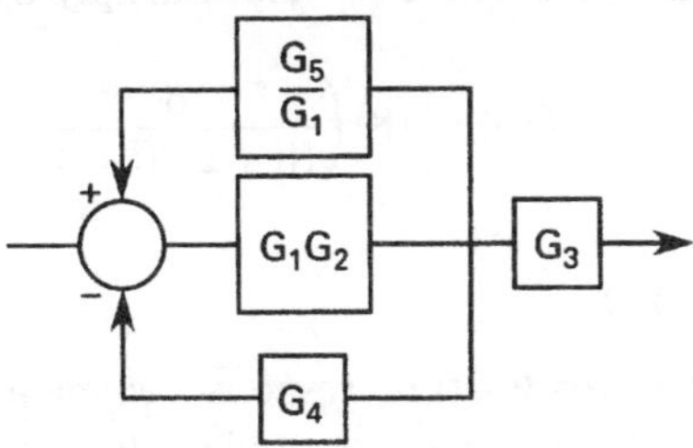

Use case 2 to combine the two feedback loops.

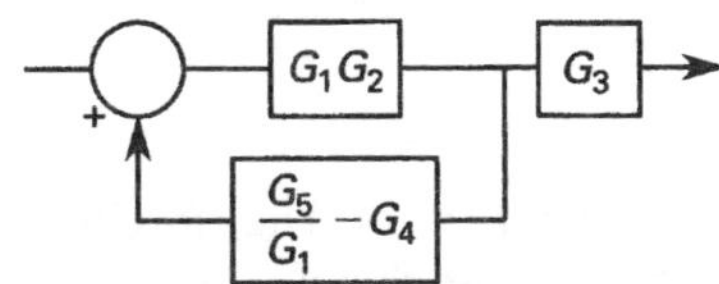

Use case 8 to move the pick-off point outside the G_3 box.

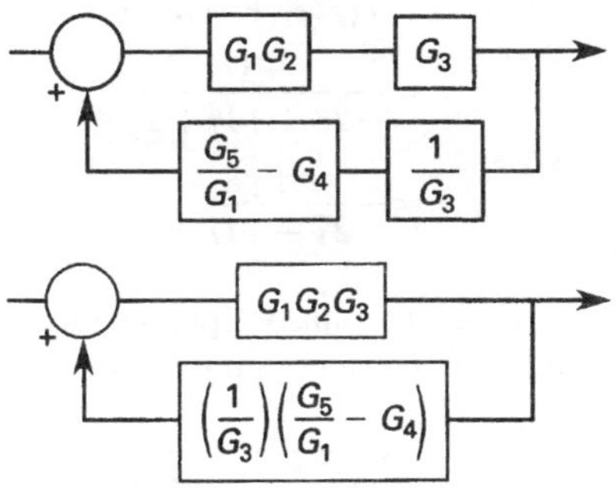

Use case 3 to determine the system gain.

$$G_{\text{loop}} = \frac{G_1G_2G_3}{1-(G_1G_2G_3)\left(\frac{1}{G_3}\right)\left(\frac{G_5}{G_1}-G_4\right)} = \frac{G_1G_2G_3}{1-G_2G_5+G_1G_2G_4}$$

7. PREDICTING SYSTEM TIME RESPONSE

The transfer function is derived without knowledge of the input and is insufficient to predict the time response of the system. The system time response depends on the form of the input function. Because the transfer function is expressed in the *s*-domain, the forcing and response functions must be also. (Laplace transforms of step, pulses, sinusoids, and other functions are evaluated in Chap. 11 and Chap. 15.)

$$R(s) = T(s)F(s) \tag{41.24}$$

The time-based response function, $r(t)$, is found by performing the inverse Laplace transform.

$$r(t) = \mathcal{L}^{-1}\Big(R(s)\Big) \tag{41.25}$$

Example 41.6

A mechanical system is acted upon by a constant force of eight units starting at $t=0$. What is the time-based response function, $r(t)$, if the transfer function is

$$T(s) = \frac{6}{(s+2)(s+4)}$$

Solution

The forcing function is a step of height 8 at $t=0$. The Laplace transform of a unit step is $1/s$. Therefore, $F(s) = 8/s$.

From Eq. 41.24,

$$R(s) = T(s)F(s) = \left(\frac{6}{(s+2)(s+4)}\right)\left(\frac{8}{s}\right) = \frac{48}{s(s+2)(s+4)}$$

The response function is found from Eq. 41.25 and a table of Laplace transforms (see App. 11.A). A product of linear terms in the denominator of $R(s)$ is equivalent to a sum of terms in $r(t)$, so

$$r(t) = \mathcal{L}^{-1}\left(\frac{48}{s(s+2)(s+4)}\right) = 6 - 12e^{-2t} + 6e^{-4t}$$

The last two terms are decaying exponentials, which represent the transient natural response. The first term does not vary with time; it is the steady-state response.

8. PREDICTING TIME RESPONSE FROM A RELATED RESPONSE

In some cases, it may be possible to use a known response to one input to determine the response to another input. For example, the impulse function is the derivative of the step function, so the response to an impulse is the derivative of the response to a step function.

9. INITIAL AND FINAL VALUES

The initial and final (steady-state) values of any function, $P(s)$, can be found from the *initial* and *final value theorems*, respectively, providing limits exist. Equation 41.26 and Eq. 41.27 are particularly valuable in determining the steady-state response (substitute $R(s)$ for $P(s)$) and the steady-state error (substitute $E(s)$ for $P(s)$).

$$\lim_{t\to 0+} p(t) = \lim_{s\to\infty}(sP(s)) \quad \text{[initial value]} \qquad 41.26$$

$$\lim_{t\to\infty} p(t) = \lim_{s\to 0}(sP(s)) \quad \text{[final value]} \qquad 41.27$$

Example 41.7

What is the final value of the response function $r(t)$ if

$$R(s) = \frac{1}{s(s+1)}$$

Solution

From Eq. 41.27, $R(s)$ is multiplied by s and the limit taken as s approaches zero.

$$r(\infty) = \lim_{s\to 0}\left(\frac{s}{s(s+1)}\right) = \lim_{s\to 0}\left(\frac{1}{s+1}\right) = 1$$

10. SPECIAL CASES OF STEADY-STATE RESPONSE

In addition to determining the steady-state response from the final value theorem (see Sec. 41.9), the steady-state response to a specific input can be easily derived from the transfer function, $T(s)$, in a few specialized cases. For example, the steady-state response function for a system acted upon by an impulse is simply the transfer function. That is, a pulse has no long-term effect on a system.

$$R(\infty) = T(s) \quad \text{[pulse input]} \qquad 41.28$$

The steady-state response for a *step input* (often referred to as a *DC input*) is obtained by substituting zero for s everywhere in the transfer function. (If the step has magnitude h, the steady-state response is multiplied by h.)

$$R(\infty) = T(0) \quad \text{[unit step input]} \qquad 41.29$$

The steady-state response for a sinusoidal input is obtained by substituting $j\omega_f$ for s everywhere in the transfer function, $T(s)$. The output will have the same frequency as the input. It is particularly convenient to perform sinusoidal calculations using phasor notation (as illustrated in Ex. 41.9).

$$R(\infty) = T(j\omega_f) \qquad 41.30$$

Example 41.8

What is the steady-state response of the system in Ex. 41.6 acted upon by a step of height 8?

Solution

Substitute zero for s in $T(s)$ and multiply by 8.

$$R(\infty) = 8T(0) = (8)\left(\frac{6}{(0+2)(0+4)}\right) = 6$$

Example 41.9

What is the steady-state response when a sinusoidal voltage of $4\sin(2t+45°)$ is applied to a system with the following transfer function?

$$T(s) = \frac{-1}{7s^2+7s+1}$$

Solution

The angular frequency of the forcing function is $\omega_f = 2$ rad/s. Substitute $j2$ for s in $T(s)$ and simplify the expression by recognizing that $j^2 = -1$.

$$\begin{aligned} T(j2) &= \frac{-1}{7(j2)^2 + 7(j2) + 1} \\ &= \frac{-1}{-28+14j+1} \\ &= \frac{-1}{-27+14j} \end{aligned}$$

Next, convert $T(j2)$ to phasor (polar) form. The magnitude and angle of the denominator are

$$\begin{aligned} \text{magnitude} &= \sqrt{(14)^2 + (-27)^2} = 30.4138 \\ \text{angle} &= 180° - \arctan\frac{14}{27} = 152.59° \end{aligned}$$

However, this is the negative reciprocal of $T(j2)$.

$$T(j2) = \frac{-1}{30.4138\angle 152.59°} = -0.03288\angle{-152.59°}$$

The forcing function expressed in phasor form is $4\angle 45°$. From Eq. 41.11, the steady-state response is

$$\begin{aligned} v(t) &= T(t)f(t) = (-0.03288\angle{-152.59°})(4\angle 45°) \\ &= -0.1315\angle{-107.6°} \end{aligned}$$

11. POLES AND ZEROS

A *pole* is a value of s that makes a function, $P(s)$, infinite. Specifically, a pole makes the denominator of $P(s)$ zero.[4] A zero of the function makes the numerator of $P(s)$ (and hence $P(s)$ itself) zero. Poles and zeros need not be real or unique; they can be imaginary and repeated within a function.

[4]Pole values are the system *eigenvalues*.

System Analysis

A *pole-zero diagram* is a plot of poles and zeros in the *s-plane*—a rectangular coordinate system with real and imaginary axes. Zero is represented as O; a pole is represented as X. Poles off the real axis always occur in conjugate pairs known as *pole pairs*.

Sometimes it is necessary to derive the function $P(s)$ from its pole-zero diagram. This will be only partially successful because repeating identical poles and zeros are not usually indicated on the diagram. Also, scale factors (scalar constants) are not shown.

Example 41.10

Draw the pole-zero diagram for the transfer function

$$T(s) = \frac{5(s+3)}{(s+2)(s^2+2s+2)}$$

Solution

The numerator is zero when $s = -3$. This is the only zero of the transfer function.

The denominator is zero when $s = -2$ and $s = -1 \pm j$. These three values are the poles of the transfer function.

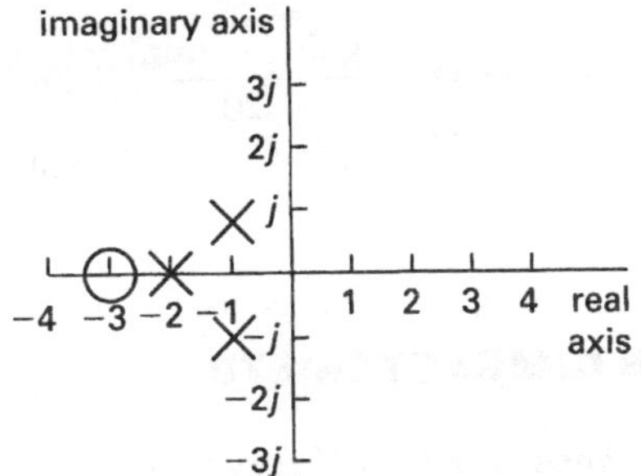

Example 41.11

A pole-zero diagram for a transfer function $T(s)$ has a single pole at $s = -2$ and a single zero at $s = -7$. What is the corresponding function?

Solution

$$T(s) = \frac{K(s+7)}{s+2}$$

The scale factor K must be determined by some other means.

12. PREDICTING SYSTEM TIME RESPONSE FROM RESPONSE POLE-ZERO DIAGRAMS

A response pole-zero diagram based on $R(s)$ can be used to predict how the system responds to a specific input. This pole-zero diagram must be based on the product $T(s)F(s)$ because that is how $R(s)$ is calculated. Plotting the product $T(s)F(s)$ is equivalent to plotting $T(s)$ and $F(s)$ separately on the same diagram. (See Fig. 41.6.)

Figure 41.6 Types of Response Determined by Pole Location

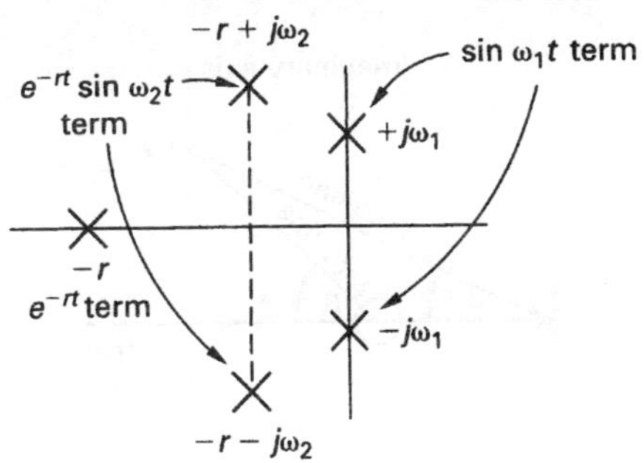

The system will experience an *exponential decay* when a single pole falls on the real axis. A pole with a value of $-r$, corresponding to the linear term $s + r$, will decay at the rate of e^{-rt}. The quantity $1/r$ is the decay *time constant*, the time for the response to achieve approximately 63% of its steady-state value. The farther left the point is located from the vertical imaginary axis, the faster the motion will die out.

Undamped sinusoidal oscillation will occur if a pole pair falls on the imaginary axis. A conjugate pole pair with the value of $\pm j\omega$ indicates oscillation with a natural frequency of ω rad/s.

Pole pairs to the left of the imaginary axis represent *decaying sinusoidal response*. The closer the poles are to the real (horizontal) axis, the slower the oscillations. The closer the poles are to the imaginary (vertical) axis, the slower the decay. The *natural frequency*, ω, of undamped oscillation can be determined from a conjugate pole pair having values of $r \pm \omega_f$.

$$\omega = \sqrt{r^2 + \omega_f^2} \qquad 41.31$$

The magnitude and phase shift can be determined for any input frequency from the pole-zero diagram with the following procedure: Locate the angular frequency, ω_f, on the imaginary axis. Draw a line from each pole (i.e., a pole-line) and zero (i.e., a zero-line) of $T(s)$ to this point. The angle of each of these lines is the angle between it and the horizontal real axis. The overall magnitude is the product of the lengths of the zero-lines divided by the product of the lengths of the pole-lines. (The scale factor must also be included because it is not shown on the pole-zero diagram.) The phase is the sum of the pole-angles less the sum of the zero-angles. (See Fig. 41.7.)

$$|R| = \frac{K\prod_z |L_z|}{\prod_p |L_p|} = \frac{K\prod_z \text{length}}{\prod_p \text{length}} \qquad 41.32$$

$$\angle R = \sum_p \alpha - \sum_z \beta \qquad 41.33$$

System Analysis

Figure 41.7 Calculating Magnitude and Phase froma Pole-Zero Diagram

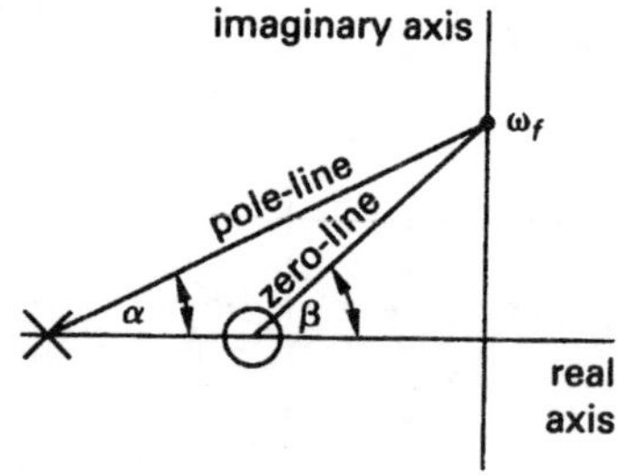

Example 41.12

What is the response of a system with the response pole-zero diagram representing $R(s) = T(s)F(s)$ shown?

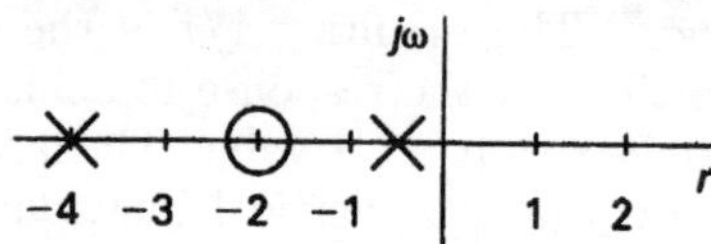

Solution

The poles are at $r = -\frac{1}{2}$ and $r = -4$. The response is

$$r(t) = C_1 e^{(-1/2)t} + C_2 e^{-4t}$$

Constants C_1 and C_2 must be found from other data.

Example 41.13

What is the system response if $T(s) = (s+2)/(s+3)$ and the input is a unit step?

Solution

The transform of a unit step is $1/s$. The response is

$$R(s) = T(s)F(s) = \frac{s+2}{s(s+3)}$$

The response pole-zero diagram is

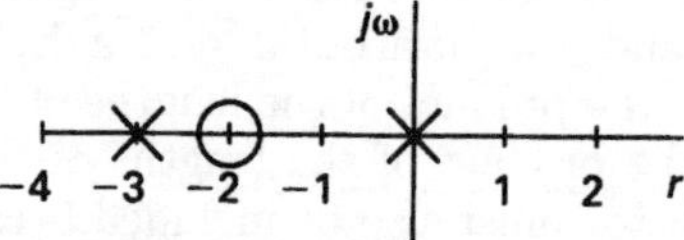

The pole at $r=0$ contributes the exponential $C_1 e^{-0t}$ (or simply C_1) to the total response. The pole at $r=-3$ contributes the term $C_2 e^{-3t}$. The total response is

$$r(t) = C_1 + C_2 e^{-3t}$$

13. FREQUENCY RESPONSE

The gain and phase angle frequency response of a system changes as the forcing frequency is varied. (The dependence of $r(t)$ on ω_f is illustrated in Ex. 41.9.) The *frequency response* is the variation in these parameters, always with a sinusoidal input. *Gain* and *phase characteristics* are plots of the steady-state gain and phase angle responses with a sinusoidal input versus frequency. While a linear frequency scale can be used, frequency response is almost always presented against a logarithmic frequency scale.

The steady-state gain response is expressed in decibels, while the steady-state phase angle response is expressed in degrees. The gain is calculated from Eq. 41.34 where $|T(s)|$ is the absolute value of the steady-state response.

$$\text{gain} = 20\log|T(j\omega)| \quad \text{[in dB]} \qquad 41.34$$

A doubling of $|T(j\omega)|$ is referred to as an *octave* and corresponds to a 6.02 dB increase. A ten-fold increase in $|T(j\omega)|$ is a *decade* and corresponds to a 20 dB increase.

$$\begin{aligned}\text{number of octaves} &= \frac{\text{gain}_2 - \text{gain}_1}{6.02} \quad \text{[in dB]}\\ &= 3.32 \times \text{number of decades}\end{aligned} \qquad 41.35$$

$$\begin{aligned}\text{number of decades} &= \frac{\text{gain}_2 - \text{gain}_1}{20} \quad \text{[in dB]}\\ &= 0.301 \times \text{number of octaves}\end{aligned} \qquad 41.36$$

14. GAIN CHARACTERISTIC

The *gain characteristic* (*M*-curve for magnitude) is a plot of the gain as ω_f is varied. It is possible to make a rough sketch of the gain characteristic by calculating the gain at a few points (pole frequencies, $\omega=0$, $\omega=\infty$, etc.). The curve will usually be asymptotic to several lines. The frequencies at which these asymptotes intersect are *corner frequencies*. The peak gain, M_p, coincides with the natural (resonant) frequency of the system.[5] Large peak gains indicate reduced stability and large overshoots. The *gain crossover point* is the frequency at which log(gain) = 0.

The *half-power points* (*cut-off frequencies*) are the frequencies for which the gain is 0.707 (i.e., $\sqrt{2}/2$) times the peak value. (This is equivalent to saying the gain is 3 dB less than the peak gain.) The *cut-off rate* is the slope of the gain characteristic in dB/octave at a half-power point. The frequency difference between the half-power points is the *bandwidth*, BW. (See Fig. 41.8.) The *closed-loop bandwidth* is the frequency range over which the closed-loop gain falls 3 dB below its value at $\omega=\omega_n$.

[5]The gain characteristic peaks when the forcing frequency equals the natural frequency. It is also said that this peak corresponds to the resonant frequency. Strictly speaking, this is true, although the gain may not actually be resonant (i.e., be infinite).

(The term bandwidth often means closed-loop bandwidth.) The *quality factor*, Q, is

$$Q = \frac{\omega_n}{\text{BW}} \quad 41.37$$

Figure 41.8 Bandwidth

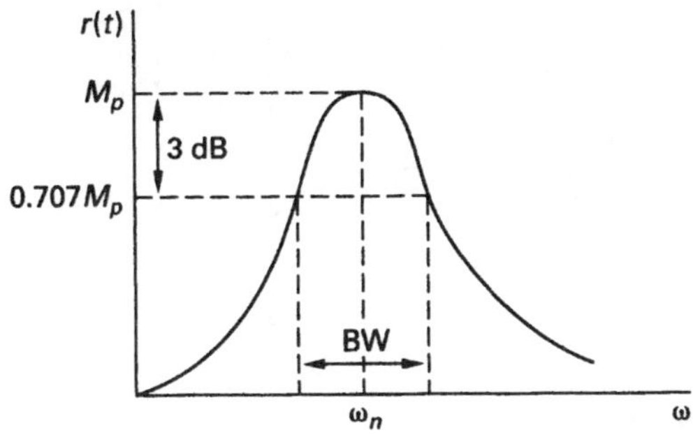

Because a low or negative gain (compared to larger parts of the curve) effectively represents attenuation, the gain characteristic can be used to distinguish between low- and high-pass filters. A low-pass filter has a large gain at low frequencies and a small gain at high frequencies. Conversely, a high-pass filter has a high gain at high frequencies and a low gain at low frequencies.

It may be possible to determine certain parameters (e.g., the natural frequency and BW) from the transfer function directly. For example, when the denominator of $T(s)$ is a single linear term of the form $s+r$, the BW will be equal to r. The BW and time constant are reciprocals.

Another important case is that in which $T(s)$ has the form of Eq. 41.38. (Compare the form of $T(s)$ to the form of Eq. 41.11.) The coefficient of the s^2 term must be unity, and ω_n must be much larger than BW so that the pole is close to the imaginary axis. The zero defined by constants a and b in the numerator is not significant.

$$T(s) = \frac{as+b}{s^2+(\text{BW})s+\omega_n^2} \quad 41.38$$

Example 41.14

What are the maximum gain, BW, upper half-power frequency, and half-power gain of a system whose transfer function is

$$T(s) = \frac{1}{s+5}$$

Solution

The steady-state response to a sinusoidal input is determined by substituting $j\omega$ for s in $T(s)$. When $\omega=0$, $T(s)$ will have a value of $1/5=0.2$, and $T(s)$ decreases thereafter. The maximum gain is therefore 0.2. The BW is 5 rad/s. Because the lower half-power frequency is implicitly 0 rad/s, the upper half-power frequency is 5 rad/s, at which point the system gain will be $(0.707)(0.2)=0.141$.

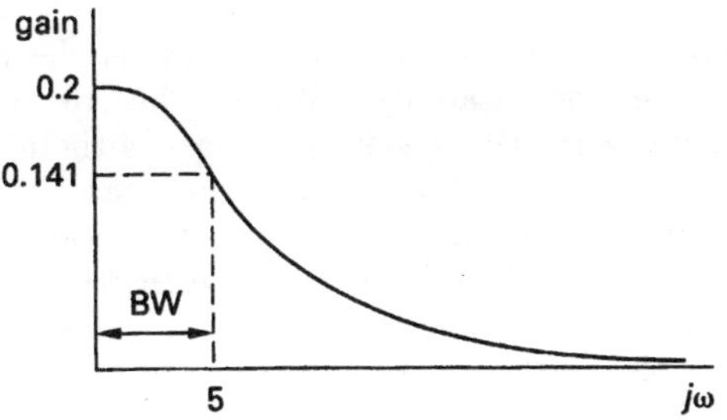

Example 41.15

Predict the natural frequency and BW for the following transfer function.

$$T(s) = \frac{s+19}{s^2+7s+1000}$$

Solution

This equation has the same form as Eq. 41.38. The BW is 7, and the natural frequency is $\sqrt{1000} = 31.62$ rad/s.

15. PHASE CHARACTERISTIC

The phase angle response will also change as the forcing frequency is varied. The *phase characteristic* (*α curve*) is a plot of the phase angle as ω_f is varied.

16. STABILITY

A stable system will remain at rest unless disturbed by external influence and will return to a rest position once the disturbance is removed. A pole with a value of $-r$ on the real axis corresponds to an exponential response of e^{-rt}. Because e^{-rt} is a decaying signal, the system is stable. Similarly, a pole of $+r$ on the real axis corresponds to an exponential response of e^{rt}. Because e^{rt} increases without limit, the system is unstable.

Because any pole to the right of the imaginary axis corresponds to a positive exponential, a *stable system* will have poles only in the left half of the s-plane. If there is an isolated pole on the imaginary axis, the response is stable. However, a conjugate pole pair on the imaginary axis corresponds to a sinusoid that does not decay with time. Such a system is considered to be unstable.

Passive systems (i.e., the homogeneous case) are not acted upon by a forcing function and are always stable. In the absence of an energy source, exponential growth cannot occur. *Active systems* contain one or more energy sources and may be stable or unstable.

There are several *frequency response (domain) analysis* techniques for determining the stability of a system, including Bode plots, root-locus diagrams, Routh stability criteria, Hurwitz tests, and Nichols charts. The term *frequency response* almost always means the steady-state response to a sinusoidal input.

The value of the denominator of $T(s)$ is the primary factor affecting stability. When the denominator approaches zero, the system increases without bound. In the typical feedback loop, the denominator is $1 \pm GH$, which can be zero only if $|GH| = 1$. It is logical, then, that most of the methods for investigating stability (e.g., Bode plots, root-locus diagrams, Nyquist analyses, and Nichols charts) investigate the value of the open-loop transfer function, GH. Because $\log 1 = 0$, the requirement for stability is that $\log GH$ must not equal 0 dB.

A negative feedback system will also become unstable if it changes to a positive feedback system, which can occur when the feedback signal is changed in phase more than 180°. Therefore, another requirement for stability is that the phase angle change must not exceed 180°.

17. BODE PLOTS

Bode plots are gain and phase characteristics for the open-loop $G(s)H(s)$ transfer function that are used to determine the *relative stability* of a system. The gain characteristic is a plot of $20 \log |G(s)H(s)|$ versus ω for a sinusoidal input. It is important to recognize that Bode plots, though similar in appearance to gain and phase frequency response charts, are used to evaluate stability and do not describe the closed-loop system response.

The *gain margin* is the number of decibels that the open-loop transfer function, $G(s)H(s)$, is below 0 dB at the *phase crossover frequency* (i.e., where the phase angle is $-180°$). (If the gain happens to be plotted on a linear scale, the gain margin is the reciprocal of the gain at the phase crossover point.) The gain margin must be positive for a stable system, and the larger it is, the more stable the system will be.

The *phase margin* is the number of degrees the phase angle is above $-180°$ at the *gain crossover point* (i.e., where the logarithmic gain is 0 dB or actual gain is 1). (See Fig. 41.9.)

In most cases, large positive gain and phase margins will ensure a stable system. However, the margins could have been measured at other than the crossover frequencies. Therefore, a Nyquist stability plot is needed to verify the absolute stability of a system.

18. ROOT-LOCUS DIAGRAMS

A *root-locus diagram* is a pole-zero diagram showing how the poles of $G(s)H(s)$ move when one of the system parameters (e.g., the gain factor) in the transfer function is varied. The diagram gets its name from the need to find the roots of the denominator (i.e., the poles). The locus of points defined by the various poles is a line or curve that can be used to predict *points of instability* or other critical operating points. A point of instability is reached when the line crosses the imaginary axis into the right-hand side of the pole-zero diagram.

Figure 41.9 Gain and Phase Margin Bode Plots

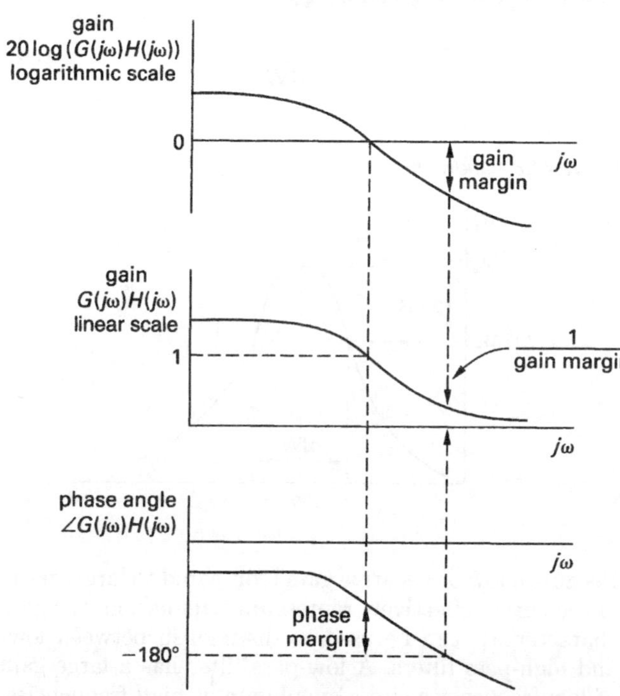

A root-locus curve may not be contiguous, and multiple curves will exist for different sets of roots. Sometimes the curve splits into two branches. In other cases, the curve leaves the real axis at *breakaway points* and continues on with constant or varying slopes approaching asymptotes. One branch of the curve will start at each open-loop pole and end at an open-loop zero.

Example 41.16

Draw the root-locus diagram for a feedback system with open-loop transfer function $G(s)H(s)$. K is a scalar constant that can be varied.

$$G(s)H(s) = \frac{Ks(s+1)(s+2)}{s(s+2) + K(s+1)}$$

Solution

The poles are the zeros of the denominator.

$$s_1, s_2 = -\tfrac{1}{2}(2+K) \pm \sqrt{1 + \tfrac{1}{4}K^2}$$

Because the second term can be either added or subtracted, there are two roots for each value of K.

System Analysis

Allowing K to vary from zero to infinity produces a root-locus diagram with two distinct branches. The first branch extends from the pole at the origin to the zero at $s=-1$. The second branch extends from the pole at $s=-2$ to $-\infty$. All poles and zeros are not shown. Because neither branch crosses into the right half, the system is stable for all values of K.

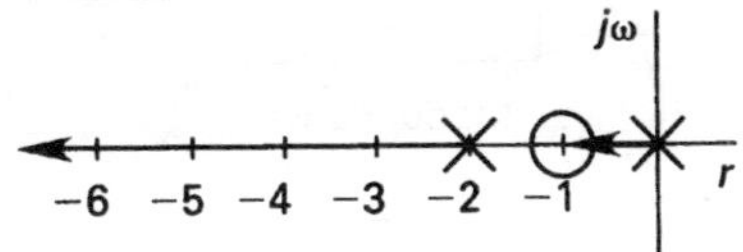

19. HURWITZ TEST

A stable system has poles only in the left half of the s-plane. These poles correspond to roots of the *characteristic equation.* The characteristic equation can be expanded into a polynomial of the form

$$a_0s^n + a_1s^{n-1} + \cdots + a_{n-1}s + a_n = 0 \quad 41.39$$

The *Hurwitz stability criterion* requires that all coefficients be present and be of the same sign (which is equivalent to requiring all coefficients to be positive); if the coefficients differ in sign, the system is unstable. If the coefficients are all alike in sign, the system may or may not be stable. The Routh test should be used in that case.

20. ROUTH CRITERION

The *Routh criterion,* like the Hurwitz test, uses the coefficients of the polynomial characteristic equation. A table (the *Routh table*) of these coefficients is formed. The Routh-Hurwitz criterion states that the number of sign changes in the first column of the table equals the number of positive (unstable) roots. Therefore, the system will be stable if all entries in the first column have the same sign.

The table is organized in the following manner.

a_0	a_2	a_4	a_6	$\cdots$
a_1	a_3	a_5	a_7	$\cdots$
b_1	b_2	b_3	b_4	$\cdots$
c_1	c_2	c_3	c_4	$\cdots$
$\vdots$	$\vdots$	$\vdots$	$\vdots$	

The remaining coefficients are calculated in the following pattern until all values are zero.

$$b_1 = \frac{a_1a_2 - a_0a_3}{a_1} \quad 41.40$$

$$b_2 = \frac{a_1a_4 - a_0a_5}{a_1} \quad 41.41$$

$$b_3 = \frac{a_1a_6 - a_0a_7}{a_1} \quad 41.42$$

$$c_1 = \frac{b_1a_3 - a_1b_2}{b_1} \quad 41.43$$

Special methods are used if there is a zero in the first column but nowhere else in that row. One of the methods is to substitute a small number, represented by ϵ or δ, for the zero and calculate the remaining coefficients as usual.

Example 41.17

Evaluate the stability of a system that has a characteristic equation of

$$s^3 + 5s^2 + 6s + C = 0$$

Solution

All of the polynomial terms are present (Hurwitz criterion), so the Routh table is

s^3	1	6
s^2	5	C
s^1	$\frac{30-C}{5}$	0
s^0	C	

To be stable, all of the entries in the first column must be positive, which requires $0 < C < 30$.

21. NYQUIST ANALYSIS

Nyquist analysis is another graphical method that is particularly useful when time delays are present in a system or when frequency response data are available. *Nyquist's stability criterion* is $N = P - Z$. P is the number of poles in the right half of the s-plane. Z is the number of zeros in the right half of the s-plane, and N is the number of encirclements (revolutions) of $1 + G(s)H(s)$ around the critical point. N may be positive, negative, or zero.

22. APPLICATION TO CONTROL SYSTEMS

A control system monitors a process and makes adjustments to maintain performance within certain acceptable limits. Feedback (see Fig. 41.10) is implicitly a part of all control systems.[6] The *controller* (*control element*) is the part of the control system that establishes the acceptable limits of performance, usually by

[6]Not all controlled systems are feedback systems. The positions of many precision devices (e.g., print heads in dot matrix printers and cutting heads on some numerically controlled machines) are controlled by precision *stepped motors.* However, unless the device has feedback (e.g., a position sensor), it will have no way of knowing if it gets out of control.

setting its own reference inputs. The controller transfer function for a proportional controller is a constant: $G_1(s) = K$.[7] The *plant (controlled system)* is the part of the system that responds to the controller. Both of these are in the forward loop. The input signal, V_i, in Fig. 41.4, is known in a control system as the *command* or *reference value.*

Figure 41.10 Typical Feedback Control System

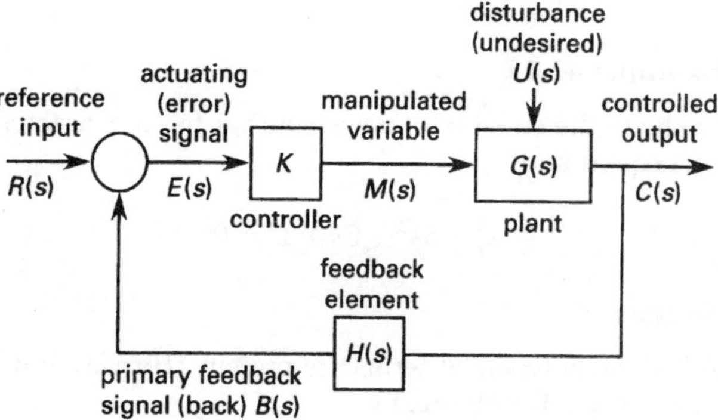

A *servomechanism* is a special type of control system in which the controlled variable is mechanical position, velocity, or acceleration. In many servomechanisms, $H(s) = 1$ (i.e., unity feedback) and it is desired to keep the output equal to the reference input (i.e., maintain a zero error function). If the input, $R(s)$, is constant, the descriptive terms *regulator* and *regulating system* are used.

The output of any given control system may not be in the assigned parameter or metric range. A *compensator*, which is a circuit designed to correct the discrepant parameter (e.g., the phase), may be added in the appropriate location of the control system to provide the necessary correction.

23. CONTROL SYSTEM TYPES/MODES

The method of controlling the desired signal depends on the level and type of desired control. The type of control action used is called the *mode of control.*

The simplest type of controller, and likely the most familiar, is the *on-off* or *bang-bang controller*. As the name implies, the system controlled is turned on when a signal goes above a given setpoint and is turned off when the signal goes below the setpoint. An example is the thermostat controller in most households. Controllers can also be made to respond to the rate that the input parameter or signal is changing. Such controllers are shown in Fig. 41.11.

[7]Sometimes the notation K_n is used for K where n is the type of the system. In a type 0 system, a constant error signal results in a constant value of the output signal. In a type 1 system, a constant error signal results in a constant rate of change of the output signal. In a type 2 system, a constant error signal produces a constant second derivative of the output variable.

Figure 41.11 Control System Types

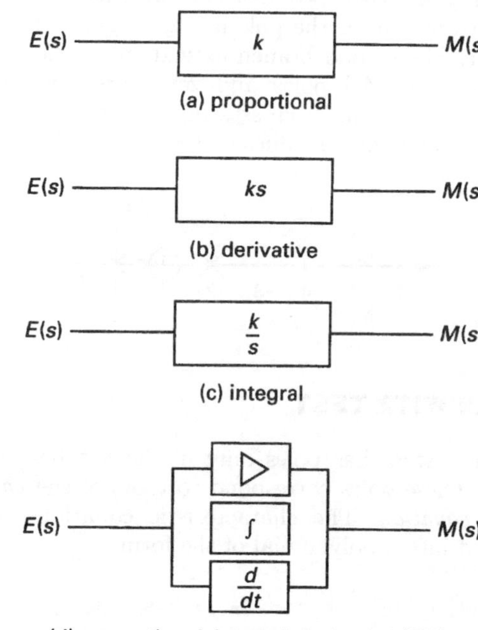

A *proportional controller* is shown in Fig. 41.11(a). Such a controller adjusts the manipulated variable so that the controlled output matches the input demand. In the case of the household thermostat so designed, the heat input to the house from the furnace would match demand, maintaining a stable temperature.

A *derivative controller* is shown in Fig. 41.11(b). As the name implies, the rate of change is the variable controlled. The advantage is the ability to respond to a rapidly changing parameter. The *integral controller* is shown in Fig. 41.11(c). The advantage is the ability to respond in a controlled fashion to rapid changes in demand. The disadvantage is a large overshoot if there is a one-time jump in demand.

Figure 41.11(d) shows a *proportional-integral-derivative (PID) controller*. The advantage of this controller is its ability to respond to complex demand signals without overshoot of the controlled parameter.

24. STATE MODEL REPRESENTATION

While the classical methods of designing and analyzing control systems are adequate for most situations, state model representations are preferred for more complex cases, particularly those with multiple inputs and outputs or when behavior is nonlinear or varies with time. This method of evaluation is almost always carried out on a digital or analog computer.

The state variables completely define the dynamic state, $x_i(t)$ (i.e., position, voltage, pressure), of the system at time t. In simple problems, the number of state variables corresponds to the number of *degrees of freedom*, n, of the system. The n state variables are written in matrix form as a state vector, **X**

$$\mathbf{X} = \begin{pmatrix} x_1 \\ x_2 \\ x_3 \\ \vdots \\ x_n \end{pmatrix} \quad \textit{41.44}$$

It is a characteristic of state models that the state vector is acted upon by a first-degree derivative operator, d/dt, to produce a differential term of order 1, $\mathbf{X}'$.

$$\mathbf{X}' = \frac{d\mathbf{X}}{dt} \quad \textit{41.45}$$

Equation 41.46 and Eq. 41.47 show the general form of a state model representation: **U** is an r-dimensional (i.e., an $r \times 1$ matrix) control vector; **Y** is an m-dimensional (i.e., an $m \times 1$ matrix) output vector; **A** is an $n \times n$ system matrix; **B** is an $n \times r$ control matrix; and **C** is an $m \times n$ output matrix. The actual unknowns are the x_i's. The y_i's, which may not be needed in all problems, are only linear combinations of the x_i's. (For example, the x's might represent spring end positions; the y's might represent stresses in the spring. Then, $y = k\Delta x$.) Equation 41.46 is the *state equation*, and Eq. 41.47 is the *response equation*.

$$\mathbf{X}' = \mathbf{AX} + \mathbf{BU} \quad \textit{41.46}$$

$$\mathbf{Y} = \mathbf{CX} \quad \textit{41.47}$$

A conventional block diagram can be modified to show the multiplicity of signals in a state model, as shown in Fig. 41.12.[8] The actual physical system does not need to be a feedback system. The form of Eq. 41.46 and Eq. 41.47 is the sole reason that a feedback diagram is appropriate.

Figure 41.12 *State Variable Diagram*

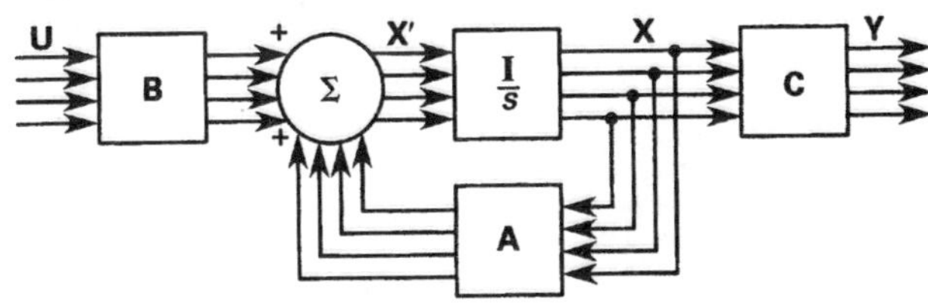

A state variable model permits only first-degree derivatives, so additional x_i state variables are used for higher-order terms (e.g., acceleration).

System controllability exists if all of the system states can be controlled by the inputs, **U**. In state model language, system controllability means that an arbitrary initial state can be steered to an arbitrary target state in a finite amount of time. *System observability* exists if the initial system states can be predicted by observing the outputs, **Y**, when the inputs, **U**, are known.[9]

Example 41.18

Write the state variable formulation of the following mechanical system's transfer function $T(s)$.

$$T(s) = \frac{7s^2 + 3s + 1}{s^3 + 4s^2 + 6s + 2}$$

Solution

Recognize the transfer function as the quotient of two terms, and multiply $T(s)$ by the dimensionless quantity x/x.

$$T(s) = \frac{Y(s)}{U(s)} = \left(\frac{7s^2 + 3s + 1}{s^3 + 4s^2 + 6s + 2}\right)\left(\frac{x}{x}\right)$$

$$Y(s) = 7s^2x + 3sx + x$$

$$U(s) = s^3x + 4s^2x + 6sx + 2x$$

$Y(s)$ and $U(s)$ represent the following differential equations.

$$y(t) = 7x''(t) + 3x'(t) + x(t)$$

$$u(t) = x'''(t) + 4x''(t) + 6x'(t) + 2x(t)$$

Make the following substitutions. ($x_4(t)$ is not needed because one level of differentiation is built into the state model.)

$$x_1(t) = x(t)$$

$$x_2(t) = x'(t) = x_1'(t)$$

$$x_3(t) = x''(t) = x_2'(t)$$

Write the first derivative variables in terms of the $x_i(t)$ to get the **A** matrix entries.

$$x_1'(t) = \ \ 0x_1(t) + 1x_2(t) + 0x_3(t) + 0$$

$$x_2'(t) = \ \ 0x_1(t) + 0x_2(t) + 1x_3(t) + 0$$

$$x_3'(t) = -2x_1(t) - 6x_2(t) - 4x_3(t) + u(t)$$

Determine the coefficients of the **C** matrix by rewriting $y(t)$ in the same variable order.

$$y(t) = 1x_1(t) + 3x_2(t) + 7x_3(t)$$

[8]The block $\mathbf{I}/s$ is a diagonal identity matrix with elements of $1/s$. This effectively is an integration operator.

[9]*Kalman's theorem*, based on matrix rank, is used to determine system controllability and observability.

System Analysis

From Eq. 41.46 and Eq. 41.47, the state variable representation of $T(s)$ is

$$\begin{bmatrix} x_1' \\ x_2' \\ x_3' \end{bmatrix} = \begin{bmatrix} 0 & 1 & 0 \\ 0 & 0 & 1 \\ -2 & -6 & -4 \end{bmatrix} \begin{bmatrix} x_1 \\ x_2 \\ x_3 \end{bmatrix} + \begin{bmatrix} 0 \\ 0 \\ 1 \end{bmatrix} [u(t)]$$

$$[y(t)] = [1\ 3\ 7] \begin{bmatrix} x_1 \\ x_2 \\ x_3 \end{bmatrix}$$

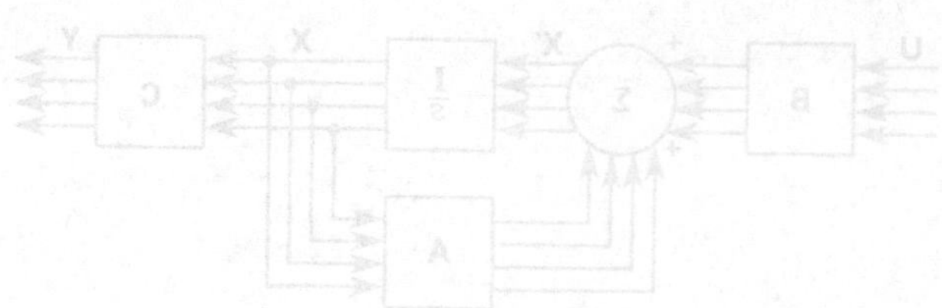

Topic VIII: Protection and Safety

Chapter

42 Protection and Safety

Nomenclature

C	capacitance	F
E	voltage	V
I	current	A
pf	power factor	–
R	resistance	Ω
V	constant or rms voltage	V
X	reactance	Ω
Z	impedance	Ω

Symbols

θ	torque angle during normal conditions	degrees
τ	time delay	s
τ	torque	N·m
ϕ	impedance angle or torque angle during referenced condition	degrees

Subscripts

adj	adjustable
f	fault
p	phase or pickup
r	ratio or relay
ref	reference

1. POWER SYSTEM STRUCTURE

The protection and safety of power systems are determined by the structures of the systems. Most power systems consist of the three elements shown in Fig. 42.1. The *control system* uses equipment to maintain a nominal voltage and frequency, as well as maintain the power system's economic operation and security in interconnected networks. The control system operates continuously on the power system to adjust operating parameters. The *protection system* operates circuit breakers and other devices, such as relays, that change the structure of the power system to protect it from abnormal situations, such as electrical faults. The protection system operates when required and is faster than the control system. The advance of computer-based electronic protection systems blurs the line between protection and control, incorporating both elements into one system; however, the functions remain the same.

Figure 42.1 *Power System Structure*

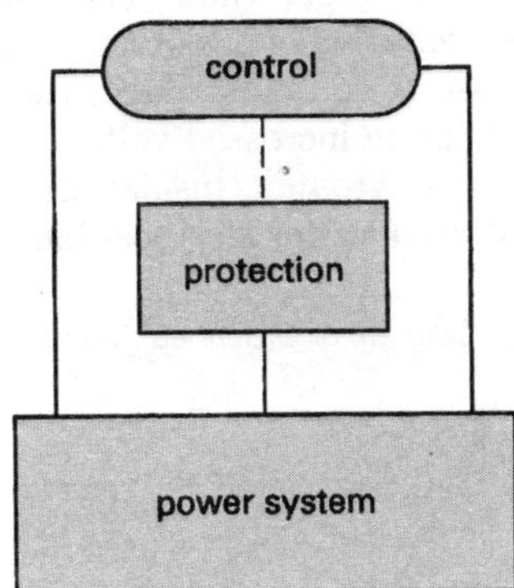

2. POWER SYSTEM GROUNDING

A power system's *grounding scheme* influences the fault current levels. As a result, the grounding scheme impacts the protection system *relaying*. (See Sec. 42.5.) In a true ungrounded system, there is zero fault current. In a real system, capacitive coupling of the feeders to ground results in a ground path if a fault occurs. The ungrounded system with a ground fault is shown in Fig. 42.2.

If the capacitance is large enough, the capacitive ground fault becomes self-sustaining and must be interrupted by circuit breakers. *Protective relays* (devices used to make or break one or more connections in an electric circuit) must sense the low level of ground-fault current before a ground fault can damage the power system. To do so,

Protection and Safety

impedance is inserted in the dashed box, as illustrated in Fig. 42.2. The impedance ensures adequate current flow that can then be sensed by the protective system, which can then react before the fault causes excessive damage.

Figure 42.2 Ungrounded System with Ground Fault

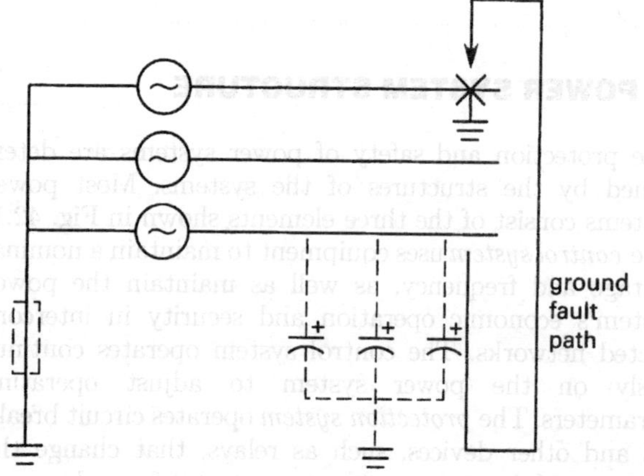

Ungrounded systems provide for high system reliability because one ground does not cause protective systems to react. A second ground fault, or the capacitive ground fault, is required to cause breakers to open and equipment to be removed from service. A disadvantage of ungrounded systems is that when a ground fault occurs on one phase, as shown in Fig. 42.2, the other two phase voltages become $\sqrt{3}$ larger than their normal values. The phasor representation of a grounded phase is shown in Fig. 42.3. Phase A voltage changes the reference on the ground resulting in increased voltage differences on phases B and C to ground. Insulation levels on the conductors must account for this possible change.

Figure 42.3 Phasor Diagram of Grounded Phase

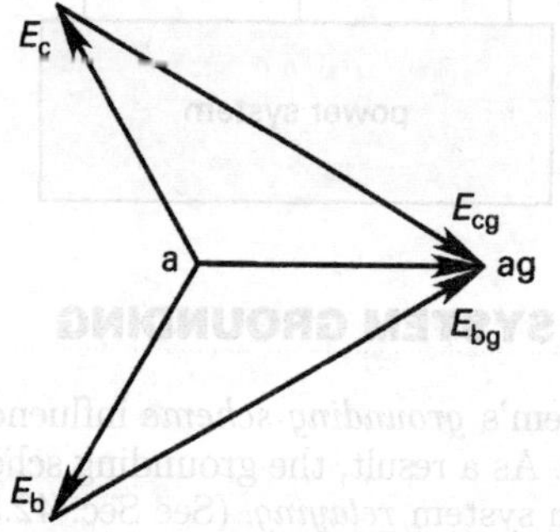

On low voltage systems, the insulation level is usually set based on lightning-induced phenomena, which are generally larger than the fault-induced voltage increase. On high voltage systems (e.g., those with voltages greater than 100 kV), fault-induced voltages are more critical. As a result, on high voltage systems, a solidly grounded neutral is used. Such grounded systems are called *effectively grounded*. These systems have high fault currents and the impedance in the dashed box shown in Fig. 42.2 is set to limit the fault current to a value that circuit breakers can safely interrupt.[1]

The system in Fig. 42.2 is called an ungrounded system even though the wye connection may be grounded to allow detection of faults. (The conductors in the distribution system are *ungrounded*.) Such a ground is called a *functional earth*. A functional earth is defined as a ground that serves a purpose other than protecting against electrical shock. The conductors supply various loads that are usually encased in metal frames. These frames must be grounded for the safety of personnel.

The ground used to provide this safety is called a *protective earth* (PE).[2]

3. POWER SYSTEM CONFIGURATIONS

The configuration of buses making up a power system will determine the available protection system options. A *radial system* is a serially connected system with a single source, as shown in Fig. 42.4. Such a system is normally limited to distribution systems (generally operating at voltages of 100 kV or less). It has the advantage of being economical to build and the disadvantage of being less reliable since a problem at any point in the network affects all downstream loads. Radial systems are also easier to protect because a fault current can only flow from the source to the fault, making it less complex to calculate. Because the system is usually remote from the source, the fault current is approximately independent of the generating capacity.

Figure 42.4 Radial Bus Configuration

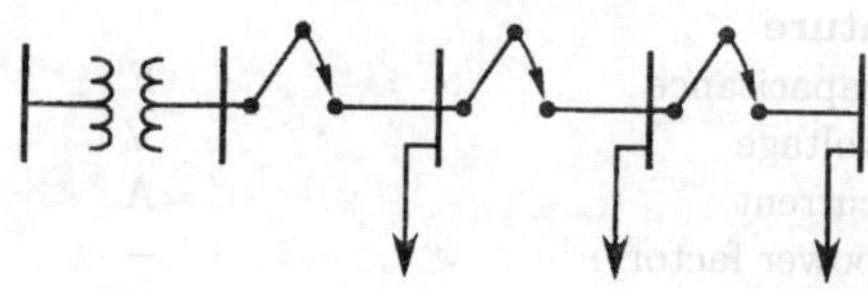

A *network system*, as shown in Fig. 42.5, is normally used for transmission and sub-transmission systems (generally operating at voltages of 100–200 kV or higher). It has increased the reliability of power delivery to loads, and the loss of a single source of power has minimal impact on reliability. However, network systems are more complicated to protect than radial systems. A fault current can occur from multiple sources and from multiple directions. Because the source may be close to the fault, the fault current may vary widely depending on location.

[1]In lower voltage networks, an alternative to ungrounded system is the *ground fault neutralizer* (GFN) known as the *Petersen coil*.
[2]International standards call an ungrounded system an IT system. The "I" means isolation and the "T" is for terra (or Earth). When the neutral is effectively grounded along with the protective earth, the system is called a *TN system*. The "T" means terra and the "N" means neutral. When the neutral is grounded, but is physically separate from the protective earth ground, the system is called a TT system, for terra-terra.

Figure 42.5 Network Bus Configuration

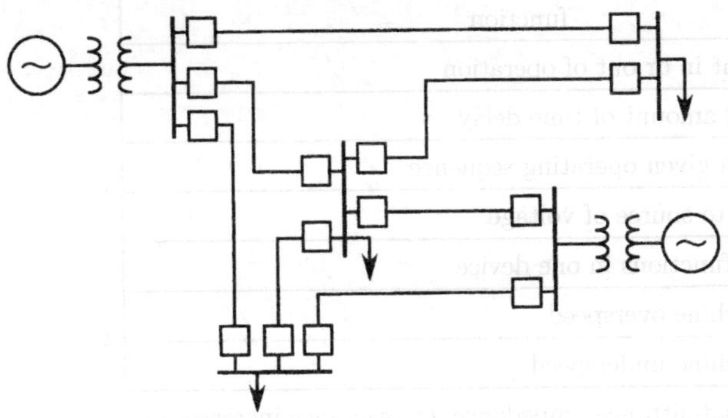

Each *substation* within a power system is named for its bus and breaker setup. Figure 42.6 shows a *single bus, single breaker substation configuration.*[3] This configuration is simple and cost effective because it uses fewer parts than other configurations. It is also the least flexible configuration and cannot undergo maintenance without de-energizing the bus and associated incoming transmission line.

Figure 42.6 Single Bus, Single Breaker Substation Configuration

A *two bus, single breaker substation configuration* is shown in Fig. 42.7. Bus 1 and bus 2 represent two separate energized circuits. This configuration allows maintenance of individual breakers without de-energizing either bus. It also lets loads be split between the buses, which increases reliability.

Figure 42.7 Two Bus, Single Breaker Substation Configuration

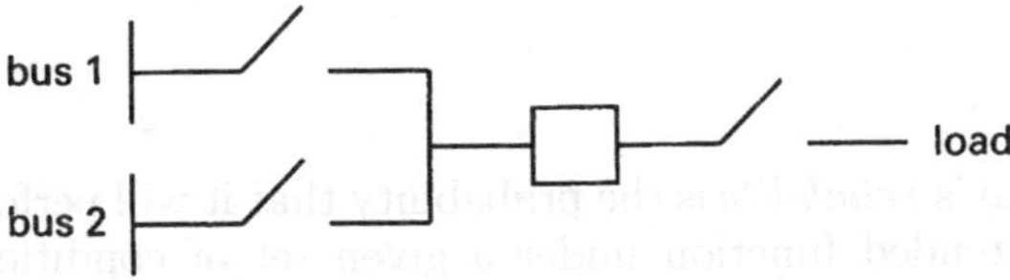

A *two bus, two breaker configuration* is shown in Fig. 42.8. This configuration allows a bus or breaker to be removed for maintenance while maintaining power to the load. A bus fault should trip only one breaker, so that the remaining breaker can maintain power to the load. A load fault will cause both breakers to trip, which is not necessarily ideal for safety; however, this is the most flexible arrangement from the standpoint of operation and maintenance.

Figure 42.8 Two Bus, Two Breaker Substation Configuration

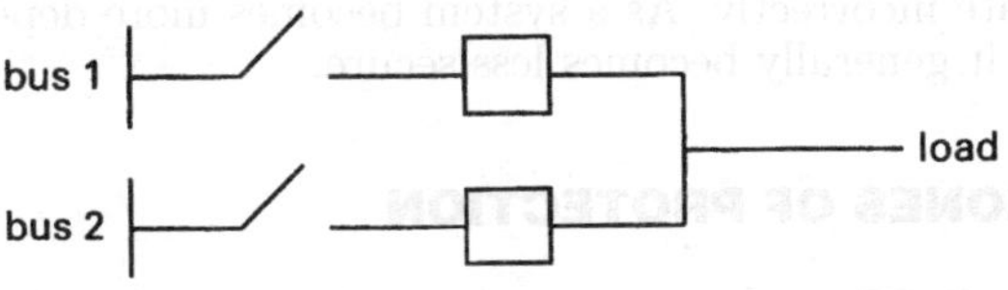

[3]The IEEE standard practice is to label a breaker with a "52" within its box.

A *breaker-and-a-half substation configuration* is shown in Fig. 42.9. This configuration has the advantages of the two bus, two breaker system, but with one-and-a-half breakers per load instead of two. With fewer breakers needed, this configuration is more economical, and it allows orderly expansion of the substation. As a result, this is the most common configuration for extra high voltage substations.

Figure 42.9 Breaker-and-a-Half Substation Configuration

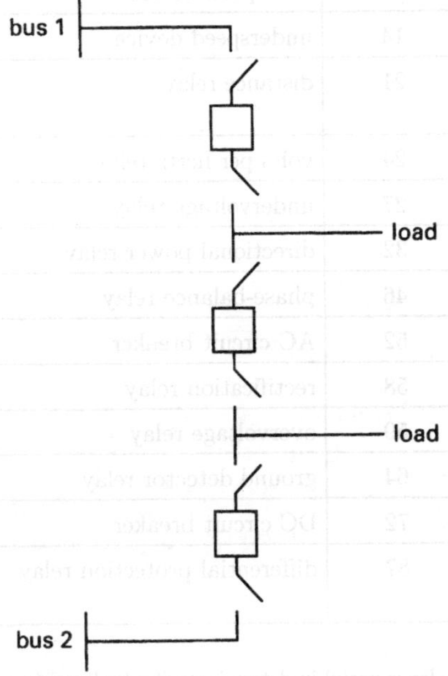

Figure 42.10 shows a *ring bus configuration.* This system is highly flexible as long as the ring is maintained so that all the breakers are working. When a breaker is out of service for any reason, the system flexibility is significantly degraded.

Figure 42.10 Ring Bus Substation Configuration

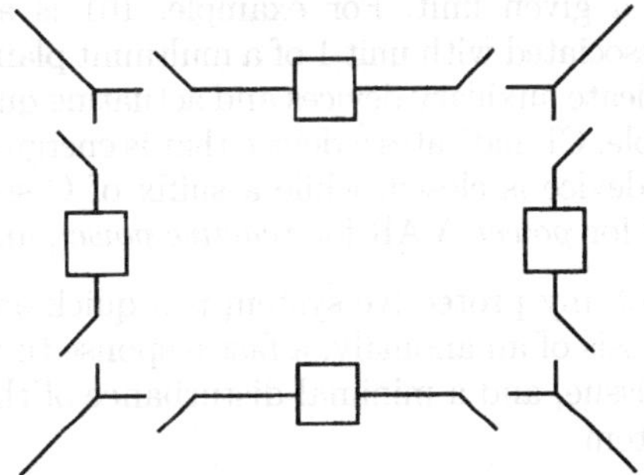

4. RELAYS

Relays are designed to determine electrical circuit operating systems and to trip circuit breaks when a fault is detected. They can be electromagnetic or electronic. The various protective functions available on a relay are given by standardized numbers in ANSI/IEEE Standard C37.2, examples of which are given in Table 42.1. Prefixes to the numbers indicate its location. For example, RE

Protection and Safety

Table 42.1 ANSI/IEEE Power System Device Function Numbers

number	title	function
1	master element	places equipment in or out of operation
2	time delay or closing relay	provides desired amount of time delay
3	interlocking relay	allows or stops a given operating sequence
6	starting circuit breaker	connects device to source of voltage
11	multifunction device	combination of functions in one device
12	overspeed device	operates on machine overspeed
14	underspeed device	operates on machine underspeed
21	distance relay	functions when admittance, impedance, or reactance increases or decreases beyond a certain value[a]
24	volts per hertz relay	operates on a given V/Hz ratio[b]
27	undervoltage relay	operates on machine undervoltage
32	directional power relay	operates on power flow in a given direction[c]
46	phase-balance relay	operates when negative phase sequence exceeds a given value[d]
52	AC circuit breaker	operates a circuit breaker
58	rectification relay	operates when a rectifier fails to conduct or block correctly
59	overvoltage relay	operates on overvoltage condition
64	ground detector relay	operates upon failure of insulation to ground
72	DC circuit breaker	operates a circuit breaker
87	differential protection relay	operates upon a quantitative difference between two or more currents or other electrical quantities

[a]A *distance relay* is useful in determining if a fault exists in a given distribution system.
[b]In variable speed devices, the volts per hertz (V/Hz) ratio is adjusted to maintain the current to a machine constant. A constant torque results in a constant speed.
[c]Reverse power on a generator can damage the machine. This relay detects the angular relationship between the current and voltage, which determines the direction of power flow.
[d]This relays checks for negative phase-sequence currents above a given value. Negative phase-sequence currents indicate unbalance phases meaning that a fault exists in the system.

stands for *remote*. Numerical prefixes indicate an association with a given unit. For example, 101 is a master element associated with unit 1 of a multiunit plant. Suffix letters indicate auxiliary devices and actuating quantities. For example, CL indicates a device that is energized when the main device is closed, while a suffix of C stands for *current*, P for *power*, VAR for *reactive power*, and so on.

The goal of any protective system is a quick and accurate diagnosis of an anomaly, a fast response time to the indicated issue, and a minimal disturbance of the entire power system.

5. RELAY RELIABILITY

The purpose of *relaying* is to remove any element that is functioning abnormally from a given power system. In general, the protective system does not prevent equipment damage. For the most part, relays operate after some damage has already occurred. Their purpose is to prevent further damage, minimize danger to personnel, and stabilize the system.

A relay's *reliability* is the probability that it will perform its intended function under a given set of conditions. The conditions can be environmental or electrical. Additionally, the assumption is made that maintenance is accomplished as specified.

A relay can be unreliable in two manners. First, a relay can fail to operate when intended. Secondly, a relay can operate when not expected. Therefore, the definition of protective systems relays must be expanded to include the requirement that they are also dependable and secure. A relay is *dependable* if there is a measure of certainty that it will operate for designed faults. A relay is *secure* if there is a measure of certainty that it will not operate incorrectly. As a system becomes more dependable, it generally becomes less secure.

6. ZONES OF PROTECTION

The reliability requirement that relays are dependable (i.e., they operate only as intended) is further defined by the zones in which they operate. A *zone* is a region of the power system and can be closed or open. A *closed zone*, also known as a *differential*, *unit*, or *absolutely selective zone* is one where all power apparatuses entering the zone are monitored at the entry points. In an *open zone*, also known as an *unrestricted*, *non-unit*, or *relatively selective zone*, monitoring does not occur, and the zone is defined by level of the fault current.

Current transformers (CTs) whose input is utilized by the relays monitoring the zones are the proximate boundaries of the zones.[4] An example of zones is shown in Fig. 42.11.

Figure 42.11 *Closed and Open Zones*

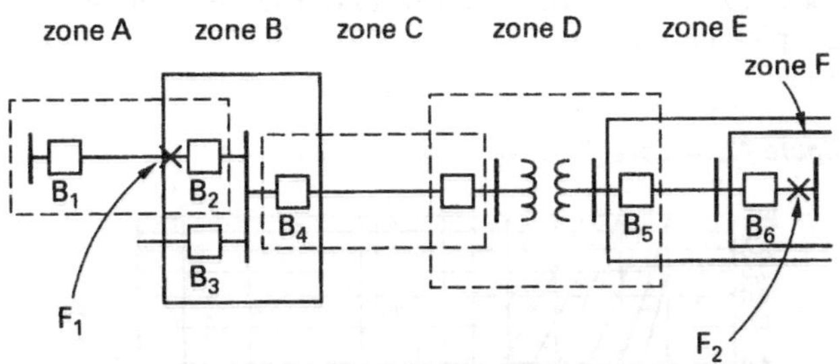

Coordination in protective systems is designed to meet the general goal of having the breaker, or other protective device, closest to the fault open first, and the breaker farthest from the fault open last. In Fig. 42.11, zones A, B, C, and D are closed zones. Zones E and F are open zones. Fault 1 occurs in two zones: A and B. As a result, breakers 1 through 4 will open, even though breakers 3 and 4 do not isolate the fault. Fault 2 occurs in an open zone. Breaker 6 is programmed to open first. If it fails to do so, breaker 5 will open after a specified delay. The setup in Fig. 42.11 represents a form of coordination.

7. RELAY SPEED

A relay senses the operating parameters of a given system. When those parameters or waveforms are distorted due to a fault, the relay must analyze the distortion to determine the relevant information and make a decision, based on internal logic parameters, regarding the fault. The time it takes for the relay to make the decision and the security of that decision is an inverse relationship. This inverse-time operating characteristic applies to all protection relays.

An *instantaneous relay* operates as soon as the logic circuitry makes the decision. No intentional delay is inserted after the decision. A *time-delay relay* has an intentional time delay inserted after the decision is made, such as would be done for backup protection. A *high-speed relay* is defined as one that operates in a specified time, such as 50 ms, or three cycles of a 60 Hz system, as given in various electrical standards. An *ultra high-speed relay* generally indicates a relay that operates in 4 ms or less, or approximately one-quarter of a cycle in a 60 Hz system, though it is not defined by standards.

8. PROTECTION SYSTEM ELEMENTS

A protection system is comprised of three main elements: transducers, relays, and breakers. *Transducers* sense the desired signal and convert it into a form usable by the relays. Current and voltage instrument transformers are examples of transducers. *Relays* monitor the input from the transducers and provide an output to operate breakers in the event of a fault. *Breakers* are designed to interrupt power and isolate sections of the power system. They come in a variety of types and interrupting mediums, such as air, oil, gas, solid dielectric, and SF_6.

When a fault occurs in a system, the AC voltage available may not be adequate to allow protective system operation. To ensure the necessary trips occur, battery backups are used to power protective system components. An *uninterruptable power supply* (UPS) is often used. A UPS ensures normal system power is provided to components unless that power is interrupted or inadequate. When such an interruption occurs, the batteries in the UPS automatically, and with minimal interruption time, provide the necessary backup power.

Two separate battery systems are sometimes used. One system powers electromechanical relays, and the other system powers solid-state relays. Electromechanical relays can produce severe transients on battery leads and potentially impact sensitive solid-state devices, which is why they are separated.

9. TRANSFORMERS

The symbol for a *current transformer* (CT) is shown in Fig. 42.12. Current transformers are used as instrument transformers (also known as *metering transformers*) or as transducers for protection circuits (i.e., relaying transformers).[5] They have low power, and ideally, high accuracy. The accuracy is measured in terms of the transformer's ability to lower the current; in other words, they transform the current without changing the expected magnitude or the phase relationship. In Fig. 42.12, the dots imply that the current leaving the secondary dot and flowing to the load is in phase with the current entering the primary dot.[6]

Figure 42.12 *Current Transformer*

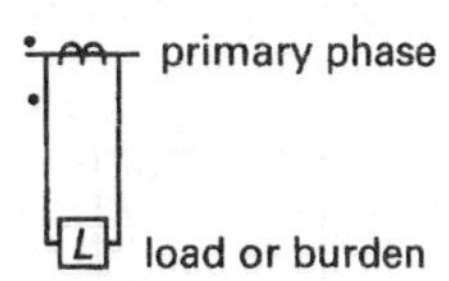

[4]Neighboring zones always overlap to ensure system protection. Faults in an overlap region can disconnect both regions, possibly even more of the system, so overlaps are as small as practical.

[5]Metering transformers are designed to operate most accurately between no-load and full-load. Relaying transformers are designed to operate accurately at expected fault conditions.

[6]In terms of voltage polarity, both dots are positive at the same time.

The *burden* is the amount of power drawn from the circuit connecting the secondary terminals of instrument transformers, usually given as apparent power in voltamperes. The term burden distinguishes power for operation of the device itself from power sensed, which is being monitored and is referred to as the load. As long as the burden is within the rating of the CT, only acceptably small errors are introduced in the phase relationship between the primary and secondary currents, and the current transformation ratio is exact. For example, a CT with a primary current angle of 30° with respect to the voltage and a ratio of 100:1 will transform a 100 A current at 30° degrees to a 1 A current at 30°. When the burden is too high, or when non-ideal conditions exist, a CT will exhibit both *phase angle errors* and *transformation errors*. These errors vary with the loading of the transformer and the power factor.

Voltage transformers are also used for metering or relaying. They have low power and, ideally, high accuracy. Accuracy is measured in terms of the ability to transform the voltage without changing the expected magnitude and phase relationship. The change in voltage output from no-load to full-load is called the *voltage regulation*.[7] The standard connections for transformers are shown in Fig. 42.13(b) and Fig. 42.13(c). The polarities are called *additive* or *subtractive*. Additive and subtractive polarities are a way to avoid the dot convention. Figure 42.13(a) shows a voltage transformer with positive polarity using the dot convention. Figure 42.13(b) shows additive polarity, and Fig. 42.13(c) shows subtractive polarity. The "H" connection represents the primary phase. The "X" connection represents the secondary phase.

Figure 42.13 *Voltage Transformer*

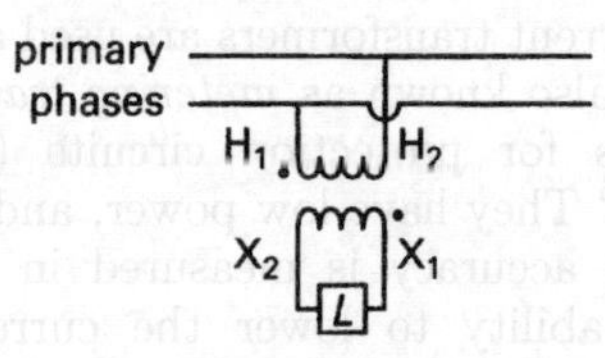

(a) voltage transformer

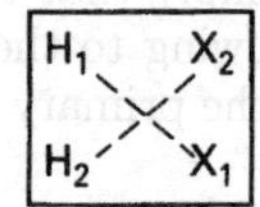

(b) additive polarity

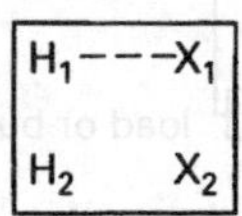

(c) subtractive polarity

[7]The phase change can be eliminated with filtering circuits.

10. RELAY TYPES

Generally when faults occur, currents increase, voltages drop, harmonics develop, and phase relationships shift. Relays, both analog and digital, can be made to sense and respond to these changes. A relay can be comprised of multiple types embedded in a single device. An explanation of some of these types follows.

Level detection relays, also called *magnitude* and *overcurrent relays*, operate when a fault exceeds a given value.[8] The *pickup setting* is the setting at which the relay operates. The properties of a typical level detection relay are shown in Fig. 42.14. The x-axis is a normalized value; that is, a ratio of the fault current to the pickup current.

Figure 42.14 *Level Detector Characteristics*

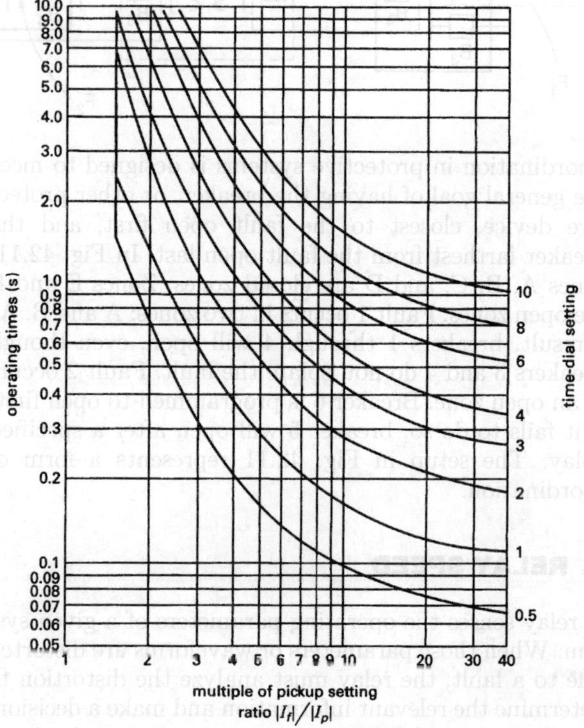

The time dial setting and the pickup setting are selected to give the desired time response in the event of a fault. The time for the relay to function is found by first dividing the fault current by the pickup setting to obtain the multiple of the pickup setting. Using this value, starting at the bottom axis, trace a straight line to the value of the time-dial setting selected (see the right axis). From this point, trace a line to the left axis to obtain the operating time. The characteristic curves are plotted on log-log scale.

[8]A *fuse* is a simple level detector.

Differential comparison relays operate when currents at the input and output of a zone are not identical. This method is very sensitive to changes and is effective in isolating faults. A sample scheme is shown in Fig. 42.15. The disadvantage to this scheme is that currents at each end of the zone can be separated by a significant distance, requiring an increased equipment cost for wiring and imposing possible time delays.

Figure 42.15 *Differential Relay Scheme*

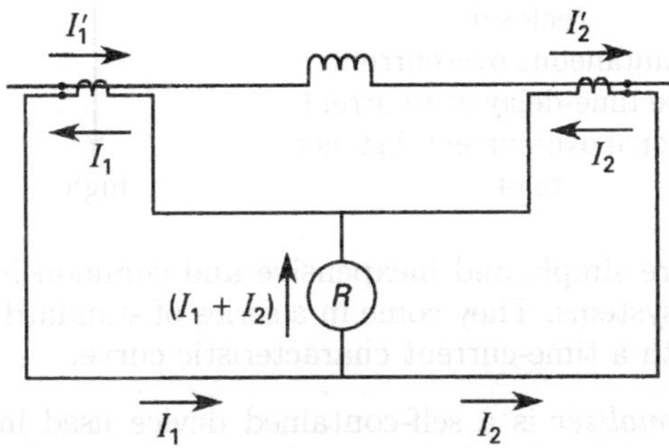

A *phase angle relay* compares the current and voltage to determine, for example, the direction of power flow. When power flow is normal, the impedance angle of the fault circuit, ϕ, is 37° for a load power factor of 0.8. When power flow reverses, the phase angle becomes 180°±37°. If a fault occurs, the phase angle becomes either $-\phi$ or $180° - \phi$. For power transmission networks, the impedance is primarily inductive and ϕ is near 90°. A phase angle relay detecting a 90° impedance angle can react under the assumption that a fault has occurred.

A *distance relay* measures the relationship between the voltage and current at a single point. Assuming standard impedance per unit distance, the relay can be adjusted for any desired distance. Since the distance relay depends on the ratio of the voltage to the current, it is also called a *ratio relay* or *impedance relay*. If the relay is not bidirectional, but rather measures faults in only one direction, it is called a *mho relay.*

Other relay types include the *pilot relay*, which operates with information received from a remote location, and the *harmonic relay*, which operates on the harmonic content of the transmission system. *Frequency relays* operate on changes in frequency.

11. UNIVERSAL RELAY EQUATION

Equation 42.1 is called the *universal relay equation.* Developed from analog relay devices, it can be used with *solid-state relays* or programmed into *computer relays.* The term θ is the torque angle in an analog system during normal conditions and is related directly to the impedance angle. The term ϕ is the torque angle in an analog system at the referenced (potentially, the fault) condition, which is also related to the impedance angle. By manipulating the constants in either an analog or digital protective system, a variety of relays can be realized. Figure 42.16 shows the relay types that can occur.[9]

$$\begin{aligned}\tau &= \tau_{\text{magnetic}} - \tau_{\text{spring}} \\ &= C_1 I^2 + C_2 V^2 + C_3 I V \sin(\theta - \phi) - \tau_{\text{spring}}\end{aligned} \qquad 42.1$$

Figure 42.16 *Universal Relay Equation Characteristics*

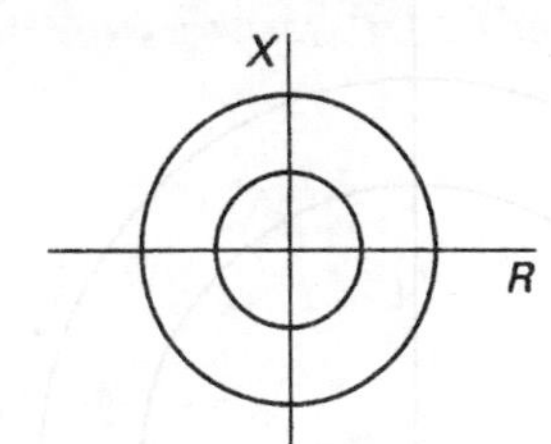

(a) impedance relay

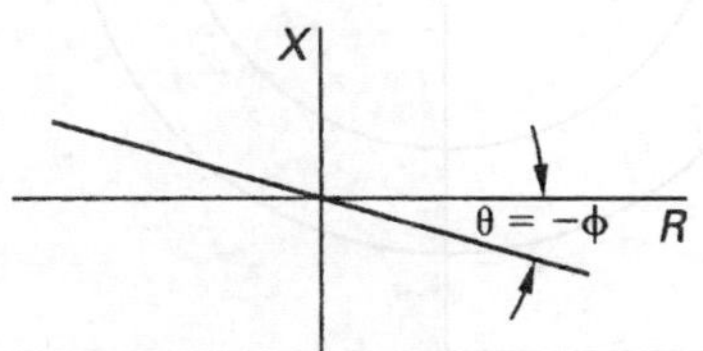

(b) directional relay

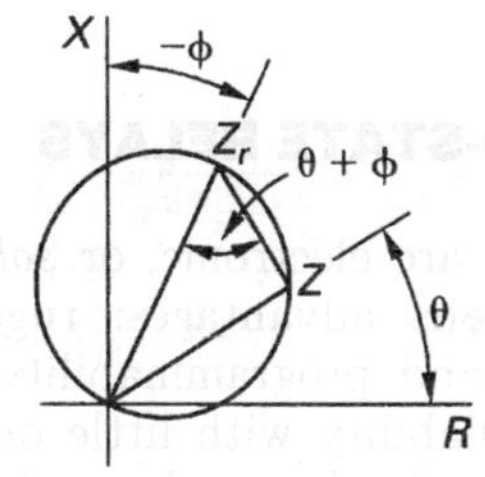

(c) siemens relay

While the *R*-*X* plane used in Fig. 42.16 appears to use negative resistance, this is not the case. A relay can be located in any number of locations within a system. The power flows, under normal conditions, toward the load. If a fault occurs upstream of the relay's location, power flow reverses, and the power will then flow from the load to the fault point. Recall that motors exposed to a fault in the system become generators feeding power to the fault from their inertial energy. Transformers do the same from the inductive energy stored in their magnetic fields.

The various zones of the *R*-*X* plane are shown in Fig. 42.17. Current and voltage in the system can be plotted on the plane to indicate phase relationships.

[9]The term Z_r is the impedance of the transmission line at the maximum response of the relay; that is, the maximum distance from the relay that will result in a response. The angle $(\theta + \phi)$ between a line connecting Z_r and the actual impedance, Z, when less than 90°, means the point lies inside the circle and a trip will result. When the angle is greater than 90°, the point lies outside the circle and the trip will be blocked.

Impedance values plot as points. Changes to the voltage-current relationships can then be analyzed as conditions shift due to faults.

Figure 42.17 R-X Diagram

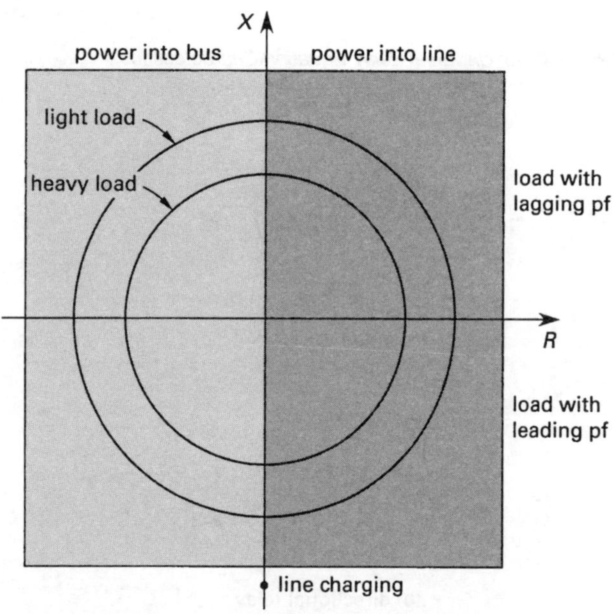

12. SOLID-STATE RELAYS

Many relays are electronic, or *solid-state.* Such relays offer numerous advantages: rugged design, ease of adjustment and programmability, accurate response, and high reliability with little or no maintenance. A solid-state level relay is shown in Fig. 42.18. A level relay is also known as an *overcurrent relay.* The signal sensed is passed through the resistor, R, rectified by the bridge network, amplified and filtered by the R-C circuit, and then compared to a reference voltage by an operational amplifier, A. If the signal exceeds the reference value, the output of the op amp increases, resulting in a protective signal after a self-imposed time delay occurs.

Figure 42.18 Solid-State Overcurrent Relay

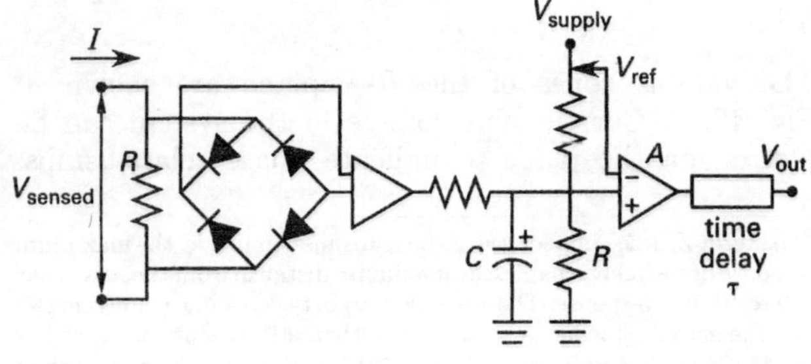

13. PROTECTIVE DEVICES FOR TRANSMISSION LINES

A wide variety of protective devices are available. Figure 42.19 lists these devices in order of complexity and cost.

Figure 42.19 Protective Devices

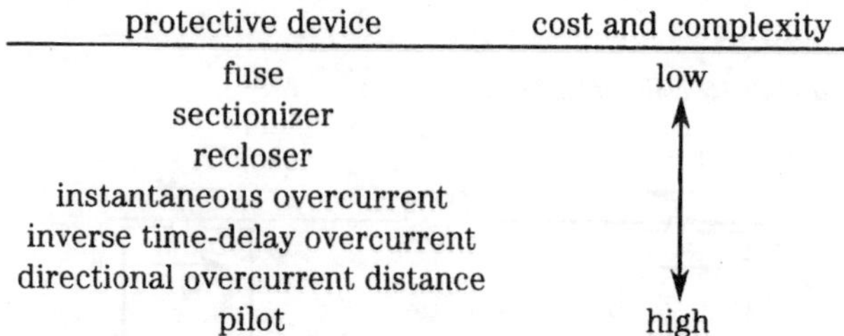

protective device	cost and complexity
fuse	low
sectionizer	
recloser	
instantaneous overcurrent	
inverse time-delay overcurrent	
directional overcurrent distance	
pilot	high

Fuses are simple and inexpensive and common in protective systems. They come in a series of standard sizes, each with a time-current characteristic curve.

A *sectionalizer* is a self-contained device used in conjunction with other devices, such as reclosers and circuit breakers, to permanently isolate a faulted portion of a distribution system. A sectionalizer does not interrupt a fault current. Instead, it counts the number of times a fault is seen (i.e., sensed) and opens when the count reaches a pre-programmed level. Only then is the circuit deenergized, usually by a recloser or circuit breaker. Sectionalizers are an economical way to isolate faults without the additional step of coordinating trip levels.

A *recloser* senses a fault and interrupts power to a system. After a preset time, the recloser attempts to reenergize the system by closing. If other protective devices have isolated the fault, the recloser remains closed.

An *overcurrent device* operates when a fault exceeds a given value.

Pilot protection refers to a communication channel between the ends of a long transmission line that enables the system to instantaneously clear faults over 100% of the line. Such communication channels include the power line itself using a *power line carrier, microwaves, fiber optics,* or *communication cables.*

Non-pilot protection refers to a system that responds only to the parameters at one end of a transmission line. If a fault occurs on the line away from the protection, a time-delay is incurred automatically and the fault is not interrupted instantaneously at both ends of the line. Such a system is also called a *graded, relatively selective,* or *non-unit* system.

14. ELECTRICAL COMPONENT PROTECTION: CROWBAR CIRCUITS

Circuits in all electrical components need some type of protection to prevent faults from damaging equipment. A variety of circuits provide protection against overvoltage conditions.

One common protective device is a *crowbar circuit*, which is used with a circuit attached to a power supply to prevent damage due to overvoltage from the power supply. By putting a short circuit or a very low resistance path across the voltage output, the crowbar circuit places a high overload on the actuating element of a circuit breaker or other protective device. Once triggered, a crowbar circuit brings the voltage between the terminals of the device to near zero, thus protecting downstream electrical components.[10] Even though the voltage is now too low to trigger it, the crowbar circuit will not return to normal operation until power is removed.

Figure 42.20 shows an example of a crowbar circuit. In this case, overvoltage on the output line is compared with the reference voltage provided by the resistor, R_{ref}.[11] When the voltage, V_{ref}, is greater than the breakdown voltage of the gated zener diode, the gated zener diode is triggered, and the triac's voltage is set by the zener's voltage. This voltage is at the appropriate level for the triac to turn on, resulting in a large current between the output and return lines due to the very low impedance of the triac.[12]

Figure 42.20 *Crowbar Example Circuit*

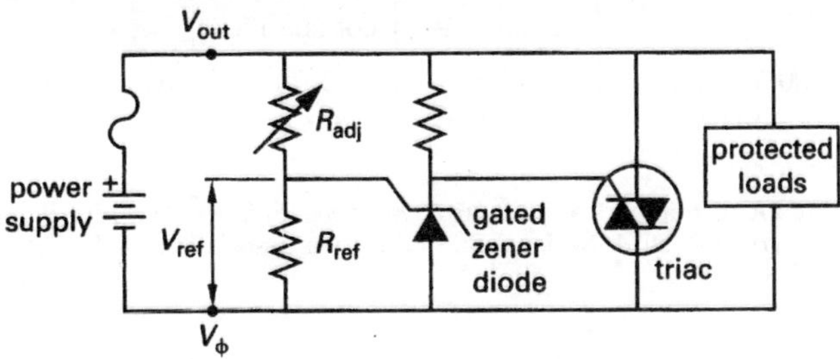

If the zener diode were used alone, the voltage between the terminals would be set at a level higher than desired. Instead, the zener diode turns on the triac, which provides an impedance path that is a short circuit between the terminals, bringing the voltage of the downstream circuitry well below the triggering voltage of the zener diode itself. In fact, the voltage is near zero; that is, near the normal voltage of the return line to the power supply. As a result, the downstream loads do not experience the overvoltage from the power supply and are protected.

[10]An actual crowbar (or large gauge wire, generically called a *jumper*) can be used during maintenance on large pieces of electrical equipment or transmission system segments. A crowbar placed across the output terminals of the power supply creates a short circuit and prevents voltage in downstream elements, which provides protection to people working on the system.

[11]While a battery is shown as the power supply in Fig. 42.20, often the source is an AC/DC converter, which does have high-voltage failure modes.

[12]See Chap. 45 for the operating characteristics of the zener diode and the triac.

15. ELECTRICAL SAFETY IN THE WORKPLACE

The ANSI-accredited standard on workplace safety and safe work practices is NFPA 70E, *Standard for Electrical Safety in the Workplace.*[13] The standard covers both shock and arc-flash hazards. These hazards are categorized using terms and risk assessments that are consistent with national and international risk standards for *personal protective equipment* (PPE).

Ensuring electrical safety in the workplace requires the following measures, which are applicable to all types of work.

- performing shock and arc-flash analysis on the electrical installation
- establishing an electrical safety program for the installation
- training maintenance and operations workers, as well as those exposed to such work
- identifying the hazards and risks associated with any particular installation task
- obtaining appropriate work permit(s)
- using proper PPE
- installing protective shields or barriers where necessary

NFPA 70E contains both *mandatory rules*, which use the words *shall* or *shall not*, and *permissive rules*, which use the words *permitted* or *not required.*[14] Also included are *informational notes* (in accordance with the NEC) that contain items for information only or provide reference material for further details.

NFPA 70E contains an Introduction labeled Art. 90 (also in accordance with the NEC), which explains the organization and purpose of the standard.

Chapter 1, Safety-Related Work Practices, contains Art. 100 through Art. 130 and covers definitions, the application of safety-related work practices, general requirements for electrical safety-related work practices, how to establish an electrically safe work condition, and work involving electrical hazards.

Chapter 2, Safety-Related Maintenance Requirements, contains Art. 200 to Art. 250 and covers general maintenance requirements, switchboards, panelboards, wiring, controller equipment, fuses, circuit breakers, rotating equipment, hazardous locations, batteries, tools, and, in Art. 250, PPE maintenance.

[13]A professional engineer should have some knowledge of NFPA 70E, given its widespread use and its impact on engineering design and operations. The explanatory sections in NFPA handbooks on the NFPA 70E are very useful.

[14] Many industry standards use *shall* to indicate requirements, *may* to indicate permitted or optional measures, and *will* to indicate an expectation of compliance that is beyond debate but for which compliance need not be proven with data.

Chapter 3, Safety Requirements for Special Equipment, contains Art. 300 to Art. 350 and covers electrolytic cells, batteries and battery rooms, lasers, power electronic equipment, and research and development (R&D) laboratories. Chapter 3 is meant to supplement or modify the requirements in Chap. 1.

The Informative Annexes contain references, calculation methods for arc flash, examples of safety programs, sample lockout/tagout procedures, guidance on the selection of PPE, work permit examples, and safety-related design requirements.

16. ARC-FLASH AND SHOCK HAZARD LEVELS

IEEE standards specify the methods used to determine arc-flash and shock hazard levels. These levels are given in terms of the energy per unit area that can be produced by any given equipment design and installation, usually in cal/cm^2. These levels are compared to the *approach boundaries* defined in NFPA 70E Art. 130 to determine the PPE required (using the guidance and tables in Annex H). The boundaries are shown in Fig. 42.21.

Figure 42.21 Approach Boundaries

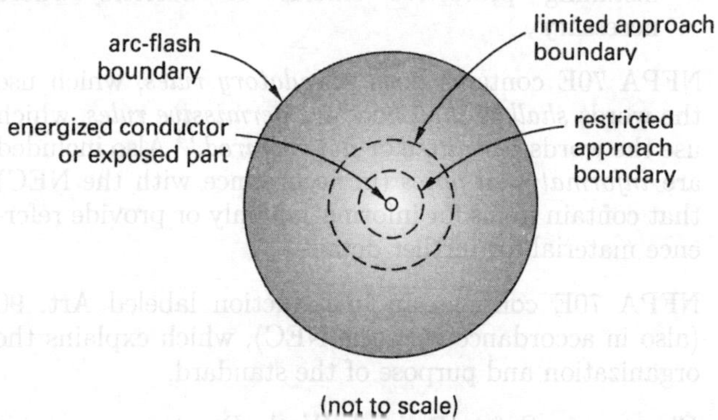

The point in the center of Fig. 42.21 represents an energized conductor or exposed part from which the boundaries are measured. There are two types of approach boundaries: *shock protection boundaries* (which are either restricted or limited) and *arc-flash boundaries.*

Shock protection boundaries vary with voltage level and whether the energized conductor is fixed (as in a panelboard) or movable (as in a high-voltage transmission line). The *limited approach boundary* is the boundary within which only qualified persons are allowed. (Unqualified persons may be allowed if supervision requirements are met.) The *restricted approach boundary* is also limited to qualified persons, but only if one of the following conditions is met.

(a) The energized conductor or part is operating at 50 V or more, and the qualified person is insulated or guarded (physically protected) from it.

(b) The energized conductor or part is operating at 50 V or more, and it is insulated from the qualified person.

(c) The qualified person is insulated from any other conductive object (for example, being insulated from the ground in a live-line, bare-handed work area).

A fixed, exposed part with an AC or DC voltage of 120 V has a limited approach boundary of 1 m (3 ft 6 in), and a restricted approach boundary of "avoid contact" rather than a specific distance. When a conductor or part such as a high-voltage transmission line is not under the control of a maintenance person, the part is considered "movable," and requires greater shock protection distances. Table 42.2, which is adapted from NFPA 70E, shows approach boundary limits for some common low voltages.

Table 42.2 Approach Boundary Limits to Energized Components

	limited approach boundary		
nominal system voltage range, phase to phase	exposed movable conductor	exposed fixed circuit part	restricted approach boundary
AC < 50 V	not specified	not specified	not specified
50 V ≤ AC ≤ 150 V	3.0 m (10 ft 0 in)	1.0 m (3 ft 6 in)	avoid contact
DC < 100 V	not specified	not specified	not specified
100 V ≤ DC ≤ 300 V	3.0 m (10 ft 0 in)	1.0 m (3 ft 6 in)	avoid contact

Adapted from NFPA 70E®-2015, *Electrical Safety in the Workplace,* copyright © 2014, National Fire Protection Association.

The arc-flash boundary is determined from arc-flash analysis, which is based on the energy level within the boundary and not on any particular voltage. Only qualified personnel with the proper PPE are allowed across the boundary. Requirements for PPE are generally divided into three ranges, according to the possible *incident energy exposure*: ≤1.2 cal/cm^2, >1.2 cal/cm^2 to ≤12 cal/cm^2, and >12 cal/cm^2. In the lowest range, protective clothing is required; that is, clothing that is nonmelting or untreated natural fiber. In the middle range, arc-rated clothing is required. In the uppermost range, additional arc-rated clothing is required, including an arc-flash suit with a hood. See Fig. 42.22 for an example of an arc-flash suit with gloves and long pants that cover the feet. All three ranges require some common protection, including safety glasses or goggles, hearing protection, and gloves (though the requirements for gloves vary). See App. 42.A for guidance on selection of PPE from NFPA 70E. If marked "AN" in App. 42.A, the item is "as needed." If marked "SR" in App. 42.A, "selection [is] required" of one of a number of items from a list. For example, when a list ending with "SR" includes heavy-duty leather gloves and rubber insulating gloves with leather protectors, one of the two choices must be used.

Protection and Safety

Figure 42.22 Arc-Flash Suit with Gloves

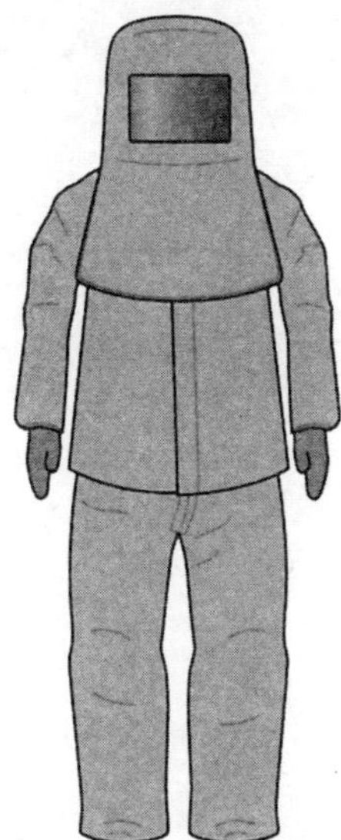

Example 42.1

Which of the three ranges of incident energy exposure require the use of hearing protection?

Solution

From NFPA 70E, the three incident energy ranges are $\leq 1.2\ \text{cal/cm}^2$, $> 1.2\ \text{cal/cm}^2$ to $< 12\ \text{cal/cm}^2$, and $> 12\ \text{cal/cm}^2$. In each of those ranges, some safety glasses, hearing protection, and gloves are required.

17. GENERATOR CAPABILITY/CONTROL/ PROTECTION COORDINATION

Generators have capability limits based on their electrical design characteristics. These characteristics include the amount of copper, the configuration and material properties of the stator and rotor, and the excitation system controlling the voltage and frequency.[15] Protection for the generator is provided by many of the devices discussed earlier in this chapter.[16] The capability of the generating unit, the limits of the excitation control system, and the generator protection must be coordinated to avoid equipment damage and power outages.[17]

[15]The frequency is nominally sensed by electrical devices and the signal representing the frequency is sent to an electrical-mechanical control system that uses the mechanical interface to control fuel flow (for diesel generators), steam input (for steam turbines), or any other input that controls the speed or frequency output, as well as the real power, of the generator.

[16]See IEEE C37.2 for details on a numbering system that has been developed and standardized for protection devices, as well as for other devices used in power and energy generation, distribution, and relaying. These numbers appear on electric schematics and one-line diagrams.

[17]The North American Electric Reliability Corporation (NERC) has standards associated with the reliability of the bulk power grid, including the coordination of generator protection and control. These standards are accredited by the American National Standards Institute (ANSI) and refer to the Institute of Electrical and Electronics Engineers (IEEE) standards. Conversely, the IEEE standards refer to the NERC standards.

Limiting the real power output (i.e., controlling the frequency) controls the mechanical power required by the generating unit. Limiting the reactive power output (i.e., controlling the voltage) controls the electromagnetic conditions within the generating unit. Both real and reactive limits maintain a generating unit (both the electrical and mechanical portions) within its temperature, structural, and other stress limits.

18. IEEE BUFF BOOK

IEEE Standard 242, *IEEE Recommended Practice for Protection and Coordination of Industrial and Commercial Power Systems* (*IEEE Buff Book*), describes the industry best practices for the selection, setup, and coordination of components of the protection system for industrial and commercial power networks. The information, often extracted from other codes, standards, and texts covers reasonable electrical abnormalities and how they affect the system as a whole. The *IEEE Buff Book* contains chapters on system protection principles; short-circuit calculations; instrument transformers; selection of relays; low- and high-voltage fuses; low-voltage circuit breakers; ground fault, conductor, motor, and transformer protection; bus and switchgear protection; overcurrent coordination; and maintenance, testing, and calibration.

19. IEEE GOLD BOOK

IEEE Standard 493, *IEEE Recommended Practice for the Design of Reliable Industrial and Commercial Power Systems* (*IEEE Gold Book*) describes the best practices for performing a comprehensive reliability analysis. The information it contains is derived primarily from extensive surveys of electrical equipment in power plants; other codes, standards, and industry texts are not referenced. The *IEEE Gold Book* is an accurate historical reference on power reliability and the associated costs.

The chapters in the *IEEE Gold Book* are on industrial and commercial power systems concepts, planning, and design; evaluating and improving reliability on existing systems; preventative maintenance; emergency and standby power; voltage sag analysis; 7×24 continuous power facilities; reliability and maintainability verification; and equipment reliability data.

20. IEEE YELLOW BOOK

IEEE Standard 902, *IEEE Guide for Maintenance, Operation, and Safety of Industrial and Commercial Power Systems* (*IEEE Yellow Book*) consists of information, often extracted from other codes, standards, and texts on safety during operation and maintenance of all sizes and types of industrial and commercial power systems. The *IEEE Yellow Book* provides guidelines for the establishment of procedures, recommended organizational structure, and maintenance strategies. It contains chapters on operating diagrams; system management;

control responsibilities and clearing procedures; maintenance strategies; electrical safety; safety programs; maintaining safe facilities; and safe work practices, tools, and test equipment.

21. IEEE BLUE BOOK

IEEE Standard 1015, *IEEE Recommended Practice for Applying Low Voltage Circuit Breakers Used in Industrial and Commercial Power Systems* (*IEEE Blue Book*) provides information necessary for the proper selection of low-voltage circuit breakers. The information, often extracted from other codes, standards, and texts, helps determine the type of breaker, appropriate ratings, requisite trips, acceptance testing, and maintenance needed. The *IEEE Blue Book* contains chapters on definitions, acronyms, and abbreviations; rating and testing; coordination; fused and special-purpose breakers; and acceptance and maintenance requirements.

Topic IX: Machinery and Devices

Chapter

43 Rotating DC Machinery

Nomenclature

a	number of parallel armature paths	–
a	ratio of transformation	–
A	area	m^2
B	magnetic flux density	T
B	susceptance	S
B, $\mathbf{B}$	magnetic field	Wb/m (T·m)
E	generated emf	V
F, $\mathbf{F}$	force	N
I, $\mathbf{I}$	current	A
k	constant	various
l, l, $\mathbf{L}$	length	m
n	rotational speed	rev/min
N	number of turns	–
p	number of poles	–
P	power	W
Q	charge	C
R	resistance	Ω
SR	speed regulation	–
t	time	s
T	torque	N·m
v	variable voltage	V
v, $\mathbf{v}$	velocity	m/s
V	constant or rms voltage	V
V	line voltage	V
VR	voltage regulation	–
z	number of conductors	–

Symbols

η	efficiency	–
θ	angle	rad
ϕ	flux	Wb
Φ	magnetic flux	Wb
ω	angular frequency	rad/s
ω_{mech}	rotational speed	rad/s

Subscripts

a	armature
cp	commutating pole
CEMF	counter-electromotive force
Cu	copper
E	emf
f	field
fl	full load
net	net field
nl	no load
rev/min	revolutions per minute
T	torque

1. INTRODUCTION

Electromagnetic devices, including sensors, relays, transformers, reactors (inductors), and rotating machines, operate using a similar set of electrical principles. These principles are applicable to each device, with each using a different application of the theory to accomplish a given task. Understanding these principles is important in understanding a wide variety of devices and machinery.

2. NOTATION METHODS

Conventional current flow is used as a standard (i.e. the most widely accepted) in electrical engineering. *Current* is said to flow from positive terminals to negative terminals, while electrons flow in the opposite direction —from negative terminals to positive terminals. A *source* delivers power and a *load* absorbs power.

Figure 43.1 shows two black boxes, each with a positive and negative terminal. Current direction is as shown. The source and load boxes are as follows.

- When current flows out of the positive terminal, the circuit element is a source; therefore, A is the source in Fig. 43.1.
- When current flows into the positive terminal, the circuit element is a load; therefore, B is the load in Fig. 43.1.

Figure 43.1 *Source and Load*

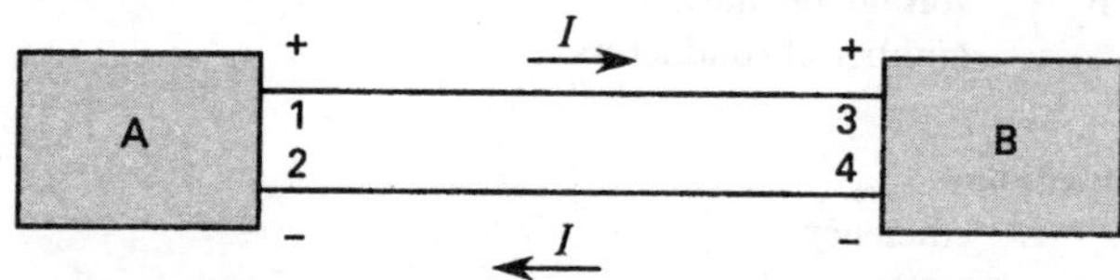

The sign notation for voltage uses *polarity markings*, as shown in Fig. 43.1. For example, from Fig. 43.1, E_{12} is the voltage between the terminals of source A with the voltage at 1 positive with respect to the voltage at 2. For load B, the same convention applies, where V_{34} indicates the voltage at 3 is positive with respect to the voltage at 4.

The *single-subscript notation* is similar, but uses one subscript and one polarity marking. When a reference node is used, the single-subscript notation can be called the *potential level representation.* The potential level representation is useful in understanding electronic circuit operation used to control and monitor motors and generators. For example, if the line between terminals 2 and 4 in Fig. 43.1 is considered the reference, then E_{12} can be represented as E_1. Similarly, V_{34} can be represented as V_3. Using Kirchhoff's voltage law (KVL) gives

$$E_{12} - V_{34} = E_1 - V_3 = 0 \qquad 43.1$$

Equation 43.1 assumes no voltage drop over the lines between the source and the load. This assumption may be used for the internals of motors and generators and short distribution systems, but cannot be used if higher accuracy is required or if excessive voltage drop will result in improper operation of the equipment.

The voltage and currents can be either AC or DC quantities; the representation remains the same. For AC quantities, the polarities, subscripts, and current directions represent an instant in time. Though the polarity and current directions change, the relationship remains, and what was a positive voltage from 1 to 2 becomes a negative voltage from 1 to 2 during the opposite half-cycle.

Example 43.1

In the following circuit, the voltages use the single-subscript notation and have the following values. What is the actual polarity of the voltages?

$$V_1 = -40\text{ V}$$
$$V_2 = -5\text{ V}$$
$$V_3 = +10\text{ V}$$

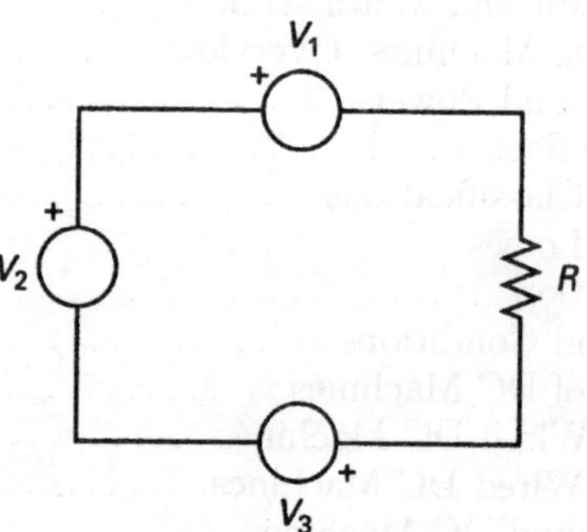

Solution

The voltages V_1 and V_2 are both given as negative. Therefore, the positive sign is on the terminal opposite. The result is as follows.

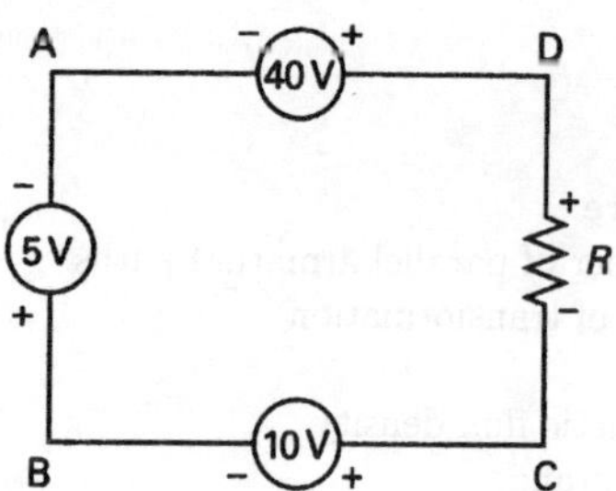

The potential across resistor R is 25 V with the polarity as shown. While it can appear that the potentials at nodes B and C have two values and that potential is 30 V, this is not the case. With single-subscript notation, the potentials are neither inherently positive nor negative. Instead, they have a value in reference to, or with respect to, another location.

Example 43.2

Consider the circuit shown with the reference level as indicated. The potentials are graphed with respect to the reference node at 1. What does the graph look like with the reference node at node 3?

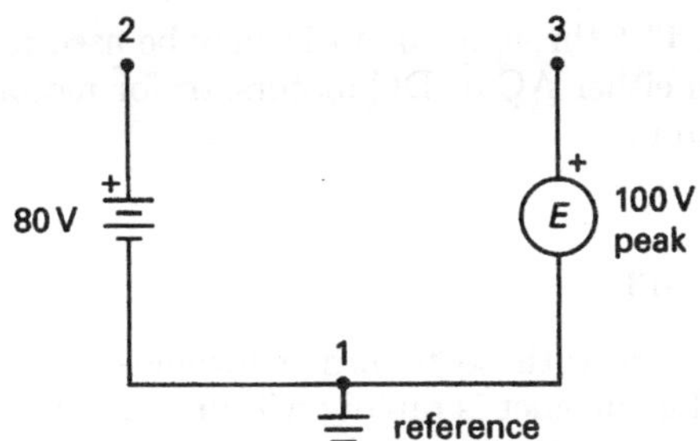

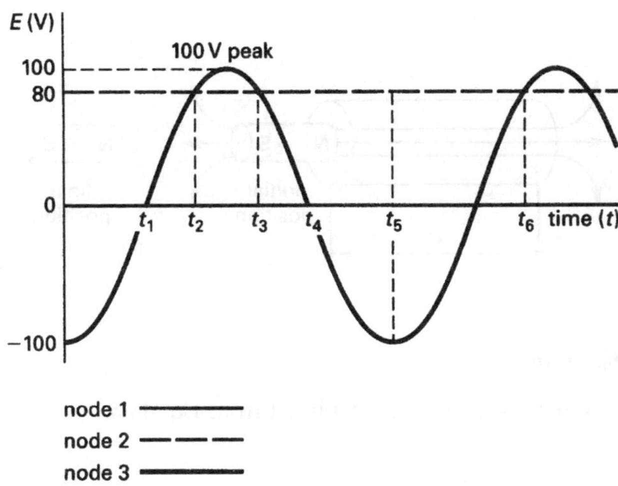

Solution

The graph is as follows.

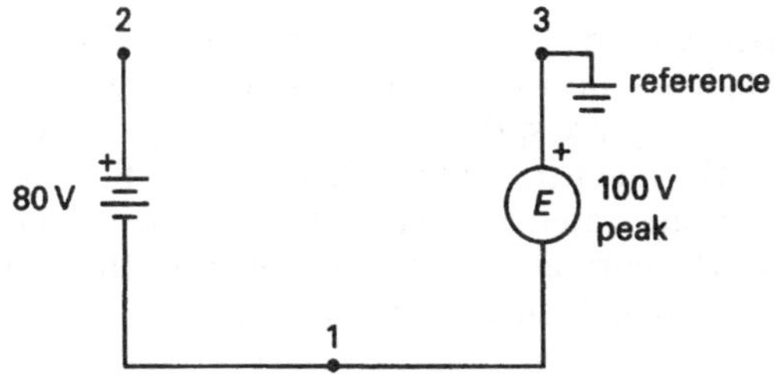

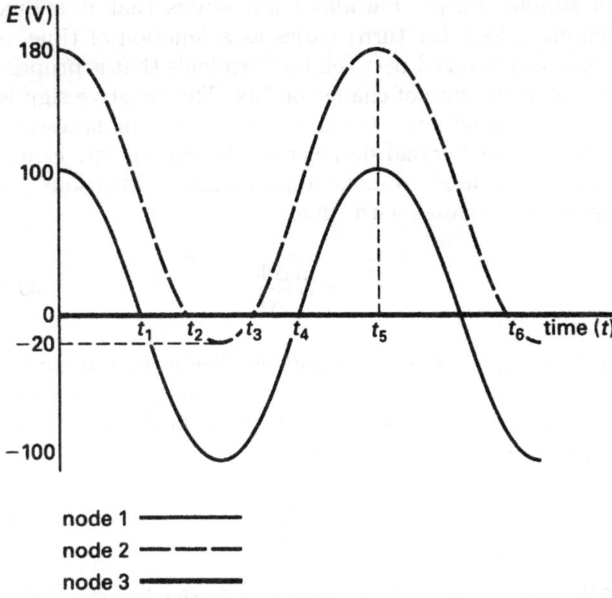

3. FARADAY'S LAW

Electric rotating machines and electromechanical devices (such as relays) consist of coupled electric and magnetic circuits. An electric circuit provides a path for the flow of electric current, and a magnetic circuit provides a path for magnetic flux. Sources of magnetic flux include permanent magnets and electric currents.

Faraday's law of electromagnetic induction describes how these coupled circuits interact. It states that the electromotive force induced in a circuit by a changing magnetic field is equal to the negative rate of change of the flux linking the circuit.

$$v = -\frac{d\phi}{dt} \tag{43.2}$$

The law refers to the rate of change for flux linking the circuit. Such flux linkage is the product of the number of turns, N, in a coil and the magnetic flux passing through the coil.

$$v = -N\frac{d\phi}{dt} \tag{43.3}$$

The rate of change can originate by two means: the flux can change with time, as indicated in Eq. 43.2 and Eq. 43.3; or, the flux can remain constant while the conductor moves with velocity **v** over a path, **l**, to "cut" the lines of flux, giving Eq. 43.4. The negative sign is not necessary since the mixed triple product provides the correct direction.

$$v = \oint (\mathbf{v} \times \mathbf{B}) \cdot d\mathbf{l} \tag{43.4}$$

In simpler terms, Faraday's law states that if a flux linking a loop (or turn) varies as a function of time, a voltage is induced between the terminals that is proportional to the rate of change of flux. The negative sign is often excluded since the direction is clear or necessary only for the original designer of the equipment. Equation 43.5 is used to make approximate calculations for fluxes that change with time.

$$v = N\frac{\Delta\Phi}{\Delta t} \qquad 43.5$$

In rotating machines, the angle is often a right angle (or close to one).[1] If the conductor is moving at constant speed at right angles to the magnetic field, the voltage induced is given by Eq. 43.6.

$$V = Blv \qquad 43.6$$

B is the magnetic flux density. l is the length of the conductor in the magnetic field. The symbol v is the velocity of the conductor through the magnetic field. If the conductor and the magnetic field are not at right angles, a sine term may be added to account for the angle between the magnetic field and the conductor.

$$V = Blv\sin\theta \qquad 43.7$$

In motors, the force is the quantity of first use. To obtain the force equivalent of Eq. 43.7, the velocity of the conductor, **v**, is replaced by the velocity of the positive charges, represented by the current, **I**.[2] This transition is accomplished by Eq. 43.8. The vector **l**, representing a unit vector in the direction of the current **I**, is often used in the place of the current vector **I**.[3]

$$\mathbf{F} = \oint \mathbf{I}\times\mathbf{B}dl = \oint I d\mathbf{l}\times\mathbf{B} = I\mathbf{L}\times\mathbf{B} \qquad 43.8$$

In terms of individual positive charges, Eq. 43.8 becomes

$$\mathbf{F} = I\mathbf{L}\times\mathbf{B} = Q\mathbf{v}\times\mathbf{B} \qquad 43.9$$

The same substitution (i.e., I for v) made in Eq. 43.6 and Eq. 43.7 results in the following force equations. Equation 43.10 determines the force felt on a conductor with length, l, at right angles to a magnetic field of flux density, B, carrying a current I. If the current (i.e., the conductor) is not at right angles to the magnetic field, then Eq. 43.11 must be used.

$$F = BlI \qquad 43.10$$

$$F = BlI\sin\theta \qquad 43.11$$

Equation 43.1 through Eq. 43.11 may be used for calculations on either AC or DC motors, or for motor action in generators.

Example 43.3

A coil of wire with 5000 turns surrounds a permanent magnet. The magnet is rapidly withdrawn, which causes the flux in the coil to change from 5×10^{-3} Wb to 1×10^{-3} Wb in 1×10^{-3} s. What voltage is induced between the terminals?

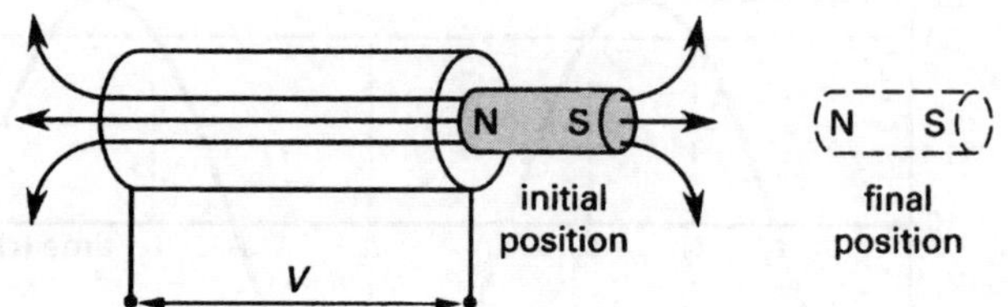

Solution

The potential can be calculated from Eq. 43.5.

$$\begin{aligned} v &= N\frac{\Delta\Phi}{\Delta t} \\ &= (5000)\left(\frac{5\times10^{-3}\ \text{Wb} - 1\times10^{-3}\ \text{Wb}}{1\times10^{-3}\ \text{s}}\right) \\ &= 20{,}000\ \text{V} \end{aligned}$$

Fireplace igniters, barbecue pit lighters, and other automatic lighters operate on the principles illustrated in Ex. 43.3. They use springs to initiate the physical movement of the magnet, rapidly changing the flux. Ignition coils use the same principal, but use a switch that removes the current quickly and collapses the magnetic field, thereby inducing the rapid change.

Example 43.4

A large generator contains conductors that have 2 m of their length (A to B) on one side within a 0.6 T magnetic field, as shown. The rotating magnetic field moves at 100 m/s. What is the voltage induced in the conductor between points A and B?

[1]The direction of the potential can be found from considering the direction of $\mathbf{v}\times\mathbf{B}$.

[2]The use of unit analysis can aid in confirming the validity of such a substitution.

[3]The vector **L** in equation Eq. 43.8 represents the total length of the conductor carrying current in the magnetic field. It is capitalized in this case only to avoid confusion with the current variable I. Where magnitudes and the italic l are used, equations are not prone to the same confusion, and the lowercase is more prominent.

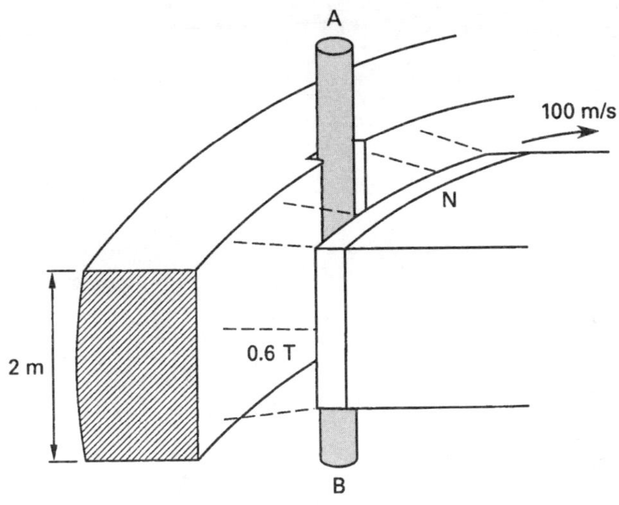

Solution

The induced voltage can be calculated from Eq. 43.6.

$$\begin{aligned} V_{AB} &= Blv = (0.6\ \text{T})(2\ \text{m})\left(100\ \frac{\text{m}}{\text{s}}\right) \\ &= 120\ \text{V} \end{aligned}$$

4. REVOLVING FIELD ROTATING MACHINE

Consider a stationary metal ring inside of which is located a rotating magnet, as shown in Fig. 43.2. The rotating magnet may be a permanent magnet or an electromagnet.

The stationary metallic ring is meant to reduce the reluctance of the magnetic circuit. The magnetic circuit runs from the north pole, across the air gap, into the ring, around the ring (in both directions), across the air gap, into the south pole, and then internally back to the north pole.

Figure 43.2 *Elementary Machine with Rotating Field*

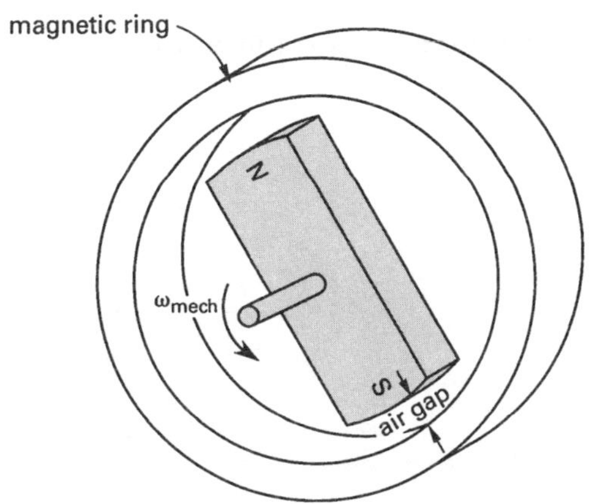

A single turn of a rectangular conductor is embedded in the ring as shown in Fig. 43.3(a). The coil is physically attached to the ring and does not rotate. With the revolving magnet included, the illustration becomes that of Fig. 43.3(b). When the revolving magnetic field is in the plane of the rectangular conductor, the induced voltage is at a maximum with the polarity as shown. (Consider this the 0° position.) Explained differently, when the poles are closest to conductor segments AB and CD, the voltage is at a maximum.

With the revolving magnet perpendicular to the plane of the rectangular conductor, the induced voltage is at a minimum (zero) with the polarity as shown in Fig. 43.3(c). (This is the 270° position.) If the rotation is 60 rev/min, the voltage between terminals A and D is as shown in Fig. 43.3(d) when considered in reference to the angle of rotation. When considered in reference to time, the voltage is as shown in Fig. 43.3(e). The advantage of the revolving field is that the electromagnet carries a much lower current through revolving connections and permanent, nonmoving connections can be made to the coil and field. Nearly all AC generators are of the revolving-field type.[4]

5. STATIONARY FIELD ROTATING MACHINE

Consider a stationary field located around a moving coil, as in Fig. 43.4. The stationary field may be a permanent magnet or an electromagnet.

Just as with the rotating field, the magnitude of the induced voltage, the polarity of the voltage, and the graph of the voltage as a function of angle or time all remain the same (assuming the same rotational speed). The principles and equations are identical. However, the difference is that within a stationary field, a direct connection cannot be made to the coil.

Two basic options exist for the connection to points A and D in Fig. 43.4. The first is a connection made with slip rings joined to each end of the coil and mounted on a shaft, but insulated from the shaft, as shown in Fig. 43.5. With this connection, when the coil is at the position shown in Fig. 43.4, the polarity of brush X is positive and brush Y is negative. When the coil segment AB of Fig. 43.4 has rotated such that it is directly under the south pole (i.e., the position where segment CD is shown), the polarity reverses. The output is identical to that shown in Fig. 43.3(d) and Fig. 43.3(e).

The second option for the connection is a *commutator*, as shown in Fig. 43.6. In its simplest form, it comprises a slip ring cut in half with an electrical insulator between

[4]This setup can be an AC generator or motor. To act as a motor, a rotating field would need to be established, and the rotor could be shorted or the resistance (now shown as the load) could act to control the current in the rotor, and thereby the speed. Physically, AC generator construction and AC motor construction differ significantly from one another, unlike construction of DC generators and motors, which can be identical.

Figure 43.3 Revolving Field

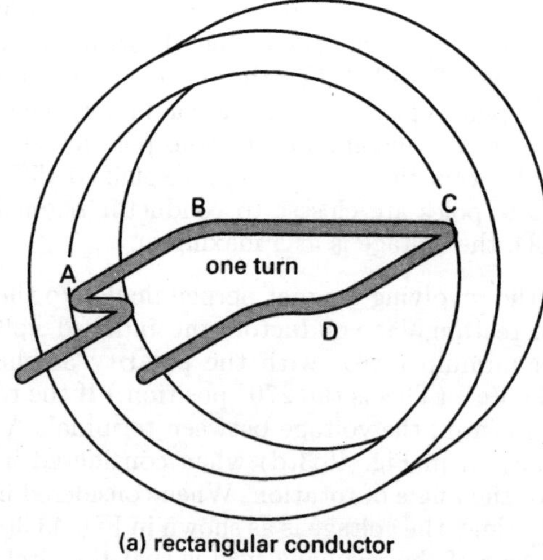

(a) rectangular conductor

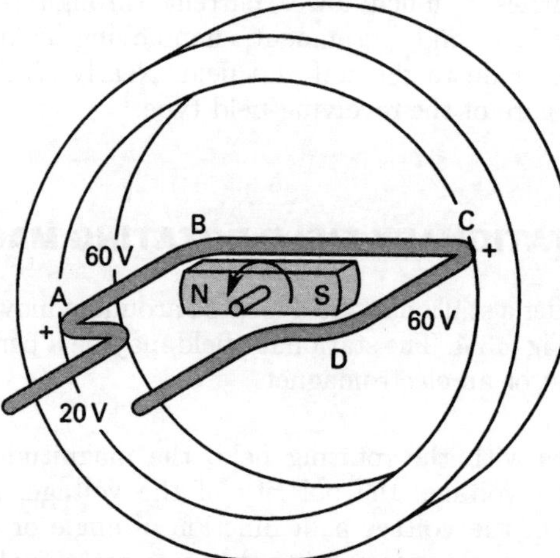

(b) revolving field at maximum induced voltage

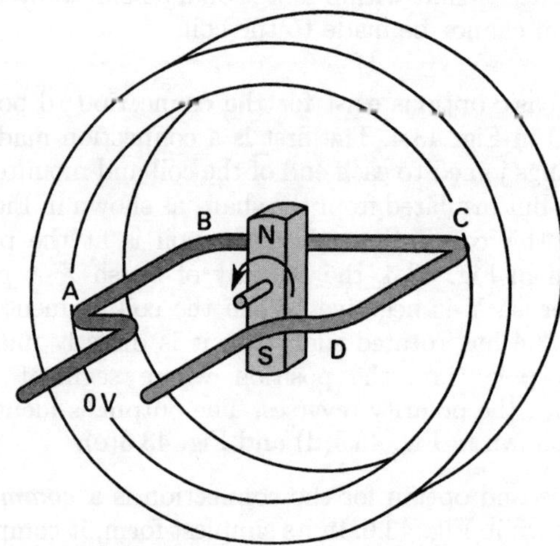

(c) revolving field at minimum (zero) induced voltage

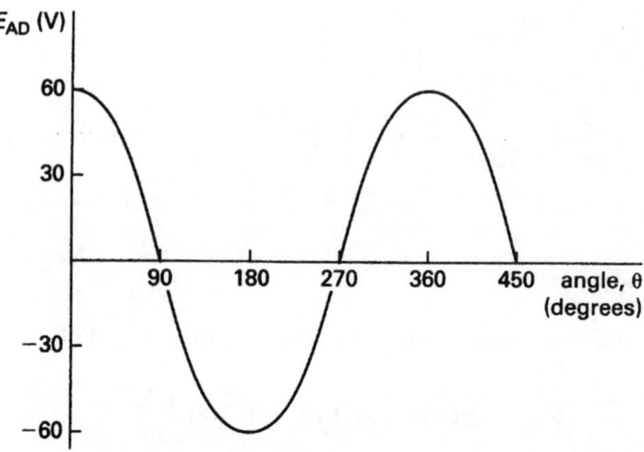

(d) induced voltage as a function of rotation angle

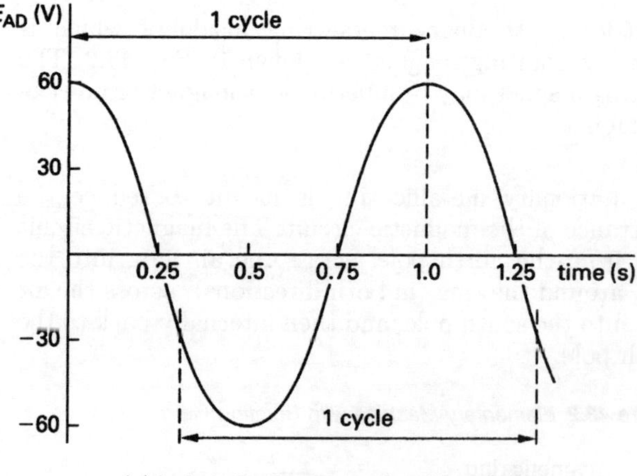

(e) induced voltage as a function of time

Figure 43.4 Elementary Machine with Stationary Field

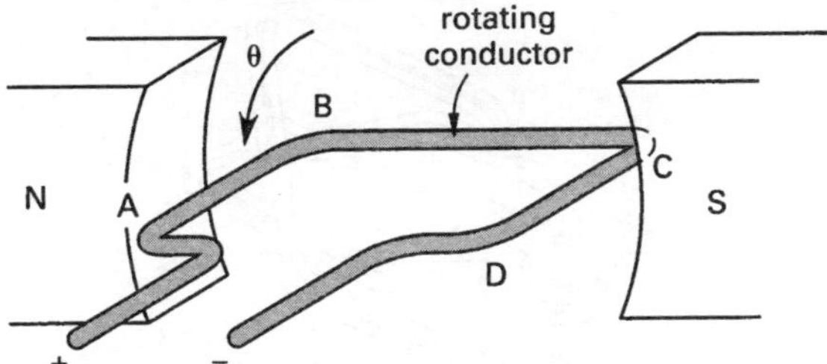

Figure 43.5 Slip Ring and Brushes Connection

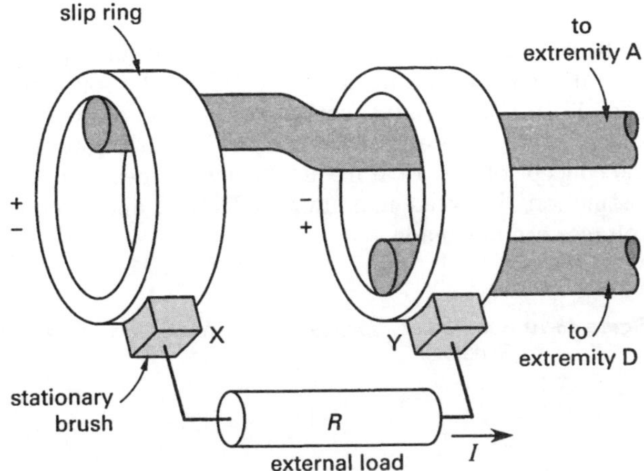

Figure 43.6 Commutator Connection

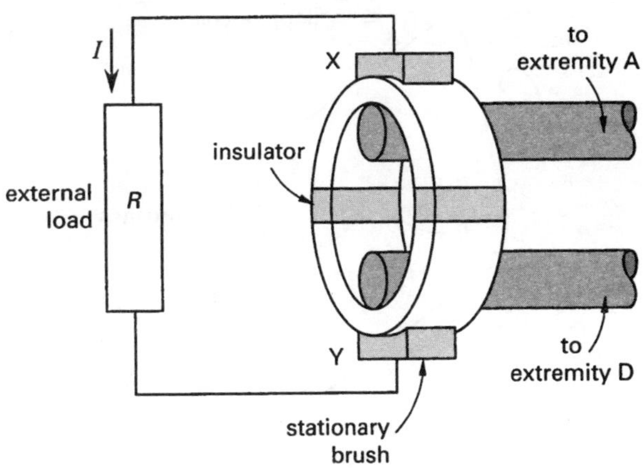

(a) commutator

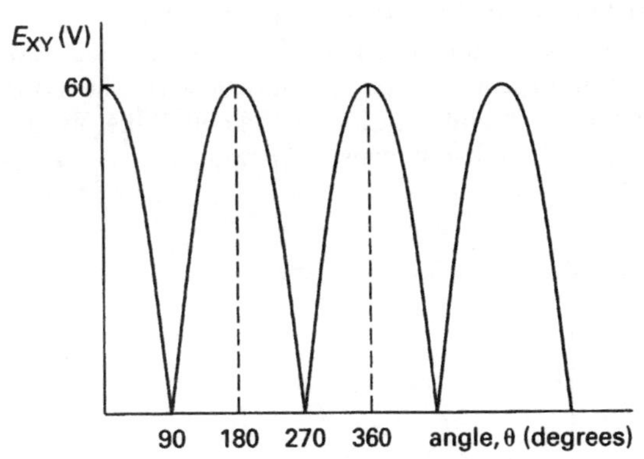

(b) commutator DC output

the two halves. One segment is connected to coil end A and the other to coil end D of Fig. 43.4, as illustrated in Fig. 43.6(a). The commutator revolves with the coil, so brush X is always connected to the segment of the coil under the north pole, which means it is always positive. The commutator rectifies the AC output to DC, as shown in Fig. 43.6(b). A commutator serves dual functions—it provides the direct connection point between the rotating portion of the machine (the rotor) to the stationary portion (the stator), and it reverses current flow mechanically.

The advantage of the stationary field is that rectification can take place on the rotating portion using a mechanical connection rather than power electronic circuitry. Nearly all DC generators are of the stationary-field type. Additionally, with this setup, the machine can be a DC generator (if the input is a mechanical torque on the rotor) or a DC motor (if the input is a DC current on the rotor).[5]

6. DC WAVESHAPE IMPROVEMENT

A *generator* is a device that produces DC potential. The actual voltage induced is sinusoidal (i.e., AC). However, brushes on *split-ring commutators* make the connection to the rotating armature and rectify the AC potential.[6] The more coils there are, the smoother the DC voltage. Two commutator segments are needed for each coil, and each coil produces its own sine wave. Because mechanical commutation is difficult unless the emf is produced in a rotating armature, DC generators are based on the simple design shown in Fig. 43.7.

DC generators suffer from several limitations. The rotating coils must be well insulated to prevent shorting with the high voltages that are induced. Structural bracing is required to counteract the large centrifugal force that results from rotating many coils of wire. It is difficult to make efficient high-voltage, high-power connections through slip rings.

The armature in a simple DC generator consists of a single coil with several turns (loops) of wire. The two ends of the coil terminate at the commutator, which

[5]To make the machine a generator, mechanical power must be applied to the rotor resulting in relative motion between the stator field and the rotor armature. To make the machine a motor, a current is applied to the rotor resulting in the interaction of two magnetic fields, which then causes motion. The construction of the machine stays the same.

[6]*Sparking* is one of the problems associated with commutators and occurs at points of low resistance. Brush resistance is nonlinear and drops as current increases.

Figure 43.7 Commutator Action

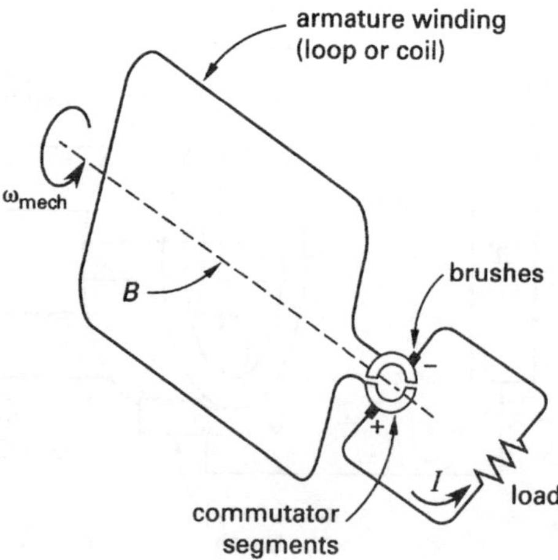

consists of a single ring split into two halves known as *segments* as is shown in Fig. 43.7. The brushes slide on the commutator and make contact with the adjacent segment every half-rotation of the coil. This produces a rectified (though not constant) potential, as shown in Fig. 43.8.

Figure 43.8 Rectified DC Voltage Induced in a Single Cell

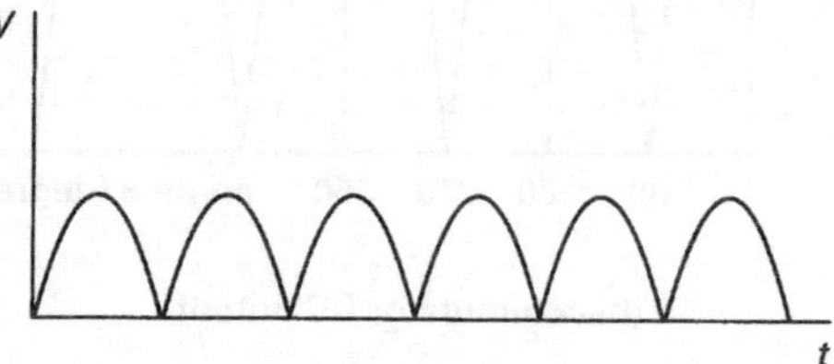

Modern DC generators contain multiple coils connected in series, an arrangement known as *closed-coil winding.* See Fig. 43.9 for a two-coil, four-segment closed-coil armature. The coils are spaced uniformly around the armature core. The single-ring commutator is divided into as many pairs of segments as there are coils. There are only two brushes, however, located on opposite sides of the commutator. Because there are many coils, as the armature rotates, the brushes always make contact with two segments of the commutator that are in nearly the same positions relative to the magnetic field. The average induced emf, $\overline{E}$, for a finite number of coils approaches the voltage induced with an infinite number of coils. N is the total number of series turns in all of the coils.

$$\overline{E} \approx E_\infty = 2\left(\frac{n}{60}\right)NAB \qquad 43.12$$

Because the coils are connected in series (in the modern closed-coil winding arrangement), the emf induced is the sum of the emfs induced in the individual coils. The voltage induced in each coil of a DC generator with multiple coils is still sinusoidal, but the terminal output is nearly constant, and not a (rectified) sinusoid. (See Fig. 43.10.) The slight variations in the voltage are known as *ripple.* Because the output is nearly constant, the concept of electrical frequency has no meaning, and distinctions between maximum, effective, and average voltages are not made.

Figure 43.9 Two-Coil, Four-Segment Closed-Coil Armature

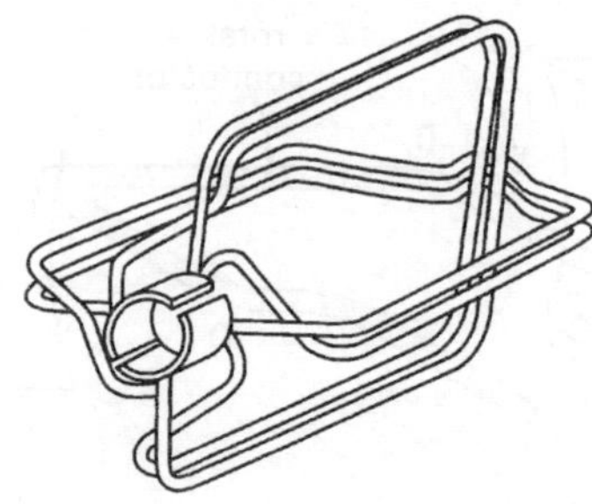

Figure 43.10 Rectified DC Voltage from a Two-Coil, Four-Segment Generator

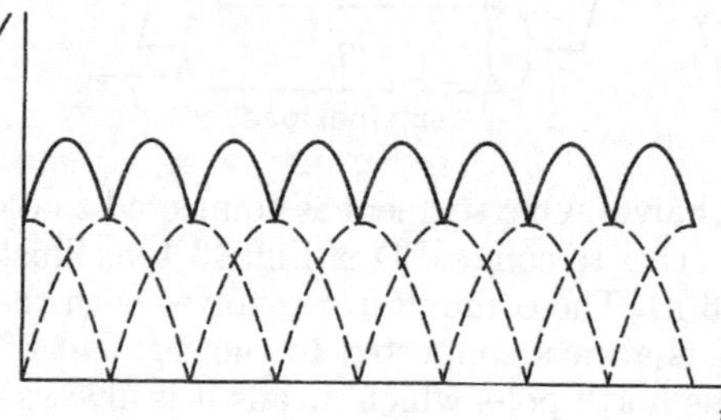

It is also possible to use an *open-coil connection* (see Fig. 43.11), though this is seldom done. Lower voltages are produced, as the average induced voltage is the maximum voltage from a single coil.

The average emf, E, for a DC machine (motor or generator) is given by Eq. 43.13. Φ is the flux per pole. For a generator, the emf is greater than the armature voltage ($E > V_a$). For a motor, the armature voltage is greater than the back emf ($V_a > E$).

$$E = \frac{Np\Phi n}{60} = \frac{zp\Phi n}{60a} = k_E\Phi n \qquad 43.13$$

$$k_E = \frac{zp}{60a} \qquad 43.14$$

More complex generators use additional coils and commutator segments, rather than the closed-coil arrangement, as shown in Fig. 43.12(a).[7] This not only more

[7]The waveform can also be improved by using electronic circuitry to filter the signal.

Figure 43.11 Two-Coil, Four-Segment Open-Coil Armature

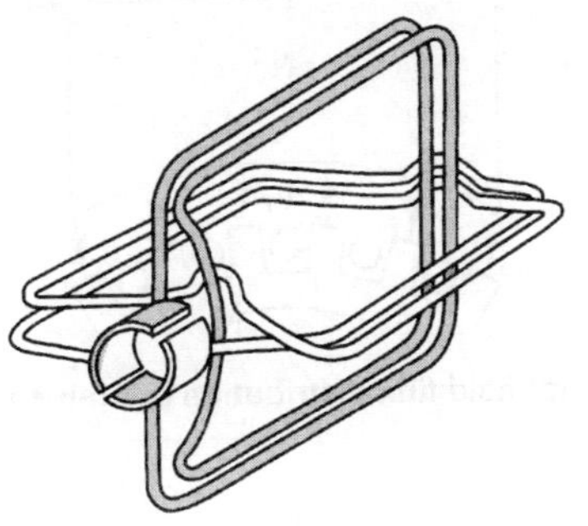

Figure 43.12 Four-Segment Commutation

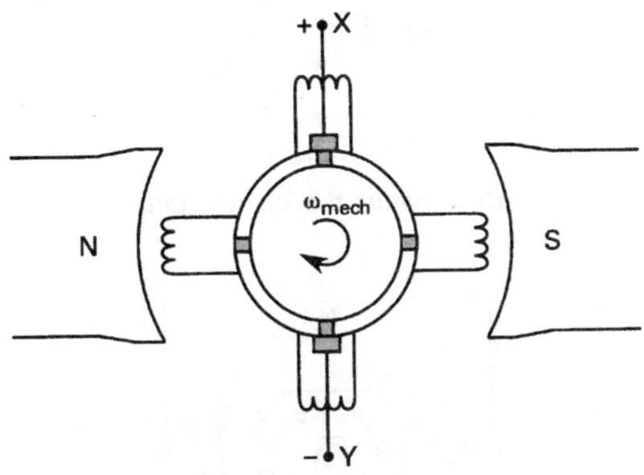

(a) four-segment windings

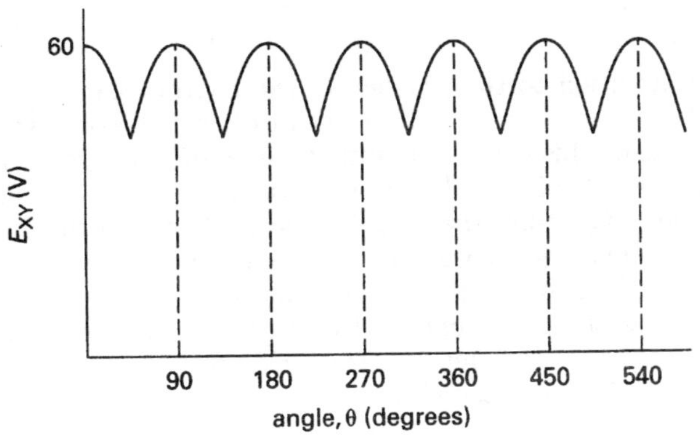

(b) four-segment output

closely approximates a steady DC, it smoothes the opposing torque (from motor action) and lowers the vibration level. Additionally, it allows lower current levels per winding while maintaining the output, overcoming some of the limitations mentioned.

7. ARMATURE REACTION

The field windings (either permanent or electromagnetic) of a generator in an AC or DC machine induce a voltage in the armature windings. The induced voltage creates current flow in the armature that results in a separate magnetic field. This armature field interacts with the original field to generate an armature reaction.[8]

Consider the magnetic field of the machine (generator) shown in Fig. 43.13. The dashed line represents the neutral plane. The magnetic field of the field windings (or permanent magnet) points from the north to the south pole, as shown in Fig. 43.13(a). The *neutral plane* is the location perpendicular to this field where windings on the armature (rotor) move parallel to the lines of force, so no voltage is induced.

In Fig. 43.13(b) the magnetic field of the armature (rotor) is shown. For a generator, the current flows as indicated. Use the cross product of the velocity, **v**, and the magnetic field, B_a, of the armature to determine the direction of the current flow in the armature windings. The right-hand rule can be used to determine the direction of the armature field.

Each field interacts leaving only one net field, $\mathbf{B}_{net}$. This net field is the vector sum of the two interacting fields. Figure 43.13(c) shows the direction. Because of this shift in the direction of the net magnetic field, the neutral plane, which is perpendicular to it, also shifts. In the case of a generator, the neutral plane shifts in the direction of rotation. For a motor, the neutral plane shifts opposite the direction of rotation.

Consider the windings in locations X and Y and compare Fig. 43.12(a) to Fig. 43.13(b). Windings shorted by the brushes at terminals X and Y had no induced voltage present in the condition of Fig. 43.12(a). This is the reason for the selected physical location of the brushes. Current is flowing via terminals X and Y even though no voltage exists across the winding associated with the shorted commutator segments. When armature reaction occurs, the neutral plane shifts, and the shorted windings now have an induced voltage as indicated in the condition of Fig. 43.13(c). The result is a voltage difference across the shorted commutator segments resulting in arcing across the insulating segment between the shorted segments, directly under the brushes. The voltage difference causes inefficient operation, noise, harmonics, and physical damage to the brushes and commutator segments.

Figure 43.14 shows a closer view of armature reaction. In Fig. 43.14(a), the stationary magnetic field, $\mathbf{B}_f$, is shown for a generator under no load. In Fig. 43.14(b),

[8]An armature reaction occurs in both AC and DC machines. The impact in DC machines is dramatic. The arcing of brushes, no longer in the neutral plane (where they would be shorting commutator segments with zero induced voltage across them), can result in immediate damage to the machine and can be seen with the naked eye. In AC machines, the interaction results in non-sinusoidal components and time harmonics in the output. The physical distribution of the conductors results in space harmonics, which are not discussed here, though the method of analysis depends heavily on Fourier series in both time and space.

Figure 43.13 Armature Reaction Fields: Vector Approach

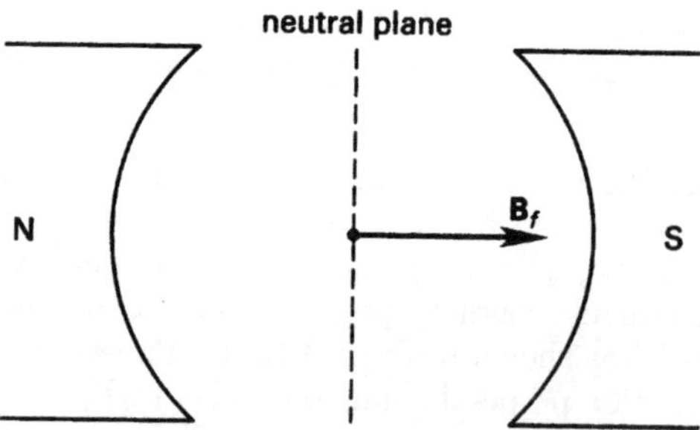

(a) magnetic field of the field windings

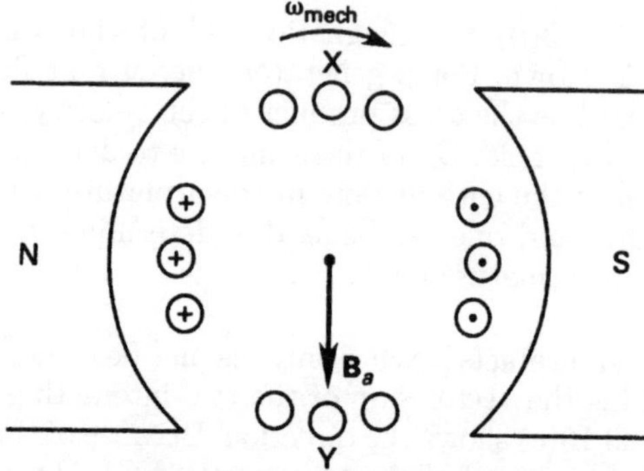

(b) magnetic field of the armature windings

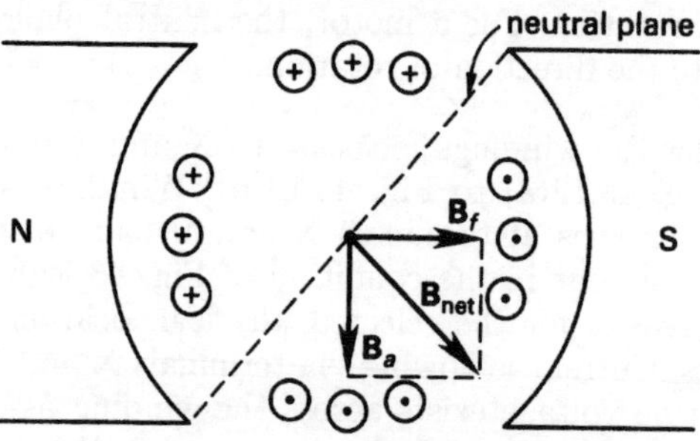

(c) net magnetic field or armature reaction

the armature (rotor) magnetic field, $\mathbf{B}_a$, is shown.[9] In Fig. 43.14(c), armature reaction, $\mathbf{B}_{net}$, is shown. The twisting of the magnetic field results in the associated movement of the neutral plane.

[9]Consider the direction of rotation in Fig. 43.13(b). The direction can be determined using the cross product of $\mathbf{v} \times \mathbf{B}$, with $\mathbf{v}$ being the velocity of the positive charges. Alternately, the right-hand rule can be used. Using the right-hand rule, the thumb is pointed in the direction of current flow on the rotor and the fingers curl in the direction of the surrounding magnetic field. For the conductors under the north pole, the magnetic field on top of the conductors is in the opposite direction of the north-south flux. The field on the bottom of the conductor is in the same direction as the north-south flux. This results in a weakening of the field above the conductors and a strengthening of the field below the conductors. The effect is a movement "up" or clockwise. If the magnetic field is mapped, it appears similar to a rubber band that has been stretched and is attempting to straighten out. This goes by the colloquial name the *rubber band theory of motor action*.

Figure 43.14 Armature Reaction Fields: Flux Approach

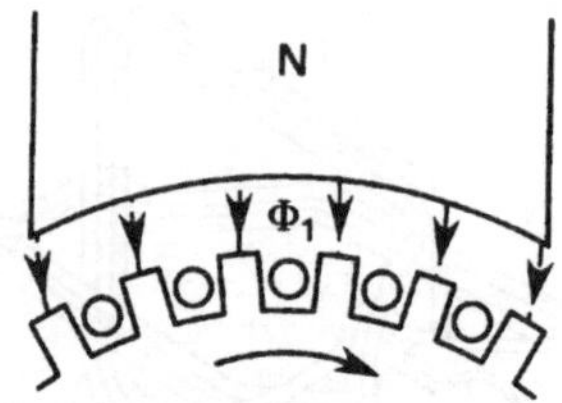

(a) field flux distribution at no-load

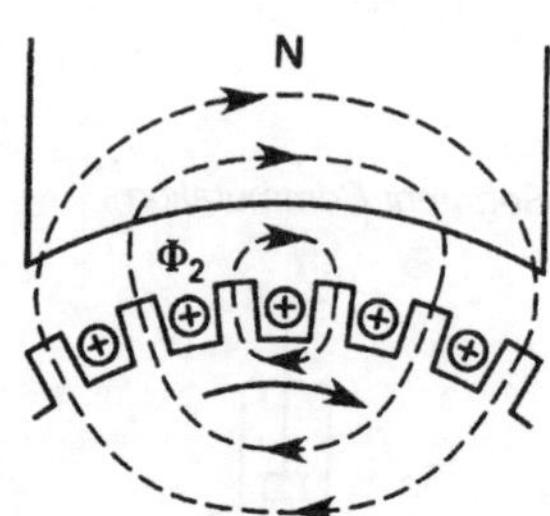

(b) armature flux distribution

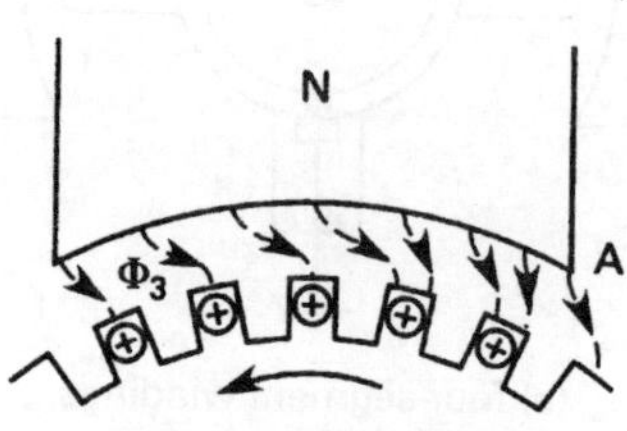

(c) armature reaction

In addition to the problem of the neutral plane shift, the pole tip at A in Fig. 43.14(c), can become saturated. This causes the flux increase on the right side of the pole piece to be less than the decrease on the left side. As a result, the total flux, Φ_3, is less than the flux at no-load, Φ_1. To prevent this potential instability, a series field (in series with the armature though wrapped on the stator) of one or two turns can be added to strengthen the shunt field (the field on the stator). This is called a ***stabilized shunt***.

8. COMMUTATING POLES

A way to correct for armature reaction is with the physical movement of the brushes to the new neutral plane. The obvious drawback to this correction is constructing such a movable device. Additionally, as the loading changes, the net magnetic field changes. The more frequent the loading changes, the more frequent the brush position movement. A more efficient approach is to use commutating poles.

Consider the individual magnetic fields of Fig. 43.15(a) and Fig. 43.15(b). It is the effect of the armature field in Fig. 43.15(b) that must be eliminated to prevent armature reaction. If a winding is placed in series with the armature, a commutating field, $\mathbf{B}_{cp}$, results, as shown in Fig. 43.15(c). This commutating field is in direct opposition to the armature field, as shown in Fig. 43.15(d). The commutating pole field cancels the shift due to the armature magnetic field, and the net result is the original magnetic field of Fig. 43.15(a). The neutral plane does not shift as it did in Fig. 43.13(c).

9. COMPENSATING WINDINGS

Large DC motors (i.e., those in the MW range) that operate with sudden acceleration, deceleration, or reversals can have large changes in armature reaction over a period of a few seconds. In such severe-duty cases, commutating poles and stabilizing windings are not enough to neutralize the armature magnetic field (i.e., the magnetomotive force, or mmf). In these cases, compensating windings are connected in series with the armature and distributed in slots cut into the pole faces of the main windings. (See Fig. 43.16.)

10. GENERATOR AND MOTOR ACTION

Recall that Eq. 43.6 ($V = Blv$) is used for a generated voltage. This equation represents the three items required for generator action; that is, for inducting a voltage.

- a magnetic field
- a conductor
- relative motion between the two

The force generated was given by Eq. 43.10 ($F = BlI$). This equation represents the three items required for motor action; that is, for the generation of a force (or for a torque when the force is applied across a moment arm, such as the radius of the rotor).

- a magnetic field
- a conductor
- current in the conductor

In a generator, a torque is applied to the shaft causing a relative motion between the conductors in the stator and the rotor, which carries the magnetic field. This is called *generator action.* Once there is relative motion, a voltage is induced in the stator windings resulting in current flow and the generation of a separate magnetic field that interacts with the first. The current flow is the desired output.

The interaction of magnetic fields can be understood in terms of *Lenz's law* (the current induced by an emf will always oppose the change that caused the emf). It can also be understood by noting that once the induced voltage causes a current flow, the conditions for motor action are met. That is, a current carrying conductor

Figure 43.15 *Commutating Poles*

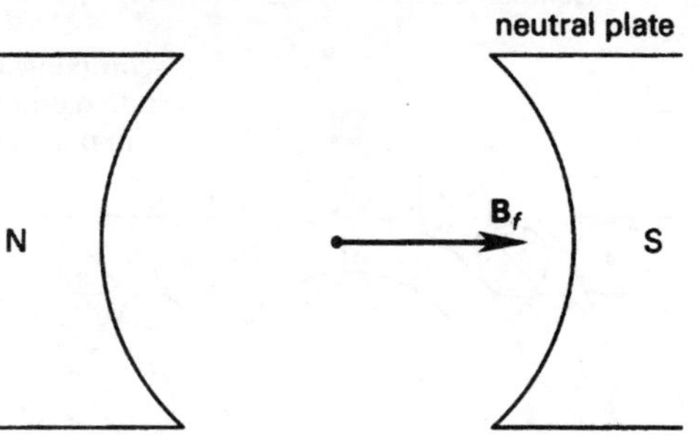

(a) magnetic field of the field windings

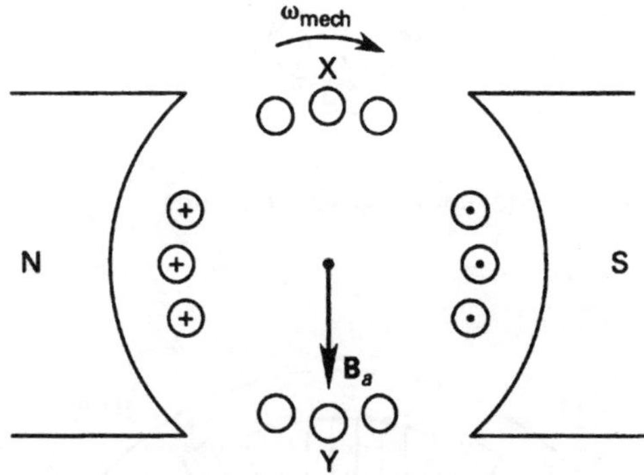

(b) magnetic field of the armature windings

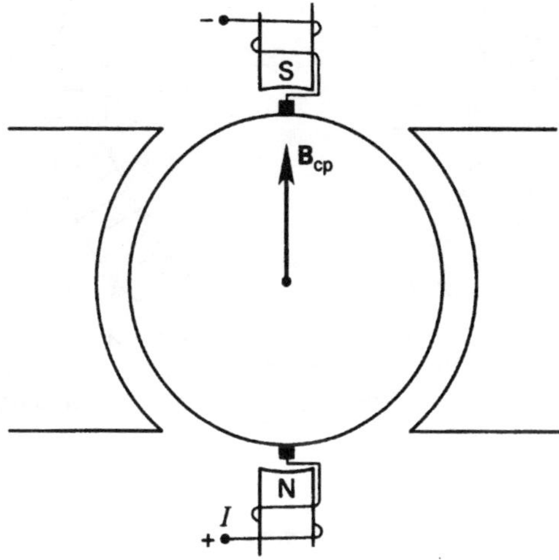

(c) commutating pole field

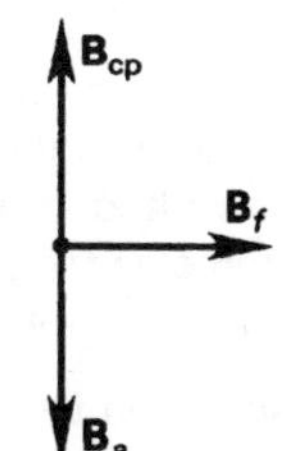

(d) net magnetic field—armature reaction commutated

Figure 43.16 Six-Pole DC Motor with Commutating Poles and Compensating Windings

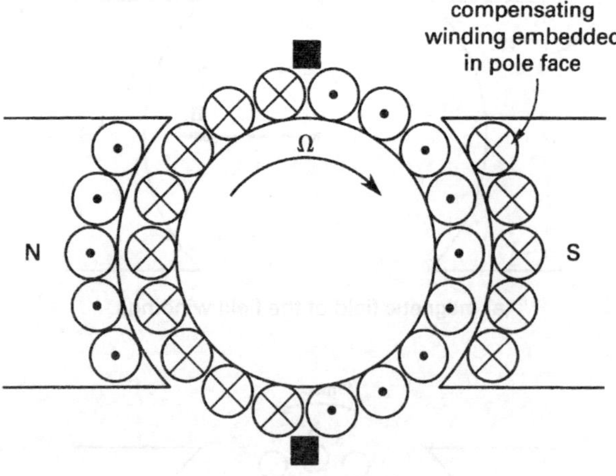

(a) compensating windings

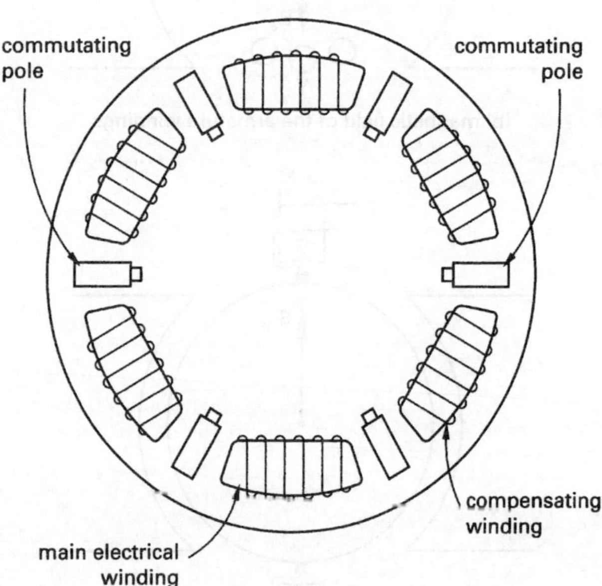

(b) DC motor stator

exists in a magnetic field. A force or torque is generated that opposes the motion of the rotor.

In a generator, generator action (the desired result) results in motor action (the undesired result). The motor action is a result of the conservation of energy. A constant input force or torque is required to continue generating an electrical output.

For a motor, the situation reverses. Motor action occurs first (the desired result), then generator action occurs (the undesirable output).

In any rotating machine there is generator action and motor action. If the generator action is the desired output, the opposing force or torque is called *counter torque.* If motor action is the desired output, the opposing voltage or current is called *counter-electromotive force* (or CEMF).

11. ROTATING MACHINES: OVERVIEW

Rotating machines are broadly categorized as AC and DC machines. Both categories include machines that use power (i.e., motors) and those that generate power (alternators and generators).

12. TORQUE AND POWER

Torque and *power* are operating parameters. It takes power to turn an alternator or generator. A motor converts electrical power into mechanical power. In the SI system, power is given in kilowatts (kW). One horsepower is equivalent to 745.7 W. The relationship between torque and power is

$$T_{\text{ft-lbf}} = \frac{5252P_{\text{horsepower}}}{n_{\text{rev/min}}} \qquad 43.15$$

$$T_{\text{N·m}} = \frac{1000P_{\text{kW}}}{\omega_{\text{mech}}} = \frac{9549P_{\text{kW}}}{n_{\text{rev/min}}} \qquad 43.16$$

There are many important torque parameters for motors. The *starting torque* (also known as *static torque, breakaway torque,* and *locked-rotor torque*) is the turning effort exerted in starting a load from rest. *Pull-up torque* (*acceleration torque*) is the minimum torque developed during the period of acceleration from rest to full speed. The *steady-state torque* must be provided to the load on a continuous basis. It establishes the temperature increase that the motor can withstand without deterioration. The *rated torque* is developed at rated speed and rated horsepower. *Breakdown torque* is the maximum torque the motor can develop without stalling (i.e., coming rapidly to a complete stop).

Equation 43.17 is the general torque expression for a rotating machine with N coils of cross-sectional area A, each carrying current I through a magnetic field of strength B.

$$T = NBAI\cos\omega t \qquad 43.17$$

13. SERVICE FACTOR

The horsepower and torque ratings listed on the nameplate of a motor can be provided on a continuous basis without overheating. Motors can be operated at slightly higher loads without exceeding a safe temperature rise, but the higher temperature has a deteriorating effect on the winding insulation. (A general rule of thumb is that a motor loses two or three hours of useful life for each hour run at the factored load.) The ratio of the safe to standard loads is the *service factor,* usually expressed as a decimal. Service factors vary from 1.15 to 1.4, with the lower values going to larger, more efficient motors.

$$\text{service factor} = \frac{\text{safe load}}{\text{nameplate load}} \tag{43.18}$$

14. MOTOR CLASSIFICATIONS

The National Electrical Manufacturers Association (NEMA) has categorized motors in several ways: *speed classification* (constant-, adjustable-, multi-, varying-speed, etc.); *service classification* (general, definite, and special purpose); and motor class. *Motor class* is a primary indicator of the maximum motor operating temperature, which, in turn, depends on the type of insulation used on the conductors. The classes are Class A, 105°C; Class B, 130°C; Class F, 155°C; and Class H, 180°C.

15. POWER LOSSES[10]

The losses for all rotating machines can be divided into four categories. *Copper losses*, P_{Cu}, are real power losses caused by wire and winding resistance. In a DC machine, copper losses occur in the armature and field windings as well as from the brush contact resistance. In an AC machine, copper losses occur in the armature and exciter field windings. There are no brush losses in an induction machine.

$$P_{\text{Cu}} = \sum I^2 R \tag{43.19}$$

Core losses, including hysteresis and eddy current losses, are constant losses that are independent of the load and, for that reason, are also known as *open-circuit* and *no-load losses.* In DC and synchronous AC machines, core losses occur in the armature iron. In induction machines, core losses occur in the stator iron.

Mechanical losses (also known as *rotational losses*) include brush and bearing friction and *windage* (air friction). (Windage is a no-load loss but is not an electrical core loss.) Mechanical losses are determined by measuring the power input at the rated speed with no load.

Stray losses are caused by nonuniform current distribution in the conductors. Stray losses are approximately 1% for DC machines and zero for AC machines.

[10]The subject of transformer losses covered in Sec. 37.2 is equivalent in concept.

Real power is only used to compute the *efficiency* of a rotating machine.

$$\eta = \frac{\text{output}}{\text{input}} = \frac{\text{output}}{\text{output} + \text{losses}} = \frac{\text{input} - \text{losses}}{\text{input}} \tag{43.20}$$

16. REGULATION

The *voltage regulation*, VR, is

$$\text{VR} = \frac{V_{\text{nl}} - V_{\text{fl}}}{V_{\text{fl}}} \times 100\% \tag{43.21}$$

The *speed regulation*, SR, is

$$\text{SR} = \frac{n_{\text{nl}} - n_{\text{fl}}}{n_{\text{fl}}} \times 100\% \tag{43.22}$$

17. NO-LOAD CONDITIONS

The meaning of the term *no load* is different for generators and motors. For unloaded shunt-wired alternators and generators (see Sec. 43.20), there is no electrical load connected across the output terminals, so although the field current flows, the line current, I, is zero. For unloaded shunt-wired motors, the work performed is zero, but line current is still drawn to keep the motor turning. All of the current is field current, however, and (neglecting mechanical losses) the armature current, I_a, is zero.

$$I = 0; \quad I_f \neq 0; \quad I_a = I_f \quad \text{[generator]} \tag{43.23}$$

$$I \neq 0; \quad I_f = I; \quad I_a = 0 \quad \text{[motor]} \tag{43.24}$$

18. TYPES OF DC MACHINES

DC machines, both generators and motors, can be connected in a number of ways, as shown in Fig. 43.17. The arrows indicate the direction of the magnetic field. If the arrows have identical directions, the magnetic fields aid one another (the magnetomotive forces of the fields are additive). If the arrows are in opposite directions, the magnetic fields oppose one another (the magnetomotive forces of the fields are subtractive).

The fields are shown as coils. Such coils have inductive and resistive components. Often variable resistors are added in series with the fields—especially the shunt field—to allow for control of the current through a given field. The current then determines the flux. The flux level controls either the generated voltage (in a generator) or the counter-electromotive force (in a motor), both of which impact the speed of the machine. In a

Figure 43.17 DC Machine Connections

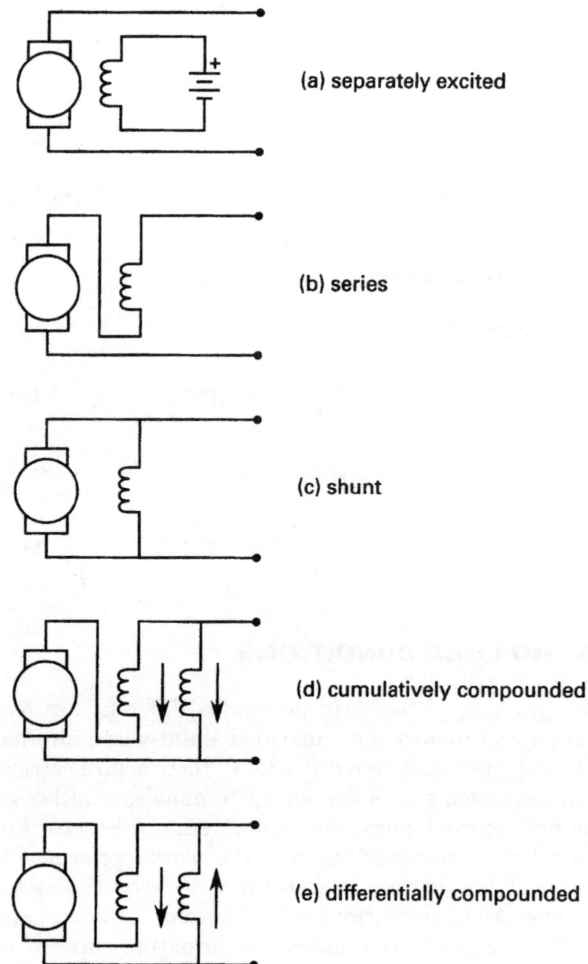

generator, an external energy input maintains the speed constant. In a motor, the speed is generally allowed to change with the load.

Example 43.5

What type of machine and compounding are shown in the following illustration?

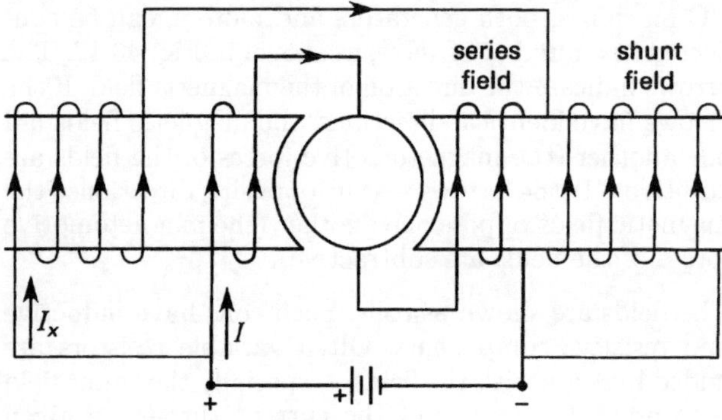

Solution

The machine has an external power source providing current to the windings, so this is a motor. A series and shunt winding are shown, which indicates the motor is either cumulatively compounded or differentially compounded. Use the direction of the arrows and the right-hand rule to determine that the series and shunt field directions (magnetomotive forces) are additive. Therefore, this is a cumulatively compounded motor.

19. SERIES-WIRED DC MACHINES

The equivalent circuit of a *series-wired DC machine* is shown in Fig. 43.18. The only components in the circuit are the field and armature resistances in series (from which the name is derived). The *brush resistance* is also considered to be in series but is often included in the armature resistance specification. The governing equations are Eq. 43.25 through Eq. 43.28. I_a is positive for a motor and negative for a generator. The magnetic flux varies with the armature current.

$$E = k'_E n\phi = k_E n I_a = V - I_a(R_a + R_f) \qquad 43.25$$

$$V = E + I_a(R_a + R_f) \qquad 43.26$$

Figure 43.18 Series-Wired DC Motor Equivalent Circuit

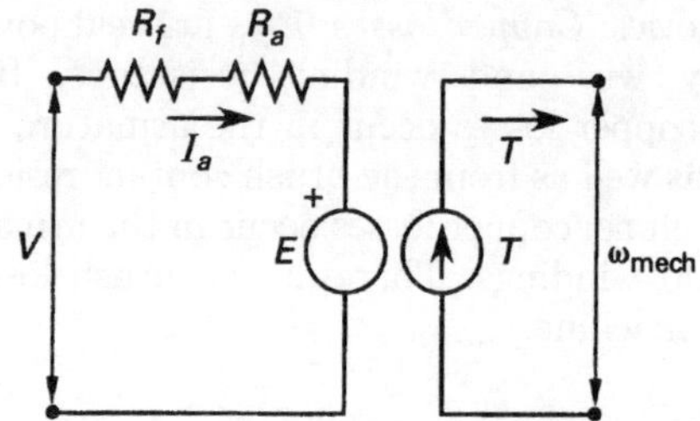

The speed, torque, and current are related by[11]

$$\frac{T_1}{T_2} = \left(\frac{I_{a,1}}{I_{a,2}}\right)^2 \approx \frac{n_1}{n_2} \qquad 43.27$$

The torque is

$$\begin{aligned} T &= k'_T \Phi I_a = k_T I_a^2 \\ &= k_T\left(\frac{V}{k_E n + R_a + R_f}\right)^2 \end{aligned} \qquad 43.28$$

For a motor, a reduction in load (torque) causes a corresponding reduction in armature current. However, because the field and armature currents are identical, the flux is reduced and the speed increases to maintain Eq. 43.26. Therefore, a series motor is not a constant-speed device. A load should never be completely removed from a running DC motor, and gears (not belts, which can slip) are the preferred method of connecting DC motors to their loads.

[11]The current, *I*, is directly related to the rotational speed, *n*. The torque, *T*, is related to the rotational speed squared, n^2, while the power, *P*, is related to the rotational speed cubed, n^3.

The back emf, E, is zero when the motor starts from rest. Therefore, the armature current, I_a, must be excessively high in order to keep Eq. 43.26 valid. For this reason, reduced voltages are required when starting, and the field resistance, R_f, is often a rheostat or switchable resistor bank.

At high speeds, the back emf, E, counteracts the applied voltage, increasing as the rotational speed increases. The *stall speed* is the speed at which Eq. 43.29 is valid.

$$E = I_a(R_a + R_f) = \frac{V}{2} \quad \text{[stall]} \tag{43.29}$$

20. SHUNT-WIRED DC MACHINES

Shunt-wired DC machines have a constant (but adjustable) magnetic field because the field current is constant. For shunt-wired motors, this results in a relatively constant speed. The magnetic coil is fed from the same line as the armature (as it is in Fig. 43.19) in a *self-excited machine*; in a *separately excited machine* the field coil is fed from another source. In Eq. 43.30 through Eq. 43.36, I_a is positive for a motor and negative for a generator.

$$E = k_E n\Phi \tag{43.30}$$

$$V = E + I_aR_a = I_fR_f \tag{43.31}$$

$$I = I_a + I_f \quad \text{[motor]} \tag{43.32}$$

$$I = I_a - I_f \quad \text{[generator]} \tag{43.33}$$

Figure 43.19 *Equivalent Circuits*

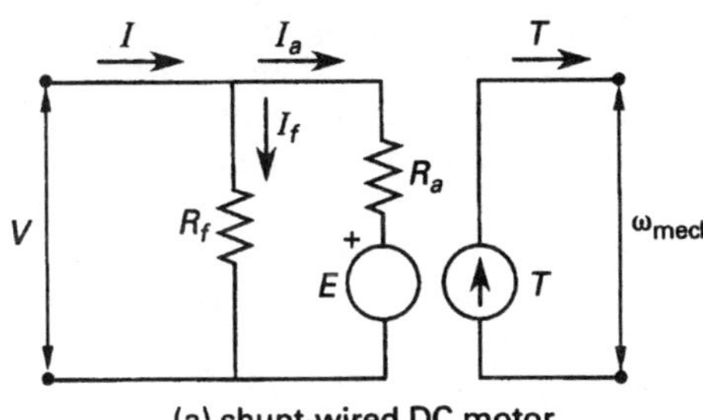

(a) shunt-wired DC motor

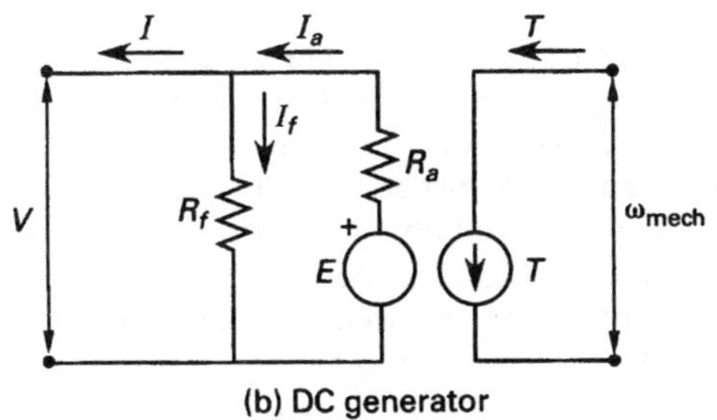

(b) DC generator

The speed-current relationship is

$$n = n_{nl} - k_nT = \frac{V - I_aR_a}{k_E\Phi} \tag{43.34}$$

The torque is

$$T = k_T\Phi I_a \tag{43.35}$$

The torque and current are proportional.

$$\frac{T_1}{T_2} = \frac{I_{a,1}}{I_{a,2}} \tag{43.36}$$

21. COMPOUND DC MACHINES

Compound DC machines have both series and shunt windings. Their performance is between those of shunt and series machines.

Example 43.6

A two-pole DC generator with a lap-wound armature is turned at 1800 rpm. There are 100 conductors between the brushes. The average magnetic flux density in the air gap between the pole faces and armature is 1.2 T. The pole faces have an area of 0.03 m^2. What is the no-load terminal voltage?

Solution

The flux per pole is

$$\Phi = BA = (1.2\ \text{T})(0.03\ \text{m}^2) = 0.036\ \text{Wb}$$

From Sec. 43.17, the term "no load" for a generator means the line current is zero. Also, $N = z/a = z/p$ for a lap-wound armature; that is, $a = p$.

$$\begin{aligned} E &= \frac{zp\Phi n}{60a} \\ &= \frac{(100)(2)(0.036\ \text{Wb})\left(1800\ \frac{\text{rev}}{\text{min}}\right)}{\left(60\ \frac{\text{s}}{\text{min}}\right)(2)} \\ &= 108\ \text{V} \end{aligned}$$

22. VOLTAGE-CURRENT CHARACTERISTICS FOR DC GENERATORS

Figure 43.20 illustrates the voltage-current characteristics for a DC generator.

Figure 43.20 DC Generator Voltage-Current Characteristics

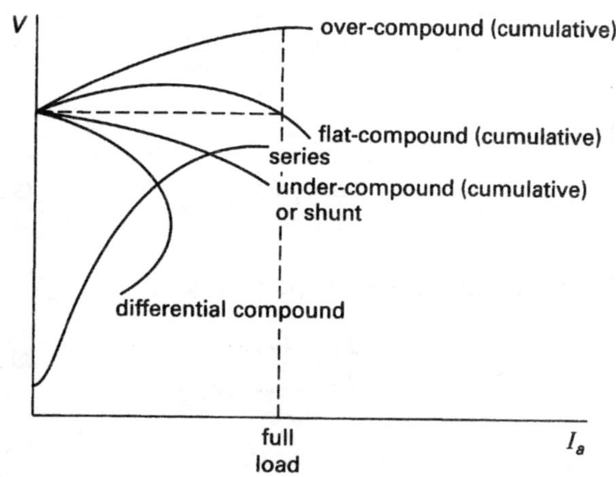

23. TORQUE CHARACTERISTICS FOR DC MOTORS

The torque produced by a DC motor is illustrated by Fig. 43.21.

Figure 43.21 DC Motor Torque Characteristics

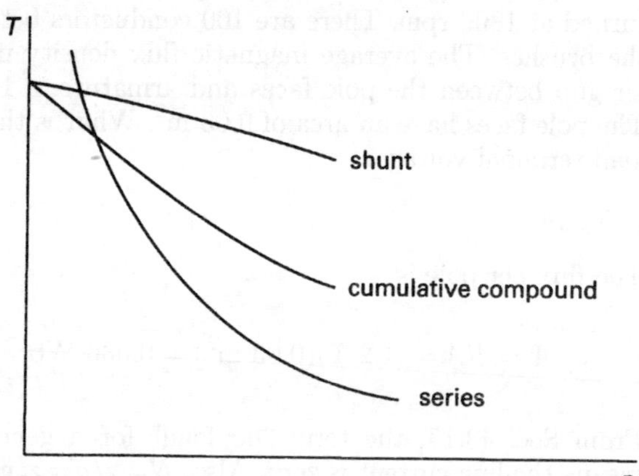

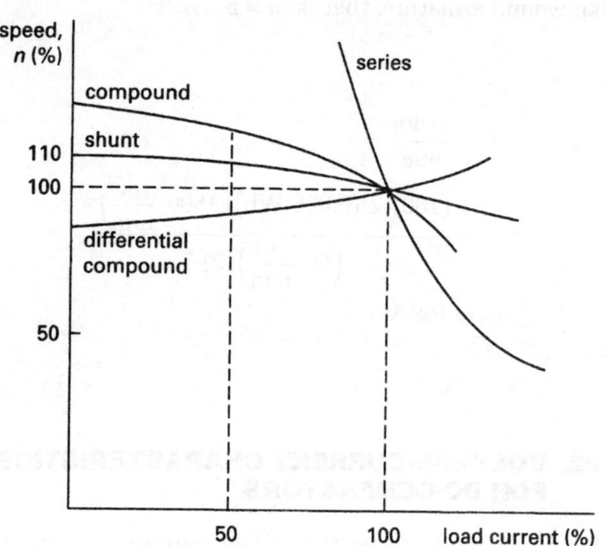

24. STARTING DC MOTORS

DC motors have very low armature resistance. At rest, there is no back emf, E. If connected across the full line voltage, the high current could damage the motor. Such motors are almost always started with a resistance in series with the armature winding.[12] An initial resistance is chosen that limits the starting current to approximately 150% of the full-load current. As the motor builds up speed, the back emf opposes the line voltage, reducing the current. The starting resistance can then be gradually reduced and removed.

Because torque depends on the current, starting torque is limited to approximately 225%, 175%, and 150% of the full-load torque for series, compound, and shunt motors, respectively, when the starting current is 150% of full-load current.

25. SPEED CONTROL FOR DC MOTORS

The speed of a DC motor can be controlled by changing the armature conditions, field conditions, or both. Changes can be made manually or automatically. *Armature control* techniques include (a) placing a variable resistance in series or parallel with the armature and (b) changing the voltage across the armature. In all control techniques, the field voltage is held constant. With a constant torque load, the DC motor speed varies in approximate proportion to armature voltage changes.

Two *field control* (*field weakening*) techniques include (a) changing the resistance of the field winding (series or shunt) and (b) changing the voltage across the field. In all field control techniques, the armature voltage is held constant. The DC motor speed increases when the flux is reduced, which happens when the field current is reduced.

To increase the rotational speed of a DC motor, the resistance of the field winding is increased. Increasing the resistance of the field winding causes a sequence of changes to the current, flux, generated emf, and torque, which changes the rotational speed of the DC motor, as shown in Fig. 43.22. Figure 43.22(a) shows the responses of variables within a DC motor, and Fig. 43.22(b) shows the sequence of changes of those variables. The magnitude of each change can be calculated using the applicable equations given in Sec. 43.12, Sec. 43.19, and Sec. 43.20. Similar principles apply when changing the voltage across the field.

DC motors can be controlled electronically by supplying either the armature or field (or both) with voltage through silicon-controlled rectifiers (SCRs). Through a feedback element or other control mechanism, rectifier output is automatically adjusted to maintain the requirements of the machine or process. This can provide constant speed, as well as control over a widely varying range of speeds and torques.

[12]Small fractional horsepower (1/4 hp or smaller) motors are an exception.

Figure 43.22 *DC Motor Speed Control*

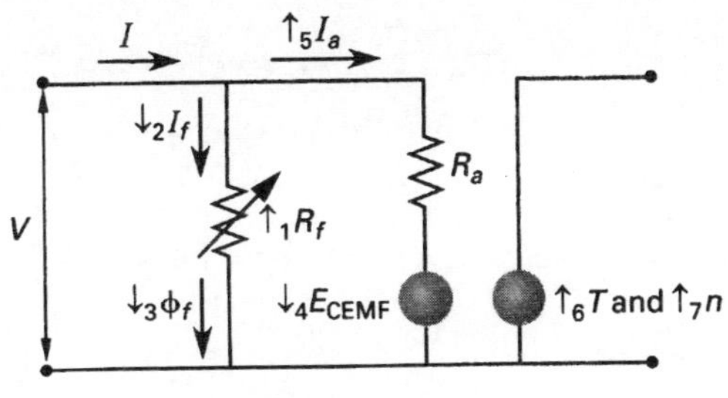

(a) variable responses

$$\uparrow_1 R_f \rightarrow \downarrow_2 I_f \rightarrow \downarrow_3 \Phi_f \rightarrow \downarrow_4 E_a \text{ or } E_{CEMF}$$
$$\rightarrow \uparrow_5 I_a \rightarrow \uparrow_6 T \text{ and } \uparrow_7 n$$

(b) sequence of changes

26. DIRECTION OF ROTATION OF DC MOTORS

A DC motor's direction of rotation can be reversed either by changing the direction of the armature's current, or by changing the direction of current in both the series and shunt fields. If both the armature current and field current directions are changed, the DC motor will continue to rotate in the same direction as before the change.

Rotating AC Machinery

Nomenclature

a	number of parallel armature paths	–
a	ratio of transformation	–
A	area	m^2
B	magnetic flux density	T
B	susceptance	S
E	energy	J
E, $\mathbf{E}$	generated emf	V
E	generated voltage	V
f	electrical frequency	Hz
G	conductance	S
I, $\mathbf{I}$	constant or rms current	A
I_B	magnetization (quadrature) current	A
I_G	in-phase component of exciting current	A
k	constant	various
l	length	m
n	rotational speed	rev/min
N	number of items	–
N	number of series armature paths	–
p	number of poles or poles per phase	–
pf	power factor	–
P	power or power loss	W
q	number of loops	–
r	radius	m
R	resistance	Ω
s	slip	–
S	apparent power	VA
SR	speed regulation	%
t	time	s
T	period	s
T	torque	N·m (ft-lbf)
V, $\mathbf{V}$	constant or rms voltage	V
V_0	generated voltage	V
VR	voltage regulation	–
X	inductance or reactance	Ω
Y	admittance	S
z	total number of conductors	–
Z	impedance	Ω

Symbols

δ	torque angle	rad
η	efficiency	–
θ	phase angle difference	rad
κ	torque conversion factor	–
ϕ	angle	rad
Φ	magnetic flux	Wb
ω	electrical frequency	rad/s
ω_{mech}	rotational speed	rad/s

Subscripts

1	equivalent stator
2	equivalent rotor
a	armature
adj	adjusted
aux	auxiliary
ave	average
b	blocked rotor
c	capacitor
Cu	copper
eff	effective
f	field
h	hysteresis
L	load
max	maximum
mech	mechanical
nl	no load
p	phase
pf	power factor
r	rotor
s	synchronous
st	stator
t	terminal or total
T	torque

1. ROTATING MACHINES

Rotating machines are broadly categorized as AC or DC machines. Both categories include machines that use power (i.e., motors) and those that generate power (alternators and generators). Most machines can be constructed in either single-phase or polyphase configurations, although single-phase machines may be outclassed in terms of economics and efficiency.

Types of small AC motors include split-phase, repulsion-induction, universal, capacitor, and series motors. Large AC motors are almost always three-phase, but it is necessary to analyze only one phase of the motor. Torque and power are divided evenly among the three phases. Machines can be wye- or delta-wired or both.

Wye connections have the following benefits.

- A neutral (ground) wire is intrinsically part of the circuit.
- Higher-order (harmonic) terms are not shorted out.
- Starting current is lower.

Nevertheless, high horsepower motors are usually run in delta. To avoid large starting currents, the motor can be started in wye and switched over to delta.

It is common to refer to line-to-line voltage as the *terminal voltage*, *V*.

$$V_p = \begin{cases} V & \text{[delta-wired]} \\ \dfrac{V}{\sqrt{3}} & \text{[wye-wired]} \end{cases} \tag{44.1}$$

$$T_p = \frac{T_t}{3} \tag{44.2}$$

$$P_p = \frac{P_t}{3} \tag{44.3}$$

$$S_p = \frac{S_t}{3} \tag{44.4}$$

2. TORQUE AND POWER

Torque and *power* are operating parameters. It takes power to turn an alternator or generator. A motor converts electrical power into mechanical power. In the SI system, power is given in kilowatts (kW). One horsepower is equivalent to 745.7 watts. The relationship between torque and power is

$$T_{\text{ft-lbf}} = \frac{5252 P_{\text{horsepower}}}{n_{\text{rpm}}} \tag{44.5}$$

$$T_{\text{N·m}} = \frac{1000 P_{\text{kW}}}{\omega_{\text{mech}}} = \frac{9549 P_{\text{kW}}}{n_{\text{rpm}}} \tag{44.6}$$

There are many important torque parameters for motors. The *starting torque* (also known as *static torque*, *breakaway torque*, and *locked-rotor torque*) is the turning effort exerted in starting a load from rest. *Pull-up torque* (*acceleration torque*) is the minimum torque developed during the period of acceleration from rest to full speed. *Pull-in torque* (as developed in synchronous motors) is the maximum torque that brings the motor back to synchronous speed. (Nominal pull-in torque is the torque that is developed at 95% of synchronous speed.) The *steady-state torque* must be provided to the load on a continuous basis. It establishes the temperature increase that the motor can withstand without deterioration. The *rated torque* is developed at rated speed and rated horsepower. The maximum torque a motor can develop at its synchronous speed is the *pull-out torque*. *Breakdown torque* is the maximum torque the motor can develop without stalling (i.e., without coming rapidly to a complete stop).

Equation 44.7 is the general torque expression for a rotating machine with *N* coils of cross-sectional area *A*, each carrying current *I* through a magnetic field of strength *B*.

$$T = NBAI\cos\omega t \tag{44.7}$$

3. SERVICE FACTOR

The horsepower and torque ratings listed on the nameplate of a motor can be provided on a continuous basis without overheating. Motors can be operated at slightly higher loads without exceeding a safe temperature rise, but the higher temperature has a deteriorating effect on the winding insulation. (A general rule of thumb is that a motor loses two or three hours of useful life for each hour run at the factored load.) The ratio of the safe to standard loads is the *service factor*, usually expressed as a decimal. Service factors vary from 1.15 to 1.4, with the lower values going to larger, more efficient motors.

$$\text{service factor} = \frac{\text{safe load}}{\text{nameplate load}} \tag{44.8}$$

4. MOTOR CLASSIFICATIONS

The National Electrical Manufacturers Association (NEMA) has categorized motors in several ways: *speed classification* (constant-, adjustable-, multi-, varying-speed, etc.); *service classification* (general, definite, and special purpose); and motor class. *Motor class* is a primary indicator of the maximum motor operating temperature, which, in turn, depends on the type of insulation used on the conductors. The motor classes are: Class A, 105°C; Class B, 130°C; Class F, 155°C; and Class H, 180°C.

5. POWER LOSSES[1]

The losses for all rotating machines can be divided into four categories. *Copper losses*, P_{Cu}, are real power losses to wire and winding resistance. In a DC machine, copper losses occur in the armature and field windings as well as from the brush contact resistance. In an AC machine, copper losses occur in the armature and exciter field windings. There are no brush losses in an induction machine.

$$P_{Cu} = \sum I^2 R \qquad 44.9$$

Core losses, including hysteresis and eddy current losses, are constant losses that are independent of the load and, for that reason, are also known as *open-circuit* and *no-load losses*. In DC and synchronous AC machines, core losses occur in the armature iron. In induction machines, core losses occur in the stator iron.

Mechanical losses (also known as *rotational losses*) include brush and bearing friction and *windage* (air friction). (Windage is a no-load loss but is not an electrical core loss.) Mechanical losses are determined by measuring the power input at rated speed and no load.

Stray losses are caused by nonuniform current distribution in the conductors. Stray losses are approximately 1% for DC machines and zero for AC machines.

Real power only is used to compute the *efficiency* of a rotating machine.

$$\begin{aligned}\eta &= \frac{\text{output}}{\text{input}} \\ &= \frac{\text{output}}{\text{output} + \text{losses}} \\ &= \frac{\text{input} - \text{losses}}{\text{input}}\end{aligned} \qquad 44.10$$

6. REGULATION

The *voltage regulation*, VR, is

$$\text{VR} = \frac{V_{\text{nl}} - V_{\text{fl}}}{V_{\text{fl}}} \times 100\% \qquad 44.11$$

The *speed regulation*, SR, is

$$\text{SR} = \frac{n_{\text{nl}} - n_{\text{fl}}}{n_{\text{fl}}} \times 100\% \qquad 44.12$$

[1]The subject of transformer losses covered in Sec. 37.3 is equivalent in concept.

7. NO-LOAD CONDITIONS

The meaning of the term *no load* is different for generators and motors. For unloaded shunt-wired alternators and generators, there is no electrical load connected across the output terminals, so although the field current flows, the line current, I, is zero. For unloaded shunt-wired motors, the work performed is zero, but line current is still drawn to keep the motor turning. All of the current is field current, however, and (neglecting mechanical losses) the armature current, I_a, is zero.

$$I = 0; \quad I_f \neq 0; \quad I_a = I_f \quad \text{[generator]} \qquad 44.13$$

$$I \neq 0; \quad I_f = I; \quad I_a = 0 \quad \text{[motor]} \qquad 44.14$$

8. PRODUCTION OF AC POTENTIAL

A potential of alternating polarity is produced by an *alternator* (*AC generator*). A permanent magnet or DC electromagnet produces a constant magnetic field. Figure 44.1 illustrates how several loops of wire can be combined into a rotating *induction coil* or *armature* to produce a continuously varying potential in a *dynamo* (*coil dynamo*). Three-phase alternators have three sets of independent windings. A single-phase AC alternator has only one set of windings.

Figure 44.1 *Elementary Two-Pole, Single-Coil Dynamo*

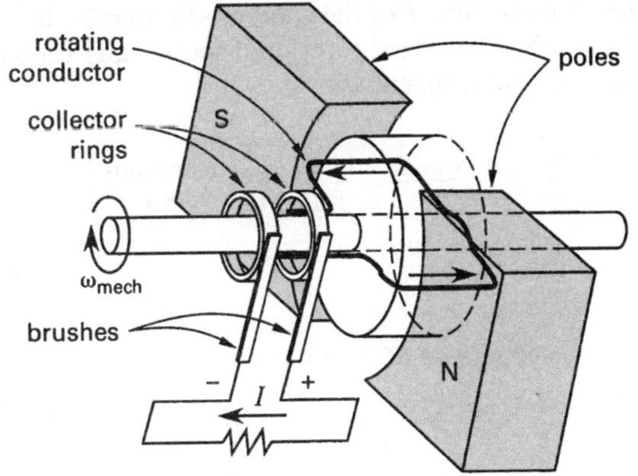

The induced voltage, E, is commonly called *electromotive force* (emf). In an elementary alternator, emf is the desired end result and is picked off by stationary brushes making contact with *slip rings* on the rotating shaft.[2] In a motor, emf is also produced but is referred to as *back emf* (*counter emf*) because it opposes the input current.

Assuming the magnetic field flux density under each pole face, B, is uniform, the maximum flux linked by a coil with N turns and area A is NAB. Because the coil rotates, the flux linkage is a function of the projected coil area.

[2]Practical alternators, described in Sec. 44.10, use slip rings to feed the field, not to transfer the generated voltage.

The instantaneous induced voltage is predicted by Faraday's law. Care must be taken to distinguish between the armature speed, n (in rpm), the angular armature speed, ω_{mech} (in rad/s), and the linear and angular voltage frequencies, f and ω (in Hz and rad/s, respectively). Note the distinction in Eq. 44.15 between the rotational speeds of the armature ($\omega_{\text{mech}} = 2\pi n/60$) and the electrical waveform ($\omega = 2\pi f$).

$$\begin{aligned} V(t) &= V_m \sin \omega t = \omega NAB \sin \omega t \\ &= \left(\frac{p}{2}\right)\omega_{\text{mech}} NAB \,\sin\left(\left(\frac{p\omega_{\text{mech}}}{2}\right)t\right) \\ &= \frac{\pi npNAB}{60}\sin\left(\left(\frac{p\omega_{\text{mech}}}{2}\right)t\right) \end{aligned} \qquad 44.15$$

$$\begin{aligned} V &= \frac{V_{\text{max}}}{\sqrt{2}} = \frac{\omega NAB}{\sqrt{2}} = \frac{p\omega_{\text{mech}}NAB}{2\sqrt{2}} \\ &= \frac{\pi npNAB}{60\sqrt{2}} \quad \text{[effective]} \end{aligned} \qquad 44.16$$

Table 44.1 summarizes the most frequently used formulas for a single-phase AC alternator.

Alternators are characterized by the number of magnetic poles, p. (Both north and south poles are counted to distinguish the quantity from the number of *pole pairs.*) The *coil pitch* (*pole pitch*) is the angle between the poles, or 360° divided by p. A two-pole alternator produces one complete sinusoidal cycle per revolution. An alternator with p poles produces $(1/2)p$ cycles per revolution. Because the armature normally turns at a constant speed known as the *synchronous speed*, n_s, the *electrical frequency*, f, of the generated potential is given by Eq. 44.18. The actual rotational speed, n, is known as the *mechanical frequency*.

$$n_s = \frac{120f}{p} = \frac{60\omega_{\text{mech}}}{2\pi} = \frac{60\omega}{\pi p} \quad \left[\begin{array}{c}\text{synchronous}\\ \text{speed}\end{array}\right] \qquad 44.17$$

$$f = \frac{1}{T} = \frac{\omega}{2\pi} = \frac{pn_s}{120} \qquad 44.18$$

Because the coils in an alternator have inductance as well as resistance (see Sec. 44.16), the rated capacity of an AC machine is stated as apparent power at some rated voltage and power factor.

Example 44.1

A four-pole alternator produces a 60 Hz potential. What is the (a) mechanical speed of the armature, (b) angular velocity of the potential, and (c) angular velocity of the armature?

Solution

(a) From Eq. 44.17, the rotational speed is

$$\begin{aligned} n = n_s &= \frac{120f}{p} = \frac{\left(120\ \frac{\text{pole·s}}{\text{min}}\right)(60\ \text{Hz})}{4\ \text{poles}} \\ &= 1800\ \text{rpm} \end{aligned}$$

(b) The angular velocity of the 60 Hz potential is

$$\omega = 2\pi f = \left(2\pi\ \frac{\text{rad}}{\text{cycle}}\right)(60\ \text{Hz}) = 377\ \text{rad/s}$$

(c) From Eq. 44.15, the angular velocity of the armature is

$$\begin{aligned} \omega_{\text{mech}} &= \frac{2\omega}{p} = \frac{2\pi n}{60} = \frac{\left(2\pi\ \frac{\text{rad}}{\text{rev}}\right)\left(1800\ \frac{\text{rev}}{\text{min}}\right)}{60\ \frac{\text{s}}{\text{min}}} \\ &= 188.5\ \text{rad/s} \end{aligned}$$

Example 44.2

The rotor of a single-phase, four-pole alternator rotates at 1200 rpm. The effective diameter and length of the 20-turn (loop) coil are 0.12 m and 0.24 m, respectively. The magnetic flux density is 1.2 T. What is the effective voltage produced?

***Table 44.1** Single-Phase AC Alternator Formulas*

	in terms of n	in terms of ω_{mech}	in terms of f	in terms of ω
n (rpm)	–	$\frac{30\,\omega_{\text{mech}}}{\pi}$	$\frac{120f}{p}$	$\frac{60\omega}{\pi p}$
ω_{mech} (mechanical rad/s)	$\frac{\pi n}{30}$	–	$\frac{4\pi f}{p}$	$\frac{2\omega}{p}$
f (Hz)	$\frac{pn}{120}$	$\frac{p\omega_{\text{mech}}}{4\pi}$	–	$\frac{\omega}{2\pi}$
ω (electrical rad/s)	$\frac{\pi pn}{60}$	$\frac{p\omega_{\text{mech}}}{2}$	$2\pi f$	–
single-phase V_{max} (V)	$\frac{\pi pnNAB}{60}$	$\frac{\omega_{\text{mech}}pNAB}{2}$	$2\pi fNAB$	ωNAB
single-phase V_{eff} (V)	$\frac{\pi pnNAB}{60\sqrt{2}}$	$\frac{\omega_{\text{mech}}pNAB}{2\sqrt{2}}$	$\frac{2\pi f\,NAB}{\sqrt{2}}$	$\frac{\omega NAB}{\sqrt{2}}$
single-phase V_{ave} (V)	$\frac{pnNAB}{30}$	$\frac{\omega_{\text{mech}}pNAB}{\pi}$	$4fNAB$	$\frac{2\omega NAB}{\pi}$

Solution

The coil area is

$$A = 2rl = (0.12\ \text{m})(0.24\ \text{m}) = 0.0288\ \text{m}^2$$

From Eq. 44.16, the effective voltage is

$$V = \frac{p\pi nNAB}{60\sqrt{2}} = \frac{(4)\left(1200\ \frac{\text{rev}}{\text{min}}\right)\left(2\pi\ \frac{\text{rad}}{\text{rev}}\right) \times (20)(0.0288\ \text{m}^2)(1.2\ \text{T})}{(2\sqrt{2})\left(60\ \frac{\text{s}}{\text{min}}\right)} = 122.8\ \text{V}$$

9. ARMATURE WINDINGS

An *armature winding* consists of several *coils* of continuous wire formed into *q loops*. Each loop in an armature contributes two *conductors*, also known as *bars*. The number of conductors, z, is

$$z = 2q \qquad 44.19$$

Voltage is induced only in conductors and inductors that are parallel to the armature shaft (i.e., that cut magnetic flux lines). The number of *series paths*, N, between positive and negative brush sets depends on how the coils are wound and connected to the commutator.

$$N = \frac{z}{a} = \frac{2q}{a} \qquad 44.20$$

The number of parallel armature paths between each pair of brushes, a, equals the number of poles, p, in a *lap-wound armature* (*simplex-lap armature* or *multiple-drum armature*).[3] (The number of poles is also equal to the number of brushes.) The coil connections are made to adjacent commutator segments. This type of winding is commonly used in DC machines and induction motors and, to a lesser extent, in AC generators.

In a *wave-wound armature* (*two-circuit* or *series-drum armature*), $a = 2$, and the coil connections are on commutator segments on opposite sides of the armature regardless of the number of poles. Only two brushes are needed, although more can be used. This configuration is commonly used in DC armatures requiring the generation or use of higher voltages than lap windings can tolerate.

A *multiplex winding* has two or more distinct coils connected in parallel. The number of parallel current paths, a, in the armature can be determined by the type of winding and the number of poles, p, as indicated in Table 44.2.

[3]Auxiliary windings are used in *shaded-pole motors*; *split-phase motors* have two separate windings; *capacitor motors* have capacitors in series with an auxiliary winding.

Table 44.2 *Multiplex Winding Current Paths*

type	lap winding	wave winding
simplex	$a=p$	$a=2$
duplex	$a=2p$	$a=4$
triplex	$a=3p$	$a=6$
quadriplex	$a=4p$	$a=8$

10. PRACTICAL ALTERNATORS

There are several reasons why large alternators are not designed with a stationary magnetic field and rotating coil as shown in Fig. 44.1.

- The rotating coils must be well insulated to prevent shorting with the high voltages that are induced.
- Structural bracing is required to counteract the large centrifugal force that results from rotating many coils of wire.
- It is difficult to make efficient high-voltage, high-power connections through slip rings.

For these reasons, practical alternators (see Fig. 44.2) reverse the locations of the field and induction coils so that the low-voltage field revolves and the high-voltage field is induced in stationary coils. The magnetic field is produced by *field windings* whose DC *magnetization current* is supplied through brushes and slip rings. Such an armature containing field coils is known as a *rotor*. The stationary induction coils are placed in slots a small distance from the rotor and are known as the *stator*. The closer the stator and rotor are, the smaller the magnetization current required.

Figure 44.2 *Practical Alternator*

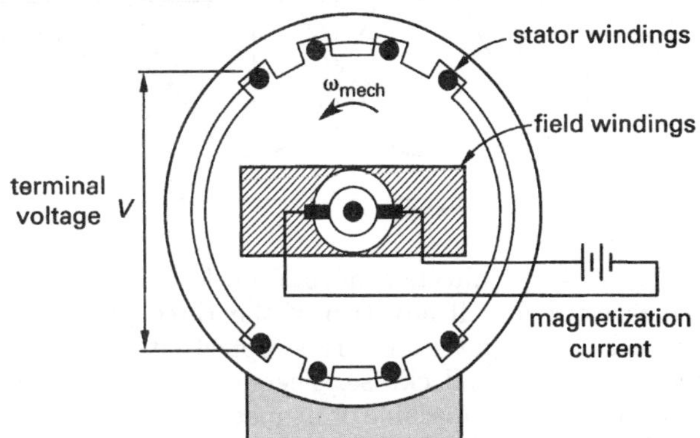

With *cylindrical rotors* (*wound rotors*), the field windings are embedded in an externally smooth-surfaced cylindrical rotor. This makes them suitable for high-speed (i.e., greater than 1800 rpm) operation because windage losses are reduced. The poles of *salient-pole rotors*, on the other hand, resemble exposed wound electromagnets. This is adequate for 1800 rpm and less.

11. ROTATING MAGNETIC FIELD

When a three-phase AC output is applied to a three-phase motor stator, the result is a magnetic field on the motor's stationary windings. The resultant magnetic field rotates with respect to those stationary windings. The speed of rotation is called the *synchronous speed*, n_s, in revolutions per minute (rpm) and is dependent on the number of poles per phase on the stator of the motor, p, and the applied frequency, f. This rotating magnetic field determines the speed of a synchronous motor and is responsible for the induced voltage, current, and torque in an induction motor rotor.

$$n_s = \frac{120f}{p} \qquad 44.21$$

A three-phase current is shown in Fig. 44.3. The seven points beneath the currents represent one complete cycle. In Fig. 44.4, each of the points is applied to a two-pole machine. The magnetic field moves 60° between points, which correlates with the 60° between subsequent poles for a rotational speed of 3600 rpm.[4] As shown in Eq. 44.21 and Fig. 44.3, the correlation between the mechanical angles between poles does not need to be the same as the electrical angles of the three-phase source. In a four-pole machine, the electrical angle between the points of Fig. 44.3 would remain 60°, while the mechanical angle between points would be 30° with a subsequent rotational speed of 1800 rpm.

Figure 44.3 *Three-Phase Currents*

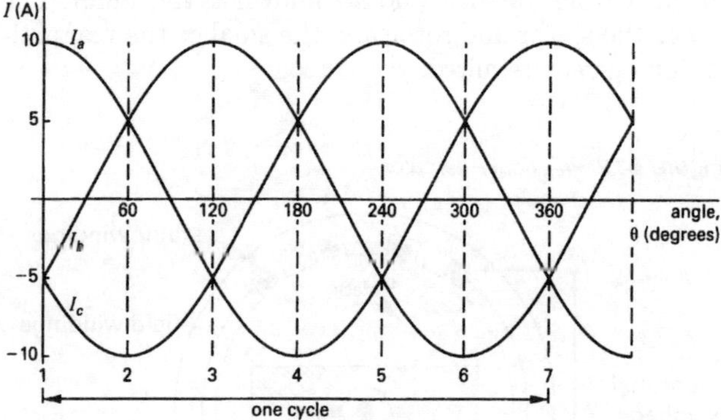

In Fig. 44.4, the direction of rotation for the magnetic field is clockwise. If any two of the three phases are reversed (the input leads are swapped), the rotation changes to counterclockwise. Therefore, to change the direction of an AC machine or its magnetic field, change the direction of the rotating field. For a three-phase machine, this is accomplished by swapping any two leads. For a single-phase machine, numerous methods exist; however, all methods involve changing the direction of the revolving magnetic field.

[4]The field moves 60° between points, or $^1/_6$ of a revolution. In one cycle, one revolution occurs. A cycle is $^1/_{60}$ of a second. Therefore, 60 revolutions occur in one second, for a synchronous speed of 3600 rpm.

12. SYNCHRONOUS MOTORS

Synchronous motors are essentially dynamo alternators operating in reverse. Alternating current is supplied to the stationary stator windings. DC current is applied to the field windings in the rotor through brushes and slip rings as in an alternator.[5] The field current interacts with the stator field, causing the armature to turn. Because the stator field frequency is fixed, the motor runs only at a single *synchronous speed* (see Eq. 44.17).

Important features of synchronous motors follow.

- They turn at constant speeds, regardless of the load. (The speed may momentarily change when the load is changed.) Stalling occurs when a motor's counter-torque is exceeded.
- The power factor can be adjusted manually without losing synchronization by varying the field current. A unity power factor occurs with *normal excitation* current. The power factor is leading (lagging) when the current is more (less) than normal, and this is known as being *over-* (*under-*) *excited.*
- They can draw leading currents and be used for power factor correction, in which case they are known as *synchronous capacitors* (*synchronous condensers*).

Because the starting torque is zero, it is necessary to bring a synchronous motor up to speed by some other means. The most common method is to include auxiliary windings in the pole faces so that the motor can be started as an induction motor. At synchronous speed, the auxiliary windings draw no current. If the motor speed becomes nonsynchronous, the auxiliary windings draw power to resynchronize the rotor.

Because synchronous motors run at only one speed, curves of horsepower (or other characteristics) versus speed are not used. Instead, motors are categorized on the basis of speed, torque, and horsepower at a specific power factor.

13. SYNCHRONOUS MACHINE EQUIVALENT CIRCUIT

Figure 44.5 illustrates a simple equivalent circuit for a synchronous machine. The vector voltage relationship is defined by Eq. 44.22 and Eq. 44.23, which introduce the equivalent *synchronous inductance*, X_s, of each phase. V_p is the phase voltage. The series armature resistance, R_a, is small and normally disregarded.

$$\begin{aligned}\mathbf{E} &= \mathbf{V}_p + (R_a + jX_s)\mathbf{I}_a \\ &\approx \mathbf{V}_p + jX_s\mathbf{I}_a \quad \text{[alternator]}\end{aligned} \qquad 44.22$$

[5]The magnetization energy can also be generated through induction. In that case, the rotor includes diodes to rectify the induced AC potential.

Figure 44.4 *Rotating Magnetic Field Flux Patterns*

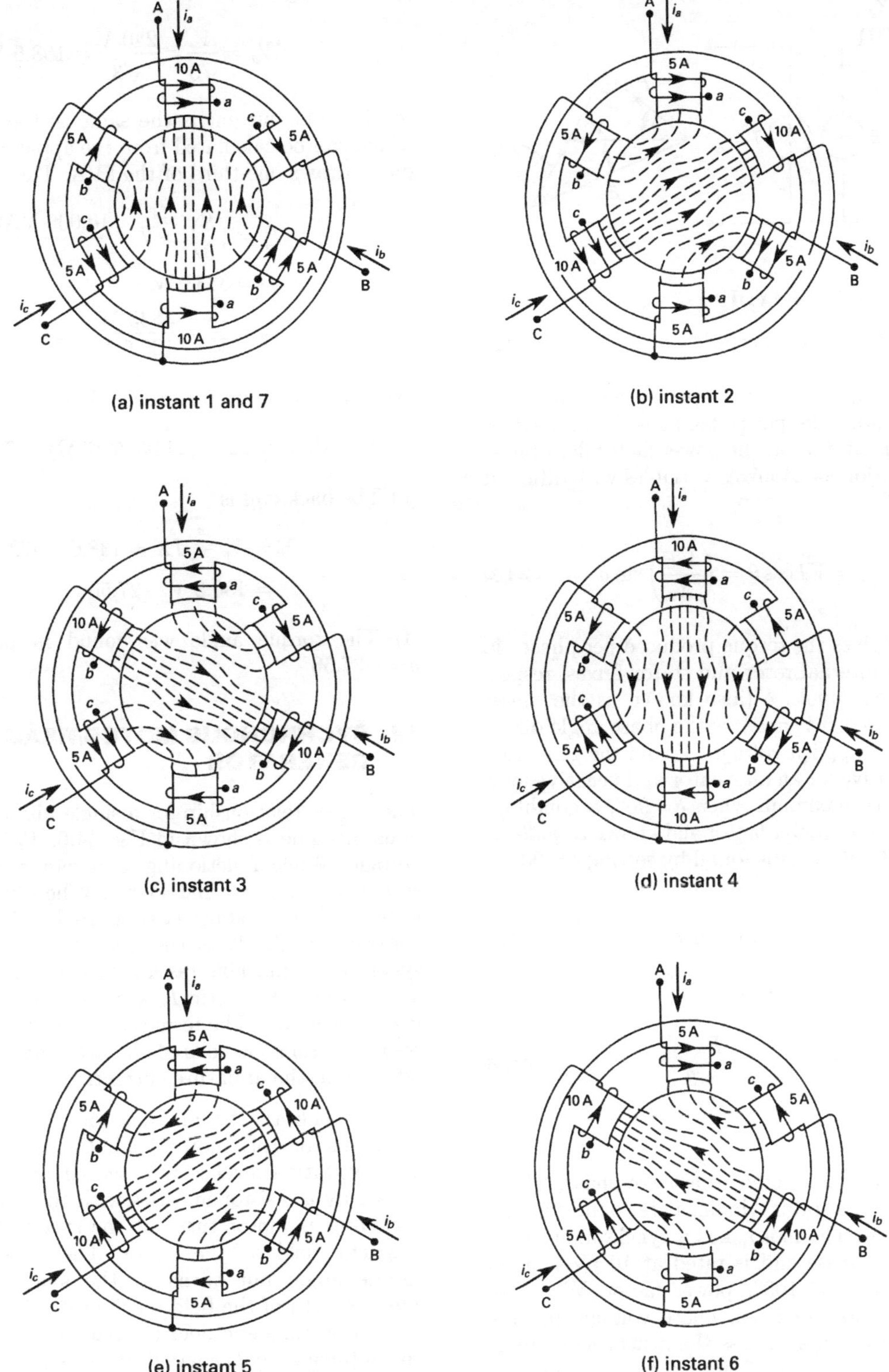

(a) instant 1 and 7

(b) instant 2

(c) instant 3

(d) instant 4

(e) instant 5

(f) instant 6

Figure 44.5 Synchronous Motor Equivalent Circuit

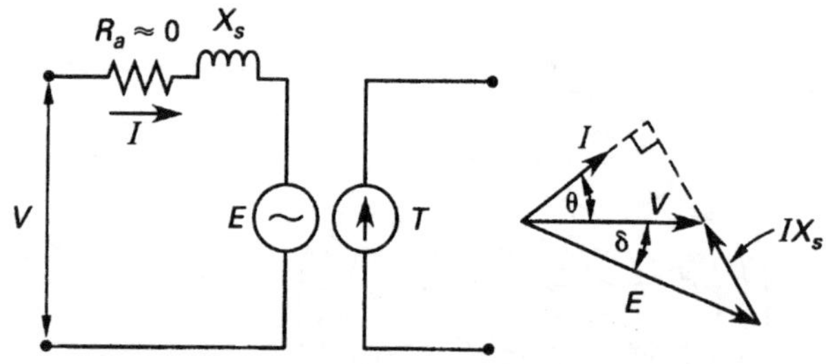

$$\mathbf{V}_p = \mathbf{E} + (R_a + jX_s)\mathbf{I}_a$$
$$\approx \mathbf{E} + jX_s\mathbf{I}_s \quad \text{[motor]} \qquad 44.23$$

Equation 44.24 gives the real power generated per phase. For a motor, the power factor is determined by E and I_f. For an alternator, the power factor determines E and I_f. (Equation 44.24 mixes variables with different units.)

$$P_p = T_p\omega_{\text{mech}} = VI\cos\theta = \left(\frac{VE}{X_s}\right)\sin\delta \qquad 44.24$$

Equation 44.25 gives the torque produced per phase. Φ_r and Φ_{st} are the internal rotor and stator fluxes, respectively. The *torque angle*, δ (also known as the *power angle* and *displacement angle*), is the phase angle difference between the applied voltage, V, and the generated emf, E. It is positive for an alternator and negative for a motor. Torque is maximum when $\delta = 90°$, a condition equivalent to a unity power factor and known as *pull-out torque*. The *pull-out power* is found by setting $\delta = 90°$ in Eq. 44.24.

$$T_p = k_T\Phi_r\Phi_{st}\sin\delta \qquad 44.25$$

The apparent power per phase is

$$S = VI = \frac{P_p}{\cos\theta} \qquad 44.26$$

Example 44.3

A six-pole motor is connected to a three-phase, 240 V (rms), 60 Hz line. Its stator windings are connected in a wye configuration. The motor has a synchronous reactance of 3 Ω per phase and is rated at 10 kVA at its synchronous speed and 100% power factor. What are the (a) synchronous speed, (b) phase voltage, (c) line current, (d) voltage drop across the synchronous reactance, (e) generated back emf, and (f) torque angle?

Solution

(a) Equation 44.17 gives the synchronous speed.

$$n_s = \frac{120f}{p} = \frac{(120)(60 \text{ Hz})}{6} = 1200 \text{ rev/min}$$

(b) Because the windings are in a wye configuration, the phase voltage is

$$V_p = \frac{V}{\sqrt{3}} = \frac{240 \text{ V}}{\sqrt{3}} = 138.6 \text{ V}$$

(c) The line current is the same as the phase current, which can be calculated from the apparent power. Each phase draws one-third of the kVA. The real power is

$$P_p = \left(\frac{S}{3}\right)\cos\theta = \frac{10{,}000 \text{ VA}}{3}$$
$$= 3333 \text{ W}$$
$$I_p = \frac{P_p}{V_p} = \frac{3333 \text{ W}}{138.6 \text{ V}} = 24.05 \text{ A}$$

(d) The voltage drop across each winding is

$$V_p = I_pX_p = (24.05 \text{ A})(3 \text{ Ω}) = 72.15 \text{ V}$$

(e) The back emf is

$$\mathbf{E} = V_p - jIX = 138.6 - j72.15$$
$$= 156.3 \text{ V}\angle{-27.50°}$$

(f) The torque angle was found in part (e) to be $\delta = -27.50°$.

14. SYCHRONOUS MOTOR/REACTIVE GENERATOR

The vector relationship for a single phase of a synchronous machine is shown in Fig. 44.6. V_t is the terminal voltage. When functioning as a motor, the voltage is supplied by an external source. When functioning as a generator, the voltage is supplied by the synchronous generator itself. V_0 is the generated voltage from the synchronous machine (as a motor or generator), shown as E earlier.[6] The term I_a is the armature current flowing in the stator. The term jI_aX_s is the inductive voltage resulting from the coils of the armature reactance, 90° offset from the armature current.

In Fig. 44.6(a), the motor is operating as a lagging power factor (pf) motor load, the underexcited case. The excitation level is represented by V_0 and is controlled by a resistor in series with the rotating field. A synchronous motor turns at the speed of the rotating magnetic field on the stator. The rotor lags by the torque angle but turns at the synchronous speed. The strength of the field on the rotor interacting with the stator magnetic field is what maintains the torque and subsequently the speed. If the excitation is lowered excessively, the magnetic field may not be able to remain synchronized with the stator depending on the motor load. If this occurs, the machine may suffer *pole slip*. Pole slippage results in excessively high induced voltages and damaging currents that may render the motor useless.

[6]In other electrical books, a generated voltage V is shown as E.

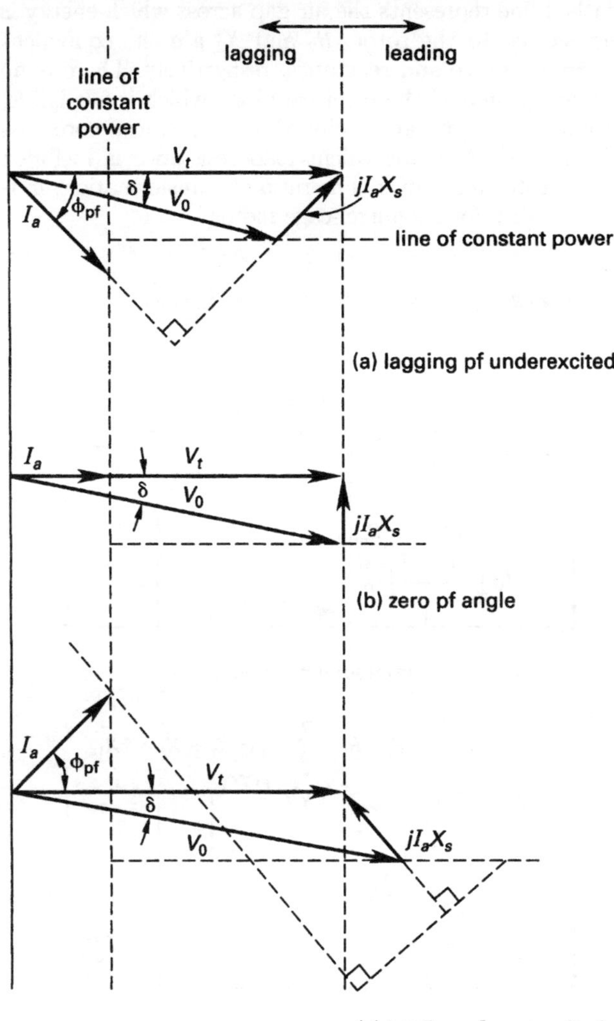

Figure 44.6 *Synchronous Motor Phasor Relationship*

In Fig. 44.6(b), the excitation level increases when the V_0 phasor increases. As long as the motor load has not increased, the excitation voltage phasor follows the line of constant power. The torque angle becomes smaller in magnitude due to the increased strength of the magnetic field on the rotor, but the overall power delivered is determined by the load, and therefore remains constant. At this point, the power factor angle of the motor is zero, which is also called the *unity power factor*.

In Fig. 44.6(c), the machine is operating as a motor with a leading power factor and excitation has increased to the point where the armature current causes the terminal voltage, as opposed to the original system source causing the terminal voltage. The motor then appears to the supplying system as a capacitive load, or a *reactive generator*. Therefore, a synchronous motor can correct the power factor of the supplying distribution system by reducing the power factor angle to zero (if the motor is capable of supplying the entire reactive load on the system). This will cause the total loading on the system source to appear resistive, minimizing the total current.[7] This occurs because the reactive current component, previously supplied by the system source, is now supplied by the motor, which is essentially acting as a large capacitor.

15. INDUCTION MOTORS

Induction motors are essentially constant-speed devices that receive power through induction—there are no brushes or slip rings. A motor can be considered as a rotating transformer secondary (the rotor) with a stationary primary (the stator). The stator field rotates at the synchronous speed given in Eq. 44.17. An emf is induced as the stator field moves past the rotor conductors. Because the rotor windings have reactance, the rotor field lags the induced emf.

In order to have a change in flux linkage, the rotor must turn at less than the synchronous speed. The difference in speed is small but essential. *Percent slip*, s, typically 2% to 5%, is the percentage difference in speed between the rotor and stator fields. Slip can be expressed as either a decimal or a fraction (e.g., 0.05 slip or 5% slip). Percent slip and *percent synchronism* are complements (i.e., add to 100%). Slip in rpm is the difference between actual and synchronous speeds.

$$s = \frac{n_s - n}{n_s} = \frac{\omega_{\text{mech},s} - \omega_{\text{mech}}}{\omega_{\text{mech},s}} \qquad 44.27$$

The stator is identical to that in an alternator or synchronous motor. In a *wound rotor*, the rotor is similar to an armature winding in a dynamo.[8] However, in a *squirrel-cage rotor* there are no wire windings at all.[9] The squirrel-cage rotor consists of copper or aluminum bars embedded in slots in the cylindrical iron core of the rotor. The ends of the bars are shorted by conductive rings, as illustrated in Fig. 44.7, to form loops. The rotor and its components, connections, and lines constitute the *rotor circuit* or *secondary circuit*.

Figure 44.7 *Squirrel-Cage Rotor Induction Motor*

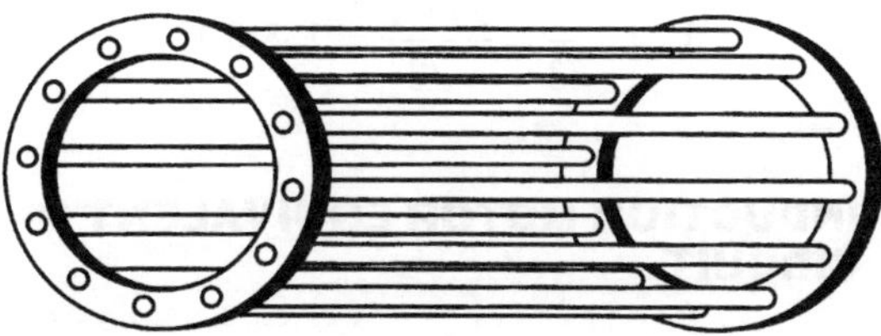

[7]Correcting the power factor means bringing it to zero so that the total loading appears resistive and the total current is at a minimum.

[8]Windings can be incorporated in the rotor to obtain better speed control or high torque.

[9]If the slip rings of a wound rotor are shorted, the motor behaves like a squirrel-cage motor.

Example 44.4

An induction motor developing 10 hp is connected to a three-phase, 240 V (rms), 60 Hz power line. The stator windings are connected in a wye configuration. The synchronous speed is 1800 rpm, but the motor turns at 1738 rpm when loaded. Its energy efficiency is 80% and power factor (pf) is 70%. Calculate the (a) slip, (b) number of poles, (c) line current drawn, and (d) phase voltage.

Solution

(a) From Eq. 44.27, the slip is

$$s = \frac{n_s - n}{n_s} = \frac{1800\ \frac{\text{rev}}{\text{min}} - 1738\ \frac{\text{rev}}{\text{min}}}{1800\ \frac{\text{rev}}{\text{min}}} = 0.03444$$

(b) The number of poles is calculated from the synchronous speed.

$$p = \frac{120f}{n_s} = \frac{(120)(60\ \text{Hz})}{1800\ \frac{\text{rev}}{\text{min}}} = 4$$

(c) The total real power input is

$$P = \frac{(10\ \text{hp})\left(745.7\ \frac{\text{W}}{\text{hp}}\right)}{0.8} = 9321\ \text{W}$$

The power per phase is

$$P_p = \frac{P}{3} = \frac{9321\ \text{W}}{3} = 3107\ \text{W}$$

Because the phase and line currents are identical in wye-connected loads (see Sec. 34.9), the line current is

$$I = I_p = \frac{P_p}{V_p(\text{pf})} = \frac{3107\ \text{W}}{\left(\frac{240\ \text{V}}{\sqrt{3}}\right)(0.70)}$$

$$= 32.03\ \text{A}$$

(d) The voltage across each phase is

$$V_p = \frac{V}{\sqrt{3}} = \frac{240\ \text{V}}{\sqrt{3}} = 138.6\ \text{V}$$

16. INDUCTION MOTOR EQUIVALENT CIRCUIT

Figure 44.8(a) illustrates the equivalent circuit for an induction motor.[10] It is very similar to the equivalent circuit for a transformer. R_1 and X_1 represent the equivalent stator resistance and reactance, respectively. G_{nl} and B_{nl} represent the stator core loss and susceptance, respectively, as determined from no-load testing. The dashed line represents the air gap across which energy is transferred to the rotor. R_2 and X_2 are the equivalent rotor resistance and reactance, respectively. There is no element to model the rotor core loss, which is negligible. Rotational losses are included in the stator core loss element, G. Any equivalent load resistance is included in the rotor resistance. The ratio of transformation, a, is taken as 1.0 for a squirrel-cage motor.[11]

Figure 44.8 *Equivalent Circuits of an Induction Motor*

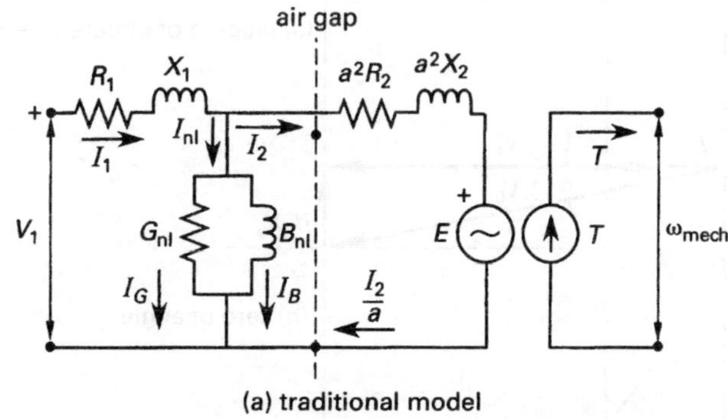

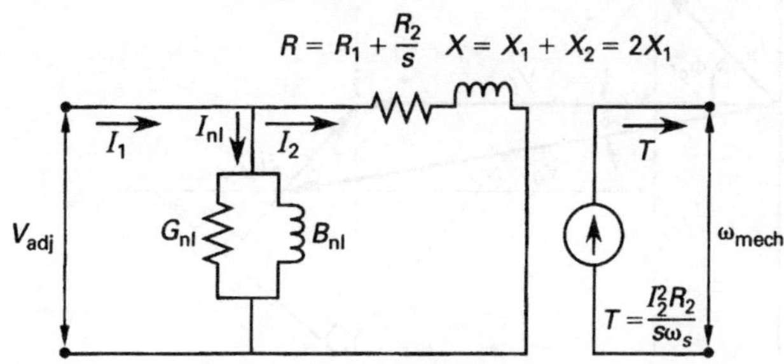

Using an adjusted voltage, V_{adj}, simplifies the model, as shown in Fig. 44.8(b). The relationship between the applied terminal voltage, V_1, and the adjusted voltage is

$$\mathbf{V}_{\text{adj}} = \mathbf{V}_1 - \mathbf{I}_{\text{nl}}(R_1 + jX_1) \qquad 44.28$$

$$V_{\text{adj}} \approx V_1 - I_{\text{nl}}\sqrt{R_1^2 + X_1^2} \qquad 44.29$$

The total series resistance per phase, R, is

$$R = R_1 + \frac{R_2}{s} \qquad 44.30$$

[10]This equivalent circuit cannot be used for a double squirrel-cage motor.

[11]With a phase-wound rotor, the brushes can be lifted from the slip rings. The ratio of transformation, a, can be determined as the ratio of applied voltage to voltage across the slip rings. This cannot be done with a squirrel-cage motor, so the equivalent circuit parameters are redefined with the ratio of transformation.

Equation 44.31 gives the torque-speed relationship predicted by this model.

$$T_p = \frac{I_2^2 R_2}{s\omega_s} = \frac{V_{\text{adj}}^2 R_2}{s\omega_s\left(\left(R_1 + \frac{R_2}{s}\right)^2 + X^2\right)} \quad 44.31$$

$$T_t = 3T_p \quad 44.32$$

17. OPERATING CHARACTERISTICS OF INDUCTION MOTORS

Under normal operating conditions, slip is small (less than 0.05) and Eq. 44.33 predicts that the torque will be directly proportional to slip, s, and inversely proportional to the rotor resistance, R_2. At low speeds, the reactive term is larger than the resistive term in Eq. 44.34.

$$T_p \approx \frac{V_{\text{adj}}^2 s}{\omega_s R_2} \quad \text{[high speed, per phase]} \quad 44.33$$

$$\approx \frac{V_{\text{adj}}^2 R_2}{s\omega_s X^2} \quad \text{[low speed, per phase]} \quad 44.34$$

The starting torque per phase is proportional to rotor winding resistance and terminal voltage and can be found by setting s equal to one in Eq. 44.34.

$$T_{\text{starting}} \approx \frac{V_{\text{adj}}^2 R_2}{\omega_s X^2} \quad \text{[starting, per phase]} \quad 44.35$$

The maximum torque (known as *breakdown torque*) is independent of the rotor circuit resistance, but the rotor resistance does affect the speed at which the maximum torque occurs. A large rotor circuit resistance only causes the maximum torque to occur at a larger slip. Maximum torque varies directly with the square of the stator voltage.

$$T_{\text{max}} = \frac{V_{\text{adj}}^2}{2\omega_s\left(R_1 + \sqrt{R_1^2 + X^2}\right)} \quad \text{[per phase]} \quad 44.36$$

$$s_{T,\text{max}} = \frac{R_2}{\sqrt{R_1^2 + X^2}} \quad 44.37$$

Figure 44.9 illustrates typical characteristic curves of an induction motor. Such curves are conventionally provided for the motor running at its optimum efficiency and power factor. Curves can be provided for polyphase or single-phase operation.

18. TESTING INDUCTION MOTORS

Performance of induction motors is evaluated in a manner analogous to transformer testing. The *no-load motor test* (*running-light test*) corresponds to an open-circuit transformer test. This test determines the values of B and G in the equivalent circuit. The *blocked-rotor test* corresponds to a closed-circuit transformer test and determines the values of R and X.

***Figure 44.9** Characteristic Curves for an Induction Motor*

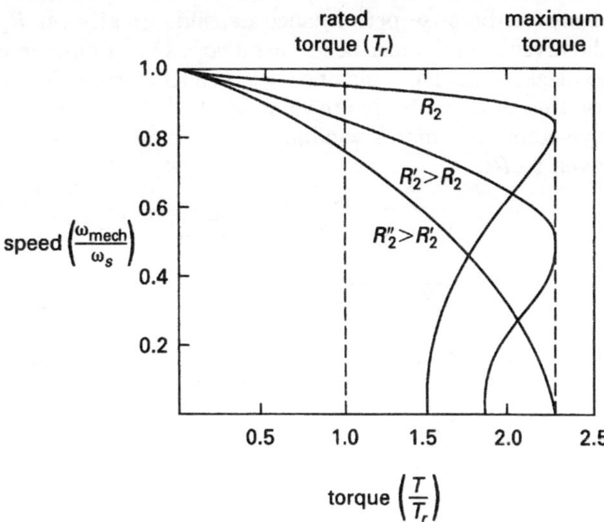

In a *no-load test*, the motor is run at the rated voltage without load. The line voltage, V, line current, I_{nl}, and power per phase, P_{nl}, are measured.

$$\text{no-load power factor} = \cos\theta_{\text{nl}} = \frac{P_{\text{nl}}}{V_{\text{adj}} I_{\text{nl}}} \quad 44.38$$

Referring to Fig. 44.8, the magnetization (quadrature) current, I_B, and the in-phase component of the exciting current, I_G, are

$$I_B = I_{\text{nl}} \sin\theta_{\text{nl}} \quad 44.39$$

$$I_G = I_{\text{nl}} \cos\theta_{\text{nl}} \quad 44.40$$

The stator parameters are

$$G_{\text{nl}} = \frac{P_{\text{nl}}}{V_{\text{adj}}^2} \quad 44.41$$

$$B_{\text{nl}} = -\sqrt{Y_{\text{nl}}^2 - G_{\text{nl}}^2} = -\sqrt{\left(\frac{I_{\text{nl}}}{V_{\text{nl}}}\right)^2 - G_{\text{nl}}^2}$$

$$\approx \frac{I_B}{V_{\text{adj}}} \quad 44.42$$

In a *blocked-rotor test*, the rotor is blocked. A (low) voltage is applied and measured so that the rated current flows. The phase current, I_b, power per phase, P_b, and phase voltage, V_b, are measured.

The phase current lags the phase voltage by the following angle.

$$\cos\theta_b = \frac{P_b}{I_b V_b} \quad 44.43$$

The stator reactance, X_1, is usually assumed to be one-half of X, the remainder being the rotor reactance. However, because performance depends greatly on R_2, the division of R must be less arbitrary. One approach is to measure the DC resistance between the terminals and use this as an approximate value of $2R_1$.[12] (This assumes wye-connected motor windings.) The remainder of R is given to R_2.

$$X = X_1 + X_2 = 2X_1 = \frac{V_b}{I_b \sin\theta_b} = \sqrt{Z^2 - R^2} \quad 44.44$$

$$R = R_1 + R_2 = \frac{P_b}{I_b^2} \quad 44.45$$

$$Z = \frac{V_b}{I_b} \quad 44.46$$

19. STARTING INDUCTION MOTORS

Induction motors draw their maximum current when starting (i.e., when slip, s, is one). The starting torque for a polyphase motor varies directly with the square of the stator voltage but also depends on the rotor resistance and reactance. (Starting torque for a single-phase induction motor is zero. Therefore, single-phase induction motors must be brought up to speed by some other means.[13]) For a given stator voltage, there is a particular rotor resistance that maximizes starting torque. Increasing or decreasing resistance from this optimum value decreases the starting torque. Starting current can be calculated from Eq. 44.47 by setting s equal to one.

$$I_{\text{starting}} = I_1 = I_{\text{nl}} + I_2 = I_{\text{nl}} + \frac{V_{\text{adj}}}{R_1 + \frac{R_2}{s} + jX} \quad 44.47$$

The *blocked- (locked-) rotor current* is the current drawn by the motor with the rotor held stationary. It is a worst-case starting current because the rotor begins to move immediately upon starting, reducing the current drawn. *Free-rotor starting current* is approximately 75% of the blocked-rotor current.

20. SPEED CONTROL FOR INDUCTION MOTORS

The rotational speed of an induction motor depends on the number of poles, line voltage, supply frequency, and rotor circuit resistance. For a given machine, the number of poles cannot be varied without excessive complexity in winding, switching, and increased manufacturing cost. Because the breakdown torque is proportional to the square of the voltage, reducing the voltage may stall the motor. As a result, voltage speed control is rarely used. Changing the supply frequency is also impractical in most instances. Introducing a resistance in series with the rotor decreases the motor speed but is possible only in wound-rotor motors.

Speed control is commonly accomplished by introducing a foreign voltage in the secondary (rotor) circuit. If the foreign voltage opposes the voltage induced in the secondary circuit, the motor speed will be reduced, and vice versa.

If two induction motors are available, one can be used to control the other by connecting them in *cascade*. The shafts are rigidly connected and the rotor and stator windings are interconnected. The two armatures can also be constructed on a single shaft.

21. POWER TRANSFER IN INDUCTION MOTORS

Figure 44.10 illustrates the power transfer in an induction motor. Equation 44.48 through Eq. 44.53 are per phase.

$$\text{input power} = V_1 I_1 \cos\theta \quad 44.48$$

$$\text{stator copper losses} = I_1^2 R_1 \quad 44.49$$

$$\text{rotor input power} = \frac{I_2^2 R_2}{s} \quad 44.50$$

$$\text{rotor copper losses} = I_2^2 R_2 \quad 44.51$$

$$\text{electrical power delivered} = I_2^2 R_2 \left(\frac{1-s}{s}\right) \quad 44.52$$

$$\text{shaft output power} = T\omega_{\text{mech}} \quad 44.53$$

Core losses are constant (see Sec. 44.5). Copper losses are proportional to the square of the delivered power.

Figure 44.10 *Induction Motor Power Transfer*

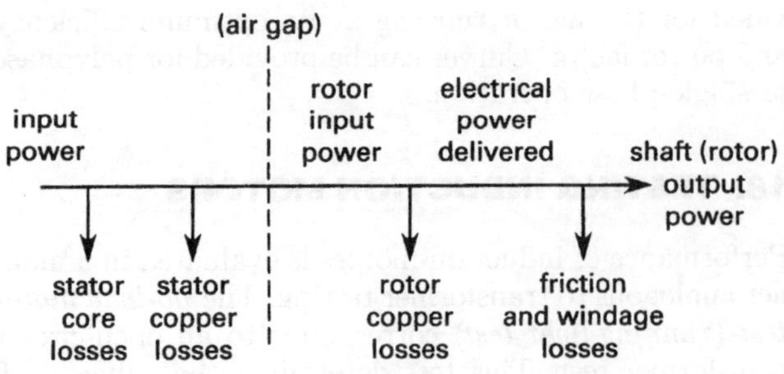

[12]This value is corrected empirically, but the correction is beyond the scope of this book.

[13]Polyphase induction motors will run but will not start on a single phase. There is a danger of current overload if synchronization is lost during single-phase operation. Protective elements (e.g., circuit breakers) are used to limit current.

Example 44.5

A 15 hp induction motor with six poles operates at 80% efficiency on a three-phase, 240 V (rms), 60 Hz line. The following losses are observed for full-load operation:

stator copper loss	540 W
friction/windage loss	975 W
core loss	675 W

What are the (a) speed and (b) torque when the motor delivers half power?

Solution

(a) The full-load output power is

$$P = (15\ \text{hp})\left(745.7\ \frac{\text{W}}{\text{hp}}\right) = 11{,}186\ \text{W}$$

The input power is

$$P_{\text{in}} = \frac{P_{\text{out}}}{\eta} = \frac{11{,}186\ \text{W}}{0.80} = 13{,}982\ \text{W}$$

The full-load rotor copper loss is

$$\begin{aligned}\text{rotor copper loss} &= 13{,}982\ \text{W} - 11{,}186\ \text{W} \\ &\quad - 540\ \text{W} - 975\ \text{W} - 675\ \text{W} \\ &= 606\ \text{W} \qquad \text{[full power]}\end{aligned}$$

The output power at half load is

$$P = \left(\tfrac{1}{2}\right)(15\ \text{hp})\left(745.7\ \frac{\text{W}}{\text{hp}}\right) = 5593\ \text{W}$$

At half load, the friction and core losses are unchanged. From Eq. 44.51 and Eq. 44.52, the ratio of actual to full-load copper losses is equal to the square of the ratio of actual to full-load output power.

$$\text{stator copper loss} = \left(\tfrac{1}{2}\right)^2(540\ \text{W}) = 135\ \text{W}$$

$$\text{rotor copper loss} = \left(\tfrac{1}{2}\right)^2(606\ \text{W}) = 152\ \text{W}$$

The rotor input power is

$$P_{\text{in}} = 5593\ \text{W} + 975\ \text{W} + 152\ \text{W} = 6720\ \text{W}$$

The synchronous speed is

$$n_s = \frac{120f}{p} = \frac{(120)(60\ \text{Hz})}{6} = 1200\ \text{rpm}$$

The slip at half power is found from Eq. 44.50.

$$s = \frac{\text{rotor copper loss}}{\text{rotor input power}} = \frac{152\ \text{W}}{6720\ \text{W}} = 0.0226$$

The speed at half load is

$$\begin{aligned}n &= n_s(1-s) = \left(1200\ \frac{\text{rev}}{\text{min}}\right)(1-0.0226) \\ &= 1173\ \text{rev/min}\end{aligned}$$

(b) The torque is found from Eq. 44.5.

$$T = \frac{5252P}{n} = \frac{(5252)\left(\tfrac{1}{2}\right)(15\ \text{hp})}{1173\ \frac{\text{rev}}{\text{min}}} = 33.6\ \text{ft-lbf}$$

22. SINGLE-PHASE MOTORS

A single-phase motor works similarly to a three-phase induction motor. The stator carries a main winding and a north-south pole set. An auxiliary winding, with the same number of poles, usually operates during the brief period of motor startup and is used to set the rotor in the desired direction. Once the direction of the rotor rotation is established, the auxiliary winding can be removed and the motor will continue to operate.[14,15]

23. SPLIT-PHASE MOTORS

A *resistance split-phase motor*, usually called a *split-phase motor*, is shown in Fig. 44.11(a). The stator winding and the auxiliary winding establish magnetic fields 90° apart. Because the auxiliary winding is made from finer wire than the stator winding (main winding), the auxiliary winding has a higher resistance and lower reactance. As a result, the main winding current lags the auxiliary winding current. For the connections shown, the result is in a counter-clockwise rotation. To change the direction of rotation, reverse the connection of either the auxiliary winding (i.e., the start winding) or the main winding (i.e., the run winding), but not both. The corresponding phasor diagram is shown in Fig. 44.11(b). The line current at startup is between 6 to 7 times the normal running current. This motor type is used where infrequent starting of low or moderate torque loads is required and a low cost is imperative.

24. CAPACITOR-START MOTORS

A *capacitor-start motor* is shown in Fig. 44.12(a). The stator and auxiliary windings establish magnetic fields 90° from one another. The auxiliary winding is made of finer wire than the stator (i.e., main) winding. The main difference between the capacitor-start and split-phase motor is that the auxiliary windings of the capacitor-start motor have nearly as many turns as the main

[14]Consider the single-phase motor as a two-phase motor at startup. Once started, the field of the rotor interacts with the stator (main) field to maintain the revolving magnetic field. This revolving field should more accurately be termed an alternating field.

[15]Two theories are used to describe the complex principle of operation, the *double revolving field theory* and the *cross-field theory*. Both result in a revolving magnetic field that sustains itself once the rotor is in motion. The theories are involved.

Figure 44.11 Split-Phase Motor

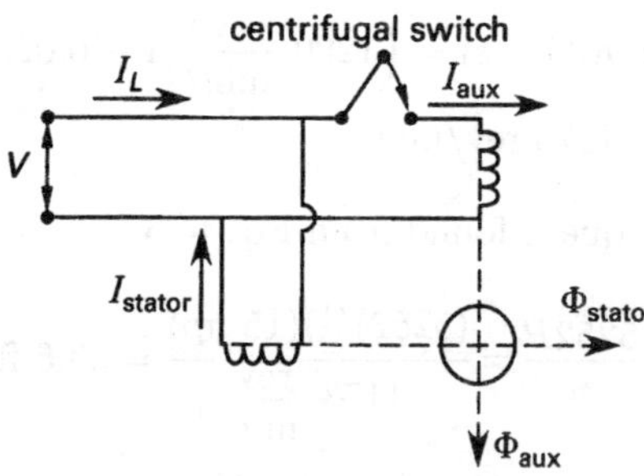

(a) startup magnetic fields

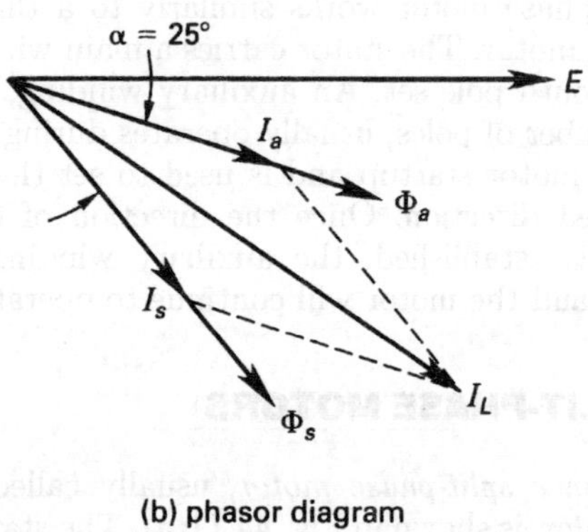

(b) phasor diagram

Figure 44.12 Capacitor-Start Motor

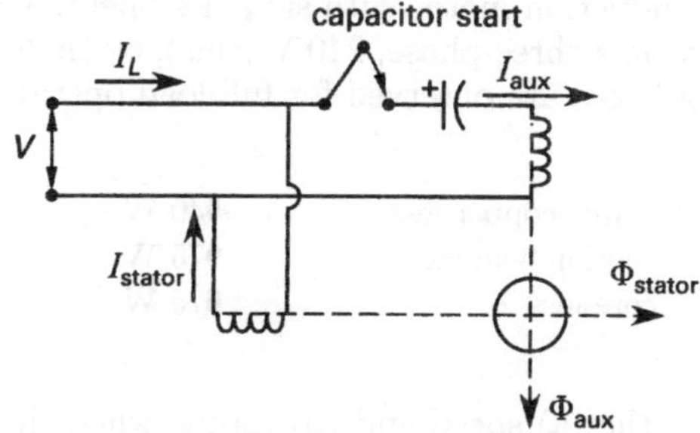

(a) startup magnetic fields

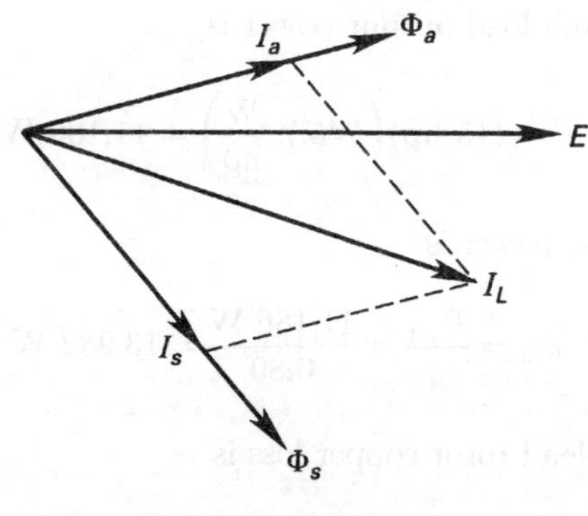

(b) phasor diagram

windings. The auxiliary winding current leads the stator current by approximately 80°. This allows for greater starting torque than the split-phase design (for the same torque, less current flows in the auxiliary winding). Therefore, the capacitor-start motor for a given output generates less heat. For the connections in Fig. 44.12(a), the rotation is counter-clockwise. To change the direction of rotation, reverse the connection of either the auxiliary winding (i.e., start winding) or the main winding (i.e., run winding), but not both. The corresponding phasor diagram is shown in Fig. 44.12(b). The line current at startup is 4 to 5 times the nominal running current. This motor type is used where high starting torques are required.

25. SINGLE-PHASE MOTOR VIBRATIONS

A *single-phase motor* receives pulsating AC power even while delivering constant mechanical power that causes vibration. This vibration occurs at twice the line frequency. In contrast, two-phase and three-phase motors do not vibrate because the total instantaneous power received from all the phases remains constant (for a constant load).

26. CAPACITOR-RUN MOTORS

A *capacitor-run motor*, also called *capacitor-start capacitor-run*, is shown in Fig. 44.13(a). The stator and auxiliary windings establish magnetic fields 90° from one another. The auxiliary winding is made of finer wire than the stator (i.e., main) winding. The primary difference between the capacitor-run and capacitor-start motors is that the auxiliary windings of the capacitor-run motor have more turns than the main windings. The auxiliary winding current leads the stator current by 90° at full load. This allows for a higher power factor and increased efficiency but a lower starting torque. For the connections shown in Fig. 44.13(a), the rotation is counter-clockwise. To change the direction of rotation, reverse the connection of either the auxiliary winding (i.e., start winding) or the main winding (i.e., run winding), but not both. The corresponding phasor diagram is shown in Fig. 44.12(b). The line current at startup is 4 to 5 times the nominal running current. This motor type is used where quiet operation is desired (e.g., hospitals or recording studios).

27. SHADED-POLE MOTORS

A *shaded-pole motor* is shown in Fig. 44.14. The main winding creates the driving magnetic flux. Some of the flux links the copper ring on pole A, some links the copper ring on pole B, and the remainder links the rotor. The copper rings have low resistance; therefore, higher currents flow (with resulting stronger magnetic fields) in the rings as opposed to the main stator structure. The flux in the copper rings lags the originating fluxes, which establishes a weak revolving magnetic field. In the

Figure 44.13 Capacitor-Run Motor

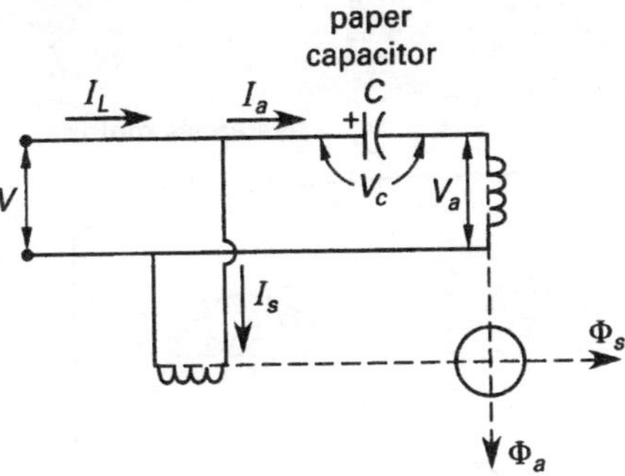

(a) startup magnetic fields

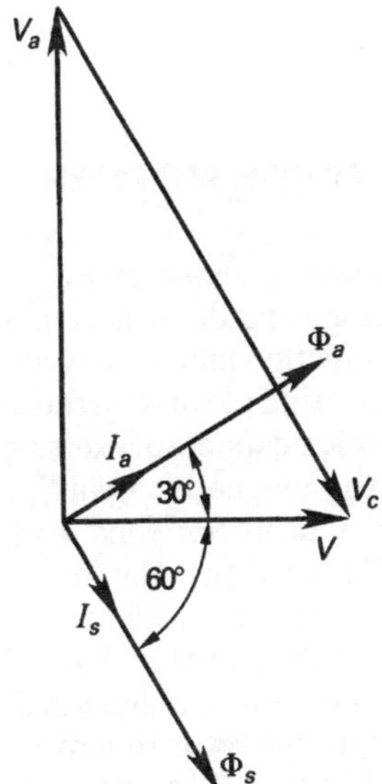

(b) phasor diagram at full load

figure, the direction of rotation is from the unshaded region to the shaded region (copper ring side) of the pole. This motor type has low starting torque, poor efficiency, and a low power factor. However, the simple and inexpensive construction makes it useful in very low power applications.

28. UNIVERSAL MOTORS

A *universal motor*, also called a *series motor*, is shown in Fig. 44.15. This is an alternating current series motor whose construction is very similar to a DC series motor. The primary difference is that the magnetic circuit in the stator is completely laminated to decrease eddy current losses. The motor can operate on AC or DC power. When operating on AC power, the magnetic fields on the stator and the rotor change with each alteration of the AC signal, thereby maintaining the relative relationship between the poles. As a result, the direction of rotation remains the same as the alternating current changes from positive to negative. Because the armature current and the flux reverse simultaneously, no revolving magnetic field exists. Therefore, the motor operates using the same principle as a DC series motor. Universal motors tend to be noisy, have relatively short lives, operate at high speeds, and have a high starting torque. They are used in vacuum cleaners, portable tools, and toys. Larger versions are used in locomotive engines.

Figure 44.14 Shaded-Pole Motor

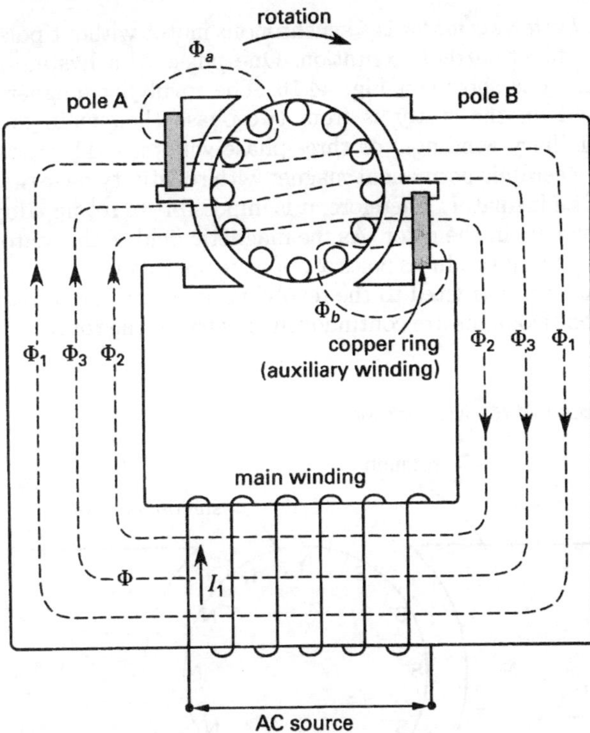

Figure 44.15 Universal Motor

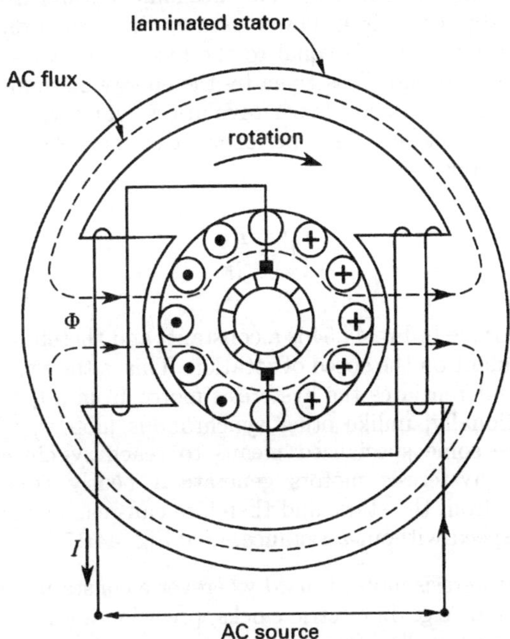

29. HYSTERESIS MOTORS

A *hysteresis motor* is a synchronous motor without poles or direct-current excitation. One phase of a hysteresis motor is shown in Fig. 44.16. The revolving magnetic field on the stator is from two-phase (i.e., main and auxiliary windings) or three-phase windings. The rotor is a ceramic permanent magnet with resistivity near that of an insulator. Therefore, it is impossible to set up eddy currents in the rotor. As the magnetic field of the stator rotates, it produces poles of opposite polarity on the rotor that are attracted to the revolving poles causing a rotation. Domains are continuously created in the rotor.

Figure 44.16 Hysteresis Motor

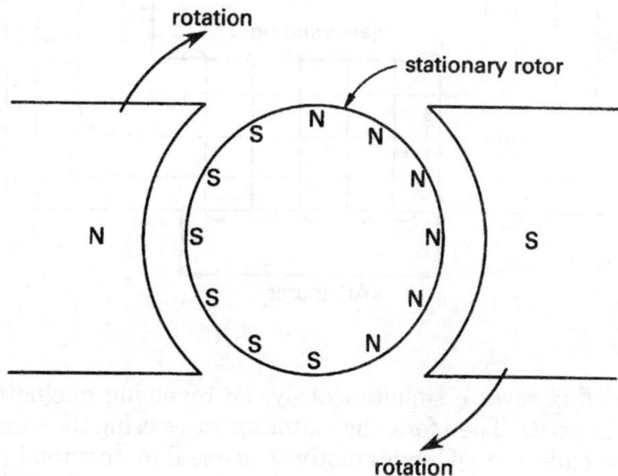

During each cycle of the stator, the domains on the rotor complete a cycle called a *hysteresis loop*. The area of the loops represents losses. These hysteresis losses are evident as heat in the rotor. The mechanical power driving the rotor comes from power used to drive the domains on the rotor, and is equal to the hysteresis losses. The resulting torque, T, is given by the energy dissipated in the rotor, E_h, in joules per revolution, which is divided by a constant that performs the necessary unit conversions.

$$T = \frac{E_h}{2\pi} \qquad 44.54$$

The torque in Eq. 44.54 is a constant, and therefore, not dependant on the speed of rotation. This is the property that differentiates the hysteresis motor from all others. Additionally, unlike other synchronous motors, which require some specialized means to reach synchronous speed, hysteresis motors generate a nearly constant torque from the start, and therefore can reach synchronous speed without assistance. (See Fig. 44.17.)

The hysteresis motor is used wherever a constant torque is desired (e.g., in electric clocks, precision audio equipment, and cooling fans).

Figure 44.17 Torque-Speed Curve Comparison: Induction versus Hysteresis

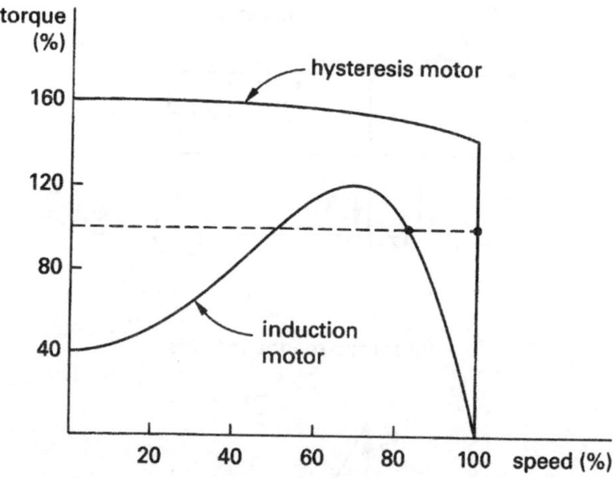

30. RELUCTANCE MOTORS

A *reluctance motor* is shown in Fig. 44.18. The rotor of a reluctance motor is made of ferromagnetic material and resembles an induction motor, only with salient poles. (See Fig. 44.19.) The principle of reluctance motors is that lines of magnetic flux are capable of exerting a force to minimize the length of the magnetic circuit (in other words, iron poles carrying a magnetic flux tend to align with one another). In Fig. 44.18(a), the lines of magnetic flux on the stator pass through a highly permeable rotor. The magnetic field on the stator (in Fig. 44.18(b)) has rotated clockwise and the lines of magnetic flux are stretched. A force is exerted on the rotor to move it into alignment, as shown in Fig. 44.18(c). If the stator magnetic field moves rapidly, the rotor operates as a standard induction motor until it reaches near synchronous speed at which point the rotor locks on to the stator field.

A reluctance motor is the most inexpensive synchronous motor type. It can operate at high or low rotating magnetic field speeds. The position of the rotor can be accurately controlled at low speed. It is well suited to variable frequency and electronic speed control mechanisms. The combination of accuracy and speed control allows reluctance motors to be used as the drive motors for control rods in nuclear reactors and other motion-control and variable-speed drive applications.

31. SPEED CONTROL FOR AC MOTORS

The speed of an AC motor can be controlled by changing the armature conditions, the field conditions, or both. The type of control is determined by the motor type. (See Sec. 44.22 through Sec. 44.30.) *Armature control* techniques include (a) varying the resistance on the rotor side using slip-rings with adjustable resistors attached, (b) controlling the voltage on the armature electronically, or (c) using motor-generator setups to provide synchronous machine excitation. *Field*

Figure 44.18 Reluctance Motor

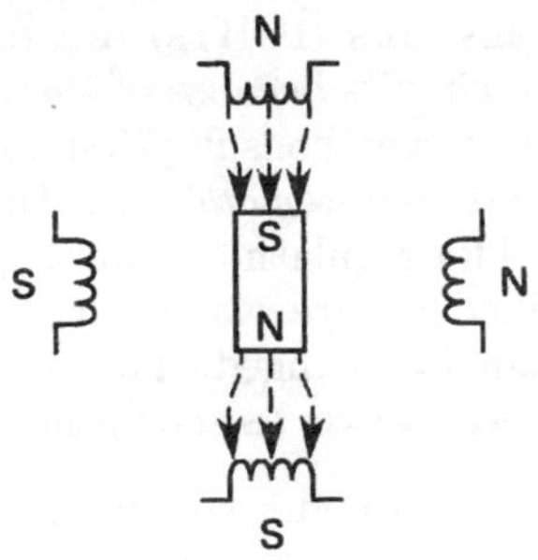

(a) rotor concentrating flux lines

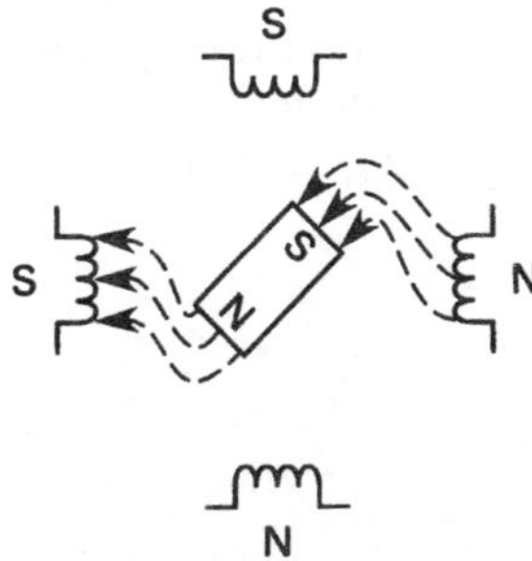

(b) rotor aligning to the strongest magnetic field

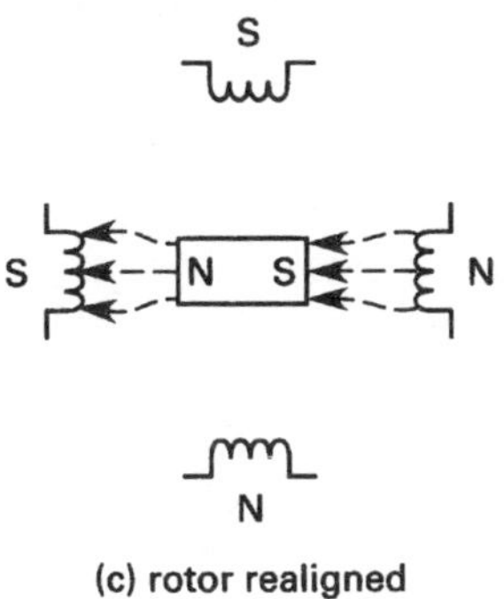

(c) rotor realigned

Figure 44.19 Reluctance Rotor

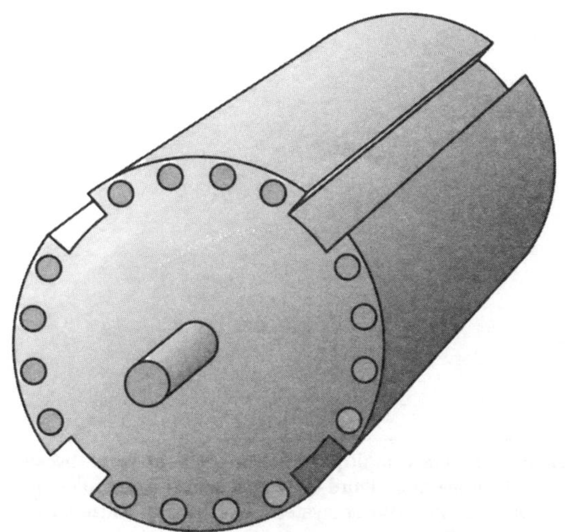

control techniques include (a) changing the number of poles, (b) controlling the voltage on the field, or (c) controlling the frequency.

Changes to the armature or field can be made manually or automatically. Automatic starters generally control the voltage per hertz, or V/Hz ratio, thereby controlling the torque delivered, up to 60 Hz. Above 60 Hz, the V/Hz ratio is adjusted to maintain a constant torque. The range above 60 Hz, where the frequency can no longer be adjusted, is called the *constant power region* or *constant torque region.*

32. VARIABLE FREQUENCY DRIVE (VFD)

A *variable speed drive* (VSD) is an electrical device whose speed varies across a considerable range as a function of the load. It is sometimes called an *adjustable speed drive* (ASD). When the frequency is controlled, the device is electronic and is called a *variable frequency drive* (VFD). Such drives change the frequency in order to change the speed of the rotational field, as shown by

$$np = 120f \qquad 44.55$$

Since the number of poles, p, is generally fixed, changing the frequency changes the speed of the rotational field, n, which varies the overall speed of the motor.

A block diagram of a VFD is shown in Fig. 44.20. A VFD converts AC input power to a controllable output which varies in frequency and produces a wave that closely approximates a sine wave.

Figure 44.20 VFD Block Diagram

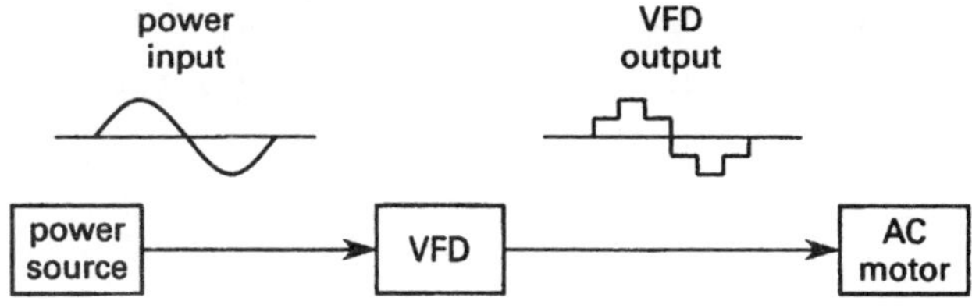

A simplified electrical schematic of a VFD is shown in Fig. 44.21. The AC power source is applied to an input rectifier section that changes the signal from AC to DC.[16] In Fig. 44.21, the VFD shown is a *six-pulse device*, since it generates three pulses on the positive portion of the cycle, and three on the negative portion. This creates a rippled DC signal in the DC bus section.

Additional pulses may be added by using a phase-shifted input and additional rectifier pairs. For example, if the supplying transformer has a delta primary and a delta/wye secondary pair, the wye secondary output is phase-shifted 30° from the corresponding voltage on the delta secondary. Using the six outputs from the dual

[16]Diodes are shown for simplicity, but *silicon-controlled rectifiers* (SCRs) or transistors can be used for more precise control.

Figure 44.21 VFD Functional Schematic

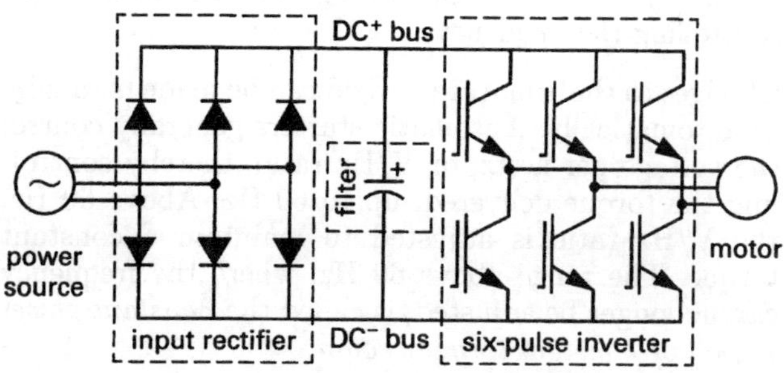

secondary, with two rectifiers for each phase, results in a twelve-pulse device. The greater the number of pulses, the more consistent the DC signal.

Even without additional pulses to smooth the signal, the DC bus section contains filtering components such as inductors, DC links, chokes, and other items to minimize ripples in the signal. However, the DC bus section's major components are capacitors, only one of which is shown in Fig. 44.21. The capacitors in the DC bus section store the energy from the pulses, maintaining the DC voltage level and delivering the energy to the inverter section.

The inverter section is controlled by circuitry that controls the firing of the individual transistors. Insulated gate bipolar transistors (IGBTs) are shown. They are common in such circuits because of their high switching speeds and low power consumption in the switching mode. IGBTs use *pulse-width modulation* (PWM), a common method for regulating power supplies, to create an easily modified square wave whose frequency and overall value can be changed. The modifications produce a variable sine wave, as shown in Fig. 44.20.

VFDs have several benefits compared to other VSDs. They reduce energy consumption, since the speed can be varied to provide only the flow required for a given application. Because the frequency at start-up is low (sometimes called a *soft start*), VFDs also produce lower levels of mechanical stress during start-up than other types of VSDs. Some designs use an *active front end*, which controls the firing of transistors. This helps minimize harmonics.[17]

[17]Harmonics that are multiples of 2 are not an issue because they cancel each other out. Third-order harmonics (multiples of 3) also cancel out in a three-phase system. The largest harmonics are the 5th, 7th, 11th, 13th, and so on.

Topic X: Electronics

45 Electronics Fundamentals

Nomenclature

A	constant	$K^{-3}\cdot m^{-3}$
A	gain	–
c	speed of light, 2.9979×10^8	m/s
C	capacitance	F
D	diffusion constant	m^2/s
E	electric field strength	V/m
E	energy	J or eV
h	Planck's constant, 6.6256×10^{-34}	J·s
i	instantaneous or variable current	A
I	effective or DC current	A
n	electron concentration	cm^{-3}
n, N	concentration	m^{-3}
p	hole concentration	cm^{-3}
q	electric unit charge, 1.6022×10^{-19}	C
r, R	resistance	Ω
T	temperature	K
v	instantaneous or variable voltage	V
V	effective or DC voltage	V

Symbols

β	CE forward current ratio or gain	–
η	empirical constant	–
κ	Boltzmann's constant, 1.3807×10^{-23}	J/K
λ	wavelength	m
μ	mobility	$m^2/V\cdot s$
σ	conductivity	S/m

Subscripts

0	at 0K
0	barrier or source value
ac	alternating current or small signal
A	acceptor
be, BE	base to emitter
B	DC or biased base
BB	base supply or base biasing
BO	breakover
BQ	base to quiescent point
c	conduction band or collector
C	collector
CB	collector to base
CC	collector supply
CE	collector to emitter
CO	cutoff
d	diffusion
D	donor, diode, or drain
e, E	emitter
f	forward (AC or instantaneous)
F	forward (DC component) or total
G	gap or gate
H	hold
i	intrinsic or input (AC or instantaneous)
I	current
L	load
m	maximum
n	electrons or n-type
p	holes or p-type
pn	p to n
r	reverse (AC or instantaneous)
R	reverse (DC component or total)
s	saturation
S	source
t	transition
T	thermal
v	valence band
V	constant or effective voltage
Z	zener
ZK	zener knee

1. OVERVIEW

Electronics involves charge motion through materials other than metals, such as vacuums, gases, or semiconductors. The focus in this chapter will be on semiconductor materials. Because an understanding of the electron structure is vital to understanding electronics, a periodic table of the elements is given in App. 45.A.

An *electronic component* is one able to amplify, control, or switch voltages or currents without mechanical or other nonelectrical commands. The charge in metals is carried by the electron, with a charge of -1.6022×10^{-19} C. In semiconductor materials, the charge is carried both by the electron and by the absence of the electron in a covalent bond, which is referred to as a *hole*[1] with a charge of $+1.6022 \times 10^{-19}$ C. The concentration of electrons, n, and holes, p, is given by the *mass action law*, Eq. 45.1.

$$n_i^2 = np \qquad 45.1$$

The term n_i is the concentration of carriers in a pure (*intrinsic*) semiconductor.[2] In intrinsic semiconductor materials, the number of electrons equals the number of holes, and the mass action law is stated as in Eq. 45.1. Nevertheless, the mass action law applies for intrinsic and extrinsic semiconductor materials. *Extrinsic* semiconductor materials are those that have had impurities deliberately added to modify their properties, normally their conductivity. For extrinsic semiconductors, the mass action law is probably better understood as indicating that the product np remains constant regardless of position in the semiconductor or doping level. Carriers, either n or p, in a semiconductor are constantly generated by thermal creation of electron-hole pairs and constantly eliminated by recombination of electron-hole pairs. The carriers generated are caused to diffuse by concentration gradients in the semiconductor, a phenomenon that does not occur in metals.

Semiconductor devices are inherently nonlinear. Nevertheless, they are commonly analyzed over ranges in which their behavior is approximately linear. Such an analysis is called *small-signal analysis*. The DC or effective value of the electrical parameters in the models used for analysis will be represented by uppercase letters. AC or instantaneous values will be represented by lowercase letters. Equivalent parameters in the models will use lowercase letters. For convenience, the lowercase letter t is omitted from functions of time. For example, $v(t)$ is written simply as v. The subscripts on currents indicate the terminals into which current flows. Subscripts on voltages indicate the terminals across which the voltage appears. Subscripts indicating biasing voltages are capitalized. For example, V_{BB} is the base biasing voltage while v_{be} is the instantaneous voltage signal applied to the base-emitter junction. A list of common designations is given in App. 45.B. *Ultra large-scale integration* (ULSI) is a term used to describe circuits with 1,000,000 or more components per chip.

Electronic components are often connected in an array known as an *integrated circuit* (IC). An integrated circuit is a collection of active and passive components on a single semiconductor substrate (*chip*) that function as a complete electronic circuit. *Small-scale integration* (SSI) involves the use of less than 100 components per chip. *Medium-scale integration* (MSI) involves between 100 and 999 components per chip. *Large-scale integration* involves between 1000 and 9999 components per chip. *Very large-scale integration* involves more than 10,000 components per chip.

The nonlinearity of electronic devices makes them attractive for use as amplifiers. An *amplifier* is a device capable of increasing the amplitude or power of a physical quantity without distorting the wave shape of the quantity. The amplifying properties of transistors are covered in this chapter. Topics peculiar to amplifiers and amplifier types are covered in Chap. 46 and Chap. 47. The application of the electronic components to a variety of circuits is extremely broad. A few examples are provided in the end sections of this chapter. Additionally, some electronic circuits are used often enough to be designed into *standard modular devices*, that is, off-the-shelf ICs with all the necessary components to accomplish a given circuit task. An understanding of the basic electrical and electronic principles for the components used in such devices will enable an understanding of the overall circuit functions of the standard modular devices and electrical/electronic circuits in general.

Example 45.1

Silicon has an intrinsic carrier concentration of 1.6×10^{10} cm^{-3}. What is the concentration of holes?

Solution

Regardless of whether the material is intrinsic or extrinsic, the law of mass action applies. However, in intrinsic materials, the concentration of electrons and the concentration of holes are equal. Equation 45.1 can be used as follows.

$$n_i^2 = np = p^2$$

$$n_i = p = 1.6 \times 10^{10}\ \text{cm}^{-3}$$

2. SEMICONDUCTOR MATERIALS

The most common semiconductor materials are silicon (Si) and germanium (Ge). Both are in group IVA of the periodic table and contain four valence electrons. A *valence electron* is one in the outermost shell of an atom. Both materials use covalent bonding to fill the outer shell of eight when they create a crystal lattice. Semiconductor materials ($N \approx 10^{10}$ to 10^{13} electrons/m^3) are slightly more conductive than insulators ($N \approx 10^7$ electrons/m^3) but less conductive than metals ($N \approx 10^{28}$ electrons/m^3). The conductivity, σ, of semiconductor materials can be made to vary from approximately 10^{-7} to 10^5 S/m.

[1]The concept of holes explains the conduction of electricity without free electrons. The hole is considered to behave as a free positive charge—quantum mechanics justifies such an interpretation. (The *Hall voltage* is experimental confirmation.) The calculation of total charge motion in semiconductors is simplified as a result. The same equations used for electron movement can be used for hole movement with a change of sign and a change of values for some terms, such as mobility.

[2]Intrinsic means natural. Pure silicon is silicon in its natural state, with no doping, though it will have naturally occurring impurities. The terms pure and intrinsic are used interchangeably.

The semiconductor crystal lattice has many defects (i.e., free electrons and their corresponding holes). The formation of free electrons and holes is driven by the temperature and is called *thermal carrier generation*. The density of these electron-hole pairs in intrinsic materials is given by

$$n_i^2 = A_0 T^3 e^{-E_{G0}/\kappa T} \qquad 45.2$$

The term A_0 is a constant independent of temperature and related to the density states at the bottom edge of the conduction band and the top edge of the valence band. T is the absolute temperature. E_{G0} is the energy gap at 0K, that is, the energy between the conduction band and the valence band. This is the energy required to break the covalent bond. The energy gap for silicon at 0K is approximately 1.21 eV. The energy gap for germanium is approximately 0.78 eV. The term κ is Boltzmann's constant, with values of 1.3807×10^{-23} J/K and 8.621×10^{-5} eV/K.[3] At a given temperature at thermal equilibrium, Eq. 45.2 becomes

$$n_i^2 = N_c N_v e^{-E_G/\kappa T} \qquad 45.3$$

The terms N_c and N_v are the effective density states in the conduction band and valence band, respectively. The energy gap for silicon at room temperature (300K) is 1.12 eV. The energy gap for germanium at room temperature is 0.80 eV.

When minor amounts of impurities called *dopants* are added, the materials are termed *extrinsic semiconductors*.[4] If the impurities added are from group IIIA, with three valence electrons, an additional hole is created in the lattice. These dopants are called *acceptors*, and semiconductors with such impurities are called *p-types*. The *majority carriers* in *p*-type semiconductors are holes, and the *minority carriers* are electrons. The majority of the charge movement takes place in the valence band. Typical dopants are indium (In) and gallium (Ga). If the impurities added are from group VA, with five valence electrons, an additional electron is provided to the lattice.[5] These dopants are called *donors*, and semiconductors with such impurities are called *n-types*. The majority carriers in *n*-type semiconductors are electrons, and the minority carriers are holes. The majority of charge movement takes place in the conduction band. Typical dopants are phosphorus (P), arsenic (As), and antimony (Sb).

The conductivity of the semiconductor is determined by the carriers. The mobility of electrons is higher than that of holes.[6] The total conductivity in any semiconductor is a combination of the movement of the electrons and holes and is given by

$$\sigma = q(n\mu_n + p\mu_p) \qquad 45.4$$

The law of mass action, Eq. 45.1, applies to extrinsic semiconductors. With the addition of dopants, the law of electrical neutrality given by Eq. 45.5 also applies. The concentration of acceptor atoms is N_A, each contributing one positive charge to the lattice. (Do not confuse N_A with Avogadro's number.) The concentration of donor atoms is N_D, each contributing one negative charge. Since neutrality is maintained, the result is[7]

$$N_A + n = N_D + p \qquad 45.5$$

In a *p*-type material, the concentration of donors is zero, ($N_D = 0$). Additionally, the concentration of holes is much greater than the number of electrons ($p \gg n$). For a *p*-type material, Eq. 45.5 can be rewritten as $N_A \approx p$. Using the law of mass action, Eq. 45.1, the concentration of electrons in a *p*-type material is

$$n = \frac{n_i^2}{N_A} \qquad 45.6$$

In an *n*-type material, the concentration of acceptors is zero ($N_A = 0$). Additionally, the concentration of electrons is much greater than the number of holes ($n \gg p$). For an *n*-type material, Eq. 45.5 can be rewritten as $N_D \approx n$. Using the law of mass action, Eq. 45.1, the concentration of holes in an *n*-type material is

$$p = \frac{n_i^2}{N_D} \qquad 45.7$$

Example 45.2

The energy gap for silicon at room temperature (300K) is 1.12 eV. The energy density states are $N_c = 2.8 \times 10^{19}$ cm^{-3} and $N_v = 1.02 \times 10^{19}$ cm^{-3}. What is the intrinsic carrier concentration?

Solution

The intrinsic carrier concentration, assuming thermal equilibrium, is given by Eq. 45.3.

$$\begin{aligned} n_i^2 &= N_c N_v e^{-E_G/\kappa T} \\ &= (2.8 \times 10^{19} \text{ cm}^{-3}) \\ &\quad \times (1.02 \times 10^{19} \text{ cm}^{-3}) e^{-1.12 \text{ eV}/(8.621\times10^{-5} \text{eV/K})(300\text{K})} \\ &= 4.452 \times 10^{19} \text{ cm}^{-6} \\ n_i &= \sqrt{4.452 \times 10^{19} \text{ cm}^{-6}} = 6.67 \times 10^9 \text{ cm}^{-3} \\ &\approx 0.7 \times 10^{10} \text{ cm}^{-3} \end{aligned}$$

This number differs slightly from the number determined experimentally and more exact calculations (1.6×10^{10} cm^{-3}).

[3]The symbology for Boltzmann's constant often varies with the units of the energy gap. If eV is used as the energy unit, Boltzmann's constant may be seen as either κ or $\overline{\kappa}$ to distinguish between eV/K and J/K.

[4]A minor amount of dopant material is on the order of 10 parts per billion.

[5]The additional electron exists because the outer shell octet is satisfied by the covalent bonding of the first four electrons in the impurity.

[6]The mobility of electrons is a factor of 2.5 higher in silicon and about 2.1 higher in germanium.

[7]Neutrality is maintained because each atom of an added impurity removes one intrinsic atom.

3. DEVICE PERFORMANCE CHARACTERISTICS

Electronic components function on some variation of *pn* junction principles. Amplifiers function on some variation of transistor principles, that is, *pnp* or *npn* junction principles. Most electronic components can be modeled as two-port devices with two variables—current and voltage—for each port. The relationship between the variables depends on the type of device. The relationship can be expressed mathematically, as for MOSFETs (*metal-oxide semiconductor field-effect transistors*), modeled in equivalent circuits, as with transistor *h*-parameters, or described graphically, as with BJT characteristic curves.

Semiconductor devices are inherently nonlinear. The performance of such devices is analyzed in a linear fashion over small portions of the characteristic curve(s). Such analysis is called *small-signal analysis.* A small signal is one that is much less than the average, that is, steady-state, value for the device, which usually is the biasing value. The models used in each linear portion of the characteristic curve are termed *piecewise linear models.* If the operation is outside the linear region or if the input signal is large compared to the average value, the device distorts the input signal. Such operation is called *nonlinear operation.*

The characteristic curve for an ideal transistor operating as a current-amplifying device is shown in Fig. 45.1. (The shape of the curve is the same for a diode, that is, a *pn* junction.)

Figure 45.1 *Typical Semiconductor Performance*

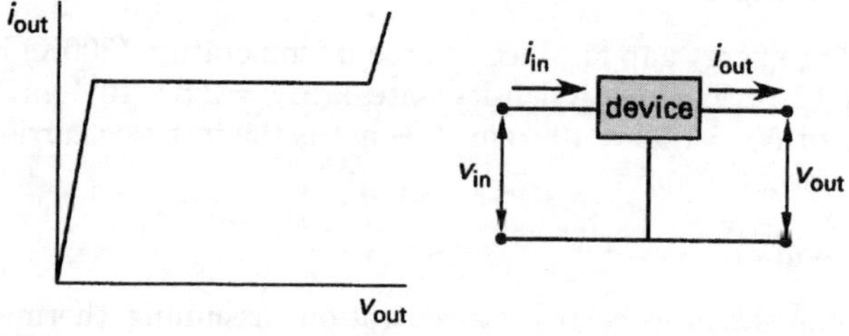

The voltage-current graph is divided into various regions known by names such as *saturation* (or *on*), *cutoff* (or *off*), *active*, *breakdown*, *avalanche*, and *pinchoff* regions. The locations of these regions depend on the type of transistor—for example, BJT or FET—and its polarity. Operation is normally in the linear active region, but applications for operation in other regions exist in digital and communications (radio frequency) applications.

4. BIAS

Bias is a DC voltage applied to a semiconductor junction to establish the operating point, also called the *quiescent* (*no-signal*) *point. Biasing* establishes the operating point with no input signal.[8] Biasing, then, is the process of establishing the DC voltages and currents (the bias) at the device's terminals when the input signal is zero (or nearly so).

For *pn* junctions, *forward bias* (or *on condition*) is the application of a positive voltage to the *p*-type material or, equivalently, the flow of current from the *p*-type to the *n*-type material. In a small semiconductor device, forward bias results in current in the milliampere range.

Reverse bias (or *off condition*) is the application of a negative voltage to the *p*-type material, or equivalently, the flow of current from the *n*-type to the *p*-type material. In a small semiconductor device, reverse bias results in current in the nanoampere range.

Self-biasing is the use of the amplifier's output voltage, rather than a separate power source, as the supply for the input bias voltage. This negative feedback control regulates the output current and voltage against variations in transistor parameters.

5. AMPLIFIERS

An *amplifier* produces an output signal from the input signal. The input and output signals can be either voltage or current. The output can be either smaller or larger (the usual case) than the input in magnitude. While most amplifiers merely scale the input voltage or current upward, the amplification process can include a sign change, a phase change, or a complete phase shift of 180°.[9] The ratio of the output to the input is known as the *gain* or *amplification factor*, *A*. A *voltage amplification factor*, A_V, and *current amplification factor*, A_I or β, can be calculated for an amplifier.

Figure 45.2 illustrates a simplified current amplifier with current amplification factor β. The additional current leaving the amplifier is provided by the bias battery, V_2.

$$i_{\text{out}} = \beta i_{\text{in}} \qquad 45.8$$

A capacitor, *C*, is placed in the output terminal to force all DC current to travel through the *load resistor*, R_L. Kirchhoff's voltage law for loop abcd is

$$V_2 = i_{\text{out}} R_L + V_{\text{ac}} = \beta i_{\text{in}} R_L + V_{\text{ac}} \qquad 45.9$$

If there is no input signal (i.e., $i_{\text{in}} = 0$), then $i_{\text{out}} = 0$ and the entire battery voltage appears across terminals ac ($V_{\text{ac}} = V_2$). If the voltage across terminals ac is zero, then the entire battery voltage appears across R_L so that $i_{\text{out}} = V_2/R_L$.

6. AMPLIFIER CLASSIFICATION

Amplifiers are classified on the basis of how much input is translated into output. A sinusoidal input signal is

[8]Bias is used both as a verb and a noun in electronics.

[9]An *inverting amplifier* is one for which $v_{\text{out}} = -A_V v_{\text{in}}$. For a sinusoidal input, this is equivalent to a phase shift of 180° (i.e., $V_{\text{out}} = A_V V_{\text{in}} \angle -180°$).

Figure 45.2 General Amplifier

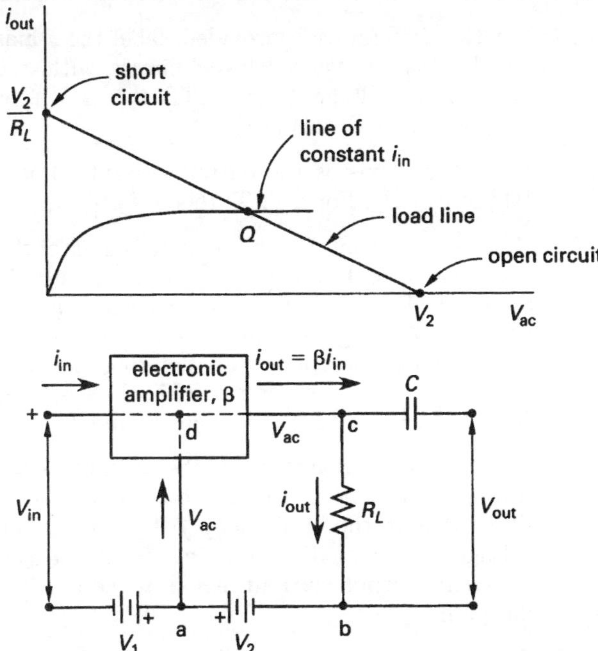

assumed. The output of an amplifier depends on the bias setting, which in turn establishes the quiescent point.

A *Class A amplifier*, as shown in Fig. 45.3, has a quiescent point in the center of the active region of the operating characteristics. Class A amplifiers have the greatest linearity and the least distortion. Load current flows throughout the full input signal cycle. Because the load resistance of a properly designed amplifier will equal the Thevenin equivalent source resistance, the maximum power conversion efficiency of an ideal Class A amplifier is 50%.

Figure 45.3 Class A Amplifier

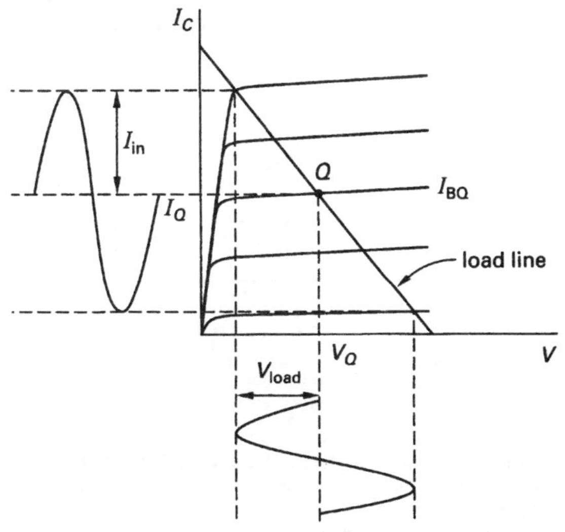

For *Class B amplifiers*, as shown in Fig. 45.4, the quiescent point is established at the cutoff point. A load current flows only if the signal drives the amplifier into its active region, and the circuit acts like an amplifying half-wave rectifier. Class B amplifiers are usually combined in pairs, each amplifying the signal in its respective half of the input cycle. This is known as *push-pull operation*. The output waveform will be sinusoidal except for the small amount of crossover distortion that occurs as the signal processing transfers from one amplifier to the other. The maximum power conversion efficiency, of an ideal Class B push-pull amplifier is approximately 78%.

Figure 45.4 Class B Amplifier

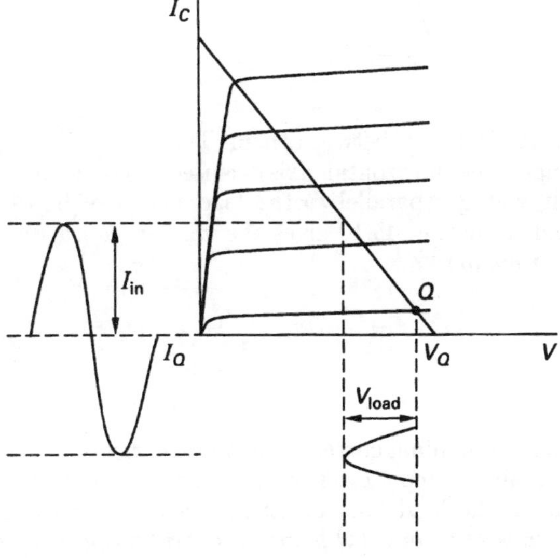

The intermediate *Class AB amplifier* has a quiescent point somewhat above cutoff but where a portion of the input signal still produces no load current. The output current flows for more than half of the input cycle. AB amplifiers are also used in push-pull circuits.

Class C amplifiers, as shown in Fig. 45.5, have quiescent points well into the cutoff region. Load current flows during less than one-half of the input cycle. For a purely resistive load, the output would be decidedly nonsinusoidal. However, if the input frequency is constant, as in radio frequency (rf) power circuits, the load can be a parallel LRC tank circuit tuned to be resonant at the signal frequency. The LRC circuit stores electrical energy, converting the output signal to a sinusoid. The power conversion efficiency of an ideal Class C amplifier is 100%.

7. LOAD LINE AND QUIESCENT POINT CONCEPT

The i_{in}-v_{out} curves of Sec. 45.5 and Sec. 45.6 illustrate how amplification occurs. The two known points, $(v_{out}, i_{out}) = (V_2, 0)$ and $(v_{out}, i_{out}) = (0, V_2/R_L)$, are plotted on the voltage-current characteristic curve. The straight

Figure 45.5 Class C Amplifier (Resistive Load)

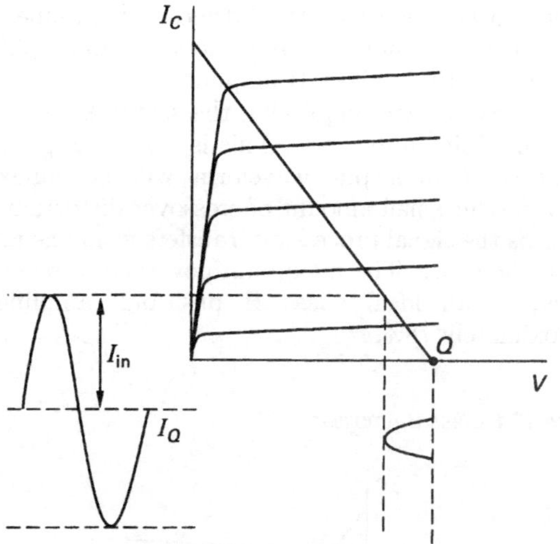

load line is drawn between them. The change in output voltage (the horizontal axis) caused by a change in input voltage (parallel to the load line) can be determined. Equation 45.10 gives the voltage *gain* (*amplification factor*).[10,11]

$$A_V = \frac{\partial v_{out}}{\partial v_{in}} \approx \frac{\Delta v_{out}}{\Delta v_{in}} \qquad 45.10$$

Usually, a nominal current (the *quiescent current*) flows in the abcd circuit even when there is no signal. The point on the load line corresponding to this current is the *quiescent point* (*Q-point* or *operating point*). It is common to represent the quiescent parameters with uppercase letters (sometimes with a subscript Q) and to write instantaneous values in terms of small changes to the quiescent conditions.

$$v_{in} = V_Q + \Delta v_{in} \qquad 45.11$$

$$v_{out} = V_{out} + \Delta v_{out} \qquad 45.12$$

$$i_{out} = I_{out} + \Delta i_{out} \qquad 45.13$$

Because it is a straight line, the load line can also be drawn if the quiescent point and any other point, usually (V_2, 0), are known.

The ideal voltage amplifier has an infinite *input impedance* (so that all of v_{in} appears across the amplifier and no current or power is drawn from the source) and zero *output impedance* so that all of the output current flows through the load resistor.

[10]Gain can be increased by increasing the load resistance, but a larger biasing battery, V_2, is required. The choice of battery size depends on the amplifier circuit devices, size considerations, and economic constraints.

[11]A *high-gain amplifier* has a gain in the tens of hundreds of thousands.

Determination of the load line for a generic transistor amplifier is accomplished through the following steps.

step 1: For the configuration provided, label the x-axis on the *output characteristic curves* with the appropriate voltage. (For a BJT, this is V_{CE} or V_{CB}. For a FET, this is V_D.)[12]

step 2: Label the y-axis as the output current. (For a BJT, this is I_C. For a FET, this is I_D.)

step 3: Redraw the circuit with all three terminals of the transistor open. Label the terminals. (For a BJT, these are base, emitter, and collector. For a FET, these are gate, source, and drain.) Label the current directions all pointing inward, toward the amplifier. (For a BJT, these are I_B, I_E, and I_C. For a FET, these are I_D and I_S.)

step 4: Perform KVL analysis in the output loop. (For a BJT, this is the collector loop. For a FET, this is the drain loop.) The transistor voltage determined is a point on the x-axis with the output current equal to zero. Plot the point.

step 5: Redraw the circuit with all three terminals of the transistor shorted. Label as in step 3.

step 6: Use Ohm's law, or another appropriate method, in the output loop to determine the current. (For the BJT, this is the collector current. For the FET, this is the drain current.) The transistor current determined is a point on the y-axis with the applicable voltage in step 1 equal to zero. Plot the point.

step 7: Draw a straight line between the two points. This is the DC load line.[13]

Example 45.3

Consider the following common emitter (CE).

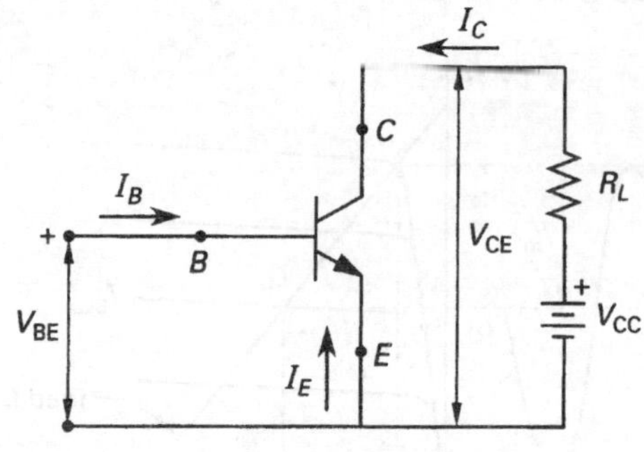

[12]The output characteristic curves for the BJT are also called the *collector characteristics* or the *static characteristics*. Figure 45.2 is an example.

[13]AC load lines are determined in the same manner, but active components, that is, inductors and capacitors, are accounted for in the analysis.

If the load resistance is 500 Ω and the collector supply voltage, V_{CC}, is 10 V, determine and draw the load line.

Solution

The x- and y-axes are drawn as shown.

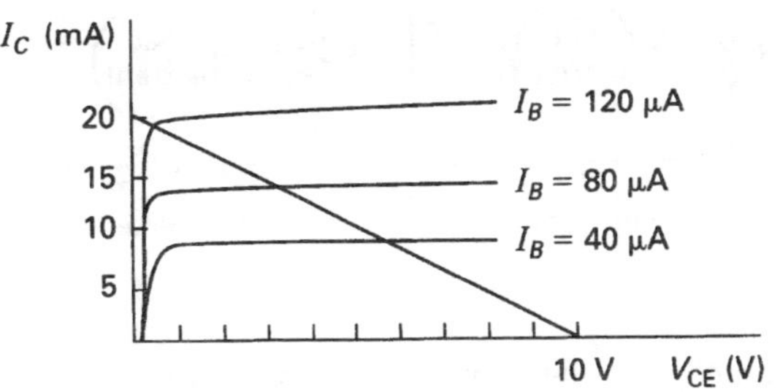

Redrawing the circuit and labeling gives

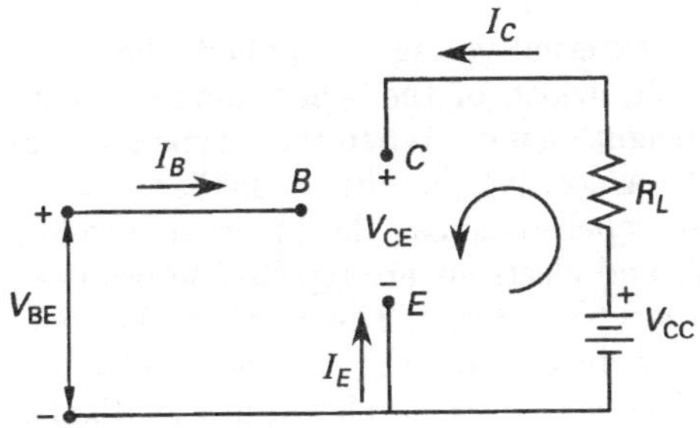

Write KVL around the indicated loop.

$$V_{CC} - I_C R_L - V_{CE} = 0$$

With the terminals open-circuited, $I_C = 0$. Substituting and rearranging gives

$$V_{CE} = V_{CC} = 10 \text{ V}$$

Plot this point (10, 0) on the x-axis. Redraw and label the circuit with the terminals shorted.

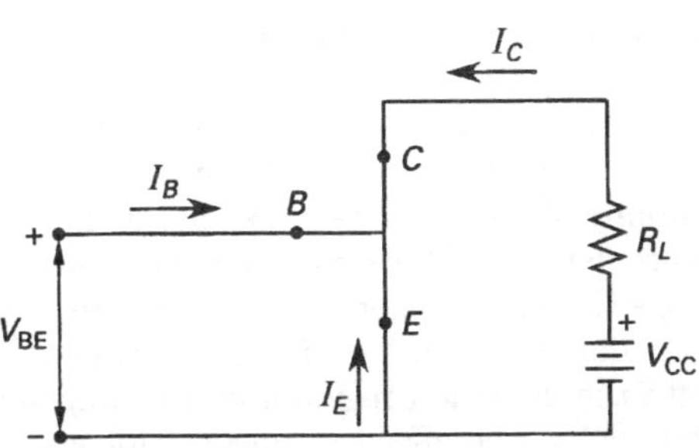

Using Ohm's law and the given values,

$$I_C = \frac{V_{CC}}{R_L} = \frac{10 \text{ V}}{500 \ \Omega} = 0.20 \text{ A} \quad (20 \text{ mA})$$

Plot this point on the y-axis. Draw a straight line between the two points. The load line is shown superimposed on the output characteristic curves in the first drawing.

8. *pn* JUNCTIONS

The *pn junction* forms the basis of diode and transistor operation. The *pn* junction is constructed of a p-type material (the anode) and an n-type material (the cathode) bonded together as shown in Fig. 45.6(a). Some of the acceptor atoms are shown as ions with a minus sign because after an impurity atom accepts an electron, it becomes negatively charged. Some of the donor atoms are shown as ions with a plus sign because after an impurity atom gives up an electron it is positively charged. (Overall, the law of charge neutrality holds, and the *pn* junction is neutral.)

The concentration gradient across the junction causes holes to diffuse to the right, and electrons to diffuse to the left. As a result, the concentration of holes on the p-side near the junction is depleted and a negative charge exists. The concentration of electrons on the n-side near the junction is depleted as well, and a positive charge exists. The result of this diffusion is shown in Fig. 45.6(b), the shape of which is determined by the level of doping.[14] The diffusion process continues until the electrostatic field set up by the charge separation is such that no further charge motion is possible. The net electric field intensity is shown in Fig. 45.6(c). The net result is a small region in which no mobile charge carriers exist. This region is called the *space-charge region*, *depletion region*, or *transition region*. Holes have a potential barrier they must overcome to move from left to right just as electrons have an energy barrier they must overcome to move from right to left, as shown in Fig. 45.6(d) and Fig. 45.6(e).

The flow of carriers caused by the concentration gradient is called the *diffusion current*, $I_{\text{diffusion}}$, also called the *recombination current* or the *injection current*. The flow of carriers caused by the established electric field is called the *drift current*, I_s, also called the *saturation current*, *thermal current*, or *reverse saturation current*.[15] The movement of carriers caused by recombination (diffusion) and drift (saturation) is continual, though at thermal equilibrium, without any applied voltage, the net current is zero.

$$I_{\text{junction}} = I_{\text{diffusion}} + I_s = 0 \quad \text{[algebraic sum]} \qquad 45.14$$

[14]A *step-graded junction* and a *linearly graded junction* are two possible types.

[15]Numerous symbols are used, among them I_o and I_{co}.

Electronics

Figure 45.6 pn Junction Characteristics

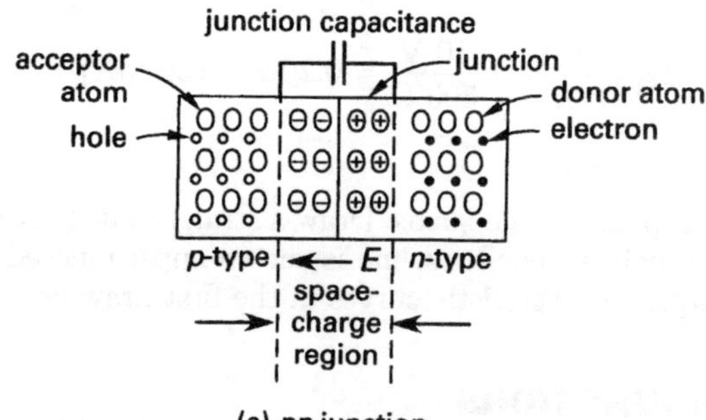

(a) *pn* junction

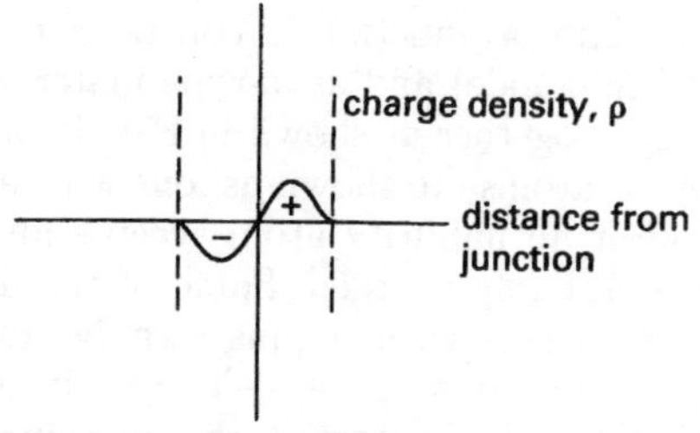

(b) charge density

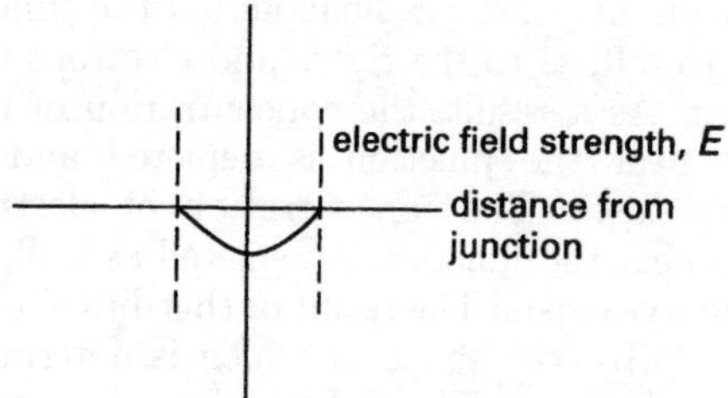

(c) electric field intensity

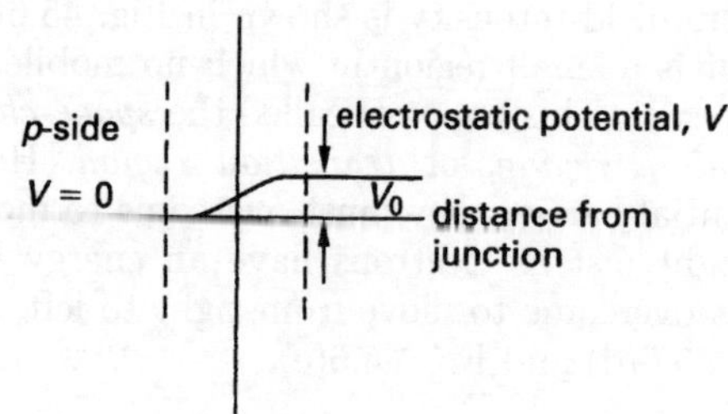

(d) electrostatic potential

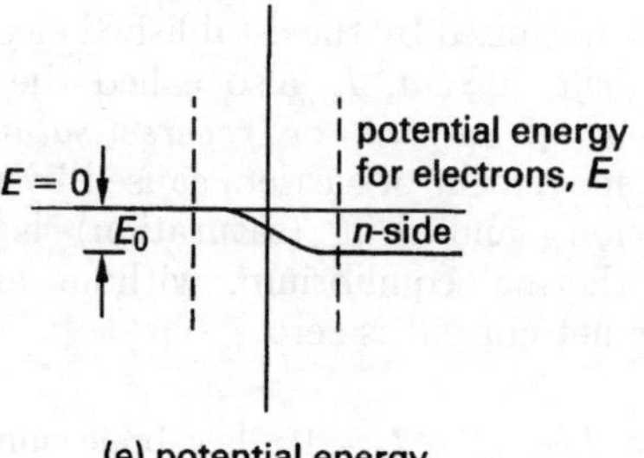

(e) potential energy

Figure 45.7 pn Junction Carrier Movement

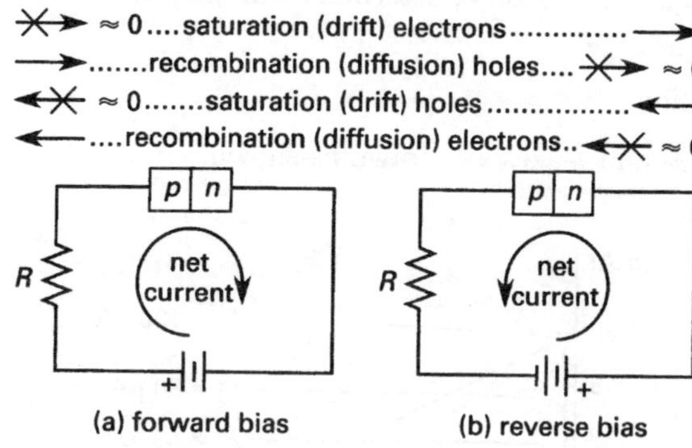

(a) forward bias (b) reverse bias

A summary of the movement of the carriers is shown in Fig. 45.7.

The width of the space-charge region at equilibrium is shown in Fig. 45.8(a).

When an external voltage is applied, that is, a biasing voltage, the width of the space-charge region and the barrier height change. When the *p*-type material is connected to a positive potential, that is, forward biased, holes are repelled across the junction into the *n*-type material, and electrons are repelled across the junction into the *p*-type material. The width of the space-charge region is reduced and the barrier height for *p* to *n* current flow is reduced as shown in Fig. 45.8(b). A DC forward bias voltage, V_F, of approximately 0.5–0.7 V for silicon and 0.2–0.3 V for germanium is required to overcome the barrier voltage. Once the barrier is overcome, the junction current increases significantly from an increase in the diffusion current. That is, holes cross the junction into the *n*-type material, where they are considered injected minority carriers. Electrons cross the junction into the *p*-type material, where they too are injected minority carriers. Because hole movement in one direction and electron movement in the opposite direction constitute a current in the same direction, the total current is the sum of the hole and electron minority currents.

When the *p*-type material is connected to a negative potential, that is, reverse biased, holes and electrons move away from the junction. The width of the space-charge region is increased and the barrier height for *p* to *n* current flow is increased as shown in Fig. 45.8(c). The process nominally stops when the holes in the *p*-type material are depleted. However, a few holes in the *n*-type material are thermally generated, as there are electrons in the *p*-type material. These minority carriers thermally diffuse into the depletion region and are swept across by the electric field. The effect is constant for a given temperature and independent of the reverse bias. This is the reverse saturation current, I_s.[16] The reverse saturation

[16]The reverse saturation current also accounts for any current leakage across the surface of the semiconductor.

Figure 45.8 *pn Junction Space-Charge Region*

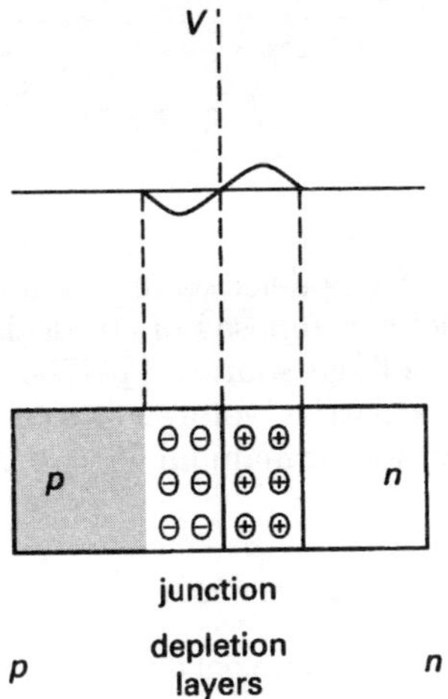

(a) no applied voltage

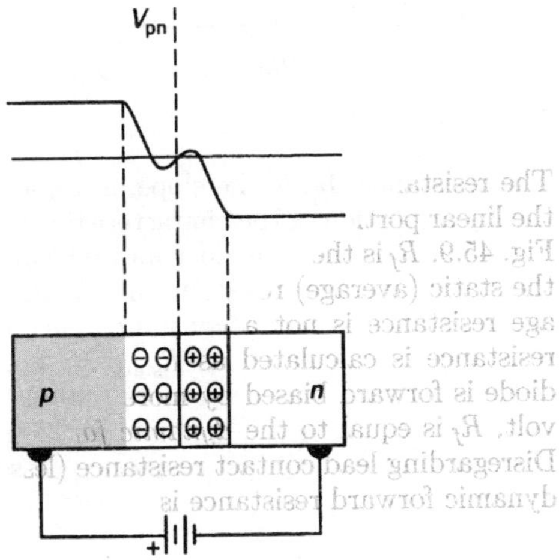

(b) forward bias

V_{pn}

p

n

+

(c) reverse bias

current is small ($\approx 10^{-9}$ A). An ideal *pn* junction, excluding the breakdown region, is governed by[17]

$$I_{\text{pn}} = I_s\left(e^{qV_{\text{pn}}/\kappa T} - 1\right) \qquad 45.15$$

Breakdown is a large, abrupt change in current for a small change in voltage. When a *pn* junction is reverse biased, the saturation current is small up to a certain reverse voltage, where it changes dramatically. Two mechanisms can cause this change. The first is *avalanche breakdown.* Avalanche occurs when thermally generated minority carriers are swept through the space-charge region and collide with ions. If they possess enough energy to break a covalent bond, an electron-hole pair is created. The same effect may occur for these newly generated carriers, resulting in an avalanche effect. Avalanche breakdown occurs in lightly doped materials at greater than 6 V reverse bias. The second breakdown mechanism is *zener breakdown.* Zener breakdown occurs through the disruption of covalent bonds caused by the strength of the electric field near the junction. No collisions are involved. The additional carriers created by the breaking covalent bonds increase the reverse current. Zener breakdown occurs in highly doped materials at less than 6 V reverse bias.[18]

9. DIODE PERFORMANCE CHARACTERISTICS

A *diode* is a two-electrode device. The diode is designed to pass current in one direction only. An *ideal diode,* approximated by a *pn* junction, has a zero voltage drop, that is, no forward resistance, and acts as a short circuit when forward biased (on). When reversed biased (off), the resistance is infinite and the device acts as an open circuit. Diode construction and theory is that of a *pn* junction. (See Sec. 45.8.)

The characteristics and symbol for a typical real semiconductor diode are shown in Fig. 45.9. The *reverse bias voltage* is any voltage below which the current is small, that is, less than 1% of the maximum rated current. The *peak inverse (reverse) voltage,* PIV or PRV, is the maximum reverse bias the diode can withstand without damage. The forward current is also limited by heating effects. Maximum forward current and peak inverse voltage for silicon diode rectifiers are approximately 600 A and 1000 V, respectively.

The ideal *pn* junction current was given in Eq. 45.15. For practical junctions (*diodes* or *rectifiers*), the equation becomes

$$I = I_s\left(e^{qV/\eta\kappa T} - 1\right) = I_s\left(e^{V/\eta V_T} - 1\right) \qquad 45.16$$

[17]The subscript "pn" is used here for clarification. Standard diode voltage and current directions are defined in Sec. 45.9, after which the subscript is no longer used.

[18]The name zener is commonly used regardless of the breakdown mechanism.

Figure 45.9 Semiconductor Diode Characteristics and Symbol

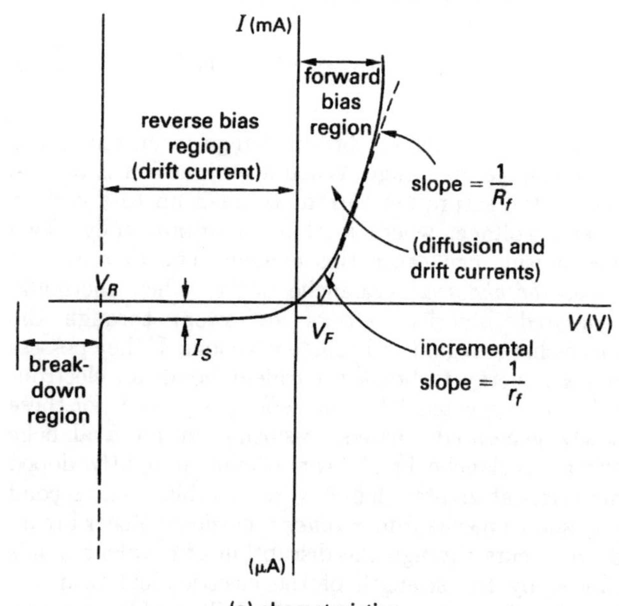

(a) characteristics

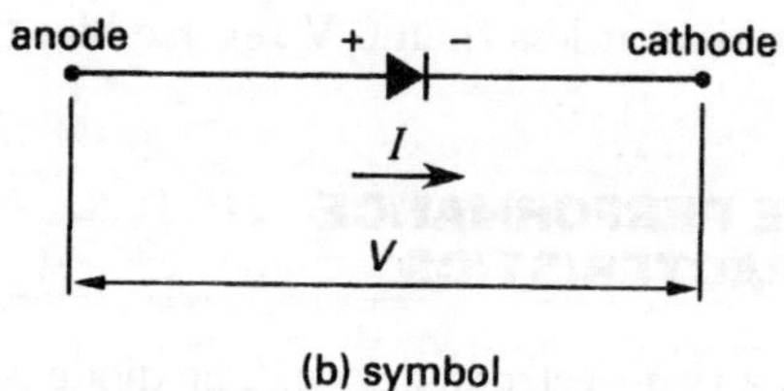

(b) symbol

Equation 45.16, which is based on the *Fermi-Dirac probability function*, is valid for all but the breakdown region (see Fig. 45.9). The term η is determined experimentally. For discrete silicon diodes, $\eta = 2$. For germanium diodes, $\eta = 1$. The saturation current, taken from any value of I with a small reverse bias (for example, between 0 and -1 V) gives $I_s \approx 10^{-9}$ A for silicon and 10^{-6} A for germanium. The term V_T represents the *voltage equivalent of temperature* and is related to the diffusion occurring at the junction.[19]

$$V_T = \frac{\kappa T}{q} = \frac{D_p}{\mu_p} = \frac{D_n}{\mu_n} \qquad 45.17$$

Boltzmann's constant is given by κ. The absolute temperature is T. An electron charge is represented by q. Diffusion constants, measured in m^2/s, are given the symbol D. The mobility, measured in $m^2/V{\cdot}s$, is given by the symbol μ.

The voltage equivalent of temperature is also known as the *thermal voltage*. The value of V_T is often quoted at *room temperature*, which can vary from 293K (20°C) to 300K (27°C) depending upon the reference used. The temperature has the effect of doubling the saturation current every 10°C. So,

$$\frac{I_{s2}}{I_{s1}} = (2)^{T_2 - T_1/10^\circ\text{C}} \qquad 45.18$$

A *real diode* can be modeled as shown in Fig. 45.10. The real diode model is composed of an ideal diode, a resistor, R_f, and a voltage source, V_F. The voltage source accounts for the barrier voltage. Typically, for silicon $V_F = 0.7$ V, and for germanium $V_F = 0.2$ V.

Figure 45.10 Diode Equivalent Circuit

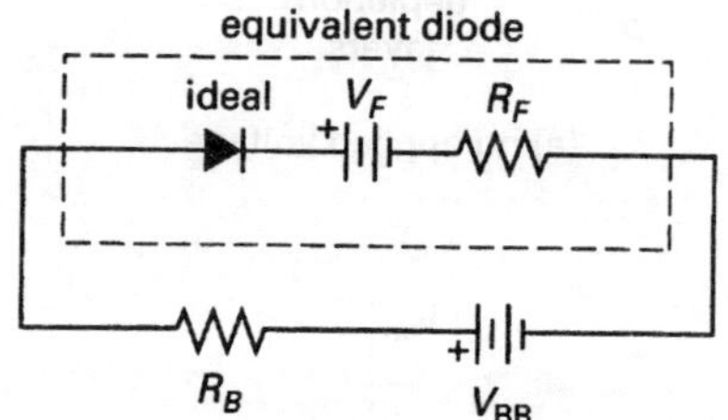

The resistance, R_f, is the slope of a line approximating the linear portion of the characteristic curve as shown in Fig. 45.9. R_f is the resistance for an ideal diode. It is not the static (average) resistance of the diode, as the average resistance is not a constant. The static (average) resistance is calculated as $R_{\text{static}} = V_D/I_D$. When the diode is forward biased by more than a few tenths of a volt, R_f is equal to the *dynamic forward resistance*, r_f. Disregarding lead contact resistance (less than 2 Ω), the dynamic forward resistance is

$$R_f = r_f = \frac{\eta V_T}{I_D} \qquad 45.19$$

A *dynamic reverse resistance*, r_r, also exists and is the inverse of the slope at a point in the reverse bias region.[20] Because the reverse current is very small, the resistance is often considered infinite. Capacitances associated with the junction are also ignored in most models of the diode. The *diffusion capacitance*, C_d, is associated with the charge stored during forward biased operation. The *transition capacitance*, C_t, is associated with the space-charge region width and is the primary capacitance of concern during reverse bias operation. (Diodes designed to take advantage of this voltage-sensitive capacitance are called *varactors*).

The model of Fig. 45.10 assumes that (1) the reverse bias current is suffciently small that the diode acts as an open circuit in that direction, (2) the reverse bias

[19]This is also called the *Einstein relationship*.

[20]Specifying the reverse current, I_{co}, is equivalent to specifying the reverse resistance.

voltage does not exceed the breakdown voltage, and (3) the switching time is instantaneous. The *switching time* is the transient that occurs from the time interval of the application of a voltage to forward (reverse) bias and the achievement of the actual condition. The switching time depends upon the speed of movement of minority carriers near the junction and the junction capacitance.

Example 45.4

What is the thermal voltage at a room temperature of 300K?

Solution

The thermal voltage is given by Eq. 45.17 as

$$V_T = \frac{\kappa T}{q} = \frac{\left(1.3807 \times 10^{-23}\ \frac{\text{J}}{\text{K}}\right)(300\text{K})}{1.6022 \times 10^{-19}\ \text{C}} = 0.026\ \text{V}$$

10. DIODE LOAD LINE

Figure 45.11 shows a forward-biased real diode in a simple circuit. V_{BB} is the *bias battery* (hence the subscripts), and R_B is a current-limiting resistor. R_f and V_f are equivalent diode parameters, not discrete components. If R_f is known in the vicinity of the operating point, the diode current, I_D, is found from Kirchhoff's voltage law. The diode voltage is found from Ohm's law: $V_D = I_D R_f$.

$$V_{BB} - V_f = I_D(R_B + R_f) \qquad 45.20$$

The diode current and voltage drop can also be found graphically from the *diode characteristic curve* (see Fig. 45.11) and the *load line*, a straight line representing the locus of points satisfying Eq. 45.20.[21] The load line is defined by two points. If the diode current is zero, all of the battery voltage appears across the diode (point $(V_{BB}, 0)$). If the voltage drop across the diode is zero, all of the voltage appears across the current-limiting resistor (point $(0,\ V_{BB}/R_B)$). (Because the diode characteristic curve implicitly includes the effects of V_f and R_f, these terms should be omitted.) The no-signal *operating point*, also known as the *quiescent point*, is the intersection of the diode characteristic curve and the load line.

The *static load line* is derived assuming there is no signal (i.e., $v_{in} = 0$). With a signal, the *dynamic load line* shifts left or right while keeping the same slope. This is

Figure 45.11 *Diode Load Line*

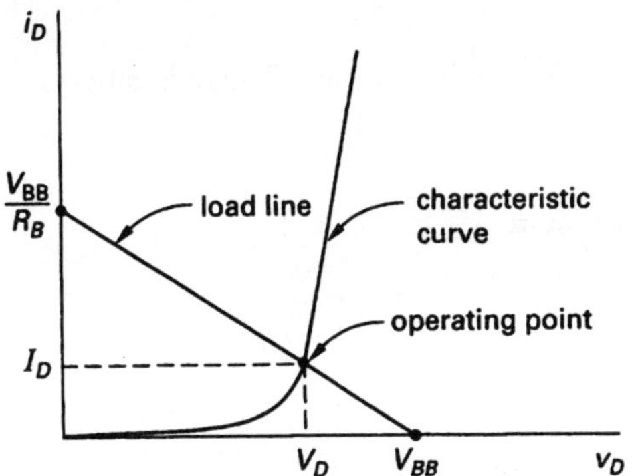

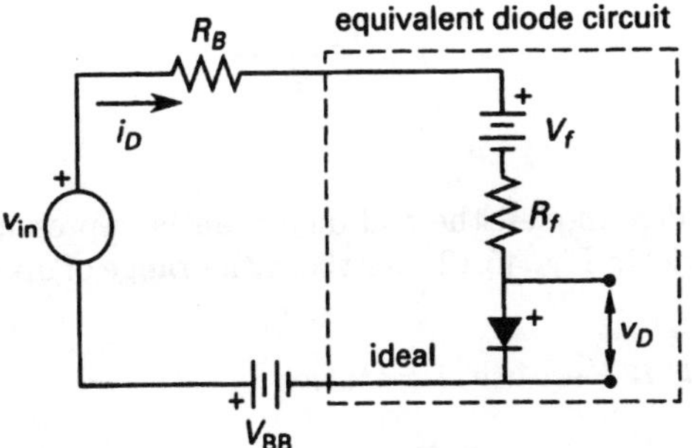

equivalent to solving Eq. 45.20 with an additional voltage source.

$$i_D = I_D + \Delta i_D = \frac{V_{BB} - V_f + v_{in}}{R_B + R_f} \qquad 45.21$$

Although presented in a slightly different manner, the method for determining the load line is similar to that given in Sec. 45.7. That is,

step 1: Open-circuit the electronic component's equivalent circuit and determine the point $(x, 0)$.

step 2: Short-circuit the electronic component's equivalent circuit and determine the point $(0, y)$.

step 3: Connect the two points.

The load line superimposed on the characteristics curve is then used to determine the operating point (*Q*-point) of the overall circuit.

11. DIODE PIECEWISE LINEAR MODEL

If the voltage applied to a diode varies over an extensive range, a piecewise linear model may be used.[22] For the real diode characteristic shown in Fig. 45.9, three regions are evident from V_R to V_F.

[21] The horizontal axis voltage is the voltage across the diode (modeled as a resistor).

[22] The model in Sec. 45.10 is for the forward bias region only and assumes no reverse current flow.

(1) *Reverse breakdown region,* $V_D < V_R$

$$I_D = \frac{V_D + V_R}{R_r} \quad [V_D \text{ and } V_R \text{ are negative}] \tag{45.22}$$

(2) *Off region,* $V_R < V_D < V_F$

$$I_D = 0 \tag{45.23}$$

(3) *Forward bias region,* $V_D > V_F$

$$I_D = \frac{V_D - V_F}{R_f} \tag{45.24}$$

Using ideal diodes, the real diode can be represented by the model in Fig. 45.12 over the entire range of operation.

Figure 45.12 Piecewise Linear Model

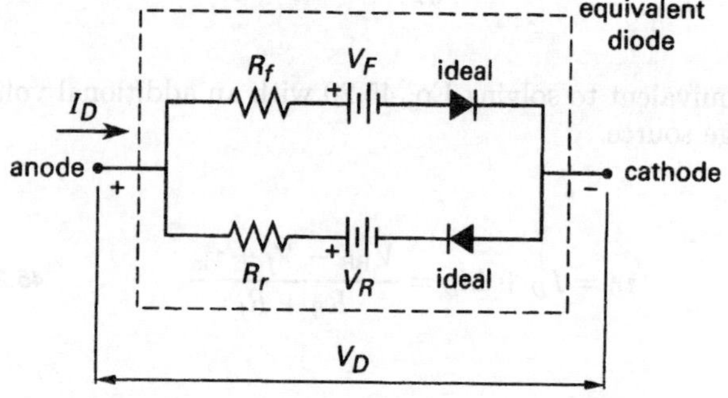

12. DIODE APPLICATIONS AND CIRCUITS

Diodes are readily integrated into *rectifier, clipping,* and *clamping circuits.* A clipping circuit cuts the peaks off of waveforms; a clamping circuit shifts the DC (average) component of the signal. Figure 45.13 illustrates the response to a sinusoid with peak voltage V_m for several simple circuits.

13. SCHOTTKY DIODES

A *Schottky diode,* also called a *barrier diode* or a *hot-carrier diode,* is a diode constructed with a metal semiconductor contact as shown in Fig. 45.14(a). The Schottky diode symbol is shown in Fig. 45.14(b). The metal semiconductor rectifying junction is similar to the *pn* junction, but the physical mechanisms are somewhat different.

In the forward direction, electrons from the lightly doped semiconductor cross into the metal anode, where

Figure 45.13 Output from Simple Diode Circuits

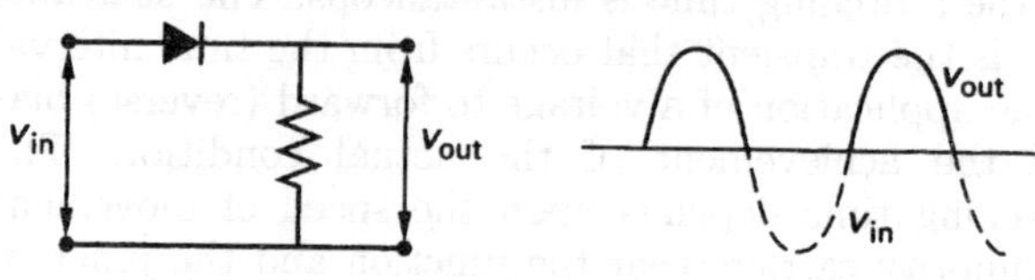

(a) half-wave rectifier

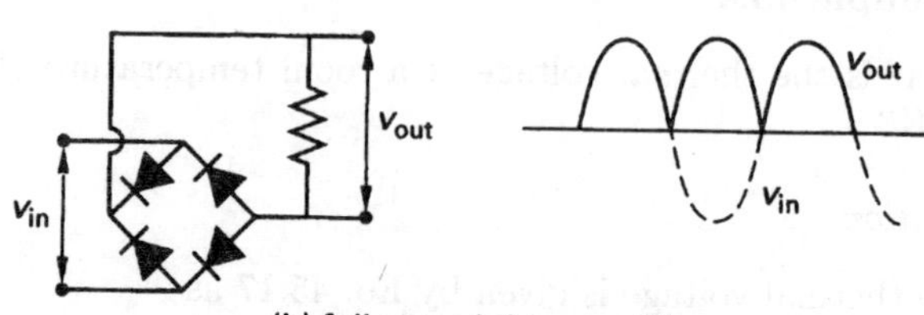

(b) full-wave bridge rectifier

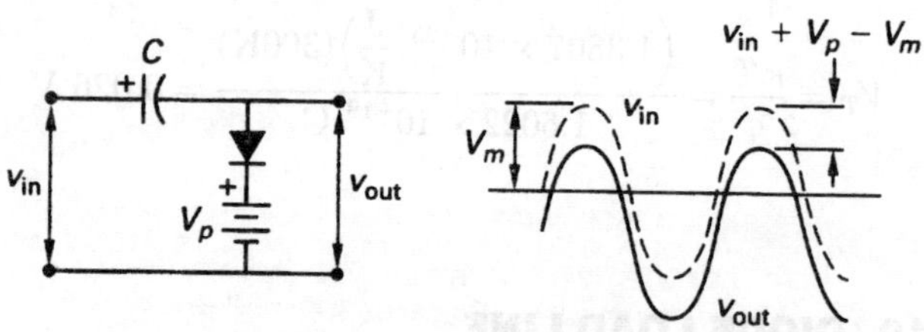

(c) clamping circuit (*C* charges to V_m)

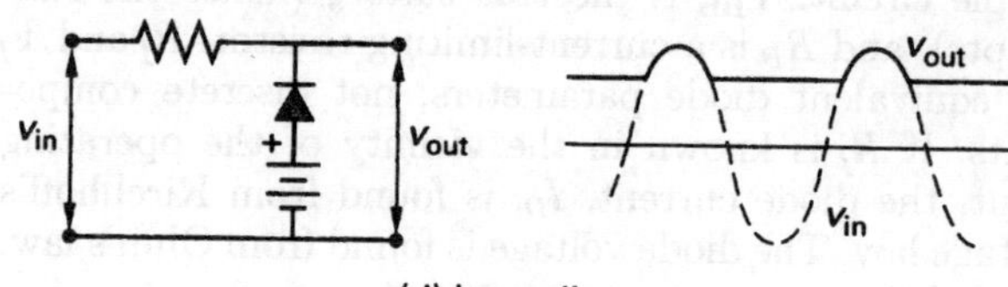

(d) base clipper

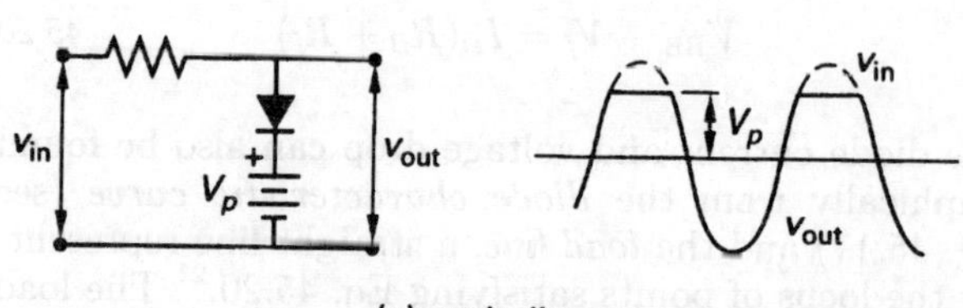

(e) peak clipper

electrons are plentiful.[23] The electrons in the metal are majority carriers, whereas in a *p*-type material they are minority carriers. Being majority carriers, they are indistinguishable from other carriers and are not stored near the junction. This means that no minority carrier population exists to move when switching occurs from forward to reverse (on to off) bias. Consequently, switching times are extremely short (approximately 10^{-12} s).

The Schottky symbol is used on any electronic component designed with a *Schottky contact,* that is, a rectifying, rather than an ohmic, contact as shown in Fig. 45.14(c).

[23]The electrons injected into the metal are above the *Fermi energy level,* determined by the Fermi-Dirac distribution of electron energies, and are called *hot carriers.*

Figure 45.14 *Schottky Diode*

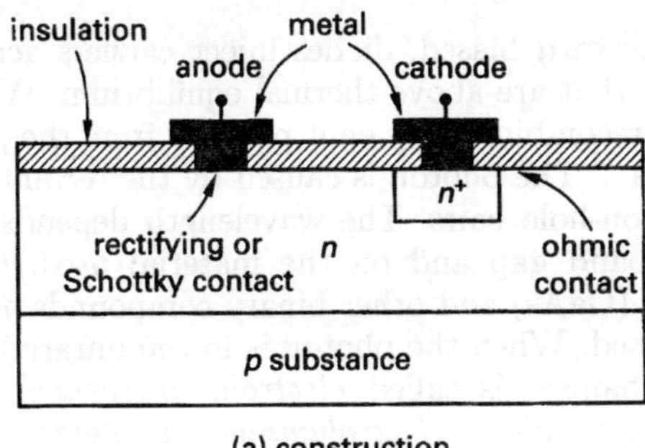

(a) construction

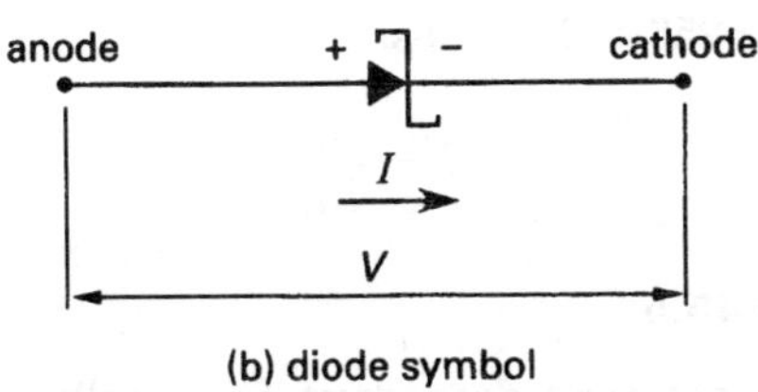

(b) diode symbol

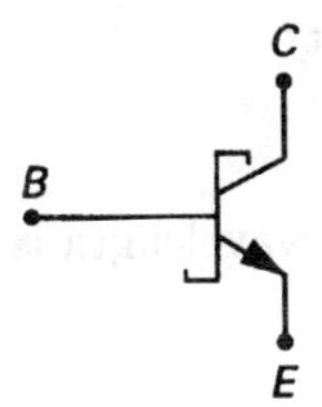

(c) Schottky contact symbol

In the figure, the Schottky contact lies between the base and the collector.

14. ZENER DIODES

A *zener diode* is a diode specifically designed to operate within the breakdown region. The construction and theory of a zener diode is similar to that of a *pn* junction, with the design allowing for greater heat dissipation capabilities. The characteristics and symbol are shown in Fig. 45.15.

When a reverse voltage, known as the *zener voltage*, V_Z (which is negative with respect to the anode), is applied, a reverse saturation current, I_Z, flows. The voltage-current relationship is nearly linear. Current flows until the reverse saturation current drops to I_{ZK} near the knee of the characteristic curve. This minimum current is the *keep-alive current*, that is, the minimum current for which the output characteristic is linear.

Because the current is large, an external resistor must be used to limit the current to within the power dissipation capability of the diode. Zener diodes are used as voltage regulating and protection devices. The regulated zener

Figure 45.15 *Zener Diode*

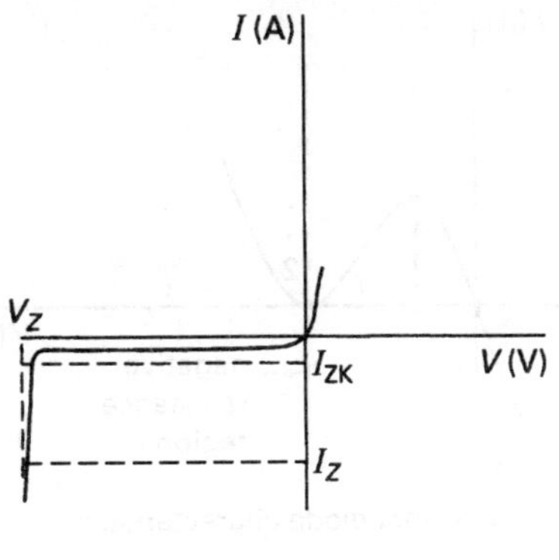

(a) characteristics

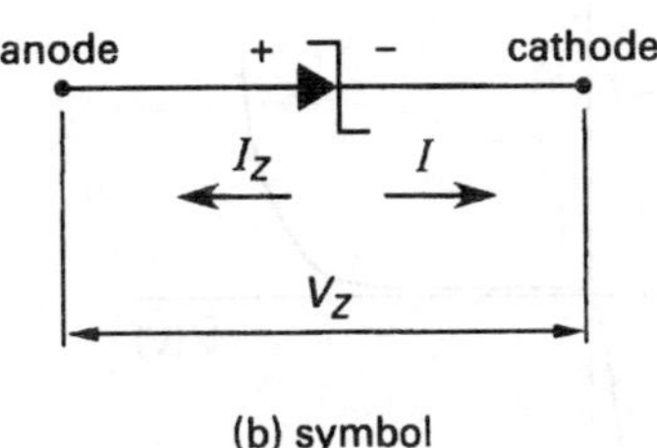

(b) symbol

voltage varies with the temperature. The *temperature coefficient* is

$$\text{tc} = \frac{\Delta V_Z}{V_Z \Delta T} \times 100\% \tag{45.25}$$

15. TUNNEL DIODES

A *tunnel diode* (*Esaki diode*) is a two-terminal device with an extremely thin potential barrier to electron flow, so that the output characteristic is dominated by the quantum-mechanical tunneling process. To make a thin potential barrier (50 Å to 100 Å), both regions of the diode are heavily doped. The amount of tunneling is limited by the electrons available in the *n*-type material or by the available empty energy states in the *p*-type material to which they can tunnel. The characteristics are shown in Fig. 45.16(a).

Point 1 in Fig. 45.16(a) corresponds to a biasing level that allows for maximum tunneling. (At this point, the minimum electron energy in the *n*-type material conduction band equals the Fermi level in the *p*-type material's valence band.) Between points 1 and 2, increases in bias voltage result in a decrease in current as available energy states are filled and the total amount of tunneling drops. This is an area of negative resistance, or negative conductance, and is the primary use for tunnel diodes. If the doping levels are slightly reduced from their typical values of $5 \times 10^{19}\ \text{cm}^{-3}$, the forward tunneling current becomes negligible and the output

Figure 45.16 Tunnel Diode

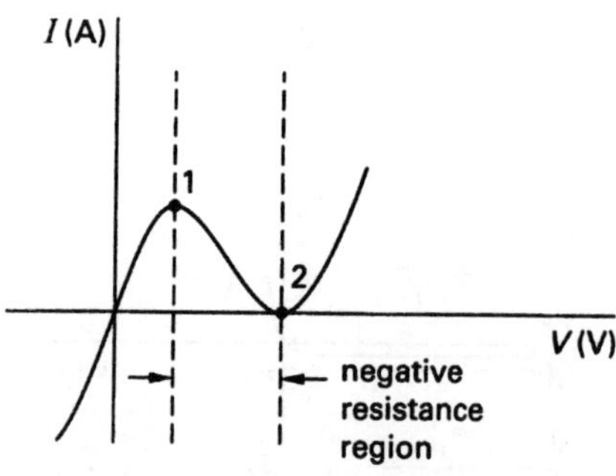

(a) tunnel diode characteristics

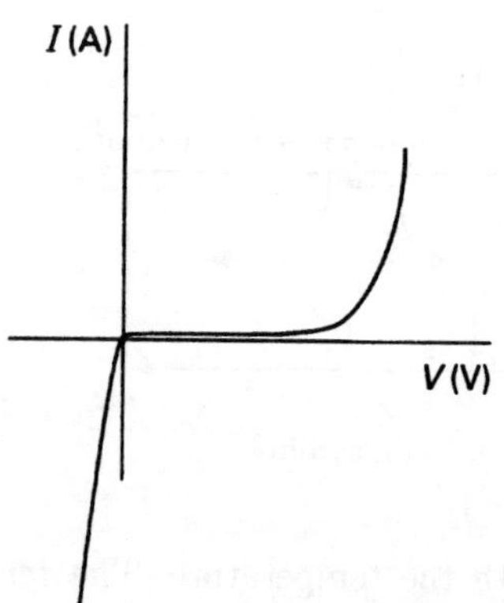

(b) backward diode

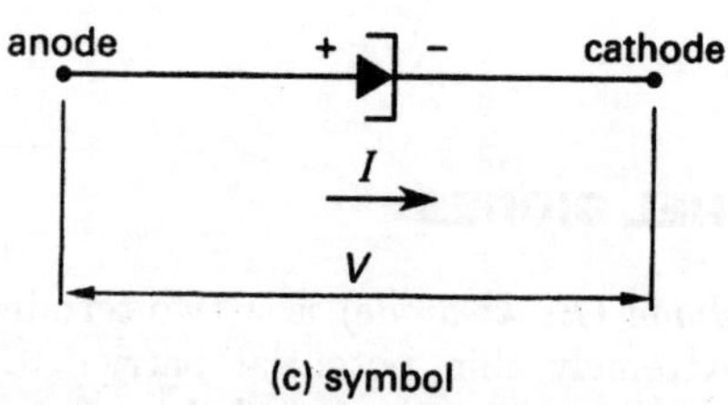

(c) symbol

characteristic becomes that of a *backward diode*, Fig. 45.16(b). The symbol for a tunnel diode is shown in Fig. 45.16(c).

16. PHOTODIODES AND LIGHT-EMITTING DIODES

If a semiconductor junction is constructed so that it is exposed to light, the incoming photons generate electron-hole pairs. When these carriers are swept from the junction by the electric field, they constitute a *photocurrent*, which is seen as an increase in the reverse saturation current. The holes generated move to the *p*-type material and the electrons move to the *n*-type material in response to the electric field that is established whenever *p*- and *n*-type semiconductors are joined (see Fig. 45.6(a)). Such devices are used as light sensors and are called *photodiodes*. When the device is designed without a biasing source, it becomes a *solar cell.*

When forward biased, diodes inject carriers across the junction that are above thermal equilibrium. When the carriers recombine, they emit photons from the *pn* junction area.[24] The photon is caused by the recombination of electron-hole pairs. The wavelength depends on the energy band gap and on the material used. Gallium arsenide (GaAs) and other binary compounds are commonly used. When the photon is in the infrared region, the mechanism is called *electroluminescence* and the diodes are called *electroluminescent diodes.* If the photons are in the visible region, the devices are called *light-emitting diodes* (LEDs). The emitted wavelength, λ, is given by Eq. 45.26. h is Planck's constant, with a value of 6.6256×10^{-34} J·s.

$$\lambda = \frac{hc}{E_G} \qquad 45.26$$

Example 45.5

The manufacturer's data sheet for a gallium arsenide diode shows a band gap energy of 1.43 eV. Will such a band gap result in emitted photons within the wavelength of visible light?

Solution

The photon emitted wavelength is given by

$$\begin{aligned}\lambda &= \frac{hc}{E_G} \\ &= \frac{(6.6256 \times 10^{-34}\ \text{J·s})\left(2.9979 \times 10^{8}\ \frac{\text{m}}{\text{s}}\right)}{(1.43\ \text{eV})\left(1.6022 \times 10^{-19}\ \frac{\text{J}}{\text{eV}}\right)} \\ &= 8.67 \times 10^{-7}\ \text{m}\end{aligned}$$

The wavelength of visible light is from approximately 8×10^{-7} m to 4×10^{-7} m. Consequently, GaAs is a good choice for a light-emitting diode.

17. SILICON-CONTROLLED RECTIFIERS

A four-layer *pnpn* device with an anode, cathode, and gate terminal is called a *silicon-controlled rectifier* (SCR) or *thyristor.*[25] A conceptual construction is shown in Fig. 45.17(a).

When the SCR is reverse biased, that is, when the anode is negative with respect to the cathode, the characteristics are similar to a reverse-biased *pn* junction as shown in Fig. 45.17(b). When the SCR is forward biased, that is, when the anode is positive with respect to the cathode, four distinct regions of operation are evident.

[24] They can also release the energy as heat, or *phonons.*

[25] A *thyristor* is defined as a transistor with thyratron-like characteristics. That is, as collector current is increased to a critical value, the alpha (common base current gain) rises above unity and results in a high-speed triggering action.

Figure 45.17 Silicon-Controlled Rectifier

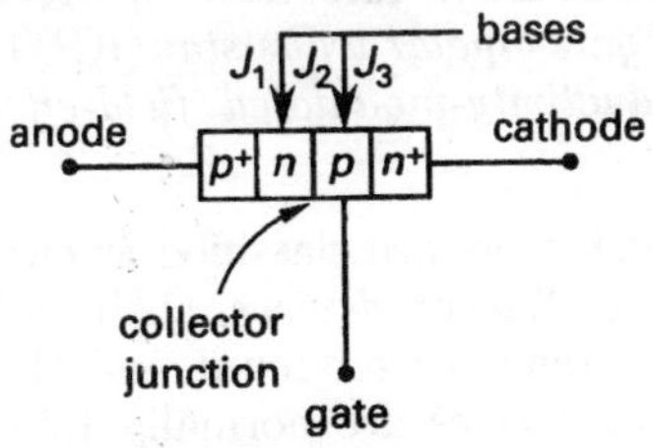

(a) conceptual construction

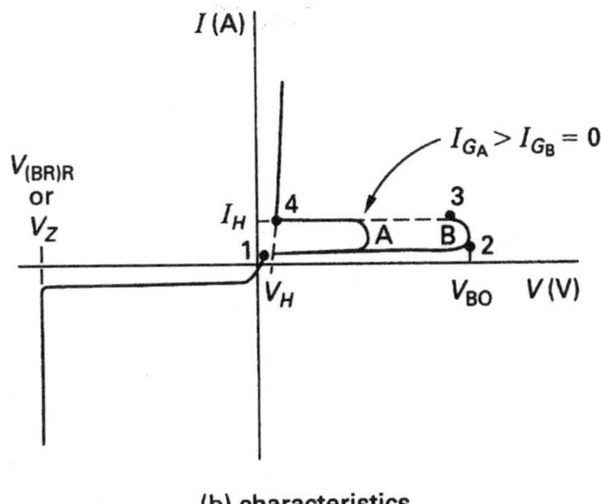

(b) characteristics

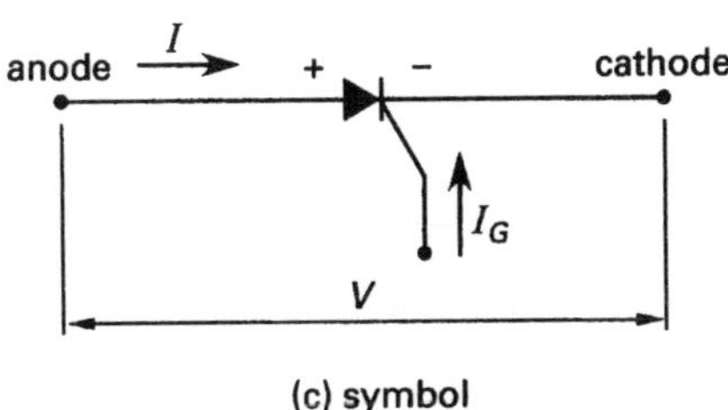

(c) symbol

From point 1 to point 2, junctions J_1 and J_3 are forward biased. Junction J_2 is reverse biased. The external voltage appears primarily across the reverse-biased junctions. The device continues to operate similarly to a reverse-biased *pn* junction. This is called the *off* or *high-impedance region.*

From point 2 to point 3, the current increases slowly to the *breakover voltage*, V_{BO}. At this point, junction J_2 undergoes breakdown and the current increases sharply.

From point 3 to point 4, the current increases as the voltage decreases. This region is called the *negative resistance region.*

From point 4, junction J_2 is forward biased. The voltage across the device is essentially that of a forward-biased *pn* junction (approximately 0.7 V). If the current through the diode is reduced by the external circuit, the diode remains on until the current falls below the *hold current*, I_H, or the *hold voltage*, V_H. Below this point, the diode switches off, to the high-impedance state.

The gate functions to increase the current at the collector junction of the *npn* transistor, J_2, which is an integral part of the SCR. By increasing the current through the reverse-biased junction, J_2, the anode current is increased. This increases the gain of the two transistors, resulting in a lowering of the forward breakover voltage. This is seen in the difference between paths A and B in Fig. 45.17(b). Consequently, for a given anode to cathode voltage, the gate can be used to turn the SCR on. Once on, however, the SCR must be reverse biased to turn off. A thyristor designed to be turned off by the gate is called a *gate turn-off thyristor.*

SCRs are used in power applications to allow small voltages and currents to control much larger electrical quantities. The symbol for an SCR is shown in Fig. 45.17(c).

Other devices with multiple *pn* junctions using the same principles are the *diac* and *triac*, shown in Fig. 45.18 and Fig. 45.19, respectively.

Most electronic components are unable to handle large amounts of current. When properly designed for power dissipation, semiconductors handling large amounts of power are called *power semiconductors.* Such devices constitute a branch of electronics called *power electronics.* Silicon-controlled rectifiers and Schottky diodes are traditional power semiconductors. Newer designs include the *high-power bipolar junction transistor*

Figure 45.18 Diac

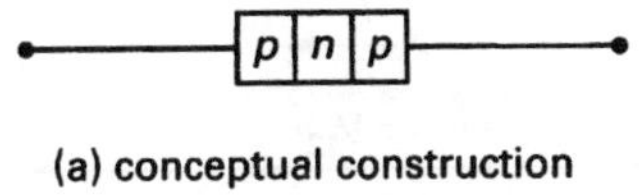

(a) conceptual construction

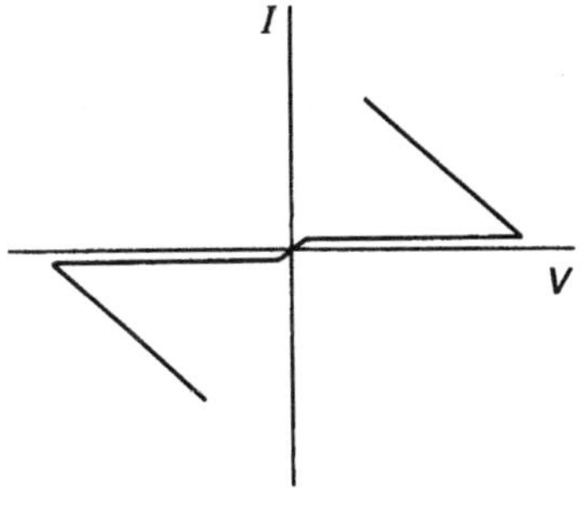

(b) characteristics

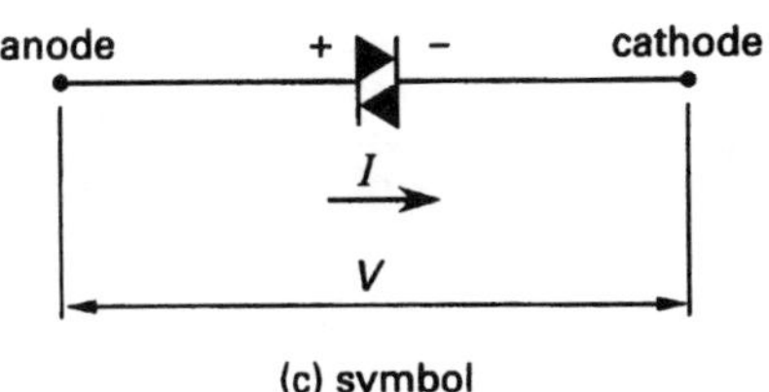

(c) symbol

Electronics

Figure 45.19 Triac

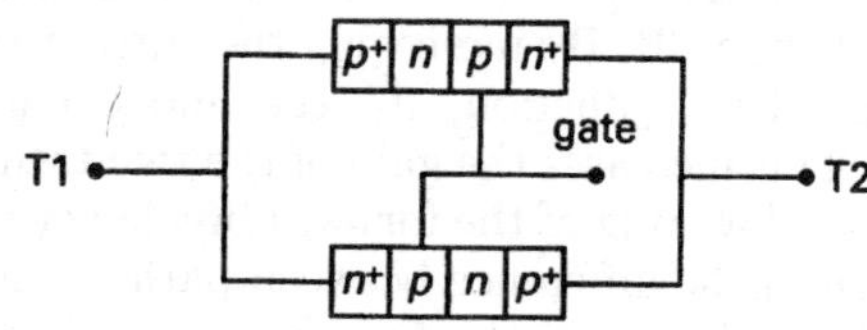

(a) conceptual construction

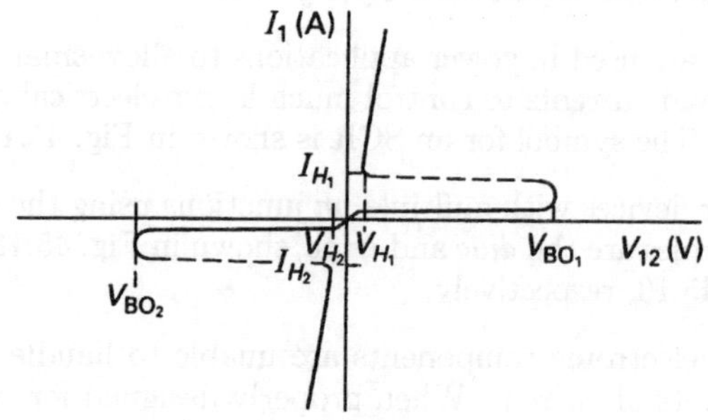

(b) characteristics

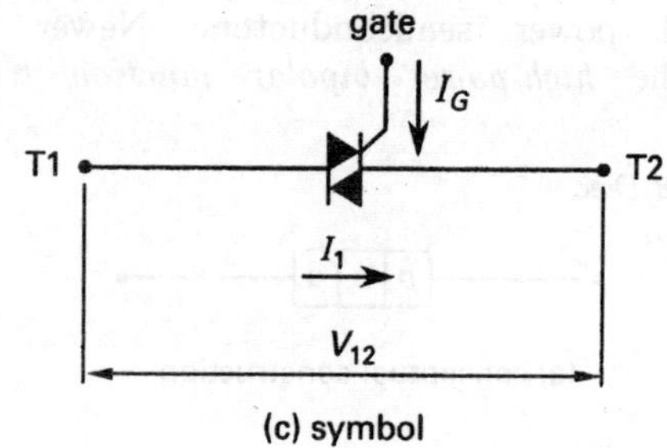

(c) symbol

(HPBT), power metal-oxide semiconductor field-effect transistor (MOSFET), *gate turn-off thyristor* (GTO), and *insulated gate bipolar transistor* (IGBT), sometimes called a *conductivity-modulated field-effect transistor* (COMFET).

Power semiconductors are classified as either trigger or control devices. *Trigger devices*, such as GTOs, start conduction by some trigger input and then behave as diodes. *Control devices* are normally BJTs and FETs used in full-range amplifiers.

46 Junction Transistors

Nomenclature

A	gain	–
C	capacitance	F
E	energy	J or eV
g	transconductance	S
g_m	mutual conductance or transconductance	S
h_f	forward transfer voltage ratio with output shorted	–
h_{fe}	CE small signal (AC) forward current transfer ratio or gain	–
h_i	input impedance with output shorted	Ω
h_{ie}	CE small signal (AC) input impedance	Ω
$h_{i,o,f,r \text{ and } e,c,b}$	hybrid parameter[1]	–
$h_{I,O,F,R \text{ and } E,C,B}$	hybrid parameter[2]	–
h_o	output admittance with input open	S
h_{oe}	CE small signal (AC) open-circuit output admittance	S
h_r	reverse transfer voltage ratio with input open	Ω
h_{re}	CE small signal (AC) reverse voltage transfer ratio or gain	–
i	instantaneous or variable current	A
I	effective or DC current	A
I_{CBO}	DC collector cutoff current, emitter open[3]	A
n	electron concentration	m^{-3} or cm^{-3}
n, N	concentration	m^{-3} or cm^{-3}
p	hole concentration	m^{-3} or cm^{-3}
P	power dissipation	W
Q	quiescent point	–
r	small-signal resistance[4]	Ω
$r_{bb'}$	base spreading resistance	Ω
$r_{b'c}$	feedback resistance	Ω
$r_{b'e}$	base input resistance	Ω
r_{ce}	output resistance	Ω
R	resistance	Ω
S	apparent power	VA
v	instantaneous or variable voltage	V
V	effective or DC voltage	V

Symbols

α	CB forward current ratio or gain	–
β	CE forward current ratio or gain	–
ω	angular frequency	rad/s

Subscripts

0	at 0K
0	barrier or source value
ac	alternating current or small signal
b	AC or small signal base
b′c	from internal base to collector
b′e	from internal base to emitter
bc, BC	base to collector
be, BE	base to emitter
B	DC or biased base
BB	base supply or base biasing
c	conduction band or collector
cb, CB	collector to base
ce, CE	collector to emitter
cf	corner frequency or cutoff frequency
co	cutoff
C	collector
CBO	collector cutoff current
CC	collector supply
CQ	Q-point collector current

[1]The first subscript refers to the AC or small signal input (i), output (o), forward (f), or reverse (r) value. The second subscript refers to the configuration, that is, e for common emitter (CE), c for common collector (CC), b for common base (CB).

[2]The first subscript refers to the DC or static input (I), output (O), forward (F), or reverse (R) value. The second subscript refers to the configuration, that is, E for common emitter (CE), C for common collector (CC), B for common base (CB).

[3]This is essentially the reverse-saturation current between the collector and the base.

[4]This is also known as the *dynamic resistance* or the *instantaneous resistance*. Any resistance represented with a lowercase letter is the small-signal value.

d	diffusion
DC	direct current or static
e	AC or small signal emitter
eb, EB	emitter to base
ec, EC	emitter to collector
E	DC or biased emitter
EE	emitter to supply voltage
f	forward (AC or instantaneous)
i	input (AC or instantaneous)
I	current or input (DC or total)
max	maximum
o	output (AC or instantaneous)
P	power
r	reverse (AC or instantaneous)
R	resistance
t	transition
T	thermal
V	constant or effective voltage

1. TRANSISTOR FUNDAMENTALS

A *transistor* is an active device comprised of semiconductor material with three electrical contacts (two rectifying and one ohmic). Two major types of transistors exist: *bipolar junction transistors* (BJTs) and *field-effect transistors* (FETs). A BJT uses the base current to control the flow of charges from the emitter to the collector. Field-effect transistors use an electric field, that is, voltage, established at the gate (equivalent to the base) to control the flow of charges from the source to the drain (equivalent to the emitter and collector). The controlling variable in a BJT is the current at the base; in a FET it is the voltage at the gate. An *npn* bipolar junction transistor (BJT) is shown in Fig. 46.1(a).

The transistor has three operating regions: *cutoff*, *saturation*, and *active* or *linear*. In the cutoff region, both the base-emitter junction and the collector-base junction are reverse biased. In the saturation region, both the base-emitter junction and the collector-base junction are forward biased. Small amounts of current injected into the base control the movement of charges in the output (from emitter to collector) in a manner similar in theory to that of an SCR, specifically, the triac (see Sec. 45.17). When operating from the cutoff to the saturation region, the transistor is a switch.

In the active, or linear, region, the base-emitter junction is forward biased and the collector-base junction is reverse biased. When so biased, the electron potential energy is of the shape shown in Fig. 46.1(b). Small amounts of current injected into the base lower the potential barrier across the emitter-base junction. As carriers move from the emitter to the base, most are swept across the narrow base by the relatively large electric field established by the reverse-biased collector-base junction. (See the large space-charge, or depletion region, shown in Fig. 46.1(b).) When operating in the active (linear) region, the transistor is an amplifier.

The symbol for BJTs is shown in Fig. 46.1(c). Positive currents are defined as those flowing into the associated terminals. Actual current flows follow the arrows.[5] Majority carriers are always injected into the base by the *emitter* (E). The thin controlling center of a transistor is the *base* (B). The majority carriers are gathered by the *collector* (C).

Figure 46.1 *Bipolar Junction Transistor*

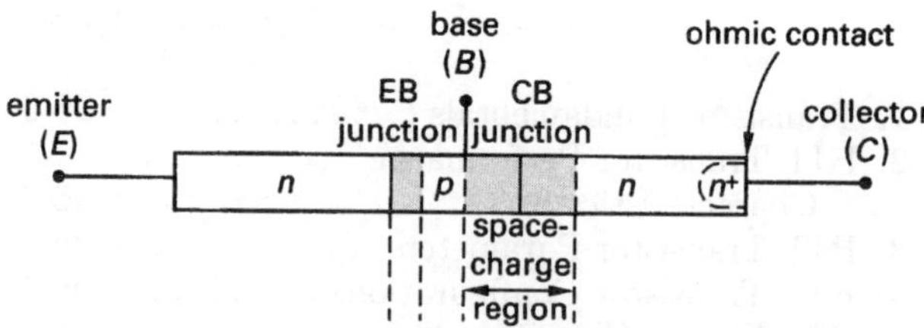

(a) conceptual construction

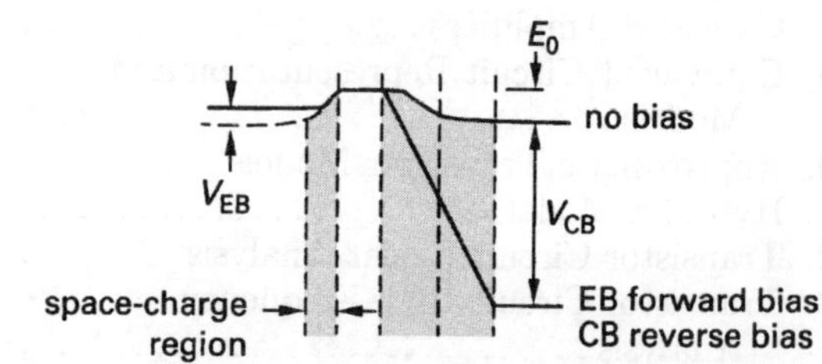

(space-charge regions shown for active region biasing)

(b) electron energy level

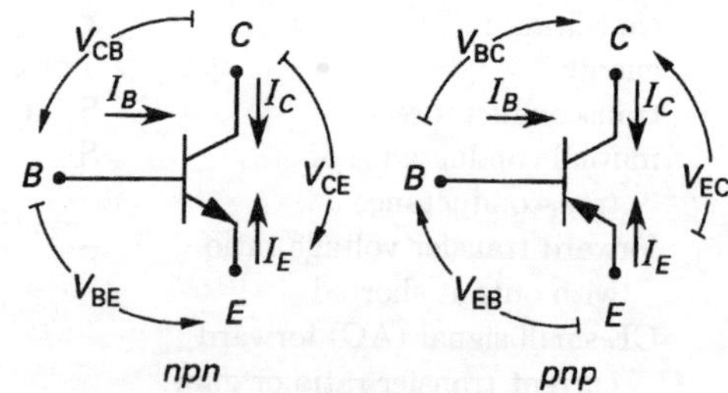

(all currents shown in positive direction)
(all voltages shown for active region biasing)

(c) symbol

When operated over all three ranges—cutoff, saturation, and active—the transistor is modeled in a piecewise linear fashion. This is *large-signal analysis*. Operated as a switch, between the cutoff and saturation regions, the transistor is modeled as an open circuit and a short circuit. This is *digital circuit operation*. Operated as an amplifier, in the active region, the transistor is modeled as equivalent parameters, usually h-parameters. This is *analog circuit operation*. Analog analysis occurs in two phases. First, the operating point, or Q-point, is determined. Determining the Q-point is a combination of determining the load line for the biasing used and setting the base current by using the values of the biasing components (see Sec. 46.5 and Sec. 46.6). This is often a

[5]That is, conventional current flow follows the arrows. Electrons flow opposite to the arrows.

reiterative process. (For a FET, the items determined are the load line and gate voltage.) Second, the incremental AC performance is determined using *small-signal analysis*, that is, using *h*-parameters to model the transistor and replacing independent sources with the appropriate models (see Sec. 46.9). Real sources are replaced with internal resistance. Ideal voltage sources are replaced with short circuits. Ideal current sources are replaced with open circuits.

2. BJT TRANSISTOR PERFORMANCE CHARACTERISTICS

When the base-emitter junction is forward biased and the collector-base junction is reverse biased, the transistor is said to be operating in the *active region.*

When the base-emitter junction is not forward biased (as when V_{BB} is zero), the base current will be nearly zero and the transistor acts like a simple switch. This is known as being *off* or *open*, and operating in the *cutoff mode.*[6] Also, the collector and emitter currents are zero when the transistor operates in the *cutoff region.* However, a very small input voltage (in place of V_{BB}) will forward bias the base-emitter junction, which, because I_B is so low, instantly forces the transistor into its saturation region. This results in a large collector current.

When the collector-emitter voltage is very low (usually about 0.3 V for silicon and 0.1 V for germanium), the transistor operates in its *saturation region.* Regardless of the collector current, the transistor operates as a closed switch (i.e., a short circuit between the collector and emitter). This is known as being *on* or *closed.*

$$V_{CE} \approx 0 \quad \text{[saturation]} \qquad 46.1$$

The BJT operating regions are shown in Fig. 46.2. The output characteristics for the common emitter and common base configurations are shown in Fig. 46.3. (A sample maximum power curve is shown as a dashed line. The *Q*-point must be to the left and below the power curve.)

Figure 46.2 *BJT Operating Regions*

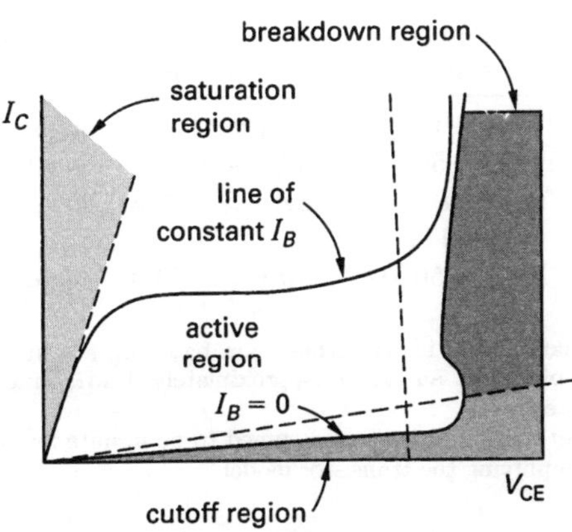

[6]Except for digital and switching applications, this condition usually results from improper selection of circuit resistances.

Figure 46.3 *BJT Output Characteristics*

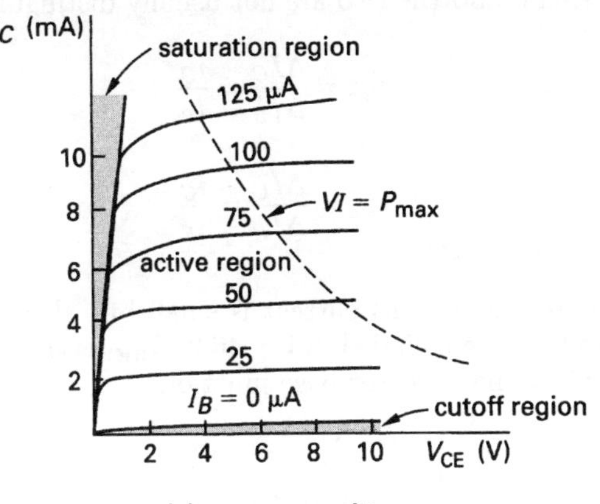

(a) common emitter

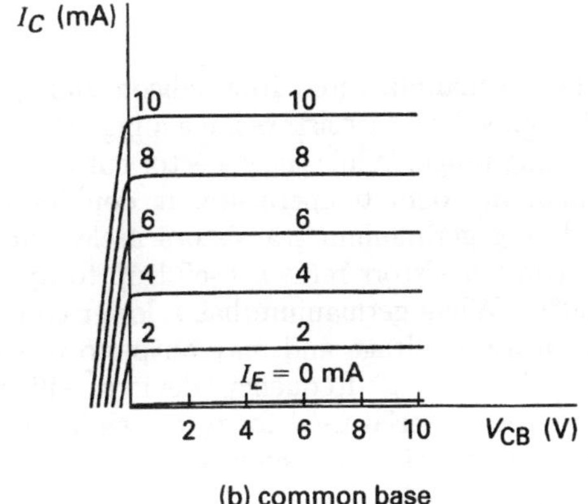

(b) common base

3. BJT TRANSISTOR PARAMETERS

Equation 46.2 is Kirchhoff's current law, taking the transistor as a node. Usually, the collector current is proportional to, and two or three orders of magnitude larger than, the base current, I_B. Because of this, a small change in base current of, for example, 1 mA, can produce a change in collector current of 100 mA. The *current (amplification) ratio*, β_{DC}, is the ratio of collector-base currents.

$$I_E = I_C + I_B \qquad 46.2$$

$$\beta_{DC} = \frac{I_C}{I_B} = \frac{\alpha_{DC}}{1 - \alpha_{DC}} \qquad 46.3$$

$$\alpha_{DC} = \frac{I_C}{I_E} = \frac{\beta_{DC}}{1 + \beta_{DC}} \qquad 46.4$$

Both α_{DC} and β_{DC} are for DC signals only. The corresponding values for small signals are designated α_{ac} and β_{ac}, respectively, and are calculated from differentials. (The difference between β_{ac} and β_{DC} is very small, and the two are not usually distinguished.)

$$\beta_{ac} = \frac{\Delta I_C}{\Delta I_B} = \frac{i_C}{i_B} \quad \text{46.5}$$

$$\alpha_{ac} = \frac{\Delta I_C}{\Delta I_E} = \frac{i_C}{i_E} \quad \text{46.6}$$

Thermal (saturation) current is small but always present and can be included in Eq. 46.2. I_{CBO} is the thermal current at the collector-base junction.

$$I_C = I_E - I_B \quad \text{46.7}$$

$$I_C = \alpha I_E - I_{CBO} \approx \alpha I_E \quad \text{46.8}$$

Transistors are manufactured from silicon and germanium, although silicon transistors have a higher temperature operating range. While the collector cutoff current is very small at room temperature, it doubles every 10°C, rendering germanium transistors useless around 100°C. Silicon transistors remain useful up to approximately 200°C. While germanium has a lower collector-emitter saturation voltage and may outperform silicon in high-speed and high-frequency devices, silicon is nevertheless the material used for most semiconductor devices and integrated circuit systems.

Figure 46.4 illustrates a family of curves for various base currents. The DC amplification factor, β_{DC}, can be found (for a wide range of V_{CE} values) by taking a point on any line in the active (horizontal line) region and calculating the ratio of the coordinates, I_C/I_B. The small-signal amplification factor, β_{ac}, is calculated as the difference in the I_C between two I_B lines divided by the differences in I_B.

Figure 46.4 *Small Signal Terms*

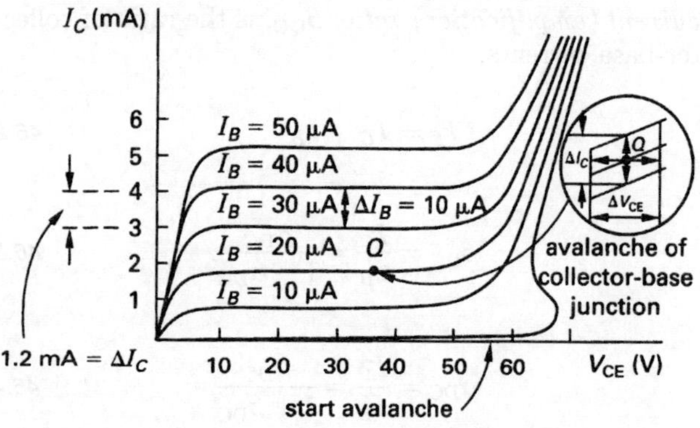

Many of the parameters used have multiple symbols. When given in manufacturers' data sheets or as equivalent parameters, *h*-parameter symbols are more common. The equivalences are given in Eq. 46.9 through Eq. 46.12.

$$\alpha_{DC} = h_{FB} \quad \text{46.9}$$

$$\alpha_{ac} = h_{fb} \quad \text{46.10}$$

$$\beta_{DC} = h_{FE} \quad \text{46.11}$$

$$\beta_{ac} = h_{fe} \quad \text{46.12}$$

4. BJT TRANSISTOR CONFIGURATIONS

There are six ways a BJT transistor can be connected in a circuit, depending on which leads serve as input and output. However, only three configurations have significant practical use. The terminal not used for either input or output is referred to as the *common terminal.* For example, in a *common emitter* circuit, the base receives the input signal and the output signal is at the collector. For the *common collector* (also known as an *emitter follower*) circuit, the input is to the base and the output is from the emitter. For a *common base* circuit, the input is to the emitter and the output is from the collector. The various transistor configurations are shown in Fig. 46.5.[7] A summary of the relative properties of the configurations is given in Table 46.1.

The common emitter configuration is arguably the most versatile and useful. It is the only configuration with a voltage and current gain that is greater than unity. The common collector is widely used as a buffer stage. That is, it is used between a high-impedance source and a low-impedance load. The common base has the fewest applications, though it is sometimes used to match a low-impedance source with a high-impedance load, or to act as a noninverting amplifier with a voltage gain greater than unity.

Table 46.1 *Comparison of Transistor Configuration Properties*

symbol[a]	common emitter[a]	common collector[a]	common base
A_V	high (−100)	low (< 1)	high (+100)
A_I	high (−50)	high (+50)	low (< 1)
R_{in}	medium (1000 Ω)	high (> 100 kΩ)	low (< 100 Ω)
R_{out}	high (≈ 50 kΩ)[b]	low (< 100 Ω)	high (≈ 2 MΩ)[b]

[a]Approximate values in parentheses are based on a source and load resistance of 3 kΩ, an h_{ie} of approximately 1 kΩ, and an h_{fe} of approximately 50.
[b]The output resistance is often assumed to be infinite for this configuration, simplifying the transistor model.

[7]Bypass capacitors are not shown.

Figure 46.5 *Transistor Configurations*

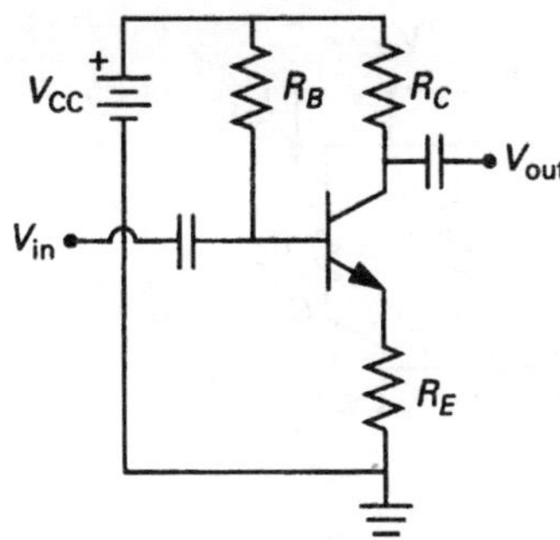

(a) common emitter

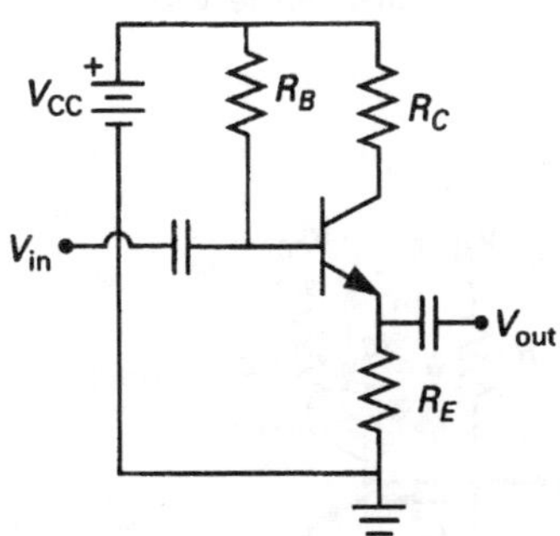

(b) common collector

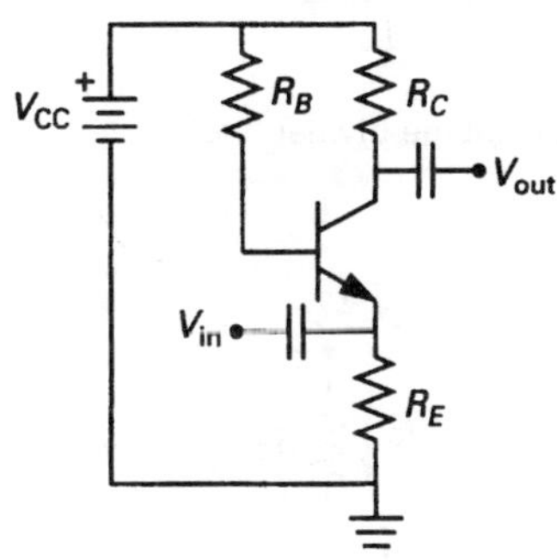

(c) common base

5. BJT BIASING CIRCUITS

To maximize the transistor's operating range when large swing-signals are expected, the quiescent point should be approximately centered in the active region. The purpose of biasing is to establish the base current and, in conjunction with the load line, the quiescent point. Figure 46.6 illustrates several typical biasing methods: fixed bias, fixed bias with feedback, self-bias, voltage-divider bias, multiple-battery bias, and switching circuit (cutoff) bias. All of the methods can be used with all three common-lead configurations and with both *npn* (shown) and *pnp* transistors.[8] It is common to omit the bias battery (shown in Fig. 46.6(a) in dashed lines) in transistor circuits.

The base current can be found by writing Kirchhoff's voltage law around the input loop, including the bias battery, the external resistances, and the V_{BE} barrier voltage that opposes the bias battery. Because the base is thin, there is negligible resistance from the base to the emitter, and V_{BE} (being less than 1 V) may be omitted as well. For the case of *fixed bias with feedback* (*fixed bias with emitter resistance*) illustrated in Fig. 46.6(b), the base current is found from

$$V_{CC} = I_B R_B + V_{BE} + I_E R_E \quad [v_{in} = 0] \qquad 46.13$$

$$\begin{aligned} I_B &= \frac{V_{CC} - V_{BE} - I_E R_E}{R_B} \quad [v_{in} = 0] \\ &= \frac{V_{CC} - V_{BE}}{R_B + \frac{\beta}{\alpha} R_E} \\ &= \frac{V_{CC} - V_{BE}}{R_B + (1+\beta) R_E} \quad [v_{in} = 0] \end{aligned} \qquad 46.14$$

The first step in designing a fixed-bias amplifier with emitter resistance is choosing the quiescent collector current, I_{CQ}. R_E is selected so that the voltage across R_E is approximately three to five times the intrinsic V_{BE} voltage (i.e., 0.3 V for germanium and 0.7 V for silicon).

$$R_E \approx \frac{3 V_{BE}}{I_{CQ}} \qquad 46.15$$

Once R_E is found, R_B can be calculated from Eq. 46.13.

The *bias stability*, with respect to any quantity M, is given by Eq. 46.16. Variable M commonly represents temperature (T), current amplification (β), collector-base cutoff (I_{CBO}), and base-emitter voltage (V_{BE}).

$$S_M = \frac{\frac{\Delta I_C}{I_{CQ}}}{\frac{\Delta M}{M}} \qquad 46.16$$

The collector-base cutoff current—essentially, the reverse saturation current—doubles with every 10°C rise in temperature, as was mentioned in Sec. 45.9 and Sec. 46.3. The thermal stability of a transistor, then, is affected by this current. The effect is self-reinforcing: as the temperature increases, the saturation current increases, which further increases the temperature. This phenomenon is called *thermal runaway*. The emitter resistor used in the biasing circuits of Fig. 46.6 helps stabilize the transistor against this trend. As the current rises, the voltage drop across the emitter resistor rises in a direction that opposes the forward biased base-emitter junction. This decreases the base current, and so the collector current increases less than it would without the self-biasing resistor R_E.

[8]The polarity of the DC supply voltages must be reversed to convert the circuits shown for use with *pnp* transistors.

Figure 46.6 DC Biasing Methods

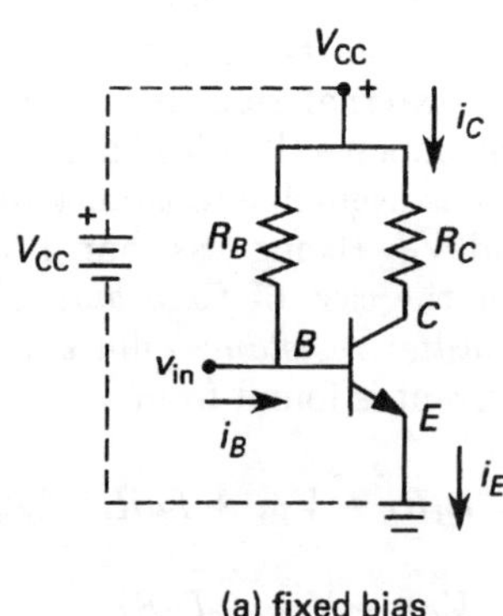

(a) fixed bias

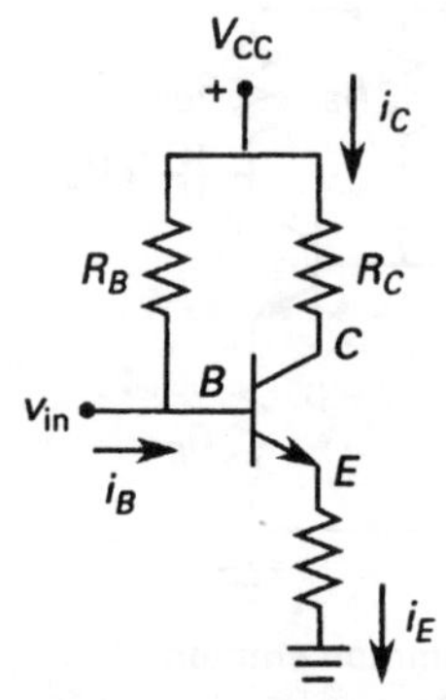

(b) fixed bias with series feedback

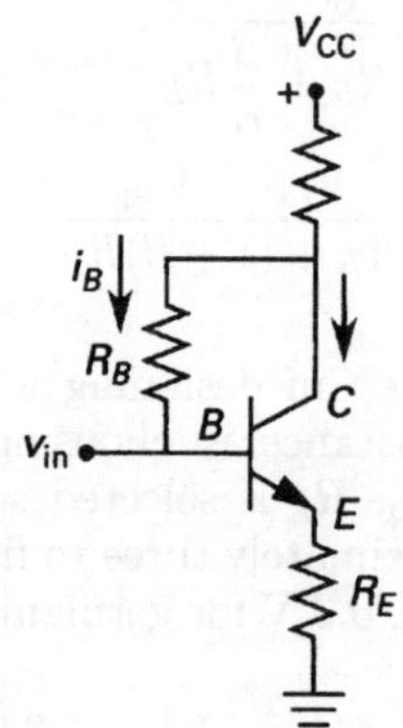

(c) self-bias with shunt feedback

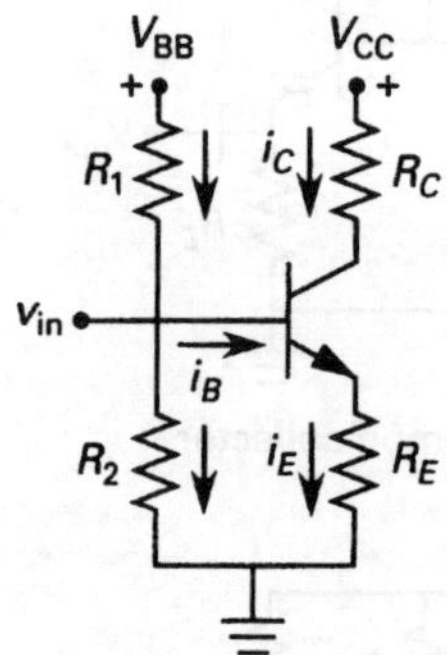

(d) voltage-divider bias

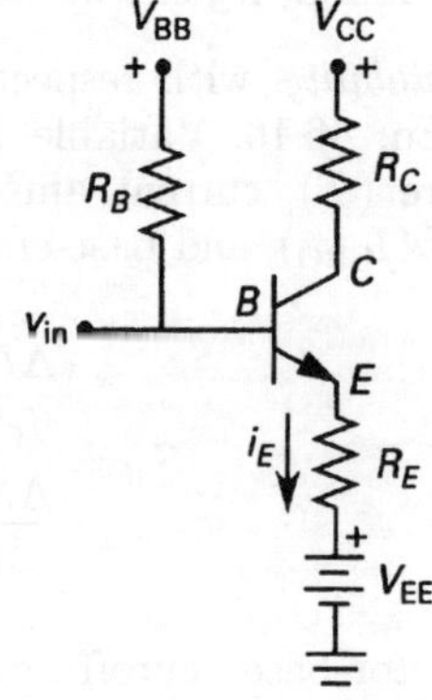

(e) multiple-battery bias

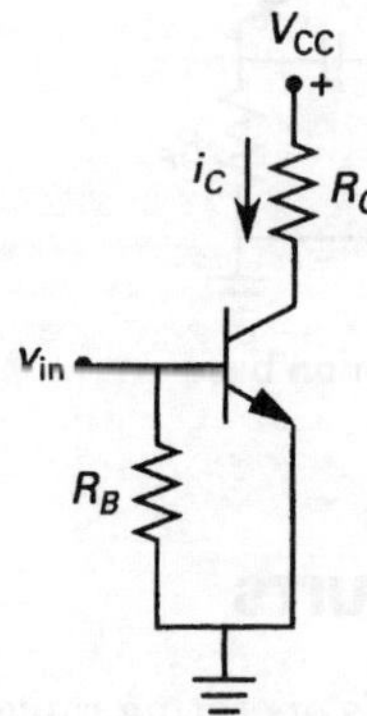

(f) cutoff bias

6. BJT LOAD LINE

Figure 46.7 illustrates part of a simple common emitter transistor amplifier circuit. The bias battery and emitter and collector resistances define the *load line.* If the emitter-collector junction could act as a short circuit (i.e., V_{CE} equals zero), the collector current would be $V_{CC}/(R_C + R_E)$. If the signal is large enough, it can completely oppose the battery-induced current, in which case the net collector current is zero and the full bias battery voltage appears across the emitter-collector junction. The intersection of the load line and the base current curve defines the *quiescent point.* The load line, base current, and quiescent point are all illustrated in Fig. 46.7.

Determination of the load line for a BJT transistor is accomplished by the following steps.

step 1: For the configuration provided, label the x-axis on the *output characteristic curves* with the appropriate voltage, V_{CE} or V_{CB}.

step 2: Label the y-axis as the output current, I_C.

step 3: Redraw the circuit with all three terminals of the transistor open. Label the terminals as the base, emitter, and collector. Label the current directions all pointing inward, that is, toward the transistor, I_B, I_E, and I_C.

step 4: Perform KVL analysis in the collector loop. The transistor voltage determined is a point on the x-axis with the output current equal to zero. Plot the point.

step 5: Redraw the circuit with all three terminals of the transistor shorted. Label as in step 3.

step 6: Use Ohm's law, or another appropriate method, in the collector loop to determine the current. The transistor current determined is a point on the y-axis with the applicable voltage in step 1 equal to zero. Plot the point.

step 7: Draw a straight line between the two points. This is the DC load line. The DC load line is used to determine the biasing and Q-point. The AC load line is determined in the same manner, but with active components, that is, inductors and capacitors, included. The AC load line is used to determine the transistor's response to small signals.

Figure 46.7 *Common Emitter Load Line and Quiescent Point*

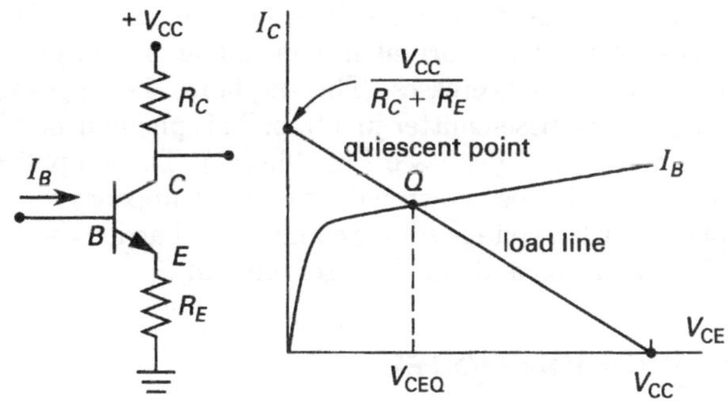

7. AMPLIFIER GAIN AND POWER

The *voltage-*, *current-*, *resistance-*, and *power-gain* are

$$A_V = \frac{\Delta V_{\text{out}}}{\Delta V_{\text{in}}} = \frac{v_{\text{out}}}{v_{\text{in}}} = \beta A_R \quad 46.17$$

$$A_I = \frac{\Delta I_{\text{out}}}{\Delta I_{\text{in}}} = \frac{i_{\text{out}}}{i_{\text{in}}} = \beta \quad 46.18$$

$$A_R = \frac{Z_{\text{out}}}{Z_{\text{in}}} = \frac{A_V}{\beta} \quad 46.19$$

$$A_P = \frac{P_{\text{out}}}{P_{\text{in}}} = \beta^2 A_R = A_I A_V \quad 46.20$$

The collector power dissipation, P_C, should not exceed the rated value. (This restriction applies to all points on the load line.) See Fig. 46.3(a) for a sample power restriction.

$$P_C = \tfrac{1}{2} I_C V_{\text{CE}} \quad \text{[rms values]} \quad 46.21$$

8. CASCADED AMPLIFIERS

Several amplifiers arranged so that the output of one is the input to the next are said to be *cascaded amplifiers.*[9] When each *amplifier stage* is properly coupled to the following, the overall gain is

$$A_{\text{total}} = A_{V,1} A_{V,2} A_{V,3} \cdots \quad 46.22$$

Capacitors are used in amplifier circuits to isolate stages and pass small signals. This is known as *capacitive coupling.*[10] A capacitor appears to a steady (DC) voltage as an open circuit. However, it appears to a small (AC) voltage as a short circuit, and so input and output signals pass through, leaving the DC portion behind.

9. EQUIVALENT CIRCUIT REPRESENTATION AND MODELS

A transistor can be modeled as any of the equivalent two-port networks. The different transistor configurations are shown as two-port networks in Fig. 46.8.

Figure 46.8 *Transistors as Two-Port Networks*

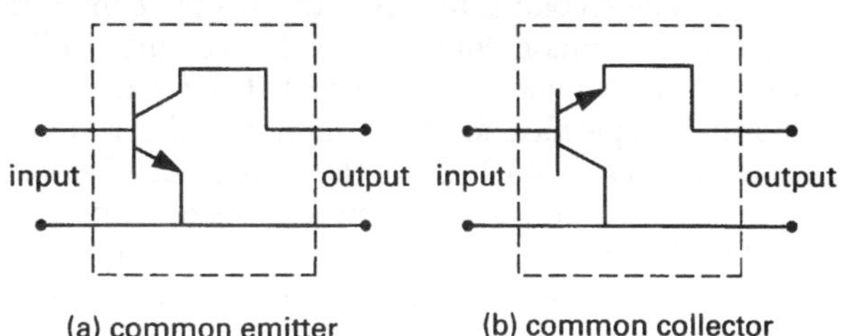

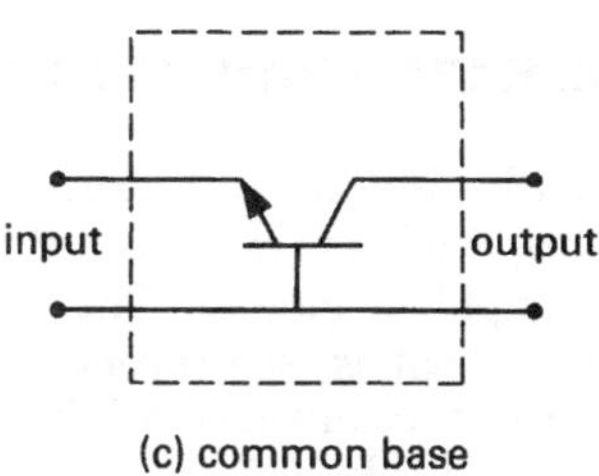

[9]A cascade amplifier should not be confused with a *cascode amplifier* (a high-gain, low-noise amplifier with two transistors directly connected in common emitter and common base configurations).
[10]The term *resistor-capacitor coupling* is also used.

Electronics

The equivalent parameters most often used are the hybrid parameters, also called the *h-parameters*. The h-parameters are defined for any two-port network in terms of Eq. 46.23 through Eq. 46.25.

$$\begin{bmatrix} v_1 \\ i_2 \end{bmatrix} = \begin{bmatrix} h_{11} & h_{12} \\ h_{21} & h_{22} \end{bmatrix} \begin{bmatrix} i_1 \\ v_2 \end{bmatrix} \qquad 46.23$$

$$v_1 = h_{11} i_1 + h_{12} v_2 \qquad 46.24$$

$$i_2 = h_{21} i_1 + h_{22} v_2 \qquad 46.25$$

Each of the h-parameters in Eq. 46.24 and Eq. 46.25 is defined specifically for transistor models in the following way.

- h_i input impedance with output shorted (Ω)
- h_r reverse transfer voltage ratio with input open (dimensionless)
- h_f forward transfer current ratio with output shorted (dimensionless)
- h_o output admittance with input open (S)

Using the h-parameters as so defined, Eq. 46.26 and Eq. 46.27 become the governing equations for *small-signal circuit models*, also called *AC incremental models*.

$$v_i = h_i i_{\text{in}} + h_r v_{\text{out}} \qquad 46.26$$

$$i_o = h_f i_{\text{in}} + h_o v_{\text{out}} \qquad 46.27$$

Normally the h-parameters are given with two subscripts. The first is defined as in Eq. 46.26 and in Eq. 46.27. The second subscript indicates the configuration, such as common emitter (e or E), common collector (c or C), or common base (b or B). The h-parameters are normally specified for the common emitter configuration only. Table 46.2 shows the equivalence between the parameters for the different transistor configurations. The defining equations for the small-signal models are given in Table 46.3. Typical values for a widely used *npn* transistor, the 2N2222A, are given in Table 46.4.

10. APPROXIMATE TRANSISTOR MODELS

The models used in Sec. 46.9 are exact. In many practical applications, sufficiently accurate results can be obtained with simplified models. The values of h_r and h_o are very small. That is, the reverse transfer voltage ratio and the output admittance are insignificant. The simplified models of Table 46.5 are obtained by ignoring these two parameters.

The simplified models assume that the output resistance is infinite; that is, $1/h_o$ is approximately infinite. (See Table 46.1.) The simplified models further assume that the reverse voltage source, $h_r v$, is negligible. The reverse

Table 46.2 *Equivalent Circuit Parameters*

symbol	common emitter	common collector	common base
h_{11}, h_{ie}	h_{ie}	h_{ic}	$\dfrac{h_{ib}}{1+h_{fb}}$
h_{12}, h_{re}	h_{re}	$1 - h_{rc}$	$\dfrac{h_{ib}h_{ob}}{1+h_{fb}} - h_{rb}$
h_{21}, h_{fe}	h_{fe}	$-1 - h_{fc}$	$\dfrac{-h_{fb}}{1+h_{fb}}$
h_{22}, h_{oe}	h_{oe}	h_{oc}	$\dfrac{h_{ob}}{1+h_{fb}}$
h_{11}, h_{ib}	$\dfrac{h_{ie}}{1+h_{fe}}$	$\dfrac{-h_{ic}}{h_{fc}}$	h_{ib}
h_{12}, h_{rb}	$\dfrac{h_{ie}h_{oe}}{1+h_{fe}} - h_{re}$	$h_{rc} - \dfrac{h_{ic}h_{oc}}{h_{fc}} - 1$	h_{rb}
h_{21}, h_{fb}	$\dfrac{-h_{fe}}{1+h_{fe}}$	$\dfrac{-1-h_{fc}}{h_{fc}}$	h_{fb}
h_{22}, h_{ob}	$\dfrac{h_{oe}}{1+h_{fe}}$	$\dfrac{-h_{oc}}{h_{fc}}$	h_{ob}
h_{11}, h_{ic}	h_{ie}	h_{ic}	$\dfrac{h_{ib}}{1+h_{fb}}$
h_{12}, h_{rc}	$1 - h_{re}$	h_{rc}	1
h_{21}, h_{fc}	$-1 - h_{fe}$	h_{fc}	$\dfrac{-1}{1+h_{fb}}$
h_{22}, h_{oc}	h_{oe}	h_{oc}	$\dfrac{h_{ob}}{1+h_{fb}}$

voltage accounts for the narrowing of the base width as the collector-base junction reverse bias increases. As the effectiveness of the base current in controlling the output is minimized, gain decreases. This explains the opposing voltage at the base-emitter junction. This phenomenon is called the *Early effect* (see Fig. 46.1(a)). The simplified models can also be shown with the input impedance, h_i, replaced with a voltage source equal to the barrier voltage (0.7 V for silicon and 0.3 V for germanium).

11. HYBRID-π MODEL

Though the h-parameter model is common, another significant model is the *hybrid-π* or *Giacoletto* model. The hybrid-π model is shown in Fig. 46.9. The terms used in the model follow.

- $r_{bb'}$ *base spreading resistance*, or the *interbase resistance*, or the *small-signal base bulk resistance* (Ω)
- $r_{b'e}$ *small-signal base input resistance* (Ω)
- $r_{b'c}$ *small-signal feedback resistance* (Ω) This resistance accounts for the Early effect.
- r_{ce} *small-signal output resistance* (Ω)
- g_m *transistor transconductance* (S)

Table 46.3 BJT Equivalent Circuits

common connection	equivalent circuit	network equations
CE; B, C, E; i_b, i_c, i_e; $v_{be} = v_{in}$; $v_{ce} = v_{out}$	CE; B, C, E; h_{ie}; i_b; i_c; $h_{fe}i_b$; $h_{re}v_{ce}$; $\frac{1}{h_{oe}}$; v_{be}; v_{ce}	common emitter $v_{be} = h_{ie}i_b + h_{re}v_{ce}$ $i_c = h_{fe}i_b + h_{oe}v_{ce}$
CC; B, C, E; i_b, i_c, i_e; $v_{bc} = v_{in}$; $v_{ec} = v_{out}$	CC; B, E, C; h_{ic}; i_b; i_e; $h_{fc}i_b$; $h_{rc}v_{ec}$; $\frac{1}{h_{oc}}$; v_{bc}; v_{ec}	common collector $v_{bc} = h_{ic}i_b + h_{rc}v_{ec}$ $i_e = h_{fc}i_b + h_{oc}v_{ec}$
CB; E, C, B; i_e, i_c, i_b; $v_{eb} = v_{in}$; v_{cb}	CB; E, C, B; h_{ib}; i_e; i_c; $h_{fb}i_e$; $h_{rb}v_{cb}$; $\frac{1}{h_{ob}}$; v_{eb}; v_{cb}	common base $v_{eb} = h_{ib}i_e + h_{rb}v_{cb}$ $i_c = h_{fb}i_e + h_{ob}v_{cb}$

Table 46.4 Typical h-Parameter Values

h-parameter	range of values
h_{ie}	$0.25 \times 10^3\ \Omega$ to $8.0 \times 10^3\ \Omega$
h_{re}	$4.0 \times 10^{-4}\ \Omega$ to $8.0 \times 10^{-4}\ \Omega$ [max values]
h_{fe}	50 to 375
h_{oc}	5.0×10^{-6} S to 200×10^{-6} S

Figure 46.9 Hybrid-π Model

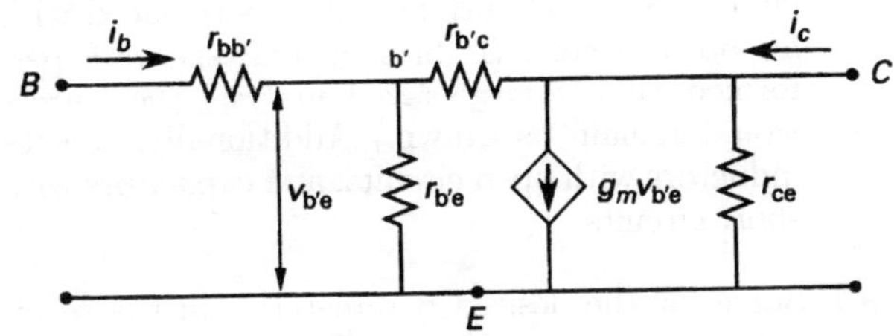

If the common emitter *h*-parameters are known, the hybrid-π parameters can be calculated from Eq. 46.28 through Eq. 46.32, in the order given.

$$g_m = \frac{|I_C|}{V_T} \quad (46.28)$$

$$r_{b'e} = \frac{h_{fe}}{g_m} \quad (46.29)$$

$$r_{bb'} = h_{ie} - r_{b'e} \quad (46.30)$$

$$r_{b'c} = \frac{r_{b'e}}{h_{re}} \quad (46.31)$$

$$g_{ce} = h_{oe} - \frac{1 + h_{fe}}{r_{b'c}} \quad (46.32)$$

12. TRANSISTOR CIRCUIT LINEAR ANALYSIS

Because the small-signal, low-frequency response of a transistor is linear, it can be obtained analytically rather

Table 46.5 BJT Simplified Equivalent Circuits

common connection	equivalent circuit	network equations
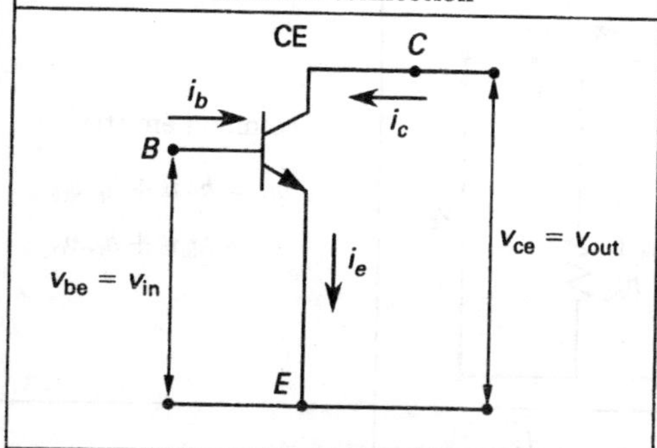 CE: B, C, E; i_b, i_c, i_e; $v_{be} = v_{in}$, $v_{ce} = v_{out}$	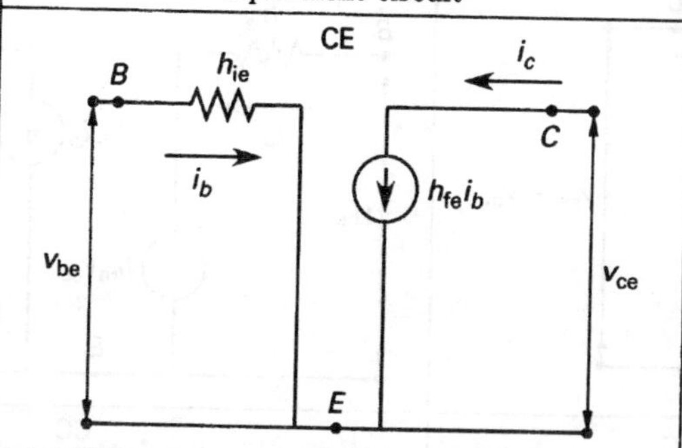 CE: B, h_{ie}, i_b, $h_{fe}i_b$, i_c, C, E; v_{be}, v_{ce}	common emitter* $v_{be} = h_{ie} i_b \approx 0.7\text{ V}$ $i_c = h_{fe} i_b$
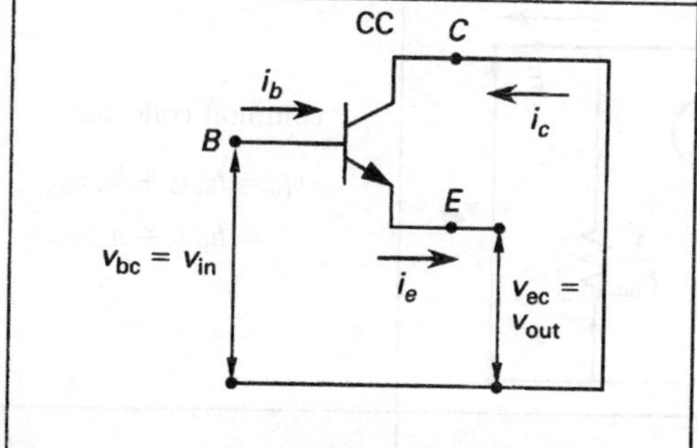 CC: B, C, E; i_b, i_c, i_e; $v_{bc} = v_{in}$, $v_{ec} = v_{out}$	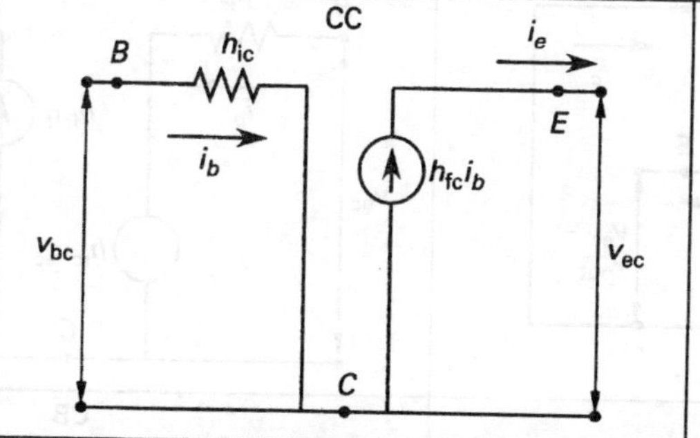 CC: B, h_{ic}, i_b, $h_{fc}i_b$, i_e, E, C; v_{bc}, v_{ec}	common collector $v_{bc} = h_{ic} i_b$ $i_e = h_{fc} i_b$
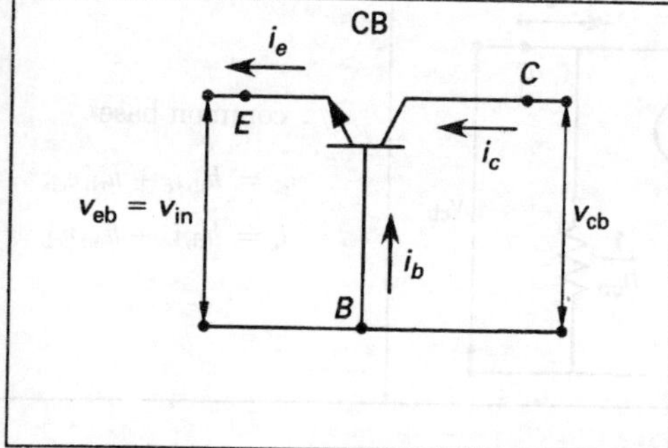 CB: E, C, B; i_e, i_c, i_b; $v_{eb} = v_{in}$, v_{cb}	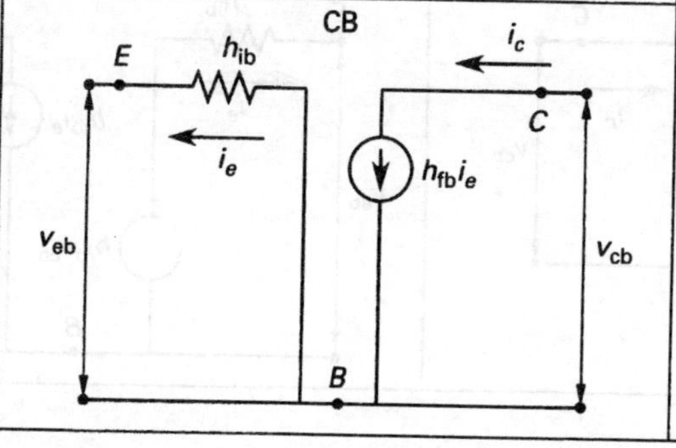 CB: E, h_{ib}, i_e, $h_{fb}i_e$, i_c, C, B; v_{eb}, v_{cb}	common base* $v_{eb} = h_{ib} i_e \approx 0.7\text{ V}$ $i_c = h_{fb} i_e$

*Germanium transistors are *pnp* types. $|v_{be}| = |v_{eb}| = 0.3$ V for germanium.

than graphically by using the models in Sec. 46.9.[11] The procedure for small-signal, or AC incremental, analysis follows.

step 1: Draw the circuit diagram. Include all external components, such as resistors, capacitors, and sources from the network.

step 2: Label the points for the base, emitter, and collector but do not draw the transistor. Maintain the points in the same relative position as in the original circuit.

step 3: Replace the transistor by the desired model. (Several models exist. Exact models are given in Sec. 46.9. Approximate models are given in Sec. 46.10. The hybrid-π model is given in Sec. 46.11.)

step 4: Because small-signal analysis is to be accomplished, only slight changes around the quiescent point are of interest. Therefore, replace each *independent* source by its internal resistance. An ideal voltage source is replaced with a short circuit. An ideal current source is replaced with an open circuit. (If biasing analysis were performed, that is, large-signal analysis, the sources would remain as drawn.) Additionally, replace inductors with open circuits, and capacitors with short circuits.

step 5: Solve for the desired parameter(s) in the resultant circuit using Kirchhoff's current and voltage laws.

[11]The *h*-parameters must be considered complex functions of the frequency to analyze high-frequency circuits, so the *h*-parameter model is not used.

Example 46.1

A transistor is used in a common emitter amplifier circuit as shown. Assume the inductor has infinite impedance and the capacitors have zero impedance to AC signals. The transistor h-parameters are

$$h_{ie} = 750\ \Omega$$
$$h_{oe} = 9.09 \times 10^{-5}\ \text{S}$$
$$h_{fe} = 184$$
$$h_{re} = 1.25 \times 10^{-4}$$

(a) If a quiescent point is wanted at $V_{CE} = 11$ V and $I_C = 10 \times 10^{-3}$ A, what should the emitter resistance, R_E, be?

For parts (b) through (g), assume that $R_E = 800\ \Omega$. (b) Draw the DC load line. (c) If the base current (i_B) is 80×10^{-6} A, what is the collector current (i_C)? (d) What is the AC circuit voltage gain? (e) What is the AC circuit current gain? (f) What is the circuit's input impedance? (g) What is the output impedance? (h) Given $v_{in} = 0.25 \sin 1400t$ V, what is v_{out}? (i) What is the purpose of the inductor, L_C?

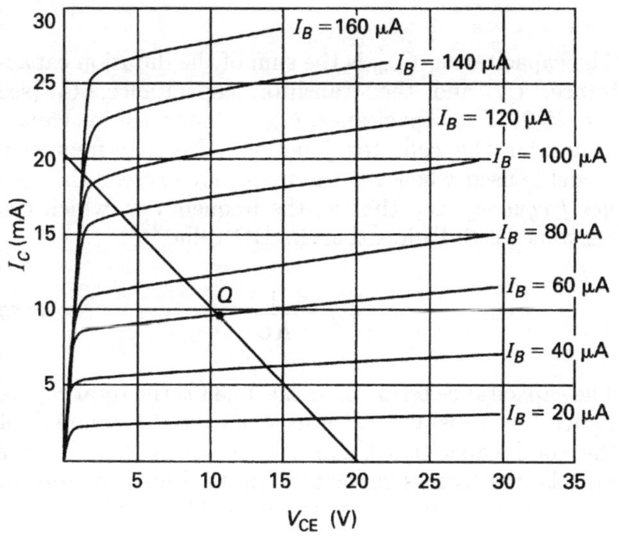

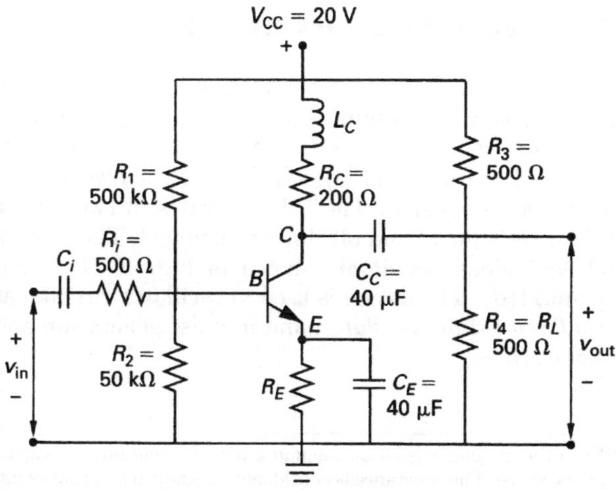

Solution

(a) Write the voltage drop in the common emitter circuit. Disregard the inductor (which passes DC signals). Use h_{fe} for h_{FE} because they are essentially the same and both are large. Kirchhoff's voltage law is

$$\alpha = \frac{\beta}{1+\beta} \approx \frac{h_{fe}}{1+h_{fe}} = \frac{184}{1+184} \approx 1$$

$$\begin{aligned} V_{CC} &= I_C R_C + I_E R_E + V_{CE} \\ &= I_C R_C + \left(\frac{I_C}{\alpha}\right) R_E + V_{CE} \\ &\approx I_C(R_C + R_E) + V_{CE} \end{aligned}$$

$$\begin{aligned} R_E &= \frac{V_{CC} - V_{CE}}{I_C} - R_C \\ &= \frac{20\ \text{V} - 11\ \text{V}}{10 \times 10^{-3}\ \text{A}} - 200\ \Omega \\ &\approx 700\ \Omega \end{aligned}$$

(b) If $I_C = 0$, then $V_{CE} = V_{CC}$. This is one point on the load line. If $V_{CE} = 0$, then

$$I_C = \frac{V_{CC}}{R_C + R_E} = \frac{20\ \text{V}}{200\ \Omega + 800\ \Omega} = 0.02\ \text{A}$$

These two points define the DC load line.

(c) From Eq. 46.3,

$$\begin{aligned} i_C &= \beta i_B = h_{fe} i_B \\ &= (184)(80 \times 10^{-6}\ \text{A}) \\ &= 0.0147\ \text{A} \quad (14.7\ \text{mA}) \end{aligned}$$

(d) To determine the AC circuit voltage gain, it is necessary to simplify the circuit. The bias battery V_{CC} is shorted out. (This is because the battery merely shifts the signal without affecting the signal swing.) Therefore, many of the resistors connect directly to ground. The inductor has infinite impedance, so R_C is disconnected. Both capacitors act as short circuits, so R_E is bypassed.

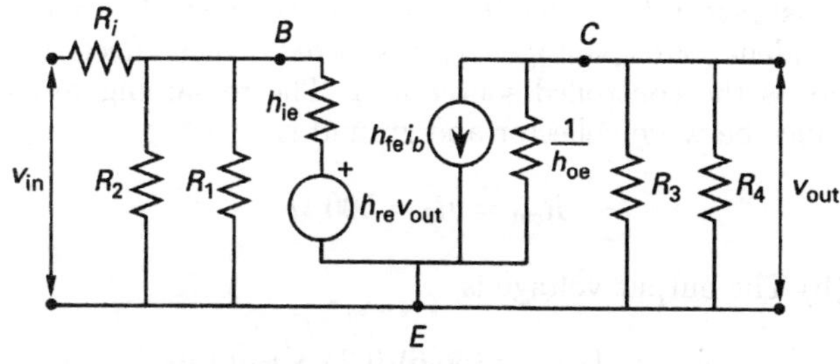

To simplify the circuit further, recognize that R_1, R_2, and $1/h_{oe}$ are very large and can be treated as infinite impedances.

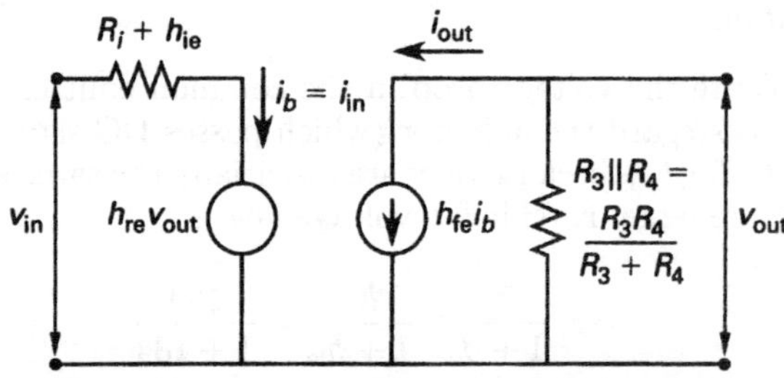

Continuing with the simplified analysis,

$$v_{out} = h_{fe}i_b\left(\frac{R_3R_4}{R_3 + R_4}\right) = (184i_b)\left(\frac{(500\ \Omega)(500\ \Omega)}{500\ \Omega + 500\ \Omega}\right)$$
$$= (46{,}000\ \Omega)i_b$$
$$v_{in} = i_b(R_i + h_{ie}) + h_{re}v_{out}$$
$$= i_b(500\ \Omega + 750\ \Omega) + (1.25 \times 10^{-4})(46{,}000\ \Omega)i_b$$
$$= (1256\ \Omega)i_b$$

To perform an exact analysis, h_{oe} and h_{re} must be considered. Convert R_i to its Norton equivalent resistance and place it in parallel with R_1, R_2, and h_{ie}. Use the current-divider concept to calculate i_b.

The voltage gain is

$$A_V = \frac{v_{out}}{v_{in}} \approx \frac{(46{,}000\ \Omega)i_b}{(1256\ \Omega)i_b} = 36.6$$

(e) The current gain is

$$A_I = \frac{i_{out}}{i_{in}} \approx \frac{h_{fe}i_b}{i_b} = 184$$

(f) The input impedance (resistance) is

$$R_{in} = \frac{v_{in}}{i_{in}} \approx \frac{(1256\ \Omega)i_b}{i_b} = 1256\ \Omega$$

(g) The output impedance is effectively the Thevenin equivalent resistance of the output circuit. The load resistance (R_4 in this instance) is removed. The independent source voltage (v_{in}) is shorted, which effectively opens the controlled source $h_{fe}i_b$. The remaining resistance between collector and ground is

$$R_{out} = R_3 = 500\ \Omega$$

(h) The output voltage is

$$v_{out} = A_V v_{in} = (36.6)(0.25\ \text{V})\sin 1400t$$
$$= (9.15\ \text{V})\sin 1400t$$

(i) There are several possible uses for the inductor. It might be included to limit voltage extremes, such as high-voltage spikes, which could damage the transistor when the amplifier is turned on or off. Alternatively, it might prevent AC current from being drawn across R_C, and, in so doing, hold V_C at a constant value.

13. TRANSISTOR CIRCUIT HIGH-FREQUENCY ANALYSIS

The *h*-parameter models are useful at low frequencies. For high frequencies, the hybrid-π model is modified to include the effects of various capacitances as shown in Fig. 46.10.

Figure 46.10 *High-Frequency Hybrid-π Model*

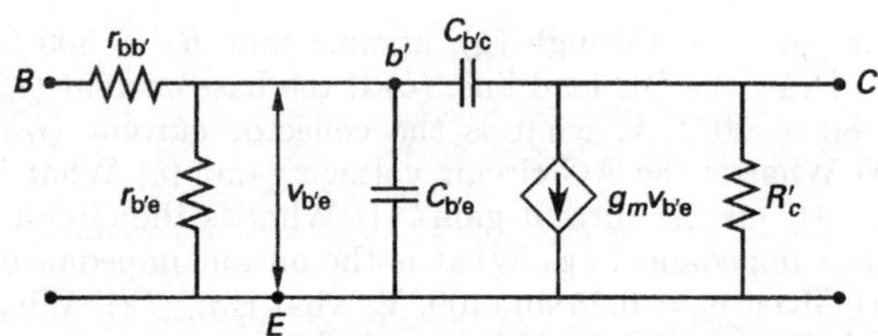

The capacitance, $C_{b'e}$, is the sum of the diffusion capacitance, C_d, and the transition capacitance, C_t (see Sec. 45.9). The capacitance, $C_{b'c}$, is the transition capacitance for the collector junction. The high-frequency model is used whenever the frequency exceeds the *corner frequency*, ω_{cf}, that is, the frequency at which the resistance is 3 dB less than its DC value.[12]

$$\omega_{cf} = \frac{1}{RC} \qquad 46.33$$

The capacitance term, C, in Eq. 46.33 is the total capacitance. The resistance, R, is as seen from the terminals of the total capacitance. (In low-frequency *h*-parameter models, the bypass capacitors are typically not shown. It is to these capacitors that Eq. 46.33 refers.)

14. UNIJUNCTION TRANSISTORS

A *unijunction transistor* is an electronic device with three contacts, two ohmic and one to a *pn* junction, as shown in Fig. 46.11(a). The output characteristics are illustrated in Fig. 46.11(b), which clearly shows a negative resistance region. The UJT operates in two states: on (low resistance) and off (high resistance). The symbol and equivalent circuit are shown in Fig. 46.11(c) and Fig. 46.11(d). The device is used in oscillator circuits, as a *UJT relaxation oscillator*, and in pulse generation and delay circuits.

[12]When the frequency is $10\omega_{cf}$, the impedance is essentially the capacitive reactance. The resistance is considered insignificant at that point.

Figure 46.11 Unijunction Transistor

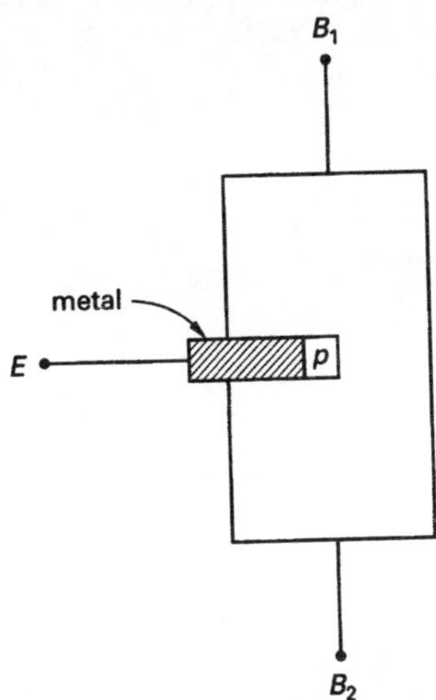

(a) conceptual construction

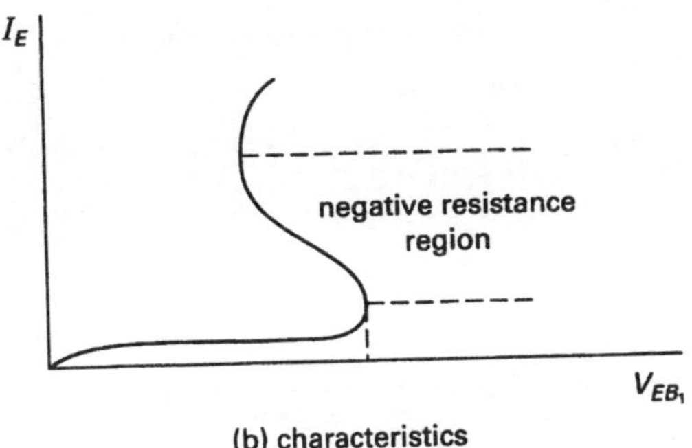

(b) characteristics

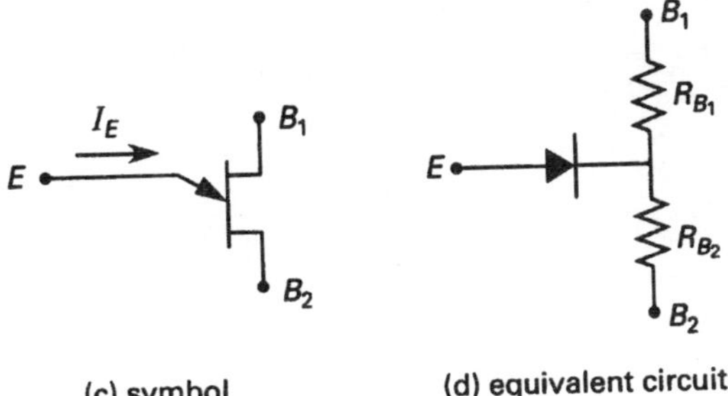

(c) symbol (d) equivalent circuit

15. DARLINGTON TRANSISTORS

A *Darlington transistor*, more commonly called a *Darlington pair*, is a current amplifier consisting of two separate transistors treated as one. A Darlington transistor has a high input impedance and a significant current gain. For example, the forward current transfer gain, h_{fe}, can be as high as 30,000. A Darlington transistor is illustrated in Fig. 46.12.

Figure 46.12 Darlington Transistor

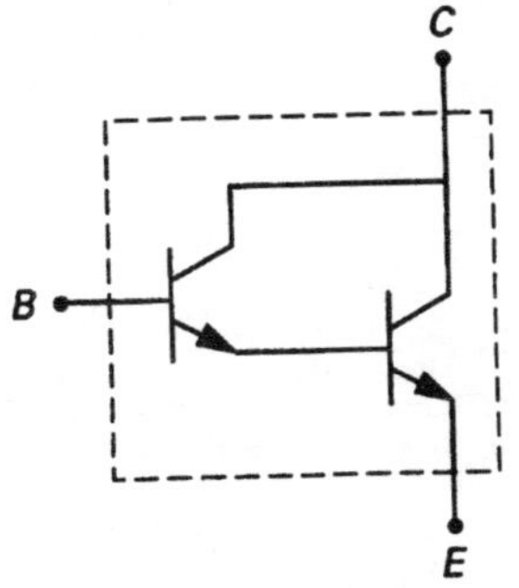

Electronics

47 Field Effect Transistors

Nomenclature

A	gain	–
C	capacitance	F
g_m	mutual conductance or transconductance	S
i	instantaneous or variable current	A
I	effective or DC current	A
I_{DSS}	zero-gate-voltage drain current	A
n	electron concentration	m^{-3}
p	hole concentration	m^{-3}
r	small-signal resistance[1]	Ω
R	resistance	Ω
v	instantaneous or variable voltage	V
V	effective or DC voltage	V
V_{DD}	drain supply voltage	V

Subscripts

CC	collector supply
d	dynamic or diffusion
ds, DS	drain to source
D	donor, diode, or drain
DSQ	drain to source at Q-point
DSS	drain to source saturation
gs, GS	gate to source
G	gap, gate
GD	gate to drain
GSQ	gate to source at Q-point
i	intrinsic or input (AC or instantaneous)
iss	common-source short-circuit input
L	load
m	mutual
oss	common-source short-circuit output
P	pinchoff or power
$P0$	pinchoff at 0 V
s	saturation
S	source
SD	source to drain
SG	source to gate
V	constant or effective voltage

[1]This is also known as the *dynamic resistance* or the *instantaneous resistance*. Any resistance represented with a lowercase letter is the small-signal value.

1. FET FUNDAMENTALS

Field-effect transistors (FETs) are bidirectional devices constructed of an *n*-type channel surrounded by a *p*-type gate, or vice versa.[2] The connections are named in a different manner than BJTs in order to distinguish the two. For a FET (BJT), the connections are the *gate* (base), *source* (emitter), and *drain* (collector). In a BJT, the base current controls the overall operation of the transistor. In a FET, the gate voltage and the corresponding electric field control the overall operation of the transistor. The various configurations and the concepts regarding biasing, load lines, and amplifier operation described for BJTs also apply to FETs.

There are two major types of FETs: the *junction field-effect transistor* (JFET) and the *metal-oxide semiconductor field-effect transistor* (MOSFET). Both types are made in *n*- and *p*-channel types, with the *n*-channel types more common. The fundamental difference between the JFET and the MOSFET is that the latter can operate in the enhancement mode (see Sec. 47.5).[3]

Variations in construction result in different names and properties. An *n*-channel MOSFET is sometimes referred to as an NMOS, and the *p*-channel MOSFET is sometimes called a PMOS. A high-power MOSFET is called an HMOS. A MOSFET with increased drain current capacity caused by its V-type structure is called a VMOS. A double-diffused MOSFET, which has replaced the VMOS except in high-frequency applications, is called a DMOS. A bipolar transistor using an insulated gate is called an *insulated gate bipolar transistor* (IGBT). An SCR using an insulated gate for control is called a *MOS-controlled thyristor* (MCT). A MOSFET with *p*-channel and *n*-channel devices on the same chip is called a *complementary MOSFET* (CMOS) and is used primarily because of its low power dissipation.

When the conventional MOSFET oxide layer is replaced by a double layer of nitride and oxide, which allows electrons to tunnel from the gate, the device is called a *metal-nitride-oxide-silicon* (MNOS) transistor. The MNOS device is used as a read-only memory (ROM) and MOS transistors are also used as memory devices. The MOS transistor is used to control the transfer of the charge from one location to the next. As a result, it is also called a *charge-transfer device* (CTD). When the charge transfer is

[2]Bidirectional indicates that the current flow direction depends on the potential between the source and the drain.

[3]MOSFETs are also much more susceptible to damage from *electrostatic discharges* (ESD) due to the thin oxide layer at the gate.

Electronics

accomplished on the circuit level by discrete MOS transistors and capacitors, it is called a *bucket-brigade device* (BBD). When the charge transfer takes place on the device level, it is called a *charge-coupled device* (CCD).

2. JFET CHARACTERISTICS

Junction field-effect transistor (JFET) construction, characteristics, and symbols are shown in Fig. 47.1.

A bidirectional channel for current flow exists between the source, *S*, and the drain, *D*. The current flow is controlled by the reverse biased *pn* junction at the gate, *G*, with the depletion width providing the control.[4] For example, when the gate is unbiased, a normal depletion zone is set up at the gate *pn* junction, as shown in Fig. 47.2(a) for an *n*-channel JFET.

Once biased, the depletion layer shifts, and the JFET operates in a relatively linear fashion in the *ohmic region*, also called the *triode region.* The ohmic depletion layer is shown in Fig. 47.2(b). The ohmic region is shown in Fig. 47.1(b). As the drain-source voltage increases, the depletion layer widens until it encompasses the drain terminal. At this point, the current is pinched off, and the JFET operates in a relatively constant manner in the *pinchoff region.* The pinchoff region depletion layer is shown in Fig. 47.2(c). The pinchoff region is shown in Fig. 47.1(b). As the drain-source voltage continues to increase, avalanche breakdown occurs and the voltage remains approximately constant as the current increases dramatically. This is known as the *breakdown* or *avalanche region* (not shown).

For a fixed value of V_{GS}, the drain-source voltage separating the resistive and pinchoff regions is the *pinchoff voltage.* As Fig. 47.1(b) shows, there is a value for V_{GS} for which no drain current flows. This is also referred to as the pinchoff voltage but is designated $V_{GS(off)}$. The drain current corresponding to the horizontal part of a curve (for a given value of V_{GS}) is the *saturation current*, represented by I_{DSS}.

The term "pinchoff voltage" and the symbol V_P are ambiguous, as the actual pinchoff voltage in a circuit depends on the gate-source voltage, V_{GS}. When V_{GS} is zero, the pinchoff voltage is represented unambiguously by V_{P0} (where the zero refers the value of V_{GS}). For other values of V_{GS},

$$V_P = V_{P0} + V_{GS} \qquad 47.1$$

Some manufacturers do not adhere to this convention when reporting the pinchoff voltage for their JFETs. They may give a value for V_{P0} and refer to it as V_P. The absence of a value for V_{GS} implies that the value given is actually V_{P0}.

[4]The source "emits" the *n*- or *p*-channel carriers, by convention. The reverse bias is therefore with respect to the gate. FETs are bidirectional, nevertheless.

Figure 47.1 *Junction Field-Effect Transistor*

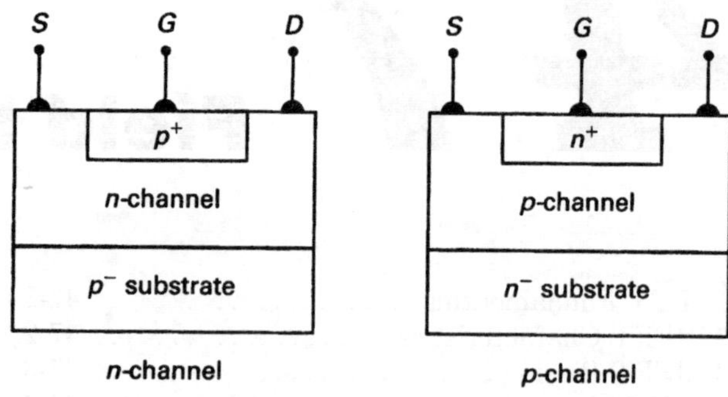

(a) conceptual construction

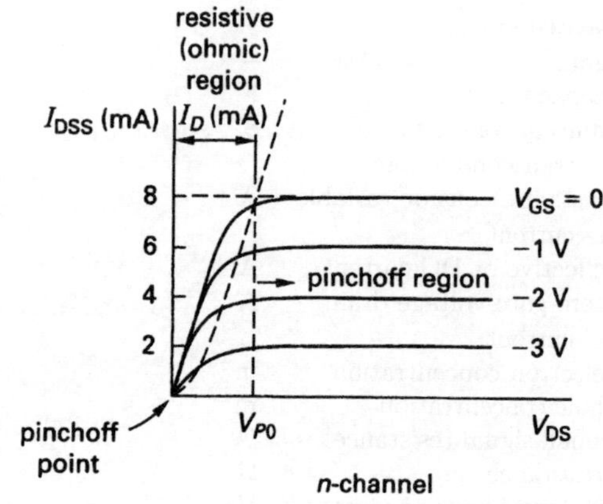

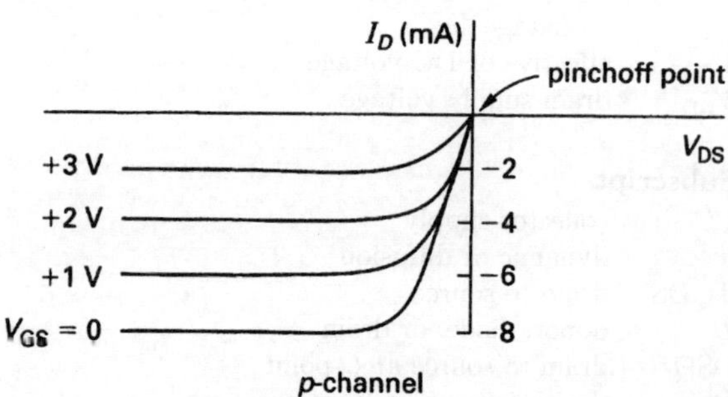

(b) characteristics

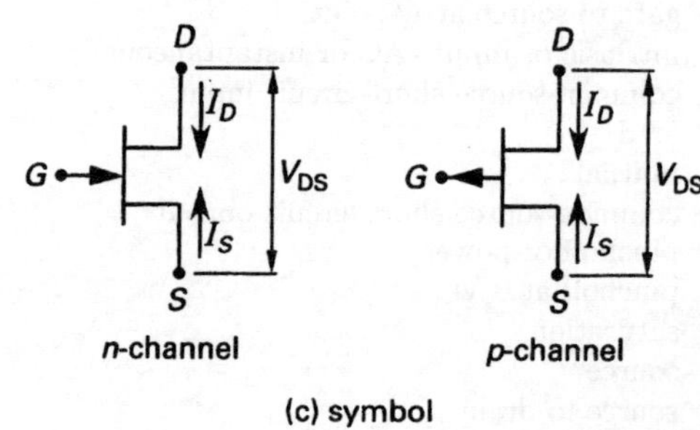

(c) symbol

Figure 47.2 *Pinchoff Theory*

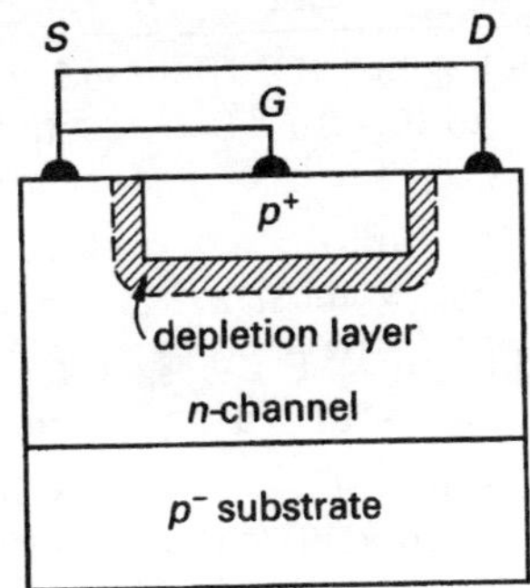

(a) unbiased depletion layer

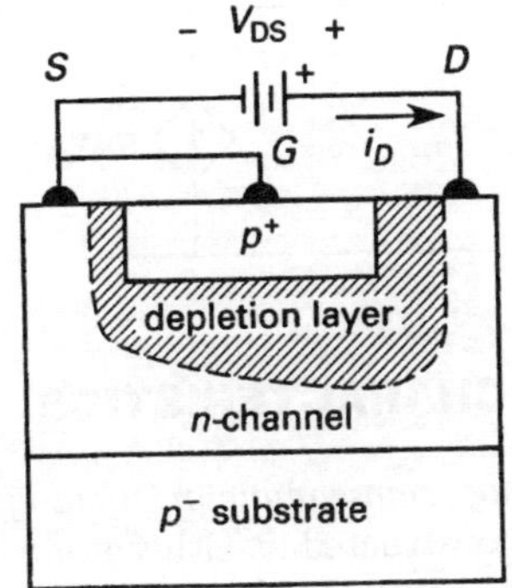

(b) ohmic region depletion layer

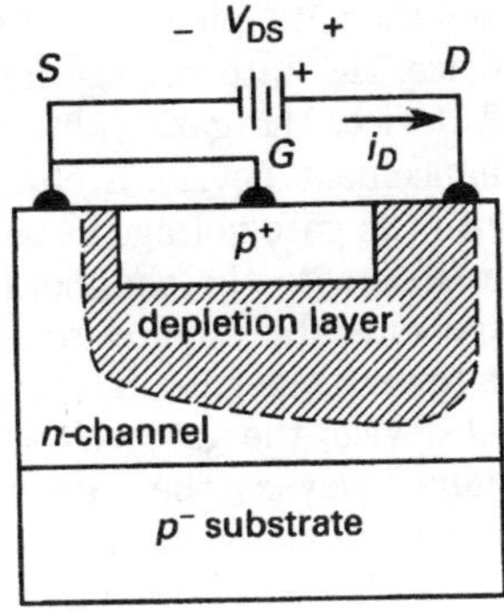

(c) pinchoff region depletion layer

Some manufacturers do not provide characteristic curves, choosing instead to indicate only I_{DSS} and $V_{GS(off)}$. The characteristic curves can be derived, as necessary, from *Shockley's equation.*

$$I_D = I_{DSS}\left(1 - \frac{V_{GS}}{V_P}\right)^2 \qquad 47.2$$

The *transconductance*, g_m, is defined for small-signal analysis by Eq. 47.3.

$$\begin{aligned} g_m &= \frac{\Delta I_D}{\Delta V_{GS}} = \frac{i_D}{v_{GS}} \\ &= \left(\frac{-2I_{DSS}}{V_P}\right)\left(1 - \frac{V_{GS}}{V_P}\right) \\ &= g_{mo}\left(1 - \frac{V_{GS}}{V_P}\right) \\ &\approx \frac{A_V}{R_{out}} \end{aligned} \qquad 47.3$$

The drain-source resistance can be obtained from the slope of the V_{GS} characteristic in Fig. 47.1(b).

$$r_d = r_{DS} = \frac{\Delta V_{DS}}{\Delta I_D} = \frac{v_{DS}}{i_D} \qquad 47.4$$

3. JFET BIASING

JFETs operate with a reverse biased gate-source junction. The quiescent point is established by choosing V_{GSQ}. I_D is then determined by Shockley's equation (see Eq. 47.2).

Figure 47.3 shows a JFET self-biasing circuit. Because the gate current is negligible, the load line equation is

$$\begin{aligned} V_{DD} &= I_S(R_D + R_S) + V_{DS} \\ &= I_D(R_D + R_S) + V_{DS} \end{aligned} \qquad 47.5$$

Figure 47.3 *Self-Biasing JFET Circuit*

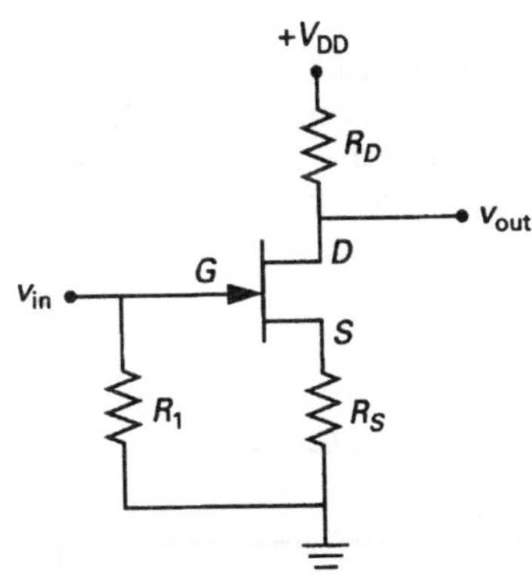

The load line for any JFET is determined using the procedures outlined in Sec. 45.7.

At the quiescent point, $V_{in} = 0$. From Kirchhoff's voltage law, around the input loop,

$$V_{GS} = -I_S R_S = -I_D R_S \qquad 47.6$$

Electronics

Example 47.1

A JFET with the characteristics shown operates as a small-signal amplifier. The supply voltage is 24 V. A quiescent bias source current of 5 mA (0.005 A) is desired at a bias voltage of $V_{DS} = 15$ V. Design a self-biasing circuit similar to Fig. 47.3.

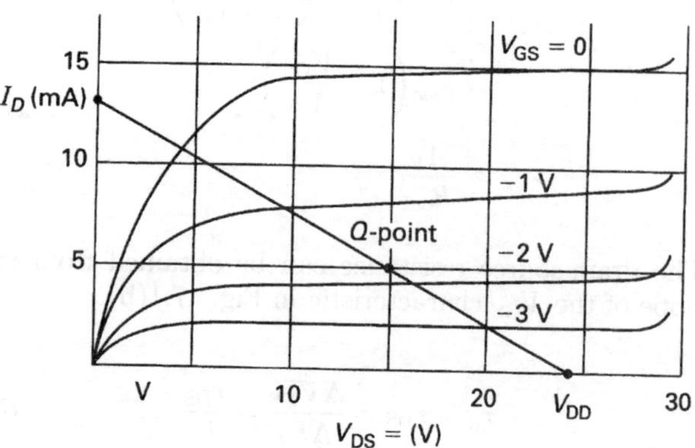

Solution

Because the gate draws negligible current, $I_D = I_S$. At the quiescent point, $V_{GS} = -1.75$ V. From Eq. 47.6,

$$R_S = \frac{-V_{GS}}{I_D} = \frac{-(-1.75 \text{ V})}{0.005 \text{ A}} = 350 \ \Omega$$

From Eq. 47.5,

$$R_D = \frac{V_{DD} - V_{DS}}{I_S} - R_S = \frac{24 \text{ V} - 15 \text{ V}}{0.005 \text{ A}} - 350 \ \Omega$$
$$= 1450 \ \Omega$$

4. FET MODELS

A field-effect transistor can be modeled as shown in Fig. 47.4. The model is valid for both the JFET and the MOSFET. Because the drain resistance is very high (see the typical values in Table 47.1), the model can be simplified further by removing r_{DS}, that is, by assuming $r_{DS} \approx \infty$.

Figure 47.4 *FET Equivalent Circuit*

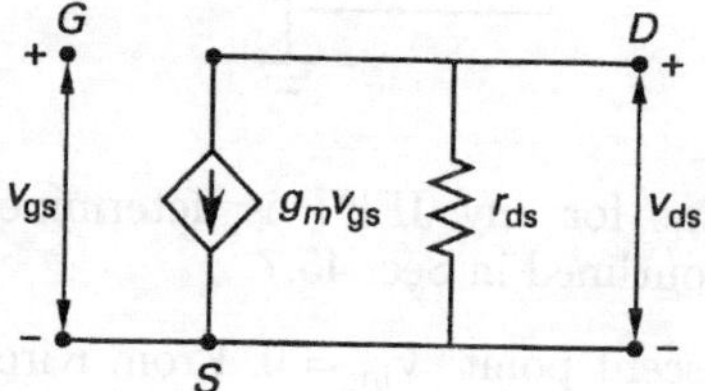

At high frequencies the various capacitances associated with the FET must be accounted for, as shown in the model in Fig. 47.5.

Table 47.1 *Typical FET Parameter Values*

parameter	JFET	MOSFET
g_m	0.1×10^{-3} S to 10×10^{-3} S	0.1×10^{-3} S to 20×10^{-3} S or greater
r_{ds}	$0.1 \times 10^{6}\ \Omega$ to $1 \times 10^{6}\ \Omega$	$1 \times 10^{3}\ \Omega$ to $50 \times 10^{3}\ \Omega$
C_{ds}	0.1×10^{-12} F to 1×10^{-12} F	0.1×10^{-12} F to 1×10^{-12} F
C_{gd}, C_{gs}	1×10^{-12} F to 10×10^{-12} F	1×10^{-12} F to 10×10^{-12} F

Figure 47.5 *FET High-Frequency Model*

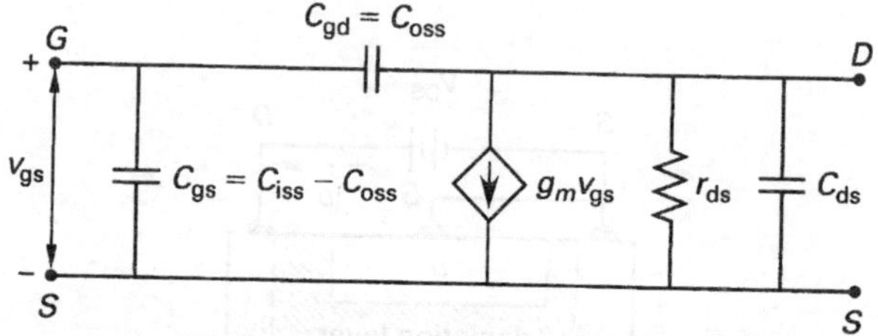

5. MOSFET CHARACTERISTICS

The *metal-oxide semiconductor field-effect transistor* (MOSFET) is constructed as either a *depletion device* or an *enhancement device* as shown in Fig. 47.6(a). In the depletion device, a channel for current exists with the gate voltage at zero. In order to control the current, a voltage applied to the gate must push channel majority carriers away from the gate, pinching off the current. For an *n*-channel device, the gate voltage must be negative. For a *p*-channel device, the gate voltage must be positive.[5] In an enhancement device, a channel for current does not exist with the gate voltage at zero, but must be created. In order to create the channel and control the current, a voltage must be applied to the gate to pull majority carriers toward the gate, increasing the current. For an *n*-channel device, the gate voltage must be positive. For a *p*-channel device, the gate voltage must be negative. A summary of these characteristics, which are similar to those for JFETs, is shown in Fig. 47.6(b). Three operating regions exist: ohmic, pinchoff, and breakdown. The various symbols used to represent MOSFETs are shown in Fig. 47.6(c) and Fig. 47.6(d). The four-terminal devices shown are IEEE standard symbols for MOSFETs with passive substrates (bulk). The three-terminal symbols are shorthand notation for active substrates (i.e., the bulk is connected to the source). The arrow of the bulk terminal of the four-terminal devices indicates the polarity of the bulk-drain, bulk-source, and bulk-channel diodes. The arrow points toward the channel for an NMOS device and away for a PMOS device. The arrow on the three-terminal devices indicates the source terminal and points in the direction of conventional (i.e., positive) current flow. The MOSFET model is identical to that for the JFET (see Sec. 47.4).

[5]The gate voltage is referenced to the source by convention, since the source is considered the emitter of majority carriers into the channel.

Figure 47.6 *Metal-Oxide Semiconductor Field-Effect Transistor*

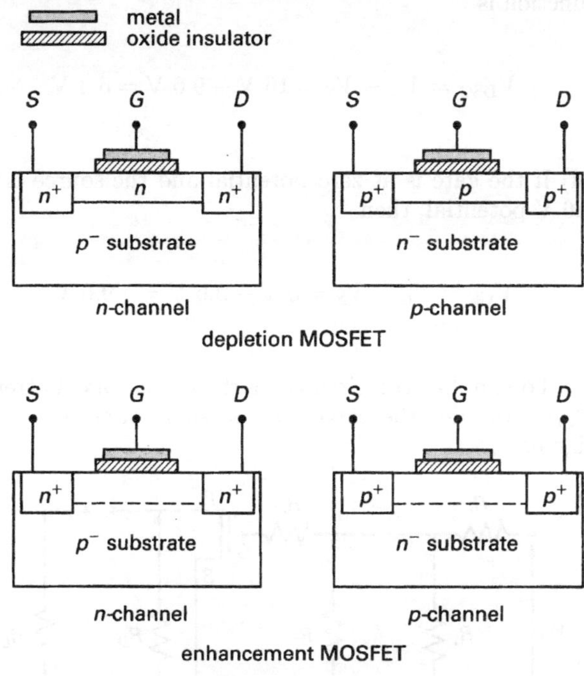

(a) conceptual construction

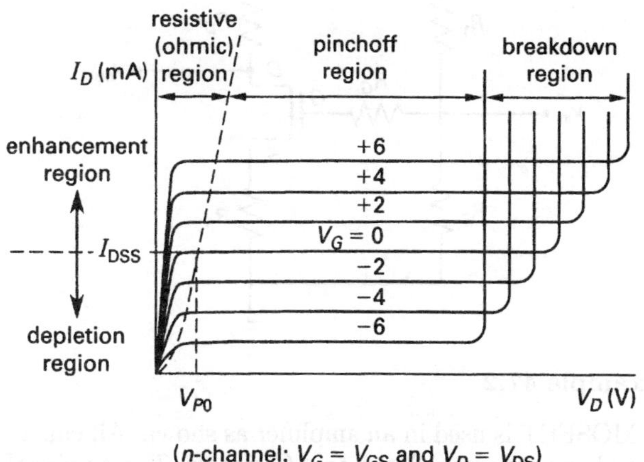

(b) depletion-mode characteristics

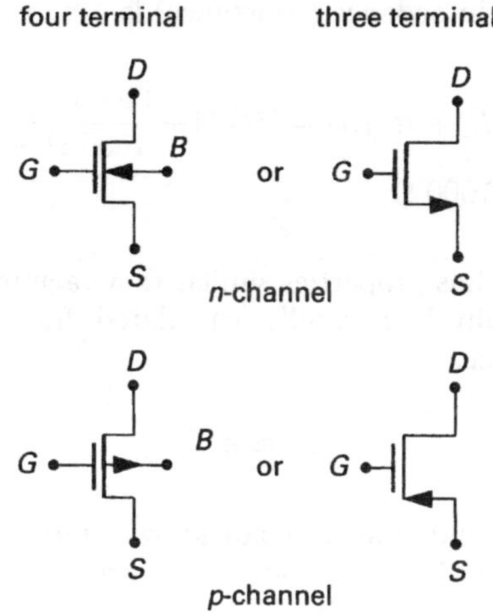

(c) depletion MOSFET symbols

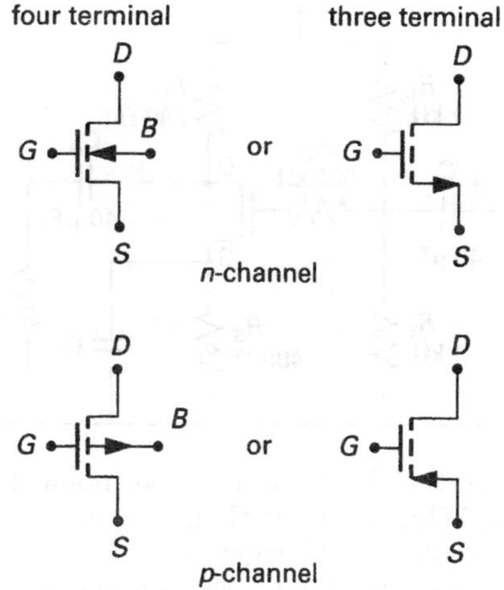

(d) enhancement MOSFET symbols

6. MOSFET BIASING

A typical MOSFET biasing circuit is shown in Fig. 47.7. Because MOSFETs do not have a *pn* junction between the gate and channel, biasing can be either forward or reverse. The polarity of the gate-source bias voltage depends on whether the transistor is to operate in the depletion mode or enhancement mode. Because no gate current flows, the resistor R_G is used merely to control the input impedance. When R_G is very large, the input impedance is essentially R_2.

Together, R_1 and R_2 form a voltage divider. Because the gate current is zero, the voltage divider is unloaded. Therefore, the gate voltage is

$$V_G = V_{DD}\left(\frac{R_2}{R_1 + R_2}\right) = V_{GS} + I_S R_S \qquad 47.7$$

The load line equation is found from Kirchhoff's voltage law and is the same as for the JFET.

$$V_{DD} = I_S(R_D + R_S) + V_{DS} \qquad 47.8$$

Electronics

Figure 47.7 *Typical MOSFET Biasing Circuit*

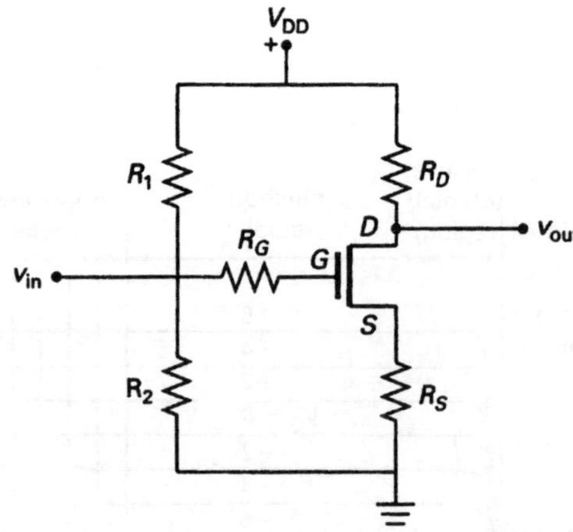

Example 47.2

A MOSFET is used in an amplifier as shown. All capacitors have zero impedance to AC signals. The no-signal drain current (I_D) is 20×10^{-3} A. The performance of the transistor is defined by

$$I_D = (14 + V_{GS})^2 \quad \text{[in A]}$$

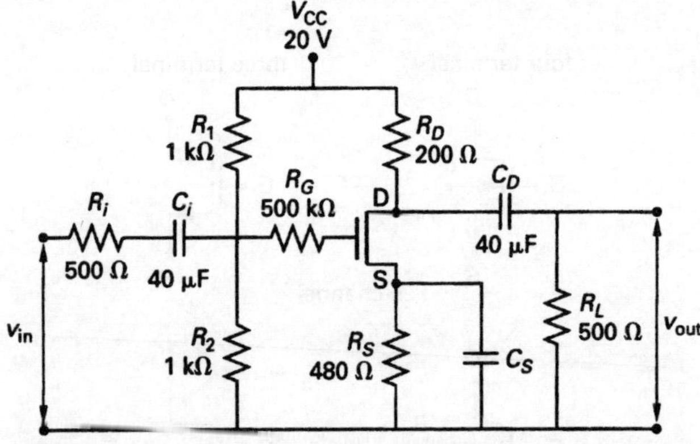

(a) What is the V_D (potential) at point D with no signal? (b) What is V_S with no signal? (c) What is V_{DSQ}? (d) If $V_G = 0$ V, what is V_{GS}? (e) What is the input impedance? (f) What is the voltage gain? (g) What is the output impedance?

Solution

(a)

$$\begin{aligned} V_D &= V_{CC} - I_D R_D \\ &= 20 \text{ V} - (20 \times 10^{-3} \text{ A})(200\ \Omega) \\ &= 16 \text{ V} \end{aligned}$$

(b) The voltage drop between the source and the ground is through R_S. Because the gate draws negligible current, the drain and source currents are the same.

$$V_S = I_S R_S = (20 \times 10^{-3} \text{ A})(480\ \Omega) = 9.6 \text{ V}$$

(c) The quiescent voltage drop across the drain-source junction is

$$V_{DSQ} = V_D - V_S = 16 \text{ V} - 9.6 \text{ V} = 6.4 \text{ V}$$

(d) If the gate is at zero potential and the source is at 9.6 V potential, then

$$V_{GS} = V_G - V_S = 0 \text{ V} - 9.6 \text{ V} = -9.6 \text{ V}$$

(e) To simplify the circuit, short out the bias battery, V_{CC}. Consider the capacitors as short circuits to AC signals.

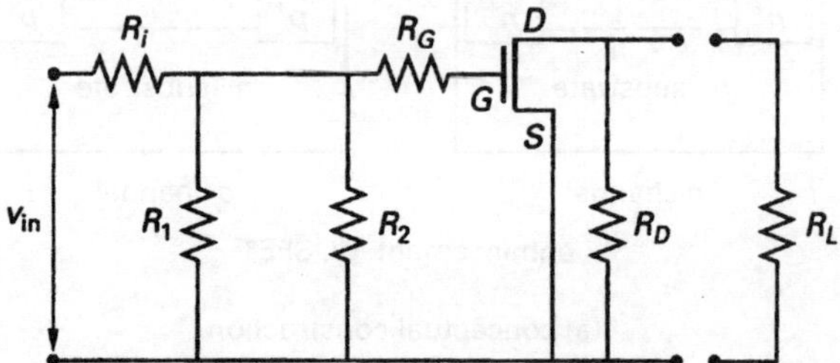

Because R_G is so large, it effectively is an open circuit. The input impedance (resistance) is

$$\begin{aligned} R_{in} &= R_i + R_1 \| R_2 = 500\ \Omega + \frac{(1000\ \Omega)(1000\ \Omega)}{1000\ \Omega + 1000\ \Omega} \\ &= 1000\ \Omega \end{aligned}$$

(f) A FET has properties similar to a vacuum tube. The voltage gain is normally calculated from the transconductance.

$$A_V \approx g_m R_{out}$$

The transconductance is not known, but it can be calculated from Eq. 47.3 and the performance equation.

$$\begin{aligned} g_m &= \frac{dI_D}{dV_{GS}} = \frac{d(14 + V_{GS})^2 \times 10^{-3}}{dV_{GS}} \\ &= (2)(14 + V_{GS}) \times 10^{-3} \end{aligned}$$

Because V_{GS} is −9.6 V at the quiescent point,

$$g_m = (2)(14 - 9.6 \text{ V}) \times 10^{-3} = 8.8 \times 10^{-3} \text{ S}$$

The resistance is a parallel combination of R_D, R_L, and r_d. The drain-source resistance, r_{ds}, is normally very large and, because it was not given in this problem, is disregarded. Then,

$$R = R_D \| R_L = \frac{R_D R_L}{R_D + R_L} = \frac{(200\ \Omega)(500\ \Omega)}{200\ \Omega + 500\ \Omega}$$

$$= 143\ \Omega$$

$$A_V = g_m R = (8.8 \times 10^{-3}\ \text{S})(143\ \Omega) = 1.26$$

This is a small gain. If advantage is not being taken of other properties possessed by the circuit, the resistances should be adjusted to increase the voltage gain.

(g) Proceeding as in the solution to Ex. 46.1, part (g), the output impedance is $R_D = 200\ \Omega$.

48 Electrical and Electronic Devices

Nomenclature

A	gain	–
BW	bandwidth	Hz
C	capacitance	F
CMRR	common-mode rejection ratio	–
D	diffusion constant	m^2/s
E	source voltage	V
f	frequency	Hz
F_m	figure of merit	rad/s
G	gain	–
GBW	gain bandwidth	rad/s
h_{fe}	CE small-signal (AC) forward current transfer ratio or gain	–
h_{FE}	CE static input (DC) forward current transfer ratio or gain	–
H	high	V
i	instantaneous or variable current	A
I	effective or DC current	A
I_{CBO}	reverse saturation current	A
I_{EBO}	DC emitter cutoff current, collector open	A
I_{ZK}	keep-alive current	A
N	fan-out (number of loads)	–
P	power	W
q	electric unit charge, 1.6022×10^{-19}	C
R	resistance	Ω
S_R	slew rate	V/s
t	time	s
T	temperature	K
v	instantaneous or variable voltage	V
V	effective or DC voltage	V
V_T	voltage equivalent of temperature	V
VR	voltage regulation	%
Z	impedance	Ω

Symbols

α	CB forward current ratio or gain	–
δ	feedback factor	–
κ	Boltzmann's constant, 1.3805×10^{-23}	J/K
ω	angular frequency	rad/s

Subscripts

0	resonance
amp	amplifier
A, B	input A, input B
b, B	base
BE	base to controller
c	AC or small-signal collector
c	control or collector
cl	current limiting
cm	common mode
co	cutoff
C	collector
C	DC or biased collector
CC	collector supply
CE	collector to emitter
d	delay
dm	differential mode
D	diode
DC	direct current
DD	drain supply
e	AC or small signal emitter
E	DC or biased emitter
EBO	emitter to base, collector open
EE	emitter supply

f	fall, feedback, or forward (AC or instantaneous)
fl	full load
F	forward (DC component or total)
GS	gate to source
i	current
L	load
m	merit
max	maximum
min	minimum
n	noise
nl	no load
p	power
r	rise
ref	reference
R	resistance
RTD	resistance temperature device
T	thermal
u	unity
v	instantaneous voltage
V	voltage
Z	zener
ZK	zener knee
ZM	zener at maximum rated current

PART 1: INTRODUCTION

An understanding of basic electronics and associated circuits is necessary for the electronics engineer who will apply this knowledge to fabricate more complex circuits from the basic building blocks. A power electrical engineer makes use of such knowledge because of the widespread use of electronic devices in instrumentation, illumination controls, battery charging automatic circuits, power electronics, variable speed drives, protective relays, transmission line monitoring, and control circuits. A computer engineer, while primarily focused on utilizing the computer to accomplish various tasks, must have an appreciation of the underlying hardware in order to take full advantage of the available capabilities.

This chapter develops the principles of the operational amplifier, various pulse circuits used for waveform shaping and logic circuits, and specific circuits with widespread applications. These principles are essential to an electrical engineer, as they provide a basis for understanding electronics and associated circuits.

PART 2: AMPLIFIERS

1. FUNDAMENTALS

An *amplifier* is any device capable of increasing the magnitude or power level of a physical quantity. The prime concern of most amplifiers is gain, A. The gain can be a voltage (A_V), current (A_i), or power gain (A_p or G_p). The input impedance (Z_{in}) and the output impedance (Z_{out}) are also parameters of concern. The *bandwidth* is the range of frequencies for which the amplifier will respond. The bandwidth (BW) is normally defined as the difference between the upper and lower frequencies at the *3 dB half-power points*, also called the *3 dB downpoints*, as shown in Fig. 48.1.

Figure 48.1 Bandwidth

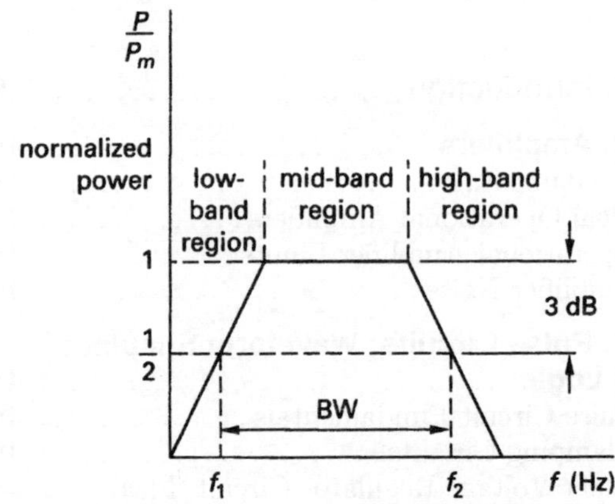

The amplifier in Fig. 48.1 is a *band-pass* amplifier. There are also *low-pass*, *high-pass*, and *notch* amplifiers, though they are more commonly called *filters*. The gain-bandwidth product is used as a *figure of merit*, F_m, measured in rad/s, for band-pass and high-pass amplifiers. (Bandwidth is sometimes measured in Hz instead of rad/s, in which case the figure of merit is also given in Hz.) The *gain-bandwidth product* (GBW) for a band-pass amplifier is

$$\text{GBW} = F_m = A_{\text{ref}}(\text{BW}) \qquad 48.1$$

The term A_{ref} is a reference gain, which is either the maximum gain or the gain at the frequency at which the gain is purely real or purely imaginary. For a low-pass amplifier, a high-frequency cutoff, ω_0, at the 3 dB downpoint is defined and the figure of merit is[1]

$$\text{GBW} = F_m = A_{\text{ref}}\omega_0 \qquad 48.2$$

The transistors described in Chap. 53, when biased in the linear region, are amplifiers. The small-signal models given are for the *mid-band region* of Fig. 48.1. In this region, the reactive elements act as short circuits and so are not included in any models.[2] In the *low-band region*, the impedance of the reactive elements, normally the coupling capacitors and emitter bypass capacitors, is significant.[3] Capacitors must be added

[1]The 3 dB downpoint occurs at the resonant frequency.

[2]In *tuned circuits*, the reactive elements must be accounted for in the mid-frequency region to ensure proper design.

[3]"Significant" means the impedance seen at the terminals of the capacitor is of the same magnitude as the impedance of the capacitor itself.

to the small-signal models in the low-band region, and the gain decreases.[4] In the *high-band region*, the impedance caused by the capacitance of the device itself becomes significant, and the gain decreases. In this manner, the device capacitances are added to the models (see Sec. 46.13 and Sec. 47.4).

Individual transistors amplify. When used in conjunction with other transistors and electronic devices, different types of amplifiers can be constructed. An amplifier with a wide range of applications is the *operational amplifier* (op amp). It takes its name from the multiple configurations with which mathematical operations may be performed. Originally designed as the major component of an analog computer, the op amp is a high-gain DC (direct-coupled or direct current) amplifier with high stability and immunity to oscillation, which uses large amounts of negative feedback. The symbols used for the op amp are shown in Fig. 48.2.

Figure 48.2 *Operational Amplifier Symbols*

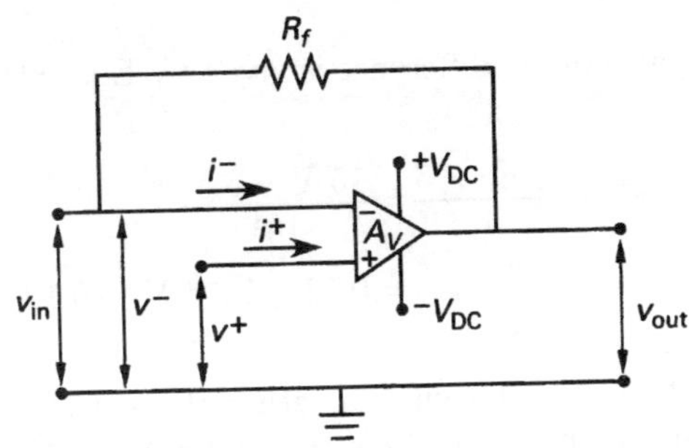

The op amp has two input terminals. The one marked "–" is the *inverting terminal*; the one marked "+" is the *noninverting terminal*. The fundamental op amp formula is

$$v_{out} = A_V(v^+ - v^-) \qquad 48.3$$

The voltages marked $+V_{DC}$ and $-V_{DC}$ are the power supplies to the op amp, and are often called the *rail voltages*. The range of v_{out} is between these two voltages.[5] In fact, the output voltage should stay 3 V from either value to avoid distortion of the output signal at rated values. The range of the input signal, derived from Eq. 48.3 and this distortion restriction, is

$$|v^+ - v^-| < \frac{V_{DC} - 3\text{ V}}{A_V} \qquad 48.4$$

When the op amp input is within the range of Eq. 48.4, the operation is linear. Typical values of op amp voltage gain, A_V, range from 10^5 to 10^8. Power supply voltages, $\pm V_{DC}$, are in the range of $\pm$10–15 V. Typical input impedances are greater than 10^5 Ω, while output impedances are very low. The internal circuits of the op amp consist of the items in the block diagram of Fig. 48.3.

Figure 48.3 *Operational Amplifier Internal Configuration*

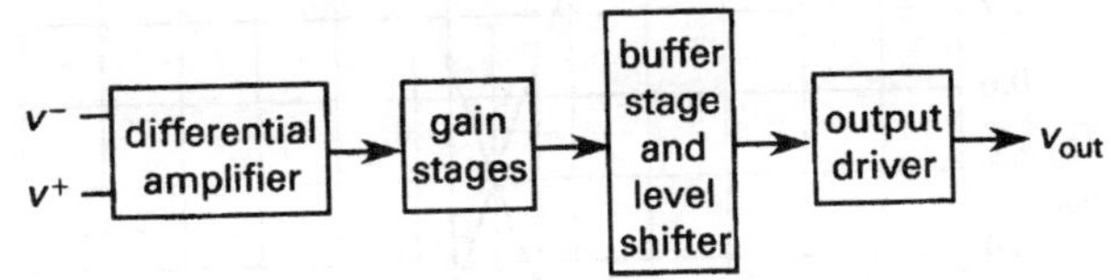

The *differential amplifier* constitutes the first stage of the op amp. A differential amplifier is shown in Fig. 48.4. The resulting output current, which is a combination of the emitter currents, is shown in Fig. 48.5. The linear region for operation restricts the voltage difference at the op amp input, as in Eq. 48.4, though here the difference is expressed in terms of the thermal voltage, V_T. Ideally, the differential amplifier amplifies only the difference between the signals. In practice, both the *differential-mode signal*, v_{dm}, and the *common-mode signal*, v_{cm}, are amplified.

$$v_{dm} = v^+ - v^- \qquad 48.5$$

$$v_{cm} = \tfrac{1}{2}(v^+ + v^-) \qquad 48.6$$

Figure 48.4 *Differential Amplifier*

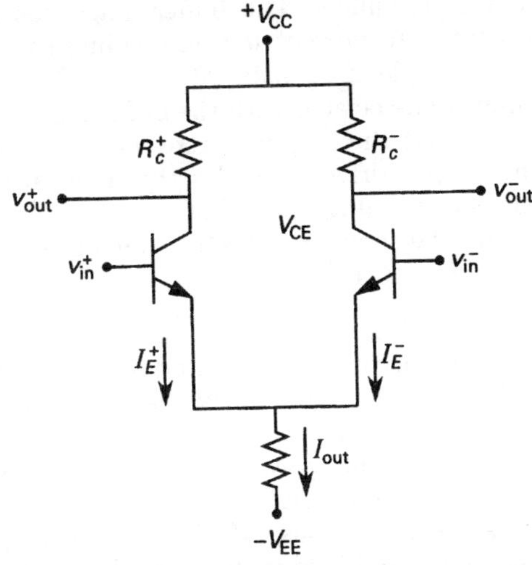

To measure how well the differential amplifier amplifies the difference signal and not the common signal, the concept of the *common-mode rejection ratio* (CMRR) is used. The CMRR is defined as

$$\text{CMRR} = \left|\frac{A_{dm}}{A_{cm}}\right| \qquad 48.7$$

[4]These models are not shown in Chap. 44. They would consist of the small-signal low-frequency model with coupling and emitter bypass capacitors added—the same circuit that would be used to determine the AC load line.

[5]Although the gain is large, the output cannot be greater than the available voltage. The rail voltages represent $\pm\infty$ for the output.

Figure 48.5 Differential Amplifier Emitter Currents (Ideal Current Source Biasing)

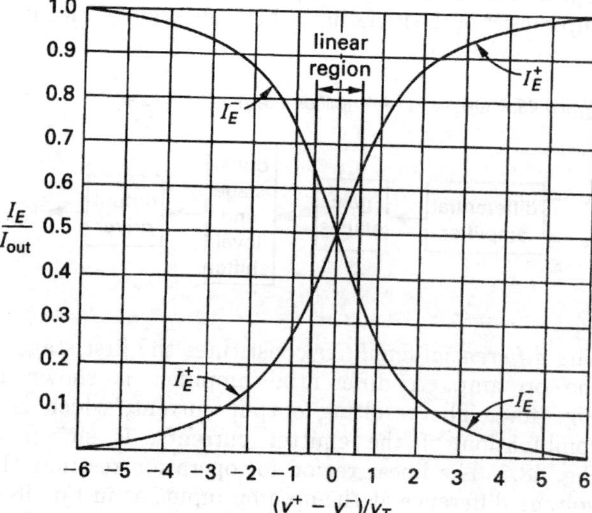

The output voltage in terms of the CMRR is

$$v_{out} = A_{dm}v_{dm}\left(1 + \left(\frac{1}{\text{CMRR}}\right)\left(\frac{v_{cm}}{v_{dm}}\right)\right) \qquad 48.8$$

The gain stages in Fig. 48.3 supply the required amplification and represent cascaded amplifiers; for example, cascaded CE amplifiers. The buffer stage has a high input resistance to prevent it from loading the output driver; for example, an *emitter follower* (CC configuration). Circuitry associated with the buffer shifts the DC bias level so that the output is zero when the input is zero. The output driver provides the necessary large-signal current or voltage gain. For example, for a large current gain, the common collector configuration could be used (see Sec. 46.4).[6]

Op amps are designed to closely approximate ideal op amp behavior. Because of this and because of their versatility of use, they are widely used in place of discrete transistor amplifiers. Op amps are used in linear systems such as voltage-to-current converters, DC instrumentation, voltage followers, and filters, and in nonlinear systems such as AC/DC converters, peak detectors, sample-and-hold systems, analog multipliers, and analog-to-digital and digital-to-analog converters, among many others. A low-frequency op amp equivalent circuit is shown in Fig. 48.6.

[6]The common collector (CC) configuration is also called the *emitter follower*. Voltage gain is approximately one. This means that a change in base voltage appears as an equal change across the load at the emitter. Consequently, the emitter follows the input signal.

Figure 48.6 Operational Amplifier Equivalent Circuit

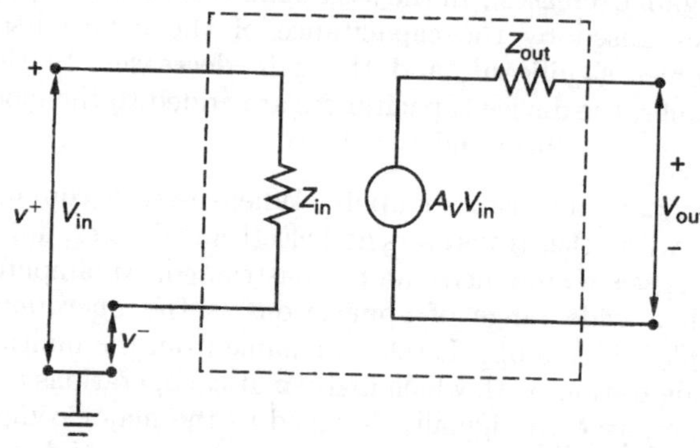

Example 48.1

An op amp has a voltage gain of 10^8. The DC power supply is ±15 V. What is the maximum input voltage difference for linear operation?

Solution

The input voltage difference is given by Eq. 48.4.

$$|v^+ - v^-| < \frac{V_{DC} - 3}{A_V} = \frac{15\ \text{V} - 3\ \text{V}}{10^8} = 0.12 \times 10^{-6}\ \text{V} \quad (0.12\ \mu\text{V})$$

The difference between the terminals of the op amp must be kept under 0.12 μV to remain under the linear region. (This is accomplished with negative feedback.) This voltage difference is so small that the terminals are essentially at the same voltage. One of the assumptions for an ideal op amp is that the voltages are the same. Consequently, if one of the terminals is connected to ground, the other terminal is also at ground potential. This situation is referred to as having a *virtual ground* at the ungrounded op amp terminal.

Example 48.2

The input impedance for the op amp in Ex. 48.1 is $10^5\ \Omega$. What is the input current?

Solution

The voltage difference was calculated as 0.12 μV. The current between the terminals is given by Ohm's law.

$$I = \frac{V}{R} = \frac{0.12 \times 10^{-6}\ \text{V}}{10^5\ \Omega} = 1.2 \times 10^{-12}\ \text{A} \quad (1.2\ \text{pA})$$

This current is so low as to be negligible. An ideal op amp implies that the input current is zero.

Example 48.3

If the gain given for Ex. 48.1 is the differential-mode gain and the CMRR is 10,000, what is the output voltage when the op amp is at the edge of linear operation and $v^+ = 50\ \mu\text{V}$?

Solution

The op amp is at the edge of linear operation for the maximum voltage difference allowed by Eq. 48.4, which was calculated in Ex. 48.1 as 0.12 μV. The output voltage is given by Eq. 48.8.

$$v_{\text{out}} = A_{\text{dm}} v_{\text{dm}} \left(1 + \left(\frac{1}{\text{CMRR}}\right)\left(\frac{v_{\text{cm}}}{v_{\text{dm}}}\right)\right)$$

The differential-mode voltage was calculated as 0.12 μV. The common-mode voltage is unknown. From Eq. 48.4 and Eq. 48.5,

$$\begin{aligned} v_{\text{dm}} &= v^+ - v^- = 0.12\ \mu\text{V} = 50\ \mu\text{V} - v^- \\ v^- &= 50\ \mu\text{V} - 0.12\ \mu\text{V} = 49.88\ \mu\text{V} \\ v_{\text{cm}} &= \tfrac{1}{2}(v^+ + v^-) \\ &= \frac{1}{2}(50\ \mu\text{V} + 49.88\ \mu\text{V}) \\ &\approx 50\ \mu\text{V} \end{aligned}$$

Substituting the calculated and given values gives

$$\begin{aligned} v_{\text{out}} &= A_{\text{dm}} v_{\text{dm}} \left(1 + \left(\frac{1}{\text{CMRR}}\right)\left(\frac{v_{\text{cm}}}{v_{\text{dm}}}\right)\right) \\ &= (10^8)(0.12 \times 10^{-6}\ \text{V})\left(1 + \left(\frac{1}{10{,}000}\right)\left(\frac{50\ \mu\text{V}}{0.12\ \mu\text{V}}\right)\right) \\ &= 12.5\ \text{V} \end{aligned}$$

This output voltage is within 3 V of the power supply voltage. Distortion could be caused by the nonlinear operation. In this case, because $A_{\text{dm}} v_{\text{dm}} = 12\ \text{V}$, the negative feedback must be increased.

2. IDEAL OPERATIONAL AMPLIFIERS

An *ideal operational amplifier* exhibits the following properties.

- $Z_{\text{in}} = \infty$
- $Z_{\text{out}} = 0$
- $A_V = \infty$
- $\text{BW} = \infty$
- $V_{\text{out}} = 0$ when $v^+ = v^-$[7,8]
- characteristics do not drift with temperature[8]

Because the actual op amp so closely approximates these conditions, it is possible to use ideal op amp analysis for most calculations. Only when high-frequency behavior is desired or circuit limitations have been calculated do the assumptions need to be discarded. A typical ideal op amp configuration with *voltage shunt feedback* is shown in Fig. 48.7.

Figure 48.7 Ideal Operational Amplifier Typical Configuration

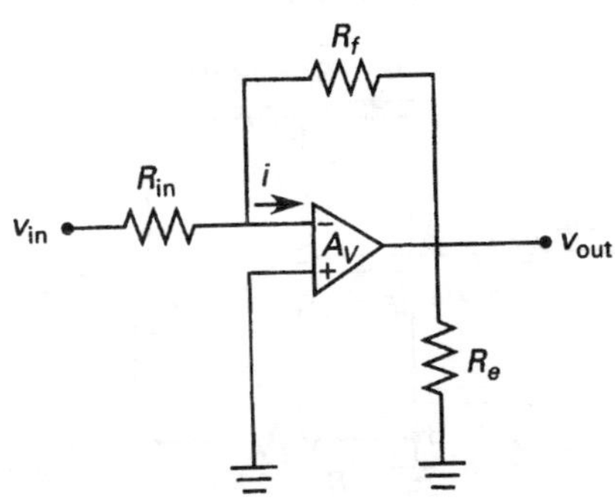

The assumptions regarding the properties of the ideal op amp result in the following practical results during analysis.

- The current to each input is zero.
- The voltage between the two input terminals is zero.
- The op amp is operating in the linear range.

The voltage difference of zero between the two terminals is called a *virtual short circuit*, or, because the positive terminal is often grounded, a *virtual ground*. The term "virtual" is used because, although the feedback from the output is used to keep the voltage difference between the terminals at zero, no current actually flows into the short circuit.

The procedure for analyzing an ideal op amp circuit follows.

step 1: Draw the circuit and label all the nodes, voltages, and currents of interest.

step 2: Write Kirchhoff's current law (KCL) at the op amp input node (normally the inverting terminal).[9]

step 3: Simplify the resulting equation using the ideal op amp assumptions. Specifically, the current into the op amp is zero and the voltage difference between the terminals is zero. The known voltage at one terminal must be the voltage at the other (e.g., if the positive terminal is at ground, the negative terminal voltage is zero).

step 4: Solve for the desired quantity or expression.

[7]This is true for the op amp itself, without feedback. When negative feedback is used, the output voltage is whatever value maintains $v^+ = v^-$.

[8]This property is sometimes omitted from lists of ideal op amp properties.

[9]Methods other than KCL may be used, using KCL incorporates all the desired quantities and minimizes errors.

Electronics

Using the procedure given, the relationships or values for any op amp circuit parameter can be found. The relationship between the output voltage and the input voltage for a variety of op amp circuits is shown in Fig. 48.8.

Example 48.4

Consider the circuit shown. The op amp is considered ideal with a 1 μV input signal. What is the current through the feedback resistor?

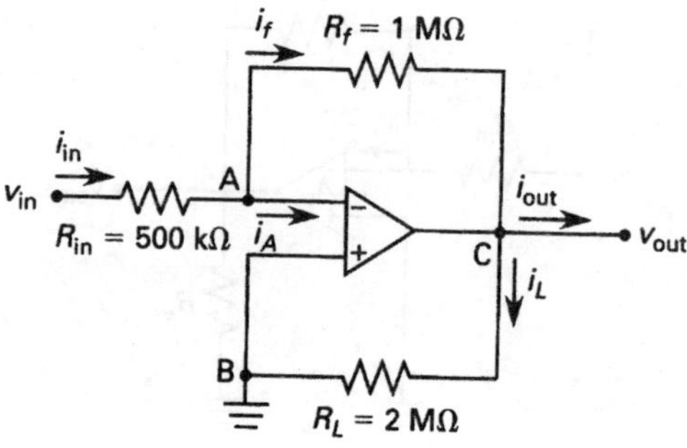

Solution

The circuit is already drawn. Writing KCL (assuming that currents out of the node are positive[10]) at the input terminal of the op amp, that is, node A, gives

$$-i_{in} + i_A + i_f = 0$$

$$i_f = i_{in} - i_A = \frac{v_{in} - V_A}{R_{in}} - i_A$$

Using the ideal op amp assumptions, the input current to the op amp is zero, that is, $i_A = 0$. Also, the voltage difference between the terminals is zero. Because the positive terminal is connected to ground, $v^+ = 0 = v^- = V_A$. Substituting gives

$$\begin{aligned} i_f = i_{in} - i_A &= \frac{v_{in} - V_A}{R_{in}} - i_A \\ &= \frac{1 \times 10^{-6}\ \text{V} - 0\ \text{V}}{500 \times 10^3\ \Omega} - 0\ \text{A} \\ &= 2 \times 10^{-12}\ \text{A} \end{aligned}$$

The negative sign indicates that the current direction is opposite that shown. Because i_A is zero, the input current is equal to 2×10^{-12} A.

Example 48.5

For the circuit in Ex. 48.4, determine the voltage gain.

[10]The current into the node, i_{in}, is shown as negative to be consistent with the assumption that all currents into the node are negative and those out are positive. Ohm's law is written as shown to be consistent with the direction of the input current given.

Solution

The voltage gain is

$$A_V = \frac{v_{out}}{v_{in}}$$

The input voltage is known. The output voltage can be calculated in a number of ways. Using a portion of KCL at node A (or Ohm's law) gives

$$i_f = \frac{v_A - v_{out}}{R_f} = \frac{0\ \text{V} - v_{out}}{1 \times 10^6\ \Omega} = 2 \times 10^{-12}\ \text{A}$$

$$\begin{aligned} v_{out} &= (-1)(2 \times 10^{-12}\ \text{A})(1 \times 10^6\ \Omega) \\ &= -2 \times 10^{-6}\ \text{V} \end{aligned}$$

Care should be taken with the negative signs. The feedback current, i_f, is positive because the direction shown in the example figure is used. Substituting gives

$$A_V = \frac{v_{out}}{v_{in}} = \frac{-2 \times 10^{-6}\ \text{V}}{1 \times 10^{-6}\ \text{V}} = -2$$

The gain is negative because the input occurs at the inverting terminal. The gain is determined by the resistors (since $|i_f| = |i_{in}|$), as can be seen by rewriting the gain formula.

$$A_V = \frac{v_{out}}{v_{in}} = \frac{-i_f R_f}{i_{in} R_{in}} = \frac{-R_f}{R_{in}}$$

Example 48.6

For the circuit in Ex. 48.4, determine the current through the load resistor.

Solution

Several possible methods exist for determining the solution. Because v_{out} is known and the direction of i_L is given, using Ohm's law at node C provides the answer directly.

$$\begin{aligned} i_L = \frac{v_{out} - v_B}{R_L} &= \frac{-2 \times 10^{-6}\ \text{V} - 0\ \text{V}}{2 \times 10^6\ \Omega} \\ &= -1 \times 10^{-12}\ \text{A} \end{aligned}$$

3. OPERATIONAL AMPLIFIER LIMITS

A real op amp has limits that must be accounted for during design or use. Those limits are summarized in this section.

Op amps have a finite bandwidth. Specifically, they have a finite gain-bandwidth product. The product of the gain and the bandwidth is essentially constant (see Eq. 48.1 and Eq. 48.2). The first-order approximation of the frequency response of an amplifier is shown in Fig. 48.9.

Figure 48.8 Operational Amplifier Circuits

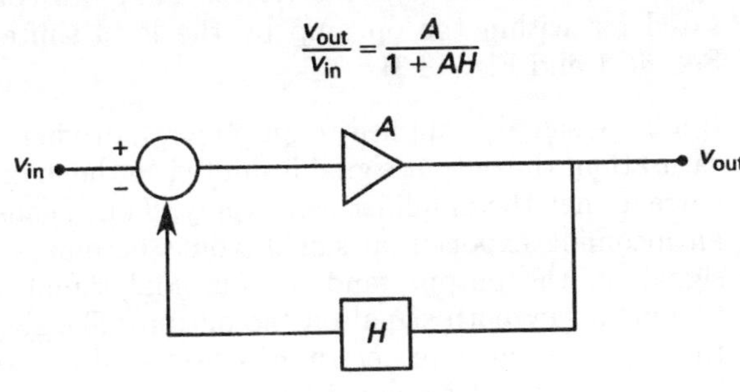

(a) feedback system

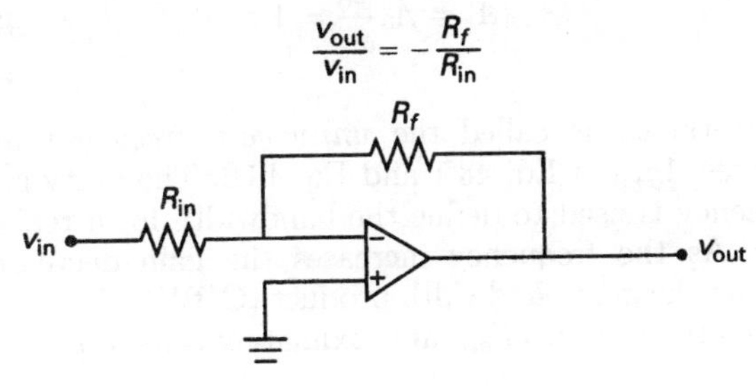

(b) inverting amplifier

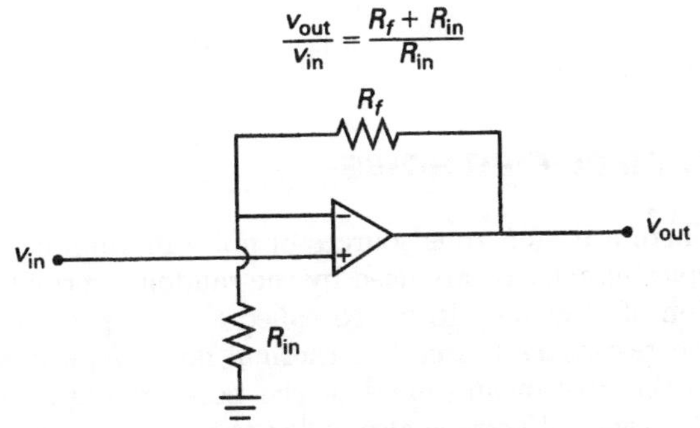

(c) noninverting amplifier

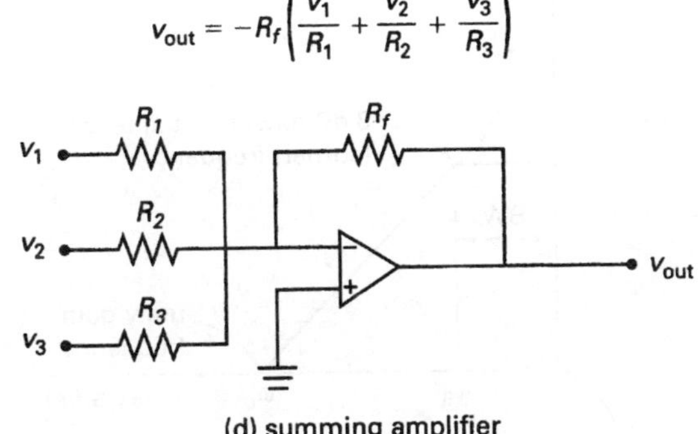

(d) summing amplifier

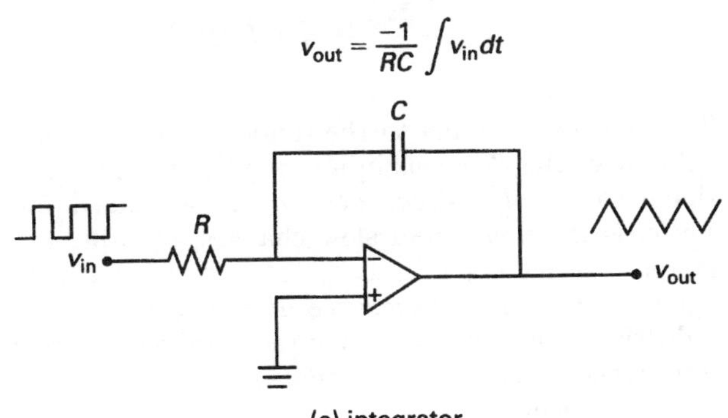

(e) integrator

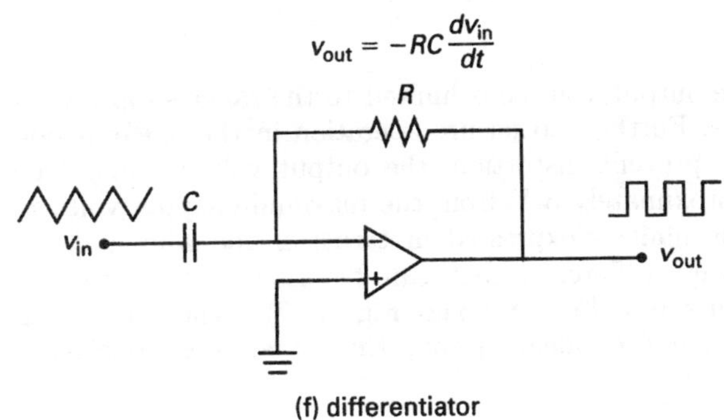

(f) differentiator

$$\frac{V_{out}}{V_{in}} = \frac{-R_f}{R_{in}(1 + j\omega R_f C)} \quad \text{[sinusoidal input]}$$

C

R_f

R_{in}

V_{in}

V_{out}

(g) low-pass filter

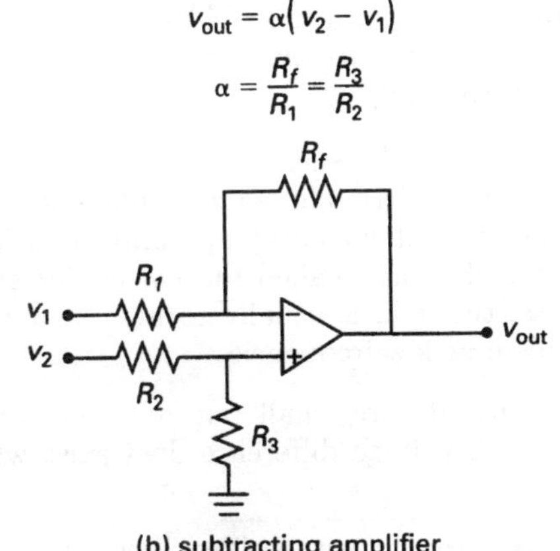

(h) subtracting amplifier

The *unity gain* is defined as

$$|A| = A_0 \frac{\omega_0}{\omega_u} = 1 \qquad 48.9$$

The term ω_u is called the *unity gain frequency* and replaces $A_0\omega_0$ in Eq. 48.1 and Eq. 48.2. The unity gain frequency is used to define the bandwidth for a real op amp. As the frequency increases the gain decreases, keeping the gain bandwidth product (GBW), also called the figure of merit (F_m), approximately constant.[11]

Figure 48.9 *Operational Amplifier Frequency Response*

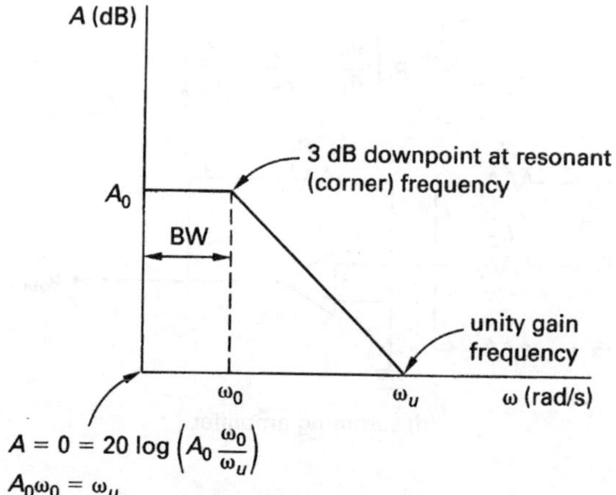

The output voltage is limited to the range supply voltages. Further, to ensure operation in the linear region and prevent distortion, the output voltage should be approximately 3 V from the maximum supply voltage. This limit is expressed in terms of the gain and the voltage difference that can be used at the terminal inputs (see Eq. 48.4 and Fig. 48.5). When operating outside the linear region, the op amp is considered *saturated.*

The op amp amplifies both the difference signal and the common signal. Amplifying the common signal results in problems, especially for precision applications. The common-mode rejection ratio (CMRR) is a measure of the op amp's ability to overcome this obstacle (see Eq. 48.7 and Eq. 48.8).

The gain is finite. As such, the output is less than expected (see Ex. 48.4). The voltage difference is determined from the fundamental op amp equation (see Eq. 48.3), which is also called the *constraint equation.* Nevertheless, the gain is usually so high that this is a minor problem at low frequencies.

The input currents are small but not zero when the input is zero. A voltage difference does exist when the input is zero. These currents and voltages are called *offset currents* and *offset voltages.* They are compensated for within the op amp by the level shifter (see Sec. 48.1 and Fig. 48.3).

If a large signal is applied to an op amp, driving it into saturation, the output signal is limited to the maximum current that the amplifier can supply. This means that an incoming exponential signal would become a ramp signal on the output, and a sinusoidal signal would become a sawtooth signal on the output. The slopes of the output ramps are determined by the *slew rate*, S_R. The slew rate is determined by the internal capacitance of the amplifier and is

$$S_R = \left.\frac{dv}{dt}\right|_{\max} \approx \frac{I_{\max}}{C} \qquad 48.10$$

4. AMPLIFIER NOISE

There are numerous sources of noise in amplifiers. The *thermal noise* is produced by the random thermal agitation of electrons. It is also called the *Johnson noise* or the *resistance noise.* The thermal noise is proportional to the root-mean-square of the noise voltage squared. The thermal noise is also called the *noise power*, P_n.

$$P_n = \frac{V^2_{\text{noise,rms}}}{4R} = \kappa T(\text{BW}) \qquad 48.11$$

The *shot noise* is due to the random variations in the velocity of the electrons in a current flow. It is proportional to $2qI$. *Low-frequency noise*, also called *flicker noise*, is the noise from slow changes, or drift, in the system parameters, caused by temperature, component aging, and so on. *Interference noise* is electromagnetic induction from sources outside the amplifier. To reduce interference, ensure the *electromagnetic compatibility* of the components.

The total noise in a given circuit is compared to the incoming signal, or power, to ensure the ability of the circuit to discern between the two. The measure of this ability is the signal-to-noise ratio (SNR).

$$\text{SNR} = 10 \log \frac{P_s}{P_n} = 20 \log \frac{V_s}{V_n} \qquad 48.12$$

Because the resistance is common to both the signal and the noise, the units of the signal power, P_s, and the noise power, P_n, can be expressed in watts or simply volts squared (in which case the noise power equals $4R\kappa T(\text{BW})$).[12] A signal-to-noise ratio of approximately 10 dB is required for the op amp output to reflect the signal.

[11]The frequency response of any amplifier circuit can be found by transforming the external electrical components into the *s* domain (see Sec. 31.10). Then obtain the gain equation using the procedures of Sec. 48.2. Finally, plot the response on a Bode plot (see Chap. 32).

[12]When ratios are involved, such as the signal-to-noise ratio, the only term remaining is the rms noise voltage squared, V^2. This is why the square of the voltage is sometimes referred to as the *noise power.*

PART 3: PULSE CIRCUITS: WAVEFORM SHAPING AND LOGIC

5. PULSE CIRCUIT FUNDAMENTALS

A *pulse circuit* is any active electrical network designed to respond to discrete pulses of current or voltage. Such circuits abound in computer and digital systems. Examples are waveform shaping circuits and logic circuits. An *active circuit*, such as a transistor circuit, is a circuit that is capable of amplifying a current or voltage signal. Waveform shaping is primarily accomplished with active nonlinear circuits. Active nonlinear elements are described in Chap. 44, though when used as switches, they are considered pulsed circuits—hence their coverage here. *Passive linear elements*, those elements that are not sources of energy (for example, resistors, capacitors, and inductors) are encountered in the design of pulse, waveform shaping, and logic circuits. *Passive nonlinear elements*, primarily diodes, are widely used for waveform shaping elements and elements of logic circuits. The delta (Dirac) function and the step function, described in Chap. 15, are used to mathematically describe the response of pulse circuits. *Pulse generators* are circuits that are designed with high-speed switches, capable of generating very short-duration pulses for physically testing the response of a network to a delta-like input.[13] The ability to recognize the overall function of a pulse, waveform, or logic circuit is a practical skill required in electrical engineering. However, detailed analysis of such circuits is a specialized skill, which requires the application of knowledge of the individual components and is beyond the scope of the general skills required in electrical engineering.

6. CLAMPING CIRCUITS

A diode represents a purely resistive passive nonlinear element. The diode can be used as a simple means of limiting voltage to a desired value. When the voltage is limited in some manner, it is *clamped.* Two types of clamping exist: limiting and clipping. A *limiter* is an electronic circuit designed to prevent a waveform from exceeding a predetermined value. A *clipper* is an electronic circuit designed to maintain a waveform at a predetermined value when the input is above a certain level. The two types of circuits are shown in Fig. 48.10 for an ideal diode, that is, a diode with no resistance and a zero voltage drop. The ideal circuits can be used for many practical calculations and for determining overall circuit response.

The diode is not an ideal device. The forward threshold voltage drop, V_F, for silicon diodes is approximately 0.7 V, while that for germanium diodes is approximately

[13]An *RC differentiator* with a switching transistor to control the application of the power is a pulse generator. The *RC* differentiator is an *RC* circuit with the output taken across the resistor.

Figure 48.10 *Ideal Diode Limiter and Clipper Circuits*

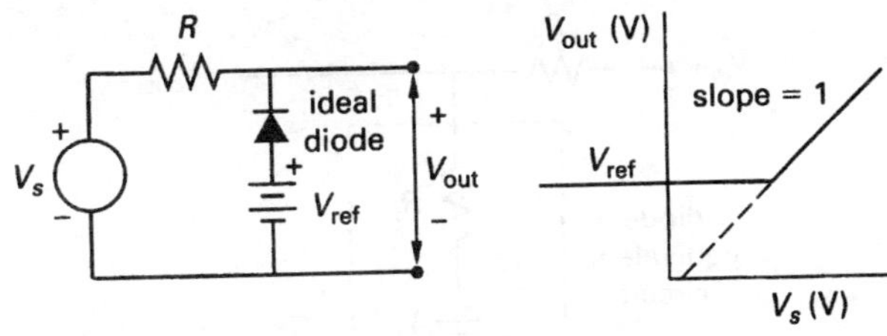

(a) shunt diode limiter circuit

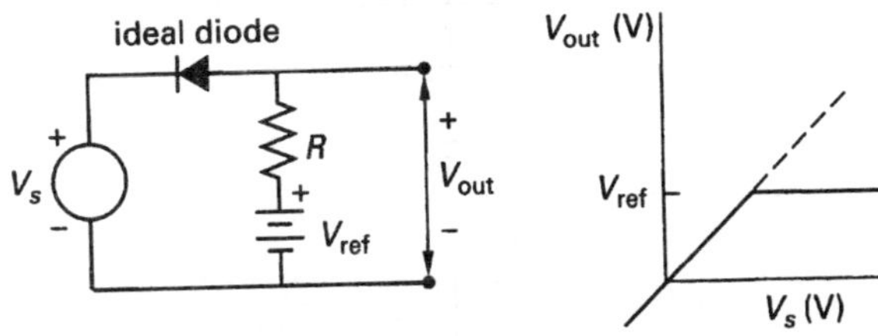

(b) series diode clipper circuit

0.2 V. Additionally, diodes exhibit some forward resistance, R_f, which accounts for the small but finite slope of the forward characteristic. (The reverse resistance or zener resistance is not a concern, because limiter and clipper circuits do not operate in the breakdown region.) The limiter and clipper circuits for an actual diode are shown in Fig. 48.11. Actual circuits are more complex than ideal circuits but produce more accurate results.

Analysis of the circuit in Fig. 48.11(a) yields Eq. 48.13 and Eq. 48.14.

$$v_{\text{out}} = \frac{R(V_{\text{ref}} - V_F)}{R + R_f} + \frac{R_f V_{\text{in}}}{R_f + R} \quad [V_{\text{in}} < V_{\text{ref}} - V_F] \qquad 48.13$$

$$V_{\text{out}} = V_{\text{in}} \quad [V_{\text{in}} > V_{\text{ref}} - V_F] \qquad 48.14$$

Equation 48.13 represents the forward-biased case. Equation 48.14 represents the reverse-biased case, ignoring the small reverse saturation current. Analysis of the circuit in Fig. 48.11(b) yields Eq. 48.15 and Eq. 48.16.

$$v_{\text{out}} = \frac{R_f V_{\text{in}}}{R_f + R} + \frac{R(V_{\text{ref}} + V_F)}{R + R_f} \quad [V_{\text{in}} > V_{\text{ref}} + V_F] \qquad 48.15$$

$$V_{\text{out}} = V_{\text{in}} \quad [V_{\text{in}} < V_{\text{ref}} + V_F] \qquad 48.16$$

Equation 48.15 represents the forward-biased case. Equation 48.16 represents the reverse-biased case, ignoring the small reverse saturation current.

The ideal limiter or clipper circuit would have R_f equal to zero, and the effect of the threshold voltage would be negligible. This type of circuit can be realized using a *precision diode.* A precision diode circuit, equivalent circuit, and characteristics are shown in Fig. 48.12. A precision diode is used to rectify very small signals: those less

Electronics

Figure 48.11 Real Diode Limiter and Clipper Circuits

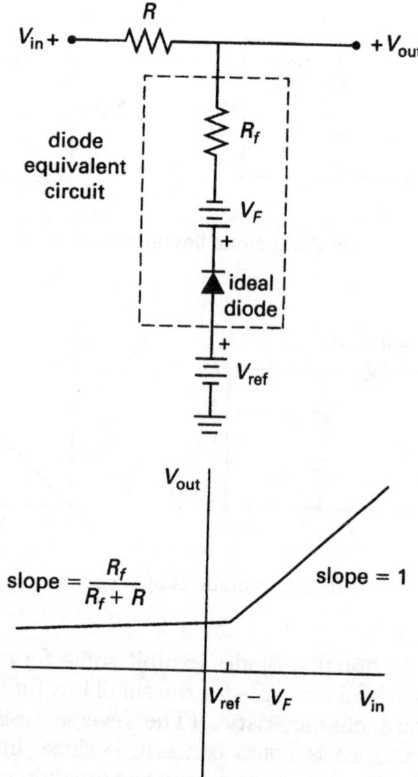

(a) limiter circuit

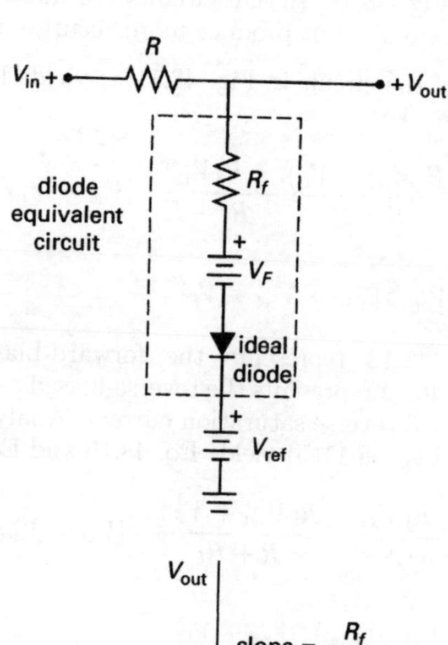

(b) clipper circuit

Figure 48.12 Precision Diode Circuit

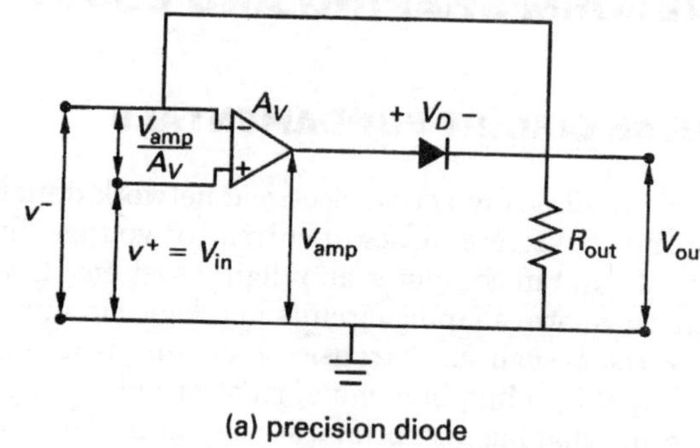

(a) precision diode

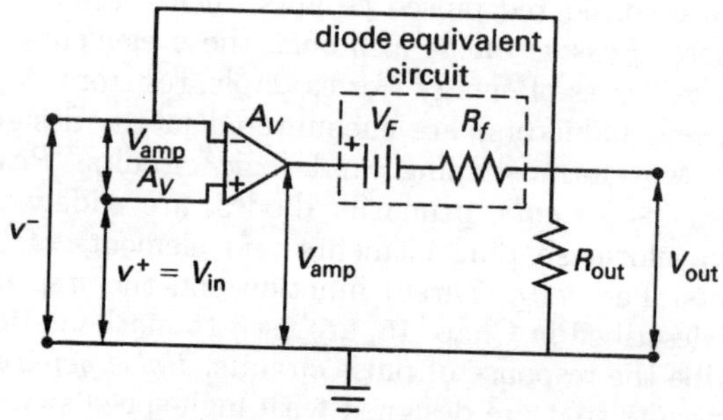

(b) forward-biased precision diode

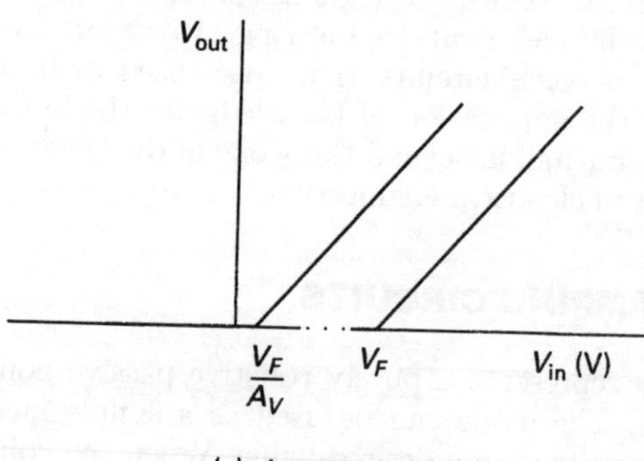

(c) characteristics

than 0.7 V for silicon or 0.2 V for germanium. In fact, the precision diode rectifies microvolt-level signals.

In Fig. 48.12(a), a diode is placed between the operational amplifier output and the feedback point of a voltage follower. For negative voltage inputs, $-v_{\text{in}}$, the op amp output voltage, V_{amp}, is negative and the diode is reverse biased.[14] The output voltage is then equal to zero because no current flows through the output resistor, R_{out}. For positive voltage inputs, $+v_{\text{in}}$, the op amp output voltage, V_{amp}, is equal to $A_V V_{\text{in}}$ until the diode becomes forward biased, that is, until the diode voltage, V_D, exceeds

[14]For proper operation, the diode must have a reverse breakdown voltage greater in magnitude than the maximum negative voltage of the op amp, that is, greater in magnitude than the negative supply voltage (not shown in the figure).

V_F. This forward-biased condition, which occurs when $A_V V_{in} = V_D$ or $V_{in} = V_D/A_V$, is shown in Fig. 48.12(b). From the forward-biased condition onward, the output voltage across R_{out} is the input voltage magnified by the op amp as shown in Fig. 48.12(c). The approximate transfer equation for the precision diode of Fig. 48.12(a) is

$$V_{out} = V_{in} - \frac{V_F}{A_V} \qquad 48.17$$

Precision diode circuits sensitive to negative voltages are shown in Fig. 48.13.

Example 48.7

Consider the precision diode in Fig. 48.12(b) with a 12 V power supply and a voltage gain of 10^6. What is the maximum voltage difference that can be amplified without distortion or saturation?

Solution

The maximum voltage difference that can be amplified is determined by the constraint equation.

$$v^+ - v^- = \frac{V_{amp}}{A_V}$$

The constraint equation is Eq. 48.17 with the diode removed, which changes V_F to V_{amp}. The maximum voltage of the amplifier is 9 V (3 V less than the maximum power supply voltage) to prevent distortion or saturation. So,

$$v^+ - v^- = \frac{V_{amp}}{A_V} = \frac{9 \text{ V}}{10^6} \quad (9\ \mu\text{V})$$

Example 48.8

Use the op amp in Ex. 48.7 with a silicon diode. What is the output voltage if the input voltage is 9 μV?

Solution

With the silicon diode, the threshold voltage is 0.7 V. Writing KCL at the negative terminal (ignoring the small current into the op amp) gives

$$\frac{v^- - 0 \text{ V}}{R_{out}} + \frac{v^- + V_F - V_{amp}}{R_f} = 0$$

$v^- = V_{out}$, and $V_F = 0.7$ V. Substituting gives

$$\frac{V_{out}}{R_{out}} + \frac{V_{out} + 0.7 \text{ V} - V_{amp}}{R_f} = 0$$

Figure 48.13 *Precision Diode Circuits Sensitive to Negative Voltage Input*

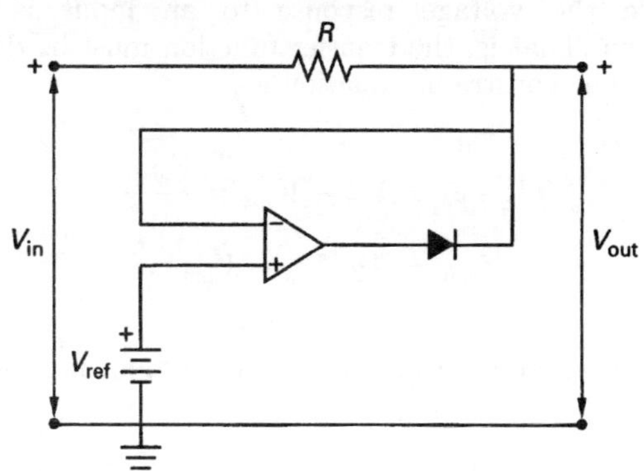

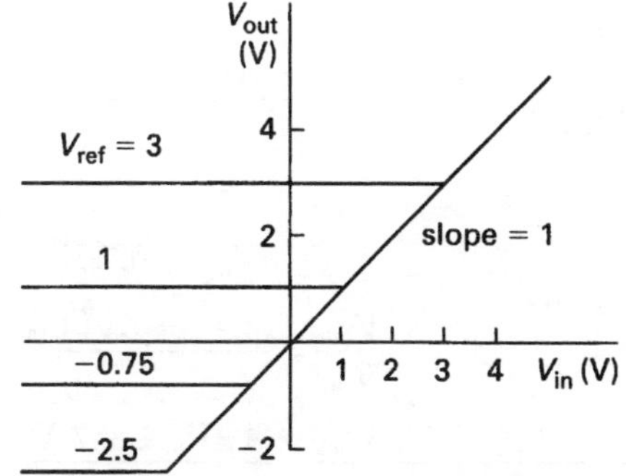

(a) precision limiter/clipper circuit

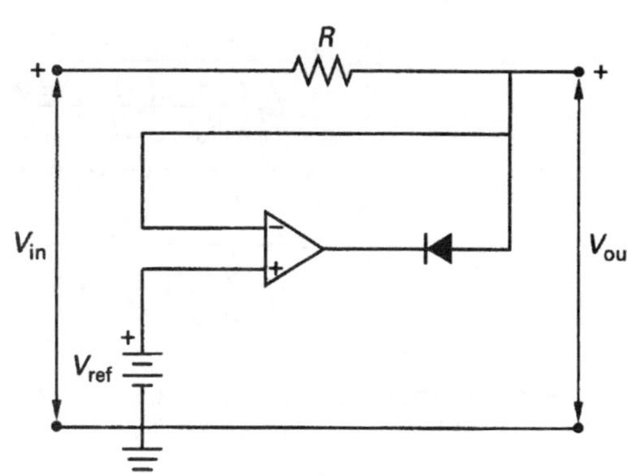

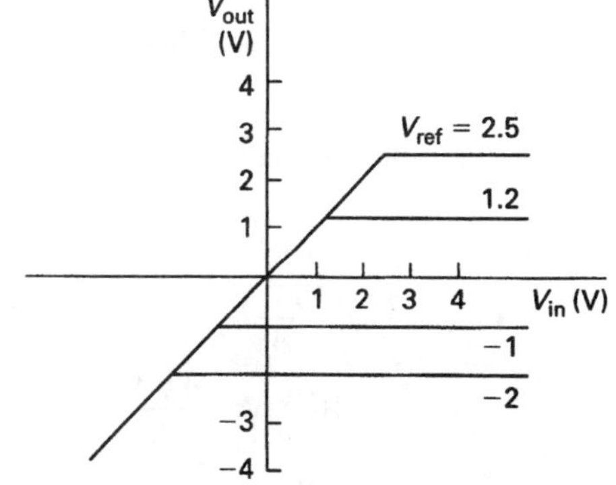

(b) precision limiter/clipper circuit sensitive to negative voltages

Electronics

The constraint equation must be used, rather than the assumption used with ideal op amps that $v^+ = v^-$, because the voltage response to an input is the unknown. That is, the transfer function must be determined. The constraint equation is

$$v^+ - v^- = V_{\text{in}} - V_{\text{out}} = \frac{V_{\text{amp}}}{A_V}$$

$$V_{\text{amp}} = A_V (V_{\text{in}} - V_{\text{out}})$$

Substituting and rearranging gives the following result.

$$\frac{V_{\text{out}}}{R_{\text{out}}} + \frac{V_{\text{out}} + 0.7\text{ V} - V_{\text{amp}}}{R_f} = 0$$

$$\frac{V_{\text{out}}}{R_{\text{out}}} + \frac{V_{\text{out}} + 0.7\text{ V} - A_V(V_{\text{in}} - V_{\text{out}})}{R_f} = 0$$

$$\frac{V_{\text{out}}}{R_{\text{out}}} = \frac{V_{\text{out}} + 0.7\text{ V} - A_V(V_{\text{in}} - V_{\text{out}})}{R_f}$$

$$\frac{V_{\text{out}}}{R_{\text{out}}} + \frac{V_{\text{out}}}{R_f} + \frac{A_V V_{\text{out}}}{R_f} = \frac{A_V V_{\text{in}} - 0.7\text{ V}}{R_f}$$

$$V_{\text{out}}\left(\frac{1}{R_{\text{out}}} + \frac{1}{R_f} + \frac{A_V}{R_f}\right) = \frac{A_V V_{\text{in}} - 0.7\text{ V}}{R_f}$$

$$V_{\text{out}} = \frac{A_V V_{\text{in}} - 0.7\text{ V}}{R_f\left(\frac{1}{R_{\text{out}}} + \frac{1}{R_f} + \frac{A_V}{R_f}\right)}$$

$$= \frac{V_{\text{in}} - \frac{0.7\text{ V}}{A_V}}{\frac{R_f}{A_V R_{\text{out}}} + \frac{1}{A_V} + 1}$$

The gain is very large. The transfer function reduces to Eq. 48.17, with 0.7 V substituted for V_F.

$$V_{\text{out}} = V_{\text{in}} - \frac{0.7\text{ V}}{A_V}$$

Substituting the input voltage of 9 μV and the gain of 10^6 gives

$$V_{\text{out}} = V_{\text{in}} - \frac{0.7\text{ V}}{A_V}$$

$$= 9 \times 10^{-6}\text{ V} - \frac{0.7\text{ V}}{10^6}$$

$$= 8.3 \times 10^{-6}\text{ V} \quad (8.3\ \mu\text{V})$$

The offset voltage of 0.7 V for silicon has been reduced by the precision diode to 0.7×10^{-6} V, or 0.7 μV, which is essentially zero.

Example 48.9

Design a circuit that has the following output characteristic.

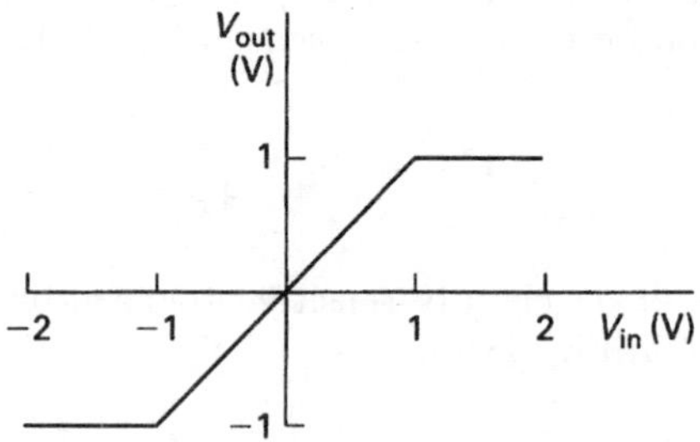

This circuit linearly transfers any signal of less than 1.0 V in magnitude, but clips all signals greater than 1.0 V in magnitude. The power supply voltage available is ±12.0 V.

Solution

A precision diode circuit set up as voltage followers exhibits the characteristics given. For the positive voltages, a clipper circuit similar to that in Fig. 48.13(b) can be used with the reference voltage at the noninverting terminal set to 1.0 V by a voltage divider. For the negative voltages, a limiter circuit similar to that in Fig. 48.13(a) can be used with the reference voltage at the noninverting terminal set to −1.0 V by a voltage divider. The resulting circuit follows.

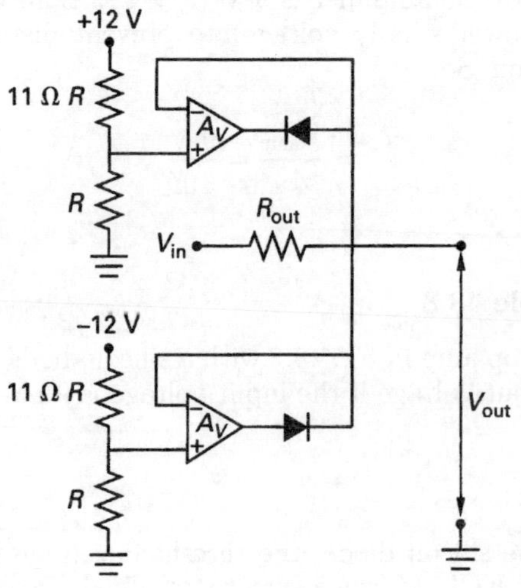

7. ZENER VOLTAGE REGULATOR CIRCUIT: IDEAL

Zener diodes operating in the breakdown region are used as voltage regulators. That is, they are used to control the load voltage (output voltage) within a narrow set of values in circuits for which the supply voltage (input voltage) varies and the load current ranges from zero to a maximum value.

An ideal zener diode voltage regulator circuit is shown in Fig. 48.14(a) with the model for the ideal zener illustrated in Fig. 48.14(b). The zener will maintain a constant voltage as long as the zener current, I_Z, is positive as shown in Fig. 48.14(c).

Figure 48.14 *Ideal Zener Diode Voltage Regulator*

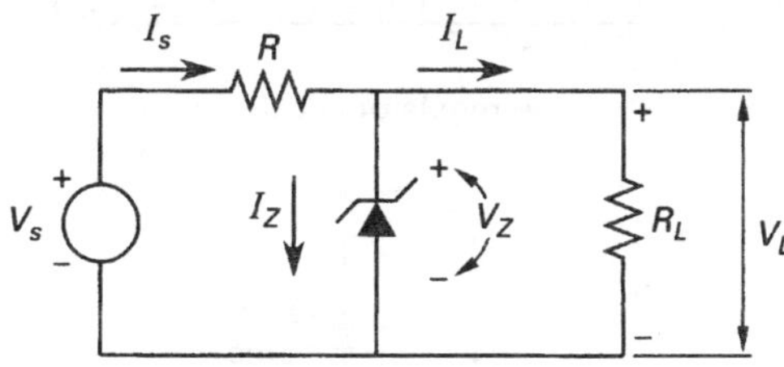

(a) regulator circuit

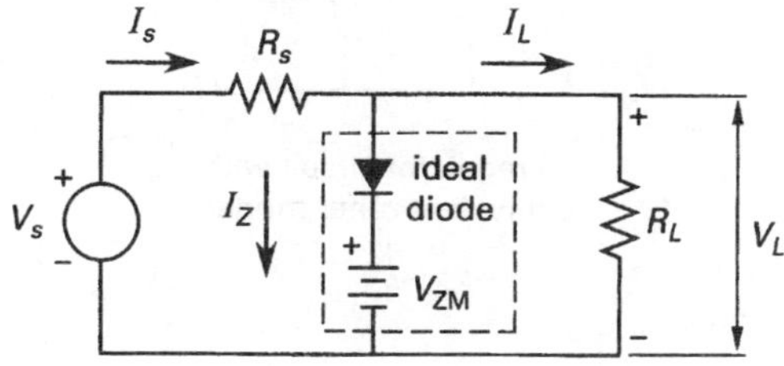

(b) regulator circuit with ideal zener model

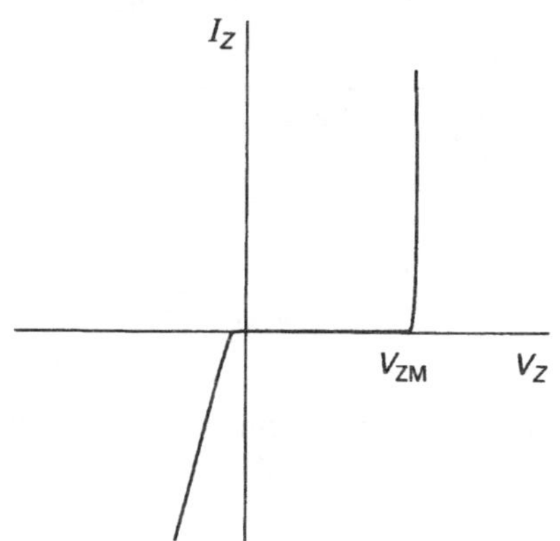

(c) ideal zener characteristics

The design restrictions can be understood from Eq. 48.18 and Eq. 48.19, or derived from Ohm's law and the application of KCL to Fig. 48.14(b).

$$I_s = \frac{V_s - V_Z}{R_s} \tag{48.18}$$

$$I_Z = I_s - I_L \tag{48.19}$$

As the load current, I_L, varies, the supply current given by Eq. 48.18 remains constant, but, according to KCL as stated in Eq. 48.19, the zener current will vary to compensate. As the load current increases the zener diode current decreases. However, to keep the ideal diode forward biased, that is, to keep the zener diode in breakdown, the zener current must be zero or positive.[15] That is, the load voltage, which equals the zener voltage, must be equal to V_{ZM}. To minimize the total power consumption, the value of R_s selected should be as small as possible while still compensating for changes in supply voltage and load current. The combination of these requirements gives

$$R_s = \frac{V_{\text{in,min}} - V_{\text{ZM}}}{I_{L,\text{max}}} \tag{48.20}$$

$$I_{Z,\text{max}} = \frac{V_{\text{in,max}} - V_{\text{ZM}}}{R_s} = \left(\frac{V_{\text{in,max}} - V_{\text{ZM}}}{V_{\text{in,min}} - V_{\text{ZM}}}\right) I_{L,\text{max}} \tag{48.21}$$

Equation 48.20 represents the requirement that at the maximum possible load current, no negative current flows in the ideal diode, with an equal sign used to minimize the value of R_s.[16] The phrase "no negative current" means that the minimum supply voltage, $V_{\text{in,min}}$, is greater than or equal to V_{ZM}. Equation 48.20 represents the maximum current drawn by the zener, setting the power requirements for the diode and the supply resistor, given by Eq. 48.22 and Eq. 48.23, respectively.

$$P_D = I_{Z,\text{max}} V_{\text{ZM}} \tag{48.22}$$

$$P_{R_s} = I^2_{Z,\text{max}} R_s \tag{48.23}$$

An ideal zener diode voltage regulator circuit is designed as follows.

step 1: Determine the supply resistance that minimizes power loss but keeps the zener operating in the breakdown region (see Eq. 48.20). This step accounts for the minimum supply (input) voltage and the maximum load (output) current.

step 2: Determine the maximum zener current drawn with the supply resistance calculated in step 1 (see Eq. 48.20). This step accounts for the maximum supply (input) voltage and the minimum load (output) current.

step 3: Determine the required diode power rating with the current calculated in step 2 (see Eq. 48.22).

step 4: Determine the required resistor power rating with the current calculated in step 2 (see Eq. 48.23).

[15]In a practical zener diode, the current must be greater than the keep-alive current, I_{ZK}.

[16]The actual requirement is that the supply resistance be less than or equal to the right-hand side of Eq. 48.20.

Example 48.10

Design a zener voltage regulator circuit, patterned after Fig. 48.14(a). The output voltage requirement is 5 V, while the input varies from 10 V to 15 V and the load current ranges from 0 A to 5 A.

Solution

Following the design steps, the supply resistance is

$$R_s = \frac{V_{\text{in,min}} - V_{\text{ZM}}}{I_{L,\text{max}}} = \frac{10\text{ V} - 5\text{ V}}{5\text{ A}} = 1\ \Omega$$

The maximum zener current drawn is

$$I_{Z,\text{max}} = \frac{V_{\text{in,max}} - V_{\text{ZM}}}{R_s} = \frac{15\text{ V} - 5\text{ V}}{1\ \Omega} = 10\text{ A}$$

The diode power rating is

$$P_D = I_{Z,\text{max}} V_{\text{ZM}} = (10\text{ A})(5\text{ V}) = 50\text{ W}$$

The resistor power rating must be at least

$$P_{R_s} = I_{Z,\text{max}}^2 R_s = (10\text{ A})^2(1\ \Omega) = 100\text{ W}$$

Ideal zener diode regulator calculations can be used as first approximations of actual circuits. The inherent assumptions are that the keep-alive current, I_{ZK}, equals zero and that the zener resistance, R_Z, is negligible (see Sec. 48.8).

8. ZENER VOLTAGE REGULATOR CIRCUIT: PRACTICAL

A practical (equivalent) zener diode voltage regulator circuit, illustrated in Fig. 48.15, differs from the ideal case in a number of ways. First, the voltages that can be used are standardized. Appendix 48.A lists some of the standard zener diodes and their respective zener impedances, Z_Z, and keep-alive currents, I_{ZK}. Second, equivalent zener diodes exhibit some resistance, R_Z, as shown in the equivalent model of Fig. 48.15(b). This resistance, R_Z, can cause the zener voltage to vary as much as 20% from the stated value at rated current. The effect is significant for zener diodes rated for 8 V or less, as the zener effect dominates. Third, the zener diode requires a small but finite current called the keep-alive current, I_{ZK}, to remain in the breakdown region. The zener current must be kept above the keep-alive current for the voltage-regulating effect to occur.

The defining equations for the equivalent circuit are derived from Fig. 48.15(b).

$$V_{\text{in}} - V_L = (I_L + I_Z)R_s \qquad 48.24$$

$$V_L = V_{\text{ZM}} + I_Z R_Z \qquad 48.25$$

Equation 48.24 and Eq. 48.25 reduce to the ideal conditions when I_Z equals zero. Because the zener current must be kept above the keep-alive current, Eq. 48.24 and

Figure 48.15 *Equivalent Zener Diode Voltage Regulator*

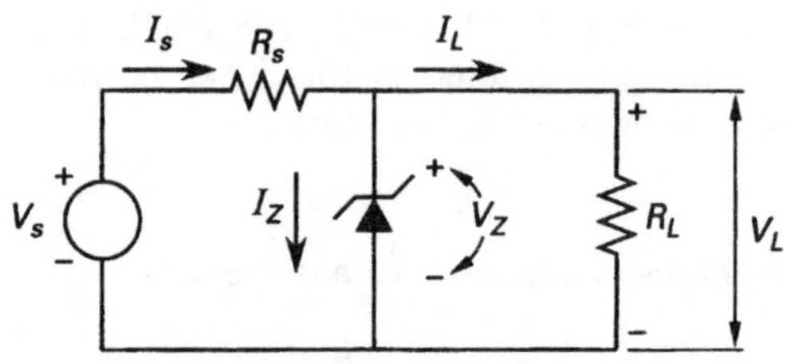

(a) regulator circuit

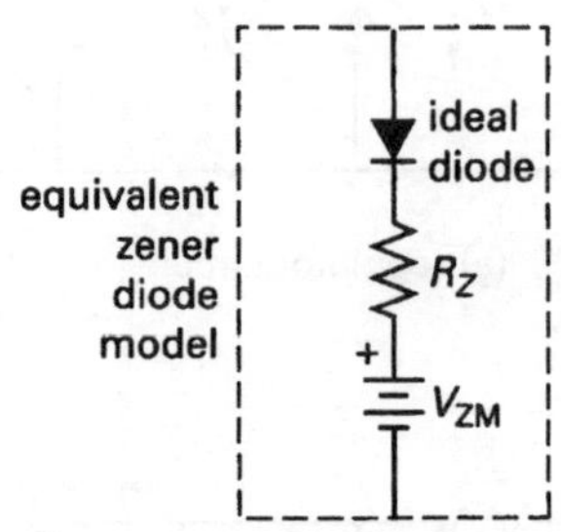

(b) regulator circuit with equivalent zener model

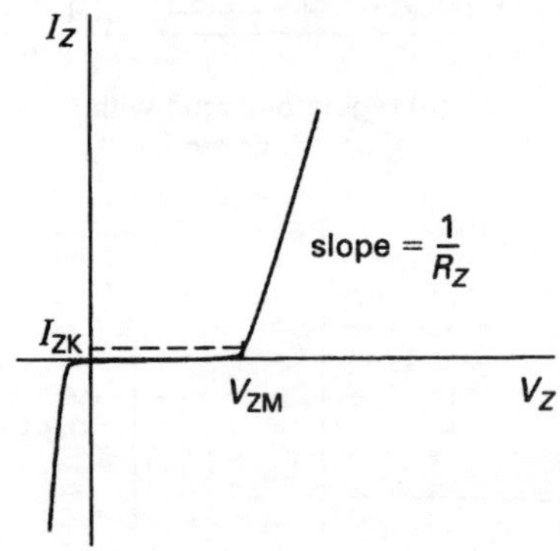

(c) characteristics

Eq. 48.25 must be solved simultaneously, which results in Eq. 48.26. (Equation 48.26 is similar to Eq. 48.20 for the ideal case, in that it is used to determine the source resistance with the input voltage at its minimum value, and the load current at its maximum value.)

$$V_L = \frac{V_{\text{ZM}} R_s + V_{\text{in}} R_Z}{R_s + R_Z} - I_L\left(\frac{R_s R_Z}{R_s + R_Z}\right) \qquad 48.26$$

The zener current is maximum when the input voltage, V_{in}, is maximum and the load current, I_L, is minimum.

$$\begin{aligned} I_{Z,\text{max}} &= \frac{V_{\text{ZM}} R_s + V_{\text{in,max}} R_Z}{R_Z (R_s + R_Z)} - \frac{V_{\text{ZM}}}{R_Z} \\ &= \frac{V_{\text{in,max}}}{R_s} - \frac{V_{\text{ZM}} R_s + V_{\text{in,max}} R_Z}{R_s (R_s + R_Z)} \end{aligned} \qquad 48.27$$

Equation 48.27 represents the maximum current drawn by the zener and sets the power requirements for the diode and the supply resistor, given by Eq. 48.28 and Eq. 48.29, respectively.

$$P_D = I_{Z,\text{max}} V_{\text{ZM}} + I^2_{Z,\text{max}} R_Z \qquad \textit{48.28}$$

$$P_{R_s} = I^2_{Z,\text{max}} R_s \qquad \textit{48.29}$$

If the load current, I_L, is allowed to exceed the maximum level, the zener current, I_Z, will decrease below the keep-alive level, I_{ZK}, and the zener diode will cease to regulate. Equation 48.26 and Eq. 48.27 will no longer be valid. The load voltage will be given by Eq. 48.30, which is derived from Eq. 48.24.

$$V_L = V_{\text{in}} - I_L R_s \qquad \textit{48.30}$$

A practical zener diode voltage regulator circuit is designed as follows.

step 1: Determine the supply resistance that minimizes power loss but keeps the zener operating in the breakdown region. Use Eq. 48.25 with the zener current, I_Z, equal to the keep-alive current, I_{ZK}, to determine the load voltage minimum that maintains the keep-alive current. Then use Eq. 48.26 with the calculated value of the minimum load voltage, the minimum input voltage, and the maximum load current to determine the source resistance. This step accounts for the minimum supply (input) voltage and the maximum load (output) current.

step 2: Determine the maximum zener current drawn with the supply resistance calculated in step 1 (see Eq. 48.27). This step accounts for the maximum supply (input) voltage and the minimum load (output) current.

step 3: Determine the required diode power rating with the current calculated in step 2 (see Eq. 48.28).

step 4: Determine the required resistor power rating with the current calculated in step 2 (see Eq. 48.29).

step 5: If desired, determine temperature effects on the zener diode at maximum expected temperature changes, and recalculate the parameters in the above steps to determine the circuit functions correctly (see Sec. 45.14).

step 6: If desired, determine the voltage regulation of the circuit. The regulation is calculated between the maximum input voltage at no load current and the minimum input voltage at maximum current.

Example 48.11

Determine the source resistance for a zener voltage regulator circuit, patterned after Fig. 48.15(b). The output voltage requirement is 5.1 V, while the input varies from 10 V to 15 V, and the load current ranges from 0 A to 5 A. The keep-alive current, I_{ZK}, equals 5 mA, and the zener resistance, R_Z, equals 0.12 Ω. (The parameters are identical to those used in Ex. 48.10 with the exception that the voltage output is 5.1 V, which is one of the standard zener voltages from App. 48.A.)

Solution

The supply resistance is determined from Eq. 48.26 by substituting the value of the minimum load voltage from Eq. 48.25. (See step 1 of the procedure in this section.) The minimum load voltage is

$$\begin{aligned} V_{L,\text{min}} &= V_{\text{ZM}} + I_{\text{ZK}} R_Z \\ &= 5.1 \text{ V} + (5 \times 10^{-3} \text{ A})(0.12 \ \Omega) \\ &= 5.1006 \text{ V} \end{aligned}$$

Substituting into Eq. 48.26, using the minimum input voltage and maximum load current, gives

$$\begin{aligned} V_L &= \frac{V_{\text{ZM}} R_s + V_{\text{in,min}} R_Z}{R_s + R_Z} - I_{L,\text{max}} \left(\frac{R_s R_Z}{R_s + R_Z} \right) \\ 5.1006 \text{ V} &= \frac{(5.1 \text{ V})(R_s) + (10 \text{ V})(0.12 \ \Omega)}{R_s + 0.12 \ \Omega} - (5 \text{ A}) \left(\frac{(R_s)(0.12 \ \Omega)}{R_s + 0.12 \ \Omega} \right) \\ R_s &= 0.98 \ \Omega \end{aligned}$$

Using a practical zener diode circuit instead of an ideal diode circuit requires a minimal source resistance change.

Example 48.12

Determine the voltage regulation for the circuit in Ex. 48.11.

Solution

The voltage regulation (step 6 of the procedure in this section) is determined by obtaining the equation for the load voltage and current. The value of R_s for the circuit to operate in the breakdown region is 0.98 Ω (from Ex. 48.11). Substituting this value into Eq. 48.26 gives the relationship between the load current and voltage.

$$\begin{aligned} V_L &= \frac{V_{\text{ZM}} R_s + V_{\text{in}} R_Z}{R_s + R_Z} - I_L \left(\frac{R_s R_Z}{R_s + R_Z} \right) \\ &= \frac{(5.1 \text{ V})(0.98 \ \Omega) + (V_{\text{in}})(0.12 \ \Omega)}{0.98 \ \Omega + 0.12 \ \Omega} - I_L \left(\frac{(0.98 \ \Omega)(0.12 \ \Omega)}{0.98 \ \Omega + 0.12 \ \Omega} \right) \\ &= 4.54 \text{ V} + 0.109 \ V_{\text{in}} - 0.107 I_L \end{aligned}$$

For an input voltage of 10 V and no-load current, the load voltage is 5.63 V. At the maximum load current of 5 A, the load voltage is 5.10 V. (If the load current goes above the maximum of 5 A, the diode ceases to regulate and the load voltage is determined by Eq. 48.30.) Repeating the calculation for an input voltage of 15 V gives the values of 6.18 V and 5.64 V for no-load current and maximum load current, respectively. Plotting the results gives the following.

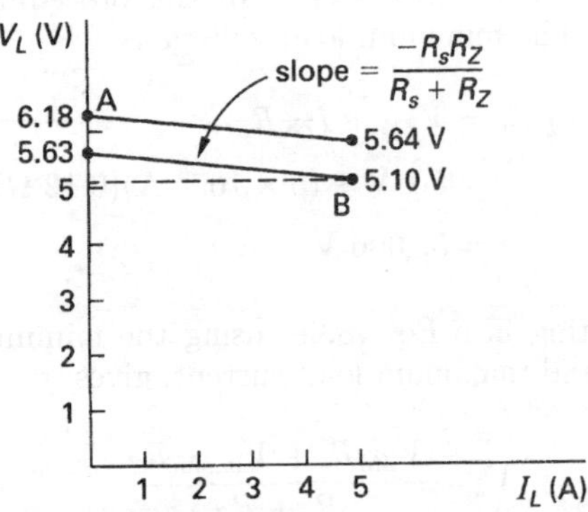

The regulation is calculated between the maximum input voltage at no-load current (point A) and the minimum input voltage at maximum current (point B).

$$\text{VR} = \frac{V_{\text{nl}} - V_{\text{fl}}}{V_{\text{fl}}} \times 100\% = \frac{6.18\ \text{V} - 5.10\ \text{V}}{5.10\ \text{V}} \times 100\%$$

$$= 21.2\%$$

9. OPERATIONAL AMPLIFIER REGULATOR CIRCUIT

Zener diodes provide voltage regulation at low-current demands. Better regulation is achieved using a precision diode, but a precision diode also lowers the operational amplifier current output. Using the zener diode as a reference for input to an op amp and passing the output to the base of a transistor improves the regulation and increases the current capability. Such an op amp regulator is shown in Fig. 48.16.

Figure 48.16 Operational Amplifier Voltage Regulator

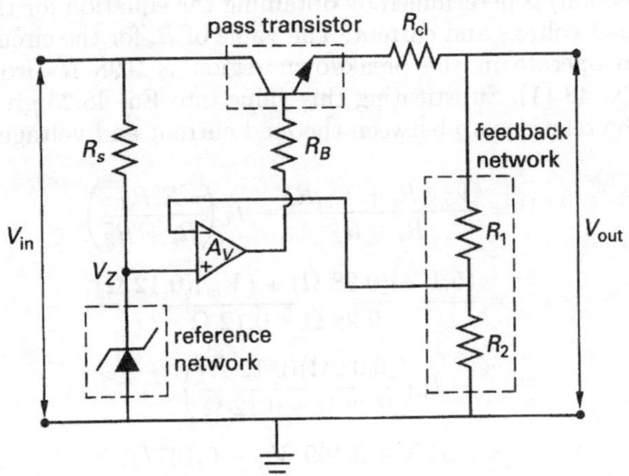

The common base transistor is called the *pass transistor* and is used to supply a current gain, α. The zener provides the reference voltage, V_Z, at the noninverting terminal of the op amp. The input to the op amp is $V_Z - \delta V_{\text{out}}$ where delta is called the *feedback factor* and is given by[17]

$$\delta = \frac{R_2}{R_1 + R_2} \qquad 48.31$$

The output voltage is approximately

$$V_{\text{out}} \approx \frac{V_Z}{\delta} \approx V_Z\left(\frac{A_V}{1 + \delta A_V}\right) \qquad 48.32$$

The supply resistance, R_s, is set to ensure that the minimum keep-alive current flows in the zener diode.

$$R_s = \frac{V_{\text{in,min}} - V_{\text{ZM}}}{I_{\text{ZK}}} - R_Z \qquad 48.33$$

The equivalent circuit for the zener diode, which shows R_Z, is shown in Fig. 48.15(b). The minimum and maximum zener voltages are

$$V_{Z,\text{min}} = V_{\text{ZM}} + I_{\text{ZK}}R_Z \qquad 48.34$$

$$V_{Z,\text{max}} = \frac{V_{\text{in,max}}R_Z + V_{\text{ZM}}R_s}{R_s + R_Z} \qquad 48.35$$

The method of design follows the same basic steps given in Sec. 48.8.

10. TRANSISTOR SWITCH FUNDAMENTALS

The bipolar junction transistor (BJT) is an active nonlinear device that, when used between its saturation and cutoff regions, becomes an electronic switch (see Sec. 46.2). A transistor circuit set up to operate as a switch is shown in Fig. 48.17(a). The pulse input that controls the switch is shown in Fig. 48.17(b). The pulses can be considered to represent logic conditions.[18] Combinations of such switches can be used to build logic circuits. The characteristics of the switch, the two operating points, and approximate models are shown in Fig. 48.17(c).

When the transistor switch is in the on state, that is, saturated, the collector current is given by KVL in the output loop as

$$I_{C,\text{sat}} = \frac{V_{\text{CC}} - V_{\text{CE,sat}}}{R_L} \qquad 48.36$$

The collector-emitter voltage, V_{CE}, is typically 0.1 V for germanium diodes and 0.2 V for silicon diodes and can be ignored for first-order calculations—hence the

[17]The *feedback factor* is sometimes symbolized β. This symbol is avoided here to prevent confusion with the CE current gain, β.

[18]Normally the logic conditions are voltage inputs. For transistor-transistor-logic (TTL) circuits, 0.0 V to 0.8 V is logic 0 and approximately 3.0 V is logic 1.

Figure 48.17 Transistor Switch

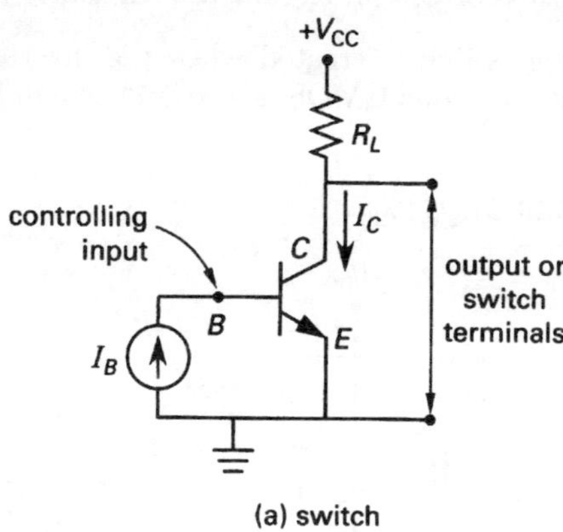

(a) switch

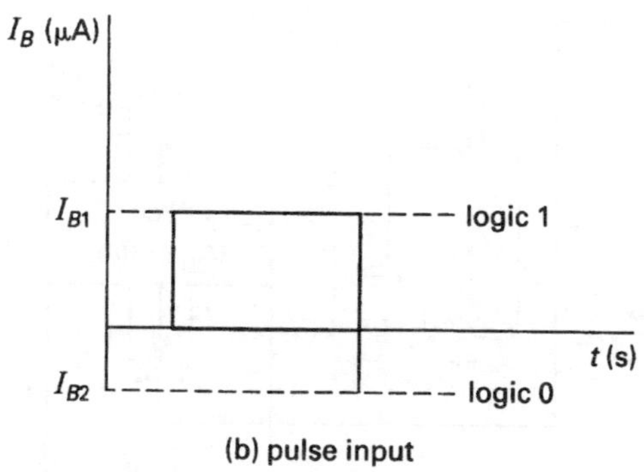

(b) pulse input

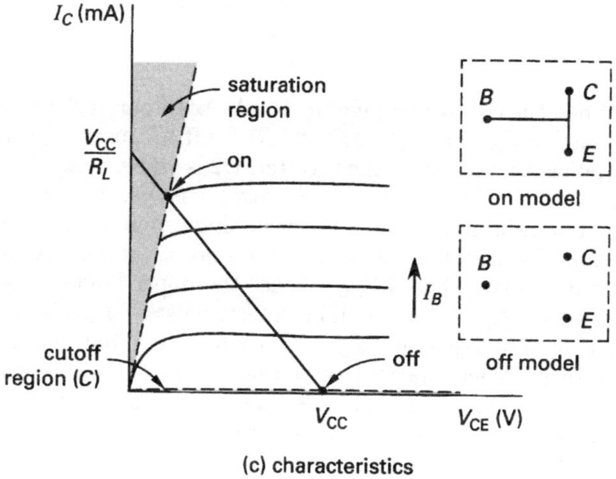

(c) characteristics

shortcircuit model of Fig. 48.17(c). The value of the load resistance must be determined so that when in saturation the condition of Eq. 48.37 is satisfied.[19]

$$I_{B1} \geq \frac{I_{C,\text{sat}}}{h_{FE}} \approx \frac{V_{CC}}{h_{FE} R_L} \qquad 48.37$$

The base current calculated by Eq. 48.37 is the minimum current required to drive the transistor into saturation, ignoring the small collector-emitter voltage. When the transistor is in saturation (on), the base-emitter and collector-base junctions are both forward biased. When the transistor is in cutoff (off), the base-emitter and collector-base junctions are both reverse biased.

[19]At low frequencies, $h_{FE} \approx h_{fe}$, as is the case for all of the small-signal parameters.

Field-effect transistors (FETs) are also used as switches. The gate provides the controlling input. In the on condition, unlike the BJT, the properties of the FET are more like that of a resistor. In the off condition, small leakage currents flow, similar to reverse saturation currents in the BJT.

11. JFET SWITCHES

Junction field-effect transistors used as switches are constructed from depletion-mode *p*-channel types as shown in Fig. 48.18.

Figure 48.18 JFET Switch

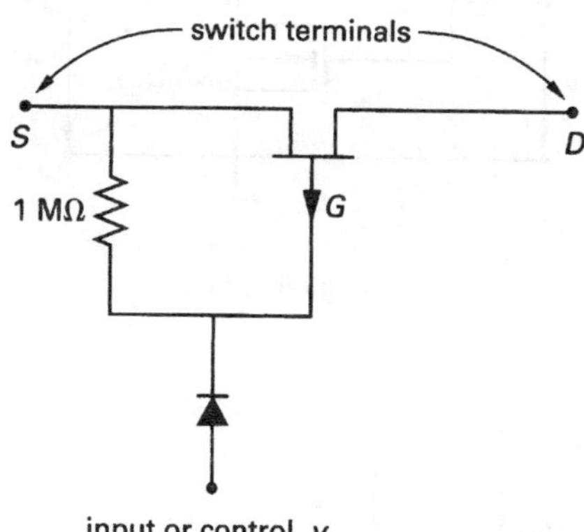

When the control voltage, v_c, is zero, the gate-source voltage, V_{GS}, is equal to zero and the switch is on. This occurs because the gate-channel *pn* junction controls the transistor by varying the depletion region caused by reverse bias (see Sec. 47.2). A positive control voltage results in a positive gate-source voltage, turning the switch off. This positive voltage is the pinchoff voltage, V_{GS}(off). The diode prevents forward bias (i.e., negative voltage) on the gate-channel *pn* junction, which would destroy the switching characteristics of the JFET.[20]

The advantage of the JFET is its low resistance while in the on condition. Disadvantages include complexity and price. Further, the JFET is restricted to switching voltages a few volts less than the control voltages, because anything larger could turn the switch on. Also, if the switching voltage becomes too large while in the on condition, the switch can turn off. That is, the JFET moves from the ohmic region to the pinchoff region (see Sec. 47.2).

[20]The current from the gate would add to that flowing from source to drain. The device would no longer function strictly as a switch controlling external current.

12. CMOS SWITCHES

When a *p*-channel MOS device and an *n*-channel MOS device are fabricated on the same chip, the device is called a *complementary MOSFET* (CMOS).[21] A CMOS device used as an *inverter* or digital switch is illustrated in Fig. 48.19 along with the approximate output characteristics.

Figure 48.19 *CMOS Inverter*

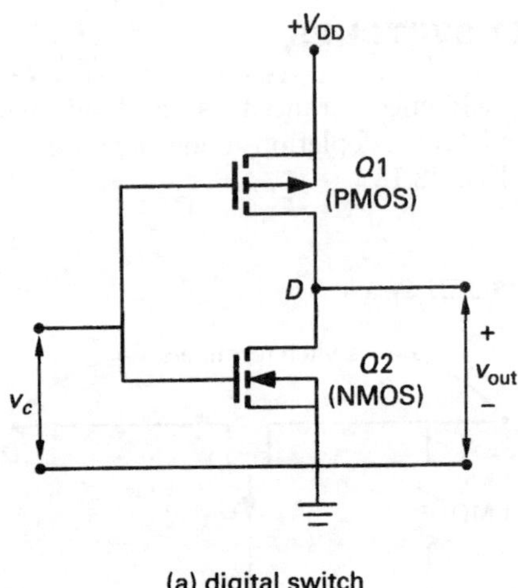

(a) digital switch

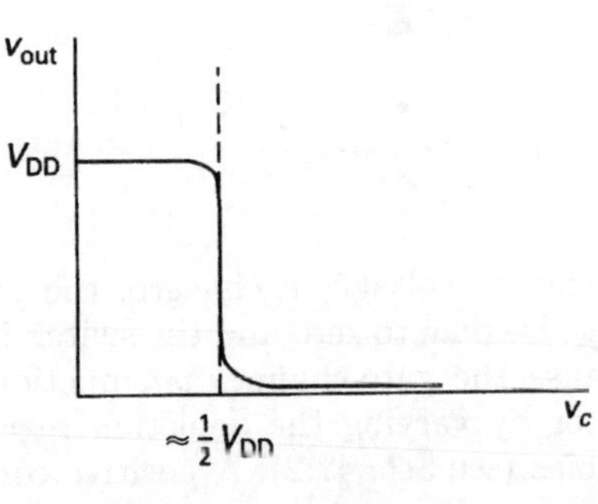

(b) characteristics

The control or input voltage, v_c, varies from 0 V (logic 0) to $+V_{DD}$ (logic 1). When the control voltage is 0 V, the gate-source voltage of $Q1$ is negative and $Q1$ is on. (The negative voltage attracts holes, enhancing the channel and allowing current flow.) The gate-source voltage of $Q2$ is zero and $Q2$ is off. (A positive voltage is required to attract electrons into the channel.) Consequently, $Q1$ can be modeled as a short circuit from source to drain, resulting in $v_{out} = +V_{DD}$ (logic 1). When the control voltage is $+V_{DD}$, the gate-source voltage of $Q1$ is zero and $Q1$ is off. The gate-source voltage of $Q2$ is positive and $Q2$ is on. Consequently, $Q2$ can be modeled as a short circuit from source to drain, resulting in $v_{out} = 0$ V (logic 0). Current flows only during the switching transient. The static current flow and static power are zero, so the power requirements of CMOS circuits are extremely low.

An analog switch is created when the inverter is combined with a second CMOS device as shown in Fig. 48.20.

Figure 48.20 *CMOS Switch*

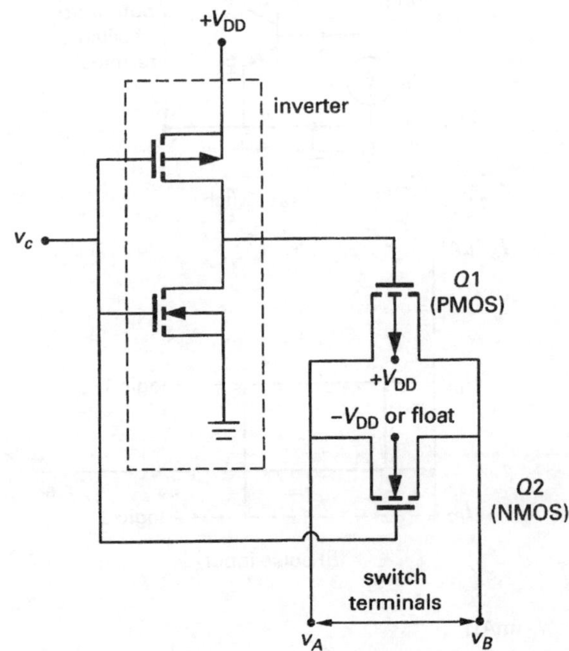

When the control voltage, v_c, is 0 V, both parallel switch channels ($Q1$ and $Q2$) are off. $Q1$ is off because $+V_{DD}$ is applied to its gate by the inverter. $Q2$ is off because 0 V is applied to its gate by the control voltage. In the off condition, restrictions on the switching terminals' voltage exist. For instance, $Q1$ will turn on if both terminals are more positive than $+V_{DD}$—hence the upper limit on the voltage. $Q2$ will turn on if both terminals are negative by an amount equal to V_{GS}—hence a lower limit on the voltage. When the control voltage, v_c, is $+V_{DD}$, $Q1$ will turn on when the switching voltage is positive enough to make V_{GS} negative enough to create the *p*-channel. $Q2$ will turn on when the switching voltage is negative or within a few volts of V_{DD}.

13. ACTIVE WAVEFORM SHAPING

Waveform shaping occurs in many circuits using nonlinear active devices, with and without storage elements (capacitors and inductors). These devices can be discrete or integrated, although integrated circuits rely primarily on capacitors rather than inductors. No single simple theory describes the behavior of all these elements. The analysis is generally done by breaking the circuit into a sequence of linear problems and combining the results.

Examples of such circuits include operational amplifiers with capacitors and inductors (integrators and differentiators, filters), waveform generators (sweep, square,

[21]The design of such circuits is known as *static CMOS implementation.*

triangular), oscillator circuits (monostable, bistable, astable), and so on.

14. LOGIC FAMILIES

Integrated circuits are classified into families based on the circuit elements they contain and the manner in which these elements are combined. *Passive logic circuits* have no transistors (i.e., they have only resistors and diodes), while *active logic circuits* do have transistors.

One of the earliest logic families is *diode-transistor logic* (DTL). *High-threshold logic* (HTL) is similar to DTL but uses higher voltages and better diodes to produce greater noise immunity. *Variable-threshold logic* (VTL) is a variation of DTL with adjustable noise immunity. *Direct-coupled transistor logic* (DCTL) is an unsophisticated connection of transistors with poor immunity to noise and low logic levels. Because there are no resistors, it has low power dissipation. *Resistor-transistor logic* (RTL) is similar to DCTL with the addition of resistors to limit current hogging. *Resistor-capacitor transistor logic* (RCTL) is a low-cost compromise between speed and power. *Transistor-transistor logic* (TTL or T^2L) is the fastest of all logic families using transistors operating in the saturated region. The fastest of all logic families using bipolar junction transistors is *emitter-coupled logic* (ECL), also known as *current-mode logic* (CML). ECL transistors do not operate in the saturation region. ECL circuits consume more power than TTL.

Metal-oxide semiconductor (MOS) circuitry contains MOSFETs. It requires few resistors and has high packing density. *Complementary construction* uses both n-channel and p-channel transistors in the same circuit. NMOS logic uses n-channel MOSFETs, while PMOS uses p-channel MOSFETs. (NMOS is approximately sixtimes slower than PMOS.) *Complementary metal-oxide semiconductor* (CMOS), also known as *complementary transistor logic* (CTL), circuits have high packing density and reliability. They are capable of higher switching speeds than MOS.

Circuitry based on bipolar transistors switches faster, offers greater current drive, and is more suitable for analog applications than CMOS circuitry.[22] On the other hand, CMOS consumes less power, dissipates less heat, and has a higher packing density. Bipolar and CMOS circuitry are combined in biCMOS integrated circuits.

15. LOGIC GATES

A *gate* performs a logical operation on one or more inputs. The inputs (labeled A, B, C, etc.) and output are limited to the values of zero or one. A listing of the output value for all possible input values is known as a *truth table*. Table 48.1 combines the symbols, names, and truth tables for the most common gates. For more information on logical operations, see Chap. 49.

[22]CMOS is equivalent to TTL in speed but slower than ECL.

16. SIMPLIFICATION OF BINARY VARIABLES

The rules of *Boolean algebra* are used to write and simplify expressions of binary variables (i.e., variables constrained to two values). The basic laws governing Boolean variables are as follows.

- *commutative*: $A+B=B+A$; $A\cdot B=B\cdot A$
- *associative*: $A+(B+C)=(A+B)+C$; $A\cdot(B\cdot C)=(A\cdot B)\cdot C$
- *distributive*: $A\cdot(B+C)=(A\cdot B)+(A\cdot C)$; $A+(B\cdot C)=(A+B)\cdot(A+C)$
- *absorptive*: $A+(A\cdot B)=A$; $A\cdot(A+B)=A$

De Morgan's theorems are

$$\overline{A+B}=\overline{A}\cdot\overline{B}$$

$$\overline{A\cdot B}=\overline{A}+\overline{B}$$

The following basic identities are used to simplify Boolean expressions.

$$\begin{array}{lll} 0+0=0 & 0\cdot 0=0 & A+0=A \\ 0+1=1 & 0\cdot 1=0 & A+1=1 \\ 1+0=1 & 1\cdot 0=0 & A+A=A \\ 1+1=1 & 1\cdot 1=1 & A+\overline{A}=1 \end{array}$$

$$\begin{array}{ll} A\cdot 0=0 & -0=1 \\ A\cdot 1=A & -1=0 \\ A\cdot A=A & -\overline{A}=A \\ A\cdot\overline{A}=0 & \end{array}$$

Switching algebra, a form of Boolean algebra, is discussed in Sec. 49.5.

Example 48.13

Simplify and write the truth tables for the following network of logic gates.

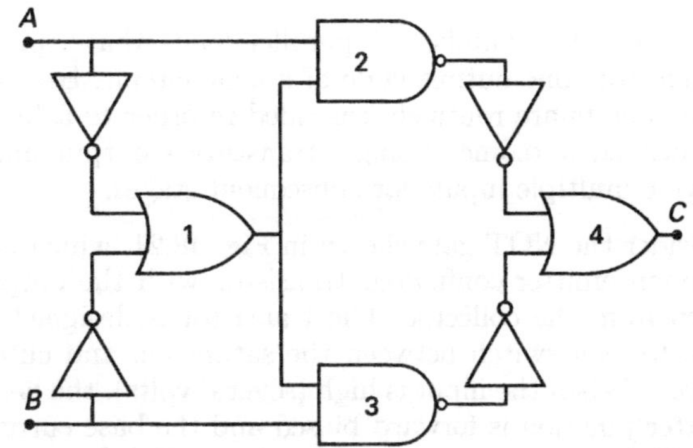

Table 48.1 Logic Gates

inputs		not	and	or	nand	nor	exclusive or
A	B	$-A$ or $\overline{A}$	AB	$A+B$	$\overline{AB}$	$\overline{A+B}$	$A \oplus B$
0	0	1	0	0	1	1	0
0	1	1	0	1	1	0	1
1	0	0	0	1	1	0	1
1	1	0	1	1	0	0	0

Solution

Determine the inputs and output of each gate in turn.

gate	inputs	output
1	$\overline{A}$, $\overline{B}$	$(\overline{A}+\overline{B})$
2	A, $(\overline{A}+\overline{B})$	$\overline{A\cdot(\overline{A}+\overline{B})}$
3	B, $(\overline{A}+\overline{B})$	$\overline{B\cdot(\overline{A}+\overline{B})}$
4	$A\cdot(\overline{A}+\overline{B})$, $B\cdot(\overline{A}+\overline{B})$	$A\cdot(\overline{A}+\overline{B})+$ $B\cdot(\overline{A}+\overline{B})$

Simplify the output of gate 4.

$A\cdot(\overline{A}+\overline{B})+B\cdot(\overline{A}+\overline{B})$	[original]
$A\cdot\overline{A}+A\cdot\overline{B}+B\cdot\overline{A}+B\cdot\overline{B}$	[distributive]
$A\cdot\overline{B}+B\cdot\overline{A}$	[since $A\cdot\overline{A}=0$]
$A\oplus B$	[definition]

The truth table is

A	B	C
0	0	0
0	1	1
1	0	1
1	1	0

17. LOGIC CIRCUIT FAN-OUT

Fan-out is the number of parallel loads that can be driven from one output node of a logic circuit. Because logic circuits are routinely cascaded in order to achieve the desired outcome, a single transistor's output must provide multiple inputs for subsequent stages.

Consider the NOT gate shown in Fig. 48.21, which is a common emitter-configured transistor with the output taken from the collector. The transistor is designed to operate as a switch between the saturation and cutoff regions. When the input is high (several volts), the base-emitter junction is forward biased and the base current is sufficient to saturate the transistor. The output is then low, nearly zero if the small collector-emitter voltage (0.2 V for silicon transistors and 0.1 V for germanium) is ignored. The input has been inverted, that is, changed from high to low—hence the designation as a NOT gate.[23]

Figure 48.21 NOT Gate

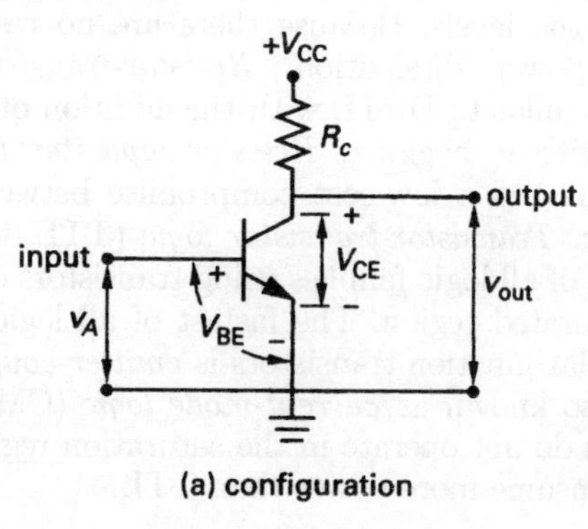

(a) configuration

input → [NOT] → output

(b) logic symbol

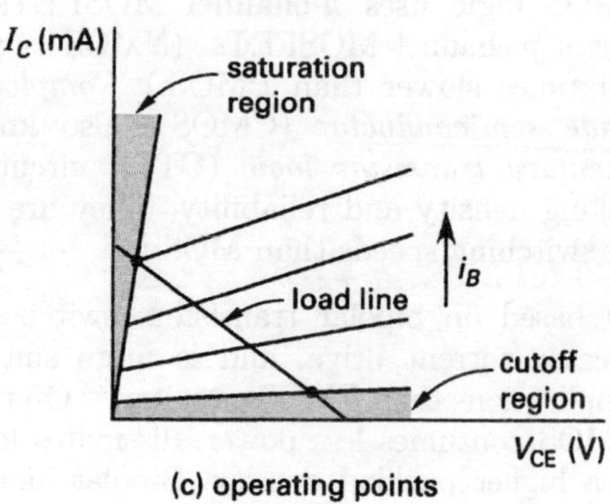

(c) operating points

V_A	V_{out}
L	H
H	L

(d) voltage truth table

[23]The logic described here and used throughout this chapter is *positive logic*, that is, logic where the high voltage is logic condition 1 and the low voltage is logic condition 0. *Negative logic* is the reverse. *Dynamic* or *pulse logic* is a system where the presence of a pulse is logic condition 1 and the absence of a pulse is logic condition 0.

In practical applications, this transistor does not operate in isolation, but instead is cascaded with other logic devices. The base current drawn while in saturation is sourced from previous stages. When in cutoff, the reverse saturation current, I_{CBO}, flowing through the collector-base junction must be sunk by previous stages. This situation is shown in Fig. 48.22.

Figure 48.22 *Fan-Out Unit Loads*

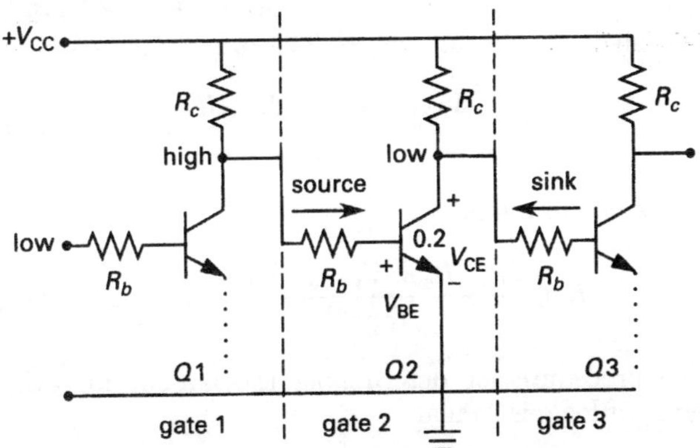

The number of stages a logic gate can source, that is, the current a stage can supply when its collector voltage is high, is called the fan-out. In Fig. 48.22, transistor $Q1$ is sourcing the current to $Q2$. The number of stages that a logic gate can sink, that is, the current the collector (or output circuit) can accept without removing the transistor from the saturated state, is also called the fan-out. In Fig. 48.22, transistor $Q2$ is sinking the current from $Q3$. A given logic family's fan-out specification is stated in terms of *unit loads*. A *unit source current* is the input current drawn by a single gate when the voltage is high. A *unit sink current* is the current drawn from the input of a single gate when the input voltage is low. The fan-out specification is the minimum number of unit loads, source or sink, the gate can provide.

Gate 1 in Fig. 48.22 sources a unit load to $Q2$. $Q1$ is cut off and appears as an open circuit.[24] The unit source current is the base current to $Q2$ given by

$$I_{B2} = \frac{V_{CC} - V_{BE2}}{R_c + R_b} \qquad 48.38$$

$Q2$ is saturated, which causes $Q3$ to be cut off. The reverse saturation current, I_{CBO}, from the collector-base junction of $Q3$ is drawn to the collector of $Q2$. (This sink current raises the voltage at the collector of $Q2$. If too much current is drawn, the voltage rise can reverse-bias the collector-base junction of $Q2$, bringing it out of saturation.) The unit sink current is I_{CBO3}. The collector current is

$$I_{C2} = \frac{V_{CC} - V_{CE,sat}}{R_c} + I_{CBO3} \qquad 48.39$$

[24]Because the threshold voltage is approximately 0.7 V for a silicon base-emitter junction, $Q1$ will be cut off at voltages of 0.6 V or less.

When a gate's function is to maintain multiple loads, the source and sink transistors can be modeled as shown in Fig. 48.23 in order to determine the fan-out.

Figure 48.23 *Fan-Out Circuts*

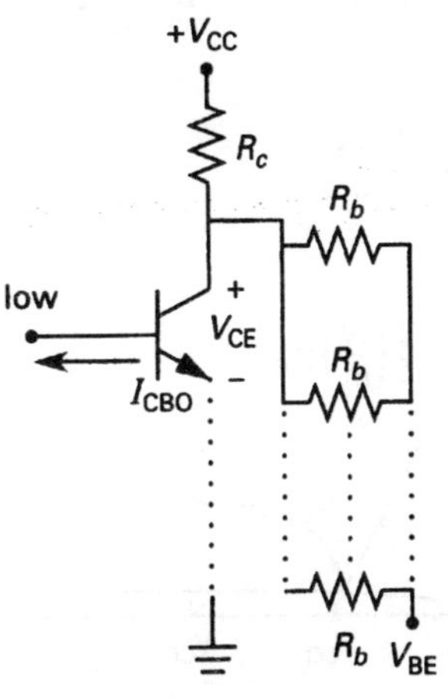

(a) source loads

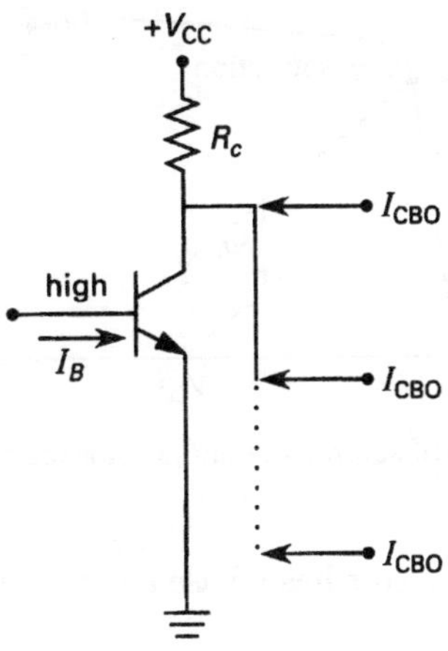

(b) sink loads

For the sourcing condition of Fig. 48.23(a), applying KCL at the collector node gives

$$\frac{V_{CE} - V_{CE}}{R_c} - I_C = N\left(\frac{V_{CE} - V_{BE}}{R_b}\right) \qquad 48.40$$

The forward base-emitter voltage required to hold the load transistors in saturation is typically ≈ 0.7 V, and the reverse saturation is I_{CBO}. The current in the collector resistor, R_c, includes the reverse saturation current, I_{CBO}, of the cutoff transistor.[25]

[25]The collector current is not shown in Eq. 48.40 as the reverse saturation current because this equation is used in a general manner to determine the load line.

The load line equation without sourced loads is shown as a dashed line in Fig. 48.24(a). The load line for a transistor sourcing multiple loads is shown as a solid line in Fig. 48.24(a) and is given by Eq. 48.40 rearranged as

$$I_C + \left(\frac{1}{R_c} + \frac{N}{R_b}\right) V_{\text{CE}} = \frac{V_{\text{CC}}}{R_c} + \left(\frac{N}{R_b}\right) V_{\text{BE}} \qquad 48.41$$

Figure 48.24 Load Lines

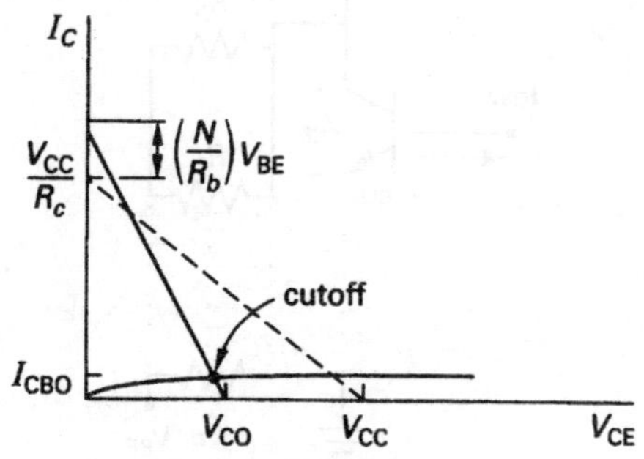

(a) cutoff: multiple source loads

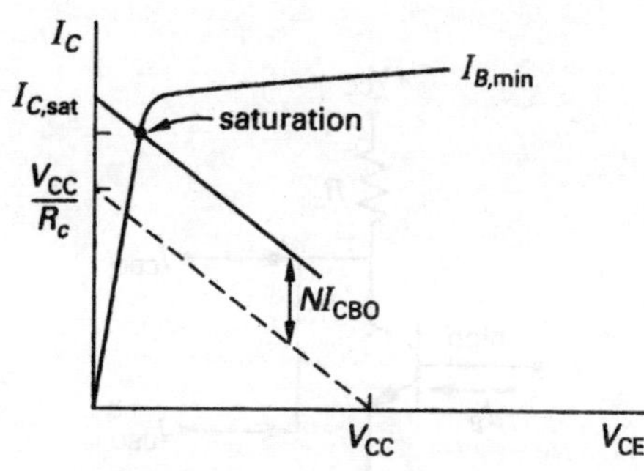

(b) saturated: multiple sink loads

Each load transistor has a base current given by

$$I_B = \frac{V_{\text{CE}} - V_{\text{BE}}}{R_b} \qquad 48.42$$

Substituting the restriction of Eq. 48.42 into the load line equation, Eq. 48.41, gives the sourcing equation for this type of logic gate.

$$I_C + \left(N + \frac{R_b}{R_c}\right) I_B = \frac{V_{\text{CC}} - V_{\text{BE}}}{R_c} \qquad 48.43$$

For the sink condition of Fig. 48.23(b), applying KCL at the collector node gives

$$I_{C,\text{sat}} = \frac{V_{\text{CC}} - V_{\text{CE,sat}}}{R_c} + NI_{\text{CBO}} \qquad 48.44$$

The load lines for the sink condition are taken directly from Eq. 48.44 and are shown in Fig. 48.24(b). Once the minimum collector saturation current is determined by Eq. 48.44 for the sink condition, which depends on the sink fan-out (N), the minimum base current required to maintain the transistor in saturation can be graphically determined from the characteristic curve as shown in Fig. 48.24(b). The base current curve that exists at the point where the load line curve for multiple sinking loads intersects the calculated collector saturated current determines the minimum base current.[26]

Once the equations for the source condition, Eq. 48.43, and sink condition, Eq. 48.44, are determined, the collector current, I_C, is set as the reverse saturation current, I_{CBO}, in the source equation and both equations are solved for the fan-out, N. The *source fan-out* is

$$N + \frac{R_b}{R_c} \leq \frac{V_{\text{CC}} - V_{\text{BE}} - I_{\text{CBO}}R_c}{I_{B,\text{min}}R_c} \qquad 48.45$$

The *sink fan-out* is

$$N \leq \frac{I_{C,\text{sat}}R_c + V_{\text{CE,sat}} - V_{\text{CC}}}{I_{\text{CBO}}R_c} \qquad 48.46$$

The procedure for determining the fan-out for other types of logic is similar.

Example 48.14

Given a transistor with the characteristic shown, $R_c = 1000\ \Omega$ and $V_{\text{CC}} = 5$ V, determine the base current necessary for a fan-out of five.

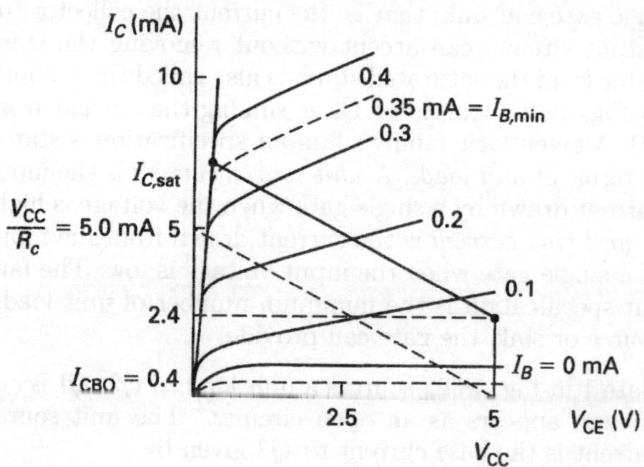

Solution

Because $I_{\text{CBO}} = 0.4$ mA, five load gates require 5×0.4 mA = 2 mA excess current. The corresponding load line is shown solid. The resulting saturation current requires a base current of about 0.35 mA. (Setting the value at 0.4 mA gives a margin of error for temperature effects and variations between transistors.)

Example 48.15

Determine the maximum base resistance, R_b, that will allow sourcing five gates from the transistor gate in Ex. 48.14.

[26]The minimum base current, $I_{B,\text{min}}$, is the base current that a sourcing gate must supply as well as the minimum base current it requires when operating in the sink condition.

Solution

Using Eq. 48.45 with $V_{BE} = 0.7$ V and $I_{B,min}$ set at 0.4 mA,

$$N + \frac{R_b}{R_c} \le \frac{V_{CC} - V_{BE} - I_{CBO}R_c}{I_{B,min}R_c}$$

$$5 + \frac{R_b}{1000\ \Omega} \le \frac{5\ \text{V} - 0.7\ \text{V} - (0.0004\ \text{A})(1000\ \Omega)}{(0.0004\ \text{A})(1000\ \Omega)}$$

$$R_{b,max} = 4750\ \Omega$$

18. LOGIC CIRCUIT DELAYS

Logic circuits do not switch instantly from cutoff to saturation, or vice versa, upon application of an input pulse. Various delays are inherent, as shown in Fig. 48.25.

Figure 48.25 *Logic Circuit Delays*

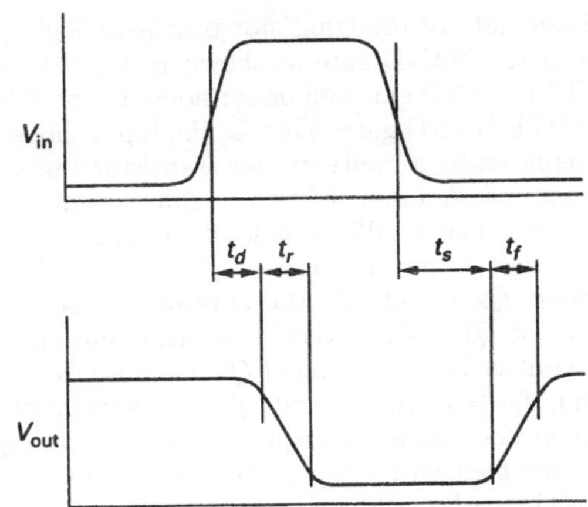

When the input voltage level (pulse) rises, a finite time, called the *delay time*, t_d, is required for the base-emitter junction of the transistor in the inverter circuit to reach the turn-on level. The time for the input voltage to rise enough for the collector output voltage to fall to the saturation level is called the *rise time*, t_r.

With the transistor in saturation, both the base-emitter and collector-base junctions are forward biased. Excess charge is stored in the base region, that is, the space-charge region is small. When the input voltage level (pulse) drops, a finite time is required to remove the excess charges in the base. This delay is called the *storage time*, t_s. Once the excess charge is removed, the collector current falls. This time is called the *fall time*, t_f.

The minimum time for a transistor to pass a pulse, that is, the time from cutoff to saturation and back to cutoff, or vice versa, is the sum of all the delays. The maximum frequency at which a transistor can operate is the reciprocal of the sum of these delays.

The sum of the delays is called the ***propagation delay*** for a given logic gate. Approximate fan-out and propagation delay values for various transistor logic families, listed left to right in the approximate order of development, are given in Table 48.2.[27]

19. RESISTOR-TRANSISTOR LOGIC (RTL)

The basic gate of the resistor-transistor logic (RTL) family is the NOR gate as shown in Fig. 48.26. The inverter used to describe fan-out in Sec. 48.17 is the basic building block. Although a ***fan-in*** of two is shown in Fig. 48.26, multiple inputs are possible. Logic 1 (high voltage) at any input saturates the associated transistor, causing the output to be connected to ground across the small collector-emitter voltage, resulting in a logic 0 (low voltage) output.

Figure 48.26 *Resistor-Transistor Logic*

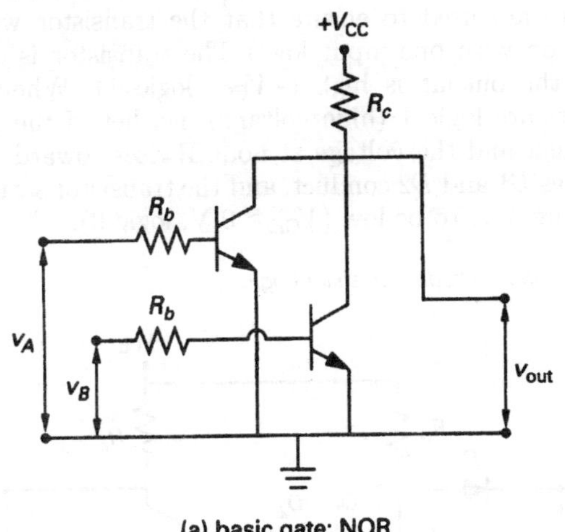

(a) basic gate: NOR

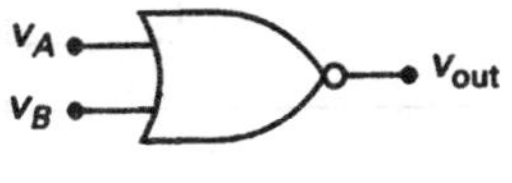

(b) logic symbol

V_A	V_B	V_{out}
L	L	H
L	H	L
H	L	L
H	H	L

(c) voltage truth table

[27]One method of designing to handle such delays is by using *clocked logic*, that is, logic controlled by a clock so that the flow of signals follows a known parameter.

Table 48.2 *Logic Family Data*

parameter	RTL	DTL	HTL	TTL	ECL	MOS	CMOS
basic gate[a]	NOR	NAND	NAND	NAND	OR-NOR	NAND	NOR OR NAND
fan-out[b]	5	8	10	10	25	20	> 50
propagation delay[c]	12	30	90	10	0.1	10	0.1

[a]Positive logic is assumed when determining the basic gate.
[b]Worst-case condition.
[c]Approximate values in nanoseconds.

20. DIODE-TRANSISTOR LOGIC (DTL)

The basic gate of the diode-transistor logic (DTL) family is the NAND gate shown in Fig. 48.27. A fan-in of two is shown, but multiple inputs are possible. When either input is logic 0 (low voltage), the associated diode, *D*, conducts, resulting in a low voltage at node B. The base current to the transistor will be logic 0 because any current flow is blocked by diodes *D*1 and *D*2. (Two diodes are used to ensure that the transistor will not turn on with one input low.) The transistor is cut off and the output is high ($+V_{CC}$, logic 1). When both inputs are logic 1 (high voltage), neither of the diodes conduct and the voltage at node B rises toward $+V_{CC}$. Diodes *D*1 and *D*2 conduct, and the transistor saturates, causing V_{out} to be low ($V_{CE} \approx 0$ V, logic 0).

Figure 48.27 *Diode-Transistor Logic*

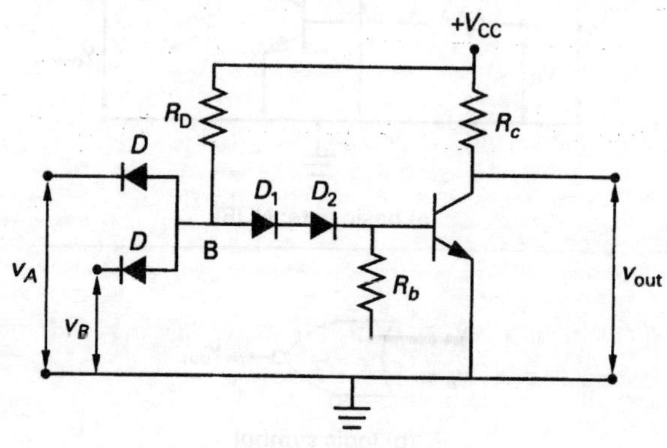

(a) basic gate: NAND

V_A V_B V_{out}

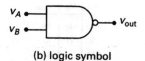

(b) logic symbol

V_A	V_B	V_{out}
L	L	H
L	H	H
H	L	H
H	H	L

(c) voltage truth table

DTL uses less current than RTL when sourcing loads, but more current when sinking loads. Zero current is required to source a load using DTL logic because of the diodes. *High-threshold logic* (HTL) is similar but uses higher voltages and better diodes to improve noise performance.

21. TRANSISTOR-TRANSISTOR LOGIC (TTL OR T²L)

The basic gate of the transistor-transistor logic (TTL) family is the NAND gate as shown in Fig. 48.28. The basic TTL NAND gate can be considered a modification of the DTL NAND gate. That is, the input diodes have been replaced by a multi-emitter transistor's base-emitter junctions. A fan-in of two is shown, but multiple inputs are possible. When at least one input is logic 0 (low voltage), the emitter of *Q*1 is forward biased. Assuming *Q*2 is cut off, the current supplied to the collector of *Q*1 is the reverse saturation current, I_{EBO}, of the emitter-base junction of *Q*2. Because this current is small, $I_{B1} > I_{EBO}/h_{FE}$ and *Q*1 is in saturation. The voltage at node B2 is then equal to the input voltage, v_A or v_B, summed with $V_{CE,sat}$, that is, about 0.4 V to 0.8 V. This voltage is not enough to forward bias *Q*2 and *Q*3, because about 0.7 V apiece is required. *Q*3 is cut off and the output is high ($+V_{CC}$, logic 1). When all the inputs are at logic 1 (high voltage), the emitter base junction of *Q*1 is reverse biased. *Q*1 is cut off and the base-collector junction is forward biased by approximately $+V_{CC}$. This voltage is high enough to forward bias the base-collector junction of *Q*1 and the base-emitter junctions of *Q*2 and *Q*3, about 2.1 V. So, the output is low ($V_{CE3} \approx 0$ V, logic 0).

TTL logic is the fastest-saturating logic gate. The fall time of TTL circuits is improved by reducing the collector current in the saturated state. This is accomplished by using a *totem-pole configuration*, also called *active pull-up*, which uses an additional transistor and diode in series with R_{C3}.[28] Additionally, the collector of *Q*3 can be left open, in which case it is called the *open-collector TTL gate*. The open-collector TTL gate is used with *wired logic* or *collector logic*, because the collector outputs are wired to the next gate's output without intervening circuitry. This allows the logic voltage level to be

[28]This configuration is called the totem pole because the additional transistor sits on top of *Q*3.

Figure 48.28 *Transistor-Transistor Logic*

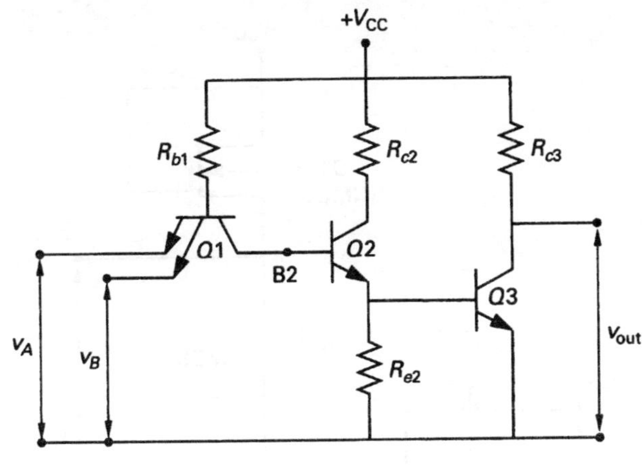

(a) basic gate: NAND

v_A
v_B
v_{out}

(b) logic symbol

v_A	v_B	v_{out}
L	L	H
H	L	H
L	H	H
H	H	L

(c) voltage truth table

changed. For example, in Fig. 48.28, the driving voltage is $+V_{CC}$. If the collector of $Q3$ is open, that is, not connected to $+V_{CC}$ via R_{c3}, then a power source can be attached to the collector of $Q3$, providing a different voltage logic, say V'_{CC}. (One can connect a +5 V logic circuit to a +12 V logic circuit without damaging the components.)

A type of logic with operational similarities to the open-collector logic is *tri-state logic.* In this case, the output of the transistor can be logic level 0, level 1, or level Z. The Z-level is a high-impedance level at which the output is completely disconnected from the rest of the circuit. This is accomplished by adding an enable (inhibit) circuit to the output, similar to the circuit used on clocked flip-flops to control sequencing (see Chap. 54). The high-impedance level allows multiple circuits to connect to the same output line or bus so that several circuits can timeshare a bus. Only the active circuit controls the bus. The others are in the Z-level state and have no impact on the voltage level, or signal, on the bus. Circuits with this tri-state logic are called tri-state buffers.

22. EMITTER-COUPLED LOGIC (ECL)

The basic gate of the emitter-coupled logic (ECL) family is the OR-NOR gate shown in Fig. 48.29. The basic building block of the family is the difference amplifier ($Q2$ and $Q3$ in Fig. 48.29). The transistors in this logic family do not saturate. Instead, they remain in the active region as the input voltages, v_A and v_B, operating above and below (logic 1 and logic 0) the reference voltage, V_{ref}. Because the emitter current through R_e remains approximately constant, when both inputs are logic 0 (i.e., below the reference voltage) the emitter currents from $Q1$ and $Q2$ are minimal and the emitter current of $Q3$ is maximum. This causes the voltage output at v_{OR} to be low (logic 0) and the voltage output at v_{NOR} to be high (logic 1). When either or both of the inputs are logic 1 (i.e., above the reference voltage) the emitter currents of $Q1$ and $Q2$ are maximum, and the emitter current of $Q3$ is minimal. This causes the voltage output at v_{OR} to be high (logic 1), and the voltage output at v_{NOR} to be low (logic 0).

Figure 48.29 *Emitter-Coupled Logic*

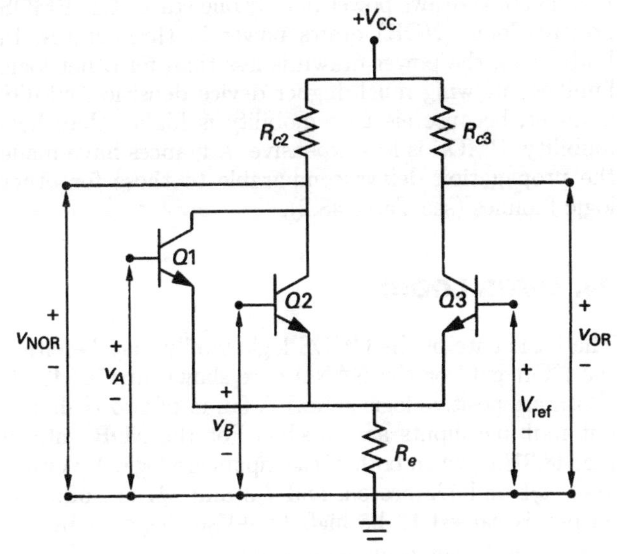

(a) basic gate: OR–NOR

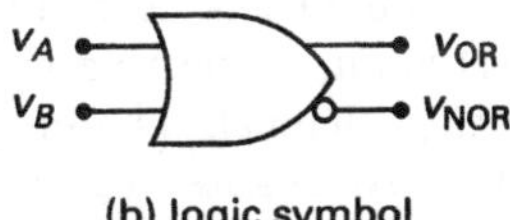

(b) logic symbol

v_A	v_B	v_{OR}	v_{NOR}
L	L	L	H
H	L	H	L
L	H	H	L
H	H	H	L

(c) voltage truth table

ECL improves on the propagation delay of TTL by not driving the transistors into saturation. This minimizes the switching time, because charge carriers do not have to be cleared from the base junction. However, the power requirements of ECL are greater than TTL (see Table 48.2). Emitter-coupled logic is also called *current-mode logic* (CML).

23. MOS LOGIC

The basic gate of the MOS logic family is the NAND gate as shown in Fig. 48.30.[29] The basic building block of the family is the inverter ($Q1$ and $Q2$ in Fig. 48.30). A fan-in of two is shown, but multiple inputs are possible. $Q1$ is always biased on by the drain power supply, $+V_{DD}$. When either input is logic 0 (low voltage), the associated transistor, $Q2$ or $Q3$, is off. No current flows and v_{out} is high ($+V_{DD}$, logic 1). When both of the inputs are logic 1 (high voltage), both transistors are on. Current flows and the output voltage is low ($v_{out} \approx 0$ V, logic 0).

MOS logic is simple and requires no external resistors or capacitors, making it well suited for realization in integrated circuit form. The NMOS positive logic shown in Fig. 48.30(a) draws power in only one state. The PMOS positive logic (NOR) draws power in three states. In both cases, the power drawn is less than for other logic families, allowing much higher device densities. NMOS is faster, because electron mobility is higher than hole mobility. PMOS is less expensive. Advances have made the propagation delays comparable to those for other logic families (see Table 48.2).

24. CMOS LOGIC

The basic gate of the CMOS logic family may be either the NOR gate or the NAND gate shown in Fig. 48.31. (Both are positive logic gates.) A fan-in of two is shown, but multiple inputs are possible. For the NOR gate in Fig. 48.31(a), when both of the inputs are logic 0 (voltage low), $Q1$ and $Q2$ are on, and $Q3$ and $Q4$ are off. The output is caused to be high ($\approx +V_{DD}$, logic 1) by the connection to the drain supply via the conducting transistors $Q1$ and $Q2$. When either or both of the inputs are logic 1 (voltage high), one or both of the upper transistors ($Q1$ and/or $Q2$) is off. One or both of the lower transistors ($Q3$ and/or $Q4$) is on, connecting the output to ground via the conducting channel. The output is low (≈ 0 V, logic 0). For the NAND gate, a logic 0 (low voltage) on one or both inputs causes one or both upper transistors ($Q1$ and/or $Q2$) to conduct, connecting the output to the drain supply voltage. The output is high ($\approx +V_{DD}$, logic 1). If both of the inputs are logic 1 (high voltage), the lower series transistors, $Q3$ and $Q4$, turn on and connect the output to ground. The output is low (≈ 0 V, logic 0).

CMOS logic requires only a single supply voltage and is relatively easy to fabricate. The only input current that flows is that required to charge the gate-channel capacitances and any leakage through the off transistor. The power supply consumption is extremely low. Because of the low power consumption, the fan-out is very high.

[29]This is the basic gate for positive logic NMOS. For positive logic PMOS, the basic gate is the NOR gate.

Figure 48.30 MOS Logic

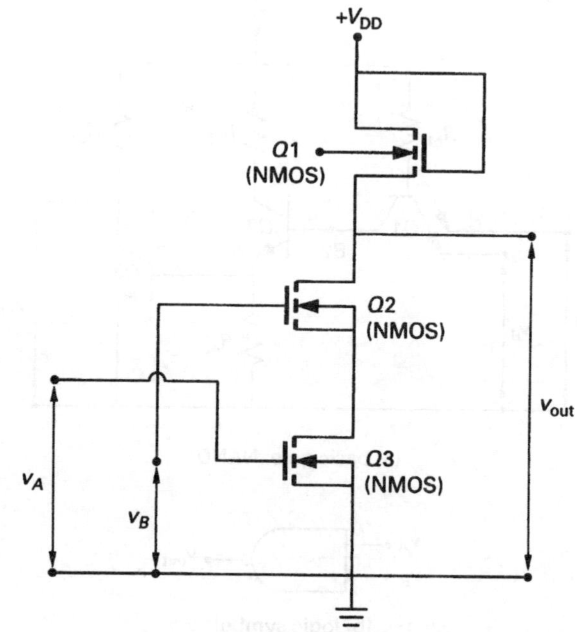

(a) basic gate: NAND

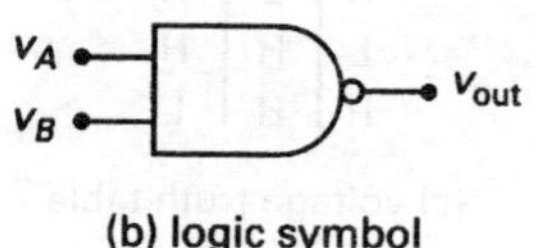

(b) logic symbol

V_A	V_B	V_{out}
L	L	H
L	H	H
H	L	H
H	H	L

(c) voltage truth table

Speeds are comparable to that for ECL (see Table 48.2). Consequently, CMOS has the best overall properties of any logic family.

25. MULTIVIBRATORS

Multivibrators are pulse generator circuits that produce rectangular-wave outputs. There are three types.

The *astable multivibrator*, also called the free-running multivibrator, has no particular stable state and generates a continuous flow of pulses without input, as shown in Fig. 48.32(a). The astable multivibrator may be used as a computer or digital system clock. The operational amplifier circuitry realization of an astable multivibrator is shown in Fig. 48.33(a).

Figure 48.31 CMOS Logic

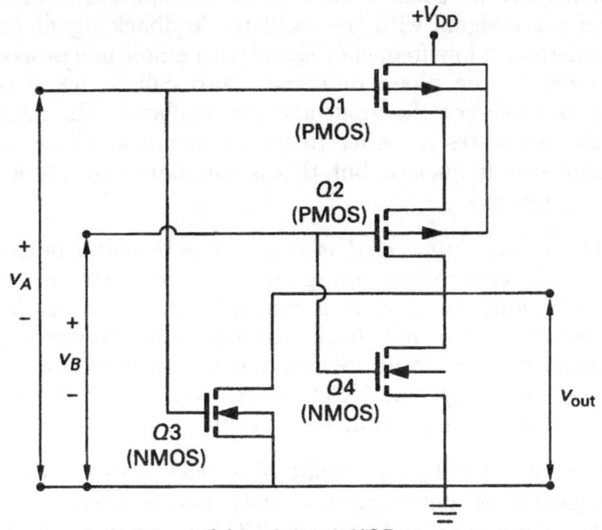

(a) basic gate: NOR

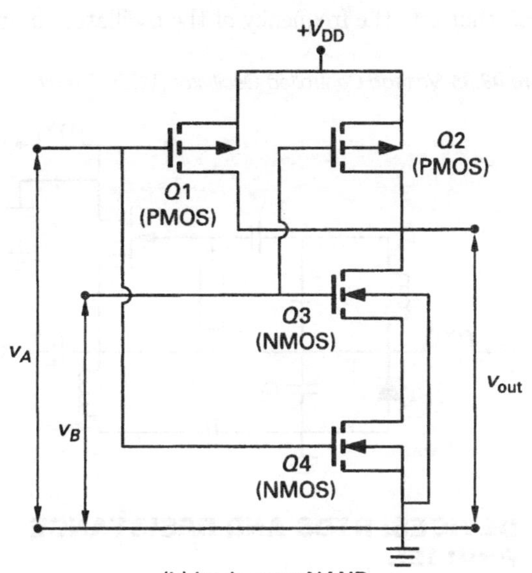

(b) basic gate: NAND

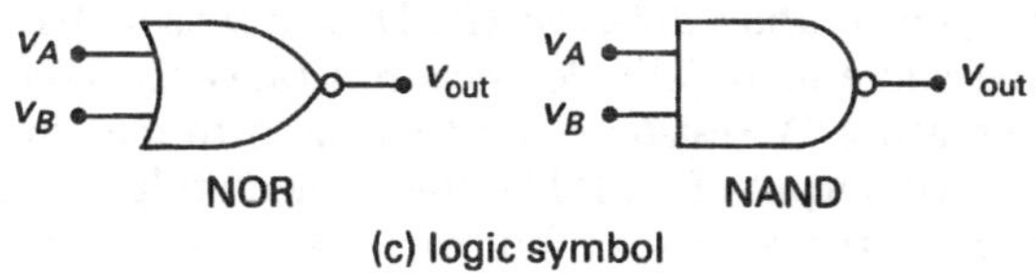

(c) logic symbol

v_A	v_B	v_{out}
L	L	H
L	H	L
H	L	L
H	H	L

NOR

v_A	v_B	v_{out}
L	L	H
L	H	H
H	L	H
H	H	L

NAND

(d) voltage truth tables

Figure 48.32 Multivibrator Types

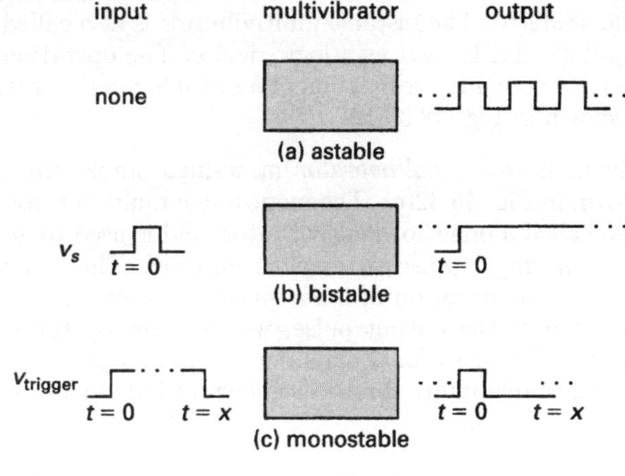

Figure 48.33 Multivibrator [Op Amp] Circuits

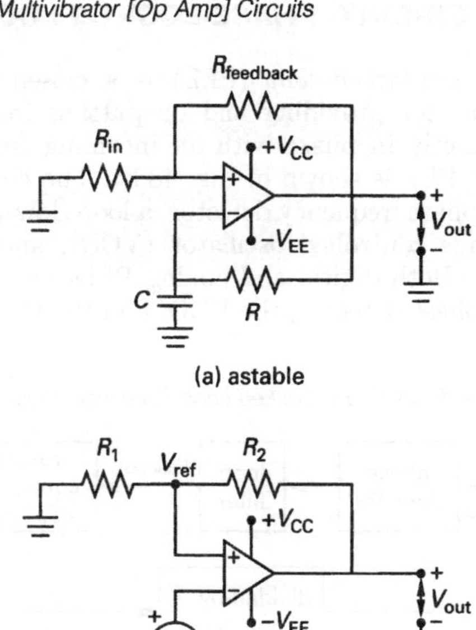

(a) astable

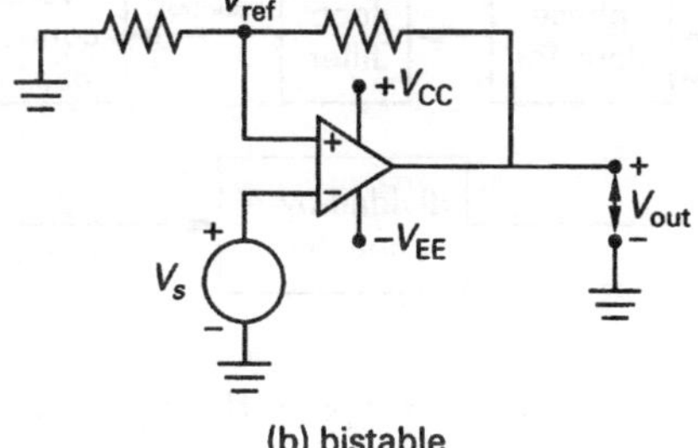

(b) bistable

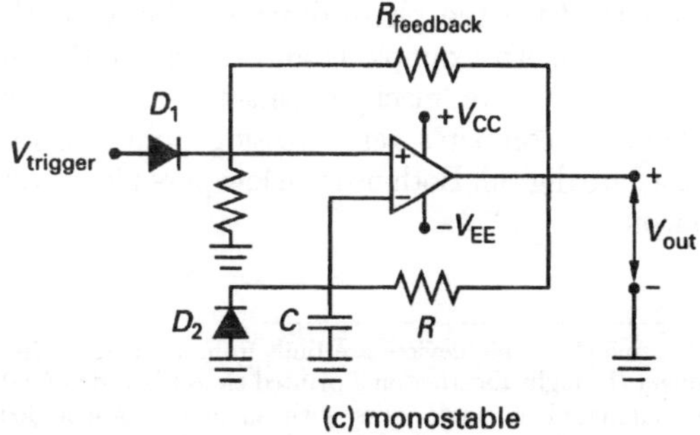

(c) monostable

The *bistable multivibrator* has two stable states and generates either a LOW or HIGH output as shown in Fig. 48.32(b). The bistable multivibrator is also called a flip-flop, and is used as a logic device. The operational amplifier circuitry realization of a bistable multivibrator is shown in Fig. 48.33(b).

The *monostable multivibrator* has a single stable state as shown in Fig. 48.32(c). The monostable multivibrator is also called a one-shot multivibrator, and is used to generate a single pulse of known duration. This occurs because the duration of input pulse is unrelated to the duration of the output pulse, which is set by the circuitry. The operational amplifier circuitry realization of a monostable multivibrator is shown in Fig. 48.33(c).

PART 4: CIRCUITS AND DEVICES[30]

26. CIRCUIT: PHASE-LOCKED LOOP (PLL)

A *phase-locked loop* (PLL) is a closed-loop control circuit for providing and outputting frequency that is exactly in phase with an incoming frequency.[31] A basic PLL is shown in Fig. 48.34. The circuit consists of a phase frequency detector, a loop (low-pass) filter, a voltage controlled oscillator (VCO), and a feedback path. Both digital and analog PLLs have in common the phase detector, the VCO, and the feedback path.

Figure 48.34 *Phase-Locked Loop Functional Block Diagram*

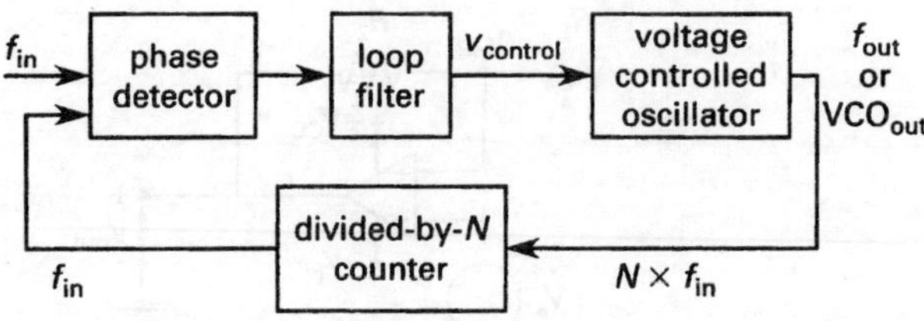

The phase detector compares the input frequency with the feedback frequency and outputs an error (or corrective) signal to the oscillator (via the loop filter) that then responds to drive the phase difference between the two inputs to zero. An example is an exclusive OR gate or a more complex state machine called a *phase frequency detector* that compares zero crossing points. Both such detectors are digital. Both use the low pass filter to smooth the output.

An example analog phase detector is the *four-quadrant multiplier* or *balance mixer* that multiplies the input reference signal with the oscillator feedback signal and generates a low-frequency signal with amplitude proportional to the phase difference (also called the *phase error*) between the input and the oscillator. The mixer also generates a higher frequency signal at twice the oscillator frequency, but this is eliminated by the low pass filter.

The *voltage controlled oscillator* uses the filter output and the controlling voltage input to generate an output frequency that is a multiple of the input frequency. Once fed back through the divide-by-N circuitry, the result drives the phase detector and filter to change the VCO voltage input until the frequencies are aligned at the output.

A digital VCO may consist of a clock source (crystal oscillator or multivibrator circuit), two counters, and a digital comparator. An analog VCO is shown in Fig. 48.35. The VCO uses the LRC *tank circuit* or *antiresonant circuit* that sets the frequency of the oscillator circuit.

Figure 48.35 *Voltage Controlled Oscillator (VCO) Circuit*

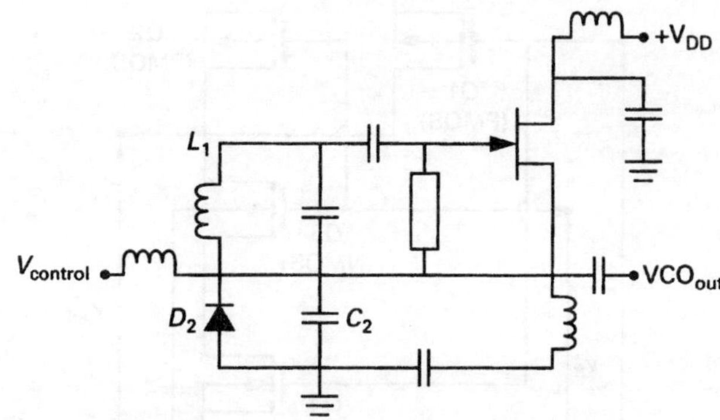

27. DEVICES: RTDS AND RESISTANCE BRIDGES

A *resistance temperature device*, also known as a *resistance temperature detector*, (RTD) is a sensor that uses changes in a material's electrical resistance to determine temperature. A *resistance bridge* is used to measure the unknown output of an RTD. Resistance bridges can use two wires, three wires, or four wires. A two-wire bridge is used when a single element of the bridge varies; for example, when the RTD is the varying element. (See Fig. 48.36.) A three-wire bridge is used when a remotely located RTD (or thermistor) could experience lead resistance and noise pickup. (See Fig. 48.37.) A four-wire bridge provides additional accuracy and compensation for lead resistance issues, although normally these improvements are not worth the additional cost. Using operational amplifiers in conjunction with these bridges results in higher output levels and improved power dissipation, while minimizing output noise, distortion, and clipping. The bridge is in balance when the voltage at the output is zero, that is, when no current flows in an indicating device attached to the output terminals.

[30]While some electronic devices are built from an array of individual components brought together on a printed circuit board (PCB), many can be purchased as single units with all functions included. Such devices are called *standard modular devices*.

[31]A mechanical analogy is tuning a piano. A tuning fork is used to provide a reference frequency. The piano string is adjusted until the beat frequency is inaudible. Another example is the manual transmission in an automobile. When being shifted, gears on the engine side and wheel side are turning at different rates. The rates must be matched before the shift is completed or a grinding sound is heard.

When balanced, the relationship between the resistances is

$$\frac{R_{L1} + R_{\text{RTD}} + R_{L2}}{R_3} = \frac{R_2}{R_4} \qquad 48.47$$

Figure 48.36 Two-Wire Resistance Bridge

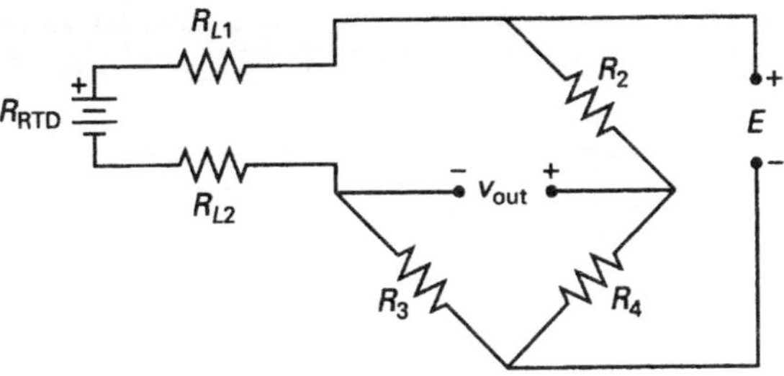

Figure 48.37 Three-Wire Resistance Bridge

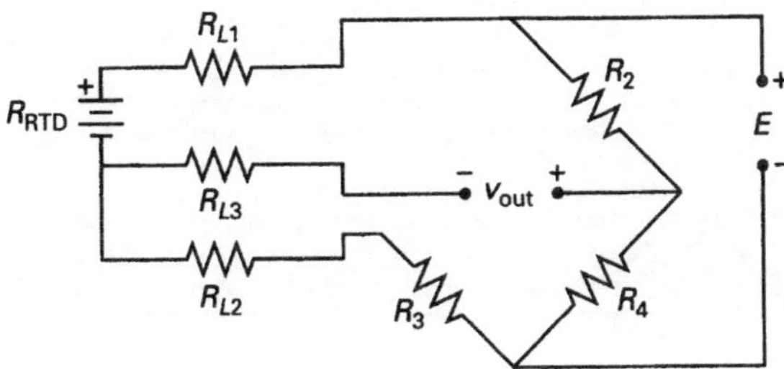

The power supply and bridge resistances are normally located within control circuitry that can be tens or hundreds of feet from the RTD. Wire resistances can be significant, and temperature-driven resistance changes in the two wires further degrade accurate detection of temperature changes. The result is that the bridge has a nonzero output voltage, which causes an imbalance in the bridge. If the imbalance is small, it can be subtracted electronically from the output signal and the correct result obtained. If a large imbalance exists, the measurement range of the RTD (or other connected instrument) may be compromised.

The three-wire bridge shown in Fig. 48.37 is symmetrical along a line that runs through the bridge output corners. The negative of the output moves electronically from the top of R_3 to the bottom of the RTD. Two lead wires, R_{L1} and R_{L2}, are on opposite arms of the bridge circuit. If the two wires are the same type and length, their resistances are equal and the bridge is in balance. Assuming both wires are run together, temperature changes in both wires will not unbalance the bridge because the effects are balanced in each wire. R_{L3} is a voltage-sensing wire and it provides the input signal at v_{out}. It carries a small current (in the microampere range in many cases) and is not in series with either bridge arm. It has no impact on the balance of the bridge circuit.

Electronics

Example 48.16

A three-wire bridge with equal wire lengths leads to a sensor. How does the system remain balanced regardless of the lead resistance value?

Solution

Simplify the resistance relationship. From Fig. 48.37, the balanced (or null) relationship is

$$\frac{R_{L1} + R_{\text{RTD}}}{R_{L2} + R_3} = \frac{R_2}{R_4}$$

$$\frac{R_{L1}\left(1 + \dfrac{R_{\text{RTD}}}{R_{L1}}\right)}{R_{L2}\left(1 + \dfrac{R_3}{R_{L2}}\right)} = \frac{R_2}{R_4}$$

Since $R_{L1} = R_{L2}$, the values cancel. The resistance of each is known as R.

$$\frac{1 + \dfrac{R_{\text{RTD}}}{R}}{1 + \dfrac{R_3}{R}} = \frac{R_2}{R_4}$$

The numerator and denominator on the left-hand side of the equation represent new resistance values, R'_{RTD} and R'_3.

$$\frac{R'_{\text{RTD}}}{R'_3} = \frac{R_2}{R_4}$$

Any changes in the value of R affect the numerator and denominator ratios equally. Only the change in resistance from the RTD impacts the balance.

Digital Logic

Nomenclature

$f(x)$	scalar function of x	–
$F(X)$	vector function of X	–
m	minterm	–
M	maxterm	–
n	number	–
N	total number of possible values	–

1. DIGITAL INFORMATION REPRESENTATION

Computers process information in digital form. The simplest representation of digital information is that of a single binary variable called a *scalar*. The notation for a scalar is x_i. A *binary scalar* takes on a value of zero or one, and represents one bit of information. The smallest information storage unit is termed a *cell*.

At the electronic device level, a zero or one is defined in terms of a voltage output specific to the logic family chosen. For transistor-transistor logic (TTL), the range of values of the zero and one, called the *state* of the cell, is shown in Fig. 49.1. In TTL, zero is represented by the voltage range 0–0.8 V, while one is represented by the voltage range 2.0–5.0 V.

Figure 49.1 *Definition of Binary States for TTL*

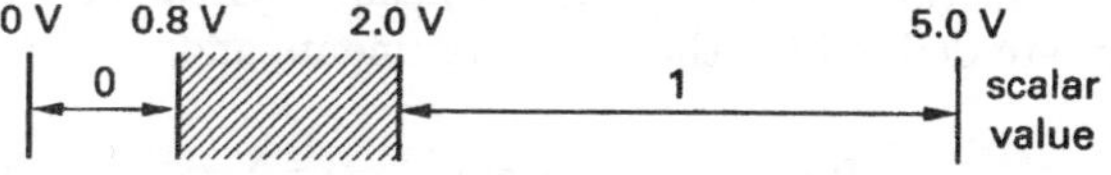

A grouping of n binary variables is called an *n-tuple* and is represented by $[x_1, x_2, x_2, \ldots, x_n]$. The value of n is called the *word length*. The cells can be listed and numbered left to right or right to left. Left to right is the normal notation. An exception occurs when representing a binary number, in which case the cells are listed right to left, from zero to $n-1$, corresponding to $2^0, 2^1, \ldots, 2^{n-1}$. Because each binary variable in the n-tuple takes on only one of two values, the total number of values that can be represented, N, is

$$N = 2^n \qquad 49.1$$

An ordered collection of cells is called a *register*. The information within each register is encoded to represent numerical, logical, or character information. The variables within the register are called *state variables* because they define the value (state) of the register (vector). Particular values within a register are called *states of the register*. For example, a 2-tuple register has four states: $[0, 0]$, $[0, 1]$, $[1, 0]$, and $[1, 1]$.

Example 49.1

A certain computer uses a 32-bit word length. What is the maximum number of possible state variables?

Solution

The maximum number of state variables is given by Eq. 49.1.

$$N = 2^n = 2^{32} = 4{,}294{,}967{,}296$$

2. COMPUTER NUMBER SYSTEMS

Several types of number systems other than base 10 find application in computer systems. These include the binary, octal, and hexadecimal number systems. These systems and their manipulation are described in Chap. 13.

3. ELECTRONIC LOGIC DEVICE LEVELS AND LIMITS

Logic one is considered to indicate a true condition. Logic zero indicates a false condition. Logic one is normally associated with the higher voltage in an electronic circuit, and logic zero is associated with the lower voltage. This type of association is called *positive logic*. When the low-voltage condition represents the true statement, the logic is called *negative logic*. In order to avoid confusion, truth tables are sometimes given in terms of high (H) and low (L) voltages, rather than one and zero.

An alternative is to use *assertion levels*. When the assertion (true) condition is the high-voltage level, the term *active high* is used. This corresponds to positive logic.

When the assertion (true) condition is the low-voltage level, the term *active low* is used

In practice, the type of device and its limits need to be considered. Depending on the circuit, the high- and low-voltage levels will be represented by a range of values. Also, the fan-in and fan-out capabilities limit the number of variable inputs and the number of logic gates that can be attached to the output. Additionally, the propagation delays must be accounted for to ensure that levels of logic have time to respond, providing the proper output before the next set of variables arrives. The propagation delays, therefore, determine the operating speed of the logic network. Timing diagrams are used to map these delays and ensure proper operation.

4. FUNDAMENTAL LOGIC OPERATIONS

Information to be processed is combined with, and compared to, other information. The logic that does this in such a manner that the output depends only on the value of the inputs is called *combinational logic.*[1] *Logic variables* and *logic constants* have one of two possible values, zero or one. By convention, zero is considered "false," and one is considered "true."

Logic functions are commonly represented in *truth tables.* Because an n-tuple can take on 2^n values, a truth table has 2^n rows. Each of these rows has an associated output that can be either zero or one. Combining these two facts gives the number of scalar functions that can be generated from n logic variables. A logic operator involving two operands is called a *binary operator.* For a two-variable input, that is, a 2-tuple, there are 16 possible functions.

$$\text{no. of scalar functions} = (2)^{2^n} \qquad 49.2$$

Not all the scalar functions are useful. In fact, only six are of primary interest. Further, all logic functions can be represented in terms of only two of these, NAND and NOR. These often-used functions are called *logic operators.* A logic operator involving a single variable or operand is called a *unary operation.* A unary function is shown in Table 49.1.

Table 49.1 *Unary Function*

x	$f_1(x)=0$	$f_2(x)=x$	$f_3(x)=\overline{x}$	$f_4(x)=1$
0	0	0	1	1
1	0	1	0	1

The function $f_2(x)$ is called the *identity function.* The function $f_3(x)$ is called the NOT, or negation, function. The symbol and truth table for the NOT function are shown in Fig. 49.2.

[1]When the output has a memory, that is, when the output is used as feedback to the input, the logic is called *sequential.*

Figure 49.2 *NOT Logic*

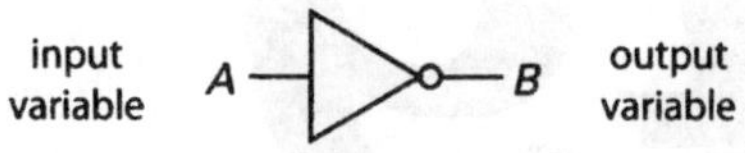

input variable A —▷o— B output variable

(a) symbol

A	B
0	1
1	0

(b) truth table

$$B = \overline{A}$$

(c) equation

Six logic operators (or gates) are used so often that they have been given their own symbols: AND, OR, XOR, NAND, NOR, and XNOR (or coincidence). The X indicates the word "exclusive," while N indicates the word "not." The logic operator AND is indicated by "·" or by "∧." The dot is commonly not used when all the variables are represented by single letters. (For example, $A \cdot B$ can be written as AB.) The logic operator OR is indicated by "+" or "∨." The small circle on the end of any gate symbol, or on the input or output, indicates the NOT function. The standard symbol for the NOT operation is $\overline{x}$. Alternate symbols for the NOT operation include $-x$ or x'. The NOT operator applied over an expression has the effect of enclosing the expression in parentheses. For example,

$$\overline{x_1 x_2} = \overline{(x_1 x_2)}$$

A circle around the AND or OR symbol indicates "exclusive" operation, ⊗ or ⊕. The fundamental logic operations and their properties are shown in Table 49.2.[2] When combinations of logic operations are to be analyzed, the following *order of precedence* is applicable. If in doubt when writing an expression, use parentheses to establish the proper order of precedence.

step 1: Evaluate those items within parentheses first.

step 2: Evaluate logic expressions from left to right, applying all instances of the NOT operations first, the AND operations second, and the OR operations last.

Two functions, f and g, are *logically equivalent* when they are defined for the same arguments and

$$f(x_n, \ldots, x_1) = g(x_n, \ldots, x_1) \qquad 49.3$$

[2]Argument inputs and results are commonly shown using capital letters. They can be scalars or vectors. Scalars have only two values. Vectors represent information and may have numerous values. The properties and laws applicable to logic variables are also applicable to logic vectors. Capital letters are sometimes reserved for logic vectors in educational texts. The symbols of A, B, C, or X, Y, Z are used for inputs and outputs with no special significance attached.

Table 49.2 Logic Operators

operator	symbol	A	B	C	equation
AND		0	0	0	$A\cdot B=C$
		0	1	0	
		1	0	0	
		1	1	1	
OR	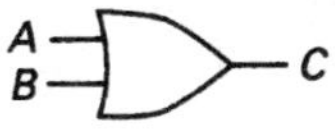	0	0	0	$A+B=C$
		0	1	1	
		1	0	1	
		1	1	1	
XOR	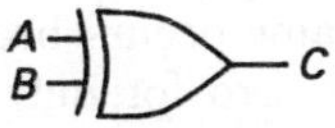	0	0	0	$A\oplus B=C$
		0	1	1	
		1	0	1	
		1	1	0	
NAND		0	0	1	$\overline{A\cdot B}=C$
		0	1	1	
		1	0	1	
		1	1	0	
NOR	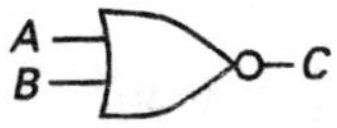	0	0	1	$\overline{A+B}=C$
		0	1	0	
		1	0	0	
		1	1	0	
XNOR or coincidence	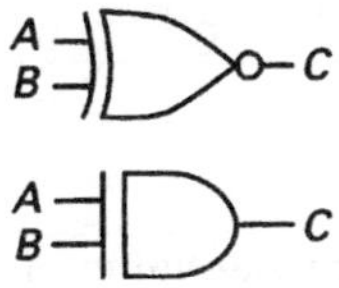	0	0	1	$A\odot B=C$ or $A\otimes B=C$
		0	1	0	
		1	0	0	
		1	1	1	

Equation 49.3 must be valid for all combinations on the n-tuple. Logical equivalence is easily verified by using a truth table for both functions and ensuring that the results are identical. Additionally, the rules of *switching algebra*, usually referred to as *Boolean algebra*, can be used to transform one of the functions into the other, which proves equivalence.

Example 49.2

If $A=1$ and $B=0$, what is the output of the following expression?

$$\overline{\left(\overline{(AB)}\,\overline{(AB)}\right)}$$

Solution

Applying the rules of precedence, the expression is reduced as follows.

$$\begin{aligned}\overline{\left(\overline{(AB)}\,\overline{(AB)}\right)} &= \overline{\left(\overline{(1\cdot 0)}\,\overline{(1\cdot 0)}\right)} = \overline{\left(\overline{(0)}\,\overline{(0)}\right)}\\ &= \overline{(1\cdot 1)}\\ &= \overline{1}\\ &= 0\end{aligned}$$

5. SWITCHING ALGEBRA

Switching algebra is a particular form of Boolean algebra, though the distinction is seldom made. (Section 48.16 has more information on Boolean algebra.) Three basic postulates define switching algebra.

- *postulate 1*: A Boolean variable has two possible values, zero and one, and these values are exclusive. That is,

$$\text{if } x=0 \text{ then } x\neq 1, \text{and}$$

$$\text{if } x=1 \text{ then } x\neq 0$$

- *postulate 2*: The NOT operation is defined as

$$\overline{0}=1 \text{ and } \overline{1}=0$$

- *postulate 3*: The logic operations AND and OR are defined as in Table 49.2.

Using these postulates, the special properties, laws, and theorems of switching algebra are developed. The properties, laws, and theorems of switching algebra are summarized in Table 49.3.

De Morgan's theorem, stated in Table 49.3, has many applications and is used when constructing logic networks from NAND and NOR logic. NAND and NOR are *universal operations*, sometimes called *complete sets.* The term "complete set" indicates that any logic expression can be represented solely by NAND logic, or solely by NOR logic. Because any logic expression is a combination of NOT, AND, and OR operations, proving that NAND logic can represent any of these three proves universality. This proof is shown in Eq. 49.4 through Eq. 49.6 for the NOT, AND, and OR operations, respectively.

$$\overline{(A\cdot A)} = \overline{A}+\overline{A} = \overline{A} \qquad 49.4$$

$$\begin{aligned}\overline{\overline{(A\cdot B)}\cdot\overline{(A\cdot B)}} &= (A\cdot B)+(A\cdot B)\\ &= A\cdot B\end{aligned} \qquad 49.5$$

$$\begin{aligned}\overline{\overline{(A\cdot A)}\cdot\overline{(B\cdot B)}} &= (A\cdot A)+(B\cdot B)\\ &= A+B\end{aligned} \qquad 49.6$$

Table 49.3 Basic Properties of Switching Algebra (Boolean algebra)

name	property	dual
special properties: 0	$0 + A = A$	$0 \cdot A = 0$
special properties: 1	$1 + A = 1$	$1 \cdot A = A$
idempotence law	$A + A = A$	$A \cdot A = A$
complementation law	$A + \overline{A} = 1$	$A \cdot \overline{A} = 0$
involution	$\overline{\overline{A}} = A$	
commutative law	$A + B = B + A$	$A \cdot B = B \cdot A$
associative law	$A + (B + C) = (A + B) + C$	$A \cdot (B \cdot C) = (A \cdot B) \cdot C$
distributive law	$A \cdot (B + C) = (A \cdot B) + (A \cdot C)$	$A + (B \cdot C) = (A + B)(A + C)$
absorption law	$A + (A \cdot B) = A$	$A \cdot (A + B) = A$
	$A + (\overline{A} \cdot B) = A + B$	$A \cdot (\overline{A} + B) = A \cdot B$
De Morgan's theorem	$\overline{(A + B)} = \overline{A} \cdot \overline{B}$	$\overline{(A \cdot B)} = \overline{A} + \overline{B}$

Using the dual of Eq. 49.4 through Eq. 49.6 proves the universality of the NOR operation. The dual of each of the items in Table 49.2 is obtained using the *principle of duality*. The dual of an expression is obtained by applying the following steps, without changing the variables involved.

step 1: If present, change zeros to ones and ones to zeros.

step 2: Change the original OR operations to AND.

step 3: Change the original AND operations to OR.

6. MINTERMS AND MAXTERMS

A *product term* is a function defined by a logical AND set of terms with either a variable, x_i, or its negation, $\overline{x_i}$. A product term where all the variables appear, but only once, is called a *minterm*. The name occurs because the minterms take on a value of one for only one of the 2^n combinations of an n-tuple, that is, for the input variables, and a value of zero for all other combinations. Expressed in terms of true/false notation, the minterms give the rows of a truth table in which the function is true. Minterms for a three-variable function are shown in Table 49.4.

Table 49.4 Minterms and Maxterms

decimal row number	binary input combinations ABC	minterm (product term)	maxterm (sum term)
0	000	$m_0 = \overline{A}\,\overline{B}\,\overline{C}$	$M_0 = A + B + C$
1	001	$m_1 = \overline{A}\,\overline{B}C$	$M_1 = A + B + \overline{C}$
2	010	$m_2 = \overline{A}B\overline{C}$	$M_2 = A + \overline{B} + C$
3	011	$m_3 = \overline{A}BC$	$M_3 = A + \overline{B} + \overline{C}$
4	100	$m_4 = A\overline{B}\,\overline{C}$	$M_4 = \overline{A} + B + C$
5	101	$m_5 = A\overline{B}C$	$M_5 = \overline{A} + B + \overline{C}$
6	110	$m_6 = AB\overline{C}$	$M_6 = \overline{A} + \overline{B} + C$
7	111	$m_7 = ABC$	$M_7 = \overline{A} + \overline{B} + \overline{C}$

A *sum term* is a function defined by a logical OR set of terms with either a variable x_i or its negation $\overline{x_i}$. A sum term where all the variables appear, but only once, is called a *maxterm*. The name occurs because the maxterms take on a value of zero for only one of the 2^n combinations of an n-tuple, that is, for the input variables, and a value of one for all other combinations. Expressed in terms of true/false notation, the maxterms give the rows of a truth table in which the function is false. Maxterms for a three-variable function are shown in Table 49.4.

Example 49.3

Consider the truth table for the two-variable function $F(A, B)$ shown. List the minterms.

A	B	F(A, B)
0	0	0
0	1	1
1	0	1
1	1	0

Solution

The minterms occur where the value of the function is true (that is, one). So, the minterms are

$$m_1 = \overline{A}B$$

$$m_2 = A\overline{B}$$

Example 49.4

For the function in Ex. 49.3, list the maxterms.

Solution

The maxterms occur where the value of the function is false (that is, zero). Therefore, the maxterms are

$$M_0 = A + B$$

$$M_3 = \overline{A} + \overline{B}$$

7. CANONICAL REPRESENTATION OF LOGIC FUNCTIONS

Any arbitrary function described by a truth table can be represented using minterms and maxterms. For example, the *minterm form* of the three-variable function in Table 49.4 is

$$F(A,B,C) = \sum_{i=0}^{7} m_i \qquad \text{49.7}$$

When Eq. 49.7 is expanded, it is referred to as the *canonical sum-of-product form* (SOP) of the function. For example,

$$\begin{aligned} F(A,B,C) &= m_0 + m_1 + m_2 + m_3 + m_4 + m_5 \\ &\quad + m_6 + m_7 \\ &= \overline{A}\,\overline{B}\,\overline{C} + \overline{A}\,\overline{B}\,C + \overline{A}\,B\,\overline{C} + \overline{A}\,B\,C \\ &\quad + A\,\overline{B}\,\overline{C} + A\,\overline{B}\,C + A\,B\,\overline{C} + A\,B\,C \end{aligned} \qquad \text{49.8}$$

The terms are called *minterms* because they have a value of one for only one of the 2^i possible values of Eq. 49.7. (These are the terms of a function that are true; that is, logic one.)

The *maxterm form* of the three-variable function in Table 49.4 is

$$F(A,B,C) = \prod_{i=0}^{7} M_i \qquad \text{49.9}$$

When Eq. 49.9 is expanded, it is referred to as the *canonical product-of-sum form* (POS) of the function. For example,

$$\begin{aligned} F(A,B,C) &= M_0M_1M_2M_3M_4M_5M_6M_7 \\ &= (A+B+C)(A+B+\overline{C}) \\ &\quad \times (A+\overline{B}+C)(A+\overline{B}+\overline{C}) \\ &\quad \times (\overline{A}+B+C)(\overline{A}+B+\overline{C}) \\ &\quad \times (\overline{A}+\overline{B}+C)(\overline{A}+\overline{B}+\overline{C}) \end{aligned} \qquad \text{49.10}$$

The terms are called *maxterms* because they have a value of zero for only one of the 2^i possible values of Eq. 49.9. (These are the terms of a function that are false; that is, logic zero.)

The term "canonical" in this context means a simple or significant form of an equation or function. Canonical forms contain all the input variables in each term. SOP and POS forms may not be the simplest expressions with regard to realizing these functions in terms of logic gates and may require minimization.

Example 49.5

Write the function in Ex. 49.3 in SOP form.

Solution

The minterms were determined in Ex. 49.3. The SOP form, using the format of Eq. 49.8 is

$$F(A,B) = \overline{A}B + A\overline{B}$$

Example 49.6

Write the function of Ex. 49.3 in POS form.

Solution

The maxterms were determined in Ex. 49.4. The POS form, using the format of Eq. 49.10 is

$$F(A,B) = (A+B)(\overline{A}+\overline{B})$$

Example 49.7

Consider the truth table shown. What are the (a) SOP and (b) POS forms of the function?

X	Y	Z	F(X, Y, Z)
0	0	0	0
0	0	1	1
0	1	0	1
0	1	1	0
1	0	0	1
1	0	1	0
1	1	0	0
1	1	1	1

Solution

(a) Minterms (logic value one) exist in rows 1, 2, 4, and 7. The SOP form is

$$\begin{aligned} F(X,Y,Z) &= m_1 + m_2 + m_4 + m_7 \\ &= \overline{X}\,\overline{Y}\,Z + \overline{X}\,Y\,\overline{Z} + X\,\overline{Y}\,\overline{Z} + X\,Y\,Z \end{aligned}$$

(b) Maxterms (logic value zero) exist in rows 0, 3, 5, and 6. The POS form is

$$\begin{aligned} F(X,Y,Z) &= M_0M_3M_5M_6 \\ &= (X+Y+Z)(X+\overline{Y}+\overline{Z}) \\ &\quad \times (\overline{X}+Y+\overline{Z})(\overline{X}+\overline{Y}+Z) \end{aligned}$$

8. CANONICAL REALIZATION OF LOGIC FUNCTIONS: SOP

Realization is the process of creating a logic diagram from the truth table for a given function. Since any scalar function can be represented in either canonical sum-of-product or product-of-sum form,[3] any function can be realized by a two-level logic circuit. The connection points are shown with a dot. Undotted line crossings indicate nonconnecting crossings. It is normally assumed that the input variables and their complements are available.

The SOP form of the function in Ex. 49.7 indicates a two-level AND-OR realization, as shown in Fig. 49.3.

Figure 49.3 *SOP Realization*

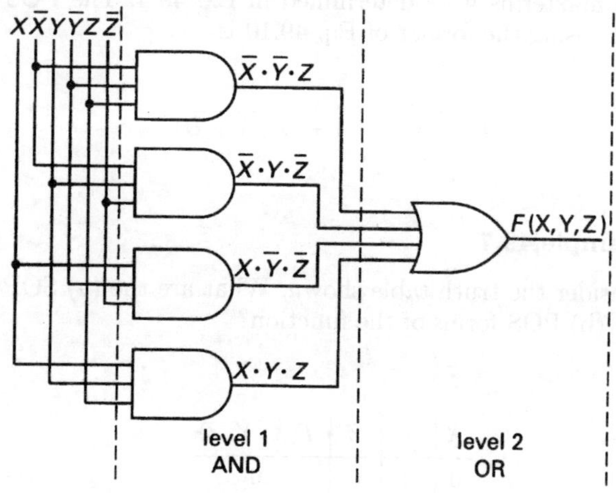

In Fig. 49.3, the AND gates are in logic level one. The OR gate is in logic level two.

The same realization can occur using only NAND gates. Consider the equivalence shown in Fig. 49.4.

Figure 49.4 *NAND Equivalence*

X	Y	Z	$\overline{X\cdot Y\cdot Z}$	$\overline{X}+\overline{Y}+\overline{Z}$
0	0	0	1	1
0	0	1	1	1
0	1	0	1	1
0	1	1	1	1
1	0	0	1	1
1	0	1	1	1
1	1	0	1	1
1	1	1	0	0

(a) truth table

X, Y, Z — $\overline{X\cdot Y\cdot Z}$ X, Y, Z — $\overline{X}+\overline{Y}+\overline{Z}$

(b) symbols

The NAND gate can be represented in two ways: the standard way or as an OR gate with the inputs inverted. Using this equivalence, the SOP form of the function in Ex. 49.7 can be realized with NAND gates only, as shown in Fig. 49.5.

Figure 49.5 *SOP NAND-NAND Realization*

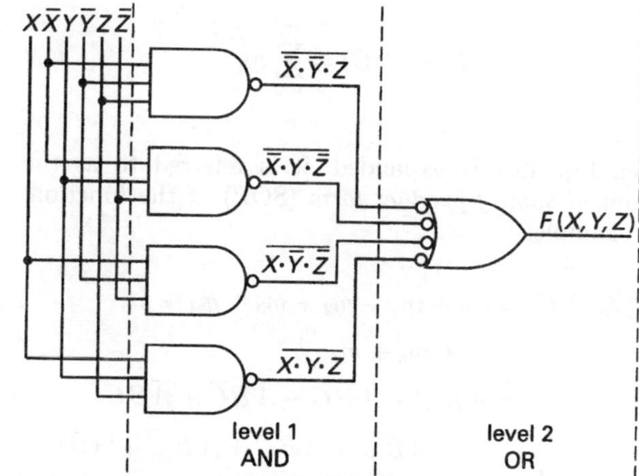

The realization in Fig. 49.5 is AND-OR. This can be seen by considering the inversions represented by the bubbles (small circles) on the output of the level-one logic to cancel with the inversions on the input to the level-two logic.[4]

9. CANONICAL REALIZATION OF LOGIC FUNCTIONS: POS

The POS form of the function in Ex. 49.7 indicates a two-level OR-AND realization, as shown in Fig. 49.6.

Figure 49.6 *POS Realization*

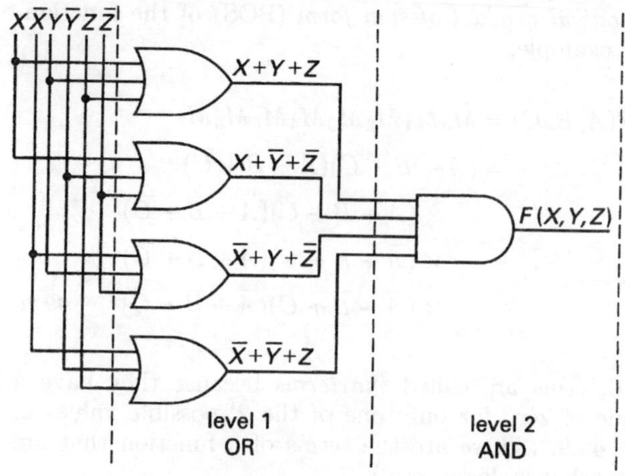

In Fig. 49.6, the OR gates are in logic level one. The AND gate is in logic level two.

[3]Canonical forms of logic realization are useful in the design of read-only memory (ROM) and programmable logic array (PLA) circuits.

[4]Figure 49.5 is not a minimum realization because the function is the XOR of *X*, *Y*, and *Z* and could be realized with two two-input XOR gates.

The same realization can occur using only NOR gates. Consider the equivalence shown in Fig. 49.7.

Figure 49.7 NOR Equivalence

X	Y	Z	$\overline{X+Y+Z}$	$\bar{X}\cdot\bar{Y}\cdot\bar{Z}$
0	0	0	1	1
0	0	1	0	0
0	1	0	0	0
0	1	1	0	0
1	0	0	0	0
1	0	1	0	0
1	1	0	0	0
1	1	1	0	0

(a) truth table

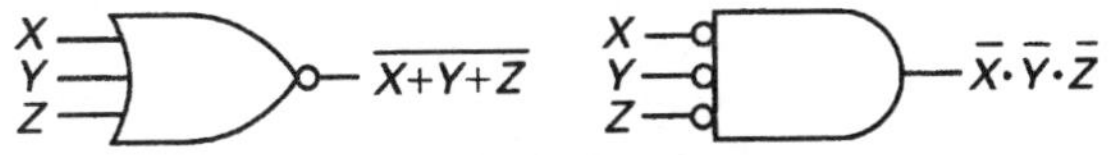

X, Y, Z — $\overline{X+Y+Z}$ X, Y, Z — $\bar{X}\cdot\bar{Y}\cdot\bar{Z}$

(b) symbols

The NOR gate can be represented in two ways: the standard way or as an AND gate with the inputs inverted. Using this equivalence, the POS form of the function in Ex. 49.7 can be realized with NOR gates only, as shown in Fig. 49.8.

Figure 49.8 POS NOR-NOR Realization

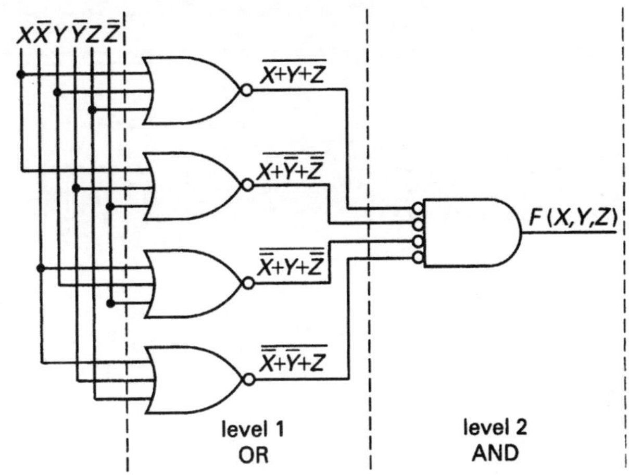

The realization in Fig. 49.8 is OR-AND. This can be understood by considering the inversions represented by the bubbles on the output of the level-one logic to cancel with the inversions on the input to the level-two logic.[5]

[5]Figure 49.8 is also not a minimum realization because the function is the XOR of X, Y, and Z and could be realized with two two-input XOR gates.

Topic XI: Special Applications

Chapter

50. Lightning Protection and Grounding
51. Illumination
52. Power System Management

50 Lightning Protection and Grounding

Nomenclature

f	frequency	Hz
l	length	m
PM	protective margin	–
PQI	protection quality index	–
V	effective or DC voltage	V

Symbols

δ	skin depth	m
μ	permeability	H/m
σ	conductivity	S/m

Subscripts

p	protection level
r	reseal level
w	withstand

1. FUNDAMENTALS

Lightning is an atmospheric electrical discharge resulting from the creation and separation of electric charges in cumulonimbus clouds. The exact mechanism causing the charge separation is unknown, with several theories explaining some, but not all, of the effects. Among these theories are the *precipitation theory*, the *convection theory*, and the theory of *graupel-ice crystal interaction with charge-reversal temperature*. The combination of all three is nominally a tripolar-electrified thundercloud as shown in Fig. 50.1.

Lightning occurs when the charge accumulation is such that the electric field between the cloud and the earth is high enough to break down the dielectric between them (i.e., the air). During the first phase of the process, local ionization occurs and discharges called *pilot streamers* create a path for the *stepped leaders* that follow.[1] The luminous stepped leaders move at 15% to 20% the speed of light, approximately 50 m at a time (see Fig. 50.1). Each step is separated by a few microseconds. The second phase occurs when the stepped leader reaches the earth or a building or power line on the earth's

[1]Stepped leaders may originate from the earth and move skyward.

Figure 50.1 *Lightning Development*

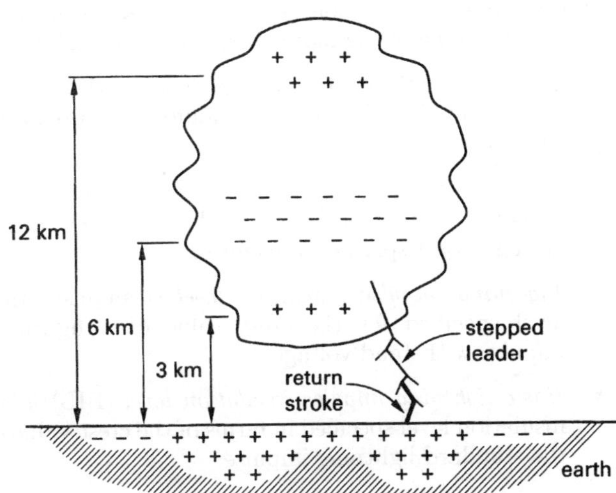

surface. When this occurs, an extremely luminous high-intensity discharge occurs between the charge centers of the earth and cloud. The discharge moves at 10% to 50% of the speed of light and results in a current ranging from several thousand amperes to 200,000 A, with the peak between 1 μs and 10 μs. This discharge is called the *return stroke* or the *lightning stroke*.[2] In the final phase, strokes may occur between the charge centers within the cloud or to the ground through the established conducting channel because of the shifting potentials within the cloud. Approximately one-third of all lightning is a single stroke. When multiple strokes occur, the mean number is three. A *stroke* is any one of a series of repeated discharges that comprise a single discharge, which is called a *lightning flash* or *strike*.

The voltage of a lightning stroke varies from 10 MV to 1000 MV. The voltage of concern to the electrical engineer, however, is that which appears on the equipment. The overvoltage on the equipment depends on the impedance and the current. Lightning strokes are considered current sources at the point of the strike. The current varies from 1 kA to 200 kA.

2. CONCEPTS AND DEFINITIONS

A power system must use a coordinated design to protect from internally generated surges, usually caused by

[2]The strokes mentioned are called *ground flashes*. Other strokes occur between clouds but are not of concern in power electrical engineering.

switching, and externally generated surges, which are usually caused by lightning. Coordinated design includes shielding, grounding, surge arresters, switching resistors, breaker timing, and surge capacitors. The first three are primarily used to protect against lightning strikes, and the last three are used to protect against switching transients. Protection devices include spark gaps and surge arresters. The main purpose of the protection is to prevent insulation breakdown, damage to equipment, and outages.

Power system protection terminology has been standardized.[3] Some of the more important terms follow.

- *Withstand voltage* is the voltage that electrical equipment can handle without failure or disruptive discharge.
- *Transient insulation level* (TIL) is an insulation level specified in terms of the crest value of the withstand voltage, for a specified waveform.
- *Lightning impulse insulation level* is an insulation level specified for the crest value of a lightning impulse withstand voltage.
- *Basic lightning impulse insulation level* (BIL) is an insulation level specified in terms of the crest value of the standard lightning impulse.
- *Standard lightning impulse* is a full impulse with a *front time* of 1.2 μs and a half-value time of 50 μs, normally termed a 1.2/50 impulse.[4] See Fig. 50.2.

Figure 50.2 *Standard Lightning Impulse*

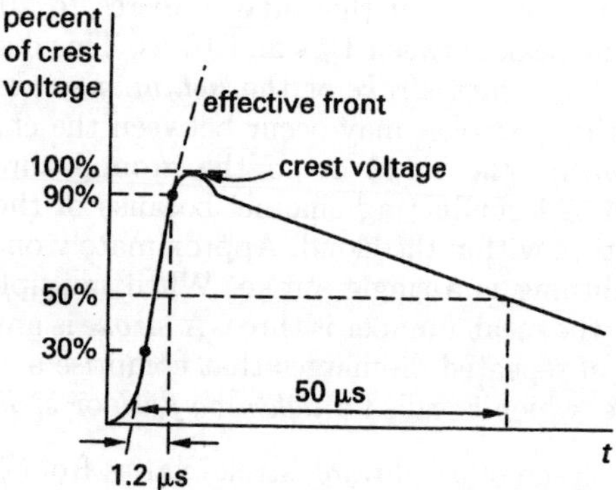

Systems of approximately 230 kV or less cannot be insulated against direct lightning strikes because of the high voltage of the lightning, on the order of several million volts. To protect such systems a combination of shielding and grounding is used as shown in Fig. 50.3.[5]

Indirect lightning strikes result in voltage surges by means of conductive coupling through the soil or by means of inductive and capacitive coupling to the lines.

[3]Refer to the American National Standards Institute (ANSI) Std. C92.1.
[4]This is taken from American National Standard Measurement of Voltage in Dielectric Tests, C68.1.
[5]The physical grounding of systems is covered in Chap. 56.

Figure 50.3 *Shielding System*

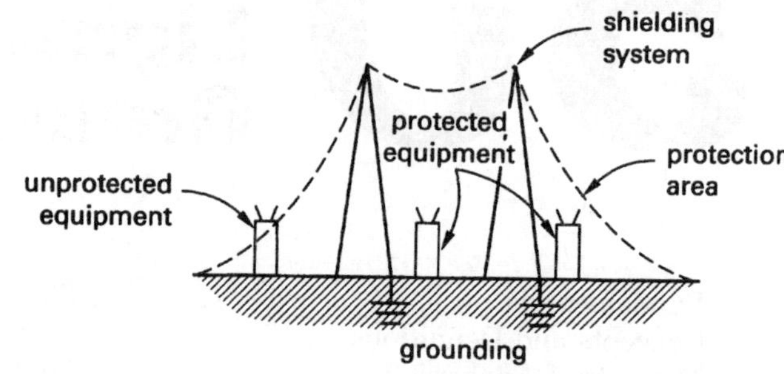

(a) equipment shielding

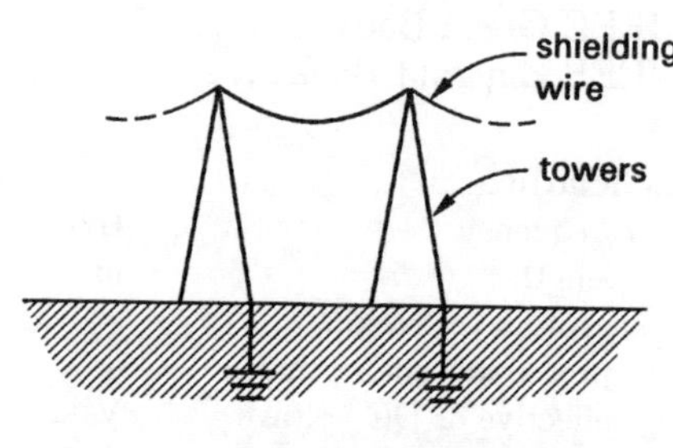

(b) line shielding

Induced voltage surges are typically less than 400 kV and can damage systems of 35 kV or less. Systems of 69 kV and more have sufficient insulation to withstand such surges. To protect such systems, surge arresters are used.

3. METHODS OF ANALYSIS

The first step in analysis is the determination of the models to use for the various components of a distribution system. The models can be lumped parameters, distributed parameters, nonlinear equations, time varying equations, or logical equations. Once the models have been determined, three primary methods of analysis are used. *Graphical analysis* is based on the representation of voltages and currents as traveling waves and the use of a *Bewley diagram* to determine the waves that exist at any given point. *Analytical analysis* consists of a systematic approach to obtaining the solution of the differential equations that are used to represent a system, usually with Laplace transforms. *Numerical analysis* transforms the differential equations into discrete time equations, which are algebraic. The algebraic equations can be interpreted as *resistive companion circuits* and analyzed using standard techniques, such as KVL or KCL. Additionally, Monte Carlo techniques are often applied to the many uncertainties in the analysis of lightning overvoltages.

4. GROUNDING MODELS

Two basic grounding models are used: low frequency and high frequency. Low-frequency models consider grounds to be purely resistive and require DC analysis.[6] High-frequency models require a complete electromagnetic analysis.[7] A general rule for determining which model to use is to compare the largest dimension of the grounding system, say length l, to the skin depth, δ, and then apply the rule of thumb given by Eq. 50.1.

$$\frac{l}{\delta} = \frac{l}{\frac{1}{\sqrt{\pi f \mu \sigma}}} \begin{cases} < 0.1 \text{ DC analysis} \\ > 0.1 \text{ electromagnetic analysis} \end{cases} \qquad 50.1$$

Grounding for protection against lightning overvoltages is in addition to individual equipment grounding and power system grounding. The relationship among the different grounding systems is shown in Fig. 50.4.

Figure 50.4 *Grounding Systems*

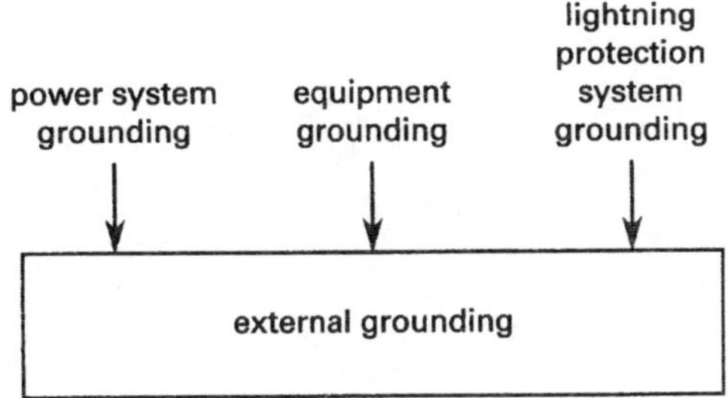

5. PROTECTIVE DEVICES

When overvoltages occur, the transient is minimized and the system is protected by connecting the overvoltage condition to the ground until the energy is dissipated. This connection must be temporary and ideally will not allow current flow at normal operating voltages. This is accomplished by the use of strategically placed *arresters*, also called *surge arresters*. The quality of surge arresters is roughly measured by a *protection quality index* (PQI) given by

$$\text{PQI} = \frac{V_r}{V_p} \qquad 50.2$$

The denominator term is the voltage *protection level*, the maximum voltage allowed by the arrester. The numerator term is the *reseal level*, the voltage level at which minimal current flow through the arrester occurs. The ideal protection quality index is one.

Basic protection is provided by *air gaps*, also known as *spark gaps*. These devices are composed of two electrodes, the shape of which varies depending upon the application, separated by a distance. When the voltage across the air gap exceeds a given value, an arc is initiated. *Gapped surge arresters* use a series of spark gaps in conjunction with a nonlinear resistor as shown in Fig. 50.5. The nonlinear resistors have low resistance at high voltages, thus allowing the surge current. As the voltage of the lightning approaches that of the system, the resistance of the nonlinear resistors increases (typically as the fourth power of the voltage), extinguishing the arc without tripping protective devices.

Figure 50.5 *Gapped Surge Arrester*

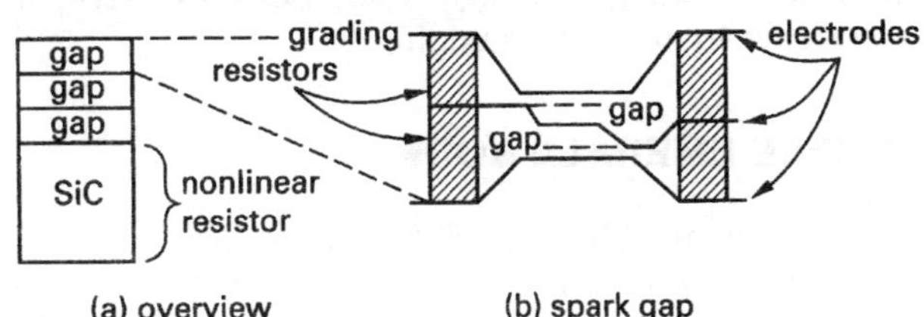

Improved protection is provided by *metal-oxide varistor arresters* (MOVs). An arrester is shown in Fig. 50.6. The device is manufactured in such a manner as to approximate an ideal protective device, that is, a device that clips the voltage level at the maximum allowed and reseals at the same level. An MOV functions like back-to-back diodes. The MOV conducts current at the voltages normally experienced by the electrical system. Heating effects that could deteriorate the device must be accounted for in the design. The performance of MOVs at high discharge currents is improved by adding a shunt gap much like the construction in Fig. 50.5(a). When constructed in such a manner, they are called *shunt-gap MOVs*.[8]

Figure 50.6 *Metal-Oxide Varistor (MOV) Arrester*

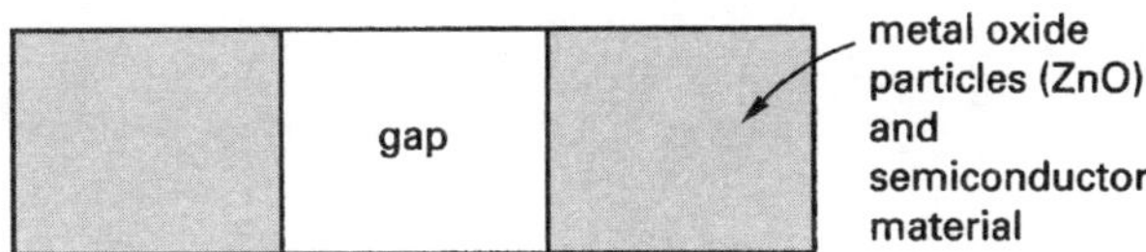

The effectiveness of surge arresters for a particular system is measured by the *protective margin* (PM) provided. The protective margin is determined by

$$\text{PM} = \frac{V_w - V_p}{V_p} \qquad 50.3$$

The term V_w is the voltage a given device can withstand before damage or insulation breakdown occurs. The term V_p is the voltage protection level provided by the arrester.[9]

[6]DC analysis is based on the *method of moments* or the *relaxation method.*

[7]Electromagnetic analysis is based on the *method of moments* or *finite-element analysis.*

[8]Testing methods for MOVs, as well as for other electrical power equipment and systems, can be found in the InterNational Electrical Testing Association (NETA) *Standard for Maintenance Testing Specifications for Electrical Power Equipment and Systems* (ANSI/NETA MTS).

[9]The protection level provided is obtained from manufacturer's nameplate data on the arrester.

Special Applications

6. IEEE GREEN BOOK

IEEE Standard 142, *IEEE Recommended Practice for Grounding of Industrial and Commercial Power Systems* (*IEEE Green Book*), covers industry best practices for grounding using information from other codes, standards, and texts. It discusses the principles and theories of grounding practices, system grounding, equipment grounding, static protection and grounding, lightning protection and grounding, connections to earth, and electronic equipment grounding.

7. IEEE EMERALD BOOK

IEEE Standard 1100, *IEEE Recommended Practice for Powering and Grounding Electronic Equipment* (*IEEE Emerald Book*), covers the best practices and principles for applying power to and grounding sensitive electronic equipment in order to ensure the power quality necessary for proper operation of the equipment. It contains information from other codes, standards, and texts concerning the design, installation, and maintenance practices for grounding systems on electronic equipment. Chapters cover general needs, fundamentals, instrumentation, site surveys and power analysis, selection of equipment and materials, design and installation practices, telecommunications, information technology, distributed computing, industrial systems, and case studies.

51 Illumination

Nomenclature

A	area	m^2 (ft^2)
b	Wien's wavelength displacement constant, 2.8978×10^{-3}	m·K
b'	Wien's spectral radiance displacement constant, 4.0956×10^{-4}	$W/sr{\cdot}m^2{\cdot}m{\cdot}K^5$
c	speed of light, 2.9979×10^8	m/s
CR	cavity ratio	–
CU	coefficient of utilization	–
d	distance	m
DF	daylight factor	–
E	energy	J
E	illuminance	lx (fc)
ERC	externally reflected component	–
f	frequency	Hz
FCR	floor cavity ratio	–
h	Planck's constant, 6.6256×10^{-34}	J·s
h, H	height	m (ft)
I	luminous intensity	cd
IRC	internally reflected component	–
K	luminous efficacy	lm/W
κ	Boltzmann's constant, 1.3807×10^{-23}	J/K
L	length or wavelength	m (ft)
L	luminance	lm
L	radiance	$W/sr{\cdot}m^2{\cdot}m$
$L(f, T)$ or L_f	spectral radiance	$W/sr{\cdot}m^2{\cdot}Hz$
$L(\lambda, T)$ or L_λ	spectral radiance	$W/sr{\cdot}m^2{\cdot}m$
LLF	light loss factor	–
m	mass	g
M	luminous or radiant exitance	lm/m^2
n	index of refraction	–
P	total radiant power per unit area	W
Q_λ	spectral radiant energy	J/m
r	radius	m
R	ratio	–
RCR	room cavity ratio	–
s	distance	m
SC	sky component	–
t	time	s
T	absolute temperature	K
T	transmittance	–
U_λ	blackbody radiant energy	J/m
v	velocity	m/s
W	width	m (ft)
WCR	well cavity ratio	–

Symbols

α	spectral absorption factor	–
ϵ	spectral emissivity	–
η	efficiency	–
θ	angle	rad
λ	wavelength	m
ν	frequency	Hz
ρ	spectral reflection factor	–
σ	Stefan-Boltzmann constant, 5.670×10^{-8}	$W/m^2{\cdot}K^4$
σ_L	luminance (or radiance) Stefan-Boltzmann constant, 1.804×10^{-8}	$W/m^2{\cdot}sr{\cdot}K^4$

τ	net transmittance	–
τ	spectral transmission factor	–
Φ	luminous flux	lm
ω	solid angle	sr

Subscripts

a	absorbed
c	controls or critical
$c\text{-}wp$	ceiling to workplane
CC	ceiling cavity
d	diffuse
D	direct
e	exitance (radiometric or radiant)
f	per unit of frequency
FC	floor cavity
g	gap
i	incident
ic	incident critical
max	maximum
RC	room cavity
s	skylights or surface
t	total
v	vision (photometric or luminous)
w	wall, well, or workplane
xh	exterior horizontal
λ	per unit of wavelength

1. HISTORY AND OVERVIEW

One of the earliest known theories of light was that it was emitted by one's eyes. Once Aristotle pointed out that we should be able to see in the dark but cannot, this theory lost standing. Sir Isaac Newton advanced the corpuscular theory in the 17th and 18th centuries from the observation that moving particles, or corpuscles, possess kinetic energy. Newton theorized that light is radiated from luminous bodies, travels in straight lines, and acts on the retina to provide a visual sensation.

Simultaneous with Newton's particle theory of light was Christiaan Huygens' wave theory of light. His theory noted that light is a product of the molecular vibration of luminous material, is transmitted through an ether (now considered nonexistent), and acts on the retina.

The two theories fought for supremacy until the 19th century when James Clerk Maxwell developed his theory of electromagnetic radiation, which seemed to settle the case for light's being a wave (results are in Chap. 25). Maxwell theorized that light is emitted from luminous material as radiant energy in the form of an electromagnetic wave, which then acts on the retina.

In the 20th century Max Planck developed quantum theory, which reasserted that light is a particle. Planck theorized that energy is emitted in discrete quantities known as *photons*, the magnitude of which is determined by the product of Planck's constant, h, and the frequency of the photon, ν (see Eq. 51.6).[1]

The particle and wave theories were unified through the efforts of Louis de Broglie and Werner Heisenberg in what is currently termed the wave-particle duality. Their theory noted that every object of mass has an associated wavelength given by

$$\lambda = \frac{h}{m\text{v}} \tag{51.1}$$

They further asserted one cannot determine whether light is a wave or a particle; that indeed, it is both.

For the purposes of lighting design, the quantum and electromagnetic theories provide the necessary theoretical background. Light is radiant energy capable of exciting the retina and producing a visual sensation. The efficiency of systems in radiating energy and the effects of that light on our visual sensations are the focus of the illumination engineer.

The International Commission on Illumination (CIE for Commision Internationale de l'Eclairage) was founded in 1913 as an information exchange for the science of lighting. The CIE was recognized by ISO (International Organization for Standards) as the standardizing body for "fundamental aspects of metrology, evaluation, and application of light and color including other radiation aspects in the optical radiation range." ANSI, as part of the ISO, utilizes and disseminates standards. IEEE, NFPA, NEC, and many other bodies issue illumination (commonly called lighting) requirements.

The NEC (*National Electrical Code*), for example, specifies illumination requirements for health care facilities in NEC Art. 517. The NEC does not, however, specify the actual lighting level, leaving that for the (National Fire Protection Association) NFPA 101® *Life Safety Code®*. Nevertheless, there is no description on how to attain the NFPA 101 code-specified levels in either the NEC or the NFPA 101. Definitive guidance may be found for illumination design (that is, calculation of the light levels required while accounting for a variety of secondary effects) in the Illuminating Engineering Society of North America's *IESNA Lighting Handbook: Reference & Application*. Chapter 10 of the *Lighting Handbook* contains the *IESNA Lighting Design Guide*, which should be the starting point for any illumination design.

The information in the present chapter covers light (electromagnetic wave) principles; general knowledge of lighting—units, terminology, and interaction with matter; optical control; and design methods and considerations. The standard lighting symbols are given in App. 51.A. Both the SI and the customary U.S. systems are used in illumination engineering. Unless otherwise indicated, the SI system is used in this chapter.

[1]The symbol for frequency in circuit theory is f. In optics and quantum mechanics, the symbol is the Greek letter nu, ν.

2. ELECTROMAGNETIC WAVES[2]

According to electromagnetic wave theory, accelerating electric charges radiate energy in the form of electromagnetic waves. This radiant energy is represented in a spectrum according to wavelength (or frequency). The electromagnetic spectrum is shown in App. 18.A. The radiant energy portion of the spectrum extends from wavelengths of 10^{-16} m to approximately 105 m. Wavelengths are normally given in meters, and frequency in hertz. Wavelengths, however, are often expressed in *microns* or *micrometers* (10^{-6} m), nanometers (10^{-9} m), and angstroms, Å (10^{-10} m). The nanometer is commonly used for the ultraviolet (UV) and visible regions, and the micrometer in the infrared (IR) region. Three regions, ultraviolet, visible, and infrared, comprise the primary electromagnetic spectrum area of concern for the illumination engineer. These regions are shown in Fig. 51.1.

Figure 51.1 *Electromagnetic Spectrum: Ultraviolet, Visible, and Infrared Radiant Energy*

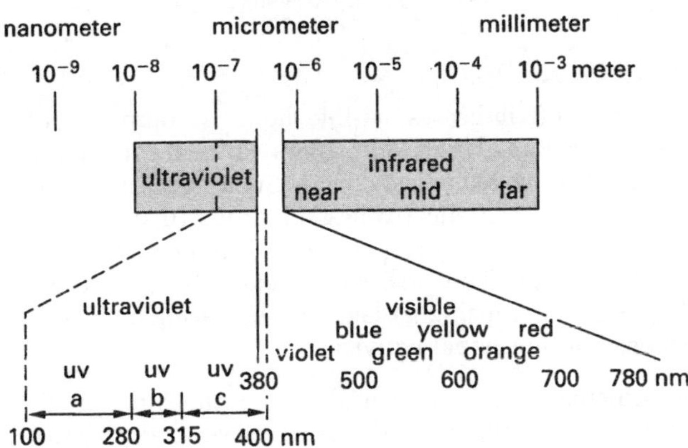

All electromagnetic waves (radiation) travel at the speed of light in a vacuum, which is approximately 186,000 mi/sec or 2.9979×10^8 m/s. The distance, s, traveled by the electromagnetic radiation, is given by $s = \nu\lambda t$, where ν is the frequency of the wave, λ is the associated wavelength, and t is the time. Setting the time, t, equal to 1 s results in the distance traveled in that same time period, which corresponds to the speed of light, c. This results in Eq. 51.2, which is valid for all types of electromagnetic radiation.

$$c = \lambda\nu \qquad 51.2$$

The product of wavelength and frequency, $\lambda\nu$, is constant, c, in a vacuum. This effect, from Einstein's special theory of relativity, results in time dilation and length contractions that can be adjusted for with *Lorentz transformations*. Such effects result in a *red shift* (that is, the visible light from an object that is rapidly receding from an observer appears in a redder region of the frequency spectrum than it would if the object were stationary, and the light from a rapidly approaching object appears bluer). The frequency shift is called the *Doppler shift* or *Doppler effect*. The *Lorentz factor*, γ, is represented in Eq. 51.3, where the common relativity symbols, γ and β, are used.[3]

$$\gamma = \frac{1}{\sqrt{1 - \left(\frac{\mathrm{v}}{c}\right)^2}} = \frac{1}{\sqrt{1 - \beta^2}} \qquad 51.3$$

When an electromagnetic wave passes through from one medium to another, the frequency remains constant while the velocity and wavelength change. In terms of wavelength, Eq. 51.2 is written as

$$\lambda = \frac{c}{\nu} \qquad 51.4$$

When the medium is not a vacuum in Eq. 51.4, the speed of light is replaced by the velocity, v, and the *index of refraction*, n (also called the *refractive index*), giving Eq. 51.5. (The index of refraction for a vacuum is equal to one, making the velocity equal to that of light, c, which brings Eq. 51.5 back to the form of Eq. 51.4.) The resulting change of speed (velocity) and wavelength bends the light (electromagnetic wave) toward the surface normal in a material denser than the original medium, and away from the normal in a less dense medium. (See Sec. 51.22 for further explanation.)

$$\lambda = \frac{\mathrm{v}n}{\nu} \qquad 51.5$$

The frequency of the electromagnetic wave remains constant in any given medium because the energy contained in the wave remains constant, a result of the conservation of energy. The wavelength and velocity must change as the medium changes. The energy of a *quantum* or *photon* is given by Eq. 51.6. The product $h\nu$ in Eq. 51.6 indicates that the energy is constant and dependent only on the frequency. The second and third portions of Eq. 51.6 combine to give Eq. 51.4, while the third and fourth portions combine to give Eq. 51.5. The net result is that light, an electromagnetic wave, retains its energy (and therefore, its frequency) as it passes through differing media. However, the light must change speed (velocity) and wavelength to do so, thereby changing direction. (Or, in terms of optics, the

[2]The topic of electromagnetic waves is often termed electromagnetic radiation. The term "radiation" has come to have negative implications, and so is often avoided. Nevertheless, there are numerous positive aspects to radiation. Indeed, radiation is necessary for human life to exist. The terms "electromagnetic wave" and "electromagnetic radiation" will be used interchangeably as the topic demands.

[3]Relativity concerns are for those dealing with illumination from celestial objects. In the design of illumination systems, the concern is with wavelength changes in mediums, as measured by the index of refraction, indicating a bending of the light toward or away from the normal to the surface.

Special Applications

light ray—a beam of light with a small cross section—is refracted.)

$$E = h\nu = \frac{hc}{\lambda} = \frac{h v n}{\lambda} \quad 51.6$$

The speeds of light (that is, electromagnetic waves) in several substances of concern are given in Table 51.1. The speed of an electromagnetic wave in air differs from that in a vacuum by less than 0.03%, so the two mediums are often considered equivalent for design purposes.

Table 51.1 *Speed of Light (Sodium D-lines at 589 nm) in Various Mediums*

medium	speed (m/s)
vacuum	2.9979×10^8
air (760 mm Hg at 0°C)	2.9972×10^8
water	2.2492×10^8
crown (optical) glass	1.9822×10^8

3. ELECTROMAGNETIC SPECTRUM: ULTRAVIOLET

The UV spectrum division shown in Fig. 51.1 is that given by the CIE. Ultraviolet light, though primarily not visible, has potential significant biological effects, as indicated in Table 51.2.

Table 51.2 *Ultraviolet Light: Biological Division*

wavelength (nm)	region effect	comments
180–220	ozone production	decomposes O_2 to O_3
220–300	bactericidal (germicidal)	destroys bacteria, fungi, and viruses
280–320	erythema	reddening of the skin
300–400	black light	used for rating the effectiveness of lamps on fluorescent materials (excluding phosphors)

As Table 51.2 indicates, UV light can have biological (physiological) effects, some of which are not desirable. By limiting lighting systems emissions to the *black light region* (300–400 nm), the negative biological effects of UV can be avoided; positive effects also exist. UV light is used by the body to generate vitamin D and to create the protective pigmentation that minimizes the negative effects.

4. ELECTROMAGNETIC SPECTRUM: VISIBLE LIGHT

The visible portion of the spectrum shown in Fig. 51.1 is that given by the CIE. The range is somewhat variable, but for practical purposes it is from 3.8×10^{-7} m to 7.8×10^{-7} m (380 nm to 780 nm). This region comprises many colors. Visible light of "many colors" is called *polychromatic light* to indicate that such light contains multiple wavelengths. When wavelengths of all or nearly all of the colors are included, the light is called *white light.* White light is not a color, but a combination of colors.

The human retina is most sensitive to light at the center of the visible spectrum, 5.55×10^{-7} m, or 555 nm. This wavelength is the basis of the unit for luminous intensity, the candela. The visual sensation from differing wavelengths in the visible spectrum is sensed as color. The approximate color sensations are given in Table 51.3.

Table 51.3 *Color Versus Wavelength*

color	wavelength (nm)
violet	380–450
blue	450–495
green	495–570
yellow	570–590
orange	590–620
red	620–750

As Table 51.3 indicates, visible light is comprised of numerous colors. Numerous colors of light, including white light, have biological effects. Such light can result in burns, alter the effectiveness of drugs (drug photosensitivity), form lesions, and cause numerous other negative effects. The positive effects include altering one's mood, minimizing jet lag, and regulating biological rhythms and hormonal activity.

The principal source of natural light is the sun, specifically the photosphere, a plasma envelope that radiates energy in the visible band. The sun's illuminance of the Earth varies with the time of day and year, weather, and other factors. The CIE specifies *standard illuminants* or *spectral power distributions* (SPDs) for use in computing colors, design, and comparison applications.[4] The preferred standard illuminate for the sun is D_{65}, which is light at a correlated color temperature of 6500K. The *correlated color temperature* (CCT) for a source is the absolute temperature of a blackbody whose chromaticity most closely resembles that of the light source (see Sec. 51.6). The chromaticity (x, y, z) values for the sun standard D_{65} are 0.33, 0.66, 0.01.[5] The x value correlates with red, the y value with green, and the z value with blue. (Hence the term "RGB" when referring to color reference systems.) The human eye

[4]Different SPDs can result in the same color. Such SPDs are called metamers.

[5]Chromaticity values x, y, z are the ratios of the tristimulus values X, Y, Z to their sum. That is, $x = X/(X + Y + Z)$, and so forth. The chromaticity values are usually specified in a graph of x and y, only given that $x + y + z = 1$. The tristimulus values define all metametric pairs (SPDs) by providing the amount of each primary color, (red, green, blue or X, Y, Z) required by a standard observer to match the color being specified.

has short, medium, and long wavelength sensors (blue, green, and red, respectively). The primarily red and green mixture of the sun's chromaticity results in the visual sensation of a yellow sun.

Other natural sources of light include *sky lighting*, caused by the Rayleigh scattering of incoming electromagnetic radiation from the sun; fire; moonlight; lightning; the aurora borealis (northern lights) and aurora australis (southern lights); and bioluminescence, which is a form of chemiluminescence in which light is produced by plants and animals through the process of oxidation.

5. ELECTROMAGNETIC SPECTRUM: INFRARED

The *infrared* (IR) portion of the spectrum, shown in Fig. 51.1, extends from 0.78×10^{-6} m (0.78 μm) to 1.0×10^{-3} m (1000 μm). The IR spectrum is arbitrarily divided as shown in Table 51.4.

Table 51.4 *Infrared Band Versus Wavelength*

IR band	wavelength (μm)
near (short wavelength)	0.78–1.4
mid (medium wavelength)	1.4–3.0
far (long wavelength)	3.0–1000

The near IR region results in the same biological effects as visible light. The mid and far IR regions can cause burns, generate cataracts, cause erythema (a different form than that from UV radiation), and other negative effects. The positive aspect is that IR radiation is useful for radiant heating.[6]

6. BLACKBODY RADIATION

A *blackbody* is an ideal body that would absorb all incident electromagnetic radiation. A blackbody is also known as a *hohlraum* or an *ideal radiator*. The term "blackbody" arises because there is no transmission or reflection of incident electromagnetic radiation. (Additionally, below approximately 1000K any visible light incident upon a blackbody is absorbed, and the body appears black.) An approximation of an ideal blackbody that was created in the laboratory is shown in Fig. 51.2. A small aperture ensures any incident radiation is likely to be absorbed before it can escape, simulating blackbody characteristics.[7]

Figure 51.2 *Laboratory Blackbody*

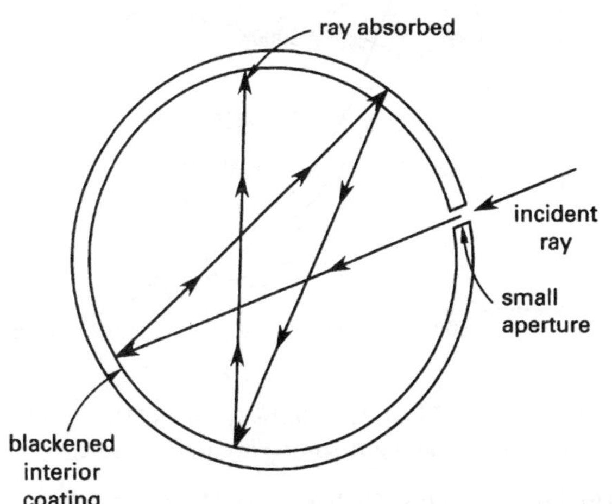

While a blackbody neither transmits nor reflects incoming electromagnetic radiation, it does absorb such radiation and emit thermal radiation. *Thermal radiation* is the energy radiated by solids, liquids, and gases in the form of electromagnetic waves as a result of their temperature. The thermal radiation is in the UV, visible, and IR regions of the electromagnetic spectrum.

By maintaining the walls of the laboratory blackbody of Fig. 51.2 at a constant temperature, and measuring the radiation emitted from the small aperture, thermal radiation curves such as those shown in Fig. 51.3 are obtained.[8] Kirchhoff's law of thermal radiation implies that the shape of the curves is dependent on the temperature only, and not the composition or material of the body. Therefore, an object's temperature may be determined from the color of its emitted thermal radiation.[9]

The concept of a blackbody is useful in the study of radiation phenomena. In the case of illumination engineering, by specifying both the magnitude of the radiation at any given wavelength and the absolute temperature (see

[6]Radiant heating is the source of warmth one feels on a sunny day, or next to a campfire on a cold night, or next to a hot substance, such as steel coming out of a furnace. Unlike the engineering of both UV and visible energy, design with IR energy is usually more concerned with the total energy deposited on a surface rather than the wavelength. Common applications of IR energy include industrial heating, drying or curing, baking in an oven, and photoreproduction. However, the wavelengths or spectral characteristics of IR are important in IR detectors and viewing devices.

[7]An interior coating of lampblack results in an absorption of approximately 97% of incident light. Polished metal surfaces absorb less than 6% of the incident light. Most surfaces are between these extremes. (Lampblack is soot formed by burning oil, coal tar, resin, or other carbonaceous substances in an insufficient supply of air. It is essentially pure carbon.)

[8]When the curves shown in Fig. 51.3 relate the power radiated by an actual object, the symbol used is Q or Q_λ. When the radiated power refers to that from a blackbody, the symbol used is U or U_λ.

[9]When an object is *red hot*, it is emitting radiation in the long wavelength portion of the visible band. As temperature increases, the glow changes color from red to blue. When the frequency is in the middle of the visible band, the object appears white, hence the term *white hot*.

Figure 51.3 *Radiated Power Versus Wavelength as Temperature Varies*

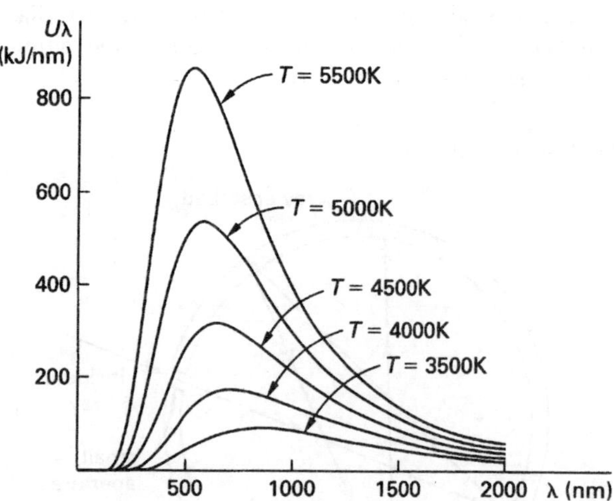

Fig. 51.3 or Fig. 51.4), the characteristics (the spectral power density) of the light source are determined.[10]

Thermal radiation is exchanged according to Eq. 51.7. α is the spectral absorption factor; ρ is the spectral reflection factor; and τ is the spectral transmission factor. All these elements are dependent on the wavelength, λ.

$$\alpha+\rho+\tau=1 \qquad 51.7$$

Kirchhoff's law of thermal radiation specifies that, at thermal equilibrium, the emissivity, ϵ, of a body (or surface) equals its spectral absorption, α. An object is a blackbody if for all frequencies Eq. 51.8 holds. (This, consistent with the definition of a black body, makes $\rho=\tau=0$.)

$$\epsilon=\alpha=1 \qquad 51.8$$

A blackbody, for a given area, radiates more total power at a given wavelength than any other light source at the same temperature.[11]

7. PLANCK RADIATION LAW

The *Planck radiation law* is an expression providing the spectral radiance, L, of a blackbody as a function of wavelength and temperature. This is also known as *Planck distribution law* or *Planck's law*. Planck's law provided the shape of the curves in Fig. 51.3. In terms of wavelength, a form of the law is given in Eq. 51.9. In terms of frequency, a form of the law is given in Eq. 51.10.[12] The two equations have different units. Equation 51.9 is the radiance per unit wavelength interval and Eq. 51.10 is radiance per unit frequency interval. The two equations are related by Eq. 51.11.

$$L(\lambda,T)=L_\lambda=\left(\frac{2hc^2}{\lambda^5}\right)\left(\frac{1}{e^{hc/\lambda\kappa T}-1}\right) \qquad 51.9$$

$$L(f,T)=L_f=\left(\frac{2hf^3}{c^2}\right)\left(\frac{1}{e^{hf/\kappa T}-1}\right) \qquad 51.10$$

$$L(\lambda,T)\,d\lambda=L(f,T)\,df \qquad 51.11$$

In the temperature range of incandescent filament lights (2000K to 3400K), with a wavelength in the visible range (380 nm to 780 nm), a simplification of the Planck radiation law (specifically, Eq. 51.9) can be used as a useful approximation (within 1%) of the spectral radiance. This is called the *Wien radiation law*. The Wien radiation law, given by Eq. 51.12, is applicable to the shaded region in Fig. 51.4 between 2000K and 3400K.

$$L(\lambda,T)=L_\lambda=\left(\frac{2hc^2}{\lambda^5}\right)e^{-hc/\lambda\kappa T} \qquad 51.12$$

8. WIEN DISPLACEMENT LAW

In Fig. 51.4, as the absolute temperature increases, the peak of the radiated power curve shifts to the left toward higher frequencies and shorter wavelengths. The law itself is similar to the Planck radiation law, but two principle corollaries to the law are the focus for the illumination engineer. The first gives the maximum wavelength for a given temperature, Eq. 51.13. (The term b in Eq. 51.13 is a constant whose current value is 2.8978×10^{-3} m·K.) Equation 51.14 gives the maximum spectral radiance for a given wavelength. (The term b' in Eq. 51.14 is a constant whose value is 4.0956×10^{-4} W/sr·m^2·m·K^5.)[13]

$$\lambda_{\text{max}}=\frac{b}{T} \qquad 51.13$$

$$L_{\text{max}}=b'T^5 \qquad 51.14$$

[10]Such a specification is also reasonably accurate in the visible region of the spectrum for tungsten filaments and other incandescent sources of light. The temperature used, however, is the *color temperature*, which is the temperature of a blackbody that produces the same color as the specified radiator (light source).

[11]The unit of luminous intensity was, until 1979, based on a blackbody operated at 2042K (the freezing point for platinum). The unit was 60 cd/m^2. The candela is now defined as the luminous intensity of a 555.016 nm source with a radiant intensity of (1/683) W/sr. Therefore, the candela is defined in terms of an electrical unit, the watt, which is much easier to measure, and not dependent upon the international temperature scale, which has changed over time.

[12]In a more exact version of the Planck radiation law, several of the terms are replaced by two constants (one for the numerator of the first fraction, the other for the numerator of the exponent) whose values are determined from experimental results.

[13]The units are shown with an m^2 and m separate instead of m^3. This is to emphasize that the radiance term (L) (with units of W/sr·m^2) is here given as spectral, that is per unit wavelength (hence units of 1/m). This also applies to L_λ and L_f in Eq. 51.9 and Eq. 51.10. Units are shown in a similar fashion in the nomenclature. The radiance, L, is dependent upon the direction of light that strikes a surface. That is, a $dA\cos\theta$ term is part of the definition. (See App. 51.A for a full listing of lighting units and terminology.)

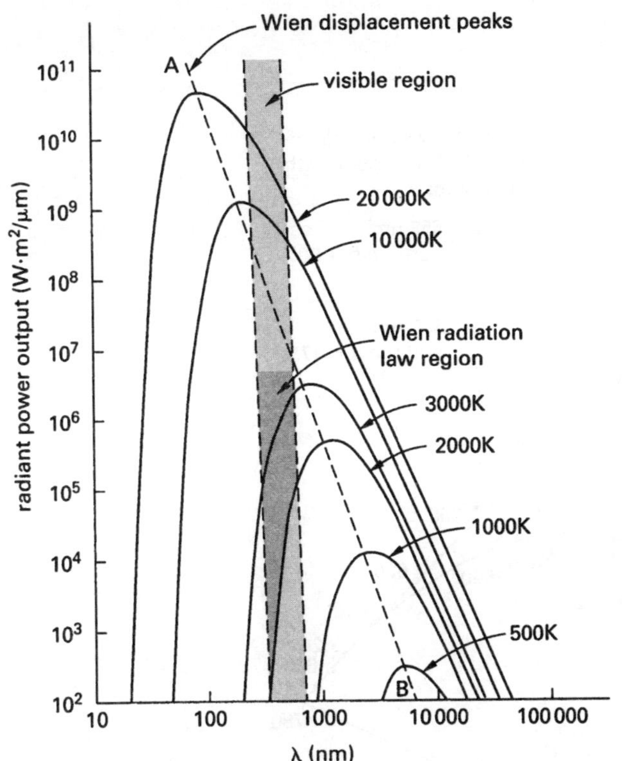

Figure 51.4 Blackbody Radiation Curves

The peaks calculated from Eq. 51.13 and Eq. 51.14 are located on the dashed line between points A and B in Fig. 51.4 (once the results are adjusted for wavelengths in micrometers as plotted in the figure).

9. STEFAN-BOLTZMANN LAW

The *Stefan-Boltzmann law* stipulates that the total radiant power per unit area of a blackbody varies as the fourth power of the absolute temperature, as given in Eq. 51.15. This is also known as the *fourth-power law* or *Stefan's law.*

$$P = \sigma A T^4 \qquad 51.15$$

In terms of the *luminous exitance*, *M*, the law is written

$$M_v = \sigma T^4 \qquad 51.16$$

In terms of the *radiance*, *L*, the law is written

$$L_e = \sigma_L T^4 \qquad 51.17$$

The Stefan-Boltzmann law applies to the entire spectrum, that is, it measures the total radiated power, not just that in the visible portion of the spectrum.

10. GRAYBODY AND SELECTIVE RADIATORS

The blackbody radiator has the most emissive power known. Other radiators must be corrected or compared to a blackbody radiator at the same temperature. The correction factor is called the *spectral emissivity*, ϵ, which is a function of the wavelength, $\epsilon(\lambda)$. The spectral emissivity is the ratio of the radiation emitted by a surface at a specified wavelength to the radiation emitted by an ideal blackbody radiator at the same wavelength and temperature. In terms of the luminous (in this case, radiant) exitance, the ratio is

$$\epsilon = \frac{M_e}{M_{\text{blackbody}}} \qquad 51.18$$

A *graybody* is an energy radiator that has a constant emissivity. That is, a graybody has a blackbody energy distribution reduced by the constant factor, ϵ, at all wavelengths. A *selective radiator* is a radiator with an emissivity that varies with wavelength. All three types of radiators are shown in Fig. 51.5.

Figure 51.5 Radiator Types

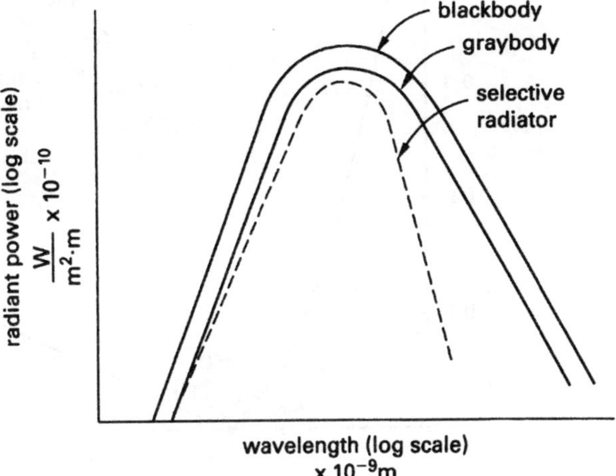

11. COLOR TEMPERATURE

The radiation characteristics of a blackbody may be determined from Planck's law by specifying only two parameters, the magnitude of the radiation at a given wavelength (L) and the absolute temperature (T). The same may be done in the visible region of the spectrum for selective radiators, such as the tungsten filaments in many incandescent sources, by using a temperature other than the actual temperature of the filament. This temperature is the color temperature. The *color temperature* is the temperature of a blackbody radiator of essentially the same color. The *distribution temperature* is the temperature of a blackbody whose relative spectral power distribution is most nearly that of the selective radiator.[14]

[14]Many different spectral power distributions (SPDs) exhibit the same color. The distribution temperature determines the specific power at a given wavelength, which is important in determining the overall effects of the light, including the physiological effects.

Figure 51.6 *CIE Chromaticity Diagram, 1951: Dominant Wavelength and Purity*

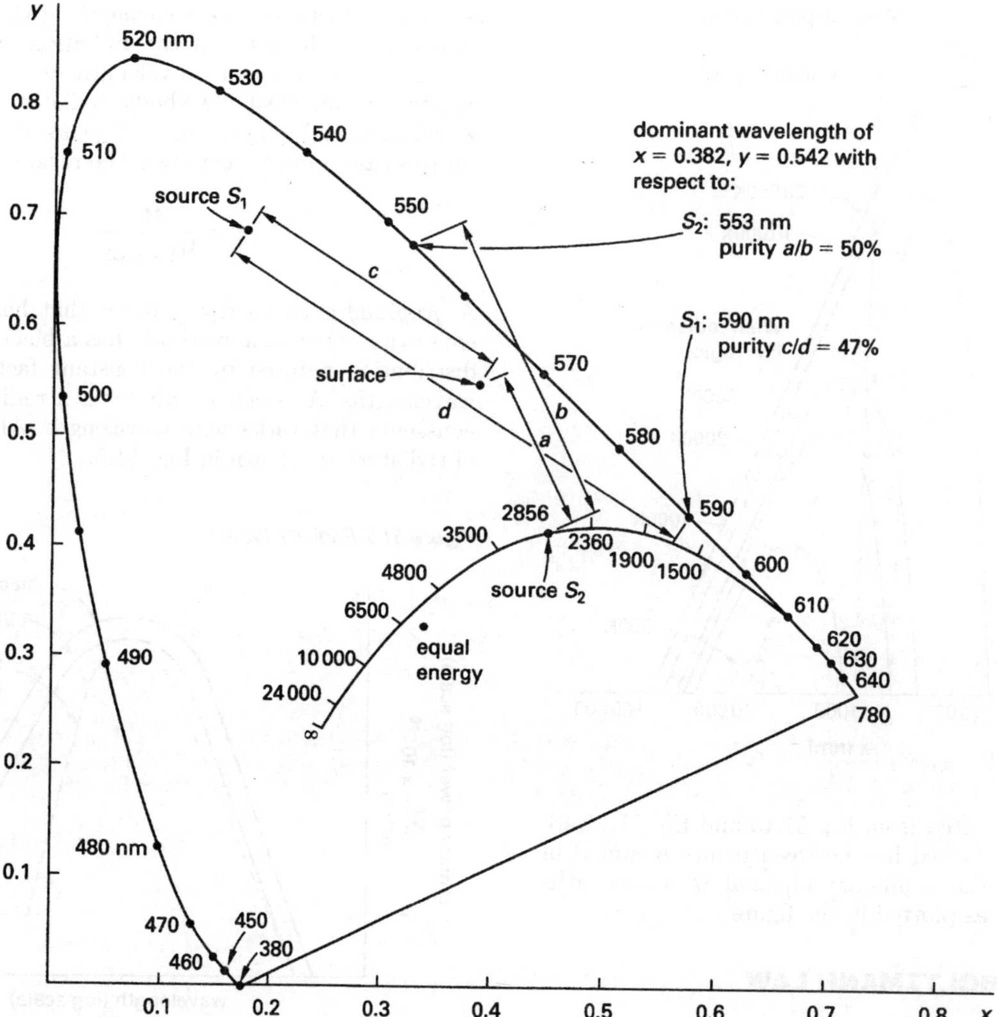

The color temperature is calculated from the chromaticity coordinates. An example of such a calculation using the (x, y) chromaticity coordinates is shown in Fig. 51.6. The details of the calculation are in the *IESNA Lighting Handbook*, and in the CIE standards. Once the wavelength is determined, the color temperature is known. Distances in the CIE (x, y) diagram (or X, Y, Z space) did not correlate well with perceived differences in color. Therefore, a CIE Uniform Color Space (UCS) was developed, as shown in Fig. 51.7. The u' and v' axes correspond to values determined by formulas for adjusting X, Y, Z space to the updated coordinate system used in the CIE Uniform Color Space (UCS).

The color temperature and distribution temperature specify the output of incandescent (hot body) sources only. For other sources, the *correlated color temperature* (CCT) is used. The CCT and the color temperature both are compared to a blackbody whose chromaticity most nearly matches that of the source. The terms simply differentiate the source. Both are commonly referred to as the color temperature.

Figure 51.7 *CIE Uniform Color Space (UCS), 1976*

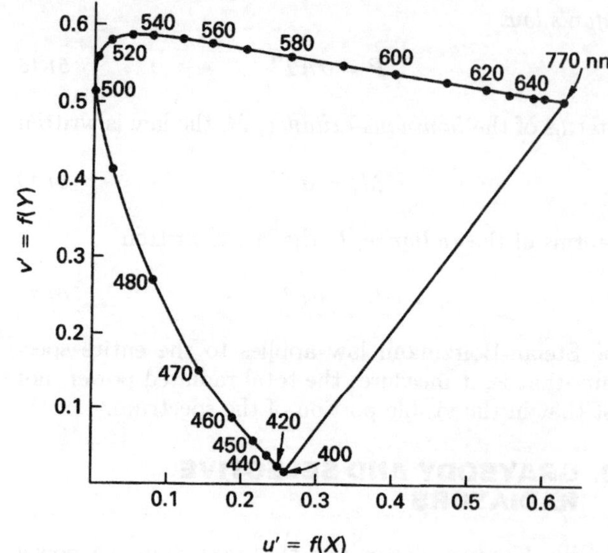

12. ATOMIC STRUCTURE: ELECTROMAGNETIC RADIATION

Electrons in orbit around the nucleus of an atom exist in various quantized energy states, or levels, that are categorized as groups called shells. Those electrons in the inner shells are not easily removed. Those in the outer shells are readily excited to higher energy levels and can also be removed from their associated atoms. These outer shell electrons are called valence electrons. When UV or visible radiation is absorbed by an outer shell electron, the electron moves to a higher energy state. (In photodiodes, this higher energy state is in the conduction band.) When the electron eventually returns to its original energy level, it emits a photon of radiation with energy given by

$$E_2 - E_1 = h\nu_{21} \qquad 51.19$$

The energy level differences can be represented as a potential difference in volts. Equation 51.19 can be put in the useful form of Eq. 51.20, in which the wavelength in nanometers is given in terms of the potential difference, in volts. In terms of the energy gap, the wavelength is given by Eq. 51.21 and is illustrated in Fig. 51.8.

$$\lambda_{\text{nm}} = \frac{1239.76\ \frac{\text{nm}}{\text{V}}}{\Delta V} \qquad 51.20$$

$$\lambda = \frac{hc}{E_g} \qquad 51.21$$

Figure 51.8 Light Emitting Diode (LED) p-n Junction

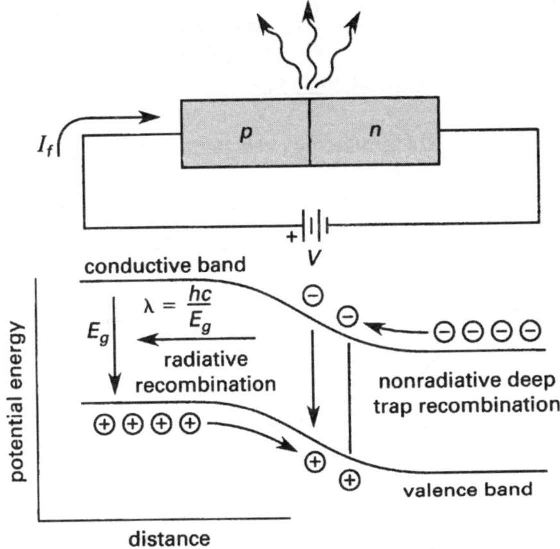

The cause of light emission is the electronic transitions from higher to lower energy states with the release of electromagnetic radiation (waves) to maintain the energy balance.[15] In spite of this, light sources are historically divided into two types, incandescent and luminescent. *Incandescent* light sources emit visible light because of their temperature. Examples are filament lamps, pyroluminescence (flames), candoluminescence (gas mantles), and carbon arc radiation. *Luminescent* light sources emit visible light caused by any factor other than their temperatures, including the following examples. Gas discharges and fluorescence, which are forms of immediate light release, are termed photoluminescence (see Fig. 51.9). Phosphorescence is a delayed release of light. Lasers emit coherent light. Electroluminescence includes lamps that may be AC capacitive or light emitting diodes. Electron excitation causes cathodoluminescence. Galvanoluminescence and chemiluminescence result from chemical changes; crystalloluminescence results from

[15]Care should be taken in the study of illumination as the terms energy and power are often used interchangeably, which may lead to errors in calculations.

Figure 51.9 Fluorescent Phenomena

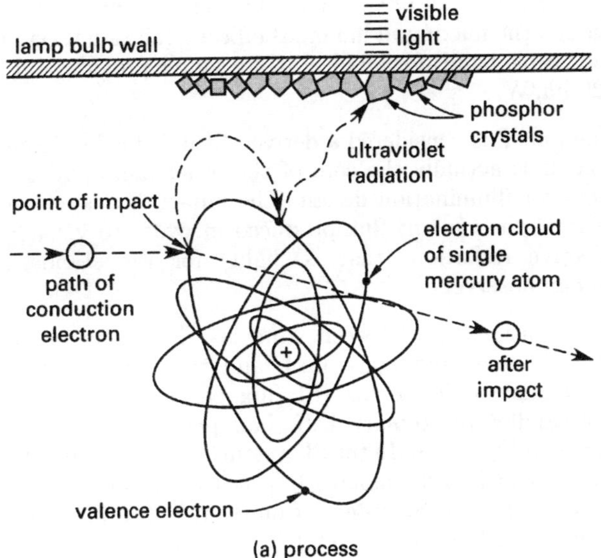

(a) process

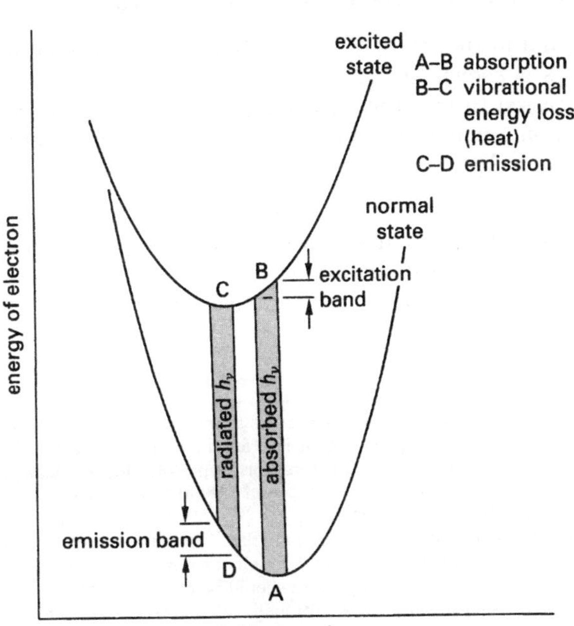

(b) typical phosphor energy diagram

Special Applications

crystallization; thermoluminescence from heat; triboluminescence from friction; sonoluminescence from ultrasonics; and radioluminescence from the interactions of α, β, γ, and X-ray particles with matter.

13. THE CANDELA AND LUMEN

The base SI unit for luminous intensity, I, is the candela (cd), which represents the luminous flux per unit solid angle, or steradian. As defined in the SI, the *candela* is the luminous intensity, in a given direction, of a source that emits a monochromatic radiation of frequency 540×10^{12} Hz, and that has a radiant intensity in that direction of 1/683 watt per steradian. The frequency in the definition correlates with a wavelength of 555 nm, which is the wavelength of maximum photopic (cone) or daylight efficiency for human vision.[16] Additionally, there is a total of 683 lm available per watt at 555 nm, though the maximum luminous efficacy (all radiation in the visible band) for an ideal white source provides only 220 lm/W.

The lumen is considered a derived SI unit for luminous flux. It is actually the core of the other units and the focus for illumination design. The lumen, lm, is a unit relating the radiant flux produced in watts to visually effective radiation (that is, light) for the standard human observer.

The relationships between candelas (luminous intensity), lumens (luminous flux), and the corresponding illuminance units of lux (SI flux per unit area) and foot-candles (customary U.S. flux per unit area) are shown in Fig. 51.10. In the SI system, at 1 m, a luminous intensity of 1 cd illuminates 1 lx over an area of 1 sr. In the customary U.S. system of units, at 1 ft, a luminous intensity of 1 cd illuminates 1 fc over an area of 1 sr. A 1 cd source provides 4π total lumens of light.

An aid in the study of illumination terms and relationships is shown in Fig. 51.11. The terms and units are discussed throughout this chapter. Additional information on luminance and illuminance conversion factors is given in App. 51.B and App. 51.C.

The *luminous efficacy*, K, is a measure of the luminous flux output emitted relative to the total source power input, and is given by Eq. 51.22.[17] The term "luminous efficiency" has been used extensively for this term,

Figure 51.10 Relationships Between Candela, Lumen, Lux, and Foot-candle

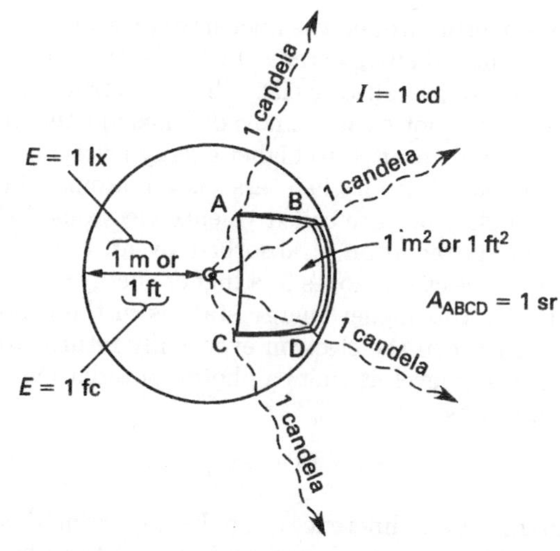

Figure 51.11 Illumination Terms and Relationships

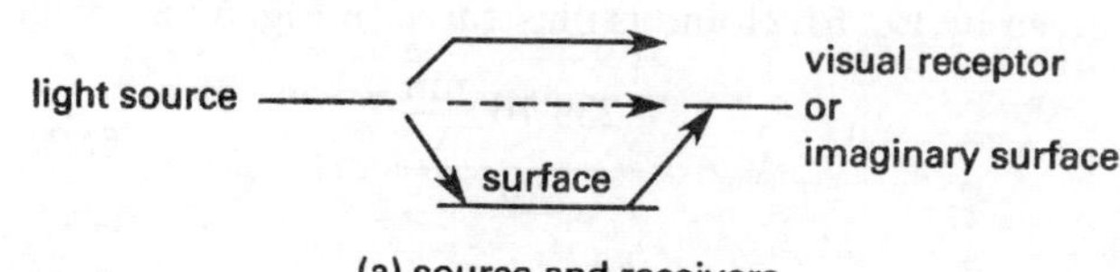

(a) source and receivers

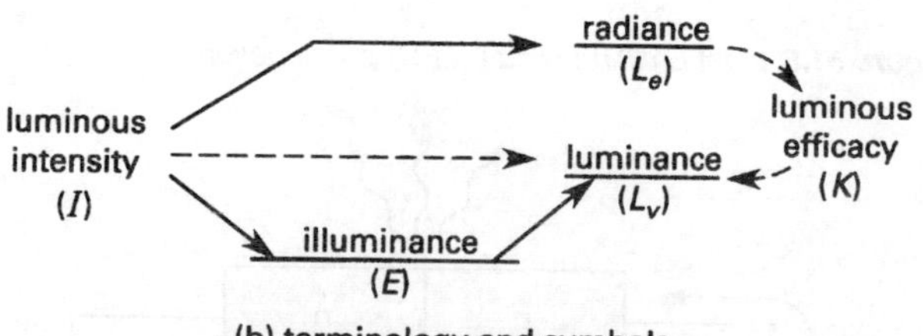

(b) terminology and symbols

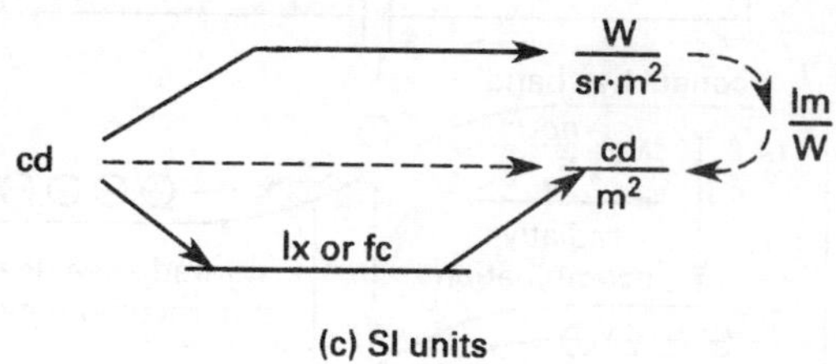

(c) SI units

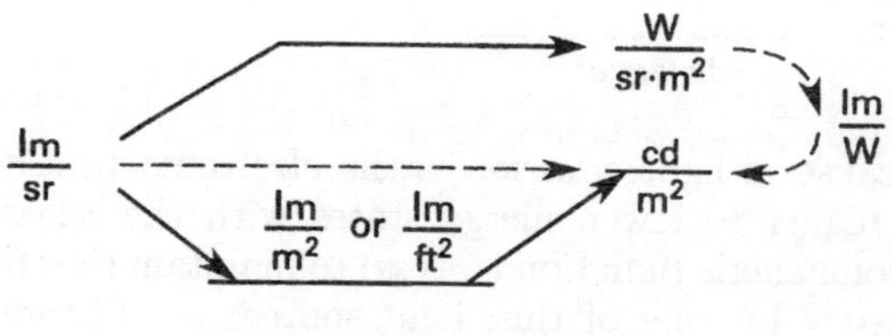

(d) decomposition of units

[16]There are two types of photoreceptors in the human eye. One is the cone or photopic receptor, responsible for daylight or bright vision. The other is the rod or scotopic receptor, responsible for nighttime or dark condition vision. The two types of receptor have different efficiencies at the same wavelength.

[17]The terms "radiant flux," "luminous flux," "radiant energy," "luminous energy," "radiant power," and "luminous power" may all be used to indicate the same quantity. Care should be taken with the units to ensure correct calculations. Additionally, if the term "radiant" or a variant is used, the power output (in watts) is the focus. If the term "luminous" or a variant is used, the visible light output (in lumens) is the focus.

Special Applications

therefore, the symbol η is used as well as K. Typical luminous efficacy values are shown in Table 51.5.

$$K = \eta = \frac{\Phi_v}{\Phi_e} = \frac{\Phi_v}{P} \qquad 51.22$$

The *radiant exitance*, M_e, is the radiant flux leaving a surface (that is, the source of the flux) in units of W/m^2, and is given by Eq. 51.23.

$$M_e = \frac{\Phi_e}{A_{\text{source}}} \qquad 51.23$$

The *luminous exitance*, M_v, (formerly called the *luminous emittance*, and roughly referred to as the brightness) is the luminous flux per unit area of the source (lm/m^2), given by Eq. 51.24.

$$M_v = \frac{\Phi_v}{A_{\text{source}}} \qquad 51.24$$

***Table 51.5** Luminous Efficacy*

power (W)	source	K (lm/W)
	candle	0.1
25/100/1000	tungsten lamp	10/16/22
100/1000	mercury arc	35/65
400	mercury fluorescent lamp	58
1000	carbon arc	60
40	fluorescent lamp	80
400/1000	metal halide lamp	85/100
400/1000	high-pressure sodium lamp	125/130
180	low-pressure sodium lamp	180
	ideal white light	220

Example 51.1

A 100 W lamp has a luminous flux of 4400 lumens. What is the lamp's luminous efficiency?

Solution

From Eq. 51.22,

$$K = \eta = \frac{\Phi_v}{P} = \frac{4400 \text{ lm}}{100 \text{ W}} = 44 \text{ lm/W}$$

Example 51.2

A tungsten filament emitting 1600 lm has a surface area of 0.35 in^2 (2.26×10^{-4} m^2). What is its brightness?

SI Solution

$$M = \frac{\Phi}{A_{\text{surface}}} = \frac{1600 \text{ lm}}{2.26 \times 10^{-4} \text{ m}^2} = 7.08 \times 10^6 \text{ lm/m}^2$$

Customary U.S. Solution

$$M = \frac{\Phi}{A_{\text{surface}}} = \frac{1600 \text{ lm}}{\dfrac{0.35 \text{ in}^2}{144 \dfrac{\text{in}^2}{\text{ft}^2}}} = 6.58 \times 10^5 \text{ lm/ft}^2$$

14. LUMINOUS INTENSITY

The *luminous intensity*, I, of a source is the flux emitted per unit solid angle from the point where the source radiated. The units are candela (cd), which are lumens per steradian (lm/sr).[18]

$$I = \frac{d\Phi}{d\omega} \qquad 51.25$$

Equation 51.26 gives the luminous intensity from a uniformly radiating point source a distance r from that source.

$$I = \frac{\Phi}{\omega} = \frac{\Phi}{\dfrac{A}{r^2}} = \frac{\Phi_t r^2}{4\pi r^2} = \frac{\Phi_t}{4\pi} \qquad 51.26$$

Example 51.3

Two lumens pass through a circular hole (diameter is 0.5 m) in a screen located 6.0 m from an omnidirectional source. (a) What is the luminous intensity of the source? (b) What luminous flux is emitted by the source?

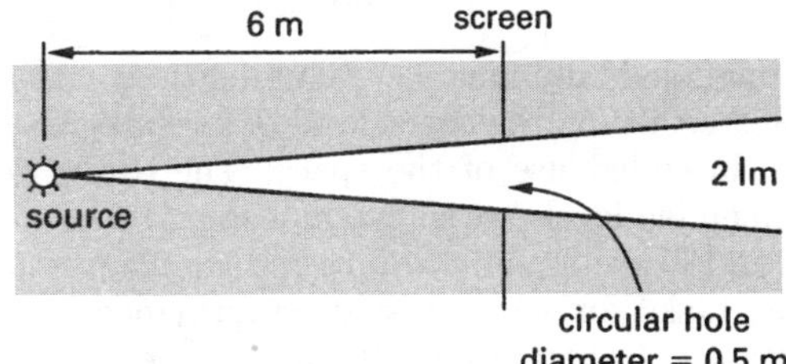

Solution

(a) The solid angle subtended by the hole is

$$\omega = \frac{A}{r^2} = \frac{\frac{\pi}{4}(0.5 \text{ m})^2}{(6 \text{ m})^2} = 0.005454 \text{ sr}$$

[18]The unit "steradian" is technically a ratio and is normally not shown (see Chap. 7). In photometry, and as approved by National Institute of Standards and Technology (NIST) for use in the SI system, the name "steradian" and the symbol "sr" are retained in expressions for units, for clarity.

Special Applications

From Eq. 51.26, the intensity is

$$I = \frac{\Phi}{\omega} = \frac{2\ \text{lm}}{0.005454\ \text{sr}} = 366.7\ \text{lm/sr}$$

(b) From Eq. 51.26, the luminous flux is

$$\Phi_t = 4\pi I = (4\pi\ \text{sr})\left(366.7\ \frac{\text{lm}}{\text{sr}}\right) = 4608\ \text{lm}$$

15. ILLUMINANCE

The *illuminance*, E, is the areal density of the luminous flux at a point on a real or imaginary surface. The units are lux (lx), which are equivalent to lumens per square meter (lm/m^2). (When referring to the radiant flux at a point on the surface, that is, the power incident to the surface, the units are watts per square meter, W/m^2, and the term *irradiance* is used. Subscripts are used to differentiate photometric from radiometric quantities only when necessary for clarification. The symbol changes from E to E_v for illuminance, and to E_e for irradiance. Additionally, the term illuminance is often called *illumination*, irradiance, and brightness.[19])

$$E = \frac{d\Phi}{dA} \qquad 51.27$$

Equation 51.28 gives the illuminance from an omnidirectional source at a spherical receptor of radius r. (The receptor, as in all illumination equations, may be a real or imaginary area at which the illuminance is of interest.)

$$E = \frac{\Phi_t}{4\pi r^2} \qquad 51.28$$

Recommended illuminance (illumination) levels are specified either by the exact level or by category, based on the intended use of the space. The categories and recommended levels are shown in Table 51.6. Categories A, B, and C are for *orientation and simple visual tasks*, that is, tasks for which visual performance is of little importance. Categories D, E, and F are for *common visual tasks* for which visual performance is important. Categories D, E, and F represent the majority of commercial, industrial, and residential lighting requirements. Category G is for *special visual tasks* for which visual performance is critical, as in hospital operating rooms. Illumination levels in general should be within 10% of the indicated values.

[19]Definitions using the SI standard are given throughout this text. Many terms, including irradiance, illumination, and brightness, are no longer recommended. However, they still occur in a variety of publications, and are used for clarity when speaking to those not familiar with illumination engineering. When in doubt, refer to ANSI/IES RP-16, *Nomenclature and Definitions for Illuminating Engineering*.

Table 51.6 *Recommended Illuminance*

category	location/condition/ magnitude	lux	fc
A	public spaces	30	3
B	simple orientation/short visits	50	5
C	cooking space/simple visual tasks	100	10
D	visual task/high contrast required/large size	300	30
E	visual task/high contrast/ small size	500	50
F	visual task/low contrast/ small size	1000	100
G	critical visual tasks	3000–10 000	300–1000

16. ILLUMINANCE: INVERSE SQUARE LAW

The illuminance on a surface varies in accordance with the *inverse square law*, given by Eq. 51.29. The inverse square law is accurate to within 1% when the distance d is at least five times the maximum dimension of the source (or luminaire). This is called the *five-times rule*.

$$E = \frac{I}{d^2} \qquad 51.29$$

When comparing the illuminance (illumination) from the same source at two different points, the inverse square law takes the form of Eq. 51.30. The impact of the inverse square law on light flux is illustrated in Fig. 51.12(a).

$$E_1 r_1^2 = E_2 r_2^2 \qquad 51.30$$

Example 51.4

A lamp radiating hemispherically and rated at 2000 lm is positioned 20 ft (6 m) above the ground. What is the illumination on a walkway directly below the lamp?

SI Solution

$$E = \frac{\Phi}{A} = \frac{\Phi}{\frac{1}{2}A_{\text{sphere}}} = \frac{2000\ \text{lm}}{\left(\frac{1}{2}\right)(4\pi)(6\ \text{m})^2} = 8.84\ \text{lx}$$

Customary U.S. Solution

$$E = \frac{\Phi}{A} = \frac{2000\ \text{lm}}{\left(\frac{1}{2}\right)(4\pi)(20\ \text{ft})^2} = 0.796\ \text{lm/ft}^2\ \text{(ft-c)}$$

Figure 51.12 Illuminance Laws

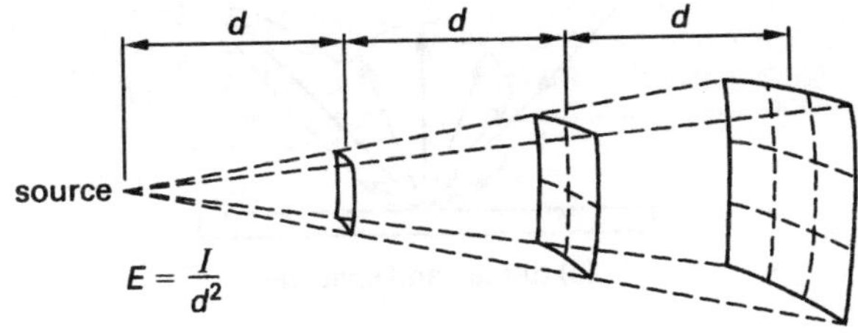

(a) inverse square law

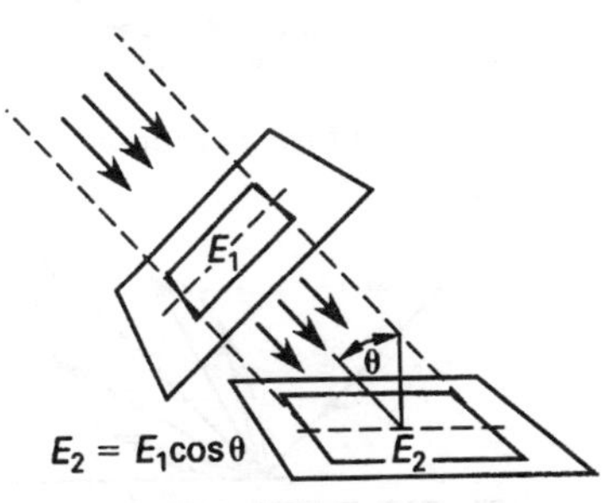

(b) Lambert's law

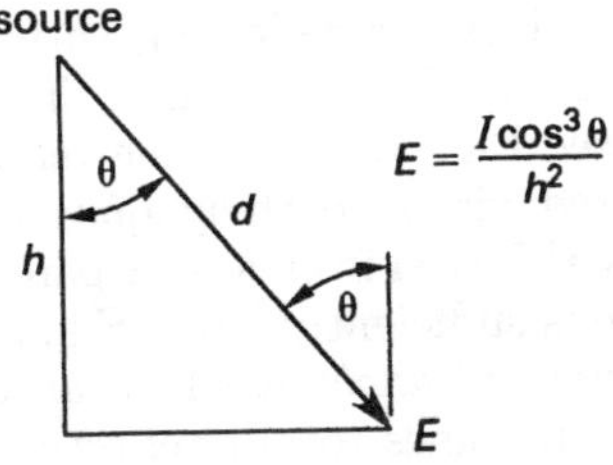

(c) cosine cubed law

17. ILLUMINANCE: LAMBERT'S LAW

The illuminance on a surface depends on the angle at which the light flux strikes the surface. *Lambert's law,* also called the *cosine law,* states that the illuminance on any surface varies with the cosine of the angle of incidence, θ, as shown in Eq. 51.31 and in Fig. 51.12(b). When combined with the inverse square law, the result is Eq. 51.32.

$$E_2 = E_1 \cos\theta \quad \text{51.31}$$

$$E = \frac{I}{d^2}\cos\theta \quad \text{51.32}$$

18. ILLUMINANCE: COSINE-CUBED LAW

When determining the illuminance level at an angle from the source, a useful extension of Lambert's cosine law is the *cosine-cubed law.* This law is derived by considering the geometry of Fig. 51.12(c), which allows the substitution of $h/\cos\theta$ for d in the inverse square law, Eq. 51.29, resulting in

$$E = \frac{I\cos^3\theta}{h^2} \quad \text{51.33}$$

19. LUMINANCE

The *luminance, L,* is the ratio of the luminous intensity, in a given direction, of an infinitesimal element of a surface containing the point, to the orthogonally projected area of the element on a plane perpendicular to the given direction. Equation 51.34 represents the concept mathematically. Figure 51.13 illustrates the concept physically. (When referring to the radiant flux at a point on the surface, that is, the power incident to the surface, the units are watts per steradian per square meter, W/sr·m², and the term *radiance* is used. Subscripts are used only to differentiate photometric from radiometric quantities, when necessary for clarification. The symbol changes from L to L_v for luminance, or to L_e for radiance.)

$$L = \frac{d^2\Phi}{d\omega dA\cos\theta} = \frac{dI}{dA\cos\theta} \quad \text{51.34}$$

Figure 51.13 Mathematical Concept of Luminance

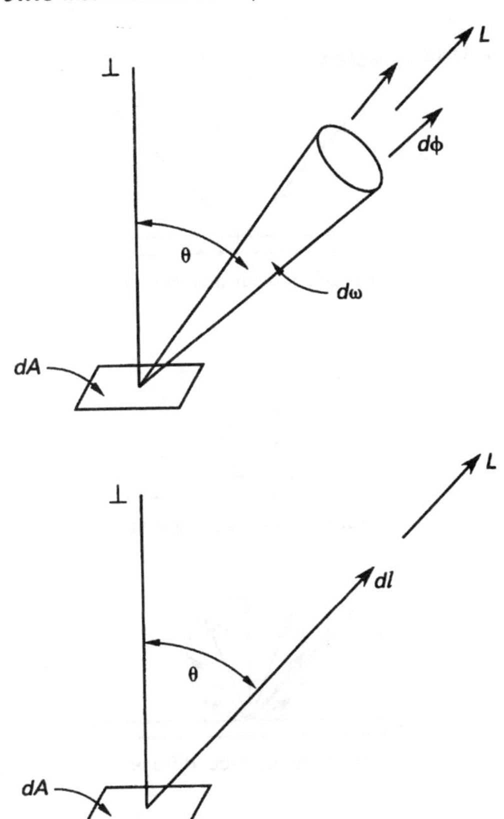

Special Applications

20. INTERACTION OF LIGHT WITH MATTER

Light travels through a vacuum as an electromagnetic wave. When the light makes contact with matter, some of the wave energy is absorbed by the matter, sometimes causing electrons to jump into higher energy states. (A polished metal surface, for example, will absorb only about 10 percent of the incident energy, reflecting the remaining 90 percent away.) Some of this absorbed energy is re-emitted when the electrons drop back to a lower energy level. Generally, the re-emitted light will not be at its original wavelength.

If the reflecting surface is smooth, the *reflection angle* for most of the light will be the same as the *incident angle* and the light is said to be *regularly reflected* (i.e., the case of *specular reflection*). If the surface is rough, however, the light will be scattered and reflected randomly (the case of *diffuse reflection*).

Examples of specular and diffuse reflections are shown in Fig. 51.14. Compound reflections are illustrated in Fig. 51.15.

The energy that is absorbed is said to be *refracted*. In the case of an *opaque material*, the refracted energy is absorbed within a very thin layer and converted to heat. However, the light is able to pass through a *transparent material* without being absorbed. Light is partially absorbed in a *translucent material*.

Figure 51.14 Reflections

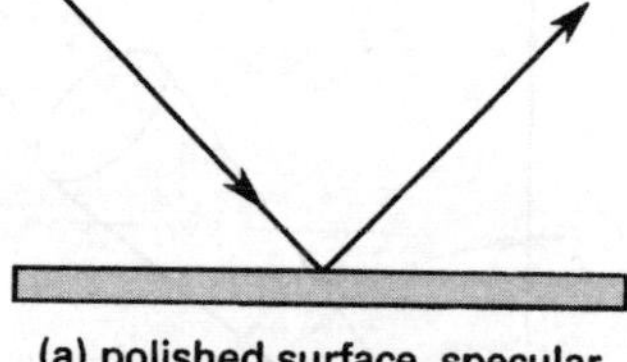

(a) polished surface, specular

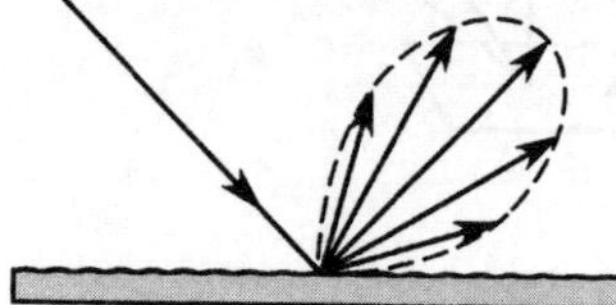

(b) rough surface, spread

(c) matte surface, diffuse

Figure 51.15 Compound Reflections

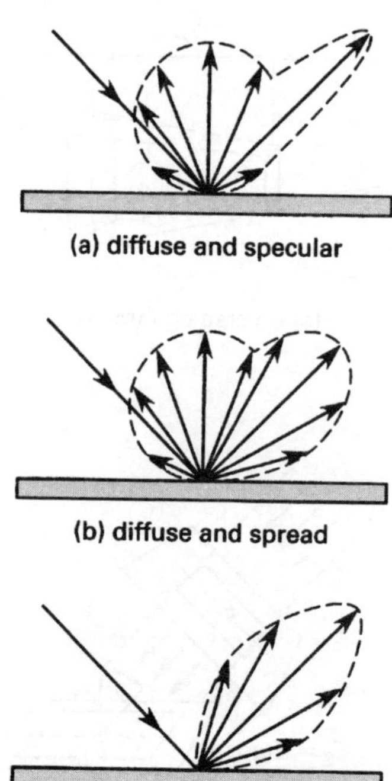

21. REFLECTION

Reflected light leaves a reflecting surface at the same angle (as measured from a normal line) it approaches the surface (see Fig. 51.16). While reflections are normally associated with smooth opaque surfaces, light can also be totally reflected from a transparent surface if the incident angle is sufficiently large. Equation 51.35 gives the *critical incident angle* at which an optically transparent surface becomes totally reflecting. Total reflecting prisms are used in place of silvered mirrors when precise reflection is required. Total internal reflection is also the principle by which *optical fibers (light pipes)* transmit light.

$$\sin \theta_c = \frac{1}{n} \qquad 51.35$$

Figure 51.16 Reflection from a Surface

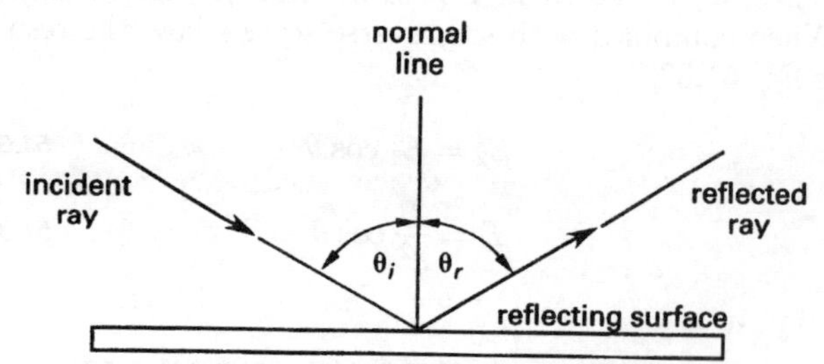

An example of the critical angle for a light ray interacting with glass (or a fiber-optic cable) is illustrated in Fig. 51.17. Control of the light radiation is vital to the output of a fiber-optic system. Therefore, fiber-optic cables are coated with a material with a very low refractive index, n, making the $\sin\theta \approx 1$.

Figure 51.17 *Critical Angle*

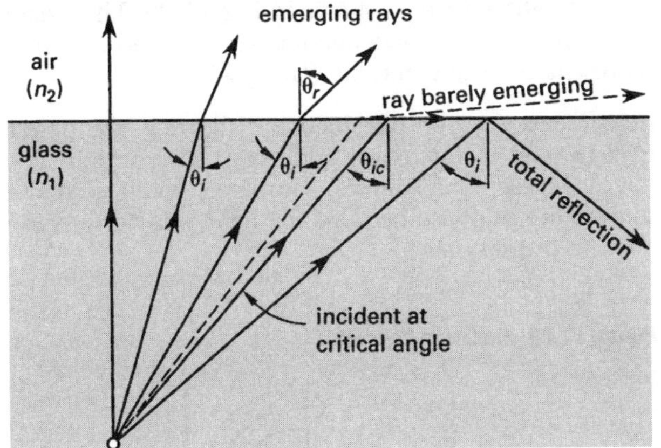

22. REFRACTION

Refraction is the bending of light as it passes from one transparent medium into another. *Snell's law*, Eq. 51.36, relates the incident and refracted angles and predicts that the light will bend *toward the normal* when it enters an optically denser material. For a vacuum (and, for practical purposes, air), $n_1 = 1$.

$$n_{\text{relative}} = \frac{n_2}{n_1} = \frac{\sin\theta_1}{\sin\theta_2} \qquad 51.36$$

If a light beam passes through a transparent medium with parallel surfaces, the emergent beam will be parallel to the incident beam, as illustrated in Fig. 51.18(a). However, the refraction due to a prism with apex angle α, illustrated in Fig. 51.18(b), is more complex. Equation 51.37 predicts the minimum *angle of refraction* (*angle of deviation*) from the original path.[20]

$$n_{\text{relative}} = \frac{\sin\frac{\alpha+\theta}{2}}{\sin\frac{\alpha}{2}} \qquad 51.37$$

Refraction causes submerged objects to appear closer to the surface than they actually are. Equation 51.38 gives the apparent depth.

$$\text{apparent depth} = \frac{\text{actual depth}}{n_{\text{relative}}} \qquad 51.38$$

[20]Different wavelengths (colors) will be reflected at slightly different angles.

Figure 51.18 *Refraction of Light*

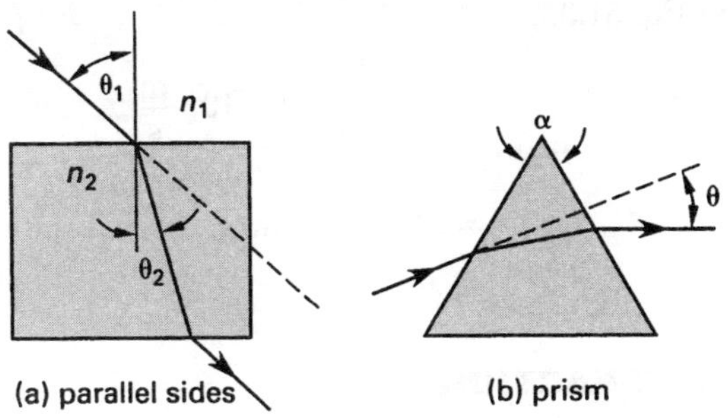

23. INDEX OF REFRACTION

While the speed of light in a vacuum is constant, it changes in different transparent media. The absolute *index of refraction* (*refractive index*), n, is the ratio of the speed of light in a vacuum (essentially the same as in air) to the speed of light in a particular medium.[21] The approximate absolute indices of refraction are listed in Table 51.7. It is not strictly constant but varies 1–2% over the visible light spectrum. This variation is disregarded, however, in simple studies.

$$n = \frac{c_{\text{vacuum}}}{c_{\text{medium}}} = \frac{3\times10^8\ \frac{\text{m}}{\text{s}}}{c_{\text{medium,m/s}}} \qquad 51.39$$

Most transparent substances including glass, water, air, and polymethyl metacrylate (Lucite™) are *isotropic*; that is, light travels at the same speed in all directions. However, some transparent crystals are *anisotropic*.

Table 51.7 *Approximate Absolute Indices of Refraction (at $\lambda = 5.893\times10^{-7}$ m)*

medium	n
air (20°C, 1 atm)	1.00029
benzene	1.50
borosilicate crown glass	1.5243
diamond	2.417
flint glass, dense	1.6555
hydrogen (20°C, 1 atm)	1.00013
ice	1.31
quartz, fused	1.46
salt	1.53
water (20°C)	1.3330

Example 51.5

At a particular frequency corresponding to red light, the refractive index of water is approximately 1.3300. What is the speed of this light in water?

[21]The *relative index of refraction*, the ratio of the speeds of light in two different media, is also encountered.

Solution

From Eq. 51.39,

$$c_{water} = \frac{c_{vacuum}}{n} = \frac{3 \times 10^8\ \frac{m}{s}}{1.3300}$$
$$= 2.26 \times 10^8\ m/s$$

24. DIFFRACTION

Diffracted light is light whose path has been changed by passing around corners or through narrow slits. As a consequence of *Huygens' principle*, each edge of a single diffraction slit (shown in Fig. 51.19) acts as a source of light waves and is capable of producing interference. For diffraction of monochromatic light by a single slit, the *m*th order reinforcements and cancellations are the same as predicted by Eq. 51.40 through Eq. 51.43.

The same equations can also be used for a *diffraction grating*, a transparent sheet containing numerous slits (scratches) spaced a distance *d* (known as the *grating space* or *grating constant*) apart and typically found in a *diffraction grating spectrometer*.[22,23] The angle at which the spectrometer must be turned to view an *m*th order image can be calculated from $\sin\theta = y/s \approx y/x$. The maximum number of orders of interference produced can be determined by setting θ equal to 90°.

$$\sin\theta = \frac{m\lambda}{d} \quad \text{[in phase]} \qquad 51.40$$

$$\sin\theta = \frac{(2m+1)\lambda}{2d} \quad \text{[out of phase]} \qquad 51.41$$

Figure 51.19 *Diffraction Around a Corner*

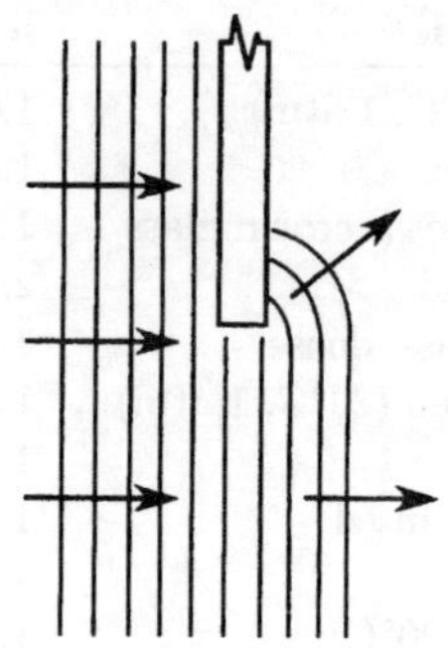

25. INTERFERENCE

Interference (also known as *cancellation*) occurs when two waves combine so that one subtracts from the other.

It is also possible for two waves to add to each other, which is known as *reinforcement* or *superposition.*

Most light sources emit light waves with random phase relationships. However, a source of parallel rays can be obtained by allowing light to pass through a narrow slit. According to *Huygens' principle* (also known as *Huygens' construction*), each point in a wave front can be considered as a source of waves. Therefore, the slit will act as a light source, as shown in Fig. 51.20. The circular lines represent the wave maxima (crests) and the spaces represent the wave minima (troughs).

Interference can be obtained by allowing the parallel rays to pass through two additional slits, which themselves act as two sources. However, these secondary sources are in phase because the light was derived from a single primary slit.

Figure 51.20 *Huygens' Principle*

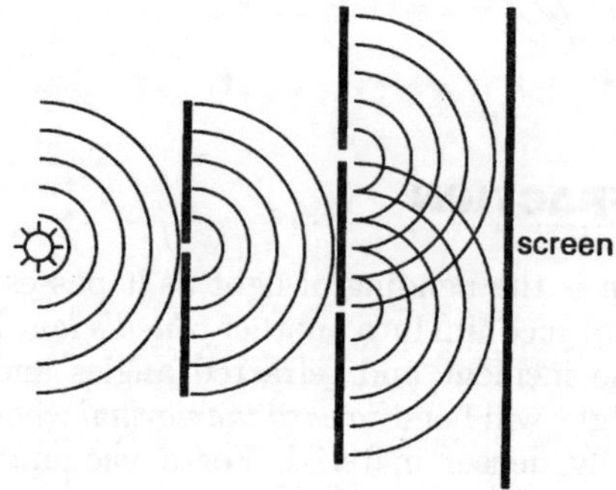

26. INTERFERENCE FROM SLITS

The visual effect on a distant screen of combining two in-phase sources will be regions of darkness and light. (The arrangement shown in Fig. 51.21 is known as *Young's experiment.*) A bright spot or band means that two wave crests or two wave troughs coincide (i.e., are in phase at that point) and thereby reinforce each other. If a trough coincides with a crest, the two sources are 180 degrees out of phase, and the result is a dark spot or band.

Figure 51.21 *Interference*

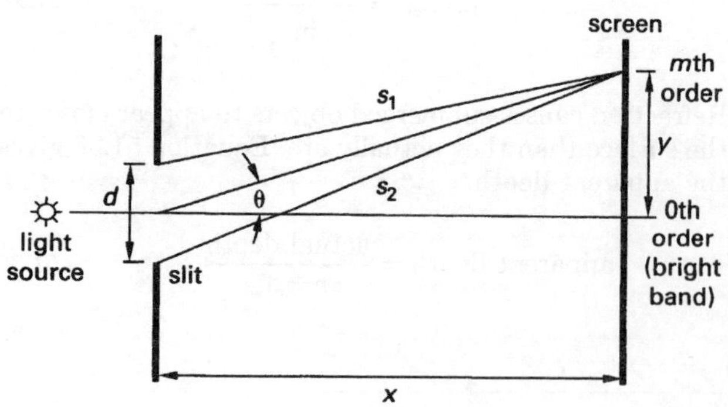

[22]The number of slits per centimeter ranges from 400 to 6000.

[23]While the number of slits (one, two, or hundreds) does not affect the position of the image, it does affect the brightness of the image.

For reinforcement (i.e., a bright band) to occur, the difference in path lengths must be a whole number, m, of wavelengths. The number m is known as the *order of interference.* The central bright band is of the zeroth order, and there are two of each mth order image, one on each side of the center. When $x \gg d$, small angle approximations are valid and $y/x = (s_2 - s_1)/d$. For the mth reinforcement (light band, maximum, etc.),

$$s_1 - s_2 = m\lambda \quad \text{[in phase]} \tag{51.42}$$

$$y = \frac{m\lambda x}{d} \quad \text{[in phase]} \tag{51.43}$$

Between the bright bands are dark bands of cancellation. For cancellation, the difference in path lengths is an odd number of half wavelengths. For the mth cancellation (dark band, minimum, etc.),

$$s_1 - s_2 = \frac{(2m+1)\lambda}{2} \quad \text{[out of phase]} \tag{51.44}$$

$$y = \frac{(2m+1)\lambda x}{2d} \quad \text{[out of phase]} \tag{51.45}$$

Example 51.6

A screen is placed 2.4 m from a sodium gas discharge tube that is emitting two beams of in-phase light ($\lambda = 5.893 \times 10^{-7}$ m) separated by 0.0005 m. What is the distance from the central image to the first reinforcement?

Solution

From Eq. 51.43 with $m = 1$,

$$y = \frac{m\lambda x}{d} = \frac{(1)(5.893 \times 10^{-7}\ \text{m})(2.4\ \text{m})}{0.0005\ \text{m}} = 0.0028\ \text{m}$$

27. INTERFERENCE FROM THIN FILMS

Interference occurs when light passes through thin films (walls of soap bubbles, layers of oil on water, etc.)[24] If the incident beam is white light (i.e., is composed of light of all colors), each wavelength will produce its own interference pattern, resulting in a rainbow effect.

Figure 51.22 shows a solid reflective surface covered by a transparent film of thickness t. Ray D is composed of a partial reflection of ray B and the remainder of ray A. Because of the 180 degree phase reversal at point G, cancellation along ray D requires that the path FGH be an integral number of wavelengths. However, the path length is also approximately equal to twice the film thickness. Equation 51.46 and Eq. 51.47 define the relationships for the mth order reinforcement and cancellation, respectively. It is essential that the wavelength in the film, λ_{film}, be used, not the free space wavelength.

$$(m + {}^1\!/_2)\lambda_{\text{film}} = 2t \quad \text{[in phase]} \tag{51.46}$$

$$m\lambda_{\text{film}} = 2t \quad \text{[out of phase]} \tag{51.47}$$

$$\lambda_{\text{film}} = \frac{\lambda_{\text{vacuum}}}{n_{\text{film}}} \tag{51.48}$$

Figure 51.22 *Interference from a Thin Film*

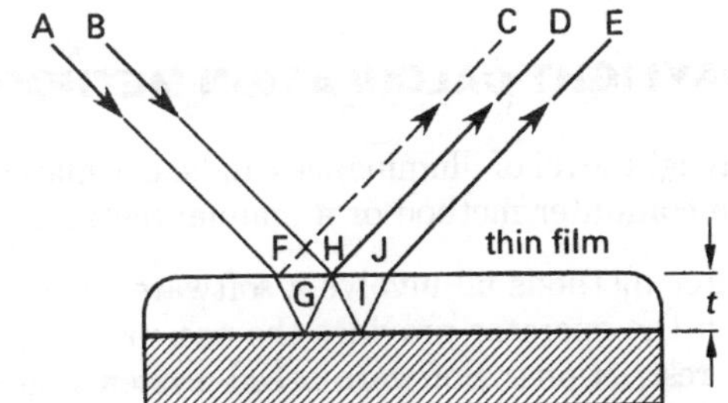

28. LIGHTING DESIGN METHODS

Various design methods exist for determining the illuminance in a general area or at particular point. These methods are dependent on the source of light, the type of lighting used, and the tasks to be accomplished in the space. Both direct and indirect (from reflection) light sources are considered in the methods.[25]

General area calculations focus on the average illuminance, considering all sources—daylight, reflections, interior lighting sources (luminaries). When daylight is part of the calculation, the *lumen method* is appropriate. If only interior lighting sources are involved, the *zonal cavity method* is appropriate. The lumen and zonal cavity methods differ in the use of factors. However, the terms "lumen" and "zonal cavity" are often used interchangeably.

If lighting for specific task accomplishment is the design goal, the *point method* is used. This method calculates the illuminance at a point or surface using variations of the laws as described in Sec. 51.16, Sec. 51.17, and Sec. 51.18. Supplementary sources, controlled separately from the main lighting source, are often incorporated into designs requiring such *task lighting.*

[24]A circular pattern of dark bands known as *Newton's rings* is created by interference in a film of thin air trapped between a flat reflecting surface and a plano-convex lens placed flat side up.

[25]Details of these calculations are extensive. If interested, Chap. 8 and Chap. 9 of the *IESNA Lighting Handbook* (Ninth Edition, 2000, Illuminating Engineering Society of North America, New York) provide the details.

29. DAYLIGHT

Daylight has unique spectra and light-level distributions that distinguish it from artificial lighting. Daylight varies daily and seasonally due to movement of the sun.[26] Variations in daylight also occur from weather and local environmental factors. As a result, the CIE has developed three spectral radiant power distributions for daylight. They are known by the symbols D_{55}, D_{65}, and D_{75}, which stand for daylight at 5500K, 6500K, and 7500K respectively (see Sec. 51.11 for an explanation of color temperature).

Using the standard power distributions of daylight, mean illuminance from the sun can be determined by factoring in the site location, the time of desired illumination, the solar position, skylight and groundlight contributions, and human response factors. Such calculations can be difficult and complicated.

30. DAYLIGHT CALCULATION METHODS

The daylight level of illuminance can be calculated using either a computer method or a manual method.

Computer methods all involve a software-based system in which the operator provides the necessary input and design restrictions. Software-based systems generally utilize the CIE standard distributions and can be customized with any number of input factors to determine the mean illuminance. Computer methods are the most accurate, and depending upon the level of detail and accuracy, can be expensive.

A manual method is appropriate for first-order calculations on direct and reflected components of light. Some examples of manual methods are the lumen method, used for both toplighting and sidelighting, and the daylight factor method.[27]

31. DAYLIGHT LUMEN METHOD

The daylight lumen method[28] uses the following steps:

step 1: Calculate exterior illuminances using the CIE standard spectral radian power distributions modified for the factors mentioned in Sec. 51.29.

step 2: Accounting for fenestration, determine the net transmittance into the room. The fenestration is defined as any opening or arrangement of openings exposing daylight. Fenestration determines the level of light available after accounting for furniture placement, reflectivity of walls and components (i.e., light loss factors), and additional factors depending upon the level of complexity desired.

step 3: *Coefficients of utilization* (CU) are applied. CUs are ratios of the interior to exterior horizontal illuminances. For the toplighting lumen method, the coefficients provide the illuminance on the workplane. For the sidelighting lumen method, coefficients are given for five predetermined points.

step 4: Calculate the product of the factors from steps 1, 2, and 3 to determine the interior illuminance.

These steps are expressed in the general equation for the lumen method, given as Eq. 51.49.

$$E_i = E_{xh}\tau_{\text{net}}(\text{CU})\left(\frac{A_s}{A_w}\right) \qquad 51.49$$

The average incident illuminance, given in lux, on the workplane is represented by E_i. The exterior horizontal illuminance available on the skylights is given as E_{xh}. The symbol τ_{net} represents the net transmittance through the skylights and light well, including loss factors for any control devices that limit the light and maintenance (i.e., losses due to dirt or other environmental factors that are limiting the light transmission). The area of the skylights and the area of the workplane are represented by A_s and A_w, respectively.

Most of the terms in Eq. 51.49 are either given in the design or can be determined from tabulated values based on design criteria. For example, E_{xh} can be calculated using one of the daylight calculation methods, or given for the design. The net transmittance is calculated by Eq. 51.50.

$$\tau_{d \text{ or } D} = (T_{d \text{ or } D})\eta_w R_a T_c(\text{LLF}) \qquad 51.50$$

The net transmittance, τ, can be either diffuse (d) or direct (D) depending upon the transmittance term used, T_d or T_D. This data is either provided by the manufacturer or it is measured. The ratio of the net skylight area to the gross skylight area, R_a, is calculated from the design data.

Figure 51.23 shows an example of a graph used to find the well efficiency, η_w. The wall reflectance is generally tabulated or given in manufacturing data. The well cavity ratio (WCR) is calculated by Eq. 51.51. The wall reflectance and the WCR comprise the data necessary to calculate well efficiency, as seen in Fig. 51.23.

$$\text{WCR} = \frac{5H(L+W)}{LW} \qquad 51.51$$

The light loss factor, LLF, is typically given in a table. An example is shown in Table 51.8.

CU is a tabulated factor and is based on the design conditions. For the toplighting method, ceiling and wall reflectance are two of the input criteria, and are

[26]The phrase "movement of the sun" is used here to reflect that many equations for lighting assume a static Earth with the sun movement as the entering argument for the equations.

[27]The results from such calculations may be required to augment the electric lighting calculations, normally the province of the electrical engineer.

[28]The lumen method is similar to the zonal method used for electric lighting. Given the similarities, the details and equations are explained in the following section on lighting calculations.

Figure 51.23 Well Efficiency

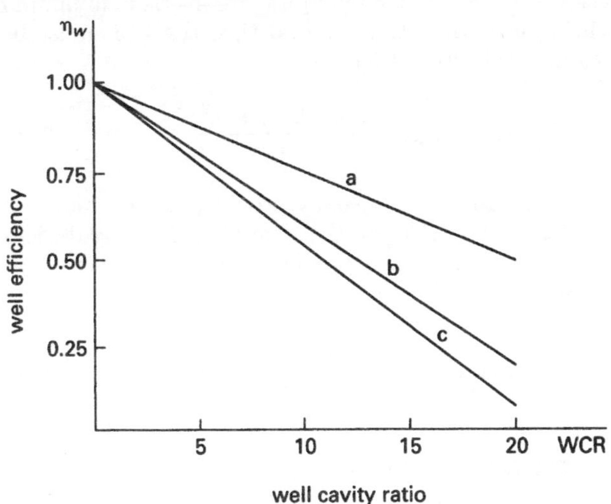

a = 80% wall reflectance
b = 60% wall reflectance
c = 40% wall reflectance

Table 51.8 Example of Light Loss Factors

location	glazing position		
	vertical	sloped	horizontal
clean	0.9	0.8	0.7
industrial	0.8	0.7	0.6
extremely dirty	0.7	0.6	0.5

provided by manufacturing (or measured) data. Another common input criteria is the room cavity ratio (RCR). The room cavity ratio is calculated using Eq. 51.52.

$$\mathrm{RCR} = \frac{5H_{c\text{-}wp}(L+W)}{LW} \qquad 51.52$$

Both Eq. 51.51 and Eq. 51.52 provide a unique ratio based on the height. This ratio is used to estimate the CU value. The larger the RCR value, the less light available in any given situation. Table 51.9 shows examples of CU values. All such tables are based on the space to mounting height ratio,[29] a lambertian distribution[30] from the skylight, and an assumed floor reflectance. Care must be taken to select the correct table for the given design constraints.

[29]This refers to the mounting height of the skylight opening from the workplane.
[30]A lambertian distribution is one following Lambert's cosine law.

Table 51.9 Coefficients of Utilization for Skylighting

ceiling reflectance (%)	RCR	wall reflectance 50%	wall reflectance 10%
80	0	1.19	1.19
	5	0.67	0.53
	10	0.52	0.40
20	0	1.04	1.04
	5	0.61	0.51
	10	0.49	0.40

The general lumen method represented by Eq. 51.49 can now be made specific for toplighting in overcast skies by Eq. 51.53, and for clear skies by Eq. 51.54.

$$E_i = E_{xh,\,\mathrm{sky}}\tau_d(\mathrm{CU})\left(\frac{A_s}{A_w}\right) \qquad 51.53$$

$$E_i = (E_{xh,\,\mathrm{sky}}\tau_d + E_{xh,\,\mathrm{sun}}\tau_D)(\mathrm{CU})\left(\frac{A_s}{A_w}\right) \qquad 51.54$$

For the lumen method for sidelighting, Eq. 51.49 is modified to eliminate the area ratio and to replace the exterior horizontal illuminance term, E_{xh}, with the term for exterior vertical illuminance on the window wall, E_{xv}. The resulting equation is shown as Eq. 51.55. The individual terms of the equation are calculated the same way as for the lumen method for toplighting. The assumptions embedded in the sidelighting method are: the ceiling cavity has a 70% reflectance, the room cavity has a 50% reflectance, and the floor cavity has a 30% reflectance.[31]

$$E_i = E_{xv}\tau_{\mathrm{net}}(\mathrm{CU}) \qquad 51.55$$

32. DAYLIGHT FACTOR METHOD

The daylight factor method is a low precision process. It is used with known illuminance distributions, such as the standard three from the CIE. The data for calculating the illuminance is given in a table. The daylight factor, DF, is a ratio of the illuminance at a point to the illuminance from the unobstructed sky. It is shown in Eq. 51.56. This unitless factor is then multiplied by the standard illuminance distribution to obtain the estimated illuminance in the space.

$$\mathrm{DF} = \mathrm{SC} + \mathrm{ERC} + \mathrm{IRC} \qquad 51.56$$

The term SC represents the sky component, ERC is the externally reflected component, and IRC is the internally reflected component.

[31]The ceiling cavity is from the top of the windows to the ceiling. The floor cavity is from the windowsill to the floor. The room cavity is the space in between the two.

33. ELECTRIC LIGHTING PRINCIPLES

Lighting calculations are approximations of complex physical processes and are designated by the intended application of the calculation. Specific applications include daylighting, luminaire design, runway design, roadway design, and security lighting. Basic equations govern the first order calculations and are a good basis for initial design. Those methods previously discussed for daylight are applicable with slight adjustments to the equations. The two most important lighting methods are the lumen method and the zonal cavity method. Both methods assume that air is nonabsorbing and nonscattering. The basic equation, applicable to both methods, is given by Eq. 51.57.

$$E_{\text{avg}} = \frac{\text{total flux onto workplace}}{\text{workplane area}} \qquad 51.57$$

34. LUMEN METHOD

The lumen method accounts for the total lumens projected by a luminaire to the workplane derated with a coefficient of utilization factor. The CU factor accounts for the efficiency of the luminaire, the distribution of the light from the luminaire (which is highly dependent upon the shape and properties of the enclosing material), and the reflective properties of the surrounding space. The average initial expected illuminance is given by Eq. 51.58.

$$E_{\text{initial}} = \frac{L_{\text{total}}(\text{CU})}{A_w} \qquad 51.58$$

The luminance emitted is represented by L with units of lumens and is determined from manufacturing data. The CU factor must be given (by the manufacturer for a given luminaire and assumed use), estimated (using reflectance calculations for the given material in a space),[32] or calculated using the zonal cavity method. A_w is the area that the designer is attempting to light to a given level. If the area is given in square meters, then the average illuminance units are lux (lumen per m^2). If the area is given in square feet, then the average illuminance units are footcandles (lumen per ft^2).

Lighting dims over time due to a variety of factors, including light loss due to dirt accumulation. To account for this loss, a light loss factor (LLF) is introduced. This is another derating factor whose values are normally tabulated. The LLF consists of recoverable losses (such as dirt accumulations) and nonrecoverable losses (such as luminaire physical degradation).[33] The designer's goal is to maintain an average minimum lighting level over time. To do this, the LLF must be included as in Eq. 51.59.

$$E_{\text{maintained}} = \frac{L_{\text{total}}(\text{CU})(\text{LLF})}{A_w} \qquad 51.59$$

The total lumens are usually provided by a number of luminaries. To calculate the number of lighting devices needed, assuming each provides the same lumens, use Eq. 51.60 and solve for N_{lights}.

$$E_{\text{maintained}} = \frac{N_{\text{lights}} L_{\text{per light}}(\text{CU})(\text{LLF})}{A_w} \qquad 51.60$$

35. CAVITY RATIOS

The amount of light exchanged between the top of a space and an area below or above it is a function of the proportions of its length, width, and height. The cavity ratio provides a single number approximating the effects. The concept is illustrated in Fig. 51.24. The zonal cavity method uses the cavity ratio to calculate coefficients of utilization.

Figure 51.24 *Zonal Cavity Terminology*

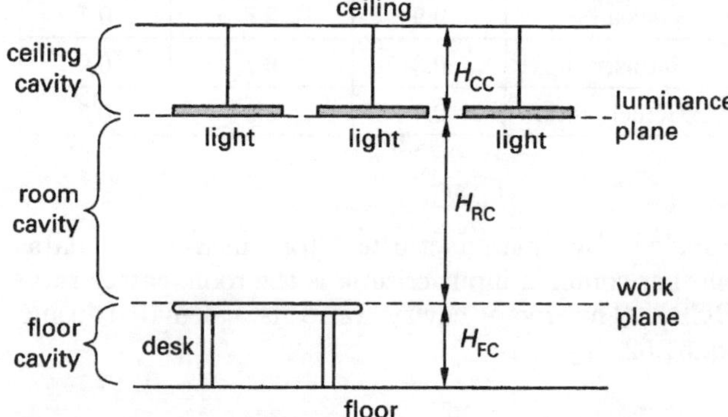

A room or space is divided into three cavities. The cavity ratio (CR) is calculated using

$$\text{CR} = \frac{5H_{xx}(L+W)}{LW} \qquad 51.61$$

When H_{xx} is replaced with H_{CC}, Eq. 51.61 gives the ceiling cavity ratio (CCR). When H_{xx} is replaced with H_{FC}, Eq. 51.61 gives the floor cavity ratio (FCR). When H_{xx} is replaced with H_{RC}, Eq. 51.61 gives the room cavity ratio (RCR).

[32]The calculation of effective cavity reflectances is not covered in this text. The value is usually given, and otherwise the equation for calculating the reflectance is provided. The use of such reflectances can enhance a lighting calculation but should be left to computer-based methods. Effective cavity reflectances are also called base reflectances for entering arguments into a zonal cavity table to calculate a coefficient of utilization.

[33]Nonrecoverable losses include degradation to the luminaire from ambient temperature effect, supply voltage level, ballast factor, and luminaire surface depreciation. Recoverable losses include degradation from room surface dirt depreciation (RSDD), lumen lamp depreciation (LLD), lamp burnout factor (LBO), and luminaire dirt depreciation (LDD).

If a room is an irregular shape, the perimeter of the room and the area of its base are used, along with an adjusted factor, which is −2.5 because the perimeter of the rectangular room in Eq. 51.61 is actually $2(L+W)$, as shown in Eq. 51.62.

$$\text{CR} = \frac{2.5 H_{xx} P}{A_{\text{base}}} \qquad 51.62$$

Using the reflectance of the appropriate surface (the ceiling, the floor, or the floor of the cavity itself—for example, the workplane of the room cavity) as the *base reflectance*, and knowing or estimating the wall reflectance, and finally calculating the appropriate cavity ratio, one has the input arguments necessary to use a zonal cavity table to determine the coefficient of utilization. See Table 51.10 for an example.[34] Select the base reflectance (either floor or ceiling), then select the wall reflectance. Move down the column, stopping at the row associated with the appropriate cavity ratio. The CU appears, as a percentage, in the intersection of the column and the row. For example, if the ceiling reflectance is 80% and the wall reflects at 50%, the CU is 71% for a cavity ratio of 0.6.

Table 51.10 *Coefficient of Utilization (%)*

base reflectance (%)	90	90	80	80
wall reflectance (%)	50	0	50	0
cavity ratio				
0.2	86	82	77	72
0.6	80	73	71	63
1.0	75	62	67	55

Special Applications

[34]Such tables also assume a length-to-width ratio for the luminaire. Care should be taken to ensure the correct table is being used for the design conditions.

52 Power System Management

Nomenclature

a	per unit growth rate	–
E	energy	J
P	power	W
t	time	s

Subscripts

0	initial
d	doubling

1. ENERGY AND POWER UNITS

Energy is the work that can be done by a force over a distance. It can result from electrical, chemical, mechanical, nuclear, and other processes. Electrically, energy is the work required to move a positive charge through an electric field. Chemically, the rearrangement of electron positions and energy levels releases energy in the form of heat. Mechanically, kinetic or potential energy results in forces that move objects over a distance. Nuclear energy is obtained from the conversion of mass directly to energy. All these sources, and others, can be used to create electrical energy through the appropriate power conversion system.

Work and energy have the same units. In colloquial terms, the energy of a body is its capacity to do work. The unit of energy in electrical engineering is the watt-second (W·s), which is equivalent to a joule (J). A more common unit is the kilowatt-hour (kWh). The relationship between the two as well as between other units is given in Table 52.1.

Power is the rate at which work is done, or energy per unit time. Common power units are the watt, where 1 W equals 1 J/s, and horsepower, where 1 hp equals 745.7 W.

$$P = \frac{E}{t} \qquad 52.1$$

***Table 52.1** Energy Equivalence*

energy	equivalent
1 W·s	1 J
1 kWh	3.6×10^6 J
1 cal	4.186 J
1 Btu	1.055×10^3 J
1 Btu	0.252×10^3 cal
1 quad	10^{15} Btu*

*Sometimes a quad will be defined as 10^{18} Btu.

2. ENERGY GROWTH RATE

Accommodation for future growth in energy usage requires knowledge of past growth and assumptions regarding that growth in the future. Energy usage, as a general rule, grows at a rate proportional to the amount of energy that is currently being used, E. The per unit growth rate is given by a.

$$\frac{dE}{dt} = aE \qquad 52.2$$

The solution to Eq. 52.2 is given in Eq. 52.3, where E_0 is the energy usage at some initial time.

$$E = E_0 e^{at} \qquad 52.3$$

The *doubling time*, or the time it takes for energy usage to double, is given by Eq. 52.4.

$$t_d = \frac{\ln 2}{a} = \frac{0.693}{a} \qquad 52.4$$

Example 52.1

Worldwide energy usage is expected to increase from 495 quad to 739 quad between 2007 and 2035, according to the Reference Case Projections published by the U.S. Energy Information Administration. What per unit growth rate is the Energy Information Administration predicting?

Solution

Use Eq. 52.3 to calculate the per unit growth rate over the 28 year period.

$$E = E_0 e^{at}$$

The future energy usage ratio is

$$\frac{E}{E_0} = e^{at}$$

Take the natural log of both sides and solve for a.

$$t = 2035 - 2007 = 28 \text{ yr}$$

$$\ln \frac{E}{E_0} = \ln e^{at} = at$$

$$a = \frac{\ln \frac{E}{E_0}}{t} = \frac{\ln \frac{739 \text{ quad}}{495 \text{ quad}}}{28 \text{ yr}}$$

$$= 0.0143 \text{ yr}^{-1} \quad (1.43\% \text{ per year})$$

3. ELECTRIC POWER ECONOMICS

The cost of electricity can be separated into three general categories of utility operations: generation, transmission, and distribution. These can be further separated into three major components: fuel, equipment, and labor.

The fuel costs depend on the fuel selected. Major sources of energy include fossil fuels (e.g., oil, coal, and natural gas); nuclear fuel (e.g., uranium and plutonium); hydroelectric power; geothermal steam; fuel cells; batteries; solar; waves; and wind. Fossil fuels develop their power from chemical conversion, nuclear from mass to energy conversion; hydroelectric from potential to kinetic energy conversion; geothermal from convection heat transfer; fuel cells and batteries from chemical conversion; solar from radiation to electric conversion; wave and wind from kinetic to mechanical conversion.

To compare electrical projects, the equipment (capital costs) must be combined with annual costs for fuel and labor. The principles involved are covered in Chap. 58.

4. ENERGY MANAGEMENT SYSTEM

An *energy management system* (EMS) is the system of components used to control a utility network. The task of an EMS is to control power generation and scheduling, network analysis, and the training of operators. Such a system includes everything from the transducers that sense the network's condition, to the computers at the control center displaying the information. This is illustrated in Fig. 52.1.

The *supervisory control and data acquisition* (SCADA) subsystem of the EMS provides for three critical functions: data acquisition, alarm display and control, and supervisory control. The supervisory processors communicate with the attached electronic devices using a serial communications protocol called *Modbus®*.

Figure 52.1 Energy Management System Block Diagram

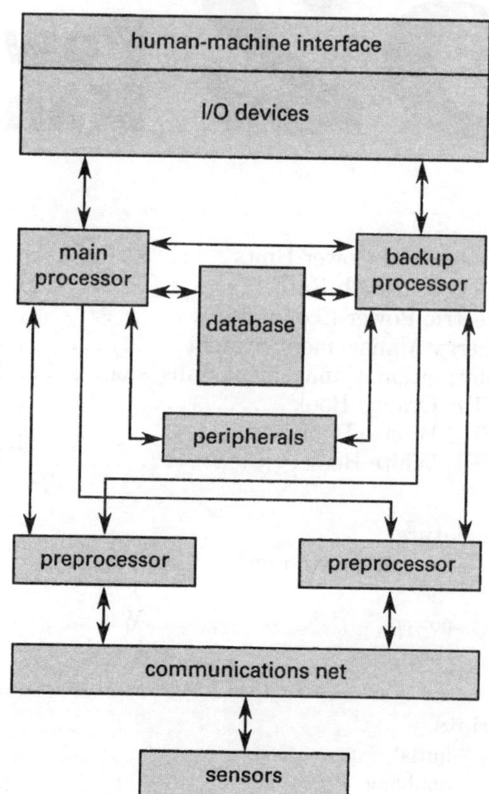

The *Modbus protocol* is a de facto standard primarily because while it is not formally adopted, it is openly published, royalty free, relatively easy to use, and robust.[1] Originally developed for communication with *programmable logic controllers* (PLCs), it has evolved for use in connecting very large numbers (i.e., hundreds) of devices in *distributed automated systems* such as SCADA. The formats in serial communications include *Remote Terminal Unit* (RTU), *American Standard Code for Information Interchange* (ASCII), and *Transmission Control Protocol/Internet Protocol* (TCP/IP). These formats determine the order and meaning of the bits in a serial message and determine how the message is decoded for use in the control system.

Data acquisition consists of collection, processing, monitoring, calculations, and configuration control. Collection includes gathering information from remote sensors. Processing converts the raw data—often analog—into *engineering units* (EU) that an operator or computer can utilize for calculation purposes. The system continuously monitors the information and uses the calculations to

[1]Modbus is an application layer protocol, level 7 of the *open systems interconnection* (OSI) model. The Modbus Organization owns and promotes the protocol.

provide alarms or automatic protective actions. Configuration control is the function that shifts transducers or monitoring points, or removes such sensors or portions of the system when errors occur.

The SCADA system can also be used for *scheduling.* This is the process whereby the system evaluates the load, load forecast, fuel cost, and the usage of each generation machine. Then, the system adjusts the loading on each machine to ensure the most efficient, cost-effective usage.

5. INFORMATION MANAGEMENT SUBSYSTEM

The *information management subsystem* (IMS) consists of the computer network and supporting devices that gather, store, and display the information used by the EMS. Advances in computers have made such systems powerful tools for fast and accurate monitoring of electrical systems. Software languages vary, but Fortran retains a high usage rate. New applications, especially those with *graphical user interfaces* (GUI), are usually written in C or C++.

6. IEEE ORANGE BOOK

IEEE Standard 446, *IEEE Recommended Practice for Emergency and Standby Power Systems for Industrial and Commercial Applications* (*IEEE Orange Book*), describes the best practices and principles for developing and installing emergency and standby power systems. It contains information, often extracted from other codes, standards, and texts, concerning the design, installation, and maintenance practices for installations designed to provide uninterruptible power free of frequency excursions and voltage transients. Chapters cover general need guidelines, generator and electric utility systems, stored energy systems, protection, grounding, maintenance, specific industry applications, and reliability improvement.

7. IEEE BRONZE BOOK

IEEE Standard 739, *IEEE Recommended Practice for Energy Management in Industrial and Commercial Facilities* (*IEEE Bronze Book*), describes the best practices and principles for designing and operating electrical systems for energy conservation. It contains information, often extracted from other codes, standards, and texts, concerning the design, installation, and maintenance practices for optimizing the efficiency of electrical, industrial, and commercial power systems. Chapters cover organization, energy cost, load management, energy management of motors, metering, and energy management of lighting.

8. IEEE WHITE BOOK

IEEE Standard 602, *IEEE Recommended Practice for Electric Systems in Health Care Facilities* (*IEEE White Book*), describes the best practices and principles for providing safe, adequate, and reliable electricity to health care facilities such as hospitals, nursing homes, residential custodial care facilities, clinics, ambulatory care centers, and medical offices. It contains information, often extracted from other codes, standards, and texts, concerning the design, installation, and maintenance practices for such facilities. Chapters cover electric loads, power distribution systems, power systems for patient care, emergency power systems, lighting, communication systems, medical equipment and instrumentation, and health care renovations.

Topic XII: Measurement and Instrumentation

53 Measurement and Instrumentation

Nomenclature

a, b	Fourier series coefficient	–
B	magnetic flux density	T
BW	bandwidth	Hz
C	compensating winding	–
C	capacitance	F
d	deflection	various
f	frequency	Hz
$f(t)$	function of time	–
H	magnetic field strength	A/m
I, $\mathbf{I}$	effective or DC current	A
k	coupling coefficient	–
κ	Boltzmann's constant, 1.3807×10^{-23}	J/K
n, N	number of turns	–
pf	power factor	–
P, $\mathbf{P}$	power	W
Q	quality factor	–
R	resistance	Ω
S, $\mathbf{S}$	apparent power	VA
t	time	s
T	temperature	K
T	period	s
T	torque	N·m
v	instantaneous voltage	V
V, $\mathbf{V}$	effective or DC voltage	V
$\mathbf{Z}$	impedance	Ω

Symbols

θ	angular displacement	rad
ϕ	angle	rad
φ	current angle $(\theta \pm \phi)$	rad

Subscripts

0	initial
ave	average
ext	external
fs	full scale
H	higher
L	lower
n	number
pf	power factor
rms	root-mean-square

1. FUNDAMENTALS

An *instrument* is a device for measuring, and sometimes for controlling or recording, a value under observation.[1] *Instrumentation* is the use of such devices for measurement, detection, observation, computation, communication, or control. Instrument outputs are given in terms of some reference value. A *reference standard* is an accepted measurement base.[2] *Precision* is the range of values of a set of measurements. An instrument's precision is a measure of the reproducibility of observations. The *accuracy* is a statement of the limits that bound the departure of the measured value from the true quantity. Accuracy includes the imprecision of the instrument and any accumulated errors from the national reference standards in the measurement itself. The accuracy of meter movements is given in terms of the percentage of error relative to the full-scale value, regardless of where the error occurs. That is, the error may occur in the low end of the indicating range, but the accuracy is nevertheless given as a percentage of the highest indicated value.

Errors can be separated into two general categories: systematic and random. A *systematic error* is an error that is reproducible and the result of some physical effect, such as a weakened magnetic field, altered resistance value, repeatable signal distortion, and the like. A *random error* is a deviation in the output caused by chance. Random errors are accidental, inconsistent in magnitude and sign, and not reproducible. The effects of errors are analyzed using statistical mathematics.

[1]Some of the instruments included in this chapter, such as an electrodynamometer instrument, have been replaced with digital instruments. The older versions remain useful for explaining the interactions that take place in the proper measurement of electrical parameters. Those interactions can be lost in the intricacies of digital circuitry.

[2]United States national reference standards are maintained by the National Institute of Standards and Technology, formerly the National Bureau of Standards (NBS).

Measurement/ Instrumentation

The *resolution* is the smallest increment that can be distinguished by the instrumentation system. The resolution is determined by the amount of deflection per unit input and is limited by the amount of signal beyond the noise. *Noise* is any portion of a signal that does not convey useful information, such as unwanted currents and voltages in an electrical device or system. Random disturbances caused by thermal agitation of electrons are a significant contributor to noise in electrical systems and are called *Johnson noise.* A measure of the Johnson noise in resistors is[3]

$$V_{\text{noise}} = \sqrt{4\kappa TR\Delta f} = \sqrt{4\kappa TR(\text{BW})} \qquad 53.1$$

The term V_{noise} is the root-mean-square (rms) equivalent voltage generated by the thermal agitation. Boltzmann's constant is κ, the absolute temperature is T, the resistance is R, and the bandwidth over which the noise occurs is Δf or BW.

Environmental factors influence the accuracy of instrumentation systems. Stray magnetic fields can combine with instruments' magnetic fields to alter the output. Ambient temperature differences between readings can result in differences caused by mechanical changes in the instrumentation or changes in resistance in the circuitry. Electrostatic charges influence the position of indicating needles. For example, the rubbing of an instrument face can move a needle with constant sensing input. Mechanical shock, vibration, and even mounting position influence the output of instrumentation and must be accounted for or protected against in order to obtain accurate indications.

Most measurement circuits are not exposed to the full voltage or current of the system monitored. The measurement circuits are isolated from the system by *instrument transformers.* An instrument transformer extends the range measured by a given meter, isolates the meter from the line, and allows the instrument circuit to be grounded. When the primary winding is in series with the line, the instrument transformer is called a *current transformer.* When the primary winding is in parallel with the line, the instrument transformer is called a *potential transformer.* Instrument transformers are tested in the same manner as power transformers.

Example 53.1

What is the expected thermal agitation noise at 20°C in a digital voltmeter using a 10 kΩ resistor and designed to operate within the frequency range of 0 Hz to 1000 Hz?

Solution

Ignoring any other resistance in the voltmeter, which will be small compared to the 10 kΩ resistance, and noting that 20°C is equal to 293K, the Johnson noise voltage is

$$\begin{aligned} V_{\text{noise}} &= \sqrt{4\kappa TR\Delta f} \\ &= \sqrt{\begin{array}{c}(4)\left(1.3807\times 10^{-23}\ \frac{\text{J}}{\text{K}}\right)(293\text{K}) \\ \times (10\times 10^{3}\ \Omega)(1000\ \text{Hz})\end{array}} \\ &= 4.02\times 10^{-7}\ \text{V} \end{aligned}$$

Such a small voltage level is of significance in only the most sensitive of measurements.

2. SIGNAL REPRESENTATION

DC signals are represented by their constant values. AC signals can be represented in a number of ways. The average value of a generic AC signal is

$$f_{\text{ave}} = \frac{1}{T}\int_{t_1}^{t_1+T} f(t)\,dt \qquad 53.2$$

The average value of a sinusoidal function is zero, so the rms method of measuring AC sinusoidal signals is required in order to use their values in electrical calculations. The rms or effective value of a generic function is

$$f_{\text{rms}} = \sqrt{\frac{1}{T}\int_{t_1}^{t_1+T} f^2(t)\,dt} \qquad 53.3$$

Because nearly all periodic functions can be represented by a Fourier series, many important nonsinusoidal signals are changed into such series for further manipulation. A Fourier series is given by

$$\begin{aligned} f(t) &= \tfrac{1}{2}a_0 + \sum_{n=1}^{\infty} a_n \cos\left(\frac{2\pi n}{T}\right)t \\ &\quad + \sum_{n=1}^{\infty} b_n \sin\left(\frac{2\pi n}{T}\right)t \end{aligned} \qquad 53.4$$

The average value of a function given by a Fourier series is

$$f_{\text{ave}} = \tfrac{1}{2}a_0 \qquad 53.5$$

The rms value of a function given by a Fourier series is

$$f_{\text{rms}} = \sqrt{\left(\tfrac{1}{2}\,a_0\right)^2 + \tfrac{1}{2}\sum_{n=1}^{\infty}(a_n^2 + b_n^2)} \qquad 53.6$$

[3]Random disturbances in mechanical systems are explained by the mechanism of *Brownian motion.* In magnetic systems, the *Barkhausen effect* explains such disturbances.

For a sinusoid, the average and rms values are given in Eq. 53.7 through Eq. 53.9. (Equation 53.7 is the average of a rectified sinusoid.)

$$V_{\text{ave}} = \frac{2}{\pi} V_p \qquad 53.7$$

$$V_{\text{rms}} = \frac{1}{\sqrt{2}} V_p \qquad 53.8$$

$$V_{\text{rms}} = 1.11 V_{\text{ave}} \qquad 53.9$$

To measure the rms value accurately in all situations, a *true rms* instrument squares the signal, computes the mean, and then takes the root of the result.[4] Less expensive instruments measure the rms value using Eq. 53.8, with the assumption that the signal is sinusoidal. These instruments measure the peak value and divide the peak by $\sqrt{2}$. This is valid only for sinusoidal signals that are also symmetrical (i.e., the frequency is stable and unchanging). If the signal varies from this ideal, if it contains harmonics, or if the regulation system allows the frequency or voltage to shift widely within the allowed control bands, the displayed value will be incorrect.

Example 53.2

A voltage waveform is represented by the truncated Fourier series,

$$\begin{aligned} v(t) &= 5 + 3\cos 100t + 0.8\cos 300t \\ &\quad + 4\sin 100t + 0.5\sin 200t \end{aligned}$$

Determine the average value of this voltage.

Solution

The average value is the zero frequency term as determined by Eq. 53.5.

$$\begin{aligned} a_0 &= (2)(5 \text{ V}) = 10 \text{ V} \\ f_{\text{ave}} &= \tfrac{1}{2} a_0 = \left(\tfrac{1}{2}\right)(10 \text{ V}) \\ &= 5 \text{ V} \end{aligned}$$

Example 53.3

For the voltage waveform given in Ex. 53.2, what is the rms voltage?

Solution

The rms value is given by Eq. 53.6.

$$\begin{aligned} f_{\text{rms}} &= \sqrt{\left(\tfrac{1}{2}a_0\right)^2 + \tfrac{1}{2}\sum_{n=1}^{\infty}\left(a_n^2 + b_n^2\right)} \\ &= \sqrt{(5.0)^2 + \left(\tfrac{1}{2}\right)\left((3)^2 + (0.8)^2 + (4)^2 + (0.5)^2\right)} \\ &= 6.16 \text{ V} \end{aligned}$$

3. MEASUREMENT CIRCUIT TYPES

Electrical parameters can be measured with *electromechanical circuits.* Such instruments use the measured parameter to generate a magnetic field that interacts with a separate magnetic field to produce the rotation of a restrained needle. Because of mechanical inertia, the instruments do not function for AC in the range of 5 Hz to 20 Hz. Because of inductive effects, the instruments do not function well above a maximum of 100 Hz to 2500 Hz, depending on the design. Because of these restrictions, the electromechanical instruments are primarily used in DC systems.

Substitution circuits compare an unknown electrical quantity with a reference. *Bridge circuits* are a simple means of determining impedance by such substitution. A *Wheatstone bridge* is shown in Fig. 53.1.

Figure 53.1 *Wheatstone Bridge*

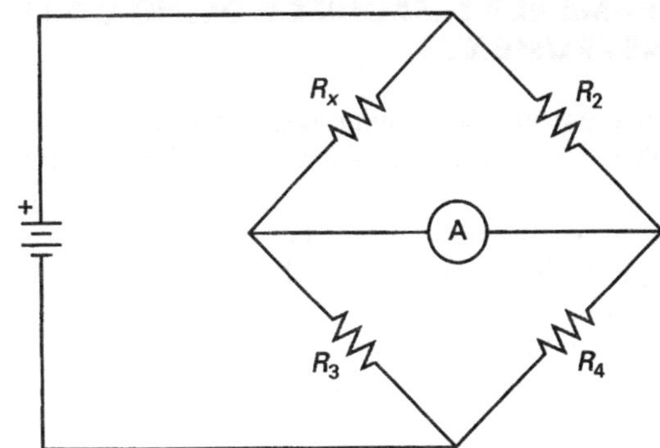

Balance exists in the bridge when no current flows through the ammeter. When balanced, the relationship between the resistors is

$$\frac{R_x}{R_3} = \frac{R_2}{R_4} \qquad 53.10$$

Typically, R_3 and R_4 are fixed resistors that can be selected to vary their ratio from 10^{-3} to 10^3. Resistor R_2 is adjustable in steps and used to balance the bridge with the unknown resistor R_x. When balance is achieved, the unknown resistance is given by Eq. 53.10. Many other

[4]ANSI/IEEE Standard 120, *IEEE Master Test Guide for Electrical Measurements in Power Circuits,* provides guidance for taking electrical measurements of power circuits.

Measurement/ Instrumentation

types of bridges are used to measure such things as inductance, capacitance, frequency, angles, and the like.[5]

Analog circuits use the mathematical relationships among electrical parameters to determine the desired parameters. Designs vary, but the instruments are used to determine such parameters as impedance, quality factor of coils (Q), the coefficient of coupling (k), magnetic parameters (H and B), frequency (f), and the like.

Digital circuits use digital processing of analog signals and analog-to-digital converters to display the results.[6] Digital instruments offer ease of use, quick response, and accurate results.

Transducer circuits are used in conjunction with transducers themselves to provide a signal of sufficient strength for use in follow-on systems.[7]

Example 53.4

A Wheatstone bridge with an R_3/R_4 ratio of 500 is balanced when the adjustable resistor has a value of 10 Ω. What is the unknown resistance?

Solution

At a balanced condition, the resistance is given by Eq. 53.10.

$$\frac{R_x}{R_3} = \frac{R_2}{R_4}$$

$$R_x = \left(\frac{R_3}{R_4}\right) R_2 = (500)(10\ \Omega) = 5000\ \Omega \quad (5\ \text{k}\Omega)$$

4. PERMANENT MAGNET MOVING-COIL INSTRUMENTS

A permanent magnet moving-coil instrument is shown in Fig. 53.2. Such an instrument is also called a *d'Arsonval movement* or a *galvanometer*. A current is passed through the coil on the rotor. The rotor is located within a magnetic field created by the permanent magnets. Because a current-carrying conductor within a magnetic field will experience a torque, the coil rotates. The torque produced is proportional to the current through the coil and the number of turns of wire in the coil ($T_{\text{coil}} \propto IN$). The spring is restrained in such a manner as to provide an opposing torque proportional to the angular displacement ($T_{\text{spring}} \propto \theta$). The pointer comes to rest when the torques are equal in magnitude.

The current necessary to cause full-scale deflection of a specific instrument is given the symbol I_{fs}. The usual range of currents is 50 μA to 10 mA. In ammeter applications, the voltage at which full-scale current flow occurs is usually in the range of voltages from 50 mV to 100 mV, with a nominal value of 50 mV used as the standard. Because this is usually more than required by the coil for I_{fs} to flow, a *swamping resistor* is placed in series with the coil.

Figure 53.2 *Permanent Magnet Moving-Coil Mechanism (d'Arsonval)*

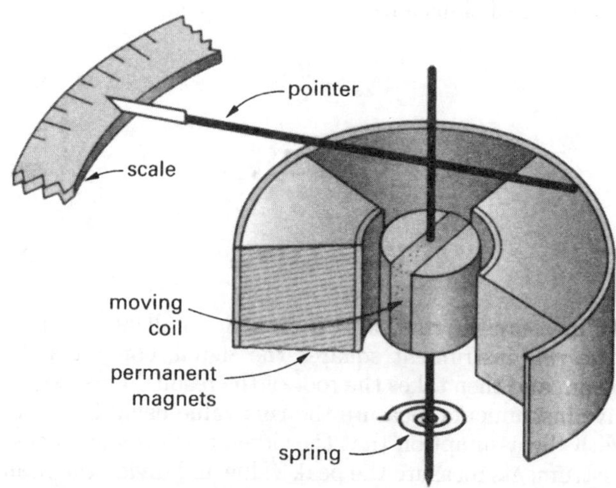

Because the permanent magnet's field is unidirectional, a change in the direction of the current in the coil reverses the torque, so the unmodified instrument is useful for detecting DC quantities only.[8] It can be made to detect AC quantities with the addition of a rectifier circuit on the coil input. When measuring AC rectified quantities, the d'Arsonval movement responds to the average voltage or current. In view of Eq. 53.9, adjusting the scale by a factor of 1.11 allows the rms value to be read directly.

The d'Arsonval movement can be designed to detect AC quantities with the use of a thermocouple. The hot junction of a thermocouple is placed in thermal, but not electrical, contact with a resistance carrying the current to be measured. The cold junction is connected to a heat sink that maintains ambient temperature. The measured current heats the wire via the resistance and subsequently heats one end of the thermocouple. A potential difference is developed between the ends of the dissimilar metals of the thermocouple proportional to the temperature difference ($I \propto \Delta T$), which is fed as input to the d'Arsonval meter movement. Because the heating effect is proportional to the square of the current, using a square-root factor on the scale output allows the rms value to be read directly. Such a *thermoelement instrument* can function up to several hundred megahertz.

5. DC VOLTMETERS

A d'Arsonval meter movement configured to perform as a DC voltmeter is shown in Fig. 53.3.

[5]Some of the named bridges are the Maxwell-Wien, Anderson, Owen, Hay, Campbell, and Schering.

[6]The digital processing is often accomplished with the use of operational amplifiers, which are covered in Chap. 48.

[7]Chapter 19 provides background on transducer principles.

[8]Specially designed d'Arsonval movements can respond to currents up to 100 Hz.

Figure 53.3 DC Voltmeter

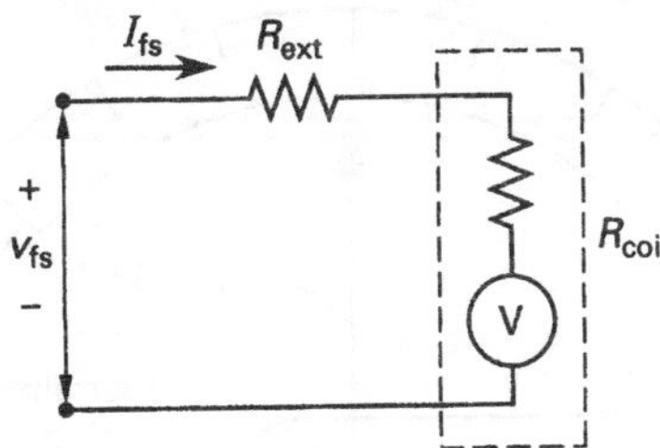

The external resistance is used to limit the current to the full-scale value, I_{fs}, at the desired full-scale voltage, V_{fs}. The electrical relationships in the voltmeter are given by Eq. 53.11.

$$\frac{1}{I_{fs}} = \frac{R_{ext} + R_{coil}}{V_{fs}} \tag{53.11}$$

The quantity $1/I_{fs}$ is fixed for a given instrument and is called the *sensitivity*. The sensitivity is measured in ohms per volt (Ω/V).

Example 53.5

Common DC voltmeters have a sensitivity of 1000 Ω/V. Given such a meter with a coil resistance of 200 Ω and a 1% accuracy, determine the external resistance necessary to use a full-scale voltage of 10 V.

Solution

The external resistance is found from Eq. 53.11 by noting that the sensitivity is $1/I_{fs}$.

$$\frac{1}{I_{fs}} = \frac{R_{ext} + R_{coil}}{V_{fs}} = 1000\ \Omega/\text{V}$$

$$\begin{aligned} R_{ext} &= V_{fs}\left(1000\ \frac{\Omega}{\text{V}}\right) - R_{coil} \\ &= (10\ \text{V})\left(1000\ \frac{\Omega}{\text{V}}\right) - 200\ \Omega \\ &= 9800\ \Omega \quad (9.8\ \text{k}\Omega) \end{aligned}$$

Example 53.6

An overall accuracy of 2% is desired for the meter in Ex. 53.5. What is the maximum error possible for the external resistance?

Solution

The maximum error introduced by the external resistance will occur at full-scale voltage. The maximum error from the movement itself is given at the full-scale voltage as 1%, regardless of where the error actually occurs. The external resistance must be a 9.8 kΩ resistor with a 1% accuracy.

Example 53.7

The manufacturer's data for a DC voltmeter lists the sensitivity as 20,000 Ω/V. What current is required to deflect the movement to the 100% value?

Solution

The 100% value is the full-scale deflection. The full-scale deflection current is found from the definition of the sensitivity.

$$\text{sensitivity} = \frac{1}{I_{fs}}$$

$$I_{fs} = \frac{1}{20{,}000\ \frac{\Omega}{\text{V}}} = 5 \times 10^{-5}\ \text{A} \quad (50\ \mu\text{A})$$

6. DC AMMETERS

A d'Arsonval meter movement configured to perform as a DC ammeter is shown in Fig. 53.4.

Figure 53.4 DC Ammeter

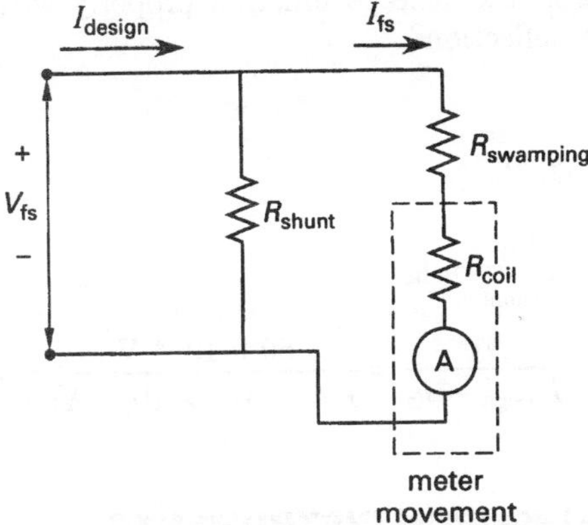

The swamping resistance is used to limit the current to the full-scale value, I_{fs}, at the desired full-scale voltage, V_{fs}. For ammeters, the standard full-scale voltage is 50 mV.[9] The electrical relationships in the ammeter are given by Eq. 53.12.

$$I_{design} = \frac{V_{fs}}{R_{shunt}} + I_{fs} \tag{53.12}$$

Example 53.8

A d'Arsonval movement is used as an ammeter with coil and swamping resistors totaling 200 Ω. The full-scale current value, I_{fs}, is 0.5 μA. If no shunt resistor is used, what percentage of the full-scale current will flow through the meter movement?

[9]The standard ammeter is designed to withstand a 50 mV voltage across the shunt with I_{fs} flowing in the movement at the desired design current flow.

Solution

Without a shunt resistor, and noting that the standard full-scale standard voltage is 50 mV, the full-scale current is

$$I_{fs} = \frac{V_{fs}}{R_{swamping} + R_{coil}} = \frac{50 \times 10^{-3}\ \text{V}}{200\ \Omega}$$
$$= 2.5 \times 10^{-4}\ \text{A}$$

The actual full-scale current should be 0.5 μA. The percentage of full-scale current flowing without the shunt resistor is

$$\%I = \left(\frac{2.5 \times 10^{-4}\ \text{A}}{0.5 \times 10^{-6}\ \text{A}}\right) \times 100\% = 50{,}000\%$$

If the full-scale current had not been specified as 0.5 μA, the calculated current of 2.5×10^{-4} A would have been the correct value for proper operation.

Example 53.9

What value of shunt resistance is required for the ammeter in Ex. 53.8 to function properly with a 1 A full-scale deflection?

Solution

Use Eq. 53.12.

$$I_{design} = \frac{V_{fs}}{R_{shunt}} + I_{fs}$$

$$R_{shunt} = \frac{V_{fs}}{I_{design} - I_{fs}} = \frac{50 \times 10^{-3}\ \text{V}}{1\ \text{A} - (0.5 \times 10^{-6}\ \text{A})} = 0.05\ \Omega$$

7. MOVING-IRON INSTRUMENTS

A *moving-iron instrument* is shown in Fig. 53.5. The instrument shown is called a *radial-vane mechanism.* A current is passed through the coil. A stationary vane and a moving vane are located within the resulting magnetic field and are similarly magnetized. Because the induced magnetic field is the same for both vanes, they repel one another. The moving vane is attached to a pointer and restraining spring. The angular displacement of the moving vane is proportional to the square of the current and the number of turns of wire in the coil ($T_{coil} \propto I^2 N$). The spring is restrained in such a manner as to provide an opposing torque proportional to the angular displacement ($T_{spring} \propto \theta$). The pointer comes to rest when the torques are equal in magnitude.

The vanes repel one another regardless of the polarity of the current or its changing value. As a result, the instrument can be used to measure AC quantities. Because the torque is proportional to the square of the current, using a square-root factor on the scale output allows the rms value to be read directly. The instrument can measure AC quantities up to approximately 500 Hz. With special compensation for eddy currents and skin effects, this range can be extended to 10 kHz.

Figure 53.5 *Radial-vane Mechanism*

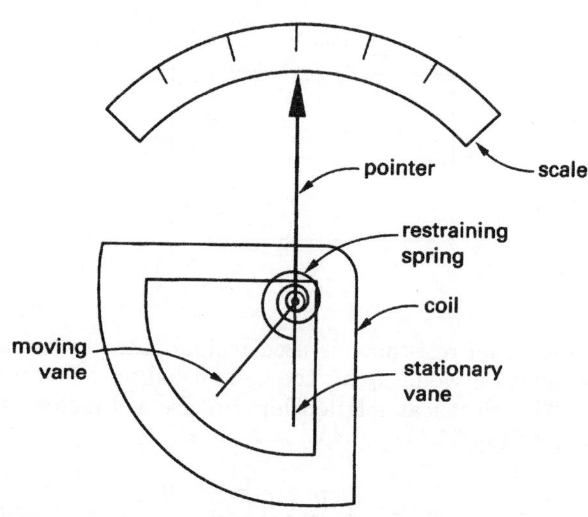

Most moving-iron instruments can also measure DC quantities. They are less useful for DC quantities because significant remnant magnetization occurs over time.

Moving-iron instruments can be used for voltmeters or ammeters using setups similar to those of d'Arsonval instruments. The sensitivity and power consumption of moving-iron instruments is between that of d'Arsonval movements and electrodynamometer movements.

8. ELECTRODYNAMOMETER INSTRUMENTS

An *electrodynamometer instrument* is shown in Fig. 53.6(a). A current is passed through a moving coil and two stationary coils. The magnetic field of the moving coil interacts with the magnetic fields of the stationary coils to produce a torque. The torque produced is proportional to the current and the number of turns of wire in the moving coil as well as the current and number of turns in the stationary coils ($T \propto I^2 N$). The spring is restrained in a manner that provides an opposing torque proportional to the angular displacement ($T_{spring} \propto \theta$). The pointer comes to rest when the torques are equal in magnitude.

Because the torque is proportional to the square of the current, using a square-root factor on the scale output allows the rms value to be read directly.[10] The instrument can measure AC quantities in the range of 25–2500 Hz. The lower limit is caused by the instrument's attempt to follow the instantaneous values of the AC quantities. The upper limit is caused by inductive reactance.

[10]This is true when the same current is flowing in all coils. That is, the device is configured so that the coils are in series.

Measurement/ Instrumentation

Figure 53.6 Electrodynamometer

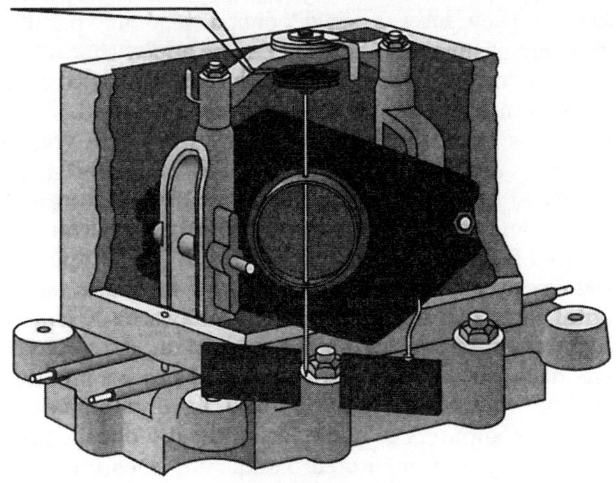

(a) mechanism

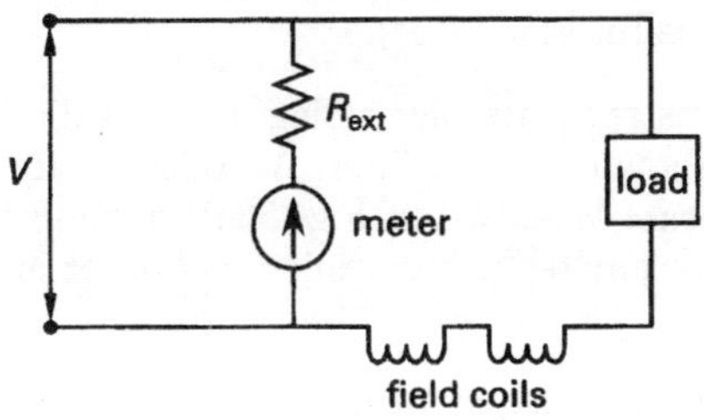

(b) configured to measure power

Though currents and voltages can be measured, the dynamometer is most commonly used to measure power. When the moving coil is placed across a load, the current in the moving coil will be proportional to the voltage across the load. When the stationary coils are placed in series with the load, the current in the stationary coils will be proportional to the current in the load. The combined magnetic field is then proportional to the instantaneous product of the voltage and the current—that is, the power. One possible configuration for measuring power is shown in Fig. 53.6(b). Corrections for the power consumption of the coils are required to ensure an accurate result.

The electrodynamometer can also be used to measure DC quantities because average DC values are equivalent to rms AC values.

The instruments can be used for voltmeters or ammeters using setups similar to those of d'Arsonval instruments. The sensitivity is higher than either d'Arsonval or moving-iron instruments, but the power consumption is higher as well.

9. POWER MEASUREMENTS

Many different types of instruments are used to measure power. All require compensation for the power they consume, inductance of the measuring coils, and phase angle corrections for any instrument transformers used. Three-phase systems require a minimum of two wattmeters to measure power. The power meter most commonly used in residential applications is the induction-type watt-hour meter shown in Fig. 53.7.

Figure 53.7 Induction-Type Watt-Hour Meter

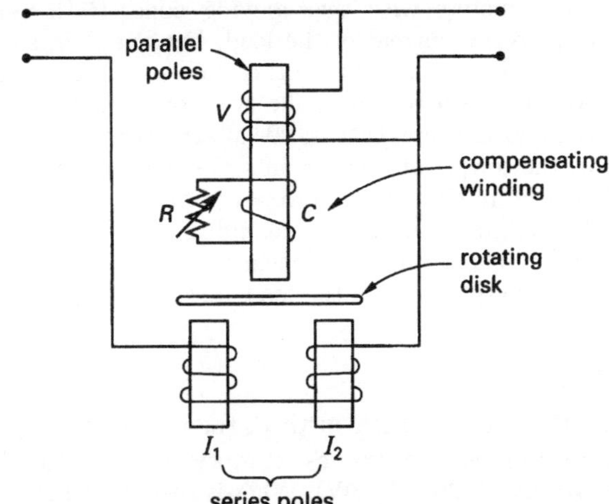

The parallel coil voltage, V, produces a magnetic flux at its pole tip that lags the applied voltage by nearly 90°, because of its high inductance and the large number of turns used in the coil. The fluxes in the series coils, I_1 and I_2, are nearly in phase with the current. All three poles induce currents in a circular disk located between the poles.[11] A torque, proportional to the power, is generated in the disk and causes it to rotate.[12] The disk speed is made proportional to the driving torque by a braking torque that is also proportional to the disk speed. This breaking torque (countertorque) arises from the interaction of the rotating disk with a permanent magnetic field (not shown) that induces eddy currents in the disk. The compensating winding, C, with adjustable resistor, R, is used to ensure that the flux from the parallel pole, V, is exactly 90° out of phase. The net result is a rotating disk speed proportional to the power delivered to the load.

10. POWER FACTOR MEASUREMENTS

The power factor, pf, is given by the ratio of the true power to the apparent power.

$$\text{pf} = \frac{P}{S} \qquad 53.13$$

For a balanced three-phase, three-wire system, results obtained using the *two-wattmeter method* can be used to determine the angle between the current and the voltage, and therefore the power factor. Let P_H be the

[11]This is the same principle as that of the induction motor.

[12]The torque is proportional to the power only if the fluxes between the series coils and the parallel coil are in quadrature at unity power factor.

larger power, and let P_L be the power measured by the other wattmeter. The desired angle is determined from

$$\tan\phi = \frac{\sqrt{3}(P_H - P_L)}{P_H + P_L} \qquad 53.14$$

Power factor meters can also be constructed from electrodynamometers with two moving coils. Both coils carry the series current of the load, but one does so in series with a resistor and the other does so in series with an inductor. Consequently, the two coils' currents are nearly in quadrature, that is, 90° apart. At unity power factor, the torque on the coil in series with the resistor will be at a maximum. At zero power factor, the torque on the coil in series with the inductor will be at a maximum. The angular displacement of the needle measures the power factor of the load.

Example 53.10

Using the two-wattmeter method, the power of a three-phase circuit is measured as 1074 kW (1074×10^3 W) and 426 kW (426×10^3 W). What is the power factor?

Solution

The angle necessary to determine the power factor comes from Eq. 53.14.

$$\begin{aligned}\tan\phi &= \frac{\sqrt{3}(P_H - P_L)}{P_H + P_L} \\ &= \frac{\sqrt{3}(1074 \times 10^3 \text{ W} - 426 \times 10^3 \text{ W})}{1074 \times 10^3 \text{ W} + 426 \times 10^3 \text{ W}} \\ &= 0.748\end{aligned}$$

The angle between the current and the voltage is

$$\phi = \arctan 0.748 = 36.8°$$

The power factor is

$$\text{pf} = \cos\phi = \cos 36.8° = 0.8$$

Given the nature of most systems, the power factor is 0.8 lagging.

11. ELECTRONIC INSTRUMENTATION

The advent of improved electronics has advanced instrumentation accuracy and rapidity. Digital instruments combine the analog circuit and sensor outputs with digital processing. Combinations of the instruments previously discussed can be used to create a *multimeter*, a device for measuring multiple electric quantities. When such an instrument remains in the analog realm, it is called a VOM meter (volt-ohm-milliamp meter). When the parameter being sensed is the primary analog portion of the circuit and the remainder is digital, the instrument is called a *digital multimeter*, *DMM* (or *DVOM*).

Such DMMs, especially when combined with a microprocessor, provide the following advantages and properties. Because they have a single digital readout, parallax errors are eliminated. Circuitry can be added that allows auto-ranging, a feature that searches for and selects the correct range for the parameter being measured. Autopolarity is a feature that can display whether a voltage agrees or disagrees with the red (positive) and black (negative) leads. Sample and hold circuitry captures the reading and continues to display it after the leads are removed from the *element under test* (*EUT*). Sample and hold circuitry is also used to digitize a waveform so the digital form can be stored. The microprocessor can be programmed to provide for self-calibration, data conversion, data processing, linearization, and process monitoring and control. Meters can also have interfaces to personal computers to allow for automatic downloading and extraction of monitored parameters. Such interfaces include IrDA (Infrared Data Association), RS-232 (Recommended Standard 232), USB (Universal Serial Bus), and IEEE-488 (commonly called the GPIB or General Purpose Interface Bus).

A Texas Instruments block diagram of a digital multimeter is shown in Fig. 53.8. A Keithley Instruments block diagram is shown in Fig. 53.9. A highly accurate digital electrometer using a microcomputer is shown in Fig. 53.10.

The operation principles of the individual electronic components shown in Fig. 53.8, Fig. 53.9, and Fig. 53.10 are discussed in detail in Chap. 48.

12. INSULATION AND GROUND TESTING

Insulation testing is conducted to ensure equipment can operate properly and personnel are protected. There are a variety of standards, such as those of IEEE or NEMA, for insulation testing. A successful insulation test will yield a high resistance value, and generally, 1 MΩ indicates satisfactory resistance. Figure 53.11 illustrates a simple, low-voltage insulation test adequate for most checks. Using a digital multimeter, the leads are attached as shown. Care must be taken not to penetrate the insulation with the positive lead, and the resistance must be checked.

If the resistance is expected to be above 1 MΩ, a more accurate insulation test will be required, and can be accomplished by raising the voltage at which the test occurs. The voltage can be increased using a *megger*. (See Fig. 53.12.) The megger operates at a higher voltage than the digital multimeter and provides more accurate resistance readings. A hand-crank generator in simpler models powers the circuit. The generator provides the power to coils A and B, which are situated on a rotor, C, all of which can pivot in response to the interaction of the magnetic fields of the coils and the permanent magnet, M. When the circuit leads, labeled *earth* and *line*, are not connected and the hand-crank is turned, current flows through coil B, generating a magnetic field. This

Figure 53.8 Digital Multimeter Overview

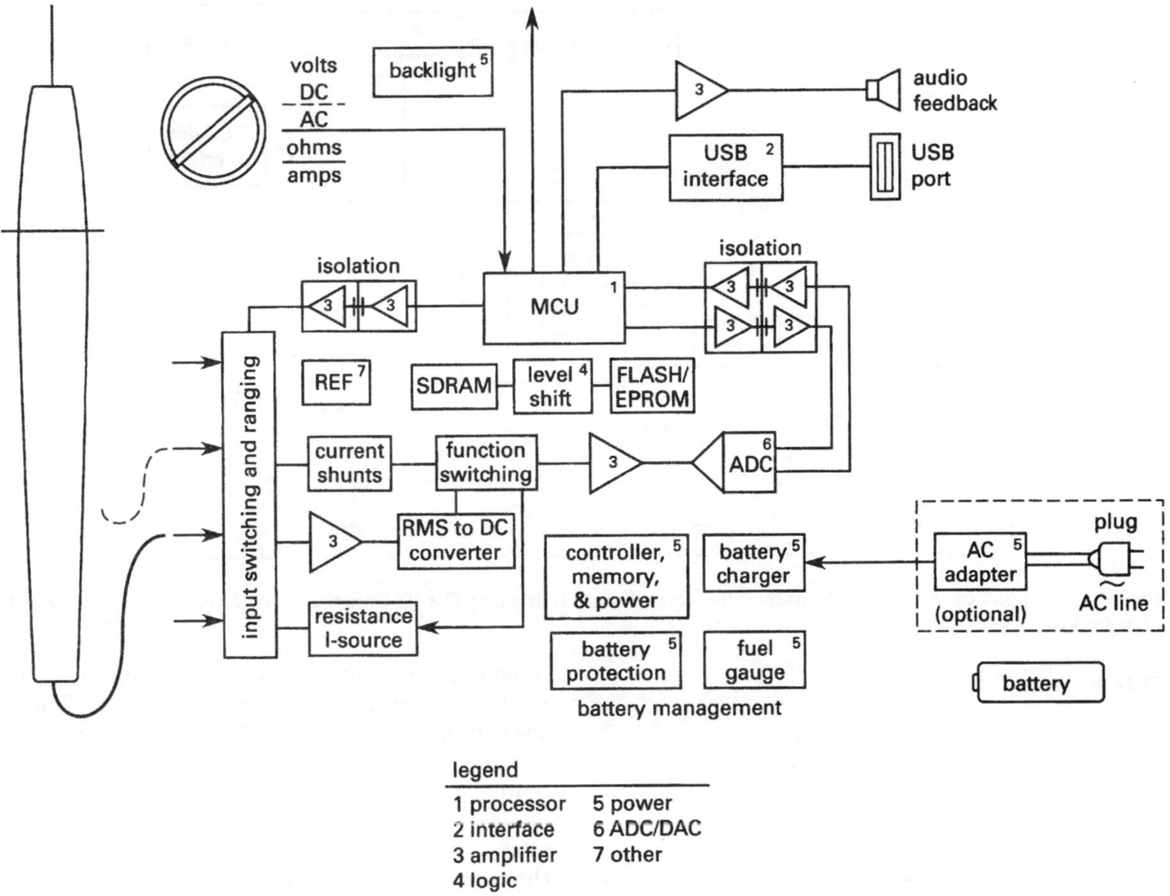

Reprinted courtesy of Texas Instruments Incorporated, ti.com, copyright © 1995-2015 by Texas Instruments Incorporated.

Figure 53.9 Keithley Instruments Block Diagram

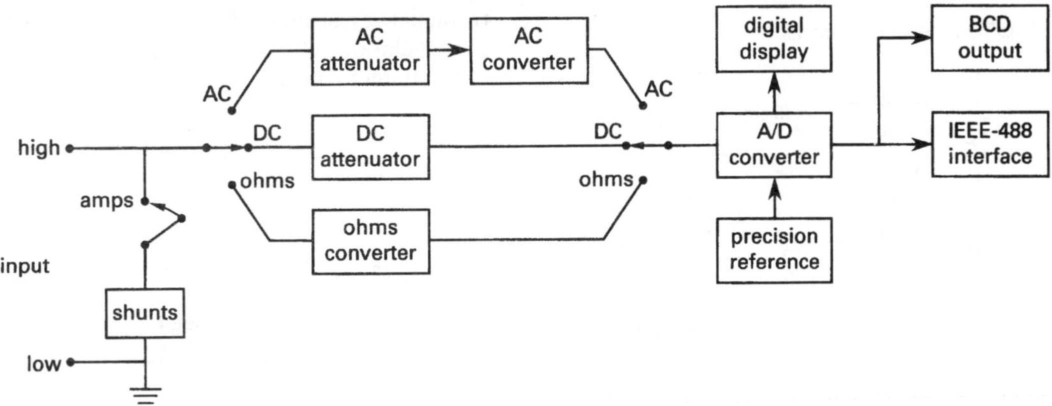

Reprinted with permission from *Electronics Engineers' Handbook*, Fourth Edition, by Donald Christiansen et al., copyright © 1997 by McGraw-Hill Companies, Inc.

magnetic field aligns itself with the permanent magnetic field of M, pivoting the rotor counterclockwise such that the pointer points to infinity. When a circuit is under test, the earth and line connections are attached to the wire and insulation. As the generator turns, current flows through both coils A and B. Coil A is a deflecting coil and generates a torque that turns the rotor clockwise, causing the needle to point to a position that represents equal torques between coils A and B. Since the current and the torque generated by coil A are dependent on the external resistance of the insulation under test, this provides a reading of that resistance.

Figure 53.10 *Typical Digital Electrometer*

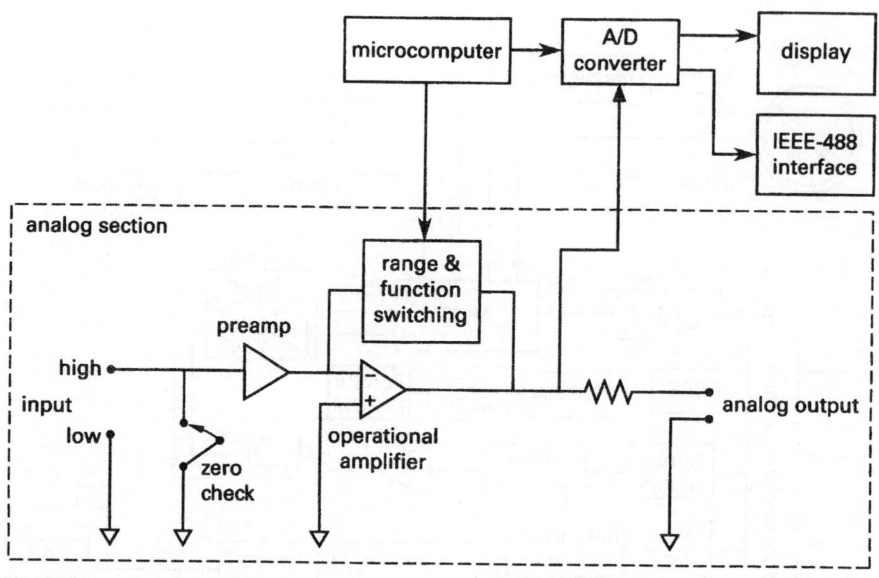

Reprinted with permission from *Electronics Engineers' Handbook*, Fourth Edition, by Donald Christiansen et al., copyright © 1997 by McGraw-Hill Companies, Inc.

Figure 53.11 *Insulation Testing*

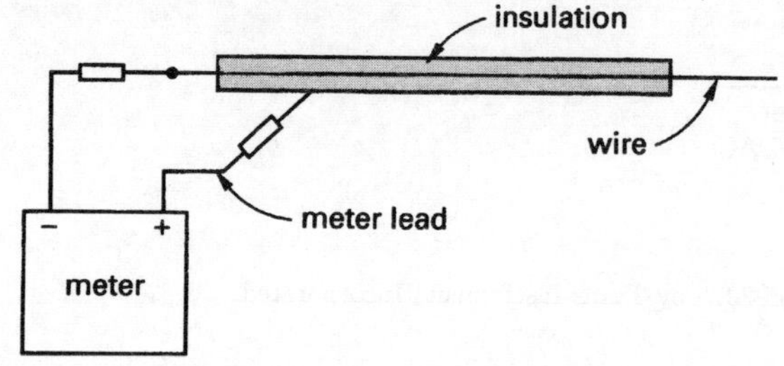

Figure 53.12 *Simplified Megger Schematic*

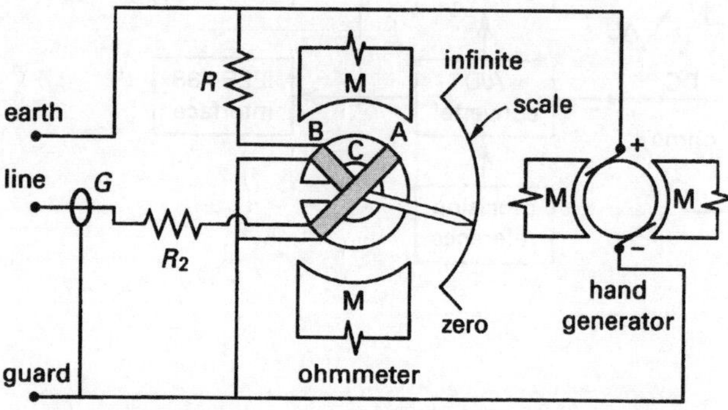

Hand-crank meggers operate on the order of 500 V. They often come equipped with a slip clutch that enables the generator to turn at a constant speed, providing consistent readings. A circuit under test should not be exposed to voltage levels that can damage the insulation. Depending on the standard and the test, this voltage level varies from one to three times the rated voltage.

If increased accuracy is desired, electronic meggers can be used. Such a megger is shown in Fig. 53.13. This is called the *leakage method* or *loss of charge method*. R represents the insulation resistance to be checked. The battery is the power supply and can be set to whatever voltage the meter is capable of generating. When the resistance is placed under test, switch A closes long enough to charge the capacitor, C, and then reopens. Switch B then immediately closes and the deflection, d_1, of the meter is noted. The process is repeated, allowing a period of time to pass before switch B is closed. The deflection, d_2, is then noted. The deflections differ by an amount equal to the charge loss on the capacitor due to the current flow through the resistance, R. The loss can be used to calculate the insulation resistance per Eq. 53.15, with the capacitance in microfarads.

$$R = \frac{t}{2.303C\log_{10}\left(\frac{d_1}{d_2}\right)} \quad \text{[in the units of M}\Omega\text{]} \qquad 53.15$$

Figure 53.13 *Leakage Method for Measuring Insulation Resistance*

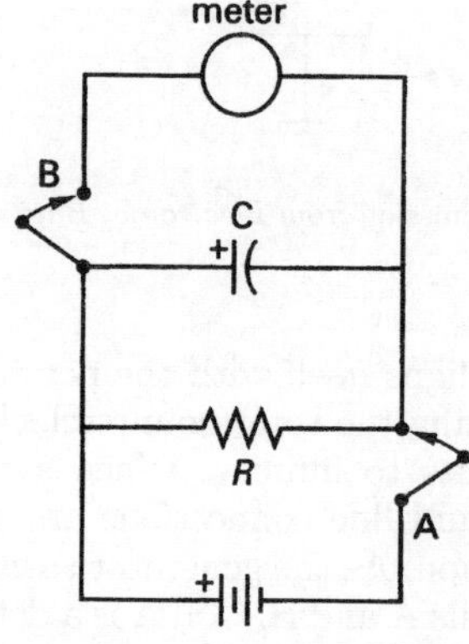

Ground testing indicates testing to determine if a conducting connection exists between an electrical circuit or equipment and the ground or some conducting body. If the connection is desired, as in a safety ground, then the result desired is a low resistance connection. If the connection is to have a specific value, as in impedance connections made on wye generators or transformers, then the desired value is the result. If a ground connection is not meant to exist, then a high value is desired. Standards exist for ground impedance values, and for checking said values. For building structures, such guidance can be found in the *National Electrical Code*.

13. MEASUREMENT STANDARDS AND CONVENTIONS

In the United States, standards and conventions for instrumentation and measurement pertaining to electric power are governed by the Institute of Electrical and Electronics Engineers (IEEE), in accordance with IEEE Standard 1459, *IEEE Standard Definitions for the Measurement of Electrical Power Quantities Under Sinusoidal, Nonsinusoidal, Balanced,* or *Unbalanced Conditions.* Internationally, these standards and conventions are governed by the International Electrotechnical Commission (IEC), in accordance with IEC Standard 62053-23, *Electricity Metering Equipment (A.C.)—Particular Requirements—Part 23: Static Meters for Reactive Energy.* The standards differ slightly—in particular, the sign convention for power factors differs[13] but both are in use in the United States. Figure 53.14 gives an overview of the sign conventions used by the standards.

Reference angles and signs related to the power factor angle are established in Sec. 27.9. Angles are measured counterclockwise, in accordance with the Cartesian coordinate system that uses the positive x-axis pointing to the right and the positive y-axis pointing upward. The reactive power for an inductive load is $+Q$, and that for a capacitive load is $-Q$.

The apparent power vector, **S**, is

$$\mathbf{S} = \mathbf{VI}^* = |V||I|\angle(\theta - \varphi) \qquad 53.16$$

The angle associated with **S**, and more specifically with the difference between **P** and **S**, is the power angle, ϕ. This is the same as the overall impedance angle, ϕ. The magnitude of the overall impedance angle is equal to the power factor angle, ϕ_{pf}, which is usually shown simply as ϕ as well.[14] This is symbolized by

$$\text{pf} = \cos\phi = \frac{\mathbf{P}}{|\mathbf{S}|} \qquad 53.17$$

If a sign is given to the power factor in Eq. 53.17, it should be the sign of the angle obtained when the angle associated with **P** is subtracted from the angle associated with **S**. The sign should be consistent with the sign assigned to **S**, and the power factor angle, ϕ_{pf}, should be equal in magnitude to the voltage angle, θ, subtracted from the current angle, φ.

The impedance angle, $|\phi_{\text{pf}}|$, or $\pm\phi$, takes its sign from

$$\mathbf{Z} = \frac{\mathbf{V}}{\mathbf{I}} \qquad 53.18$$

The sign for impedance is positive for lagging loads and negative for leading loads and may not match the sign of the power angle or power factor angle.

The impedance angle is positive for lagging currents and negative for leading currents, as shown in Fig. 53.14.[15] Lagging currents are in quadrants I and II, and leading currents are in quadrants III and IV. For example, Fig. 53.15 shows the voltage/current relationship for a lagging load. Setting the voltage as the reference at 0° as shown in Fig. 53.15(a), the current lags the voltage as in an inductive load. The current shown in Fig. 53.15(a) is associated with the power triangle of quadrant I in Fig. 53.14 even though the current is in quadrant IV. When current is made the reference, as in Fig. 53.15(b), the lagging current is in quadrant I, which is consistent with quadrant I of Fig. 53.14. Figure 53.15(b) and Fig. 53.14 illustrate that the power factor angle and the power angle are in the same quadrant and are equal in magnitude.

The voltage/current relationship for a leading load is shown in Fig. 53.16. Setting the voltage as the reference at 0° as shown in Fig. 53.16(a), the current leads the voltage as in a capacitive load. The current shown in Fig. 53.16(a) is associated with the power triangle of quadrant IV in Fig. 53.14 even though the current is in quadrant I. When current is made the reference, as in Fig. 53.16(b), the leading current is in quadrant IV, which is consistent with quadrant IV in Fig. 53.14. Figure 53.16(b) and Fig. 53.14 illustrate that the power factor angle and the power angle are in the same quadrant and are equal in magnitude.

When the capacitive load voltage and current are reflected, they move from quadrant IV to quadrant II, as shown in Fig. 53.16(c).[16] This reflection of reactive power must be absorbed by the generator or source. The reflected current and reactive power come from the capacitive load, but the current becomes a lagging current. This is consistent with the concept in power system engineering of a capacitor as a generator of $+Q$[17]. The capacitors in a given circuit provide the reactive

[13]An understanding of both conventions is important, given the wide number of instruments that use either or both conventions. Particular meters must be set during installation for the desired configuration.

[14]See Sec. 27.19 for more on angular relationships and the power factor and Sec. 27.22 for more on power triangles.

[15]A positive impedance angle indicates the current is delayed (with voltage as a reference at time zero). The impedance occurs before the current flows. For a negative impedance angle, the current flows and then impedance occurs.

[16]During one half-cycle, the $-Q$ is absorbed by the capacitor and stored in the electric field. During the next half-cycle, the Q is returned to the generator or source as $+Q$, with a lagging current that must be absorbed by the generator or source. Generators deliver or receive power, while loads import or export power, depending on the system used by the manufacturer.

[17]See Sec. 44.14 for a description of capacitance as a generator of $+Q$.

Figure 53.14 Power Factor Sign Convention

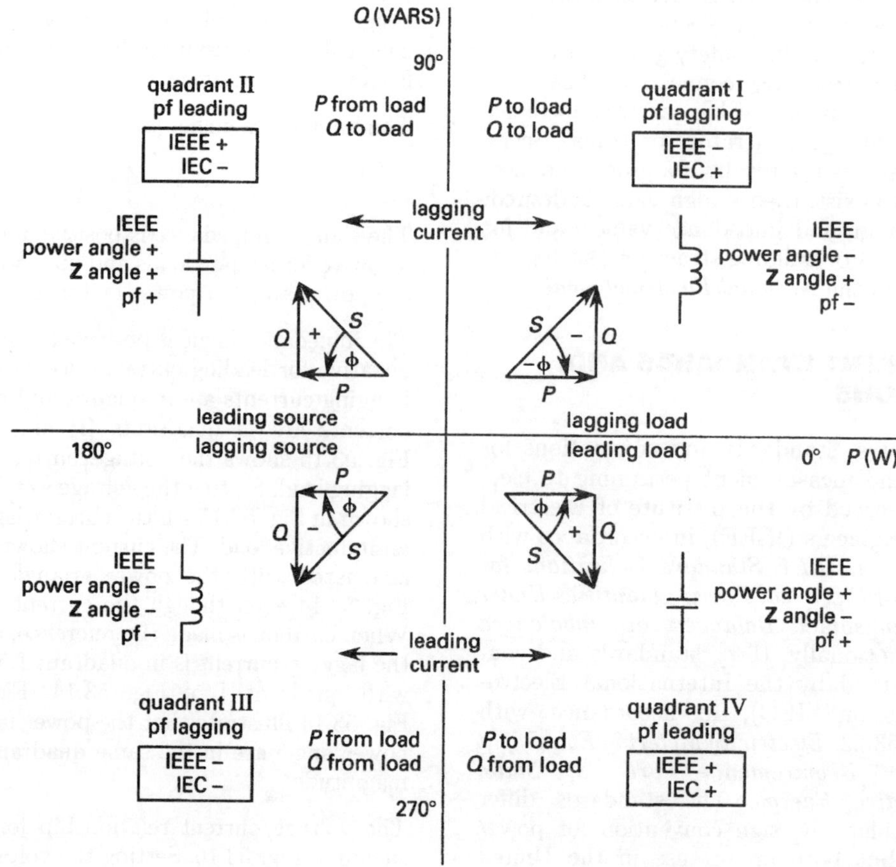

Figure 53.15 Lagging Relationships

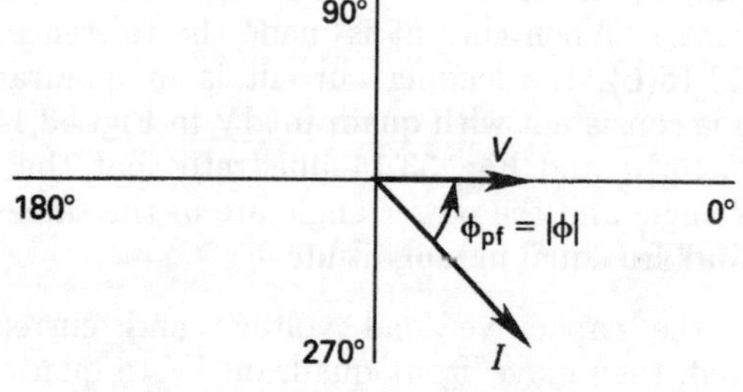

(a) lagging load with V as reference

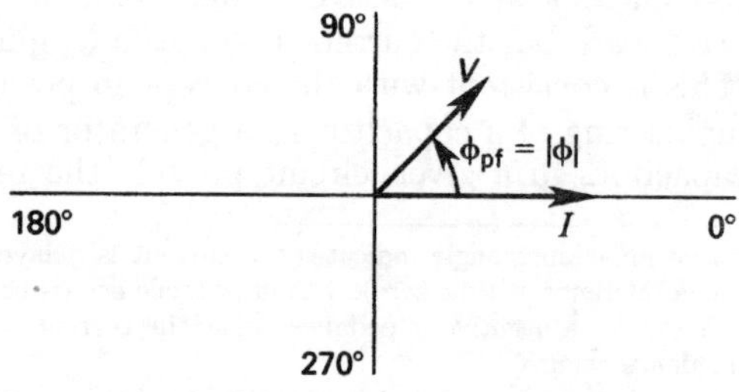

(b) lagging load with I as reference

load, $+Q$, that is required by the inductors in the same circuit.

Combining the concepts discussed in this section shows that the angle associated with the apparent power, which is the difference between the voltage angle θ and the current angle, φ, should have the same sign as the power factor and the same relationship defined by the impedance angle in order to remain consistent with the formulas used in this section. Using power as the reference, as shown in Fig. 53.14, in each quadrant the IEEE standard has the same sign for the power angle and the power factor. The only variable sign that is consistent between the top quadrants and between the bottom quadrants is the impedance angle. The impedance angle is positive when there is lagging current (in quadrants I and II) and negative when there is leading current (in quadrants III and IV).

In this way, the IEEE convention is consistent with the electrical formulas. A negative power factor is applied to lagging loads, and a positive power factor is applied to leading loads.[18]

Figure 53.17 shows another way to visualize the sign of a power factor in relation to loads. On an analog power factor meter, the term "leading" is associated with

[18]This is somewhat counterintuitive. The reactive lagging loads are positive, so one would expect the power factor to be positive, but it is in fact negative.

Figure 53.16 Leading Relationships

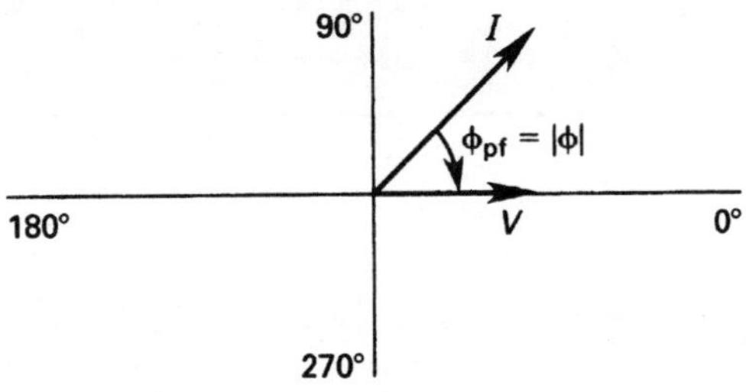

(a) leading load with V as reference

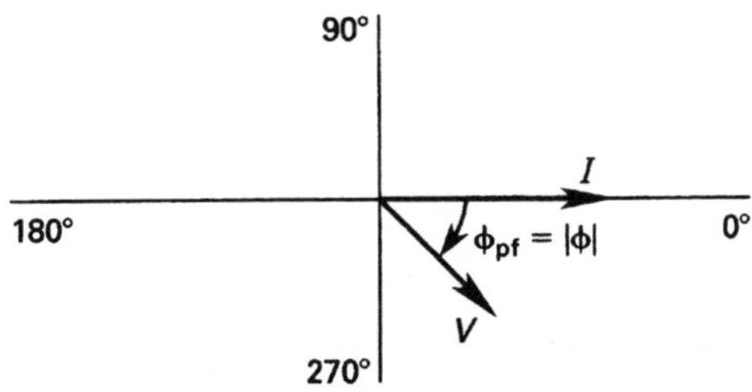

(b) leading load with I as reference

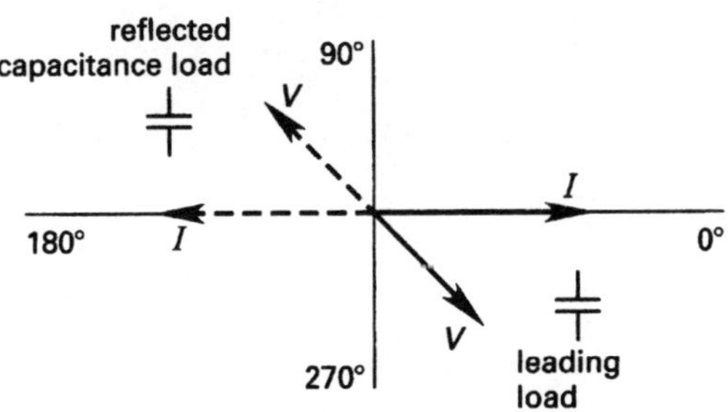

(c) leading load reflection of power export

Figure 53.17 Analog Power Factor Meter

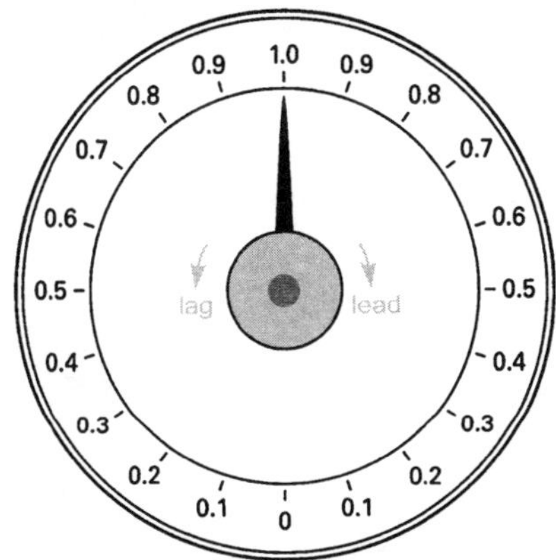

positive values that move the meter to the right in the Cartesian coordinate system. The term "lagging" is associated with negative values that move the meter to the left in the Cartesian coordinate system.

Voltage/current relationships for lagging loads and leading loads may also be described in terms of an electromagnetic wave. In Fig. 53.14, the capacitor as a load is shown in quadrant IV, where it absorbs a leading current. When the current wave reflects from the load, the current angle shifts 180° to quadrant II where the current now lags the voltage. The capacitor provides $+Q$ to the circuit, as shown by the direction of Q in the power triangle in quadrant II. For consistency with the signs used by the IEEE standard and the conventions used for capacitors and inductors, Fig. 53.14 shows the capacitor symbol in quadrant II and the inductor in quadrant III. Some manufacturer's documentation that discusses the IEEE sign convention shows inductors in quadrants I and II and capacitors in quadrants III and IV to be consistent with the lagging and leading current labels. However, this is inconsistent with the positive and negative values given in the IEEE standard for the power angle, the impedance angle, and the power factor angle.

By contrast, the IEC convention uses a positive power factor if the power, **P**, is delivered to the load (or imported by the load), as in quadrants I and IV of Fig. 53.14. If the power, **P**, is from the load (exported by the load, or received by the generator), the power factor is negative, as in quadrants II and III.[19]

Representations of either or both the IEEE and IEC power factor sign conventions are found in manufacturers' documentation for instrumentation and for protective relaying. The standard or combination of standards being used must be ascertained before instrumentation is installed or the data from it are analyzed.

Example 53.11

A transmission system has the instrumentation shown. If IEEE standards are used, what are (a) the sign of the power factor on the load meter for the load shown, (b) the terms used when the power flow is A to B, and (c) the terms used when the power flow is C to D?

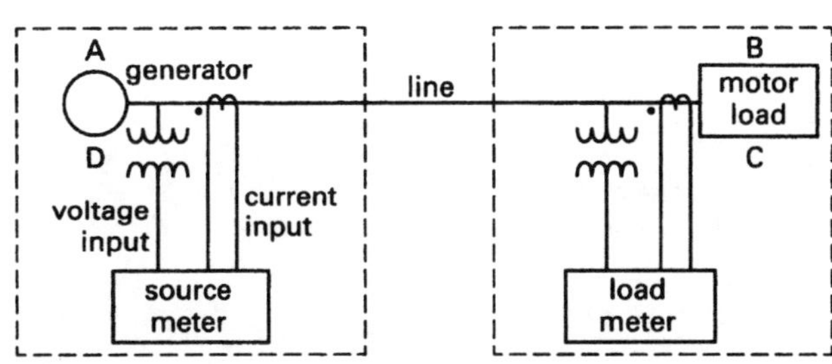

[19]By IEC convention, positive values indicate normal power flow, while negative values indicate a reverse power flow, or potentially a fault.

Solution

(a) According to IEEE standards, a motor load will be primarily inductive, so the load will be in quadrant I of Fig. 53.14. The sign for this inductive load is negative. The sign on the IEEE source meter will also be negative. If an IEC standard meter is used, the signs would technically differ (see the IEC standard shown in quadrants I and III of Fig. 53.14), but IEC meters are often set to provide readings that match those of the load meter.

(b) The term for A is *delivered.* The term for B is *imported.*

(c) The term for C is *exported.* The term for D is *received.*

Topic XIII: Electrical Materials

Chapter

54 Electrical Materials

1. OVERVIEW

The electrical properties of materials impact all engineering designs. An understanding of the properties and potential reactions of chosen materials within the component parts and the surrounding environment is necessary both to ensure proper operation over the service lifetime of the components and to minimize long-term effects on the environment.

The fundamental properties of conducting, insulating and dielectric, semiconductor, magnetic, electron emitting, and radiation-emitting materials are covered throughout this book. Materials handbooks are available that provide specific values for these properties for elements and compounds under various conditions. A periodic table with a summary of materials' properties is given in App. 54.A. A table of relative atomic weights is given in App. 54.B.

2. TYPES OF MATERIALS

Materials may be grouped into three main types: metals, polymers (that is, plastics), and ceramics.[1] *Metals* are materials characterized by high thermal and electrical conductivity. They are opaque and can generally be polished to a high luster. Further, they are normally heavy and deformable. Metals are located to the left of the metal-to-nonmetal transition staircase shown in App. 54.A. Elements adjacent to the staircase exhibit characteristics of both metals and nonmetals and are sometimes called *metalloids* or *semiconductors*. Further, in some texts, they are called metals and in others they are called semiconductors. Items to the right of the staircase are called *nonmetals*.

Polymers, or *plastics*, are materials characterized by low thermal and electrical conductivity. They are generally transparent or translucent and poor reflectors of light. Finally, they are normally flexible and subject to deformation. Polymers are composed of those elements listed in Table 54.1.

Table 54.1 *Elements in Polymers*

element atomic number	element name
1	hydrogen
6	carbon
7	nitrogen
8	oxygen
9	fluorine
14	silicon

Ceramics are compounds that contain metallic and nonmetallic elements. Concrete, rocks, glass, and nuclear fuel (UO_2) are all ceramics, to name but a few. Ceramics have extremely low thermal and electrical conductivity and for this reason are also called *refractory materials*. Indeed, they are highly chemically resistant (inert), can be opaque or translucent, and are very hard and brittle. Ceramics are composed of metallic elements combined with carbon, nitrogen, oxygen, silicon, or phosphorous, or some combination thereof.

3. CONDUCTING MATERIALS

A *conductor* is a wire, cable, or other body or medium that is suitable for carrying an electric current. While no specific dividing lines exist between conductors, semiconductors, and insulators, conductors have a conductivity on the order of 10^6 S/m.

Copper is used extensively in electrical applications because of its high electrical and thermal conductivity and malleability. Aluminum is used when additional conductor strength is required. Silver has the highest conductivity at room temperature. Gold has conductivity similar to aluminum, as well as excellent corrosion and oxidation resistance. Platinum, a precious metal, is used primarily in resistance thermometers and thermocouples. Palladium is similar to platinum and less expensive. Nickel or nickel-coated conductors are used at temperatures up to 300°C with no enhanced corrosion at defective areas. Tungsten has a very high melting point and is erosion resistant, finding use in voltage

[1]Glass is sometimes listed as a type of material. "Glass," however, is a term applied to noncrystalline materials with specific expansion/contraction characteristics. For example, below a temperature known as the *glass temperature*, further rearrangement of atoms in glass ceases and contraction results from smaller thermal vibrations—rather than the formation of a metallic crystalline structure. Glasses have short-range order with an absence of long-range order. Glass is probably best described as a ceramic.

Electrical Materials

regulators, repetitive low-current applications, and heating wires or filaments for lamps. Carbon has two crystalline forms: graphite and diamond. Carbon (that is, amorphous carbon) or graphite (that is, crystalline carbon) is used as an electric sliding contact in motor brushes. The diamond structure is used where high heat conduction is desired.

4. INSULATING AND DIELECTRIC MATERIALS

An *insulator* is a device or material having a high electrical resistance (i.e., a low conductivity). A *dielectric* is an electrically insulating material or a material in which an electric field can be maintained with minimum dissipation of power. Insulators generally have a conductivity on the order of 10^{-12} S/m.

Plastics are widely used because of their low cost, ability to be molded into many shapes, and good mechanical properties. Ceramics, because of their very low electrical conductivity, also find many uses in electrical insulating applications. Oils, gases, and papers are inexpensive insulators. Mica, with its stability, arc resistance, and high dielectric strength, is another common insulator.

5. MAGNETIC MATERIALS

A *magnetic material* is generally defined as a material exhibiting ferromagnetism. Magnetic materials are classified into two broad categories: *Nonretentive* or *soft* materials are low-loss materials. *Retentive* or *hard* materials are high-loss, high-retentivity, high-energy product materials.

The soft magnetic materials are used in transformers, motors and generators, cores, and other electromagnetic apparatuses. Iron is the most common soft magnetic material. Iron-silicon, also called *electrical steel*, is a soft material with very low core and eddy current losses and very high permeability. Iron-cobalt alloys are used where high saturation is required. Other alloy combinations are used in special applications.

The hard magnetic materials are used as permanent magnets. Hard materials are composed of a fine structure created during various heat treatments. High-carbon steels with tungsten, chromium, cobalt, aluminum, or vanadium in a martensitic structure are common. Precipitation-hardened alloys also form a fine structure. In this category are *alnico* (aluminum, nickel, and cobalt), *cunife* (copper, nickel, and iron), and *vicalloy* (vanadium, cobalt, and iron), among others.[2]

6. RADIATION-EMITTING MATERIALS

Materials at excited energy levels emit electrons, photons of light, or other radiation. These emissions have numerous electrical applications.

Metal filaments in incandescent lamps emit light when heated. Tungsten is the most common material used in this application. *Discharge lamps*, also called *fluorescent lamps*, use material in vapor form.[3] The noble gases, sodium, and mercury are the materials of choice. *Phosphors* are organic or inorganic luminescent materials.

Electrical Materials

[2]Ordered alloys and fine-particle magnets are also retentive.
[3]When materials emit light almost instantaneously after excitation, they are called *fluorescent* materials. When the emission is delayed, they are called *phosphorescent* materials.

Topic XIV: Codes and Standards

Chapter

55 Biomedical Electrical Engineering

1. OVERVIEW

Biomedical engineering is the application of engineering technology to medicine. Electrical and electronic engineering play a large role in biomedical engineering. Physiological monitoring, diagnostic instruments, patient internal imaging, implant devices, artificial organs, and computer data processing of biological information are but a few of the areas of concern.

The theory of operation of the instruments and equipment used is described throughout this book and is consistent with standard techniques of electrical analysis. The difference in electrical engineering for medical purposes is that the technology interfaces with the human body. This interface requires stringent controls to ensure electrical safety, radiation safety, reliability, and biocompatibility.

Electrical sources may often touch the human body but not transfer electrical energy to it, either directly or by radiating energy. Nonengineering personnel and patients will handle and/or operate the equipment. They must be protected, and the equipment must withstand such usage. The electrical components must also be protected from the biological environment to function as designed. The components must be highly reliable and fail-safe for obvious reasons. Biocompatibility is an absolute necessity to prevent interactions with body tissue and ensure nontoxicity. Inert materials are used, usually metals and polymers. An understanding of materials is required because the fabrication, molding, and surface condition of electrical components affect the body's reaction to them.

The safety and effectiveness of medical devices in the United States are the responsibility of the Food and Drug Administration (FDA) Center for Devices and Radiological Health (CDRH). Most biomedical devices require the approval of the CDRH and are subject to the general requirements of the Federal Food, Drug, and Cosmetic Act (FD&C). The requirements of the FD&C Act are documented in Title 21 of the *Code of Federal Regulations*, Chapter 1, Parts 800 through 1299.

The FDA recognizes the standards of numerous organizations in the design and use of biomedical devices. These include the standards of the Association for the Advancement of Medical Information (AAMI), International Electrotechnical Commission (IEC), International Organization for Standardization (ISO), American Society for Testing and Materials (ASTM), American Institute of Ultrasound in Medicine (AIUM), Institute of Electrical and Electronics Engineers (IEEE), National Electrical Manufacturers Association (NEMA), and Underwriters Laboratories, Inc. (UL).

2. SENSORS

Sensors are used to convert biomedical quantities into electrical signals, which can then be monitored and analyzed. Examples of biomedical sensor applications are given in Table 55.1.

Table 55.1 *Biomedical Sensor Applications*[*]

quantity sensed	device or principle used	biomedical application(s)
electric potential	electrodes	nerve or muscle activity (e.g., heart beats)
electric impedance	bridge	respiration rate
magnetic field	permeability	blood flow
displacement	capacitive	orthopedics
flow	ultrasonic Doppler	body fluid flow
image	X ray	shape or organ motion

[*]The devices and the applications listed are by no means the only possible choices.

Sensor size and weight are dependent upon the application and the patient. The size is often small, requiring high sensitivity and minimal power consumption.

Biocompatibility is often ensured by isolating the sensor from the biological environment or by using relatively inert elements or compounds whose interactions with

Codes and Standards

the human body are minimal. For example, the surfaces of insulated electrodes are coated with tantalum or titanium oxides, Ta_2O_3 or TiO_2, to provide good stability and minimal chemical interactions. Additionally, silver and platinum electrodes, Ag-AgCl or Pt-PtIr, are widely used in health care.

3. IMPLANTABLE STIMULATORS

The application of electricity to the human body for medical purposes existed well prior to modern medicine. One of the first major benefits was to the human cardiovascular system with the advent of commercial pacemakers. Implants are now used to treat deafness (for example, the cochlear implant), paralyzed muscles, spinal cord pain, and epilepsy, among many other conditions. Implants also monitor blood flow, glucose levels, oxygen concentration, and the delivery of drugs.

Electrical potential exists across the membranes of living cells. The human nervous system detects and changes these potentials, and now implantable stimulators can do the same. When the implantable stimulator is equipped with a telemetry device to allow external monitoring, it is called an *implantable stimulator telemetry* (IST).

4. MONITORS

Monitors are used to observe numerous physiologic variables and display the patient's status, provide a historical record, and perform elementary logical decisions regarding care. Some of the possible variables that can be monitored are given in Table 55.2.

***Table 55.2** Monitored Physiologic Variables*

vital signs	bioelectric signals	chemical concentrations
temperature	electrocardiogram (ECG)	oxygen
heart rate	electroencephalogram (EEG)*	carbon dioxide
blood pressure		pH
respiration rate		glucose

*measurement of electric signals on the scalp that arise from brain activity, indicative of the state of the central nervous system

A general monitoring system block diagram is shown in Fig. 55.1. Other than the interface with the human system, all the electrical and electronic principles are identical to those in other electrical engineering applications.

5. APPLICATIONS

The types of electrical and electronic equipment in the field of medicine are broad in scope but similar in principle to other such electrical engineering applications.

***Figure 55.1** General Medical Monitoring System*

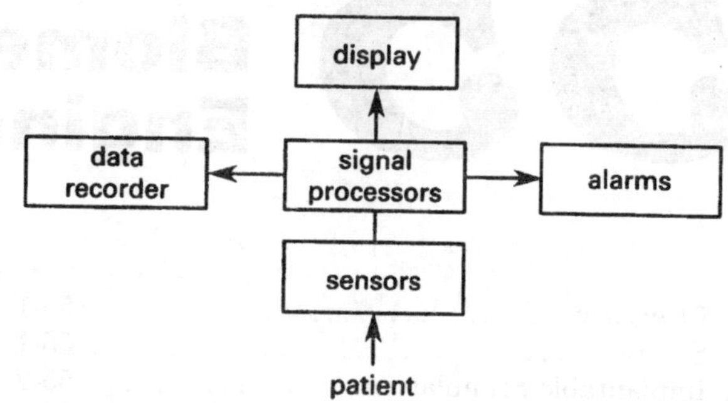

They include radiology, magnetic resonance imaging (MRI), and ultrasound equipment. Support systems that replace or augment human organs, such as artificial limbs, artificial hearts, dialysis machines, insulin delivery systems, and blood-detoxification systems, rely on electrical and electronic components. Finally, an entire array of electrical devices exists to analyze the many samples taken from a patient to aid in the diagnostic and recovery processes.

6. SHOCK AND BURNS

The effect of electric current on the human body varies according to the individual, the frequency of oscillation, the path of the current through the body, and the environmental conditions. Nevertheless, the observed effects on large populations have been documented, investigated, and used to identify general ranges the average current at which certain effects occur have been determined. Figure 55.2 illustrates some general the current ranges and their resulting effects for 60 Hz. More specific ranges are provided in standards.[1]

***Figure 55.2** Human Physiological Reactions*

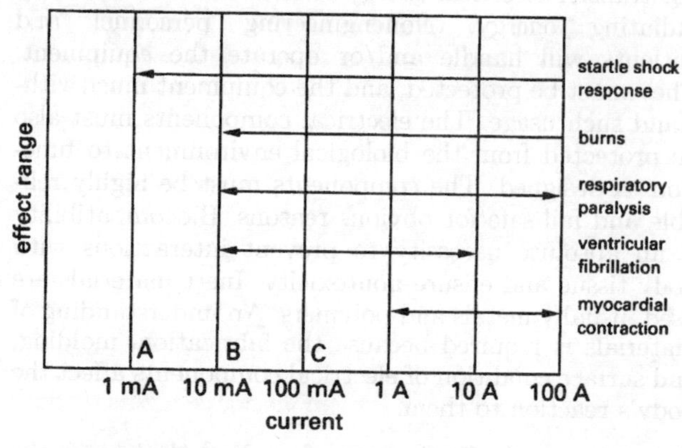

[1] One of the international standards on this subject is IEC 60479-2. An American standard is ANSI/UL C101.

The label points A, B, and C are meant as memory aids. Level A (1 mA) is an average for the perception of current flow in most people. Level B (10 mA) is an average representing the current level resulting in an inability to let go of an electric line or item once it is in one's grasp.[2] Level C (100 mA) represents the point where various effects occur that may result in death. In addition to individual and environmental variables, these also vary by person, as mentioned, but also by the type of current, be it a static discharge, DC, 60 Hz AC, 400 Hz AC, or higher.

Electric current models for the human body vary in complexity. However, a simple, useful rule of thumb is to consider the body as a single resistor with a resistance of 500 Ω. A 500 Ω body with a level C current of 100 mA (enough to potentially cause death) can be driven by as little as 50 V. Most safety standards require workers to use additional protection during electrical work involving 50 V or greater.

Ground fault circuit interrupters (GFCIs) are designed to prevent exceeding the let-go level and to prevent ventricular fibrillation. Underwriters Laboratory (UL) Standard 943 GFCI Class A devices are rated for 5 mA (60 Hz) to prevent let go from becoming an issue. The UL standard considers 35 mA (60 Hz) to be the level at which ventricular fibrillation becomes an issue. As a result, GFCIs are designed to trip fast enough to prevent exceeding that level of total rms current. Even so, at 120 V, a 500 Ω body can experience enough current to result in the startle or shock sensation.

7. SHOCK PROTECTION

Electrical equipment requires an installation designed to minimize shock hazards. Common installation design methods include grounding, bonding, insulation of electrical devices, and the use of protective devices (such as GFCIs), isolation transformers, and circuit protective devices (such as circuit breakers, fuses, and voltage or current limiters) to limit exposure to higher levels of current.

[2]Standards are moving toward 5 mA as the appropriate current for the let go level. The level of 10 mA is used here because it is easier to remember in conjunction with the others and is above the threshold.

56 National Electrical Code

Nomenclature

A	area	m^2
d	diameter	m
F	factor	–
I	effective or DC current	A
l	length	m
P	power	W
pf	power factor	–
R	resistance	Ω
V	effective or DC voltage	V

Subscripts

l per unit length

1. HISTORY AND OVERVIEW

The first practical incandescent light bulb and lighting system were developed by Thomas Alva Edison in 1879. That same year, the National Association of Fire Engineers met to discuss standards for electrical installations. By 1895 there were five separate codes in the United States. In 1896, a national meeting was held between various concerned organizations and by 1897 the *National Electrical Code* (NEC) (hereafter referred to as the "Code") was adopted. The Code gradually came under the sponsorship and control of the National Fire Protection Association (NFPA). The Code is officially endorsed by the American National Standards Institute (ANSI), and the NFPA *National Electrical Code* committee responsible for the Code is known as ANSI Standards Committee C1. The Code is the most widely adopted code in the United States and the most widely accepted set of electrical safety requirements in the world. Local codes in the United States normally adopt the Code en masse and then add supplemental requirements that are deemed necessary, or remove and change items to ensure local applicability or uniqueness.

The Code generally applies to the installation of electrical conductors and equipment within or on public and private buildings up to and including the connection to the electrical supply. Its general purpose is to prevent fires and explosions by providing guidance for proper electrical installations.

The Code consists of an introduction followed by nine chapters, which are further subdivided into articles. The outline of the Code follows. Chapter numbers are given in parentheses. Chapter 1 through Chap. 4 generally apply to all situations except as modified by Chap. 5 through Chap. 7.

- Introduction
- (1) General
- (2) Wiring and Protection
- (3) Wiring Methods and Materials
- (4) Equipment for General Use
- (5) Special Occupancies
- (6) Special Equipment
- (7) Special Conditions
- (8) Communications Systems
- (9) Tables
- Annex A: Product Safety Standards
- Annex B: Application Information for Ampacity Calculation
- Annex C: Conduit and Tubing Fill Tables
- Annex D: Examples
- Annex E: Types of Construction
- Annex F: Critical Operations Power Systems
- Annex G: SCADA
- Annex H: Administration and Enforcement

- Annex I: Recommended Torque Tables
- Annex J: ADA Standards

In this chapter, article numbers greater than 90 indicate the article or article number in the Code and are for ease in referencing the Code directly. Concepts covered and requirements mentioned focus on dwelling units and voltages less than 1000 V. The code uses a mixture of English engineering units and SI units.

2. INTRODUCTION

The Introduction to the Code contains Article 90 and explains that the purpose of the Code is the "practical safeguarding of persons and property from hazards arising from the use of electricity." The intended coverage or applicability of the Code is illustrated in Fig. 56.1.

The Code is meant to be a legal document that can be implemented and interpreted by local governmental bodies that have jurisdiction over electrical installations. Most do just that, with some additions and changes for local needs. A *permissive rule* in the Code is one that is allowed but not required (i.e., an alternative). The term "shall be permitted" is used to indicate these alternatives. Any *informational note* used in the Code is for informational purposes only and does not represent a requirement. Factory-installed internal wiring is not required to be inspected if the equipment has been listed by a qualified electrical testing laboratory that requires installation in accordance with the Code. When metric measurements are used, they conform to the SI System. Most of the values in the Code remain in the English Engineering System, with approximate metric equivalents given. Footnotes are also used to provide the necessary SI conversion factors.

The Code addresses the principles of protection and safety contained in the International Electrotechnical Commission (IEC) Standard 60364-1, *Electrical Installation of Buildings*, so knowledge of the Code is widely applicable. The Code only applies to the parts of a utility power generation plant that are not an integral part of the generating plant, such as offices.

While the Code primarily uses SI units, it accounts for the use of customary U.S. units. The Code defines conversion between units as either hard or soft. *Hard conversion* is a change in dimensions that may result in a new quantity that is not interchangeable with the original. A *soft conversion* is a change in the description of the measurement but not in the actual dimensions, making the part interchangeable. For example, a soft conversion of 1/2 in is 12.7 mm, while a hard conversion is 13 mm.

3. GENERAL

The "General" portion of the Code contains Art. 100, "definitions" and Art. 110, "requirements for electrical installations."

The definitions are those required for the proper application of the Code and are not all-encompassing. Generally understood terms that can be found in a standard dictionary are not included. Also not included are electrical terms that are not unique to the Code, such as "volt." Some of the terms that are used in a distribution system and found in Art. 100 are shown in Fig. 56.2. The following definitions are also important to understand when reading the Code.

Ampacity is the maximum current, in amperes, that a conductor can carry continuously under the conditions of use without exceeding its temperature rating. A *continuous load* is expected to operate at maximum current for three hours or more. *Continuous duty* is operation at a substantially constant load for an indefinite period of time. In a *nonlinear load*, the steady-state current wave shape does not follow the wave shape of the applied voltage. *Approved* indicates acceptability to the authority having jurisdiction (AHJ). *Bonding* is a process that establishes reliable electrical continuity.

A *feeder circuit* is from the service equipment, the source of a separately derived system (such as a

Figure 56.1 NEC Coverage

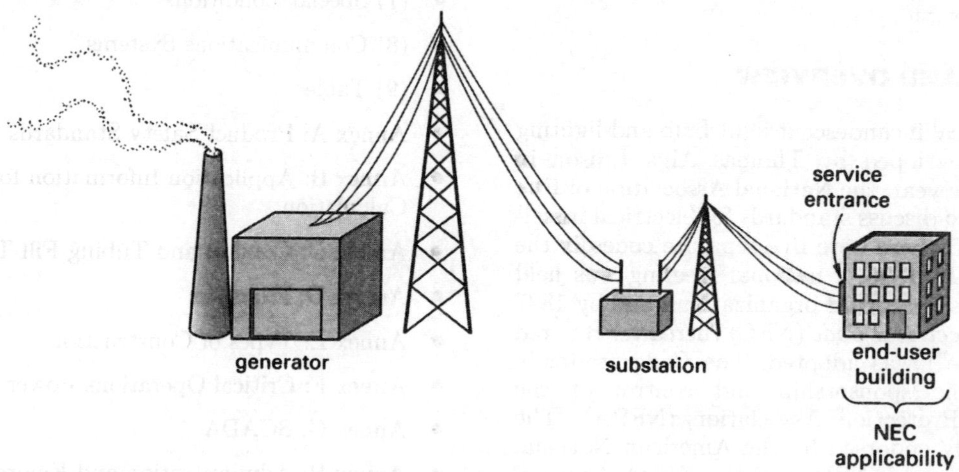

Codes and Standards

Figure 56.2 *Typical Distribution System*

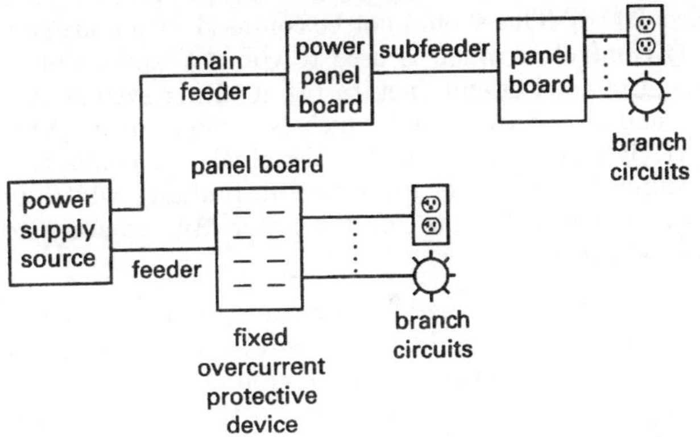

transformer), and the final branch-circuit overcurrent device. A *branch circuit* is from the final overcurrent device protecting the circuit and the outlet(s). When a branch circuit supplies a load other than an outlet, a descriptive name is added, such as *appliance branch circuit* or *general-purpose branch circuit* (which supplies receptacles or outlets for lighting and appliances), or *individual branch circuit* (which supplies utilization equipment).

An *equipment grounding conductor* (EGC) is the conductive path installed to connect normally non-current-carrying metal parts of equipment that are not intended to carry current during normal operation to the system's *grounded conductor*, which is intentionally grounded. A *grounding electrode* is the conducting object through which a connection to earth is made, while a *grounding electrode conductor* is the conductor that connects the system grounded conductor to the grounding electrode.[1]

A *damp location* is protected from the weather but subject to moderate degrees of moisture. By contrast, a *wet location* is subject to saturation with water or other liquids.

A *surge-protective device* (SPD) limits transient voltages by diverting or limiting surge currents and is capable of repetitive operation. *Overcurrent* is defined as any current in excess of rated current. Overcurrent includes overload, short circuit, and ground fault. An *overload* is defined as the operation of equipment in excess of its normal full-load rating or of a conductor in excess of its rated ampacity, which after sufficient time, causes damage or overheating.

NEC Art. 110, "Requirements for Electrical Installation," notes that conductor size will be given in *American Wire Gage* (AWG) or in circular mils. A comparison of AWG diameters to other systems is given in App. 56.A. Use Eq. 56.1 and Eq. 56.2 to convert between diameters and cross-sectional areas.

$$A_{\text{cmil}} = \left(\frac{d_{\text{in}}}{0.001}\right)^2 \qquad 56.1$$

$$A_{\text{cmil}} = 1.273 \times 10^6 A_{\text{in}^2} \qquad 56.2$$

Wire gage sizes up to 4/0 are listed as AWG in the Code. Beyond that, the size is listed in circular mils (cmil) or, more likely, thousands of circular mils (kcmil). The metric standard for marking wire size is square millimeters (mm^2). The conversion is

$$A_{\text{cmil}} = 1973.53 A_{\text{mm}^2} \qquad 56.3$$

Temperature limitations are imposed and ratings coordinated so that conductors and their terminations are rated for the same operating temperature for circuits rated at 100 A or less [NEC Sec. 110.14(C)(1)]. Required working space within and around electrical installations, safety issues, and illumination of such spaces fill out the remaining items in NEC Art. 110.

Enclosure types designed to protect electrical components are listed in NEC Table 110.28. This information is expanded upon in a variety of standards, including those of Underwriters Laboratories (UL), the International Electrotechnical Commission (IEC), the Institute of Electrical and Electronics Engineers (IEEE), the National Electrical Manufacturers Association (NEMA), and others. Care should be taken when selecting an enclosure. Names such as driptight (2, 5, 12, 12K, 13), watertight (4, 4X, 6, 6P), and rainproof (3R) apply to entirely different enclosures, which cannot be substituted for each other. The environment of the space in which the equipment is located is the determining factor for selecting the enclosure type.

Example 56.1

A wire of unknown gage is measured. The diameter is found to be 0.2057 cm. What is the American Wire Gage (AWG) designation?

Solution

The tables for wire gage comparisons are generally in English Engineering System units, as is App. 56.A. Converting the diameter gives

$$d = (0.2057 \text{ cm})\left(\frac{1 \text{ in}}{2.54 \text{ cm}}\right) = 0.0810 \text{ in}$$

[1]Unfortunately, these names and definitions are very similar. In general usage, the *green wire* in a house is referred to as the *ground* and the *white wire* as the *neutral*. However, in the NEC, both wires contain some form of "ground" in their definitions. To further complicate matters, the NEC defines a *grounded conductor* as any circuit conductor that is intentionally grounded, which includes both the white and green wires. The NEC also defines the *neutral conductor* as the conductor connected to the neutral point of a system that is intended to carry current under normal conditions, which includes the white wire (*grounded conductor*), the *grounding electrode conductor*, and the *grounding electrode*. So to relate the NEC's usage to generic usage, the grounded conductor is the neutral (the white wire). While "grounded conductor" could be used to describe the equipment grounding conductor (the green wire) since it also is intentionally grounded, in general usage the green wire is not called the grounded conductor. In the *NEC Handbook*, the neutral is referred to as the *grounded service conductor* to differentiate it from the equipment grounding conductor. To summarize the connections, the grounding electrode conductor connects the green and white wires (which are connected only at the power panel and nowhere else in the dwelling unit) to the grounding electrode, which is generally outside buried in the earth. So the connection goes from the system grounded conductor(s) (or grounded service conductor and equipment grounding conductors) to the grounding electrode conductor to the grounding electrode.

Codes and Standards

From App. 56.A, this diameter correlates with AWG 12. The generic term for it is "twelve gage wire."

4. WIRING AND PROTECTION

NEC Chap. 2, "Wiring and Protection," contains Art. 200 (which covers grounded conductors), Art. 210 (branch circuits), Art. 215 (feeders), Art. 220 (branch circuit, feeder, and service calculations), Art. 225 (outside branch circuits and feeders), Art. 230 (services), Art. 240 (overcurrent protection), Art. 250 (grounding), Art. 280 (surge arresters), and Art. 285 (surge-protective devices).

The term "nominal" is used throughout the Code. Actual power supply voltage specifications, high and low, are set by ANSI standards. Nominal voltage is the desired voltage of the power source and is given in multiples of 120 V. Rated voltage is the desired voltage for operating equipment and is given in multiples of 115 V. Operating equipment is called *utilization equipment* and includes any device that utilizes electric energy for electronic, electromechanical, chemical, heating, lighting, or another similar purpose [NEC Art. 100]. Four-wire wye-connected systems normally list the type of connection next to the voltage for phase-to-phase voltage and give the phase-to-neutral voltage after a slash (i.e., 208Y/120 or 480Y/277). Examples of common wye and delta four-wire systems are illustrated in Fig. 56.3.

Figure 56.3 *Four-Wire System Nomenclature*

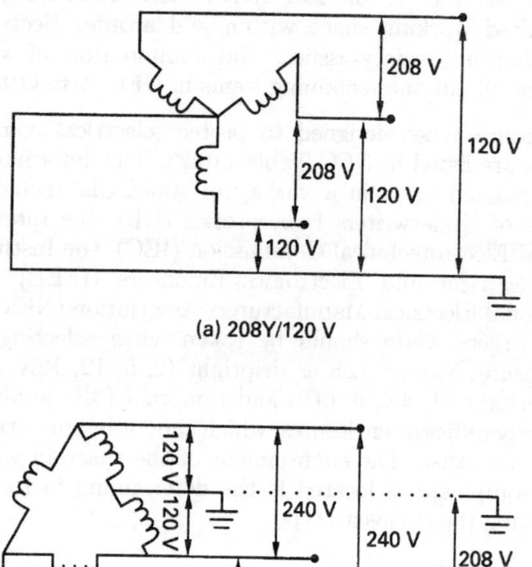

5. WIRING AND PROTECTION: GROUNDED CONDUCTORS

A *grounded conductor*, which is a conductor that is intentionally grounded (for example, the neutral of a system), is identified by a white or natural gray outer finish or by three continuous white or gray stripes on insulation that is any color except green [NEC Sec. 200.6]. This should not be confused with a *grounding conductor*, which is used to connect equipment or the grounded circuit (neutral) of a wiring system to a grounding electrode and which is colored green [NEC Art. 100 and Art. 250.119]. Terminals for connecting grounded conductors must be substantially white, or have the word "white" or the letter "W" adjacent to the terminal.

The important distinction between a grounded conductor (called the neutral or white wire) and an equipment grounding conductor (called the ground, safety ground, equipment ground, or green wire) is that the neutral is designed to carry current during normal operations, while the ground wire should only carry current during fault conditions.

6. WIRING AND PROTECTION: BRANCH CIRCUITS

A *branch circuit* consists of the circuit conductors between the final overcurrent device protecting the circuit and the outlet. Such circuits are covered in NEC Art. 210. However, if the branch circuit supplies only motor loads, it is covered in NEC Art. 430. A cross-reference for articles relating to branch circuits for other special conditions is given in NEC Table 210.2.

A *ground-fault circuit interrupter* (GFCI) is a device that de-energizes a circuit within an established period of time when the current to ground exceeds some predetermined value, which is less than that required to operate the overcurrent protective device of the supply circuit [NEC Art. 100]. The concept of such a device is illustrated in Fig. 56.4.

Figure 56.4 *Ground-Fault Circuit Interrupter (GFCI)*

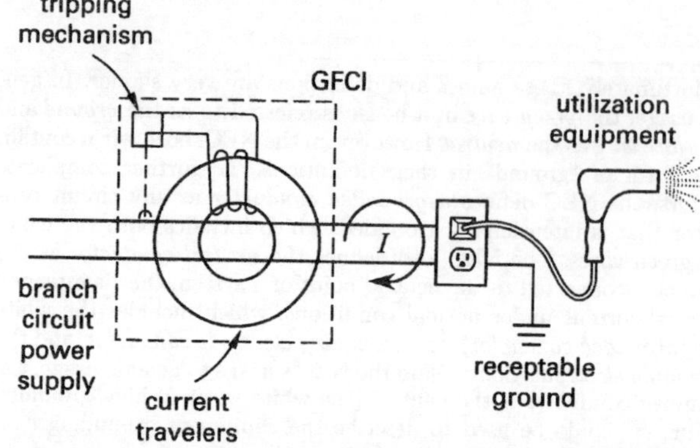

As long as the current is balanced in the lines running through the transformer core, no induced voltage is felt, no current flows in the tripping mechanism, and the GFCI contact remains shut. When the current is unbalanced due to a ground fault, the unbalanced magnetic fields produced induce a net voltage in the transformer, current flows in the tripping mechanism, and the GFCI contact opens, protecting any person holding the offending equipment. GFCIs are required to be installed outdoors and in dwelling-unit bathrooms, garages, crawl spaces, unfinished basements, kitchens, wet bar sinks, and boat houses [NEC Sec. 210.8(A)].

While outlets in certain locations require only GFCIs for safety, the 15 A and 20 A branch circuits supplying outlets located in family rooms, dining rooms, living rooms, parlors, dens, bedrooms, and other rooms in the home are required to have arc-fault circuit interrupter (AFCI) protection. Breakers supplying such circuits have AFCI protection built in as an integral part of the breakers. Detailed information is in NEC Sec. 210.12.

The number of branch circuits required in a given building is calculated per NEC Sec. 210.11. A dwelling unit is also required to have two or more 20 A small-appliance branch circuits beyond that calculated, one 20 A branch circuit for laundry, and one 20 A branch circuit for each bathroom [NEC Sec. 210.11(C)]. The conductors that supply branch circuits, before any adjustment is made or derating is allowed, must be able to carry 100% of the noncontinuous load as well as 125% of the continuous load [NEC Sec. 210.19(A)(1)]. The *overcurrent protective device* (OCPD) selected must meet identical requirements [NEC Sec. 210.20(A)]. The allowable conductor ampacities are found in App. 56.B [NEC Table 310.15(B)(16)].

A single receptacle on a branch circuit must have an ampere rating that is not less than the branch circuit. If two or more receptacles are used, NEC Table 210.21(B)(3) gives their allowed values [NEC Sec. 210.21(B)]. The rating of a branch circuit is based on the overcurrent protective device (OCPD) rating regardless of the actual load or the conductor ampacity. A 15 A or 20 A branch circuit can supply lighting, utilization equipment, or a combination of both. Any single cord-and-plug-connected device cannot exceed 80% of the branch circuit ampere rating. In addition, the rating of utilization equipment that is fastened in place, other than lighting, cannot exceed 50% of the branch circuit ampere rating [NEC Sec. 210.23(A)(2)]. NEC Table 210.24 summarizes the branch circuit requirements.

Receptacles are spaced in dwellings so that no point along the floor of any wall space is more than 1.8 m (6 ft) from a receptacle. (A distance of 12 ft can exist between receptacles.) Receptacles that are higher than 1.68 m (5.5 ft) above the floor, part of a lighting system or appliance, or in the floor located within cupboards or cabinets, do not count toward the requirement [NEC Sec. 210.52]. On countertops, the receptacle spacing is no greater than 600 mm (2 ft). Receptacles attached and controlled by switches are lighting outlets, which have separate requirements from general receptacles [NEC Sec. 210.70].

Example 56.2

A given grocery store branch circuit is meant to carry a continuous lighting load of 8 A. In addition, three intermittent loads designed for display stands are fastened in place and require 2 A each when operating. What is the rating of the overcurrent protective device (OCPD) on the branch circuit?

Solution

By NEC Sec. 210.20(A), the overcurrent device must be rated at 100% of the noncontinuous load and 125% of the continuous load.

$$\begin{aligned}\text{OCPD} &= (1.00)(2\text{ A} + 2\text{ A} + 2\text{ A}) + (1.25)(8\text{ A}) \\ &= 16\text{ A minimum}\end{aligned}$$

The standard ratings for fuses and fixed-trip circuit breakers are given in NEC Sec. 240.6 as 15 A, 20 A, 25 A, 30 A, and so on. Therefore, a 20 A OCPD will suffice.

Utilization equipment that is "fastened in place" cannot, and does not in this case, exceed 50% of the branch circuit ampere rating [NEC Sec. 210.23(A)(2)]. That is, 6 A is less than 50% of 20 A.

7. WIRING AND PROTECTION: FEEDERS

A *feeder* consists of all circuit conductors between the service equipment, the source of a separately derived system, or another power supply source and the final branch circuit overcurrent device (see Fig. 56.2) [NEC Art. 100]. The capacity of the conductors that constitute feeders, before any adjustment or derating allowed, must be 100% of the noncontinuous load plus 125% of the continuous load [NEC Sec. 215.2]. The *overcurrent protective device* (OCPD) selected must meet identical requirements [NEC Sec. 215.3]. Several special conditions for sizing the feeders also apply, including the exception that allows dwelling unit feeders carrying the full current of the service conductors to be the same size as those conductors. Though not required, a feeder is recommended to have no more than a 3% voltage drop, with a 5% voltage drop on both feeders and branch circuit conductors to the farthest outlet [NEC Sec. 215.2(A)(1)(b), Informational Note 2]. Tables for calculating voltage drops based on conductor properties when used in DC or AC circuits are given in App. 56.E, Conductor Properties, and App. 56.F, Conductor AC Properties [NEC Chap. 9, Tables 8 and 9, respectively].

For two-wire circuits, DC or AC, or three-wire single-phase AC circuits, with a power factor equal to one (purely resistive), the voltage drop is given by

$$V_{\text{drop}} = IR_l 2l = \frac{2lIR}{1000\text{ ft}} \qquad 56.4$$

The number 2 in Eq. 56.4 is used because the voltage drop is felt over the entire circuit length. The farthest equation to the right is used only if the resistance per unit length, R_l, is given in "per 1000 ft" quantities (as found in NEC Chap. 9, Tables 8 and 9). Many manufacturers in the United States list conductor information in units of 1000 ampere-feet.

Example 56.3

A 480Y/277, 80 kW noncontinuous load with a 0.8 pf consists of less than 50% ballast-type lighting.[2] The feeder length is 110 m. Applying the 3% recommendation from NEC Sec. 215.2(A)(1)(b) Informational Note No. 2, what is the maximum recommended voltage drop for every 1000 ampere-foot?

Solution

To determine the line current drawn on a noncontinuous basis, and noting that the voltage given is that of a three-phase four-wire system, use[3]

$$P = \sqrt{3} I V (\text{pf})$$

$$I = \frac{P}{\sqrt{3} V (\text{pf})} = \frac{80 \times 10^3 \text{ W}}{(\sqrt{3})(480 \text{ V})(0.8)} = 120.3 \text{ A}$$

The total number of ampere-feet is

$$\begin{aligned} \text{ampere-feet} &= (120.3 \text{ A})(110 \text{ m})\left(3.281 \ \frac{\text{ft}}{\text{m}}\right) \\ &= 43{,}417 \text{ A-ft} \quad (43.4 \times 1000 \text{ A-ft}) \end{aligned}$$

The reason for specifying "noncontinuous" is to emphasize that the ampacity is to be calculated at 100%, not 125%. Regardless of the load, continuous or noncontinuous, the voltage drop is calculated using the 100% ampacity. Nevertheless, the conductor must be sized to handle 125% ampacity according to Code standards. The OCPD must be set at 125% ampacity as well.

Since this system has a neutral, the voltage drop of concern is the phase-to-neutral drop.

$$\begin{aligned} V_{\text{drop}} &= (\text{allowed drop})(\text{line-to-neutral voltage}) \\ &= (0.03)(277 \text{ V}) \\ &= 8.31 \text{ V} \end{aligned}$$

The maximum voltage drop per 1000 A-ft is then

$$\begin{aligned} V_{\text{drop}} &= \frac{8.31 \text{ V}}{(43.4 \times 1000 \text{ A-ft})} \\ &= 0.192 \text{ V/1000 A-ft} \end{aligned}$$

This value could then be used with a manufacturer's table of conductors to determine which conductor size meets the voltage drop recommendations. Voltage drop specifications on conductor size are in addition to other requirements and must be coordinated to determine the most limiting case. If the conductor size is increased to lower the voltage drop, the overcurrent protection for the wiring also increases.

Tables for conductor properties and voltage drops are often given in terms of line-to-neutral values while the recommendation for percentage voltage drop in the Code refers to the line voltage. Calculations should normally be accomplished using line-to-neutral voltages (i.e., phase voltages), and then conversions applied to obtain line quantities and the percent voltage drop. The line current is used instead of the line-to-neutral values since this is the current causing the voltage drop in the conductor cable, be it a branch circuit or feeder conductor.

8. WIRING AND PROTECTION: BRANCH CIRCUIT, FEEDER, AND SERVICE CALCULATIONS

NEC Art. 220 contains the requirements for calculations involving branch circuits, feeders, and service loads. Examples that are representative of the requirements are given in App. D of the Code.

The voltages used in calculations are standardized as 120, 120/240, 208Y/120, 240, 347, 480Y/277, 480, 600Y/347, and 600 V [NEC Sec. 220.5(A)]. Loads are determined in terms of volt-amperes (VA) or kilovolt-amperes (kVA). This is considered equivalent to equipment power ratings, which are given in watts (W) or kilowatts (kW). This concept of equivalence is used since the power factor is unknown and the conductors will carry the volt-ampere load (i.e., the apparent power), and so should be rated for such. Fractions of less than 0.5 A are dropped.

Branch circuit lighting loads are determined by the square footage of a building, using the unit loads given in NEC Sec. 220.12. Dwelling units' lighting loads are calculated based on 3 VA per square foot. Additional receptacle loads depend on the equipment attached [NEC Sec. 220.14]. Receptacle outlets are calculated based on 180 VA per outlet minimum, with a minimum of 90 VA per receptacle for outlets with four or more receptacles [NEC Sec. 220.14(I)].

Lighting loads that have ballasts (i.e., fluorescent lamps), transformers, or autotransformers are based on the ampere rating of the units and not the watt ratings [NEC Sec. 220.18(B)]. These types of devices are termed *nonlinear loads* and affect the sizing of the neutral conductor [NEC Sec. 310.15(B)(5)].

Requirements for various types of loads, including general lighting, motors, and small appliances, are given in NEC Sec. 220. Demand factors that allow less than the full-load wattage, that is, volt-amperes, to be used in determining the size of conductors and overcurrent devices are also given. Optional calculations for those dwelling units with a total connected load greater than 100 A are described in NEC Sec. 220.82.

[2]If the ballast-type lighting constitutes greater than 50% of the load, the neutral conductor would be considered a current-carrying conductor and would have to be sized according to the calculations in the example problem [NEC Sec. 310.15(B)(5)(b)].

[3]Only four-wire systems have a voltage specified after the slash, for example, 208Y/120, 240/120, 480Y/277, and 4160Y/2400.

Example 56.4

For the circuit shown, what is the total VA load?

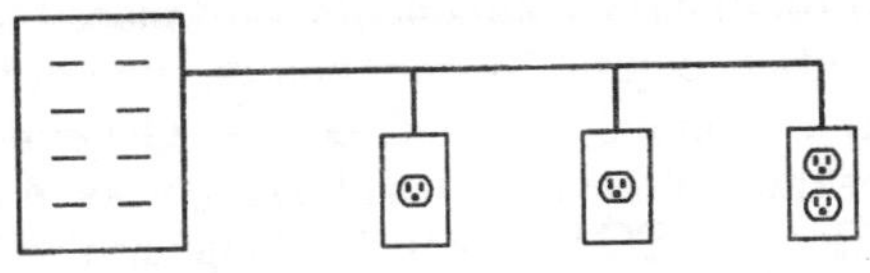

Solution

The minimum load is 180 VA [NEC Sec. 220.14(I)]. The total load is

$$\begin{aligned} VA_{total} &= (\text{number of outlets})\left(180\ \frac{VA}{\text{outlet}}\right) \\ &= (3)\left(180\ \frac{VA}{\text{outlet}}\right) \\ &= 540\ VA \end{aligned}$$

9. WIRING AND PROTECTION: OVERCURRENT PROTECTION

Article 240 of the Code describes the overcurrent protection requirements and the electrical devices that accomplish them. One of the first sections of Part I lists specific types of equipment and the applicable article containing the overcurrent requirements. NEC Table 240.3 is a useful starting point in any determination of overcurrent requirements.

Conductors are protected in accordance with their ampacities [NEC Sec. 240.4]. Standard ampere ratings for fuses and circuit breakers are given in NEC Sec. 240.6 as 15, 20, 25, 30, 35, 40, 45, 50, and so on. Thermal devices not designed to interrupt short circuits cannot be used for overcurrent protection for short circuits or grounds [NEC Sec. 240.9]. The remainder of the article discusses the location of overcurrent devices, tap requirements, fuses, and circuit breakers.

Most circuit breakers have an instantaneous trip. Those that do not must use one of the following equivalent means as required by NEC Sec. 240.87: zone-selective interlocking, differential relaying, or energy-reducing maintenance switching with a local status indicator. With *zone-selective interlocking*, two or more circuit breakers communicate with one another so that a short circuit or ground fault is cleared by the breaker closest to the fault in a minimum amount of time. *Differential relaying* measures current at the ends of a zone to ensure that the current entering the zone equals the current leaving the zone. If the currents do not, a fault has occurred, and the relay sends a trip signal to the appropriate breakers. With *energy-reducing maintenance switching*, a technician manually sets a circuit breaker to remove any intentional delays, so that during the maintenance action any fault will result in an instantaneous trip. All of these methods minimize *let-through energy* (measured in units of It^2, called *energy units*, $A{\cdot}s^2$) that could result in an arc flash.

10. WIRING AND PROTECTION: GROUNDING

Article 250 of the Code describes general grounding and bonding in Part I, circuit and system grounding in Part II, grounding electrode requirements in Part III, and equipment grounding in Part VI.

Electrical systems are connected to the earth in a way that will limit voltage from lightning strikes, line surges, or other transients [NEC Sec. 250.4(A)(2)]. Electrical systems are also grounded in order to maintain the voltage at a stable level relative to ground to ensure connected equipment is subjected to a maximum potential difference set by the designer—for example, three-phase four-wire systems (see NEC Sec. 250.26). Equipment is grounded as well, the primary goal being the safety of people in contact with the equipment. The overall requirements of the fault current path and their organization in the Code are given in NEC Sec. 250.4(B)(4), while specific equipment requirements are cross-referenced in NEC Sec. 250.3, which is a good starting point for finding grounding information.

The grounding electrode requirements are described in NEC Sec. 250.50. Electrically continuous water pipes in contact with the earth for 3.0 m (10 ft) or more are used as grounding electrodes [NEC Sec. 250.52(A)(1)]. Grounding electrodes 6.0 m (20 ft) long are allowed to be encased in concrete with a minimum thickness of 50 mm (2 in), if the electrodes are within the concrete and near the foundation, which must be in direct contact with the earth [NEC Sec. 250.52(A)(3)]. Rod- and pipe-type electrodes must be 2.44 m (8 ft) long [NEC Sec. 250.52(A)(5)]. The sizing of grounding electrode conductors for AC systems is given in NEC Table 250.66.

Cord-and-plug-connected equipment is generally grounded unless double-insulated [NEC Sec. 250.114]. Major appliance equipment within dwellings must be grounded [NEC Sec. 250.114(3)]. Insulated equipment grounding conductors are green, or green with one or more yellow stripes [NEC Sec. 250.119]. Grounding terminals use green screws or nuts, the word "green" or "ground", the symbols "G" or "GR", or the grounding symbol [NEC Sec. 250.126]. The sizing of equipment grounding conductors is given in Table 56.1 [NEC Table 250.122].

Surge-protective devices (SPDs) protect systems against electrical surges. NEC Art. 285 lists the requirements for SPDs. The term SPD is used in place of two older terms. Previously, if the surge device was connected on the supply side of a disconnecting means, it was referred to as a *surge arrester* (now known as a type 1 SPD). If the surge device was on the load side, it was called a *transient voltage surge suppressor* (TVSS) (now known as a type 2 SPD). Both terms are still in general use.

***Table 56.1** Equipment Grounding (NEC Table 250.122)*

rating or setting of automatic overcurrent device in circuit ahead of equipment, conduit, etc., not exceeding (amperes)	size (AWG or kcmil)	
	copper	aluminum or copper-clad aluminum*
15	14	12
20	12	10
60	10	8
100	8	6
200	6	4
300	4	2
400	3	1
500	2	1/0
600	1	2/0
800	1/0	3/0
1000	2/0	4/0
1200	3/0	250
1600	4/0	350
2000	250	400
2500	350	600
3000	400	600
4000	500	750
5000	700	1200
6000	800	1200

(Note: Where necessary to comply with NEC Sec. 250.4(A)(5) or (B)(4), the equipment grounding conductor must be sized larger than this table.)

*See installation restrictions in Sec. 250.120 of the *National Electrical Code®* (NEC).

Codes and Standards

11. WIRING METHODS AND MATERIALS

NEC Chap. 3, "Wiring Methods and Materials," contains Art. 300, "General Requirements for Wiring Methods and Materials" and Art. 310 "Conductors for General Wiring," several articles on conduit, tubing, and raceways, and many others. The requirements regarding switchboards and panelboards are also in this portion of the Code.

The wiring methods covered in the Code are for all wiring installations, but do not include internal equipment wiring [NEC Sec. 300.1(B)]. All conductors in the same circuit should be contained together [NEC Sec. 300.3(B)]. In general, conductors for systems over 1000 V are not mixed with those for systems under 1000 V [NEC Sec. 300.3(C)(2)]. Conductors, as well as raceways and cables, must be protected from damage [NEC Sec. 300.4]. Directly buried conduit must be at least 450 mm (18 in) from the surface [NEC Sec. 300.5(D)(1)].

Concerns over corrosion, temperature, wet locations, and the proper securing of raceways and cables are covered in NEC Sec. 300.6 to Sec. 300.14. Box, conduit bodies, and fittings requirements are found in NEC Sec. 300.15. These components are placed at splice points, outlet points, switch points, junction points, termination points, and pull points. General requirements for the number and size of conductors in raceways are specified in NEC Sec. 300.17, along with an informational note that describes where information on specific types of installations can be found in the NEC. Raceways are installed per NEC Sec. 300.18.

The support required for vertically run conductors is covered in NEC Sec. 300.19, with the required spacing listed in NEC Table 300.19(A). Such support can consist of clamps, junction boxes, or a change in the cable direction of at least 90 degrees, for a distance of not less than twice the diameter of the cable. Any other method of equal effectiveness is also acceptable.

Preventing the spread of fire in hollow spaces or ducts is addressed in NEC Sec. 300.21, which is often done with *fire-stops*. *Fire-stops* are used to close open parts of a building to prevent the spread of fire. Openings around electrical penetrations that go into or through walls must be fire-stopped using approved methods that maintain the fire resistance rating.[4]

Conductor ampacities are determined using App. 56.B [NEC Table 310.15(B)(16)] and numerous other NEC tables given in Sec. 310.15. The loads used to determine the ampacities are calculated in accordance with Art. 220 of the Code. *Derating*, or lowering the allowed ampere rating, of a conductor can occur due to temperature limitations of the terminal connections [NEC Sec. 110.14(C)]; coordination with system overcurrent protection [NEC Sec. 240.4]; the ambient operating environment [NEC Table 310.15(B)(2)(a) and (b)]; carrying more than three current-carrying conductors in a raceway or cable [NEC Sec. 310.15(B)(3)(a) and NEC Table 310.15(B)(3)(a)]; and a variety of other conditions [NEC Annex B]. Under engineering supervision, conductor ampacities may be calculated using the formula given in NEC Sec. 310.15(C). (See NEC Annex B for sample calculations using the formula.)

Conductor application and insulation codes, such as THHN for heat-resistant thermoplastic, are given in NEC Table 310.104(A). The appropriate letter designations are explained, the type of insulation is specified, and the design applications are listed.

NEC Art. 312 covers cabinets, cutout boxes, and meter socket enclosures. Cabinets and cutout boxes are electrical enclosures with swinging doors. They are designed to accommodate metering devices, instrument transformers, and myriad kinds of control equipment.

[4]Additional information regarding fire-stops and methods of testing can be found in the UL *Guide Information for Electrical Equipment: The White Book*; ASTM E814 *Standard Test Method for Fire Tests of Penetration Firestop Systems*; ANSI/UL 1479 *Standard for Fire Tests of Penetration Firestops*; and various other standards.

Electric utility metering requirements are standardized by both the NEC and the Electric Utility Service Equipment Requirements Committee (EUSERC).

NEC Art. 314 covers the requirements of outlet, device, pull, and junction boxes, in addition to those of conduit bodies, fittings, and handhole enclosures. One of the more practical aspects of this article is NEC Table 314.16(A), which establishes the maximum number of conductors to be placed in any given volume of metal box. The *box fill calculations* are guided by NEC Art. 314.16(B)(1) through (5), which describes how to count the wires, clamps, support fittings, and other items that may be located in any given box. While no volume allowance is necessary for a splice made by a wire nut, or locknut, each spliced wire counts as a separate wire, whereas a wire passing into and out of the box unspliced counts as a single wire.

Armored cable is covered in NEC Art. 320. Flat cable, seen frequently in offices, is covered in NEC Art. 322 through 324. Other cable types are covered in NEC Art. 326 through Art. 340. Conduit types are covered in NEC Art. 342 through Art. 362. The more common types of conduit are Intermediate Metal Conduit (IMC), which is lighter and thinner than Rigid Metal Conduit (RMC) yet interchangeable with it, and Electrical Metallic Tubing (EMT), which is a light, unthreaded version of conduit. All are designed for the physical protection of wiring.

The remaining articles focus on the numerous methods for routing wires, including gutters, busways, cable buses, concrete floor raceways, metal wireways, nonmetallic wireways, strut-type channel raceways, surface metal raceways, underfloor raceways, cable trays, knob-and-tube wiring, messenger-supported wiring, open wiring, and outdoor wiring.

Example 56.5

A single AWG 10 TW copper conductor is considered for use in an environment expected to have an ambient temperature of 100°F. What is the maximum ampacity?

Solution

A TW cable sized at AWG 10, located in the 60°C column of App. 56.B, has an ampacity of 30 A before derating. The correction factor given at the bottom of the table for 100°F (38°C) is 0.82. The allowable ampacity is

$$\begin{aligned} A_{\text{total}} &= A_{\text{rated}} F_{\text{correction}} \\ &= (30\ \text{A})(0.82) \\ &= 24.6\ \text{A} \end{aligned}$$

12. EQUIPMENT FOR GENERAL USE

NEC Chap. 4, "Equipment for General Use," contains Art. 400 (which covers flexible cords and cables), Art. 402 (fixture wires), Art. 404 (switches), Art. 406 (receptacles), Art. 408 (switchboards and panelboards), Art. 409 (industrial control panels), Art. 410 (luminaires), Art. 411 (low-voltage lighting systems), Art. 422 (appliances), Art. 424 (space-heating equipment), Art. 426 (outdoor deicing and snow melting equipment), Art. 427 (pipeline and vessel heating), Art. 430 (motors, motor circuits, and controllers),[5] Art. 440 (air conditioning and refrigeration), Art. 445 (generators), Art. 450 (transformers), Art. 455 (phase converters), Art. 460 (capacitors), Art. 470 (resistors and reactors), Art. 480 (storage batteries), and Art. 490 (equipment over 1000 V).

Flexible cords and cables are covered by NEC Art. 400, and descriptions of the different cable types are given in NEC Table 400.4. The allowable ampacity of flexible cords and cables is given in NEC Table 400.5(A)(1) and Table 400.5(A)(2), depending upon the cable type. The notes below each table place restrictions on the use and application of the values. In addition, these tables are based on an ambient temperature of 30°C. If the ambient temperature is something other than 30°C, the ampacity is corrected using the factors provided for power cables in NEC Sec. 310.15(B)(16), which in turn directs the reader to NEC Table 310.15(B)(2)(a) or Table 310.15(B)(2)(b). If there are more than three current-carrying conductors in each cable, additional restrictions apply per NEC Table 400.5(A)(3). Per NEC Sec. 400.9, such cords and cables are generally used without splices, but AWG 14 may be spliced in accordance with NEC Sec. 110.14(B). Wires for fixtures are covered in NEC Art. 402, and switches in NEC Art. 404. Switches should not interrupt the grounded conductor [NEC Sec. 404.2(B)].

Receptacles, cords, and plugs are covered in NEC Art. 406. Receptacle and plug configuration is controlled by the National Electrical Manufacturers Association (NEMA) in ANSI/NEMA WD 6 *Wiring Devices—Dimensional Specifications.* (See App. 56.G.) Receptacles using an isolated ground for electrical noise in order to minimize electromagnetic interference (EMI) are identified by an orange triangle on the face of the receptacle [NEC Sec. 406.3(D)].

Switchboards and panelboards are covered in NEC Art. 408, which also references several additional articles. The additional articles are listed in NEC Sec. 408.2, including NEC Art. 240 (which covers overcurrent protection), Art. 250 (grounding and bonding), Art. 312 (cabinets), Art. 314 (outlet and junction boxes), and Art. 406 (receptacles). NEC Sec. 408.3(E) describes the phase arrangement within a switchboard, switchgear, or panelboard as a-b-c from front to back, top to bottom, or left to right as viewed from the front. NEC Sec. 408.18(A) requires that switchboards be given 900 mm (3 ft) of clearance from the top of the switchboard to any combustible ceiling. Wire bending space is

[5]NEC Art. 430 covers motors. NEC Fig. 430.1 shows common motor circuit components cross-referenced to the articles defining them, while NEC Table 430.5 includes cross-references to other articles containing specific motor equipment requirements.

mentioned in NEC Sec. 408.55, but the required dimensions are given in NEC Sec. 312.6(B).

Panelboards containing only power circuit components, control circuit components (e.g., push buttons, pilot lights, and selector switches), or a combination of such components are called industrial control panels and are covered in NEC Art. 409.

Luminaires, lampholders, and lamps are covered in NEC Art. 410. Lighting systems operating at less than 30 V are covered in NEC Art. 411. Electrical appliances used in any occupancy are covered in NEC Art. 422, although motor-operated appliances are considered to be covered by NEC Art. 430 (see NEC Sec. 422.3). The covered appliances include kitchen and cooking equipment, central heating equipment, storage-type water heaters, waste disposals, dishwashers, and trash compactors. Fixed, electric space-heating equipment has its own set of requirements in NEC Art. 424.

NEC Art. 430 covers motors, motor circuits, and controllers. The article is divided into 14 parts indicated by Roman letters: general (I), motor circuit conductors (II), overload protection (III), branch-circuit protection (IV), feeder circuit protection (V), control circuits (VI), controllers (VII), motor control centers (MCCs) (VIII), disconnecting means (IX), adjustable speed drives (X), systems over 1000 V (XI), protection of live parts (XII), grounding (XIII), and tables (XIV).

NEC Art. 430 must be understood in terms of the history of rating motors by their horsepower output. Circuits supplying a motor must be rated for the input conditions to the motor, which include the losses involved and the power factor. In NEC Sec. 430, Part XIV, Table 430.247 through Table 430.250 utilize industry-wide input ampere ratings that are compared to the nameplate horsepower ratings to ensure adequate sizing of conductors and overprotection.

In NEC Sec. 430, part I, Fig. 430.1 shows the contents of NEC Art. 430 in graphical form in order to focus on a particular portion of the motor circuitry from the feeder through the motor itself. NEC Table 430.5 lists cross-referenced articles about various motor applications with special requirements beyond those in NEC Art. 430.

The ampacity and motor rating are determined from the motor nameplate, which is then cross referenced against the ampacity required using NEC Table 430.247, Table 430.248, Table 430.249, or Table 430.250 as applicable. If the motor nameplate lists a full-load ampere rating, that ampere rating is considered the guiding value (regardless of what the horsepower rating says). The horsepower rating determined from the NEC tables by crossing the nameplate ampere rating is considered the horsepower rating of the motor (regardless of what the horsepower rating says). So-called torque motors use the locked-rotor, full-load current to determine what conductor size is necessary [NEC Sec. 430.6(B)]. Locked-rotor letter codes are provided in NEC Table 430.7(B). The conductors supplying a motor must be rated for 125% of the motor full-load rating per NEC Sec. 430.22 of part II of Sec. 430. NEC Sec. 430.27 requires conductors to comply with Art. 480.8 and Art. 460.9 when capacitors are installed in motor circuits.

Motors greater than 1 hp are protected against overloads using a separate overload device [NEC Sec. 430.32(A)(1)] at 125% of full-load current if their service factor is 1.15 or greater and the temperature rise is 40°C or higher.[6] All other motors are protected at the 115% level.

In NEC Sec. 430, part III, thermal protectors, which are integral to the motor, have higher settings that are controlled by NEC Sec. 430.32(A)(2). Motors greater than 1500 hp [NEC Sec. 430.32(A)(4)] and those less than 1 hp [NEC Sec. 430.32(B) and Sec. 430.32(D)] have different requirements.[7] Motors that are combined with other loads on a branch circuit are covered in NEC Sec. 430, Part III.

The requirements for short-circuit and ground-fault protection are in NEC Sec. 430, part IV for branch circuits and NEC Sec. 430, part V for feeder circuits. Short circuits and ground faults must be cleared quickly to prevent or minimize damage. This is done with fuses, time-delay fuses, instantaneous-trip circuit breakers, or inverse-time circuit breakers. NEC Table 430.52 lists the maximum ratings or settings for the indicated devices, depending upon the motor type. Where the required settings do not correspond to the standard sizes or ratings of fuses, nonadjustable circuit breakers, thermal protective devices, or adjustable circuit breakers, the next higher size that does not exceed the standard ampere rating is allowed [NEC Sec. 430.52(C)(1) Exception No. 1].

Motor control circuits can be tapped from the load side of a motor circuit. When they are, the governing requirements for control circuits in NEC Sec. 430, part VI apply. The controller itself is governed by NEC Sec. 430, part VII. NEC Sec. 430, part VIII covers motor control centers (MCCs) installed for the control of motors, lighting, and power circuits. Such MCCs are considered service equipment.

Adjustable speed drives (ASDs) are covered by NEC Sec. 430, part X, although the requirements of part I through IX still apply unless modified or supplemented by part X. ASDs can operate at a lower than rated speed, with less cooling, and at higher temperatures. Conductor sizing and overload protection become critical in this case, and in general use an integral thermal protector [NEC Sec. 430.126].

[6]The *service factor* indicates the level at which a motor may operate continuously without damage. NEMA defines the service factor as a multiple that, when applied to the rated horsepower, indicates a permissible horsepower loading that may be carried under the conditions specified for that service factor. For a service factor of 1.15, that would be 115% of full load. The temperature rise is the rise above ambient temperature that the motor can handle when operating at 100% load.

[7]Motors less than 1 hp are sometimes called *fractional horsepower motors.*

Systems over 1000 V are covered in NEC Sec. 430, part XI. Protection of live parts is covered in NEC Sec. 430, part XII. Grounding is covered in NEC Sec. 430, part XIII. NEC Sec. 430, part XIV contains tables of the most used two-pole and four-pole induction motors.

NEC Art. 440 focuses on air conditioning and refrigeration equipment. It describes the special conditions necessary for circuits supplying hermetic refrigerant motor-compressors. This article applies only if a hermetic unit is involved. The requirements listed in NEC Art. 430 apply to the units discussed in NEC Art. 440. In such units, the *branch-circuit selection current* is found on the nameplate. This value should be used in the selection of conductors, disconnecting means, controllers, and short-circuit and ground-fault protective devices. The branch-circuit selection current will always be larger than the *rated-load current*, which is the current of the motor-compressor running at rated load, rated voltage, and rated frequency. Cross-references to related articles are provided in NEC Table 440.3(D).

NEC Art. 445 focuses on generators that are used for fire pumps, emergency systems, standby systems, and critical operations power systems. The article covers the requirements for the generators themselves, while the systems mentioned are covered in NEC Chap. 7, "Special Conditions."

NEC Art. 450 covers transformers and transformer vaults, with the exception of current transformers and several other special-use transformers.[8] The overcurrent protection for the primary and/or secondary of a transformer is listed in NEC Table 450.3(A) and Table 450.3 (B). Autotransformers are covered in NEC Sec. 450.4.

Phase converters that convert single-phase power to three-phase power are covered in NEC Art. 455. Capacitors, whose ability to store energy poses an additional safety hazard, are covered in NEC Art. 460. Capacitors rated for use under 1000 V must discharge to 50 V within one minute of disconnection from the source [NEC Sec. 460.6]. The conductors supplying the capacitors must be rated for at least 135% of rated current because capacitors are built to a 15% tolerance and the current drawn varies with the applied voltage. The special requirements for resistors and reactors (i.e., inductors) are covered in NEC Art. 470.

Storage batteries, which provide backup power in many systems, are covered in Art. 480. NEC Art. 490 covers systems over 1000 V, and NEC Table 490.24 lists the minimum clearance from live parts at this increased voltage.

[8]Because of this exception, the requirement to short-circuit an unused current transformer (CT) secondary is in NEC Art. 110.23. This requirement is designed to ensure the safety of personnel working in the area of an unloaded CT, which can develop extremely high voltages on the secondary when it is open-circuited because the primary flux is determined by the current level in the primary (which is never open-circuited).

13. SPECIAL OCCUPANCIES

NEC Chap. 5, "Special Occupancies" covers hazardous locations [NEC Art. 500] and temporary installations [NEC Art. 590]. These articles contain information on a variety of specialty locations with specific hazards and corresponding requirements for improving safety. Some specialty locations discussed are health care facilities [NEC Art. 517]; theaters [NEC Art. 520]; recreational vehicles and recreational vehicle parks [NEC Art. 551]; and marinas [NEC Art. 555].

NEC Art. 500, "Hazardous (Classified) Locations, Classes I, II, and III, Division 1 and 2," uses the terms *hazardous* and *classified* interchangeably. In NEC Art. 500, *nonincendive* is used for items that are not capable of igniting a flammable gas-air, vapor-air, or dust-air mixture under intended operating conditions where an arc or thermal effect can be produced. Standards relating to the design of circuitry for use in hazardous locations are listed in NEC Art. 500.4(B) Informational Note No. 1. The NFPA has several additional standards for hazardous operations beyond those listed in the NEC.[9]

Transformers and capacitors containing combustible liquid must be installed in approved vaults for class I locations [NEC Sec. 501.100(A)(1)]. The requirements for meters, instruments, relays, motors, wiring, lighting, and other equipment are found in NEC Chap. 5, according to the classification of the space in which the equipment is used.

Definitions of and requirements for intrinsically safe systems are specified in NEC Art. 504.

14. SPECIAL EQUIPMENT

NEC Chap. 6, "Special Equipment," contains information specific to equipment rather than to occupancy. The chapter includes electric signs [NEC Art. 600], manufactured wiring systems [NEC Art. 604], office furnishings [NEC Art. 605], electric vehicle charging [NEC Art. 625], information technology (IT) equipment [NEC Art. 645], industrial machinery [NEC Art. 670], solar photovoltaic (PV) systems [NEC Art. 690], fuel cell systems [NEC Art. 692], small wind electric systems [NEC Art. 694], and fire pumps [NEC Art. 695].

NEC Art. 604 covers *manufactured wiring systems*, which are defined as systems containing assembled component parts that cannot be inspected at the manufacturing site without damaging or destroying the assembly. Manufactured wiring systems are used for the connection of luminaires, utilization equipment, continuous plug-in busways, and other devices.

Information technology (IT) equipment is covered in NEC Art. 645. A *critical operations data center* is defined as an IT equipment system that requires

[9]For example, NFPA 85 *Boiler and Combustion Systems Hazards Code*.

continuous operation due to public safety, emergency management, national security, or business continuity [NEC Sec. 645.2]. Conductors for IT must have an ampacity of 125% of the total connected load [NEC Sec. 645.5(A)]. Power cords for IT equipment cannot be longer than 15 ft [NEC Sec. 645.5(B)(1)]. Restrictions on the cable types used under raised floors can be found in NEC Table 645.5(E)(6). Abandoned supply circuits and interconnecting cables should be removed unless they are in a raceway [NEC Sec. 645.5(G)]. Sensitive electronic equipment is covered in NEC Art. 647. An *integrated electrical system* is a system in a unitized segment of an industrial system in which an orderly shutdown is required and only qualified persons service the system [NEC Art. 685].

Solar photovoltaic (PV) systems are covered in NEC Art. 690. Their components are identified in NEC Fig. 690.1(b). The voltage rating of cables, disconnecting means, and settings of overcurrent devices is determined by the open-circuit voltage of the solar cell modules as corrected by NEC Table 690.7, which accounts for changes in voltage due to temperature. Correction factors for conductors used in PV systems are given in NEC Table 690.31(E).

Fire pump systems, which are discussed in NEC Art. 695, have very specialized requirements that differ in many respects from standard requirements for other systems.

15. SPECIAL CONDITIONS

NEC Chap. 7, "Special Conditions," contains information specific to a condition that must be met, as opposed to information on a specialized occupancy or equipment. The chapter discusses emergency systems [NEC Art. 700], legally required standby systems [NEC Art. 701], optional standby systems [NEC Art. 702], interconnected electric power production sources [NEC Art. 705], circuits and equipment operating at less than 50 V [NEC Art. 705], critical operations power systems (COPS) [NEC Art. 708], remote-control, signaling, power-limited circuits, and instrumentation tray cable (ITC) [NEC Art. 725], fire alarm systems [NEC Art. 760], and optical fiber cables and raceways [NEC Art. 770].

16. COMMUNICATIONS SYSTEMS

NEC Chap. 8, "Communications Systems," portion of the Code covers communication circuits [NEC Art. 800], radio and television equipment, including community-based systems [NEC Art. 810 and Art. 820], network-powered broadband communication systems [NEC Art. 830], and premises-powered broadband communication systems [NEC Art. 840].

A pictorial representation of some of the bonding and grounding terms is given in the informational notes for NEC Fig. 800(a) and Fig. 800(b). Communication cables may have their own overcurrent protection, provided they meet specific requirements that the wire itself act as a fuse [NEC Art. 800.90(A)(1)]. Uses of various communications wires and cables are listed in NEC Table 800.154(a).

NEC Art. 830 covers network-powered broadband communication systems that come into many households via a *network interface unit* (NIU) and provide audio, video, data, and interactive services. These systems are classified as either low- or medium-power based on their source as described in NEC Table 830.15. The bonding and grounding conductors should be as short as practicable but should not exceed 6.0 m (20 ft) [NEC Sec. 830.100(A)(4)]. Correct cable use is specified in NEC Table 830.154(a). Optical broadband systems are covered in NEC Art. 840.

17. TABLES

NEC Chap. 9 provides tables used when installing systems that must comply with the Code. These tables are referenced elsewhere in the NEC. Some of the more commonly used tables are NEC Table 8, "Conductor Properties" and Table 9, "Alternating-Current Resistance and Reactance for 600-Volt Cables, 3-Phase, 60-Hz, 75°C (167°F)—Three Single Conductors in Conduit."

18. ANNEXES

There are ten NEC Annexes. Annex A cross-references product standards. Annex B gives examples of ampacity calculations. Annex C contains fill tables that list, for example, the maximum number of wires in a given conduit. Annex D has examples of applications of various NEC articles. Annex E lists construction types. Annex F has reliability information for critical power systems. Annex G provides information on supervisory control and data acquisition (SCADA) systems. Annex H has enforcement information. Annex I contains recommended torque tables. Annex J gives ADA standards for accessible design.

57 National Electrical Safety Code

1. OVERVIEW

The *National Electrical Safety Code* (NESC) was initiated in 1913 at the National Bureau of Standards. It is now under the sponsorship and control of the Institute of Electrical and Electronic Engineers (IEEE) and is published as a voluntary IEEE standard. The NESC (hereafter referred to as the "Safety Code") is officially endorsed by the American National Standards Institute (ANSI), and the committee responsible for the Safety Code is known as *ANSI Standards Committee* C2. While voluntary, the Safety Code is often adopted by each individual state, specifically by the state authority with jurisdiction over utilities.

The Safety Code generally applies to "practical safeguarding of persons during the installation, operation, and maintenance of electric supply and communications lines, and associated equipment." Its general purpose is to safeguard utility persons and the general public.

The Safety Code consists of General Sections followed by four parts, all of which are further subdivided into sections and then rules. The outline of the Safety Code follows. Part numbers (single digits) and Section numbers (two digits) are given in parentheses. Rules are three digit numbers.

General Sections

(01) Introduction

(02) Definitions

(03) References

(09) Grounding

Part 1 Rules for the Installation and Maintenance of Electric Supply Stations and Equipment

(10) Purpose and Scope

(11) Protective Arrangements in Electric Supply Stations

(12) Installation and Maintenance of Equipment

(13) Rotating Equipment

(14) Storage Batteries

(15) Transformers and Regulators

(16) Conductors

(17) Circuit Breakers, Reclosers, Switches, and Fuses

(18) Switchgear and Metal-Enclosed Bus

(19) Surge Arrestors

Part 2 Safety Rules for the Installation and Maintenance of Overhead Electric Supply and Communication Lines

(20) Purpose, Scope, and Application of Rules

(21) General Requirements

(22) Relations between Various Classes of Lines and Equipment

(23) Clearance

(24) Grades of Construction

(25) Loading for Grades B and C

(26) Strength Requirements

(27) Line Insulation

Part 3 Safety Rules for the Installation and Maintenance of Underground Electric Supply and Communication Lines

(30) Purpose, Scope, and Application of Rules

(31) General Requirements Applying to Underground Lines

(32) Underground Conduit Systems

(33) Supply Cable

(34) Cable in Underground Structures

(35) Direct Buried Cable

(36) Risers

(37) Supply Cable Terminations

(38) Equipment

(39) Installation in Tunnels

Part 4 Rules for the Operation of Electric Supply and Communication Lines and Equipment

(40) Purpose and Scope

(41) Supply and Communications Systems—Rules for Employers

(42) General Rules for Employees

(43) Additional Rules for Communications Employees

(44) Additional Rules for Supply Employees

Appendices

(A) Uniform System of Clearances

(B) Uniform Clearance Calculations for Conductors Under Ice and Wind Conditions

(C) Example Applications for Rule 250C Tables 250-2 and 250-3

(D) Determining Maximum Anticipated Per-Unit Overvoltage Factor (T) at the Worksite

(E) Bibliography

The Safety Code parts separate into the general areas as shown in Fig. 57.1. In this chapter, numbers in brackets, if used, indicate the rule number in the Safety Code and are meant for ease in referencing the Safety Code directly. Specific limits may be found in the latest code. Though metric is the official system for the Safety Code, a mixture of customary U.S. engineering units and SI units is used.

2. INTRODUCTION

Section 01 of the General Sections is titled "Introduction to the National Electrical Safety Code." Rule 010

Figure 57.1 *NESC Areas of Concern*

part 1

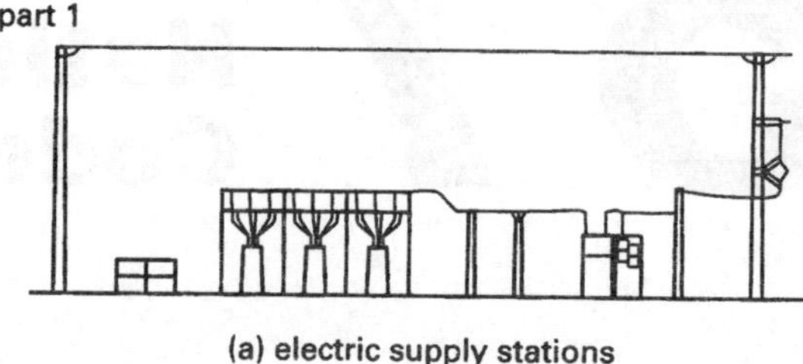

(a) electric supply stations

part 2

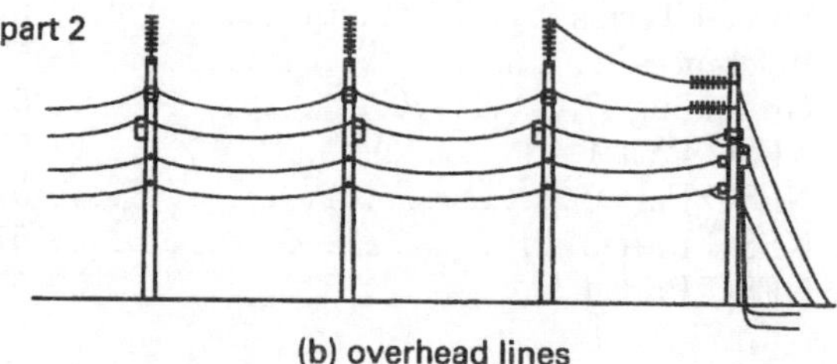

(b) overhead lines

part 3

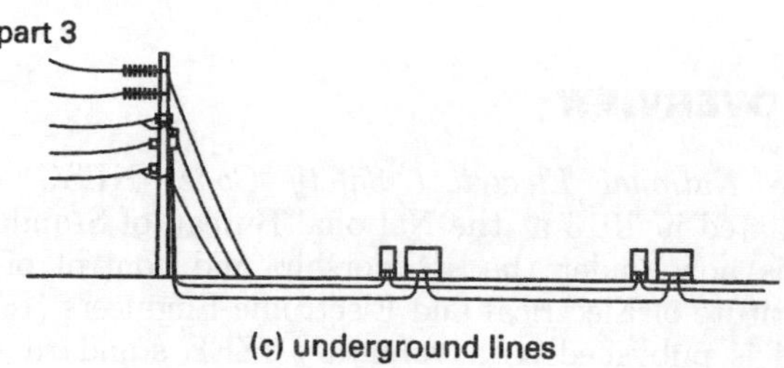

(c) underground lines

part 4

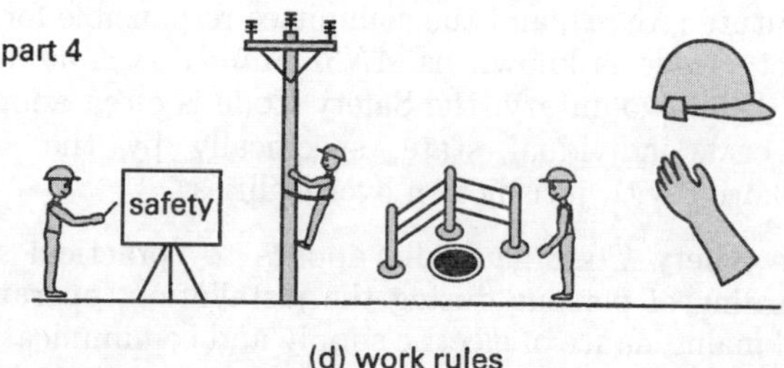

(d) work rules

specifies the purpose as the safeguarding of persons during installation, operation, or maintenance. The Safety Code is meant to contain the basic guidelines for safety. It covers electric power, which the Safety Code refers to as electric "supply," lines and associated equipment, as well as communication lines and equipment. Communications refers to, but is not limited to, telephone and cable television.

Rule 011 covers the scope of the code. The area of coverage for the Safety Code is shown in Fig. 57.2. The exact demarcation between the NEC and the NESC is shown in Fig. 57.3.

In addition to covering different areas of responsibility, the NESC and NEC differ as indicated in Table 57.1.

Rule 012 requires the use of the Safety Code. If an item or situation is not covered in the Safety Code, then "accepted

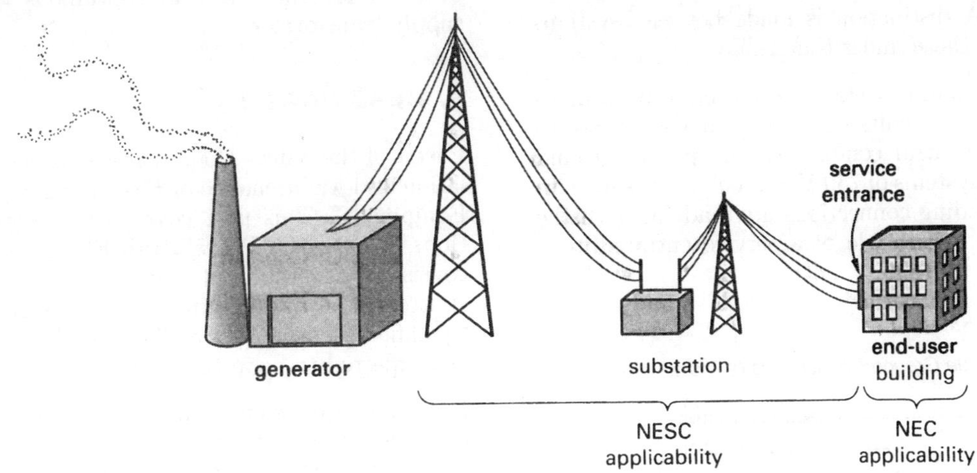

Figure 57.2 NESC Coverage

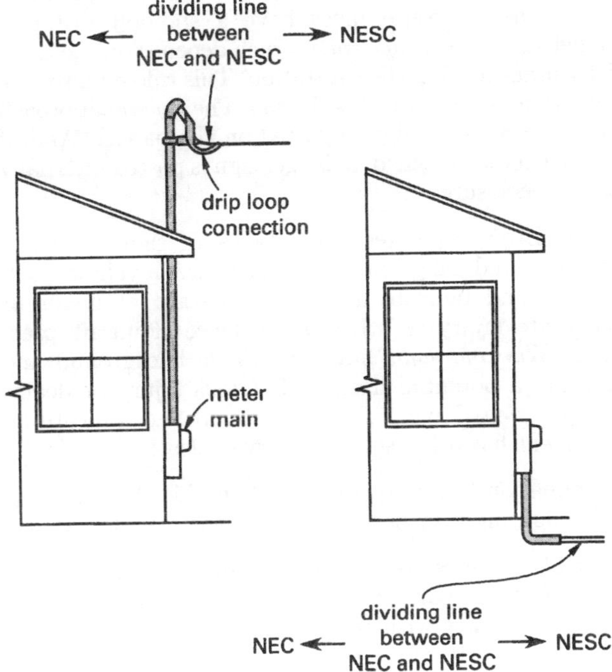

Figure 57.3 NEC/NESC Division of Responsibility

Table 57.1 NEC and NESC Comparison

NEC	NESC
published by NFPA	published by IEEE
code book for building industry	code book for utilities
legal authority by state, city, or local law	legal authority by state utility commissions
revised every 3 yr	revised every 5 yr
used by engineers and electricians	used by engineers and utility linemen

good practice" is allowed. Rule 013 requires all new installations to comply with the Safety Code. Rule 014 provides the conditions for a waiver of requirements.

Rule 015 defines "shall," "should," and "RECOMMENDATION." The term "shall" indicates a rule is mandatory and must be met. The term "should" indicates a rule that is applied normally, and is practical for the given situation. The term "RECOMMENDATION" is used to indicate a rule that is desired but not mandatory. Additionally, Rule 015 specifies that "NOTES" and "EXAMPLES" are not mandatory. However, "Footnotes" carry the force of the table or rule with which they are associated. Interpretation of the scope of "RECOMMENDATION," "NOTE," and "EXCEPTION" is also provided in this rule.

Rule 016 specifies that metric units are the primary units. Customary U.S. units are shown in parenthesis. Rule 017 specifies that values in the Safety Code are nominal and tolerances may exist in other standards.

3. SPECIAL TERMS

Section 02 of the General Sections is titled "Definitions of Special Terms." If a term is not defined in this section, then IEEE Standard 100 is to be used. If the term cannot be found in either, then a standard dictionary is to be used.

4. REFERENCES

Section 03 of the General Sections is titled "References." Any reference in this section is considered part of the Safety Code. If the reference is listed in Appendix B of the code, it is there for information only.

5. GROUNDING

Section 09 of the General Sections is titled "Grounding Methods for Electric Supply and Communications

Facilities." This entire section is devoted to the requirements for properly grounding electrical systems to ensure safety. A distinction is made between systems over 750 V and those under that value.

Rules in the grounding section require lighting loads to use wye and delta circuits that are grounded or else to use a common neutral conductor. The proper ground connections for systems of 750 V and below are shown in Fig. 57.4. Grounding connections are made at both the source and line (load) side of a service entrance to a building.

Figure 57.4 *Allowable Ground Connections (≤ 750 V)*

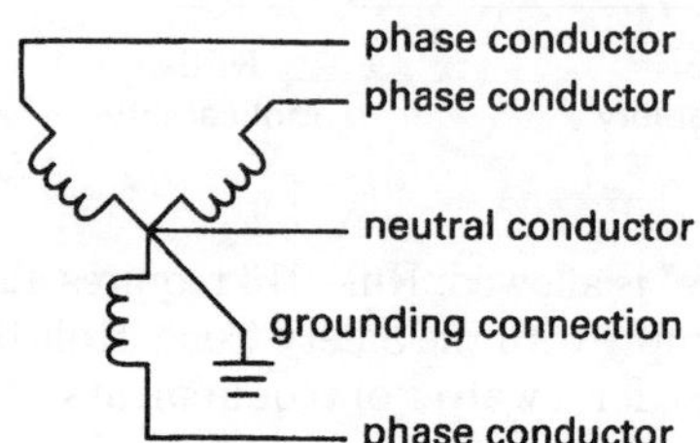

wye connected, 3-phase, 4-wire system (e.g., 120/208 V, 3ϕ (phase), 4 W, and 277/480 V, 3ϕ (phase), 4 W)

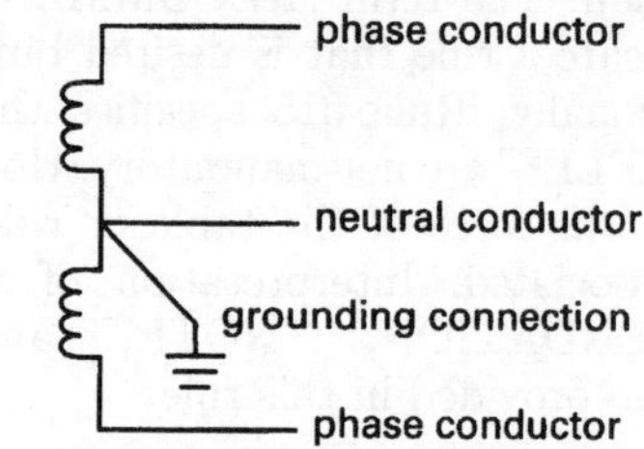

single-phase, 3-wire system (e.g., 120/240 V, 1ϕ (phase), 3 W)

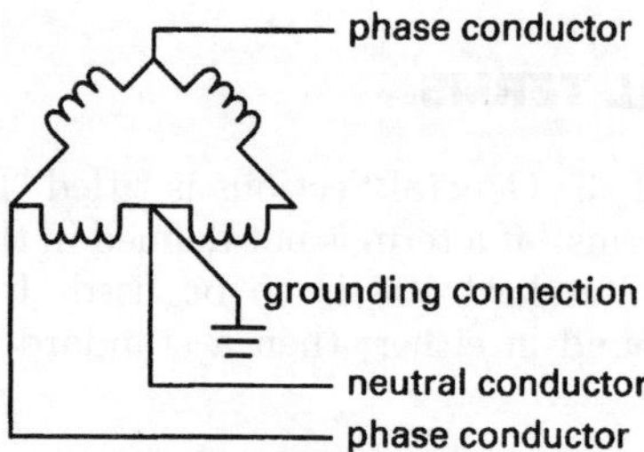

single-phase, 3-wire system on a delta connected 3-phase, 4-wire system (e.g., 120/240 V, 1ϕ (phase), 3 W)

The ampacity of grounding conductors is also specified in this section. The main specification is that fault current of a given time duration cannot melt or damage the grounding conductor. The "given time" is determined by the time-current characteristics of the fault protective device (e.g., the fuse, recluse, relay, or breaker) and the system impedance. The ampacity comes from manufacturers' *short-circuit withstand charts.* If the appropriate value is not determined, the grounding conductor must be sized for the full load continuous current of the supply transformer.

6. NESC PART 1

Part 1 of the Safety Code is titled "Rules for the Installation and Maintenance of Electric Supply Stations and Equipment." This part covers the electric supply stations as shown in Fig. 57.1(a). Electric supply stations consist of generating stations, substations, and switch stations. The rules assume such areas are accessible to "qualified personnel" only. That is, the general public is not expected to be in the area.

Rule 110 requires enclosures for supply stations to consist of fences, screens, partitions, or walls. The electrical components within the enclosure are required to meet clearance requirements to ensure personnel are not likely to come into contact with live electrical components. The enclosure is generally required to be 7 ft high and to limit access. Additionally, the electrical equipment inside the enclosure must have a standoff distance, called *safety clearance zone*, which depends on the highest voltage level in the substation. This rule requires the signage as specified in ANSI Z535. The general approach is that a warning sign is posted on the barrier. And, if the barrier is breached, a danger sign is posted internally to the enclosure.

ANSI Z535 stipulates three types of signs: caution, warning, and danger. *Caution signs* have a yellow background and indicate a potential hazard of minor or moderate injury or other consequence of unsafe practices. *Warning signs* have an orange background and indicate a potential hazard of serious injury or death. *Danger signs* have a red background and indicate an imminent hazard of serious injury or death.

Illumination levels are given in Rule 111, although portable lighting may be used to meet the specified levels.

Rule 112 requires floors to be even and nonslip. Obstructions must be painted, marked, or posted with safety signs. Stairways with three steps do not require handrails; those with four or more steps do require such handrails.

Rule 113 requires exits to be free of obstructions. In addition, double exits are required if an accident can make a single exit inaccessible.

Rule 123 requires protective grounding in accordance with IEEE Standard 80.

Rule 124 requires live parts over 150 V to be guarded. Guarding implies a physical barrier. Additionally, if the electrical potential of a given electrical part is unknown, the vertical clearance must be at least 8 ft 6 in.

Working space, that is, horizontal clearance, is specified in Rule 125. The space required is dependent on the voltage level. The requirements differ for voltages greater than 600 V and voltages of 600 V and less. For

600 V or less, at least 7 ft of head clearance and 30 in of width are the minimum. The working spaces in NESC Table 125-1 are for exposed energized parts, and do not apply if the parts are de-energized during maintenance or servicing.

Rule 131 requires that motors do not restart automatically after a power outage if such starting creates a hazard for personnel. This means that such motors should have *low-voltage protection* (LVP) that requires an operator to physically restart the motor following an outage. *Low-voltage release* (LVR) is a protective scheme that allows automatic restart without interaction with a human operator. Such LVR schemes are allowed if warning signals and time delays are incorporated.

Rule 140 governs the installation and use of batteries. The main thrust of the rule, and associated rules, is that hydrogen accumulation must be kept below "explosive mixtures" though the code does not specify an explosive percentage. (The hydrogen percentage that is explosive is approximately 4%.)

Conductors are covered in Section 16 of Part 1, in Rules 160–164. In general, the conductors must meet four conditions as follows. Conductors must

- be suitable for the intended use
- be suitable for the environmental location in which they are used
- have a voltage rating (i.e., insulation rating) adequate for the intended use
- have an ampacity (i.e., current rating) adequate for the intended use

According to Rule 174 fuses of more than 150 V to ground, or carrying more than 60 A, must be classified as disconnecting, or be arranged to be able to disconnect from the power source, or be removed with insulated handles.

Surge arrestors are common in the utility industry and are used to protect equipment from switching surges and lightning strikes. Arrestors are mentioned in Section 19, Rules 190–193. The basic requirements, however, come from IEEE Standard C62.1 and C62.11.

7. NESC PART 2

Part 2 of the Safety Code is titled "Safety Rules for the Installation and Maintenance of Overhead Electric Supply and Communication Lines." This part covers overhead lines as shown in Fig. 57.1(b). The requirements in Part 2 focus on spacing (center-to-center measurements) between conductors, clearances (surface-to-surface measurements), and the strength of construction.

Accessibility for servicing and maintenance is a concern of this part. The accessibility issue is covered in Rule 213 and throughout Part 2 under the topics "climbing space," "working space," "working facilities," and "clearance between conductors."

A wide variety of clearances are required depending on the situation. However, two important arguments come from the sag and tension chart and the effects of weather conditions, called *loadings*, on the overhead lines. The *sag and tension chart* uses the type and temperature of the cable, the span length, and the weather conditions (including ice accumulation and wind speed) as inputs. The output is the maximum expected sag and tension on the overhead line that is then used to determine the construction grade. The sag and tension chart details are in Rule 230. The clearances required by Rule 230 vary with regions of the country, with Zone 1, 2, and 3 corresponding to weather designations in Rule 250 or heavy, medium, and light, respectively.

The standard weather conditions for U.S. areas are in Rule 250. The classification is light, medium, and heavy. The southern parts of the United States are in the light category, the northwestern and central portions are in the medium category, and the northeastern portions are in the heavy category.

The resultant construction grades required are in Rules 240–243, and are Grade B (the highest grade), Grade C (the medium grade), and Grade N (the lowest grade). The strength requirements for each construction grade are in Rules 260–264.

8. NESC PART 3

Part 3 of the Safety Code is titled "Safety Rules for the Installation and Maintenance of Underground Electric Supply and Communication Lines." This part covers underground installations as shown in Fig. 57.1(c). The requirements of Part 3 focus on underground installations for supply (power) and communications (telephone and cable television, and others). There is overlap at the underground riser and overhead pole between Part 2 and Part 3.

Accessibility is a concern of this part. The accessibility issue is covered in Rule 313 and throughout Part 3 under the topics working space, working facilities, and clearance between conductors.

Definitions are in Rule 320, and are shown in Fig. 57.5.

Surprisingly, the Safety Code does not specify conduit burial depths but instead defers to "accepted good practice," which generally means that the requirements of the NEC are applicable.

Rule 320 requires that conduit running longitudinally under a road should be installed on the shoulder, and if not practical, under one lane only. This rule also requires that supply (power) and communications lines be separated as shown in Table 57.2.

Figure 57.5 Underground Installation Technology

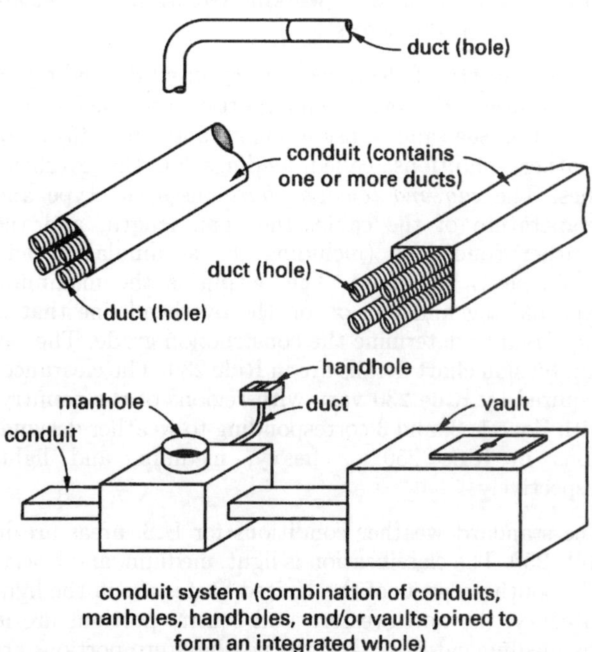

Table 57.2 Underground Supply (Power) and Communication Conduit Separation Requirements

burial method	separation requirement (in)
concrete	3
masonry	4
tamped earth	12

Table 57.3 Supply (Power) and Communication Cable Separation in Joint-Use Manholes and Vaults

supply cable phase-to-phase voltage (V)	communication cable minimum separation (in)
0–15,000	6
15,001–50,000	9
50,001–120,000	12
120,001 and greater	24

Manhole dimensions and requirements are in Rule 323. Cable separation requirements in manholes and vaults are covered in Rule 341. Rule 341 requires communications cables to be a minimum distance from supply (power) cables depending upon their voltage as given in Table 57.3.

The Safety Code does contain burial (depth) requirements for directly buried (that is, non-conduit encased) supply (power) cables, but not communication cables. These depth requirements are shown in Fig. 57.6(a), whose source is Rule 352. The depth varies with the voltage and is measured surface to surface and not center to center. Nevertheless, communications cables must meet the general burial requirements of Rule 352. Directly buried lighting cable is an exception as is shown in Fig. 57.6(b).

Figure 57.6 Underground Installation Technology

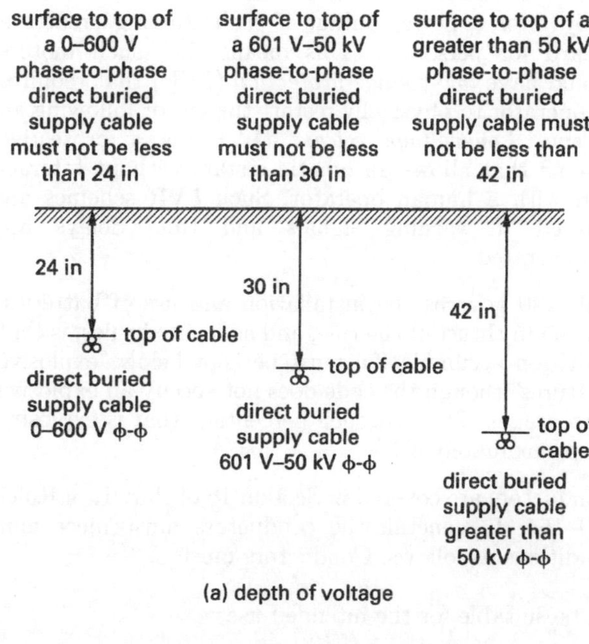

(a) depth of voltage

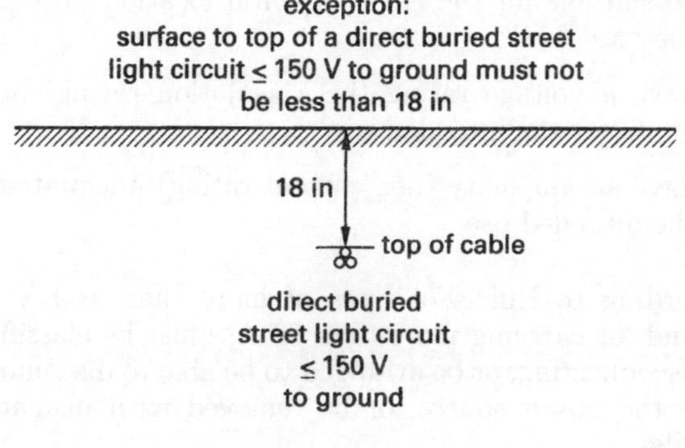

(b) lighting exception

Rules 353 and 354 are similar. Both levy radial separation requirements from underground structures, including sewer, water, natural gas, fuel, and steam lines. These rules also apply to separation from building foundations. Rule 353 covers the specifications for 12 in or greater radial separation. If the 12 in separation cannot be maintained, Rule 354 applies or the cables must be installed in a conduit system, in which case Rule 320 (covered earlier) would apply. The intent of Rule 353 is to allow for maintenance on either system without interference with the other. The intent of Rule 354 is to prevent interference between supply (power) and communication cables.

Risers bring underground cabling to the vault or pole. Riser requirements are in Rules 360–363. The supply cable terminations located in the same area are covered by Rules 370–374. The location of equipment, both on the surface and in vaults is covered in Section 38, Rules 380–385. Many of the separation distances are not specified in Section 38, instead the Safety Code defers to accepted good practice. Here again, the NEC provides a guide to much of that good practice. The Safety Code does require that supply and communications equipment be located horizontally at least 4 ft from fire hydrants, though 3 ft is allowed where conditions do not permit the 4 ft separation.

9. NESC PART 4

Part 4 of the Safety Code is titled "Rules for the Operation of Electric Supply and Communications Lines and Equipment." This part covers work rules, as indicated in Fig. 57.1(d). Part 5 focuses on practical *work rules* designed to safeguard both the employees performing the work and the general public. The work rules of the Safety Code have been coordinated with Occupation Safety and Health Administration (OSHA) regulations related to electric supply and communications systems.

Topic XV: Professional Practice

Chapter

58. Engineering Economic Analysis
59. Engineering Law
60. Engineering Ethics
61. Electrical Engineering Frontiers
62. Engineering Licensing in the United States

58 Engineering Economic Analysis

Nomenclature

A	annual amount	\$
B	present worth of all benefits	\$
BV_j	book value at end of the jth year	\$
C	cost or present worth of all costs	\$
d	declining balance depreciation rate	decimal
D	demand	various
D	depreciation	\$
DR	present worth of after-tax depreciation recovery	\$
e	constant inflation rate	decimal
$\mathcal{E}$	expected value	various
E_0	initial amount of an exponentially growing cash flow	\$
EAA	equivalent annual amount	\$
EUAC	equivalent uniform annual cost	\$
f	federal income tax rate	decimal
F	forecasted quantity	various
F	future worth	\$
g	exponential growth rate	decimal
G	uniform gradient amount	\$
i	effective interest rate	decimal
$i\%$	effective interest rate	%
i'	effective interest rate corrected for inflation	decimal
j	number of years	–
k	number of compounding periods per year	–

m	an integer	–
n	number of compounding periods or years in life of asset	–
p	probability	decimal
P	present worth	\$
r	nominal rate per year (rate per annum)	decimal per unit time
ROI	return on investment	\$
ROR	rate of return	decimal per unit time
s	state income tax rate	decimal
S_n	expected salvage value in year n	\$
t	composite tax rate	decimal
t	time	years (typical)
T	a quantity equal to $\frac{1}{2}n(n+1)$	–
TC	tax credit	\$
z	a quantity equal to $\dfrac{1+i}{1-d}$	decimal

Symbols

α	smoothing coefficient for forecasts	–
ϕ	effective rate per period	decimal

Subscripts

0	initial
j	at time j
n	at time n
t	at time t

1. IRRELEVANT CHARACTERISTICS

In its simplest form, an *engineering economic analysis* is a study of the desirability of making an investment.[1] The decision-making principles in this chapter can be applied by individuals as well as by companies. The nature of the spending opportunity or industry is not important. Farming equipment, personal investments, and multimillion dollar factory improvements can all be evaluated using the same principles.

Similarly, the applicable principles are insensitive to the monetary units. Although *dollars* are used in this chapter, it is equally convenient to use pounds, yen, or euros.

Finally, this chapter may give the impression that investment alternatives must be evaluated on a year-by-year basis. Actually, the *effective period* can be defined as a day, month, century, or any other convenient period of time.

2. MULTIPLICITY OF SOLUTION METHODS

Most economic conclusions can be reached in more than one manner. There are usually several different analyses that will eventually result in identical answers.[2] Other than the pursuit of elegant solutions in a timely manner, there is no reason to favor one procedural method over another.[3]

3. PRECISION AND SIGNIFICANT DIGITS

The full potential of electronic calculators will never be realized in engineering economic analyses. Considering that calculations are based on estimates of far-future cash flows and that unrealistic assumptions (no inflation, identical cost structures of replacement assets, etc.) are routinely made, it makes little sense to carry cents along in calculations.

The calculations in this chapter have been designed to illustrate and review the principles presented. Because of this, greater precision than is normally necessary in everyday problems may be used. Though used, such precision is not warranted.

Unless there is some compelling reason to strive for greater precision, the following rules are presented for use in reporting final answers to engineering economic analysis problems.

- Omit fractional parts of the dollar (i.e., cents).
- Report and record a number to a maximum of four significant digits unless the first digit of that number is 1, in which case, a maximum of five significant digits should be written. For example,

\$49	not	\$49.43
\$93,450	not	\$93,453
\$1,289,700	not	\$1,289,673

4. NONQUANTIFIABLE FACTORS

An engineering economic analysis is a quantitative analysis. Some factors cannot be introduced as numbers into the calculations. Such factors are known as *non-quantitative factors*, *judgment factors*, and *irreducible factors*. Typical nonquantifiable factors are

- preferences
- political ramifications
- urgency
- goodwill
- prestige
- utility
- corporate strategy
- environmental effects
- health and safety rules

[1]This subject is also known as *engineering economics* and *engineering economy*. There is very little, if any, true economics in this subject.

[2]Because of round-off errors, particularly when factors are taken from tables, these different calculations will produce slightly different numerical results (e.g., \$49.49 versus \$49.50). However, this type of divergence is well known and accepted in engineering economic analysis.

[3]This does not imply that approximate methods, simplifications, and rules of thumb are acceptable.

Professional Practice

- reliability
- political risks

Since these factors are not included in the calculations, the policy is to disregard the issues entirely. Of course, the factors should be discussed in a final report. The factors are particularly useful in breaking ties between competing alternatives that are economically equivalent.

5. YEAR-END AND OTHER CONVENTIONS

Except in short-term transactions, it is simpler to assume that all receipts and disbursements (cash flows) take place at the end of the year in which they occur.[4] This is known as the *year-end convention*. The exceptions to the year-end convention are initial project cost (purchase cost), trade-in allowance, and other cash flows that are associated with the inception of the project at $t = 0$.

On the surface, such a convention appears grossly inappropriate since repair expenses, interest payments, corporate taxes, and so on seldom coincide with the end of a year. However, the convention greatly simplifies engineering economic analysis problems, and it is justifiable on the basis that the increased precision associated with a more rigorous analysis is not warranted (due to the numerous other simplifying assumptions and estimates initially made in the problem).

There are various established procedures, known as *rules* or *conventions*, imposed by the Internal Revenue Service on U.S. taxpayers. An example is the *half-year rule*, which permits only half of the first-year depreciation to be taken in the first year of an asset's life when certain methods of depreciation are used. These rules are subject to constantly changing legislation and are not covered in this book. The implementation of such rules is outside the scope of engineering practice and is best left to accounting professionals.

6. CASH FLOW DIAGRAMS

Although they are not always necessary in simple problems (and they are often unwieldy in very complex problems), *cash flow diagrams* can be drawn to help visualize and simplify problems having diverse receipts and disbursements.

The following conventions are used to standardize cash flow diagrams.

- The horizontal (time) axis is marked off in equal increments, one per period, up to the duration (or *horizon*) of the project.
- Two or more transfers in the same period are placed end to end, and these may be combined.
- Expenses incurred before $t = 0$ are called *sunk costs*. Sunk costs are not relevant to the problem unless they have tax consequences in an after-tax analysis.
- *Receipts* are represented by arrows directed upward. *Disbursements* are represented by arrows directed downward. The arrow length is proportional to the magnitude of the cash flow.

[4]A *short-term transaction* typically has a lifetime of five years or less and has payments or compounding that are more frequent than once per year.

Example 58.1

A mechanical device will cost $20,000 when purchased. Maintenance will cost $1000 each year. The device will generate revenues of $5000 each year for five years, after which the salvage value is expected to be $7000. Draw and simplify the cash flow diagram.

Solution

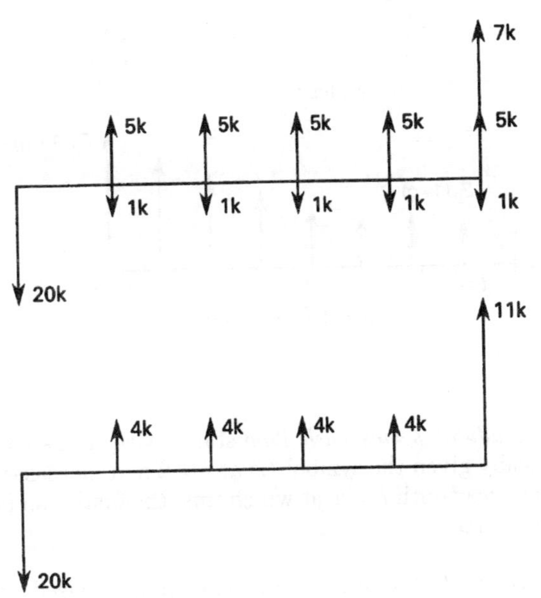

7. TYPES OF CASH FLOWS

To evaluate a real-world project, it is necessary to present the project's cash flows in terms of standard cash flows that can be handled by engineering economic analysis techniques. The standard cash flows are single payment cash flow, uniform series cash flow, gradient series cash flow, and the infrequently encountered exponential gradient series cash flow. (See Fig. 58.1.)

A *single payment cash flow* can occur at the beginning of the time line (designated as $t = 0$), at the end of the time line (designated as $t = n$), or at any time in between.

The *uniform series cash flow* consists of a series of equal transactions starting at $t = 1$ and ending at $t = n$. The symbol A is typically given to the magnitude of each individual cash flow.[5]

Professional Practice

[5]The cash flows do not begin at $t = 0$. This is an important concept with all of the series cash flows. This convention has been established to accommodate the timing of annual maintenance (and similar) cash flows for which the year-end convention is applicable.

Figure 58.1 Standard Cash Flows

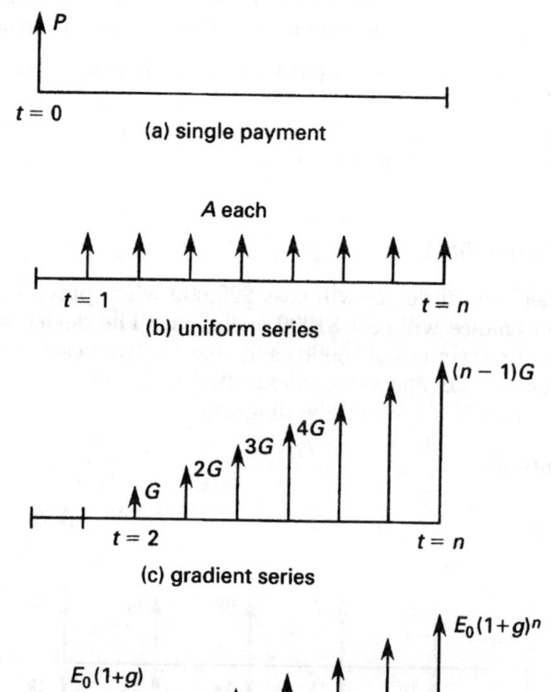

The *gradient series cash flow* starts with a cash flow (typically given the symbol G) at $t = 2$ and increases by G each year until $t = n$, at which time the final cash flow is $(n - 1)G$.

An *exponential gradient series cash flow* is based on a phantom value (typically given the symbol E_0) at $t = 0$ and grows or decays exponentially according to the following relationship.[6]

$$\text{amount at time } t = E_t = E_0(1+g)^t \quad [t = 1, 2, 3, \ldots, n] \qquad 58.1$$

In Eq. 58.1, g is the *exponential growth rate*, which can be either positive or negative. Exponential gradient cash flows are rarely seen in economic justification projects assigned to engineers.[7]

[6]By convention, for an exponential cash flow series: The first cash flow, E_0, is at $t = 1$, as in the uniform annual series. However, the first cash flow is $E_0(1 + g)$. The cash flow of E_0 at $t = 0$ is absent (i.e., is a *phantom cash flow*).

[7]For one of the few discussions on exponential cash flow, see *Capital Budgeting*, Robert V. Oakford, The Ronald Press Company, New York, 1970.

8. TYPICAL PROBLEM TYPES

There is a wide variety of problem types that, collectively, are considered to be engineering economic analysis problems.

By far, the majority of engineering economic analysis problems are *alternative comparisons*. In these problems, two or more mutually exclusive investments compete for limited funds. A variation of this is a *replacement/retirement analysis*, which is repeated each year to determine if an existing asset should be replaced. Finding the percentage return on an investment is a *rate of return problem*, one of the alternative comparison solution methods.

Investigating interest and principal amounts in loan payments is a *loan repayment problem*. An *economic life analysis* will determine when an asset should be retired. In addition, there are miscellaneous problems involving economic order quantity, learning curves, break-even points, product costs, and so on.

9. IMPLICIT ASSUMPTIONS

Several assumptions are implicitly made when solving engineering economic analysis problems. Some of these assumptions are made with the knowledge that they are or will be poor approximations of what really will happen. The assumptions are made, regardless, for the benefit of obtaining a solution.

The most common assumptions are the following.

- The year-end convention is applicable.
- There is no inflation now, nor will there be any during the lifetime of the project.
- Unless otherwise specifically called for, a before-tax analysis is needed.
- The effective interest rate used in the problem will be constant during the lifetime of the project.
- Nonquantifiable factors can be disregarded.
- Funds invested in a project are available and are not urgently needed elsewhere.
- Excess funds continue to earn interest at the effective rate used in the analysis.

This last assumption, like most of the assumptions listed, is almost never specifically mentioned in the body of a solution. However, it is a key assumption when comparing two alternatives that have different initial costs.

For example, suppose two investments, one costing \$10,000 and the other costing \$8000, are to be compared at 10%. It is obvious that \$10,000 in funds is available, otherwise the costlier investment would not be under consideration. If the smaller investment is chosen, what is done with the remaining \$2000? The last assumption yields the answer: The \$2000 is "put to work" in some investment earning (in this case) 10%.

10. EQUIVALENCE

Industrial decision makers using engineering economic analysis are concerned with the magnitude and timing of a project's cash flow as well as with the total profitability of that project. In this situation, a method is required to compare projects involving receipts and disbursements occurring at different times.

By way of illustration, consider \$100 placed in a bank account that pays 5% effective annual interest at the end of each year. After the first year, the account will have grown to \$105. After the second year, the account will have grown to \$110.25.

Assume that you will have no need for money during the next two years, and any money received will immediately go into your 5% bank account. Then, which of the following options would be more desirable?

option A: \$100 now

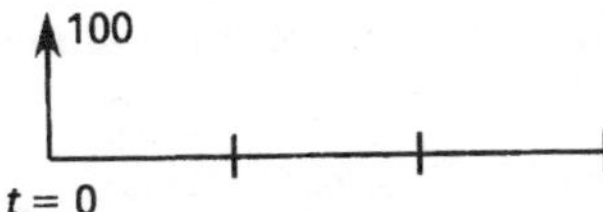

option B: \$105 to be delivered in one year

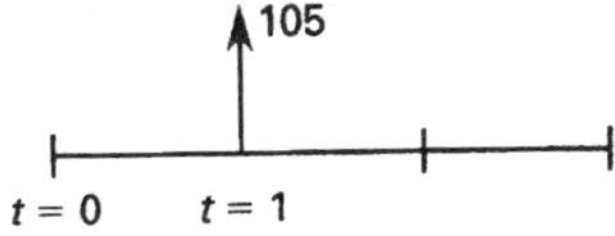

option C: \$110.25 to be delivered in two years

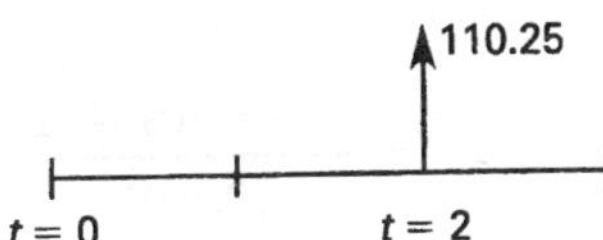

As illustrated, none of the options is superior under the assumptions given. If the first option is chosen, you will immediately place \$100 into a 5% account, and in two years the account will have grown to \$110.25. In fact, the account will contain \$110.25 at the end of two years regardless of the option chosen. Therefore, these alternatives are said to be *equivalent*.

Equivalence may or may not be the case, depending on the interest rate, so an alternative that is acceptable to one decision maker may be unacceptable to another. The interest rate that is used in actual calculations is known as the *effective interest rate*.[8] If compounding is once a year, it is known as the *effective annual interest rate*. However, effective quarterly, monthly, daily, and so on, interest rates are also used.

[8]The adjective *effective* distinguishes this interest rate from other interest rates (e.g., nominal interest rates) that are not meant to be used directly in calculating equivalent amounts.

The fact that \$100 today grows to \$105 in one year (at 5% annual interest) is an example of what is known as the *time value of money* principle. This principle simply articulates what is obvious: Funds placed in a secure investment will increase to an equivalent future amount. The procedure for determining the present investment from the equivalent future amount is known as *discounting*.

11. SINGLE-PAYMENT EQUIVALENCE

The equivalence of any present amount, P, at $t = 0$, to any future amount, F, at $t = n$, is called the *future worth* and can be calculated from Eq. 58.2.

$$F = P(1+i)^n \qquad 58.2$$

The factor $(1 + i)^n$ is known as the single payment *compound amount factor* and has been tabulated in App. 58.B for various combinations of i and n.

Similarly, the equivalence of any future amount to any present amount is called the *present worth* and can be calculated from Eq. 58.3.

$$P = F(1+i)^{-n} = \frac{F}{(1+i)^n} \qquad 58.3$$

The factor $(1 + i)^{-n}$ is known as the *single payment present worth factor*.[9]

The interest rate used in Eq. 58.2 and Eq. 58.3 must be the effective rate per period. Also, the basis of the rate (annually, monthly, etc.) must agree with the type of period used to count n. Therefore, it would be incorrect to use an effective annual interest rate if n was the number of compounding periods in months.

Example 58.2

How much should you put into a 10% (effective annual rate) savings account in order to have \$10,000 in five years?

Solution

This problem could also be stated: What is the equivalent present worth of \$10,000 five years from now if money is worth 10% per year?

$$P = F(1+i)^{-n} = (\$10{,}000)(1+0.10)^{-5}$$
$$= \$6209$$

The factor 0.6209 would usually be obtained from the tables.

[9]The *present worth* is also called the *present value* and *net present value*. These terms are used interchangeably and no significance should be attached to the terms *value*, *worth*, and *net*.

Professional Practice

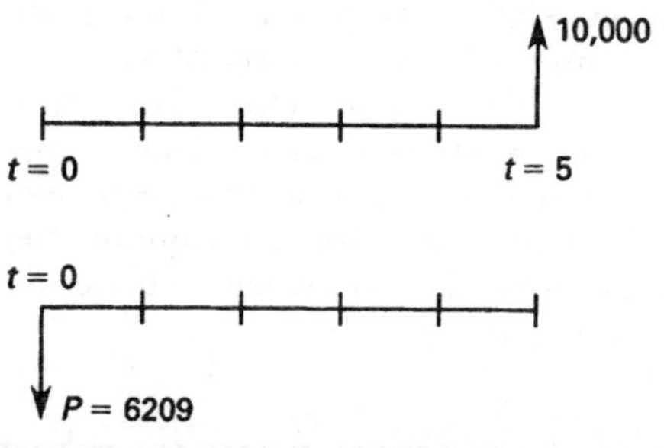

12. STANDARD CASH FLOW FACTORS AND SYMBOLS

Equation 58.2 and Eq. 58.3 may give the impression that solving engineering economic analysis problems involves a lot of calculator use, and, in particular, a lot of exponentiation. Such calculations may be necessary from time to time, but most problems are simplified by the use of tabulated values of the factors.

Rather than actually writing the formula for the compound amount factor (which converts a present amount to a future amount), it is common convention to substitute the standard functional notation of $(F/P, i\%, n)$. Therefore, the future value in n periods of a present amount would be symbolically written as

$$F = P(F/P, i\%, n) \quad 58.4$$

Similarly, the present worth factor has a functional notation of $(P/F, i\%, n)$. The present worth of a future amount n periods hence would be symbolically written as

$$P = F(P/F, i\%, n) \quad 58.5$$

Values of these *cash flow* (*discounting*) *factors* are tabulated in App. 58.B. There is often initial confusion about whether the (F/P) or (P/F) column should be used in a particular problem. There are several ways of remembering what the functional notations mean.

One method of remembering which factor should be used is to think of the factors as conditional probabilities. The conditional probability of event **A** given that event **B** has occurred is written as $p\{\mathbf{A}|\mathbf{B}\}$, where the given event comes after the vertical bar. In the standard notational form of discounting factors, the given amount is similarly placed after the slash. What you want comes before the slash. (F/P) would be a factor to find F given P.

Another method of remembering the notation is to interpret the factors algebraically. The (F/P) factor could be thought of as the fraction F/P. Algebraically, Eq. 58.4 would be

$$F = P\left(\frac{F}{P}\right) \quad 58.6$$

This algebraic approach is actually more than an interpretation. The numerical values of the discounting factors are consistent with this algebraic manipulation. The (F/A) factor could be calculated as $(F/P) \times (P/A)$. This consistent relationship can be used to calculate other factors that might be occasionally needed, such as (F/G) or (G/P). For instance, the annual cash flow that would be equivalent to a uniform gradient may be found from

$$A = G(P/G, i\%, n)(A/P, i\%, n) \quad 58.7$$

Formulas for the compounding and discounting factors are contained in Table 58.1. Normally, it will not be necessary to calculate factors from the formulas. For solving most problems, App. 58.B is adequate.

Example 58.3

What factor will convert a gradient cash flow ending at $t = 8$ to a future value at $t = 8$? (That is, what is the $(F/G, i\%, n)$ factor?) The effective annual interest rate is 10%.

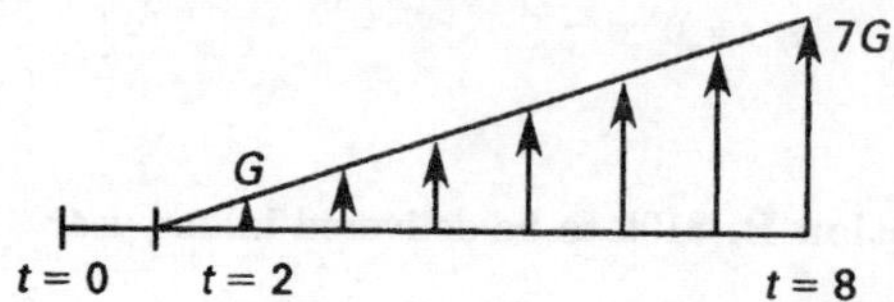

Solution

method 1:

From Table 58.1, the $(F/G, i\%, n)$ factor is

$$\begin{aligned}(F/G, 10\%, 8) &= \frac{(1+i)^n - 1}{i^2} - \frac{n}{i} \\ &= \frac{(1+0.10)^8 - 1}{(0.10)^2} - \frac{8}{0.10} \\ &= 34.3589\end{aligned}$$

method 2:

The tabulated values of (P/G) and (F/P) in App. 58.B can be used to calculate the factor.

$$\begin{aligned}(F/G, 10\%, 8) &= (P/G, 10\%, 8)(F/P, 10\%, 8) \\ &= (16.0287)(2.1436) \\ &= 34.3591\end{aligned}$$

The (F/G) factor could also have been calculated as the product of the (A/G) and (F/A) factors.

Professional Practice

Table 58.1 Discount Factors for Discrete Compounding

factor name	converts	symbol	formula
single payment compound amount	P to F	$(F/P, i\%, n)$	$(1+i)^n$
single payment present worth	F to P	$(P/F, i\%, n)$	$(1+i)^{-n}$
uniform series sinking fund	F to A	$(A/F, i\%, n)$	$\dfrac{i}{(1+i)^n-1}$
capital recovery	P to A	$(A/P, i\%, n)$	$\dfrac{i(1+i)^n}{(1+i)^n-1}$
uniform series compound amount	A to F	$(F/A, i\%, n)$	$\dfrac{(1+i)^n-1}{i}$
uniform series present worth	A to P	$(P/A, i\%, n)$	$\dfrac{(1+i)^n-1}{i(1+i)^n}$
uniform gradient present worth	G to P	$(P/G, i\%, n)$	$\dfrac{(1+i)^n-1}{i^2(1+i)^n}-\dfrac{n}{i(1+i)^n}$
uniform gradient future worth	G to F	$(F/G, i\%, n)$	$\dfrac{(1+i)^n-1}{i^2}-\dfrac{n}{i}$
uniform gradient uniform series	G to A	$(A/G, i\%, n)$	$\dfrac{1}{i}-\dfrac{n}{(1+i)^n-1}$

13. CALCULATING UNIFORM SERIES EQUIVALENCE

A cash flow that repeats each year for n years without change in amount is known as an *annual amount* and is given the symbol A. As an example, a piece of equipment may require annual maintenance, and the maintenance cost will be an annual amount. Although the equivalent value for each of the n annual amounts could be calculated and then summed, it is more expedient to use one of the uniform series factors. For example, it is possible to convert from an annual amount to a future amount by use of the (F/A) factor.

$$F = A(F/A, i\%, n) \quad 58.8$$

A *sinking fund* is a fund or account into which annual deposits of A are made in order to accumulate F at $t = n$ in the future. Since the annual deposit is calculated as $A = F(A/F, i\%, n)$, the (A/F) factor is known as the *sinking fund factor*. An *annuity* is a series of equal payments, A, made over a period of time.[10] Usually, it is necessary to "buy into" an investment (a bond, an insurance policy, etc.) in order to ensure the annuity. In the simplest case of an annuity that starts at the end of the first year and continues for n years, the purchase price, P, is

$$P = A(P/A, i\%, n) \quad 58.9$$

The present worth of an *infinite (perpetual) series* of annual amounts is known as a *capitalized cost*. There is no $(P/A, i\%, \infty)$ factor in the tables, but the capitalized cost can be calculated simply as

$$P = \frac{A}{i} \quad [i \text{ in decimal form}] \quad 58.10$$

Alternatives with different lives will generally be compared by way of *equivalent uniform annual cost* (EUAC). An EUAC is the annual amount that is equivalent to all of the cash flows in the alternative. The EUAC differs in sign from all of the other cash flows. Costs and expenses expressed as EUACs, which would normally be considered negative, are actually positive. The term *cost* in the designation EUAC serves to make clear the meaning of a positive number.

Example 58.4

Maintenance costs for a machine are \$250 each year. What is the present worth of these maintenance costs over a 12-year period if the interest rate is 8%?

[10] An annuity may also consist of a lump sum payment made at some future time. However, this interpretation is not considered in this chapter.

Professional Practice

Solution

Using Eq. 58.9,

$$P = A(P/A, 8\%, 12) = (-\$250)(7.5361)$$
$$= -\$1884$$

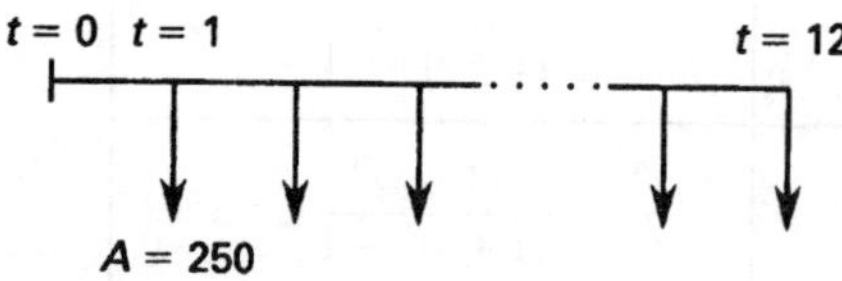

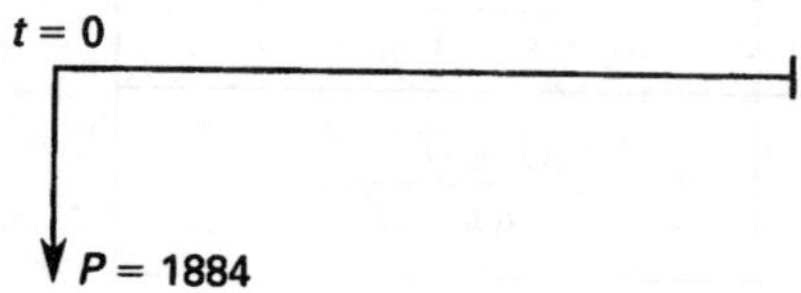

14. FINDING PAST VALUES

From time to time, it will be necessary to determine an amount in the past equivalent to some current (or future) amount. For example, you might have to calculate the original investment made 15 years ago given a current annuity payment.

Such problems are solved by placing the $t = 0$ point at the time of the original investment, and then calculating the past amount as a P value. For example, the original investment, P, can be extracted from the annuity, A, by using the standard cash flow factors.

$$P = A(P/A, i\%, n) \qquad 58.11$$

The choice of $t = 0$ is flexible. As a general rule, the $t = 0$ point should be selected for convenience in solving a problem.

Example 58.5

You currently pay $250 per month to lease your office phone equipment. You have three years (36 months) left on the five-year (60-month) lease. What would have been an equivalent purchase price two years ago? The effective interest rate per month is 1%.

Solution

The solution of this example is not affected by the fact that investigation is being performed in the middle of the horizon. This is a simple calculation of present worth.

$$P = A(P/A, 1\%, 60)$$
$$= (-\$250)(44.9550)$$
$$= -\$11,239$$

15. TIMES TO DOUBLE AND TRIPLE AN INVESTMENT

If an investment doubles in value (in n compounding periods and with $i\%$ effective interest), the ratio of current value to past investment will be 2.

$$F/P = (1+i)^n = 2 \qquad 58.12$$

Similarly, the ratio of current value to past investment will be 3 if an investment triples in value. This can be written as

$$F/P = (1+i)^n = 3 \qquad 58.13$$

It is a simple matter to extract the number of periods, n, from Eq. 58.12 and Eq. 58.13 to determine the *doubling time* and *tripling time*, respectively. For example, the doubling time is

$$n = \frac{\log 2}{\log(1+i)} \qquad 58.14$$

When a quick estimate of the doubling time is needed, the *rule of 72* can be used. The doubling time is approximately $72/i$.

The tripling time is

$$n = \frac{\log 3}{\log(1+i)} \qquad 58.15$$

Equation 58.14 and Eq. 58.15 form the basis of Table 58.2.

***Table 58.2** Doubling and Tripling Times for Various Interest Rates*

interest rate (%)	doubling time (periods)	tripling time (periods)
1	69.7	110.4
2	35.0	55.5
3	23.4	37.2
4	17.7	28.0
5	14.2	22.5
6	11.9	18.9
7	10.2	16.2
8	9.01	14.3
9	8.04	12.7
10	7.27	11.5
11	6.64	10.5
12	6.12	9.69
13	5.67	8.99
14	5.29	8.38
15	4.96	7.86
16	4.67	7.40
17	4.41	7.00
18	4.19	6.64
19	3.98	6.32
20	3.80	6.03

Professional Practice

16. VARIED AND NONSTANDARD CASH FLOWS

Gradient Cash Flow

A common situation involves a uniformly increasing cash flow. If the cash flow has the proper form, its present worth can be determined by using the *uniform gradient factor*, (P/G, i%, n). The uniform gradient factor finds the present worth of a uniformly increasing cash flow that starts in year two (not in year one).

There are three common difficulties associated with the form of the uniform gradient. The first difficulty is that the initial cash flow occurs at $t = 2$. This convention recognizes that annual costs, if they increase uniformly, begin with some value at $t = 1$ (due to the year-end convention) but do not begin to increase until $t = 2$. The tabulated values of (P/G) have been calculated to find the present worth of only the increasing part of the annual expense. The present worth of the base expense incurred at $t = 1$ must be found separately with the (P/A) factor.

The second difficulty is that, even though the factor (P/G, i%, n) is used, there are only $n - 1$ actual cash flows. It is clear that n must be interpreted as the *period number* in which the last gradient cash flow occurs, not the number of gradient cash flows.

The sign convention used with gradient cash flows may seem confusing. If an expense increases each year (as in Ex. 58.5), the gradient will be negative, since it is an expense. If a revenue increases each year, the gradient will be positive. In most cases, the sign of the gradient depends on whether the cash flow is an expense or a revenue.[11] Figure 58.2 illustrates positive and negative gradient cash flows.

Example 58.6

Maintenance on an old machine is $100 this year but is expected to increase by $25 each year thereafter. What is the present worth of five years of the costs of maintenance? Use an interest rate of 10%.

Solution

In this problem, the cash flow must be broken down into parts. (The five-year gradient factor is used even though there are only four nonzero gradient cash flows.)

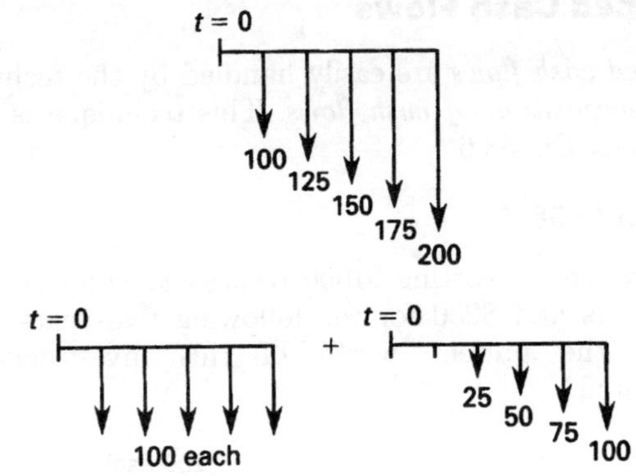

Therefore, the present worth is

$$\begin{aligned}P &= A(P/A, 10\%, 5) + G(P/G, 10\%, 5)\\ &= (-\$100)(3.7908) - (\$25)(6.8618)\\ &= -\$551\end{aligned}$$

Figure 58.2 *Positive and Negative Gradient Cash Flows*

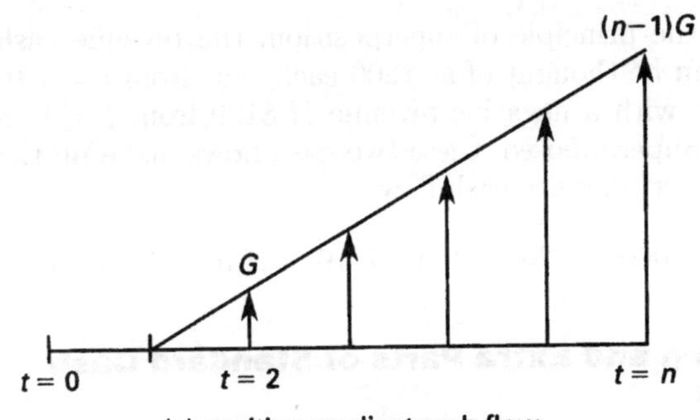

(a) positive gradient cash flow
$P = G(P/G)$

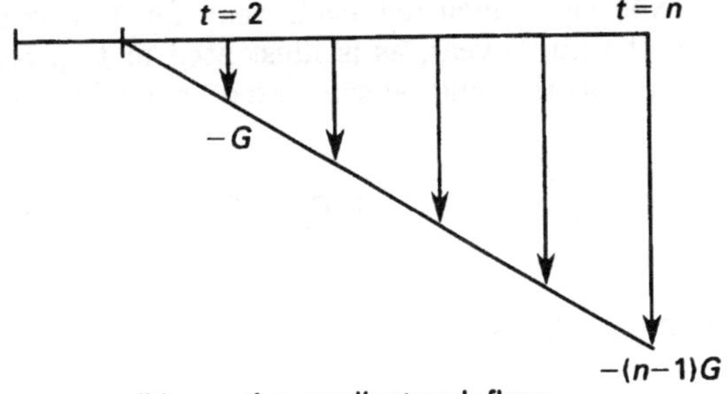

(b) negative gradient cash flow
$P = -G(P/G)$

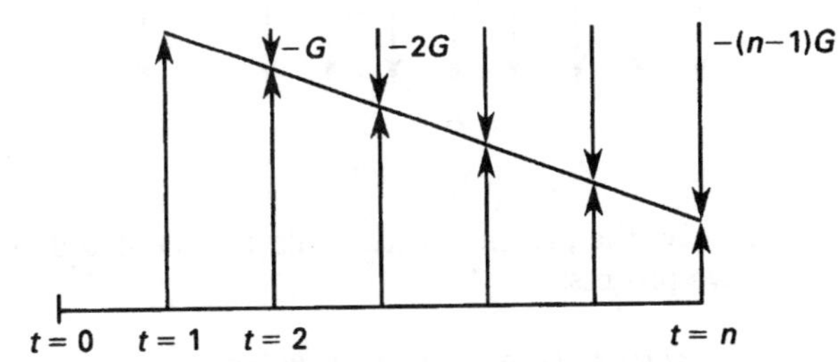

(c) decreasing revenue incorporating a negative gradient
$P = A(P/A) - G(P/G)$

[11]This is not a universal rule. It is possible to have a uniformly decreasing revenue as in Fig. 58.2(c). In this case, the gradient would be negative.

Professional Practice

Stepped Cash Flows

Stepped cash flows are easily handled by the technique of *superposition of cash flows.* This technique is illustrated by Ex. 58.6.

Example 58.7

An investment costing \$1000 returns \$100 for the first five years and \$200 for the following five years. How would the present worth of this investment be calculated?

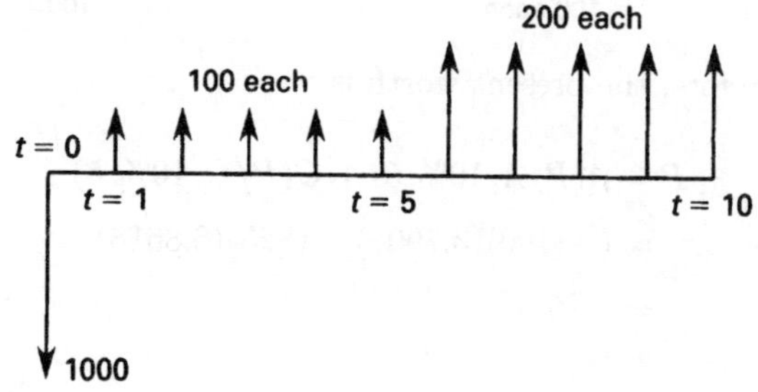

Solution

Using the principle of superposition, the revenue cash flow can be thought of as \$200 each year from $t = 1$ to $t = 10$, with a negative revenue of \$100 from $t = 1$ to $t = 5$. Superimposed, these two cash flows make up the actual performance cash flow.

$$P = -\$1000 + (\$200)(P/A, i\%, 10) - (\$100)(P/A, i\%, 5)$$

Missing and Extra Parts of Standard Cash Flows

A missing or extra part of a standard cash flow can also be handled by superposition. For example, suppose an annual expense is incurred each year for ten years, except in the ninth year, as is illustrated in Fig. 58.3. The present worth could be calculated as a subtractive process.

$$P = A(P/A, i\%, 10) - A(P/F, i\%, 9) \qquad 58.16$$

Figure 58.3 *Cash Flow with a Missing Part*

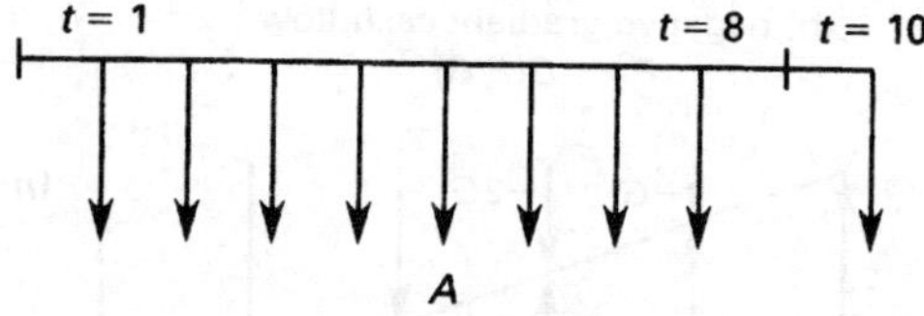

Alternatively, the present worth could be calculated as an additive process.

$$P = A(P/A, i\%, 8) + A(P/F, i\%, 10) \qquad 58.17$$

Delayed and Premature Cash Flows

There are cases when a cash flow matches a standard cash flow exactly, except that the cash flow is delayed or starts sooner than it should. Often, such cash flows can be handled with superposition. At other times, it may be more convenient to shift the time axis. This shift is known as the *projection method.* Example 58.7 illustrates the projection method.

Example 58.8

An expense of \$75 is incurred starting at $t = 3$ and continues until $t = 9$. There are no expenses or receipts until $t = 3$. Use the projection method to determine the present worth of this stream of expenses.

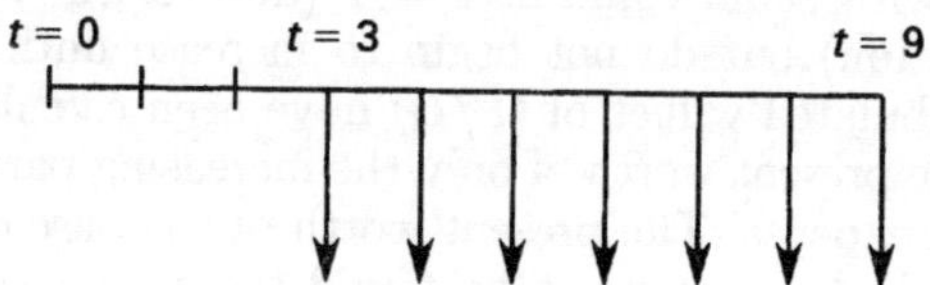

Solution

Determine a cash flow at $t = 2$ that is equivalent to the entire expense stream. If $t = 0$ was where $t = 2$ actually is, the present worth of the expense stream would be

$$P' = (-\$75)(P/A, i\%, 7)$$

P' is a cash flow at $t = 2$. It is now simple to find the present worth (at $t = 0$) of this future amount.

$$P = P'(P/F, i\%, 2) = (-\$75)(P/A, i\%, 7)(P/F, i\%, 2)$$

Cash Flows at Beginnings of Years: The Christmas Club Problem

This type of problem is characterized by a stream of equal payments (or expenses) starting at $t = 0$ and ending at $t = n - 1$, as shown in Fig. 58.4. It differs from the standard annual cash flow in the existence of a cash flow at $t = 0$ and the absence of a cash flow at $t = n$. This problem gets its name from the service provided by some savings institutions whereby money is automatically deposited each week or month (starting immediately, when the savings plan is opened) in order to accumulate money to purchase Christmas presents at the end of the year.

Figure 58.4 *Cash Flows at Beginnings of Years*

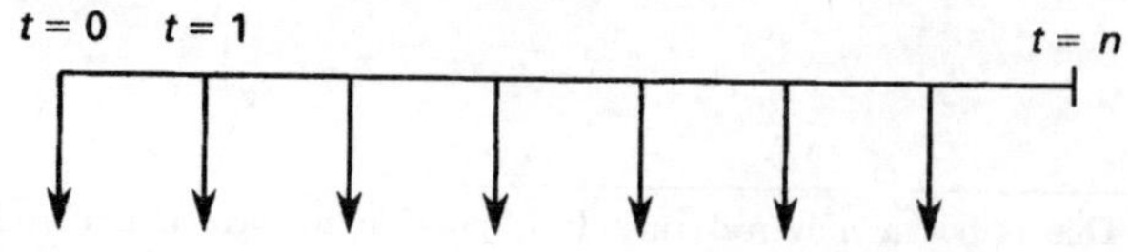

Professional Practice

It may seem that the present worth of the savings stream can be determined by directly applying the (P/A) factor. However, this is not the case, since the Christmas Club cash flow and the standard annual cash flow differ. The Christmas Club problem is easily handled by superposition, as illustrated by Ex. 58.8.

Example 58.9

How much can you expect to accumulate by $t = 10$ for a child's college education if you deposit \$300 at the beginning of each year for a total of ten payments?

Solution

The first payment is made at $t = 0$ and that there is no payment at $t = 10$. The future worth of the first payment is calculated with the (F/P) factor. The absence of the payment at $t = 10$ is handled by superposition. This "correction" is not multiplied by a factor.

$$\begin{aligned} F &= (\$300)(F/P, i\%, 10) + (\$300)(F/A, i\%, 10) - \$300 \\ &= (\$300)(F/A, i\%, 11) - \$300 \end{aligned}$$

17. THE MEANING OF PRESENT WORTH AND *i*

If \$100 is invested in a 5% bank account (using annual compounding), you can remove \$105 one year from now; if this investment is made, you will receive a *return on investment* (ROI) of \$5. The cash flow diagram, as shown in Fig. 58.5, and the present worth of the two transactions are as follows.

$$\begin{aligned} P &= -\$100 + (\$105)(P/F, 5\%, 1) \\ &= -\$100 + (\$105)(0.9524) \\ &= 0 \end{aligned}$$

Figure 58.5 *Cash Flow Diagram*

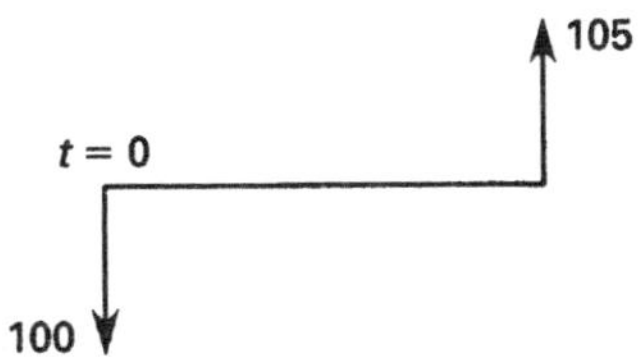

The present worth is zero even though you will receive a \$5 return on your investment.

However, if you are offered \$120 for the use of \$100 over a one-year period, the cash flow diagram, as shown in Fig. 58.6, and present worth (at 5%) would be

$$\begin{aligned} P &= -\$100 + (\$120)(P/F, 5\%, 1) \\ &= -\$100 + (\$120)(0.9524) \\ &= \$14.29 \end{aligned}$$

Figure 58.6 *Cash Flow Diagram*

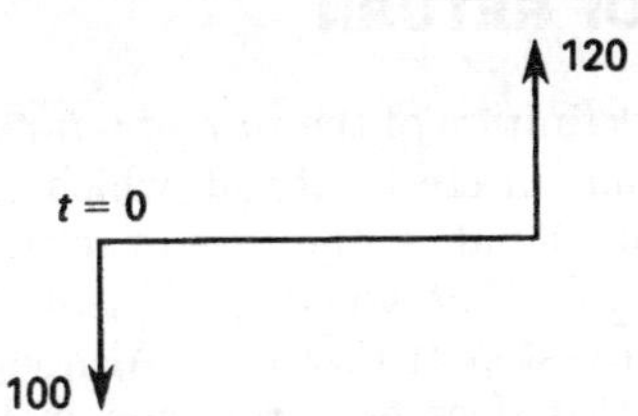

Therefore, the present worth of an alternative is seen to be equal to the equivalent value at $t = 0$ of the increase in return above that which you would be able to earn in an investment offering $i\%$ per period. In the previous case, \$14.29 is the present worth of (\$20 – \$5), the difference in the two ROIs.

The present worth is also the amount that you would have to be given to dissuade you from making an investment, since placing the initial investment amount along with the present worth into a bank account earning $i\%$ will yield the same eventual return on investment. Relating this to the previous paragraphs, you could be dissuaded from investing \$100 in an alternative that would return \$120 in one year by a $t = 0$ payment of \$14.29. Clearly, (\$100 + \$14.29) invested at $t = 0$ will also yield \$120 in one year at 5%.

Income-producing alternatives with negative present worths are undesirable, and alternatives with positive present worths are desirable because they increase the average earning power of invested capital. (In some cases, such as municipal and public works projects, the present worths of all alternatives are negative, in which case, the least negative alternative is best.)

The selection of the interest rate is difficult in engineering economics problems. Usually, it is taken as the average rate of return that an individual or business organization has realized in past investments. Alternatively, the interest rate may be associated with a particular level of risk. Usually, i for individuals is the interest rate that can be earned in relatively *risk-free investments*.

18. SIMPLE AND COMPOUND INTEREST

If \$100 is invested at 5%, it will grow to \$105 in one year. During the second year, 5% interest continues to be accrued, but on \$105, not on \$100. This is the principle of *compound interest:* The interest accrues interest.[12]

If only the original principal accrues interest, the interest is said to be *simple interest.* Simple interest is rarely encountered in long-term engineering economic analyses, but the concept may be incorporated into short-term transactions.

[12]This assumes, of course, that the interest remains in the account. If the interest is removed and spent, only the remaining funds accumulate interest.

19. EXTRACTING THE INTEREST RATE: RATE OF RETURN

An intuitive definition of the *rate of return* (ROR) is the effective annual interest rate at which an investment accrues income. That is, the rate of return of a project is the interest rate that would yield identical profits if all money were invested at that rate. Although this definition is correct, it does not provide a method of determining the rate of return.

It was previously seen that the present worth of a \$100 investment invested at 5% is zero when $i = 5\%$ is used to determine equivalence. Therefore, a working definition of rate of return would be the effective annual interest rate that makes the present worth of the investment zero. Alternatively, rate of return could be defined as the effective annual interest rate that will discount all cash flows to a total present worth equal to the required initial investment.

It is tempting, but impractical, to determine a rate of return analytically. It is simply too difficult to extract the interest rate from the equivalence equation. For example, consider a \$100 investment that pays back \$75 at the end of each of the first two years. The present worth equivalence equation (set equal to zero in order to determine the rate of return) is

$$P = 0 = -\$100 + (\$75)(1+i)^{-1} + (\$75)(1+i)^{-2} \quad 58.18$$

Solving Eq. 58.18 requires finding the roots of a quadratic equation. In general, for an investment or project spanning n years, the roots of an nth-order polynomial would have to be found. It should be obvious that an analytical solution would be essentially impossible for more complex cash flows. (The rate of return in this example is 31.87%.)

If the rate of return is needed, it can be found from a trial-and-error solution. To find the rate of return of an investment, proceed as follows.

step 1: Set up the problem as if to calculate the present worth.

step 2: Arbitrarily select a reasonable value for i. Calculate the present worth.

step 3: Choose another value of i (not too close to the original value), and again solve for the present worth.

step 4: Interpolate or extrapolate the value of i that gives a zero present worth.

step 5: For increased accuracy, repeat steps 2 and 3 with two more values that straddle the value found in step 4.

A common, although incorrect, method of calculating the rate of return involves dividing the annual receipts or returns by the initial investment. (See Sec. 58.54.) However, this technique ignores such items as salvage, depreciation, taxes, and the time value of money. This technique also is inadequate when the annual returns vary.

It is possible that more than one interest rate will satisfy the zero present worth criteria. This confusing situation occurs whenever there is more than one change in sign in the investment's cash flow.[13] Table 58.3 indicates the numbers of possible interest rates as a function of the number of sign reversals in the investment's cash flow.

Table 58.3 *Multiplicity of Rates of Return*

number of sign reversals	number of distinct rates of return
0	0
1	0 or 1
2	0, 1, or 2
3	0, 1, 2, or 3
4	0, 1, 2, 3, or 4
m	0, 1, 2, 3, ..., $m-1$, m

Difficulties associated with interpreting the meaning of multiple rates of return can be handled with the concepts of external investment and external rate of return. An *external investment* is an investment that is distinct from the investment being evaluated (which becomes known as the internal investment). The *external rate of return*, which is the rate of return earned by the external investment, does not need to be the same as the rate earned by the internal investment.

Generally, the multiple rates of return indicate that the analysis must proceed as though money will be invested outside of the project. The mechanics of how this is done are not covered here.

Example 58.10

What is the rate of return on invested capital if \$1000 is invested now with \$500 being returned in year 4 and \$1000 being returned in year 8?

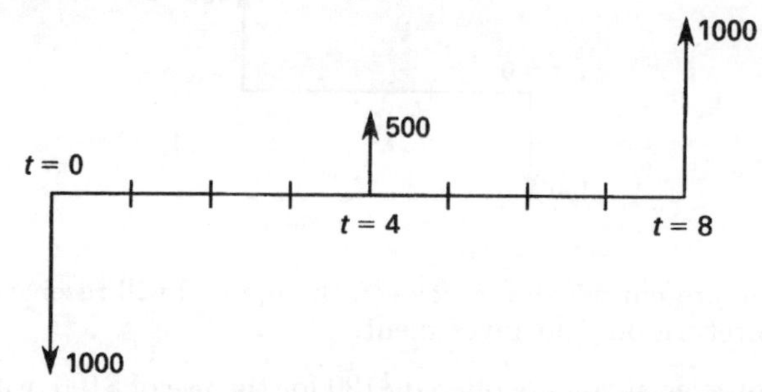

[13]There will always be at least one change of sign in the cash flow of a legitimate investment. (This excludes municipal and other tax-supported functions.) At $t = 0$, an investment is made (a negative cash flow). Hopefully, the investment will begin to return money (a positive cash flow) at $t = 1$ or shortly thereafter. Although it is possible to conceive of an investment in which all of the cash flows were negative, such an investment would probably be classified as a *hobby*.

Solution

First, set up the problem as a present worth calculation.

Try $i = 5\%$.

$$\begin{aligned} P &= -\$1000 + (\$500)(P/F, 5\%, 4) \\ &\quad + (\$1000)(P/F, 5\%, 8) \\ &= -\$1000 + (\$500)(0.8227) + (\$1000)(0.6768) \\ &= \$88 \end{aligned}$$

Next, try a larger value of i to reduce the present worth. If $i = 10\%$,

$$\begin{aligned} P &= -\$1000 + (\$500)(P/F, 10\%, 4) \\ &\quad + (\$1000)(P/F, 10\%, 8) \\ &= -\$1000 + (\$500)(0.6830) + (\$1000)(0.4665) \\ &= -\$192 \end{aligned}$$

Using simple interpolation, the rate of return is

$$\begin{aligned} \text{ROR} &= 5\% + \left(\frac{\$88}{\$88 + \$192}\right)(10\% - 5\%) \\ &= 6.57\% \end{aligned}$$

A second iteration between 6% and 7% yields 6.39%.

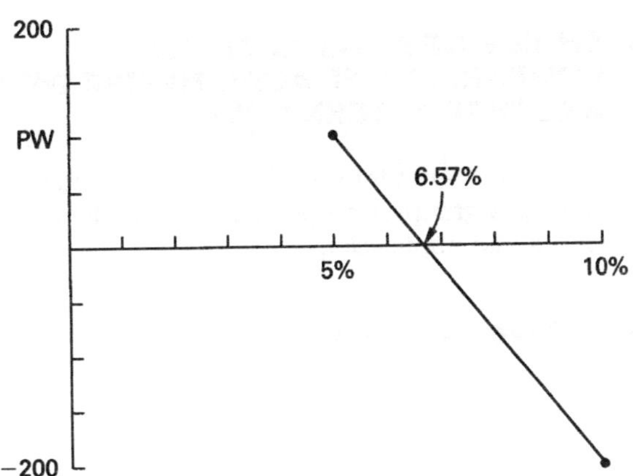

Example 58.11

A biomedical company is developing a new drug. A venture capital firm gives the company \$25 million initially and \$55 million more at the end of the first year. The drug patent will be sold at the end of year 5 to the highest bidder, and the biomedical company will receive \$80 million. (The venture capital firm will receive everything in excess of \$80 million.) The firm invests unused money in short-term commercial paper earning 10% effective interest per year through its bank. In the meantime, the biomedical company incurs development expenses of \$50 million annually for the first three years. The drug is to be evaluated by a government agency and there will be neither expenses nor revenues during the fourth year. What is the biomedical company's rate of return on this investment?

Solution

Normally, the rate of return is determined by setting up a present worth problem and varying the interest rate until the present worth is zero. Writing the cash flows, though, shows that there are two reversals of sign: one at $t = 2$ (positive to negative) and the other at $t = 5$ (negative to positive). Therefore, there could be two interest rates that produce a zero present worth. (In fact, there actually are two interest rates: 10.7% and 41.4%.)

time	cash flow (millions)
0	+25
1	+55 − 50 = +5
2	−50
3	−50
4	0
5	+80

However, this problem can be reduced to one with only one sign reversal in the cash flow series. The initial \$25 million is invested in commercial paper (an *external investment* having nothing to do with the drug development process) during the first year at 10%. The accumulation of interest and principal after one year is

$$(25)(1 + 0.10) = 27.5$$

This 27.5 is combined with the 5 (the money remaining after all expenses are paid at $t = 1$) and invested externally, again at 10%. The accumulation of interest and principal after one year (i.e., at $t = 2$) is

$$(27.5 + 5)(1 + 0.10) = 35.75$$

This 35.75 is combined with the development cost paid at $t = 2$.

The cash flow for the development project (the internal investment) is

time	cash flow (millions)
0	0
1	0
2	35.75 − 50 = −14.25
3	−50
4	0
5	+80

There is only one sign reversal in the cash flow series. The *internal rate of return* on this development project is found by the traditional method to be 10.3%. This is different from the rate the company can earn from investing externally in commercial paper.

Professional Practice

20. RATE OF RETURN VERSUS RETURN ON INVESTMENT

Rate of return (ROR) is an effective annual interest rate, typically stated in percent per year. *Return on investment* (ROI) is a dollar amount. *Rate of return* and *return on investment* are not synonymous.

Return on investment can be calculated in two different ways. The accounting method is to subtract the total of all investment costs from the total of all net profits (i.e., revenues less expenses). The time value of money is not considered.

In engineering economic analysis, the return on investment is calculated from equivalent values. Specifically, the present worth (at $t = 0$) of all investment costs is subtracted from the future worth (at $t = n$) of all net profits.

When there are only two cash flows, a single investment amount and a single payback, the two definitions of return on investment yield the same numerical value. When there are more than two cash flows, the returns on investment will be different depending on which definition is used.

21. MINIMUM ATTRACTIVE RATE OF RETURN

A company may not know what effective interest rate, i, to use in engineering economic analysis. In such a case, the company can establish a minimum level of economic performance that it would like to realize on all investments. This criterion is known as the *minimum attractive rate of return* (MARR). Unlike the effective interest rate, i, the minimum attractive rate of return is not used in numerical calculations.[14] It is used only in comparisons with the rate of return.

Once a rate of return for an investment is known, it can be compared to the minimum attractive rate of return. To be a viable alternative, the rate of return must be greater than the minimum attractive rate of return.

The advantage of using comparisons to the minimum attractive rate of return is that an effective interest rate, i, never needs to be known. The minimum attractive rate of return becomes the correct interest rate for use in present worth and equivalent uniform annual cost calculations.

22. TYPICAL ALTERNATIVE-COMPARISON PROBLEM FORMATS

With the exception of some investment and rate of return problems, the typical problem involving engineering economics will have the following characteristics.

- An interest rate will be given.
- Two or more alternatives will be competing for funding.

[14]Not everyone adheres to this rule. Some people use "minimum attractive rate of return" and "effective interest rate" interchangeably.

- Each alternative will have its own cash flows.
- It will be necessary to select the best alternative.

23. DURATIONS OF INVESTMENTS

Because they are handled differently, short-term investments and short-lived assets need to be distinguished from investments and assets that constitute an infinitely lived project. Short-term investments are easily identified: a drill press that is needed for three years or a temporary factory building that is being constructed to last five years.

Investments with perpetual cash flows are also (usually) easily identified: maintenance on a large flood control dam and revenues from a long-span toll bridge. Furthermore, some items with finite lives can expect renewal on a repeated basis.[15] For example, a major freeway with a pavement life of 20 years is unlikely to be abandoned; it will be resurfaced or replaced every 20 years.

Actually, if an investment's finite lifespan is long enough, it can be considered an infinite investment because money 50 or more years from now has little impact on current decisions. The $(P/F, 10\%, 50)$ factor, for example, is 0.0085. Therefore, one dollar at $t = 50$ has an equivalent present worth of less than one penny. Since these far-future cash flows are eclipsed by present cash flows, long-term investments can be considered finite or infinite without significant impact on the calculations.

24. CHOICE OF ALTERNATIVES: COMPARING ONE ALTERNATIVE WITH ANOTHER ALTERNATIVE

Several methods exist for selecting a superior alternative from among a group of proposals. Each method has its own merits and applications.

Present Worth Method

When two or more alternatives are capable of performing the same functions, the superior alternative will have the largest present worth. The *present worth method* is restricted to evaluating alternatives that are mutually exclusive and that have the same lives. This method is suitable for ranking the desirability of alternatives.

Example 58.12

Investment A costs \$10,000 today and pays back \$11,500 two years from now. Investment B costs \$8000 today and pays back \$4500 each year for two years. If an interest rate of 5% is used, which alternative is superior?

[15]The term *renewal* can be interpreted to mean replacement or repair.

Solution

$$\begin{aligned} P(\text{A}) &= -\$10{,}000 + (\$11{,}500)(P/F, 5\%, 2) \\ &= -\$10{,}000 + (\$11{,}500)(0.9070) \\ &= \$431 \end{aligned}$$

$$\begin{aligned} P(\text{B}) &= -\$8000 + (\$4500)(P/A, 5\%, 2) \\ &= -\$8000 + (\$4500)(1.8594) \\ &= \$367 \end{aligned}$$

Alternative A is superior and should be chosen.

Capitalized Cost Method

The present worth of a project with an infinite life is known as the *capitalized cost* or *life-cycle cost.* Capitalized cost is the amount of money at $t = 0$ needed to perpetually support the project on the earned interest only. Capitalized cost is a positive number when expenses exceed income.

In comparing two alternatives, each of which is infinitely lived, the superior alternative will have the lowest capitalized cost.

Normally, it would be difficult to work with an infinite stream of cash flows since most economics tables do not list factors for periods in excess of 100 years. However, the (A/P) discounting factor approaches the interest rate as n becomes large. Since the (P/A) and (A/P) factors are reciprocals of each other, it is possible to divide an infinite series of equal cash flows by the interest rate in order to calculate the present worth of the infinite series. This is the basis of Eq. 58.19.

$$\text{capitalized cost} = \text{initial cost} + \frac{\text{annual costs}}{i} \qquad 58.19$$

Equation 58.19 can be used when the annual costs are equal in every year. If the operating and maintenance costs occur irregularly instead of annually, or if the costs vary from year to year, it will be necessary to somehow determine a cash flow of equal annual amounts (EAA) that is equivalent to the stream of original costs.

The equal annual amount may be calculated in the usual manner by first finding the present worth of all the actual costs and then multiplying the present worth by the interest rate (the (A/P) factor for an infinite series). However, it is not even necessary to convert the present worth to an equal annual amount since Eq. 58.20 will convert the equal amount back to the present worth.

$$\begin{aligned} \text{capitalized cost} &= \text{initial cost} + \frac{\text{EAA}}{i} \\ &= \text{initial cost} + \begin{array}{l}\text{present worth}\\ \text{of all expenses}\end{array} \end{aligned} \qquad 58.20$$

Example 58.13

What is the capitalized cost of a public works project that will cost \$25,000,000 now and will require \$2,000,000 in maintenance annually? The effective annual interest rate is 12%.

Solution

Worked in millions of dollars, from Eq. 58.19, the capitalized cost is

$$\begin{aligned} \text{capitalized cost} &= 25 + (2)(P/A, 12\%, \infty) \\ &= 25 + \frac{2}{0.12} \\ &= 41.67 \end{aligned}$$

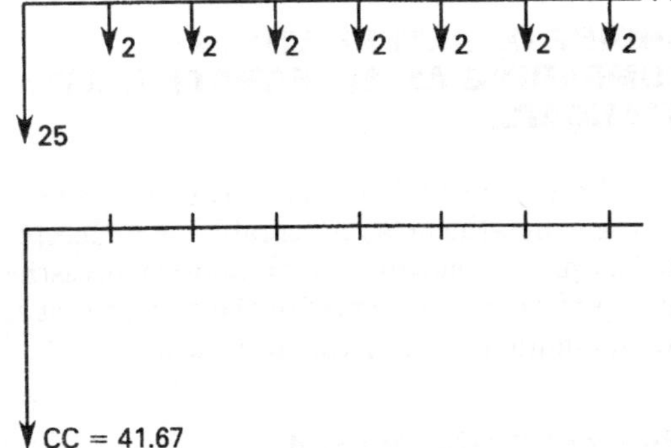

Annual Cost Method

Alternatives that accomplish the same purpose but that have unequal lives must be compared by the *annual cost method.*[16] The annual cost method assumes that each alternative will be replaced by an identical twin at the end of its useful life (infinite renewal). This method, which may also be used to rank alternatives according to their desirability, is also called the *annual return method* or *capital recovery method.*

Restrictions are that the alternatives must be mutually exclusive and repeatedly renewed up to the duration of the longest-lived alternative. The calculated annual cost is known as the *equivalent uniform annual cost* (EUAC) or just *equivalent annual cost.* Cost is a positive number when expenses exceed income.

Example 58.14

Which of the following alternatives is superior over a 30-year period if the interest rate is 7%?

	alternative A	alternative B
type	brick	wood
life	30 years	10 years
initial cost	\$1800	\$450
maintenance	\$5/year	\$20/year

[16]Of course, the annual cost method can be used to determine the superiority of assets with identical lives as well.

Professional Practice

Solution

$$\begin{aligned}\text{EUAC(A)} &= (\$1800)(A/P, 7\%, 30) + \$5 \\ &= (\$1800)(0.0806) + \$5 \\ &= \$150 \\ \text{EUAC(B)} &= (\$450)(A/P, 7\%, 10) + \$20 \\ &= (\$450)(0.1424) + \$20 \\ &= \$84\end{aligned}$$

Alternative B is superior since its annual cost of operation is the lowest. It is assumed that three wood facilities, each with a life of 10 years and a cost of $450, will be built to span the 30-year period.

25. CHOICE OF ALTERNATIVES: COMPARING AN ALTERNATIVE WITH A STANDARD

With specific economic performance criteria, it is possible to qualify an investment as acceptable or unacceptable without having to compare it with another investment. Two such performance criteria are the benefit-cost ratio and the minimum attractive rate of return.

Benefit-Cost Ratio Method

The *benefit-cost ratio* method is often used in municipal project evaluations where benefits and costs accrue to different segments of the community. With this method, the present worth of all benefits (irrespective of the beneficiaries) is divided by the present worth of all costs. The project is considered acceptable if the ratio equals or exceeds 1.0, that is, if $B/C \geq 1.0$.

When the benefit-cost ratio method is used, disbursements by the initiators or sponsors are *costs*. Disbursements by the users of the project are known as *disbenefits*. It is often difficult to determine whether a cash flow is a cost or a disbenefit (whether to place it in the numerator or denominator of the benefit-cost ratio calculation).

Regardless of where the cash flow is placed, an acceptable project will always have a benefit-cost ratio greater than or equal to 1.0, although the actual numerical result will depend on the placement. For this reason, the benefit-cost ratio method should not be used to rank competing projects.

The benefit-cost ratio method of comparing alternatives is used extensively in transportation engineering where the ratio is often (but not necessarily) written in terms of annual benefits and annual costs instead of present worths. Another characteristic of highway benefit-cost ratios is that the route (road, highway, etc.) is usually already in place and that various alternative upgrades are being considered. There will be existing benefits and costs associated with the current route. Therefore, the *change* (usually an increase) in benefits and costs is used to calculate the benefit-cost ratio.[17]

$$B/C = \frac{\Delta \begin{array}{c}\text{user}\\\text{benefits}\end{array}}{\Delta \begin{array}{c}\text{investment}\\\text{cost}\end{array} + \Delta\,\text{maintenance} - \Delta \begin{array}{c}\text{residual}\\\text{value}\end{array}} \quad \text{58.21}$$

The change in *residual value* (*terminal value*) appears in the denominator as a negative item. An increase in the residual value would decrease the denominator.

Example 58.15

By building a bridge over a ravine, a state department of transportation can shorten the time it takes to drive through a mountainous area. Estimates of costs and benefits (due to decreased travel time, fewer accidents, reduced gas usage, etc.) have been prepared. Should the bridge be built? Use the benefit-cost ratio method of comparison.

	millions
initial cost	40
capitalized cost of perpetual annual maintenance	12
capitalized value of annual user benefits	49
residual value	0

Solution

If Eq. 58.21 is used, the benefit-cost ratio is

$$B/C = \frac{49}{40 + 12 - 0} = 0.942$$

Since the benefit-cost ratio is less than 1.00, the bridge should not be built.

If the maintenance costs are placed in the numerator (per Ftn. 17), the benefit-cost ratio value will be different, but the conclusion will not change.

$$B/C_{\text{alternate method}} = \frac{49 - 12}{40} = 0.925$$

Rate of Return Method

The minimum attractive rate of return (MARR) has already been introduced as a standard of performance against which an investment's actual *rate of return*

[17]This discussion of highway benefit-cost ratios is not meant to imply that everyone agrees with Eq. 58.21. In *Economic Analysis for Highways* (International Textbook Company, Scranton, PA, 1969), author Robley Winfrey took a strong stand on one aspect of the benefits versus disbenefits issue: highway maintenance. According to Winfrey, regular highway maintenance costs should be placed in the numerator as a subtraction from the user benefits. Some have called this mandate the *Winfrey method.*

Professional Practice

(ROR) is compared. If the rate of return is equal to or exceeds the minimum attractive rate of return, the investment is qualified. This is the basis for the *rate of return method* of alternative selection.

Finding the rate of return can be a long, iterative process. Usually, the actual numerical value of rate of return is not needed; it is sufficient to know whether or not the rate of return exceeds the minimum attractive rate of return. This *comparative analysis* can be accomplished without calculating the rate of return simply by finding the present worth of the investment using the minimum attractive rate of return as the effective interest rate (i.e., $i = \text{MARR}$). If the present worth is zero or positive, the investment is qualified. If the present worth is negative, the rate of return is less than the minimum attractive rate of return.

26. RANKING MUTUALLY EXCLUSIVE MULTIPLE PROJECTS

Ranking of multiple investment alternatives is required when there is sufficient funding for more than one investment. Since the best investments should be selected first, it is necessary to place all investments into an ordered list.

Ranking is relatively easy if the present worths, future worths, capitalized costs, or equivalent uniform annual costs have been calculated for all the investments. The highest ranked investment will be the one with the largest present or future worth, or the smallest capitalized or annual cost. Present worth, future worth, capitalized cost, and equivalent uniform annual cost can all be used to rank multiple investment alternatives.

However, neither rates of return nor benefit-cost ratios should be used to rank multiple investment alternatives. Specifically, if two alternatives both have rates of return exceeding the minimum acceptable rate of return, it is not sufficient to select the alternative with the highest rate of return.

An *incremental analysis*, also known as a *rate of return on added investment study*, should be performed if rate of return is used to select between investments. An incremental analysis starts by ranking the alternatives in order of increasing initial investment. Then, the cash flows for the investment with the lower initial cost are subtracted from the cash flows for the higher-priced alternative on a year-by-year basis. This produces, in effect, a third alternative representing the costs and benefits of the added investment. The added expense of the higher-priced investment is not warranted unless the rate of return of this third alternative exceeds the minimum attractive rate of return as well. The choice criterion is to select the alternative with the higher initial investment if the incremental rate of return exceeds the minimum attractive rate of return.

An incremental analysis is also required if ranking is to be done by the benefit-cost ratio method. The incremental analysis is accomplished by calculating the ratio of differences in benefits to differences in costs for each possible pair of alternatives. If the ratio exceeds 1.0, alternative 2 is superior to alternative 1. Otherwise, alternative 1 is superior.[18]

$$\frac{B_2 - B_1}{C_2 - C_1} \geq 1 \quad \text{[alternative 2 superior]} \qquad 58.22$$

27. ALTERNATIVES WITH DIFFERENT LIVES

Comparison of two alternatives is relatively simple when both alternatives have the same life. For example, a problem might be stated: "Which would you rather have: car A with a life of three years, or car B with a life of five years?"

However, care must be taken to understand what is going on when the two alternatives have different lives. If car A has a life of three years and car B has a life of five years, what happens at $t = 3$ if the five-year car is chosen? If a car is needed for five years, what happens at $t = 3$ if the three-year car is chosen?

In this type of situation, it is necessary to distinguish between the length of the need (the *analysis horizon*) and the lives of the alternatives or assets intended to meet that need. The lives do not have to be the same as the horizon.

Finite Horizon with Incomplete Asset Lives

If an asset with a five-year life is chosen for a three-year need, the disposition of the asset at $t = 3$ must be known in order to evaluate the alternative. If the asset is sold at $t = 3$, the salvage value is entered into the analysis (at $t = 3$) and the alternative is evaluated as a three-year investment. The fact that the asset is sold when it has some useful life remaining does not affect the analysis horizon.

Similarly, if a three-year asset is chosen for a five-year need, something about how the need is satisfied during the last two years must be known. Perhaps a rental asset will be used. Or, perhaps the function will be "farmed out" to an outside firm. In any case, the costs of satisfying the need during the last two years enter the analysis, and the alternative is evaluated as a five-year investment.

If both alternatives are "converted" to the same life, any of the alternative selection criteria (present worth method, annual cost method, etc.) can be used to determine which alternative is superior.

Finite Horizon with Integer Multiple Asset Lives

It is common to have a long-term horizon (need) that must be met with short-lived assets. In special instances, the horizon will be an integer number of asset lives. For example, a company may be making a

[18]It goes without saying that the benefit-cost ratios for all investment alternatives by themselves must also be equal to or greater than 1.0.

12-year transportation plan and may be evaluating two cars: one with a three-year life, and another with a four-year life.

In this example, four of the first car or three of the second car are needed to reach the end of the 12-year horizon.

If the horizon is an integer number of asset lives, any of the alternative selection criteria can be used to determine which is superior. If the present worth method is used, all alternatives must be evaluated over the entire horizon. (In this example, the present worth of 12 years of car purchases and use must be determined for both alternatives.)

If the equivalent uniform annual cost method is used, it may be possible to base the calculation of annual cost on one lifespan of each alternative only. It may not be necessary to incorporate all of the cash flows into the analysis. (In the running example, the annual cost over three years would be determined for the first car; the annual cost over four years would be determined for the second car.) This simplification is justified if the subsequent asset replacements (renewals) have the same cost and cash flow structure as the original asset. This assumption is typically made implicitly when the annual cost method of comparison is used.

Infinite Horizon

If the need horizon is infinite, it is not necessary to impose the restriction that asset lives of alternatives be integer multiples of the horizon. The superior alternative will be replaced (renewed) whenever it is necessary to do so, forever.

Infinite horizon problems are almost always solved with either the annual cost or capitalized cost method. It is common to (implicitly) assume that the cost and cash flow structure of the asset replacements (renewals) are the same as the original asset.

28. OPPORTUNITY COSTS

An *opportunity cost* is an imaginary cost representing what will not be received if a particular strategy is rejected. It is what you will lose if you do or do not do something. As an example, consider a growing company with an existing operational computer system. If the company trades in its existing computer as part of an upgrade plan, it will receive a *trade-in allowance.* (In other problems, a *salvage value* may be involved.)

If one of the alternatives being evaluated is not to upgrade the computer system at all, the trade-in allowance (or, salvage value in other problems) will not be realized. The amount of the trade-in allowance is an opportunity cost that must be included in the problem analysis.

Similarly, if one of the alternatives being evaluated is to wait one year before upgrading the computer, the *difference in trade-in allowances* is an opportunity cost that must be included in the problem analysis.

29. REPLACEMENT STUDIES

An investigation into the retirement of an existing process or piece of equipment is known as a *replacement study.* Replacement studies are similar in most respects to other alternative comparison problems: An interest rate is given, two alternatives exist, and one of the previously mentioned methods of comparing alternatives is used to choose the superior alternative. Usually, the annual cost method is used on a year-by-year basis.

In replacement studies, the existing process or piece of equipment is known as the *defender.* The new process or piece of equipment being considered for purchase is known as the *challenger.*

30. TREATMENT OF SALVAGE VALUE IN REPLACEMENT STUDIES

Since most defenders still have some market value when they are retired, the problem of what to do with the salvage arises. It seems logical to use the salvage value of the defender to reduce the initial purchase cost of the challenger. This is consistent with what would actually happen if the defender were to be retired.

By convention, however, the defender's salvage value is subtracted from the defender's present value. This does not seem logical, but it is done to keep all costs and benefits related to the defender with the defender. In this case, the salvage value is treated as an opportunity cost that would be incurred if the defender is not retired.

If the defender and the challenger have the same lives and a present worth study is used to choose the superior alternative, the placement of the salvage value will have no effect on the net difference between present worths for the challenger and defender. Although the values of the two present worths will be different depending on the placement, the difference in present worths will be the same.

If the defender and the challenger have different lives, an annual cost comparison must be made. Since the salvage value would be "spread over" a different number of years depending on its placement, it is important to abide by the conventions listed in this section.

There are a number of ways to handle salvage value in retirement studies. The best way is to calculate the cost of keeping the defender one more year. In addition to the usual operating and maintenance costs, that cost includes an opportunity interest cost incurred by not selling the defender, and also a drop in the salvage value if the defender is kept for one additional year. Specifically,

$$\begin{aligned}\text{EUAC (defender)} = {} & \text{next year's maintenance costs} \\ & + i(\text{current salvage value}) \\ & + \text{current salvage} \\ & - \text{next year's salvage}\end{aligned} \qquad 58.23$$

Professional Practice

It is important in retirement studies not to double count the salvage value. That is, it would be incorrect to add the salvage value to the defender and at the same time subtract it from the challenger.

Equation 58.23 contains the difference in salvage value between two consecutive years. This calculation shows that the defender/challenger decision must be made on a year-by-year basis. One application of Eq. 58.23 will not usually answer the question of whether the defender should remain in service indefinitely. The calculation must be repeatedly made as long as there is a drop in salvage value from one year to the next.

31. ECONOMIC LIFE: RETIREMENT AT MINIMUM COST

As an asset grows older, its operating and maintenance costs typically increase. Eventually, the cost to keep the asset in operation becomes prohibitive, and the asset is retired or replaced. However, it is not always obvious when an asset should be retired or replaced.

As the asset's maintenance cost is increasing each year, the amortized cost of its initial purchase is decreasing. It is the sum of these two costs that should be evaluated to determine the point at which the asset should be retired or replaced. Since an asset's initial purchase price is likely to be high, the amortized cost will be the controlling factor in those years when the maintenance costs are low. Therefore, the EUAC of the asset will decrease in the initial part of its life.

However, as the asset grows older, the change in its amortized cost decreases while maintenance cost increases. Eventually, the sum of the two costs reaches a minimum and then starts to increase. The age of the asset at the minimum cost point is known as the *economic life* of the asset. The economic life generally is less than the length of need and the technological lifetime of the asset, as shown in Fig. 58.7.

Figure 58.7 *EUAC Versus Age at Retirement*

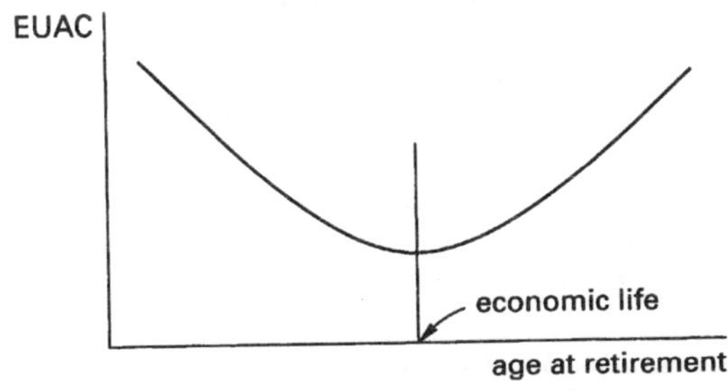

The determination of an asset's economic life is illustrated by Ex. 58.16.

Example 58.16

Buses in a municipal transit system have the characteristics listed. In order to minimize its annual operating expenses, when should the city replace its buses if money can be borrowed at 8%?

initial cost of bus: $120,000

year	maintenance cost	salvage value
1	35,000	60,000
2	38,000	55,000
3	43,000	45,000
4	50,000	25,000
5	65,000	15,000

Solution

The annual maintenance is different each year. Each maintenance cost must be spread over the life of the bus. This is done by first finding the present worth and then amortizing the maintenance costs. If a bus is kept for one year and then sold, the annual cost will be

$$\begin{aligned}\text{EUAC}(1) &= (\$120{,}000)(A/P, 8\%, 1) \\ &\quad + (\$35{,}000)(A/F, 8\%, 1) \\ &\quad - (\$60{,}000)(A/F, 8\%, 1) \\ &= (\$120{,}000)(1.0800) + (\$35{,}000)(1.000) \\ &\quad - (\$60{,}000)(1.000) \\ &= \$104{,}600\end{aligned}$$

If a bus is kept for two years and then sold, the annual cost will be

$$\begin{aligned}\text{EUAC}(2) &= \big(\$120{,}000 + (\$35{,}000)(P/F, 8\%, 1)\big) \\ &\quad \times (A/P, 8\%, 2) \\ &\quad + (\$38{,}000 - \$55{,}000)(A/F, 8\%, 2) \\ &= \big(\$120{,}000 + (\$35{,}000)(0.9259)\big)(0.5608) \\ &\quad + (\$38{,}000 - \$55{,}000)(0.4808) \\ &= \$77{,}296\end{aligned}$$

If a bus is kept for three years and then sold, the annual cost will be

$$\begin{aligned}\text{EUAC}(3) &= \big(\$120{,}000 + (\$35{,}000)(P/F, 8\%, 1) \\ &\quad + (\$38{,}000)(P/F, 8\%, 2)\big)(A/P, 8\%, 3) \\ &\quad + (\$43{,}000 - \$45{,}000)(A/F, 8\%, 3) \\ &= \big(\$120{,}000 + (\$35{,}000)(0.9259) \\ &\quad + (\$38{,}000)(0.8573)\big)(0.3880) \\ &\quad - (\$2000)(0.3080) \\ &= \$71{,}158\end{aligned}$$

This process is continued until the annual cost begins to increase. In this example, EUAC(4) is $71,700. Therefore, the buses should be retired after three years.

Professional Practice

32. LIFE-CYCLE COST

The *life-cycle cost* of an alternative is the equivalent value (at $t = 0$) of the alternative's cash flow over the alternative's lifespan. Since the present worth is evaluated using an effective interest rate of i (which would be the interest rate used for all engineering economic analyses), the life-cycle cost is the same as the alternative's present worth. If the alternative has an infinite horizon, the life-cycle cost and capitalized cost will be identical.

33. CAPITALIZED ASSETS VERSUS EXPENSES

High expenses reduce profit, which in turn reduces income tax. It seems logical to label each and every expenditure, even an asset purchase, as an expense. As an alternative to this *expensing the asset*, it may be decided to capitalize the asset. *Capitalizing the asset* means that the cost of the asset is divided into equal or unequal parts, and only one of these parts is taken as an expense each year. Expensing is clearly the more desirable alternative, since the after-tax profit is increased early in the asset's life.

There are long-standing accounting conventions as to what can be expensed and what must be capitalized.[19] Some companies capitalize everything—regardless of cost—with expected lifetimes greater than one year. Most companies, however, expense items whose purchase costs are below a cutoff value. A cutoff value in the range of $250–500, depending on the size of the company, is chosen as the maximum purchase cost of an expensed asset. Assets costing more than this are capitalized.

It is not necessary for a large corporation to keep track of every lamp, desk, and chair for which the purchase price is greater than the cutoff value. Such assets, all of which have the same lives and have been purchased in the same year, can be placed into groups or *asset classes*. A group cost, equal to the sum total of the purchase costs of all items in the group, is capitalized as though the group was an identifiable and distinct asset itself.

34. PURPOSE OF DEPRECIATION

Depreciation is an *artificial expense* that spreads the purchase price of an asset or other property over a number of years.[20] Depreciating an asset is an example of capitalization, as previously defined. The inclusion of depreciation in engineering economic analysis problems will increase the after-tax present worth (profitability) of an asset. The larger the depreciation, the greater will be the profitability. Therefore, individuals and companies eligible to utilize depreciation want to maximize and accelerate the depreciation available to them.

Although the entire property purchase price is eventually recognized as an expense, the net recovery from the expense stream never equals the original cost of the asset. That is, depreciation cannot realistically be thought of as a fund (an annuity or sinking fund) that accumulates capital to purchase a replacement at the end of the asset's life. The primary reason for this is that the depreciation expense is reduced significantly by the impact of income taxes, as will be seen in later sections.

35. DEPRECIATION BASIS OF AN ASSET

The *depreciation basis* of an asset is the part of the asset's purchase price that is spread over the *depreciation period*, also known as the *service life*.[21] Usually, the depreciation basis and the purchase price are not the same.

A common depreciation basis is the difference between the purchase price and the expected salvage value at the end of the depreciation period. That is,

$$\text{depreciation basis} = C - S_n \quad 58.24$$

There are several methods of calculating the year-by-year depreciation of an asset. Equation 58.24 is not universally compatible with all depreciation methods. Some methods do not consider the salvage value. This is known as an *unadjusted basis*. When the depreciation method is known, the depreciation basis can be rigorously defined.[22]

36. DEPRECIATION METHODS

Generally, tax regulations do not allow the cost of an asset to be treated as a deductible expense in the year of purchase. Rather, portions of the depreciation basis must be allocated to each of the n years of the asset's depreciation period. The amount that is allocated each year is called the *depreciation*.

[19]For example, purchased vehicles must be capitalized; payments for leased vehicles can be expensed. Repainting a building with paint that will last five years is an expense, but the replacement cost of a leaking roof must be capitalized.

[20]In the United States, the tax regulations of Internal Revenue Service (IRS) allow depreciation on almost all forms of *business property* except land. The following types of property are distinguished: *real* (e.g., buildings used for business), *residential* (e.g., buildings used as rental property), and *personal* (e.g., equipment used for business). Personal property does *not* include items for personal use (such as a personal residence), despite its name. *Tangible personal property* is distinguished from *intangible property* (goodwill, copyrights, patents, trademarks, franchises, agreements not to compete, etc.).

[21]The *depreciation period* is selected to be as short as possible within recognized limits. This depreciation will not normally coincide with the *economic life* or *useful life* of an asset. For example, a car may be capitalized over a depreciation period of three years. It may become uneconomical to maintain and use at the end of an economic life of nine years. However, the car may be capable of operation over a useful life of 25 years.

[22]For example, with the Accelerated Cost Recovery System (ACRS) the *depreciation basis* is the total purchase cost, regardless of the expected salvage value. With declining balance methods, the depreciation basis is the purchase cost less any previously taken depreciation.

Various methods exist for calculating an asset's depreciation each year.[23] Although the depreciation calculations may be considered independently (for the purpose of determining book value or as an academic exercise), it is important to recognize that depreciation has no effect on engineering economic analyses unless income taxes are also considered.

Straight Line Method

With the *straight line method*, depreciation is the same each year. The depreciation basis ($C - S_n$) is allocated uniformly to all of the n years in the depreciation period. Each year, the depreciation will be

$$D = \frac{C - S_n}{n} \qquad 58.25$$

Constant Percentage Method

The *constant percentage method*[24] is similar to the straight line method in that the depreciation is the same each year. If the fraction of the basis used as depreciation is $1/n$, there is no difference between the constant percentage and straight line methods. The two methods differ only in what information is available. (With the straight line method, the life is known. With the constant percentage method, the depreciation fraction is known.)

Each year, the depreciation will be

$$\begin{aligned} D &= (\text{depreciation fraction})(\text{depreciation basis}) \\ &= (\text{depreciation fraction})(C - S_n) \end{aligned} \qquad 58.26$$

Sum-of-the-Years' Digits Method

In *sum-of-the-years' digits* (SOYD) depreciation, the digits from 1 to n inclusive are summed. The total, T, can also be calculated from

$$T = \tfrac{1}{2}n(n+1) \qquad 58.27$$

The depreciation in year j can be found from Eq. 58.28. Notice that the depreciation in year j, D_j, decreases by a constant amount each year.

$$D_j = \frac{(C - S_n)(n - j + 1)}{T} \qquad 58.28$$

Double Declining Balance Method[25]

Double declining balance[26] (DDB) depreciation is independent of salvage value. Furthermore, the book value never stops decreasing, although the depreciation decreases in magnitude. Usually, any book value in excess of the salvage value is written off in the last year of the asset's depreciation period. Unlike any of the other depreciation methods, double declining balance depends on accumulated depreciation.

$$D_{\text{first year}} = \frac{2C}{n} \qquad 58.29$$

$$D_j = \frac{2\left(C - \sum\limits_{m=1}^{j-1} D_m\right)}{n} \qquad 58.30$$

Calculating the depreciation in the middle of an asset's life appears particularly difficult with double declining balance, because all previous years' depreciation amounts seem to be required. It appears that the depreciation in the sixth year, for example, cannot be calculated unless the values of depreciation for the first five years are calculated. However, this is not true.

Depreciation in the middle of an asset's life can be found from the following equations. (d is known as the *depreciation rate*.)

$$d = \frac{2}{n} \qquad 58.31$$

$$D_j = dC(1 - d)^{j-1} \qquad 58.32$$

Statutory Depreciation Systems

In the United States, property placed into service in 1981 and thereafter must use the *Accelerated Cost Recovery System* (ACRS), and after 1986, the *Modified Accelerated Cost Recovery System* (MACRS) or other statutory method. Other methods (straight line, declining balance, etc.) cannot be used except in special cases.

Property placed into service in 1980 or before must continue to be depreciated according to the method originally chosen (e.g., straight line, declining balance, or sum-of-the-years' digits). ACRS and MACRS cannot be used.

Under ACRS and MACRS, the cost recovery amount in the jth year of an asset's cost recovery period is

---

[23] This discussion gives the impression that any form of depreciation may be chosen regardless of the nature and circumstances of the purchase. In reality, the IRS tax regulations place restrictions on the higher-rate (accelerated) methods, such as declining balance and sum-of-the-years' digits methods. Furthermore, the *Economic Recovery Act of 1981* and the *Tax Reform Act of 1986* substantially changed the laws relating to personal and corporate income taxes.

[24] The *constant percentage method* should not be confused with the declining balance method, which used to be known as the *fixed percentage on diminishing balance method*.

[25] In the past, the *declining balance method* has also been known as the *fixed percentage of book value* and *fixed percentage on diminishing balance method*.

[26] Double declining balance depreciation is a particular form of *declining balance depreciation*, as defined by the IRS tax regulations. Declining balance depreciation includes 125% declining balance and 150% declining balance depreciations that can be calculated by substituting 1.25 and 1.50, respectively, for the 2 in Eq. 58.29.

Professional Practice

calculated by multiplying the initial cost by a factor. (See Table 58.4.)

$$D_j = C \times \text{factor} \tag{58.33}$$

Table 58.4 *Representative ACRS and MACRS Depreciation Factors**

	recovery period, n		
	3 yr	5 yr	10 yr
year j	ACRS/MACRS	ACRS/MACRS	ACRS/MACRS
1	0.25/0.3333	0.15/0.2000	0.08/0.100
2	0.38/0.4445	0.22/0.3200	0.14/0.180
3	0.37/0.1481	0.21/0.1920	0.12/0.1440
4	0/0.0741	0.21/0.1152	0.10/0.1152
5		0.21/0.1152	0.10/0.0922
6		0/0.0576	0.10/0.0737
7			0.09/0.0655
8			0.09/0.0655
9			0.09/0.0656
10			0.09/0.0655
11			0/0.0328

*MACRS values are for the "half-year" convention. This table gives typical values only. Because these factors are subject to continuing revision, they should not be used without consulting an accounting professional.

The initial cost used is not reduced by the asset's salvage value for ACRS and MACRS calculations. The factor used depends on the asset's cost recovery period. Such factors are subject to continuing legislation changes. Current tax publications should be consulted before using this method.

Production or Service Output Method

If an asset has been purchased for a specific task and that task is associated with a specific lifetime amount of output or production, the depreciation may be calculated by the fraction of total production produced during the year. The depreciation is not expected to be the same each year.

$$D_j = (C - S_n)\left(\frac{\text{actual output in year } j}{\text{estimated lifetime output}}\right) \tag{58.34}$$

Sinking Fund Method

The *sinking fund method* is seldom used in industry because the initial depreciation is low. The formula for sinking fund depreciation (which increases each year) is

$$D_j = (C - S_n)(A/F, i\%, n)(F/P, i\%, j-1) \tag{58.35}$$

Disfavored Methods

Three other depreciation methods are mentioned here, not because they are currently accepted or in widespread use, but because they are still occasionally encountered in the literature.[27]

The *sinking fund plus interest on first cost* depreciation method, like the following two methods, is an attempt to include the *opportunity interest cost* on the purchase price with the depreciation. That is, the purchasing company not only incurs an annual expense from the drop in book value, but it also loses the interest on the purchase price. The formula for this method is

$$D = (C - S_n)(A/F, i\%, n) + Ci \tag{58.36}$$

The *straight line plus interest on first cost* method is similar. Its formula is

$$D = \left(\frac{1}{n}\right)(C - S_n) + Ci \tag{58.37}$$

The *straight line plus average interest method* assumes that the opportunity interest cost should be based on the book value only, not on the full purchase price. Because the book value changes each year, an average value is used. The depreciation formula is

$$D = \left(\frac{C - S_n}{n}\right)\left(1 + \frac{i(n+1)}{2}\right) + iS_n \tag{58.38}$$

Example 58.17

An asset is purchased for \$9000. Its estimated economic life is ten years, after which it will be sold for \$200. Find the depreciation in the first three years using the straight line (SL), double declining balance (DDB), and sum-of-the-years' digits (SOYD) depreciation methods.

Solution

SL:

$$D = \frac{C - S_n}{n} = \frac{\$9000 - \$200}{10} = \$880 \text{ each year}$$

DDB:

$$D_1 = \frac{2C}{n} = \frac{(2)(\$9000)}{10} = \$1800 \text{ in year 1}$$

$$D_j = \frac{2\left(C - \sum_{m=1}^{j-1} D_m\right)}{n}$$

$$D_2 = \frac{(2)(\$9000 - \$1800)}{10} = \$1440 \text{ in year 2}$$

$$D_3 = \frac{(2)(\$9000 - \$3240)}{10} = \$1152 \text{ in year 3}$$

[27]The three disfavored depreciation methods should not be used in the usual manner (e.g., in conjunction with the income tax rate). These methods are attempts to calculate a more accurate annual cost of an alternative, but sometimes they give misleading answers. Their use is not recommended, and they are included in this chapter only for the sake of completeness.

$$\text{SOYD:}\quad T = \tfrac{1}{2}n(n+1) = \left(\tfrac{1}{2}\right)(10)(11) = 55$$

$$D_1 = \left(\tfrac{10}{55}\right)(\$9000 - \$200) = \$1600 \text{ in year 1}$$

$$D_2 = \left(\tfrac{9}{55}\right)(\$8800) = \$1440 \text{ in year 2}$$

$$D_3 = \left(\tfrac{8}{55}\right)(\$8800) = \$1280 \text{ in year 3}$$

37. ACCELERATED DEPRECIATION METHODS

An *accelerated depreciation method* is one that calculates a depreciation amount greater than a straight line amount. Double declining balance and sum-of-the-years' digits methods are accelerated methods. The ACRS and MACRS methods are explicitly accelerated methods. Straight line and sinking fund methods are not accelerated methods.

Use of an accelerated depreciation method may result in unexpected tax consequences when the depreciated asset or property is disposed of. Professional tax advice should be obtained in this area.

38. BOOK VALUE

The difference between original purchase price and accumulated depreciation is known as *book value.*[28] At the end of each year, the book value (which is initially equal to the purchase price) is reduced by the depreciation in that year.

It is important to properly synchronize depreciation calculations. It is difficult to answer the question, "What is the book value in the fifth year?" unless the timing of the book value change is mutually agreed upon. It is better to be specific about an inquiry by identifying when the book value change occurs. For example, the following question is unambiguous: "What is the book value at the end of year 5, after subtracting depreciation in the fifth year?" or "What is the book value after five years?"

Unfortunately, this type of care is seldom taken in book value inquiries, and it is up to the respondent to exercise reasonable care in distinguishing between beginning-of-year book value and end-of-year book value. To be consistent, the book value equations in this chapter have been written in such a way that the year subscript, j, has the same meaning in book value and depreciation calculations. That is, BV_5 means the book value at the end of the fifth year, after five years of depreciation, including D_5, have been subtracted from the original purchase price.

[28]The balance sheet of a corporation usually has two asset accounts: the *equipment account* and the *accumulated depreciation account.* There is no book value account on this financial statement, other than the implicit value obtained from subtracting the accumulated depreciation account from the equipment account. The book values of various assets, as well as their original purchase cost, date of purchase, salvage value, and so on, and accumulated depreciation appear on detail sheets or other peripheral records for each asset.

There can be a great difference between the book value of an asset and the *market value* of that asset. There is no legal requirement for the two values to coincide, and no intent for book value to be a reasonable measure of market value.[29] Therefore, it is apparent that book value is merely an accounting convention with little practical use. Even when a depreciated asset is disposed of, the book value is used to determine the consequences of disposal, not the price the asset should bring at sale.

The calculation of book value is relatively easy, even for the case of the declining balance depreciation method.

For the straight line depreciation method, the book value at the end of the jth year, after the jth depreciation deduction has been made, is

$$\text{BV}_j = C - \frac{j(C - S_n)}{n} = C - jD \qquad 58.39$$

For the sum-of-the-years' digits method, the book value is

$$\text{BV}_j = (C - S_n)\left(1 - \frac{j(2n+1-j)}{n(n+1)}\right) + S_n \qquad 58.40$$

For the declining balance method, including double declining balance (see Ftn. 26), the book value is

$$\text{BV}_j = C(1-d)^j \qquad 58.41$$

For the sinking fund method, the book value is calculated directly as

$$\text{BV}_j = C - (C - S_n)(A/F, i\%, n)(F/A, i\%, j) \qquad 58.42$$

Of course, the book value at the end of year j can always be calculated for any method by successive subtractions (i.e., subtraction of the accumulated depreciation), as Eq. 58.43 illustrates.

$$\text{BV}_j = C - \sum_{m=1}^{j} D_m \qquad 58.43$$

Figure 58.8 illustrates the book value of a hypothetical asset depreciated using several depreciation methods. Notice that the double declining balance method initially produces the fastest write-off, while the sinking fund method produces the slowest write-off. Also, the book value does not automatically equal the salvage value at the end of an asset's depreciation period with the double declining balance method.[30]

[29]Common examples of assets with great divergences of book and market values are buildings (rental houses, apartment complexes, factories, etc.) and company luxury automobiles (Porsches, Mercedes, etc.) during periods of inflation. Book values decrease, but actual values increase.

[30]This means that the straight line method of depreciation may result in a lower book value at some point in the depreciation period than if double declining balance is used. A *cut-over* from double declining balance to straight line may be permitted in certain cases. Finding the *cut-over point,* however, is usually done by comparing book values determined by both methods. The analytical method is complicated.

Figure 58.8 Book Value with Different Depreciation Methods

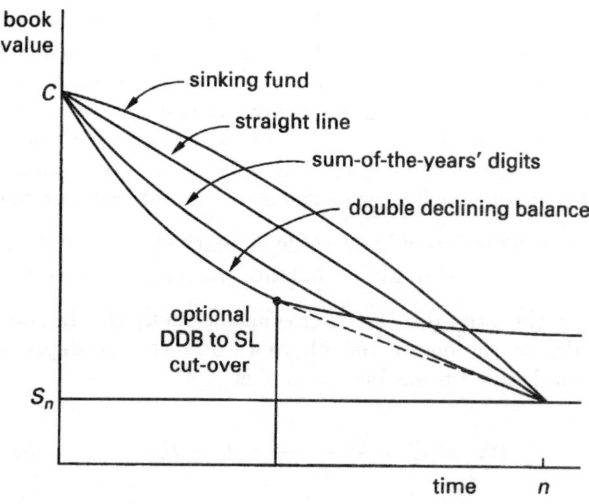

Example 58.18

For the asset described in Ex. 58.17, calculate the book value at the end of the first three years if sum-of-the-years' digits depreciation is used. The book value at the beginning of year 1 is $9000.

Solution

From Eq. 58.43,

$$\begin{aligned} BV_1 &= \$9000 - \$1600 = \$7400 \\ BV_2 &= \$7400 - \$1440 = \$5960 \\ BV_3 &= \$5960 - \$1280 = \$4680 \end{aligned}$$

39. AMORTIZATION

Amortization and depreciation are similar in that they both divide up the cost basis or value of an asset. In fact, in certain cases, the term "amortization" may be used in place of the term "depreciation." However, depreciation is a specific form of amortization.

Amortization spreads the cost basis or value of an asset over some base. The base can be time, units of production, number of customers, and so on. The asset can be tangible (e.g., a delivery truck or building) or intangible (e.g., goodwill or a patent).

If the asset is tangible, if the base is time, and if the length of time is consistent with accounting standards and taxation guidelines, then the term "depreciation" is appropriate. However, if the asset is intangible, if the base is some other variable, or if some length of time other than the customary period is used, then the term "amortization" is more appropriate.[31]

[31]From time to time, the U.S. Congress has allowed certain types of facilities (e.g., emergency, grain storage, and pollution control) to be written off more rapidly than would otherwise be permitted in order to encourage investment in such facilities. The term "amortization" has been used with such write-off periods.

Example 58.19

A company purchases complete and exclusive patent rights to an invention for $1,200,000. It is estimated that once commercially produced, the invention will have a specific but limited market of 1200 units. For the purpose of allocating the patent right cost to production cost, what is the amortization rate in dollars per unit?

Solution

The patent should be amortized at the rate of

$$\frac{\$1{,}200{,}000}{1200 \text{ units}} = \$1000 \text{ per unit}$$

40. DEPLETION

Depletion is another artificial deductible operating expense, designed to compensate mining organizations for decreasing mineral reserves. Since original and remaining quantities of minerals are seldom known accurately, the *depletion allowance* is calculated as a fixed percentage of the organization's gross income. These percentages are usually in the 10–20% range and apply to such mineral deposits as oil, natural gas, coal, uranium, and most metal ores.

41. BASIC INCOME TAX CONSIDERATIONS

The issue of income taxes is often overlooked in academic engineering economic analysis exercises. Such a position is justifiable when an organization (e.g., a nonprofit school, a church, or the government) pays no income taxes. However, if an individual or organization is subject to income taxes, the income taxes must be included in an economic analysis of investment alternatives.

Assume that an organization pays a fraction, f, of its profits to the federal government as income taxes. If the organization also pays a fraction s of its profits as state income taxes and if state taxes paid are recognized by the federal government as tax-deductible expenses, then the composite tax rate is

$$t = s + f - sf \qquad 58.44$$

The basic principles used to incorporate taxation into engineering economic analyses are the following.

- Initial purchase expenditures are unaffected by income taxes.
- Salvage revenues are unaffected by income taxes.
- Deductible expenses, such as operating costs, maintenance costs, and interest payments, are reduced by the fraction t (i.e., multiplied by the quantity $(1 - t)$).

Professional Practice

- Revenues are reduced by the fraction t (i.e., multiplied by the quantity $(1 - t)$).
- Since tax regulations allow the depreciation in any year to be handled as if it were an actual operating expense, and since operating expenses are deductible from the income base prior to taxation, the after-tax profits will be increased. If D is the depreciation, the net result to the after-tax cash flow will be the addition of tD. Depreciation is multiplied by t and added to the appropriate year's cash flow, increasing that year's present worth.

For simplicity, most engineering economics practice problems involving income taxes specify a single income tax rate. In practice, however, federal and most state tax rates depend on the income level. Each range of incomes and its associated tax rate are known as *income bracket* and *tax bracket*, respectively. For example, the state income tax rate might be 4% for incomes up to and including \$30,000, and 5% for incomes above \$30,000. The income tax for a taxpaying entity with an income of \$50,000 would have to be calculated in two parts.

$$\begin{aligned} \text{tax} &= (0.04)(\$30{,}000) + (0.05)(\$50{,}000 - \$30{,}000) \\ &= \$2200 \end{aligned}$$

Income taxes and depreciation have no bearing on municipal or governmental projects since municipalities, states, and the U.S. government pay no taxes.

Example 58.20

A corporation that pays 53% of its profit in income taxes invests \$10,000 in an asset that will produce a \$3000 annual revenue for eight years. If the annual expenses are \$700, salvage after eight years is \$500, and 9% interest is used, what is the after-tax present worth? Disregard depreciation.

Solution

$$\begin{aligned} P &= -\$10{,}000 + (\$3000)(P/A, 9\%, 8)(1 - 0.53) \\ &\quad - (\$700)(P/A, 9\%, 8)(1 - 0.53) \\ &\quad + (\$500)(P/F, 9\%, 8) \\ &= -\$10{,}000 + (\$3000)(5.5348)(0.47) \\ &\quad - (\$700)(5.5348)(0.47) + (\$500)(0.5019) \\ &= -\$3766 \end{aligned}$$

42. TAXATION AT THE TIMES OF ASSET PURCHASE AND SALE

There are numerous rules and conventions that governmental tax codes and the accounting profession impose on organizations. Engineers are not normally expected to be aware of most of the rules and conventions, but occasionally it may be necessary to incorporate their effects into an engineering economic analysis.

Tax Credit

A *tax credit* (also known as an *investment tax credit* or *investment credit*) is a one-time credit against income taxes.[32] Therefore, it is added to the after-tax present worth as a last step in an engineering economic analysis. Such tax credits may be allowed by the government from time to time for equipment purchases, employment of various classes of workers, rehabilitation of historic landmarks, and so on.

A tax credit is usually calculated as a fraction of the initial purchase price or cost of an asset or activity.

$$\text{TC} = \text{fraction} \times \text{initial cost} \qquad 58.45$$

When the tax credit is applicable, the fraction used is subject to legislation. A professional tax expert or accountant should be consulted prior to applying the tax credit concept to engineering economic analysis problems.

Since the investment tax credit reduces the buyer's tax liability, a tax credit should be included only in after-tax engineering economic analyses. The credit is assumed to be received at the end of the year.

Gain or Loss on the Sale of a Depreciated Asset

If an asset that has been depreciated over a number of prior years is sold for more than its current book value, the difference between the book value and selling price is taxable income in the year of the sale. Alternatively, if the asset is sold for less than its current book value, the difference between the selling price and book value is an expense in the year of the sale.

Example 58.21

One year, a company makes a \$5000 investment in a historic building. The investment is not depreciable, but it does qualify for a one-time 20% tax credit. In that same year, revenue is \$45,000 and expenses (exclusive of the \$5000 investment) are \$25,000. The company pays a total of 53% in income taxes. What is the after-tax present worth of this year's activities if the company's interest rate for investment is 10%?

Solution

The tax credit is

$$\text{TC} = (0.20)(\$5000) = \$1000$$

[32]Strictly, *tax credit* is the more general term, and applies to a credit for doing anything creditable. An *investment tax credit* requires an investment in something (usually real property or equipment).

This tax credit is assumed to be received at the end of the year. The after-tax present worth is

$$\begin{aligned} P &= -\$5000 + (\$45{,}000 - \$25{,}000)(1 - 0.53) \\ &\quad \times (P/F, 10\%, 1) + (\$1000)(P/F, 10\%, 1) \\ &= -\$5000 + (\$20{,}000)(0.47)(0.9091) \\ &\quad + (\$1000)(0.9091) \\ &= \$4455 \end{aligned}$$

43. DEPRECIATION RECOVERY

The economic effect of depreciation is to reduce the income tax in year j by tD_j. The present worth of the asset is also affected: The present worth is increased by $tD_j(P/F, i\%, j)$. The after-tax present worth of all depreciation effects over the depreciation period of the asset is called the *depreciation recovery* (DR).[33]

$$\text{DR} = t\sum_{j=1}^{n} D_j(P/F, i\%, j) \qquad 58.46$$

There are multiple ways depreciation can be calculated, as summarized in Table 58.5. *Straight line depreciation recovery* from an asset is easily calculated, since the depreciation is the same each year. Assuming the asset has a constant depreciation of D and depreciation period of n years, the depreciation recovery is

$$\text{DR} = tD(P/A, i\%, n) \qquad 58.47$$

$$D = \frac{C - S_n}{n} \qquad 58.48$$

Sum-of-the-years' digits depreciation recovery is also relatively easily calculated, since the depreciation decreases uniformly each year.

$$\text{DR} = \left(\frac{t(C - S_n)}{T}\right)\big(n(P/A, i\%, n) - (P/G, i\%, n)\big) \qquad 58.49$$

Finding *declining balance depreciation recovery* is more involved. There are three difficulties. The first (the apparent need to calculate all previous depreciations in order to determine the subsequent depreciation) has already been addressed by Eq. 58.32.

The second difficulty is that there is no way to ensure (that is, to force) the book value to be S_n at $t = n$. Therefore, it is common to write off the remaining book value (down to S_n) at $t = n$ in one lump sum. This assumes $\text{BV}_n \geq S_n$.

The third difficulty is that of finding the present worth of an *exponentially decreasing cash flow*. Although the proof is omitted here, such exponential cash flows can be handled with the *exponential gradient factor*, (P/EG).[34]

$$(P/EG, z-1, n) = \frac{z^n - 1}{z^n(z-1)} \qquad 58.50$$

$$z = \frac{1+i}{1-d} \qquad 58.51$$

Then, as long as $\text{BV}_n > S_n$, the declining balance depreciation recovery is

$$\text{DR} = tC\left(\frac{d}{1-d}\right)(P/EG, z-1, n) \qquad 58.52$$

Example 58.22

For the asset described in Ex. 58.17, calculate the after-tax depreciation recovery with straight line and sum-of-the-years' digits depreciation methods. Use 6% interest with 48% income taxes.

Solution

SL:

$$\begin{aligned} \text{DR} &= (0.48)(\$880)(P/A, 6\%, 10) \\ &= (0.48)(\$880)(7.3601) \\ &= \$3109 \end{aligned}$$

SOYD: The depreciation series can be thought of as a constant \$1600 term with a negative \$160 gradient.

$$\begin{aligned} \text{DR} &= (0.48)(\$1600)(P/A, 6\%, 10) \\ &\quad - (0.48)(\$160)(P/G, 6\%, 10) \\ &= (0.48)(\$1600)(7.3601) \\ &\quad - (0.48)(\$160)(29.6023) \\ &= \$3379 \end{aligned}$$

The ten-year (P/G) factor is used even though there are only nine years in which the gradient reduces the initial \$1600 amount.

Example 58.23

What is the after-tax present worth of the asset described in Ex. 58.20 if straight line, sum-of-the-years' digits, and double declining balance depreciation methods are used?

Solution

Using SL, the depreciation recovery is

$$\begin{aligned} \text{DR} &= (0.53)\left(\frac{\$10{,}000 - \$500}{8}\right)(P/A, 9\%, 8) \\ &= (0.53)\left(\frac{\$9500}{8}\right)(5.5348) \\ &= \$3483 \end{aligned}$$

[33] Since the depreciation benefit is reduced by taxation, depreciation cannot be thought of as an annuity to fund a replacement asset.

[34] The (P/A) columns in App. 58.B can be used for (P/EG) as long as the interest rate is assumed to be $z - 1$.

Using SOYD, the depreciation recovery is calculated as follows.

$$T = \left(\tfrac{1}{2}\right)(8)(9) = 36$$

$$\text{depreciation base} = \$10{,}000 - \$500 = \$9500$$

$$D_1 = \left(\tfrac{8}{36}\right)(\$9500) = \$2111$$

$$G = \left(\tfrac{1}{36}\right)(\$9500) = \$264$$

$$\begin{aligned}\text{DR} &= (0.53)\Big((\$2111)(P/A, 9\%, 8) \\ &\quad - (\$264)(P/G, 9\%, 8)\Big) \\ &= (0.53)\Big((\$2111)(5.5348) \\ &\quad - (\$264)(16.8877)\Big) \\ &= \$3830\end{aligned}$$

Using DDB, the depreciation recovery is calculated as follows.[35]

$$d = \frac{2}{8} = 0.25$$

$$z = \frac{1 + 0.09}{1 - 0.25} = 1.4533$$

$$(P/EG, z-1, n) = \frac{(1.4533)^8 - 1}{(1.4533)^8(0.4533)} = 2.095$$

From Eq. 58.52,

$$\begin{aligned}\text{DR} &= (0.53)\left(\frac{(0.25)(\$10{,}000)}{0.75}\right)(2.095) \\ &= \$3701\end{aligned}$$

The after-tax present worth, neglecting depreciation, was previously found to be −$3766.

The after-tax present worths, including depreciation recovery, are

SL: $P = -\$3766 + \$3483 = -\$283$

SOYD: $P = -\$3766 + \$3830 = \$64$

DDB: $P = -\$3766 + \$3701 = -\$65$

44. OTHER INTEREST RATES

The *effective interest rate per period*, i (also called *yield* by banks), is the only interest rate that should be used in equivalence equations. The interest rates at the top of the factor tables in App. 58.B are implicitly all effective interest rates. Usually, the period will be one year, hence the name *effective annual interest rate*. However, there are other interest rates in use as well.

The term *nominal interest rate*, r (*rate per annum*), is encountered when compounding is more than once per year. The nominal rate does not include the effect of compounding and is not the same as the effective rate. And, since the effective interest rate can be calculated from the nominal rate only if the number of compounding periods per year is known, nominal rates cannot be compared unless the method of compounding is specified. The only practical use for a nominal rate per year is for calculating the effective rate per period.

45. RATE AND PERIOD CHANGES

If there are k compounding periods during the year (two for semiannual compounding, four for quarterly compounding, twelve for monthly compounding, etc.) and the nominal rate is r, the *effective rate per compounding period* is

$$\phi = \frac{r}{k} \qquad \text{58.53}$$

The effective annual rate, i, can be calculated from the effective rate per period, ϕ, by using Eq. 58.54.

$$\begin{aligned}i &= (1 + \phi)^k - 1 \\ &= \left(1 + \frac{r}{k}\right)^k - 1\end{aligned} \qquad \text{58.54}$$

Sometimes, only the effective rate per period (e.g., per month) is known. However, that will be a simple problem since compounding for n periods at an effective rate per period is not affected by the definition or length of the period.

The following rules may be used to determine which interest rate is given.

- Unless specifically qualified, the interest rate given is an annual rate.
- If the compounding is annual, the rate given is the effective rate. If compounding is other than annual, the rate given is the nominal rate.

The effective annual interest rate determined on a *daily compounding basis* will not be significantly different than if *continuous compounding* is assumed.[36] In the case of continuous (or daily) compounding, the discounting factors can be calculated directly from the nominal interest rate and number of years, without having to find the effective interest rate per period. Table 58.6 can be used to determine the discount factors for continuous compounding.

[35]This method should start by checking that the book value at the end of the depreciation period is greater than the salvage value. In this example, such is the case. However, the step is not shown.

[36]The number of *banking days in a year* (250, 360, etc.) must be specifically known.

Table 58.5 Depreciation Calculation Summary

method	depreciation basis	depreciation in year j, D_j	book value after jth depreciation, BV_j	present worth of after-tax depreciation recovery, DR	supplementary formulas
straight line (SL)	$C - S_n$	$\frac{C - S_n}{n}$ (constant)	$C - jD$	$tD(P/A, i\%, n)$	
constant percentage	$C - S_n$	fraction × $(C - S_n)$ (constant)	$C - jD$	$tD(P/A, i\%, n)$	
sum-of-the-years' digits (SOYD)	$C - S_n$	$\frac{(C - S_n) \times (n - j + 1)}{T}$	$(C - S_n) \times \left(1 - \frac{j(2n+1-j)}{n(n+1)}\right) + S_n$	$\frac{t(C - S_n)}{T} \times (n(P/A, i\%, n) - (P/G, i\%, n))$	$T = \frac{1}{2}n(n+1)$
double declining balance (DDB)	C	$dC(1-d)^{j-1}$	$C(1-d)^j$	$tC\left(\frac{d}{1-d}\right) \times (P/EG, z-1, n)$	$d = \frac{2}{n};\ z = \frac{1+i}{1-d}$ $(P/EG, z-1, n) = \frac{z^n - 1}{z^n(z-1)}$
sinking fund (SF)	$C - S_n$	$(C - S_n) \times (A/F, i\%, n) \times (F/P, i\%, j-1)$	$C - (C - S_n) \times (A/F, i\%, n) \times (F/A, i\%, j)$	$\frac{t(C - S_n)(A/F, i\%, n)}{1+i}$	
accelerated cost recovery system (ACRS/MACRS)	C	$C \times$ factor	$C - \sum_{m=1}^{j} D_m$	$t\sum_{j=1}^{n} D_j(P/F, i\%, j)$	
units of production or service output	$C - S_n$	$(C - S_n) \times \left(\frac{\text{actual output in year } j}{\text{lifetime output}}\right)$	$C - \sum_{m=1}^{j} D_m$	$t\sum_{j=1}^{n} D_j(P/F, i\%, j)$	

Table 58.6 Discount Factors for Continuous Compounding (n is the number of years)

symbol	formula
$(F/P, r\%, n)$	e^{rn}
$(P/F, r\%, n)$	e^{-rn}
$(A/F, r\%, n)$	$\frac{e^r - 1}{e^{rn} - 1}$
$(F/A, r\%, n)$	$\frac{e^{rn} - 1}{e^r - 1}$
$(A/P, r\%, n)$	$\frac{e^r - 1}{1 - e^{-rn}}$
$(P/A, r\%, n)$	$\frac{1 - e^{-rn}}{e^r - 1}$

Example 58.24

A savings and loan offers a nominal rate of 5.25% compounded daily over 365 days in a year. What is the effective annual rate?

Solution

method 1: Use Eq. 58.54.

$$r = 0.0525,\ k = 365$$

$$i = \left(1 + \frac{r}{k}\right)^k - 1 = \left(1 + \frac{0.0525}{365}\right)^{365} - 1 = 0.0539$$

Professional Practice

method 2: Assume daily compounding is the same as continuous compounding.

$$\begin{aligned} i &= (F/P, r\%, 1) - 1 \\ &= e^{0.0525} - 1 \\ &= 0.0539 \end{aligned}$$

Example 58.25

A real estate investment trust pays \$7,000,000 for an apartment complex with 100 units. The trust expects to sell the complex in ten years for \$15,000,000. In the meantime, it expects to receive an average rent of \$900 per month from each apartment. Operating expenses are expected to be \$200 per month per occupied apartment. A 95% occupancy rate is predicted. In similar investments, the trust has realized a 15% effective annual return on its investment. Compare to those past investments the expected present worth of this investment when calculated assuming (a) annual compounding (i.e., the year-end convention), and (b) monthly compounding. Disregard taxes, depreciation, and all other factors.

Solution

(a) The net annual income will be

$$\begin{aligned} &(0.95)(100 \text{ units})\left(\frac{\$900}{\text{unit-mo}} - \frac{\$200}{\text{unit-mo}}\right)\left(12\ \frac{\text{mo}}{\text{yr}}\right) \\ &\quad = \$798{,}000/\text{yr} \end{aligned}$$

The present worth of ten years of operation is

$$\begin{aligned} P &= -\$7{,}000{,}000 + (\$798{,}000)(P/A, 15\%, 10) \\ &\quad + (\$15{,}000{,}000)(P/F, 15\%, 10) \\ &= -\$7{,}000{,}000 + (\$798{,}000)(5.0188) \\ &\quad + (\$15{,}000{,}000)(0.2472) \\ &= \$713{,}000 \end{aligned}$$

(b) The net monthly income is

$$\begin{aligned} &(0.95)(100 \text{ units})\left(\frac{\$900}{\text{unit-mo}} - \frac{\$200}{\text{unit-mo}}\right) \\ &\quad = \$66{,}500/\text{mo} \end{aligned}$$

Equation 58.54 is used to calculate the effective monthly rate, ϕ, from the effective annual rate, $i = 15\%$, and the number of compounding periods per year, $k = 12$.

$$\begin{aligned} \phi &= (1+i)^{1/k} - 1 \\ &= (1+0.15)^{1/12} - 1 \\ &= 0.011715 \quad (1.1715\%) \end{aligned}$$

The number of compounding periods in ten years is

$$n = (10 \text{ yr})\left(12\ \frac{\text{mo}}{\text{yr}}\right) = 120 \text{ mo}$$

The present worth of 120 months of operation is

$$\begin{aligned} P &= -\$7{,}000{,}000 + (\$66{,}500)(P/A, 1.1715\%, 120) \\ &\quad + (\$15{,}000{,}000)(P/F, 1.1715\%, 120) \end{aligned}$$

Since table values for 1.1715% discounting factors are not available, the factors are calculated from Table 58.1.

$$\begin{aligned} (P/A, 1.1715\%, 120) &= \frac{(1+i)^n - 1}{i(1+i)^n} \\ &= \frac{(1+0.011715)^{120} - 1}{(0.011715)(1+0.011715)^{120}} \\ &= 64.261 \end{aligned}$$

$$\begin{aligned} (P/F, 1.1715\%, 120) &= (1+i)^{-n} = (1+0.011715)^{-120} \\ &= 0.2472 \end{aligned}$$

The present worth over 120 monthly compounding periods is

$$\begin{aligned} P &= -\$7{,}000{,}000 + (\$66{,}500)(64.261) \\ &\quad + (\$15{,}000{,}000)(0.2472) \\ &= \$981{,}357 \end{aligned}$$

46. BONDS

A *bond* is a method of long-term financing commonly used by governments, states, municipalities, and very large corporations.[37] The bond represents a contract to pay the bondholder specific amounts of money at specific times. The holder purchases the bond in exchange for specific payments of interest and principal. Typical municipal bonds call for quarterly or semiannual interest payments and a payment of the *face value of the bond* on the *date of maturity* (end of the bond period).[38] Due to the practice of discounting in the bond market, a bond's face value and its purchase price generally will not coincide.

In the past, a bondholder had to submit a coupon or ticket in order to receive an interim interest payment. This has given rise to the term *coupon rate*, which is the nominal annual interest rate on which the interest payments are made. Coupon books are seldom used with modern bonds, but the term survives. The coupon rate determines the magnitude of the semiannual (or otherwise) interest payments during the life of the bond. The bondholder's own effective interest rate should be used for economic decisions about the bond.

Actual *bond yield* is the bondholder's actual rate of return of the bond, considering the purchase price, interest payments, and face value payment (or, value realized

[37]In the past, 30-year bonds were typical. Shorter term 10-year, 15-year, 20-year, and 25-year bonds are also commonly issued.

[38]A *fully amortized bond* pays back interest and principal throughout the life of the bond. There is no balloon payment.

if the bond is sold before it matures). By convention, bond yield is calculated as a nominal rate (rate per annum), not an effective rate per year. The bond yield should be determined by finding the effective rate of return per payment period (e.g., per semiannual interest payment) as a conventional rate of return problem. Then, the nominal rate can be found by multiplying the effective rate per period by the number of payments per year, as in Eq. 58.54.

Example 58.26

What is the maximum amount an investor should pay for a 25-year bond with a $20,000 face value and 8% coupon rate (interest only paid semiannually)? The bond will be kept to maturity. The investor's effective annual interest rate for economic decisions is 10%.

Solution

For this problem, take the compounding period to be six months. Then, there are 50 compounding periods. Since 8% is a nominal rate, the effective bond rate per period is calculated from Eq. 58.53 as

$$\phi_{\text{bond}} = \frac{r}{k} = \frac{8\%}{2} = 4\%$$

The bond payment received semiannually is

$$(0.04)(\$20{,}000) = \$800$$

10% is the investor's effective rate per year, so Eq. 58.54 is again used to calculate the effective analysis rate per period.

$$0.10 = (1+\phi)^2 - 1$$
$$\phi = 0.04881 \quad (4.88\%)$$

The maximum amount that the investor should be willing to pay is the present worth of the investment.

$$P = (\$800)(P/A, 4.88\%, 50) + (\$20{,}000)(P/F, 4.88\%, 50)$$

Table 58.1 can be used to calculate the following factors.

$$(P/A, 4.88\%, 50) = \frac{(1+i)^n - 1}{i(1+i)^n} = \frac{(1+0.0488)^{50} - 1}{(0.0488)(1.0488)^{50}} = 18.600$$

$$(P/F, 4.88\%, 50) = \frac{1}{(1+i)^n} = \frac{1}{(1+0.0488)^{50}} = 0.09233$$

Then, the present worth is

$$P = (\$800)(18.600) + (\$20{,}000)(0.09233) = \$16{,}727$$

47. PROBABILISTIC PROBLEMS

If an alternative's cash flows are specified by an implicit or explicit probability distribution rather than being known exactly, the problem is *probabilistic.*

Probabilistic problems typically possess the following characteristics.

- There is a chance of loss that must be minimized (or, rarely, a chance of gain that must be maximized) by selection of one of the alternatives.
- There are multiple alternatives. Each alternative offers a different degree of protection from the loss. Usually, the alternatives with the greatest protection will be the most expensive.
- The magnitude of loss or gain is independent of the alternative selected.

Probabilistic problems are typically solved using annual costs and expected values. An *expected value* is similar to an *average value* since it is calculated as the mean of the given probability distribution. If cost 1 has a probability of occurrence, p_1, cost 2 has a probability of occurrence, p_2, and so on, the expected value is

$$\mathcal{E}\{\text{cost}\} = p_1(\text{cost 1}) + p_2(\text{cost 2}) + \cdots \qquad 58.55$$

Example 58.27

Flood damage in any year is given according to the following table. What is the present worth of flood damage for a ten-year period? Use 6% as the effective annual interest rate.

damage	probability
0	0.75
$10,000	0.20
$20,000	0.04
$30,000	0.01

Solution

The expected value of flood damage in any given year is

$$\mathcal{E}\{\text{damage}\} = (0)(0.75) + (\$10{,}000)(0.20) + (\$20{,}000)(0.04) + (\$30{,}000)(0.01) = \$3100$$

The present worth of ten years of expected flood damage is

$$\text{present worth} = (\$3100)(P/A, 6\%, 10) = (\$3100)(7.3601) = \$22{,}816$$

Example 58.28

A dam is being considered on a river that periodically overflows and causes $600,000 damage. The damage is essentially the same each time the river causes flooding. The project horizon is 40 years. A 10% interest rate is being used.

Three different designs are available, each with different costs and storage capacities.

design alternative	cost	maximum capacity
A	$500,000	1 unit
B	$625,000	1.5 units
C	$900,000	2.0 units

The National Weather Service has provided a statistical analysis of annual rainfall runoff from the watershed draining into the river.

units annual rainfall	probability
0	0.10
0.1–0.5	0.60
0.6–1.0	0.15
1.1–1.5	0.10
1.6–2.0	0.04
2.1 or more	0.01

Which design alternative would you choose assuming the dam is essentially empty at the start of each rainfall season?

Solution

The sum of the construction cost and the expected damage should be minimized. If alternative A is chosen, it will have a capacity of 1 unit. Its capacity will be exceeded (causing $600,000 damage) when the annual rainfall exceeds 1 unit. Therefore, the expected value of the annual cost of alternative A is

$$\begin{aligned}\mathcal{E}\{\text{EUAC(A)}\} &= (\$500{,}000)(A/P, 10\%, 40) \\ &\quad + (\$600{,}000)(0.10 + 0.04 + 0.01) \\ &= (\$500{,}000)(0.1023) + (\$600{,}000)(0.15) \\ &= \$141{,}150\end{aligned}$$

Similarly,

$$\begin{aligned}\mathcal{E}\{\text{EUAC(B)}\} &= (\$625{,}000)(A/P, 10\%, 40) \\ &\quad + (\$600{,}000)(0.04 + 0.01) \\ &= (\$625{,}000)(0.1023) + (\$600{,}000)(0.05) \\ &= \$93{,}938\end{aligned}$$

$$\begin{aligned}\mathcal{E}\{\text{EUAC(C)}\} &= (\$900{,}000)(A/P, 10\%, 40) \\ &\quad + (\$600{,}000)(0.01) \\ &= (\$900{,}000)(0.1023) + (\$600{,}000)(0.01) \\ &= \$98{,}070\end{aligned}$$

Alternative B should be chosen.

48. FIXED AND VARIABLE COSTS

The distinction between fixed and variable costs depends on how these costs vary when an independent variable changes. For example, factory or machine production is frequently the independent variable. However, it could just as easily be vehicle miles driven, hours of operation, or quantity (mass, volume, etc.). Examples of fixed and variable costs are given in Table 58.7.

Table 58.7 *Summary of Fixed and Variable Costs*

fixed costs
- rent
- property taxes
- interest on loans
- insurance
- janitorial service expense
- tooling expense
- setup, cleanup, and tear-down expenses
- depreciation expense
- marketing and selling costs
- cost of utilities
- general burden and overhead expense

variable costs
- direct material costs
- direct labor costs
- cost of miscellaneous supplies
- payroll benefit costs
- income taxes
- supervision costs

If a cost is a function of the independent variable, the cost is said to be a *variable cost.* The change in cost per unit variable change (i.e., what is usually called the *slope*) is known as the *incremental cost.* Material and labor costs are examples of variable costs. They increase in proportion to the number of product units manufactured.

If a cost is not a function of the independent variable, the cost is said to be a *fixed cost.* Rent and lease payments are typical fixed costs. These costs will be incurred regardless of production levels.

Some costs have both fixed and variable components, as Fig. 58.9 illustrates. The fixed portion can be determined by calculating the cost at zero production.

An additional category of cost is the *semivariable cost.* This type of cost increases stepwise. Semivariable cost structures are typical of situations where *excess capacity* exists. For example, supervisory cost is a stepwise function of the number of production shifts. Also, labor cost for truck drivers is a stepwise function of weight (volume) transported. As long as a truck has room left (i.e., excess capacity), no additional driver is needed. As soon as the truck is filled, labor cost will increase.

49. ACCOUNTING COSTS AND EXPENSE TERMS

The accounting profession has developed special terms for certain groups of costs. When annual costs are

Figure 58.9 Fixed and Variable Costs

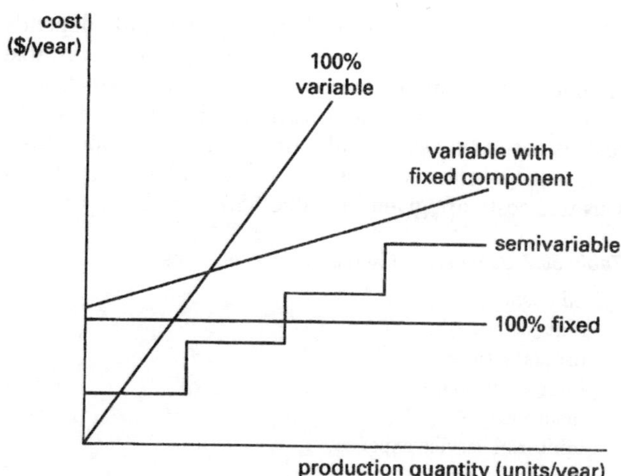

incurred due to the functioning of a piece of equipment, they are known as *operating and maintenance* (O&M) *costs.* The annual costs associated with operating a business (other than the costs directly attributable to production) are known as *general, selling, and administrative* (GS&A) *expenses.*

Direct labor costs are costs incurred in the factory, such as assembly, machining, and painting labor costs. *Direct material costs* are the costs of all materials that go into production.[39] Typically, both direct labor and direct material costs are given on a per-unit or per-item basis. The sum of the direct labor and direct material costs is known as the *prime cost.*

There are certain additional expenses incurred in the factory, such as the costs of factory supervision, stock-picking, quality control, factory utilities, and miscellaneous supplies (cleaning fluids, assembly lubricants, routing tags, etc.) that are not incorporated into the final product. Such costs are known as *indirect manufacturing expenses* (IME) or *indirect material and labor costs.*[40] The sum of the per-unit indirect manufacturing expense and prime cost is known as the *factory cost.*

Research and development (R&D) *costs* and *administrative expenses* are added to the factory cost to give the *manufacturing cost* of the product.

Additional costs are incurred in marketing the product. Such costs are known as *selling expenses* or *marketing expenses.* The sum of the selling expenses and manufacturing cost is the *total cost* of the product. Figure 58.10 illustrates these terms.[41]

The distinctions among the various forms of cost (particularly with overhead costs) are not standardized.

Figure 58.10 Costs and Expenses Combined

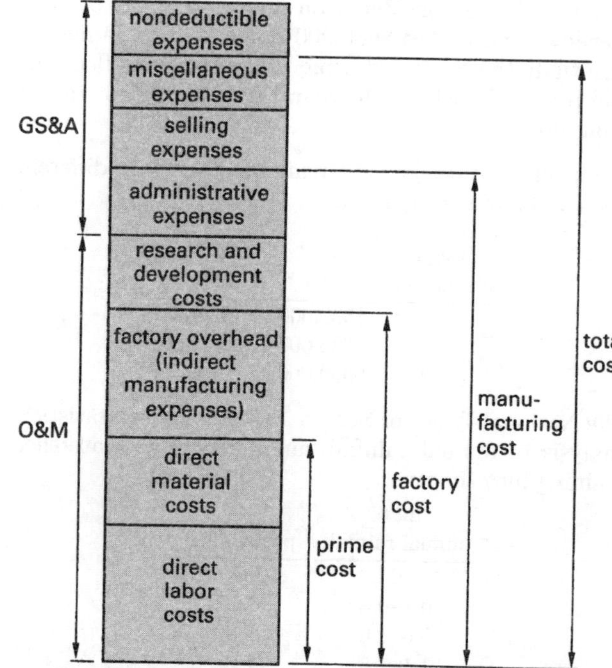

Each company must develop a classification system to deal with the various cost factors in a consistent manner. There are also other terms in use (e.g., *raw materials, operating supplies, general plant overhead*), but these terms must be interpreted within the framework of each company's classification system. Table 58.8 is typical of such classification systems.

50. ACCOUNTING PRINCIPLES

Basic Bookkeeping

An accounting or *bookkeeping system* is used to record historical financial transactions. The resultant records are used for product costing, satisfaction of statutory requirements, reporting of profit for income tax purposes, and general company management.

Bookkeeping consists of two main steps: recording the transactions, followed by categorization of the transactions.[42] The transactions (receipts and disbursements) are recorded in a *journal* (*book of original entry*) to complete the first step. Such a journal is organized in a simple chronological and sequential manner. The transactions are then categorized (into interest income, advertising expense, etc.) and posted (i.e., entered or written) into the appropriate *ledger account.*[43]

[39]There may be problems with pricing the material when it is purchased from an outside vendor and the stock on hand derives from several shipments purchased at different prices.

[40]The *indirect material and labor costs* usually exclude costs incurred in the office area.

[41]*Total cost* does not include income taxes.

[42]These two steps are not to be confused with the *double-entry bookkeeping method.*

[43]The two-step process is more typical of a *manual bookkeeping system* than a computerized *general ledger system.* However, even most computerized systems produce reports in journal entry order, as well as account summaries.

Table 58.8 *Typical Classification of Expenses*

direct labor expenses
- machining and forming
- assembly
- finishing
- inspection
- testing

direct material expenses
- items purchased from other vendors
- manufactured assemblies

factory overhead expenses (indirect manufacturing expenses)
- supervision
- benefits
  - pension
  - medical insurance
  - vacations
- wages overhead
  - unemployment compensation taxes
  - social security taxes
  - disability taxes
- stock-picking
- quality control and inspection
- expediting
- rework
- maintenance
- miscellaneous supplies
  - routing tags
  - assembly lubricants
  - cleaning fluids
  - wiping cloths
  - janitorial supplies
- packaging (materials and labor)
- factory utilities
- laboratory
- depreciation on factory equipment

research and development expenses
- engineering (labor)
- patents
- testing
- prototypes (material and labor)
- drafting
- O&M of R&D facility

administrative expenses
- corporate officers
- accounting
- secretarial/clerical/reception
- security (protection)
- medical (nurse)
- employment (personnel)
- reproduction
- data processing
- production control
- depreciation on nonfactory equipment
- office supplies
- office utilities
- O&M of offices

selling expenses
- marketing (labor)
- advertising
- transportation (if not paid by customer)
- outside sales force (labor and expenses)
- demonstration units
- commissions
- technical service and support
- order processing
- branch office expenses

miscellaneous expenses
- insurance
- property taxes
- interest on loans

nondeductible expenses
- federal income taxes
- fines and penalties

The ledger accounts together constitute the *general ledger* or *ledger*. All ledger accounts can be classified into one of three types: *asset accounts, liability accounts*, and *owners' equity accounts.* Strictly speaking, income and expense accounts, kept in a separate journal, are included within the classification of owners' equity accounts.

Together, the journal and ledger are known simply as "the books" of the company, regardless of whether bound volumes of pages are actually involved.

Balancing the Books

In a business environment, *balancing the books* means more than reconciling the checkbook and bank statements. All accounting entries must be posted in such a way as to maintain the equality of the *basic accounting equation*,

$$\text{assets} = \text{liability} + \text{owner's equity} \qquad 58.56$$

In a *double-entry bookkeeping system*, the equality is maintained within the ledger system by entering each transaction into two balancing ledger accounts. For example, paying a utility bill would decrease the cash account (an asset account) and decrease the utility expense account (a liability account) by the same amount.

Transactions are either *debits* or *credits*, depending on their sign. Increases in asset accounts are debits; decreases are credits. For liability and equity accounts, the opposite is true: Increases are credits, and decreases are debits.[44]

Cash and Accrual Systems[45]

The simplest form of bookkeeping is based on the *cash system*. The only transactions that are entered into the journal are those that represent cash receipts and disbursements. In effect, a checkbook register or bank deposit book could serve as the journal.

During a given period (e.g., month or quarter), expense liabilities may be incurred even though the payments for those expenses have not been made. For example, an invoice (bill) may have been received but not paid. Under the *accrual system*, the obligation is posted into the appropriate expense account before it is paid.[46]

[44]There is a difference in sign between asset and liability accounts. An increase in an expense account is actually a decrease. The accounting profession, apparently, is comfortable with the common confusion that exists between debits and credits.

[45]There is also a distinction made between cash flows that are known and those that are expected. It is a *standard accounting principle* to record losses in full, at the time they are recognized, even before their occurrence. In the construction industry, for example, losses are recognized in full and projected to the end of a project as soon as they are foreseeable. Profits, on the other hand, are recognized only as they are realized (typically, as a percentage of project completion). The difference between cash and accrual systems is a matter of *bookkeeping*. The difference between loss and profit recognition is a matter of *accounting convention*. Engineers seldom need to be concerned with the accounting tradition.

[46]The expense for an item or service might be accrued even *before* the invoice is received. It might be recorded when the purchase order for the item or service is generated, or when the item or service is received.

Analogous to expenses, under the accrual system, income will be claimed before payment is received. Specifically, a sales transaction can be recorded as income when the customer's order is received, when the outgoing invoice is generated, or when the merchandise is shipped.

Financial Statements

Each period, two types of corporate financial statements are typically generated: the *balance sheet* and *profit and loss* (P&L) *statement*.[47] The profit and loss statement, also known as a *statement of income and retained earnings*, is a summary of sources of *income* or *revenue* (interest, sales, fees charged, etc.) and *expenses* (utilities, advertising, repairs, etc.) for the period. The expenses are subtracted from the revenues to give a *net income* (generally, before taxes).[48] Figure 58.11 illustrates a simplified profit and loss statement.

Figure 58.11 *Simplified Profit and Loss Statement*

revenue			
interest	2000		
sales	237,000		
returns	(23,000)		
net revenue		216,000	
expenses			
salaries	149,000		
utilities	6000		
advertising	28,000		
insurance	4000		
supplies	1000		
net expenses		188,000	
period net income			28,000
beginning retained earnings			63,000
net year-to-date earnings				91,000

The *balance sheet* presents the *basic accounting equation* in tabular form. The balance sheet lists the major categories of assets and outstanding liabilities. The difference between asset values and liabilities is the *equity*, as defined in Eq. 58.56. This equity represents what would be left over after satisfying all debts by liquidating the company.

Figure 58.12 is a simplified balance sheet.

[47]Other types of financial statements (*statements of changes in financial position, cost of sales statements*, inventory and asset reports, etc.) also will be generated, depending on the needs of the company.

[48]Financial statements also can be prepared with percentages (of total assets and net revenue) instead of dollars, in which case they are known as *common size financial statements.*

Figure 58.12 *Simplified Balance Sheet*

ASSETS			
current assets			
cash	14,000		
accounts receivable	36,000		
notes receivable	20,000		
inventory	89,000		
prepaid expenses	3000		
total current assets		162,000	
plant, property, and equipment			
land and buildings	217,000		
motor vehicles	31,000		
equipment	94,000		
accumulated depreciation	(52,000)		
total fixed assets		290,000	
total assets			452,000
LIABILITIES AND OWNERS' EQUITY			
current liabilities			
accounts payable	66,000		
accrued income taxes	17,000		
accrued expenses	8000		
total current liabilities		91,000	
long-term debt			
notes payable	117,000		
mortgage	23,000		
total long-term debt		140,000	
owners' and stockholders' equity			
stock	130,000		
retained earnings	91,000		
total owners' equity		221,000	
total liabilities and owners' equity			452,000

There are several terms that appear regularly on balance sheets.

- *current assets:* cash and other assets that can be converted quickly into cash, such as accounts receivable, notes receivable, and merchandise (inventory). Also known as *liquid assets.*
- *fixed assets:* relatively permanent assets used in the operation of the business and relatively difficult to convert into cash. Examples are land, buildings, and equipment. Also known as *nonliquid assets.*
- *current liabilities:* liabilities due within a short period of time (e.g., within one year) and typically paid out of current assets. Examples are accounts payable, notes payable, and other accrued liabilities.
- *long-term liabilities:* obligations that are not totally payable within a short period of time (e.g., within one year).

Analysis of Financial Statements

Financial statements are evaluated by management, lenders, stockholders, potential investors, and many other groups for the purpose of determining the *health of the company.* The health can be measured in terms of *liquidity* (ability to convert assets to cash quickly), *solvency* (ability to meet debts as they become due), and *relative risk* (of which one measure is *leverage*—the portion of total capital contributed by owners).

The analysis of financial statements involves several common ratios, usually expressed as percentages. The following are some frequently encountered ratios.

- *current ratio:* an index of short-term paying ability.

$$\text{current ratio} = \frac{\text{current assets}}{\text{current liabilities}} \qquad 58.57$$

- *quick* (or *acid-test*) *ratio:* a more stringent measure of short-term debt-paying ability. The *quick assets* are defined to be current assets minus inventories and prepaid expenses.

$$\text{quick ratio} = \frac{\text{quick assets}}{\text{current liabilities}} \qquad 58.58$$

- *receivable turnover:* a measure of the average speed with which accounts receivable are collected.

$$\text{receivable turnover} = \frac{\text{net credit sales}}{\text{average net receivables}} \qquad 58.59$$

- *average age of receivables:* number of days, on the average, in which receivables are collected.

$$\text{average age of receivables} = \frac{365}{\text{receivable turnover}} \qquad 58.60$$

- *inventory turnover:* a measure of the speed with which inventory is sold, on the average.

$$\text{inventory turnover} = \frac{\text{cost of goods sold}}{\text{average cost of inventory on hand}} \qquad 58.61$$

- *days supply of inventory on hand:* number of days, on the average, that the current inventory would last.

$$\text{days supply of inventory on hand} = \frac{365}{\text{inventory turnover}} \qquad 58.62$$

- *book value per share of common stock:* number of dollars represented by the balance sheet owners' equity for each share of common stock outstanding.

$$\text{book value per share of common stock} = \frac{\text{common shareholders' equity}}{\text{number of outstanding shares}} \qquad 58.63$$

- *gross margin:* gross profit as a percentage of sales. (Gross profit is sales less cost of goods sold.)

$$\text{gross margin} = \frac{\text{gross profit}}{\text{net sales}} \qquad 58.64$$

- *profit margin ratio:* percentage of each dollar of sales that is net income.

$$\text{profit margin} = \frac{\text{net income before taxes}}{\text{net sales}} \qquad 58.65$$

- *return on investment ratio:* shows the percent return on owners' investment.

$$\text{return on investment} = \frac{\text{net income}}{\text{owners' equity}} \qquad 58.66$$

- *price-earnings ratio:* indication of relationship between earnings and market price per share of common stock, useful in comparisons between alternative investments.

$$\text{price-earnings} = \frac{\text{market price per share}}{\text{earnings per share}} \qquad 58.67$$

51. COST ACCOUNTING

Cost accounting is the system that determines the cost of manufactured products. Cost accounting is called *job cost accounting* if costs are accumulated by part number or contract. It is called *process cost accounting* if costs are accumulated by departments or manufacturing processes.

Cost accounting is dependent on historical and recorded data. The unit product cost is determined from actual expenses and numbers of units produced. Allowances (i.e., budgets) for future costs are based on these historical figures. Any deviation from historical figures is called a *variance.* Where adequate records are available, variances can be divided into *labor variance* and *material variance.*

When determining a unit product cost, the direct material and direct labor costs are generally clear-cut and easily determined. Furthermore, these costs are 100% variable costs. However, the indirect cost per unit of product is not as easily determined. Indirect costs (*burden, overhead,* etc.) can be fixed or semivariable costs. The amount of indirect cost allocated to a unit will depend on the unknown future overhead expense as well as the unknown future production (*vehicle size*).

A typical method of allocating indirect costs to a product is as follows.

step 1: Estimate the total expected indirect (and overhead) costs for the upcoming year.

step 2: Determine the most appropriate vehicle (basis) for allocating the overhead to production. Usually, this vehicle is either the number of units expected to be produced or the number of direct hours expected to be worked in the upcoming year.

step 3: Estimate the quantity or size of the overhead vehicle.

Professional Practice

step 4: Divide expected overhead costs by the expected overhead vehicle to obtain the unit overhead.

step 5: Regardless of the true size of the overhead vehicle during the upcoming year, one unit of overhead cost is allocated per unit of overhead vehicle.

Once the prime cost has been determined and the indirect cost calculated based on projections, the two are combined into a *standard factory cost* or *standard cost*, which remains in effect until the next budgeting period (usually a year).

During the subsequent manufacturing year, the standard cost of a product is not generally changed merely because it is found that an error in projected indirect costs or production quantity (vehicle size) has been made. The allocation of indirect costs to a product is assumed to be independent of errors in forecasts. Rather, the difference between the expected and actual expenses, known as the *burden (overhead) variance*, experienced during the year is posted to one or more *variance accounts*.

Burden (overhead) variance is caused by errors in forecasting both the actual indirect expense for the upcoming year and the overhead vehicle size. In the former case, the variance is called *burden budget variance*; in the latter, it is called *burden capacity variance*.

Example 58.29

A company expects to produce 8000 items in the coming year. The current material cost is \$4.54 each. Sixteen minutes of direct labor are required per unit. Workers are paid \$7.50 per hour. 2133 direct labor hours are forecasted for the product. Miscellaneous overhead costs are estimated at \$45,000.

Find the per-unit (a) expected direct material cost, (b) direct labor cost, (c) prime cost, (d) burden as a function of production and direct labor, and (e) total cost.

Solution

(a) The direct material cost was given as \$4.54.

(b) The direct labor cost is

$$\left(\frac{16 \text{ min}}{60 \frac{\text{min}}{\text{hr}}}\right)\left(\frac{\$7.50}{\text{hr}}\right) = \$2.00$$

(c) The prime cost is

$$\$4.54 + \$2.00 = \$6.54$$

(d) If the burden vehicle is production, the burden rate is \$45,000/8000 = \$5.63 per item.

If the burden vehicle is direct labor hours, the burden rate is \$45,000/2133 = \$21.10 per hour.

(e) If the burden vehicle is production, the total cost is

$$\$4.54 + \$2.00 + \$5.63 = \$12.17$$

If the burden vehicle is direct labor hours, the total cost is

$$\$4.54 + \$2.00 + \left(\frac{16 \text{ min}}{60 \frac{\text{min}}{\text{hr}}}\right)\left(\frac{\$21.10}{\text{hr}}\right) = \$12.17$$

Example 58.30

The actual performance of the company in Ex. 58.27 is given by the following figures.

actual production: 7560

actual overhead costs: \$47,000

What are the burden budget variance and the burden capacity variance?

Solution

The burden capacity variance is

$$\$45{,}000 - (7560)(\$5.63) = \$2437$$

The burden budget variance is

$$\$47{,}000 - \$45{,}000 = \$2000$$

The overall burden variance is

$$\$47{,}000 - (7560)(\$5.63) = \$4437$$

The sum of the burden capacity and burden budget variances should equal the overall burden variance.

$$\$2437 + \$2000 = \$4437$$

52. COST OF GOODS SOLD

Cost of goods sold (COGS) is an accounting term that represents an inventory account adjustment.[49] Cost of goods sold is the difference between the starting and ending inventory valuations. That is,

$$\begin{aligned} \text{COGS} &= \text{starting inventory valuation} \\ &\quad - \text{ending inventory valuation} \end{aligned} \qquad 58.68$$

Cost of goods sold is subtracted from *gross profit* to determine the *net profit* of a company. Despite the fact that cost of goods sold can be a significant element in the profit equation, the inventory adjustment may not be made each accounting period (e.g., each month) due to the difficulty in obtaining an accurate inventory valuation.

[49]The cost of goods sold inventory adjustment is posted to the COGS *expense account*.

Professional Practice

With a *perpetual inventory system*, a company automatically maintains up-to-date inventory records, either through an efficient stocking and stock-releasing system or through a *point of sale* (POS) *system* integrated with the inventory records. If a company only counts its inventory (i.e., takes a *physical inventory*) at regular intervals (e.g., once a year), it is said to be operating on a *periodic inventory system*.

Inventory accounting is a source of many difficulties. The inventory value is calculated by multiplying the quantity on hand by the standard cost. In the case of completed items actually assembled or manufactured at the company, this standard cost usually is the manufacturing cost, although factory cost also can be used. In the case of purchased items, the standard cost will be the cost per item charged by the supplying vendor. In some cases, delivery and transportation costs will be included in this standard cost.

It is not unusual for the elements in an item's inventory to come from more than one vendor, or from one vendor in more than one order. Inventory valuation is more difficult if the price paid is different for these different purchases. There are four methods of determining the cost of elements in inventory. Any of these methods can be used (if applicable), but the method must be used consistently from year to year. The four methods are as follows.

- *specific identification method:* Each element can be uniquely associated with a cost. Inventory elements with serial numbers fit into this costing scheme. Stock, production, and sales records must include the serial number.
- *average cost method:* The standard cost of an item is the average of (recent or all) purchase costs for that item.
- *first-in, first-out* (FIFO) *method:* This method keeps track of how many of each item were purchased each time and the number remaining out of each purchase, as well as the price paid at each purchase. The inventory system assumes that the oldest elements are issued first.[50] Inventory value is a weighted average dependent on the number of elements from each purchase remaining. Items issued no longer contribute to the inventory value.
- *last-in, first-out* (LIFO) *method:* This method keeps track of how many of each item were purchased each time and the number remaining out of each purchase, as well as the price paid at each purchase.[51] The inventory value is a weighted average dependent on the number of elements from each purchase remaining. Items issued no longer contribute to the inventory value.

[50]If all elements in an item's inventory are identical, and if all shipments of that item are agglomerated, there will be no way to guarantee that the oldest element in inventory is issued first. But, unless *spoilage* is a problem, it really does not matter.

[51]See previous footnote.

53. BREAK-EVEN ANALYSIS

Special Nomenclature

a	incremental cost to produce one additional item (also called *marginal cost* or *differential cost*)
C	total cost
f	fixed cost that does not vary with production
p	incremental value (price)
Q	quantity sold
Q^*	break-even quantity sold
R	total revenue

Break-even analysis is a method of determining when the value of one alternative becomes equal to the value of another. A common application is that of determining when costs exactly equal revenue. If the manufactured quantity is less than the break-even quantity, a loss is incurred. If the manufactured quantity is greater than the break-even quantity, a profit is made. (See Fig. 58.13.)

Figure 58.13 Break-Even Quantity

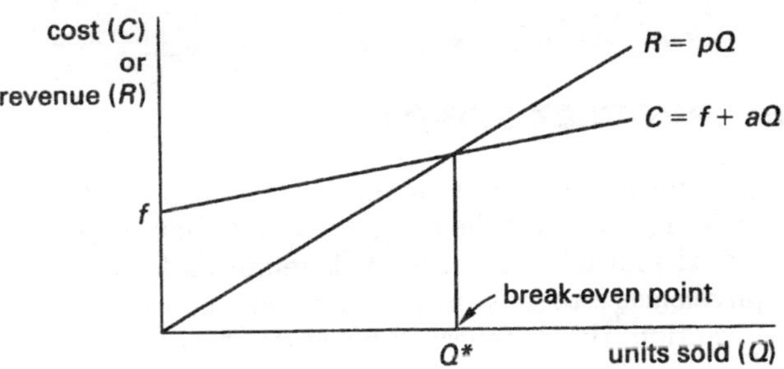

Assuming no change in the inventory, the *break-even point* can be found by setting costs equal to revenue ($C = R$).

$$C = f + aQ \quad 58.69$$

$$R = pQ \quad 58.70$$

$$Q^* = \frac{f}{p - a} \quad 58.71$$

An alternative form of the break-even problem is to find the number of units per period for which two alternatives have the same total costs. Fixed costs are to be spread over a period longer than one year using the equivalent uniform annual cost (EUAC) concept. One of the alternatives will have a lower cost if production is less than the break-even point. The other will have a lower cost for production greater than the break-even point.

Example 58.31

Two plans are available for a company to obtain automobiles for its salesmen. How many miles must the cars

Professional Practice

be driven each year for the two plans to have the same costs? Use an interest rate of 10%. (Use the year-end convention for all costs.)

plan A: Lease the cars and pay $0.15 per mile.

plan B: Purchase the cars for $5000. Each car has an economic life of three years, after which it can be sold for $1200. Gas and oil cost $0.04 per mile. Insurance is $500 per year.

Solution

Let x be the number of miles driven per year. Then, the EUAC for both alternatives is

$$\begin{aligned}\text{EUAC(A)} &= 0.15x \\ \text{EUAC(B)} &= \$0.04x + \$500 + (\$5000)(A/P, 10\%, 3) \\ &\quad - (\$1200)(A/F, 10\%, 3) \\ &= \$0.04x + \$500 + (\$5000)(0.4021) \\ &\quad - (\$1200)(0.3021) \\ &= \$0.04x + \$2148\end{aligned}$$

Setting EUAC(A) and EUAC(B) equal and solving for x yields 19,527 miles per year as the break-even point.

54. PAY-BACK PERIOD

The *pay-back period* is defined as the length of time, usually in years, for the cumulative net annual profit to equal the initial investment. It is tempting to introduce equivalence into pay-back period calculations, but by convention, this is generally not done.[52]

$$\text{pay-back period} = \frac{\text{initial investment}}{\text{net annual profit}} \qquad 58.72$$

Example 58.32

A ski resort installs two new ski lifts at a total cost of $1,800,000. The resort expects the annual gross revenue to increase by $500,000 while it incurs an annual expense of $50,000 for lift operation and maintenance. What is the pay-back period?

Solution

From Eq. 58.71,

$$\text{pay-back period} = \frac{\$1{,}800{,}000}{\frac{\$500{,}000}{\text{yr}} - \frac{\$50{,}000}{\text{yr}}} = 4 \text{ years}$$

[52]Equivalence (i.e., interest and compounding) generally is not considered when calculating the "pay-back period." However, if it is desirable to include equivalence, then the term *pay-back period* should not be used. Other terms, such as *cost recovery period* or *life of an equivalent investment,* should be used. Unfortunately, this convention is not always followed in practice.

55. MANAGEMENT GOALS

Depending on many factors (market position, age of the company, age of the industry, perceived marketing and sales windows, etc.), a company may select one of many production and marketing strategic goals. Three such strategic goals are

- maximization of product demand
- minimization of cost
- maximization of profit

Such goals require knowledge of how the dependent variable (e.g., demand quantity or quantity sold) varies as a function of the independent variable (e.g., price). Unfortunately, these three goals are not usually satisfied simultaneously. For example, minimization of product cost may require a large production run to realize economies of scale, while the actual demand is too small to take advantage of such economies of scale.

If sufficient data are available to plot the independent and dependent variables, it may be possible to optimize the dependent variable graphically. (See Fig. 58.14.) Of course, if the relationship between independent and dependent variables is known algebraically, the dependent variable can be optimized by taking derivatives or by use of other numerical methods.

Figure 58.14 Graphs of Management Goal Functions

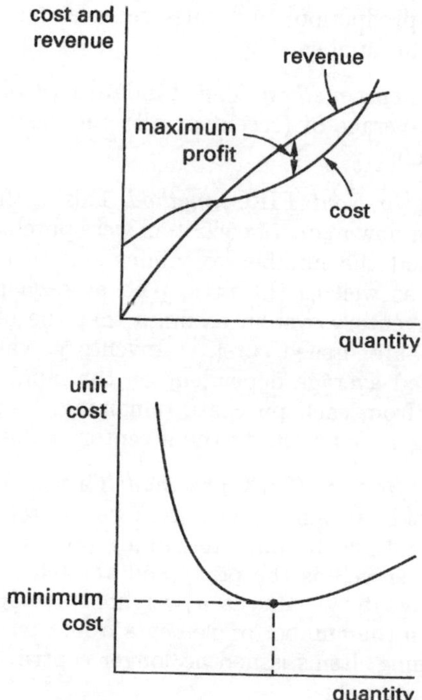

56. INFLATION

It is important to perform economic studies in terms of *constant value dollars.* One method of converting all cash flows to constant value dollars is to divide the flows by some annual *economic indicator* or price index.

If indicators are not available, cash flows can be adjusted by assuming that inflation is constant at a decimal rate, e, per year. Then, all cash flows can be converted to $t = 0$ dollars by dividing by $(1 + e)^n$, where n is the year of the cash flow.

An alternative is to replace the effective annual interest rate, i, with a value corrected for inflation. This corrected value, i', is

$$i' = i + e + ie \qquad 58.73$$

This method has the advantage of simplifying the calculations. However, precalculated factors are not available for the non-integer values of i'. Therefore, Table 58.1 must be used to calculate the factors.

Example 58.33

What is the uninflated present worth of a $2000 future value in two years if the average inflation rate is 6% and i is 10%?

Solution

$$\begin{aligned} P &= \frac{F}{(1+i)^n(1+e)^n} \\ &= \frac{\$2000}{(1+0.10)^2(1+0.06)^2} \\ &= \$1471 \end{aligned}$$

Example 58.34

Repeat Ex. 58.33 using Eq. 58.73.

Solution

$$\begin{aligned} i' &= i + e + ie \\ &= 0.10 + 0.06 + (0.10)(0.06) = 0.166 \end{aligned}$$

$$P = \frac{F}{(1+i')^2} = \frac{\$2000}{(1+0.166)^2} = \$1471$$

57. CONSUMER LOANS

Special Nomenclature

BAL_j	balance after the jth payment
j	payment or period number
LV	principal total value loaned (cost minus down payment)
N	total number of payments to pay off the loan
PI_j	jth interest payment
PP_j	jth principal payment
PT_j	jth total payment
ϕ	effective rate per period (r/k)

Many different arrangements can be made between a borrower and a lender. With the advent of creative financing concepts, it often seems that there are as many variations of loans as there are loans made. Nevertheless, there are several traditional types of transactions. Real estate or investment texts, or a financial consultant, should be consulted for more complex problems.

Simple Interest

Interest due does not compound with a *simple interest loan*. The interest due is merely proportional to the length of time that the principal is outstanding. Because of this, simple interest loans are seldom made for long periods (e.g., more than one year). (For loans less than one year, it is commonly assumed that a year consists of 12 months of 30 days each.)

Example 58.35

A $12,000 simple interest loan is taken out at 16% per annum interest rate. The loan matures in two years with no intermediate payments. How much will be due at the end of the second year?

Solution

The interest each year is

$$PI = (0.16)(\$12{,}000) = \$1920$$

The total amount due in two years is

$$PT = \$12{,}000 + (2)(\$1920) = \$15{,}840$$

Example 58.36

$4000 is borrowed for 75 days at 16% per annum simple interest. There are 360 banking days per year. How much will be due at the end of 75 days?

Solution

$$\begin{aligned} \text{amount due} &= \$4000 + (0.16)\left(\frac{75 \text{ days}}{360 \dfrac{\text{days}}{\text{bank yr}}}\right)(\$4000) \\ &= \$4133 \end{aligned}$$

Loans with Constant Amount Paid Toward Principal

With this loan type, the payment is not the same each period. The amount paid toward the principal is constant, but the interest varies from period to period.

Professional Practice

(See Fig. 58.15.) The equations that govern this type of loan are

$$BAL_j = LV - j(PP) \quad 58.74$$

$$PI_j = \phi(BAL)_{j-1} \quad 58.75$$

$$PT_j = PP + PI_j \quad 58.76$$

$$PP = \frac{LV}{N} \quad 58.77$$

$$N = \frac{LV}{PP} \quad 58.78$$

$$LV = (PP + PI_1)(P/A, \phi, N) - PI_N(P/G, \phi, N) \quad 58.79$$

$$1 = \left(\frac{1}{N} + \phi\right)(P/A, \phi, N) - \left(\frac{\phi}{N}\right)(P/G, \phi, N) \quad 58.80$$

Figure 58.15 *Loan with Constant Amount Paid Toward Principal*

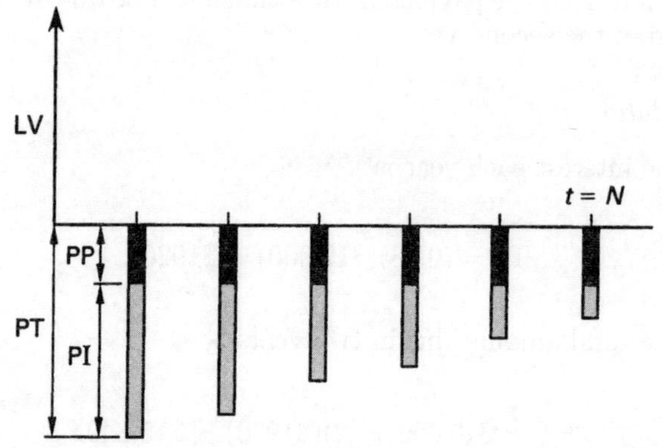

Example 58.37

A $12,000 six-year loan is taken from a bank that charges 15% effective annual interest. Payments toward the principal are uniform, and repayments are made at the end of each year. Tabulate the interest, total payments, and the balance remaining after each payment is made.

Solution

The amount of each principal payment is

$$PP = \frac{LV}{N} = \frac{\$12{,}000}{6} = \$2000$$

At the end of the first year (before the first payment is made), the principal balance is $12,000 (i.e., BAL_0 = $12,000). From Eq. 58.64, the interest payment is

$$PI_1 = \phi(BAL)_0 = (0.15)(\$12{,}000) = \$1800$$

The total first payment is

$$PT_1 = PP + PI = \$2000 + \$1800 = \$3800$$

The following table is similarly constructed.

j	BAL_j	PP_j	PI_j	PT_j
	(in dollars)			
0	12,000	–	–	–
1	10,000	2000	1800	3800
2	8000	2000	1500	3500
3	6000	2000	1200	3200
4	4000	2000	900	2900
5	2000	2000	600	2600
6	0	2000	300	2300

Direct Reduction Loans

This is the typical "interest paid on unpaid balance" loan. The amount of the periodic payment is constant, but the amounts paid toward the principal and interest both vary. (See Fig. 58.16.)

$$BAL_{j-1} = PT\left(\frac{1 - (1+\phi)^{j-1-N}}{\phi}\right) \quad 58.81$$

$$PI_j = \phi(BAL)_{j-1} \quad 58.82$$

$$PP_j = PT - PI_j \quad 58.83$$

$$BAL_j = BAL_{j-1} - PP_j \quad 58.84$$

$$N = \frac{-\ln\left(1 - \frac{\phi(LV)}{PT}\right)}{\ln(1+\phi)} \quad 58.85$$

Figure 58.16 *Direct Reduction Loan*

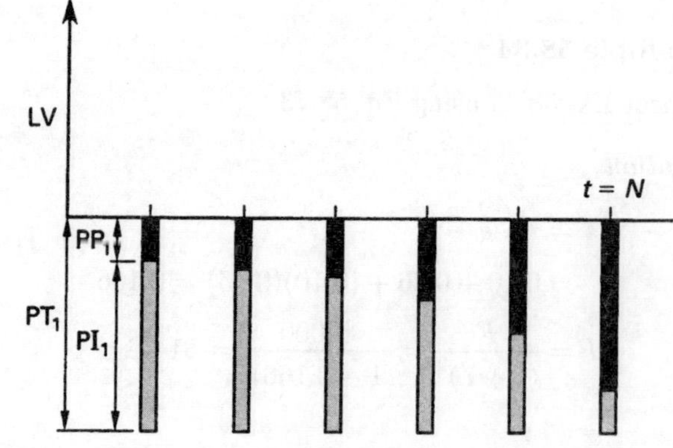

Equation 58.85 calculates the number of payments necessary to pay off a loan. This equation can be solved with effort for the total periodic payment (PT) or the initial value of the loan (LV). It is easier, however, to use the $(A/P, I\%, n)$ factor to find the payment and loan value.

$$PT = LV(A/P, \phi\%, N) \quad 58.86$$

If the loan is repaid in yearly installments, then i is the effective annual rate. If the loan is paid off monthly, then i should be replaced by the effective rate per month

Professional Practice

(ϕ from Eq. 58.54). For monthly payments, N is the number of months in the loan period.

Example 58.38

A \$45,000 loan is financed at 9.25% per annum. The monthly payment is \$385. What are the amounts paid toward interest and principal in the 14th period? What is the remaining principal balance after the 14th payment has been made?

Solution

The effective rate per month is

$$\begin{aligned}\phi &= \frac{r}{k} = \frac{0.0925}{12}\\ &= 0.0077083\ldots \quad \text{[use 0.007708]}\end{aligned}$$

From Eq. 58.74,

$$\begin{aligned}N &= \frac{-\ln\left(1 - \frac{\phi(\text{LV})}{\text{PT}}\right)}{\ln(1+\phi)}\\ &= \frac{-\ln\left(1 - \frac{(0.007708)(45{,}000)}{385}\right)}{\ln(1+0.007708)}\\ &= 301\end{aligned}$$

From Eq. 58.71,

$$\begin{aligned}\text{BAL}_{14-1} &= \text{PT}\left(\frac{1-(1+\phi)^{14-1-N}}{\phi}\right)\\ &= (\$385)\left(\frac{1-(1+0.007708)^{14-1-301}}{0.007708}\right)\\ &= \$44{,}476.39\end{aligned}$$

From Eq. 58.71,

$$\begin{aligned}\text{PI}_{14} &= \phi(\text{BAL})_{14-1}\\ &= (0.007708)(\$44{,}476.39)\\ &= \$342.82\end{aligned}$$

From Eq. 58.72,

$$\text{PP}_{14} = \text{PT} - \text{PI}_{14} = \$385 - \$342.82 = \$42.18$$

Therefore, using Eq. 58.73, the remaining principal balance is

$$\begin{aligned}\text{BAL}_{14} &= \text{BAL}_{14-1} - \text{PP}_{14}\\ &= \$44{,}476.39 - \$42.18\\ &= \$44{,}434.21\end{aligned}$$

Direct Reduction Loans with Balloon Payments

This type of loan has a constant periodic payment, but the duration of the loan is insufficient to completely pay back the principal (i.e, the loan is not fully amortized). Therefore, all remaining unpaid principal must be paid back in a lump sum when the loan matures. This large payment is known as a *balloon payment*.[53] (See Fig. 58.17.)

Figure 58.17 Direct Reduction Loan with Balloon Payment

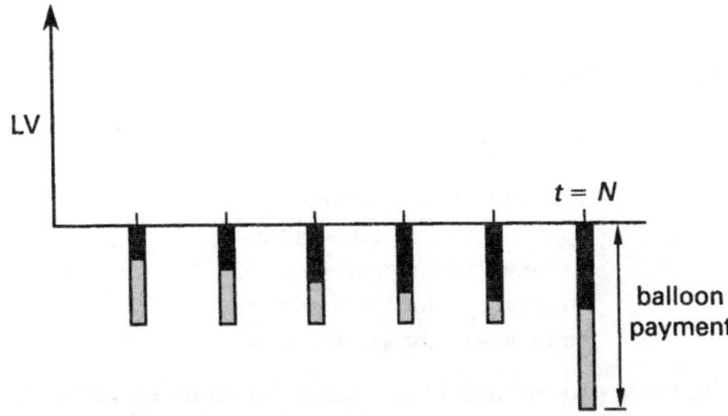

Equation 58.69 through Eq. 58.71 also can be used with this type of loan. The remaining balance after the last payment is the balloon payment. This balloon payment must be repaid along with the last regular payment calculated.

58. FORECASTING

There are many types of forecasting models, although most are variations of the basic types.[54] All models produce a *forecast* (F_{t+1}) of some quantity (*demand* is used in this section) in the next period based on actual measurements (D_j) in current and prior periods. All of the models also try to provide *smoothing* (or *damping*) of extreme data points.

Forecasts by Moving Averages

The method of *moving average forecasting* weights all previous demand data points equally and provides some smoothing of extreme data points. The amount of smoothing increases as the number of data points, n, increases.

$$F_{t+1} = \frac{1}{n}\sum_{m=t+1-n}^{t} D_m \qquad 58.87$$

[53]The term *balloon payment* may include the final interest payment as well. Generally, the problem statement will indicate whether the balloon payment is inclusive or exclusive of the regular payment made at the end of the loan period.

[54]For example, forecasting models that take into consideration steady (linear), cyclical, annual, and seasonal trends are typically variations of the exponentially weighted model. A truly different forecasting tool, however, is *Monte Carlo simulation*, which predicts trajectories for variables based on assumed interactions.

Professional Practice

Forecasts by Exponentially Weighted Averages

With *exponentially weighted forecasts*, the more current (most recent) data points receive more weight. This method uses a *weighting factor* (α), also known as a *smoothing coefficient*, which typically varies between 0.01 and 0.30. An initial forecast is needed to start the method. Forecasts immediately following are sensitive to the accuracy of this first forecast. It is common to choose $F_0 = D_1$ to get started.

$$F_{t+1} = \alpha D_t + (1-\alpha)F_t \qquad 58.88$$

59. LEARNING CURVES

Special Nomenclature

b	learning curve constant
n	total number of items produced
R	decimal learning curve rate (2^{-b})
T_1	time or cost for the first item
T_n	time or cost for the nth item

The more products that are made, the more efficient the operation becomes due to experience gained. Therefore, direct labor costs decrease.[55] Usually, a *learning curve* is specified by the decrease in cost each time the cumulative quantity produced doubles. If there is a 20% decrease per doubling, the curve is said to be an 80% learning curve (i.e., the *learning curve rate*, R, is 80%).

Then, the time to produce the nth item is

$$T_n = T_1 n^{-b} \qquad 58.89$$

The total time to produce units from quantity n_1 to n_2 inclusive is approximately given by Eq. 58.79. T_1 is a constant, the time for item 1, and does not correspond to n unless $n_1 = 1$.

$$\int_{n_1}^{n_2} T_n\,dn \approx \left(\frac{T_1}{1-b}\right)\left(\left(n_2+\tfrac{1}{2}\right)^{1-b} - \left(n_1-\tfrac{1}{2}\right)^{1-b}\right) \qquad 58.90$$

The *average time per unit* over the production from n_1 to n_2 is the above total time from Eq. 58.79 divided by the quantity produced, $(n_2 - n_1 + 1)$.

$$T_{\text{ave}} = \frac{\displaystyle\int_{n_1}^{n_2} T_n\,dn}{n_2 - n_1 + 1} \qquad 58.91$$

Table 58.9 lists representative values of the *learning curve constant*, b. For learning curve rates not listed in the table, Eq. 58.81 can be used to find b.

$$b = \frac{-\log_{10}R}{\log_{10}(2)} = \frac{-\log_{10}R}{0.301} \qquad 58.92$$

[55] Learning curve reductions apply only to direct labor costs. They are not applied to indirect labor or direct material costs.

Table 58.9 Learning Curve Constants

learning curve rate, R	b
0.70 (70%)	0.515
0.75 (75%)	0.415
0.80 (80%)	0.322
0.85 (85%)	0.234
0.90 (90%)	0.152
0.95 (95%)	0.074

Example 58.39

A 70% learning curve is used with an item whose first production time is 1.47 hr. (a) How long will it take to produce the 11th item? (b) How long will it take to produce the 11th through 27th items?

Solution

(a) From Eq. 58.78,

$$T_{11} = T_1 n^{-b} = (1.47\text{ hr})(11)^{-0.515} = 0.428\text{ hr}$$

(b) The time to produce the 11th item through 27th item is given by Eq. 58.79.

$$T \approx \left(\frac{T_{11}}{1-b}\right)\left(\left(n_{27}+\tfrac{1}{2}\right)^{1-b} - \left(n_{11}-\tfrac{1}{2}\right)^{1-b}\right)$$

$$T \approx \left(\frac{1.47\text{ hr}}{1-0.515}\right)\left((27.5)^{1-0.515} - (10.5)^{1-0.515}\right) = 5.643\text{ hr}$$

60. ECONOMIC ORDER QUANTITY

Special Nomenclature

a	constant depletion rate (items/unit time)
h	inventory storage cost ($/item-unit time)
H	total inventory storage cost between orders ($)
K	fixed cost of placing an order ($)
Q	order quantity (original quantity on hand)
t^*	time at depletion

The *economic order quantity* (EOQ) is the order quantity that minimizes the inventory costs per unit time. Although there are many different EOQ models, the simplest is based on the following assumptions.

- Reordering is instantaneous. The time between order placement and receipt is zero, as shown in Fig. 58.18.
- Shortages are not allowed.
- Demand for the inventory item is deterministic (i.e., is not a random variable).
- Demand is constant with respect to time.
- An order is placed when the inventory is zero.

Professional Practice

Figure 58.18 Inventory with Instantaneous Reorder

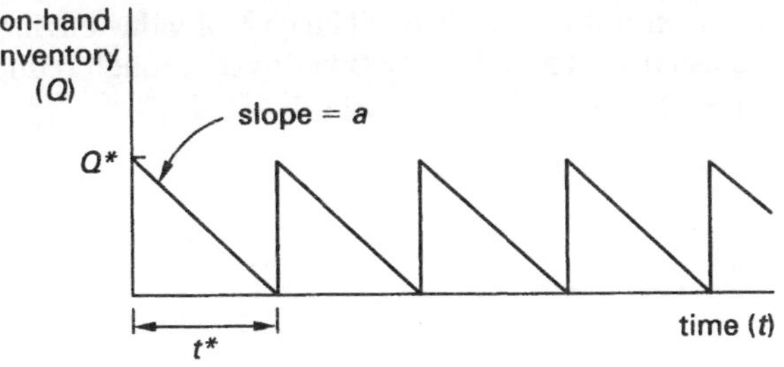

Figure 58.19 Types of Sensitivity

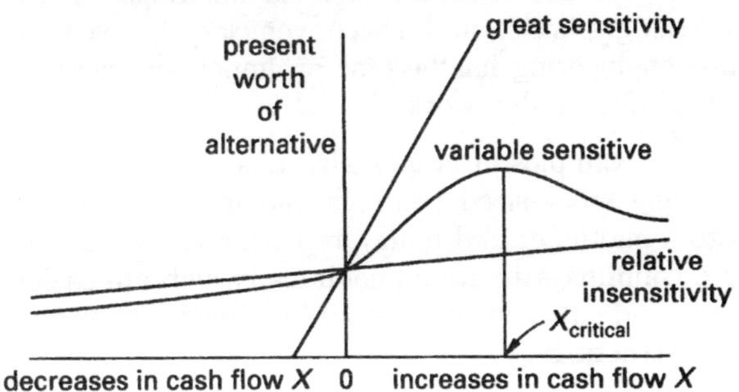

If the original quantity on hand is Q, the stock will be depleted at

$$t^* = \frac{Q}{a} \quad 58.93$$

The total inventory storage cost between t_0 and t^* is

$$H = \tfrac{1}{2}Qht^* = \frac{Q^2h}{2a} \quad 58.94$$

The total inventory and ordering cost per unit time is

$$C_t = \frac{aK}{Q} + \frac{hQ}{2} \quad 58.95$$

C_t can be minimized with respect to Q. The economic order quantity and time between orders are

$$Q^* = \sqrt{\frac{2aK}{h}} \quad 58.96$$

$$t^* = \frac{Q^*}{a} \quad 58.97$$

61. SENSITIVITY ANALYSIS

Data analysis and forecasts in economic studies require estimates of costs that will occur in the future. There are always uncertainties about these costs. However, these uncertainties are insufficient reason not to make the best possible estimates of the costs. Nevertheless, a decision between alternatives often can be made more confidently if it is known whether or not the conclusion is sensitive to moderate changes in data forecasts. Sensitivity analysis provides this extra dimension to an economic analysis.

The sensitivity of a decision is determined by inserting a range of estimates for critical cash flows and other parameters. If radical changes can be made to a cash flow without changing the decision, the decision is said to be *insensitive* to uncertainties regarding that cash flow. However, if a small change in the estimate of a cash flow will alter the decision, that decision is said to be very *sensitive* to changes in the estimate. If the decision is sensitive only for a limited range of cash flow values, the term *variable sensitivity* is used. Figure 58.19 illustrates these terms.

An established semantic tradition distinguishes between risk analysis and uncertainty analysis. *Risk analysis* addresses variables that have a known or estimated probability distribution. In this regard, statistics and probability theory can be used to determine the probability of a cash flow varying between given limits. On the other hand, *uncertainty analysis* is concerned with situations in which there is not enough information to determine the probability or frequency distribution for the variables involved.

As a first step, sensitivity analysis should be applied one at a time to the dominant factors. Dominant cost factors are those that have the most significant impact on the present value of the alternative.[56] If warranted, additional investigation can be used to determine the sensitivity to several cash flows varying simultaneously. Significant judgment is needed, however, to successfully determine the proper combinations of cash flows to vary. It is common to plot the dependency of the present value on the cash flow being varied in a two-dimensional graph. Simple linear interpolation is used (within reason) to determine the critical value of the cash flow being varied.

62. VALUE ENGINEERING

The *value* of an investment is defined as the ratio of its return (performance or utility) to its cost (effort or investment). The basic object of *value engineering* (VE, also referred to as *value analysis*) is to obtain the maximum per-unit value.[57]

Value engineering concepts often are used to reduce the cost of mass-produced manufactured products. This is done by eliminating unnecessary, redundant, or superfluous features, by redesigning the product for a less expensive manufacturing method, and by including features for easier assembly without sacrificing utility and

[56]In particular, engineering economic analysis problems are sensitive to the choice of effective interest rate, *i*, and to accuracy in cash flows at or near the beginning of the horizon. The problems will be less sensitive to accuracy in far-future cash flows, such as salvage value and subsequent generation replacement costs.

[57]Value analysis, the methodology that has become today's value engineering, was developed in the early 1950s by Lawrence D. Miles, an analyst at General Electric.

function.[58] However, the concepts are equally applicable to one-time investments, such as buildings, chemical processing plants, and space vehicles. In particular, value engineering has become an important element in all federally funded work.[59]

Typical examples of large-scale value engineering work are using stock-sized bearings and motors (instead of custom manufactured units), replacing rectangular concrete columns with round columns (which are easier to form), and substituting custom buildings with prefabricated structures.

Value engineering is usually a team effort. And, while the original designers may be on the team, usually outside consultants are utilized. The cost of value engineering is usually returned many times over through reduced construction and life-cycle costs.

[58]Some people say that value engineering is "the act of going over the plans and taking out everything that is interesting."
[59]U.S. Government Office of Management and Budget Circular A-131 outlines value engineering for federally funded construction projects.

59 Engineering Law[1]

1. FORMS OF COMPANY OWNERSHIP

There are three basic forms of company ownership in the United States: (a) sole proprietorship, (b) partnership, and (c) corporation.[2] Each of these forms of ownership has advantages and disadvantages.

2. SOLE PROPRIETORSHIPS

A *sole proprietorship* (*single proprietorship*) is the easiest form of ownership to establish. Other than the necessary licenses, filings, and notices (which apply to all forms of ownership), no legal formalities are required to start business operations. A sole proprietor (the owner) has virtually total control of the business and makes all supervisory and management decisions.

[1]This chapter is not intended to be a substitute for professional advice. Law is not always black and white. For every rule there are exceptions. For every legal principle, there are variations. For every type of injury, there are numerous legal precedents. This chapter covers the superficial basics of a small subset of U.S. law affecting engineers.

[2]The discussion of forms of company ownership in Sec. 59.2, Sec. 59.3, and Sec. 59.4 applies equally to service-oriented companies (e.g., consulting engineering firms) and product-oriented companies.

Legally, there is no distinction between the sole proprietor and the sole proprietorship (the business). This is the greatest disadvantage of this form of business. The owner is solely responsible for the operation of the business, even if the owner hires others for assistance. The owner assumes personal, legal, and financial liability for all acts and debts of the company. If the company debts remain unpaid, or in the event there is a legal judgment against the company, the owner's personal assets (home, car, savings, etc.) can be seized or attached.

Another disadvantage of the sole proprietorship is the lack of significant organizational structure. In times of business crisis or trouble, there may be no one to share the responsibility or to help make decisions. When the owner is sick or dies, there may be no way to continue the business.

There is also no distinction between the incomes of the business and the owner. Therefore, the business income is taxed at the owner's income tax rate. Depending on the owner's financial position, the success of the business, and the tax structure, this can be an advantage or a disadvantage.[3]

3. PARTNERSHIPS

A *partnership* (also known as a *general partnership*) is ownership by two or more persons known as *general partners.* Legally, this form is very similar to a sole proprietorship, and the two forms of business have many of the same advantages and disadvantages. For example, with the exception of an optional *partnership agreement,* there are a minimum of formalities to setting up business. The partners make all business and management decisions themselves according to an agreed-upon process. The business income is split among the partners and taxed at the partners' individual tax rates.[4] Continuity of the business is still a problem since most partnerships are automatically dissolved upon the withdrawal or death of one of the partners.[5]

One advantage of a partnership over a sole proprietorship is the increase in available funding. Not only do more partners bring in more start-up capital, but the

[3]To use a simplistic example, if the corporate tax rates are higher than the individual tax rates, it would be *financially* better to be a sole proprietor because the company income would be taxed at a lower rate.

[4]The percentage split is specified in the partnership agreement.

[5]Some or all of the remaining partners may want to form a new partnership, but this is not always possible.

Professional Practice

resource pool may make business credit easier to obtain. Also, the partners bring a diversity of skills and talents.

Unless the partnership agreement states otherwise, each partner can individually obligate (i.e., *bind*) the partnership without the consent of the other partners. Similarly, each partner has personal responsibility and liability for the acts and debts of the partnership company, just as sole proprietors do. In fact, each partner assumes the *sole* responsibility, not just a proportionate share. If one or more partners are unable to pay, the remaining partners shoulder the entire debt. The possibility of one partner having to pay for the actions of another partner must be considered when choosing this form of business ownership.

A *limited partnership* differs from a general partnership in that one (or more) of the partners is silent. The *limited partners* make a financial contribution to the business and receive a share of the profit but do not participate in the management and cannot bind the partnership. While *general partners* have unlimited personal liabilities, limited partners are generally liable only to the extent of their investment.[6] A written partnership agreement is required, and the agreement must be filed with the proper authorities.

4. CORPORATIONS

A corporation is a legal entity (i.e., a legal person) distinct from the founders and owners. The separation of ownership and management makes the corporation a fundamentally different business form than a sole proprietorship or partnership, with very different advantages and disadvantages.

A corporation becomes legally distinct from its founders upon formation and proper registration. Ownership of the corporation is through shares of stock, distributed to the founders and investors according to some agreed-upon investment and distribution rule. The founders and investors become the stockholders (i.e., owners) of the corporation. A *closely held* (*private*) *corporation* is one in which all stock is owned by a family or small group of co-investors. A *public corporation* is one whose stock is available for the public-at-large to purchase.

There is no mandatory connection between ownership and management functions. The decision-making power is vested in the executive officers and a *board of directors* that governs by majority vote. The stockholders elect the board of directors which, in turn, hires the executive officers, management, and other employees. Employees of the corporation may or may not be stockholders.

Disadvantages (at least for a person or persons who could form a partnership or sole proprietorship) include the higher corporate tax rate, difficulty and complexity of formation (some states require a minimum number of persons on the board of directors), and additional legal and accounting paperwork.

However, since a corporation is distinctly separate from its founders and investors, those individuals are not liable for the acts and debts of the corporation. Debts are paid from the corporate assets. Income to the corporation is not taxable income to the owners. (Only the salaries, if any, paid to the employees by the corporation are taxable to the employees.) Even if the corporation were to go bankrupt, the assets of the owners would not ordinarily be subject to seizure or attachment.

A corporation offers the best guarantee of continuity of operation in the event of the death, incapacitation, or retirement of the founders since, as a legal entity, it is distinct from the founders and owners.

5. LIMITED LIABILITY ENTITIES

A variety of other legal entities have been established that blur the lines between the three traditional forms of business (i.e., proprietorship, partnership, and corporation). The *limited liability partnership*, LLP, extends a measure of corporate-like protection to professionals while permitting partnership-like personal participation in management decisions. Since LLPs are formed and operated under state laws, actual details vary from state to state. However, most LLPs allow the members to participate in management decisions, while not being responsible for the misdeeds of other partners. As in a corporation, the debts of the LLP do not become debts of the members.[7] The *double taxation* characteristic of traditional corporations is avoided, as profits to the LLP flow through to the members.

For engineers and architects (as well as doctors, lawyers, and accountants), the *professional corporation*, PC, offers protection from the actions (e.g., malpractice) of other professionals within a shared environment, such as a design firm. While a PC does not shield the individual from responsibility for personal negligence or malpractice, it does permit the professional to be associated with a larger entity, such as a partnership of other PCs, without accepting responsibility for the actions of the other members. In that sense, the protection is similar to that of an LLP. Unlike a traditional corporation, a PC may have a board of directors consisting of only a single individual, the professional.

The *limited liability company*,[8] LLC, also combines advantages from partnerships and corporations. In an LLC, the members are shielded from debts of the LLC

[6]That is, if the partnership fails or is liquidated to pay debts, the limited partners lose no more than their initial investments.

[7]Depending on the state, the shield may be complete or limited. It is common that the protection only applies to negligence-related claims, as opposed to intentional tort claims, contract-related obligations, and day-to-day operating expenses such as rent, utilities, and employees.

[8]LLC does not mean *limited liability corporation*. LLCs are not corporations.

while enjoying the pass-through of all profits.[9] Like a partnership or shareholder, a member's obligation is limited to the *membership interest* in (i.e., contribution to) the LLC. LLCs are directed and controlled by one or more managers who may also be members. A variation of the LLC specifically for design, medical, and other professionals is the *professional limited liability company*, PLLC.

The traditional corporation, as described in Sec. 59.4, is referred to as a *Subchapter C corporation* or "*C corp.*"[10] A variant is the *S corporation* ("*S corp*") which combines characteristics of the C corporation with pass-through for taxation. S corporations can be limited or treated differently than C corporations by state and federal law.

6. PIERCING THE CORPORATE VEIL

An individual operating as a corporation, LLP, LLC, or PC entity may lose all protection if his or her actions are fraudulent, or if the court decides the business is an "alter ego" of the individual. Basically, this requires the business to be run as a business. Business and personal assets cannot be intermingled, and business decisions must be made and documented in a business-like manner. If operated fraudulently or loosely, a court may assign liability directly to an individual, an action known as *piercing the corporate veil*.

7. AGENCY

In some contracts, decision-making authority and right of action are transferred from one party (the owner, or *principal*) who would normally have that authority to another person (the *agent*). For example, in construction contracts, the engineer may be the agent of the owner for certain transactions. Agents are limited in what they can do by the scope of the agency agreement. Within that scope, however, an agent acts on behalf of the principal, and the principal is liable for the acts of the agent and is bound by contracts made in the principal's name by the agent.

Agents are required to execute their work with care, skill, and diligence. Specifically, agents have *fiduciary responsibility* toward their principal, meaning that agent must be honest and loyal. Agents are liable for damages resulting from a lack of diligence, loyalty, and/or honesty. If the agents misrepresented their skills when obtaining the agency, they can be liable for breach of contract or fraud.

[9]LLCs enjoy *check the box taxation*, which means they can elect to be taxed as sole proprietorships, partnerships, or corporations.

[10]The reference is to subchapter C of the Internal Revenue Code.

8. GENERAL CONTRACTS

A *contract* is a legally binding agreement or promise to exchange goods or services.[11] A written contract is merely a documentation of the agreement. Some agreements must be in writing, but most agreements for engineering services can be verbal, particularly if the parties to the agreement know each other well.[12] Written contract documents do not need to contain intimidating legal language, but all agreements must satisfy three basic requirements to be enforceable (binding).

- There must be a clear, specific, and definite *offer* with no room for ambiguity or misunderstanding.
- There must be some form of conditional future *consideration* (i.e., payment).[13]
- There must be an *acceptance* of the offer.

There are other conditions that the agreement must meet to be enforceable. These conditions are not normally part of the explicit agreement but represent the conditions under which the agreement was made.

- The agreement must be *voluntary* for all parties.
- All parties must have *legal capacity* (i.e., be mentally competent, of legal age, not under coercion, and uninfluenced by drugs).
- The purpose of the agreement must be *legal*.

For small projects, a simple *letter of agreement* on one party's stationery may suffice. For larger, complex projects, a more formal document may be required. Some clients prefer to use a *purchase order*, which can function as a contract if all basic requirements are met.

Regardless of the format of the written document—letter of agreement, purchase order, or standard form—a contract should include the following features.[14]

- introduction, preamble, or preface indicating the purpose of the contract
- name, address, and business forms of both contracting parties
- signature date of the agreement

[11]Not all agreements are legally binding (i.e., enforceable). Two parties may agree on something, but unless the agreement meets all of the requirements and conditions of a contract, the parties cannot hold each other to the agreement.

[12]All states have a *statute of frauds* that, among other things, specifies what types of contracts must be in writing to be enforceable. These include contracts for the sale of land, contracts requiring more than one year for performance, contracts for the sale of goods over $500 in value, contracts to satisfy the debts of another, and marriage contracts. Contracts to provide engineering services do not fall under the statute of frauds.

[13]Actions taken or payments made prior to the agreement are irrelevant. Also, it does not matter to the courts whether the exchange is based on equal value or not.

[14]*Construction contracts* are unique unto themselves. Items that might also be included as part of the *contract documents* are the agreement form, the general conditions, drawings, specifications, and addenda.

- effective date of the agreement (if different from the signature date)
- duties and obligations of both parties
- deadlines and required service dates
- fee amount
- fee schedule and payment terms
- agreement expiration date
- standard boilerplate clauses
- signatures of parties or their agents
- declaration of authority of the signatories to bind the contracting parties
- supporting documents

9. STANDARD BOILERPLATE CLAUSES

It is common for full-length contract documents to include important *boilerplate clauses.* These clauses have specific wordings that should not normally be changed, hence the name "boilerplate." Some of the most common boilerplate clauses are paraphrased here.

- Delays and inadequate performance due to war, strikes, and acts of God and nature are forgiven (*force majeure*).
- The contract document is the complete agreement, superseding all prior verbal and written agreements.
- The contract can be modified or canceled only in writing.
- Parts of the contract that are determined to be void or unenforceable will not affect the enforceability of the remainder of the contract (*severability*). Alternatively, parts of the contract that are determined to be void or unenforceable will be rewritten to accomplish their intended purpose without affecting the remainder of the contract.
- None (or one, or both) of the parties can (or cannot) assign its (or their) rights and responsibilities under the contract (*assignment*).
- All notices provided for in the agreement must be in writing and sent to the address in the agreement.
- Time is of the essence.[15]
- The subject headings of the agreement paragraphs are for convenience only and do not control the meaning of the paragraphs.
- The laws of the state in which the contract is signed must be used to interpret and govern the contract.
- Disagreements shall be arbitrated according to the rules of the American Arbitration Association.
- Any lawsuits related to the contract must be filed in the county and state in which the contract is signed.
- Obligations under the agreement are unique, and in the event of a breach, the defaulting party waives the defense that the loss can be adequately compensated by monetary damages (*specific performance*).
- In the event of a lawsuit, the prevailing party is entitled to an award of reasonable attorneys' and court fees.[16]
- Consequential damages are not recoverable in a lawsuit.

10. SUBCONTRACTS

When a party to a contract engages a third party to perform the work in the original contract, the contract with the third party is known as a *subcontract.* Whether or not responsibilities can be subcontracted under the original contract depends on the content of the *assignment clause* in the original contract.

11. PARTIES TO A CONSTRUCTION CONTRACT

A specific set of terms has developed for referring to parties in consulting and construction contracts. The *owner* of a construction project is the person, partnership, or corporation that actually owns the land, assumes the financial risk, and ends up with the completed project. The *developer* contracts with the architect and/or engineer for the design and with the contractors for the construction of the project. In some cases, the owner and developer are the same, in which case the term *owner-developer* can be used.

The *architect* designs the project according to established codes and guidelines but leaves most stress and capacity calculations to the *engineer.*[17] Depending on the construction contract, the engineer may work for the architect, or vice versa, or both may work for the developer.

Once there are approved plans, the developer hires *contractors* to do the construction. Usually, the entire construction project is awarded to a *general contractor.* Due to the nature of the construction industry, separate *subcontracts* are used for different tasks (electrical, plumbing, mechanical, framing, fire sprinkler installation, finishing, etc.). The general contractor who hires all of these different *subcontractors* is known as the *prime contractor* (or *prime*). (The subcontractors can also work directly for the owner-developer, although this is less

[15]Without this clause in writing, damages for delay cannot be claimed.

[16]Without this clause in writing, attorneys' fees and court costs are rarely recoverable.

[17]On simple small projects, such as wood-framed residential units, the design may be developed by a *building designer.* The legal capacities of building designers vary from state to state.

common.) The prime contractor is responsible for all acts of the subcontractors and is liable for any damage suffered by the owner-developer due to those acts.

Construction is managed by an agent of the owner-developer known as the *construction manager*, who may be the engineer, the architect, or someone else.

12. STANDARD CONTRACTS FOR DESIGN PROFESSIONALS

Several professional organizations have produced standard agreement forms and other standard documents for design professionals.[18] Among other standard forms, notices, and agreements, the following standard contracts are available.[19]

- standard contract between engineer and client
- standard contract between engineer and architect
- standard contract between engineer and contractor
- standard contract between owner and construction manager

Besides completeness, the major advantage of a standard contract is that the meanings of the clauses are well established, not only among the design professionals and their clients but also in the courts. The clauses in these contracts have already been litigated many times. Where a clause has been found to be unclear or ambiguous, it has been rewritten to accomplish its intended purpose.

13. CONSULTING FEE STRUCTURE

Compensation for consulting engineering services can incorporate one or more of the following concepts.

- *lump-sum fee:* This is a predetermined fee agreed upon by client and engineer. This payment can be used for small projects where the scope of work is clearly defined.
- *cost plus fixed fee:* All costs (labor, material, travel, etc.) incurred by the engineer are paid by the client. The client also pays a predetermined fee as profit. This method has an advantage when the scope of services cannot be determined accurately in advance. Detailed records must be kept by the engineer in order to allocate costs among different clients.
- *per diem fee:* The engineer is paid a specific sum for each day spent on the job. Usually, certain direct expenses (e.g., travel and reproduction) are billed in addition to the per diem rate.
- *salary plus:* The client pays for the employees on an engineer's payroll (the salary) plus an additional percentage to cover indirect overhead and profit plus certain direct expenses.
- *retainer:* This is a minimum amount paid by the client, usually in total and in advance, for a normal amount of work expected during an agreed-upon period. None of the retainer is returned, regardless of how little work the engineer performs. The engineer can be paid for additional work beyond what is normal, however. Some direct costs, such as travel and reproduction expenses, may be billed directly to the client.
- *percentage of construction cost:* This method, which is widely used in construction design contracts, pays the architect and/or the engineer a percentage of the final total cost of the project. Costs of land, financing, and legal fees are generally not included in the construction cost, and other costs (plan revisions, project management labor, value engineering, etc.) are billed separately.

14. MECHANIC'S LIENS

For various reasons, providers and material, labor, and design services to construction sites may not be promptly paid or even paid at all. Such providers have, of course, the right to file a lawsuit demanding payment, but due to the nature of the construction industry, such relief may be insufficient or untimely. Therefore, such providers have the right to file a *mechanic's lien* (also known as a *construction lien, materialman's lien, supplier's lien,* or *laborer's lien*) against the property. Although there are strict requirements for deadlines, filing, and notices, the procedure for obtaining (and removing) such a lien is simple. The lien establishes the supplier's security interest in the property. Although the details depend on the state, essentially the property owner is prevented from transferring title (i.e., selling) the property until the lien has been removed by the supplier. The act of filing a lawsuit to obtain payment is known as "perfecting the lien." Liens are perfected by forcing a judicial foreclosure sale. The court orders the

[18]There are two main sources of standardized construction and design agreements: EJCDC and AIA. Consensus documents, known as *ConsensusDOCS*, for every conceivable situation have been developed by the *Engineers Joint Contracts Documents Committee*, EJCDC. EJCDC includes the American Society of Civil Engineers (ASCE), the American Council of Engineering Companies (ACEC), National Society of Professional Engineers' (NSPE's) Professional Engineers in Private Practice Division, Associated General Contractors of America (AGC), and more than fifteen other participating professional engineering design, construction, owner, legal, and risk management organizations, including the Associated Builders and Contractors; American Subcontractors Association; Construction Users Roundtable; National Roofing Contractors Association; Mechanical Contractors Association of America; and National Plumbing, Heating-Cooling Contractors Association. The American Institute of Architects, AIA, has developed its own standardized agreements in a less collaborative manner. Though popular with architects, AIA provisions are considered less favorable to engineers, contractors, and subcontractors who believe the AIA documents assign too much authority to architects, too much risk and liability to contractors, and too little flexibility in how construction disputes are addressed and resolved.

[19]The Construction Specifications Institute (CSI) has produced standard specifications for materials. The standards have been organized according to a UNIFORMAT structure consistent with ASTM Standard E1557.

property sold, and the proceeds are used to pay off any lien-holders.

15. DISCHARGE OF A CONTRACT

A contract is normally discharged when all parties have satisfied their obligations. However, a contract can also be terminated for the following reasons:

- mutual agreement of all parties to the contract
- impossibility of performance (e.g., death of a party to the contract)
- illegality of the contract
- material breach by one or more parties to the contract
- fraud on the part of one or more parties
- failure (i.e., loss or destruction) of consideration (e.g., the burning of a building one party expected to own or occupy upon satisfaction of the obligations)

Some contracts may be dissolved by actions of the court (e.g., bankruptcy), passage of new laws and public acts, or a declaration of war.

Extreme difficulty (including economic hardship) in satisfying the contract does not discharge it, even if it becomes more costly or less profitable than originally anticipated.

16. TORTS

A *tort* is a civil wrong committed by one person causing damage to another person or person's property, emotional well-being, or reputation.[20] It is a breach of the rights of an individual to be secure in person or property. In order to correct the wrong, a civil lawsuit (*tort action* or *civil complaint*) is brought by the alleged injured party (the *plaintiff*) against the *defendant*. To be a valid *tort action* (i.e., lawsuit), there must have been injury (i.e., damage). Generally, there will be no contract between the two parties, so the tort action cannot claim a breach of contract.[21]

Tort law is concerned with compensation for the injury, not punishment. Therefore, tort awards usually consist of general, compensatory, and special damages and rarely include punitive and exemplary damages. (See Sec. 59.17 for definitions of these damages.)

17. BREACH OF CONTRACT, NEGLIGENCE, MISREPRESENTATION, AND FRAUD

A *breach of contract* occurs when one of the parties fails to satisfy all of its obligations under a contract. The breach can be *willful* (as in a contractor walking off a construction job) or *unintentional* (as in providing less than adequate quality work or materials). A *material breach* is defined as nonperformance that results in the injured party receiving something substantially less than or different from what the contract intended.

Normally, the only redress that an *injured party* has through the courts in the event of a breach of contract is to force the breaching party to provide *specific performance*—that is, to satisfy all remaining contract provisions and to pay for any damage caused. Normally, *punitive damages* (to punish the breaching party) are unavailable.

Negligence is an action, willful or unwillful, taken without proper care or consideration for safety, resulting in damages to property or injury to persons. "Proper care" is a subjective term, but in general it is the diligence that would be exercised by a reasonably prudent person.[22] Damages sustained by a negligent act are recoverable in a tort action. (See Sec. 59.13.) If the plaintiff is partially at fault (as in the case of *comparative negligence*), the defendant will be liable only for the portion of the damage caused by the defendant.

Punitive damages are available, however, if the breaching party was fraudulent in obtaining the contract. In addition, the injured party has the right to void (nullify) the contract entirely. A *fraudulent act* is basically a special case of *misrepresentation* (i.e., an intentionally false statement known to be false at the time it is made). Misrepresentation that does not result in a contract is a tort. When a contract is involved, misrepresentation can be a breach of that contract (i.e., *fraud*).

Unfortunately, it is extremely difficult to prove *compensatory fraud* (i.e., fraud for which damages are available). Proving fraud requires showing *beyond a reasonable doubt* (a) a reckless or intentional misstatement of a material fact, (b) an intention to deceive, (c) it resulted in misleading the innocent party to contract, and (d) it was to the innocent party's detriment.

For example, if an engineer claims to have experience in designing steel buildings but actually has none, the court might consider the misrepresentation a fraudulent action. If, however, the engineer has some experience, but an insufficient amount to do an adequate job, the engineer probably will not be considered to have acted fraudulently.

[20]The difference between a *civil tort* (*lawsuit*) and a *criminal lawsuit* is the alleged injured party. A *crime* is a wrong against society. A criminal lawsuit is brought by the state against a defendant.

[21]It is possible for an injury to be both a breach of contract and a tort. Suppose an owner has an agreement with a contractor to construct a building, and the contract requires the contractor to comply with all state and federal safety regulations. If the owner is subsequently injured on a stairway because there was no guardrail, the injury could be recoverable both as a tort and as a breach of contract. If a third party unrelated to the contract was injured, however, that party could recover only through a tort action.

[22]Negligence of a design professional (e.g., an engineer or architect) is the absence of a *standard of care* (i.e., customary and normal care and attention) that would have been provided by other engineers. It is highly subjective.

Professional Practice

18. STRICT LIABILITY IN TORT

Strict liability in tort means that the injured party wins if the injury can be proven. It is not necessary to prove negligence, breach of explicit or implicit warranty, or the existence of a contract (*privity of contract*). Strict liability in tort is most commonly encountered in product liability cases. A defect in a product, regardless of how the defect got there, is sufficient to create strict liability in tort.

Case law surrounding defective products has developed and refined the following requirements for winning a strict liability in tort case. The following points must be proved.

- The product was defective in manufacture, design, labeling, and so on.
- The product was defective when used.
- The defect rendered the product unreasonably dangerous.
- The defect caused the injury.
- The specific use of the product that caused the damage was reasonably foreseeable.

19. MANUFACTURING AND DESIGN LIABILITY

Case law makes a distinction between *design professionals* (architects, structural engineers, building designers, etc.) and manufacturers of consumer products. Design professionals are generally consultants whose primary product is a design service sold to sophisticated clients. Consumer product manufacturers produce specific product lines sold through wholesalers and retailers to the unsophisticated public.

The law treats design professionals favorably. Such professionals are expected to meet a *standard of care* and skill that can be measured by comparison with the conduct of other professionals. However, professionals are not expected to be infallible. In the absence of a contract provision to the contrary, design professionals are not held to be guarantors of their work in the strict sense of legal liability. Damages incurred due to design errors are recoverable through tort actions, but proving a breach of contract requires showing negligence (i.e., not meeting the standard of care).

On the other hand, the law is much stricter with consumer product manufacturers, and perfection is (essentially) expected of them. They are held to the standard of strict liability in tort without regard to negligence. A manufacturer is held liable for all phases of the design and manufacturing of a product being marketed to the public.[23]

Prior to 1916, the court's position toward product defects was exemplified by the expression *caveat emptor* ("let the buyer beware").[24] Subsequent court rulings have clarified that "...a manufacturer is strictly liable in tort when an article [it] places on the market, knowing that it will be used without inspection, proves to have a defect that causes injury to a human being."[25]

Although all defectively designed products can be traced back to a design engineer or team, only the manufacturing company is usually held liable for injury caused by the product. This is more a matter of economics than justice. The company has liability insurance; the product design engineer (who is merely an employee of the company) probably does not. Unless the product design or manufacturing process is intentionally defective, or unless the defect is known in advance and covered up, the product design engineer will rarely be punished by the courts.[26]

20. DAMAGES

An injured party can sue for *damages* as well as for specific performance. Damages are the award made by the court for losses incurred by the injured party.

- *General* or *compensatory damages* are awarded to make up for the injury that was sustained.
- *Special damages* are awarded for the direct financial loss due to the breach of contract.
- *Nominal damages* are awarded when responsibility has been established but the injury is so slight as to be inconsequential.
- *Liquidated damages* are amounts that are specified in the contract document itself for nonperformance.
- *Punitive* or *exemplary damages* are awarded, usually in tort and fraud cases, to punish and make an example of the defendant (i.e., to deter others from doing the same thing).
- *Consequential damages* provide compensation for indirect losses incurred by the injured party but not directly related to the contract.

[23]The reason for this is that the public is not considered to be as sophisticated as a client who contracts with a design professional for building plans.

[24]1916, *MacPherson v. Buick*. MacPherson bought a Buick from a car dealer. The car had a defective wheel, and there was evidence that reasonable inspection would have uncovered the defect. MacPherson was injured when the wheel broke and the car collapsed, and he sued Buick. Buick defended itself under the ancient *prerequisite of privity* (i.e., the requirement of a face-to-face contractual relationship in order for liability to exist), since the dealer, not Buick, had sold the car to MacPherson, and no contract between Buick and MacPherson existed. The judge disagreed, thus establishing the concept of *third party liability* (i.e., manufacturers are responsible to consumers even though consumers do not buy directly from manufacturers).

[25]1963, *Greenman v. Yuba Power Products*. Greenman purchased and was injured by an electric power tool.

[26]The engineer can expect to be discharged from the company. However, for strategic reasons, this discharge probably will not occur until after the company loses the case.

21. INSURANCE

Most design firms and many independent design professionals carry *errors and omissions insurance* to protect them from claims due to their mistakes. Such policies are costly, and for that reason, some professionals choose to "go bare."[27] Policies protect against inadvertent mistakes only, not against willful, knowing, or conscious efforts to defraud or deceive.

[27]Going bare appears foolish at first glance, but there is a perverted logic behind the strategy. One-person consulting firms (and perhaps, firms that are not profitable) are "judgment-proof." Without insurance or other assets, these firms would be unable to pay any large judgments against them. When damage victims (and their lawyers) find this out in advance, they know that judgments will be uncollectable. So, often the lawsuit never makes its way to trial.

60 Engineering Ethics

1. CREEDS, CODES, CANONS, STATUTES, AND RULES

It is generally conceded that an individual acting on his or her own cannot be counted on to always act in a proper and moral manner. Creeds, statutes, rules, and codes all attempt to complete the guidance needed for an engineer to do "... the correct thing."

A *creed* is a statement or oath, often religious in nature, taken or assented to by an individual in ceremonies. For example, the *Engineers' Creed* adopted by the National Society of Professional Engineers (NSPE) in 1954 is[1]

> As a professional engineer, I dedicate my professional knowledge and skill to the advancement and betterment of human welfare.
>
> I pledge...
>
> ...to give the utmost of performance;
>
> ...to participate in none but honest enterprise;
>
> ...to live and work according to the laws of man and the highest standards of professional conduct;
>
> ...to place service before profit, the honor and standing of the profession before personal advantage, and the public welfare above all other considerations.
>
> In humility and with need for Divine Guidance, I make this pledge.

A *code* is a system of nonstatutory, nonmandatory canons of personal conduct. A *canon* is a fundamental belief that usually encompasses several rules. For example, the code of ethics of the American Society of Civil Engineers (ASCE) contains seven canons.

1. Engineers shall hold paramount the safety, health, and welfare of the public and shall strive to comply with the principles of sustainable development in the performance of their professional duties.
2. Engineers shall perform services only in areas of their competence.
3. Engineers shall issue public statements only in an objective and truthful manner.
4. Engineers shall act in professional matters for each employer or client as faithful agents or trustees and shall avoid conflicts of interest.
5. Engineers shall build their professional reputation on the merit of their service and shall not compete unfairly with others.
6. Engineers shall act in such a manner as to uphold and enhance the honor, integrity, and dignity of the engineering profession, and shall act with zero tolerance for bribery, fraud, and corruption.
7. Engineers shall continue their professional development throughout their careers and shall provide opportunities for the professional development of those engineers under their supervision.

A *rule* is a guide (principle, standard, or norm) for conduct and action in a certain situation. A *statutory rule* is enacted by the legislative branch of a state or federal government and carries the weight of law. Some U.S. engineering registration boards have statutory *rules of professional conduct.*

2. PURPOSE OF A CODE OF ETHICS

Many different sets of *codes of ethics* (*canons of ethics, rules of professional conduct,* etc.) have been produced by various engineering societies, registration boards, and other organizations.[2] The purpose of these ethical guidelines is to guide the conduct and decision making of engineers. Most codes are primarily educational. Nevertheless, from time to time they have been used

[1] The *Faith of an Engineer* adopted by the Accreditation Board for Engineering and Technology (ABET), formerly the Engineer's Council for Professional Development (ECPD), is a similar but more detailed creed.

[2] All of the major engineering technical and professional societies in the United States (ASCE, IEEE, ASME, AIChE, NSPE, etc.) and throughout the world have adopted codes of ethics. Most U.S. societies have endorsed the *Code of Ethics of Engineers* developed by the Accreditation Board for Engineering and Technology (ABET). The National Council of Examiners for Engineering and Surveying (NCEES) has developed its *Model Rules of Professional Conduct* as a guide for state registration boards in developing guidelines for the professional engineers in those states.

Professional Practice

by the societies and regulatory agencies as the basis for disciplinary actions.

Fundamental to ethical codes is the requirement that engineers render faithful, honest, professional service. In providing such service, engineers must represent the interests of their employers or clients and, at the same time, protect public health, safety, and welfare.

There is an important distinction between what is legal and what is ethical. Many actions that are legal can be violations of codes of ethical or professional behavior.[3] For example, an engineer's contract with a client may give the engineer the right to assign the engineer's responsibilities, but doing so without informing the client would be unethical.

Ethical guidelines can be categorized on the basis of who is affected by the engineer's actions—the client, vendors and suppliers, other engineers, or the public at large.[4]

3. ETHICAL PRIORITIES

There are frequently conflicting demands on engineers. While it is impossible to use a single decision-making process to solve every ethical dilemma, it is clear that ethical considerations will force engineers to subjugate their own self-interests. Specifically, the ethics of engineers dealing with others need to be considered in the following order, from highest to lowest priority.

- society and the public
- the law
- the engineering profession
- the engineer's client
- the engineer's firm
- other involved engineers
- the engineer personally

4. DEALING WITH CLIENTS AND EMPLOYERS

The most common ethical guidelines affecting engineers' interactions with their employer (the *client*) can be summarized as follows.[5]

- An engineer should not accept assignments for which he/she does not have the skill, knowledge, or time.
- An engineer must recognize his/her own limitations, and use associates and other experts when the design requirements exceed his/her ability.
- The client's interests must be protected. The extent of this protection exceeds normal business relationships and transcends the legal requirements of the engineer-client contract.
- An engineer must not be bound by what the client wants in instances where such desires would be unsuccessful, dishonest, unethical, unhealthy, or unsafe.
- Confidential client information remains the property of the client and must be kept confidential.
- An engineer must avoid conflicts of interest and should inform the client of any business connections or interests that might influence his/her judgment. An engineer should also avoid the *appearance* of a conflict of interest when such an appearance would be detrimental to the profession, the client, or the engineer.
- The engineer's sole source of income for a particular project should be the fee paid by the client. An engineer should not accept compensation in any form from more than one party for the same services.
- If the client rejects the engineer's recommendations, the engineer should fully explain the consequences to the client.
- An engineer must freely and openly admit to the client any errors made.

All courts of law have required an engineer to perform in a manner consistent with normal professional standards. This is not the same as saying an engineer's work must be error-free. If an engineer completes a design, has the design and calculations checked by another competent engineer, and an error is subsequently shown to have been made, the engineer may be held responsible, but the engineer will probably not be considered negligent.

5. DEALING WITH SUPPLIERS

Engineers routinely deal with manufacturers, contractors, and vendors (*suppliers*). In this regard, engineers have great responsibility and influence. Such a relationship requires that engineers deal justly with both clients and suppliers.

An engineer will often have an interest in maintaining good relationships with suppliers since this often leads to future work. Nevertheless, relationships with suppliers must remain highly ethical. Suppliers should not be encouraged to feel that they have any special favors coming to them because of a long-standing relationship with the engineer.

[3]Whether the guidelines emphasize ethical behavior or professional conduct is a matter of wording. The intention is the same: to provide guidelines that transcend the requirements of the law.

[4]Some authorities also include ethical guidelines for dealing with the employees of an engineer. However, these guidelines are no different for an engineering employer than they are for a supermarket, automobile assembly line, or airline employer. Ethics is not a unique issue when it comes to employees.

[5]These general guidelines contain references to contractors, plans, specifications, and contract documents. This language is common, though not unique, to the situation of an engineer supplying design services to an owner-developer or architect. However, most of the ethical guidelines are general enough to apply to engineers in industry as well.

The ethical responsibilities relating to suppliers are listed as follows.

- An engineer must not accept or solicit gifts or other valuable considerations from a supplier during, prior to, or after any job. An engineer should not accept discounts, allowances, commissions, or any other indirect compensation from suppliers, contractors, or other engineers in connection with any work or recommendations.
- An engineer must enforce the plans and specifications (i.e., the *contract documents*), but must also interpret the contract documents fairly.
- Plans and specifications developed by an engineer on behalf of the client must be complete, definite, and specific.
- Suppliers should not be required to spend time or furnish materials that are not called for in the plans and contract documents.
- An engineer should not unduly delay the performance of suppliers.

6. DEALING WITH OTHER ENGINEERS

Engineers should try to protect the engineering profession as a whole, to strengthen it, and to enhance its public stature. The following ethical guidelines apply.

- An engineer should not attempt to maliciously injure the professional reputation, business practice, or employment position of another engineer. However, if there is proof that another engineer has acted unethically or illegally, the engineer should advise the proper authority.
- An engineer should not review someone else's work while the other engineer is still employed unless the other engineer is made aware of the review.
- An engineer should not try to replace another engineer once the other engineer has received employment.
- An engineer should not use the advantages of a salaried position to compete unfairly (i.e., moonlight) with other engineers who have to charge more for the same consulting services.
- Subject to legal and proprietary restraints, an engineer should freely report, publish, and distribute information that would be useful to other engineers.

7. DEALING WITH (AND AFFECTING) THE PUBLIC

In regard to the social consequences of engineering, the relationship between an engineer and the public is essentially straightforward. Responsibilities to the public demand that the engineer place service to humankind above personal gain. Furthermore, proper ethical behavior requires that an engineer avoid association with projects that are contrary to public health and welfare or that are of questionable legal character.

- An engineer must consider the safety, health, and welfare of the public in all work performed.
- An engineer must uphold the honor and dignity of his/her profession by refraining from self-laudatory advertising, by explaining (when required) his/her work to the public, and by expressing opinions only in areas of knowledge.
- When an engineer issues a public statement, he/she must clearly indicate if the statement is being made on anyone's behalf (i.e., if anyone is benefitting from his/her position).
- An engineer must keep his/her skills at a state-of-the-art level.
- An engineer should develop public knowledge and appreciation of the engineering profession and its achievements.
- An engineer must notify the proper authorities when decisions adversely affecting public safety and welfare are made.[6]

8. COMPETITIVE BIDDING

The ethical guidelines for dealing with other engineers presented here and in more detailed codes of ethics no longer include a prohibition on *competitive bidding.* Until 1971, most codes of ethics for engineers considered competitive bidding detrimental to public welfare, since cost-cutting normally results in a lower quality design.

However, in a 1971 case against NSPE that went all the way to the U.S. Supreme Court, the prohibition against competitive bidding was determined to be a violation of the Sherman Antitrust Act (i.e., it was an unreasonable restraint of trade).

The opinion of the Supreme Court does not *require* competitive bidding—it merely forbids a prohibition against competitive bidding in NSPE's code of ethics. The following points must be considered.

- Engineers and design firms may individually continue to refuse to bid competitively on engineering services.
- Clients are not required to seek competitive bids for design services.
- Federal, state, and local statutes governing the procedures for procuring engineering design services, even those statutes that prohibit competitive bidding, are not affected.

[6]This practice has come to be known as *whistle-blowing.*

Professional Practice

- Any prohibitions against competitive bidding in individual state engineering registration laws remain unaffected.
- Engineers and their societies may actively and aggressively lobby for legislation that would prohibit competitive bidding for design services by public agencies.

61 Electrical Engineering Frontiers

1. ARTIFICIAL INTELLIGENCE

Artificial intelligence (AI) is the property of machines capable of reason by which they are capable of learning functions normally associated with human intelligence. Improvements in integrated circuits and computer technology have enabled concurrent advances in the application of artificial intelligence to real-world engineering devices and projects.

2. CAD/CAM

CAD is an acronym for *computer-aided design* or *drafting*.[1] CAM is an acronym for *computer-aided manufacturing*. The "CA" portion of CAD/CAM denotes the principal components, such as the hardware and software (e.g., AutoCAD). The "D" for drafting indicates the graphic language that uses lines, symbols, and words to describe components. The standards for the line conventions used in drafting are governed by the American National Standards Institute (ANSI). Advances in computer capacity and speed, driven by advances in integrated circuit (IC) technology and software improvements, have greatly improved the efficiency and accuracy of computerized designs.

3. COMMUNICATION

Communication capability is now both global and individualistic in nature and use. Electronic engineering improvements and advances in space technology are leading to worldwide connectivity.

4. INTEGRATED CIRCUITS

The size of integrated circuits continues to shrink as design complexity increases. Wafer and chip sizes have increased while individual feature size has decreased. According to *Moore's Law*, the number of transistors that can be placed inexpensively on an integrated circuit doubles approximately every two years. Such increases have driven advances in manufacturing automation, clean-room engineering, robotics, and so on.

The analysis of such circuits has also become more automated with SPICE and PSPICE. SPICE is the *simulation program with integrated circuit emphasis*, while PSPICE is a similar version for personal computers.

5. ENGINEERING MANAGEMENT

As many aspects of life evolve technologically, engineering management challenges grow and change. When and where to apply technological advances, and whether it is necessary to apply some advances at all, are but a few of the questions managers and individuals face. An understanding of the underlying principles of engineering and the capacities and limitations of the devices, as well as an appreciation for the human component, is vital to the decision-making process.

6. SUPERCONDUCTIVITY

Superconductivity is a property of certain materials such that at low temperatures their electrical resistivity vanishes. Resistance occurs due to thermal scattering of electrons and due to scattering by lattice impurities. These types of resistivity are called *thermal resistivity* and *residual resistivity*, respectively.[2] At absolute zero, the total electrical resistivity is residual and minimal—that is, very near zero. This, however, is not the phenomenon of superconductivity. Below a transition temperature, called the *critical temperature*, certain materials exhibit a *superconducting state* in which electron interactions fundamentally change, and resistance goes to zero.

Advances in materials science have led to compounds that have transition temperatures greater than 100K—a great increase from the first single-digit transition temperatures.

[1]The acronym is sometimes given as CADD.

[2]The sum of the two resistivities gives the total resistivity according to *Matthiessen's rule*.

7. QUANTUM ELECTRODYNAMICS

Quantum electrodynamics (QED), also called the *quantum theory of light* or the *quantum theory of radiation*, successfully incorporates relativity into the explanation of electromagnetic (i.e., radiation) interactions with charged matter. This theory unites the quantum mechanical- and relativity-based view of such interactions. Further, the theory synthesizes the wave and particle views.

QED theory serves as the prototype for the quantum theory used to explain *electroweak* interactions. *Electroweak theory* is the explanation of electromagnetic and weak nuclear interactions as the same phenomena.[3] With the addition of the electroweak theory, QED theory explains all physical phenomena of the world except those involving gravity or the nuclear force.

The force particle in electromagnetic interactions is the *photon* (sometimes called the *virtual photon*). The force particles for the weak interactions are the *Z* and *W bosons*, also known as *weakons* or *intermediate vector bosons*. The force particle for the strong nuclear force is the *gluon*. The hypothesized force particle for the gravitational force is the *graviton*.

As understanding of electrical phenomena at the quantum level deepens, the capacity, speed, and types of electrical devices increase. Research is under way in an attempt to create quantum mechanical computers.

8. EDUCATION

To fully appreciate engineering advances and apply them effectively, people must understand the basics of science and engineering, which begins with a foundation of mathematics and reading comprehension. The challenge of educating people throughout their lives is enormous, and requires technical and managerial skill. The ability to explain complex subjects and processes in logical terms applicable to the targeted audience, in both verbal and written formats, is a vital skill necessary for the advancement of technology and society itself.

[3]With this combination, three general forces remain: the *electroweak force*, the *strong nuclear* (or *color*) *force*, and the *gravitational force*. A theory combining the electroweak force and strong nuclear force is called the *grand unified theory* (GUT). A theory combining all three, that is, the GUT with gravitational theory, is referred to as the *theory of everything*.

62 Engineering Licensing in the United States

1. ABOUT LICENSING

Engineering licensing (also known as *engineering registration*) in the United States is an examination process by which a state's *board of engineering licensing* (typically referred to as the "engineers' board" or "board of registration") determines and certifies that an engineer has achieved a minimum level of competence.[1] This process is intended to protect the public by preventing unqualified individuals from offering engineering services.

Most engineers in the United States do not need to be licensed.[2] In particular, most engineers who work for companies that design and manufacture products are exempt from the licensing requirement. This is known as the *industrial exemption*, something that is built into the laws of most states.[3]

Nevertheless, there are many good reasons to become a licensed engineer. For example, you cannot offer consulting engineering services in any state unless you are licensed in that state. Even within a product-oriented corporation, you may find that employment, advancement, and managerial positions are limited to licensed engineers.

Once you have met the licensing requirements, you will be allowed to use the titles *Professional Engineer* (PE), *Registered Engineer* (RE), and *Consulting Engineer* (CE) as permitted by your state.

Although the licensing process is similar in each of the 50 states, each has its own licensing law. Unless you offer consulting engineering services in more than one state, however, you will not need to be licensed in the other states.

2. THE U.S. LICENSING PROCEDURE

The licensing procedure is similar in all states. You will take two written examinations. The full process requires you to complete two applications, one for each of the two examinations. The first examination is the *Fundamentals of Engineering* (FE) *examination*, formerly known (and still commonly referred to) as the *Engineer-In-Training* (EIT) *examination*.[4] This examination covers basic subjects from all of the mathematics, physics, chemistry, and engineering courses you took during your university years.

The second examination is the *Professional Engineering* (PE) *examination*, also known as the *Principles and Practices* (P&P) *examination*. This examination covers only the subjects in your engineering discipline (e.g., civil, mechanical, electrical, and others).

The actual details of licensing qualifications, experience requirements, minimum education levels, fees, and examination schedules vary from state to state. Contact your state's licensing board for more information. You will find contact information (websites, telephone numbers, email addresses, etc.) for all U.S. state and territorial boards of registration at **ppi2pass.com/stateboards**.

3. NATIONAL COUNCIL OF EXAMINERS FOR ENGINEERING AND SURVEYING

The *National Council of Examiners for Engineering and Surveying* (NCEES) in Clemson, South Carolina, writes, prints, distributes, and scores the national FE and PE examinations.[5] The individual states purchase the examinations from NCEES and administer them in a uniform, controlled environment dictated by NCEES.

[1]Licensing of engineers is not unique to the United States. However, the practice of requiring a degreed engineer to take an examination is not common in other countries. Licensing in many countries requires a degree and may also require experience, references, and demonstrated knowledge of ethics and law, but no technical examination.

[2]Less than one-third of the degreed engineers in the United States are licensed.

[3]Only one or two states have abolished the industrial exemption. There has always been a lot of "talk" among engineers about abolishing it, but there has been little success in actually trying to do so. One of the reasons is that manufacturers' lobbies are very strong.

[4]The terms *engineering intern* (EI) and *intern engineer* (IE) have also been used in the past to designate the status of an engineer who has passed the first exam. These uses are rarer but may still be encountered in some states.

[5]National Council of Examiners for Engineering and Surveying, P.O. Box 1686, Clemson, SC 29633, (800) 250-3196. ncees.org.

4. UNIFORM EXAMINATIONS

Although each state has its own licensing law and is, theoretically, free to administer its own exams, none does so for the major disciplines. All states have chosen to use the NCEES exams. The exams from all the states are sent to NCEES to be graded. Each state adopts the cut-off passing scores recommended by NCEES. These practices have led to the term *uniform examination.*

5. RECIPROCITY AMONG STATES

With minor exceptions, having a license from one state will not permit you to practice engineering in another state. You must have a professional engineering license from each state in which you work. Most engineers do not work across state lines or in multiple states, but some do. Luckily, it is not too difficult to get a license from every state you work in once you have a license from one of them.

All states use the NCEES examinations. If you take and pass the FE or PE examination in one state, your certificate or license will be honored by all of the other states. Upon proper application, payment of fees, and proof of your license, you will be issued a license by the new state. Although there may be other special requirements imposed by a state, it will not be necessary to retake the FE or PE examinations.[6] The issuance of an engineering license based on another state's licensing is known as *reciprocity* or *comity.*

6. APPLYING FOR THE FE EXAMINATION

The FE exam is administered at approved Pearson VUE testing centers. Registration is open year-round and can be completed online through an NCEES account.[7] Registration fees may be paid online. Once a candidate receives notification of eligibility from NCEES, the exam can be scheduled through an NCEES account. A letter from Pearson VUE confirming the exam location and date will be sent via email.

Whether or not applying for and taking the exam is the same as applying for a state's FE certificate depends on the state. In most cases, a candidate might take the exam without a state board ever knowing about it. In fact, as part of the NCEES online exam application process, candidates have to agree to the following statement:

> Passage of the FE exam alone does not ensure certification as an engineer intern or engineer-in-training in any U.S. state or territory. To obtain certification, you must file an application with an engineering licensing board and meet that board's requirements.

After a state's education requirements have been met and a candidate is ready to obtain an FE (EIT, IE, etc.) certificate, he or she must apply and pay an additional fee to that state. In some cases, passing an additional nontechnical exam related to professional practice in that state may be required. Actual procedures will vary from state to state.

Approximately 7–10 days after the exam, a candidate will receive an email notification that the exam results are ready for viewing through the NCEES account. This notification may include information about additional state-specific steps in the process.

All states make special accommodations for candidates who are physically challenged or who have religious or other special needs. Needs must be communicated to Pearson VUE well in advance of the examination day.

7. APPLYING FOR THE PE EXAMINATION

As with the FE exam, applying for the PE exam includes the steps of registering with NCEES, registering with a state, paying fees to that state, paying fees to NCEES, sending documentation to that state, and notifying that state and NCEES of what has been done. The sequence of these steps varies from state to state.[8]

In some states, the first thing a candidate does is reserve a seat for the upcoming exam through an NCEES account. In other states, the first thing a candidate does is to register with his or her state. In some states, registering with NCEES is the very last thing done, and a candidate may not even be able to notify NCEES until the entire application has been approved by his or her state. Some states use a third-party testing organization with whom a candidate must register. In some states, all fees are paid directly to the state. In other states, all fees are paid to NCEES. In some states, fees are paid to both. In some states, NCEES forms are included with the state application; in other states, state forms are included with the NCEES application. Both the states and NCEES have their own application deadlines, and a state's deadlines will not necessarily coincide with NCEES deadlines.

If a candidate has special physical needs or requirements, these must be coordinated directly with the state. The method and deadline for requesting special accommodations depends on the state. A candidate should keep copies of all forms and correspondence with NCEES and the state.

[6] For example, California requires all civil engineering applicants to pass special examinations in seismic design and surveying in addition to their regular eight-hour PE exams. Licensed engineers from other states only have to pass these two special exams. They do not need to retake the PE exam.

[7] PPI is not associated with NCEES.

[8] Each state board of registration (licensure board) is the most definitive source of information about the application process. A summary of each state's application process is also listed on the NCEES website, ncees.org.

Professional Practice

All states require an application containing a detailed work history. The application form may be available online to print out, or the state may have it mailed. All states require submitted evidence of qualifying engineering experience in the form of written corroborative statements (typically referred to as "professional references") provided and signed by supervising engineers, usually other professional engineers. The nature and format of these applications and corroborative statements depend on the state. An application is not complete until all professional references have been received by the state. The state will notify a candidate if one or more references have not been received in a timely manner.

Approximately 2–3 weeks before the exam, NCEES will email the candidate a link to the Exam Authorization Admittance notice, which should be printed out. This document must be presented along with a valid, government-issued photo ID in order to enter the exam site.

The examination results will be released by NCEES to each state 8–10 weeks after the exam. Soon thereafter, states will notify candidates via mail, email, or web access, or alternatively, NCEES may notify candidates via email that results are accessible from an NCEES account. If a state-specific examination was taken, but was not written or graded by NCEES, the state will notify a candidate separately of those results.

8. EXAMINATION DATES

The FE examination is administered eight months out of the year: January, February, April, May, July, August, October, and November. There are multiple testing dates within each of those months. No exams are administered in March, June, September, or December.

The PE examination is administered twice a year (usually in mid-April and late October), on the same weekends in all states. For a current exam schedule, check **ppi2pass.com/CE**. Click on the Exam FAQs link.

9. FE EXAMINATION FORMAT

The FE exam is a computer-based test that contains 110 multiple-choice questions given over two consecutive sessions (sections, parts, etc.). Each session contains approximately 55 multiple-choice questions that are grouped together by knowledge area (subject, topic, etc.). The subjects are not explicitly labeled, and the beginning and ending of the subjects are not noted. No subject spans the two exam sessions. That is, if a subject appears in the first session of the exam, it will not appear in the second.

The exam is six hours long and includes an 8-minute tutorial, a 25-minute break, and a brief survey at the conclusion of the exam. Questions from all undergraduate technical courses appear in the examination, including mathematics, physics, and engineering.

The FE exam is essentially entirely in SI units. There may be a few non-SI problems and some opportunities to choose between identical SI and non-SI problems.

The FE exam is a limited-reference exam. NCEES provides its own reference handbook for use in the examination. Personal books or notes may not be used, though candidates may bring an approved calculator. A reusable, erasable notepad and compatible writing instrument to use for scratchwork are provided during the exam.

10. PE EXAMINATION FORMAT

There are three separate electrical examinations: Power, Electrical and Electronics, and Computer Engineering. You will choose which exam you will take the day that you register for the exam.

Each exam consists of two four-hour sessions separated by a one-hour lunch period. The material covered in the morning session is very similar to the material covered in the afternoon. Both sessions will require an understanding of the basics as well as a detailed study of the appropriate topics contained within your specific discipline. This is an open-book exam. Although some states have a few restrictions, generally you may bring in the books of your choice. Visit **ppi2pass.com/eefaq** to get the most up-to-date exam details.

The exam requires the use of a scientific calculator. NCEES has a very limited list of calculators that they will allow into the exam room. Check the PPI website (**ppi2pass.com/calculators**) for a current list of permissible devices.

The questions on the PE exam will be multiple choice. Most questions will be unique; that is, they will be scenario-based with one question per scenario. There may be some multi-part questions, where one problem statement (i.e., scenario) is followed by two or three questions. The multi-part questions will nevertheless be independent; that is, you will not have to answer the first question correctly in order to get the second or third correct (they do not "cascade"). A total of 40 questions are expected in the morning, and another 40 in the afternoon. You must answer them all to get full credit. The subjects likely to be tested on each exam are given in Table 1 of the Introduction to this book.

You may be required to call upon nonacademic knowledge that stems from practical experience. The exam may include terminology used in the field without explanation. You may be asked to display current knowledge (within the last five years) of your chosen field.

To simplify the problems, you may want to use shortcuts known to practicing engineers. Additionally, the exam may require the knowledge of use of codes such as the *National Electrical Code* (NEC), the Code of Federal Regulations (CFR), Occupational Safety and Health Administration (OSHA) rules, and various computer codes and standards.

The electrical engineering field primarily uses SI units (the mks, or metric, system). Accordingly, SI units are used extensively on the exam. There are a few exceptions, however. For example, many of the questions relating to the National Electrical Code and wiring in general will utilize U.S. units. Additionally, magnetic questions may use the cgs system.

Topic XVI: Support Material

Appendices

APPENDIX 1.A
Conversion Factors

multiply	by	to obtain
acres	0.40468	hectares
	43,560.0	square feet
	1.5625×10^{-3}	square miles
ampere-hours	3600.0	coulombs
angstrom units	3.937×10^{-9}	inches
	1×10^{-4}	microns
astronomical units	1.496×10^{8}	kilometers
atmospheres	76.0	centimeters of mercury
atomic mass unit	9.3149×10^{8}	electron-volts
	1.4924×10^{-10}	joules
	1.6605×10^{-27}	kilograms
BeV (also GeV)	1×10^{9}	electron-volts
Btu	3.93×10^{-4}	horsepower-hours
	778.2	foot-pounds
	1055.1	joules
	2.931×10^{-4}	kilowatt hours
	1.0×10^{-5}	therms
Btu/hr	0.2161	foot-pounds/sec
	3.929×10^{-4}	horsepower
	0.2931	watts
bushels	2150.4	cubic inches
calories, gram (mean)	3.9683×10^{-3}	Btu (mean)
centares	1.0	square meters
centimeters	1×10^{-5}	kilometers
	1×10^{-2}	meters
	10.0	millimeters
	3.281×10^{-2}	feet
	0.3937	inches
chains	792.0	inches
coulombs	1.036×10^{-5}	faradays
cubic centimeters	0.06102	cubic inches
	2.113×10^{-3}	pints (U.S. liquid)
cubic feet	0.02832	cubic meters
	7.4805	gallons
cubic feet/min	62.43	pounds H_2O/min
cubic feet/sec	448.831	gallons/min
	0.64632	millions of gallons per day
cubits	18.0	inches
days	86,400.0	seconds
degrees (angle)	1.745×10^{-2}	radians
degrees/sec	0.1667	revolutions/min
dynes	1×10^{-5}	newtons
electron-volts	1.0735×10^{-9}	atomic mass units
	1×10^{-9}	BeV (also GeV)
	1.60218×10^{-19}	joules
	1.78266×10^{-36}	kilograms
	1×10^{-6}	MeV
faradays/sec	96,485	amperes
fathoms	6.0	feet
feet	30.48	centimeters
	0.3048	meters
	1.645×10^{-4}	miles (nautical)
	1.894×10^{-4}	miles (statute)
feet/min	0.5080	centimeters/sec
feet/sec	0.592	knots
	0.6818	miles/hr
foot-pounds	1.285×10^{-3}	Btu
	5.051×10^{-7}	horsepower-hours
	3.766×10^{-7}	kilowatt-hours
foot-pound/sec	4.6272	Btu/hr
	1.818×10^{-3}	horsepower
	1.356×10^{-3}	kilowatts
furlongs	660.0	feet
	0.125	miles (statute)
gallons	0.1337	cubic feet
	3.785	liters
gallons H_2O	8.3453	pounds H_2O
gallons/min	8.0208	cubic feet/hr
	0.002228	cubic feet/sec
GeV (also BeV)	1×10^{9}	electron-volts
grams	1×10^{-3}	kilograms
	3.527×10^{-2}	ounces (avoirdupois)
	3.215×10^{-2}	ounces (troy)
	2.205×10^{-3}	pounds
hectares	2.471	acres
	1.076×10^{5}	square feet
horsepower	2545.0	Btu/hr
	42.44	Btu/min
	550	foot-pounds/sec
	0.7457	kilowatts
	745.7	watts
horsepower-hours	2545.0	Btu
	1.976×10^{-6}	foot-pounds
	0.7457	kilowatt-hours
hours	4.167×10^{-2}	days
	5.952×10^{-3}	weeks
inches	2.540	centimeters
	1.578×10^{-5}	miles
inches, H_2O	5.199	pounds force/ft^2
	0.0361	psi
	0.0735	inches, mercury
inches, mercury	70.7	pounds force/ft^2
	0.491	pounds force/in^2
	13.60	inches, H_2O
joules	6.705×10^{9}	atomic mass units
	9.478×10^{-4}	Btu
	1×10^{7}	ergs
	6.2415×10^{18}	electron-volts
	1.1127×10^{-17}	kilograms

(continued)

Support Material

APPENDIX 1.A *(continued)*
Conversion Factors

multiply	by	to obtain
kilograms	6.0221×10^{26}	atomic mass units
	5.6096×10^{35}	electron-volts
	8.9875×10^{16}	joules
	2.205	pounds
kilometers	3281.0	feet
	1000.0	meters
	0.6214	miles
kilometers/hr	0.5396	knots
kilowatts	3412.9	Btu/hr
	737.6	foot-pounds/sec
	1.341	horsepower
kilowatt-hours	3413.0	Btu
knots	6076.0	feet/hr
	1.0	nautical miles/hr
	1.151	statute miles/hr
light years	5.9×10^{12}	miles
links (surveyor)	7.92	inches
liters	1000.0	cubic centimeters
	61.02	cubic inches
	0.2642	gallons (U.S. liquid)
	1000.0	milliliters
	2.113	pints
MeV	1×10^6	electron-volts
meters	100.0	centimeters
	3.281	feet
	1×10^{-3}	kilometers
	5.396×10^{-4}	miles (nautical)
	6.214×10^{-4}	miles (statute)
	1000.0	millimeters
microns	1×10^{-6}	meters
miles (nautical)	6076	feet
	1.853	kilometers
	1.1516	miles (statute)
miles (statute)	5280.0	feet
	1.609	kilometers
	0.8684	miles (nautical)
miles/hr	88.0	feet/min
milligrams/liter	1.0	parts/million
milliliters	1×10^{-3}	liters
millimeters	3.937×10^{-2}	inches
newtons	1×10^5	dynes
ounces	28.349527	grams
	6.25×10^{-2}	pounds
ounces (troy)	1.09714	ounces (avoirdupois)
parsecs	3.086×10^{13}	kilometers
	1.9×10^{13}	miles

multiply	by	to obtain
pascal-sec	1000	centipoise
	10	poise
	0.02089	pound force-sec/ft^2
	0.6720	pound mass/ft-sec
	0.02089	slug/ft-sec
pints (liquid)	473.2	cubic centimeters
	28.87	cubic inches
	0.125	gallons
	0.5	quarts (liquid)
poise	0.002089	pound-sec/ft^2
pounds	0.4536	kilograms
	16.0	ounces
	14.5833	ounces (troy)
	1.21528	pounds (troy)
pounds/ft^2	0.006944	pounds/in^2
pounds/in^2	2.308	feet, H_2O
	27.7	inches, H_2O
	2.037	inches, mercury
	144	pounds/ft^2
quarts (dry)	67.20	cubic inches
quarts (liquid)	57.75	cubic inches
	0.25	gallons
	0.9463	liters
radians	57.30	degrees
	3438.0	minutes
revolutions	360.0	degrees
revolutions/min	6.0	degrees/sec
rods	16.5	feet
	5.029	meters
rods (surveyor)	5.5	yards
seconds	1.667×10^{-2}	minutes
square meters/sec	1×10^6	centistokes
	10.76	square feet/sec
	1×10^4	stokes
slugs	32.174	pounds mass
stokes	0.0010764	square feet/sec
tons (long)	1016.0	kilograms
	2240.0	pounds
	1.120	tons (short)
tons (short)	907.1848	kilograms
	2000.0	pounds
	0.89287	tons (long)
watts	3.4129	Btu/hr
	1.341×10^{-3}	horsepower
yards	0.9144	meters
	4.934×10^{-4}	miles (nautical)
	5.682×10^{-4}	miles (statute)

Support Material

APPENDIX 1.B
Common SI Unit Conversion Factors

multiply	by	to obtain
AREA		
circular mil	506.7	square micrometer
square foot	0.0929	square meter
square kilometer	0.3861	square mile
square meter	10.764	square foot
	1.196	square yard
square micrometer	0.001974	circular mil
square mile	2.590	square kilometer
square yard	0.8361	square meter
ENERGY		
Btu (international)	1.0551	kilojoule
erg	0.1	microjoule
foot-pound	1.3558	joule
horsepower-hour	2.6485	megajoule
joule	0.7376	foot-pound
	0.10197	meter·kilogram force
kilogram·calorie (international)	4.1868	kilojoule
kilojoule	0.9478	Btu
	0.2388	kilogram·calorie
kilowatt·hour	3.6	megajoule
megajoule	0.3725	horsepower-hour
	0.2778	kilowatt-hour
	0.009478	therm
meter·kilogram force	9.8067	joule
microjoule	10.0	erg
therm	105.506	megajoule
FORCE		
dyne	10.0	micronewton
kilogram force	9.8067	newton
kip	4448.2	newton
micronewton	0.1	dyne
newton	0.10197	kilogram force
	0.0002248	kip
	3.597	ounce force
	0.2248	pound force
ounce force	0.2780	newton
pound force	4.4482	newton
HEAT		
Btu/ft^2-hr	3.1546	watt/m^2
Btu/hr-ft^2-°F	5.6783	watt/m^2·°C
Btu/ft^3	0.0373	megajoule/m^3
Btu/ft^3-°F	0.06707	megajoule/m^3·°C
Btu/hr	0.2931	watt
Btu/lbm	2326	joule/kg
Btu/lbm-°F	4186.8	joule/kg·°C
Btu-in/hr-ft^2-°F	0.1442	watt/meter·°C
joule/kg	0.000430	Btu/lbm
joule/kg·°C	0.0002388	Btu/lbm-°F
megajoule/m^3	26.839	Btu/ft^3
megajoule/m^3·°C	14.911	Btu/ft^3-°F
watt	3.4121	Btu/hr
watt/m·°C	6.933	Btu-in/hr-ft^2-°F

(continued)

Support Material

APPENDIX 1.B *(continued)*
Common SI Unit Conversion Factors

multiply	by	to obtain
HEAT *(continued)*		
watt/m^2	0.3170	Btu/ft^2-hr
watt/m^2·°C	0.1761	Btu/hr-ft^2-°F
LENGTH		
angstrom	0.1	nanometer
foot	0.3048	meter
inch	25.4	millimeter
kilometer	0.6214	mile
	0.540	mile (nautical)
meter	3.2808	foot
	1.0936	yard
micrometer	1.0	micron
micron	1.0	micrometer
mil	0.0254	millimeter
mile	1.6093	kilometer
mile (nautical)	1.852	kilometer
millimeter	0.0394	inch
	39.370	mil
nanometer	10.0	angstrom
yard	0.9144	meter
MASS		
grain	64.799	milligram
gram	0.0353	ounce (avoirdupois)
	0.03215	ounce (troy)
kilogram	2.2046	pound mass
	0.068522	slug
	0.0009842	ton (long—2240 lbm)
	0.001102	ton (short—2000 lbm)
milligram	0.0154	grain
ounce (avoirdupois)	28.350	gram
ounce (troy)	31.1035	gram
pound mass	0.4536	kilogram
slug	14.5939	kilogram
ton (long—2240 lbm)	1016.047	kilogram
ton (short—2000 lbm)	907.185	kilogram
PRESSURE		
bar	100.0	kilopascal
inch, H_2O (20°C)	0.2486	kilopascal
inch, Hg (20°C)	3.3741	kilopascal
kilogram force/cm^2	98.067	kilopascal
kilopascal	0.01	bar
	4.0219	inch, H_2O (20°C)
	0.2964	inch, Hg (20°C)
	0.0102	kilogram force/cm^2
	7.528	millimeter, Hg (20°C)
	0.1450	pound force/in^2
	0.009869	standard atmosphere (760 torr)
	7.5006	torr
millimeter, Hg (20°C)	0.13332	kilopascal
pound force/in^2	6.8948	kilopascal
standard atmosphere (760 torr)	101.325	kilopascal
torr	0.13332	kilopascal

(continued)

Support Material

APPENDIX 1.B *(continued)*
Common SI Unit Conversion Factors

multiply	by	to obtain
POWER		
Btu (international)/hr	0.2931	watt
foot-pound/sec	1.3558	watt
horsepower	0.7457	kilowatt
kilowatt	1.341	horsepower
	0.2843	ton of refrigeration
meter·kilogram force/sec	9.8067	watt
ton of refrigeration	3.517	kilowatt
watt	3.4122	Btu (international)/hr
	0.7376	foot-pound/sec
	0.10197	meter·kilogram force/sec
TEMPERATURE		
Celsius	$\frac{9}{5}$°C + 32°	Fahrenheit
Celsius	°C + 273.15	Kelvin
Fahrenheit	$\frac{5}{9}$(°F − 32°)	Celsius
Kelvin	$\frac{9}{5}$	Rankine
Kelvin	K − 273.15	Celsius
Rankine	$\frac{5}{9}$	Kelvin
TORQUE		
gram force·centimeter	0.098067	millinewton·meter
kilogram force·meter	9.8067	newton·meter
millinewton	10.197	gram force·centimeter
newton·meter	0.10197	kilogram force·meter
	0.7376	foot-pound
	8.8495	inch-pound
foot-pound	1.3558	newton·meter
inch-pound	0.1130	newton·meter
VELOCITY		
feet/sec	0.3048	meters/sec
kilometers/hr	0.6214	miles/hr
meters/sec	3.2808	feet/sec
	2.2369	miles/hr
miles/hr	1.60934	kilometers/hr
	0.44704	meters/sec
VISCOSITY		
centipoise	0.001	pascal·sec
centistoke	1×10^{-6}	square meter/sec
pascal·sec	1000	centipoise
square meter/sec	1×10^{6}	centistoke
VOLUME (capacity)		
cubic centimeter	0.06102	cubic inch
cubic foot	28.3168	liter
cubic inch	16.3871	cubic centimeter
cubic meter	1.308	cubic yard
cubic yard	0.7646	cubic meter
gallon (U.S. fluid)	3.785	liter
liter	0.2642	gallon (U.S. fluid)
	2.113	pint (U.S. fluid)
	1.0567	quart (U.S. fluid)
	0.03531	cubic foot
milliliter	0.0338	ounce (U.S. fluid)
ounce (U.S. fluid)	29.574	milliliter

(continued)

APPENDIX 1.B *(continued)*
Common SI Unit Conversion Factors

multiply	by	to obtain
VOLUME (capacity) *(continued)*		
pint (U.S. fluid)	0.4732	liter
quart (U.S. fluid)	0.9464	liter
VOLUME FLOW (gas-air)		
cubic meter/sec	2119	standard cubic foot/min
liter/sec	2.119	standard cubic foot/min
microliter/sec	0.000127	standard cubic foot/hr
milliliter/sec	0.002119	standard cubic foot/min
	0.127133	standard cubic foot/hr
standard cubic foot/min	0.0004719	cubic meter/sec
	0.4719	liter/sec
	471.947	milliliter/sec
standard cubic foot/hr	7866	microliter/sec
	7.8658	milliliter/sec
VOLUME FLOW (liquid)		
gallon/hr (U.S. fluid)	0.001052	liter/sec
gallon/min (U.S. fluid)	0.06309	liter/sec
liter/sec	951.02	gallon/hr (U.S. fluid)
	15.850	gallon/min (U.S. fluid)

Support Material

APPENDIX 8.A
Mensuration of Two-Dimensional Areas

Nomenclature

A	total surface area
b	base
c	chord length
d	distance
h	height
L	length
p	perimeter
r	radius
s	side (edge) length, arc length
θ	vertex angle, in radians
ϕ	central angle, in radians

Triangle

equilateral

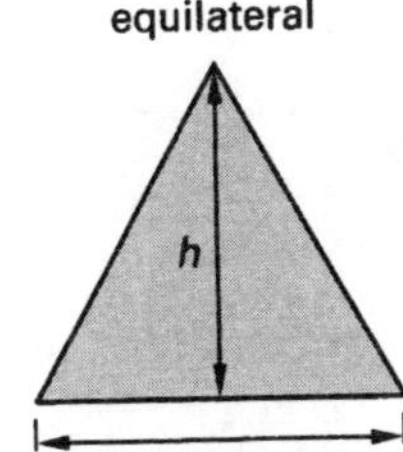

$$A = \tfrac{1}{2}bh = \frac{\sqrt{3}}{4}b^2$$

$$h = \frac{\sqrt{3}}{2}b$$

right

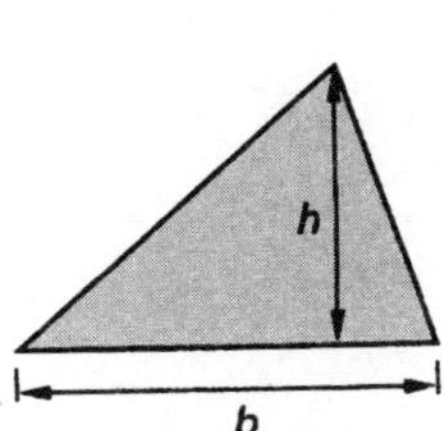

$$A = \tfrac{1}{2}bh$$

$$H^2 = b^2 + h^2$$

oblique

$$A = \tfrac{1}{2}bh$$

Circle

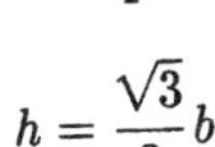

$$p = 2\pi r$$

$$A = \pi r^2 = \frac{p^2}{4\pi}$$

Circular Segment

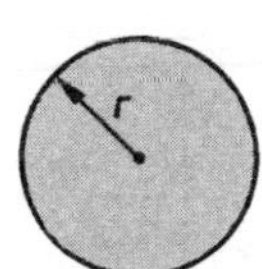

$$A = \tfrac{1}{2}r^2(\phi - \sin\phi)$$

$$\phi = \frac{s}{r} = 2\left(\arccos\frac{r-d}{r}\right)$$

$$c = 2r\sin\left(\frac{\phi}{2}\right)$$

Circular Sector

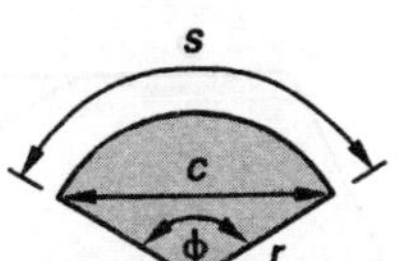

$$A = \tfrac{1}{2}\phi r^2 = \tfrac{1}{2}sr$$

$$\phi = \frac{s}{r}$$

$$s = r\phi$$

$$c = 2r\sin\left(\frac{\phi}{2}\right)$$

Parabola

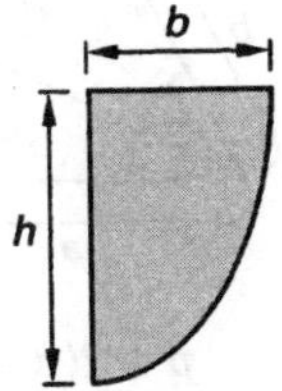

$$A = \tfrac{2}{3}bh$$

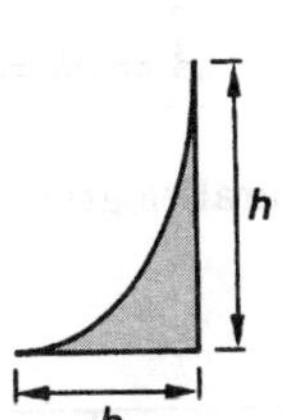

$$A = \tfrac{1}{3}bh$$

Ellipse

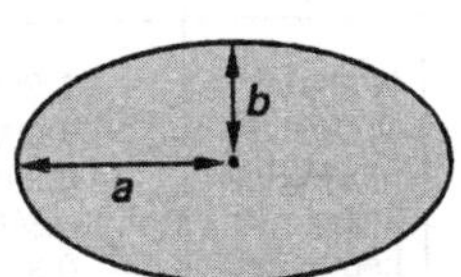

$$A = \pi ab$$

$$p \approx 2\pi\sqrt{\tfrac{1}{2}(a^2 + b^2)} \quad \left[\begin{array}{c}\text{Euler's}\\ \text{upper bound}\end{array}\right]$$

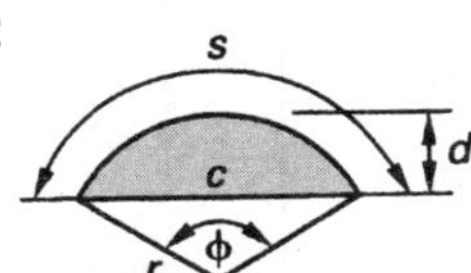

(continued)

Support Material

APPENDIX 8.A *(continued)*
Mensuration of Two-Dimensional Areas

Trapezoid

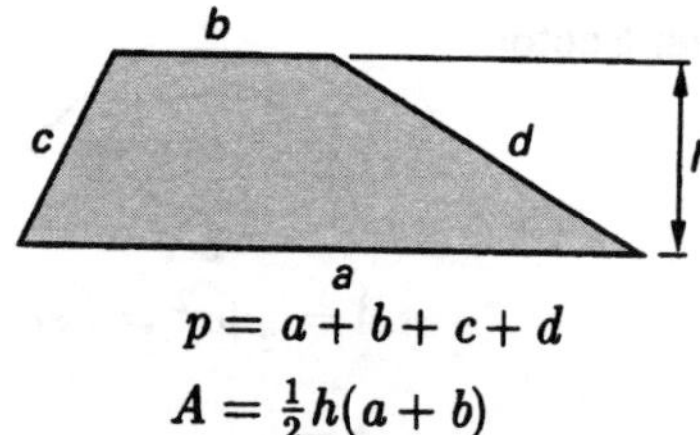

$$p = a + b + c + d$$

$$A = \tfrac{1}{2}h(a + b)$$

If $c = d$, the trapezoid is isosceles.

Parallelogram

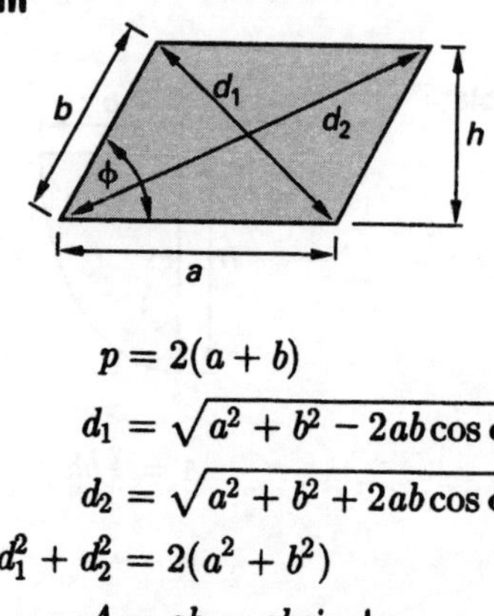

$$p = 2(a + b)$$

$$d_1 = \sqrt{a^2 + b^2 - 2ab\cos\phi}$$

$$d_2 = \sqrt{a^2 + b^2 + 2ab\cos\phi}$$

$$d_1^2 + d_2^2 = 2(a^2 + b^2)$$

$$A = ah = ab\sin\phi$$

If $a = b$, the parallelogram is a rhombus.

Regular Polygon (*n* equal sides)

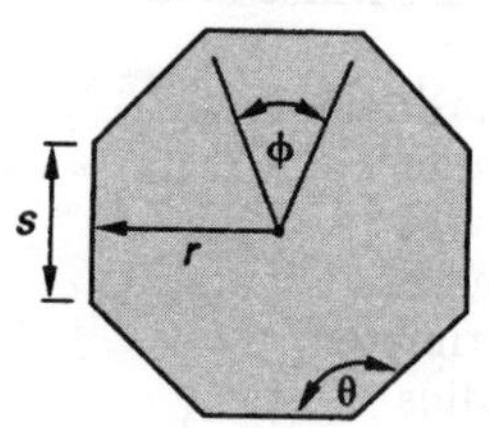

$$\phi = \frac{2\pi}{n}$$

$$\theta = \frac{\pi(n-2)}{n} = \pi - \phi$$

$$p = ns$$

$$s = 2r\tan\frac{\theta}{2}$$

$$A = \tfrac{1}{2}nsr$$

regular polygons

sides	name	area (A) when diameter of inscribed circle = 1	area (A) when side = 1	radius (r) of circumscribed circle when side = 1	length (L) of side when radius (r) of circumscribed circle = 1	length (L) of side when perpendicular to circle = 1	perpendicular (p) to center when side = 1
3	triangle	1.299	0.433	0.577	1.732	3.464	0.289
4	square	1.000	1.000	0.707	1.414	2.000	0.500
5	pentagon	0.908	1.720	0.851	1.176	1.453	0.688
6	hexagon	0.866	2.598	1.000	1.000	1.155	0.866
7	heptagon	0.843	3.634	1.152	0.868	0.963	1.038
8	octagon	0.828	4.828	1.307	0.765	0.828	1.207
9	nonagon	0.819	6.182	1.462	0.684	0.728	1.374
10	decagon	0.812	7.694	1.618	0.618	0.650	1.539
11	undecagon	0.807	9.366	1.775	0.563	0.587	1.703
12	dodecagon	0.804	11.196	1.932	0.518	0.536	1.866

APPENDIX 8.B
Mensuration of Three-Dimensional Volumes

Nomenclature

A	surface area
b	base
h	height
r	radius
R	radius
s	side (edge) length
V	internal volume

Sphere

$$V = \tfrac{4}{3}\pi r^3 = \tfrac{4}{3}\pi\left(\frac{d}{2}\right)^3 = \tfrac{1}{6}\pi d^3$$

$$A = 4\pi r^2$$

Right Circular Cone (excluding base area)

$$V = \tfrac{1}{3}\pi r^2 h = \tfrac{1}{3}\pi\left(\frac{d}{2}\right)^2 h = \tfrac{1}{12}\pi d^2 h$$

$$A = \pi r\sqrt{r^2 + h^2}$$

Right Circular Cylinder (excluding end areas)

$$V = \pi r^2 h$$

$$A = 2\pi r h$$

Spherical Segment (spherical cap)

Surface area of a spherical segment of radius r cut out by an angle θ_0 rotated from the center about a radius, r, is

$$A = 2\pi r^2(1 - \cos\theta_0)$$

$$\omega = \frac{A}{r^2} = 2\pi(1 - \cos\theta_0)$$

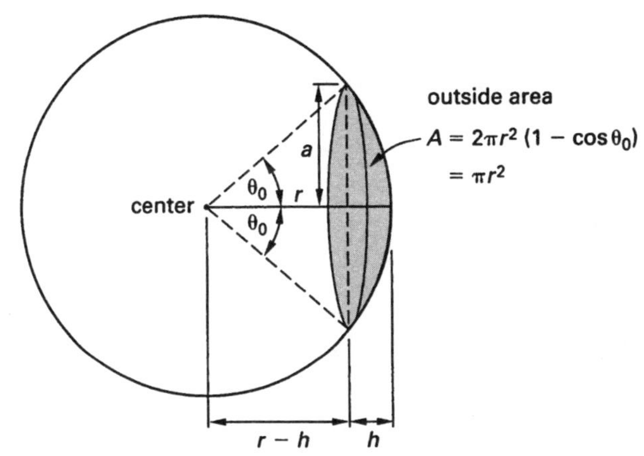

$$V_{cap} = \tfrac{1}{6}\pi h(3a^2 + h^2)$$

$$= \tfrac{1}{3}\pi h^2(3r - h)$$

$$a = \sqrt{h(2r - h)}$$

Paraboloid of Revolution

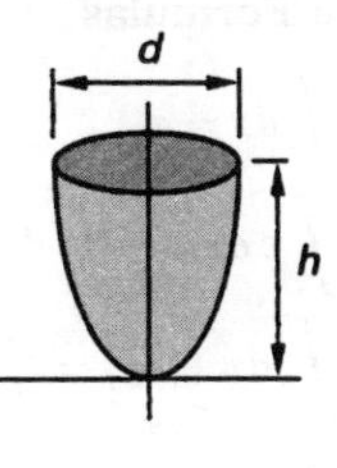

$$V = \tfrac{1}{8}\pi d^2 h$$

Torus

$$A = 4\pi^2 R r$$

$$V = 2\pi^2 R r^2$$

Regular Polyhedra (identical faces)

name	number of faces	form of faces	total surface area	volume
tetrahedron	4	equilateral triangle	$1.7321s^2$	$0.1179s^3$
cube	6	square	$6.0000s^2$	$1.0000s^3$
octahedron	8	equilateral triangle	$3.4641s^2$	$0.4714s^3$
dodecahedron	12	regular pentagon	$20.6457s^2$	$7.6631s^3$
icosahedron	20	equilateral triangle	$8.6603s^2$	$2.1817s^3$

The radius of a sphere inscribed within a regular polyhedron is

$$r = \frac{3V_{polyhedron}}{A_{polyhedron}}$$

Support Material

APPENDIX 10.A
Abbreviated Table of Indefinite Integrals
(In each case, add a constant of integration. All angles are measured in radians.)

General Formulas

1. $\int dx = x$

2. $\int c\,dx = c\int dx$

3. $\int (dx + dy) = \int dx + \int dy$

4. $\int u\,dv = uv - \int v\,du$ $\left[\begin{array}{c}\text{integration by parts; } u \text{ and } v \text{ are}\\ \text{functions of the same variable}\end{array}\right]$

Algebraic Forms

5. $\int x^n\,dx = \frac{x^{n+1}}{n+1} \quad [n \neq -1]$

6. $\int x^{-1}\,dx = \int \frac{dx}{x} = \ln|x|$

7. $\int (ax+b)^n\,dx = \frac{(ax+b)^{n+1}}{a(n+1)} \quad [n \neq -1]$

8. $\int \frac{dx}{ax+b} = \frac{1}{a}\ln(ax+b)$

9. $\int \frac{x\,dx}{ax+b} = \frac{1}{a^2}\Big(ax+b-b\ln(ax+b)\Big)$

10. $\int \frac{x\,dx}{(ax+b)^2} = \frac{1}{a^2}\left(\frac{b}{ax+b} + \ln(ax+b)\right)$

11. $\int \frac{dx}{x(ax+b)} = \frac{1}{b}\ln\left(\frac{x}{ax+b}\right)$

12. $\int \frac{dx}{x(ax+b)^2} = \frac{1}{b(ax+b)} + \frac{1}{b^2}\ln\left(\frac{x}{ax+b}\right)$

13. $\int \frac{dx}{x^2+a^2} = \frac{1}{a}\arctan\left(\frac{x}{a}\right)$

14. $\int \frac{dx}{a^2-x^2} = \frac{1}{a}\operatorname{arctanh}\left(\frac{x}{a}\right)$

15. $\int \frac{x\,dx}{ax^2+b} = \frac{1}{2a}\ln(ax^2+b)$

16. $\int \frac{dx}{x(ax^n+b)} = \frac{1}{bn}\ln\left(\frac{x^n}{ax^n+b}\right)$

17. $\int \frac{dx}{ax^2+bx+c} = \frac{1}{\sqrt{b^2-4ac}}\ln\left(\frac{2ax+b-\sqrt{b^2-4ac}}{2ax+b+\sqrt{b^2-4ac}}\right) \quad [b^2 > 4ac]$

18. $\int \frac{dx}{ax^2+bx+c} = \frac{2}{\sqrt{4ac-b^2}}\arctan\left(\frac{2ax+b}{\sqrt{4ac-b^2}}\right) \quad [b^2 < 4ac]$

19. $\int \sqrt{a^2-x^2}\,dx = \frac{x}{2}\sqrt{a^2-x^2} + \frac{a^2}{2}\arcsin\left(\frac{x}{a}\right)$

20. $\int x\sqrt{a^2-x^2}\,dx = -\frac{1}{3}(a^2-x^2)^{3/2}$

21. $\int \frac{dx}{\sqrt{a^2-x^2}} = \arcsin\left(\frac{x}{a}\right)$

22. $\int \frac{x\,dx}{\sqrt{a^2-x^2}} = -\sqrt{a^2-x^2}$

APPENDIX 11.A
Laplace Transforms
(See also App. 15.G.)

$f(t)$	$\mathcal{L}(f(t))$	$f(t)$	$\mathcal{L}(f(t))$
$\delta(t)$ [unit impulse at $t=0$]	1	$1-\cos at$	$\dfrac{a^2}{s(s^2+a^2)}$
$\delta(t-c)$ [unit impulse at $t=c$]	e^{-cs}	$\cosh at$	$\dfrac{s}{s^2-a^2}$
1 or u_0 [unit step at $t=0$]	$\dfrac{1}{s}$	$t\cos at$	$\dfrac{s^2-a^2}{(s^2+a^2)^2}$
u_c [unit step at $t=c$]	$\dfrac{e^{-cs}}{s}$	t^n [n is a positive integer]	$\dfrac{n!}{s^{n+1}}$
t [unit ramp at $t=0$]	$\dfrac{1}{s^2}$	e^{at}	$\dfrac{1}{s-a}$
rectangular pulse, magnitude M, duration a	$\left(\dfrac{M}{s}\right)(1-e^{-as})$	$e^{at}\sin bt$	$\dfrac{b}{(s-a)^2+b^2}$
triangular pulse, magnitude M, duration $2a$	$\left(\dfrac{M}{as^2}\right)(1-e^{-as})^2$	$e^{at}\cos bt$	$\dfrac{s-a}{(s-a)^2+b^2}$
sawtooth pulse, magnitude M, duration a	$\left(\dfrac{M}{as^2}\right)(1-(as+1)e^{-as})$	$e^{at}t^n$ [n is a positive integer]	$\dfrac{n!}{(s-a)^{n+1}}$
sinusoidal pulse, magnitude M, duration π/a	$\left(\dfrac{Ma}{s^2+a^2}\right)(1+e^{-\pi s/a})$	$1-e^{-at}$	$\dfrac{a}{s(s+a)}$
$\dfrac{t^{n-1}}{(n-1)!}$	$\dfrac{1}{s^n}$	$e^{-at}+at-1$	$\dfrac{a^2}{s^2(s+a)}$
$\sin at$	$\dfrac{a}{s^2+a^2}$	$\dfrac{e^{-at}-e^{-bt}}{b-a}$	$\dfrac{1}{(s+a)(s+b)}$
$at-\sin at$	$\dfrac{a^3}{s^2(s^2+a^2)}$	$\dfrac{(c-a)e^{-at}-(c-b)e^{-bt}}{b-a}$	$\dfrac{s+c}{(s+a)(s+b)}$
$\sinh at$	$\dfrac{a}{s^2-a^2}$	$\dfrac{1}{ab}+\dfrac{be^{-at}-ae^{-bt}}{ab(a-b)}$	$\dfrac{1}{s(s+a)(s+b)}$
$t\sin at$	$\dfrac{2as}{(s^2+a^2)^2}$	$t\sinh at$	$\dfrac{2as}{(s^2-a^2)^2}$
$\cos at$	$\dfrac{s}{s^2+a^2}$	$t\cosh at$	$\dfrac{s^2+a^2}{(s^2-a^2)^2}$

APPENDIX 12.A
Areas Under the Standard Normal Curve
(0 to z)

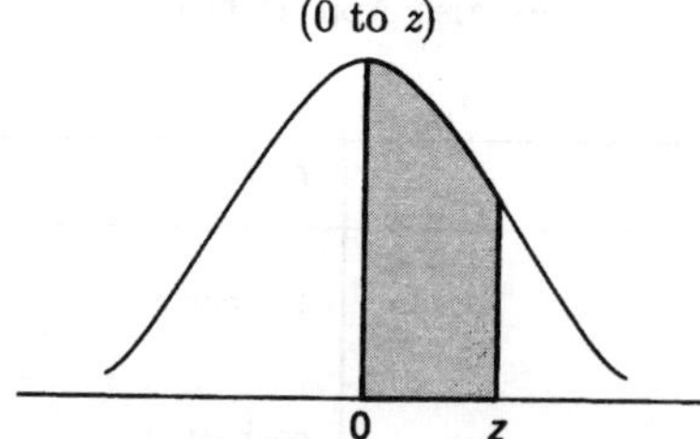

z	0	1	2	3	4	5	6	7	8	9
0.0	0.0000	0.0040	0.0080	0.0120	0.0160	0.0199	0.0239	0.0279	0.0319	0.0359
0.1	0.0398	0.0438	0.0478	0.0517	0.0557	0.0596	0.0636	0.0675	0.0714	0.0754
0.2	0.0793	0.0832	0.0871	0.0910	0.0948	0.0987	0.1026	0.1064	0.1103	0.1141
0.3	0.1179	0.1217	0.1255	0.1293	0.1331	0.1368	0.1406	0.1443	0.1480	0.1517
0.4	0.1554	0.1591	0.1628	0.1664	0.1700	0.1736	0.1772	0.1808	0.1844	0.1879
0.5	0.1915	0.1950	0.1985	0.2019	0.2054	0.2088	0.2123	0.2157	0.2190	0.2224
0.6	0.2258	0.2291	0.2324	0.2357	0.2389	0.2422	0.2454	0.2486	0.2518	0.2549
0.7	0.2580	0.2612	0.2642	0.2673	0.2704	0.2734	0.2764	0.2794	0.2823	0.2852
0.8	0.2881	0.2910	0.2939	0.2967	0.2996	0.3023	0.3051	0.3078	0.3106	0.3133
0.9	0.3159	0.3186	0.3212	0.3238	0.3264	0.3289	0.3315	0.3340	0.3365	0.3389
1.0	0.3413	0.3438	0.3461	0.3485	0.3508	0.3531	0.3554	0.3577	0.3599	0.3621
1.1	0.3643	0.3665	0.3686	0.3708	0.3729	0.3749	0.3770	0.3790	0.3810	0.3830
1.2	0.3849	0.3869	0.3888	0.3907	0.3925	0.3944	0.3962	0.3980	0.3997	0.4015
1.3	0.4032	0.4049	0.4066	0.4082	0.4099	0.4115	0.4131	0.4147	0.4162	0.4177
1.4	0.4192	0.4207	0.4222	0.4236	0.4251	0.4265	0.4279	0.4292	0.4306	0.4319
1.5	0.4332	0.4345	0.4357	0.4370	0.4382	0.4394	0.4406	0.4418	0.4429	0.4441
1.6	0.4452	0.4463	0.4474	0.4484	0.4495	0.4505	0.4515	0.4525	0.4535	0.4545
1.7	0.4554	0.4564	0.4573	0.4582	0.4591	0.4599	0.4608	0.4616	0.4625	0.4633
1.8	0.4641	0.4649	0.4656	0.4664	0.4671	0.4678	0.4686	0.4693	0.4699	0.4706
1.9	0.4713	0.4719	0.4726	0.4732	0.4738	0.4744	0.4750	0.4756	0.4761	0.4767
2.0	0.4772	0.4778	0.4783	0.4788	0.4793	0.4798	0.4803	0.4808	0.4812	0.4817
2.1	0.4821	0.4826	0.4830	0.4834	0.4838	0.4842	0.4846	0.4850	0.4854	0.4857
2.2	0.4861	0.4864	0.4868	0.4871	0.4875	0.4878	0.4881	0.4884	0.4887	0.4890
2.3	0.4893	0.4896	0.4898	0.4901	0.4904	0.4906	0.4909	0.4911	0.4913	0.4916
2.4	0.4918	0.4920	0.4922	0.4925	0.4927	0.4929	0.4931	0.4932	0.4934	0.4936
2.5	0.4938	0.4940	0.4941	0.4943	0.4945	0.4946	0.4948	0.4949	0.4951	0.4952
2.6	0.4953	0.4955	0.4956	0.4957	0.4959	0.4960	0.4961	0.4962	0.4963	0.4964
2.7	0.4965	0.4966	0.4967	0.4968	0.4969	0.4970	0.4971	0.4972	0.4973	0.4974
2.8	0.4974	0.4975	0.4976	0.4977	0.4977	0.4978	0.4979	0.4979	0.4980	0.4981
2.9	0.4981	0.4982	0.4982	0.4983	0.4984	0.4984	0.4985	0.4985	0.4986	0.4986
3.0	0.4987	0.4987	0.4987	0.4988	0.4988	0.4989	0.4989	0.4989	0.4990	0.4990
3.1	0.4990	0.4991	0.4991	0.4991	0.4992	0.4992	0.4992	0.4992	0.4993	0.4993
3.2	0.4993	0.4993	0.4994	0.4994	0.4994	0.4994	0.4994	0.4995	0.4995	0.4995
3.3	0.4995	0.4995	0.4996	0.4996	0.4996	0.4996	0.4996	0.4996	0.4996	0.4997
3.4	0.4997	0.4997	0.4997	0.4997	0.4997	0.4997	0.4997	0.4997	0.4997	0.4998
3.5	0.4998	0.4998	0.4998	0.4998	0.4998	0.4998	0.4998	0.4998	0.4998	0.4998
3.6	0.4998	0.4998	0.4999	0.4999	0.4999	0.4999	0.4999	0.4999	0.4999	0.4999
3.7	0.4999	0.4999	0.4999	0.4999	0.4999	0.4999	0.4999	0.4999	0.4999	0.4999
3.8	0.4999	0.4999	0.4999	0.4999	0.4999	0.4999	0.4999	0.4999	0.4999	0.4999
3.9	0.5000	0.5000	0.5000	0.5000	0.5000	0.5000	0.5000	0.5000	0.5000	0.5000

APPENDIX 15.A
Gamma Function Values*

$$\text{Values of } \Gamma(\alpha) = \int_0^\infty e^{-t} t^{\alpha-1} dt;\ \Gamma(\alpha+1) = \alpha\Gamma(\alpha)$$

α	$\Gamma(\alpha)$	α	$\Gamma(\alpha)$	α	$\Gamma(\alpha)$	α	$\Gamma(\alpha)$
1.00	1.00000	1.25	0.90640	1.50	0.88623	1.75	0.91906
1.01	0.99433	1.26	0.90440	1.51	0.88659	1.76	0.92137
1.02	0.98884	1.27	0.90250	1.52	0.88704	1.77	0.92376
1.03	0.98355	1.28	0.90072	1.53	0.88757	1.78	0.92623
1.04	0.97844	1.29	0.89904	1.54	0.88818	1.79	0.92877
1.05	0.97350	1.30	0.89747	1.55	0.88887	1.80	0.93138
1.06	0.96874	1.31	0.89600	1.56	0.88964	1.81	0.93408
1.07	0.96415	1.32	0.89464	1.57	0.89049	1.82	0.93685
1.08	0.95973	1.33	0.89338	1.58	0.89142	1.83	0.93969
1.09	0.95546	1.34	0.89222	1.59	0.89243	1.84	0.94261
1.10	0.95135	1.35	0.89115	1.60	0.89352	1.85	0.94561
1.11	0.94740	1.36	0.89018	1.61	0.89468	1.86	0.94869
1.12	0.94359	1.37	0.88931	1.62	0.89592	1.87	0.95184
1.13	0.93993	1.38	0.88854	1.63	0.89724	1.88	0.95507
1.14	0.93642	1.39	0.88785	1.64	0.89864	1.89	0.95838
1.15	0.93304	1.40	0.88726	1.65	0.90012	1.90	0.96177
1.16	0.92980	1.41	0.88676	1.66	0.90167	1.91	0.96523
1.17	0.92670	1.42	0.88636	1.67	0.90330	1.92	0.96877
1.18	0.92373	1.43	0.88604	1.68	0.90500	1.93	0.97240
1.19	0.92089	1.44	0.88581	1.69	0.90678	1.94	0.97610
1.20	0.91817	1.45	0.88566	1.70	0.90864	1.95	0.97988
1.21	0.91558	1.46	0.88560	1.71	0.91057	1.96	0.98374
1.22	0.91311	1.47	0.88563	1.72	0.91258	1.97	0.98768
1.23	0.91075	1.48	0.88575	1.73	0.91466	1.98	0.999171
1.24	0.90852	1.49	0.88595	1.74	0.91683	1.99	0.99581
						2.00	1.00000

*For large positive values of t, $\Gamma(t)$ approximates Stirling's asymptotic series.

$$\Gamma(t) \approx t^t e^{-t} \sqrt{\frac{2\pi}{t}} \left(1 + \frac{1}{12t} + \frac{1}{288t^2} - \frac{139}{51{,}840t^3} - \frac{2{,}488{,}320t^4}{571} + \cdots\right)$$

APPENDIX 15.B
Bessel Functions J_0 and J_1

x	$J_0(x)$	$J_1(x)$	x	$J_0(x)$	$J_1(x)$	x	$J_0(x)$	$J_1(x)$
0.0	1.0000	0.0000	5.0	−0.1776	−0.3276	10.0	−0.2459	0.0435
0.1	0.9975	0.0499	5.1	−0.1443	−0.3371	10.1	−0.2490	0.0184
0.2	0.9900	0.0995	5.2	−0.1103	−0.3432	10.2	−0.2496	−0.0066
0.3	0.9776	0.1483	5.3	−0.0758	−0.3460	10.3	−0.2477	−0.0313
0.4	0.9604	0.1960	5.4	−0.0412	−0.3453	10.4	−0.2434	−0.0555
0.5	0.9385	0.2423	5.5	−0.0068	−0.3414	10.5	−0.2366	−0.0789
0.6	0.9120	0.2867	5.6	0.0270	−0.3343	10.6	−0.2276	−0.1012
0.7	0.8812	0.3290	5.7	0.0599	−0.3241	10.7	−0.2164	−0.1224
0.8	0.8463	0.3688	5.8	0.0917	−0.3110	10.8	−0.2032	−0.1422
0.9	0.8075	0.4059	5.9	0.1220	−0.2951	10.9	−0.1881	−0.1603
1.0	0.7652	0.4401	6.0	0.1506	−0.2767	11.0	−0.1712	−0.1768
1.1	0.7196	0.4709	6.1	0.1773	−0.2559	11.1	−0.1528	−0.1913
1.2	0.6711	0.4983	6.2	0.2017	−0.2329	11.2	−0.1330	−0.2039
1.3	0.6201	0.5220	6.3	0.2238	−0.2081	11.3	−0.1121	−0.2143
1.4	0.5669	0.5419	6.4	0.2433	−0.1816	11.4	−0.0902	−0.2225
1.5	0.5118	0.5579	6.5	0.2601	−0.1538	11.5	−0.0677	−0.2284
1.6	0.4554	0.5699	6.6	0.2740	−0.1250	11.6	−0.0446	−0.2320
1.7	0.3980	0.5778	6.7	0.2851	−0.0953	11.7	−0.0213	−0.2333
1.8	0.3400	0.5815	6.8	0.2931	−0.0652	11.8	0.0020	−0.2323
1.9	0.2818	0.5812	6.9	0.2981	−0.0349	11.9	0.0250	−0.2290
2.0	0.2239	0.5767	7.0	0.3001	−0.0047	12.0	0.0477	−0.2234
2.1	0.1666	0.5683	7.1	0.2991	0.0252	12.1	0.0697	−0.2157
2.2	0.1104	0.5560	7.2	0.2951	0.0543	12.2	0.0908	−0.2060
2.3	0.0555	0.5399	7.3	0.2882	0.0826	12.3	0.1108	−0.1943
2.4	0.0025	0.5202	7.4	0.2786	0.1096	12.4	0.1296	−0.1807
2.5	−0.0484	0.4971	7.5	0.2663	0.1352	12.5	0.1469	−0.1655
2.6	−0.0968	0.4708	7.6	0.2516	0.1592	12.6	0.1626	−0.1487
2.7	−0.1424	0.4416	7.7	0.2346	0.1813	12.7	0.1766	−0.1307
2.8	−0.1850	0.4097	7.8	0.2154	0.2014	12.8	0.1887	−0.1114
2.9	−0.2243	0.3754	7.9	0.1944	0.2192	12.9	0.1988	−0.0912
3.0	−0.2601	0.3391	8.0	0.1717	0.2346	13.0	0.2069	−0.0703
3.1	−0.2921	0.3009	8.1	0.1475	0.2476	13.1	0.2129	−0.0489
3.2	−0.3202	0.2613	8.2	0.1222	0.2580	13.2	0.2167	−0.0271
3.3	−0.3443	0.2207	8.3	0.0960	0.2657	13.3	0.2183	−0.0052
3.4	−0.3643	0.1792	8.4	0.0692	0.2708	13.4	0.2177	0.0166
3.5	−0.3801	0.1374	8.5	0.0419	0.2731	13.5	0.2150	0.0380
3.6	−0.3918	0.0955	8.6	0.0146	0.2728	13.6	0.2101	0.0590
3.7	−0.3992	0.0538	8.7	−0.0125	0.2697	13.7	0.2032	0.0791
3.8	−0.4026	0.0128	8.8	−0.0392	0.2641	13.8	0.1943	0.0984
3.9	−0.4018	−0.0272	8.9	−0.0653	0.2559	13.9	0.1836	0.1165
4.0	−0.3971	−0.0660	9.0	−0.0903	0.2453	14.0	0.1711	0.1334
4.1	−0.3887	−0.1033	9.1	−0.1142	0.2324	14.1	0.1570	0.1488
4.2	−0.3766	−0.1386	9.2	−0.1367	0.2174	14.2	0.1414	0.1626
4.3	−0.3610	−0.1719	9.3	−0.1577	0.2004	14.3	0.1245	0.1747
4.4	−0.3423	−0.2028	9.4	−0.1768	0.1816	14.4	0.1065	0.1850
4.5	−0.3205	−0.2311	9.5	−0.1939	0.1613	14.5	0.0875	0.1934
4.6	−0.2961	−0.2566	9.6	−0.2090	0.1395	14.6	0.0679	0.1999
4.7	−0.2693	−0.2791	9.7	−0.2218	0.1166	14.7	0.0476	0.2043
4.8	−0.2404	−0.2985	9.8	−0.2323	0.0928	14.8	0.0271	0.2066
4.9	−0.2097	−0.3147	9.9	−0.2403	0.0684	14.9	0.0064	0.2069

APPENDIX 15.C
Properties of Fourier Series for Periodic Signals

periodic signal	Fourier series coefficients
$\left.\begin{array}{l} x(t) \\ y(t) \end{array}\right\}$ periodic with period T_0	a_k b_k
$Ax(t)+By(t)$	Aa_k+Bb_k
$x(t-t_0)$	$a_k e^{-jk(2\pi/T_0)t_0}$
$e^{jM(2\pi/T_0)t}x(t)$	a_{k-M}
$x^*(t)$	a^*_{-k}
$x(-t)$	a_{-k}
$x(\alpha t)$, $\alpha > 0$ [periodic with period T_0/α]	a_k
$\int_{T_0} x(\tau)y(t-\tau)d\tau$	$T_0 a_k b_k$
$x(t)y(t)$	$\sum_{l=-\infty}^{+\infty} a_l b_{k-l}$
$\frac{dx(t)}{dt}$	$jk\frac{2\pi}{T_0}a_k$
$\int_{-\infty}^{t} x(t)\,dt$ $\left[\begin{array}{c}\text{finite-valued and periodic} \\ \text{only if } a_0 = 0\end{array}\right]$	$\left(\frac{1}{jk(2\pi/T_0)}\right)a_k$
$x(t)$ real	$\begin{cases} a_k = a^*_{-k} \\ \mathcal{Re}\{a_k\} = \mathcal{Re}\{a_{-k}\} \\ \mathcal{Im}\{a_k\} = -\mathcal{Im}\{a_{-k}\} \\ \|a_k\| = \|a_{-k}\| \\ \angle a_k = -\angle a_{-k} \end{cases}$
$x_e(t) = \text{even}\{x(t)\}$ [$x(t)$ real]	$\mathcal{Re}\{a_k\}$
$x_o(t) = \text{odd}\{x(t)\}$ [$x(t)$ real]	$j\mathcal{Im}\{a_k\}$

Parseval's relation for periodic signals

$$\frac{1}{T_0}\int_{T_0} |x(t)|^2\,dt = \sum_{k=-\infty}^{+\infty} |a_k|^2$$

Support Material

APPENDIX 15.D
Properties of Fourier Transform for Aperiodic Signals

aperiodic signal	Fourier transform
$x(t)$	$X(\omega)$
$y(t)$	$Y(\omega)$
$ax(t) + by(t)$	$aX(t) + bY(t)$
$x(t - t_0)$	$e^{-j\omega t_0}X(\omega)$
$e^{j\omega_0 t}x(t)$	$X(\omega - \omega_0)$
$x^*(t)$	$X^*(-\omega)$
$x(-t)$	$X(-\omega)$
$x(at)$	$\frac{1}{\lvert a\rvert}X\left(\frac{\omega}{a}\right)$
$x(t) * y(t)$	$X(\omega)Y(\omega)$
$x(t)y(t)$	$\frac{1}{2\pi}X(\omega) * Y(\omega)$
$\frac{d}{dt}x(t)$	$j\omega X(\omega)$
$\int_{-\infty}^{t} x(t)\,dt$	$\frac{1}{j\omega}X(\omega) + \pi X(0)\delta(\omega)$
$tx(t)$	$j\frac{d}{d\omega}X(\omega)$
$x(t)$ real	$\begin{cases} X(\omega) = X^*(-\omega) \\ \mathcal{R}e\{X(\omega)\} = \mathcal{R}e\{X(-\omega)\} \\ \mathcal{I}m\{X(\omega)\} = -\mathcal{I}m\{X(-\omega)\} \\ \lvert X(\omega)\rvert = \lvert X(-\omega)\rvert \\ \angle X(\omega) = -\angle X(-\omega) \end{cases}$
$x_e(t) = \text{even}\{x(t)\}$ [$x(t)$ real]	$\mathcal{R}e\{X(\omega)\}$
$x_o(t) = \text{odd}\{x(t)\}$ [$x(t)$ real]	$j\mathcal{I}m\{X(\omega)\}$

duality

$$f(u) = \int_{-\infty}^{+\infty} g(v)e^{juv}\,dv$$

$$g(t) \overset{\mathcal{F}}{\leftrightarrow} f(\omega)$$

$$f(t) \overset{\mathcal{F}}{\leftrightarrow} 2\pi g(-\omega)$$

Parseval's relation for aperiodic signals

$$\int_{-\infty}^{+\infty} \lvert x(t)\rvert^2\,dt = \frac{1}{2\pi}\int_{-\infty}^{+\infty} \lvert X(\omega)\rvert^2\,d\omega$$

Support Material

APPENDIX 15.E
Fourier Transform Pairs

signal	Fourier transform
$\sum_{k=-\infty}^{+\infty} a_k e^{jk\omega_0 t}$	$2\pi \sum_{k=-\infty}^{+\infty} a_k \delta(\omega - k\omega_0)$
$e^{j\omega_0 t}$	$2\pi\delta(\omega - \omega_0)$
$\cos \omega_0 t$	$\pi\Big(\delta(\omega - \omega_0) + \delta(\omega + \omega_0)\Big)$
$\sin \omega_0 t$	$\frac{\pi}{j}\Big(\delta(\omega - \omega_0) + \delta(\omega + \omega_0)\Big)$
$x(t) = 1$	$2\pi\delta(\omega)$
Periodic square wave $x(t) = \begin{cases} 1, & \lvert t\rvert < T_1 \\ 0, & T_1 < \lvert t\rvert \le \frac{T_0}{2} \end{cases}$ and $x(t + T_0) = x(t)$	$\sum_{k=-\infty}^{+\infty} \frac{2 \sin k\omega_0 T_1}{k} \delta(\omega - k\omega_0)$
$\sum_{k=-\infty}^{+\infty} \delta(t - nT)$	$\frac{2\pi}{T} \sum_{k=-\infty}^{+\infty} \delta\left(\omega - \frac{2\pi k}{T}\right)$
$x(t) = \begin{cases} 1, & \lvert t\rvert < T_1 \\ 0, & \lvert t\rvert > T_1 \end{cases}$	$2T_1 \operatorname{sinc} \frac{\omega T_1}{\pi} = \frac{2 \sin \omega T_1}{\omega}$
$\frac{W}{\pi} \operatorname{sinc} \frac{Wt}{\pi} = \frac{\sin Wt}{\pi t}$	$X(\omega) = \begin{cases} 1, & \lvert\omega\rvert < W \\ 0, & \lvert\omega\rvert > W \end{cases}$
$\delta(t)$	1
$u(t)$	$\frac{1}{j\omega} + \pi\delta(\omega)$
$\delta(t - t_0)$	$e^{-j\omega t_0}$
$e^{-at}u(t),\ \mathcal{R}e\{a\} > 0$	$\frac{1}{a + j\omega}$
$te^{-at}u(t),\ \mathcal{R}e\{a\} > 0$	$\frac{1}{(a + j\omega)^2}$
$\frac{t^{n-1}}{(n-1)!} e^{-at}u(t),$ $\mathcal{R}e\{a\} > 0$	$\frac{1}{(a + j\omega)^n}$

Support Material

APPENDIX 15.F
Properties of Laplace Transform

signal	transform	ROC (region of convergence)
$x(t)$	$X(s)$	R
$x_1(t)$	$X_1(s)$	R_1
$x_2(t)$	$X_2(s)$	R_2
$ax_1(t) + bx_2(t)$	$aX_1(s) + bX_2(s)$	at least $R_1 \cap R_2$
$x(t - t_0)$	$e^{-st_0}X(s)$	R
$e^{s_0 t}x(t)$	$X(s - s_0)$	shifted version of R [i.e., s is in the ROC if $(s - s_0)$ is in R]
$x(at)$	$\frac{1}{\lvert a\rvert}X\left(\frac{s}{a}\right)$	"scaled" ROC [i.e., s is in the ROC if s/a is in the ROC of $X(s)$]
$x_1(t) * x_2(t)$	$X_1(s)X_2(s)$	at least $R_1 \cap R_2$
$\frac{d}{dt}x(t)$	$sX(s)$	at least R
$-tx(t)$	$\frac{d}{ds}X(s)$	R
$\int_{-\infty}^{t} x(\tau)\,d\tau$	$\frac{1}{s}X(s)$	at least $R \cap \{\mathcal{R}e\{s\} > 0\}$

Support Material

APPENDIX 15.G
Laplace Transforms
(See also App. 11.A.)

transform pair	signal	transform	ROC (region of convergence)
1	$\delta(t)$	1	all s
2	$u(t)$	$\frac{1}{s}$	$\mathcal{R}e\{s\} > 0$
3	$-u(-t)$	$\frac{1}{s}$	$\mathcal{R}e\{s\} < 0$
4	$\frac{t^{n-1}}{(n-1)!}u(t)$	$\frac{1}{s^n}$	$\mathcal{R}e\{s\} > 0$
5	$-\frac{t^{n-1}}{(n-1)!}u(-t)$	$\frac{1}{s^n}$	$\mathcal{R}e\{s\} < 0$
6	$e^{-\alpha t}u(t)$	$\frac{1}{s+\alpha}$	$\mathcal{R}e\{s\} > \alpha$
7	$-e^{-\alpha t}u(-t)$	$\frac{1}{s+\alpha}$	$\mathcal{R}e\{s\} < -\alpha$
8	$\frac{t^{n-1}}{(n-1)!}e^{-\alpha t}u(t)$	$\frac{1}{(s+\alpha)^n}$	$\mathcal{R}e\{s\} > -\alpha$
9	$-\frac{t^{n-1}}{(n-1)!}e^{-\alpha t}u(-t)$	$\frac{1}{(s+\alpha)^n}$	$\mathcal{R}e\{s\} < -\alpha$
10	$\delta(t-T)$	e^{-sT}	all s
11	$(\cos\omega_0 t)u(t)$	$\frac{s}{s^2+\omega_0^2}$	$\mathcal{R}e\{s\} > 0$
12	$(\sin\omega_0 t)u(t)$	$\frac{\omega_0}{s^2+\omega_0^2}$	$\mathcal{R}e\{s\} > 0$
13	$(e^{-\alpha t}\cos\omega_0 t)u(t)$	$\frac{s+\alpha}{(s+\alpha)^2+\omega_0^2}$	$\mathcal{R}e\{s\} > -\alpha$
14	$(e^{-\alpha t}\sin\omega_0 t)u(t)$	$\frac{\omega_0}{(s+\alpha)^2+\omega_0^2}$	$\mathcal{R}e\{s\} > -\alpha$

Support Material

APPENDIX 15.H
Properties of Discrete-Time Fourier Series for Periodic Signals

periodic signal	Fourier series coefficients
$\left.\begin{matrix} x(n) \\ y(n) \end{matrix}\right\}$ periodic with period N	$\left.\begin{matrix} a_k \\ b_k \end{matrix}\right\}$ periodic with period N
$Ax[n] + By[n]$	$Aa_k + Bb_k$
$x[n - n_0]$	$a_k e^{-jk(2\pi/N)n_0}$
$e^{jM(2\pi/N)n}x[n]$	a_{k-M}
$x^*[n]$	a^*_{-k}
$x^*[-n]$	a_{-k}
$x_{(m)}[n] = \begin{cases} x[n/m] \text{ if } n \text{ is a multiple of } m \\ 0 \text{ if } n \text{ is not a multiple of } m \end{cases}$ (periodic with mN)	$\frac{1}{m}a_k$ [viewed as periodic with period mN]
$\sum_{r=\langle N\rangle} x[r]y[n-r]$	Na_kb_k
$x[n]y[n]$	$\sum_{l-\langle N\rangle} a_l b_{k-1}$
$x[n] - x[n-1]$	$(1 - e^{-jk(2\pi/N)})a_k$
$\sum_{k=-\infty}^{n} x[k]$ [finite-valued and periodic only if $a_0 = 0$]	$\left(\frac{1}{1 - e^{-jk(2\pi/N)}}\right)a_k$
$x[n]$ real	$\begin{cases} a_k = a^*_{-k} \\ \mathcal{R}e\{a_k\} = \mathcal{R}e\{a_{-k}\} \\ \mathcal{I}m\{a_k\} = -\mathcal{I}m\{a_{-k}\} \\ \lvert a_k\rvert = \lvert a_{-k}\rvert \\ \angle a_k = -\angle a_{-k} \end{cases}$
$x_e[n] = \text{even}\{x[n]\}$ [$x[n]$ real]	$\mathcal{R}e\{a_k\}$
$x_o[n] = \text{odd}\{x[n]\}$ [$x[n]$ real]	$j\mathcal{I}m[a_k]$

Parseval's relation for periodic signals

$$\frac{1}{N}\sum_{n=\langle N\rangle} |x[n]|^2 = \sum_{k=\langle N\rangle} |a_k|^2$$

Support Material

APPENDIX 15.I

Properties of Discrete-Time Fourier Transform for Aperiodic Signals

aperiodic signal	Fourier transform
$x(n)$ $y(n)$	$\left.\begin{matrix} X(\Omega) \\ Y(\Omega) \end{matrix}\right\}$ periodic with period 2π
$ax[n] + by[n]$	$aX(\Omega) + bY(\Omega)$
$x[n - n_0]$	$e^{-j\Omega n_0} X(\Omega)$
$e^{j\Omega_0 n} x[n]$	$X(\Omega - \Omega_0)$
$x^*[n]$	$X^*(-\Omega)$
$x[-n]$	$X(-\Omega)$
$x_{(k)}[n] = \begin{cases} x[n/k], & \text{if } n \text{ is a multiple of } k \\ 0, & \text{if } n \text{ is not a multiple of } k \end{cases}$	$X(k\Omega)$
$x[n] * y[n]$	$X(\Omega)\,Y(\Omega)$
$x[n]y[n]$	$\frac{1}{2\pi}\int_{2\pi} X(\theta)\,Y(\Omega - \theta)\,d\theta$
$x[n] - x[n-1]$	$1 - e^{-j\Omega} X(\Omega)$
$\sum_{k=-\infty}^{n} x[k]$	$\frac{1}{1 - e^{j\Omega}} X(\Omega) + \pi X(0) \sum_{k=-\infty}^{+\infty} \delta(\Omega - 2\pi k)$
$nx[n]$	$j\frac{dX(\Omega)}{d\Omega}$
$x[n]$ real	$\begin{cases} X(\Omega) = X^*(-\Omega) \\ \mathcal{R}e\{X(\Omega)\} = \mathcal{R}e\{X(-\Omega)\} \\ \mathcal{I}m\{X(\Omega)\} = -\mathcal{I}m\{X(-\Omega)\} \\ \lvert X(\Omega)\rvert = \lvert X(-\Omega)\rvert \\ \angle X(\Omega) = -\angle X(-\Omega) \end{cases}$
$x_e[n] = \text{even}\{x[n]\}$ $[x[n]$ real$]$	$\mathcal{R}e\{X(\Omega)\}$
$x_o[n] = \text{odd}\{x[n]\}$ $[x[n]$ real$]$	$j\mathcal{I}m\{X(\Omega)\}$

Parseval's relation for aperiodic signals

$$\sum_{n=-\infty}^{+\infty} \lvert x[n]\rvert^2 = \frac{1}{2\pi}\int_{2\pi} \lvert X(\Omega)\rvert^2 d\Omega$$

Support Material

APPENDIX 15.J
Properties of *z*-Transforms

sequence	transform	ROC (region of convergence)
$x[n]$	$X(z)$	R_x
$x_1[n]$	$X_1(z)$	R_1
$x_2[n]$	$X_2(z)$	R_2
$ax_1[n] + bx_2[n]$	$aX_1(z) + bX_2(z)$	at least the intersection of R_1 and R_2
$x[n - n_0]$	$z^{-n_0}X(z)$	R_x except for the possible addition or deletion of the origin
$e^{j\Omega_0 n}x[n]$	$X(e^{-j\Omega_0}z)$	R_x
$z_0^n x[n]$	$X\left(\frac{z}{z_0}\right)$	$z_0 R_x$
$a^n x[n]$	$X(a^{-1}z)$	scaled version of R_x (i.e., $\lvert a\rvert R_x =$ the set of points $\{\lvert a\rvert z\}$ for z in R_x)
$x[-n]$	$X(z^{-1})$	inverted R_x (i.e., $R_x^{-1} =$ the set of points z^{-1} where z is in R_x)
$w[n] = \begin{cases} x[r], & n = rk \\ 0, & n \neq rk \text{ for some } r \end{cases}$	$X(z^k)$	$R_x^{1/k}$ (i.e., the set of points $z^{1/k}$ where z is in R_x)
$x_1[n]x_2[n]$	$X_1(z)X_2(z)$	at least the intersection of R_1 and R_2
$nx[n]$	$-z\frac{dX(z)}{dz}$	R_x except for the possible addition or deletion of the origin
$\sum_{k=-\infty}^{n} x[k]$	$\frac{1}{1 - z^{-1}}X(z)$	at least the intersection of R_x and $\lvert z\rvert > 1$

Support Material

APPENDIX 15.K
z-Transforms

$f(t)$	$F(z)$
$\delta(t)$	1
$\delta(t-mT)$	$\frac{1}{z^m}$
$H(t)$	$\frac{z}{z-1}$
t	$\frac{Tz}{(z-1)^2}$
t^2	$\frac{T^2 z(z+1)}{(z-1)^3}$
t^3	$\frac{T^3 z(z^2+4z+1)}{(z-1)^4}$
t^n	$(-1)^n \lim_{x\to 0} \frac{\delta^n}{\delta_{x^n}} \left(\frac{z}{z-e^{-xT}}\right)$
$1-a^{\Omega\tau}$	$\frac{z(1-a^{\Omega\tau})}{(z-1)(z-a^{\Omega\tau})}$
$a^{\Omega\tau}$	$\frac{z}{z-a^{\Omega\tau}}$
$ta^{\Omega\tau}$	$\frac{Tza^{\Omega\tau}}{(z-a^{\Omega\tau})^2}$
$t^2 a^{\Omega\tau}$	$\frac{T^2 a^{\Omega\tau} z(z+a^{\Omega\tau})}{(z-a^{\Omega\tau})^3}$
$\sin \Omega\tau$	$\frac{z \sin \Omega T}{z^2-2z\cos \Omega T+1}$
$\cos \Omega t$	$\frac{z(z-\cos \Omega T)}{z^2-2z\cos \Omega T+1}$
$\sinh \Omega t$	$\frac{z \sinh \Omega T}{z^2-2z\cosh \Omega T+1}$
$\cosh \Omega t$	$\frac{z(z-\cosh \Omega T)}{z^2-2z\cosh \Omega T+1}$
$e^{-at}\sin \Omega t$	$\frac{ze^{-a\tau}\sin \Omega T}{z^2-2ze^{-\tau}\cos \Omega T+e^{-2a\tau}}$
$e^{-at}\cos \Omega t$	$\frac{z(z-e^{-a\tau}\cos \Omega T)}{z^2-2ze^{-\tau}\cos \Omega T+e^{-2a\tau}}$
$e^{-at}\sinh \Omega t$	$\frac{ze^{-a\tau}\sinh \Omega T}{z^2-2ze^{-a\tau}\cosh \Omega T+e^{-2a\tau}}$
$e^{-at}\cosh \Omega t$	$\frac{z(z-e^{-a\tau}\cosh \Omega T)}{z^2-2ze^{-a\tau}\cosh \Omega T+e^{-2a\tau}}$

APPENDIX 17.A

Room Temperature Properties of Silicon, Germanium, and Gallium Arsenide (at 300K)

property	Si	Ge	GaAs
atoms/cm^3	5.0×10^{22}	4.42×10^{22}	2.21×10^{22}
atomic weight	28.08	72.6	144.63
breakdown field	$\approx 3 \times 10^5$ V/cm	$\approx 10^5$ V/cm	$\approx 4 \times 10^5$ V/cm
crystal structure	diamond	diamond	zincblende
density	2.328 g/cm^3	5.3267 g/cm^3	5.32 g/cm^3
dielectric constant	11.8	16	10.9
effective density of states			
in conduction band, N_c	2.8×10^{19} cm^{-3}	1.04×10^{19} cm^{-3}	4.7×10^{17} cm^{-3}
in valence band, N_v	1.02×10^{19} cm^{-3}	6.1×10^{18} cm^{-3}	7.0×10^{18} cm^{-3}
effective mass m^*/m_0 (electrons)			
longitudinal	0.98 kg	1.64 kg	1.98 kg
transverse	0.19 kg	0.082 kg	0.37 kg
effective mass, m^*/m_0 (holes)			
light	0.16 kg	0.044 kg	0.12 kg
heavy	0.49 kg	0.28 kg	0.5 kg
electron affinity, x	4.05 V	4.0 V	4.07 V
energy gap	1.12 eV	0.803 eV	1.43 eV
intrinsic carrier concentration	1.6×10^{10} cm^{-3}	2.5×10^{13} cm^{-3}	1.1×10^7 cm^{-3}
lattice constant	0.543086 nm	0.565748 nm	0.56534 nm
linear coefficient of thermal expansion, $\delta = \frac{\Delta L}{L\Delta T}$	2.6×10^{-6} K^{-1}	5.8×10^{-6} K^{-1}	5.9×10^{-6} K^{-1}
melting point	1420°C	937°C	1238°C
minority-carrier lifetime	2.5×10^{-3} s	10^{-3} s	$\approx 10^{-8}$ s
mobility (drift)			
electrons, μ_n	1500 cm^2/V·s	3900 cm^2/V·s	8500 cm^2/V·s
holes, μ_p	600 cm^2/V·s	1900 cm^2/V·s	400 cm^2/V·s
specific heat	0.7 J/g·K	0.31 J/g·K	0.35 J/g·K
thermal conductivity	1.45 W/cm·K	0.64 W/cm·K	0.46 W/cm·K
thermal diffusivity	0.9 cm^2/s	0.36 cm^2/s	0.44 cm^2/s
work function	4.8 V	4.4 V	4.7 V

Support Material

APPENDIX 18.A
Electromagnetic Spectrum

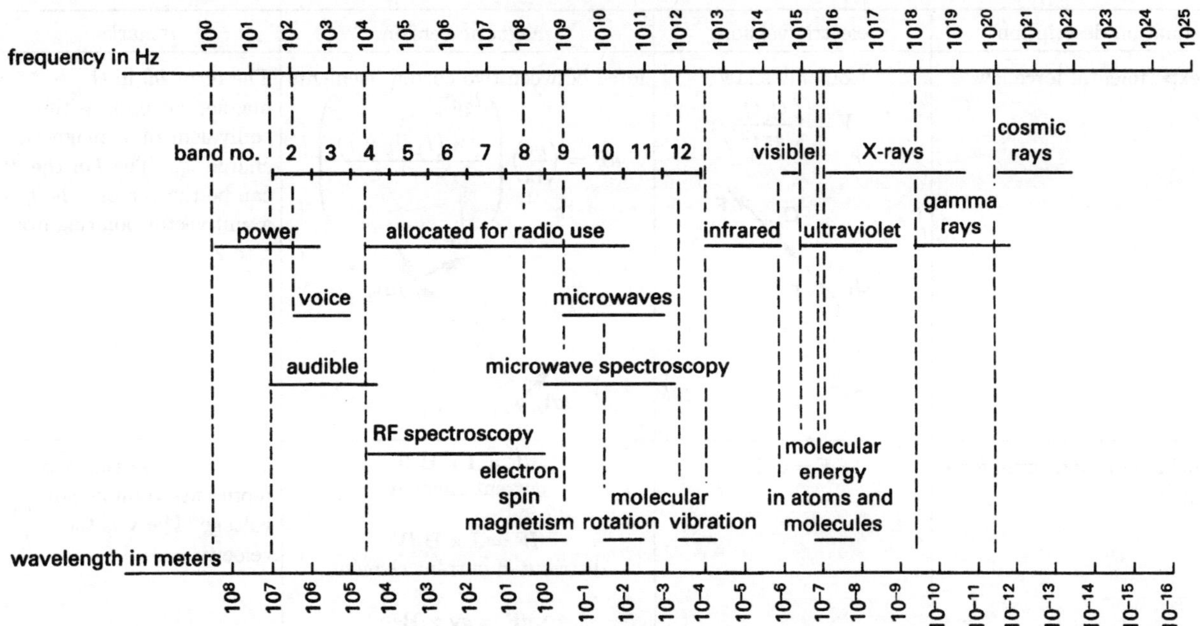

Support Material

APPENDIX 25.A
Comparison of Electric and Magnetic Equations

equation description	electric version	magnetic version	remarks
experimental force law	Coulomb's law $\mathbf{F} = \left(\frac{Q_1 Q_2}{4\pi r^2}\right)\mathbf{r}$ 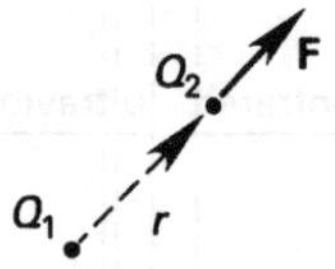	force between two current elements $d\mathbf{F} = \left(\frac{\mu_0}{4\pi}\right)\left(\frac{I_2 d\mathbf{l}_2 \times (I_1 d\mathbf{l}_1 \times r)}{r^2}\right)$ 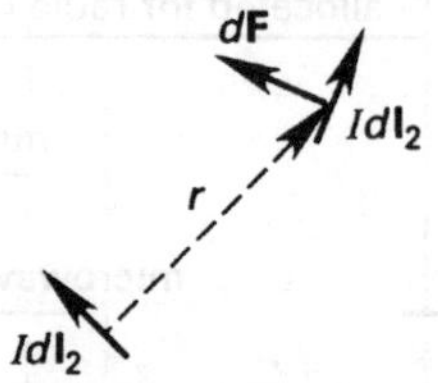	The term $Id\mathbf{l}$ in the magnetic column is the equivalent of a "magnetic charge" q_m. The I or the $d\mathbf{l}$ can be the vector. The $\mathbf{r}$ is a unit vector pointing from 1 to 2.
field definitions from force law	$\mathbf{F} = Q\mathbf{E}$	$d\mathbf{F} = \mathbf{I} \times \mathbf{B} dl$ current element $d\mathbf{F} = \mathbf{J} \times \mathbf{B} dV$ distributed current element $d\mathbf{F} = q\mathbf{v} \times \mathbf{B}$ moving charge	The V used in this row represents volume, not voltage. The $\mathbf{v}$ is the velocity.
general force law	$\mathbf{F} = q(\mathbf{E} + \mathbf{v} \times \mathbf{B})$ $d\mathbf{F} = (\rho\mathbf{E} + \mathbf{J} \times \mathbf{B})dV$ where $dQ = \rho dV$		The V in this row represents the volume, not voltage. The $\mathbf{v}$ is the velocity.
definition of scalar and vector potential	$E = -\nabla V$	$\mathbf{B} = \nabla \times \mathbf{A}$	$\mathbf{A}$ is the magnetic vector potential.
Poisson's equation for the potential function	$\nabla^2 V = -\frac{\rho}{\epsilon}$	$\nabla^2 \mathbf{A} = -\mu_0 \mathbf{J}$	From a knowledge of the charge distribution, the potential can be found and then the $\mathbf{E}$ and $\mathbf{B}$ fields determined.
Gauss's law enclosing charge and Ampere's law enclosing current	$\iint \mathbf{D} \cdot d\mathbf{A} = \iiint \rho dV = Q$ $\nabla \cdot \mathbf{D} = \rho$	$\oint H \cdot dl = I$ $\nabla \times \mathbf{H} = \mathbf{J}$	The V in this row represents volume.
constitutive relations	$\mathbf{D} = \epsilon\mathbf{E}$ $\mathbf{D} = \epsilon_0\mathbf{E} + \mathbf{P}$	$\mathbf{B} = \mu\mathbf{H}$ $\mathbf{B} = \mu_0\mathbf{H} + \mu_0\mathbf{M}$	The second set of equations is always valid. The first set assumes the medium is linear and isotropic.
definitions of relative permittivity and permeability	$\epsilon_r = \frac{\epsilon}{\epsilon_0}$ $\epsilon_0 = 8.854 \times 10^{-12}$ F/m	$\mu_r = \frac{\mu}{\mu_0}$ $\mu_0 = 4\pi \times 10^{-7}$ H/m	

(continued)

Support Material

APPENDIX 25.A *(continued)*
Comparison of Electric and Magnetic Equations

equation description	electric version	magnetic version	remarks
capacitance and inductance of a field cell	$\epsilon_0 = \frac{C}{l}$	$\mu_0 = \frac{L}{l}$	Field cells are a construct designed to represent free space in terms of a parallel plate capacitor and an inductor. This capacitance and inductance exist regardless of the presence of an electric or magnetic field.
capacitance and inductance	$C = \frac{Q}{V}$	$L = \frac{\Lambda}{I}$	Λ is the flux linkage.
energy density of a field	$U = \frac{1}{2}\epsilon E^2$	$U = \frac{1}{2}\mu H^2$	Both energy and momentum are carried by a field.
energy stored by capacitance and inductance	$W = \frac{1}{2}CV^2$	$W = \frac{1}{2}LI^2$	
electromotive and magnetomotive force with sources present	$\oint \mathcal{E} \cdot d\mathbf{l} = \mathcal{E} = V$	$\oint \mathbf{H} \cdot d\mathbf{l} = NI = F_m = V_m$	The $\mathcal{E}$ is the emf, not the permittivity. Without sources present, both line integrals are equal to zero.
dipole moments	$\mathbf{p} = q\mathbf{d}$ 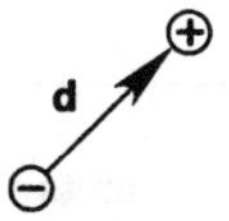	$\mathbf{m} = I\mathbf{A}$ 	
dipole torque	$\mathbf{T} = \mathbf{p} \times \mathbf{E}$	$\mathbf{T} = \mathbf{m} \times \mathbf{B}$	This torque occurs due to the dipole being immersed in an external **E** or **B** field.
dipole potential energy	$W = -\mathbf{p} \cdot \mathbf{E}$	$W = -\mathbf{m} \cdot \mathbf{B}$	

Support Material

APPENDIX 27.A
Impedance of Series-Connected Circuit Elements

circuit	impedance $Z = R + jX$ (Ω)	magnitude of impedance $\lvert Z\rvert = \sqrt{R^2 + X^2}$ (Ω)	phase angle $\theta = \arctan\frac{X}{R}$ (radians)	admittance $Y = \frac{1}{Z}$ (S)
R	R	R	0	$\frac{1}{R}$
L	$j\omega L$	ωL	$+\frac{\pi}{2}$	$-j\left(\frac{1}{\omega L}\right)$
C	$-j\left(\frac{1}{\omega C}\right)$	$\frac{1}{\omega C}$	$-\frac{\pi}{2}$	$j\omega C$
R_1 R_2	$R_1 + R_2$	$R_1 + R_2$	0	$\frac{1}{R_1 + R_2}$
M, L_1 L_2	$j\omega(L_1 + L_2 \pm 2M)$	$\omega(L_1 + L_2 \pm 2M)$	$+\frac{\pi}{2}$	$-j\left(\frac{1}{\omega(L_1 + L_2 \pm 2M)}\right)$
C_1 C_2	$-j\left(\frac{1}{\omega}\right)\left(\frac{C_1 + C_2}{C_1 C_2}\right)$	$\frac{1}{\omega}\left(\frac{C_1 + C_2}{C_1 C_2}\right)$	$-\frac{\pi}{2}$	$j\omega\left(\frac{C_1 C_2}{C_1 + C_2}\right)$
R L	$R + j\omega L$	$\sqrt{R^2 + \omega^2 L^2}$	$\arctan\frac{\omega L}{R}$	$\frac{R - j\omega L}{R^2 + \omega^2 L^2}$
R C	$R - j\left(\frac{1}{\omega C}\right)$	$\sqrt{\frac{\omega^2 C^2 R^2 + 1}{\omega^2 C^2}}$	$\arctan\frac{1}{\omega RC}$	$\frac{\omega^2 C^2 R + j\omega C}{\omega^2 C^2 R^2 + 1}$
L C	$j\left(\omega L - \frac{1}{\omega C}\right)$	$\omega L - \frac{1}{\omega C}$	$\pm\frac{\pi}{2}$	$-j\left(\frac{\omega C}{\omega^2 LC - 1}\right)$
R L C	$R + j\left(\omega L - \frac{1}{\omega C}\right)$	$\sqrt{R^2 + \left(\omega L - \frac{1}{\omega C}\right)^2}$	$\arctan\frac{\omega L - \frac{1}{\omega C}}{R}$	$\frac{R - j\left(\omega L - \frac{1}{\omega C}\right)}{R^2 + \left(\omega L - \frac{1}{\omega C}\right)^2}$

Support Material

APPENDIX 27.B
Impedance of Parallel-Connected Circuit Elements

circuit	impedance, $Z = R + jX$ (Ω)	magnitude of impedance, $\lvert Z\rvert = \sqrt{R^2 + X^2}$ (Ω)
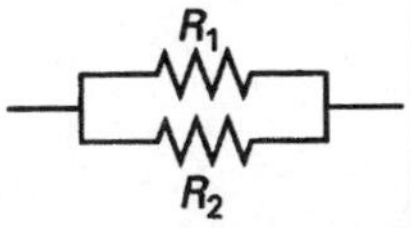 R_1, R_2	$\dfrac{R_1 R_2}{R_1 + R_2}$	$\dfrac{R_1 R_2}{R_1 + R_2}$
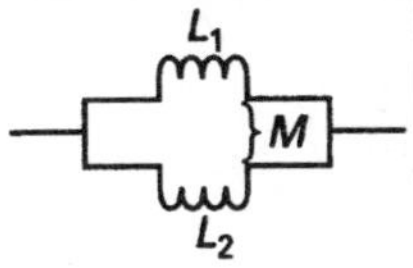 L_1, M, L_2	$+j\omega\left(\dfrac{L_1 L_2 - M^2}{L_1 + L_2 \mp 2M}\right)$	$\omega\left(\dfrac{L_1 L_2 - M^2}{L_1 + L_2 \mp 2M}\right)$
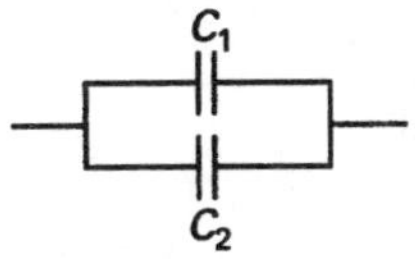 C_1, C_2	$-j\left(\dfrac{1}{\omega(C_1 + C_2)}\right)$	$\dfrac{1}{\omega(C_1 + C_2)}$
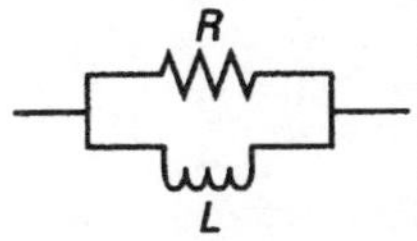 R, L	$\dfrac{\omega^2 L^2 R + j\omega L R^2}{\omega^2 L^2 + R^2}$	$\dfrac{\omega L R}{\sqrt{\omega^2 L^2 + R^2}}$
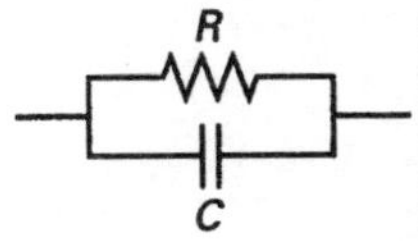 R, C	$\dfrac{R - j\omega R^2 C}{1 + \omega^2 R^2 C^2}$	$\dfrac{R}{\sqrt{1 + \omega^2 R^2 C^2}}$
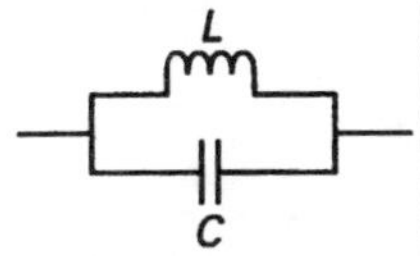 L, C	$j\left(\dfrac{\omega L}{1 - \omega^2 LC}\right)$	$\dfrac{\omega L}{1 - \omega^2 LC}$
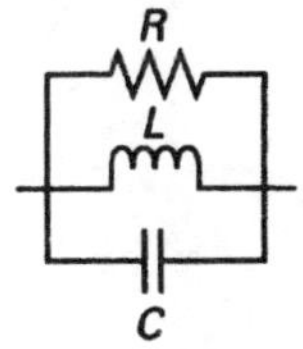 R, L, C	$\dfrac{\frac{1}{R} - j\left(\omega C - \frac{1}{\omega L}\right)}{\left(\frac{1}{R}\right)^2 + \left(\omega C - \frac{1}{\omega L}\right)^2}$	$\dfrac{1}{\sqrt{\left(\frac{1}{R}\right)^2 + \left(\omega C - \frac{1}{\omega L}\right)^2}}$
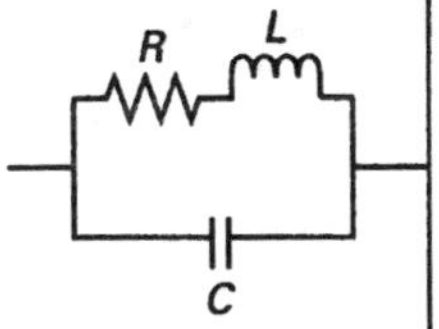 R, L, C	$\dfrac{\frac{R}{\omega^2 C^2} - j\left(\frac{R^2}{\omega C} + \frac{L}{C}\left(\omega L - \frac{1}{\omega C}\right)\right)}{R^2 + \left(\omega L - \frac{1}{\omega C}\right)^2}$	$\dfrac{\sqrt{\left(\frac{R}{\omega^2 C^2}\right)^2 + \left(\frac{R^2}{\omega C} + \left(\frac{L}{C}\right)\left(\omega L - \frac{1}{\omega C}\right)\right)^2}}{R^2 + \left(\omega L - \frac{1}{\omega C}\right)^2}$
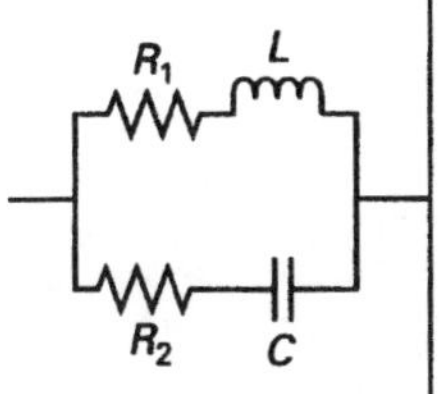 R_1, L, R_2, C	$\dfrac{R_1 R_2 (R_1 + R_2) + \omega^2 L^2 R_2 + \frac{R_1}{\omega^2 C^2}}{(R_1 + R_2)^2 + \left(\omega L - \frac{1}{\omega C}\right)^2} + j\left(\dfrac{\omega R_2^2 L - \frac{R_1^2}{\omega C} - \left(\frac{L}{C}\right)\left(\omega L - \frac{1}{\omega C}\right)}{(R_1 + R_2)^2 + \left(\omega L - \frac{1}{\omega C}\right)^2}\right)$	$\sqrt{\left(\dfrac{R_1 R_2 (R_1 + R_2) + \omega^2 L^2 R_2 + \frac{R_1}{\omega^2 C^2}}{(R_1 + R_2)^2 + \left(\omega L - \frac{1}{\omega C}\right)^2}\right)^2 + \left(\dfrac{\omega L R_2^2 - \frac{R_1^2}{\omega C} - \frac{L}{C}\left(\omega L - \frac{1}{\omega C}\right)}{(R_1 + R_2)^2 + \left(\omega L - \frac{1}{\omega C}\right)^2}\right)^2}$

(continued)

Support Material

APPENDIX 27.B *(continued)*
Impedance of Parallel-Connected Circuit Elements

circuit	phase angle, $\theta = \arctan \frac{X}{R}$ (radians)	admittance, $Y = \frac{1}{Z}$ (S)
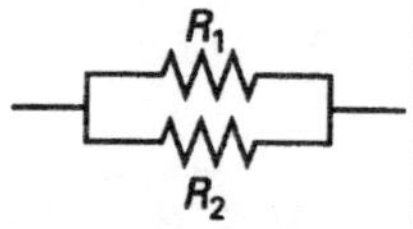 R_1 parallel R_2	0	$\frac{R_1 + R_2}{R_1 R_2}$
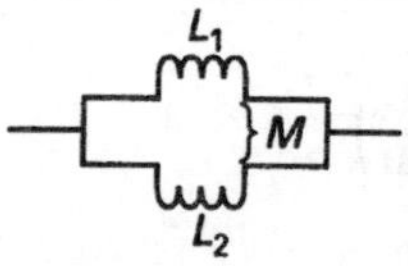 L_1 parallel L_2, mutual M	$+\frac{\pi}{2}$	$-j\left(\frac{1}{\omega}\right)\left(\frac{L_1 + L_2 \mp 2M}{L_1 L_2 - M^2}\right)$
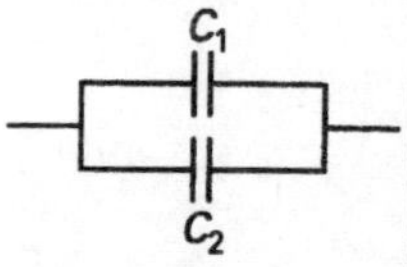 C_1 parallel C_2	$-\frac{\pi}{2}$	$+j\omega(C_1 + C_2)$
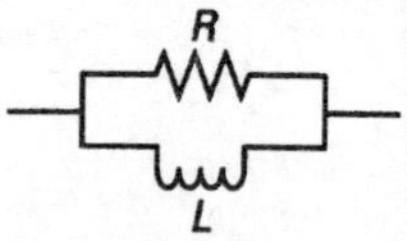 R parallel L	$\arctan \frac{R}{\omega L}$	$\frac{\omega L - jR}{\omega L R}$
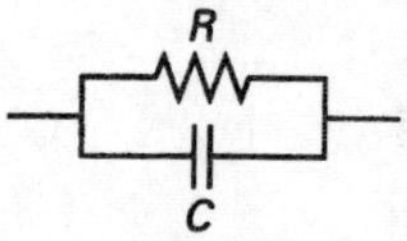 R parallel C	$\arctan(-\omega R C)$	$\frac{1}{R} + j\omega C$
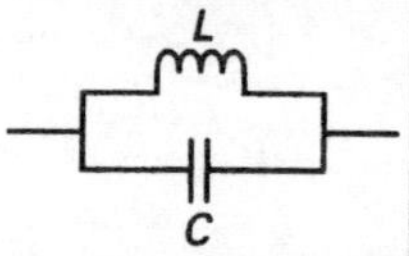 L parallel C	$\pm\frac{\pi}{2}$	$-j\left(\frac{1 - \omega^2 L C}{\omega L}\right)$
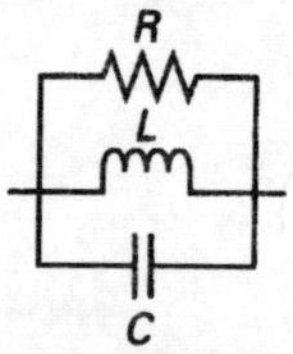 R, L, C parallel	$\arctan R\left(\frac{1}{\omega L} - \omega C\right)$	$\frac{1}{R} + j\left(\omega C - \frac{1}{\omega L}\right)$
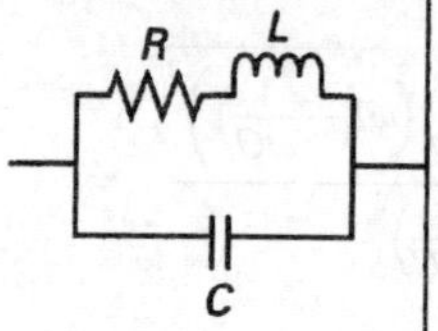 R–L series, parallel C	$\arctan\left(\frac{\frac{R^2}{\omega C} + \left(\frac{L}{C}\right)\left(\omega L - \frac{1}{\omega C}\right)}{\frac{R}{\omega^2 C^2}}\right)$	$\frac{R + j\omega(R^2 C - L + \omega^2 L^2 C)}{R^2 + \omega^2 L^2}$
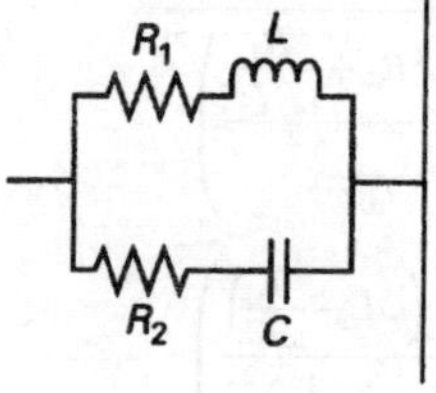 R_1–L series, parallel R_2–C series	$\arctan\left(\frac{\omega L R^2 - \frac{R_1^2}{\omega C} - \left(\frac{L}{C}\right)\left(\omega L - \frac{1}{\omega C}\right)}{R_1 R_2 (R_1 + R_2) + \omega^2 L^2 R_2 + \frac{R_1}{\omega^2 C^2}}\right)$	$\frac{R_1 + \omega^2 R_1 R_2 C^2 (R_1 + R_2) + \omega^4 L^2 C^2 R_2}{(R_1^2 + \omega^2 L^2)(\omega^2 R_2^2 C^2 + 1)} + j\left(\frac{\omega\left(R_1^2 C - L + \omega^2 L C (L - R_2^2 C)\right)}{(R_1^2 + \omega^2 L^2)(\omega^2 R_2^2 C^2 + 1)}\right)$

Support Material

APPENDIX 29.A
Two-Port Parameter Conversions

	$[z]$	$[y]$	$[h]$	$[g]$	$[TL]$	$[TLI]$
$[z]$	$\begin{bmatrix} z_{11} & z_{12} \\ z_{21} & z_{22} \end{bmatrix}$	$\begin{bmatrix} \frac{y_{22}}{\lvert y\rvert} & -\frac{y_{12}}{\lvert y\rvert} \\ -\frac{y_{21}}{\lvert y\rvert} & \frac{y_{11}}{\lvert y\rvert} \end{bmatrix}$	$\begin{bmatrix} \frac{\lvert h\rvert}{h_{22}} & \frac{h_{12}}{h_{22}} \\ -\frac{h_{22}}{h_{22}} & \frac{1}{h_{22}} \end{bmatrix}$	$\begin{bmatrix} \frac{1}{g_{11}} & -\frac{g_{12}}{g_{11}} \\ \frac{g_{21}}{g_{11}} & \frac{\lvert g\rvert}{g_{11}} \end{bmatrix}$	$\begin{bmatrix} \frac{A}{C} & \frac{\lvert TL\rvert}{C} \\ \frac{1}{C} & \frac{D}{C} \end{bmatrix}$	$\begin{bmatrix} \frac{\delta}{\gamma} & \frac{1}{\gamma} \\ \frac{\lvert TLI\rvert}{\gamma} & \frac{\alpha}{\gamma} \end{bmatrix}$
$[y]$	$\begin{bmatrix} \frac{z_{22}}{\lvert z\rvert} & -\frac{z_{12}}{\lvert z\rvert} \\ -\frac{z_{21}}{\lvert z\rvert} & \frac{z_{11}}{\lvert z\rvert} \end{bmatrix}$	$\begin{bmatrix} y_{11} & y_{12} \\ y_{21} & y_{22} \end{bmatrix}$	$\begin{bmatrix} \frac{1}{h_{11}} & -\frac{h_{12}}{h_{11}} \\ \frac{h_{21}}{h_{11}} & \frac{\lvert h\rvert}{h_{11}} \end{bmatrix}$	$\begin{bmatrix} \frac{\lvert g\rvert}{g_{22}} & \frac{g_{12}}{g_{22}} \\ -\frac{g_{21}}{g_{22}} & \frac{1}{g_{22}} \end{bmatrix}$	$\begin{bmatrix} \frac{D}{B} & -\frac{\lvert TL\rvert}{B} \\ -\frac{1}{B} & \frac{A}{B} \end{bmatrix}$	$\begin{bmatrix} \frac{\alpha}{\beta} & -\frac{1}{\beta} \\ -\frac{\lvert TLI\rvert}{\beta} & \frac{\delta}{\beta} \end{bmatrix}$
$[h]$	$\begin{bmatrix} \frac{\lvert z\rvert}{z_{22}} & \frac{z_{21}}{z_{22}} \\ -\frac{z_{21}}{z_{22}} & \frac{1}{z_{22}} \end{bmatrix}$	$\begin{bmatrix} \frac{1}{y_{11}} & -\frac{y_{12}}{y_{11}} \\ \frac{y_{21}}{y_{11}} & \frac{\lvert y\rvert}{y_{11}} \end{bmatrix}$	$\begin{bmatrix} h_{11} & h_{12} \\ h_{21} & h_{22} \end{bmatrix}$	$\begin{bmatrix} \frac{g_{22}}{\lvert g\rvert} & -\frac{g_{12}}{\lvert g\rvert} \\ -\frac{g_{21}}{\lvert g\rvert} & \frac{g_{11}}{\lvert g\rvert} \end{bmatrix}$	$\begin{bmatrix} \frac{B}{D} & \frac{\lvert TL\rvert}{D} \\ -\frac{1}{D} & \frac{C}{D} \end{bmatrix}$	$\begin{bmatrix} \frac{\beta}{\alpha} & \frac{1}{\alpha} \\ \frac{\lvert TLI\rvert}{\alpha} & \frac{\gamma}{\alpha} \end{bmatrix}$
$[g]$	$\begin{bmatrix} \frac{1}{z_{11}} & -\frac{z_{12}}{z_{11}} \\ \frac{z_{21}}{z_{11}} & \frac{\lvert z\rvert}{z_{11}} \end{bmatrix}$	$\begin{bmatrix} \frac{\lvert y\rvert}{y_{22}} & \frac{y_{12}}{y_{22}} \\ -\frac{y_{21}}{y_{22}} & \frac{1}{y_{22}} \end{bmatrix}$	$\begin{bmatrix} \frac{h_{22}}{\lvert h\rvert} & -\frac{h_{12}}{\lvert h\rvert} \\ -\frac{h_{21}}{\lvert h\rvert} & \frac{h_{11}}{\lvert h\rvert} \end{bmatrix}$	$\begin{bmatrix} g_{11} & g_{12} \\ g_{21} & g_{22} \end{bmatrix}$	$\begin{bmatrix} \frac{C}{A} & -\frac{\lvert TL\rvert}{A} \\ \frac{1}{A} & \frac{B}{A} \end{bmatrix}$	$\begin{bmatrix} \frac{\gamma}{\delta} & -\frac{1}{\delta} \\ \frac{\lvert TLI\rvert}{\delta} & -\frac{\beta}{\delta} \end{bmatrix}$
$[TL]$	$\begin{bmatrix} \frac{z_{11}}{z_{21}} & \frac{\lvert z\rvert}{z_{21}} \\ \frac{1}{z_{21}} & \frac{z_{22}}{z_{21}} \end{bmatrix}$	$\begin{bmatrix} -\frac{y_{22}}{y_{21}} & -\frac{1}{y_{21}} \\ -\frac{\lvert y\rvert}{y_{21}} & -\frac{y_{11}}{y_{21}} \end{bmatrix}$	$\begin{bmatrix} -\frac{\lvert h\rvert}{h_{21}} & -\frac{h_{11}}{h_{21}} \\ -\frac{h_{22}}{h_{21}} & -\frac{1}{h_{21}} \end{bmatrix}$	$\begin{bmatrix} \frac{1}{g_{21}} & \frac{g_{22}}{g_{21}} \\ \frac{g_{21}}{g_{21}} & \frac{\lvert g\rvert}{g_{21}} \end{bmatrix}$	$\begin{bmatrix} A & B \\ C & D \end{bmatrix}$	$\begin{bmatrix} \frac{\delta}{\lvert TLI\rvert} & \frac{\beta}{\lvert TLI\rvert} \\ \frac{\gamma}{\lvert TLI\rvert} & \frac{\alpha}{\lvert TLI\rvert} \end{bmatrix}$
$[TLI]$	$\begin{bmatrix} \frac{z_{22}}{z_{12}} & \frac{\lvert z\rvert}{z_{12}} \\ \frac{1}{z_{12}} & \frac{z_{11}}{z_{12}} \end{bmatrix}$	$\begin{bmatrix} -\frac{y_{11}}{y_{12}} & -\frac{1}{y_{12}} \\ -\frac{\lvert y\rvert}{y_{12}} & -\frac{y_{22}}{y_{12}} \end{bmatrix}$	$\begin{bmatrix} \frac{1}{h_{12}} & \frac{h_{11}}{h_{12}} \\ \frac{h_{22}}{h_{12}} & \frac{\lvert h\rvert}{h_{12}} \end{bmatrix}$	$\begin{bmatrix} -\frac{\lvert g\rvert}{g_{12}} & -\frac{g_{22}}{g_{12}} \\ -\frac{g_{11}}{g_{12}} & -\frac{1}{g_{12}} \end{bmatrix}$	$\begin{bmatrix} \frac{D}{\lvert TL\rvert} & \frac{B}{\lvert TL\rvert} \\ \frac{C}{\lvert TL\rvert} & \frac{A}{\lvert TL\rvert} \end{bmatrix}$	$\begin{bmatrix} \alpha & \beta \\ \gamma & \delta \end{bmatrix}$

Note: $|x| = \det[x] = \det\begin{bmatrix} x_{11} & x_{12} \\ x_{21} & x_{22} \end{bmatrix} = x_{11}x_{22} - x_{12}x_{21}$, where x is any of the listed parameters, z, y, h, g, TL, or TLI.

Support Material

APPENDIX 30.A
Resonant Circuit Formulas

<table>
<tr><th>unknown quantity</th><th>symbol</th><th>units</th><th>series</th><th>parallel</th></tr>
<tr><td rowspan="4">resonant frequency</td><td rowspan="2">f_0</td><td rowspan="2">Hz</td><td colspan="2">$\frac{1}{2\pi\sqrt{LC}}$</td></tr>
<tr><td>$\frac{QR}{2\pi L}=\frac{1}{2\pi QRC}$</td><td>$\frac{R}{2\pi QL}=\frac{Q}{2\pi RC}$</td></tr>
<tr><td rowspan="2">ω_0</td><td rowspan="2">$\frac{\text{rad}}{\text{s}}$</td><td colspan="2">$\frac{1}{\sqrt{LC}}$</td></tr>
<tr><td>$\frac{QR}{L}=\frac{1}{QRC}$</td><td>$\frac{R}{QL}=\frac{Q}{RC}$</td></tr>
<tr><td rowspan="4">bandwidth</td><td rowspan="2">BW</td><td rowspan="2">Hz</td><td colspan="2">$f_2-f_1=\frac{f_0}{Q}$</td></tr>
<tr><td>$\frac{R}{2\pi L}=\frac{1}{2\pi Q^2 RC}$</td><td>$\frac{1}{2\pi RC}=\frac{R}{2\pi Q^2 L}$</td></tr>
<tr><td rowspan="2">BW</td><td rowspan="2">$\frac{\text{rad}}{\text{s}}$</td><td colspan="2">$\omega_2-\omega_1=\frac{\omega_0}{Q}$</td></tr>
<tr><td>$\frac{R}{L}=\frac{1}{Q^2 RC}$</td><td>$\frac{1}{CR}=\frac{R}{Q^2 L}$</td></tr>
<tr><td rowspan="2">quality factor</td><td rowspan="2">Q</td><td rowspan="2">–</td><td>$\frac{1}{R}\sqrt{\frac{L}{C}}$</td><td>$R\sqrt{\frac{C}{L}}$</td></tr>
<tr><td>$\frac{\omega_0 L}{R}=\frac{1}{\omega_0 RC}$</td><td>$\omega_0 RC=\frac{R}{\omega_0 L}$</td></tr>
<tr><td rowspan="4">lower half-power point</td><td rowspan="2">f_1</td><td rowspan="2">Hz</td><td colspan="2">$f_0-\frac{\text{BW}}{2}=f_0\left(1-\frac{1}{2Q}\right)$</td></tr>
<tr><td>$f_0-\frac{R}{4\pi L}$</td><td>$f_0-\frac{1}{4\pi CR}$</td></tr>
<tr><td rowspan="2">ω_1</td><td rowspan="2">$\frac{\text{rad}}{\text{s}}$</td><td colspan="2">$\omega_0-\frac{\text{BW}}{2}=\omega_0\left(1-\frac{1}{2Q}\right)$</td></tr>
<tr><td>$\omega_0-\frac{R}{2L}$</td><td>$\omega_0-\frac{1}{2CR}$</td></tr>
<tr><td rowspan="4">upper half-power point</td><td rowspan="2">f_2</td><td rowspan="2">Hz</td><td colspan="2">$f_0+\frac{\text{BW}}{2}=f_0\left(1+\frac{1}{2Q}\right)$</td></tr>
<tr><td>$f_0+\frac{R}{4\pi L}$</td><td>$f_0+\frac{1}{4\pi CR}$</td></tr>
<tr><td rowspan="2">ω_2</td><td rowspan="2">$\frac{\text{rad}}{\text{s}}$</td><td colspan="2">$\omega_0+\frac{\text{BW}}{2}=\omega_0\left(1+\frac{1}{2Q}\right)$</td></tr>
<tr><td>$\omega_0+\frac{R}{2L}$</td><td>$\omega_0+\frac{1}{2CR}$</td></tr>
</table>

APPENDIX 38.A

Bare Aluminum Conductors, Steel Reinforced (ACSR)
Electrical Properties of Multi-Layer Sizes

				resistance					phase-to-neutral (60 Hz) reactance at 1 ft spacing	
					AC[b,c] (60 Hz)					
code word	area[d] (kcmil)	stranding (Al/St)	number of aluminum layers	DC[a] 20°C (Ω/mile)	25°C (Ω/mile)	50°C (Ω/mile)	75°C (Ω/mile)	GMR (ft)	inductive, X_a (Ω/mile)	capacitive, X'_a (MΩ-mile)
waxwing	266.8	18/1	2	0.3398	0.347	0.382	0.416	0.0197	0.477	0.109
partridge	266.8	26/7	2	0.3364	0.344	0.377	0.411	0.0217	0.465	0.107
ostrich	300	26/7	2	0.2993	0.306	0.336	0.366	0.0230	0.458	0.106
merlin	336.4	18/1	2	0.2693	0.276	0.303	0.330	0.0221	0.463	0.106
linnet	336.4	26/7	2	0.2671	0.273	0.300	0.327	0.0244	0.451	0.104
oriole	336.4	30/7	2	0.2650	0.271	0.297	0.324	0.0255	0.445	0.103
chickadee	397.5	18/1	2	0.2279	0.234	0.257	0.279	0.0240	0.452	0.103
ibis	397.5	26/7	2	0.2260	0.231	0.254	0.277	0.0265	0.441	0.102
lark	397.5	30/7	2	0.2243	0.229	0.252	0.274	0.0277	0.435	0.101
pelican	477	18/1	2	0.1899	0.195	0.214	0.233	0.0263	0.411	0.100
flicker	477	24/7	2	0.1889	0.194	0.213	0.232	0.0283	0.432	0.0992
hawk	477	26/7	2	0.1883	0.193	0.212	0.231	0.0290	0.430	0.0988
hen	477	30/7	2	0.1869	0.191	0.210	0.229	0.0304	0.424	0.0980
osprey	556.5	18/1	2	0.1629	0.168	0.184	0.200	0.0284	0.432	0.0981
parakeet	556.5	24/7	2	0.1620	0.166	0.183	0.199	0.0306	0.423	0.0969
dove	556.5	26/7	2	0.1613	0.166	0.182	0.198	0.0313	0.420	0.0965
eagle	556.5	30/7	2	0.1602	0.164	0.180	0.196	0.0328	0.415	0.0957
peacock	605	24/7	2	0.1490	0.153	0.168	0.183	0.0319	0.418	0.0957
squab	605	26/7	2	0.1485	0.153	0.167	0.182	0.0327	0.415	0.0953
teal	605	30/19	2	0.1475	0.151	0.166	0.181	0.0342	0.410	0.0944
kingbird	636	18/1	2	0.1420	0.147	0.162	0.175	0.0301	0.425	0.0951
rook	636	24/7	2	0.1417	0.146	0.160	0.174	0.0327	0.415	0.0950
grosbeak	636	26/7	2	0.1411	0.145	0.159	0.173	0.0335	0.412	0.0946
swift	636	36/1	3	0.1410	0.148	0.162	0.176	0.0300	0.426	0.0964
egret	636	30/19	2	0.1403	0.144	0.158	0.172	0.0351	0.406	0.0937
flamingo	666.6	24/7	2	0.1352	0.139	0.153	0.166	0.0335	0.412	0.0943
crow	715.5	54/7	3	0.1248	0.128	0.141	0.153	0.0372	0.399	0.0920
starling	715.5	26/7	2	0.1254	0.129	0.142	0.154	0.0355	0.405	0.0928
redwing	715.5	30/19	2	0.1248	0.128	0.141	0.153	0.0372	0.399	0.0920
coot	795	36/1	3	0.1146	0.119	0.130	0.142	0.0335	0.412	0.0932
cuckoo	795	24/7	2	0.1135	0.118	0.128	0.140	0.0361	0.403	0.0917
drake	795	26/7	2	0.1129	0.117	0.128	0.139	0.0375	0.399	0.0912
mallard	795	30/19	2	0.1122	0.116	0.127	0.138	0.0392	0.393	0.0904
tern	795	45/7	3	0.1143	0.119	0.130	0.141	0.0352	0.406	0.0925
condor	795	54/7	3	0.1135	0.117	0.129	0.140	0.0368	0.401	0.0917
crane	874.5	54/7	3	0.1030	0.107	0.117	0.127	0.0387	0.395	0.0902
ruddy	900	45/7	3	0.1008	0.106	0.115	0.125	0.0374	0.399	0.0907
canary	900	54/7	3	0.1002	0.104	0.114	0.124	0.0392	0.393	0.0898
corncrake	954	20/7	2	0.0950	0.099	0.109	0.118	0.0378	0.396	0.0898
rail	954	45/7	3	0.09526	0.0994	0.109	0.118	0.0385	0.395	0.0897
towhee	954	48/7	3	0.0950	0.099	0.108	0.118	0.0391	0.393	0.0896
redbird	954	24/7	2	0.0945	0.098	0.108	0.117	0.0396	0.392	0.0890
cardinal	954	54/7	3	0.09452	0.0983	0.108	0.117	0.0404	0.389	0.0890
ortolan	1033.5	45/7	3	0.08798	0.0922	0.101	0.110	0.0401	0.390	0.0886
curlew	1033.5	54/7	3	0.08728	0.0910	0.0996	0.108	0.0420	0.385	0.0878
bluejay	1113	45/7	3	0.08161	0.0859	0.0939	0.102	0.0416	0.386	0.0874
finch	1113	54/19	3	0.08138	0.0851	0.0931	0.101	0.0436	0.380	0.0867
bunting	1192.5	45/7	3	0.07619	0.0805	0.0880	0.0954	0.0431	0.382	0.0864
grackle	1192.5	54/19	3	0.07600	0.0798	0.0872	0.0947	0.0451	0.376	0.0856

(continued)

Support Material

APPENDIX 38.A *(continued)*
Bare Aluminum Conductors, Steel Reinforced (ACSR)
Electrical Properties of Multi-Layer Sizes

				resistance					phase-to-neutral (60 Hz) reactance at 1 ft spacing	
					AC[b,c] (60 Hz)					
code word	area[d] (kcmil)	stranding (Al/St)	number of aluminum layers	DC[a] 20°C (Ω/mile)	25°C (Ω/mile)	50°C (Ω/mile)	75°C (Ω/mile)	GMR (ft)	inductive, X_a (Ω/mile)	capacitive, X'_a (MΩ-mile)
bittem	1272	45/7	3	0.07146	0.0759	0.0828	0.0898	0.0445	0.378	0.0855
pheasant	1272	54/19	3	0.07122	0.0751	0.0820	0.0890	0.0466	0.372	0.0847
dipper	1351.5	45/7	3	0.06724	0.0717	0.0783	0.0848	0.0459	0.374	0.0846
martin	1351.5	54/19	3	0.06706	0.0710	0.0775	0.0840	0.0480	0.368	0.0838
bobolink	1431	45/7	3	0.06352	0.0681	0.0742	0.0804	0.0472	0.371	0.0837
plover	1431	54/19	3	0.06332	0.0673	0.0734	0.0796	0.0495	0.365	0.0829
nuthatch	1510.5	45/7	3	0.06017	0.0649	0.0706	0.0765	0.0485	0.367	0.0829
parrot	1510.5	54/19	3	0.06003	0.0641	0.0699	0.0757	0.0508	0.362	0.0821
lapwing	1590	45/7	3	0.05714	0.0620	0.0674	0.0729	0.0498	0.364	0.0822
falcon	1590	54/19	3	0.05699	0.0611	0.0666	0.0721	0.0521	0.358	0.0814
chukar	1780	84/19	4	0.05119	0.0561	0.0609	0.0658	0.0534	0.355	0.0803
mockingbird	2034.5	72/7	4	0.04488	0.0507	0.0549	0.0591	0.0553	0.348	0.0788
bluebird	2156	84/19	4	0.04229	0.0477	0.0516	0.0555	0.0588	0.344	0.0775
kiwi	2167	72/7	4	0.04228	0.0484	0.0522	0.0562	0.0570	0.348	0.0779
thrasher	2312	76/19	4	0.03960	0.0454	0.0486	0.0528	0.0600	0.343	0.0767
joree	2515	76/19	4	0.03643	0.0428	0.0459	0.0491	0.0621	0.338	0.0756

[a]Direct current (DC) resistance is based on 16.946 Ω-cmil/ft (61.2% IACS) at 20°C for nominal aluminum area of the conductors and 129.64 Ω-cmil/ft (8% IACS) at 20°C for the nominal steel area, with standard increments for stranding. ASTM B232.
[b]Alternating current (AC) resistance is based on the resistance corrected for temperature using 0.00404 1/°C as the temperature coefficient of resistivity for aluminum and 0.0029 1/°C for steel and for skin effect.
[c]The effective AC resistance of 3-layer ACSR increases with current density due to core magnetization.
[d]kcmil ≡ 1000 circular mils, also referred to as "MCM." Multiply circular mils by 5.066×10^{-10} m^2.

Used with permission from the *Aluminum Electrical Conductor Handbook*, reaffirmed 1998 by the Aluminum Association, 1989.

Support Material

APPENDIX 39.A
Smith Chart
Impedance or Admittance Coordinates

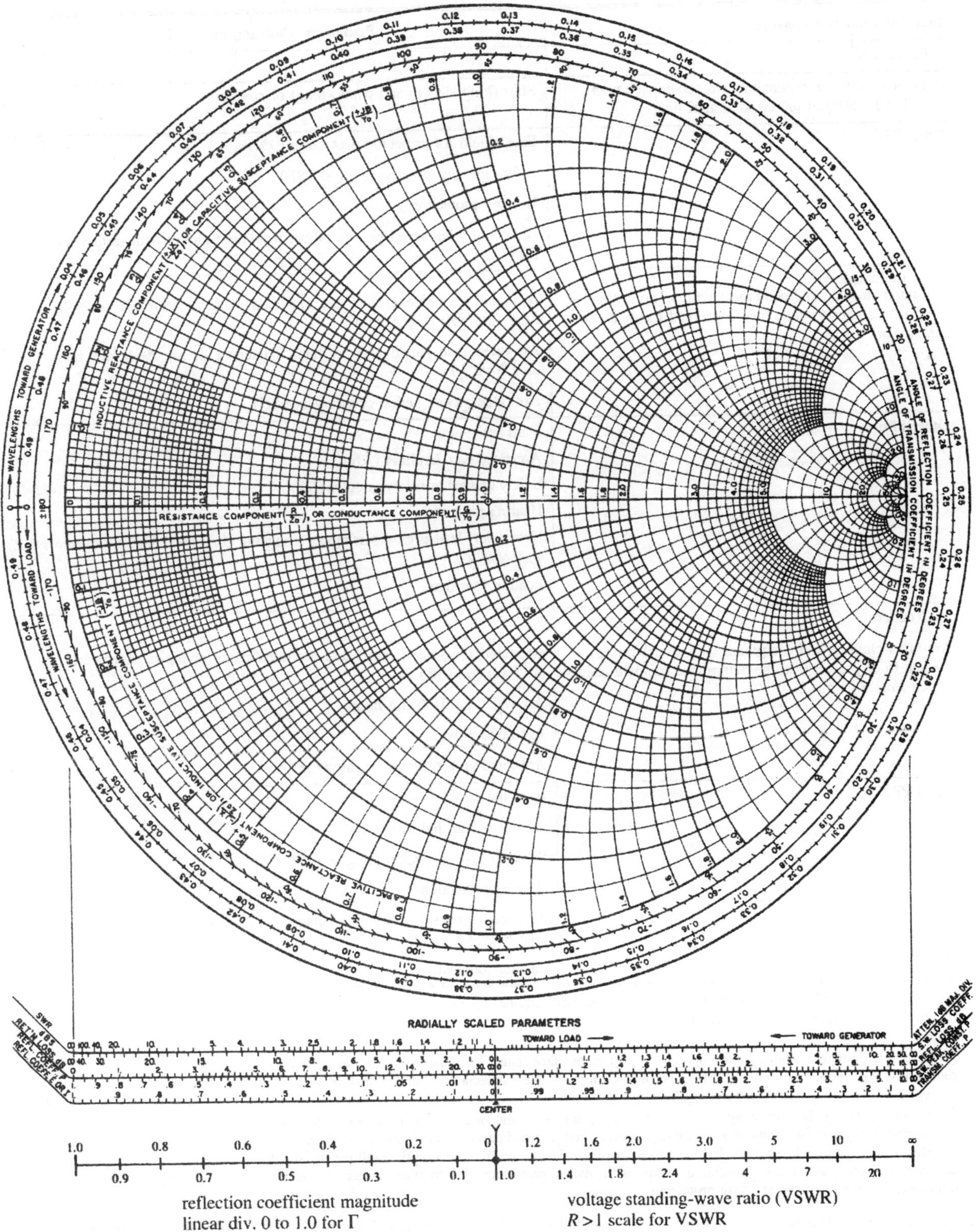

reflection coefficient magnitude
linear div. 0 to 1.0 for Γ

voltage standing-wave ratio (VSWR)
$R > 1$ scale for VSWR

Support Material

APPENDIX 42.A

Guidance on Selection of Protective Clothing and Other Protective Equipment (PPE)

Table H.3(b) Guidance on Selection of Arc-Rated Clothing and Other PPE for Use When Incident Energy Exposure Is Determined

Incident Energy Exposure	Protective Clothing and PPE
≤ 1.2 cal/cm²	
Protective clothing, nonmelting (in accordance with ASTM F 1506) or untreated natural fiber	Shirt (long sleeve) and pants (long) or coverall
Other PPE	Face shield for projectile protection (AN) Safety glasses or safety goggles (SR) Hearing protection Heavy-duty leather gloves or rubber insulating gloves with leather protectors (AN)
> 1.2 to 12 cal/cm² Arc-rated clothing and equipment with an arc rating equal to or greater than the determined incident energy *(See Note 3.)*	Arc-rated long-sleeve shirt and arc-rated pants or arc-rated coverall or arc flash suit (SR) *(See Note 3.)* Arc-rated face shield and arc-rated balaclava or arc flash suit hood (SR) *(See Note 1.)* Arc-rated jacket, parka, or rainwear (AN)
Other PPE	Hard hat Arc-rated hard hat liner (AN) Safety glasses or safety goggles (SR) Hearing protection Heavy-duty leather gloves or rubber insulating gloves with leather protectors (SR) *(See Note 4.)* Leather footwear
> 12 cal/cm² Arc-rated clothing and equipment with an arc rating equal to or greater than the determined incident energy *(See Note 3.)*	Arc-rated long-sleeve shirt and arc-rated pants or arc-rated coverall and/or arc flash suit (SR) Arc-rated arc flash suit hood Arc-rated gloves Arc-rated jacket, parka, or rainwear (AN)
Other PPE	Hard hat Arc-rated hard hat liner (AN) Safety glasses or safety goggles (SR) Hearing protection Arc-rated gloves or rubber insulating gloves with leather protectors (SR) *(See Note 4.)* Leather footwear

AN: As needed [in addition to the protective clothing and PPE required by 130.5(C)(1)].
SR: Selection of one in group is required by 130.5(C)(1).
Notes:
(1) Face shields with a wrap-around guarding to protect the face, chin, forehead, ears, and neck area are required by 130.7(C)(10)(c). For full head and neck protection, use a balaclava or an arc flash hood.
(2) All items not designated "AN" are required by 130.7(C).
(3) Arc ratings can be for a single layer, such as an arc-rated shirt and pants or a coverall, or for an arc flash suit or a multi-layer system consisting of a combination of arc-rated shirt and pants, coverall, and arc flash suit.
(4) Rubber insulating gloves with leather protectors provide arc flash protection in addition to shock protection. Higher class rubber insulating gloves with leather protectors, due to their increased material thickness, provide increased arc flash protection.

APPENDIX 45.A
Periodic Table of the Elements
(referred to carbon-12)

The Periodic Table of Elements (Long Form)

The number of electrons in filled shells is shown in the column at the extreme left; the remaining electrons for each element are shown immediately below the symbol for each element. Atomic numbers are enclosed in brackets. Atomic weights (rounded, based on carbon-12) are shown above the symbols. Atomic weight values in parentheses are those of the isotopes of longest half-life for certain radioactive elements whose atomic weights cannot be precisely quoted without knowledge of origin of the element.

metals (I A–II B; transition metals: III B–II B) | nonmetals (III A–0)

periods	I A	II A	III B	IV B	V B	VI B	VII B	VIII	VIII	VIII	I B	II B	III A	IV A	V A	VI A	VII A	0
1 0	1.00794 H[1] 1																	4.00260 He[2] 2
2 2	6.941 Li[3] 1	9.01218 Be[4] 2											10.811 B[5] 3	12.0107 C[6] 4	14.0067 N[7] 5	15.9994 O[8] 6	18.9984 F[9] 7	20.1797 Ne[10] 8
3 2,8	22.9898 Na[11] 1	24.3050 Mg[12] 2											26.9815 Al[13] 3	28.0855 Si[14] 4	30.9738 P[15] 5	32.065 S[16] 6	35.453 Cl[17] 7	39.948 Ar[18] 8
4 2,8	39.0983 K[19] 8,1	40.078 Ca[20] 8,2	44.9559 Sc[21] 9,2	47.867 Ti[22] 10,2	50.9415 V[23] 11,2	51.9961 Cr[24] 13,1	54.9380 Mn[25] 13,2	55.845 Fe[26] 14,2	58.9332 Co[27] 15,2	58.6934 Ni[28] 16,2	63.546 Cu[29] 18,1	65.38 Zn[30] 18,2	69.723 Ga[31] 18,3	72.64 Ge[32] 18,4	74.9216 As[33] 18,5	78.96 Se[34] 18,6	79.904 Br[35] 18,7	83.798 Kr[36] 18,8
5 2,8,18	85.4678 Rb[37] 8,1	87.62 Sr[38] 8,2	88.9059 Y[39] 9,2	91.224 Zr[40] 10,2	92.9064 Nb[41] 12,1	95.96 Mo[42] 13,1	(98) Tc[43] 14,1	101.07 Ru[44] 15,1	102.906 Rh[45] 16,1	106.42 Pd[46] 18	107.868 Ag[47] 18,1	112.411 Cd[48] 18,2	114.818 In[49] 18,3	118.710 Sn[50] 18,4	121.760 Sb[51] 18,5	127.60 Te[52] 18,6	126.904 I[53] 18,7	131.293 Xe[54] 18,8
6 2,8,18	132.905 Cs[55] 18,8,1	137.327 Ba[56] 18,8,2	* (57-71)	178.49 Hf[72] 32,10,2	180.948 Ta[73] 32,11,2	183.84 W[74] 32,12,2	186.207 Re[75] 32,13,2	190.23 Os[76] 32,14,2	192.217 Ir[77] 32,15,2	195.084 Pt[78] 32,17,1	196.967 Au[79] 32,18,1	200.59 Hg[80] 32,18,2	204.383 Tl[81] 32,18,3	207.2 Pb[82] 32,18,4	208.980 Bi[83] 32,18,5	(209) Po[84] 32,18,6	(210) At[85] 32,18,7	(222) Rn[86] 32,18,8
7 2,8,18,32	(223) Fr[87] 18,8,1	(226) Ra[88] 18,8,2	† (89-103)	(265) Rf[104] 32,10,2	(268) Db[105] 32,11,2	(271) Sg[106] 32,12,2	(272) Bh[107] 32,13,2	(270) Hs[108] 32,14,2	(276) Mt[109] 32,15,2	(281) Ds[110] 32,17,1	(280) Rg[111] 32,18,1	(285) Cn[112] 32,18,2						

*lathanide series	138.905 La[57] 18,9,2	140.116 Ce[58] 20,8,2	140.908 Pr[59] 21,8,2	144.242 Nd[60] 22,8,2	(145) Pm[61] 23,8,2	150.36 Sm[62] 24,8,2	151.964 Eu[63] 25,8,2	157.25 Gd[64] 25,9,2	158.925 Tb[65] 27,8,2	162.500 Dy[66] 28,8,2	164.930 Ho[67] 29,8,2	167.259 Er[68] 30,8,2	168.934 Tm[69] 31,8,2	173.054 Yb[70] 32,8,2	174.967 Lu[71] 32,9,2
†actinide series	(227) Ac[89] 18,9,2	232.038 Th[90] 18,10,2	231.036 Pa[91] 20,9,2	238.029 U[92] 21,9,2	(237) Np[93] 23,8,2	(244) Pu[94] 24,8,2	(243) Am[95] 25,8,2	(247) Cm[96] 25,9,2	(247) Bk[97] 26,9,2	(251) Cf[98] 28,8,2	(252) Es[99] 29,8,2	(257) Fm[100] 30,8,2	(258) Md[101] 31,8,2	(259) No[102] 32,8,2	(262) Lr[103] 32,9,2

APPENDIX 45.B
Semiconductor Symbols and Abbreviations*

a, A—anode
b, B—base
b_{fs}—common-source small-signal forward transfer susceptance
b_{is}—common-source small-signal input susceptance
b_{os}—common-source small-signal output susceptance
b_{rs}—common-source small-signal reverse transfer susceptance
c, C—collector
C_{cb}—collector-base interterminal capacitance
C_{ce}—collector-emitter interterminal capacitance
C_{ds}—drain-source capacitance
C_{du}—drain-substrate capacitance
C_{eb}—emitter-base interterminal capacitance
C_{ibo}—common-base open-circuit input capacitance
C_{ibs}—common-base short-circuit input capacitance
C_{ieo}—common-emitter open-circuit input capacitance
C_{ies}—common-emitter short-circuit input capacitance
C_{iss}—common-source short-circuit input capacitance
C_{obo}—common-base open-circuit output capacitance
C_{obs}—common-base short-circuit output capacitance
C_{oeo}—common-emitter open-circuit output capacitance
C_{oes}—common-emitter short-circuit output capacitance
C_{oss}—common-source short-circuit output capacitance
C_{rbs}—common-base short-circuit reverse transfer capacitance
C_{res}—common-collector short-circuit reverse transfer capacitance
C_{rss}—common-source short-circuit reverse transfer capacitance
C_{tc}—collector depletion-layer capacitance
C_{te}—emitter depletion-layer capacitance
d, D—drain
e, E—emitter
η—intrinsic standoff ratio
f_{hfb}—common-base small-signal short-circuit forward current transfer ratio cutoff frequency
f_{hfc}—common-collector small-signal short-circuit forward current transfer ratio cutoff frequency
f_{hfe}—common-emitter small-signal short-circuit forward current transfer ratio cutoff frequency
f_{max}—maximum frequency of oscillation
f_T—transition frequency (frequency at which common-emitter small-signal forward current transfer ratio extrapolates to unity)
g, G—gate
g_{fs}—common-source small-signal forward transfer conductance
g_{is}—common-source small-signal input conductance
g_{MB}—common-base static transconductance
g_{MC}—common-collector static transconductance
g_{ME}—common-emitter static transconductance
g_{os}—common-source small-signal output conductance
g_{rs}—common-source small-signal reverse transfer conductance
G_{pb}—common-base small-scale insertion power gain
G_{PB}—common-base large-signal insertion power gain
G_{pc}—common-collector small-signal insertion power gain
G_{PC}—common-collector large-signal insertion power gain
G_{pe}—common-emitter small-signal insertion power gain
G_{PE}—common-emitter large-signal insertion power gain
G_{pg}—common-gate small-signal insertion power gain
G_{ps}—common-source small-signal insertion power gain
G_{tb}—common-base small-signal transducer power gain
G_{TB}—common-base large-signal transducer power gain
G_{tc}—common-collector small-signal transducer power gain
G_{TC}—common-collector large-signal transducer power gain
G_{te}—common-emitter small-signal transducer power gain
G_{TE}—common-emitter large-signal transducer power gain
G_{tg}—common-gate small-signal transducer power gain
G_{ts}—common-source small-signal transducer power gain
h_{fb}—common-base small-signal short-circuit forward current transfer ratio
h_{FB}—common-base static forward current transfer ratio
h_{fc}—common-collector small-signal short-circuit forward current transfer ratio
h_{FC}—common-collector static forward current transfer ratio
h_{fe}—common-emitter small-signal short-circuit forward current transfer ratio
h_{FE}—common-emitter static forward current transfer ratio
h_{FEL}—inherent large-signal forward current transfer ratio
h_{ib}—common-base small-signal short-circuit input impedance
h_{IB}—common-base static input resistance
h_{ic}—common-collector small-signal short-circuit input impedance
h_{IC}—common-collector static input resistance
h_{ie}—common-emitter small-signal short-circuit input impedance
h_{IE}—common-emitter static input resistance
$h_{ie(imag)}$—imaginary part of common-emitter small-signal short-circuit input impedance
$h_{ie(real)}$—real part of common-emitter small-signal short-circuit input impedance
h_{ob}—common-base small-signal open-circuit output admittance
h_{oc}—common-collector small-signal open-circuit output admittance
h_{oe}—common emitter small-signal open-circuit output admittance
$h_{oe(imag)}$—imaginary part of common-emitter small-signal open-circuit output admittance
$h_{oe(real)}$—real part of common-emitter small-signal open-circuit output admittance
h_{rb}—common-base small-signal open-circuit reverse voltage transfer ratio
h_{re}—common-collector small-signal open-circuit reverse voltage transfer ratio
h_{ro}—common-emitter small-signal open-circuit reverse voltage transfer ratio

(continued)

APPENDIX 45.B *(continued)*
Semiconductor Symbols and Abbreviations*

i_B—instantaneous total value of base-terminal current
i_C—instantaneous total value of collector-terminal current
i_E—instantaneous total value of emitter-terminal current
i_R—instantaneous total reverse current
I_0—average forward current, 180° conduction angle, 60 Hz half sine wave
I_b—alternating component (rms value) of base-terminal current
I_B—base-terminal DC current
$I_{B2(\text{mod})}$—interbase modulated current
I_{BEV}—base cutoff current, DC
I_c—alternating component (rms value) of collector-terminal current
I_C—collector-terminal DC current
I_{CBO}—collector cutoff current (DC), emitter open
I_{CEO}—collector cutoff current (DC), base open
I_{CER}—collector cutoff current (DC), specified resistance between base and emitter
I_{CES}—collector cutoff current (DC), base shorted to emitter
I_{CEV}—collector cutoff current, DC, specified voltage between base and emitter
I_{CEX}—collector cutoff current, DC, specified circuit between base and emitter
I_D—drain current, DC
$I_{D,\text{off}}$—drain cutoff current
$I_{D,\text{on}}$—on-state drain current
I_{DSS}—zero-gate-voltage drain current
I_e—alternating component (rms value) of emitter-terminal current
I_E—emitter-terminal DC current
I_{EBO}—emitter cutoff current (DC), collector open
I_{ERR}—emitter reverse current
$I_{\text{EC,ofs}}$—emitter-collector offset current
I_{ECS}—emitter cutoff current (DC), base short-circuited to collector
$I_{\text{EIE2,off}}$—emitter cutoff current
I_f—alternating component of forward current (rms value)
I_F—for voltage-regulator and voltage-reference diodes: DC forward current. For signal diodes and rectifier diodes: DC forward current (no alternating component).
I_F—instantaneous total forward current
$I_{F,\text{AV}}$—forward current, DC (with alternating component)
I_{FM}—maximum (peak) total forward current
$I_{F,\text{OV}}$—forward current, overload
I_{FRM}—maximum (peak) forward current, repetitive
I_{FRMS}—total rms forward current
I_{FSM}—maximum (peak) forward current, surge
I_G—gate current, DC
I_{GF}—forward gate current
I_{GR}—reverse gate current
I_{GSS}—reverse gate current, drain short-circuited to source
I_{GSSF}—forward gate current, drain short-circuited to source
I_{GSSR}—reverse gate current, drain short-circuited to source
I_I—inflection-point current
I_P—peak-point current
I_r—alternating component of reverse current (rms value)
I_R—for voltage-regulator and voltage-reference diodes: DC reverse current. For signal diodes and rectifier diodes: DC reverse current (no alternating component).
$I_{R,\text{AV}}$—reverse current, DC (with alternating component)
I_{RM}—maximum (peak) total reverse current
I_{RRM}—maximum (peak) reverse current, repetitive
$I_{R,\text{RMS}}$—total rms reverse current
I_{RSM}—maximum (peak) surge reverse current
I_S—source current, DC
I_{SDS}—zero-gate-voltage source current
$I_{S,\text{off}}$—source cutoff current
I_V—valley-point current
I_Z—regulator current, reference current, DC
I_{ZK}—regulator current, reference current (DC near breakdown knee)
I_{ZM}—regulator current, reference current (DC maximum rated current)
$\text{Im}(h_{\text{oe}})$—imaginary part of common-emitter small-signal short-circuit input impedance
$\text{Im}(h_{\text{oe}})$—imaginary part of common-emitter small-signal open-circuit output admittance
k, K—cathode
L_c—conversion loss
M—figure of merit
NF_o—overall noise figure
NR_o—output noise ratio
p_{BE}—instantaneous total power input to base, common emitter
p_{CB}—instantaneous total power input to collector, common base
p_{CE}—power input (DC) to collector, common emitter
p_{EB}—instantaneous total power input to emitter, common base
p_F—instantaneous total forward power dissipation
p_{ib}—common-base small-signal input power
p_{ic}—common collector small-signal input power
p_{ie}—common-emitter small-signal input power
p_{ob}—common-base small-signal output power
p_{oc}—common-collector small-signal output power
p_{oe}—common-emitter small-signal output power
p_R—instantaneous total reverse power dissipation
p_T—nonreactive power input, instantaneous total, to all terminals
P_{BE}—power input (DC) to base, common emitter
P_{CB}—power input (DC) to collector, common base
P_{CE}—instantaneous total power input to collector, common emitter
P_{EB}—power input (DC) to emitter, common base
P_F—forward power dissipation, DC (no alternating component)
$P_{F,\text{AV}}$—forward power dissipation, DC (with alternating component)
P_{FM}—maximum (peak) total forward power dissipation
P_{IB}—common-base large-signal input power
P_{IC}—common-collector large-signal input power
P_{IE}—common-emitter large-signal input power

(continued)

Support Material

APPENDIX 45.B *(continued)*
Semiconductor Symbols and Abbreviations*

P_{OB}—common-base large-signal output power
P_{OC}—common-collector large-signal output power
P_{OE}—common-emitter large-signal output power
P_R—reverse power dissipation, DC (no alternating component)
$P_{R(AV)}$—reverse power dissipation, DC (with alternating component)
P_{RM}—maximum (peak) total reverse power dissipation
P_T—total nonreactive power input to all terminals
Q_S—stored charge
r'_b, C_c—collector-base time constant
r_{BB}—interbase resistance
$r_{CE,sat}$—saturation resistance, collector to emitter
$r_{ds,on}$—small-signal drain-source on-state resistance
$r_{DS,on}$—static drain-source on-state resistance
$r_{ele2,on}$—small-signal emitter-emitter on-state resistance
r_i—dynamic resistance at inflection point
R_θ—thermal resistance
$R_{\theta CA}$—thermal resistance, case to ambient
$R_{\theta JA}$—thermal resistance, junction to ambient
$R_{\theta JC}$—thermal resistance, junction to case
$\text{Re}(h_{ie})$—real part of common-emitter small-signal short-circuit input impedance
$\text{Re}(h_{oe})$—real part of common-emitter small-signal open-circuit output admittance
s, S—source
t_d—delay time
$t_{d,off}$—turn-off delay time
$t_{d,on}$—turn-on delay time
t_f—fall time
t_{fr}—forward recovery time
t_{off}—turn-off time
t_{on}—turn-on time
t_p—pulse time
t_r—rise time
t_{rr}—reverse recovery time
t_s—storage time
t_w—pulse average time
T_j—junction temperature
T_{stg}—storage temperature
T_A—ambient temperature or free-air temperature
T_C—case temperature
TSS—tangential signal sensitivity
u, U—bulk (substrate)
v_{bc}—instantaneous value of alternating component of base-collector voltage
v_{be}—instantaneous value of alternating component of base-emitter voltage
v_{BR}—breakdown voltage (instantaneous total)
v_{cb}—instantaneous value of alternating component of collector-base voltage
v_{ce}—instantaneous value of alternating component of collector-emitter voltage
v_{eb}—instantaneous value of alternating component of emitter-base voltage
v_{ec}—instantaneous value of alternating component of emitter-collector voltage
v_F—instantaneous total forward voltage
v_R—instantaneous total reverse voltage
V_{B2B1}—interbase voltage
V_{BB}—base supply voltage, DC
V_{BC}—average or DC voltage, base to collector
V_{BE}—average or DC voltage, base to emitter
V_{BR}—breakdown voltage, DC
$V_{BR,CBO}$—collector-base breakdown voltage, emitter open
$V_{BR,CEO}$—collector-emitter breakdown voltage, base open
$V_{BR,CER}$—collector-emitter breakdown voltage, resistance between base and emitter
$V_{BR,CES}$—collector-emitter breakdown voltage, base shorted to emitter
$V_{BR,CEV}$—collector-emitter breakdown voltage, specified voltage between base and emitter
$V_{BR,CEX}$—collector-emitter breakdown voltage, specified circuit between base and emitter
$V_{BR,E1E2}$—emitter-emitter breakdown voltage
$V_{BR,EBO}$—emitter-base breakdown voltage, collector open
$V_{BR,ECO}$—emitter-collector breakdown voltage, base open
$V_{BR,GSS}$—gate-source breakdown voltage
$V_{BR,GSSF}$—forward gate-source breakdown voltage
$V_{BR,GSSR}$—reverse gate-source breakdown voltage
V_{CB}—average or DC voltage, collector to base
V_{CBO}—collector-base voltage, DC, emitter open
$V_{CB,fl}$—collector-base DC open-circuit voltage (floating potential)
V_{CC}—collector supply voltage, DC
V_{CE}—average or DC voltage, collector to emitter
$V_{CE,fl}$—collector-emitter DC open-circuit voltage (floating potential)
V_{CEO}—collector-emitter voltage (DC), base open
$V_{CE,ofs}$—collector-emitter offset voltage
V_{CER}—collector-emitter voltage (DC), resistance between base and emitter
V_{CES}—collector-emitter voltage (DC), base shorted to emitter
$V_{CE,at}$—collector-emitter DC saturation voltage
V_{CEV}—collector-emitter voltage, DC, specified voltage between base and emitter
V_{CEX}—collector-emitter voltage, DC, specified circuit between base and emitter
V_{DD}—drain supply voltage, DC
V_{DG}—drain-gate voltage
V_{DS}—drain-source voltage
$V_{DS,on}$—drain-source on-state voltage
V_{DU}—drain-substrate voltage
V_{ED}—average or DC voltage, emitter to base
$V_{EB,fl}$—emitter-base DC open-circuit voltage (floating potential)
V_{EBO}—emitter-base voltage, DC, collector open
$V_{EB,sat}$—emitter saturation voltage
V_{EC}—average or DC voltage, emitter to collector
$V_{EC,fl}$—emitter-collector DC open-circuit voltage (floating potential)
$V_{EC,ofs}$—emitter-collector offset voltage
V_{EE}—emitter supply voltage, DC
V_f—alternating component of forward voltage (rms value)

(continued)

Support Material

APPENDIX 45.B *(continued)*
Semiconductor Symbols and Abbreviations*

$V_{F,\mathrm{AV}}$—forward voltage, DC (with alternating component)
V_{FM}—maximum (peak) total forward voltage
$V_{F,\mathrm{RMS}}$—total rms forward voltage
V_{GG}—gate supply voltage, DC
V_{GS}—gate-source voltage
V_{GSF}—forward gate-source voltage
$V_{\mathrm{GS,off}}$—gate-source cutoff voltage
V_{GSR}—reverse gate-source voltage
$V_{\mathrm{GS,th}}$—gate-source threshold voltage
V_{GU}—gate-substrate voltage
V_I—inflection-point voltage
V_{OB1}—base-1 peak voltage
V_P—peak-point voltage
V_{PP}—projected peak-point voltage
V_r—alternating component of reverse voltage (rms value)
V_R—for voltage-regulator and voltage-reference diodes: DC reverse voltage. For signal diodes and rectifier diodes: DC reverse voltage (no alternating component).
$V_{R,\mathrm{AV}}$—reverse voltage, DC (with alternating component)
V_{RM}—maximum (peak) total reverse voltage
V_{RRM}—repetitive peak reverse voltage
$V_{R,\mathrm{RMS}}$—total rms reverse voltage
V_{RSM}—nonrepetitive peak reverse voltage
V_{RT}—reach-through voltage
V_{RWM}—working peak reverse voltage
V_{SS}—source supply voltage, DC
V_{SU}—source-substrate voltage
V_{TO}—threshold voltage
V_V—valley-point voltage
V_Z—regulator voltage, reference voltage, DC
V_{ZM}—regulator voltage, reference voltage (DC at maximum rated current)
VF—for voltage-regulator and voltage-reference diodes: DC forward voltage. For signal diodes and rectifier diodes: DC forward voltage (no alternating component).
y_{fb}—common-base small-signal short-circuit forward transfer admittance
y_{fc}—common-collector small-signal short-circuit forward transfer admittance
y_{fe}—common-emitter small-signal short-circuit forward transfer admittance
y_{fs}—common-source small-signal short-circuit forward transfer admittance
$y_{\mathrm{fs,imag}}$—common-source small-signal forward transfer susceptance
$y_{\mathrm{fs,real}}$—common-source small-signal forward transfer conductance
y_{ib}—common-base small-signal short-circuit input admittance
y_{ic}—common-collector small-signal short-circuit input admittance
y_{ie}—common-emitter small-signal short-circuit input admittance
$y_{\mathrm{ie,imag}}$—imaginary part of small-signal short-circuit input admittance (common emitter)
$y_{\mathrm{ie,real}}$—real part of small-signal short-circuit input admittance (common emitter)
y_{is}—common-source small-signal short-circuit input admittance
$y_{\mathrm{is,imag}}$—common-source small-signal input susceptance
$y_{\mathrm{is,real}}$—common-source small-signal input conductance
y_{ob}—common-base small-signal short-circuit output admittance
y_{oc}—common-collector small-signal short-circuit output admittance
y_{oe}—common-emitter small-signal short-circuit output admittance
$y_{\mathrm{oe,imag}}$—imaginary part of small-signal short-circuit output admittance (common emitter)
$y_{\mathrm{oe,real}}$—real part of small-signal short-circuit output admittance (common emitter)
y_{os}—common-source small-signal short-circuit output admittance
$y_{\mathrm{os,imag}}$—common-source small-signal output susceptance
$y_{\mathrm{os,real}}$—common-source small-signal output conductance
y_{rb}—common-base small-signal short-circuit reverse transfer admittance
y_{rc}—common-collector small-signal short-circuit reverse transfer admittance
y_{re}—common-emitter small-signal short-circuit reverse transfer admittance
y_{rs}—common-source small-signal short-circuit reverse transfer admittance
$y_{\mathrm{rs,imag}}$—common-source small-signal reverse transfer susceptance
$y_{\mathrm{rs,real}}$—common-source small-signal reverse transfer conductance
z_{if}—intermediate-frequency impedance
z_m—modulator-frequency load impedance
z_{rf}—radio-frequency impedance
z_v—video impedance
z_z—regulator impedance, reference impedance (small signal at I_z)
z_{zk}—regulator impedance, reference impedance (small signal at I_{ZK})
$Z_{\theta,t}$—transient thermal impedance
$Z_{\theta\mathrm{JA},t}$—junction-to-case transient thermal impedance
$Z_{\theta\mathrm{JC},t}$—junction-to-case transient thermal impedance
Z_{zm}—regulator impedance, reference impedance (small signal at I_{ZM})

*Recommended by the Joint Electron Device Engineering Council of the Electronic Industries Association and the National Electrical Manufacturers Associaton for use in semiconductor device data sheets and specifications.

APPENDIX 48.A
Standard Zener Diodes

zener voltage (nominal)	rated power 500 mW		1 W		5 W		50 W	
	Z_Z[a] (Ω)	I_{ZK}[b] (mA)	Z_Z[a] (Ω)	I_{ZK}[b] (mA)	Z_Z[a] (Ω)	I_{ZK}[b] (mA)	Z_Z[a] (Ω)	I_{ZK}[b] (mA)
2.4	30	1.0						
2.7	30	1.0						
3.0	29	1.0						
3.3	28	1.0	10	1.0	3.0	1.0		
3.6	24	1.0	10	1.0	2.5	1.0		
3.9	23	1.0	9.0	1.0	2.0	1.0	0.16	5.0
4.3	22	1.0	9.0	1.0	2.0	1.0	0.16	5.0
4.7	19	1.0	8.0	1.0	2.0	1.0	0.12	5.0
5.1	17	1.0	7.0	1.0	1.5	1.0	0.12	5.0
5.6	11	1.0	5.0	1.0	1.0	1.0	0.12	5.0
6.2	7.0	1.0	2.0	1.0	1.0	1.0	0.14	5.0
6.8	5.0	1.0	1.5	1.0	1.0	1.0	0.16	5.0
7.5	5.5	0.5	1.5	1.0	1.5	1.0	0.24	5.0
8.2	6.5	0.5	4.5	0.5	1.5	1.0	0.4	5.0
8.7					2.0	1.0		
9.1	7.5	0.5	5.0	0.5	2.0	1.0	0.5	5.0
10	8.5	0.25	7.0	0.25	2.0	1.0	0.6	5.0
11	9.5	0.25	8.0	0.25	2.5	1.0	0.8	5.0
12	11.5	0.25	9.0	0.25	2.5	1.0	1.0	5.0
13	13	0.25	10	0.25	2.5	1.0	1.1	5.0
14					2.5	1.0	1.2	5.0
15	16	0.25	14	0.25	2.5	1.0	1.4	5.0
16	17	0.25	16	0.25	2.5	1.0	1.6	5.0
17					2.5	1.0	1.8	5.0
18	21	0.25	20	0.25	2.5	1.0	2.0	5.0
19					3.0	1.0	2.2	5.0
20	25	0.25	22	0.25	3.0	1.0	2.4	5.0
22	29	0.25	23	0.25	3.5	1.0	2.5	5.0
24	33	0.25	25	0.25	3.5	1.0	2.6	5.0
25					4.0	1.0	2.7	5.0
27	41	0.25	35	0.25	5.0	1.0	2.8	5.0
28					6.0	1.0		
30	49	0.25	40	0.25	8.0	1.0	3.0	5.0
33	58	0.25	45	0.25	10	1.0	3.2	5.0
36	70	0.25	50	0.25	11	1.0	3.5	5.0
39	80	0.25	60	0.25	14	1.0	4.0	5.0
43	93	0.25	70	0.25	20	1.0	4.5	5.0
45							4.5	5.0
47	105	0.25	80	0.25	25	1.0	5.0	5.0
50							5.0	5.0
51	125	0.25	95	0.25	27	1.0	5.2	5.0
56	150	0.25	110	0.25	35	1.0	6.0	5.0
60					40	1.0		
62	185	0.25	125	0.25	42	1.0	7.0	5.0
68	230	0.25	150	0.25	44	1.0	8.0	5.0
75	270	0.25	175	0.25	45	1.0	9.0	5.0
82	330	0.25	200	0.25	65	1.0	11	5.0
87					75	1.0		
91	400	0.25	250	0.25	75	1.0	15	5.0
100			350	0.25	90	1.0	20	5.0
105							25	5.0

(continued)

APPENDIX 48.A *(continued)*
Standard Zener Diodes

zener voltage (nominal)	rated power: 500 mW		1 W		5 W		50 W	
	Z_Z[a] (Ω)	I_{ZK}[b] (mA)	Z_Z[a] (Ω)	I_{ZK}[b] (mA)	Z_Z[a] (Ω)	I_{ZK}[b] (mA)	Z_Z[a] (Ω)	I_{ZK}[b] (mA)
110					125	1.0	30	5.0
120			550	0.25	170	1.0	40	5.0
130			700	0.25	190	1.0	50	5.0
140					230	1.0	60	5.0
150			1000	0.25	330	1.0	75	5.0
160			1100	0.25	350	1.0	80	5.0
170					380	1.0		
175							85	5.0
180			1200	0.25	430	1.0	90	5.0
190					450	1.0		
200			1500	0.25	480	1.0	100	5.0

[a]Z_Z is the zener impedance.
[b]I_{ZK} is the zener keep-alive or knee current.

APPENDIX 51.A

Units, Symbols, and Defining Equations for Fundamental Photometric and Radiometric Quantities

symbol	definition	units	unit symbols
α	absorptance	–	–
	$\alpha = \Phi_a/\Phi_i$		
E	lluminance ("illumination" is deprecated)	(SI) lux	$lx = lm/m^2$
E	irradiance	(SI) watt per square meter	W/m^2
E_e (radiometric)	$E = d\Phi/dA$	(cgs) watt per square centimeter	W/cm^2
E_v (photometric)	photometric illuminance	(cgs) phot	$ph = lm/cm^2$
	$E = d\Phi/dA$	(U.S.) footcandle	$fc = lm/ft^2$
ϵ	emissivity, spectral-total hemispherical	–	–
	$\epsilon = M/M_{\text{blackbody}}$	–	–
$\epsilon(\theta,\phi,T)$	emissivity, spectral-total directional	–	–
	$\epsilon(\theta,\phi,T) = L(T)/L_{\text{blackbody}}(T)$		
$\epsilon(\theta,\phi,\lambda,T)$	emissivity, spectral directional	–	–
	$\epsilon(\theta,\phi,\lambda,T) = L(\theta,\phi,\lambda,T)/L_{\lambda,\text{blackbody}}(\lambda,T)$		
$\epsilon(\lambda,T)$	emissivity, spectral hemispherical	–	–
	$\epsilon(\lambda,T) = M_\lambda(\lambda,T)/M_{\lambda,\text{blackbody}}(\lambda,T)$		
	$\varepsilon(\lambda,T) = M_\lambda(\lambda,T)/M_{\lambda,\text{blackbody}}(\lambda,T)$		
I	luminous intensity	candela (lumen per steradian)	cd
I	radiant intensity	(SI) watt per steradian	W/sr
I_e (radiometric)	$I = d\Phi/d\omega$		
I_v (photometric)	(candle power)		
	$I = d\Phi/d\omega$		
K	luminous efficacy	lumen per watt	lm/W
	$K = \Phi_v/\Phi_e$		
L	luminance	candela per unit area	e.g., cd/in^2
L	radiance	(SI) watt per steradian per square meter	$W/sr{\cdot}m^2$
L_e (radiometric)	$L = d^2\Phi/d\omega\, dA\cos\theta = dI/dA\cos\theta$	watt per steradian per square centimeter	$W/sr{\cdot}cm^2$
L_v (photometric)	$I = d^2\Phi/d\omega\, dA\cos\theta = dI/dA\cos\theta$	stib	$sb = cd/cm^2$
		(SI) nit	$nt = cd/m^2$
		footlambert (deprecated)	$fL = cd/\pi\text{-}ft^2$
		lambert (deprecated)	$L = cd/\pi{\cdot}cm^2$
		apostib (deprecated)	$asb = cd/\pi{\cdot}m^2$
λ	wavelength used as subscript or argument of a spectral function, to specify a wavelength, as in Q_λ for spectral concentration or $K(\lambda)$ for a function of wavelength, especially when quantities are restricted to narrow wavelength bands		
M	luminous exitance	lumen per square meter	lm/m^2
M	radiant exitance	(SI) watt per square meter	W/m^2
M_e (radiometric)	radiant flux density at a surface	(cgs) watt per square centimeter	W/cm^2
	radiant emittance (deprecated)		
	$M = d\Phi/dA$		

(continued)

Support Material

APPENDIX 51.A *(continued)*
Units, Symbols, and Defining Equations for Fundamental Photometric and Radiometric Quantities

symbol	definition	units	unit symbols
M_v (photometric)	luminous flux density at a surface	lumen per square foot	lm/ft^2
	luminous emittance (deprecated)		
	$M = d\Phi/dA$		
Φ	luminous flux	(SI) lumen	lm
Φ	radiant flux	(SI) watt	W
Φ_a	absorbed flux	–	–
Φ_i	incident flux	–	–
Φ_r	reflected flux	–	–
Φ_t	transmitted flux	–	–
Φ_e (radiometric)	$\Phi = dQ/dt$	(cgs) erg per second	erg/s
Φ_v (photometric)	$\Phi = d\Phi/dt$		
Q	luminous energy (quantity of light)	(SI) lumen-second (talbot)	lm·s
Q	radiant energy	(SI) joule	J
Q_e (radiometric)		(cgs) erg	erg
		calorie	cal
		kilowatt-hour	kW·h
Q_v (photometric)	$\int_{360}^{800} K(\lambda) Q_{e\lambda} d\lambda$	lumen-hour	lm·h
τ	transmittance	–	–
	$\tau = \Phi_t/\Phi_i$		
w	radiant energy density	(SI) joule per cubic meter	J/m^3
w_e (radiometric)	$w = dQ/dV$	(cgs) erg per cubic centimeter	erg/cm^3
V	luminous efficiency	–	–
	$V = K/K_{\text{max}}$		

APPENDIX 51.B
Luminance Conversion Factors

multiply	by	to obtain
apostilb (international)	10	mL (deprecated)
apostilb (German Hefner)	11.11	mL (deprecated)
blondel	10	mL (deprecated)
cd/m^2 or nt	0.000645	cd/in^2
cd/m^2 or nt	0.0929	cd/ft^2
cd/m^2 or nt	0.001	stilb or cd/cm^2
cd/m^2 or nt	0.2919	fL (deprecated)
cd/m^2 or nt	0.3142	mL (deprecated)
cd/in^2	1550	cd/m^2 or nt
cd/in^2	144	cd/ft^2
cd/in^2	0.155	stilb or cd/cm^2
cd/in^2	452	fL (deprecated)
cd/in^2	487	mL (deprecated)
cd/ft^2	10.76	cd/m^2 or nt
cd/ft^2	0.00694	cd/in^2
cd/ft^2	0.00108	stilb or cd/cm^2
cd/ft^2	3.142	fL (deprecated)
cd/ft^2	3.382	mL (deprecated)
lambert (L)	0.001	mL (deprecated)
stilb or cd/cm^2	10,000	candela/m^2 or nt
stilb or cd/cm^2	6.45	cd/in^2
stilb or cd/cm^2	929	cd/ft^2
stilb or cd/cm^2	2919	fL (deprecated)
stilb or cd/cm^2	3142	mL (deprecated)
fL (deprecated)	3.426	cd/m^2 or nt
fL (deprecated)	0.00221	cd/in^2
fL (deprecated)	0.3183	cd/ft^2
fL (deprecated)	0.00034	stilb or cd/cm^2
fL (deprecated)	1.076	mL (deprecated)
mL (deprecated)	3.183	cd/m^2 or nt
mL (deprecated)	0.00205	cd/in^2
mL (deprecated)	0.2957	cd/ft^2
mL (deprecated)	0.00032	stilb or cd/cm^2
mL (deprecated)	0.929	fL (deprecated)
mL (deprecated)	1000	L

APPENDIX 51.C
Illuminance Conversion Factors

multiply	by	to obtain
erg	10^7	watt-second
footcandles	10.76	lx or lm/m^2 or metercandle
footcandles	0.00108	phot or lm/cm^2
footcandles	1.076	milliphot
lumen	1/683	light-watt
lumen-hour	60	lumen-minutes
light-watt	683	lumen
lux	0.0929	footcandles
lux	0.0001	phot or lm/cm^2
lux	0.1	milliphot
phot	929	footcandle or lm/ft^2
phot	10,000	lx or lm/m^2 or metercandle
phot	1000	milliphot
milliphot	0.929	footcandle or lm/ft^2
milliphot	10	lx or lm/m^2 or metercandle
milliphot	0.001	phot or lm/cm^2
watt-second	10^{-7}	erg

Support Material

APPENDIX 54.A
Periodic Table: Materials' Properties Summary

Atomic radius decreases (increases from group VIIA to group 0); boiling and freezing point increases to group VB and then decreases; density increases up to group 8 (elements 28, 46, 78) and then decreases; electronegativity and ionization energy increase; ionic radius decreases (increases across groups VIA and VIIA).

atomic radius increases; electronegativity decreases; ionization energy decreases; ionic radius increases

boiling and freezing point increases (except for rare earths and groups IB through VIA); density increases

s block: IA–IIA; *d* block: IIIB–IIB; *p* block: IIIA–0

IA	IIA		IIIB	IVB	VB	VIB	VIIB	VIII	VIII	VIII	IB	IIB	IIIA	IVA	VA	VIA	VIIA	0
1																		2
3	4												5	6	7	8	9	10
11	12												13	14	15	16	17	18
19	20		21	22	23	24	25	26	27	28	29	30	31	32	33	34	35	36
37	38		39	40	41	42	43	44	45	46	47	48	49	50	51	52	53	54
55	56	*	71	72	73	74	75	76	77	78	79	80	81	82	83	84	85	86
87	88	**	103	104	105	106	107	108	109	110	111	112	113	114	115	116	117	118

f block

lanthanides *	57	58	59	60	61	62	63	64	65	66	67	68	69	70
actinides **	89	90	91	92	93	94	95	96	97	98	99	100	101	102

coding

- alkali metals
- alkaline-earth metals
- transition metals } rare earths
- inner transition metals } rare earths
- halogens
- noble gases

legend

metal	solid
semi-metal (single underline)	liquid
nonmetal (double underline)	gas
(staircase line)	metal-to-nonmetal transition staircase

Note: Columns are *groups*; rows are *periods*; elements are listed by atomic number; helium (element 2) is part of the *s* block; grey atomic numbers indicate temporary names and relatively unknown properties for newly discovered elements; The *spdf* blocks refer to the orbitals, filling; the states of the elements are based on STP conditions.

APPENDIX 54.B

Table of Relative Atomic Weights
(referred to carbon-12)

name	symbol	atomic number	atomic weight	name	symbol	atomic number	atomic weight
actinium	Ac	89	–	meitnerium	Mt	109	–
aluminum	Al	13	26.9815	mendelevium	Md	101	–
americium	Am	95	–	mercury	Hg	80	200.59
antimony	Sb	51	121.760	molybdenum	Mo	42	95.96
argon	Ar	18	39.948	neodymium	Nd	60	144.242
arsenic	As	33	74.9216	neon	Ne	10	20.1797
astatine	At	85	–	neptunium	Np	93	237.048
barium	Ba	56	137.327	nickel	Ni	28	58.693
berkelium	Bk	97	–	niobium	Nb	41	92.906
beryllium	Be	4	9.0122	nitrogen	N	7	14.0067
bismuth	Bi	83	208.980	nobelium	No	102	–
bohrium	Bh	107	–	osmium	Os	76	190.23
boron	B	5	10.811	oxygen	O	8	15.9994
bromine	Br	35	79.904	palladium	Pd	46	106.42
cadmium	Cd	48	112.411	phosphorus	P	15	30.9738
calcium	Ca	20	40.078	platinum	Pt	78	195.084
californium	Cf	98	–	plutonium	Pu	94	–
carbon	C	6	12.0107	polonium	Po	84	–
cerium	Ce	58	140.116	potassium	K	19	39.0983
cesium	Cs	55	132.9054	praseodymium	Pr	59	140.9077
chlorine	Cl	17	35.453	promethium	Pm	61	–
chromium	Cr	24	51.996	protactinium	Pa	91	231.0359
cobalt	Co	27	58.9332	radium	Ra	88	–
copernicium	Cn	112	–	radon	Rn	86	226.025
copper	Cu	29	63.546	rhenium	Re	75	186.207
curium	Cm	96	–	rhodium	Rh	45	102.9055
darmstadtium	Ds	110	–	roentgenium	Rg	111	–
dubnium	Db	105	–	rubidium	Rb	37	85.4678
dysprosium	Dy	66	162.50	ruthenium	Ru	44	101.07
einsteinium	Es	99	–	rutherfordium	Rf	104	–
erbium	Er	68	167.259	samarium	Sm	62	150.36
europium	Eu	63	151.964	scandium	Sc	21	44.956
fermium	Fm	100	–	seaborgium	Sg	106	–
fluorine	F	9	18.9984	selenium	Se	34	78.96
francium	Fr	87	–	silicon	Si	14	28.0855
gadolinium	Gd	64	157.25	silver	Ag	47	107.868
gallium	Ga	31	69.723	sodium	Na	11	22.9898
germanium	Ge	32	72.64	strontium	Sr	38	87.62
gold	Au	79	196.9666	sulfur	S	16	32.065
hafnium	Hf	72	178.49	tantalum	Ta	73	180.94788
hassium	Hs	108	–	technetium	Tc	43	–
helium	He	2	4.0026	tellurium	Te	52	127.60
holmium	Ho	67	164.930	terbium	Tb	65	158.925
hydrogen	H	1	1.00794	thallium	Tl	81	204.383
indium	In	49	114.818	thorium	Th	90	232.038
iodine	I	53	126.90447	thulium	Tm	69	168.934
iridium	Ir	77	192.217	tin	Sn	50	118.710
iron	Fe	26	55.845	titanium	Ti	22	47.867
krypton	Kr	36	83.798	tungsten	W	74	183.84
lanthanum	La	57	138.9055	uranium	U	92	238.0289
lawrencium	Lr	103	–	vanadium	V	23	50.942
lead	Pb	82	207.2	xenon	Xe	54	131.293
lithium	Li	3	6.941	ytterbium	Yb	70	173.054
lutetium	Lu	71	174.9668	yttrium	Y	39	88.906
magnesium	Mg	12	24.305	zinc	Zn	30	65.38
manganese	Mn	25	54.9380	zirconium	Zr	40	91.224

APPENDIX 56.A
Wire (and Sheet-Metal Gage) Diameters (Thickness) in Inches

gage no.	(1) American Wire Gage (AWG)*	(2) steel wire gage	(3) Birmingham iron wire (Stubs)	(4) Stubs steel wire gage	(5) British standard wire gage	(6) steel music wire gage	(7) manufacturer's standard gage	(8) American zinc	(9) Birmingham
15/0	–	–	–	–	–	–	–	–	1.0000
14/0	–	–	–	–	–	–	–	–	0.9583
13/0	–	–	–	–	–	–	–	–	0.9167
12/0	–	–	–	–	–	–	–	–	0.8750
11/0	–	–	–	–	–	–	–	–	0.8333
10/0	–	–	–	–	–	–	–	–	0.7917
9/0	–	–	–	–	–	–	–	–	0.7500
8/0	–	–	–	–	–	–	–	–	0.7083
7/0	–	0.4900	–	–	0.5000	–	0.5000	–	0.6666
6/0	–	0.4615	–	–	0.4640	–	0.4687	–	0.6250
5/0	–	0.4305	–	–	0.4320	–	0.4375	–	0.5883
4/0	**0.4600**	0.3938	0.454	–	0.4000	–	0.4062	–	0.5416
3/0	**0.4100**	0.3625	0.425	–	0.3720	–	0.3750	–	0.5000
2/0	**0.3650**	0.3310	0.380	–	0.3480	0.0087	0.3437	–	0.4452
1/0	**0.3250**	0.3065	0.340	–	0.3240	0.0039	0.3125	–	0.3964
1	**0.2890**	0.2830	0.300	0.227	0.3000	0.0098	0.2812	0.002	0.3532
2	**0.2580**	0.2625	0.284	0.219	0.2760	0.0106	0.2656	0.004	0.3147
3	**0.2290**	0.2437	0.259	0.212	0.2520	0.0114	0.2500	0.006	0.2804
4	**0.2040**	0.2253	0.238	0.207	0.2320	0.0122	0.2344	0.008	0.2500
5	**0.1820**	0.2007	0.220	0.204	0.2120	0.0138	0.2187	0.010	0.2225
6	**0.1620**	0.1920	0.203	0.201	0.1920	0.0157	0.2035	0.012	0.1981
7	**0.1440**	0.1770	0.180	0.199	0.1760	0.0177	0.1875	0.014	0.1764
8	**0.1280**	0.1620	0.165	0.197	0.1600	0.0197	0.1719	0.016	0.1570
9	**0.1140**	0.1483	0.148	0.194	0.1440	0.0216	0.1562	0.018	0.1398
10	**0.1020**	0.1350	0.134	0.191	0.1280	0.0236	0.1406	0.020	0.1250
11	**0.0910**	0.1205	0.120	0.188	0.1160	0.0260	0.1250	0.024	0.1113
12	**0.0810**	0.1055	0.109	0.185	0.1040	0.0283	0.1094	0.028	0.0991
13	**0.0720**	0.0915	0.095	0.182	0.0920	0.0305	0.0937	0.032	0.0882
14	**0.0640**	0.0800	0.083	0.180	0.0800	0.0323	0.0821	0.036	0.0785
15	**0.0570**	0.0720	0.072	0.178	0.0720	0.0342	0.0703	0.040	0.0699
16	**0.0510**	0.0625	0.065	0.175	0.0640	0.0362	0.0625	0.045	0.0625
17	**0.0450**	0.0540	0.058	0.172	0.0560	0.0382	0.0562	0.050	0.0556
18	**0.0400**	0.0475	0.049	0.168	0.0480	0.0400	0.0500	0.055	0.0495
19	**0.0360**	0.0410	0.042	0.164	0.0400	0.0420	0.0437	0.060	0.0440
20	**0.0320**	0.0348	0.035	0.161	0.0360	0.0440	0.0375	0.070	0.0392
21	**0.0285**	0.0317	0.032	0.157	0.0320	0.0460	0.0344	0.080	0.0349
22	**0.0253**	0.0286	0.028	0.155	0.0280	0.0480	0.0312	0.090	0.0312
23	**0.0226**	0.0258	0.025	0.153	0.0240	0.0510	0.0281	0.100	0.0278
24	**0.0210**	0.0230	0.022	0.151	0.0220	0.0550	0.0250	0.125	0.0247
25	**0.0179**	0.0204	0.020	0.148	0.0200	0.0590	0.0219	0.250	0.0220
26	**0.0159**	0.0181	0.018	0.146	0.0180	0.0630	0.0187	0.375	0.0196
27	**0.0142**	0.0173	0.016	0.143	0.0164	0.0670	0.0172	0.500	0.0175
28	**0.0126**	0.0162	0.014	0.139	0.0148	0.0710	0.0156	1.000	0.0156
29	**0.0113**	0.0150	0.013	0.134	0.0136	0.0740	0.0141	–	0.0139
30	**0.0100**	0.0140	0.012	0.127	0.0124	0.0780	0.0125	–	0.0123

(continued)

Support Material

APPENDIX 56.A *(continued)*
Wire (and Sheet-Metal Gage) Diameters (Thickness) in Inches

gage no.	(1) American Wire Gage (AWG)*	(2) steel wire gage	(3) Birmingham iron wire (Stubs)	(4) Stubs steel wire gage	(5) British standard wire gage	(6) steel music wire gage	(7) manufacturer's standard gage	(8) American zinc	(9) Birmingham
31	**0.0089**	0.0132	0.010	0.120	0.0116	0.0820	0.0109	–	0.0110
32	**0.0080**	0.0128	0.009	0.115	0.0108	0.0860	0.0101	–	0.0098
33	**0.0071**	0.0118	0.008	0.112	0.0100	–	0.0094	–	0.0087
34	**0.0063**	0.0104	0.007	0.110	0.0092	–	0.0086	–	0.0077
35	**0.0056**	0.0095	0.005	0.108	0.0084	–	0.0078	–	0.0069
36	**0.0050**	0.0090	0.004	0.106	0.0076	–	0.0070	–	0.0061
37	**0.0045**	0.0085	–	0.103	0.0068	–	0.0066	–	0.0054
38	**0.0040**	0.0080	–	0.101	0.0060	–	0.0062	–	0.0048
39	**0.0035**	0.0075	–	0.099	0.0052	–	–	–	0.0043
40	**0.0031**	0.0070	–	0.097	0.0048	–	–	–	0.0039
41	–	0.0066	–	0.095	0.0044	–	–	–	0.0034
42	–	0.0062	–	0.092	0.0040	–	–	–	0.0031
43	–	0.0060	–	0.088	0.0036	–	–	–	0.0027
44	–	0.0058	–	0.085	0.0032	–	–	–	0.0024
45	–	0.0055	–	0.081	0.0028	–	–	–	0.00215
46	–	0.0052	–	0.079	0.0024	–	–	–	0.0019
47	–	0.0050	–	0.077	0.0020	–	–	–	0.0017
48	–	0.0048	–	0.075	0.0016	–	–	–	0.0015
49	–	0.0046	–	0.072	0.0012	–	–	–	0.00135
50	–	0.0044	–	0.069	0.0010	–	–	–	0.0012
51	–	–	–	–	–	–	–	–	0.0011
52	–	–	–	–	–	–	–	–	0.00095

*The American wire gage sizes have been rounded off to the usual limits of commercial accuracy.

Support Material

APPENDIX 56.B
Conductor Ampacities
(NEC Table 310.15(B)(16))

Allowable ampacities of insulated conductors rated 0 V through 2000 V, 60°C through 90°C (140°F through 194°F), not more than three current-carrying conductors in raceway, cable, or earth (directly buried), based on ambient temperature of 30°C (86°F).[a]

size	temperature rating of conductor (see NEC Table 310.104(A))						size
	60°C (140°F)	75°C (167°F)	90°C (194°F)	60°C (140°F)	75°C (167°F)	90°C (194°F)	
AWG or kcmil	types TW, UF	types RHW, THHW, THW, THWN, XHHW, USE, ZW	types TBS, SA, SIS, FEP, FEPB, MI, RHH, RHW-2, THHN, THHW, THW-2, THWN-2 USE-2, XHH, XHHW, XHHW-2, ZW-2	types TW,UF	types RHW, THHW, THW, THWN, XHHW, USE	types TBS, SA SIS, THHN, THHW, THW-2 THWN-2, RHH, RHW-2, USE-2, XHH, XHHW, XHHW-2, ZW-2	AWG or kcmil
	copper			aluminum or copper-clad aluminum			
18[b]	–	–	14	–	–	–	–
16[b]	–	–	18	–	–	–	–
14[b]	20	20	25	–	–	–	–
12[b]	25	25	30	20	20	25	12[b]
10[b]	30	35	40	25	30	35	10[b]
8	40	50	55	30	40	45	8
6	55	65	75	40	50	60	6
4	70	85	95	55	65	75	4
3	85	100	110	65	75	85	3
2	95	115	130	75	90	100	2
1	110	130	150	85	100	115	1
1/0	125	150	170	100	120	135	1/0
2/0	145	175	195	115	135	150	2/0
3/0	165	200	225	130	155	175	3/0
4/0	195	230	260	150	180	205	4/0
250	215	255	290	170	205	230	250
300	240	285	320	190	230	255	300
350	260	310	350	210	250	280	350
400	280	335	380	225	270	305	400
500	320	380	430	260	310	350	500
600	355	420	475	285	340	385	600
700	385	460	520	310	375	420	700
750	400	475	535	320	385	435	750
800	410	490	555	330	395	450	800
900	435	520	585	355	425	480	900
1000	455	545	615	375	445	500	1000
1250	495	590	665	405	485	545	1250
1500	520	625	705	435	520	585	1500
1750	545	650	735	455	545	615	1750
2000	560	665	750	470	560	630	2000

[a]See App. 56.C (NEC Table 310.15(B)(2)(a)) for the ampacity correction factors where the ambient temperature is other than 30°C (86°F).
[b]See Sec. 240.4(D) of the *National Electrical Code* (NEC) for conductor overcurrent protection limitations.

APPENDIX 56.C
Ambient Temperature Correction Factors
(NEC Table 310.15(B)(2)(a))

ambient temp. (°C)	correction factors: For ambient temperatures other than 26–30°C (86°F), multiply the allowable ampacities specified in the ampacity tables by the appropriate factor shown. 60°C	75°C	90°C	ambient temp. (°F)
26–30	1.00	1.00	1.00	78–86
31–35	0.91	0.94	0.96	87–95
36–40	0.82	0.88	0.91	96–104
41–45	0.71	0.82	0.87	105–113
46–50	0.58	0.75	0.82	114–122
51–55	0.41	0.67	0.76	123–131
56–60	–	0.58	0.71	132–140

Reprinted with permission from NFPA 70®-2014, *National Electrical Code®*, Copyright © 2013, National Fire Protection Association, Quincy, MA. This reprinted material is not the complete and official position of the NFPA on the referenced subject, which is represented only by the standard in its entirety.

Support Material

APPENDIX 56.D
Ambient Temperature Correction Factors Based on 40°C (104°F)
(NEC Table 310.15(B)(2)(b))

ambient temp. (°C)	For ambient temperatures other than 36–40°C (96–104°F), multiply the allowable ampacities shown by the appropriate factor shown.						ambient temp. (°F)
	60°C	75°C	90°C	150°C	200°C	250°C	
10 or less	1.58	1.36	1.26	1.13	1.09	1.07	50 or less
11–15	1.50	1.31	1.22	1.11	1.08	1.06	51–59
16–20	1.41	1.25	1.18	1.09	1.06	1.05	60–68
21–25	1.32	1.2	1.14	1.07	1.05	1.04	69–77
26–30	1.22	1.13	1.10	1.04	1.03	1.02	78–86
31–35	1.12	1.07	1.05	1.02	1.02	1.01	87–95
36–40	1.00	1.00	1.00	1.00	1.00	1.00	96–104
41–45	0.87	0.93	0.95	0.98	0.98	0.99	105–113
46–50	0.71	0.85	0.89	0.95	0.97	0.98	114–122
51–55	0.50	0.76	0.84	0.93	0.95	0.96	123–131
56–60	–	0.65	0.77	0.90	0.94	0.95	132–140
61–65	–	0.53	0.71	0.88	0.92	0.94	141–149
66–70	–	0.38	0.63	0.85	0.90	0.93	150–158
71–75	–	–	0.55	0.83	0.88	0.91	159–167
76–80	–	–	0.45	0.80	0.87	0.90	168–176
81–90	–	–	–	0.74	0.83	0.87	177–194
91–100	–	–	–	0.67	0.79	0.85	195–212
101–110	–	–	–	0.60	0.75	0.82	213–230
111–120	–	–	–	0.52	0.71	0.79	231–248
121–130	–	–	–	0.43	0.66	0.76	249–266
131–140	–	–	–	0.30	0.61	0.72	267–284
141–160	–	–	–	–	0.50	0.65	285–320
161–180	–	–	–	–	0.35	0.58	321–356
181–200	–	–	–	–	–	0.49	357–392
201–225	–	–	–	–	–	0.35	393–437

APPENDIX 56.E
Conductor Properties
(NEC Chapter 9, Table 8)

size (AWG or kcmil)	area, mm²	area, circular mils	conductors: stranding quantity	stranding diameter mm	stranding diameter in	overall diameter mm	overall diameter in	overall area mm²	overall area in²	direct current resistance at 75°C (167°F): copper, uncoated Ω/km	copper, uncoated Ω/1000 ft	copper, coated Ω/km	copper, coated Ω/1000 ft	aluminum Ω/km	aluminum Ω/1000 ft
18	0.823	1620	1	–	–	1.02	0.040	0.823	0.001	25.5	7.77	26.5	8.08	42.0	12.8
18	0.823	1620	7	0.39	0.015	1.16	0.046	1.06	0.002	26.1	7.95	27.7	8.45	42.8	13.1
16	1.31	2580	1	–	–	1.29	0.051	1.31	0.002	16.0	4.89	16.7	5.08	26.4	8.05
16	1.31	2580	7	0.49	0.019	1.46	0.058	1.68	0.003	16.4	4.99	17.3	5.29	26.9	8.21
14	2.08	4110	1	–	–	1.63	0.064	2.08	0.003	10.1	3.07	10.4	3.19	16.6	5.06
14	2.08	4110	7	0.62	0.024	1.85	0.073	2.68	0.004	10.3	3.14	10.7	3.26	16.9	5.17
12	3.31	6530	1	–	–	2.05	0.081	3.31	0.005	6.34	1.93	6.57	2.01	10.45	3.18
12	3.31	6530	7	0.78	0.030	2.32	0.092	4.25	0.006	6.50	1.98	6.73	2.05	10.69	3.25
10	5.261	10380	1	–	–	2.588	0.102	5.26	0.008	3.984	1.21	4.148	1.26	6.561	2.00
10	5.261	10380	7	0.98	0.038	2.95	0.116	6.76	0.011	4.070	1.24	4.226	1.29	6.679	2.04
8	8.367	16510	1	–	–	3.264	0.128	8.37	0.013	2.506	0.764	2.579	0.786	4.125	1.26
8	8.367	16510	7	1.23	0.049	3.71	0.146	10.76	0.017	2.551	0.778	2.653	0.809	4.204	1.28
6	13.30	26240	7	1.56	0.061	4.67	0.184	17.09	0.027	1.608	0.491	1.671	0.510	2.652	0.808
4	21.15	41740	7	1.96	0.077	5.89	0.232	27.19	0.042	1.010	0.308	1.053	0.321	1.666	0.508
3	26.67	52620	7	2.20	0.087	6.60	0.260	34.28	0.053	0.802	0.245	0.833	0.254	1.320	0.403
2	33.62	66360	7	2.47	0.097	7.42	0.292	43.23	0.067	0.634	0.194	0.661	0.201	1.045	0.319
1	42.41	83690	19	1.69	0.066	8.43	0.332	55.80	0.087	0.505	0.154	0.524	0.160	0.829	0.253
1/0	53.49	105600	19	1.89	0.074	9.45	0.372	70.41	0.109	0.399	0.122	0.415	0.127	0.660	0.201
2/0	67.43	133100	19	2.13	0.084	10.62	0.418	88.74	0.137	0.3170	0.0967	0.329	0.101	0.523	0.159
3/0	85.01	167800	19	2.39	0.094	11.94	0.470	111.9	0.173	0.2512	0.0766	0.2610	0.0797	0.413	0.126
4/0	107.2	211600	19	2.68	0.106	13.41	0.528	141.1	0.219	0.1996	0.0608	0.2050	0.0626	0.328	0.100
250	127	–	37	2.09	0.082	14.61	0.575	168	0.260	0.1687	0.0515	0.1753	0.0535	0.2778	0.0847
300	152	–	37	2.29	0.090	16.00	0.630	201	0.312	0.1409	0.0429	0.1463	0.0446	0.2318	0.0707
350	177	–	37	2.47	0.097	17.30	0.681	235	0.364	0.1205	0.0367	0.1252	0.0382	0.1984	0.0605
400	203	–	37	2.64	0.104	18.49	0.728	268	0.416	0.1053	0.0321	0.1084	0.0331	0.1737	0.0529
500	253	–	37	2.95	0.116	20.65	0.813	336	0.519	0.0845	0.0258	0.0869	0.0265	0.1391	0.0424
600	304	–	61	2.52	0.099	22.68	0.893	404	0.626	0.0704	0.0214	0.0732	0.0223	0.1159	0.0353
700	355	–	61	2.72	0.107	24.49	0.964	471	0.730	0.0603	0.0184	0.0622	0.0189	0.0994	0.0303
750	380	–	61	2.82	0.111	25.35	0.998	505	0.782	0.0563	0.0171	0.0579	0.0176	0.0927	0.0282
800	405	–	61	2.91	0.114	26.16	1.030	538	0.834	0.0528	0.0161	0.0544	0.0166	0.0868	0.0265
900	456	–	61	3.09	0.122	27.79	1.094	606	0.940	0.0470	0.0143	0.0481	0.0147	0.0770	0.0235
1000	507	–	61	3.25	0.128	29.26	1.152	673	1.042	0.0423	0.0129	0.0434	0.0132	0.0695	0.0212
1250	633	–	91	2.98	0.117	32.74	1.289	842	1.305	0.0338	0.0103	0.0347	0.0106	0.0554	0.0169
1500	760	–	91	3.26	0.128	35.86	1.412	1011	1.566	0.02814	0.00858	0.02814	0.00883	0.0464	0.0141
1750	887	–	127	2.98	0.117	38.76	1.526	1180	1.829	0.02410	0.00735	0.02410	0.00756	0.0397	0.0121
2000	1013	–	127	3.19	0.126	41.45	1.632	1349	2.092	0.02109	0.00643	0.02109	0.00662	0.0348	0.0106

Notes:

1. These resistance values are valid *only* for the parameters as given. Using conductors having coated strands, different stranding type, and, especially, other temperatures changes the resistance.
2. Formula for temperature change: $R_2 = R_1(1 + \alpha(T_2 - 75°))$ where $\alpha_{Cu} = 0.00323$ 1/°C, $\alpha_{Al} = 0.00330$ 1/°C at 75°C.
3. Conductors with compact and compressed stranding have about 9% and 3%, respectively, smaller bare conductor diameters than those shown. See Table 5A for actual compact cable dimensions.
4. The IACS conductivities used: bare copper = 100%, aluminum = 61%.
5. Class B stranding is listed as well as solid for some sizes. Its overall diameter and area is that of its circumscribing circle.

Informational Note: The construction information is per NEMA WC/70-2009 or ANSI/UL 1581-2011. The resistance is calculated per National Bureau of Standards Handbook 100, dated 1966, and Handbook 109, dated 1972.

APPENDIX 56.F
Conductor AC Properties
(NEC Chapter 9, Table 9)
Alternating-Current Resistance and Reactance for 600-Volt Cables, 3-Phase, 60 Hz, 75°C (167°F)—
Three Single Conductors in Conduit

size (AWG or kcmil)	resistance to neutral per kilometer (Ω/km) resistance to neutral per 1000 ft (Ω/1000 ft)														size (AWG or kcmil)
	X_L (reactance) for all wires		alternating current resistance for uncoated copper wires			alternating current resistance for aluminum wires			effective Z at 0.85 pf for uncoated copper wires			effective Z at 0.85 pf for aluminum wires			
	PVC, aluminum conduits	steel conduit	PVC conduit	aluminum conduit	steel conduit	PVC conduit	aluminum conduit	steel conduit	PVC conduit	aluminum conduit	steel conduit	PVC conduit	aluminum conduit	steel conduit	
14	0.190 0.058	0.240 0.073	10.2 3.1	10.2 3.1	10.2 3.1	— —	— —	— —	8.9 2.7	8.9 2.7	8.9 2.7	— —	— —	— —	14
12	0.177 0.054	0.223 0.068	6.6 2.0	6.6 2.0	6.6 2.0	10.5 3.2	10.5 3.2	10.5 3.2	5.6 1.7	5.6 1.7	5.6 1.7	9.2 2.8	9.2 2.8	9.2 2.8	12
10	0.164 0.050	0.207 0.063	3.9 1.2	3.9 1.2	3.9 1.2	6.6 2.0	6.6 2.0	6.6 2.0	3.6 1.1	3.6 1.1	3.6 1.1	5.9 1.8	5.9 1.8	5.9 1.8	10
8	0.171 0.052	0.213 0.065	2.56 0.78	2.56 0.78	2.56 0.78	4.3 1.3	4.3 1.3	4.3 1.3	2.26 0.69	2.26 0.69	2.30 0.70	3.6 1.1	3.6 1.1	3.6 1.1	8
6	0.167 0.051	0.210 0.064	1.61 0.49	1.61 0.49	1.61 0.49	2.66 0.81	2.66 0.81	2.66 0.81	1.44 0.44	1.48 0.45	1.48 0.45	2.33 0.71	2.36 0.72	2.36 0.72	6
4	0.157 0.048	0.197 0.060	1.02 0.31	1.02 0.31	1.02 0.31	1.67 0.51	1.67 0.51	1.67 0.51	0.95 0.29	0.95 0.29	0.98 0.30	1.51 0.46	1.51 0.46	1.51 0.46	4
3	0.154 0.047	0.194 0.059	0.82 0.25	0.82 0.25	0.82 0.25	1.31 0.40	1.35 0.41	1.31 0.40	0.75 0.23	0.79 0.24	0.79 0.24	1.21 0.37	1.21 0.37	1.21 0.37	3
2	0.148 0.045	0.187 0.057	0.62 0.19	0.66 0.20	0.66 0.20	1.05 0.32	1.05 0.32	1.05 0.32	0.62 0.19	0.62 0.19	0.66 0.20	0.98 0.30	0.98 0.30	0.98 0.30	2
1	0.151 0.046	0.187 0.057	0.49 0.15	0.52 0.16	0.52 0.16	0.82 0.25	0.85 0.26	0.82 0.25	0.52 0.16	0.52 0.16	0.52 0.16	0.79 0.24	0.79 0.24	0.82 0.25	1
1/0	0.144 0.044	0.180 0.055	0.39 0.12	0.43 0.13	0.39 0.12	0.66 0.20	0.69 0.21	0.66 0.20	0.43 0.13	0.43 0.13	0.43 0.13	0.62 0.19	0.66 0.20	0.66 0.20	1/0
2/0	0.141 0.043	0.177 0.054	0.33 0.10	0.33 0.10	0.33 0.10	0.52 0.16	0.52 0.16	0.52 0.16	0.36 0.11	0.36 0.11	0.36 0.11	0.52 0.16	0.52 0.16	0.52 0.16	2/0
3/0	0.138 0.042	0.171 0.052	0.253 0.077	0.269 0.082	0.259 0.079	0.43 0.13	0.43 0.13	0.43 0.13	0.289 0.088	0.302 0.092	0.308 0.094	0.43 0.13	0.43 0.13	0.46 0.14	3/0
4/0	0.135 0.041	0.167 0.051	0.203 0.062	0.220 0.067	0.207 0.063	0.33 0.10	0.36 0.11	0.33 0.10	0.243 0.074	0.256 0.078	0.262 0.080	0.36 0.11	0.36 0.11	0.36 0.11	4/0
250	0.135 0.041	0.171 0.052	0.171 0.052	0.187 0.057	0.177 0.054	0.279 0.085	0.295 0.090	0.282 0.086	0.217 0.066	0.230 0.070	0.240 0.073	0.308 0.094	0.322 0.098	0.33 0.10	250
300	0.135 0.041	0.167 0.051	0.144 0.044	0.161 0.049	0.148 0.045	0.233 0.071	0.249 0.076	0.236 0.072	0.194 0.059	0.207 0.063	0.213 0.065	0.269 0.082	0.282 0.086	0.289 0.088	300
350	0.131 0.040	0.164 0.050	0.125 0.038	0.141 0.043	0.128 0.039	0.200 0.061	0.217 0.066	0.207 0.063	0.174 0.053	0.190 0.058	0.197 0.060	0.240 0.073	0.253 0.077	0.262 0.080	350
400	0.131 0.040	0.161 0.049	0.108 0.033	0.125 0.038	0.115 0.035	0.177 0.054	0.194 0.059	0.180 0.055	0.161 0.049	0.174 0.053	0.184 0.056	0.217 0.066	0.233 0.071	0.240 0.073	400
500	0.128 0.039	0.157 0.048	0.089 0.027	0.105 0.032	0.095 0.029	0.141 0.043	0.157 0.048	0.148 0.045	0.141 0.043	0.157 0.048	0.164 0.050	0.187 0.057	0.200 0.061	0.210 0.064	500
600	0.128 0.039	0.157 0.048	0.075 0.023	0.092 0.028	0.082 0.025	0.118 0.036	0.135 0.041	0.125 0.038	0.131 0.040	0.144 0.044	0.154 0.047	0.167 0.051	0.180 0.055	0.190 0.058	600
750	0.125 0.038	0.157 0.048	0.062 0.019	0.079 0.024	0.069 0.021	0.095 0.029	0.112 0.034	0.102 0.031	0.118 0.036	0.131 0.040	0.141 0.043	0.148 0.045	0.161 0.049	0.171 0.052	750
1000	0.121 0.037	0.151 0.046	0.049 0.015	0.062 0.019	0.059 0.018	0.075 0.023	0.089 0.027	0.082 0.025	0.105 0.032	0.118 0.036	0.131 0.040	0.128 0.039	0.138 0.042	0.151 0.046	1000

Notes:
1. These values are based on the following constants: UL-Type RHH wires with Class B stranding, in cradled configuration. Wire conductivities are 100% IACS copper and 61% IACS aluminum, and aluminum conduit is 45% IACS. Capacitive reactance is ignored, since it is negligible at these voltages. These resistance values are valid only at 75°C (167°F) and for the parameters as given, but are representative for 600-volt wire types operating at 60 Hz.
2. *Effective Z* is defined as $R \cos \theta + X \sin \theta$, where θ is the power factor angle of the circuit. Multiplying current by effective impedance gives a good approximation for line-to-neutral voltage drop. Effective impedance values shown in this table are valid only at 0.85 power factor. For another circuit power factor (pf), effective impedance (Z_e) can be calculated from R and X_L values given in this table as follows: $Z_e = R \times \text{pf} + X_L \sin(\arccos \text{pf})$.

APPENDIX 56.G

Plug and Receptacle Configurations

specific purpose

	DESCRIPTION	NEMA NUMBER	15 AMPERE		30 AMPERE		50 AMPERE	
			RECEPTACLE	PLUG	RECEPTACLE	PLUG	RECEPTACLE	PLUG
MIDGET LOCKING	125V, 2 POLE, 2 WIRE	ML1	ML1-15R	ML1-15P				
MIDGET LOCKING	125V, 2 POLE, 3 WIRE GROUNDING	ML2	ML2-15R	ML2-15P				
MIDGET LOCKING	125/250V, 3 POLE, 3 WIRE	ML3	ML3-15R	ML3-15P				
FSL CONFIGURATIONS	28V DC, 2 POLE, 3 WIRE GROUNDING	FSL1			FSL1-30R	FSL1-30P		
FSL CONFIGURATIONS	120V, 400HZ, 2 POLE, 3 WIRE GROUNDING	FSL2			FSL2-30R	FSL2-30P		
FSL CONFIGURATIONS	120V, 400 HZ, 3-PHASE 3 POLE, 4 WIRE GROUNDING	FSL3			FSL3-30R	FSL3-30P		
FSL CONFIGURATIONS	120/208V, 3 Ø Y, 400 HZ 4 POLE, 5 WIRE GROUNDING	FSL4			FSL4-30R	FSL4-30P		
MARINE SHIP-TO-SHORE	125V, 2 POLE, 3 WIRE GROUNDING	SS1					SS1-50R	SS1-50P
MARINE SHIP-TO-SHORE	125/250V, 3 POLE, 4 WIRE GROUNDING	SS2					SS2-50R	SS2-50P
TRAVEL TRAILER	120V AC, 2 POLE, 3 WIRE GROUNDING	TT			TT-30R	TT-30P		

(continued)

APPENDIX 56.G *(continued)*

Plug and Receptacle Configurations

non-locking

DESCRIPTION		NEMA NUMBER	15 AMPERE RECEPTACLE	15 AMPERE PLUG	20 AMPERE RECEPTACLE	20 AMPERE PLUG	30 AMPERE RECEPTACLE	30 AMPERE PLUG	50 AMPERE RECEPTACLE	50 AMPERE PLUG	60 AMPERE RECEPTACLE	60 AMPERE PLUG
2-POLE 2-WIRE	125V	1	1-15R	1-15P POLARIZED / NON POLAR.		1-20P MATES WITH 5-20R		1-30P MATES WITH 5-30R				
	250V	2		2-15P MATES WITH 6-15R	2-20R	2-20P	2-30R	2-30P				
	277V AC	3										
	600V	4										
2-POLE 3-WIRE GROUNDING	125V	5	5-15R	5-15P	5-20R	5-20P	5-30R	5-30P	5-50R	5-50P		
	125V	5ALT			5ALT-20R							
	250V	6	6-15R	6-15P	6-20R	6-20P	6-30R	6-30P	6-50R	6-50P		
	250V	6ALT			6ALT-20R							
	277V AC	7	7-15R	7-15P	7-20R	7-20P	7-30R	7-30P	7-50R	7-50P		
	347V AC	24	24-15R	24-15P	24-20R	24-20P	24-30R	24-30P	24-50R	24-50P		
	480V AC	8										
	600V AC	9										
3-POLE 3-WIRE	125 / 250V	10			10-20R	10-20P	10-30R	10-30P	10-50R	10-50P		
	3 Ø 250V	11	11-15R	11-15P	11-20R	11-20P	11-30R	11-30P	11-50R	11-50P		
	3 Ø 480V	12										
	3 Ø 600V	13										
3-POLE 4-WIRE GROUNDING	125 / 250V	14	14-15R	14-15P	14-20R	14-20P	14-30R	14-30P	14-50R	14-50P	14-60R	14-60P
	3 Ø 250V	15	15-15R	15-15P	15-20R	15-20P	15-30R	15-30P	15-50R	15-50P	15-60R	15-60P
	3 Ø 480V	16										
	3 Ø 600V	17										
4-POLE 4-WIRE	3 Ø Y 120 / 208V	18	18-15R	18-15P	18-20R	18-20P	18-30R	18-30P	18-50R	18-50P	18-60R	18-60P
	3 Ø Y 277 / 480V	19										
	3 Ø Y 347 / 600V	20										
4-POLE 5-WIRE GROUNDING	3 Ø Y 120 / 208V	21										
	3 Ø Y 277 / 480V	22										
	3 Ø Y 347 / 600V	23										

(continued)

Support Material

APPENDIX 56.G *(continued)*
Plug and Receptacle Configurations
locking

DESCRIPTION		NEMA NUMBER	15 AMPERE RECEPTACLE	15 AMPERE PLUG	20 AMPERE RECEPTACLE	20 AMPERE PLUG	30 AMPERE RECEPTACLE	30 AMPERE PLUG	50 AMPERE RECEPTACLE	50 AMPERE PLUG	60 AMPERE RECEPTACLE	60 AMPERE PLUG
2-POLE 2-WIRE	125V	1	L 1-15R	L 1-15P								
	250V	2			L 2-20R	L 2-20P						
	277V AC	3										
	600V	4										
2-POLE 3-WIRE GROUNDING	125V	5	L 5-15R	L 5-15P	L 5-20R	L 5-20P	L 5-30R	L 5-30P	L 5-50R	L 5-50P	L 5-60R	L 5-60P
	250V	6	L 6-15R	L 6-15P	L 6-20R	L 6-20P	L 6-30R	L 6-30P	L 6-50R	L 6-50P	L 6-60R	L 6-60P
	277V AC	7	L 7-15R	L 7-15P	L 7-20R	L 7-20P	L 7-30R	L 7-30P	L 7-50R	L 7-50P	L 7-60R	L 7-60P
	347V AC	24			L24-20R	L24-20P						
	480V AC	8			L 8-20R	L 8-20P	L 8-30R	L 8-30P	L 8-50R	L 8-50P	L 8-60R	L 8-60P
	600V AC	9			L 9-20R	L 9-20P	L 9-30R	L 9-30P	L 9-50R	L 9-50P	L 9-60R	L 9-60P
3-POLE 3-WIRE	125 / 250V	10			L10-20R	L10-20P	L10-30R	L10-30P				
	3 Ø 250V	11	L11-15R	L11-15P	L11-20R	L11-20P	L11-30R	L11-30P				
	3 Ø 480V	12			L12-20R	L12-20P	L12-30R	L12-30P				
	3 Ø 600V	13					L13-30R	L13-30P				
3-POLE 4-WIRE GROUNDING	125 / 250V	14			L14-20R	L14-20P	L14-30R	L14-30P	L14-50R	L14-50P	L14-60R	L14-60P
	3 Ø 250V	15			L15-20R	L15-20P	L15-30R	L15-30P	L15-50R	L15-50P	L15-60R	L15-60P
	3 Ø 480V	16			L16-20R	L16-20P	L16-30R	L16-30P	L16-50R	L16-50P	L16-60R	L16-60P
	3 Ø 600V	17					L17-30R	L17-30P	L17-50R	L17-50P	L17-60R	L17-60P
4-POLE 4-WIRE	3 Ø Y 120 / 208V	18			L18-20R	L18-20P	L18-30R	L18-30P				
	3 Ø Y 277 / 480V	19			L19-20R	L19-20P	L19-30R	L19-30P				
	3 Ø Y 347 / 600V	20			L20-20R	L20-20P	L20-30R	L20-30P				
4-POLE 5-WIRE GROUNDING	3 Ø Y 120 / 208V	21			L21-20R	L21-20P	L21-30R	L21-30P	L21-50R	L21-50P	L21-60R	L21-60P
	3 Ø Y 277 / 480V	22			L22-20R	L22-20P	L22-30R	L22-30P	L22-50R	L22-50P	L22-60R	L22-60P
	3 Ø Y 347 / 600V	23			L23-20R	L23-20P	L23-30R	L23-30P	L23-50R	L23-50P	L23-60R	L23-60P

APPENDIX 58.A
Standard Cash Flow Factors

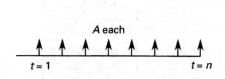

multiply	by	to obtain
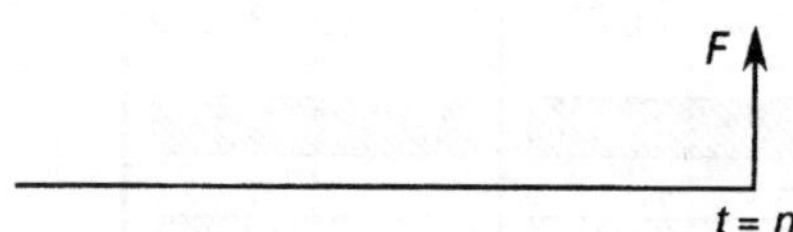 F	$P = F(1+i)^{-n}$ $(P/F, i\%, n)$	P
P	$F = P(1+i)^{n}$ $(F/P, i\%, n)$	F
A	$P = A\left(\frac{(1+i)^n - 1}{i(1+i)^n}\right)$ $(P/A, i\%, n)$	P 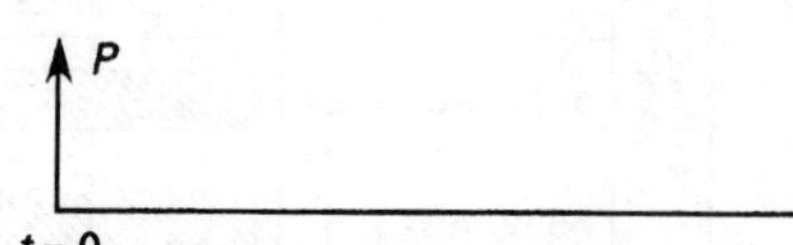
P	$A = P\left(\frac{i(1+i)^n}{(1+i)^n - 1}\right)$ $(A/P, i\%, n)$	A 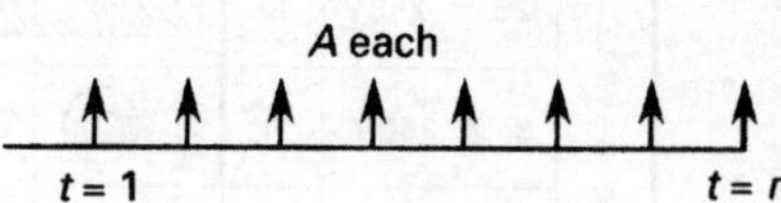
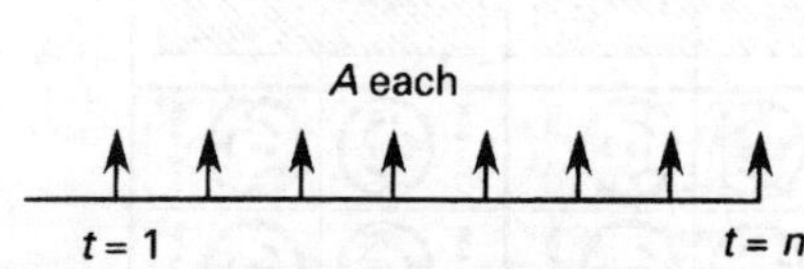 A	$F = A\left(\frac{(1+i)^n - 1}{i}\right)$ $(F/A, i\%, n)$	F 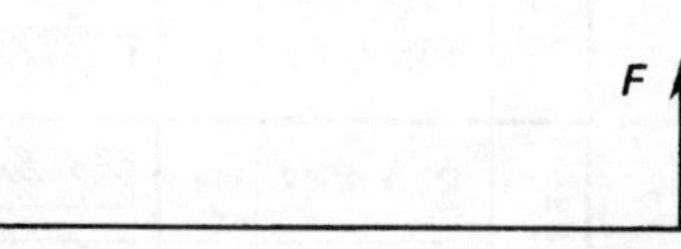
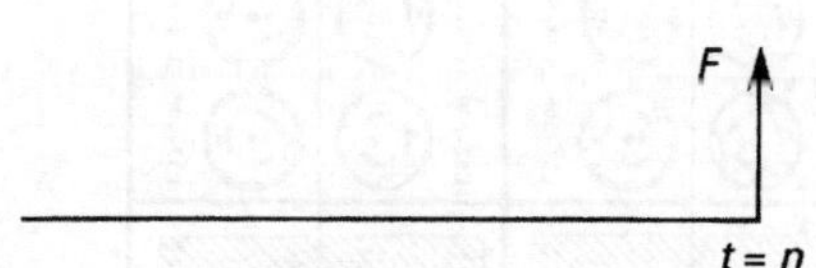 F	$A = F\left(\frac{i}{(1+i)^n - 1}\right)$ $(A/F, i\%, n)$	A 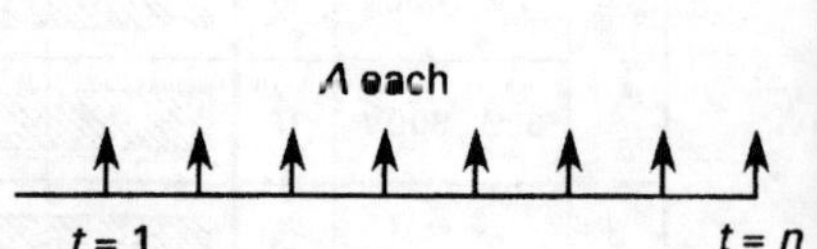
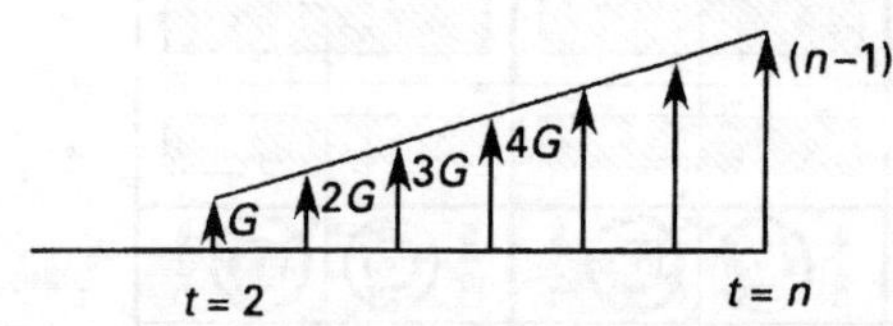 G	$P = G\left(\frac{(1+i)^n - 1}{i^2(1+i)^n} - \frac{n}{i(1+i)^n}\right)$ $(P/G, i\%, n)$	P 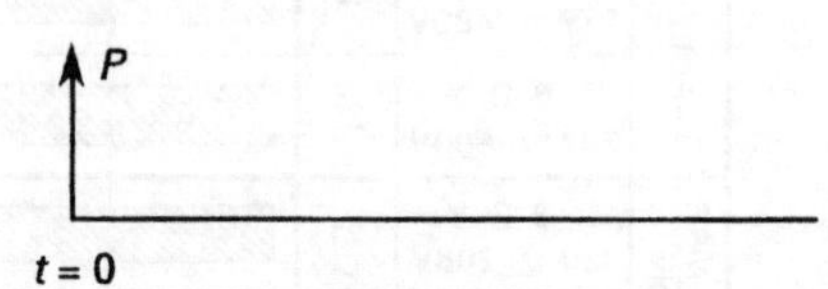
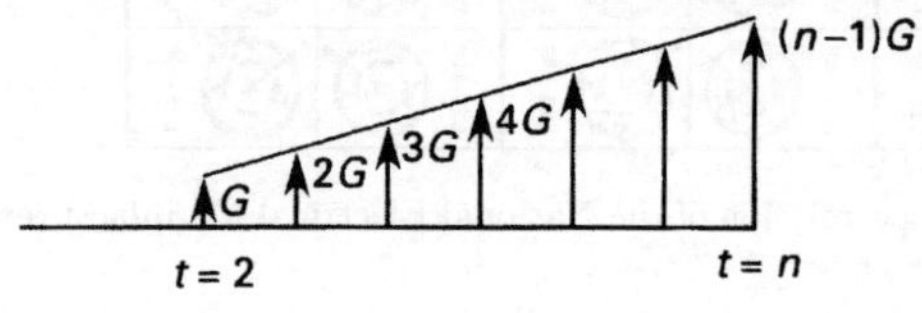 G	$A = G\left(\frac{1}{i} - \frac{n}{(1+i)^n - 1}\right)$ $(A/G, i\%, n)$	A 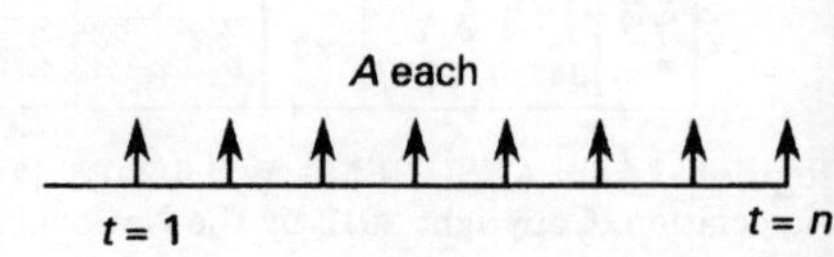

APPENDIX 58.B
Cash Flow Equivalent Factors

$i = 0.50\%$

n	P/F	P/A	P/G	F/P	F/A	A/P	A/F	A/G	n
1	0.9950	0.9950	0.0000	1.0050	1.0000	1.0050	1.0000	0.0000	1
2	0.9901	0.9851	0.9901	1.0100	2.0050	0.5038	0.4988	0.4988	2
3	0.9851	2.9702	2.9604	1.0151	3.0150	0.3367	0.3317	0.9967	3
4	0.9802	3.9505	5.9011	1.0202	4.0301	0.2531	0.2481	1.4938	4
5	0.9754	4.9259	9.8026	1.0253	5.0503	0.2030	0.1980	1.9900	5
6	0.9705	5.8964	14.6552	1.0304	6.0755	0.1696	0.1646	2.4855	6
7	0.9657	6.8621	20.4493	1.0355	7.1059	0.1457	0.1407	2.9801	7
8	0.9609	7.8230	27.1755	1.0407	8.1414	0.1278	0.1228	3.4738	8
9	0.9561	8.7791	34.8244	1.0459	9.1821	0.1139	0.1089	3.9668	9
10	0.9513	9.7304	43.3865	1.0511	10.2280	0.1028	0.0978	4.4589	10
11	0.9466	10.6770	52.8526	1.0564	11.2792	0.0937	0.0887	4.9501	11
12	0.9419	11.6189	63.2136	1.0617	12.3356	0.0861	0.0811	5.4406	12
13	0.9372	12.5562	74.4602	1.0670	13.3972	0.0796	0.0746	5.9302	13
14	0.9326	13.4887	86.5835	1.0723	14.4642	0.0741	0.0691	6.4190	14
15	0.9279	14.4166	99.5743	1.0777	15.5365	0.0694	0.0644	6.9069	15
16	0.9233	15.3399	113.4238	1.0831	16.6142	0.0652	0.0602	7.3940	16
17	0.9187	16.2586	128.1231	1.0885	17.6973	0.0615	0.0565	7.8803	17
18	0.9141	17.1728	143.6634	1.0939	18.7858	0.0582	0.0532	8.3658	18
19	0.9096	18.0824	160.0360	1.0994	19.8797	0.0553	0.0503	8.8504	19
20	0.9051	18.9874	177.2322	1.1049	20.9791	0.0527	0.0477	9.3342	20
21	0.9006	19.8880	195.2434	1.1104	22.0840	0.0503	0.0453	9.8172	21
22	0.8961	20.7841	214.0611	1.1160	23.1944	0.0481	0.0431	10.2993	22
23	0.8916	21.6757	233.6768	1.1216	24.3104	0.0461	0.0411	10.7806	23
24	0.8872	22.5629	254.0820	1.1272	25.4320	0.0443	0.0393	11.2611	24
25	0.8828	23.4456	275.2686	1.1328	26.5591	0.0427	0.0377	11.7407	25
26	0.8784	24.3240	297.2281	1.1385	27.6919	0.0411	0.0361	12.2195	26
27	0.8740	25.1980	319.9523	1.1442	28.8304	0.0397	0.0347	12.6975	27
28	0.8697	26.0677	343.4332	1.1499	29.9745	0.0384	0.0334	13.1747	28
29	0.8653	26.9330	367.6625	1.1556	31.1244	0.0371	0.0321	13.6510	29
30	0.8610	27.7941	392.6324	1.1614	32.2800	0.0360	0.0310	14.1265	30
31	0.8567	28.6508	418.3348	1.1672	33.4414	0.0349	0.0299	14.6012	31
32	0.8525	29.5033	444.7618	1.1730	34.6086	0.0339	0.0289	15.0750	32
33	0.8482	30.3515	471.9055	1.1789	35.7817	0.0329	0.0279	15.5480	33
34	0.8440	31.1955	499.7583	1.1848	36.9606	0.0321	0.0271	16.0202	34
35	0.8398	32.0354	528.3123	1.1907	38.1454	0.0312	0.0262	16.4915	35
36	0.8356	32.8710	557.5598	1.1967	39.3361	0.0304	0.0254	16.9621	36
37	0.8315	33.7025	587.4934	1.2027	40.5328	0.0297	0.0247	17.4317	37
38	0.8274	34.5299	618.1054	1.2087	41.7354	0.0290	0.0240	17.9006	38
39	0.8232	35.3531	649.3883	1.2147	42.9441	0.0283	0.0233	18.3686	39
40	0.8191	36.1722	681.3347	1.2208	44.1588	0.0276	0.0226	18.8359	40
41	0.8151	36.9873	713.9372	1.2269	45.3796	0.0270	0.0220	19.3022	41
42	0.8110	37.7983	747.1886	1.2330	46.6065	0.0265	0.0215	19.7678	42
43	0.8070	38.6053	781.0815	1.2392	47.8396	0.0259	0.0209	20.2325	43
44	0.8030	39.4082	815.6087	1.2454	49.0788	0.0254	0.0204	20.6964	44
45	0.7990	40.2072	850.7631	1.2516	50.3242	0.0249	0.0199	21.1595	45
46	0.7950	41.0022	886.5376	1.2579	51.5758	0.0244	0.0194	21.6217	46
47	0.7910	41.7932	922.9252	1.2642	52.8337	0.0239	0.0189	22.0831	47
48	0.7871	42.5803	959.9188	1.2705	54.0978	0.0235	0.0185	22.5437	48
49	0.7832	43.3635	997.5116	1.2768	55.3683	0.0231	0.0181	23.0035	49
50	0.7793	44.1428	1035.6966	1.2832	56.6452	0.0227	0.0177	23.4624	50
51	0.7754	44.9182	1074.4670	1.2896	57.9284	0.0223	0.0173	23.9205	51
52	0.7716	45.6897	1113.8162	1.2961	59.2180	0.0219	0.0169	24.3778	52
53	0.7677	46.4575	1153.7372	1.3026	60.5141	0.0215	0.0165	24.8343	53
54	0.7639	47.2214	1194.2236	1.3091	61.8167	0.0212	0.0162	25.2899	54
55	0.7601	47.9814	1235.2686	1.3156	63.1258	0.0208	0.0158	25.7447	55
60	0.7414	51.7256	1448.6458	1.3489	69.7700	0.0193	0.0143	28.0064	60
65	0.7231	55.3775	1675.0272	1.3829	76.5821	0.0181	0.0131	30.2475	65
70	0.7053	58.9394	1913.6427	1.4178	83.5661	0.0170	0.0120	32.4680	70
75	0.6879	62.4136	2163.7525	1.4536	90.7265	0.0160	0.0110	34.6679	75
80	0.6710	65.8023	2424.6455	1.4903	98.0677	0.0152	0.0102	36.8474	80
85	0.6545	69.1075	2695.6389	1.5280	105.5943	0.0145	0.0095	39.0065	85
90	0.6383	72.3313	2976.0769	1.5666	113.3109	0.0138	0.0088	41.1451	90
95	0.6226	75.4757	3265.3298	1.6061	121.2224	0.0132	0.0082	43.2633	95
100	0.6073	78.5426	3562.7934	1.6467	129.3337	0.0127	0.0077	45.3613	100

(continued)

APPENDIX 58.B *(continued)*
Cash Flow Equivalent Factors

$i = 0.75\%$

n	P/F	P/A	P/G	F/P	F/A	A/P	A/F	A/G	n
1	0.9926	0.9926	0.0000	1.0075	1.0000	1.0075	1.0000	0.0000	1
2	0.9852	1.9777	0.9852	1.0151	2.0075	0.5056	0.4981	0.4981	2
3	0.9778	2.9556	2.9408	1.0227	3.0226	0.3383	0.3308	0.9950	3
4	0.9706	3.9261	5.8525	1.0303	4.0452	0.2547	0.2472	1.4907	4
5	0.9633	4.8894	9.7058	1.0381	5.0756	0.2045	0.1970	1.9851	5
6	0.9562	5.8456	14.4866	1.0459	6.1136	0.1711	0.1636	2.4782	6
7	0.9490	6.7946	20.1808	1.0537	7.1595	0.1472	0.1397	2.9701	7
8	0.9420	7.7366	26.7747	1.0616	8.2132	0.1293	0.1218	3.4608	8
9	0.9350	8.6716	34.2544	1.0696	9.2748	0.1153	0.1078	3.9502	9
10	0.9280	9.5996	42.6064	1.0776	10.3443	0.1042	0.0967	4.4384	10
11	0.9211	10.5207	51.8174	1.0857	11.4219	0.0951	0.0876	4.9253	11
12	0.9142	11.4349	61.8740	1.0938	12.5076	0.0875	0.0800	5.4110	12
13	0.9074	12.3423	72.7632	1.1020	13.6014	0.0810	0.0735	5.8954	13
14	0.9007	13.2430	84.4720	1.1103	14.7034	0.0755	0.0680	6.3786	14
15	0.8940	14.1370	96.9876	1.1186	15.8137	0.0707	0.0632	6.8606	15
16	0.8873	15.0243	110.2973	1.1270	16.9323	0.0666	0.0591	7.3413	16
17	0.8807	15.9050	124.3887	1.1354	18.0593	0.0629	0.0554	7.8207	17
18	0.8742	16.7792	139.2494	1.1440	19.1947	0.0596	0.0521	8.2989	18
19	0.8676	17.6468	154.8671	1.1525	20.3387	0.0567	0.0492	8.7759	19
20	0.8612	18.5080	171.2297	1.1612	21.4912	0.0540	0.0465	9.2516	20
21	0.8548	19.3628	188.3253	1.1699	22.6524	0.0516	0.0441	9.7261	21
22	0.8484	20.2112	206.1420	1.1787	23.8223	0.0495	0.0420	10.1994	22
23	0.8421	21.0533	224.6682	1.1875	25.0010	0.0475	0.0400	10.6714	23
24	0.8358	21.8891	243.8923	1.1964	26.1885	0.0457	0.0382	11.1422	24
25	0.8296	22.7188	263.8029	1.2054	27.3849	0.0440	0.0365	11.6117	25
26	0.8234	23.5422	284.3888	1.2144	28.5903	0.0425	0.0350	12.0800	26
27	0.8173	24.3595	305.6387	1.2235	29.8047	0.0411	0.0336	12.5470	27
28	0.8112	25.1707	327.5416	1.2327	31.0282	0.0397	0.0322	13.0128	28
29	0.8052	25.9759	350.0867	1.2420	32.2609	0.0385	0.0310	13.4774	29
30	0.7992	26.7751	373.2631	1.2513	33.5029	0.0373	0.0298	13.9407	30
31	0.7932	27.5683	397.0602	1.2607	34.7542	0.0363	0.0288	14.4028	31
32	0.7873	28.3557	421.4675	1.2701	36.0148	0.0353	0.0278	14.8636	32
33	0.7815	29.1371	446.4746	1.2796	37.2849	0.0343	0.0268	15.3232	33
34	0.7757	29.9128	472.0712	1.2892	38.5646	0.0334	0.0259	15.7816	34
35	0.7699	30.6827	498.2471	1.2989	39.8538	0.0326	0.0251	16.2387	35
36	0.7641	31.4468	524.9924	1.3086	41.1527	0.0318	0.0243	16.6946	36
37	0.7585	32.2053	552.2969	1.3185	42.4614	0.0311	0.0236	17.1493	37
38	0.7528	32.9581	580.1511	1.3283	43.7798	0.0303	0.0228	17.6027	38
39	0.7472	33.7053	608.5451	1.3383	45.1082	0.0297	0.0222	18.0549	39
40	0.7416	34.4469	637.4693	1.3483	46.4465	0.0290	0.0215	18.5058	40
41	0.7361	35.1831	666.9144	1.3585	47.7948	0.0284	0.0209	18.9556	41
42	0.7306	35.9137	696.8709	1.3686	49.1533	0.0278	0.0203	19.4040	42
43	0.7252	36.6389	727.3297	1.3789	50.5219	0.0273	0.0198	19.8513	43
44	0.7198	37.3587	758.2815	1.3893	51.9009	0.0268	0.0193	20.2973	44
45	0.7145	38.0732	789.7173	1.3997	53.2901	0.0263	0.0188	20.7421	45
46	0.7091	38.7823	821.6283	1.4102	54.6898	0.0258	0.0183	21.1856	46
47	0.7039	39.4862	854.0056	1.4207	56.1000	0.0253	0.0178	21.6280	47
48	0.6986	40.1848	886.8404	1.4314	57.5207	0.0249	0.0174	22.0691	48
49	0.6934	40.8782	920.1243	1.4421	58.9521	0.0245	0.0170	22.5089	49
50	0.6883	41.5664	953.8486	1.4530	60.3943	0.0241	0.0166	22.9476	50
51	0.6831	42.2496	988.0050	1.4639	61.8472	0.0237	0.0162	23.3850	51
52	0.6780	42.9276	1022.5852	1.4748	63.3111	0.0233	0.0158	23.8211	52
53	0.6730	43.6006	1057.5810	1.4859	64.7859	0.0229	0.0154	24.2561	53
54	0.6680	44.2686	1092.9842	1.4970	66.2718	0.0226	0.0151	24.6898	54
55	0.6630	44.9316	1128.7869	1.5083	67.7688	0.0223	0.0148	25.1223	55
60	0.6387	48.1734	1313.5189	1.5657	75.4241	0.0208	0.0133	27.2665	60
65	0.6153	51.2963	1507.0910	1.6253	83.3709	0.0195	0.0120	29.3801	65
70	0.5927	54.3046	1708.6065	1.6872	91.6201	0.0184	0.0109	31.4634	70
75	0.5710	57.2027	1917.2225	1.7514	100.1833	0.0175	0.0100	33.5163	75
80	0.5500	59.9944	2132.1472	1.8180	109.0725	0.0167	0.0092	35.5391	80
85	0.5299	62.6838	2352.6375	1.8873	118.3001	0.0160	0.0085	37.5318	85
90	0.5104	65.2746	2577.9961	1.9591	127.8790	0.0153	0.0078	39.4946	90
95	0.4917	67.7704	2807.5694	2.0337	137.8225	0.0148	0.0073	41.4277	95
100	0.4737	70.1746	3040.7453	2.1111	148.1445	0.0143	0.0068	43.3311	100

(continued)

APPENDIX 58.B *(continued)*
Cash Flow Equivalent Factors

$i = 1.00\%$

n	P/F	P/A	P/G	F/P	F/A	A/P	A/F	A/G	n
1	0.9901	0.9901	0.0000	1.0100	1.0000	1.0100	1.0000	0.0000	1
2	0.9803	1.9704	0.9803	1.0201	2.0100	0.5075	0.4975	0.4975	2
3	0.9706	2.9410	2.9215	1.0303	3.0301	0.3400	0.3300	0.9934	3
4	0.9610	3.9020	5.8044	1.0406	4.0604	0.2563	0.2463	1.4876	4
5	0.9515	4.8534	9.6103	1.0510	5.1010	0.2060	0.1960	1.9801	5
6	0.9420	5.7955	14.3205	1.0615	6.1520	0.1725	0.1625	2.4710	6
7	0.9327	6.7282	19.9168	1.0721	7.2135	0.1486	0.1386	2.9602	7
8	0.9235	7.6517	26.3812	1.0829	8.2857	0.1307	0.1207	3.4478	8
9	0.9143	8.5660	33.6959	1.0937	9.3685	0.1167	0.1067	3.9337	9
10	0.9053	9.4713	41.8435	1.1046	10.4622	0.1056	0.0956	4.4179	10
11	0.8963	10.3676	50.8067	1.1157	11.5668	0.0965	0.0865	4.9005	11
12	0.8874	11.2551	60.5687	1.1268	12.6825	0.0888	0.0788	5.3815	12
13	0.8787	12.1337	71.1126	1.1381	13.8093	0.0824	0.0724	5.8607	13
14	0.8700	13.0037	82.4221	1.1495	14.9474	0.0769	0.0669	6.3384	14
15	0.8613	13.8651	94.4810	1.1610	16.0969	0.0721	0.0621	6.8143	15
16	0.8528	14.7179	107.2734	1.1726	17.2579	0.0679	0.0579	7.2886	16
17	0.8444	15.5623	120.7834	1.1843	18.4304	0.0643	0.0543	7.7613	17
18	0.8360	16.3983	134.9957	1.1961	19.6147	0.0610	0.0510	8.2323	18
19	0.8277	17.2260	149.8950	1.2081	20.8109	0.0581	0.0481	8.7017	19
20	0.8195	18.0456	165.4664	1.2202	22.0190	0.0554	0.0454	9.1694	20
21	0.8114	18.8570	181.6950	1.2324	23.2392	0.0530	0.0430	9.6354	21
22	0.8034	19.6604	198.5663	1.2447	24.4716	0.0509	0.0409	10.0998	22
23	0.7954	20.4558	216.0660	1.2572	25.7163	0.0489	0.0389	10.5626	23
24	0.7876	21.2434	234.1800	1.2697	26.9735	0.0471	0.0371	11.0237	24
25	0.7798	22.0232	252.8945	1.2824	28.2432	0.0454	0.0354	11.4831	25
26	0.7720	22.7952	272.1957	1.2953	29.5256	0.0439	0.0339	11.9409	26
27	0.7644	23.5596	292.0702	1.3082	30.8209	0.0424	0.0324	12.3971	27
28	0.7568	24.3164	312.5047	1.3213	32.1291	0.0411	0.0311	12.8516	28
29	0.7493	25.0658	333.4863	1.3345	33.4504	0.0399	0.0299	13.3044	29
30	0.7419	25.8077	355.0021	1.3478	34.7849	0.0387	0.0287	13.7557	30
31	0.7346	26.5423	377.0394	1.3613	36.1327	0.0377	0.0277	14.2052	31
32	0.7273	27.2696	399.5858	1.3749	37.4941	0.0367	0.0267	14.6532	32
33	0.7201	27.9897	422.6291	1.3887	38.8690	0.0357	0.0257	15.0995	33
34	0.7130	28.7027	446.1572	1.4026	40.2577	0.0348	0.0248	15.5441	34
35	0.7059	29.4086	470.1583	1.4166	41.6603	0.0340	0.0240	15.9871	35
36	0.6989	30.1075	494.6207	1.4308	43.0769	0.0332	0.0232	16.4285	36
37	0.6920	30.7995	519.5329	1.4451	44.5076	0.0325	0.0225	16.8682	37
38	0.6852	31.4847	544.8835	1.4595	45.9527	0.0318	0.0218	17.3063	38
39	0.6784	32.1630	570.6616	1.4741	47.4123	0.0311	0.0211	17.7428	39
40	0.6717	32.8347	596.8561	1.4889	48.8864	0.0305	0.0205	18.1776	40
41	0.6650	33.4997	623.4562	1.5038	50.3752	0.0299	0.0199	18.6108	41
42	0.6584	34.1581	650.4514	1.5188	51.8790	0.0293	0.0193	19.0424	42
43	0.6519	34.8100	677.8312	1.5340	53.3978	0.0287	0.0187	19.4723	43
44	0.6454	35.4555	705.5853	1.5493	54.9318	0.0282	0.0182	19.9006	44
45	0.6391	36.0945	733.7037	1.5648	56.4811	0.0277	0.0177	20.3273	45
46	0.6327	36.7272	762.1765	1.5805	58.0459	0.0272	0.0172	20.7524	46
47	0.6265	37.3537	790.9938	1.5963	59.6263	0.0268	0.0168	21.1758	47
48	0.6203	37.9740	820.1460	1.6122	61.2226	0.0263	0.0163	21.5976	48
49	0.6141	38.5881	849.6237	1.6283	62.8348	0.0259	0.0159	22.0178	49
50	0.6080	39.1961	879.4176	1.6446	64.4632	0.0255	0.0155	22.4363	50
51	0.6020	39.7981	909.5186	1.6611	66.1078	0.0251	0.0151	22.8533	51
52	0.5961	40.3942	939.9175	1.6777	67.7689	0.0248	0.0148	23.2686	52
53	0.5902	40.9844	970.6057	1.6945	69.4466	0.0244	0.0144	23.6823	53
54	0.5843	41.5687	1001.5743	1.7114	71.1410	0.0241	0.0141	24.0945	54
55	0.5785	42.1472	1032.8148	1.7285	72.8525	0.0237	0.0137	24.5049	55
60	0.5504	44.9550	1192.8061	1.8167	81.6697	0.0222	0.0122	26.5333	60
65	0.5237	47.6266	1358.3903	1.9094	90.9366	0.0210	0.0110	28.5217	65
70	0.4983	50.1685	1528.6474	2.0068	100.6763	0.0199	0.0099	30.4703	70
75	0.4741	52.5871	1702.7340	2.1091	110.9128	0.0190	0.0090	32.3793	75
80	0.4511	54.8882	1879.8771	2.2167	121.6715	0.0182	0.0082	34.2492	80
85	0.4292	57.0777	2059.3701	2.3298	132.9790	0.0175	0.0075	36.0801	85
90	0.4084	59.1609	2240.5675	2.4486	144.8633	0.0169	0.0069	37.8724	90
95	0.3886	61.1430	2422.8811	2.5735	157.3538	0.0164	0.0064	39.6265	95
100	0.3697	63.0289	2605.7758	2.7048	170.4814	0.0159	0.0059	41.3426	100

(continued)

APPENDIX 58.B *(continued)*
Cash Flow Equivalent Factors

$i = 1.50\%$

n	P/F	P/A	P/G	F/P	F/A	A/P	A/F	A/G	n
1	0.9852	0.9852	0.0000	1.0150	1.0000	1.0150	1.0000	0.0000	1
2	0.9707	1.9559	0.9707	1.0302	2.0150	0.5113	0.4963	0.4963	2
3	0.9563	2.9122	2.8833	1.0457	3.0452	0.3434	0.3284	0.9901	3
4	0.9422	3.8544	5.7098	1.0614	4.0909	0.2594	0.2444	1.4814	4
5	0.9283	4.7826	9.4229	1.0773	5.1523	0.2091	0.1941	1.9702	5
6	0.9145	5.6972	13.9956	1.0934	6.2296	0.1755	0.1605	2.4566	6
7	0.9010	6.5982	19.4018	1.1098	7.3230	0.1516	0.1366	2.9405	7
8	0.8877	7.4859	25.6157	1.1265	8.4328	0.1336	0.1186	3.4219	8
9	0.8746	8.3605	32.6125	1.1434	9.5593	0.1196	0.1046	3.9008	9
10	0.8617	9.2222	40.3675	1.1605	10.7027	0.1084	0.0934	4.3772	10
11	0.8489	10.0711	48.8568	1.1779	11.8633	0.0993	0.0843	4.8512	11
12	0.8364	10.9075	58.0571	1.1956	13.0412	0.0917	0.0767	5.3227	12
13	0.8240	11.7315	67.9454	1.2136	14.2368	0.0852	0.0702	5.7917	13
14	0.8118	12.5434	78.4994	1.2318	15.4504	0.0797	0.0647	6.2582	14
15	0.7999	13.3432	89.6974	1.2502	16.6821	0.0749	0.0599	6.7223	15
16	0.7880	14.1313	101.5178	1.2690	17.9324	0.0708	0.0558	7.1839	16
17	0.7764	14.9076	113.9400	1.2880	19.2014	0.0671	0.0521	7.6431	17
18	0.7649	15.6726	126.9435	1.3073	20.4894	0.0638	0.0488	8.0997	18
19	0.7536	16.4262	140.5084	1.3270	21.7967	0.0609	0.0459	8.5539	19
20	0.7425	17.1686	154.6154	1.3469	23.1237	0.0582	0.0432	9.0057	20
21	0.7315	17.9001	169.2453	1.3671	24.4705	0.0559	0.0409	9.4550	21
22	0.7207	18.6208	184.3798	1.3876	25.8376	0.0537	0.0387	9.9018	22
23	0.7100	19.3309	200.0006	1.4084	27.2251	0.0517	0.0367	10.3462	23
24	0.6995	20.0304	216.0901	1.4295	28.6335	0.0499	0.0349	10.7881	24
25	0.6892	20.7196	232.6310	1.4509	30.0630	0.0483	0.0333	11.2276	25
26	0.6790	21.3986	249.6065	1.4727	31.5140	0.0467	0.0317	11.6646	26
27	0.6690	22.0676	267.0002	1.4948	32.9867	0.0453	0.0303	12.0992	27
28	0.6591	22.7267	284.7958	1.5172	34.4815	0.0440	0.0290	12.5313	28
29	0.6494	23.3761	302.9779	1.5400	35.9987	0.0428	0.0278	12.9610	29
30	0.6398	24.0158	321.5310	1.5631	37.5387	0.0416	0.0266	13.3883	30
31	0.6303	24.6461	340.4402	1.5865	39.1018	0.0406	0.0256	13.8131	31
32	0.6210	25.2671	359.6910	1.6103	40.6883	0.0396	0.0246	14.2355	32
33	0.6118	25.8790	379.2691	1.6345	42.2986	0.0386	0.0236	14.6555	33
34	0.6028	26.4817	399.1607	1.6590	43.9331	0.0378	0.0228	15.0731	34
35	0.5939	27.0756	419.3521	1.6839	45.5921	0.0369	0.0219	15.4882	35
36	0.5851	27.6607	439.8303	1.7091	47.2760	0.0362	0.0212	15.9009	36
37	0.5764	28.2371	460.5822	1.7348	48.9851	0.0354	0.0204	16.3112	37
38	0.5679	28.8051	481.5954	1.7608	50.7199	0.0347	0.0197	16.7191	38
39	0.5595	29.3646	502.8576	1.7872	52.4807	0.0341	0.0191	17.1246	39
40	0.5513	29.9158	524.3568	1.8140	54.2679	0.0334	0.0184	17.5277	40
41	0.5431	30.4590	546.0814	1.8412	56.0819	0.0328	0.0178	17.9284	41
42	0.5351	30.9941	568.0201	1.8688	57.9231	0.0323	0.0173	18.3267	42
43	0.5272	31.5212	590.1617	1.8969	59.7920	0.0317	0.0167	18.7227	43
44	0.5194	32.0406	612.4955	1.9253	61.6889	0.0312	0.0162	19.1162	44
45	0.5117	32.5523	635.0110	1.9542	63.6142	0.0307	0.0157	19.5074	45
46	0.5042	33.0565	657.6979	1.9835	65.5684	0.0303	0.0153	19.8962	46
47	0.4967	33.5532	680.5462	2.0133	67.5519	0.0298	0.0148	20.2826	47
48	0.4894	34.0426	703.5462	2.0435	69.5652	0.0294	0.0144	20.6667	48
49	0.4821	34.5247	726.6884	2.0741	71.6087	0.0290	0.0140	21.0484	49
50	0.4750	34.9997	749.9636	2.1052	73.6828	0.0286	0.0136	21.4277	50
51	0.4680	35.4677	773.3629	2.1368	75.7881	0.0282	0.0132	21.8047	51
52	0.4611	35.9287	796.8774	2.1689	77.9249	0.0278	0.0128	22.1794	52
53	0.4543	36.3830	820.4986	2.2014	80.0938	0.0275	0.0125	22.5517	53
54	0.4475	36.8305	844.2184	2.2344	82.2952	0.0272	0.0122	22.9217	54
55	0.4409	37.2715	868.0285	2.2679	84.5296	0.0268	0.0118	23.2894	55
60	0.4093	39.3803	988.1674	2.4432	96.2147	0.0254	0.0104	25.0930	60
65	0.3799	41.3378	1109.4752	2.6320	108.8028	0.0242	0.0092	26.8393	65
70	0.3527	43.1549	1231.1658	2.8355	122.3638	0.0232	0.0082	28.5290	70
75	0.3274	44.8416	1352.5600	3.0546	136.9728	0.0223	0.0073	30.1631	75
80	0.3039	46.4073	1473.0741	3.2907	152.7109	0.0215	0.0065	31.7423	80
85	0.2821	47.8607	1592.2095	3.5450	169.6652	0.0209	0.0059	33.2676	85
90	0.2619	49.2099	1709.5439	3.8189	187.9299	0.0203	0.0053	34.7399	90
95	0.2431	50.4622	1824.7224	4.1141	207.6061	0.0198	0.0048	36.1602	95
100	0.2256	51.6247	1937.4506	4.4320	228.8030	0.0194	0.0044	37.5295	100

(continued)

APPENDIX 58.B *(continued)*
Cash Flow Equivalent Factors

$i = 2.00\%$

n	P/F	P/A	P/G	F/P	F/A	A/P	A/F	A/G	n
1	0.9804	0.9804	0.0000	1.0200	1.0000	1.0200	1.0000	0.0000	1
2	0.9612	1.9416	0.9612	1.0404	2.0200	0.5150	0.4950	0.4950	2
3	0.9423	2.8839	2.8458	1.0612	3.0604	0.3468	0.3268	0.9868	3
4	0.9238	3.8077	5.6173	1.0824	4.1216	0.2626	0.2426	1.4752	4
5	0.9057	4.7135	9.2403	1.1041	5.2040	0.2122	0.1922	1.9604	5
6	0.8880	5.6014	13.6801	1.1262	6.3081	0.1785	0.1585	2.4423	6
7	0.8706	6.4720	18.9035	1.1487	7.4343	0.1545	0.1345	2.9208	7
8	0.8535	7.3255	24.8779	1.1717	8.5830	0.1365	0.1165	3.3961	8
9	0.8368	8.1622	31.5720	1.1951	9.7546	0.1225	0.1025	3.8681	9
10	0.8203	8.9826	38.9551	1.2190	10.9497	0.1113	0.0913	4.3367	10
11	0.8043	9.7868	46.9977	1.2434	12.1687	0.1022	0.0822	4.8021	11
12	0.7885	10.5753	55.6712	1.2682	13.4121	0.0946	0.0746	5.2642	12
13	0.7730	11.3484	64.9475	1.2936	14.6803	0.0881	0.0681	5.7231	13
14	0.7579	12.1062	74.7999	1.3195	15.9739	0.0826	0.0626	6.1786	14
15	0.7430	12.8493	85.2021	1.3459	17.2934	0.0778	0.0578	6.6309	15
16	0.7284	13.5777	96.1288	1.3728	18.6393	0.0737	0.0537	7.0799	16
17	0.7142	14.2919	107.5554	1.4002	20.0121	0.0700	0.0500	7.5256	17
18	0.7002	14.9920	119.4581	1.4282	21.4123	0.0667	0.0467	7.9681	18
19	0.6864	15.6785	131.8139	1.4568	22.8406	0.0638	0.0438	8.4073	19
20	0.6730	16.3514	144.6003	1.4859	24.2974	0.0612	0.0412	8.8433	20
21	0.6598	17.0112	157.7959	1.5157	25.7833	0.0588	0.0388	9.2760	21
22	0.6468	17.6580	171.3795	1.5460	27.2990	0.0566	0.0366	9.7055	22
23	0.6342	18.2922	185.3309	1.5769	28.8450	0.0547	0.0347	10.1317	23
24	0.6217	18.9139	199.6305	1.6084	30.4219	0.0529	0.0329	10.5547	24
25	0.6095	19.5235	214.2592	1.6406	32.0303	0.0512	0.0312	10.9745	25
26	0.5976	20.1210	229.1987	1.6734	33.6709	0.0497	0.0297	11.3910	26
27	0.5859	20.7069	244.4311	1.7069	35.3443	0.0483	0.0283	11.8043	27
28	0.5744	21.2813	259.9392	1.7410	37.0512	0.0470	0.0270	12.2145	28
29	0.5631	21.8444	275.7064	1.7758	38.7922	0.0458	0.0258	12.6214	29
30	0.5521	22.3965	291.7164	1.8114	40.5681	0.0446	0.0246	13.0251	30
31	0.5412	22.9377	307.9538	1.8476	42.3794	0.0436	0.0236	13.4257	31
32	0.5306	23.4683	324.4035	1.8845	44.2270	0.0426	0.0226	13.8230	32
33	0.5202	23.9886	341.0508	1.9222	46.1116	0.0417	0.0217	14.2172	33
34	0.5100	24.4986	357.8817	1.9607	48.0338	0.0408	0.0208	14.6083	34
35	0.5000	24.9986	374.8826	1.9999	49.9945	0.0400	0.0200	14.9961	35
36	0.4902	25.4888	392.0405	2.0399	51.9944	0.0392	0.0192	15.3809	36
37	0.4806	25.9695	409.3424	2.0807	54.0343	0.0385	0.0185	15.7625	37
38	0.4712	26.4406	426.7764	2.1223	56.1149	0.0378	0.0178	16.1409	38
39	0.4619	26.9026	444.3304	2.1647	58.2372	0.0372	0.0172	16.5163	39
40	0.4529	27.3555	461.9931	2.2080	60.4020	0.0366	0.0166	16.8885	40
41	0.4440	27.7995	479.7535	2.2522	62.6100	0.0360	0.0160	17.2576	41
42	0.4353	28.2348	497.6010	2.2972	64.8622	0.0354	0.0154	17.6237	42
43	0.4268	28.6616	515.5253	2.3432	67.1595	0.0349	0.0149	17.9866	43
44	0.4184	29.0800	533.5165	2.3901	69.5027	0.0344	0.0144	18.3465	44
45	0.4102	29.4902	551.5652	2.4379	71.8927	0.0339	0.0139	18.7034	45
46	0.4022	29.8923	569.6621	2.4866	74.3306	0.0335	0.0135	19.0571	46
47	0.3943	30.2866	587.7985	2.5363	76.8172	0.0330	0.0130	19.4079	47
48	0.3865	30.6731	605.9657	2.5871	79.3535	0.0326	0.0126	19.7556	48
49	0.3790	31.0521	624.1557	2.6388	81.9406	0.0322	0.0122	20.1003	49
50	0.3715	31.4236	642.3606	2.6916	84.5794	0.0318	0.0118	20.4420	50
51	0.3642	31.7878	660.5727	2.7454	87.2710	0.0315	0.0115	20.7807	51
52	0.3571	32.1449	678.7849	2.8003	90.0164	0.0311	0.0111	21.1164	52
53	0.3501	32.4950	696.9900	2.8563	92.8167	0.0308	0.0108	21.4491	53
54	0.3432	32.8383	715.1815	2.9135	95.6731	0.0305	0.0105	21.7789	54
55	0.3365	33.1748	733.3527	2.9717	98.5865	0.0301	0.0101	22.1057	55
60	0.3048	34.7609	823.6975	3.2810	114.0515	0.0288	0.0088	23.6961	60
65	0.2761	36.1975	912.7085	3.6225	131.1262	0.0276	0.0076	25.2147	65
70	0.2500	37.4986	999.8343	3.9996	149.9779	0.0267	0.0067	26.6632	70
75	0.2265	38.6771	1084.6393	4.4158	170.7918	0.0259	0.0059	28.0434	75
80	0.2051	39.7445	1166.7868	4.8754	193.7720	0.0252	0.0052	29.3572	80
85	0.1858	40.7113	1246.0241	5.3829	219.1439	0.0246	0.0046	30.6064	85
90	0.1683	41.5869	1322.1701	5.9431	247.1567	0.0240	0.0040	31.7929	90
95	0.1524	42.3800	1395.1033	6.5617	278.0850	0.0236	0.0036	32.9189	95
100	0.1380	43.0984	1464.7527	7.2446	312.2323	0.0232	0.0032	33.9863	100

(continued)

APPENDIX 58.B *(continued)*
Cash Flow Equivalent Factors

$i = 3.00\%$

n	P/F	P/A	P/G	F/P	F/A	A/P	A/F	A/G	n
1	0.9709	0.9709	0.0000	1.0300	1.0000	1.0300	1.0000	0.0000	1
2	0.9426	1.9135	0.9426	1.0609	2.0300	0.5226	0.4926	0.4926	2
3	0.9151	2.8286	2.7729	1.0927	3.0909	0.3535	0.3235	0.9803	3
4	0.8885	3.7171	5.4383	1.1255	4.1836	0.2690	0.2390	1.4631	4
5	0.8626	4.5797	8.8888	1.1593	5.3091	0.2184	0.1884	1.9409	5
6	0.8375	5.4172	13.0762	1.1941	6.4684	0.1846	0.1546	2.4138	6
7	0.8131	6.2303	17.9547	1.2299	7.6625	0.1605	0.1305	2.8819	7
8	0.7894	7.0197	23.4806	1.2668	8.8923	0.1425	0.1125	3.3450	8
9	0.7664	7.7861	29.6119	1.3048	10.1591	0.1284	0.0984	3.8032	9
10	0.7441	8.5302	36.3088	1.3439	11.4639	0.1172	0.0872	4.2565	10
11	0.7224	9.2526	43.5330	1.3842	12.8078	0.1081	0.0781	4.7049	11
12	0.7014	9.9540	51.2482	1.4258	14.1920	0.1005	0.0705	5.1485	12
13	0.6810	10.6350	59.4196	1.4685	15.6178	0.0940	0.0640	5.5872	13
14	0.6611	11.2961	68.0141	1.5126	17.0863	0.0885	0.0585	6.0210	14
15	0.6419	11.9379	77.0002	1.5580	18.5989	0.0838	0.0538	6.4500	15
16	0.6232	12.5611	86.3477	1.6047	20.1569	0.0796	0.0496	6.8742	16
17	0.6050	13.1661	96.0280	1.6528	21.7616	0.0760	0.0460	7.2936	17
18	0.5874	13.7535	106.0137	1.7024	23.4144	0.0727	0.0427	7.7081	18
19	0.5703	14.3238	116.2788	1.7535	25.1169	0.0698	0.0398	8.1179	19
20	0.5537	14.8775	126.7987	1.8061	26.8704	0.0672	0.0372	8.5229	20
21	0.5375	15.4150	137.5496	1.8603	28.6765	0.0649	0.0349	8.9231	21
22	0.5219	15.9369	148.5094	1.9161	30.5368	0.0627	0.0327	9.3186	22
23	0.5067	16.4436	159.6566	1.9736	32.4529	0.0608	0.0308	9.7093	23
24	0.4919	16.9355	170.9711	2.0328	34.4265	0.0590	0.0290	10.0954	24
25	0.4776	17.4131	182.4336	2.0938	36.4593	0.0574	0.0274	10.4768	25
26	0.4637	17.8768	194.0260	2.1566	38.5530	0.0559	0.0259	10.8535	26
27	0.4502	18.3270	205.7309	2.2213	40.7096	0.0546	0.0246	11.2255	27
28	0.4371	18.7641	217.5320	2.2879	42.9309	0.0533	0.0233	11.5930	28
29	0.4243	19.1885	229.4137	2.3566	45.2189	0.0521	0.0221	11.9558	29
30	0.4120	19.6004	241.3613	2.4273	47.5754	0.0510	0.0210	12.3141	30
31	0.4000	20.0004	253.3609	2.5001	50.0027	0.0500	0.0200	12.6678	31
32	0.3883	20.3888	265.3993	2.5751	52.5028	0.0490	0.0190	13.0169	32
33	0.3770	20.7658	277.4642	2.6523	55.0778	0.0482	0.0182	13.3616	33
34	0.3660	21.1318	289.5437	2.7319	57.7302	0.0473	0.0173	13.7018	34
35	0.3554	21.4872	301.6267	2.8139	60.4621	0.0465	0.0165	14.0375	35
36	0.3450	21.8323	313.7028	2.8983	63.2759	0.0458	0.0158	14.3688	36
37	0.3350	22.1672	325.7622	2.9852	66.1742	0.0451	0.0151	14.6957	37
38	0.3252	22.4925	337.7956	3.0748	69.1594	0.0445	0.0145	15.0182	38
39	0.3158	22.8082	349.7942	3.1670	72.2342	0.0438	0.0138	15.3363	39
40	0.3066	23.1148	361.7499	3.2620	75.4013	0.0433	0.0133	15.6502	40
41	0.2976	23.4124	373.6551	3.3599	78.6633	0.0427	0.0127	15.9597	41
42	0.2890	23.7014	385.5024	3.4607	82.0232	0.0422	0.0122	16.2650	42
43	0.2805	23.9819	397.2852	3.5645	85.4839	0.0417	0.0117	16.5660	43
44	0.2724	24.2543	408.9972	3.6715	89.0484	0.0412	0.0112	16.8629	44
45	0.2644	24.5187	420.6325	3.7816	92.7199	0.0408	0.0108	17.1556	45
46	0.2567	24.7754	432.1856	3.8950	96.5015	0.0404	0.0104	17.4441	46
47	0.2493	25.0247	443.6515	4.0119	100.3965	0.0400	0.0100	17.7285	47
48	0.2420	25.2667	455.0255	4.1323	104.4084	0.0396	0.0096	18.0089	48
49	0.2350	25.5017	466.3031	4.2562	108.5406	0.0392	0.0092	18.2852	49
50	0.2281	25.7298	477.4803	4.3839	112.7969	0.0389	0.0089	18.5575	50
51	0.2215	25.9512	488.5535	4.5154	117.1808	0.0385	0.0085	18.8258	51
52	0.2150	26.1662	499.5191	4.6509	121.6962	0.0382	0.0082	19.0902	52
53	0.2088	26.3750	510.3742	4.7904	126.3471	0.0379	0.0079	19.3507	53
54	0.2027	26.5777	521.1157	4.9341	131.1375	0.0376	0.0076	19.6073	54
55	0.1968	26.7744	531.7411	5.0821	136.0716	0.0373	0.0073	19.8600	55
60	0.1697	27.6756	583.0526	5.8916	163.0534	0.0361	0.0061	21.0674	60
65	0.1464	28.4529	631.2010	6.8300	194.3328	0.0351	0.0051	22.1841	65
70	0.1263	29.1234	676.0869	7.9178	230.5941	0.0343	0.0043	23.2145	70
75	0.1089	29.7018	717.6978	9.1789	272.6309	0.0337	0.0037	24.1634	75
80	0.0940	30.2008	756.0865	10.6409	321.3630	0.0331	0.0031	25.0353	80
85	0.0811	30.6312	791.3529	12.3357	377.8570	0.0326	0.0026	25.8349	85
90	0.0699	31.0024	823.6302	14.3005	443.3489	0.0323	0.0023	26.5667	90
95	0.0603	31.3227	853.0742	16.5782	519.2720	0.0319	0.0019	27.2351	95
100	0.0520	31.5989	879.8540	19.2186	607.2877	0.0316	0.0016	27.8444	100

(continued)

APPENDIX 58.B *(continued)*
Cash Flow Equivalent Factors

$i = 4.00\%$

n	P/F	P/A	P/G	F/P	F/A	A/P	A/F	A/G	n
1	0.9615	0.9615	0.0000	1.0400	1.0000	1.0400	1.0000	0.0000	1
2	0.9246	1.8861	0.9246	1.0816	2.0400	0.5302	0.4902	0.4902	2
3	0.8890	2.7751	2.7025	1.1249	3.1216	0.3603	0.3203	0.9739	3
4	0.8548	3.6299	5.2670	1.1699	4.2465	0.2755	0.2355	1.4510	4
5	0.8219	4.4518	8.5547	1.2167	5.4163	0.2246	0.1846	1.9216	5
6	0.7903	5.2421	12.5062	1.2653	6.6330	0.1908	0.1508	2.3857	6
7	0.7599	6.0021	17.0657	1.3159	7.8983	0.1666	0.1266	2.8433	7
8	0.7307	6.7327	22.1806	1.3686	9.2142	0.1485	0.1085	3.2944	8
9	0.7026	7.4353	27.8013	1.4233	10.5828	0.1345	0.0945	3.7391	9
10	0.6756	8.1109	33.8814	1.4802	12.0061	0.1233	0.0833	4.1773	10
11	0.6496	8.7605	40.3772	1.5395	13.4864	0.1141	0.0741	4.6090	11
12	0.6246	9.3851	47.2477	1.6010	15.0258	0.1066	0.0666	5.0343	12
13	0.6006	9.9856	54.4546	1.6651	16.6268	0.1001	0.0601	5.4533	13
14	0.5775	10.5631	61.9618	1.7317	18.2919	0.0947	0.0547	5.8659	14
15	0.5553	11.1184	69.7355	1.8009	20.0236	0.0899	0.0499	6.2721	15
16	0.5339	11.6523	77.7441	1.8730	21.8245	0.0858	0.0458	6.6720	16
17	0.5134	12.1657	85.9581	1.9479	23.6975	0.0822	0.0422	7.0656	17
18	0.4936	12.6593	94.3498	2.0258	25.6454	0.0790	0.0390	7.4530	18
19	0.4746	13.1339	102.8933	2.1068	27.6712	0.0761	0.0361	7.8342	19
20	0.4564	13.5903	111.5647	2.1911	29.7781	0.0736	0.0336	8.2091	20
21	0.4388	14.0292	120.3414	2.2788	31.9692	0.0713	0.0313	8.5779	21
22	0.4220	14.4511	129.2024	2.3699	34.2480	0.0692	0.0292	8.9407	22
23	0.4057	14.8568	138.1284	2.4647	36.6179	0.0673	0.0273	9.2973	23
24	0.3901	15.2470	147.1012	2.5633	39.0826	0.0656	0.0256	9.6479	24
25	0.3751	15.6221	156.1040	2.6658	41.6459	0.0640	0.0240	9.9925	25
26	0.3607	15.9828	165.1212	2.7725	44.3117	0.0626	0.0226	10.3312	26
27	0.3468	16.3296	174.1385	2.8834	47.0842	0.0612	0.0212	10.6640	27
28	0.3335	16.6631	183.1424	2.9987	49.9676	0.0600	0.0200	10.9909	28
29	0.3207	16.9837	192.1206	3.1187	52.9663	0.0589	0.0189	11.3120	29
30	0.3083	17.2920	201.0618	3.2434	56.0849	0.0578	0.0178	11.6274	30
31	0.2965	17.5885	209.9556	3.3731	59.3283	0.0569	0.0169	11.9371	31
32	0.2851	17.8736	218.7924	3.5081	62.7015	0.0559	0.0159	12.2411	32
33	0.2741	18.1476	227.5634	3.6484	66.2095	0.0551	0.0151	12.5396	33
34	0.2636	18.4112	236.2607	3.7943	69.8579	0.0543	0.0143	12.8324	34
35	0.2534	18.6646	244.8768	3.9461	73.6522	0.0536	0.0136	13.1198	35
36	0.2437	18.9083	253.4052	4.1039	77.5983	0.0529	0.0129	13.4018	36
37	0.2343	19.1426	261.8399	4.2681	81.7022	0.0522	0.0122	13.6784	37
38	0.2253	19.3679	270.1754	4.4388	85.9703	0.0516	0.0116	13.9497	38
39	0.2166	19.5845	278.4070	4.6164	90.4091	0.0511	0.0111	14.2157	39
40	0.2083	19.7928	286.5303	4.8010	95.0255	0.0505	0.0105	14.4765	40
41	0.2003	19.9931	294.5414	4.9931	99.8265	0.0500	0.0100	14.7322	41
42	0.1926	20.1856	302.4370	5.1928	104.8196	0.0495	0.0095	14.9828	42
43	0.1852	20.3708	310.2141	5.4005	110.0124	0.0491	0.0091	15.2284	43
44	0.1780	20.5488	317.8700	5.6165	115.4129	0.0487	0.0087	15.4690	44
45	0.1712	20.7200	325.4028	5.8412	121.0294	0.0483	0.0083	15.7047	45
46	0.1646	20.8847	332.8104	6.0748	126.8706	0.0479	0.0079	15.9356	46
47	0.1583	21.0429	340.0914	6.3178	132.9454	0.0475	0.0075	16.1618	47
48	0.1522	21.1951	347.2446	6.5705	139.2632	0.0472	0.0072	16.3832	48
49	0.1463	21.3415	354.2689	6.8333	145.8337	0.0469	0.0069	16.6000	49
50	0.1407	21.4822	361.1638	7.1067	152.6671	0.0466	0.0066	16.8122	50
51	0.1353	21.6175	367.9289	7.3910	159.7738	0.0463	0.0063	17.0200	51
52	0.1301	21.7476	374.5638	7.6866	167.1647	0.0460	0.0060	17.2232	52
53	0.1251	21.8727	381.0686	7.9941	174.8513	0.0457	0.0057	17.4221	53
54	0.1203	21.9930	387.4436	8.3138	182.8454	0.0455	0.0055	17.6167	54
55	0.1157	22.1086	393.6890	8.6464	191.1592	0.0452	0.0052	17.8070	55
60	0.0951	22.6235	422.9966	10.5196	237.9907	0.0442	0.0042	18.6972	60
65	0.0781	23.0467	449.2014	12.7987	294.9684	0.0434	0.0034	19.4909	65
70	0.0642	23.3945	472.4789	15.5716	364.2905	0.0427	0.0027	20.1961	70
75	0.0528	23.6804	493.0408	18.9453	448.6314	0.0422	0.0022	20.8206	75
80	0.0434	23.9154	511.1161	23.0498	551.2450	0.0418	0.0018	21.3718	80
85	0.0357	24.1085	526.9384	28.0436	676.0901	0.0415	0.0015	21.8569	85
90	0.0293	24.2673	540.7369	34.1193	827.9833	0.0412	0.0012	22.2826	90
95	0.0241	24.3978	552.7307	41.5114	1012.7846	0.0410	0.0010	22.6550	95
100	0.0198	24.5050	563.1249	50.5049	1237.6237	0.0408	0.0008	22.9800	100

(continued)

APPENDIX 58.B *(continued)*
Cash Flow Equivalent Factors

$i = 5.00\%$

n	P/F	P/A	P/G	F/P	F/A	A/P	A/F	A/G	n
1	0.9524	0.9524	0.0000	1.0500	1.0000	1.0500	1.0000	0.0000	1
2	0.9070	1.8594	0.9070	1.1025	2.0500	0.5378	0.4878	0.4878	2
3	0.8638	2.7232	2.6347	1.1576	3.1525	0.3672	0.3172	0.9675	3
4	0.8227	3.5460	5.1028	1.2155	4.3101	0.2820	0.2320	1.4391	4
5	0.7835	4.3295	8.2369	1.2763	5.5256	0.2310	0.1810	1.9025	5
6	0.7462	5.0757	11.9680	1.3401	6.8019	0.1970	0.1470	2.3579	6
7	0.7107	5.7864	16.2321	1.4071	8.1420	0.1728	0.1228	2.8052	7
8	0.6768	6.4632	20.9700	1.4775	9.5491	0.1547	0.1047	3.2445	8
9	0.6446	7.1078	26.1268	1.5513	11.0266	0.1407	0.0907	3.6758	9
10	0.6139	7.7217	31.6520	1.6289	12.5779	0.1295	0.0795	4.0991	10
11	0.5847	8.3064	37.4988	1.7103	14.2068	0.1204	0.0704	4.5144	11
12	0.5568	8.8633	43.6241	1.7959	15.9171	0.1128	0.0628	4.9219	12
13	0.5303	9.3936	49.9879	1.8856	17.7130	0.1065	0.0565	5.3215	13
14	0.5051	9.8986	56.5538	1.9799	19.5986	0.1010	0.0510	5.7133	14
15	0.4810	10.3797	63.2880	2.0789	21.5786	0.0963	0.0463	6.0973	15
16	0.4581	10.8378	70.1597	2.1829	23.6575	0.0923	0.0423	6.4736	16
17	0.4363	11.2741	77.1405	2.2920	25.8404	0.0887	0.0387	6.8423	17
18	0.4155	11.6896	84.2043	2.4066	28.1324	0.0855	0.0355	7.2034	18
19	0.3957	12.0853	91.3275	2.5270	30.5390	0.0827	0.0327	7.5569	19
20	0.3769	12.4622	98.4884	2.6533	33.0660	0.0802	0.0302	7.9030	20
21	0.3589	12.8212	105.6673	2.7860	35.7193	0.0780	0.0280	8.2416	21
22	0.3418	13.1630	112.8461	2.9253	38.5052	0.0760	0.0260	8.5730	22
23	0.3256	13.4886	120.0087	3.0715	41.4305	0.0741	0.0241	8.8971	23
24	0.3101	13.7986	127.1402	3.2251	44.5020	0.0725	0.0225	9.2140	24
25	0.2953	14.0939	134.2275	3.3864	47.7271	0.0710	0.0210	9.5238	25
26	0.2812	14.3752	141.2585	3.5557	51.1135	0.0696	0.0196	9.8266	26
27	0.2678	14.6430	148.2226	3.7335	54.6691	0.0683	0.0183	10.1224	27
28	0.2551	14.8981	155.1101	3.9201	58.4026	0.0671	0.0171	10.4114	28
29	0.2429	15.1411	161.9126	4.1161	62.3227	0.0660	0.0160	10.6936	29
30	0.2314	15.3725	168.6226	4.3219	66.4388	0.0651	0.0151	10.9691	30
31	0.2204	15.5928	175.2333	4.5380	70.7608	0.0641	0.0141	11.2381	31
32	0.2099	15.8027	181.7392	4.7649	75.2988	0.0633	0.0133	11.5005	32
33	0.1999	16.0025	188.1351	5.0032	80.0638	0.0625	0.0125	11.7566	33
34	0.1904	16.1929	194.4168	5.2533	85.0670	0.0618	0.0118	12.0063	34
35	0.1813	16.3742	200.5807	5.5160	90.3203	0.0611	0.0111	12.2498	35
36	0.1727	16.5469	206.6237	5.7918	95.8363	0.0604	0.0104	12.4872	36
37	0.1644	16.7113	212.5434	6.0814	101.6281	0.0598	0.0098	12.7186	37
38	0.1566	16.8679	218.3378	6.3855	107.7095	0.0593	0.0093	12.9440	38
39	0.1491	17.0170	224.0054	6.7048	114.0950	0.0588	0.0088	13.1636	39
40	0.1420	17.1591	229.5452	7.0400	120.7998	0.0583	0.0083	13.3775	40
41	0.1353	17.2944	234.9564	7.3920	127.8398	0.0578	0.0078	13.5857	41
42	0.1288	17.4232	240.2389	7.7616	135.2318	0.0574	0.0074	13.7884	42
43	0.1227	17.5459	245.3925	8.1497	142.9933	0.0570	0.0070	13.9857	43
44	0.1169	17.6628	250.4175	8.5572	151.1430	0.0566	0.0066	14.1777	44
45	0.1113	17.7741	255.3145	8.9850	159.7002	0.0563	0.0063	14.3644	45
46	0.1060	17.8801	260.0844	9.4343	168.6852	0.0559	0.0059	14.5461	46
47	0.1009	17.9810	264.7281	9.9060	178.1194	0.0556	0.0056	14.7226	47
48	0.0961	18.0772	269.2467	10.4013	188.0254	0.0553	0.0053	14.8943	48
49	0.0916	18.1687	273.6418	10.9213	198.4267	0.0550	0.0050	15.0611	49
50	0.0872	18.2559	277.9148	11.4674	209.3480	0.0548	0.0048	15.2233	50
51	0.0831	18.3390	282.0673	12.0408	220.8154	0.0545	0.0045	15.3808	51
52	0.0791	18.4181	286.1013	12.6428	232.8562	0.0543	0.0043	15.5337	52
53	0.0753	18.4934	290.0184	13.2749	245.4990	0.0541	0.0041	15.6823	53
54	0.0717	18.5651	293.8208	13.9387	258.7739	0.0539	0.0039	15.8265	54
55	0.0683	18.6335	297.5104	14.6356	272.7126	0.0537	0.0037	15.9664	55
60	0.0535	18.9293	314.3432	18.6792	353.5837	0.0528	0.0028	16.6062	60
65	0.0419	19.1611	328.6910	23.8399	456.7980	0.0522	0.0022	17.1541	65
70	0.0329	19.3427	340.8409	30.4264	588.5285	0.0517	0.0017	17.6212	70
75	0.0258	19.4850	351.0721	38.8327	756.6537	0.0513	0.0013	18.0176	75
80	0.0202	19.5965	359.6460	49.5614	971.2288	0.0510	0.0010	18.3526	80
85	0.0158	19.6838	366.8007	63.2544	1245.0871	0.0508	0.0008	18.6346	85
90	0.0124	19.7523	372.7488	80.7304	1597.6073	0.0506	0.0006	18.8712	90
95	0.0097	19.8059	377.6774	103.0347	2040.6935	0.0505	0.0005	19.0689	95
100	0.0076	19.8479	381.7492	131.5013	2610.0252	0.0504	0.0004	19.2337	100

Support Material

(continued)

APPENDIX 58.B *(continued)*
Cash Flow Equivalent Factors

$i = 6.00\%$

n	P/F	P/A	P/G	F/P	F/A	A/P	A/F	A/G	n
1	0.9434	0.9434	0.0000	1.0600	1.0000	1.0600	1.0000	0.0000	1
2	0.8900	1.8334	0.8900	1.1236	2.0600	0.5454	0.4854	0.4854	2
3	0.8396	2.6730	2.5692	1.1910	3.1836	0.3741	0.3141	0.9612	3
4	0.7921	3.4651	4.9455	1.2625	4.3746	0.2886	0.2286	1.4272	4
5	0.7473	4.2124	7.9345	1.3382	5.6371	0.2374	0.1774	1.8836	5
6	0.7050	4.9173	11.4594	1.4185	6.9753	0.2034	0.1434	2.3304	6
7	0.6651	5.5824	15.4497	1.5036	8.3938	0.1791	0.1191	2.7676	7
8	0.6274	6.2098	19.8416	1.5938	9.8975	0.1610	0.1010	3.1952	8
9	0.5919	6.8017	24.5768	1.6895	11.4913	0.1470	0.0870	3.6133	9
10	0.5584	7.3601	29.6023	1.7908	13.1808	0.1359	0.0759	4.0220	10
11	0.5268	7.8869	34.8702	1.8983	14.9716	0.1268	0.0668	4.4213	11
12	0.4970	8.3838	40.3369	2.0122	16.8699	0.1193	0.0593	4.8113	12
13	0.4688	8.8527	45.9629	2.1329	18.8821	0.1130	0.0530	5.1920	13
14	0.4423	9.2950	51.7128	2.2609	21.0151	0.1076	0.0476	5.5635	14
15	0.4173	9.7122	57.5546	2.3966	23.2760	0.1030	0.0430	5.9260	15
16	0.3936	10.1059	63.4592	2.5404	25.6725	0.0990	0.0390	6.2794	16
17	0.3714	10.4773	69.4011	2.6928	28.2129	0.0954	0.0354	6.6240	17
18	0.3503	10.8276	75.3569	2.8543	30.9057	0.0924	0.0324	6.9597	18
19	0.3305	11.1581	81.3062	3.0256	33.7600	0.0896	0.0296	7.2867	19
20	0.3118	11.4699	87.2304	3.2071	36.7856	0.0872	0.0272	7.6051	20
21	0.2942	11.7641	93.1136	3.3996	39.9927	0.0850	0.0250	7.9151	21
22	0.2775	12.0416	98.9412	3.6035	43.3923	0.0830	0.0230	8.2166	22
23	0.2618	12.3034	104.7007	3.8197	46.9958	0.0813	0.0213	8.5099	23
24	0.2470	12.5504	110.3812	4.0489	50.8156	0.0797	0.0197	8.7951	24
25	0.2330	12.7834	115.9732	4.2919	54.8645	0.0782	0.0182	9.0722	25
26	0.2198	13.0032	121.4684	4.5494	59.1564	0.0769	0.0169	9.3414	26
27	0.2074	13.2105	126.8600	4.8223	63.7058	0.0757	0.0157	9.6029	27
28	0.1956	13.4062	132.1420	5.1117	68.5281	0.0746	0.0146	9.8568	28
29	0.1846	13.5907	137.3096	5.4184	73.6398	0.0736	0.0136	10.1032	29
30	0.1741	13.7648	142.3588	5.7435	79.0582	0.0726	0.0126	10.3422	30
31	0.1643	13.9291	147.2864	6.0881	84.8017	0.0718	0.0118	10.5740	31
32	0.1550	14.0840	152.0901	6.4534	90.8898	0.0710	0.0110	10.7988	32
33	0.1462	14.2302	156.7681	6.8406	97.3432	0.0703	0.0103	11.0166	33
34	0.1379	14.3681	161.3192	7.2510	104.1838	0.0696	0.0096	11.2276	34
35	0.1301	14.4982	165.7427	7.6861	111.4348	0.0690	0.0090	11.4319	35
36	0.1227	14.6210	170.0387	8.1473	119.1209	0.0684	0.0084	11.6298	36
37	0.1158	14.7368	174.2072	8.6361	127.2681	0.0679	0.0079	11.8213	37
38	0.1092	14.8460	178.2490	9.1543	135.9042	0.0674	0.0074	12.0065	38
39	0.1031	14.9491	182.1652	9.7035	145.0585	0.0669	0.0069	12.1857	39
40	0.0972	15.0463	185.9568	10.2857	154.7620	0.0665	0.0065	12.3590	40
41	0.0917	15.1380	189.6256	10.9029	165.0477	0.0661	0.0061	12.5264	41
42	0.0865	15.2245	193.1732	11.5570	175.9505	0.0657	0.0057	12.6883	42
43	0.0816	15.3062	196.6017	12.2505	187.5076	0.0653	0.0053	12.8446	43
44	0.0770	15.3832	199.9130	12.9855	199.7580	0.0650	0.0050	12.9956	44
45	0.0727	15.4558	203.1096	13.7646	212.7435	0.0647	0.0047	13.1413	45
46	0.0685	15.5244	206.1938	14.5905	226.5081	0.0644	0.0044	13.2819	46
47	0.0647	15.5890	209.1681	15.4659	241.0986	0.0641	0.0041	13.4177	47
48	0.0610	15.6500	212.0351	16.3939	256.5645	0.0639	0.0039	13.5485	48
49	0.0575	15.7076	214.7972	17.3775	272.9584	0.0637	0.0037	13.6748	49
50	0.0543	15.7619	217.4574	18.4202	290.3359	0.0634	0.0034	13.7964	50
51	0.0512	15.8131	220.0181	19.5254	308.7561	0.0632	0.0032	13.9137	51
52	0.0483	15.8614	222.4823	20.6969	328.2814	0.0630	0.0030	14.0267	52
53	0.0456	15.9070	224.8525	21.9387	348.9783	0.0629	0.0029	14.1355	53
54	0.0430	15.9500	227.1316	23.2550	370.9170	0.0627	0.0027	14.2402	54
55	0.0406	15.9905	229.3222	24.6503	394.1720	0.0625	0.0025	14.3411	55
60	0.0303	16.1614	239.0428	32.9877	533.1282	0.0619	0.0019	14.7909	60
65	0.0227	16.2891	246.9450	44.1450	719.0829	0.0614	0.0014	15.1601	65
70	0.0169	16.3845	253.3271	59.0759	967.9322	0.0610	0.0010	15.4613	70
75	0.0126	16.4558	258.4527	79.0569	1300.9487	0.0608	0.0008	15.7058	75
80	0.0095	16.5091	262.5493	105.7960	1746.5999	0.0606	0.0006	15.9033	80
85	0.0071	16.5489	265.8096	141.5789	2342.9817	0.0604	0.0004	16.0620	85
90	0.0053	16.5787	268.3946	189.4645	3141.0752	0.0603	0.0003	16.1891	90
95	0.0039	16.6009	270.4375	253.5463	4209.1042	0.0602	0.0002	16.2905	95
100	0.0029	16.6175	272.0471	339.3021	5638.3681	0.0602	0.0002	16.3711	100

(continued)

APPENDIX 58.B *(continued)*
Cash Flow Equivalent Factors

$i = 7.00\%$

n	P/F	P/A	P/G	F/P	F/A	A/P	A/F	A/G	n
1	0.9346	0.9346	0.0000	1.0700	1.0000	1.0700	1.0000	0.0000	1
2	0.8734	1.8080	0.8734	1.1449	2.0700	0.5531	0.4831	0.4831	2
3	0.8163	2.6243	2.5060	1.2250	3.2149	0.3811	0.3111	0.9549	3
4	0.7629	3.3872	4.7947	1.3108	4.4399	0.2952	0.2252	1.4155	4
5	0.7130	4.1002	7.6467	1.4026	5.7507	0.2439	0.1739	1.8650	5
6	0.6663	4.7665	10.9784	1.5007	7.1533	0.2098	0.1398	2.3032	6
7	0.6227	5.3893	14.7149	1.6058	8.6540	0.1856	0.1156	2.7304	7
8	0.5820	5.9713	18.7889	1.7182	10.2598	0.1675	0.0975	3.1465	8
9	0.5439	6.5152	23.1404	1.8385	11.9780	0.1535	0.0835	3.5517	9
10	0.5083	7.0236	27.7156	1.9672	13.8164	0.1424	0.0724	3.9461	10
11	0.4751	7.4987	32.4665	2.1049	15.7836	0.1334	0.0634	4.3296	11
12	0.4440	7.9427	37.3506	2.2522	17.8885	0.1259	0.0559	4.7025	12
13	0.4150	8.3577	42.3302	2.4098	20.1406	0.1197	0.0497	5.0648	13
14	0.3878	8.7455	47.3718	2.5785	22.5505	0.1143	0.0443	5.4167	14
15	0.3624	9.1079	52.4461	2.7590	25.1290	0.1098	0.0398	5.7583	15
16	0.3387	9.4466	57.5271	2.9522	27.8881	0.1059	0.0359	6.0897	16
17	0.3166	9.7632	62.5923	3.1588	30.8402	0.1024	0.0324	6.4110	17
18	0.2959	10.0591	67.6219	3.3799	33.9990	0.0994	0.0294	6.7225	18
19	0.2765	10.3356	72.5991	3.6165	37.3790	0.0968	0.0268	7.0242	19
20	0.2584	10.5940	77.5091	3.8697	40.9955	0.0944	0.0244	7.3163	20
21	0.2415	10.8355	82.3393	4.1406	44.8652	0.0923	0.0223	7.5990	21
22	0.2257	11.0612	87.0793	4.4304	49.0057	0.0904	0.0204	7.8725	22
23	0.2109	11.2722	91.7201	4.7405	53.4361	0.0887	0.0187	8.1369	23
24	0.1971	11.4693	96.2545	5.0724	58.1767	0.0872	0.0172	8.3923	24
25	0.1842	11.6536	100.6765	5.4274	63.2490	0.0858	0.0158	8.6391	25
26	0.1722	11.8258	104.9814	5.8074	68.6765	0.0846	0.0146	8.8773	26
27	0.1609	11.9867	109.1656	6.2139	74.4838	0.0834	0.0134	9.1072	27
28	0.1504	12.1371	113.2264	6.6488	80.6977	0.0824	0.0124	9.3289	28
29	0.1406	12.2777	117.1622	7.1143	87.3465	0.0814	0.0114	9.5427	29
30	0.1314	12.4090	120.9718	7.6123	94.4608	0.0806	0.0106	9.7487	30
31	0.1228	12.5318	124.6550	8.1451	102.0730	0.0798	0.0098	9.9471	31
32	0.1147	12.6466	128.2120	8.7153	110.2182	0.0791	0.0091	10.1381	32
33	0.1072	12.7538	131.6435	9.3253	118.9334	0.0784	0.0084	10.3219	33
34	0.1002	12.8540	134.9507	9.9781	128.2588	0.0778	0.0078	10.4987	34
35	0.0937	12.9477	138.1353	10.6766	138.2369	0.0772	0.0072	10.6687	35
36	0.0875	13.0352	141.1990	11.4239	148.9135	0.0767	0.0067	10.8321	36
37	0.0818	13.1170	144.1441	12.2236	160.3374	0.0762	0.0062	10.9891	37
38	0.0765	13.1935	146.9730	13.0793	172.5610	0.0758	0.0058	11.1398	38
39	0.0715	13.2649	149.6883	13.9948	185.6403	0.0754	0.0054	11.2845	39
40	0.0668	13.3317	152.2928	14.9745	199.6351	0.0750	0.0050	11.4233	40
41	0.0624	13.3941	154.7892	16.0227	214.6096	0.0747	0.0047	11.5565	41
42	0.0583	13.4524	157.1807	17.1443	230.6322	0.0743	0.0043	11.6842	42
43	0.0545	13.5070	159.4702	18.3444	247.7765	0.0740	0.0040	11.8065	43
44	0.0509	13.5579	161.6009	19.6285	266.1200	0.0738	0.0038	11.9237	44
45	0.0476	13.6055	163.7559	21.0025	285.7493	0.0735	0.0035	12.0360	45
46	0.0445	13.6500	165.7584	22.4726	306.7518	0.0733	0.0033	12.1435	46
47	0.0416	13.6916	167.6714	24.0457	329.2244	0.0730	0.0030	12.2463	47
48	0.0389	13.7305	169.4981	25.7289	353.2701	0.0728	0.0028	12.3447	48
49	0.0363	13.7668	171.2417	27.5299	378.9990	0.0726	0.0026	12.4387	49
50	0.0339	13.8007	172.9051	29.4570	406.5289	0.0725	0.0025	12.5287	50
51	0.0317	13.8325	174.4915	31.5190	435.9860	0.0723	0.0023	12.6146	51
52	0.0297	13.8621	176.0037	33.7253	467.5050	0.0721	0.0021	12.6967	52
53	0.0277	13.8898	177.4447	36.0861	501.2303	0.0720	0.0020	12.7751	53
54	0.0259	13.9157	178.8173	38.6122	537.3164	0.0719	0.0019	12.8500	54
55	0.0242	13.9399	180.1243	41.3150	575.9286	0.0717	0.0017	12.9215	55
60	0.0173	14.0392	185.7677	57.9464	813.5204	0.0712	0.0012	13.2321	60
65	0.0123	14.1099	190.1452	81.2729	1146.7552	0.0709	0.0009	13.4760	65
70	0.0088	14.1604	193.5185	113.9894	1614.1342	0.0706	0.0006	13.6662	70
75	0.0063	14.1964	196.1035	159.8760	2269.6574	0.0704	0.0004	13.8136	75
80	0.0045	14.2220	198.0748	224.2344	3189.0627	0.0703	0.0003	13.9273	80
85	0.0032	14.2403	199.5717	314.5003	4478.5761	0.0702	0.0002	14.0146	85
90	0.0023	14.2533	200.7042	441.1030	6287.1854	0.0702	0.0002	14.0812	90
95	0.0016	14.2626	201.5581	618.6697	8823.8535	0.0701	0.0001	14.1319	95
100	0.0012	14.2693	202.2001	867.7163	12381.6618	0.0701	0.0001	14.1703	100

Support Material

(continued)

APPENDIX 58.B *(continued)*
Cash Flow Equivalent Factors

$i = 8.00\%$

n	P/F	P/A	P/G	F/P	F/A	A/P	A/F	A/G	n
1	0.9259	0.9259	0.0000	1.0800	1.0000	1.0800	1.0000	0.0000	1
2	0.8573	1.7833	0.8573	1.1664	2.0800	0.5608	0.4808	0.4808	2
3	0.7938	2.5771	2.4450	1.2597	3.2464	0.3880	0.3080	0.9487	3
4	0.7350	3.3121	4.6501	1.3605	4.5061	0.3019	0.2219	1.4040	4
5	0.6806	3.9927	7.3724	1.4693	5.8666	0.2505	0.1705	1.8465	5
6	0.6302	4.6229	10.5233	1.5869	7.3359	0.2163	0.1363	2.2763	6
7	0.5835	5.2064	14.0242	1.7138	8.9228	0.1921	0.1121	2.6937	7
8	0.5403	5.7466	17.8061	1.8509	10.6366	0.1740	0.0940	3.0985	8
9	0.5002	6.2469	21.8081	1.9990	12.4876	0.1601	0.0801	3.4910	9
10	0.4632	6.7101	25.9768	2.1589	14.4866	0.1490	0.0690	3.8713	10
11	0.4289	7.1390	30.2657	2.3316	16.6455	0.1401	0.0601	4.2395	11
12	0.3971	7.5361	34.6339	2.5182	18.9771	0.1327	0.0527	4.5957	12
13	0.3677	7.9038	39.0463	2.7196	21.4953	0.1265	0.0465	4.9402	13
14	0.3405	8.2442	43.4723	2.9372	24.2149	0.1213	0.0413	5.2731	14
15	0.3152	8.5595	47.8857	3.1722	27.1521	0.1168	0.0368	5.5945	15
16	0.2919	8.8514	52.2640	3.4259	30.3243	0.1130	0.0330	5.9046	16
17	0.2703	9.1216	56.5883	3.7000	33.7502	0.1096	0.0296	6.2037	17
18	0.2502	9.3719	60.8426	3.9960	37.4502	0.1067	0.0267	6.4920	18
19	0.2317	9.6036	65.0134	4.3157	41.4463	0.1041	0.0241	6.7697	19
20	0.2145	9.8181	69.0898	4.6610	45.7620	0.1019	0.0219	7.0369	20
21	0.1987	10.0168	73.0629	5.0338	50.4229	0.0998	0.0198	7.2940	21
22	0.1839	10.2007	76.9257	5.4365	55.4568	0.0980	0.0180	7.5412	22
23	0.1703	10.3711	80.6726	5.8715	60.8933	0.0964	0.0164	7.7786	23
24	0.1577	10.5288	84.2997	6.3412	66.7648	0.0950	0.0150	8.0066	24
25	0.1460	10.6748	87.8041	6.8485	73.1059	0.0937	0.0137	8.2254	25
26	0.1352	10.8100	91.1842	7.3964	79.9544	0.0925	0.0125	8.4352	26
27	0.1252	10.9352	94.4390	7.9881	87.3508	0.0914	0.0114	8.6363	27
28	0.1159	11.0511	97.5687	8.6271	95.3388	0.0905	0.0105	8.8289	28
29	0.1073	11.1584	100.5738	9.3173	103.9659	0.0896	0.0096	9.0133	29
30	0.0994	11.2578	103.4558	10.0627	113.2832	0.0888	0.0088	9.1897	30
31	0.0920	11.3498	106.2163	10.8677	123.3459	0.0881	0.0081	9.3584	31
32	0.0852	11.4350	108.8575	11.7371	134.2135	0.0875	0.0075	9.5197	32
33	0.0789	11.5139	111.3819	12.6760	145.9506	0.0869	0.0069	9.6737	33
34	0.0730	11.5869	113.7924	13.6901	158.6267	0.0863	0.0063	9.8208	34
35	0.0676	11.6546	116.0920	14.7853	172.3168	0.0858	0.0058	9.9611	35
36	0.0626	11.7172	118.2839	15.9682	187.1021	0.0853	0.0053	10.0949	36
37	0.0580	11.7752	120.3713	17.2456	203.0703	0.0849	0.0049	10.2225	37
38	0.0537	11.8289	122.3579	18.6253	220.3159	0.0845	0.0045	10.3440	38
39	0.0497	11.8786	124.2470	20.1153	238.9412	0.0842	0.0042	10.4597	39
40	0.0460	11.9246	126.0422	21.7245	259.0565	0.0839	0.0039	10.5699	40
41	0.0426	11.9672	127.7470	23.4625	280.7810	0.0836	0.0036	10.6747	41
42	0.0395	12.0067	129.3651	25.3395	304.2435	0.0833	0.0033	10.7744	42
43	0.0365	12.0432	130.8998	27.3666	329.5830	0.0830	0.0030	10.8692	43
44	0.0338	12.0771	132.3547	29.5560	356.9496	0.0828	0.0028	10.9592	44
45	0.0313	12.1084	133.7331	31.9204	386.5056	0.0826	0.0026	11.0447	45
46	0.0290	12.1374	135.0384	34.4741	418.4261	0.0824	0.0024	11.1258	46
47	0.0269	12.1643	136.2739	37.2320	452.9002	0.0822	0.0022	11.2028	47
48	0.0249	12.1891	137.4428	40.2106	490.1322	0.0820	0.0020	11.2758	48
49	0.0230	12.2122	138.5480	43.4274	530.3427	0.0819	0.0019	11.3451	49
50	0.0213	12.2335	139.5928	46.9016	573.7702	0.0817	0.0017	11.4107	50
51	0.0197	12.2532	140.5799	50.6537	620.6718	0.0816	0.0016	11.4729	51
52	0.0183	12.2715	141.5121	54.7060	671.3255	0.0815	0.0015	11.5318	52
53	0.0169	12.2884	142.3923	59.0825	726.0316	0.0814	0.0014	11.5875	53
54	0.0157	12.3041	143.2229	63.8091	785.1141	0.0813	0.0013	11.6403	54
55	0.0145	12.3186	144.0065	68.9139	848.9232	0.0812	0.0012	11.6902	55
60	0.0099	12.3766	147.3000	101.2571	1253.2133	0.0808	0.0008	11.9015	60
65	0.0067	12.4160	149.7387	148.7798	1847.2481	0.0805	0.0005	12.0602	65
70	0.0046	12.4428	151.5326	218.6064	2720.0801	0.0804	0.0004	12.1783	70
75	0.0031	12.4611	152.8448	321.2045	4002.5566	0.0802	0.0002	12.2658	75
80	0.0021	12.4735	153.8001	471.9548	5886.9354	0.0802	0.0002	12.3301	80
85	0.0014	12.4820	154.4925	693.4565	8655.7061	0.0801	0.0001	12.3772	85
90	0.0010	12.4877	154.9925	1018.9151	12723.9386	0.0801	0.0001	12.4116	90
95	0.0007	12.4917	155.3524	1497.1205	18701.5069	0.0801	0.0001	12.4365	95
100	0.0005	12.4943	155.6107	2199.7613	27484.5157	0.0800	0.0000	12.4545	100

Support Material

(continued)

APPENDIX 58.B *(continued)*
Cash Flow Equivalent Factors

$i = 9.00\%$

n	P/F	P/A	P/G	F/P	F/A	A/P	A/F	A/G	n
1	0.9174	0.9174	0.0000	1.0900	1.0000	1.0900	1.0000	0.0000	1
2	0.8417	1.7591	0.8417	1.1881	2.0900	0.5685	0.4785	0.4785	2
3	0.7722	2.5313	2.3860	1.2950	3.2781	0.3951	0.3051	0.9426	3
4	0.7084	3.2397	4.5113	1.4116	4.5731	0.3087	0.2187	1.3925	4
5	0.6499	3.8897	7.1110	1.5386	5.9847	0.2571	0.1671	1.8282	5
6	0.5963	4.4859	10.0924	1.6771	7.5233	0.2229	0.1329	2.2498	6
7	0.5470	5.0330	13.3746	1.8280	9.2004	0.1987	0.1087	2.6574	7
8	0.5019	5.5348	16.8877	1.9926	11.0285	0.1807	0.0907	3.0512	8
9	0.4604	5.9952	20.5711	2.1719	13.0210	0.1668	0.0768	3.4312	9
10	0.4224	6.4177	24.3728	2.3674	15.1929	0.1558	0.0658	3.7978	10
11	0.3875	6.8052	28.2481	2.5804	17.5603	0.1469	0.0569	4.1510	11
12	0.3555	7.1607	32.1590	2.8127	20.1407	0.1397	0.0497	4.4910	12
13	0.3262	7.4869	36.0731	3.0658	22.9534	0.1336	0.0436	4.8182	13
14	0.2992	7.7862	39.9633	3.3417	26.0192	0.1284	0.0384	5.1326	14
15	0.2745	8.0607	43.8069	3.6425	29.3609	0.1241	0.0341	5.4346	15
16	0.2519	8.3126	47.5849	3.9703	33.0034	0.1203	0.0303	5.7245	16
17	0.2311	8.5436	51.2821	4.3276	36.9737	0.1170	0.0270	6.0024	17
18	0.2120	8.7556	54.8860	4.7171	41.3013	0.1142	0.0242	6.2687	18
19	0.1945	8.9501	58.3868	5.1417	46.0185	0.1117	0.0217	6.5236	19
20	0.1784	9.1285	61.7770	5.6044	51.1601	0.1095	0.0195	6.7674	20
21	0.1637	9.2922	65.0509	6.1088	56.7645	0.1076	0.0176	7.0006	21
22	0.1502	9.4424	68.2048	6.6586	62.8733	0.1059	0.0159	7.2232	22
23	0.1378	9.5802	71.2359	7.2579	69.5319	0.1044	0.0144	7.4357	23
24	0.1264	9.7066	74.1433	7.9111	76.7898	0.1030	0.0130	7.6384	24
25	0.1160	9.8226	76.9265	8.6231	84.7009	0.1018	0.0118	7.8316	25
26	0.1064	9.9290	79.5863	9.3992	93.3240	0.1007	0.0107	8.0156	26
27	0.0976	10.0266	82.1241	10.2451	102.7231	0.0997	0.0097	8.1906	27
28	0.0895	10.1161	84.5419	11.1671	112.9682	0.0989	0.0089	8.3571	28
29	0.0822	10.1983	86.8422	12.1722	124.1354	0.0981	0.0081	8.5154	29
30	0.0754	10.2737	89.0280	13.2677	136.3076	0.0973	0.0073	8.6657	30
31	0.0691	10.3428	91.1024	14.4618	149.5752	0.0967	0.0067	8.8083	31
32	0.0634	10.4062	93.0690	15.7633	164.0370	0.0961	0.0061	8.9436	32
33	0.0582	10.4644	94.9314	17.1820	179.8003	0.0956	0.0056	9.0718	33
34	0.0534	10.5178	96.6935	18.7284	196.9823	0.0951	0.0051	9.1933	34
35	0.0490	10.5668	98.3590	20.4140	215.7108	0.0946	0.0046	9.3083	35
36	0.0449	10.6118	99.9319	22.2512	236.1247	0.0942	0.0042	9.4171	36
37	0.0412	10.6530	101.4162	24.2538	258.3759	0.0939	0.0039	9.5200	37
38	0.0378	10.6908	102.8158	26.4367	282.6298	0.0935	0.0035	9.6172	38
39	0.0347	10.7255	104.1345	28.8160	309.0665	0.0932	0.0032	9.7090	39
40	0.0318	10.7574	105.3762	31.4094	337.8824	0.0930	0.0030	9.7957	40
41	0.0292	10.7866	106.5445	34.2363	369.2919	0.0927	0.0027	9.8775	41
42	0.0268	10.8134	107.6432	37.3175	403.5281	0.0925	0.0025	9.9546	42
43	0.0246	10.8380	108.6758	40.6761	440.8457	0.0923	0.0023	10.0273	43
44	0.0226	10.8605	109.6456	44.3370	481.5218	0.0921	0.0021	10.0958	44
45	0.0207	10.8812	110.5561	48.3273	525.8587	0.0919	0.0019	10.1603	45
46	0.0190	10.9002	111.4103	52.6767	574.1860	0.0917	0.0017	10.2210	46
47	0.0174	10.9176	112.2115	57.4176	626.8628	0.0916	0.0016	10.2780	47
48	0.0160	10.9336	112.9625	62.5852	684.2804	0.0915	0.0015	10.3317	48
49	0.0147	10.9482	113.6661	68.2179	746.8656	0.0913	0.0013	10.3821	49
50	0.0134	10.9617	114.3251	74.3575	815.0836	0.0912	0.0012	10.4295	50
51	0.0123	10.9740	114.9420	81.0497	889.4411	0.0911	0.0011	10.4740	51
52	0.0113	10.9853	115.5193	88.3442	970.4908	0.0910	0.0010	10.5158	52
53	0.0104	10.9957	116.0593	96.2951	1058.8349	0.0909	0.0009	10.5549	53
54	0.0095	11.0053	116.5642	104.9617	1155.1301	0.0909	0.0009	10.5917	54
55	0.0087	11.0140	117.0362	114.4083	1260.0918	0.0908	0.0008	10.6261	55
60	0.0057	11.0480	118.9683	176.0313	1944.7921	0.0905	0.0005	10.7683	60
65	0.0037	11.0701	120.3344	270.8460	2998.2885	0.0903	0.0003	10.8702	65
70	0.0024	11.0844	121.2942	416.7301	4619.2232	0.0902	0.0002	10.9427	70
75	0.0016	11.0938	121.9646	641.1909	7113.2321	0.0901	0.0001	10.9940	75
80	0.0010	11.0998	122.4306	986.5517	10950.5741	0.0901	0.0001	11.0299	80
85	0.0007	11.1038	122.7533	1517.9320	16854.8003	0.0901	0.0001	11.0551	85
90	0.0004	11.1064	122.9758	2335.5266	25939.1842	0.0900	0.0000	11.0726	90
95	0.0003	11.1080	123.1287	3593.4971	39916.6350	0.0900	0.0000	11.0847	95
100	0.0002	11.1091	123.2335	5529.0408	61422.6755	0.0900	0.0000	11.0930	100

(continued)

APPENDIX 58.B *(continued)*
Cash Flow Equivalent Factors

$i = 10.00\%$

n	P/F	P/A	P/G	F/P	F/A	A/P	A/F	A/G	n
1	0.9091	0.9091	0.0000	1.1000	1.0000	1.1000	1.0000	0.0000	1
2	0.8264	1.7355	0.8264	1.2100	2.1000	0.5762	0.4762	0.4762	2
3	0.7513	2.4869	2.3291	1.3310	3.3100	0.4021	0.3021	0.9366	3
4	0.6830	3.1699	4.3781	1.4641	4.6410	0.3155	0.2155	1.3812	4
5	0.6209	3.7908	6.8618	1.6105	6.1051	0.2638	0.1638	1.8101	5
6	0.5645	4.3553	9.6842	1.7716	7.7156	0.2296	0.1296	2.2236	6
7	0.5132	4.8684	12.7631	1.9487	9.4872	0.2054	0.1054	2.6216	7
8	0.4665	5.3349	16.0287	2.1436	11.4359	0.1874	0.0874	3.0045	8
9	0.4241	5.7590	19.4215	2.3579	13.5795	0.1736	0.0736	3.3724	9
10	0.3855	6.1446	22.8913	2.5937	15.9374	0.1627	0.0627	3.7255	10
11	0.3505	6.4951	26.3963	2.8531	18.5312	0.1540	0.0540	4.0641	11
12	0.3186	6.8137	29.9012	3.1384	21.3843	0.1468	0.0468	4.3884	12
13	0.2897	7.1034	33.3772	3.4523	24.5227	0.1408	0.0408	4.6988	13
14	0.2633	7.3667	36.8005	3.7975	27.9750	0.1357	0.0357	4.9955	14
15	0.2394	7.6061	40.1520	4.1772	31.7725	0.1315	0.0315	5.2789	15
16	0.2176	7.8237	43.4164	4.5950	35.9497	0.1278	0.0278	5.5493	16
17	0.1978	8.0216	46.5819	5.0545	40.5447	0.1247	0.0247	5.8071	17
18	0.1799	8.2014	49.6395	5.5599	45.5992	0.1219	0.0219	6.0526	18
19	0.1635	8.3649	52.5827	6.1159	51.1591	0.1195	0.0195	6.2861	19
20	0.1486	8.5136	55.4069	6.7275	57.2750	0.1175	0.0175	6.5081	20
21	0.1351	8.6487	58.1095	7.4002	64.0025	0.1156	0.0156	6.7189	21
22	0.1228	8.7715	60.6893	8.1403	71.4027	0.1140	0.0140	6.9189	22
23	0.1117	8.8832	63.1462	8.9543	79.5430	0.1126	0.0126	7.1085	23
24	0.1015	8.9847	65.4813	9.8497	88.4973	0.1113	0.0113	7.2881	24
25	0.0923	9.0770	67.6964	10.8347	98.3471	0.1102	0.0102	7.4580	25
26	0.0839	9.1609	69.7940	11.9182	109.1818	0.1092	0.0092	7.6186	26
27	0.0763	9.2372	71.7773	13.1100	121.0999	0.1083	0.0083	7.7704	27
28	0.0693	9.3066	73.6495	14.4210	134.2099	0.1075	0.0075	7.9137	28
29	0.0630	9.3696	75.4146	15.8631	148.6309	0.1067	0.0067	8.0489	29
30	0.0573	9.4269	77.0766	17.4494	164.4940	0.1061	0.0061	8.1762	30
31	0.0521	9.4790	78.6395	19.1943	181.9434	0.1055	0.0055	8.2962	31
32	0.0474	9.5264	80.1078	21.1138	201.1378	0.1050	0.0050	8.4091	32
33	0.0431	9.5694	81.4856	23.2252	222.2515	0.1045	0.0045	8.5152	33
34	0.0391	9.6086	82.7773	25.5477	245.4767	0.1041	0.0041	8.6149	34
35	0.0356	9.6442	83.9872	28.1024	271.0244	0.1037	0.0037	8.7086	35
36	0.0323	9.6765	85.1194	30.9127	299.1268	0.1033	0.0033	8.7965	36
37	0.0294	9.7059	86.1781	34.0039	330.0395	0.1030	0.0030	8.8789	37
38	0.0267	9.7327	87.1673	37.4043	364.0434	0.1027	0.0027	8.9562	38
39	0.0243	9.7570	88.0908	41.1448	401.4478	0.1025	0.0025	9.0285	39
40	0.0221	9.7791	88.9525	45.2593	442.5926	0.1023	0.0023	9.0962	40
41	0.0201	9.7991	89.7560	49.7852	487.8518	0.1020	0.0020	9.1596	41
42	0.0183	9.8174	90.5047	54.7637	537.6370	0.1019	0.0019	9.2188	42
43	0.0166	9.8340	91.2019	60.2401	592.4007	0.1017	0.0017	9.2741	43
44	0.0151	9.8491	91.8508	66.2641	652.6408	0.1015	0.0015	9.3258	44
45	0.0137	9.8628	92.4544	72.8905	718.9048	0.1014	0.0014	9.3740	45
46	0.0125	9.8753	93.0157	80.1795	791.7953	0.1013	0.0013	9.4190	46
47	0.0113	9.8866	93.5372	88.1975	871.9749	0.1011	0.0011	9.4610	47
48	0.0103	9.8969	94.0217	97.0172	960.1723	0.1010	0.0010	9.5001	48
49	0.0094	9.9063	94.4715	106.7190	1057.1896	0.1009	0.0009	9.5365	49
50	0.0085	9.9148	94.8889	117.3909	1163.9085	0.1009	0.0009	9.5704	50
51	0.0077	9.9226	95.2761	129.1299	1281.2994	0.1008	0.0008	9.6020	51
52	0.0070	9.9296	95.6351	142.0429	1410.4293	0.1007	0.0007	9.6313	52
53	0.0064	9.9360	95.9679	156.2472	1552.4723	0.1006	0.0006	9.6586	53
54	0.0058	9.9418	96.2763	171.8719	1708.7195	0.1006	0.0006	9.6840	54
55	0.0053	9.9471	96.5619	189.0591	1880.5914	0.1005	0.0005	9.7075	55
60	0.0033	9.9672	97.7010	304.4816	3034.8164	0.1003	0.0003	9.8023	60
65	0.0020	9.9796	98.4705	490.3707	4893.7073	0.1002	0.0002	9.8672	65
70	0.0013	9.9873	98.9870	789.7470	7887.4696	0.1001	0.0001	9.9113	70
75	0.0008	9.9921	99.3317	1271.8954	12708.9537	0.1001	0.0001	9.9410	75
80	0.0005	9.9951	99.5606	2048.4002	20474.0021	0.1000	0.0000	9.9609	80
85	0.0003	9.9970	99.7120	3298.9690	32979.6903	0.1000	0.0000	9.9742	85
90	0.0002	9.9981	99.8118	5313.0226	53120.2261	0.1000	0.0000	9.9831	90
95	0.0001	9.9988	99.8773	8556.6760	85556.7605	0.1000	0.0000	9.9889	95
100	0.0001	9.9993	99.9202	13780.6123	137796.1234	0.1000	0.0000	9.9927	100

(continued)

APPENDIX 58.B *(continued)*
Cash Flow Equivalent Factors

$i = 12.00\%$

n	P/F	P/A	P/G	F/P	F/A	A/P	A/F	A/G	n
1	0.8929	0.8929	0.0000	1.1200	1.0000	1.1200	1.0000	0.0000	1
2	0.7972	1.6901	0.7972	1.2544	2.1200	0.5917	0.4717	0.4717	2
3	0.7118	2.4018	2.2208	1.4049	3.3744	0.4163	0.2963	0.9246	3
4	0.6355	3.0373	4.1273	1.5735	4.7793	0.3292	0.2092	1.3589	4
5	0.5674	3.6048	6.3970	1.7623	6.3528	0.2774	0.1574	1.7746	5
6	0.5066	4.1114	8.9302	1.9738	8.1152	0.2432	0.1232	2.1720	6
7	0.4523	4.5638	11.6443	2.2107	10.0890	0.2191	0.0991	2.5515	7
8	0.4039	4.9676	14.4714	2.4760	12.2997	0.2013	0.0813	2.9131	8
9	0.3606	5.3282	17.3563	2.7731	14.7757	0.1877	0.0677	3.2574	9
10	0.3220	5.6502	20.2541	3.1058	17.5487	0.1770	0.0570	3.5847	10
11	0.2875	5.9377	23.1288	3.4785	20.6546	0.1684	0.0484	3.8953	11
12	0.2567	6.1944	25.9523	3.8960	24.1331	0.1614	0.0414	4.1897	12
13	0.2292	6.4235	28.7024	4.3635	28.0291	0.1557	0.0357	4.4683	13
14	0.2046	6.6282	31.3624	4.8871	32.3926	0.1509	0.0309	4.7317	14
15	0.1827	6.8109	33.9202	5.4736	37.2797	0.1468	0.0268	4.9803	15
16	0.1631	6.9740	36.3670	6.1304	42.7533	0.1434	0.0234	5.2147	16
17	0.1456	7.1196	38.6973	6.8660	48.8837	0.1405	0.0205	5.4353	17
18	0.1300	7.2497	40.9080	7.6900	55.7497	0.1379	0.0179	5.6427	18
19	0.1161	7.3658	42.9979	8.6128	63.4397	0.1358	0.0158	6.8375	19
20	0.1037	7.4694	44.9676	9.6463	72.0524	0.1339	0.0139	6.0202	20
21	0.0926	7.5620	46.8188	10.8038	81.6987	0.1322	0.0122	6.1913	21
22	0.0826	7.6446	48.5543	12.1003	92.5026	0.1308	0.0108	6.3514	22
23	0.0738	7.7184	50.1776	13.5523	104.6029	0.1296	0.0096	6.5010	23
24	0.0659	7.7843	51.6929	15.1786	118.1552	0.1285	0.0085	6.6406	24
25	0.0588	7.8431	53.1046	17.0001	133.3339	0.1275	0.0075	6.7708	25
26	0.0525	7.8957	54.4177	19.0401	150.3339	0.1267	0.0067	6.8921	26
27	0.0469	7.9426	55.6369	21.3249	169.3740	0.1259	0.0059	7.0049	27
28	0.0419	7.9844	56.7674	23.8839	190.6989	0.1252	0.0052	7.1098	28
29	0.0374	8.0218	57.8141	26.7499	214.5828	0.1247	0.0047	7.2071	29
30	0.0334	8.0552	58.7821	29.9599	241.3327	0.1241	0.0041	7.2974	30
31	0.0298	8.0850	59.6761	33.5551	271.2926	0.1237	0.0037	7.3811	31
32	0.0266	8.1116	60.5010	37.5817	304.8477	0.1233	0.0033	7.4586	32
33	0.0238	8.1354	61.2612	42.0915	342.4294	0.1229	0.0029	7.5302	33
34	0.0212	8.1566	61.9612	47.1425	384.5210	0.1226	0.0026	7.5965	34
35	0.0189	8.1755	62.6052	52.7996	431.6635	0.1223	0.0023	7.6577	35
36	0.0169	8.1924	63.1970	59.1356	484.4631	0.1221	0.0021	7.7141	36
37	0.0151	8.2075	63.7406	66.2318	543.5987	0.1218	0.0018	7.7661	37
38	0.0135	8.2210	64.2394	74.1797	609.8305	0.1216	0.0016	7.8141	38
39	0.0120	8.2330	64.6967	83.0812	684.0102	0.1215	0.0015	7.8582	39
40	0.0107	8.2438	65.1159	93.0510	767.0914	0.1213	0.0013	7.8988	40
41	0.0096	8.2534	65.4997	104.2171	860.1424	0.1212	0.0012	7.9361	41
42	0.0086	8.2619	65.8509	116.7231	964.3595	0.1210	0.0010	7.9704	42
43	0.0076	8.2696	66.1722	130.7299	1081.0826	0.1209	0.0009	8.0019	43
44	0.0068	8.2764	66.4659	146.4175	1211.8125	0.1208	0.0008	8.0308	44
45	0.0061	8.2825	66.7342	163.9876	1358.2300	0.1207	0.0007	8.0572	45
46	0.0054	8.2880	66.9792	183.6661	1522.2176	0.1207	0.0007	8.0815	46
47	0.0049	8.2928	67.2028	205.7061	1705.8838	0.1206	0.0006	8.1037	47
48	0.0043	8.2972	67.4068	230.3908	1911.5898	0.1205	0.0005	8.1241	48
49	0.0039	8.3010	67.5929	258.0377	2141.9806	0.1205	0.0005	8.1427	49
50	0.0035	8.3045	67.7624	289.0022	2400.0182	0.1204	0.0004	8.1597	50
51	0.0031	8.3076	67.9169	323.6825	2689.0204	0.1204	0.0004	8.1753	51
52	0.0028	8.3103	68.0576	362.5243	3012.7029	0.1203	0.0003	8.1895	52
53	0.0025	8.3128	68.1856	406.0273	3375.2272	0.1203	0.0003	8.2025	53
54	0.0022	8.3150	68.3022	454.7505	3781.2545	0.1203	0.0003	8.2143	54
55	0.0020	8.3170	68.4082	509.3206	4236.0050	0.1202	0.0002	8.2251	55
60	0.0011	8.3240	68.8100	897.5969	7471.6411	0.1201	0.0001	8.2664	60
65	0.0006	8.3281	69.0581	1581.8725	13173.9374	0.1201	0.0001	8.2922	65
70	0.0004	8.3303	69.2103	2787.7998	23223.3319	0.1200	0.0000	8.3082	70
75	0.0002	8.3316	69.3031	4913.0558	40933.7987	0.1200	0.0000	8.3181	75
80	0.0001	8.3324	69.3594	8658.4831	72145.6925	0.1200	0.0000	8.3241	80
85	0.0001	8.3328	69.3935	15259.2057	127151.7140	0.1200	0.0000	8.3278	85
90	0.0000	8.3330	69.4140	26891.9342	224091.1185	0.1200	0.0000	8.3300	90
95	0.0000	8.3332	69.4263	47392.7766	394931.4719	0.1200	0.0000	8.3313	95
100	0.0000	8.3332	69.4336	83522.2657	696010.5477	0.1200	0.0000	8.3321	100

(continued)

APPENDIX 58.B *(continued)*
Cash Flow Equivalent Factors

$i = 15.00\%$

n	P/F	P/A	P/G	F/P	F/A	A/P	A/F	A/G	n
1	0.8696	0.8696	0.0000	1.1500	1.0000	1.1500	1.0000	0.0000	1
2	0.7561	1.6257	0.7561	1.3225	2.1500	0.6151	0.4651	0.4651	2
3	0.6575	2.2832	2.0712	1.5209	3.4725	0.4380	0.2880	0.9071	3
4	0.5718	2.8550	3.7864	1.7490	4.9934	0.3503	0.2003	1.3263	4
5	0.4972	3.3522	5.7751	2.0114	6.7424	0.2983	0.1483	1.7228	5
6	0.4323	3.7845	7.9368	2.3131	8.7537	0.2642	0.1142	2.0972	6
7	0.3759	4.1604	10.1924	2.6600	11.0668	0.2404	0.0904	2.4498	7
8	0.3269	4.4873	12.4807	3.0590	13.7268	0.2229	0.0729	2.7813	8
9	0.2843	4.7716	14.7548	3.5179	16.7858	0.2096	0.0596	3.0922	9
10	0.2472	5.0188	16.9795	4.0456	20.3037	0.1993	0.0493	3.3832	10
11	0.2149	5.2337	19.1289	4.6524	24.3493	0.1911	0.0411	3.6549	11
12	0.1869	5.4206	21.1849	5.3503	29.0017	0.1845	0.0345	3.9082	12
13	0.1625	5.5831	23.1352	6.1528	34.3519	0.1791	0.0291	4.1438	13
14	0.1413	5.7245	24.9725	7.0757	40.5047	0.1747	0.0247	4.3624	14
15	0.1229	5.8474	26.9630	8.1371	47.5804	0.1710	0.0210	4.5650	15
16	0.1069	5.9542	28.2960	9.3576	55.7175	0.1679	0.0179	4.7522	16
17	0.0929	6.0472	29.7828	10.7613	65.0751	0.1654	0.0154	4.9251	17
18	0.0808	6.1280	31.1565	12.3755	75.8364	0.1632	0.0132	5.0843	18
19	0.0703	6.1982	32.4213	14.2318	88.2118	0.1613	0.0113	5.2307	19
20	0.0611	6.2593	33.5822	16.3665	102.4436	0.1598	0.0098	5.3651	20
21	0.0531	6.3125	34.6448	18.8215	118.8101	0.1584	0.0084	5.4883	21
22	0.0462	6.3587	35.6150	21.6447	137.6316	0.1573	0.0073	5.6010	22
23	0.0402	6.3988	36.4988	24.8915	159.2764	0.1563	0.0063	5.7040	23
24	0.0349	6.4338	37.3023	28.6252	184.1678	0.1554	0.0054	5.7979	24
25	0.0304	6.4641	38.0314	32.9190	212.7930	0.1547	0.0047	5.8834	25
26	0.0264	6.4906	38.6918	37.8568	245.7120	0.1541	0.0041	5.9612	26
27	0.0230	6.5135	39.2890	43.5353	283.5688	0.1535	0.0035	6.0319	27
28	0.0200	6.5335	39.8283	50.0656	327.1041	0.1531	0.0031	6.0960	28
29	0.0174	6.5509	40.3146	57.5755	377.1697	0.1527	0.0027	6.1541	29
30	0.0151	6.5660	40.7526	66.2118	434.7451	0.1523	0.0023	6.2066	30
31	0.0131	6.5791	41.1466	76.1435	500.9569	0.1520	0.0020	6.2541	31
32	0.0114	6.5905	41.5006	87.5651	577.1005	0.1517	0.0017	6.2970	32
33	0.0099	6.6005	41.8184	100.6998	664.6655	0.1515	0.0015	6.3357	33
34	0.0086	6.6091	42.1033	115.8048	765.3654	0.1513	0.0013	6.3705	34
35	0.0075	6.6166	42.3586	133.1755	881.1702	0.1511	0.0011	6.4019	35
36	0.0065	6.6231	42.5872	153.1519	1014.3457	0.1510	0.0010	6.4301	36
37	0.0057	6.6288	42.7916	176.1246	1167.4975	0.1509	0.0009	6.4554	37
38	0.0049	6.6338	42.9743	202.5433	1343.6222	0.1507	0.0007	6.4781	38
39	0.0043	6.6380	43.1374	232.9248	1546.1655	0.1506	0.0006	6.4985	39
40	0.0037	6.6418	43.2830	267.8635	1779.0903	0.1506	0.0006	6.5168	40
41	0.0032	6.6450	43.4128	308.0431	2046.9539	0.1505	0.0005	6.5331	41
42	0.0028	6.6478	43.5286	354.2495	2354.9969	0.1504	0.0004	6.5478	42
43	0.0025	6.6503	43.6317	407.3870	2709.2465	0.1504	0.0004	6.5609	43
44	0.0021	6.6524	43.7235	468.4950	3116.6334	0.1503	0.0003	6.5725	44
45	0.0019	6.6543	43.8051	538.7693	3585.1285	0.1503	0.0003	6.5830	45
46	0.0016	6.6559	43.8778	619.5847	4123.8977	0.1502	0.0002	6.5923	46
47	0.0014	6.6573	43.9423	712.5224	4743.4824	0.1502	0.0002	6.6006	47
48	0.0012	6.6585	43.9997	819.4007	5456.0047	0.1502	0.0002	6.6080	48
49	0.0011	6.6596	44.0506	942.3108	6275.4055	0.1502	0.0002	6.6146	49
50	0.0009	6.6605	44.0958	1083.6574	7217.7163	0.1501	0.0001	6.6205	50
51	0.0008	6.6613	44.1360	1246.2061	8301.3737	0.1501	0.0001	6.6257	51
52	0.0007	6.6620	44.1715	1433.1370	9547.5798	0.1501	0.0001	6.6304	52
53	0.0006	6.6626	44.2031	1648.1075	10980.7167	0.1501	0.0001	6.6345	53
54	0.0005	6.6631	44.2311	1895.3236	12628.8243	0.1501	0.0001	6.6382	54
55	0.0005	6.6636	44.2558	2179.6222	14524.1479	0.1501	0.0001	6.6414	55
60	0.0002	6.6651	44.3431	4383.9987	29219.9916	0.1500	0.0000	6.6530	60
65	0.0001	6.6659	44.3903	8817.7874	58778.5826	0.1500	0.0000	6.6593	65
70	0.0001	6.6663	44.4156	17735.7200	118231.4669	0.1500	0.0000	6.6627	70
75	0.0000	6.6665	44.4292	35672.8680	237812.4532	0.1500	0.0000	6.6646	75
80	0.0000	6.6666	44.4364	71750.8794	478332.5293	0.1500	0.0000	6.6656	80
85	0.0000	6.6666	44.4402	144316.6470	962104.3133	0.1500	0.0000	6.6661	85
90	0.0000	6.6666	44.4422	290272.3252	1935142.1680	0.1500	0.0000	6.6664	90
95	0.0000	6.6667	44.4433	583841.3276	3892268.8509	0.1500	0.0000	6.6665	95
100	0.0000	6.6667	44.4438	1174313.4507	7828749.6713	0.1500	0.0000	6.6666	100

(continued)

APPENDIX 58.B *(continued)*
Cash Flow Equivalent Factors

$i = 20.00\%$

n	P/F	P/A	P/G	F/P	F/A	A/P	A/F	A/G	n
1	0.8333	0.8333	0.0000	1.2000	1.0000	1.2000	1.0000	0.0000	1
2	0.6944	1.5278	0.6944	1.4400	2.2000	0.6545	0.4545	0.4545	2
3	0.5787	2.1065	1.8519	1.7280	3.6400	0.4747	0.2747	0.8791	3
4	0.4823	2.5887	3.2986	2.0736	5.3680	0.3863	0.1863	1.2742	4
5	0.4019	2.9906	4.9061	2.4883	7.4416	0.3344	0.1344	1.6405	5
6	0.3349	3.3255	6.5806	2.9860	9.9299	0.3007	0.1007	1.9788	6
7	0.2791	3.6046	8.2551	3.5832	12.9159	0.2774	0.0774	2.2902	7
8	0.2326	3.8372	9.8831	4.2998	16.4991	0.2606	0.0606	2.5756	8
9	0.1938	4.0310	11.4335	5.1598	20.7989	0.2481	0.0481	2.8364	9
10	0.1615	4.1925	12.8871	6.1917	25.9587	0.2385	0.0385	3.0739	10
11	0.1346	4.3271	14.2330	7.4301	32.1504	0.2311	0.0311	3.2893	11
12	0.1122	4.4392	15.4667	8.9161	39.5805	0.2253	0.0253	3.4841	12
13	0.0935	4.5327	16.5883	10.6993	48.4966	0.2206	0.0206	3.6597	13
14	0.0779	4.6106	17.6008	12.8392	59.1959	0.2169	0.0169	3.8175	14
15	0.0649	4.6755	18.5095	15.4070	72.0351	0.2139	0.0139	3.9588	15
16	0.0541	4.7296	19.3208	18.4884	87.4421	0.2114	0.0114	4.0851	16
17	0.0451	4.7746	20.0419	22.1861	105.9306	0.2094	0.0094	4.1976	17
18	0.0376	4.8122	20.6805	26.6233	128.1167	0.2078	0.0078	4.2975	18
19	0.0313	4.8435	21.2439	31.9480	154.7400	0.2065	0.0065	4.3861	19
20	0.0261	4.8696	21.7395	38.3376	186.6880	0.2054	0.0054	4.4643	20
21	0.0217	4.8913	22.1742	46.0051	225.0256	0.2044	0.0044	4.5334	21
22	0.0181	4.9094	22.5546	55.2061	271.0307	0.2037	0.0037	4.5941	22
23	0.0151	4.9245	22.8867	66.2474	326.2369	0.2031	0.0031	4.6475	23
24	0.0126	4.9371	23.1760	79.4968	392.4842	0.2025	0.0025	4.6943	24
25	0.0105	4.9476	23.4276	95.3962	471.9811	0.2021	0.0021	4.7352	25
26	0.0087	4.9563	23.6460	114.4755	567.3773	0.2018	0.0018	4.7709	26
27	0.0073	4.9636	23.8353	137.3706	681.8528	0.2015	0.0015	4.8020	27
28	0.0061	4.9697	23.9991	164.8447	819.2233	0.2012	0.0012	4.8291	28
29	0.0051	4.9747	24.1406	197.8136	984.0680	0.2010	0.0010	4.8527	29
30	0.0042	4.9789	24.2628	237.3763	1181.8816	0.2008	0.0008	4.8731	30
31	0.0035	4.9824	24.3681	284.8516	1419.2579	0.2007	0.0007	4.8908	31
32	0.0029	4.9854	24.4588	341.8219	1704.1095	0.2006	0.0006	4.9061	32
33	0.0024	4.9878	24.5368	410.1863	2045.9314	0.2005	0.0005	4.9194	33
34	0.0020	4.9898	24.6038	492.2235	2456.1176	0.2004	0.0004	4.9308	34
35	0.0017	4.9915	24.6614	590.6682	2948.3411	0.2003	0.0003	4.9406	35
36	0.0014	4.9929	24.7108	708.8019	3539.0094	0.2003	0.0003	4.9491	36
37	0.0012	4.9941	24.7531	850.5622	4247.8112	0.2002	0.0002	4.9564	37
38	0.0010	4.9951	24.7894	1020.6747	5098.3735	0.2002	0.0002	4.9627	38
39	0.0008	4.9959	24.8204	1224.8096	6119.0482	0.2002	0.0002	4.9681	39
40	0.0007	4.9966	24.8469	1469.7716	7343.8578	0.2001	0.0001	4.9728	40
41	0.0006	4.9972	24.8696	1763.7259	8813.6294	0.2001	0.0001	4.9767	41
42	0.0005	4.9976	24.8890	2116.4711	10577.3553	0.2001	0.0001	4.9801	42
43	0.0004	4.9980	24.9055	2539.7653	12693.8263	0.2001	0.0001	4.9831	43
44	0.0003	4.9984	24.9196	3047.7183	15233.5916	0.2001	0.0001	4.9856	44
45	0.0003	4.9986	24.9316	3657.2620	18281.3099	0.2001	0.0001	4.9877	45
46	0.0002	4.9989	24.9419	4388.7144	21938.5719	0.2000	0.0000	4.9895	46
47	0.0002	4.9991	24.9506	5266.4573	26327.2863	0.2000	0.0000	4.9911	47
48	0.0002	4.9992	24.9581	6319.7487	31593.7436	0.2000	0.0000	4.9924	48
49	0.0001	4.9993	24.9644	7583.6985	37913.4923	0.2000	0.0000	4.9935	49
50	0.0001	4.9995	24.9698	9100.4382	45497.1908	0.2000	0.0000	4.9945	50
51	0.0001	4.9995	24.9744	10920.5258	54597.6289	0.2000	0.0000	4.9953	51
52	0.0001	4.9996	24.9783	13104.6309	65518.1547	0.2000	0.0000	4.9960	52
53	0.0001	4.9997	24.9816	15725.5571	78622.7856	0.2000	0.0000	4.9966	53
54	0.0001	4.9997	24.9844	18870.6685	94348.3427	0.2000	0.0000	4.9971	54
55	0.0000	4.9998	24.9868	22644.8023	113219.0113	0.2000	0.0000	4.9976	55
60	0.0000	4.9999	24.9942	56347.5144	281732.5718	0.2000	0.0000	4.9989	60
65	0.0000	5.0000	24.9975	140210.6469	701048.2346	0.2000	0.0000	4.9995	65
70	0.0000	5.0000	24.9989	348888.9569	1744439.7847	0.2000	0.0000	4.9998	70
75	0.0000	5.0000	24.9995	868147.3693	4340731.8466	0.2000	0.0000	4.9999	75

(continued)

APPENDIX 58.B *(continued)*
Cash Flow Equivalent Factors

$i = 25.00\%$

n	P/F	P/A	P/G	F/P	F/A	A/P	A/F	A/G	n
1	0.8000	0.8000	0.0000	1.2500	1.0000	1.2500	1.0000	0.0000	1
2	0.6400	1.4400	0.6400	1.5625	2.2500	0.6944	0.0444	0.4444	2
3	0.5120	1.9520	1.6640	1.9531	3.8125	0.5123	0.2623	0.8525	3
4	0.4096	2.3616	2.8928	2.4414	5.7656	0.4234	0.1734	1.2249	4
5	0.3277	2.6893	4.2035	3.0518	8.2070	0.3718	0.1218	1.5631	5
6	0.2621	2.9514	5.5142	3.8147	11.2588	0.3383	0.0888	1.8683	6
7	0.2097	3.1611	6.7725	4.7684	15.0735	0.3163	0.0663	2.1424	7
8	0.1678	3.3289	7.9469	5.9605	19.8419	0.3004	0.0504	2.3872	8
9	0.1342	3.4631	9.0207	7.4506	25.8023	0.2888	0.0388	2.6048	9
10	0.1074	3.5705	9.9870	9.3132	33.2529	0.2801	0.0301	2.7971	10
11	0.0859	3.6564	10.8460	11.6415	42.5661	0.2735	0.0235	2.9663	11
12	0.0687	3.7251	11.6020	14.5519	54.2077	0.2684	0.0184	3.1145	12
13	0.0550	3.7801	12.2617	18.1899	68.7596	0.2645	0.0145	3.2437	13
14	0.0440	3.8241	12.8334	22.7374	86.9495	0.2615	0.0115	3.3559	14
15	0.0352	3.8593	13.3260	28.4217	109.6868	0.2591	0.0091	3.4530	15
16	0.0281	3.8874	13.7482	35.5271	138.1085	0.2572	0.0072	3.5366	16
17	0.0225	3.9099	14.1085	44.4089	173.6357	0.2558	0.0058	3.6084	17
18	0.0180	3.9279	14.4147	55.5112	218.0446	0.2546	0.0046	3.6698	18
19	0.0144	3.9424	14.6741	69.3889	273.5558	0.2537	0.0037	3.7222	19
20	0.0115	3.9539	14.8932	86.7362	342.9447	0.2529	0.0029	3.7667	20
21	0.0092	3.9631	15.0777	108.4202	429.6809	0.2523	0.0023	3.8045	21
22	0.0074	3.9705	15.2326	135.5253	538.1011	0.2519	0.0019	3.8365	22
23	0.0059	3.9764	15.3625	169.4066	673.6264	0.2515	0.0015	3.8634	23
24	0.0047	3.9811	15.4711	211.7582	843.0329	0.2512	0.0012	3.8861	24
25	0.0038	3.9849	15.5618	264.6978	1054.7912	0.2509	0.0009	3.9052	25
26	0.0030	3.9879	15.6373	330.8722	1319.4890	0.2508	0.0008	3.9212	26
27	0.0024	3.9903	15.7002	413.5903	1650.3612	0.2506	0.0006	3.9346	27
28	0.0019	3.9923	15.7524	516.9879	2063.9515	0.2505	0.0005	3.9457	28
29	0.0015	3.9938	15.7957	646.2349	2580.9394	0.2504	0.0004	3.9551	29
30	0.0012	3.9950	15.8316	807.7936	3227.1743	0.2503	0.0003	3.9628	30
31	0.0010	3.9960	15.8614	1009.7420	4034.9678	0.2502	0.0002	3.9693	31
32	0.0008	3.9968	15.8859	1262.1774	5044.7098	0.2502	0.0002	3.9746	32
33	0.0006	3.9975	15.9062	1577.7218	6306.8872	0.2502	0.0002	3.9791	33
34	0.0005	3.9980	15.9229	1972.1523	7884.6091	0.2501	0.0001	3.9828	34
35	0.0004	3.9984	15.9367	2465.1903	9856.7613	0.2501	0.0001	3.9858	35
36	0.0003	3.9987	15.9481	3081.4879	12321.9516	0.2501	0.0001	3.9883	36
37	0.0003	3.9990	15.9574	3851.8599	15403.4396	0.2501	0.0001	3.9904	37
38	0.0002	3.9992	15.9651	4814.8249	19255.2994	0.2501	0.0001	3.9921	38
39	0.0002	3.9993	15.9714	6018.5311	24070.1243	0.2500	0.0000	3.9935	39
40	0.0001	3.9995	15.9766	7523.1638	30088.6554	0.2500	0.0000	3.9947	40
41	0.0001	3.9996	15.9809	9403.9548	37611.8192	0.2500	0.0000	3.9956	41
42	0.0001	3.9997	15.9843	11754.9435	47015.7740	0.2500	0.0000	3.9964	42
43	0.0001	3.9997	15.9872	14693.6794	58770.7175	0.2500	0.0000	3.9971	43
44	0.0001	3.9998	15.9895	18367.0992	73464.3969	0.2500	0.0000	3.9976	44
45	0.0000	3.9998	15.9915	22958.8740	91831.4962	0.2500	0.0000	3.9980	45
46	0.0000	3.9999	15.9930	28698.5925	114790.3702	0.2500	0.0000	3.9984	46
47	0.0000	3.9999	15.9943	35873.2407	143488.9627	0.2500	0.0000	3.9987	47
48	0.0000	3.9999	15.9954	44841.5509	179362.2034	0.2500	0.0000	3.9989	48
49	0.0000	3.9999	15.9962	56051.9386	224203.7543	0.2500	0.0000	3.9991	49
50	0.0000	3.9999	15.9969	70064.9232	280255.6929	0.2500	0.0000	3.9993	50
51	0.0000	4.0000	15.9975	87581.1540	350320.6161	0.2500	0.0000	3.9994	51
52	0.0000	4.0000	15.9980	109476.4425	437901.7701	0.2500	0.0000	3.9995	52
53	0.0000	4.0000	15.9983	136845.5532	547378.2126	0.2500	0.0000	3.9996	53
54	0.0000	4.0000	15.9986	171056.9414	684223.7658	0.2500	0.0000	3.9997	54
55	0.0000	4.0000	15.9989	213821.1768	855280.7072	0.2500	0.0000	3.9997	55
60	0.0000	4.0000	15.9996	652530.4468	2610117.7872	0.2500	0.0000	3.9999	60

(continued)

Support Material

APPENDIX 58.B *(continued)*
Cash Flow Equivalent Factors

$i = 30.00\%$

n	P/F	P/A	P/G	F/P	F/A	A/P	A/F	A/G	n
1	0.7692	0.7692	0.0000	1.3000	1.0000	1.3000	1.0000	0.000	1
2	0.5917	1.3609	0.5917	1.6900	2.3000	0.7348	0.4348	0.434	2
3	0.4552	1.8161	1.5020	2.1970	3.9900	0.5506	0.2506	0.827	3
4	0.3501	2.1662	2.5524	2.8561	6.1870	0.4616	0.1616	1.178	4
5	0.2693	2.4356	3.6297	3.7129	9.0431	0.4106	0.1106	1.490	5
6	0.2072	2.6427	4.6656	4.8268	12.7560	0.3784	0.0784	1.765	6
7	0.1594	2.8021	5.6218	6.2749	17.5828	0.3569	0.0569	2.006	7
8	0.1226	2.9247	6.4800	8.1573	23.8577	0.3419	0.0419	2.215	8
9	0.0943	3.0190	7.2343	10.6045	32.0150	0.3312	0.0312	2.396	9
10	0.0725	3.0915	7.8872	13.7858	42.6195	0.3235	0.0235	2.551	10
11	0.0558	3.1473	8.4452	17.9216	56.4053	0.3177	0.0177	2.683	11
12	0.0429	3.1903	8.9173	23.2981	74.3270	0.3135	0.0135	2.795	12
13	0.0330	3.2233	9.3135	30.2875	97.6250	0.3102	0.0102	2.889	13
14	0.0254	3.2487	9.6437	39.3738	127.9125	0.3078	0.0078	2.968	14
15	0.0195	3.2682	9.9172	51.1859	167.2863	0.3060	0.0060	3.034	15
16	0.0150	3.2832	10.1426	66.5417	218.4722	0.3046	0.0046	3.089	16
17	0.0116	3.2948	10.3276	86.5042	285.0139	0.3035	0.0035	3.134	17
18	0.0089	3.3037	10.4788	112.4554	371.5180	0.3027	0.0027	3.171	18
19	0.0068	3.3105	10.6019	146.1920	483.9734	0.3021	0.0021	3.202	19
20	0.0053	3.3158	10.7019	190.0496	630.1655	0.3016	0.0016	3.227	20
21	0.0040	3.3198	10.7828	247.0645	820.2151	0.3012	0.0012	3.248	21
22	0.0031	3.3230	10.8482	321.1839	1067.2796	0.3009	0.0009	3.264	22
23	0.0024	3.3254	10.9009	417.5391	1388.4635	0.3007	0.0007	3.278	23
24	0.0018	3.3272	10.9433	542.8008	1806.0026	0.3006	0.0006	3.289	24
25	0.0014	3.3286	10.9773	705.6410	2348.8033	0.3004	0.0004	3.297	25
26	0.0011	3.3297	11.0045	917.3333	3054.4443	0.3003	0.0003	3.305	26
27	0.0008	3.3305	11.0263	1192.5333	3971.7776	0.3003	0.0003	3.310	27
28	0.0006	3.3312	11.0437	1550.2933	5164.3109	0.3002	0.0002	3.315	28
29	0.0005	3.3317	11.0576	2015.3813	6714.6042	0.3001	0.0001	3.318	29
30	0.0004	3.3321	11.0687	2619.9956	8729.9855	0.3001	0.0001	3.321	30
31	0.0003	3.3324	11.0775	3405.9943	11349.9811	0.3001	0.0001	3.324	31
32	0.0002	3.3326	11.0845	4427.7926	14755.9755	0.3001	0.0001	3.326	32
33	0.0002	3.3328	11.0901	5756.1304	19183.7681	0.3001	0.0001	3.327	33
34	0.0001	3.3329	11.0945	7482.9696	24939.8985	0.3000	0.0000	3.328	34
35	0.0001	3.3330	11.0980	9727.8604	32422.8681	0.3000	0.0000	3.329	35
36	0.0001	3.3331	11.1007	12646.2186	42150.7285	0.3000	0.0000	3.330	36
37	0.0001	3.3331	11.1029	16440.0841	54796.9471	0.3000	0.0000	3.331	37
38	0.0000	3.3332	11.1047	21372.1094	71237.0312	0.3000	0.0000	3.331	38
39	0.0000	3.3332	11.1060	27783.7422	92609.1405	0.3000	0.0000	3.331	39
40	0.0000	3.3332	11.1071	36118.8648	120392.8827	0.3000	0.0000	3.332	40
41	0.0000	3.3333	11.1080	46954.5243	156511.7475	0.3000	0.0000	3.332	41
42	0.0000	3.3333	11.1086	61040.8815	203466.2718	0.3000	0.0000	3.332	42
43	0.0000	3.3333	11.1092	79353.1460	264507.1533	0.3000	0.0000	3.332	43
44	0.0000	3.3333	11.1096	103159.0898	343860.2993	0.3000	0.0000	3.332	44
45	0.0000	3.3333	11.1099	134106.8167	447019.3890	0.3000	0.0000	3.333	45
46	0.0000	3.3333	11.1102	174338.8617	581126.2058	0.3000	0.0000	3.333	46
47	0.0000	3.3333	11.1104	226640.5202	755465.0675	0.3000	0.0000	3.333	47
48	0.0000	3.3333	11.1105	294632.6763	982105.5877	0.3000	0.0000	3.333	48
49	0.0000	3.3333	11.1107	383022.4792	1276738.2640	0.3000	0.0000	3.333	49
50	0.0000	3.3333	11.1108	497929.2230	1659760.7433	0.3000	0.0000	3.333	50

(continued)

Support Material

APPENDIX 58.B *(continued)*
Cash Flow Equivalent Factors

$i = 40.00\%$

n	P/F	P/A	P/G	F/P	F/A	A/P	A/F	A/G	n
1	0.7143	0.7143	0.0000	1.4000	1.0000	1.4000	1.0000	0.000	1
2	0.5102	1.2245	0.5102	1.9600	2.4000	0.8167	0.4167	0.416	2
3	0.3644	1.5889	1.2391	2.7440	4.3600	0.6294	0.2294	0.779	3
4	0.2603	1.8492	2.0200	3.8416	7.1040	0.5408	0.1408	1.092	4
5	0.1859	2.0352	2.7637	5.3782	10.9456	0.4914	0.0914	1.358	5
6	0.1328	2.1680	3.4278	7.5295	16.3238	0.4613	0.0613	1.581	6
7	0.0949	2.2628	3.9970	10.5414	23.8534	0.4419	0.0419	1.766	7
8	0.0678	2.3306	4.4713	14.7579	34.3947	0.4291	0.0291	1.918	8
9	0.0484	2.3790	4.8585	20.6610	49.1526	0.4203	0.0203	2.042	9
10	0.0346	2.4136	5.1696	28.9255	69.8137	0.4143	0.0143	2.141	10
11	0.0247	2.4383	5.4166	40.4957	98.7391	0.4101	0.0101	2.221	11
12	0.0176	2.4559	5.6106	56.6939	139.2348	0.4072	0.0072	2.284	12
13	0.0126	2.4685	5.7618	79.3715	195.9287	0.4051	0.0051	2.334	13
14	0.0090	2.4775	5.8788	111.1201	275.3002	0.4036	0.0036	2.372	14
15	0.0064	2.4839	5.9688	155.5681	386.4202	0.4026	0.0026	2.403	15
16	0.0046	2.4885	6.0376	217.7953	541.9883	0.4018	0.0018	2.426	16
17	0.0033	2.4918	6.0901	304.9135	759.7837	0.4013	0.0013	2.444	17
18	0.0023	2.4941	6.1299	426.8789	1064.6971	0.4009	0.0009	2.457	18
19	0.0017	2.4958	6.1601	597.6304	1491.5760	0.4007	0.0007	2.468	19
20	0.0012	2.4970	6.1828	836.6826	2089.2064	0.4005	0.0005	2.476	20
21	0.0009	2.4979	6.1998	1171.3556	2925.8889	0.4003	0.0003	2.482	21
22	0.0006	2.4985	6.2127	1639.8978	4097.2445	0.4002	0.0002	2.486	22
23	0.0004	2.4989	6.2222	2295.8569	5737.1423	0.4002	0.0002	2.490	23
24	0.0003	2.4992	6.2294	3214.1997	8032.9993	0.4001	0.0001	2.492	24
25	0.0002	2.4994	6.2347	4499.8796	11247.1990	0.4001	0.0001	2.494	25
26	0.0002	2.4996	6.2387	6299.8314	15747.0785	0.4001	0.0001	2.495	26
27	0.0001	2.4997	6.2416	8819.7640	22046.9099	0.4000	0.0000	2.496	27
28	0.0001	2.4998	6.2438	12347.6696	30866.6739	0.4000	0.0000	2.497	28
29	0.0001	2.4999	6.2454	17286.7374	43214.3435	0.4000	0.0000	2.498	29
30	0.0000	2.4999	6.2466	24201.4324	60501.0809	0.4000	0.0000	2.498	30
31	0.0000	2.4999	6.2475	33882.0053	84702.5132	0.4000	0.0000	2.499	31
32	0.0000	2.4999	6.2482	47434.8074	118584.5185	0.4000	0.0000	2.499	32
33	0.0000	2.5000	6.2487	66408.7304	166019.3260	0.4000	0.0000	2.499	33
34	0.0000	2.5000	6.2490	92972.2225	232428.0563	0.4000	0.0000	2.499	34
35	0.0000	2.5000	6.2493	130161.1116	325400.2789	0.4000	0.0000	2.499	35
36	0.0000	2.5000	6.2495	182225.5562	455561.3904	0.4000	0.0000	2.499	36
37	0.0000	2.5000	6.2496	255115.7786	637786.9466	0.4000	0.0000	2.499	37
38	0.0000	2.5000	6.2497	357162.0901	892902.7252	0.4000	0.0000	2.499	38
39	0.0000	2.5000	6.2498	500026.9261	1250064.8153	0.4000	0.0000	2.499	39
40	0.0000	2.5000	6.2498	700037.6966	1750091.7415	0.4000	0.0000	2.499	40
41	0.0000	2.5000	6.2499	980052.7752	2450129.4381	0.4000	0.0000	2.500	41
42	0.0000	2.5000	6.2499	1372073.8853	3430182.2133	0.4000	0.0000	2.500	42
43	0.0000	2.5000	6.2499	1920903.4394	4802256.0986	0.4000	0.0000	2.500	43
44	0.0000	2.5000	6.2500	2689264.8152	6723159.5381	0.4000	0.0000	2.500	44
45	0.0000	2.5000	6.2500	3764970.7413	9412424.3533	0.4000	0.0000	2.500	45

Index

B

INDEX - C

D

E

INDEX - E

F

G

H

J

INDEX - J

K

L

INDEX - L

INDEX - O

Q

R

INDEX - R

S

INDEX - S

T

INDEX - T

U

V

W

Y

Z

ENGINEERING CONVERSIONS

(5 significant digits, rounded)

(Atmospheres are standard; Btus are international thermal; calories are gram-calories; gallons are U.S. liquid; horsepower are international; kilocalories are international thermal; miles are statute; pounds are avoirdupois.)

multiply	by	to obtain
acre	43,560	ft^2
Å (angstrom)	1.0×10^{-10}	m
atm	1.01325	bar
atm	76.0	cm Hg (0°C)
atm	33.899	ft H_2O (4°C)
atm	406.783	in H_2O (4°C)
atm	29.921	in Hg (0°C)
atm	101.325	kPa
atm	14.696	lbf/in^2
bar	0.98692	atm
bar	100	kPa
bar	14.504	lbf/in^2
bar	0.1	MPa
Btu	778.17	ft-lbf
Btu	3.9301×10^{-4}	hp-hr
Btu	0.25200	kcal
Btu	1.0551	kJ
Btu	2.9307×10^{-4}	kW-hr
Btu	1.0×10^{-5}	therm
Btu/hr	0.21616	ft-lbf/sec
Btu/hr	3.9301×10^{-4}	hp
Btu/hr	2.9307×10^{-4}	kW
Btu/lbm	2.3260	kJ/kg
Btu/lbm-°R	4.1868	kJ/kg·K
cal (see kcal)	0.001	kcal
cm (see m)	0.01	m
cm (see m)	0.03281	ft
cm (see m)	0.3937	in
cm Hg (0°C)	0.013158	atm
cm^3/g	1.0×10^{-3}	m^3/kg
cP (see P)	0.01	P
eV	1.6022×10^{-22}	kJ
ft	0.30480	m
ft	1.8939×10^{-4}	mi
ft^2	2.2957×10^{-5}	acre
ft H_2O (4°C)	0.029500	atm
ft H_2O (4°C)	0.43353	lbf/in^2
ft^3	7.4805	gal
ft^3/sec	448.83	gal/min
ft-lbf	1.2851×10^{-3}	Btu
ft-lbf	0.0013558	kJ
ft-lbf	3.7661×10^{-7}	kW-hr
ft-lbf	1.3558	N·m
ft-lbf/sec	4.6262	Btu/hr
ft/min	0.00508	m/s
g/cm^3	1000	kg/m^3
g/cm^3	62.428	lbm/ft^3
gal	0.13368	ft^3
gal	3.7854	L
gal	0.0037854	m^3
gal/min	0.0022280	ft^3/sec
gr (grain)	1.4286×10^{-4}	lbm
hp	2544.4	Btu/hr
hp	550	ft-lbf/sec
hp	33,000	ft-lbf/min
hp	0.74570	kW
hp-hr	2544.4	Btu
in	0.02540	m
in H_2O (4°C)	2.4583×10^{-3}	atm
in Hg (0°C)	0.033421	atm
in Hg (0°C)	0.49115	lbf/in^2
J (see kJ)	0.001	kJ
J	9.4782×10^{-4}	Btu
J	6.2415×10^{18}	eV
J	0.73756	ft-lbf
J	1.0	N·m
J/s	1.0	W
kcal	3.9683	Btu
kcal	4.1868	kJ

multiply	by	to obtain
kg	2.2046	lbm
kg/m^3	0.062428	lbm/ft^3
kip	4.4482	kN
kip	1000	lbf
kJ	0.94782	Btu
kJ	6.2415×10^{21}	eV
kJ	737.56	ft-lbf
kJ	0.23885	kcal
kJ	2.7778×10^{-4}	kW-hr
kJ/kg	0.42992	Btu/lbm
kJ/kg·K	0.23885	Btu/lbm-°R
km	3280.8	ft
km	0.62137	mi
km/hr	0.62137	mi/hr
kN	0.22481	kip
kPa	9.8692×10^{-3}	atm
kPa	0.01	bar
kPa	0.14504	lbf/in^2
kPa	1.4504×10^{-4}	ksi
ksi	6894.8	kPa
kW	3412.1	Btu/hr
kW	0.94782	Btu/sec
kW	737.56	ft-lbf/sec
kW	1.3410	hp
kW-hr	3412.1	Btu
kW-hr	2.6552×10^{6}	ft-lbf
kW-hr	3600.0	kJ
L	0.035135	ft^3
L	61.024	in^3
L	0.26417	gal
L	0.001	m^3
L/s	2.1189	ft^3/min
L/s	15.850	gal/min
lbf	4.4482	N
lbf/in^2	0.068046	atm
lbf/in^2	0.068948	bar
lbf/in^2	2.3067	ft H_2O (4°C)
lbf/in^2	2.0360	in Hg (0°C)
lbf/in^2	6.8948	kPa
lbm	7000	gr (grain)
lbm	0.45359	kg
lbm/ft^3	0.016018	g/cm^3
lbm/ft^3	16.018	kg/m^3
m	1.0×10^{10}	Å (angstrom)
m	3.2808	ft
m	39.370	in
m	1.0×10^{6}	μm (micron)
m^3	264.17	gal
mg/L	8.3454	lbm/MG
m/s	196.85	ft/min
mi	5280	ft
mi	1.6093	km
N	0.22481	lbf
N·m	0.73756	ft-lbf
N·m	1.0	J
P (poise)	1	N·s/m^2
Pa (see kPa)	0.001	kPa
Pa (see kPa)	1.4504×10^{-4}	lbf/in^2
therm	1.0×10^{5}	Btu
ton (cooling)	12,000	Btu/hr
μm (micron)	1.0×10^{-6}	m
W (see kW)	0.001	kW
W (see kW)	3.4121	Btu/hr
W (see kW)	0.73756	ft-lbf/sec
W (see kW)	1.3410×10^{-3}	hp
W (see kW)	1.0	J/s